ARCHITECTURAL DRAFTING & DESIGN

ARCHITECTURAL DRAFTING & DESIGN

Fourth Edition

ALAN JEFFERIS ■ DAVID MADSEN

Delmar
Thomson Learning™

Africa • Australia • Canada • Denmark • Japan • Mexico • New Zealand • Philippines • Puerto Rico • Singapore • Spain • United Kingdom • United States

NOTICE TO THE READER

Delmar Staff
Business Unit Director: Alar Elken
Executive Editor: Sandy Clark
Acquisitions Editor: Michael Kopf
Developmental Editor: John Fisher
Editorial Assistant: Jasmine Hartman
Executive Marketing Manager: Maura Theriault
Channel Manager: Mona Caron
Marketing Coordinator: Paula Collins
Executive Production Manager: Mary Ellen Black
Production Manager: Larry Main
Project Editor: Christopher Chien
Art Director: Mary Beth Vought

Cover: Photo courtesy Eric Brown Design Group, Inc.

Printed in the United States of America
7 8 9 10 XXX 05 04 03

For more information, contact Delmar at 3 Columbia Circle, PO Box 15015, Albany, New York 12212-5015; or find us on the World Wide Web at http://www.delmar.com

Asia
Thomson Learning
60 Albert Street, #15-01
Albert Complex
Singapore 189969

Australia/New Zealand
Nelson/Thomson Learning
102 Dodds Street
South Melbourne, Victoria 3205
Australia

Canada
Nelson/Thomson Learning
1120 Birchmont Road
Scarborough, Ontario
Canada M1K 5G4

International Headquarters
Thomson Learning
International Division
290 Harbor Drive, 2nd Floor
Stamford, CT 06902-7477
USA

Japan
Thomson Learning
Palaceside Building 5F
1-1-1 Hitotsubashi, Chiyoda-ku
Tokyo 100 0003
Japan

Latin America
Thomson Learning
Seneca, 53
Colonia Polanco
11560 Mexico D. F. Mexico

Spain
Thomson Learning
Calle Magallanes, 25
28015-Madrid
Espana

UK/Europe/Middle East
Thomson Learning
Berkshire House
168-173 High Holborn
London
WC1V 7AA United Kingdom

Thomas Nelson & Sons Ltd.
Nelson House
Mayfield Road
Walton-on-Thames
KT 12 5PL United Kingdom

Library of Congress Cataloging-in-Publication Data
Jefferis, Alan
Architectural drafting and design / Alan Jefferis, David A. Madsen.—4th ed.
p. cm.
Includes index.
ISBN 0-7668-1546-3
1. Architectural drawing. 2. Architectural design. I. Madsen, David A. II. Title.
NA2700 .J44 2000
720′.28′4—dc21

00-029488

CONTENTS

SECTION 10

SECTION 11

SECTION 12

GENERAL CONSTRUCTION SPECIFICATIONS 809

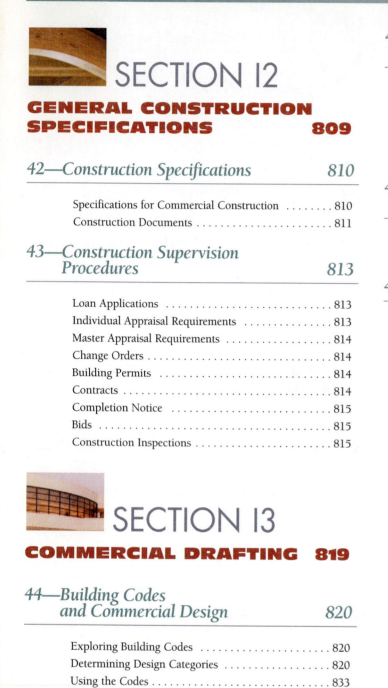

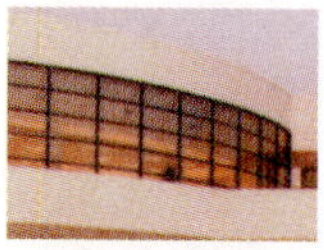

SECTION 13

COMMERCIAL DRAFTING 819

PREFACE

Architectural Drafting and Design is a practical, comprehensive textbook that is easy to use and understand. The content may be used as presented, by following a logical sequence of learning activities for residential and light commercial architectural drafting and design, or the chapters may be rearranged to accommodate alternate formats for traditional or individualized instruction.

APPROACH

Practical

Architectural Drafting and Design provides a practical approach to architectural drafting as it relates to current common practices. The emphasis on standardization is an excellent and necessary foundation of drafting training as well as implementing a common approach to drafting nationwide. After students become professional drafters, this text will serve as a valuable desk reference.

Realistic

Chapters contain professional examples, illustrations, step-by-step layout techniques, drafting problems, and related tests. The examples demonstrate recommended drafting presentation with actual architectural drawings used for reinforcement. The correlated text explains drafting techniques and provides useful information for skill development. Step-by-step layout methods provide a logical approach to beginning and finishing complete sets of working drawings.

Practical Approach to Problem Solving

The professional architectural drafter's responsibility is to convert architects', engineers', and designers' sketches and ideas into formal drawings. The text explains how to prepare formal drawings from design sketches by providing the learner with the basic guidelines for drafting layout, and minimum design and code requirements in a knowledge-building format; one concept is learned before the next is introduced. The concepts and skills learned from one chapter to the next allow students to prepare complete sets of working drawings in residential and light commercial drafting. Problem assignments are presented in order of difficulty and in a manner that provides students with a wide variety of architectural drafting experiences.

The problems are presented as preliminary designs or design sketches in a manner that is consistent with actual architectural office practices. It is not enough for students to duplicate drawings from given assignments; they must be able to think through the process of drawing development with a foundation of how drawing and construction components are implemented. The goals and objectives of each problem assignment are consistent with

recommended evaluation criteria based on the progression of learning activities. The drafting problems and tests recommend that work be done using drafting skills on actual drafting materials with either professional manual or computer drafting equipment. A problem solution or test answer should be accurate and demonstrate proper drafting technique.

FEATURES OF THE TEXT

Applications

Special emphasis has been placed on providing realistic drafting problems. Problems are presented as design sketches or preliminary drawings in a manner that is consistent with industry practices. The problems have been supplied by architectural designers and architects. Each problem solution is based on the step-by-step layout procedures provided in the chapter discussions.

Problems are given in order of complexity so students may be exposed to a variety of drafting experiences. Problems require students to go through the same thought and decision-making processes that a professional drafter faces daily, including scale and paper size selection, view layout, dimension placement, section placement, and many other activities. Problems may be solved using manual or computer drafting, as determined by individual course guidelines. Chapter tests provide complete coverage of each chapter and may be used for student evaluation or as study questions.

Illustrations

Drawings and photos are used liberally throughout this text to amplify the concepts presented. Full color treatment

QUESTIONS

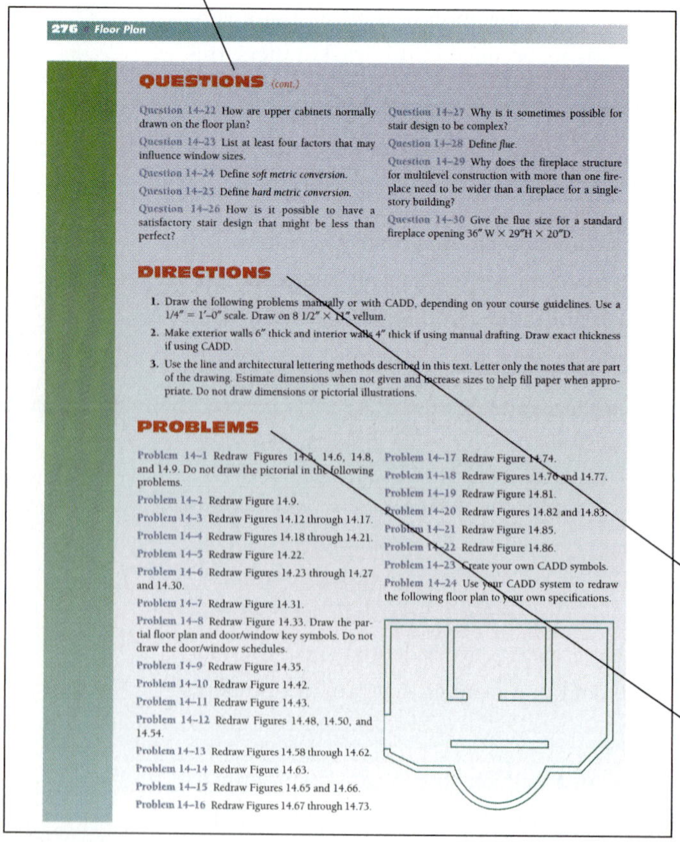

276 Floor Plan

QUESTIONS (cont.)

Question 14–22 How are upper cabinets normally drawn on the floor plan?

Question 14–23 List at least four factors that may influence window sizes.

Question 14–24 Define *soft metric conversion.*

Question 14–25 Define *hard metric conversion.*

Question 14–26 How is it possible to have a satisfactory stair design that might be less than perfect?

Question 14–27 Why is it sometimes possible for stair design to be complex?

Question 14–28 Define *flue.*

Question 14–29 Why does the fireplace structure for multilevel construction with more than one fireplace need to be wider than a fireplace for a single-story building?

Question 14–30 Give the flue size for a standard fireplace opening 36″ W × 29″H × 20″D.

DIRECTIONS

1. Draw the following problems manually or with CADD, depending on your course guidelines. Use a 1/4″ = 1′–0″ scale. Draw on 8 1/2″ × 11″ vellum.
2. Make exterior walls 6″ thick and interior walls 4″ thick if using manual drafting. Draw exact thickness if using CADD.
3. Use the line and architectural lettering methods described in this text. Letter only the notes that are part of the drawing. Estimate dimensions when not given and increase sizes to help fill paper when appropriate. Do not draw dimensions or pictorial illustrations.

PROBLEMS

Problem 14–1 Redraw Figures 14.5, 14.6, 14.8, and 14.9. Do not draw the pictorial in the following problems.

Problem 14–2 Redraw Figure 14.9.

Problem 14–3 Redraw Figures 14.12 through 14.17.

Problem 14–4 Redraw Figures 14.18 through 14.21.

Problem 14–5 Redraw Figure 14.22.

Problem 14–6 Redraw Figures 14.23 through 14.27 and 14.30.

Problem 14–7 Redraw Figure 14.31.

Problem 14–8 Redraw Figure 14.33. Draw the partial floor plan and door/window key symbols. Do not draw the door/window schedules.

Problem 14–9 Redraw Figure 14.35.

Problem 14–10 Redraw Figure 14.42.

Problem 14–11 Redraw Figure 14.43.

Problem 14–12 Redraw Figures 14.48, 14.50, and 14.54.

Problem 14–13 Redraw Figures 14.58 through 14.62.

Problem 14–14 Redraw Figure 14.63.

Problem 14–15 Redraw Figures 14.65 and 14.66.

Problem 14–16 Redraw Figures 14.67 through 14.73.

Problem 14–17 Redraw Figure 14.74.

Problem 14–18 Redraw Figures 14.76 and 14.77.

Problem 14–19 Redraw Figure 14.81.

Problem 14–20 Redraw Figures 14.82 and 14.83.

Problem 14–21 Redraw Figure 14.85.

Problem 14–22 Redraw Figure 14.86.

Problem 14–23 Create your own CADD symbols.

Problem 14–24 Use your CADD system to redraw the following floor plan to your own specifications.

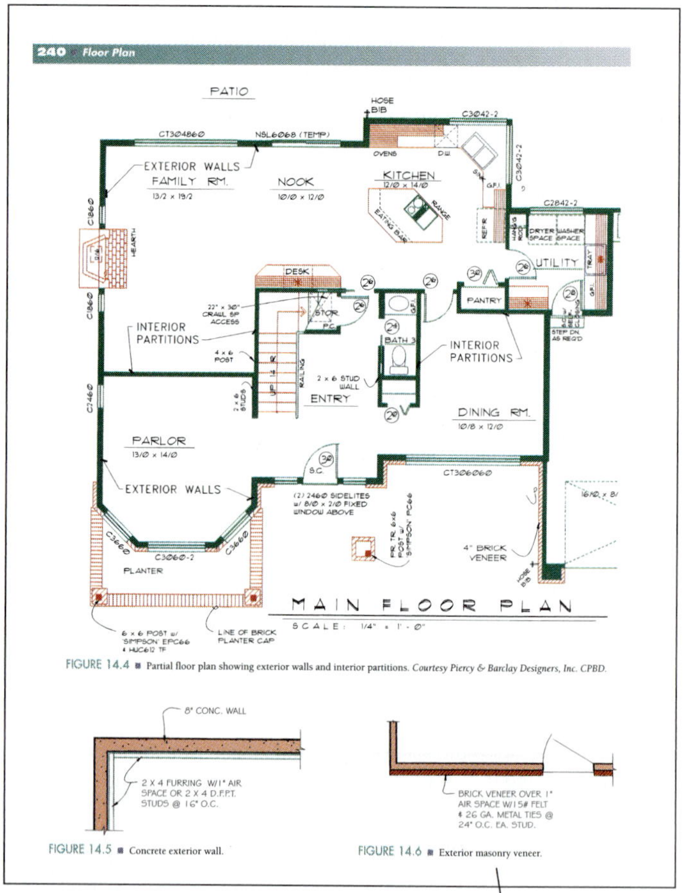

FIGURE 14.4 ■ Partial floor plan showing exterior walls and interior partitions. *Courtesy Piercy & Barclay Designers, Inc. CPBD.*

FIGURE 14.5 ■ Concrete exterior wall.

FIGURE 14.6 ■ Exterior masonry veneer.

DIRECTIONS

PROBLEMS

ILLUSTRATIONS

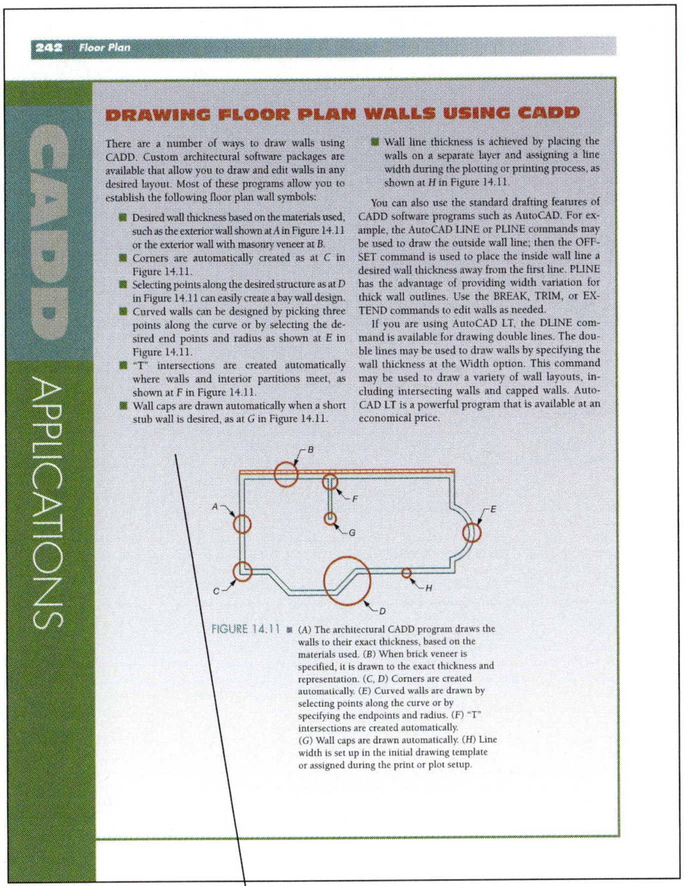

242 Floor Plan

CADD APPLICATIONS

DRAWING FLOOR PLAN WALLS USING CADD

There are a number of ways to draw walls using CADD. Custom architectural software packages are available that allow you to draw and edit walls in any desired layout. Most of these programs allow you to establish the following floor plan wall symbols:

- Desired wall thickness based on the materials used, such as the exterior wall shown at *A* in Figure 14.11 or the exterior wall with masonry veneer at *B*.
- Corners are automatically created as at *C* in Figure 14.11.
- Selecting points along the desired structure as at *D* in Figure 14.11 can easily create a bay wall design.
- Curved walls can be designed by picking three points along the curve or by selecting the desired end points and radius as shown at *E* in Figure 14.11.
- "T" intersections are created automatically where walls and interior partitions meet, as shown at *F* in Figure 14.11.
- Wall caps are drawn automatically when a short stub wall is desired, as at *G* in Figure 14.11.
- Wall line thickness is achieved by placing the walls on a separate layer and assigning a line width during the plotting or printing process, as shown at *H* in Figure 14.11.

You can also use the standard drafting features of CADD software programs such as AutoCAD. For example, the AutoCAD LINE or PLINE commands may be used to draw the outside wall line; then the OFFSET command is used to place the inside wall line a desired wall thickness away from the first line. PLINE has the advantage of providing width variation for thick wall outlines. Use the BREAK, TRIM, or EXTEND commands to edit walls as needed.

If you are using AutoCAD LT, the DLINE command is available for drawing double lines. The double lines may be used to draw walls by specifying the wall thickness at the Width option. This command may be used to draw a variety of wall layouts, including intersecting walls and capped walls. AutoCAD LT is a powerful program that is available at an economical price.

FIGURE 14.11 ■ (*A*) The architectural CADD program draws the walls to their exact thickness, based on the materials used. (*B*) When brick veneer is specified, it is drawn to the exact thickness and representation. (*C, D*) Corners are created automatically. (*E*) Curved walls are drawn by selecting points along the curve or by specifying the endpoints and radius. (*F*) "T" intersections are created automatically. (*G*) Wall caps are drawn automatically. (*H*) Line width is set up in the initial drawing template or assigned during the print or plot setup.

CADD APPLICATIONS

STAIR TYPE	IRC
Straight stair	
Max. rise	7 3/4″ (197 mm)
Min. run	10″ (254 mm)
Min. headroom	6′–8″ (2032 mm)
Min. tread width	3′–0″ (914 mm)
Handrail height	34″ (864 mm) min./38″ (965 mm) max.
Guardrail height	36″ (914 mm) min.
Winders	
Min. tread depth	6″ (153 mm) min. @ edge/10″ (254 mm) @ 12″ (305 mm)
Spiral	
Min. width	26″ (660 mm)
Min. tread depth	7 1/2″ (190 mm) @ 12″ (305 mm)
Max. rise	9 1/2″ (241 mm)
Min. headroom	6′–6″ (1982 mm)

FIGURE 37.4 ■ Basic stair dimensions according to the International Residential Code.

CONSTRUCTION TECHNIQUES

enhances the clarity. Abundant step-by-step illustrations take students through the drafting process and clarify the detailed stages. This new edition also incorporates many CADD-drawn illustrations.

Computer-Aided Design Drafting (CADD)

CADD is presented as a valuable tool that is treated no differently from manual drafting tools. The complete discussion of computer graphics introduces the workstation environment, terminology, drafting techniques, and sample drawings. While individual course guidelines may elect to solve architectural drafting problems using either computer or manual drafting equipment, the concepts remain the same; the only difference is the method of presentation.

An Online Companion (www.delmarlearning.com/resources.asp) for this text contains drawing files of the floor plans for chapters 14, 15, and 16.

Code and Construction Techniques

The 2000 International Residential Code, and the 2000 International Building Code are introduced and compared throughout the text as they relate to specific instructions and applications. Construction techniques differ throughout the country. This text clearly acknowledges the difference in construction methods and introduces the student to the format used to make complete sets of working drawings for each method of construction. Students may learn to prepare drawings from each construction method or, more commonly, for the specific construction techniques that are used in their locality. The problem assignments are designed to provide drawings that involve a variety of construction alternatives.

Construction Documents

Documents for bill of materials, contracts, and specifications are provided as PDF files on the Online Companion. These

ADDITIONAL READINGS

Chapter 14 Additional Reading

The following web sites can be used as a resource to help you keep current with changes in building materials.

ADDRESS	COMPANY OR ORGANIZATION
www.abracadata.com	Abracadata
www.andersonwindows.com	Anderson Windows
www.aristokraft.com	Arisokoft (Cabinets)
www.autodesk.com	Autodesk, Inc.-Architectural Desktop Software
www.beamvac.com	Beam Central Cleaning Systems

pertain to content coverage in chapters 42 and 43, and can be used to complete problems in these chapters.

Additional Reading

An Additional Reading section is offered at the end of selected chapters to provide users with information on additional resources related to chapter topics. This is a list of web sites for companies or organizations that offer services or materials related to the chapter content.

FEATURES OF THE NEW EDITION

- Revised order of topics to reflect the development of architectural drawings.
- Updated code material based on 2000 International Residential Code and the International Building Code.
- The Online Companion for this book contains AutoCAD drawing files of floor plans from chapters 14, 15, and 16. Students may also purchase fully functional student priced versions of current release Autodesk Student Portfolio software-AutoCAD, AutoCAD LT®, and Architectural Desktop™ at www.cadcampus.com or their college bookstore.
- Additional framing methods, including advanced framing techniques, rigid foam panels, timber construction, concrete masonry construction, steel framing, and modular construction.
- Additional framing material, including laminated beams, open web joist, laminated veneer lumber, oriented strand board, waferboard, vapor barriers, expanded plywood coverage, and expanded truss information.
- Expanded coverage of architectural line work, conservation design, barrier-free design, site design, contour layout, grading plan layout, site profile, and electrical, plumbing and HVAC plans.
- Updated coverage of CADD in architecture.
- Expanded coverage of loads resisted by a structure, including the stresses from hurricanes, floods, tornadoes, and earthquakes.
- Revised standards for joist and rafter selection based on 2000 IRC wood values and tables for sizing engineered joists and rafters.
- Expanded coverage of concrete slab foundations, including prestressed and posttensioned slabs.
- Expanded coverage of wall sections and stock details.
- Additional photos showing current commercial building methods.
- Drawing check lists are provided at the end of the drafting related chapters. This gives the student a list of features that should be included in the drawing. The student can check off each item when completed. The drawing contains all of the required elements when everything is

checked off. This also provides the instructor with a valuable check list to use when evaluating student work.

Coverage of commercial construction is completely covered in the Delmar publication Commercial Drafting and Detailing. The book features development of structures using wood, timber, masonry, steel, and concrete. Emphasis will be placed on drafting and detailing of the drawings required for the architectural and structural drawings.

ORGANIZING YOUR COURSE

Architectural drafting is the primary emphasis of many technical drafting curricula, whereas other programs offer only an exploratory course in this field. This text is appropriate for either application, as its content reflects the common elements in an architectural drafting curriculum.

Prerequisites

An interest in architectural drafting, plus basic arithmetic, written communication, and reading skills are the only prerequisites required. Basic drafting skills and layout techniques are presented as appropriate. Students with an interest in architectural drafting who begin using this text will end with the knowledge and skills required to prepare complete sets of working drawings for residential and light commercial architectural drafting.

Fundamental Through Advanced Coverage

This text may be used in an architectural drafting curriculum that covers the basics of residential architecture in a one- two- or three-semester sequence. In this application, students use the chapters directly associated with the preparation of a complete set of working drawings for a residence, where the emphasis is on the use of fundamental skills and techniques. The balance of the text may remain as a reference for future study or as a valuable desk reference.

The text may also be used in the comprehensive architectural drafting program where a four- to six-semester sequence of residential and light commercial architectural drafting and design is required. In this application, students may expand on the primary objective of preparing a complete set of working drawings for the design of residential and light commercial projects with the coverage of any one or all of the following areas: systems drafting, energy-efficient construction techniques, solar and site orientation design applications, heating and cooling thermal performance calculations, structural load calculations, and presentation drawings.

Section Length

Chapters are presented in individual learning segments that begin with elementary concepts and build until each chapter provides complete coverage of every topic. Instructors may choose to present lectures in short, 15-minute discussions or divide each chapter into 40- to 50-minute lectures.

Drafting Equipment and Materials

Identification and use of a manual and computer-aided drafting equipment is given. Students will require an inventory of equipment available for use as listed in the chapters. Professional drafting materials are explained, and it is recommended that students prepare problem solutions using actual drafting materials.

SUPPLEMENTS

E.Resource

This is an educational resource that creates a truly electronic classroom. It is a CD-ROM containing tools and instructional resources that enrich your classroom and make instructor's preparation time shorter. The elements of e.resource link directly to the text and tie together to provide a unified instructional system. With e.resource you can spend your time teaching, not preparing to teach. ISBN: 0-7668-1547-1.

Features contained in e.resource include:

- *Syllabus:* Lesson plans created by chapter. You have the option of using these lesson plans with your own course information. www.worldclasslearning.com.
- *Chapter Hints:* Objectives and teaching hints that provide the basis for a lecture outline that helps you to present concepts and material. Key points and concepts can be graphically highlighted for student retention.
- *PowerPoint® Presentation:* These slides provide the basis for a lecture outline that helps you to present concepts and material. Key points and concepts can be graphically highlighted for student retention.
- *Exam View Computerized Test Bank:* Over 800 questions of varying levels of difficulty are provided in

true/false and multiple choice formats so you can assess student comprehension.

- *CADD Drawing Files:* Drawing files for many of the plans in the text are listed in chapter folders. You can modify these plans to create new illustrations or chapter problems, or you can make them available to students who desire to work within AutoCAD.

Video and Animation Resources: These AVI files graphically depict the execution of key concepts and commands in drafting, design, and AutoCAD and let you bring multi-media presentations into the classroom.

Workbook

We have developed a workbook to correlate with Architectural Drafting and Design. This workbook is correlated with the main text. The plentiful problem assignment take the beginning architectural drafting student from the basics of line and lettering techniques to drawing a complete set of residential plans. Problems may be done manually or on CADD. Over 30 problems refer to AutoCAD 2000 drawing files that can be downloaded from our web site www.autodeskpress. com/resources/olc/index.asp to be used to complete the problems.

The workbook also provides the student with a series of complete residential plans presented as architectural design problems. The plans range from simple to complex and give the student an opportunity to vary the design elements of the home. Problems may be done on CADD or manually.

Advanced coverage includes a variety of complete light commercial drafting and design projects. The types of projects include multifamily residential, tilt-up concrete, steel, and heavy timber construction. This information leads the student into the comprehensive drafting and design projects. ISBN 0-7668-1548-X

Solutions Manual

A solutions manual is available with answers to end of chapter review questions and solutions to end of chapter problems. Solutions are also provided for the Workbook problems. ISBN 0-7668-2860-3

Videos

Two video sets, containing four 20-minute tapes each, are available. The videos correspond to the topics addressed in the text:

- Set #1, ISBN 0–7668–3094–2
- Set #2, ISBN 0–7668–3095–0

Video sets are also available on Interactive video CD-ROM.

- Set #1, ISBN 0–7668–3116–7
- Set #2, ISBN 0–7668–3117–5

WebTutor™

WebTutor on WebCT™ and Blackboard™ online resources provide content rich, web based teaching and learning that reinforces and clarifies complex concepts:

- WebTutor on WebCT, ISBN 1–4018–1390–9
- WebTutor on Blackboard, ISBN 1–4018–1391–7

ACKNOWLEDGMENTS

We would like to thank and acknowledge the many professionals who reviewed the manuscript to help us publish our architectural drafting text. A special acknowledgment is due the instructors who reviewed the chapters in detail.

REVIEWERS

Tom Bledsaw
ITT Technical Institute, Indianapolis, IN

Walter B. Cheever
South Central Technical College, N. Mankato, MN

Margaret Ann Jeffries
Pellissippi State Technical Community College

Tom Kane
Pueblo Community College, Florence, CO

David LaRue
ITT Technical Institute, Strongsville, OH

Jeff Plant
Salt Lake Community College, Salt Lake City, UT

Joe Ramos
Mt. San Antonio College, Walnut, CA

Joey Spillyards
Red Rocks Community College, Lakewood, CO

Roy Trouerbach
University of Advancing Computer Technology, Tempe, AZ

Tony Whitus
Tennessee Technology Center, Crossville, TN

CONTRIBUTING COMPANIES

The quality of this text is also enhanced by the support and contributions from architects, designers, engineers, and vendors. The list of contributors is extensive and acknowledgement is given at each illustration. The following individuals and companies gave an extraordinary amount of support with technical information and art for this edition

Dan Kovac
Pierce & Barclay Designers, Inc.

Alan Mascord
Alan Mascord Design Associates, Inc.

Charles Talcott
Home Planners, Inc.

Ken Smith
Ken Smith Architect and Associates, Inc.

Bob Ringland
Southland Corporation

Richard Wallace
Southern Forest Production Association

Evan Terry, P.C.
Architecture and Planning

Catherine Dale
Graphisoft U.S., Inc.

Deborah Parker Wong
Graphisoft Press Relations

Toni Garretson
Dawn Ostrowski
Calcomp

Eric Silver
Eagle Point Software

Catherine Reynolds
Frametech Technologies Corporation

Carol Parcels
Arel Lucas
Sandy Gramley
Hewlett Packard

Jessie O. Kempter
IBM Worldwide Image Library

Tamara McKinney Berry
Tri Star Computer

Deana J. Croussore
Kent Kuffner
Water Furnace International Inc.

Chris McElroy
Pella Corporation

Michael A. Cassidy
The Construction Specifications Institute

Kelly Malone
Visio Corporation

Walter Stein
Home Building Plan Service, Inc.

Steve Webb
Lennox Industries, Inc.

Marilyn H. LeMoine
American Plywood Association

Richard Branham
National Concrete Masonry Association

Glen Becker
ABTCO Inc.

W.A. Erdos
Alsode, Inc.

David Commery
Barentine, Bates and Lee, AIA

Pamela Allsebrook
California Redwood Association

Eric Brown
Eric S. Brown, Design Group

Barbara Catlow
Dryvit Systems, Inc.

Meredith Brandes
Elk Roofing

Peggy Nila & Cheryl Melendez
International Conference of Building Officials

D. Tracy Ward
Kew Corporation, Specialty Architectural Services

Kathleen M. Arnelt
Louisana-Pacific Corporation

Richard R. Chapman & Robert C. Guzikowski
Simpson Strong-Tie company, Inc.

Kate Cammack & Kerry Vanden Heuvel
Stephen Fuller, Incorporated

Misty Schymtzik
Trus Joist MacMillan

Havlin Kemp
VLMK Consulting Engineers

Eric C. Wilson
Western Wood Products Association

Special thanks goes to the International Code Council (ICC) for allowing us to use text from the International Residential Code/2000:

Portions of this publication reproduce text from the International Residential Code/2000. Copyright 2000, with the permission of the publisher, the International Conference of Building Officials, under license from the International Code Council Inc., Falls Church, Virginia. The 2000 International

Art work for title pages and section openers in the text were provided courtesy of the following companies:

Half title page:	California Redwood Association
Title page:	Elk Roofing
Section I, II, IV:	Eric Brown Design Group, Inc.
Section III, V:	PhotoDisc
Section VI:	John Black
Section VII, XII:	Design Basics, Inc.
Section VIII:	Southern Forest Product Association
Section IX:	Leroy Cook
Section X:	Western Wood Products Assn.
Section XI:	Graphics Software, Inc.
Section XIII:	Planned Expansion Group PC

Approximately 180 illustrations are reproduced from Engineering Drawing and Design, by David Madsen, Shumaker, and Stewart, from Delmar Publishers. Photos were provided by Leroy Cook, Krista Herbel, Janice Jefferis, Joe Jones, Denise Taylor, and Tim Taylor. Last but not least, we greatly appreciate the efforts of Teresa Jefferis and Connie Wilmon for their work with the step-by-step drawings in each chapter.

TO THE STUDENT

Architectural Drafting and Design is designed for you, the student. The development and format of the presentation have been tested in both conventional and individualized classroom instruction. The information presented is based on architectural standards, drafting room practice, and trends in the drafting industry. This text is the only architectural drafting reference that you will need. Use the text as a learning tool while in school, and take it along as a desk reference when you enter the profession. The amount of written text is complete but kept to a minimum. Examples and illustrations are used extensively. Drafting is a graphic language, and most drafting students learn best by observation of examples. Here are a few helpful hints.

1. *Read the text.* The text content is intentionally designed for easy reading. Sure, it doesn't read the same as an exciting short story, but it does give the facts in as few, easy-to-understand words as possible. Don't pass up the reading, because the content will help you to understand the drawings clearly.
2. *Look carefully at the examples.* The figure examples are presented in a manner that is consistent with drafting standards. Look at the examples carefully in an attempt to understand the intent of specific applications. If you are able to understand why something is done a certain way, it will be easier for you to apply those concepts to the drawing problems and, later, to the job. Drafting is a precise technology based on rules and guidelines. The goal of a drafter is to prepare drawings that are easy to interpret. There will always be situations when rules must be altered to handle a unique situation. Then you will have to rely on judgment based on your knowledge of accepted standards. Drafting is often like a puzzle; there may be more than one way to solve a problem.
3. *Use the text as a reference.* Few drafters know everything about drafting standards, techniques, and concepts, so always be ready to use the reference if you need to verify how a specific application is handled. Become familiar with the definitions and use of technical terms. It would be difficult to memorize everything noted in this text, but after considerable use of the concepts, architectural drafting applications should become second nature.
4. *Learn each concept and skill before you continue to the next.* The text is presented in a logical learning sequence. Each chapter is designed for learning development, and chapters are sequenced so that drafting knowledge grows from one chapter to the next. Problem assignments are presented in the same learning sequence as the chapter content and also reflect progressive levels of difficulty.
5. *Practice.* Development of good manual and computer drafting skills depends to a large extent on practice. Some individuals have an inherent talent for manual drafting, and some people are readily compatible with computers. If you fit into either group, great! If you don't, then practice is all you may need. Practice manual drafting skills to help improve the quality of your drafting presentation, and practice communicating and working with a computer. A good knowledge of drafting practice is not enough if the manual skills are not satisfactory. When the computer is used, however, most manual skills are not needed.
6. *Use sketches or preliminary drawings.* When you are drawing manually or with a computer, the proper use of a sketch or preliminary drawing can save a lot of time in the long run. Prepare a layout sketch or preliminary layout for each problem. This will give you a chance to organize thoughts about drawing scale, view selection, dimension and note placement, and paper size. After you become a drafting veteran, you may be able to design a sheet layout in your head, but until then, you will be sorry if you don't use sketches.
7. *Use professional equipment and materials.* For the best possible learning results and skill development, use the professional drafting equipment, supplies, and materials that are recommended.

Alan Jefferis
David A. Madsen

SECTION 1

Introduction to Architectural Drafting

1 CHAPTER

Professional Architectural Careers, Office Practice, and Opportunities

INTRODUCTION

As you begin working with this text, you are opening the door to many exciting careers. Each career in turn has many different opportunities within it. Whether your interest lies in theoretical problem solving, artistic creations, or working with your hands creating something practical, a course in residential architecture will help prepare you to satisfy that interest. An architectural drafting class can lead to a career as a drafter, designer, architect, or engineer. Once you have mastered the information and skills presented in this text, you will be prepared for each of these fields as well as many others.

DRAFTER

A drafter is the person who creates the drawings and details for another person's creations. It is the drafter's responsibility to use the proper line and lettering quality and to lay out properly the required drawings necessary to complete a project. Such a task requires a great attention to detail as the drafter redraws the supervisor's sketches. Because the drafter could possibly be working for several architects or engineers within an office, the drafter must be able to get along well with others.

The Beginning Drafter

Your job as a beginning or junior drafter will generally consist of making corrections to drawings rendered by others. There may not be a lot of mental stimulation to making changes, but it is a very necessary job. It is also a good introduction to the procedures and quality standards within an office.

As your line and lettering quality improve, your responsibilities will be expanded. If you're working at a firm that uses computers, you'll need to become proficient using the firm's computer standards and any special menus and LISP routines needed to work efficiently. No matter what tools are used to create the drawings, typically your supervisor will give you a sketch and expect you to draw the required drawing. Figure 1–1 shows an engineer's sketch. Figure 1–2 shows the drawing created by a drafter. As you gain an understanding of the drawings that you are making and gain confidence in your ability, the

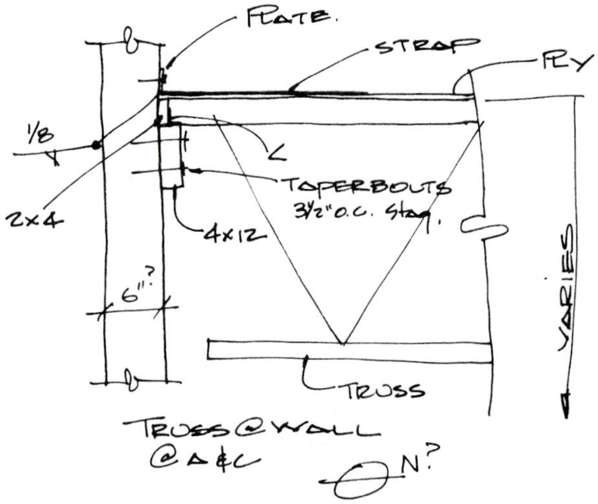

FIGURE 1–1 ■ A sketch is usually given to a junior drafter to follow for a first drawing. The drafter can find information for completing the drawing by examining similar jobs in the office.

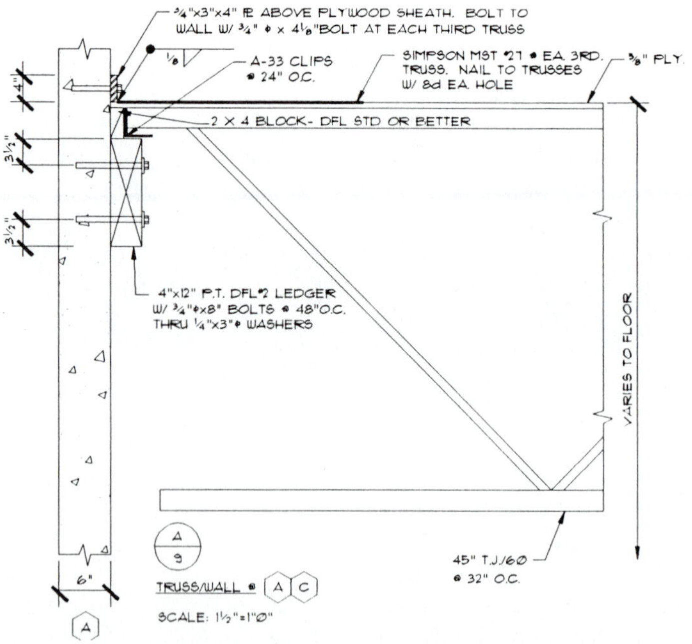

FIGURE 1–2 ■ A detail drawn by a drafter using the sketch shown in Figure 1–1.

sketches that you are given generally will become more simplified. Eventually your supervisor may just refer you to a similar drawing and expect you to be able to make the necessary adjustments to fit it to the new application.

The decisions involved in making drawings without sketches require the drafter to have a good understanding of what is being drawn. This understanding does not come just from a textbook. To advance as a drafter and become a leader on the drawing team will require you to become an effective manager of your time. This would include the ability to determine what drawings will need to be created, selected from a stock library and edited, and to estimate the time needed to complete these assignments and meet deadlines established by the team captain, the client, the lending institution or the building department. An even better way to gain an understanding of what you are drafting is to spend time working at a construction site so that you understand what a craftsman must do as a result of what you have drawn. To become a good team leader you will also need to develop skills that promote a sense of success among your teammates. Although it is against the law to discriminate on the basis of race, color, religion, gender, sexual orientation, age, marital status, or disability, moving beyond the law and creating a friendly and productive work environment is a critical skill for a team leader.

Depending on the size of the office where you work, you may also spend a lot of your time as a beginning drafter editing stock details, running prints, making deliveries, obtaining permits, and doing other such office chores. Don't get the idea that a drafter does only the menial chores around an office. But you do need to be prepared, as you go to your first drafting job, to do things other than drafting.

The Experienced Drafter

Although your supervisor may prepare the basic design for a project, experienced drafters are expected to make decisions about construction design. These might include determining structural sizes and connection methods for intersecting beams, drawing renderings, visiting job sites, and supervising beginning drafters. The types of drawings that you will be working on are also affected as you gain experience. Instead of drawing site plans, cabinet elevations, or roof plans or revising existing details, an experienced drafter and team leader may be working on the floor and foundation plans, elevations, and sections. Probably your supervisor will still make the initial design drawings but will pass these drawings on to you as soon as a client approves the preliminary drawings.

In addition to drafting, you may work with the many city and state building departments that govern your work. This will require you to do research in the codes that govern the building industry. You will also need to become familiar with vendors' catalogs. The most common is the Sweet's Catalog. Sweet's, as it is known, is a series of books that contain product information on a wide variety of building products. Sweet's is also available on CDs that are updated quarterly. Information in these catalogs is listed by manufacturer, trade name, and type of product.

Employment Opportunities

Drafters can find employment in firms of all sizes. Designers, architects, and engineers all require entry-level and advanced drafters to help produce their drawings. Many drafters are also employed by suppliers of architectural equipment. This work might include drawing construction details for a steel fabricator, making layout drawings for a cabinet shop, or designing ductwork for a heating and air-conditioning installer. Many manufacturing companies hire drafters with an architectural background to help draw and sometimes sell a product or draw installation diagrams for instruction booklets or sales catalogs. Drafters are also employed by many government agencies. These jobs include working in planning, utility, or building departments; survey crews; or other related municipal jobs.

Educational Requirements

In order to get your first drafting job, you will need a solid education, good line and lettering quality, a good understanding of basic computer skills, and the ability to *sell* yourself to an employer. Good linework and lettering will come as a result of practice. CAD skills must include a thorough understanding of drawing and editing commands, as well as the ability to quickly decide which option is best for the given situation. This ability will also come with practice. If you work full time editing details, you'll quickly become proficient on deciding the best commands to use. The education required for a drafter can range from one or more years in a high school drafting program to a diploma from a one-year accredited technical school to a degree from a two-year college program, and all the way to a master's degree in architecture.

Helpful areas of study for an entry-level drafter would include math, writing, and drawing. The math required ranges from simple addition to calculus. Although the drafter may spend most of the day adding dimensions expressed in feet and inches, a knowledge of advanced math will be helpful for solving many building problems. You'll often be required to use basic math skills to determine quantities, areas, and volume. Many firms provide their clients with a list of materials required to construct the project. Areas will need to be calculated to determine the size of each room, required areas to meet basic building code requirements, the loads on a structural member, or the size of the structure. Writing skills will also be very helpful. As a drafter, you are often required to complete the paperwork that accompanies any set of plans, such as permits, requests for variances, written specifications, or environmental impact reports. In addition to standard drafting classes, classes in photography, art, surveying, and construction could be helpful to the drafter.

DESIGNER

The meaning of the term *designer* varies from state to state. Many states restrict its use by requiring those calling themselves designers to have had formal training and to have passed

a competency test. A designer's responsibilities are very similar to those of an experienced drafter and are usually based on both education and experience. A designer is usually the coordinator of a team of drafters. The designer may work under the direct supervision of an architect, an engineer, or both and supervise the work schedule of the drafting team.

In addition to working in a traditional architectural office setting, designers often have their own office practice in which they design residential, multifamily, and some types of light commercial buildings. State laws vary regarding the types and sizes of buildings a designer may work on without the stamp of an architect or engineer. Students wishing more information about this career can obtain information from:

American Institute of Building Design (AIBD)
991 Post Road East
Westport, CT 06880
1-800-366-2423
Web site: www.aibd.org
E-mail: aibdnat@aol.com

Students can also contact the American Design and Drafting Association, the U.S. Department of Labor, or the U.S. Office of Education.

FIGURE 1–3 ■ Architects use their training to blend the needs and wishes of the client with the site, materials, and financial realities. *Courtesy of the California Redwood Assoc. Robert Corna architect, photo by Balthazar Korab.*

ARCHITECT

An architect performs the tasks of many professionals, including designer, artist, project manager, and construction supervisor. Few architects work full time in residential design. Although many architects design some homes, most devote their time to commercial construction projects such as schools, offices, and hospitals. An architect is responsible for the design of a structure and for the way the building relates to the environment. The architect often serves as a coordinator on a project to ensure that all aspects of the structure blend together to form a pleasing relationship. This coordinating includes working with the client, the contractors, and a multitude of engineering firms that may be working on the project. Figure 1–3 shows a home designed by an architect to blend the needs and wishes of the client with the site, materials, and financial realities.

Use of the term *architect* is legally restricted to individuals who have been licensed by the state where they practice. Obtaining a license typically requires three years to complete a masters program, three years as an intern, and two years to complete the registration exam process. Some states allow a designer to take the licensing test through practical work experience. Although standards vary for each state, five to seven years of experience under the direct supervision of a licensed architect or engineer is usually required.

Positions in an architectural firm include:

Technical staff—Consulting engineers such as mechanical, electrical, and structural engineers; landscape architects; interior designers; CAD operators; and drafters.

Intern—Unlicensed architectural graduates with less than three years of experience. An intern's responsibilities typically include developing design and technical solutions under the supervision of an architect.

Architect I—Licensed architect with three to five years of experience. An Architect I's job description typically includes responsibility for a specific portion of a project within the parameters set by a supervisor.

Architect II—Licensed architect with six to eight years of experience. An Architect II's job description typically includes responsibility for the daily design and technical development of a project.

Architect III—Licensed architect with eight to ten years of experience. An Architect III's job description typically includes responsibility for the management of major projects.

Manager—Licensed architect with more than ten years of experience. A Manager's responsibilities typically include management of several projects, project teams, and client contacts, as well as project scheduling and budgeting.

Associate—Senior management architect, but not an owner in the firm. This person is responsible for major departments and their functions.

Principal—Owner/partner in an architectural firm.

Education

High school and two-year college students can prepare for a degree program by taking classes in fine arts, math, science, and social science. Many two-year drafting programs offer drafting classes that can be used for credit in four- or five-year architectural programs. A student planning to transfer to a four-year program should verify with the new college which classes can be transferred.

Fine arts classes such as drawing, sketching, design, and art along with architectural history will help the future architect de-

velop an understanding of the cultural significance of structures and help transform ideas into reality. Math and science, including algebra, geometry, trigonometry, and physics, will provide a stable base for the advanced structural classes that will be required. Sociology, psychology, cultural anthropology, and classes dealing with human environments will help develop an understanding of the people who will use the structure. Because architectural students will need to read, write, and think clearly about abstract concepts, preparation should also include literature and philosophy courses.

In addition to formal study, students should discuss with local architects the opportunities and possible disadvantages that may await them in pursuing the study and practice of architecture.

Areas of Study

The study of architecture is not limited to the design of buildings. Although the architectural curriculum typically is highly structured for the first two years of study, students begin to specialize in an area of interest during the third year of the program.

Students in a bachelor's program may choose courses leading to a degree in several different areas of architecture such as urban planning, landscape architecture, and interior architecture. Urban design is the study of the relationship among the components within a city. Interior architects work specifically with the interior of a structure to ensure that all aspects of the building will be functional. Landscape architects specialize in relating the exterior of a structure to the environment. Students wishing further information on training or other related topics can contact:

American Institute of Architects
1735 New York Avenue, NW
Washington, DC 20006
202-626-7300
Web site: www.aiaonline.com

American Society of Landscape Architects
636 Eye Street NW
Washington, DC 20001-3736
202-898-2444
Web site: www.asla.org

FIGURE 1–4 ■ The structural engineering team is responsible for determining the size of materials to resist the loads and stress that a building will face. *Courtesy David Jefferis.*

ENGINEER

The term *engineer* covers a wide variety of professions. In the construction fields, structural engineers are the most common, although many jobs exist for electrical, mechanical, and civil engineers. Structural engineers typically specialize in the design of structures built of steel or concrete, similar to Figure 1–4. Many directly supervise drafters and designers in the design of multifamily and light commercial structures.

Electrical engineers work with architects and structural engineers and are responsible for the design of lighting and communication systems. They supervise the design and installation of specific lighting fixtures, telephone services, and requirements for computer networking.

Mechanical engineers are also an instrumental part of the design team. They are responsible for the sizing and layout of heating, ventilation, and air-conditioning systems (HVAC) and plan how treated air will be routed throughout the project. They work with the project architect to determine the number of occupants of the completed building and the heating and cooling load that will be generated.

Civil engineers are responsible for the design and supervision of a wide variety of construction projects, such as highways, bridges, sanitation facilities, and water treatment plants. They are often directly employed by construction companies to oversee the construction of large projects and to verify that the specifications of the design architects and engineers have been carried out.

As with the requirements for becoming an architect, a license is required to function as an engineer. The license can be applied

for after several years of practical experience, or after obtaining a bachelor's degree and three years of practical experience.

Success in any of the engineering fields requires high proficiency in math and science. As with the requirements for becoming an architect, an engineer must have five years of education at an accredited college or university, followed by successful completion of a state-administered examination. Certification can also be accomplished by training under a licensed engineer and then successfully completing the examination. Additional information can be obtained about engineering by writing to:

American Society of Civil Engineers
1015 15th Street NW, Suite 600
Washington, DC 20005
1-800-548-2723
Web site: www.asce.org

RELATED FIELDS

So far, only opportunities that are similar because they involve their drawing, design, and creativity have been covered. In addition to these careers, there are many related careers that require an understanding of drafting principles. These include model maker, illustrator, specification writer, inspector, and construction-related trades.

Model Maker

In addition to presentation drawings, many architectural offices use models of a building or project to help convey design concepts. Models such as the one shown in Figure 1–5 are often used as a public display to help gain support for large projects. Model makers need basic drafting skills to help interpret the plans required to build the actual project. Model makers may be employed within a large architectural firm or for a company that only makes models for architects.

Illustrator

Many drafters, designers, and architects have the basic skills to draw architectural renderings. Very few, though, have the expertise to make this type of drawing rapidly. Most illustrators have a background in art. By combining artistic talent with a basic understanding of architectural principles, the illustrator is able to produce drawings that show a proposed structure realistically. Figure 1–6 shows a drawing that was prepared by an architectural illustrator. Section XI provides an introduction to presentation drawings.

Specification Writer

Specifications are written instructions for methods, materials, and quality of construction. Written specifications are introduced in Section XII.

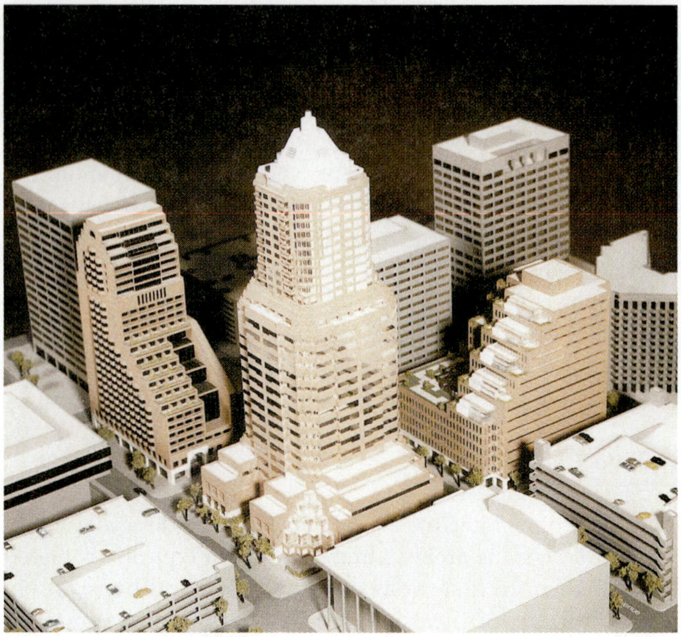

FIGURE 1–5 ■ Models such as this one of the KOIN Center in Portland, Oregon, are often used in public displays to convey design ideas. *Courtesy KOIN Center, Olympia & York Properties (Oregon) Inc.*

A specification writer must have a thorough understanding of the construction process and have a good ability to read plans. Generally, a writer will have had classes in technical writing at the two-year-college level.

Inspector

Building departments require that plans and the construction process be inspected to ensure that the required codes for public safety have been met. A plan examiner must be licensed by the state, to certify minimum understanding of the construction process. In most states, there are different levels of examiners. An experienced drafter or designer may be able to qualify as a low-level or residential-plan inspector. Generally, a degree in engineering or architecture is required to advance to an upper-level position.

The construction that results from the plans must also be inspected. Depending on the size of the building department, the plan examiner may also serve as the building inspector. In large building departments, one group inspects plans and another inspects construction. To be a construction inspector requires an exceptionally good understanding of codes limitations, print reading, and construction methods. Each of these skills has its roots in a beginning drafting class.

Jobs in Construction

Many drafters are employed directly by construction companies. The benefits of this type of position have already been discussed. These drafters typically not only do drafting but also work part time in the field. Many drafters give up their jobs for

one of the high-paying positions in the construction industry. The ease of interpreting plans as a result of a background in drafting is of great benefit to any construction worker.

DESIGN BASICS

Designing a home for a client can be an exciting but difficult process. Very rarely can an architect sit down and create a design that meets the needs of the client perfectly. The time required to design a home can range from a few days to several months. It is important for you to understand the design process and the role the drafter plays in it. This requires an understanding of basic design, financial considerations, and common procedures of design. In the rest of this chapter, the terms *architect* and *designer* can be thought of as synonymous.

Typically the designing and much of the drafting will already be done before a project is given to the junior drafter. As experience and confidence are gained, the drafter enters the design process at earlier stages.

Financial Considerations

Both designers and drafters need to be concerned with costs. Finances influence decisions made by the drafter about framing methods and other structural considerations. Often the advanced drafter must decide between methods that require more materials with less labor and those that require fewer materials but more labor. These are decisions of which the owner may never be aware, but they can make the difference in the house's being affordable or not.

The designer makes the major financial decisions that affect the cost and size of the project. The designer needs to determine the client's budget at the beginning of the design process and work to keep the project within these limits.

Through past experience and contact with builders, the designer should be able to make an accurate estimate of the cost of a finished house. This estimate is often made on a square footage basis in the initial stages of design. For instance, in some areas a modest residence can be built for approximately $65 per square foot. In other areas this same house may cost approximately $120 per square foot. A client wishing to build a 2500 sq ft house could expect to pay between $160,000 and $300,000, depending on where it will be built. Keep in mind that these are estimates. A square footage price tells very little about the home. In the design stage, square footage estimates help set parameters for the design. The estimate is based on typical cost for previous clients. The price of materials such as lumber, concrete, and roofing will vary throughout the year depending on supply and demand. Basic materials will typically account for approximately 50 percent of the project cost. Finished materials are the part of the estimate that will cause wide variation in the cost. If you have ever been in a home supply store, you know that a toilet can be purchased for between $28 and $300. Every item in the house will have a range of possible prices. The final cost of the project is determined once the house is completely drawn and a list of materials is prepared. With a list of materials, contractors are able to make accurate decisions about cost.

The source of finances can also affect the design process. Certain lending institutions may require some drawings that the local building department may not. Federal Housing Administration (FHA) and Veterans Administration (VA) loans often require extra forms, drawings, and specifications that need to be taken into account in the initial design stages.

The Client

Most houses are not designed for one specific family. In order to help keep costs down, houses are often built in subdivisions and designed to appeal to a wide variety of people. Often one basic plan can be built with several different options, thus saving the contractor the cost of paying for several different complete plans. Some families may make minor changes to an existing plan. These modified *stock plans* allow the prospective home buyer a chance to have a personalized design at a cost far below that of custom-drawn plans. See Figure 1–6. If finances allow, or if a stock plan cannot be found to meet their needs, a family can have a plan custom-designed.

FIGURE 1–6 ■ Stock plans are designed to appeal to a wide variety of buyers. *Courtesy Design Basics, Inc. Home Plan Design Service.*

The Design Process

The design of a residence can be divided into several stages. These generally include initial contact, preliminary design studies, room planning, initial working drawings, final design considerations, completion of working drawings, permit procedures, and job supervision.

Initial Contact

A client often approaches the designer to obtain background information on him or her. Design fees, schedules, and the compatibility of personalities are but a few of the basic questions to be answered. This initial contact may take place by telephone or a personal visit. The questions asked are important to both the designer and the client. The client needs to pick a designer who can work within budget and time limitations. Drawing fees are another important consideration in choosing a designer. Design fees vary based on the type of project and the range of services to be provided. Fees are generally based on hourly rates, price per square feet of construction, on a percentage of construction cost, or a combination of these methods. Square footage prices vary based on the size of the project and the area of the country, but must be set to cover design and drafting time, overhead, and profit margin. New drafters are often shocked to learn that their office bills clients at a rate that is often three to four times their hourly rate of pay for drafting time. It is important to remember, that this billing price must include supervision time, overhead, state and federal taxes and hopefully a profit for the office. Design fees for architectural services of a custom house range from five to fifteen percent of the total construction cost. The amount varies based on the services provided, the work load of the office, and the local economy. The designer needs to screen clients to determine if the client's needs fit within the office schedule.

Once an agreement has been reached, the preliminary design work can begin. This selection process usually begins with the signing of a contract to set guidelines that identify which services are to be provided and when payment is expected. This is also the time when the initial criteria for the project will be determined. Generally, clients have a basic size and a list of specifics in mind, a sketch of proposed floor plans, and a file full of pictures of items that they would like in their house. During this initial phase of design it is important to become familiar with the lifestyle of the client as well as the site where the house will be built.

Preliminary Design Studies

Once a thorough understanding of the client's lifestyle, design criteria, and financial limits has been developed, the preliminary studies can begin. These include research with the building and zoning departments that govern the site, investigation of the site, and discussions with any board of review that may be required. Once this initial research has been done, preliminary design studies can be started. The preliminary drawings usually take two stages: bubble drawings and scaled sketches.

Bubble drawings are freehand sketches used to help determine room locations and relationships. See Figure 1–7. It is during this stage of the preliminary design that consideration is given to the site and energy efficiency of the home.

Once a satisfactory layout has been sketched, these shapes are transformed into scaled sketches. Figure 1–8 shows a preliminary floor plan. Usually several sets of sketches are developed to explore different design possibilities. Consideration is given to building code regulations and room relationships and sizes. After the design options are explored, the designer selects a plan to prepare for the client. This could be the point when a drafter first becomes involved in the design process. Depending on the schedule of the designer, a senior drafter might prepare the refined preliminary drawings, which can be seen in Figure 1–9.

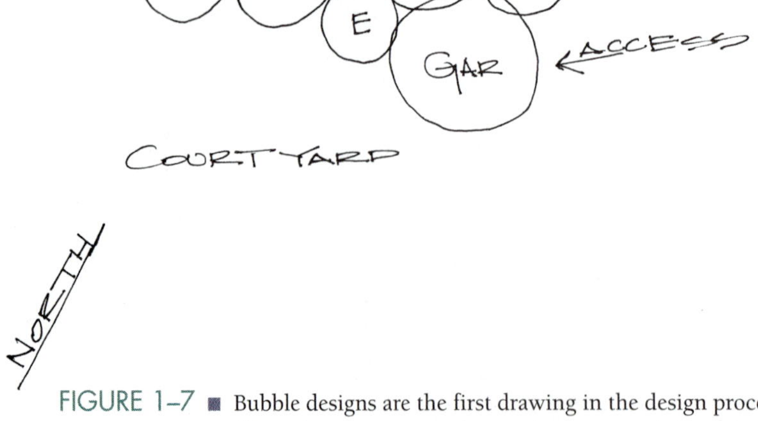

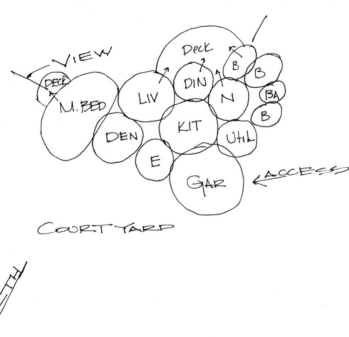

FIGURE 1–7 ■ Bubble designs are the first drawing in the design process. These drawings are used to explore room relationships.

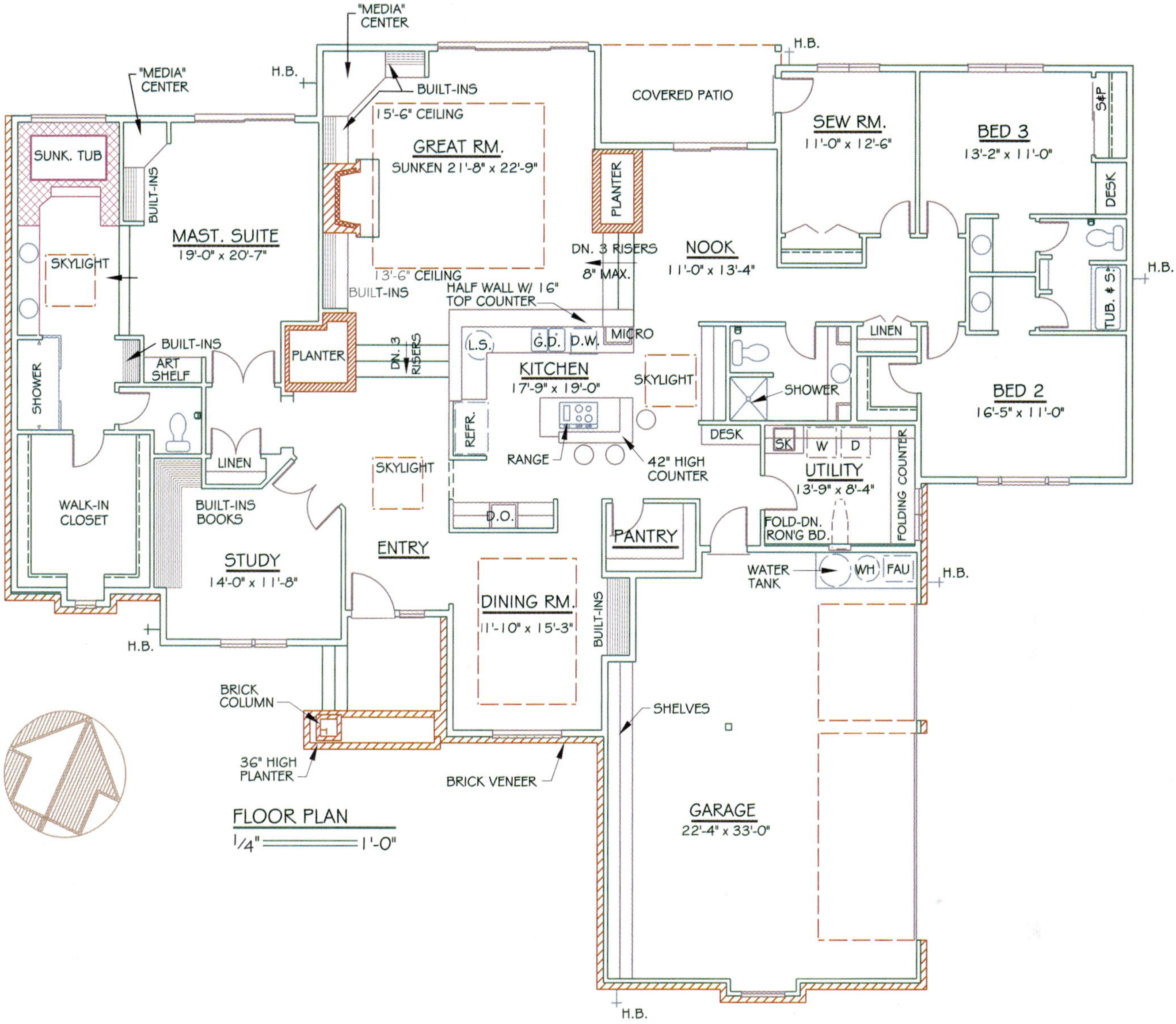

FIGURE 1–8 ■ Bubble drawings are converted to scale drawings so that basic sizes can be determined.

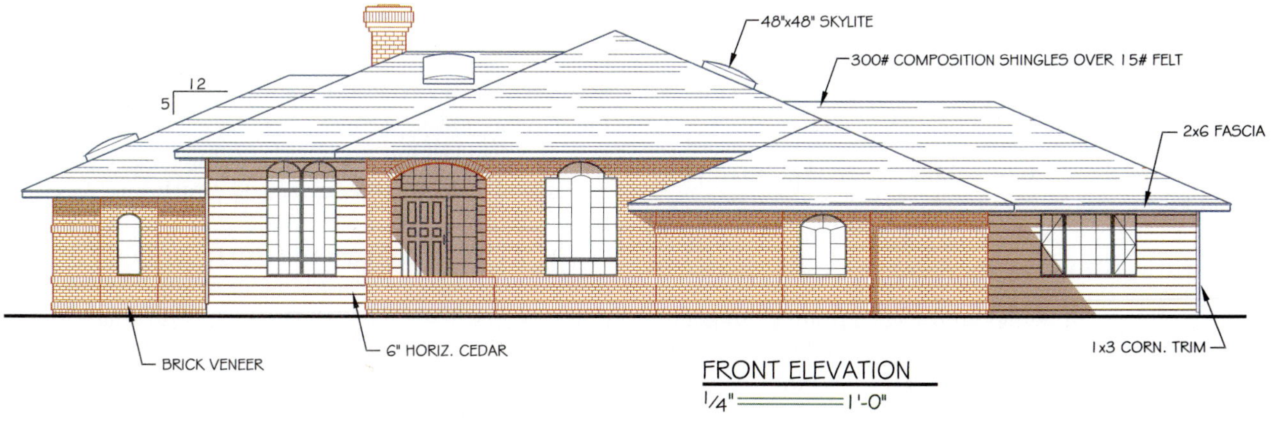

FIGURE 1–9 ■ Based on the preliminary floor plans, the front elevation is drawn to explore design options.

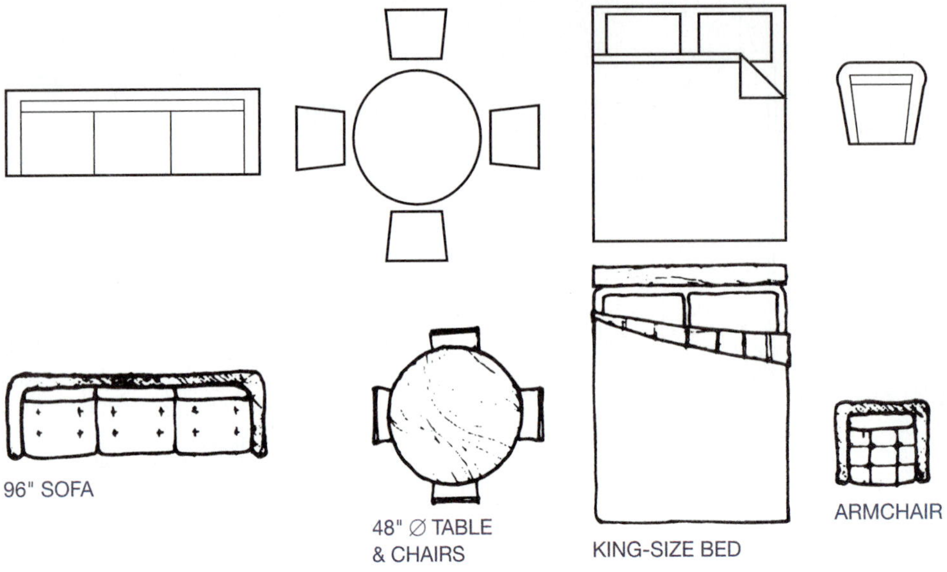

FIGURE 1–10 ■ Furniture is often drawn on preliminary floor plans to show how the space within a room can be utilized.

These would include floor plans and an elevation. These drawings are then presented to the client. Changes and revisions are made at this time. Once the plans are approved by the client, the preliminary drawings are ready to be converted to design drawings.

Room Planning Room usage must be considered throughout the design process. Books such as *Architectural Graphic Standards* and vendor catalogs are used by many professionals to verify standard sizes. Chapter 8 will introduce major design concepts to be considered. Typically the designer will have talked with the owners about how each room will be used and what types of furniture will be included. Occasionally, placement of a family heirloom will dictate the entire layout. When this is the case, the owner should specify the size requirements for a particular piece of furniture.

When placing furniture outlines to determine the amount of space in a room, the drafter can use transfer shapes known as rub-ons, templates, freehand sketching methods, or blocks if working with a CAD program. Rub-ons provide a fast method of representing furniture, but they can also prove to be expensive. Furniture can be drawn with a template or sketched. Figure 1–10 shows examples of the outlines of common pieces of furniture. Figure 1–11 shows a preliminary floor plan with the furniture added.

Initial Working Drawings

With the preliminary drawings approved, a drafter can begin to lay out the working drawings. The procedure will vary with each office, but generally each of the drawings required for the project will be started. These would include the foundation, site, roof, electrical, cabinet, and framing plans. At this stage the drafter must rely on past experience for drawing the size of beams and other structural members. Beams and other structural material will be located, but exact sizes are usually not determined until the entire project has been laid out.

Final Design Considerations

Once the drawings have been laid out, the designer will generally meet with the client several times again to get information on flooring, electrical needs, cabinets, and other finish materials. This conference will result in a set of marked drawings that the drafter will use to complete the working drawings.

Completion of Working Drawings

The complexity of the residence will determine which drawings are required. Most building departments require a site plan, a floor plan, a foundation plans, elevations, and one cross section as the minimum drawing to get a building permit. On a complicated plan, a wall framing plan, a roof framing plan, a grading plan, and construction details may be required. Depending on the lending institution, interior elevations, cabinet drawings, and finish specifications may be required.

The skills of the drafter will determine his or her participation in preparing the working drawings. As an entry-level drafter, you will often be given the job of making corrections on existing drawings or drawing, for example, site plans or cabinets. As you gain skill you will be given more drawing responsibility. With increased ability, you will start to share in the design responsibilities.

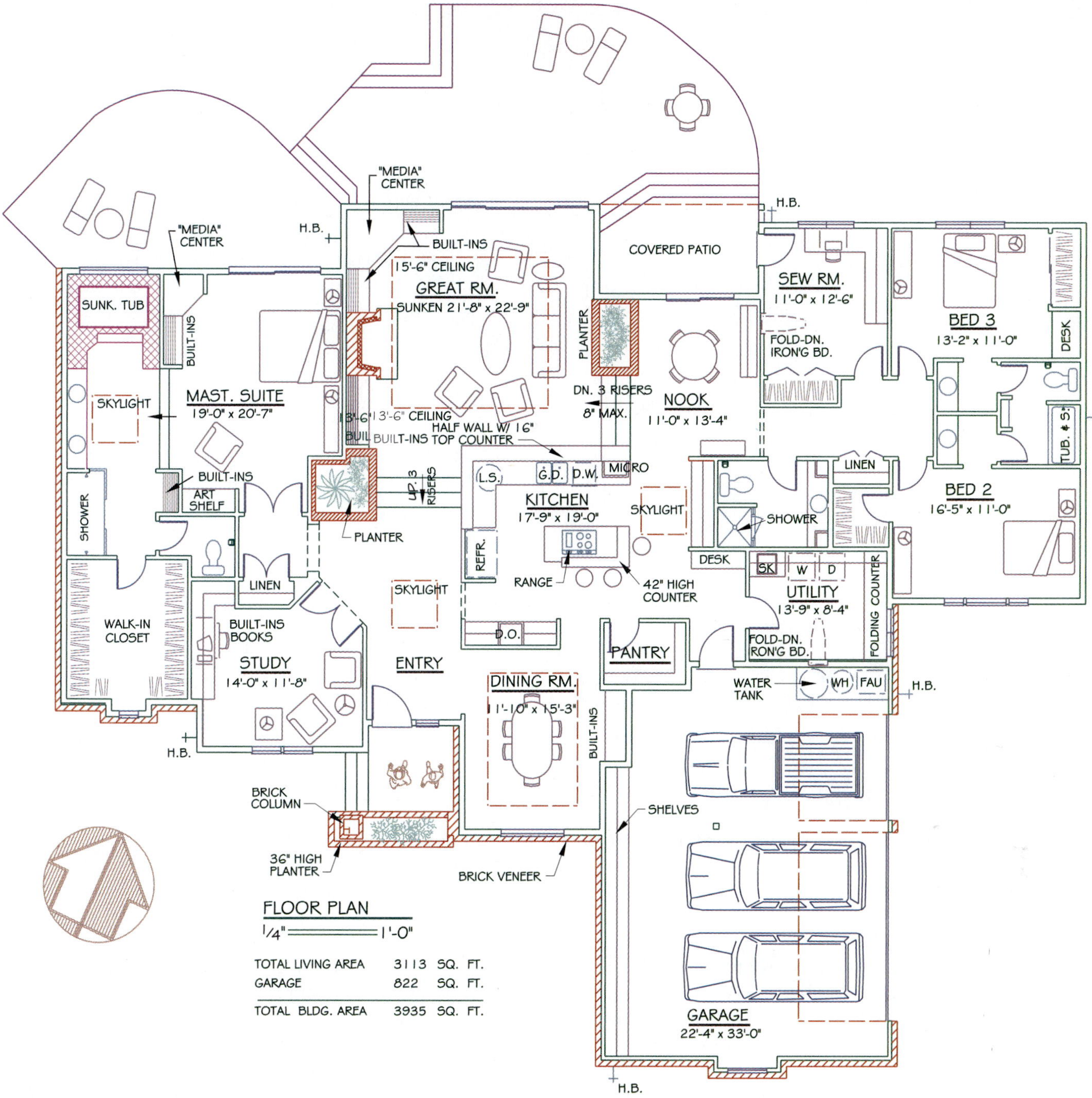

FIGURE 1–11 ■ The final floor plan shows the owner's changes, with furniture added.

Working Drawings. The drawings that will be provided vary for each residence and within each office. Figures 1–12 through 1–17 show what are typically considered the *architectural drawings.* Figures 1–18 through 1–20 show the *electrical drawings.* These are drawings that show finishing materials. Figures 1–21 through 1–24 show what are typically called the *structural drawings.* Many architectural firms number each page based on the type of drawing on that page. Pages would be represented by:

A—Architectural

S—Structural

M—Mechanical

E—Electrical

P—Plumbing

Figure 1–12 shows the site plan that was required for the residence in Figure 1–8. Typically the site plan is completed by

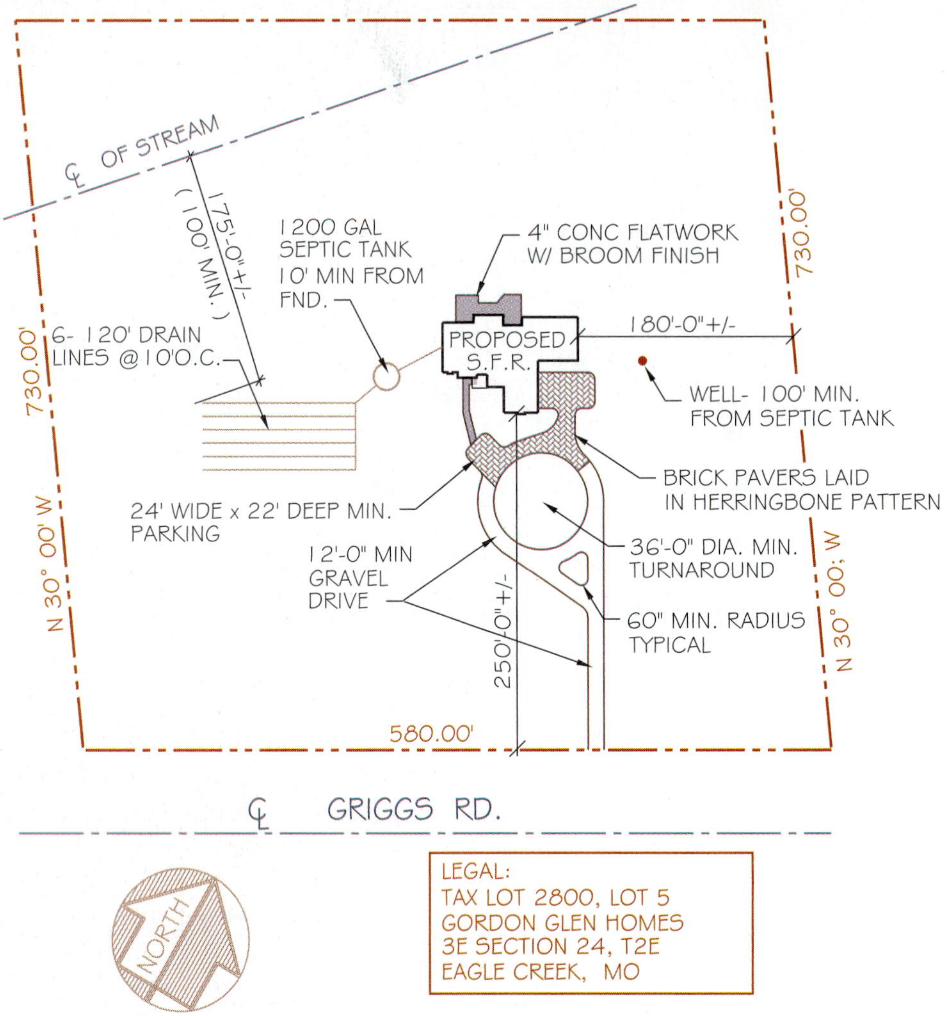

FIGURE 1–12 ■ A site plan is used to show how the structure relates to the site. When the site is relatively flat, grading information is placed on this drawing.

a junior drafter from a sketch provided by the senior drafter or designer. A grading plan is not required for every residence. Because this home is to be built on a fairly level site, the elevations are indicated at each corner of the structure. Figure 1–13 shows a site plan for a hillside home. Because of the large amount of soil to be excavated for the lower floor, the grading plan shown in Figure 1–14 was provided. The junior drafter may draw the base drawing showing the structure and site plan and the designer or senior drafter will usually complete the grading information. Typically the owner is responsible for hiring a surveyor, who will provide the topography and site map, although the designer may coordinate the work between the two offices. Section III will provide insight into how these two drawings are developed.

Using the preliminary drawing that was presented in Figure 1–9, the working elevations can be completed. Depending on the complexity of the structure, they may be completed by either the junior or the senior drafter. Figure 1–15 shows the working elevations for this structure. Section VII will provide information needed to complete working elevations, and Section XI will introduce presentation drawings.

Figures 1–16 and 1–17 show the completed plan views of this structure. Figure 1–16 shows the view from flying over the structure. The roof plan is often completed by the junior drafter using sketches provided by the senior drafter. Section VI will provide information on roof plans. The plan view represented in Figure 1–17*a* is usually completed by the senior drafter. These drawings provide information related to the room arrangements and interior finishes. Junior drafters also work on the plan views by making corrections or adding notes, which are typically placed on a marked-up set of plans provided by the designer or senior drafter. Section V will provide information for completing floor plans.

Figure 1–18 shows the electrical plan. Depending on the complexity of the residence, electrical information may be placed directly on the floor plan. A junior drafter typically completes the electrical drawings by working from marked-up prints provided by the designer. Chapter 17 will provide information for completing the electrical plans.

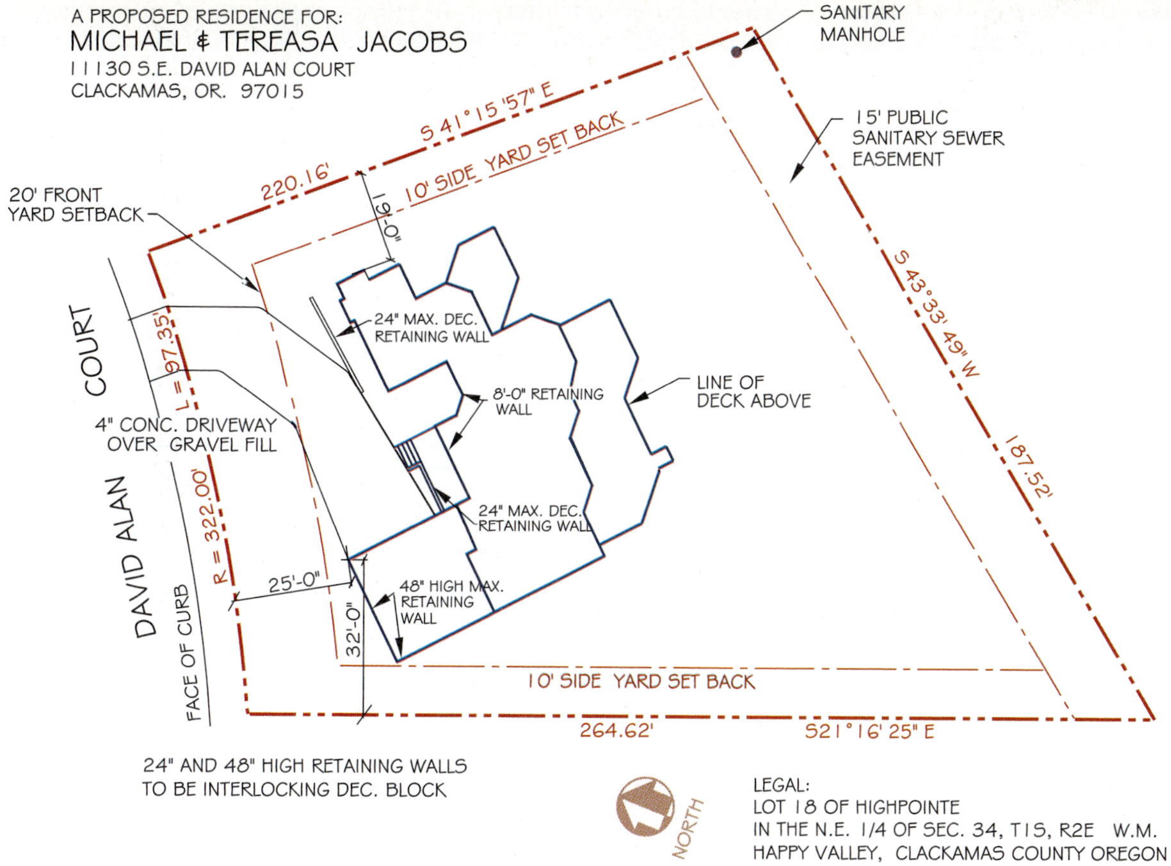

FIGURE 1–13 ■ If grading is extensive, the site plan is separated from the grading plan.

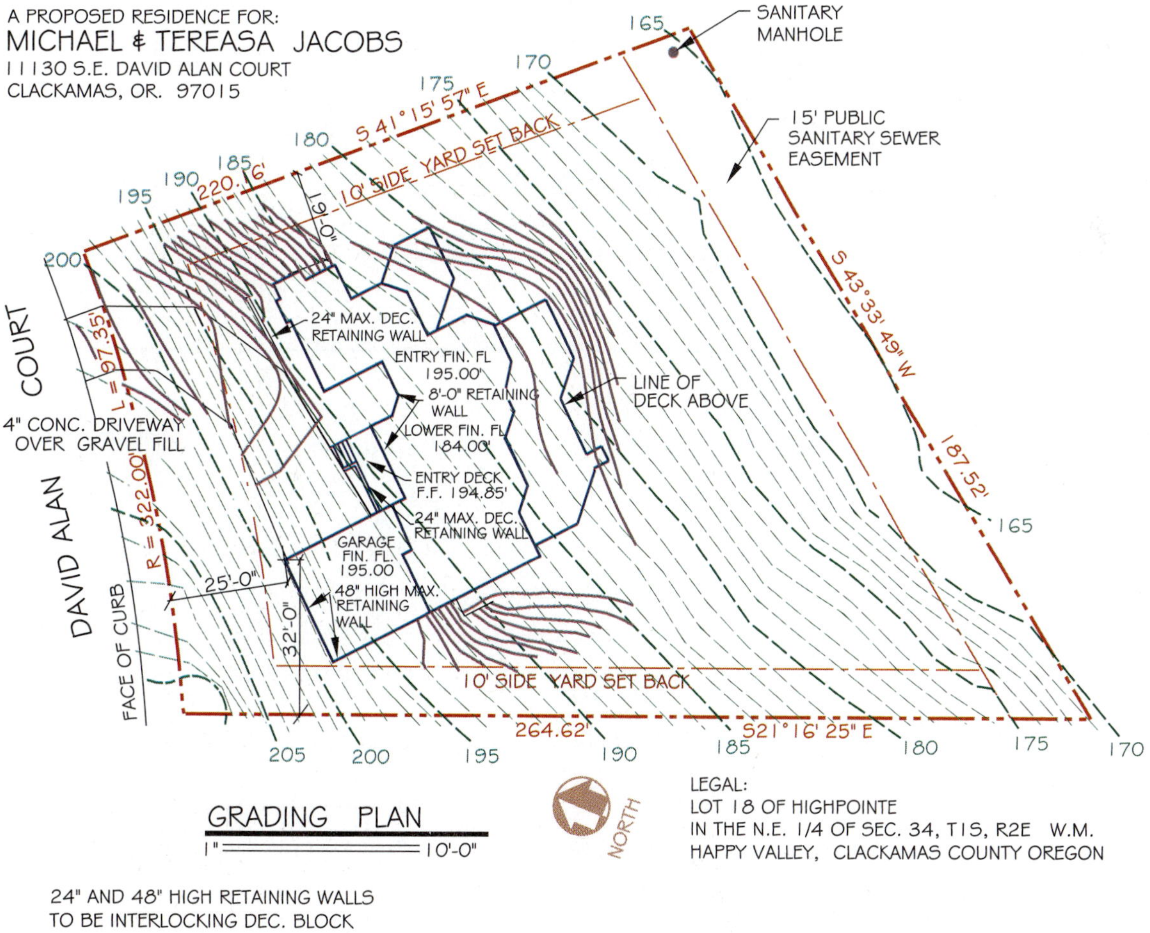

FIGURE 1–14 ■ A grading plan is used to define new and existing grades and cut and fill banks that will be created as soil is relocated.

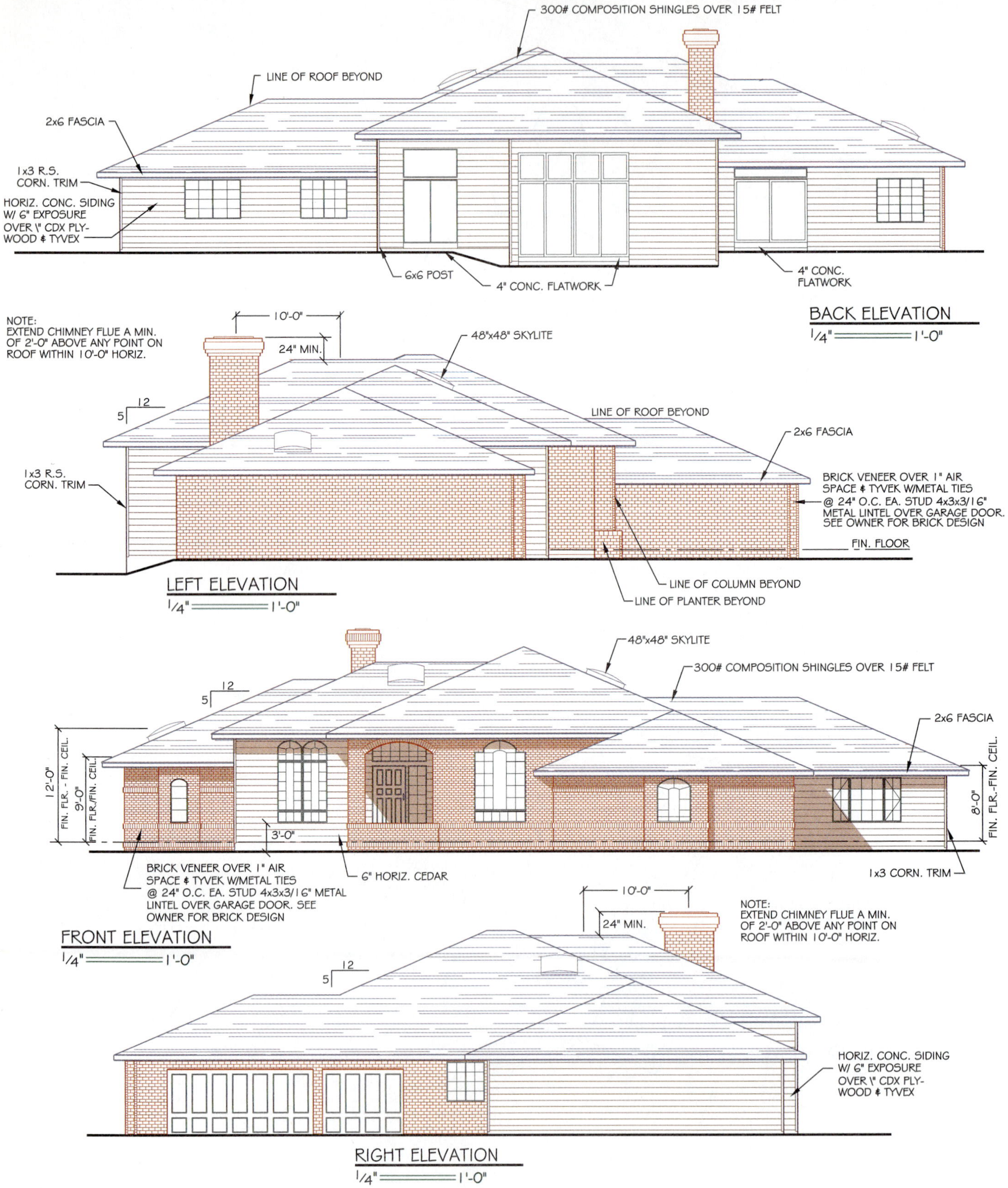

FIGURE 1–15 ■ The working elevations provide a view of each side of the structure and show the shape as well as the exterior materials.

HIP SCHEDULE

A	2x12
B	2-2x12
C	5 1/8" x 10 1/2" GLULAM
D	2-2x10
E	3 1/8" x 12" GLULAM
F	5 1/8 x 13 1/2 GLULAM
G	3 1/2" x 11 7/8" MICROLLAM

ROOF PLAN

1/4" = 1'-0"

ROOF FRAMING NOTES:

1. ALL HIPS, RIDGES & VALLEYS TO BE 2x10 UNLESS NOTED.
2. ALL RAFTERS TO BE 2x8 DFL #2 @ 24" O.C. UNLESS NOTED.

1. USE 1/2 "CCX PLY @ ALL EXPOSED EAVES.
2. ROOF VENTS MUST HAVE AN AREA EQUAL TO 1/150 OF THE ATTIC AREA.
3. SKYLITES TO BE DOUBLE DOMED PLASTIC BY VELEX OR EQUAL.
4. ALL ROOF PITCH ARE 5/12 UNLESS NOTED.
5. 300# COMPOSITION SHINGLES OVER 15# FELT.

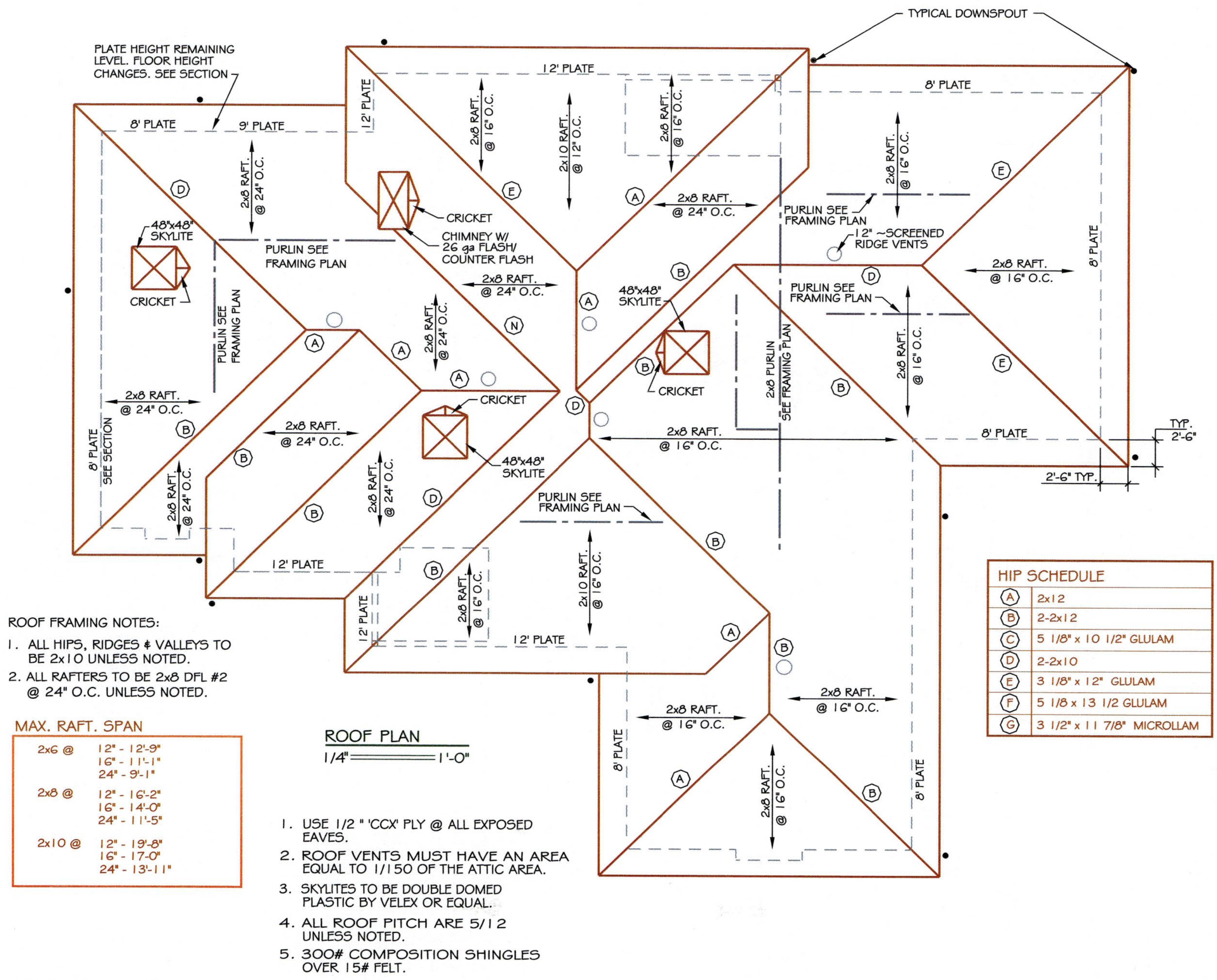

MAX. RAFT. SPAN

2x6 @	12" - 12'-9"
	16" - 11'-1"
	24" - 9'-1"
2x8 @	12" - 16'-2"
	16" - 14'-0"
	24" - 11'-5"
2x10 @	12" - 19'-8"
	16" - 17'-0"
	24" - 13'-1"

FIGURE 1–16 ■ The roof plan shows the shape of the roof structure as well as vents and drains.

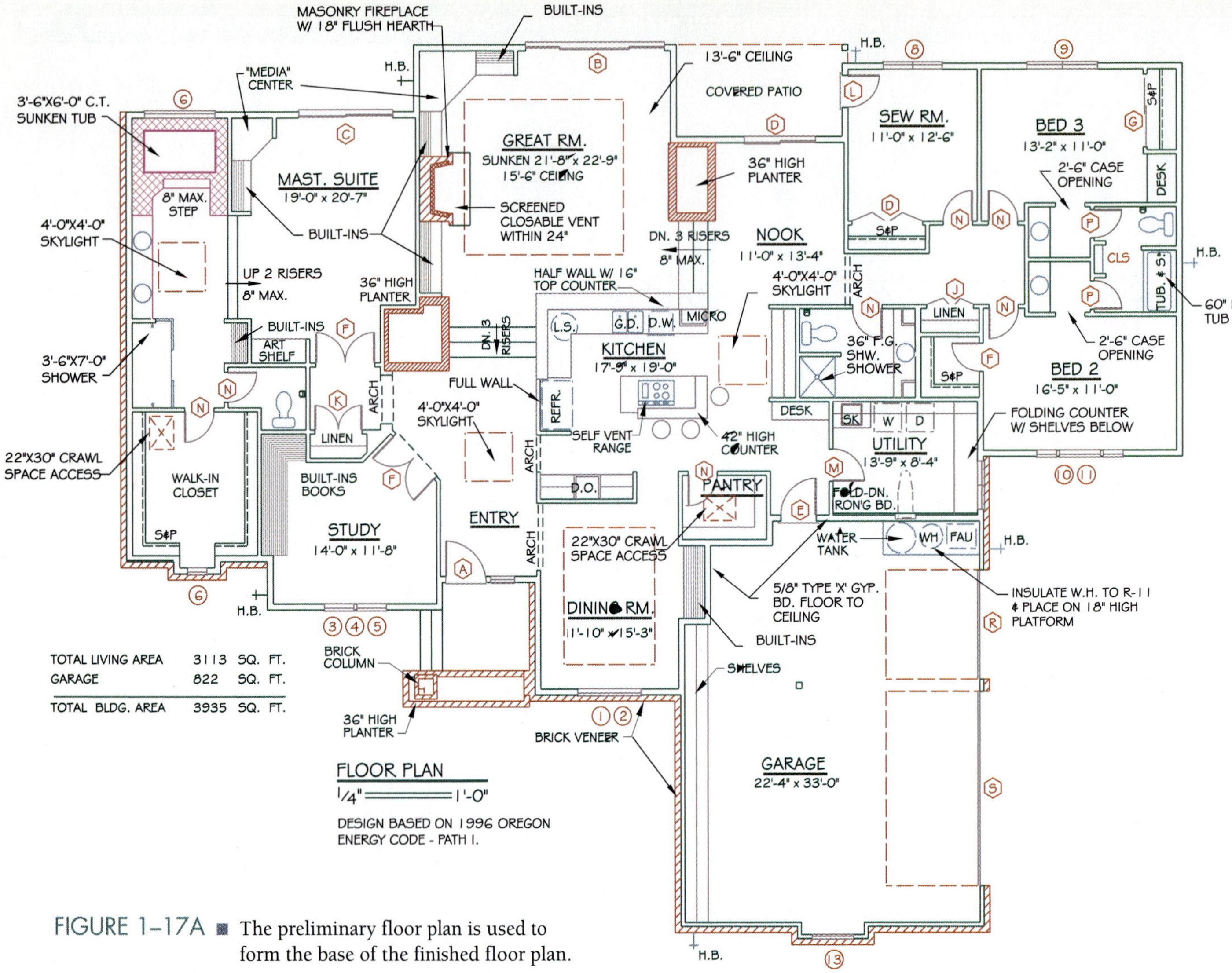

FIGURE 1–17A ■ The preliminary floor plan is used to form the base of the finished floor plan.

DOOR SCHEDULE

SYM.	SIZE	TYPE	QUAN.
A	3'-6" x 8'-0"	S.C.,R.P., W/ 24" SIDELIGHT W/ SQ. TRANSOM ABOVE	1
B	12'-0" x 8'-0"	SL. GLASS W/ 12'-0" x 3'-0" SQ. TRANSOM ABOVE	1
C	8'-0" x 6'-8"	SL. FRENCH DRS. W/ 12" SQ. TRANSOM ABOVE	1
D	6'-0" x 6'-8"	SL. GLASS DRS. W/6'-0" x 3'-0" SQ. TRANSOM ABOVE	1
E	3'-0" x 6'-8"	SELF CLOSING, M.I.	1
F	PR 2'-6" X 6'-8"	HOLLOW CORE	2
G	6'-0" x 6'-8"	SL. MIRROR	1
H	6'-0" x 6'-8"	BIFOLD	1
J	4'-0" x 6'-8"	BIFOLD	1
K	PR 2'-0" x 6'-8"	HOLLOW CORE	1
L	2'-8" X 6'-8"	S.C., M.I.	1
M	2'-8" x 6'-8"	HOLLOW CORE	1
N	2'-6" x 6'-8"	HOLLOW CORE	7
P	2'-4" x 6'-8"	HOLLOW CORE	3
Q	16'-0" x 8'-0"	OVERHEAD, GARAGE	1
R	9'-0" x 8'-0"	OVERHEAD, GARAGE	1

DOOR NOTES:

1. FRONT DOOR TO BE RATED AT 0.54 OR LESS, VERIFY DOOR STYLE WITH OWNER.
2. EXTERIOR DOORS IN HEATED WALLS TO BE U 0.20 OR LESS.
3. DOORS THAT EXCEED 50% GLASS ARE TO BE U 0.40 OR LESS.
4. ALL GLASS WITHIN 18" OF DOORS TO BE TEMPERED.

WINDOW SCHEDULE

SYM.	SIZE	TYPE	QUAN.
1	5'-6" x 7-'6"	ARCHED TOP	1
2	5'-6" x 1-0"	AWNING	1
3	2'-6" x 5'-0"	PICTURE	2
4	2'-6" x 1'-0"	AWNING	2
5	2'-6" x 2'-6"	ARCHED TOP	2
6	1'-8" x 4'-6"	ARCHED TOP	1
7	6'-0" x 4'-6"	HOR. SLIDING	1
8	5'-0" x 4'-0"	HOR. SLIDING	1
9	6'-0" x 4'-0"	HOR. SLIDING	1
10	1'-8" x 4'-6"	CASEMENT	2
11	3'-6" x 4'-6"	PICTURE	1
12	4'-0" x 3'-0"	PICTURE	1
13	3'-6" x 4'-6"	ARCHED TOP	1

WINDOW NOTES:

1. ALL WINDOWS TO BE VINYL FRAME, MILGARD OR BETTER
2. ALL WINDOWS TO BE U-.40 MIN.
3. ALL BEDROOM WINDOWS TO BE WITH IN 44" OF FIN. FLOOR
4. GLASS WITHIN 18" OF DOORS TO BE TEMPERED.
5. ALL WINDOW HDRS. FOR 8' & 9' CEIL. HEIGHTS TO BE SET AT 6'-8". ALL HDRS FOR 12' CEIL. TO BE SET AT 10'-8", UNLESS OTHERWISE NOTED.
6. VERIFY HEIGHT OF WINDOW #7 WITH TUB HEIGHT

GENERAL NOTES:

1. STRUCTURAL DESIGNS ARE BASED ON 2000 I.R.C AND 1996 OREGON ENERGY CODE PATH 1.
2. ALL PENETRATIONS IN TOP OR BOTTOM PLATES FOR PLUMBING OR ELECTRICAL RUNS TO BE SEALED. SEE ELECTRICAL PLANS FOR ADDITIONAL SPECIFICATIONS.
3. PROVIDE 1/2" WATERPROOF GYP. BD. AROUND ALL TUBS, SHOWERS, AND SPAS.
4. VENT DRYER AND ALL FANS TO OUTSIDE AIR THRU DAMPERED VENTS.
5. INSULATE WATER HEATER TO R-11. PROVIDE 18" HIGH PLATFORM FOR GAS APPLIANCES.
6. PROVIDE 1/4" COLD WATER LINE TO REFRIGERATOR.
7. BRICK VENEER TO BE PLACED OVER 1" AIR SPACE, AND TYVEK W/ METAL TIES @ 24" O.C. EA. STUD.

FIGURE 1–17B ■ Notes and schedules are typically displayed near the floor plan.

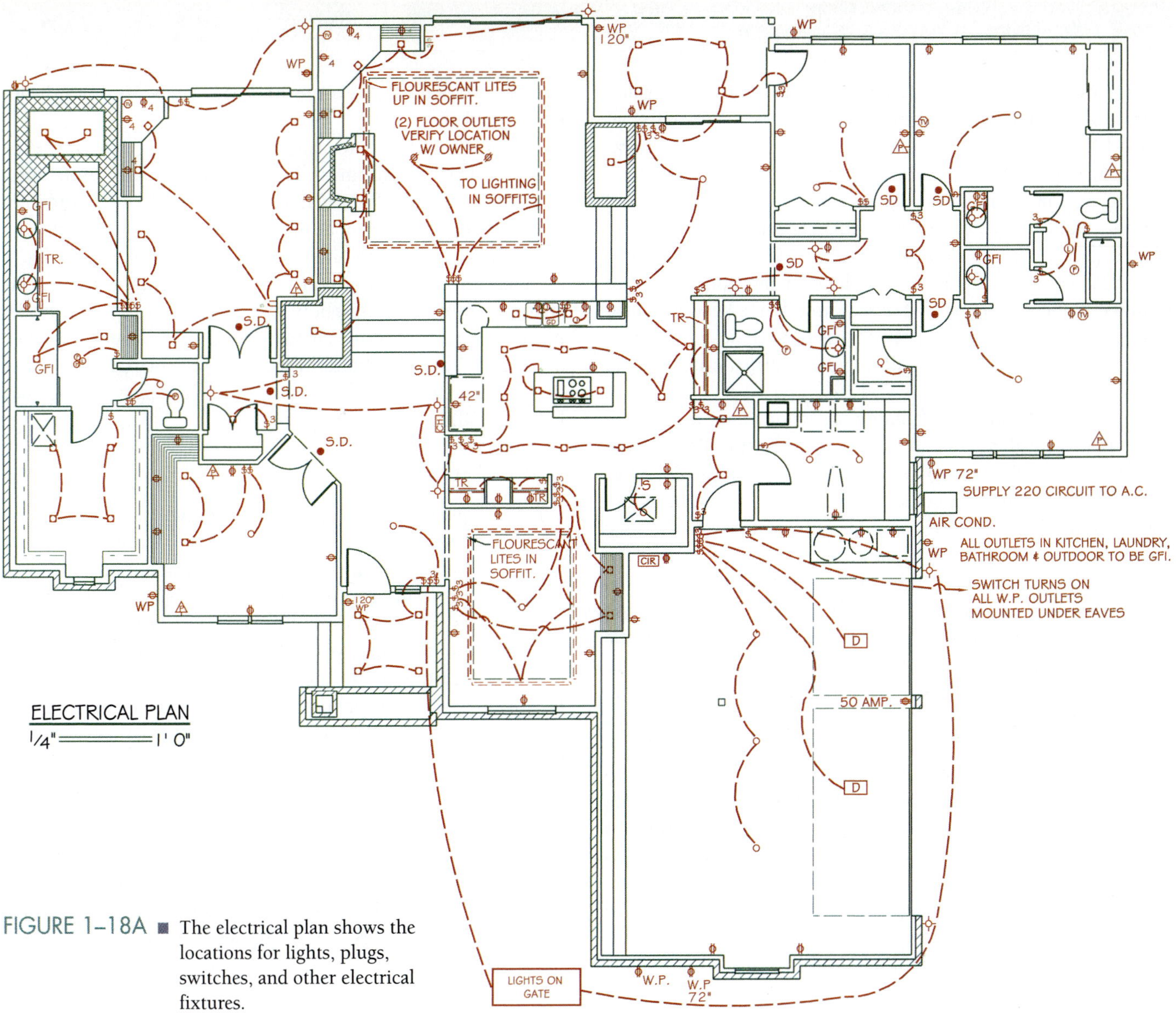

FIGURE 1–18A ■ The electrical plan shows the locations for lights, plugs, switches, and other electrical fixtures.

ELECTRICAL NOTES:

1. ALL GARAGE AND EXTERIOR PLUGS & LIGHT FIXTURES TO BE ON GFCI CIRCUIT.
2. ALL KITCHEN PLUGS AND LIGHT FIXTURES TO BE ON GFCI CIRCUIT.
3. PROVIDE A SEPARATE CIRCUIT FOR MICROWAVE OVEN.
4. PROVIDE A SEPARATE CIRCUIT FOR PERSONAL COMPUTER. VERIFY LOCATION WITH OWNER.
5. VERIFY ALL ELECTRICAL LOCATIONS W/ OWNER.
6. EXTERIOR SPOTLIGHTS TO BE ON PHOTO-ELECTRIC CELL W/TIMER.
7. ALL RECESSED LIGHTS IN EXT. CEIL. TO BE INSULATION COVER RATED.
8. ELECTRICAL OUTLET PLATE GASKETS SHALL BE INSTALLED ON RECEPTACLE, SWITCH, AND ANY OTHER BOXES IN EXTERIOR WALL.
9. PROVIDE THERMOSTATICALLY CONTROLLED FAN IN ATTIC W/ MANUAL OVERRIDE (VERIFY LOCATION W/ OWNER).
10. ALL FANS TO VENT TO OUTSIDE AIR. ALL FAN DUCTS TO HAVE AUTOMATIC DAMPERS.
11. HOT WATER TANKS TO BE INSULATED TO R-11 MINIMUM.
12. INSULATE ALL HOT WATER LINES TO R-4 MINIMUM. PROVIDE AN ALTERNATE BID TO INSULATE ALL PIPES FOR NOISE CONTROL.
13. PROVIDE 6 SQ. FT. OF VENT FOR COMBUSTION AIR TO OUTSIDE AIR FOR FIREPLACE CONNECTED DIRECTLY TO FIREBOX. PROVIDE FULLY CLOSEABLE AIR INLET.
14. HEATING TO BE ELECTRIC HEAT PUMP. PROVIDE BID FOR SINGLE UNIT NEAR GARAGE OR FOR A UNIT EACH FLOOR (IN ATTIC).
15. INSULATE ALL HEATING DUCTS IN UNHEATED AREAS TO R-11. ALL HVAC DUCTS TO BE SEALED AT JOINTS AND CORNERS.

ELECTRICAL LEGEND

Symbol	Description	Symbol	Description
	110 CONVENIENCE OUTLET		CEILING MOUNTED LITE FIXTURE
GFI	110 C.O. GROUND FAULT INTERRUPTER		CAN CEILING LITE FIXTURE
WP	110 WATER PROOF		WALL MOUNTED LITE
	110 HALF HOT		RECESSED LITE FIXTURE
	JUNCTION BOX	P.C	LITE ON PULL CHORD
	220 OUTLET		SPOT LITES
$	SINGLE POLE SWITCH		48" SURFACE MOUNTED FLOURESCANT LITE FIXTURE
$3	THREE-WAY SWITCH	S	STEREO SPEAKER
L H F	LITE, HEATER, & FAN		PHONE OUTLET
S.D.	SMOKE DETECTOR	TV	CABLE T.V. OUTLET
V	VACUUME		

FIGURE 1–18B ■ Notes and legends are typically displayed near an electrical plan.

Figures 1–19 through 1–21 represent the framing plans and notes associated with these drawings. The notes shown with Figure 1–19 are standard notes that specify the nailing for all framing connections. Depending on the complexity of the structure, framing information may be placed on the floor plan. Because of the information to be shown regarding seismic and wind problems, separate framing plans have been provided. Typically these drawings would be completed by the designer and the senior drafters. Notice in Figure 1–21*b* that notes have been provided to specify the minimum code standards used to design the residence. Sections VI and VIII will provide information needed to complete the framing plans.

Figures 1–22 and 1–23 contain the section views required to show the vertical relationships of materials specified on the framing plans. Office practice varies greatly on what drawings will be provided and the scale and detail that will be shown in the sections. Section X will provide information for drawing sections. Sections are typically drawn by the designer and the senior drafter and completed by the junior drafters.

Figure 1–24 shows what is typically the final drawing in a set of residential plans, although the foundation plan is one of the first drawings used at the construction site. Because of the importance of this drawing, it is typically completed by the designer or the senior drafter, although junior drafters may work on corrections or add notes to the drawing.

As you progress through this text, you will be exposed to each type of working drawing. It is important to understand that a drafter would rarely draw one complete drawing and then go on to another drawing. Because the drawings in a set of plans are so interrelated, often one drawing is started, and then another drawing is started so that relationships between the two can be studied before the first drawing is completed. For instance, floor plans may be laid out, and then a section may be drawn to work out the relationships between floors or any headroom problems.

When the plans are completely drawn, they must be checked. Dimensions must be checked carefully and cross-referenced from one floor plan to another and to the foundation plan. Bearing points from beams must be followed from the roof down through the foundation system. Perhaps one of the hardest jobs for a new drafter is coordinating the drawings. As changes are made throughout the design process, the drafter must be sure that those changes are reflected on all affected drawings. When the drafter has completed checking the plans, the drafting supervisor will again review the plans before they leave the office.

Permit Procedures

When the plans are complete, the owner will ask several contractors to estimate the cost of construction. Once a contractor is selected to build the house, a construction permit is obtained. Although the drafter is sometimes responsible for obtaining the permits, this is usually done by the owner or contractor. The process for obtaining a permit varies depending on the local building department and the complexity of the drawings. Permit reviews generally take several weeks. Once the review is complete, either a permit is issued or required changes are made to the plans.

Job Supervision

In residential construction, job supervision is rarely done when working for a designer, but it is quite often provided when an architect has drawn the plans. Occasionally a problem at the job site will require the designer or drafter to go to the site to help find a solution.

CAULKING NOTES:

CAULKING REQUIREMENTS BASED ON 1992 OREGON RESIDENTIAL ENERGY CODE

1. SEAL THE EXTERIOR SHEATHING @ CORNERS, JOINTS, DOOR AND WINDOW, AND FOUNDATION SILLS W/ SILICONE CAULKING.

2. CAULK THE FOLLOWING OPENINGS W/ EXPANDED FOAM OR BACKER RODS. POLYURETHANE, ELASTOMERIC COPOLYMER, SILCONIZED ACRYLIC LAYTEX CAULKS MAY ALSO BE USED WHERE APPROPRIATE.

ANY SPACE BETWEEN WINDOW AND DOOR FRAMES.

BETWEEN ALL EXTERIOR WALL SOLE PLATES AND PLY SHEATHING.

ON TOP OF RIM JOIST PRIOR TO PLYWOOD FLOOR APPLICATION

WALL SHEATHING TO TOP PLATE.

JOINTS BETWEEN WALL AND FOUNDATION

JOINTS BETWEEN WALL AND ROOF

JOINTS BETWEEN WALL PANELS

AROUND OPENINGS FOR DUCTS, PLUMBING, ELECTRICAL, TELEPHONE

AND GAS LINES IN CEILINGS, WALLS AND FLOORS. ALL VOIDS AROUND

PIPING RUNNING THROUGH FRAMING OR SHEATHING TO BE PACKED

ALTERNATE ATTACHMENTS

TABLE NO R 402.3A (1)

NOMINAL THICKNESS	DESCRIPTION (1,2) OF FASTENERS & LENGTH	SPACING OF FASTENERS	
		EDGES	INTERMEDIATE SUPPORTS
5/16"	.097-.009 NAIL 1\ STAPLE 15 GA. 1 3/8"	6"	12"
3/8"	STAPLE 15 GA. 1 3/8"	6	12
	.097- .099 NAIL 1 \	4	10
15/32" & 1/2"	STAPLE 15 GA. 1 1/2"	6	12"
	.097-.099 NAIL 1 5/8	3	6"
19/32" & 5/8"	.113 NAIL 1]" STAPLE 15 & 16 GA. 1}	6	12"
	.097-.099 NAIL 1 7/8"	3"	6"
23/32" & 3/4"	STAPLE 14 GA. 1 3/4"	6	12
	STAPLE 15 GA. 1 3/4"	5	10
	.097-.099 NAIL 2 7/8"	3	6
1"	STAPLE 14 GA. 2"	6	10"
	.113 NAIL 2 1/4" STAPLE 15 GA. 2"	4"	8"
	.097-.099 NAIL 2 1/8"	3"	6"

FLOOR UNDERLAYMENT:
PLYWOOD, HARDBOARD, PARTICLEBOARD

l" & 5/16"	.097-.099 NAIL 1 1/2 STAPLE 15 & 16 GA. 1 1/4"	6"	12"
	.080 NAIL 1 1/4"	5"	10"
	STAPLE 18 GA. 3/16 CROWN 7/8"		63"
3/8"	STAPLE 15 & 16 GA 1 3/8" .097-.099 NAIL 1 1/2"	6"	12"
	.080 NAIL 1 3/8"	5"	10"
1/2"	.113 NAIL 1 7/8" STAPLE 15 & 16 GA. 1 1/2"	6"	12"
	.097-.099 NAIL 1 3/4"	5"	6"

1. NAIL IS A GENERAL DESCRIPTION AND MAY BE T-HEAD, MODIFIED ROUND HEAD, OR ROUND HEAD.

2. STAPLES SHALL HAVE A MINIMUM CROWN WIDTH OF 7/16" O.D. EXCEPT AS NOTED.

3. NAILS OR STAPLES SHELL BE SPACED AT NOT MORE THAN 6" O.C. AT ALL SUPPORTS WHERE SPANS ARE 48" OR GREATER. NAILS OR STAPLES SHALL BE SPACED AT NOT MORE THAN 10" O.C. AT INTERMEDIATE SUPPORTS FOR FLOORS.

FASTENER SCHEDULE

DESCRIPTION OF BUILDING MATERIAL	NUMBER & TYPE OF FASTENERS (1,2,3,5,)
1 JOIST TO SILL OR GIRDER, TOE NAIL	3-8d
2 BRIDGING TO JOIST, TOENAIL EA. END	2-8d
3 1 x 6 (25 X 150) SUBFLOOR OR LESS TO EACH JOIST, FACE NAIL	2-8d
4 WIDER THAN 1 x 6 (25 x 150) SUBFLOOR TO EACH JOIST, FACE NAIL	3-8d
5 2" (50) SUBFLOOR TO JOIST OR GIRDER BLIND AND FACE NAIL	2-16d
6 SOLE PLATE TO JOIST OR BLOCKING FACE NAIL	16d @ 16" (406mm) O.C.
SOLE PLATE TO JOIST OR BLOCKING AT BRACED WALL PANELS	3- 16d PER 16" (406mm) O.C.
7 TOP OR SOLE PLATE TO STUD, END NAIL	2-16d
8 STUD TO SOLE PLATE, TOE NAIL	4-8d, TOENAIL OR 2-16d END NAIL
9 DOUBLE STUDS, FACE NAIL	16d @ 24" (610mm) O.C.
10 DOUBLE TOP PLATE, FACE NAIL	16d @ 16" (406mm) O.C.
DOUBLE TOP PLATE, LAP SPLICE	8-16d
11 BLOCKING BTWN. JOIST OR RAFTERS TO TOP PLATE, TOENAIL	3- 8d
12 RIM JOIST TO TOP PLATE, TOENAIL	8d @ 6" (152mm) O.C.
13 TOP PLATES, TAPS & INTERSEC-TIONS, FACE NAIL	2-16d
14 CONTINUED HEADER, TWO PIECES	16d @ 16" (406mm) O.C. ALONG EACH EDGE
15 CEILING JOIST TO PLATE, TOE NAIL	3-8d
16 CONTINUOUS HEADER TO STUD, TOE NAIL	4-8d
17 CEILING JOIST, LAPS OVER PARTITIONS, FACE NAIL	3-16d
18 CEILING JOIST TO PARALLEL RAFTERS, FACE NAIL	3-16d
19 RAFTERS TO PLATE, TOE NAIL	3-8d
20 1" (25mm) BRACE TO EA. STUD & PLATE FACE NAIL	2-8d
21 1 x 8 (25 x 203mm) SHEATHING OR LESS TO EACH BEARING, FACE NAIL	2-8D
22 WIDER THAN 1 x 8 (25 X 203mm) SHEATHING TO EACH BEARING, FACE NAIL	3-8d
23 BUILT-UP CORNER STUDS	16d @ 24" (610 mm) O.C.
24 BUILT-UP GIRDER AND BEAMS	20d @ 32" (813 mm) O.C. @ TOP/BOTTOM & STAGGER 2-20d @ ENDS & @ EACH SPLICE
25 2" PLANKS	2-16d AT EACH BEARING
26 WOOD STRUCTURAL PANELS AND PARTICLEBOARD:	2
SUBFLOOR, ROOF AND WALL SHEATHING (TO FRAMING) 1" =25.4mm)	
1/2" OR LESS	6d[3]
19/32 - 3/4"	8d OR 6d[5]
7/8 - 1"	8d[3]
1 1/8 - 1 1/4"	10d[4] OR 8d[5]
COMBINATION SUBFLOOR — UNDER-LAYMENT (TO FRAMING) 1" = 25.4 mm)	
3/4 AND LESS	8d[5]
7/8" - 1"	8d[5]
1 1/8 - 1 1/4"	10d[4] OR 8d[5]
27. PANEL SIDING (TO FRAMING):	
1/2" (13 mm)	6d[6]
5/8" (16 mm)	8d[6]
28. FIBERBOARD SHEATHING: 7	
1/2" (13 mm)	No. 11 ga.
	6d[4]
	No. 16 ga.[9]
25/32" (20 mm)	No. 11 ga.[8]
	6d[4]
	No. 16 ga.[9]
29. INTERIOR PANELING	
1/4" (6.4 mm)	4d[10]
3/8" (9.5 mm)	4d[11]

1. - COMMON OR BOX NAILS MAY BE USED EXCEPT WHERE OTHERWISE STATED.

2. - NAILS SPACED @ 6" (152 mm) ON CENTER @ EDGES, 12" INTERMEDIATE SUPPORTS EXCEPT 6" (152 mm) AT ALL SUPPORTS WHERE SPANS ARE 48" 1220 mm OR MORE. FOR NAILING OF WOOD STRUCTURAL PANEL AND PARTICLEBOARD DIAPHRAGMS AND SHEAR WALLS, REFER TO SECTION 2314.3. NAIL FOR WALL SHEATHING MAY BE COMMON, BOX OR CASING. (305 mm) AT

3. - COMMON OR DEFORMED SHANK.

4. - COMMON.

5. - DEFORMED SHANK..

6. - CORROSION-RESISTANT SIDING OR CASING NAILS

7. - FASTENERS SPACED 3" (76 mm) O.C. AT EXTERIOR EDGES AND 6" (152 mm) O.C. AT INTERMEDIATE SUPPORTS.

8. - CORROSION-RESISTANT ROOFING NAILS W/ 7/16" ~ (11 mm) HEAD & 1 1/2" (38 mm) LENGTH FOR 1/2" (13 mm) SHEATHING AND 1 3/4" (44 mm) LENGTH (FOR 25/32" (20 mm) SHEATHING CONFORMING TO THE REQUIREMENTS OF SECTION 2325.1.

9. - CORROSION-RESISTANT STAPLES WITH NOMINAL 7/16" (11 mm) CROWN AND 1 1/8" (29 mm) LENGTH FOR 1/2" (13 mm) SHEATHING AND 1 1/2" (38 mm) LENGTH FOR 25/32" (20 mm) SHEATHING CONFORMING TO THE REQUIREMENTS OF SECTION 2325.1.

10. - PANEL SUPPORTS @ 16" (406 mm) O.C. 20" (508 mm) IF STRENGTH AXIS IN LONG DIRECTION OF THE PANEL, UNLESS OTHERWISE MARKED CASING OR FINISH NAILS SPACED 6" (152 mm) ON PANEL EDGES, 12" (305 mm) AT INTERMEDIATE SUPPORTS.

11. - PANEL SUPPORTS @ 24" (610 mm). CASING OR FINISH NAILS SPACED 6" (152 mm) ON PANEL EDGES, 12" (305 mm) AT INTERMEDIATE SUPPORTS.

INSULATION NOTES:

INSULATION BASED ON PATH # 1 OF 1996 OREGON RESIDENTIAL ENERGY CODE.

FIGURE 1–19 ■ Many offices use one or more sheets to include all standard framing notes and building specifications.

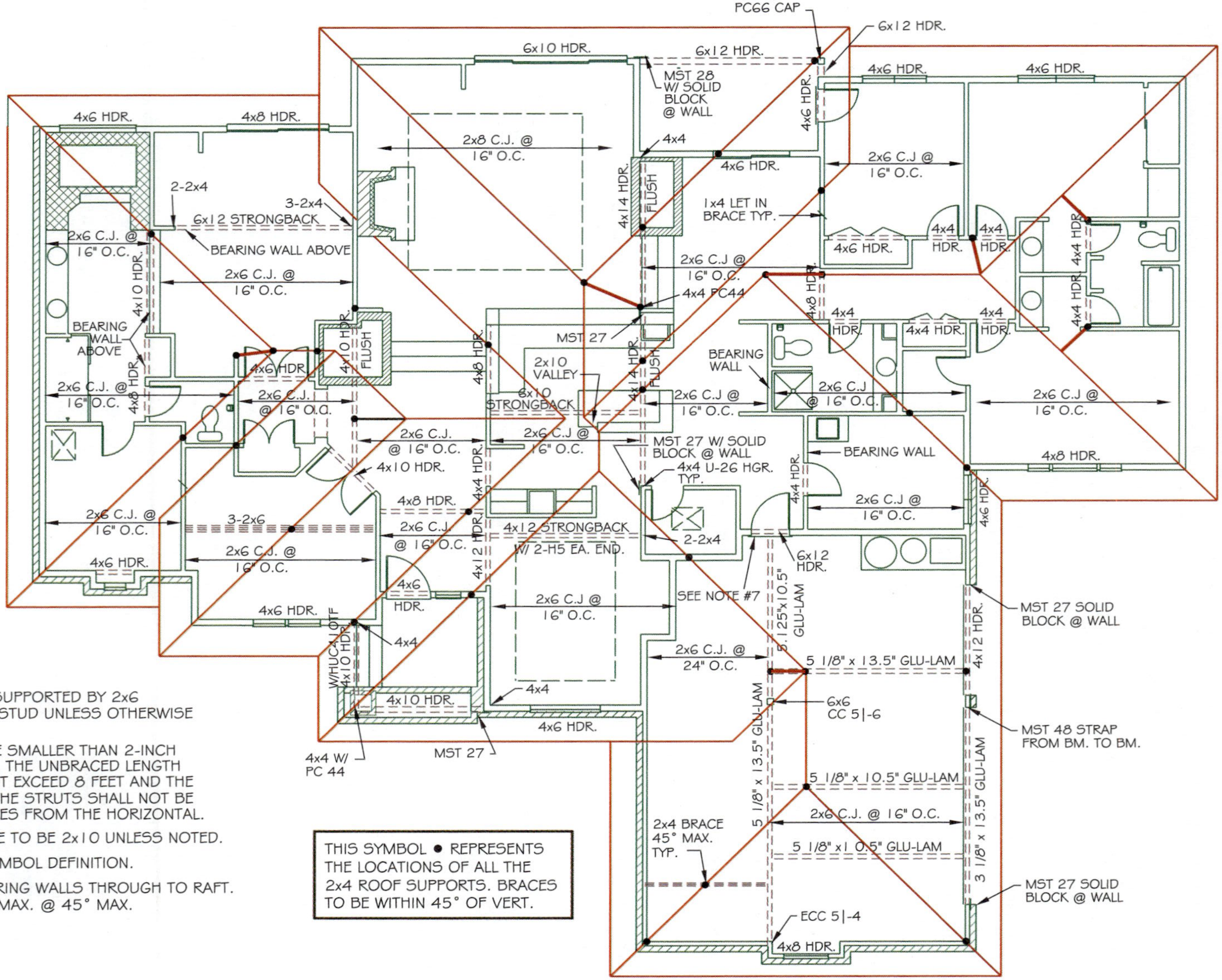

NOTES:

1. ALL HEADERS TO BE SUPPORTED BY 2x6 TRIMMERS AND KING STUD UNLESS OTHERWISE NOTED.
2. STRUTS SHALL NOT BE SMALLER THAN 2-INCH BY 4-INCH MEMBERS. THE UNBRACED LENGTH OF STRUTS SHALL NOT EXCEED 8 FEET AND THE MINIMUM SLOPE OF THE STRUTS SHALL NOT BE LESS THAN 45 DEGREES FROM THE HORIZONTAL.
3. HIP, VALLEY AND RIDGE TO BE 2x10 UNLESS NOTED.
4. SEE SHEET 8A FOR SYMBOL DEFINITION.
5. EXTEND INTERIOR BEARING WALLS THROUGH TO RAFT. USE 2x4 @ 24" O.C. MAX. @ 45° MAX.

FIGURE 1–20 ■ Because of the complicated roof structure, the material used to frame the roof is separated from the material displayed on the roof plan (Figure 1–16). The roof plan shows the material used to frame the roof. The roof framing plan is used to display the structural materials used to transfer roof loads into the walls.

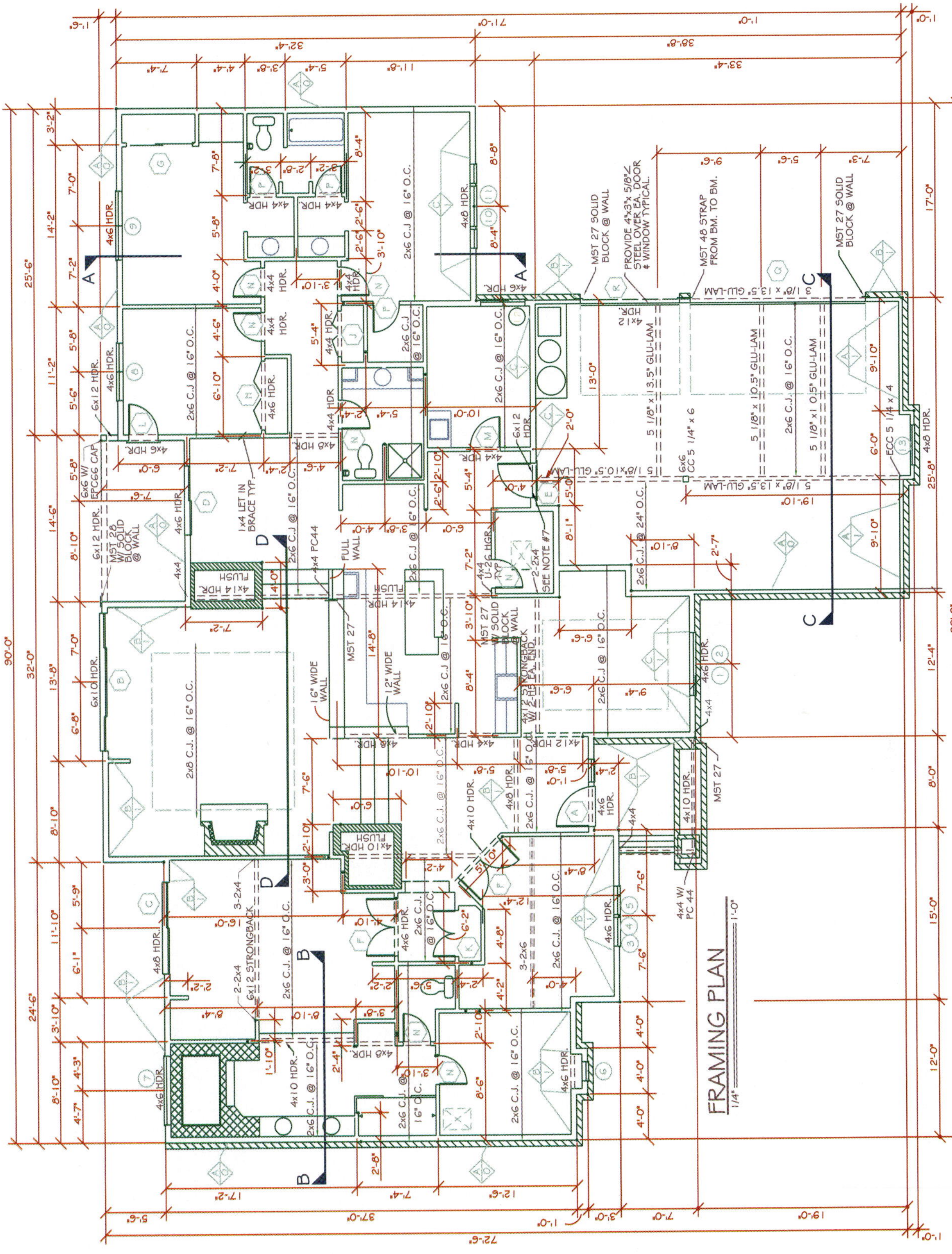

FIGURE 1–21A ■ In addition to showing materials needed to frame the structure, materials needed to resist wind, seismic, and other forces of nature are shown.

NOTES:

1. ALL FRAMING LUMBER TO BE D.F.L. #2 MIN.
2. FRAME ALL EXTERIOR WALLS W/ 2X6 STUDS @ 16" O.C. ALL EXTERIOR HEADERS TO BE 4X6 D.F.L. #2 UNLESS NOTED.
3. ALTERNATE: FOR ALL BMS. SMALLER THAN 4x10 USE 2X6 NAILER AT THE BOTTOM OF ALL 4x HEADERS @ EXTERIOR WALLS. BACK HEADER W/ 2" RIGID INSULATION.
4. ALL SHEAR PANELS TO BE 1/2" PLY NAILED W/8 d'S @ 4" O.C. @ EDGE BLOCKING AND 6 d'S @ 8" O.C. @ FIELD UNLESS NOTED.
5. ALL METAL CONNECTORS TO BE BY SIMPSON CO. OR OTHER.
6. ALL ANGLES ON INTERNAL WALLS TO BE 45° UNLESS OTHERWISE SPECIFIED.
7. USE 4x6 POST W/ EPC 5 1/4 TOP CAP TO 4x6 POST & PC 46 BASE CAP UPSIDE/DOWN TO HDR OVER GARAGE/HALL DOOR.
8. 1x4 DIAG. LET IN BRACE @ 45° MAX. TYPICAL.
9. BLOCK ALL WALLS OVER 10'-0" HIGH AT MID HEIGHT.
10. SEE SHEET 8A FOR SYMBOL DEFINITIONS.

SHEAR WALL SCHEDULE			
MARK	WALL COVER	EDGE NAIL'G	FIELD NAIL'G
A	15/32" C-D EXTR. PLYWD. (1) SIDE -UNBLOCKED	8d @ 6" O.C.	8d @ 12" O.C.
B	15/32" C-D EXTR. PLYWD. (1) SIDE - BLOCKED EDGES	8d @ 6" O.C.	8d @ 12" O.C.
C	15/32" C-D EXTR. PLYWD. (1) SIDE - BLOCKED EDGE	8d @ 4" O.C.	8d @ 12" O.C.

HOLD DOWN SCHEDULE			
MARK	HOLD DOWN	CONN. MEMBER	DETAIL
0	NON-REQ'D	———	———
1	SIMPSON PAHD42	8d @ 6" O.C.	

FIGURE 1–21B ■ Notes specifying minimum building standards are shown near the framing plan.

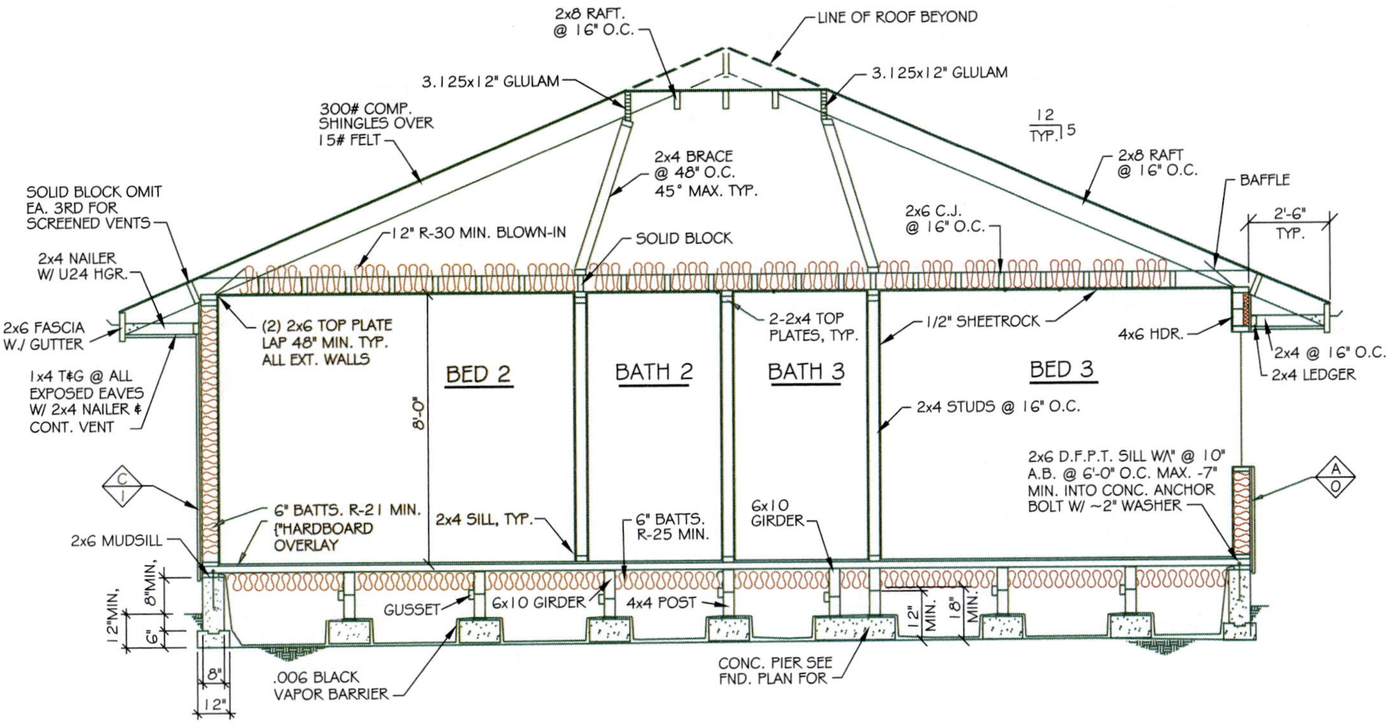

SECTION A-A

3/8" = 1'-0"

NOTES:

1. SEE FND. PLAN FOR GIRDER AND PIER SIZES.

2. SEE SHEET 8A FOR SYMBOL DEFINITIONS.

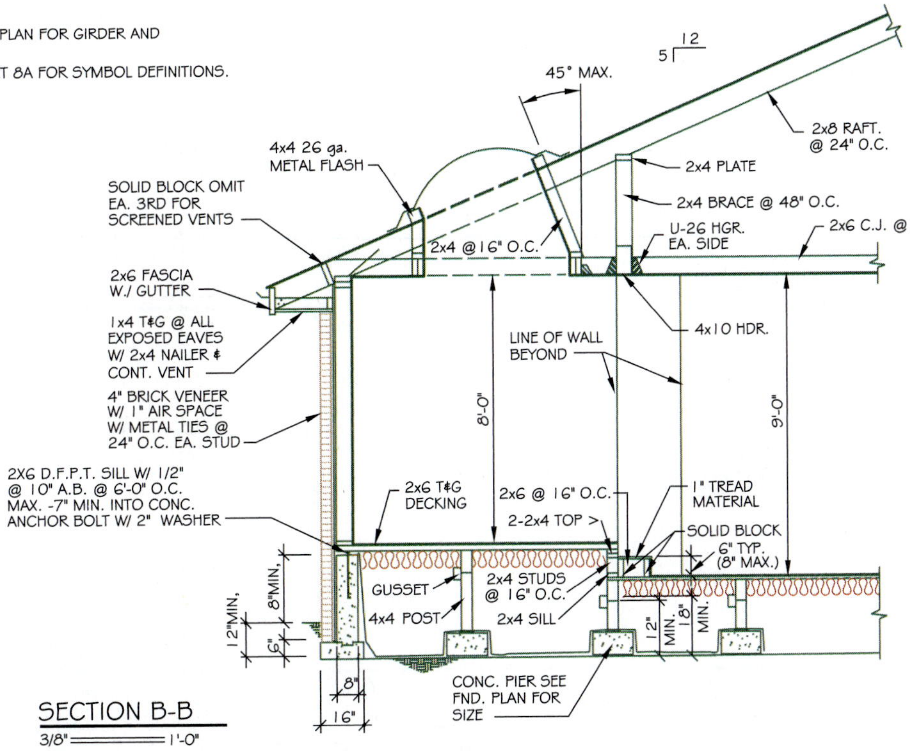

SECTION B-B

3/8" = 1'-0"

NOTES:

1. SEE SECTION A-A FOR BALANCE OF NOTES.

FIGURE 1–22 ■ Sections are used to show the vertical relationships of the structural members. A section or detail is used to represent each major shape within the structure.

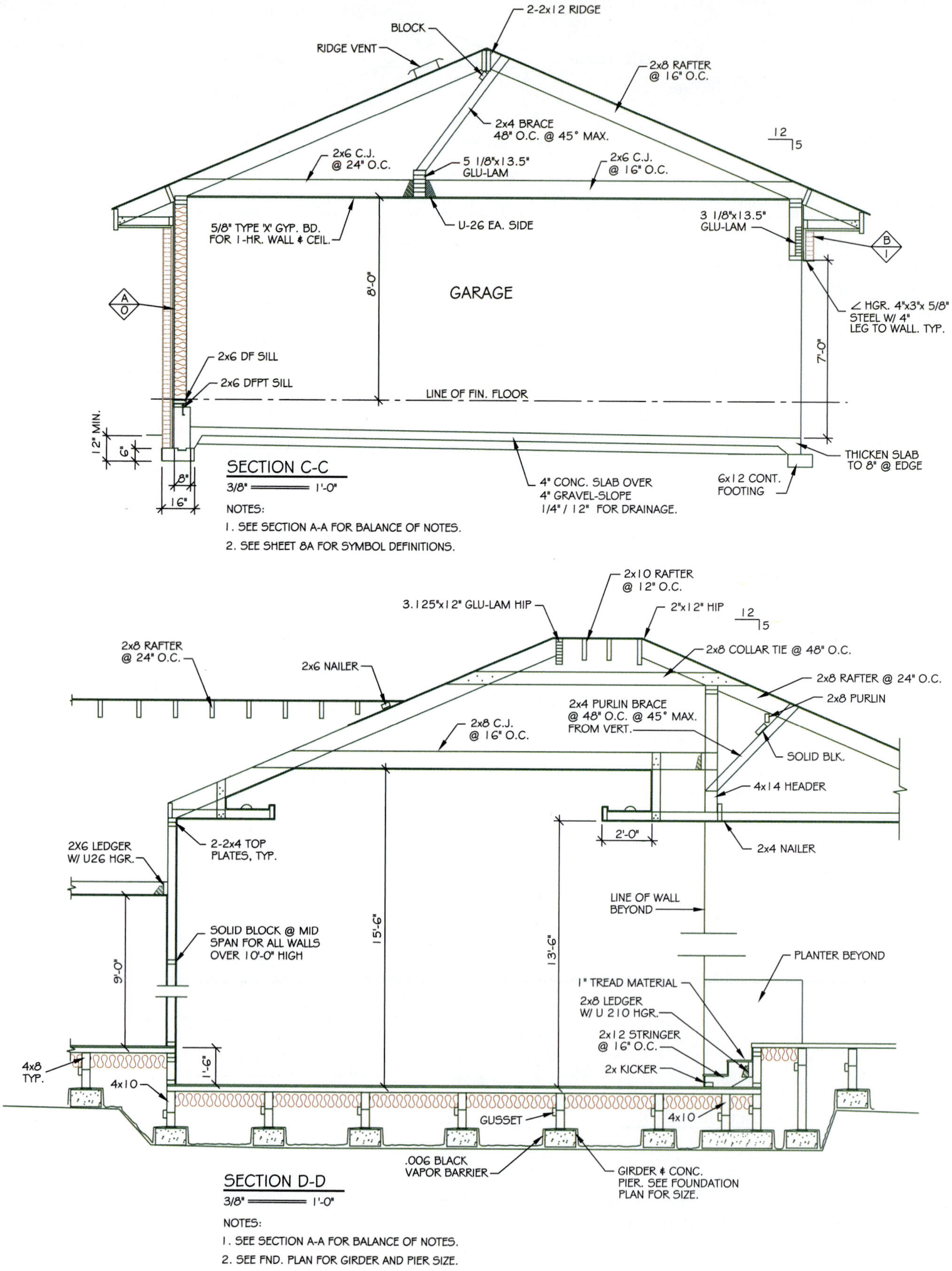

FIGURE 1–23 ■ Each section is referenced to the framing plan with a viewing plane line to show where the structure is "cut" and in which direction the viewer is looking.

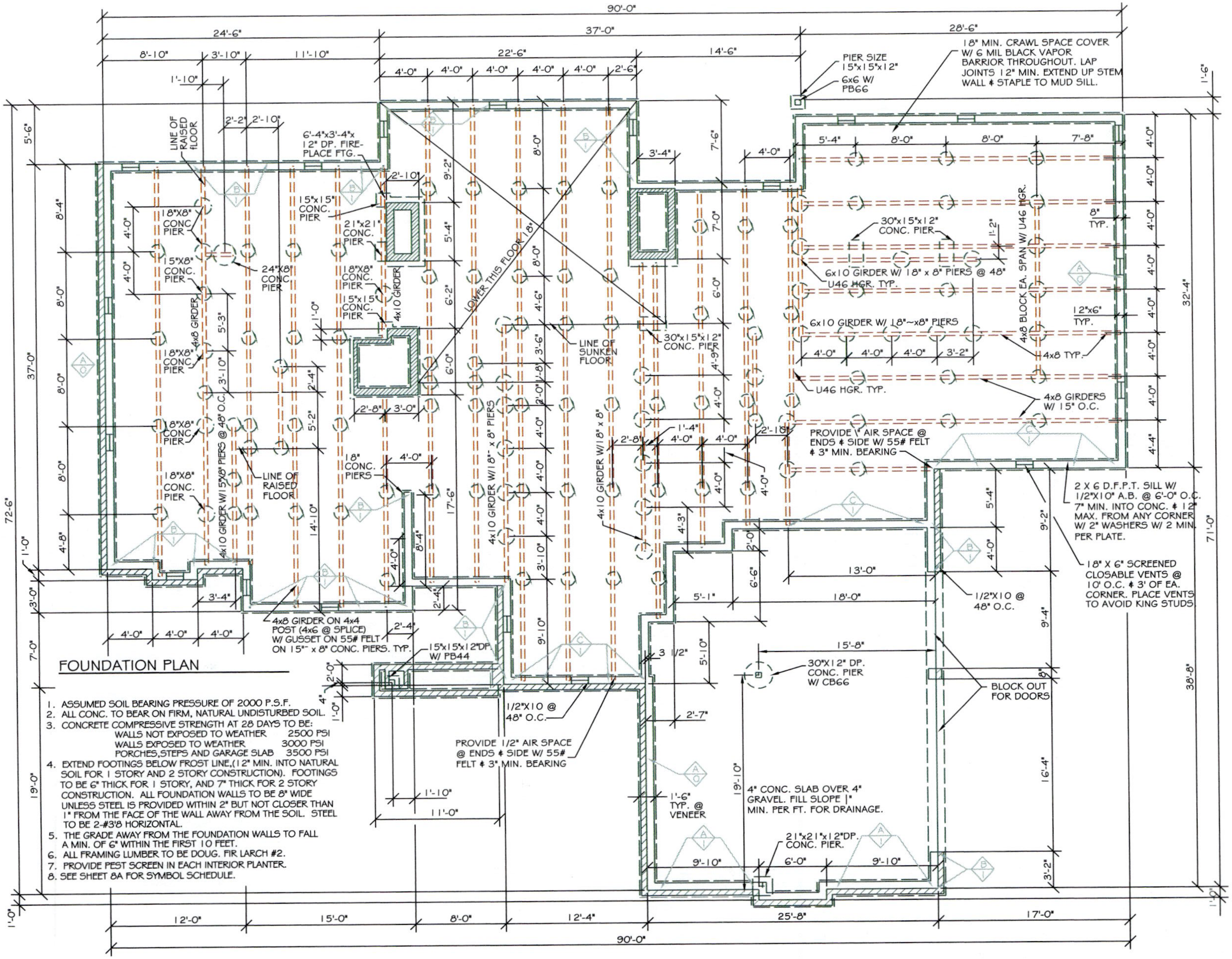

FIGURE 1–24 ■ The foundation plan shows how each of the loads will be supported. The foundation not only resists the loads caused by gravity but also resists the loads from seismic forces, winds, and flooding.

Chapter 1 Additional Reading

The following Web sites can be used as a resource to help you keep current with changes in the building industry.

ADDRESS	COMPANY OR ORGANIZATION
www.afsonl.com	Architects First Source for Products
www.builderonline.com	Builder Magazine
www.dreamhomesmagazine.com	Dream Homes
www.hbrnet.com	Home Building and Remodeling Network
www.nahbrc.org	National Association of Home Builders
www.residentialarchitect.com	Residential Architect Online (Bulletin board and buyers guide)
www.sweets.com	Sweet's System On Line Building Product Information. Sweet's is a reference for many of the leading products used throughout the construction industry. This site contains reference material and a Web site for each listing. This is one of the most useful sites you can visit!

CHAPTER 1

Professional Architectural Careers, Office Practice, and Opportunities Test

DIRECTIONS

Answer the problems with short complete statements or drawings as needed on an 8 1/2 × 11 sheet of notebook paper, as follows:

1. Letter your name, chapter number, and the date at the top of the sheet.
2. Letter the question number and provide the answer. You do not need to write out the question. Answers may be prepared on a word processor if course guidelines allow this.

PROBLEMS

Problem 1–1 List five types of work that a junior drafter might be expected to perform.

Problem 1–2 What three skills are usually required of a junior drafter, for advancement?

Problem 1–3 What types of drawings should a junior drafter expect to prepare?

Problem 1–4 Describe what the junior drafter might be given to assist in making drawings.

Problem 1–5 List four sources of written information that a drafter will need to be able to use.

Problem 1–6 List and briefly describe different careers in which drafting would be helpful.

Problem 1–7 What is the purpose of a bubble drawing?

Problem 1–8 Why do you think furniture placement should be considered in the preliminary design process?

Problem 1–9 What would be the minimum drawings required to get a building permit?

Problem 1–10 List five additional drawings that may be required for a complete set of house plans in addition to the five basic drawings.

Problem 1–11 List and describe the steps of the design process.

Problem 1–12 What are the functions of the drafter in the design process?

Problem 1–13 Following the principles of this chapter, prepare a bubble sketch for a home with the following specifications:

a. 75′ × 120′ lot with a street on the north side of the lot
b. a gently sloping hill to the south
c. south property line is 75′ long
d. 40 ft oak trees along the south property line
e. 3 bedrooms, 2 1/2 baths, living, dining, with separate eating area off kitchen
f. exterior style as per your choice

Explain why you designed the house that you did.

2 CHAPTER

Architectural Drafting Equipment

INTRODUCTION

Manual drafting is the term used to describe traditional pencil or ink drafting. Manual drafting tools and equipment are available from a number of vendors of professional drafting supplies. For accuracy and durability, always purchase high-quality equipment.

DRAFTING SUPPLIES

Whether equipment is purchased in a kit or by the individual tool, the following items are the minimum normally needed for manual drafting:.

- One mechanical lead holder with 4H, 2H, H, and F grade leads. This pencil and lead assortment allows for individual flexibility of line and lettering control, or for use in sketching. (The lead holder and leads may be considered optional if automatic pencils are used.)
- One 0.3-mm automatic drafting pencil with 4H, 2H, and H leads.
- One 0.5-mm automatic drafting pencil with 4H, 2H, and F leads.
- One 0.7-mm automatic drafting pencil with 2H, H, and F leads.
- One 0.9-mm automatic drafting pencil with H, F, and HB leads. Automatic pencils and leads may be considered optional if lead holders and leads are used.
- Lead sharpener (not needed if the lead holder is omitted).
- 6″ bow compass.
- Dividers.
- Eraser.
- Erasing shield.
- 30°–60° triangle.
- 45° triangle.
- Irregular curve.
- Adjustable triangle (optional).
- Scales:
  1. Triangular architect's scale.
  2. Triangular civil engineer's scale.
- Drafting tape.
- Architectural floor plan template, 1/4″ = 1′–0″ (for residential plans).
- Circle template (small holes).
- Lettering guide.
- Sandpaper sharpening pad.
- Dusting brush.

Not all of the items listed here are used in every school or workplace. Additional equipment and supplies may be needed depending on the specific application, for example:

- Technical pen set.
- Drafting ink.
- Lettering template.
- Assorted architectural drafting templates.

DRAFTING FURNITURE

Tables

Drafting tables are generally sized by the dimensions of their tops. Standard table-top sizes range from 24 × 36″ (610 × 915 mm) to 42 × 84″ (1067 × 2134 mm).

The features to look for in a good-quality professional table include:

- One-hand tilt control.
- One-hand or foot height control.
- Ability to position the board vertically.
- Electrical outlet.
- Drawer for tools and/or drawings.

Most offices commonly cover drafting-table tops with specially designed smooth surfaces. The material, usually vinyl, provides the proper density for effective use under normal drafting conditions. Drafting tape is commonly used to adhere drawings to the table top, although some drafting tables have magnetized tops and use magnetic strips to attach drawings.

Chairs

Good drafting chairs have these characteristics:

- Padded or contoured seat design.
- Height adjustment.
- Footrest.
- Fabric that allows air to circulate.
- Sturdy construction.

DRAFTING PENCILS AND LEADS

Mechanical Pencils

The term *mechanical pencil* refers to a pencil that requires a piece of lead to be manually inserted. Special sharpeners are available to help maintain a conical point.

Automatic Pencils

The term *automatic pencil* refers to a pencil with a lead chamber that, at the push of a button or tab, advances the lead from the chamber to the writing tip. Automatic pencils are designed to hold leads of one width that do not need to be sharpened. These pencils are available for leads 0.3, 0.5, 0.7, and 0.9 mm thick. Many drafters have several automatic pencils, each with a different grade of lead hardness, used for a specific application.

Lead Grades

Several different grades, or hardnesses, of leads are used for linework. Your selection depends on the amount of pressure you apply and other technique factors. You should experiment until you identify the leads that give you the best line quality. Leads commonly used for thick lines range from 2H to F; leads for thin lines range from 4H to H, depending on individual preference. Construction lines for layout and guidelines are very lightly drawn with a 6H or 4H lead. Figure 2.1 shows the different lead grades. Generally, leads used for sketching are softer than those used for making formal lines. Good sketching leads are: H, F, or HB.

9H 8H 7H 6H 5H 4H	3H 2H H F HB B	2B 3B 4B 5B 6B 7B
HARD	MEDIUM	SOFT
4H AND 6H ARE COMMONLY USED FOR CONSTRUCTION LINES.	HAND 2H ARE COMMON LEAD GRADES USED FOR LINE WORK. H AND F WOULD BE FOR LETTERING AND SKETCHING.	THESE GRADES ARE FOR ART WORK. THEY ARE TOO SOFT TO KEEP A SHARP POINT AND THEY SMUDGE EASILY.

FIGURE 2.1 ■ Range of lead grades.

Polyester and Special Leads

Polyester leads, also known as *plastic leads,* are for drawing on polyester drafting film, often called by its trade name, *Mylar.* Plastic leads come in grades equivalent to F, 2H, 4H, and 5H and are usually labeled with an "S," as in 2S. Some companies make a combination lead for use on both vellum and polyester film. See Chapter 3 for information about the use of polyester leads, and the techniques involved.

Colored leads have special uses. Red or yellow lead is commonly used to make corrections. Red or yellow lines appear black on a print. Blue lead can be used on an original drawing for information that is not to show on a print, such as notes, lines, and outlines of areas to be corrected. Blue lines do not appear on the print when the original is run through a Diazo machine but will reproduce in the photocopy process if not very lightly drawn. Some drafters use light blue lead for layout work and guidelines.

Basic Pencil Motions

A mechanical pencil must be kept sharp with a slightly rounded point. A point that is too sharp will break easily, but a point that is too dull will produce fuzzy lines. Keep the pencil aligned with the drafting instrument and tilted at an angle of approximately 45° in the direction you are drawing. Always pull the pencil. If you are right-handed, it is recommended that you draw from left to right and from bottom to top. Left-handed people draw from right to left. It is very important to keep lines uniform in width. This is done by keeping the pencil a uniform sharpness and by rotating the pencil as you draw. This rotation will help keep the lead from having flat spots that cause the line to vary in width and become fuzzy. Even pressure is also needed to help ensure a uniform line weight. Your lines need to be crisp and dark. Keep practicing, and you will be able to draw high-quality lines.

Keep your automatic pencil straight from side to side and tilted about 40° with the direction of travel. Try not to tilt the lead under or away from the straightedge. Automatic pencils do not require rotation, although some drafters feel that better lines are made by rotating the pencil. Provide enough pressure to make a line dark and crisp. Figure 2.2 shows some basic pencil motions.

TECHNICAL PENS AND ACCESSORIES

Technical Pens

Technical pens produce excellent inked lines. These pens function on a capillary action; a needle acts as a valve to allow ink to flow from a storage cylinder through a small tube, which is designed to meter the ink so a specific line width is created. Figure 2.3 shows a comparison of some of the different line widths available with technical pens.

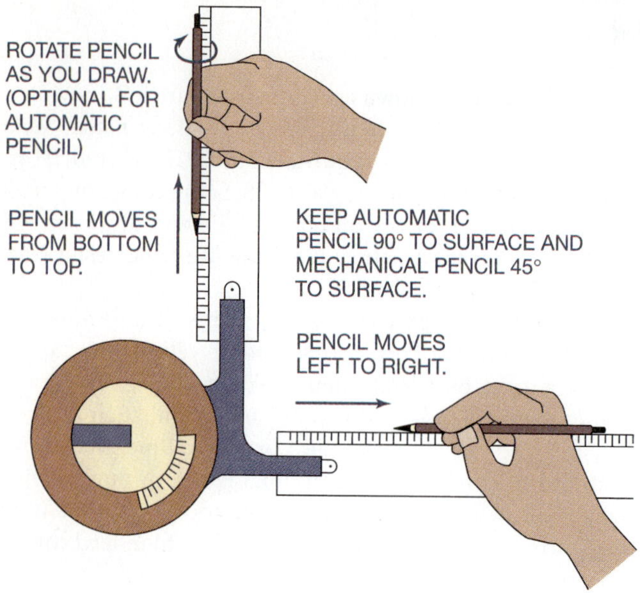

FIGURE 2.2 ■ Basic pencil motions.

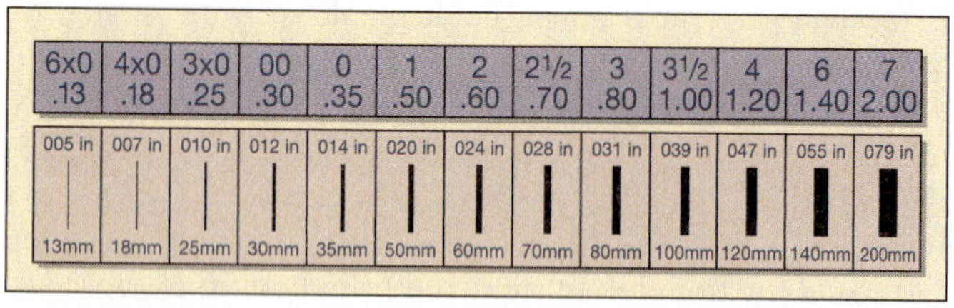

6x0 .13
4x0 .18
3x0 .25
00 .30
0 .35
1 .50
2 .60
2½ .70
3 .80
3½ 1.00
4 1.20
6 1.40
7 2.00
005 in 13mm
007 in 18mm
010 in 25mm
012 in 30mm
014 in 35mm
020 in 50mm
024 in 60mm
028 in 70mm
031 in 80mm
039 in 100mm
047 in 120mm
055 in 140mm
079 in 200mm

FIGURE 2.3 ■ Technical pen line widths. *Courtesy Koh-I-Noor, Inc.*

Technical pens may be used with templates to make circles, arcs, and symbols. Compass adapters to hold technical pens are also available. Technical pen tips are designed to fit into scribers for use with lettering guides. This concept is discussed further in Chapter 5.

Pen Cleaning

Some symptoms to look for when pens need cleaning are listed below:

- Ink constantly creates a drop at the pen tip.
- Ink tends to flow out around the tip holder.
- The point plunger does not activate properly. If you do not hear and feel the plunger when you shake the pen, it is probably clogged with thick or dried ink.
- When you are drawing a line, ink does not flow freely.
- Ink flow starts with difficulty.

The ink level in the reservoir should be kept between one-quarter to three-quarters full.

Read the cleaning instructions that come with the pen you purchase. Some pens require disassembly for cleaning, while others should not be taken apart. Pens should be cleaned before each filling or before being stored for a long period of time. Clean the technical pen nib, cartridge, and body separately in warm water or a special cleaning solution.

Ink

Drafting inks should be opaque or have a matte or semiflat black finish that will not reflect light. The ink should reproduce without hot spots or line variation. Drafting ink should have excellent adhesion properties for use on paper or film. Certain inks are recommended for use on film in order to avoid peeling, chipping, or cracking. Inks recommended for use in technical pens also have nonclogging characteristics.

ERASERS AND ACCESSORIES

The common shapes of erasers are rectangular and stick. The stick eraser works best in small areas. Select an eraser that is recommended for the particular material used.

Erasers Used for Drafting

White Pencil Eraser

The white pencil eraser is used for removing pencil marks from paper. It contains a small amount of pumice (an abrasive material) but can be used as a general-purpose pencil eraser.

Pink Pencil Eraser

A pink pencil eraser contains less pumice than the white pencil eraser. It can be used to erase pencil marks from most kinds of office paper without doing any damage to the surface.

Soft-Green or Soft-Pink Pencil Eraser

The soft-green and soft-pink erasers contain no pumice and are designed for use on even the most delicate types of paper without doing damage to the surface. The color difference is maintained only to satisfy long-established individual preferences.

Kneaded Eraser

Kneaded erasers are soft, stretchable, and nonabrasive. They can be formed into any size you want. Kneaded erasers are used for erasing pencil and charcoal; they are also used for shading. In addition, kneaded erasers are handy for removing smudges from a drawing and for cleaning a drawing surface.

Vinyl Eraser

The vinyl or plastic eraser contains no abrasive material. This eraser is used on polyester drafting film to erase plastic lead or ink.

India Ink Erasing Machine Refill

The India ink refill eraser is for removing India ink from vellum or film. It contains a solvent that dissolves the binding agents in

the ink and the vinyl, then removes the ink. This eraser does not work well with pencil lead.

Erasing Tips

In erasing, the idea is to remove an unwanted line or letter, not the surface of the paper or polyester film. Erase only hard enough to remove the unwanted line. However, you must bear down enough to eliminate the line completely. If the entire line does not disappear, ghosting results. A *ghost* is a line that should have been eliminated but still shows on a print. Lines that have been drawn so hard as to make a groove in the drawing sheet can cause a ghost too. To remove ink from vellum, use a pink or green eraser, or an electric eraser. Work the area slowly. Do not apply too much pressure or erase in one spot too long, or you will go through the paper. On polyester film, use a vinyl eraser and/or a moist cotton swab. The inked line usually comes off easily, but use caution. If you destroy the mat surface of either a vellum or film sheet, you will not be able to redraw over the erased area.

Electric Erasers

Professional manual drafters use electric erasers, available in cord and cordless models. When working with an electric eraser, you do not need to use very much pressure because the eraser operates at high speed. The purpose of the electric eraser is to remove unwanted lines quickly. Use caution; these erasers can also remove paper quickly! Eraser refills for electric units are available in all types.

Erasing Shield

Erasing shields are thin metal or plastic sheets with a number of differently shaped holes. They are used to erase small, unwanted lines or areas. For example, if you have a corner overrun, place one of the slots of the erasing shield over the area to be removed while covering the good area.

Cleaning Agents

Special eraser particles are available in a shaker-top can or a pad. Both types are used to sprinkle the eraser particles on a drawing to help reduce smudging and to keep the drawing and your equipment clean. The particles also help float triangles, straightedges, and other drafting equipment to reduce line smudging. Use this material sparingly, since too much of it can cause your lines to become fuzzy. Cleaning powders are not recommended for use on ink drawings or on polyester film.

Dusting Brush

Use a dusting brush to remove eraser particles from your drawing. Doing so helps reduce the possibility of smudges. Avoid using your hand to brush away eraser particles, because the hand tends to cause messy smudges. A clean dry cloth works better than your hand for removing eraser particles, but a brush is preferred.

DRAFTING INSTRUMENTS

Kinds of Compasses

Compasses are used to draw circles and arcs. A compass is especially useful for large circles. However, using a compass can be time-consuming. Use a template, whenever possible, to make circles or arcs quickly.

There are several basic types of drafting compasses:

1. *Drop-bow compass,* mostly used for drawing small circles. The center rod contains the needle point and remains stationary while the pencil, or pen, leg revolves around it.
2. *Center-wheel bow compass,* most commonly used by professional drafters. This compass operates on the screw-jack principle by turning the large knurled center wheel, as shown in Figure 2.4.
3. *Beam compass,* a bar with an adjustable needle and a pencil or pen attachment for swinging large arcs or circles. Also available is a beam that is adaptable to the bow compass. Such an adapter works only on bow compasses that have a removable break point, not on the fixed point models.

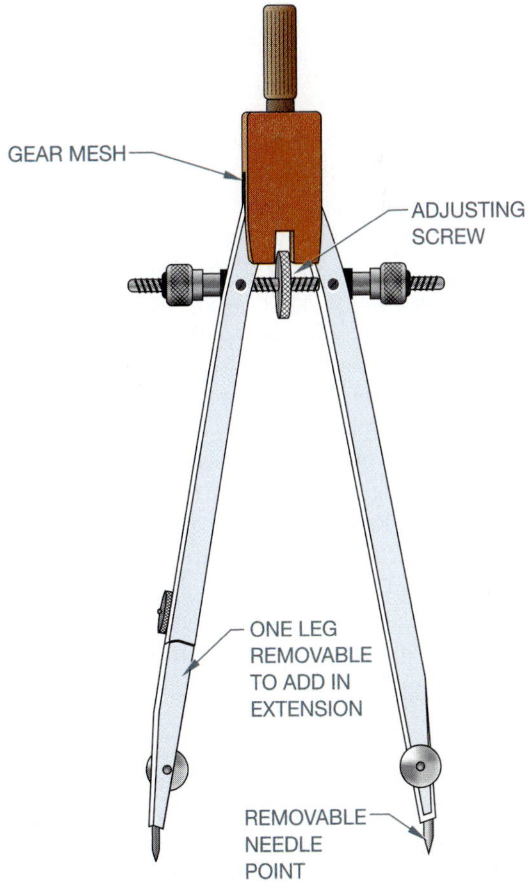

FIGURE 2.4 ■ Bow compass and its parts.

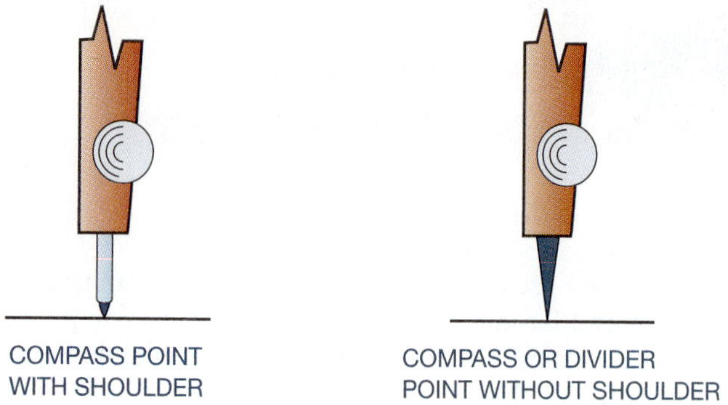

FIGURE 2.5 ■ Common compass points.

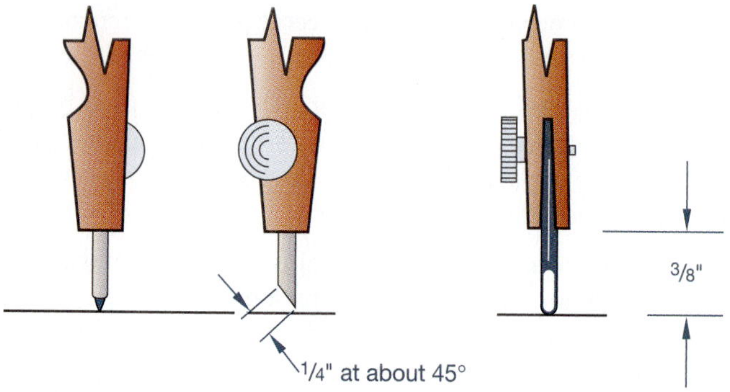

FIGURE 2.6 ■ Properly sharpened and aligned elliptical compass lead.

Compass Use

Keep both the compass needle point and the lead point sharp. The points are removable for easy replacement. The better compass needle points have a shoulder, which helps keep the point from penetrating the paper more than necessary. Compare the needle points in Figure 2.5.

The compass lead should, in most cases, be one grade softer than the lead you use for straight lines because less pressure is used on a compass than a pencil. Keep the compass lead sharp. An elliptical point is commonly used with the bevel side inserted away from the needle leg. Keep the lead and the point equal in length. Figure 2.6 shows properly aligned and sharpened points on a compass.

Use a sandpaper block to sharpen the elliptical point. Be careful to keep the graphite residue away from your drawing and off your hands. Remove excess graphite from the point with a tissue or cloth after sharpening. Sharpen the lead often.

Some drafters prefer to use a conical compass point. This is the same point used in a mechanical pencil. If you want to try this point, sharpen a piece of lead in a mechanical pencil, then transfer it to your compass.

If you are drawing a number of circles from the same center, you will find that the compass point causes an ugly hole in your drawing sheet. Reduce the chance of making such a hole by placing a couple of pieces of drafting tape at the center point for protection. There are small plastic circles available for just this purpose. Place one at the center point; then pierce the plastic with your compass.

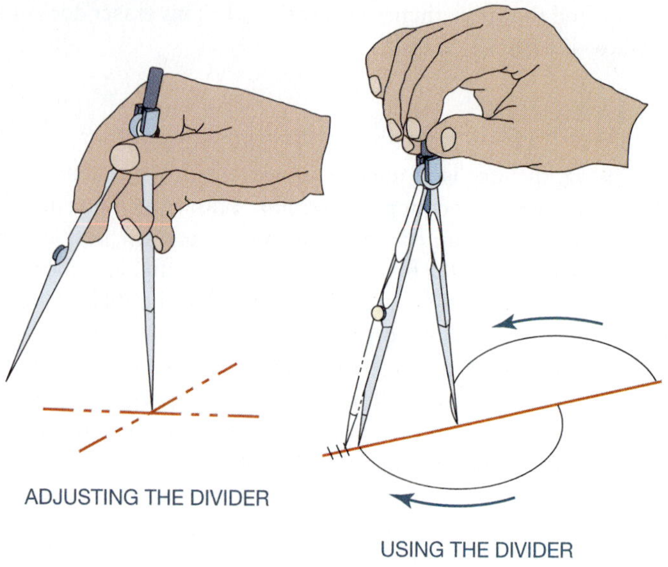

FIGURE 2.7 ■ Using the divider.

Dividers

Dividers are used to transfer dimensions or to divide a distance into a number of equal parts.

Some drafters prefer to use a bow divider because the center wheel provides the ability to make fine adjustments easily. Also, the setting remains more stable than with standard friction dividers.

A good divider should not be too loose or too tight. It should be easily adjustable with one hand. In fact, you should control dividers with one hand as you lay out equal increments or transfer dimensions from one feature to another. Figure 2.7 shows how the dividers should be handled.

Proportional Dividers

Proportional dividers are used to reduce or enlarge an object without mathematical calculations or scale manipulations. The center point of the dividers is set at the correct point for the proportion you want. Then measure the original size line with one side of the proportional dividers, and the other side automatically determines the new reduced or enlarged size.

Parallel Bar

The parallel bar slides up and down the drafting board on cables that are attached to pulleys mounted at the top and bottom corners of the table (see Figure 2.8). The function of the parallel bar is to allow you to draw horizontal lines. Vertical lines and angles are made with triangles in conjunction with the parallel bar (Figure 2.9). Angled lines are made with standard or adjustable triangles.

The parallel bar is commonly found in architectural drafting offices because architectural drawings are frequently very large. Architects and architectural drafters often need to draw straight lines the full length of their boards, and the parallel bar

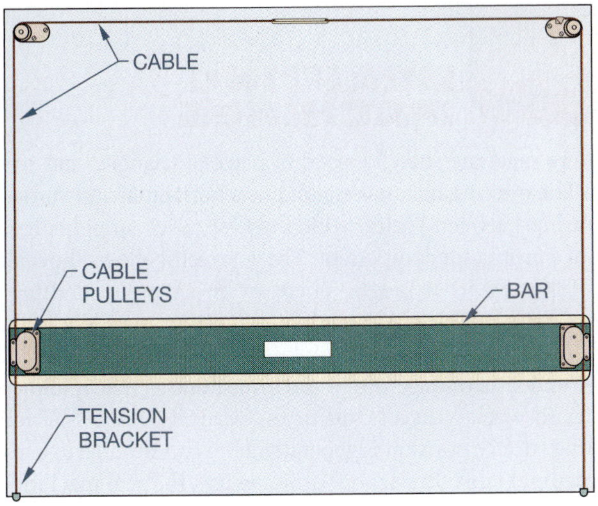

FIGURE 2.8 ■ Parallel bar.

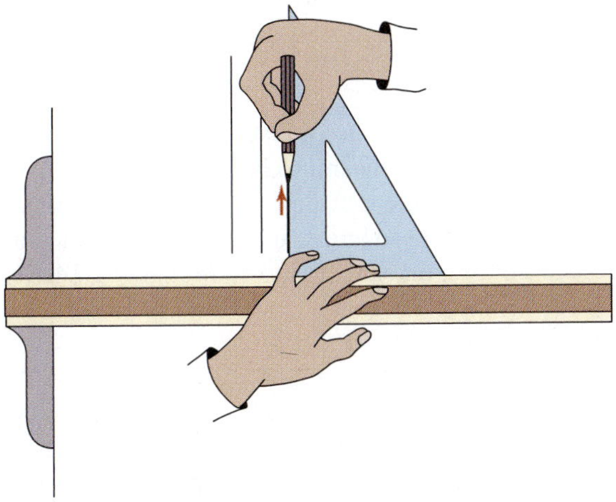

FIGURE 2.9 ■ Drawing vertical lines with the parallel bar and triangle.

is ideal for such lines. The parallel bar may be used with the board in an inclined position. Triangles and the parallel bar must be held securely when used in any position. Professional units have a brake that may be used to lock the parallel bar in place during use.

Triangles

There are two standard triangles. One has angles of 30°–60°–90° and is known as the 30°–60° triangle. The other has angles of 45°–45°–90° and is called the 45° triangle.

Triangles may be used as a straightedge to connect points for drawing lines. Triangles are used individually or in combination to draw angled lines in 15° increments (Figure 2.10). Also available are adjustable triangles with built-in protractors that are used to make angles of any degree.

Architectural Templates

Templates are plastic sheets that have standard symbols cut through them for tracing. There are as many standard architec-

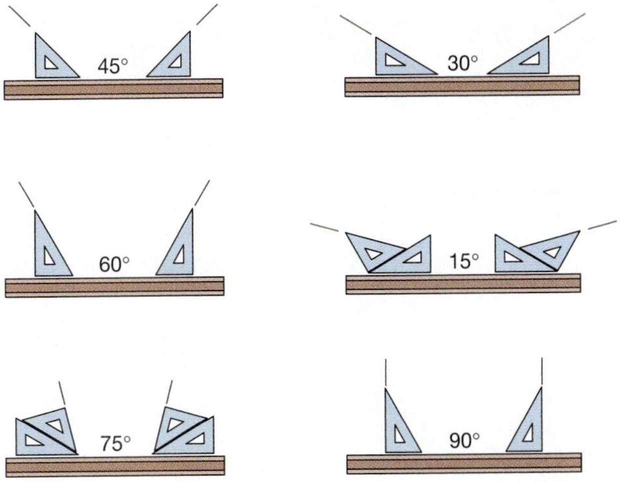

FIGURE 2.10 ■ Angles that may be made with the 30°–60° and 45° triangles, individually or in combination.

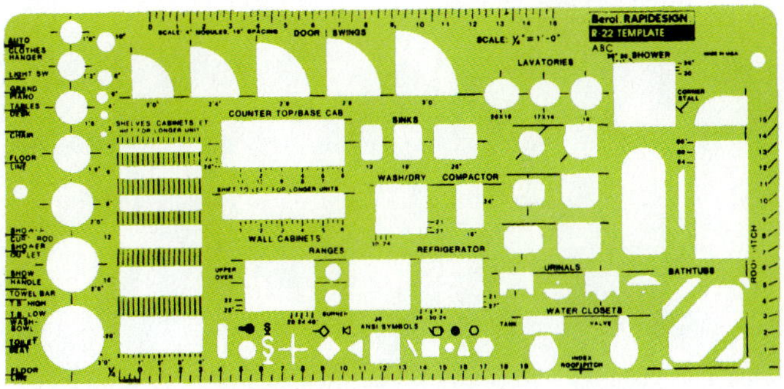

FIGURE 2.11 ■ Architectural floor plan symbol template. *Courtesy Berol Corporation.*

tural templates as there are architectural symbols to draw. Templates are commonly scaled to match a particular application, such as floor plan symbols.

One of the most common architectural templates used is the residential floor plan template (Figure 2.11). This template is typically available at a scale of 1/4″ = 1′–0″ (1:50 metric), which matches the scale usually used for residential floor plans. Features that are often found on a floor plan template include circles, door swings, sinks, bathroom fixtures, kitchen appliances, and electrical symbols. Other floor plan templates are available for both residential and commercial applications. Additional architectural templates are introduced in chapters where their specific applications are discussed.

Always use a template when you can. Templates save time, are very accurate, and help standardize features on drawings. For most applications the template should be used only after the layout has been established. For example, floor plan symbols are placed on the drawing after walls, openings, and other main floor plan features have been established.

When using a template, carefully align the desired template feature with the layout lines of the drawing. Hold the template firmly in position while you draw the feature. For best results, try to keep your pencil or pen perpendicular to the drawing surface. When using a template and a technical pen, keep the pen perpendicular to the drawing sheet. Some templates have risers built in to keep the template above the drawing sheet. Without

FIGURE 2.12 ■ Using a template with built-in risers. *Courtesy Chartpak.*

FIGURE 2.13 ■ Irregular or French curves. *Courtesy the C-Thru® Ruler Company.*

this feature, there is a risk of ink running under a template that is flat against the drawing. If your template does not have risers, purchase and add template lifters, use a few layers of tape placed on the underside of the template (although tape does not always work well), or place a second template with a large circle under the template you are using. See Figure 2.12.

Irregular Curves

Irregular curves are commonly called French curves. These curves have no constant radii. A selection of irregular curves is shown in Figure 2.13. Also available are ship's curves, which become progressively larger and, like French curves, have no constant radii. They are used for layout and development of ships' hulls. Flexible curves are made of plastic and have a flexible core. They may be used to draw almost any curve desired by bending and shaping.

Irregular curves are used for a variety of applications in architectural drafting. Some drafters use the curves to draw floor plan electrical runs or leader lines that connect notes to features on the drawing.

DRAFTING MACHINES

Drafting machines may be used in place of triangles and parallel bars. The drafting machine maintains a horizontal and vertical relationship between scales, which also serve as straightedges, by way of a protractor component. The protractor allows the scales to be set quickly at any angle. There are two types of drafting machines: arm and track. Although both types are excellent tools, the track machine has generally replaced the arm machine in industry. A major advantage of the track machine is that it allows the drafter to work with a board in the vertical position. A vertical drafting surface position is generally more comfortable to use than a horizontal table. For school or home use, the arm machine may be a good economical choice. When you order a drafting machine, the size specified should relate to the size of the drafting board. For example, a 37 1/2 × 60″ machine would fit the same size table.

Arm Drafting Machine

The arm drafting machine is compact and less expensive than a track machine. The arm machine clamps to a table and through an elbow-like arrangement of supports allows the drafter to position the protractor head and scales anywhere on the board.

The arm drafting machine is shown in Figure 2.14.

Track Drafting Machines

A track drafting machine has a traversing arm that moves left and right across the table, and a head unit that moves up and down the traversing arm. There is a locking device for both the head and the traversing arm. The shape and placement of the controls of a track machine vary with the manufacturer, although most brands have the same operating features and procedures. Figure 2.15 shows the component parts of a track drafting machine.

As with the arm machines, track drafting machines have a vernier head that allows you to measure angles accurately to 5 minutes of arc (5′). Remember, 1 degree equals 60 minutes (1° = 60′) and 1 minute equals 60 seconds (1′ = 60″).

FIGURE 2.14 ■ Arm-type drafting machine. *Courtesy VEMCO America, Inc.*

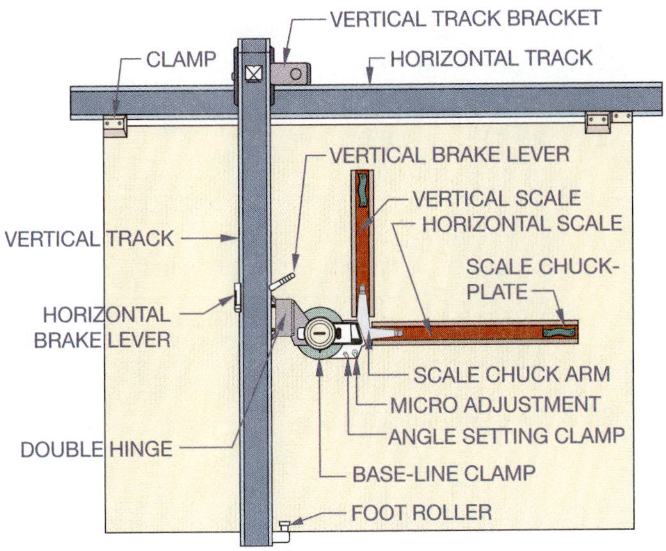

FIGURE 2.15 ■ Track drafting machine and its parts. *Courtesy Mutoh America, Inc.*

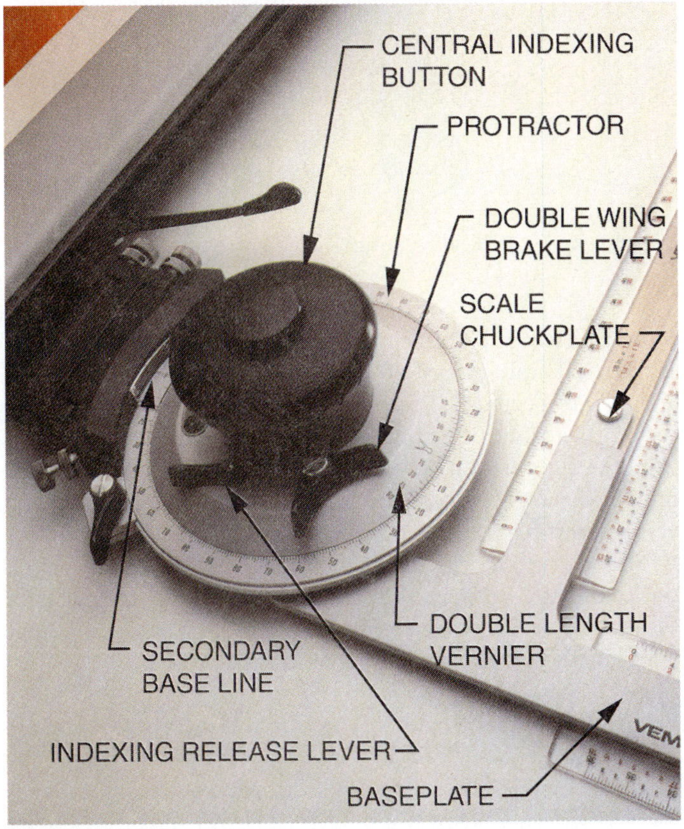

FIGURE 2.16 ■ Drafting machine head, controls, and parts. *Courtesy Vemco.*

Controls and Machine Head Operation

The drafting machine head contains the controls for horizontal, vertical, and angular movement. Although each brand has similar features, controls may be found in different places on different brands. See Figure 2.16.

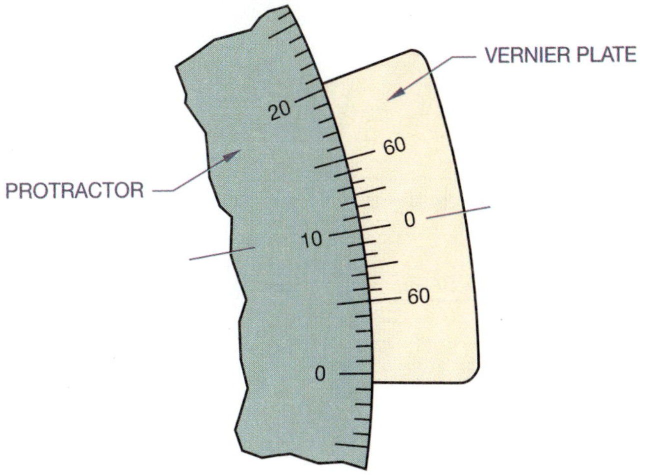

FIGURE 2.17 ■ Vernier plate and protractor showing a reading of 10°.

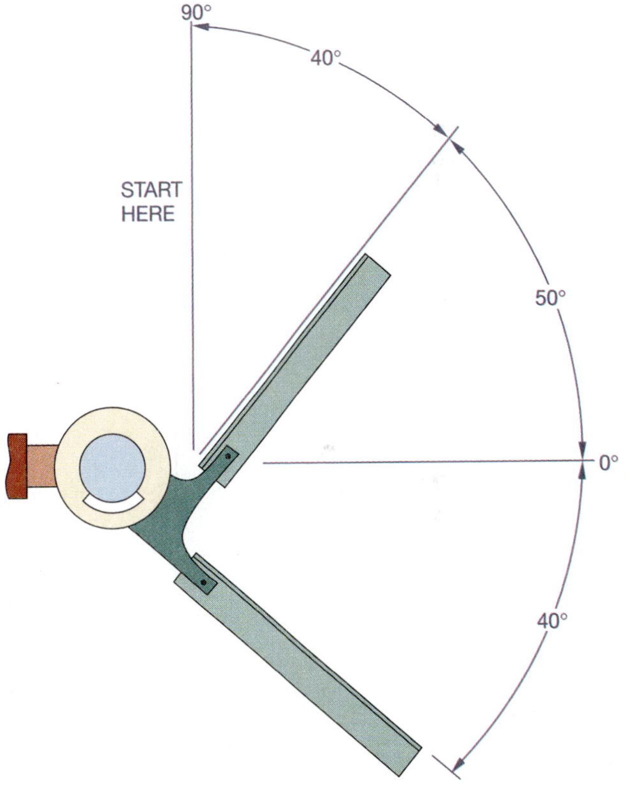

FIGURE 2.18 ■ Angle measurement.

To operate the drafting machine protractor head, place your hand on the handle, and with your thumb, depress the index thumb piece. Doing so allows the head to rotate. Each increment marked on the protractor is 1°, with a label every 10° (Figure 2.17). As the vernier plate (the small scale numbered from 0 to 60) moves past the protractor, the zero on the vernier aligns with the angle that you wish to read. Figure 2.17 shows a reading of 10°. As you rotate the handle, notice that the head automatically locks every 15°. To move the protractor past the 15° increment, you must again depress the index thumb piece. (Figure 2.16 shows the index thumb piece.)

With the protractor head rotated 40° clockwise, the machine is in the position shown in Figure 2.18. The vernier plate at the

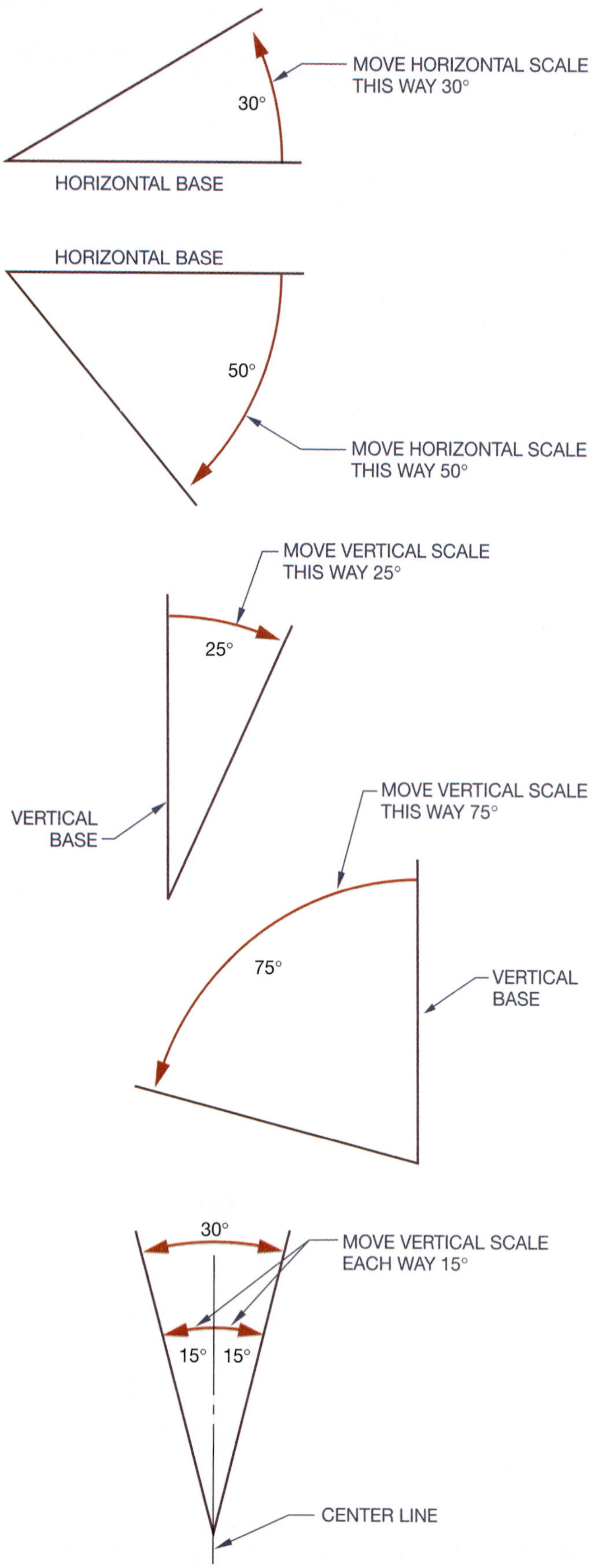

FIGURE 2.19 ■ Angle measurements reference from horizontal or vertical.

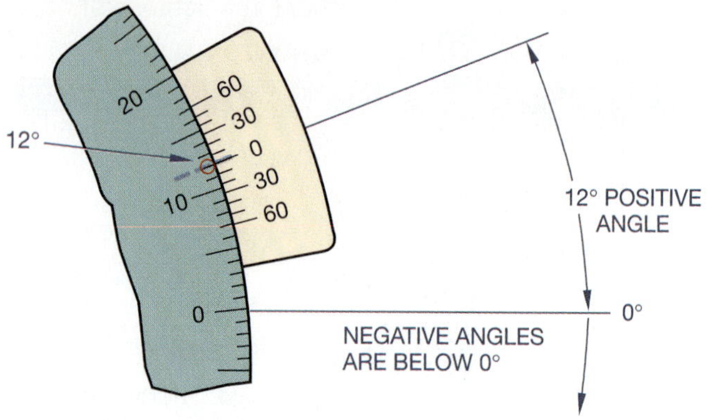

FIGURE 2.20 ■ Measuring full degrees.

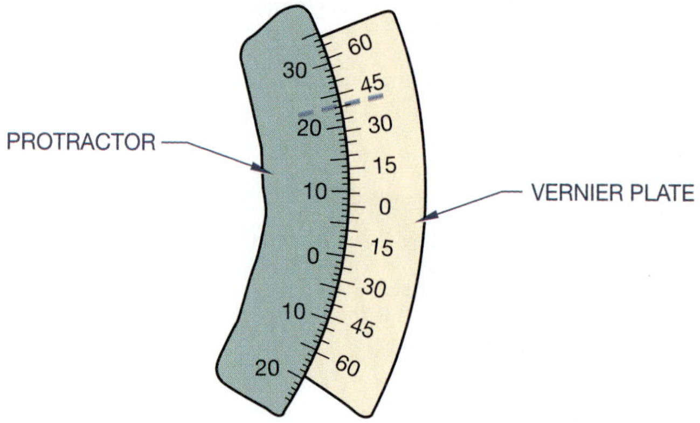

FIGURE 2.21 ■ Reading positive angles with the vernier.

protractor reads 40°, which means that both the horizontal and the vertical scale moved 40° from their original position at 0° and 90°, respectively. The horizontal scale reads directly from the protractor starting from 0°. The vertical scale reading begins from the 90° position. The key to measuring angles is to determine if the angle is to be measured from the horizontal or the vertical starting point. See the examples in Figure 2.19.

Measuring full degree increments is easy: you simply match the zero mark on the vernier plate with a full degree mark on the protractor. See the reading of 12° in Figure 2.20. The vernier scale allows you to measure angles as accurately as 5′.

Reading and Setting Angles with the Vernier

To read an angle other than a full degree, we will assume that the vernier scale is set at a *positive angle,* as shown in Figure 2.21. Positive angles read upward, and negative angles read downward from zero. Each mark on the vernier scale represents 5′. First see that the angle to be read is between 7° and 8°. Then find the 5′ mark on the *upper* half of the vernier (the direction in which the scale has been turned) that is most closely aligned with a full degree on the protractor. In this example, it is the 40′ mark. Add the minutes to the degree just past. The correct reading, then, is 7°40′.

The procedure for reading *negative* angles is the same, except that you read the minute marks on the *lower* half of the vernier.

SCALES

Scale Shapes

There are four basic scale shapes (Figure 2.22). The two-bevel scales are also available with chuckplates for use with standard arm or track drafting machines. These machine scales have typical calibrations; and some have no scale reading, for use as a straightedge alone. Drafting machine scales are purchased by designating the length needed—12, 18, or 24″—and the scale calibration, such as metric, engineer's full scale in tenths and half scale in twentieths or architect's scale 1/4″ = 1′–0″ and 1/2″ = 1′–0″ Many other scales are also available.

The triangular scale is commonly used because it offers many scale options and it is easy to handle.

Scale Notation

The scale of a drawing is usually noted in the title block or below the view that differs in scale from that given in the title block. Drawings are scaled so the object represented can be illustrated clearly on standard sizes of paper. It would be difficult, for example, to draw a house full size; thus, a scale that reduces the size of such large objects must be used:

The scale selected depends on:

- Actual size of the structure.
- Amount of detail to be shown.
- Paper size selected.
- Amount of dimensions and notes required.
- Common practice that regulates certain scales.

The following scales and their notation are frequently used on architectural drawings.

1/8″ = 1′–0″	3/8″ = 1′–0″	1 1/2″ = 1′–0″
3/32″ = 1′–0″	1/2″ = 1′–0″	3″ = 1′–0″
3/16″ = 1′–0″	3/4″ = 1′–0″	12″ = 1′–0″
1/4″ = 1′–0″	1″ = 1′–0″	

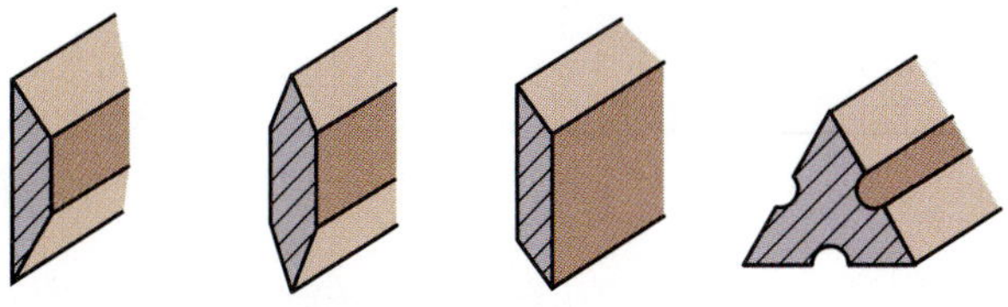

FIGURE 2.22 ■ Scale shapes.

Some scales used in civil drafting are noted as follows:

1″ = 10′	1″ = 50′	1″ = 300′
1″ = 20′	1″ = 60′	1″ = 400′
1″ = 30′	1″ = 100′	1″ = 500′
1″ = 40′	1″ = 200′	1″ = 600′

Metric Scale

According to the American National Standards Institute (ANSI), the SI (International System of Units) linear unit commonly used on drawings is the millimeter. On drawings where all dimensions are in either inches or millimeters, individual identification of units is not required. However, the drawing shall contain a note stating: UNLESS OTHERWISE SPECIFIED, ALL DIMENSIONS ARE IN INCHES [or MILLIMETERS as applicable]. Where some millimeters are shown on an inch-dimensioned drawing, the millimeter value should be followed by the symbol mm. Where some inches are shown on a millimeter-dimensioned drawing, the inch value should be followed by the abbreviation IN.

Metric symbols are as follows:

millimeter = mm	dekameter = dam
centimeter = cm	hectometer = hm
decimeter = dm	kilometer = km
meter = m	

Some metric-to-metric equivalents are the following:

10 millimeters = 1 centimeter
10 centimeters = 1 decimeter
10 decimeters = 1 meter
10 meters = 1 dekameter
10 dekameters = 1 kilometer

Some metric-to-U.S. customary equivalents are the following:

1 millimeter = 0.03937 inch
1 centimeter = 0.3937 inch
1 meter = 39.37 inches
1 kilometer = 0.6214 mile

Some U.S. customary-to-metric equivalents are the following:

1 mile = 1.6093 kilometers = 1609.3 meters
1 yard = 914.4 millimeters = 0.9144 meter
1 foot = 304.8 millimeters = 0.3048 meter
1 inch = 25.4 millimeters = 0.0254 meter

To convert inches to millimeters, multiply inches by 25.4.

One advantage of metric scales is that any scale is a multiple of 10; therefore, reductions and enlargements are easily performed. In most cases, no mathematical calculations should be required when using a metric scale. Whenever possible, select a direct reading scale. If no direct reading scale is available, use multiples of other scales. To avoid the possibility of error, avoid multiplying or dividing metric scales by anything but multiples of ten. Some metric scale calibrations are shown in Figure 2.23. See "Using Metric" in Chapter 15.

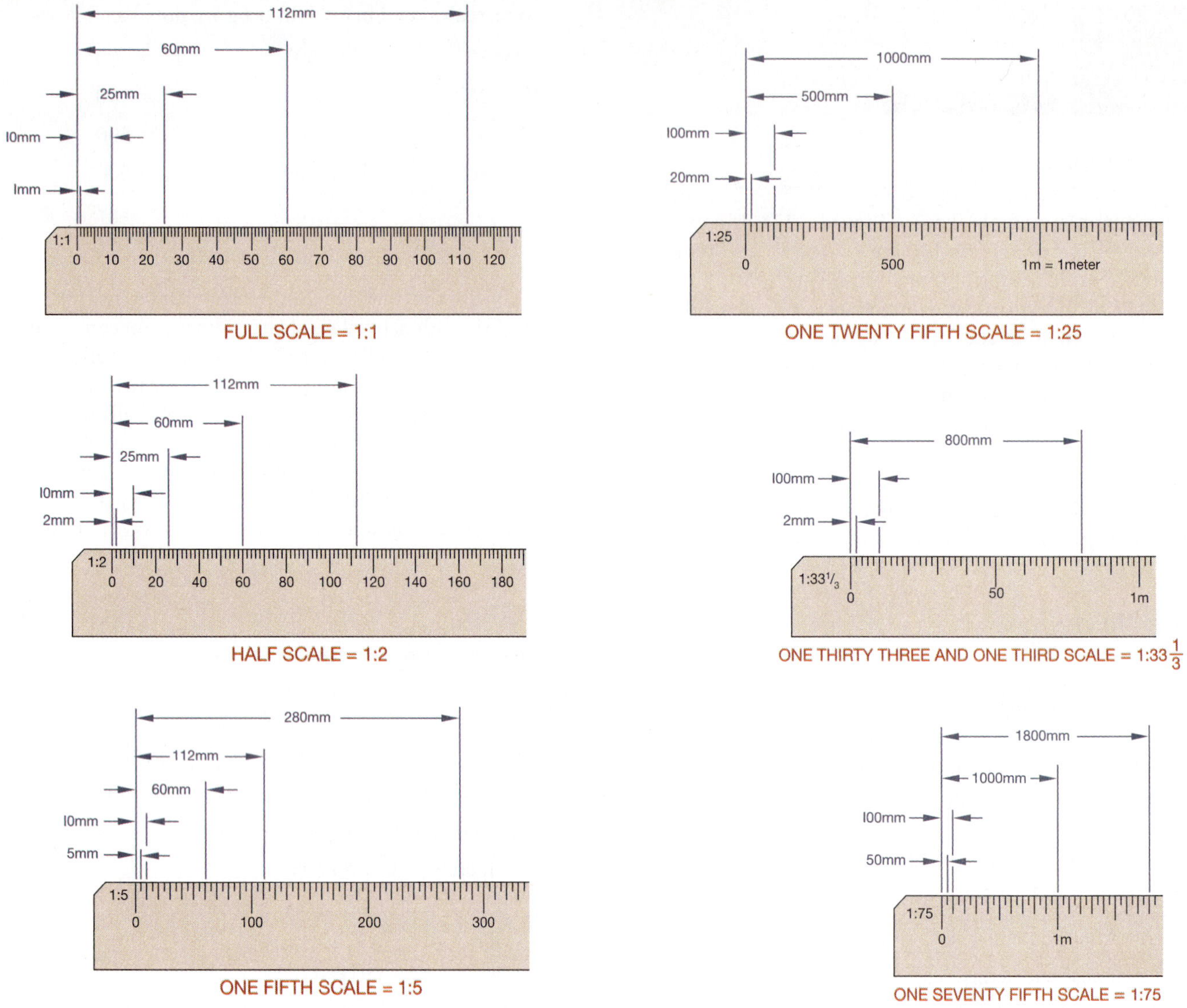

Metric Scales	METRIC SCALES THAT MATCH ARCHITECT AND CIVIL SCALES	
	Inch-to-Feet Architectural Scales	Civil Engineering Scales
1:1	FULL SCALE, 1:1	
1:5	Close to 3" = 1'–0"	
1:10	Between 1" = 1'–0" and 1 1/2" = 1'–0"	
1:20	Between 1/2" = 1'–0" and 3/4" = 1'–0"	
1:50	Close to 1/4" = 1'–0"	
1:100	Close to 1/8" = 1'–0"	
1:200	Close to 1/16" = 1'–0"	
1:500		Close to 1" = 40'
1:1000		Close to 1" = 80'

FIGURE 2.23 ■ Common metric scale calibrations and measurements.

Architect's Scale

The triangular architect's scale contains 11 different scales. On ten of them, each inch represents a foot and is subdivided into multiples of 12 parts to represent inches and fractions of an inch. The eleventh scale is the full scale with a 16 in the margin. The 16 means that each inch is divided into 16 parts, and each part is equal to 1/6 inch. Figure 2.24 shows an example of the full architect's scale; Figure 2.25 shows the fraction calibrations.

Look at the architect's scale examples in Figure 2.26. Note the form in which scales are expressed on a drawing. The scale is expressed as an equation of the drawing size in inches or fractions of an inch to one foot. For example: 3″ = 1′–0″, 1/2″ = 1′–0″, or 1/4″ = 1′–0″. The architect's scale commonly has scales running in both directions along an edge. Be careful when reading a scale from left to right so as not to confuse its calibrations with the scale that reads from right to left.

When selecting scales for preparing a set of architectural drawings, there are several factors that must be considered. Some parts of a set of drawings are traditionally drawn at a certain scale. For example:

- Floor plans for most residential structures are drawn at 1/4″ = 1′–0″.
- Although some applications may require a larger or smaller scale, commercial buildings are often drawn at 1/8″ = 1′–0″ when they are too large to draw at the same scale as a residence.
- Exterior elevations are commonly drawn at 1/4″ = 1′–0″, although some architects prefer to draw the front or most important elevation at 1/4″ = 1′–0″ and the balance of the elevations at 1/8″ = 1′–0″.
- Construction details and cross sections may be drawn at larger scales to help clarify specific features. Some cross sections can be drawn at 1/4″ = 1′–0″ with clarity, but complex cross sections require a scale of 3/8″ = 1′–0″ or 1/2″ = 1′–0″ to be clear.
- Construction details may be drawn at any scale between 1/2″ = 1′–0″ and 3″ = 1′–0″, depending on the amount of information presented.

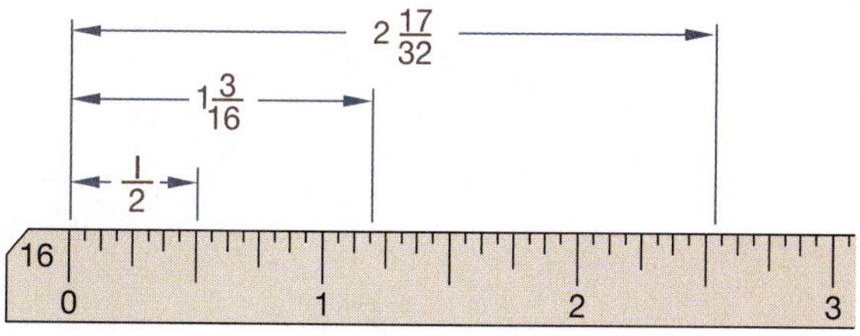

FIGURE 2.24 ■ Full (1:1), or 12″ = 1′–0″, architect's scale.

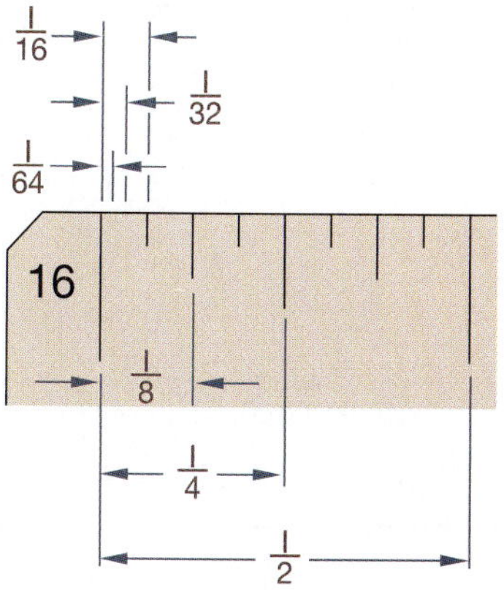

FIGURE 2.25 ■ Enlarged view of an architect's 16 scale.

Civil Engineer's Scale

The architectural drafter may need to use the civil engineer's scale to draw site plans, maps, or subdivision plats or to read existing land documents. Most land-related plans such as construction site plans are drawn using the civil engineer's scale, although some are drawn with an architect's scale. Common architect's scales used are 1/8″ = 1′–0″, 3/32″ = 1′–0″, and 1/16″ = 1′–0″. Common civil engineer's scales used for site plans are 1″ = 10′, 1″ = 50′, and 1″ = 100′. The triangular civil engineer's scale contains six scales, one on each of its sides. The civil engineer's scales are calibrated in multiples of ten. The scale margin specifies the scale represented on a particular edge. Table 2.1 shows some of the many scale options available when you are using the civil engineer's scale. Keep in mind that any multiple of ten is available with this scale.

The 10 scale is used in civil drafting for scales of 1″ = 10′ or 1″ = 100′ and so on. See Figure 2.27. The 20 scale is used for scales such as 1″ = 2′, 1″ = 20′, or 1″ = 200′.

The remaining scales on the civil engineer's scale may be used in a similar fashion. The 50 scale is popular in civil drafting for drawing plats of subdivisions. Figure 2.28 shows scales that are commonly used on the civil engineer's scale, with sample measurements.

PROTRACTORS

When a drafting machine is not used or when combining triangles does not provide enough flexibility for drawing angles, a protractor is the instrument to use. Protractors can be used to measure angles to within 1/2° or even closer if you have a good eye.

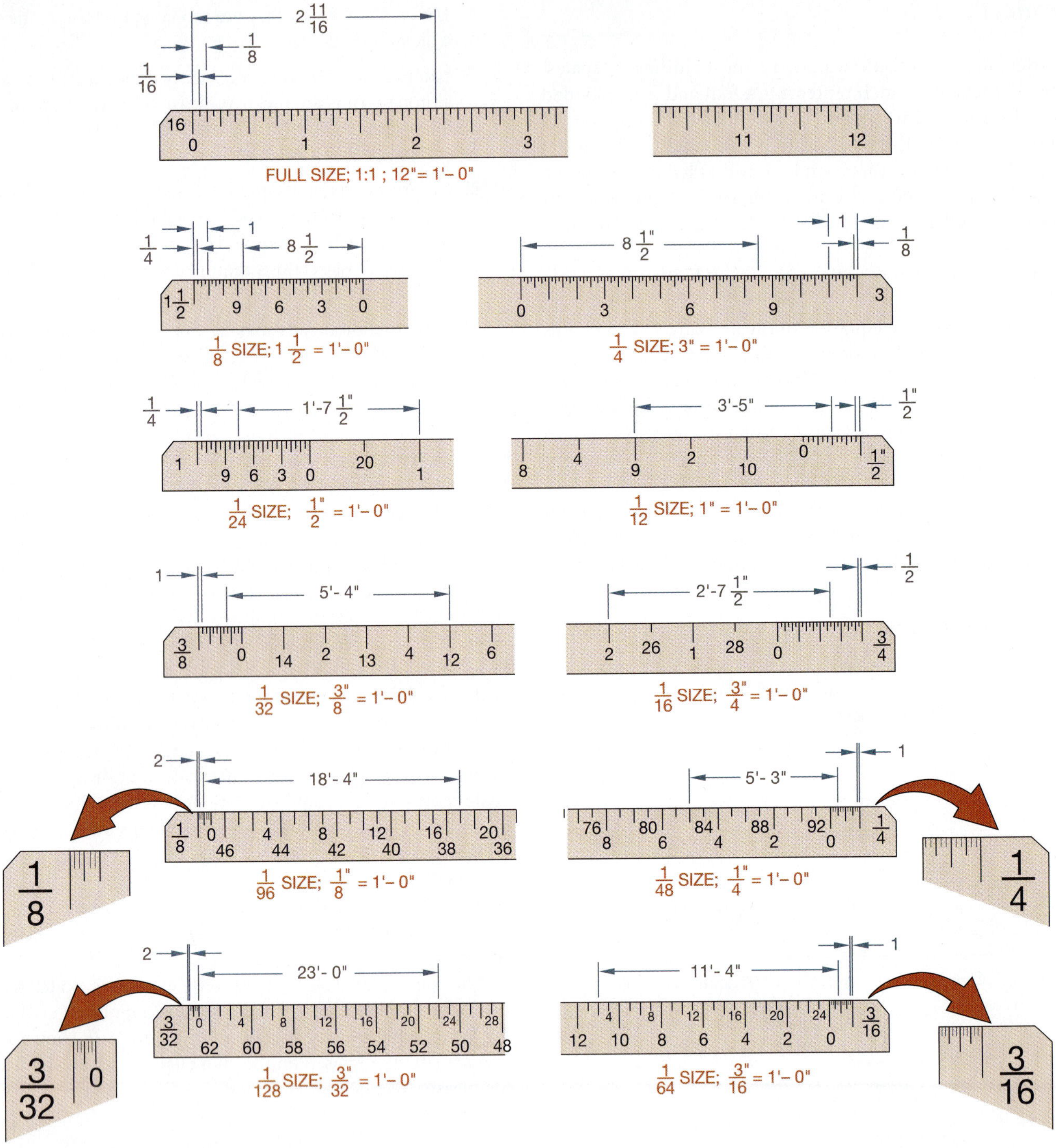

FIGURE 2.26 ■ Architect's scale combinations.

Divisions	Ratio	Scales Used with This Division			
10	1:1	1" = 1"	1" = 1'	1" = 10'	1" = 100'
20	1:2	1" = 2"	1" = 2'	1" = 20'	1" = 200'
30	1:3	1" = 3"	1" = 3'	1" = 30'	1" = 300'
40	1:4	1" = 4"	1" = 4'	1" = 40'	1" = 400'
50	1:5	1" = 5"	1" = 5'	1" = 50'	1" = 500'
60	1:6	1" = 6"	1" = 6'	1" = 60'	1" = 600'

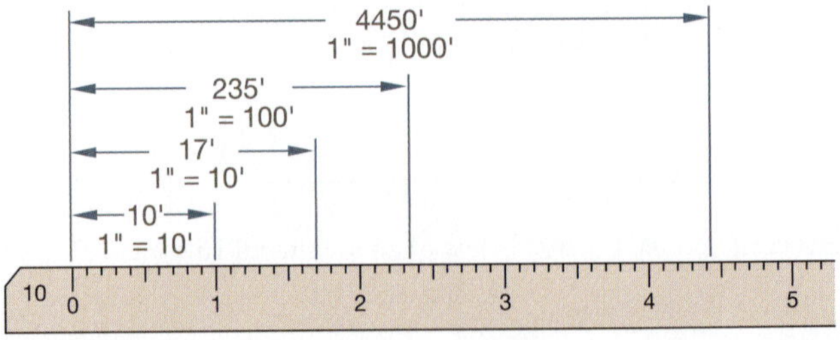

FIGURE 2.27 ■ Civil engineer's scale, units of ten.

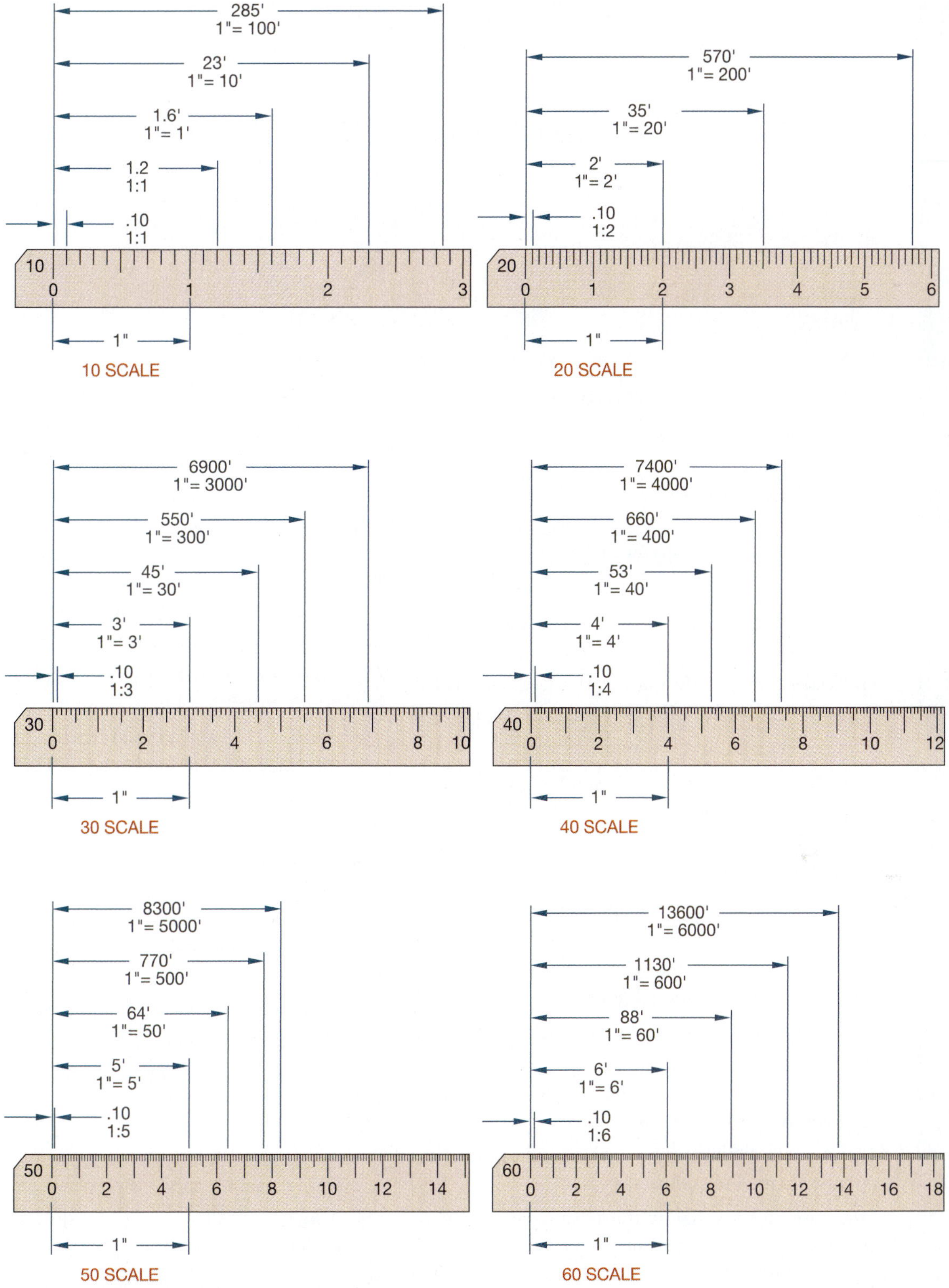

FIGURE 2.28 ■ Civil engineer's scales and sample measurements at different scales.

CHAPTER 2

Architectural Drafting Equipment Test

DIRECTIONS

Answer the questions with short, complete statements or drawings as needed on 8 1/2 × 11 lined paper, as follows:

1. Letter your name, Chapter 2 Test, and the date at the top of the sheet.
2. Letter the question number and provide the answer. You do not need to write out the question.

Answers may be prepared on a word processor if course guidelines allow this.

QUESTIONS

Question 2–1 Describe a mechanical drafting pencil.

Question 2–2 Describe an automatic drafting pencil.

Question 2–3 List three lead sizes available for automatic pencils.

Question 2–4 Identify how graphite leads are labeled.

Question 2–5 Describe in a short complete paragraph how the mechanical pencil is properly used.

Question 2–6 Describe in a short complete paragraph how the automatic pencil is properly used.

Question 2–7 List three characteristics recommended for technical pen inks.

Question 2–8 Describe one advantage and one disadvantage of using an electric eraser.

Question 2–9 Name the type of compass most commonly used by professional drafters.

Question 2–10 Why should templates be used whenever possible?

Question 2–11 Why should a compass point with a shoulder be used when possible?

Question 2–12 Identify two lead points that are commonly used in compasses.

Question 2–13 What is the advantage of placing drafting tape or plastic at the center of a circle before using a compass?

Question 2–14 A good divider should not be too loose or too tight. Why?

Question 2–15 List two uses for dividers.

Question 2–16 At what number of degree increments do most drafting machine heads automatically lock?

Question 2–17 How many minutes are there in 1°? How many seconds in 1′?

Question 2–18 Explain the uses of and differences between the soft-green and soft-pink eraser.

Question 2–19 Name the type of eraser that is recommended for use on polyester drafting film.

Question 2–20 Provide a short complete paragraph explaining proper erasing technique.

Question 2–21 Name at least five features found on a residential floor plan template.

Question 2–22 Name the piece of drafting equipment used to draw curves that have no constant radii.

Question 2–23 Describe the parallel bar and how it is used.

Question 2–24 Explain why the parallel bar is commonly used for manual drafting in the architectural office.

Question 2–25 In combination, the 30°–60° and 45° triangles may be used to make what range of angles?

Question 2–26 Show how the following scales are noted on an architectural drawing: 1/8″, 1/4″, 1/2″, 3/4″, 1 1/2″, and 3″.

Question 2–27 Identify a civil engineer's scale that is commonly used for subdivision plats.

Question 2–28 What scale is usually used for drawing residential floor plans?

Question 2–29 List three factors that influence the selection of a scale for a drawing.

Question 2–30 Name the scale that is commonly used, and why.

Question 2–31 To convert inches to millimeters, multiply inches by ______.

Question 2–32 List an advantage of using metric scales.

(continued)

QUESTIONS *(cont.)*

Determine the length of the following lines using an architect's, civil engineer's, and a metric scale. The scales to use when measuring are given above each line. The architect's scale is first; the civil engineer's scale is second; and the metric scale is third.

Question 2–33 a. Scale: 1/8″ = 1′–0″ b. 1″ = 10′–0″ c. 1:1 (metric)

Question 2–34 a. Scale: 1/4″ = 1′–0″ b. 1″ = 20′–0″ c. 1:2 (metric)

Question 2–35 a. Scale: 3/8″ = 1′–0″ b. 1″ = 30′–0″ c. 1:5 (metric)

Question 2–36 a. Scale: 1/2″ = 1′–0″ b. 1″ = 40′–0″ c. 1:10 (metric)

Question 2–37 a. Scale: 3/4″ = 1′–0″ b. 1″ = 50′–0″ c. 1:20 (metric)

Question 2–38 a. Scale: 1″ = 1′–0″ b. 1″ = 60′–0″ c. 1:50 (metric)

Question 2–39 a. Scale: 1 1/2″ = 1′–0″ b. 1″ = 100′ c. 1:75 (metric)

Question 2–40 a. Scale: 3″ = 1′–0″ b. 1″ = 200′ c. 1:75 (metric)

3 CHAPTER

Drafting Media and Reproduction Methods

INTRODUCTION

This chapter covers the types of media that are used for the drafting and the reproduction of architectural drawings. Drawings are generally prepared on precut drafting sheets with printed graphic designs for company borders and title blocks. Drawing reproductions or copies are made for use in the construction process so the original drawings are not damaged. Alternative materials and reproduction methods are discussed as they relate to common practices.

PAPERS AND FILMS

Selection of Drafting Media

Several factors influence the choice and use of drafting media: durability, smoothness, erasability, dimensional stability, transparency, and cost.

Durability should be considered if the original drawing will have a great deal of use. Originals may tear or wrinkle, and the images become difficult to see if the drawings are used often.

Smoothness relates to how the medium accepts linework and lettering. The material should be easy to draw on so the image is dark and sharp without excessive effort on the part of the drafter.

Erasability is important because errors need to be corrected and changes frequently made. When images are erased, ghosting should be kept to a minimum. *Ghosting* is the residue when lines are not completely removed. These unsightly ghost images reproduce in a print. Materials that have good erasability are easy to clean up.

Dimensional stability is the quality of the medium that maintains size despite the effects of atmospheric conditions such as heat and cold. Some materials are more dimensionally stable than others.

Transparency is one of the most important characteristics of drawing media. The diazo reproduction method requires light to pass through the material. The final goal of a drawing is good reproduction, so the more transparent the material is, the better is the reproduction, assuming that the image drawn is of professional quality. Transparency is not a factor with photocopying.

Cost may influence the selection of drafting media. When the best reproduction, durability, dimensional stability, smoothness, and erasability are important, there may be few cost alternatives. If drawings are to have normal use and reproduction quality, then the cost of the drafting media may be kept to a minimum. The following discussion of available materials will help you evaluate cost differences.

Vellum

Vellum is drafting paper that is specially designed to accept pencil or ink. Lead on vellum is the most common combination used for manual drafting.

Vellum is the least expensive material having good smoothness and transparency. Use vellum originals with care. Drawings made on vellum that require a great deal of use could deteriorate; vellum is not as durable a material as others. Also, some brands are more erasable than others. Affected by humidity and other atmospheric conditions, vellum generally is not as dimensionally stable as other materials.

Polyester Film

Polyester film, also known by its trade name, Mylar®, is a plastic material that is more expensive than vellum but offers excellent dimensional stability, erasability, transparency, and durability. Drawing on polyester film is best accomplished using ink or special polyester leads. Do not use regular graphite leads; they smear easily. Drawing techniques that drafters use with polyester leads are similar to those used with graphite leads except that polyester lead is softer and feels like a crayon.

Film is available with a single or double mat surface. *Mat* is surface texture. Double-mat film has texture on both sides so drawing can be done on either side if necessary. Single-mat film, the most common, has a slick surface on one side. When using polyester film, be very careful not to damage the mat by erasing. Erase at right angles to the direction of your lines, and do not use too much pressure. This helps minimize damage to the mat surface. Once the mat is destroyed (removed), the surface will not accept ink or pencil. Also, be careful not to get moisture on the polyester film surface. Moisture from your hands can cause ink to skip across the material.

Normal handling of drawing film is bound to soil it. Inked lines applied over soiled areas do not adhere well and in time will chip off or flake. It is always good practice to keep the film clean. Soiled areas can be cleaned effectively with special film cleaner.

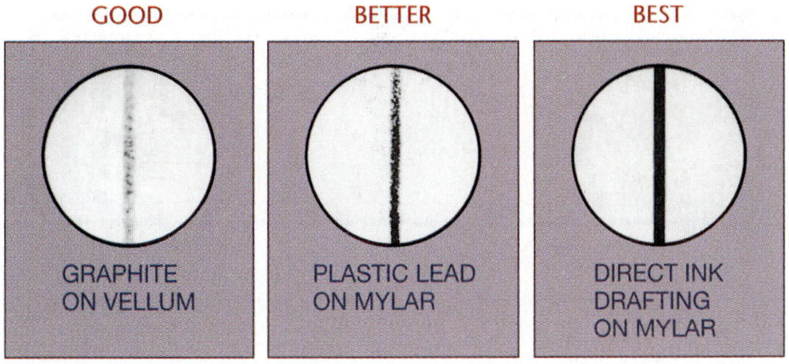

FIGURE 3.1 ■ A magnified comparison of graphite on vellum, plastic lead on Mylar, and ink on Mylar. *Courtesy Koh-I-Noor, Inc.*

Reproductions

The one thing most designers, engineers, architects, and drafters have in common is that their finished drawings are made to be reproduced. The goal of every professional is to produce drawings of the highest quality that give the best possible prints when they are reproduced.

Reproduction is the most important factor in the selection of media for drafting. The primary combination that achieves the best reproduction is the blackest, most opaque lines or images on the most transparent base or material. Each of the materials mentioned makes good prints if the drawing is well done. Some products have better characteristics than others. It is up to the individual or company to determine the combination that works best for its needs and budget. The question of reproduction is especially important when sepias must be made. (Sepias, discussed later in this chapter, are second- or third-generation originals.)

Figure 3.1 shows a magnified view of graphite on vellum, plastic lead on polyester film, and ink on polyester film. Judge for yourself which combination provides the best reproduction. As you can see from the figure, the best reproduction is achieved with a crisp, opaque image on transparent material. If your original drawing is not of good quality, it will not get better on the print.

SHEET SIZES, TITLE BLOCKS, AND BORDERS

Every professional drawing has a title block. Standards have been developed for the information put into the title block and on the sheet next to the border. A standardized drawing is easier to read and file than drawings that do not follow a standard format.

Sheet Sizes

Drafting materials are available in standard sizes, which are determined by manufacturers' specifications. Paper and polyester film may be purchased in cut sheets or in rolls. Architectural drafting offices generally use cut sheet sizes, in inches, of 18 × 24″, 24 × 36″, 28 × 42″, 30 × 42″, 30 × 48″, or 36 × 48″. Sheet sizes used in architectural drafting are often different, but they can be the same as sheet sizes recommended by the American National Standards Institute (ANSI). This depends on the practice followed in a specific architectural office. Table 3–1 shows the standard ANSI recommended sheet sizes.

TABLE 3–1 STANDARD ANSI INCH DRAWING SHEET SIZES

DIMENSIONS (INCHES)	SHEET SIZES
A	81/2 × 11 or 11 × 81/2
B	17 × 11
C	22 × 17
D	34 × 22
E	44 × 34
F	40 × 28

Metric sheet sizes vary from 420 × 594 mm to 841 × 1189 mm. Roll sizes vary in width from 18 to 48″. Metric roll sizes range from 297 to 420 mm wide. The International Organization for Standardization (ISO) has established standard metric drawing sheet sizes (Table 3–2).

TABLE 3–2 STANDARD METRIC DRAWING SHEET SIZES

SHEET SIZE	DIMENSIONS (MM)	DIMENSIONS (INCHES)
A0	1189 × 841	46.8 × 33.1
A1	841 × 594	33.1 × 23.4
A2	594 × 420	23.4 × 16.5
A3	420 × 297	16.5 × 11.7
A4	297 × 210	11.7 × 8.3

Zoning

Some companies use a system of numbers along the top and bottom margins and letters along the left and right margins. This is called *zoning*. Zoning allows a drawing to be read like a road map. For example, the reader can refer to the location of a specific item as D-4, which means the item can be found on or near the intersection of D across and 4 up or down. Zoning may be found on some architectural drawings, although it is more commonly used in mechanical drafting.

Architectural Drafting Title Blocks

Title blocks and borders are normally preprinted on drawing paper or polyester film to help reduce drafting time and cost.

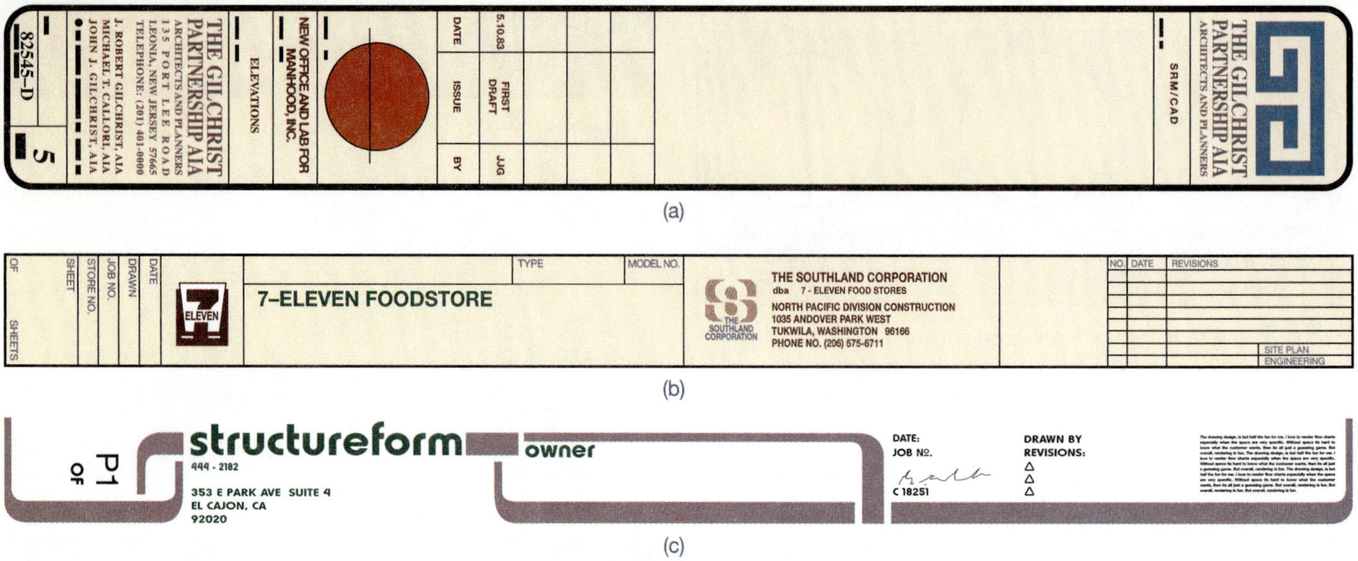

FIGURE 3.2 ■ Sample architectural title blocks. (a) *Courtesy Summagraphics Corporation;* (b) *courtesy Southland Corporation;* (c) *courtesy Structureform.*

Drawing borders are thick lines that go around the entire sheet. Top, bottom, and right-side border lines are usually between 3/8 and 1/2″ away from the paper edge; the left border may be between 3/4 and 1 1/2″ away. This extra-wide left margin allows for binding drawing sheets.

Preprinted architectural drawing title blocks are generally placed along the right side of the sheet, although some companies place them across the bottom of the sheet. Each company uses a slightly different title block design, but the same general information is found in almost all blocks:

1. *Drawing number.* This may be a specific job or file number for the drawing.
2. *Company name, address, and phone number.*
3. *Project or client.* This is an identification of the project by company or client, project title, or location.
4. *Drawing name.* This is where the title of the drawing may be placed—for example, MAIN FLOOR PLAN or ELEVATIONS. Most companies leave this information off the title block and place it on the face of the sheet below the drawing.
5. *Scale.* Some company title blocks provide space for the drafter to fill in the general scale of the drawing. Any view or detail on the sheet that differs from the general scale must have the scale identified below the view title, and both must be placed directly below the view.
6. *Drawing or sheet identification.* Each sheet is numbered in relation to the entire set of drawings. For example, if the complete set of drawings has eight sheets, then consecutive sheets are numbered 1 of 8, 2 of 8, 3 of 8, and so on, to 8 of 8.
7. *Date.* The date noted is the one on which the drawing or project is completed.
8. *Drawn by.* This is where the drafter, designer, or architect who prepared this drawing places his or her initials or name.
9. *Checked by.* This is the identification of the individual who approves the drawing for release.
10. *Architect or designer.* Most title blocks provide for the identification of the individual who designed the structure.
11. *Revisions.* Many companies provide a revision column where changes in the drawing are identified and recorded. After a set of drawings is released for construction, there may be some need for changes. When this is necessary, there is usually a written request for a change from the contractor, owner, or architect. The change is then implemented on the drawings that are affected. Where changes are made on the face of the drawing, a circle with a revision number accompanies the change. This revision number is keyed to a place in the drawing title block where the revision number, the revision date, the initials of the individual making the change, and an optional brief description of the revision are located. For further record of the change, the person making the change may also fill out a form, called a *change notice.* This change notice contains complete information about the revision for future reference.

Figure 3.2 shows different architectural title blocks. Notice the similarities and differences among title blocks.

HOW TO FOLD PRINTS

Architectural drawings are generally created on large sheets. It is necessary to fold the architectural prints properly when they must be mailed or filed in a standard file cabinet. Folding large prints is much like folding a road map. Folding is done in a pattern of bends that puts the title block and sheet identification on the front. Using the proper method to fold prints also aids in unfolding and refolding prints. Figure 3.3 shows how large prints are folded.

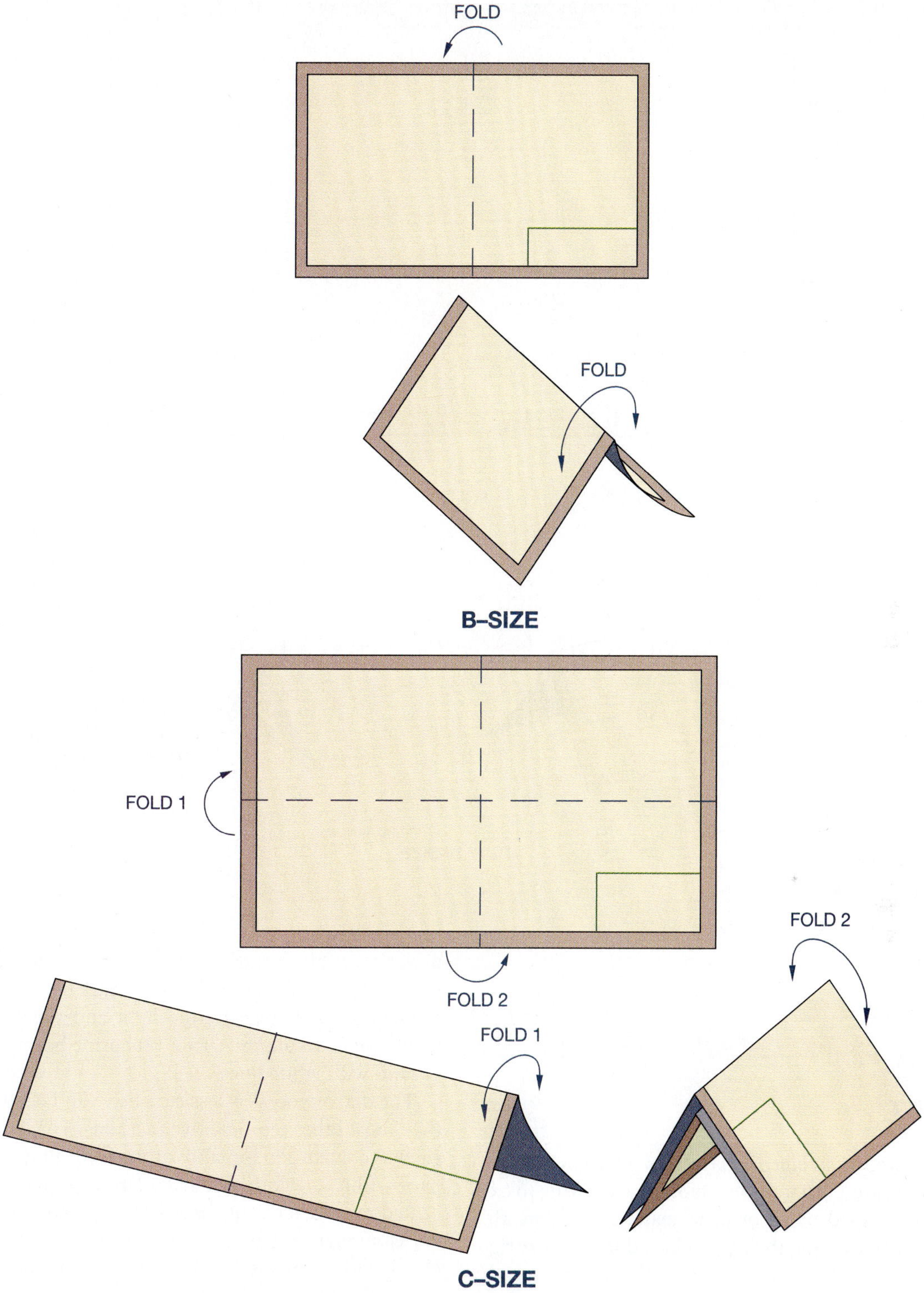

FIGURE 3.3 ■ How to fold B-size, C-size, and D-size prints.

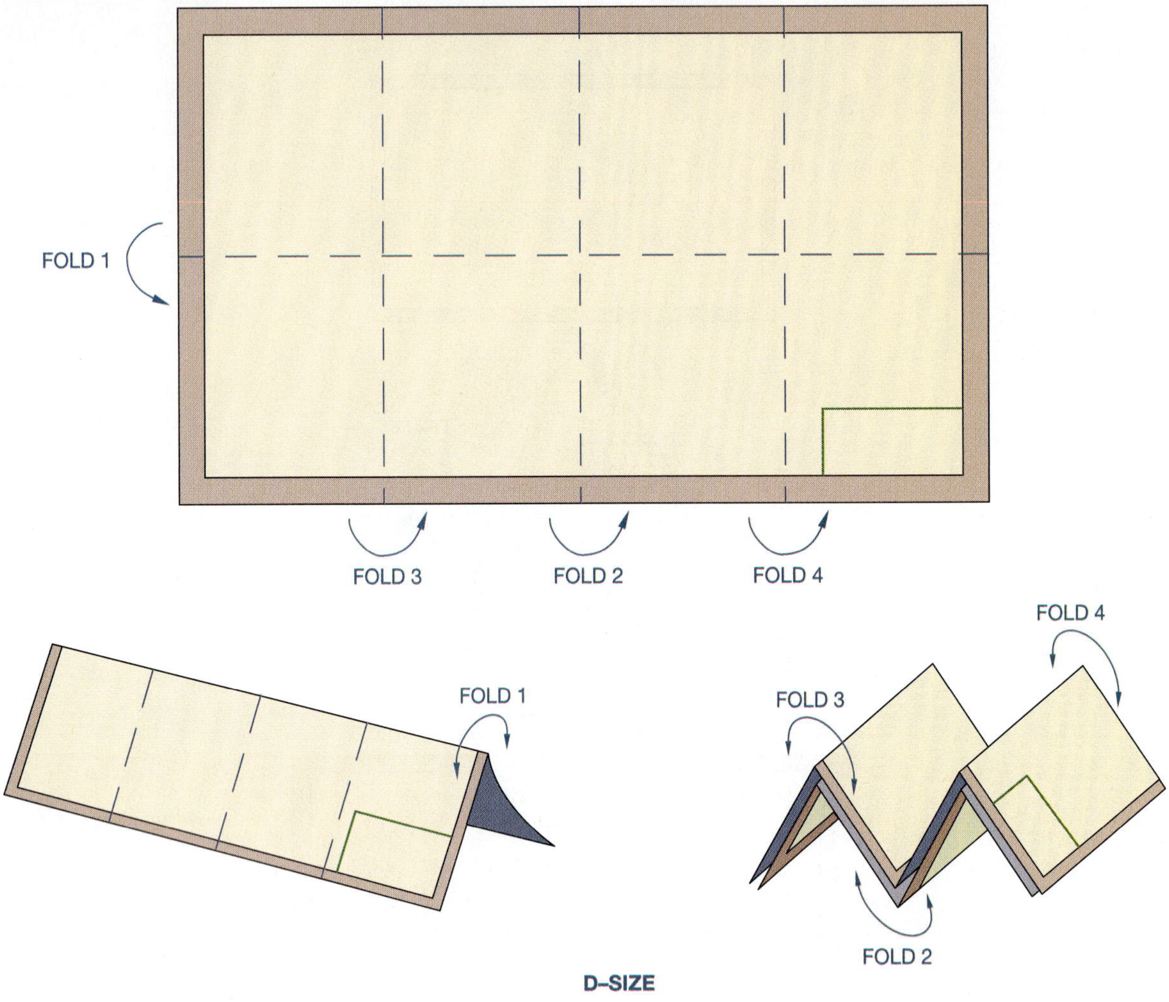

FIGURE 3.3 ■ *(continued)*

DIAZO REPRODUCTION

Diazo Printer

Diazo prints, also known as *blue-line prints,* are made by passing an ultraviolet light through a translucent original drawing to expose a chemically coated paper or print material underneath. The light does not go through the dense black lines on the original drawing; thus the chemical coating on the paper beneath the lines is not exposed. The print material is then exposed to ammonia vapor, which activates the remaining chemical coating to produce blue, black, or brown lines on a white or clear background. The print that results is a diazo or blue-line print, not a blueprint. *Blueprint,* however, is now a generic term that is used to refer to diazo prints, even though they are not true blueprints.

Diazo is a direct print process, and the print remains relatively dry. There are other processes available that require a moist development. For example, blueprinting, an old method of making prints, uses a light to expose sensitized paper placed under an original drawing. The prints are developed in a water wash, which turns the background dark blue. The lines from the original are not fixed by the light and wash out, leaving the white paper to show. A true blueprint has a dark blue background with white lines.

The diazo process is less expensive and less time-consuming than most other reproductive methods. The diazo printer is designed to make good-quality prints from all types of translucent paper, film, or cloth originals. Diazo prints may be made on coated roll stock or cut sheets. The operation of all diazo printers is similar, but be sure to read the instruction manual or check with someone familiar with a particular machine if you are using a machine that is new to you.

Diazo Printer Speed

The operating speed of the machine is determined by the speed of the diazo material and the transparency of the original. For example, polyester film may be run faster than vellum. The speed of the diazo material also influences the quality of the print. Diazo materials that are classified as slow produce a higher-quality print than fast materials. Some materials, such as fast-speed blue-line print paper, require a

high-speed or fast setting. Other materials, such as some sepias, require a low-speed or slow setting.

Testing the Printer

Before you run a print using a complete sheet, run a test strip to determine the quality of the print you will get. Cut a sheet of diazo material into strips; then use a strip to get the proper speed setting. You know you have the proper speed setting when the lines and letters are dark blue and the background is white. Once you are satisfied with the quality of the test strip, you are ready to run a full sheet to make a print.

Making a Diazo Print

To make a diazo print, place the diazo material on the feedboard, coated (yellow) side up. Position the original drawing, image side up, on top of the diazo material, making sure to align the leading edges. See Figure 3.4.

Using fingertip pressure from both hands, push the original and diazo material into the machine. The light will pass through the original and expose the sensitized diazo material, except where images (lines and lettering) exist on the original. When the exposed diazo material and the original drawing emerge from the printer section, remove the original and carefully feed the diazo material into the developer section, as shown in Figure 3.5.

In the developer section the sensitized material that remains on the diazo paper activates with ammonia vapor, and blue lines form. (Some diazo materials make black or brown lines.) Special materials are also available to make other colored lines, or to make transfer sheets and transparencies.

Storage of Diazo Materials

Diazo materials are light-sensitive, so they should always be kept in a dark place until ready for use. A long exposure to room light causes the diazo chemicals to deteriorate and reduces the quality of your print. If you notice brown or blue edges around unexposed diazo material, it is getting old. To repeat, then: Always keep diazo material tightly stored in the shipping package and in a dark place,

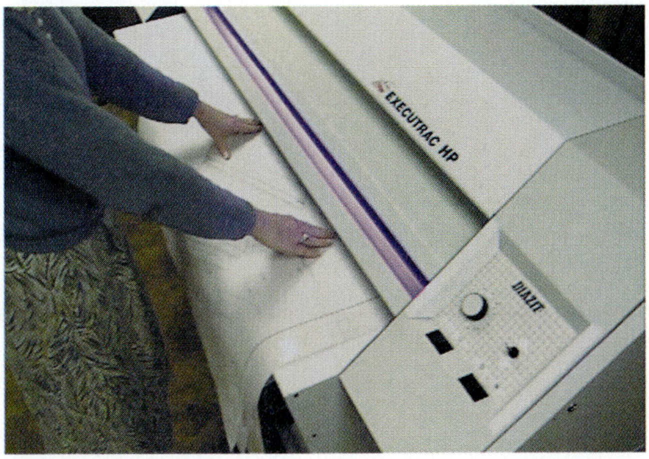

FIGURE 3.4 ■ Feeding the original drawing and diazo material into the machine. *Courtesy Ozalid Corporation.*

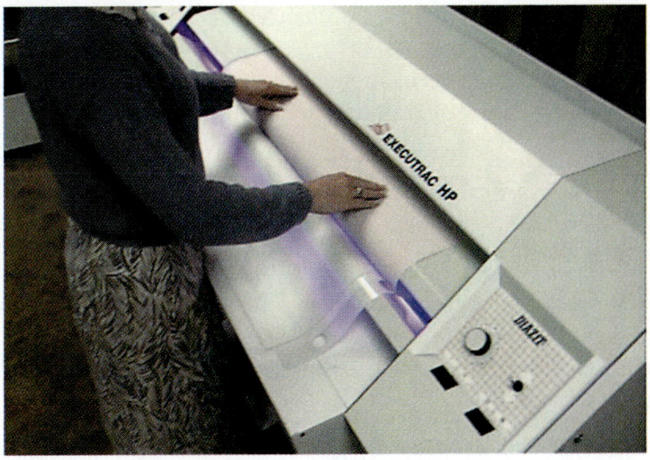

FIGURE 3.5 ■ Feeding exposed sensitized diazo material into the developer section. *Courtesy Ozalid Corporation.*

such as a drawer. Some people prefer to keep all diazo material in a small, dark refrigerator. This preserves the chemical for a long time. Old diazo prints, or spoiled diazo paper should not be placed with recycled paper. The chemicals on the paper do not need to be treated as hazardous waste and can be disposed of as trash.

Sepias

Sepias are diazo materials that are used to make secondary originals. A secondary, or second-generation, original is a print of an original drawing that can be used as an original. Changes can be made on the secondary original while the original drawing remains intact. Sepias are also used when originals are required at more than one company location. To make a secondary original, the diazo process is performed with sepia. Sepia prints normally form dark brown lines on a translucent background; therefore, they are sometimes called brown-lines.

Sepias are made in reverse (reverse sepias) so corrections or changes can be made on the mat side (face side) and erasures can be made on the sensitized side.

Sepia materials are available in paper or polyester. Paper sepia is similar to vellum; polyester sepia is polyester film. Drawing changes can be made on sepia paper with pencil or ink, and on sepia polyester film with polyester lead or ink.

Computer-aided drafting is rapidly taking the place of sepia, because drawings are easily changed on the computer and a new original is easily created.

Safety Precautions with Diazo

Ammonia

Ammonia has a strong, unmistakable odor. You may detect it, at times, while operating your diazo printer. Diazo printers are designed and built to provide safe operation free from ammonia exposure. Some machines have ammonia filters; others require outside exhaust fans. When print quality begins to deteriorate and it has been determined that the print paper is in good condition, the ammonia may be old. Ammonia bottles should be changed periodically, as determined by the quality of prints.

When the ammonia bottle is changed, it is generally time to change the filter also.

In handling bottles of ammonia or replacing a bottle supplying a diazo printer, precautions are required.

Eye Protection Avoid direct contact of ammonia with your eyes. Always wear safety goggles or equivalent eye protection when handling ammonia containers directly or handling ammonia supply systems.

Ammonia and Filter Avoid any direct contact between ammonia or the ammonia filter and your skin or clothing. Ammonia and ammonia residue on the filter can cause uncomfortable irritation and burns.

Ammonia Fumes The disagreeable odor of ammonia is usually sufficient to prevent breathing harmful concentrations of ammonia vapors. Avoid prolonged periods of breathing close to open containers of ammonia or where a strong, pungent odor is present. The care, handling, and storage of ammonia containers should be in accordance with the suppliers' instructions and all applicable regulations. Although ammonia may weaken to the point of not being suitable for making blue prints, it still emits a harmful vapor. Most municipalities allow spent containers of ammonia to be flushed down the waste water system when diluted with water per the distributors instructions. Because of the ammonia fumes released when it is diluted, a better practice is to recycle the spent ammonia to the distributor when new ammonia is delivered

First Aid If ammonia is spilled on your skin, promptly wash with plenty of water, and remove your clothing if necessary, to flush affected areas adequately. If your eyes are affected, irrigate with water as quickly as possible for at least 15 minutes, and, if necessary, consult a physician.

Anyone overcome by ammonia fumes should be removed to fresh air at once. Apply artificial respiration, preferably with the aid of oxygen if breathing is labored or has stopped. Obtain medical attention at once in the event of eye contact or burns to the nose or throat, or if the person is unconscious.

Ultraviolet Light

Under the prescribed operating instructions, there is no exposure to the ultraviolet rays emitted from illuminated printing lamps. However, to avoid possible eye damage, no one should attempt to look directly at the illuminated lamps under any circumstances.

PHOTOCOPY REPRODUCTION

The photocopy engineering-size printer makes prints up to 24″ wide and up to 25′ long from originals up to 36″ wide by 25′ long. Prints can be made on bond paper, vellum, polyester film, colored paper, or other translucent materials. Reproduction capabilities also include reduction and enlargement on some models.

Almost any large original can be converted into a smaller-sized reproducible print for distribution, inclusion in manuals, or more convenient handling. Also, a random collection of mixed-scale drawings can be enlarged or reduced and converted to one standard scale and format. Reproduction clarity is so good that halftone illustrations (photographs) and solid or fine line work have excellent resolution and density.

MICROFILM

In many companies original drawings are filed in drawers by drawing number. When a drawing is needed, the drafter finds the original, removes it, and makes a copy. This process works well, although, depending on the size of the company or the number of drawings generated, storage often becomes a problem. Sometimes an entire room is needed for storage cabinets. Another problem occurs when originals are used over and over. They often become worn and damaged, and old vellum becomes yellowed and brittle. Also, in case of a fire or some other kind of destruction, originals may be lost and endless hours of drafting may vanish. For these and other reasons some companies are using microfilm for the storage and reproduction of original drawings.

CHAPTER 3

Drafting Media and Reproduction Test

DIRECTIONS

Answer the questions with short, complete statements or drawings as needed on $8\frac{1}{2} \times 11$ lined paper, as follows:

1. Letter your name, Chapter 3 Test, and the date at the top of the sheet.
2. Letter the question number and provide the answer. You do not need to write out the question.

Answers may be prepared on a word processor if course guidelines allow this.

QUESTIONS

Question 3–1 List five factors in the choice and use of drafting materials.

Question 3–2 Why is transparency so important in reproducing copies using the diazo process?

Question 3–3 Describe vellum.

Question 3–4 Describe polyester film.

Question 3–5 What is another name for polyester film?

Question 3–6 Define mat, and describe the difference between single and double mat.

Question 3–7 Which of the following combinations would yield the best reproduction: graphite on vellum, plastic lead on polyester film, or ink on polyester film?

Question 3–8 What are the primary elements that give the best reproduction?

Question 3–9 Identify three standard sheet sizes that may commonly be used by architectural offices.

Question 3–10 Describe zoning.

Question 3–11 List six elements that normally may be found in a standard architectural title block.

Question 3–12 Identify two different locations where the scale may be located on an architectural drawing.

Question 3–13 What are the results of a properly folded print?

Question 3–14 Briefly describe the revision process and its steps.

Question 3–15 What is another name for a diazo print?

Question 3–16 Is a diazo print the same as a blueprint? Explain.

Question 3–17 Describe the diazo process.

Question 3–18 How does the speed of the diazo printer affect the print?

Question 3–19 Describe how to run a test strip, and explain the importance of this test.

Question 3–20 What should a good-quality diazo print look like?

Question 3–21 Describe the characteristics of slightly exposed or old diazo materials.

Question 3–22 Define sepias and their use.

Question 3–23 Why are sepias made in reverse (reverse sepia)?

Question 3–24 Discuss safety precautions in the handling of ammonia.

Question 3–25 Describe the recommended first aid for the following accidents: ammonia spilled on the skin, ammonia in the eyes, inhaling excess ammonia vapor.

Question 3–26 List four advantages of photocopying over the diazo process.

Question 3–27 Give at least two reasons for using microfilm.

Question 3–28 What is the big advantage of microfilm?

QUESTIONS *(cont.)*

Question 3–29 Given the architectural title block shown in the illustration for this question, identify elements a–h.

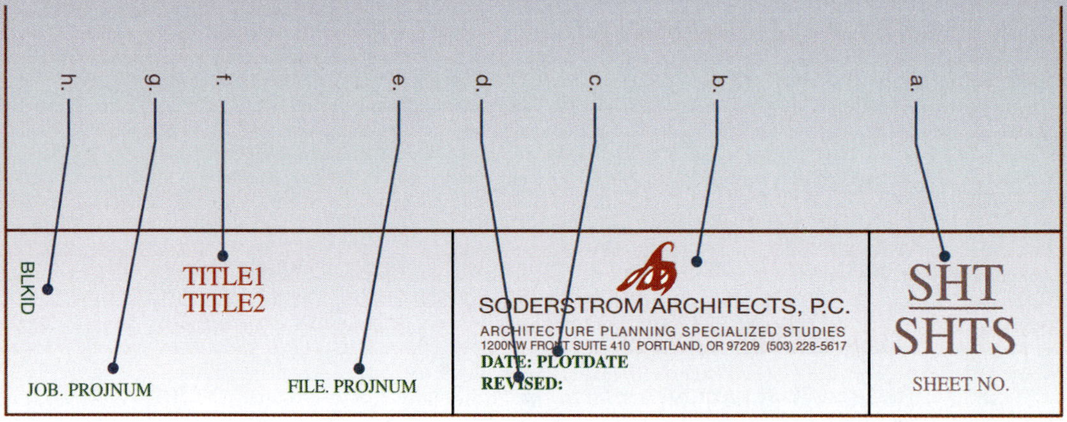

CHAPTER 4

Sketching and Orthographic Projection

INTRODUCTION

Sketching is freehand drawing, that is, drawing without the aid of drafting equipment. Sketching is convenient, since only paper, pencil, and an eraser are needed. There are a number of advantages to freehand sketching. Sketching is fast visual communication. The ability to make an accurate sketch quickly can often be an asset in communicating with people at work or at home. Especially for technical concepts, a sketch may be the best form of communication. Most drafters prepare a preliminary sketch to help organize thoughts and minimize errors on the final drawing. The computer-aided drafter usually prepares a sketch on graph paper to help establish the coordinates for drawing components. Some drafters use sketches to help record the stages of progress until a final design is ready for formal drawings.

TOOLS AND MATERIALS

The pencil should have a soft lead; a common number 2 pencil works fine. A mechanical pencil with a soft lead such as H, F, or HB is good. An automatic 0.7- or 0.9-mm pencil with F or HB lead is also good. Soft pencil leads allow for maximum control and variation of line weight. The pencil lead should not be sharp. A slightly rounded pencil point is best, because it is easy to control and will not break easily. Different thickness of line, if needed, can be drawn by changing the amount of pressure that you apply to the pencil. The quality of the paper is not critical either. A good sketching paper is newsprint, although almost any kind of paper works. Paper with a surface that is not too smooth is best. Many architectural designs have been created on a napkin at a lunch table. Sketching paper should not be taped down to the table. The best sketches are made when you are able to move the paper to the most comfortable drawing position. Some people make horizontal lines better than vertical lines. If this is your situation, move the paper so vertical lines become horizontal. Such movement of the paper may not always be possible, so it does not hurt to keep practicing all forms of lines for best results.

SKETCHING STRAIGHT LINES

Lines should be sketched lightly in short connected segments, as shown in Figure 4.1. If you sketch one long stroke in one continuous movement, your arm tends to make the line curved rather than straight. If you make a dark line, you may have to erase it if you make an error, whereas if you draw a light line, often there is no need to erase an error, because it does not show up very much when the final lines are darkened to your satisfaction.

Use the following procedure to sketch a horizontal straight line with the dot-to-dot method:

STEP 1 Mark the starting and ending positions, as in Figure 4.2. The letters A and B are only for instruction. All you need are the points.

STEP 2 Without actually touching the paper with the pencil point, make a few trial motions between the marked points to adjust your eye and hand to the anticipated line.

STEP 3 Sketch very light lines between the points by moving the pencil in short, light strokes, 2 to 3″ long. With each stroke, attempt to correct the most obvious defects of the preceding stroke so that the finished light lines will be relatively straight. Look at Figure 4.3.

STEP 4 Darken the finished line with a dark, distinct, uniform line directly on top of the light line. Usually the darkness can be obtained by pressing on the pencil. See Figure 4.4.

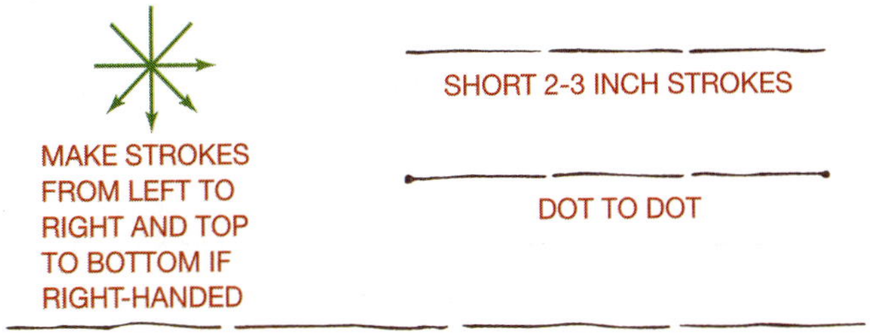

SHORT 2-3 INCH STROKES
MAKE STROKES FROM LEFT TO RIGHT AND TOP TO BOTTOM IF RIGHT-HANDED
DOT TO DOT

FIGURE 4.1 ■ Sketching short line segments.

FIGURE 4.2 ■ Step 1: Dot-to-dot.

FIGURE 4.3 ■ Step 3: Short, light strokes.

FIGURE 4.4 ■ Step 4: Darken to finish the line.

SKETCHING CIRCULAR LINES

Figure 4.5 shows the parts of a circle. Three sketching techniques are used in making a circle: trammel, hand compass, and nail and string.

Trammel Method

A *trammel* is an instrument used for making circles. In this example, the instrument is a strip of paper.

STEP 1 To sketch a 6-in.-diameter circle, tear a strip of paper approximately 1″ wide and longer than the radius, 3″ On the strip of paper, mark an approximate 3″ radius with tick marks, such as A and B in Figure 4.6.

STEP 2 Sketch a straight line representing the diameter at the place where the circle is to be located. On the sketched line, locate with a dot the center of the circle. Use the marks on the trammel to mark the other end of the radius line, as shown in Figure 4.7. Place the trammel next to the sketched line, being sure point B on the trammel is aligned with the center of the circle.

STEP 3 Pivot the trammel at point B, making tick marks at point A as you go, as shown in Figure 4.8, until you have a complete circle, as shown in Figure 4.9.

STEP 4 Lightly sketch the circumference over the tick marks to complete the circle, as shown in Figure 4.9.

STEP 5 Darken the circle, as shown in Figure 4.9. You can darken the circle directly over the tick marks if you are in a hurry.

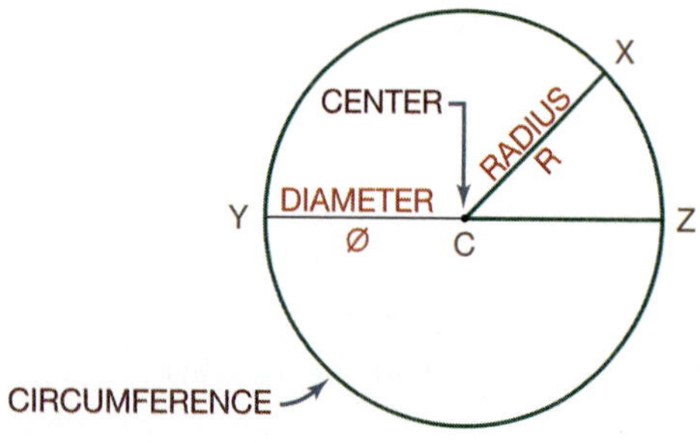

FIGURE 4.5 ■ Parts of a circle.

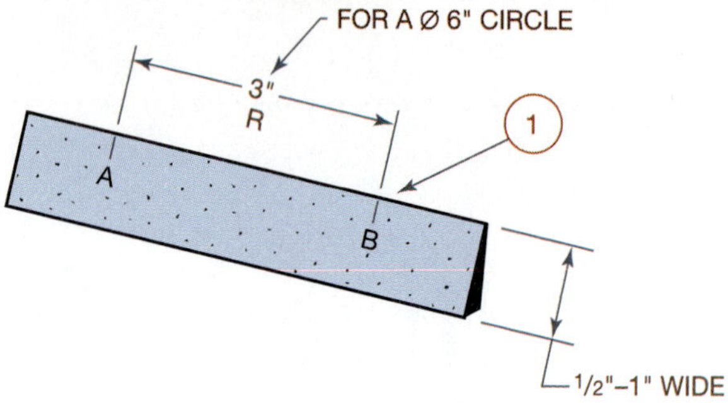

FIGURE 4.6 ■ Step 1: Making a trammel.

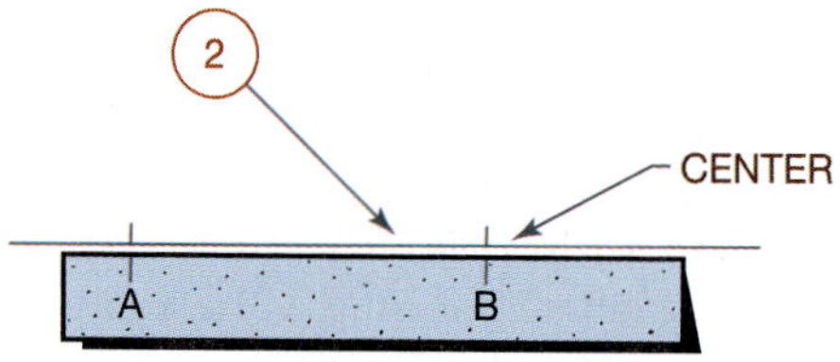

FIGURE 4.7 ■ Step 2: Marking the radius with a trammel.

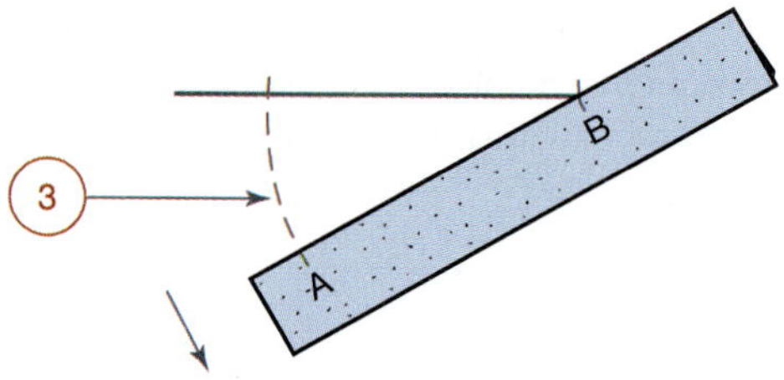

FIGURE 4.8 ■ Step 3: Sketching the circle with the trammel.

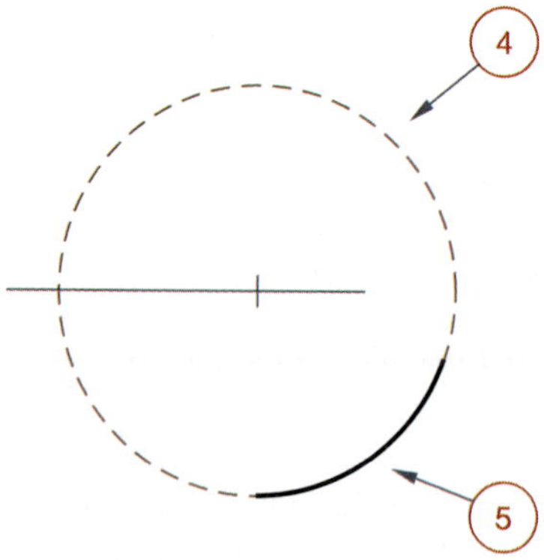

FIGURE 4.9 ■ Step 4: Completing the circle.

Hand-Compass Method

The hand-compass method is a quick and fairly accurate way to sketch circles, although it takes some practice.

STEP 1 Be sure your paper is free to rotate completely around 360°. Remove anything from the table that might stop such a rotation.

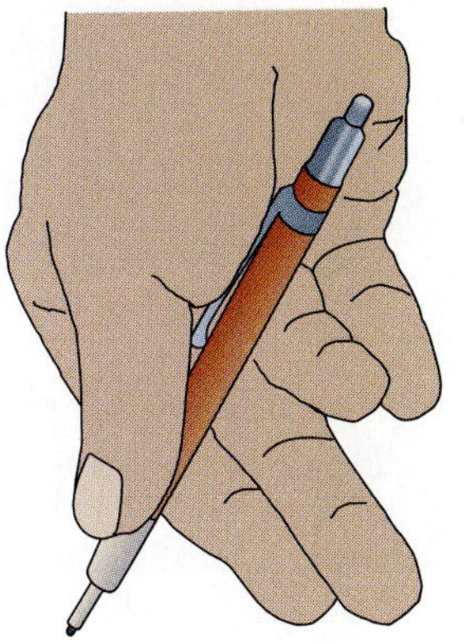

FIGURE 4.10 ■ Step 2: Holding the pencil for the hand compass method.

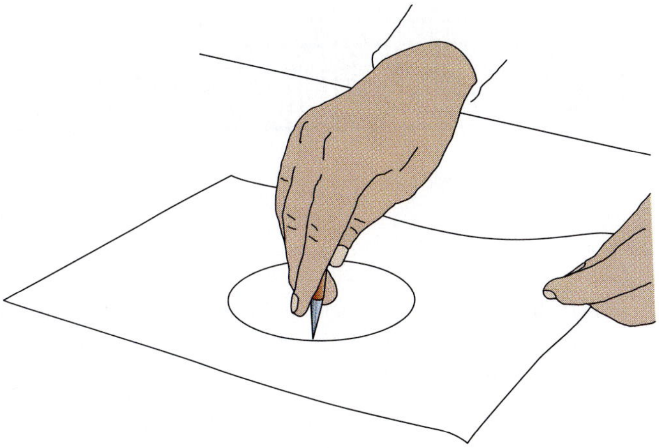

FIGURE 4.11 ■ Steps 4 and 5: Sketching a circle with the hand compass method.

STEP 2 To use your hand and a pencil as a compass, place the pencil in your hand between your thumb and the upper part of your index finger so that your index finger becomes the *compass point* and the pencil becomes the *compass lead*. The other end of the pencil rests in your palm, as shown in Figure 4.10.

STEP 3 Determine the approximate radius by adjusting the distance between your index finger and the pencil point. Now, with the radius established, place your index finger on the paper at the proposed center of the circle.

STEP 4 Using your radius, keep your hand and pencil point in one place while you rotate the paper with your other hand. Try to keep the radius steady as you rotate the paper. Look at Figure 4.11.

STEP 5 You can perform Step 4 very lightly and then go back and darken the circle or, if you have had a lot of practice, you may be able to draw a dark circle as you go. See Figure 4.11.

Nail-and-String Method

Another method used to sketch large circles is to tie a pencil and a pin together with a string. The distance between the pencil and pin is the radius of the circle. Use this method when a large circle is to be sketched, since the other methods may not work as well. This method is used, with a nail driven at the center and a string connected to a pencil, for drawing large circles at a construction site.

MEASUREMENT LINES AND PROPORTIONS

In sketching an object, all the lines that make it up are related to each other by size and direction. In order for a sketch to communicate accurately and completely, it should be proportional to the object. The actual size of the sketch depends on the paper size and how large you want the sketch to be. The sketch should be large enough to be clear, but the proportions of the features are more important than the size of the sketch.

Look at the lines in Figure 4.12. How long is line 1? How long is line 2? Answer these questions without measuring either line; instead, relate each line to the other. For example, line 1 could be described as half as long as line 2, or line 2 as twice as long as line 1. Now you know how long each line is relative to the other (proportion). You do not know how long either line is in terms of a measured scale, but no scale is used for sketching, so this is not a concern. Whatever line you decide to sketch first determines the scale of the drawing. This first line sketched is called the *measurement line*. You relate all the other lines in your sketch to the first line. This is one of the secrets in making a sketch look like the object being sketched.

The second thing you must know about the relationship of the two lines in the example is their direction and position relative to each other. Do they touch each other? Are they parallel, perpendicular, or at some other angle to each other? When you look at a line, ask yourself the following questions (for this example use the two lines given in Figure 4.13):

1. How long is the second line?
   Answer: Line 2 is about three times as long as line 1.
2. In what direction and position is the second line relative to the first line?
   Answer: Line 2 touches the lower end of line 1 with about a 90° angle between them.

FIGURE 4.12 ■ Measurement lines.

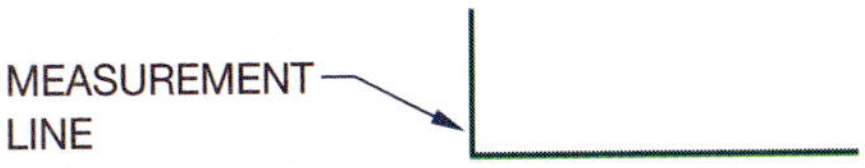

FIGURE 4.13 ■ Measurement line.

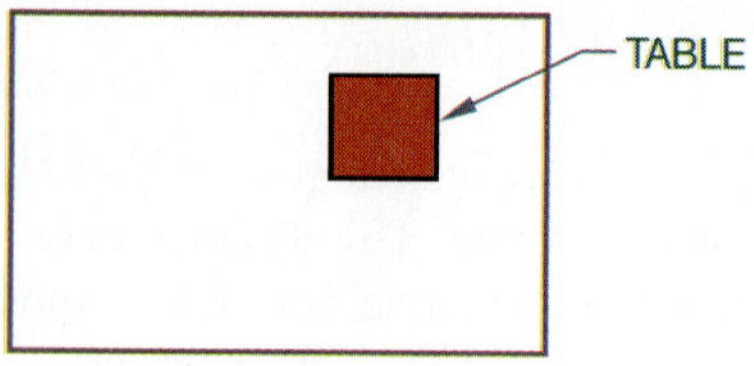

FIGURE 4.14 ■ Spatial proportions.

Carrying this concept a step further, a third line can relate to the first line or the second line, and so forth. Again, the first line drawn (measurement line) sets the scale for the entire sketch.

This idea of relationship can also apply to spaces. In Figure 4.14, the location of a table in a room can be determined by spatial proportions. A typical verbal location for the table in this floor plan might be as follows: "The table is located about one-half the table width from the top of the floor plan or about two table widths from the bottom, and about one table width from the right side or about three table widths from the left side of the floor plan."

Using Your Pencil to Establish Measurements

Your pencil can be a handy tool for establishing measurements on a sketch. When you are sketching an object that you can hold, use your pencil as a ruler. Place the pencil next to the feature to be sketched and determine the length by aligning the pencil tip at one end of the feature and marking the other end on the pencil with your thumb. The other end may also be identified by a specific contour or mark on the pencil. Then transfer the measurement to your sketch. The two distances are the same.

A similar technique may be used to sketch a distant object. For example, to sketch a house across the street, hold your pencil at arm's length and align it with a feature of the house, such as the width. With the pencil point at one end of the house, place your thumb on the pencil, marking the other end of the house. Transfer this measurement to your sketch in the same orientation as it was taken from the house. Repeat this technique for all of the house features until the sketch is complete. If you keep your pencil at arm's length and your arm straight, each measurement will have the same accuracy and proper proportions.

Block Technique

Any illustration of an object can be surrounded overall with a rectangle or any other geometric shapes. See Figure 4.15. Before starting a sketch, visualize the object inside a rectangle. Then use the measurement-line technique with the rectangle, or block, to help determine the shape and proportions of your sketch.

Procedures in Sketching

You can sketch any object, such as an elevation, as you look at an actual house or a floor plan and think about it. To use the measurement and block techniques, follow these steps:

STEP 1 When you start to sketch an object, try to visualize the object surrounded with a rectangle overall. Sketch this rectangle first with very light lines. Sketch it in the proper proportion with the measurement-line technique, as shown in Figure 4.16.

STEP 2 Cut sections out or away using proper proportions as measured by eye. Use light lines, as shown in Figure 4.17.

STEP 3 Finish the sketch by darkening the outlines desired for the finished sketch. See Figure 4.18.

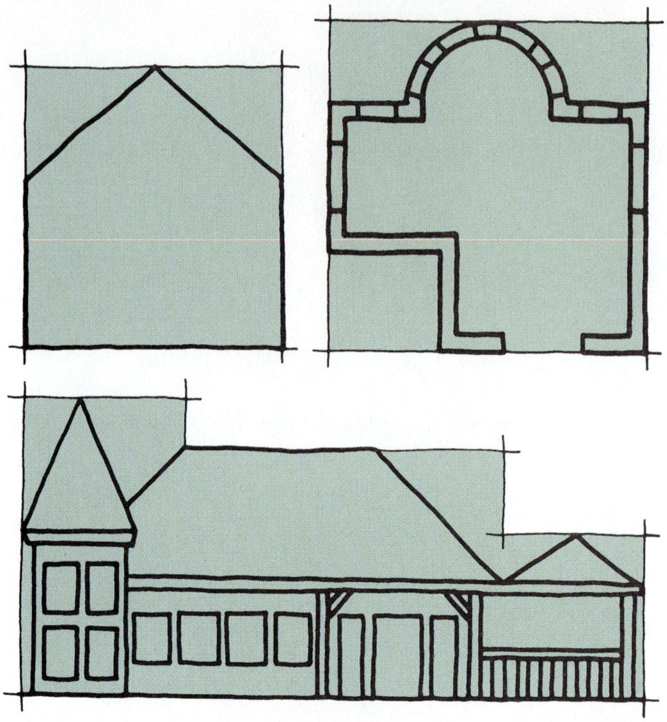

FIGURE 4.15 ■ Block technique.

Irregular Shapes

By using a frame of reference or an extension of the block method, irregular shapes can be sketched easily to their correct proportions. Follow these steps to sketch the free-form swimming pool shown in Figure 4.19.

STEP 1 Enclose the object in a lightly constructed box. See Figure 4.20.

STEP 2 Sketch several evenly spaced horizontal and vertical lines, forming a grid, as shown in Figure 4.21. If you are sketching an object already drawn, draw your reference lines on top of the object lines to establish a frame of reference. Make a photocopy if the original may not be used. If you are sketching an object directly, you have to visualize these reference lines on the object you sketch.

STEP 3 On your sketch, correctly locate a proportioned box similar to the one established on the original drawing or object, as shown in Figure 4.22.

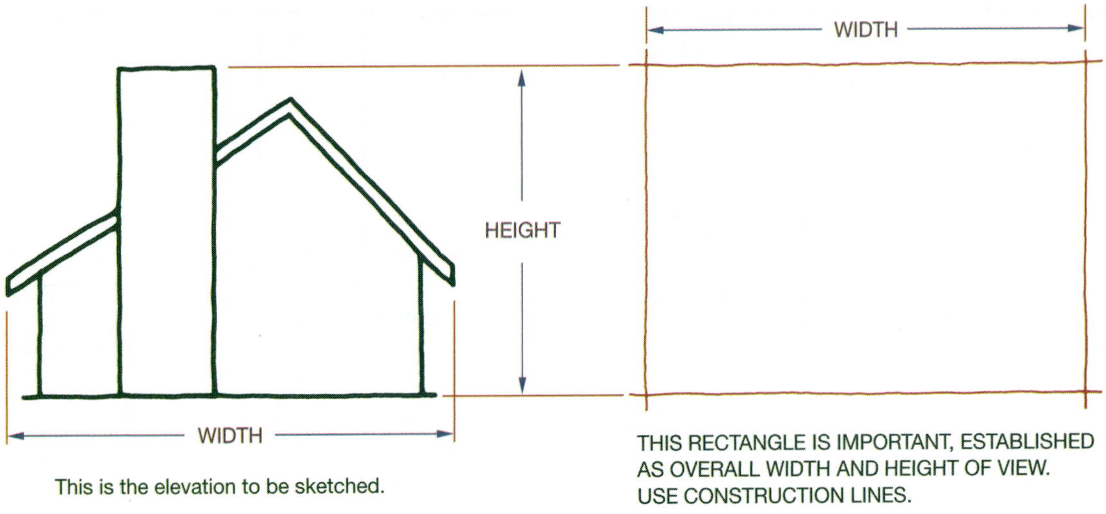
WIDTH
HEIGHT
WIDTH
This is the elevation to be sketched.
THIS RECTANGLE IS IMPORTANT, ESTABLISHED AS OVERALL WIDTH AND HEIGHT OF VIEW. USE CONSTRUCTION LINES.

FIGURE 4.16 ■ Step 1: Sketch the block.

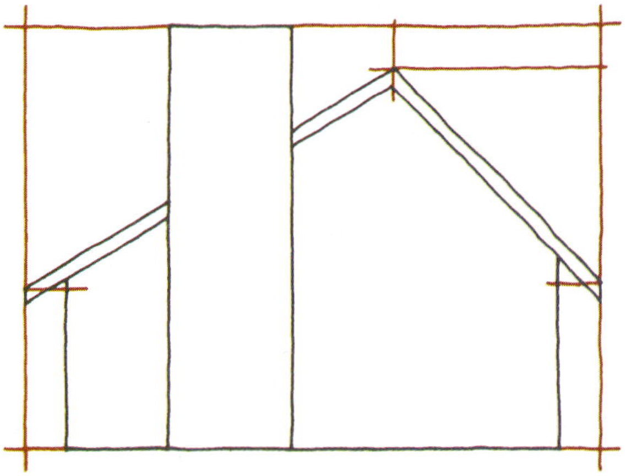

FIGURE 4.17 ■ Step 2: Block technique. Cut the sections out and lightly sketch the shapes.

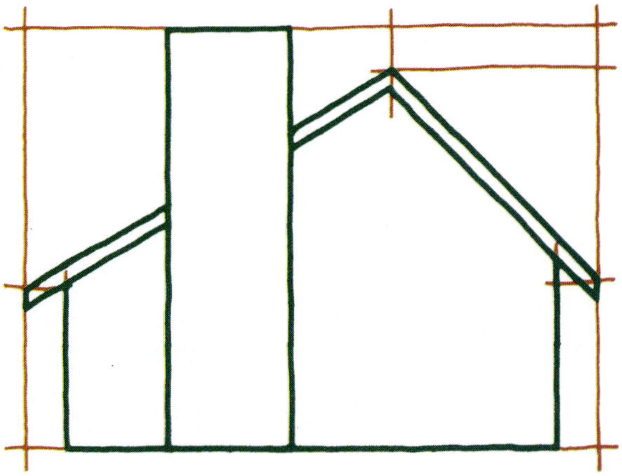

FIGURE 4.18 ■ Step 3: Darken the view.

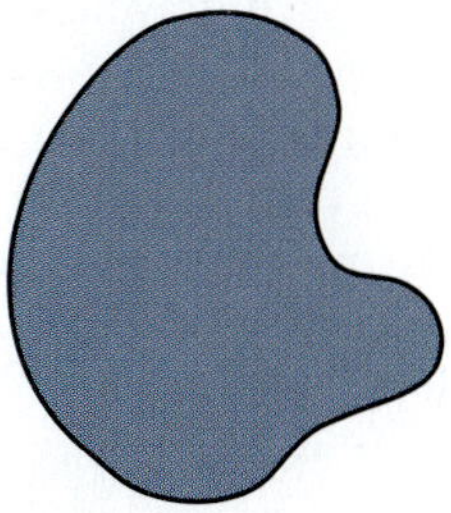

FIGURE 4.19 ■ Free-form swimming pool.

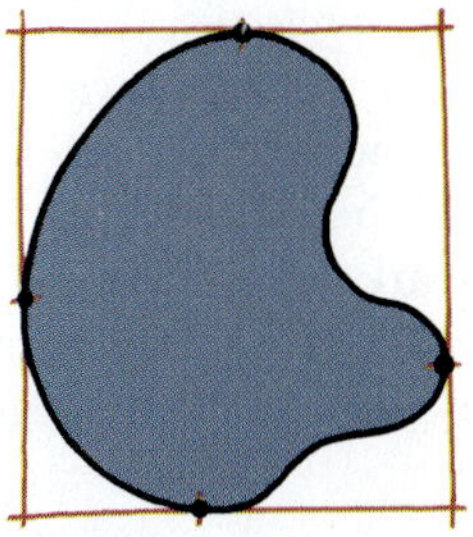

FIGURE 4.20 ■ Step 1: Imaginary box.

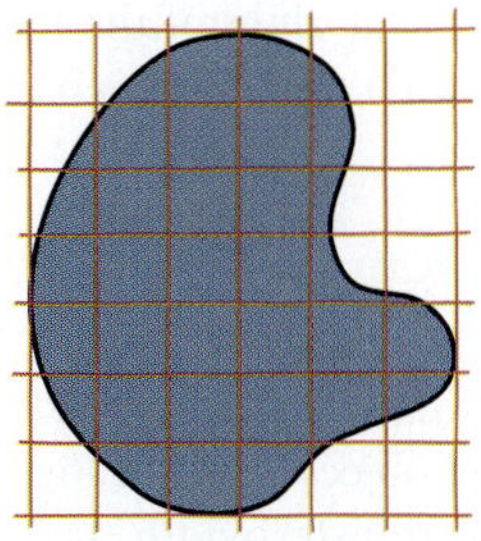

FIGURE 4.21 ■ Step 2: Evenly spaced grid sketched over the swimming pool.

FIGURE 4.22 ■ Steps 3 and 4: Proportioned box with a regular grid.

STEP 4 Using the drawn box as a frame of reference, include the grid lines in correct proportion, as shown in Figure 4.22.

STEP 5 Using the grid, sketch small, irregular arcs and lines that match the lines of the original, as in Figure 4.23.

STEP 6 Darken the outline for a complete proportioned sketch, as shown in Figure 4.24.

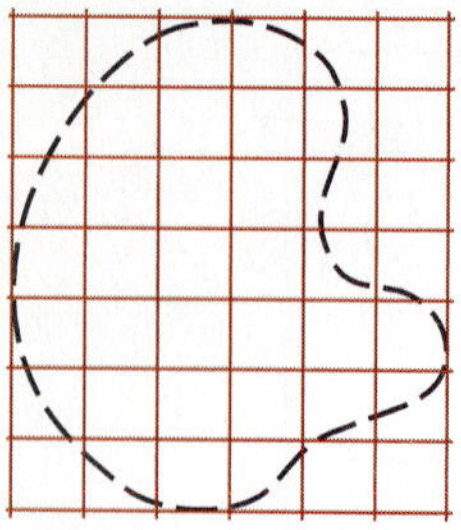

FIGURE 4.23 ■ Step 5: Sketching the shape using a regular grid.

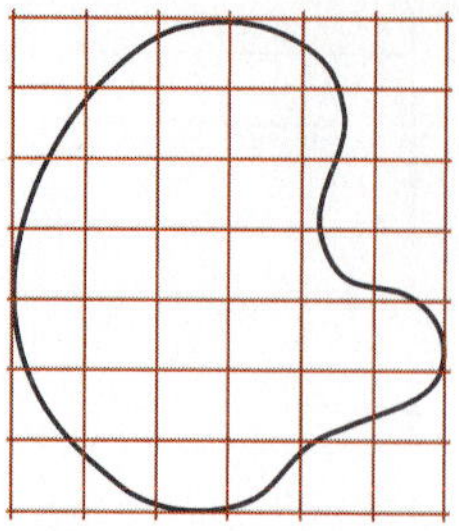

FIGURE 4.24 ■ Step 6: Darkening the object completely.

MULTIVIEW SKETCHES

A *multiview,* or *multiview projection,* is also known as an *orthographic projection.* In architectural drafting such drawings are referred to as elevation views. *Elevation views* are two-dimensional views of an object (a house, for example) that are established by a line of sight perpendicular (at a 90° angle) to the surface of the object. When you make multiview sketches, follow a methodical order. Learning to make these sketches will help you later as you begin to prepare elevation drawings.

Multiview Alignment

To keep your drawing in a standard form, sketch the front view in the lower left portion of the paper, the top view directly above the front view, and the right-side view to the right of the front view. See Figure 4.25. The views needed may differ depending on the object.

Multiview Sketching Technique

STEP 1 Sketch and align the proportional rectangles for the front, top, and right side of the object given in Figure 4.26. Sketch a 45° line to help transfer width dimensions. The 45° line is established by projecting the width from the top view across and the width from the right-side view up until the lines intersect, as shown in Figure 4.27.

STEP 2 Complete the shapes within the blocks, as shown in Figure 4.28.

STEP 3 Darken the lines of the object, as in Figure 4.29. Remember to keep the views aligned for ease of sketching and understanding.

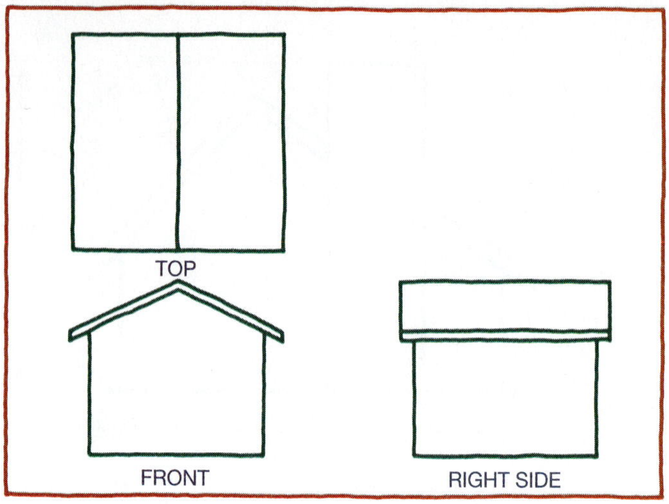

FIGURE 4.25 ■ Multiviews.

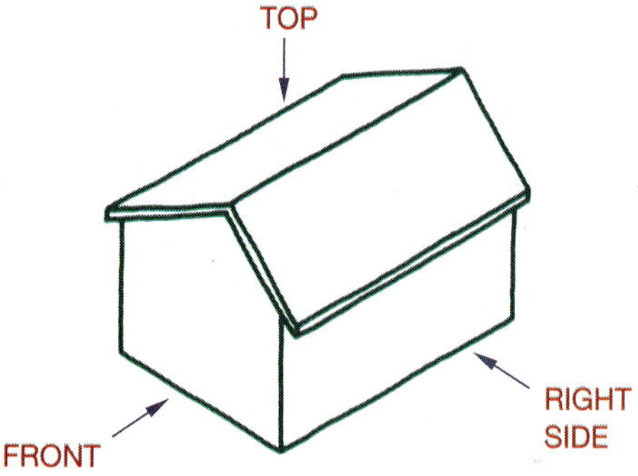

FIGURE 4.26 ■ Pictorial view.

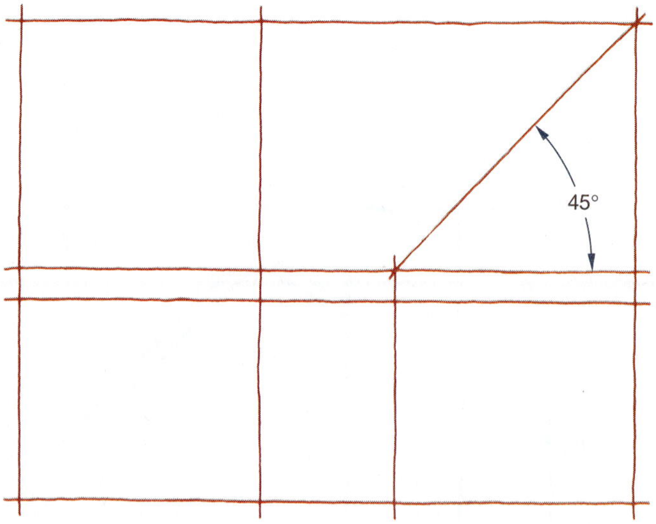

FIGURE 4.27 ■ Step 1: Block out views and establish a 45° line.

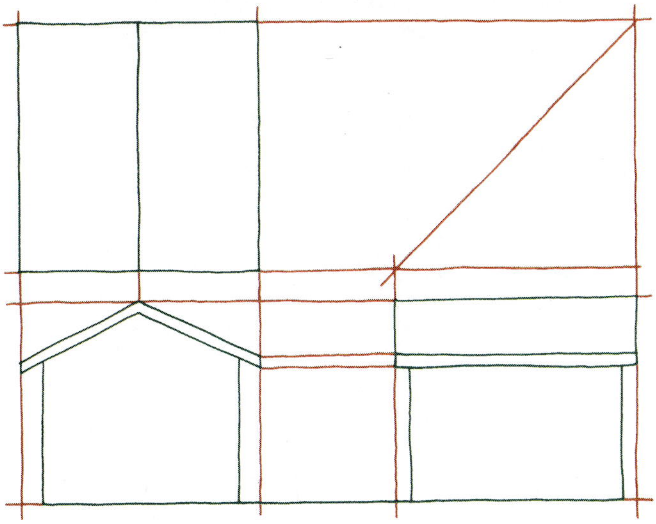

FIGURE 4.28 ■ Step 2: Block out shapes.

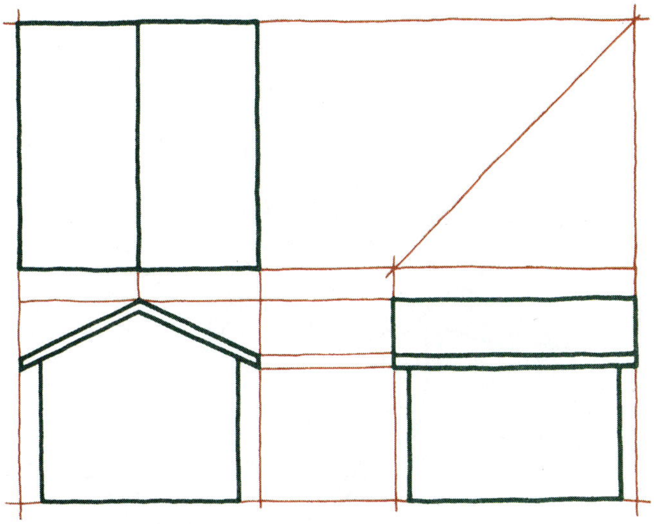

FIGURE 4.29 ■ Step 3: Darken all lines.

ISOMETRIC SKETCHES

Isometric sketches provide a three-dimensional pictorial representation of an object, such as the shape of a building. Isometric sketches are easy to draw and are fairly realistic. An isometric sketch tends to represent objects as they appear to the eye. Such a sketch helps us visualize an object because three sides of the object are shown in a single three-dimensional view.

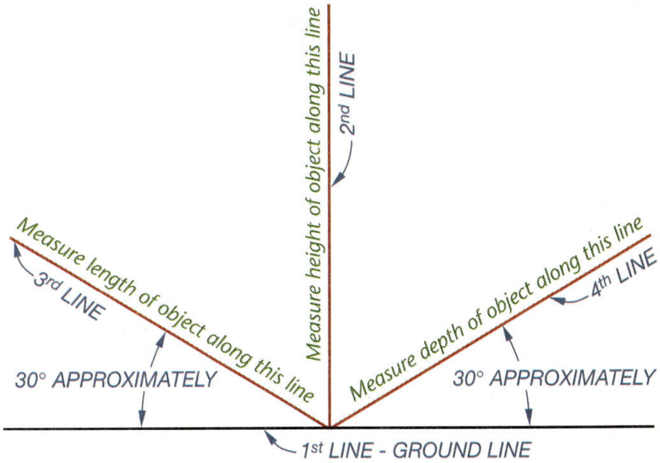

FIGURE 4.30 ■ Isometric axes.

Establishing Isometric Axes

To establish isometric axes, you need four beginning lines: a horizontal reference line, two 30° angular lines, and one vertical line. Draw them as very light construction lines. Look at Figure 4.30.

STEP 1 Sketch a horizontal reference line (consider this the ground-level line).

STEP 2 Sketch a vertical line perpendicular to the ground line somewhere near its center. This vertical line is used to measure height.

STEP 3 Sketch two 30° angular lines, each starting at the intersection of the first two lines, as shown in Figure 4.30.

Making an Isometric Sketch

The steps in making an isometric sketch from a multiview drawing or a real object are as follows:

STEP 1 Select an appropriate view of the object to use as a front view, or study the front view of the multiview drawing.

STEP 2 Determine the best position in which to show the object.

STEP 3 Begin your sketch by setting up the isometric axes as just described. See Figure 4.30.

STEP 4 Using the measurement-line technique, draw a rectangular box to the correct proportion, which could surround the object to be drawn. Use the object shown in Figure 4.31 as an example. Imagine the rectangular box. Begin to sketch the box by marking off the width at any convenient length (measurement line), as in Figure 4.32. Next, estimate and mark the length and height

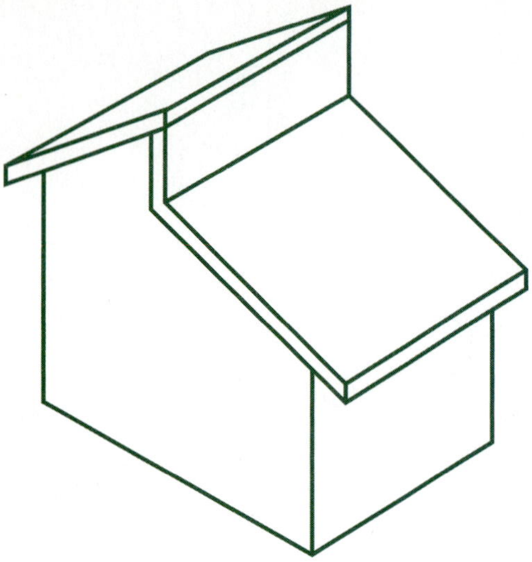

FIGURE 4.31 ■ Given structure.

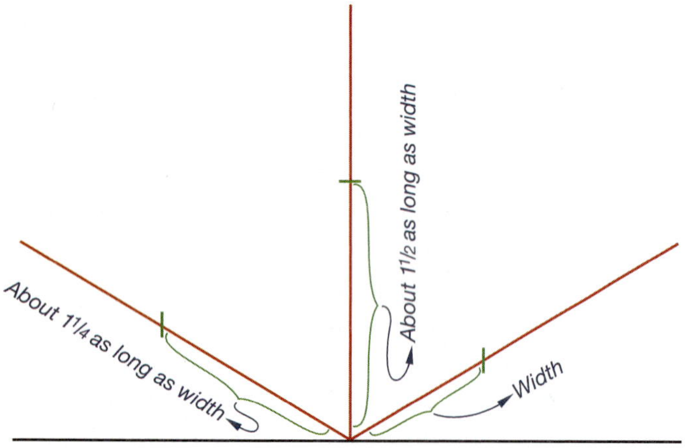

FIGURE 4.32 ■ Step 4: Lay out length, width, and height.

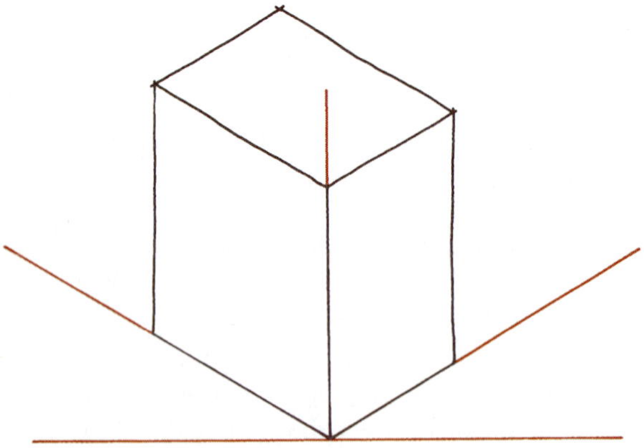

FIGURE 4.33 ■ Step 4: Sketch the three-dimensional box.

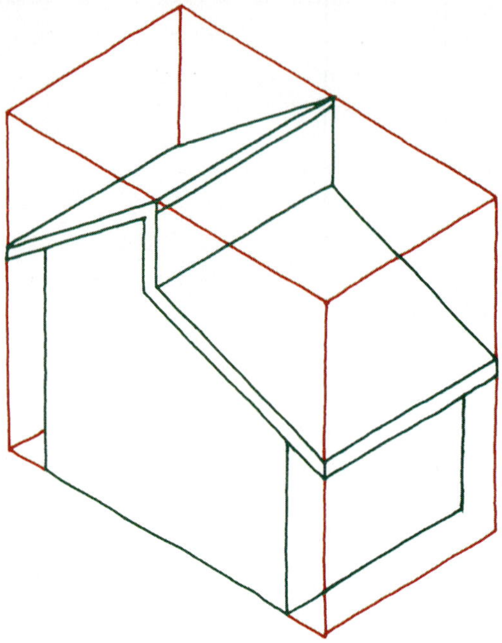

FIGURE 4.34 ■ Step 5: Sketch the features of the structure.

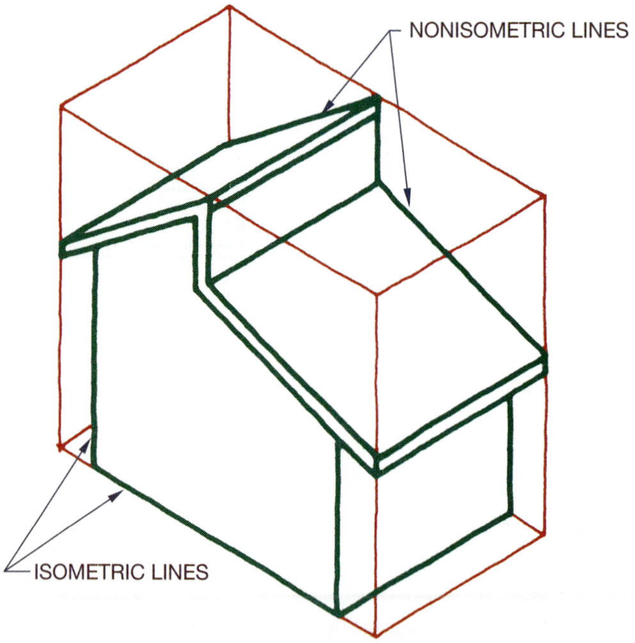

FIGURE 4.35 ■ Step 6: Darken the structure.

relative to the measurement line. See Figure 4.33. Sketch the three-dimensional box by using lines parallel to the original axis lines. See Figure 4.33. The box must be sketched correctly; otherwise the rest of your sketch will be out of proportion. All lines drawn in the same direction must be parallel.

STEP 5 Lightly sketch the features that define the details of the object. When you estimate distances on the rectangular box, you will find that the features of the object are easier to sketch in correct proportion than they would be if you tried to draw them without the box. See Figure 4.34.

STEP 6 To finish the sketch, darken all the outlines, as in Figure 4.35.

Nonisometric Lines

Isometric lines are lines that are on or parallel to one of the three original axes. All other lines are nonisometric lines. Isometric lines can be measured to true length. Nonisometric lines appear either longer or shorter than they actually are. See Figure 4.35.

You can measure and draw nonisometric lines by connecting their endpoints. Find the endpoints of the nonisometric lines by measuring along isometric lines. To locate where nonisometric lines should be placed, you have to relate to an isometric line.

Sketching Isometric Circles

Circles and arcs appear as ellipses in isometric views. To sketch isometric circles and arcs correctly, you need to know the relationship between circles and the faces, or planes, of an isometric cube. Depending on which plane the circle is to appear in, isometric circles look like one of the ellipses shown in Figure 4.36. The angle at which the ellipse (isometric circle) slants is determined by the surface on which the circle is to be sketched.

Four-Center Method

The four-center method of sketching an isometric ellipse is simple, but care must be taken to form the ellipse arcs properly so that the ellipse does not look distorted.

STEP 1 Draw an isometric cube (box) similar to Figure 4.37.

STEP 2 On each surface of the box draw line segments that connect the 120° corners to the centers of the opposite sides. See Figure 4.38.

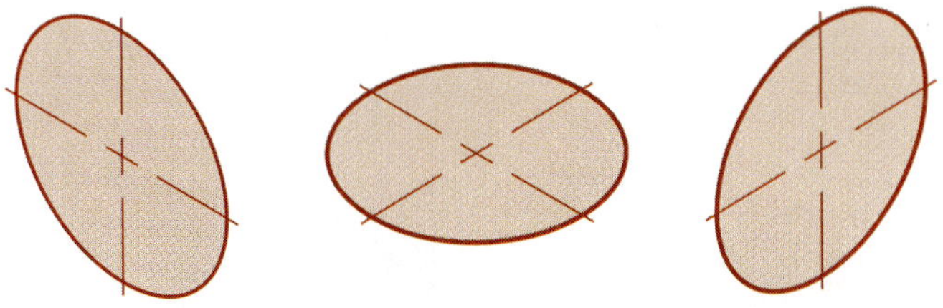

FIGURE 4.36 ■ Isometric circles.

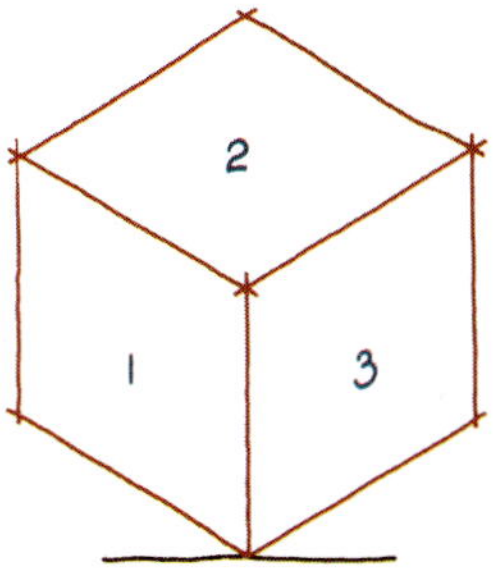

FIGURE 4.37 ■ Step 1: Isometric cube.

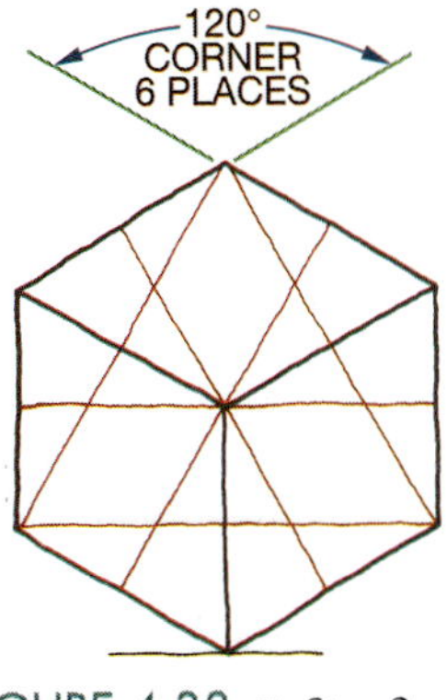

FIGURE 4.38 ■ Step 2: Four-center construction of an isometric ellipse.

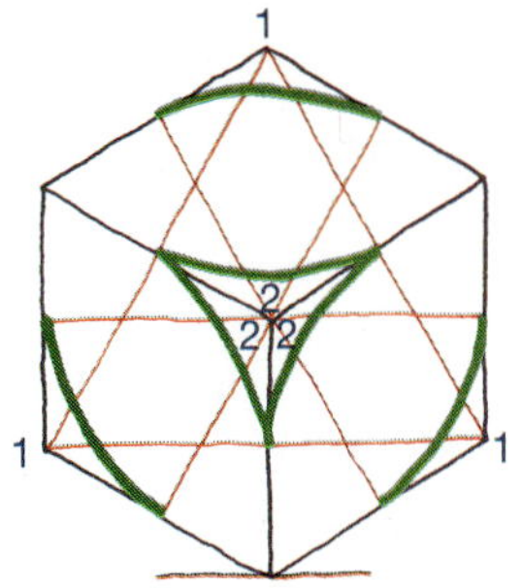

FIGURE 4.39 ■ Step 3: Sketch arcs from points 1 and 2.

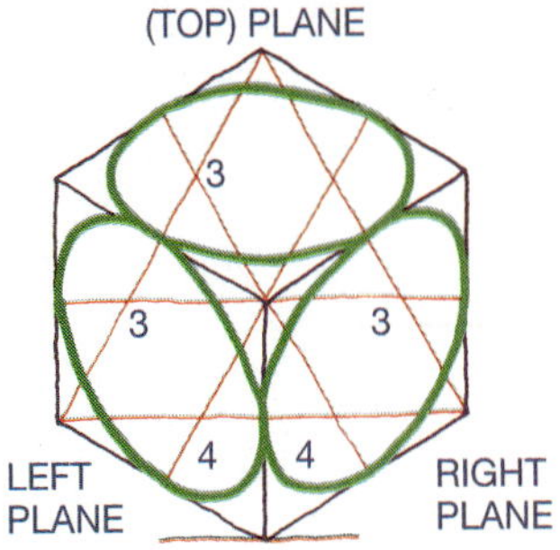

FIGURE 4.40 ■ Step 4: Sketch arcs from points 3 and 4.

STEP 3 With points 1 and 2 as the centers, sketch arcs that begin and end at the centers of the opposite sides on each isometric surface. See Figure 4.39.

STEP 4 On each isometric surface, with points 3 and 4 as the centers, complete the isometric ellipses by sketching arcs that meet the arcs sketched in step 3. See Figure 4.40.

Sketching Isometric Arcs

Sketching isometric arcs is similar to sketching isometric circles. First, block out the overall configuration of the object; then establish the centers of the arcs; finally, sketch the arc shapes. Remember that isometric arcs, just like isometric circles, must lie in the proper plane and have the correct shape.

ORTHOGRAPHIC PROJECTION

Orthographic projection is any projection of features of an object onto an imaginary plane called a plane of projection. The projection of the features of the object is made by lines of sight that are perpendicular to the plane of projection. When a surface of the object is parallel to the plane of projection, the surface appears in its true size and shape on that plane. In Figure 4.41, the plane of projection is parallel to the surface of the object. The line of sight (projection from the object) from the object is perpendicular to the plane of projection. Notice also that the object appears three-dimensional (it has width, depth, and height), while the view on the plane of projection is two-dimensional (it has only width and height). In situations where the plane of projection is not parallel to the surface of the object, the resulting orthographic view is foreshortened, or shorter than the true length. See Figure 4.42.

MULTIVIEW PROJECTION

Multiview projection establishes two or more views of an object as projected on two or more planes by using orthographic projection techniques. The result of multiview projection is a multiview drawing. Multiview drawings represent the shape of an

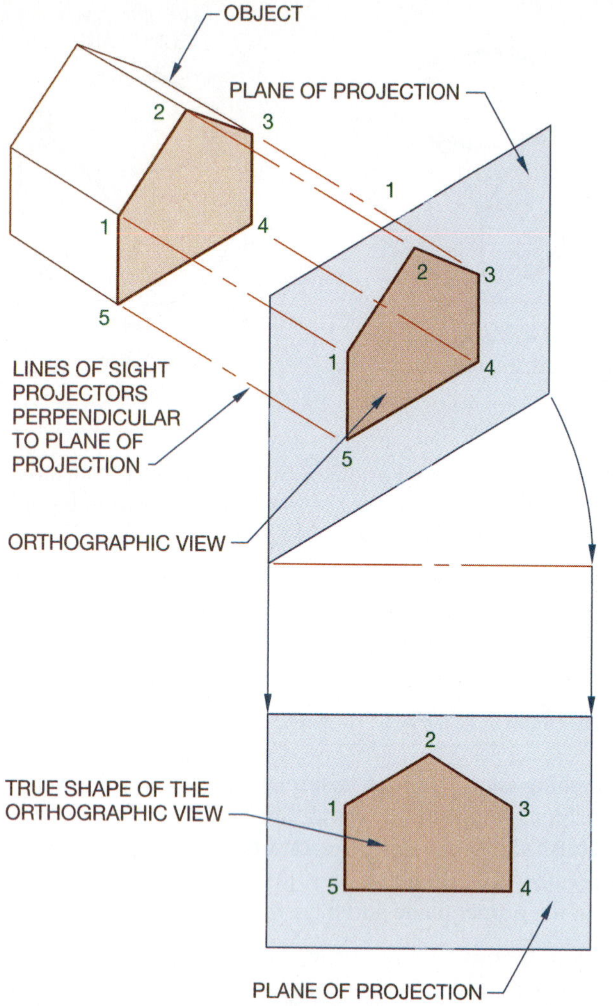

FIGURE 4.41 ■ Orthographic projection to form orthographic view.

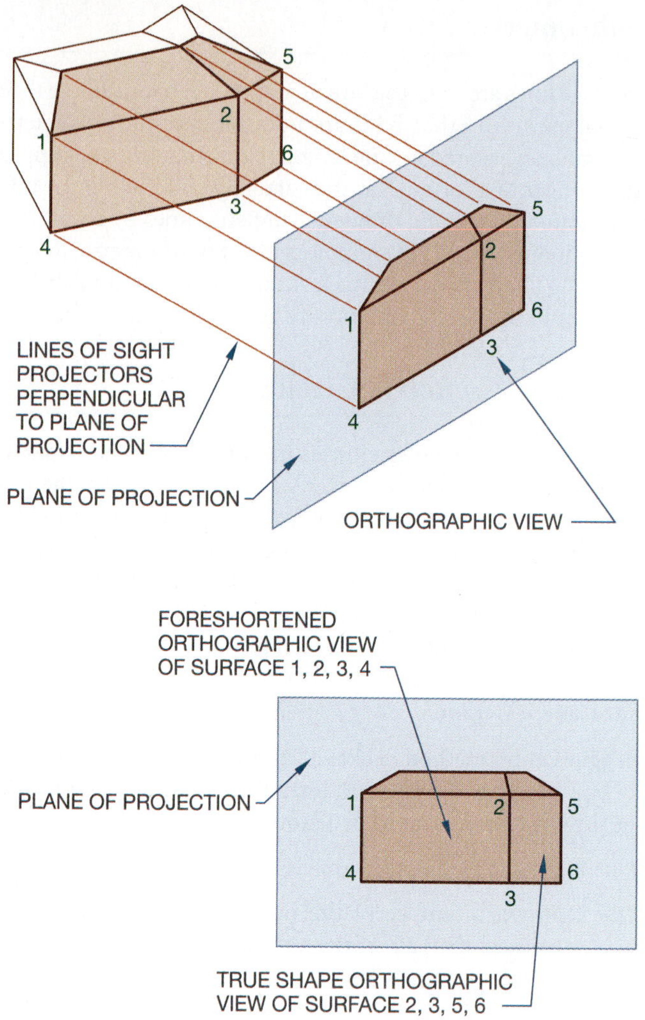

FIGURE 4.42 ■ Projection of a foreshortened orthographic surface.

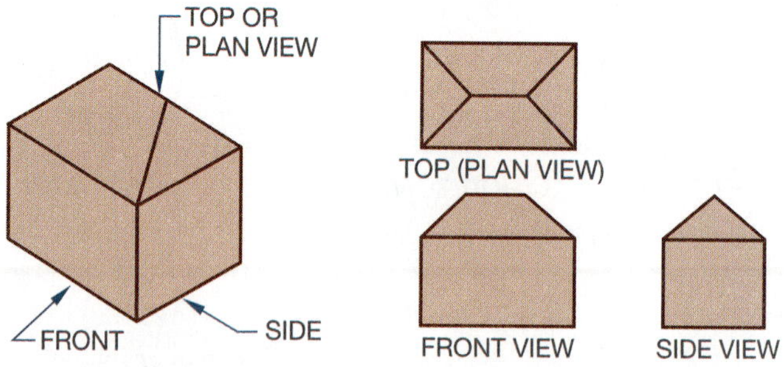

FIGURE 4.43 ■ Comparison of a pictorial and a multiview presentation.

object using two or more views. Consideration should be given to the choice and number of views so, when possible, the surfaces of the object are shown in their true size and shape.

Elevations As Multiviews

It is often easier to visualize a three-dimensional drawing of a structure than it is to visualize a two-dimensional drawing. In architectural drafting, however, it is common to prepare construction drawings showing two-dimensional exterior views of a structure that provide representations of exterior materials or interior views of features such as kitchen and bath cabinets. These drawings are referred to as *elevations*. The method used to draw elevations is multiview projection. A more detailed discussion of elevation drawing is found in Chapters 22 and 23. Figure 4.43 shows an object represented by a three-dimensional drawing, called a pictorial drawing, and three two-dimensional views.

Glass Box

If you place the object in Figure 4.43 in a glass box so that the sides of the box are parallel to the major surfaces of the object, you can project the surfaces of the object onto the sides of the glass box and create multiviews (Figure 4.44). Imagine that the sides of the glass box are the planes of projection previously described. Look at Figure 4.45. If you look at all sides of the glass box, you see six views: front, top, right side, left side, bottom, and rear. Now unfold the glass box as if the corners were hinged about the front view (except the rear view, which is attached to the left-side view), as demonstrated in Figure 4.45. These hinge lines are commonly called *fold lines*.

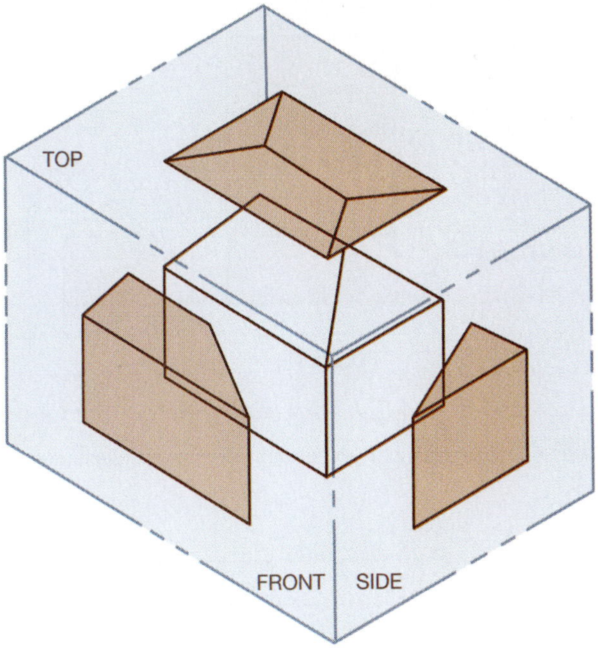

FIGURE 4.44 ■ Glass box.

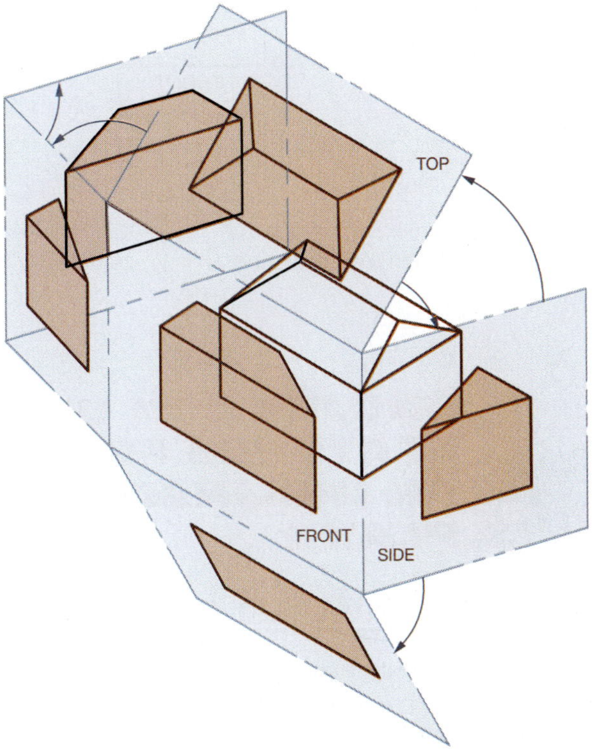

FIGURE 4.45 ■ Unfolding the glass box at hinge lines, also called fold lines.

Completely unfold the glass box onto a flat surface, and you have the six views of an object represented in multiview. Figure 4.46 shows the glass box unfolded. Notice also that the views are labeled: front, top, right side, left side, rear, and bottom. It is in this arrangement that the views are always found when using multiviews. Analyze Figure 4.47 in detail so you see the features that are common among the views. Knowing how to identify the

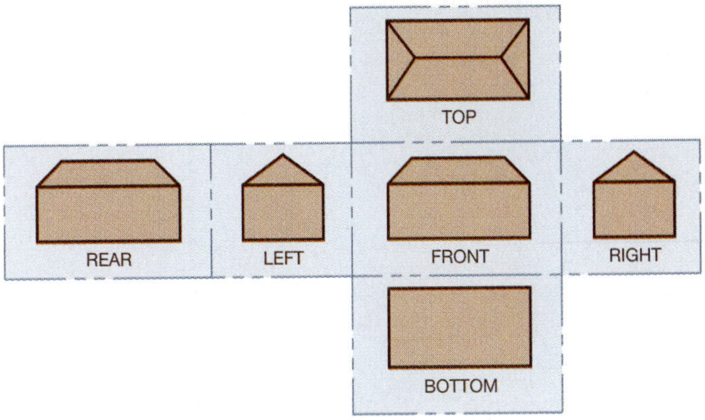

FIGURE 4.46 ■ Glass box unfolded.

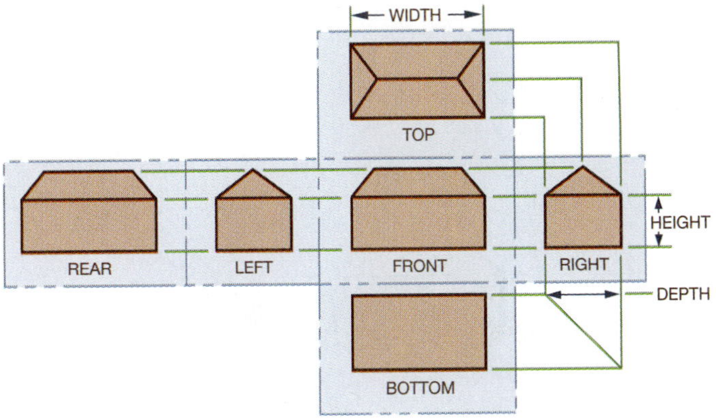

FIGURE 4.47 ■ View alignment.

features of an object that are common among views will aid you later in the visualization of elevations.

Notice how the views are aligned in Figure 4.47. The top view is directly above and the bottom view is directly below the front view. The left side is directly to the left, while the right side is directly to the right of the front view. This alignment allows the drafter to project features from one view to the next to help establish each view.

Now look closely at the relationship of the front, top, and right-side views. A similar relationship exists using the left-side view. Figure 4.48 shows a 45° projection line established by projecting the fold or reference line (hinge) between the front and side view up, and the fold line between the front and top view over. All of the features established on the top view can be projected to the 45° line and then down onto the side view. This is possible because the depth dimension is seen in both the top and the side views. The reverse is also true. Features from the side view may be projected to the 45° line and then over to the top view.

The transfer of features in Figure 4.48 using the 45° line can also be accomplished by using a compass with one leg at the intersection of the horizontal and vertical fold lines. The compass establishes the relationship between the top and side views, as shown in Figure 4.49.

Another method for transferring the size of features from one view to the next is the use of dividers to transfer distances from the fold line at the top view to the fold line at the side view. The

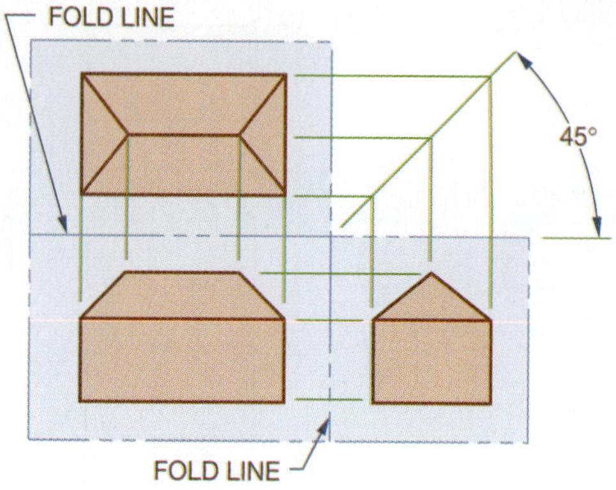

FIGURE 4.48 ■ Establishing a 45° projection line.

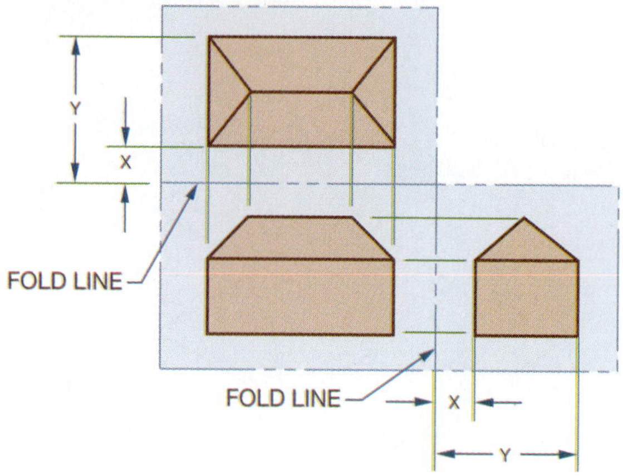

FIGURE 4.50 ■ Using dividers to transfer view projections.

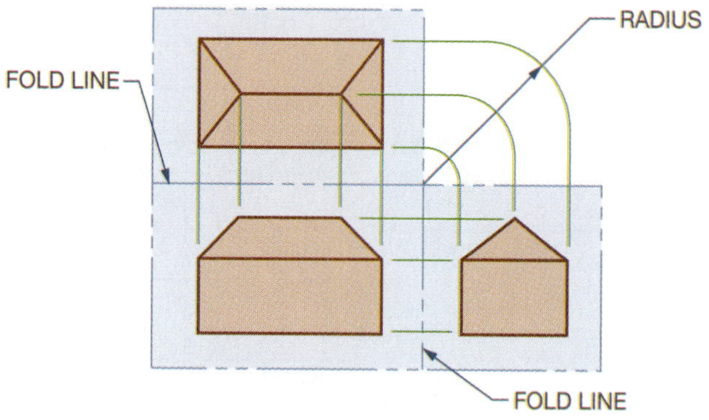

FIGURE 4.49 ■ Projection with a compass.

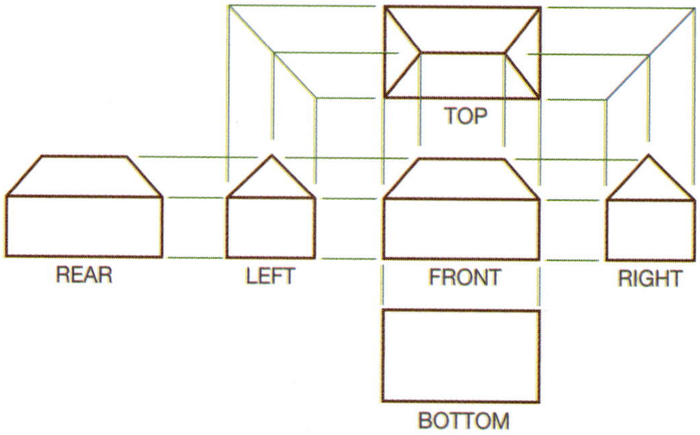

FIGURE 4.51 ■ Multiview orientation.

relationship between the fold lines and the two views is shown in Figure 4.50.

The front view is usually most important, since it is the one from which the other views are established. There is always one common dimension between adjacent views. For example, the width is common between the front and top views and the height between the front and side views. This knowledge allows you to relate information from one view to another. Take one more look at the relationship of the six views, as shown in Figure 4.51.

Selection of Views

There are six primary views that you may select to describe a structure completely. In architectural drafting, the front, left side, right side, and rear views are used as elevations to describe the exterior appearance of a structure completely. Elevation drawings are discussed in more detail in Chapters 22 and 23. The top view is called the roof plan view. This view shows the roof of the structure and provides construction information and dimensions. Roof plans are discussed in detail in Chapters 20 and 21. The bottom view is not used in architectural drafting.

PROJECTION OF FEATURES FROM AN INCLINED PLANE

Rectangular Features on an Inclined Plane

When a rectangular feature such as a skylight projects out of a sloped roof, the intersection of the skylight with the roof appears as a line when the roof also appears as a line. This intersection may then be projected onto adjacent views, as shown in Figure 4.52.

Circles on an Inclined Plane

When the line of sight in a view is perpendicular to a circle, such as a round window, the window appears round, as shown in Figure 4.53. When a circle is projected onto an inclined surface, such as a round skylight projected onto a sloped roof, the view of the inclined circle is elliptical. See Figure 4.54.

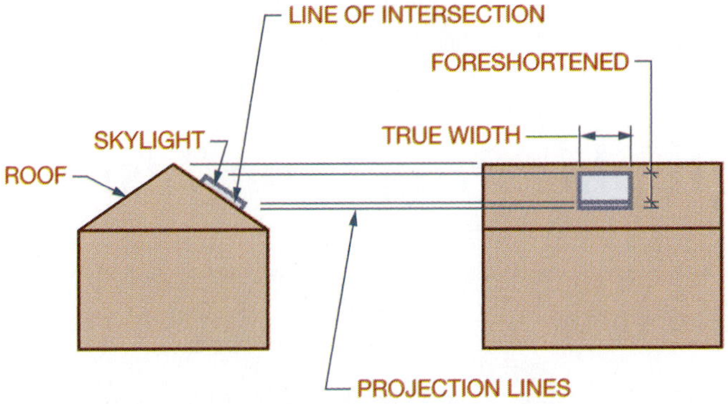

FIGURE 4.52 ■ Rectangular features on an inclined plane.

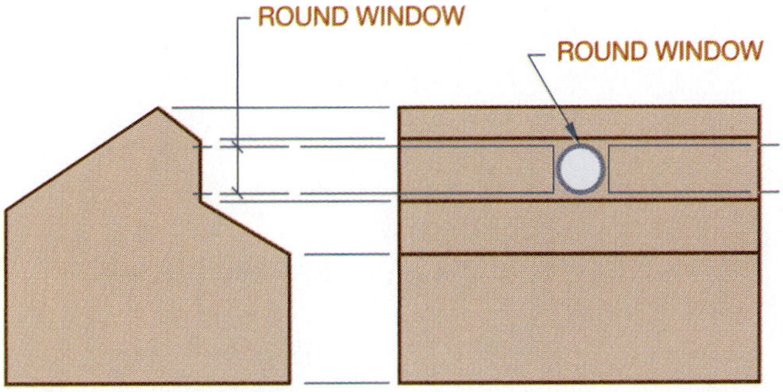

FIGURE 4.53 ■ Round window is a circle when the line of sight is perpendicular.

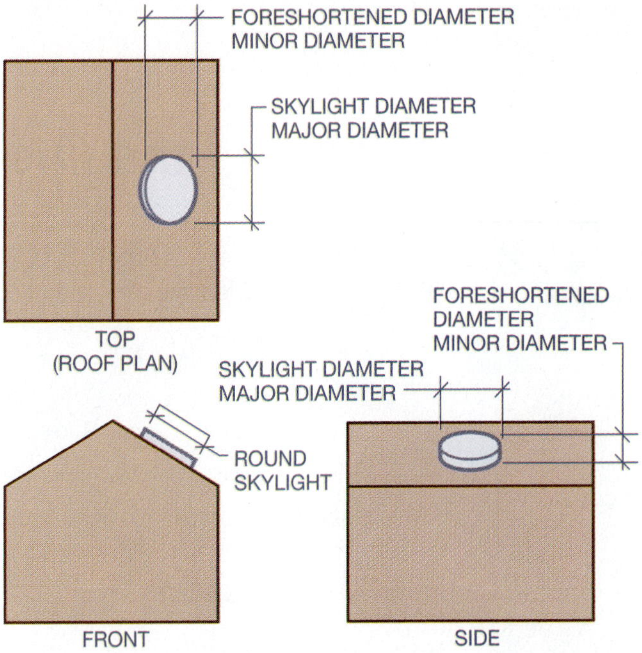

FIGURE 4.54 ■ Circle projected onto an inclined surface appears as an ellipse.

CHAPTER 4

Sketching and Orthographic Projection Test

DIRECTIONS

Answer the questions with short, complete statements or drawings as needed on 8 1/2 × 11 notebook paper as follows:

1. Letter your name, Chapter 4 Test, and the date at the top of the sheet.
2. Letter the question number and provide the answer. You do not need to write out the question.

Answers may be prepared on a word processor if course guidelines allow this.

QUESTIONS

Question 4–1 Define sketching.

Question 4–2 How are sketches useful in computer-aided drafting?

Question 4–3 Describe the proper sketching tools.

Question 4–4 Should paper for sketching be taped to the drafting board or table. Why or why not?

Question 4–5 What kind of problem can occur if a long, straight line is drawn without moving the hand?

Question 4–6 What type of paper should be used for sketching?

Question 4–7 Briefly describe a method that can be used to sketch irregular shapes.

Question 4–8 Define an isometric sketch.

Question 4–9 What is the difference between an isometric line and a nonisometric line?

Question 4–10 What do proportions have to do with sketching techniques?

Question 4–11 Define orthographic projection.

Question 4–12 What is the relationship between the orthographic plane of projection and the projection lines from the object or structure?

Question 4–13 When is a surface foreshortened in an orthographic view?

Question 4–14 How many principal multiviews of an object are possible?

Question 4–15 Give at least two reasons why the multiviews of an object are aligned in a specific format.

Question 4–16 In architectural drafting, what are the exterior front, right side, left side, and rear views also called?

Question 4–17 If a round window appears as a line in the front view and the line of sight is perpendicular to the window in the side view, what shape is the window in the side view?

Question 4–18 If a round skylight is positioned on a 5/12 roof slope and appears as a line in the front view, what shape will the skylight be in the side view?

Question 4–19 Briefly describe the trammel method for sketching a circle.

Question 4–20 Why must the paper be free to rotate when you are using the hand-compass method for sketching a circle?

Question 4–21 What is the distance from the center of a circle to the circumference called?

Question 4–22 Name the distance that goes all the way across a circle and passes through the center.

Question 4–23 Describe an easy way to sketch a 14″ circle on paper and a 6′-6″ in. circle at a construction site.

Question 4–24 In a short but complete paragraph, discuss the importance and use of measurement lines and proportions in sketching objects.

Question 4–25 Briefly describe how the block technique works for making sketches of objects.

Question 4–26 Name two basic applications for multiviews in architectural drafting.

Question 4–27 In multiview projection, what is the top view commonly called?

Question 4–28 Name the four views that are commonly used as elevations to describe the exterior appearance of a structure.

Question 4–29 Why should a soft lead and slightly rounded pencil point be used when sketching?

Question 4–30 Briefly describe how you would use your pencil to establish measurements if you are sketching a house across the street.

DIRECTIONS

Use proper sketching materials and techniques to solve the following sketching problems, on 8 1/2 × 11 bond paper or newsprint. Use very lightly sketched construction lines for all layout work. Darken the lines of the object, but do not erase the layout lines.

Additional problems are available on the CD accompanying this text.

PROBLEMS

Problem 4–1 Sketch the front view of your home or any other local single-family residence using the block technique. Use the measurement line method to approximate proper proportions.

Problem 4–2 Use the box method to sketch a circle with a diameter of approximately 4 in. Sketch the same circle using the trammel and hand-compass methods.

Problem 4–3 Find an object with an irregular shape, such as a French (irregular) curve, and sketch a two-dimensional view using the grid method. Sketch the object to correct proportions without measuring.

Problem 4–4 Use the same structure you used for Problem 4-1, or a different structure, to prepare an isometric sketch.

Problem 4–5 Use the same structure you used for Problem 4-4 to sketch a front and right-side view.

Problem 4–6 Use a scale of 1/4″ = 1′-0″ to draw a 38° acute angle with one side horizontal and both sides 8′– 6″ long.

Problem 4–7 Given the sketch for this problem of a swimming pool, spa, and patio, resketch these elements large enough to fill most of an 8 1/2″ × 11 sheet of paper.

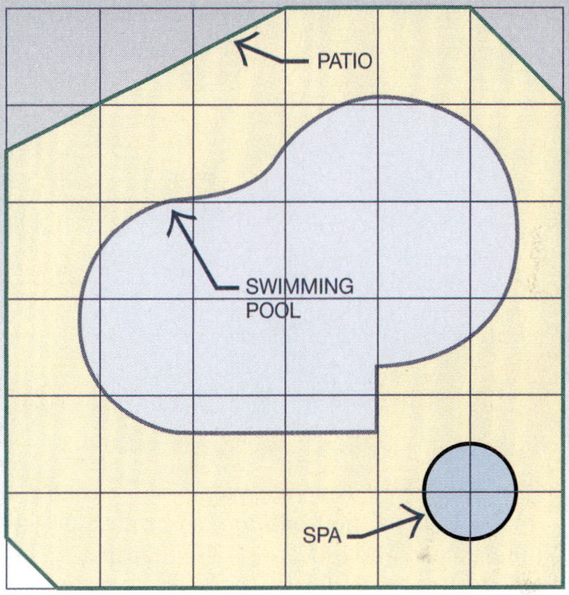

(continued)

Problem 4–8 Given the top and side views shown in the sketch for this problem, redraw these views and draw the missing front view, filling most of an 8 1/2 × 11 sheet of paper.

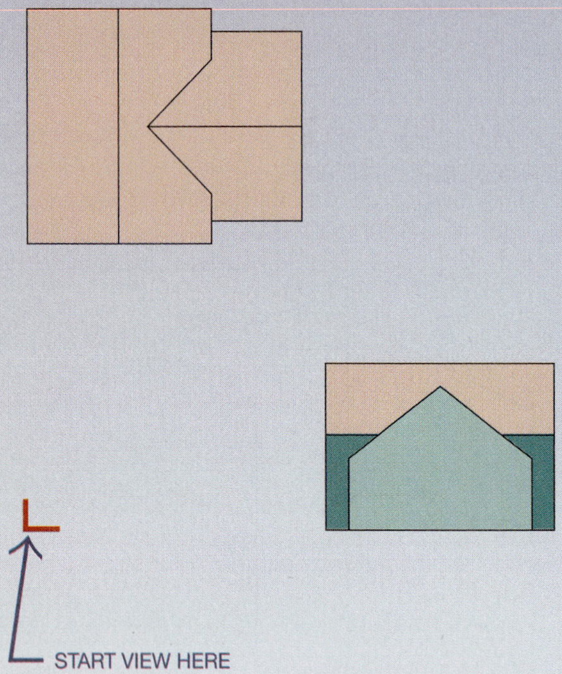

Problem 4–9 Given the pictorial sketch for this problem, draw the front, top, and right-side views, filling most of an 8 1/2 × 11 sheet of paper.

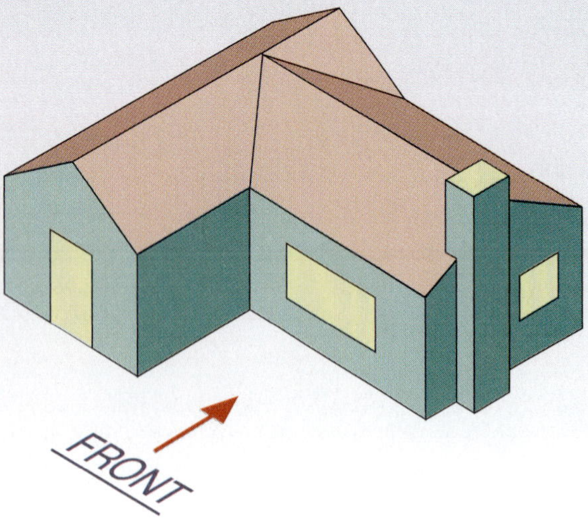

Problem 4–10 Given the three views of the house shown in the sketches for this problem, sketch an isometric view, filling most of an 8 1/2 × 11 sheet of paper.

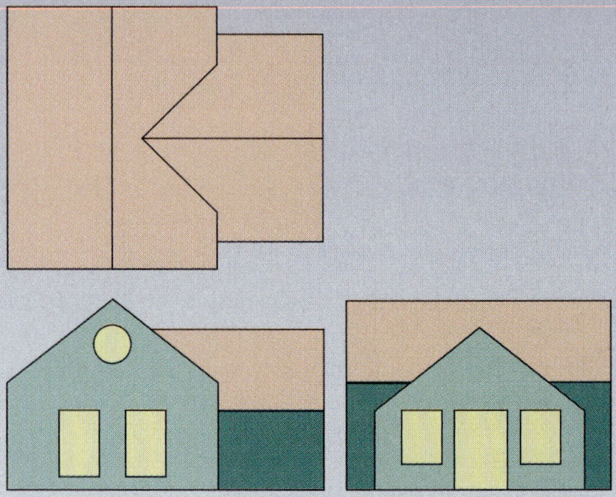

CHAPTER 5

Architectural Lines and Lettering

INTRODUCTION

Drafting is a universal graphic language that uses lines, symbols, dimensions, and notes to describe a structure to be built. Lines and lettering on a drawing must be of a quality that reproduces clearly. Properly drawn lines are dark, crisp, sharp, and of a uniform thickness. There should be no variation in darkness, only a variation in thickness known as line contrast. Certain lines may be drawn thick to stand out clearly from other information on the drawing. Other lines are drawn thin. Thin lines are not necessarily less important than thick lines, but they may be subordinate for identification purposes. Recommended line thicknesses are much more defined in mechanical drafting than in architectural drafting. Architectural drafting line standards have traditionally been more flexible, and the creativity of the individual drafter or company may influence line and lettering applications. This is not to say that each drafter can do anything that he or she desires when preparing an architectural drawing. This chapter presents some general guidelines, styles, and techniques that are recommended for architectural drafters. Practices using computer-aided design and drafting (CADD) are covered in Chapter 6 and applied to specific content in later chapters.

TYPES OF LINES

Lines are the primary method of displaying images on architectural drawings. The lines must accurately and clearly represent the drawing content. There are a variety of line types found on drawings. Each type of line conveys a meaning in the way it is represented and its placement on the drawing. There are basically two widths of lines commonly found on architectural drawings. The purpose of different line widths is to make certain lines stand out more than others. These thicker lines are meant to be more dominant than other lines. They may not be any more important, but they are the first lines that are intended to be seen by the viewer. For example, when we look at a floor plan, the wall lines and related features should be the main focus at first glance. Other lines, such as dimension lines, are equally important, but their appearance is subordinate to the lines used to create the plan. This discussion will introduce you to each of the lines commonly used in architectural drafting. Figure 5.1 shows each type of architectural line and its desired width.

Construction Lines and Guidelines

Construction lines are used for laying out a drawing. They are drawn very lightly so they will not reproduce and will not be mistaken for any other lines on the drawing. Construction lines are drawn with very little pressure using a 4H to 6H pencil, and if drawn properly will not need to be erased. Use construction lines for all preliminary work. Construction lines are drawn on a separate layer when CADD is used. This layer can be turned on or off as needed. CADD layers are discussed in Chapter 6.

Guidelines are like construction lines and will not reproduce when properly drawn. Guidelines are drawn to guide your lettering. For example, if lettering on a drawing is 1/8″ (3 mm) high, then the lightly drawn guidelines are placed 1/8″ (3 mm) apart. Guidelines are not used with CADD.

Some drafters prefer to use a light-blue lead rather than a graphite lead for all construction and guidelines. Light-blue lead will not reproduce in a diazo printer and is usually cleaner than graphite. Blue-line will reproduce in a photocopy machine unless it is drawn very lightly.

Object Lines

In architectural drafting the outline lines—or *object lines*, as they are commonly called—are a specific thickness so that they stand out from other lines, as they form the outline of views. Object lines are used to define the outline and characteristic features of architectural plan components, but the method of presentation may differ slightly from one office to another. The following techniques may be alternatives for object line presentation:

1. One popular technique is to enhance certain drawing features so that they stand out clearly from outer items on the drawing. For example, the outline of floor plan walls and partitions, or beams in a cross section, may be drawn thicker than other lines so that they are more apparent than the other lines on the drawing. These thicker lines may be drawn with a mechanical pencil or 0.7- or 0.9-mm automatic pencil using a 2H, H, or F lead. See Figure 5.2*a*. This technique may also use light shading to accentuate the walls of a floor plan. Dark shading would not be used, because it would nullify the thicker outline lines.
2. Another technique is for all lines of the drawing to be the same thickness. This method does not differentiate one type of line from another, except that construction lines

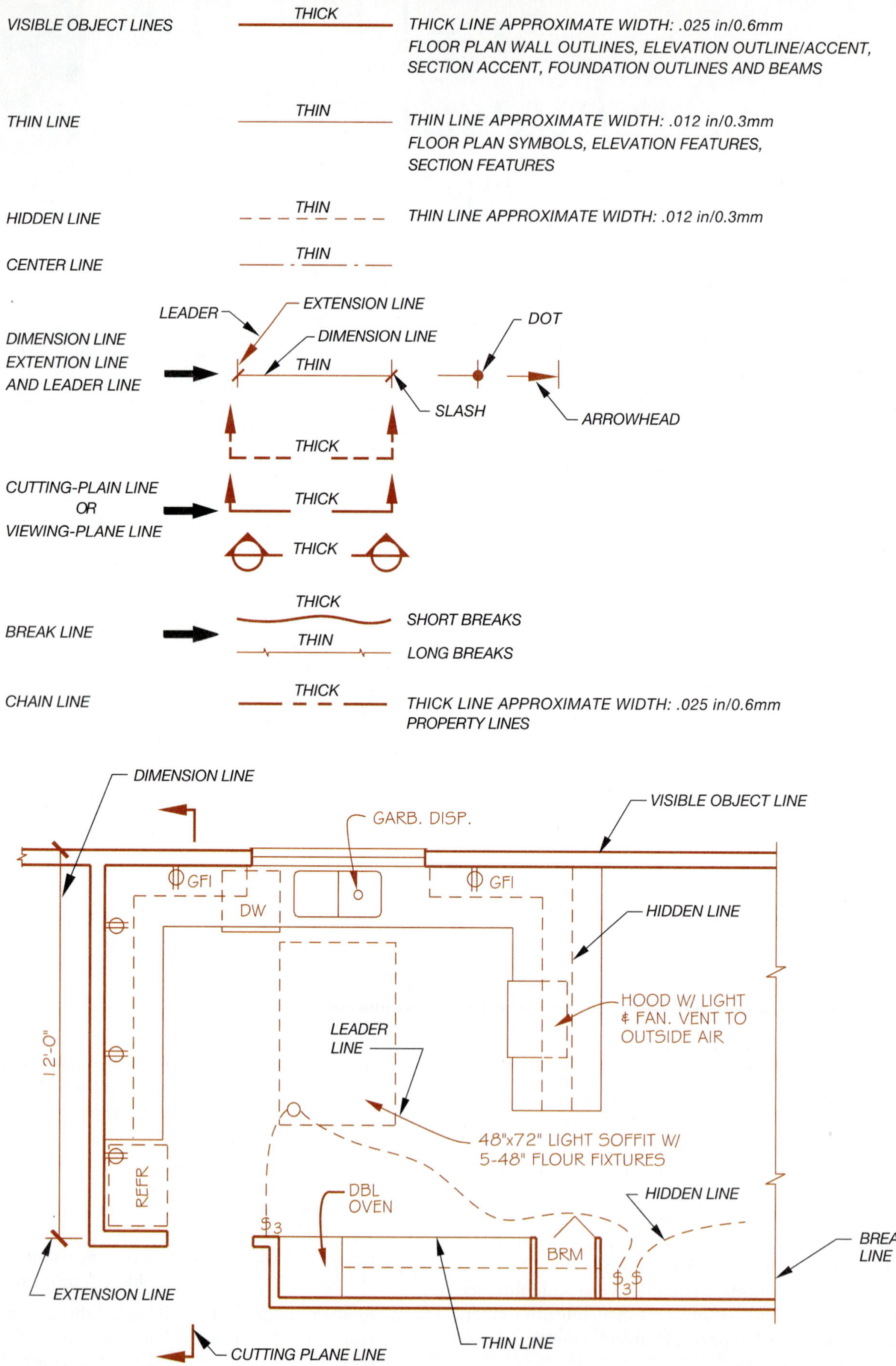

FIGURE 5.1 ■ Recommended architectural line styles and weights, with an example as used in a partial floor plan.

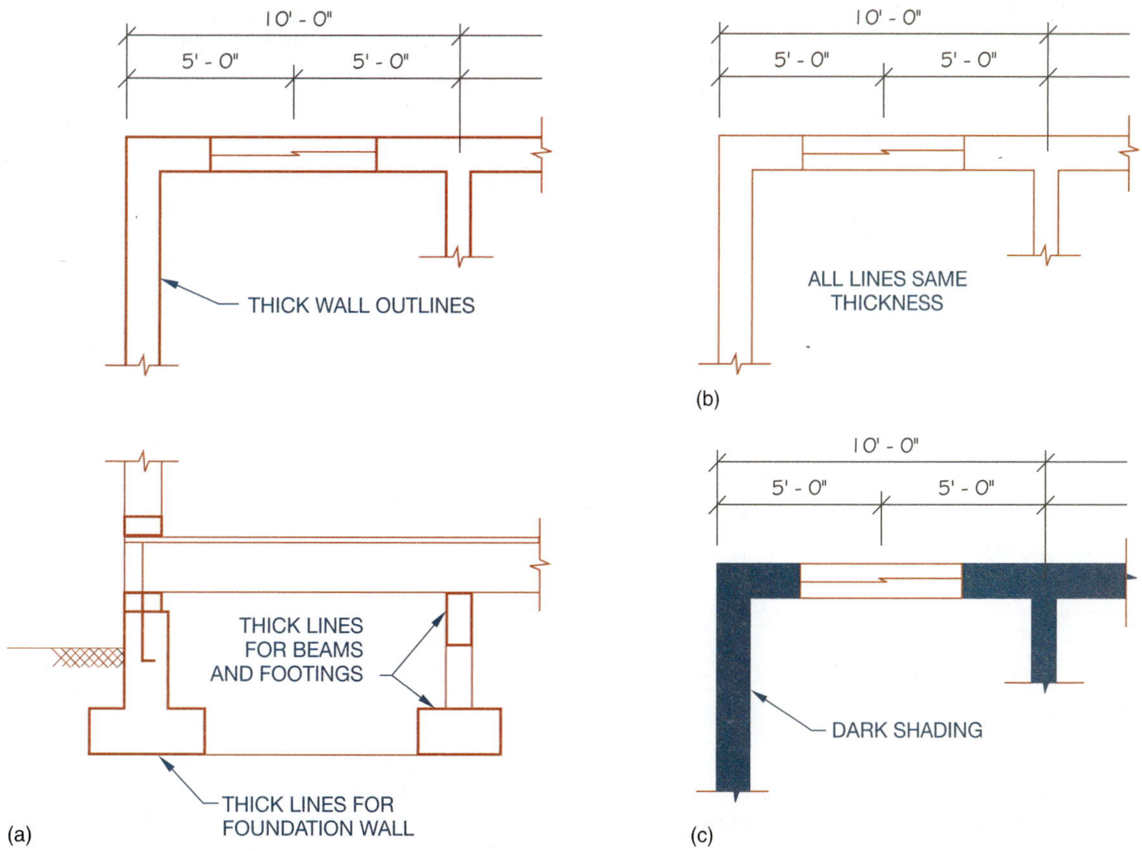

FIGURE 5.2 ■ (a) Thick outlines. (b) All lines the same thickness. (c) Accent with shading.

are always very lightly drawn. The idea of this technique is to make all lines medium-thick to save drafting time. The drafter uses a lead that works best for him or her, although a mechanical pencil or 0.5-mm automatic pencil with 2H or H lead is popular. See Figure 5.2*b*. This technique may use dark shading to accentuate features such as walls in floor plans, as shown in Figure 5.2*c*. The idea is to get all lines dark and crisp. If the lines are fuzzy, they will not reproduce well.

3. Some object lines may not be drawn as thick as others. This idea may be confusing at first. When you are creating an architectural drawing, you need to think about what features you want to be most visual. On a floor plan, for example, the outlines of the walls might be drawn thick to stand out clearly from other features. However, the outlines of cabinets, doors, and other objects might be drawn thin even though they are objects and their outlines might be considered object lines. The partial floor plan in Figure 5.1 shows this variation between object lines. While this practice is common, the confusion can be compounded for the beginner, because some variations may exist between offices. The best thing for you to do is spend as much time as you can looking at sets of architectural drawings with the idea of observing how lines are displayed. Also, quickly become familiar with the methods preferred by your instructor and the company where you work.

Dashed Lines

In mechanical drafting dashed lines are called *hidden lines*. In architectural drafting, dashed lines may also be considered hidden lines, since they are used to show drawing features that are not visible in the view or plan. These dashed features may also be subordinate to the main emphasis of the drawing.

Dashed lines vary slightly from one office to the next. These lines are thin and generally drawn about 1/8 to 3/8″ (3–10 mm) in length with a space of 1/16 to 1/8″ (1.5–3 mm) between dashes. The dashes should be kept uniform in length on the drawing, for example, all 1/4″ (6 mm) with equal spaces. Dashed lines are thin, and the spacing should be by eye, never measured. Drawing dashed lines well takes practice. Recommended leads are a 0.5-mm automatic pencil with 2H or H lead, or a sharp mechanical pencil with 4H, 2H, or H lead. Examples of dashed-line representations include beams (Figure 5.3); headers (Figure 5.4); and upper kitchen cabinets, under-counter appliances (dishwasher), and electrical circuit runs (Figure 5.5). These concepts are discussed further in later chapters.

Extension and Dimensions Lines

Extension lines show the extent of a dimension, and dimension lines show the length of the dimension and terminate at the related extension lines with slashes, arrowheads, or dots. The dimension numeral in feet and inches is placed above and near

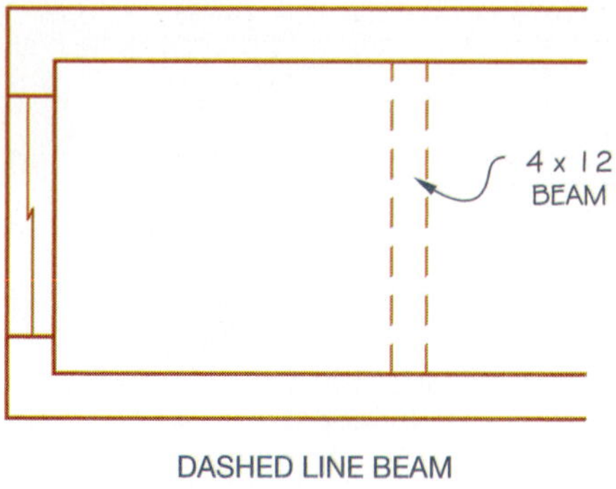

FIGURE 5.3 ■ Dashed line beam.

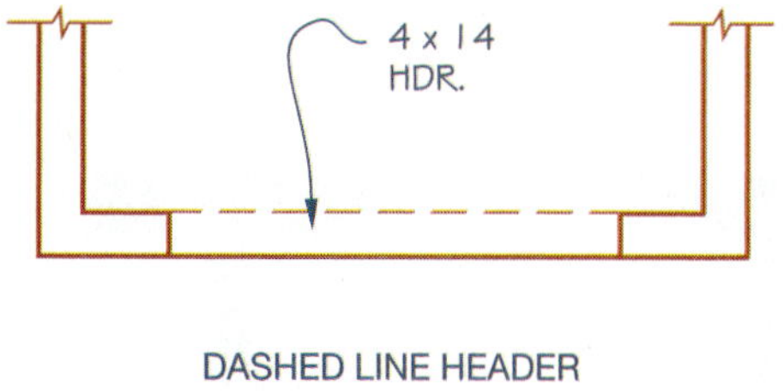

FIGURE 5.4 ■ Dashed line header.

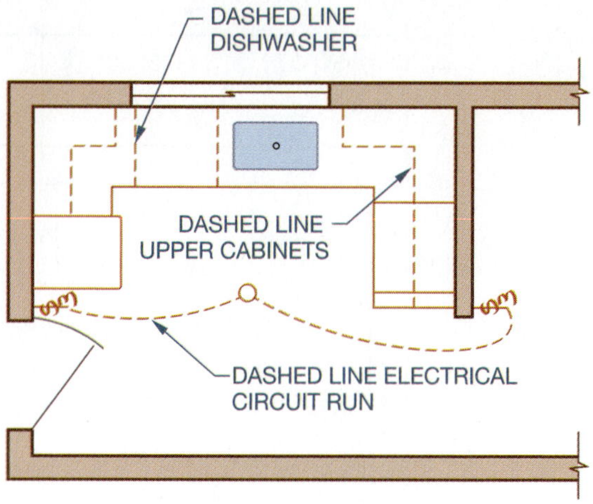

FIGURE 5.5 ■ Dashed lines for upper kitchen cabinets, dishwasher, and electrical circuit run.

the center of the solid dimension line. Extension lines generally start a short distance, such as 1/16″, away from the feature being dimensioned and run 1/8″ beyond the last dimension line. When you are dimensioning to a feature such as the center of a window, the center line becomes an extension line. Figure 5.6 shows several dimension and extension lines. Further discussion and examples are provided in later chapters. Extension and dimension lines are generally thin, dark, crisp lines that may be drawn with a sharp mechanical pencil or 0.5-mm automatic pencil using 4H, 2H, or H lead, depending on the amount of pressure you use.

Leader Lines

Leader lines are also thin, sharp, crisp lines. These lines are used to connect notes to related features on a drawing. Leader lines may be drawn freehand or with an irregular curve. Do them freehand if you can do a good job; but if they are not smooth, use an irregular curve. The leader should start from the vertical center at the beginning or end of a note and be terminated with an arrowhead at the feature. Some companies prefer that leaders be straight lines beginning with a short shoulder and angling to the feature. Figure 5.7 shows several examples.

Break Lines

The two types of break lines are the long break line and the short break line. The long break line is normally associated with architectural drafting. The break symbol is generally drawn free-

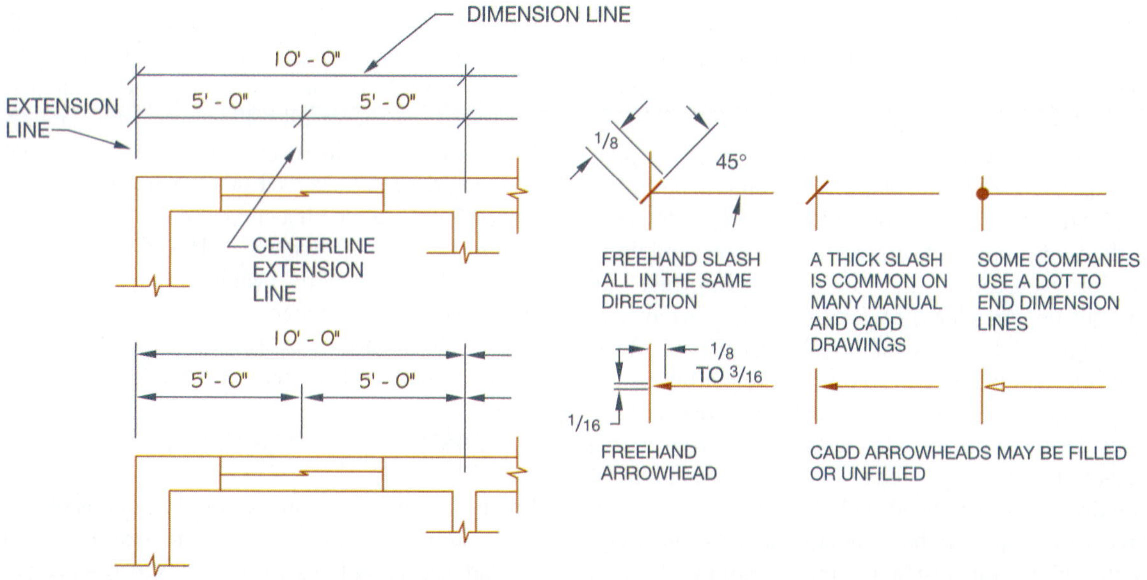

FIGURE 5.6 ■ Dimension and extension lines.

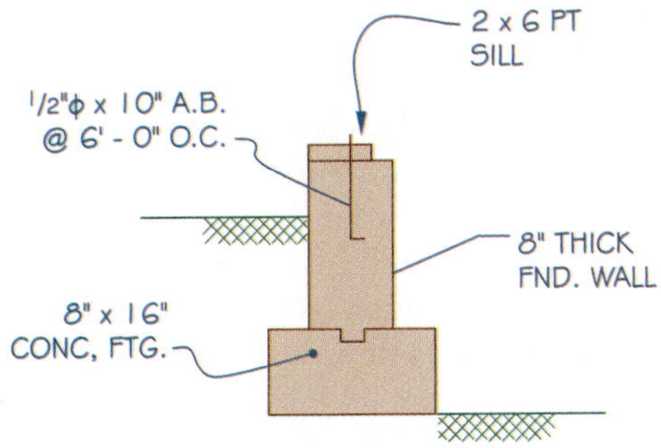

FIGURE 5.7 ■ Sample leader lines. The style for leaders used should be the same throughout the drawing.

hand. Break lines are used to terminate features on a drawing when the extent of the feature has been clearly defined. Figure 5.8*a* shows several examples. The short break line may be found on some architectural drawings. This line, as shown in Figure 5.8*a*, is an irregular line drawn freehand and may be used for a short area. Keep in mind that it is not as common as the long break line. Breaks in cylindrical objects such as steel bars and pipes are shown in Figure 5.8*b*.

LINE TECHNIQUES

When you are working on vellum or polyester film with pencil, polyester lead, or ink, use these basic techniques to make the completion of the drawing easier:

1. Prepare a sketch to help organize your thoughts before beginning the formal drawing.
2. Do all layout work using construction lines. If you make an error, it is easy to correct.
3. Begin the formal drawing by making all horizontal lines from the top of the sheet to the bottom. Try to avoid going back over lines that have been drawn any more than is necessary.
4. If you are right-handed, draw all vertical lines from right to left. If you are left-handed, draw vertical lines from right to left.
5. Place all symbols on the drawing. Try to work from one side of the sheet to the other.
6. Do all lettering last. Place a clean piece of paper under your hand when lettering to avoid smudging lines or perspiring on the drawing.
7. Most important, keep your hands and equipment clean.

Pencil Line Methods on Vellum

To see if your lines are dark and crisp enough, turn your drawing over and hold it up to the light or put it on a light table. The lines should have a dark, consistent density. Problems to look for are lines that you can see through or that have rough or fuzzy edges. You can also run a diazo copy or a diazo test strip to check line quality. If your lines are fuzzy on a properly run print, the quality is not dark and crisp enough. The following hints may help you to make a proper line:

1. Use a lead of the proper hardness. A lead that is too hard will not make a dark line without a lot of extra work. A lead that is too soft causes fuzzy lines.
2. Draw with the automatic pencil perpendicular with the sheet. Draw with the mechanical pencil tilted about 45° in the direction you are drawing. See Figure 5.9.

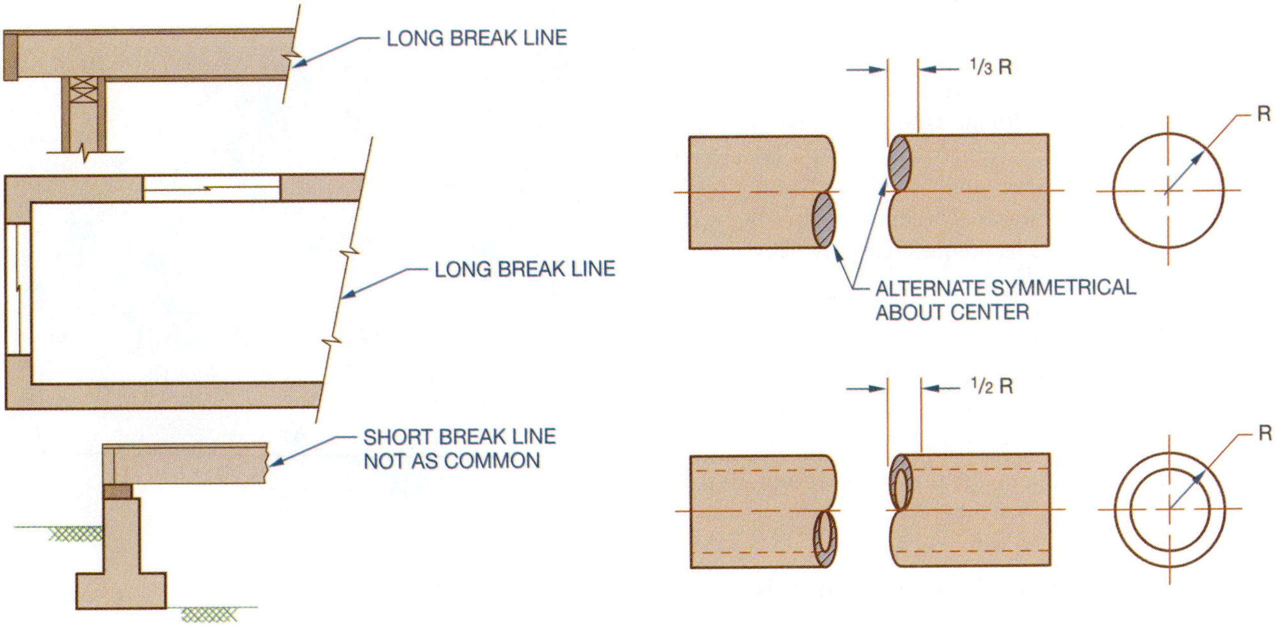

FIGURE 5.8 ■ (a) Solid break lines. (b) Cylindrical break lines.

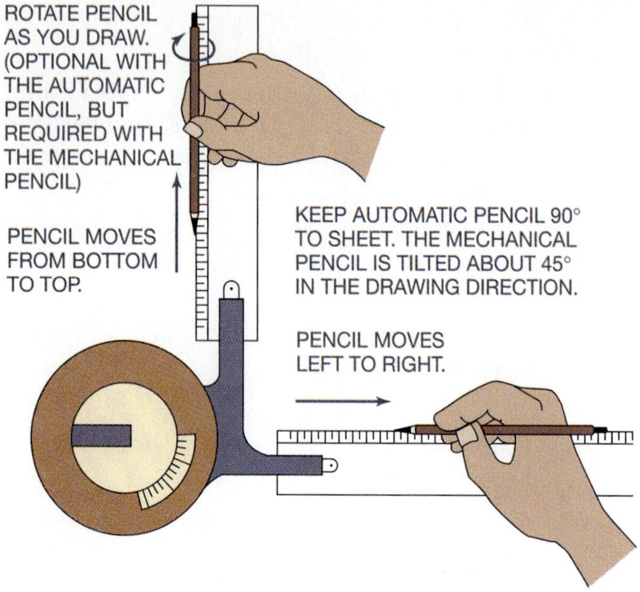

ROTATE PENCIL AS YOU DRAW. (OPTIONAL WITH THE AUTOMATIC PENCIL, BUT REQUIRED WITH THE MECHANICAL PENCIL)
PENCIL MOVES FROM BOTTOM TO TOP.
KEEP AUTOMATIC PENCIL 90° TO SHEET. THE MECHANICAL PENCIL IS TILTED ABOUT 45° IN THE DRAWING DIRECTION.
PENCIL MOVES LEFT TO RIGHT.

FIGURE 5.9 ■ Basic pencil motions.

3. When using an automatic pencil, you are not required to rotate it, although some drafters feel that rotation does help. Always rotate the mechanical pencil as you draw. This helps keep the lead uniformly sharp. See Figure 5.9.
4. Use enough pressure on the pencil. The amount of pressure depends on the individual. The suggested leads are only a recommendation. You may need to experiment with different lead hardnesses and pressures to find the right combination. Too much pressure can engrave the paper or break the point; not enough pressure may make fuzzy lines.
5. Some drafters may need to go over lines more than once to make them dark. Try to avoid this, as it slows you down and can cause an unwanted double line.

Polyester Lead Methods on Polyester Film

Some companies use polyester lead on film to help improve lines and lettering quality without using ink. Polyester lead is faster to use than ink and, in general, produces a better print than graphite on vellum. After you have worked with graphite on vellum, the use of polyester lead is similar to drawing with a crayon. Here are some basic techniques that may help:

1. Always draw on the mat (textured) side of the film. Some films have double mat.
2. Draw a single line in one direction. Retracing a line in both directions deposits a double line, which will smear and damage the mat.
3. Draft with a light touch. Drafting films require up to 40 percent less pressure than other media. Smearing and embossing can be reduced with less pressure.
4. Erase with a vinyl eraser. If an electric eraser is used, be very careful not to destroy the mat surface.

Inking Methods on Vellum or Polyester Film

When inking on either vellum or film, be sure to ink on the mat surface. On vellum the inking surface has a watermark, printed label, or title block and border. Inking can be easy if you remember that ink is wet until it turns a dull color. If the ink is shiny, do not move your equipment over it. Follow the same recommended procedures as previously discussed. Following are some helpful hints to make inking easier.

1. When using technical pens, hold the pen perpendicular to the vellum or polyester film. Move the pen at a constant speed that is not too fast. Do not slow at the end of a line, since this may widen the line or cause a drop of ink to form at the end of a line. Do not apply any pressure to the pen. Allow the pen to flow easily. See Figure 5.10.
2. Care should be taken to provide a space between the instrument edge and the ink so that the ink does not flow under the instrument and smear. Drafting machine scales have long been manufactured with edge relief for inking. Now templates, triangles, and other devices are being made with ink risers. Other ways to keep instruments away from inked lines include adhesive template lifters or template risers, which are long plastic strips that fit on the template edges. If these are unavailable, it is possible to place a second template with a larger opening under the template being used.
3. Periodically check pens for leaks around the tip or a drop of ink at the end of the tip. Have a piece of tissue paper or cloth available to help keep the tip free of ink drops.
4. Keep technical pens clean and the reservoir between one-quarter and three-quarters full.
5. Shake a technical fountain pen to get the ink started, but do not shake the pen over your drawing.

FIGURE 5.10 ■ Inking with a technical pen.
Courtesy Koh-I-Noor, Inc.

DRAWING LINES WITH CADD

There are a variety of ways to draw lines using a CADD system. Drawing line segments is the simplest form of drafting on CADD. Each line is drawn between two points. This is referred to as *point entry.* The command name is LINE. Lines are drawn in CADD using one or a combination of the point entry methods.

Cartesian Coordinate System

The *Cartesian coordinate* point entry system, also referred to as *rectangular coordinates,* is based on selecting points that are a given distance from the horizontal and vertical axes, known, respectively, as the *X* and *Y* axes. The intersection of the axes is called the *origin.* This is where $X = 0$ and $Y = 0$, or 0, 0. Points are then identified by their distance from 0, 0. For example a point $X = 2$, $Y = 2$ is two units to the right and two units above the origin. This point is entered in CADD as 2, 2.

The rectangular coordinate system is divided into quadrants. Points in the upper right quadrant are *X, Y.* Points in the upper left quadrant are $-X$, *Y.* Points in the lower left quadrant are $-X$, $-Y$, and points in the lower right quadrant are *X*, $-Y$. Figure 5.11 shows the Cartesian coordinate system. Generally, the origin (0, 0) is in the lower left corner of the drawing limits (*limits* refers to the size of the CADD drawing area). You can relate this to sheet size. The point entry methods associated with the Cartesian coordinate system are absolute coordinate, relative coordinate, polar coordinate, and picking points with the screen cursor.

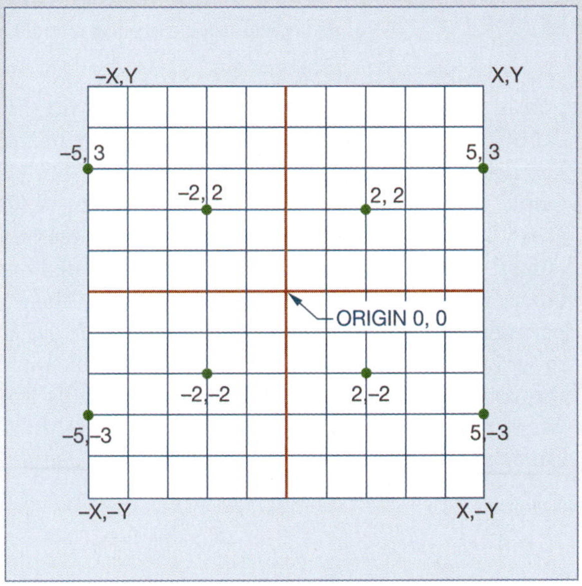

FIGURE 5.11 ■ Cartesian coordinate system.

Using the Absolute Coordinate System

Points located using the *absolute coordinate* system are always measured from the origin (0, 0). To draw the lines shown in Figure 5.12, you enter these absolute coordinates:

- First point: 2, 1
- Second point: 2, 4
- Third point: 4, 4

Using the Relative Coordinate System

Relative coordinates are always located from the previous point, or relative to the last point. To use this point entry method, you need to enter the @ symbol before entering the coordinates. The @ symbol is used in AutoCAD for this point entry system. (Other CAD software may use other command entries. Check your users' guide for these applications.) This tells the computer to place the next point a given distance from the previous point. To draw the lines shown in Figure 5.13, you enter these relative coordinates:

- First point: 2, 1
- Second point: @0, 2
- Third point: @2, 0
- Fourth point: @0, 2
- Fifth point: @−2, 0

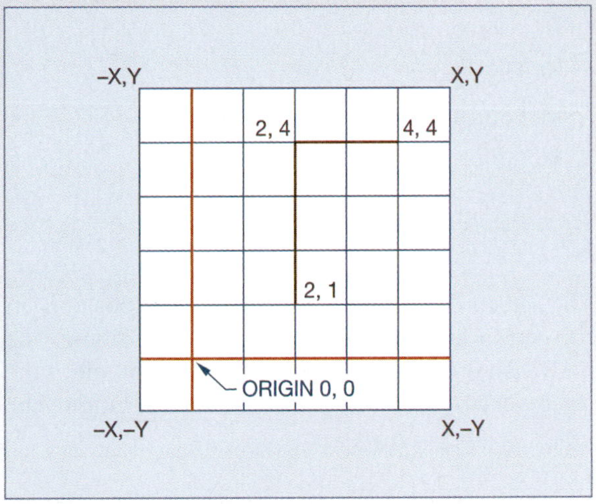

FIGURE 5.12 ■ Points located using the absolute coordinate method.

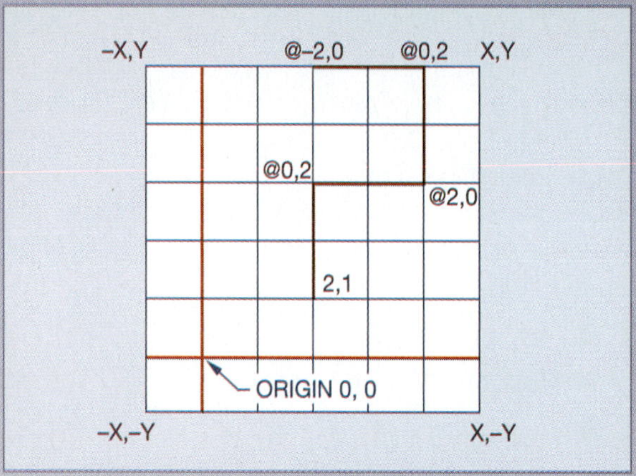

FIGURE 5.13 ■ Points located using the relative coordinate method.

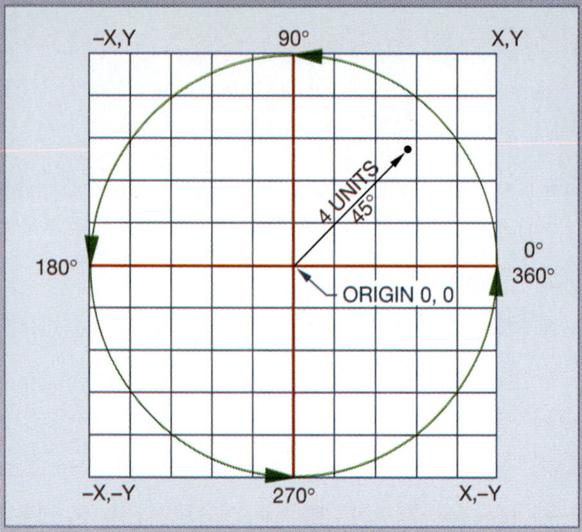

FIGURE 5.14 ■ Angular relationship of points from the origin in the Cartesian coordinate system.

Notice the fifth point. This point is @−2, 0 because it is negative two units along the *X* axis from the previous point.

Using the Polar Coordinate System

Polar coordinates are located from the previous point using a distance and angle. The angular relationship of points from the origin in the Cartesian coordinate system is shown in Figure 5.14. Also shown is a point that is four units and 45° from the origin. With the polar coordinate method, the symbols @ and < are used in AutoCAD. The symbol @ means from the previous point. The symbol < establishes an angular measurement to follow. (Other command entries may be used with other CAD programs. Check your users' guide for these applications.) To draw the lines shown in Figure 5.15, you enter these polar coordinates:

- First point: 2, 1
- Second point: @2 < 90
- Third point: @2 < 0
- Fourth point: @2 < 90
- Fifth point: @2 < 180

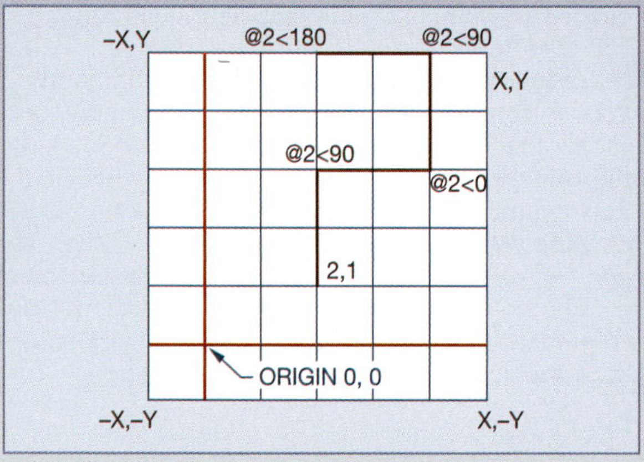

FIGURE 5.15 ■ Points located using the polar method.

Picking Points with the Screen Cursor

The screen cursor may be moved to any position on the screen and a point picked at that position. The CADD system has a screen grid that lets you move the screen cursor accurately to any desired location. The grid is generally a pattern of equally spaced dots. In AutoCAD, the grid is set up using the GRID command. You determine spacing based on the drawing. For example, a floor plan may have the grid dots spaced 12″ apart. In AutoCAD the SNAP command forces the cursor to "snap" to these dots or any designated distance. The point coordinates are displayed in the upper right corner of the screen as you move the cursor to various positions on the screen. This is known as the *coordinate display,* and it constantly tells you where the cursor is located in relationship to the origin or to the previous point, as desired.

LETTERING

Information on drawings that cannot be represented graphically by lines may be presented by lettered dimensions, notes, and titles. It is extremely important that these lettered items be exact, reliable, and entirely legible in order for the user to have confidence in them and never have any hesitation as to their meaning. Poor lettering will ruin an otherwise good drawing.

Following is some terminology commonly associated with lettering:

Composition refers to the spacing, layout, and appearance of the lettering.

Justify means to align the text. Several lines of text that are left-justified are aligned along the left side.

Lettering is the term used to describe the traditional handmade letters and numbers on a drawing. *Text* is the term for lettering that is done using computer-aided drafting. The term *annotation* is commonly used in architectural CADD applications to refer any text, notes, dimensions, and text symbols on a drawing.

A *font* is a complete assortment of any one size and style of lettering or text.

Text *style* is a set of text characters, such as font, height, width, and angle.

Rules for Metric Applications

- Unit names are lowercase—even those derived from proper names (for example, millimeter, meter, kilogram, kelvin, newton, and pascal).
- Use vertical text for unit symbols (abbreviations). Use lowercase text, such as mm (millimeter), m (meter), and kg (kilogram), unless the unit name is derived from a proper name, as in K (kelvin), N (newton), or Pa (pascal).
- Leave a space between a numeral and a symbol (for example, 55 kg, 24 m, 38 °C).
- Do not leave a space between a unit symbol and its prefix (for example, kg, *not* k g).
- Do not use plural unit symbols (for example, 5 kg, *not* 5 kgs).
- Use the plural of spelled-out metric measurements (for example, 125 meters).
- Use *either* names or symbols for units; do not mix them. Symbols are preferred on drawings. Millimeters (mm) are assumed on architectural drawings unless something else is specified. See Appendix A for other metric guidelines.

Single-Stroke Lettering

The standard lettering that has been used for generations by drafters is vertical single-stroke letters. The term *single stroke* comes from the fact that each letter is made up of single straight or curved line elements that makes it easy to draw and clear to read. There are uppercase and lowercase letters, although the industry has traditionally used uppercase lettering. Figure 5.16 shows the recommended strokes that are used to form architectural letters and numbers.

Architectural lettering styles vary, but there are similarities. Figure 5.17 shows typical architectural lettering.

As a beginning drafter, you would do well to be conservative. Too much flair may cause unnatural lettering. One important point regarding lettering technique is to make letters consistent. Do not make a letter one way one time and another way the next time. Also, keep letters vertical and always use guidelines. Entry-level drafters should pay particular attention to the style used at the architectural firm where they are employed and match it. A professional drafter should be able to letter rapidly with clarity and neatness.

Lettering Size

Minimum lettering height should be 1/8″ (3 mm). Some companies use 5/32″ (4 mm). All dimension numerals, notes, and other lettered information should be the same height except for titles, drawing numbers, and other captions. Titles and subtitles, for example, may be 3/16 or 1/4″ (4 or 6 mm) high. The height of numbers in fractions should be the same height as the other lettering associated with the fraction. The fraction bar can be placed horizontally at midheight of the guidelines, or it can be placed diagonally. The exact practice depends on the preference of the architectural office. Some offices allow fraction numbers to be smaller than the other lettering, but for best readability, this is not preferred. Verify the specific requirements with your instructor or employer.

Lettering Legibility

Lettering should be dark, crisp, and sharp so that it will be reproducible. The composition of letters in words and the space between words in sentences should be such that the individual letters are uniformly spaced, with approximately equal background areas. To achieve such spacing usually requires that letters such as I, N, or S be spaced slightly farther apart than L, A, or W. A minimum recommended space between letters is approximately 1/16″ (1.5 mm). The space between words in a note or title should be about the same as the height of the letters. When a note is made up of more than one sentence, the space between separate notes should be twice the height of the letters.

For lettering notes, sentence, or dimensions that require more than one line, the space between the lines should be one-half to full height of the letters. Notes should be lettered horizontally on the sheet, or, in some cases where space dictates, notes may be lettered vertically so they read from the right side of the sheet. For some applications, some companies may prefer that lettering be done using a lettering template or guide.

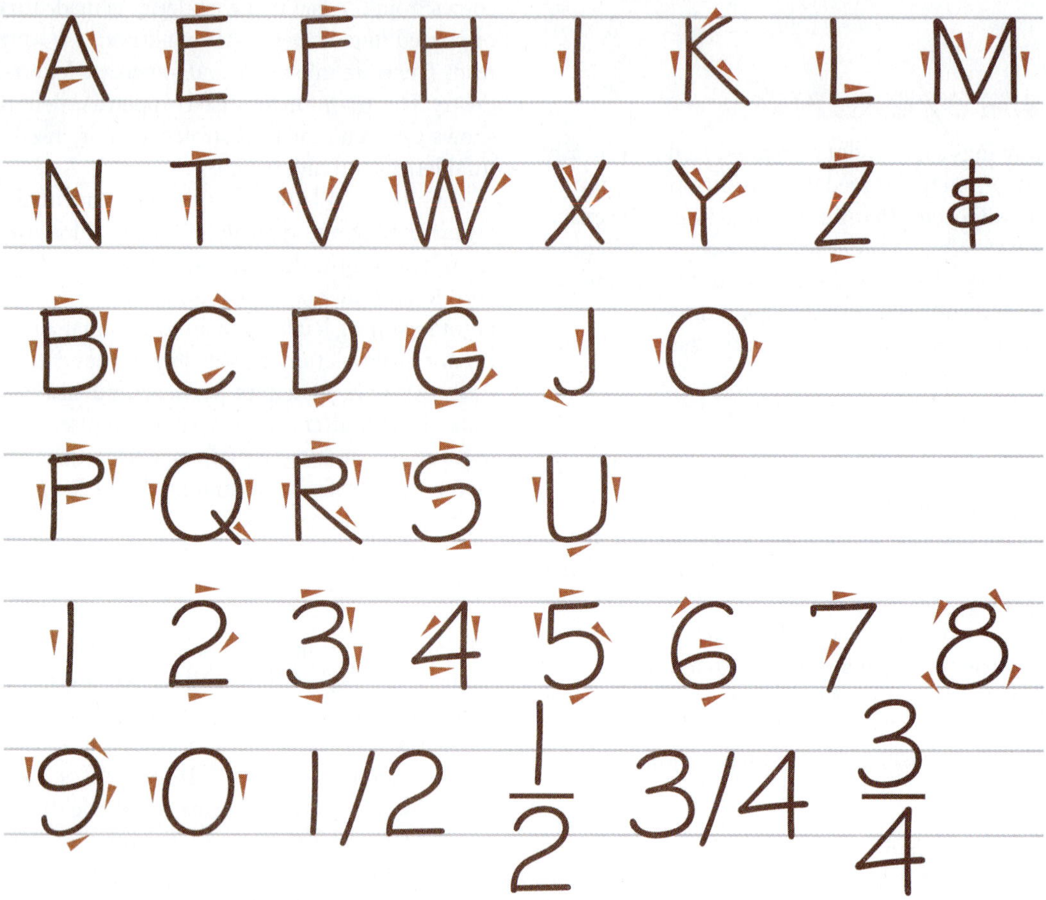

FIGURE 5.16 ■ Vertical straight element letters, curved element letters, numerals, and fractions.

FIGURE 5.17 ■ Examples of architectural lettering.

Hints for Professional Lettering

Vertical freehand lettering is the standard for architectural drafting. The ability to make good-quality lettering quickly is important. A common concern of employers hiring entry-level drafters is their ability to do high-quality lettering and linework. The following suggestions may help establish good lettering skills:

1. Always use guidelines, which are lightly drawn horizontal or vertical lines spaced equal to the height of the letters. Guidelines should be so light that they do not reproduce.
2. Use a slightly rounded point on a mechanical pencil lead or a 0.5-mm automatic pencil with H, F, or HB lead. Automatic pencils are usually easy to control for effective lettering and do not require sharpening.
3. Protect the drawing by resting your hand on a clean protective sheet placed over the drawing. This helps prevent smearing and smudging.
4. *Lettering composition* means spacing letters so that background areas look the same. This is done by eye, and there is no substitute for experience. In general, space vertical-line letters farther apart than angled or curved-line letters.
5. If your letters are wiggly or if you are nervous, try using a straightedge for vertical strokes, as shown in Figure 5.18. Making each letter rapidly also helps some people. This tends to eliminate wiggly letters. Another option, if the problem continues, is to try using a softer lead; also, be sure that the lead does not extend too far out of the pencil tip.
6. Make sure your hand and arm are comfortable on the board.

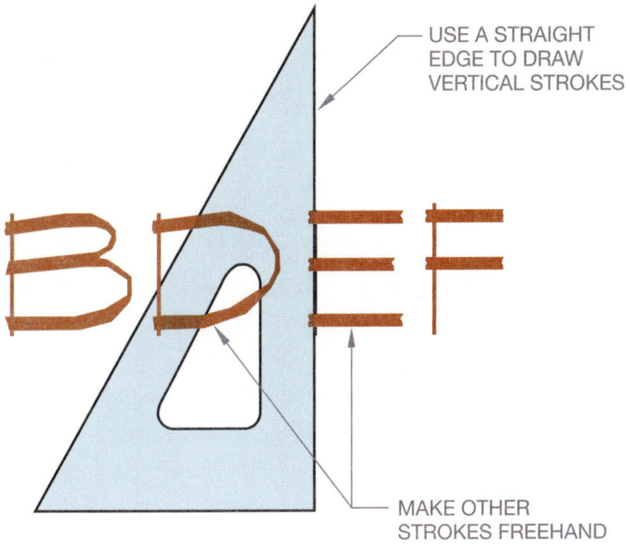

FIGURE 5.18 ■ Using a straightedge to draw the vertical strokes of letters.

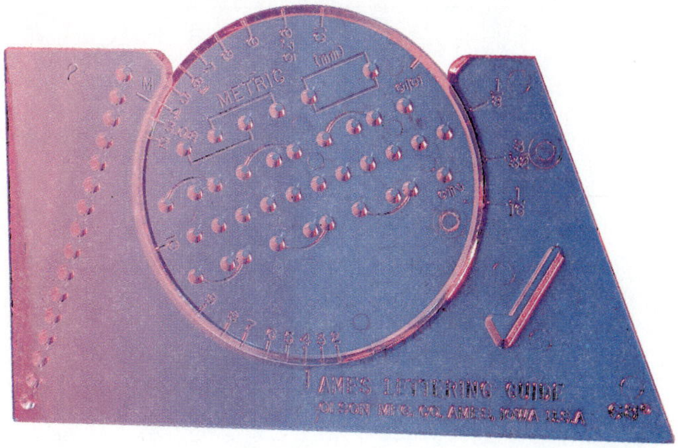

FIGURE 5.19 ■ Ames Lettering Guide. Courtesy Olson Manufacturing Co.

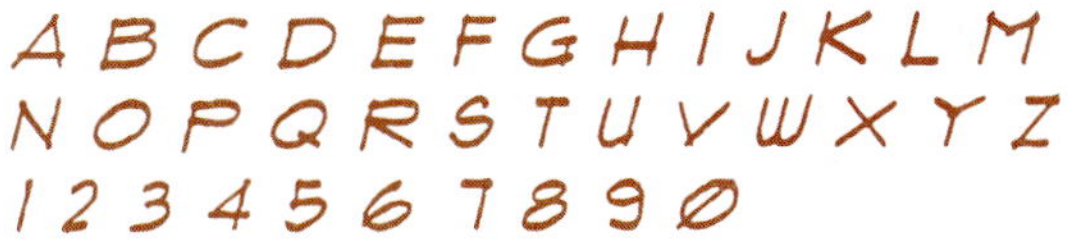

FIGURE 5.20 ■ Slanted letters (uppercase) and numbers.

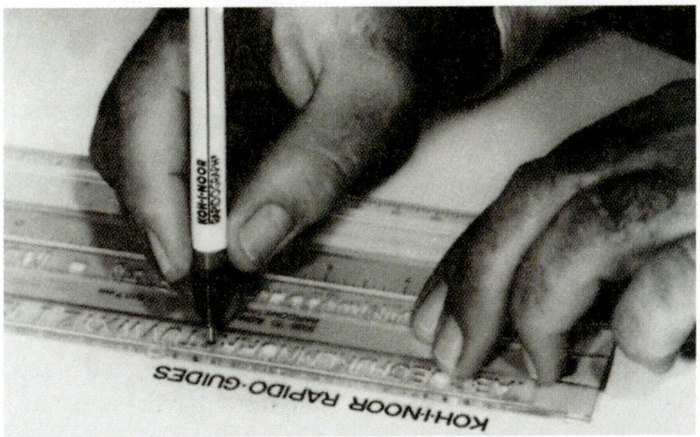

FIGURE 5.21 ■ Lettering guide template in use. *Courtesy Koh-I-Noor, Inc.*

Making Lettering Guidelines

A commonly used device for making guidelines is the Ames Lettering Guide. It is possible to draw guidelines and sloped lines for lettering from 1/16 to 2″ (1.5–51 mm) in height. Disk numbers from 10 to 2 give the height of letters in thirty-seconds of an inch. If 1/4″ high letters are required, simply rotate the disk so the 8 (8/32″ = 1/4″) is at the index mark on the bottom of the frame. See Figure 5.19. Instructions for use should be included when you purchase the guide. Metric guidelines may also be drawn.

Slanted or vertical guidelines can be drawn easily to help keep the letters vertical or slanted.

The numbers and the set of six holes to the left of the disk relate to metric heights for guidelines. This column of six holes offers the drafter the option of spacing guidelines equally (right brackets) or at half space (left brackets).

Other guideline lettering aids for equidistant spacing of lines have parallel slots ranging in width from 1/16 to 1/4″ (1.5 to 6 mm). These lettering guideline aids are not as complex as the Ames Lettering Guide, but they are not as versatile or as flexible for drawing guidelines.

Slanted Lettering

Some companies prefer *slanted lettering*. The general slant of these letters is 68°. Structural drafting is one field where slanted lettering may be found. Figure 5.20 shows slanted uppercase letters.

Lettering Guide Templates

Some drafting applications may be done using lettering guides. Standard lettering guide templates are available with vertical letters and numerals ranging in lettering height from 3/32 to 3/8″ (1–10 mm). Lettering guides are also available in many other styles, in either uppercase or lowercase. Figure 5.21 shows a lettering guide in use.

Mechanical Lettering Equipment

Mechanical lettering equipment is available in kits with templates for letters and numerals in a wide range of sizes. A complete lettering equipment kit includes a scriber plus templates, tracing pins, and lettering pens. Figure 5.22 shows the component parts of a lettering equipment set. Mechanical lettering equipment sets generally contain instructions for use. You will need to practice to become good at using this equipment.

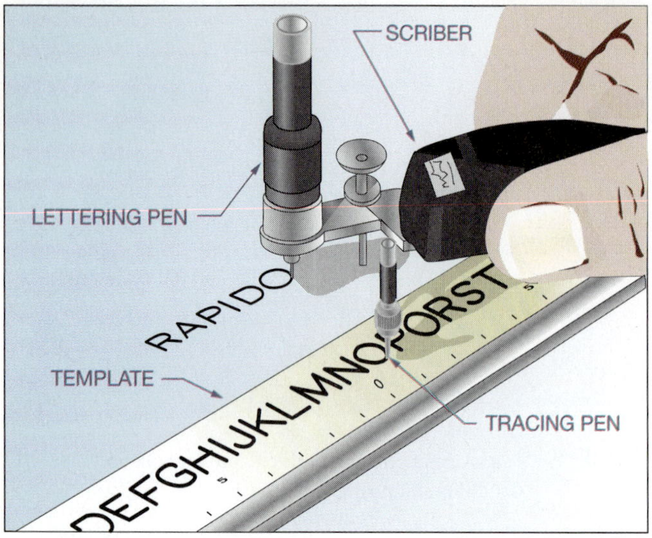

FIGURE 5.22 ■ Components of a lettering equipment set. *Courtesy Koh-I-Noor, Inc.*

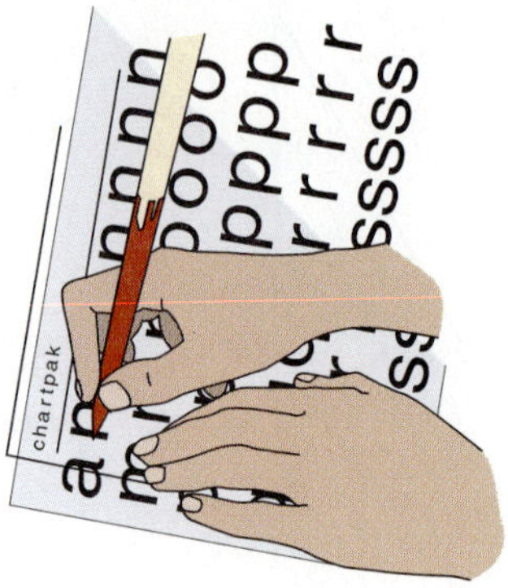

FIGURE 5.23 ■ Using rub-on letters. *Courtesy Chartpak.*

Machine Lettering

Lettering machines are available that produce a variety of fonts, styles, and sizes to prepare drawing titles, labels, or special headings. These features are especially useful for making display letters, cover sheets, and drawing titles. Most machines use a typewriter keyboard for quick preparation of lettering. A personal computer may also interface with a lettering machine to increase speed and provide additional flexibility. Lettering machines prepare strips of lettering on clear adhesive-backed tape for placement on drawing originals. The tape is also available in a variety of colors for special displays and presentation drawings.

Transfer Lettering

A large variety of transfer lettering fonts, styles, and sizes are available on sheets. These transfer letters may be used in any combination to prepare drawing titles, labels, or special headings. They may be used to improve the quality of a presentation drawing or for titles on all drawings.

Transfer letters may be purchased as vinyl sheets; individual letters are removed from a sheet and placed on the drawing in the desired location. Transfer letters are also available on sheets; letters are placed on the drawing by rubbing with a burnishing tool (Figure 5.23). Rub-on letters, as they are often called, offer high quality that is excellent for titles and special displays.

Special drawing transfers are also manufactured that can be used on presentation drawings. For example, a wide assortment of home furnishings or plant and office products may be purchased for use in preparing layout drawings. Other special items are scales, callouts, trees, landscaping items, representations of people, and transportation figures.

Drawing aids in the form of transfer tapes may be used to prepare borders, line drawings, or special symbols.

Some vendors prepare custom transfer templates for architectural customers. Such transfer templates may be used as standard drawing details.

Another type of transfer lettering is called *sticky back*. In this process, all lettering is done in a word processing or CADD file and then printed onto adhesive-backed film or paper. The sticky back material is then adhered to the drawing or the presentation.

LETTERING WITH CADD

Lettering with a CADD system is one of the easiest tasks associated with computer-aided drafting. It is just a matter of deciding on the font and style, and then locating the text where it is needed.

The term *annotation* is commonly used in architectural CADD applications to refer to any text, notes, dimensions, and text symbols on a drawing.

The CADD drafter often rejoices when the time comes to place text and notes on the drawing, because no freehand lettering is involved. The computer places text of a consistent shape and size on a drawing in any number of styles or fonts. The TEXT command is one of several that can be found in a section of a CADD menu labeled text, or text attributes. The drafter is also able to specify the height, width, and slant angle of characters (letters and numbers). Most systems maintain a certain size of text, called the *default size*, that is used if you do not specify one. The term *default* refers to any value that is maintained by the computer for a command or function that has variable parameters. The default text height may be 1/8″ (3 mm), but you can change it.

CADD Drawing Scale Factors

You may need to scale your text height relative to the drawing scale. For example, text that is 1/8″ high would be too small to see if you are working on a floor plan at a 1/4″ = 1′–0″ scale. This is because 1/8″ at a 1/4″ = 1′–0″ scale is very small. You need to size the text height in relation to the drawing scale. If you want the text on the final drawing to be 1/8″ high, then you need to set the text height at 6″ at a 1/4″ = 1′–0″ scale. Six inches at this scale is actually 1/8″. To do this properly, you need to determine the drawing scale factor. The *scale factor* establishes the relationship between the drawing scale and the desired text height. This should be determined when you are setting up the drawing as part of a prototype drawing. A *prototype drawing* is the basis for starting a drawing. It contains all of the standard elements that you need in the drawing format—for example, a border, title block, text style, and scale factor, and drawing default values. A prototype is also referred to as a *template* in CADD terminology. The architectural drawing to be plotted at a 1/4″ = 1′–0″ scale has a scale factor of 48, calculated as follows:

1/4″ = 1′–0″ (1/4″ = 0.25″)
0.25″ = 12″
12/.25 = 48

If your drawing is in millimeters and the scale is 1:1, the drawing scale factor may be converted to inches with the formula 1″ = 25.4 mm. Therefore, the scale factor is 25.4. When the metric drawing is 1:2, the scale factor is 2 × 25.4 = 50.8. For floor-plan drawings scaled at 1:50 metric, the scale factor is 50 × 25.4 = 1270.

CADD Text Styles

Many CADD systems have a variety of lettering fonts and styles. The drafter can select a style simply by pressing a menu-pad command or symbol, or by typing a command at the keyboard. Figure 5.24*a* shows some of the styles and sizes of characters that can be used in CADD. The size and styles used are dictated by the nature of the drawing. Some CADD lettering font styles include symbols such as the mapping symbols shown in Figure 5.24*b*.

Locating Text on the Drawing

The process of locating text has not changed. You still need to decide where to locate dimensions and notes,

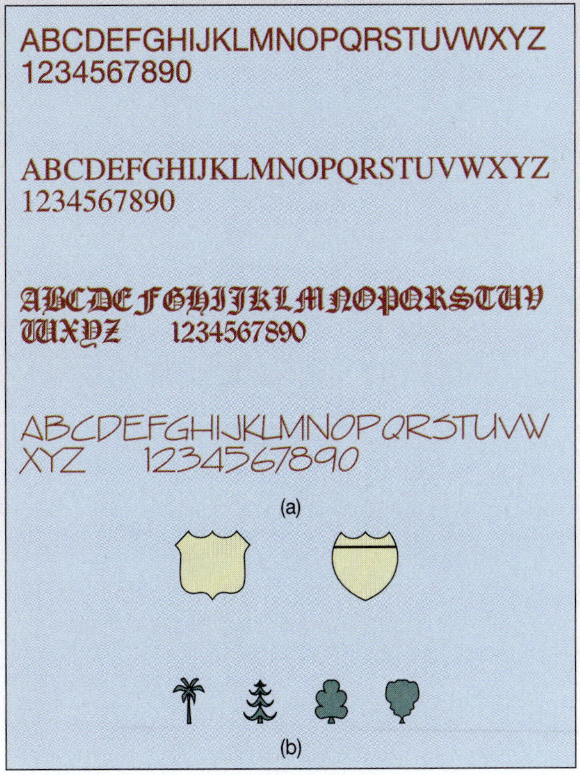

FIGURE 5.24 ■ (a) Some text character styles and sizes that can be used with CADD. (b) Special CADD mapping symbols.

but with CADD the process of placing notes is a bit more technical. Most CADD systems provide for keyboard entry of location and size coordinates, thus allowing you to locate the text accurately. Picking the location with the cursor can also place text. The location where text is started is called the *insertion point*. Text can be located by establishing any one of several starting places on the text.

Figure 5.25 shows an example of several points on the text that can be used for location purposes. Not all CADD systems use all the points shown, but most systems have a command known as TEXT that is used for placing written information on the drawing. Some systems may allow you to locate text between two points. The computer calculates the size of each letter to enable the text to fit in the desired space.

The first decision regarding text is to determine its height, width, and slant angle. Most CADD systems maintain a default text size that is used by the computer if you forget or decide not to change it. The rotation angle direction or text path is also determined by the drafter. The rotation angle is the angle from horizontal that the text is on. An example of rotation angle is shown in Figure 5.26. Text located on a horizontal line has a direction of 0°, and text that reads from the bottom up vertically has a rotation angle of 90°.

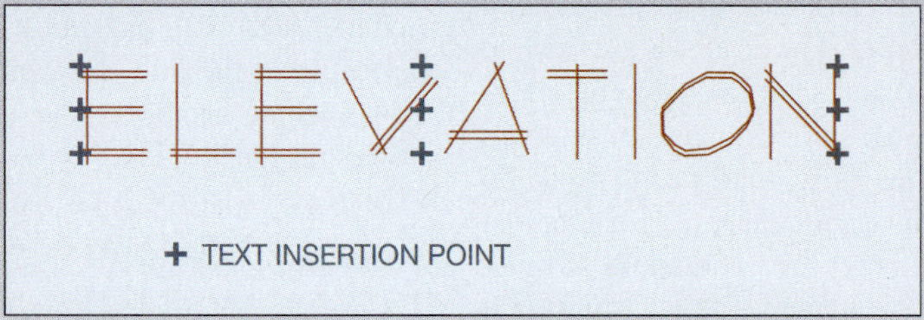

FIGURE 5.25 ■ Some points on text that can be used for location purposes.

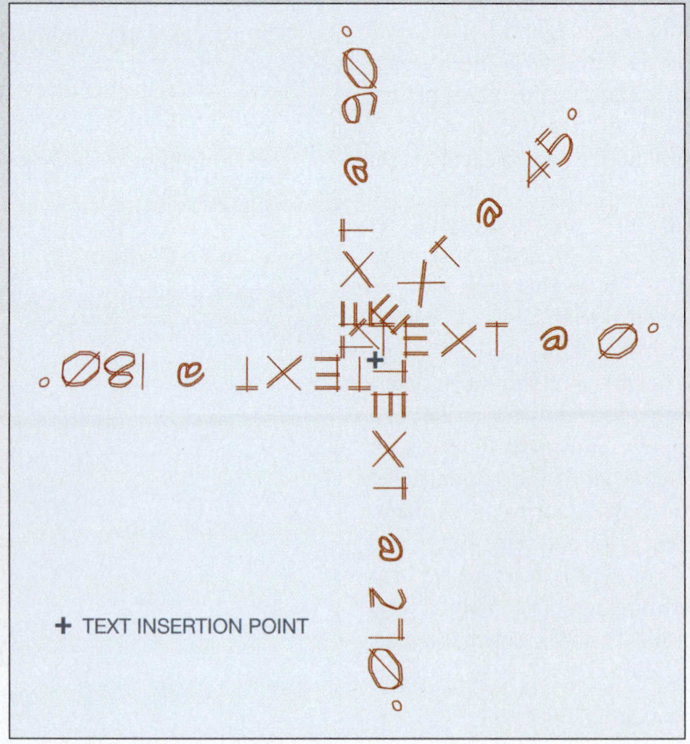

FIGURE 5.26 ■ Text rotation angles.

CHAPTER 5

Architectural Lines and Lettering Test

DIRECTIONS

Answer the questions with short complete statements or drawings as needed on 8 1/2 × 11 notebook paper, as follows:

1. Letter your name, Chapter 5 Test, and the date at the top of the sheet.
2. Letter the question number and provide the answer. You do not need to write out the question.

Answers may be prepared on a word processor if course guidelines allow this.

QUESTIONS

Question 5–1 Is there any recommended variation in line darkness, or are all properly drawn lines the same darkness?

Question 5–2 What are construction lines used for, and how should they be drawn?

Question 5–3 Discuss line uniformity and line contrast.

Question 5–4 Define guidelines.

Question 5–5 What is the recommended thickness of outlines?

Question 5–6 Identify two items that dashed lines represent on a drawing.

Question 5–7 Describe a situation in which an extension line is also the center line of a feature.

Question 5–8 Extension lines are thin lines that are used for what purpose?

Question 5–9 Where should extension lines begin relative to the object and end relative to the last dimension line?

Question 5–10 Describe leaders.

Question 5–11 Describe and show an example of three methods used to terminate dimension lines at extension lines.

Question 5–12 How does architectural linework differ from mechanical drafting linework?

Question 5–13 What is the advantage of drawing certain outlines thicker than other lines?

Question 5–14 If all lines on a drawing are the same thickness, how can walls and partitions, for example, be represented so that they stand out clearly on a floor plan?

Question 5–15 Describe the proper technique to use for drawing lines with a technical pen.

Question 5–16 Describe proper architectural lettering.

Question 5–17 What are the minimum recommended lettering heights?

Question 5–18 How should letters within words be spaced?

Question 5–19 What is the recommended space between words in a note?

Question 5–20 What is the recommended horizontal space between horizontal lines of lettering?

Question 5–21 When should guidelines be used for lettering on a drawing?

Question 5–22 Why are guidelines necessary for freehand lettering?

Question 5–23 Describe the recommended leads and points used in mechanical and automatic pencils for freehand lettering.

Question 5–24 Identify a method to help avoid smudging a drawing when lettering.

Question 5–25 For lettering fractions, what is the recommended relationship of the fraction division line?

Question 5–26 List two manual methods that can be used to make guidelines rapidly.

Question 5–27 Why should the mechanical lettering template be placed along a straightedge when lettering?

Question 5–28 Identify an advantage of using lettering guides.

(continued)

QUESTIONS *(cont.)*

Question 5–29 Describe a use of lettering machines.

Question 5–30 Describe four uses of transfer materials.

Question 5–31 The Cartesian coordinate system is also called what?

Question 5–32 What is the basis of the Cartesian coordinate system?

Question 5–33 Where is the origin?

Question 5–34 How are points in the absolute coordinate system located relative to the origin?

Question 5–35 How is an absolute coordinate specified that is 4 units on the *X* axis and 3 units on the Y axis?

Question 5–36 Describe how relative coordinate points are located.

Question 5–37 If the first point is located at 2,2, then how do you specify the next point, using relative coordinates, if it is to be 2 units directly above the first point?

Question 5–38 Describe how polar coordinate points are located.

Question 5–39 If the first point is located at 2,2, how do you specify the next point, using polar coordinates, if it is to be 2 units directly above the first point?

Question 5–40 Explain how a CADD system might have commands and options that allow you to pick points accurately with the cursor.

Question 5–41 What is the function of the coordinate display on a CADD system?

Question 5–42 Define font.

Question 5–43 Define annotation.

Question 5–44 What does the term *default* mean when applied to CADD?

Question 5–45 Explain the function of a prototype drawing.

Question 5–46 What is another name for a prototype?

Question 5–47 Discuss the purpose of drawing scale factors when this is related to text height.

Question 5–48 What drawing scale factor should be used if you want text 1/8″ high on a drawing with a 1/4″ = 1′–0″ scale? Show your calculations.

Question 5–49 Give at least three advantages of doing lettering with a computer-aided drafting system.

Question 5–50 Explain how it might be possible to have CADD text that looks like high-quality architectural freehand lettering.

DIRECTIONS

1. Read all instructions carefully before you begin.
2. Use an 8 1/2 × 11 vellum or bond paper drawing sheet for each drawing or lettering exercise.
3. Use the architect's scale of 1/4″ = 1″–0″ for each drawing.
4. Draw the floor plan described in Problem 5–1 twice. On the first drawing, use thick lines for the walls and thin lines for extension, dimension, and symbol lines. On the second drawing, make all lines the same thickness. Each drawing will represent a line thickness technique discussed in this chapter.
5. Use guidelines for all lettering. Using architectural lettering, letter the title FLOOR PLAN; center it below the drawing in 1/4″ high letters. In 1/8″ high letters and centered below the title, letter the following: SCALE: 1/4″ = 1′–0″. Letter your name and all other notes and dimensions using 1/8″ high architectural lettering.

6. Lightly lay out the drawing using construction lines. When you are satisfied with the layout, darken all lines.
7. For the lettering exercise, use 1/8″ guidelines with 1/8″ space between the lines.
8. Leave a 3/4″ margin around the lettering exercise sheet.
9. Make a diazo print or photocopy of your original drawings or lettering exercise sheet as specified by your instructor.
10. Submit your copies and originals for evaluation and grading.
11. Problems may be completed manually or with a CADD system, or both, depending on your specific course objectives and instructions. Confirm this with your instructor.

PROBLEMS

Problem 5–1 Lines and lettering. Given the following information, draw the garage floor plan:

1. Overall dimensions are 24′–0″ × 24′–0″. Dimensions are measured to the outside of walls.
2. Make all walls 4″ thick.
3. Center a 16′–0″ wide garage door in the front wall.
4. Using dashed lines, draw a 4 × 12 header over the garage door and label it 4 × 12 HEADER with a leader.
5. Label the garage floor 4″ CONC OVER 4″ GRAVEL FILL. Use a leader.

Problem 5–2 Lettering practice. Using uppercase architectural lettering with proper guidelines, letter the following statement as instructed:

YOUR NAME
MOST ARCHITECTURAL DRAWINGS THAT ARE NOT MADE WITH CADD EQUIPMENT ARE LETTERED USING VERTICAL FREEHAND LETTERING. THE QUALITY OF THE FREEHAND LETTERING GREATLY AFFECTS THE APPEARANCE OF THE ENTIRE DRAWING. MANY ARCHITECTURAL DRAFTERS LETTER WITH PENCIL ON VELLUM OR POLYESTER LEAD ON POLYESTER FILM. LETTERING IS COMMONLY DONE WITH A SOFT, SLIGHTLY ROUNDED LEAD IN A MECHANICAL PENCIL OR A 0.5-MILLIMETER LEAD IN AN AUTOMATIC PENCIL. LETTERS ARE MADE BETWEEN VERY LIGHTLY DRAWN GUIDELINES. GUIDELINES ARE PARALLEL AND SPACED AT A DISTANCE EQUAL TO THE HEIGHT OF THE LETTERS. GUIDELINES ARE REQUIRED TO HELP KEEP ALL LETTERS THE SAME UNIFORM HEIGHT. THE SPACE BETWEEN LINES OF LETTERING MAY BE BETWEEN HALF TO EQUAL THE HEIGHT OF THE LETTERS. ALWAYS USE GUIDELINES. WHEN LETTERING FREEHAND. LEARN TO RELAX SO THAT THE STROKES FOR EACH LETTER FLOW SMOOTHLY. WHEN YOUR HAND BECOMES TIRED, REST FOR A WHILE BEFORE BEGINNING AGAIN.

(continued)

PROBLEMS *(cont.)*

Problem 5–3 Given the partial floor plan shown for this problem, identify the line types labeled *a* through *k*.

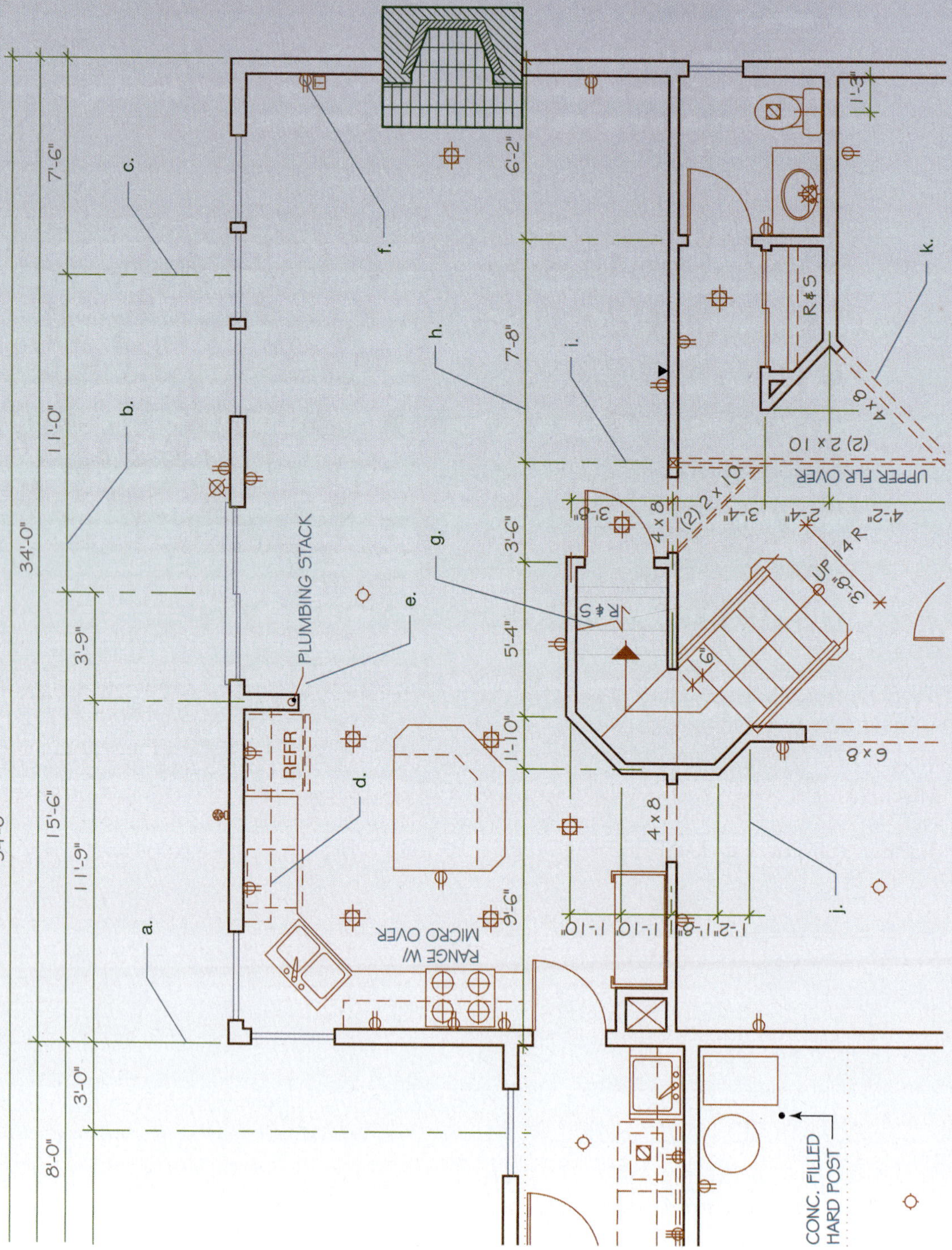

CHAPTER 6

Computer-Aided Design and Drafting in Architecture

INTRODUCTION

In Chapter 2 you learned about the traditional manual drafting workstation with modern conveniences such as an adjustable drafting table and a drafting machine or parallel bar. Also available to enhance your speed and accuracy are automatic pencils, electric erasers, and templates of all kinds. As a professional drafter, designer, or architect, you may continue to see the traditional equipment in many offices. However, drafting is changing. The concepts and theories are the same, but the tools are different. Traditional manual drafting, is rapidly converting to computer-aided design and drafting (CADD). CADD is now available in drafting and architecture schools and is being used in architectural firms.

TERMINOLOGY OF COMPUTER-AIDED DESIGN AND DRAFTING

Attribute: Text or numeric information attached to a symbol or entity on a computer-aided drawing.

Background drawing: Base drawing in the CADD layers.

Bit: Smallest amount of information a computer can read—either 0 or 1.

Byte: Eight bits that make up letters or numbers.

CADD: Computer-aided design and drafting. CAD is often used to refer to computer-aided design or computer-aided drafting.

Command: Operator-supplied information that results in a task performed by the computer.

Compatible: Term used for two or more computers or software programs that can work together. *IBM compatible* means that non-IBM computers work just as IBM computers.

Clone: A replica of another computer. Clones are often referred to as "IBM clones" because they are very similar to IBM computers. Different models of clones are generally interchangeable. Clones are normally less expensive than the models they are patterned after. Clones are also known as *compatibles*.

Cursor: Cross, space, box, or other image on the screen that moves when the user moves the pointing device, such as a mouse, puck, or stylus.

Customize: To alter a software program to perform specific tasks. The standard AutoCAD program, for example, may be customized to have special screen, pull-down, puck, or tablet menus that may be customized for drafting special applications or symbols.

Data: Collection of computer information.

Database: Collection of data stored in computer form.

Default value: One or more values established by the program. Alternatives must be selected, or default values take precedence. For example, the default value for character height may be 0.125″. The computer will always output 0.125″ lettering unless the operator changes the size.

Digitizer tablet: Input device that converts graphic data (points) into *X*, *Y* coordinates for the computer to use. The tablet usually is equipped with an electronic cursor (sometimes called a puck) or a stylus.

Directory: Device used to make hard or floppy disks work like a file cabinet. The main directory might be a design and drafting program such as AutoCAD; the subdirectories are used to store drawings such as ARCH (architectural), ELEC (electrical), or PLUMB (plumbing).

Disk drive: Used to write data to, and read from, a floppy disk.

Display: Information placed on the monitor or screen.

DOS: Acronym for disk operating system, generally meaning the operating system for IBM PCs and IBB compatibles.

Edit: To change an existing or new drawing.

Entity: Geometric element or single item of data on a computer-aided drawing–for example, a line, a circle, or text.

File: Group of computer information. Individual CADD drawings are referred to as drawing files.

File server: Computer used for storing and transmitting information in a network system.

Floppy disk: Thin disk coated with magnetic material used to store data. The most common floppy disk sizes is now 3 1/2″. Your computer may also have a high-density 5 1/4″ disk drive for compatibility with older systems.

Hard copy: Drawing produced on paper or film.

Hard drive: The part of the computer where information is stored. Also known as a hard disk.

Hardware: Workstation components, such as the computer, monitor, digitizer, keyboard, and plotter.

Input: Information given to the computer by the operator.

Joystick: Control, similar to that used on video games. When shifted, it moves a crosshair target on the screen. Used for digitizing information. Another type on some units consists of two thumbwheels, which, when moved, control the *X* and *Y* axes of the crosshairs on the screen.

Laser: Acronym for *light amplification from the stimulated emission of radiation.* A laser light beam is focused very tightly and used for a variety of commercial purposes, such as laser printers.

Layers: In CADD, details of a design or different drafting information might be separated on layers, that is, one type of information over the other. Layers are generally different colors and have their own names for clarity. Layers may be kept together, or individual layers may be turned off or on as needed.

LISP: Programming language; LISP stands for *list programming.* LISP is used to customize design and drafting software such as AutoCAD.

Load: To move a program from storage to the computer.

Memory: There are two kinds of memory. (1) Random-access memory (RAM): In the computer, a program goes from disk to RAM, which is made up of computer chips. Information is held in RAM until it is stored on the disk or power is removed. (2) Read-only memory (ROM): Computer chip with fixed data that are not lost when power is removed.

Menu: Group of commands that are related in some way and allow the operator to order items needed to make a drawing.

Modem: Device that lets computers and other electronic equipment communicate through telephone lines.

Mouse: Pointing device with a roller ball underneath. When the mouse is moved on a flat surface, it moves the screen cursor. The mouse also has a button or buttons for issuing commands or for picking items.

Network: Connection system that allows computers to communicate electronically with each other or with other peripherals.

Operating system: Software that controls the basic operation of the computer. Often referred to as OS.

Optical disk: Commonly known as compact disk or CD, these are a popular storage device. They can store up to 4 GB on a single disk.

Output: Information given to the operator by the computer.

PC: Acronym for *personal computer;* the trademark name of the first IBM personal computer.

Peripheral: Devices outside the computer, such as the keyboard, monitor, digitizer, or plotter.

Program: System that runs the computer and a set of instructions.

Puck: Pointing device that works with a digitizer and normally has multiple buttons used to issue commands or pick items from the tablet menu. Movement of the puck on the digitizer moves the screen cursor correspondingly.

Reference file: Drawing file that can be displayed as a background but cannot be edited.

Relocate: Command that allows the operator to change the origin of a drawing or a portion of a drawing.

Soft copy: Image on the screen or monitor.

Software: Programs or instructions that run the computer.

Stylus: Pointing device that works with a digitizer to pick points or items from the tablet menu. Movement of the puck on the digitizer moves the screen cursor correspondingly.

Symbol: Collection of CADD entities stored under a single name and available for use on other drawings. Referred to as a *block* or *wblock* in AutoCAD.

Text: Letters and numbers in a computer file.

Windows: Graphic system developed by Microsoft Corporation and similar to the Macintosh (by Apple). The operator uses symbols used on the screen to issue commands, rather than typing everything at the keyboard. This is an easy system, and it allows the user to integrate several activities at the same time (called *multitasking*).

Zip disk: A high-capacity, high-performance storage medium, Zip disks come in 100 MB and 250 MB formats. They are good for expanding your hard drive storage capacity and for backup.

Other, more specific terminology is defined throughout this textbook where related to the application.

DESIGNING

Architectural designers continue to make sketches even as the computer waits for input. The traditional method of making balloons to determine a preliminary room arrangement continues to be a common practice. Designers still create sketches in an effort to establish design ideas. The computer can help in the design process, but individual creativity takes place before the computer is turned on.

The CADD system can perform a number of design functions in a manner that often exceeds expectations. Plan components, such as room arrangements, can be stored in a design file; this allows the designer to call on a series of these components and rearrange them in a new design. Another aid to design is that a given plan can be quickly reduced or enlarged in size, with all components remaining in proportion. One particular room can be changed while the balance of the design stays the same. The design capability of drawing layers allows the architect to prepare a preliminary layout on one layer while changing and rearranging components on another layer.

Not only is creativity enhanced with CADD, but the repetitive aspects of design are handled more effectively. For example, in the design of an apartment or condominium complex, an initial unit can be designed, and then any number of the same units can be attached to one another as desired, very quickly. Alternative designs can also be made. Some units can be expanded or reduced in size, or alternatives for bedrooms and baths can be implemented quickly.

Design Coordination

CADD software programs offer a wide variety of applications that make the design and drafting process efficient. One example is a concept referred to as the Virtual Building™. With regard to computer usage, the term *virtual* refers to something that appears to have the properties of a real or actual object or experience. The Virtual Building stores your work in a single building file where the information remains integrated, up-to-date, and easy to manage. Working from the Virtual Building file, you can create and modify buildings in two-dimensional (2D) and three-dimensional (3D) views and derive sections, elevations, details, and other working drawings easily. Because the Virtual Building is integrated, changes are automatically updated in all views, saving time and reducing errors along the way. Architects and designers can share their work directly across a local network or the Intranet and Internet. *Intranet* links computers within a company or organization, while *Internet* is a worldwide network of communication between computers.

The Virtual Building allows coordination between the activities and elements of a project, as shown in Figure 6.1:

- Complete plans, sections, and elevations.
- Architectural and structural details.
- Window, door, and finish schedules with elevation and plan symbols.
- Component lists with quality calculations for estimating and building management.
- Photorealistic renderings for presentation and marketing.
- Virtual reality (VR) to show clients or to display on a Website. *Virtual reality* refers to a world that appears to be a real or actual world, having many of the properties of a real world. The VR world often appears and feels so real that it almost is real. This is where the computer is used to simulate environments, including the inside and outside of buildings; sound; and touch.
- Walk-through or fly-through animation. *Walk-through* can be described as a camera in a computer program that is set up like a person walking through a building, around a building, or through a landscape. *Fly-through* is similar, but the camera is like a helicopter flying over the area. Fly-through is generally not used to describe a tour through a building. Walk-through or fly-through has the effect of a computer-generated movie in which the computer images represent the real architecture; or it is like a VR presentation in which the computer images turn or move as you turn your head in the desired direction. Realistic renderings, animations, and VR are excellent tools for showing a client how a building will look inside and out. At this stage, design ideas can be created and changes made easily.

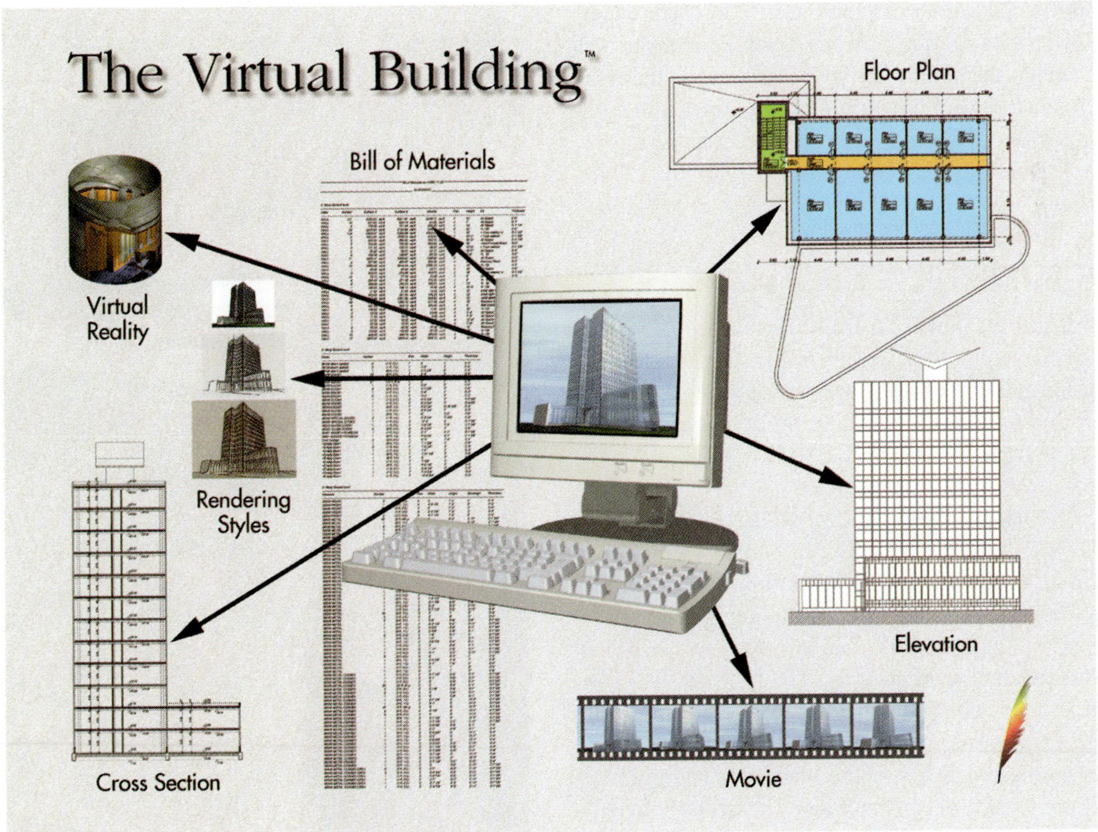

FIGURE 6.1 ■ The Virtual Building covers every facet of the architectural process communication, and collaboration. *ArchiCAD®* IMAGES COURTESY OF GRAPHISOFT®.

- Sun studies for any location, date, and time.
- Reference drawings for sharing within the company and for sharing with consultants in other companies.

Productivity

There is agreement among users that CADD increases a drafter's productivity over traditional manual drafting methods; estimates range from a twofold increase to a threefold increase. In fact, though, any increase in productivity depends on the task, the system, and how quickly employees learn to use CADD. Productivity is also related directly to the amount of time a company has had CADD in use, and to employees' acceptance of the technology and their experience with it. Most companies can expect more productivity after the users become comfortable with the capabilities of the equipment and software.

For many duties, CADD does multiply productivity several times; this is especially true of redundant and time-consuming tasks, which CADD performs much faster and more accurately than manual techniques. A great advantage of CADD is that it increases the time available to designers and drafters for creativity by reducing the time they spend on the actual preparation of drawings. Some of the business tasks related to architectural design and drafting, such as analyzing construction costs, writing specifications, computing, taking materials inventories, scheduling time, and storing information, are also done better and faster on a computer.

With some units, a designer can look at several design alternatives at one time. The drafter draws in layers; one layer may be the plan perimeter, the next layer fixtures, the next electrical, and the next dimensions and notes. Each layer appears in a different color for easy comparison.

FIGURE 6.2 ■ An architectural office with computer workstations and a manual drafting table in use. *Photo courtesy of Hewlett-Packard Company.*

THE CADD WORKSTATION

The CADD workstation is different from the traditional station because the tools have changed. For example, the new table is flat and holds a computer, monitor, keyboard, and mouse or digitizer. A printer and a plotter are usually on or available to the workstation (see Figures 6.2 and 6.3).

Computer, Monitor, Keyboard, and Mouse

There are two basic computer-aided workstations. Some companies may have a large, centrally located computer, generally in a special room or area. This *network computer,* as it is called, can serve several independent workstations. The network is a connection system that lets computers communicate electronically with each other, or to printers, plotters, and other devices. The network may be connected by way of the Internet to computers at other companies or for a variety of services. The network may be set up to operate each workstation fully, or the network can support workstations. When computers are connected in an office environment, they are referred to as a *local area network* (LAN).

FIGURE 6.3 ■ A computer system being used in an architectural office. *Photo courtesy of Hewlett-Packard Company.*

FIGURE 6.4 ■ Computer workstations operating in a network environment. *Photo courtesy of Hewlett-Packard Company.*

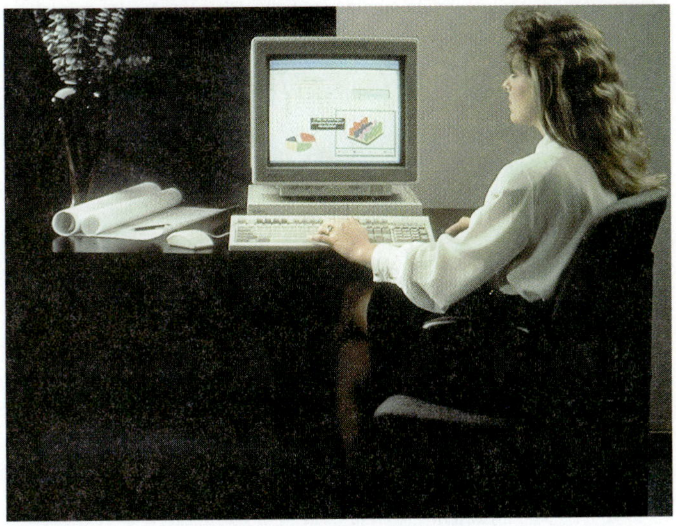

FIGURE 6.5 ■ A mouse, keyboard, and monitor in use. *Photo courtesy of Hewlett-Packard Company.*

It is common for individual workstations to be able to operate on a stand-alone basis. This means the workstation computer and its software may be used independently of the network computer. In this application, the network computer is often used, in part, to store drawing files and to access software programs. This is an advantage, because if there is a breakdown in the network computer—a situation known as *downtime*—the workstation can continue to operate on its own without expensive delays. (Downtime is also necessary periodically for repair and maintenance.) Figure 6.4 shows computer workstations operating in a network environment. Each person in the team can work on different parts of the same project, and also come together with other team members for discussions and problem solving related to the entire project.

Whichever way a company uses CADD, one result is the same: the computer is the core of the CADD workstation. The monitor is similar to a TV screen, and resting below it or attached to it is the keyboard. The monitor displays alphanumeric information (characters and numbers) from input given at the keyboard. *Input* is any data or information that the operator puts into the computer. The monitor also displays graphics (lines, symbols, or pictures) as input from the keyboard, mouse, or digitizer.

Figure 6.5 shows a mouse, keyboard, and monitor in use. The computer's central processing unit (CPU) is often placed in a convenient location under the table.

Architectural drawings are often very complex. For this reason, the monitors used in an architectural office are often larger than those in a business office or a home. A monitor is often referred to as a *screen*. The quality of the screen's graphics display is very important. Display quality is referred to as *resolution*. It is measured by total number of *pixels*, or *picture elements*. A typical resolution is 800 × 600. This means that the display has 800 pixels horizontally and 600 pixels vertically. A high-resolution monitor may have 1280 × 1024. If you see three numbers, such as 1024 × 768 × 256, the last number indicates how many colors can be displayed. A flat panel monitor takes up very little space on the desktop and gives a crisp, clear image without glare. Figure 6.6 shows a computer, keyboard, mouse, and flat panel monitor.

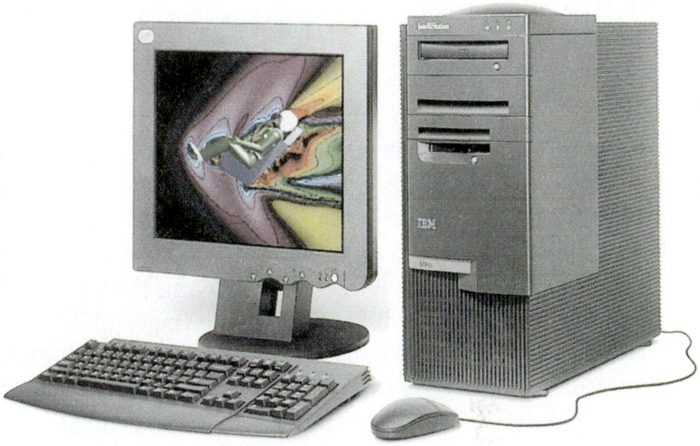

FIGURE 6.6 ■ A computer, keyboard, mouse, and flat panel monitor. *Courtesy of International Business Machines Corporation.*

The keyboard is a primary tool for providing input to your computer. Commands can be issued to the computer, calculations can be made using the number keys, and text can be entered using the alphanumeric and number keys. Fast typing is normally not critical for CADD work, but some companies do require a certain proficiency. The traditional keyboard is the 101-key flat keyboard, but the 104-key ergonomic keyboard should be considered if repetitive strain is a concern. Keyboarding can cause physical problems that are discussed later in this chapter.

A *pointing device* is used to move the screen cursor or crosshairs to a desired location. The *cursor* is an on-screen symbol (usually an arrow or a box), and *crosshairs* are a horizontal

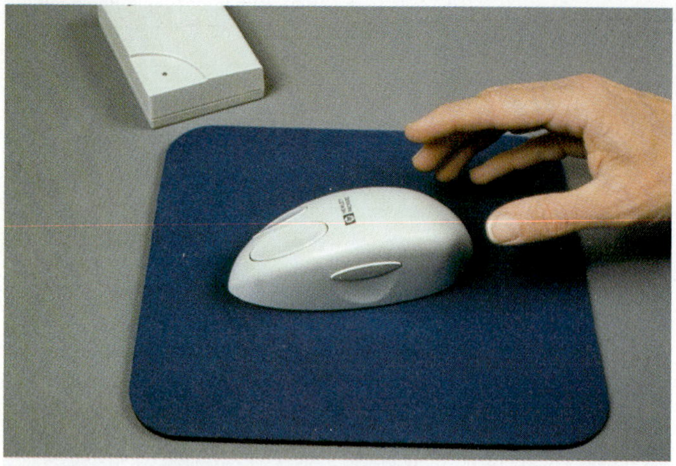

FIGURE 6.7 ■ Wireless mouse. *Photo courtesy of Hewlett-Packard Company.*

and a vertical line. The *mouse* that you have seen in the previous illustrations is the most common pointing device. It has a small ball on the underside that rolls as you move it across a flat surface. Sensors inside the mouse send messages to the software that operates the screen cursor or crosshairs. A mouse typically has two or three buttons. When the cursor is in the desired position, the pick button can be pressed to cause the desired input to the computer. A mouse is commonly connected to the computer by a wire, but a wireless mouse is shown in Figure 6.7.

A trackball is also a popular pointing device; it is like a mouse that is turned over. Rather than moving the entire device, the user moves only the ball, with the fingertips. Trackballs are easy to use and are convenient for people with limited hand movement.

Three-dimensional (3D) pointing devices are available that allow you to move objects on the screen in any direction. These devices commonly have fingertip control similar to a trackball. The 3D pointing device is used for modeling and animation projects.

Digitizer

A *digitizer* is an electronically sensitized drafting board. The digitizer may be adjacent to the monitor or built into the unit. Digitizers may be equipped with a penlike instrument called a *stylus* or a handheld device called a *cursor* that contains a crosshair lens and command buttons. Input is displayed on the monitor from the keyboard in the form of alphanumeric information or as graphics by providing the coordinates of the ends of a line or some other geometric shape. The digitizer, however, allows the operator to perform graphics tasks more rapidly and with greater ease. Provided with a list of symbols, lines, or words known as a *menu,* the operator can instantly input information on the monitor. The related commands are called a menu because each symbol or element can be ordered for immediate display with a touch of the stylus or cursor. Figure 6.8 shows a digitizer used to establish coordinate points for drawing lines.

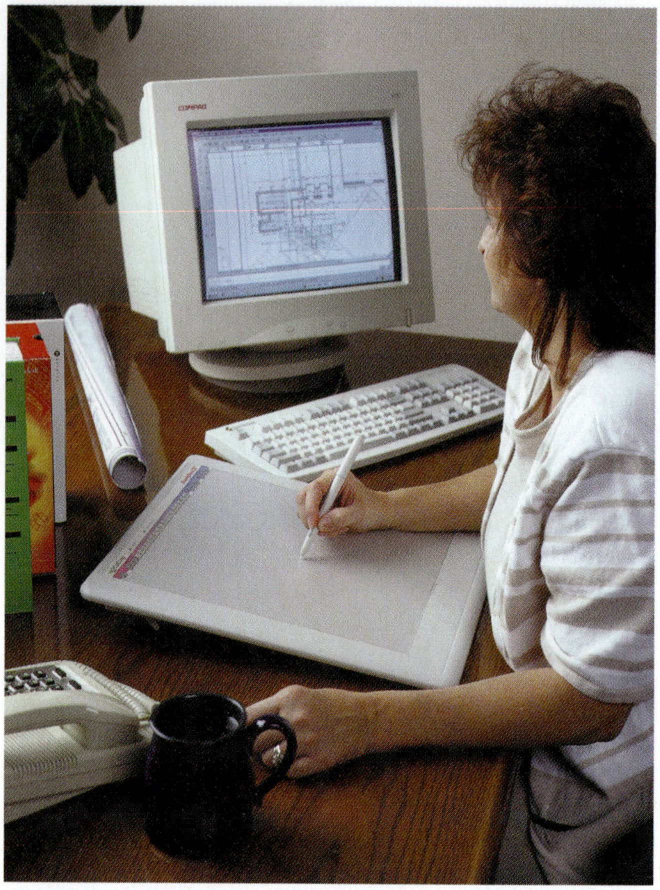

FIGURE 6.8 ■ Digitizer used to establish coordinate points for drawing lines. *Photo courtesy GTCO Calcomp Peripherals Inc.*

Another method used to input graphics is to place a sketch over the digitizer surface or to work from a sketch or series of commands next to the digitizer. Here the stylus or cursor is used to establish coordinate points or geometric shapes by a touch on the active surface of the digitizer. A *stylus* is similar in appearance to a pen. It is used for transferring information to a computer by writing, moving, or picking on an electronic tablet. The position of the command is transmitted from the electronic sensing device inside the digitizer to the monitor. For example, press the stylus at the beginning and endpoints of a line, and the coordinates of these endpoints are transmitted to the computer so that a line is shown as output. *Output* is any information the computer has to show or tell the operator. A CADD operator can digitize an entire drawing in this manner. Figure 6.9 shows a digital tablet with a 16-button programmable mouse and a stylus. This product offers accuracy and the optional use of a mouse or stylus for ergonomic flexibility.

Data Storage Devices

A typical CADD system has one or two disk drives for *floppy diskettes,* which are referred to as *disks.* These removable disks are convenient for storing and transferring data to and from your computer. The most commonly used disk is the 3 1/2″

FIGURE 6.9 ■ Digital tablet with a 16-button programmable mouse and a stylus. *Photo courtesy CalComp.*

diskette, which stores up to 1.44 megabytes (MB) of data files. The 3 1/2″ disk is in a hard case that makes it easy to transport. Your computer may also have a 5 1/4″ floppy disk drive for compatibility with older systems. The 5 1/4″ floppy disk is less durable and less reliable, and holds less data, than the 3 1/4″ disk.

Computers normally have *hard drives* that are nonremovable. The hard drives are used to store program files and to store your work. Hard drives are available in many sizes, but large hard drives are recommended for CADD applications. One *gigabyte* (GB, 1 billion bytes) or larger hard drives are common, because the operating system and the CADD program can themselves occupy over 100 MB or more of disk space.

Optical disk drives are commonly called *compact disk* (CD) drives. One CD can store up to 4 GB. Some optical disks are *rewritable*. This means that you can store, erase, and replace files as needed. A common optical disk drive is the *CD-ROM*. ROM stands for read-only memory. You can access data from a CD-ROM, but you cannot store additional data. A CD-ROM is important in modern computers, because CADD programs are shipped on CDs.

Another way to store data is the high-capacity disk drive. Such drives give you not only high-capacity but also high-performance storage. These drives are available in 40, 100 and 250 MB, and 1 and 2 GB drives. Also available are 3, 5, 7, and 10 GB tape drives. These products provide you with large movable storage capabilities. High-capacity disks are commonly used to store one or more projects or individual customer accounts. They are also used to back up the hard drive and to archive files or to share large files or many files with coworkers or other company sites. To *back up* means to make a copy of all important software and files in case something happens to the original. The high-capacity drive can also be used to store items that you do not want to fill up the hard drive.

Plotter and Printers

Once a drawing has been created on the computer and a desired monitor image has been established, an actual drawing may be needed. This drawing is called a hard copy. A *hard copy* is actually any printout of a document that you created on the computer. The printer or plotter is the piece of peripheral equipment that produces a hard copy.

FIGURE 6.10 ■ High-resolution monochrome inkjet plotter in use. *Photo courtesy of Hewlett-Packard Company.*

Pen plotters use liquid ink or felt or roller-tip pens to reproduce the computer image on plotter bond paper, vellum, or polyester film. Liquid-ink pens, similar in construction to technical fountain pens, produce high-quality drawings. These pens are available in disposable and refillable models. Pen plotters produce high-quality plots, but a complex drawing can take an hour or more to reproduce. Pen plotters are being replaced by inkjet plotters.

Inkjet plotters produce high-quality plots very fast. An inkjet plotter sprays droplets of ink onto the paper to form dot-matrix images. Print quality is measured in *dots per inch* (dpi). This is also referred to as *resolution*. Inkjet plotters can produce resolutions of 720 dpi or higher. Figure 6.10 shows a high-resolution monochrome inkjet plotter in use. *Monochrome* means that the plotter uses only one color. Inkjet plotters are available for the reproduction of small- or large-format drawings and can be networked for access from all of the computers in the group or company. Figure 6.11 shows a desktop

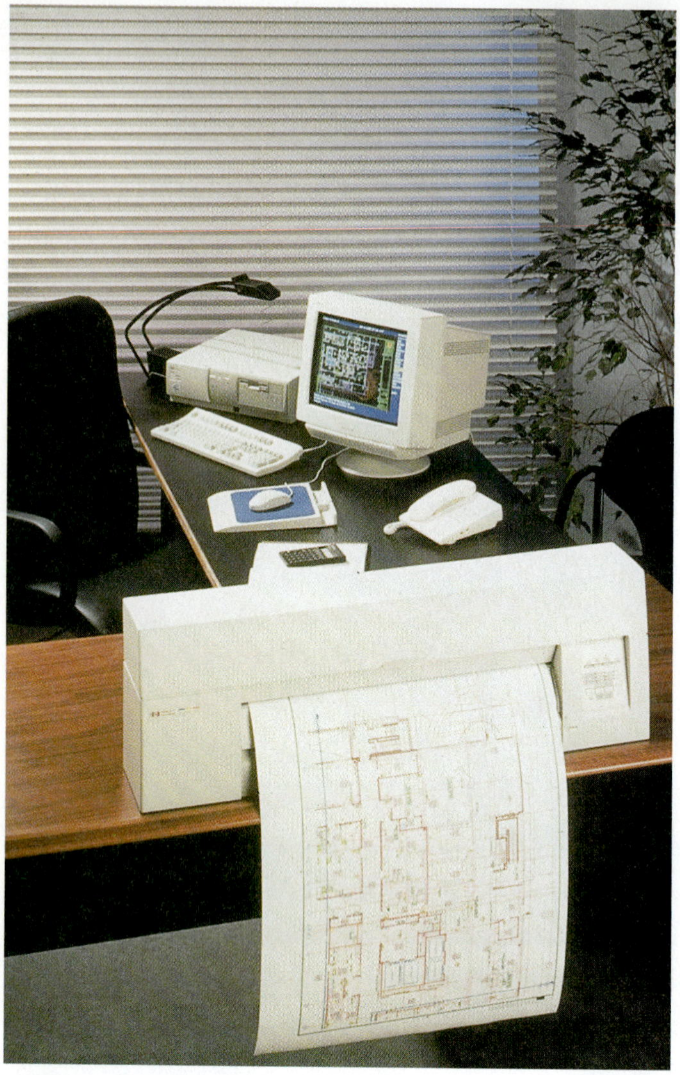

FIGURE 6.11 ■ A desktop color inkjet plotter. *Photo courtesy of Hewlett-Packard Company.*

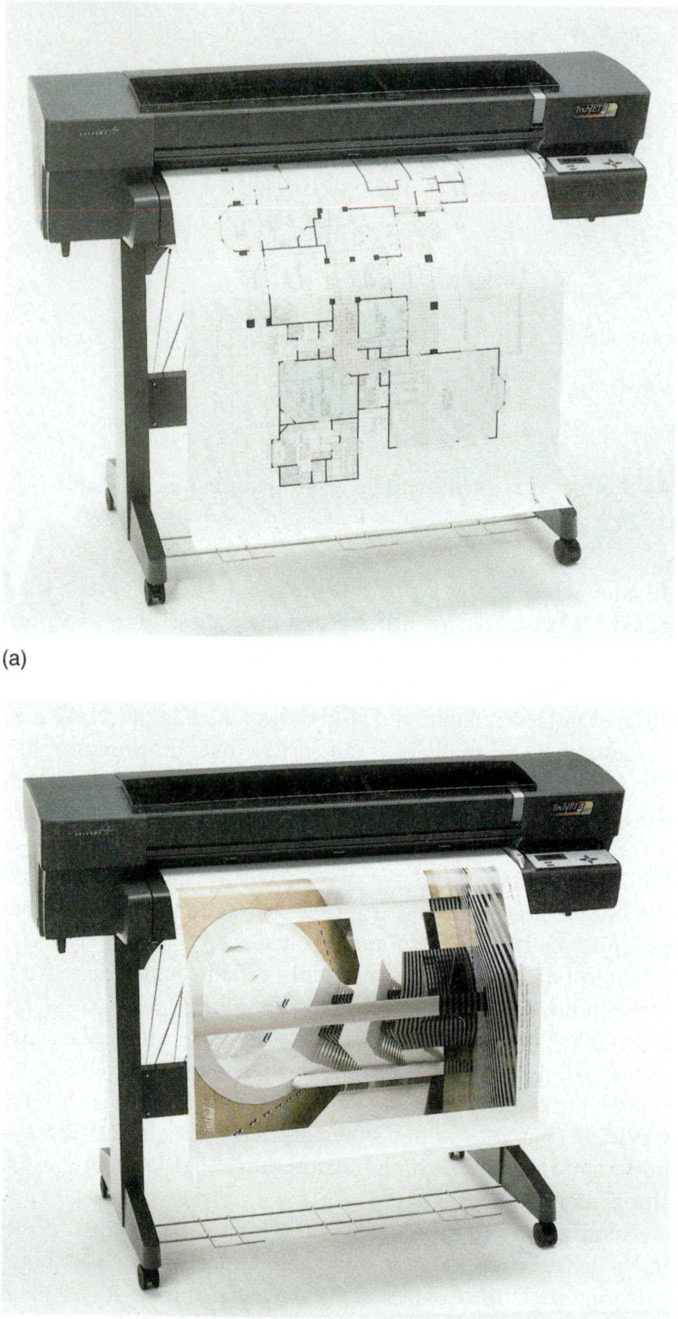

(a)

(b)

FIGURE 6.12 ■ (a) A color inkjet plotter displaying a floor plan with color used to accent features for a client presentation. (b) A color architectural rendering being printed on a color inkjet plotter. *Photos courtesy CalComp.*

color inkjet plotter. Some applications require photorealistic reproduction. Figure 6.12*a* shows a color inkjet plotter displaying a floor plan with color used to accent features for a client presentation. Figure 6.12*b* shows a color architectural rendering being printed on a color inkjet plotter.

Also available are *electrostatic plotters*, used where a large volume of plots must be made. Electrostatic plotters produce plots faster and more quietly than pen plotters. The electrostatic process uses a line of closely spaced electrically charged wire nibs to produce dots on coated paper. Pressure or heat is then used to attach ink permanently to the charged dot. These plotters are about ten times faster than pen plotters.

Two basic types of printers are impact and nonimpact printers. *Impact printers* operate by striking a ribbon to transfer ink from the ribbon to the paper. Two well-known members of this family are the daisy-wheel printer and the dot-matrix printer. The daisy-wheel printer produces characters similar to a typewriter and is considered a letter-quality printer. The dot-matrix printer creates images composed of tiny dots. Usually, a dot-matrix printer that can strike 24 dots at one time (a 24-pin printer) is considered a letter-quality printer. Many companies use dot-matrix printers to run economical check prints. Nonimpact printers are thermal, electrostatic, inkjet, and laser printers.

FIGURE 6.13 ■ A universal laser printer-fax-copier-scanner machine in use. *Photo courtesy of Hewlett-Packard Company.*

The thermal printer uses tiny heat elements to burn dot-matrix characters into treated paper. The best-quality image comes from the laser printer. A laser draws lines on a revolving plate that is charged with a high voltage. The laser light causes the plate to discharge. An ink toner then adheres to the laser-drawn images, and the ink is bonded to the paper by pressure or heat.

Many modern products are highly versatile. For example, some machines can print, copy, scan, and fax. Scanning is explained later in this chapter. The term *fax* is short for *facsimile transmission*. A *facsimile* is a duplicate or copy. A fax machine transmits a copy of an image and sends it over telephone lines to a receiving fax machine. Figure 6.13 shows a universal laser printer-fax-copier-scanner machine.

Digitizing Existing Drawings

Digitizing is a method of transferring drawing information from a digitizer into a computer. A digitizing tablet may be used for existing drawings. Digitizer tablets range in size from 6 by 6 in. to 44 by 60 in. Most school and industry CAD departments use 11- by 11-in. digitizer tablets to input commands from standard and custom tablet menus. In most cases companies do not have the time to convert existing drawings to CAD, because the CAD systems are heavily used for new drawings. Therefore, an increasing number of businesses digitize existing drawings for other companies. Figure 6.8 shows a digitizer tablet in use.

Optical Scanning of Existing Drawings

Scanning is a method of reproducing existing drawings in the form of computer drawings. Scanners work much the same way as taking a photograph of a drawing. One advantage of scanning over digitizing existing drawings is that the entire drawing, including dimensions, symbols, and text, is transferred to the computer. A disadvantage is that some scanned drawings require a lot of editing to make them presentable. The scanning process picks up visual graphic information from the drawing. What appears to be a dimension, for example, is only a graphic representation of the dimension picked up from the original drawing. What you get is an exact duplicate of the existing drawing. However, for the dimensional information to be technically accurate, the dimensions must be redrawn and edited on the computer. There are software packages that make this editing process quick and easy.

After the drawing is scanned, the image is sent to a raster converter that translates information to digital or vector format. A *raster* is an electron beam that generates a matrix of pixels. *Pixels* make up the drawing image on the computer. A raster editor is then used to display the image for changes.

Companies involved in scanning can often reproduce existing drawings more efficiently than companies that digitize existing drawings.

Computer Networks

Computer networks were introduced briefly earlier in this chapter. It can be very difficult to control files in an architectural office where there is more than one CADD workstation. When drawings are changed, it is important that every person in the office have the latest revision available for access. If two or more workstations have a copy of the same project, then editing can happen at each workstation. This can lead to different versions of the same project, which can become a serious problem. This situation can be difficult to manage unless the computers are connected in a network. A *network* is a communication or connection system that allows each computer to be linked to other computers, to printers and plotters, to data storage devices, and to other equipment.

A network system that connects the computers and peripherals within an office or company is called a *local area network* (LAN). A network system that connects computers all over the world is called a *wide area network* (WAN). A typical CADD network uses one computer with a large storage space as a file server. The file server provides a central location for all of the company's drawing and document files, so only one working copy of each file exists. All of the other computers can be connected to the file server to provide access to the stored files. Networks provide various degrees of security for files, and a network can be set up to allow only one person at a time to work on the file. In addition to file control, networks provide communication capabilities through the use of e-mail (electronic mail) on the Intranet or Internet, security for sensitive files, and accurate logs of time spent working on projects. Figure 6.14 shows a typical architectural office network system.

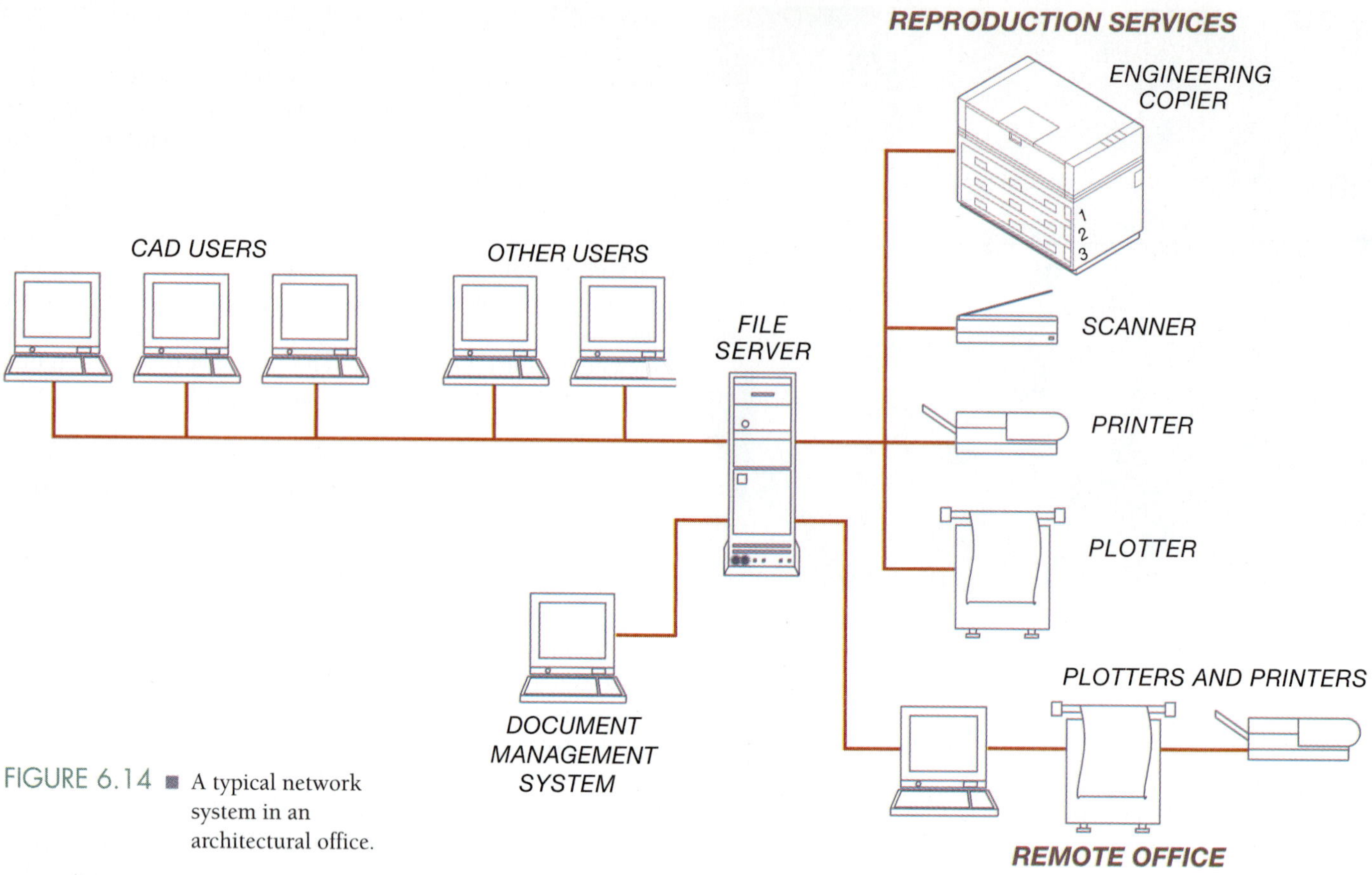

FIGURE 6.14 ■ A typical network system in an architectural office.

MOBILE COMPUTERS

So far, we have discussed the typical architectural office computer workstation and its peripherals. *Mobile computers*, commonly called *laptop* or *notebook* computers, are small and light enough to use on your lap. These computers are primarily intended for use when you are traveling (rather than at a workstation), because they fold up like a suitcase and are battery-operated or can be plugged into a wall outlet. Modern mobile computers are powerful and have features that are found in typical office PCs. Many architects and designers travel to local or distant project sites. The mobile computer allows communication with the office at such times. Design changes can be made in the field, while traveling, or while in direct communication with a client at the client's location. Figure 6.15 shows a powerful mobile computer with a 14.1-in. flat panel screen.

FIGURE 6.15 ■ A powerful mobile computer with a 14.1″ flat panel screen. *Courtesy of International Business Machines Corporation.*

ERGONOMICS

Ergonomics is the study of a worker's relationship to physical and psychological environments. There is concern about the effect of the CADD working environment on the individual worker. Some studies have found that people should not work at a computer workstation for longer than about four hours without a break. Physical problems can develop when someone is working at a poorly designed CADD workstation; these problems can range from injury to eyestrain. The most common type of injury is referred to as *repetitive strain injury* (RSI), but it also has other names, which include *repetitive movement injury* (RMI), *cumulative trauma disorder* (CTD),

and *occupational overuse syndrome* (OOS). One of the most common effects is *carpal tunnel syndrome* (CTS), which develops when inflamed muscles trap the nerves that run through your wrists. Symptoms of CTS include tingling, numbness, and possibly loss of sensation. An RSI usually develops slowly and can affect many parts of your body. You should be concerned if you experience different degrees and types of discomfort that seem to be aggravated by computer usage.

Many factors can help reduce the possibility of RSI. These factors are outlined below.

Ergonomic Workstations

Figure 6.16 shows an ergonomically designed workstation. In general, a workstation should be designed so that you sit with your feet flat on the floor, your calves should be perpendicular to the floor and your thighs parallel to the floor. Your back should be straight, your forearms should be parallel to the floor, and your wrists should be straight. For some people, the keyboard should either be adjustable or separate from the computer, to provide more flexibility. The keyboard should be positioned, and arm or wrist supports may be used to reduce elbow and wrist tension. Also, when the keys are depressed, a slight sound should be heard, to assure the user that the key has made contact. Ergonomically designed keyboards are available.

The monitor is a concern. It should be 18 to 28″, or approximately one arm's length, away from your head. The screen should be adjusted to 15° to 30° below your horizontal line of sight. Eyestrain and headache can be a problem with extended use. If the position of the monitor is adjustable, you can tilt or turn the screen to reduce glare from overhead or adjacent lighting. Some users have found that a small amount of background light is helpful. Monitor manufacturers offer large, flat, nonglare screens that help reduce eyestrain. Some CADD users have suggested changing screen background and text colors weekly to give variety and reduce eyestrain.

The chair should be designed for easy adjustments, to give you optimum comfort. It should be comfortably padded. Your back should be straight or up to 10° back, your feet should be flat on the floor, and your elbow-to-hand movement should be horizontal when you are using the keyboard, mouse, or digitizer. The mouse or digitizer puck should be close to the monitor so that movement is not strained and equipment use is flexible. You should not have to move a great deal to look directly over the cursor to activate commands.

Positive Work Habits

In addition to an ergonomically designed workstation, your own personal work habits can contribute to a healthy environment. Try to concentrate on good posture until it becomes second nature. Keeping your feet flat on the floor helps improve posture. Try to keep your stress level low, because increased stress can contribute to tension, which may aggravate physical problems. Take breaks periodically to help reduce muscle fatigue and tension. You should consult with your doctor for further advice and recommendations.

Exercise

If you feel the pain and discomfort that can be associated with computer use, some stretching exercises can help. Figure 6.17 shows some exercises that can help reduce workstation-related problems. Some people have also had success with yoga, biofeedback, and massage. Again, you should consult with your doctor for further advice and recommendations.

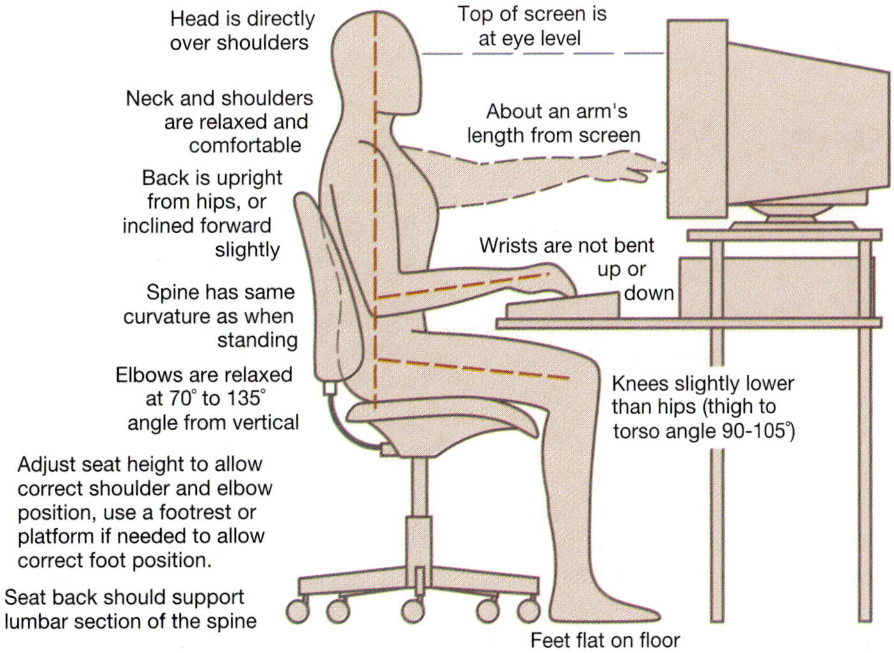

FIGURE 6.16 ■ An ergonomically designed workstation.

Sitting at a computer workstation for several hours in a day can produce a great deal of muscle tension and physical discomfort. You should do these stretches throughout the day, whenever you are feeling tension (mental or physical).

As you stretch, you should breathe easily—inhale through your nose, exhale through your mouth. Do not force any stretch, do not bounce, and stop if the stretch becomes painful. The most benefit is realized if you relax, stretch slowly, and *feel* the stretch. The stretches shown here take about 2 1/2 minutes to complete.

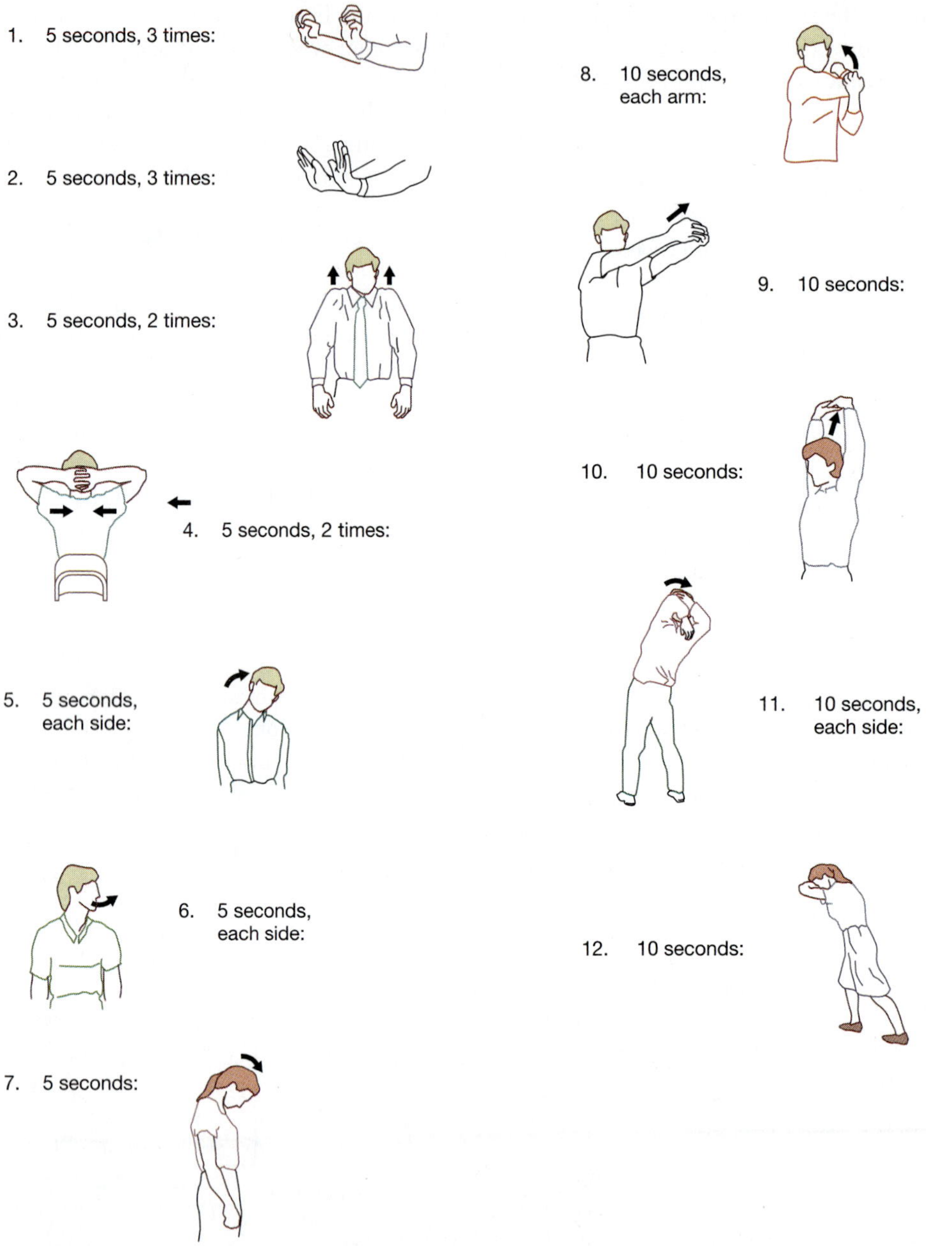

FIGURE 6.17 ■ Stretches that can be used at a computer workstation to help avoid repetitive stress injury.

The Plotter

The plotter makes some noise and is best located in a separate room next to the workstation. Some companies put the plotter in a central room, with small office workstations around it. Others prefer to have plotters near the individual workstations, which may be surrounded by acoustical partition walls or partial walls.

Other Factors

The CADD working environment may be different from traditional drafting rooms, in that air conditioning and ventilation should be designed to accommodate the computers and equipment. Carpets should be antistatic. Noise should be kept to a minimum. If smoking is permitted, the room should have special ventilation. Smoke or other contaminants can affect computer equipment.

Drafters who are required to do tedious, repetitive tasks on the computer could develop symptoms of fatigue. Consideration should be given to allowing CADD employees to do a variety of jobs or take periodic breaks.

ARCHITECTURAL CADD SYMBOLS

Many companies have a manager who is responsible for setting up the architectural CADD system and standards. This person and the staff often create custom architectural CADD menus and symbol libraries. The systems generally include custom screen, cursor, pull-down, icon, and tablet menus. Every time a new symbol or drawing detail is made, it can be saved for future use. In this way, it is never necessary to draw anything more than once with CADD. For example, if a designer or drafter draws a structural detail, as shown in Figure 6.18, this detail is saved for future use. A copy of the symbol or detail is kept in a symbols manual, and information about it is recorded.

Although every company saves symbols for future use, many companies purchase their own core architectural programs from a software developer. These are complete architectural CADD programs that are ready to use out of the box. Some of these packages provide commands for the design and layout of walls, symbols, stairs, roofs, structural grids, and plot plan symbols. Figure 6.19

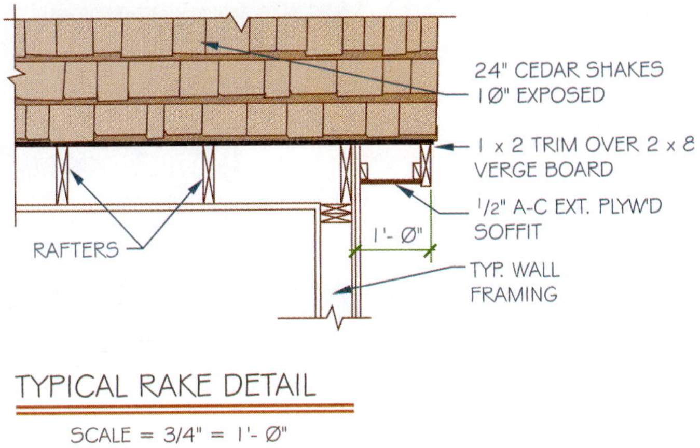

FIGURE 6.18 ■ Drawing a structural detail on CADD and saving it for future use. *Courtesy Piercy & Barclay Designers, Inc. CPBD.*

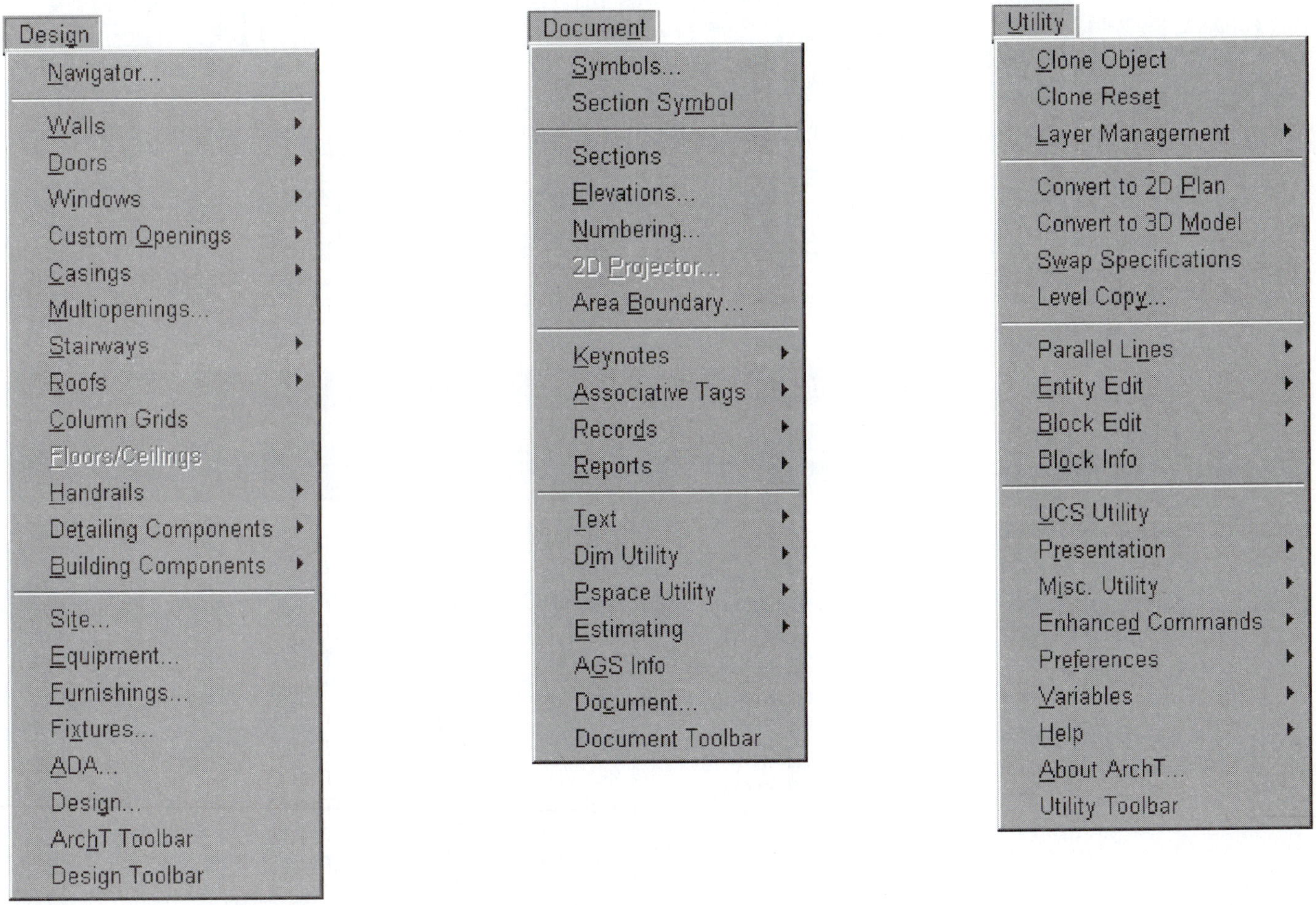

(a)

FIGURE 6.19 ■ Samples of: (a) pull-down menus, (b) toolbar menus, (c) tablet menu, (d) automatic layer editing dialog box. *Courtesy of Visio Corporation. (continued)*

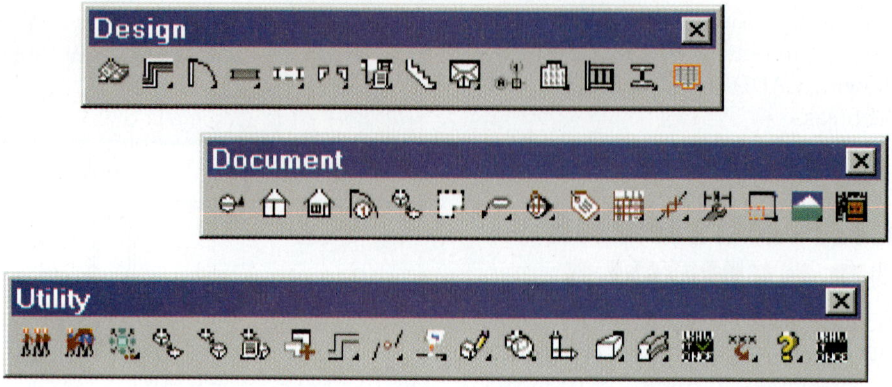

(b)

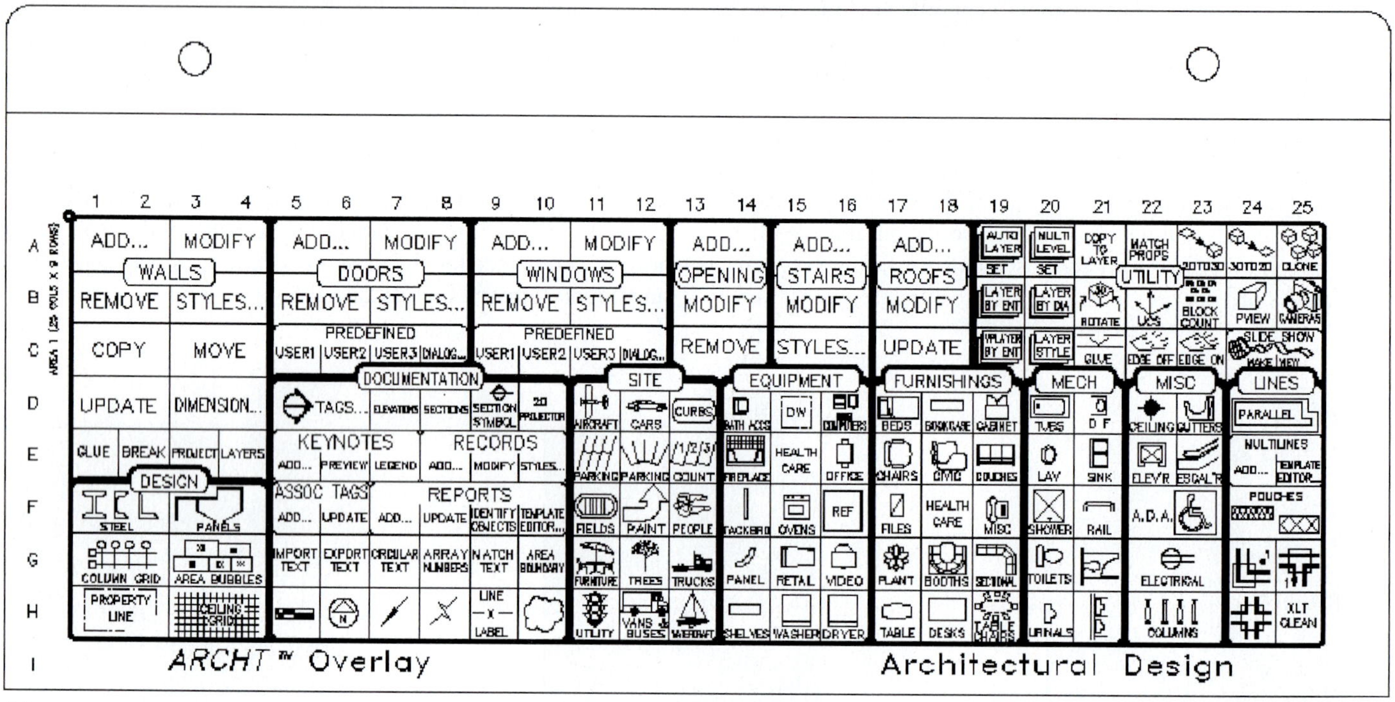

(c)

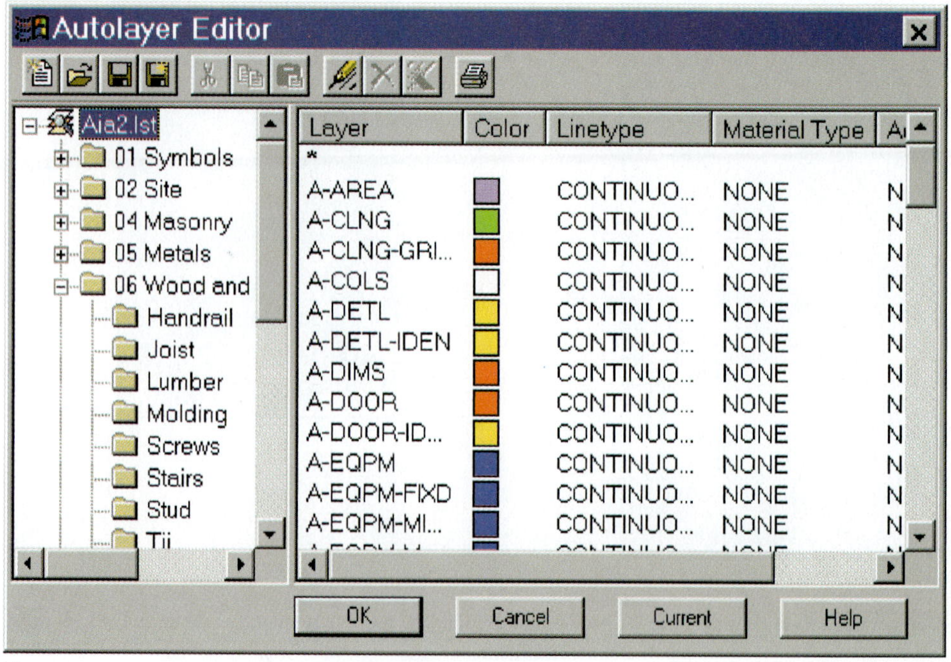

FIGURE 6.19 ■ Continued

(d)

shows samples of pull-down, icon, and tablet menus available from an architectural software developer. Drawings can be created in two or three dimensions and switched back and forth at any time. These packages may also let you render and animate three-dimensional models. The packages often have standard layer systems based on the American Institute of Architects (AIA) format. The dialog box in Figure 6.19*d* shows an automatic layer management and editing system available in an architectural software product.

Drawing Architectural Symbols

Symbols are drawn on plans by placing the puck or stylus pick on the table menu box, icon menu, or pull-down menu that contains the desired symbol. For example, if an exterior door is to be drawn, the door location is digitized and the direction of rotation is specified. To do this, pick the symbol from the appropriate menu; the symbol appears on the monitor, where you drag it into place and pick the insertion point. Each symbol has an established insertion point, as shown in Figure 6.20. Figure 6.21 shows an architectural

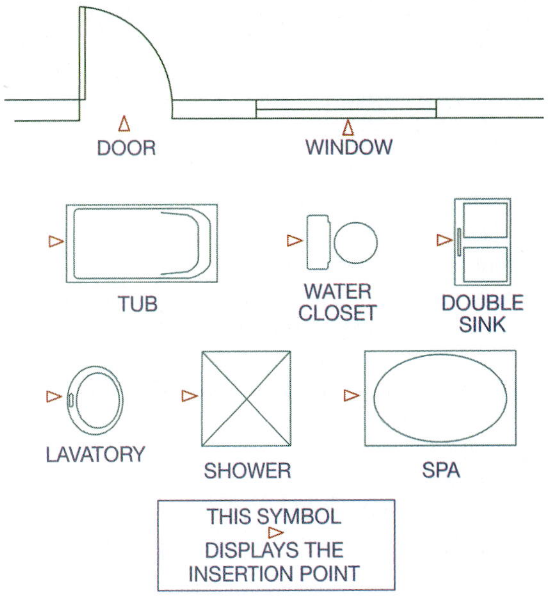

FIGURE 6.20 ■ Insertion points for several architectural symbols.

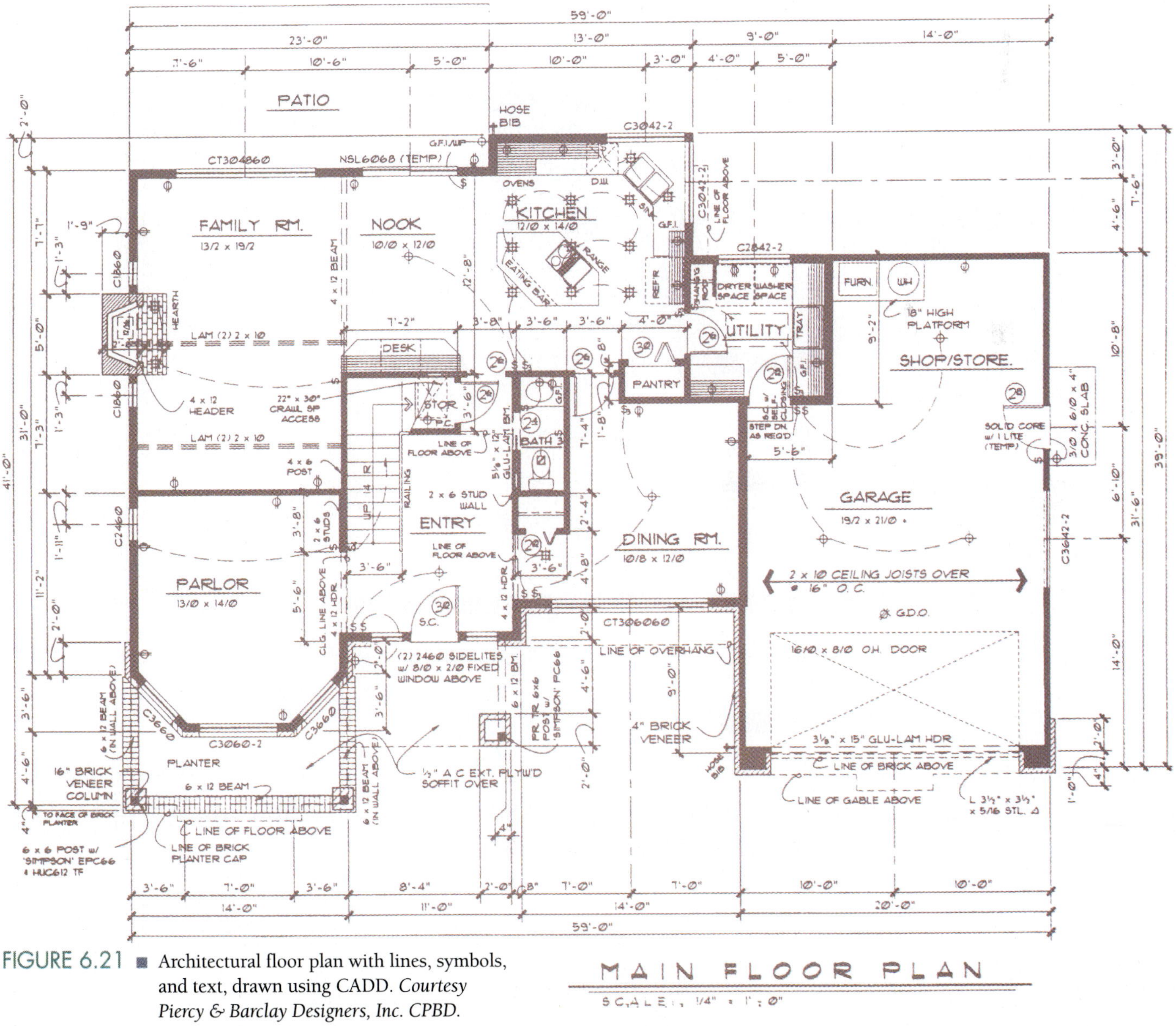

FIGURE 6.21 ■ Architectural floor plan with lines, symbols, and text, drawn using CADD. *Courtesy Piercy & Barclay Designers, Inc. CPBD.*

floor plan, with text and its many lines and symbols, drawn using CADD.

Creating Your Own CADD Symbols

You can create your own CADD symbols if a customized architectural menu is not available. The symbols that you can make are called *blocks* or *wblocks* in AutoCAD. You can design an entire group of symbols for inserting into architectural drawings. As you create these symbols, prepare a symbols library or catalog, where each symbol is represented for reference. The symbol catalog should include the symbol name, the insertion point, and other important information about the symbol.

DRAWING ON CADD LAYERS

Drawings are made in layers, each perfectly aligned with the others. Each layer contains its own independent information. Layers may be reproduced individually, in combination, or together as one composite drawing. Figure 6.22 shows how layers may be used to share information among drawings.

Some architectural CADD systems automatically set drawing elements on separate layers; others require that you create your own layering system. These layers are part of the drawing and may be turned on or off as needed. In addition to the layer identification system, many companies set up other layer systems based on their own project needs. For example, layers may be identified using a system in which elements are divided into three groups: major group, minor group, and basic group. The major group contains the type of drawing, such as "Floor Plan." The minor group identifies the elements on the drawing, such as "Electrical." The basic group labels the type of information, such as "Lines" or "Text." Here is an example of this system of layer identification: FL1ELECL, which means first floor (FL1), electrical (ELEC), and lines (L). This is a typical layout of the three groups:

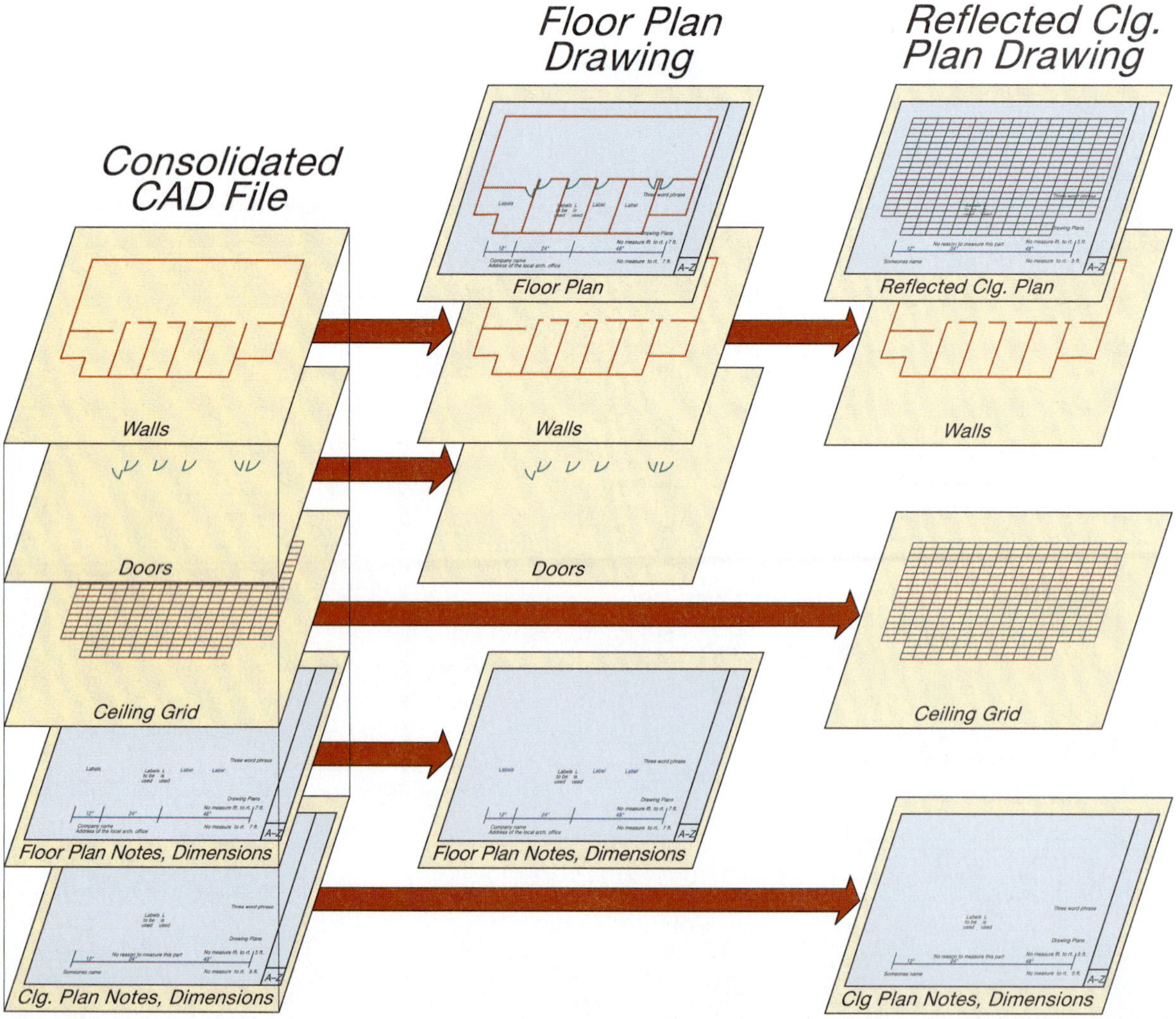

FIGURE 6.22 ■ An example of using layers to share information.

MAJOR GROUP	MINOR GROUP	BASIC GROUP
BSMT	HVAC	LINE
FL1	ELEC	TXT
FL2	PLUMB	DIM
GAR	WALLS	INSERT
SITE	DOORS	HATCH
SCHED	WINDOWS	
NOTES	APPL	
FDN		
ELEV		
ROOF		
REFLECT		

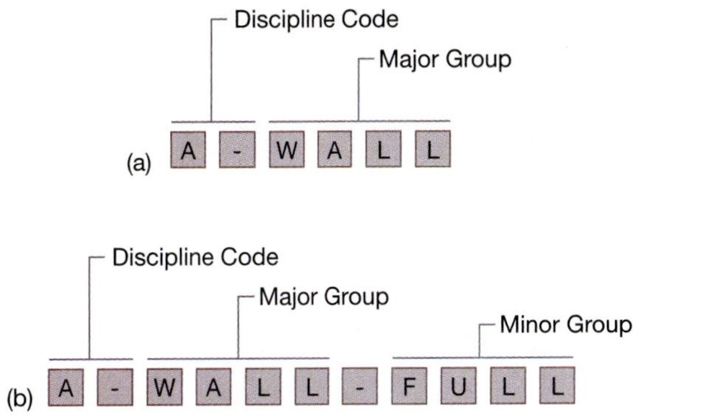

FIGURE 6.23 ■ Examples of AIA-recommended format for simple layer names showing use of a) major group only and b) major and minor group.

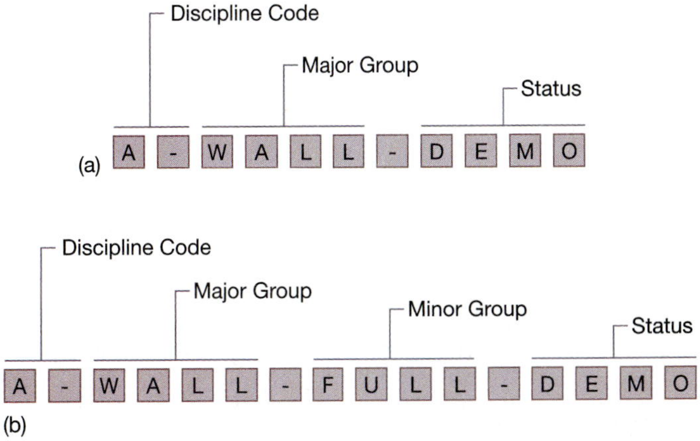

FIGURE 6.24 ■ Examples of AIA-recommended format for layer names showing use of a) major group and status field and b) major group, minor group, and status field.

Layer Names

Layer names should be kept short, such as FDN for "foundation." Layer identification may also contain numbers specifying the pen width used to produce the plot. For example, 1 = 0.25 mm, 2 = 0.35 mm, 3 = 0.5 mm, and 4 = 0.7 mm pen width.

The AIA recommends the use of layer names that contain two to four parts. Refer to the second edition of the AIA standard, CAD Layer Guidelines, for complete information. The AIA-recommended layering system has a discipline code, a major group, a minor group, and a status group of information in the layer names. The discipline code is identified by a letter, followed by a hyphen, and describes the project element, as follows:

- A Architecture, interiors, and facilities
- S Structural
- M Mechanical
- P Plumbing
- F Fire protection
- E Electrical
- C Civil engineering and site work
- L Landscape architecture

The major groups reduce the layer names to type of information found in the discipline code such as doors, windows, walls, ceilings, furniture, or equipment. A minor group may also be added to the layer name for further definition. Layers might have minor groups such as NOTE, SYMB (for "symbols"), DIMS (for "dimensions"), or TITLB (for "title block information"). In the recommended format, the major group is followed by a dash, and the minor group is used. This format also provides for status characters that are entered after the minor group. These status entries are for special project requirements that relate to new and existing construction in remodeling projects. An example of a simple layer name with only the discipline code and major group identified is shown in Figure 6.23a, and a major and minor group layer name is shown in Figure 6.23b. The AIA layer name system with major group and status field is shown in Figure 6.24a, and the layer name system with major group, minor group, and status field is shown in Figure 6.24b.

Each chapter provides recommended AIA layer names that are related to the layers used in the topics covered. The complete layer naming system recommended by the AIA is displayed in Appendix B.

Layer Systems in Use

A big advantage of CADD is the ability to send drawing files on disk to different subcontractors involved in a project. This is important in commercial applications, because the architect often creates the building design and has others design the electrical, plumbing, and mechanical systems. The mechanical system encompasses heating, ventilating, and air conditioning. To

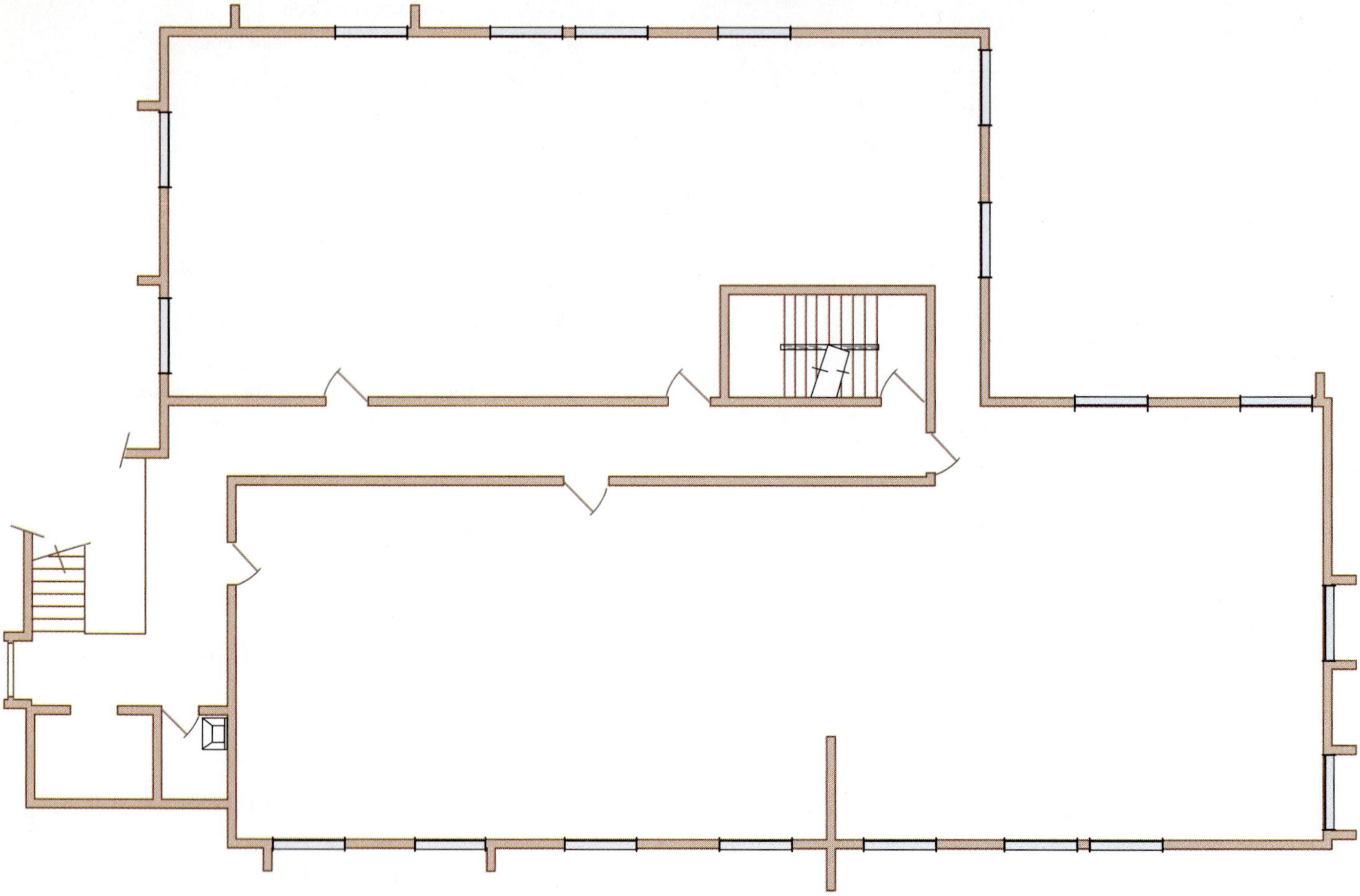

FIGURE 6.25 ■ Base layer sheet for exterior and support structure is unchanged through the project. *Courtesy Structureform.*

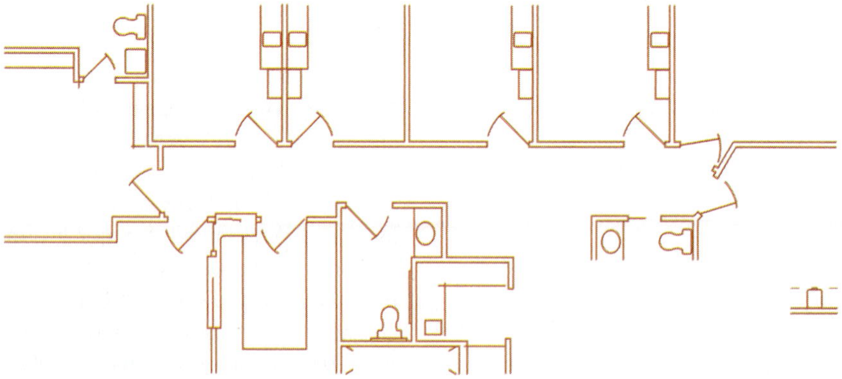

FIGURE 6.26 ■ Interior design of nonbearing partitions may be changed as desired using layers. *Courtesy Structureform.*

make this system work effectively, the architect prepares a base drawing, which represents the floor plan and its bearing walls and partitions (Figure 6.25). A disk copy, modem, or network communication of this drawing is sent to the interior design department to create the layers containing the nonbearing facility layout, which includes the room layout with fixtures and appliances (Figure 6.26). The CADD drawing files then go to the plumbing, electrical, and HVAC contractors for design of their related elements. Figure 6.27 shows the plumbing layer for the previous building. A composite drawing of all layers is shown in Figure 6.28. Each layer is generally set up in its own color on the computer screen.

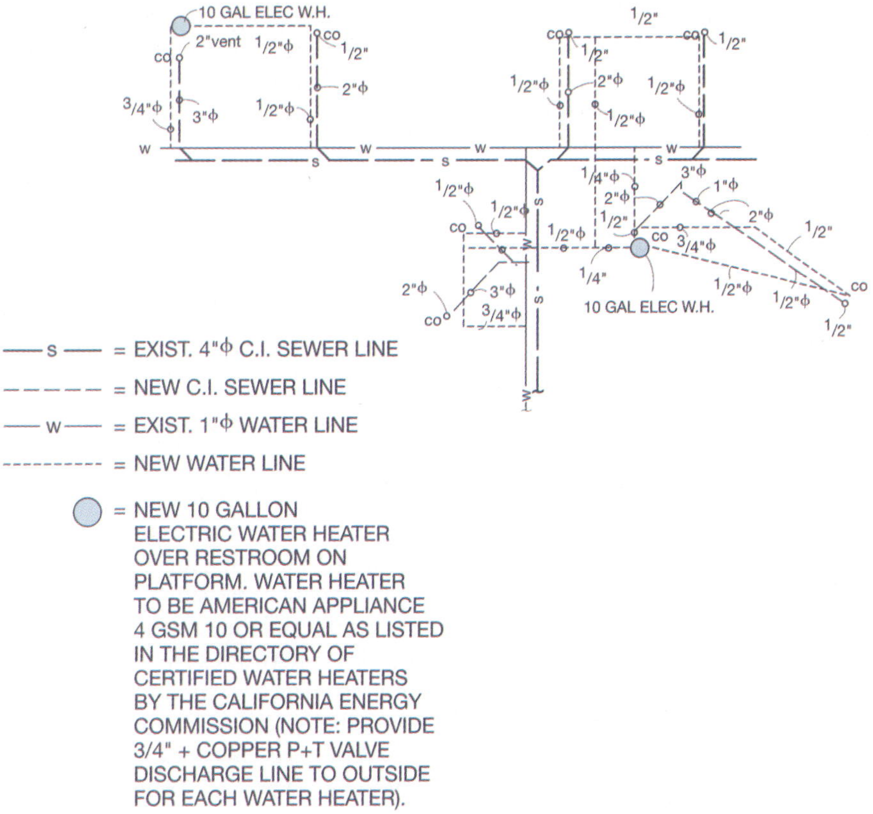

FIGURE 6.27 ■ Plumbing layer. *Courtesy Structureform.*

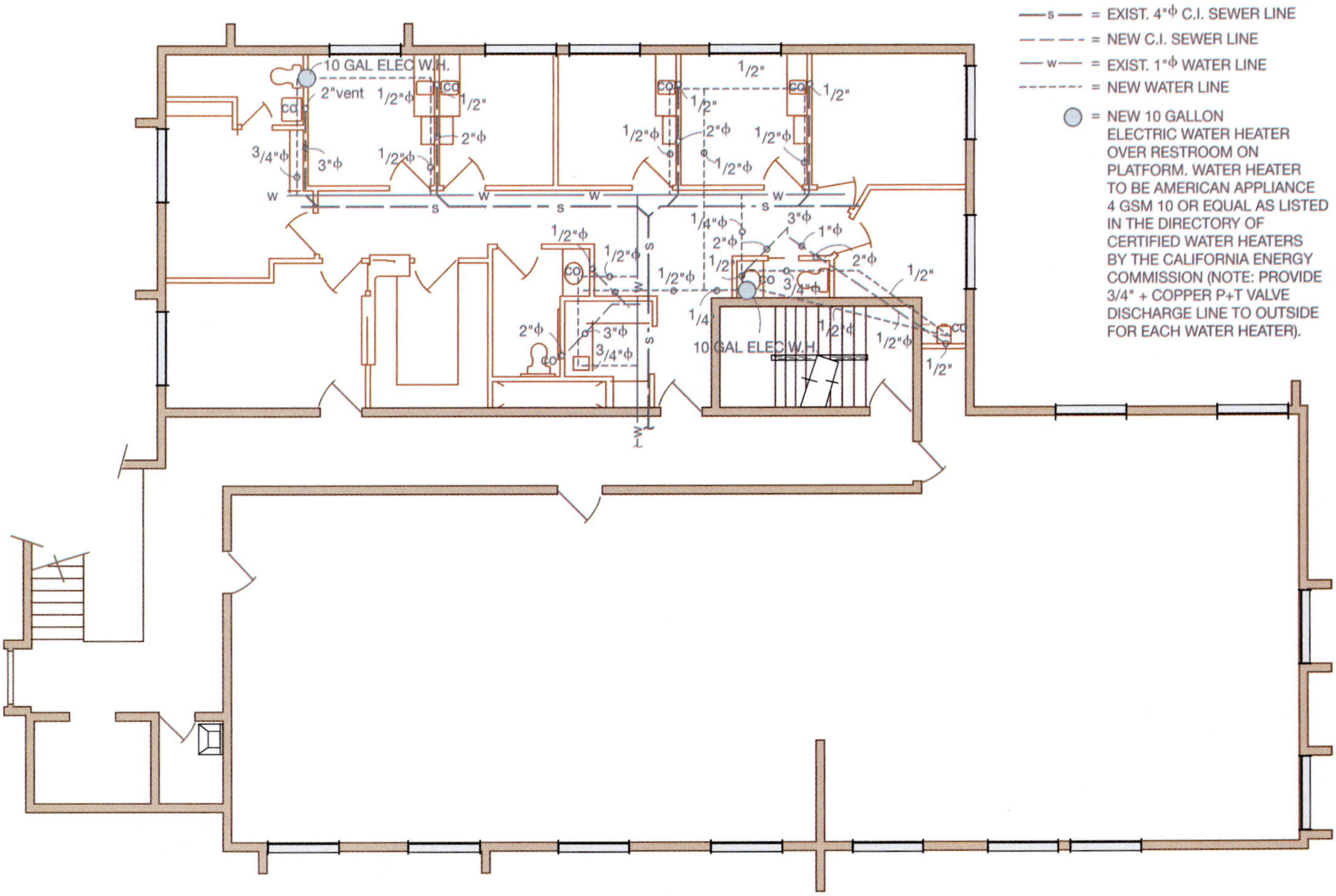

FIGURE 6.28 ■ Composite of base layer, interior design layer, and overlays. Courtesy Structureform.

CADD layers provide these advantages:

- Reduced drafting time.
- Improved project coordination; this is achieved by sharing information that is common to different elements of the project or common to several floors in a multistory building.
- Ease of preparing design alternatives; the same drawing can be shared by turning layers on or off as needed for display.

WEB-BASED ARCHITECTURAL COLLABORATION

Architecture is beginning to see the development of technologies that will allow teamwork over the World Wide Web (WWW). This technology has major implications for the evolution of the practice. Architecture is a cooperative process involving multiple disciplines such as structural, mechanical, and electrical engineering; interior design; and facility management. One of the most time-consuming and risky aspects of the practice is document management. Hundreds of drawings and documents can be issued during the design process. All of these documents can be in various states of completion. The coordination of this information is a big task. Web-based coordination uses the power of sophisticated database management systems in conjunction with the Internet to simplify and streamline document management and team communications. These products allow architects, designers, drafters, product suppliers, building contractors, and owners to communicate and coordinate throughout the entire project regardless of where each participant is located.

This changing technology also has implications for the way design projects are produced. Web-based coordination allows for the increased use of outsourcing. *Outsourcing* means sending parts of a project out to subcontractors for completion. Outsourcing to multiple resources requires consistency in procedure and similarity in tools. This interactive teamwork of the future requires clear measurement of progress. The collaboration tools now being developed give architects a higher level of communication and control over the production process. Threaded E-mail will allow issue tracking throughout the project. The outsourcing model of the future is based on defined tasks rather than hours worked. These discrete tasks are based on interrelated logical steps rather than drawing sheets. Drawings move forward on the basis of their relationship to other drawings and the available information. The development of drawings and documents is treated as stepped tasks or slices of time. Outsource personnel will be independent contractors responsible for managing their own time rather than punching a clock. These professional subcontractors can be found all over the world, and they are connected through the WWW. Through this huge network of potential workers, it is possible to find professionals who produce high-quality drawings on schedule and economically.

Web-based collaboration will also affect the relationship of the architect and the client. Because the work product is visible on the Internet, the owner can view its progress at any time. Progress on the documents is obvious, and the owner can involve more people in reviewing the work. This model of practice enhances the owner's involvement in the process and improves the sense of overall teamwork.

CHAPTER 6

Computer-Aided Design and Drafting in Architecture Test

DIRECTIONS

Answer the questions with short, complete statements or drawings as needed on 8 1/2 × 11 notebook paper, as follows:

1. Place your name, Chapter 6 Test, and the date at the top of the sheet.
2. Letter the question number and provide the proper answer. You do not need to write out the question.

Use a CADD workstation or word processor, if available, to prepare the answers to the test questions.

QUESTIONS

Question 6–1 What does the abbreviation CADD mean?

Question 6–2 Define a microcomputer in general terms.

Question 6–3 Identify three factors that influence an increase in productivity with CADD.

Question 6–4 Identify five tasks, not related to drafting, that may be performed by a computer.

Question 6–5 Describe how drawing layers can be prepared on a computer graphics system.

Question 6–6 Describe three ways architectural design functions are improved with a CADD system.

Question 6–7 Define ergonomics.

Question 6–8 Describe three potential disadvantages of CADD.

Question 6–9 Describe the following components of a computer-aided drafting workstation:

a. Computer
b. Monitor
c. Digitizer
d. Plotter

Question 6–10 Define the following CADD terminology:

a. Hardware
b. Software
c. Program
d. Commands
e. Soft copy
f. Default value
g. Disk drive
h. Storage
i. Floppy disk
j. Menu
k. Hard copy
l. File
m. Network
n. PC
o. Puck
p. Mouse
q. Stylus
r. Windows
s. Attribute
t. Font
u. LISP
v. Entity
w. Compatible
x. Text
y. Layers
z. Backup

Question 6–11 What does WWW stand for?

Question 6–12 Define outsourcing.

Question 6–13 Define virtual reality.

Question 6–14 Explain the difference between the Intranet and the Internet.

Question 6–15 Discuss the similarities and difference between walk-through and fly-through.

Question 6–16 Name the most common type of injury that can develop at a poorly designed workstation.

(continued)

QUESTIONS *(cont.)*

Question 6–17 Identify at least three factors related to positive work habits that can contribute to a healthy environment.

Question 6–18 What is a clone?

Question 6–19 Explain the difference between a cursor and crosshairs.

Question 6–20 Briefly discuss each of the following data storage devices:

a. Disk
b. Hard drive
c. Optical disk
d. High-capacity disk drive

CADD PROBLEMS

Write a report of 250 or fewer words on one or more of the following topics:

Problem 6–1 Productivity.

Problem 6–2 Ergonomics.

Problem 6–3 CADD workstation.

Problem 6–4 Customizing your own architectural symbols.

Problem 6–5 Architectural program by a software developer.

Problem 6–6 CADD layers.

Problem 6–7 Digitizing existing drawings.

Problem 6–8 Scanning existing drawings.

Problem 6–9 Virtual Building™

Problem 6–10 Virtual reality.

Problem 6–11 Repetitive strain injury.

Problem 6–12 Ergonomic workstation.

Problem 6–13 Positive work habits.

Problem 6–14 Data storage devices.

Problem 6–15 Computer networks.

Problem 6–16 Mobile computers.

Problem 6–17 Design coordination.

Problem 6–18 Web-based architectural collaboration.

SECTION 2

Residential Design

7 CHAPTER

Building Codes and Interior Design

INTRODUCTION

Building codes are intended to protect the public by establishing minimum standards of safety. Although landowners often assume they can build whatever they want because they own the property, building codes are intended to protect others from such shortsighted individuals. Consider the stories you have seen in the national news media about structures that have been damaged by high winds, raging floodwaters, hurricanes, tornadoes, mud slides, earthquakes, and fire. Although they do not make the national news, hundreds of other structures become uninhabitable because of inadequate foundation design, failed members from poor design, rot, or termite infestation. Building codes are designed to protect consumers by providing minimum guidelines for construction and inspection of a structure to prevent fire, structural collapse, and general deterioration. Many other aspects of building construction, such as electrical wiring, heating equipment, and sanitary facilities, represent a potential hazard to occupants. Building codes are enforced as a safeguard against these risks.

The regulation of buildings can be traced through recorded history for over 4,000 years. In early America, George Washington and Thomas Jefferson encouraged building regulations as minimum standards for health and safety. Building codes are now used throughout most of the United States to regulate issues related to fire, structural ability, health, security, and energy conservation. Architects, engineers, and contractors also rely on building codes to help regulate the use of new materials and technology. Each major building code provides testing facilities for new materials, to develop safe, cost-effective, timely construction methods. In addition to the testing facilities of each major code, several major material suppliers have their own testing facilities. Following is a list of several major testing labs, quality assurance and inspection agencies, and their Web sites:

- www.apawood.org — APA The Engineered Wood Association
- www.csa.ca — Canadian Standards Association
- www.etl.go.jp — ETL Testing Services Ltd.
- www.osha.slc.gov — Factory Mutual Research Corporation
- www.erols.com/hpva — Hardwood Plywood & Veneer Association
- www.itsqs.com/hk — Inchcape Testing Services Inc.
- www.lasapprovals.org — International Approval Services, Inc.
- www.nahbrc.org — NAHB Research Center
- www.ul.com — Underwriters Laboratories Inc.
- www.wwpa.org — Western Wood Products Association

Although most areas of the country understand the benefits of regulating construction, each city, county, or state has the authority to write its own building codes. Some areas choose to have no building codes. Some areas have codes, but compliance is voluntary. Other areas have building codes but exempt residential construction. This variation in attitudes has produced over 2,000 different codes in the United States. Trying to design a structure can be quite challenging when one is faced with so many possibilities.

NATIONAL CODES

Most states have adopted one of three national codes:

- www.bocai.org — The 1999 BOCA (Building Officials and Code Administrators International, Inc.)
- www.sbcci.org — The 1999 SBC (Standard Building Code) written by the SBCCI (Southern Building Code Congress International, Inc.)
- www.icbo.org — The 1997 UBC (Uniform Building Code) written by the ICBO (International Conference of Building Officials).

Figure 7.1 shows the current building code used by each state. It is important to remember that each municipality and state has the right to adopt all, or a portion, of the indicated code.

INTERNATIONAL BUILDING CODE

In 1972, in order to serve designers better, BOCA, SBCCI, and ICBO joined forces to form CABO (the Council of American Building Officials) to create a national residential code. In 1994, in another attempt to create a national code, the International Code Council (ICC–www.intlcode.org) was formed. Because many regulations affecting construction come from the federal government, in 1997 CABO and ICC merged in an attempt to

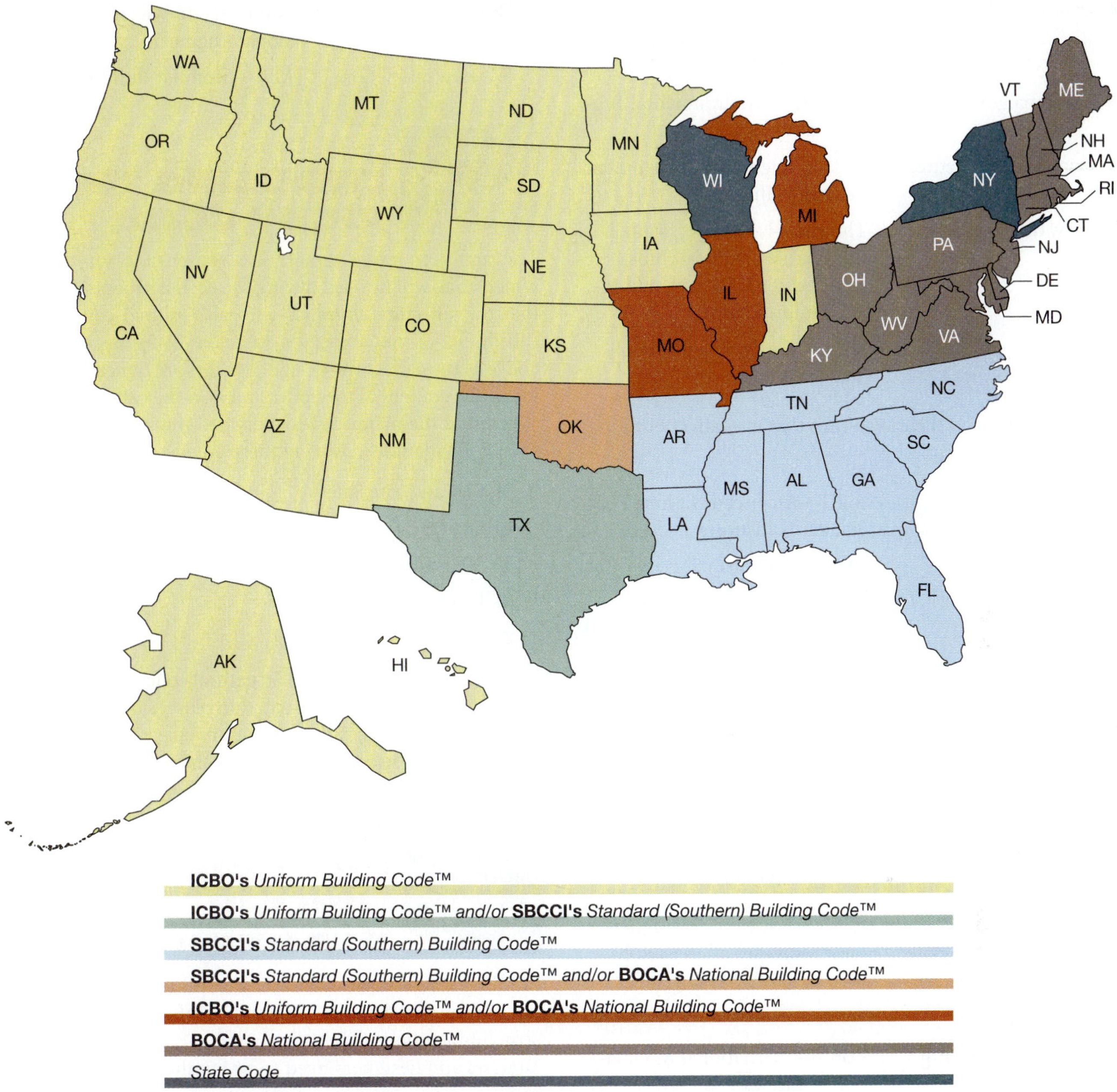

FIGURE 7.1 ■ Current building codes, by state. Because each municipality has the right to adopt all or a portion of the indicated code, and the development of the 2000 International Residential Code, you should verify with the governing agency what code is to be used as the design standard.

provide an agency that could channel the many regulations to the state level. The goal of ICC is to develop a single set of comprehensive, coordinated national codes that will eliminate disparities among the three major codes. It is hoped that one building code will aid in the design of structures located in different regions of the country from the design firm.

Note

In the hope that this code will become the national standard, information in this text will be based on the 2000 International Residential Code (IRC) for One- and Two-Family Dwelling Code published by ICC.

In addition to these national codes, the Department of Housing and Urban Development (HUD), the Federal Housing Authority (FHA) and the Americans with Disabilities Act (ADA); each publish guidelines for minimum property standards for residential construction.

Other United States and Federal Agencies related to construction include:

- www.access-board.gov — The Access Board—U.S. Architectural and Transportation Barriers Compliance Board

- www.eren.doe.gov   DOE—Department of Energy's Energy Efficiency and Renewable Energy Network
- www.epa.gov   EPA—U.S. Environmental Protection Agency.
- www.nara.gov/fedreg   Federal Register—Government Printing Office.
- www.thomas.voc.gov   Thomas—Federal legislative information, status of bills, Congressional Record, and related information.
- www.nist.gov   NIST—National Institute of Standards and Technology.

Other agencies and their Web sites that may affect building codes in your area include:

- www.ansi.org   ANSI—American National Standards Institute
- www.asce.org   ASCE—American Society of Civil Engineers
- www.ashrae.org   ASHRAE—American Society of Heating, Refrigerating, and Air Conditioning Engineers.
- www.astm.org   ASTM—American Society for Testing and Materials
- www.cabo.org   CABO—Council of American Building Officials
- www.nfpa.org   NFPA—National Fire Protection Association

In addition to these regulatory agencies, which influence construction, several related organizations are often used by designers to ensure the quality of materials. These agencies and their Web sites include:

- www.asfconline.org   IAFC—International Association of Fire Chiefs
- www.iec.ch   IEC—International Electrotechnical Commission
- www.ieee.org   IEEE—Institute of Electrical and Electronics Engineers
- www.iso.ch   ISO—International Standards Organization
- www.nahb.com   NAHB—National Association of Home Builders
- www.ncma.org   NCMA—National Concrete Masonry Association
- www.nibs.org   NIBS—National Institute of Building Sciences
- www.nmhc.org   NMHC—National Institute Multi-Housing Council
- www.nspe.org   NSPE—National Society of Professional Engineers
- www.mbinet.org   MBI—Modular Building Institute

CHOOSING THE RIGHT CODE

The architect and engineer are responsible for determining which code will be used during the design process and ensuring that the structure complies with all required codes. Although the drafter is not expected to make decisions regarding the codes as they affect the design, drafters are typically expected to know the content of the codes and how they affect construction.

Each of the major codes is divided into similar sections that specify regulations covering these areas:

- Fire and Life Safety
- Structural
- Mechanical
- Electrical
- Plumbing

The Fire and Life Safety and Structural codes have the largest effect on the design and construction of a residence. The effect of the codes on residential construction will be discussed throughout this text. Chapter 45 examines the effects of the UBC and IBC on commercial construction.

BASIC DESIGN CRITERIA FOR BUILDING PLANNING

Building codes have their major influence on construction methods rather than on design. (The influence of codes on construction methods will be discussed in Sections, IV, VIII and X.) Drafters and designers need to be familiar with several areas in order to meet minimum design standards, including basic design criteria. Areas of the code such as climatic and geographical design criteria will affect the materials used to resist forces such as wind, snow, and earthquakes. These forces and how they affect a residence will be discussed in Chapter 27. Keep in mind that the following discussion is only an introduction to building codes. As a drafter or a designer you will need to become familiar with and constantly use the building code that governs your area.

Occupancy

Building codes divide structures into different categories or occupancies. Houses are a Group R-1 occupancy. This occupancy includes hotels, apartments, convents, single-family dwellings, and lodging houses with more than ten inhabitants. The R occupancy is further divided into three categories. Single-family

dwellings and apartments with fewer than ten inhabitants are defined as R-3. This designation allows the least restrictive type of construction and fire rating group to be used.

In addition to the division of structures into different occupancy ratings, the space within a home or dwelling unit is subdivided into habitable and nonhabitable space. A room is considered *habitable space* when it is used for sleeping, living, cooking, or dining purposes. *Nonhabitable spaces* include closets, pantries, bath or toilet rooms, hallways, utility rooms, storage spaces, garages, darkrooms, and other similar spaces.

Location on the Property

Exterior walls of a type R occupancy cannot be located within 3′ (914 mm) of the property lines unless special provisions are met. Typically, an exterior wall that is built within 3′ of the property line must be made from materials that will resist a fire for one hour. This is known as a one-hour fire rating. Using 5/8″ type X gypsum board on each side of the wall is a common method of achieving a one-hour wall. Many municipalities require any walls built into the minimum side yard to be of one-hour construction. Alternatives based on the siding materials used are available in the building code. Openings such as doors or windows are not allowed in a wall with a fire separation distance less than 3′ (914 mm). Projections such as a roof or chimney cannot project more than 12″ (305 mm) into areas where openings are permitted.

Accessibility

Although this is not required for type R-3 occupancies, BOCA, SBCCI, and ICBO have each added information dealing with accessibility. Accessibility requirements affect type R-1 occupancies (hotels, apartments, and private homes with more than ten inhabitants—six or more for SBCCI). The following requirements do apply in R-3 occupancies having four or more dwelling units in one structure. R-1 units are divided in type A and type B units. Type A units are to be designed and constructed to meet ICC/ANSI A117.1–1998 standards. This publication is printed by the American National Standards Institute, Inc. (ANSI) with the cooperation of BOCA, SBCCI, and ICBO. Major sections of this standard include accessible routes, general site and building elements, plumbing elements and fixtures, communication elements and fixtures, special rooms and spaces, and built-in furnishing and equipment. These sections are applicable to R-1 occupancies. Chapter 10 of the ANSI guidelines applies specifically to type A units. Major guidelines apply to bathroom and kitchen design. The main consideration in bathroom design is wheelchair access. See Figure 7.2. A major consideration in kitchen design is access to work areas. Figure 7.3 shows counter requirements.

Type B units are to be designed in accordance with chapter 11 of the governing code. Consideration is given in each code to accessible routes; operating controls for electrical, environmental, intercom, and security controls; doorways; kitchens; and bath design.

Accessible Route

An accessible route is the walking surface from the exterior access through the residence. The exterior access (the front door) cannot be located in a bedroom. At least one route is required to connect all spaces that are part of the dwelling unit. If only one route is provided, it cannot pass through a bathroom, closet, or similar space. The UBC allows that one of the three following exceptions is not required to be on the access route:

- A raised floor area in a portion of a living, dining, or sleeping area;
- A sunken floor area in a portion of a living, dining, or sleeping area; or
- A mezzanine that does not have plumbing fixtures or an enclosed habitable space

The access route is required to have a minimum width of 36″ (914 mm) except at doors. Changes in floor level greater than 1/2″ (12.7 mm) are not allowed unless a ramp, elevator, or wheelchair lift is provided. Ramps cannot have a slope greater than 1:48. The only exception to these elevation requirements is at an exterior door that leads to a deck, patio, or balcony. Here the lower exterior floor surface is allowed to be a maximum of 4″ (102 mm) below the finished interior floor surface.

Operating Controls

This portion of the building code regulates placement of and access to controls for electrical, environmental, security, and intercom controls. Guidelines for placement of controls can be seen in Figure 7.4. The control should be centrally located in a clear floor space measuring a minimum of 30″ × 48″ (762 × 1219 mm), and should be no more than 48″ (1219 mm) and no less than 15″ (381 mm) from the finished floor. Exceptions to these locations include:

- Electrical receptacles serving a dedicated use (i.e., a 110 c.o. for a refrigerator)
- Appliance-mounted controls or switches
- A single receptacle located above a portion of a countertop uninterrupted by a sink or appliance (this need not be accessible as long as one receptacle is provided)
- Floor electrical receptacles
- Plumbing fixture controls

Doors

All doors are required to provide a minimum clear opening of 32″ (815 mm), as shown in Figure 7.5. Although maneuvering space is not required on the dwelling unit side of the door, good design dictates that space be provided for turning a wheelchair. Figure 7.6 shows required clear access for type A units.

For exterior swinging doors, a threshold cannot exceed a maximum height of 1/2″ (12.7 mm). Sliding exterior doors are allowed to have a beveled threshold with a maximum height of 3/4″ (19 mm). The bevel cannot exceed 1 vertical unit /2 horizontal units (50% slope).

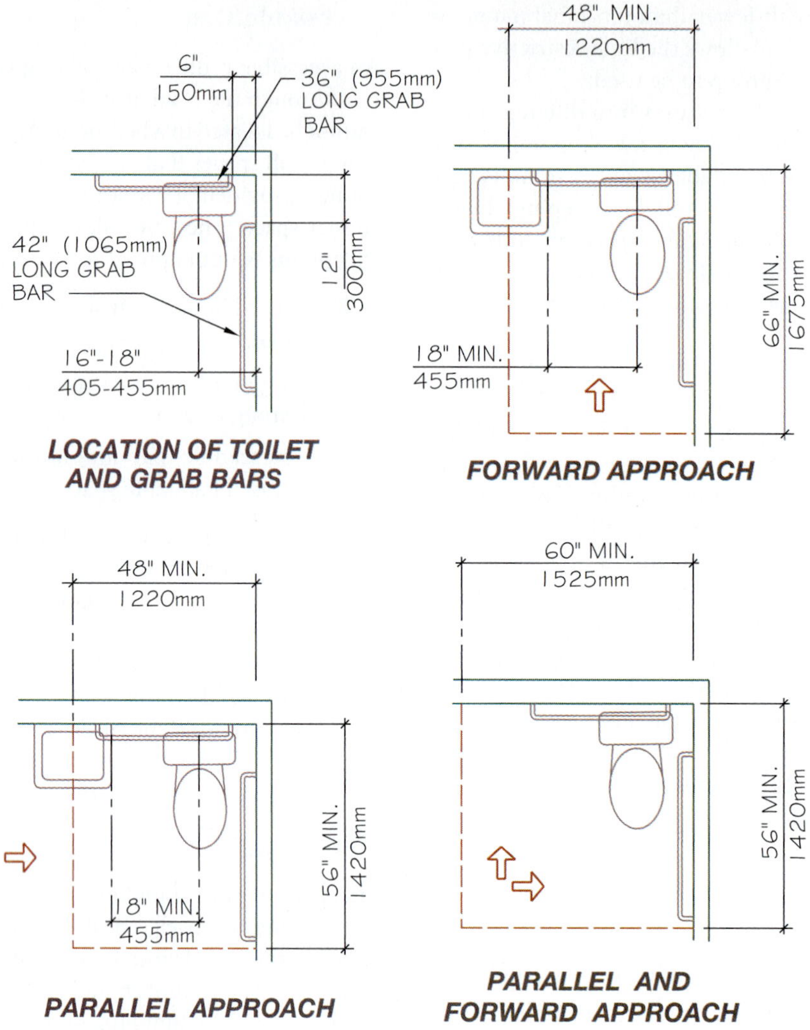

FIGURE 7.2 ■ Water closet clearance in type A dwelling units.

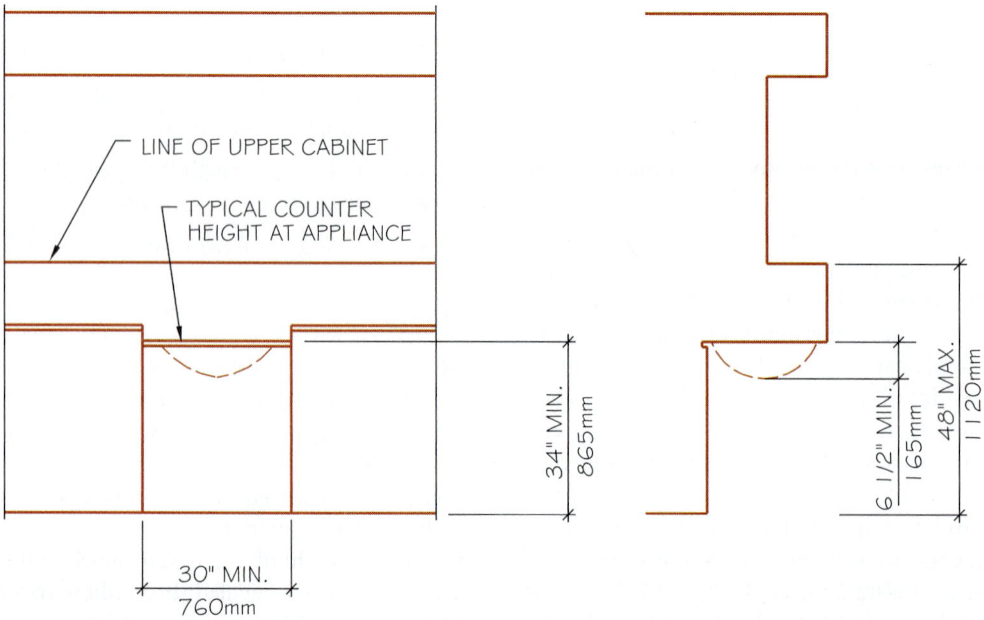

FIGURE 7.3 ■ A major consideration of kitchen guidelines is access to work areas.

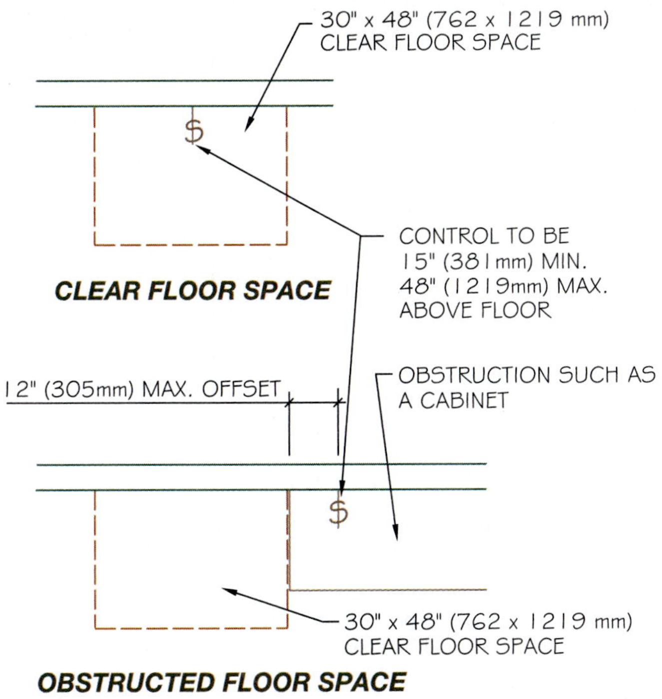

FIGURE 7.4 ■ A clear floor area of 30″ × 48″ (762 × 1219 mm) must be provided to access controls.

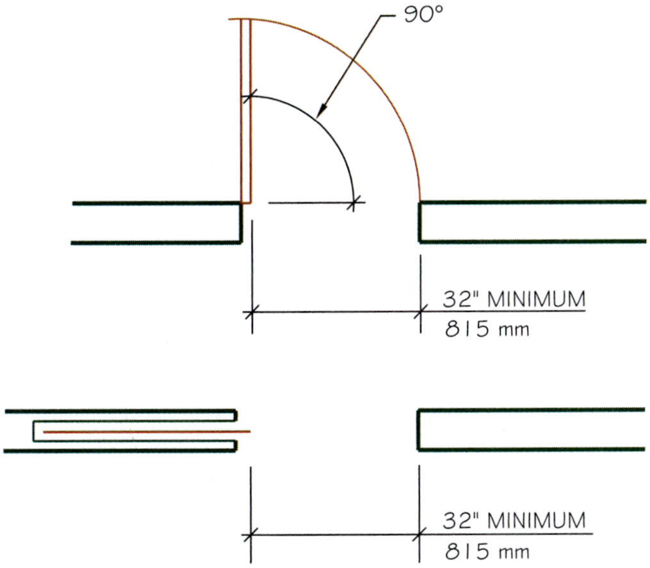

FIGURE 7.5 ■ Doorways must have a minimum clear opening of 32″ (815 mm) measured between the face of the door when open 90° and the stop. A maximum tolerance of −1/4″ (6.4 mm) is allowed.

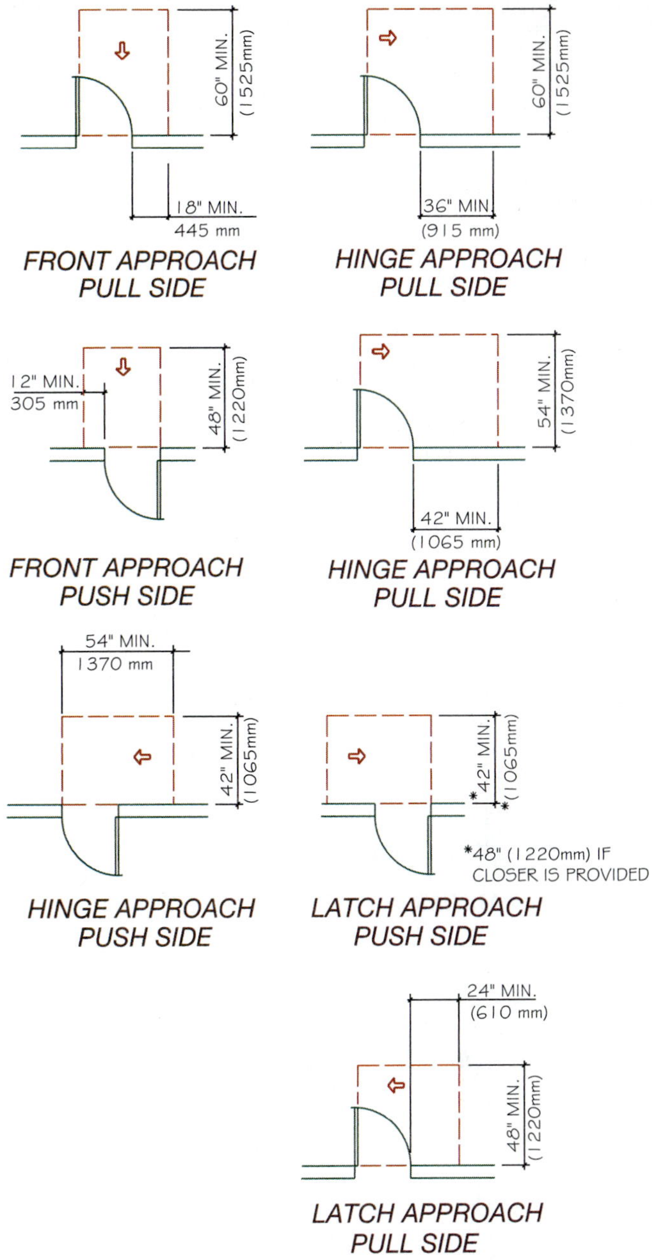

FIGURE 7.6 ■ Clear floor spaces required for type A units. For exterior swinging doors, a threshold cannot exceed a maximum height of 1/2″ (12.7 mm). Sliding exterior doors are allowed to have a beveled threshold with a maximum height of 3/4″ (19 mm). The bevel cannot exceed 1 vertical units/2 horizontal units (50% slope).

Kitchens

Figure 7.7 shows the required minimum clearances for kitchen base cabinet placement. A clear floor area of 30″ × 48″ (762 × 1219 mm) is required for the cooktop, dishwasher, freezer, oven, range, refrigerator, sink, and trash compactor when provided.

Toilet and Bathing Facilities

Figure 7.8 shows a typical bathroom layout. Doors are not allowed to swing into the clear floor space required for any fixture unless a clear floor space of at least 30″ × 48″ (762 × 1219 mm) is provided beyond the swing of the door. This space allows a person to enter, close the door, and then move to a fixture. If this space is provided, the door may swing into the required space

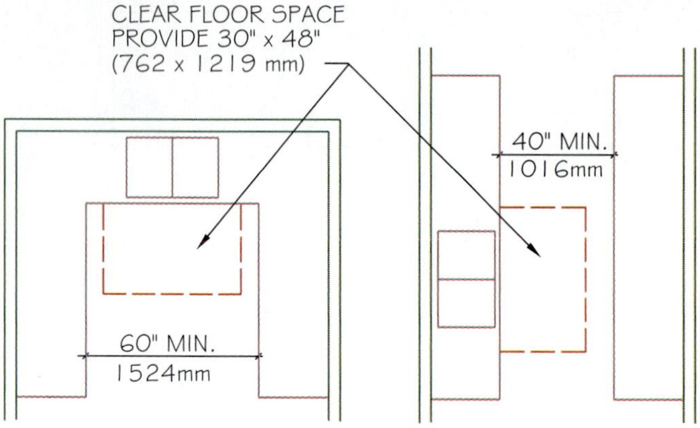

FIGURE 7.7 ■ Clear floor spaces required between cabinets in a kitchen.

FIGURE 7.8 ■ A typical bathroom layout (left) is often too small to provide accessibility. Doors are not allowed to swing into the clear floor space required for any fixture unless a 30″ × 48″ (762 × 1219 mm) minimum clear floor space is provided beyond the swing of the door.

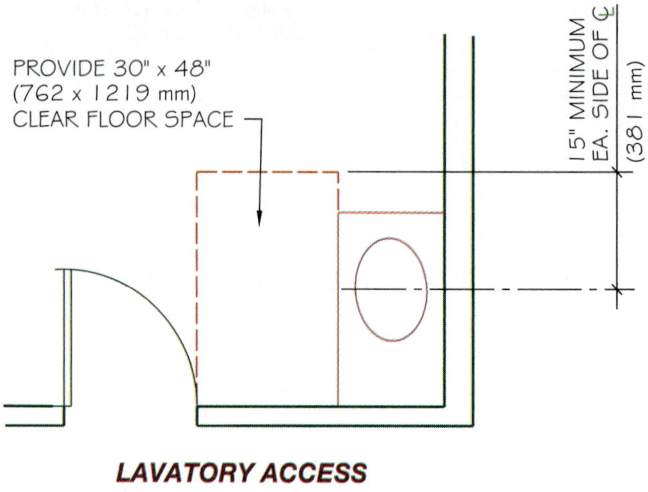

FIGURE 7.9 ■ Required space for a lavatory. A water closet must be at least 18″ (457 mm) from the centerline of the fixture and 15″ (381 mm) on the other side when located between a bathtub or lavatory. When it is located by a wall, a distance of 18″ (457 mm) must be provided from the centerline of the fixture to the wall.

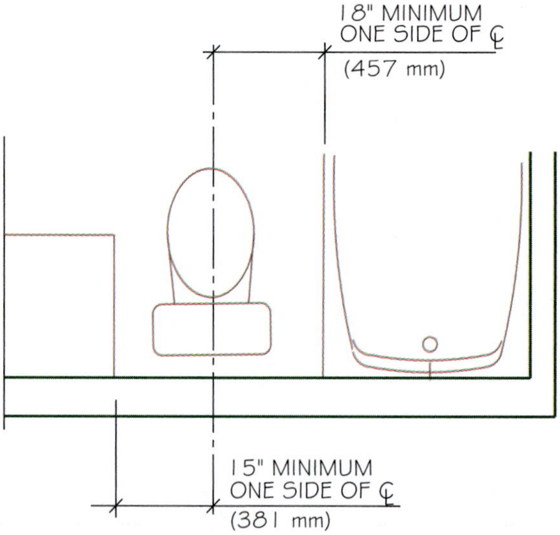

FIGURE 7.10 ■ A water closet must be 18″ (457 mm) from the centerline of the fixture and 15″ (381 mm) minimum on the other side when located between a bathtub or lavatory. When it is located by a wall, a distance of 18″ (457 mm) must be provided from the centerline of the fixture to the wall.

for each bath fixture. Figure 7.9 shows the required space for a lavatory. As shown in Figure 7.10, a water closet must have a minimum of 18″ (457 mm) from the centerline of the fixture and a minimum of 15″ (381 mm) on the other side when located between a bathtub or lavatory. When it is located by a wall, a distance of 18″ (457 mm) must be provided from the centerline of the fixture to the wall. Figure 7.11 shows access methods and clear floor space required for a toilet. For each access method, vanities or lavatories located on a wall behind the water closet are permitted to overlap the clear floor space. Where a tub and/or shower is provided, it must comply with Figure 7.12. If a separate shower and tub are provided, the guidelines apply to

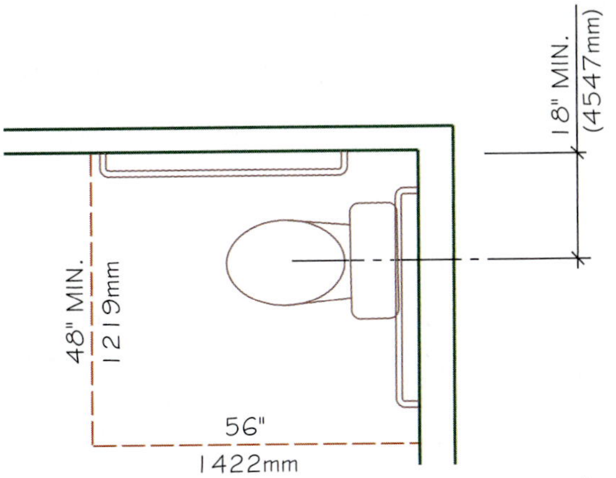

PARALLEL APPROACH

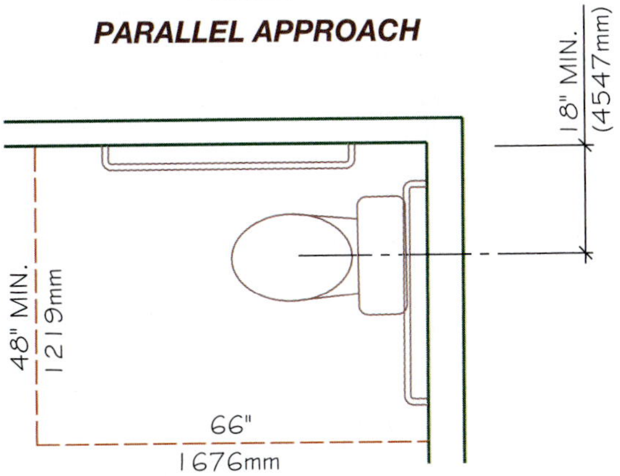

FORWARD APPROACH

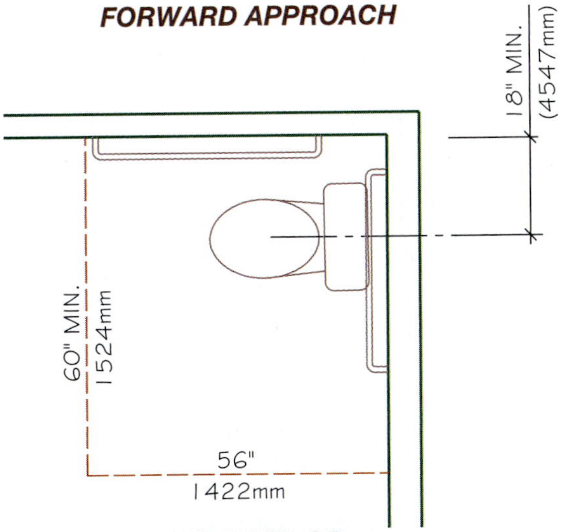

PARALLEL OR FORWARD APPROACH

FIGURE 7.11 ■ Access methods and clear floor space required for a toilet. For each access method, vanities or lavatories located on a wall behind the water closet are permitted to overlap the clear floor space.

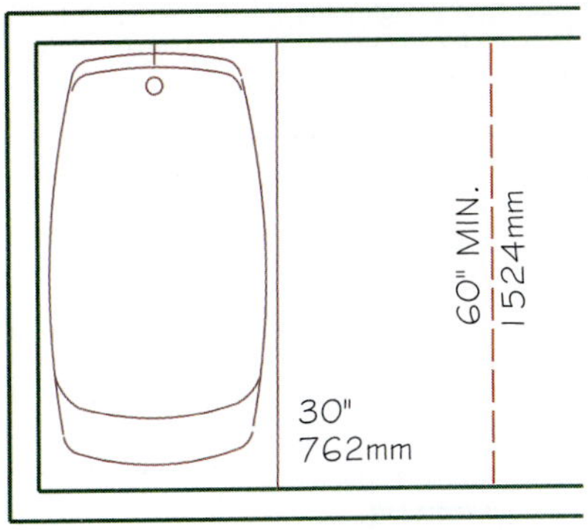

PARALLEL APPROACH

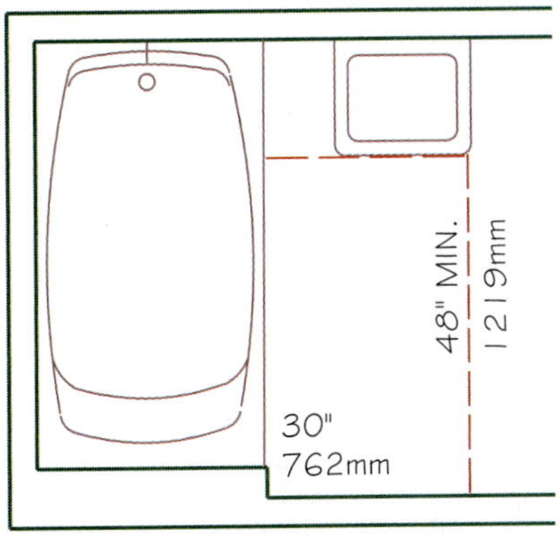

PARALLEL APPROACH

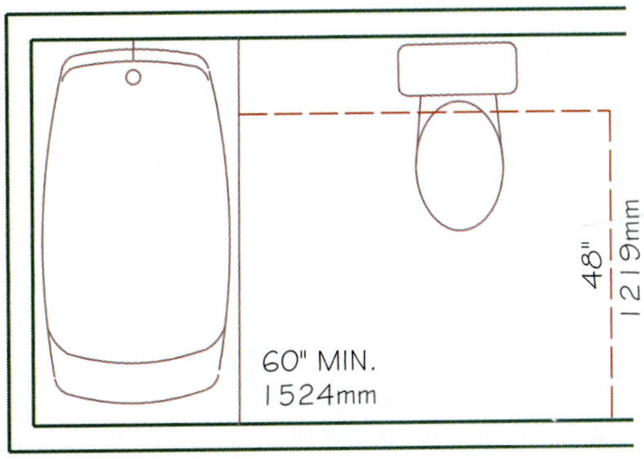

FORWARD APPROACH

FIGURE 7.12 ■ Where a tub and/or shower is provided, it must have a clear floor area of 30″ × 60″ (762 × 1524 mm). A lavatory placed at the control end of the tub may extend into the clearance if a 30″ × 48″ (762 × 1219 mm) clear area is maintained. When the forward approach is used, the floor space must be 48″ × 60″ (1219 × 1524 mm). A toilet placed at the control end of the tub may be placed in the clear floor area as long as the minimum requirements for the toilet are still met. Access to the toilet can be achieved using the parallel or the parallel/forward approach.

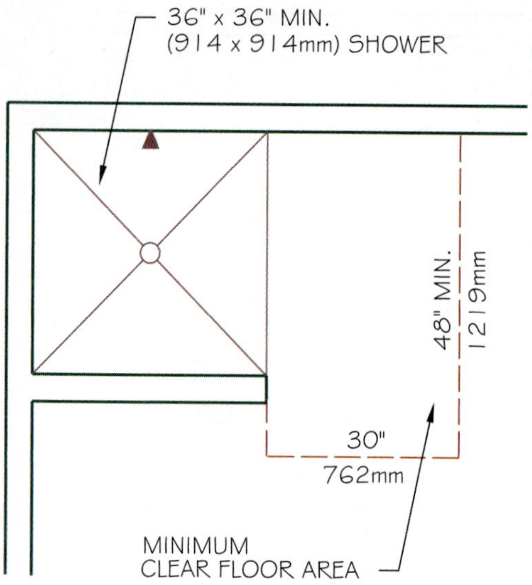

FIGURE 7.13 ■ If a separate shower and tub are provided, the guidelines apply to only one of the fixtures. If only a shower is provided, minimum clear floor requirements must be met.

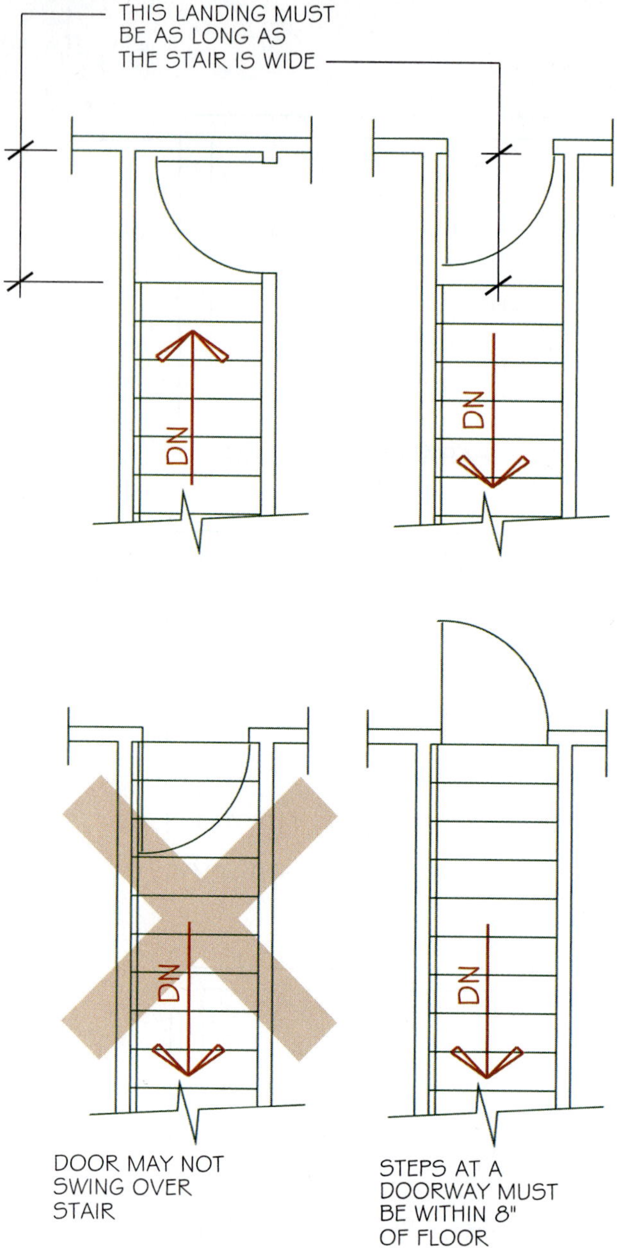

FIGURE 7.14 ■ Placement of doors near stairs. Landings by doors must be within 1 1/2″ (38 mm) of the door threshold. When the door does not swing over the landing, the landing can be a maximum of 8″ (197 mm) below the threshold.

only one of the fixtures. Figure 7.13 shows the clear floor requirements when only a shower is provided.

Exit Facilities

The major subjects to be considered in this area of the code are access doors, emergency egress (exits), and stairs.

Doors

Each dwelling unit, as a residence is referred to in the codes, must have a minimum of one door that is at least 36″ (914 mm) wide and 6′–8″ tall (2032 mm). Any hallway adjacent to that door must also be a minimum of 36″ (914 mm) clear. All other door sizes may be determined by the architect. Notice that the codes do not mandate that the front door be 36″ (914 mm) wide; only that one exit door needs to have this dimension. Chapter 14 provides guidelines for determining door sizes throughout a residence.

A floor or landing must be provided on each side of an exterior door. The landing must be within 8″ (197 mm) from the top of the door threshold provided that the door does not swing over the landing. The IRC requires the landing to be a maximum distance of 1 1/2″ (38 mm) from the top of door threshold. A minimum of a 3′ × 3′ (914 mm × 914 mm) landing is required on each side of the required egress door. Interior landings are to be level. Exterior landings may have a maximum slope of 1/4″/12″ (2% slope). This often becomes a design problem when trying to place a door near a stairway. Figure 7.14 shows some common door and landing problems and their solutions.

Emergency Egress Openings

The term *egress* is used in most building codes to specify areas of access or exits. It is typically used in reference to doors, windows, and hallways. Windows are a major consideration in designing exits. Emergency egress is required in every sleeping room and in every basement with habitable space. Escape may be made through a door or window that opens directly into a public street, alley, yard, or exit court. The emergency escape must be operable from the inside without the use of any keys or tools. The sill of all emergency escape bedroom windows must be within 44″ (1118 mm) of the floor. Windows used for emergency egress must have a minimum net clear area of 5.7 sq ft (0.530 m^2). The net clear opening area must have a minimum width of 20″ (508 mm) and a minimum height of 24″

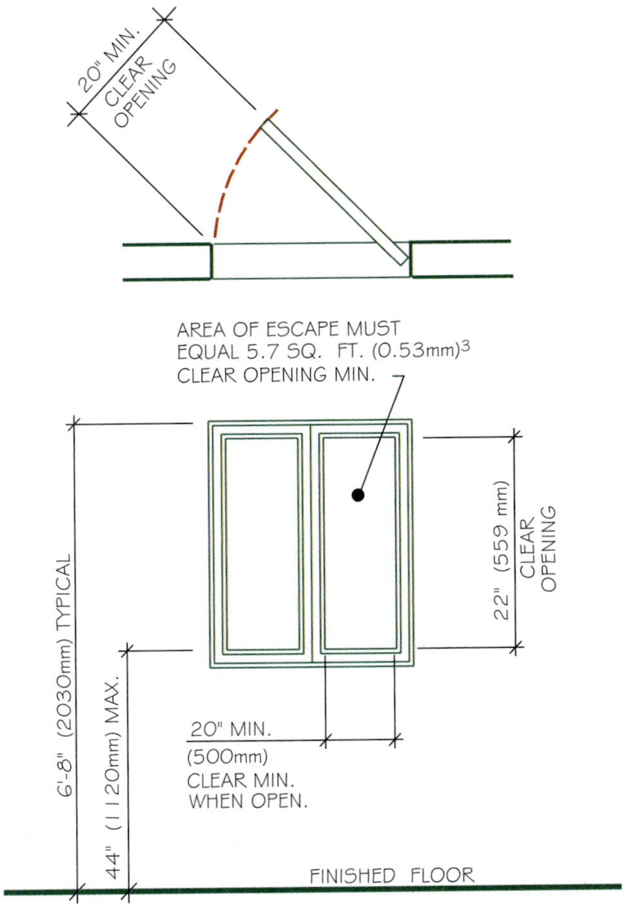

FIGURE 7.15 ■ Minimum emergency escape window opening sizes. A means of escape is required for all sleeping units. The IRC also requires escape windows in any basement with habitable space.

(559 mm). This opening gives occupants in each sleeping area a method of escape in case of a fire. See Figure 7.15.

When a habitable space is in a basement, the room must be provided with a rescue window that has a window well. Emergency escape windows with a finished sill height below the surrounding ground elevation must have a window well that allows the window to be fully opened. The window well is required to have a clear opening of 9 sq ft (0.84 m^2) with a minimum horizontal projection and width of 36″ (914 mm) when the window is fully open. If the window well has a depth of more than 44″ (1118 mm), the well must have an approved permanently fixed ladder or stair that can be accessed when the window is fully opened. The ladder or stair cannot encroach more than 6″ (152 mm) into the required well dimensions and must have rungs that have a minimum inside width of 12″ (305 mm). The rungs must be a minimum of 3″ (76mm) from the wall and must be spaced at a maximum distance of 18″ (457 mm) O.C. vertically for the full height of the window well.

Smoke Detectors

Closely related to emergency egress is the use of smoke alarms to provide an opportunity for safe exit through early detection of fire and smoke. IRC requires that a smoke alarm be installed in each sleeping room, as well as at a point centrally located in a corridor that provides access to the bedrooms. For a one-level residence, a smoke detector must be located at the start of every hall that serves a bedroom, as well as in each sleeping room. For multilevel homes, a smoke alarm is required on every floor, including the basement. The smoke alarm should be located over the stair leading to the upper level. In split-level homes, the smoke alarm is required only on the upper level if the lower level is less than one full story below the upper level. If a door separates the levels of a split-level home, a smoke alarm is required on each level. A smoke alarm is required on the lower level if a sleeping unit is located on that level. Smoke alarms should not be placed in or near kitchens or fireplaces because a small amount of smoke can set off a false alarm.

Smoke alarms must be located within 12″ (305 mm) of the ceiling or mounted on the ceiling. Most codes require alarms to be connected to electrical wiring with a battery-powered backup system. Smoke alarms must be interconnected so that if one alarm is activated, all will sound.

Halls

Hallways must be a minimum of 36″ (914 mm) wide. This is very rarely a design consideration because hallways are often laid out to be 42″ (1067 mm) wide or wider to create an open feeling and to enhance accessibility from room to room.

Stairs

Stairs often can dictate the layout of an entire structure. Because of their importance in the design process, stairs must be considered early in the design stage. For a complete description of stair construction see Chapter 37. Minimum code requirements for stairs can be seen in Figure 7.16. Following the minimum standards would result in stairs that are extremely steep, very narrow, and have very little room for foot placement. Good design practice would provide stairs with a width of between 36 and 42″ (914–1067 mm) for ease of movement. A common tread width is between 10 and 10 1/2″ (254–267 mm) with a rise of about 7 1/2″ (191 mm). Figure 7.17 shows the difference between the minimum stair layout and some common alternatives. The rise of a stairway can be no less than 4″ (102 mm). Within any flight of stairs, the largest step run cannot exceed the smallest step by 3/8″ (9.5 mm). The difference between the largest and smallest rise cannot exceed 3/8″ (9.5 mm).

Winding stairs may be used if a minimum width of 10″ (254 mm) is provided at a point not more than 12″ (305 mm) from the side of the treads. The narrowest portion of a winding stairway may not be less than 6″ (152 mm) wide at any point.

When a spiral stair is the only stair serving an upper floor area, it can be difficult to move furniture from one floor to another. Spiral stair treads must provide a clear walking width of 26″ (660 mm) measured from the outer edge of the support column to the inner edge of the handrail. A run of 7 1/2″ (190 mm) must be provided within 12″ (305 mm) of the narrowest part of the tread. The rise for a spiral stair must be sufficient to provide 6′–6″ (1982 mm)

INTERNATIONAL RESIDENTIAL CODE MINIMUM STAIR GUIDELINES		
STANDARD	IRC	
Width (minimum)	*36″	*914 mm
Rise (maximum)	7 3/4″	196 mm
Tread (minimum)	10″	254 mm

*Minimum clear width above the permitted handrail and below the required headroom height. Below the handrail, 31.5″ (787 mm) minimum width with handrail on one side, 27″ (698) (698 mm) minimum width with handrail on each side.

FIGURE 7.16 ■ Basic design values for stairs.

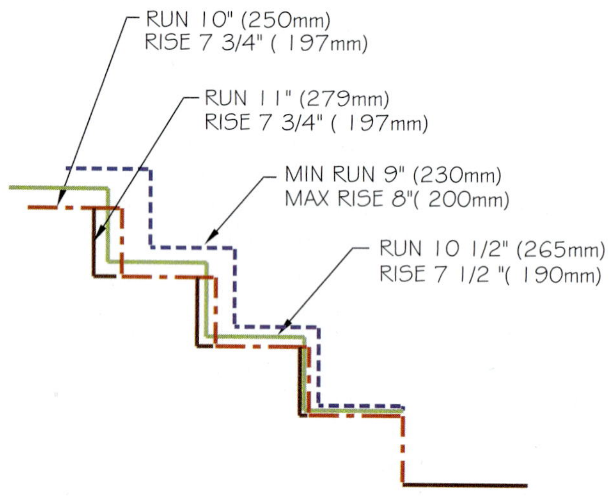

FIGURE 7.17 ■ Stair layouts comparing minimum run and maximum rise based on common practice.

headroom, but no riser may exceed 9 1/2″ (241 mm), and all risers must be equal. Circular stairways must have a minimum tread run of 11″ (279 mm) measured at a point not more than 12″ (305 mm) from the narrow edge. At no point can the run be less than 6″ (152 mm).

Headroom over stairs must also be considered as the residence is being designed. Stairs are required to have 6′–8″ (2032 mm) minimum headroom. Headroom of 6′–6″ (1982 mm) is allowed for spiral stairs. This headroom can have a great effect on wall placement on the upper floor over the stairwell. Figure 7.18 shows some alternatives in wall placement around stairs.

All stairways with three or more risers are required to have at least one smooth handrail that extends the entire length of the stair. The rail must be placed on the open side of the stairs and must be between 34″ and 38″ (864–965 mm) above the front edge of the stair. Handrails are required to be 1 1/2″ (38 mm) from the wall but may not extend into the required stair width by more than 3 1/2″ (89 mm).

A guardrail must be provided at changes in floor or ground elevation that exceed 30″ (762 mm). Guardrails are required to be 36″ (914 mm) high. Railings must be constructed so that a 4″ (102 mm) diameter sphere cannot pass through any opening in the rail. The triangular opening formed by the bottom of the rail and the stairs must be constructed so that a 6″ (152 mm) sphere cannot pass through the opening. Horizontal rails or other ornamental designs that form a ladder effect cannot be used.

Room Dimensions

Room dimension requirements affect the size and ceiling height of rooms. Every dwelling unit is required to have at least one room with a minimum of 120 sq ft (11.2 m^2) of floor area. Other habitable rooms except kitchens are required to have a minimum of 70 sq ft (6.5 m^2) and shall not be less than 7′ (2134 mm) in any direction.

These code requirements rarely affect home design. One major code requirement affecting room size governs the space allowed for a toilet, which is typically referred to as a water closet. A space 30″ (762 mm) wide must be provided for water closets. A distance of 21″ (533 mm) is required in front of a toilet. This can often affect the layout of a bathroom.

Habitable rooms, hallways, corridors, bathrooms, toilet rooms, laundry rooms and basements must have a minimum ceiling height of 7′–0″ (2134 mm). This lowered ceiling can often be used for lighting or for heating ducts. Beams that are spaced at a maximum distance of 48″ (1219 mm) O.C. are allowed to extend 6″ (152 mm) below the required ceiling height. In rooms that have a sloping ceiling, the minimum ceiling must be maintained in at least half of the room. The balance of the room height may slope to 5′ (1524 mm) minimum. Any part of a room that has a sloping ceiling less than 5′ (1524 mm) high, or 7′–0″ (2134 mm) for furred ceilings, may not be included as habitable square footage. See Figure 7.19.

Light, Ventilation, Heating, and Sanitation Requirements

The light and ventilation requirements of building codes have a major effect on window size and placement. The heating and sanitation requirements are so minimal that they rarely affect the design process.

The codes covering light and ventilation have a broad impact on the design of the house. Many preliminary designs often show entire walls of glass to take advantage of beautiful surroundings. At the other extreme, some houses have very little glass and thus no view but very little heat loss. Building codes affect both types of designs.

All habitable rooms must have natural light provided by windows. Kitchens are habitable rooms but are not required to meet the light and ventilation requirements. Windows for other habitable rooms are required to open directly into a street, public alley, or yard on the same building site. Windows may open into an enclosed structure such as a porch as long as the area is at least 65% open. Required windows may also be provided by skylights.

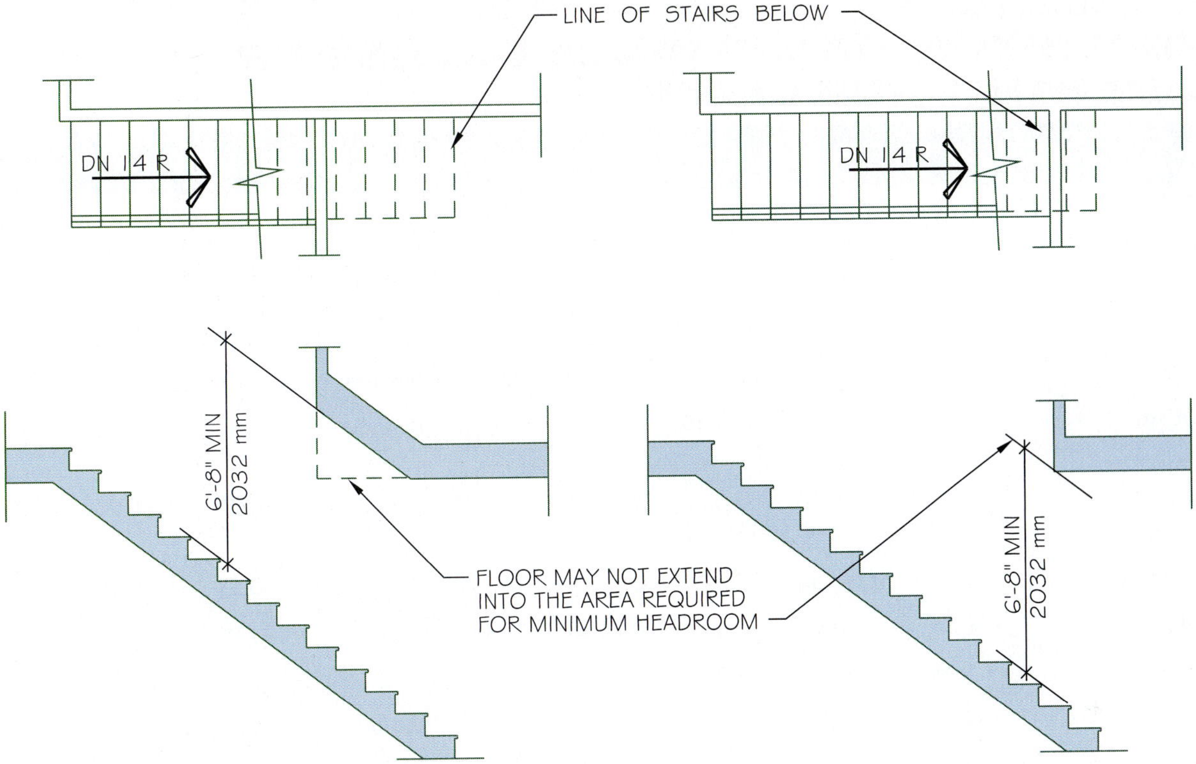

FIGURE 7.18 ■ Wall placement over stairs.

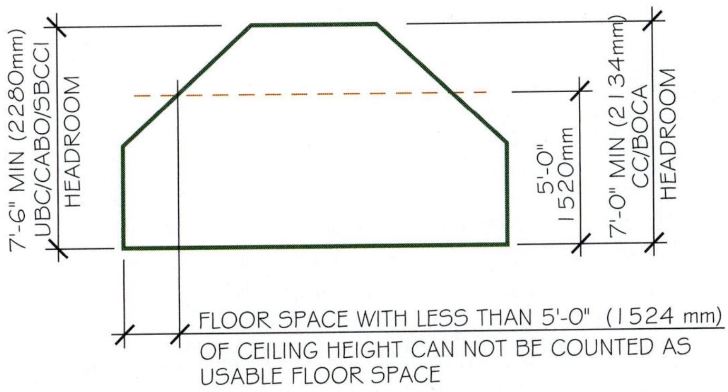

FIGURE 7.19 ■ Minimum ceiling heights for habitable rooms.

IRC requires all habitable rooms to have a glazing area equal to 8% of the room's floor area, and that 1/2 of the area used to provide light also be openable to provide ventilation. A bedroom that is 9 × 10′ (2743 × 3050 mm) is required to have a window with a glass area of 7.2 sq ft (0.66 m^2) to meet IRC standards. Another limit to the amount of glass area would be in locations that are subject to strong winds or earthquakes. Because of the lateral movement created by winds and earthquakes, window and wall areas must be carefully proportioned to provide lateral stability. Chapter 30 will discuss lateral design.

Two alternatives to the light and ventilation requirements are typically allowed in all habitable rooms. Mechanical ventilation and lighting equipment can be used in place of openable windows in habitable rooms except bedrooms. Glazing can be eliminated except when required for emergency egress when artificial light capable of producing 6 footcandles (6.46 lux) over the area of the room at a height of 30″ (762 mm) above the floor is provided.

Ventilation can be eliminated when an approved mechanical ventilation system capable of producing 0.35 air changes per hour is provided for the room. Ventilation is also waived if a whole-house mechanical ventilation system capable of producing 15 cfm (7.08 L/s) per occupant is provided. Occupancy is based on 2 per the first bedroom and 1 for each additional bedroom.

The standards for emergency egress require bedrooms to have a window even if mechanical light and ventilation are provided. The second method allows floor areas of two adjoining rooms to be considered as one if half of the area of the common wall is open and unobstructed. This opening must also be equal to one-tenth of the floor area of the interior room or 25 sq ft (2.3 m^2), whichever is greater.

Although considered nonhabitable, bathrooms and laundry rooms must be provided with an openable window. The window must be a minimum of 3 sq ft (0.279 m^2) of which one-half must open. A fan that provides five air changes per hour can be used instead of a window. These fans must be vented to outside air.

Alternatives

Many municipalities are developing their own code variations to meet the needs of light and code restrictions. The 1996 Oregon Residential Energy Code, for example, allows nine different design alternatives to develop an energy-efficient structure. Rather than strict percentages for windows, variable areas are allowed depending on the quality of insulation, the framing system used, and the quality of windows to be installed. Figure 7.20 shows part of this energy code.

Oregon Residential Energy Code

PRESCRIPTIVE COMPLIANCE PATHS FOR RESIDENTIAL BUILDINGS

PRESCRIPTIVE COMPLIANCE PATHS FOR RESIDENTIAL BUILDINGS

Building Components	PATH 1	PATH 2 Sun Tempered[4]	PATH 3	PATH 4 Sun Tempered[4]	PATH 5	PATH 6 Sun Tempered[4]	PATH 7 Sun Tempered[4]	PATH 8 House Size Limited[5]	PATH 9 Log Homes/ Solid Timber
Maximum Allowable Window Area[6]	No Limit	No Limit	No Limit	No Limit	No Limit	No Limit	No Limit	12%	No Limit
Window Class[7]	U=0.40	U=0.40	U=0.50	U=0.50	U=0.60	U=0.60	U=0.60	U=0.40	U=0.40
Doors, Other Than Main Entry	U=0.20	U=0.20	U=0.20	U=0.20	U=0.20	U=0.20	U=0.20	U=0.20	U=0.54
Main Entry Door, maximum 24 sq. ft.	U=0.54	U=0.54	U=0.20	U=0.20	U=0.20	U=0.20	U=0.54	U=0.20	U=0.54
Wall Insulation	R-21[9]	R-15	R-21A[8]	R-15A[8]	R-24A[8]	R-21A[8]	R-21A[8]	R-15	—[3]
Underfloor Insulation	R-25	R-21	R-25	R-21	R-30	R-21	R-25	R-21	R-30
Flat Ceilings	R-38	R-49	R-49A[8]	R-38	R-49A[8]	R-49A[8]	R-49A[8]	R-49	R-49
Vaulted Ceilings[10,11]	R-30	R-30	R-30	R-38	R-38	R-38	R-38	R-38	R-38
Skylight Class[7]	U=0.50	U=0.50	U=0.50	U=0.50	U=0.50	U=0.50	U=0.50	U=0.50	U=0.50
Skylight Area[12]	<2%	<2%	<2%	<2%	<2%	<2%	<2%	<2%	<2%
Basement Walls	R-21	R-21	R-21	R-21	R-21	R-21	R-21	R-21	R-21
Slab Floor Edge Insulation	R-15	R-15	R-15	R-15	R-15	R-15	R-15	R-15	R-15
Forced Air Duct Insulation	R-8	R-8	R-8	R-8	R-8	R-8	R-8	R-8	R-8

Notes:

[1] Path 1 is based on cost-effectiveness. Paths 2-7 are based on energy equivalence with Path 1. Cost-effectiveness of Paths 2-9 not evaluated.

[2] As allowed in current Chapter 13, section 1306, thermal performance of a component may be adjusted provided that overall heat loss does not exceed the total resulting from conformance to the required U-value standards. Calculations to document equivalent heat loss shall be performed using the procedure and approved U-values contained in Table 13-B.

[3] R-values used in this table are nominal, for the insulation only and not for the entire assembly. The wall component for Path 9 shall be a minimum solid log or timber wall thickness of 3.5 inches (88.9 mm).

[4] The sun-tempered house shall have one lot line which borders on a street oriented within 30 degrees of true east-west and 50 percent or more of the total glazing area for the heated space on the south elevation. An approved alternate to street orientation based on solar design and access shall be accepted by the building official.

[5] Path 8 applies only to homes with less than 1,500 sq. ft. (139 m^2) heated floor space and glazing area less than 12 percent of heated space floor area.

[6] Reduced window area may not be used as a trade-off criterion for thermal performance of any component, except as noted in Table 13-B.

[7] Window and skylight U-values shall not exceed the number listed. U-values may also be listed as "class" on some windows and skylights (i.e., CL40 is the same as U = 0.40).

[8] A = advanced frame construction as defined in Section 1307.1.4 for walls and Section 1307.1.5 for ceilings.

[9] R-19 Advanced Frame or 2×4 wall with rigid insulation may be substituted if total nominal insulation R-value is 18.5 or greater.

[10] Partially vaulted ceilings and ceilings totaling not more than 150 sq. ft. (139m^2) in area for dormers, bay windows or similar architectural features may be reduced to not less than R-21. When reduced, the cavity shall be filled (except for required ventilation spaces) and a 0.5 perm (dry cup) vapor retarder installed.

[11] Vaulted area, unless insulated to R-38, may not exceed 50 percent of the total heated space floor area.

[12] Skylight area is a percentage of the heated space floor area. Any glazing in the roof/ceiling assembly above the conditioned space shall be considered a skylight.

FIGURE 7.20 ■ Many municipalities make additions to their building codes to cover specific concerns. This table is part of the 1996 Oregon Residential Energy Code. Varied paths allow for varied construction materials, glazing, and insulation to provide an energy-efficient home.

Heating

The heating requirements for a residence are very minimal. IRC requires a heating unit capable of producing and maintaining a room temperature of 68° F, or 20° C at a point 3′ (914 mm) above the floor for all habitable rooms. Chapter 19 provides information on sizing heating units.

Sanitation

Code requirements for sanitation rarely affect the design of a structure. Each residence is required to have a toilet, a sink, and a tub or shower, but not all are required to have hot and cold water. All plumbing fixtures must be connected to a sanitary sewer or an approved private sewage disposal system. The room containing the toilet must be separated from the food preparation area by a tight-fitting door. Chapter 14 provides further information regarding the layout of each fixture, and Chapter 18 provides information regarding plumbing plans.

CLIMATIC AND GEOGRAPHIC DESIGN CRITERIA

The major influence of the IRC on a structure has to do with structural components. Chapters covering foundations, wall construction, wall coverings, floors, roof-ceiling construction, roof coverings, and chimney and fireplace construction are provided to specify minimum construction standards. To use the information in the codes as it relates to each area of construction requires some basic climatic and geographic information about the area where the structure will be constructed. This includes knowledge of various factors of the building site:

Roof live load

Roof snow load

Wind pressure

Seismic condition by zone

Susceptibility to damage by weathering

Susceptibility to damage by termites

Susceptibility to damage by decay

Frost line depth

Winter design temperature for heating facilities

Information for each of these categories can be obtained from a table in the applicable building code or from the governing building department. Building codes specify how a building can be constructed to resist the forces from wind pressure, seismic activity, freezing, decay, and insects. Each of these areas will be discussed in detail in succeeding chapters. In the initial design stage, an understanding of these factors will determine what types of materials will be required.

CHAPTER 7

Building Codes and Interior Design Test

DIRECTIONS

Answer the questions with short complete statements or drawings as needed on 8 1/2 × 11 notebook paper, as follows:

1. Letter your name and the date at the top of the sheet.
2. Letter the question number and provide the answer. You do not need to write out the question. Answers may be prepared on a word processor if course guidelines allow this.
3. Answers should be based on the code that governs your area, unless otherwise noted.

QUESTIONS

Answers to all questions should be based on the code that governs your area.

Question 7–1 What is the minimum required size for an entry door?

Question 7–2 List the five major building codes used throughout the United States, starting with the code that governs your area.

Question 7–3 All habitable rooms must have what percent of floor area in glass for natural light?

Question 7–4 What is the required width for hallways?

Question 7–5 List the minimum width for residential stairs.

Question 7–6 What is the minimum ceiling height for a kitchen?

Question 7–7 What is the maximum height that bedroom windows can be above the finish floor?

Question 7–8 Toilets must have a space how many inches wide?

Question 7–9 Are bathrooms in a single-family residence required to be handicapped-accessible?

Question 7–10 List three sanitation requirements for a residence.

Question 7–11 What are the area limitations of spiral stairs?

Question 7–12 For a bedroom that is 10′ × 12′, what area is required to provide the minimum ventilation requirements?

Question 7–13 List the minimum square footage required for habitable rooms.

Question 7–14 For a sloping ceiling, what is the lowest height allowed for usable floor area?

Question 7–15 What is the minimum size opening for an emergency egress?

Question 7–16 What is the minimum required size for an exterior door?

Question 7–17 Visit the Web site of the building department that governs residential construction in your area. Determine the following current design criteria:

a. Roof live loads
b. Wind pressure
c. Risk of weathering
d. Frost line depth
e. Snow loads
f. Seismic zone
g. Risk of termites

Question 7–18 Visit the Web site of the agency that regulates your state's building code. List and describe any code changes currently under consideration.

Question 7–19 Visit the Web site of one of the testing labs and obtain information related to a class drawing project.

Question 7–20 Visit the Web site of one of the regulatory agencies that influence construction in your area and describe your findings.

CHAPTER 8

Room Relationships and Sizes

INTRODUCTION

Architecture probably has more amateur experts than any other field. Wide exposure to houses causes many people to feel that they can design their own house. Because they know what they like, many people attempt to draw their own plans, only to have them rejected by building departments. Designing can be done by almost anyone. Designing a home to meet a variety of family needs takes more knowledge.

THE FLOOR PLAN

The floor plan is the first drawing that is started as a home is designed for a family. The floor plan must be designed to harmonize with the site and the inhabitants. Planning for the site is introduced in Chapters 11 through 13. The balance of this chapter will introduce both the design process and basic requirements for a home.

Clients typically provide the designer with many pictures collected from years of planning, which can be very helpful in planning for the family's needs and lifestyle. Many architectural firms have a written questionnaire to provide insight into how the structure will be used. Important design considerations include the following factors:

- The number of inhabitants
- The ages and sex of children
- Future plans to add onto the dwelling
- A list of general family activities to be done in the home
- Entertainment habits
- Desired number of bedrooms and bathrooms
- Kitchen appliances desired
- Planned length of stay in the residence
- Live-in guests or requirements for people with disabilities
- The budget for the home
- The style of the home

It is helpful to have clients write out a specific list of requirements including: minimum number of rooms, minimum size of each room, and a listing of how each room is to be furnished. The written list is an excellent way to have the clients agree on the design criteria before the design team spends hours pursuing unwanted features. Providing minimum room sizes is also much more helpful than allowing a client to request "three big bedrooms." An understanding of how the room is to be furnished will also help the designer be sure that the owner has realistic expectations for the size of the room. It is also helpful to have clients provide a "wish list" of items for their residence. These items can be eliminated if the budget is exceeded, but it may be possible to provide them.

Once this information has been provided, the design process described in Chapter 1 can be started. The bubble drawings and preliminary sketches will need to integrate the living, sleeping, and service areas, as well as movement among these areas. Consideration must also be given to the style of the home. Single-level and multilevel homes pose different design challenges. This chapter will explore the challenges of placing the living, sleeping, and service areas in various styles of homes.

LIVING AREA

The living area can be considered the living, dining, and family rooms; den; and nook. These are the rooms or areas of a house where family and friends will spend most of their leisure time. The rooms of the living area should be clustered together near the entry to allow easy access for guests.

Entries

Entries serve as a transition point between areas of the home. A well-planned home will have at least two points of entry—the main entry, to draw guests into the home, and a service entry that is used by the family for access between the garage, yards, and service areas.

Main Entry

The main entry, as in Figure 8.1, provides an outside focal point to draw guests to the front door and serves as a hub for traffic to the living areas. A raised ceiling is a common method of accenting the front door. Figure 8.2 shows an example of an inviting entry created by using an arbor and good landscaping. Both examples clearly define access to the front door. The entry should be kept proportional to the entire structure and should be made of compatible materials. The

FIGURE 8.1 ■ The main entry should provide a focal point from the outside of the home to draw guests to the front door. *Courtesy Janice Jefferis.*

FIGURE 8.2 ■ An inviting entry created using an arbor and good landscaping to guide guests to the main entry. *Courtesy David Jefferis.*

FIGURE 8.3 ■ The recessed, covered entry provides protection from the weather and blends well with the rest of the home. Once inside, guests can see through the foyer to the fireplace in the great room. Direct access is provided to the dining area and den, with close access to the coat closet, the stairs, and a hallway that serves the balance of the home. *Courtesy Design Basics, Inc.*

size of the entry will be influenced by the number of entry doors and their size. Single entry doors are generally 3′ (900 mm) wide. Larger homes may have doors that are 42″. (1050 mm) or 48″ (1200 mm) wide. Double doors are typically 5′ (1500 mm) or 6′ (1800 mm) wide, and one or both panels can be operable. Single and double doors are usually 6′–8″ (2000 mm) high but are also available in 8′ (2400 mm) and custom heights.

A second major consideration in planning the entry is providing protection from the weather. Depending on your area of the country, this might include direct sunlight (in arid climates), rain, snow, and wind. A well-planned entry will provide a covered access into the home as well as a windbreak. The windbreak can often be achieved by recessing the entry area. The depth of the recess should be kept proportional to the width and height of the entry. See Figure 8.3.

Foyer

The ideal entry will open to a foyer rather than directly into one of the living areas. The foyer serves as a place either for welcoming guests or saying final good-byes. It also provides a place to put on or remove coats or other weather-related clothing. It should provide access to a closet for seasonal clothing as well as space for guests' coats. The foyer also provides access to each area of the home. This might include a hall leading to the sleeping area, a stair leading to an upper or lower floor, and access to the living room or dining areas. The foyer should be proportional to the rest of the home, but it should also be of sufficient size to hold several people at a time. The size will be based on the client's needs and personal preferences, as well as the size and cost of the home. In a small home, a foyer 48″ (1200 mm)

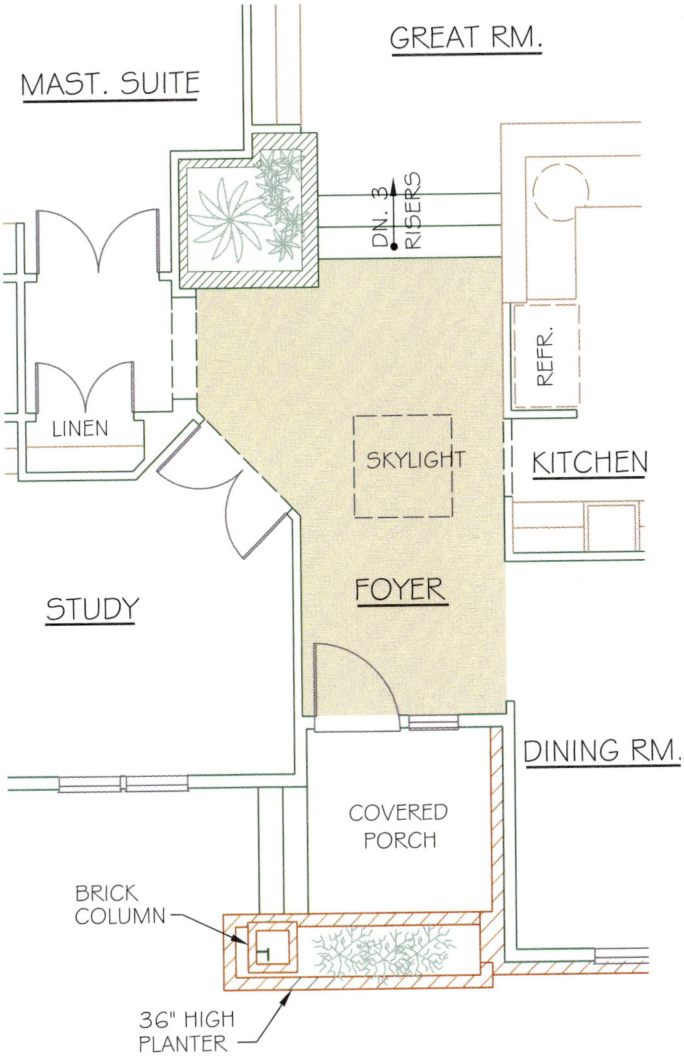

FIGURE 8.4 ■ The foyer should create a warm, inviting feeling and enhance traffic flow throughout the house. *Courtesy Stark/Jefferis Designs.*

wide may be tolerable, although 6′ × 6′ (1500 × 1500 mm) is still considered small. In a larger home, a room 8′ to 12′ (2400 to 3600 mm) wide would be more appropriate. The height of the foyer should be considered as the size is determined. A small foyer with tall ceiling will create a feeling of being in an elevator shaft. Remember, the goal of the foyer is to create a warm, inviting feeling and to enhance traffic flow throughout the home. See Figure 8.4.

Service Entry

The service entry links the garage, kitchen, utility room, bathrooms, and patios or decks. In many homes, although guests typically use the main entry, the family's main access is through the service entry. Depending on the size and location of the home, the service entry may serve as a mudroom. The service entry will be discussed later in this chapter.

Living Room

The purpose of the living room will vary, depending on the size of the home and the preferences of the homeowner. Points to consider in planning the living room include how the room will be used, how many people will use it, how often it will be used, and the type and size of furniture will be placed in it. Some families use the living room as a place for quiet conversation and depend on other living areas for noisier activities. In homes designed to appeal to a wide variety of interests, the formal living room is giving way to a large multipurpose room.

If both a living room and a family room are provided, the living room is typically used to entertain guests or, for quiet conversations. Figure 8.5 shows a formal living room. If the home has no family room, the living room will also need to provide room for recreation, hobbies, and relaxation.

The size of the living room should be determined by the number of people who will use it, the furniture (see Figure 8.6*a*) that is intended for it, and the budget for the entire construction project. Rectangular rooms are typically easier to plan for furniture arrangements. The closer a room is to being square, the harder furniture placement becomes. The minimum width needed for common activities is 12′–14′ (3660–4270 mm). A room 13′ × 18′ (3960 × 5500 mm) serves for most furniture arrangements and allows for moving furniture easily. Figure 8.6*b* lists common sizes of living room furniture.

Within the room, an area approximately 9′ (2740 mm) in diameter should be provided for the primary seating area. This area is often arranged to take advantage of a view or centered on a fireplace. In smaller homes, the living room can be made to

FIGURE 8.5 ■ A formal living room provides an area for quiet conversation away from the noise of other living areas. In larger homes, the living room is often set apart from the rest of the living areas to create a formal area. *Courtesy Eric S. Brown Design Group Inc. Photo by Michael Lowry Photography.*

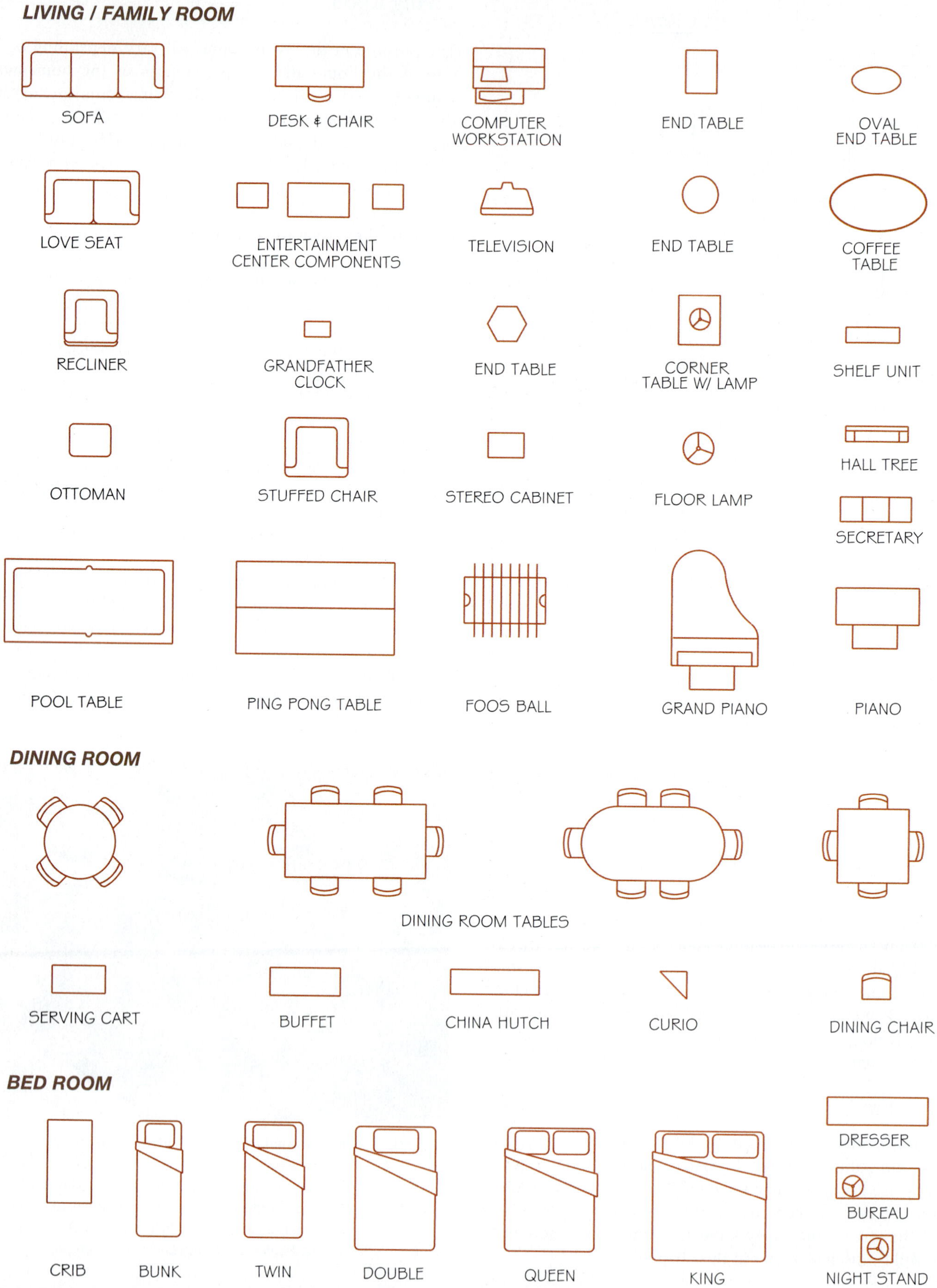

FIGURE 8.6A ■ Furniture typically found in living areas.

COMMON FURNITURE SIZES

LIVING ROOM AND FAMILY ROOM

Sofas	34 × 78 34 × 90 34 × 96 34 × 102	End Tables	18 × 18 18 × 24 18 × 30 18 × 36 24 × 24	Arm Chairs	18 × 21 18 × 24 22 × 24 28 × 32 32 × 34
Sectional Pieces	26 × 30 26 × 53 26 × 62		26 × 26 24 hex 26 hex	Recliners	30 × 29 to 66
				Piano	24 × 56
Love Seats	32 × 50 30 × 66	Coffee Tables	18 × 36 20 × 52 20 × 60 20 × 75	Grand Piano	58 × 75

DINING ROOM

Oval Tables	54 × 76	Hutch	18 × 36 × 72
Round Tables	30 32 36 42 48	Chairs	18 × 18 18 × 20
		Tea Cart	16 × 24 18 × 33
Rectangular Tables	30 × 34 36 × 44 to 108 42 × 48 to 108		

BEDROOM

Beds:		Dressers	18 × 30 to 82
Cribs	30 × 54		
Bunk	30 × 75	Desks	24 × 30 24 × 35
Twin	39 × 75		24 × 42 32 × 42
Double	54 × 75		32 × 48 32 × 60
Queen	60 × 80		
King	76 × 84	Night Stands	12 × 15 15 × 15 to 21

FIGURE 8.6B ■ Common furniture sizes.

FIGURE 8.7A ■ If the living room is open to other living areas, a casual feeling is created. *Courtesy Eric S. Brown Design Group Inc. Photo by Oscar Thompson.*

FIGURE 8.7B ■ In addition to being open, this room creates an elegant room in which to entertain. *Courtesy Eric S. Brown Design Group Inc. Photo by Oscar Thompson.*

seem larger if it is attached to the dining area as in Figure 8.7*a*. In larger residences, the living room is often set apart from other living areas to give it a more formal effect.

The living room is usually placed near the entry so that guests may enter it without having to pass through the rest of the house. The living room should also be placed so that the rest of the home can be accessed without having to pass through it. Many designers place the living room one or more steps below the level of the balance of the home, as in Figure 8.8, to enhance a feeling of separation and to discourage through traffic. The location of the room should also take energy efficiency into consideration. If the room is used primarily in the evening, it should be placed on the west side of the residence so it will have natural

FIGURE 8.8 ■ Many designers place the living room one or more steps below the level of the rest of the home to enhance a feeling of separation and to discourage through traffic. *Courtesy Michael Jefferis.*

FIGURE 8.9 ■ A casual living room that draws the outside in. Living areas of this residence are close to each other for easy entertaining, but separated for noise control. *Courtesy Eric S. Brown Design Group Inc. Photo by Laurence Taylor.*

light at that time. In cooler climates, a southwestern orientation will provide the most hours of sunlight; in warmer climates a northwestern orientation will help the room take advantage of evening light while offering some protection from the low summer sun. The overhang of roofs, wall projections, type and amount of glazing, and landscaping can all be used to control the effects on the structure.

Figure 8.9 shows a casual living room that is open to both the rest of the home and to the yard.

Family Room

The family room is probably the most used area in the house. A multipurpose area, it is used for such activities as watching television, informal entertaining, sorting laundry, playing pool, or eating. With such a wide variety of possible uses, planning the family room can be quite difficult. The room needs to be separated from the living room but still be close enough to have easy access from the living and dining rooms for entertaining. Figure 8.10 shows a possible layout of living areas providing close but separate rooms. The family room needs to be near the kitchen, since it is often used as an eating area. Placing the family room adjacent to the kitchen combines the two most used rooms into one area called a country kitchen. It is also common to have the family room near the service areas of the residence.

Sizes can vary greatly depending on the design criteria for the residence. When the family room is designed for a specific family, its size can be planned to meet their needs. When the room is being designed for a house where the owners are unknown, it must be of sufficient size to meet a variety of needs. An area of about 13′ × 16′ (3960 × 4900 mm) should be the minimum. This is a small room but would provide sufficient area for most activities. If a wood stove or fireplace is in the family room, the room should be large enough to allow for the heat from the stove. Often in an enclosed room where air does not circulate well, the area around a stove or fireplace is too warm to sit near comfortably.

For most families, space should be provided for an entertainment center storing the television, the stereo system, and related equipment. Figure 8.11 shows a pleasing arrangement of a fireplace, a television, and entertainment units.

Dining Room

Dining areas, depending on the size and atmosphere of the residence, can be treated in several ways. The dining area is often part of, or adjoining, the living area, as in Figure 8.12. For a more formal eating environment, the dining area will be near but separate from the living room area. See Figures 8.13*a* and 8.13*b*. The two areas are usually adjoining so that guests may go easily from one to the other without passing through other areas of the house. The dining room should also be near the kitchen for easy serving of meals but without providing a direct view into the work areas of the kitchen. Because of the need to be by the living room and kitchen, the dining room is often placed between these two areas or in a corner between the two rooms, as in Figure 8.14.

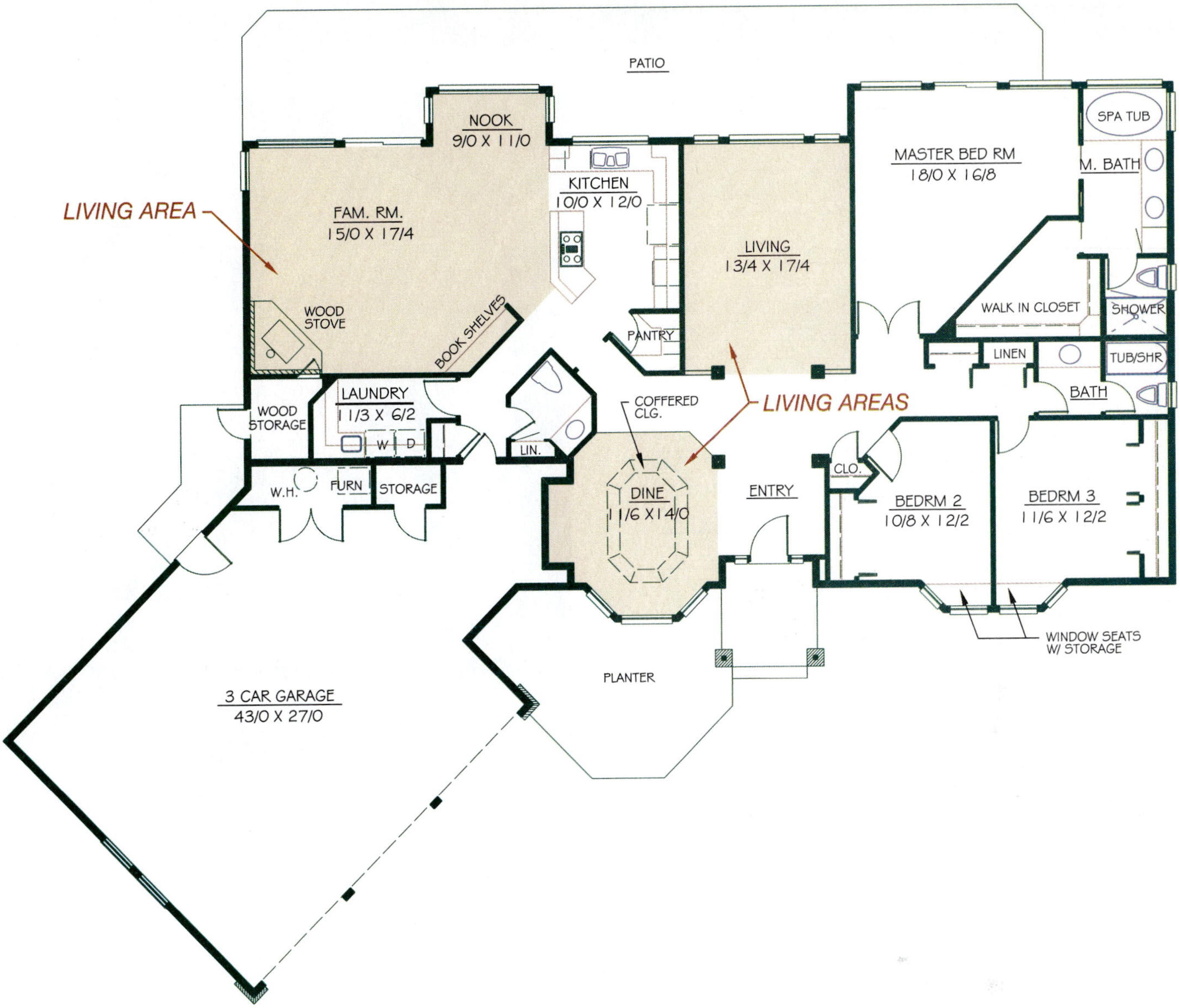

FIGURE 8.10 ■ Living areas of this residence are close to each other for entertaining, but separated for noise control. *Courtesy Wally Griener, SunRidge Designs, CPBD.*

FIGURE 8.11 ■ A pleasing arrangement of fireplace, television, and storage units should be part of every family room.

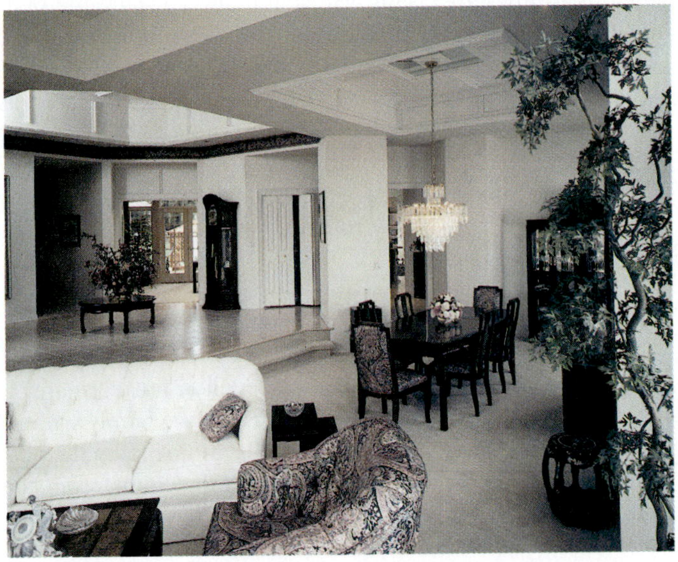

FIGURE 8.12 ■ The dining area can be combined with other living areas to create a spacious environment. *Courtesy Piercy & Barclay Designers, CPBD.*

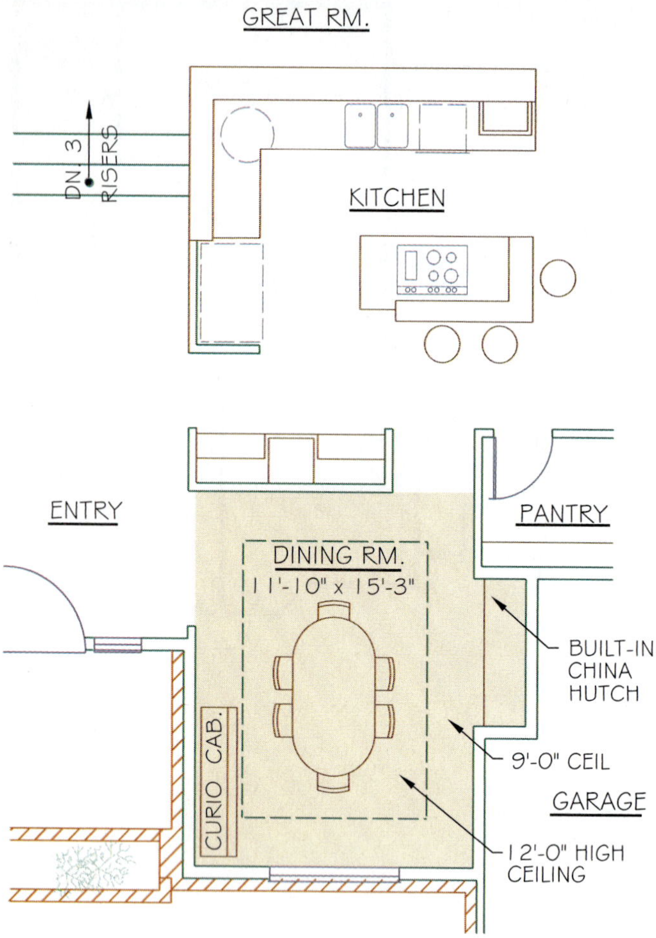

FIGURE 8.13A ■ Separating the dining room from the other living areas creates a more formal eating area. *Courtesy Stark/Jefferis Designs.*

FIGURE 8.13B ■ Although open to other rooms, a formal atmosphere is created by enclosing the dining area with columns and arches. *Courtesy Eric S. Brown Design Group Inc. Photo by Laurence Taylor.*

A casual dining room can be as small as 9′ × 11′ if it is open to another area. This size would allow for a table seating four to six people and a small storage hutch. A formal dining room should be about 11′ × 14′ (3350 × 4300 mm). This will allow room for a hutch or china cabinet as well as the table and chairs. Allow for a minimum of 32″ (810 mm) from the table edge to any wall or furniture for placement of chairs and a minimal passage area. A space of approximately 42″ (1067 mm) will allow room for walking around a chair when it is occupied.

Nook

When space and finances will allow, a nook or breakfast area is often included in the design. This is where the family will eat most meals and snacks (Figure 8.15). The dining room then becomes an area for formal eating only. Both the dining room and the nook need to be near the kitchen. If possible, the nook should also be near the family room.

Where space is at a premium, these areas are often placed together in one room called a grand or great room (Figure 8.16). In larger homes a patio or deck may be enclosed to provide a solarium (Figure 8.17).

Den/Study/Office

Although the name may vary, many families plan for a room for quiet reading and study, as in Figure 8.18. This room is typically located off the entry and near the living room. The den often serves as a buffer between the living and sleeping areas of a residence. In many tract houses, the den is used as a spare bedroom or guest room.

If the room is to be a true office within the home, the entrance should be directly off the entry. An entry separate from the residence is ideal. Entry to the office can be provided from a covered porch but should be distinctive enough so that clients will know where to enter. The size of the office depends on the equipment to be used and the number of clients to be accommodated. It should have room for a desk, computer work space, storage areas for books and files, and a separate area for phones, fax, and photocopying machines.

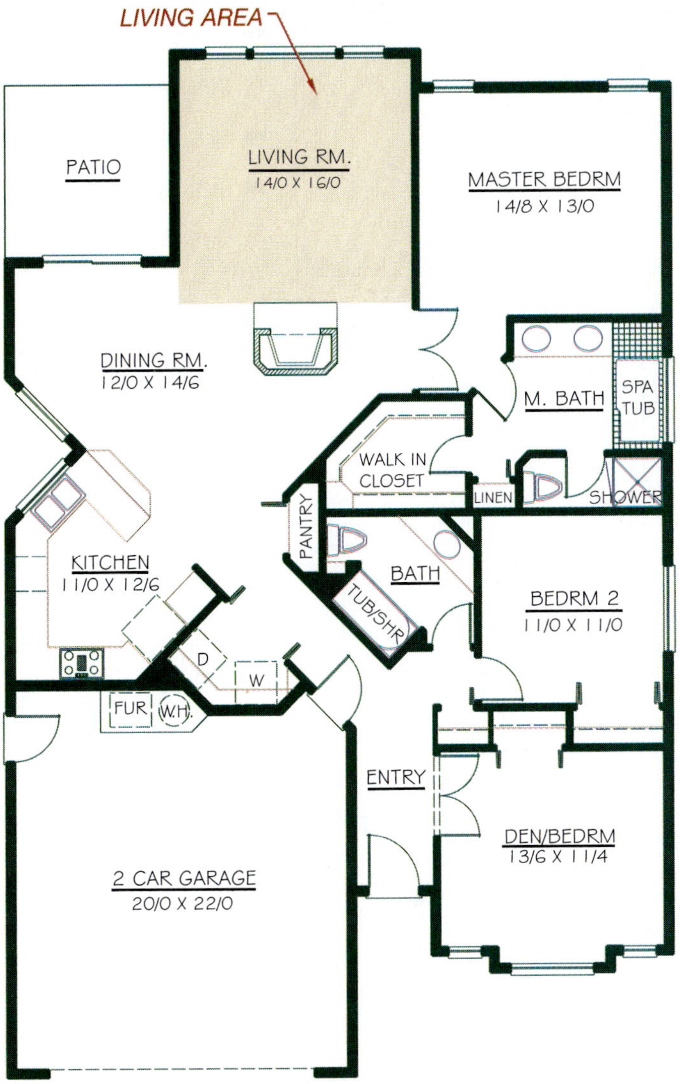

FIGURE 8.14 ■ Combining the dining, family and living areas provides ample space for entertaining. *Courtesy Wally Griener, SunRidge Designs, CPBD.*

FIGURE 8.15 ■ A nook is a casual area for family dining. *Courtesy Marvin Windows and Doors.*

FIGURE 8.16 ■ A grand room is a combination of a living, dining, and family room. *Courtesy National Oak Flooring Association.*

Home Theaters

Although not included in most tract homes, personal theaters or media rooms are becoming increasingly popular in many high-end custom homes. What used to be unfinished space, or a bonus room above a garage, can be made into a home theater. Home theaters are custom-designed rooms designed, as the name implies, for high-quality viewing of movies. The home designer will work closely with media consultants to meet the needs of the family and to help plan the locations of sound equipment. Figure 8.19 shows four common room layouts for home theaters with surround sound. Points to consider will be the size and type of the television, the number of viewers, and the type of seating. A wall-mounted digital television will require far less room than a projection television or even a big-screen standard TV. While most home theaters are designed for eight to twelve people, larger-capacity rooms are occasionally found. Sofas, recliners, and easy chairs make for casual, comfortable seating. Theater-type seating is also available. For larger seating groups, an inclined or stepped floor, should be part of the design. No matter how small the room, space should be provided for snack preparation. Items that may be included are a popcorn machine, a counter with a microwave, and a small refrigerator or wet bar.

FIGURE 8.17 ■ A sunroom or solarium can provide added square footage at a relatively low construction cost. *Courtesy Velux America, Inc.*

FIGURE 8.18 ■ Many homes provide a room that can be used as an office, study or guest bedroom. *Courtesy Eric S. Brown Design Group Inc. Photo by Laurence Taylor.*

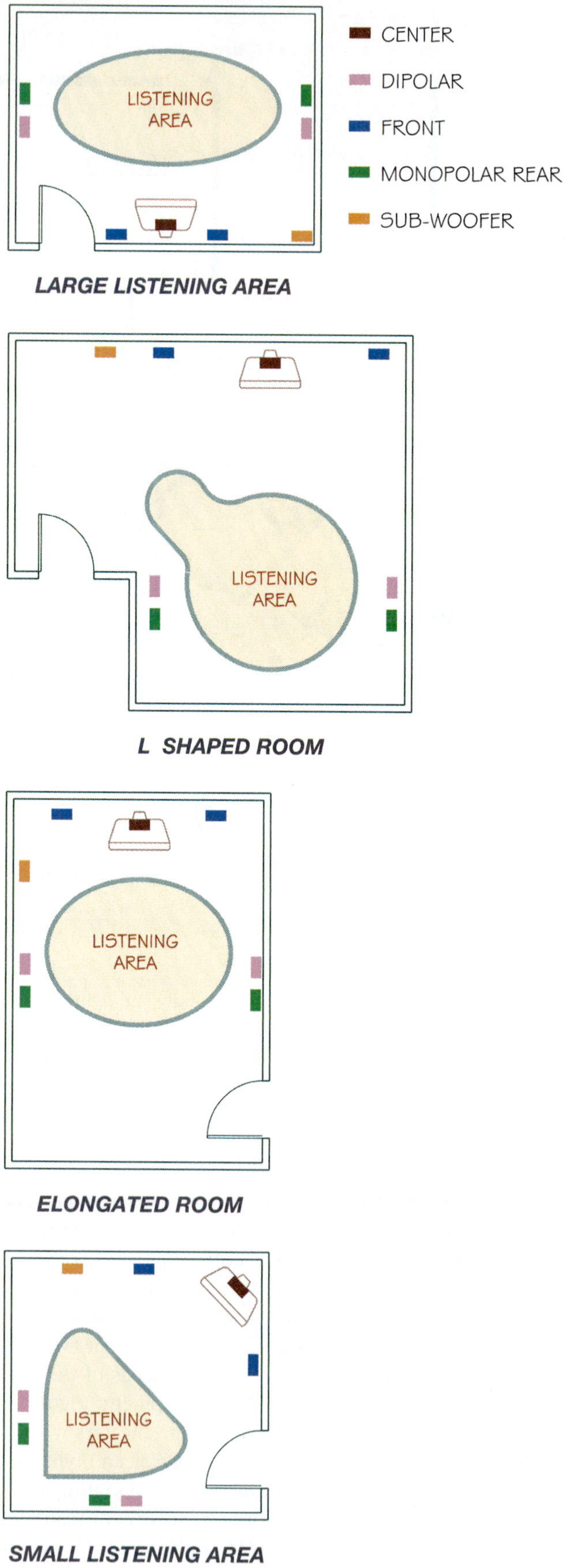

FIGURE 8.19 ■ Four common room layouts for a home theater.

SLEEPING AREA

The age, sex, and number of children will determine how many bedrooms will be required. Preteens can share a bedroom if space is limited. Teens of the same sex have even been known to survive sharing a room together, although this situation often results in distinctive markings of territory during tense times. The ideal situation is to provide a separate bedroom for each child. Each room should have space for sleeping, relaxation, study, storage, and dressing.

The sleeping area consists of the bedrooms. These rooms should be placed away from the noise of the living and service areas and out of the normal traffic patterns. The number of bedrooms will vary depending on the size of the family and the ages of the children. Even in a family with no children, a minimum of two rooms should be provided. A second bedroom also greatly increases the value of the home for resale. Most homes have a minimum of three bedrooms; and four bedrooms and a den are a common option for many subdivision homes. The extra room provides space for guests, or it can double as a craft, sewing, or hobby room. Many custom homes plan for a space about the size of a bedroom to be used as a multipurpose room. An exercise or equipment room, a game room, and a darkroom are common types of multipurpose rooms located near the bedrooms.

The arrangement of bedrooms will vary greatly depending on the needs of the family. Common arrangements include placing all the bedrooms together, or placing the master bedroom separate from the bedrooms for children. It is also becoming common to plan a bedroom-living unit for long-term care of a live-in relative.

Bedrooms are generally located with access from a hallway for privacy from living areas and to be near bathrooms, as in Figures 8.20*a* and 8–20*b*. Care must be taken to keep the bathroom plumbing away from bedroom walls. One way to accomplish this is to place a closet between the bedroom and the bath. If plumbing must be placed in a bedroom wall, use insulation to help control noise.

Bedrooms

Bedrooms function best on the southeast side of the house. This location will bring morning sunlight to the rooms. When a two-level layout is used, bedrooms are often placed on the upper level, away from the living areas. Not only does this arrangement provide a quiet sleeping area, but bedrooms can often be heated by the natural convection of heat rising from the living area. Another option is placing the bedrooms in a

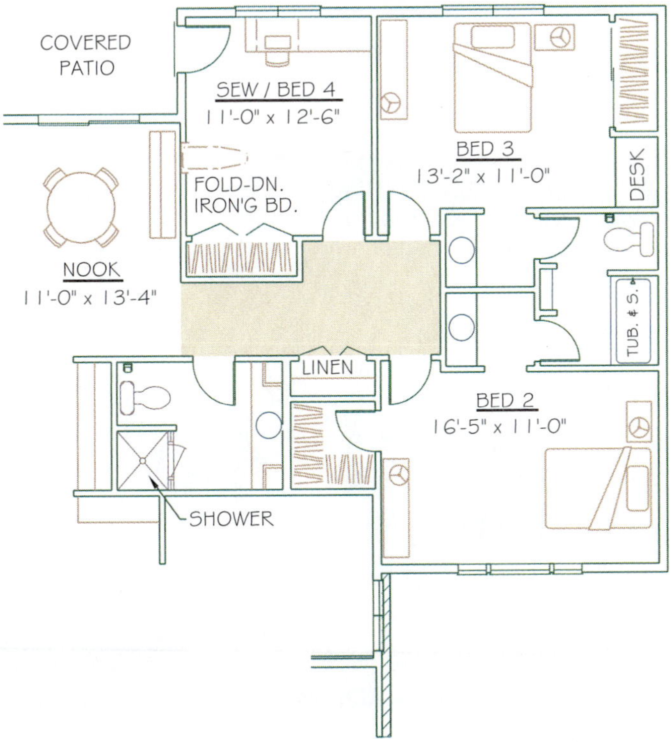

FIGURE 8.20A ■ Bedrooms should be located with access from a hallway for privacy from the living areas and be near bathrooms. *Courtesy Stark/Jefferis Designs.*

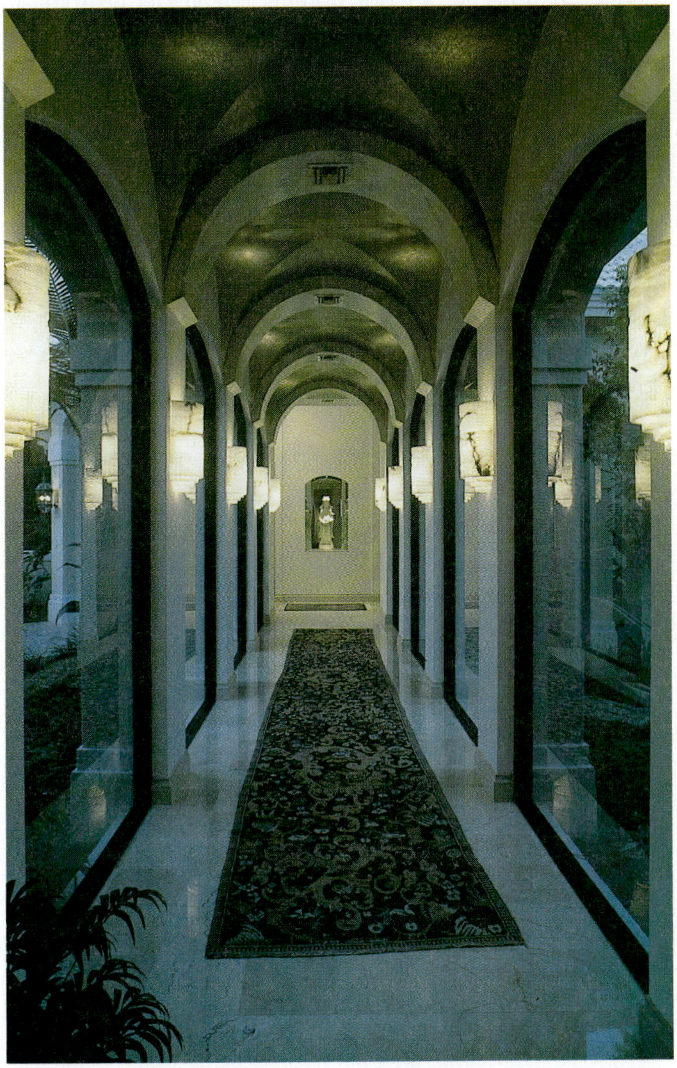

FIGURE 8.20B ■ A hallway can provide an interesting transition between rooms. *Courtesy Eric S. Brown Design Group Inc. Photo by Oscar Thompson.*

daylight basement. Care must be taken in basement bedrooms to provide direct emergency exits to the outside. An advantage of a bedroom in a daylight basement is the cool sleeping environment that the basement provides.

Sizes of bedrooms vary greatly depending on the size of the home, the age of the occupants, and the furniture to be placed in a bedroom. The IRC national building code requires a minimum of 70 sq ft (6.5 m^2). Homes financed by FHA loans are required to provide a minimum of 100 sq ft (9.3 m^2). Spare or children's bedrooms are often as small as 9′ × 10′ (2740 × 3050 mm). This might be adequate for a sleeping area, but not if the room is also to be used as a study or play area. Each bedroom should have enough space for a single bed, a small bedside table, and a dresser. Clients often specify enough space for a twin or double bed. Plan for a minimum of 24″ (600 mm) on each side of a bed, where space to walk is required. A space of approximately 36″ (900 mm) should be provided between dressers and any obstruction to allow for opening of drawers or doors. This requires a space of about 12′ × 14′ (3600 × 4300 mm) plus closet areas. Minimum closet space will be covered in the next section.

More than just a room for sleeping, the master bedroom in many custom homes is a retreat from the day's troubles. The master suite, as it is often referred to, serves as a bedroom, sitting area, and bathing area. A well-designed master suite, as in Figure 8.21, provides spacious room for a queen- or king-sized bed, plus room for a fireplace, sitting area, direct access to the wardrobe and bathing areas, and access to a private deck or balcony. Figure 8.22 shows another example of a master suite.

The master bedroom should be at least 12′ × 14′ (3658 ×4267 mm) plus closet areas. An area of 13′ × 16′ (4000 × 4900 mm) will make a spacious bedroom suite, although additional room is required for a sitting or reading area (Figure 8.23). A window seat can often be used to provide sit-

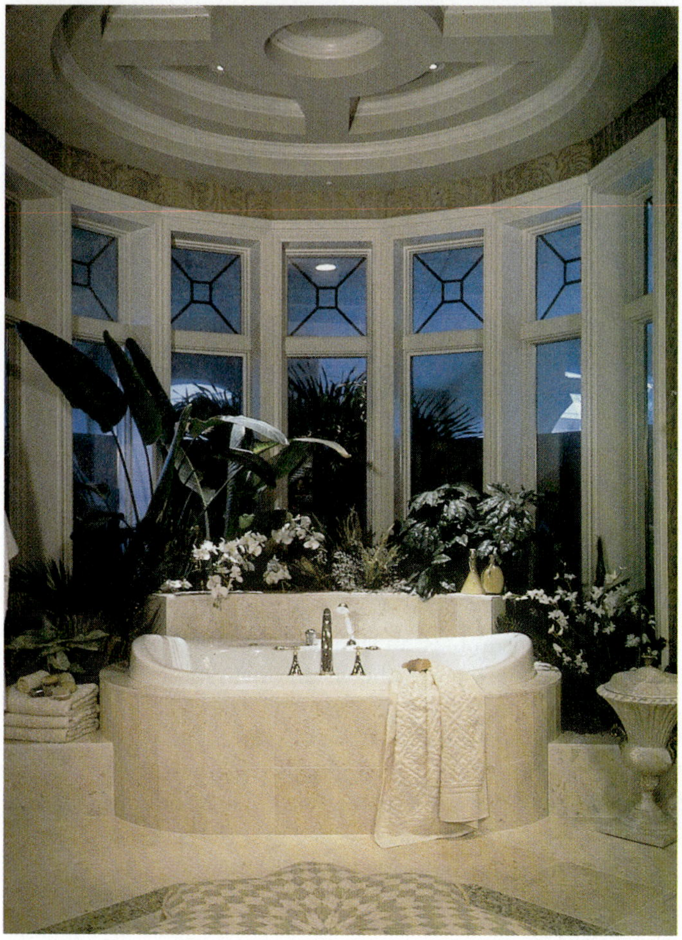

FIGURE 8.22 ■ A spa or soaking tub is often a part of the master suite. *Courtesy Eric S. Brown Design Group Inc. Photo by Laurence Taylor.*

FIGURE 8.21 ■ A well-designed master suite provides spacious room for a queen- or king-size bed, a fireplace, a sitting area, direct access to the wardrobe and bathing areas, and access to a private deck or balcony. *Courtesy Janice Jefferis.*

FIGURE 8.23 ■ A master suite often includes a sitting area for quiet reading or conversation. *Courtesy Marvin Windows and Doors.*

ting room without greatly increasing the square footage of the room. Try to arrange bedrooms so that at least two walls can be used for bed placement. This is especially important in the master bedroom to allow for periodic furniture movement.

Window planning is very important in a bedroom. Although morning sun is desirable, space should be planned for bed placement so that sunlight will not shine directly on it. Especially important in planning windows for a bedroom is to allow for emergency egress from the bedroom. (See Chapter 7 for a review of emergency window requirements.) For upper-level bedrooms, try to provide a roof, porch, or balcony as an escape route from a window for young children. A flexible escape ladder can be used for older children.

Closets

Building codes do not require bedroom closets. FHA recommends a length of 48″ (1200 mm) of closet space for males and 72″ (1800 mm) for females. Six feet (1800 mm) should be con-

FIGURE 8.24 ■ An area containing shelves, baskets and drawers increases storage without increasing square footage. *Courtesy Denise Taylor.*

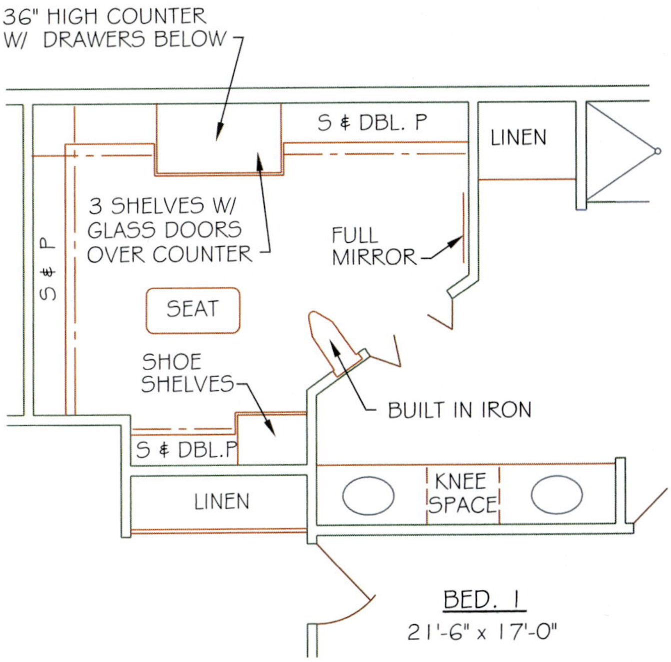

FIGURE 8.25 ■ A well-planned wardrobe provides ample storage space for hanging and folded clothes and easy access to the bed and bathrooms.

sidered the practical minimum for resale if a traditional shelf-and-pole storage system is to be used for storage. Space can be reduced if closet organizers are used. Minimum closet depth is typically considered to be 24″ (600 mm). A depth of 30″ (750 mm) is adequate to keep clothing from becoming wrinkled.

If a closet must be small, a double-pole system can be used to double the storage space or a premanufactured closet organizer can be used to provide storage. Closets can often be used as a noise buffer between rooms or located in what would be considered wasted space in areas with sloping ceilings. Master bedrooms often will have a walk-in closet. It should be a minimum of 6′ × 6′ (1800 × 1800 mm) to provide adequate space for clothes storage. A closet of 6′ × 8′ (1800 × 2400 mm) provides much better access to all clothes. Providing multiple rods for hanging pants, shirts, skirts, and blouses can increase storage. A single pole should be provided for hanging dresses and seasonal coats. An area containing shelves, baskets, and drawers, as in Figure 8.24, is also desirable if space permits. Figure 8.25 shows a well-planned wardrobe area for a master suite. Many clients request space for a freestanding armoire to provide additional storage. Space must also be provided near the bedrooms for linen storage. A space 2′ (610 mm) wide and 18″ (450 mm) deep would be minimal.

SERVICE AREA

Bath, kitchen, utility rooms, and the garage are each considered a part of the service area. Notice that three of the four areas have plumbing. Because of the plumbing and the services that each

provides, an attempt should be made to keep the service areas together. Another consideration in placing the service area is noise. Each of these four areas tends to have noises that will interrupt activities of the living and sleeping areas.

Bathrooms

To provide privacy, bathrooms are often reached by a short hallway, apart from living areas. A house with only one bathroom must have it located for easy access from both the living and sleeping areas. Access to the bathroom should not require having to pass through the living or sleeping areas.

Options for bathrooms include half-bath, three-quarter bath, full bath, and bathroom suite. A *half-bath* has a lavatory and a water closet (a toilet). A *three-quarter* bath combines a shower with a toilet and a lavatory. A *full bath* has a lavatory, a toilet, and a tub or a combination tub-and-shower unit. A *bathroom suite* typically includes the features of a full bath with a separate tub and shower. The tub is often an enlarged tub or spa (Figure 8.26). If the tub is to be raised, skidproof steps should be provided. If windows are placed around a tub or spa, the glass is required to be tempered. The wardrobe, dressing area, and a sitting area are also usually part of, or adjoining, the bathroom suite. Figure 8.27 shows a plan view for a master suite. Many custom homes feature a full bath with an adjoining exercise room, sauna, or steam room.

The style of the house affects the number and locations of bathrooms. A single-story residence typically has one bathroom for the master bedroom and a second to serve the living room and the other bedrooms. A two-story house typically has two full bathrooms upstairs, with a minimum of a half-bath downstairs. Multilevel homes should have a bath on each level containing bedrooms or a half-bath on each level with any type of living area.

When a residence has two or more baths, they are often placed back-to-back or above each other to reduce plumbing costs. A full bath is typically provided near the master bedroom, with a separate bathroom to serve the balance of the bedrooms. If space and budget allow, a bathroom for each bedroom is common. For families with young children, a tub is usually a must. Many bathrooms that are designed to be shared place the sinks in one room, with the toilet and tub in an adjoining room (Figure 8.28). For homes designed to appeal to a wide variety of families, a combination of a tub and shower is often used. Homes designed for outside activities often have a three-quarter bath near the utility room or combine the laundry facilities with a bathroom to create what is often called a mudroom.

FIGURE 8.26 ■ A master bedroom often includes a soaking tub with a separate tub and shower. *Courtesy Sara Jefferis.*

Kitchen

A kitchen is used through most of a family's waking hours. It not only serves for meal preparation but often includes areas for eating, working, and laundry (see Figure 8.29). It needs to be located close to the dining areas so that serving meals does

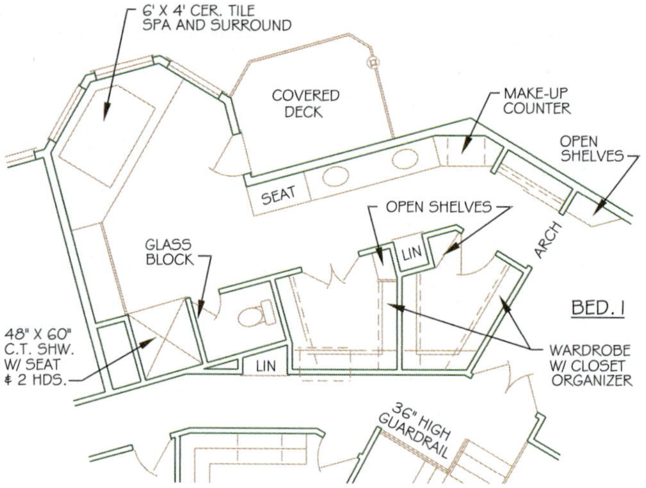

FIGURE 8.27 ■ Master bath suite.

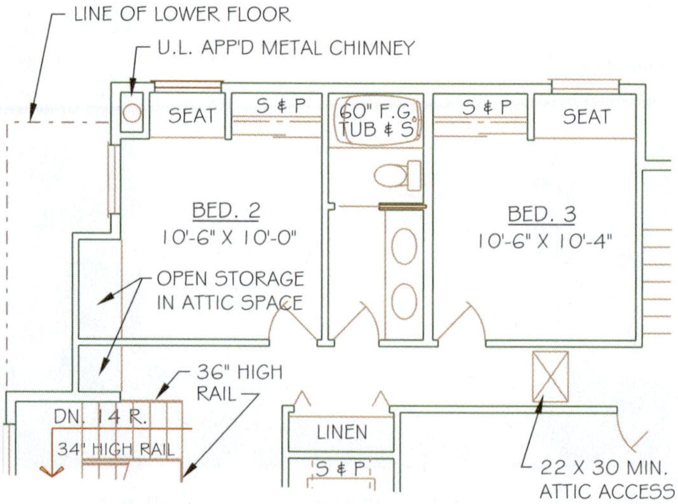

FIGURE 8.28 ■ Many bedrooms share a bathroom. When the lavatories are separated from the bathing area, two people can use the room at once.

FIGURE 8.29 ■ The kitchen is used for meal preparation and is often a place for eating and a social hub at family gatherings and parties. *Courtesy Eric S. Brown Design Group Inc. Photo by Laurence Taylor.*

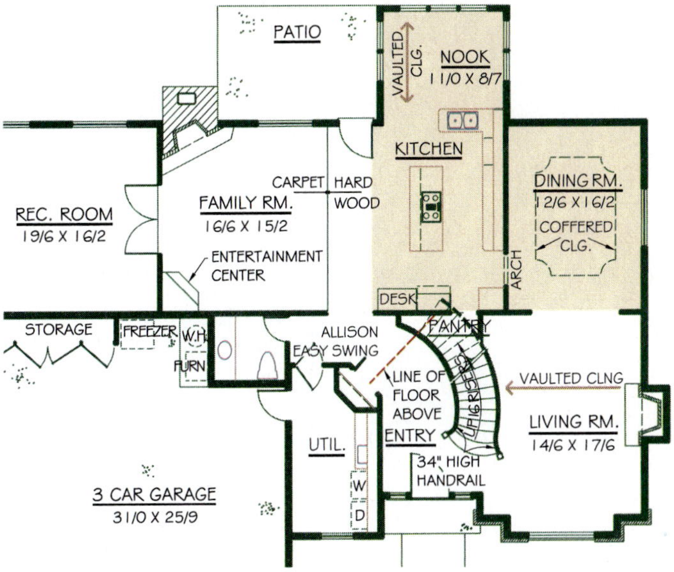

FIGURE 8.30 ■ For efficient living, the kitchen must be near the dining, living, family, and utility areas and the service entry. If traffic must flow through the kitchen, it should not pass through the work triangle. *Courtesy Wally Griener, SunRidge Designs, CPBD.*

not require extra steps. The kitchen should also be near the family room. This will allow those preparing meals to be part of family activities. When possible, avoid placing the kitchen in the southwest corner of the home. This location will receive the greatest amount of natural sunlight, which could easily cause the kitchen to overheat. Since the kitchen creates its own heat, try to place it in a cooler area of the house unless venting and shading precautions are taken. One advantage of a western placement is the natural sunlight available in the late evening.

In a house for a family with young children, a kitchen with a view of indoor and outdoor play areas is a valuable asset. This will allow for the supervision of playtimes and control of traffic in and out of the house.

The kitchen is often closely related to the utility area. Because these are the two major workstations in the home, a kitchen close to the utility room can save valuable time and energy as daily chores are done. It is also helpful to place the kitchen near the garage or carport. This location will allow groceries to be unloaded easily (Figure 8.30).

Kitchen Work Areas

Not only must the kitchen location be given much thought, but the space within the kitchen also demands great consideration. Perhaps the greatest challenge facing the designer is to create a workable layout in a small kitchen. Layout within the kitchen includes the relationship of the appliances to the work areas. The main work areas of a kitchen are those for storage, preparation, and cleaning. Each can be seen in the plan in Figure 8.31.

Storage Area

The storage area consists of the refrigerator, the freezer, and cabinet space for food and utensils. Most families prefer to have a refrigerator in the kitchen, with a separate freezer in the laundry or garage. Storage for cans and dried foods is typically in base cabinets 24″ (600 mm) wide or in a pantry unit, similar to Figure 8.32. Upper units 12″ (300 mm) deep are also typically used for dishes and nonperishable foods. A minimum counter surface 18″. (450 mm) wide should be provided next to the refrigerator to facilitate access to it (see Figure 8.33). This area can also be useful for preparing snacks if the width is increased to approximately 36″ (900 mm). As the size is increased, this counter can also be used to store appliances, such as mixers and blenders in an appliance garage similar to the one shown in Figure 8.34. This is a useful area for a breadboard and a drawer for silverware storage. Additional drawer space could be used for mixing bowls, baking pans, and other cooking utensils. Enclosed cabinet space should also be provided for storage of cookbooks, writing supplies, and telephone books, similar to Figure 8.35.

Preparation Areas

The main components of the preparation areas are the sink, cooking units, and a clear counter workspace. Clear counter workspace should be placed near the storage area, sink, and cooking areas with a minimum of one counter that provides approximately 48″ (1200 mm) of work space. Each work area should have a counter at least 18″ (450 mm) wide. Specific requirements will be introduced as each appliance is discussed. Large kitchens often have a small vegetable sink in the preparation area and a larger sink for cleaning utensils. A minimum size for a vegetable sink is 16″ × 16″ (400 × 400 mm).

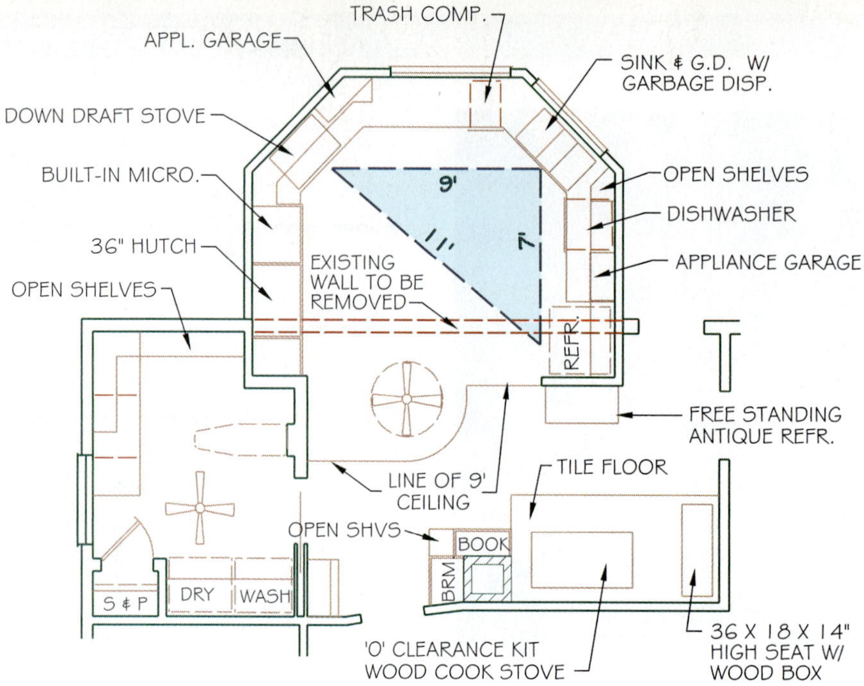

FIGURE 8.31 ■ Plan for a kitchen added to a home built in 1910. The kitchen was designed to provide the convenience of the new millennium but still blend in with a historic home. A well-planned kitchen will provide preparation areas, storage areas, and cleaning areas separated by adequate counter space. These areas are generally not specified on the floor plan except to denote specific appliances required for each area. Related information is specified on the interior elevations (Chapter 24). Figures 8–32 through 8–37 show features of this kitchen.

FIGURE 8.32A ■ A pantry should be provided in or near the kitchen to provide storage for dry and canned goods. This walk-in closet, located between the kitchen and garage, uses shelves, bins, and pullout drawers for long-term storage. *Courtesy Janice Jefferis.*

FIGURE 8.32B ■ Storage cabinet located in the kitchen. The cabinet has pullout shelves and drawers for storage of foods for daily use. Because the cabinet is located just to the left of the baking center, the lower drawers provide convenient storage for large pans and other baking utensils. *Courtesy Janice Jefferis.*

FIGURE 8.33 ■ The refrigerator is used for the short-term cold storage of food. A counter 18″ (450 mm) wide should be provided next to the refrigerator. *Courtesy Janice Jefferis.*

FIGURE 8.34 ■ An appliance garage can be placed between the countertop and the upper cabinet to store small appliances that are used on a regular basis. *Courtesy Janice Jefferis.*

The cooking appliances are usually a stove that includes both an oven and surface heating units or separate appliances for baking and cooking. Most kitchens also provide space for a microwave oven that can be part of a built-in oven, be mounted below the upper cabinets, or be placed on the counter. A minimum counter space of 18″ (450 mm) should be placed on each side of a stove or oven unit to prevent burns and provide temporary storage while cooking. Figure 8.36 shows a cooking area.

Cleaning Center

The cleaning center typically includes the sink, garbage disposal, and dishwasher, as in the arrangement shown in Figure

FIGURE 8.35 ■ Cookbooks can be stored and antique kitchen tins can be displayed in this storage cabinet. *Courtesy Janice Jefferis.*

FIGURE 8.36 ■ The cooking center of this home includes a downdraft stove, a microwave, and preparation areas on each side of the stove. Storage includes open upper cabinets for spices, hanging storage for common utensils, and enclosed upper cabinets for cooking supplies. The counter provides space for storage of baking supplies. Lower cabinets feature breadboards on each side of the stove, and drawers and doors for storage. The drawers on the right side of the stove have false fronts for the display of dry food items such as lentils, beans and corn. *Courtesy Janice Jefferis.*

8.37. The sink and the surrounding cabinets and counter space are the most heavily used work area of the kitchen, so this area should be centrally located.

Most clients prefer a double sink rather than a single sink. A typical double sink is 32″ × 21″ (800 × 530 mm), but 36″ (900 mm) and 42″ (1070 mm) wide units are also available. Double sinks are also available, at 90 degrees to each other for use in a corner. If the sink is connected to a public sewer, a

FIGURE 8.37 ■ The kitchen sink is the hub of the cleaning center. Located near windows providing a view of the outside entertaining areas, the sink is flanked by a trash compactor on the left and a dishwasher on the right. Storage for dishes is provided above the dishwasher. The drawers provide storage for eating utensils and other basic kitchen supplies. *Courtesy Janice Jefferis.*

garbage disposal is typically installed to one side of the sink to eliminate wet garbage. A garbage disposal should not be used on sinks connected to septic tanks. A minimum area 36″ (900 mm) wide should be provided on one side of the sink for stacking dirty dishes, and about 24″–30″ (600–700 mm) on the other side for clean dishes. The dishwasher can be placed on either side of the sink, depending on the client's wishes. An upper cabinet should be provided near the dishwasher to store dishes.

The Work Triangle

The mere fact that a kitchen is big and has all the latest appliances does not make it efficient. Designing a functional kitchen requires careful consideration of the relationship between the work areas. This relationship, referred to as the *work triangle,* is a key aspect of kitchen ergonomics. The work triangle is formed by drawing lines from the centers of the storage, preparation, and cleaning areas. This triangle outlines the main traffic area required to prepare a meal. Food will be taken from the refrigerator, cleaned at the sink, and cooked at the microwave or stove, and then leftovers will be returned to the refrigerator. Using this work pattern and considering the work triangle, good design placed the work centers at approximately equally spaced points of the triangle connected by counters. General rules for efficient kitchen design include:

- Always place work space between each workstation of the triangle.
- No side of the work triangle should be less than 4′ (1200 mm) or greater than 7′ (2100 mm).
- The sum of the sides of the work triangle should be at least 15′ (4500 mm) but not more than 22′ (6700 mm).

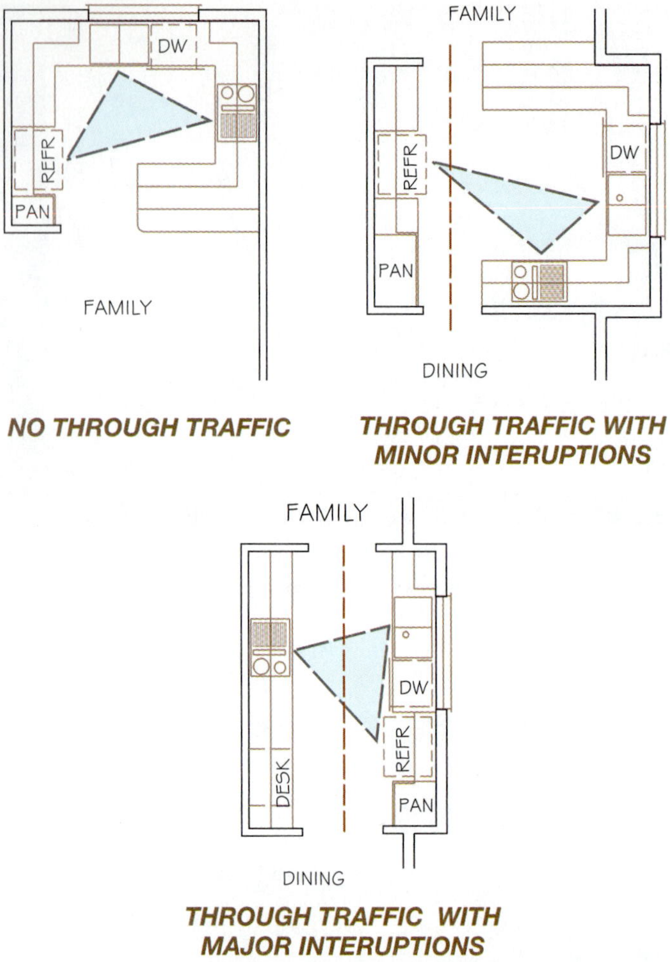

FIGURE 8.38 ■ Traffic flow should never disrupt the working area of a kitchen; plan *a* accomplishes this. Plan *b* allows traffic through the kitchen, but it will not disrupt work. Plan *C* shows traffic that will be very disruptive.

- Never arrange rooms so that traffic is required to pass through the work triangle.

Keep in mind that these are only guidelines, and that a large kitchen consists of more than just three appliances or workstations. The triangle may need to be modified to accommodate doors or to preserve a view. Appliances such as convection ovens, grills, microwave ovens, multiple sinks, and even multiple refrigerators will all affect the arrangement of work areas and appliances.

In addition to traffic within the kitchen, the relationship of the kitchen to other rooms also needs to be considered. If care is not taken in the design process the kitchen can become a hallway. The kitchen needs to be in a central location, but traffic must flow around the kitchen, not through it. Figure 8.38 shows the traffic pattern in three different kitchens.

Common Counter Arrangements

Designers typically use one of six common counter arrangements to define the work triangle: straight (one wall), L-shape,

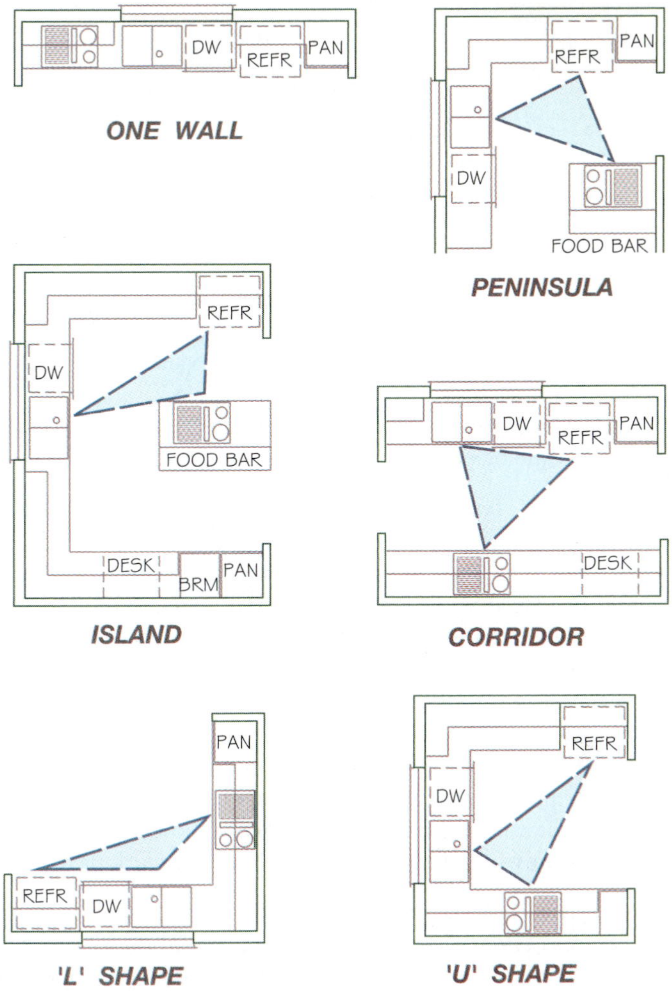

FIGURE 8.39 ■ The sum of the three sides of the work triangle must be between 15′–22′ (4600–6700 mm) for efficient work pattern.

corridor, U-shape, peninsula, and island. See Figure 8.39. Each arrangement presents its own design challenges.

Straight The straight-line or single-wall kitchen layout is ideal when appliances must be placed in a very small space. This type of layout is often used in a small apartment, a mother-in-law unit, or a recreation room that is on a different floor from the main kitchen. While this arrangement requires few steps to move between appliances, it also provides a very limited amount of cabinet space and work areas. This arrangement can be improved, however, by using a movable cart or cutting block to expand the work triangle.

Corridor or Galley The corridor or galley arrangement places all of the cabinets on two parallel walls. This arrangement is a great improvement over a straight-line kitchen because it allows convenient storage, provides ample counter space in a small area, and minimizes walking distance in the triangle. The distance between the cabinet faces should be a minimum of 48″ (1200 mm). A width of 54″–64″ (1350–1600 mm) will allow two or more cooks to use the kitchen simultaneously. This arrangement is appropriate for a small home but should not be used in larger homes because of the possibility of poor traffic flow. A galley kitchen is usually placed between two living areas. Unless an alternative route is provided, the galley kitchen will become a thoroughfare for room-to-room traffic.

L-Shape An L-shape cabinet arrangement is suitable for both small and large kitchens, but it loses efficiency as the size of the kitchen increases. Cabinets are placed on two adjacent walls with two workstations placed along one wall and the third workstation placed on the remaining wall. The L enables efficient travel between workstations that are located close to the right angle, but the arrangement loses its effectiveness as the legs of the L become longer. This type of cabinet arrangement eliminates traffic through the work area and is well suited to great rooms.

U-Shape The U-shape cabinet arrangement provides an efficient layout with easy access between workstations. It also eliminates through traffic unless a door is placed in one of the walls. Many designers consider the U shape ideal for a large kitchen, but it can result in hard-to-reach storage areas in the corners. The U arrangement may also seem confining in a small kitchen. A minimum space of 60″ (1500 mm) should be used between cabinet faces. As the distance increases, an island or peninsula can be added to improve the cabinet arrangement.

Peninsula A peninsula arrangement provides ample cabinet and workspace by adding one additional leg to an L-shape or U-shape kitchen. The peninsula can be used as work space, as the location for a work center, as a food bar, or as a combination of these features.

Island An island can be added to any of the other five cabinet arrangements. The island can be used to provide added counter space or work space. A range or cooktop, or a small sink, is often placed in the island. A minimum of 42″ (1050 mm) should be provided between the island and other counters to allow for traffic flow around the island. If the island contains an appliance such as a range, 48″ (1200 mm) should be the minimum distance provided between counters, to allow for traffic flow when the oven door is open.

Counter and Cabinet Sizes The standard kitchen counter is 36″ (900 mm) high. For custom homes, the height may be adjusted to meet the physical needs of the owners, but it is important to remember that radical changes from the norm may greatly limit resale value. Common cabinet dimensions listed in table on page 144.

If a food bar is provided in a peninsula or island, consideration must be given to the height and width of the eating counter. Common heights include 30″, 36″, and 42″ (750, 900, and 1050 mm). Widths range from 12″–18″ (300–450 mm). Figure 8.40 shows an example of each type of food bar located in a peninsula. If a counter 30″ (750 mm) high is provided, chairs can be used with it. The other heights will require stools for seating. When the food bar is the same height as the other counters, the counter area will seem much larger than when two different heights are used. A width of 12″ (300 mm) will provide sufficient space for eating. If the food bar is higher than the other kitchen cabinets, it can be used as a visual shield to hide clutter in the kitchen.

CABINET DIMENSIONS		
	OVERHEAD CABINETS	BASE CABINETS
Standard depth	(12″ (300 mm)	24″ (600 mm)
Maximum depth	18″ (450 mm)	36″ (900 mm)
Height	30″–33″ (750–825 mm) when placed above countertop 12″–18″ (300–450 mm) when placed above an appliance	36″ (900 mm) floor to countertop
Width	9″–48″ (225–1200 mm). Stock cabinets are available in 3″ (75 mm) increments	9″–48″ (225–1200 mm). Stock cabinets are available in 3″. (75 mm) increments.
Location	15″ (375 mm) above counter 24″ (600 mm) above sink 30″ (750 mm) above range	

When the height of the counter is different from the food bar, a width of 15″–18″ (375–450 mm) should be provided.

Arrangement of Appliances

Many homeowners prefer the sink to be located in front of a window, but a location with a view into other living areas is also popular. Placement of the sink at a window allows for supervision of outdoor activities and provides a source of light at this workstation. Because of its placement at a window, the sink is often located first as the kitchen is being planned. The sink should also be placed to promote easy movement between work areas. The sink should be near the cooking units and the refrigerator to facilitate the preparation of fresh foods. Provide a minimum of 48″ (1200 mm) between counters to accommodate someone working at the sink and another person walking behind him or her. Try to avoid placing the sink and dishwasher on different counters, even if they would be just around the corner from each other. Such a layout often leads to accidents from water dripping onto the floor. If the sink and dishwasher are located at right angles, plan the dishwasher location so that when the door is open it will not interfere with the sink or the work space of any other appliances. Provide a loading space with a minimum width of

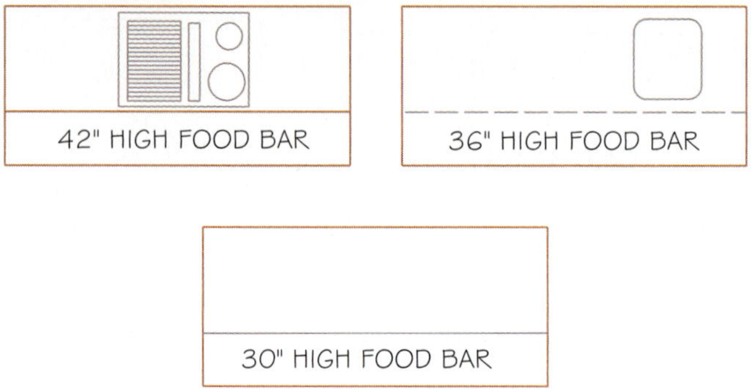

FIGURE 8.40 ■ Common arrangements of food bars.

20″ (500 mm) between the sink and the dishwasher. An upper cabinet should be provided near the dishwasher to store dishes.

Refrigerator Typically a space 36″ (900 mm) wide is provided for the refrigerator. The refrigerator should be placed near the service entry to ease unloading of groceries. The refrigerator should be placed at the end of a counter, so that it will not divide the counter into small work spaces. If it must be placed near a cabinet corner, allow a minimum of 12″ (300 mm) in the corner for access to the back of the counter. The location must also be convenient for access to the sink for food preparation and to the cooking areas for food storage. The refrigerator should be placed within 60″–72″ (1500–1800 mm) of both the sink and the cooking center. If possible, it should be placed so that the door will not block traffic through the work triangle.

Stove The cooking units should be located near the daily eating area. The stove should be placed so that the person using it will not be standing in the path of traffic flowing through the kitchen. This will reduce the chance of hot utensils being knocked from the stove. Stoves should be placed so that there is approximately 18″ (457 mm) of counter space between the stove and the end of the counter to prevent burns to people passing by. Stoves should not be placed within 18 in. (457 mm) of an interior cabinet corner. This precaution will allow the oven door to be open and still allow access to the interior cabinet. A walkway 48″ (1200 mm) wide should be provided to allow someone to walk behind a person working at the stove.

Cooktop A cooktop and oven should be within three or four steps of each other if separate units are to be used. Cooktops are available in units 30, 36, 42, and 48″ (760, 900, 1070, and 1200 mm) wide. Oven units are typically 27, 30, or 33″ (680, 760, or 840 mm) wide. The bottom of the oven is typically 30″ (760 mm) above the floor. Avoid placing a stove next to a refrigerator, a trash compactor, or a storage area for produce or breads.

Microwave A microwave unit may be placed above the oven unit, but the height of each unit above the floor should be considered in regard to the height of the clients. A microwave mounted approximately 48″ (1200 mm) above the floor and near the daily eating area often proves to be very practical.

Breadboards In addition to the major kitchen appliances, care must be given to other details such as breadboards, counter work spaces, and specialized storage areas. The breadboard should be

FIGURE 8.41 ■ A wine cellar can be located in an insulated room near the kitchen or family room, or in a cool area of the basement. *Courtesy of Krista Herbel.*

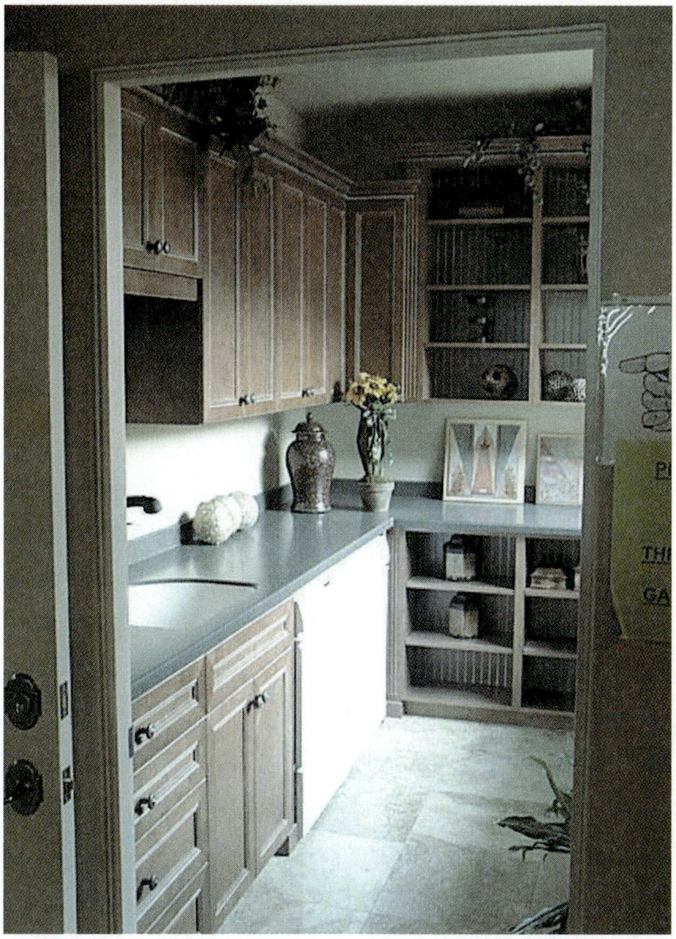

FIGURE 8.42 ■ A small utility room near the kitchen can be used for additional kitchen storage as well as for washing, drying, and folding clothes. *Courtesy of Tereasa Jefferis.*

placed near the sink and the stove but not in a corner. Ideally a minimum of 5″ (1500 mm) of counter space can be placed between appliances to allow for food preparation. Specialized storage often needs to be considered to meet clients' needs. These needs will be covered as cabinets are explained in Section VII.

Wine Cellar

For most kitchens a cabinet or an under-cabinet rack is provided for wine. In custom homes, a special room may be requested for storing wine. Although this is referred to as a wine cellar, it could be as simple as a small insulated closet near the kitchen or family room. Figure 8.41 shows a small wine cellar that also provides a quiet sitting area.

Utility Room

The room that is called a utility room or mudroom should be planned to include space for washing, drying, folding, mending, ironing, and storing clothes. Additional space is often used for long-term storage of dry and canned food, as well as for a freezer. Figure 8.42 shows a small utility room. In a well-planned clothing center, a means of hanging wet clothing should be provided in addition to the standard clothes dryer. A counter for folding and storing clean clothes as well as a closet for hanging clothes as they come out of the dryer should also be included. A built-in ironing board should also be provided. Figure 8.43 shows common items found in a laundry room.

Utility rooms are often placed near either the bedrooms or the kitchen area. There are advantages to both locations. Placing the utility room near the bath and sleeping areas puts the washer and dryer near the primary source of laundry. Care must be taken to insulate the sleeping area from the noise of the washer and dryer.

Placing the utilities near the kitchen allows for a much better traffic flow between the two major work areas of the home. With the utility room near the kitchen, space can often be provided in the utility room for additional kitchen storage. See Figure 8.9. In smaller homes, the laundry facilities may be enclosed in a closet near the kitchen. In warm climates, laundry equipment may be placed in the garage or carport. In homes with a basement, laundry facilities are often placed near the water heater source. Many home owners find that laundry rooms in a basement are too separated from the living areas. If bedrooms are on the upper floor,

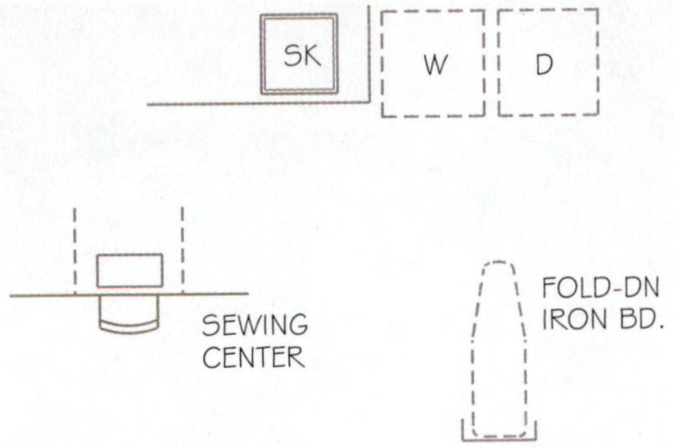

FIGURE 8.43 ■ Common utility room appliances.

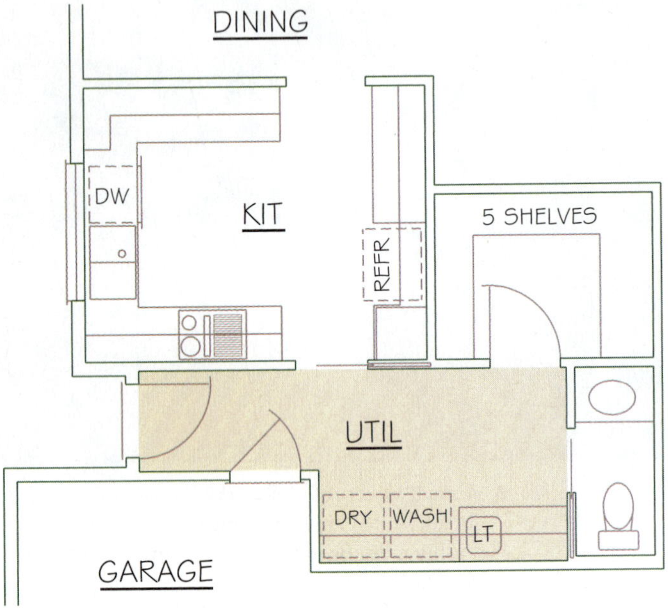

FIGURE 8.44 ■ The utility room may serve as a service entry where dirty clothes can be removed as the home is entered.

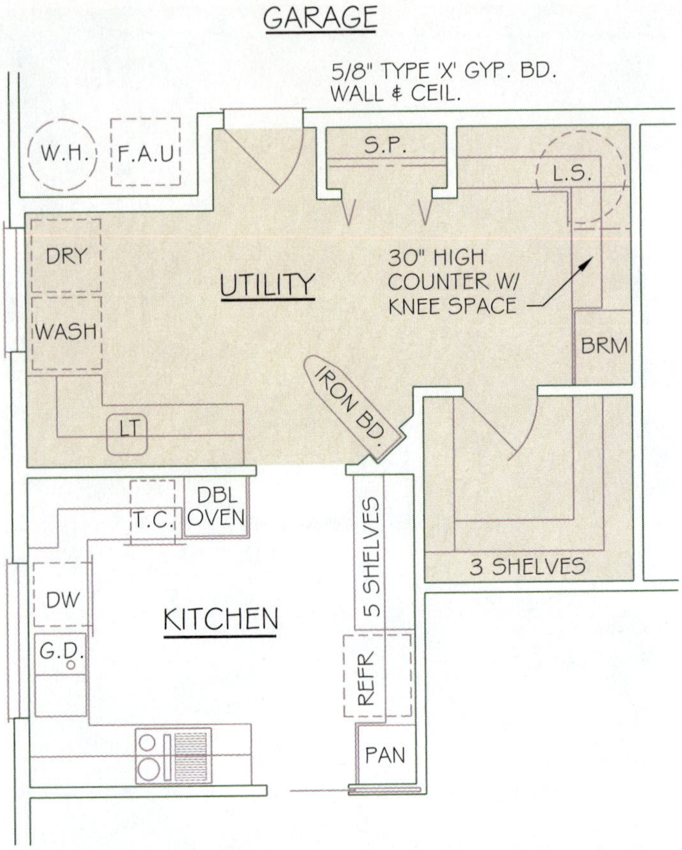

FIGURE 8.45 ■ This utility room provides a spacious layout for chores.

a laundry chute to the utility room can be a nice convenience. The utility room often has a door leading to the exterior. This allows the utility room to function as a mudroom. Entry can be made from the outside directly into the mudroom, where dirty clothes can be removed. This allows for cleanup near the service entry and helps keep the rest of the house clean. Figure 8.44 shows this type of utility layout. Another common use for a utility room is to provide an area for sewing and ironing (Figure 8.45).

Garage or Carport

Believe it or not, some people actually park their car in the garage. For those who do not, the garage often becomes a storage area, a second family room, or a place for the water heater and furnace. The minimum space for a single car is 11′ × 20′ (3400 × 6100 mm). A space for two cars should be 21′ × 21′ (6400 × 6400 mm). These sizes will allow minimal room to open doors and to walk around the car. Additional space should also be included for a workbench. A garage 22′ × 24′ provides for parking, storage, and room to walk around. Many custom homes include enclosed parking for three cars. Consideration is often given for parking a boat, truck, camper, or recreational vehicle.

Space should be planned for the storage of a lawn mower and other yard equipment. Additional space must be provided if a post is located near the center of the garage to support an upper floor. It will also be helpful if a required post can be located off center from front to back so that it will not interfere with the opening of car doors.

The garage location is often dictated by the site. Although access and size are important, the garage should be designed to blend with the residence. Many homeowners prefer that the garage doors not be seen from the street as the home is being approached. Clients usually have a preference for either one large single door or smaller doors for each parking space. A single door 8′ wide × 7′ high (2400 × 2100 mm) is common. A door 9′ (2700 mm) wide is the smallest that should be used for a truck with wide side mirrors, or for most boat trailers.

A double door is typically 16′ (4800 mm) wide. Aside from the owner's preference, posts required to support upper floors may influence the type of door to be used. Typically, the space between doors will be based on the amount of space needed inside the garage to allow car doors to open without hitting anything. When multiple single doors are to be used, consideration should be given to lateral bracing for the front of the garage. Al-

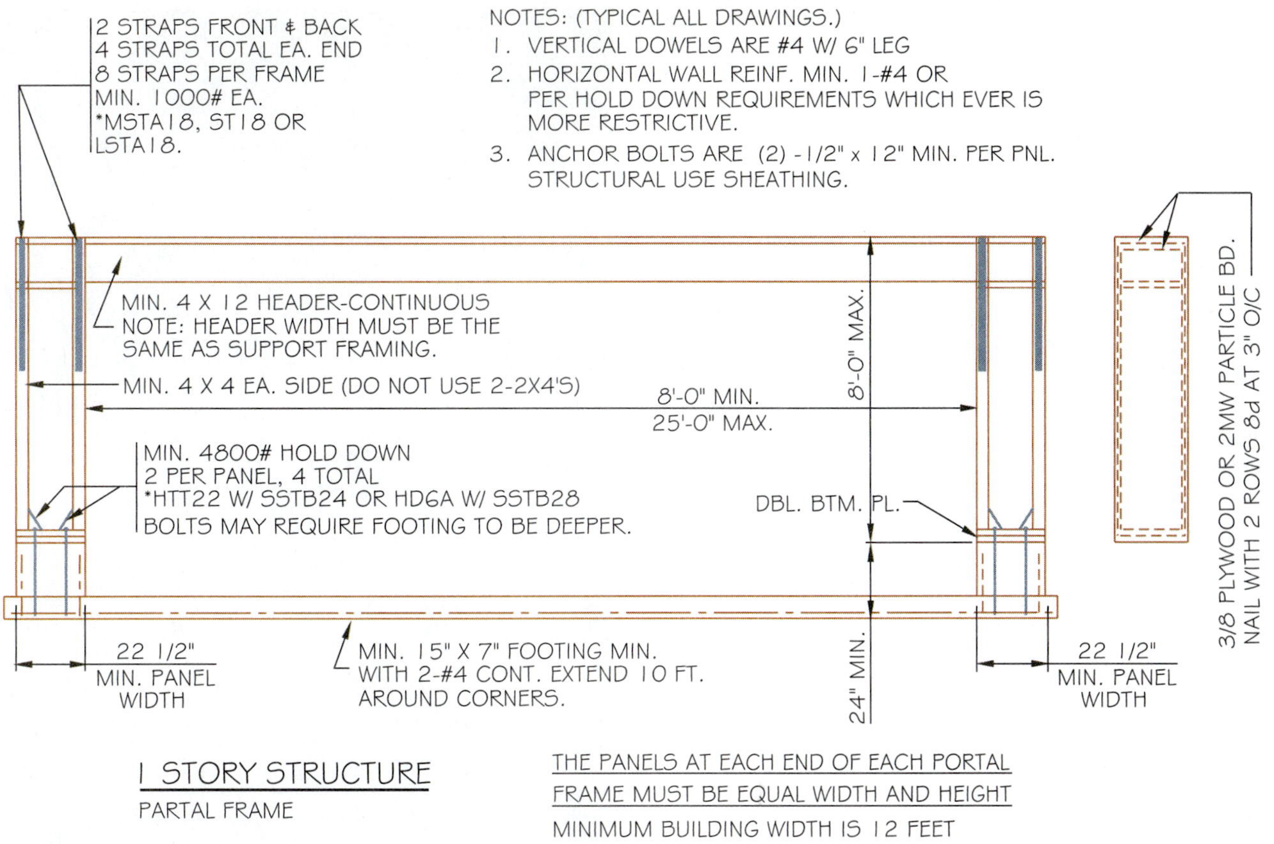

FIGURE 8.46 ■ Minimum opening requirements for a portal frame.

though a good engineer can make any structure stand up, keeping the space between doors to a minimum and increasing the size of the walls at each outer edge of the garage can minimize the cost of keeping the front wall rigid. Figure 8.46 shows the minimum sizes for a lateral bracing method, referred to as a portal frame. Chapter 30 will explore lateral bracing in more detail. As a general guideline, do not let structural considerations dictate the initial design. A door 32 in. (800 mm) wide should be installed to provide access to the yard.

In areas where cars do not need to be protected from the weather, a carport can be an inexpensive alternative to a garage. Provide lockable storage space on one side of the carport if possible. This can usually be provided between the supports at the exterior end.

TRAFFIC PATTERNS

Interior Traffic

A key aspect of any design is the traffic flow between areas of the residence. *Traffic flow* is the route that people follow as they move from one area to another. Often this means hallways, but areas of a room are also used to aid circulation. Rooms should be of sufficient size to allow circulation through the room without disturbing the use of the room. By careful arrangements of doors, traffic can flow around furniture rather than through conversation or work areas.

Codes that cover the width of hallways were covered in Chapter 7. Good design practice will provide a width of 36″ to 48″ (900–1200 mm) for circulation pathways. A space of 48″ (1200 mm) is appropriate for pathways that are used frequently, such as those that connect main living areas. Pathways used less frequently can be smaller. In addition to hallways, the entries of a residence are important in controlling traffic flow.

The foyer is typically the first access to the residence for guests. In cold climates, many homes have an enclosure that leads to the foyer, to prevent heat loss. The foyer should provide access to the living areas, the sleeping area, and the service areas of the house. The size of the foyer will vary. Provide ample room to open a door completely and allow it to swing out of the way of people entering the home. A closet should be provided near the front door for storage of outdoor coats and sweaters. Provide access to the closet that will not be blocked by opening the front door.

In addition to considering traffic flow between rooms, it is important to determine how the levels of a home will relate to each other. The ideal location for stairs is off a central hallway. Many floor plans locate the stairs in or near the main entry to provide an attractive focal point in the entry, as well as convenient access to upper areas. Stairs should be located for convenient flow from each floor level and should not require access through another room. The location will also be influenced by how often the stairs will be used. Stairs serving mechanical

equipment in a basement will be used less frequently than stairs connecting the family room to other living areas.

Traffic Flow Between Interior and Exterior

Just as important as how the major interior areas relate to each other is the traffic flow between each of these areas and exterior areas. A well-planned home will expand the inside living areas to outside areas such as courtyards, patios, decks, balconies, and porches, as shown in Figure 8.47. Sunrooms and solariums are interior areas that combine interior and exterior features. Exterior structures that enhance outdoor living include gazebos and arbors. Each should be planned so that it blends with the overall design of the home.

A sunroom or solarium is a common way to bring the outside into the interior living areas. A sunroom is a multipurpose room, similar to the one shown in Figure 8.48, that can be used as a reading room, a breakfast room, or a music room, or as extra space for entertaining while providing transition between the interior and exterior spaces. A solarium is typically associated with passive solar heating, but the room also provides an excellent location for casual living and may include a lap pool, a spa, or an area for indoor gardening. Each of these uses will require that ventilation, sun orientation, and views are carefully planned. Chapter 10 will introduce materials to help in planning a sunroom or solarium. Figure 8.49 shows a solarium used to provide added room for entertainment and to enclose a spa.

Most of the terms describing exterior areas are strictly defined and regulated by building codes, but the use of these terms by the general public will vary slightly. IRC defines a *court* as an exterior space that is at grade level, is enclosed on three or more sides by walls or a building, and is open and unobstructed to the sky. Figure 8.50 shows a home with an entry court. The courtyard provides an excellent screen, allowing privacy for outdoor living in the front yard and a secure play area for young children. Courtyards can be used as an extension of the living, dining, and family rooms by providing a patio or deck and landscaped areas.

A *patio* is a ground-level exterior entertaining area made of concrete, stone, brick, or treated wood. The patio in Figure 8.51*a* provides good access to the interior, ample areas of sunny and shaded seating, and room for pool activities. The patio in Figure 8.51*b* is designed to provide an elegant extension of the living areas to the outside.

Balconies and decks are elevated floors. Building codes define a *deck* as an exterior floor supported on at least two opposing sides by adjoining structures, posts, or piers. Figure 8.52 shows a covered deck supported by columns on the upper floor of a home. A *balcony* is an above-ground deck that projects from a wall or building with no additional supports. Figure 8.53 shows a home with two upper balconies. The balcony on the left cantilevers, allowing access to the garage. The enclosed balcony on the right cantilevers over the entry to provide a protected area for outside activities.

Although a *porch* is not defined by building codes, a common perception is that it is an enclosed patio or deck, as in Figure 8.54*a*. In some areas of the country, the term *enclosed* would refer to walls, windows and doors, and a roofed area. In warmer

FIGURE 8.47 ■ A well-planned home will expand the interior living areas into outside areas such as courtyards, decks, and balconies. *Courtesy California Redwood Association. Tobey Zinkhan, Architects; photo by Barbeau Engh.*

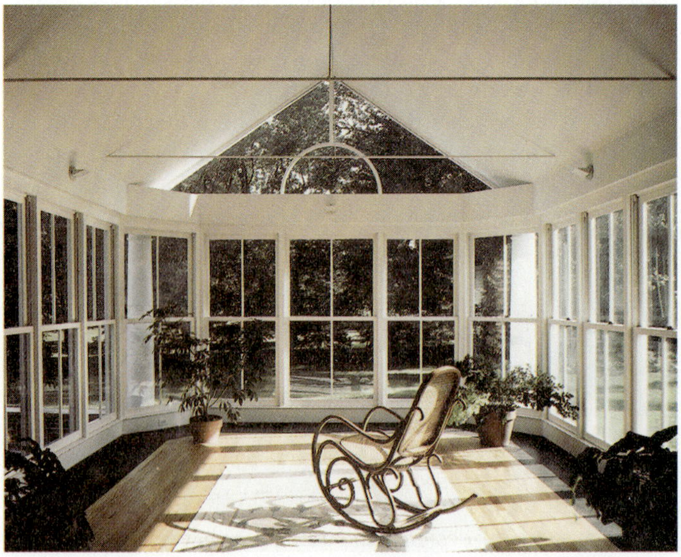

FIGURE 8.48 ■ A sunroom is a multipurpose room that can be used as a reading room, a breakfast room, or a music room, or as extra space for entertaining while providing transition between interior and exterior spaces. *Courtesy Marvin Windows and Doors.*

FIGURE 8.49 ■ A solarium can provide an excellent location for casual living and may include a lap pool, a spa, or an area for indoor gardening. *Courtesy Sunbuilt Solar Products.*

FIGURE 8.50 ■ A courtyard provides an excellent screen, giving privacy and allowing the living, dining, and family rooms to be extended to a patio or deck and landscaped areas. *Courtesy Janice Jefferis.*

FIGURE 8.51A ■ A *patio* is a ground-level exterior entertaining area made of concrete, stone, brick, or treated wood. This patio provides good access to the interior of the home, ample areas of sunny and shaded seating, and room for pool activities. *Courtesy Georgia Pacific Corp.*

FIGURE 8.51B ■ A patio designed to provide an elegant extension of the living areas to the outside. *Courtesy Eric S. Brown Design Group Inc. Photo by Laurence Taylor.*

areas of the country, *enclosed* may include only a roof or awning-type covering, as in Figure 8.54*b*, to protect the areas from sunlight, insects, or other natural elements. The usefulness of an outdoor area may be greatly increased if it is enclosed by screens, plastic, or glass panels. Gazebos and arbors are free-standing structures that provide protection from the elements. Common uses include protected outdoor eating areas, covered pools or spas (as in Figure 8.55*a*), and shaded garden living.

Most families will require one or more outdoor entertaining areas. To determine which areas are to be included in a home, needs, function, location, orientation, climate, and size must be considered. Key questions to be considered with the client include: When will the area be used? Who will use it? What activities are planned for the area? Will traffic flow through the area? Areas that may need to be provided include a play area for

FIGURE 8.52 ■ What many would consider the balcony or veranda of this historic southern home is considered a *deck* by building codes. IRC defines a deck as an exterior floor that is supported on at least two opposing sides by adjoining structures, posts, or piers. *Courtesy Eric S. Brown Design Group Inc. Photo by Oscar Thompson.*

FIGURE 8.53 ■ Building codes define a *balcony* as an above-ground deck that projects from a wall or building with no additional supports. *Courtesy California Redwood Association. Architect Charles J. Trevisan; photo by William Helsel.*

children; seating for dining, quiet activities, or lounging in the sun; and space for outdoor games such as Ping Pong, shuffleboard, and tennis. A patio or deck that will be used for outdoor eating will have different size requirements if a pool or spa is to be included. A deck will seem smaller than a patio of the same size because of the addition of a rail. Any deck, patio, or balcony is required to have a guardrail when the finished floor height is 30″ (750 mm) or more above the finish grade or floor below. A balcony as small as 36″ (900 mm) wide can be used to place a lawn chair in a sunny spot for one person. A minimum balcony width of 48″ (1200 mm) should be provided for two or more small chairs, to allow access. The minimum space for patios and

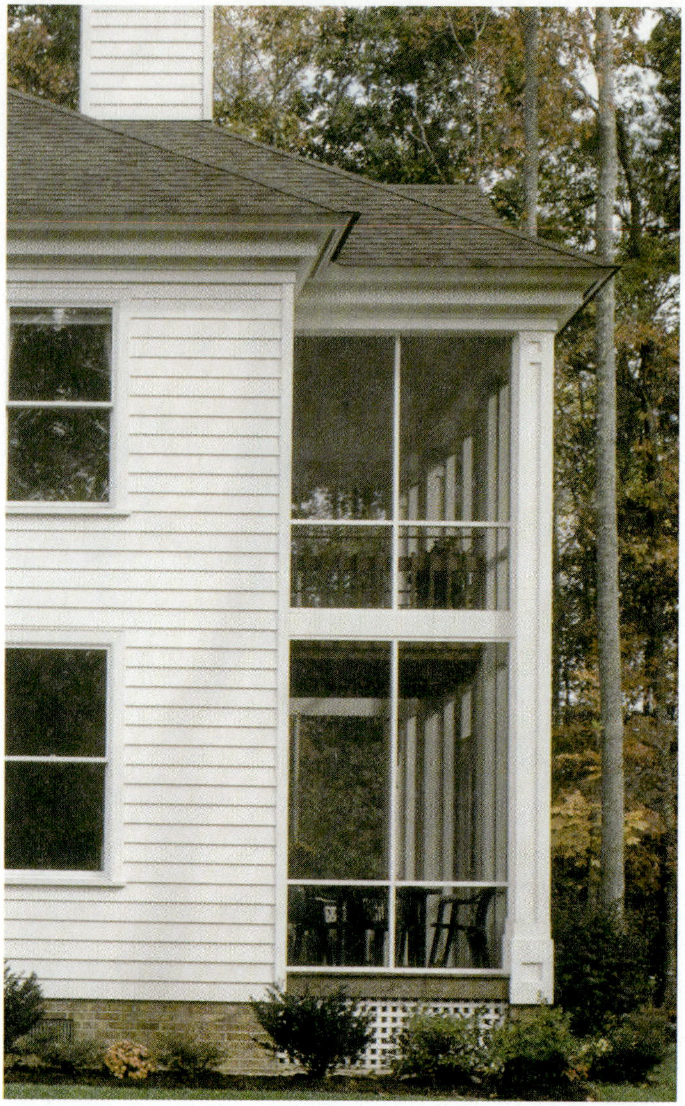

FIGURE 8.54A ■ An enclosed porch can provide outdoor living protected from insects and weather. *Courtesy California Redwood Association.*

FIGURE 8.54B ■ A covered porch provides an excellent area for outdoor activities. *Courtesy Elk Roofing.*

FIGURE 8.55A ■ Freestanding structures can be used as a wind block and to provide shade for outdoor activities. *Courtesy California Redwood Association, Designer/Builder Bryan Hays; photo by Ernest Braun.*

FIGURE 8.55B ■ Covered patios and creative landscaping can be used to beautify outdoor living areas. *Courtesy Eric S. Brown Design Group Inc. Photo by Oscar Thompson.*

decks that will include a seating and traffic area is 10′ × 14′ (3050 × 4200 mm).

Once the function is determined, the location of the deck or patio must be considered. A patio or deck located near main living areas will need to incorporate traffic flow in addition to the space required for its own functions. Exterior areas located near a bedroom or bathroom suite should have some type of screen or landscaping to provide privacy. Figure 8.56 shows how landscaping can be incorporated into a hillside site to provide a sense of privacy and beauty.

A designer must consider the orientation of the outdoor area and the local climate. A balcony or deck may be placed on the west side of a home in a northern region to take advantage of the sun. In a southern region, designing for shade will be of great importance. Just as important as the movement of the sun will be prevailing wind directions and the view. Chapter 11 will discuss methods of designing homes to enhance the relationship of the home to terrain, view, sun, wind, and sound.

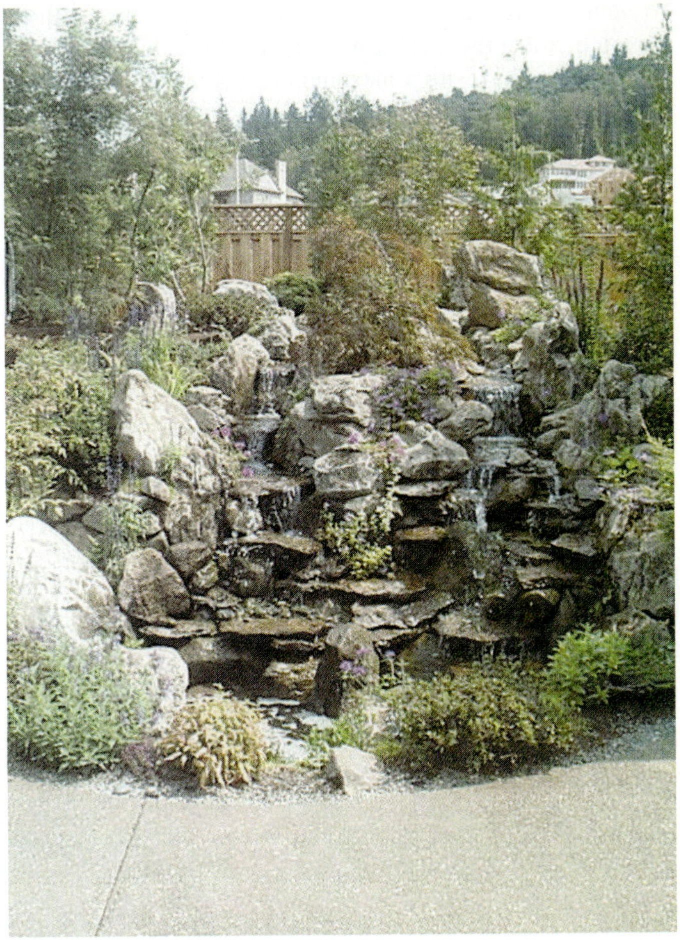

FIGURE 8.56 ■ Landscaping can be an effective way to enhance outdoor living. *Courtesy Janice Jefferis.*

PROVIDING UNIVERSAL ACCESSIBILITY

Well-planned, attractive, accessible housing is essential for millions of people with disabilities—those with temporary injuries who walk with the aid of crutches or a walker or those who depend on a wheelchair for mobility. A house can be made accessible with just a few minor design changes based on the Americans with Disabilities Act (ADA), 1997, and the guidelines of the International Code Council.

Exterior Access

An accessible home has a level site for parking with paved walkways from parking areas to the main entry. A parking space 9′ (2700 mm) clear should be provided with a minimum walkway of 42″ (1070 mm) between cars. A minimum ceiling height of

9′ (2896 mm) is required for raised-top vans. Any changes in elevation between the parking area and the front door can be no more than a 1/12, with a maximum rise of 30″ (760 mm) per run. Inclined ramps should be covered with a nonskid surface to provide sure footing for those walking and to help keep a wheelchair from going out of control. Ramps longer than 30″ (9100 mm) or higher than 30″ (760 mm) should have a landing 48″ (1200 mm) wide at the midpoint. Provide handrails that are 29″ to 32″ (740–810 mm) high. Guidelines for ramp access are shown in Figure 8.57.

Interior Access

Beyond the scope of the codes presented in Chapter 7, a commonsense approach to design can greatly aid the quality of a home. Doors with a 32″ (813 mm) clear opening should be provided throughout the residence to ease movement in a wheelchair or with a walker or crutches.

Hinges are available that swing the door out of the doorway for maximum clearance. Hallways with a width of 42″ (1070 mm) will provide convenient traffic flow, but 48″ (1200 mm) will make turning from the hall to doorways much easier. Doors with a lever action handle or latches are typically easier to open than round doorknobs. Windows should be within 24″ (600 mm) of the floor and should not require more than 8 lb of pressure to open.

An area 60″ (1500 mm) square will allow for a 360° turn in a wheelchair. An area of this size should be provided in each room. Floors covered with hardwood or tile provide excellent traction, although a low-pile carpet is also suitable. Avoid using small area rugs that move easily. Figure 8.58 shows minimum turning alternatives.

Work Areas

To provide a suitable kitchen work area, countertops should be 30–32″ (760–810 mm), high with pull-out cutting boards. Stoves should have all controls on a front panel rather than top-mounted. A wall oven should be mounted between 30 and 42″ (762–1067 mm) above the floor. A side-by-side frost-free refrigerator with door-mounted water and ice dispensers provides easy access.

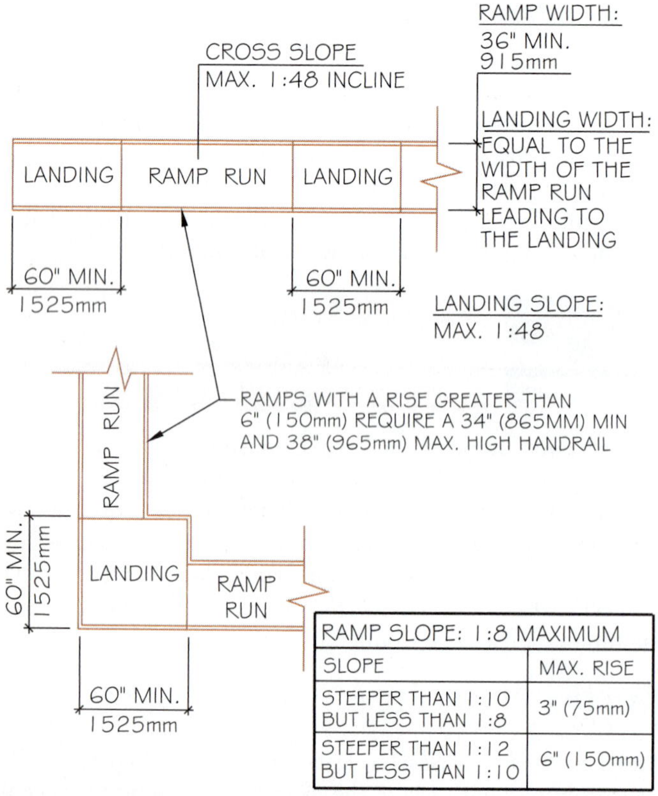

RAMP SLOPE: 1:8 MAXIMUM	
SLOPE	MAX. RISE
STEEPER THAN 1:10 BUT LESS THAN 1:8	3" (75mm)
STEEPER THAN 1:12 BUT LESS THAN 1:10	6" (150mm)

FIGURE 8.57 ■ Guidelines for designing ramp access, based on ICC/ANSI A117.1-1998 standards.

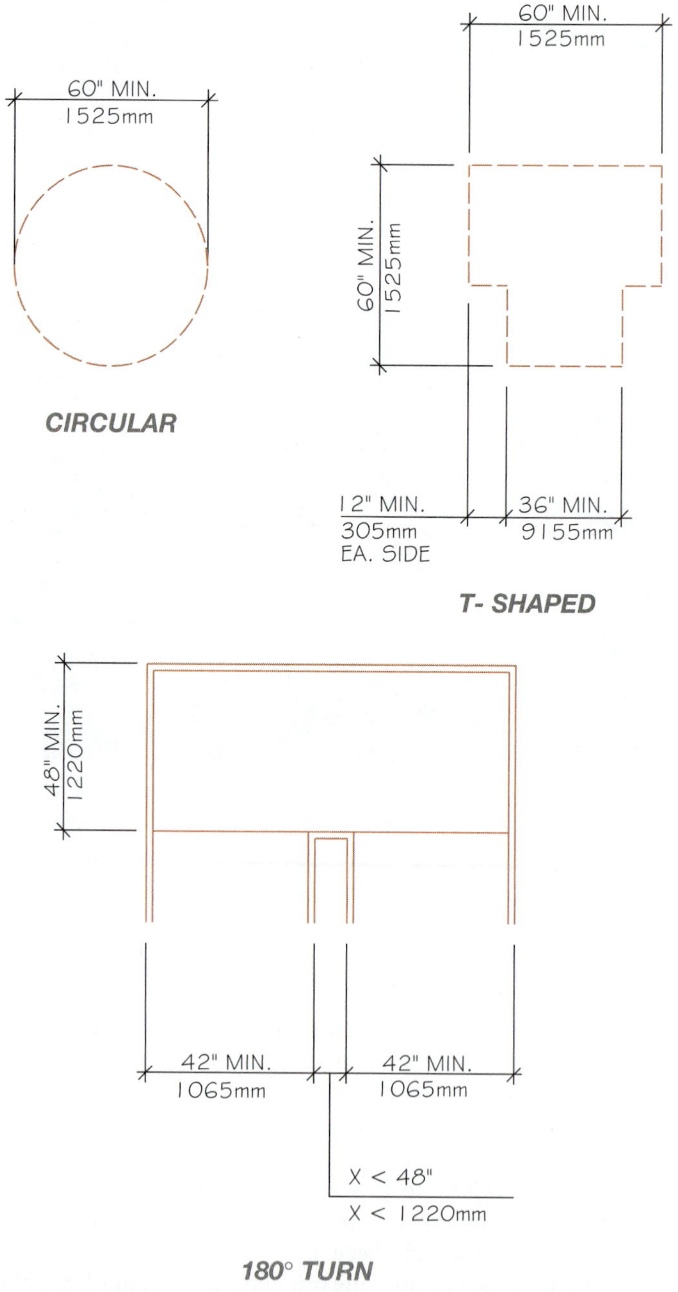

FIGURE 8.58 ■ Guidelines for turning radii, based on ICC/ANSI A117.1-1998 standards.

Bathrooms

Doors that swing out will provide the most usable space in a small room such as a bathroom. Vanities should have a roll-under countertop with a lowered or tilted mirror. If possible, reinforce areas of each wall where a grab bar will be used with 3/4" (20-mm) plywood backing. Grab bars should be provided next to toilets and inside a tub or shower. Each tub and shower should have nonskid floor surfaces. Showers with a seat and a handheld showerhead will provide added safety for people with limited strength. A roll-in shower with no curb should be provided for clients using a wheelchair.

Chapter 8 Additional Reading

One of the best ways to improve your ability to design pleasing homes is to look at successful designs. Excellent sources for review include magazines, television, and the Internet. Magazines used by designers include:

Architect's Journal	*Design Times*
Architectural Digest	*DesignLine (AIBD)*
Better Homes and Gardens	*Home*
Country Living	*House Beautiful*
Country Home	*Metropolitan Home*
Design Architecture	

If you have access to cable or satellite television, HGTV and TLC provides an excellent selection of programs that explore interior design, including:

Awesome Interiors	*Homes by Design*
Decorating Cents	*Interiors by Design*
Decorating with Style	*Kitchen Designs*
Designing for the Sexes	*Landscape Smart*
Dream Builders	*Room by Room*
Garden Architecture	*This Small Space*

Information about each program can be obtained by contacting www.hgtv.com. Other Web sites that promote interior design include:

www.aibd.org (American Institute of Building Design)

www.aham.org (Association of Home Appliance Manufacturers)

www.iida.com (International Interior Design Association)

www.kcma.org (Kitchen Cabinet Manufacturers)

www.maplefloor.org (Maple Flooring Manufacturers Association, Inc.)

www.marble-institute.com (Marble Institute of America)

www.nspi.org (National Spa and Pool Institute)

www.glasswebsite.com/nsa (National Sunroom Association)

The best way to improve your skills is to visit open homes. Walking through a wide variety of homes will help you visualize the materials you will be drawing. As you walk through homes, it may be helpful to take measurements of each room and develop a list of sizes that you find pleasing or confining. Photographing homes and their decor is an additional method to help you remember fine details of interior design. Photos of each area of a home can be a valuable aid in your future, as you help clients define their taste.

CHAPTER 8

Room Relationships and Sizes Test

DIRECTIONS

Answer the questions with short, complete statements or drawings as needed on an 8 1/2 × 11 notebook sheet, as follows:

1. Letter your name, Chapter 8 Test, and the date at the top of the sheet.
2. Letter the question number and provide the answer. You do not need to write out the question.

Answers may be prepared on a word processor, if course guidelines allow this.

QUESTIONS

Question 8–1 What are the three main areas of a home?

Question 8–2 How much closet area should be provided for each bedroom?

Question 8–3 What rooms should the kitchen be near? Why?

Question 8–4 What are the functions of a utility room?

Question 8–5 Give the standard size of a two-car garage.

Question 8–6 List five functions of a family room.

Question 8–7 What are the advantages of placing bedrooms on an upper floor?

Question 8–8 What are the advantages of placing bedrooms on a lower floor?

Question 8–9 List four design criteria to consider in planning a dining room.

Question 8–10 What are the service areas of the home?

Question 8–11 List and describe two types of entries.

Question 8–12 Using the Internet, research five sources of information about the design of a home theater.

Question 8–13 Describe the use of a foyer in a home.

Question 8–14 Describe the difference between a formal and a casual living room.

Question 8–15 What should be considered in placing a living room?

Question 8–16 You are designing a home for a family with no children. The clients insist on having only one bedroom. How would you counsel them?

Question 8–17 Visit several open homes in your area and take pictures or obtain floor plans of a minimum of five different master bedroom suites.

Question 8–18 Visit several different home supply stores or use the Internet to obtain information on several different closet storage options.

Question 8–19 Redesign Figure 8.14 to allow access to the dining room without passing through the living room or kitchen.

Question 8–20 Use the Internet to visit builder organizations such as the National Association of Home Builders. Research current issues affecting home building in your region.

CHAPTER 9

Exterior Design Factors

INTRODUCTION

The design of a house does not stop once the room arrangements have been determined. The exterior of the residence must also be considered. Often a client has a certain style in mind that will dictate the layout of the floor plan. In this chapter, ideas will be presented to help you better understand the design process. To design a structure properly, consideration must be given to the site, the style and shape of the floor plan, and exterior styles.

SITE CONSIDERATIONS

Several site factors affect the design of a house. Among the most important are the neighborhood and access to the lot. For a complete description of each item, see Section III.

Neighborhood

In the initial planning of a residence, the neighborhood must be considered. It is extremely poor judgment to design a $500,000 residence in a neighborhood of $150,000 houses. This is not to say the occupants will not be able to coexist with their neighbors, but the house will have poor resale value because of the lower value of the houses in the rest of the neighborhood. The style of the houses in the neighborhood should also be considered. Not all houses should look alike, but some unity of design can help keep the value of all the properties in the neighborhood high.

Review Boards

To help keep the values of the neighborhood uniform, many areas have architectural control committees. These are review boards made up of residents who determine what may or may not be built. Although once found only in the most exclusive neighborhoods, review boards are now common in undeveloped subdivisions, recreational areas, and retirement areas. These boards often set standards for minimum square footage, height limitations, and the type and color of siding and roofing materials. A potential homeowner or designer is usually required to submit preliminary designs to the review board showing floor plans and exterior elevations.

Access

Site access can have a major effect on the design of the house. The narrower the lot, the more access will affect the location of the entry and the garage. Figure 9.1 shows typical access and garage locations for a narrow lot with access from one side. Usually only a straight driveway is used on interior lots, because of space restrictions.

When a plan is being developed for a corner lot, there is much more flexibility in garage and house placement. To enhance livability, some municipalities are moving away from the layouts shown in Figure 9.1. Traditional layout has produced what is referred to as a *snout house,* meaning a home that is dominated by a view of the garage. In a design where the home is dominated by the garage, the main entrance to the home is often secondary to the entrance for cars, and the driveway often dominates the front yard. In an attempt to eliminate barriers between homes and enhance visibility, building covenants may call for:

- At least one main entrance to the house that meets one of the following requirements.
  The main entrance can be no further than 6′–0″ (1800 mm) behind the longest wall of the house that faces the street.
  The main entry must face the street, or be at an angle of up to 45°.
- At least 15% of the area of the street-facing facade of the home must be windows.

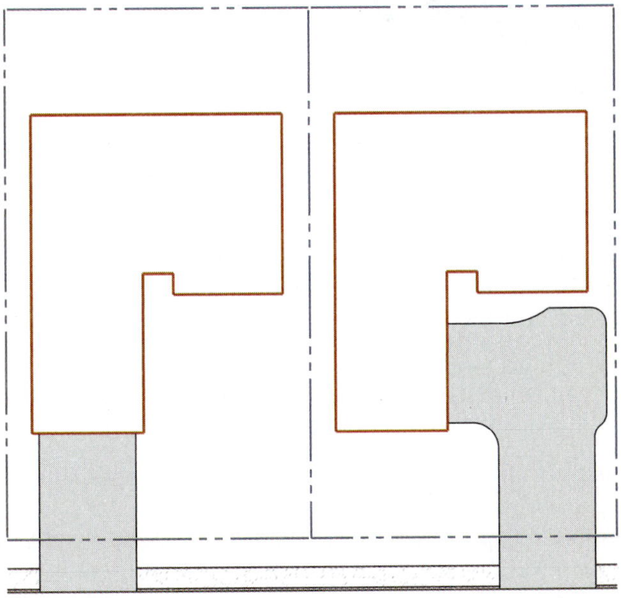

FIGURE 9.1 ■ Access to an inner lot is limited by the street and garage location.

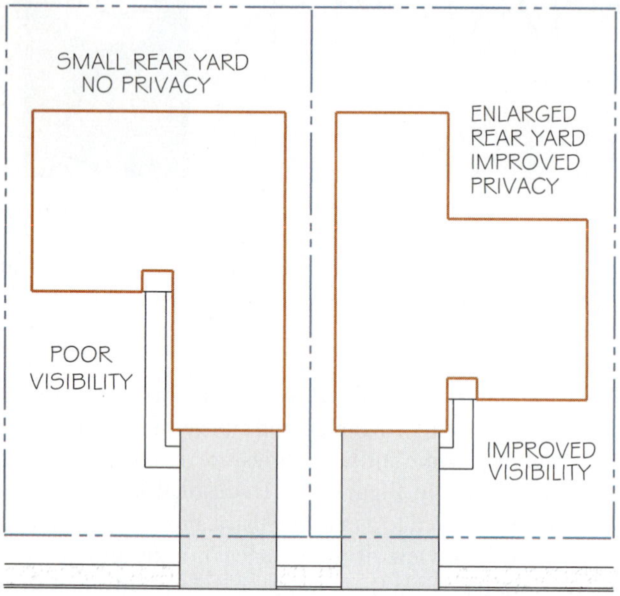

FIGURE 9.2 ■ By limiting the distance that the garage can project from the balance of the home, improved visibility is provided increasing friendlier neighborhood atmosphere.

FIGURE 9.3 ■ A pleasing relationship between the residence and the garage. *Courtesy of Design Basics. Photo by Ken Alan.*

- The length of the garage wall facing the street may not be greater than 50% of the length of the entire facade of the home.
- A garage wall that faces a street may be no closer to the street property line than the longest street-facing wall of the home. (See Figure 9.2.)

Figure 9.3 shows an example of a pleasing relationship between the garage and the balance of the home.

Another popular way to make a neighborhood more livable is to remove the garage altogether from the front of the site. Many planned areas in Florida, Maryland, Oregon, and Tennessee have moved the garage to the rear of the site so that no driveway or parking is provided on the entry or front side of the home. Automobile access is provided to the rear of the lot by an alley.

FIGURE 9.4 ■ On larger sites, the entrance to the garage can often be moved away from the main entry. *Courtesy ABTco Building Products.*

When a residence is being planned for a rural site, weather and terrain can affect access. Studying weather patterns at the site will help reveal areas of the lot that may be inaccessible during parts of the year because of poor water drainage or drifted snow. The shape of the land will determine where the access to the house can be placed. Figure 9.4 shows a garage carefully blended into a home on a large site.

ELEMENTS OF DESIGN

The elements of design are the tools the designer uses to create a structure that will be both functional and pleasing to the eye. These tools are line, form, color, and texture.

Line

Line provides a sense of direction or movement in the design of a structure and helps relate it to the site and the natural surroundings. Lines may be curved, horizontal, vertical, or diagonal and can accent or disguise features of a structure. Curved lines in a design tend to provide a soft, graceful feeling. Curves are often used in decorative arches, curved walls, and round windows and doorways. Figure 9.5 shows how curved surfaces can be used to accent a structure.

Horizontal lines, typically seen in long roof or floor and in balconies and siding patterns, can be used to minimize the height and maximize the width of a structure, as seen in Figure 9.6. Horizontal surfaces often create a sense of relaxation and peacefulness. Vertical lines create an illusion of height; they lead the eye upward and tend to provide a sense of strength and stability. Vertical lines are often used in columns, windows, trim, and siding patterns (Figure 9.7). Diagonal lines can often be used to create a sense of transition. Diagonal lines are typically used in rooflines and siding patterns (Figure 9.8).

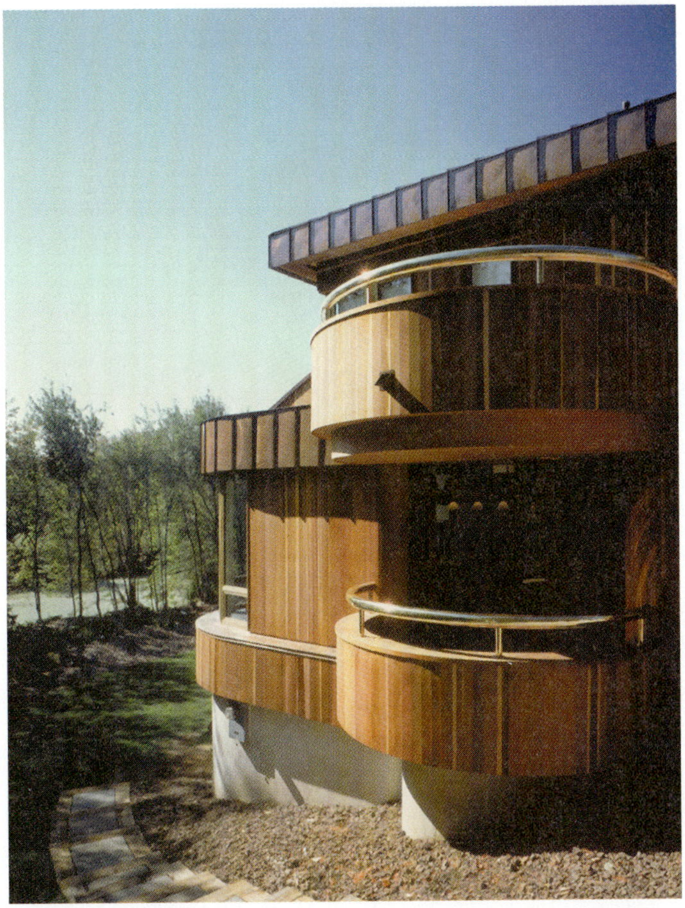

FIGURE 9.5 ■ Curved lines are used to provide smooth transitions and accent a structure. *Courtesy California Redwood Association.*

FIGURE 9.6 ■ Horizontal lines can be used to accent the length, or hide the height, of a structure. *Courtesy Western Red Cedar Lumber Association.*

FIGURE 9.7 ■ Vertical lines can be used to accent the height of a structure. *Courtesy Louisiana Pacific Corporation.*

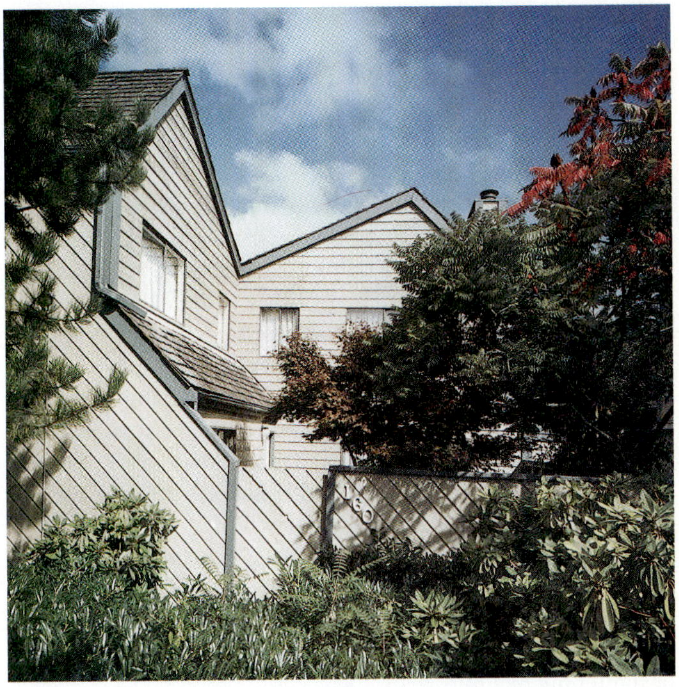

FIGURE 9.8 ■ Diagonal lines are often used to enhance a design.

Form

Lines are used to produce forms or shapes. Rectangles, squares, circles, ovals, and ellipses are the most common shapes found in structures. These shapes are typically three-dimensional, and the proportions between them are important to design. For the best results, the form of a structure should be dictated by its function. Forms are typically used to accent specific features of a structure. Figure 9.9 shows how form can be used to break up the length of a residence. Form also is used to create a sense of security. For example, large columns provide a greater sense of stability than thinner columns.

FIGURE 9.9 ■ Forms such as rectangles, circles, and ovals are used to provide interest. *Courtesy Louisiana Pacific Corporation.*

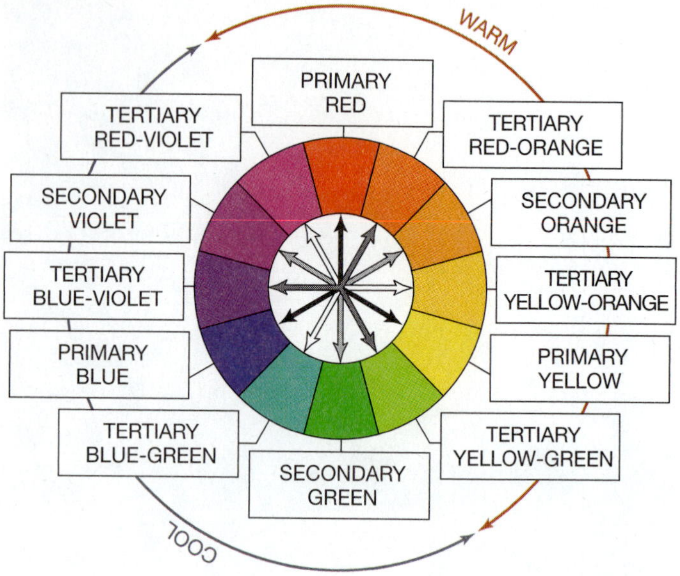

FIGURE 9.10 ■ Colors are primary, secondary, or tertiary. These categories are then divided into warm and cool colors.

Color

Color is an integral part of interior design and decorating and helps distinguish exterior materials and accent shapes. A pleasing blend of colors creates a dramatic difference in the final appearance of any structure. Color is described by the terms *hue, value,* and *intensity.*

Hue represents what you typically think of as the color. Colors are categorized as primary, secondary, or tertiary on a color wheel similar to Figure 9.10. Primary colors—red, yellow, and blue—cannot be created from any other color. All other colors are made by mixing, darkening, or lightening primary colors. Secondary colors—orange, green, and violet—are made from an equal combination of two primary colors. Mixing a primary color with a secondary color will produce a tertiary or intermediate color—for example, red-orange, yellow-orange, yellow-green, blue-green, blue-violet, and red-violet. Mixing each of the primary colors in equal amounts will produce black.

Mixing black with a color will darken the color, producing a shade of the original color. White has no color pigment and gray is a mixture of black and white. Adding white to a color lightens the original color, producing a tint. *Value* is the darkening or lightening of a hue.

Intensity is the brightness or strength of a specific color. A color is brightened as its purity is increased by removing neutralizing factors. Sports cars often have high-intensity colors. A color is softened by adding the color that is directly opposite it on the color wheel. Low-intensity colors such as mint green are used to create a calming effect.

Colors are also classified as warm or cool. Colors seen in warm objects, such as the reds and oranges of burning coals, are warm colors. Warm colors tend to make objects appear larger or closer than they really are. Blues, greens, and violet are cool colors. These colors often make objects appear farther away. Most designers use neutral colors for the major portions of a home and rely on the trim and roofing for accent colors. Figure 9.11 shows an exclusive home designed for wide appeal. Color can be added through landscaping and interior design features.

FIGURE 9.11 ■ Most designers use neutral colors for major portions of the home and rely on the trim and roofing for accent colors. Color can also be added through landscaping and interior design. *Courtesy Eric S. Brown Design Group Inc. Photo by Laurence Taylor.*

Texture

Texture, which refers to the roughness or smoothness of an object, is an important factor in selecting materials to complete a structure. Rough surfaces tend to create a feeling of strength and security—examples include concrete masonry and rough-sawn wood. Rough surfaces also give an illusion of reduced height. Resawn wood, plastic, glass, and most metals have a very smooth surface and create a sense of luxury. A smooth surface tends to give an illusion of increased height; also, it reflects more light and makes colors seem brighter.

FIGURE 9.12 ■ Repetitive features lead the eye from form to form. *Courtesy Monier.*

FIGURE 9.13 ■ Formal balance places features evenly along an imaginary centerline. *Courtesy Aaron Jefferis.*

PRINCIPLES OF DESIGN

Line, form, color, and texture are the tools of design. The principles of design affect how these tools are used to create an aesthetically pleasing structure. The basic principles to be considered are rhythm, balance, proportion, and unity.

Rhythm

Most people can usually recognize rhythm in music. The beat of a drum is a repetitive element that sets the foot tapping. In design, a repetitive element provides rhythm and leads the eye through the design from one place to another in an orderly fashion. Rhythm can also be created by a gradual change in materials, shape, and color. Gradation in materials could be from rough to smooth, gradation in shape from large to small, and gradation in color from dark to light. Rhythm can also be created with a pattern that appears to radiate outward from a central point. A consistent pattern of shapes, sizes, or material can create a house that is pleasing to the eye as well as provide a sense of ease for the inhabitants and convey a sense of equilibrium. Figure 9.12 shows elements of rhythm in a structure.

Balance

Balance is the relationship between the various areas of the structure and an imaginary centerline. Balance may be formal or informal. Formal balance is symmetrical; one side of the structure matches the opposite side in size. The residence in Figure 9.13 is an example of formal balance. Its two sides are similar in mass. Informal balance is nonsymmetrical; in this case, balance can be achieved by placing shapes of different sizes in various positions around the imaginary centerline. This type of balance can be seen in Figure 9.14.

FIGURE 9.14 ■ Informal balance is achieved by moving the centerline away from the mathematical center of a structure and altering the sizes and shapes of objects on each side of the centerline. *Avallon Estate, Plan No. 128 © 1988 courtesy Stephen Fuller Home Gallery, Ltd.*

Proportion

Proportion is related to both size and balance. It can be thought of in terms of size, as in Figure 9.15, where one exterior area is compared with another. Many current design standards are based on the designs of the ancient Greeks. Rectangles using the proportions 2:3, 3:5, and 5:8 are generally considered very pleasing. Proportion can also be thought of in terms of how a residence relates to its environment.

Proportions must also be considered inside the house. A house with a large living and family area needs to have the rest of the structure in proportion. In large rooms the height must also be considered. A 24′ × 34′ (7300 × 10,900 mm) family room is too large to have flat ceiling 8′ high. This standard-height ceiling would be out of proportion to the room size. A ceiling 10′ high, or a vaulted ceiling, would be much more in keeping with the size of the room.

Unity

Unity relates to rhythm, balance, and proportion. Unity ties a structure together with a common design or decorating pattern. Similar features that relate to each other can give a sense of well-being. Avoid adding features to a building that appear to be just there or tacked on. See Figure 9.16.

As an architectural drafter, you should be looking for these basic elements in residences that you see. Walk through houses at every opportunity to see how other designers have used these basics of design. Many magazines are available that feature interior and exterior home designs. Study the photos for pleasing relationships and develop a scrapbook of styles and layouts that are pleasing to you. This will provide valuable resource material as you advance in your drafting career and become a designer or architect.

FIGURE 9.15 ■ Shapes that are related by size and shape create a pleasing flow from form to form. *Photo courtesy Saundra Powell.*

FIGURE 9.16 ■ Unity is a blend of varied shapes and sizes to create a pleasing appearance. *Courtesy California Redwood Association. Waters, Cluts, and O'Brien Architects; photo by Saari Forai.*

FLOOR PLAN STYLES

Many clients come to a designer with specific ideas about the kind and number of levels they want in a house. Some clients want a home with only one level so that no stairs will be required. For other clients, the levels in the house are best determined by the topography of the lot. Common floor plan layouts are single, split-level, daylit basement, two-story, dormer, and multilevel.

Single-Level

The single-level house (Figure 9.17) has become one of the most common styles. It is a standard of many builders because it provides stair-free access to all rooms, which makes it attractive to people with limited mobility. It is also preferred by many homeowners because it is easy to maintain and can be used with a variety of exterior styles.

Split-Level

The split-level plan (Figure 9.18) is an attempt to combine the features of a one- and a two-story residence. This style is best

FIGURE 9.17 ■ A single-level residence is popular because it allows stair-free access to each area. *Courtesy Eric S. Brown Design Group Inc.*

FIGURE 9.18 ■ Homes with split floor levels are ideal for gently sloping sites. *Courtesy LeRoy Cook.*

suited to sloping sites, which allow one area of the house to be two stories and another area to be one story. Many clients like the reduced number of steps from one level to another that is found in the split level design. The cost of construction is usually greater for a split-level plan than for a single-level structure of the same size because of the increased foundation cost that the sloping lot will require.

Split-level plans may be split from side to side or front to back. In side-to-side split-levels, the front entrance and the main living areas are typically located on one level, with the sleeping area located over the garage and service areas. In front-to-rear splits, the front of the residence is typically level with the street, with the rear portion stepping to match the contour of the building site.

Daylight Basement

The style of house called *daylight basement* could be either a one-story house over a basement or garage, or two complete living levels. This style of house is well suited for a sloping lot. From the high side of the lot, the house will appear to be a one-level structure. From the low side of the lot, both levels of the structure can be seen. See Figure 9.19.

Two-Story

A two-story house (Figure 9.20) provides many options for families that don't mind stairs. Living and sleeping areas can be easily separated, and a minimum of land will be used for the building site. On a sloping lot, depending on the access, the living area can be on either the upper or lower level. The most popular feature of a two-level layout is that it provides the maximum building area at a lower cost per square foot than other styles of houses. This saving results because less material is used on the foundation, exterior walls, and roof.

FIGURE 9.19 ■ A home with a daylight basement allows living areas on the lower level to have access to the exterior on one or more side of the home, while looking like a one-level home on the opposite side. *Courtesy Janice Jefferis.*

Dormer

The dormer style provides for two levels, with the upper level usually about half of the square footage of the lower floor. This floor plan is best suited to an exterior style that incorporates a steep roof (Figure 9.21). The dormer level is formed in what would have been the attic area. The dormer has many of the same economic features as a two-story home.

FIGURE 9.20 ■ Two-level homes provide the maximum square footage of living area using the minimum amount of the lot. *Courtesy ABT Co. Building Products.*

FIGURE 9.21 ■ Dormer-style home.

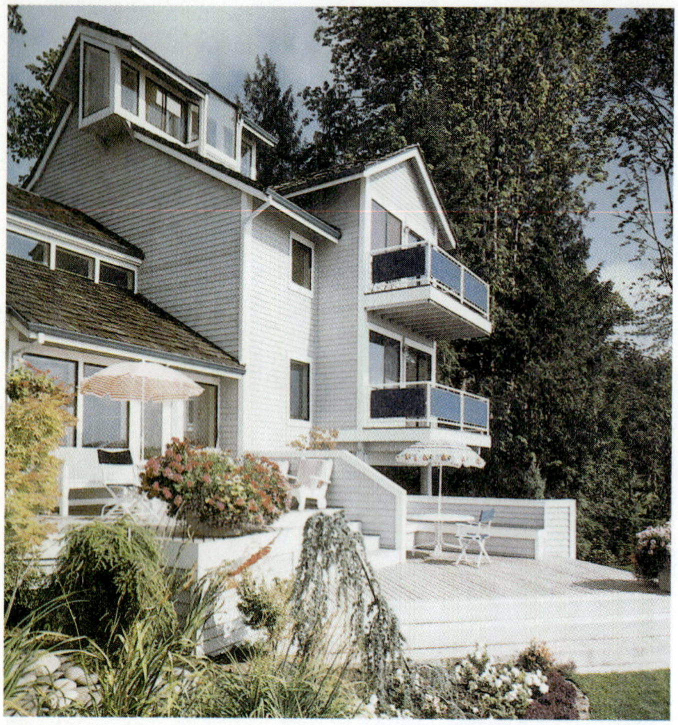

FIGURE 9.22 ■ Multilevel homes provide endless possibilities for floor level arrangements on sloping sites.

Multilevel

With a multilevel layout, the possibilities for floor levels are endless. Site topography as much as the owners' living habits will dictate this style. Figure 9.22 shows a multilevel layout. The cost for this type of home exceeds all other styles because of the problems of excavation, foundation construction, and roof intersections.

EXTERIOR STYLES

Exterior style is often based on past styles of houses. Some early colonists lived in lean-to shelters called wigwams, formed with poles and covered with twigs woven together and then covered with mud or clay. Log cabins were introduced by Swedish settlers in Delaware and soon were used throughout the colonies by the poorer settlers. Many companies now sell plans and precut kits for assembling log houses. Many of the colonial houses were similar in style to the houses that had been left behind in Europe. Construction methods and materials were varied because of the weather and building materials available.

Colonial influence can still be seen in houses that are built to resemble the Georgian, saltbox, garrison, Cape Cod, federal, Greek revival, and southern colonial styles of houses. Other house styles from the colonial period include influences of the English, Dutch, French, and Spanish. Other popular styles include farmhouse, ranch, Victorian, and contemporary.

FIGURE 9.23 ■ Georgian-style homes have formal balance, with the entry on the centerline. *Courtesy Jordan Jefferis.*

FIGURE 9.24 ■ A saltbox residence has a two-level front and a one-level rear. *Courtesy Home Planners, Inc.*

Georgian

The Georgian style is a good example of a basic style that was modified throughout the colonies in response to available material and weather. This style is named for the kings of England who were in power when it flourished. The Georgian style follows the classical principles of design used by the ancient Greeks. The principles of form and symmetry can be seen throughout the structure but are most evident in the front elevation. The front entry is centered on the wall, and equally spaced windows are placed on each side. The front entry is usually covered with a columned porch, and the doorway trimmed with carved wood detailing. In the South, much of the facade is built of brick. In northern states, wood siding is the major covering. Other common exterior materials are stucco and stone. An example of the Georgian style can be seen in Figure 9.23.

Saltbox

One of the most common modifications of the Georgian style is the saltbox. The saltbox maintained the symmetry of the Georgian style but omitted much of the detailing. A saltbox is typically a two-story structure at the front but tapers to one story at the rear. Window areas usually have shutters to provide protection from winter winds. Figure 9.24 shows features of saltbox styling in a residence.

Vanderhorst, Plan No. 421
©1994 Stephen Fuller American Home Gallery, Ltd.

FIGURE 9.25 ■ A garrison-style home has an upper level that extends over the lower level. *Vanderhorst, Plan No. 421 © 1994 courtesy Stephen Fuller Home Gallery, Ltd.*

Amesbury Grove, Plan No. 350
©1994 Stephen Fuller American Home Gallery, Ltd.

FIGURE 9.26 ■ A Cape Cod home features the formal balance of other styles but adds dormers and shutters. *Amesbury Grove, Plan no. 350 © 1994 courtesy Stephen Fuller Home Gallery, Ltd.*

Garrison

The garrison style combines saltbox and Georgian style with the construction methods of log buildings. The garrison was originally modeled after the lookout structures of early forts. The upper level extends past the lower level; in a fort, this made the walls harder to scale. Originally, heavy timbers were used to support the overhang and were usually carved. Figure 9.25 shows the features of the garrison-style residence.

Cape Cod

Cape Cod style is typically one level with a steep roof, which allows an upper-floor level to be formed throughout the center of the house. Dormers are typically placed on the front side of the roof to make the upper floor area habitable. Windows are placed symmetrically around the door and have shutters on the lower level. An example of Cape Cod styling can be seen in Figure 9.26. An offshoot of Cape Cod style is Cape Ann, which has many of the same features as a Cape Cod house but is covered by a gambrel roof.

Federal

The Federal style of the late 1700s combines Georgian with classical Roman and Greek styles to form a very dignified style. Federal homes were built of wood or brick. Major additions to the Georgian home include a high, covered entry porch or portico supported on Greek-style columns centered over the front door. The door typically has arched trim, and windows are capped with a projected pediment. Figure 9.27 shows common pediment styles found throughout American architecture. Figure 9.28 shows an example of a Federal-style residence.

Greek Revival

Greek revival homes are built using classic proportions and decorations of classical Greek architecture. Typically, these homes are large, rectangular, and very boxlike. Decoration is added by a two-story portico with a low, sloped gable roof supported on Greek columns centered on the residence. Figure 9.29 shows an example of Greek revival architecture.

FIGURE 9.27 ■ Pediment styles.

FIGURE 9.28 ■ Federal style combines features of Georgian and classical Roman and Greek architecture.

FIGURE 9.29 ■ Greek revival features classic Greek proportions and ornamentation.

Southern Colonial

Southern colonial homes are similar to the Georgian style, with symmetrical features. The southern colonial style usually has a flat, covered porch, which extends the length of the house to protect the windows from the summer sun. Figure 9.30 shows an example.

English

English-style houses are fashioned after houses that were built in England prior to the early 1800s. These houses feature an unsymmetrical layout and walls that are usually constructed of stone, brick, or heavy timber and plaster. Window glass typically has a diamond-shape rather than the more traditional rectangle. Figure 9.31 shows common features of English Tudor styling. The Elizabethan-style home is influenced by the English and Dutch as well as Gothic styling. These homes shared the half-timber styling of Tudor homes but were typically irregularly shaped.

FIGURE 9.30 ■ Southern colonial or plantation-style homes reflect classic symmetry. They typically have a large covered porch to provide protection from the sun.

FIGURE 9.31 ■ English Tudor combines half-timber with brick, stone, and plaster. *Courtesy Janice Jefferis.*

Dutch

The Dutch colonial style has many of the features of homes already described. The major difference is the roof shape. Dutch colonial style features a gambrel roof, also known as a barn roof. This roof is made of two levels. The lower level is usually very steep and serves as the walls for the second floor of the structure. The upper area of the roof is the more traditional gable roof. An example of Dutch colonial style can be seen in Figure 9.32. The gambrel roof will be described in Section VI.

French

French colonial styling is also a matter of roof design, which is similar to the gambrel, with its steep lower roof. French colonial styling uses a hipped or mansard roof to hide the upper floor

FIGURE 9.32 ■ Dutch colonial homes often feature a gambrel roof. *Courtesy Mike Jefferis.*

FIGURE 9.33 ■ French colonial homes often hide the upper floor behind a mansard roof. *Courtesy David Jefferis.*

area. A mansard roof is basically an angled wall and will be discussed in Section VI. Figure 9.33 shows an example of French colonial styling.

Later influences of the French are visible in the single-level French manor. Originally found in the northern states, these homes are usually a rectangle with a smaller wing on each side. The roof is usually a mansard to hide the upper floor, but hip roofs were also used, as seen in Figure 9.34. Another common French style is shown in Figure 9.35.

The French Normandy style first became popular in the 1600s. These homes were typically multilevel and framed with brick, stone or wood, and plaster with half-timber decoration. The roof is typically a gable, although some are now built with a hip roof. A circular turret is typically found near the center of the home. Figure 9.36 shows an example of a French Normandy residence.

In the southern states, the French Normandy style has been modified into the French plantation style. This home typically is two full floors with a wraparound porch and is covered with a hip roof. A third floor is a common component, with dormers added for light and ventilation. Figure 9.37 shows a French plantation home. This style was modified around New Orleans into what is now known as the Louisiana French style. The size of the balconies is diminished, but the supporting columns and rails have become fancier.

Whitehall, Plan No. 179
©1989 Stephen Fuller American Home Gallery, Ltd.

FIGURE 9.34 ■ A French manor typically features a wing on each side of a central rectangle. *Whitehall, Plan no. 179 © 1989 courtesy Stephen Fuller Home Gallery, Ltd.*

FIGURE 9.35 ■ French influence can be seen in the roof, trim, and design of this home. *Westminster Estate, Plan no. 181, © 1989 courtesy Stephen Fuller Home Gallery, Ltd.*

FIGURE 9.36 ■ French Normandy homes typically feature a turret, which houses the stairs leading to an upper level. *Courtesy Ludowici Celadon Company.*

FIGURE 9.37 ■ This southern home reflects French features in the upper deck covered by a hip roof. *Lafayette Hall, Plan no. 356, © 1995 courtesy Stephen Fuller Home Gallery, Ltd.*

FIGURE 9.38 ■ Spanish-style homes typically have low-sloping tile roofs, arches, and window grills. *Courtesy of Michael Jefferis.*

FIGURE 9.39 ■ Farmhouse style features a simple structure with a covered porch. *Broadwings, Plan no. 329 © 1992 courtesy Stephen Fuller Home Gallery, Ltd.*

FIGURE 9.40 ■ Ranch-style homes feature a one-level elongated floor plan covered by a low, sloping roof. *Courtesy Piercy & Barclay Designers, Inc. CPBD.*

Spanish

Spanish colonial buildings were constructed of adobe or plaster and were usually one story. Arches and tiled roofs are two of the most common features of Spanish, or mission-style, architecture. Timbers are often used to frame a flat or very low-pitched roof. Windows with grills or spindles and balconies with wrought-iron railings are also common features. Figure 9.38 shows an example of Spanish style architecture.

Farmhouse

The farmhouse style makes use of two-story construction and is usually surrounded by a covered porch. Trim and detail work, common in many other styles of architecture, are rarely found in farmhouse style. Figure 9.39 shows an example of this style.

Ranch

The ranch style of construction comes from the Southwest. This style is usually defined by a one-story, rambling layout, which is made possible because of the mild climate and plentiful land on which such houses are built. The roof is typically low-pitched, with a large overhang to block the summer sun. The major exterior material is usually stucco or adobe. Figure 9.40 shows the ranch style.

Victorian

The Victorian and Queen Anne styles from the late 1800s are still being copied in many parts of the country. These styles feature irregularly shaped floor plans and very ornate detailing throughout the residence. Victorian houses often borrow elements from many other styles of architecture, including partial mansard roofs, arched windows, and towers. Exterior materials include a combination of wood and brick. Wrought iron is often used. Figure 9.41 shows an example of the Victorian style.

Contemporary

It is important to remember that a client may like the exterior look of one of the traditional styles, but very rarely would the traditional floor plan of one of those houses be desired. Quite often the floor plans of older houses produced small rooms with

FIGURE 9.41 ■ Victorian homes typically have very ornate, irregular shapes. *Courtesy Janice Jefferis.*

FIGURE 9.42 ■ Many contemporary homes use clean lines with little or no trim to provide an attractive structure. *Design by Jack Smuckler, A.I.A.; photo by Jerry Swanson. Courtesy California Redwood Association.*

poor traffic flow. A designer must take the best characteristics of a particular style of house and work them into a plan that will best suit the needs of the owner.

Contemporary, or modern, does not denote any special style of house. Some houses are now being designed to meet a wide variety of needs, and others reflect the lifestyle of the owner. Figures 9.42 through 9.47 show examples of the wide variety of contemporary houses.

FIGURE 9.43 ■ Many contemporary homes feature a portico at the entry door to provide a warm, inviting effect. *Courtesy Georgia Pacific Corporation.*

FIGURE 9.44 ■ Many contemporary homes are open from room to room and from the interior to the exterior. *Courtesy Eric S. Brown Design Group Inc. Photo by Oscar Thompson.*

FIGURE 9.45 ■ Although this elegant residence reflects no particular style, it has pleasing proportions, symmetry, and balance. *Courtesy Eric S. Brown Design Group Inc. Photo by Oscar Thompson.*

FIGURE 9.46 ■ Many homes are now being constructed that have a common property line. *Courtesy Western Red Cedar Lumber Association.*

FIGURE 9.47 ■ Subterranean construction is energy-efficient in severe climates. *Courtesy Weather Shield Mfg., Inc.*

CHAPTER 9

Exterior Design Factors Test

DIRECTIONS

Answer the questions with short complete statements or drawings as needed on an 8 1/2 × 11 notebook sheet, as follows:

1. Letter your name, Chapter 9 Test, and the date at the top of the sheet.
2. Letter the question number and provide the answer. You do not need to write out the question.
3. Answers may be prepared on a word processor if course guidelines allow this.

QUESTIONS

Question 9–1 Explain how the neighborhood can influence the type of house that will be built.

Question 9–2 Describe how lines can be used in the design of a structure.

Question 9–3 Sketch a simple two-bedroom house in two of the following shapes: L, U, T, or V.

Question 9–4 List four functions of a review board.

Question 9–5 Describe four floor plan styles, and explain the benefits of each.

Question 9–6 What factors make a two-story house more economical to build than a single-story house of similar size?

Question 9–7 What house shape is the most economical to build?

Question 9–8 Photograph or sketch examples of the following historical styles found in your community:

a. Dutch colonial
b. Garrison
c. Saltbox
d. Victorian

Question 9–9 What forms are most typically seen in residential design?

Question 9–10 List and define the three terms used to describe color.

Question 9–11 Identify the primary and secondary colors.

Question 9–12 List the major features of the following styles:

a. Ranch
b. Tudor
c. Cape Cod
d. Spanish

Question 9–13 How is a tint created?

Question 9–14 Photograph or sketch three contemporary houses that have no apparent historic style.

Question 9–15 What period of architecture influenced the Georgian style?

Question 9–16 Sketch a Dutch colonial house.

Question 9–17 Sketch or photograph a house in your community built with a traditional influence, and explain which styles this house has copied.

Question 9–18 What are some of the drawbacks of a traditional style?

Question 9–19 What are the common proportions of classical Greek design?

Question 9–20 What style of residence has two full floors with a wraparound porch covered with a hip roof?

10 CHAPTER

Conservation and Environmental Design and Construction

INTRODUCTION

Conservation design uses different technologies to improve construction methods and provide alternative energy sources. Architectural designers, building contractors, and home buyers today are concerned about quality of life on the earth and in the home. Improving energy efficiency is a good way to reduce oil dependence and address environmental concerns. Throughout North America, private, local, state, and national agencies sponsor projects such as green building, earth smart, sustainable building, and EnviroHome (Canada). The goal of these programs is to provide incentives and education for designers and builders in an effort to create buildings that will be good for the earth today and into the future. The terms *environmentally friendly* and *environmentally sound* refer to designing and constructing buildings with renewable materials such as wood from managed forests, earthen materials, and recycled products. These buildings are healthier, because nontoxic and low-toxicity materials are used. Energy from renewable resources—such as geothermal, solar, photovoltaic, or wind energy—is also taken into consideration in protecting the environment.

ENERGY-EFFICIENT DESIGN

Energy-conscious individuals have been experimenting with energy-efficient construction for decades. Their goal has been to reduce home heating and cooling costs. In recent years several formal studies have been done around the country, sponsored by various private and governmental agencies that are, as a result of rising energy costs, committed to conservation. The experimental programs have had the following goals:

- Conserve natural resources.
- Save the environment.
- Preserve the earth's ozone layer.
- Create a better and healthier living environment.
- Meet consumers' demand for more economical living.
- Evaluate realistic material and construction alternatives that may be used to alter building codes in the future.

Today's home buyers are concerned about energy-efficient design. They expect energy-efficient design and construction to be a part of the total home package, and, in fact, such design and construction can be achieved without a great deal of additional cost. For example, consumers look for construction that uses high-quality insulation and air infiltration barriers. Such construction increases comfort while reducing energy costs.

Home builders around the country have realized the advantages of energy-efficient construction for both consumers and themselves. An energy-efficient home is easier to sell than an ordinary construction.

In several controlled situations where energy-efficient construction techniques were implemented to determine if there would be a significant reduction in energy consumption, the results clearly show a saving. A number of techniques can be used that do not cost much more than standard construction, although at the other extreme some energy-conserving construction techniques are complex and labor- and material-intensive.

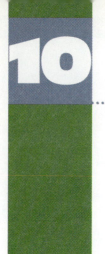

ENERGY CODES

The Model Energy Code was originally developed jointly by Building Officials and Code Administrators International, Inc. (BOCA), International Conference of Building Officials (ICBO), National Conference of States on Building Codes and Standards (NCSBCS), and Southern Building Code Congress International, Inc. (SBCCI) under a contract funded by the U.S. Department of Energy (DOE). The Model Energy Code sets minimum requirements for the design of new buildings and additions to existing buildings. BOCA has also established the National Energy Conservation Code, which is updated every three years. The purpose of the code is to regulate the design and construction of the exterior envelope and selection of HVAC, service water heating, electrical distribution and illuminating systems, and equipment required for effective use of energy in buildings for human occupancy. The *exterior envelope* is made up of the elements of a building that enclose conditioned (heated and cooled) spaces through which thermal energy transfers to or from the exterior.

ENERGY-EFFICIENT CONSTRUCTION

No matter what system of construction or architectural style is used, energy efficiency can be a part of the construction process. The goal of energy-efficient construction is to decrease dependency on the heating and cooling system. This is best done by the use of framing techniques, caulking, vapor retarders, radiant barriers and insulation.

Caulking

Caulking normally consists of filling small seams in the siding or the trim to reduce air drafts. In energy-efficient construction, caulking is added during construction at the following places:

- Exterior joints around the window and door frames.
- Joints between wall cavities and window and door frames.
- Joints between wall and foundation.
- Joints between wall and roof.
- Joints between wall panels.
- Penetrations or utility services through exterior walls, floors, and roofs.
- All other openings that may cause air leakage. Check with local building officials for a complete list.

Additional energy-efficient construction techniques related to specific construction trades are discussed in Chapters 17, 18, and 19. Figure 10.1 shows typical areas where caulking can be added. These beads of caulk keep air from leaking between joints in construction materials. Figure 10.2 shows the general caulking notes that may be found on a drawing. Caulking may seem like extra work for a small effect, but it is well worth the effort. It involves minimal expense for material and labor.

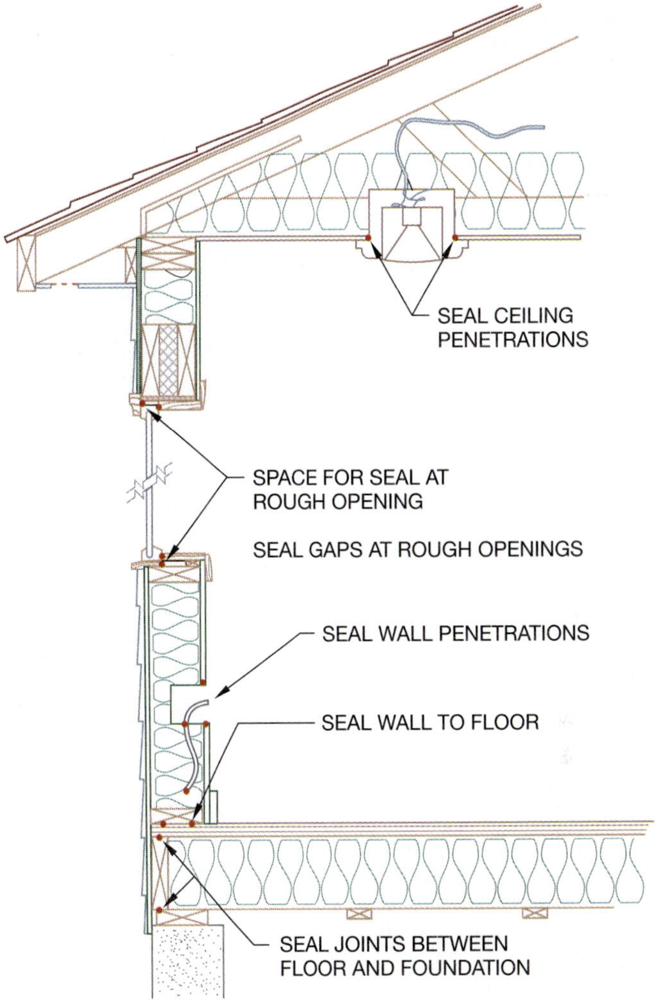

FIGURE 10.1 ■ Typical areas where caulking should be added. *Courtesy Oregon Residential Energy Code.*

CAULKING NOTES:

1. SEAL THE EXTERIOR SHEATHING @ CORNERS JOINTS DOOR AND WINDOW AND FOUNDATION SILLS WITH SILICONE CAULKING

2. CAULK THE FOLLOWING OPENINGS W/ EXPANDED FOAM OR BACKER RODS, POLYURETHANE, ELASTOMERIC COPOLYMER SILCONIZED ACRYLIC LAYTEX CAULKS MAY ALSO BE USED WHERE APPROPRIATE.

ANY SPACE BETWEEN WINDOW AND DOOR FRAMES.

BETWEEN ALL EXTERIOR WALL SOLE PLATES AND PLY SHEATHING

ON TOP OF RIM JOIST PRIOR TO PLYWOOD FLOOR APPLICATION

WALL SHEATHING TO TOP PLATE.

JOINTS BETWEEN WALL AND FOUNDATION

JOINTS BETWEEN WALL AND ROOF

JOINTS BETWEEN WALL PANELS

AROUND OPENINGS FOR DUCTS, PLUMBING, ELECTRICAL, TELEPHONE, AND GAS LINES IN CEILINGS, WALLS, AND FLOORS. ALL VOIDS AROUND PIPING RUNNING THROUGH FRAMING OR SHEATHING TO BE PACKED W/ GASKETING OR OAKUM TO PROVIDE A DRAFT FREE BARRIER.

FIGURE 10.2 ■ General caulking notes that may be found on a set of architectural drawings.

Vapor Retarders

For an energy-efficient system, air tightness is critical. The ability to eliminate air infiltration through small cracks is imperative if heat loss is to be minimized. Vapor retarders are a very effective method of decreasing heat loss. Most building codes require 6-mil-thick plastic to be placed over the earth in the crawl space. Many energy-efficient construction methods add a continuous vapor retarder to the walls. This added vapor retarder is designed to keep exterior moisture from the walls and insulation. Figure 10.3 shows the effect of water vapor on insulated walls with and without a vapor retarder.

Under normal conditions, wall and ceiling insulation with a foil face on the interior side allows small amounts of air to leak in at each seam. To eliminate this leakage, a continuous vapor retarder can be installed. To be effective, the vapor retarder must be lapped and sealed to keep air from penetrating through the seams in the plastic.

The vapor retarder can be installed in the ceiling, walls, and floor system for effective air control. Figure 10.4 shows three

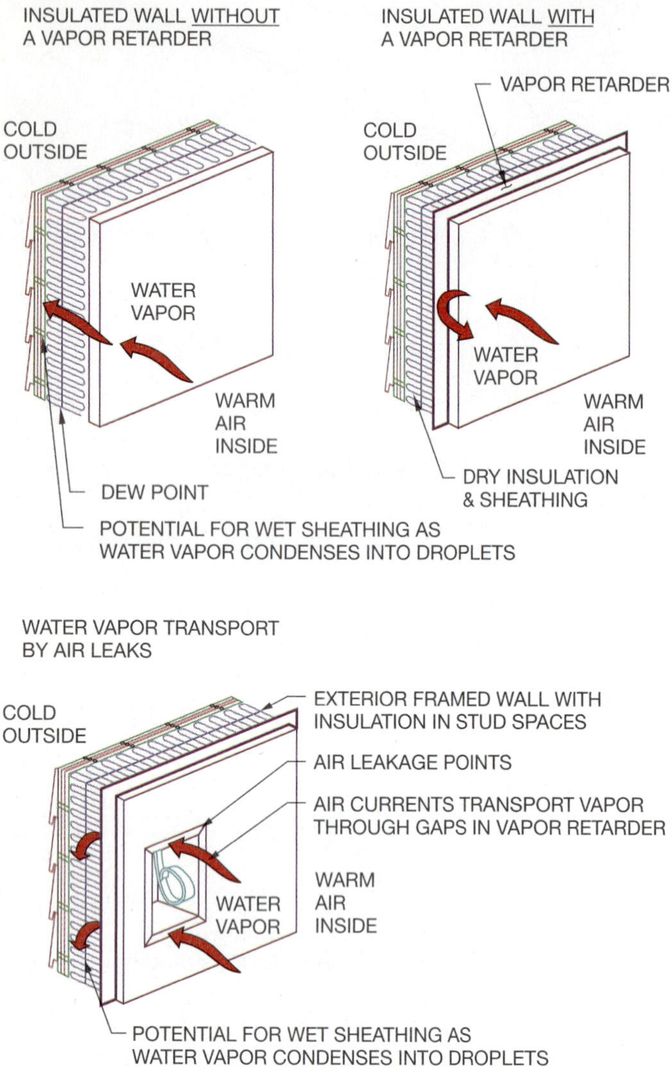

FIGURE 10.3 ■ Effect of water vapor on insulated walls with and without a vapor retarder. *Courtesy Oregon Residential Energy Code.*

different ceiling applications. Each is designed to help prevent small amounts of heated air from escaping to the attic. All the effort required to keep the vapor retarder intact at the seams must be continued wherever an opening in the wall or ceiling is required.

The cost of materials for a vapor retarder is low compared with the overall cost of the project. The expense of the vapor retarder comes in the labor to install and to maintain its seal. Great care is required by the entire construction crew to maintain the barrier.

Caution should be exercised when building a completely airtight structure with increased insulation and vapor retarders. This type of construction may cause problems with the quality of air inside. These potential problems may be countered with the installation of an air-to-air exchanger. A complete discussion of air contaminants and air-to-air exchangers is found in Chapter 19. A heating, ventilating, and air-conditioning (mechanical) engineer should be consulted when these structures are being designed. Garages need screened openings through exterior walls at or near the floor level with a clear area of screen not less than 60 square inches per motor vehicle to be accommodated.

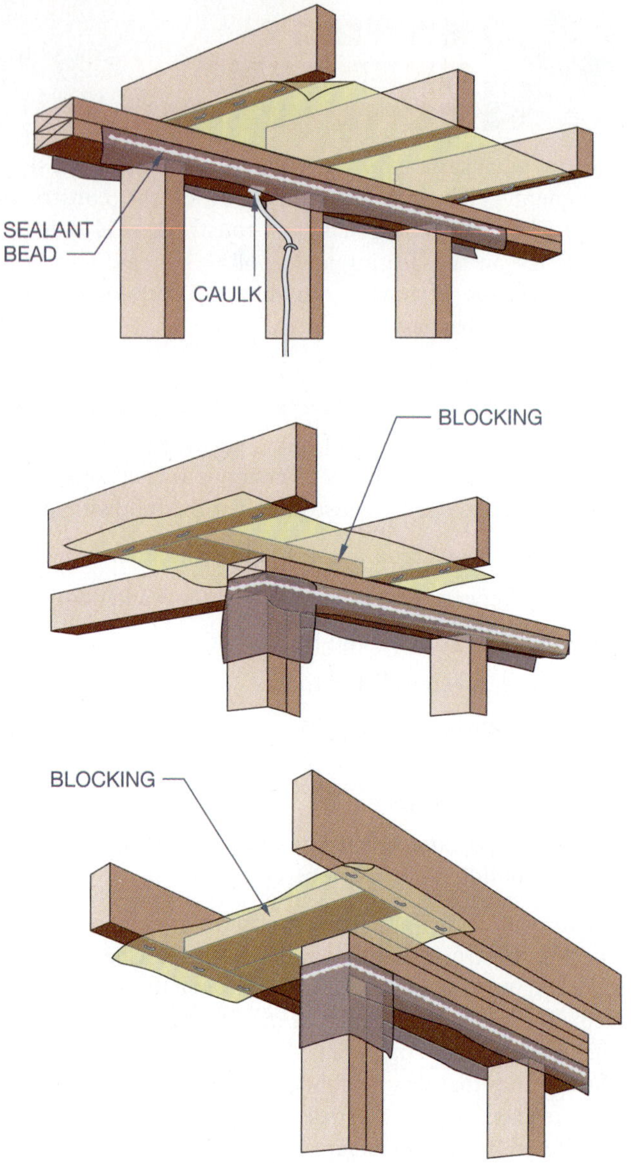

FIGURE 10.4 ■ Ceiling application of continuous vapor retarders. *Courtesy Oregon Department of Energy.*

Radiant Barriers

Radiant barriers stop heat from radiating through the attic; they help to reduce attic temperatures in hot climates. Radiant barriers are made of aluminum foil with a backing. Radiation-control roof coatings are also available to reduce roof temperature in hot climates.

Insulation

Insulation saves energy costs and makes the home comfortable; properly insulated walls, ceilings, and floors stay warmer in winter and cooler in summer. This helps maintain a uniform temperature throughout the house. Properly installed insulation can also reduce noise. Various types of insulation are available, but any type must be installed properly for the best efficiency. Figure 10.5 shows the locations where insulation is typically installed.

Many energy-efficient construction methods depend on added insulation to help reduce air infiltration and heat loss. Some systems require not only adding more insulation to the structure but also adding more framing material to contain the insulation.

Insulation acts to reduce the amount of heat lost through walls, ceilings, and floors during the winter and helps keep heat from entering the building during the summer. The ability of materials to slow heat transfer is called *thermal resistance*. The R-value of a material is a measure of thermal resistance to heat flow. The higher the R-value, the greater the insulating ability. Insulation is critical in helping reduce heat loss, but insulation may not stop air infiltration due to leaks. These leaks must be stopped with proper caulking, as previously discussed. As noted above, insulation must be installed properly for maximum efficiency.

How much insulation a home should have depends on local and national codes, climate, energy costs, budget, and personal preference. The building codes give minimum required insulation levels. The U.S. Department of Energy (DOE) provides recommended insulation levels, by zip codes, for walls, ceilings, floors, and foundations. Typically, R-38 is recommended in the ceiling, with R-49 used in cold climates. R-25 insulation is recommended for use in floors for most of the country. Some warm climates do not require floor insulation. Wall insulation levels vary between R-11 to R-21, again depending on the climate. The DOE recommendations are a good guide, but always confirm the minimum insulation required by the codes governing your area. In general, better energy efficiency can be gained by using more insulation. Adding insulation during construction is easier and less expensive than adding it later.

Types of Insulation

Common insulation materials include fiberglass, rock wool, cellulose, urethane foam, and recycled cotton. Insulation is available in the form of loose fill, blanket, rigid board, and expanding spray foam.

Loose-Fill Insulation

Loose-fill insulation is fibers or granules made from cellulose, fiberglass, rock wool, cotton, or other materials. Loose-fill insulation is normally blown into areas and cavities with special equipment. A benefit of this insulation is that it conforms to and completely fills the space where it is applied. This helps eliminate air spaces and increases efficiency. Some applications mix loose-fill fibers with a spray adhesive to cover areas before drywall is installed. This is a special advantage for irregular and hard-to-reach areas. Some applications use a vapor barrier, a net, or a temporary frame over the studs to hold the insulation while it is blown into the wall cavity.

Loose-fill insulation must be installed correctly, without excess air in the cavity. Some manufactures guarantee that the insulation will have a specific R-value when installed to a given thickness.

Loose-fill insulation settles over a period of time, leaving an uninsulated area at the top of wall cavities. Settling is reduced

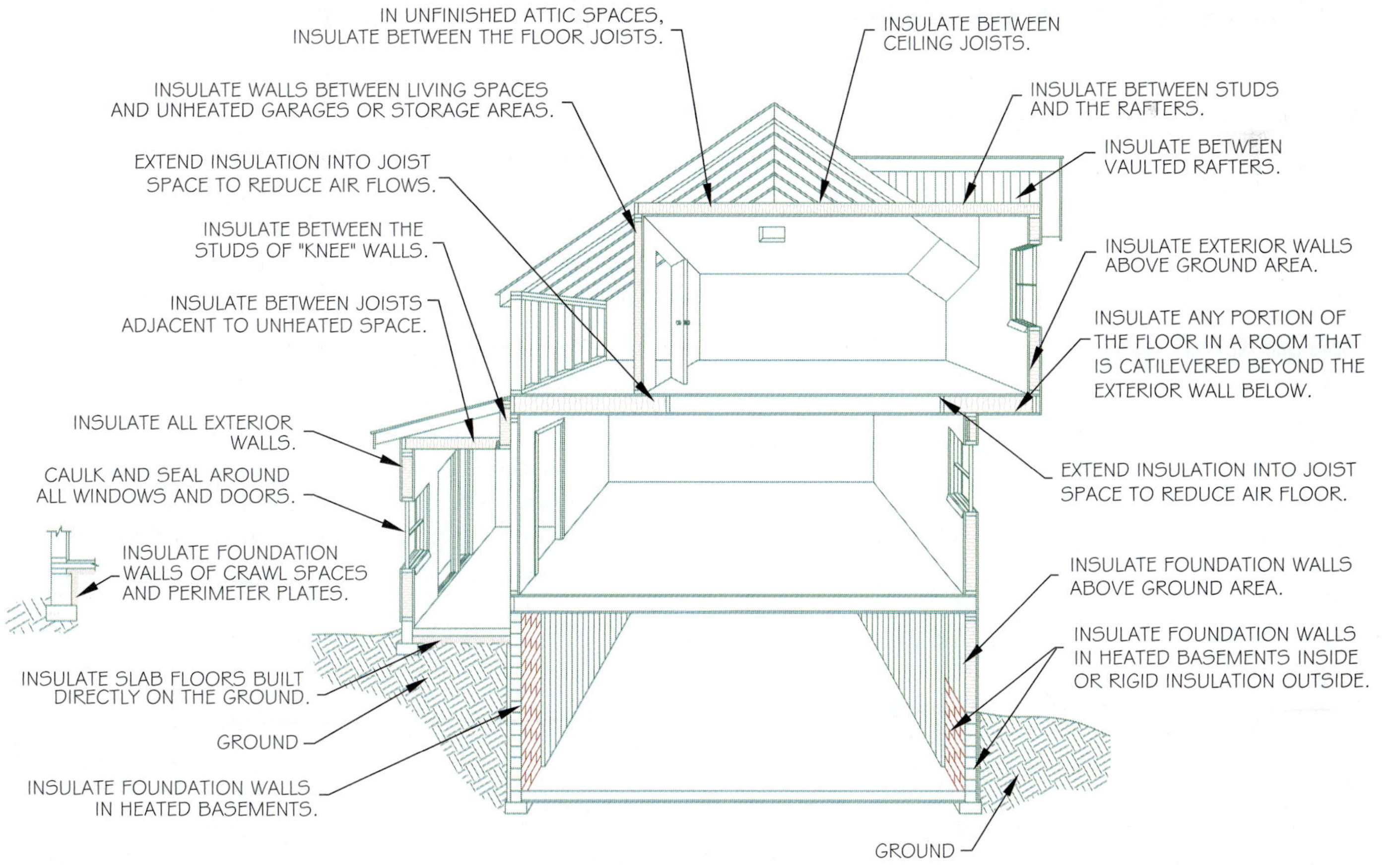

FIGURE 10.5 ■ Locations where residential insulation is typically installed.

or avoided by installing higher-than-normal density. This technique, called *dense-pack,* also reduces air leakage.

Batts and Blanket Insulation

Batts are strips and *blankets* are roll insulation made of fiberglass or cotton fibers. This insulation is available 16 to 24 in. wide for placement between standard framing member spacing. The thickness of the insulation determines its R-value. Batts and blankets are available with or without vapor-retarded facings. Careful installation of batts and blanket insulation is important to avoid gaps that reduce their effectiveness. Be sure the insulation is tight to the framing members and is not compacted.

The inside of basement walls can be insulated with batts or blanket by building a frame with pressure-treated two-by-fours against the wall, or by attaching pressure furring strips to the wall and placing the insulation between the frame or furring members.

Rigid Insulation

Rigid insulation is made from fibrous material or plastic foams and pressed or extruded into sheets of varying thickness. Rigid insulation provides high R-values per inch of thickness, sound insulation, some structural strength, and air sealing. Rigid foam insulation is used in foundations, under siding to increase R-value, and in tight places such as vaulted ceilings where other materials would be too thick. Foam insulation must be covered with finish material for fire safety. A common use is under concrete slabs and on the outside of basement walls. Any insulation exposed aboveground must be covered with protective material such as plastic, treated wood, plaster, or concrete.

A type of construction called *stress skin walls* replaces conventional framing with structural sheathing glued to both sides of foam core panels. A recycling technology uses agricultural byproducts such as wheat straw compressed between strandboard.

Insulated structural concrete walls are being used for foundations. These walls use rigid foam blocks that are stacked and filled with concrete and reinforcing steel. These foundations provide structural qualities and excellent insulation.

Foam-in-Place Insulation

Foam-in-place, also referred to as *spray-in* foam insulation, provides high R-values per inch and air sealing. These materials require special equipment to spray or extrude into place. They are sprayed into open wall and ceiling cavities and hard-to-access areas, where they expand to fill the space. Any excess foam is trimmed away when it hardens. A product called Icynene is being used in energy-efficient construction to get high R-values and airtight seals; it also provides low toxicity for a healthier environment.

Insulation Values and Applications

Table 10–1 shows the R-value per inch and application of different insulation materials.

TABLE 10–1 INSULATION VALUES AND APPLICATION

INSULATION TYPE	R-VALUE/IN.	APPLICATION	ADVANTAGES
Loose-fill Fiberglass Rock wool Cellulose	 2.2–2.9 2.2–2.9 3.1–3.7	Anyplace the frame is covered on both sides, such as finished walls and vaulted ceilings. Unfinished attic floors and hard-to-reach places.	The only insulation that can be used in closed cavities. Easy to use in irregular shaped areas and around obstructions. Dense-pack provides air seal.
Batts and blanket Fiberglass Cotton	 2.9–3.8 3–3.7	All unfinished walls, floors, and attics fitted between framing.	Standard stud and joist spacing with few obstacles.
Foam board Expanded Polystyrene Extruded Polystyrene Polylsocyanurate Polyurethane Phenolic Fiberglass board	 3.9–4.2 5 5.6–7 5.6–7 4.4–8.2 3–4	Basement masonry walls. Exterior walls under siding, and under roofing.	High insulating values for little thickness. Insulates over framing and cavities.
Spray-in foam Polyurethane Icyene	 5.6–6.2 3.6–4.3	Unfinished walls, attics, and floors.	Air sealing and high insulating value.

TABLE 10–2 SAMPLE ENERGY CODE INSULATION REQUIREMENTS

BUILDING COMPONENTS	INSULATION CODE
Walls	R-21
Flat ceilings	R-38
Vaulted ceilings	R-30 not to exceed 50% of the floor space unless insulated to R-38
Under floor	R-25
Basement walls	R-21
Slab floor edge	R-15
Forced air ducts	R-8
Windows	U = 0.40, CL40
Main entry door	U = 0.54, maximum 24 sq ft
Other exterior doors	U = 0.20
Skylights	U = 0.50, CL50, not to exceed 2% of heated floor space

Insulated Windows

There are a variety of ways to insulate windows. One common method is to use two or more panes of glass with a sealed space between. This is referred to as *insulating glass*. Another technique is to fill the sealed space with argon, a colorless gas that raises the window's insulating value. *Low-E*, or *low-emissivity*, glass is an additional technology that improves windows' energy efficiency; it has a transparent coating that acts as a thermal mirror, which increases insulating value, blocks heat from the sun, and reduces fading of objects inside the house. Window frames made of wood or a combination of materials that have a thermal brake installed are an advantage in terms of energy efficiency.

Whereas insulation is rated by R-value, windows are rated by U-value. R is the reciprocal of the U-value: 1/U = R. *U-value* is defined as the coefficient of heat transfer expressed as BTUH sq ft/°F of surface area. BTUH is heat loss calculated in BTUs per hour. *BTU* means British Thermal Unit, a measure of heat. The lower the U-value, the more the insulating value. The U-value of windows can also be identified by *window class* (CL). For example, a window with a value of U = 0.40 is CL40.

Insulation Requirements in Energy Codes

As stated earlier, insulation requirements differ around the country. Requirements are stricter in colder climates than in warmer climates. The U.S. Department of Energy has recommendations based on climate zones. Always check with your local and state energy codes for specific information. Table 10–2 gives sample requirements.

Figure 10.6 shows general insulation notes that may be found in a set of plans designed for energy-efficient construction.

INSULATION NOTES:

1. INSULATE ALL EXTERIOR HEATED WALLS W/ HIGH DENSITY FIBERGLASS BATT INSULATION R-21 MIN. W/ PAPER FACE. INSULATE EXTERIOR WALLS PRIOR TO INSULATION OF TUB / SHOWER UNITS.
2. INSULATE ALL FLAT CEILINGS W/ 12" R-38 FIBERGLASS BATT INSULATION / NO PAPER FACE REQUIRED.
3. INSULATE ALL VAULTED CEILINGS TO W/ 10 1/4' HIGH DENSITY PAPER FACED FIBERGLASS BATTS R-30 MIN. W/ 2 MIN. A R SPACE ABOVE.
4. INSULATE ALL WOOD FLOORS W/ 8' FIBERGLASS BATTS R-25 MIN. W/ PAPER FACE OR 1 PERM FORMULATED VAPOR RETARDED INSTALLED ABOVE FLOOR DECKING PRIOR TO INSTALLING FINISHED FLOORING. INSTALL PLUMBING ON HEATED SIDE OF INSULATION.
5. INSULATE CONCRETE SLAB FLOORS BENEATH HEATED ROOMS W/ 3" EXTRUDED POLYSTYRENE R-15 MIN. X 24" WIDE AT ALL SLAB EDGES IN CONTACT WITH UNHEATED EARTH.
   - PROVIDE ALT. BID FOR R-15 RIGID INSULATION ON EXTERIOR FACE OF STEM WALL FROM TOP OF SLAB TO FOUNDATION BTM. PROVIDE FLASHING AND PROTECTION BOARD AT EXPOSED INSULATION ABOVE GRADE.
6. INSULATE BASEMENT WALLS W/ R-21 HIGH DENSITY BATTS IN 2 X 4 FURRING WALL. PROVIDE ALTERNATE BID FOR R-21 RIGID INSULATION ON EXTERIOR FACE OF WALL W/ FLASHING AND PROTECTION BOARD FOR EXPOSED INSULATION ABOVE GRADE.
7. COVER THE EXTERIOR FACE OF ALL EXTERIOR HEATED WALLS W/ TYVEK VAPOR BARRIER. LAP ALL JOINTS 6" MIN. AND TAPE ALL JOINTS.
8. WEATHER STRIP THE ATTIC AND CRAWL ACCESS DOORS. INSULATE ATTIC ACCESS DOOR TO R-38.
9. SET ALL MUDSILLS FOR HEATED WALLS ON NONPORUS SILL SEAL.
10. INSULATE ALL HEATING DUCTS IN UNHEATED AREAS TO R-8 INSULATION TO HAVE A FLAME SPREAD RATING OF 50 MAX. W/ AND A SMOKE DEVELOPMENT RATING OF 100 MAX. ALL DUCT SEAMS TO BE SEALED.

FIGURE 10.6 ■ General insulation notes that could be found on a set of architectural drawings. *Courtesy Residential Designs.*

SOLAR ENERGY DESIGN

Sunlight becomes solar energy when it is transferred to a medium that has the capacity or ability to provide useful heat. This useful solar energy may be regulated in order to heat water or the inside of a building or create power to run electrical utilities. Most areas of the earth receive about 60 percent direct

sunlight each year, and in very clear areas, up to 80 percent of the annual sunlight is available for use as solar energy. When the sun's rays reach the earth, air and the things on the earth become heated. Certain dense materials such as concrete can absorb more heat than less dense materials such as wood. During the day the dense materials absorb and store solar energy. Then at night, with the source of energy gone, the stored energy is released in the form of heat. Some substances, such as glass, absorb thermal radiation while transmitting light. This concept, in part, is what makes solar heating possible. Solar radiation enters a structure through a glass panel and warms the surfaces of the interior areas. The glass keeps the heat inside by absorbing the radiation.

Two basic residential and commercial uses for solar energy are heating spaces and hot water. Other uses are industrial and include drying materials such as lumber, masonry, or crops. Solar energy has also been used for nearly a century with desalinization plants to provide freshwater from either mineral or saltwater.

Solar space heating systems are passive, active, or a combination of the two. *Passive,* or architectural, systems use no mechanical devices to retain, store, or radiate solar heat. *Active,* or mechanical, systems do use mechanical devices to absorb, store, and use solar heat.

The potential reduction of fossil fuel energy consumption can make solar heat an economical alternative. A number of factors contribute to the effective use of solar energy. Among them are energy-efficient construction techniques and full insulation of a building to reduce heat loss and air infiltration.

An auxiliary heating system is often used as a backup or supplemental system in conjunction with solar heat. The amount of heat needed from the auxiliary system depends on the effectiveness of the solar system. Both the primary and the supplemental system should be professionally engineered for optimum efficiency and comfort.

A southern exposure, when available, provides the best site orientation for solar construction. A perfect solar site allows the structure to have an unobstructed southern exposure. See Chapter 11.

Room placement is another factor to consider for taking advantage of solar heat. Living areas, such as the living and family rooms, should be on the south side of the house, while inactive rooms, such as bedrooms, laundries, and baths, should be located on the north side, if a cooler environment is desirable. If possible, the garage should be placed on the north, northeast, or northwest side of the home. A garage can act as an effective barrier insulating the living areas from cold exterior elements. The inactive rooms, including the bedrooms, baths, laundry, pantry, and garage, shelter the living area from the northern exposure. Masonry walls also allow the house to be built into a slope on its northern side, or berms to be used as shelter from the elements.

Another energy-efficient element of design is an *air-lock entry,* known as a *vestibule.* This is an entry that provides a hall or chamber between an exterior and interior door to the building. The vestibule should be designed so that the interior and exterior doors will not be open at the same time. The distance between doors should be at least 7′, to force the occupants to close one door before they reach the other. The main idea behind an air-lock entry is to provide a chamber that is always closed to the living area by a door. When the exterior door is opened, the air lock loses heat, but the heat loss is confined to the small space of the vestibule and the warm air of the living area is exposed to minimum heat loss.

Living with Solar Energy Systems

Solar energy and energy-efficient construction require a commitment to conserving energy. Each individual must evaluate cost against potential savings and become aware of the responsibility of living with energy conservation.

Solar systems can be designed that provide some heat from the sun and require little or no involvement from the occupant to assist the process. Such a minimal energy-saving design is worth the effort in many cases. Active solar space heating systems are available that can provide a substantial amount of needed heat energy. These systems are automatic and also do not generally require involvement from the homeowner, although he or she should be aware of operation procedures and maintenance schedules. On the other hand, some passive solar heating systems require much participation by the occupant. For example, a mechanical shade that must be maneuvered by the homeowner can be used to block the summer sun's rays from entering southern-exposure windows. During the winter months, when the sun is heating the living area the heat should be retained as long as possible. In the morning the homeowner needs to open shutters or drapes to allow sunlight to enter and heat the rooms. In the early evening, before the heat begins to radiate out of the house, the occupant should close the shutters or draperies to keep the heat within.

Codes and Solar Rights

Building permits are generally required for the installation of active solar systems or the construction of passive solar systems. Some installations also require plumbing and electrical permits. Verify the exact requirements for solar installations with local building officials. During the initial planning process, always check the local zoning ordinances to determine the feasibility of the installation. For example, many areas restrict dwelling height. If the planned solar system encroaches on this zoning rule, then a different approach or a variance to the restriction should be considered.

The individual's access to sunlight is not always guaranteed. A solar home may be built in an area that has excellent solar orientation, and then a few years later, a tall structure may be built across the street that blocks the sun. A neighbor's trees may grow tall and reduce sunlight. Determine the possibility of such a problem before construction begins. Some local zoning ordinances, laws, or even deed restrictions do protect the individual's right to *light.* In the past, laws generally provided the right to receive light from above the property but not from across neighboring land. This situation is changing in many areas of the country.

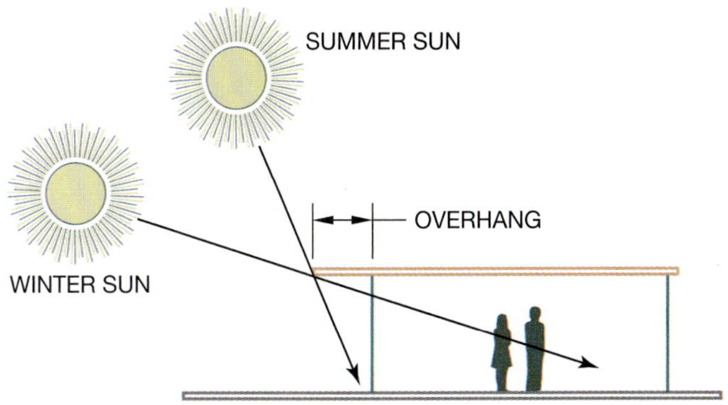

FIGURE 10.7 ■ Effective overhang.

Roof Overhang

Since the sun's angle changes from season to season—it is lower on the horizon in the winter and higher in the summer—an overhang can shield a major glass area from the heat of the summer sun and also allow the lower winter sun to help warm the home. Figure 10.7 shows an example of how a properly designed overhang can aid in the effective use of the sun's heat.

An overhang that provides about 100 percent of shading at noon on the longest day of the year can be calculated with a formula that divides the window height by a factor determined in relationship to latitude:

$$\text{Overhang} = \frac{\text{Windowsill height}}{\text{F}}$$

NORTH LATITUDE	F	NORTH LATITUDE	F
28°	8.4	44°	2.4
32°	5.2	48°	2.0
36°	3.8	52°	1.7
40°	3.0	56°	1.4

Calculate the recommended southern overhang for a location at 36° latitude and provide for a window height of 6 ft 8 in.

$$\text{Overhang (OH)} = \frac{6'\ 8''.\ \text{(Window height)}}{3.8\ \text{(F at 36° latitude)}}$$

$$= \frac{6.6}{3.8} = 1.8 \approx 1'\ 10''$$

The recommended overhang is greater for a more northerly latitude.

Overhang protection can be constructed in ways other than a continuation of the roof structure; there is a great deal of flexibility in architectural design. An awning, porch cover, or trellis may be built to serve the same function as an overhang. Alternative methods of shading from summer heat and exposing window areas to winter sun can be achieved with mechanical devices. These movable devices require that the occupant be aware of the need for shade or heat at different times of the year. Figure 10.8 shows some options.

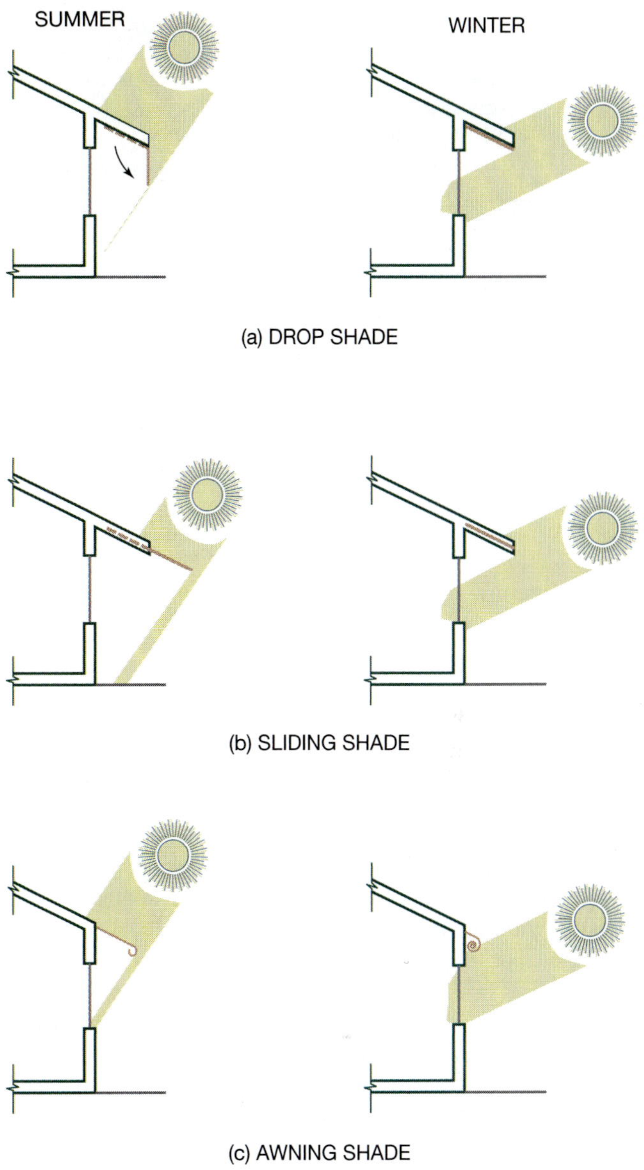

FIGURE 10.8 ■ Optional shading devices.

Passive Solar Systems

In passive solar architecture, the structure is designed so that the sun directly warms the interior. A passive solar system allows sunlight to enter the structure and be absorbed into a structural mass. The stored heat then warms the living space. Shutters or curtains are used to control the amount of sunlight entering the home, and vents help control temperature. In passive solar construction, the structure is the system. The amount of material needed to store heat depends on the amount of sunlight, the desired temperature within the structure, and the ability of the material to store heat. Materials such as water, steel, concrete, and masonry have good heat capacity; wood does not. Several passive

FIGURE 10.9 ■ South-facing glass provides direct gain. *Courtesy Residential Designs.*

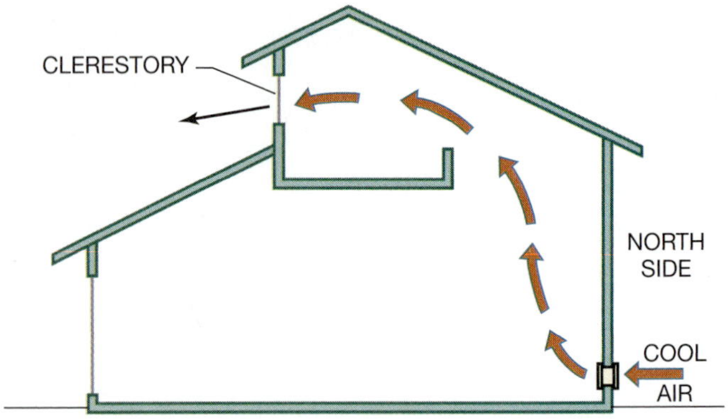

FIGURE 10.10 ■ Clerestory circulation.

solar architectural methods have been used individually and together, including south-facing glass, thermal storage walls, roof ponds, solariums, and envelope construction.

South-Facing Glass

Direct solar gain is a direct gain in heat created by the sun. This direct gain is readily created in a structure through south-facing windows. Large window areas facing south can provide up to 60 percent of a structure's heating needs when the windows are insulated at night with tight-fitting shutters or insulated curtains. When the window insulation is not used at night, the heat gain during the day is quickly lost. The sun's energy must heat a dense material in order to be retained. Floors and walls are often covered with or made of materials other than traditional wood and plasterboard. Floors constructed of or covered with tile, brick, or concrete, and special walls made of concrete or masonry or containing water tubes, can provide heat storage for direct gain applications. Figure 10.9 shows a typical application of direct-gain south-facing glass.

Clerestory windows can be used to provide light and direct solar gain to a second-floor living area and increase the total solar heating capacity of a house. These clerestory units can also be used to help ventilate a structure during the summer months, when the need for cooling may be greater than the need for heating. Look at Figure 10.10.

Skylights can be used effectively for direct solar gain. When skylights are placed on a south-sloping roof, they provide needed direct gain during the winter. In the summer, these units can cause the area to overheat unless additional care is taken to provide for ventilation or a shade cover. Some manufacturers have skylights that open and may provide sufficient ventilation. Figure 10.11 shows an application of openable skylights.

FIGURE 10.11 ■ Skylights. *Courtesy Velux-America, Inc.*

Thermal Storage Walls

Thermal storage walls can be constructed of any good heat-absorbing material such as concrete, masonry, or water-filled cylinders. The storage wall can be built inside and next to a large southern-exposed window or group of windows. The wall receives and stores energy during the day and releases the heat slowly at night.

The *Trombe wall,* designed by a French scientist, Dr. Felix Trombe, is a commonly used thermal storage wall. The Trombe wall is a massive dark-painted masonry or concrete wall situated a few inches inside and next to south-facing glass. The sun heats the air between the wall and the glass. The heated air rises and enters the room through vents at the top of the wall. At the same

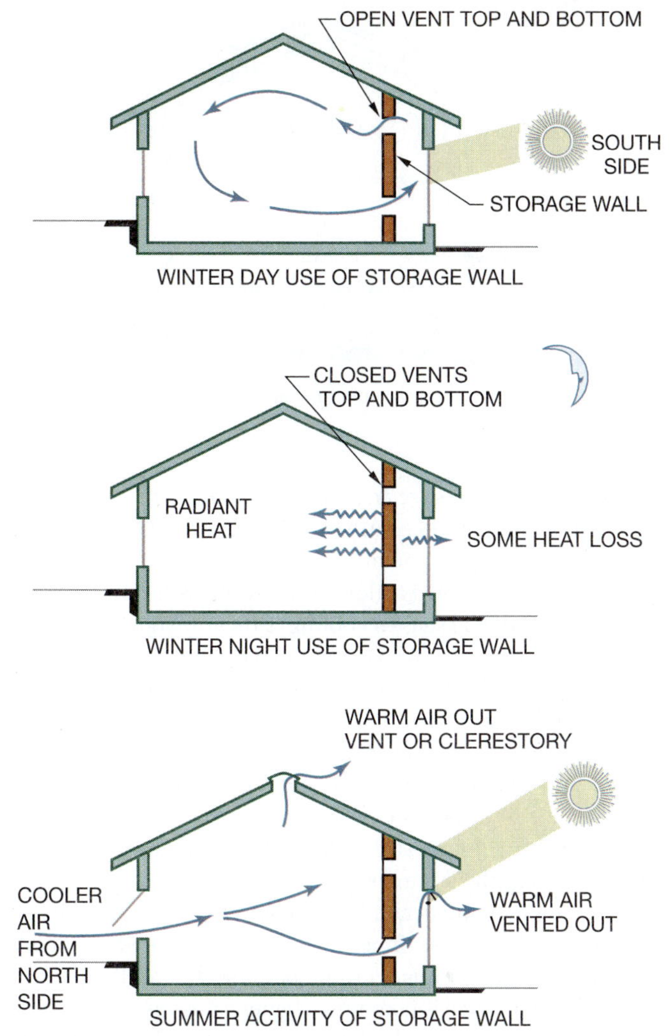

FIGURE 10.12 ■ Thermal storage wall.

time, cool air from the floor level of adjacent rooms is pulled in through vents at the bottom of the wall. The vents in the Trombe wall must be closable to avoid losing the warm air from within the structure through the windows at night. The heat absorbed in the wall during the day radiates back into the room during the night hours. The Trombe wall also acts to cool the structure during summer. This happens when warm air rises between the wall and glass and is vented to the outside. The air currents thus created work to pull cooler air from an open north-side window or vent. Figure 10.12 shows an example of the thermal storage wall.

Some passive solar structures use water as the storage mass: large vertical water-filled tubes or drums painted dark to absorb heat are installed as the Trombe wall. The water functions as a medium to store heat during the day and release heat at night.

Roof Ponds

Roof ponds are used occasionally in residential architecture, although they are more common in commercial construction. A roof pond is usually constructed of containers filled with antifreeze and water on a flat roof. The water is heated during the winter days, and then at night the structure is covered with in-

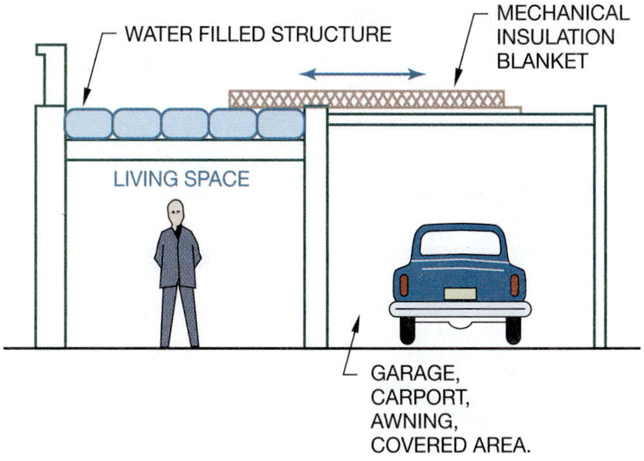

FIGURE 10.13 ■ Roof ponds.

sulation, which allows the absorbed heat to radiate into the living space. This process functions in reverse during the summer. The water-filled units are covered with insulation during the day and uncovered at night to allow any stored heat to escape. To assist the radiation of heat at night, the structure should be constructed of a good thermal-conducting material such as steel. Figure 10.13 shows an example of the roof pond system.

Solariums

A *solarium*, sometimes called a sunroom or solar greenhouse, is designed to be built on the south side of a house next to the living area. Solariums are a nice part of the living area when designed into a home. The additional advantage of the solarium is the greenhouse effect. These rooms allow plants to grow well all year.

The theory behind the solarium is that it will absorb a great amount of solar energy and transmit it to the balance of the structure. A thermal mass may also be used in conjunction with the solarium. Heat from the solarium can be circulated throughout the entire house by natural convection or through a forced-air system. The circulation of hot air during the winter and cool air during the summer in the solar greenhouse functions as in a Trombe wall.

The solarium can overheat during long, hot summer days. This potential problem can be reduced by mechanical ventilators or a mechanical humidifier. Exterior shading devices are often advantageous; a cover can be rolled down over the greenhouse glass area. The use of landscaping with southern deciduous trees, as discussed in Chapter 11, is a suggested alternative. Figure 10.14 shows how a solarium operates to provide solar heat and circulate summer cooling.

Envelope Design

Envelope design is based on the idea of constructing an envelope or continuous cavity around the perimeter of a structure. A solarium is built on the south side with insulating double-pane glass to act as a solar collector. During the winter, when solar energy is required to heat the structure, the warm air that is

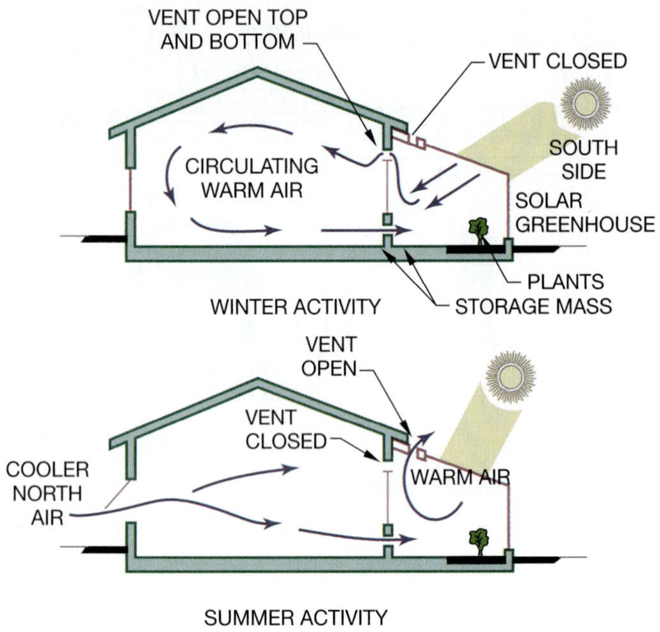

FIGURE 10.14 ■ Solarium construction.

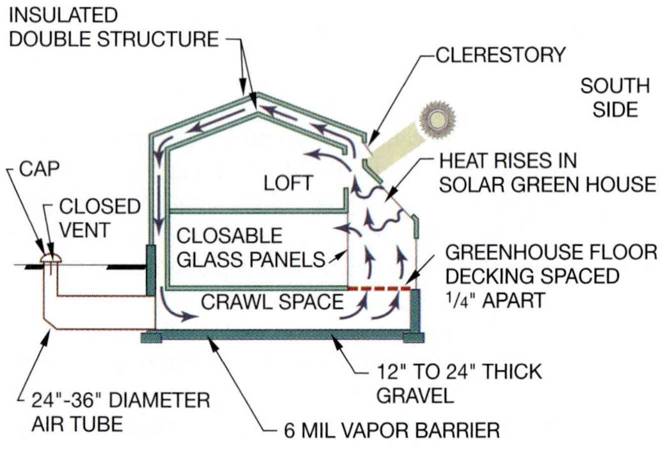

FIGURE 10.15 ■ Winter envelope activity.

created in the solarium rises, and the convection currents cause the heated air to flow around the structure through the envelope cavity. Look at Figure 10.15.

The envelope crawl space floor is a storage mass. The rock stores heat during the day and releases the energy slowly at night. This functions in reverse during the summer to help cool the cavity and the structure. The greenhouse floor acts as a vent, with floor decking spaced to allow air currents to pass through and provide the continuous current. The heated air in the envelope cavity helps keep the living space at a constant temperature. A properly designed envelope structure can completely eliminate the need for a backup heating system. Some envelope dwellings use a wood- or coal-burning appliance for backup heat; others use small forced-air systems.

Summer heat is often a concern in solar design, although an envelope can cool effectively when the sun is high in the sky.

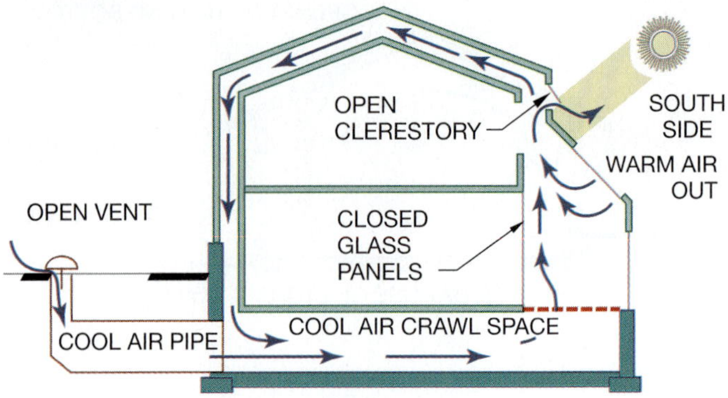

FIGURE 10.16 ■ Summer envelope activity.

Deciduous shade trees are an added advantage for summer cooling when planted on the south side. The cooling process is a reverse of the heating operation. The clerestory windows are opened, and heated air is allowed to escape as it rises. Another design feature that assists in cooling is the underground pipe or tube that is constructed in connection with the cool crawl space. The underground temperature remains about 55° F, and when the vent to the underground pipe is open to outside air ventilation, a convection current is again created. This time the air is cool and flows under the cool crawl space and around the envelope. This same system can be used in conjunction with a solar greenhouse. Figure 10.16 shows how the envelope design can cool during the summer.

Insulation is an important factor in envelope design, with both the outer shell and the inner shell being insulated. Generally R-19 insulation is adequate on the outer shell, and R-11 insulation on the inner shell is common. Higher values generally do not add to the efficiency of the structure. Foundation walls at the crawl space perimeter and floor should be adequately insulated.

The passive envelope system is an effective method of reducing energy consumption and creating an excellent living environment. Standard construction techniques are generally used, although the cost of two walls and roof structures is a factor. Another problem is the framing and installation of windows and openings in the double walls of the envelope. Research local and national building codes when considering the envelope design. Factors that should be discussed include fire safety. Codes often require that specific safety precautions be taken to damper the cavity mechanically in case of fire. The same air currents that carry the air around the envelope also allow flames to engulf the structure quickly.

If proper precautions are not taken, a fire in the home can turn the envelope into a giant wooden flue, and the inside of the house will ultimately become the firebox and proliferate the flames into an all-fuel fireplace. National codes in general specify fire blocking at the floor and ceiling plates and require the designer to submit alternatives for the envelope design. Most alternatives include the addition of smoke-activated dampers in the cavity at the floor and ceiling lines plus lining the air chamber with drywall. The liner may improve the efficiency of the structure but add tremendously to the cost. Some envelope designs have proved better at cooling in the summer than heating

FIGURE 10.17 ■ Typical applications of solar hot water collectors. *Courtesy Lennox Industries, Inc.*

in the winter, and this may be a plus. However, the popularity of the envelope design is declining because of the added cost compared with other systems. The potential energy saving may not be much better than that achieved with excellent energy-efficient construction.

Active Solar Systems

Active solar systems for space heating use collectors to gather heat from the sun, which is transferred to a fluid. Fans, pumps, valves, and thermostats move the heated fluid from the collectors to an area of heat storage. The heat collected and stored is then distributed to the structure. Heat in the system is transported to the living space by insulated ducts, which are similar to the ducts in a conventional forced-air heating system. An active solar heating system generally requires a backup heating system that is capable of handling the entire heating needs of the building. Some backup systems are forced-air systems that use the same ducts as the solar system. The backup system may be heated by electricity, natural gas, or oil. Some people rely on wood- or coal-burning appliances for supplemental heat, but local codes should be checked. Active solar systems are commonly adapted to existing homes or businesses that normally use a forced-air heating system.

Solar Collectors

As the name implies, solar collectors catch sunlight and convert this light to heat. Well-designed solar collectors are nearly 100 percent efficient.

The number of solar collectors needed to provide heat for a given structure depends on the size of the structure and the volume of heat needed. These are the same kinds of determinations made when sizing a conventional heating system. Collectors are commonly placed in rows on a roof or on the ground next to the structure, as shown in Figure 10.17. Architects and designers are integrating the solar panels into the total design in an effort to make them less obvious. The best placement requires an unobstructed southern exposure. The angle of the solar collectors should be in proper relationship to the angle of the sun during winter months, when the demand for heat is greatest. Verify the collector angle that is best suited for a specific location. Some collectors are designed to be positioned at 60° from horizontal so that winter sun hits at about 90° to the collector. While this is the ideal situation, a slight change from this angle of up to 15° may not alter the efficiency very much. Other collectors are designed to obtain solar heat by direct and reflected sunlight. Reflecting, or focusing, lenses help concentrate the light on the collector surface. Figure 10.18 shows an example of tilt.

Storage

During periods of sunlight, active solar collectors transfer heat energy to a storage area and then to the living space. After the demand for heat is met, the storage facilities allow the heat to be contained for use when solar activity is reduced, such as at night or during cloud cover.

The kind of storage facilities used depends on the type of collector system used. There are water, rock, and chemical storage systems. Water storage is excellent, since water has a high capacity for storing heat. Water in a storage tank is used to absorb heat from a collector. When the demand for heat exceeds the collector's output, the hot water from the storage tank is pumped into a radiator, or through a water-to-air heat exchanger for dispersement to the forced-air system. Domestic hot water may be provided by a water-to-water heat exchanger in the storage tank.

Rock storage is often used when air (rather than water) is the fluid in the collectors. Heated air flows over a rock storage bed. The rock absorbs some heat from the air while the balance is distributed to the living space. When the solar gain is minimal, cooler air passes over the rock storage, where it absorbs heat and is distributed by fans to the living space. Domestic hot water may also be provided by an air-to-water heat exchanger situated in the rock storage or in the hot air duct leading from the collector.

Chemicals used in collectors and in storage facilities absorb large amounts of heat at low temperatures. Many of these systems claim to absorb heat on cloudy days or even when the

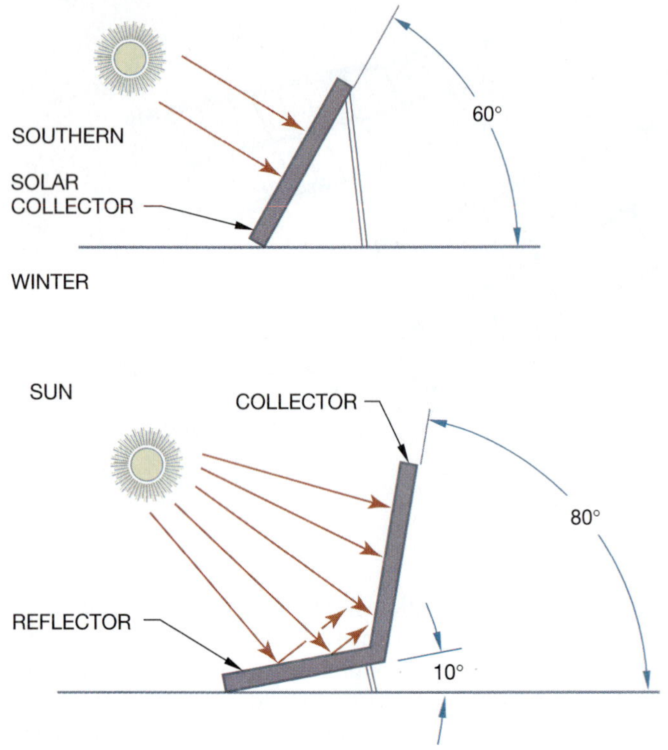

FIGURE 10.18 ■ Solar collector tilt.

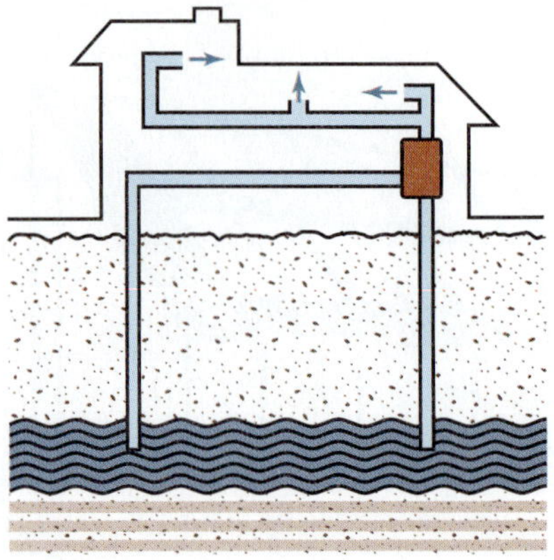

FIGURE 10.19 ■ Groundwater system. *Courtesy Solar Oriented Environmental Systems, Inc.*

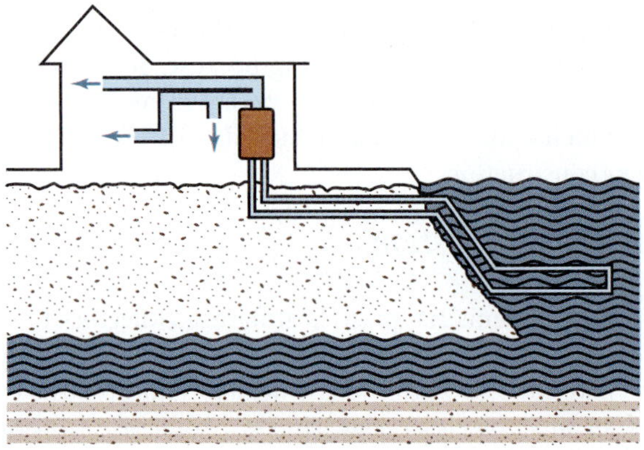

FIGURE 10.20 ■ Geothermal freon exchanger in a lake. *Courtesy Solar Oriented Environmental Systems, Inc.*

collectors are snow-covered. Chemicals with very low freezing temperatures can actually absorb heat during winter months, when the outside temperature is low.

Solar Architectural Concrete Products

Solar collectors may be built into a patio, driveway, tennis court, pool deck, or tile roof.

Solar architectural concrete products make the most versatile solar collector on the market. It can be used as a driveway, sidewalk, patio, pool deck, or roofing material, or as the side surface of a building wall or fence. The finished surface can resemble cobblestone, brick, or roof tiles. Pigments are added to the specially formulated material, giving it a lasting color that is pleasing to the eye. These solar products are completely different from all other solar collectors in that they add to the aesthetic value of the property. They are manufactured from a specially formulated mixture that is strong, dense, very conductive, and waterproof. Solar architectural cement products absorb and collect heat from the sun and outside air and transfer the heat into water, glycol, or another heat-transferring fluid passing through imbedded tubes.

Geothermal Systems

Geothermal heating and cooling equipment is designed to use the constant, moderate temperature of the ground to provide space heating and cooling , or domestic hot water, by placing heat exchangers in the ground, or in wells, lakes, rivers, or streams. The basic types of geothermal systems are water- and refrigerant-based. The *water-based system* has closed-loop and open-loop options. The *closed-loop system* has water or water-antifreeze fluid pumped through polyethylene tubes. The *open-loop system* has water pumped from a well or reservoir through a heat exchanger, and then discharged into a drainage ditch, a field tile, a reservoir or another well. The *refrigerant-based system* is also known as direct exchange where refrigerant flows in copper tubing around a heat exchanger. Direct exchange represents only a small number of geothermal systems and the Air Conditioning and Refrigeration Institute does not recognize these.

A geothermal system operates by pumping groundwater from a supply well and then circulating it through a heat exchanger, where either heat or cold is transferred by a refrigerant. The water, which has undergone only a temperature change, is then returned through a discharge well back to the strata, as shown in Figure 10.19. Lakes, rivers, ponds, streams, and swimming pools may be alternate sources of water. Rather than pumping the water up to the heat exchanger, a geothermal exchanger can be inserted into a lake, river, or other natural body of water to extract heat, as shown in Figure 10.20. To cool, the geothermal heat pump extracts heat from within the structure

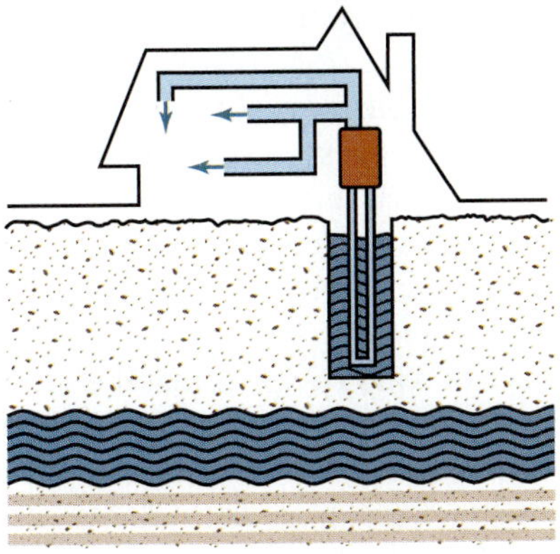

FIGURE 10.21 ■ Geothermal freon exchanger in a well. *Courtesy Solar Oriented Environmental Systems, Inc.*

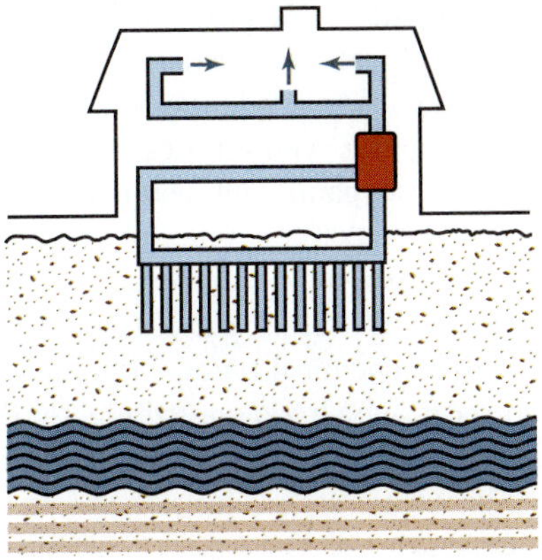

FIGURE 10.23 ■ Ground loop system. *Courtesy Solar Oriented Environmental Systems, Inc.*

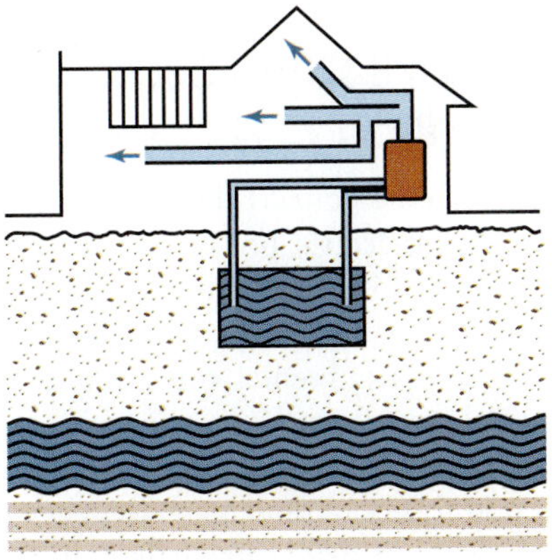

FIGURE 10.22 ■ Solar-assisted system. *Courtesy Solar Oriented Environmental Systems, Inc.*

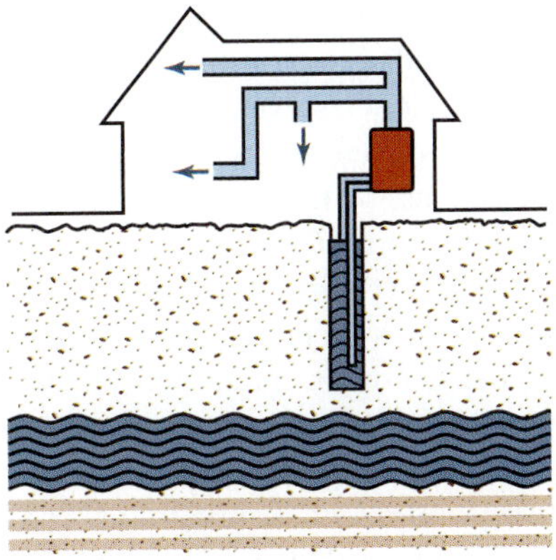

FIGURE 10.24 ■ Vertical dry system. *Courtesy Solar Oriented Environmental Systems, Inc.*

and transfers it to the fluid, where the heat dissipates into the ground as it circulates.

Another alternative is to insert a geothermal freon exchanger directly into a well, where the natural convection of the groundwater temperature, assisted by the normal flow of underground water in the strata, is used to transfer the hot or cold temperature that is required. Sometimes the thermal transfer must be assisted by forcing the freon to circulate through tubes within the well by a low-horsepower pump. See Figure 10.21.

A geothermal system may assist a solar system that uses water as a heat storage and transfer medium. After the water heated by solar collectors has given up enough heat to reduce its temperature to below 100°F, this water becomes the source for operating the geothermal system. See Figure 10.22.

When an adequate supply of water is unavailable, the ground, which always maintains a constant temperature below the frost line, can be used to extract either heat or cold. The thermal extraction is done through the use of a closed-loop system consisting of polyethylene tubing filled with a glycol solution and circulated through the geothermal system. One such system is shown in Figure 10.23. Another method is to use a vertical dry-hole well, which is sealed to enable it to function as a closed-loop system, as shown in Figure 10.24.

The geothermal system is mechanically similar to a conventional heat pump, except that it uses available water to cool the refrigerant or to extract heat, as opposed to using 90° F air for cooling or 20 to 40° F air for heating. The difference between air source and ground or water source is that air temperature can change much more than ground or water temperatures. This gives the ground or water source heat pump more capacity and makes it more efficient. Water can store great amounts of geothermal energy because of its high *specific heat* that is the amount of energy required to raise the temperature of any substance 1° F. The specific heat of air is only 0.018, so it can absorb and release only 1/50 of the energy that water can. To produce a given amount of heat, fifty times more air volume than water is required to pass through the unit.

Photovoltaic Modules

Photovoltaic technology has been developed and refined, and photovoltaic modules are now powering thousands of

installations worldwide. Solar photovoltaic systems are now producing millions of watts of electricity to supply power for remote cabins, homes, railroad signals, water pumps, telecommunications stations, and even utilities. Uses for this technology are continuing to expand, and new installations are being constructed.

Photovoltaic cells turn light into electricity. The word *photovoltaic* is derived from the Greek *photo,* "light"; and *voltaic,* "to produce electricity by chemical action." Photons strike the surface of a silicon wafer, which is a semiconductor diode, to stimulate the release of mobile electric charges that can be guided into a circuit to become a useful electric current.

Photovoltaic modules produce direct current (DC) electricity. This type of power is useful for many applications and for charging storage batteries. When alternating current (AC) is required, DC can be changed to AC by an inverter. Some photovoltaic systems are designed to use the energy produced immediately, as is often the case in water-pumping installations. If the energy produced by solar electric systems is not to be used immediately, or if an energy reserve is required for use when sunlight is not available, the energy must be stored. The most common storage devices are batteries.

Although both DC and AC systems can stand alone, AC systems can also be connected to a utility grid. During times of peak power usage, the system can draw on the grid for extra electricity if needed. At other times, such a system may actually return extra power to the grid. In most areas of the country the utility (grid) is required to purchase excess power.

Correct site selection is vitally important. The solar modules should be situated where they receive maximum exposure to direct sunlight for the longest period of time every day. Other considerations are distance from the load (appliance); shade from trees or buildings, which changes with the seasons; and accessibility.

The future looks bright for solar technologies. The costs and efficiencies of high-technology systems are improving. As homeowners watch their heating and cooling costs rise and become more concerned about shortages of oil and gas, solar heating becomes an attractive alternative. In preparing a preliminary design, solar alternatives should be considered. A qualified solar engineer should evaluate the site and recommend solar design alternatives. Additional assistance is usually available from state or national Departments of Energy.

Wind Energy

Wind energy is a form of solar energy. Winds are caused by the uneven heating of the atmosphere by the sun, the irregularities of the earth's surface, and the earth's rotation. Wind turbines capture energy from the wind. A *wind turbine* is turned by the wind passing over propellers that power an electric generator, creating a supply of electricity. Large-scale applications use several technologically advanced turbines grouped together in what is referred to as a *wind farm.* This wind power plant generates electricity that is fed into a local utility for distribution to customers. Small-scale wind turbines are also available for home or business use. The wind speed in an area determines the effectiveness of possible power generation by wind. Good wind areas have an annual minimum average wind speed of 13 miles per hour. Contact your state or the U.S. Department of Energy to determine if your area is a good location for wind power.

The advantage of wind energy is that the source is a free, renewable, clean resource. However, you must have a good location where wind is available, the initial cost is a factor to consider, there is some noise from the turbines, and mortality of birds flying into the turbines is a concern. The initial cost is balanced out over time when compared with the cost of fuels used for other systems. It is possible that wind power could produce 10% of the United States' energy needs by 2050.

HYDROELECTRIC POWER

Hydroelectric generators convert the energy from falling water into electricity. Hydroelectric power is a renewable resource, which supplies about 15% of the United States' electricity needs and about 60% in Canada. While hydroelectric resources are abundant, there is environmental concern about the blockage by dams of natural river flow.

Hydroelectric power for residential and business use is dependent on a good source of flowing water. The two key factors are the vertical distance that water falls that is referred to as *head,* and flow rate. The site should be evaluated and tested by a hydrology engineer to determine its feasibility. Local codes and environmental regulations must be considered before planning a hydroelectric project. Power produced from hydroelectric sources is clean, and this makes it an excellent consideration when compared with power plants fired by fossil fuels. However, other environmental impacts, such as fish passage, must be considered.

HEALTHY ARCHITECTURAL DESIGN AND CONSTRUCTION

Constructing a healthy building involves issues related to construction techniques, construction materials, and heating, ventilating, and air-conditioning design. Many of the proper construction methods have been discussed throughout this chapter. Techniques that result in energy-efficient construction are important, because the goal is to design and build an airtight structure. The methods used to reduce outside air infiltration range from the selection and proper installation of insulation and caulking products to the installation of continuous vapor retarders.

The risks involved in building an airtight structure involve air quality and range from stale air to a buildup of harmful contaminants. These issues can be addressed with a well-designed and properly installed heat recovery and ventilation system. Sources of pollutants and methods used to deal with these problems are discussed in Chapter 19, with complete coverage of heating, ventilating, and air conditioning (HVAC) systems. The selection of building materials and products is also an important consideration. Avoid manufactured wood products such as fiberboard, particleboard, and plywood. Use water-based caulking, paint, and finishes. Keep carpeting to a minimum, or eliminate it completely for very sensitive clients. An alternative to synthetic carpeting is natural fiber products such as wool and cotton. Combustion appliances (use gas fuel) such as furnaces, ranges, water heaters, and clothes dryers must be properly ventilated, and combustion must be complete. An electric or heat pump furnace; electric water heater, range, and dryer; or sealed combustion appliances should be considered. Avoid treated wood products in the construction of the building envelope, although they may be considered for outside use, as in decks. In regions where treated wood is commonly used for termite and rot control, steel frame construction is a consideration. Steel frame construction has been commonly used in commercial structures and is becoming a serious consideration for residences.

The following is a checklist of features and products that can be considered for designing a healthy structure:

- Design and build with high-quality insulation, caulking, and vapor retarders.
- Use airtight well-insulated wood or metal frame windows.
- Use unbleached plasterboard and plaster rather than chemically produced sheetrock and materials.
- Install a properly designed heat recovery ventilation system, also called an air-to-air heat exchanger.
- Install a high-performance particle air filtration system.
- Use low-toxicity ceramic tiles and hardwood floors or natural fiber carpets rather than synthetic carpets.
- Use wood decking such as tongue-and-groove rather than plywood and particleboard. Low-formaldehyde products can be considered.
- Consider low-toxicity foam insulation or other nonchemical insulation products.
- Use water-base, solvent-free, low-toxicity paints and finishes.
- Construct a tightly sealed foundation to help keep out radon gas and moisture. Radon gas is discussed in Chapter 19.
- Use a heat pump such as a geothermal unit that produces no indoor pollutants.
- Install gas appliances and fireplaces that use only outdoor air and exhaust to the outside without any downdraft.
- Install an electrical system that is designed to reduce stray voltage. Some people are sensitive to power surges.
- Use a sealed and well-balanced metal duct system. Avoid fiberglass insulation ducts. Metal ducts can be insulated on the outside.
- Install a central vacuum system that vents particles out of the house into a filtered container in the garage.
- Install a garage ventilation system.
- Use ceramic roofing.

Chapter 10 Additional Reading

The following Web sites can be used as a resource to help you keep current with changes in building materials.

ADDRESS	COMPANY OR ORGANIZATION
www.ase.org	Alliance to Save Energy
www.ases.org	American Solar Energy Society
www.certainteed.com	Certainteed (Insulation)
www.dap.com	Dap (caulking)
www.doe.gov	U.S. Department of Energy
www.osisealants.com	OSI Sealants, Inc.
www.seia.org	Solar Energy Industry Association
www.solstice.crest.org	Sustainable Energy
www.usgbc.org	U.S. Green Building Council

CHAPTER 10

Conservation and Environmental Design and Construction Test

DIRECTIONS

Answer the questions with short, complete statements or drawings as needed on an 8 1/2 by 11 sheet as follows:

1. Letter your name, Chapter 10 Test, and the date at the top of the sheet.
2. Letter the question number and provide the answer. You do not need to write out the question.
3. Answers may be prepared on a word processor if course guidelines allow this.

QUESTIONS

Question 10–1 List three reasons that energy-efficient construction techniques are becoming important.

Question 10–2 Describe at least one factor that may be a disadvantage of energy-efficient construction.

Question 10–3 What is the primary goal of energy-efficient design and construction?

Question 10–4 Identify at least one goal of conservation design.

Question 10–5 Why is it possible to use energy-efficient design in any architectural style?

Question 10–6 Why is it possible to implement some energy-efficient design in every new home?

Question 10–7 List two site factors that contribute to energy-efficient design.

Question 10–8 List two room layout factors that contribute to energy-efficient design.

Question 10–9 List two window placement factors that contribute to energy-efficient design.

Question 10–10 List two factors that influence a reduction in air infiltration into a structure.

Question 10–11 Identify at least two goals of an energy-efficient construction system.

Question 10–12 List three general materials for energy-efficient construction.

Question 10–13 What is caulking? Where is it typically used?

Question 10–14 What is the thickness for a vapor barrier in a crawl space?

Question 10–15 List the typical R value for the following uses based on the DOE recommendations:

a. Flat ceilings
b. Floors
c. Walls

Question 10–16 What are the two basic residential and commercial uses for solar heat?

Question 10–17 Describe passive and active solar heating.

Question 10–18 List and describe three factors that influence a good solar design.

Question 10–19 Discuss the concept of right to light.

Question 10–20 Define each of the following solar systems and make a sketch to show how it functions:

a. South-facing glass
b. Clerestory windows
c. Thermal storage wall
d. Roof ponds
e. Solarium
f. Envelope design

Question 10–21 Define *solar collector.*

Question 10–22 What are solar architectural concrete products, and how do they function?

Question 10–23 Describe three applications of geothermal heating and cooling systems.

Question 10–24 Define the function of photovoltaic cells.

Question 10–25 What are radiant barriers, and what is their function?

Question 10–26 Describe the function of insulation.

Question 10–27 List at least four items that influence how much insulation a home should have.

QUESTIONS *(cont.)*

Question 10–28 What does R-value measure?

Question 10–29 The ability of material to slow heat transfer is called what?

Question 10–30 What does R mean?

Question 10–31 Briefly describe loose-fill insulation.

Question 10–32 Give at least one advantage of loose-fill insulation.

Question 10–33 Provide a brief description of batts and blanket insulation.

Question 10–34 Why is careful installation of batts and blanket insulation important?

Question 10–35 Give at least three advantages of rigid insulation.

Question 10–36 Identify at least three places where rigid insulation is commonly used.

Question 10–37 Give at least two advantages of foam in place insulation.

Question 10–38 Name the only insulation that can be used in closed cavities.

Question 10–39 What is the R-value range per inch of thickness for fiberglass batts and blanket insulation?

Question 10–40 Identify a foam-in-place insulation material that gives high R-values and airtight seals and provides low-toxicity.

Question 10–41 Describe insulated windows.

Question 10–42 What is the advantage of using argon gas in insulated windows?

Question 10–43 What is Low-E glass and what does it do?

Question 10–44 What is the class of a window that has a U-value of 0.40?

Question 10–45 Give the insulation code for the following building components based on sample energy code requirements:

a. Walls
b. Flat ceilings
c. Vaulted ceilings
d. Under floor
e. Basement walls
f. Slab floor edge
g. Forced air ducts
h. Windows
i. Front entry door
j. Other exterior doors
k. Skylights

Question 10–46 What is the recommended overhang for 36° north latitude?

Question 10–47 What is the recommended overhand for 48° north latitude?

Question 10–48 Define *direct solar gain.*

Question 10–49 Give two advantages of wind power generation.

Question 10–50 List at least ten features and/or products that can be considered for designing a healthy home.

11 CHAPTER

Site Orientation

INTRODUCTION

Site orientation is the placement of a structure on the property with certain environmental and physical factors taken into consideration. Site orientation is one of the preliminary factors that an architect or designer takes into consideration when beginning the design process. The specific needs of the occupants—such as individual habits, perceptions, and aesthetic values—are important and need to be considered. The designer or architect has the responsibility of putting together the values of the owner and other factors that influence the location of the structure when designing the home or business. In some cases site orientation is predetermined. For example, in a residential subdivision where all of the lots are 50′ × 100′ (1524 by 3048 m), the street frontage clearly dictates the front of the house. The property line setback requirements do not allow much flexibility. In such a case, site planning has a minimal influence on the design. This chapter presents factors that may influence site orientation: terrain, view, solar, wind, and sound.

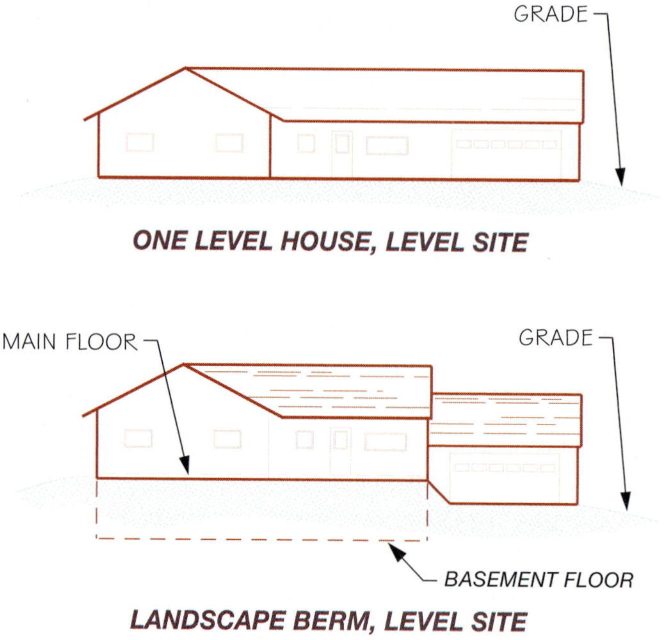

FIGURE 11.1 ■ Level site.

TERRAIN ORIENTATION

The *terrain* is the characteristic of the land on which the proposed structure will be placed. Terrain affects the type of structure to be built. A level construction site is a natural location for a single-level or two-story home. Some landscape techniques on a level site may allow for alternatives.

For example, the excavation material plus extra topsoil could be used to construct an earth *berm*, which is a mound or built-up area. The advantage of the berm is to help reduce the apparent height of a secondary story or to add earth insulation to part of the structure. See Figure 11.1. Construction on any site requires that a slope be graded away from the house. This is to help ensure that normal rainwater will drain away.

Sloped sites are a natural location for multilevel or daylight basement homes. A single-level home is a poor choice on a sloped site because of the extra construction cost for excavation or building up the foundation. See Figure 11.2. Some builders have taken advantage of very steep construction sites by designing homes on stilts. This requires careful geological and structural engineering to ensure safe construction. Figure 11.3 shows how terrain can influence housing types. Subterranean construction has gained popularity, owing to some economical advantages in energy consumption. The terrain of the site is an important factor to consider in the implementation of subterranean designs. See Figure 11.4.

VIEW ORIENTATION

In many situations future homeowners purchase a building site even before they begin the home design. In a large number of these cases, the people are buying a view. The view may be of mountains, city lights, a lake, the ocean, or even a golf course. These view sites are usually more expensive than comparable sites without a view. The architect's obligation to the client in this situation is to provide a home design that optimizes the view. Actually, it is best to provide an environment that allows the occupants to feel as though they are part of the view. Figure 11.5 shows a dramatic example of a view as part of the total environment.

View orientation may conflict with the advantages of other orientation factors such as solar or wind orientation. When a client pays a substantial amount to purchase a view site, such a trade-off may be necessary. When the view dictates that a large glass surface face a wind-exposed nonsolar orientation, some energy-saving alternatives should be considered. Provide as much solar-related glass as possible. Use small window surfaces in other areas to help minimize heat loss. Use triple-glazed windows in the exposed view surface as well as in the

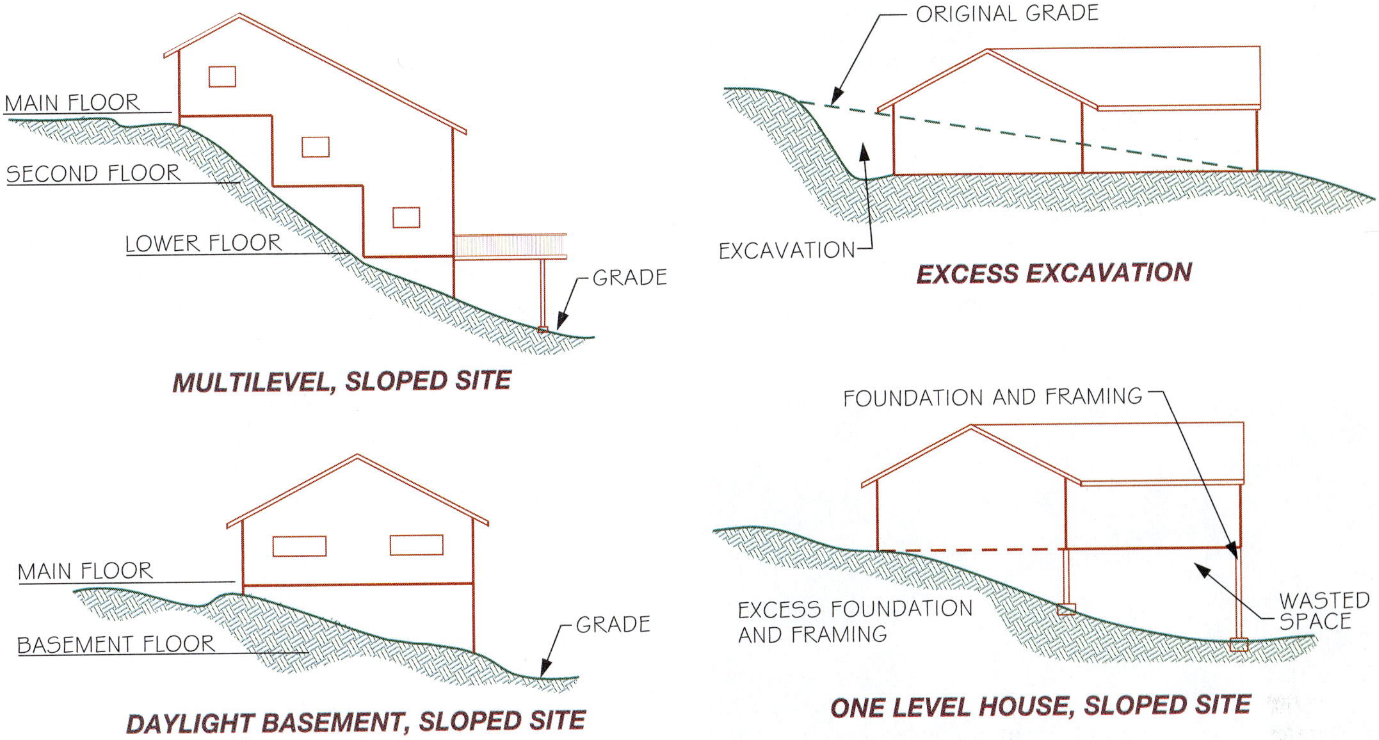

FIGURE 11.2 ■ Sloped sites.

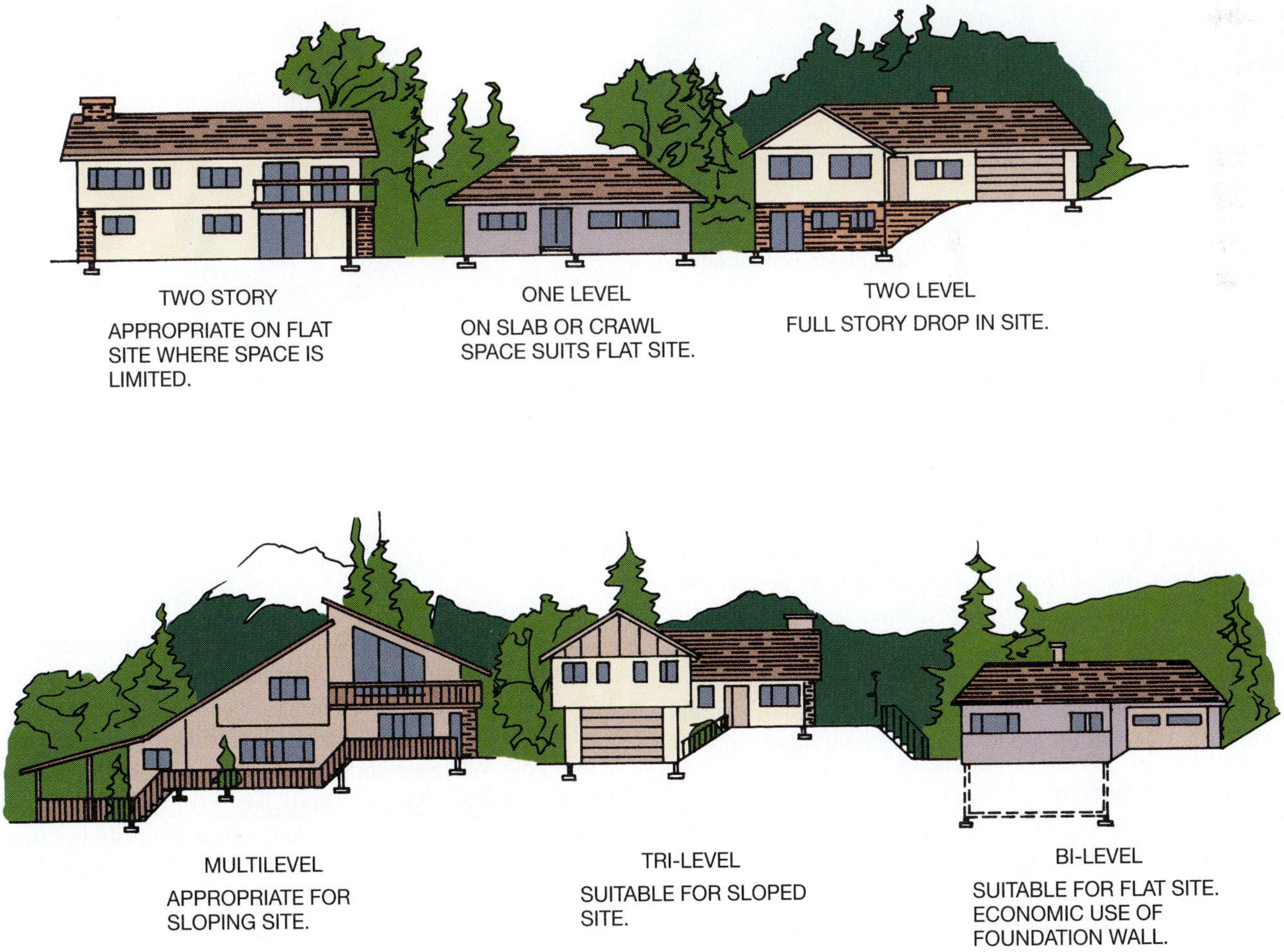

FIGURE 11.3 ■ Terrain determines housing types.

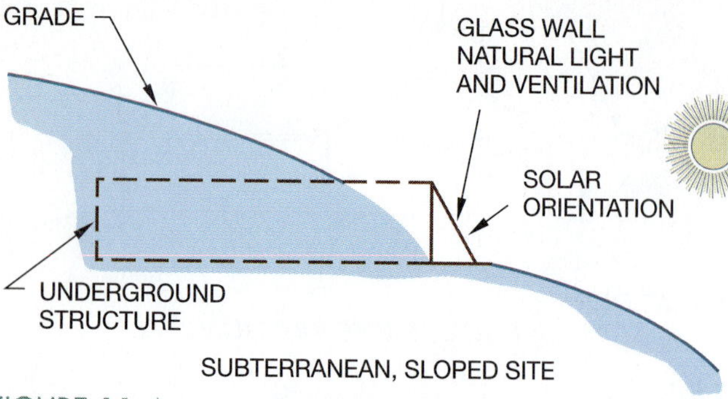

FIGURE 11.4 ■ Subterranean construction site.

FIGURE 11.5 ■ View orientation. *Courtesy of Pella® Windows and Doors.*

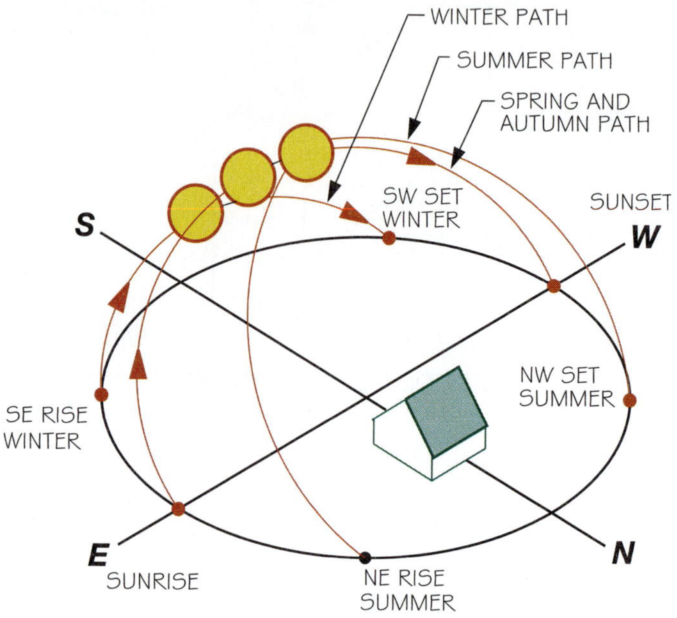

FIGURE 11.6 ■ Southern exposure.

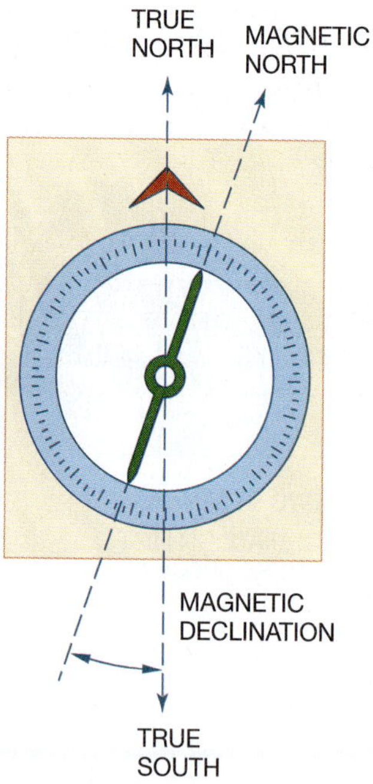

FIGURE 11.7 ■ Magnetic declination.

balance of the structure if economically possible. Insulate to stop air leakage.

SOLAR ORIENTATION

The sun is an important factor in home orientation. Sites with a solar orientation allow for excellent exposure to the sun. There should not be obstacles such as tall buildings, evergreen trees, or hills that have the potential to block the sun. Generally a site located on a south slope has these characteristics.

When a site has southern exposure that allows solar orientation, a little basic astronomy can contribute to proper placement of the structure. Figure 11.6 shows how a southern orientation relative to the sun's path provides the maximum solar exposure.

Establishing South

A perfect solar site allows the structure to have unobstructed southern exposure. When a site has this potential, true south should be determined. This determination should be established in the preliminary planning stages. Other factors that contribute to orientation, such as view, also may be taken into consideration at this time. If view orientation requires that a structure be turned slightly away from south, it is possible that the solar potential will not be significantly reduced.

True south is determined by a line from the North Pole to the South Pole. When a compass is used to establish north-south, the compass points to *magnetic north*. Magnetic north is not the same as true north. The difference between true north and magnetic north is known as the *magnetic declination*. Figure 11.7 shows the compass relationship between true north-south and magnetic north. Magnetic declination differs throughout the country. Figure 11.8 shows a map of the United States with lines that represent the magnetic declination at different locations.

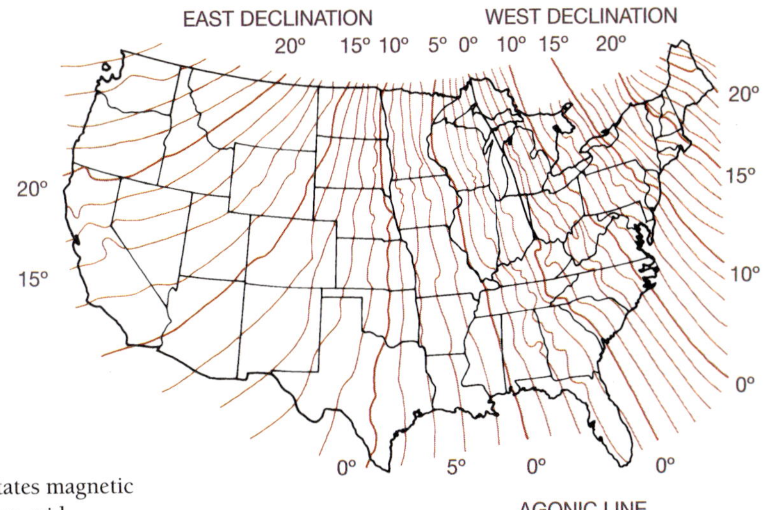

FIGURE 11.8 ■ United States magnetic declination grid.

A magnetic declination of 18° east, which occurs in northern California, means that the compass needle points 18° to the east of true north, or 18° to the west of true south. In this location, as you face toward magnetic south, true south will be 18° to the left.

Solar Site Planning Tools

Instruments are available that calculate solar access and demonstrate shading patterns for any given site throughout the year. *Solar access* refers to the availability of direct sunlight to a structure or construction site.

Some instruments provide accurate readings for the entire year at any time of day, in clear or cloudy weather.

Solar Site Location

A solar site in a rural or suburban location where there may be plenty of space to take advantage of a southern solar exposure allows the designer a great deal of flexibility. Some other factors, however, may be considered when selecting an urban solar site. Select a site where zoning restrictions have maximum height requirements. This prevents future neighborhood development from blocking the sun from an otherwise good solar orientation. Avoid a site where large coniferous trees hinder the full potential of the sun exposure. Sites that have streets running east-west and lots 50′ × 100′ (1524 × 3048 m) provide fairly limited orientation potential. Adjacent homes can easily block the sun unless southern exposure is possible. Figure 11.9 shows how the east-west street orientation can provide the maximum solar potential. If the client loves trees but also wants solar orientation, consider a site where the home could be situated so southern exposure can be achieved with coniferous trees to the north and deciduous trees on the south side. The coniferous trees can effectively block wind without interfering with the solar orientation. The deciduous trees provide shade from the hot summer sun. In the winter, when these trees have lost their leaves, the winter sun exposure is not substantially reduced. Figure 11.10 shows a potential solar site with trees.

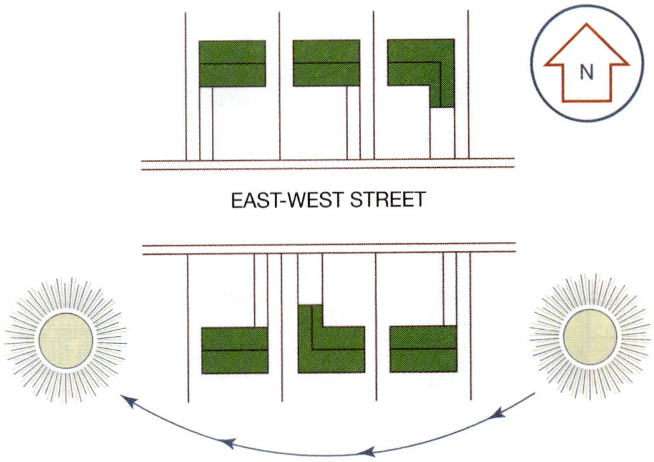

FIGURE 11.9 ■ East-west street orientation.

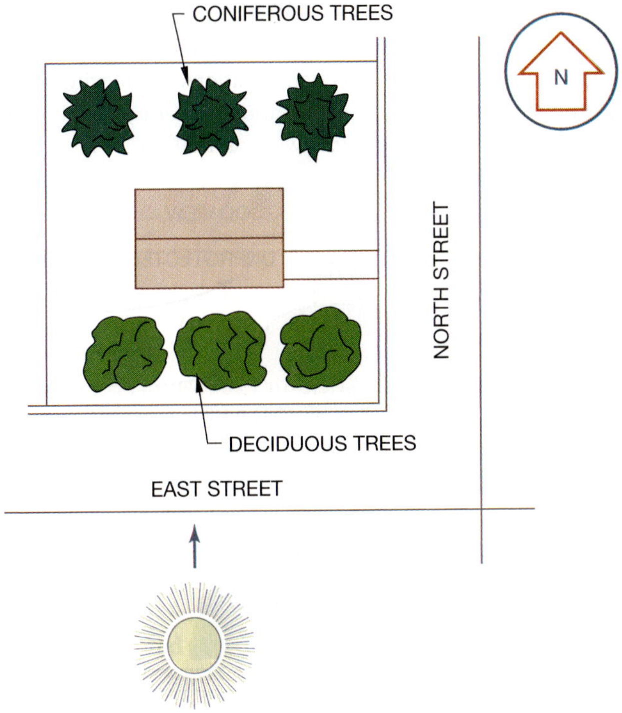

FIGURE 11.10 ■ Solar orientation with trees.

The
wir
exa
tha
sou
erly
ver
enc
of v
in t
hav
V
orie
con
amp
win
the
C
ano
goo
to ta

Sit

Win
alm
mat
E
ing t
Selec
is pr
of a
of a
the

FIGU

SECTION 3

Site Plan

12 CHAPTER

Legal Descriptions and Site Plan Requirements

LEGAL DESCRIPTIONS

Virtually every piece of property in the United States is described for legal purposes; such descriptions are referred to as *legal descriptions*. Every legal description is unique and cannot be confused with any other property. Legal descriptions of properties are filed in local jurisdictions, generally county or parish courthouses. Legal descriptions are public records and may be reviewed at any time.

This section deals with site plan characteristics and requirements. A site is an area of land generally one plot or construction lot in size. The term site is synonymous with *plot* and *lot*. A *plat* is a map of part of a city or township showing some specific area, such as a subdivision made up of several individual lots. There are usually many sites or plots in a plat. Some dictionary definitions, however, do not differentiate between *plot* and *plat*.

There are three basic types of legal descriptions: metes and bounds, rectangular survey system, and lot and block.

Metes and Bounds System

Metes, or measurements, and *bounds*, or boundaries, may be used to identify the perimeters of any property. This is referred to as the *metes and bounds system*.The metes are measured in feet, yards, rods (rd), or surveyor's chains (ch). There are 3′ (914 mm) in 1 yard, 5.5 yards or 16.5′ (5029 mm) in one rod, and 66′ (20,117 mm) in one surveyor's chain. The boundaries may be a street, fence, or river. Boundaries are also established as *bearings*, directions with reference to one quadrant of the compass. There are 360° in a circle or compass, and each quadrant has 90°. Degrees are divided into minutes and seconds. There are 60 minutes (60′) in 1° and 60 seconds (60″) in 1 minute. Bearings are measured clockwise or counterclockwise from north or south. For example, a reading 45° from north to west is labeled N 45° W. See Figure 12.1. If a bearing reading requires great accuracy, fractions of a degree are used. For example, S 30° 20′ 10″ E reads from south 30 degrees 20 minutes 10 seconds to east.

The metes and bounds land survey begins with a *monument*, known as the *point-of-beginning* (*POB*). This point is a fixed location and in times past has been a pile of rocks, a large tree, or an iron rod driven into the ground. Figure 12.2 shows an example of a site plan that is laid out using metes and bounds, and the legal description for the plot is given.

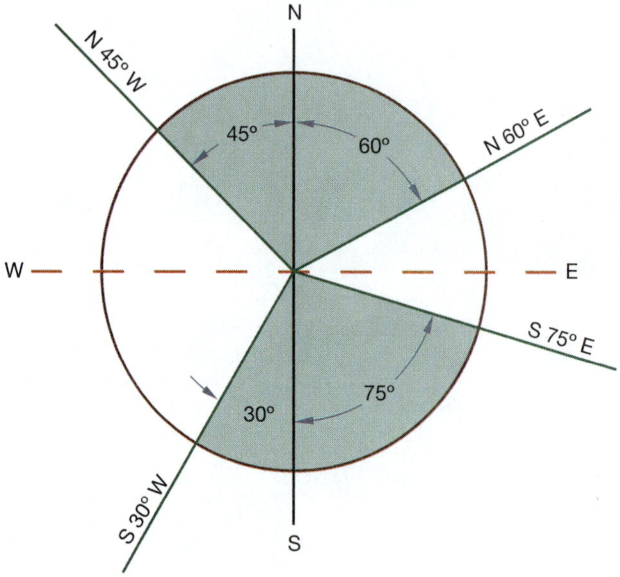

FIGURE 12.1 ■ Bearings.

Rectangular Survey System

The states in an area of the United States starting with the western boundary of Ohio to the Pacific Ocean, and including some southeastern states were described as public land states. Within this area the U.S. Bureau of Land Management devised a system for describing land known as the *rectangular survey system*.

Parallels of latitude and meridians of longitude were used to establish areas known as *great land surveys*. Lines of *latitude*, also called *parallels*, are imaginary parallel lines running east and west. Lines of *longitude*, also called *meridians*, are imaginary lines running north and south. The point of beginning of each great land survey is where two basic reference lines cross. The lines of latitude, or parallels, are termed the *baselines*, and the lines of longitude, or meridians, are called *principal meridians*. There are 31 sets of these lines in the continental United States, with 3 in Alaska. At the beginning, the principal meridians were numbered, and the numbering system ended with

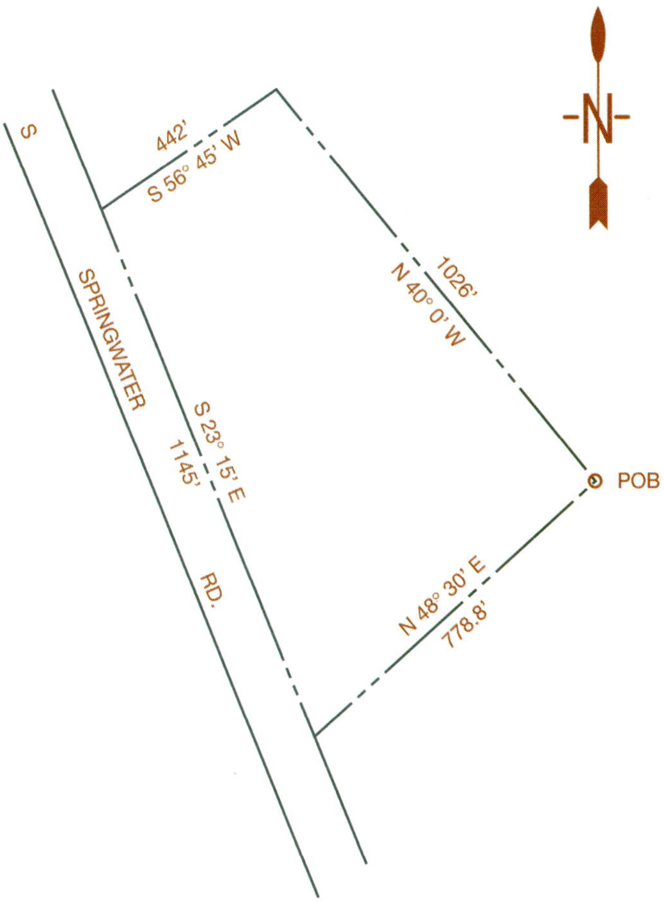

Beginning at a point 1200′ north 40° 0′ west from the southeast corner of the Asa Stone Donation Land Claim No. 49, thence north 40° 0′ west 1026′ to a pipe, thence south 56° 45′ west 442′ chains to center of road, thence south 23° 15′ east 1145′, thence north 48° 30′ east 778.8′ to place of beginning.

FIGURE 12.2 ■ Metes and bounds plot plan and legal description for the plot.

the sixth principal meridian passing through Nebraska, Kansas, and Oklahoma. The remaining principal meridians were given local names. The meridian through one of the last great land surveys near the West Coast is named the Willamette Meridian because of its location in the Willamette Valley of Oregon. The principal meridians and baselines of the great land surveys are shown in Figure 12.3.

Townships

The great land surveys were broken down into smaller surveys known as *townships* and *sections*. The baselines and meridians were divided into blocks called *townships*. Each township measures 6 miles square. The townships are numbered by tiers running north-south. The tier numbering system is established either north or south of a principal baseline. For example, the fourth tier south of the baseline is labeled Township Number 4 South, abbreviated T. 4 S. Townships are also numbered according to vertical meridians, known as *ranges*. Ranges are established either east or west of a principal meridian. The third range east of the principal meridian is called Range Number 3 East, abbreviated R. 3 E. Now combine T. 4 S. and R. 3 E. to locate a township or a piece of land 6 miles by 6 miles or a total of 36 square miles. Figure 12.4 shows the township just described.

Sections

To further define the land within a township 6 miles square, the area was divided into units 1 mile square, called *sections*. Sections in a township are numbered from 1 to 36. Section 1 always begins in the upper right corner, and consecutive numbers are arranged as shown in Figure 12.5. The legal descriptions of land can be carried one stage further. For example, Section 10 in the township given would be described as Sec. 10, T. 4 S., R. 3 E. This is an area of land 1 mile square. Sections are divided into acres. One acre equals 43,560 sq′, and one section of land contains 640 acres.

In addition to dividing sections into acres, sections are divided into quarters, as shown in Figure 12.6. The northeast one-quarter of Section 10 is a 160-acre piece of land described as NE 1/4, Sec. 10, T. 4 S., R. 3 E. When this section is keyed to a specific meridian, it can be only one specific 160-acre area. The section can be broken further by dividing each quarter into quarters, as shown in Figure 12.7. If the SW 1/4 of the NE 1/4 of Section 10 were the desired property, then you would have 40 acres known as SW 1/4, NE 1/4, Sec. 10, T. 4 S., R. 3 E. The complete rectangular system legal description of a 2.5-acre piece of land in Section 10 reads: SW 1/4, SE 1/4, SE 1/4, SE 1/4 Sec. 10, T. 4 N., R. 8 W. of the San Bernardino Meridian, in the County of Los Angeles, State of California. See Figure 12.8.

The rectangular survey system may be used to describe very small properties by continuing to divide a section of a township. Often the township section's legal description may be used to describe the location of the point of beginning of a metes and bounds legal description, especially when the surveyed land is an irregular site or plot within the rectangular survey system.

Lot and Block System

The *lot and block* legal description system can be derived from either the metes and bounds or the rectangular system. Generally when a portion of land is subdivided into individual building sites, the subdivision is established as a legal plot and recorded as such in the local county records. The subdivision is given a name and broken into blocks of lots. A subdivision may have several blocks, each divided into a series of lots. Each lot may be 50′ × 100′, for example, depending on the zoning requirements of the specific area. Figure 12.9 shows an example of a typical lot and block system. A typical lot and block legal description might read: LOT 14, BLOCK 12, LINCOLN PARK NO. 3, CITY OF SALEM, STATE. This lot is the shaded area in Figure 12.9 (page 202).

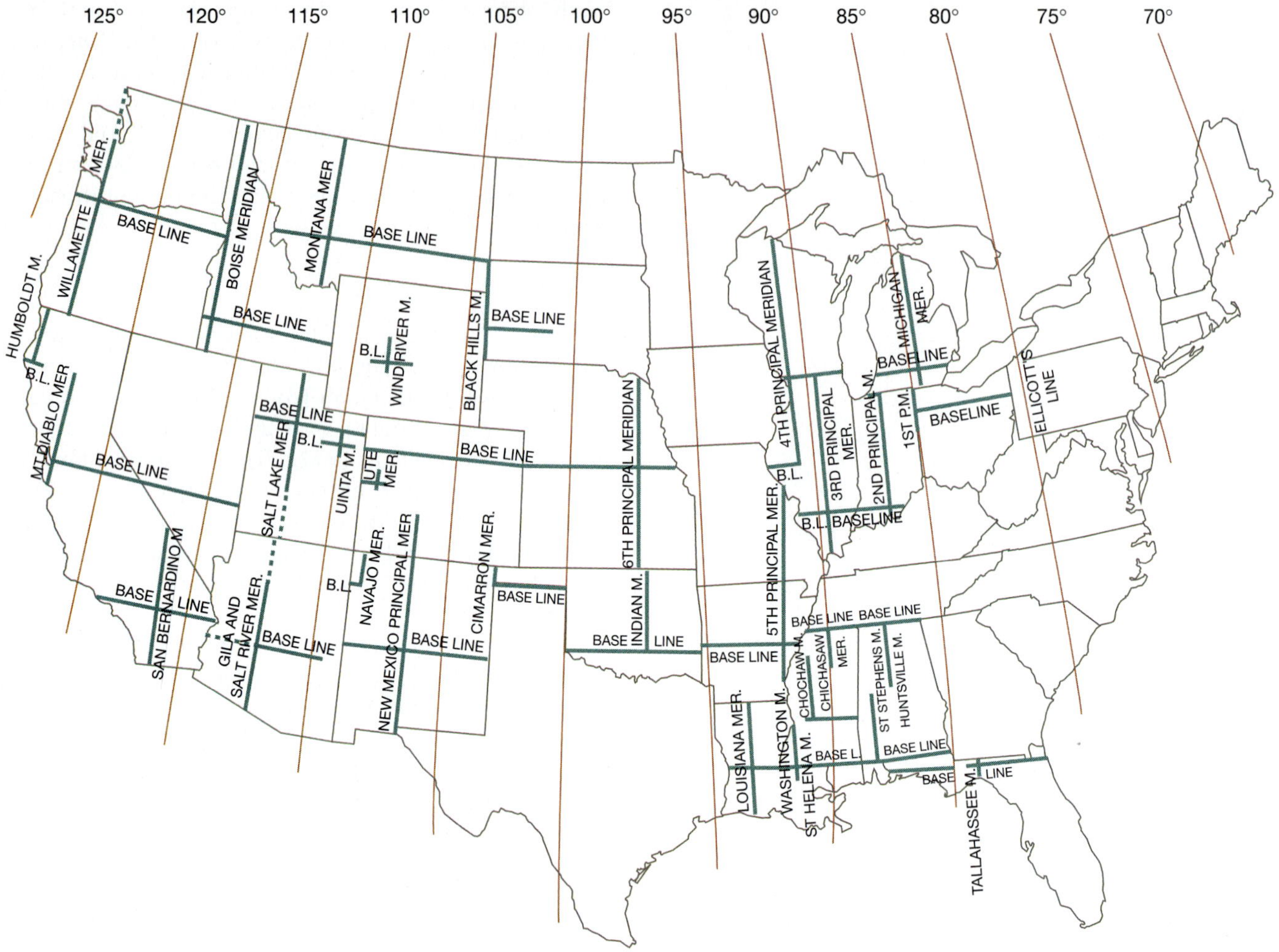

FIGURE 12.3 ■ Principal meridians and baselines of the great land survey (not including Alaska).

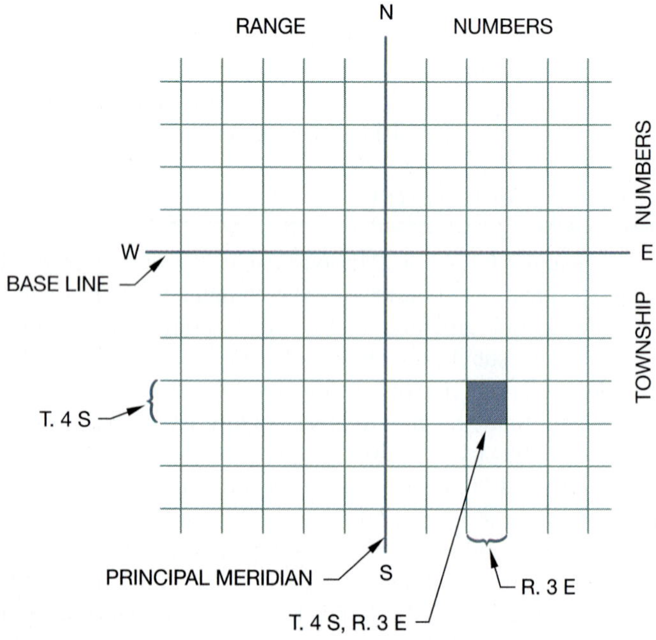

FIGURE 12.4 ■ Townships.

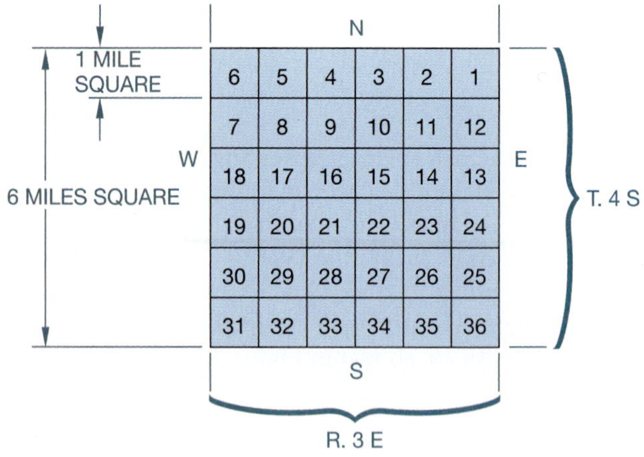

FIGURE 12.5 ■ Sections.

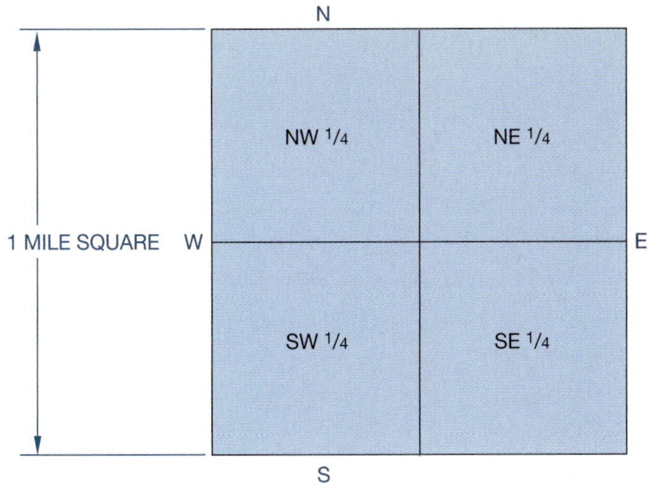

FIGURE 12.6 ■ Section quarters.

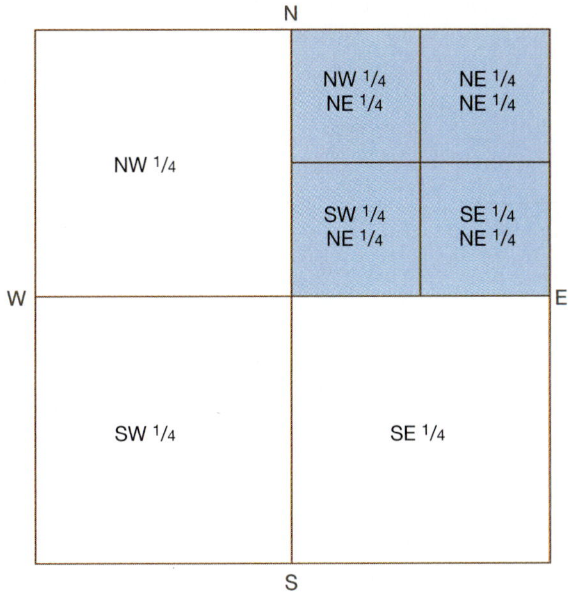

FIGURE 12.7 ■ Dividing a quarter section.

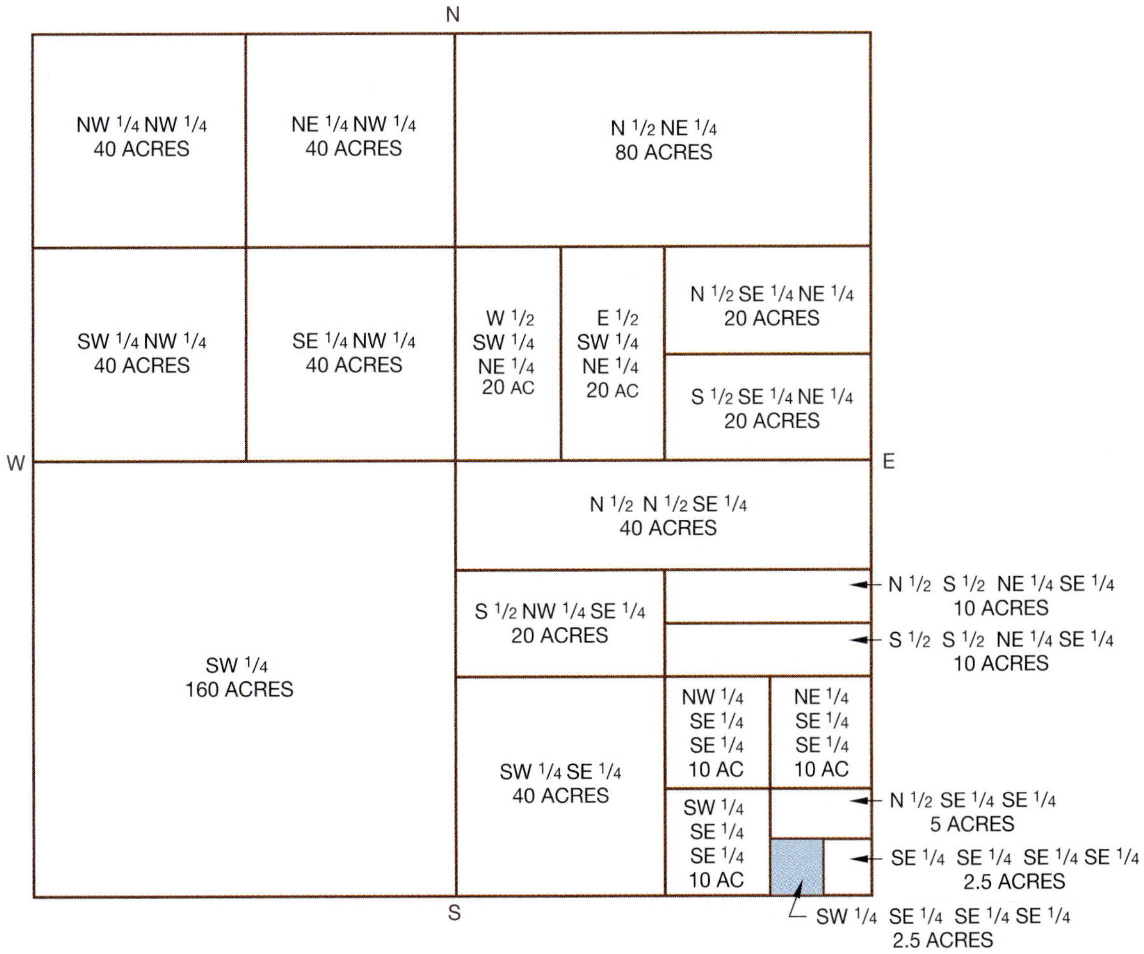

FIGURE 12.8 ■ Sample divisions of a section.

FIGURE 12.9 ■ Part of a lot and block subdivision. *Courtesy GLADS Program.*

SITE PLAN REQUIREMENTS

A *site plan*, also known as a *plot* or *lot plan*, is a map of a piece of land that may be used for any number of purposes. Site plans may show a proposed construction site for a specific property. Sites may show topography with contour lines, or the numerical value of land elevations may be given at certain locations. Site plans are also used to show how a construction site will be excavated and are then known as *grading plans*. Site plans can be drawn to serve any number of required functions; all have some similar characteristics, which include showing the following:

- A legal description of the property based on a survey
- Property line bearings and directions
- North direction
- Roads and easements
- Utilities
- Elevations
- Map scale

Plats are maps that are used to show an area of a town or township. They show several or many lots and may be used by a developer to show a proposed subdivision of land.

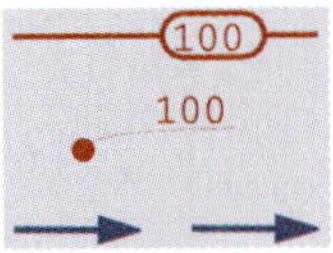

TOPOGRAPHY

Topography is a physical description of land surface showing its variation in elevation, known as relief, and locating other features. Surface relief can be shown with graphic symbols that use shading methods to accent the character of land, or the differences in elevations can be shown with contour lines. Site plans that require surface relief identification generally use *contour lines*. These lines connect points of equal elevation and help show the general lay of the land.

A good way to visualize the meaning of contour lines is to look at a lake or ocean shoreline. When the water is high during the winter or at high tide, a high-water line establishes a contour at that level. As the water recedes during the summer or at low tide, a new lower-level line is obtained. This new line represents another contour. The high-water line goes all around the lake at one level, and the low-water line goes all around the

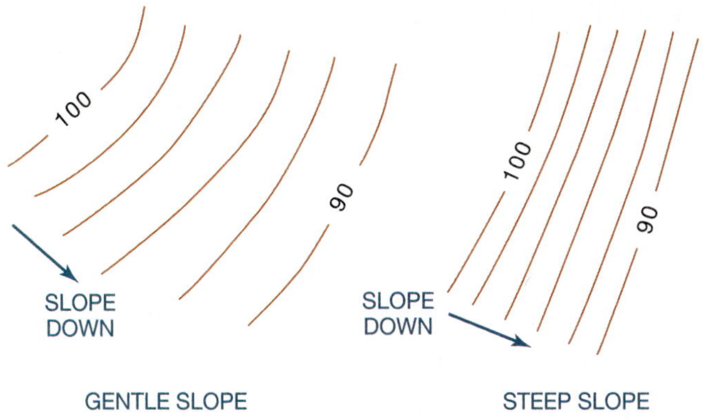

FIGURE 12.10 ■ Contour lines showing both gentle and steep slopes.

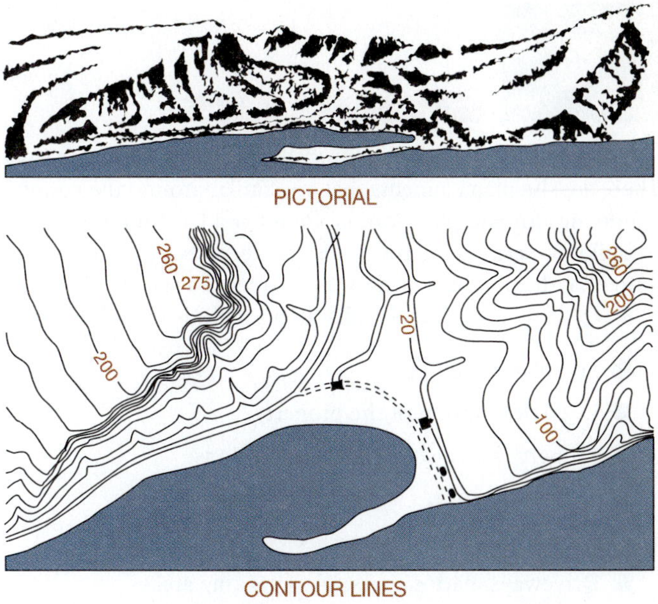

FIGURE 12.11 ■ Land relief pictorial and contour lines. *Courtesy U.S. Department of the Interior, Geological Survey.*

lake at another level. These two lines represent *contours,* or lines of equal elevation. The vertical distance between contour lines is known as *contour interval.* When the contour lines are far apart, the contour interval shows relatively flat or gently sloping land. When the contour lines are close together, the contour interval shows land that is much steeper. Contour lines are broken periodically, and the numerical value of the contour elevation above sea level is inserted. Figure 12.10 shows sample contour lines. Figure 12.11 shows a graphic example of land relief in pictorial form and contour lines of the same area.

Figure 12.12 shows a site with contour lines. Site plans do not always require contour lines showing topography. Verify the requirements with the local building codes. In most instances the only contour-related information required is property corner elevations, street elevation at a driveway, and the elevation of the finished floor levels of the structure. Additionally, slope may be defined and labeled with an arrow.

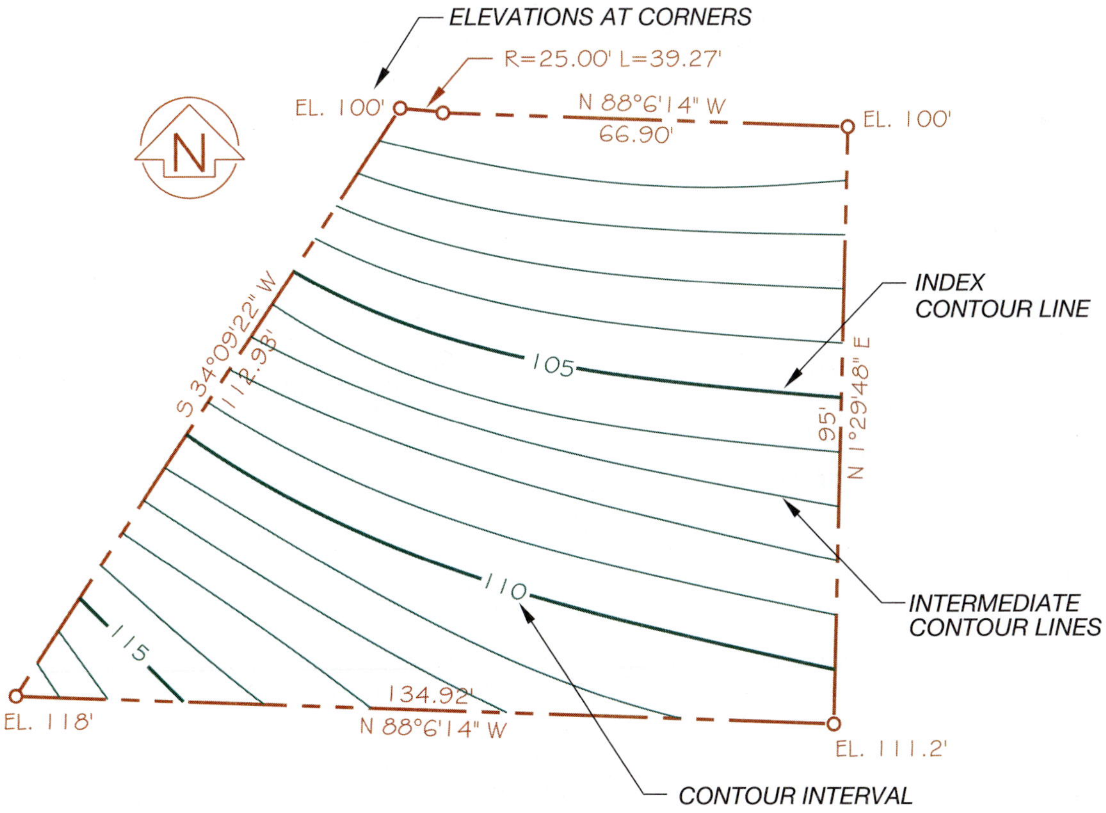

FIGURE 12.12 ■ Site plan with contour lines.

SITE PLANS

Site plan requirements vary by jurisdiction: city, county, or state. There are site plan elements that are similar around the country. Guidelines for site plans can be found at a local building office or building permit department. Some agencies, for example, require that the plot plan be drawn on paper of a specific size, such as 8 1/2″ × 14″. Typical site plan items include the following:

- Site plan scale.
- Legal description of the property.
- Property line bearings and dimensions.
- North direction.
- Existing and proposed roads.
- Driveways, patios, walks, and parking areas.
- Existing and proposed structures.
- Public or private water supply.
- Public or private sewage disposal.
- Location of utilities.
- Rain and footing drains, and storm sewers or drainage.
- Topography, including contour lines or land elevations at lot corners, street centerline, driveways, and floor elevations.
- Setbacks—The minimum distance from the property lines to the front, rear, and sides of the structure.
- Specific items on adjacent properties may be required.
- Existing and proposed trees may be required.

Figure 12.13 shows a site plan layout that is used as an example at a local building department. Figure 12.14 shows a basic site plan for a proposed residential addition.

The method of sewage disposal is generally an important item shown on a site plan drawing. There are a number of alternative methods of sewage disposal, including public sewers and private systems. Chapter 18 gives more details of sewage disposal methods. The site plan representation of a public sewer connection is shown in Figure 12.15. A private septic sewage disposal system is shown in a site plan in Figure 12.16.

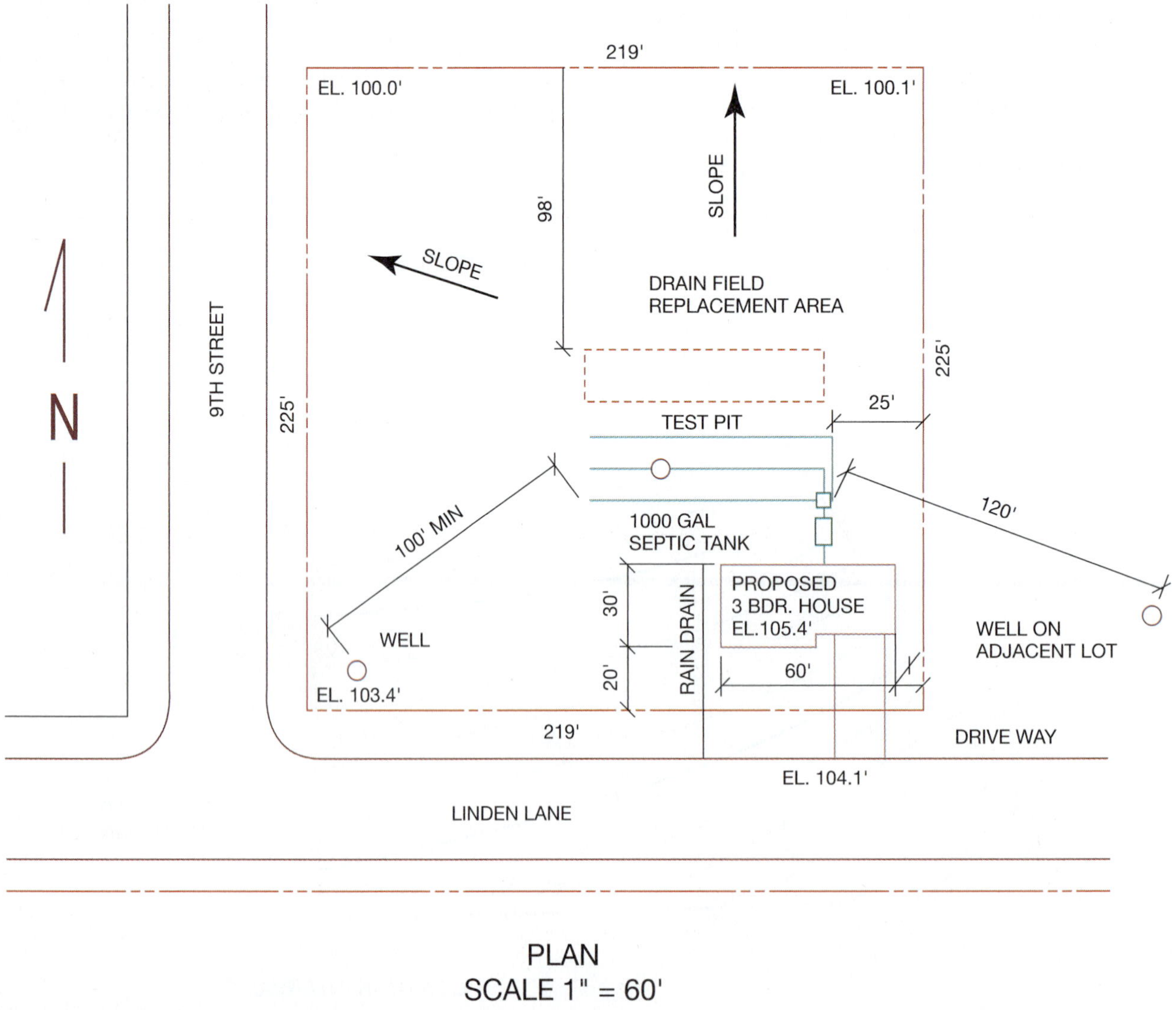

FIGURE 12.13 ■ Recommended typical site plan layout.

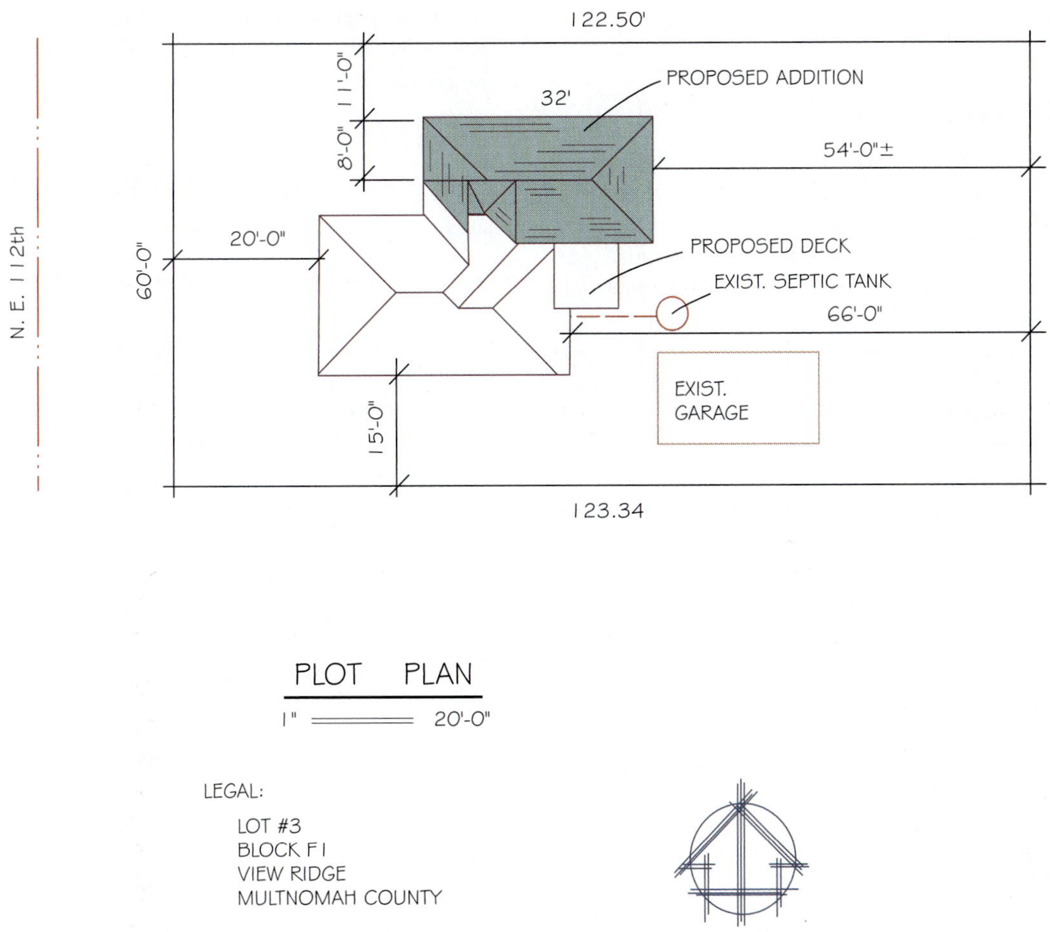

FIGURE 12.14 ■ Sample site plan showing existing home and proposed addition.

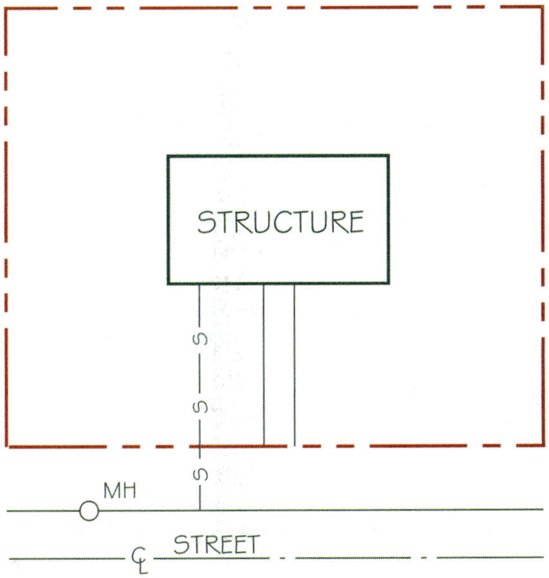

FIGURE 12.15 ■ Plot plan showing a public sewer connection.

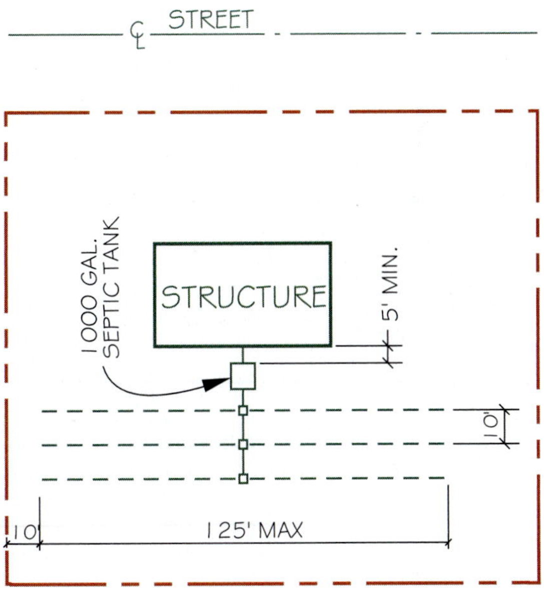

FIGURE 12.16 ■ Plot plan showing a private septic sewage system.

GRADING PLAN

Grading plans are construction drawings that generally show existing and proposed topography. The outline of the structure may be shown with elevations at each building corner and the elevation given for each floor level. Figure 12.17 shows a detailed grading plan for a residential construction site. Notice that the legend identifies symbols for existing and finished contour lines. This particular grading plan provides retaining walls and graded slopes to accommodate a fairly level construction site from the front of the structure to the extent of the rear yard. The finished slope represents an embankment that establishes the relationship of the proposed contour to the existing contour. This particular grading plan also shows a proposed irrigation and landscaping layout.

Grading plan requirements may differ from one location to the next. Some grading plans may also show a cross section through the site at specified intervals or locations to evaluate the contour more fully. This cross section is called a *profile*. A profile through the grading plan in Figure 12.17 is shown in Figure 12.18.

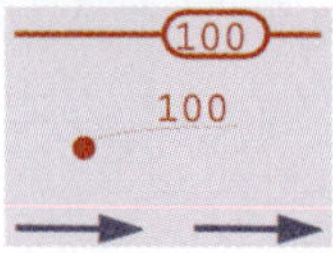

SITE ANALYSIS PLAN

In areas or conditions where zoning and building permit applications require a design review, a site analysis plan may be required. The site analysis provides the basis for the proper design relationship of the proposed development to the site and to adjacent properties. The degree of detail of the site analysis is generally appropriate to the scale of the proposed project. A site analysis plan, shown in Figure 12.19, often includes the following:

- A vicinity map showing the location of the property in relationship to adjacent properties, roads, and utilities.

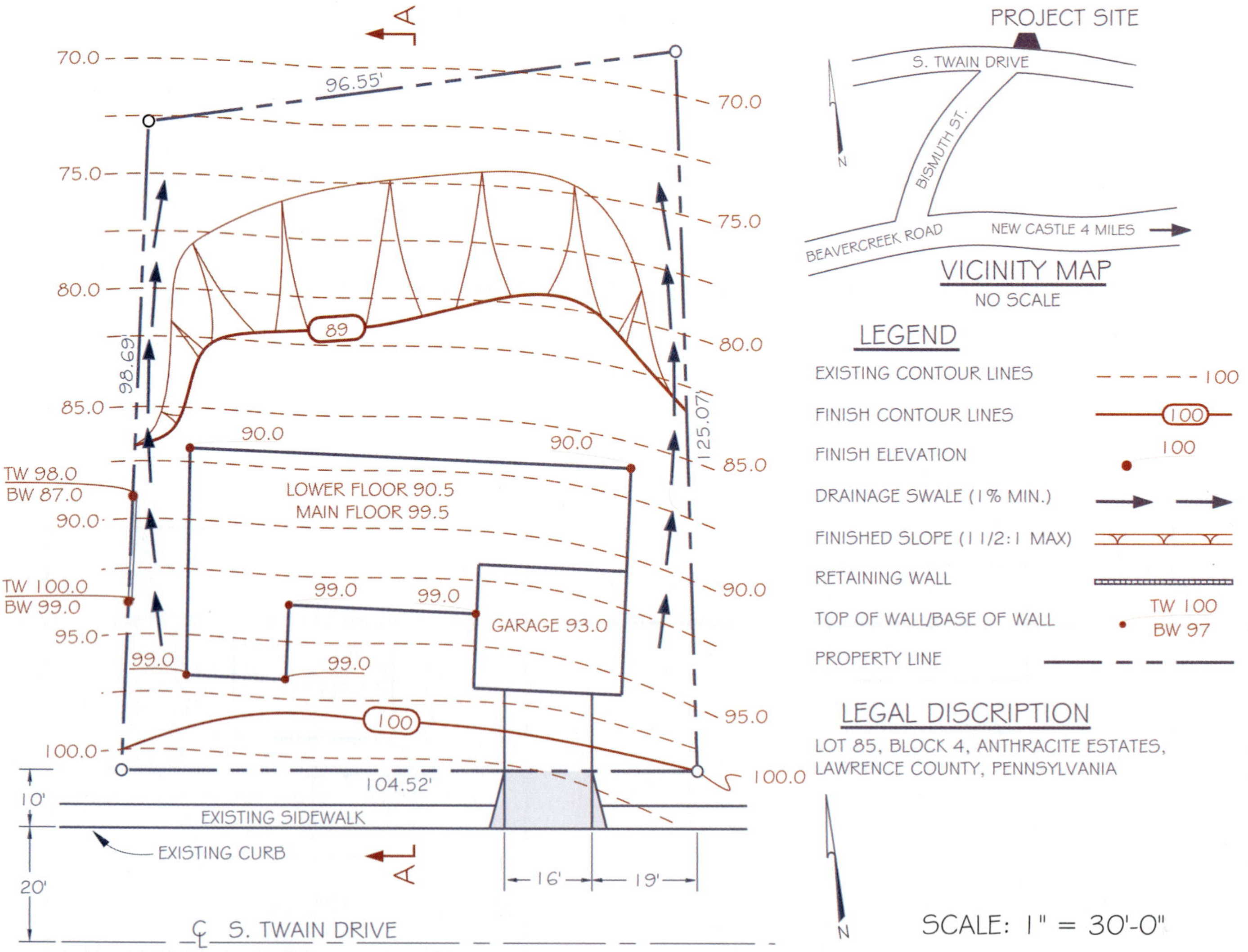

FIGURE 12.17 ■ Grading plan.

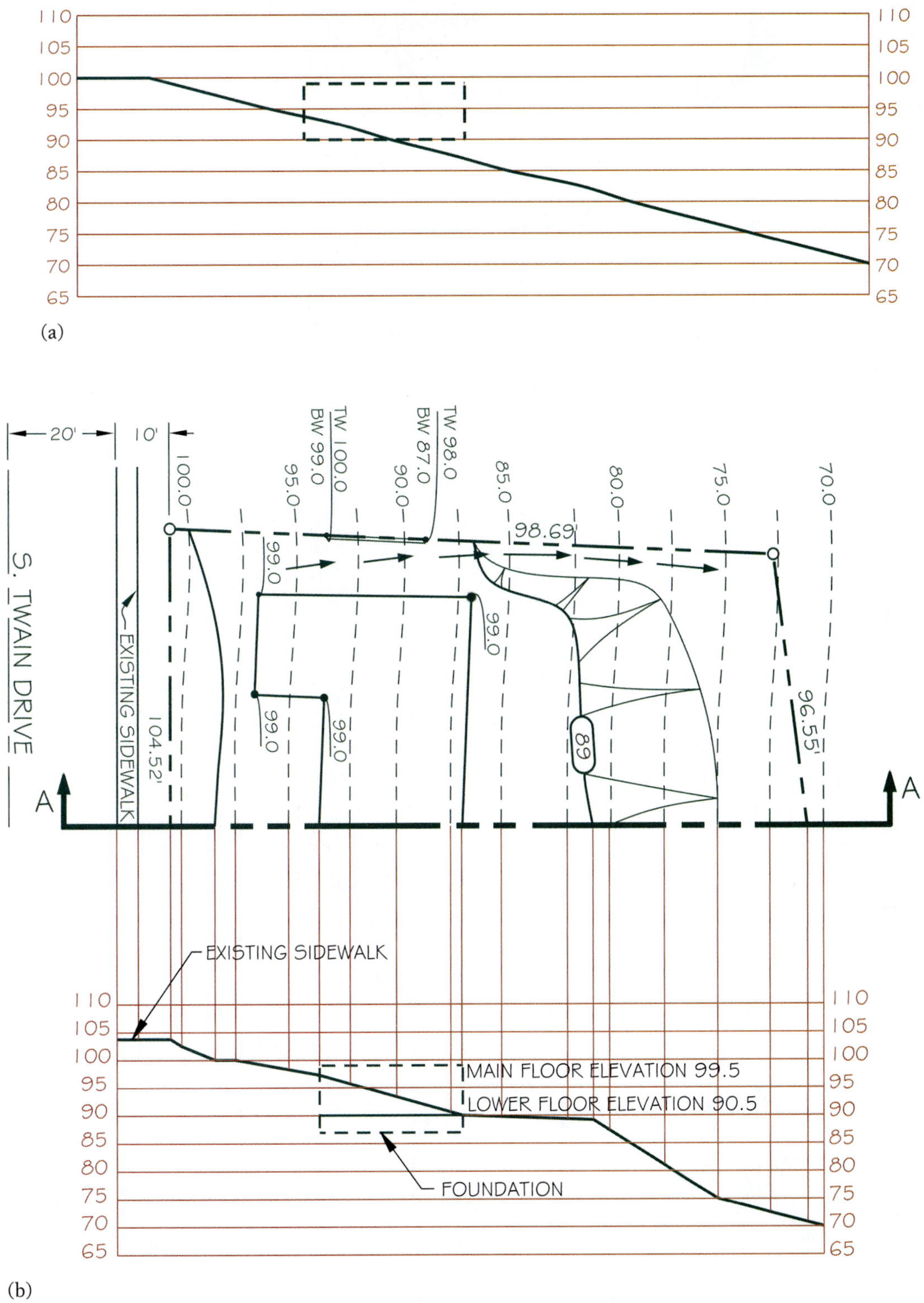

FIGURE 12.18 ■ Constructing a profile from the grading plan in Figure 12.17.

- Site features, such as existing structures and plants on the property and adjacent property.
- Scale.
- North direction.
- Property boundaries.
- Slope shown by contour lines or cross sections, or both.
- Plan legend.
- Traffic patterns.
- Solar site information if solar application is intended.
- Pedestrian patterns.

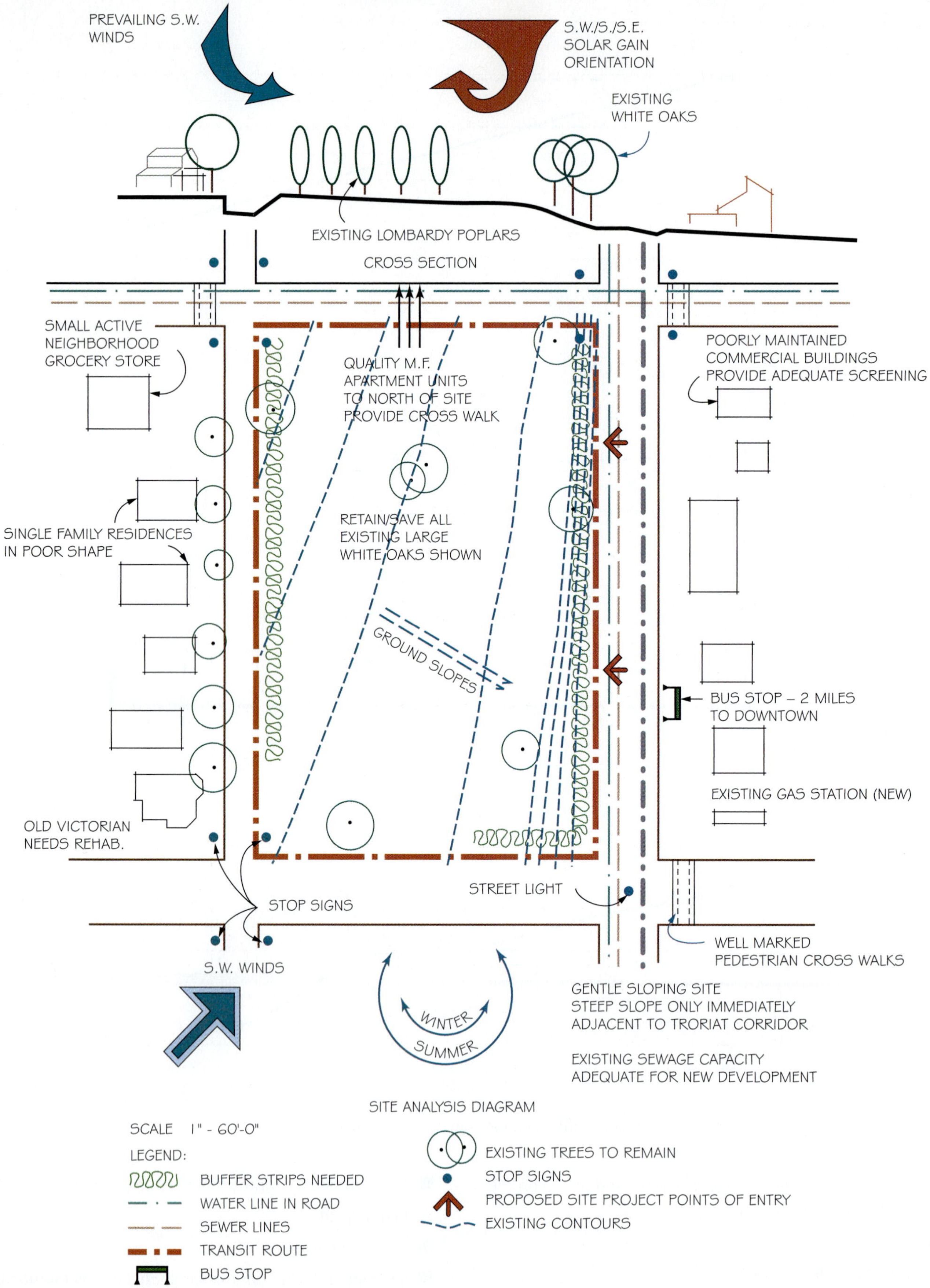
PREVAILING S.W. WINDS
S.W./S./S.E. SOLAR GAIN ORIENTATION
EXISTING WHITE OAKS
EXISTING LOMBARDY POPLARS
CROSS SECTION
SMALL ACTIVE NEIGHBORHOOD GROCERY STORE
QUALITY M.F. APARTMENT UNITS TO NORTH OF SITE PROVIDE CROSS WALK
POORLY MAINTAINED COMMERCIAL BUILDINGS PROVIDE ADEQUATE SCREENING
RETAIN/SAVE ALL EXISTING LARGE WHITE OAKS SHOWN
SINGLE FAMILY RESIDENCES IN POOR SHAPE
GROUND SLOPES
BUS STOP – 2 MILES TO DOWNTOWN
EXISTING GAS STATION (NEW)
OLD VICTORIAN NEEDS REHAB.
STREET LIGHT
STOP SIGNS
WELL MARKED PEDESTRIAN CROSS WALKS
S.W. WINDS
WINTER
SUMMER
GENTLE SLOPING SITE STEEP SLOPE ONLY IMMEDIATELY ADJACENT TO TRORIAT CORRIDOR
EXISTING SEWAGE CAPACITY ADEQUATE FOR NEW DEVELOPMENT
SITE ANALYSIS DIAGRAM
SCALE 1" - 60'-0"
LEGEND:
BUFFER STRIPS NEEDED
WATER LINE IN ROAD
SEWER LINES
TRANSIT ROUTE
BUS STOP
EXISTING TREES TO REMAIN
STOP SIGNS
PROPOSED SITE PROJECT POINTS OF ENTRY
EXISTING CONTOURS

FIGURE 12.19 ■ Site analysis plan. *Courtesy Planning Department, Clackamas County, Oregon.*

SUBDIVISION PLANS

Local requirements for subdivisions should be confirmed because procedures vary, depending on local guidelines and zoning rules.

Some areas have guidelines for minor subdivisions that differ from major subdivisions. The plat required for a minor subdivision may include the following:

1. Legal description.
2. Name, address, and telephone number of applicant.
3. Parcel layout with dimensions.
4. Direction of north.
5. All existing roads and road widths.
6. Number identification of parcels, such as Parcel 1, Parcel 2.
7. Location of well or proposed well, or name of water district.
8. Type of sewage disposal (septic tank or public sanitary sewers). Name of sewer district.
9. Zoning designation.
10. Size of parcel(s) in square feet or acres.
11. Slope of ground. (Arrows pointing downslope.)
12. Setbacks of all existing buildings, septic tanks, and drainfields from new property lines.
13. All utility and drainage easements.
14. Any natural drainage channels. Indicate direction of flow and whether drainage is seasonal or year-round.
15. Map scale.
16. Date.
17. Building permit application number, if any.

Figure 12.20 shows a typical subdivision of land with three proposed parcels.

A major subdivision may require more detailed plats than a minor subdivision. Some of the items that may be included on the plat or in a separate document are the following:

1. Name, address, and telephone number of the property owner, applicant, and engineer or surveyor.
2. Source of water.
3. Method of sewage disposal.
4. Existing zoning.
5. Proposed utilities.
6. Calculations justifying the proposed density.
7. Name of the major partitions or subdivision.

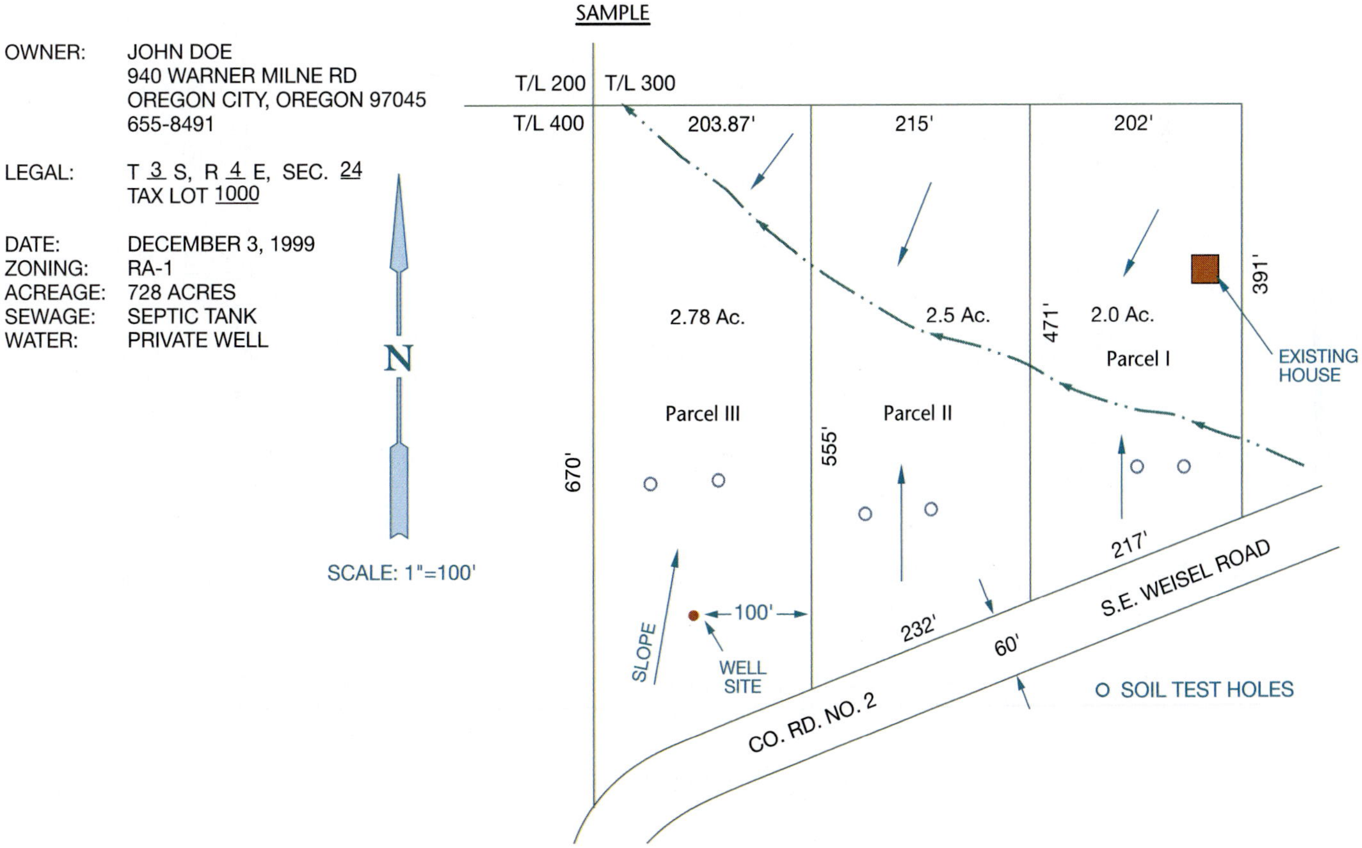

FIGURE 12.20 ■ Subdivision with three proposed lots. *Courtesy Planning Department, Clackamas County, Oregon.*

8. Date the drawing was made.
9. Legal description.
10. North arrow.
11. Vicinity sketch showing location of the subdivision.
12. Identification of each lot or parcel and block by number.
13. Gross acreage of property being subdivided or partitioned.
14. Dimensions and acreage of each lot or parcel.
15. Streets abutting the plat, including name, direction of drainage, and approximate grade.
16. Streets proposed, including names, approximate grades, and radius of curves.
17. Legal access to subdivision or partition other than public road.
18. Contour lines at 2′ interval for slopes of 10 percent or less, 5′ interval if slopes exceed 10 percent.
19. Drainage channels, including width, depth, and direction of flow.
20. Locations of existing and proposed easements.
21. Location of all existing structures, driveways, and pedestrian walkways.
22. All areas to be offered for public dedication.
23. Contiguous property under the same ownership, if any.
24. Boundaries of restricted areas, if any.
25. Significant vegetative areas such as major wooded areas or specimen trees.

Figure 12.21 shows an example of a small major subdivision plat.

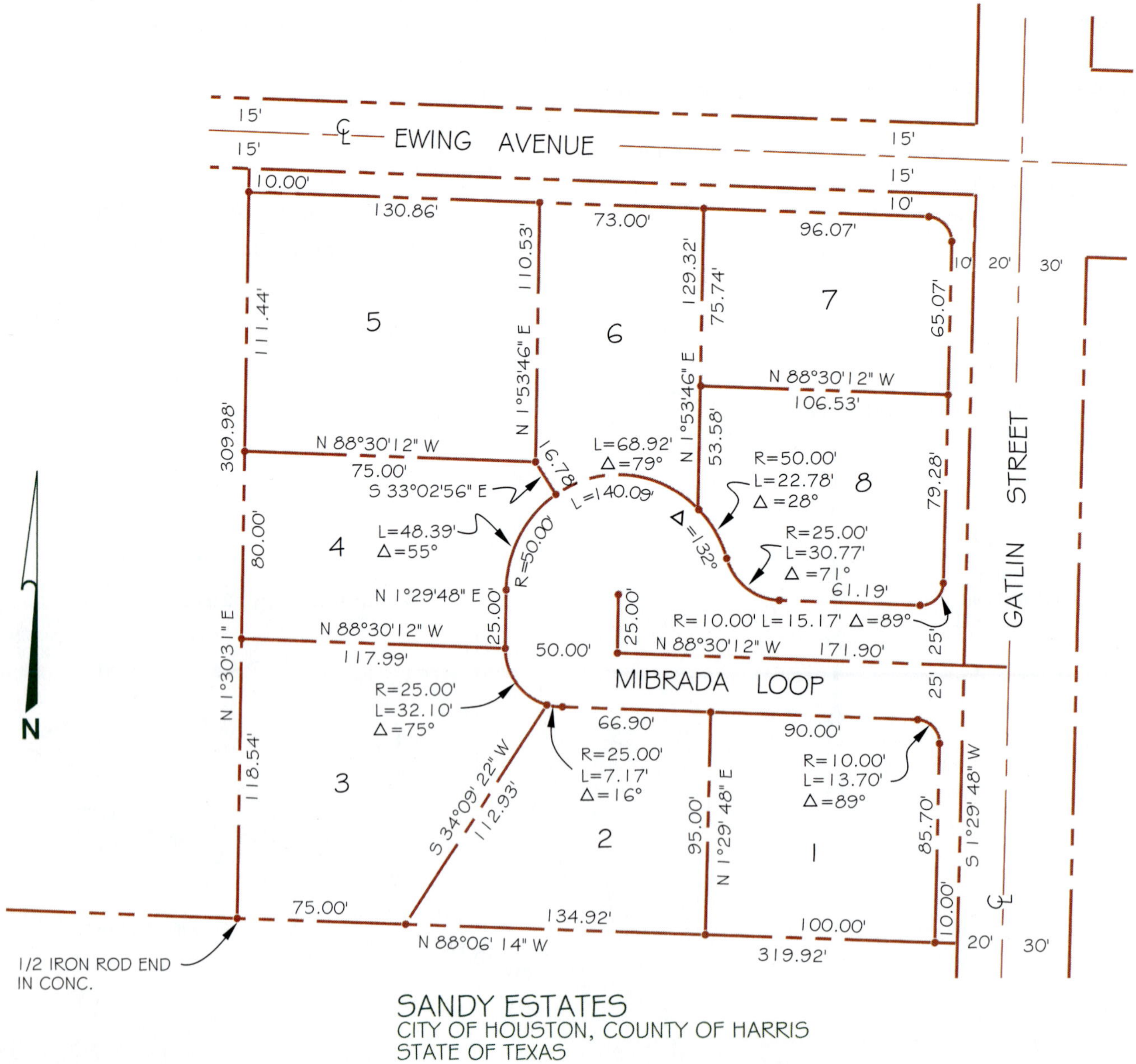

FIGURE 12.21 ■ Small major subdivision plat.

PLANNED UNIT DEVELOPMENT

A creative and flexible approach to land development is a *planned unit development*. Planned unit developments may include such uses as residential areas, recreational areas, open spaces, schools, libraries, churches, or convenient shopping facilities. Developers involved in these projects must pay particular attention to the impact on local existing developments. Generally the plats for these developments must include all of the same information shown on a subdivision plat, plus:

1. Detailed vicinity map, as shown in Figure 12.22.
2. Land use summary.
3. Symbol legend.
4. Special spaces such as recreational and open spaces or other unique characteristics.

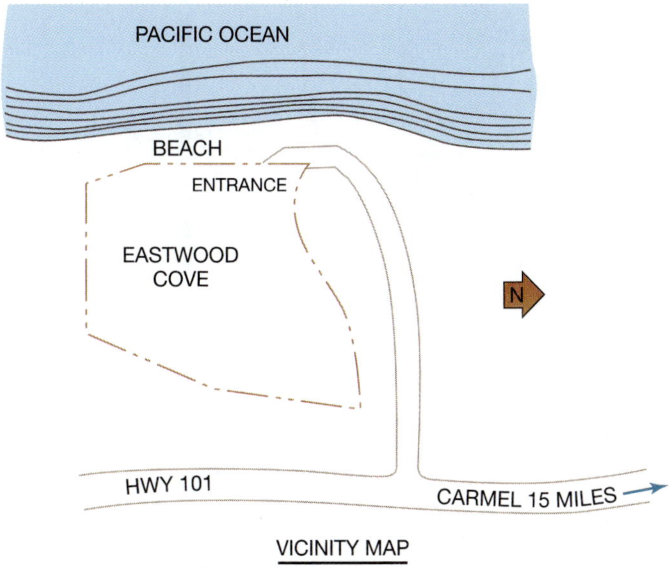

FIGURE 12.22 ■ Vicinity map.

Figure 12.23 shows a typical planned unit development plan. These plans, like any proposed site plan, may require changes before the final drawings are approved for development.

There are several specific applications for a site plan. The applied purpose of each is different, although the characteristics of each type of site plan may be similar. Local districts have guidelines for the type of plan required. Be sure to evaluate local guidelines before preparing a site plan for a specific purpose. To get a plot plan accepted, prepare it in strict accordance with the requirements.

A variety of plot plan–related templates and CADD programs are available that can help make the preparation of these plans easier.

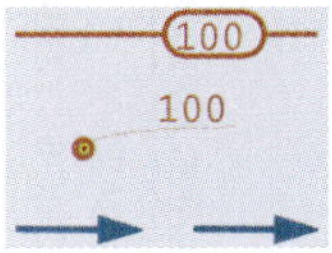

METRICS IN SITE PLANNING

The recommended metric values used in the design and drafting of site plans based on surveying, excavating, paving and concrete construction are as follows:

	QUANTITY	UNIT AND SYMBOL
Surveying	Length	meter (m) and kilometer (km)
	Area	square meter (m^2), hectare (ha), and square kilometer (km^2)
	Plane angle	degree (°), minute (′), second (″), and percent (%)
Excavating	Length	millimeter (mm) and meter (m)
	Volume	cubic meter (m^3)
Paving	Length	millimeter (mm) and meter (m)
	Volume	cubic meter (m^3)
Concrete	Length	millimeter (mm) and meter (m)
	Area	square meter (m^2)
	Volume	cubic meter (m^3)

TYPICAL SECTION
UNOBSTRUCTED OCEAN VIEW
UNOBSTRUCTED OCEAN VIEW
100'
LAND USE SUMMARY
ZONE R1
GROSS AREA 7.63 ACRES
NET AREA 705 ACRES
ROADS PACKING 119 ACRES
STRUCTURES (44 UNITS) 44 ACRES
RECREATIONAL 4 OPEN SPACES 350 ACRES
INTERFACING 1.92 ACRES
DENSITY 12.24 UNITS/ACRE
PACIFIC OCEAN
BEACH
ENTRANCE
GATEHOUSE PHONE
RECREATIONAL SPACE
OPEN SPACE
OPEN SPACE
OPEN SPACE
OPEN SPACE
OPEN SPACE
OPEN SPACE
OPEN SPACE
OPEN SPACE
25
50
75
100
125
150
150
100
75
50
25
N
0 50 100 200
EASTWOOD COVE
DEVELOPMENT PLAN
LEGEND
PROPERTY BOUNDARY
CONTOURS
NEW FOOT PATHS
PROPOSED ROADS
EXISTING TREES
ONE FOURPLEX

FIGURE 12.23 ■ Planned unit development plan.

CHAPTER 12

Legal Descriptions and Site Plan Requirements Test

DIRECTIONS

Answer the questions with short complete statements or drawings as needed on an 8 1/2″ × 11″ sheet of notebook paper, as follows:

1. Letter your name, Chapter 12 Test, and the date at the top of the sheet.
2. Letter the question number and provide the answer. You do not need to write out the question.
3. Answers may be prepared on a word processor if course guidelines allow this.

QUESTIONS

Question 12–1 Define *site.*

Question 12–2 What is the difference between the terms *site, plot* and *lot?*

Question 12–3 Name the three basic types of legal descriptions.

Question 12–4 How many total degrees are in a compass?

Question 12–5 How many minutes are in 1 degree?

Question 12–6 How many seconds are in 1 minute?

Question 12–7 Define *bearing.*

Question 12–8 Give the bearing of a property line positioned 60° from north toward west.

Question 12–9 Give the bearing of a property line positioned 20° from south toward east.

Question 12–10 Define *baselines.*

Question 12–11 Define *principal meridians.*

Question 12–12 Give the dimensions of a township.

Question 12–13 How many acres are in a section?

Question 12–14 How many square feet are in 1 acre?

Question 12–15 How are sections in a township numbered?

Question 12–16 How many acres are in a quarter section?

Question 12–17 Give the partial legal description of 2.5 acres of land in the farthest corner of the northwest corner of section 16, township six north, range four west.

Question 12–18 List at least six items that are required on a site plan.

Question 12–19 Define *topography.*

Question 12–20 Define *contour lines.*

Question 12–21 What is the purpose of a grading plan?

Question 12–22 Under what conditions is a site analysis plan normally required?

Question 12–23 What are setbacks?

Question 12–24 Name the two main categories of sewage disposal systems.

Question 12–25 Identify the type of legal description that is used for most major subdivisions.

13 CHAPTER

Site Plan Layout

INTRODUCTION

Site plans may be drawn on media ranging in size from 8 1/2″ × 11″ up to 34″ × 44″ depending on the purpose of the plan and the guidelines of the local government agency or lending institution requiring the plan. Many local jurisdictions recommend that site plans be drawn on a sheet 8 1/2″ × 14″.

Before you begin the site plan layout, there is some important information that you need. This information can often be found in the legal documents for the property, the surveyor's map, the local assessor's office, or the local zoning department. Figure 13.1 is a plat from a surveyor's map that can be used as a guide to prepare the site plan. The scale of the surveyor's plat may vary, although in this case it is 1″ = 200′ (1:1000 metric). The plot plan to be drawn may have a scale ranging from 1″ = 10′ (1:50 metric) to 1″ = 200′. The factors that influence the scale include the following:

- Sheet size.
- Plot size.
- Amount of information required.
- Amount of detail required.

Additional information that should be determined before the site plan can be completed usually includes the following:

- Legal description.
- North direction.
- All existing roads, utilities, water, sewage disposal, drainage, and slope of land.
- Zoning information, including front, rear, and side yard setbacks.

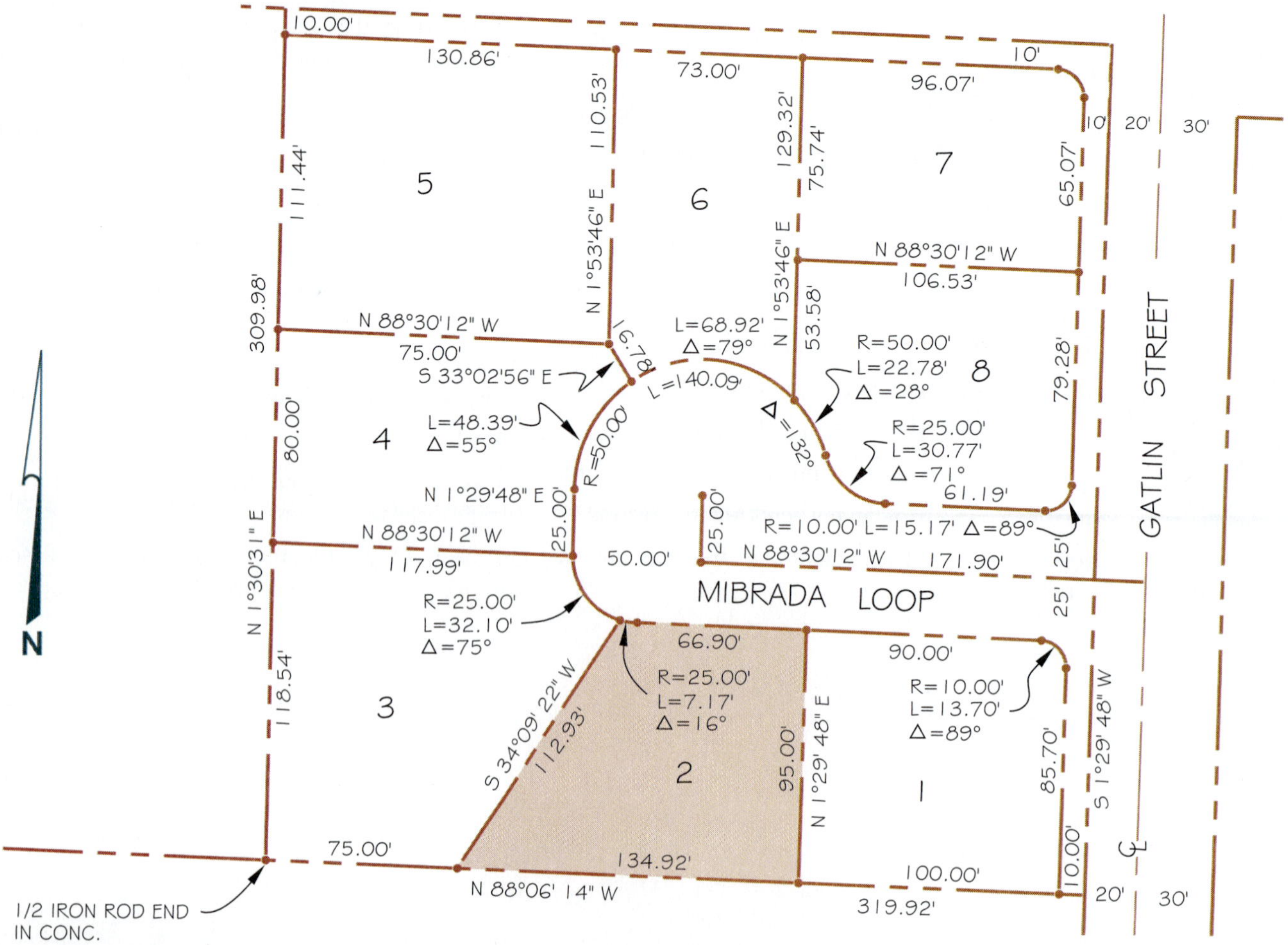

FIGURE 13.1 ■ Plat from a surveyor's map.

- Size of proposed structures.
- Elevations at property corners, driveway at street, or contour elevations.

SITE DESIGN CONSIDERATIONS

Several issues that can effect the quality of the construction project and adjacent properties should be considered during site design. Some of these factors are outlined below:

- Provide a minimum driveway slope of 1/4″ per foot. The maximum slope depends on surface conditions and local requirements.
- Provide a minimum lawn slope of 1/4″ per foot if possible.
- Single-car driveways should be a minimum of 10′ (3 m) wide, and double-car driveways should be a minimum of 18′ (5.5 m) wide but can taper to 10′ wide at the entrance. Any reduction of driveway width should be centered on the garage door. A turning apron is preferred when space permits. This allows the driver to back into the parking apron and then drive forward into the street, which is safer than backing into the street.
- The minimum turning radius for a driveway should be 15′ (4.5 m). The turning radius for small cars can be less, but more should be considered for trucks. A turning radius of 20′ (6 m) is preferred if space permits. Figure 13.2 shows a variety of driveway layouts for you to use as examples. The dimensions are given as commonly recommended minimums for small to standard-size cars. Additional room should be provided if available.
- Provide adequate room for the installation of and future access to water, sewer, and electrical utilities.
- Do not build over established easements. An *easement* is the right-of-way for access to property and for the purpose of construction and maintenance of utilities.
- Follow basic grading rules, which include not grading on adjacent property. Do not slope the site so as to cause water drainage onto adjacent property. Slope the site away from the house. Adequate drainage of at least 10′ away from the house is recommended.
- Identify all trees that are to remain on the site after construction.
- Establish retaining walls where needed to minimize the slope, control erosion, and level portions of the site.

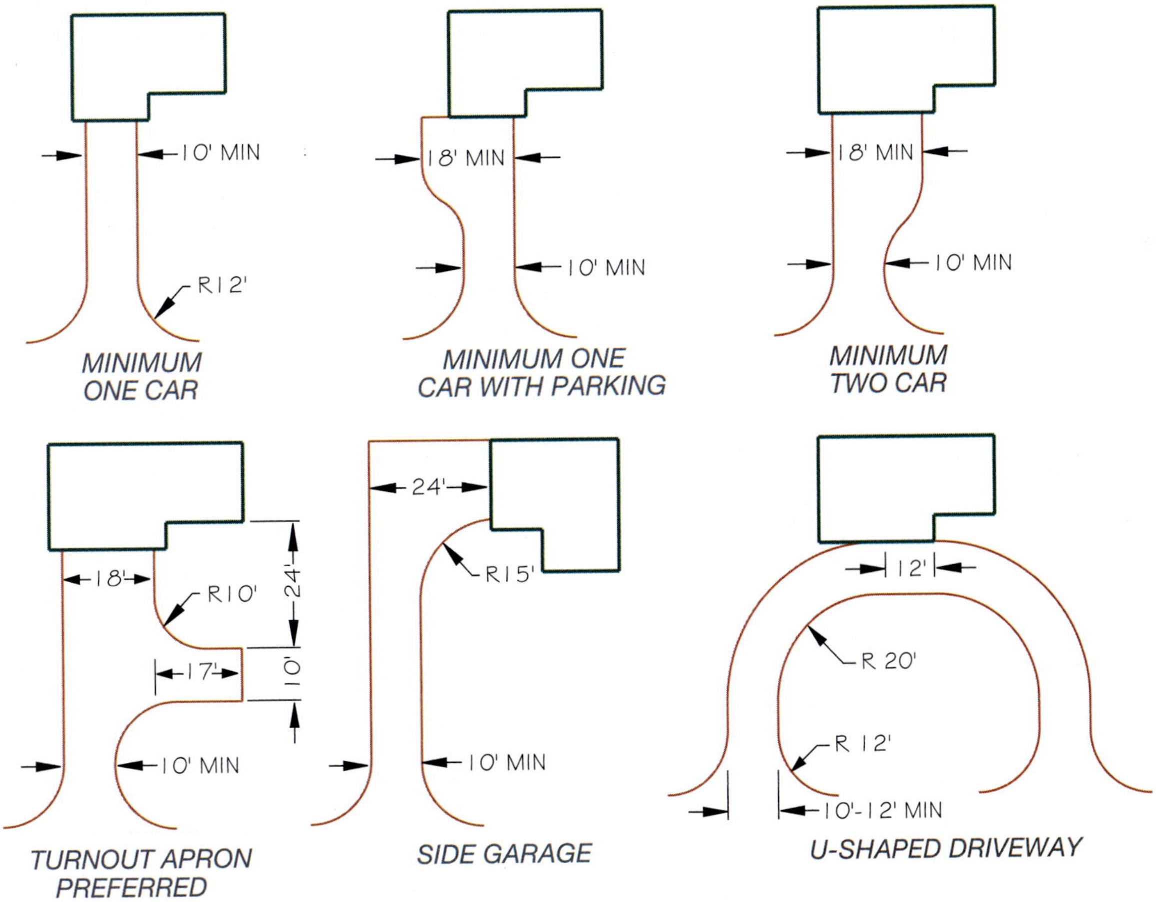

FIGURE 13.2 ■ Typical driveway layout options.

PROPERTY LINES

Laying Out Property Lines

In Chapter 12, you were introduced to the methods for describing properties. Look at Figure 13.1 and notice that the property lines of each lot and the boundaries of the plat are labeled with distances and bearings. An example is the west property line of lot 2, which is 112.93′ for the length and S34°09′22″W for the bearing. This property line is set up to be drawn as shown in Figure 13.3A. After the property lines are drawn, they are labeled with the distance and bearing, as shown in Figure 13.3B.

Many plats have property lines that curve, such as the lines around the cul-de-sac in Figure 13.1. For example, the largest curve is labeled R = 50.00′ L = 140.09′. R is the radius of the curve, and L is the length of the curve. There are three sublengths in this curve. L = 48.39′ is for lot 4, L = 68.92′ is for lot 6, and L = 22.78′ is for lot 8. Figure 13.4A shows the setup for drawing this curve using the radius and arc lengths. Plats also typically show a Delta angle for curves, which is represented by the symbol Δ. The *Delta* angle is the included angle of the curve. The *included angle* is the angle formed between the center and the endpoints of the arc, as shown in Figure 13.4B. Figure 13.4C shows the final labeling of the curve.

CAD Applications

Generally the plat or legal description provides metes and bounds coordinates for property line boundaries. For example, the south property line for Lot 2 has the coordinates given with the bearing and distance of N 88°06′ 14″ W and 134.92′. CADD systems that are used for either general or architectural applications have a surveyor's units setting. When the CADD system is set to draw lines based on surveyor's units, the south property line of Lot 2 is drawn with this computer prompt: @134.92′ < N 88d06′ 14″ W. Notice that the degree symbol (°) is replaced with the lowercase letter d when entering a bearing at the computer prompt.

Most CADD programs have options for drawing arcs that allow you to create the curves in Figure 13.4. The commands typically allow you to draw arcs with a combination of radius, arc length, and included angle. This provides you with the flexibility needed to use the R, L, and Δ information provided on the plat or in the surveyor's notes.

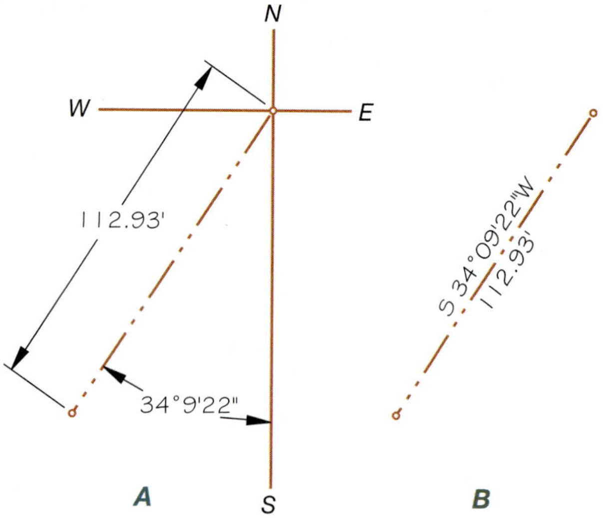

FIGURE 13.3 ■ Drawing and labeling a property line.

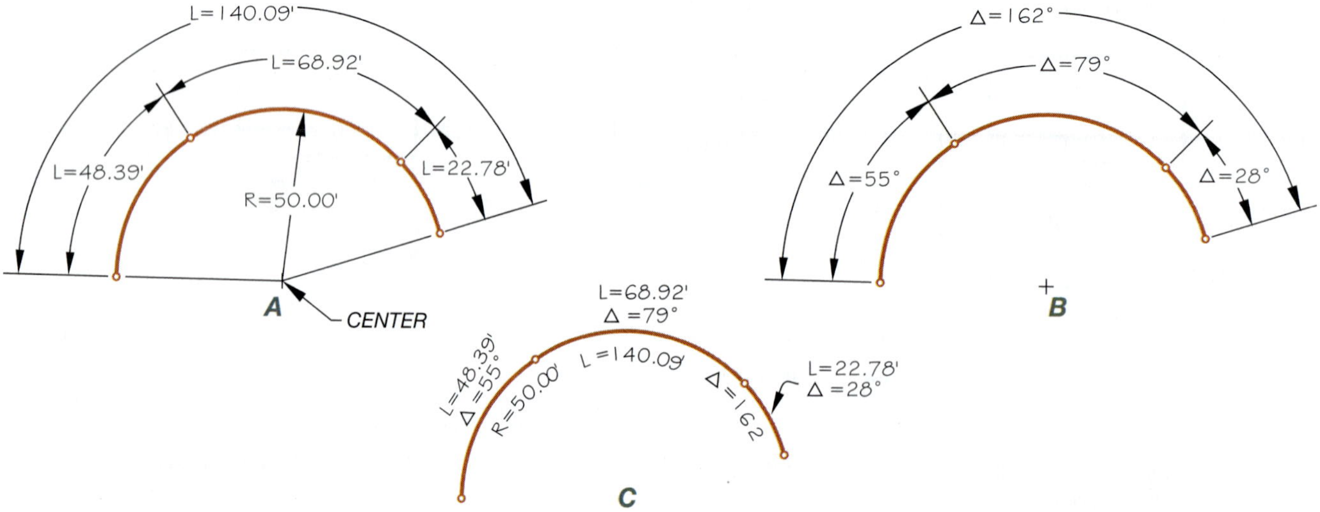

FIGURE 13.4 ■ Drawing and labeling curved property lines.

STEPS IN SITE PLAN LAYOUT

Follow these steps to draw a site plan.

STEP 1 Select the paper size. In this case the size is 8 1/2″ × 11″. Evaluate the plot to be drawn. Lot 2 of Sandy Estates, shown in Figure 13.2, will be used. Determine the scale to use by considering how the longest dimension (134.92′) fits on the sheet. Always try to leave at least a 1/2″ margin around the sheet.

STEP 2 Use the given plat as an example to lay out the proposed site plan. If a plat is not available, then the site plan can be laid out from the legal description by establishing the boundaries using the bearings and dimensions in feet. Lay out the entire plot plan using construction lines. If errors are made, the construction lines are very easy to erase, or use CADD layers. See Figure 13.5.

STEP 3 Lay out the proposed structure using construction lines. The proposed structure in this example is 60′ long and 36′ wide. The house can be drawn on the site plan with or without showing the roof. A common practice is to draw only the outline of the floor plan. In some cases, the roof overhang may be considered in the setback. In these situations, the house outline is drawn as a dashed line under the roof and dimensions are given for the house and the overhang. The front setback is 25′, and the east side is 15′. Lay out all roads, driveways, walks, and utilities. Be sure the structure is inside or on the minimum setback requirements. *Setbacks* are imaginary boundaries beyond which the structure may not be placed. Think of them as property line offsets that are established by local regulations. Minimum setbacks may be confirmed with local zoning regulations. The minimum setbacks for this property are 25′ front, 10′ sides, and 35′ back. Look at Figure 13.6.

STEP 4 Darken all property lines, structures, roads, driveways, walks, and utilities, as shown in Figure 13.7. Some drafters use a thick line or shading for the structure.

STEP 5 Add dimensions and contour lines (if any) or elevations. The property line dimensions are generally placed on the inside of the line in decimal feet, and the bearing is placed on the outside of the line. The dimensions locating and giving the size of the structure are commonly in feet and inches or in decimal feet. Try to keep the amount of extension and dimension lines to a minimum on the site plan. One way to do this is to dimension directly to the house and place size dimensions inside the house outline. Add all labels, including the road name, property dimensions and bearings (if used), utility names, walks, and driveways, as shown in Figure 13.8.

STEP 6 Complete the plot plan by adding the north arrow, the legal description, title, scale, client's name, and other title block information. Figure 13.9 shows a complete plot plan.

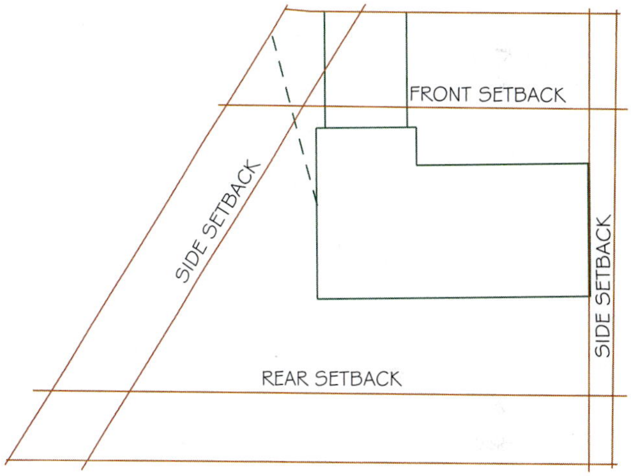

FIGURE 13.6 ■ Step 3: Lay out the structures, roads, driveways, walks, and utilities. Be sure the structure is on or within the minimum required setbacks.

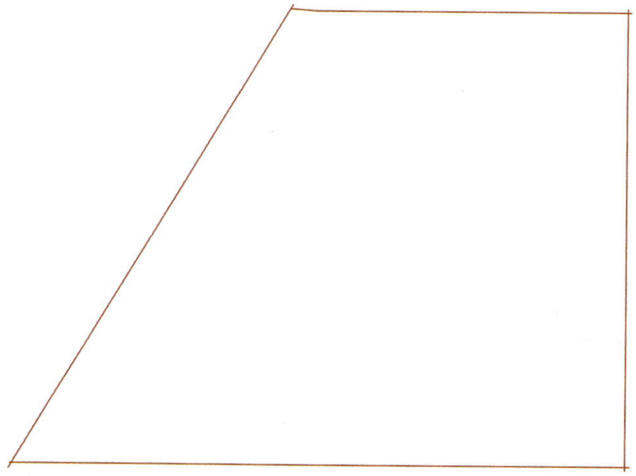

FIGURE 13.5 ■ Step 2: Lay out the plot plan property lines.

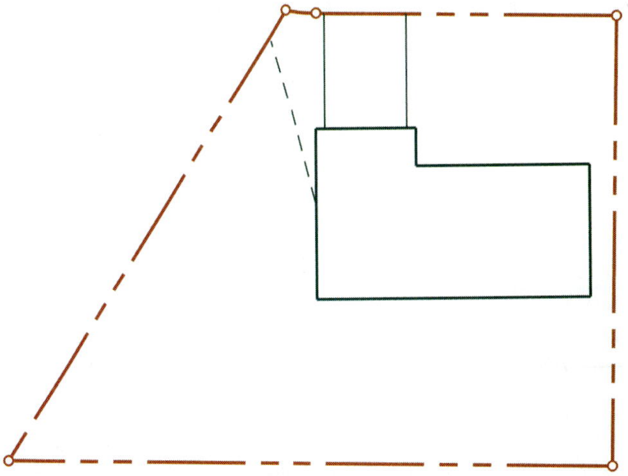

FIGURE 13.7 ■ Step 4: Darken boundary lines, structures, roads, driveways, walks, and utilities.

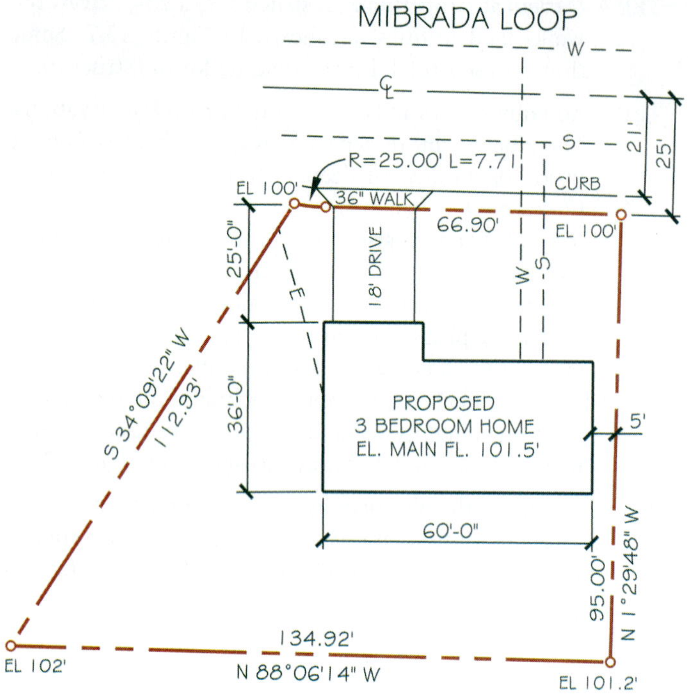

FIGURE 13.8 ■ Step 5: Add dimensions and elevations; then label all roads, driveways, walks and utilities.

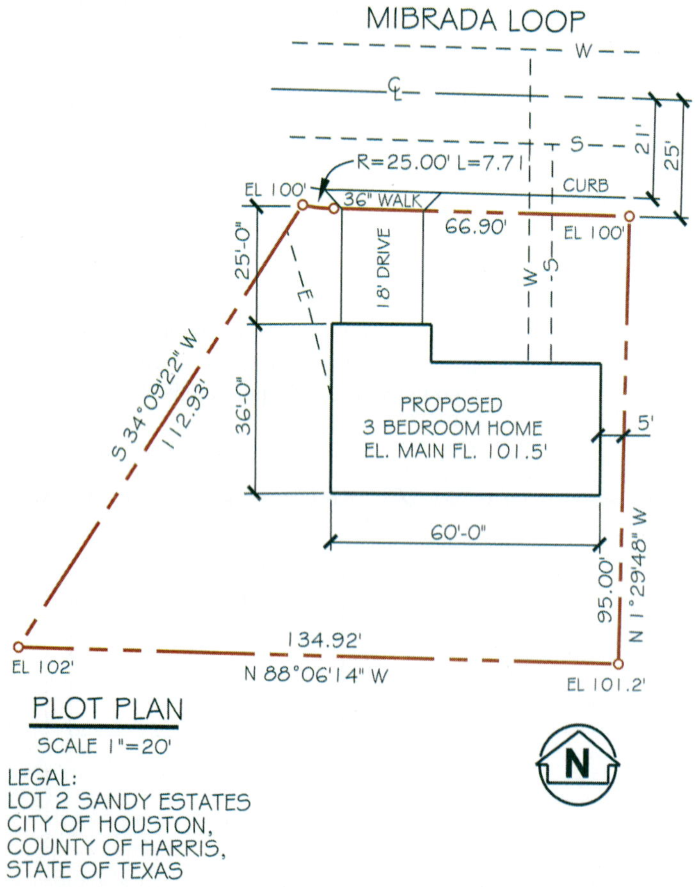

FIGURE 13.9 ■ Step 6: Complete the plot plan. Add title, scale, north arrow, legal description, and other necessary information, such as the owner's name if required.

SITE PLAN DRAWING CHECKLIST

Check off the items in the following list as you work on the basic site plan, to be sure that you have included all of the necessary details. Site plans for special applications may require additional information. Refer to this chapter for features found in special plans such as grading plan, subdivision plan, site analysis plan, planned unit development, and commercial plan.

- Site plan title and scale.
- Property legal description.
- Property line dimensions and bearings.
- North arrow.
- Existing and proposed roads with the elevation at the center of roads.
- Driveways, patios, decks, walks, and parking areas.
- Existing and proposed structures with floor level elevations.
- Public or private water supplies.
- Public or private sewage disposal.
- Location of utilities.
- Rain and footing drains and storm sewer or drainage.
- Topography, including contour lines or elevations at property corners, street centerline, driveways, and floor elevations.
- Front, side, and rear setbacks dimensioned and in compliance with zoning.
- Specific items on adjacent properties, if required, such as existing structures, water supply, sewage disposal, trees, or water features.
- Existing and proposed trees may be required.

DRAWING CONTOUR LINES

Contour lines represent intervals of equal elevation, as explained in Chapter 12. The following discussion shows you how to lay out contour lines on a construction site. Surveyors can use various methods to establish elevations at points on the ground. This information is recorded in field notes. You then use the field notes to plot the contour lines. This discussion will explain the grid survey method. Another common technique is the *control point survey*, which establishes elevations that are recorded on a map. You then lay out the elevations for the contour lines based on the given elevation points. The *radial survey* is also a common method, used for locating property corners, structures, natural features, and elevation points. This system establishes control points using a process called *radiation* in which measurements are taken from a survey instrument located at a base known as a *transit station*. From the transit station, a series of angular and distance measurements are

established to specific points on the ground. You then establish the property lines, land features, and contours based on these points.

Using the Grid Survey to Draw Contour Lines

A *grid survey* divides the site into a pattern similar to a checkerboard. Stakes are driven into the ground at each grid intersection. The surveyor then establishes an elevation at each stake and records this information in field notes. The spacing of the stakes depends on the land area and the topography. The stakes may be placed in a grid 10′, 20′, 50′, or 100′ apart, for example. An example is shown in Figure 13.10. This grid has lines spaced 20 feet apart. The vertical lines are labeled with letters, and the horizontal lines are labeled with numbers that are called *stations*. The station numbers are in two parts: for example, 0 + 20. The first number is hundreds of feet. Zero represents 0 hundreds, or 0′. The first number is followed by a plus (+) sign. This means that you add the first number to the second number to get the actual distance to this station. The second number is in tens of feet. For example, 0′ + 20′ = 20′, or 1′ + 20′ = 120′. The intersection of vertical line B and station 0 + 40 is identified as station B-0 + 40. The field notes record the elevation at each station, as shown in Table 13–1. For example, the elevation at A-0 + 80 is 105′.

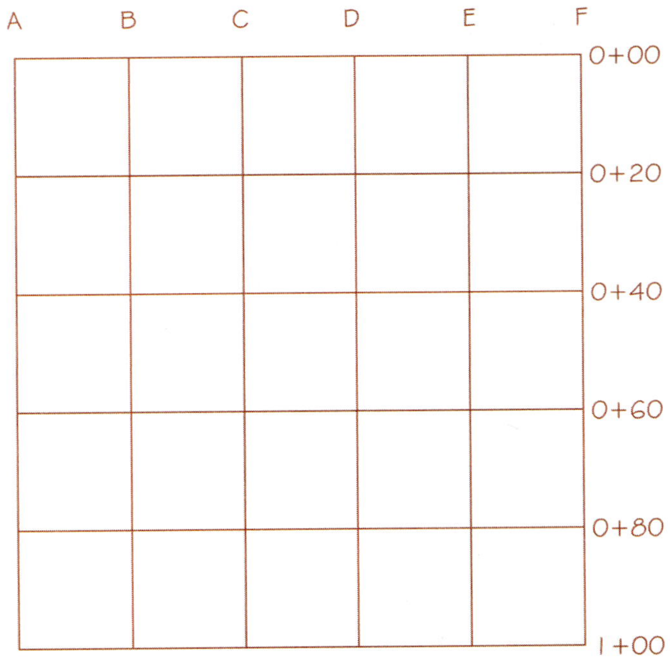

FIGURE 13.10 ■ Draw a grid at a desired scale. Label the vertical lines with letters and the horizontal lines with station numbers, as shown in this example.

TABLE 13–1 GRID SURVEY FIELD NOTES

STATION	ELEVATION	STATION	ELEVATION
A-0 + 00	101	D-0 + 00	95
A-0 + 20	104	D-0 + 20	97
A-0 + 40	108	D-0 + 40	99
A-0 + 60	112	D-0 + 60	105
A-0 + 80	105	D-0 + 80	100
A-1 + 00	102	D-1 + 00	94
B-0 + 00	100	E-0 + 00	94
B-0 + 20	102	E-0 + 20	96
B-0 + 40	105	E-0 + 40	98
B-0 + 60	110	E-0 + 60	100
B-0 + 80	104	E-0 + 80	95
B-1 + 00	100	E-1 + 00	92
C-0 + 00	98	F-0 + 00	92
C-0 + 20	100	F-0 + 20	93
C-0 + 40	102	F-0 + 40	95
C-0 + 60	108	F-0 + 60	98
C-0 + 80	102	F-0 + 80	90
C-1 + 00	95	F-1 + 00	85

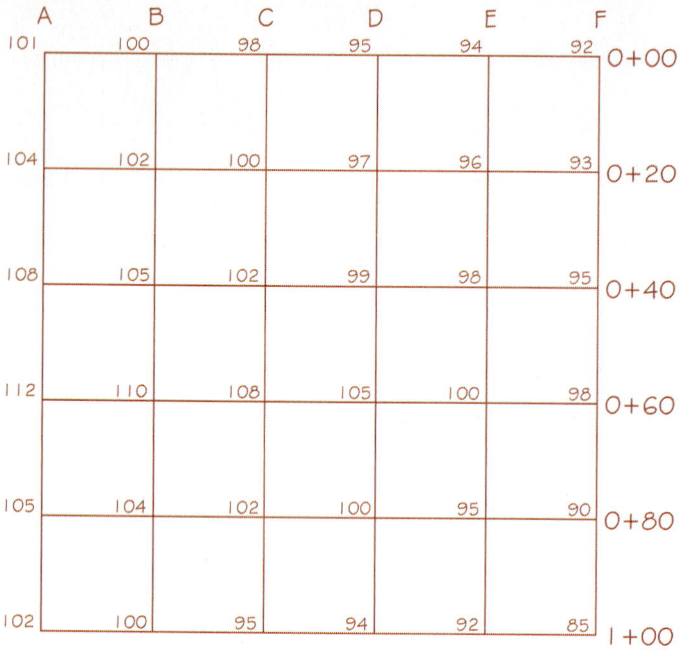

FIGURE 13.11 ■ Use the field notes in Table 13–1 to label the elevation at each grid intersection.

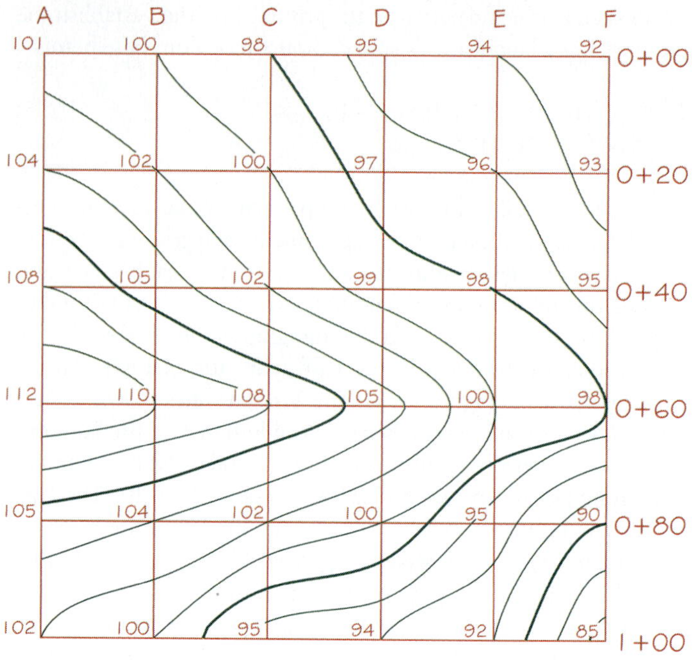

FIGURE 13.12 ■ Connect the points at the elevations for each contour interval.

Steps in Drawing Contour Lines

Contour lines are drawn from grid survey field notes using the following steps:

STEP 1 Draw a grid at a desired scale similar to Figure 13.10. Use construction lines if you are drafting manually, or establish a GRID or CONSTRUCTION layer if you are using CADD. Label the vertical and horizontal grid lines.

STEP 2 Use the field notes to label the elevation at each grid intersection. The elevations, labeled on the grid in Figure 13.11, are based on the field notes found in Table 13–1.

STEP 3 Determine the desired contour interval and connect the points on the grid that represent contour lines at this interval. The contour interval for the grid in Figure 13.11 is 2′. This means that you will establish contour lines every 2′, such as 92, 94, 96 and 98.

STEP 3 Establish the contour lines at the desired contour interval by picking points on the grid that represent the desired elevations. If a grid intersection elevation is 100, then this is the exact point for the 100′ elevation on the contour line. If the elevation at one grid intersection is 98′ and the next is 102′, then you estimate the location of 100′ between the two points. This establishes another 100′ elevation point.

STEP 4 When you have identified the points for all of the 100′ elevations, then connect the points to create the contour line. See Figure 13.12. Do this for the elevation at each contour interval.

FIGURE 13.13 ■ Darken the contour lines. The intermediate lines are thin, and the index contour lines are thick. The index contour lines are broken and labeled with the elevation of the contour line.

STEP 5 Darken the contour lines. The intermediate lines are thin. The index contour lines are broken and labeled with the elevation and are generally drawn thicker than intermediate lines, as shown in Figure 13.13.

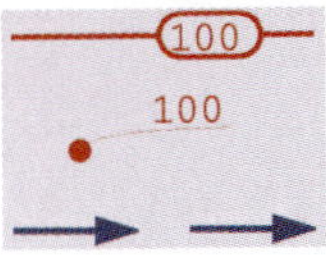

DRAWING PROFILES

A *profile* is a vertical section of the surface of the ground, and/or of underlying earth that is taken along any desired fixed line. The profile of a construction site is usually through the building excavation location, but more than one profile can be drawn as needed. The profile for road construction is normally placed along the center line. Profiles are drawn from the contour lines at the section location. The contour map and its related profile are commonly referred to as the plan and profile. Profiles have uses such as showing road grades and site excavation. Projecting directly from the desired cut location on the contour map following these steps creates the profile:

STEP 1 Draw a straight line on the contour map at the location of the desired profile. See Figure 13.14.

STEP 2 Set up the profile vertical scale. The horizontal scale is the same as the map, because you are projecting directly from the map. The vertical scale can be the same, or it can be exaggerated. Exaggerated scales are used to give a clearer representation of the contour when needed. Establish a vertical scale increment that is above the maximum elevation and one that is below the minimum elevation. Figure 13.15 shows how the vertical is set up. Notice that the profile is projected 90° from the profile line on the map.

STEP 3 Project a line from the location where every contour line crosses the profile line on the contour map. See Figure 13.16.

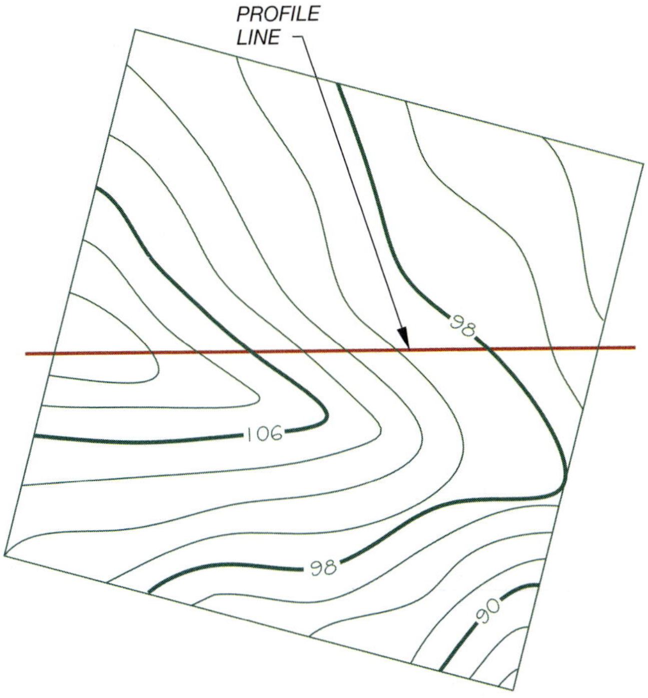

FIGURE 13.14 ■ Draw a straight line on the contour map at the location of the desired profile.

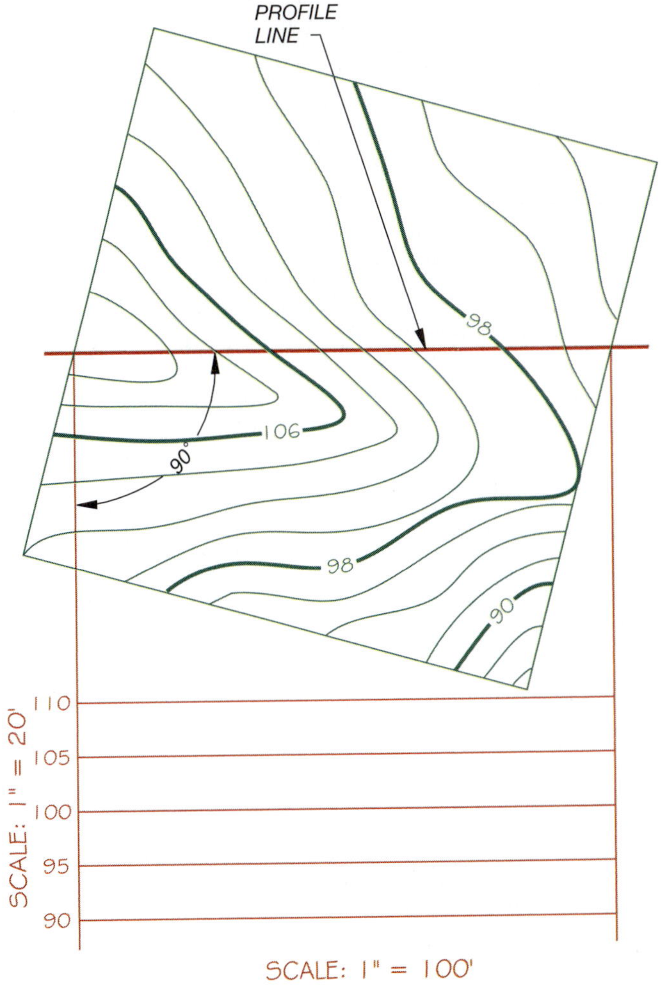

FIGURE 13.15 ■ Project the profile 90° from the start of the profile line. Set up the vertical scale. Label the vertical scale and the elevation of each contour along the vertical scale. Draw a horizontal line at each contour interval. Label the horizontal scale.

STEP 4 Draw the profile by connecting the points where every two vertical and horizontal lines of the same elevation intersect, as shown in Figure 13.17.

DRAWING THE GRADING PLAN

If you have a specific location on a site where a level excavation must take place for the proposed construction, you can lay out a grading plan. The *grading plan* shows the elevation of the site after excavation. This plan shows where areas need to be cut and filled. This is referred to as *cut and fill.* Figure 13.18 shows the site plan location for the desired level excavation. The following steps can be used to draw the grading plan for a level construction site:

STEP 1 Determine the *angle of repose,* which is the slopes of cut and fill from the excavation site measured in feet of horizontal run to feet of vertical rise. One unit of rise to one

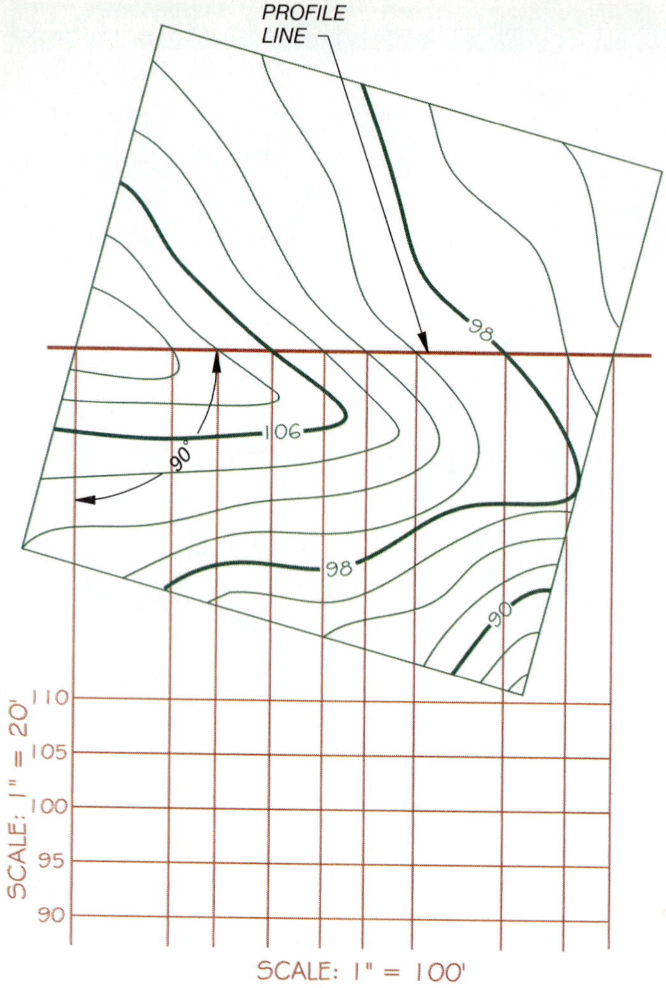

FIGURE 13.16 ■ Project a line 90° from the location where every contour line crosses the profile line.

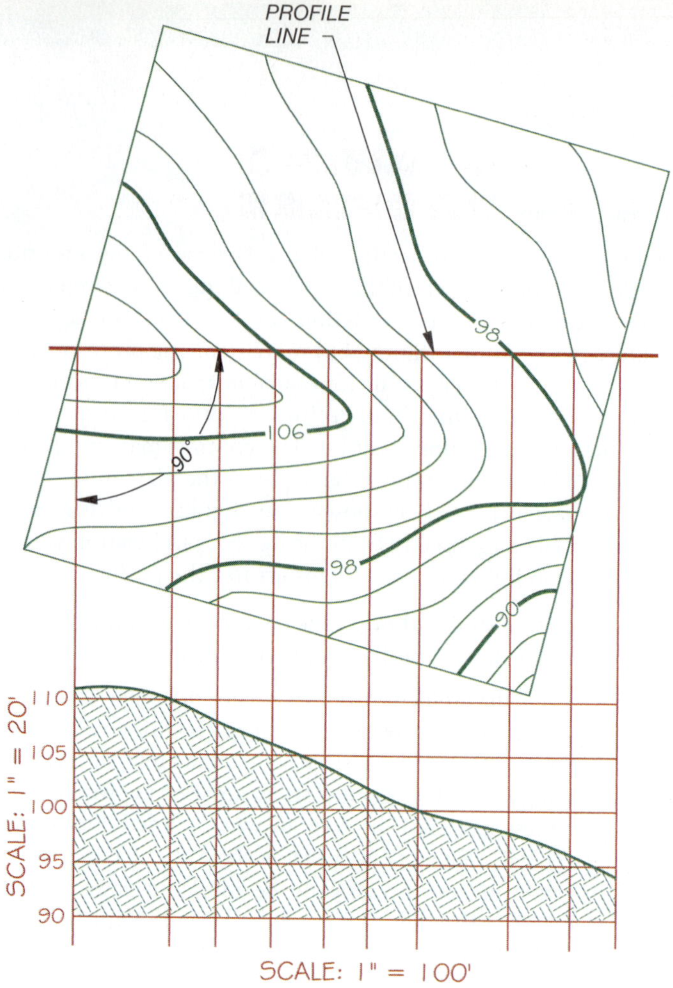

FIGURE 13.17 ■ Draw the profile by connecting the points where every two vertical and horizontal lines of the same elevation intersect.

unit of run is specified as 1:1. See Figure 13.19. The actual angle of repose for cuts and fills is normally determined by approved soils engineering or engineering geology reports. The slope of cut surfaces shall be no steeper than is safe for the intended use and shall not exceed 1 unit vertical in 2 units horizontal (1:2). Alternative designs may be allowed if soils engineering and/or engineering geology reports state that the site has been investigated and give an opinion that a cut at a steeper slope will be stable and not create a hazard to property. Fill slopes shall not be constructed on natural slopes steeper than 1 unit vertical in 2 units horizontal (1:2). The ground surface shall be prepared to receive fill by removing vegetation, previous unstable fill material, topsoil, and other unsuitable materials to provide a bond with the new fill. Other requirements include soil engineering where stability, steeper slopes, and heights are issues. Soil engineering can require benching the fill into sound material and specific drainage and construction methods. A *bench* is a fairly level step excavated into the earth material on which fill is placed.

STEP 2 Draw parallel lines around the excavation site, with each line representing the elevation at the cut and fill. If the angle of repose is 1:1 and the contour interval is 2′, then these parallel lines are 2′ from the level excavation site.

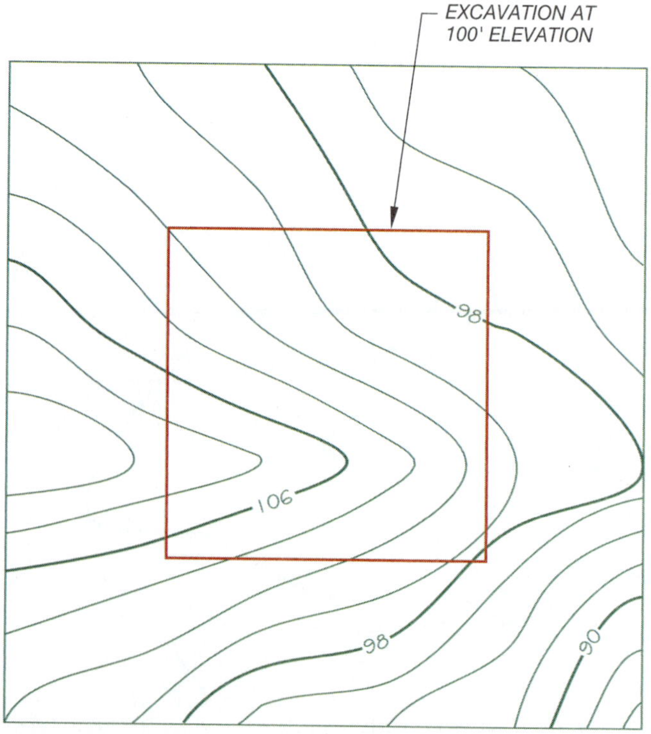

FIGURE 13.18 ■ Site plan location for the desired level excavation at 100′.

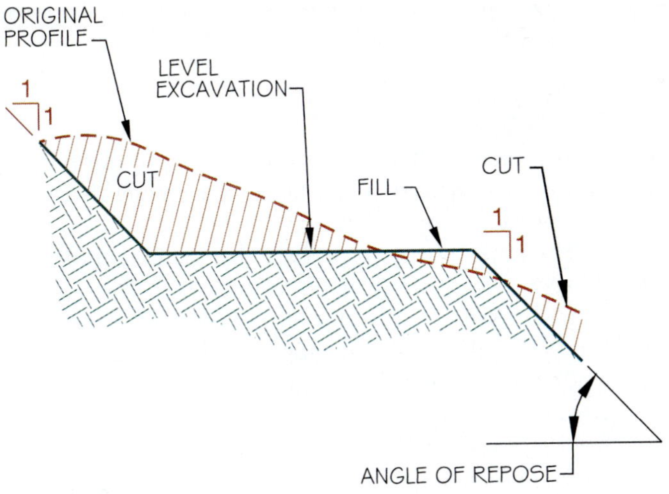

FIGURE 13.19 ■ Angle of repose.

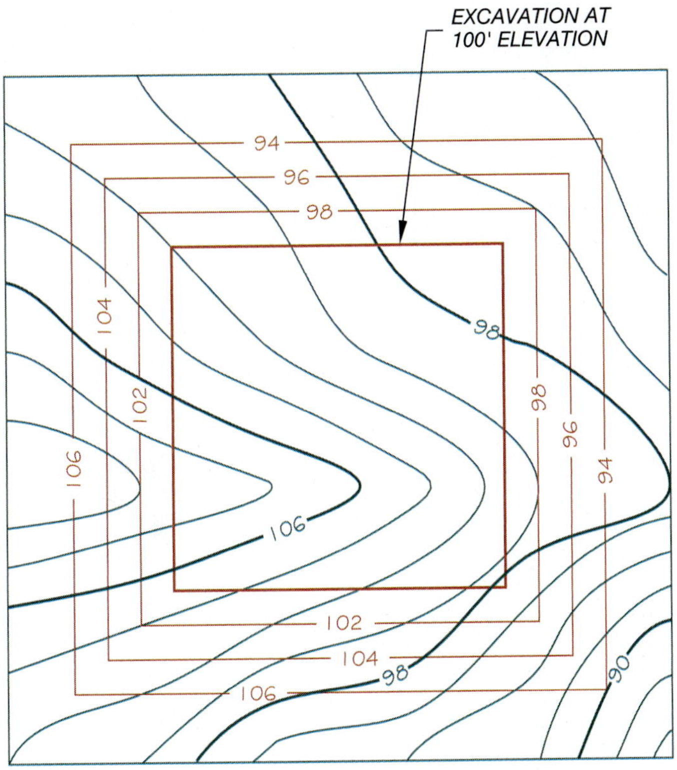

FIGURE 13.20 ■ Draw parallel lines around the excavation site, with each line representing the elevation of the cut and fill.

This is determined by multiplying *rise times contour interval* ($1 \times 2 = 2$ in this example). Look at Figure 13.20.

STEP 3 The elevation of the excavation is 100′. Elevations above this are considered cuts, and elevations below this are fills. To establish the cut and fills, mark where the elevations of the parallel lines around the excavation match the corresponding elevations of the contour lines, as shown in Figure 13.21.

STEP 4 Connect the points established in step 3, as shown in Figure 13.22. The cut and fill areas can be shaded or left unshaded.

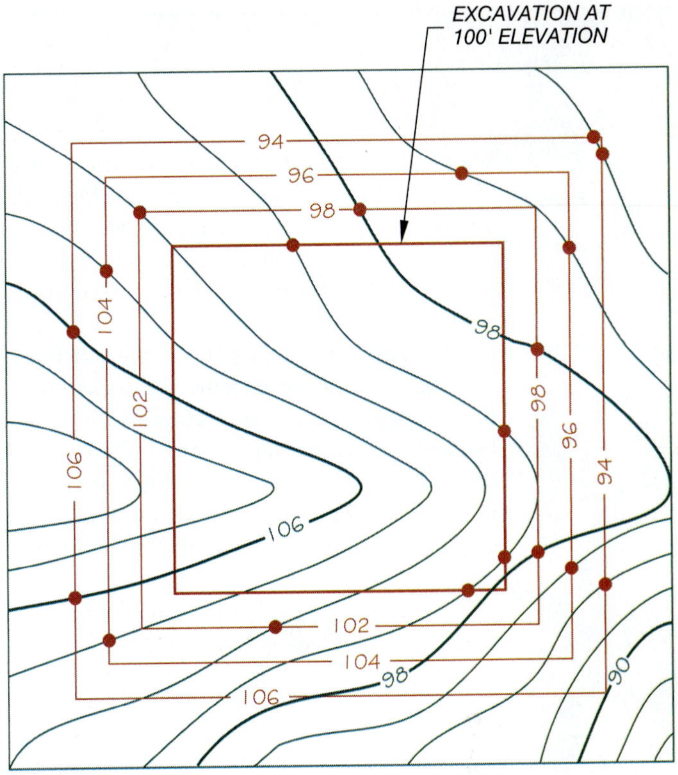

FIGURE 13.21 ■ To establish the cut and fill, mark where the elevations of the parallel lines around the excavation match the corresponding elevations of the contour lines.

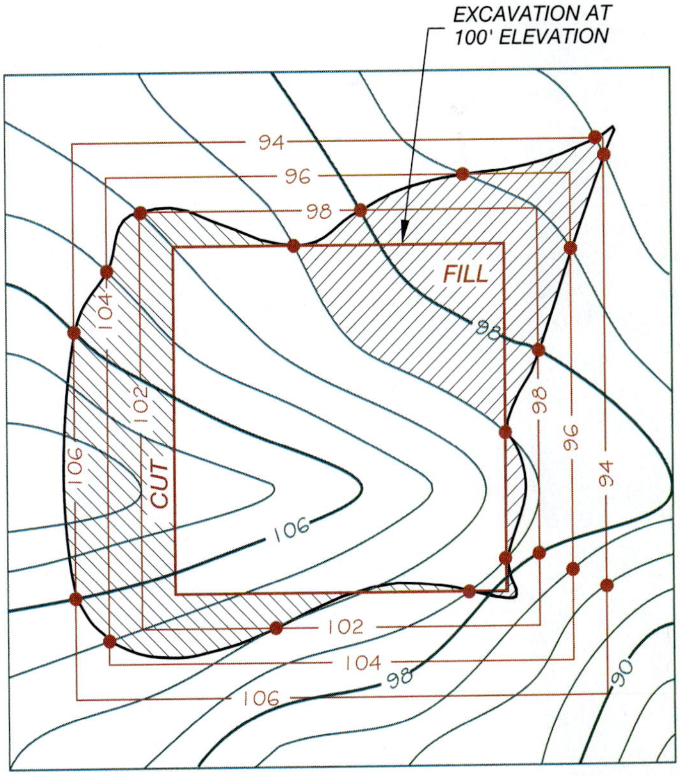

FIGURE 13.22 ■ Complete the cut and fill drawing by connecting the points. The cut and fill areas can be labeled, and they can be shaded or left unshaded.

CADD APPLICATIONS

USING CADD TO DRAW PLOT PLANS

There are CADD software packages that may easily be customized to assist in drawing site plans. Also available are complete CADD mapping packages that allow you to draw topographic maps, terrain models, grading plans, and land profiles. It all depends on the nature of your business and how much power you need in the CADD mapping program. One of the benefits of CADD over manual drafting is accuracy. For example, you can draw a property boundary line by giving the length and bearing. The computer automatically draws the line, then labels the length and bearing. Continue by entering information from the surveyor's notes to draw the entire property boundary in just a few minutes. Such features increase the speed and accuracy of drawing site plans. A residential site plan, drawn using CADD mapping software, is shown in Figure 13.23.

The needs of the commercial site plan are a little different from the residential requirements. The commercial CADD site plan package should use the same features as the residential application and, additionally, have the ability to design street and parking lot layouts. The commercial CADD drafter uses features from symbol libraries, including utility symbols; street, curb, and gutter designs; landscaping; parking lot layouts; titles; and scales. A commercial site plan is shown in Figure 13.24.

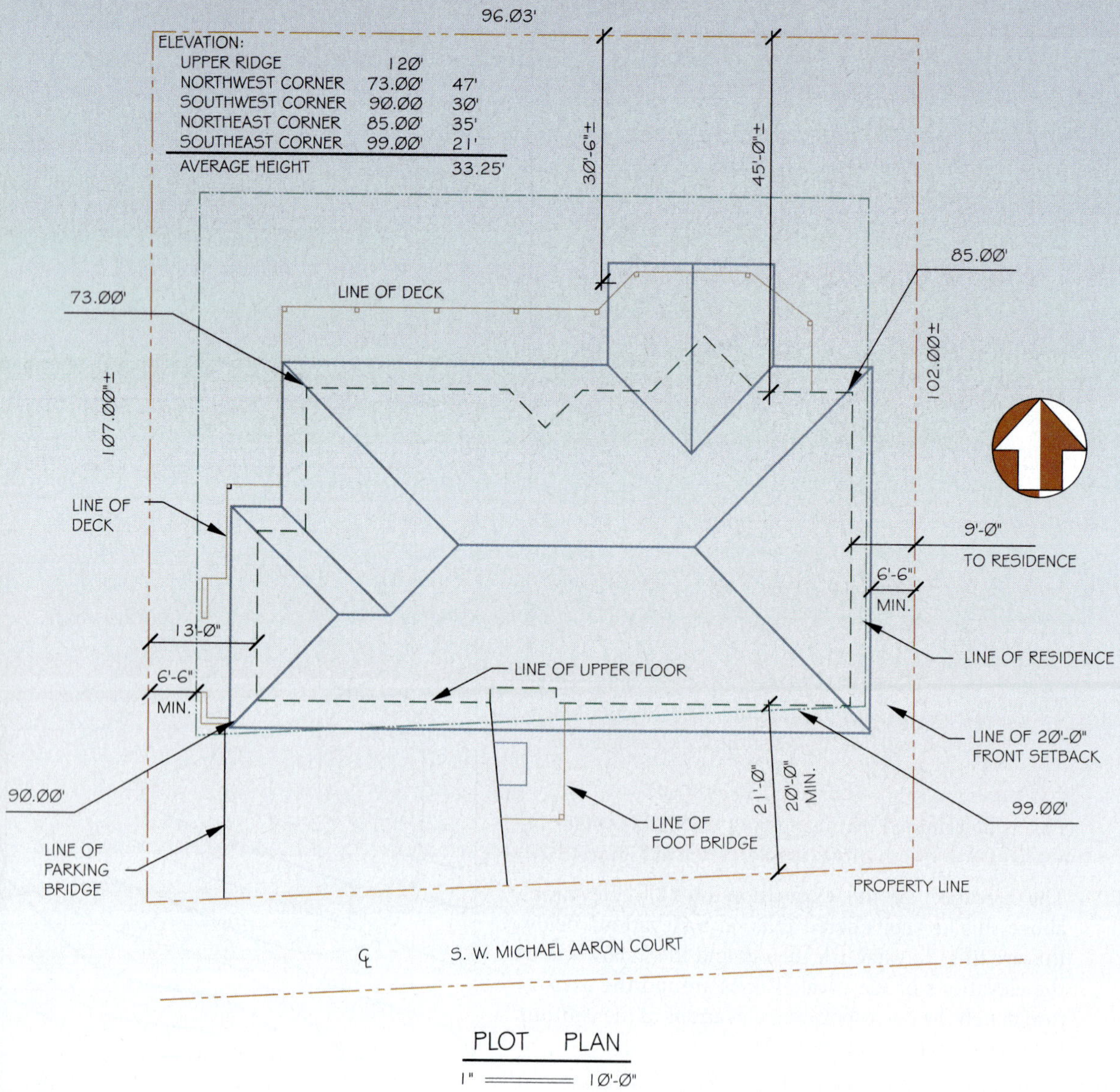

FIGURE 13.23 ■ A CADD-drawn site plan.

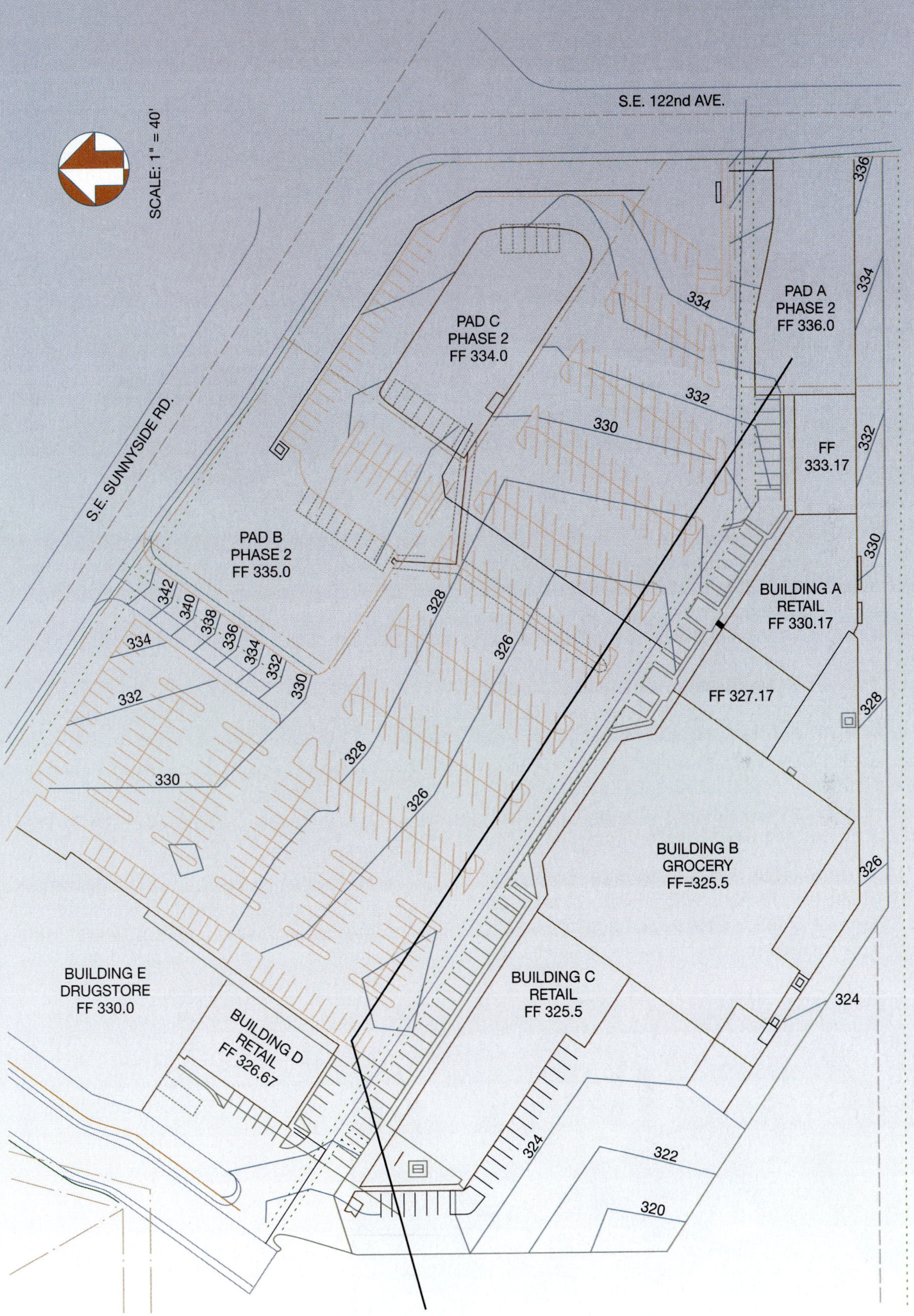

FIGURE 13.24 ■ CADD drawn commercial site plan. Courtesy Saderstrum Architects, PC.

DEVELOPING A CADD TERRAIN MODEL

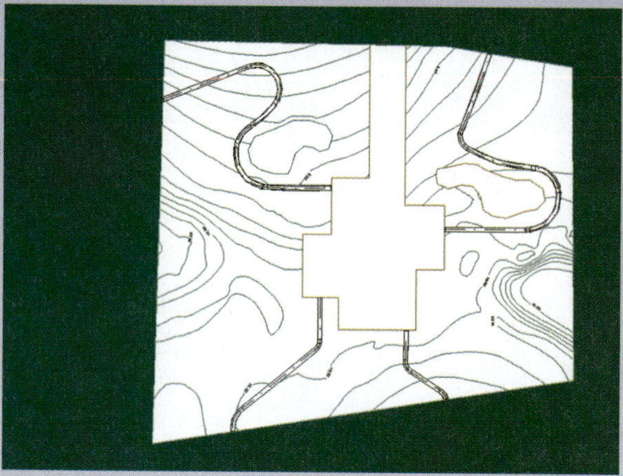

FIGURE 13.25 ■ A topographic site plan to be used for developing a terrain model. *ArchiCAD® images courtesy of Graphisoft®*, planned by architects Csaba Szakál and Geörgy Péchy.

FIGURE 13.26 ■ The CADD program recognizes the contour lines from the site plan in Figure 13.25 and creates a 3D wireframe model (a) or a 3D terrain model (b). *ArchiCAD® images courtesy of Graphisoft®*, planned by architects Csaba Szakál and Geörgy Péchy.

CADD programs are available that allow you to include site work, landscaping, roads, driveways, retaining walls, and construction excavation as part of the project. A first step is to build a terrain model from the survey data or from a topographic site plan. Figure 13.25 shows an example of a topographic site plan to be used. The CADD program recognizes the contour lines or survey data and creates a 3D model in wireframe or as a 3D-terrain model, as shown in Figure 13.26. The terrain model can be viewed from any angle to help you fully visualize the contours of the site. Now the terrain model can be used for any of the following design applications:

- Define borders and property lines.
- Find the elevation at any point, contour line, surface object, or feature and modify the elevation to determine how this affects the model.
- Modify the terrain model by editing the contour shapes.
- Define and display building excavation sites, roads, and other features.
- Show and calculate cut and fill requirements.
- Display the model in plan or 3D view.

When the site is designed as desired, the 3D rendering of the home can be placed on the site, as shown in Figure 13.27. This is an excellent way to demonstrate how a client's home will look when it is finished.

FIGURE 13.27 ■ When the site is designed as desired, a 3D rendering of the house can be placed on the site. *ArchiCAD® images courtesy of Graphisoft®*, planned by architects Csaba Szakál and Geörgy Péchy.

CADD SITE PLAN SOFTWARE

CADD site plan software is available from developers who have a variety of architectural products to simplify your design and drafting practice. Figure 13.28*a* shows a site plan tablet menu and Figure 13.28*b* shows icon menus that were used to help create the site plan displayed.

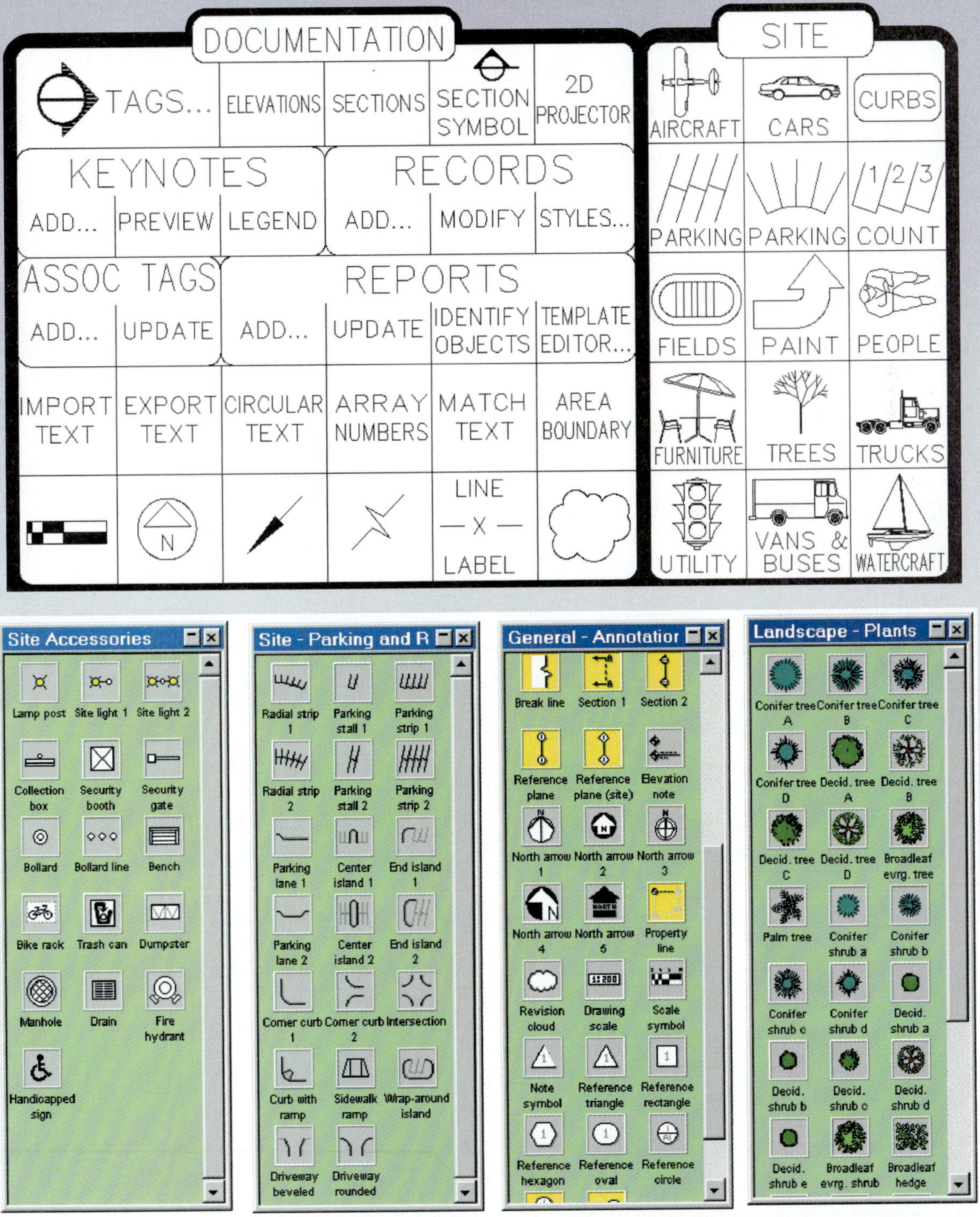

FIGURE 13.28 ■ (a) Documentation and site plan tablet menus. (b) Site and landscape plan icon menus and a site plan that has been created with the aid of these CADD applications. *Courtesy Visio Corporation.*

CADD APPLICATIONS

CADD LAYERS FOR SITE PLAN DRAWINGS

The American Institute of Architects (AIA) CAD Layer Guidelines establish the heading Civil Engineering and Site Work as the major group for CADD layers related to site plans. The following are some of the recommended CADD layer names for site plan applications:

LAYER NAME	DESCRIPTION
C-PROP	Property lines and survey benchmarks.
C-TOPO	Proposed contour lines and elevations.
C-BLDG	Proposed building footprints. The term *footprint* is often used to describe the area directly below a structure.
C-PKNG	Parking lots.
C-ROAD	Roads.
C-STRM	Storm drainage, catch basins and manholes, C-COMM Site communications systems, such as telephone.
C-WATR	Domestic water.
C-FIRE	Fire protection hydrants, connections.
C-NGAS	Natural gas systems.
CSSWR	Sanitary sewer systems.
C-ELEV	Elevations.
C-SECT	Sections.
C-DETL	Details.
C-PSIT	Site plan.
C-PELC	Site electrical systems plan.
C-PUTL	Site utility plan.
C-PGRD	Grading plan.
C-PPAV	Paving plan.
C-P***	Other site, landscape.

Landscape plans also have CADD layer designations, with the major heading Landscape Architecture. The following are some of the commonly used layer names:

LAYER NAME	DESCRIPTION
L-PLNT	Plant and landscape materials.
L-IRRG	Irrigation systems.
L-WALK	Walks and steps.

CHAPTER 13

Site Plan Layout Test

DIRECTIONS

Answer the questions with short complete statements or drawings as needed on an 8 1/2″ × 11″ sheet of notebook paper, as follows:

1. Letter your name, Chapter 13 Test, and the date at the top of the sheet.
2. Letter the question number and provide the answer. You do not need to write out the question.

Answers may be prepared using a word processor or CADD if course guidelines allow this.

QUESTIONS

Question 13–1 What influences the size of the drawing sheet recommended for site plans?

Question 13–2 What four factors influence scale selection in drawing a site plan?

Question 13–3 List five elements of information that should be determined before starting a site plan drawing.

Question 13–4 Why are construction lines helpful in site plan layout?

Question 13–5 Information used to prepare a site plan may come from one or more of several available sources. List four possible sources.

Question 13–6 List at least six types of information that should be determined before the site plan can be completed.

Question 13–7 Define *setbacks*.

Question 13–8 If you are using a CADD system to draw a site plan, what would be a typical prompt in drawing a 134.92′ property line with a N88°06′ 14″W bearing?

Question 13–9 Is it true or false that one of the advantages of CADD over manual drafting is accuracy?

Question 13–10 Identify at least one difference between residential and commercial site plans.

Question 13–11 Define *contour lines*.

Question 13–12 Describe the control point survey.

Question 13–13 Describe the radial survey.

Question 13–14 What is the transit station?

Question 13–15 Explain how a grid survey is set up.

Question 13–16 What is the distance to station point 0 + 40?

Question 13–17 What is the distance to station point 1 + 60?

Question 13–18 Explain the difference between index and intermediate contour lines.

Question 13–19 Define *profile*.

Question 13–20 Why is the vertical profile scale exaggerated?

Question 13–21 In addition to the elevation of the excavation site, what does the grading plan show?

Question 13–22 Define *angle of repose*.

Question 13–23 What are field notes?

Question 13–24 Why is the horizontal scale of the profile the same as the map scale?

Question 13–25 In a grid survey, what is used to label the vertical lines?

(continued)

DIRECTIONS

1. Use an 8 1/2″ × 14″ drawing sheet unless otherwise specified by your instructor.
2. Select an appropriate scale.
3. Minimum front setback is 25′-0″.
4. Minimum rear yard setback is 25′-0″.
5. Minimum side yard setback is 7′-0″.
6. Select, or have your instructor assign, one or more of the plot plan sketches or drawings that follow.

PROBLEMS

Problems 13–1 through 13–7 Begin the selected site plan problem(s) by drawing the given information. The site plan will then be completed after you have selected a residential design project(s) in Chapter 16. Two evaluations are recommended:

- Site plan without structures, drives, and utilities.
- Complete site plan after selection and placement of structures, drives, and utilities. Structures selected in Chapter 15.

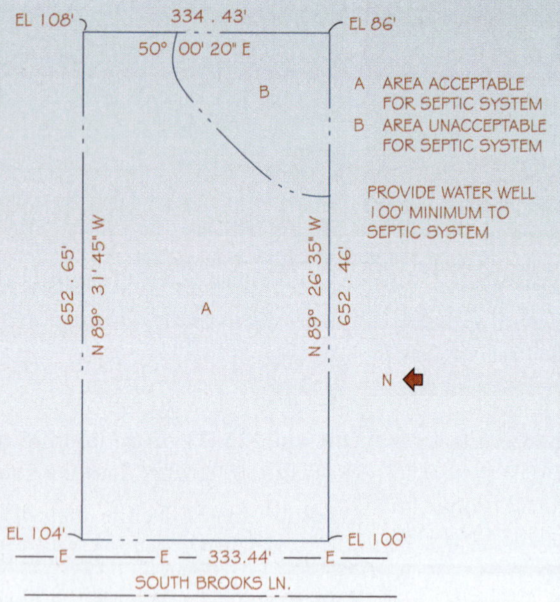

TAX LOT 2300, LOT 12 CLARKES ESTATES SECTION 17, T. 4 S., R. 3E., SALT LAKE MERIDIAN, TOOELE COUNTY, UTAH

PROBLEM 13–1

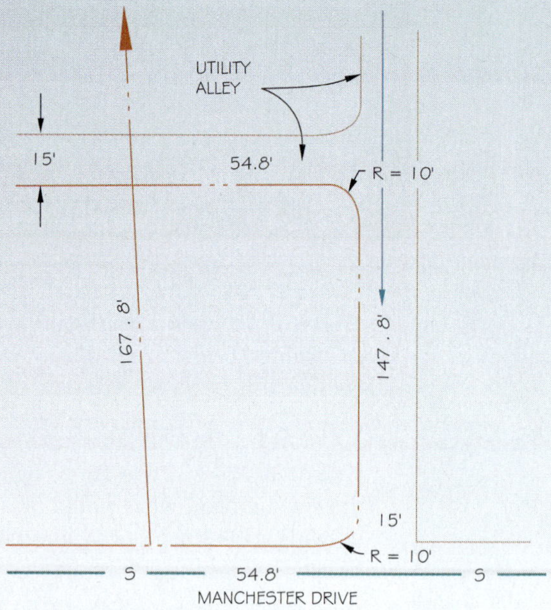

LOT 17, BLOCK 3, PLAT OF GARTHWICK, YOUR CITY, COUNTY, STATE

PROBLEM 13–2

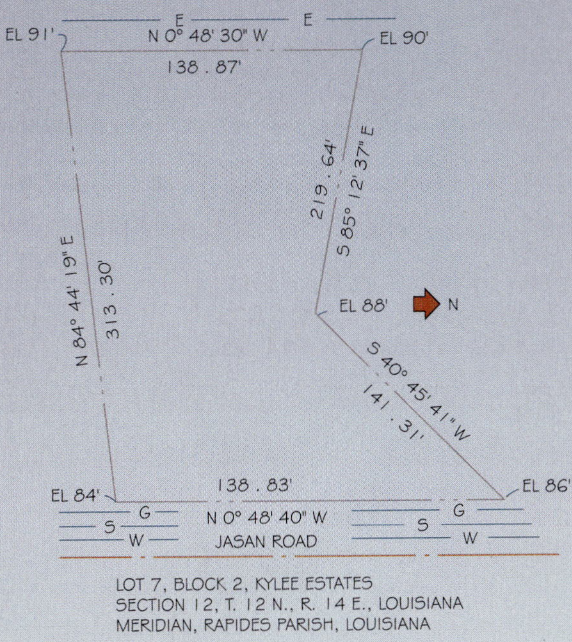

PROBLEM 13–3

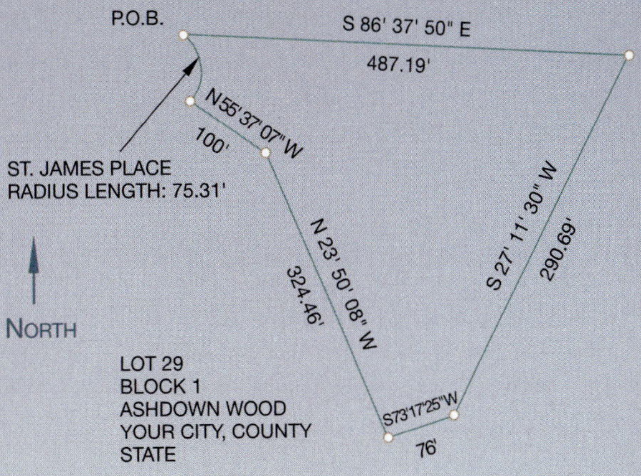

PROBLEM 13–5

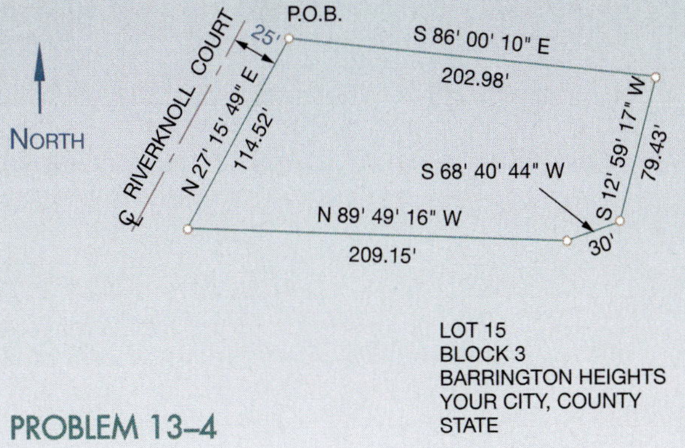

PROBLEM 13–4

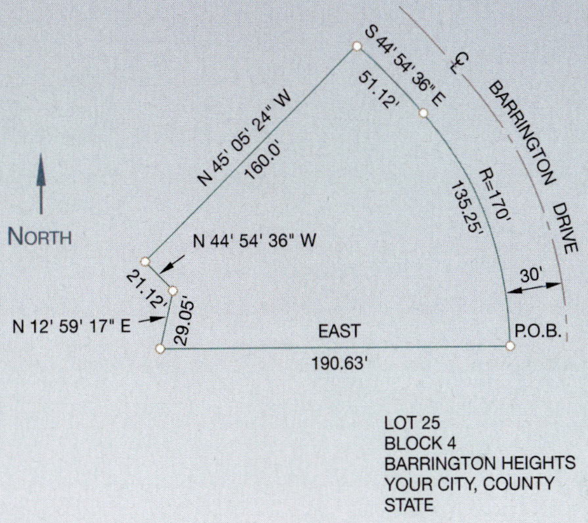

PROBLEM 13–6

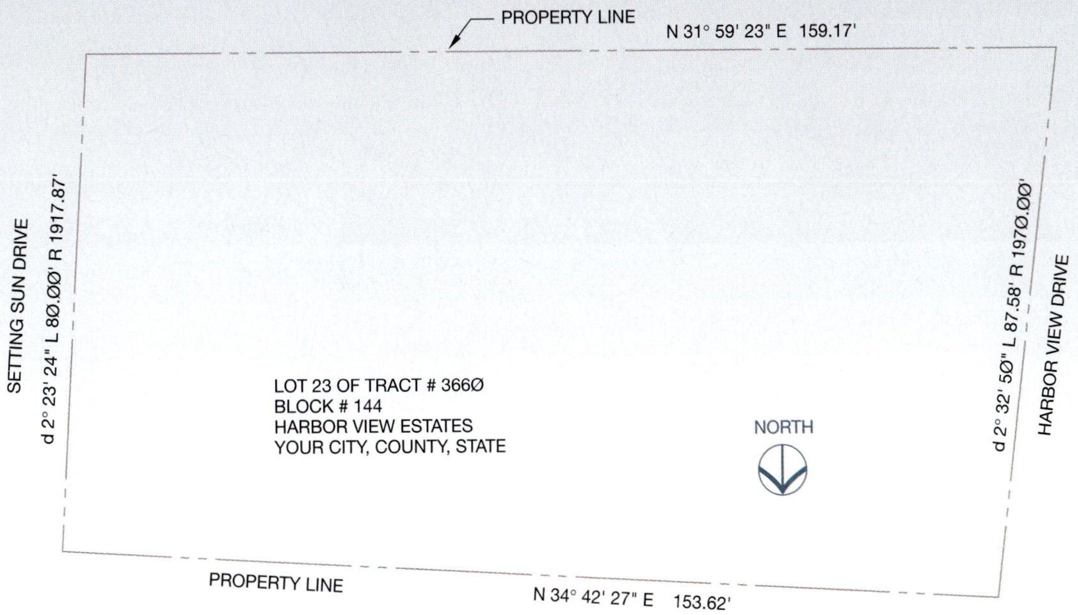

PROBLEM 13–7

(continued)

Problem 13–8 Draw the complete site plan shown.

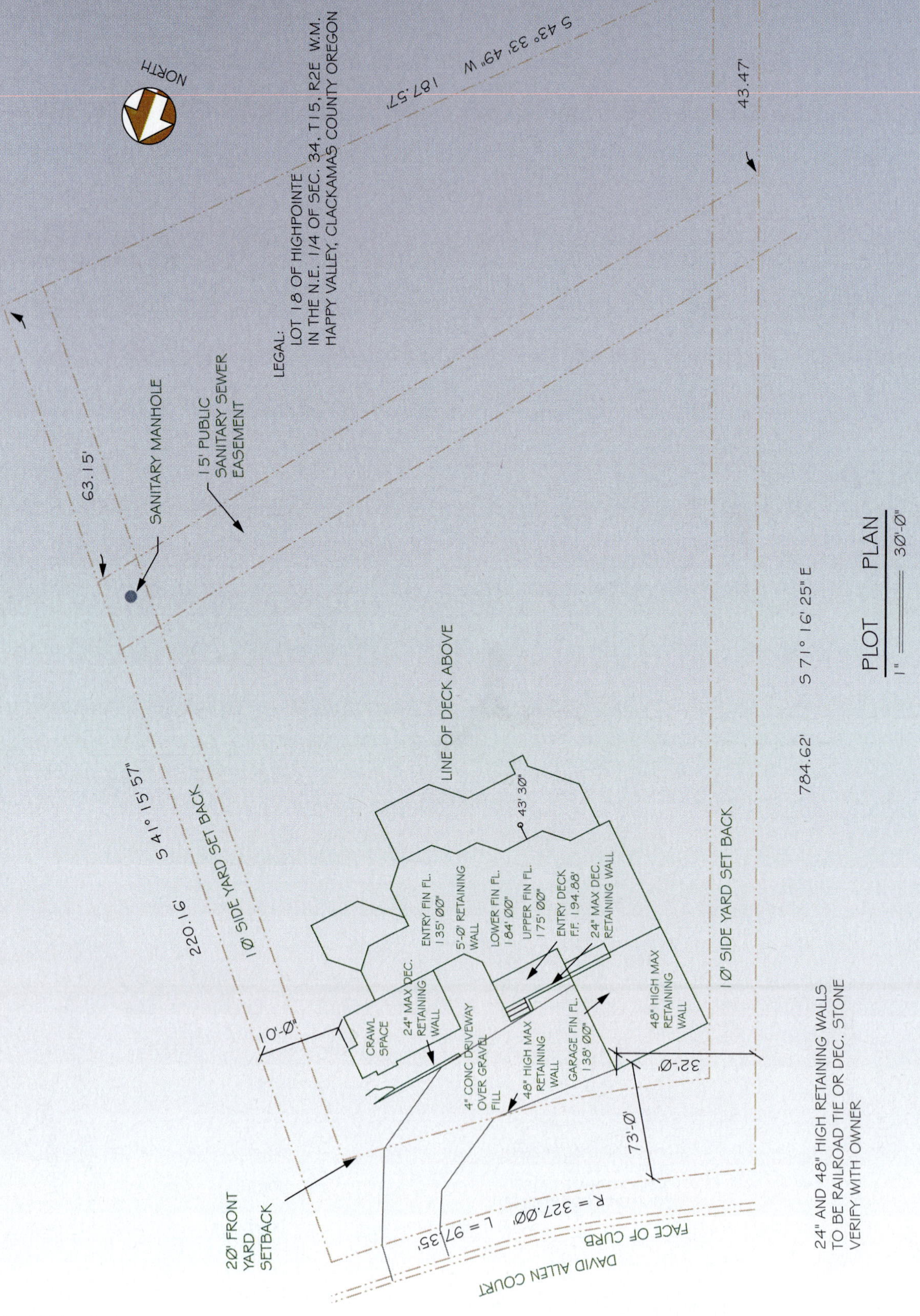
NORTH
S 43° 33' 49" W
187.57'
43.47'
LEGAL:
LOT 18 OF HIGHPOINTE
IN THE N.E. 1/4 OF SEC. 34, T1S, R2E W.M.
HAPPY VALLEY, CLACKAMAS COUNTY OREGON
63.15'
SANITARY MANHOLE
15' PUBLIC
SANITARY SEWER
EASEMENT
LINE OF DECK ABOVE
PLOT PLAN
1" = 30'-0"
S 71° 16' 25" E
784.62'
S 41° 15' 57"
220.16'
10' SIDE YARD SET BACK
43' 30"
ENTRY FIN FL.
135' 00"
5'-0" RETAINING
WALL
LOWER FIN FL.
184' 00"
UPPER FIN FL.
175' 00"
ENTRY DECK
F.F. 194.88'
24" MAX DEC.
RETAINING WALL
10' SIDE YARD SET BACK
CRAWL
SPACE
24" MAX DEC.
RETAINING
WALL
48" HIGH MAX
RETAINING
WALL
10'-0"
4" CONC DRIVEWAY
OVER GRAVEL
FILL
48" HIGH MAX
RETAINING
WALL
GARAGE FIN FL.
138' 00"
32'-0"
73'-0"
24" AND 48" HIGH RETAINING WALLS
TO BE RAILROAD TIE OR DEC. STONE
VERIFY WITH OWNER
20' FRONT
YARD
SETBACK
R = 327.00'
L = 97.35'
FACE OF CURB
DAVID ALLEN COURT

PROBLEM 13–8

Problem 13–9 Continue the drawing started in Problem 13–8 by drawing the complete grading plan using the layout for this problem as a guide.

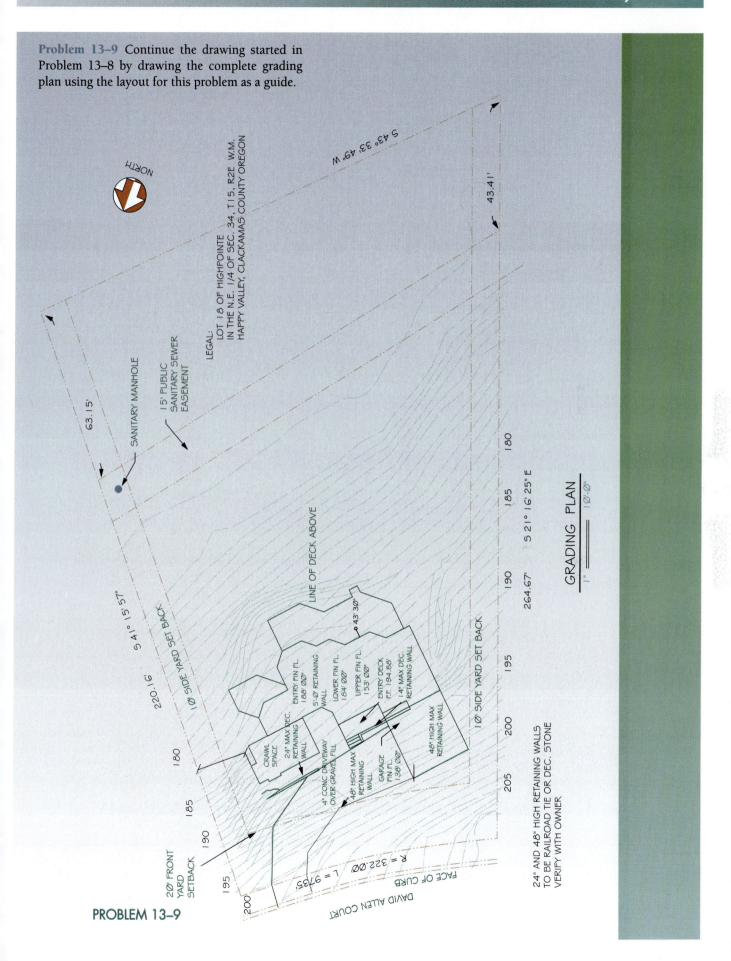

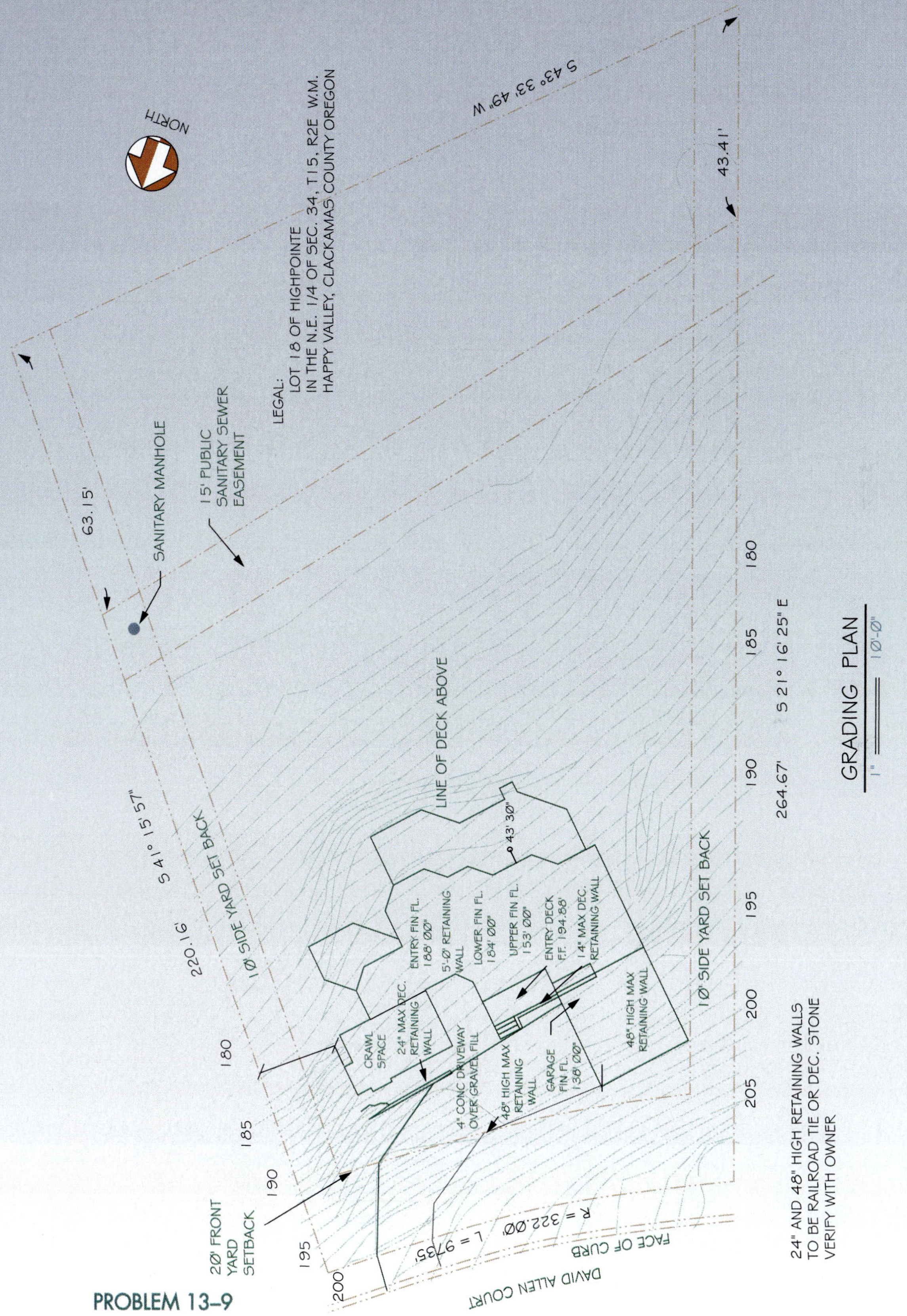

PROBLEM 13–9

SECTION 4

Floor Plan

14 CHAPTER

Floor-Plan Symbols

INTRODUCTION

The floor plans for a proposed home provide the future homeowner with the opportunity to evaluate the design in terms of livability and suitability for the needs of the family. In addition, the floor plans quickly communicate the overall construction requirements to the builder. Symbols are used on floor plans to describe items that are associated with living in the home, such as doors, windows, cabinets, and plumbing fixtures. Other symbols are more closely related to the construction of the home, such as electrical circuits and material sizes and spacing. One of the most important concerns of the drafter is to combine carefully all of the symbols, notes, and dimensions on the floor plan so the plan is easy to read and uncluttered.

Floor plans that are easy to read are also easy to build, since there is less of a possibility of construction errors than with unorganized, cluttered plans. Figure 14.1 shows a typical complete floor plan.

The floor plan is a representation of an imaginary horizontal cut made approximately 4′ (1220 mm) above the floor line. The relationship of the structure to the floor plan is shown in Figure 14.2. This example shows projecting the bottom of the wall onto the floor plan. Residential floor plans are generally drawn at a scale of 1/4″ = 1′–0″ (1:50 metric). Architectural templates are available with a wide variety of floor-plan symbols at the proper scale.

In addition to knowing the proper symbols to use, architects, designers, and architectural drafters should be familiar with standard products that the symbols represent. Products used in the structures they design such as plumbing fixtures, appliances, windows, and doors are usually available from local vendors. Occasionally some special items must be ordered far enough in advance to ensure delivery to the job site on time. By becoming familiar with the standard products, you will also become familiar with their cost. Knowing costs allows you to know if the product you are considering using is affordable within the owner's budget.

One of the best sources of standard product information is Sweet's catalogs. Sweet's publishes several sets of catalogs in different categories, and the set you use depends on the kind of building you are designing. For residential designs, you would most likely use Sweet's *Home Building and Remodeling* catalog. Sweet's, as it is known, is a set of books or CD-ROMs containing building products manufacturers' catalogs arranged by category.

WALL SYMBOLS

Professional Perspective

There are a variety of opinions regarding how walls are drawn on a floor plan. Practices range from drawing all exterior walls thicker than interior walls, for contrast, to drawing all exterior and interior walls the same thickness for convenience and because with manual drafting it is difficult to draw exact thicknesses. A common practice with CADD applications is to draw the floor-plan walls the exact thickness of the construction materials. This is because the CADD system is extremely accurate for drafting. Some architectural CADD programs ask you to specify the exact construction materials and then set up the wall symbols to match. CADD programs are also valuable for establishing information from the drawings. For example, a command such as DISTANCE can be used to determine the exact dimension between walls for the length of a cabinet to be built in that location. This is another reason for making the drawing accurate.

When the thickness of walls is drawn to the exact construction dimensions, the material used establishes these values. Exterior walls can be built using 2 × 4 or 2 × 6 studs. *Studs* are the vertical construction members used for framing walls and partitions, and the nominal size of the construction member before milling is 2 × 4 or 2 × 6. The milling process makes the finished member 1 1/2″ × 3 1/2″ and 1 1/2″ × 5 1/2″. The metric equivalents of these lumber dimensions are 40 × 90 mm and 40 × 150 mm. Exterior and interior construction materials are then applied to the studs. Typical exterior and interior construction is shown in Figure 14.3.

This text recommends drawing exterior and interior walls with contrasting thickness, such as 6″ exterior and 4″ interior, in manual drafting; but the exact wall thickness is recommended when a CADD system is used. Any actual practice should be confirmed with your instructor or company, since either may have a preferred practice. The samples in this text are typically drawn using CADD with text that closely matches high-quality architectural hand lettering, in an effort to maintain quality and demonstrate the best possible manual drafting representation.

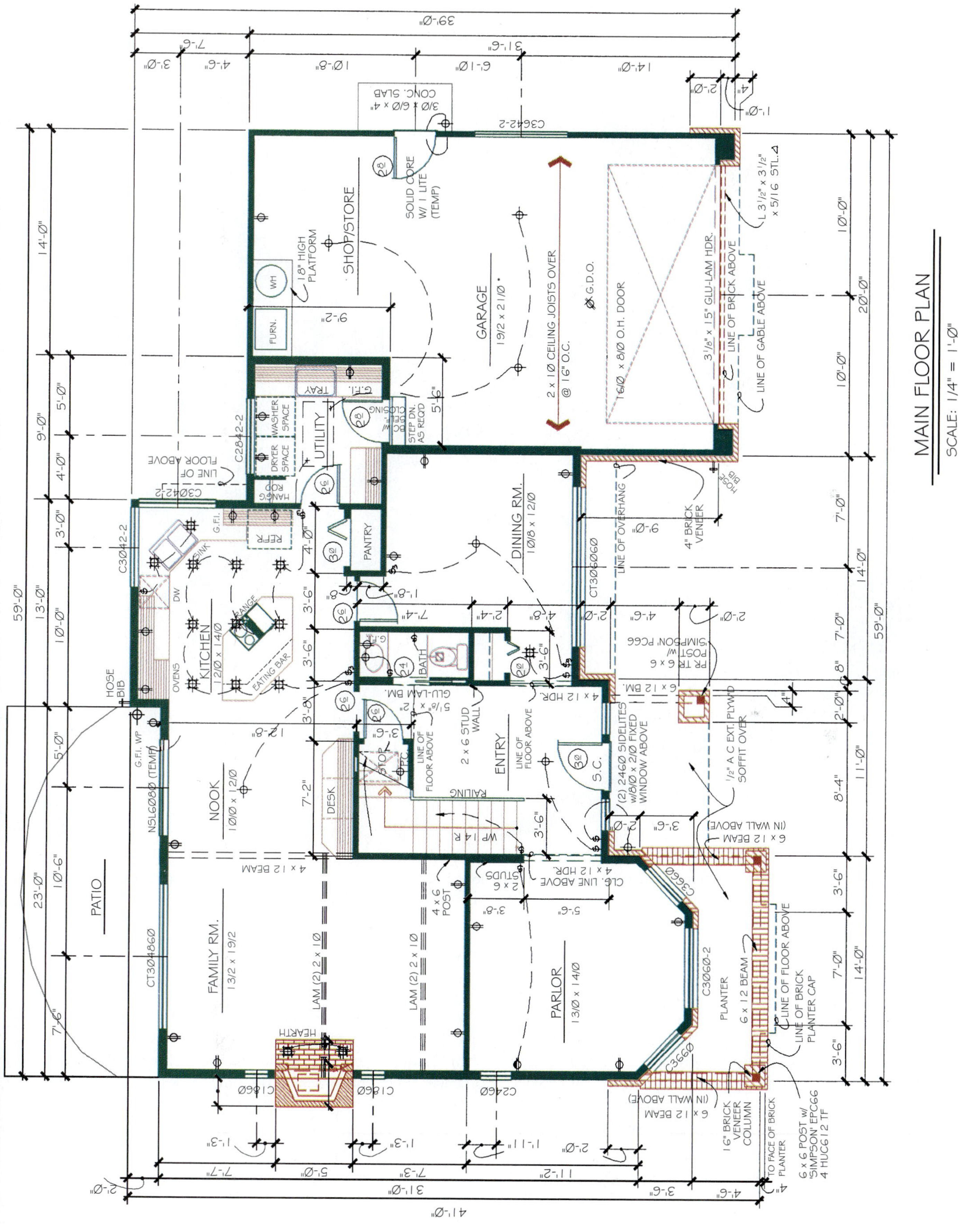

FIGURE 14.1 ■ Complete floor plan. *Courtesy Piercy & Barclay Designer, Inc.*

FIGURE 14.2 ■ The floor plan is a representation of an imaginary horizontal cut made approximately 4′ (1220 mm) above the floor line. This example shows projecting the bottom of the walls onto the floor plan.

Exterior Walls

Exterior wood-frame walls are generally drawn 6″ thick at a 1/4″ = 1′–0″ (1:50 metric) scale for manual drafting or 5″ for CADD. See Figure 14.4. When 2 × 6 (50 × 150 mm) studs are used, the exterior walls are drawn the same thickness or thicker for manual drafting, or exactly 7″ with CADD.

The wall thickness depends on the type of construction. If the exterior walls are to be concrete or masonry construction with wood framing to finish the inside surface, they are drawn substantially thicker, as shown in Figure 14.5. Exterior frame walls with masonry veneer construction applied to the outside surface are drawn an additional 5″ (approx. 125 mm) thick with the masonry veneer represented as 4″ (100 mm) thick over a 1″ (25 mm) air space and 6″ (150 mm) wood frame walls, as seen in Figure 14.5.

Interior Partitions

Interior walls, known as *partitions,* are frequently drawn 4″ (100 mm) thick. Drafters commonly draw the partitions 4″

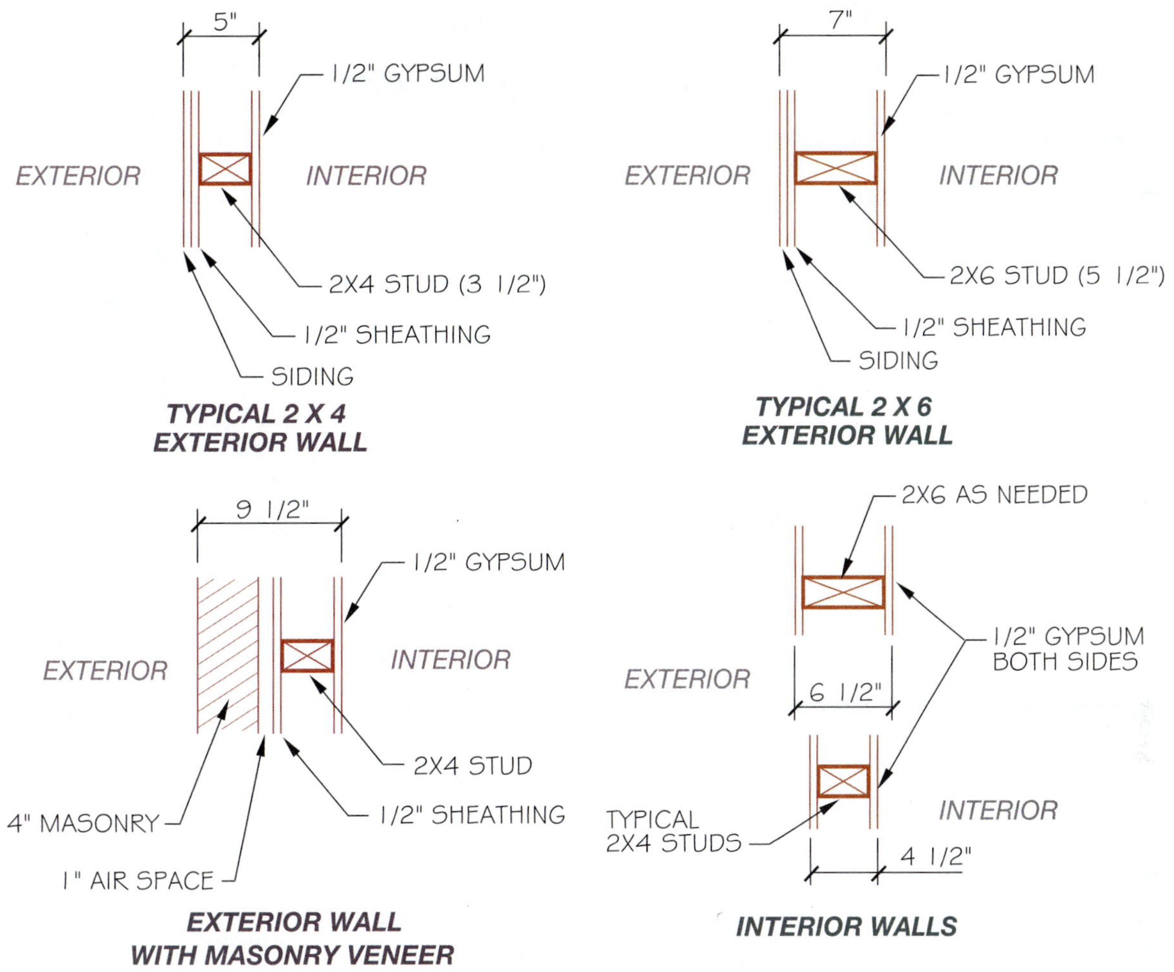

FIGURE 14.3 ■ Typical exterior and interior construction.

(100 mm) thick for manual drafting or the exact 4 1/2″ (114 mm) with CADD. The 4 1/2″ (115 mm) dimension is the 2 × 4 (50 × 100 mm) stud with 1/2″ (12 mm) drywall on each side, as shown in Figure 14.3. The dimension is established with this math: 3 1/2″ + 1/2″ + 1/2″ = 4 1/2″. *Drywall* refers to any materials used for interior stud covering that do not need to be mixed with water before application. The terms *wallboard, plasterboard,* and *gypsum* are also used. Gypsum is commonly used to make drywall. When gypsum is heated and the water removed, the resulting product is plaster of paris. This is covered by paper to make the drywall sheets. Walls with 2 × 6 (40 × 150 mm) studs are generally used behind a water closet (toilet). This wall, commonly known as an *8″ nominal wall,* is used to house a soil stack that is 4″ in diameter. The *soil stack* is a vertical soil pipe carrying the discharge from the toilet fixture. The soil stack is too big to be placed in the cavity of a 2 × 4 stud wall. This applies if the water closet is on an upper floor where the soil stack must run down the wall below. There is no need for the increase in stud size if the water closet is on the main floor, because the soil stack connects directly into the sewer line below the slab or crawl space. There is also a *soil vent* that runs up the wall and vents out the roof. The soil vent, also called the *stack vent* or *waste vent,* allows vapor to escape and ventilates the system. The soil vent can be in a standard 2 × 4 stud wall, because it normally measures 1 1/2″ in diameter. Occasionally masonry veneer is used on interior walls and is drawn in a manner similar to the exterior application shown in Figure 14.6. In many architectural offices wood-frame exterior and interior walls are drawn the same thickness to save time. The result is a wall representation that clearly communicates the intent.

Wall Shading and Material Indications

Several methods are used to shade walls so the walls stand out clearly from the rest of the drawing. Wall shading is also referred to as *poché,* a French word that means the art of outlining large letters or areas and filling the centers. Wall shading should be the last drafting task performed.

Wall shading should be done on the back of the drawing, although some drafters shade on the front, a practice that could cause the drawing to become smudged easily. When shading is done on the back of the sheet, blank paper should be placed on

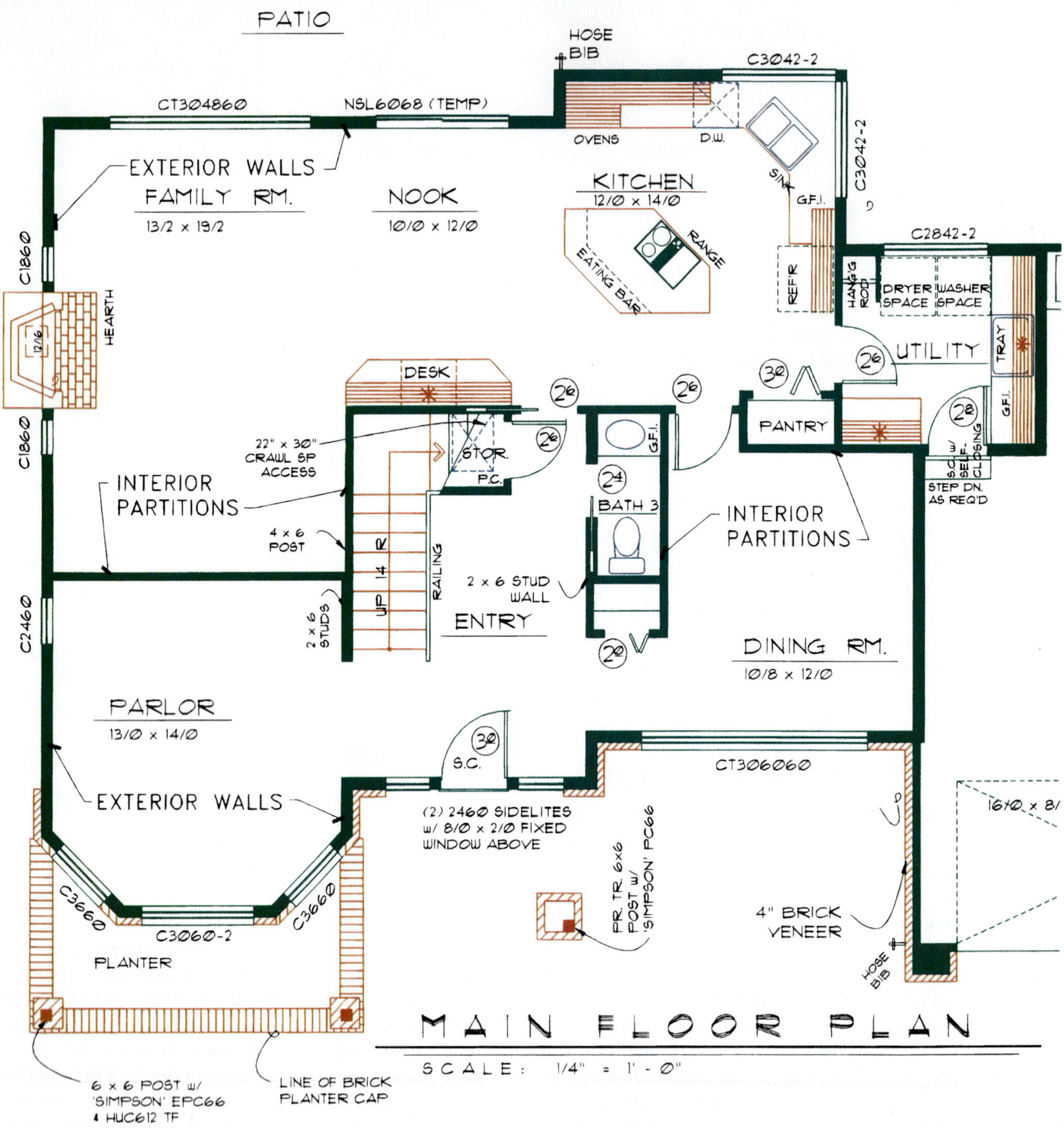

FIGURE 14.4 ■ Partial floor plan showing exterior walls and interior partitions. *Courtesy Piercy & Barclay Designers, Inc. CPBD.*

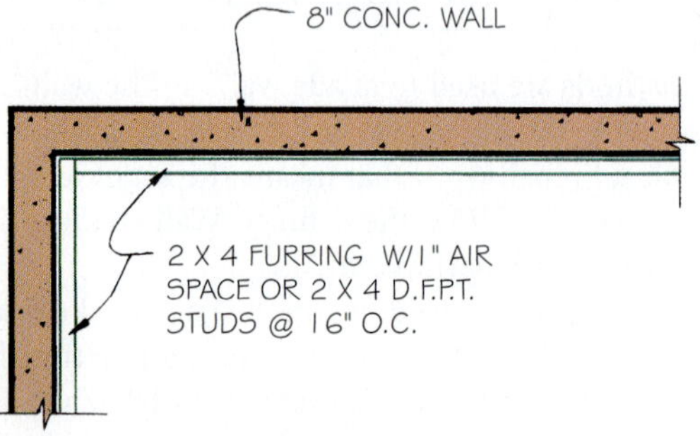

FIGURE 14.5 ■ Concrete exterior wall.

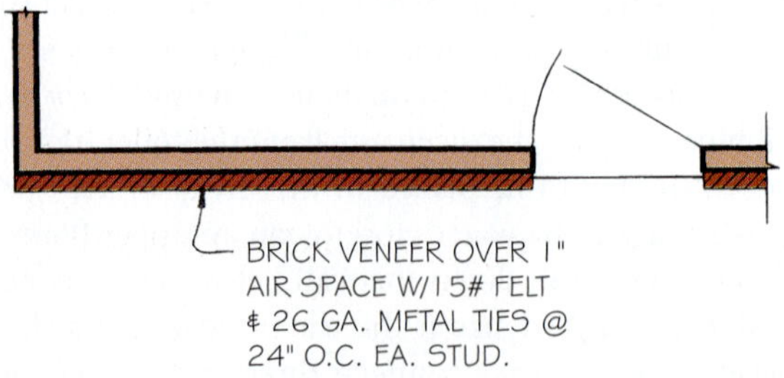

FIGURE 14.6 ■ Exterior masonry veneer.

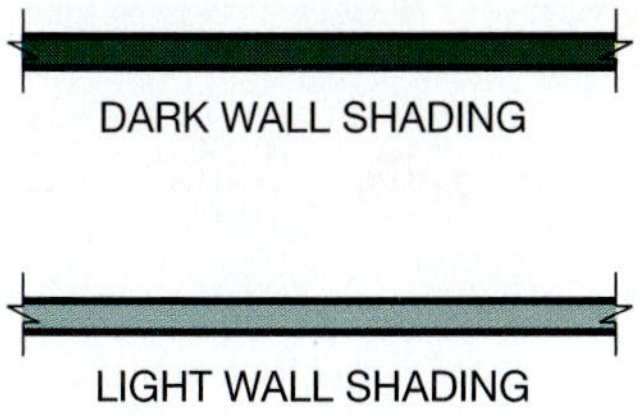

FIGURE 14.7 ■ Wall shading.

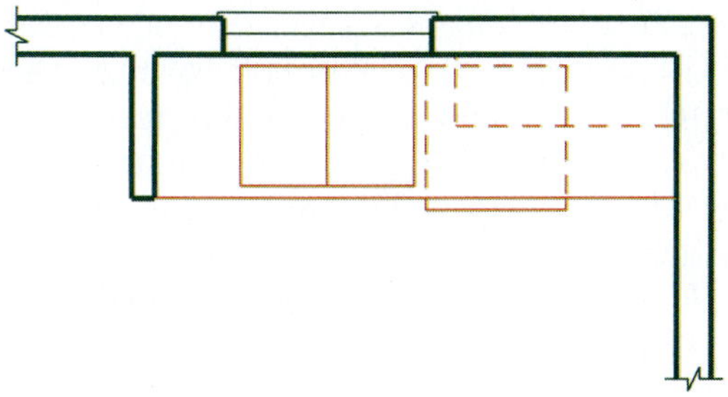

FIGURE 14.8 ■ Thick lines used on walls and partitions.

the drawing table to avoid transferring lines from the drawing image to the drafting table. Some drafters shade walls very dark for accent, while others shade the walls lightly for a more subtle effect, as shown in Figure 14.7. Wall shading should be done with a soft (F or HB) pencil lead. Other wall shading techniques include closely spaced thin lines, wood grain effect, and the use of colored pencils.

Office practice that requires light wall shading may also use thick wall lines to help accent the walls and partitions so they stand out from other floor-plan features. Thick wall lines may be achieved with an H or F lead in a 0.7-mm automatic pencil. Drafters using a mechanical pencil should work with a slightly rounded H or F lead. CADD drawings create different line weights by placing each drawing element on its own layer. The wall layer is then set up to print or plot as thick lines. CADD programs also offer the flexibility of leaving the walls open or filled as desired. With some CADD programs that are designed for architectural applications, you can select either option. With a standard CADD program, you can use commands that are designed to fill areas. Figure 14.8 shows how walls and partitions appear when outlined with thick lines. This method is not used when dark wall shading is used, since the darker shading results in walls and partitions that stand out clearly from other floor-plan features. Some architectural drafters prefer to draw walls and partitions unshaded with a thick outline.

Partial Walls

Partial walls are used as room dividers in situations where an open environment is desired—for example, guardrails at balconies or next to a flight of stairs. Partial walls require a minimum height above the floor of 36″ and are often capped with wood or may have decorative spindles that connect to the ceiling. Partial walls can be differentiated from other walls by wood

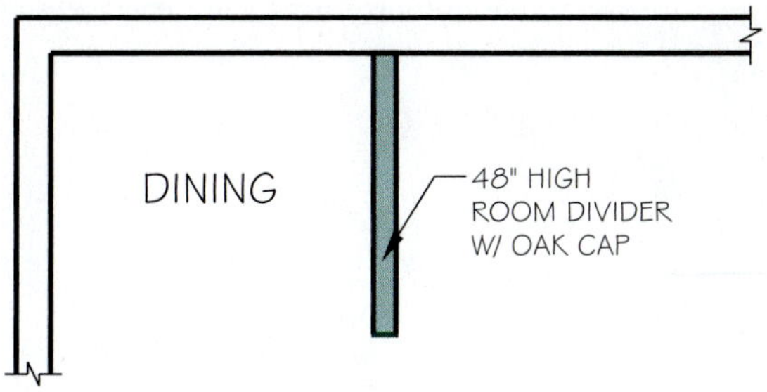

FIGURE 14.9 ■ Partial wall used as a partition wall or room divider.

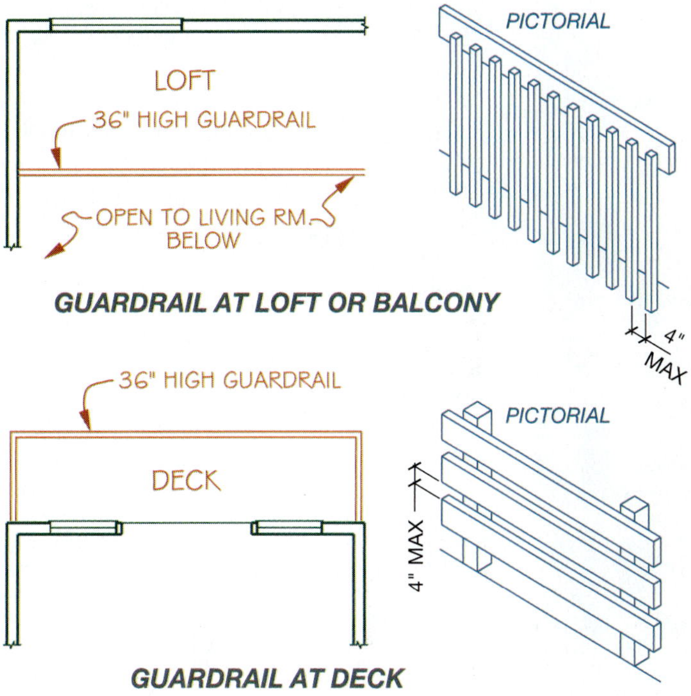

FIGURE 14.10 ■ Guardrail representations.

grain or very light shading and should be defined with a note that specifies the height, as shown in Figure 14.9.

Guardrails

Guardrails are used for safety at balconies, lofts, stairs, and decks over 30″ (762 mm) above the next lower level. Residential guardrails, according to IRC (International Residential Code), should be noted on floor plans as being at least 36″ (915 mm) above the floor and may include another note specifying that the intermediate rails should not have more than a 4″ (100 mm) open space. The minimum space helps ensure that small children will not fall through. Decorative guardrails are also used as room dividers—especially at a sunken area or to create an open effect between two rooms. Guardrails are commonly made of either wood or wrought iron. Creatively designed guardrails add a great deal to the aesthetics of the interior design. Figure 14.10 shows guardrail designs and how they are drawn on a floor plan.

CADD APPLICATIONS

DRAWING FLOOR PLAN WALLS USING CADD

There are a number of ways to draw walls using CADD. Custom architectural software packages are available that allow you to draw and edit walls in any desired layout. Most of these programs allow you to establish the following floor plan wall symbols:

- Desired wall thickness based on the materials used, such as the exterior wall shown at *A* in Figure 14.11 or the exterior wall with masonry veneer at *B*.
- Corners are automatically created as at *C* in Figure 14.11.
- Selecting points along the desired structure as at *D* in Figure 14.11 can easily create a bay wall design.
- Curved walls can be designed by picking three points along the curve or by selecting the desired end points and radius as shown at *E* in Figure 14.11.
- "T" intersections are created automatically where walls and interior partitions meet, as shown at *F* in Figure 14.11.
- Wall caps are drawn automatically when a short stub wall is desired, as at *G* in Figure 14.11.
- Wall line thickness is achieved by placing the walls on a separate layer and assigning a line width during the plotting or printing process, as shown at *H* in Figure 14.11.

You can also use the standard drafting features of CADD software programs such as AutoCAD. For example, the AutoCAD LINE or PLINE commands may be used to draw the outside wall line; then the OFFSET command is used to place the inside wall line a desired wall thickness away from the first line. PLINE has the advantage of providing width variation for thick wall outlines. Use the BREAK, TRIM, or EXTEND commands to edit walls as needed.

If you are using AutoCAD LT, the DLINE command is available for drawing double lines. The double lines may be used to draw walls by specifying the wall thickness at the Width option. This command may be used to draw a variety of wall layouts, including intersecting walls and capped walls. AutoCAD LT is a powerful program that is available at an economical price.

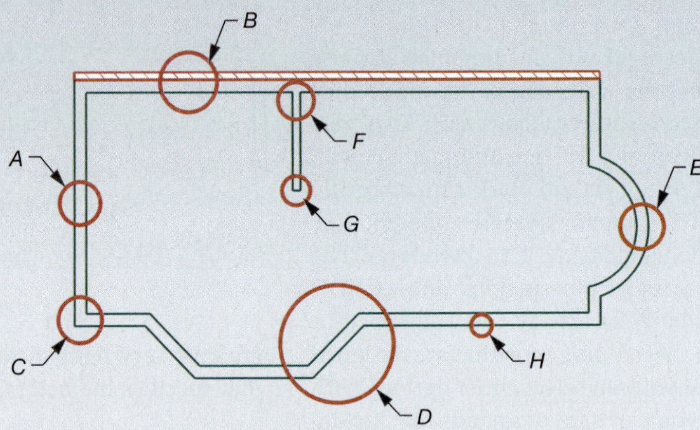

FIGURE 14.11 ■ (*A*) The architectural CADD program draws the walls to their exact thickness, based on the materials used. (*B*) When brick veneer is specified, it is drawn to the exact thickness and representation. (*C, D*) Corners are created automatically. (*E*) Curved walls are drawn by selecting points along the curve or by specifying the endpoints and radius. (*F*) "T" intersections are created automatically. (*G*) Wall caps are drawn automatically. (*H*) Line width is set up in the initial drawing template or assigned during the print or plot setup.

DOOR SYMBOLS

Exterior doors are drawn on the floor plan with the sill shown on the outside of the house. The sill is commonly drawn projected about 1/16″, although it may be drawn flush depending on individual company standards. See Figure 14.12. The main entry door is usually 3′–0″ (915 mm) wide. An exterior door from a garage or utility room is usually 2′–8″ (813 mm) wide, although, 3′–0″ is good if space permits. Exterior doors are typically solid wood or hollow metal with insulation. Doors may be either smooth, slab, or have decorative panels on the surface or in their construction. Doors are typically 6′–8″ (2032 mm) high, although 8′–0″ (2438 mm) doors are available. Refer to Chapter 7 for a review of code requirements for exterior doors.

The interior door symbol, as shown in Figure 14.13, is drawn without a sill. Interior doors should swing into the room being entered and against a wall. Interior doors are typically placed within 3″ (80 mm) of a corner, or at least 2′ from the corner. This allows for furniture placement behind the door if desired. Common interior door sizes are:

2′–8″: Utility rooms

2′–8″ to 2′–6″: Bedrooms, dens, family rooms, and dining rooms

2′–6″ to 2′–4″: Bathrooms

2′–4″ to 2′–0″: Closets (or larger for bifold and bipass doors)

Metric door sizes vary from 600 mm to 1000 mm depending on use and location.

The Americans with Disabilities Act (ADA) specifies an opening 36″ (915 mm) wide for wheelchair access. This is a clear, unobstructive dimension measured to the edge of the door at 90°, if it does not open 180°. The larger-size door is normally used in more expensive homes with wider halls; smaller-size doors are usually used in homes where space is at a premium. Interior doors are often flush but may have raised panels or glass panels.

Pocket doors are commonly used when space for a door swing is limited, as in a small room. Look at Figure 14.14. Pocket doors should not be placed where the pocket is in an exterior wall or there is interference with plumbing or electrical wiring. With regards to size, pocket doors should follow the same guidelines as interior swinging doors. Pocket doors are more expensive to purchase and install than standard interior doors because the pocket door frame must be built while the house is being framed. The additional cost is often worth the results, however.

A common economical wardrobe uses the bipass door, as shown in Figure 14.15. Bipass doors normally range in width from 4′–0″ (1200 mm) through 12′–0″ (3660 mm) in 1′ intervals. Pairs of doors are usually adequate for widths up to 8′. Three panels are usually provided for doors between 8′ and 10′ wide, and four panels are normal for doors wider than 10′. Bifold wardrobe or closet doors are used when complete access

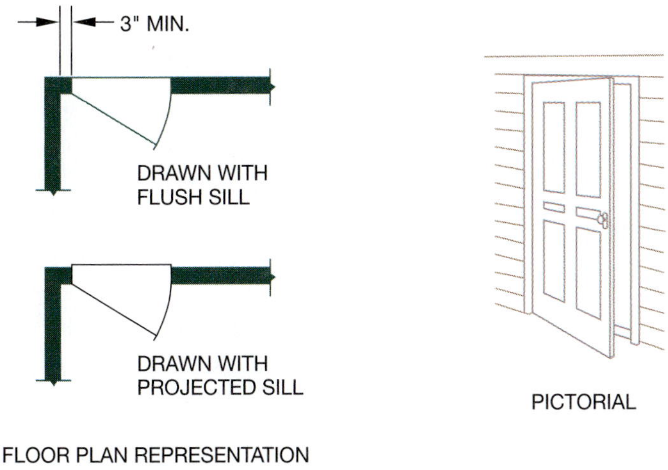

FIGURE 14.12 ■ Exterior door.

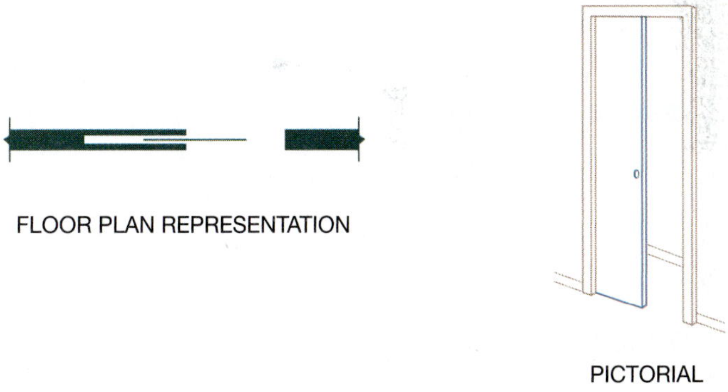

FIGURE 14.14 ■ Pocket door.

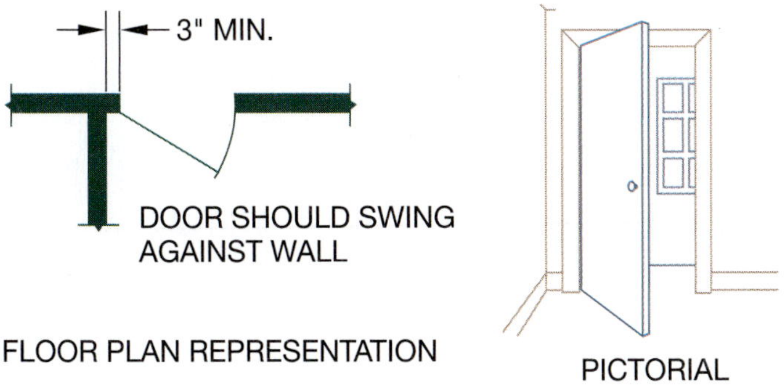

FIGURE 14.13 ■ Interior door.

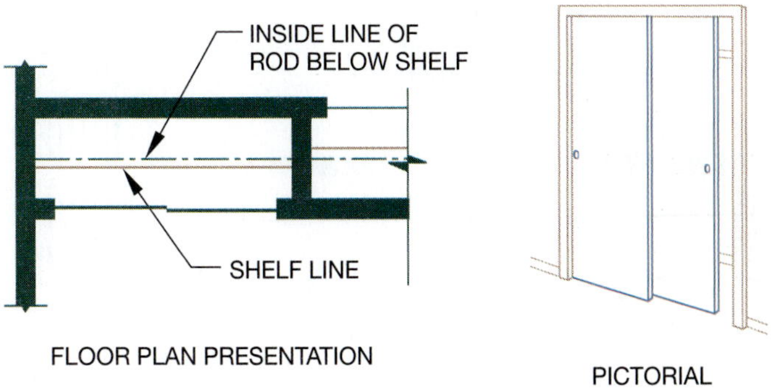

FIGURE 14.15 ■ Bipass door.

to the closet is required. Sometimes bifold doors are used on a utility closet that houses a washer and dryer or other utilities. See Figure 14.16. Bifold wardrobe doors often range in width from 4′–0″ through 9′–0″, in 6″ intervals.

Double-entry doors are common where a large formal foyer design requires a more elaborate entry than can be achieved by one door. The floor-plan symbol for double-entry and French doors is the same; therefore, the door schedule should clearly identify the type of doors to be installed. The term *French doors* normally refers to double exterior or interior doors that have glass panels and swing into a room. See Figure 14.17.

Glass sliding doors are made with wood or metal frames and tempered glass for safety. Figure 14.18 shows the floor-plan symbols for representing a flush and a projected exterior sill. These doors are used to provide glass areas and are excellent for access to a patio or deck. Glass sliding doors typically range in width from 5′–0″ through 12′–0″ at 1′ intervals. Common sizes are 6′–0″ and 8′–0″.

French doors are used in place of glass sliding doors when a more traditional door design is required. Glass sliding doors are associated with contemporary design and do not take up as much floor space as French doors. French doors may be purchased with wood mullions and muntins (the upright and bar partitions respectively) between the glass panes or with one large glass pane and a removable decorative grill for easy cleaning. See Figure 14.17. French doors range in width from 2′–4″ through 3′–6″ in 2″ increments. Doors may be used individually, in pairs, or in groups of threes or fours. The three- and four-panel doors typically have one or more fixed panels, which should be specified.

Double-acting doors are often used between a kitchen and eating area, to swing in either direction for easy passage. See Figure 14.19. Common sizes for pairs of doors range from 2′–6″ through 4′–0″ wide in 2″ increments.

Dutch doors are used when it is desirable to have a door that may be half open and half closed. The top portion may be opened and used as a pass-through. Look at Figure 14.20. Dutch doors range in width from 2′–6″ through 3′–6″ in 2″ increments.

Accordion doors may be used at closets or wardrobes, or they are often used as room dividers where an openable partition is

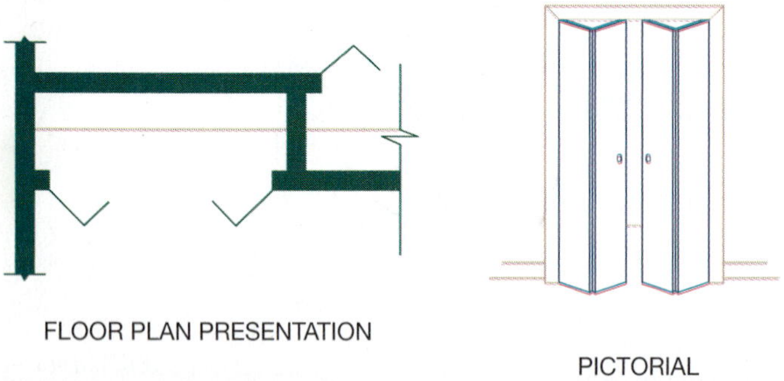

FIGURE 14.16 ■ Bifold door.

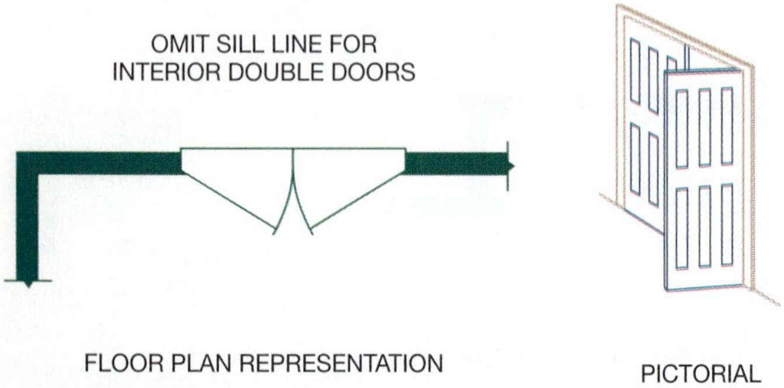

FIGURE 14.17 ■ Double-entry or French doors.

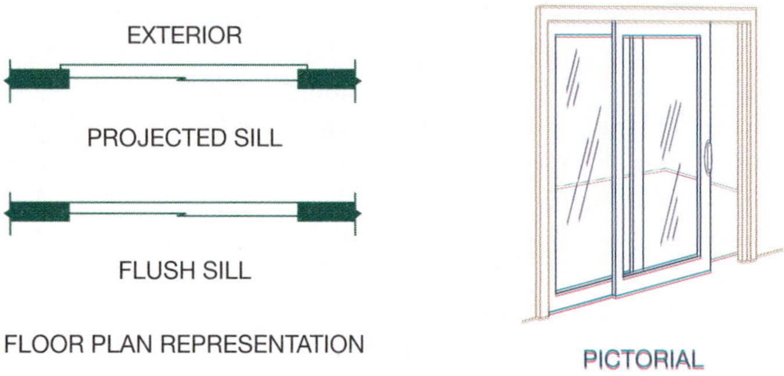

FIGURE 14.18 ■ Glass sliding door.

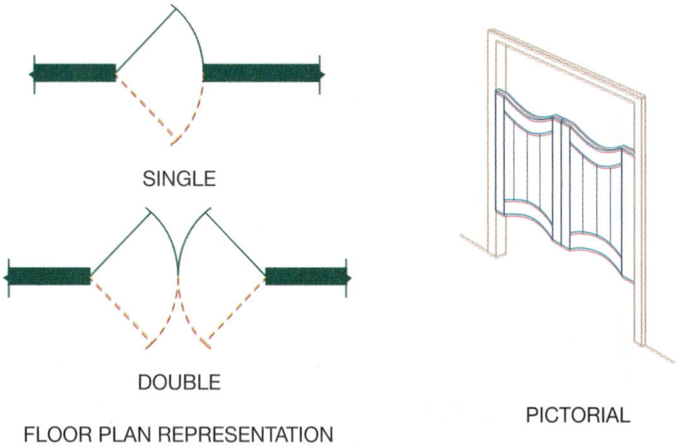

FIGURE 14.19 ■ Double-acting doors.

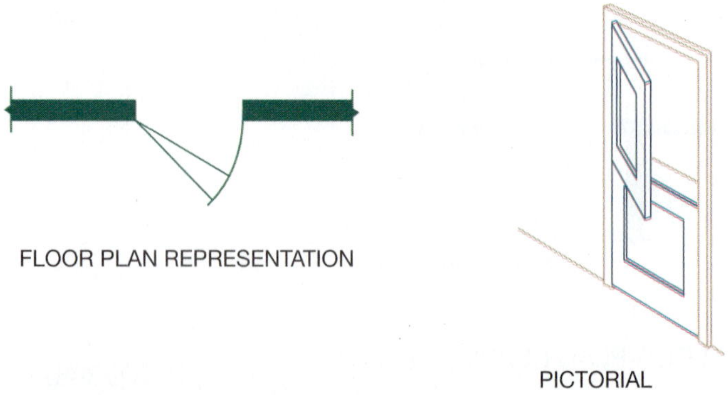

FIGURE 14.20 ■ Dutch doors.

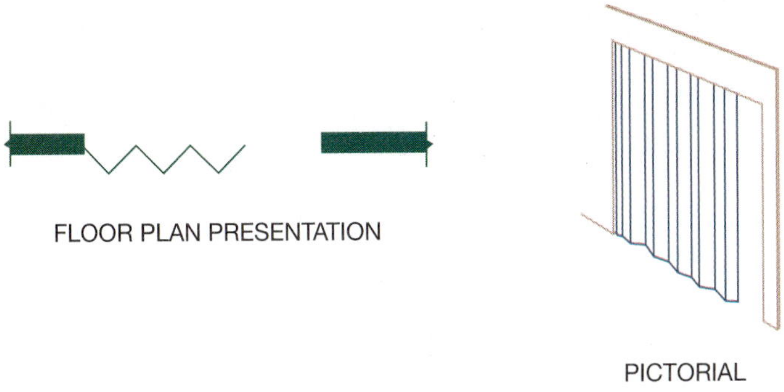

FIGURE 14.21 ■ Accordion door.

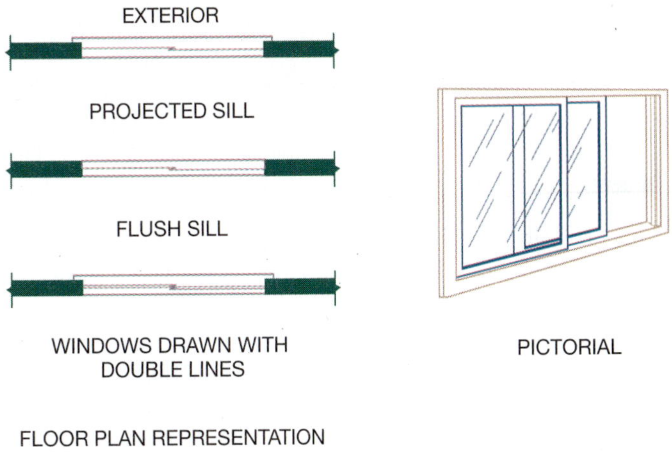

FIGURE 14.23 ■ Horizontal sliding window.

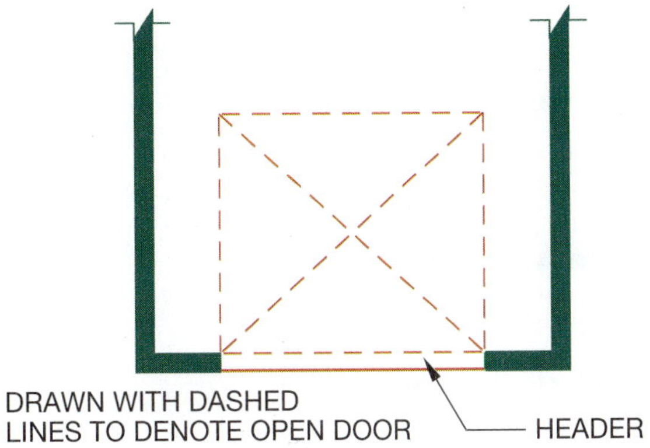

FIGURE 14.22 ■ Overhead garage door.

needed. Figure 14.21 shows the floor-plan symbol. Accordion doors range in width from 4′–0″ through 12′–0″ in 1′ increments. Commercial doors are available for bigger openings. Standard metric door sizes for all large openings such as sliding glass or double French doors vary from 1500 mm to 3600 mm.

The floor-plan symbol for an overhead garage door is shown in Figure 14.22. The dashed lines show the size and extent of the garage door when open. The extent of the garage door is typically shown when the door interferes with something on the ceiling. The garage door header is normally shown as a hidden line, and the size is labeled with a note. A *header* is a horizontal structural member that supports the ends of joists or rafters and the load over an opening.

Garage doors range in width from 8′–0″ through 18′–0″. An 8′–0″ door is a common width for a single-car. A size of 9′–0″ or 10′–0″ is common for a single door and accommodates a pickup truck or large van. A door 16′–0″ is a common double-car width. Doors are typically 7′–0″ high, although doors 8′–0″ or 10′–0″ high are common for campers or recreational vehicles. Standard metric sizes vary from 2400 mm to 5400 mm.

Window Symbols

Window symbols are drawn with a sill on the outside and inside. A sliding window, shown in Figure 14.23, is a popular 50 percent openable window. Notice that windows may be drawn with exterior sills projected or flush, and the glass pane may be drawn with single or double lines. The method used should be consistent throughout the plan and should be determined by the preference of the specific architectural office. Many offices draw all windows with a projected sill and one line to represent the glass; they then specify the type of window in the window schedule.

Windows typically come in widths that range from 2′–0″ through 12′–0″ at intervals of about 6″. Vinyl frame windows generally fall within the range of these nominal sizes, but sizes of wood frame windows are often different and should be confirmed with the manufacturer's specifications. Metric window sizes also vary and should be confirmed.

The location of a window in the house and the way the window opens have an effect on the size. A window between 6′–0″ and 12′–0″ wide is common for a living, dining, or family room. To let the occupants take advantage of a view while sitting, windows in these rooms are normally between 4′–0″ and 5′–0″ tall.

Windows in bedrooms are typically 3′–0″ to 6′–0″ wide. The depth often ranges from 3′–6″ to 4′–0″. The type of window used in the bedroom is important because of emergency egress requirements in most codes. See Chapter 7 for a review.

Kitchen windows are often between 3′–0″ and 5′–0″ wide and between 3′–0″ and 3′–6″ tall. Wide windows are nice to have in a kitchen for the added light they provide, but some of the upper cabinets may be eliminated. If the tops of windows are at the normal 6′–8″ height, windows deeper than 3′–6″ interfere with the countertops.

Bathroom windows often range between 2′–0″ and 3′–0″ wide, with an equal depth. A longer window with less depth is often specified if the window is to be located in a shower area. Most bathroom windows have obscure glass.

Types of Windows

Casement windows may be 100 percent openable and are best used where extreme weather conditions require a tight seal when the window is closed, although these windows are in

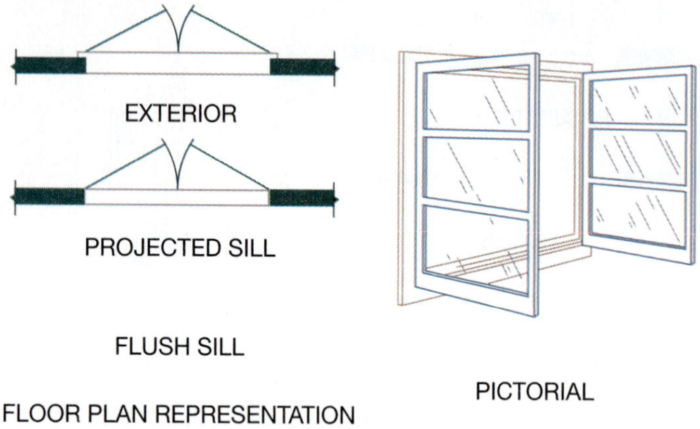

FIGURE 14.24 ■ Casement window.

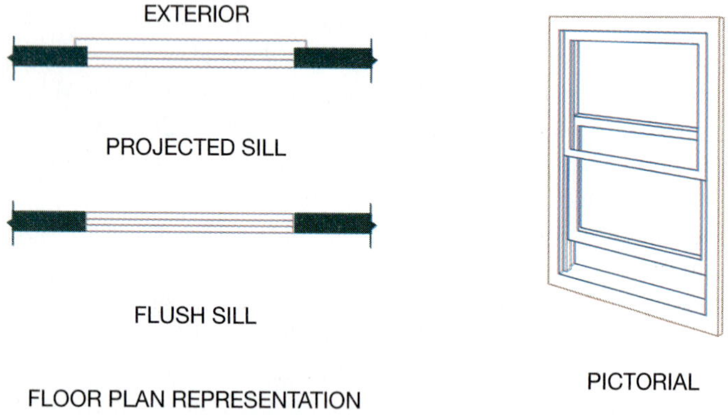

FIGURE 14.25 ■ Double-hung window.

common use everywhere. See Figure 14.24. Casement windows may be more expensive than sliding units.

The traditional double-hung window has a bottom panel that slides upward, as shown in Figure 14.25. Double-hung wood-frame windows are designed for energy efficiency and are commonly used in traditional as opposed to contemporary architectural designs. Double-hung windows are typically taller than they are wide. Single windows usually range in width from 1′–6″ through 4′–0″. It is very common to have double-hung windows grouped together in pairs of two or more.

Awning windows are often used in basements or below a fixed window to provide ventilation. Another common use places awning windows between two different roof levels; this provides additional ventilation in vaulted rooms. These windows would then be opened with a long pole connected to the opening device. These windows are hinged at the top and swing outward, as shown in Figure 14.26. Hopper windows are drawn in the same manner; however, they hinge at the bottom and swing inward. Jalousie windows are used when a louvered effect is desired, as seen in Figure 14.27.

Fixed windows are popular when a large, unobstructed area of glass is required to take advantage of a view or to allow for solar heat gain. Figure 14.28 shows the floor-plan symbol of a fixed window.

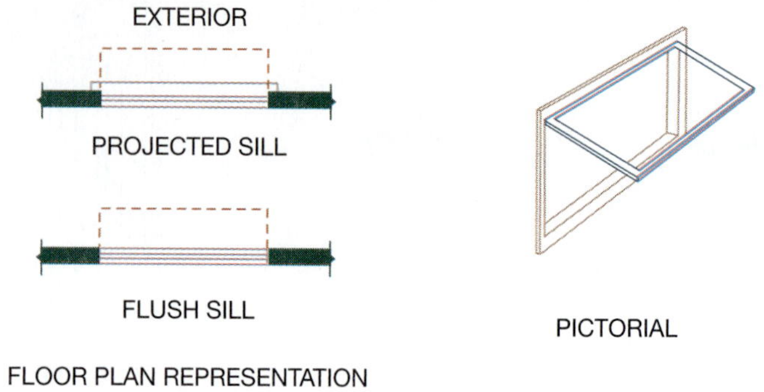

FIGURE 14.26 ■ Awning window.

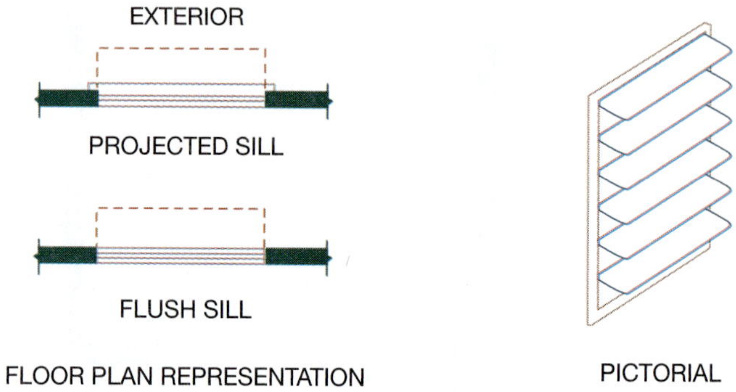

FIGURE 14.27 ■ Jalousie window.

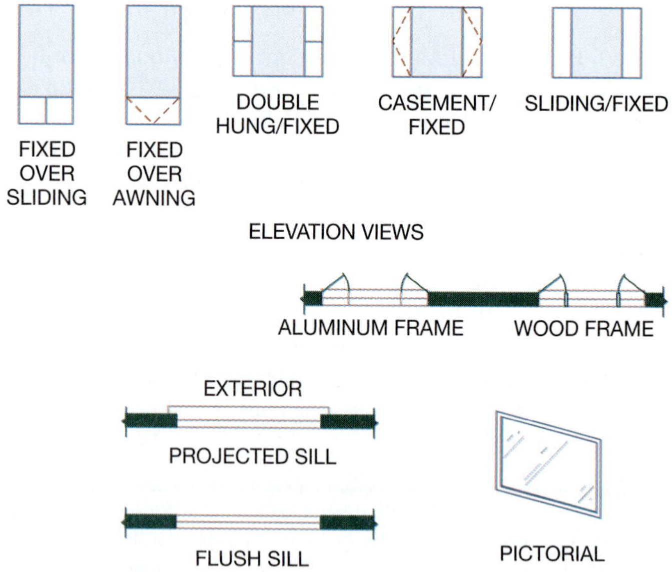

FIGURE 14.28 ■ Options for fixed windows. Fixed part of combination windows is shaded for reference.

Bay windows are often used when a traditional style is desired. Figure 14.29 shows the representation of a bay. Usually the sides are drawn at a 45° or 30°. The depth of the bay is often between 18″ and 24″. The total width of a bay is limited by the size of the center window, which is typically either a fixed panel, double-hung, or casement window. Large bays can have more than one window at the center and sides. Bays can be either premanufactured or built at the job site. Bay windows are

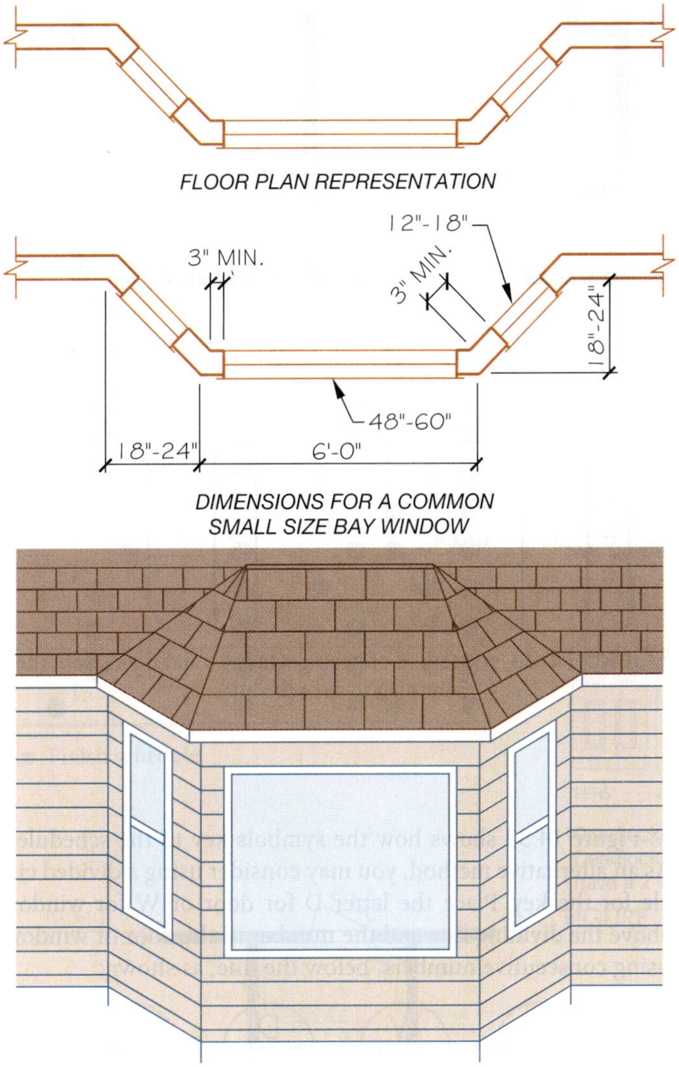

FIGURE 14.29 ■ Bay window.

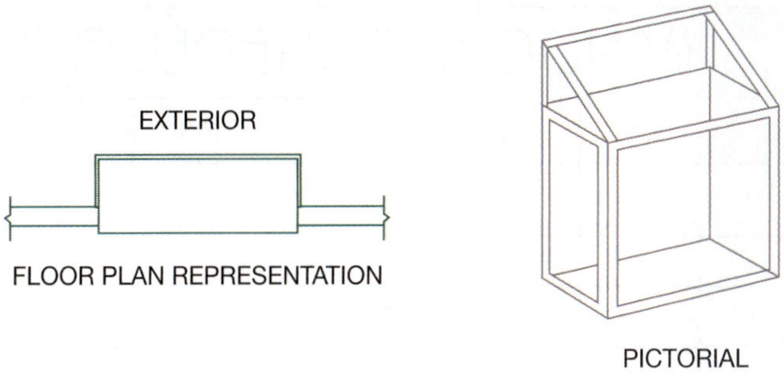

FIGURE 14.30 ■ Garden window.

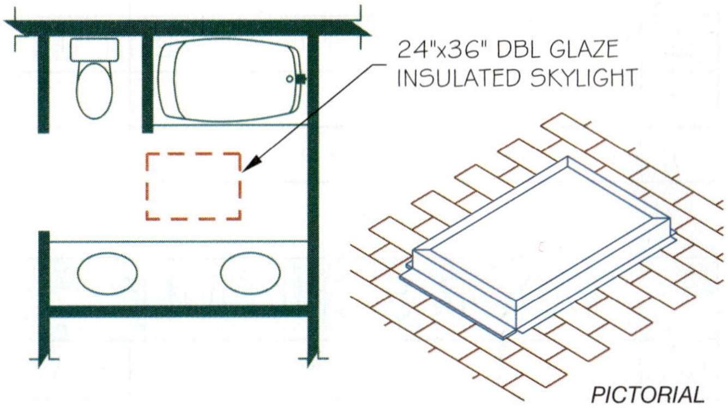

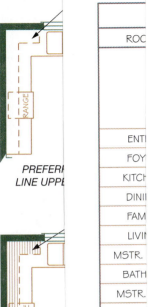

FIGURE 14.31 ■ Skylight representation.

designed for a specific size and location. Bay windows commonly extend from floor to ceiling. The roof of the house can be extended down to cover the bay if adequate ceiling height is available. The bay commonly has its own roof structure, which often adds to the quality of the home's exterior design. The bay window structure can also be framed out with a bench seat if it does not extend to the floor. The actual design and size depend on the purpose and the available space.

A garden window, as shown in Figure 14.30, is a popular style for utility rooms or kitchens. Garden windows usually project between 12″ and 18″ from the residence. Depending on the manufacturer, either the side or the top panels open.

Skylights

When additional daylight is desirable in a room, or to let natural light enter an interior room, consider using a skylight. Skylights are available fixed or openable. They are made of plastic in a dome shape or flat tempered glass. Tempered double-pane insulated skylights are energy-efficient, do not cause any distortion of view, and generally are not more expensive than plastic skylights. Figure 14.31 shows how a skylight is represented in the floor plan.

SCHEDULES

Numbered symbols used on the floor plan key specific items to charts known as *schedules*. Schedules are used to describe items such as doors, windows, appliances, materials, fixtures, hardware, and finishes. See Figure 14.32. Schedules help keep drawings clear of unnecessary notes, since the details of an item are off the drawing or on another sheet.

Content of Schedules

There are many different ways to set up a schedule, but it may include any or all of the following information about the product:

- Manufacturer's name
- Product name
- Model number
- Type
- Quantity
- Size
- Rough opening size
- Color

Placement of quantities on schedules varies among offices. Some companies do not include quantities on the schedules,

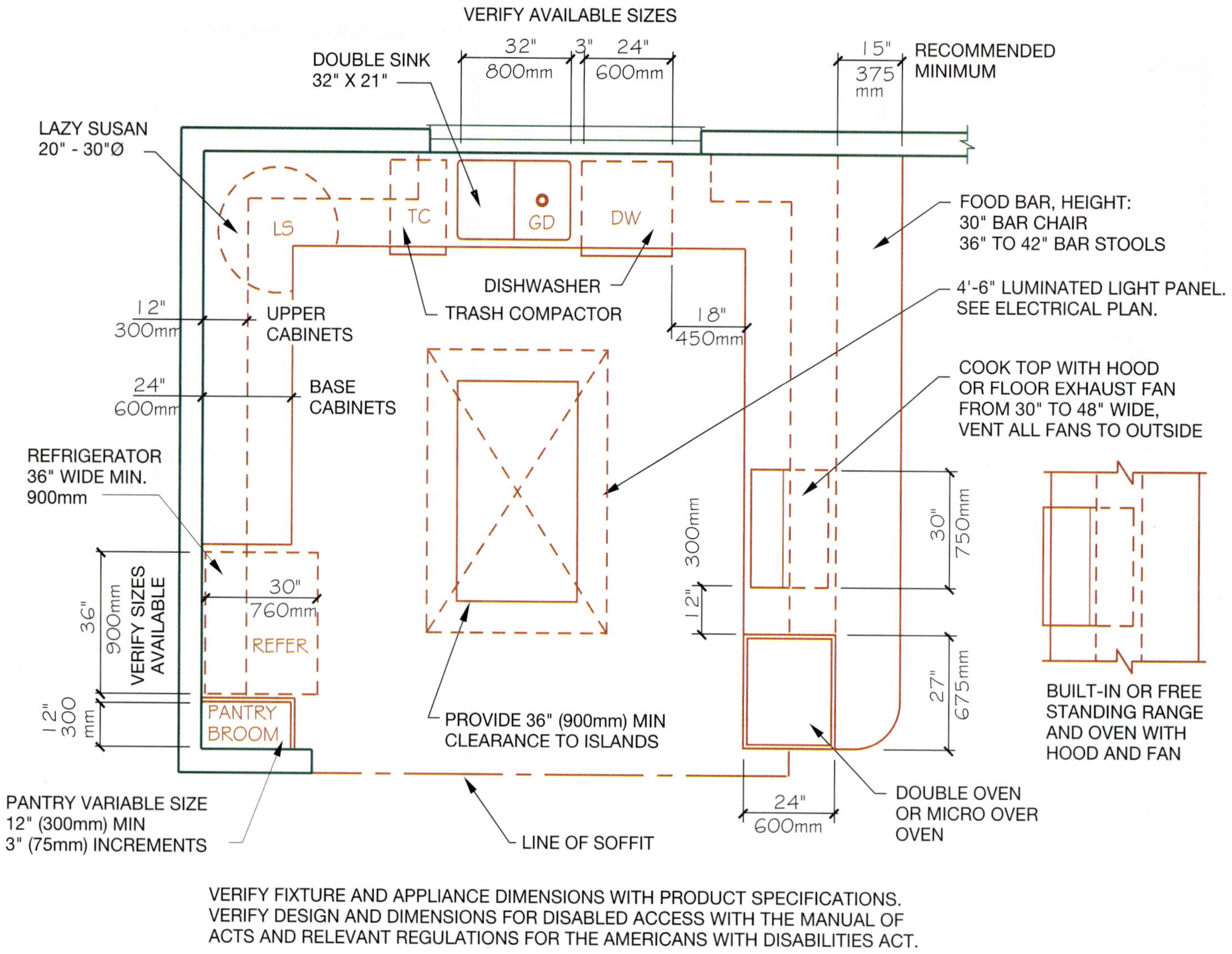

FIGURE 14.42 ■ Standard sizes of kitchen cabinets, fixtures, and appliances.

Verify the availability of products and manufacturers' specifications as you design and draw the floor plan. Figure 14.45 shows the floor-plan representation of a 60″ × 72″ raised bathtub in a solarium. Often, owing to cost considerations, the space available for bathrooms is minimal. Figure 14.46 shows some minimum sizes to consider. The *Manual of Acts and Relevant Regulations,* published for the ADA, specifies design and dimensions for people with disabilities. For example, wheelchair access requires up to 60″ (1524 mm) minimum width, depending on the design used.

Utility Rooms

The utility room can be a very important room in the home, and its placement should be based on the family's needs. The design location of the utility room was discussed in Chapter 7.

The symbols for the clothes washer, dryer, and laundry tray are shown in Figure 14.47. The symbols for clothes washers and dryers may be drawn with dashed lines if these items are not part of the construction contract. Laundry utilities may be placed in a closet when only minimum space is available, as shown in Figure 14.48. Ironing boards may be built into the laundry room wall or attached to the wall surface, as shown in Figure 14.49. Provide a note to the electrician if power is required to the unit, or provide an adjacent outlet.

The furnace and water heater are sometimes placed together in a location central to the house. They may be placed in a closet, as shown in Figure 14.50. These utilities may also be placed in a separate room, the basement, or the garage.

Wardrobes and closets, used for clothes and storage, are utilitarian in nature. Bedroom closets should be labeled CLOSET or WARDROBE. Other closets may be labeled

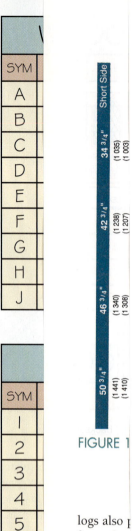

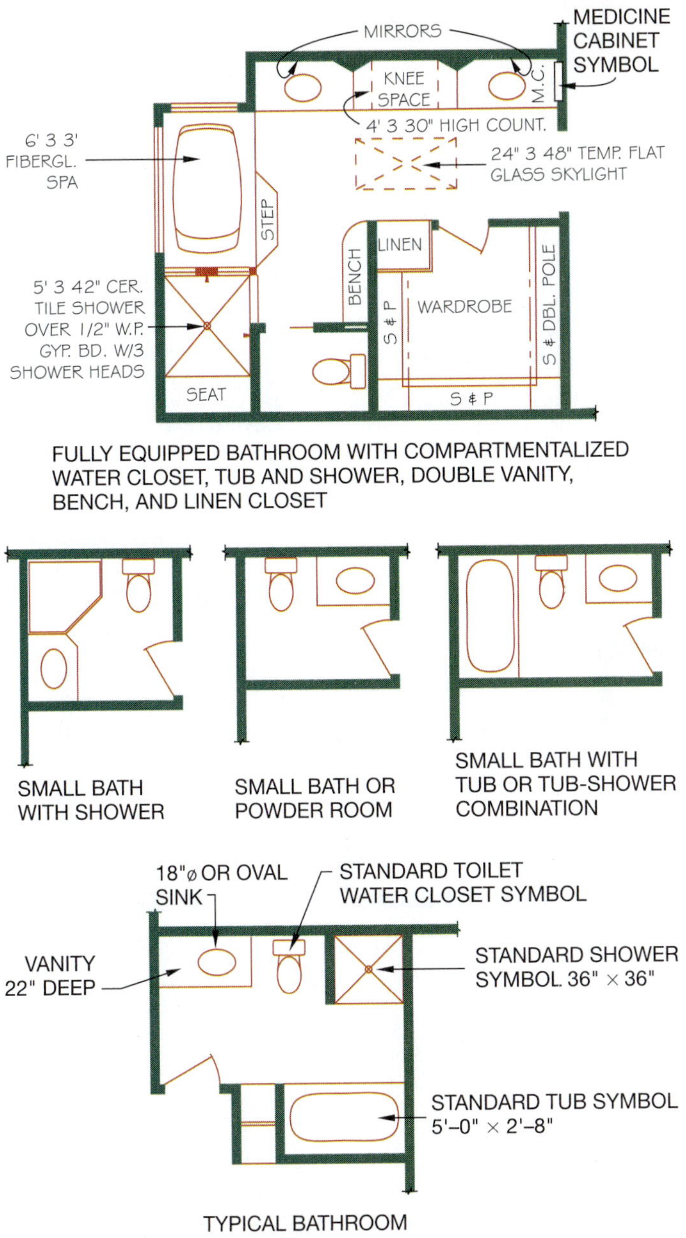

FIGURE 14.43 ■ Floor-plan symbols for layout of bathroom cabinets and fixtures.

FIGURE 14.44 ■ Unique bathroom products. *Courtesy Kohler Co.*

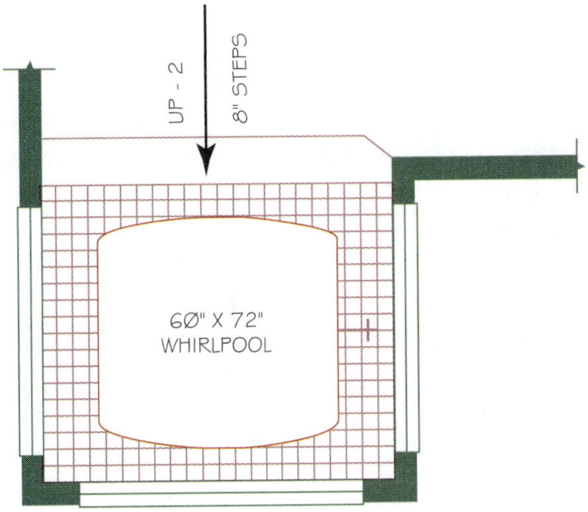

FIGURE 14.45 ■ Raised bathtub and solarium.

ENTRY CLOSET, LINEN CLOSET, BROOM CLOSET, and so forth. Wardrobe or guest closets should be provided with a shelf and pole. The shelf and pole may be drawn with a thin line to represent the shelf and a centerline to show the pole, as in Figure 14.51. Some drafters used two dashed lines to symbolize the shelf and pole. The rod is placed directly below the shelf with enough space to pass the hanger through easily. The height of the shelf can vary, depending on the function of the closet. For example, a shelf height of 5′–0″ (1524 mm) is fine for shirts, pants, and dresses. A shelf height of 6′–0″ (1825 mm) is preferred for long gowns. Wardrobe closets that are designed for children have shelf and rods at lower heights that vary depending on the children's ages. Some wardrobe closets are designed with dual shelf and rod combinations or shelf and rod over a set of drawers. Closet packages are available from companies that customize the wardrobe closet to meet the specific needs of family members.

When a laundry room is below the bedroom area, provide a chute from a convenient area near the bedrooms through the ceiling and into a cabinet directly in the utility room. The

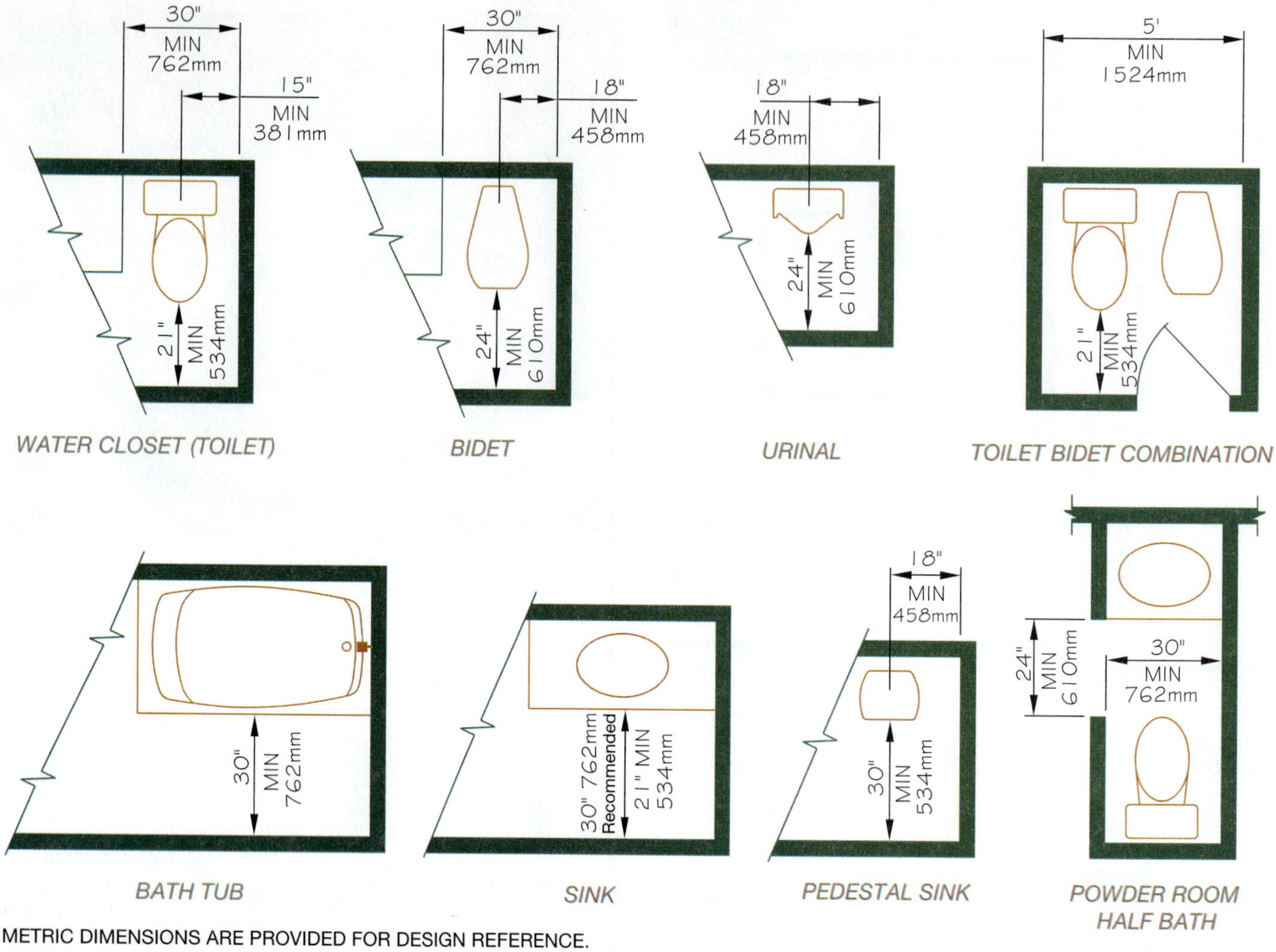

METRIC DIMENSIONS ARE PROVIDED FOR DESIGN REFERENCE.

VERIFY FIXTURE SIZES WITH PRODUCT SPECIFICATIONS.

VERIFY DESIGN AND DIMENSIONS FOR DISABLED ACCESS WITH THE MANUAL OF ACTS AND RELEVANT REGULATIONS FOR THE AMERICANS WITH DISABILITIES ACT. AND LOCAL CODE REQUIREMENTS.

FIGURE 14.46 ■ Minimum bath spaces. All dimensions are IRC minimums.

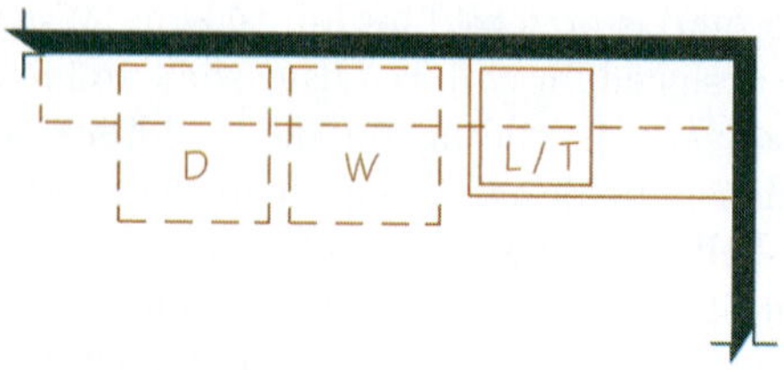

FIGURE 14.47 ■ Washer, dryer, and laundry tray.

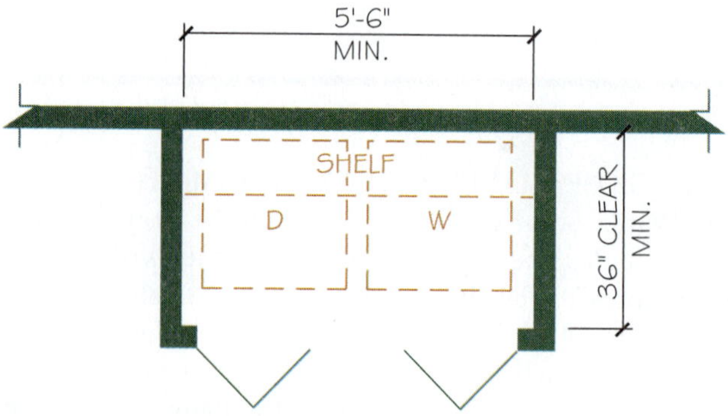

FIGURE 14.48 ■ Minimum washer and dryer space.

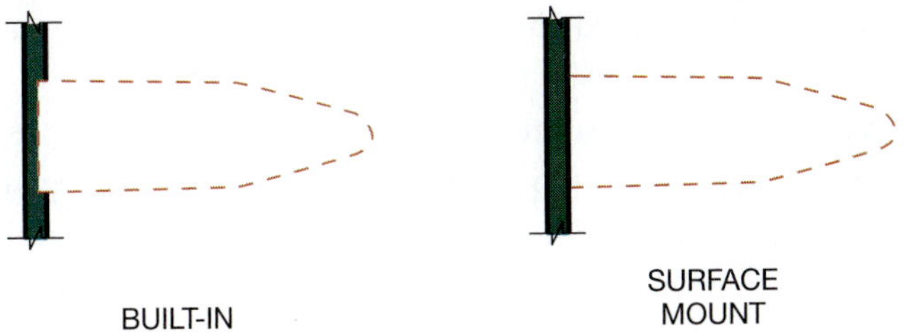

FIGURE 14.49 ■ Ironing boards.

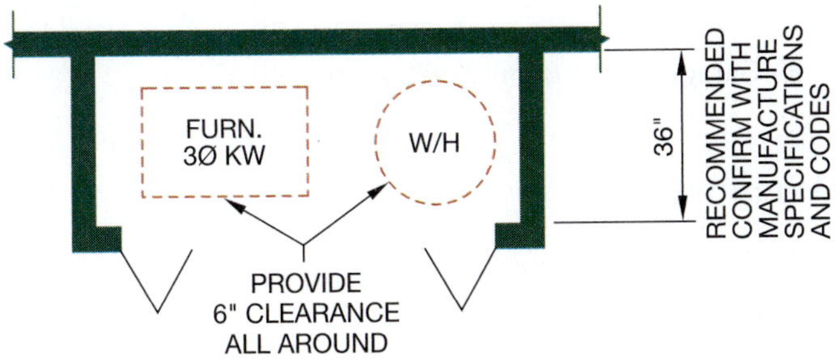

FIGURE 14.50 ■ Furnace and water heater.

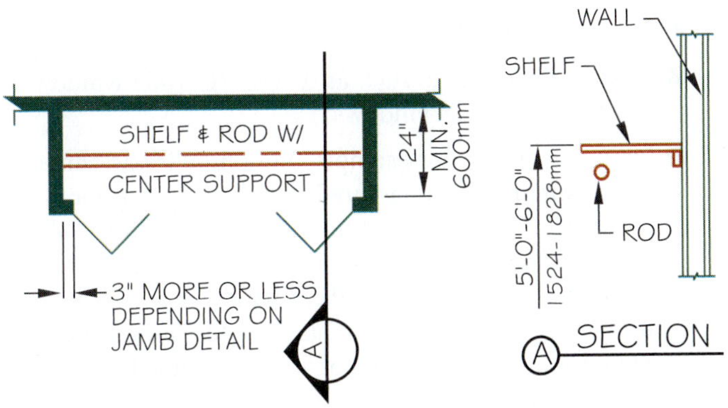

FIGURE 14.51 ■ Standard wardrobe closet and a walk-in wardrobe.

cabinet should be above or next to the clothes washer. Figure 14.52 shows a laundry chute noted in the floor plan.

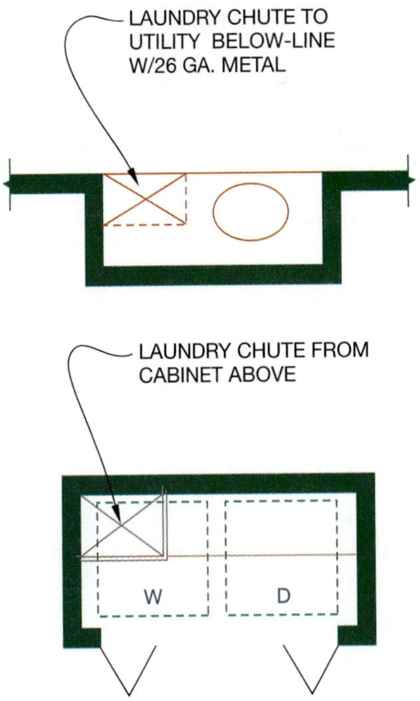

FIGURE 14.52 ■ Laundry chute.

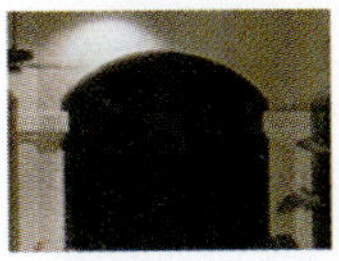

COMMON SIZES OF ARCHITECTURAL FEATURES

All walls and edges of masonry are thick lines. All other lines are thin. The refrigerator, water heater, washer and dryer, dishwasher, trash compactor, built-in range/oven, and upper cabinets are hidden lines. Closet poles are 15″ (380 mm) from the back wall and are drawn using center lines. Closet shelves are solid lines 12″ (300 mm) from the back wall.

The following features and product dimensions are based on common sizes and recommended minimums. Dimensions may vary depending on space available and manufacturers' specifications. Establish the final design on the basis of the client's needs and budget. Always determine the exact dimensions during the design process by using manufacturers' specifications, and then confirm during construction on the basis of the actual product. Metric values are given where appropriate. Metric values are not provided for products that vary depending on the manufacturer.

Room Components

- Exterior heated walls: 6″ manual drafting or exact dimensions with CADD. See Figure 14.3.
- Exterior unheated walls: (garage/shop) 4″ manual drafting or exact dimensions with CADD. See Figure 14.3.
- Interior: 4″ manual drafting or exact dimensions with CADD. See Figure 14.3.
- Interior plumbing below upper floor water closet: 6″ manual drafting or exact dimensions with CADD. See Figure 14.3.
- Hallways: 36″ (900 mm) minimum clearance.
- Entry hallways: 42″–60″ (1060–1500 mm) minimum.
- Bedroom closets: 24″ (600 mm) minimum depth; 48″ (1200 mm) minimum length.
- Linen closets: 14″–24″ (350–600 mm) deep (not over 30″, 760 mm).
- Base cabinets: 24″ (600 mm) deep; 15–18″ (380–450 mm) deep bar; 24″ (600 mm) deep island with 12″ (300 mm) minimum at each side of cooktop; 36″ (900 mm) minimum from island to cabinet (for passage, 48″ (1200 mm) is better on sides with an appliance).

- Upper cabinets: 12″ (300 mm) deep.
- Washer/dryer space: 36″ (900 mm) deep, 5′–6″ (1675 mm) long minimum.
- Stairways: 36″ (900 mm) minimum wide; 10.5″–12″ (260–300 mm) tread.
- Fireplace: See pages 264–268.

Plumbing

- Kitchen sink: double, 32″ × 21″; triple, 42″ × 21″.
- Vegetable sink and/or bar sink: 16″ × 16″ or 16″ × 21″.
- Laundry sink: 21″ × 21″.
- Bathroom sink: 19″ × 16″, oval; provide 9″ from edge to wall and about 12″ (300 mm) between two sinks; 36″ (900 mm) minimum length.
- Water closet space: 30″ (760 mm) wide (minimum); 24″ (600 mm) clearance in front.
- Tub: 32″ × 60″ or 32″ × 72″.
- Shower: 36″ square; 42″ square; or combinations of 36″, 42″, 48″, and 60″ for fiberglass; any size for ceramic tile.
- Washer/dryer: 2′–4″ square (approximately).

Appliances

- Forced air unit: gas, 18″ square (minimum) with 6″ space all around; cannot go under stair; electric, 24″ × 30″ (same space requirements as gas).
- Water heater: gas, 18–22″ diameter; cannot go under stair.
- Refrigerator: 36″ wide space; approximately 27″ deep; 4″ from wall; 1″ from base cabinet.
- Stove/cooktop: 30″ × 21″ deep.
- Built-in oven: 27″ × 24″ deep.
- Dishwasher: 24″ × 24″; place near sink for precleaning and near upper cabinet to ease putting away dishes.
- Trash compactor: 15″ × 24″ deep; near sink, away from stove.
- Broom/pantry: 12″ (300 mm) minimum × 24″ (600 mm) deep; width increasing by 3″ increments.
- Desk: 30″ × 24″ (760 × 600 mm) deep (minimum); out of main work triangle.
- Built-in vacuum: 24″–30″ diameter.

Doors

- Entry: 36″ × 6′–8″; 42″ × 8′.
- Sliders or French: 5′, 6′, 8′ (double); 9′, 10′ (triple); 12′ (four-panel).
- Garage, utility, kitchen, and bedrooms on custom houses: 2′–8″, 2′–10″, or 3′–0″.
- Bedrooms and bathrooms of nice houses: 2′–6″ and 2′–8″.
- Closets and wardrobes: 2′–4″, 2′–6″, 2′–8″, 2′–10″, 3′–0″, 4′–0″, 5′–0″, 6′–0″, 7′–0″, 8′–0″, 9′–0″, 10′–0″, 11′–0″ or 12′–0″ depending on the size of the closet or wardrobe and the type of door used.
- Bathrooms: 2′–4″, and 2′–6″.
- Garage car door: 8′ × 7′, 9′ × 8′, 16′ × 7′, and 18′ × 7′.
- Garage people door: 2′–8″, 2′–10″, 3′–0″.

Windows

The following factors should be considered, because they may influence window sizes:

- Architectural style. For example, contemporary style might use large fixed windows, but traditional style might use smaller double-hung windows.
- Amount of light desired.
- Desired appearance, such as one 8′–10′ wide window or four to six 2′ wide windows.
- Type of window. For example, aluminum frame windows are commonly manufactured in nominal even measurements such as 24″, 36″, and 48″ in 12″ modules. Wood frame window dimensions vary, depending on the manufacturer. Also, sliding and fixed windows are often sized differently from casement and double-hung windows. Dimensions of these types of windows must be based on the manufacturer's specifications.
- Window location and function. For example, bedroom windows require emergency egress as explained in Chapter 7. When the kitchen window is placed in a cabinet wall, its height must be selected to maintain the window at or above the base cabinet height.
- Living, family: 8′–10′.
- Dining, den: 6′–8′.
- Bedrooms: 4′–6′.
- Kitchens: 3′–5′.
- Bathrooms: 2′–3′.
- Sliding: 4′, 5′, 6′, 8′, 10′, 12′.
- Single hung: 24″, 30″, 36″, 42″.
- Casement: same as sliding.
- Fixed/awning: 24″, 30″, 36″, 42″, 48″.
- Fixed/sliding: 24″, 30″, 36″, 42″, 48″.
- Picture: 4′, 5′, 6′, 8′.
- Bay: 8′–10′ total; sides, 18′–24″ wide.

FLOOR-PLAN MATERIALS

Finish Materials

Finish materials used in construction may be identified on the floor plan with notes, characteristic symbols, or key symbols that relate to a finish schedule. The key symbol is also placed on the finish schedule next to the identification of the type of finish needed at the given location on the floor plan. Finish schedules may also be set up on a room-by-room basis. When flooring finish is identified, the easiest method is to label the material directly under room designation as shown in Figure 14.53. Other methods include using representative material symbols and a note to describe the finish materials as shown in Figure 14.54.

Drafters with an artistic flair may draw floor-plan symbols freehand. Other drafters may use adhesive material symbols that are available in sheet form.

FIGURE 14.53 ■ Room labels with floor finish material noted.

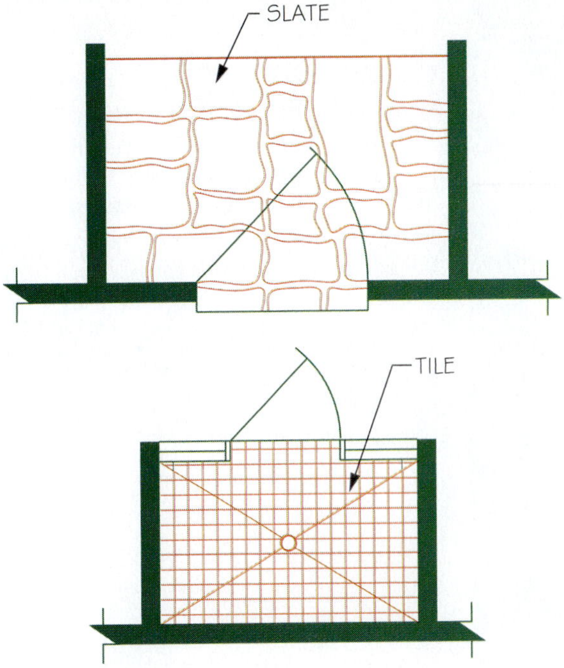

FIGURE 14.54 ■ Finish material symbols.

CADD APPLICATIONS

CADD programs provide for easy drawing of material symbols. Software such as AutoCAD has a command that allows you to place HATCH patterns. These patterns are available in a variety of material symbols for placement on plan and sectional views. Custom architectural programs also have a variety of material symbol patterns that can be easily added to your drawing. For example, one product has an architectural CD that contains nearly 3000 2-D, 3-D and front elevation symbols for AutoCAD architectural drafting. The CD symbols are divided into a logical directory structure so you can easily find the desired symbol. Figure 14.55 shows a portion of a drawing with CADD floor plan symbols added.

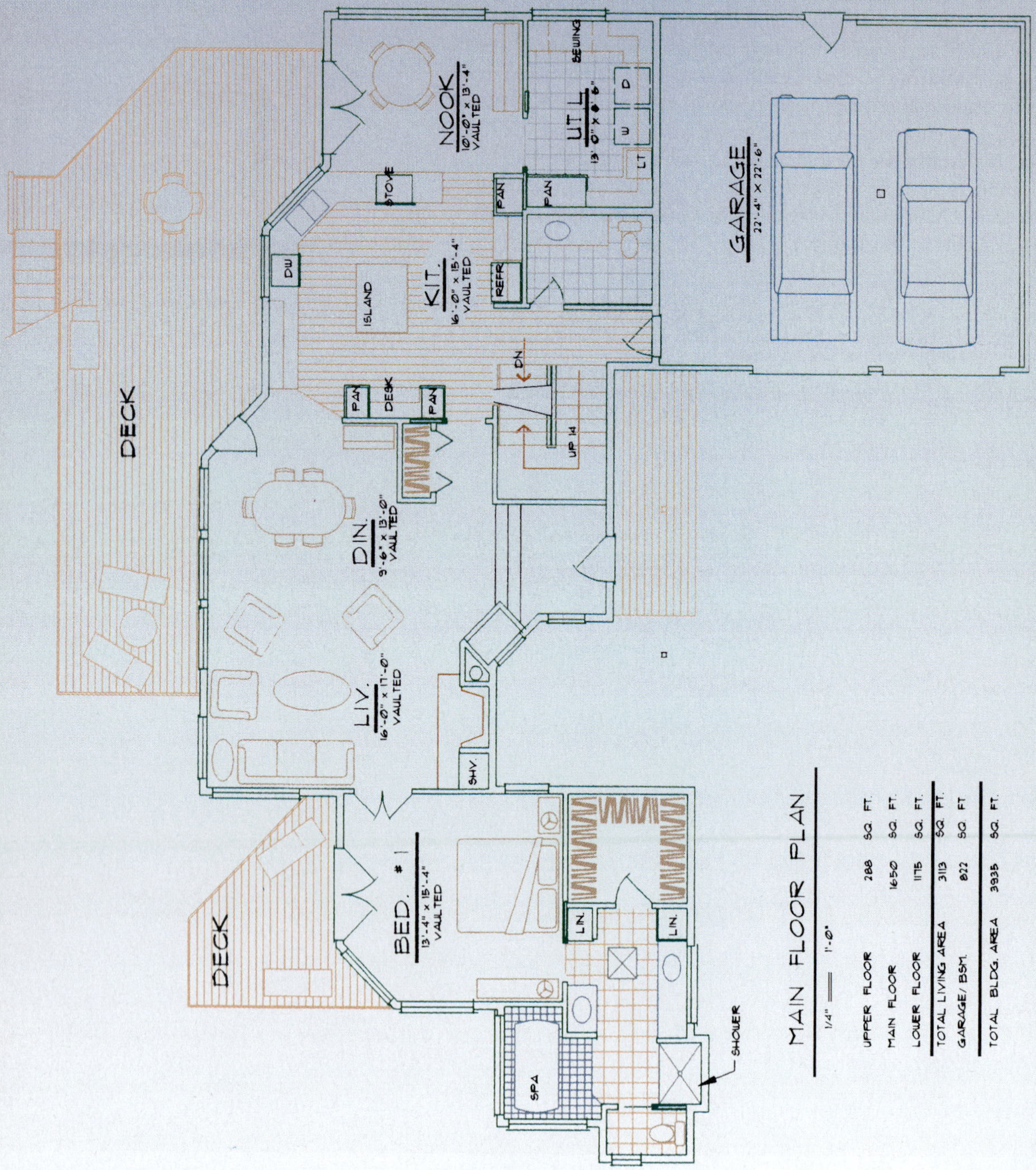

FIGURE 14.55 ■ CADD drawing with floor plan finish materials used.

Structural Materials

Structural materials are identified on floor plans with notes and symbols or in framing plans. This is discussed in detail in Chapter 31.

STAIRS

Stair Planning

This section covers the design and drafting of the floor-plan representation of stairs. The discussion will give you basic information about stair design and codes and show you different types of stairs and how they are drawn. You will also be able to make calculations for placement of stairs on the floor plan. When you learn about drawing sections and details, Chapter 37 will provide you with complete information on stair construction and layout.

Metric dimensions are given for design reference. Metric values may vary, depending on the method of calculation. The calculations can be soft or hard conversion. A *soft metric conversion,* the preferred metric construction standard, rounds dimensions off to convenient metric modules. For example, stairs 36″ wide are 900 mm when soft conversion is used. Inches are converted to millimeters with the formula 25.4 × inches = mm. So 25.4 × 36″ = 914.4 mm. This is rounded off using soft conversion to 900 mm, because 900 is in even metric modules. A *hard metric conversion* is calculated close to or exactly the same as the inch equivalent. A hard conversion of 36″ would be 25.4 × 36″ = 914.4, rounded off to 914 mm. This formula may be used in a situation where the code requires a guardrail 36″ high, because 914 mm is safely close to 36″.

The design of a multistory home is often made complex by the need to plan stairs. Figure 14.56 shows several common flights of stairs. Stairs should be conveniently located for easy access. Stair components and construction are shown in plan view and sectional view in Figure 14.57. Properly designed stairs have several important characteristics based on IRC (International Residential Code) requirements:

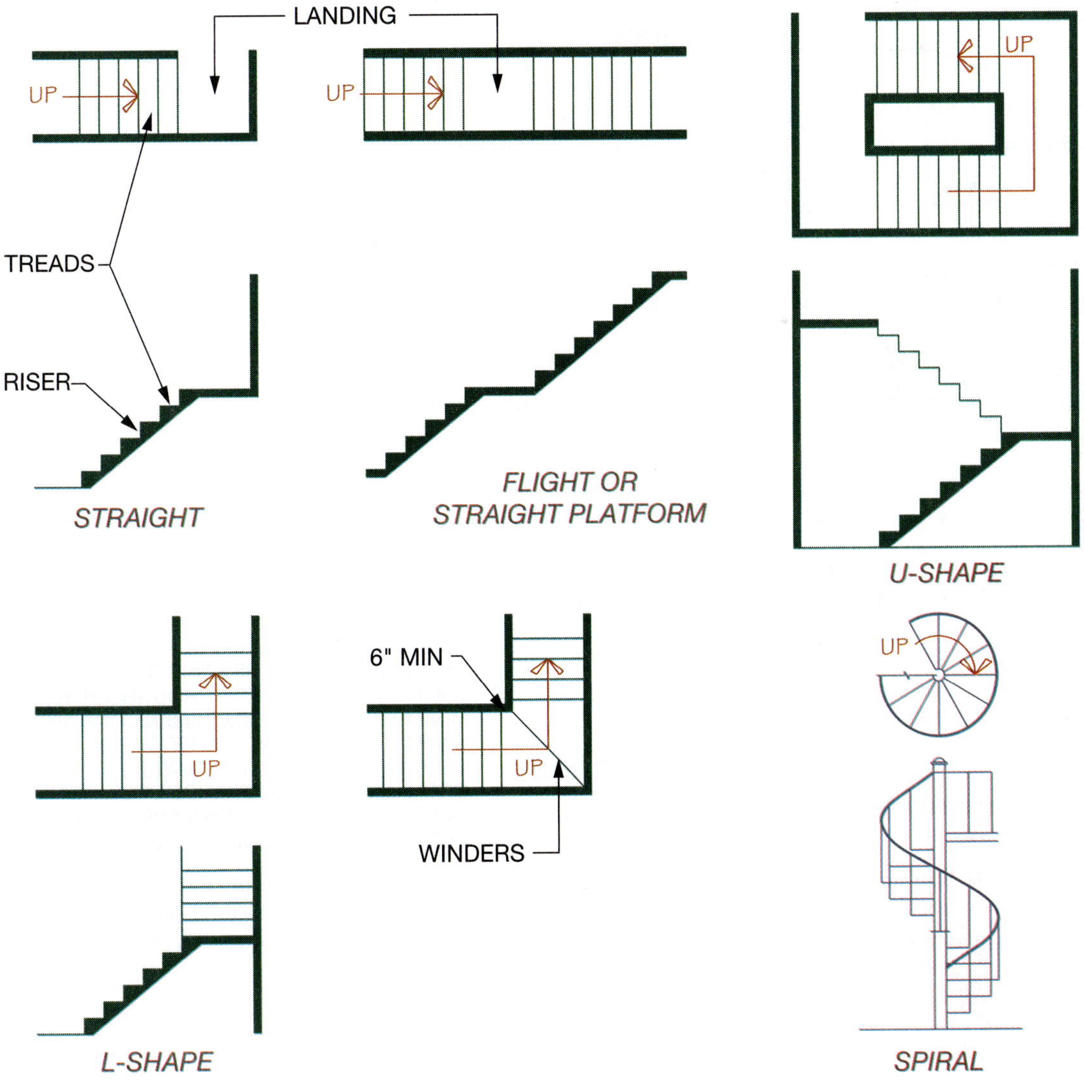

FIGURE 14.56 ■ Stair types.

24″, or PROVIDE SCREENED CLOSABLE VENT TO OUTSIDE AIR WITHIN 24″ OF FIREBOX.

Gas-Burning Fireplace

Natural gas may be provided to the fireplace, either for starting the wood fire or for fuel to provide flames on artificial logs. The gas supply should be noted on the floor plan, as shown in Figure 14.72.

Built-In Masonry Barbecue

When a home has a fireplace in a room next to the dining room, nook, or kitchen, the masonry structure may also incorporate a built-in barbecue. The floor-plan representation for a barbecue is shown in Figure 14.73. A built-in barbecue may be purchased as a prefabricated unit that is set into the masonry structure surrounding the fireplace. There may be a gas or electricity supply to the barbecue as a source of heat for cooking. As an alternative, the barbecue unit may be built into the exterior structure of a fireplace for outdoor cooking. The barbecue may also be installed separately from a fireplace.

Multilevel Fireplaces

In two- or three-story homes, it is common to have a fireplace structure that is constructed up through all levels with a chimney exiting at the roof. In this situation, a fireplace can be placed at each floor if the design is appropriate. The fireplace on the first floor is drawn similar to the example in Figure 14.65, except that additional width is needed for the flues of all stacked fireplaces to extend to the top of the chimney. A *flue* is a heat-resistant, incombustible enclosed passageway in a chimney to carry combustion products from the fireplace. The flue lining within a chimney is made out of square, rectangular, or round firebrick or fireclay products or round double or triple-insulated steel for zero-clearance fireplaces. There must be a separate flue for each fireplace. This means that the fireplace structure and the chimney need to be sized large enough to house the fireplaces and the flues running up in the same structure. Figure 14.74 shows the floor-plan drawing of a fireplace at the main floor and the second floor, and the chimney beyond the second floor. If a third-floor fireplace is used, an additional flue would be added to the structure. The size of each flue and the dimensions of the fireplace structure and the chimney should be given. The chimney above the second floor can taper to a smaller size than the fireplace structure, but it needs to be large enough to house the flues with minimum masonry material surrounding. Installation of zero-clearance flues for metal fireplaces should be based on the manufacturer's specifications. Flues must be designed to provide proper ventilation and draft for the fireplace. *Draft* is a current of air and gases through the fireplace and chimney. The size of the flue is determined by the size of the fireplace opening and the chimney height. The following table

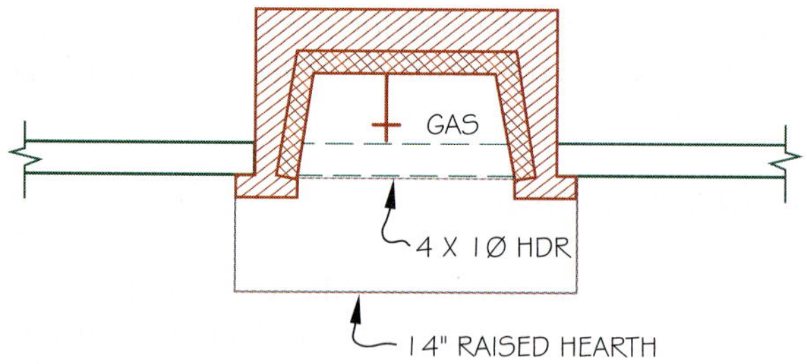

FIGURE 14.72 ■ Gas supply to fireplace.

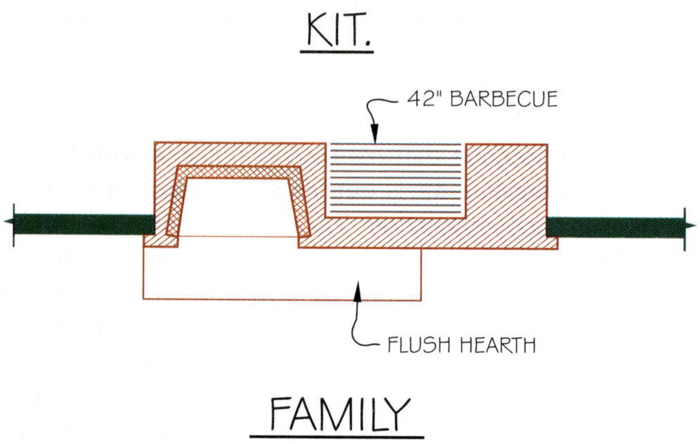

FIGURE 14.73 ■ Barbecue in fireplace.

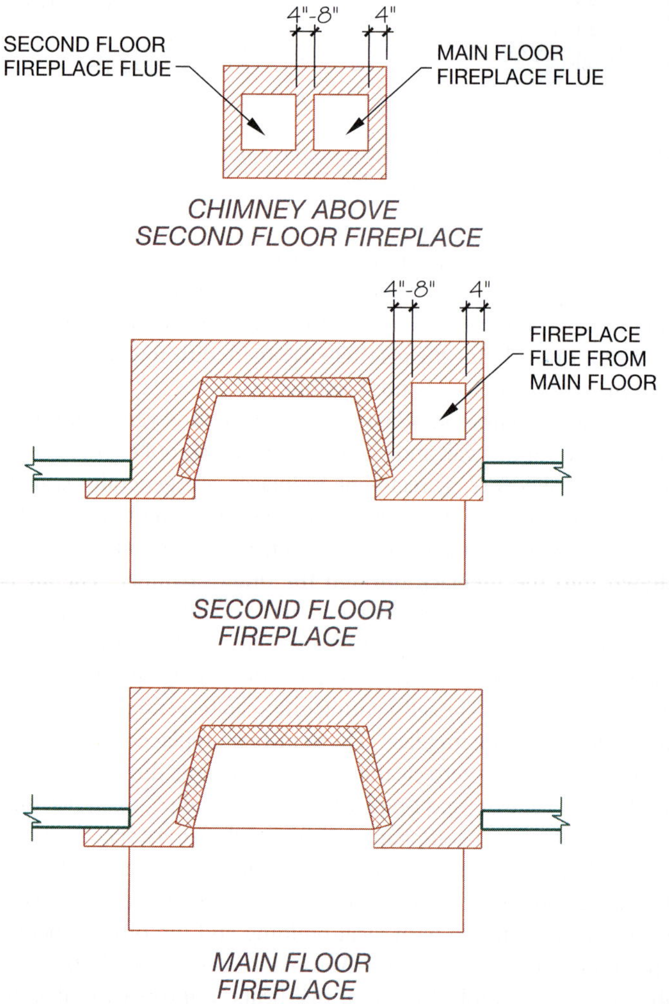

FIGURE 14.74 ■ Floor-plan drawing of a fireplace at the main floor, at the second floor, and beyond the second floor.

FIREPLACE OPENING WIDTH	OPENING HEIGHT	DEPTH	FLUE SIZE
24″ (600 mm)	24″ (600 mm)	18″ (430 mm)	8″ × 12″ (200 × 300 mm)
32″ (800 mm)	27″ (680 mm)	20″ (500 mm)	12″ × 12″ (300 × 300 mm)
36″ (900 mm)	29″ (730 mm)	20″ (500 mm)	12″ × 12″ (300 × 300 mm)
40″ (800 mm)	27″ (680 mm)	20″ (500 mm)	12″ × 12″ (300 × 300 mm)
48″ (1200 mm)	32″ (800 mm)	24″ (600 mm)	16″ × 16″ (400 × 400 mm)
60″ (1500 mm)	37″ (940 mm)	24″ (600 mm)	16″ × 20″ (400 × 500 mm)

gives general guidelines for standard fireplace openings and chimney heights. Flues for custom designs and multistory chimney heights should be confirmed with the masonry contractor, masonry supply, or fireplace manufacturer.

SOLID FUEL–BURNING APPLIANCES

Solid fuel–burning appliances are such items as airtight stoves, freestanding fireplaces, fireplace stoves, room heaters, zero-clearance fireplaces, antique stoves, and fireplace inserts for existing masonry fireplaces. This discussion shows the floor-plan representation and minimum distance requirements for typical installations of some approved appliances.

Appliances that comply with nationally recognized safety standards may be noted on the floor plan with wording such as: "IRC APPROVED WOOD STOVE AND INSTALLATION REQUIRED VERIFY ACTUAL INSTALLATION REQUIREMENTS WITH VENDORS' SPECIFICATIONS AND LOCAL FIRE MARSHAL OR BUILDING CODE GUIDELINES."

General Rules

Verify these rules with the local fire marshal or building code guidelines.

Floor Protection

Combustible floors must be protected. Floor protection material shall be noncombustible, with no cracks or holes, and strong enough not to crack, tear, or puncture with normal use. Materials commonly used are brick, stone, tile, or metal.

Wall Protection

Wall protection is critical whenever solid fuel-burning units are designed into a structure. Direct application of noncombustible materials will not provide adequate protection. When solid fuel–burning appliances are installed at recommended distances to combustible walls, a 7″ airspace is necessary between the wall and floor to the noncombustible material, plus a bottom opening for air intake and a top opening for air exhaust to provide positive air change behind the structure. This helps reduce superheated air next to combustible material (Figure 14.75). Noncombustible materials include brick, stone, or tile over cement asbestos board. Minimum distances to walls should be verified in regard to vendors' specifications and local requirements.

Combustion Air

Combustion air is generally required as a screened closable vent installed within 24″ of the solid fuel–burning appliance.

Air Pollution

Some local areas have initiated guidelines that help control air pollution from solid fuel–burning appliances. The installation

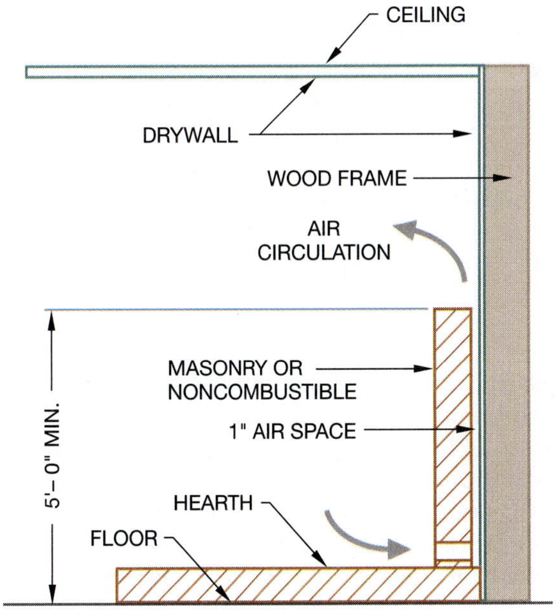

FIGURE 14.75 ■ Air circulation around wall protection.

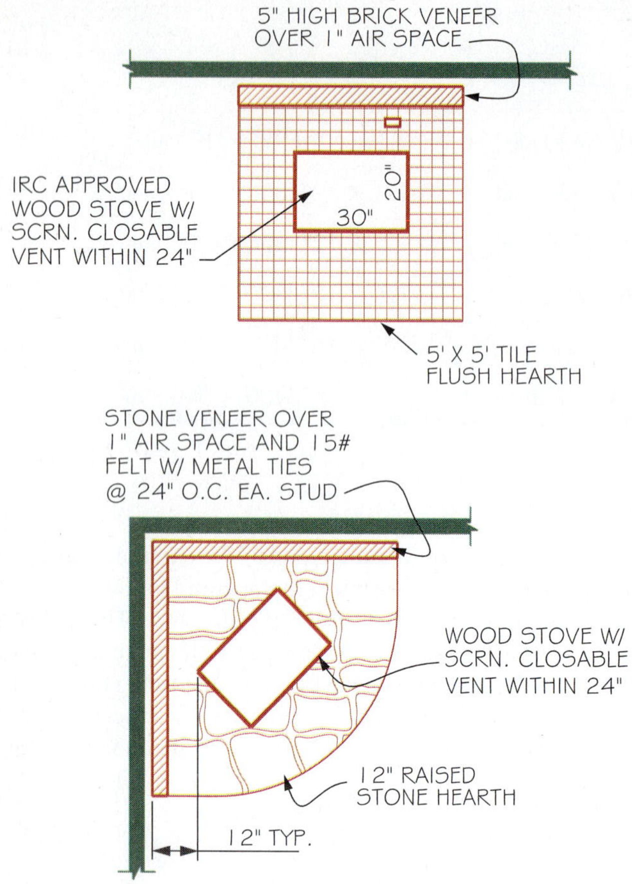

FIGURE 14.76 ■ Woodstove installation.

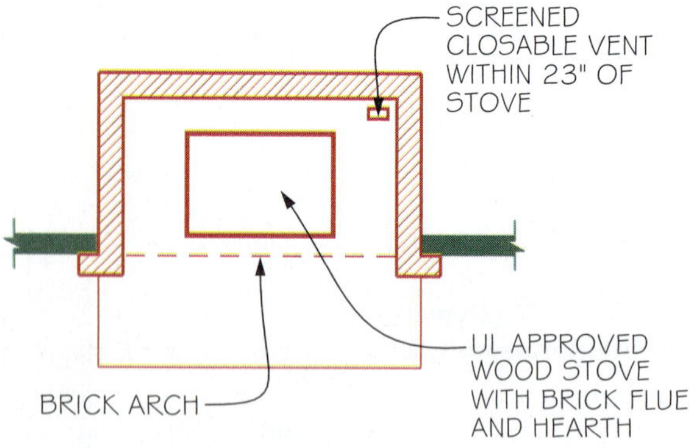

FIGURE 14.77 ■ Masonry alcove for solid-fuel burning stove.

of a catalytic converter or other devices may be required. Check with local regulations.

Figure 14.76 shows floor-plan representations of common woodstove installations.

A current trend in housing is to construct a masonry alcove within which a solid fuel unit is installed. The floor-plan layout for a typical masonry alcove is shown in Figure 14.77.

FIGURE 14.78 ■ Room titles.

FIGURE 14.79 ■ Room title with room sizes noted.

ROOM TITLES

Rooms are labeled with a name on the floor plan. Generally the lettering is larger than that used for other notes and dimensions. Room titles may be lettered using letters 3/16″ or 1/4″ (5 or 7 mm) high, as shown in Figure 14.78. The interior dimensions of the room may be lettered under the title, as shown in Figure 14.79.

OTHER FLOOR-PLAN SYMBOLS

Hose Bibb

A *hose bibb* is an outdoor water faucet to which a garden hose may be attached. Hose bibbs should be placed at locations convenient for watering lawns or gardens and for washing a car. The floor-plan symbol for a hose bibb is shown in Figure 14.80. This symbol may be shown on a separate plumbing plan, as discussed in Chapter 18. This often depends on the complexity of the structure.

Concrete Slab

Concrete slabs used for patio walks, garages, or driveways may be noted on the floor plan. A typical example would be 4″-THICK CONCRETE WALK. Concrete slabs used for the floor of a garage should slope toward the front, or to a floor drain in cold climates, to allow water to drain. The amount of slope is 1/8″/ft minimum. A garage slab is noted in Figure 14.81. Calling for a slight slope on patios and driveway aprons is also common.

Attic and Crawl Space Access

Access is necessary to attics and crawl spaces. The crawl access may be placed in any convenient location such as a closet or

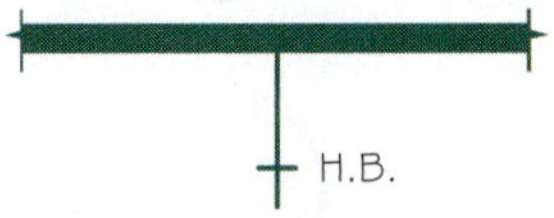

FIGURE 14.80 ■ Hose bibb symbol.

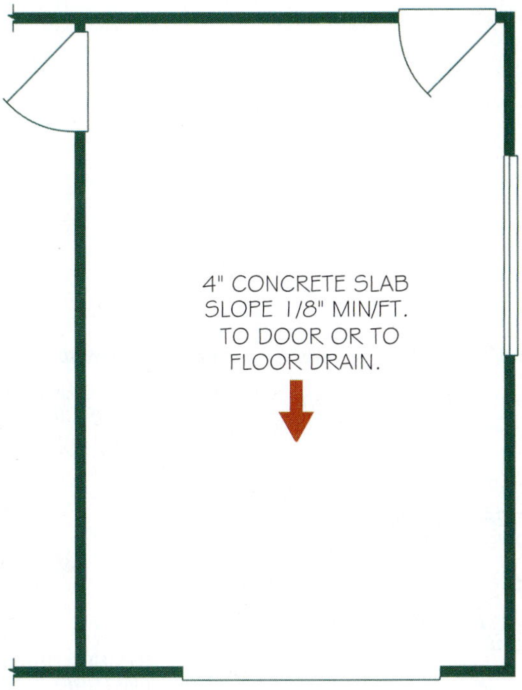

FIGURE 14.81 ■ Garage slab note.

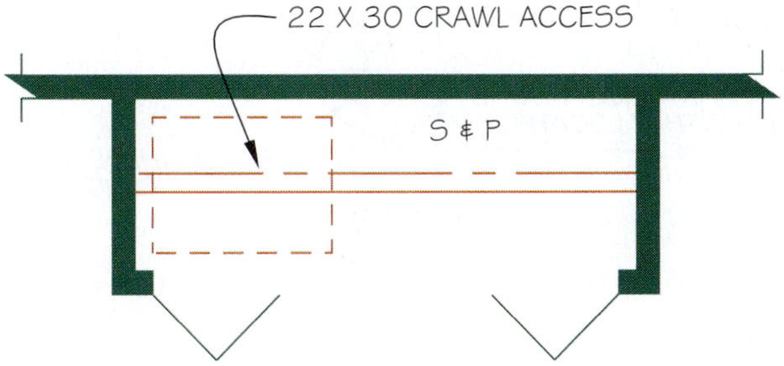

FIGURE 14.82 ■ Crawl space access note.

hallway. The crawl access should be 22″ × 30″ (560 × 762 mm) if located in the floor, as shown in Figure 14.82.

Crawl space access may be shown on the foundation plan when it is constructed through the foundation wall. When it is in the foundation wall, the access can be 24″ × 18″ (610 × 460 mm).

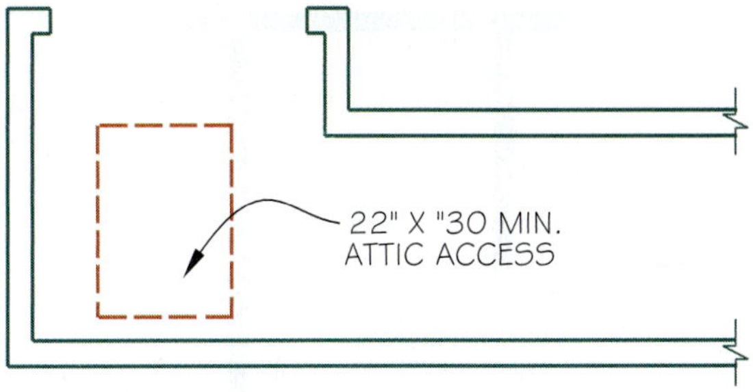

FIGURE 14.83 ■ Attic access symbol and related note.

The minimum size of the attic access is 22″ × 30″ (560 × 762 mm), and it must have 30″ (760 mm) unobstructed headroom above as you enter the attic. The attic access is commonly placed at the end of a hallway, in a walk-in wardrobe closet, or in a utility room. Figure 14.83 shows an attic access symbol and related note. The attic access may include a fold-down ladder if the attic is to be used for storage. A minimum of 30″ (762 mm) must be provided above the attic access.

Floor Drains

A floor drain is shown on the floor-plan symbol for a shower seen in Figure 14.84. Floor drains should be used in any location where water could accumulate on the floor, such as the laundry room, bathroom, or garage. The easiest application for a floor drain is in a concrete slab floor. Although drains may be designed in any type of floor construction, this is much more difficult in a wood frame. Figure 14.85 shows a floor drain in a utility room.

Cross-Section Symbol

The location on the floor plan where a cross section is taken is identified with symbols known as *cutting-plane lines.* These symbols are discussed in detail in Chapter 35.

Figure 14.86 shows options for the cutting-plane line symbol. The method used depends on your school or office practice and the drawing complexity.

Floor plans are a key element in a complete set of architectural drawings. Clients are interested in floor plans so they can see how their future home or business is laid out. As you have seen in this chapter, a large variety of symbols and drawing techniques go into preparing a floor plan. The challenge to the architectural drafter is to be sure to include all the necessary symbols and notes needed for construction and yet make the plan easy to read.

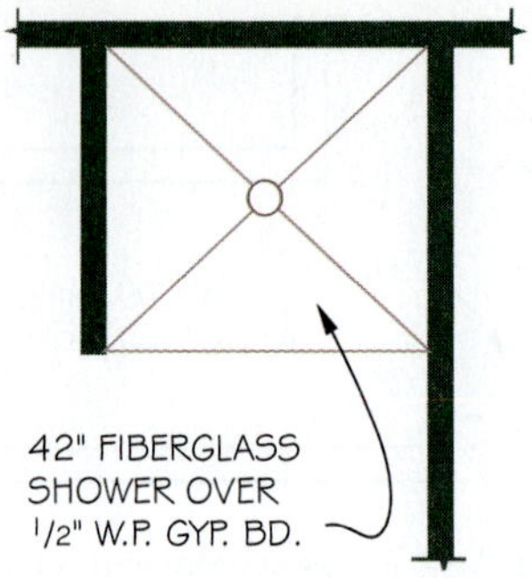

FIGURE 14.84 ■ Shower symbol showing floor drain.

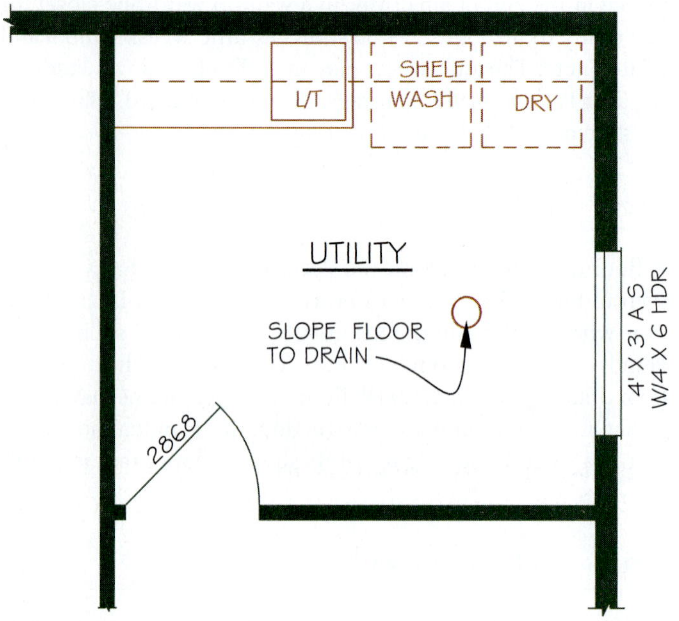

FIGURE 14.85 ■ Floor drain symbol (symbol may also be square).

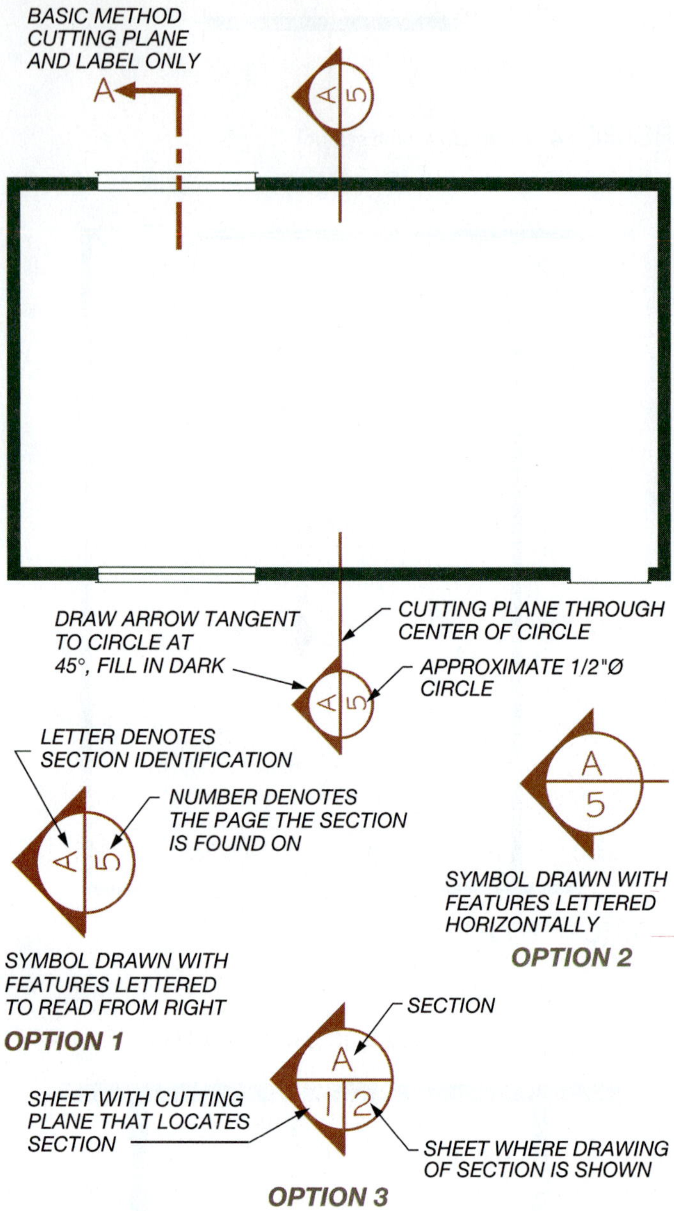

FIGURE 14.86 ■ Options for the cutting-plane line symbol.

CADD FLOOR-PLAN SYMBOLS

Various architectural CADD software packages are available that provide complete floor-plan symbol library templates. These architectural CADD programs usually include applications for windows, doors, structural symbols, electrical, furniture, appliance, stairs, and site-plan symbols. Also available are titles, drawing and editing functions, complete dimensioning capabilities, and pictorial drawing applications. The CADD commands provide (1) residential and commercial symbols for single insertion and (2) commercial symbols for multiple insertion. The multiple-insertion symbols are based on parametric design, in which you provide specifications for several variables and the CADD system automatically draws the units. For example, if you design a restroom for a hotel, all you do is select vanity sinks for multiple insertion, and you are given several sink types to choose from. You continue by specifying the sink size, the number of units, and the distance between walls. You also specify the cabinet width, height, and the backsplash dimensions. Figure 14.87 shows the multiple sink arrangement previously described.

Drawing stairs is also easy with the CADD system. Pick the start point of the stairs, specify the number of steps to the landing, give the stair width and direction, and give the total rise; the program automatically calculates the rise of each step. The program asks you to provide handrails with or without balusters and provides several options of handrail ends. After you have provided all the required information, the stair is automatically drawn, as shown in Figure 14.88.

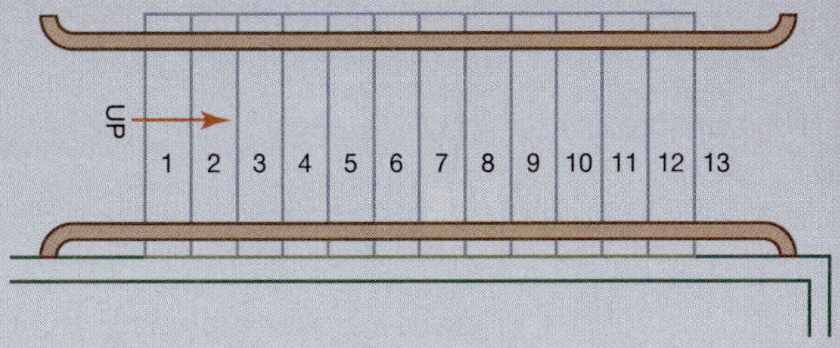

FIGURE 14.88 ■ Stairs are drawn automatically from information provided.

In most cases, the symbol you select is "dragged" onto the screen cursor at the symbol-insertion point. The term *drag* defines the CADD command that allows you to move the symbol to a desired location, where it is fixed in place on the drawing. If you decide at a later time that you do not like the selected position, you can use commands such as MOVE and ROTATE to reposition the symbol; you might also use ERASE to get rid of it altogether. Figure 14.89 shows a symbol being dragged into position and displays insertion points of several common symbols. The insertion point refers to a convenient point on the symbol that is used when placing the symbol on the drawing.

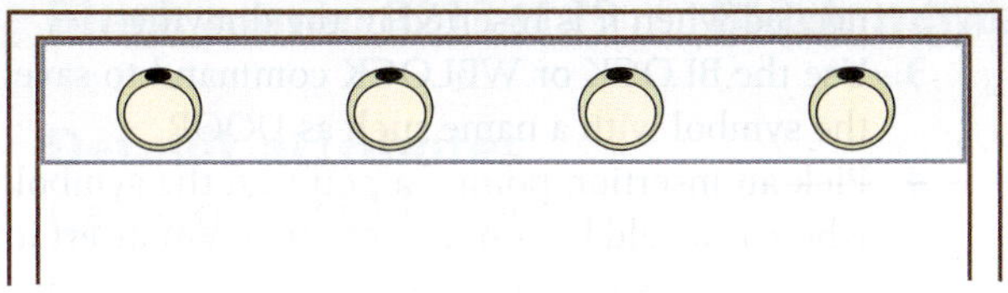
FIGURE 14.87 ■ Multiple sink arrangement automatically placed with CADD.

Manufacturers' Symbols Libraries. Manufacturers of most architectural products such as doors, windows, plumbing fixtures, and appliances have CADD symbols libraries available for design and drafting. *Symbols libraries* are a collection of related shapes, views, symbols, and attributes that may be used repeatedly. *Attributes,* also known as *tags,* go along with the architectural symbols to provide identification and numerical values for the products. Attributes are also referred to as *notes.* There is more than the simple insertion of a symbol. The symbols also carry information that may be used as a bill of materials or compiled from the drawing into a bill of materials. Any part or all of the attributes may be displayed on the drawing or kept invisible. Manufacturers' symbol libraries are usually free to architects, designers, and builders. They are available to help the user select a product, make a specification, prepare drawings, and tabulate bills of material. Most of these products allow you to create drawings quickly in plan view, elevation, pictorial, and construction details. *Elevations* are drawings looking at the front and sides of the product. *Pictorial drawings* display the product three-dimensionally. Figure 14.90 shows what the computer screen looks like when displaying a window from a manufacturer's symbol library. The left side of the screen shows a pictorial image, the upper right display is an elevation, and the lower right shows the plan view. Figure 14.91 displays the computer screen image of a detail drawing from a manufacturer's symbol library.

In addition to the CADD symbol library, manufacturers provide a printed reference library. This library allows you to see what is available, and it provides the block name and default attributes. A *block* is a symbol that can be inserted into a drawing full-size or scaled, and rotated if necessary. A *default* is a value that is maintained by the computer until you change it. Figure 14.92 shows an example of several items found in

Chapter 14 Additional Reading

The following web sites can be used as a resource to help you keep current with changes in building materials.

ADDRESS	COMPANY OR ORGANIZATION
www.abracadata.com	Abracadata
www.andersonwindows.com	Anderson Windows
www.aristokraft.com	Arisokoft (Cabinets)
www.autodesk.com	Autodesk, Inc.-Architectural Desktop Software
www.beamvac.com	Beam Central Cleaning Systems
www.builtinvacuum.com	M.D. Manufacturing
www.crestlineonline.com	Crestline Windows and Doors
www.eaglepoint.com	EaglePoint Software, Inc.-Arch T and Solid Builder software
www.generalshale.com	General Shale Brick
www.graphisoft.com	Graphisoft-ArchiCAD software
www.heatilator.com	Heatilator (Fireplaces)
www.heatnglo.com	Heat-N-Glo (Fireplaces)
www.homecrestcab.com	HomeCrest (Cabinets)
www.jacuzzi.com	Jacuzzi Whirlpool bath
www.kcma.org	Kitchen Cabinet Manufacturers
www.kitchenaid.com	KitchenAid Home Appliances
www.kraftmaid.com	KraftMaid Cabinetry, Inc.
www.marvin.com	Marvin Windows and Doors
www.nutone.com	Nutone
www.pella.com	Pella
www.plan-publishers.com	Plan Publishers, Inc.
www.schultestorage.com	Schulte Storage Systems
www.ssina.com	The Stainless Steel Information Center
www.superiorfireplace.com	Superior The Fireplace Company
www.waterfurnace.com	Water Furnace International, Inc.
www.whirlpool.com	Whirlpool Home Appliances
www.winecelarinnovations.com	Wine Cellar Innovations

CHAPTER 14

Floor-Plan Symbols Test

DIRECTIONS

Answer the questions with short, complete statements or drawings as needed on an 8 1/2″ × 11″ sheet of notebook paper as follows:

1. Letter your name, Floor-Plans Symbol Test, and the date at the top of the sheet.
2. Letter the question number and provide the answer. You do not need to write out the question.
3. Answers may be prepared on a word processor if course guidelines allow this.

QUESTIONS

Question 14–1 Exterior walls for a wood-frame residence are usually drawn how thick?

Question 14–2 Interior walls are commonly drawn how thick?

Question 14–3 How does the floor-plan symbol for an interior door differ from the symbol for an exterior door?

Question 14–4 What are the recommended spaces for the following?

a. wardrobe closet depth
b. water closet compartment
c. stair width
d. fireplace hearth depth

Question 14–5 Describe an advantage of using schedules for windows and doors.

Question 14–6 Sketch the following floor-plan symbols:

a. pocket door
b. bifold closet door
c. casement window
d. sliding window
e. skylight
f. ceiling beam
g. single-run stairs, up
h. hose bibb

Question 14–7 Letter the note that would properly identify a wall 48″ high.

Question 14–8 What is the required minimum height of guardrails?

Question 14–9 Provide the appropriate note used to identify a guardrail at a balcony overlooking the living room.

Question 14–10 What door would you recommend for use when space for a door swing is not available?

Question 14–11 Name the window that you would recommend when a 100 percent openable unit is required.

Question 14-12 What does the note 6040 next to a window on the floor plan mean?

Question 14–13 What is the minimum crawl space or attic access required?

Question 14–14 What is the abbreviation for garbage disposal?

Question 14–15 When should the clothes washer and dryer be shown with dashed lines on the floor plan?

Question 14–16 Identify in short, complete statements four factors to consider in stair planning.

Question 14–17 Show the note used to identify the steps at a two-step sunken living room.

Question 14–18 What is the amount of slope for a concrete garage floor and why should it have a slope?

Question 14–19 If you are using CADD to draw a floor plan, what dimension will you use to draw the interior walls to their exact width if 2 × 4 studs are used with 1/2″ gypsum on each side? Show the calculations.

Question 14–20 If you are using CADD to draw a floor plan, what dimension will you use to draw the exterior walls to their exact width if 2 × 4 studs are used with 1/2″ gypsum on the inside and 1/2″ sheathing and siding on the outside? Show the calculations.

Question 14–21 Define *header.*

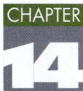

QUESTIONS *(cont.)*

Question 14–22 How are upper cabinets normally drawn on the floor plan?

Question 14–23 List at least four factors that may influence window sizes.

Question 14–24 Define *soft metric conversion.*

Question 14–25 Define *hard metric conversion.*

Question 14–26 How is it possible to have a satisfactory stair design that might be less than perfect?

Question 14–27 Why is it sometimes possible for stair design to be complex?

Question 14–28 Define *flue.*

Question 14–29 Why does the fireplace structure for multilevel construction with more than one fireplace need to be wider than a fireplace for a single-story building?

Question 14–30 Give the flue size for a standard fireplace opening 36″ W × 29″H × 20″D.

DIRECTIONS

1. Draw the following problems manually or with CADD, depending on your course guidelines. Use a 1/4″ = 1′–0″ scale. Draw on 8 1/2″ × 11″ vellum.
2. Make exterior walls 6″ thick and interior walls 4″ thick if using manual drafting. Draw exact thickness if using CADD.
3. Use the line and architectural lettering methods described in this text. Letter only the notes that are part of the drawing. Estimate dimensions when not given and increase sizes to help fill paper when appropriate. Do not draw dimensions or pictorial illustrations.

PROBLEMS

Problem 14–1 Redraw Figures 14.5, 14.6, 14.8, and 14.9. Do not draw the pictorial in the following problems.

Problem 14–2 Redraw Figure 14.9.

Problem 14–3 Redraw Figures 14.12 through 14.17.

Problem 14–4 Redraw Figures 14.18 through 14.21.

Problem 14–5 Redraw Figure 14.22.

Problem 14–6 Redraw Figures 14.23 through 14.27 and 14.30.

Problem 14–7 Redraw Figure 14.31.

Problem 14–8 Redraw Figure 14.33. Draw the partial floor plan and door/window key symbols. Do not draw the door/window schedules.

Problem 14–9 Redraw Figure 14.35.

Problem 14–10 Redraw Figure 14.42.

Problem 14–11 Redraw Figure 14.43.

Problem 14–12 Redraw Figures 14.48, 14.50, and 14.54.

Problem 14–13 Redraw Figures 14.58 through 14.62.

Problem 14–14 Redraw Figure 14.63.

Problem 14–15 Redraw Figures 14.65 and 14.66.

Problem 14–16 Redraw Figures 14.67 through 14.73.

Problem 14–17 Redraw Figure 14.74.

Problem 14–18 Redraw Figures 14.76 and 14.77.

Problem 14–19 Redraw Figure 14.81.

Problem 14–20 Redraw Figures 14.82 and 14.83.

Problem 14–21 Redraw Figure 14.85.

Problem 14–22 Redraw Figure 14.86.

Problem 14–23 Create your own CADD symbols.

Problem 14–24 Use your CADD system to redraw the following floor plan to your own specifications.

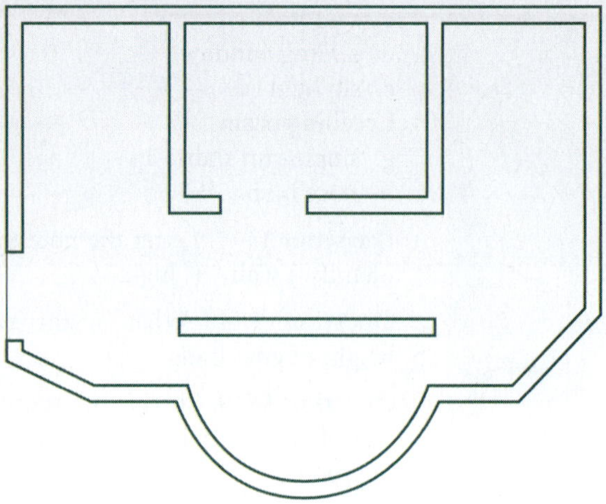

CHAPTER 15

Floor-Plan Dimensions and Notes

ALIGNED DIMENSIONS

The dimensioning system most commonly used in architectural drafting is known as *aligned dimensioning*. With this system, dimensions are placed in line with the dimension lines and read from the bottom or right side of the sheet. Dimension numerals are centered on and placed above the solid dimension lines. Figure 15.1 shows a floor plan that has been dimensioned using the aligned dimensioning system.

FLOOR-PLAN DIMENSIONS

Basic Dimensioning Concepts

You should place dimensions so that the drawing does not appear crowded. However, this is often difficult because of the great number of dimensions that must be placed on an architectural drawing. When placing dimensions, space dimension lines a minimum of 3/8″ (10 mm) from the object and from each other. If there is room, 1/2″ (13 mm) is preferred. Some drafters use 1″ (26 mm) or more if space is available. Other drafters place the first dimension line 1″ (26 mm) away from the plan and space additional dimensions equally 1/2″ (13 mm) or 3/4″ (20 mm) apart. The minimum recommended spacing of dimensions is shown in Figure 15.2. Regardless of the distance chosen, be consistent so that dimension lines are evenly spaced.

Extension and dimension lines are drawn thin and dark so that they will not detract from the overall appearance and balance of the drawing. Dimension lines terminate at extension lines with dots, arrowheads, or slash marks that are each drawn in the same direction. See Figure 15.3.

Dimension numerals are drawn 1/8″ (3 mm) high with the aid of guidelines. The dimension units used are feet and inches for all lengths over 12″. Inches and fractions are used for units less than 12″. Foot units are followed by the symbol (′), and inch units shown by the symbol (″). Some drawing dimensions may not use the foot and inch symbols if company standards are to omit them.

Metric dimensions are given in millimeters (mm), as discussed later in this chapter; however, metric sizes vary, depending on hard or soft conversion. Soft conversion is the preferred method; this is explained in Chapter 14 and again later.

Exterior Dimensions

The overall dimensions on frame construction are understood to be given to the outside of the stud frame of the exterior walls. The reason for locating dimensions on the outside of the stud frame is that the frame is established first, and windows, doors, and partitions are usually put in place before sheathing and other wall-covering material is applied. Another construction option is for the exterior dimensions to be to the outside face of the exterior sheathing. This depends on the preferred construction method. Figure 15.4*A* shows an example of the construction used where the exterior dimension is to the outside face of studs, as compared with the outside face of the sheathing in Figure 15.4*B*. Your floor plan dimensioning practice is the same either way. The contractor determines the actual construction method.

The first line of dimensions on the plan is the smallest distance from the exterior wall to the center of windows, doors, and partition walls. The second line of dimensions generally gives the distance from the outside walls to partition centers. The third line of dimensions is usually the overall distance between two exterior walls. See Figure 15.5. This method of applying dimensions eliminates the need for workers to add dimensions at the job site and reduces the possibility of error.

This textbook shows dimensions placed from the outside face of exterior stud walls to the center of interior walls and partitions to the center of doors and windows. Many architectural offices, however, use a different dimensioning system. They place dimensions from the outside and inside face of studs between exterior and interior walls, followed by dimensions from the outside face of exterior studs to the center of doors and windows, as shown in Figure 15.5. Both methods provide an overall dimension as the last dimension.

Interior Dimensions

Interior dimensions locate all interior partitions and features in relation to exterior walls. Figure 15.6 shows some common interior dimensions. Notice how they relate to the outside walls. When dimensioning interior features, ask yourself if you have provided workers with enough dimensions to build the house. The contractor should not have to guess where a wall or feature should be located, nor should workers have to use a scale to try to locate items on a plan.

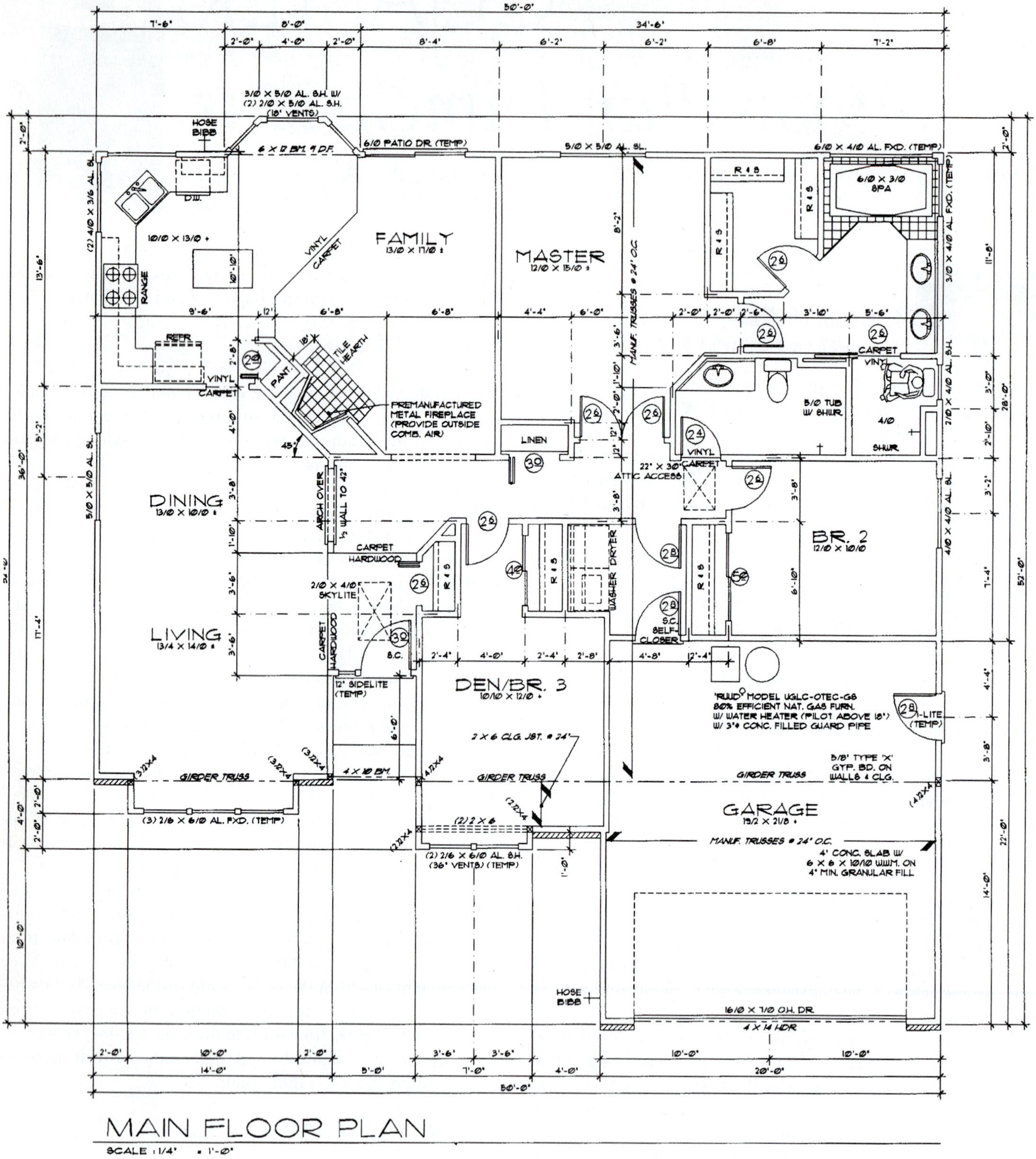

FIGURE 15.1 ■ Aligned dimensions on a floor plan. *Courtesy Alan Mascord Design Associates.*

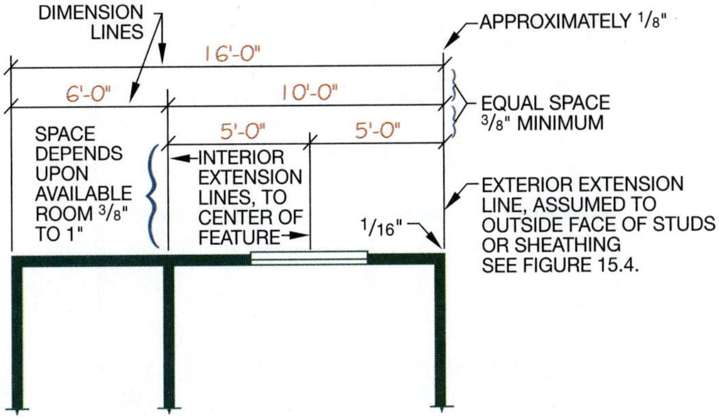

FIGURE 15.2 ■ Recommended spacing of dimension lines.

Dimensions are generally given by locating interior walls and partitions from the outside face of exterior studs to their centers. Dimensions between interior walls are from center to center. *Stub walls*—walls that do not go all the way across the room—are dimensioned by giving their length. Windows and doors are dimensioned to their centers, unless the location is assumed. Some offices practice dimensioning all features to the face of studs rather than to the centers, as previously discussed. Figure 15.6 shows examples of dimensioning interior features.

Standard Features

Some interior features that are considered to be standard sizes may not require dimensions. Figure 15.7 shows a situation in which the drafter elected to dimension the depth of the pantry even though it is directly next to a refrigerator, which has an assumed depth of 30″ (762 mm). Notice that the refrigerator width is given because there are many different widths available. The base cabinet is not dimensioned; such cabinets are typically 24″ (600 mm) deep. When there is any doubt, it is better to apply a dimension. The refrigerator and other appliances may be drawn with dashed lines or marked with the abbreviation NIC (not in contract) if the unit will not be supplied according to the contract for the home being constructed. Follow the practice of the architect's office in which you work.

Other situations in which dimensions may be assumed are when a door is centered between two walls as at the end of a hallway, and when a door enters a room and the minimum distance from the wall to the door is assumed. See the examples in Figure 15.8. Some dimensions may be provided in the form of a note for standard features, as seen in Figure 15.9. The walls around a shower need not be dimensioned when a note, 36″ (900 mm) square shower, defines the inside dimensions. The shower must be located and actual product dimensions verified during construction.

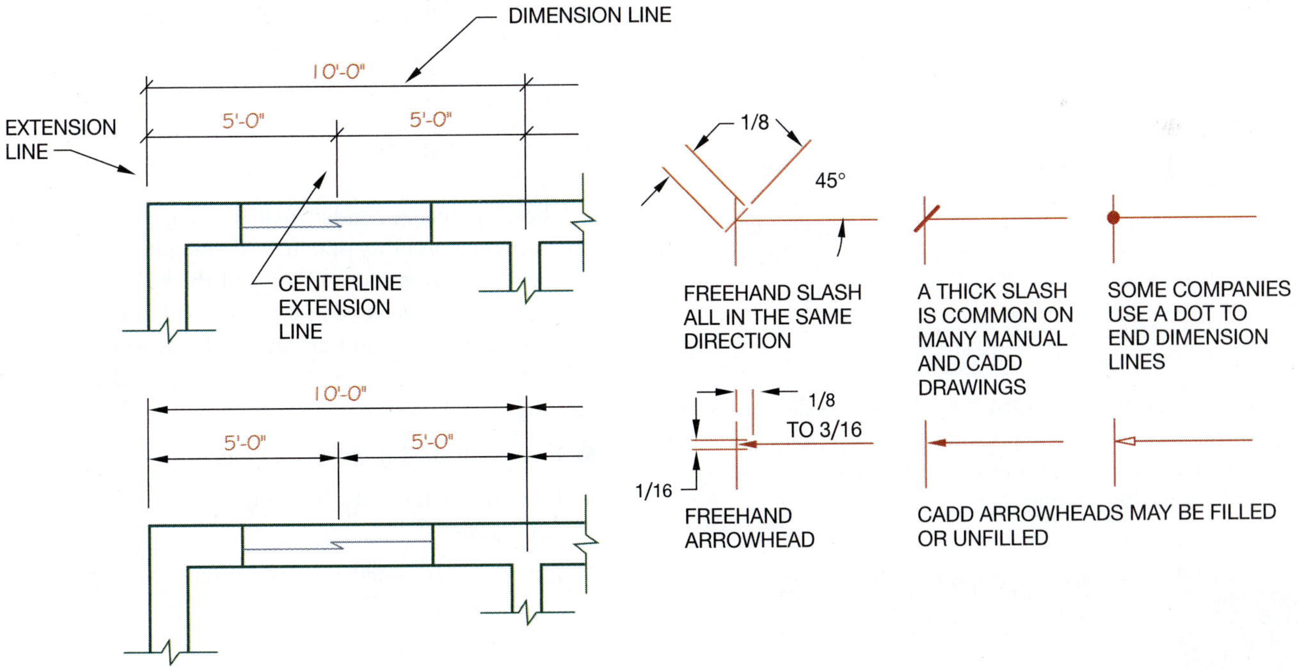

FIGURE 15.3 ■ Methods for terminating dimension lines.

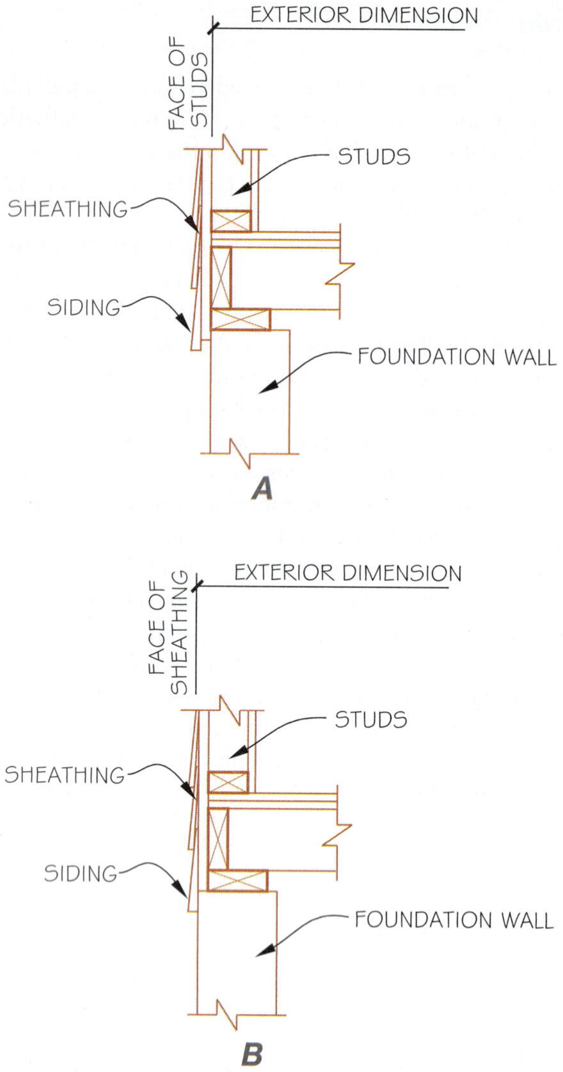

FIGURE 15.4 ■ (*A*) An example of the construction used when exterior dimensions are to the outside face of studs: (*B*) Construction used when exterior dimensions are to the outside face of sheathing.

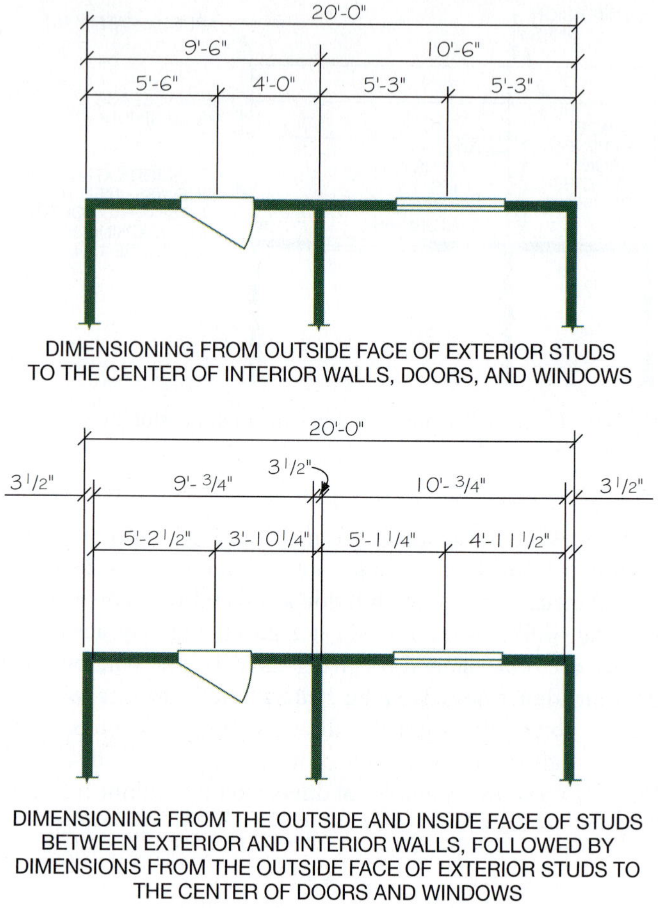

FIGURE 15.5 ■ Placing exterior dimensions.

One of the best ways to learn how experienced drafters lay out dimensions is to study and evaluate existing plans. Metric dimensions may vary depending on hard or soft conversion, as discussed later in this chapter.

COMMON SIZES OF ARCHITECTURAL FEATURES

All walls, edges of brick, and brick fireplaces are thick lines. All other lines are thin. The refrigerator, water heater, washer and dryer, dishwasher, and trash compactor are dashed lines. Closet poles are 12″ (300 mm) from the back wall and are center lines. Closet shelves are solid lines 15″ (330 mm) from the back wall.

Room Components

The following recommended wall dimensions are based on common manual drafting practices. If you are using CADD, it is recommended that you draw the wall thickness to their exact measurements, as discussed in Chapter 14. For example, an exterior wall with 2 × 4 stud framing would measure 3 1/2″ (stud) + 1/2″ (interior drywall) + 1/2″ (sheathing) + 1/2 to 1″ (exterior siding). See Figure 14.3.

- Exterior heated walls: 6″ (150 mm).
- Exterior unheated finished walls: 5″ (130 mm) (garage/shop).
- Exterior unheated walls interior unfinished: 5″ (garage/shop)
- Interior: 4″ (100 mm).
- Interior plumbing (toilet): 6″ (150 mm).
- Hallways: 36″ (900 mm) minimum clearance.
- Entry hallways: 42″–60″ (1060–1500 mm).
- Bedroom closets: 24″ (600 mm) minimum depth; 48″ (1200 mm) length.

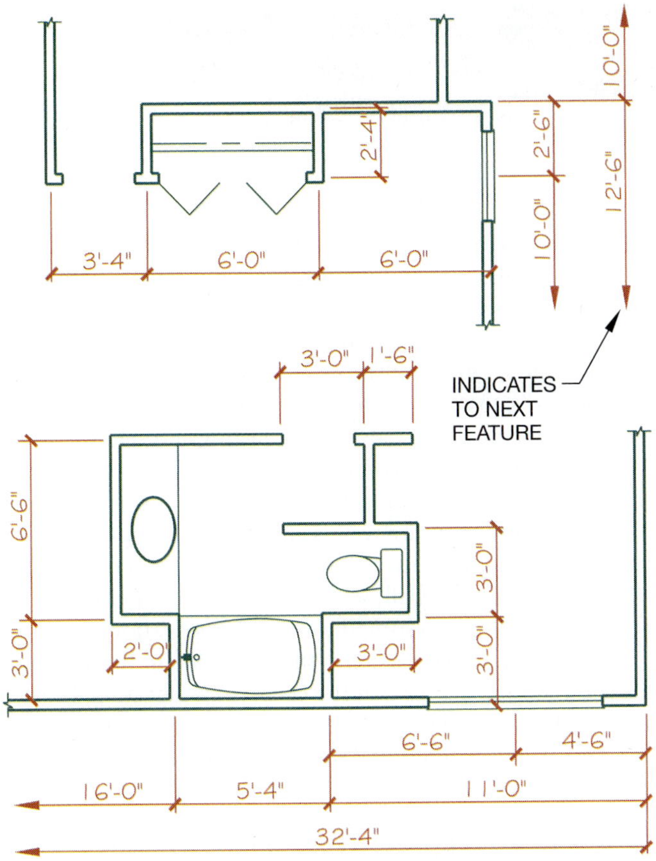

FIGURE 15.6 ■ Placing interior dimensions.

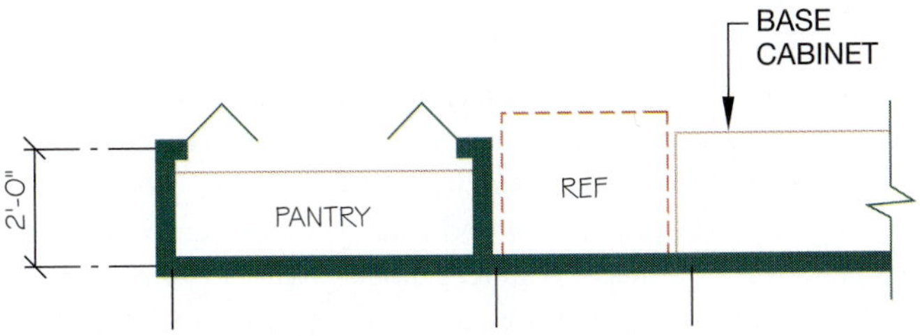

FIGURE 15.7 ■ Assumed dimensions.

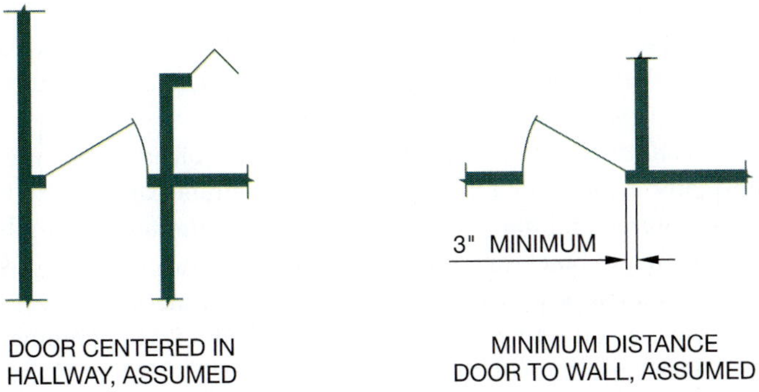

FIGURE 15.8 ■ Assumed location of features without dimensions given.

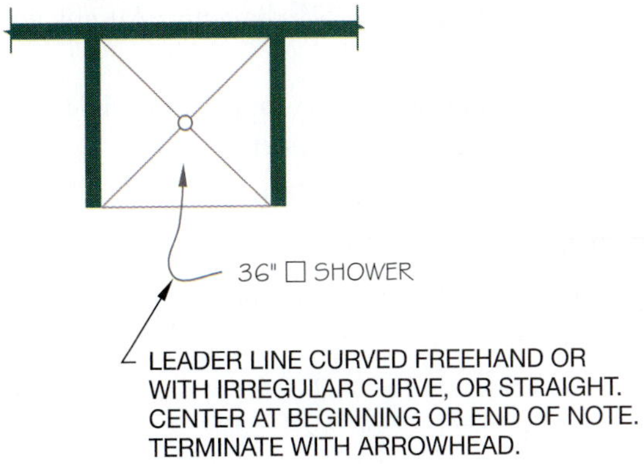

FIGURE 15.9 ■ Standard features dimensioned with a specific note.

- Linen closets: 14″–24″ (350–600 mm) deep (not over 30″ (760 mm)).
- Base cabinets: 24″ (600 mm) deep; 15″–18″ (380–460 mm) wide bar; 24″ (600 mm) deep island with 12″ (300 mm) minimum at each side of cooktop; 36″ (900 mm) minimum from island to cabinet (for passage, 48″ (1200 mm) is better on sides with an appliance).
- Upper cabinets: 12″ (600 mm) deep.
- Washer/dryer space: 36″ (900 mm) deep, 5′ × 6″ (1680 mm) long minimum.
- Stairways: 36″ (900 mm) minimum wide; 9″ (230 mm) minimum to 12″ (300 mm) tread. Review Chapter 14 for detailed information.
- Fireplace.

Plumbing

- Kitchen sink: double, 32″ × 21″; triple, 42″ × 21″.
- Vegetable sink and/or bar sink: 16″ × 16″ or; 16″ × 21″.
- Laundry sink: 21″ × 21″.
- Bathroom sink: 19″ × 16″, oval; provide 9″ from edge to wall and about 12″ between two sinks; 36″ minimum length.
- Toilet space: 30″ wide (minimum); 24″ clearance in front.
- Tub: 32″ × 60″ or 32″ × 72″.
- Shower: 36″ square; 42″ square; or combinations of 36″, 42″, 48″, and 60″ for fiberglass; any size for ceramic tile.
- Washer/dryer: 2′–4″ square (approximately).

Metric values of above products vary, depending on manufacturer, and are generally in even modules.

Appliances

- Forced air unit: gas, 18″ square (minimum) with 6″ space all around; cannot go under stair; electric, 24″ × 30″ (same space requirements as gas).

- Water heater: gas, 18–22″ diameter; cannot go under stair.
- Refrigerator: 36″ wide space; approximately 27″ deep; 4″ from wall; 4″ from base cabinet.
- Stove/cooktop: 30″ × 21″ deep.
- Built-in oven: 27″ × 24″ deep.
- Dishwasher: 24″ × 24″; place near upper cabinet to ease putting away dishes.
- Trash compactor: 15″ × 24″ deep; near sink, away from stove.
- Broom/pantry: 12″ minimum × 24″ deep, increasing by 3″ increments.
- Desk: 30″ × 24″ deep (minimum); out of main work triangle.
- Built-in vacuum: 24–30″ diameter.

Metric values of above products vary, depending on manufacturer, but are generally in even modules.

Doors

- Entry: 36″ × 6′–8″; 42″ × 8′.
- Sliders or French: 5′, 6′, 8′ (double); 9′, 10′ (triple); 12′ (four-panel).
- Garage, utility, kitchen, and bedrooms on custom houses: 2′–8″.
- Bedrooms and bathrooms of nice houses: 2′–6″.
- Bathrooms closets: 2′–4″.
- Garage 8′ × 7′, 9′ × 8′, 16′ × 7′, and 18′ × 7′.

Windows

- Living, family: 8′–10′.
- Dining, den: 6′–8′.
- Bedrooms: 4′–6′.
- Kitchens: 3′–5′.
- Bathrooms: 2′–3′.
- Sliding: 4′, 5′, 6′, 8′, 10′, 12′.
- Single-hung: 24″, 30″, 36″, 42″.
- Casement: same as sliding.
- Fixed/awning: 24″, 30″, 36″, 42″, 48″.
- Fixed/sliding: 24″, 30″, 36″, 42″, 48″.
- Picture 4′, 5′, 6′, 8′.
- Bay: 8′–10′ total; sides, 18–24″ wide.

DIMENSIONING FLOOR-PLAN FEATURES

Masonry Veneer

The discussion thus far has provided examples of floor-plan dimensioning for wood-framed construction. Other methods of residential construction include masonry veneer, structural masonry, concrete block, and solid concrete. Masonry veneer construction is the application of thin (4″) masonry, such as stone or brick, to the exterior of a wood-framed structure. This kind of construction provides a long-lasting, attractive exterior. Masonry veneer may also be applied to interior frame partitions where the appearance of stone or brick is desired. Brick or stone may cover an entire wall that contains a fireplace, for example. Occasionally either material may be applied extending to the lower half of a wall, with another material extending to the ceiling for a contrasting decorative effect. Masonry veneer construction is dimensioned on the floor plan in the same way as wood framing, except that the veneer is dimensioned and labeled with a note describing the product. See Figure 15.10. This construction requires foundation support, discussed in Chapters 32–34.

Concrete-Block and Structural Masonry Construction

Concrete blocks are made in standard sizes and may be solid or have hollow cavities. Concrete block may be used to construct exterior or interior walls of residential or commercial structures. Some concrete blocks have a textured or sculptured surface to provide a pleasant exterior appearance. Most concrete-block construction must be covered, such as by masonry veneer, for a finished look. Some structures use concrete block for the exterior bearing walls and wood-framed construction for interior partitions. Standard concrete blocks are 8″ (190 mm) wide.

Dimensioning concrete-block construction is different from wood-framed construction in that each wall, partition, and window and door opening is dimensioned to the edge of the feature, as shown in Figure 15.11. While this method is most common, some drafters do prefer to dimension the concrete-block structure in the same way as wood-framed construction. To do so, they must add specific information in notes or section views about wall thicknesses and openings. Some practices use the abbreviation MO to identify masonry openings.

Structural masonry, also referred to as *reinforced masonry,* is normally a blend of materials that are manufactured at high temperatures and generally have higher compressive strength than concrete block or poured concrete. Structural masonry is attractive in appearance and does not normally need to be covered with wood frame or other masonry products. In combination with certain insulation, the standard 4 5/8″ reinforced ma-

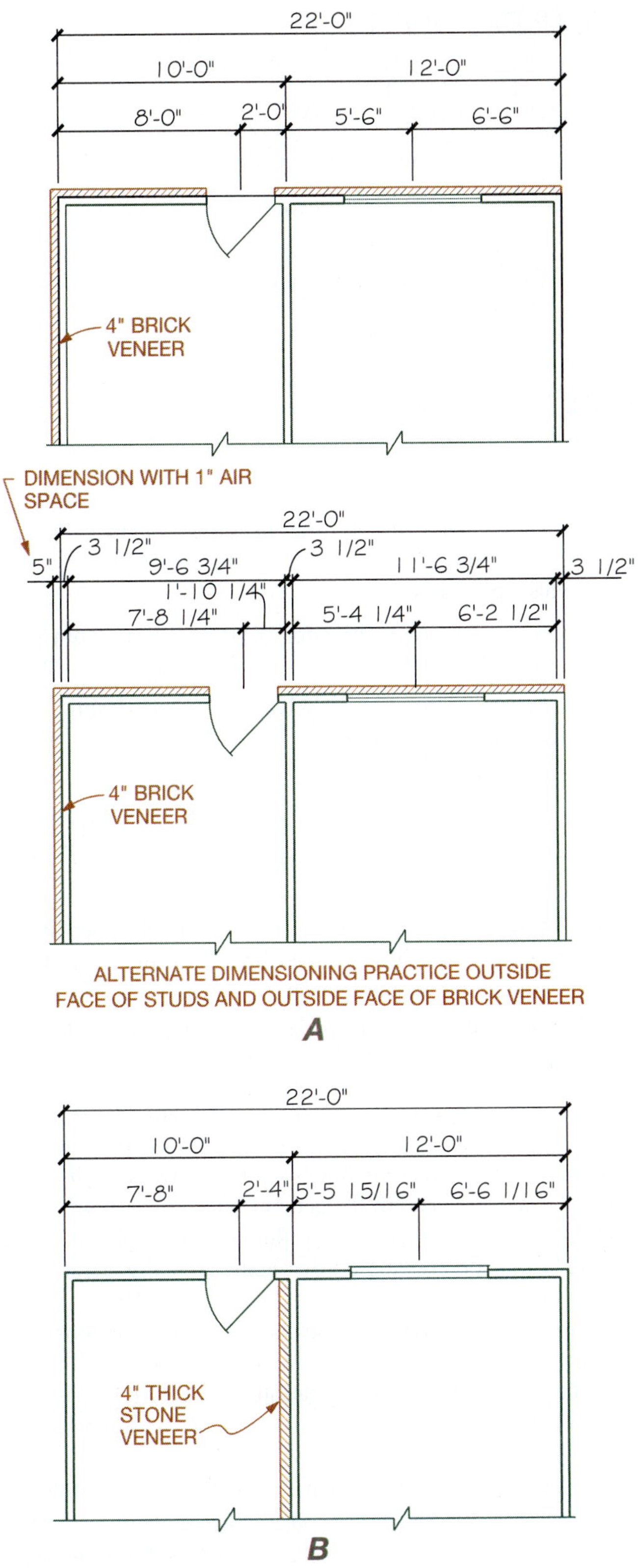

FIGURE 15.10 ■ Dimensioning (*A*) brick exterior veneer; (*B*) interior stone veneer.

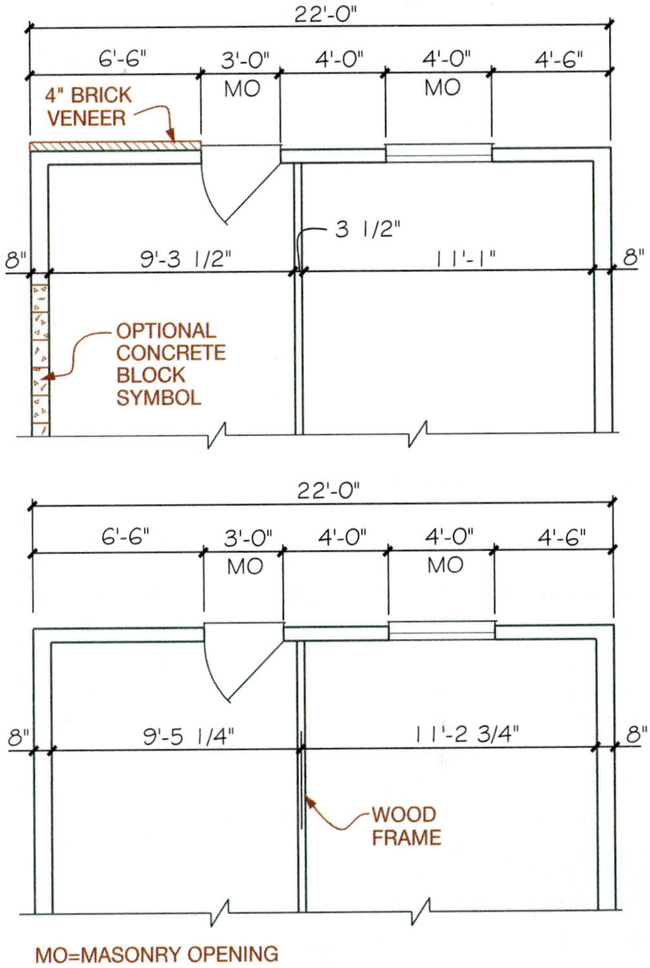

FIGURE 15.11 ■ Dimensioning concrete-block construction.

sonry wall can have R-values up to 20. Structural masonry floor plan drawings are similar to those for concrete block, except that the standard unit is 4 5/8″ wide rather than 8″ wide.

Solid Concrete Construction

Solid concrete construction is used in residential and commercial structures. In residences it is mostly limited to basements and subterranean homes. Concrete is poured into forms that mold the mixture to the shape desired. When the concrete cures (hardens), the forms are removed and the structure is complete, except for any surface preparation that may be needed.

Masonry veneer may be placed on either side of concrete walls for a good appearance. Wood framing and typical interior finish materials may also be added to the inside of concrete walls. The wood framing attached to concrete walls is in the form of furring, which is usually nailed in place with concrete nails. *Furring* is wood strips which are fastened to the studs or interior walls and ceilings and on which wall materials are attached. Furring is used on masonry walls to provide airspace

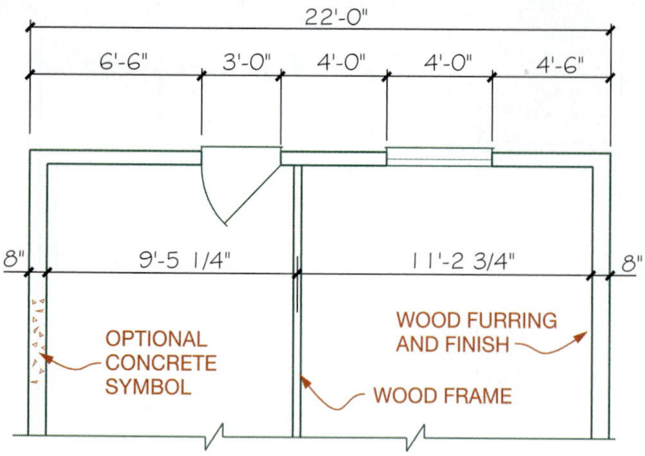

FIGURE 15.12 ■ Dimensioning solid concrete construction.

GENERAL NOTES:

1. ALL PENETRATIONS IN TOP OR BOTTOM PLATES FOR PLUMBING OR ELECTRICAL RUNS TO BE SEALED. SEE ELECTRICAL PLANS FOR ADDITIONAL SPECIFICATIONS.
2. PROVIDE 1/2 " WATER PROOF GYPSUM BOARD AROUND ALL TUBS, SHOWERS, AND SPAS.
3. VENT DRYER AND ALL FANS TO OUTSIDE AIR THRU VENT WITH DAMPER.
4. INSULATE WATER HEATER TO R-11. W.H. IN GARAGE TO BE ON 18" HIGH PLATFORM.
5. PROVIDE 1-HOUR FIREWALL BETWEEN GARAGE AND RESIDENCE WITH 5/8" TYPE 'X' GYPSUM BOARD FROM FLOOR TO BOTTOM OF SHEATHING.

FIGURE 15.13 ■ Some typical general notes.

and a nailing surface for attaching finish wall materials. Furring attached to masonry is pressure-treated or foundation-grade wood or galvanized steel. Standard interior finish materials such as drywall or paneling may be added to the furring. Solid concrete construction is typically dimensioned the same as concrete-block construction. See Figure 15.12.

More information about masonry and concrete construction in a residence is found in Chapters 32 through 34.

NOTES AND SPECIFICATIONS

Notes on plans are either specific or general. *Specific notes* relate to specific features within the floor plan, such as the header size over a window opening. A specific note is often connected to a feature with a leader line. Specific notes are also called *local notes*, since they identify isolated items. Common information that may be specified in the form of local notes is:

1. Window schedule. Place all windows in a schedule. Group by type (sliding, fixed, awning, etc.) from largest to smallest. Window size and type can also be placed directly on the floor plan.
2. Door schedule. Place all doors in schedule. Group by type (solid core, slab, sliding, bifold, etc.) from largest to smallest. Door size and type can be placed directly on the floor plan.
3. Place room names in the center of all habitable rooms (1/4″ [6 mm] lettering) with the interior size below (1/8″ [3 mm] lettering).
4. Label appliances such as furnace, water heater, dishwasher, compactor, stove, and refrigerator. Items that can be distinguished by shape, such as toilets and sinks, do not need to be identified.
5. Label all tubs, showers, or spas, giving size, type and material.
6. Label fireplace or solid-fuel-burning appliance. Specify vent within 24″ (610 mm), hearth, U.L. approved materials, and wood box.
7. Label stairs, giving direction of travel, number of risers, and rail height.
8. Label all closets with shelves, S&P (shelf and pole), linen, broom, or pantry.
9. Specify attic and crawl access openings.
10. Designate 1-hour firewall between garage and residence with 5/8 (16 mm) type X gypsum board from floor to ceiling.
11. On multilevel structures, call out "line of upper floor," balcony above, line of lower floor, or other projections of one level beyond another.
12. Verify codes and construction methods for additional local notes.

General notes apply to the drawing overall rather than to specific items. General notes are commonly lettered in the *field* of the drawing. The field is any open area that surrounds the main views. A common location for general notes is the lower right or left corner of the sheet. Notes should not be placed closer than 1/2″ from the drawing border. Some typical general notes are seen in Figure 15.13.

Some specific notes that are too complex or take up too much space may be lettered with the general notes and keyed to the floor plan with a short identification such as the phrase see NOTE #1 or SEE NOTE 1, or with a number within a symbol such as ① or ⚠1.

Written specifications are separate notes that identify the quality, quantity, or type of materials and fixtures to be used in the entire project. Specifications for construction are prepared in a format different from drawing sheets. Specifications may be printed in a format that categorizes each phase of the construction and indicates the precise methods and materials to be used. Architects and designers may publish specifications for a house so that the client knows exactly what the home will contain, even including the color and type of paint. This information

OBJECT	ME
Construction modules	100 600
Standard brick	90 ×
Mortar joints	10
Concrete block	200 ›
Sheet metal	tenths
Drywall, plywood, and rigid insulation	1200 1200
Batt insulation	400 600
Door height	2050 2100
Door width	750 800 850 900 o 1000
Cut glass	mm

According to the Brick Institute of Americ
Modular Construction of Clay and Conc
building module. Many common brick si
within a 100 mm vertical module by usir
Owing to the size of concrete blocks, ho

also sets a standard that allows contractors to prepare construction estimates on an equal basis.

Lending institutions require material specifications to be submitted with plans when builders apply for financing. Generally each lender has a form to be used that supplies a description of all construction materials and methods along with a cost analysis of the structure. More information about specifications is found in Chapters 43 and 44.

USING METRIC

The unit of measure commonly used is the millimeter (mm). Meters (m) are used for large site plans and civil engineering drawings. Metric dimensioning is based on the International System of Units (SI). Canada is one country that uses metric dimensions.

When materials are purchased from the United States, it is often necessary to make a *hard conversion* to metric units. This means that the typical inch units are converted directly to metric. For example, a 2 × 4 that is milled to 1 1/2″ × 3 1/2″ converts to 38 × 89 mm. To convert from the imperial inch to millimeters, use the formula 25.4 × inches = millimeters.

The preferred method of metric dimensioning is called *soft conversion*. This means that the lumber is milled directly to metric units. The 2 × 4 lumber is 40 × 90 mm using a soft conversion. This method is much more convenient to use for drawing plans and measuring in construction. When plywood thickness is measured in metric units, 5/8″ thick equals 17 mm, and 3/4″ thick equals 20 mm. The length and width of plywood also change, from 48″ × 96″ to 1200 × 2400 mm. Modules for architectural design and construction in the United States are typically 12″, 16″, or 24″. In countries using metric measurement, the dimensioning module is 100 mm. For example, construction members may be spaced 24″ on center (OC) in the United States, while the spacing in Canada is 600 mm OC; and the spacing between studs is 16″ OC in the United States and 400 mm OC in Canada. These metric modules allow 1200 × 2400 mm plywood to fit exactly on center with the construction members. Interior dimensions are also designed in 100-mm increments. For example, the kitchen base cabinet measures 600 mm deep.

Expressing Metric Units on a Drawing

In placing metric dimensions on a drawing, all dimensions specified with dimension lines are in millimeters and the millimeter symbol (mm) is omitted. When more than one dimension is quoted, the millimeter symbol (mm) is placed only after the last dimension. For example, the size of a plywood sheet reads 1200 × 2400 mm, or the size and length of a wood stud reads 40 × 90 × 2400 mm. The millimeter symbol is omitted in the notes associated with a drawing, except when referring to a single dimension, such as the thickness of material or the spacing of members. For example, a note might read 90 × 1200 BEAM, while the reference to material thickness of 12 mm GYPSUM or the spacing of joists of 400 mm OC places the millimeter symbol after the size (OC in the abbreviation for On Center).

Rules for Writing Metric Symbols and Names

- Unit names are lowercase, even those derived from proper names (for example, millimeter, meter, kilogram, kelvin, newton, and pascal).
- Use vertical text for unit symbols. Use lowercase text, such as mm (millimeter), m (meter), and kg (kilogram), unless the unit name is derived from a proper name as in K (kelvin), N (newton), or Pa (pascal).
- Use lowercase for values less than 10^3, and use uppercase for values greater than 10^3.
- Leave a space between a numeral and symbol, 55 kg, 24 m, 38° C. Do *not* close up the space like this: 55 kg, 24 m, 38°C.
- Do *not* leave a space between a unit symbol and its prefix—for example, use kg not k g.
- Do *not* use the plural of unit symbols—for example, use 55 kg, *not* 55 kgs.
- The plural of a metric name should be used, such as 125 meters.
- Do not mix unit names and symbols; use one or the other. Symbols are preferred on drawings where necessary. Millimeters (mm) are assumed on architectural drawings unless otherwise specified.

Metric Scales

In Chapter 2 metric scales were introduced with respect to the use of the metric scale as a drafting tool. When drawings are produced in metric, floor plans, elevations, and foundation plans are generally drawn at a scale of 1:50 rather than the 1/4″ = 1′–0″ scale used in the imperial system. Larger-scale drawings, such as those showing construction details, are often drawn at a scale of 1:5. Small-scale drawings, such as plot plans, may be drawn at a scale of 1:500. Figure 15.14 shows a floor plan completely drawn using metric dimensioning. The preferred method of soft conversion is used in this figure, and the metric scale is 1:50.

Metric Standards and Specifications

Refer to Appendix A for a complete list and examples of metric standards and specifications for architectural design and construction materials.

Metric Construction Dimensions

The chart on page 287 gives a comparison of metric and inch construction modules. These are not a direct conversion of metric and inches, but a relationship between the metric and inch standard based on soft conversion.

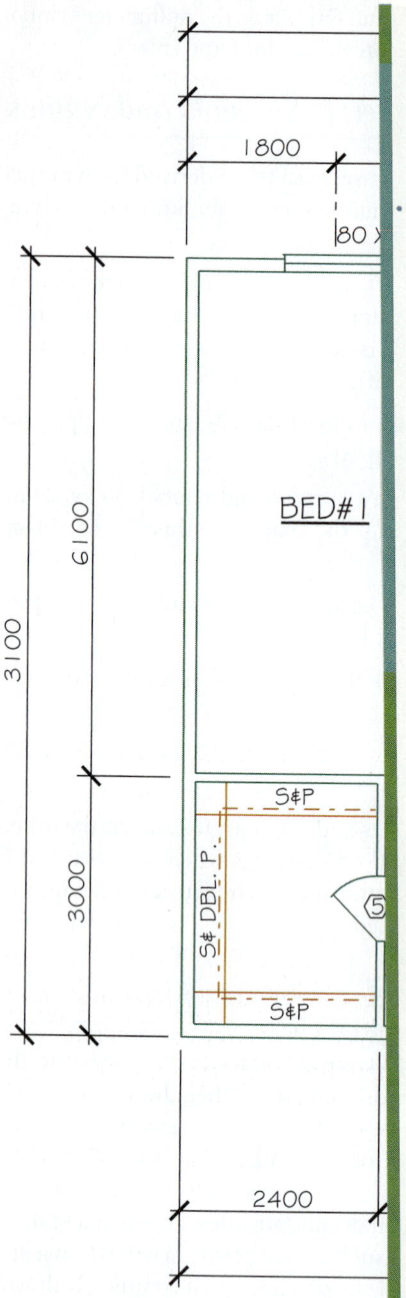

FIGURE 15.14 ■ Floor plan drawn us

CHAPTER 15

Floor-Plan Dimensions and Notes Test

DIRECTIONS

Answer the questions with short complete statements or drawings as needed on an 8 1/2″ × 11″ sheet of notebook paper, as follows:

1. Letter your name, Floor Plan Dimensions and Notes Test, and the date at the top of the sheet.
2. Letter the question number and provide the answer. You do not need to write out the question.

Answers may be prepared on a word processor if course guidelines allow this.

QUESTIONS

Question 15–1 Define *aligned dimensioning.*

Question 15–2 Why should dimension lines be spaced evenly and to avoid crowding?

Question 15–3 Should extension and dimension lines be drawn thick or thin?

Question 15–4 Show an example of how dimension numerals that are less than 1′ are lettered.

Question 15–5 Show an example of how dimension numerals that are greater than 1′ are lettered.

Question 15–6 Are the overall dimensions on frame construction given to the outside of the stud frame at exterior walls?

Question 15–7 Describe the dimensional information provided in each line when three lines of dimensions are used.

Question 15–8 Describe and give an example of a specific note.

Question 15–9 What is another name for a specific note?

Question 15–10 Describe and give an example of a general note.

Question 15–11 Give at least two examples of when it might be acceptable for a dimension to be assumed.

Question 15–12 Define *soft metric conversion.*

Question 15–13 Define *hard metric conversion.*

Question 15–14 Identify the metric conversion method preferred in Canada.

Question 15–15 Convert the following inch dimensions to metric using soft conversion:

36″ base cabinet	12″ upper cabinet
2 × 4 stud	1/2″ × 4′ × 8′ plywood

Question 15–16 Give the dimension that you would use to draw a wall thickness using CADD if the wall is constructed with the following material:

2 × 6 studs, 1/2″ drywall, 1/2″ exterior sheathing, 3/4″ siding.

Question 15–17 Give the standard wall thickness for concrete block construction and structural masonry construction.

Question 15–18 What formula would you use to convert inches to millimeters?

Question 15–19 What are the typical units found on a metric architectural drawing?

Question 15–20 On a metric drawing, is the metric symbol used to identify the units when the dimensions are specified with dimension lines?

PROBLEMS

1. Draw the following problems using pencil on 8 1/2″ × 11″ vellum or with CADD.
2. Use architectural lettering.
3. Prepare lines and dimensions as specified in your course guidelines, or select your own preferred technique as discussed in this text.

Problem 15–1 Draw an example of dimensioning a wood-framed structure on a floor plan. Show at least one interior partition, one exterior door, and one window.

Problem 15–2 Show an example of how masonry veneer is dimensioned on a floor plan when used with wood-framed construction.

Problem 15–3 Show an example of a floor plan constructed with concrete block. Show at least one wood-frame interior partition, one exterior door, and one window.

Problem 15–4 Show an example of a floor plan constructed with solid concrete exterior walls. Show at least one interior wood-frame partition, one wall with interior wood furring, one exterior door, and one window.

Problem 15–5 Redraw the portion of Figure 15.14 that contains the main bathroom above the stairs and bedroom number 2.

16 CHAPTER

Floor-Plan Layout

INTRODUCTION

Residential plans are commonly drawn on 17″ × 22″, 18″ × 24″, 22″ × 34″, or 24″ × 36″ drawing sheets. This discussion explains floor-plan layout techniques using 17″ × 22″ vellum. The layout methods may be used with any sheet size. A complete floor plan is shown in Figure 16.1. This floor plan is shown without the electrical layout and the structural elements. Designing and drawing the electrical plan are discussed in Chapter 17, and calculating and drawing the structural members are covered in Chapters 25 through 31. The step-by-step method used to draw this floor plan provides you with a method for laying out any floor plan. As you progress through the step-by-step discussion of the floor-plan layout, refer back to previous chapters to review specific floor-plan symbols and dimensioning techniques. Keep in mind that these layout techniques represent a suggested typical method used to establish a complete floor plan. You may alter these layout steps to suit your individual preference as your skills develop and your knowledge increases.

The step-by-step floor plan layout guidelines can be used for manual drafting or computer-aided design and drafting. Steps used to layout a drawing manually is often similar to the practice used for CADD. The following steps will give you commonly used methods, with each step followed by conditions that can affect the CADD layout when appropriate.

LAYING OUT A FLOOR PLAN

Before you begin the floor-plan layout, be sure that your hands and equipment are clean, tape your drawing sheet down very tightly, and align and square the scales of your drafting machine if you use one.

STEP 1 Determine the working area of the paper to be used. This is the distance between borders. In this exercise, the 17″ × 22″ vellum has a 16″ vertical and 9 1/2″ horizontal distance between borders. Verify optional sheet sizes with a review of Chapter 3.

The final CADD drawing will be printed or plotted within the selected sheet size. Determining this in advance is important, and for many companies the sheet size is a standard used in the office. If the office allows for flexibility in selecting sheet size, there is more flexibility with CADD. This is because the boundaries of your CADD drawing can change at any time. The final drawing can also be plotted on any desired sheet size.

STEP 2 Determine the drawing area, which is approximately the area that the floor plan plus dimensions requires when completely drawn. Residential floor plans are drawn to a scale of 1/4″ = 1′–0″. The house to be drawn is 60′ long and 36′ wide. At a scale of 1/4″ = 1′–0″ the house will be drawn 15″ × 9″ (60′ ÷ 4 = 15″ by 36′ ÷ 4 = 9″). This size does not take into consideration any area needed for dimensions. However, it appears that in this case there is enough space for dimensions because the house size is much less than the working area. You should consider at least 2″ on each side of the house for dimensions. When the plan does not fit within the working area, you need to increase the sheet size or reduce the scale. You need to confirm this decision with your instructor or company, because a standard sheet size may be used.

The floor-plan scale is *not* normally reduced, as floor plans are usually drawn at 1/4″ to 1′–0″, except for some commercial applications where 1/8 = 1′–0″ or 3/16″ = 1′–0″ may be used. Floor plans drawn in metric usually use a 1:50 scale.

CADD drawings are created actual size. You still need to think ahead about the sheet size that will be used for your final plot, but this is not as critical as with manual drafting. You have flexibility and can change the sheet size throughout the CADD drawing. The boundary for a CADD drawing is called the *limits*. There are 4′ in every 1″ of the drawing if you want the final drawing to be plotted at a scale of 1/4″ = 1′–0″. If you use the working area of 19 1/2″ × 16″ described in Step 1, the multiplying this by 4 gives you limits of (4)19 1/2 × (4)16 = 78′ × 64′. Now you can draw the floor plan actual size within the 78′ × 64′ limits that you have established.

STEP 3 Center the drawing within the working area using these calculations:

Working area length = 19.5″
Drawing area length = − 15.0″
4.5″

4.5″ ÷ 2 = 2.25″

2.25″ = Space on each side of floor plan

Working area height = 16″
Drawing area height = − 9″
7″

7″ ÷ 2 = 3.5″

3.5″ = Space above and below floor plan

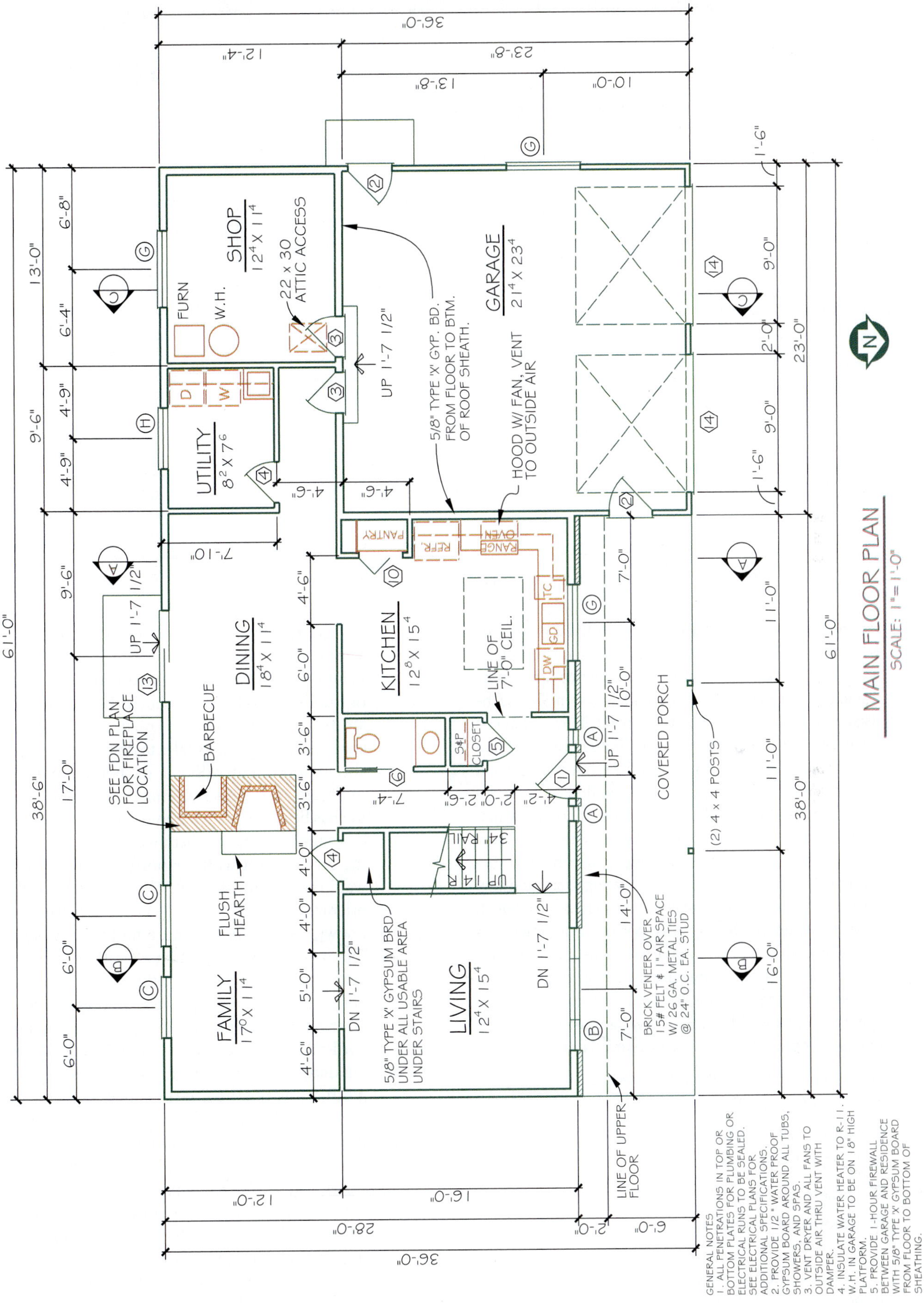

FIGURE 16.1 ■ Complete floor plan.

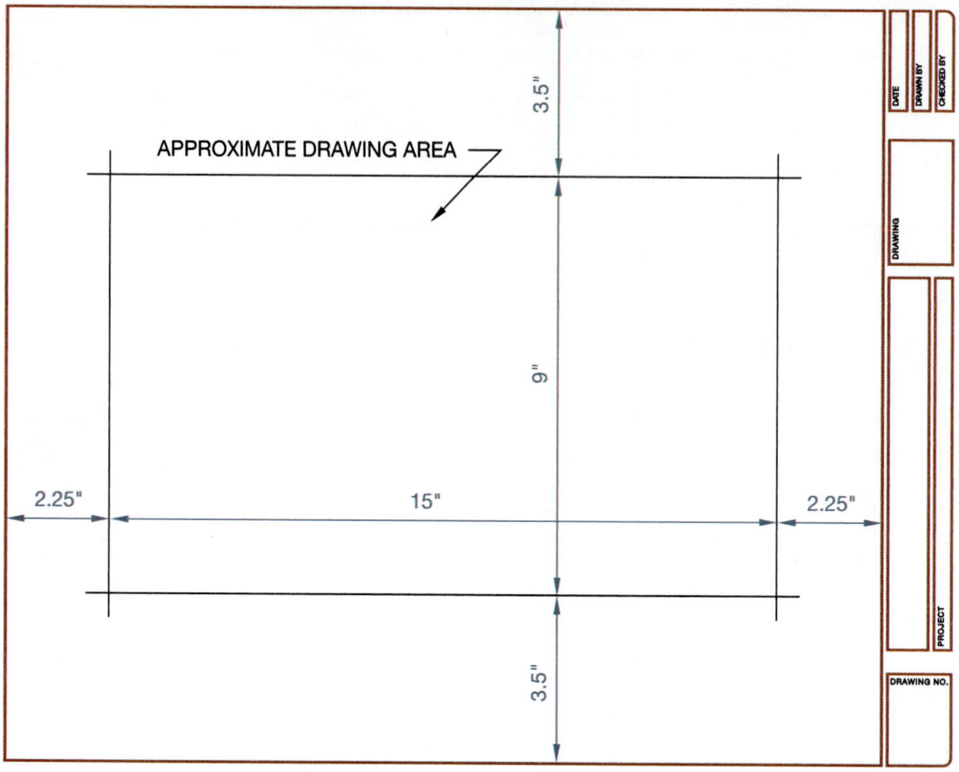

FIGURE 16.2 ■ Approximate drawing area centered within the working area.

Figure 16.2 shows the drawing area centered within the working area. Use construction lines drawn very lightly with a 4H or 3H lead to outline the drawing area. Some drafters use a 6H lead, but extreme caution should be used with hard leads such as 6H and 4H. These leads can easily pierce through paper. Some drafters prefer to use light blue layout lead, because it does not reproduce if used properly. Caution should be used if your drawings are to be reproduced with an engineering photocopy machine. These machines are generally sensitive to all lines. This method of centering a drawing is important for beginning drafters. It can be quite frustrating to start a drawing and find that there is not enough room to complete it. More experienced drafters may not use this process to this extent, although they make some quick calculations to determine where the drawing should be located: In actual practice, the drawing does not have to be perfectly centered, but a well-balanced drawing is important. Several factors contribute to a well-balanced drawing:

- Actual size of the required drawing
- Scale of the drawing
- Amount of detail
- Size of drawing sheet
- Dimensions needed
- Amount of general and local notes and schedules required

Care in centering the drawing on the sheet is not as important with CADD, because the drawing can be moved at any time for a more convenient location within the limits.

STEP 4 Before you start drawing, the orientation of the house or building on the sheet should be considered. There are two standard ways to orient the building on the sheet. A common method is to draw the building on the sheet with the entry door at the bottom of the sheet, as in Figure 16.1. This is normally done when the building has been designed for construction on any site to be determined later. Another method is to place the building on the sheet with the north direction pointing toward the top of the sheet. This is possible if the north orientation of the building on the construction site is known. When this is done, a north arrow is placed on the drawing in a convenient location, which may be near the drawing title. Sometimes the north orientation pointing toward the top of the sheet does not allow for a convenient positioning of the building parallel to the borderlines. In this case, it is possible to orient the building with the front entry door at the bottom and a north arrow pointing in the proper compass direction relative to the building if it were on the construction site. Whether the north arrow is placed on the floor plan drawing or not, it is placed on the site plan. This drawing generally accompanies a complete set of drawings and can be referred to for compass orientation.

Lay out all exterior walls 6″ (150 mm) thick within the drawing area using construction lines. Construction lines are very lightly drawn with a 4H or 3H lead and easy to erase if you make an error. If properly drawn, however,

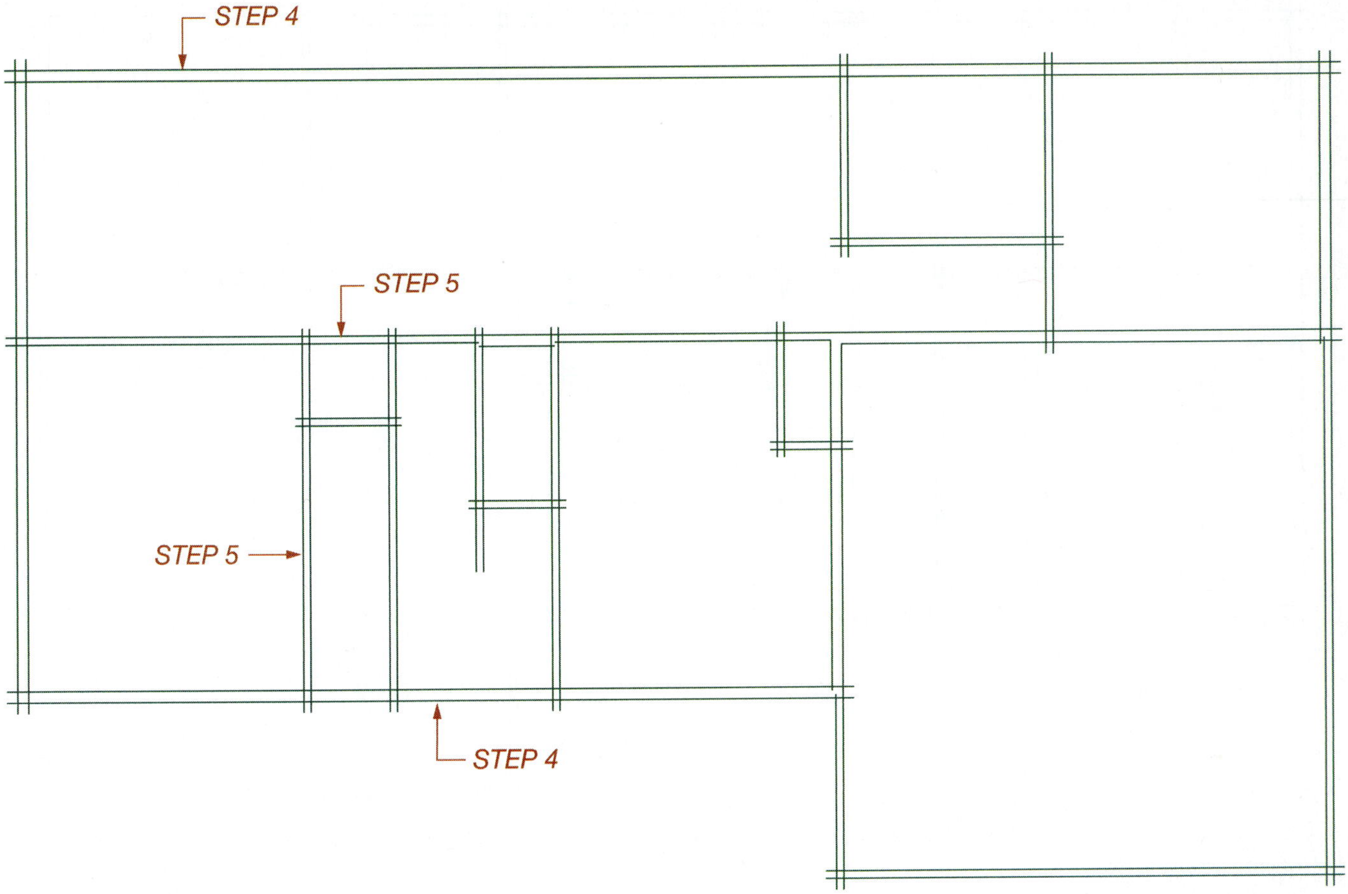

FIGURE 16.3 ■ Steps 4 and 5: Lay out exterior walls and interior partitions.

construction lines need not be erased. Properly drawn construction lines are too light to reproduce using a diazo print, although they may show as very light lines on a photocopy reproduction. Some drafters use a light-blue pencil lead for preliminary layout work because the light blue does not reproduce. Use your 1/4″ = 1′–0″ architect's scale for the floor plan layout. Look at Figure 16.3.

Floor plan walls should be drawn to their exact size if CADD is used. This was explained in detail in Chapter 14. Exterior walls might be drawn if you use 5 1/4″ if 2 × 4 studs + 1/2″ drywall + 1/2″ sheathing + 3/4″ siding. The CADD floor plan should also be drawn on the appropriate layer. The American Institute of Architects (AIA) CAD Layer Guidelines recommended using the A-WALL layer name for architectural walls. This was discussed in Chapter 6. Any necessary construction work can be put on a layer that can be turned off before plotting. This is any nonplot information and construction lines; identified with the minor group −NPLT. For example, construction lines on the architectural floor plans might have the layer name A-FLOR-NPLT.

STEP 5 Lay out all interior walls 4″ (100 mm) thick within the floor plan using construction lines. See Figure 16.3.

Floor-plan walls should be drawn to their exact size if CADD is used. Interior walls might be drawn 4 1/2″ wide if 2 × 4 studs + (2)1/2″ drywall are used. Also use the A-WALL layer unless your instructor or company specifies otherwise.

STEP 6 Block out all doors and windows in their proper locations using construction lines, as shown in Figure 16.4. Be sure that doors and windows are centered within the desired areas.

Doors and windows are inserted at this time if you are using CADD. When using CADD, you will need to either create door and window symbols for insertion or use a customized architectural program. If you insert your own doors and windows, you will need to use commands such as TRIM to remove the unwanted wall lines where the doors are located, and for windows as needed. Architectural programs are available that have any desired door and window combination available for insertion in the drawing. Some of these programs automatically create an inventory of the items that you insert and create the door and window schedules or other material lists. Doors can be drawn on a layer such as A-DOOR and windows on the A-GLAZ layer.

STEP 7 Using construction lines, draw all cabinets. Base kitchen cabinets are 36″ (900 mm), upper cabinets 12″ (300 mm), and bath vanities 22″ (560 mm) deep, as shown in Figure 16.5.

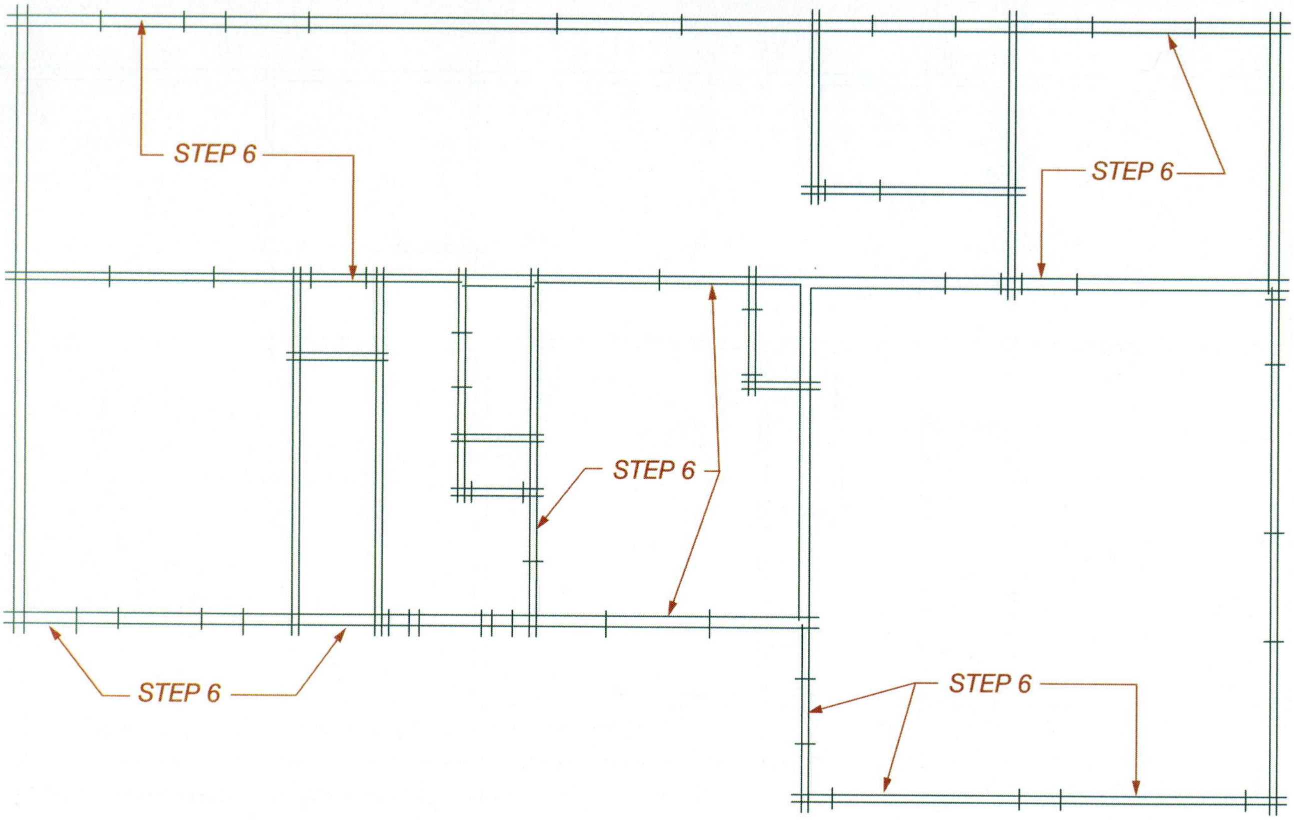

FIGURE 16.4 ■ Step 6: Block out all doors and windows. Doors and windows would be inserted at this time when using CADD.

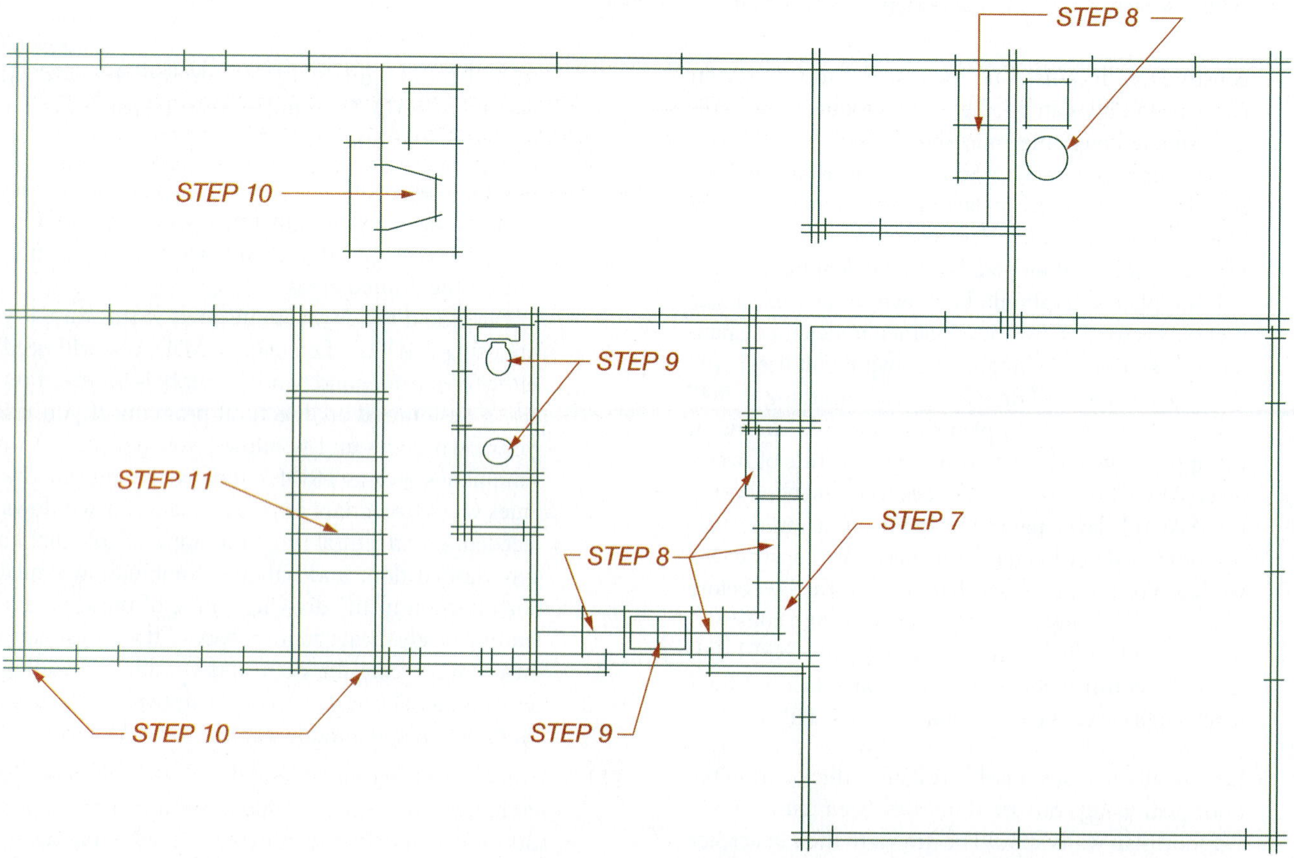

FIGURE 16.5 ■ Steps 7 through 11: Lay out the cabinets, appliances, utilities, plumbing fixtures, fireplace, and stairs.

Cabinets are drawn with CADD by using typical commands such as LINE or are inserted as symbols. Programs that have cabinet symbols available are parametric, which means that you can specify any desired dimensions or shape. Any change to a parametric design automatically updates the entire design to match the change. Cabinets can be placed a layer with casework and manufactured cabinets if they are built by a cabinetmaker and delivered to the project. *Casework* is all of the components that make up the cabinets. The layer name for this application is A-FLOR-CASE, which stands for architectural-floor-casework. If the cabinets are built on the job, they are referred to as *field-built* cabinets and are architectural woodwork. In this case, the layer name is A-FLOR-WDWK, for woodwork.

STEP 8 Draw all appliances and utilities using construction lines. Draw the refrigerator 36″ × 30″, range 30″ wide, dishwasher 24″ wide, furnace 24″ × 30″, water heater 18″ in diameter, and the washer and dryer 30″ × 30″ each. See Figure 16.5. Metric values vary depending on manufacturer and soft or hard conversion.

Appliances and utilities are generally inserted as symbols when CADD is used. Appliances can be placed on the A-FLOR-APPL layer, which stands for architectural-floor-appliances. The water heater can be on the plumbing layer as domestic hot and cold water systems with the P-DOMW layer name.

STEP 9 Draw all plumbing fixtures, sinks, and toilets, as seen in Figure 16.5.

When CADD is used, plumbing fixtures are generally inserted as symbols. Plumbing fixtures can be placed on the plumbing layer as fixtures with the layer name P-FIXT.

STEP 10 Use construction lines to draw a 5′-wide single-face fireplace with a 36″-wide opening and an 18″ × 5′ hearth. Include a 30″ × 18″ barbecue. Add 4″ brick veneer to the front wall, as shown in Figure 16.5.

You can create your own CADD fireplace symbols for insertion or use a custom architectural program. Programs of this type often have automatic fireplace and chimney utilities that allow you to create fireplaces with multiple design options, such as opening, hearth, mantels, and single- or multistory designs. This was discussed in Chapter 14. The fireplace can be placed on an architectural fixed-equipment layer such as A-EQPM-FIXD or a modified floor layer such as A-FLOR-FPLC.

STEP 11 Lightly draw the stairs representation, providing several 10 1/2″-wide treads to a long break line with the necessary handrails or guardrails. See Figure 16.5.

Standard CADD programs can be used to draw stairs by drawing one tread line and then making multiple copies with a command such as ARRAY; or you can create your own stair symbols for insertion. Custom architectural programs are available that automatically calculate and draw straight, L-shaped, U-shaped, and circular or spiral stairs based on the dimensional information and specifications that you enter. This was discussed in Chapter 15. The stairs can be placed on an architectural floor plan layer as stairs using the layer name A-FLOR-STRS. The handrails and guardrails can be on their own layer, A-FLOR-HRAL.

STEP 12 Check your accuracy; then darken all the construction lines that you have drawn. Add all door and window floor-plan symbols. Draw lines and symbols from top to bottom and from left to right if you are right-handed or from right to left if left-handed. To minimize smudging, try to avoid passing your hands and equipment over darkened lines. Darken walls with a mechanical pencil or a 0.7- or 0.9-mm automatic pencil with H or F lead. For inked drawings or CADD, use a 0.7-mm plotter setting to plot walls. All cabinets, appliances, fixtures, and other items are drawn with a 0.5-mm pencil or plotted line width. Lead grades are only suggestions. Look at Figure 16.6. At this point, the floor plan is ready to use as a preliminary drawing for client approval.

This step would have already been completed if you were using CADD.

STEP 13 Lay out all exterior dimension lines using construction lines. Place an overall dimension line on all sides of the residence. The second row of dimensions should go from exterior face to exterior face of major jogs. A third line should be used to go from the exterior face of exterior walls to the center of interior walls. There is a common exception: dimension to the exterior face of the walls between the garage and the residence because of the change in foundation materials. A line should go from each wall to the center of each door and window. There is also a common exception here: you may not need to locate the doors in the garage on the floor plan. These are dimensioned on the foundation plan. The garage doors can be located on the floor plan for reference. Confirm the preferred practice with your company. See Chapter 15 for a complete review of common dimensioning practices and alternatives. Do not place dimension numerals on at this time. Add symbols for door and window schedules. See Figure 16.7.

The dimensions are easy to place and they are very accurate when using CADD. This was discussed in Chapter 15. Dimensions are placed on a separate layer. Dimensions are part of most drawings. For this reason, the dimension layer name is usually an extension modifying the layer name where the dimensions are found. For example, the dimension layer for the

Problem 16–21 *Courtesy SunRidge Design.*

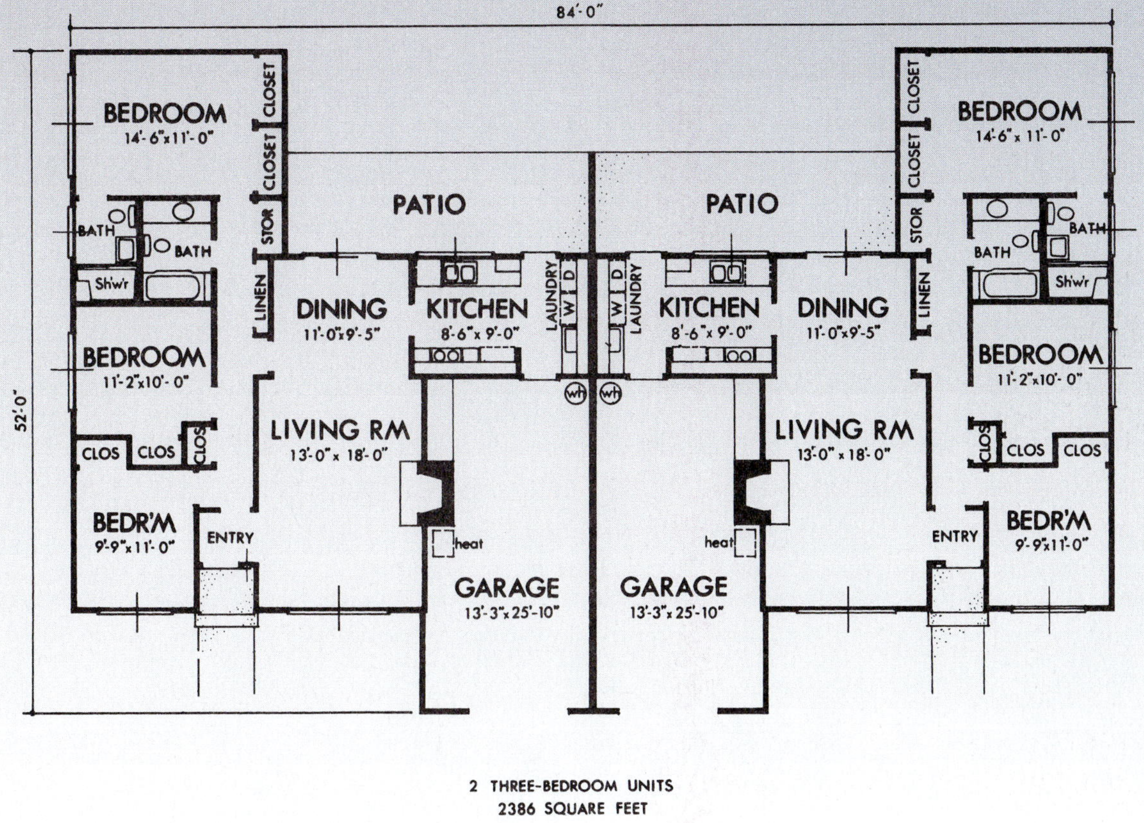

Problem 16–22 *Courtesy Home Building Plan Service, Inc.*

SECTION 5

Supplemental Floor Plan Drawings

17 CHAPTER

Electrical Plans

INTRODUCTION

The electrical plans display all of the circuits and systems to be used by the electrical contractor during installation. Electrical installation for new construction occurs in these three phases.

- Temporary—the installation of a temporary underground or overhead electrical service near the construction site and close to the final meter location.
- Rough in—when the electrical boxes and wiring are installed. Rough-in happens after the structure is framed and covered with roofing. The electrical meter and permanent service are hooked up.
- Finish—the installation of the light fixtures, outlets and covers, and appliances. This is one of the last construction phases.

Electrical plans may be placed on the floor plan with all the other symbols, information, and dimensions (as discussed in Chapters 14 through 16). This is a common practice on simple floor plans where the addition of the electrical symbols does not overcomplicate the drawing. Electrical plans are also drawn on a separate sheet, which displays the floor-plan walls and key symbols, such as doors, windows, stairs, fireplaces, cabinets, and room labels. This chapter takes the latter approach but provides discussion on how both methods might be used. CADD provides an excellent tool for creating electrical plans after the floor plans have been drawn. Floor plan layers that are not needed may be turned off or frozen while the electrical layer is turned on to create the electrical plan. It is easy to create a separate drawing from the key elements of the base drawing in this manner.

Following are definitions of typical terms related to electrical plans and construction. A basic understanding of terminology is important before beginning the electrical drawing.

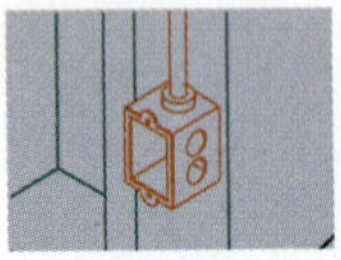

ELECTRICAL TERMS AND DEFINITIONS

It is important to be familiar with key electrical terminology to help you understand this chapter and communicate effectively.

Ampere: A measurement of electrical current flow. Referred to by its abbreviation: *amp* or *amps*.

Boxes: A box equipped with clamps, used to terminate a conduit. Also known as an *outlet box*. Connections are made in the box, and a variety of covers are available for finish electrical. A premanufactured box or casing is installed during electrical rough-in to house the switches, outlets, and fixture mounting. See Figure 17.1 and *electrical work*.

Breaker: An electric safety switch that automatically opens a circuit when excessive amperage occurs. Also referred to as a *circuit breaker.*

Circuit: The various conductors, connections, and devices found in the path of electrical flow from the source through the components and back to the source.

Conductor: A material that permits the free motion of electricity. Copper is a common conductor in architectural wiring.

Conduit: A metal or fiber pipe or tube used to enclose one or more electrical conductors.

Distribution panel: Where the conductor from the meter base is connected to individual circuit breakers, which are connected to separate circuits for distribution to various locations throughout the structure. Also known as a *panel.*

Electrical work: The installation of the wiring and fixtures for a complete residential or commercial electrical system. The installation of the wiring is referred to as the *rough-in.* Rough-in takes place after the framing is completed and the structure is *dried-in.* Dried-in refers to installing the roof or otherwise making the building dry. The light fixtures, outlets, and all other final electrical work are done when the construction is nearly complete. This is referred to as the *finish electrical*. The rough-in and finish electrical make up the electrical work.

Ground: An electrical connection to the earth by means of a rod.

Junction box: A box which protects electrical wiring splices in conductors or joints in runs. The box has a removable cover for easy access. See Figure 17.1.

Lighting outlet: An electrical outlet that is intended for the direct connection of a lighting fixture. See Figure 17.1.

Meter: An instrument used to measure electrical quantities. The electrical meter for a building is where the power enters and is monitored for the electrical utility.

Meter base: The mounting base on which the electrical meter is attached. It contains all of the connections and clamps.

Outlet: An electrical connector used to plug in devices. A *duplex outlet,* with two outlets, is the typical wall plug.

Switch leg: The electrical conductor from a switch to the electrical device being controlled.

Volt: The unit of measure for electrical force.

Watt: A unit measure of power.

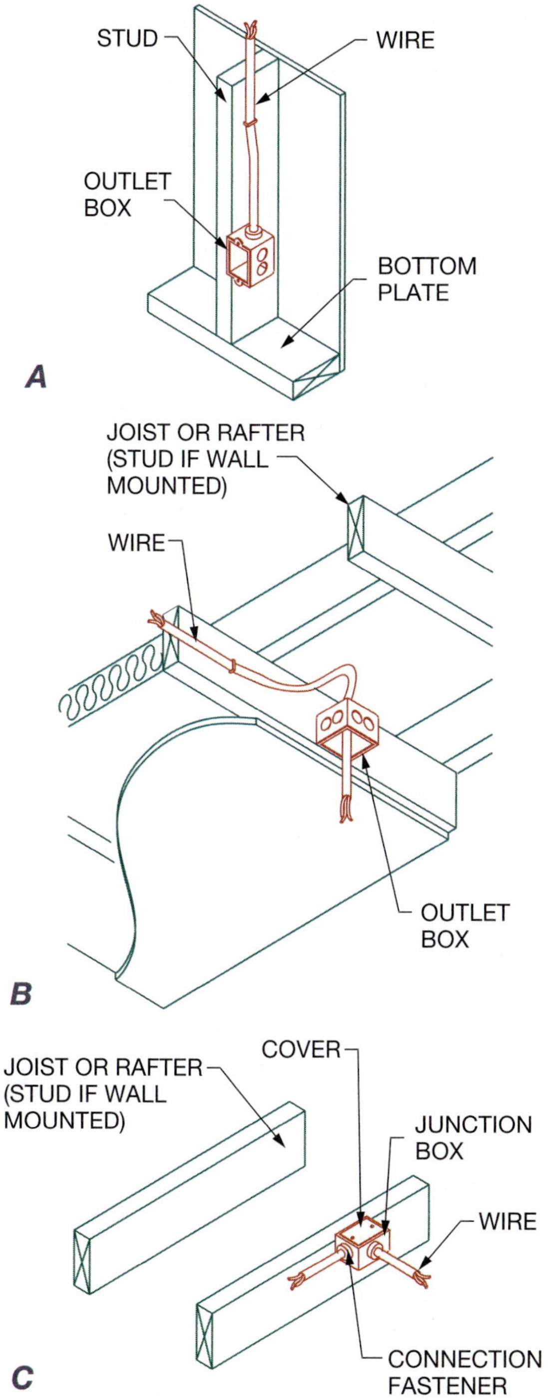

FIGURE 17.1 ■ Installation of outlets and junction box. (*A*) Outlet box. (*B*) Lighting outlet. (*C*) Junction box.

ELECTRICAL CIRCUIT DESIGN

The design of the electrical circuits in a home is important because of the number of electrical appliances that make modern living enjoyable. Discuss with the client any anticipated needs for electric energy in the home. Such a discussion includes the intended use of each of the rooms and the potential placement of furniture in them. Try to design the electrical circuits so there are enough outlets and switches at convenient locations. National Electrical Code requirements dictate the size of some circuits and the placement of certain outlets and switches within the home.

In addition to convenience and code requirements, you have to consider cost. A client may want outlets 4′ (1220 mm) apart in each room, but the construction budget may not allow such a luxury. Try to establish budget guidelines and then work closely with the client to achieve desired results. While switches, convenience outlets, and electrical wire and boxes may not seem to cost very much individually, their installation in great quantity could be costly. In addition, attractive and useful light fixtures required in most rooms can also be very expensive.

Code Requirements

Common codes to consider when designing electrical circuits based on convenience and national and local codes include the following:

- Duplex convenience outlets (wall plugs) should be a maximum of 12′ (3657 mm) apart. Closer spacing is desirable, although economy is a factor.
- Duplex outlets should be no more than 6′ (1820 mm) from an opening.
- Each wall over 2′ (610 mm) in length must have an outlet.
- Consider furniture layout, if possible, so duplex outlets do not become inaccessible behind large pieces of furniture.
- Place a duplex outlet next to or behind a desk or table where light is needed or near a probable reading area.
- Place a duplex outlet near a fireplace, for an adjacent table light, for a vacuum, or for fireplace maintenance.
- Place a duplex outlet in a hallway, for a vacuum.
- Duplex outlets should be placed close together in kitchens and installed where fixed and portable appliances will be located for use. Outlets are required every 4′ (1200 mm) of counter space, no more than 2′ (600 mm) from a corner, end of counter, or appliance. Some designs include an outlet in a pantry for use with portable appliances.
- Kitchen, bathroom, laundry, and outdoor circuits are required to be protected by a ground-fault circuit interrupter (GFI or GFCI), which trips a circuit breaker when there is any unbalance in the circuit current. The standard breaker condition is 0.005 amp fault in 1/40 second. This protection is required for any electric fixture within 6′ of water. Fixtures and appliances produce a potential hazard for electrocution.
- Outlets in garages and outbuildings must be GFCI.
- All exterior outlets must be waterproof and GFCI.
- Each bath sink or makeup table should have a duplex outlet.

- Each enclosed bath or laundry (utility) room should have an exhaust fan. Rooms with openable windows do not require fans, although fans are desirable.
- A lighting outlet with light fixture is necessary outside all entries and exterior doors.
- A waterproof duplex outlet should be placed at a convenient exterior location, such as a patio. This exterior outlet must have a GFCI if within direct grade level access.
- Ceiling lights are common in children's bedrooms for easy illumination. Master bedrooms may have a switched duplex outlet for a bedside lamp.
- The kitchen may have a light over the sink, although in a very well-lighted kitchen, this may be an extra light.
- Bathroom lights are commonly placed above the mirror at the vanity and located in any area that requires additional light. In a large bath, a recessed light should be placed in a shower or above a tub.
- Place switches in a manner that provides the easiest control of the lights.
- Locate lights in large closets or any alcove or pantry that requires light.
- Light stairways well, and provide a switch at each level.
- Place lights and outlets in garages or shops in relationship to their use. For example, a welder or shop equipment requires 220-volt outlets.
- Place exterior lights properly to illuminate walks, drives, patios, decks, and other high-use areas.

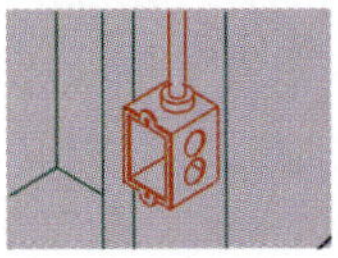

ENERGY CONSERVATION

Energy efficiency is a very important consideration in today's home design and construction market. There are a number of steps that can easily help contribute to energy savings for the homeowner at little additional cost. Some of these items will have to be explained by specific or general notes or included in construction specifications; other methods must be defined with detail drawings. Energy-efficient considerations related to electrical design include the following:

- Keep electrical outlets and recessed appliances or panels to a minimum at exterior walls. Any units recessed into an exterior wall can eliminate or severely compress the insulation and reduce its insulating value.
- Fans or other systems exhausting air from the building should be provided with back draft or automatic dampers to limit air leakage.
- Timed switches or humidistats should be installed on exhaust fans to control unnecessary operation.
- Electrical wiring is often located in exterior walls at a level convenient for electricians to run wires. This practice causes insulation to become compacted, and a loss of insulating value results. Wires could be run along the bottom of the studs at the bottom plate. Preparing the studs is easily accomplished during framing by cutting a V groove in the stud bottoms while the studs are piled at the site after delivery.
- Select energy-efficient appliances such as a self-heating dishwasher or a high-insulation water heater. Evaluate the vendor's energy statement before purchase.
- Use energy-saving fluorescent lighting fixtures where practical, such as in the kitchen, laundry, utility, garage, or shop.
- Fully insulate above and around recessed lighting fixtures. Verify code and vendors' specifications for this practice.
- Carefully caulk and seal around all light and convenience outlets. Also, caulk and seal where electrical wires penetrate the top and bottom plates.
- Use recessed lights that are IC (insulation cover) rated. This allows you to insulate around and over the recessed light to help avoid heat loss.

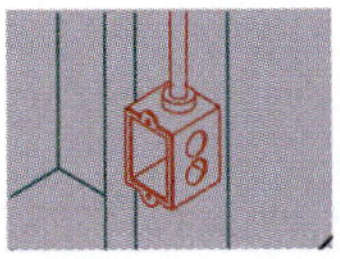

HOME AUTOMATION

Many modern homes are being built with automation systems. This technology is rapidly changing and requires that you continuously research the available products. The installation of home automation systems allows the builder to sell the same-square-footage home for more and increases marketability. *Automation* is a method or process of controlling and operating mechanical devices by other than human power. Such operations include the computerized control of the heating, ventilating, air conditioning, landscape sprinkling systems, lighting, and security systems. Among the many home automation systems available are the following:

- Entertainment centers, to provide an in-home theater. These are fully automatic and contain the best technology in picture and sound performance. For example, pressing one button automatically dims the lights, opens the curtain, and begins the movie.
- Computerized programming of house functions from a personal computer. These systems allow the computer user to set and monitor a variety of electrical circuits throughout the home, including security, sound, yard watering, cooking, and lighting.
- Residential elevators, to transport people between floors who require a wheelchair or have other disabilities.

The design and drafting of the home automation system may be done by the product supplier or the architect. Electrical symbols and specific notes are placed on the floor plan or on separate drawings used for construction purposes.

Structured Wiring

As home computer use and Internet access become common in nearly every home, structured wiring systems are part of the electrical work needed in the architectural design. *Structured wiring systems* are high-speed voice and data lines and video cables wired to a central service location. These wires and cables optimize the speed and quality of various communication signals coming into and going out of the house. This kind of wiring is typically used in commercial construction projects and is the future trend for residential design. In a system of this type, each electrical outlet, telephone jack, or computer port has a dedicated line back to the central service location. The central service location allows the wires to be connected as needed for a network configuration or for dedicated wiring from the outside. High-quality structured wiring systems use network connectors and parallel circuits, because conventional outlets and series circuits degrade the communication signal. A *parallel circuit* is an electrical circuit that contains two or more paths for the electricity or signal to flow from a common source. A *series circuit* is a circuit that supplies electricity or a signal to a number of devices connected so that the same current passes through each device in completing its path to the source. Structured wiring systems allow the use of fax, multiline telephone, and computers at the same time. Additional applications include digital satellite system (DSS), digital broadcast system (DBS), stereo audio, and closed-circuit security systems.

ELECTRICAL DESIGN CONSIDERATIONS

The discussion and many examples in this chapter provide you with the minimum requirements for the electrical design. You need to meet minimum code requirements while also considering the needs and preferences of the occupants. One way to design the electrical layout is to pretend that you are living in a home, or working in a commercial building, and plan a system based on your convenience and daily activities. The following list provides some of the basic design features to consider when creating the electrical plan.

Entry: Every entry shall have at least one lighting outlet controlled by a switch on the inside of the building. There shall be at least one weatherproof ground-fault circuit interrupter outlet (GFCI).

Entry foyer: There is generally a light fixture centered in the entry foyer, with a single switch or two switches if necessary. Depending on the size of the foyer, there may be additional ceiling or wall-mounted lighting outlets. At least one duplex convenience outlet, with more spaced 12′ apart if needed.

Patios and porches: Patios and porches are generally designed with an entry and should have the same requirements. Additionally, these areas need adequate lighting. One lighting outlet per 150 square feet should be considered. The number of duplex convenience outlets in an area depends on its use. A light and weatherproof GFCI outlet should be placed at a sink, serving counter, or outside cooking area.

Living area: Living rooms and great rooms should have at least one switch-controlled duplex convenience outlet. Additional outlets are needed for convenience and to meet code requirements. There is generally no central ceiling lighting outlet, but there are often accent lights such as recessed ceiling or wall lights.

Dining room: Dining rooms and eating nooks should have a switch-controlled centrally located ceiling light. Outlets are needed for convenience and to meet code requirements.

Kitchen: The kitchen is a place where lighting and convenience outlets are very important. There should be switch-controlled centrally located lighting. Many designs use ceiling fluorescent fixtures or adequately spaced recessed ceiling lights. An additional switch-controlled light is often placed over the sink and other work areas. Duplex convenience outlets should be placed at least 4′ apart, no more than 2′ from a fixture or appliance, and with one at each end of an island or peninsula. The space between outlets along a wall should not exceed 24″ at each wall counter space of 12″ or more in depth. Each appliance should have an outlet. This includes refrigerator, range, hood light and fan, oven, microwave, dishwasher, and trash compactor. There should also be a small appliance outlet in every kitchen, pantry, breakfast room, dining room, or similar area. A switch-controlled garbage disposal may be used. Custom kitchens can also have under- and over-cabinet lighting and portable appliance compartments or garages. An exhaust fan that is vented to the outside is required. The exhaust fan can be in combination with a hood and light over the range or adjacent to the range.

Bedrooms: Bedrooms often have a switch-controlled central ceiling light. Bedrooms may have a recessed ceiling light next to a wardrobe closet. The master bedroom may use a switch-controlled duplex convenience outlet rather than the ceiling light. Additional duplex convenience outlets should be spaced by code requirements or better. Walk-in wardrobes should have a switch-controlled ceiling light.

Bathrooms: The bathrooms have at least one lighting outlet in the ceiling or wall above the sink mirror. There should be one light over each sink and a GFCI outlet next to each sink and at 48″ above the floor and not more than 36″ from the edge of the fixture. Additional convenience outlets can be placed around the room to meet personal requirements. Outlets shall not be installed face-up in the work surface or cabinet top in a bathroom sink location. If fixtures are used in an enclosed shower or spa, they should be vapor-proof. Ceiling lights are often recessed and should be placed in areas such as over the water closet, tub, spa, or sauna as needed for adequate lighting. An exhaust fan that is vented to the outside is required, unless an openable window is used. Some owners prefer an exhaust fan even if an openable window is provided.

Laundry, utility room: The laundry room should have at least one switch-controlled ceiling light, with other lights considered over work areas. Washer and dryer outlets are required along with adequate duplex convenience outlets. The appliance outlet shall not be more than 6′ from the appliance location.

Office; recreation or hobby room: An office or other auxiliary room should have ceiling lighting placed over planned work areas, and there should be duplex convenience outlets placed as needed for the intended purpose and to meet codes. If computers are used, dedicated outlets should be considered.

Hallways: Ceiling lights should be placed no more than 12′ apart, with a switch controlling them from each end of the hall. Duplex convenience outlets should be at least 12′ apart.

Stairs: Place at least one ceiling light at each floor or landing and have additional lights as needed to light the stairs adequately. Convenience outlets are not required but may be considered, especially if landings are involved.

Garage: At least one ceiling outlet for every two cars, but one light per bay is better. Use at least one weatherproof GFCI duplex convenience outlet of 125-volt, single phase, 15–20 amps. Provide additional outlets and lights as needed for work areas and other conveniences. An outlet and controls should be wired for every electric garage door opener.

Outdoor outlets: At least one weatherproof outlet shall be installed at the front and back of each home with direct access from grade and not exceed 6′–6″ above grade.

Crawl spaces: All outlets in crawl spaces shall be GFCI, 125-volt, single phase, and 15 or 20 amps.

Smoke detectors: Smoke detectors are required at each sleeping area and on each additional story and basement of a living unit. The smoke detectors are generally placed in the ceiling immediately outside the entrances to bedrooms or above a stairs leading to adjacent bedrooms or living areas. The type of smoke detectors can be indicated in the specifications. Smoke detectors that are wired directly into the electrical system are preferred, and some installations use additional battery backup smoke detectors.

Telephones: Careful consideration should be given to the locations where telephones are typically used. Common locations include the kitchen, master bedroom, laundry room, and office. Telephone jacks should be placed in every desired location or rough-wired for future installations. *Rough-wire* means to run wiring in the walls during construction for possible future hookup and use. Another term for *rough-wired* is *prewired.* Wiring during construction is always less expensive and easier than wiring after the building is occupied.

Television, cable, stereo, security: Wiring for planned television, stereo, and security systems should be designed into the project. These items should also be prewired for installation during or after construction. The type of television wiring depends on the hookup to an antenna, cable, or satellite. The location of television outlets depend on the needs of the occupants, but outlets are generally provided in the living room, family room, recreation room, and master bedroom, and in other rooms as needed. Stereo installations can be wired separately for sound throughout the home or associated with the cable television. The sound system can be connected to a central system with AM/FM, CD, and other devices and can be part of an intercom system. The security system should be wired throughout the building and should be designed in cooperation with a security expert to provide the best possible installation needed by the owner. Security systems can be internal or connected to a monitoring station such as a private security provider or public police department.

Computers and peripherals: Wiring for computers and peripherals is becoming an important part of the electrical system, as explained earlier in this chapter. *Peripherals* are items outside of the main computer box. These items can include printers, modems, and fax machines. Refer to the earlier discussion covering home automation and structured wiring systems.

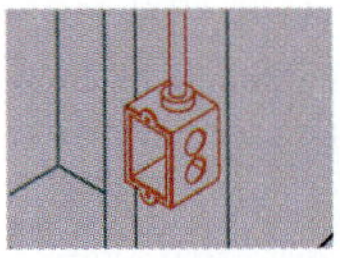

UNIVERSAL ELECTRICAL INSTALLATIONS

The design of the electrical system to accommodate people of all ages and possible disabilities should be considered. The following are some possible considerations.

Switch locations: Place light switches 2′–6″ (760 mm) above the floor for easy use by children and people in wheelchairs. Additionally, provide an adequate turning radius next to the switch for wheelchair movement. More switches than what is considered normal should be considered for access by disabled people. Switches should be the type that operate by touch or may even be sound or motion-activated and on a timer if necessary. A master switch that controls all of the lights in a home might be located in a convenient and commonly used place, such as in the master bedroom. Place the distribution panel in an easily accessible location and install the panel at a height lower than normal, such as 4′–6;″ (1320 mm) for wheelchair access.

Convenience outlets: Bathroom vanity outlets should be placed in the side wall for easy access. Place outlets on the front or top surface of the base kitchen cabinets in convenient locations. Provide more than the normal number of outlets. For example, a spacing of 8′ (2400 mm) is preferred over the code requirement of 12′ (3600 mm). Outlets should be at least 15″ (380 mm) above the floor rather than the normal 12″ (300 mm). Determine the owner's need for special medical equipment, communication systems, and emergency alarm circuits. Provide a special outlet in a ventilated area for a homeowner to charge an electric wheelchair battery.

Lighting: Provide specifications for wall-mounted lights to be placed within reach and ceiling fixtures that pull down for easy access from a wheelchair. Extra lighting may be considered,

especially in locations where a disabled person might have difficulty seeing, such as a shower, bathroom, and kitchen. Provide extra lighting around stairs and landings. Additional lighting should be considered as occupants grow older. Lights can be controlled with dimmer switches as needed to provide varying intensity. Fluorescent lights should be avoided when designing a home for people with epilepsy, Alzheimer's disease, hyperactivity, hearing aids, and cataracts. Specify natural light tubes when fluorescent lights are used. Track lighting should be considered when it is desirable to add or remove lights or change the angle of lights.

Communications: Provide more than the normal number of telephone jacks when designing a home for disabled people. Also provide additional television and audio outlets if they are preferred by the owner. An intercom system is recommended for communication within the home.

ELECTRICAL SYMBOLS

Electrical symbols are used to show the lighting arrangement desired in the home. This includes all switches, fixtures, and outlets. Lighting fixture templates are available with a variety of fixture shapes for drafting convenience. Switch symbols are generally lettered freehand unless CADD is used.

All electrical symbols should be drawn with circles 1/8″ (3 mm) in diameter, as seen in Figure 17.2. The electrical layout should be subordinate to the plan and should not clutter or detract from the other information. All lettering for switches and other notes should be 1/8″ (3 mm) high, depending on space requirements and office practice.

Switch symbols are generally drawn perpendicular to the wall and are placed to read from the right side or bottom of the sheet. Look at Figure 17.3. Also, notice that the switch relay should intersect the symbol at right angles to the wall, or the relay may begin next to the symbol. Verify the preference of your instructor or employer, and do not mix methods.

Figure 17.4 shows several typical electrical installations with switches to light outlets. The switch leg or electrical circuit line may be drawn with an irregular (French) curve, or as a curved dash line when CADD is used.

When a fixture with special characteristics is required, such as a specific size, location, or any other specification, a local note that briefly describes the situation may be applied next to the outlet, as shown in Figure 17.5.

Figure 17.6 shows some common errors related to the placement and practice of electrical floor plan layout.

Figure 17.7 shows some examples of maximum spacing recommended for installation of wall outlets.

Figure 17.8 shows some examples of typical electrical layouts. Figure 17.9 shows a bath layout. Figure 17.10 shows a typical kitchen electrical layout.

ELECTRICAL WIRING SPECIFICATIONS

The following specifications describing the installation of the service entrance and meter base for a residence have been adapted from information furnished by Home Building Plan Service.

Service Entrance and Meter Base Installation

Before the installation of the service entrance and meter base, the following questions must be resolved:

1. *What is the service capacity to be installed?* For capacity of service, the average single-family residence should be equipped with a 200-amp service entrance. If heating, cooking, water heating, and similar heavy loads are supplied by energy sources other than electricity, the service entrance conductors may be sized for 100 amps, provided local codes are not violated. Some local codes require a minimum capacity of 200 amps for each single dwelling. Local inspection authorities and the power company serving the area can be of assistance if difficulty is encountered in sizing the service entrance.
2. *Where is the service entrance to be located?* The service entrance location depends on whether the power company serves the houses in the area from an overhead or underground distribution system. The underground service entrance is the most desirable from the standpoint of both aesthetics and reliability. However, if the power company services the area from an overhead pole line, an underground service entrance may be impractical, prohibitively expensive, or even impossible.
3. *Where will the meter base be located?* The meter base must always be mounted so the meter socket is on the exterior of the house, preferably on the side wall of a garage. The meter base should not be installed on the front of the house for aesthetic reasons. If the house does not have a garage, the meter base should be located on an exterior side wall as near to the distribution panel as practical.
4. *Where will the distribution panel be located?* The preferred location for the distribution panel is on an inside garage wall, close to the heaviest electrical loads, such as a range, clothes dryer, or electric furnace. The panel should be mounted flush unless the location does not permit this. If the house does not have a garage, the panel should be located in a readily accessible area such as a utility room or kitchen wall.

In some localities the electrical power billing structure requires two meters. For example, lighting and general-use energy may be billed at a rate different from energy used for space and heating water. In such a situation, provide two meter bases and two distribution panels to suit requirements.

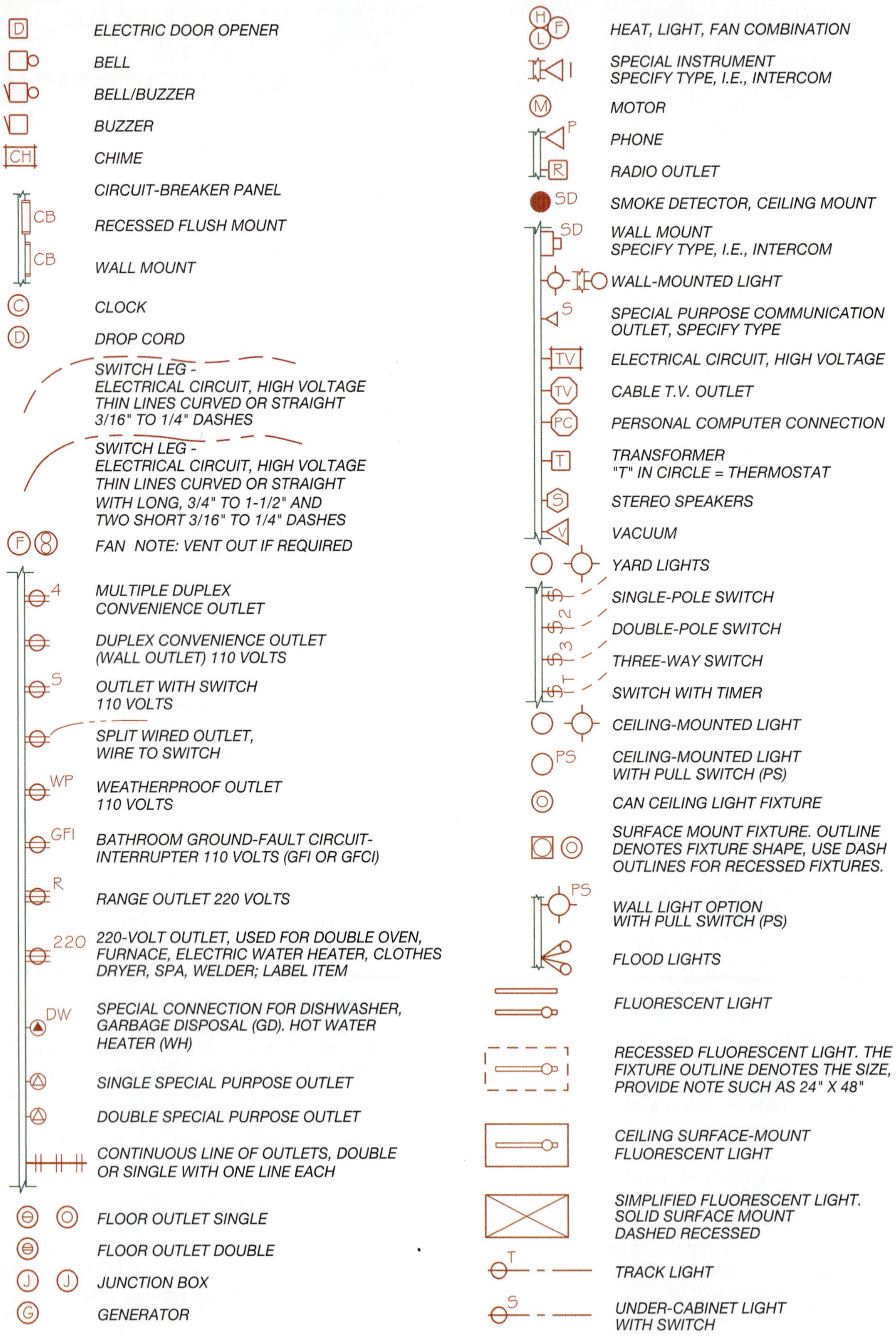

FIGURE 17.2 ■ Common electrical symbols.

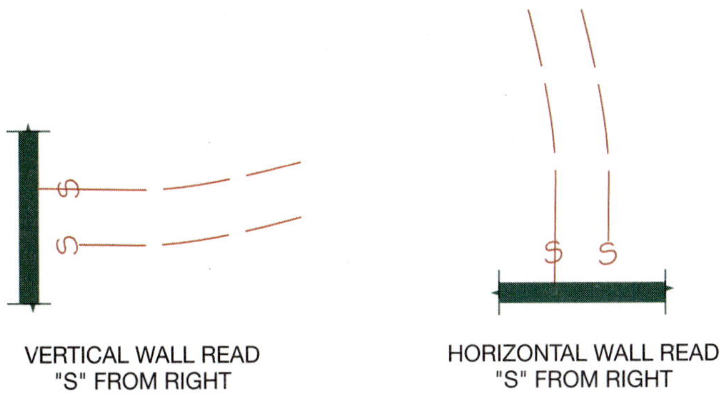

FIGURE 17.3 ■ Placement of switch symbols.

METRICS IN ELECTRICAL INSTALLATIONS

Electrical conduit designations are expressed in millimeters. *Electrical conduit* is a metal or fiber pipe or tube used to enclose a single or several electrical conductors. Electrical conduit is produced in decimal inch dimensions and is identified in nominal inch sizes. *Nominal size* is referred to as the conventional size; for example, a 0.500″ pipe has a 1/2″ nominal size. The actual size of a conduit will remain in inches but will be labeled in metric:

INCH	METRIC (mm)	INCH	METRIC (mm)
1/2	16	2 1/2	63
3/4	21	3	78
1	27	3 1/2	91
1 1/4	35	4	103
1 1/2	41	5	129
2	53	6	155

Existing American Wire Gage (AWG) sizes will remain the same without a metric conversion. The diameter of wires conforms to various gaging systems. The AWS is one system for the designation of wire sizes.

STEPS IN DRAWING THE ELECTRICAL PLAN

The electrical plan is prepared as a separate floor-plan drawing, labeled ELECTRICAL PLAN, for complex homes, or combined with all other floor-plan information for basic one-story homes. When drawing the electrical plan with all other floor-plan information, follow the next steps after placing everything on the drawing, as explained in Chapter 16. If you

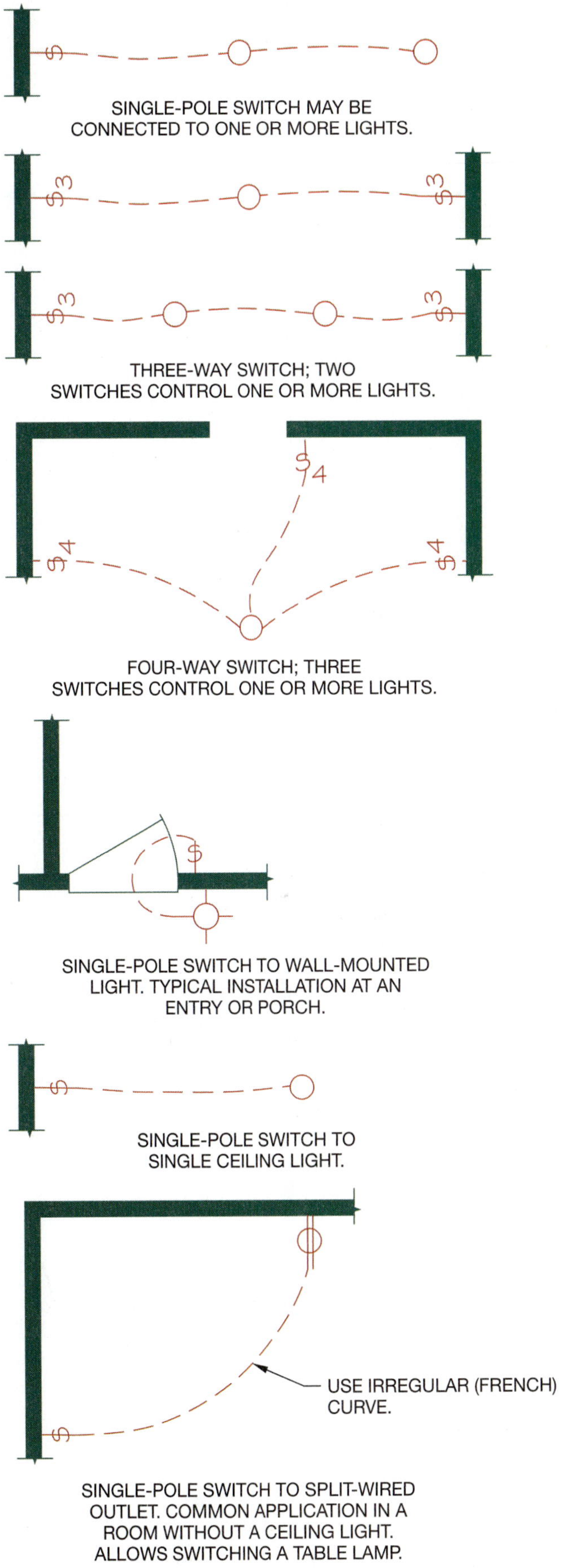

SINGLE-POLE SWITCH MAY BE
CONNECTED TO ONE OR MORE LIGHTS.
THREE-WAY SWITCH; TWO
SWITCHES CONTROL ONE OR MORE LIGHTS.
FOUR-WAY SWITCH; THREE
SWITCHES CONTROL ONE OR MORE LIGHTS.
SINGLE-POLE SWITCH TO WALL-MOUNTED
LIGHT. TYPICAL INSTALLATION AT AN
ENTRY OR PORCH.
SINGLE-POLE SWITCH TO
SINGLE CEILING LIGHT.
USE IRREGULAR (FRENCH)
CURVE.
SINGLE-POLE SWITCH TO SPLIT-WIRED
OUTLET. COMMON APPLICATION IN A
ROOM WITHOUT A CEILING LIGHT.
ALLOWS SWITCHING A TABLE LAMP.

FIGURE 17.4 ■ Typical electrical installations.

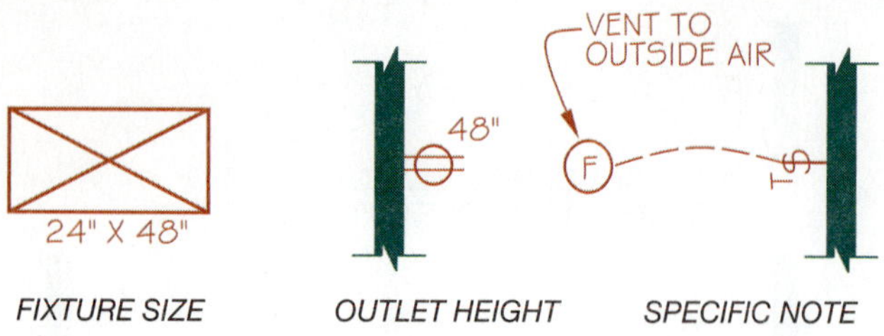

FIGURE 17.5 ■ Special notes for electrical fixtures.

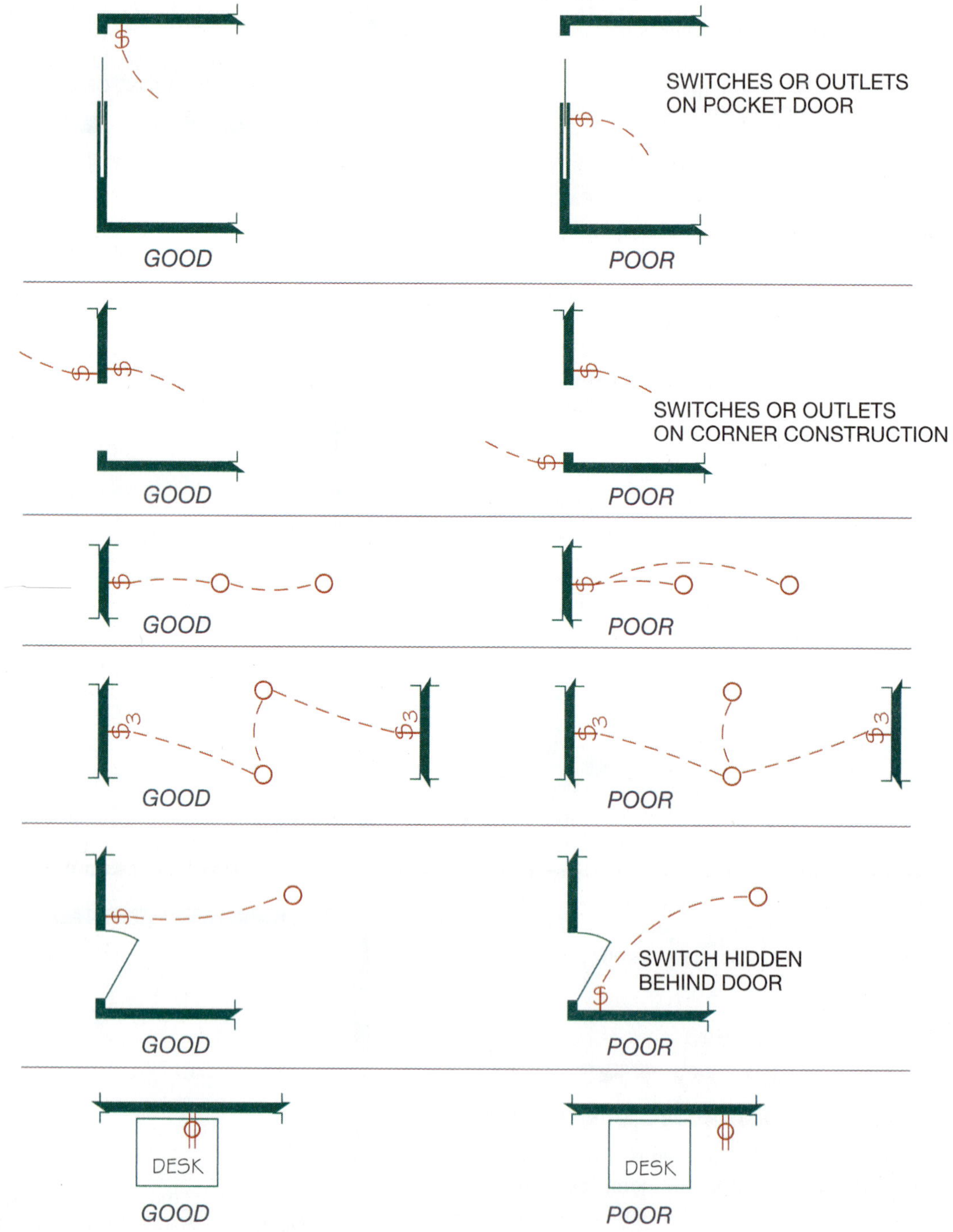

FIGURE 17.6 ■ Good and poor electrical layout techniques.

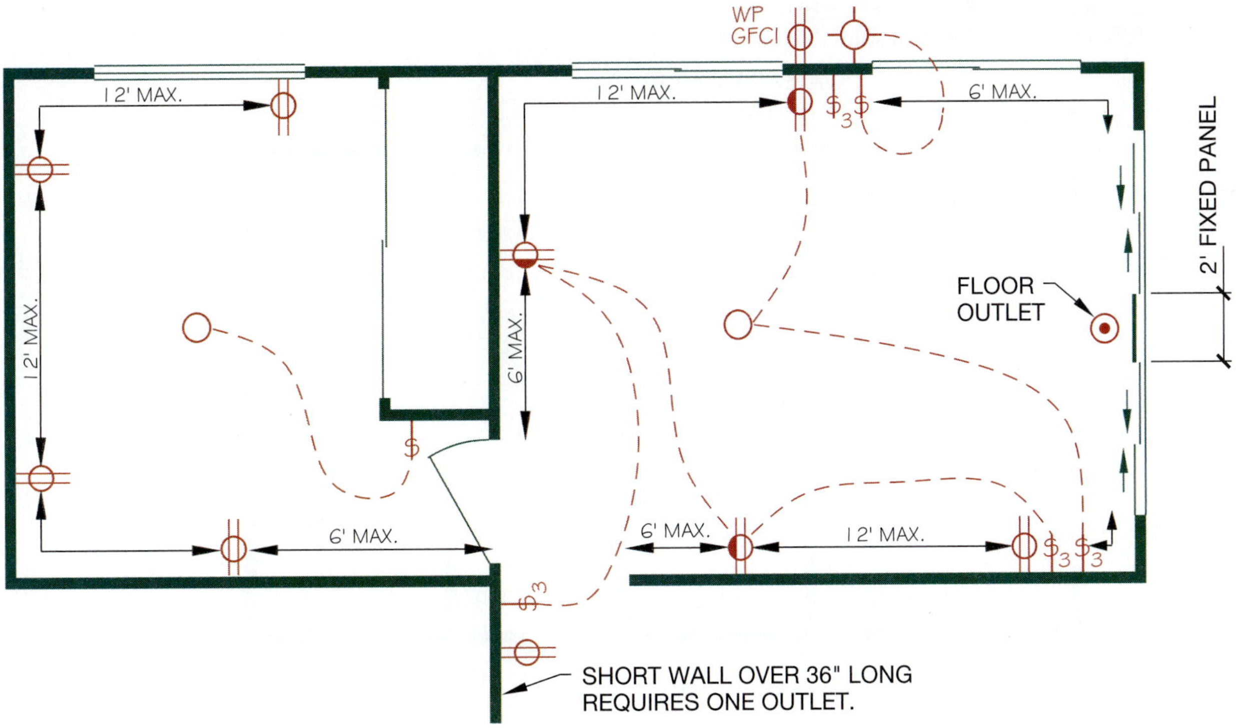

FIGURE 17.7 ■ Maximum spacing requirements.

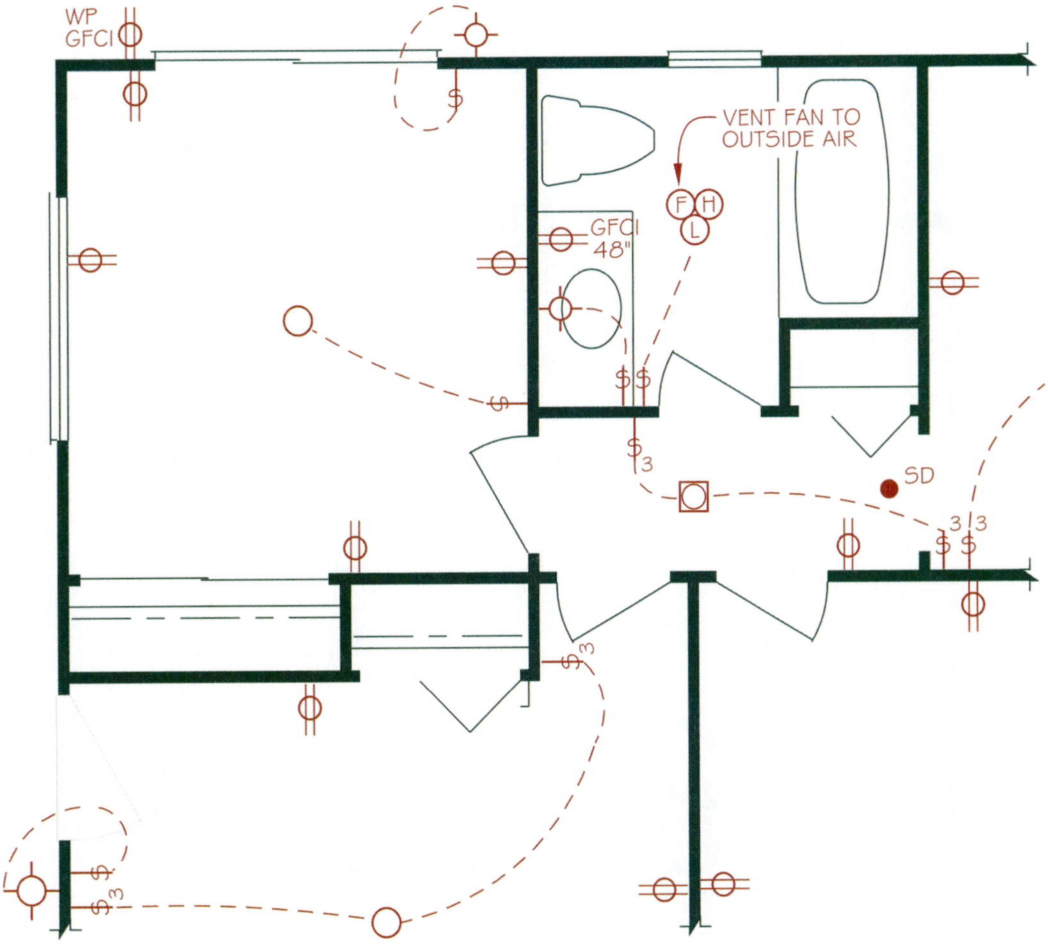

FIGURE 17.8 ■ Typical electrical layouts.

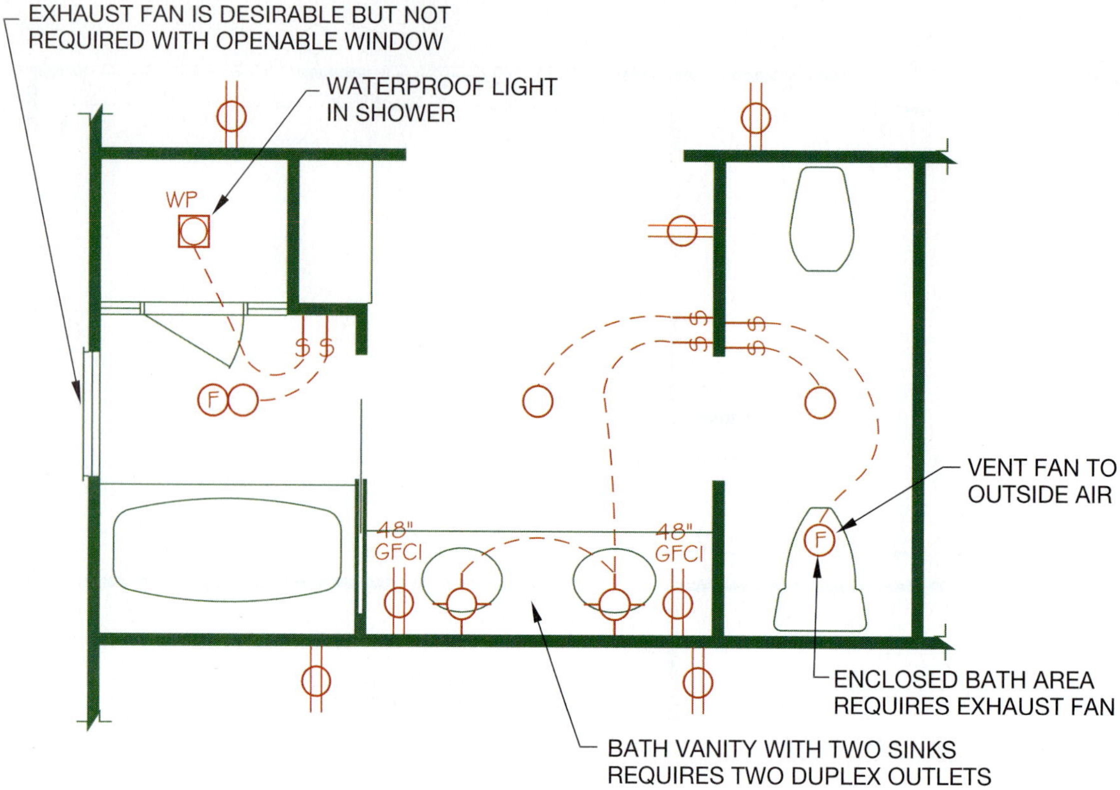

FIGURE 17.9 ■ Bath electrical layout.

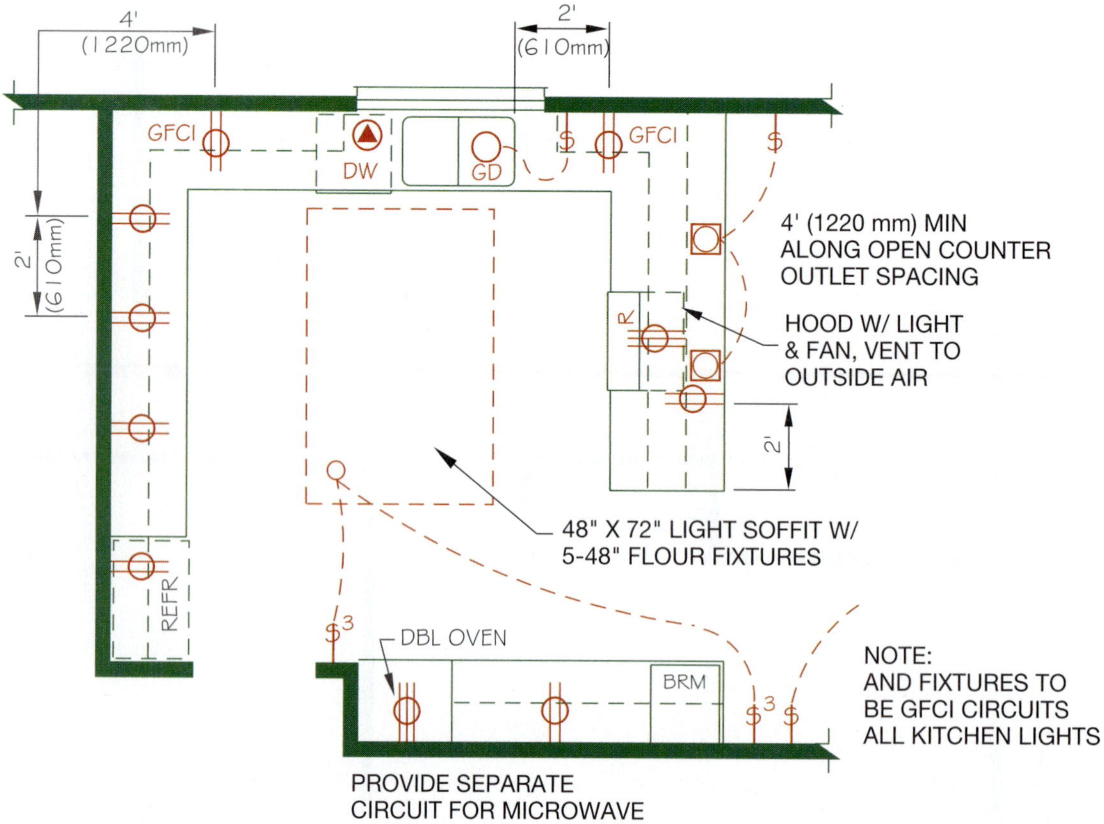

FIGURE 17.10 ■ Kitchen electrical layout.

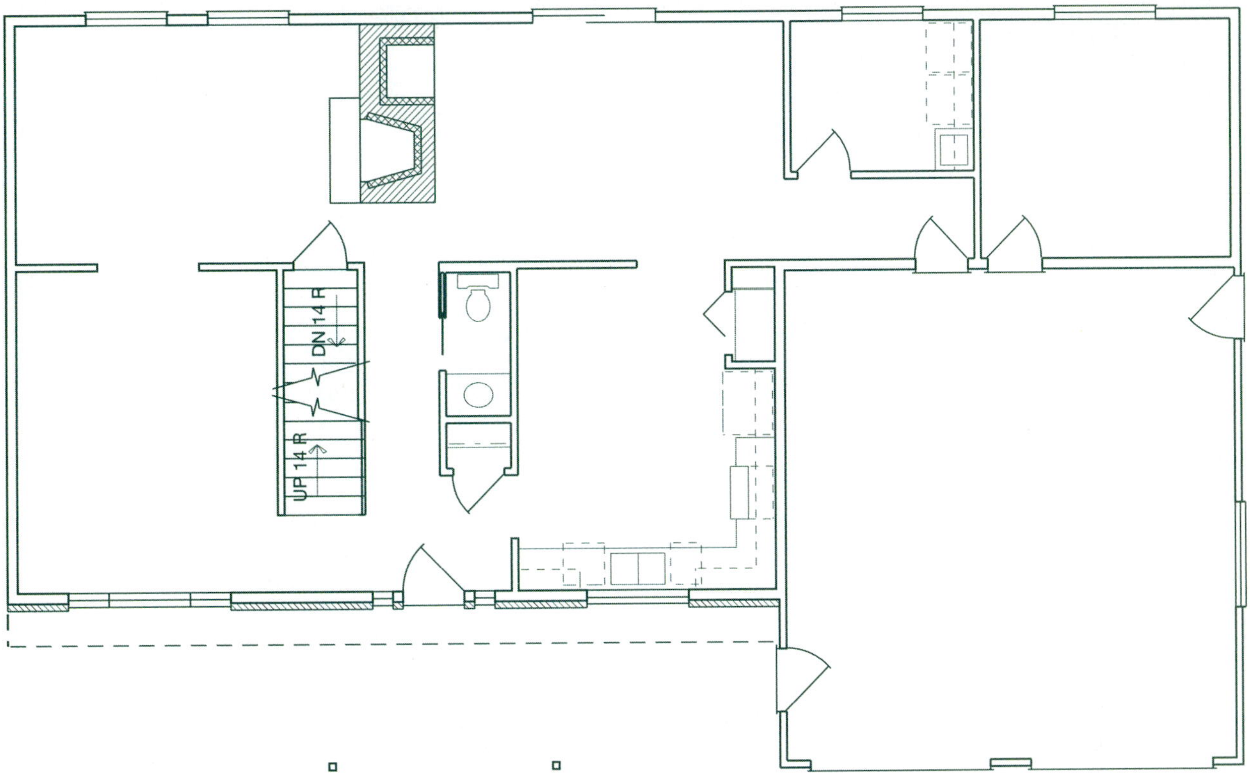

FIGURE 17.11 ■ Base sheet for the floor plan used in Chapter 16.

are creating a separate electrical plan, draw the plan as described in the following steps or use the CADD layering system. To do this, first begin with a base drawing of the floor plan, with all of the walls, doors, windows, stairs, cabinets, and the fireplace; the basic room titles are optional. A base drawing for the residence that was first used in Chapter 16 is shown in Figure 17.11.

STEP 1 Letter all switch locations and draw all light fixture locations. Use a 1/8″ diameter circle (6″ at a scale of 1/4″ = 1′–0″) for the light fixture symbols. If electrical symbols are too large, they distract from the appearance of the drawing. If they are too small, they can be difficult to read. CADD electrical symbol libraries provide quick insertion of these symbols. Look at Figure 17.12. When a CADD system is used, each symbol and feature on the electrical plan can be placed on an individual layer. If you are using the American Institute of Architects (AIA) layering system, the following layer names can be applied to your electrical plan:

LAYER NAME	DESCRIPTION
E-LITE	Lighting
E-POWR	Power
E-CTRL	Electric control systems
E-GRND	Ground system
E-AUXL	Auxiliary systems
E-LTNG	Lighting protection system
E-FIRE	Fire alarm system
E-COMM	Telephones and communication systems
E-DATA	Data systems
E-SOUN	Sound system
E-TVAN	Television antenna system
E-CCTV	Closed-circuit television
E-NURS	Nurse call system
E-SERT	Security system
E-PGNG	Paging system
E-BELL	Bell system
E-CLOCK	Clock system

STEP 2 Draw electrical circuits or switch legs from switches to fixtures using a dashed arc line. For manual drafting, use an irregular curve and a sharp mechanical pencil or a 0.5-mm automatic pencil with 2H or H lead. For CADD applications, use either a dashed or a hidden line type and the ARC command. Figure 17.12 shows Step 2.

STEP 3 Place all electrical outlets, such as duplex convenience, range, and television. Figure 17.12 shows the placement of items for Step 3.

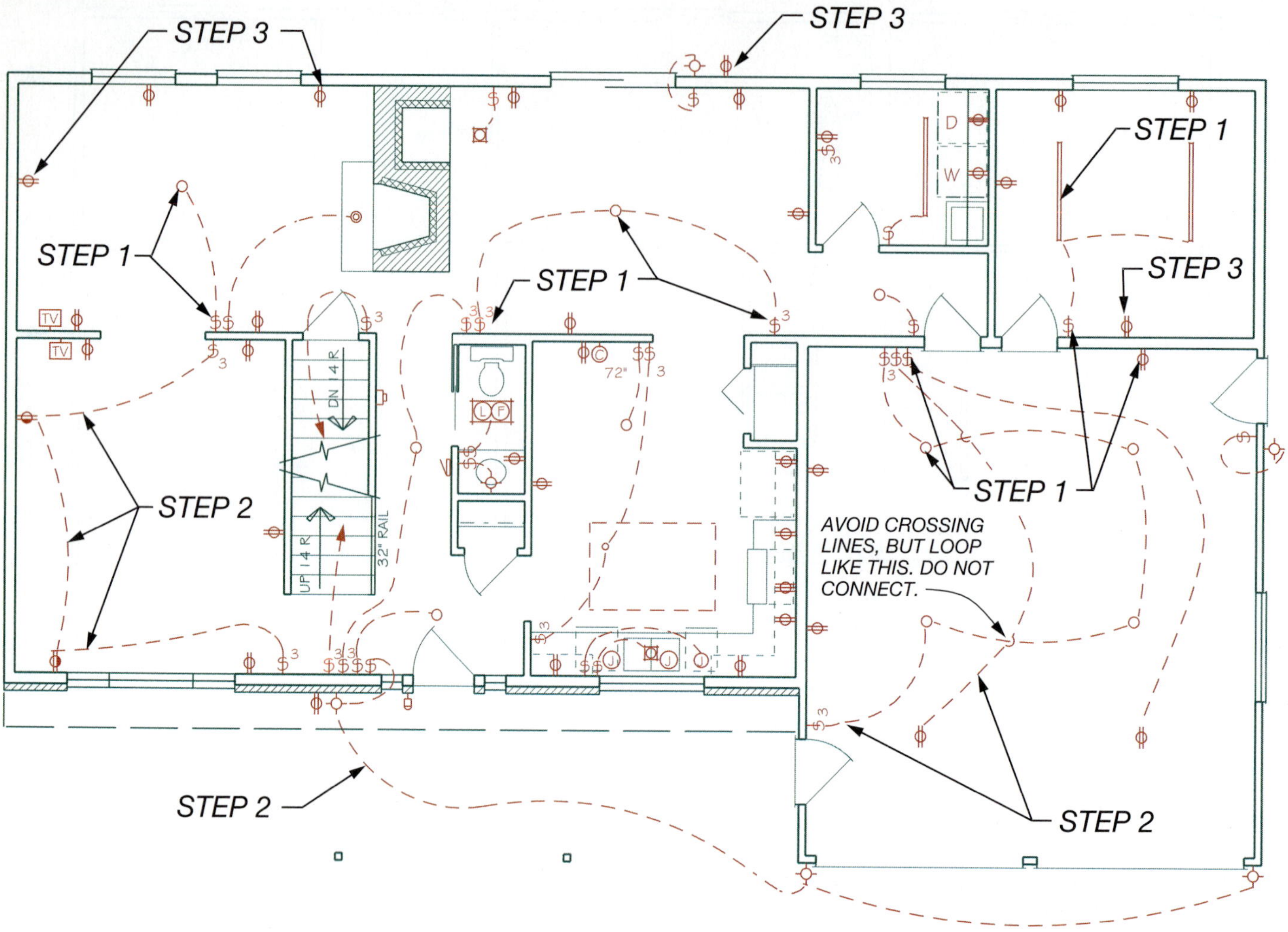

FIGURE 17.12 ■ Step 1: Placing the switches and lighting outlets. Step 2: Drawing the switch legs. Step 3: Placing the outlets. Note: Stairs to basement are included in this plan. If there is no basement, the stairs would be omitted, and a light with switch would be used in the storage area.

STEP 4 Letter all notes, and the drawing title and scale, as shown in Figure 17.13. Figure 17.14 shows the second floor electrical plan for the sample home, and Figure 17.15 shows the basement electrical plan for the sample home.

ELECTRICAL PLAN DRAWING CHECKLIST

Check off the items in the following list as you work on the electrical plan, to be sure that you have included everything necessary.

- Switch locations labeled with proper placement and identification, such as three-way, four-way, dimmer.
- Inside and outside light fixture locations.
- Recessed lights where used.
- Fluorescent light fixtures in 24″ (600 mm) modules where used.
- Switch legs drawn with dashed lines.
- Duplex convenience outlets located based on code requirements.
- Kitchen and bath GFCI outlets.
- Exterior duplex convenience outlets, weatherproof and GFCI.
- Outlets for refrigerator, microwave on a separate circuit, garbage disposal with switch, dishwasher, trash compactor, clothes washer.
- Utility and appliance outlets: furnace, clothes dryer, range, ovens, and water heater shown and labeled. 200-V outlets for electric appliances as needed, or gas and 110-V outlets.

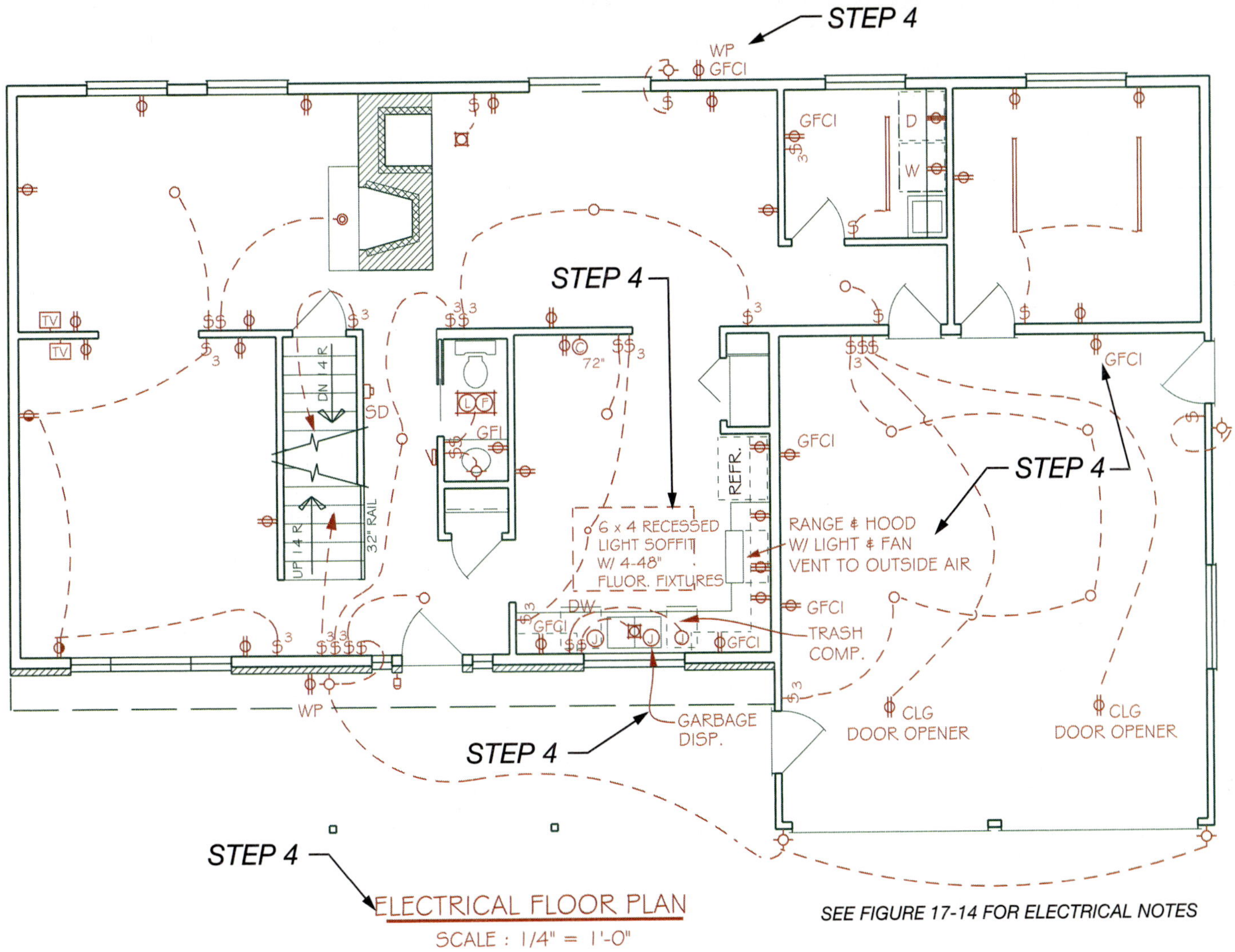

FIGURE 17.13 ■ Lettering the notes and drawing title.

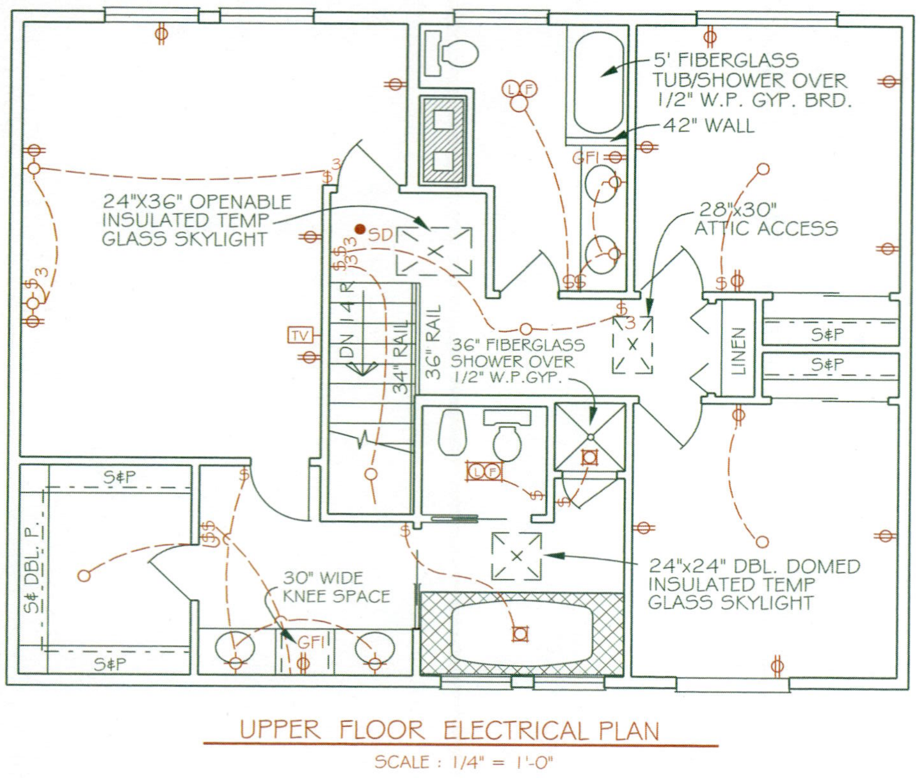

ELECTRICAL NOTES:

1. ALL GARAGE AND EXTERIOR PLUGS AND LIGHT FIXTURES TO BE ON GFCI CIRCUIT.
2. ALL KITCHEN PLUGS AND LIGHT FIXTURES TO BE ON GFCI CIRCUIT.
3. PROVIDE A SEPARATE CIRCUIT FOR MICROWAVE OVEN.
4. PROVIDE A SEPARATE CIRCUIT FOR PERSONAL COMPUTER. VERIFY LOCATION WITH OWNER.
5. VERIFY ALL ELECTRICAL LOCATIONS W/ OWNER.
6. EXTERIOR SPOTLIGHTS TO BE ON PHOTO-ELECTRIC CELL W/ TIMER.
7. ALL RECESSED LIGHTS IN EXTERIOR CEILINGS TO BE INSULATION COVER RATED.
8. ELECTRICAL OUTLET PLATE GASKETS SHALL BE INSTALLED ON RECEPTACLE, SWITCH, AND ANY OTHER BOXES IN EXTERIOR WALL.
9. PROVIDE THERMOSTATICALLY CONTROLLED FAN IN ATTIC WITH MANUAL OVERRIDE (VERIFY LOCATION W/ OWNER).
10. ALL FANS TO VENT TO OUTSIDE AIR. ALL FAN DUCTS TO HAVE AUTOMATIC DAMPERS.
11. HOT WATER TANKS TO BE INSULATED TO R-11 MINIMUM.
12. INSULATE ALL HOT WATER LINES TO R-4 MINIMUM. PROVIDE AN ALTERNATE BID TO INSULATE ALL PIPES FOR NOISE CONTROL.
13. PROVIDE 6 SQ. FT. OF VENT FOR COMBUSTION AIR TO OUTSIDE AIR FOR FIREPLACE CONNECTED DIRECTLY TO FIREBOX. PROVIDE FULLY CLOSEABLE AIR INLET.
14. HEATING TO BE ELECTRIC HEAT PUMP. PROVIDE BID FOR SINGLE UNIT NEAR GARAGE OR FOR A UNIT EACH FLOOR (IN ATTIC) .
15. INSULATE ALL HEATING DUCTS IN UNHEATED AREAS TO R-11. ALL HVAC DUCTS TO BE SEALED AT JOINTS AND CORNERS.

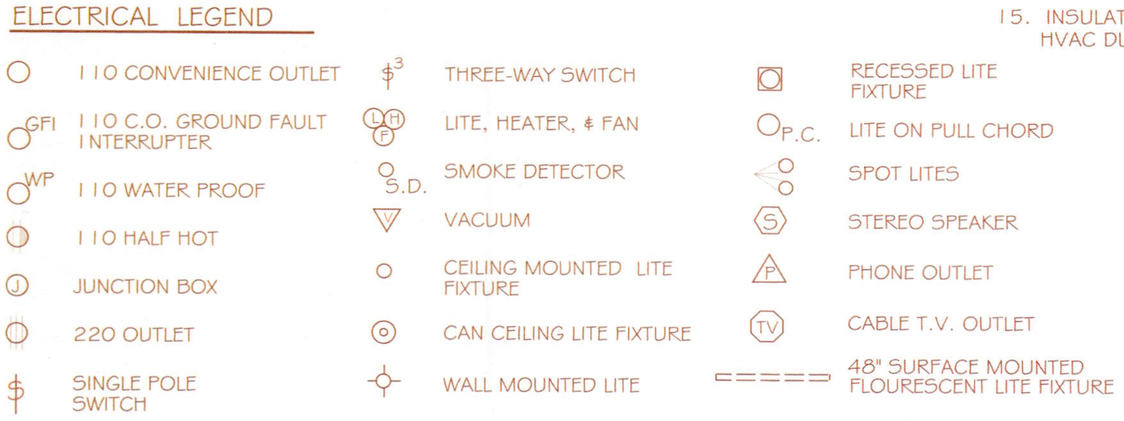
ELECTRICAL LEGEND
110 CONVENIENCE OUTLET
GFI 110 C.O. GROUND FAULT INTERRUPTER
WP 110 WATER PROOF
110 HALF HOT
JUNCTION BOX
220 OUTLET
SINGLE POLE SWITCH
THREE-WAY SWITCH
LITE, HEATER, & FAN
S.D. SMOKE DETECTOR
VACUUM
CEILING MOUNTED LITE FIXTURE
CAN CEILING LITE FIXTURE
WALL MOUNTED LITE
RECESSED LITE FIXTURE
P.C. LITE ON PULL CHORD
SPOT LITES
STEREO SPEAKER
PHONE OUTLET
CABLE T.V. OUTLET
48" SURFACE MOUNTED FLOURESCENT LITE FIXTURE

FIGURE 17.14 ■ The second floor electrical plan for the sample home used in this book.

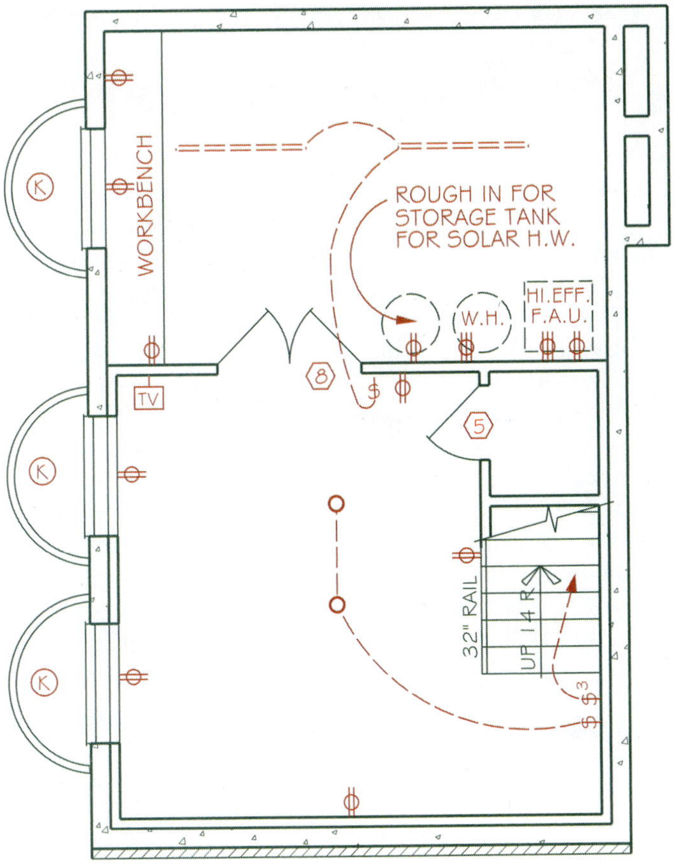

FIGURE 17.15 ■ The basement electrical plan for the sample home used in this book.

- Garage door junction boxes with switches for automatic door openers.
- Television, telephone, separate telephone Internet jack, and computer outlets shown and labeled on separate circuits.
- Vacuum system outlets located and labeled.
- Circuit panel location.
- Door buzzer and chime location.
- Smoke detectors at each bedroom or sleeping area.
- Review all specific notes.
- General notes.
- Drawing title and scale.
- Title block information.

DRAWING ELECTRICAL SYMBOLS WITH CADD

In most cases, electrical symbols are drawn on separate CADD layers. The floor plan serves as the base sheet. So, by turning the electrical layers ON you have a composite drawing with the floor plan, electrical symbols, and electrical layout. You also have the flexibility to turn the electrical layer OFF when you want to display the floor plan without electrical. Many companies provide two separate drawings for residential and commercial projects: the floor plan completely dimensioned and the floor plan with only the walls and electrical layout. All you have to do is turn ON or OFF specific layers as needed to display what you want. Architectural CADD programs have symbols available for you to select from pull-down menus, icon menus, dialogue boxes, and tablet menus. These programs provide you with maximum variety and flexibility to create a high-quality electrical plan in an efficient manner. Figure 17.16 shows some of the electrical menus available from a custom architectural CADD software program. Programs of this type often have the architectural layer system already created. For example, when you select an electrical symbol to insert in your drawing, the symbol is automatically inserted on the correct layer. Some general CADD programs can be customized. This allows you to create the symbols that are needed for specific drawing applications. When these symbols are available, you simply select the desired symbol from the menu and drag the symbol into position on the drawing. Parametric design plays an important role in applications such as ceiling lighting grids. You specify the room width, length, and grid size; the computer then automatically draws the entire ceiling grid. Figure 17.17 shows a complete floor plan with electrical symbols.

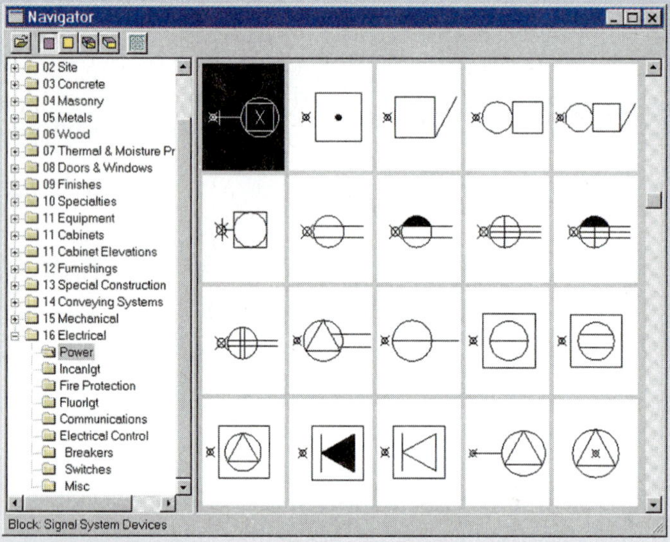

A.

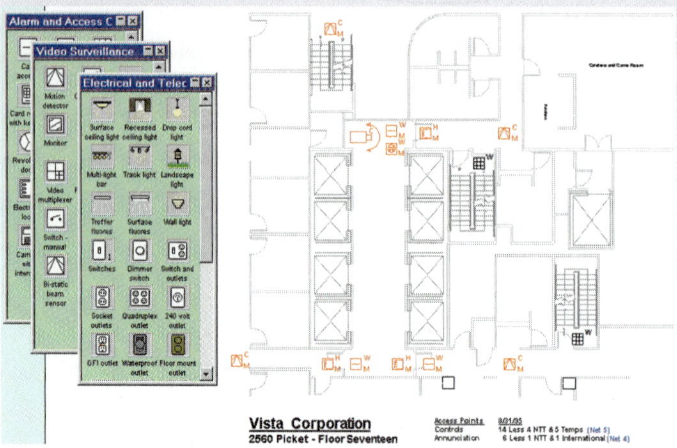

B.

FIGURE 17.16 ■ (a) Electrical signal system devices icon menu; (b) electrical symbol button menus and floor plan application. *Courtesy Visio Corporation.*

CADD APPLICATIONS

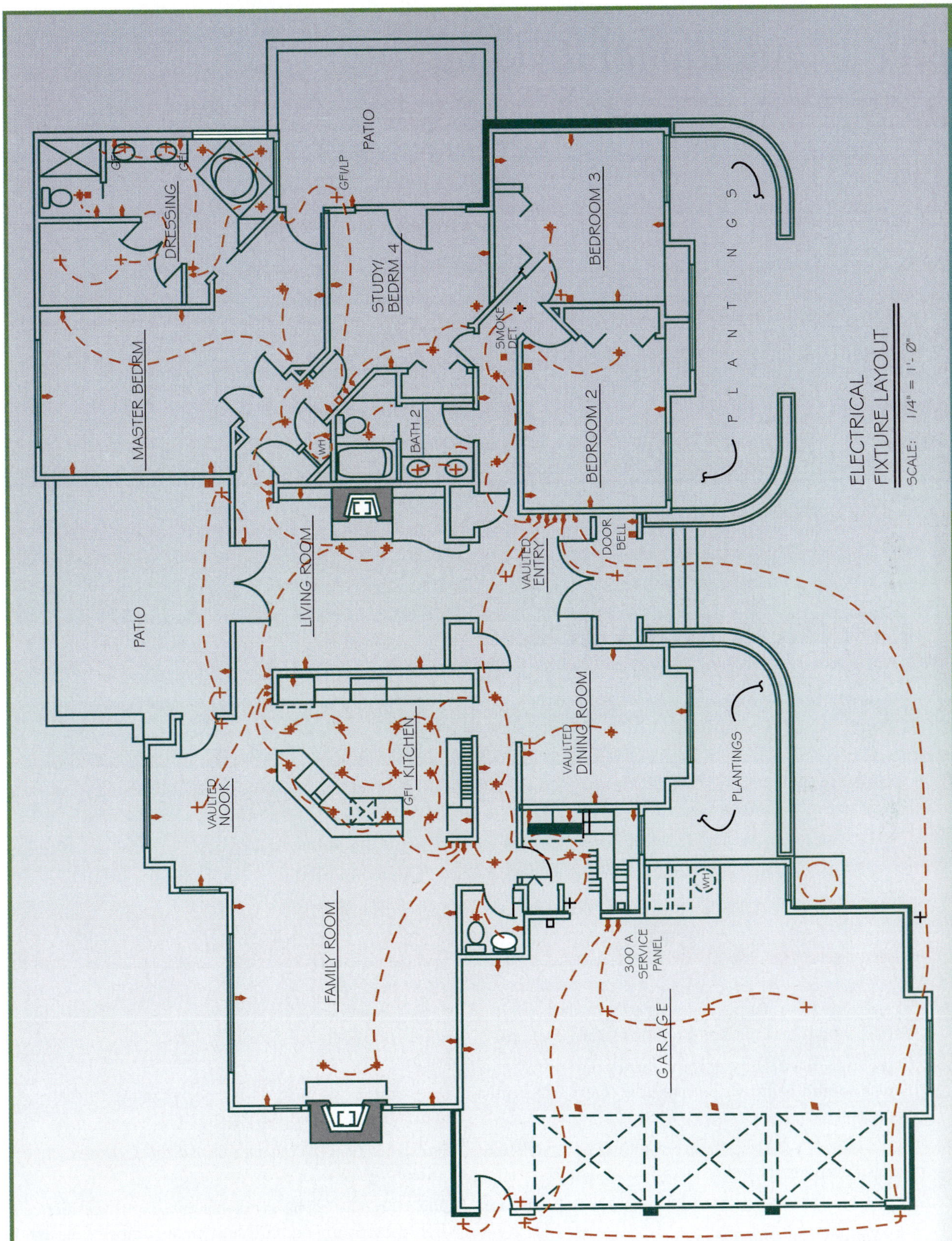

FIGURE 17.17 ■ Complete floor plan with electrical symbols. *Courtesy Piercy & Barclay Designers, Inc. CPBD.*

Chapter 17 Additional Reading

The following Web sites can be used as a resource to help you keep current with changes in electrical materials.

ADDRESS	COMPANY OR ORGANIZATION
www.avenow.com	Audio Video Environments
www.beamvac.com	Beam Central Cleaning Systems
www.broan.com	Broan (Bath Fans)
www.leviton.com	Leviton (Lighting controls)
www.lucent.com	Lucent Technologies (Wiring systems)
www.nutone.com	Nutone (Central Cleaning Systems, Intercoms)
www.onqtech.com	On Q Home Wiring Systems
www.squared.com	SquareD (Electrical systems)

CHAPTER 17

Electrical Plans Test

DIRECTIONS

Answer the questions with short, complete statements or drawings as needed on an 8 1/2″ × 11″ sheet of notebook paper, as follows:

1. Letter your name, Electrical Plans Test, and the date at the top of the sheet.
2. Letter the question number and provide the answer. You do not need to write out the question.

Answers may be prepared on a word processor if course guidelines allow this.

QUESTIONS

Question 17–1 Duplex convenience outlets in living areas should be a maximum of how many feet apart?

Question 17–2 Duplex convenience outlets in living areas should be no more than how many feet from a corner?

Question 17–3 Describe at least four energy-efficient considerations related to electrical design.

Question 17–4 Draw the proper floor-plan symbol for:

a. Duplex convenience outlet
b. Range outlet
c. Recessed circuit breaker panel
d. Phone
e. Light
f. Wall-mounted light
g. Single-pole switch
h. Simplified fluorescent light fixture
i. Fan

Question 17–5 Why is it poor practice to place a switch behind a door swing?

Question 17–6 What is a GFI duplex convenience outlet?

Question 17–7 The average single-family residence should be equipped with how many amps of electrical service?

Question 17–8 Define *outlet box*.

Question 17–9 When does rough-in of electrical wiring take place?

QUESTIONS *(cont.)*

Question 17–10 Define *outlet.*

Question 17–11 Define *lighting outlet.*

Question 17–12 What is the code requirement for the spacing of outlets in a kitchen?

Question 17–13 Give two possible abbreviations for the ground-fault circuit interrupter.

Question 17–14 Define *structured wiring systems.*

Question 17–15 Where are smoke detectors required?

PROBLEMS

Problem 17–1 Draw a typical floor-plan representation of the following on 8 1/2″ × 11″ vellum in pencil or using CADD:

a. Three-way switch controlling three ceiling-mounted lighting fixtures.
b. Four-way switch to one ceiling-mounted lighting fixture.
c. Single-pole switch to wall-mounted light fixture.
d. Single-pole switch to split-wired duplex convenience outlet.

Problem 17–2 Draw a typical small bathroom layout with tub, water closet, one sink vanity, one wall-mounted light over sink with single-pole switch at door, GFCI duplex convenience outlet next to sink, ceiling-mounted light-fan-heat unit with timed switch at door, proper vent note for fan.

Problem 17–3 Draw a typical kitchen layout with double light above sink, dishwasher, range with oven below and proper vent note, refrigerator, centered ceiling-recessed fluorescent lighting fixture, adequate GFCI duplex convenience outlets, and garbage disposal.

Problem 17–4 Use the drawing of the house that you started in the floor-plan problem from Chapter 16 and do one of the following as directed by your instructor:

a. If you are using manual drafting, design an electrical floor-plan for the house.
b. If you are using CADD, create the appropriate electrical plan layer or layers, and design the electrical plan on these layers.
c. If you drew the entire floor plan with the intention of entering the electrical plan on the same drawing, do so now.

18 CHAPTER

Plumbing Plans

INTRODUCTION

There are two classifications of piping: industrial and residential. Industrial piping is used to carry liquids and gases used in the manufacture of products. Steel pipe with welded or threaded connections and fittings is used in heavy construction.

Residential piping is called plumbing and carries fresh water, gas, or liquid and solid waste. The pipe used in plumbing may be made of copper, plastic, galvanized steel, or cast iron.

Copper pipes have soldered joints and fittings and are used for carrying hot or cold water. Plastic pipes have glued joints and fittings and are used for vents and for carrying fresh water or solid waste. Many contractors are replacing copper pipe with plastic piping for both hot and cold water. One example is a plastic pipe with the chemical name of polybutylene (PB), also known as *poly pipe*. Some concerns have been addressed about the strength and life expectancy of PB pipe in construction. Plastic polyvinyl chloride (PVC) pipe has been used effectively for cold water installations. Corrosion-resistant plastic piping is available in a thermoplastic with the chemical name postchlorinated polyvinyl chloride (CPVC). While metal piping loses heat, CPVC pipe retains insulation and saves energy. CPVC piping lasts longer than copper pipe because it is corrosion-resistant, it maintains water purity even under severe conditions, and it does not cause condensation, as copper does. Plastic pipe is considered quieter than copper pipe, and it costs less to buy and install.

Steel pipe is used for large-distribution water piping and for natural gas installations. Steel pipe is joined by threaded joints and fittings or grooved joints. The steel pipe used for water is galvanized. *Galvanized pipe* is steel pipe that has been cleaned and dipped in a bath of molten zinc. The steel pipe used for natural gas applications is protected with a coat of varnish. This pipe is commonly referred to as *black pipe* because of its color. Steel pipe is strong, rugged, and fairly inexpensive. However, it is more expensive than plastic and copper pipe, and labor costs for installation are generally higher, because of the threaded joints.

Corrugated stainless steel tubing (CSST) is also used for natural gas piping. CSST is a flexible piping system that is easier and less expensive to install than black pipe. This type of pipe comes in rolls that allow the plumbing contractor to run pipe through walls and under floors nearly as easily as electric wire. The flexible piping system is easy to cut with traditional pipe cutters and has easy-to-assemble fittings and fixtures.

Stainless steel piping is commonly used in chemical, pollution control, pharmaceutical, and food industries because of its resistance to corrosion. Stainless steel is also sometimes used for water piping in large institutional and commercial buildings.

Cast-iron pipe is commonly used to carry solid and liquid waste as the sewer pipe that connects a structure with a local or regional sewer system. Cast-iron pipe may also be used for the drain system throughout the structure to help reduce water flow noise substantially in the pipes. It is more expensive than plastic pipe but may be worth the price if a quiet plumbing system is desired.

Residential plans may not require a complete plumbing plan. The need for a complete plumbing plan should be verified with the local building code. In most cases, the plumbing requirements can be clearly provided on the floor plan in the form of symbols for fixtures and notes for specific applications or conditions. The plumbing fixtures are drawn in their proper locations on the floor plans at a scale of 1/4″ = 1′–0″. Templates with a large variety of floor-plan plumbing symbols are available.

Other plumbing items to be added to the floor plan include floor drains, vent pipes, and sewer or water connections. Floor drains are shown in their approximate location, with a note identifying size, type, and slope to drain. Vent pipes are shown in the wall where they are to be located and labeled by size. Sewer and water service lines are located in relationship to the position in which these utilities enter the home. The service lines are commonly found on the plot plan. In the situation described here, where a very detailed plumbing layout is not provided, the plumbing contractor is required to install plumbing of a quality and in a manner that meet local code requirements and are economical. Figure 18.1 shows plumbing fixture symbols in plan, frontal, and profile view.

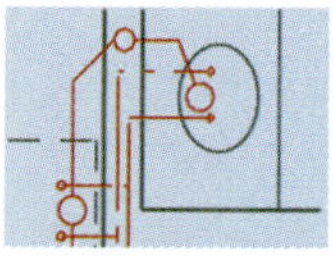

PLUMBING TERMS AND DEFINITIONS

To understand the contents of this chapter, and to communicate effectively, it is important for you to understand key plumbing terminology.

Cleanout: A fitting with a removable plug that is placed in plumbing drainage pipe lines to allow access for cleaning out the pipe.

Drain: Any pipe that carries wastewater in a building drainage system.

Fitting: A standard pipe part such as a coupling, elbow, reducer, tee, and union; used for joining two or more sections of pipe together.

Hose bibb: A faucet that is used to attach a hose.

Lavatory: A fixture that is designed for washing hands and face, usually found in a bathroom.

Main: The primary supply pipe, also called *water main* or *sewer main,* depending on its purpose.

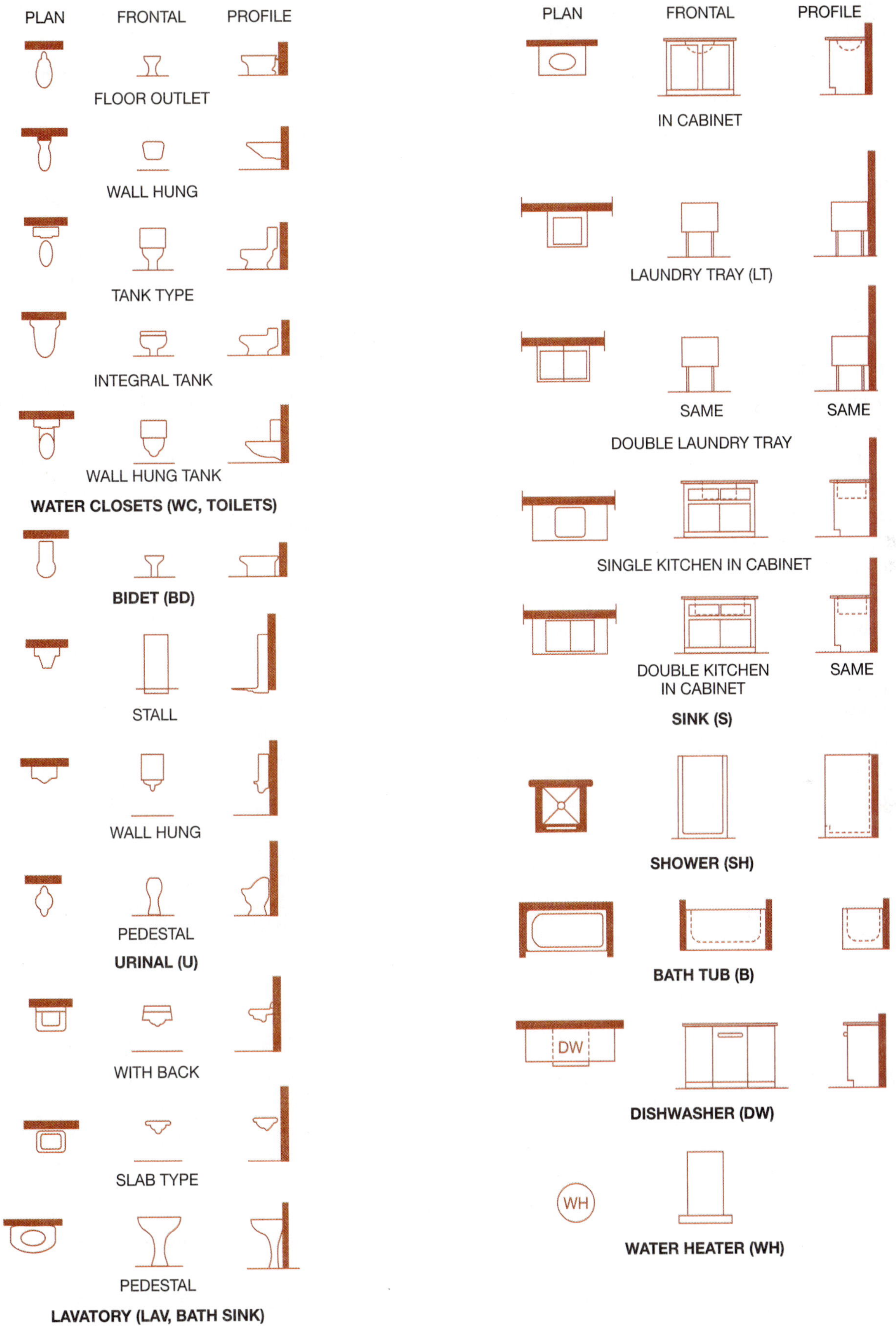

FIGURE 18.1 ■ Common plumbing fixture symbols.

Plumbing fixture: A unit used to contain and discharge waste. Examples of fixtures are sinks, lavatories, showers, tubs, and water closets.

Plumbing system: The plumbing system of a building has these elements:

- Water supply pipes.
- Fixtures and fixture traps.
- Soil, waste, and vent pipes.
- Drain and sewer.
- Storm water drainage.

Plumbing wall: The walls in a building where plumbing pipes are installed.

Potable water: Drinking water, which is free from impurities.

Riser: A water supply pipe that extends vertically one story or more to carry water to fixtures.

Rough-in: Installation of the plumbing system before the installation of fixtures.

Run: The portion of a pipe or fitting continuing in a straight line in the direction of flow in which it is connected.

Sanitary sewer: A sewer that carries sewage without any storm, surface, or groundwater.

Sewer: A pipe, normally underground, that carries wastewater and refuse.

Soil pipe: A pipe that carries the discharge of water closets or other similar fixtures.

Soil stack: A vertical pipe that extends one or more floors and carries discharge of water closets and other similar fixtures.

Stack: A general term referring to any vertical pipe for soil waste or vent piping.

Storm sewer: A sewer used for carrying groundwater, rainwater, surface water, or other nonpolluting waste.

Trap: A vented fitting that provides a liquid seal to prevent the emission of sewer gases without affecting the flow of sewage or wastewater.

Valve: A fitting that is used to control the flow of fluid or gas.

Vanity: A bathroom lavatory fixture that is freestanding or in a cabinet.

Vent pipe: The pipe installed to ventilate the building drainage system and to prevent drawing liquid out of traps and stopping back pressure.

Waste pipe: A pipe that carries only liquid waste free of fecal material.

Waste stack: A vertical pipe that runs one or more floors and carries the discharge of fixtures other than water closets and similar fixtures.

Water closet: A water-flushing plumbing fixture, such as a toilet, that is designed to receive and discharge human excrement. This term is sometimes used to mean the compartment where the fixture is located.

Water distributing pipe: A pipe that carries water from the service to the point of use.

Water heater: An appliance used for heating and storing hot water.

Water main: See *Main.*

Water meter: A device used to measure the amount of water that goes through the water service.

Water service: The pipe from the water main or other supply to the water-distributing pipes.

SIZING OF PLUMBING PIPING

Plumbing must be sized enough to allow fixtures to operate properly and for proper draining and venting. The size of water supply piping is based on these conditions:

- Amount of water needed.
- Supply pressure.
- Pipe length.
- Number of stories to be supplied.
- Flow pressure needed at the farthest point from the source.

The size of drainage piping is based on standards established for the type of fixture and the average amount of waste that can be discharged through the fixture in a given amount of time. The size of vent pipes is based on the number of drainage fixture units that drain into the waste portion of the vent stack.

The following chart lists common minimum plumbing pipe sizes for residential installations. Pipe sizes are also given in the discussions and examples throughout this chapter. Actual plumbing pipe sizes should be confirmed with local, state, and national codes that apply to your area.

ENERGY CONSERVATION

Energy-conserving methods of construction can be applied to the plumbing installation. If such methods are used, they can be applied in the form of specific or general notes on detailed drawings or in construction specifications. Energy-conserving methods include:

- Insulating all exposed hot water pipes. Cold water pipes should be insulated in climates where freezing is a problem.
- Running water pipes in insulated spaces where possible.
- Keeping water pipes out of exterior walls where practical.

FIXTURE OR PIPE SIZE	WATER SIZE	MINIMUM FIXTURE TRAP AND DRAIN SIZE	MINIMUM VENT SIZE
Distributing pipe	3/4″		
Bathroom group*	1/2″		
Plus one or more fixtures	3/4″		
Two fixtures	1/2″		
Three fixtures	3/4″		
Hose bibb	1/2″		
Plus one or more fixtures	3/4″		
Water closet (toilet)	3/8″	3″	2″
Bathtub	1/2″	1 1/2″	1 1/4″
Shower	1/2″	2″	1 1/4″
Lavatory	3/8″	1 1/4″	1 1/4″
Kitchen sink	1/2″	1 1/2″	1 1/4″
Laundry tray	1/2″	1 1/2″	1 1/4″
Clothes washer	1/2″	1 1/2″	1 1/4″
Bidet	3/8″	1 1/2″	1 1/4″
Dishwasher	3/8″	1 1/2″	1 1/4″

*The typical bathroom group consists of a lavatory, tub with shower, and toilet.

- Placing thermosiphon traps in hot water pipes to reduce heat loss from excess hot water in the pipes.
- Locating the water heater in a heated space and insulating it well.
- Selecting low-flow showerheads.
- Putting flow restrictors in faucets.
- Completely caulking plumbing pipes where they pass through the plates.
- Sealing and covering drain penetrations in floors.
- Sealing all wall penetrations.

UNIVERSAL PLUMBING INSTALLATIONS

Plumbing systems and fixtures can be designed to accommodate a variety of needs for people of all ages, and people with disabilities. Design considerations include the following.

Sinks

Provide a sink that is approximately 6″ (150 mm) deep, to be easily reachable from a wheelchair. The cabinet below the sink can be recessed for easier access. The drain can have an auxiliary control that is mounted in a close, convenient place on the countertop. More than one sink is a possibility; this allows placement at different heights. Stainless steel sinks might be a consideration for durability. Bathroom sinks that are either pedestal-based or wall-mounted can help provide easy access for wheelchairs. Wall-mounted sinks should have extra support. The height of the sink should be approximately 34″ (860 mm) for convenient wheelchair access. The water pipes should be concealed behind the sink. Single-lever or automatic faucets can be provided that allow ease of operation and better reach. The faucets can be mounted to the side rather than (as with most installations) in the back. Gooseneck faucets allow for easier access than regular units. A sink can also be provided in the water closet compartment for convenience and privacy.

Water Closet

The water closet compartment should be designed to provide easy access and maneuverability when a wheelchair is in use. There should be at least 42″ (1070 mm) of clear space provided to the side of the toilet. A toilet designed for wheelchair access should be wall-mounted and have an elongated bowl that is between 15″ and 19″ (380–480 mm) high. The toilet seat should be specified with a front opening. Some wheelchairs are designed for bathroom use and should be coordinated with the shower and toilet access. In some designs, a toilet inside the shower can be considered. Some homeowners may want a toilet with warm water for washing and a dryer, or a bidet for wash-

ing. The toilet flush lever should be specified on the access side and may have an extension for convenient use.

Bathtubs

A bathtub can be specified with a height to match the wheelchair for easier access. Single-lever controls should be considered. A seat should be designed into the length of the bathtub. This allows a person to transfer from wheelchair to seat before entering the tub. The tub should be at least 60″ (1524 mm) long with a minimum 15″ (380 mm) seat. The clear floor space in front of the tub should be 30″ (760 mm) minimum. A slant at the back of the tub and on the seat allows for easier access to the tub and serves as a headrest. Select a bathtub or design a tub area that offers easy access and comfort and has a slip-resistant bottom and handles for getting in and out.

Showers

Showers are generally easier to access and use than bathtubs. The shower should be designed for wheelchair access by providing a minimum length and width of 60″ × 30″ (1524 × 760 mm). The shower should have rails and have a handheld faucet with temperature and surge controls to help maintain a constant water temperature and flow rate. The shower should have a slight ramp for easy access and have a nonslip bottom. A shower should be designed with a convenient seat when the owner does not use a shower wheelchair. Shower controls should be carefully placed at the shower entrance. Controls can be specified that start the shower and maintain the temperature automatically.

PLUMBING SCHEDULES

Plumbing schedules are similar to door, window, and finish schedules. Schedules provide specific information regarding plumbing equipment, fixtures, and supplies. The information is condensed in a chart so that the floor plan is not unnecessarily crowded. Figure 18.2 shows a typical plumbing fixture schedule.

PLUMBING FIXTURE SCHEDULE

LOCATION	ITEM	MANUFACTURER	REMARKS
MASTER BATH	36" F.G. SHR. COLOR BIDET C.I. SR. COLOR LAV. COLOR W. C.	HYTEC K-4868 K-2904 K-3402-PBR	M2620 BRASS K1940 BRASS M4625 BRASS PLAS. SEAT
BATH #2	KEG STYLE TUB C.I. COLOR PED. LAV. COLOR W.C.	KOHLER KOHLER K3402-PBR	M2850 BRASS M4625 BRASS PLAS. SEAT
BATH #3	URINAL F.G. SHOWER C.I. LAVS.	K-4980 HYTEC K2904	M2620 BRASS M4625 BRASS
KITCHEN	C.I. 3 HOLE SINK	K5960	M7531 BRASS
WATER HTR.	82 GAL. ELEC.	MORFIO	P & T VALVE
UTILITY	F.G. LAUN. TRAY	24 X21	D2121 BRASS

FIGURE 18.2 ■ Plumbing fixture schedule.

Other schedules may include specific information regarding floor drains, water heaters, pumps, boilers, or radiators. These schedules generally key specific items to the floor plan with complete information describing size, manufacturer, type, and other specifications as appropriate.

Description of Plumbing Fixtures and Materials

Mortgage lenders may require a complete description of materials for the structure. Part of the description often includes a plumbing section in which certain plumbing specifications are described, as shown in Figure 18.3.

PLUMBING DRAWINGS

Plumbing drawings are usually not on the same sheet as the complete floor plan. The only plumbing items shown on the floor plans are fixtures, as previously noted. The drafter or designer prepares an accurate outline of the floor plan showing all walls, partitions, doors, windows, plumbing fixtures, and utilities. This outline can be used as a preliminary drawing for the client's approval and can also be used to prepare a sepia or photocopy secondary original. This secondary original allows the drafter to save the original drawing for the complete floor plan. The secondary original saves drafting time, since the drafter will not have to prepare another outline for the plumbing drawing. When the drawings are done with CADD, the base drawing is the floor plan, and layers are set up to display the plumbing symbols and piping layout when needed.

Plumbing drawings may be prepared by a drafter in an architectural office or in conjunction with a plumbing contractor. In some situations, when necessary, plumbing contractors or mechanical engineers work up their own rough sketches or field drawings. Plumbing drawings are made up of lines and symbols that show how liquids, gases, or solids are transported to various locations in the structure. Plumbing lines and features are drawn thicker than wall lines, so they are clearly distinguishable. Lateral lines are commonly broken when they cross over other lines. Symbols identify types of pipes, fittings, valves, and other components of the system. Sizes and specifications are provided in local or general notes. Figure 18.4 shows some typical plumbing symbols. Certain abbreviations commonly used in plumbing drawings are shown in Figure 18.5.

WATER SYSTEMS

Water supply to a structure begins at a water meter for public systems or at a water storage tank for private well systems. The water supply to the home or business, known as the *main line,* is generally 1″ plastic pipe. This size may vary in relation to the service needed. The plastic main line joins a copper line within a few feet of the structure. The rest of the water system piping

PLUMBING

FIXTURE	NUMBER	LOCATION	MAKE	MANUFACTURER'S FIXTURE INDENTIFICATION NUMBER	SIZE	COLOR
Sink						
Lavatory						
Water closet						
Bathtub						
Shower over tub △						
Stall shower △						
Laundry trays						

△☐ Curtain rod △☐ Door ☐ Shower pan: material ______

Water Supply: ☐ public; ☐ community system; ☐ individual (private) system.★

Sewage disposal: ☐ public; ☐ community system; ☐ individual (private) system.★

★ *Show and describe individual system in complete detail in separate drawings and specifications according to requirements.*

House drain (inside): ☐ cast iron; ☐ tile; ☐ other ______ House sewer (outside): ☐ cast iron; ☐ tile; ☐ other ______

Water piping: ☐ galvanized steel; ☐ copper tubing; ☐ other ______ Sill cocks, number ______

Domestic water heater: type ______; make and model ______; heating capacity ______ ______ gph 100° rise. Storage tank: material ______; capacity ______ gallons.

Gas service: ☐ utility company; ☐ liq. pet. gas; other ______ Gas piping: ☐ cooking: ☐ house heating.

Footing drains connected to: ☐ storm sewer; ☐ sanitary sewer; ☐ dry well Sump pump; make and model ______ ______; capacity ______; discharges into ______

FIGURE 18.3 ■ Form for description of plumbing materials.

is usually copper pipe, although plastic pipe is increasing in popularity for cold water. The 1″ main supply often changes to 3/4″ pipe where a junction is made to distribute water to various specific locations. From the 3/4″ distribution lines, 1/2″ pipe usually supplies water to specific fixtures, for example, the kitchen sink. Figure 18.6 shows a typical installation from the water meter of a house with distribution to a kitchen. The water meter location and main line representation are generally shown on the plot plan. Verify local codes regarding the use of plastic pipe and the water meter location.

There is some advantage to placing plumbing fixtures back-to-back when possible. This practice saves materials and labor costs. Also, in designing a two-story structure, placing plumbing fixtures one above the other is economical. If the functional design of the floor plan clearly does not allow for such economy measures, then good judgment should be used in the placement of plumbing fixtures so the installation of the plumbing is physically possible. Figure 18.7 shows a back-to-back bath situation. If the design allows, another common installation may be a bath and laundry room next to each other, to provide a common plumbing wall. See Figure 18.8. If a specific water temperature is required, that specification can be applied to the hot water line, as shown in Figure 18.9.

Water Heaters

Water heaters should be placed on a platform with an overflow tray. Water heaters with nonrigid water connections and over 4′ (1219 mm) in height shall be anchored or strapped to the building. If the water heater is located in the garage, it shall be protected from impact by automobiles with a steel pipe embedded in concrete placed in front. Fuel combustion water heaters shall not be installed in sleeping rooms, bathrooms, clothes closets, or closets or other confined spaces opening into a bedroom or bathroom.

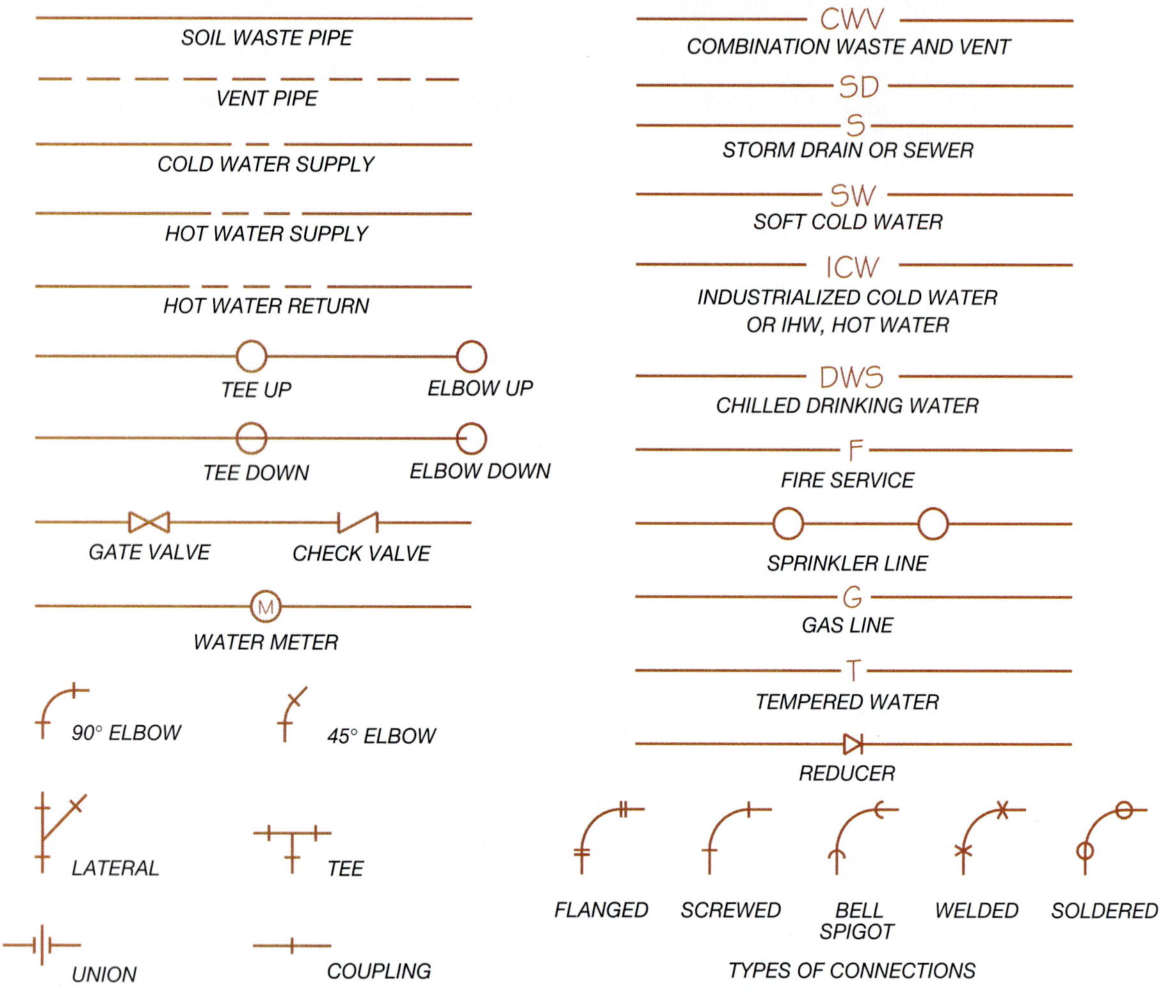

FIGURE 18.4 ■ Typical plumbing pipe symbols.

CURRENT		MCS	
CW	Cold-water supply	WC	Water closet (toilet)
HW	Hot-water supply	LA	Lavatory (bath sink)
HWR	Hot-water return	B	Bathrub
HB	Hose bibb	S	Sink
CO	Cleanout	U	Urinal
DS	Downspout	SH	Shower
RD	Rain drain	DF	Drinking fountain
FD	Floor drain	WH	Water heater
SD	Shower drain	DW	Dishwasher
CB	Catch basin	BD	Bidet
MH	Manhole	GD	Garbage disposal
VTR	Vent thru roof	CW	Clothes washer

FIGURE 18.5 ■ Some typical plumbing abbreviations.

Recommended water heater sizes vary, depending on the number of bedrooms, number of bathrooms, and type of water heater, such as electric, gas, or oil-fired. Size specifications for water heaters may also vary, depending on the fuel type and manufacturer, but the following chart gives some basic size guidelines:

NUMBER OF BATHS	NUMBER OF BEDROOMS	WATER HEATER SIZE, GALLONS		
		GAS	OIL	ELECTRIC
1 to 1 1/2	1	20	30	20
	2	30	30	30
	3	30	30	40
2 to 2 1/2	2	30	30	40
	3	40	30	50
	4	40	30	50
	5	50	30	66
3 to 3 1/2	3	40	30	50
	4	50	30	66
	5	50	30	66
	6	50	40	80

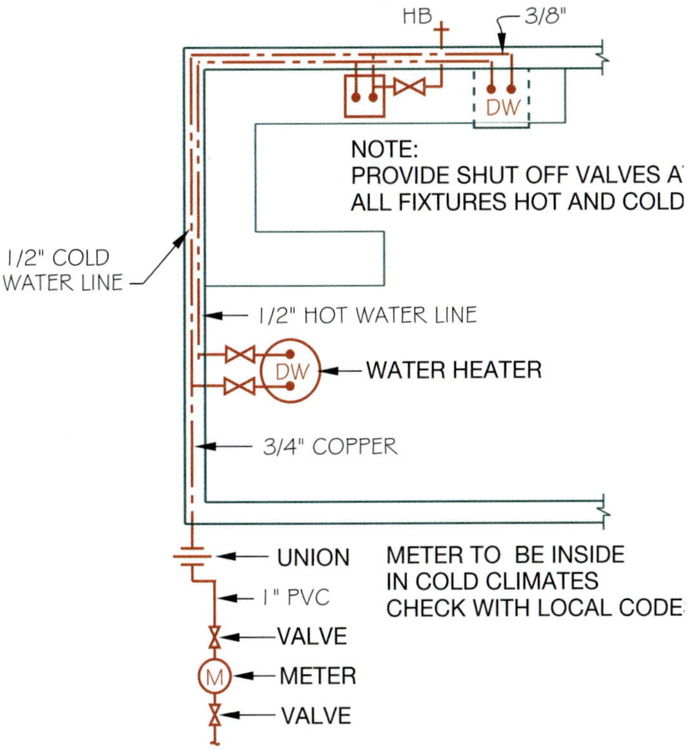

FIGURE 18.6 ■ Partial water supply system.

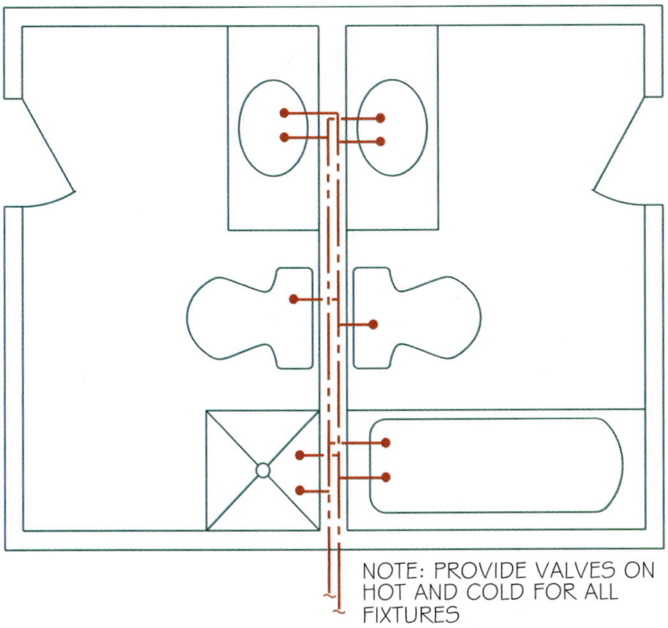

FIGURE 18.7 ■ Common plumbing wall with back-to-back baths.

Hose Bibbs

Hose bibbs are faucets that are used to attach a hose. Outside hose bibbs require a separate valve that allows the owner to turn off the water to the hose bibb during freezing weather. The separate valve must be located inside a heated area and is generally under a cabinet where a sink is located, such as in the kitchen or bathroom. This provides easy access to the

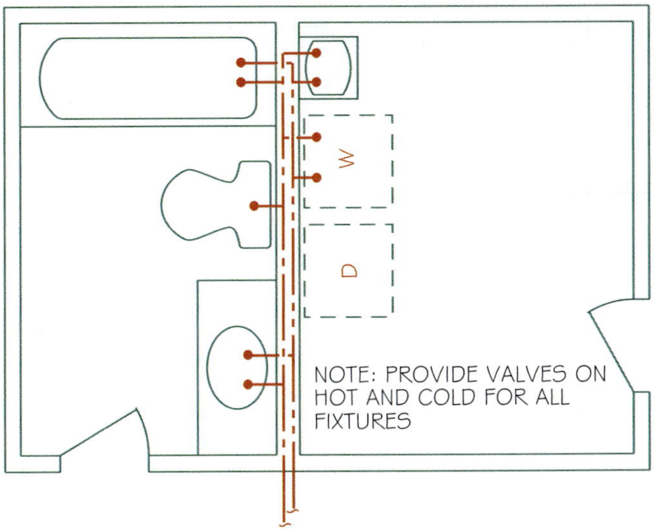

FIGURE 18.8 ■ Common plumbing wall with a bath and laundry back-to-back.

FIGURE 18.9 ■ Hot water temperature.

valve. Frost proof hose bibbs do not require a separate inside valve, but the stem must extend through the building insulation into an open heated or semiconditioned space to avoid freezing.

DRAINAGE AND VENT SYSTEMS

The *drainage system* provides for the distribution of solid and liquid waste to the sewer line. Drainage pipes are required to have a minimum slope of 1/4″ per foot. The *vent system* allows for a continuous flow of air through the system so that gases and odors may dissipate and bacteria do not have an opportunity to develop. The vent system also protects drawing liquid out of the traps and stops back pressure. These pipes throughout the house are generally made of PVC plastic, although the pipe from the house to the concrete sewer pipe is commonly 3″ or 4″ cast iron. Drainage and vent systems, like water systems, are drawn with thick lines using symbols, abbreviations, and notes. Figure 18.10 shows a sample drainage vent system. Figure 18.11 shows a house plumbing plan.

ISOMETRIC PLUMBING DRAWINGS

Isometric drawings may be used to provide a three-dimensional representation of a plumbing layout, sometimes called a *plumbing riser diagram*. Isometric plumbing templates and CADD programs are available to make the job easier. Especially for a two-story structure, an isometric drawing provides an easy-to-understand pictorial representation. Figure 18.12 is an isometric drawing of

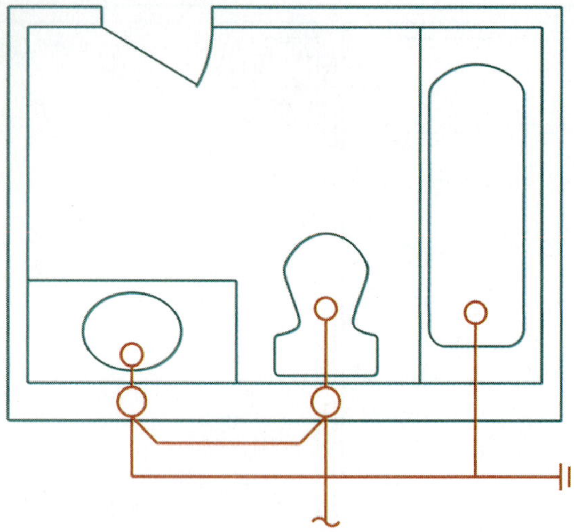

FIGURE 18.10 ■ Drainage and vent drawing.

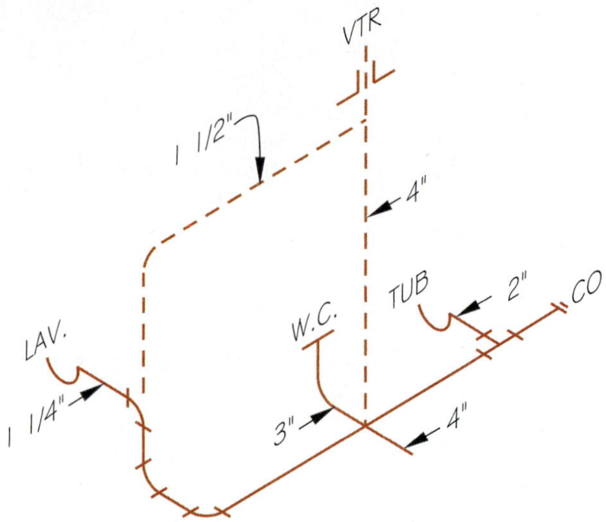

FIGURE 18.12 ■ Single-line isometric drawing of a drainage and vent system.

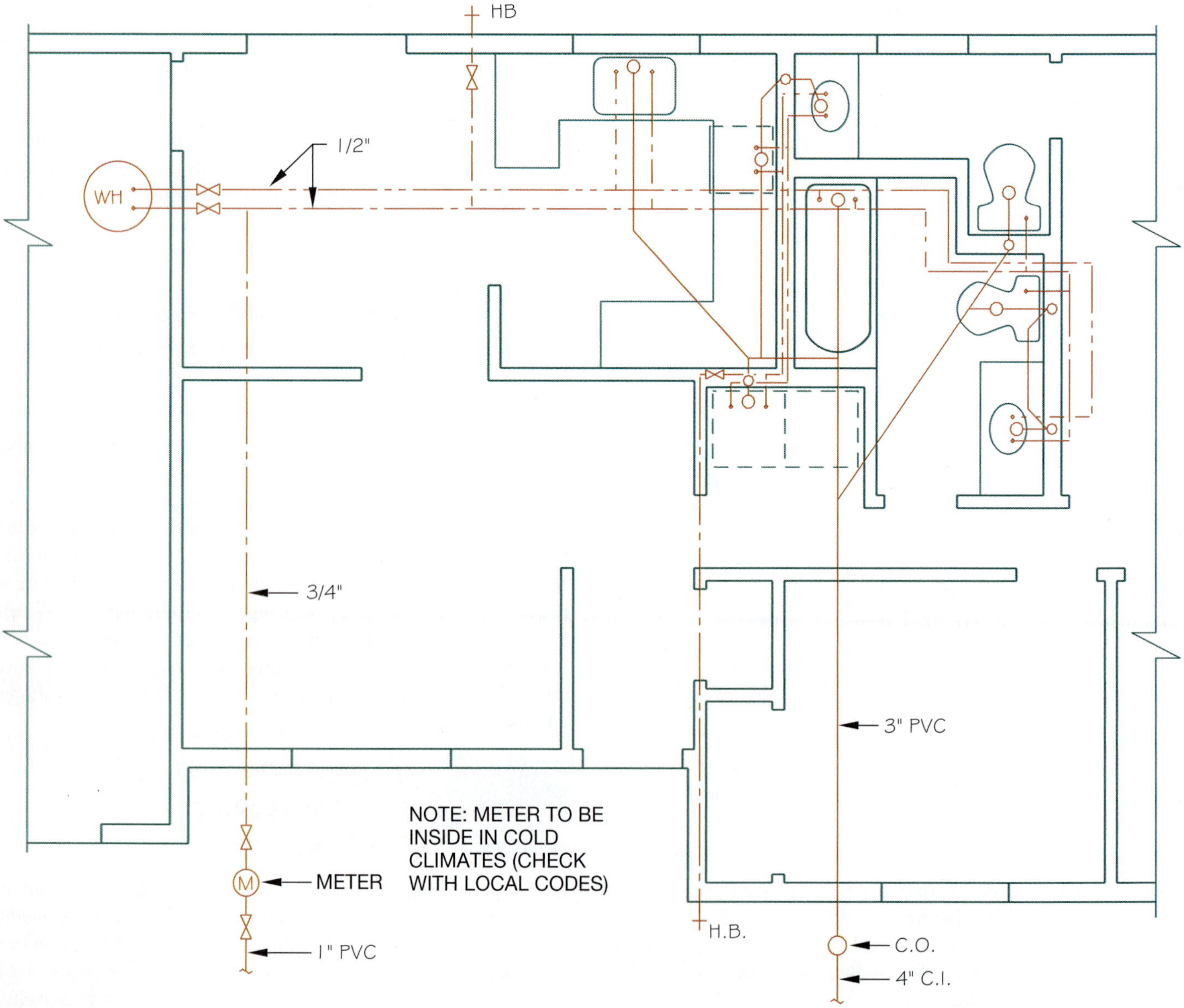

FIGURE 18.11 ■ Residential plumbing plan.

the system shown in plan view in Figure 18.10. Figure 18.13 shows a detailed isometric drawing of a typical drain, waste, and vent system. Figure 18.14 shows a single-line isometric drawing of the detailed isometric drawing in Figure 18.13. Figure 18.15 shows a detailed isometric drawing of a hot and cold water supply system. Figure 18.16 shows a single-line isometric drawing of that same system.

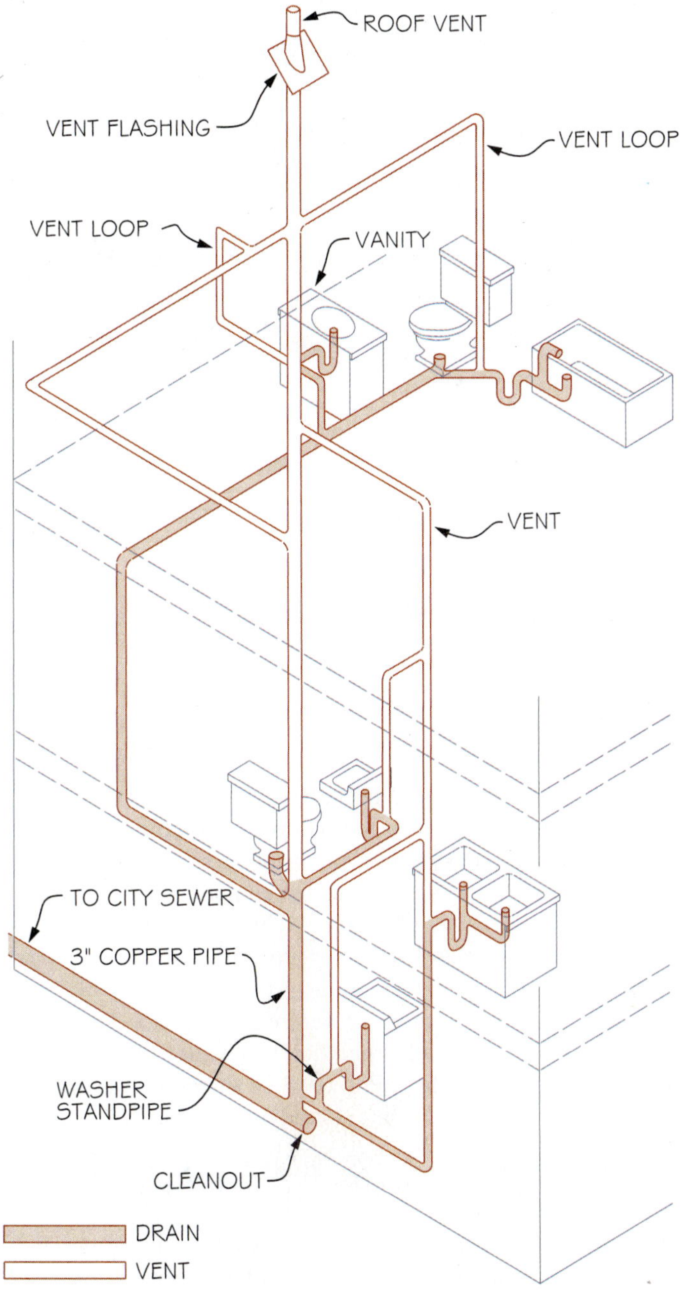

FIGURE 18.13 ■ Detailed isometric drawing of drain, waste, and vent system.

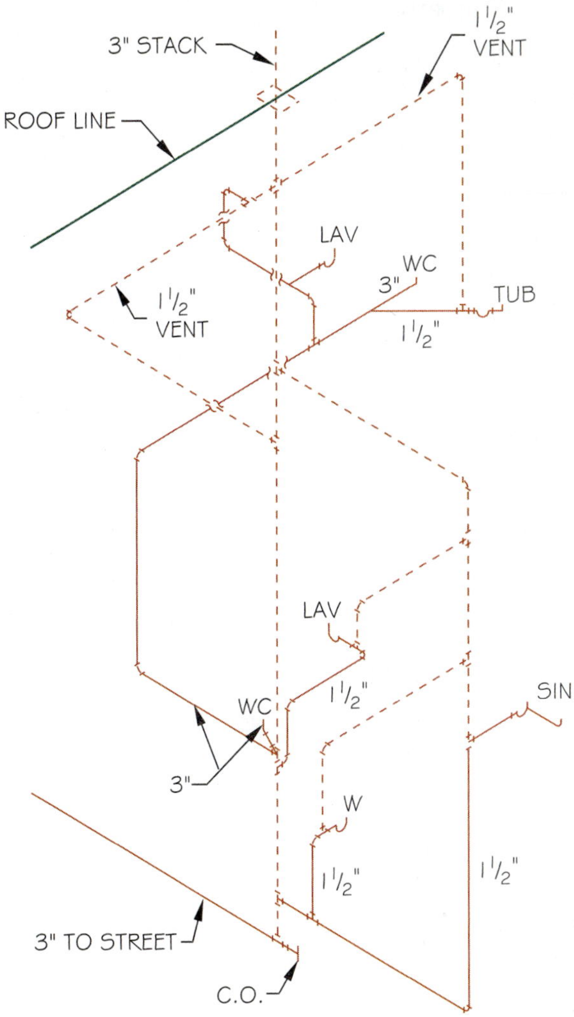

FIGURE 18.14 ■ Single-line isometric drawing of drain, waste, and vent system.

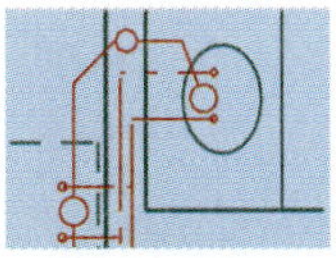

SOLAR HOT WATER

Solar hot water collectors are available for new or existing residential, commercial, or industrial installations. Solar collectors may be located on a roof, on a wall, or on the ground. Figure 18.17 shows a typical application of solar collectors. Solar systems vary in efficiency. The number of collectors needed to provide heat to a structure depends on the size of the structure and the volume of heat needed.

The flat-plate collector is the heart of a solar system. Its main parts are the transparent glass cover, absorber plate, flow tubes, and insulated enclosure. Piping and wiring drawings for a complete solar hot water heating system are available through the manufacturer.

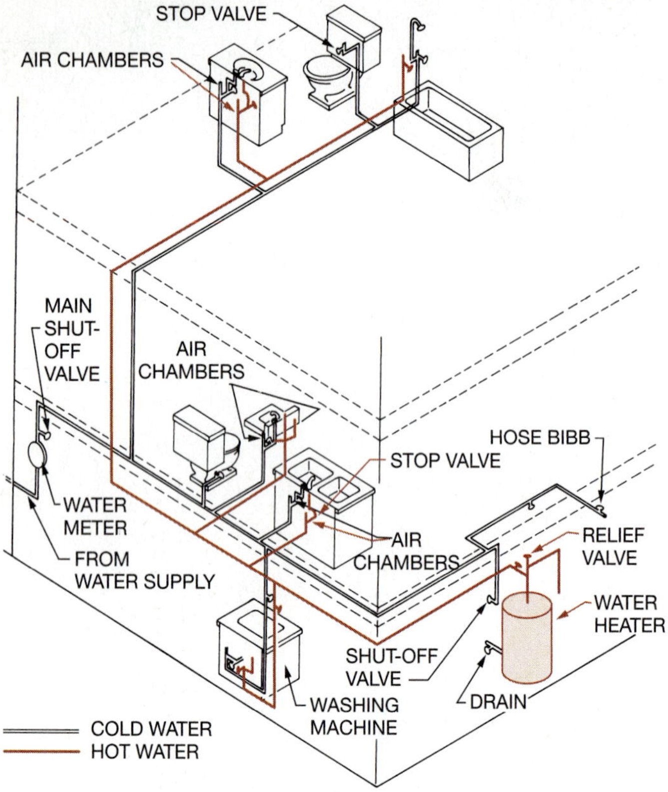

FIGURE 18.15 ■ Detailed isometric drawing of hot and cold water system.

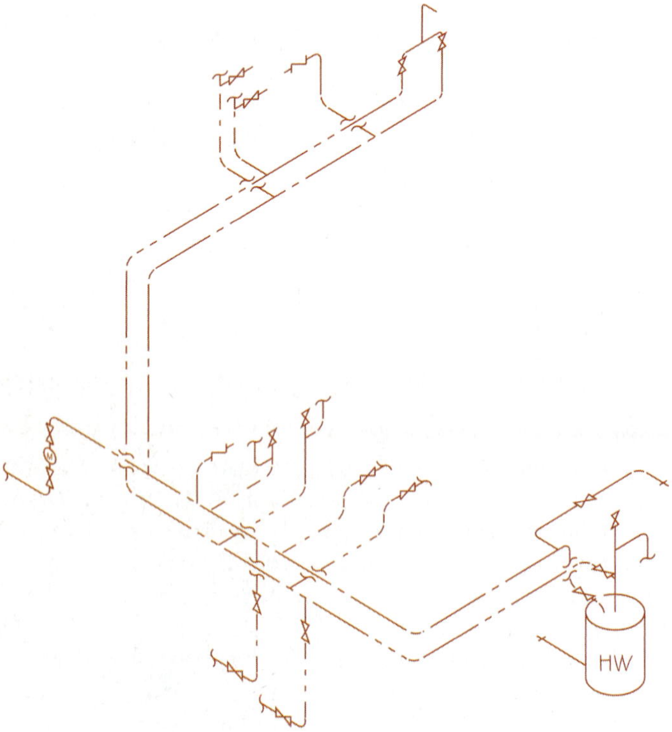

FIGURE 18.16 ■ Single-line isometric drawing of hot and cold water system.

FIGURE 18.17 ■ Typical application of solar hot water collectors. *Courtesy Lennox Industries, Inc.*

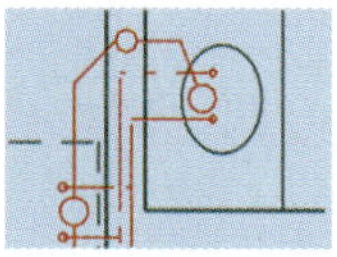

SEWAGE DISPOSAL

Public Sewers

Public sewers are available in and near most cities and towns. Public sewers are generally located under the street or an easement next to the construction site. In some situations the sewer line may have to be extended to accommodate the addition of another home or business in a newly developed area. The cost of this extension may be the responsibility of the developer. Usually the public sewer line is under the street, so the new construction has a line from the structure to the sewer. The cost of this construction usually includes installation, street repair, sewer tap, and permit fees. Figure 18.18 shows a sewer connection and the plan view usually found on the plot plan.

Private Sewage Disposal: A Septic System

A *septic system* consists of a storage tank and an absorption field and operates as follows. Solid and liquid waste enters the septic tank, where it is stored and begins to decompose into sludge. Liquid material, or effluent, flows from the tank outlet and is

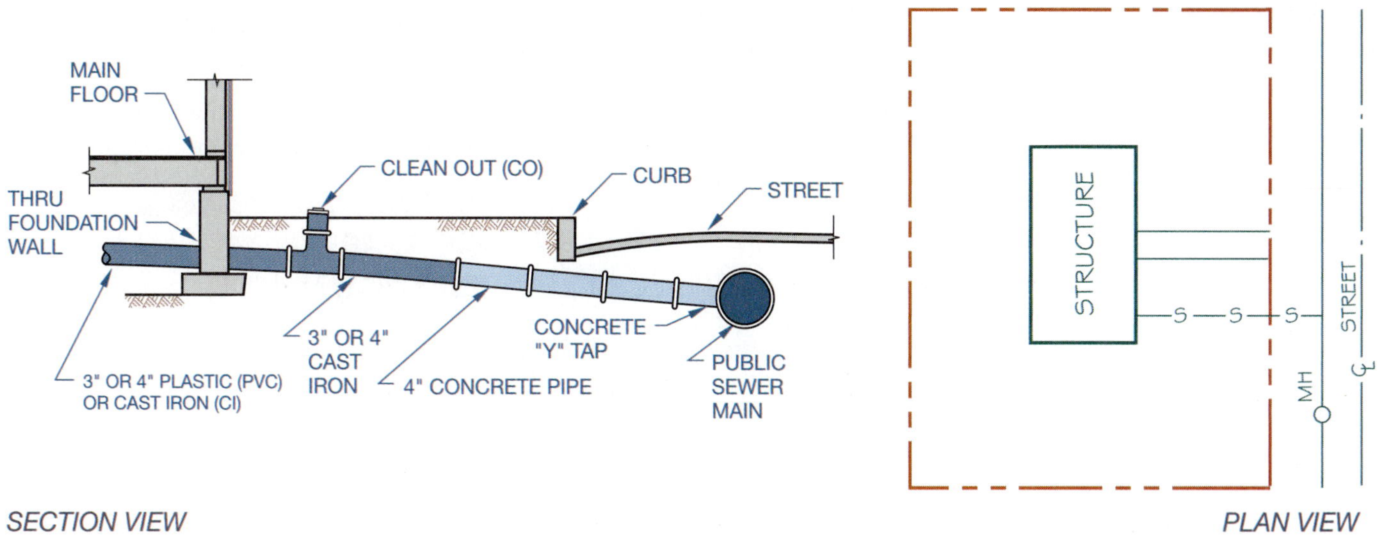

FIGURE 18.18 ■ Public sewer system.

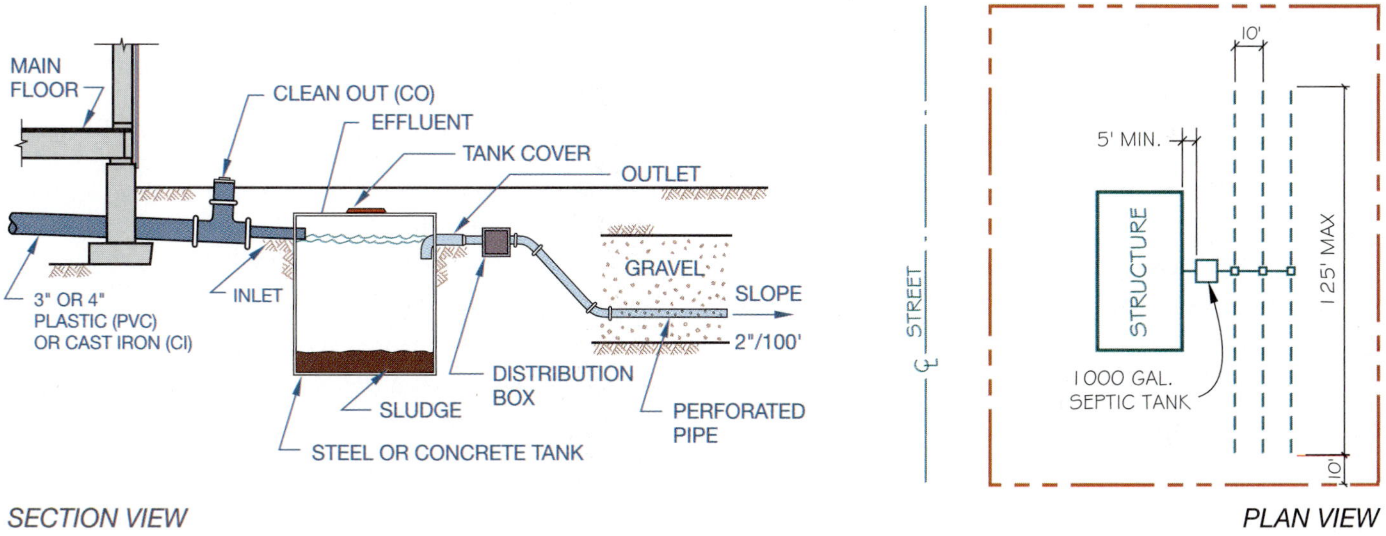

FIGURE 18.19 ■ Septic sewer system.

dispersed into a soil-absorption field, or drain field (also known as *leach lines*). When the solid waste has effectively decomposed, it also dissipates into the soil absorption field. The owner should use a recommended chemical to work as a catalyst for complete decomposition of solid waste. Septic tanks may become overloaded in a period of up to ten years and may require pumping.

The characteristics of the soil must be verified for suitability for a septic system by a soil feasibility test, also known as a *percolation test.* This test, performed by a soil scientist or someone from the local government, determines if the soil will accommodate a septic system. The test should also identify certain specifications that should be followed for installation. The Veterans Administration (VA) and the Federal Housing Administration (FHA) require a minimum of 240′ of field line, or more if the soil feasibility test shows it is necessary. Verify these dimensions with local building officials. When the soil characteristics do not allow a conventional system, there may be some alternatives such as a sand filter system, which filters the effluent through a specially designed sand filter before it enters the soil absorption field. Check with your local code officials before calling for such a system. Figure 18.19 shows a typical serial septic system. The serial system allows for one drain field line to fill before the next line is used. The drain field lines must be level and must follow the contour of the land perpendicular to the slope. The drain field should be at least 100′ from a water well, but verify the distance with local codes. There is usually no minimum distance to a public water supply.

METRICS IN PLUMBING

Pipe is made of a wide variety of materials identified by trade names; the nominal sizes are related only loosely to actual dimensions. For example, a 2″ galvanized pipe has an inside diameter of about 2 1/8″ but is called 2″ pipe for convenience. Since few pipe products have even inch dimensions that match

their specifications, there is no reason to establish even metric sizes. Metric values established by the International Organization for Standardization relate nominal pipe sizes (NPS) in inches to metric equivalents, referred to as *diameter nominal* (DN). The following equivalents relate to all plumbing, natural gas, heating, oil, drainage, and miscellaneous piping used in buildings and civil works projects:

NPS (in.)	DN (mm)	NPS (in.)	DN (mm)
1/8	6	8	200
3/16	7	10	250
1/4	8	12	300
3/8	10	14	350
1/2	15	16	400
5/8	18	18	450
3/4	20	20	500
1	25	24	600
1 1/4	32	28	700
1 1/2	40	30	750
2	50	42	800
2 1/2	65	36	900
3	80	40	1000
3 1/2	90	44	1100
4	100	48	1200
4 1/2	115	52	1300
5	125	56	1400
6	150	60	1500

In giving a metric pipe size, it is recommended that DN precede the metric value. For example, the conversion of a 2 1/2″ pipe to metric is DN65.

The standard thread for thread pipe is the National Standard Taper Pipe Threads (NPT). The thread on 1/2″ pipe reads 1/2 –14NPT, where 14 is threads per inch. The metric conversion affects only the nominal pipe size—1/2 in this case. The conversion of the 1/2–14NPT pipe thread to metric is DN15–14NPT.

ADDING PLUMBING INFORMATION TO THE FLOOR PLAN

Most residential floor plan drawings prepared by the architect or designer place the plumbing symbols and notes with all other information. Plumbing symbols such as the sinks and water closets aid in reading and understanding the drawing. Plumbing notes are generally minimal. The plumbing symbols and notes may also be placed on a separate drawing or on plumbing layers if CADD is used. Figure 18.20 shows the base floor plan drawing with the plumbing symbols and notes displayed in color.

COMMERCIAL PLUMBING DRAWINGS

In most cases, plumbing drawings are found as an individual component of the complete set of plans for a commercial building. The architect or mechanical engineer prepares the plumbing drawings over the floor-plan layer. This method keeps the drawing clear of any unwanted information and makes it easier for the plumbing contractor to read the print. An example is shown in Figure 18.21. Notice the floor-plan outline with unnecessary detail and information omitted. The plumbing plan shown in color is the only other information provided. As you can see, the plumbing drawing may be fairly complex.

Common Commercial Plumbing Symbols

Displaying the fixture is the most important aspect of a residential plumbing drawing. In fact, usually the actual piping is left off the residential plan. Commercial plumbing drawings are different; they must clearly show the pipe lines and fittings as symbols, with related information given in specific and general notes. Some common plumbing symbols are shown in Figure 18.22. A variety of valve symbols may be used in a plumbing print. A *valve* is a device used to regulate the flow of a gas or liquid. The symbols for common valves are shown in Figure 18.23. The direction in which the pipe is running is also important to plumbing drawings. When a pipe is rising, the end of the pipe is shown as if it were cut through, and the inside is crosshatched with section lines. When a pipe is turned down, or running away from you, the symbol appears as if you are looking at the outside of the pipe. Figure 18.24 shows several common symbols in side view, in a rising pipe view, and in a view with the pipe turning down.

Pipe Size and Elevation Shown on the Drawing

The size of the pipe may be shown with a leader and a note, or with an area of the pipe expanded to provide room for the size to be given, as shown in Figure 18.25.

Some plumbing installations provide the elevation of a run of pipe or the elevation of a pipe fixture when a specific location is required. *Elevation* is the height of a feature from a known base, usually given as 0 (zero elevation). The elevation of a pipe or fitting is given in feet and inches from the base—for example, EL 24′–6″, where EL is the abbreviation for elevation and 24′–6″ is the height from the zero elevation base reference. Elevation may be noted on the print in relationship to a construction member such as TOC ("top of concrete") or TOS ("top of steel"). Elevation may also be related to a location on the pipe, such as BOP ("bottom of pipe") or CL ("center line of pipe").

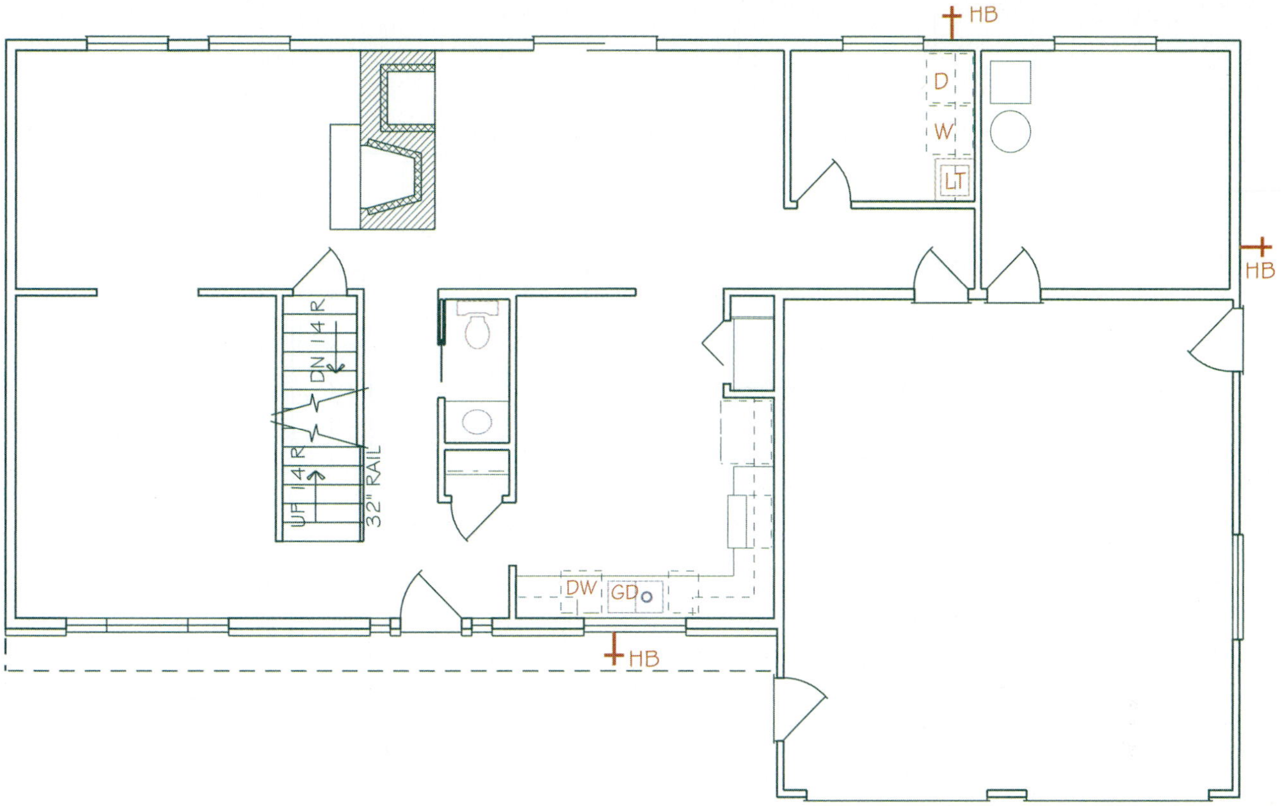

FIGURE 18.20 ■ Base floor plan with plumbing layer shown in color.

Special Symbols, Schedules, and Notes on a Plumbing Drawing

As with any other drawing, plumbing information may be provided in the form of special symbols, schedules, and general notes. Figure 18.22 shows special symbols in a *legend,* and an example of a plumbing connection schedule and a general notes. Architectural and mechanical engineering offices often place a legend on the drawing for reference. The legend displays symbols used on the drawing. The legend and many general notes are often saved as a *block* for insertion in any drawing when needed.

PLUMBING PLAN DRAWING CHECKLIST

Check off the items in the following list as you work on the plumbing plan to be sure that you have included all of the necessary items.

- Sinks: kitchen sink, laundry tray, vanity sinks.
- Dishwasher.
- Clothes washer.
- Garbage disposal.
- Toilets, bidet, urinal.
- Hose bibbs.
- Complete plumbing schedules as needed.
- Review all specific notes.
- General notes.
- Drawing title and scale.
- Title block information.

Water supply system:

- Use proper symbols.
- All pipe sizes specified.
- Water line from main with valve, meter, valve union to house or building.
- Hot water line.
- Cold water line.
- All valves shown.
- Break lines that cross over other lines.

Drainage and vent system:

- All pipe sizes specified.
- Drainage and vent lines drawn.
- Fixtures labeled or symbols shown.

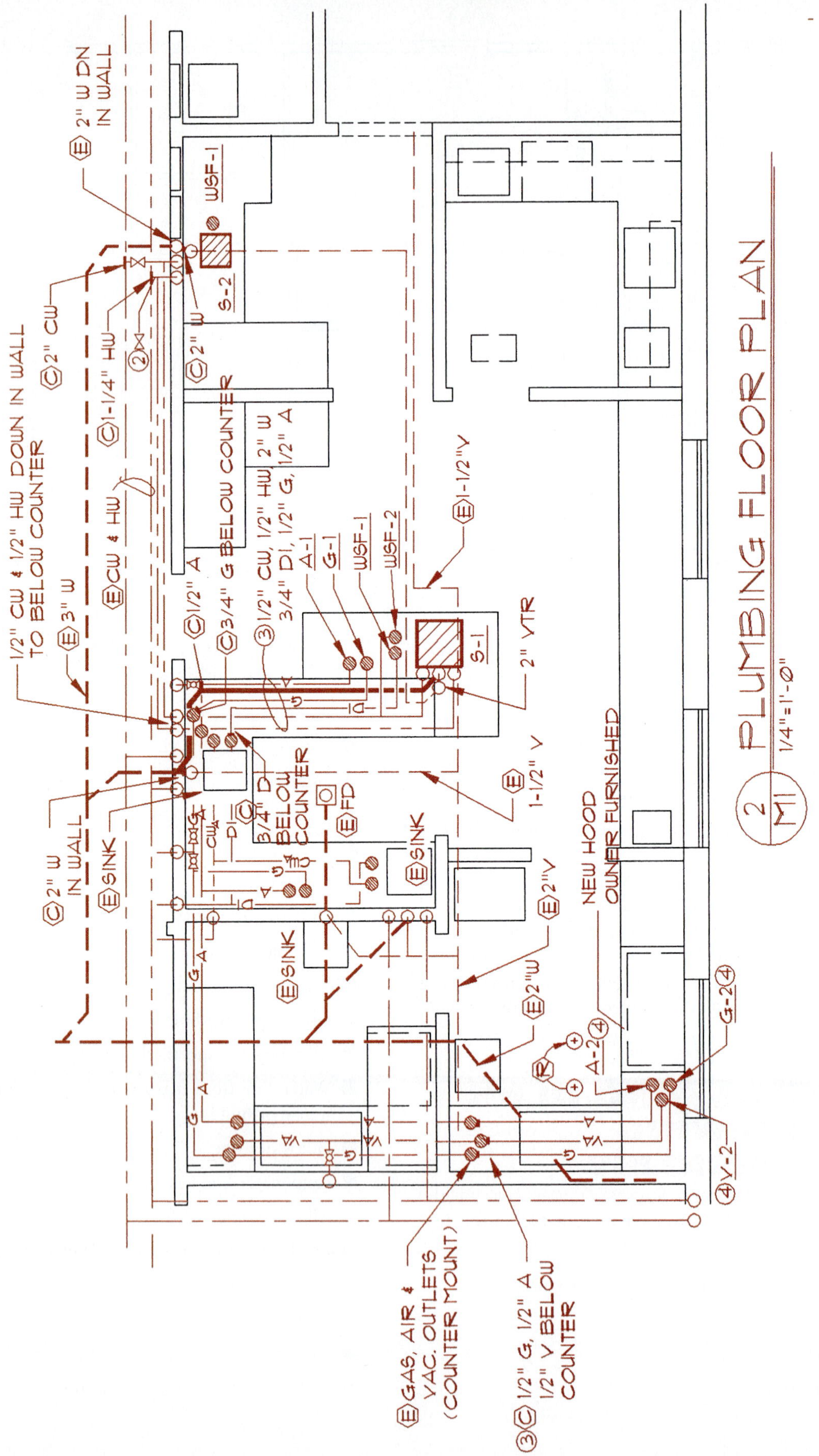

FIGURE 18.21 ■ Architectural floor plan with the plumbing layout shown on a color CADD layer. *Courtesy System Design Consultants.*

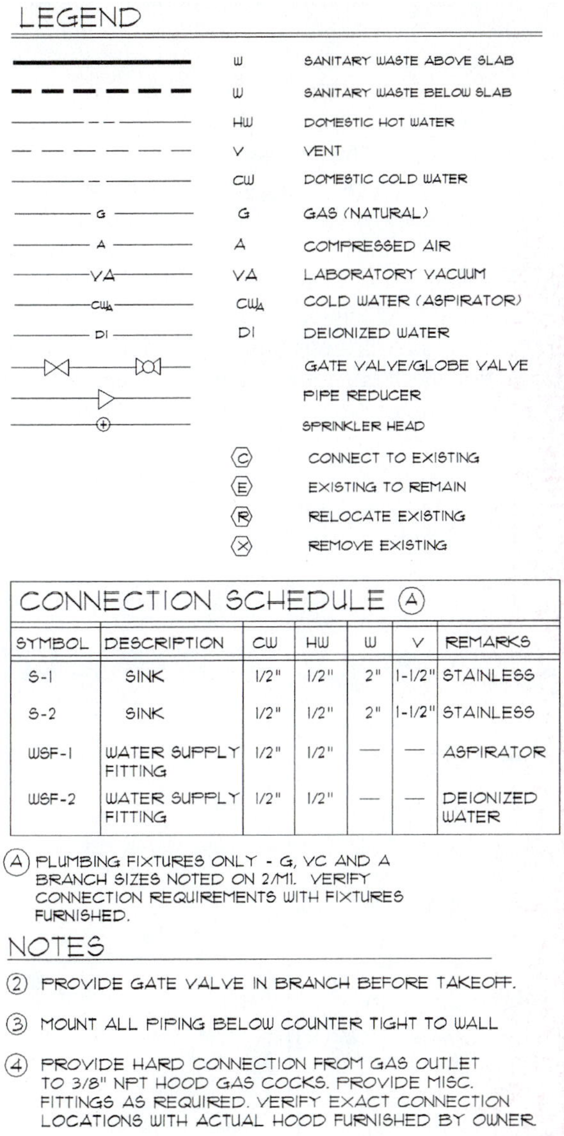

CONNECTION SCHEDULE (A)

SYMBOL	DESCRIPTION	CW	HW	W	V	REMARKS
S-1	SINK	1/2"	1/2"	2"	1-1/2"	STAINLESS
S-2	SINK	1/2"	1/2"	2"	1-1/2"	STAINLESS
WSF-1	WATER SUPPLY FITTING	1/2"	1/2"	—	—	ASPIRATOR
WSF-2	WATER SUPPLY FITTING	1/2"	1/2"	—	—	DEIONIZED WATER

(A) PLUMBING FIXTURES ONLY - G, VC AND A BRANCH SIZES NOTED ON 2/M1. VERIFY CONNECTION REQUIREMENTS WITH FIXTURES FURNISHED.
NOTES
(2) PROVIDE GATE VALVE IN BRANCH BEFORE TAKEOFF.
(3) MOUNT ALL PIPING BELOW COUNTER TIGHT TO WALL
(4) PROVIDE HARD CONNECTION FROM GAS OUTLET TO 3/8" NPT HOOD GAS COCKS. PROVIDE MISC. FITTINGS AS REQUIRED. VERIFY EXACT CONNECTION LOCATIONS WITH ACTUAL HOOD FURNISHED BY OWNER.

FIGURE 18.22 ■ Some common plumbing symbols, a schedule, and general notes. *Courtesy System Design Consultants.*

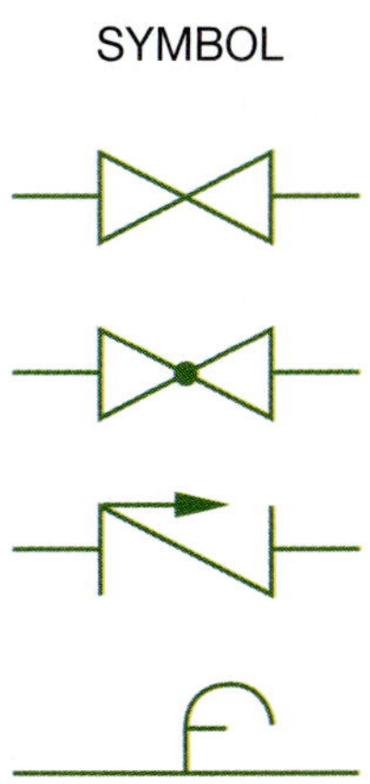

FIGURE 18.23 ■ Common valve symbols.

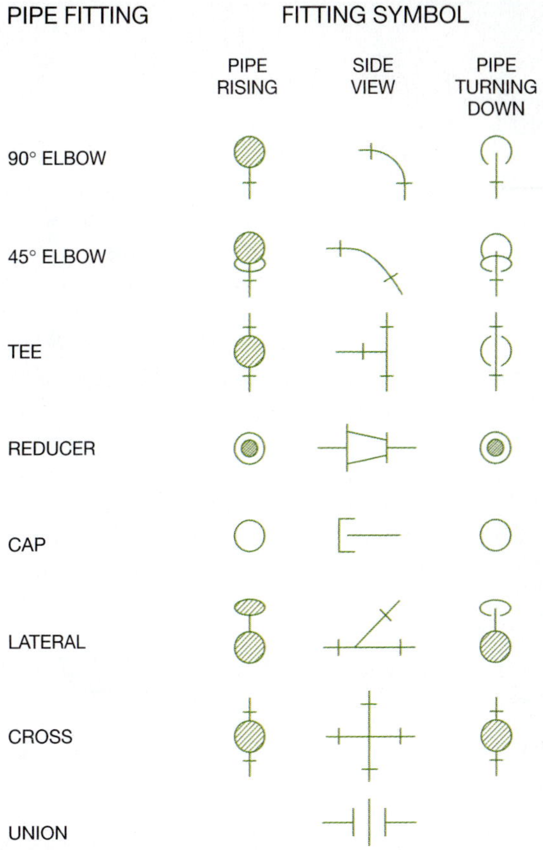

FIGURE 18.24 ■ Several common piping symbols in side view, in a rising pipe view, and in a view with the pipe turning down.

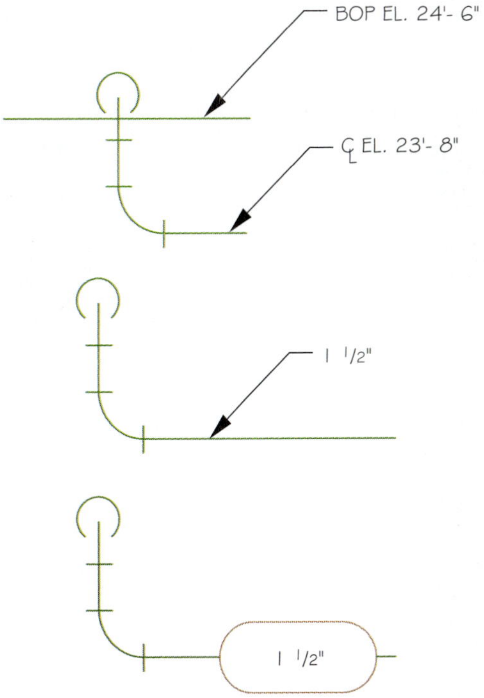

FIGURE 18.25 ■ Specifying the pipe size and elevation.

DRAWING PLUMBING SYMBOLS WITH CADD

In residential architecture, plumbing drawings usually amount to placing floor-plan symbols such as sinks, water closets, and tubs. In commercial architecture applications, a separate plumbing drawing shows the complete plumbing layout with symbols and pipes on a separate CADD layer. The floor plan serves as the base sheet. So, by turning the plumbing layer ON, you have a composite drawing with the floor plan, plumbing symbols, and plumbing layout shown. You also have the flexibility to turn the plumbing layer OFF when you want to display the floor plan without plumbing. Many companies show two separate drawings for commercial projects: (1) the floor plan completely dimensioned and (2) the floor-plan walls and plumbing layout. This is easy to do, because all you have to do is turn ON or OFF specific layers as needed to display what you want. Some architectural CADD programs have menus with a symbols library for plumbing applications, as shown in Figure 18.26. You select the desired symbol from the menu and then drag the symbol into position on the drawing. Parametric design plays an important role in applications such as multiple water closet stalls in a commercial building rest room. You specify the distance between walls or the width of each stall, the number of stalls, the stall length, and the door-swing direction; the computer then automatically draws the entire series of rest room stalls.

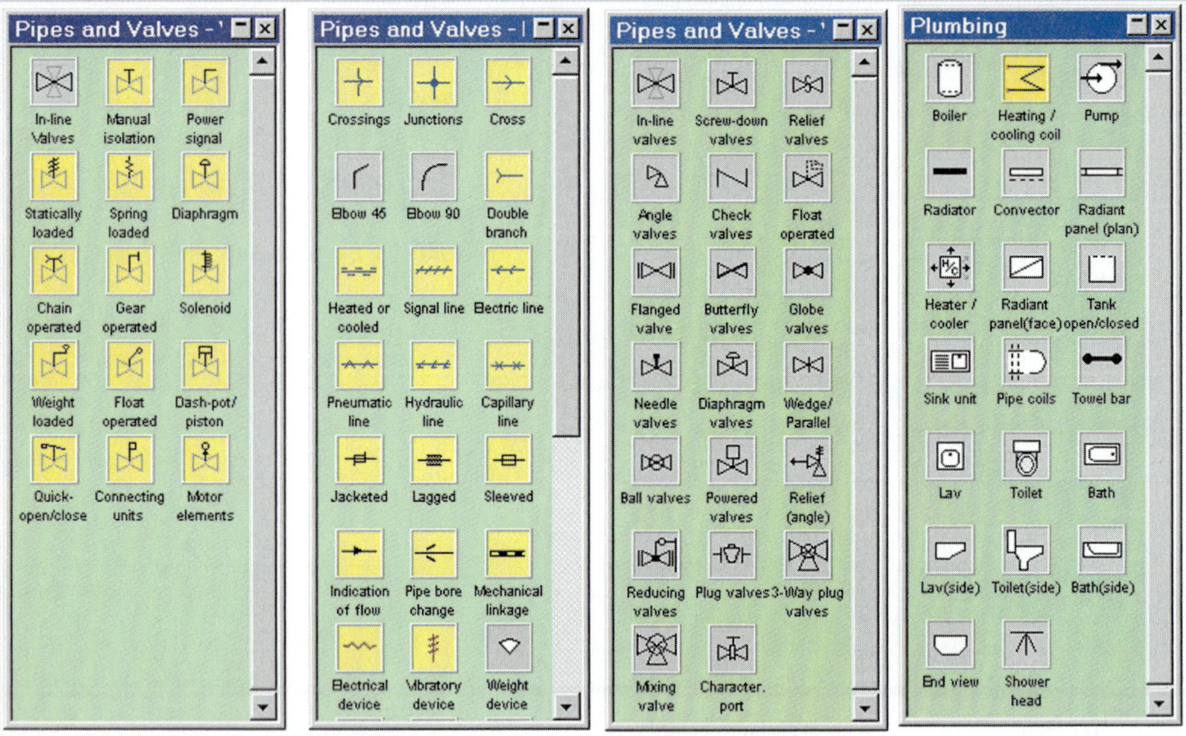

FIGURE 18.26 ■ CADD symbols library for plumbing and piping applications. *Courtesy Visio Corporation.*

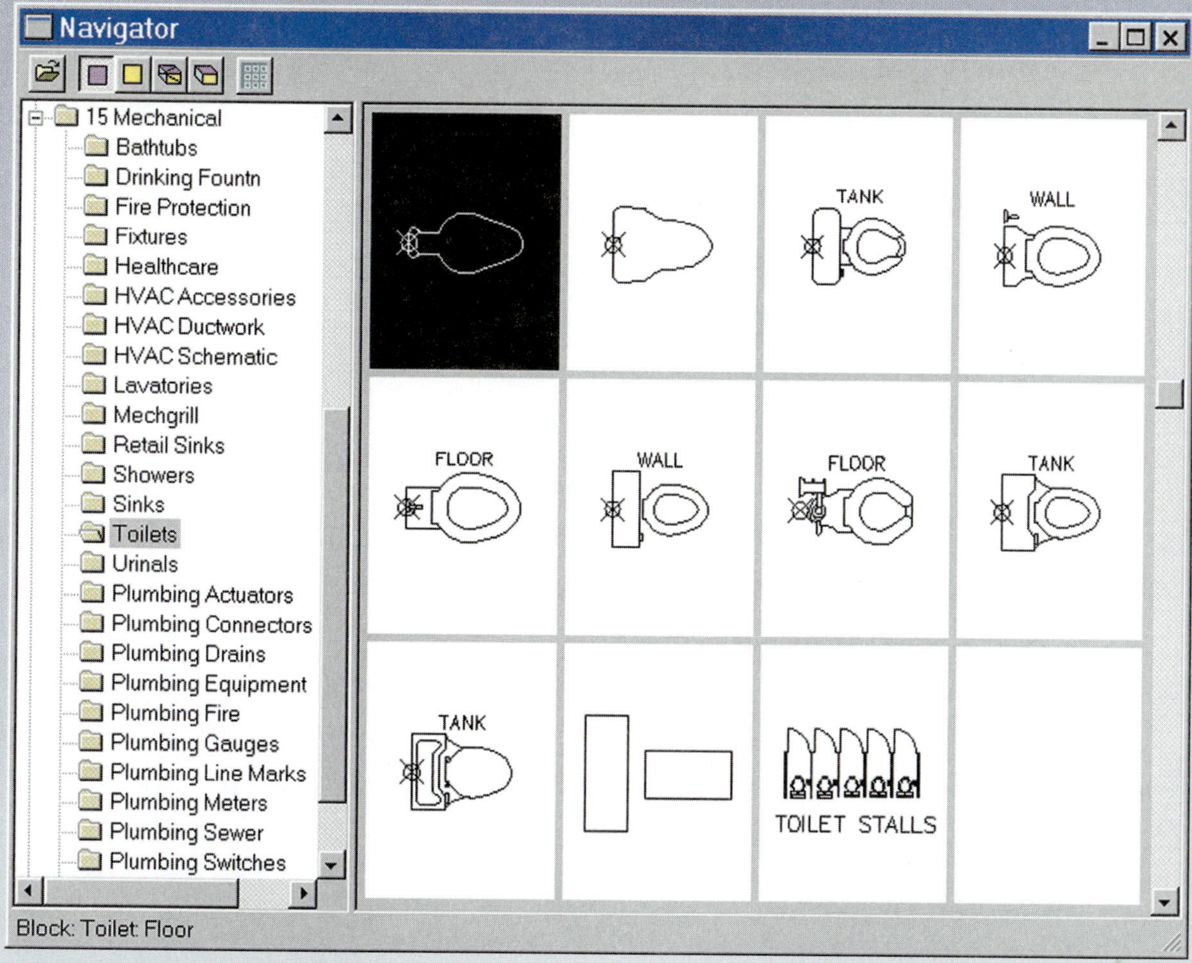

FIGURE 18.26 ■ *(Continued)*

CADD PLUMBING LAYERS

CADD plumbing layers have a P designation for the major group identification and then four letter abbreviations for the minor group when the American Institute of Architects (AIA) layer guidelines are used. The following list gives recommended layer names:

LAYER NAME	LAYER DESCRIPTION	LAYER NAME	LAYER DESCRIPTION
P-ACID	Acid, alkaline, and oil waste systems	P-ELEV	Plumbing elevations
P-DOMW	Domestic hot and cold water systems	P-SECT	Plumbing sections
P-SANR	Sanitary drainage	P-DETL	Plumbing details
P-STRM	Storm drainage system	P-PPLM	Plumbing plan
P-EQPM	Miscellaneous plumbing equipment	P-SCHD	Plumbing schedules
P-FIXT	Plumbing fixtures		

Chapter 18 Additional Reading

The following Web sites can be used as a resource to help you keep current with changes in plumbing materials.

ADDRESS	COMPANY OR ORGANIZATION
www.americanstandard.com	American Standard
www.aquaglass.com	Aqua Glass (Bathing fixtures)
www.flowguardgold.com	B F Goodrich
www.nibco.com	Nibco (Flow Control)

CHAPTER 18

Plumbing Plans Test

DIRECTIONS

Answer the questions with short, complete statements or drawings as needed on an 8 1/2″ × 11″ sheet of notebook paper, as follows:

1. Letter your name, Plumbing Plans Test, and the date at the top of the sheet.
2. Letter the question number and provide the answer. You do not need to write out the question.

Answers may be prepared on a word processor if course guidelines allow this.

QUESTIONS

Question 18–1 Floor-plan plumbing symbols are generally drawn at what scale?

Question 18–2 Identify at least four methods that can contribute to energy-efficient plumbing.

Question 18–3 Define the following plumbing abbreviations:

a. CO
b. FD
c. VTR
d. WC
e. WH
f. SH

Question 18–4 What information is required on a plumbing schedule?

Question 18–5 Are plumbing drawings required by all contractors? Why or why not?

Question 18–6 List at least one advantage and one disadvantage of solar hot water systems.

Question 18–7 Briefly explain how public sewer and private septic systems differ.

Question 18–8 What is the name for steel pipe that has been cleaned and dipped in molten zinc?

Question 18–9 Give the name of a steel pipe, protected with a coat of varnish, that is used for natural gas.

Question 18–10 Briefly describe CSST.

Question 18–11 Why is stainless steel pipe often used in chemical, pollution control, pharmaceutical, and food industries?

Question 18–12 Define *soil pipe.*

Question 18–13 Give the proper term for a vented fitting that provides a liquid seal to prevent the emission of sewer gases without affecting the flow of sewage or waste water.

Question 18–14 Define *vent pipe.*

Question 18–15 List at least four factors that influence the sizing of water supply pipes.

PROBLEMS

Use 8 1/2″ × 11″ vellum and pencil or a CADD system to prepare the following drawings, unless otherwise specified.

Problem 18–1 Draw the following plumbing fixture symbols in plan view:

a. Water closet
b. Bathtub 5′–0″ long × 2′–6″ wide
c. Shower 3′–0″ × 3′–0″
d. Urinal

Problem 18–2 Draw the following plumbing piping symbols:

a. Hot water supply
b. Cold water supply
c. Soil waste pipe
d. Gate valve
e. 90° elbow
f. Tee

Problem 18–3 Make a single-line isometric drawing of a typical kitchen sink piping diagram.

Problem 18–4 Make a single-line isometric drawing of a typical drain, waste, and vent system.

Problem 18–5 Make a floor-plan plumbing drawing of a typical back-to-back bath arrangement.

Problem 18–6 Draw a partial single line water supply system (see Figure 18.6).

Problem 18–7 Draw a typical bathroom drainage and vent system. See Figure 18.10.

Problem 18–8 Continue the floor plan(s) that you started in Chapter 16 by adding any additional symbols and notes related to the plumbing, such as hose bibbs, washer, laundry trays, sinks, and water heater.

19 CHAPTER

Heating, Ventilating, and Air-Conditioning

THE BOCA NATIONAL ENERGY CONSERVATION CODE REQUIREMENTS

The Building Officials and Code Administrators (BOCA) International updates the National Energy Conservation Code every three years. The purpose of the code is to regulate the design and construction of the exterior envelope and selection of heating, ventilating, and air-conditioning (HVAC); service water heating; electrical distribution and illuminating systems; and equipment required for effective use of energy in buildings for human occupancy. The *exterior envelope* is made up of elements of a building that enclose conditioned (heated and cooled) spaces through which thermal energy transfers to or from the exterior.

The following are general code requirements for some common heating and cooling equipment and duct systems. Specific applications and installations must be confirmed with the exact code regulations and manufacturers' specifications.

Heating and Cooling Equipment

- Heating and cooling equipment located in a garage shall be protected from impact by automobiles. This is generally done by placing a steel pipe embedded in concrete in front of the equipment.
- Fixed appliances shall be fastened in place.
- Heating and cooling equipment shall be located and with clearances to allow access for servicing and replacement of the largest piece of equipment. This overrides minimum requirements.
- A door or access at least 20″ (508 mm) or large enough to remove the largest piece of equipment shall be provided to access the equipment location. An unobstructed working space not less than 30″ (762 mm) wide and high shall be provided next to the control side of the equipment.
- When equipment is located in an attic, a passageway at least 22″ (559 mm) wide by 30″ (762 mm) high shall be provided from the attic access to the equipment and controls.
- An electrical outlet and lighting fixture shall be provided near the equipment and shall have a switch at the doorway or access.
- Equipment located in a crawl space shall have access and a passageway not less than 20″ (508 mm) wide by 30″ (762 mm) high.
- Equipment located on the ground shall be supported on a concrete slab and shall be at least 3″ (76 mm) above the ground.
- Suspended equipment shall be at least 6″ (153 mm) from the ground.
- When equipment is placed in an excavated (dug-out) area, there shall be at least 12″ (305 mm) on the sides and 30″ (762 mm) in front of the controls. Excavations deeper than 12″ (305 mm) shall have a concrete or masonry wall extending 4″ (102 mm) above the adjoining ground.
- Equipment installed outdoors shall be listed and labeled for such use.
- Fuel-burning warm-air furnaces shall not be installed in a room designed for use as a storage closet. Furnaces located in a bedroom or bathroom shall be installed in a sealed enclosure so that combustion air will not be taken from the living space. *Combustion air* is air provided at a burner for proper combustion.
- Direct-vent furnaces are not required to be in an enclosure.
- Fuel-burning warm-air furnaces shall be supplied with combustion air.
- Return air shall be taken from inside the dwelling and can be mixed with outside air. Return air cannot be taken from kitchens, bathrooms, garages, or other dwelling units. Outside air shall not be taken from within 10′ (3048 mm) of an appliance or plumbing vent outlet that is located less than 3′ (914 mm) above the air inlet.

Duct Systems

- Equipment connected to duct systems shall have a 250°F (121°C) temperature limit control.
- Galvanized steel ducts shall be ASTM A 525 specification.
- Metal ducts shall have the following thickness.
  Round or rectangular enclosed: 14″ or less, 0.016″ thick; over 14″, 0.019″ thick.

Rectangular exposed: 14″ or less, 0.019″ thick; over 14″, 0.022″ thick.

- Factory-made ducts shall be approved for installation and have UL label and compliance. UL is the abbreviation for Underwriters' Laboratories, Inc., which tests and labels products for approved uses.
- Gypsum products can be used as ducts if the air temperature does not exceed 125°F (52°C) and there is no condensation.
- Spaces between structural members can be used as return ducts if they are isolated from other spaces with tight-fitting sheet metal or wood at least 2″ (51 mm) thick.
- Underground duct systems shall be constructed of approved concrete, clay, metal, or plastic.
- Duct coverings and linings shall have a flame-spread rating not greater than 25 and a smoke-development rating not more than 50, and shall withstand a test temperature of at least 250°F (121°C).
- Insulated ducts shall have a minimum thermal resistance value of R-4.2. Insulate ducts in unconditioned areas.
- Exterior ducts shall have a weatherproof covering.
- Joints and seams shall be made substantially airtight.
- Metal ducts shall be supported by 18-gage metal straps or 12-gage galvanized wire spaced no more than 10′ (3048 mm) apart and must be at least 4″ (102 mm) from the ground. Nonmetallic ducts shall be supported to the manufacturer's specifications.
- Crawl spaces used as a supply plenum shall not have gas lines or plumbing cleanouts, shall be clean of loose combustible material, and shall be airtight; the ground shall be covered with a 4 mill (0.102 mm) moisture barrier; and the space shall be formed by materials with a flame spread rating of no more than 200. A *plenum* is an enclosed duct or area that is connected to a number of branch ducts.

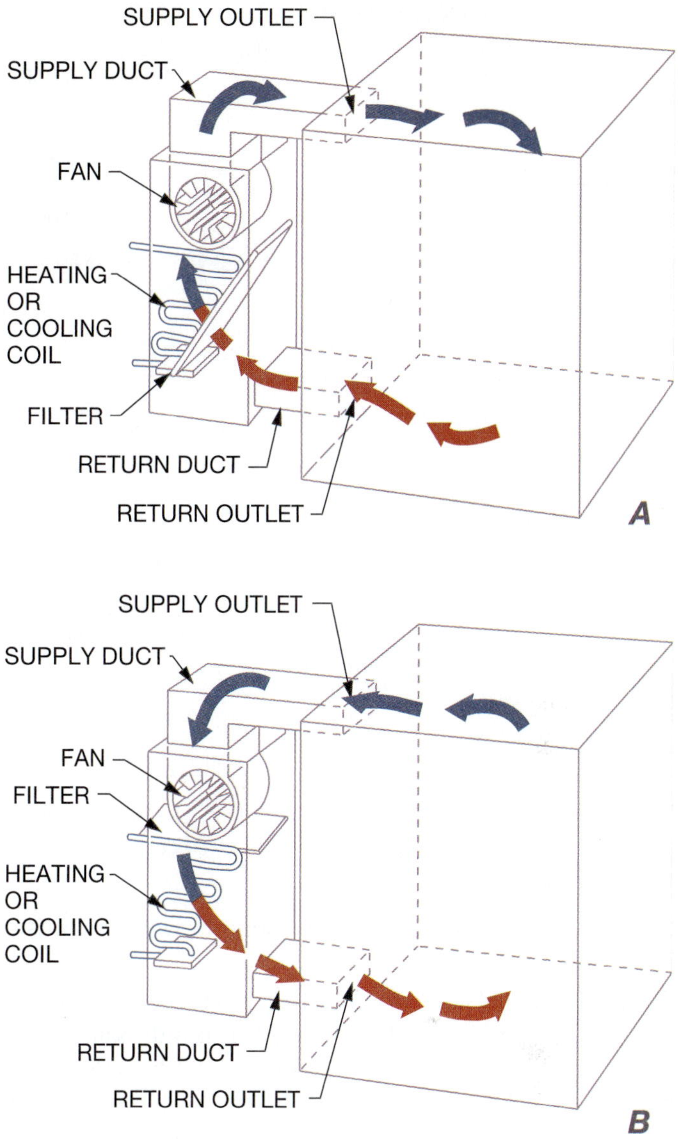

FIGURE 19.1 ■ (*A*) Downdraft forced-air system heated air cycle. (*B*) Updraft forced-air system heated air cycle.

CENTRAL FORCED-AIR SYSTEMS

One of the most common systems for heating and air-conditioning circulates the air from living spaces through or around heating or cooling devices. A fan forces the air into sheet metal or plastic pipes called *ducts*, which connect to openings called *diffusers*, or *air supply registers*. Warm air or cold air passes through the ducts and registers to enter the rooms, heating or cooling them as needed.

Air then flows from the room through another opening into the return duct, or return air register. The return duct directs the air from the rooms over the heating or cooling device. If warm air is required, the return air is passed over the surface of either a combustion chamber (the part of a furnace where fuel is burned) or a heating coil. If cool air is required, the return air passes over the surface of a cooling coil. Finally, the conditioned air is picked up again by the fan and the air cycle is repeated. Figure 19.1 shows the air cycle in a forced-air system.

Heating Cycle

If the air cycle just described is used for heating, the heat is generated in a furnace. Furnaces for residential heating produce heat by burning fuel oil or natural gas, or by using electric heating coils or heat pumps. If the heat comes from burning fuel oil or natural gas, the combustion (burning) takes place inside a combustion chamber. The air to be heated does not enter the combustion chamber but absorbs heat from the outer surface of the chamber. The gases given off by combustion are vented through a chimney. In an electric furnace, the air to be heated is passed directly over the heating coils. This type of furnace does not require a chimney.

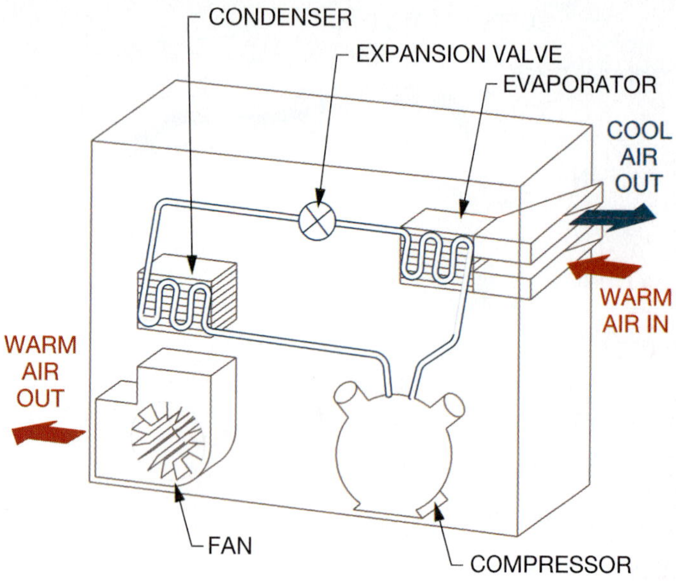

FIGURE 19.2 ■ Schematic diagram of refrigeration cycle. *From Lang,* Principles of Air Conditioning, *3d ed., Delmar Publishers Inc.*

Cooling Cycle

If the air from the room is to be cooled, it is passed over a cooling coil. The most common type of residential cooling system is based on two principles:

1. As liquid changes to vapor, it absorbs large amounts of heat.
2. The boiling point of a liquid can be changed by changing the pressure applied to the liquid. This is the same as saying that the temperature of a liquid can be raised by increasing its pressure and lowered by reducing its pressure.

Common refrigerants boil (change to a vapor) at very low temperatures, some as low as 21°F (6°C) below zero.

The principal parts of a refrigeration system are the cooling coil (evaporator), the compressor, the condenser, and the expansion valve. Figure 19.2 shows a diagram of the cooling cycle. The cooling cycle operates as follows. Warm air from the ducts is passed over the evaporator. As the cold liquid refrigerant moves through the evaporator coil, it picks up heat from the warm air. As the liquid picks up heat, it changes to a vapor. The heated refrigerant vapor is then drawn into the compressor, where it is put under high pressure. This raises the temperature of the vapor even more.

Next, the high-temperature, high-pressure vapor passes to the condenser, where the heat is removed. This is done by blowing air over the coils of the condenser. As the condenser removes heat, the vapor changes to a liquid. It is still under high pressure, however. From the condenser, the refrigerant flows to the expansion valve. As the liquid refrigerant passes through the valve, the pressure is reduced; this lowers the temperature of the liquid still further so that it is ready to pick up more heat.

The cold low-pressure liquid then moves to the evaporator. The pressure in the evaporator is low enough to allow the refrigerant to boil again and absorb more heat from the air passing over the coil of the evaporator.

Forced-Air Heating Plans

Complete plans for the heating system may be needed in applying for a building permit or a mortgage, depending on the requirements of the local building jurisdiction or the lending agency. If a complete heating layout is required, it can be prepared by the architectural drafter, a heating contractor, or a mechanical engineer.

When forced-air electric, gas, or oil heating systems are used, the warm-air outlets and return air locations may be shown, as in Figure 19.3. Notice that the registers are normally placed in front of a window so warm air is circulated next to the coldest part of the room. As the warm air rises, a circulation action is created as the air goes through the complete heating cycle. Cold-air returns are often placed in the ceiling or floor of a central location.

A complete forced-air heating plan shows the size, location, and number of Btu dispersed to the rooms from the warm-air supplies. Btu stands for *British thermal unit,* a measure of heat. The location and size of the cold-air return and the location, type, and output of the furnace are also shown.

The warm-air registers are sized in inches—for example, 4″ × 12″. The size of the duct is also given, as shown in Figure 19.4. The note 20/24 means a 20″ × 24″ register; a number 8 next to a duct means an 8″-diameter duct. This same system may be used as a central cooling system when cool air is forced from an air conditioner through the ducts and into the rooms. WA means warm air, and RA is return air. CFM is cubic feet per minute, the rate of airflow.

Providing Duct Space

When ducted heating and cooling systems are used, the location of the ducts often becomes a serious consideration. The ducts are placed in a crawl space or attic when possible. When ducts cannot be confined to a crawl space or attic, they must be run inside the occupied areas of the home or building. The designer and building contractor try to conceal ducts when they must be placed in locations where they could be seen. There are several ways to conceal ducts.

When they run parallel to structural members, ducts can be placed within the space created by the structural members. This is referred to as running the ducts in the *joist space, rafter space,* or *stud space;* these terms identify the type of construction members used. This works when the duct size is equal to or smaller than the size of the construction members. In the case of return air ducts, the construction members and enclosing materials can be used as the duct plenum. When this can be done, no extra framing needs to be done to conceal the ducts. When ducts must be exposed to occupied areas, they are normally enclosed by framing covered with gypsum or other finish material. When ducts are run in an area such as an unfinished basement, it is possible to leave them exposed.

When possible, ducts are run in the ceiling of a hallway, because they can be framed and finished with a lowered ceiling. The

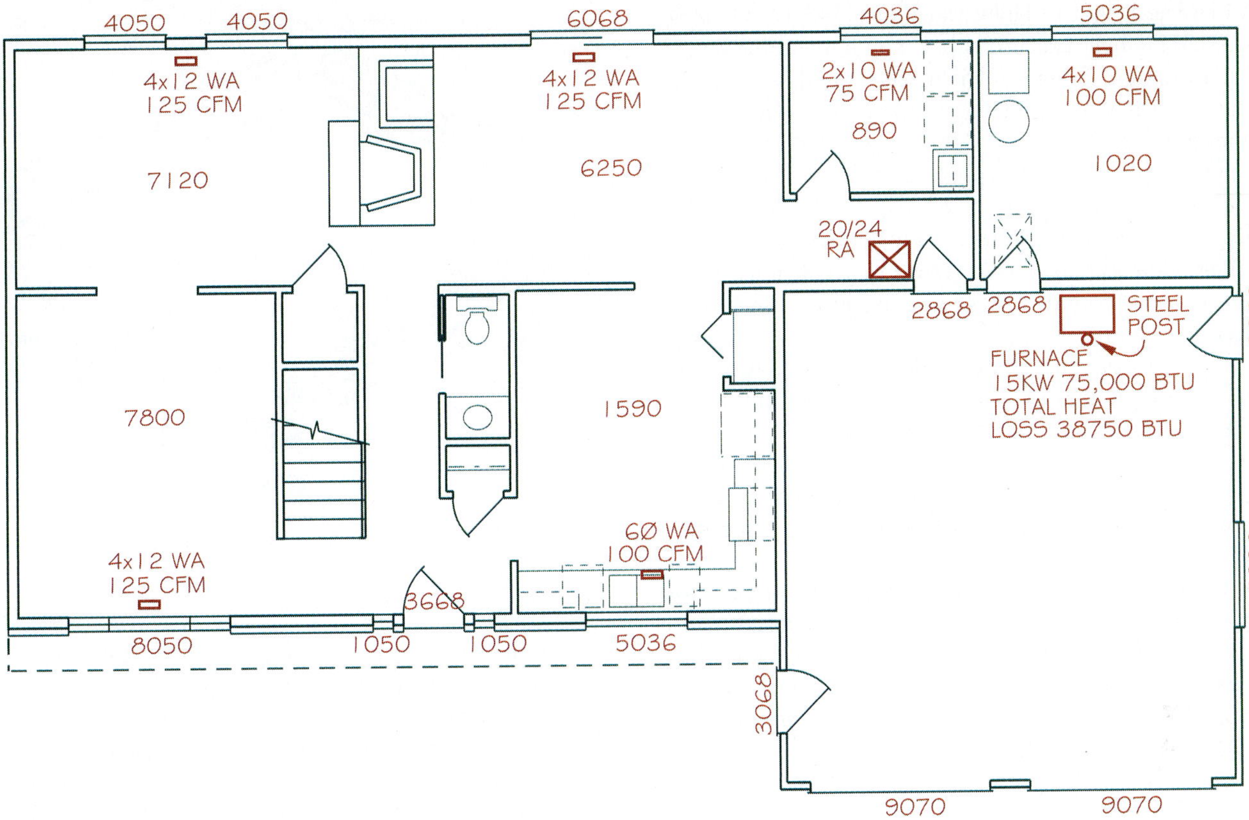

FIGURE 19.3 ■ Simplified forced-air plan.

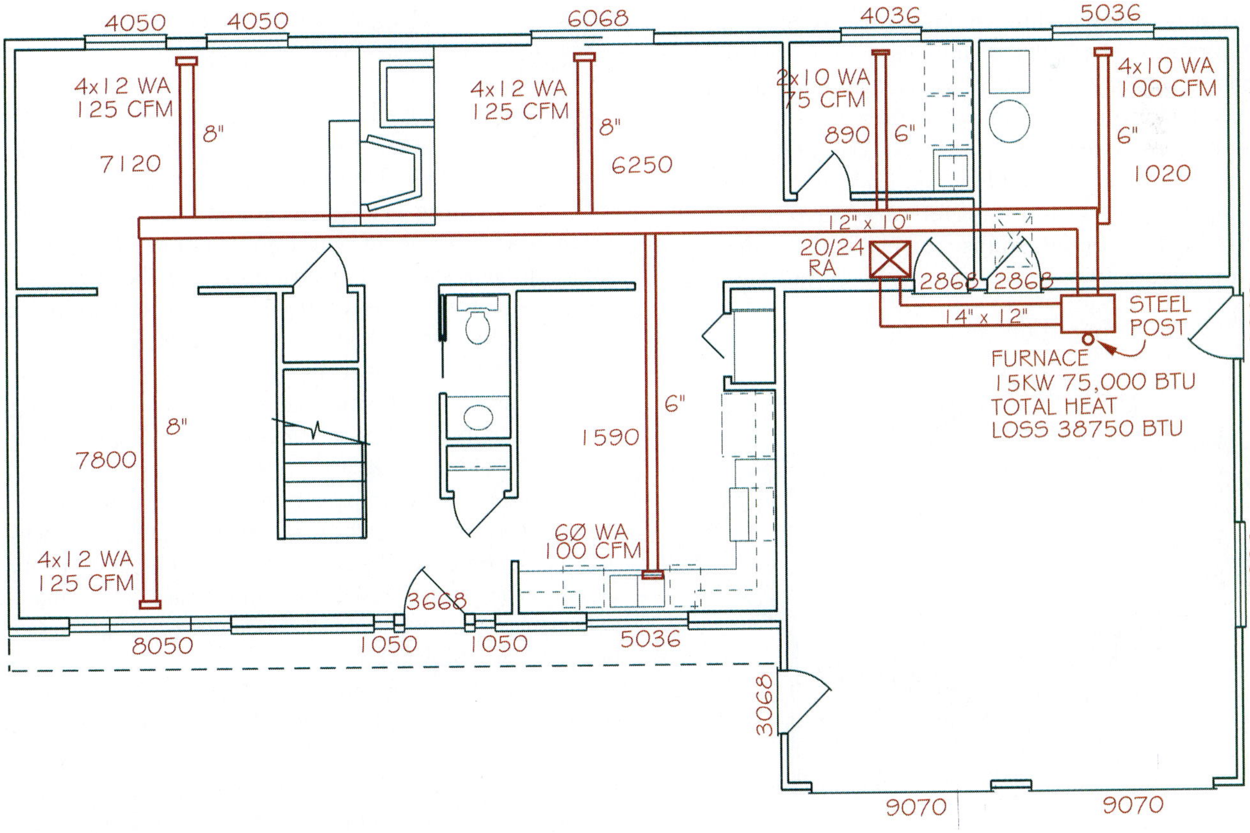

FIGURE 19.4 ■ Detailed forced-air plan.

minimum ceiling height in hallways can be 7′–0″ (2134 mm). Bathrooms and kitchens can also have a minimum ceiling height of 7′–0″ (2134 mm), though this is normally undesirable. The typical ceiling height in habitable rooms is approximately 8′ (2440 mm), but a ceiling can be as low as 7′–6″ (2286 mm) for at least 50 percent of the room, with no portion less than 7′–0″ (2134 mm). This information is valuable, because it tells you how low ceilings can be framed down to provide space for ducts. When running ducts along a ceiling creates a problem with ceiling height, the wall framing needs to be higher to give enough floor-to-ceiling height.

When ducts must be run between floors, they need to be run in an easily concealed location, such as in a closet or in the stud space. The stud space can be used for ducts that are 3 1/2″ deep for 2×4 studs, 5 1/2″ deep for 2×6 studs, or 7 1/2″″ deep for 2×8 studs. If the duct can be run up through a closet, then it can be framed in and covered in a chase. A *chase* is a continuous recessed area built to carry or conceal ducts, pipes or other construction products. If the duct cannot be located in a convenient place for concealment, then it might need to be framed into the corner of a room; although, this is normally not preferred. The framing for a chase is shown on the floor plan as a wall surrounding the duct to be concealed and a note is usually placed that indicates the use, such as CHASE FOR 22 × 24RETURN DUCT.

HOT WATER SYSTEMS

In a hot water system, the water is heated in an oil- or gas-fired boiler and then circulated through pipes to radiators or convectors in the rooms. The boiler is supplied with water from the fresh water supply for the house. The water is circulated around the combustion chamber, where it absorbs heat.

In one kind of system, one pipe leaves the boiler and runs through the rooms of the building and back to the boiler. In this one-pipe system, the heated water leaves the supply, is circulated through the outlet, and is returned to the same pipe, as shown in Figure 19.5. In a two-pipe system, two pipes run

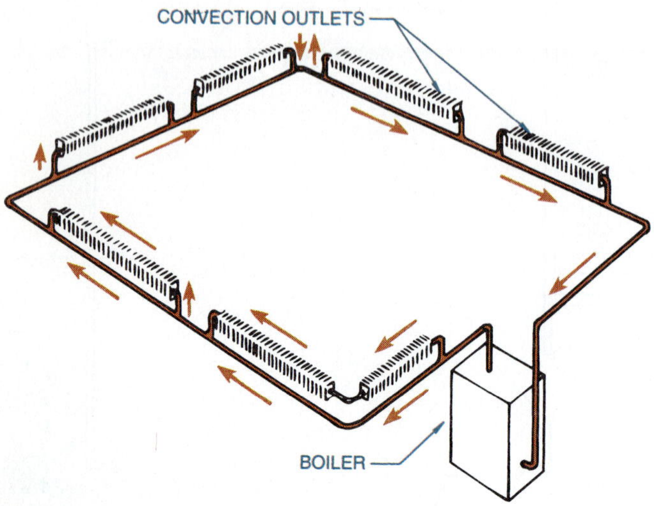

FIGURE 19.5 ■ One-pipe hot water system.

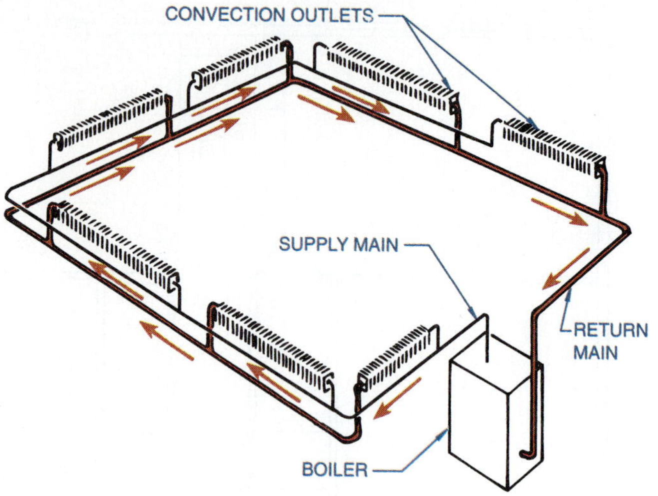

FIGURE 19.6 ■ Two-pipe hot water system.

throughout the building. One pipe supplies heated water to all of the outlets. The other is a return pipe, which carries the water back to the boiler for reheating, as seen in Figure 19.6.

Hot water systems use a pump, called a *circulator,* to move the water through the system. The water is kept at a temperature between 150°F and 180°F in the boiler. When heat is needed, the thermostat starts the circulator, which supplies hot water to the convectors in the rooms.

HEAT PUMP SYSTEMS

A *heat pump* is a forced-air central heating and cooling system. It operates using a compressor and a circulating refrigerant system. Heat is extracted from the outside air and pumped inside the structure. The heat pump supplies up to three times as much heat per year for the same amount of electrical consumption as a standard electric forced-air heating system. In comparison, this can result in a 30 to 50 percent annual energy saving. In the summer the cycle is reversed, and the unit operates as an air conditioner. In this mode, the heat is extracted from the inside air and pumped outside. On the cooling cycle the heat pump also acts as a dehumidifier. Figure 19.7 shows how a heat pump works.

Residential heat pumps vary in size from 2 to 5 tons. Some vendors carry two-stage heat pumps that alternate—between 3 and 5 tons, for example. During minimal demand, the more efficient 3-ton phase is used; the 5-ton phase is operable during peak demand. Each ton of rating removes approximately 12,000 Btu per hour (Btuh) of heat.

Residential or commercial structures that, because of their size or design, cannot be uniformly heated or cooled from one compressor may require a split system with two or more compressors. The advantage of such an arrangement may be that two smaller units can more effectively control the needs of two zones within the structure.

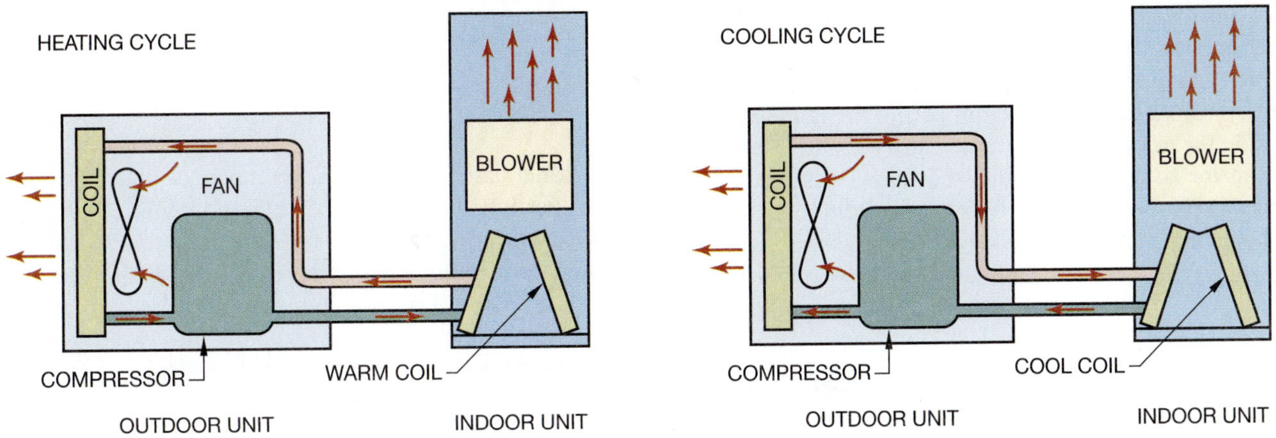

FIGURE 19.7 ■ Heat pump heating and cooling cycle. *Courtesy Lennox Industries, Inc.*

Other features, such as an air cleaner, a humidifier, or an air freshener, can be added to the system. In general, the initial cost of the heat pump will be over twice as much as a conventional forced-air electric heat system. However, the advantages and long-range energy savings make it a significant option for heating. Keep in mind that if cooling is a requirement, the cost difference is less.

The total heat pump system uses an outside compressor, an inside blower to circulate air, a backup heating coil, and a complete duct system. Heat pump systems move a large volume of air; therefore, the return air and supply ducts must be adequately sized. Figure 19.8 shows the relationship between the compressor and the balance of the system. The compressor should be placed in a location where some noise will not cause a problem. The compressor should be placed on a concrete slab about 36″ × 48″ × 6″ in size in a location that will allow adequate service access. To avoid transmitting vibration, do not connect the slab to the structure. Show or note the concrete slab on the foundation plan. Verify the vendor's specifications. Heat pumps may not be as efficient in some areas of the country as in other areas because of annual low or high temperatures. Verify the product efficiency with local vendors.

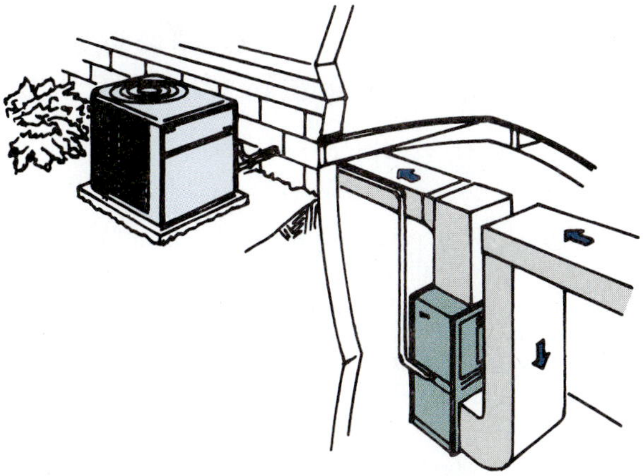

FIGURE 19.8 ■ Heat pump compressor. *Courtesy Lennox Industries, Inc.*

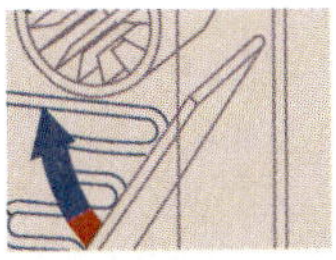

ZONE CONTROL SYSTEMS

A zoned heating system requires one heater and one thermostat per room. No ductwork is required, and only the heaters in occupied rooms need be turned on.

One of the major differences between a zonal and a central system is flexibility. A zonal heating system allows the occupant to determine how many rooms are heated, how much energy is used, and how much money is spent on heat. A zonal system allows the home to be heated to the family's needs; a central system requires using all the heat produced. If the airflow is restricted, the efficiency of the central system is reduced. There is also a 10 to 15 percent heat loss through ductwork in central systems.

Regardless of the square footage in a house, its occupants normally use less than 40 percent of the entire area on a regular basis. A zoned system is very adaptable to heating the 40 percent of the home that is occupied: automatic controls allow for night setback, day setback, and unheated areas. The homeowner can save as much as 60 percent on energy costs through controlled heating systems.

There are typically two types of zone heaters: baseboard and fan. Baseboard heaters have been the most popular type for zoned heating system for the past several decades. They are used in many different climates and under various operating conditions. No ducts, motors, or fans are required. Baseboard units have an electric heating element, which causes a convection current as the air around the unit is heated. The heated air rises into the room and is replaced by cooler air that falls to the floor. Baseboard heaters should be placed on exterior walls under or next to windows or other openings. These units do project a few inches into the room at floor level. Some homeowners do not care for this obstruction, in comparison with the floor duct vent of a central system or the recessed wall-mounted fan heater. Furniture arrangements should be a factor in locating any heating element.

Fan heaters are generally mounted in a wall recess. A resistance heater is used to generate heat, and a fan circulates the heat into the room. These units should be placed to circulate the

warmed air in each room adequately. Avoid placing the heaters on exterior walls, as a recessed unit reduces or eliminates insulation. The fan in these units does cause some noise.

Split systems using zonal heat in part of the home and central heat in the rest of the structure are also possible. This is also an option for additions to homes that have a central system. Zoned heat can be used effectively in an addition so that the central system is not overloaded and need not be replaced.

RADIANT HEAT

Radiant heating and cooling systems provide a comfortable environment by controlling surface temperatures and minimizing excessive air motion within the space. Warm ceiling panels are effective for winter heating because they warm the floor surfaces and glass surfaces by direct transfer of radiant energy. The surface temperature of well-constructed and properly insulated floors is 2°F to 3°F (−16.4°C) above the ambient air temperature, and the inside surface temperature of glass is increased significantly. *Ambient temperature* is the temperature of the air surrounding a heating or cooling device. As a result of these heated surfaces, downdrafts are minimized to the point where no discomfort is felt.

Radiant heat systems yield operating cost savings of 20 to 50 percent annually, compared with conventional convective systems. This saving is accomplished through lower thermostat settings and through the superior, cost-effective design inherent in radiant heating products.

It is generally accepted that 3 to 4 percent of the energy cost is saved for each degree the thermostat is lowered. Users of surface-mounted radiant panels take advantage of this fact in two ways. First, they are able to achieve comfort at a lower ambient air temperature, normally 60°F to 64°F (16.6°C) as compared with convection heating air temperatures of 68°F to 72°F (21°C). Second, they are able to practice comfortable day and night temperature setback of about 58°F in areas used frequently, 55°F (12.8°C) in areas used occasionally, and 50°F (10°C) in areas used seldom.

Radiant heat may be achieved with oil- or gas-heated hot water piping in the floor or ceiling, to electric coils, wiring, or elements either in or above the ceiling gypsum board, and transferred to metal radiator panels generally mounted by means of a bracket about 1″ below the ceiling surface.

Zone radiant heating panels are an alternative when space is limited or the location of built-in zone heaters may be a factor.

THERMOSTATS

A *thermostat* is an automatic mechanism for controlling the amount of heating or cooling given by a central or zonal heating or cooling system. The thermostat floor-plan symbol is shown in Figure 19.9.

The location of the thermostat is important to the proper functioning of the system. For zoned heating or cooling units, thermostats may be placed in each room, or a central thermostat panel that controls each room may be placed in a convenient location. For central heating and cooling systems, there may be one or more thermostats, depending on the layout of the system or the number of units required to service the structure. For example, a very large home or office building may have a split system that divides the structure into two or more zones, each individual zone with its own thermostat.

FIGURE 19.9 ■ Thermostat floor-plan symbol.

Several factors contribute to the effective placement of the thermostat for a central system. A good common location is near the center of the structure and close to a return air duct for a central forced-air system. The air entering the return air duct is usually temperate, thus causing little variation in temperature on the thermostat. A key to successful thermostat placement is to find a stable location where an average temperature reading can be achieved. There should be no drafts that would adversely affect temperature settings. The thermostat should not be placed in a location where sunlight or a heat register causes an unreliable reading. The same rationale suggests that a thermostat not be placed close to an exterior door, where temperatures can change quickly. Thermostats should be placed on inside partitions rather than on outside walls, where a false temperature reading could also be obtained. Avoid placing the thermostat near stairs or a similar traffic area, where significant bouncing or shaking could cause the mechanism to alter the actual reading.

Energy consumption can be reduced by controlling thermostats in individual rooms. Central panels are available that make it easy to lower or raise temperatures in any room where each panel switch controls one remote thermostat.

Programmable microcomputer thermostats effectively help reduce the cost of heating or cooling. Some units automatically switch from heat to cool while minimizing temperature deviation from the setting under varying load conditions. These computers can be used to alter heating and cooling temperature settings automatically for different days of the week or different months of the year.

HEAT RECOVERY AND VENTILATION

Sources of Pollutants

Air pollution in a structure is a principal reason for installing a heat recovery and ventilation system. A number of sources contribute to an unhealthy environment within a home or business:

- Moisture in the form of relative humidity can cause structural damage as well as health problems, such as respiratory problems. Sources of relative humidity include the atmosphere, steam from cooking and showers, and individuals, who can produce up to 1 gallon (3.8 liters) of water vapor per day.

- Incomplete combustion from gas-fired appliances or wood-burning stoves and fireplaces can generate a variety of pollutants, including carbon monoxide, aldehydes, and soot.
- Humans and pets can transmit bacterial and viral diseases through the air.
- Tobacco smoke contributes chemical compounds to the air; these can affect both smokers and nonsmokers alike.
- Formaldehyde, when present, is considered a factor in the cause of eye irritation, certain diseases, and respiratory problems. Formaldehyde is found in carpets, furniture, and the glue used in construction materials, such as plywood and particle board, as well as some insulation products.
- Radon is a naturally occurring radioactive gas that breaks down into compounds that are carcinogenic (cancer-causing) when large quantities are inhaled over a long period of time. Radon may be more apparent in a structure that contains a large amount of concrete or in certain areas of the country. Radon can be monitored scientifically at a nominal cost, and barriers can be built that help reduce concern about radon contamination.
- Household products such as those available in aerosol spray cans and craft materials such as glues and paints can contribute a number of toxic pollutants.

Air-to-Air Heat Exchangers

Government energy agencies, architects, designers, and contractors around the country have been evaluating construction methods that are designed to reduce energy consumption. Some of the tests have produced superinsulated, vapor-barrier-lined, airtight structures. The result has been a dramatic reduction in heating and cooling costs; however, the air quality in these houses has been significantly reduced and may even be harmful to health. In essence, the structure does not breathe, and stale air and pollutants have no place to go. A technology has emerged from this problem: an air-to-air heat exchanger. In the past the air in a structure was exchanged in leakage through walls, floors, and ceilings and around openings. Although this random leakage was no insurance that the building was properly ventilated, it did ensure a certain amount of heat loss. Now, with the concern for energy conservation, it is clear that the internal air quality of a home or business cannot be left to chance.

An *air-to-air heat exchanger* is a heat recovery and ventilation device that pulls polluted, stale warm air from the living space and transfers the heat in that air to fresh, cold air being pulled into the house. Heat exchangers do not produce heat; they only exchange heat from one airstream to another. The heat transfer takes place in the core of the heat exchanger, which is designed to avoid mixing the two airstreams and to ensure that indoor pollutants are expelled. Moisture in the stale air condenses in the core and is drained from the unit. Figure 19.10 shows the function and basic components of an air-to-air heat exchanger.

The recommended minimum effective air change rate is 0.5 air change per hour (ACH). Codes in some areas of the country have established a rate of 0.7 ACH. The American Society of Heating, Refrigeration, and Air Conditioning Engineers, Inc. (ASHRAE), recommends ventilation levels based on the amount of air entering a room. The recommended amount of air entering most rooms is 10 cubic feet (10 cu′) per minute (CFM). The rate for kitchens is 100 CFM; for bathrooms, 50 CFM. Mechanical exhaust devices vented to outside air should be added to kitchens and baths to maintain the recommended air exchange rate.

The minimum heat exchanger capacity needed for a structure can be determined easily. Assume a 0.5 ACH rate in a 1,500 sq′ single-level, energy-efficient house and follow these steps:

STEP 1 Determine the total floor area in square feet. Use the outside dimensions of the living area only.

$$30' \times 50' = 1{,}500 \text{ sq}'$$

STEP 2 Determine the total volume within the house in cubic ft. by multiplying the total floor area by the ceiling height.

$$1{,}500 \text{ sq}' \times 8' = 12{,}000 \text{ cu}'$$

STEP 3 Determine the minimum exchanger capacity in CFM by first finding the capacity in cubic feet per hour (CFH). To do this, multiply the house volume in cubic feet (CUFT) by the ventilation rate in air changes per hour (ACH).

$$12{,}000 \text{ CUFT} \times 0.5 \text{ ACH} = 6{,}000 \text{ CFH}$$

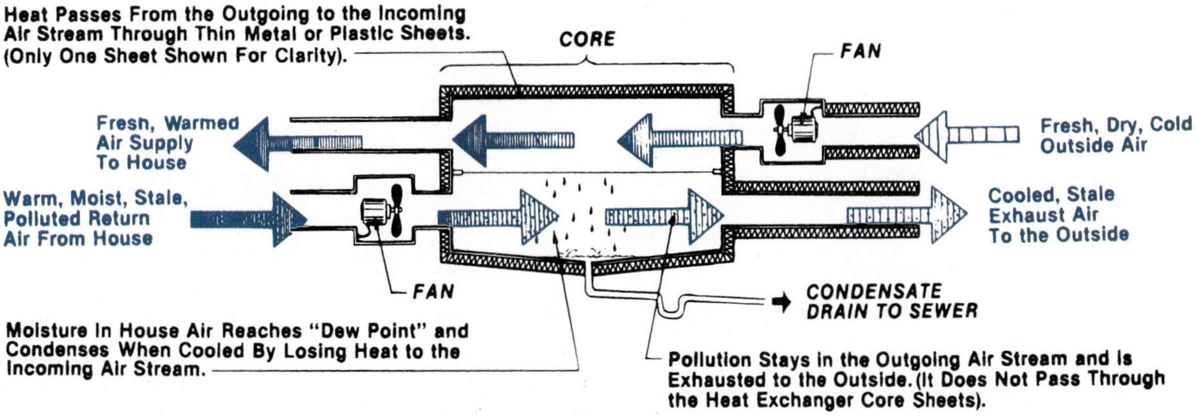
Heat Passes From the Outgoing to the Incoming Air Stream Through Thin Metal or Plastic Sheets. (Only One Sheet Shown For Clarity).
CORE
FAN
Fresh, Warmed Air Supply To House
Fresh, Dry, Cold Outside Air
Warm, Moist, Stale, Polluted Return Air From House
Cooled, Stale Exhaust Air To the Outside
FAN
CONDENSATE DRAIN TO SEWER
Moisture In House Air Reaches "Dew Point" and Condenses When Cooled By Losing Heat to the Incoming Air Stream.
Pollution Stays In the Outgoing Air Stream and Is Exhausted to the Outside. (It Does Not Pass Through the Heat Exchanger Core Sheets).

FIGURE 19.10 ■ Components and function of air-to-air heat exchanger. *Courtesy U.S. Department of Energy.*

STEP 4 Convert the CFH rate to cubic feet per minute (CFM) by dividing the CFH rate by 60 minutes.

$$6{,}000 \text{ CFH} \div 60 \text{ min} = 100 \text{ CFM}$$

There is a percentage of capacity loss due to mechanical resistance that should be considered by the system designer.

Most homeowners are not knowledgeable about potential air pollution and how internal air may be adequately controlled by ventilation. Therefore, the architect designing an energy-efficient home must plan the proper ventilation. The ducts, intake, and exhaust for such a system should be located on the advice of a ventilation consultant or a knowledgeable HVAC contractor.

Exhaust Systems: Code Requirements

Exhausts are part of the HVAC system. Exhaust systems are required to remove odors, steam, moisture, and pollutants from inside a building. The following are some general code requirements that need to be considered during the design process.

- Range hoods shall be vented to the outside by a single-wall galvanized stainless steel or copper duct. This duct shall have a smooth inner surface, be substantially airtight, and have a backdraft damper.
- Metal or noncombustible range hoods cannot be any closer than 24″ (610 mm) to the range.
- Clothes dryer vents shall be independent of all other systems and shall carry the moisture outside. The vent cannot be connected with screws that extend into the vent. Exhaust vents shall have a backdraft damper. The vent duct shall be rigid metal 0.016″ (0.406 mm) thick with a smooth inside and joints running in the direction of the airflow. Approved flexible duct can be used but cannot be concealed within the construction.
- A 4″ (102 mm) clothes dryer duct cannot be longer than 25′ (7620 mm) from the dryer to the wall or roof vent and shall have a full opening exhaust hood. The total length shall be reduced by 2′–6″ (762 mm) for each 45° bend and 5′ (1524 mm) for each 90° bend.
- When gas clothes dryers are installed in a closet, they must be approved for such use, and no other fuel-burning appliance can be in the same closet.

CENTRAL VACUUM SYSTEMS

Central vacuum systems have a number of advantages over portable vacuum cleaners.

1. Cost: some systems cost no more than a major appliance.
2. Increased resale value of home.
3. Removal of dirt too heavy for most portable units.
4. Exhaust of dirt and dust out of the house or office.

FIGURE 19.11 ■ Vacuum wall outlet. *Courtesy Vacu-maid, Inc.*

FIGURE 19.12 ■ Vacuum outlet floor-plan symbol.

5. No motor unit or electric cords; however, there are often long hoses.
6. No noise.
7. Savings in cleaning time.
8. Ability to vary vacuum pressure.

A well-designed system requires only a few inlets to cover an entire home or business, including exterior use. The hose plugs into a wall outlet and the vacuum is ready for use, as shown in Figure 19.11. The hose is generally lightweight; and without a portable unit to carry, it is often easier to vacuum in hard-to-reach places. The central canister empties easily to remove dust and debris from the house or business. The floor plan symbol for vacuum cleaner outlets is shown in Figure 19.12. The central unit may be shown in the garage or storage area as a circle labeled CENTRAL VACUUM SYSTEM.

HVAC SYMBOLS

There are over a hundred HVAC symbols that may be used in residential and commercial heating plans. Only a few of the symbols are typically used in residential HVAC drawings. Figure 19.13 shows some common HVAC symbols. Sheet metal conduit template and heating and air-conditioning templates are timesaving devices that are used to help improve the quality of drafting for HVAC systems. Custom HVAC CADD programs are also available to help improve drafting productivity. CADD applications are discussed later.

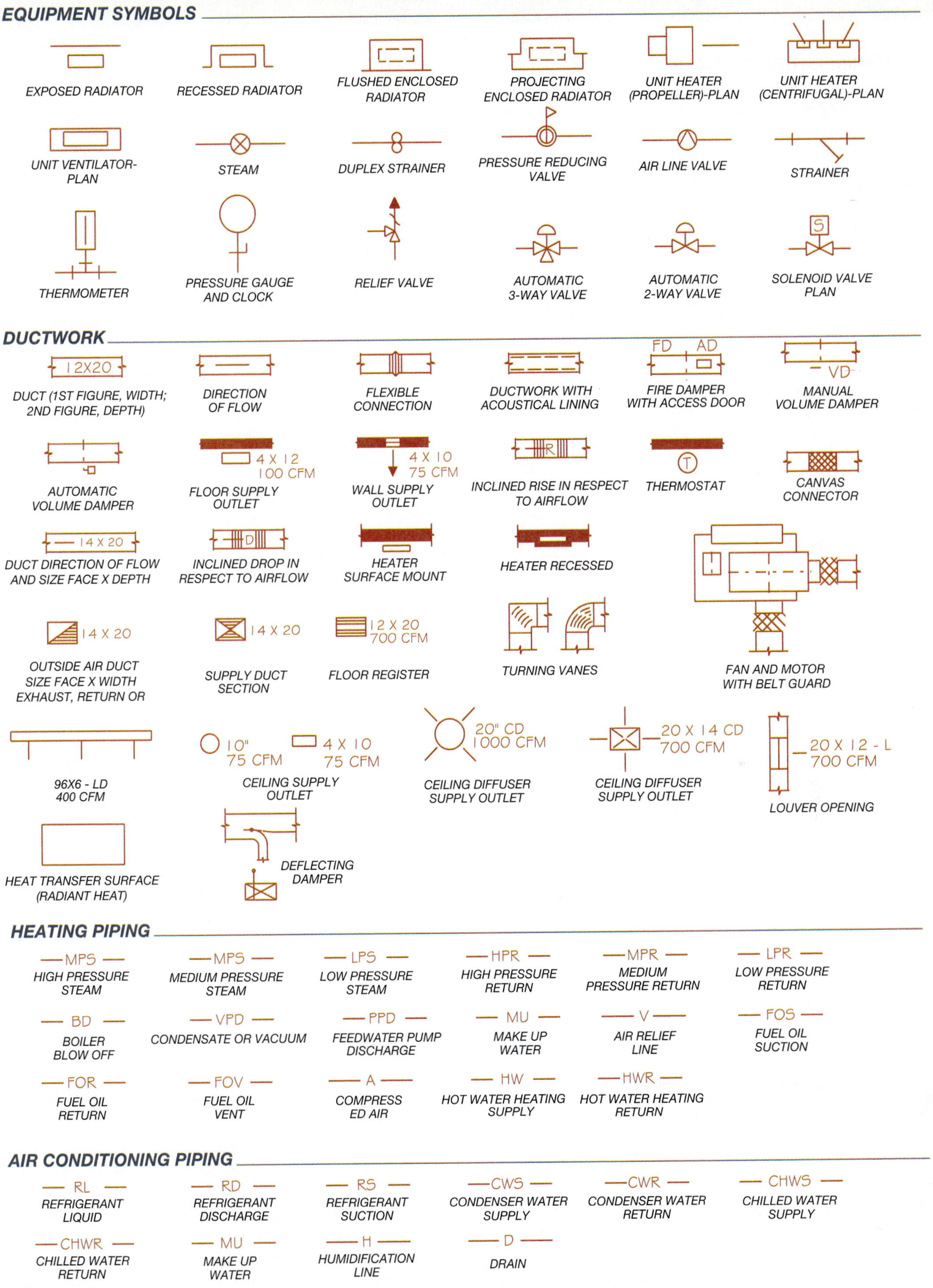
EQUIPMENT SYMBOLS
EXPOSED RADIATOR
RECESSED RADIATOR
FLUSHED ENCLOSED RADIATOR
PROJECTING ENCLOSED RADIATOR
UNIT HEATER (PROPELLER)-PLAN
UNIT HEATER (CENTRIFUGAL)-PLAN
UNIT VENTILATOR-PLAN
STEAM
DUPLEX STRAINER
PRESSURE REDUCING VALVE
AIR LINE VALVE
STRAINER
THERMOMETER
PRESSURE GAUGE AND CLOCK
RELIEF VALVE
AUTOMATIC 3-WAY VALVE
AUTOMATIC 2-WAY VALVE
S
SOLENOID VALVE PLAN
DUCTWORK
12X20
DUCT (1ST FIGURE, WIDTH; 2ND FIGURE, DEPTH)
DIRECTION OF FLOW
FLEXIBLE CONNECTION
DUCTWORK WITH ACOUSTICAL LINING
FD AD
FIRE DAMPER WITH ACCESS DOOR
VD
MANUAL VOLUME DAMPER
AUTOMATIC VOLUME DAMPER
4 X 12
100 CFM
FLOOR SUPPLY OUTLET
4 X 10
75 CFM
WALL SUPPLY OUTLET
R
INCLINED RISE IN RESPECT TO AIRFLOW
T
THERMOSTAT
CANVAS CONNECTOR
14 X 20
DUCT DIRECTION OF FLOW AND SIZE FACE X DEPTH
D
INCLINED DROP IN RESPECT TO AIRFLOW
HEATER SURFACE MOUNT
HEATER RECESSED
FAN AND MOTOR WITH BELT GUARD
14 X 20
OUTSIDE AIR DUCT SIZE FACE X WIDTH EXHAUST, RETURN OR
14 X 20
SUPPLY DUCT SECTION
12 X 20
700 CFM
FLOOR REGISTER
TURNING VANES
96X6 - LD
400 CFM
10"
75 CFM
4 X 10
75 CFM
CEILING SUPPLY OUTLET
20" CD
1000 CFM
CEILING DIFFUSER SUPPLY OUTLET
20 X 14 CD
700 CFM
CEILING DIFFUSER SUPPLY OUTLET
20 X 12 - L
700 CFM
LOUVER OPENING
HEAT TRANSFER SURFACE (RADIANT HEAT)
DEFLECTING DAMPER
HEATING PIPING
MPS
HIGH PRESSURE STEAM
MPS
MEDIUM PRESSURE STEAM
LPS
LOW PRESSURE STEAM
HPR
HIGH PRESSURE RETURN
MPR
MEDIUM PRESSURE RETURN
LPR
LOW PRESSURE RETURN
BD
BOILER BLOW OFF
VPD
CONDENSATE OR VACUUM
PPD
FEEDWATER PUMP DISCHARGE
MU
MAKE UP WATER
V
AIR RELIEF LINE
FOS
FUEL OIL SUCTION
FOR
FUEL OIL RETURN
FOV
FUEL OIL VENT
A
COMPRESS ED AIR
HW
HOT WATER HEATING SUPPLY
HWR
HOT WATER HEATING RETURN
AIR CONDITIONING PIPING
RL
REFRIGERANT LIQUID
RD
REFRIGERANT DISCHARGE
RS
REFRIGERANT SUCTION
CWS
CONDENSER WATER SUPPLY
CWR
CONDENSER WATER RETURN
CHWS
CHILLED WATER SUPPLY
CHWR
CHILLED WATER RETURN
MU
MAKE UP WATER
H
HUMIDIFICATION LINE
D
DRAIN

FIGURE 19.13 ■ Common HVAC symbols.

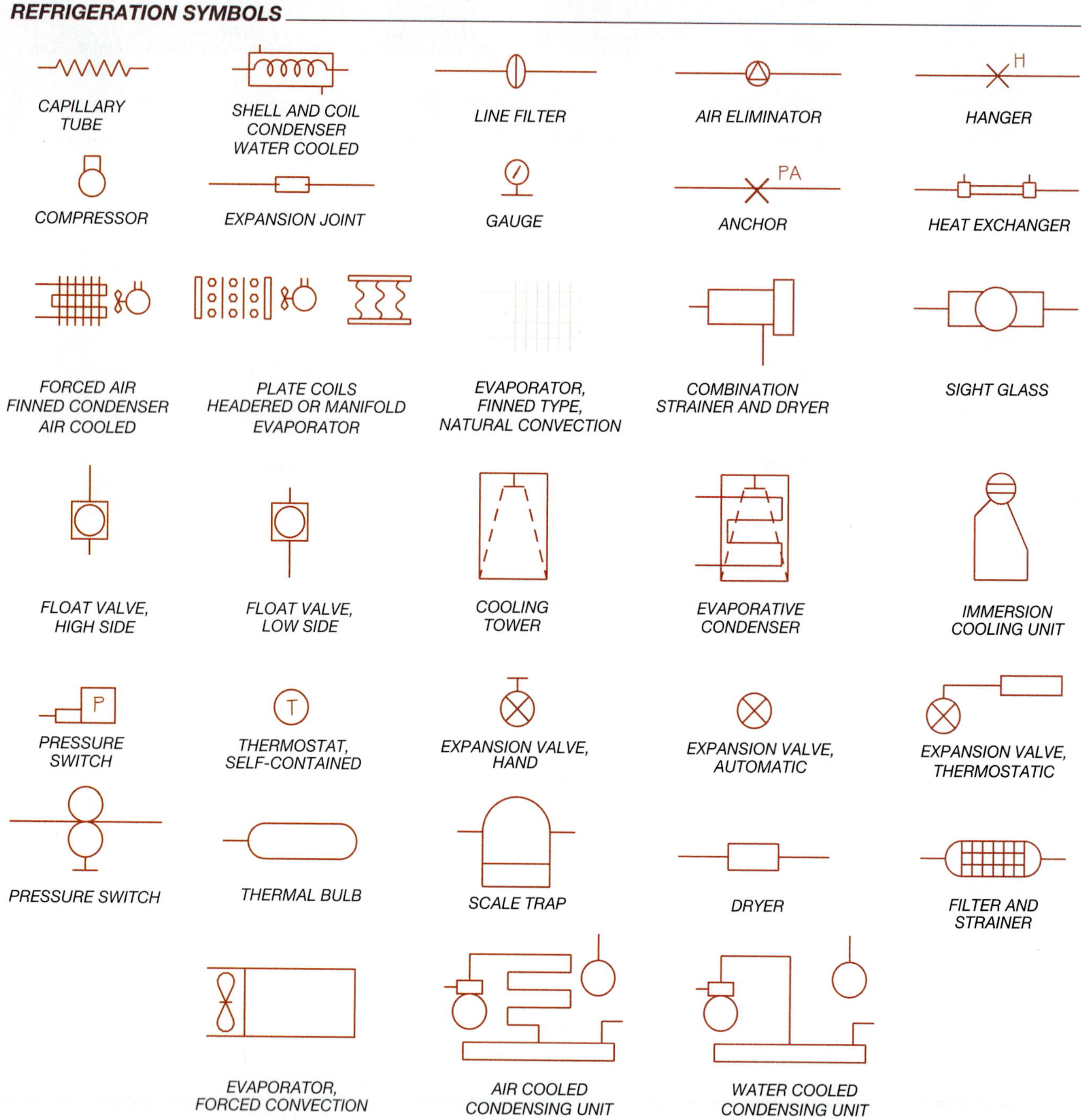

FIGURE 19.13 ■ *(continued)*

UNIVERSAL HVAC DESIGN

Following are a few issues that can be considered in designing a building for disabled people.

- **Thermostats:** The thermostat should be placed between 36″ and 48″ (915–1220 mm) high. Use the higher distance or a protective cover if children are present. Thermostats are available with graduations printed in Braille or large digital numbers for sight-impaired people. Some are available with clicks so the user can feel or hear the movement.
- **Heating systems:** Electric forced-air furnaces or heat pumps may be preferred by disabled people. However, place floor registers in places out of the traffic pattern or place them 90° to the wall so that wheelchairs can cross easily. Electric radiant heat is a consideration for people with allergies: install an electronic air filter if forced-air systems are used. An auxiliary ceiling heater with a timer may be placed in a bathroom to heat the space.

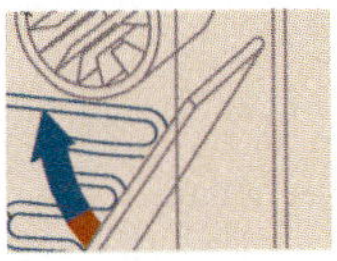

THERMAL CALCULATIONS FOR HEATING/ COOLING

Development of Methodology for Estimating Heat Loss

The basic methodology for estimating residential heat loss today has been in use since the early 1900s. Historically, its primary use was to calculate the design heat load of houses in order to estimate the size of gas and oil heating systems required. Because home designers, builders, and heating system installers did not want to receive complaints from cold homeowners, they commonly designed the heating system for worst-case weather conditions, with a bias toward overestimating the design heat load to ensure that the furnace would never be too small. Consequently, many gas and oil furnaces were oversize.

As early as 1915 an engineer for a gas utility began modifying the design heat load with a degree-day method to estimate annual energy consumption. Oil companies also began using this method to predict when to refill their customers' tanks. Data from this period, for houses with little or no insulation, indicated that the proportionality between annual heating energy requirements and average outside temperature began at 65°F (18°C) in residential buildings. This was the beginning of the 65°F base for degree days. Studies made by the American Gas Association up to 1932 and by the National District Heating Association in 1932 also indicated that a 65°F base was appropriate for houses of the period.

Experience in the 1950s and early 1960s with electrically heated homes indicated that the traditional degree-day procedure was overestimating annual heating loads. This was due primarily to tighter, better-insulated houses with balance temperatures below 65°F, and to more appliances. These later studies led to a modified degree-day procedure incorporating a modifying factor, C. This C factor compensates for such things as higher insulation levels and more heat-producing appliances in the house. The high insulation levels found in homes built to current codes, and those that have been retrofitted with insulation, cause even lower balance temperatures than those of houses built in the 1950s and 1960s. (This history has been adapted from *Standard Heat Loss Methodology*, Bonneville Power Administration.)

Terminology

Btu (British thermal unit): A unit of measure determined by the amount of heat required to raise 1 pound of water 1 degree Fahrenheit. Heat loss is calculated in Btu per hour (Btuh).

Compass point: The relationship of the structure to compass orientation. In referring to cooling calculations it is important to evaluate the amount of glass in each wall as related to compass orientation. This is due to the differences effected by solar gain.

Duct loss: Heat loss through ductwork in an unheated space, which has an effect on total heat loss. Insulating those ducts helps increase efficiency.

Grains: The amount of moisture in the air (grains of moisture in 1 cubic foot). A grain is a unit of weight. Air at different temperatures and humidities holds different amounts of moisture. The effect of this moisture content becomes more of a factor in the southern and midwestern states than in other parts of the country.

Heat transfer multiplier: The amount of heat that flows through square foot of the building surface. The product depends on the type of surface and whether it is applied to heating or cooling.

Indoor temperature: The indoor design temperature is generally 60 to 70°F.

Indoor wet bulb: Relates to the use of the wet-bulb thermometer inside. A wet-bulb thermometer is one in which the bulb is kept moistened and is used to determine humidity level.

Infiltration: The inward flow of air through a porous wall or crack. In a loosely constructed home, infiltration substantially increases heat loss. Infiltration around windows and doors is calculated in CFM per linear foot of crack. Window infiltration greater than 0.5 CFM per linear foot of crack is excessive.

Internal heat gain: Heat gain associated with factors such as heat transmitted from appliances, lights, other equipment, and occupants.

Latent load: The effects of moisture entering the structure from the outside by humidity infiltration or from the inside, produced by people, plants, and daily activities such as cooking, showers, and laundry.

Mechanical ventilation: Amount of heat loss through mechanical ventilators such as range hood fans or bathroom exhaust fans.

Outdoor temperature: Related to average winter and summer temperatures for a local area. If the outdoor winter design temperature is 20°F, this means that the temperature during the winter is 20° or higher 97 1/2 percent of the time. If the outdoor summer design temperature in an area is 100°F, this means that the temperature during the summer is 100° or less 97 1/2 percent of the time. Figures for each area of the country have been established by ASHRAE. Verify the outdoor temperature with your local building department or heating contractor.

Outdoor wet bulb: Relates to the use of a wet-bulb thermometer outside. A wet-bulb thermometer is one in which the bulb is kept moistened and is used to determine humidity level.

R factor: Resistance to heat flow. The more resistance to heat flow, the higher the R value. For example, 3 1/2″ of mineral wool insulation has a value of R-11; 6″ of the same insulation has a value of R-19. R is equal to the reciprocal of the U factor, 1/U = R.

Sensible load calculations: Load calculations associated with temperature change that occurs when a structure loses or gains heat.

Temperature difference: The indoor temperature less the outdoor temperature.

U factor: The coefficient of heat transfer expressed in Btuh sq. ft./°F of surface area.

Steps in Filling Out the Residential Heating Data Sheet

Figure 19.14 is a completed Residential Heating Data Sheet. The two-page data sheet is divided into several categories with calculations resulting in total heat loss for the structure, shown in Figure 19.15. The large numbers alongside the categories refer to the following steps, used in completing the form. Notice that the calculations are rounded off to the nearest whole unit.

STEP 1 *Outdoor temperature.* Use the recommended outdoor design temperature for your area. The area selected for this problem is Dallas, Texas, with an outdoor design temperature of 22°F, which has been rounded off to 20°F to make calculations simple to understand. The proper calculations would interpolate the tables for a 22°F outdoor design temperature.

STEP 2 *Indoor temperature:* 70°F.

STEP 3 *Temperature difference:*

$$70° - 20° = 50°\text{F}$$

STEP 4 *Movable glass windows:* Select double glass; find area of each window (frame length × width); then combine for total: 114 sq′.

STEP 5 *Btuh heat loss:* Using 50° design temperature difference, find 46 approximate heat transfer multiplier:

$$46 \times 114 = 5244 \text{ Btuh}$$

STEP 6 *Sliding glass doors:* Select double glass; find total area of 34 sq′.

STEP 7 *Btuh heat loss:*

$$34 \times 48 = 1632 \text{ Btuh}$$

STEP 8 *Doors:* Weatherstripped solid wood; 2 doors at 39 sq. ft.

STEP 9 *Btuh heat loss:*

$$39 \times 30 = 1170 \text{ Btuh}$$

STEP 10 *Walls:* Excluding garage; perimeter in running feet 132 × ceiling height 8′ = gross wall area 1056 sq′. Subtract window and door areas − 187 sq′ = net wall area 869 sq′.

STEP 11 *Frame wall:* No masonry wall above or below grade. When there is masonry wall, subtract the square footage of masonry wall from total wall for the net frame wall. Fill in net amount on approximate insulation value; 869 sq′ frame wall, R-13 (given).

STEP 12 *Btuh heat loss:*

$$869 \times 3.5 = 3042 \text{ Btuh}$$

STEPS 13–14 No masonry above grade in this structure.

STEPS 15–16 No masonry below grade in this structure.

STEP 17 Heat loss subtotal: Add together items 5, 7, 9, and 12; 11,088 Btuh. Transfer the amount to the top of page 2 of the form.

STEP 18 Ceilings: R-30 insulation (given), same square footage as floor plan 976 sq′ (given).

STEP 19 *Btuh heat loss:*

$$976 \times 1.6 = 1562 \text{ Btuh}$$

STEP 20 *Floor over an unconditioned space:* R-19 insulation (given), 97 sq′.

STEP 21 *Btuh heat loss:*

$$976 \times 2.6 = 2538 \text{ Btuh}$$

STEPS 22–23 *Basement floor:* Does not apply to this house.

STEPS 24–25 *Concrete slab without perimeter system:* Does not apply to this house.

STEPS 26–27 *Concrete slab with perimeter system:* Does not apply to this house.

STEP 28 *Infiltration:* 976 sq′ floor × 8′ ceiling height = 7808 cu′

$$0.40 \times 7808 \text{ cu}' \div 60 = 52 \text{ CFM}$$

Infiltration = mechanical ventilation

CFM = fresh air intake.

STEP 29 *Btuh heat loss:*

$$52 \times 55 = 2860 \text{ Btuh}$$

STEP 30 *Heat loss subtotal:* Add items 17, 19, 21, and 29; 20,908 Btuh.

STEP 31 *Duct loss:* R-4 insulation (given); add 15 percent to item 30:

$$0.15 \times 20{,}923 = 3138$$

When no ducts are used, exclude this item.

STEP 32 *Total heat loss:* Add items 30 and 31; 24,046 Btuh.

Steps in Filling Out the Residential Cooling Data Sheet

Figure 19.16 is a completed residential cooling data sheet. The two-page data sheet is divided into several categories, with calculations resulting in total sensible and latent heat gain for the structure, shown in Figure 19.15, used for heat loss calculations. The large numbers by each category refer to the following steps used in completing the form.

JOB NAME:		DATE
ADDRESS:		
1 OUTDOOR TEMP: 20°	2 INDOOR TEMP: 70°	3 TEMP. DIFFERENCE: 50°

	MOVABLE GLASS WINDOWS	SQUARE FEET	DESIGN TEMPERATURE DIFFERENCE														BTUH HEAT LOSS	
4			30	35	40	45	50	55	60	65	70	75	80	85	90	95		
			HEAT TRANSFER MULTIPLIER															
	SINGLE GLASS		39	45	52	58	65	71	78	84	90	97	103	110	116	123		
	SINGLE GLASS W/STORM		21	25	28	31	35	38	42	45	49	52	56	59	63	66		
	DOUBLE GLASS	114	28	32	37	41	46	50	55	60	64	69	73	78	82	87	5244	5
	DOUBLE GLASS W/STORM		16	19	21	24	27	29	32	35	37	40	42	45	48	50		

	SLIDING GLASS DOORS	SQUARE FEET	DESIGN TEMPERATURE DIFFERENCE														BTUH HEAT LOSS	
6			30	35	40	45	50	55	60	65	70	75	80	85	90	95		
			HEAT TRANSFER MULTIPLIER															
	SINGLE GLASS		42	48	55	62	69	76	83	90	97	104	110	117	124	131		
	SINGLE GLASS W/STORM		22	26	29	33	37	40	44	48	51	55	59	62	66	70		7
	DOUBLE GLASS	34	29	34	39	43	48	53	58	63	67	72	77	82	87	91	1632	

	DOORS	SQUARE FEET	DESIGN TEMPERATURE DIFFERENCE														BTUH HEAT LOSS	
			30	35	40	45	50	55	60	65	70	75	80	85	90	95		
			HEAT TRANSFER MULTIPLIER															
8	SOLID WOOD		31	36	41	46	51	56	62	67	72	77	82	87	92	97		
	SOLID WOOD**	39	18	21	24	27	30	33	36	39	42	45	47	50	53	56	1170	9
	METAL URETHANE		23	27	30	34	38	42	45	49	53	57	60	64	68	72		
	METAL URETHANE**		13	16	18	20	22	25	27	29	31	33	36	38	40	42		

**Weatherstripped or Storm

10	RUNNING FEET		132
	CEILING HEIGHT	X	8
WALLS	GROSS WALL	=	1056
	WINDOWS & DOOR AREAS	–	187
	NET WALL AREA		869

	FRAME WALL	SQUARE FEET	DESIGN TEMPERATURE DIFFERENCE														BTUH HEAT LOSS	
			30	35	40	45	50	55	60	65	70	75	80	85	90	95		
			HEAT TRANSFER MULTIPLIER															
	NO INSULATION		8	10	11	12	14	15	17	18	19	21	22	23	25	26		
	R-11, 3" INSULATION		2.7	3.1	3.6	4.0	4.5	4.9	5.4	5.8	6.3	6.7	7.2	7.6	8.1	8.5		
11	R-13, 3-1/2" INSULATION	869	2.1	2.4	2.8	3.2	3.5	3.8	4.2	4.6	4.9	5.3	5.6	5.9	6.3	6.6	3042	12
	R-13 + 1" POLYSTYRENE		1.8	2.1	2.4	2.7	3.0	3.3	3.6	3.9	4.2	4.5	4.8	5.1	5.4	5.7		
	R-19 + 1/2" POLYSTYRENE		1.6	1.9	2.2	2.5	2.8	3.0	3.3	3.6	3.8	4.1	4.4	4.7	4.9	5.2		

	MASONRY WALL ABOVE GRADE	SQUARE FEET	DESIGN TEMPERATURE DIFFERENCE														BTUH HEAT LOSS	
			30	35	40	45	50	55	60	65	70	75	80	85	90	95		
			HEAT TRANSFER MULTIPLIER															
	NO INSULATION		16	18	21	23	26	28	31	33	36	38	41	44	46	49		
	R-5, 1" INSULATION		4.3	5.0	5.8	6.5	7.2	7.9	8.6	9.4	10.1	10.8	11.5	12.2	13.0	13.7		14
13	R-11, 3" INSULATION		2.3	2.7	3.1	3.5	3.8	4.2	4.6	5.0	5.4	5.8	6.2	6.5	6.9	7.3		
	R-19, 6" INSULATION		1.4	1.7	1.9	2.2	2.4	2.6	2.9	3.1	3.4	3.6	3.8	4.1	4.3	4.6		

	MASONRY WALL BELOW GRADE	SQUARE FEET	DESIGN TEMPERATURE DIFFERENCE														BTUH HEAT LOSS	
			30	35	40	45	50	55	60	65	70	75	80	85	90	95		
			HEAT TRANSFER MULTIPLIER															
	NO INSULATION		4.4	5.1	5.9	6.6	7.3	8.1	8.8	9.6	10.3	11.0	11.8	12.5	13.2	14.0		16
15	R-5, 1" INSULATION		2.6	3.0	3.5	3.9	4.3	4.8	5.2	5.7	6.1	6.5	7.0	7.4	7.8	8.3		
	R-11, 3" INSULATION		1.8	2.1	2.4	2.7	3.0	3.3	3.6	3.9	4.2	4.5	4.8	5.1	5.4	5.7		
	R-19, 6" INSULATION		1.2	1.4	1.6	1.8	2.0	2.2	2.4	2.6	2.8	3.0	3.2	3.4	3.6	3.8		

HEAT LOSS SUBTOTAL	11088	17

FIGURE 19.14 ■ Completed residential heating data sheet form. *Courtesy Lennox Industries, Inc.*

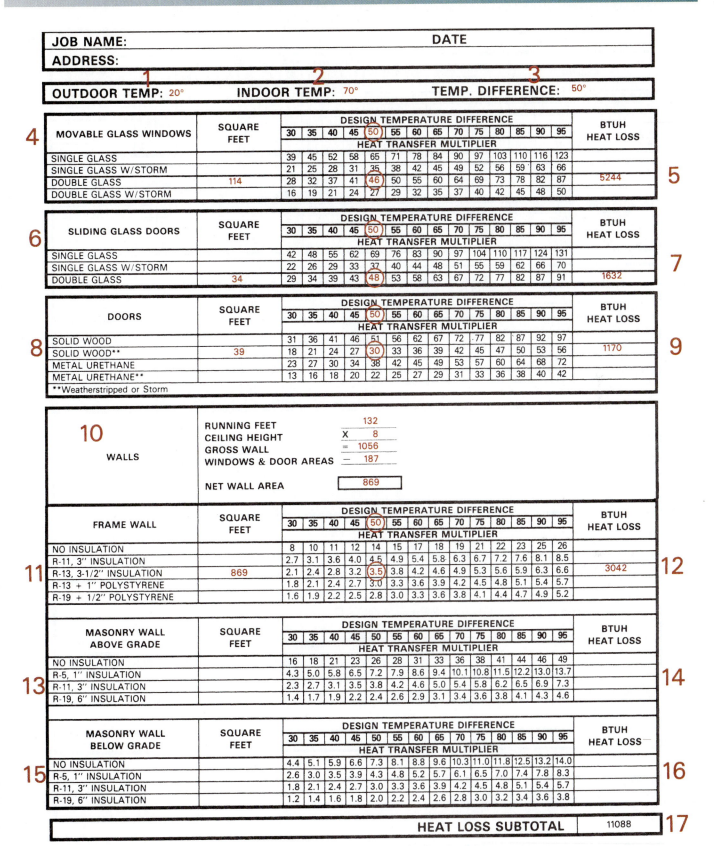

Heat Loss Subtotal from Page 1 | 11088 | 17

CEILING	SQUARE FEET	DESIGN TEMPERATURE DIFFERENCE														BTUH HEAT LOSS
		30	35	40	45	50	55	60	65	70	75	80	85	90	95	
		HEAT TRANSFER MULTIPLIER														
NO INSULATION		18	21	24	27	30	33	36	39	42	45	48	51	54	57	
R-11, 3" INSULATION		2.6	3.1	3.5	4.0	4.4	4.8	5.3	5.7	6.2	6.6	7.0	7.5	7.9	8.4	
18 R-19, 6" INSULATION		1.6	1.9	2.1	2.4	2.6	2.9	3.2	3.4	3.7	4.0	4.2	4.5	4.8	5.0	
R-30, 10" INSULATION	976	1.0	1.2	1.3	1.5	1.6	1.8	2.0	2.1	2.3	2.5	2.6	2.8	3.0	3.1	1562 19
R-38, 12" INSULATION		0.8	0.9	1.0	1.2	1.3	1.4	1.6	1.7	1.8	2.0	2.1	2.2	2.3	[illegible].5	

FLOOR OVER AN UNCONDITIONED SPACE	SQUARE FEET	DESIGN TEMPERATURE DIFFERENCE														BTUH HEAT LOSS
		30	35	40	45	50	55	60	65	70	75	80	85	90	95	
		HEAT TRANSFER MULTIPLIER														
NO INSULATION		10	11	13	14	16	17	19	21	22	24	25	27	28	30	
20 R-11, 3" INSULATION		2.4	2.8	3.2	3.6	4.0	4.4	4.8	5.2	5.6	6.0	6.4	6.8	7.2	7.6	
R-19, 6"INSULATION	976	1.6	1.8	2.1	2.3	2.6	2.9	3.1	3.4	3.6	3.9	4.2	4.4	4.7	4.9	2538 21
R-30, 10"INSULATION		1.1	1.3	1.5	1.7	1.8	2.0	2.2	2.4	2.6	2.8	3.0	3.1	3.3	3.5	

22 BASEMENT FLOOR	SQUARE FEET	DESIGN TEMPERATURE DIFFERENCE														BTUH HEAT LOSS 23
		30	35	40	45	50	55	60	65	70	75	80	85	90	95	
		HEAT TRANSFER MULTIPLIER														
BASEMENT FLOOR		0.8	1.0	1.1	1.3	1.4	1.5	1.7	1.8	2.0	2.1	2.2	2.4	2.5	2.7	

CONCRETE SLAB WITHOUT PERIMETER SYSTEM	LINEAR FOOT	DESIGN TEMPERATURE DIFFERENCE														BTUH HEAT LOSS
		30	35	40	45	50	55	60	65	70	75	80	85	90	95	
24		HEAT TRANSFER MULTIPLIER														
NO EDGE INSULATION		25	29	33	37	41	45	49	53	57	61	65	69	73	77	25
1" EDGE INSULATION		13	15	17	19	21	23	25	27	29	31	33	35	37	39	
2" INSULATION		6.3	7.4	8.4	9.4	10.5	11.5	12.6	13.6	14.7	15.8	16.8	17.8	18.9	20.0	

CONCRETE SLAB WITH PERIMETER SYSTEM	LINEAR FOOT	DESIGN TEMPERATURE DIFFERENCE														BTUH HEAT LOSS
		30	35	40	45	50	55	60	65	70	75	80	85	90	95	
26		HEAT TRANSFER MULTIPLIER														
NO EDGE INSULATION		57	67	76	86	95	105	114	124	133	143	152	162	171	181	27
1" EDGE INSULATION		34	40	46	52	57	63	69	74	80	86	91	97	103	109	
2" EDGE INSULATION		28	33	37	42	47	51	56	61	65	70	75	79	84	89	

An additional infiltration load is calculated only if the home is loosely constructed or when window infiltration is greater than .5 CFM per linear foot of crack.

28 INFILTRATION/ VENTILATION

976 FLOOR SQ FT. x 8 CEILING HEIGHT = 7808 CUBIC FT

0.40 x 7808 CUBIC FT ÷ 60 = 52 CFM

MECHANICAL VENTILATION CFM = FRESH AIR INTAKE

	CFM	DESIGN TEMPERATURE DIFFERENCE														BTUH HEAT LOSS
		30	35	40	45	50	55	60	65	70	75	80	85	90	95	
		HEAT TRANSFER MULTIPLIER														
INFILTRATION	52	33	39	44	50	55	61	66	72	77	83	88	94	99	105	2860 29
MECHANICAL VENTILATION	52	33	39	44	50	55	61	66	72	77	83	88	94	99	105	2860

HEAT LOSS SUBTOTAL	20908	30

31 DUCT LOSS	BTUH HEAT LOSS
R-4, 1" Flexible Blanket Insulation: ADD 15% (.15)	3138 31
R-7, 2" Flexible Blanket Insulation: ADD 10% (.10)	

TOTAL HEAT LOSS	24046	32

NOTE: All Heat Transfer Multipliers from ACCA Manual "J" Sixth Edition.

FIGURE 19.14 ■ *(continued)*

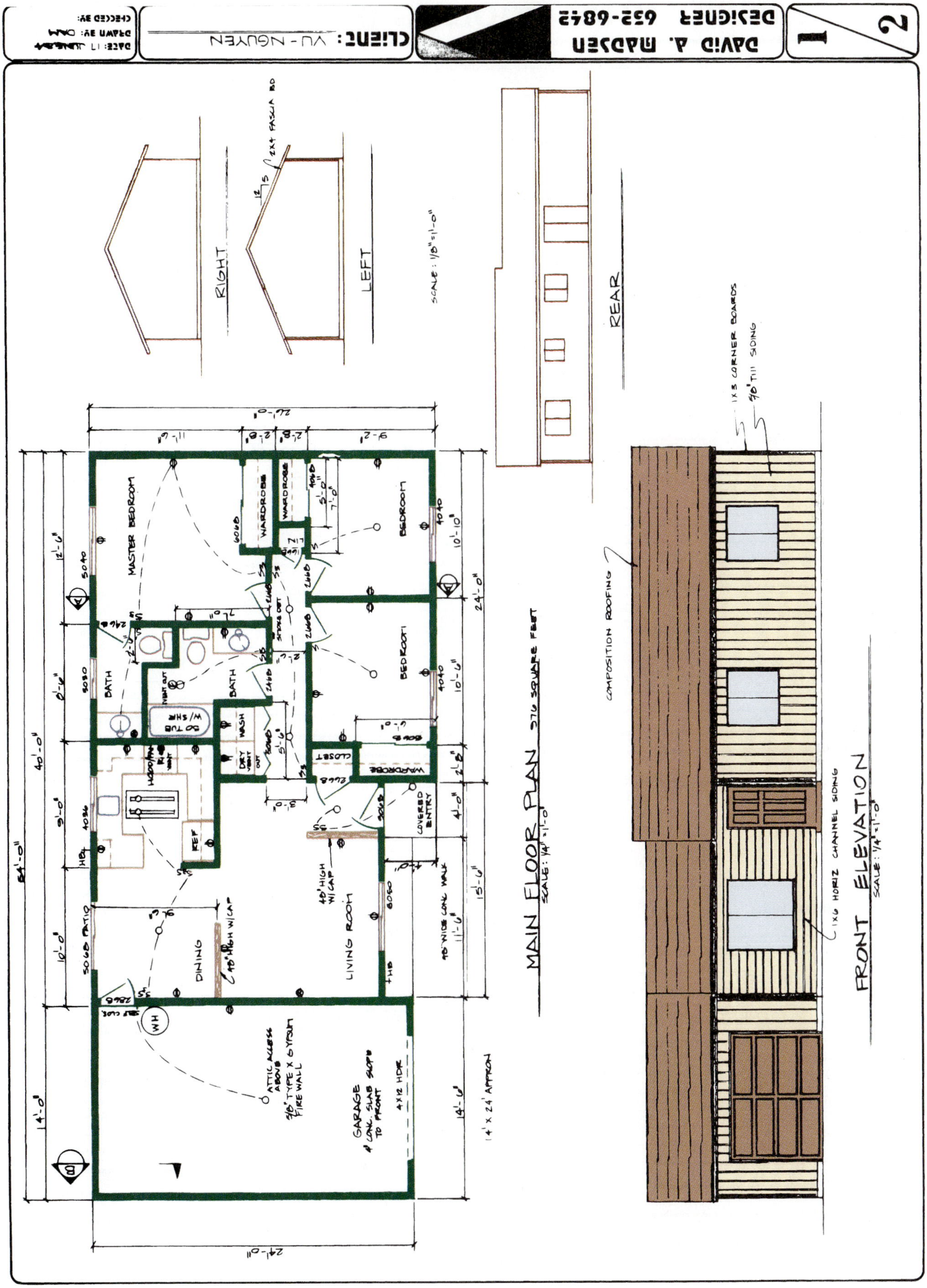

FIGURE 19.15 ■ Sample floor plan for heat loss and heat gain calculations. *Courtesy Madsen Designs.*

RESIDENTIAL COOLING DATA SHEET

JOB NAME:	DATE
ADDRESS:	

1	2	3
OUTDOOR TEMP: 100°	INDOOR TEMP: 70°	TEMP DIFFERENCE: 30°

SENSIBLE LOAD CALCULATIONS

	GLASS NO SHADE		SINGLE						DOUBLE						TRIPLE							
			DESIGN TEMPERATURE DIFFERENCE																			
	COMPASS POINT	GLASS AREA SQ. FEET	10	15	20	25	30	35	10	15	20	25	30	35	10	15	20	25	30	35	BTUH HEAT GAIN	
			HEAT TRANSFER MULTIPLIER																			
4	N	81	25	29	33	37	41	45	20	22	24	26	28	30	15	16	18	19	20	21	2268	5
6	NE & NW		55	60	65	70	75	80	50	52	54	56	58	60	37	38	40	41	42	44		7
8	E & W		80	85	90	95	100	105	70	72	74	76	78	80	55	56	58	59	60	62		9
10	SE & SW		70	74	78	82	86	90	60	62	64	66	68	70	47	49	51	52	53	54		11
12	S	67	40	44	48	52	56	60	35	37	39	41	43	45	26	27	29	31	32	33	2881	13

	GLASS INSIDE SHADE		SINGLE						DOUBLE						TRIPLE							
			DESIGN TEMPERATURE DIFFERENCE																			
	COMPASS POINT	GLASS AREA SQ. FEET	10	15	20	25	30	35	10	15	20	25	30	35	10	15	20	25	30	35	BTUH HEAT GAIN	
			HEAT TRANSFER MULTIPLIER																			
14	N		15	19	23	27	31	35	15	17	19	21	23	25	10	12	14	16	17	19		15
16	NE & NW		35	39	43	47	51	55	30	32	34	36	38	40	22	24	26	28	30	31		17
18	E & W		50	54	58	62	66	70	45	47	49	51	53	55	35	36	38	40	42	44		19
20	SE & SW		40	44	48	52	56	60	35	37	39	41	43	45	29	30	32	34	36	38		21
22	S		25	29	33	37	41	45	20	22	24	26	28	30	16	18	20	22	24	26		23

	DOORS	SQUARE FEET	DESIGN TEMPERATURE DIFFERENCE						BTUH HEAT GAIN	
			10	15	20	25	30	35		
			HEAT TRANSFER MULTIPLIER							
	SOLID WOOD		6.3	8.6	10.9	13.2	14.4	15.5		
24	SOLID WOOD **	39	4.2	5.7	7.3	8.8	9.6	10.4	374	25
	METAL URETHANE		2.6	3.5	4.5	5.4	5.9	6.4		
	METAL URETHANE **		2.2	3.0	3.8	4.6	5.0	5.4		

** Weatherstripped or Storm

26	WALLS		
		RUNNING FEET	132
		CEILING HEIGHT	X 8
		GROSS WALL	= 1056
		WINDOWS & DOOR AREAS	– 187
		NET WALL AREA	869

	FRAME WALL	SQUARE FEET	DESIGN TEMPERATURE DIFFERENCE						BTUH HEAT GAIN	
			10	15	20	25	30	35		
			HEAT TRANSFER MULTIPLIER							
	NO INSULATION		3.7	5.0	6.4	7.8	8.5	9.1		
	R-11, 3" INSULATION		1.2	1.7	2.1	2.6	2.8	3.0		
27	R-13, 3-1/2" INSULATION	869	1.1	1.5	1.9	2.3	2.5	2.7	2173	28
	R-13 + 1" POLYSTYRENE		0.8	1.1	1.4	1.7	1.8	2.0		
	R-19 + 1/2" POLYSTYRENE		0.7	1.0	1.3	1.6	1.7	1.8		

29	MASONRY WALL ABOVE GRADE	SQUARE FEET	DESIGN TEMPERATURE DIFFERENCE						BTUH HEAT GAIN	30
			10	15	20	25	30	35		
			HEAT TRANSFER MULTIPLIER							
	NO INSULATION		3.2	5.8	8.3	10.9	12.2	13.4		
	R-5, 1" INSULATION		0.9	1.6	2.3	3.1	3.5	3.8		
	R-11, 3" INSULATION		0.5	0.9	1.3	1.6	1.8	2.0		
	R-19, 6" INSULATION		0.3	0.5	0.8	1.0	1.2	1.3		

SENSIBLE HEAT GAIN SUBTOTAL	7696	31

FIGURE 19.16 ■ Completed residential cooling data sheet form. *Courtesy Lennox Industries, Inc.*

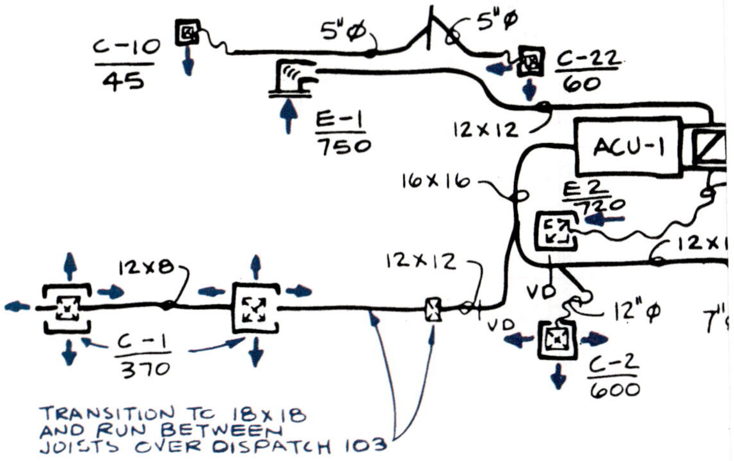

FIGURE 19.17 ■ Single-line HVAC plan engineer's sketch.

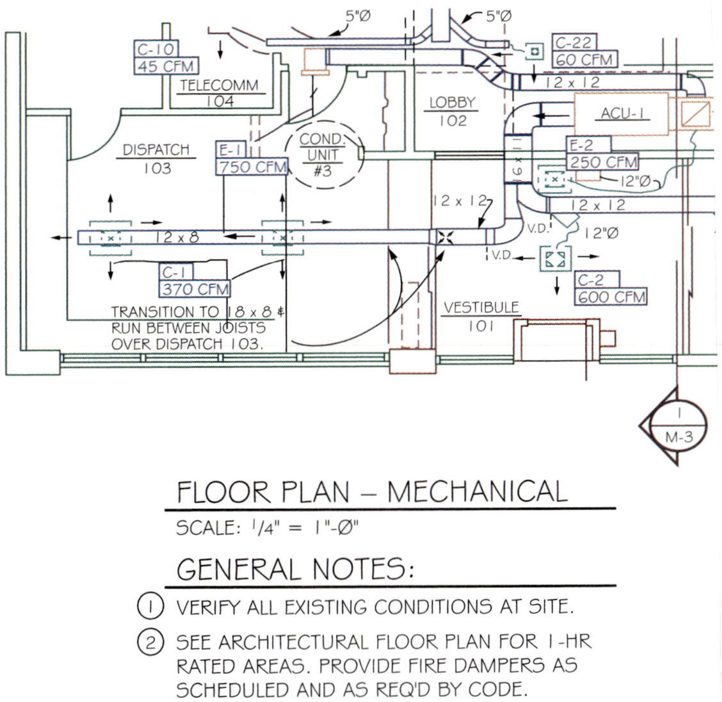

FIGURE 19.18 ■ HVAC plan for the engineer's sketch shown in Figure 1
Robert Evenson Associates AIA Architects.

sketches or submit data and calculations to a design drafter, who prepares design sketches or final drawings. Drafters without design experience work from engineering or design sketches to prepare formal drawings. A single-line engineer's sketch is shown in Figure 19.17. The next step in the HVAC design is for the drafter to convert the rough sketch into a preliminary drawing. This preliminary drawing goes back to the engineer and architect for verification and corrections or changes. The final step in the design process is for the drafter to implement the design changes on the preliminary drawing to establish the final HVAC drawing. The final HVAC drawing is shown in Figure 19.18.

dr

	BTUH	
Sensible Heat Gain Subtotal from Page 1	7696	31

	CEILING	SQUARE FEET	DESIGN TEMPERATURE DIFFERENCE 10	15	20	25	30	35	BTUH HEAT GAIN	
			HEAT TRANSFER MULTIPLIER							
	No Insulation		14.9	17.0	19.2	21.4	22.5	23.6		
	R-11, 3" Insulation		2.8	3.2	3.7	4.1	4.3	4.5		
	R-19, 6" Insulation		1.8	2.1	2.3	2.6	2.8	2.9		
32	R-30, 10" Insulation	976	1.1	1.3	1.5	1.6	1.7	1.8	1659	33
	R-38, 12" Insulation		0.9	1.0	1.1	1.3	1.3	1.4		

	FLOOR OVER UNCONDITIONED SPACE	SQUARE FEET	DESIGN TEMPERATURE DIFFERENCE 10	15	20	25	30	35	BTUH HEAT GAIN	
			HEAT TRANSFER MULTIPLIER							
	No Insulation		1.9	3.9	5.8	7.7	8.7	9.6		
	CARPET FLOOR NO INSULATION		1.3	2.5	3.8	5.1	5.7	6.3		
	R-11, 3" INSULATION		0.4	0.8	1.3	1.7	1.9	2.1		
34	R-19, 6" INSULATION	976	0.3	0.5	0.8	1.1	1.2	1.3	1171	35
	R-30, 10" INSULATION		0.2	0.4	0.6	0.7	0.8	0.9		

36 INFILTRATION/VENTILATION

976 FLOOR SQ FT. x 8 CEILING HEIGHT = 7808 CUBIC FT

0.40 x 7808 CUBIC FT ÷ 60 = 52 CFM

MECHANICAL VENTILATION CFM – FRESH AIR INTAKE

	CFM	DESIGN TEMPERATURE DIFFERENCE 10	15	20	25	30	35	BTUH HEAT GAIN	
		HEAT TRANSFER MULTIPLIER							
INFILTRATION	52	11.0	16.5	22.0	27.0	32.0	38.0	1664	
MECHANICAL VENTILATION	52	11.0	16.5	22.0	27.0	32.0	38.0	1664	37

	INTERNAL HEAT GAIN	BTUH HEAT GAIN	
38	Number of People 4 × 300	1200	38
	Kitchen Allowance	1,200	

SENSIBLE HEAT GAIN SUBTOTAL	15054	39

	DUCT GAIN	BTUH HEAT GAIN	
40	R-4, 1" Flexible Blanket Insulation: ADD 15% (.15)	2258	40
	R-7, 2" Flexible Blanket Insulation: ADD 10% (.10)		

TOTAL SENSIBLE HEAT GAIN	17312	41

LATENT LOAD CALCULATIONS

42

Conditions	Outdoor Wet Bulb	Indoor Wet Bulb	Grains
Wet	80	62.5	50
Medium	75	62.5	35
Medium Dry	70	62.5	20
Dry	65	62.5	0

Based on 75°F Indoor Dry Bulb at 50% RH.

43	LATENT LOAD INFILTRATION 0.68 × 35 Grains × 52 Infiltration CFM	1238	43
44	LATENT LOAD-VENTILATION 0.68 × 35 Grains × 52 Ventilation CFM	1238	44
45	LATENT LOAD-PEOPLE Number of People 4 x 230	920	45

TOTAL LATENT HEAT GAIN	3396	46
TOTAL SENSIBLE AND LATENT HEAT GAIN	20716	47

NOTE: All Heat Transfer Multipliers from ACCA Manual "J" Sixth Edition and for a medium outdoor daily temperature range.

Litho U.S.A.

FIGURE 19.16 ■ *(continued)*

STEP 1 *Outdoor temperature:* 100°F.

STEP 2 *Indoor temperature:* 70°F.

STEP 3 *Temperature difference.* $100° - 70° = 30°$F.

STEPS 4–13 Glass, no shade, double-glazed.

STEP 4 *North glass:* Including sliding glass door; 81 sq. ft.

STEP 5 *Btuh heat gain:*

$$81 \times 28 = 2268 \text{ Btuh}$$

STEPS 6–7 *NE and NW glass:* None; house faces N, E, W, S.

STEPS 8–9 *East and west glass:* None in this plan.

STEPS 10–11 *SE and SW glass:* None; house faces N, E, W, S.

STEP 12 South glass: 67 sq. ft.

STEP 13 *Btuh heat loss:*

$$67 \times 43 = 2881 \text{ Btuh}$$

STEPS 14–23 Applies to glass with inside shade. This house is sized without considering inside shade. If items 14–23 are used, omit items 4–13.

STEP 24 *Doors:* Solid wood weather-stripped, 2 doors; 39 sq. ft.

STEP 25 *Btuh heat gain:*

$$39 \times 9.6 = 374 \text{ Btuh}$$

STEP 26 Walls (excluding garage walls):

$$132' \times 8' = 1056 \text{ sq. ft.} - 187 \text{ sq. ft.} = 869 \text{ sq. ft. net wall area}$$

STEP 27 *Frame wall:* 869 sq. ft.

STEP 28 *Btuh heat gain:*

$$869 \times 2.5 = 2173 \text{ Btuh}$$

STEPS 29–30 *Masonry wall above grade:* None in this house.

STEP 31 *Sensible heat gain subtotal:* Add together items 5, 13, 25, 28. 7696 Btuh heat gain. Transfer the amount to the top of page 2 of the form.

STEP 32 *Ceiling:* R-30 insulation; 976 sq. ft.

STEP 33 *Btuh heat gain:*

$$976 \times 1.7 = 1659 \text{ Btuh}$$

STEP 34 *Floor over unconditioned space:* R-19 insulation; 976 sq. ft.

STEP 35 *Btuh heat gain:*

$$976 \times 1.2 = 1171 \text{ Btuh}$$

STEP 36 *Infiltration/ventilation:*

$$976 \text{ sq. ft.} \times 8' = 7808 \text{ cu. ft.}$$

$$0.40 \times 7808 \text{ cu. ft.} \div 60 = 52 \text{ CFM}$$

STEP 37 *Btuh heat gain:*

$$52 \times 32 = 1664 \text{ Btuh}$$

CADD APPLICATIONS

while you draw, generating a complete bill of materials. The systems that offer the greatest flexibility and productivity are designed as a parametric package. This program type allows you to set the design parameters that you want to use. Then the computer automatically draws and details according to these settings. As you draw, information is placed with each fitting, including type of fitting, CFM, and gauge. The CADD menus and an HVAC plan are shown in Figure 19.31.

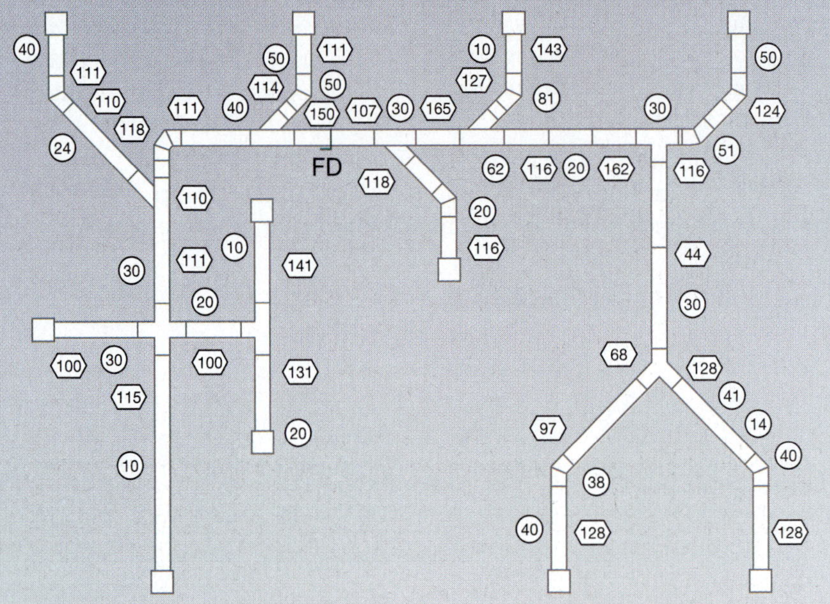

FIGURE 19.30 ■ Complete HVAC CADD layout.

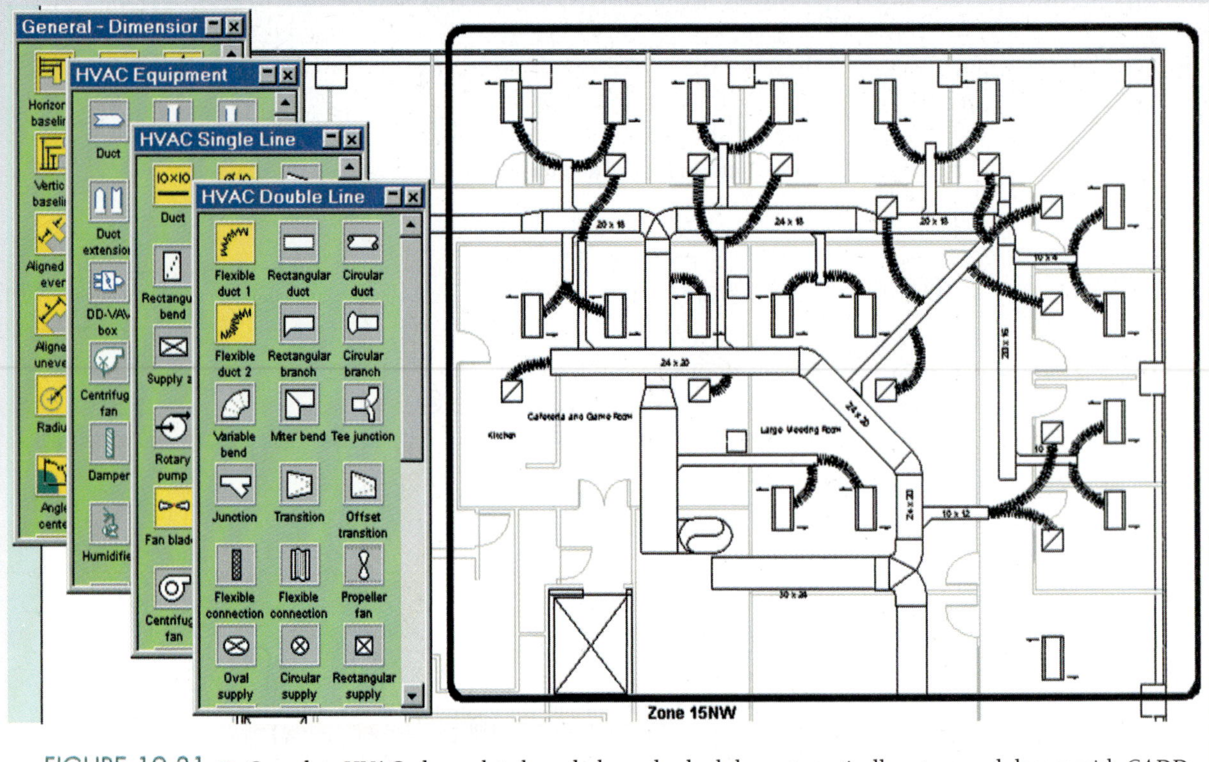

FIGURE 19.31 ■ Complete HVAC plan, related symbols, and schedules automatically set up and drawn with CADD systems. HVAC menus and a sample HVAC plan drawn using a custom CAD program. *Courtesy Visio Corporation.*

CADD APPLICATIONS

HVAC PICTORIALS

CADD applications often make pictorial representations much easier to implement, especially when the HVAC program allows direct conversion from the plan view to the pictorial. The CADD pictorials, known as *graphic models,* may be used to view the HVAC system from any angle or orientation. Some CADD programs automatically analyze the layout for obstacles where an error in design may result in a duct that does not have a clear path. One of the biggest advantages of a CADD system occurs when changes are made to the HVAC plan. These changes are corrected simultaneously on all drawings, schedules, and lists of materials. Figure 19.32 shows a CADD-generated perspective of HVAC duct routing.

CADD LAYERS FOR HVAC DRAWINGS

The American Institute of Architects (AIA) CAD Layer Guidelines establish the heading "Mechanical" as the major group identification for HVAC-related CADD layers. Following are some of the recommended CADD layer names for HVAC applications:

LAYER NAME	DESCRIPTION
M-CHIM	Prefabricated chimneys
M-CMPA	Compressed air systems
M-CONT	Controls and instrumentation
M-DUST	Dust and fume collection systems
M-ENER	Emergency management systems
M-EXHS	Exhaust systems
M-FUEL	Fuel system piping
M-HVAC	HVAC system
M-HOTW	Hot water heating system
M-CWTR	Chilled water system
M-REFG	Refrigeration systems
M-STEM	Steam systems
M-ELEV	Elevations
M-SECT	Sections
M-DETL	Details
M-SCHD	Schedules and title block sheets

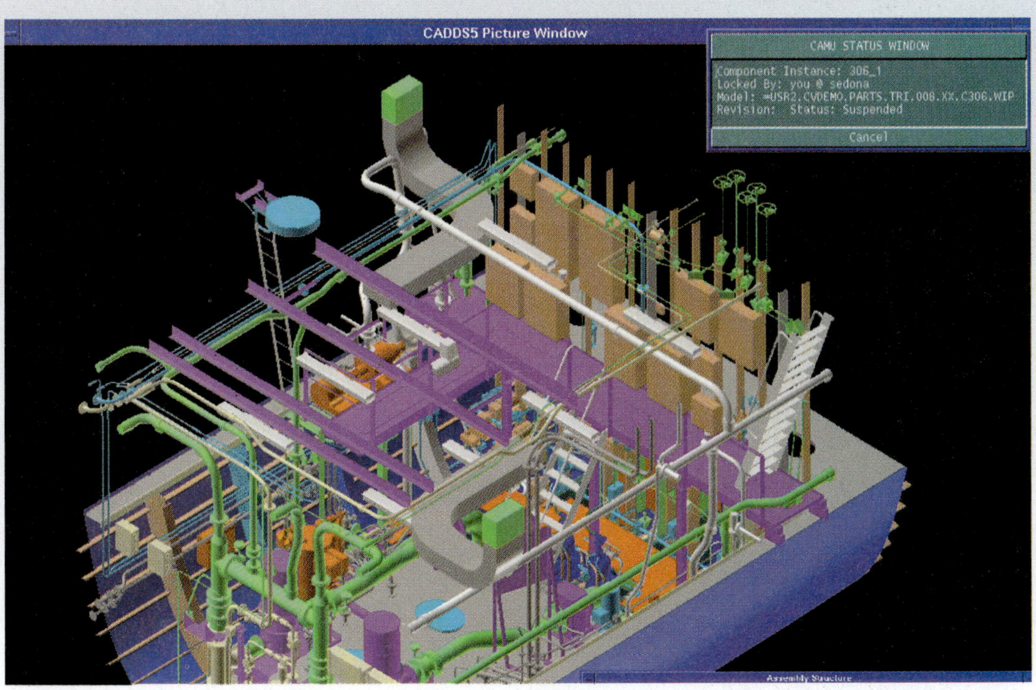

FIGURE 19.32 ■ CADD-generated perspective of HVAC duct routing. *Courtesy Computervision Corporation.*

Chapter 19 Additional Reading

The following Web sites can be used as a resource to help you keep current with changes in building materials.

ADDRESS	COMPANY OR ORGANIZATION
www.carrier.com	Carrier
www.lennox.com	Lennox
www.trane.com	Trane

CHAPTER 19

Heating, Ventilating, and Air-Conditioning Test

DIRECTIONS

Answer the questions with short, complete statements or drawings as needed on an 8 1/2″ × 11″ sheet of notebook paper, as follows:

1. Letter your name, HVAC Test, and the date at the top of the sheet.
2. Letter the question number and provide the answer. You do not need to write out the question.

Answers may be prepared on a word processor if course guidelines allow this.

QUESTIONS

Question 19–1 Describe the following heating and cooling systems:

a. Central forced air
b. Hot water
c. Heat pump
d. Zoned heating

Question 19–2 Describe a type of zoned heating called radiant heat.

Question 19–3 List two advantages and two disadvantages of zonal heat as compared with central forced-air heating.

Question 19–4 A heat pump may supply up to how many times as much heat per year for the same amount of electrical consumption as a standard electric forced-air system?

Question 19–5 Discuss four factors that influence the placement of a thermostat.

Question 19–6 Describe five sources that can contribute to an unhealthy living environment.

Question 19–7 Discuss the function of an air-to-air heat exchanger.

Question 19–8 List five advantages of a central vacuum system.

Question 19–9 What does the size of heating and cooling equipment have to do with providing access clearances for servicing and replacement?

Question 19–10 Define *combustion air.*

Question 19–11 List the rooms or locations where return air cannot be accessed.

Question 19–12 What does the abbreviation UL stand for?

Question 19–13 Give the minimum thermal resistance value for insulated ducts.

Question 19–14 Define *plenum.*

Question 19–15 Describe the purpose of schedules in an HVAC drawing.

PROBLEMS

Problem 19–1 Calculate the minimum heat exchanger capacity for one of the floor plans found in Floor Plan Problems at the end of Chapter 16. Use the formulas established earlier in the Air-to-Air Heat Exchangers section of this chapter.

Problem 19–2 Using this problem's residential heating engineering sketch of main floor plan and basement, do the following on vellum of appropriate size:

1. Make a formal double-line HVAC floor plan layout at a scale of 1/4″ = 1′–0″.
2. Approximate the location of undimensioned items such as windows.
3. Use thin lines for the floor plan and thick lines for the heating equipment and duct runs.

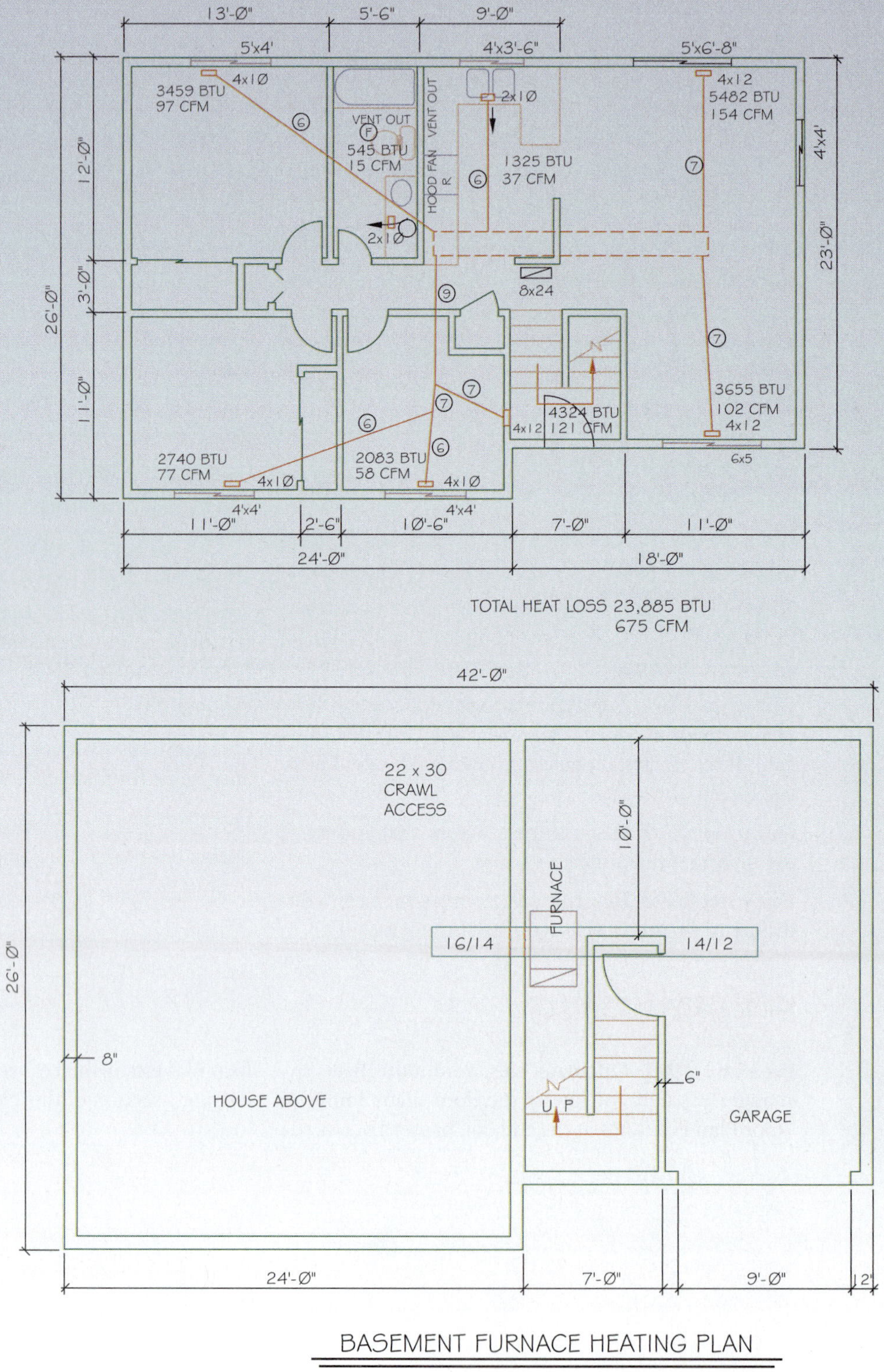

PROBLEM 19–2

Problem 19–3 Using this problem's residential air-to-air heat exchanger ducting engineering sketch of a basement floor plan, do the following on vellum of appropriate size (one B or C size sheet is recommended):

1. Make a formal single-line air-to-air heat exchanger floor plan layout at a scale of 1/4″ = 1′–0″.
2. Approximate the location of undimensioned items such as doors.
3. Use thin lines for the floor plan and use thick lines for the air-to-air heat exchanger equipment and duct runs.

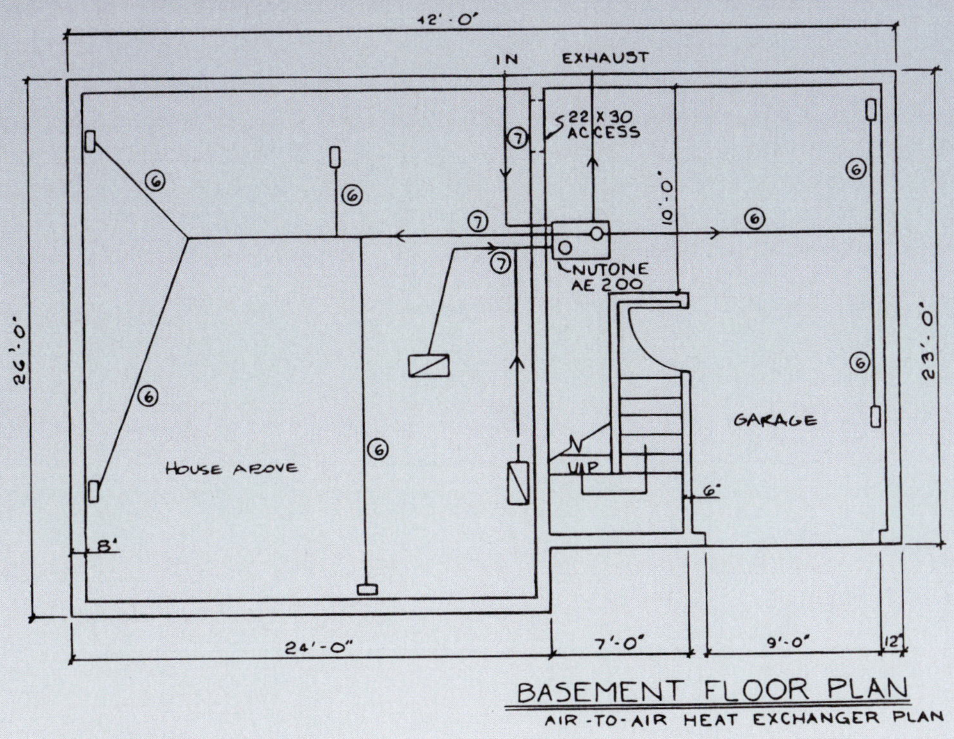

PROBLEM 19–3

Problem 19–4 Using the plan that you have been drawing from Chapter 16, design and draw your own simplified HVAC plan.

Problem 19–5 Using the plan that you have been drawing from Chapter 16, design and draw your own double line HVAC plan.

Problem 19–6 Using the plan that you have been drawing from Chapter 16, design and draw your own double line air-to-air heat exchanger plan.

Problem 19–7 Prepare heat loss and gain calculations for one of the floor plans found at the end of Chapter 16 Floor Plan Problems. You may use the floor plan of your choice, or your instructor will assign a floor plan. Use the residential heating data sheets and the residential cooling data sheets found in the CD for this text. Complete the data sheets according to the method used in CD instructions to solve for the total winter heat loss and summer heat gain. Use the following criteria unless your instructor specifies otherwise:

1. Outdoor temperature as recommended for your area.
2. Indoor temperature: 70°F.
3. Double glass windows and glass doors.
4. Urethane insulated metal exterior doors, weather-stripped.
5. Ceiling height: 8′–0″.
6. Ceiling insulation: R-30.
7. Frame wall insulation: R-19 + 1/2″ polystyrene.
8. Insulation in floors over unconditioned space: R-19.
9. Duct insulation: R-4.
10. Masonry above and below grade as required in plan used.
11. Basement floor or concrete slab construction as required in plan used. Verify use with your instructor.
12. Assume entry foyer door faces south to establish window orientation.
13. Assume glass with inside shade.
14. Number of occupants is optional. Select the family size you desire or count the bedrooms and add one. For example, a three-bedroom home plus one equals four people.
15. Use latent load conditions typical for your local area. Select wet, medium, medium dry, or dry as appropriate. Consult your instructor.

Problem 19–8 Given the following:

1. HVAC floor plan engineering layout at approximately 1/16″ = 1′–0″. The engineer's layout is rough, so round off dimensions to the nearest convenient units. For example, if the dimension you scale reads 24′–3″, round off to 24′–0″. The floor plan will not require dimensioning; therefore, the representation is more important than the specific dimensions.
2. Related schedules.
3. Engineer's sketch for exhaust hood.

Do the following on vellum of appropriate size (C or D size is recommended; all required items will fit on one sheet with careful planning):

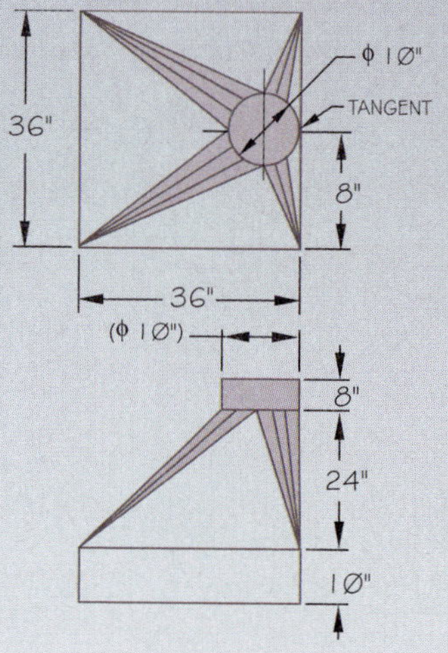

EXHAUST HOOD DETAIL

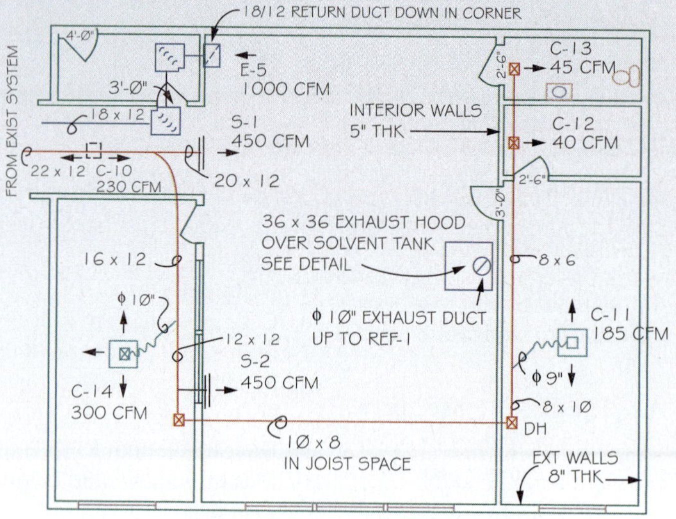

PROPOSED ADDITION HVAC PLAN

PROBLEM 19–8

SCHEDULES

CEILING OUTLET SCHEDULE

SYMBOL	SIZE	CFM	DAMPER TYPE	PANEL SIZE
C-10	9 × 9	230	Key-operated	12 × 12
C-11	8 × 8	185	Key-operated	12 × 12
C-12	6 × 6	40	Key-operated	12 × 12
C-13	6 × 6	45	Key-operated	12 × 12
C-14	6 × 18	300	Fire damper	24 × 24

SUPPLY GRILL SCHEDULE

SYMBOL	SIZE	CFM	LOCATION	DAMPER TYPE
S-1	20 × 8	450	High wall	Key operation
S-2	20 × 8	450	High wall	External operation

EXHAUST GRILL SCHEDULE

SYMBOL	SIZE	CFM	LOCATION	DAMPER TYPE
E-5	18 × 24	1000	Low wall	No damper

ROOF EXHAUST FAN SCHEDULE

SYMBOL	AREA SERVED	CFM	FAN SPECIFICATIONS
REF-1	Solvent Tank	900	1/4 HP, 12″ Nonspark wheel, 1050 Maximum, outlet velocity

1. Make a formal double-line HVAC floor plan layout at a scale of 1/4″ = 1′–0″. (Note: You measured the given engineer's sketch at 1/16″ = 1′–0″.) Now you convert the established dimensions to a formal drawing at 1/4″ = 1′–0″. Approximate the location of the HVAC duct runs and equipment in proportion to the presentation on the sketch.
2. Prepare correlated schedules in the space available. Set up the schedules in a manner similar to the examples below.
3. Make a detail drawing of the exhaust hood, either scaled or unscaled. Make the detail large enough to clearly show the features. Refer to the detail drawings below for examples.

SECTION 6

Roof Plans

20 CHAPTER

Roof Plan Components

INTRODUCTION

The design of the roof must be considered long before the roof plan is drawn. The architect or designer will typically design the basic shape of the roof as the floor plan and elevations are drawn in the preliminary design stage. This does not mean that the designer plans the entire structural system for the roof during the initial stages, but the general shape and type of roofing material to be used will be planned. By examining the structure in Figure 20.1, you can easily see how much impact the roof design has on the structure. Often the roof can present a larger visible surface area than the walls. In addition to aesthetic considerations, the roof can also be used to provide rigidity in a structure when wall areas are filled with glass, as seen in Figure 20.2. To ensure that the roof will meet the designer's criteria, a roof plan is usually drawn by the drafter to provide construction information. In order to draw the roof plan, a drafter should understand types of roof plans, various pitches, common roof shapes, and common roof materials.

TYPES OF ROOF PLANS

The plan that is drawn of the roof area may be either a *roof plan* or a *roof framing plan.* For some types of roofs a roof drainage plan may also be drawn. Roof framing and drainage plans will be discussed in Chapter 30.

FIGURE 20.1 ■ The shape of the roof can play an important role in the design of the structure as seen in this home by the architect Robert Roloson. *Photo Courtesy Elk Roofing.*

FIGURE 20.2 ■ In addition to aesthetic considerations, the roof is often used to resist wind and seismic forces when walls of the structure contain large amounts of glass. *Courtesy of the California Redwood Association, Robert Corna architect, photo by Balthazar Korab.*

Roof Plans

A roof plan is used to show the shape of the roof. Materials such as the roofing material, vents and their location, and the type of underlayment are also typically specified on the roof plan, as seen in Figure 20.3. Roof plans are typically drawn at a scale smaller than the scale used for the floor plan. A scale of 1/8″ = 1′–0″ or 1/16″ = 1′–0″ is commonly used for a roof plan. A roof plan is typically drawn on the same sheet as the exterior elevations.

Roof Framing Plans

Roof framing plans are usually required for complicated residential roof shapes and for most commercial projects. A roof framing plan shows the size and direction of the construction

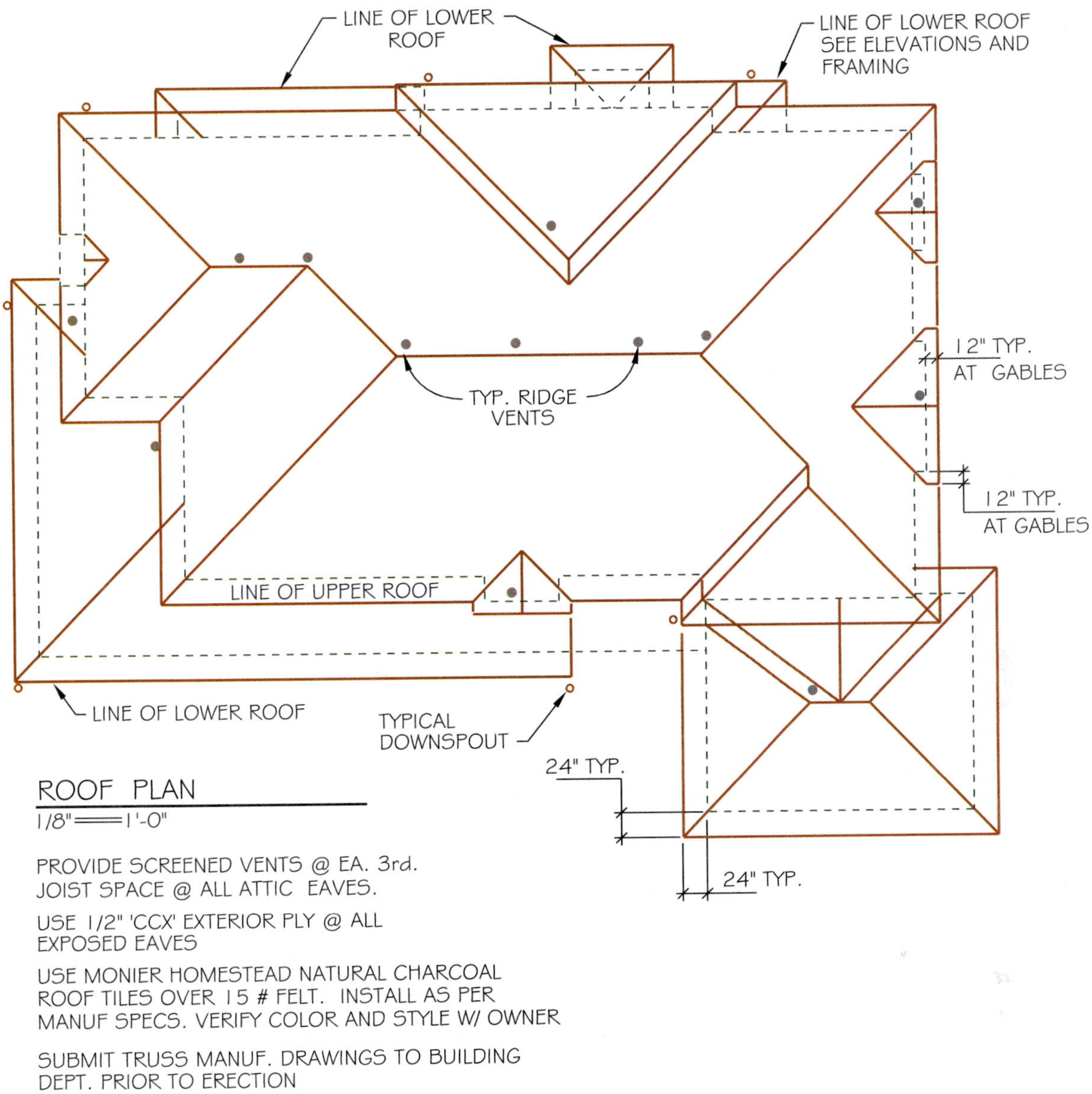

FIGURE 20.3 ■ A roof plan is drawn to show the shape of the roof.

members that are required to frame the roof. Figure 20.4 shows an example of a roof framing plan. On very complex projects, every framing member is shown, as seen in Figure 20.5. Framing plans will be discussed further in Chapter 30.

ROOF PITCHES

Roof pitch, or *slope,* is a description of the angle of the roof that compares the horizontal run and the vertical rise. The slope, shown when the elevations and sections are drawn, will be discussed in Chapters 23 and 26. The intersections that result from various roof pitches must be shown on the roof plan. In order to plot the intersection between two roof surfaces correctly, the drafter must understand how various roof pitches are drawn. Figure 20.6 shows how the pitch can be visualized. The drafter can plot the roof shape using this method for any pitch. Adjustable triangles for plotting roof angles are available and save the time of having to measure the rise and run of a roof. The roof pitch can also be drawn if the drafter knows the proper angle that a certain pitch represents. Knowing that a 4/12 roof equals 18 1/2° allows the drafter to plot the correct angle without having to plot the layout. Figure 20.7 shows angles for common roof pitches.

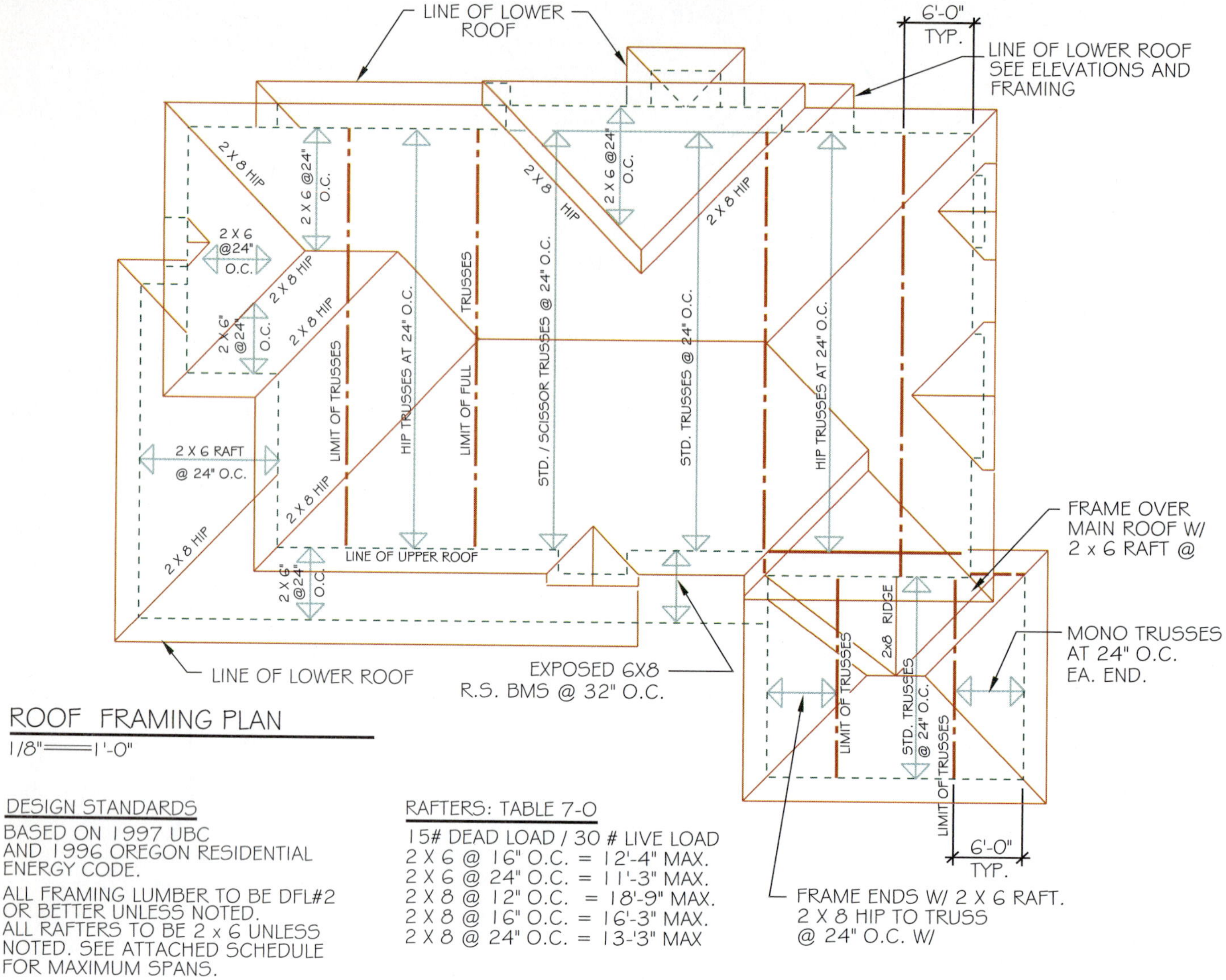

FIGURE 20.4 ■ A roof framing plan is used to show the framing members for the roof.

ROOF SHAPES

By changing the roof pitch, the designer can change the shape of the roof. Common roof shapes include flat, shed, gable, A-frame, gambrel, hip, Dutch hip, and mansard. See Chapter 26 for a complete discussion of roof framing terms.

Flat Roofs

The flat roof is a very common style in areas with little rain or snow. In addition to being used in residential construction, the flat roof is typically used on commercial structures to provide a platform for heating and other mechanical equipment. The flat roof is economical to construct because ceiling joists are eliminated and rafters are used to support both the roof and ceiling loads. Figure 20.8 shows the materials commonly used to frame a flat roof. Figure 20.9 shows how a flat roof could be represented on the roof plan.

Often the flat roof has a slight pitch in the rafters. A pitch of 1/4″ per foot (2 percent slope) is often used to help prevent water from ponding on the roof. As water flows to the edge, a metal diverter is usually placed at the eave to prevent dripping at walkways. A flat roof will often have a *parapet,* or false wall, surrounding the perimeter of the roof. Figure 20.10 provides an example of a parapet wall. This wall can be used for decoration or for protection of mechanical equipment. When used, it must be shown on the roof plan.

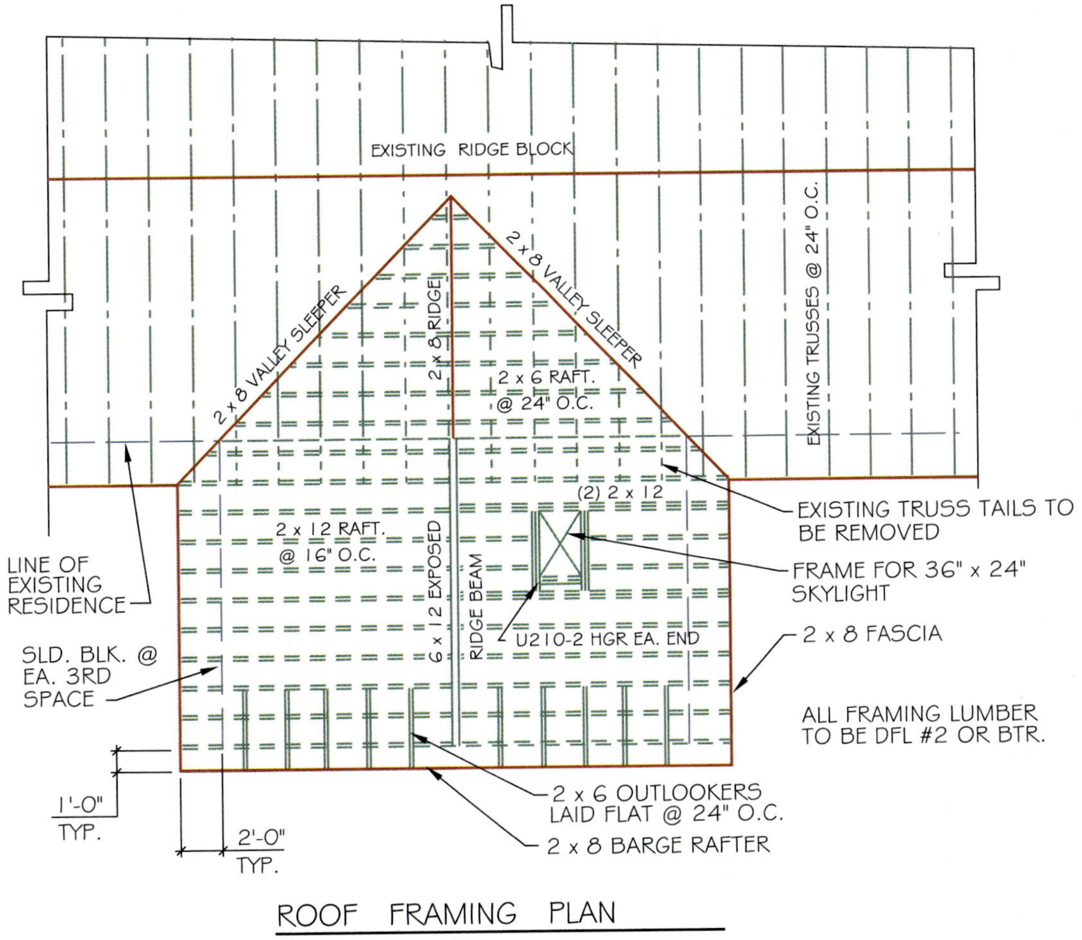

FIGURE 20.5 ■ For complicated roofs, a roof framing plan may be drawn to show the size and location of every structural member.

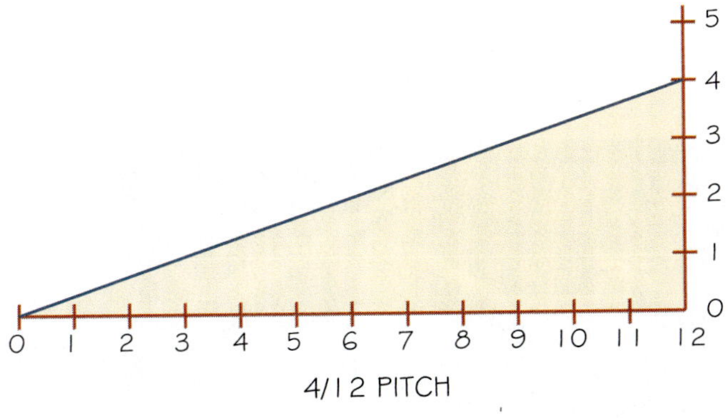

FIGURE 20.6 ■ In determining the roof slope, the angle is expressed as a comparison of equal units. Units may be inches, feet, meters, etc., as long as the horizontal and vertical units are of equal length.

COMMON ANGLES FOR DRAWING ROOF PITCHES

ROOF PITCH	ANGLE
1/12	4°–30′
2/12	9°–30′
3/12	15°–0′
4/12	18°–30′
5/12	22°–30′
6/12	26°–30′
7/12	30°–0′
8/12	33°–45′
9/12	37°–0′
10/12	40°–0′
11/12	42°–30′
12/12	45°–0′

FIGURE 20.7 ■ Common roof pitches and angles. Angles shown are approximate and are to be used for drawing purposes only.

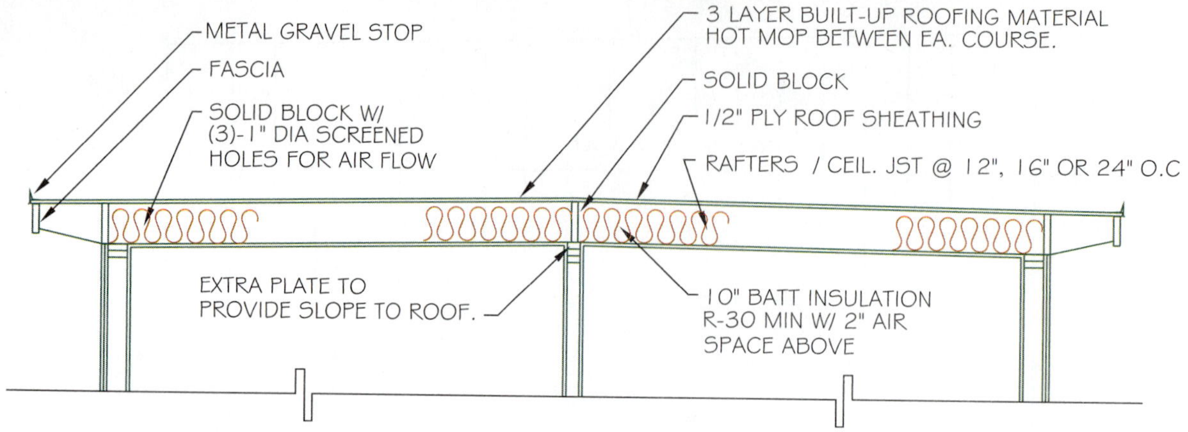

FIGURE 20.8 ■ Common construction components of a flat roof.

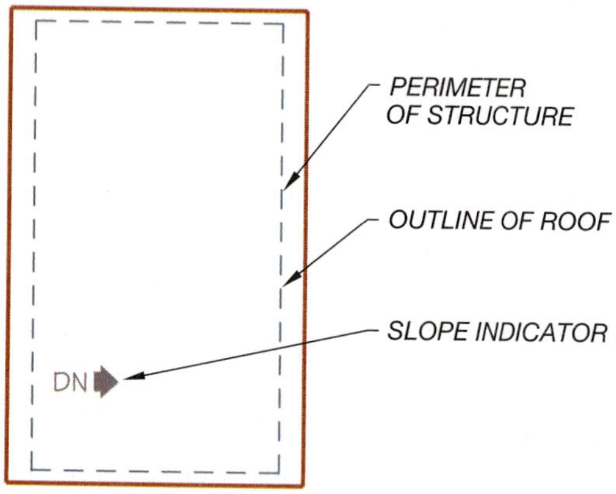

FIGURE 20.9 ■ Flat roof in plan view.

Shed Roofs

The shed roof, as seen in Figure 20.11, offers the same simplicity and economical construction methods as a flat roof but does not have the drainage problems associated with a flat roof. Figure 20.12 shows construction methods for shed roofs. The shed roof may be constructed at any pitch. The roofing material and aesthetic considerations are the only factors limiting the pitch. Drawn in plan view, the shed roof will resemble the flat roof, as seen in Figure 20.13.

Gable Roofs

A gable roof is one of the most common roof types in residential construction. As seen in Figure 20.14, it uses two shed roofs that meet to form a ridge between the support walls. Figure 20.15 shows the construction of a gable roof system. The gable can be constructed at any pitch, with the choice of

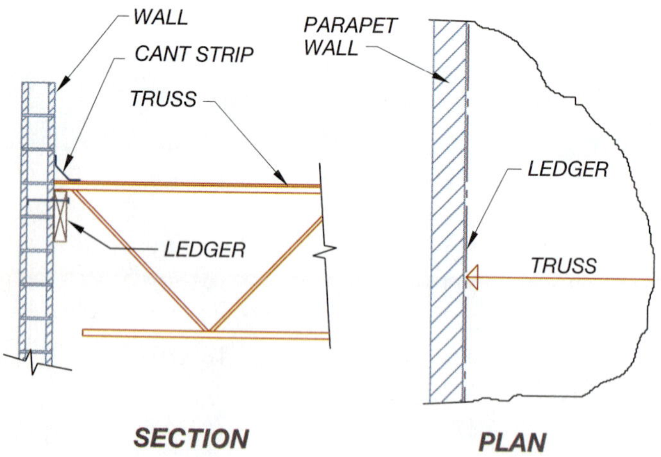

FIGURE 20.10 ■ A parapet wall is often placed around a flat roof to hide mechanical roof equipment. The thickness of the wall should be represented on the roof plan.

FIGURE 20.11 ■ Many contemporary homes combine flat and shed roofs to create a pleasing design. *Courtesy LeRoy Cook.*

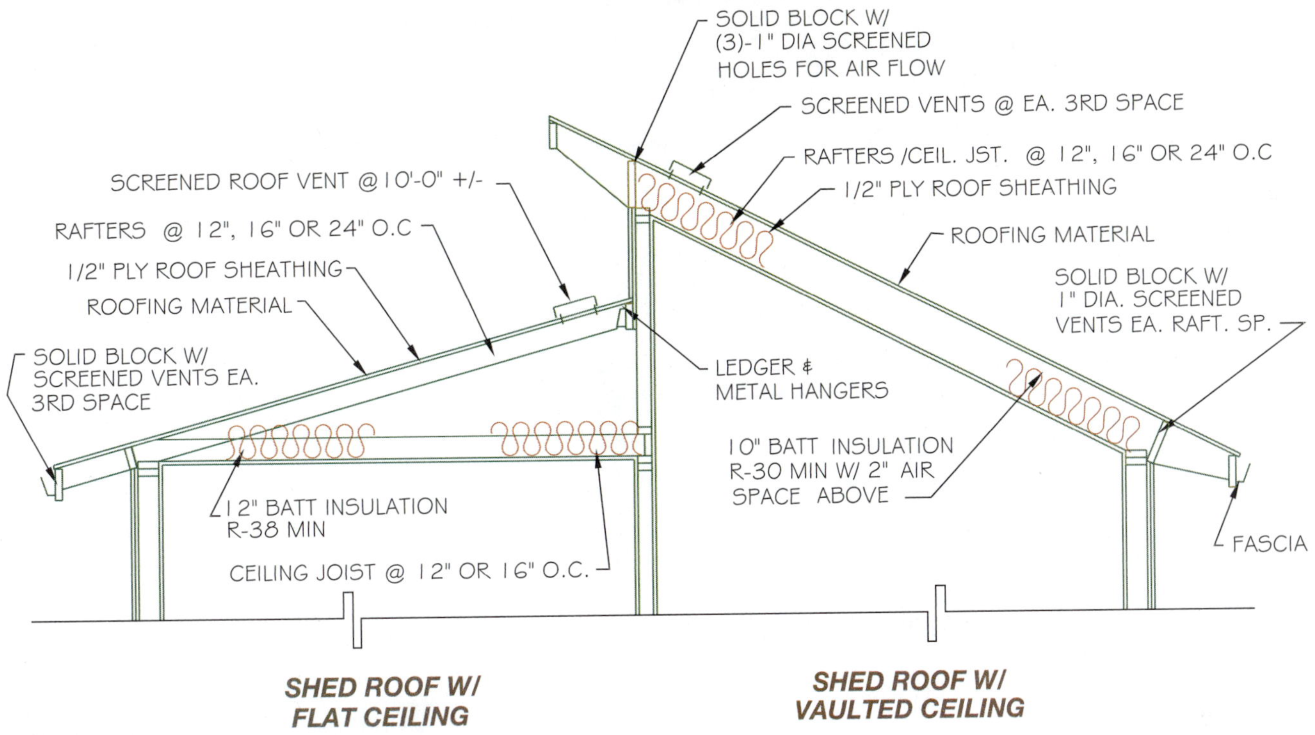

FIGURE 20.12 ■ Common construction components of shed roofs.

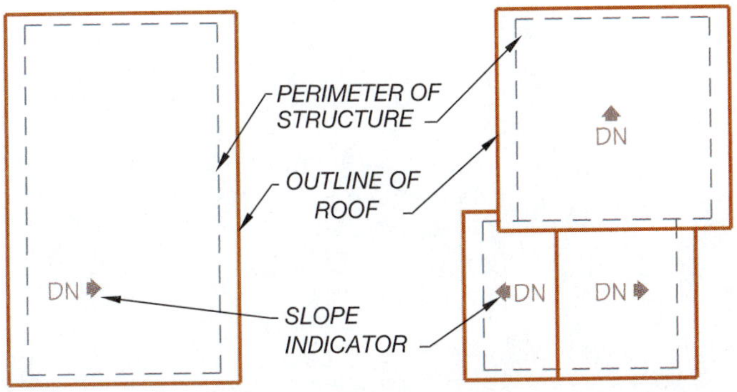

FIGURE 20.13 ■ Shed roof shapes in plan view.

FIGURE 20.14 ■ A gable roof is composed of two intersecting planes that form a peak (the ridge) between the planes. *Courtesy LeRoy Cook.*

pitch limited only by the roofing material and the effect desired. A gable roof is often used on designs seeking a traditional appearance and formal balance. Figure 20.16 shows how a gable roof is typically represented in plan view. Many plans use two or more gables at 90° angles to each other. The intersections of gable surfaces are called either hips or valleys. Typically, the valley and hip are specified on the roof plan.

A-Frame Roofs

An A-frame is a method of framing walls, as well as a system of framing roofs. An A-frame structure uses rafters to form its supporting walls, as shown in Figure 20.17. The structure gets its name from the letter A that is formed by the roof and floor systems. See Figure 20.18. The roof plan for an A-frame is very similar to the plan for a gable roof. However, the framing materials are usually quite different. Figure 20.19 shows how an A-frame can be represented on the roof plan.

Gambrel Roofs

A gambrel roof can be seen in Figure 20.20. The gambrel roof is a traditional shape that dates back to the colonial period. Figure 20.21 shows construction methods for a gambrel roof. The lower level is covered with a steep roof surface, which

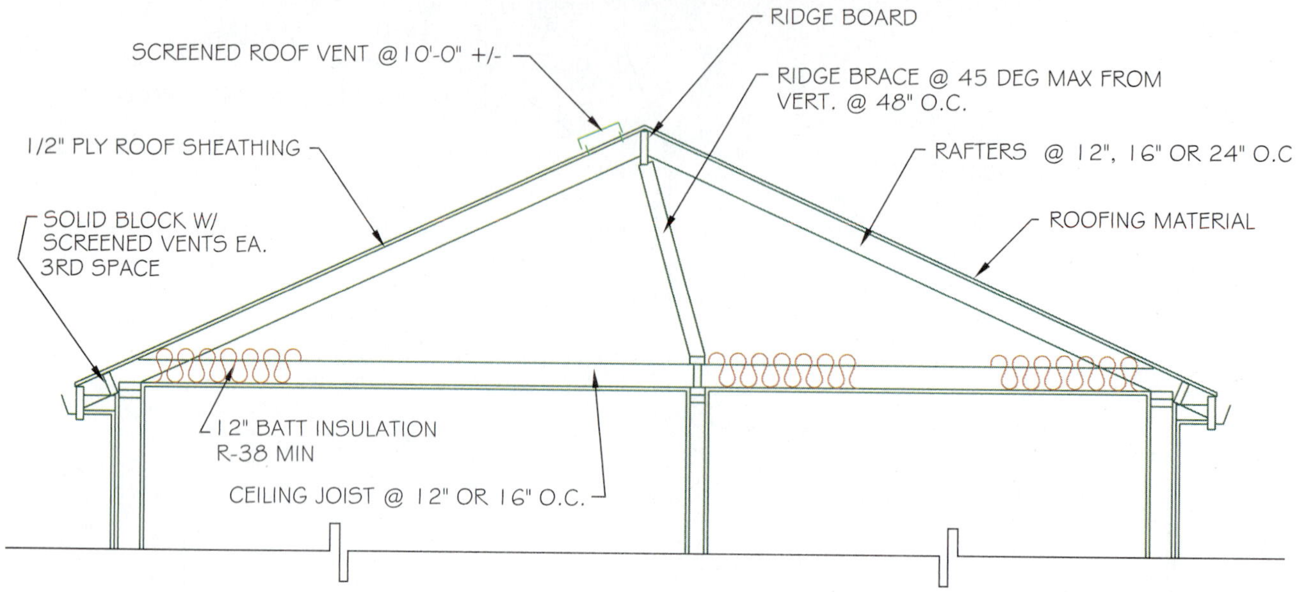

FIGURE 20.15 ■ Common construction components of a gable roof.

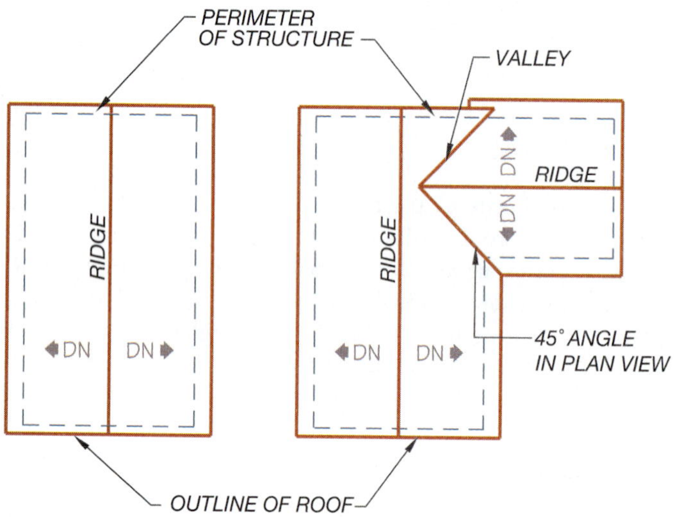

FIGURE 20.16 ■ Gable roof in plan view.

connects into the upper roof system with a slighter pitch. By covering the lower level with roofing material rather than siding, the structure appears shorter than it actually is. This roof system can also reduce the cost of siding materials by using less expensive roofing materials. Figure 20.22 shows a plan view of a gambrel roof.

Hip Roofs

The hip roof (Figure 20.23) is a traditional shape that can be used to help eliminate some of the roof mass and create a structure with a smaller appearance. A hip roof has many similarities to a gable roof but has four surfaces instead of

FIGURE 20.17 ■ An A-frame uses a steep roof to form the walls of the upper level. *Courtesy Janice Jefferis.*

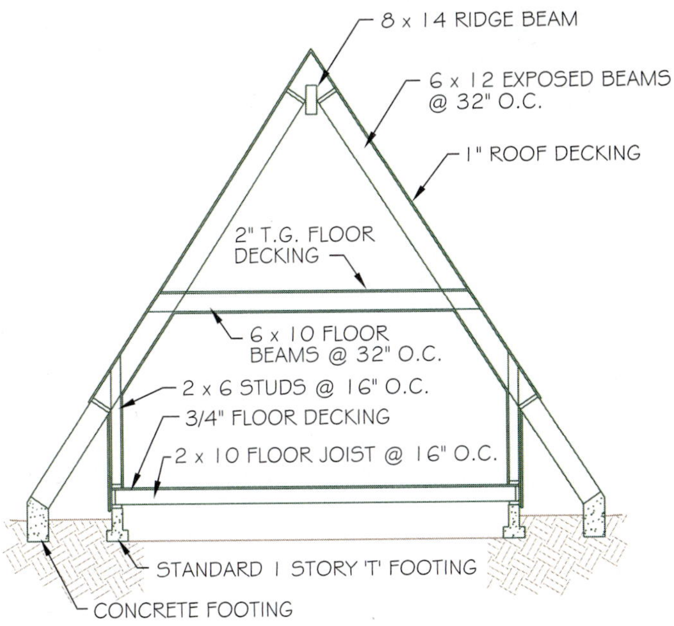

FIGURE 20.18 ■ Common components of A-frame construction.

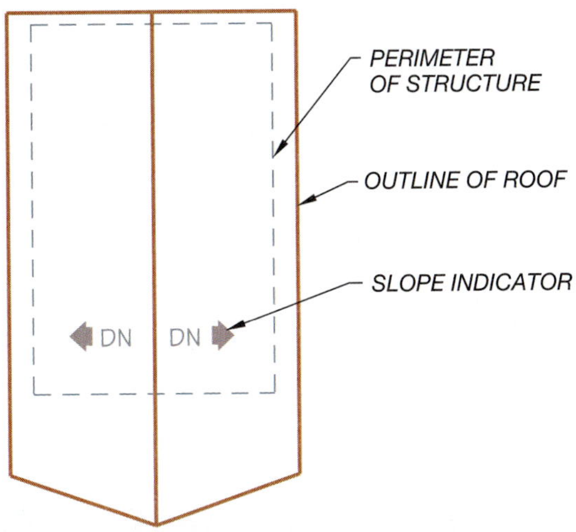

FIGURE 20.19 ■ A-frame in plan view.

FIGURE 20.20 ■ The gambrel roof is often used to enhance the traditional appearance of a residence. *Courtesy Michael Jefferis.*

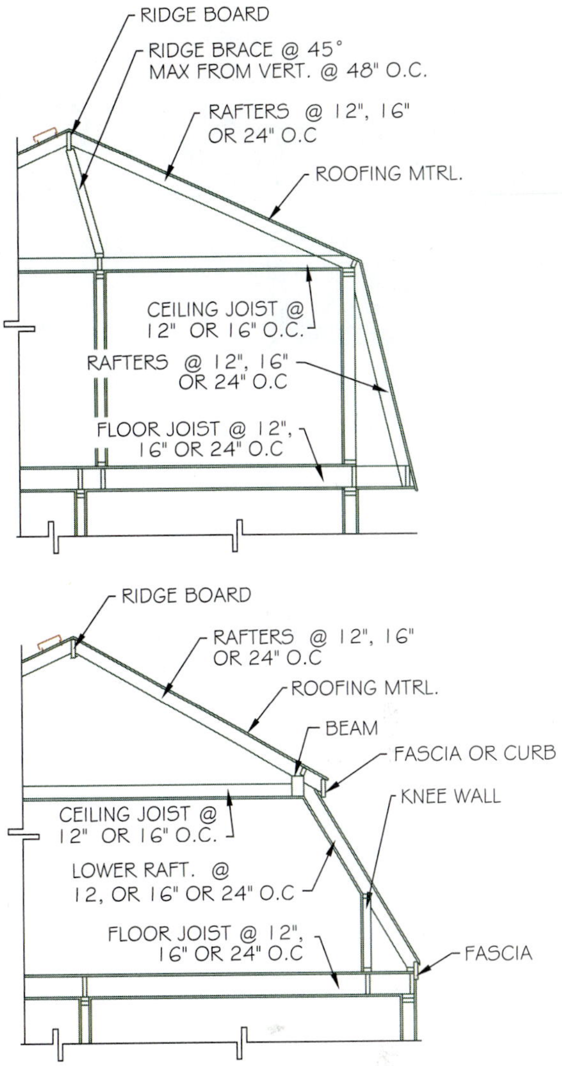

FIGURE 20.21 ■ A gambrel roof can be constructed with or without a fascia or curb between the upper and lower roofs.

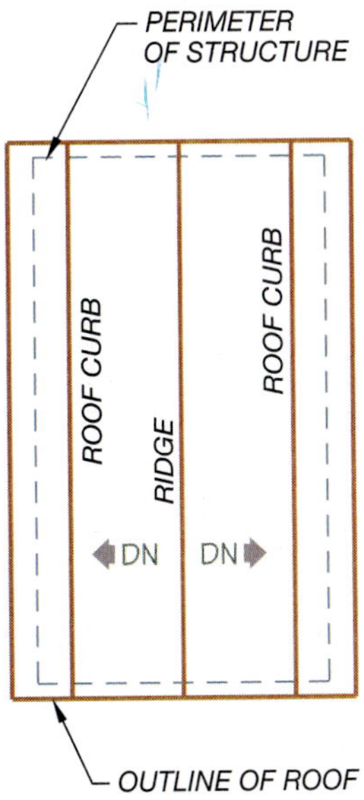

FIGURE 20.22 ■ Gambrel roof in plan view.

FIGURE 20.23 ■ A hip roof is made of three or more intersecting planes. This home uses a combination of hip roofs with a gable roof over the entry area to create a pleasing roof structure. *Courtesy LeRoy Cook.*

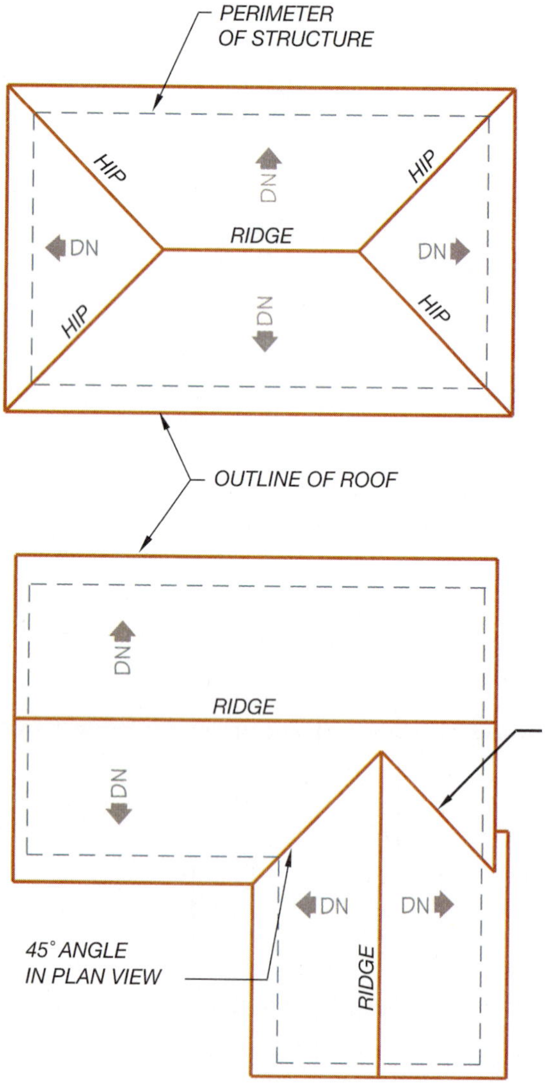

FIGURE 20.24 ■ Hips and valleys in plan view.

two. The intersection between surfaces is called a *hip*. If built on a square structure, the hips will come together to form a point. If built on a rectangular structure, the hips will form two points with a ridge spanning the distance between them. When hips are placed over an L- or T-shaped structure, an interior intersection will be formed; this is called a valley. The valley of a hip roof is the same as the valley of a gable roof. Hips and valleys can be seen in plan view as shown in Figure 20.24.

Dutch Hip Roofs

The Dutch hip roof is a combination of a hip and a gable roof. See Figure 20.25. The center section of the roof is framed using a method similar to a gable roof. The ends of the roof are framed with a partial hip that blends into the gable. A small wall (gable end wall) is formed between the hip and the gable roofs, as seen in Figure 20.26. On the roof plan, the shape, distance, and wall location must be shown, as in the plan in Figure 20.27.

Mansard Roofs

The mansard roof is similar to a gambrel roof but has the angled lower roof on all four sides rather than just two. A mansard roof is often used as a parapet wall to hide mechanical equipment on the roof or to help hide the height of the upper level of a structure. An example can be seen in Figure 20.28. Mansard roofs can be constructed in many different ways. Figure 20.29 shows

FIGURE 20.25 ■ A Dutch hip is a combination of a hip and a gable roof. *Courtesy CertainTeed Roofing.*

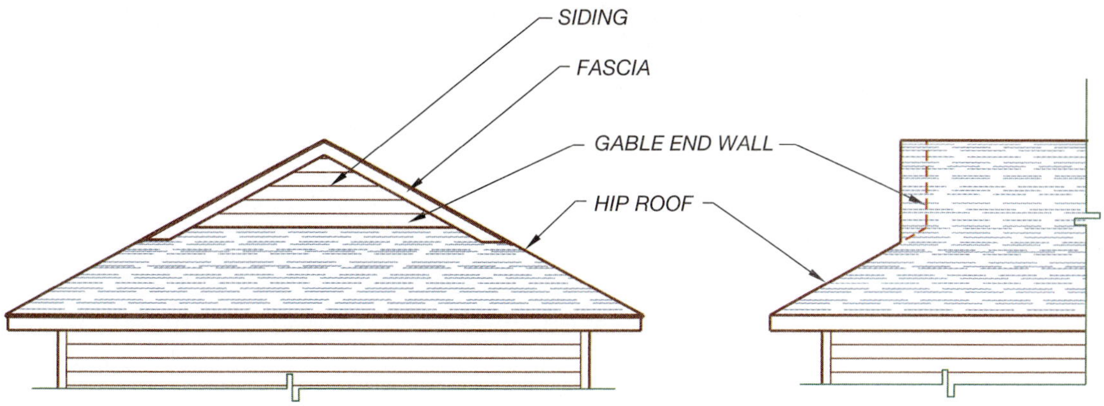

FIGURE 20.26 ■ A wall is formed between the hip and gable roof.

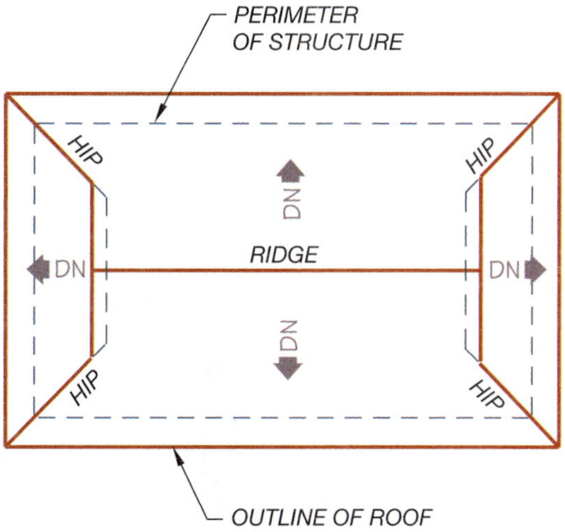

FIGURE 20.27 ■ Dutch hip roof in plan view.

FIGURE 20.28 ■ Mansard roofs are used to help disguise the height of a structure. *Courtesy Ken Stead.*

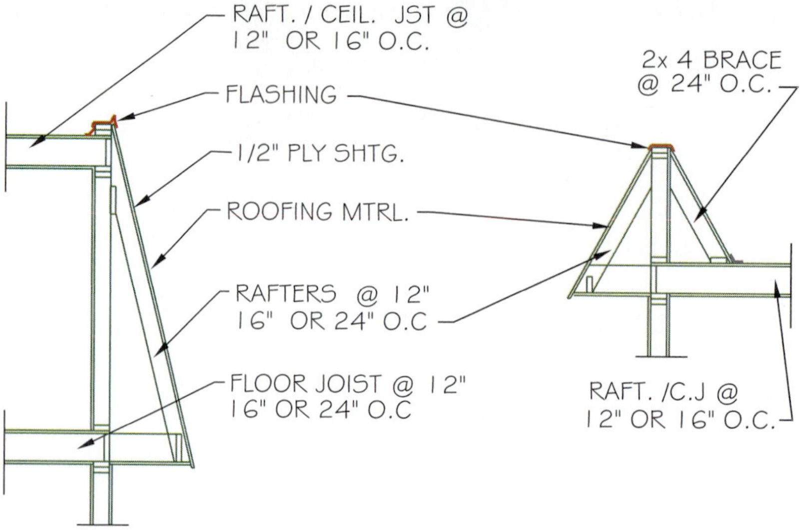

FIGURE 20.29 ■ Common methods of constructing a mansard roof.

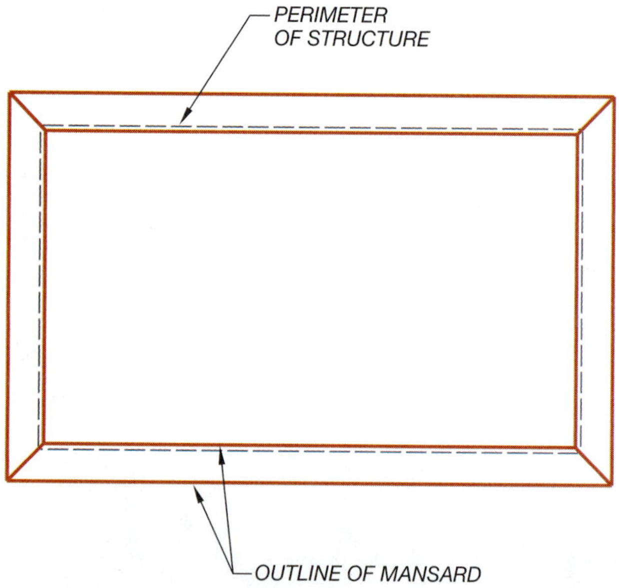

FIGURE 20.30 ■ Mansard in plan view.

FIGURE 20.31 ■ Dormers allow windows to be added to attic areas. *Courtesy LeRoy Cook.*

two common methods of constructing a mansard roof. The roof plan for a mansard roof will resemble the plan shown in Figure 20.30.

Dormers

A dormer is an opening framed in the roof to allow for window placement. Figure 20.31 shows a dormer that has been added to provide light and ventilation to rooms in what would have been attic space. Dormers are most frequently used on traditional roofs such as the gable or hip. Figure 20.32 shows one of the many ways that dormers can be constructed. Dormers are usually shown on the roof plan as seen in Figure 20.33.

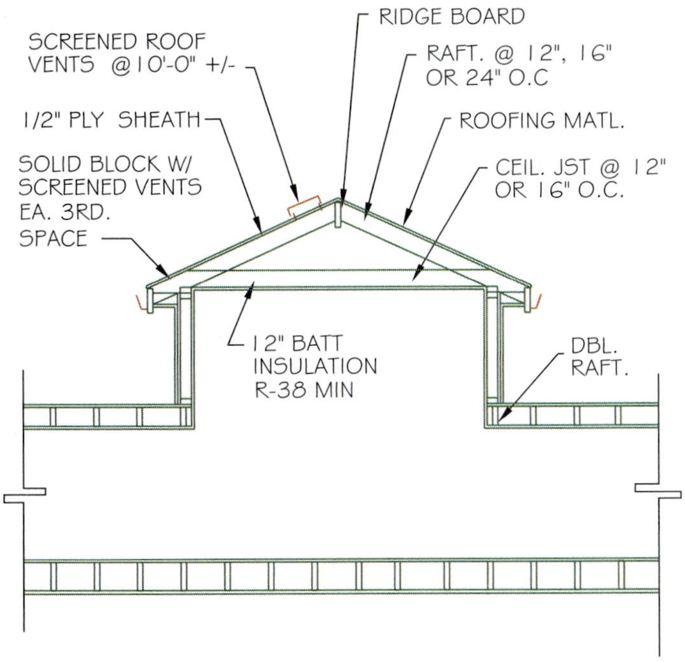

FIGURE 20.32 ■ Typical components of dormer construction.

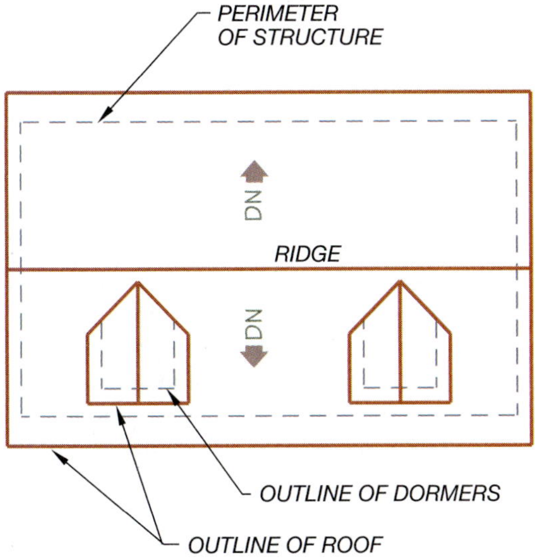

FIGURE 20.33 ■ Dormers in plan view.

ROOFING MATERIALS

The material to be used on the roof depends on pitch, exterior style, the cost of the structure, and the weather. Common roofing materials include built-up roofing, composition and wood shingles, clay and cement tiles, and metal panels. In ordering or specifying these materials, the term *square* is often used. A square is used to describe an area of roofing that covers 100 sq. ft. (9.3 m^2). The drafter will need to be aware of the weight per square and the required pitch as the plan is being drawn. The weight of the roofing material will affect the size of the framing members all the way down to the foundation level. The material will also affect the required pitch and the appearance that results from the selected pitch.

Built-Up Roofing

Built-up roofing of felt and asphalt is used on flat or low-sloped roofs below a 3/12 pitch. When the roof has a low pitch, water will either pond or drain very slowly. To prevent water from leaking into a structure, built-up roofing is used because it has no seams. On a residence, a built-up roof may consist of three alternate layers of felt and hot asphalt placed over solid roof decking. The decking is usually plywood. In commercial uses, a four- or five-layer roof is used to provide added durability. Gravel is often used as a finishing layer to help cover the felt. On roofs with a pitch over 2/12, course rocks 2″ or 3″ (50–75 mm) in diameter are used for protecting the roof and for appearance. When built-up roofs are to be specified on the roof plan, the note should include the number of layers, the material to be used, and the size of the finishing material. A typical note would be:

- 3 LAYER BUILT UP ROOF WITH HOT ASPHALTIC EMULSION BTWN. LAYERS WITH 1/4″ (6 mm) PEA GRAVEL.

Other roofing materials suitable for low-sloped (1/4 /12 minimum pitch) roofs and typical specifications include:

- MODIFIED BITUMEN—MODIFIED BITUMEN SHEET ROOFING BY JOHN MANVILLE OR EQUAL OVER 2 LAYERS OF UNDERLAYMENT PER ASTM D226 TYPE I CEMENTED TOGETHER.
- SINGLE-PLY THERMOPLASTIC—THERMOPLASTIC SINGLE-PLY ROOF SYSTEM BY SARNAFIL OR EQUAL INSTALLED PER ASTM D4434.
- SPRAYED POLYURETHANE FOAM—SPF ROOFING BY MAINLAND INDUSTRIAL COATINGS, INC. APPLIED PER ASTM 1029.
- LIQUID APPLIED COATING—GREENSEAL LIQUID WATERPROOFING MEMBRANE OR EQUAL INSTALLED PER MANUF. SPECS.

Each material can be applied to a roof with minimum pitch of 1/4 /12. Mineral surface roll roofing can be used on roofs with a minimum pitch of 1/12.

Shingles

Asphalt, fiberglass, and wood are the most typical types of shingles used as roofing materials. Most building codes and manufacturers require a minimum roof pitch of 4/12 with an underlayment of one layer of 15-lb. felt. Asphalt and fiberglass shingles can be laid on roofs as low as 2/12 if two layers of 15-lb felt are laid under the shingles and if the shingles are sealed. Wood shingles must usually be installed on roofs having a pitch of at least 3/12. Asphalt and fiberglass are similar in appearance and application.

Asphalt shingles come in a variety of colors and patterns. Also known as composition shingles, they are typically made

FIGURE 20.34 ■ Composition or three-tab shingles are a common roofing material on high sloped roofs. *Courtesy Elk Roofing.*

of fiberglass backing and covered with asphalt and a filler with a coating of finely crushed particles of stone. The asphalt waterproofs the shingle, and the filler provides fire protection. The standard shingle is a three-tab rectangular strip weighing 235 lb per square. The upper portion of the strip is coated with self-sealing adhesive and is covered by the next row of shingles. The lower portion of a three-tab shingle is divided into three flaps that are exposed to the weather. See Figure 20.34.

Composition shingles are also available in random width and thickness to give the appearance of cedar shakes. These shingles weigh approximately 300 lb per square. Both types of shingles can be used in a variety of conditions on roofs having a minimum slope of 2/12. The lifetime of shingles varies from twenty- to forty-year guarantees. See Figure 20.35. Asbestos cement shingles are also available; they weigh approximately 560 lb per square, depending on the manufacturer and the pattern used.

Shingles are typically specified on drawings in note form listing the material, the weight, and the underlayment. The color and manufacturer may also be specified. This information is often omitted in residential construction to allow the contractor to purchase a suitable brand at the best cost. A typical call-out would be:

- 235-lb composition shingles over 15-lb felt.
- 300-lb composition shingles over 15-lb felt.
- Architect 80 "Driftwood" class A fiberglass shingles by Genstar with 5 5/8″ exposure over 15-lb felt underlayment with thirty-year warranty.

Wood is also used for shakes and shingles. Wood shakes are thicker than shingles and are also more irregular in their texture. See Figure 20.36. Wood shakes and shingles are generally installed on roofs with a minimum pitch of 3/12 using a base layer of 15-lb felt. An additional layer of 15 lb × 18″ (457 mm) wide felt is also placed between each courses or layers of shingles. Wood shakes and shingles can be installed over solid or spaced sheathing. The weather, material availability, and labor practices affect the type of underlayment used.

Depending on the area of the country, shakes and shingles are usually made of cedar, redwood, or cypress. They are also produced in various lengths. When shakes or shingles are specified on the roof plan, the note should usually include the thickness, the material, the exposure, the underlayment, and the type of sheathing. A typical specification for wood shakes would be:

- Med. cedar shakes over 15# felt w/15# × 18″ wide felt between each course. Lay with 10 1/2″ exposure.

Other materials such as Masonite and metal are also used to simulate shakes. Metal is sometimes used for roof shingles on roofs with a 3/12 or greater pitch. Metal provides a durable, fire-resistant roofing material. Metal shingles are usually installed using the same precautions applied to asphalt shingles. Metal is typically specified on the roof plan in a note listing the manufacturer, type of shingle, and underlayment.

FIGURE 20.35 ■ 300-lb composition shingles are made with tabs of random width and length. *Courtesy Elk Roofing.*

FIGURE 20.36 ■ Cedar shakes are a rustic but elegant roofing material. *Courtesy Tim Taylor.*

Clay and Cement Tiles

Tile is the material most often used for homes on the high end of the price scale or where the risk of fire is extreme. Although tile may cost twice as much as the better grades of asphalt shingle, it offers a lifetime guarantee. Tile is available in a variety of colors, materials, and patterns. Clay, concrete, and metal are the most common materials. See Figure 20.37*a* and 20.37*b*.

Roof tiles are manufactured in both curved and flat shapes. Curved tiles are often called Spanish tiles and come in a variety of curved shapes and colors. Flat, or barr, tiles are also produced in many different colors and shapes. Tiles are installed on roofs having a pitch of 2 1/2/12 or greater. Tiles can be placed over either spaced or solid sheathing. If solid sheathing is used, wood strips are generally added on top of the sheathing to support the tiles.

When tile is to be used, special precautions must be taken with the design of the structure. Tile roofs weigh between 850 and 1,000 lb per square. These weights require rafters, headers, and other supporting members to be larger than normally required for other types of roofing material. Tiles are generally specified on the roof plan in a note, which lists the manufacturer, style, color, weight, fastening method, and underlayment. A typical note on the roof plan might be:

- Monier burnt terra cotta mission's roof tile over 15# felt and 1 × 3 skip sheathing. Use a 3″ minimum head lap and install as per manufacturer's specifications.

FIGURE 20.37A ■ Tile is an excellent choice for a roofing material because of its durability. *Courtesy Tim Taylor.*

FIGURE 20.37B ■ Many tile patterns are made to simulate Spanish clay tiles. *Courtesy Tim Taylor.*

Metal Panels

Metal roofing panels often provide savings because of the speed and ease of installation. Metal roof panels provide a water- and fireproof material that comes with a warranty for a protected period that can range from twenty to fifty years. Panels are typically produced in either 22- or 24-gage metal in widths of either 18″ or 24″. See Figure 20.38. The length of the panel can be specified to meet the needs of the roof in lengths up to 40′. Metal roofing panels typically weigh between 50 and 100 lb per square. Metal roofs are manufactured in many colors and patterns and can be used to blend with almost any material. Steel, stainless steel, aluminum, copper, and zinc alloys are most typically used for metal roofing. Steel panels are heavier and more durable than other metals but must be covered with a protective coating to provide protection from rust and corrosion. A baked-on acrylic coating typically provides both color and weather protection. Stainless steel does not rust or corrode but is more expensive than steel. Stainless steel weathers to a natural matte-gray finish. Aluminum is extremely lightweight and does not rust. Finish coatings are similar to those used for steel. Copper has been used for centuries as a roofing material. Copper roofs weather to a blue-green color and do not rust. In specifying metal roofing on the roof plan, the note should include the manufacturer, the pattern, the material, the underlayment, and the trim and flashing. A typical note would be:

- Amer-X-9 ga. 36″ wide, kodiak brown metal roofing by American Building Products or equal. Install over 15# felt as per manuf. specs.

FIGURE 20.38 ■ Metal is often selected for its durability and pleasing appearance. *Courtesy LeRoy Cook.*

ROOF VENTILATION AND ACCESS

As the roof plan is drawn, the drafter must determine the size of the attic space. The attic is the space formed between the ceiling joists and the rafters. The attic space must be provided with vents that are covered with 1/8″ (3.2 mm) screen mesh. These vents must have an area equal to 1/150 of the attic area. This area can be reduced to 1/300 of the attic area if a vapor barrier is provided on the heated side of the attic floor or if half of the required vents but not more than 80% of the vents are placed in the upper half of the roof area.

The method used to provide the required vents varies throughout the country. Vents may be placed in the gabled end walls near the ridge. This allows the roof surface to remain vent-free. In some areas, a continuous vent is placed in the eaves, or a vent may be placed in each third rafter space. The drafter needs to specify the proper area of vents that are required and the area in which they are to be placed.

The drafter must also specify how to get into the attic space. The actual opening into the attic is usually shown on the floor plan, but its location must be considered when the roof plan is being drawn. The minimum size of the access opening is 22″ × 30″ (560 × 760 mm) with 30″ (760 mm) minimum of headroom. While planning the roof shape, the drafter must find a suitable location for the attic access that meets both code and aesthetic requirements. The access should be placed where it can be easily reached but not where it will visually dominate a space. Avoid placing the access in areas such as the garage; areas with high moisture content, such as bathrooms and utility rooms; or in bedrooms that will be used by young children. Hallways usually provide an area to place the access that is easily accessible but not a focal point of the structure.

Chapter 20 Additional Reading

The following Web sites can be used as a resource to help you keep current with changes in roof materials.

ADDRESS	COMPANY OR ORGANIZATION
www.asphaltroofing.org	Asphalt Roofing Manufacturers
www.calredwood.org	California Redwood Association
www.cedarbureau.org	Cedar Shake & Shingle Bureau
www.certainteed.org	Certainteed Corporation Asphalt Shingles
www.elkcorp.com/index.htm	Elk Corporation Asphalt Shingles
www.gaf.com	GAF Corporation Asphalt Shingles
www.jm.com	Johns Manville (roofing)
www.lpcorp.com	Louisiana Pacific (hardboard and wood siding)
www.ludowici.com	Ludowici Roof Tile
www.mca-tile.com	MCA, Inc. (tile roofing)
www.malarkey-rfg.com	Malarkey Corporation High Wind Asphalt Shingles
www.monier.com	Monier Lifetile Concrete Roofing
www.owenscorning.com/ow	Owens-Corning Corporation Asphalt Shingles
www.riei.org	Roofing Industry Educational Institute
www.solatube.com	Solatube (skylights)
www.spri.org	Single Ply Roofing Institute
www.stone-slate.com	Slate/Select Inc.
www.sunoptic.com	Sunoptics (skylights)
www.velux.com	Velux (skylights)
www.zappone.com	Zappone Manufacturing (copper shingles)

CHAPTER 20

Roof Plan Components Test

DIRECTIONS

1. Letter your name, Chapter 20 Test, and the date at the top of the sheet.
2. Letter the question number and provide the answer. You do not need to write out the question.
3. Do all lettering with vertical uppercase architectural letters. If the answer requires line work, use proper drafting tools and technique.

Answers may be prepared on a word processor if course guidelines allow this.

QUESTIONS

Question 20–1 List and describe three different types of roof plans.

Question 20–2 In describing roof pitch, what do the numbers 4/12 represent?

Question 20–3 What angle represents a 6/12 pitch?

Question 20–4 Is a surface built at a 28° angle from vertical a wall or a roof?

Question 20–5 What are two advantages of using a flat roof?

Question 20–6 What is the major disadvantage of using a flat roof?

Question 20–7 List three traditional roof shapes.

Question 20–8 Sketch and define the difference between a hip and a Dutch hip roof.

Question 20–9 What are the two uses for a mansard roof?

Question 20–10 List two common weights for asphalt or fiberglass shingles.

Question 20–11 What are two common shapes of clay roof tiles?

Question 20–12 What advantage do metal roofing panels have over other roofing materials?

Question 20–13 What is the minimum headroom required at the attic access?

Question 20–14 What is the minimum size of an attic access opening?

Question 20–15 What type of roof is both a roof system and a framing system?

21 CHAPTER

Roof Plan Layout

INTRODUCTION

The design of a roof plan is considered early in the design process of a structure. The designer will often draw a preliminary roof plan to coordinate key design elements of the floor roof and elevations, and to project the preliminary elevations. The actual drawing of the roof plan for the working drawings can be completed once the design has been finalized. To complete the roof plan, a drafter must be familiar with the correct lines and symbols that are used to represent roofing material and the type of roof framing system to be used. This chapter will examine the lines and symbols that are used to represent the roof shape, structural materials, nonstructural materials and dimensions. Roof framing systems that will be discussed in this chapter include gable, hip and Dutch hip drawings.

ROOF SHAPE

The overhang that forms the outline of the roof is usually drawn with bold solid lines. The size of the overhang varies depending on the pitch of the roof and the amount of shade desired. As seen in Figure 21.1, the steeper the roof pitch, the smaller the overhang that is required to shade an area. If you are drawing a home for a southern location, an overhang large enough to protect glazing from direct sunlight is usually desirable. In northern areas, the overhang is usually restricted to maximize the amount of sunlight received during winter months. Figure 21.2 compares the effect of an overhang at different times of the year. Another important consideration regarding the size of the overhang is how the view from windows will be affected. As the angle of the roof is increased, the size of the overhang may need to be decreased so that the eave will not extend into the line of sight from a window. The designer will need to base the size of the overhang on the roof pitch so view is not hindered.

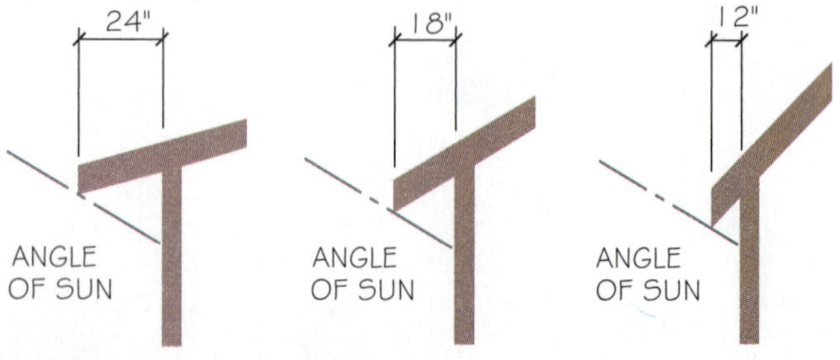

FIGURE 21.1 ■ The steeper the roof pitch, the more shadow will be cast. Typically the overhang is decreased as the pitch is increased.

Lines representing walls are drawn with thin dashed lines. Changes in the shape of the roof such as ridges, hips, Dutch hips, and valleys are also drawn with solid lines. Each can be seen in Figure 21.3. These changes in shape can be drawn easily when the pitches are equal. Because the ridge is centered between the two bearing walls, it can be located by measuring the distance between the two supporting walls and dividing that distance in half. A faster method of locating the ridge is to draw two 45° angles, as shown in Figure 21.4. The ridge extends through the intersection of the two angles. When the roof

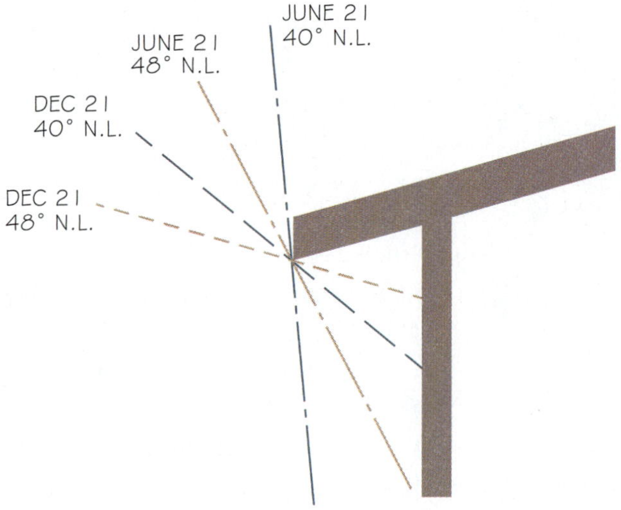

FIGURE 21.2 ■ The overhang blocks different amounts of sunlight at different times of the year. Notice the difference in the sun's angles at 40° and 48° north latitude.

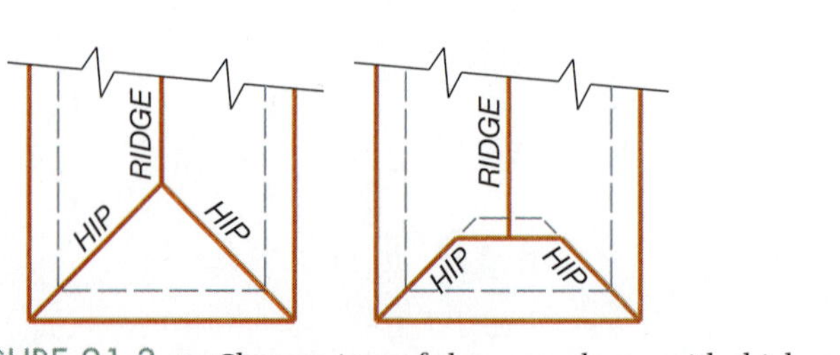

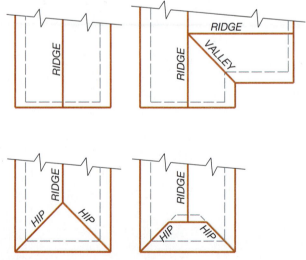

FIGURE 21.3 ■ Changes in roof shape are shown with thick, solid lines.

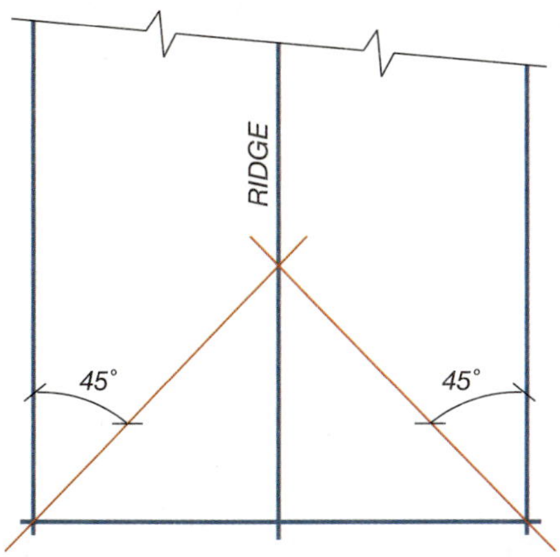

FIGURE 21.4 ■ If the slope of each plane is equal, and the supporting walls are equal in height, the ridge will be located halfway between the two supporting walls. The ridge can be located on the roof plan by selecting the midpoint, or by drawing two lines at 45° from the intersection of the walls. The ridge will extend through the intersection of the two lines.

pitches are unequal, the drafter must plot the proper pitches to determine the location of the ridge. The ridge location can be determined by drawing a partial section to reflect the angle of the roof. See Figure 21.5.

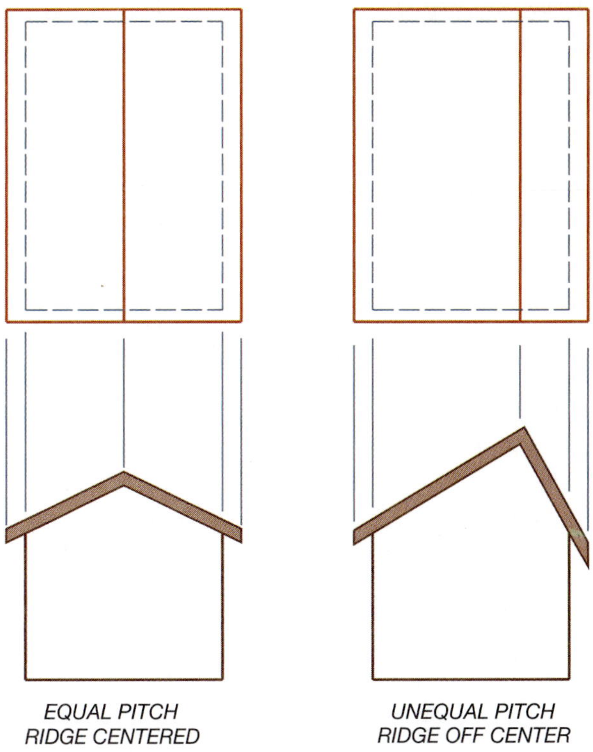

FIGURE 21.5 ■ Roof pitches affect the intersection of each surface.

Hip Roofs

If a hip roof is to be drawn, the hips, or external corners of the roof, can be drawn in a manner similar to that used to locate the ridge. The hips represent the intersection between two roof planes and are represented by a line drawn at an angle that is one-half of the angle formed between the two supporting walls. For walls that are perpendicular, an internal corner or valley is formed. The valley will be drawn at an angle of 45° between the two intersecting walls, as seen in Figure 21.6. Keep in mind that the 45° line represents the intersection between two equally pitched roofs in plan view only. This angle has nothing to do with the slope of the roof. The angle that represents the slope or pitch of the roof can be shown only in an elevation of the structure. Drawing elevations will be introduced in Chapters 22 and 23.

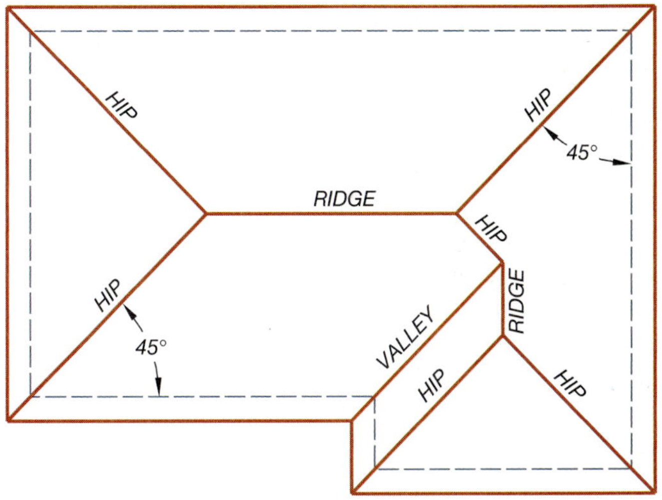

FIGURE 21.6 ■ A valley (interior corner) or a hip (exterior corner) is formed between two intersecting roof planes. When each plane is framed from walls that are perpendicular to each other and the slope of each roof is identical, the valley is drawn at a 45° angle.

Dutch Hip Roofs

A Dutch hip is drawn by first lightly drawing a hip, as shown in Figure 21.7. Once the hip is drawn, determine the location of the gable wall. The wall is usually located over a framing member, spaced at 24″ (600 mm) on center. The wall is typically located 48″ or 72″ (1200–1800 mm) from the exterior wall. With the wall located, the overhang can be drawn in a manner similar to a gable roof. The overhang line will intersect the hip lines to form the outline of the Dutch hip.

Hips and Valleys

When two perpendicular roofs intersect each other, a valley will be formed at 45° to the walls. The drafter must consider both the distance between supporting walls and the pitch to determine how the roof intersections are represented on the plan. Notice in Figure 21.8*a* that the shape of the roof changes dramatically as the width between the walls is changed. When one roof is taller than another, the change can be made by using a wall similar to the example on the left in Figure 21.8*b*, or by extending the lower plane as in the example on the right in Figure 21.8*b*. Remember that when the pitches are equal, the wider the distances are between the walls, the higher the roof will be.

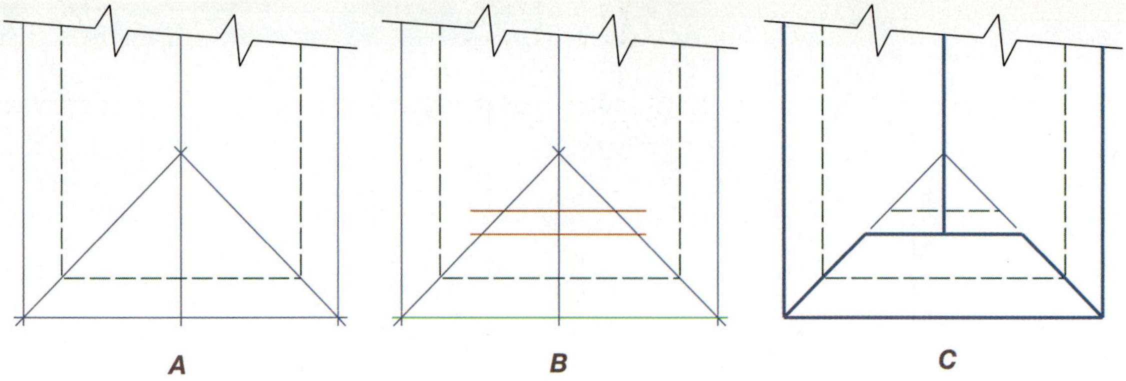

FIGURE 21.7 ■ Layout of a Dutch hip roof: (*A*) Establish the ridge and hip locations; (*B*) locate the gable end wall; (*C*) use bold lines to complete the roof.

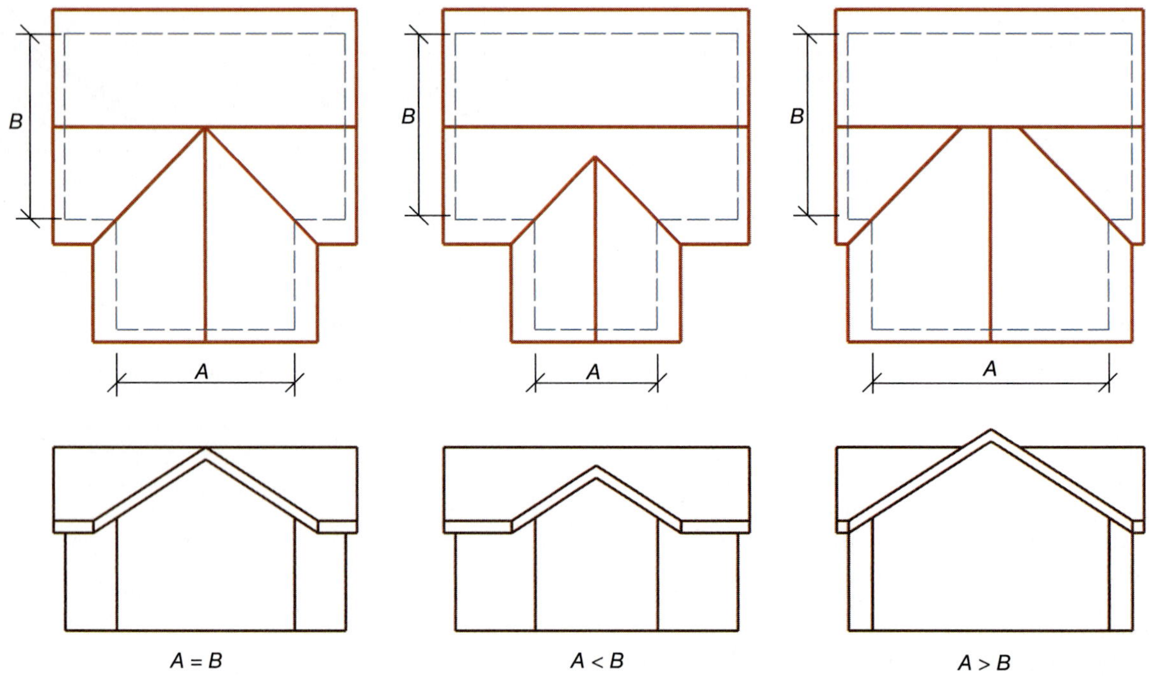

FIGURE 21.8(A) ■ Although the basic shape remains the same, the roof plan and elevations change as the distance between walls varies.

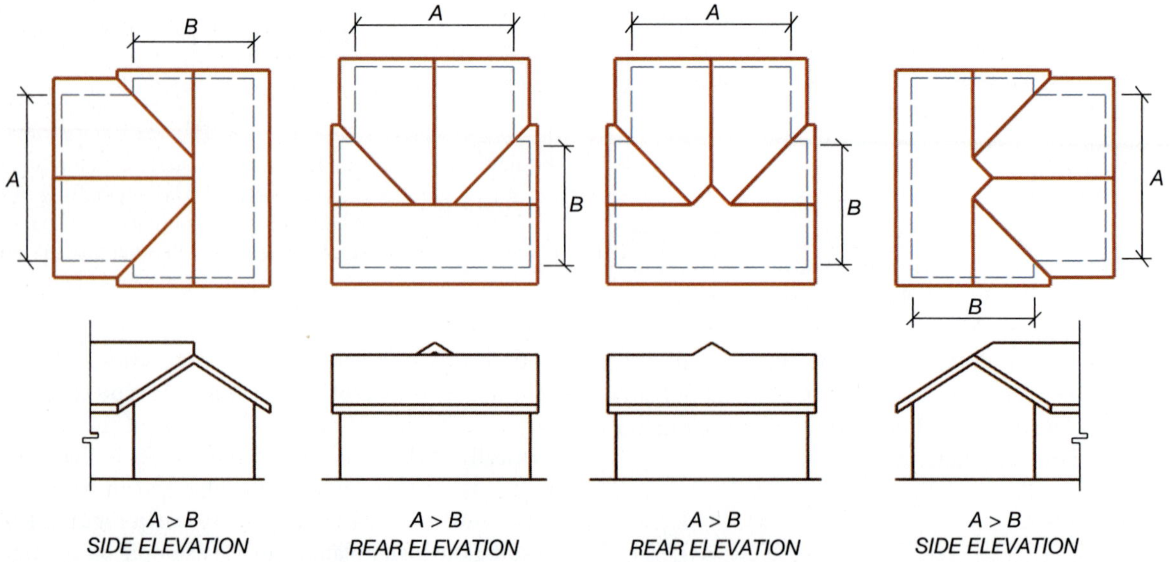

FIGURE 21.8(B) ■ When roof heights are unequal, the transition can be made by extending the lower roof pitch until it lines with the upper roof (right) or by allowing a gable roof to be formed between the two roofs.

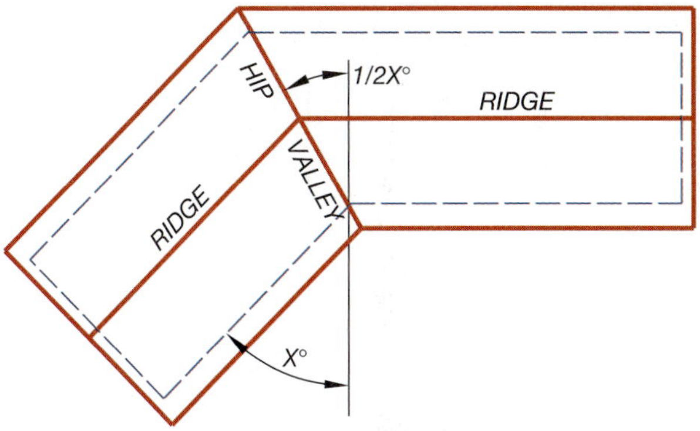

FIGURE 21.9 ■ When walls are equally spaced but not perpendicular, the roof surfaces occur along a line, as shown.

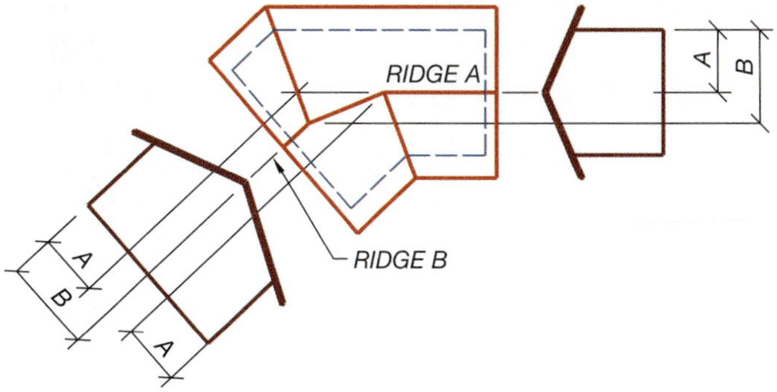

FIGURE 21.11 ■ The true roof shape can be seen by drawing sections of the roof. Project the heights represented on the sections onto the roof.

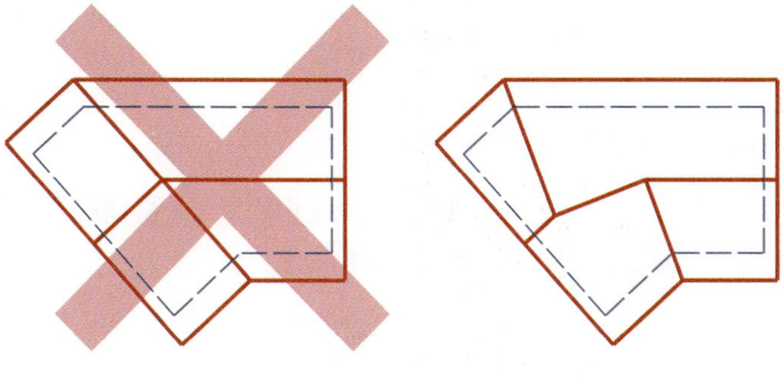

FIGURE 21.10 ■ When walls are unequally spaced, a drafter may be tempted to draw the roof as shown at *A*. The actual intersection can be seen at *B*. Remember that each valley is drawn at an angle equal to half of the angle between the supporting walls.

When the walls are equally spaced but not perpendicular, the hip, valley, and ridge intersection will occur along a line, as shown in Figure 21.9. With unequal distances between the supporting walls, the hips, valleys, and ridge can be drawn as shown in Figure 21.10. Such a drawing is often a difficult procedure for an inexperienced drafter. The process is simplified if you draw partial sections by the roof plan, as shown in Figure 21.11. By comparing the heights and distances in the sections, you can often visualize the intersections better.

Bay Projections

If a bay is to be included on the floor plan, special consideration will be required to draw the roof covering it. Figure 21.12 shows the steps required to lay out a bay roof. The layout of the roof can be eased by squaring the walls of the bay. Lay out the line of the ridge from the hips and valleys created by the intersecting rectangles. The true hips over the bay are drawn with a line from the intersection of the roof overhangs, which extends up to the end of the ridge.

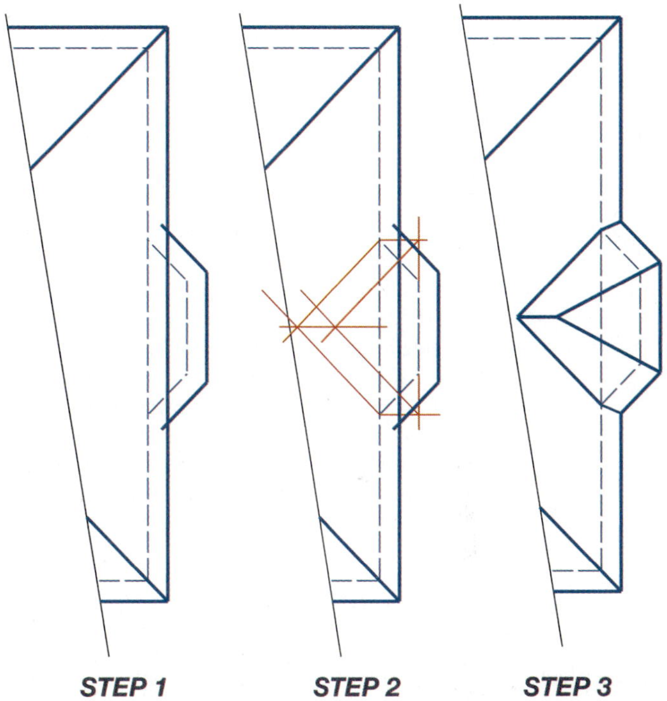

FIGURE 21.12 ■ The roof to cover a bay can be represented by drawing the outline of the bay and the desired overhangs parallel to the bay walls (step 1). Draw construction lines to represent the roof that would be created if the bay were a rectangle. This will establish the ridge and the intersection of the ridge with the main roof (step 2). Drawing the hips created by a rectangular bay will also establish the point where the true hips intersect the ridge. The true hips will start at the intersection of the overhangs, pass through the intersection of the bay walls, and end at the ridge (step 3).

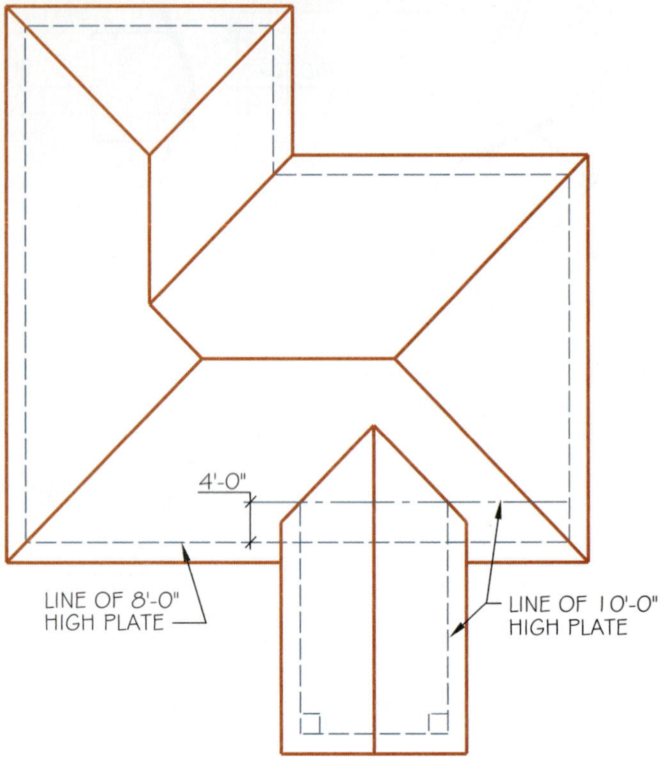

FIGURE 21.13 ■ The slope of the roof must be known to draw the intersection between roofs built on walls of differing heights. With a 6/12 pitch, a horizontal distance of 48″ (1200 mm) would be required to reach the starting height of the upper roof (6″ per foot × 4′ = 24″ rise). By projecting the line of the upper walls to a line 48″ in from the lower wall, the intersection of the two roof planes can be determined. A valley will be formed between the two intersecting roofs starting at the intersection of the wall and the line representing the rise.

Roof Intersections with Varied Wall Heights

Another feature common to roof plans is the intersection between two roof sections of different heights. Figure 21.13 shows how an entry portico framed with a 10′ (3000 mm) wall would intersect with a residence framed with walls 8′ (2400 mm) high. The roof pitch must be known to determine where the intersection between the two roofs will occur. In this example a 6/12 pitch was used. At this pitch, the lower roof must extend 4′ (1200 mm) before it will be 10′ (3000 mm) high. A line was drawn 48″ (1200 mm) from the wall representing the horizontal distance. The valley between the two roofs will occur where the lines representing the higher walls intersect the line representing the horizontal distance.

STRUCTURAL MATERIAL

The type of plan to be drawn will affect the method used to show the structural material. Many offices draw a separate roof framing plan. (See Figure 20.4). Chapter 27 explains in detail

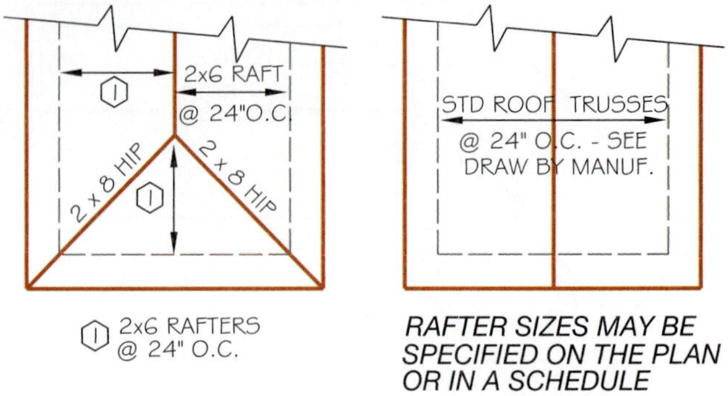

FIGURE 21.14 ■ Rafters or trusses can be represented on the roof plan by a thin line showing the proper direction and span.

how to represent structural members on a roof framing plan. Chapter 30 will discuss drawing framing plans. If framing is to be drawn on the roof plan, the rafters or trusses can be represented as shown in Figure 21.14.

NONSTRUCTURAL MATERIAL

Vents, chimneys, skylights, solar panels, diverters, cant strips, slope indicators, and downspouts are the most common nonstructural materials that will need to be shown on the roof plan. Vents are typically placed as close to the ridge as possible, usually on the side of the roof that is the least visible. The size of the vent varies with each manufacturer, but an area equal to 1/300 of the attic area must be provided for ventilation. Vents can typically be represented by a 12″ (300 mm) diameter circle or a 12″ (300 mm) square placed at approximately 10′–0″ (3000 mm) o.c. Figure 21.15 shows how ridge vents are represented.

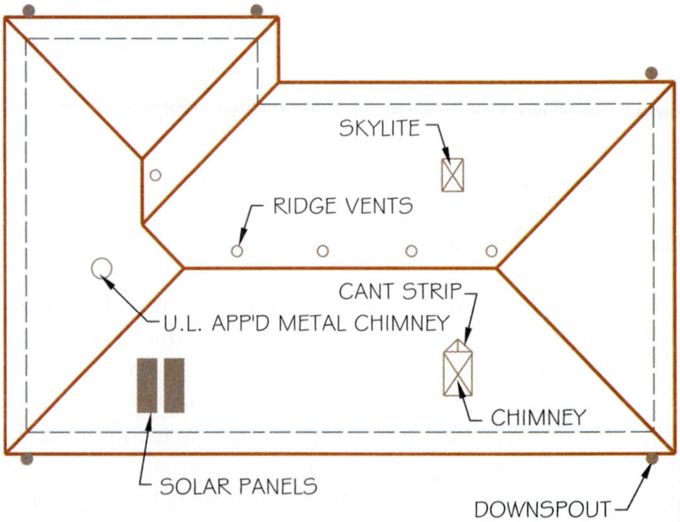

FIGURE 21.15 ■ Nonstructural materials shown on the roof plan.

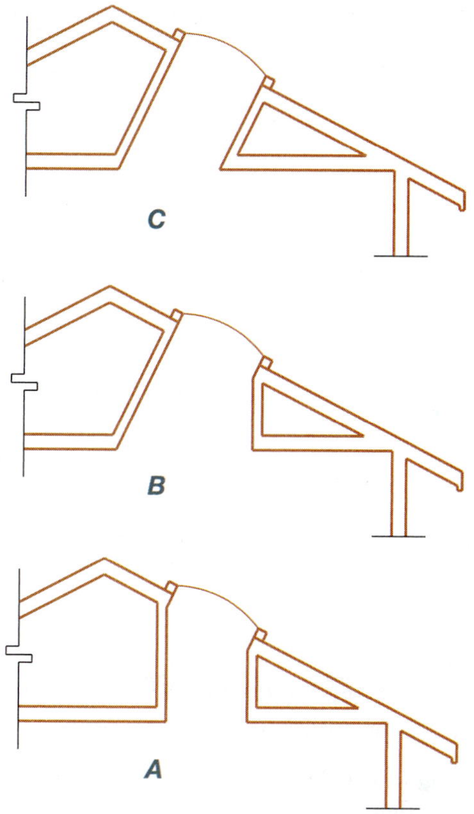

FIGURE 21.16 ■ By altering the angle of the well (the enclosed space connecting the skylight to the ceiling), the location in the ceiling can be altered. Walls can be framed perpendicular to the ceiling (*A*), with one side perpendicular to both the roof and the ceiling (*B*), and perpendicular to the roof framing (*C*).

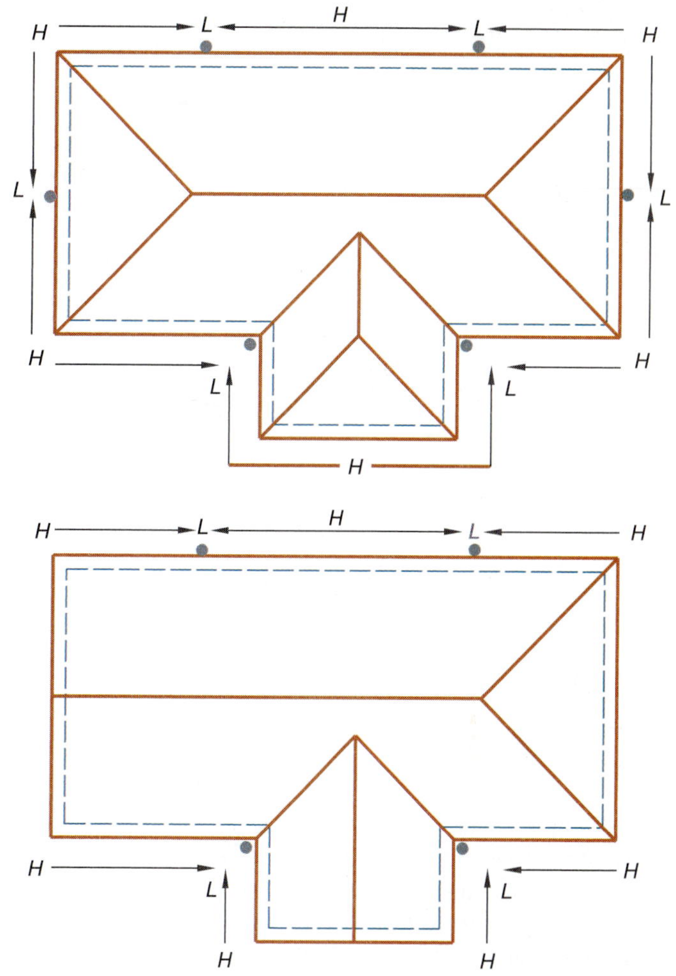

FIGURE 21.17 ■ The location of gutters and downspouts will be determined by the type of roof and amount of rainfall to be drained. Gutters must be sloped to the downspouts, which should be placed so that no view is blocked.

The method used to represent the chimney depends on the chimney material. Two common methods of representing a chimney can be seen in Figure 21.15. The metal chimney can be represented by a 14″ (300 mm) diameter circle. The masonry chimney can range from a minimum of 16″ (400 mm) square, up to matching the size used to represent the fireplace on the floor plan. The location of skylights can usually be determined from their location on the floor plan. As seen in Figure 21.16, the opening in the roof for the skylight does not have to align directly with the opening in the ceiling. The skylight is connected to the ceiling by an enclosed area, called a *chase* or *well.* By adjusting the angle of the chase, the size and location of the skylight can be adjusted. If solar panels are to be represented, the size, angle, and manufacturer should be specified.

The amount of rainfall determines the need for gutters and downspouts. In semiarid regions, a metal strip called a *diverter* can be placed on the roof to route runoff on the roof from doorways. The runoff should be diverted to an area of the roof where it can drop to the ground and be adequately diverted from the foundation. In wetter regions, gutters should be provided to collect and divert the water collected by the roof. Local codes will specify if the drains must be connected to storm sewers, private drywell, or a splash block. The downspouts that bring the roof runoff to ground level can be represented by approximately a 3″ (70 mm) circle or square. Care should be taken to keep downspouts out of major lines of sight. Each downspout can typically drain approximately 20′ (6100 mm) of roof on each side of the downspout, allowing the downspouts to be spaced at approximately 40′ (12,200 mm) intervals along the eave. This spacing can be seen in Figure 21.17. The distance between downspouts will vary depending on the amount and rate of rainfall and should be verified with local manufacturers. Common methods of showing downspouts can be seen in Figure 21.15. The saddle is a small gable built behind the chimney to divert water away from the chimney, as seen in Figure 21.18.

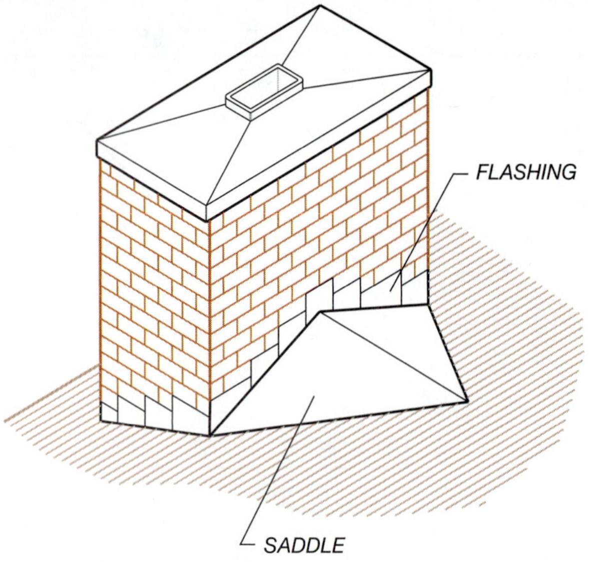

FIGURE 21.18 ■ A saddle is used to divert water around the chimney.

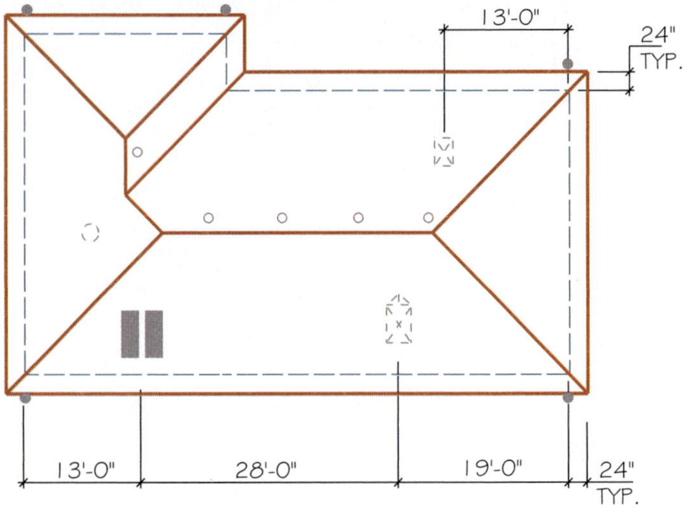

FIGURE 21.19 ■ Dimensions should be placed by using leader and extension lines or in a note.

DIMENSIONS

The roof plan requires very few dimensions. Typically only the overhangs and openings are dimensioned. These may even be specified in note form rather than with dimensions. Figure 21.19 shows how dimensions are placed on roof and framing plans. When drawing framing plans for commercial projects, beam locations are often dimensioned to help in estimating the materials required.

NOTES

As with the other drawings, notes on the roof or framing plans can be divided into general and local notes. General notes might include the following:

- Vent notes
- Sheathing information
- Roof covering
- Eave sheathing
- Pitch

Material that should be specified in local notes includes the following:

- Skylight type, size, and material
- Chimney caps
- Solar panel type and size
- Cant strips and saddles

In addition to these notes, the drafter should also place a title and scale on the drawing.

DRAWING GABLE ROOF PLANS

The following instructions are for the roof plan to accompany the residence that was used in Chapter 16. The plan will be drawn using a gable roof. Use construction lines for steps 1 through 6. Each step can be seen in Figure 21.20. Use the dimensions on the floor plan to determine all sizes.

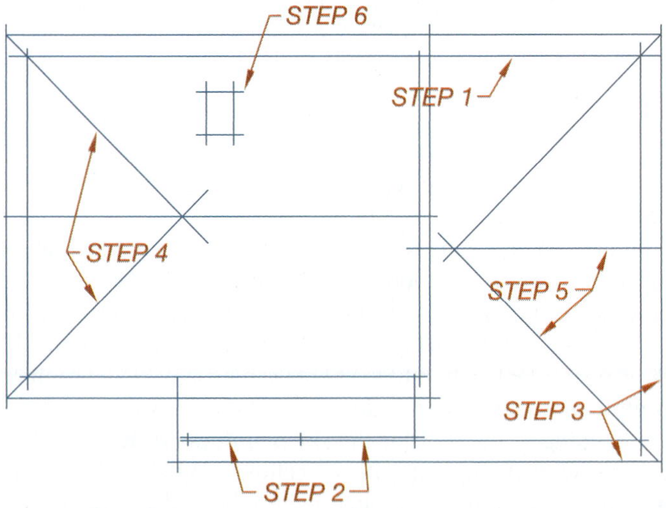

FIGURE 21.20 ■ The layout of the roof plan should be done using construction lines.

STEP 1 Draw the perimeter walls.

STEP 2 Locate any supports required for covered porches.

STEP 3 Draw the limits of the overhangs. Unless your instructor provides different instructions, use 1′–0″ (305 mm) overhangs for the gable end walls and 2′–0″ (610 mm) overhangs for the eaves.

STEP 4 Locate and draw the ridge or ridges.

STEP 5 Locate and draw any hips and valleys required by design.

STEP 6 Locate the chimney.

STEP 7 Using the line quality described in this chapter, draw the materials of steps 1 through 6. Draw the outline of the upper roof, and then work down to lower levels. Your drawing should now resemble Figure 21.21.

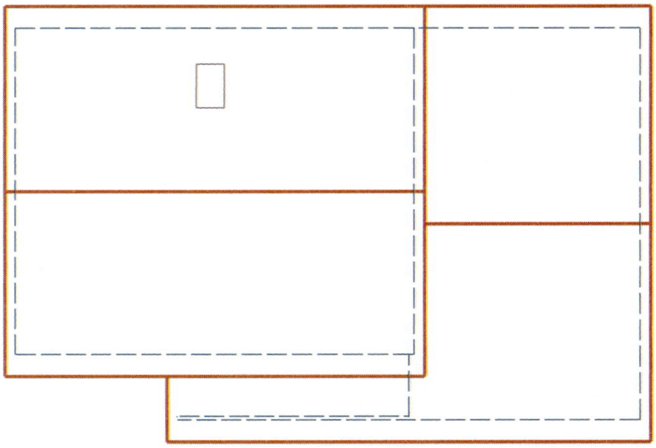

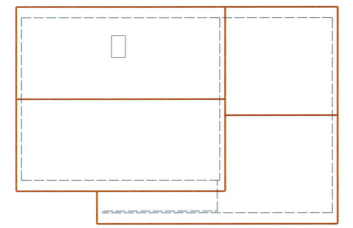

FIGURE 21.21 ■ Drawing the roof plan using finished quality lines.

Use the line quality described in this chapter to draw steps 8 through 13. Each step can be seen in Figure 21.22.

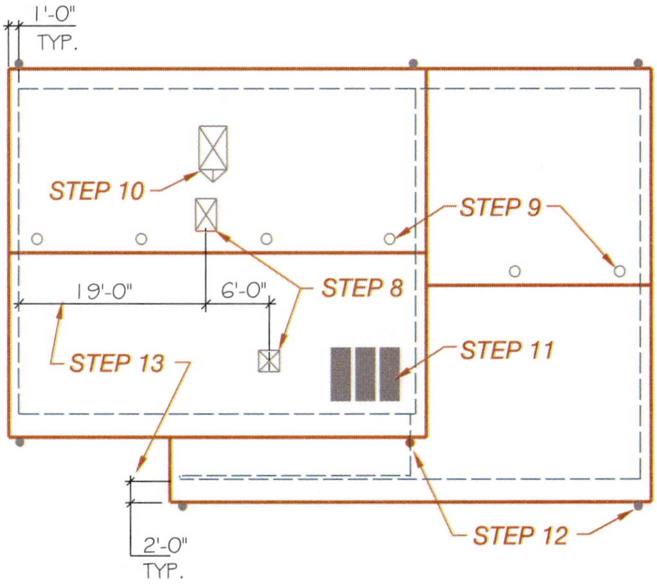

FIGURE 21.22 ■ Drawing the nonstructural material.

STEP 8 Draw any skylights that are specified on the floor plans.

STEP 9 Calculate the area of the attic and determine the required number of vents. Assume that 12″ (305 mm) round vents will be used. Draw the required vents on a surface of the roof that will make the vents least visible.

STEP 10 Draw the saddle by the chimney if a masonry chimney or wood chase was used.

STEP 11 Draw solar panels if required.

STEP 12 Draw the downspouts.

STEP 13 Add dimensions for the overhangs and skylights.

STEP 14 Label all materials using local and general notes, a title, and a scale. Your drawing should now resemble Figure 21.23.

STEP 15 Evaluate your drawing for completeness, and make any minor revisions required before giving your drawing to your instructor.

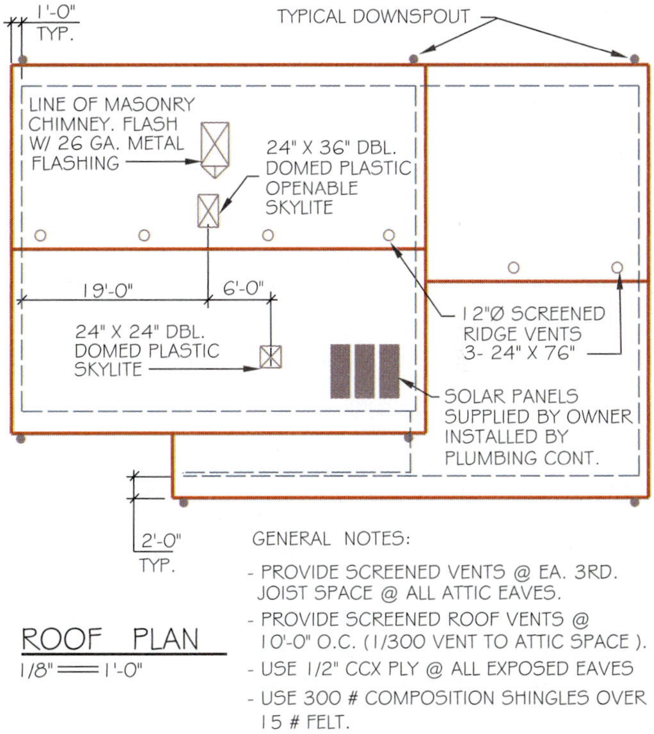

FIGURE 21.23 ■ The size of the overhangs and the location of all openings in the roof should be dimensioned. The roof plan is completed by adding the required local and general notes to specify all roofing materials.

DRAWING HIP ROOF PLANS

The step-by-step process can be used for other styles of roofs. A hip roof can be drawn by completing the following steps. Although drawing roof plans may prove frustrating, remember that lines will always be vertical, horizontal, or at a 45° angle. Another helpful hint for drawing a hip roof is to remember that three lines are always required to represent an intersection of hips, valleys, and ridges. Use construction lines for steps 1–6. Each step can be seen in Figure 21.24. Use the dimensions on the floor plan to determine all sizes.

STEP 1 Draw the perimeter walls.

STEP 2 Locate all supports required for any covered porches.

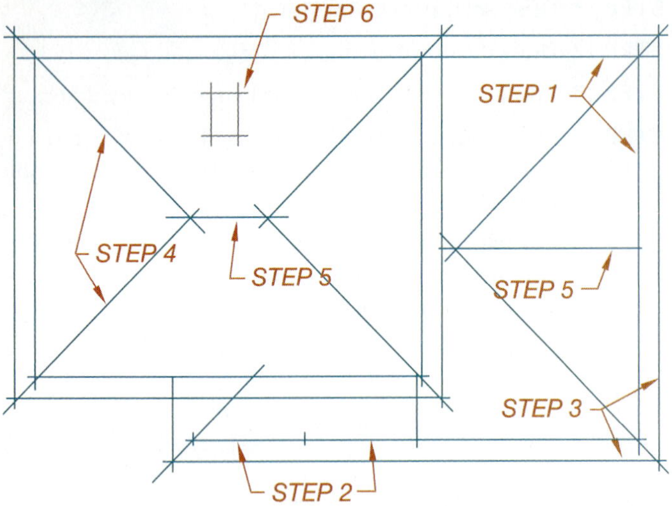

FIGURE 21.24 ■ The layout of a hip roof plan should be done with construction lines.

STEP 3 Draw the limits of the overhangs. Unless your instructor provides different instructions, assume 24″ (600 mm) overhangs.

STEP 4 Locate and draw all hips and valleys.

STEP 5 Locate and draw all ridges.

STEP 6 Locate the chimney if required.

STEP 7 Using the line quality described in this chapter, draw the materials of steps 1 through 6. Draw the upper roof, and then work down to lower roof levels. When complete, your drawing should resemble Figure 21.25.

Use the line quality described in this chapter to draw Steps 8 through 13. When complete, your drawing should resemble Figure 21.26.

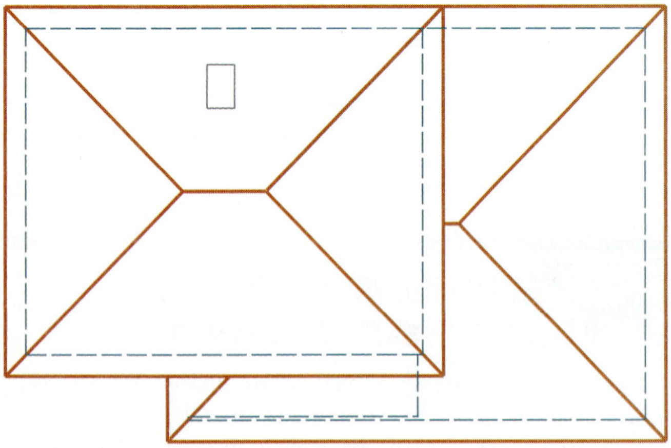

FIGURE 21.25 ■ A hip roof plan drawn with finished line quality.

STEP 8 Draw any skylights that are specified on the floor plan.

STEP 9 Calculate the area of the attic, and determine the required number of vents. Assume that 12″ (300 mm) round vents at 10′–0″ o.c. will be used.

STEP 10 Draw a saddle for any chimneys.

STEP 11 Draw solar panels if required.

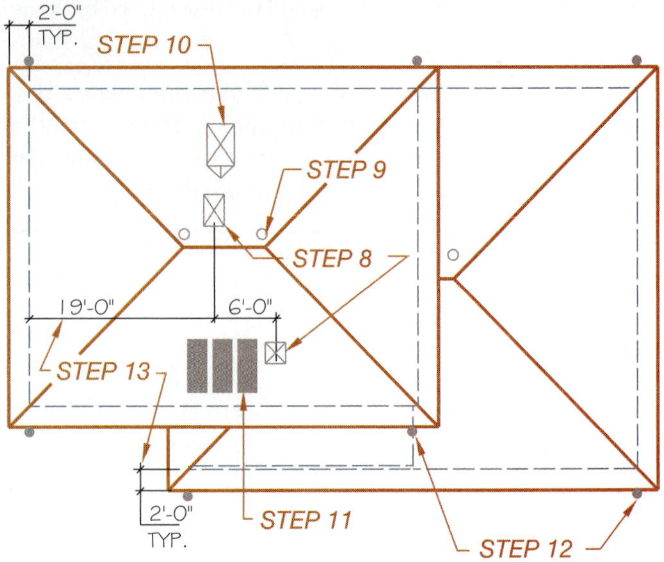

FIGURE 21.26 ■ All openings in the roof, such as the chimney, skylights, and materials to be mounted on the roof, should be identified. The overhangs and the location of all openings in the roof should be dimensioned.

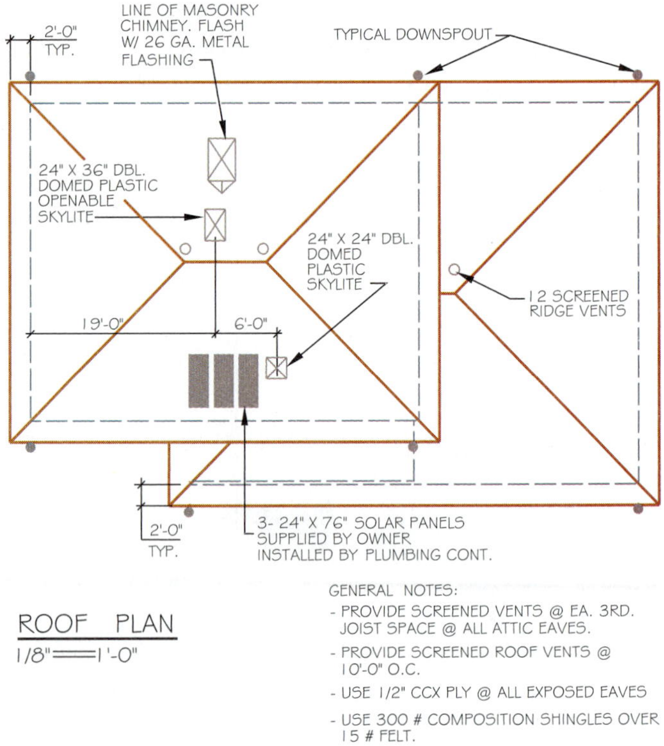

FIGURE 21.27 ■ The roof plan is completed by adding the required local and general notes to specify all roofing materials.

STEP 12 Draw downspouts.

STEP 13 Provide dimensions to locate any roof openings and all overhangs.

STEP 14 Label all materials using local and general notes, a title, and a scale. Your drawing should now resemble Figure 21.27.

STEP 15 Evaluate your drawing for completeness, and make any minor revisions required before giving your drawing to your instructor.

DRAWING DUTCH HIP ROOF PLANS

The step-by-step process can be used for other styles of roofs. A Dutch hip roof can be drawn by completing the following steps. Use construction lines for steps 1 through 6. Each step can be seen in Figure 21.28. Use the dimensions on the floor plan to determine all sizes.

STEP 1 Draw the perimeter walls.

STEP 2 Locate supports required for any covered porches.

STEP 3 Draw the limits of the overhangs. Unless your instructor provides different instructions, assume 24″ (600 mm) overhangs for eaves and 12″ (300 mm) at gable end walls.

STEP 4 Locate and draw all hips, valleys, and Dutch hip locations. Hips and valleys are located following methods used to draw a hip roof. The location of a Dutch hip is determined by the framing method. Typically the gable end wall, which is formed between the hip and the gable roof, is located approximately 6′–0″ (1800 mm) from the end wall. Chapter 30 will explain framing considerations that can affect the location of the Dutch hip gable end walls.

STEP 5 Locate and draw all ridges.

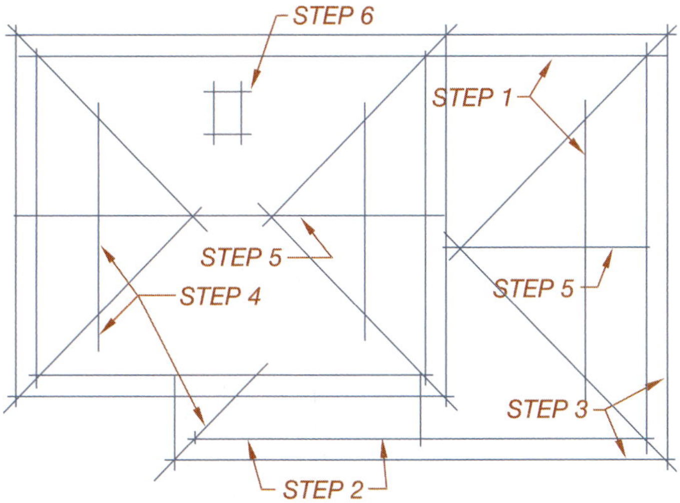

FIGURE 21.28 ■ The layout of a Dutch hip roof plan should be done with construction lines.

STEP 6 Locate the chimney if required.

STEP 7 Using line quality described in this chapter, draw the materials of steps 1 through 6. Draw the upper roof, and then work down to lower roof levels. When complete, your drawing should resemble Figure 21.29.

Use the line quality described in this chapter to draw Steps 8 through 13. When complete, your drawing should resemble Figure 21.30.

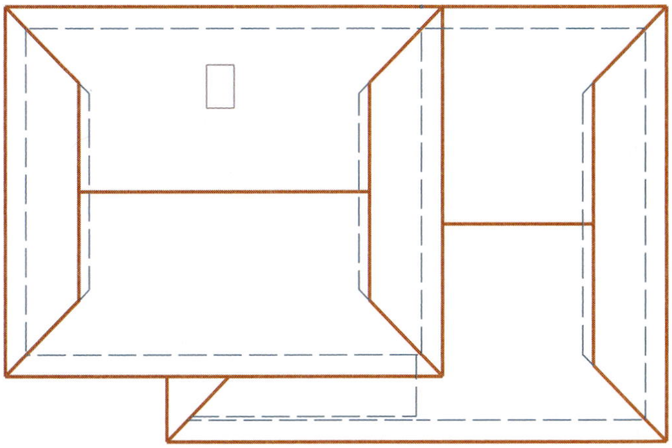

FIGURE 21.29 ■ A Dutch hip roof plan drawn with finished line quality.

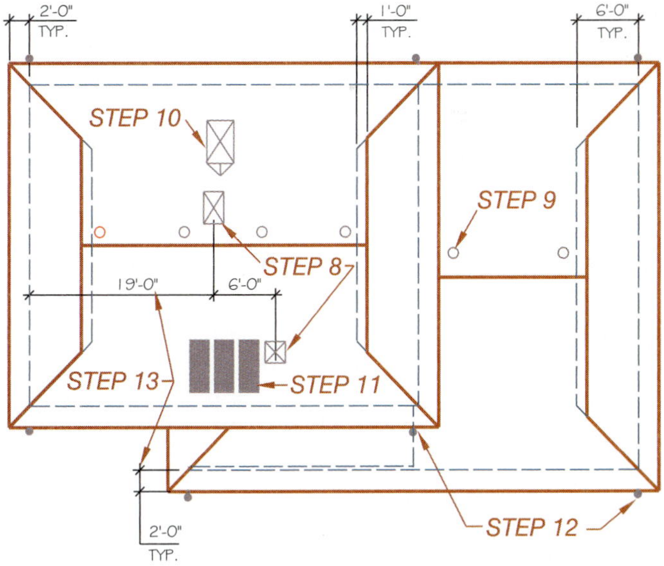

FIGURE 21.30 ■ All openings in the roof, such as the chimney, skylights, and materials to be mounted on the roof, should be identified. The overhangs and the locations of all openings in the roof should be dimensioned.

STEP 8 Draw any skylights that are specified on the floor plan.

STEP 9 Calculate the area of the attic and determine the required number of vents. Assume that 12″ (300 mm) round vents at 10′–0″ o.c. will be used.

STEP 10 Draw a saddle for any chimneys.

STEP 11 Draw solar panels if required.

STEP 12 Draw downspouts.

STEP 13 Provide dimensions to locate any roof openings and all overhangs.

STEP 14 Label all materials using local and general notes, a title, and a scale. Your drawing should now resemble Figure 21.31.

STEP 15 Evaluate your drawing for completeness and make any minor revisions required before giving it to your supervisor.

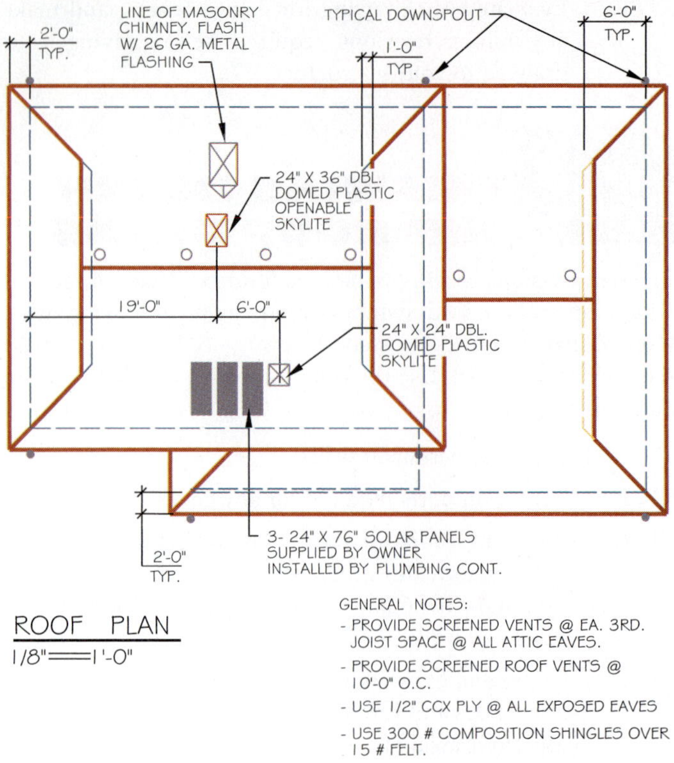

FIGURE 21.31 ■ The roof plan is completed by adding the required local and general notes to specify all roofing materials.

CADD APPLICATIONS

DRAWING ROOF PLANS WITH CADD

A roof plan can be drawn using a CADD program following similar procedures that are used for a manual drawing. The drawing can be created in the file containing the floor plan.

Start by freezing all floor-plan material except the wall layout. Create layers to contain the roof plan using the ROOF prefix. Layers such as ROOFWALL, ROOFLINE, ROOFTEXT, ROOFDIMEN, or A.I.A. layer names can be used to keep the roof information separate from the floor-plan information and to ease plotting. Draw the outline of the residence on the ROOFWALL layer, and then freeze the walls of the floor plan. The plan can now be completed using the appropriate step-by-step process for the required roof type. Figure 21.32 shows an example of a roof plan drawn by CADD. When the roof plan is completed, a copy can be saved as a WBLOCK with a title ROOF. This drawing eventually can be inserted into and plotted with the drawing file that will contain the elevations. To ease development of the elevations of a structure, a copy of the roof plan is usually left as an overlay of the FLOOR drawing file.

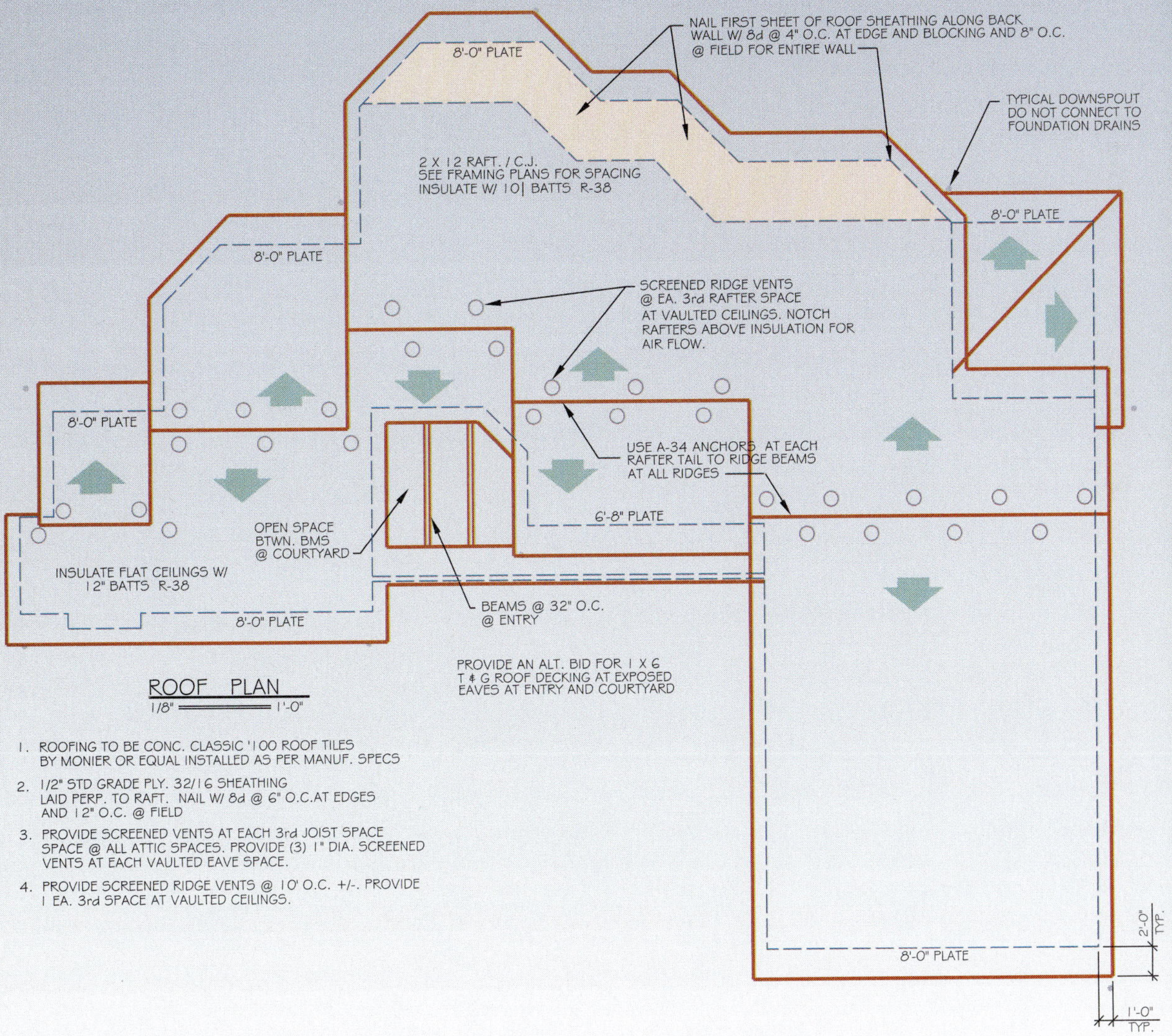

FIGURE 21.32 ■ CADD-drawn roof plan.

CHAPTER 21

Roof Plan Layout Problems

DIRECTIONS

1. Using a print of your floor plan drawn for problems from Chapter 16, draw a roof plan for your house. Design a roof system appropriate for your area. Sketch the layout you will use and have it approved by your instructor prior to starting your drawing.
2. Place the plan on the same sheet as the elevations if possible. If a new sheet is required, place the drawing so that other drawings can be put on the same sheet.
3. When you have completed your drawing, turn in a copy to your instructor for evaluation.

SECTION 7

Elevations

22 CHAPTER

Introduction to Elevations

INTRODUCTION

Elevations are an essential part of the design and drafting process. The elevations are a group of drawings that show the exterior of a building. An elevation is an orthographic drawing that shows one side of a building. In true orthographic projection, the elevations would be displayed as shown in Figure 22.1*a*. The true projection is typically modified as shown in Figure 22.1*b* to ease viewing. No matter how they are displayed, it is important to realize that between each elevation projection and the plan view is an imaginary 90° fold line. An imaginary 90° fold line also exists between elevations in Figure 22.1*b*. Elevations are drawn to show exterior shapes and finishes, as well as the vertical relationships of the building levels. By using the elevations, sections, and floor plans, the exterior shape of a building can be determined.

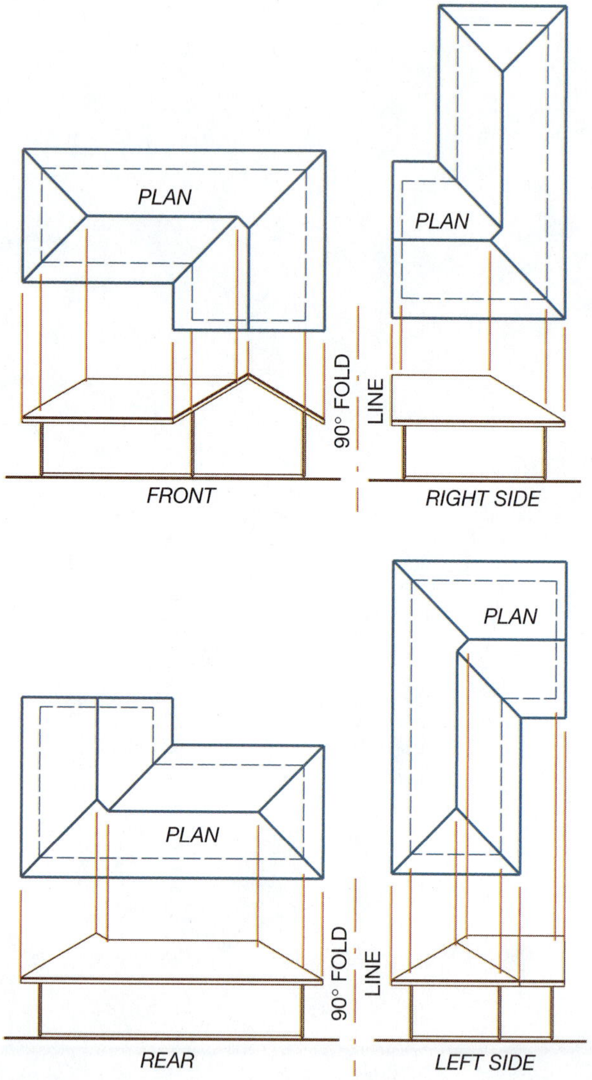

FIGURE 22.1(B) ■ The placement of elevations is usually altered to ease viewing. Group elevations so that a 90° rotation exists between views.

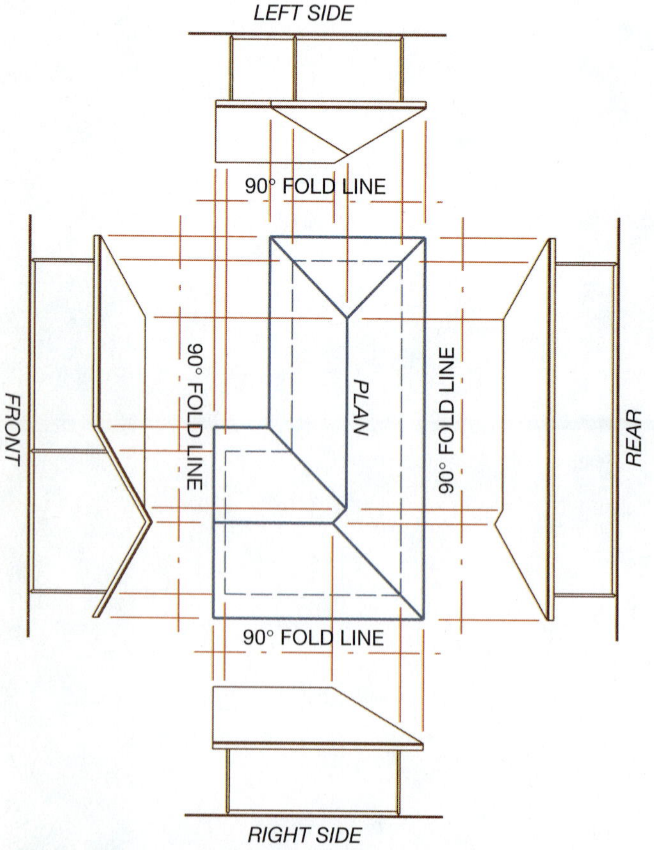

FIGURE 22.1(A) ■ Elevations are orthographic projections showing each side of a structure.

REQUIRED ELEVATIONS

Typically, four elevations will be required to show the features of a building. On a simple building, only three elevations will be needed, as seen in Figure 22.2. When drawing a building with an irregular shape, parts of the house may be hidden. An elevation of each different surface should be drawn as shown in Figure 22.3. If a building has walls that are not at 90° to

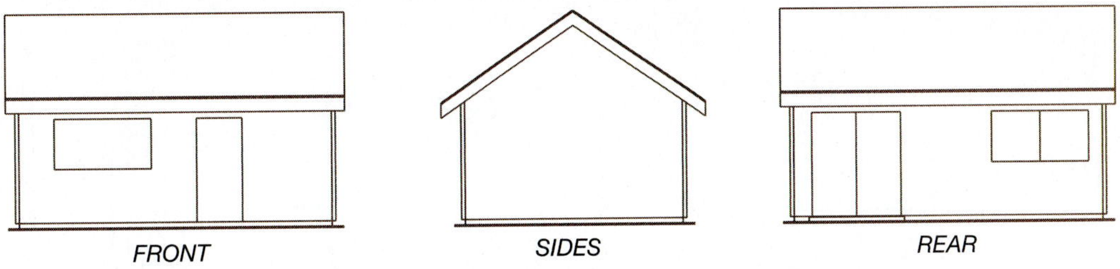

FIGURE 22.2 ■ Elevations are used to show the exterior shape and material of a building. For a simple structure, only three views are required.

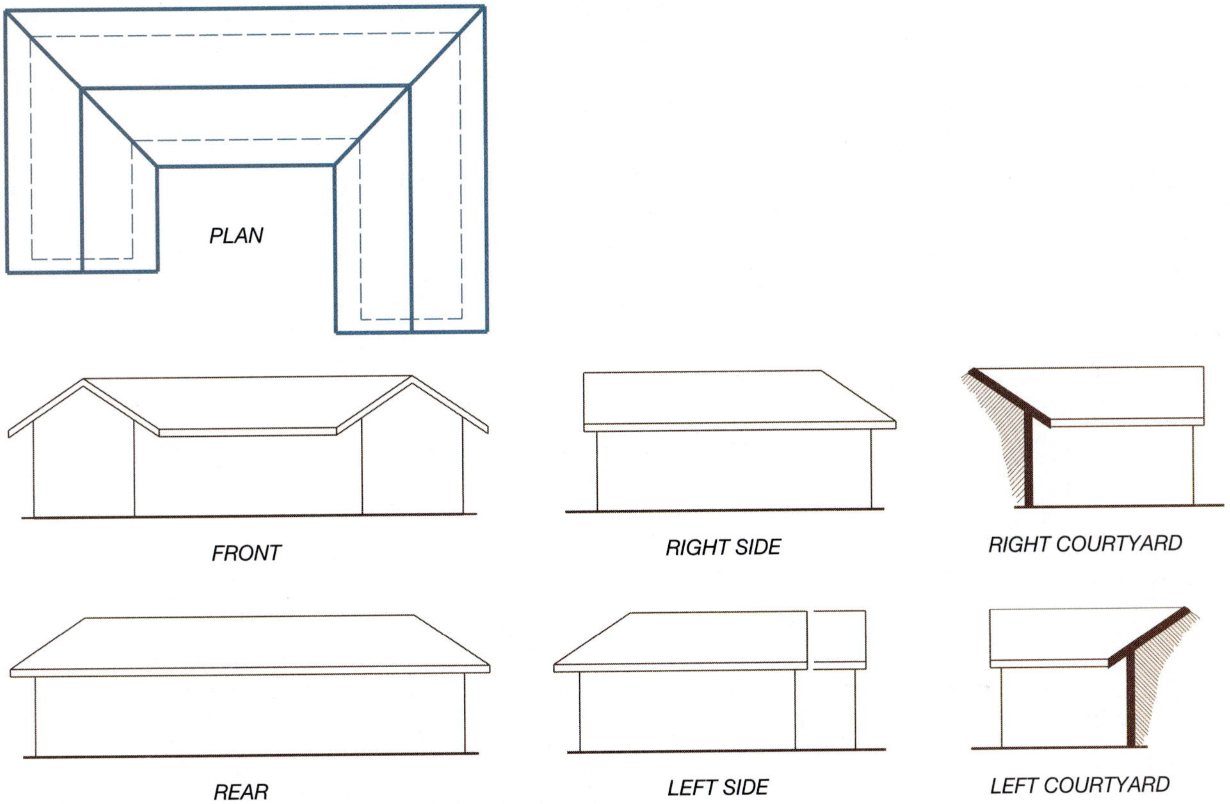

FIGURE 22.3 ■ Plans of irregular shapes often require an elevation of each surface.

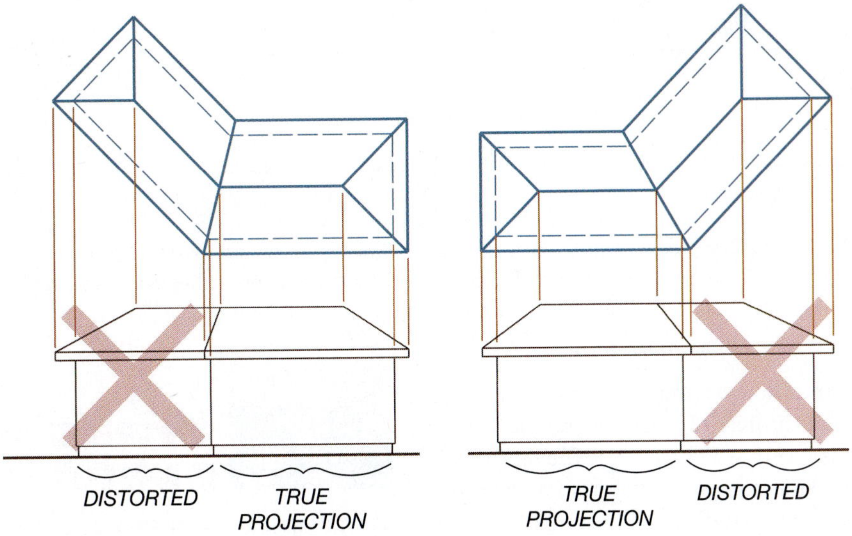

FIGURE 22.4 ■ Using true orthographic projection methods with a plan of irregular shapes will result in a distortion of part of the view.

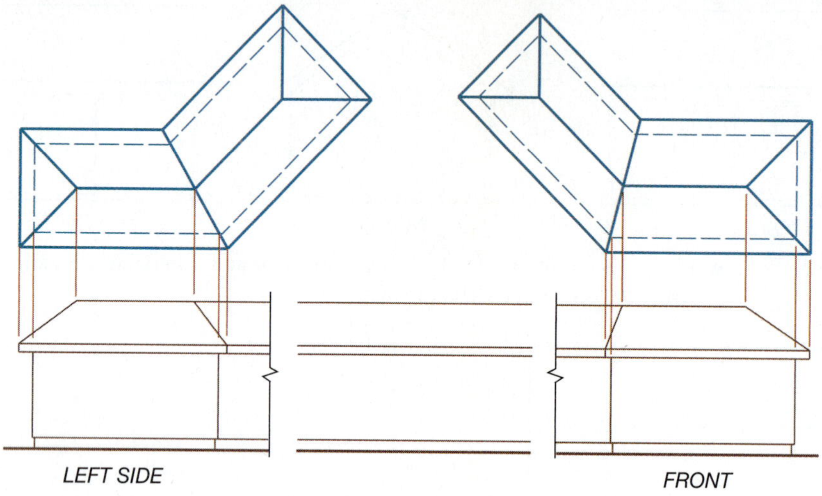

FIGURE 22.5(A) ■ If walls are not at right angles to each other, expanded elevations should be drawn.

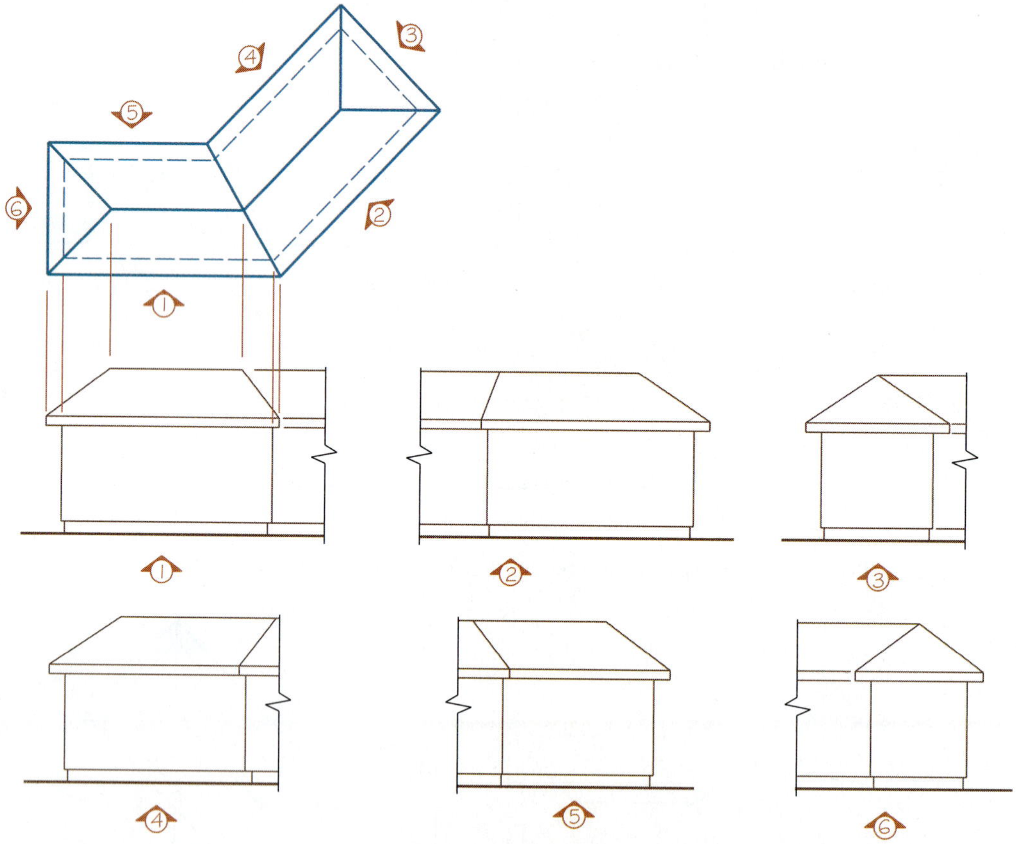

FIGURE 22.5(B) ■ A common practice when drawing an irregular shaped structure is to draw an elevation of each surface. Each elevation is then given a reference number to tie the elevation to the floor plan.

each other, a true orthographic drawing could be very confusing. In the orthographic projection, part of the elevation will be distorted, as can be seen in Figure 22.4. Elevations of this type of building are usually expanded so that a separate elevation of each face is drawn, similar to Figure 22.5*a*. Figure 22.5*b* shows the complete layout for a structure with an irregular shape.

TYPES OF ELEVATIONS

Elevations can be drawn as either presentation drawings or working drawings. Presentation drawings were introduced in Chapter 1 and will be covered in depth in Chapter 39. These

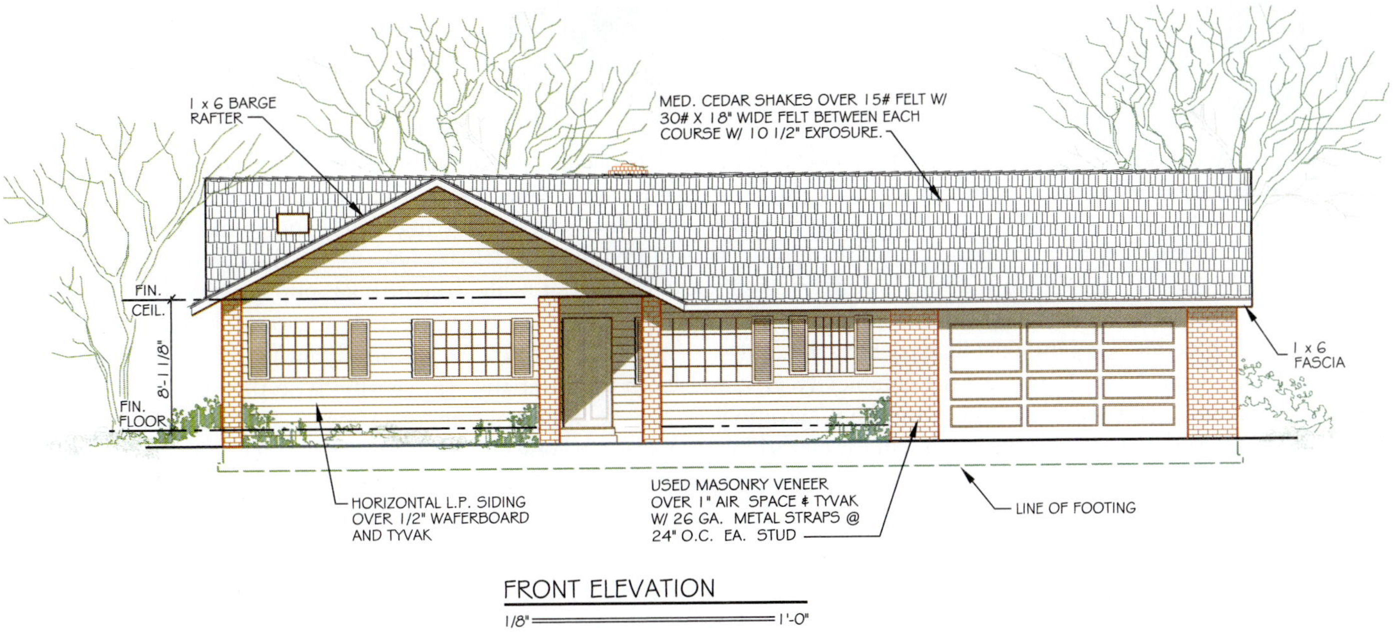

FIGURE 22.6 ■ A presentation elevation is a highly detailed drawing used to show the exterior shapes and material to be used. Shades, shadows, and landscaping are usually added to enhance the drawing.

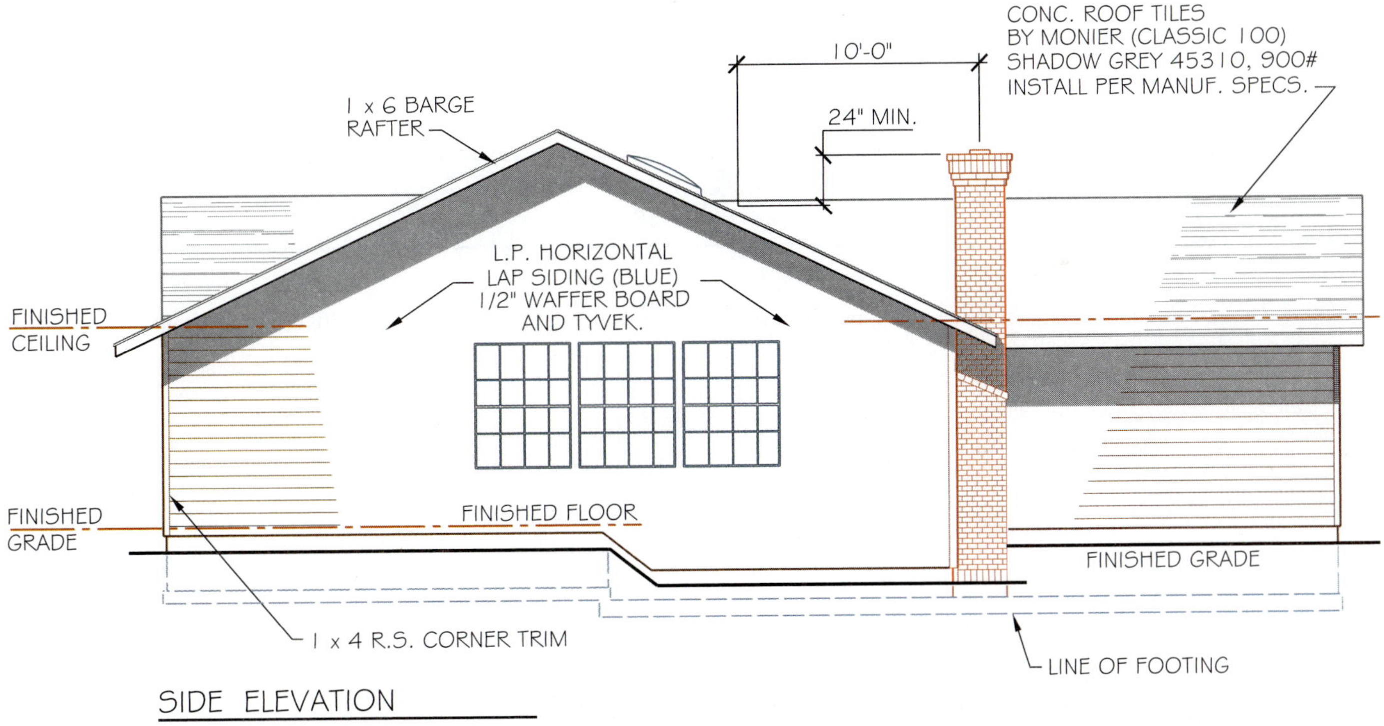

FIGURE 22.7 ■ Working elevations contain less finish detail but still show the shape of a structure accurately. *Courtesy April Muilenburg.*

drawings are part of the initial design process and may range from sketches to very detailed drawings intended to help the owner and lending institution understand the basic design concepts. See Figure 22.6.

Working elevations are part of the working drawings and provide information for the building team. They include information on roofing, siding, openings, chimneys, land shape, and sometimes even the depth of footings, as shown in Figure 22.7. The floor plans and elevations provide the information for the contractor to determine surface areas. Once surface areas are known, exact quantities of material can be determined. The elevations are also necessary in calculating heat loss, as described in Chapter 19. The elevations are used to determine the surface area of walls and wall openings for the required heat loss formulas.

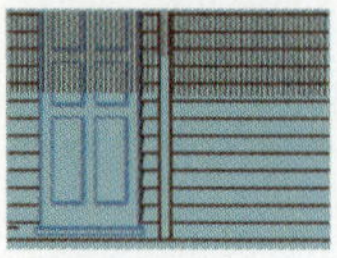

ELEVATION SCALES

Elevations are typically drawn at the same scale as the floor plan. For most plans, this means a scale of 1/4″ = 1′–0″ will be used. This allows for the elevations to be projected directly from the floor plans. Some floor plans for multifamily and commercial projects may be laid out at a scale of 1/16″ = 1′–0″ or even as small as 1/32″ = 1′–0″. When a scale of 1/8″ = 1′–0″ or less is used, generally very little detail appears in the drawings, as in Figure 22.8. Depending on the complexity of the project or the amount of space on a page, the front elevation may be drawn at 1/4″ = 1′–0″ and the balance of the elevations at a smaller scale.

ELEVATION PLACEMENT

It is usually the drafter's responsibility to plan the layout for drawing the elevations. The layout will depend on the scale to be used, size of drawing sheet, and number of drawings required. Because of size limitations, the elevations are not usually laid out in the true orthographic projection of front, side, rear, side. A common method of layout for four elevations can be seen in Figure 22.9. This layout places a side elevation by both the front and rear elevations so that true vertical heights may be projected from one view to another. If the drawing paper is not long enough for this placement, the layout shown in Figure 22.10 may be used. This arrangement is often used when

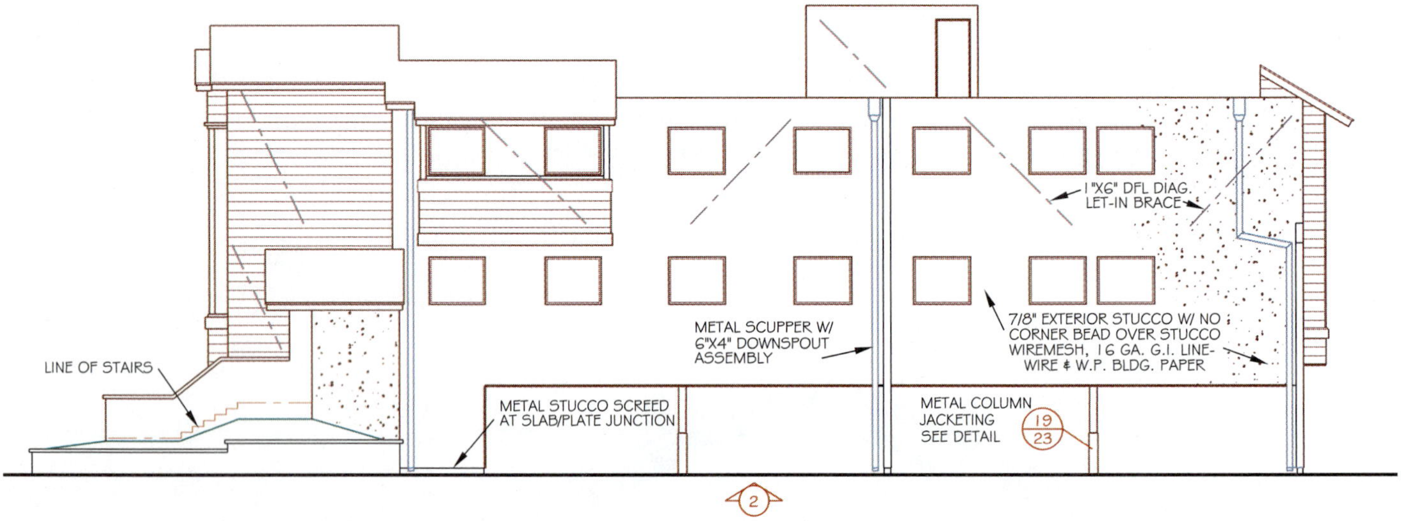

FIGURE 22.8 ■ Elevations of large structures are typically drawn at a small scale such as 1/16″ = 1′–0″, with very little detail shown. *Courtesy Kenneth D. Smith, and Associates, Inc.*

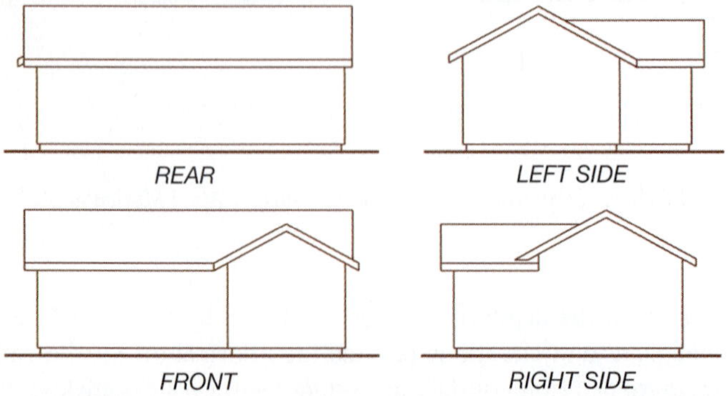

FIGURE 22.9 ■ A common method of elevation layout is to place a side elevation beside both the front and the rear elevations. This allows for heights to be directly transferred from one view to the other.

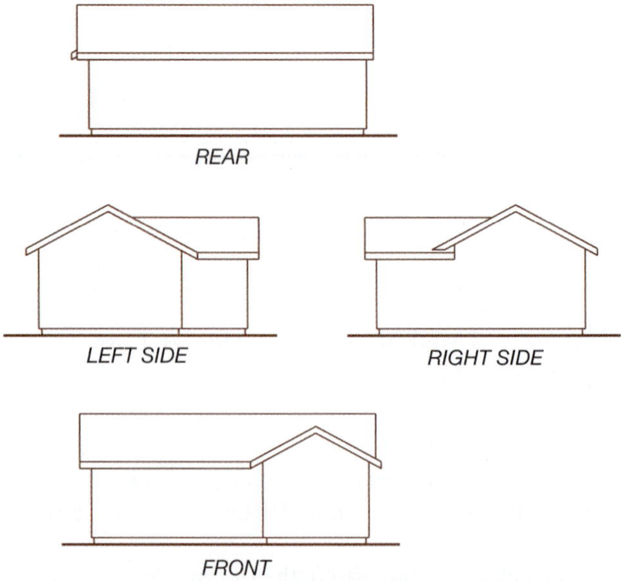

FIGURE 22.10 ■ An alternative elevation arrangement is to place the two shortest elevations side by side.

the elevations are placed next to the floor plan to conserve space. A drawback of this layout is that the heights established on the side elevations cannot be quickly transferred. A print of one elevation will have to be taped next to the location of the new elevation so that heights can be transferred.

SURFACE MATERIALS IN ELEVATION

The materials that are used to protect the building from the weather need to be shown on the elevations. This information will be considered in four categories: roofing, wall coverings, doors, and windows. Additional considerations are rails, shutters, eave vents, and chimneys.

Roofing Materials

Several common materials are used to protect the roof. Among the most frequently used are asphalt shingles, wood shakes and shingles, clay and concrete tiles, metal sheets, and built-up roofing materials, which were introduced in Chapter 20. It is important for the architectural drafter to have an idea of what each material looks like so that it can be drawn in a realistic style. It is also important to remember that the elevations are meant for the framing crew. The framer's job will not be made easier by seeing every shingle drawn or other techniques that are used on presentation drawings and renderings. The elevations need to have materials represented clearly and quickly.

Asphalt Shingles

Asphalt shingles come in many colors and patterns. Asphalt shingles are typically drawn using the method seen in Figure 22.11. Notice that line weights and lengths vary. Lines can be placed by using a back-and-forth method with varied pressure. When placed using AutoCAD, lines representing shingles will resemble Figure 22.12.

Wood Shakes and Shingles

Figure 22.13 shows a roof protected with wood shakes. Other materials such as Masonite are used to simulate wood shakes. Shakes and Masonite create a jagged surface at the ridge and the edge. These types of materials are often represented as shown in Figure 22.14. When placed using AutoCAD, lines representing shingles will resemble Figure 22.15.

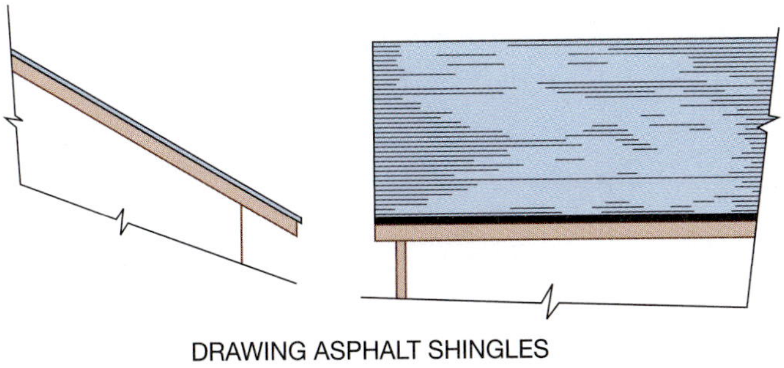

FIGURE 22.11 ■ Drawing asphalt shingles manually.

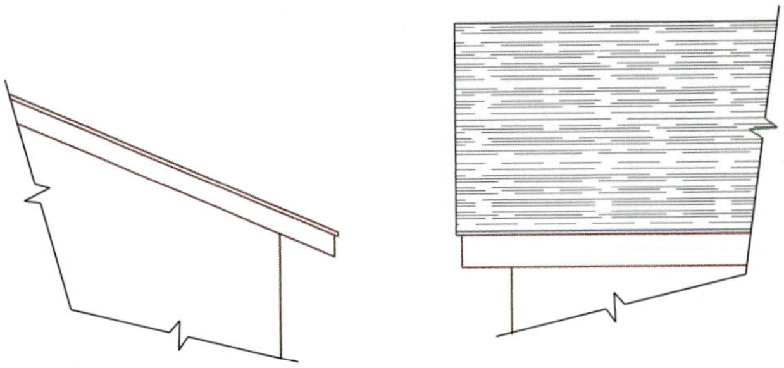

FIGURE 22.12 ■ The AR-RROOF hatch pattern of AutoCAD can be used to represent composition shingles.

FIGURE 22.13 ■ Wood shakes and shingles have much more texture than asphalt shingles.

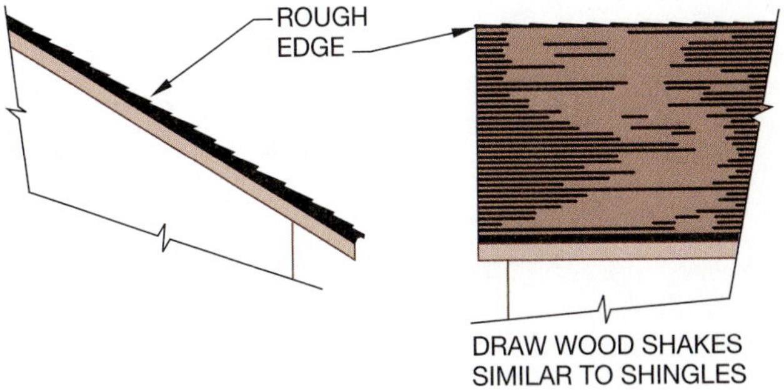

FIGURE 22.14 ■ Drawing wood shakes and shingles manually.

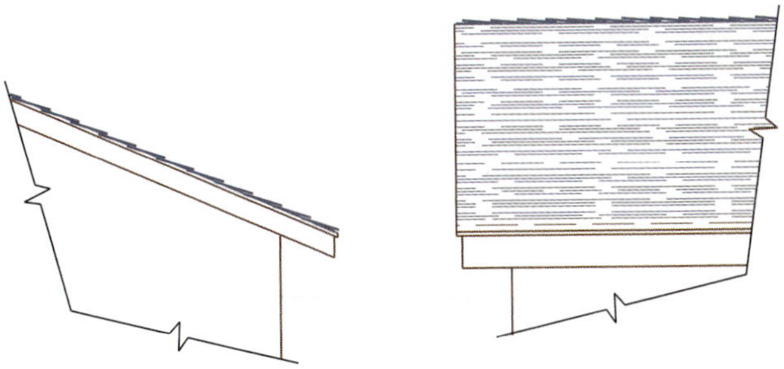

FIGURE 22.15 ■ Shakes and shingles can be represented using AR-RROOF or AR-RSKE. Notice that an inclined line is added to the eave and to the ridge to represent the contour of the roofing material.

Tile

Concrete, clay, or a lightweight simulated tile material presents a very rugged surface at the ridge and edge, as well as many shadows throughout the roof. Figure 22.16 shows flat tiles. This type of tile is usually drawn in a manner similar to Figure 22.17*a* and Figure 22.17*b*. Figure 22.18 shows a roof with Spanish tile. This type of tile is often drawn as shown in Figure 22.19*a* and Figure 22.19*b*. Rub-on films are available for some types of tile roofs. Figure 22.20 shows an elevation where the drafter used a rub-on film to represent the tile rather than drawing each tile.

Metal

Metal shingles are usually drawn in a manner similar to asphalt shingles.

Built-Up Roofs

Because of the low pitch and the lack of surface texture, built-up roofs are usually outlined and left blank. Occasionally a built-up roof will be covered with rock 2″ or 3″ (50–75 mm) in diameter. The drawing technique for this roof can be seen in Figure 22.21.

FIGURE 22.16 ■ Flat tiles are used in many parts of the country because of their low maintenance, durability, and resistance to the forces of nature. *Courtesy LeRoy Cook.*

FIGURE 22.18 ■ Curved or Spanish tiles are a traditional roofing material of Spanish and Mediterranean style homes. *Courtesy Jordan Jefferis.*

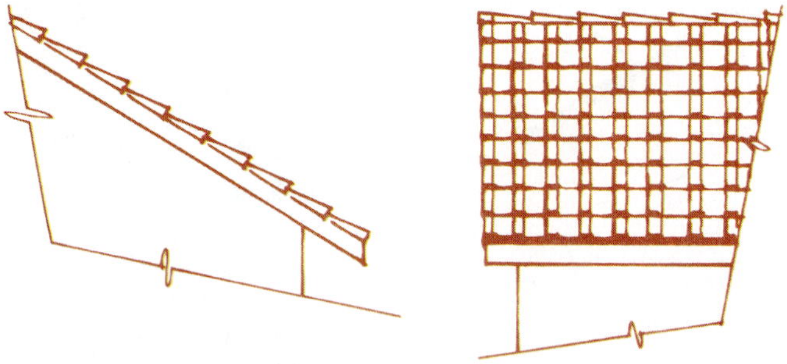

FIGURE 22.17(A) ■ Flat tiles are drawn manually using methods similar to drawing shakes.

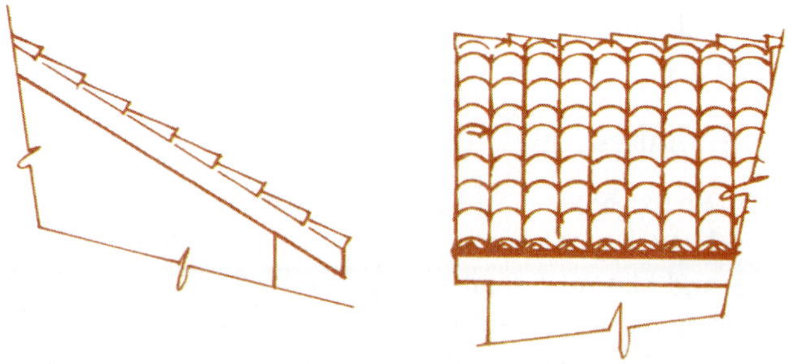

FIGURE 22.19(A) ■ Drawing curved tiles manually.

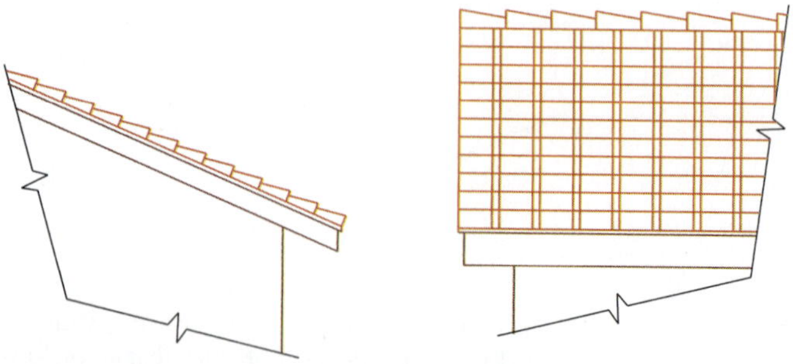

FIGURE 22.17(B) ■ Representing tile using CADD may require a hatch pattern to be created.

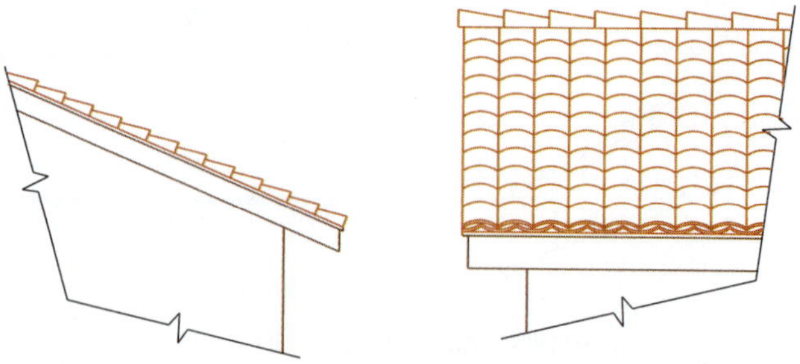

FIGURE 22.19(B) ■ Representing Spanish tiles using CADD.

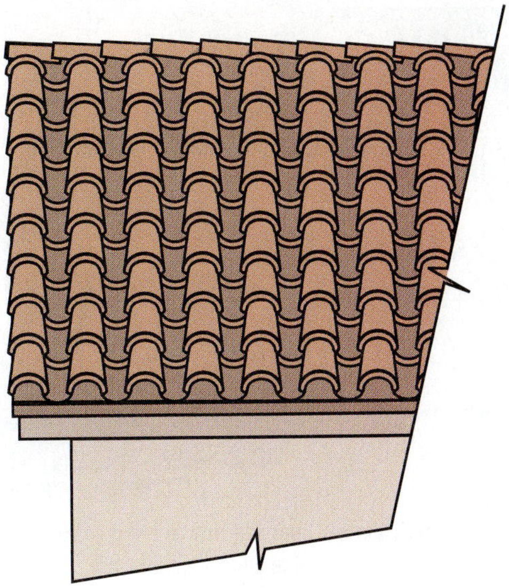

FIGURE 22.20 ■ Tiles may be drawn in elevation by using rub-on film. *Courtesy Chartpak.*

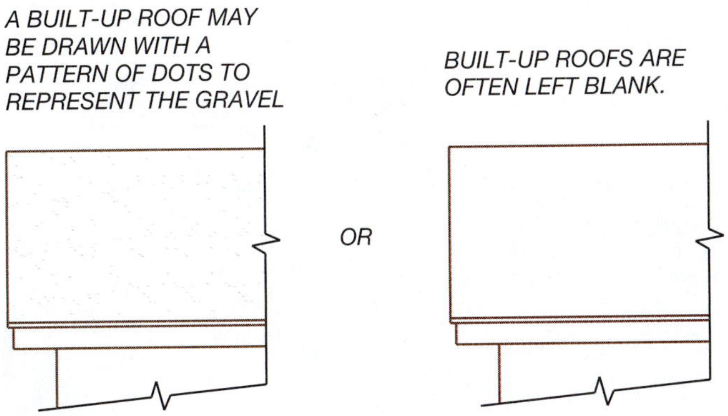
A BUILT-UP ROOF MAY BE DRAWN WITH A PATTERN OF DOTS TO REPRESENT THE GRAVEL
OR
BUILT-UP ROOFS ARE OFTEN LEFT BLANK.

FIGURE 22.21 ■ Drawing built-up roofs.

Skylights

Skylights may be made of either flat glass or domed plastic. Although they come in a variety of shapes and styles, skylights usually resemble the model shown in Figure 22.22. Depending on the pitch of the roof, skylights may or may not be drawn. On very low-pitched roofs a skylight may be unrecognizable. On roofs over 3/12 pitch the shape of the skylight can usually be drawn without creating confusion. Unless the roof is very steep, a rectangular skylight will appear almost square. The flatter the roof, the more distortion there will be in the size of the skylight. Figure 22.23 shows common methods of drawing both flat-glass and domed skylights.

Wall Coverings

Exterior wall coverings are usually made of wood, wood substitutes, masonry, metal, plaster or stucco. Each has its own distinctive look in elevation.

FIGURE 22.22 ■ Double-domed plastic skylights are a common feature of many roofs. *Courtesy Duralite.*

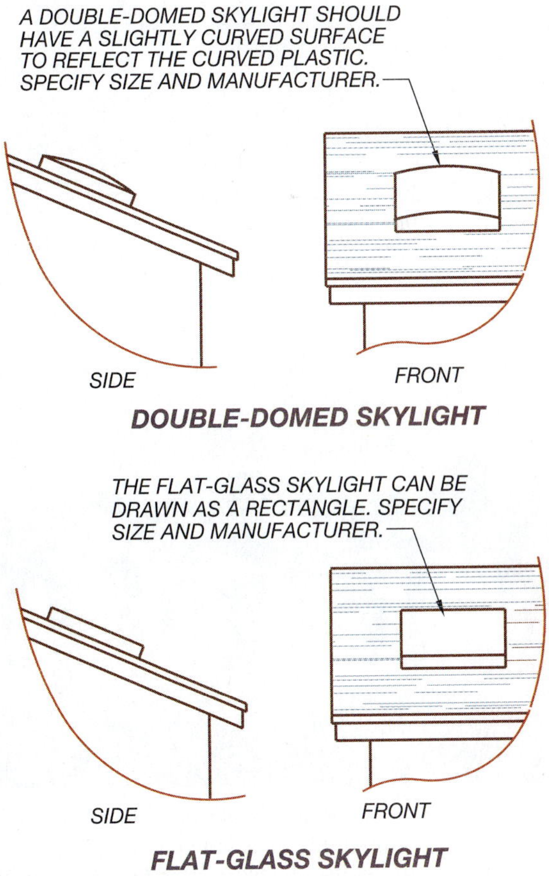
A DOUBLE-DOMED SKYLIGHT SHOULD HAVE A SLIGHTLY CURVED SURFACE TO REFLECT THE CURVED PLASTIC. SPECIFY SIZE AND MANUFACTURER.
SIDE
FRONT
DOUBLE-DOMED SKYLIGHT
THE FLAT-GLASS SKYLIGHT CAN BE DRAWN AS A RECTANGLE. SPECIFY SIZE AND MANUFACTURER.
SIDE
FRONT
FLAT-GLASS SKYLIGHT

FIGURE 22.23 ■ Representing domed and flat-glass skylights on elevations.

Wood

Wood siding can be installed in large sheets or in individual pieces. Plywood sheets are a popular wood siding because of their low cost and ease of installation. Individual pieces of wood provide an attractive finish but usually cost more than plywood. This higher cost results from differences in material and the labor to install each individual piece.

Plywood siding can have many textures, finishes, and patterns. Textures and finishes are not shown on the elevations but may be specified in a general note. Patterns in the plywood are usually shown. The most common patterns in plywood are T-1-11 (Figure 22.24), board on board (Figure 22.25), board and batten (Figure 22.26), and plain or rough-cut plywood (Figure 22.27). Figure 22.28 shows methods for drawing each type of siding.

FIGURE 22.26 ■ Batt-on-board siding is used to highlight the horizontal siding of this home.

FIGURE 22.24 ■ T-1-11 plywood is a common siding. *Courtesy Ryan McFadden.*

FIGURE 22.27 ■ Rough-cut plywood is used on the sides of these structures to provide contrast to the bevel cedar siding used on the front surfaces. *Courtesy Western Red Cedar Lumber Association.*

FIGURE 22.25 ■ Redwood board-on-board siding is used to protect this home from the elements. Design by James Bischoff of Callister, Gately, Heckmann & Bischoff, Photo by Charles Callister, Jr. *Courtesy California Redwood Association.*

Lumber siding comes in several types and can be laid in many patterns. Common lumber for siding is cedar, redwood, pine, fir, spruce, and hemlock. Common styles of lumber siding are tongue and groove, bevel, and channel. Each can be seen in Figures 22.29 through 22.31. Figure 22.32 shows common shapes of wood siding. Each of these materials can be installed in a vertical, horizontal, or diagonal position. The material and type of siding must be specified in a general note on the elevations. The pattern in which the siding is to be installed must be shown on the elevations in a manner similar to Figure 22.33. The type of siding and the position in which it is laid will affect how the siding appears

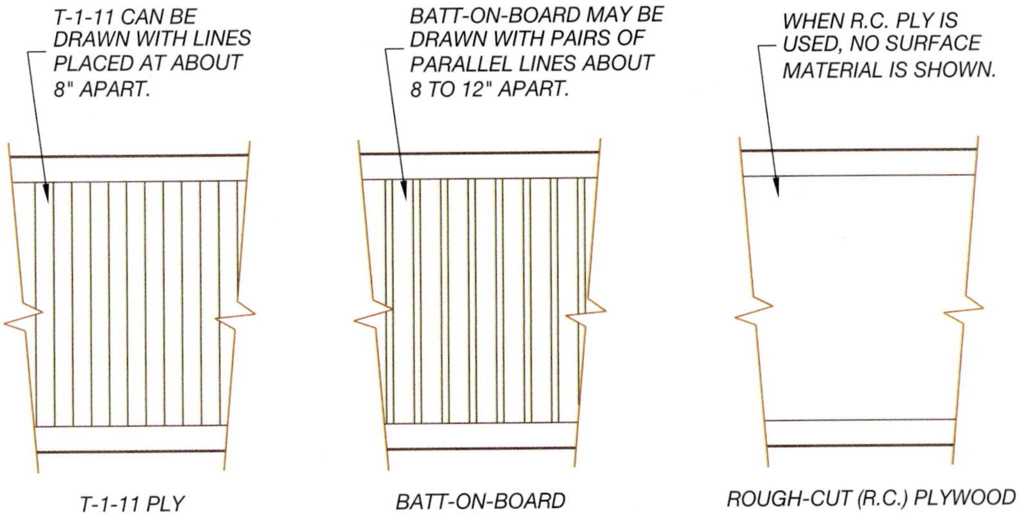
T-1-11 CAN BE DRAWN WITH LINES PLACED AT ABOUT 8" APART.
BATT-ON-BOARD MAY BE DRAWN WITH PAIRS OF PARALLEL LINES ABOUT 8 TO 12" APART.
WHEN R.C. PLY IS USED, NO SURFACE MATERIAL IS SHOWN.
T-1-11 PLY
BATT-ON-BOARD
ROUGH-CUT (R.C.) PLYWOOD

FIGURE 22.28 ■ Drawing plywood siding in elevation.

NORDSTROM & CO.
3050 BROADWAY

FIGURE 22.29 ■ Vertical siding is a common natural material for protecting a structure from the elements. Designed by Architects Knudson-Williams. *Photo by T.S. Gordon. Courtesy California Redwood Association.*

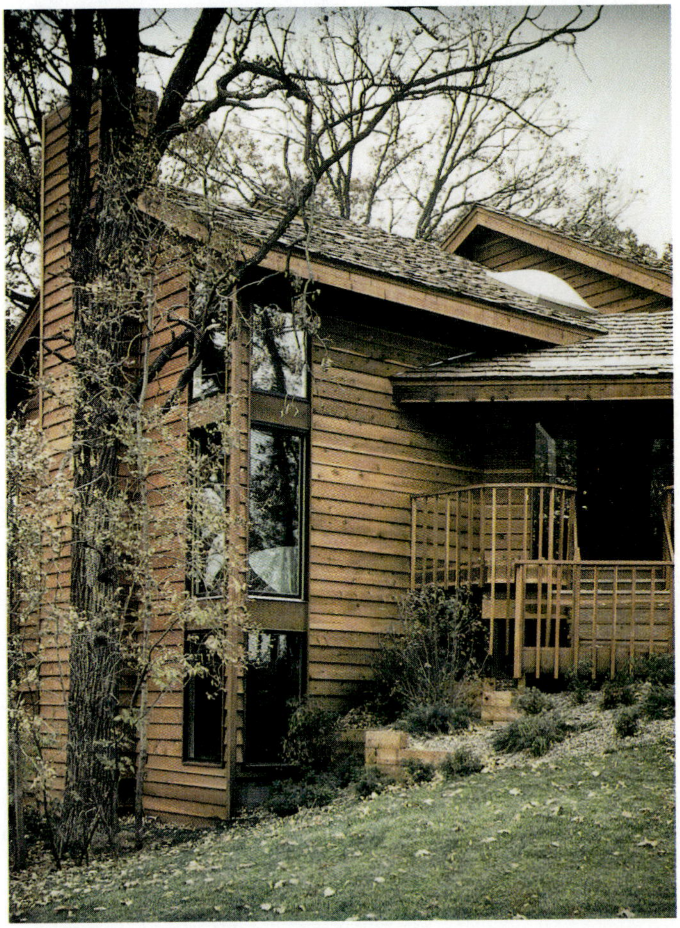

FIGURE 22.30 ■ Beveled redwood siding is used on this home to create a pleasing blend of the home with the site. Designed and built by Cramer-Weir Co. *Photo by Saari & Forrai Photography. Courtesy California Redwood Association.*

at a corner. Figure 22.34 shows two common methods of corner treatment.

Wood shingles similar to Figure 22.35 can either be installed individually or in panels. Shingles are often drawn as shown in Figure 22.36.

Wood Substitutes

Hardboard, fiber cement, aluminum, and vinyl siding can be produced to resemble lumber siding. Figure 22.37 shows a home finished with hardboard siding. Hardboard siding is generally installed in large sheets similar to plywood but often has more detail than plywood or lumber siding. It is typically

FIGURE 22.31 ■ Finger-jointed redwood siding is used to create the lower walls of the project by William Zimmerman, architect. *Photo by Wayne LeNouve, Courtesy California Redwood Association.*

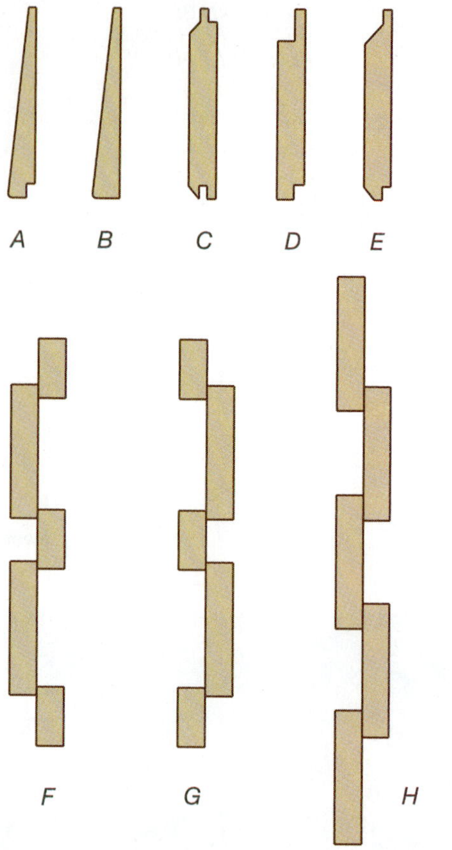

FIGURE 22.32 ■ Common types of siding: (*A*) bevel; (*B*) rabbeted; (*C*) tongue and groove; (*D*) channel shiplap; (*E*) V shiplap; (*F, G, H*) these types can have a variety of appearances, depending on the width of the boards and battens being used.

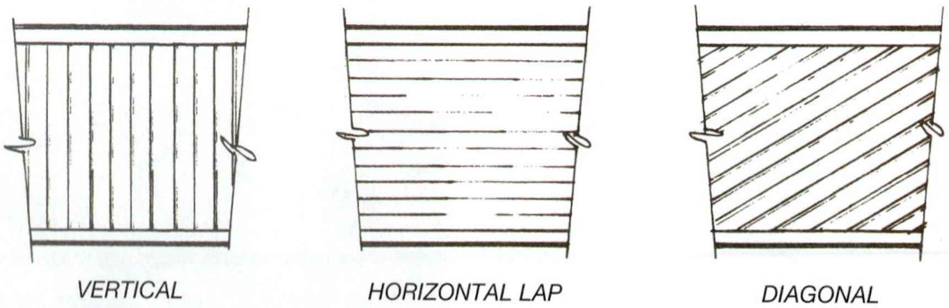

FIGURE 22.33(A) ■ The type of siding and the position in which it is to be installed must be shown on the elevations.

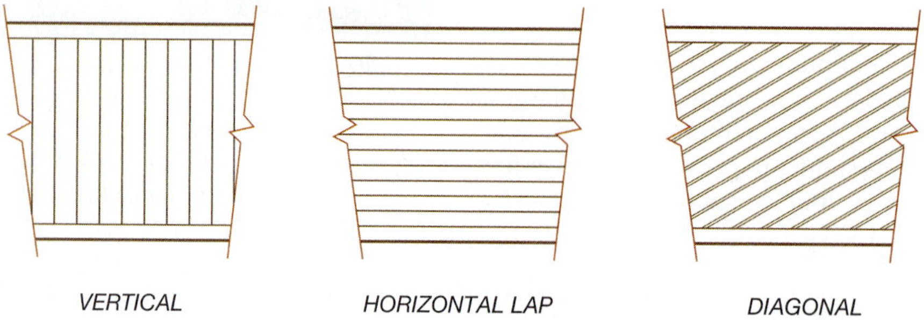

FIGURE 22.33 ■ Representing siding using CADD.

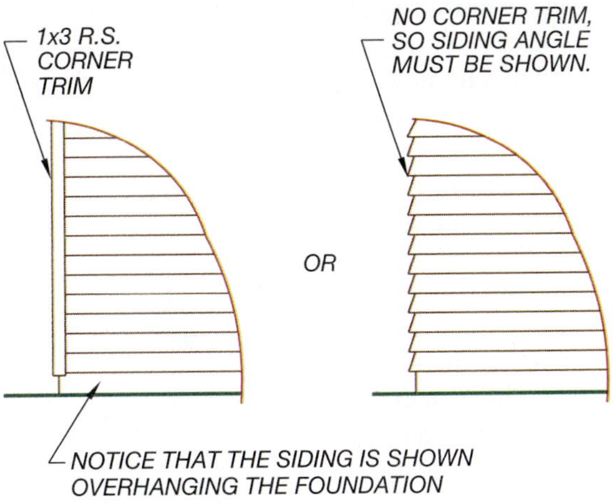

FIGURE 22.34 ■ Common methods of corner treatment to be shown on elevations.

FIGURE 22.35 ■ Shingles are often used as a siding material to provide a casual or rustic finish. *Courtesy ABTco, Inc.*

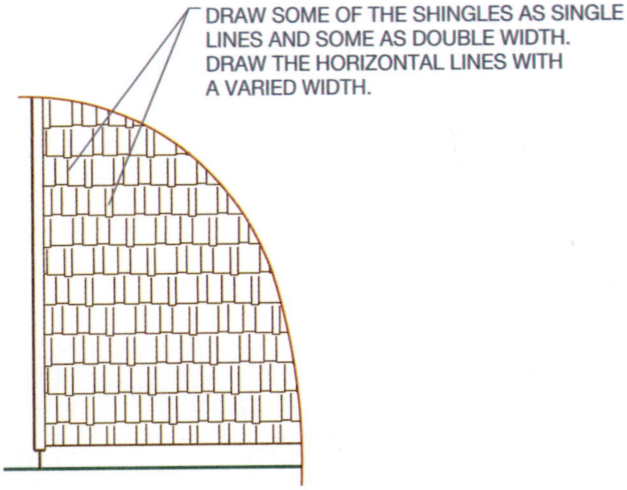

DRAW SOME OF THE SHINGLES AS SINGLE
LINES AND SOME AS DOUBLE WIDTH.
DRAW THE HORIZONTAL LINES WITH
A VARIED WIDTH.

FIGURE 22.36 ■ The AR-RSHKE hatch pattern can be used to represent shingle siding.

FIGURE 22.37 ■ Hardboard can be used as a siding material if precautions are taken to protect against moisture.

drawn using methods similar to those used for drawing lumber sidings. Each of the major national wood distributors has also developed siding products made from wood by-products that resemble individual pieces of beveled siding. Strands of wood created during the milling process are saturated with a water-resistant resin binder and compressed under extreme heat and pressure. The exterior surface typically has an embossed finish to resemble the natural surface of cedar. Most engineered lap sidings are primed to provide protection from moisture prior to installation.

Fiber cement siding products are becoming an increasingly popular wood substitute because of their durability. Engineered to be resistant to moisture, cold, insects, salt air and fire, fiber cement products such as DuraPress by ABTco and Hardiplank by James Hardie are being used in many areas of the country as an alternative to wood and plaster products. Available in widths of 6 1/2″, 7 1/2″ and 9 1/2″ (165, 190, and 240 mm) or 4′ × 8′ (100 × 200 mm), sheets, fiber cement products can reproduce smooth or textured wood patterns as well as cedar plywood panels and stucco. Products are installed similar to their wood counterparts and typically offer a fifty-year warranty. Figure 22.38 shows an example of fiber cement bevel siding.

Aluminum and vinyl sidings also resemble lumber siding in appearance, as shown in Figure 22.39. Aluminum and vinyl sidings are drawn similar to their lumber counterpart.

FIGURE 22.38 ■ Fiber cement siding provides excellent protection from moisture, insects, and other natural elements. *Courtesy ABTco, Inc.*

FIGURE 22.39 ■ Vinyl and aluminum sidings are manufactured to resemble their wood counterparts. *Courtesy ABTco, Inc.*

Masonry

Masonry finishes may be made of brick, concrete block, or stone.

Brick is used on many homes, like the one in Figure 22.40, because of its beauty and durability. Bricks come in a variety of sizes, patterns, and textures. When drawing elevations, the drafter must represent the pattern of the bricks on the drawing and the material and texture in the written specifications. A common method for drawing bricks is shown in Figure 22.41. Although bricks are not usually drawn exactly to scale, the proportions of the brick must be maintained. Figure 22.42 shows common patterns drawn using CADD.

FIGURE 22.40 ■ Brick is used in many traditional designs because of its elegant appearance and its resistance to natural forces.

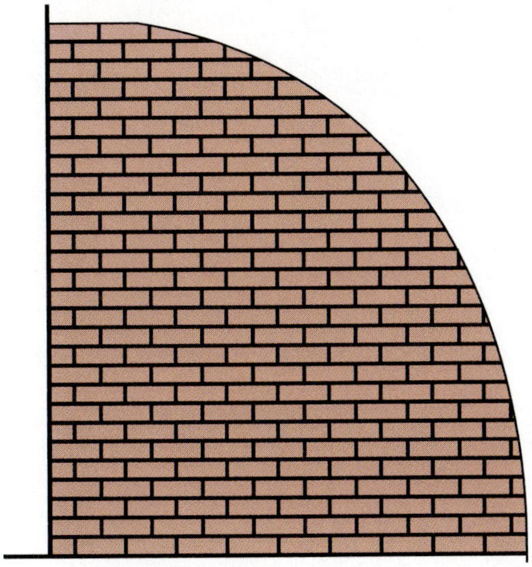

FIGURE 22.41 ■ Drawing brick in elevation.

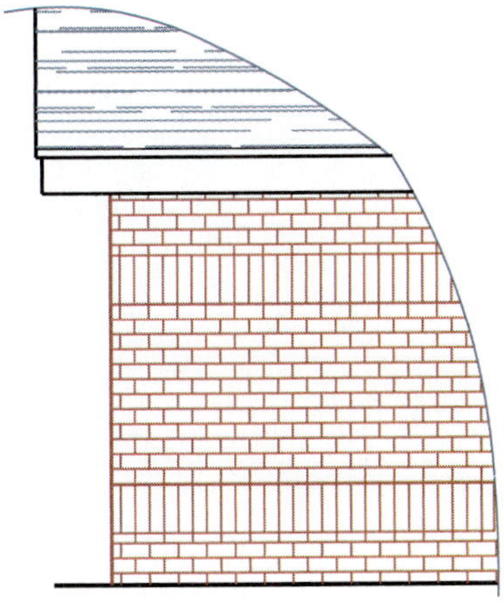

FIGURE 22.42 ■ Drawing brick using CADD.

Concrete block is often used as a weather-resistant material for above- and-below-grade construction. Figure 22.43 shows two examples of concrete block used to form aboveground walls. A drafter should show size, pattern, and texture on the elevation when drawing concrete blocks. Figure 22.44 shows methods of representing some of the various concrete blocks.

Stone is often used to provide a charming, traditional style that is extremely weather-resistant. Figure 22.45 shows an example of stone used to accent a home. Stone or rock finishes also come in a wide variety of sizes and shapes, and are laid in a variety of patterns, as shown in Figure 22.46. Stone or rock may be natural or artificial. Both appear the same when drawn in elevation. When showing stone and rock surfaces in elevation, the drafter must be careful to represent the irregular shape, as in Figure 22.47.

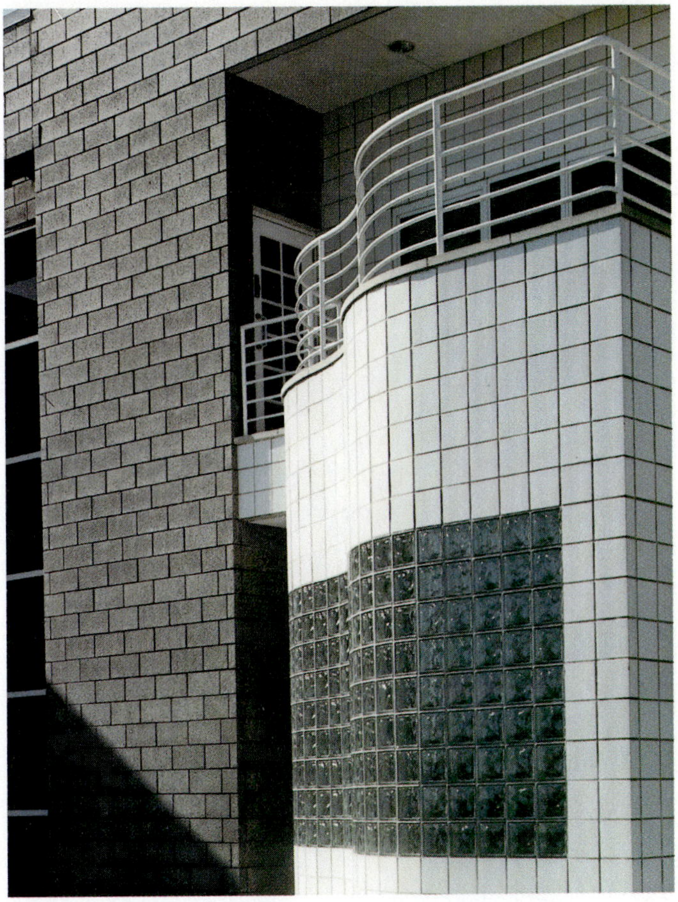

FIGURE 22.43 ■ Concrete block is used in many areas to provide a long-lasting, energy-efficient building material.

FIGURE 22.44 ■ Representing 8″ × 8″ × 16″ concrete blocks in elevation.

Metal

Although primarily a roofing material, metal can be used as an attractive wall covering. Drawing metal in elevation uses a method similar to drawing lumber siding.

Plaster or Stucco

Although primarily used in areas with little rainfall, plaster or stucco can be found throughout the country. Figure 22.48

FIGURE 22.45 ■ Stone is used on many traditional home styles. *Courtesy LeRoy Cook.*

FIGURE 22.46 ■ Stone is used in a variety of shapes and patterns. *Courtesy Stucco Stone Products, Inc., Napa.*

shows an example of a common stucco pattern. No matter what the pattern, stucco is typically drawn as shown in Figure 22.49.

Similar in appearance to stucco or plaster are exterior insulation and finishing systems (EIFS) such as Dryvit. Typically a rigid insulation board is used as a base for a fiberglass-reinforced base coat. A weather-resistant colored finish coat is then applied by trowel to seal the structure. Shapes made out of insulation board can be added wherever three-dimensional details are desired. Figure 22.50 shows a home sealed with Dryvit.

Doors

On elevations, doors are drawn to resemble the type of door specified in the door schedule. The drafter should be careful not to try to reproduce an exact likeness of a door. This is especially true of entry doors and garage doors. Typically these doors have a decorative pattern on them. It is important to show this pattern, but do not spend time trying to reproduce the exact pattern. Since the door is manufactured, the drafter is wasting time on details that add nothing to the plan. Figure 22.51 shows the layout of a raised-panel door, and Figure 22.52 shows how other common types of doors can be drawn.

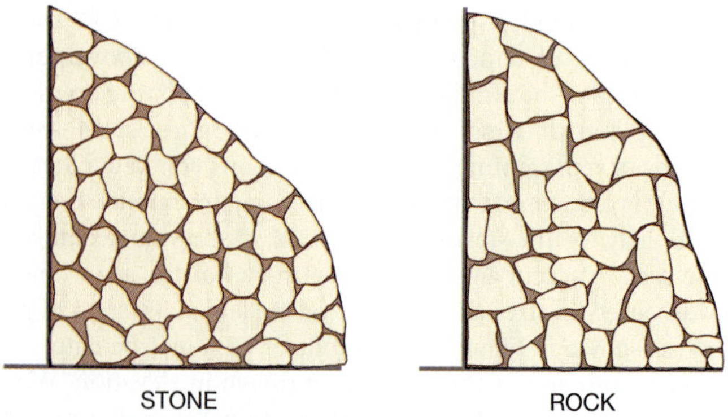

FIGURE 22.47 ■ Drawing stone and rock surfaces in elevation.

FIGURE 22.48 ■ Stucco and plaster can be installed in many different patterns and colors to provide a durable finish. *Courtesy Western Red Cedar Lumber Association.*

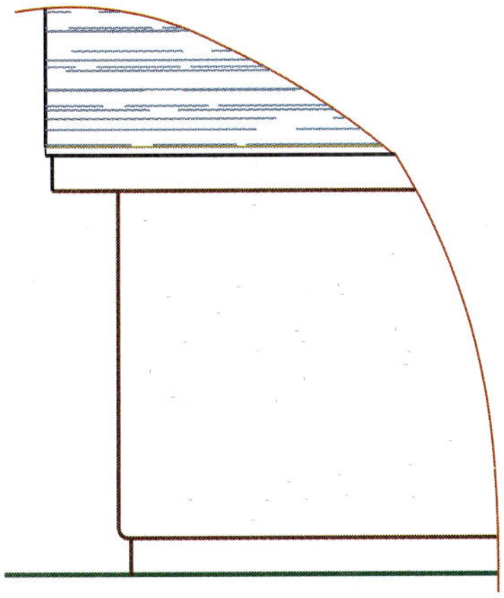

FIGURE 22.49 ■ Drawing stucco or plaster in elevation.

FIGURE 22.50 ■ EIFS offers the look of stucco while providing added insulation and durability. *Courtesy Dryvit.*

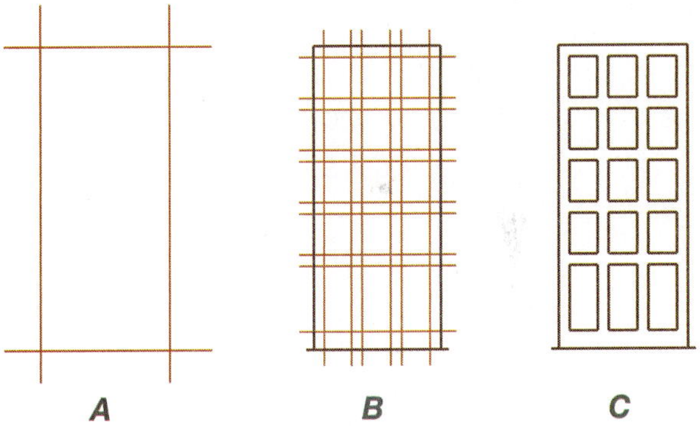

FIGURE 22.51 ■ Layout steps for drawing a raised-panel door.

Windows

The same precautions about drawing needless details for doors should be taken in drawing windows. Care must be given to the frame material. Wooden frames are wider than metal frames. When the elevations are started in the preliminary stages of design, the drafter may not know what type of frames will be used. In this case, the drafter should draw the windows in the most typical usage for the area. Figure 22.53 shows the layout steps for an aluminum sliding window. Figure 22.54 shows how other common types of windows can be drawn.

Rails

Rails can be solid to match the wall material or open. Open rails are typically made of wood or wrought iron. Although spacing varies, vertical rails are required to be no more than 4″ (100 mm) clear. Rails are often drawn as shown in Figure 22.55. Although it is not necessary to measure the location of each vertical, care must be taken to space the rails evenly.

Shutters

Shutters are sometimes used as part of the exterior design and must be shown on the elevations. Figure 22.56 shows a typical shutter and how it could be drawn. Although the spacing should be uniform, the drafter should try to lay out the shutter by eye rather than measure each louver.

Eave Vents

Drawing eave vents is similar to drawing shutters. Figure 22.57 shows common methods of drawing eave vents.

FIGURE 22.52 ■ Common doors represented in elevation.

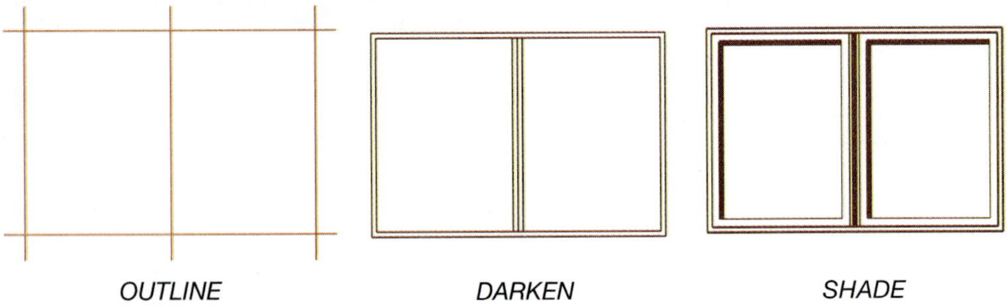

FIGURE 22.53 ■ Layout steps for a sliding window.

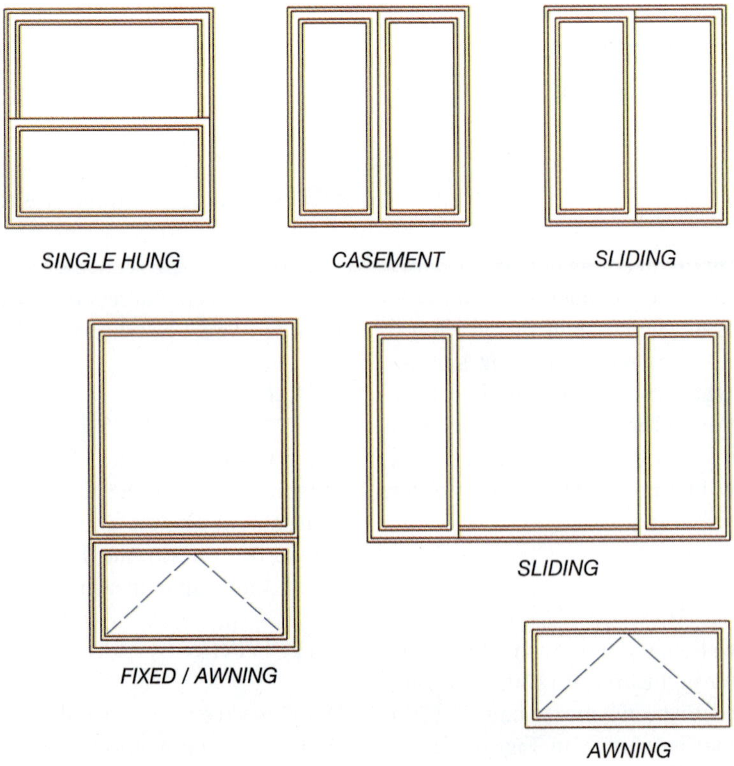

FIGURE 22.54 ■ Representing common window shapes on elevations.

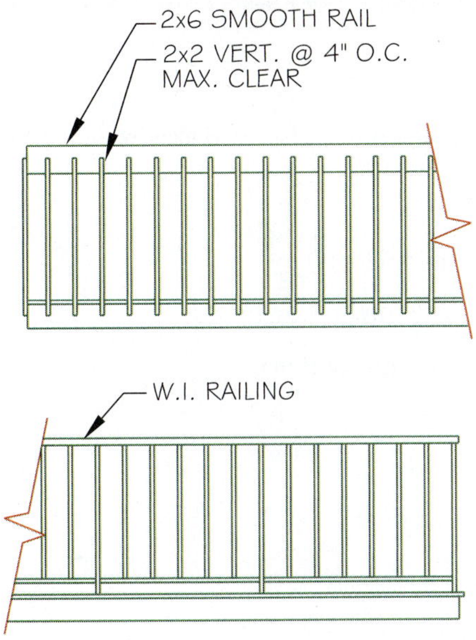

FIGURE 22.55 ■ Common types of railings.

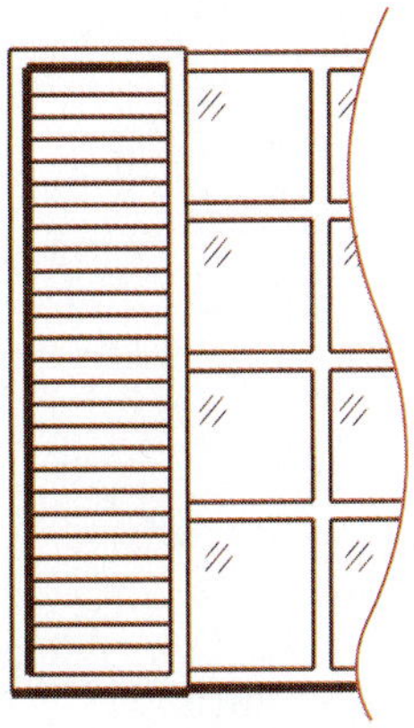

FIGURE 22.56 ■ Representing shutters in elevation.

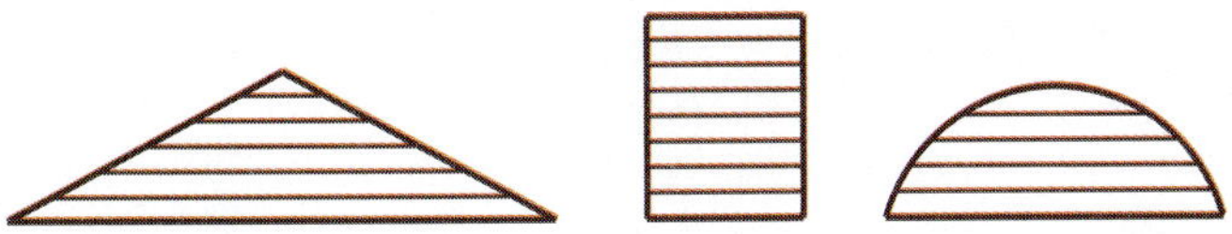

FIGURE 22.57 ■ Common methods of representing attic vents on elevations.

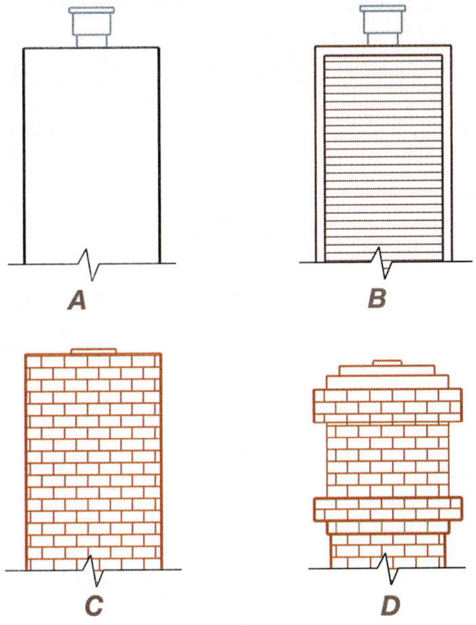

FIGURE 22.58 ■ Common methods of drawing chimneys: (*A*) stucco chase with metal cap; (*B*) horizontal wood siding with metal cap; (*C*) common brick; (*D*) decorative masonry.

Chimney

Several different methods can be used to represent a chimney. Figure 22.58 shows examples of wood and masonry chimneys.

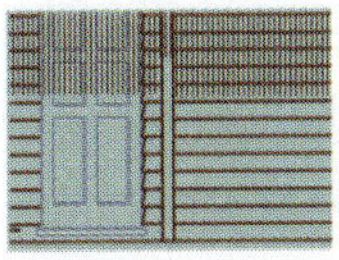

CADD ELEVATION SYMBOLS

Drawing exterior and interior elevation features with CADD is easy. Most architectural CADD programs have a variety of elevation symbols representing doors, windows, and materials. For example, pick a DOOR or WINDOW symbol from a tablet menu library; a screen menu will display your options. You then pick doors such as entrance, flush, or French. You can also add trim and doorknobs. When you pick WINDOW from the tablet menu library, you get options such as sliders, casement, double-hung, awning, and fixed. You can customize your window symbols with divided lights, arrows, reflection marks, solid black glass, or hinge marks.

Chapter 22 Additional Reading

The following web sites can be used as a resource to help you keep current with changes in roofing and exterior siding materials.

ADDRESS	COMPANY OR ORGANIZATION
www.abtco.com	Abtco, Inc. (hardboard and vinyl siding products)
www.alcoahomes	Alcoa Building Products (aluminum and vinyl siding products)
www.alsco.com	Alsco Building Products (vinyl and aluminum siding)
www.alside.com	Alside (vinyl siding, windows, trim)
www.ambrico.com	American Brick
www.brickinfo.org	The Brick Institute of America
www.canamould.com	Canamould (exterior moldings)
www.cemplank.com	Chemplank (fiber-cement siding)
www.culturedstone.com	Cultured Stone
www.apawood	APA - The Engineered Wood Association
www.cwc.ca	Canadian Wood Council
www.caststone.org	Cast Stone Institute
www.cedarbureau.org	Cedar Shake and Shingle Bureau
www.eifsfacts.com	EIFS Industry Members Association
www.fabral.com	Fabral (metal roofing and siding)
www.gp.com	Georgia Pacific (siding)
www.shakertown.com	Hardie Building Products (cedar shingles)
www.jameshardie.com	James Hardie Building Products
www.nailite.com	Nailite International (replica brick, stone, shake products)
www.napcobuildingmaterials.com	NAPCO Building Specialties
www.norandex.com	Norandex (vinyl siding)
www.reynoldsbp.com	Reynolds Building Products
www.senergyeifs.com	Senergy Inc. (EIFS)
www.vinylinfo.org	The Vinyl Institute
www.wrcla.org	Western Red Cedar Lumber Association
www.vinylsiding.com	Wolverine (metal siding)

CHAPTER 22

Introduction to Elevations Test

DIRECTIONS

Answer the questions with short, complete statements or drawings as needed.

1. Letter your name and the date at the top of the sheet.
2. Letter the question number, and provide the answer. You do not need to write out the question.

QUESTIONS

Question 22–1 Under what circumstances could a drafter be required to draw only three elevations?

Question 22–2 When are more than four elevations required?

Question 22–3 What are the goals of an elevation?

Question 22–4 What is the most common scale for drawing elevations?

Question 22–5 Would an elevation that has been drawn at a scale of 1/16″ = 1′–0″ require the same methods to draw the finishing materials as an elevation that has been drawn at some larger scale?

Question 22–6 Describe two methods of transferring the heights of one elevation to another.

Question 22–7 Describe two different methods of showing concrete tile roofs.

Question 22–8 What are the two major types of wood siding?

Question 22–9 In drawing a home with plywood siding, how should the texture of the wood be expressed?

Question 22–10 Sketch the pattern most typically used for brickwork.

Question 22–11 What problems is the drafter likely to encounter when drawing stone?

Question 22–12 Sketch the way wood shingles appear when one is looking at the gable end of a roof.

Question 22–13 Give the major consideration for drawing doors in elevation.

Question 22–14 What are the two most common materials used for rails?

Question 22–15 In drawing stucco, how should the pattern be expressed?

23 CHAPTER

Elevation Layout and Drawing Techniques

INTRODUCTION

Drawing elevations is one of the most enjoyable projects in architecture. As a student, though, you should keep in mind that a beginning drafter very rarely gets to draw the main elevations. The architect usually draws the main elevations in the preliminary stage of design, and then a senior drafter gets to finish the elevations. As a beginner you may be introduced to elevations at your first job by making corrections resulting from design changes. Not a lot of creativity is involved in corrections, but it is a start. As you display your ability, you'll most likely be involved in earlier stages of the drawing process. Elevations are drawn in three steps: layout, drawing, and lettering. Usually you will be given the preliminary design elevation or the designer's sketch to use as a guide.

LAYOUT

The size of drawing sheet you'll be using will affect the scale, placement, and layout of the elevations. The drawing sheet should match the size used for the floor plan. Because C-sized paper is so commonly used in educational settings, these instructions will assume that the drawing is to be made on a C-size sheet. Usually two elevations cannot be placed side by side on C-sized material. Thus, you will not be able to use a side elevation for projecting heights onto the front or rear elevations. The following instructions will be given for drawing the front elevation at a scale of 1/4″ = 1′–0″ and the rear and both side elevations at a scale of 1/8″ = 1′–0″. Using two scales will require more steps than when all of the elevations are drawn at one scale.

The layout process can be divided into three stages: overall-shape layout, roof layout, and layout of openings. Use construction lines for each stage.

Overall-Shape Layout

STEP 1 With a clean sheet of drawing paper in place, tape down a print of the floor and roof plans to use as a guide for all horizontal measurements. If a separate roof plan has not been drawn, draw the outline of the roof shape on the print of the floor plan. Assume 24″ (600 mm) overhangs and a 12″ (300 mm) gable end wall overhang. Each horizontal measurement can be projected as shown in Figure 23.1.

See Figure 23.2 for steps 2 through 8.

STEP 2 Lay out a baseline to represent the finish grade.

STEP 3 Project construction lines down from the floor plan to represent the edges of the house.

STEP 4 Establish the finish floor line. See Figure 23.3 for help in determining minimum measurements. This home will have a concrete slab that is 8″ (200 mm) above finish grade.

STEP 5 Measure up from the finished floor line, and draw a line to represent the ceiling level, typically 8′–0″ (2400 mm).

STEP 6 Measure up the depth of the floor joist to establish the finished floor level of the second floor. If the floor joist size has not been determined, assume that 10″ (250 mm) depth (2 × 10s) will be used.

STEP 7 Measure up 8′–0″ (2400 mm) from the second finished floor line, and draw a line to represent the upper ceiling level.

STEP 8 Measure up from each finished floor line 6′–8″ (2000 mm), and draw a line to represent the tops of the windows and doors.

Roof Layout

Because a side elevation is not being drawn beside the front elevation, there is no easy way to determine the true roof height. If the sections have been drawn before the elevations, the height from the ceiling to the ridge could be measured on one of the section drawings and then transferred to the elevations. Although this might save time, it also might duplicate an error. If the roof was incorrect on the section drawing, the error will be repeated if dimensions are simply transferred from section to elevation.

The best procedure to determine the height of the roof in the front view is to lay out a side view lightly. This can be done in the middle of what will become the front view. An example can

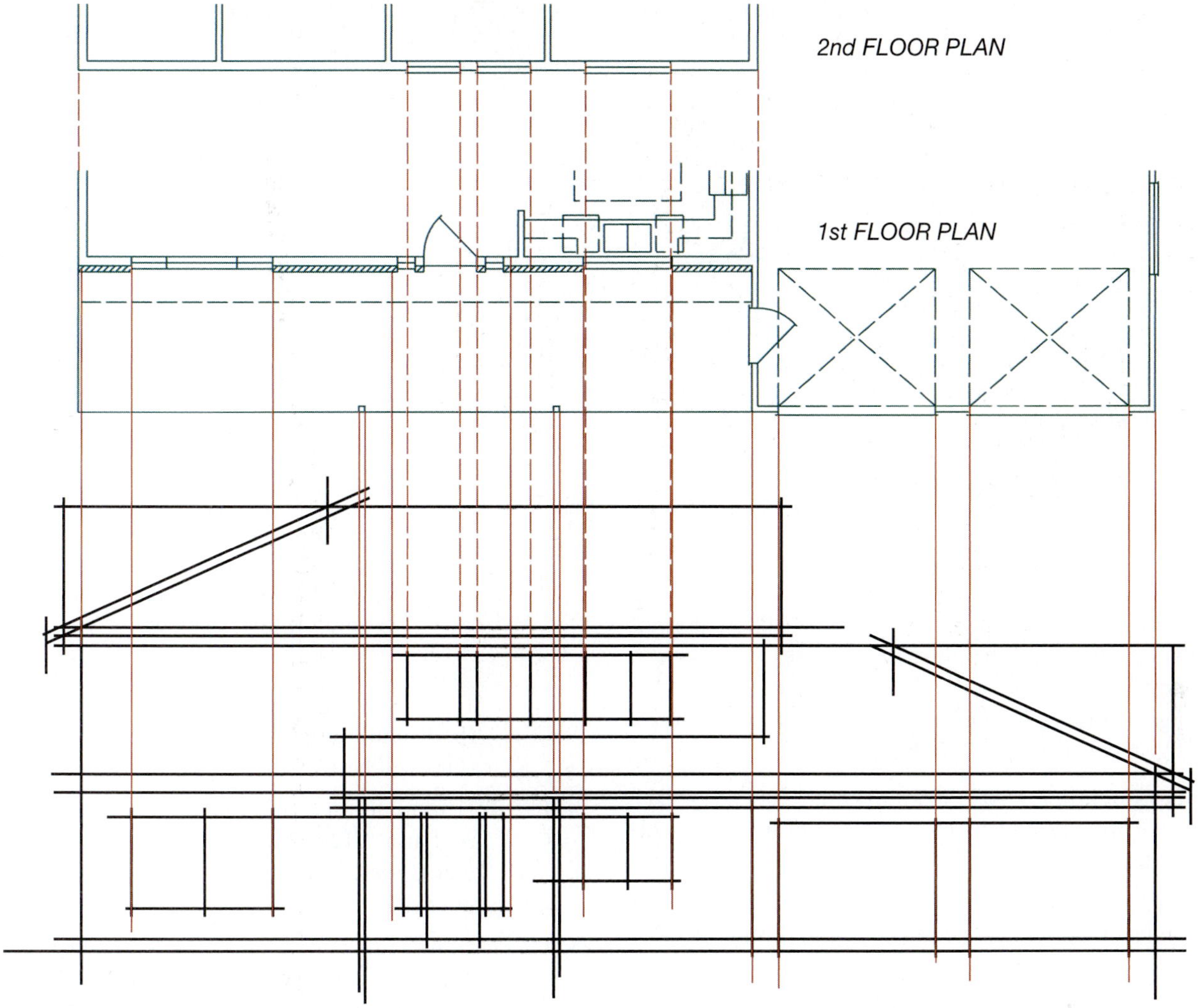

FIGURE 23.1 ■ With a print of the floor plans taped above the area where the elevations will be drawn, horizontal distances can be projected down to the elevation.

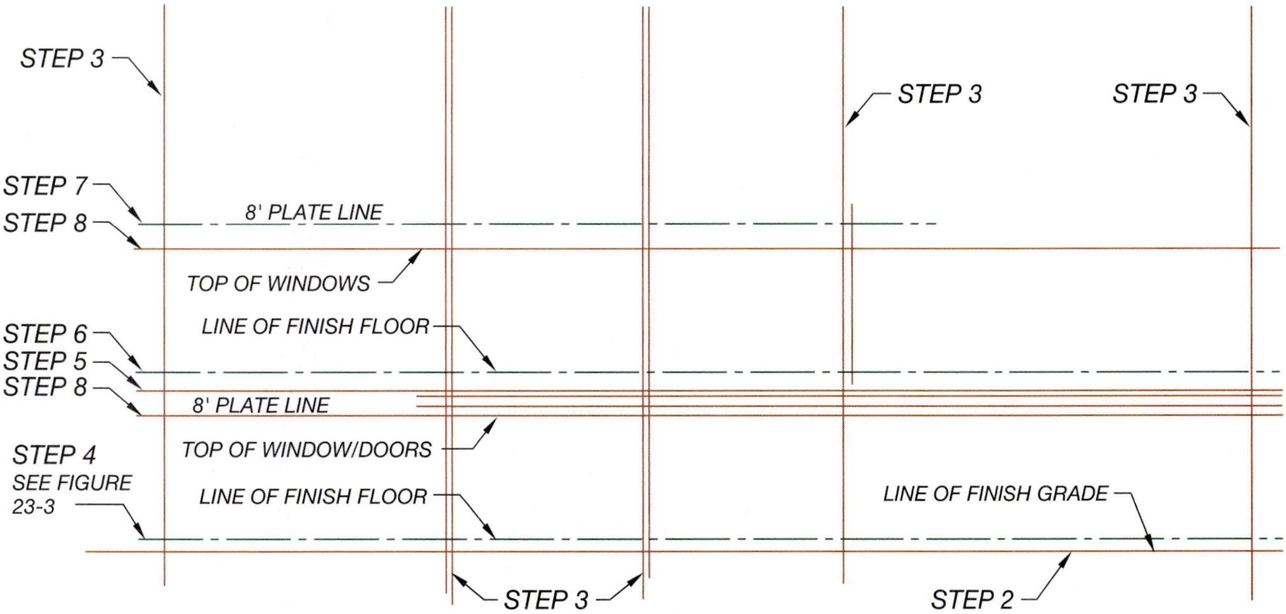

FIGURE 23.2 ■ Layout of the overall shape of the front elevation.

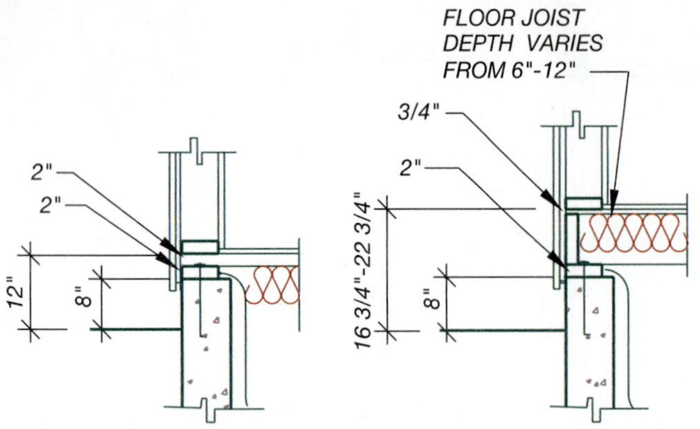

FIGURE 23.3 ■ The height of the finish floor above grade will be determined by the type of the floor framing system. Exact framing sizes are often rounded up to the nearest inch in elevation drawings.

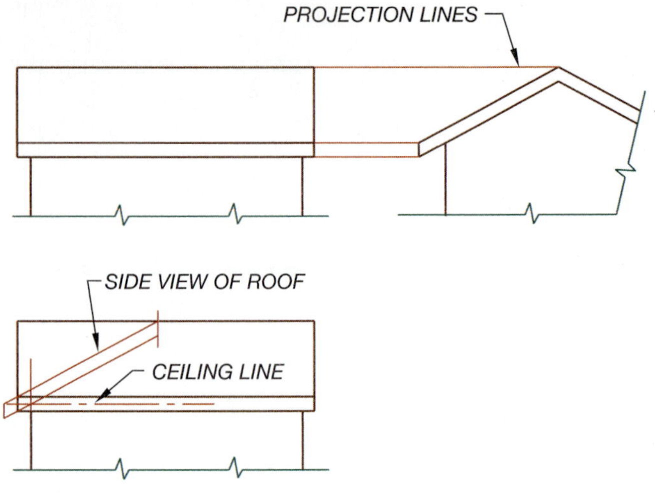

FIGURE 23.4 ■ When elevations are drawn side by side, heights can be projected with construction lines. When only one elevation is drawn, a side elevation will need to be drawn in the middle of what will become the front elevation.

be seen in Figure 23.4. For this home, three different roofs must be drawn: over the two-story area, garage, and porch. It would be best to lay out the porch roof last, since it is a projection of the garage roof. The upper or garage roof can be drawn by following steps 9 through 19 in Figure 23.5.

STEP 9 Lay out a line 1′–0″ (300 mm) from the end of each wall to represent the edge of the roof for each gable end wall overhang.

STEP 10 Lay out a line 2′–0″ (600 mm) from one of the end walls to represent the overhang projection.

STEP 11 Lay out a line to represent the ridge. Because the lower floor is 36′ wide (11,000 mm), the ridge will be 18′ (5500 mm) from the front wall. For the upper roof, the ridge will be 15′ (4570 mm) from the front wall.

STEP 12 Determine the roof slope. For this home, draw the roof at a 5/12 pitch. A 5/12 slope equals 22 1/2 degrees. See Section 6 for a complete explanation of pitches.

STEP 13 Draw a line at the desired angle to represent the roof. This line will represent the bottom of the truss or rafter. See Figure 23.6.

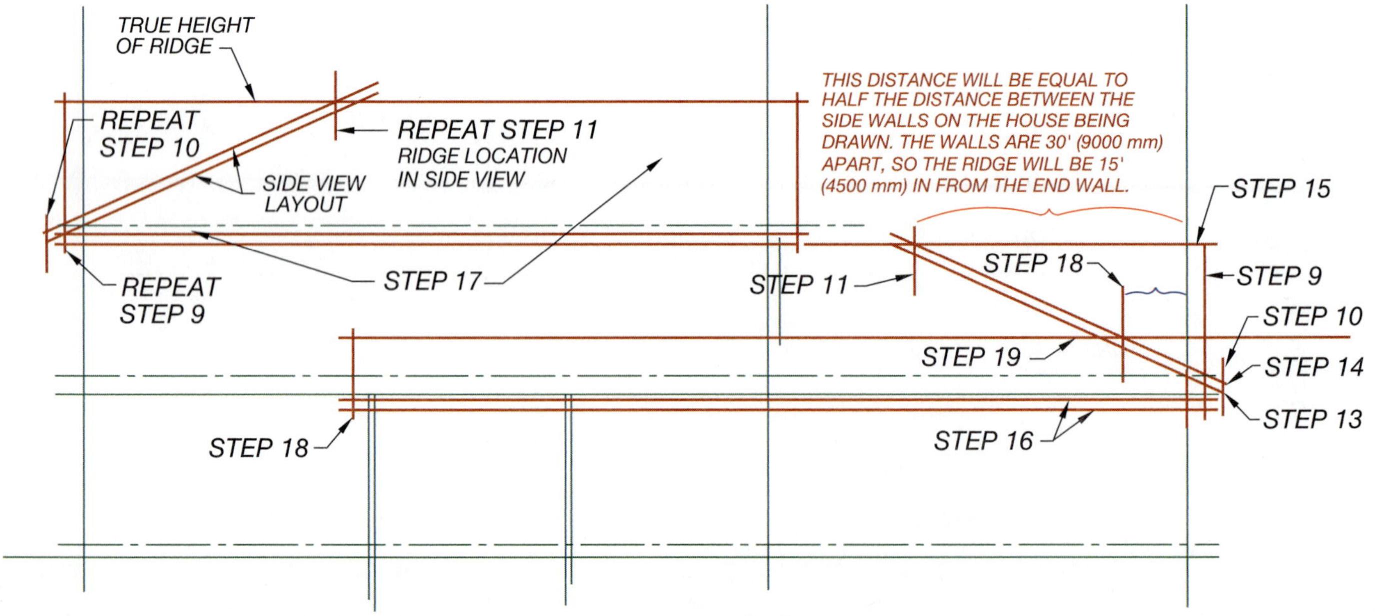

FIGURE 23.5 ■ Layout of the roof shapes. When looking at a roof from the front, the true height cannot be determined without drawing a side elevation. If the roof pitch and the ridge location are known, the true height can be drawn quickly and accurately.

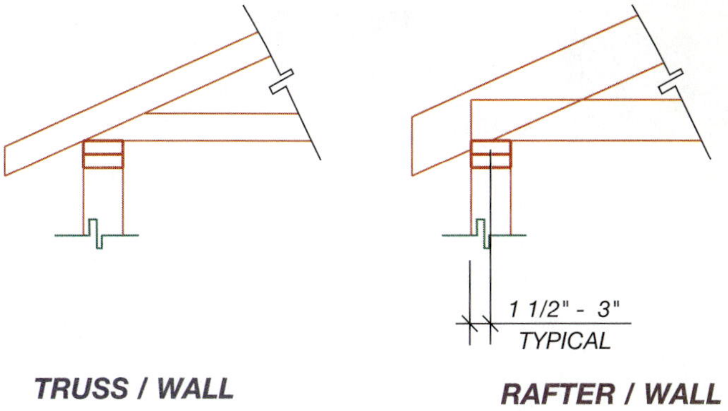

FIGURE 23.6 ■ The layout of the roof incline will vary slightly, depending on whether trusses or rafters are used to frame the roof. The incline of a truss passes through the intersection of the ceiling and walls. Because the rafter must have at least 1 1/2″ of bearing surface, the intersection of the rafter and wall must be adjusted slightly. If the notch in the rafter exceeds 3″, the rafter may not be able to support its design loads. Future chapters will explore minimum framing requirements.

STEP 14 Measure up 6″ (150 mm), and draw a line parallel to the line just drawn, which will represent the top of the rafter. On the elevations, it is not important that the exact depth of the truss or rafter be represented.

STEP 15 Establish the top of the roof. The true height of the roof can be measured where the line from step 14 intersects the line from step 11.

STEP 16 Establish the bottom of the eave. This can be determined from the intersection of the line from step 10 with the lines of steps 13 and 14.

STEP 17 Lay out the upper roof using the same procedure: drawing a partial side elevation and projecting exact heights.

The porch roof will be an extension of the garage roof, so much of the layout work will have been done. Only the left end and the upper limits of the roof must be determined. Because the ends of the upper and garage roof extend 12″ (300 mm), this distance will be maintained. The upper edge of the roof will be formed by the intersection of the roof and the wall for the upper floor. This height can be determined by following steps 18 and 19.

STEP 18 Place a line on the slope lines (the lines from steps 13 and 14) to represent the distance from the garage wall to the front wall on the upper floor plan.

STEP 19 At the intersection of the lines from steps 14 and 18, project a line to represent the top of the roof.

You now have the basic shape of the walls and roof blocked out. The only items to be drawn now are the door and window openings.

Layout of Openings

See Figure 23.7 for steps 20 through 23.

STEP 20 Lay out the width of all doors, windows, and skylights by projecting their location from the floor plan.

STEP 21 Lay out the bottom of the windows and doors. The doors should extend to the line representing the finished floor. Determine the depth of the windows from

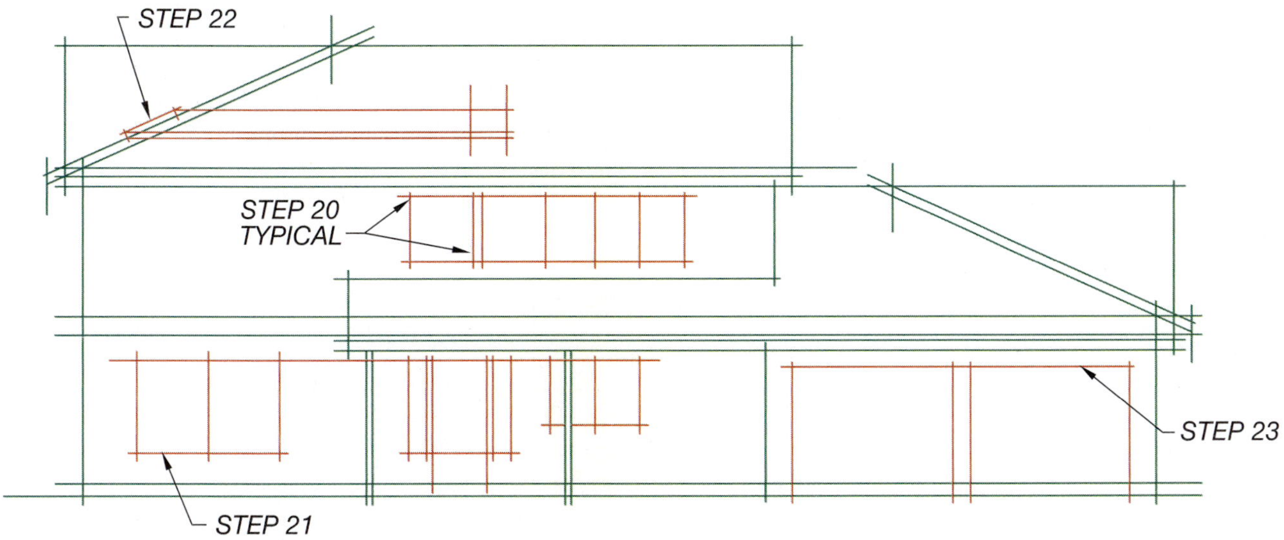

FIGURE 23.7 ■ The widths of doors, windows, and skylights can be projected from the floor and roof plans into the elevations. Heights can be determined using the window schedule.

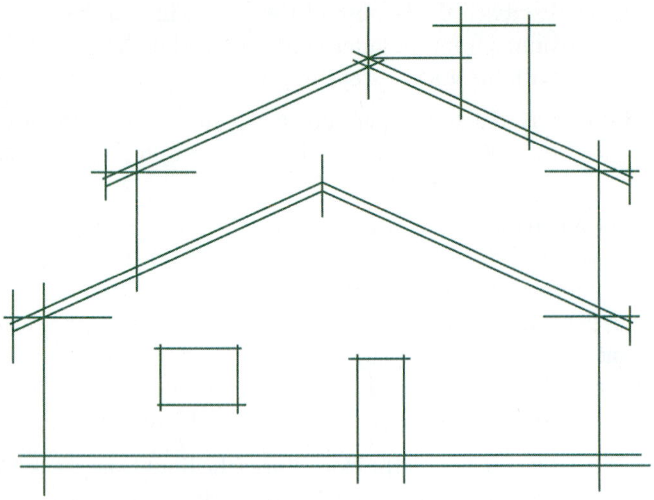

FIGURE 23.8 ■ The layout of a side elevation is similar to the layout of the front elevation.

the window schedule, and draw a line to represent the correct depth.

STEP 22 Determine the height of the skylights. This must be done by measuring the true size of the skylights on the line from step 14, and then projecting this height to the proper location.

STEP 23 Draw the shape of the garage doors. *Remember that the height of these doors is measured from the line representing the ground, not from the finished floor level.*

You now have the front elevation blocked out. The same steps can be used to draw the right and left side and rear elevations. The major difference in laying out these elevations is that the widths of walls, windows, and doors cannot be projected from the floor plan. To lay out these items, you will need to rely on the dimensions of the floor plan.

Start by drawing either side elevation. The side elevation can be drawn by following the procedure for the layout of the front elevation. Figure 23.8 shows a completed side elevation. Once this is drawn, use the side elevation to establish the heights for the other side and the rear elevation, as shown in Figure 23.9.

DRAWING FINISHED-QUALITY LINES

After all of the elevations are drawn with construction lines, they must be completed with finished-quality lines. An H and 2H lead will provide good contrast for many of the materials. Ink and graphite will also produce a very nice effect. Methods of drawing items required in elevations were discussed in Chapter 22. Line quality cannot be overstressed. By using varied line widths and density, a more realistic elevation can be created.

A helpful procedure for completing the elevations is to begin to darken the lines at the top of the page first and work down the page to the bottom. You will be using a soft lead, that tends to smear easily. If the lower elevations are drawn first, cover them with a sheet of paper when they are complete to prevent smearing.

Another helpful procedure for completing the elevations is to start at the top of one elevation and work toward the ground level. Also start by drawing surfaces that are in the foreground and work toward the background of the elevation. This procedure will help cut down on erasing background items for items that are in the foreground. Figure 23.10 shows a completed front elevation. It can be drawn by following steps 24 through 28.

STEP 24 Draw the outline of all roofs.

STEP 25 Draw the roofing material.

STEP 26 Draw all windows and doors.

STEP 27 Draw all corner trim and posts.

STEP 28 Draw the siding.

At this point your front elevation is completely drawn. The same procedures can be used to finish the rear and side elevations, shown in Figure 23.11. Note that only a small portion of the siding has been shown in these elevations. It is common practice to have the front elevation highly detailed and have the other elevations show just the minimum information for the construction crew.

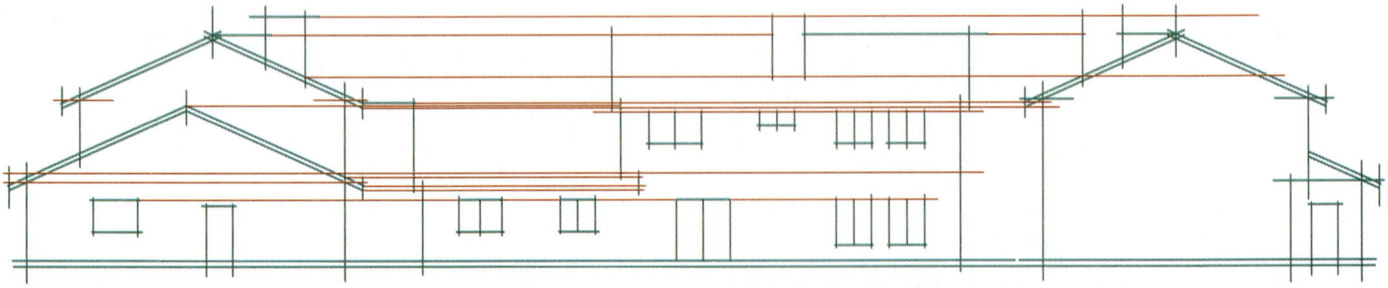

FIGURE 23.9 ■ With a side elevation drawn, heights for the rear and other side can be projected quickly.

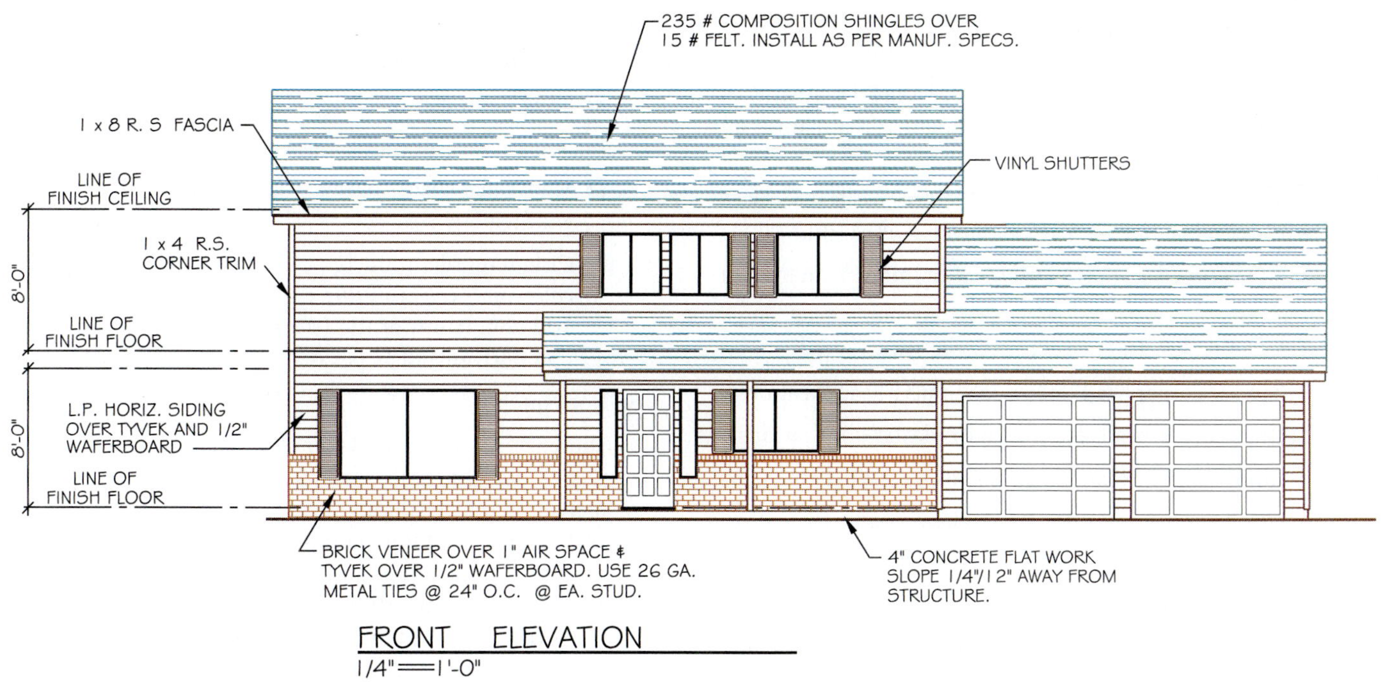

FIGURE 23.10 ■ Drawing the front elevation using finished-quality lines. Note that features in the background such as the chimney are often omitted for clarity.

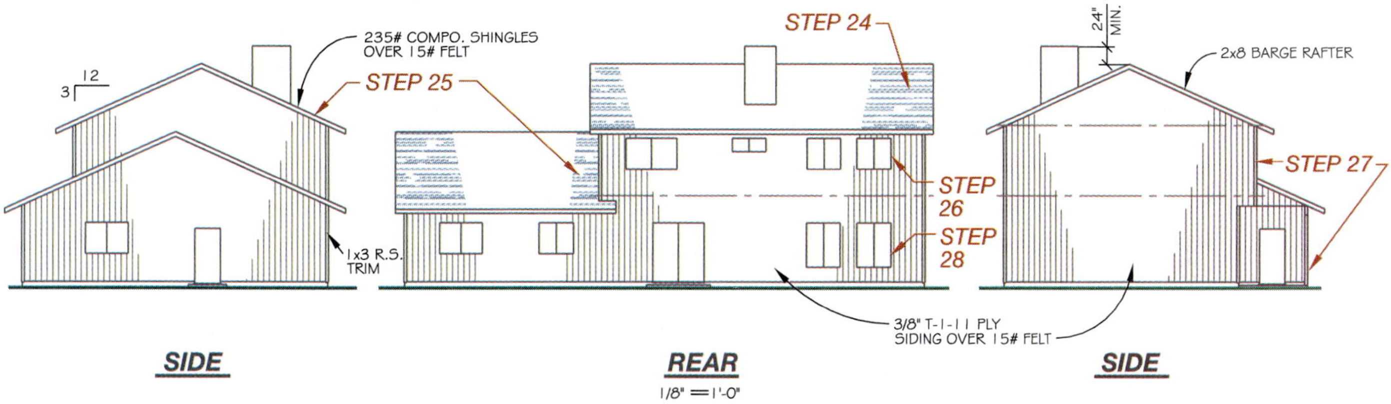

FIGURE 23.11 ■ Drawing small-scale elevations does not require drawing all of the finishing materials. The same procedures used on the front elevation can be used on the side and rear elevations.

LETTERING

With the elevations completely drawn, the material must now be specified. Chapter 26 will discuss each of the materials to be specified in the notes that must be placed on the exterior elevations. If the elevations are all on one page, a material needs to be specified on only one elevation. Use Figures 23–10 and 23–11 as guides for placing notes. The notes needed include the following:

1. Siding. Generally the kind of siding, its thickness, and the backing material must be specified. For each of the following siding materials, the OSB or plywood may be provided or omitted, according to regional preferences. When the OSB or plywood is omitted, the siding is installed directly over the vapor barrier and studs. Tyvek and 15# felt can be used interchangeably with the trade names of other vapor barriers. Siding examples include the following:

- Horiz. siding over 1/2″ (13 mm) OSB and Tyvek.
- Horiz. siding over 3/8″ (10 mm) plywood and 15 # felt.
- Hardiplank lap siding over . . .
- Hardie Shingleside w/ 8″ max. exposure over . . .
- LP horiz. lap siding over . . .
- 5/8″ (15.8 mm) T-1-11 siding w/grooves @ 4″ (100 mm) o.c. over . . .
- 1″ (25 mm) ext. stucco over 15# felt w/ 26 ga. linewire at 24″ (600 mm) o.c. w/stucco wire mesh, and no corner bead over . . .

- CD-EIFS exterior insulation and finishing system by Dryvit. Install as per manuf. specifications.
- Monterey Sand, Nova IV vinyl siding by Alside or equal installed over insulation board as per manuf. instructions.
- Brick veneer over 15# felt over 1″ (25 mm) air space with 26 ga. metal ties @ 24″ (600 mm) o.c. @ ea. stud.

2. Corner and decorative trims. Common trim might include:
   - 1 × 4 (25 × 100 mm) R.S. corner trim.
   - Provide corner bead [for stucco or plaster applications].
   - Omit corner bead [for stucco or plaster applications].
3. Chimney height *(2′ minimum above any roof within 10′).*
4. Flatwork would include any decks or concrete patios. Include rail material, vertical material, and spacing such as:
   - 2 × 6 (50 × 150 mm) dfl. smooth rail w/ 2 × 2 (50 × 50 mm) vert. @ 4″ (100 mm) max. clear.
   - 6 × 6 (150 × 150 mm) decorative post.
   - 2 × 6 (50 × 150 mm) cedar decking laid flat w/ 1/4″ (6 mm) gap between.
   - W.I. railing w/ vert. @ 4″ (100 mm) clear max.
   - 4″ (100 mm) concrete flatwork.
5. Posts and headers.
6. Fascias and barge rafters.
7. Roof material. For example:
   - 235# composition shingles over 15# felt.
   - 300# composition shingles over 15# felt.
   - Med. cedar shakes over 15# felt w/ 30# × 18″ (450 mm) wide felt btwn. ea. course w/ 10 1/2″ (270 mm) exposure.
   - Concrete tile roof [*give manufacturer's name, style, color, and weight of tile;* e.g., Chestnut brown, Classic 100 concrete roof tile (950# per sq.) by Monier over 15# felt]. Install as per manuf. specs.
8. Roof pitch.
9. Dimensions. The use of dimensions on elevations varies greatly from office to office. Some companies place no dimensions on elevations, using the sections to show all vertical heights. Other companies show the finished floor lines on the elevations and dimension floor-to-ceiling levels. Unless your instructor tells you otherwise, no dimensions will be placed on the elevations.

ALTERNATIVE ELEVATIONS

Many homes such as those that will be constructed in a subdivision are designed to be built more than once, and they require alternative facades. Figures 23–12 and 23–13 show

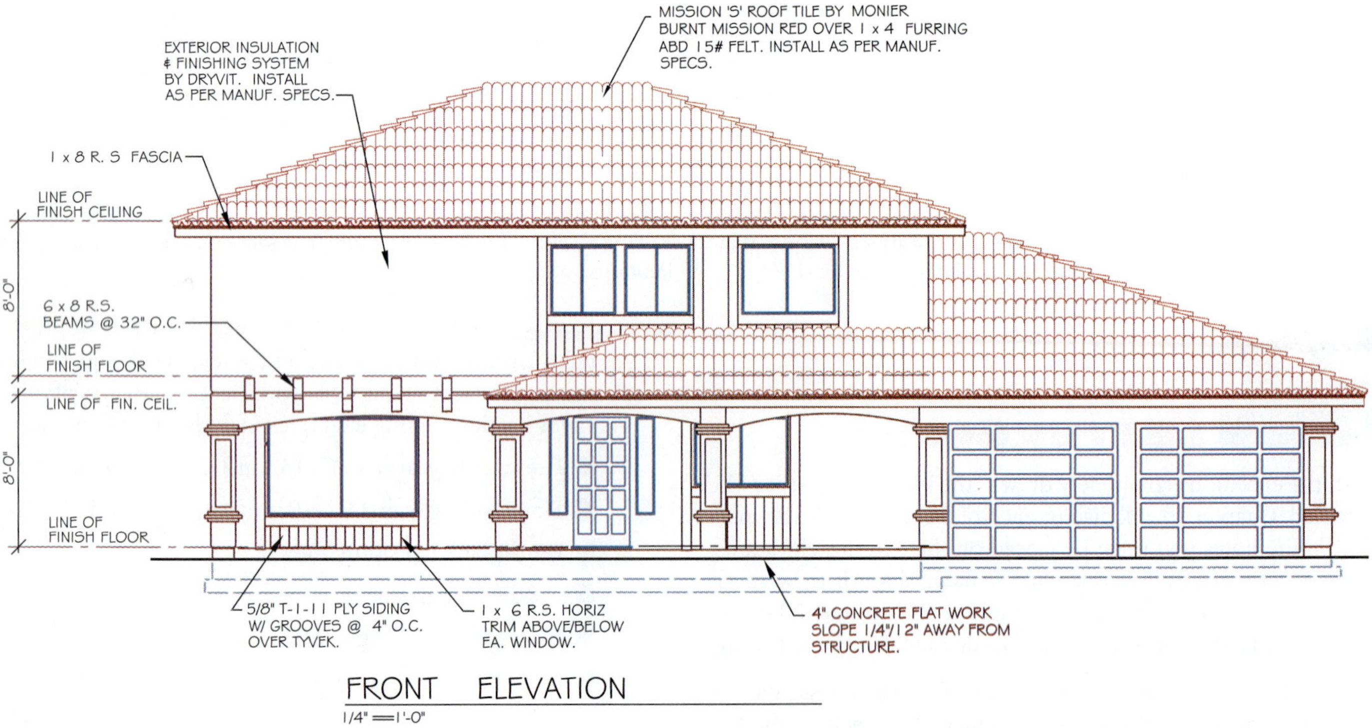

FIGURE 23.12 ■ An elevation with a hip roof can be drawn using the process described in Chapters 20 and 21. Locate the ridge height, assuming that a gable roof will be used. Once the ridge height is determined, the guidelines used to locate the ridge can be used to form the hips.

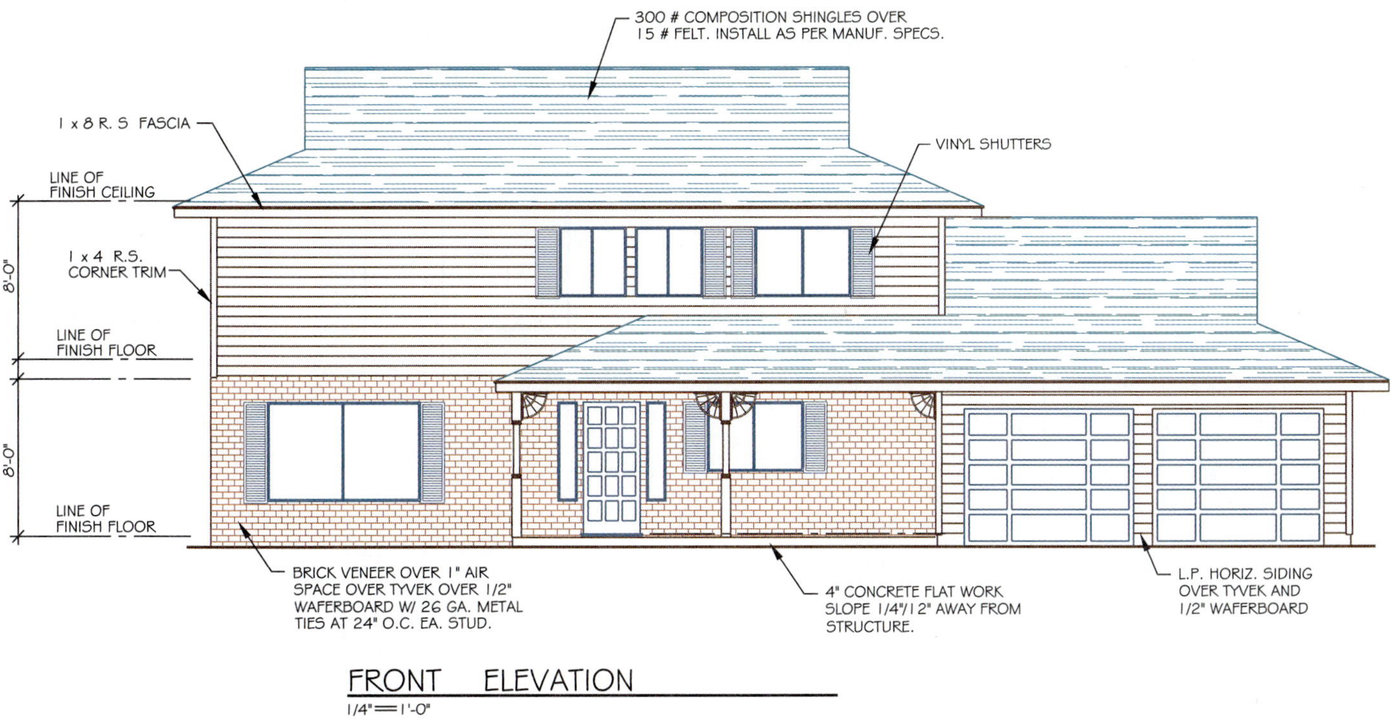

FIGURE 23.13 ■ An elevation with a Dutch hip roof can be drawn using the process described in Chapters 20 and 21. Locate the ridge height, assuming that a gable roof will be used. Once the ridge height is determined, the guideline used to locate the ridge can be used to form the hips. The location of the Dutch hip can be projected from the roof plan.

examples of alternative elevations that were drawn using the same step-by-step process as the plan in Figure 23.10. These plans could be completed by using the information for the floor and roof plans. If they are drawn using manual drafting, much of the information such as wall and opening locations can be traced from the original elevation. By using CADD, information common to all three drawings can be placed on a base layer, with information specific to each roof plan placed on separate layers. The elevations for the gable, hip, and Dutch hip roofs can be stored in one drawing file or separated for ease of plotting.

DRAWING ROOF INTERSECTIONS

The roof shape can be projected from the roof plan or from an overlay of the roof placed on the floor plan. Figures 23–4 through 23–7 provide examples for projecting a simple roof. Gable roofs often have irregular-size overhangs, overhangs of varied shapes, and altered rooflines and roof intersections that must be represented on the elevations. Figure 23.14 shows a small overhang at the gable end wall and a larger overhang where the truss or rafter tails are perpendicular to the support wall. The elevation was created by drawing the roof on the right side first to establish the incline of the roof. The intersection of the 1′ and 2′ overhangs was established by projecting their location in the roof plan down to the incline in the elevation. A similar process can be used by projecting heights from one elevation to another. Figure 23.15 shows a portion of a roof plan with a 12″ wide overhang at the gable end wall and a 24″ wide overhang. Once the inclined plane is drawn, the heights of the 1′ and 2′ overhangs can be projected into other elevations.

Projecting the intersection of the chimney with the roof is another matter that often frustrates new drafters. To project the true height of a chimney easily, the height must be drawn in an elevation where the angle of the roof can be seen. Figure 23.16 shows the process. The chimney must be 24″ (600 mm) above any portion of the structure that is within 10′ (3000 mm) of the chimney. The height is determined by measuring 10′ (3000 mm) from the center of the flue, and then measuring 24″ (600 mm) above this point. Once the height of the chimney above the roof has been determined, it can be projected into other elevations where the roof pitch can't be seen. Remember that the inner face of the chimney does not align with the outer face of the structure.

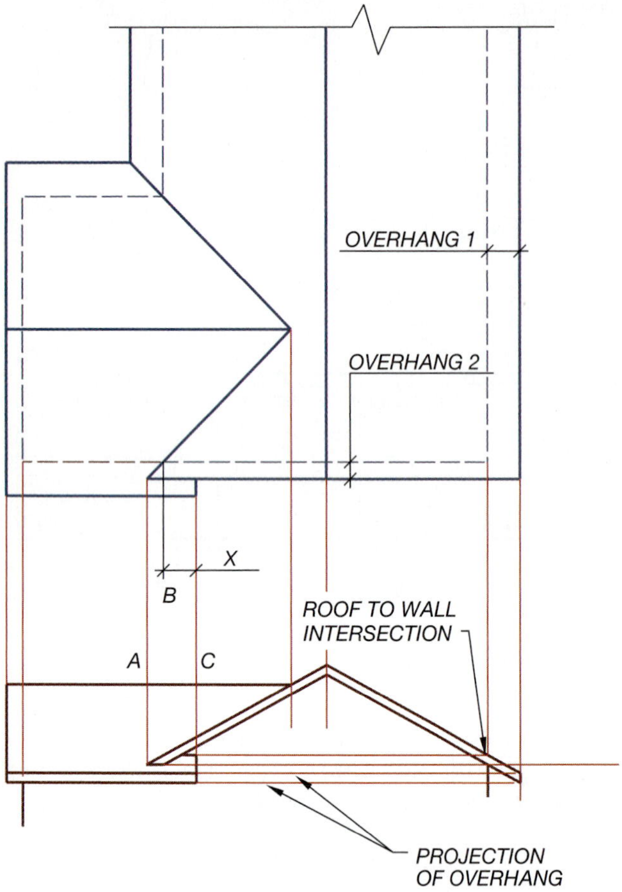

FIGURE 23.14 ■ The intersection of different-sized overhangs can be projected from the roof plan to the elevations. Once the incline of the roof is established, the small overhang (represented by line *A*) can be projected down until it intersects the roof angle. The distance that the lower roof extends past the upper roof (*A* to *C*) will vary, depending on the design. *X* represents the distance between where the roof intersects the outline of the wall and the end of the eave. This overhang (*X*) is usually equal to overhang 1 or 2.

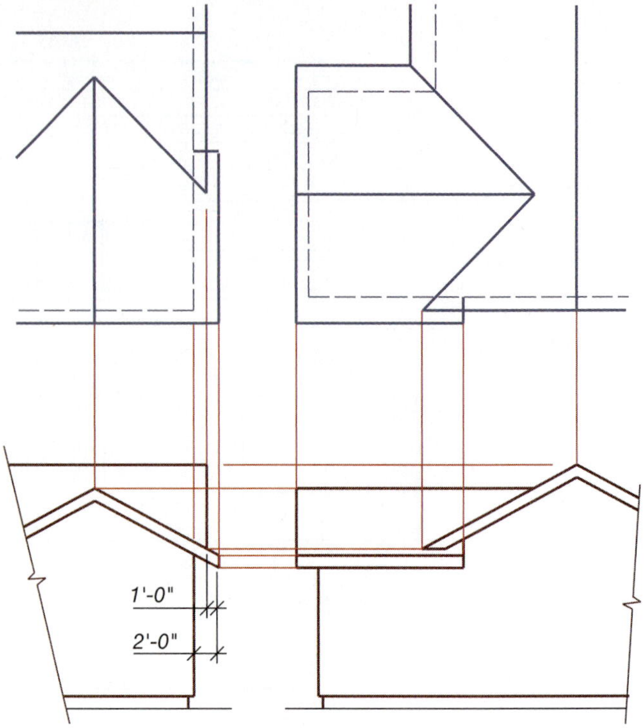

FIGURE 23.15 ■ The intersection of different-sized overhangs can be projected from one elevation to another.

Figure 23.17*a* shows a gable roof with a modified Dutch hip. The gable wall of a full Dutch hip roof is usually set back several feet from the wall of the structure. Figure 23.17*b* shows methods of projecting the overhangs into an elevation. A similar projection method is used to project eaves that appear to wrap around the corner of a structure, as in Figure 23.18*a*. The overhangs can be projected as shown in Figure 23.18*b*. Figure 23.19*a* shows a gable modified with a small hip. Figure 23.19*b* shows the process for projecting the hip into the elevation.

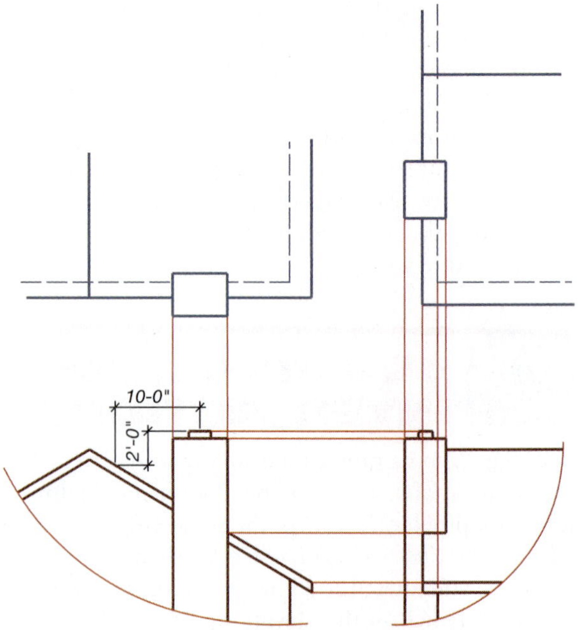

FIGURE 23.16 ■ The intersection of the chimney with the roof.

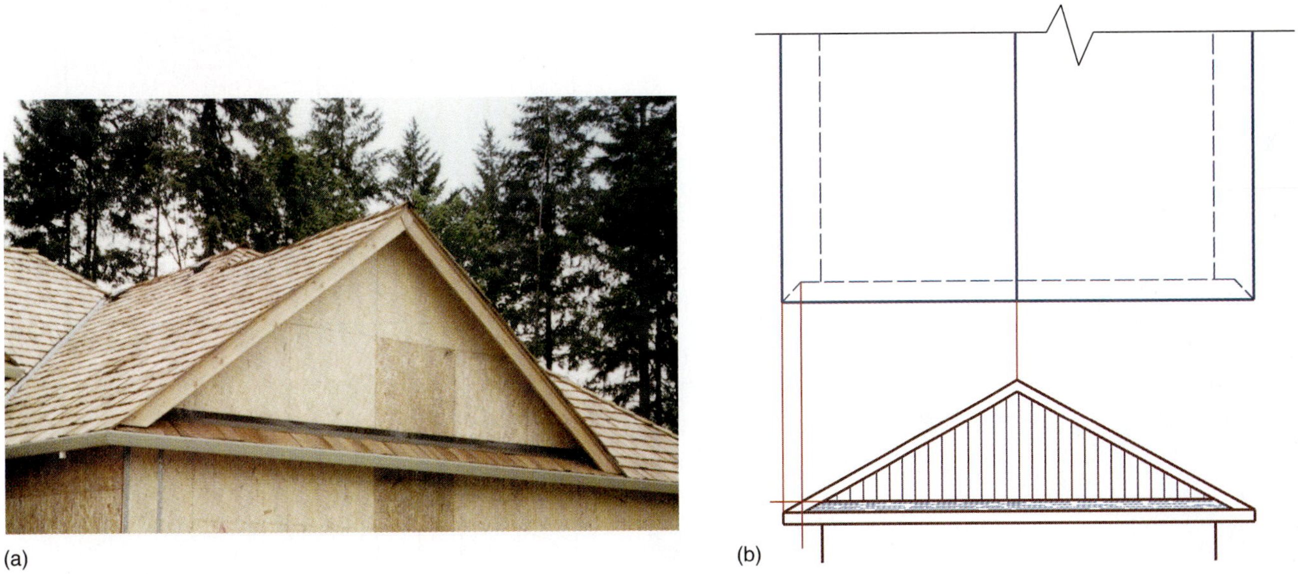

(a) (b)

FIGURE 23.17 ■ (a) Gable roof with an eave similar to a Dutch hip roof. (b) Projection of a gable roof with a continuous eave.

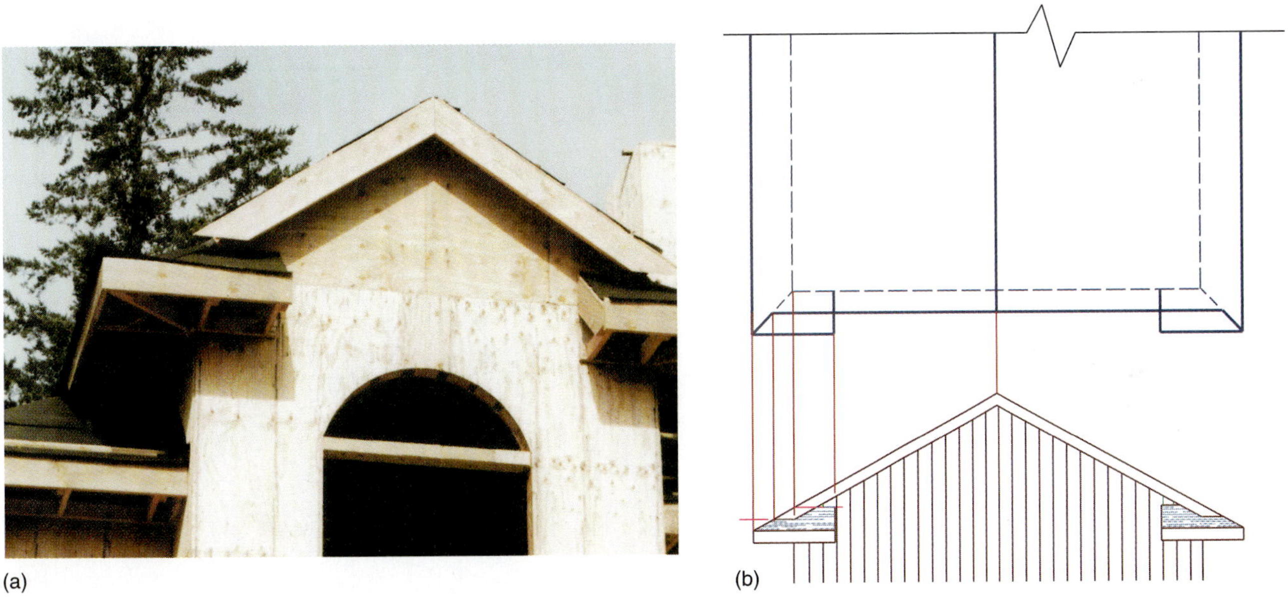

(a) (b)

FIGURE 23.18 ■ (a) Gable roof with corner eaves. (b) Representing corner eaves.

DRAWING IRREGULAR SHAPES

Not all plans that will be drawn have walls constructed at 90° to each other. When the house has an irregular shape, a floor plan and a roof plan will be required to draw the elevations. The process is similar to the process just described but will require more patience in projecting all of the roof and wall intersections. The projection of an irregularly shaped home can be seen in Figure 23.20. See Section 6 for an explanation of how the roof plan is drawn. Once the roof plan has been determined, it can be drawn on a print of the floor plan, and the intersections of the roof can be projected to the elevation.

PROJECTING GRADES

Grade is used to describe the shape of the finish ground. Projecting ground levels is required on elevations when a structure is being constructed on a hillside. Figure 23.21 shows the layout of a hillside residence. To complete this elevation a site, floor, and roof plan were used to project locations. Usually the grading plan and the floor plan are drawn at different scales. The easiest method of projecting the ground level onto the elevation is to measure the distance on the site plan from one contour line to the next. This distance can then be marked on the floor plan and projected onto the elevation.

(a)

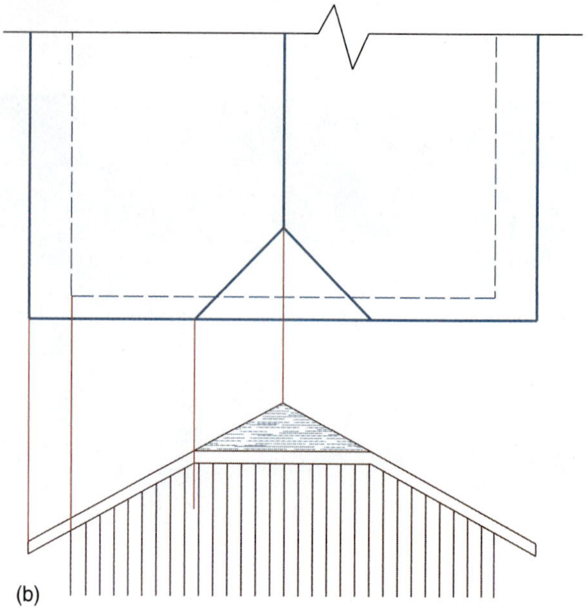

(b)

FIGURE 23.19 ■ (a) Gable roof with a partial hip. (b) Representing a gable roof with a partial hip.

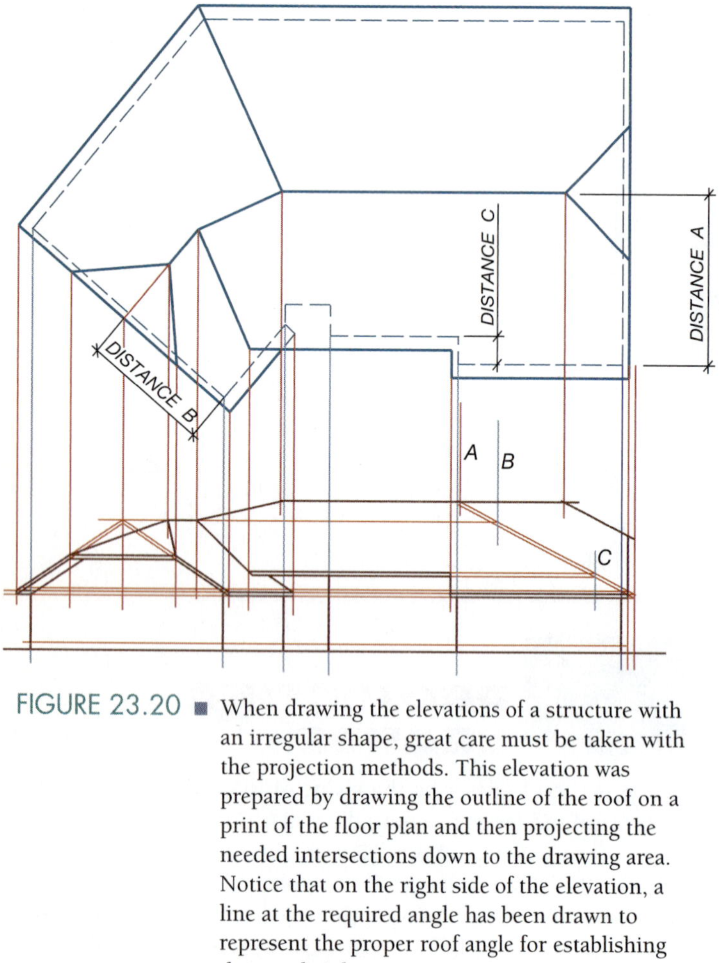

FIGURE 23.20 ■ When drawing the elevations of a structure with an irregular shape, great care must be taken with the projection methods. This elevation was prepared by drawing the outline of the roof on a print of the floor plan and then projecting the needed intersections down to the drawing area. Notice that on the right side of the elevation, a line at the required angle has been drawn to represent the proper roof angle for establishing the true heights.

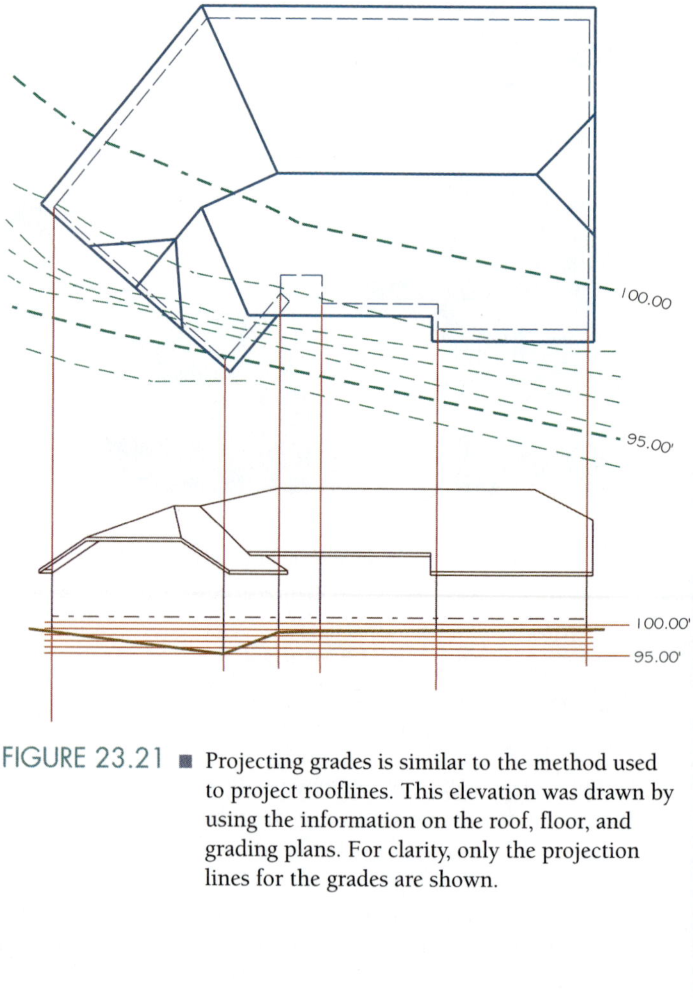

FIGURE 23.21 ■ Projecting grades is similar to the method used to project rooflines. This elevation was drawn by using the information on the roof, floor, and grading plans. For clarity, only the projection lines for the grades are shown.

DRAWING ELEVATIONS WITH CADD

Drawing exterior elevations with CADD is similar to techniques used to draw elevations manually. With manual drafting, you project lines from the floor plan to create lines for the elevation drawing. This was shown in Figure 23.1. With CADD, the same thing is done using a computer.

Create layers for the elevations with prefixes that clearly identify the contents. Layer titles such as ELEVWALLS, ELEVLAYOUT, ELEVROOF, ELEVWIN, ELEVSIDING, or ELEVTEXT will be useful for keeping track of features as the elevations are created.

After the floor plan is drawn, freeze all layers that are not relevant to the exterior of the residence. Make a *block* of the floor plan. A block is a custom symbol or drawing that can be transferred between drawings and displayed at any time. Set the drawing limits large enough so that the block of the floor can be inserted and rotated; in this way, each elevation can be projected from the floor plan using techniques similar to manual methods (Figure 23.22).

Once the roof and wall locations have been projected, the locations of doors and windows can be determined. Windows and doors representing specific styles can be created and stored as blocks in a library and then inserted as needed. Trees and bushes can also be created as blocks and inserted as needed. Once the design is completed, the elevations can be moved into the desired position for presentation and plotting. Figure 23.23 shows an example of elevations created using AutoCAD.

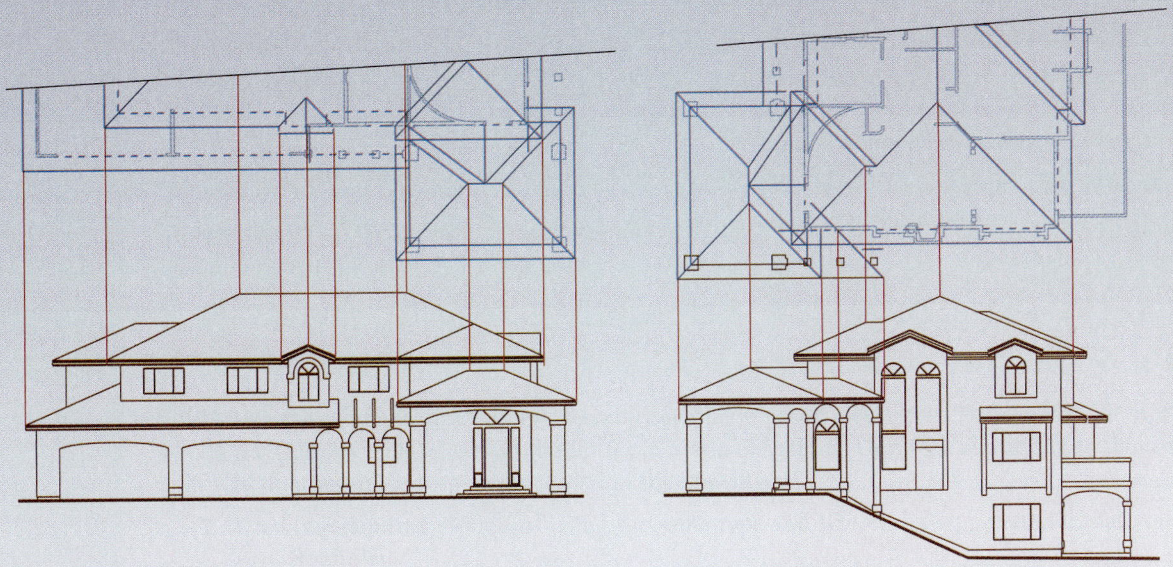

FIGURE 23.22 ■ Use a block of the relevant parts of the floor and roof plans to line out each of the exterior elevations. Once the shapes have been projected, doors, windows, and other elements can be added. The completed elevations can be moved into any position for plotting. *Courtesy of Residential Designs.*

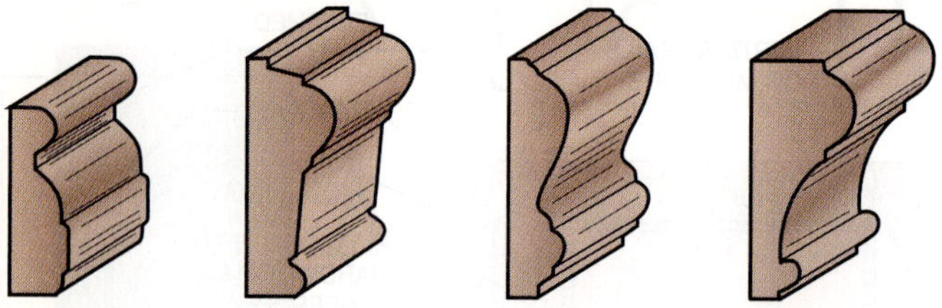

FIGURE 24.7 ■ Standard chair rails. *Courtesy Hillsdale Pozzi Co.*

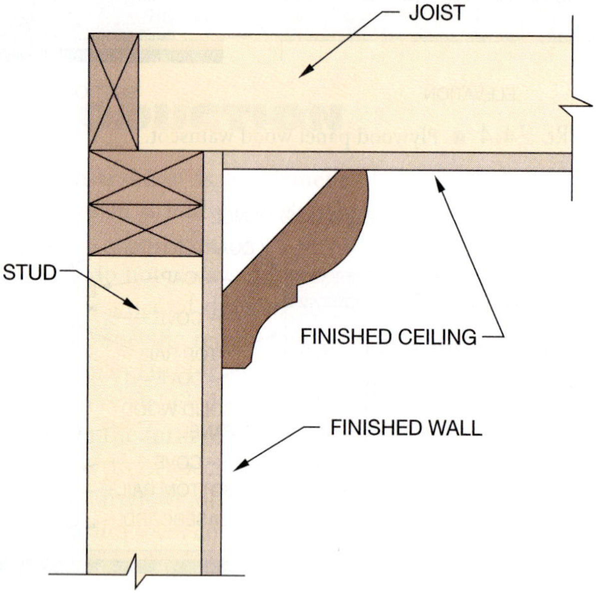

FIGURE 24.8 ■ Individual-piece cornice.

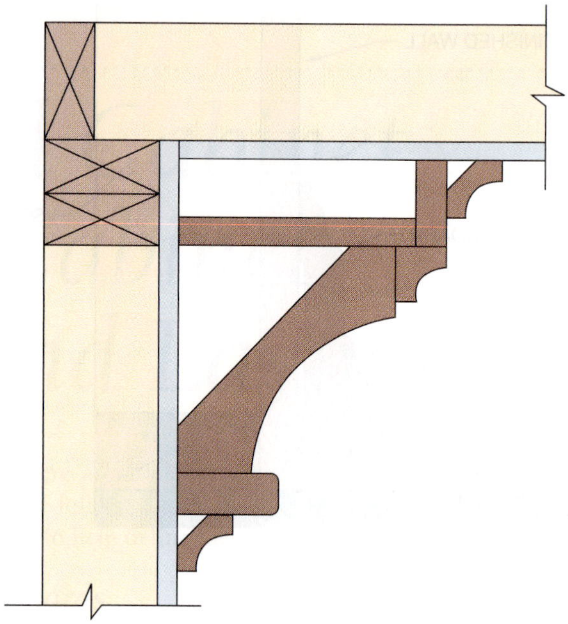

FIGURE 24.9 ■ Multipiece cornice.

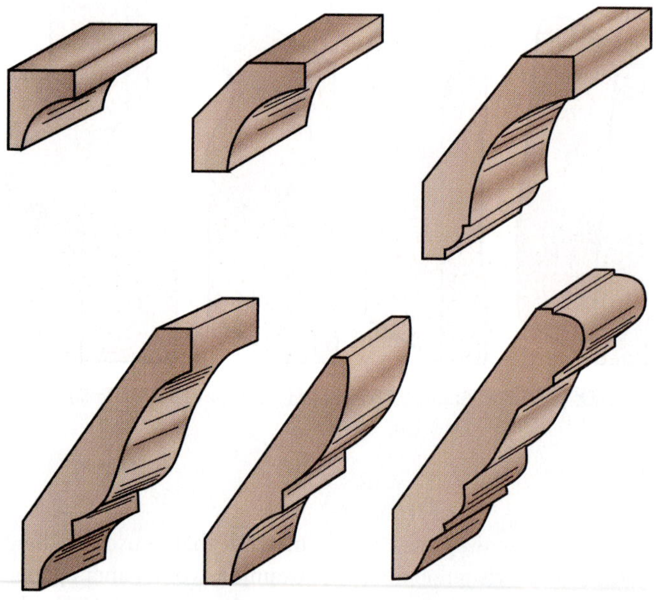

FIGURE 24.10 ■ Standard cornice (cove and ceiling) moldings. *Courtesy Hillsdale Sash and Door Co.*

rail is an excellent division between two different materials or wall textures. Figure 24.7 shows sample chair rail moldings.

Cornice

The *cornice* is decorative trim placed in the corner where the wall meets the ceiling. A cornice may be a single shaped wood member called *cove* or *crown* molding, as seen in Figure 24.8; or a more elaborate structure made up of several individual wood members, as shown in Figure 24.9. Cornice boards are not commonly used; they traditionally fit into specific types of architectural styles, such as English Tudor, Victorian, or colonial. Figure 24.10 shows some standard cornice moldings. In most construction, where contemporary architecture or cost saving is important, wall-to-ceiling corners are left square.

Casings

Casings are the members that are used to trim around windows and doors. Casings are attached to the window or doorjamb (frame) and to the adjacent wall, as shown in Figure 24.11. Casings may be decorative to match other moldings or plain to serve the functional purpose of covering the space between the doorjamb or window jamb and the wall. Figure 24.12 shows a variety of standard casings.

Mantels

The *mantel* is an ornamental shelf or structure that is built above a fireplace opening, as seen in Figure 24.13. Mantel designs vary with individual preference. Mantels may be made of masonry as part of the fireplace structure, or ornate decorative wood moldings, or even a rough-sawn length of lumber bolted

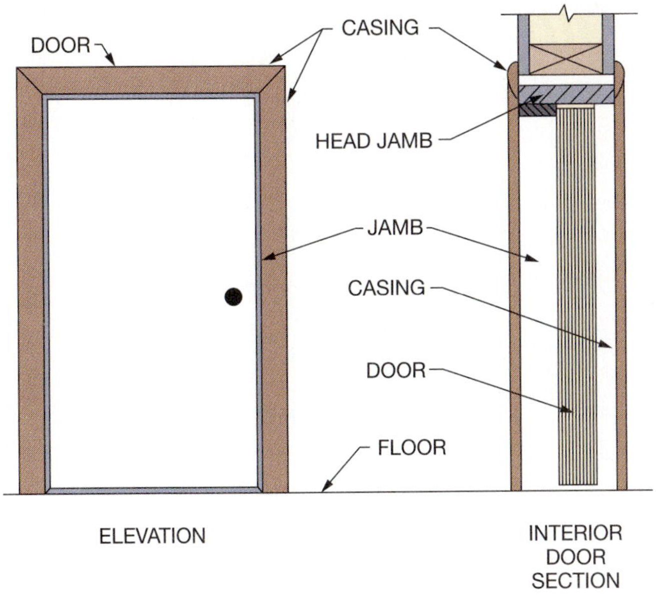

FIGURE 24.11 ■ Casings.

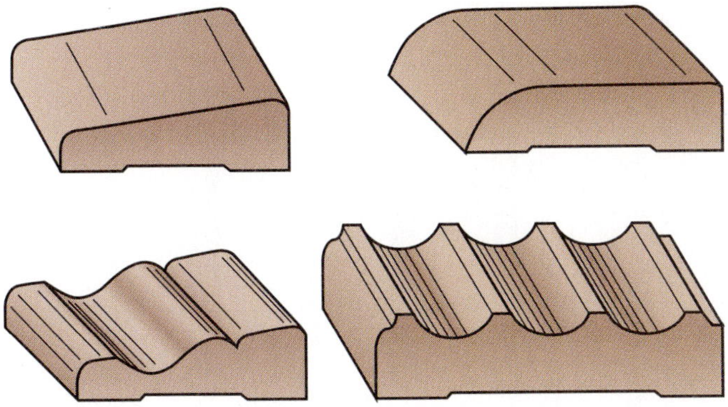

FIGURE 24.12 ■ Standard casings. *Courtesy Hillsdale Sash and Door Co.*

FIGURE 24.13 ■ Mantel. *Courtesy Pella Corporation.*

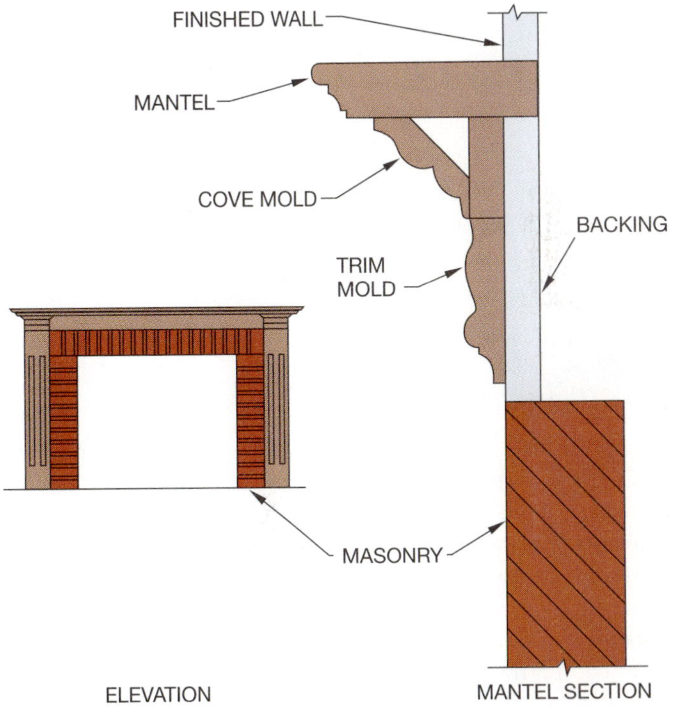

FIGURE 24.14 ■ Traditional mantel.

to the fireplace face. Figure 24.14 shows a traditional mantel application.

Bookshelves

Bookshelves and display shelves may have simple construction with metal brackets and metal or wood shelving, or they may be built detailed as fine furniture. Bookshelves are commonly found in the den, library, office, or living room. Shelves that are designed to display items other than books are also found in almost any room of the house. A common application is placing shelves on each side of a fireplace, as in Figure 24.13. Shelves are also used for functional purposes in storage rooms, linen closets, laundry rooms, and any other location where additional storage is needed.

Railings

Railings are used for safety at stairs, landings, decks, and open balconies where people could fall. Rails are recommended at any rise that measures 24″ (610 mm) or is three or more stair risers. Verify the size with local codes. Railings may also be used as decorative room dividers or for special accents. Railings may be built enclosed or open and constructed of wood or metal. Enclosed rails are often the least expensive to build because they require less detailed labor than open rails. A decorative wood cap is typically used to trim the top of enclosed railings. See Figure 24.15.

Open railings can be one of the most attractive elements of interior design. They may be as detailed as the designer's or craftsperson's imagination. Detailed open railings built of exotic hardwoods can be one of the most expensive items in the structure. Figure 24.16 shows some standard railing components. *Newels* are the posts that support the rail at the ends and at landings. *Balusters* are placed between the newels.

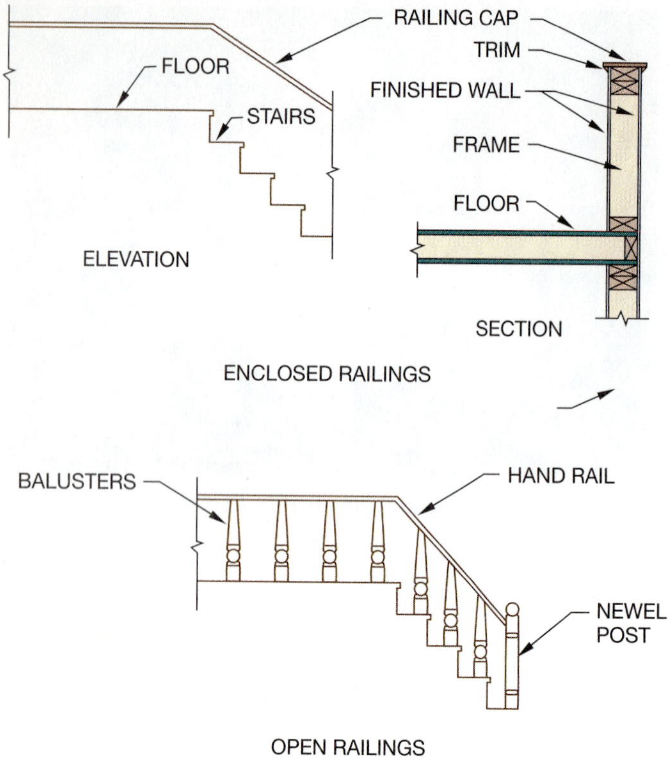

FIGURE 24.15 ■ Railings.

CABINETS

One of the most important items for buyers of a new home is the cabinets. The quality of cabinetry can vary greatly. Cabinet designs may reflect individual taste, and a variety of styles are available. Cabinets are used for storage and as furniture. The most obvious locations for cabinets are the kitchen and bath.

The design and arrangement of kitchen cabinets have been the object of numerous studies. Over the years design ideas have resulted in attractive and functional kitchens. Kitchen cabinets have two basic elements: the base cabinet and the upper cabinet. See Figure 24.17. Drawers, shelves, cutting boards, pantries, and appliance locations must be carefully considered.

Bathroom cabinets are called vanities, linen cabinets, and medicine cabinets. See Figure 24.18. Other cabinetry may be found throughout a house, as in utility or laundry rooms, for storage. For example, a storage cabinet above a washer and dryer is common.

Cabinet Types

There are as many cabinet styles and designs as the individual can imagine. However, there are only two general types of cabinets, based on their method of construction. These are modular, or prefabricated, cabinets; and custom cabinets.

Modular Cabinets

The term *modular* refers to prefabricated cabinets because they are constructed in specific sizes called modules. Modular cabinets are best used when a group of modules can be placed side by side in a given space. If a little more space is available, then pieces of wood, called *filler,* are spliced between modules. Many brands of modular cabinets are well crafted and very attractive. Most modular cabinet vendors offer different door styles, wood species, or finish colors. Modular cabinets are sized in relationship to standard or typical applications, although many modular

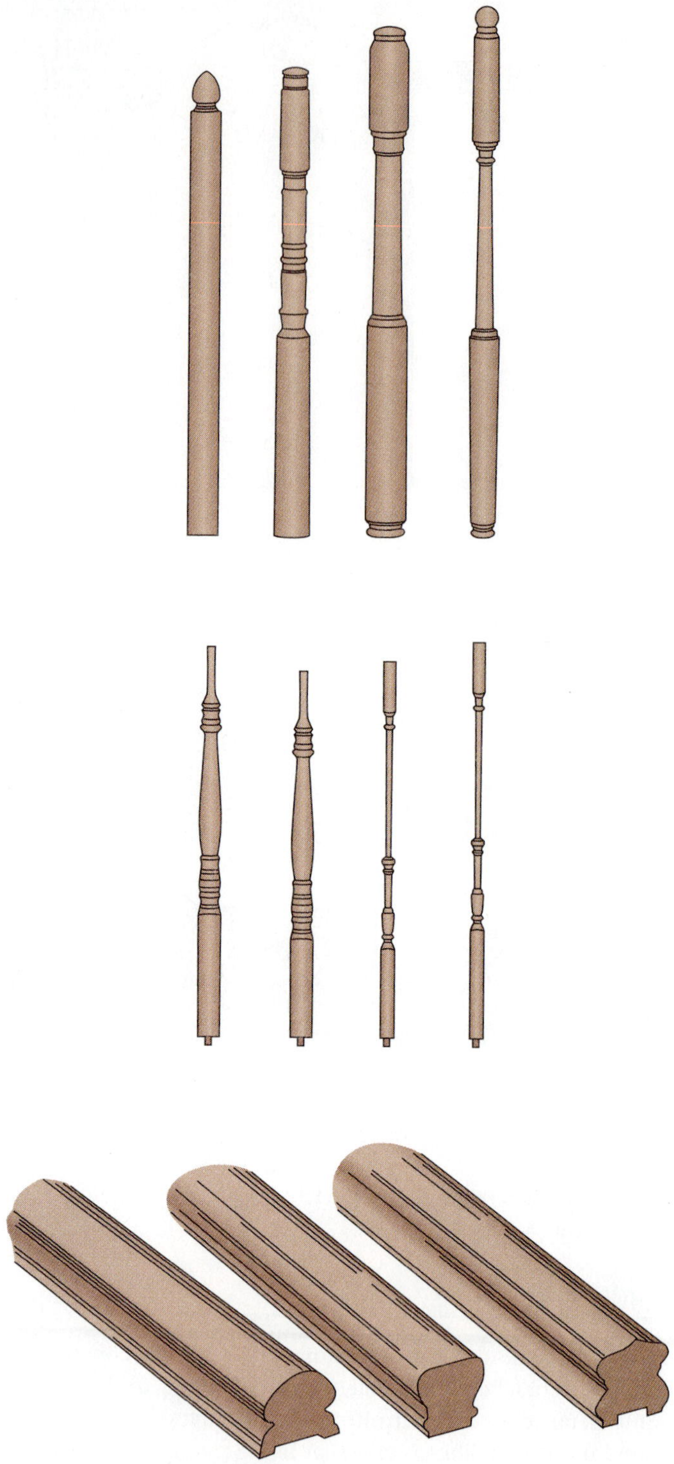

FIGURE 24.16 ■ Examples of standard railing components. *Courtesy Hillsdale Pozzi Co.*

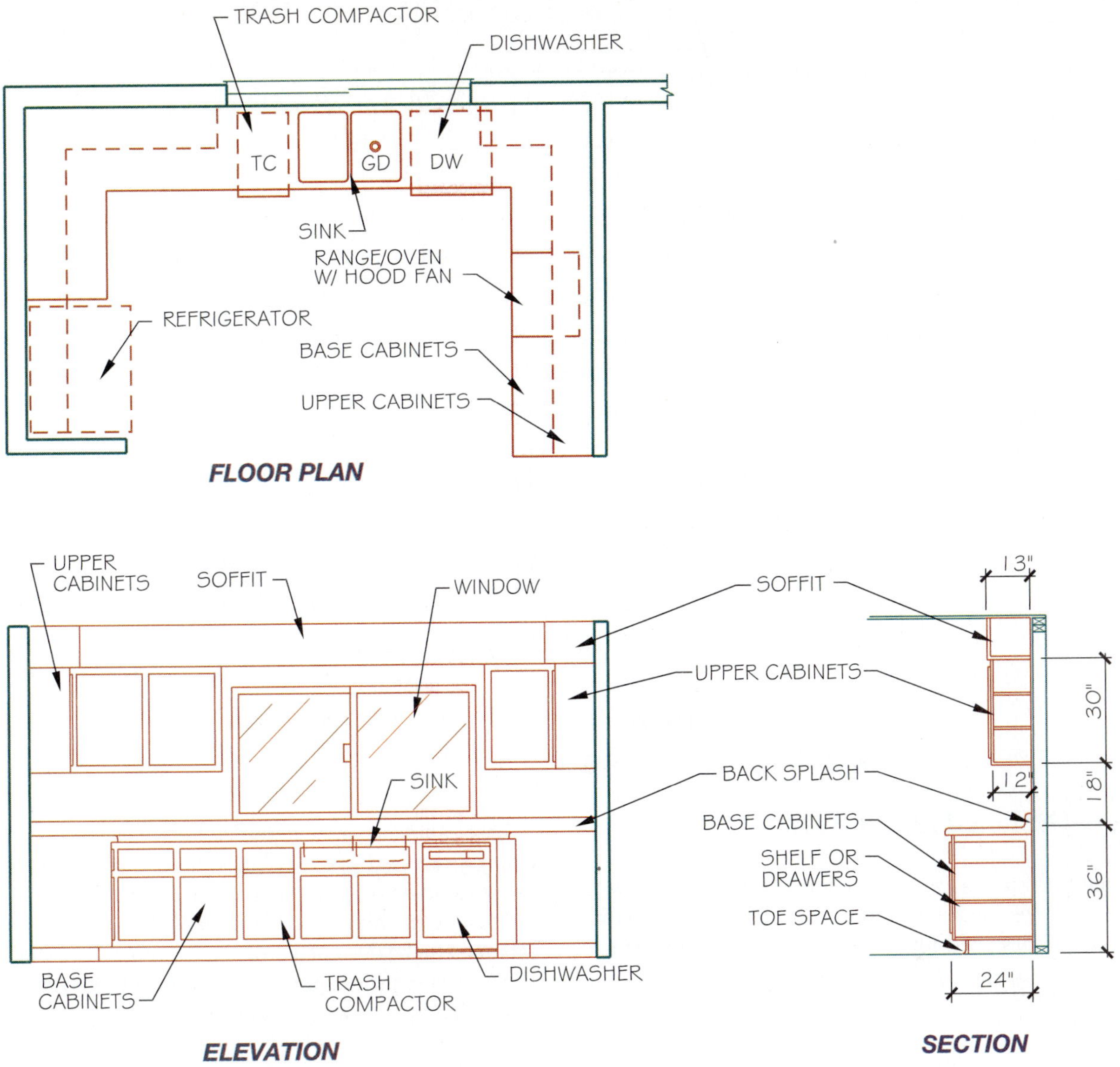

FIGURE 24.17 ■ Components of kitchen cabinets.

cabinet manufacturers can make components that will fit nearly every design situation. Modular cabinets are often manufactured nationally and delivered in modules. See Figure 24.19.

Custom Cabinets

The word *custom* means "made to order." Custom cabinets are generally fabricated locally in a shop to the specifications of an architect or designer, and after construction they are delivered to the job site and installed in large sections.

One of the advantages of custom cabinets is that their design is limited only by the imagination of the architect and the cabinet shop. Custom cabinets may be built for any situation, such as for any height, space, type of exotic hardwood, type of hardware, on geometric shape, or for any other design criteria.

Custom cabinet shops estimate a price based on the architectural drawings. The actual cabinets are constructed from measurements taken at the job site after the building has been framed. In most cases, if cabinets are ordered as soon as framing is completed, the cabinets should be delivered on time.

Cabinet Options

Cabinet designs available from either custom or modular cabinet manufacturers include the following:

- Various door styles, materials, and finishes.
- Self-closing hinges.
- A variety of drawer slides, rollers, and hardware.
- Glass cabinet fronts for a more traditional appearance.
- Wooden range hoods.
- Specially designed pantries, appliance hutches, and a lazy Susan or other corner cabinet design for efficient storage.
- Bath, linen storage, and kitchen specialties.

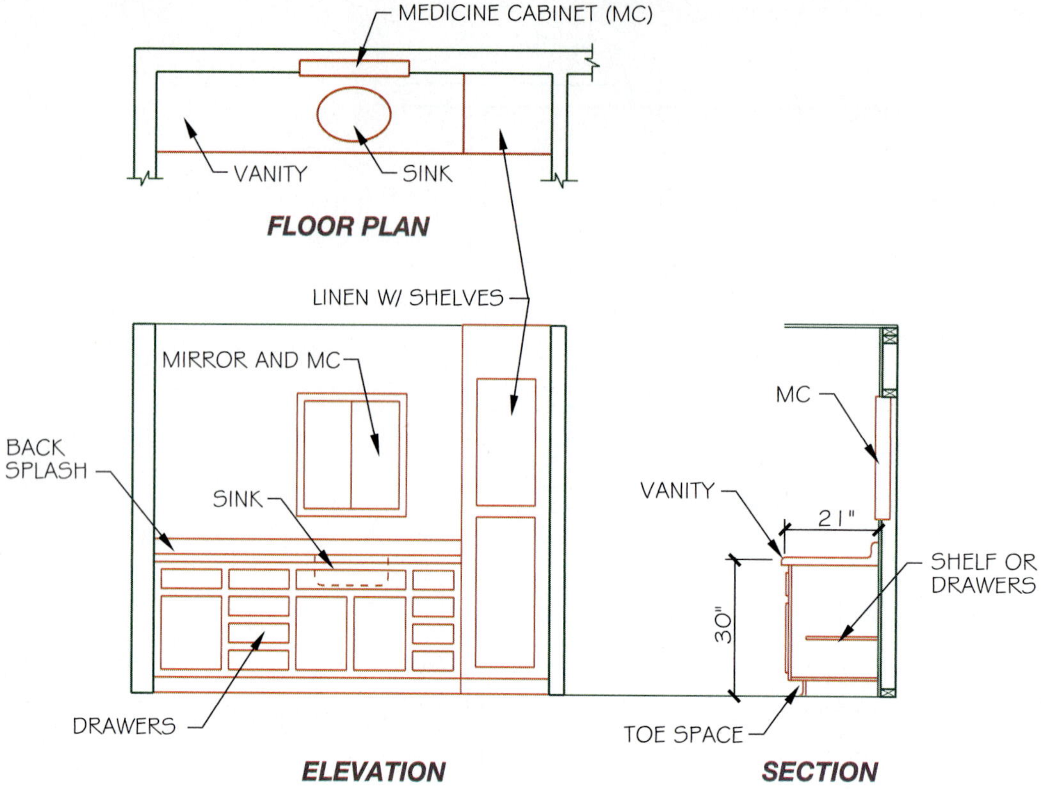

FIGURE 24.18 ■ Components of bathroom cabinets.

Cabinets Designed for Universal Use

Kitchen and bathroom access can be improved for people with disabilities by applying a few design considerations. Whether the cabinets are designed for wheelchair access or for people who need to sit, the same possibilities exist.

Kitchen Design

The kitchen cabinets can be designed specifically for easy access, or they can be adjustable or removable for future needs. A versatile kitchen has at least one counter at a height of 32″ (800 mm) for convenient wheelchair access. A lower cabinet or pullout shelf might be considered for heavy work that requires leaning over or down. The standard base cabinet height is 36″ (900 mm), but a height of 42″ (1060 mm) can be considered for very tall people. The range, sink, and workspace next to the refrigerator should be connected and on the same level to make it easier for people to slide heavy items from one workstation to the next. This design is possible in a straight, L-shaped, or U-shaped kitchen. The standard kitchen shape can be modified with angled or curved corners to provide easier access to the corners. Placing the sink between the refrigerator and the cooking center allows for the most convenient food preparation. This may be altered on the basis of the owner's preference.

Some designs have two kitchens when both standing and wheelchair-using members of the same family actively do work. A second kitchen can be added later if enough space and rough plumbing and electrical work are considered during the design process. This second kitchen might become useful at a later time in life or if there is a need to care for aging family members.

Base cabinets can be designed for wheelchair access by providing a recessed space 27″ (686 mm) high under the cabinet. These cabinets should also have a larger-than-normal toe space 10″ (254 mm) high by 6″ (152 mm) deep. Some of the upper cabinets can be designed down to the top of base cabinets. This allows access up to approximately 46″ (1168 mm) for wheelchair use. These lower upper cabinets can be installed without doors for easier access to the contents. The clear space in front of cabinets should be at least 48″ (1220 mm), but more is preferred.

The eating area should be next to the kitchen for convenience, or provide a passageway between the kitchen and dining room. Part of the kitchen cabinet can be movable to allow for transporting items to and from the eating area.

Additional storage is a consideration in designing a kitchen for people who cannot go shopping frequently. Storage should also be compartmentalized for convenient organization, and provide a lazy Susan in the corners for access to under-counter items. Wire racks placed on the back of doors are accessible, and wire shelves and drawers make it easier to see the contents. Pantries are beneficial for storage in any kitchen design. When full-extension pullout shelves are added to the pantry, items at the back of the shelves become easy to reach. All kitchen drawers should have full-extension guides for easy access. Touch latches on cabinet doors allow the doors to be opened easily. For people in wheelchairs, sliding cabinet doors allow easier access to the cabinet than swinging doors.

Many modern appliances are designed with easy-to-see and easy-to-use controls near the front. Double-door refrigerators provide easy access to the inside, and one door with ice and water dispensers can be specified.

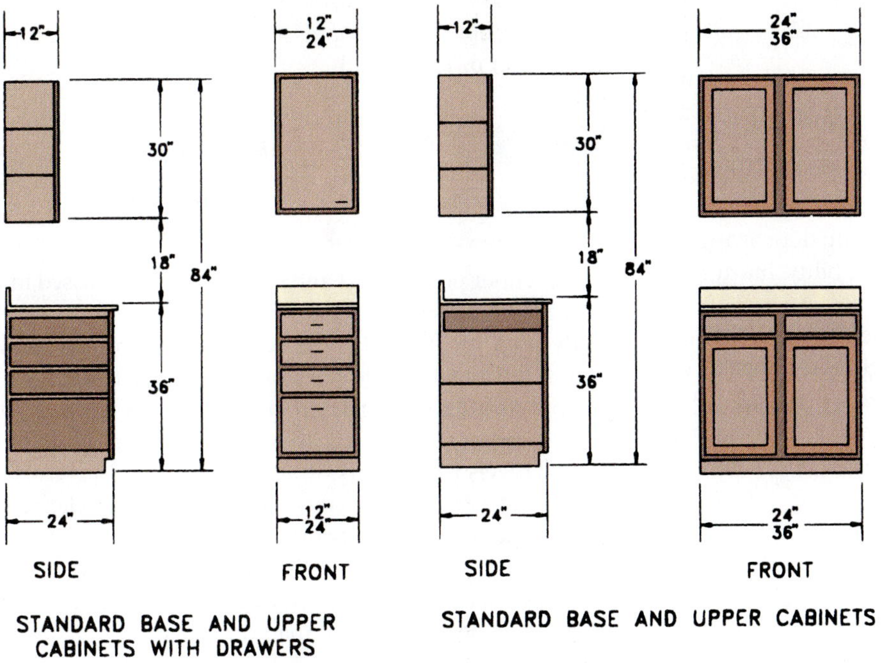

METRIC CONVERSION	
INCHES	MILLIMETERS
12	300
15	380
18	450
24	600
30	760
36	900
48	1200
72	1800
84	2200

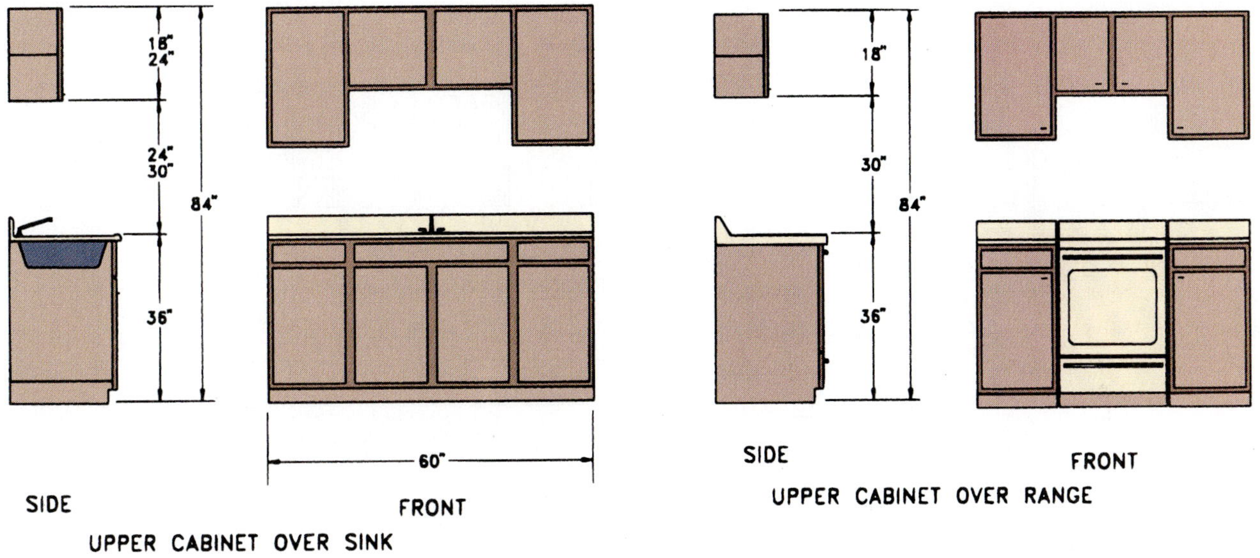

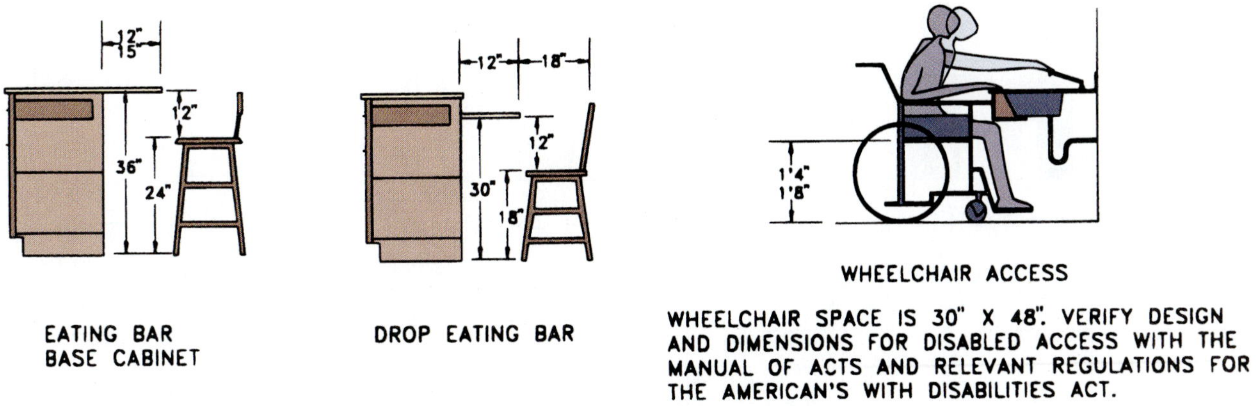

FIGURE 24.19 ■ Standard cabinet dimensions. Note that all metric equivalents are soft conversions.

Bathroom Design

The bathroom vanity can be designed with a recessed space underneath for sitting or wheelchair access. The vanity cabinet height should be 28-32″ (710-860 mm). Other cabinet design features that were discussed in the previous recommendations about kitchen design can be considered: for example, the toe kick space, storage options, drawers guides, and door swings. A wall-mounted or pedestal sink is also a possibility, but it does not provide counter space.

Grab bars should be placed to assist movement from the vanity to the shower and the water closet. The grab bars must be designed and installed to support the weight of a person.

Benches can be designed in the bathroom in locations where sitting or transferring from a wheelchair may be convenient.

Laundry Room Design

Laundry rooms designed with base cabinets next to the clothes dryer make folding convenient. With regard to height and cabinet recessing, the same considerations as previously discussed apply. Foldout ironing boards are available that can be installed in a drawer or in the wall.

CABINET ELEVATIONS AND LAYOUT

Cabinet floor plan layouts and symbols were discussed in Chapter 14. Cabinet elevations or exterior views are developed directly from the floor-plan drawings. The purpose of the elevations is to show how the exterior of the cabinets will look when completed and to give general dimensions, notes, and specifications. Cabinet elevations may be as detailed as the architect or designer feels is necessary.

In Figure 24.20 the cabinet elevations are very clear and well done without detail or artwork that is not specifically necessary to construct the cabinets. In Figure 24.21 the cabinet elevations are drawn more artistically. Neither set of cabinet drawings is necessarily more functional than the other.

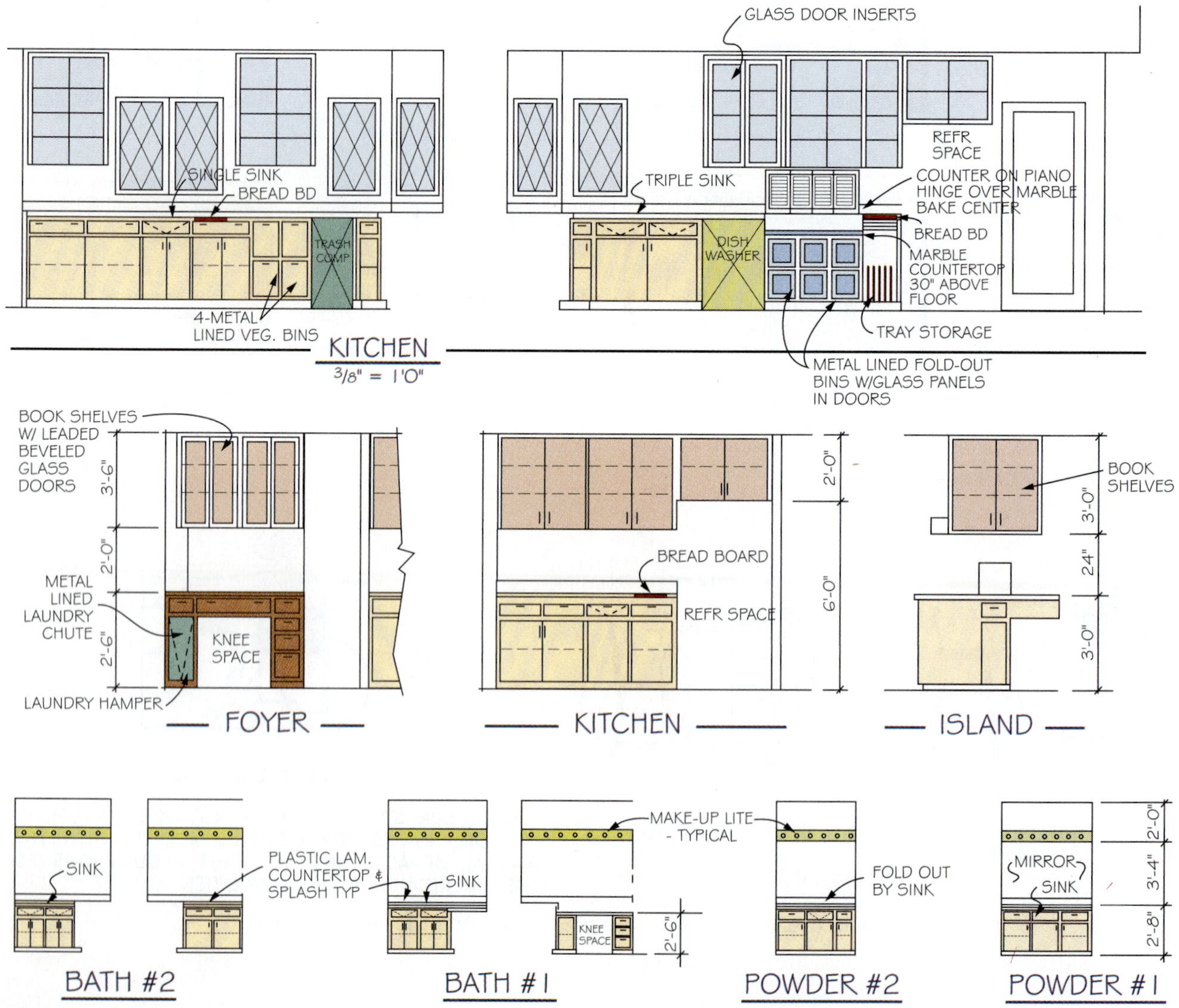

FIGURE 24.20 ■ Simplified cabinet elevations. *Courtesy Residential Designs, AIBD.*

FIGURE 24.21 ■ Detailed cabinet elevations. *Courtesy Madsen Designs.*

Keying Cabinet Eleations to Floor Plans

Several methods may be used to key the cabinet elevations to the floor plan. In Figure 24.20, the designer keyed the cabinet elevations to the floor plans with room titles such as KITCHEN ELEVATION or BATH ELEVATION. In Figure 24.21, the designer used an arrow with a letter inside to correlate the elevation to the floor plan. For example, in Figure 24.22 the E and F arrows pointing to the vanities are keyed to letters E and F that appear below the vanity elevation in Figure 24.21.

In the commercial drawing shown in Figure 24.23, notice that the floor plan has identification numbers within a combined circle and arrow. The symbols, similar to the one shown in Figure 24.24, are pointing to various areas that correlate

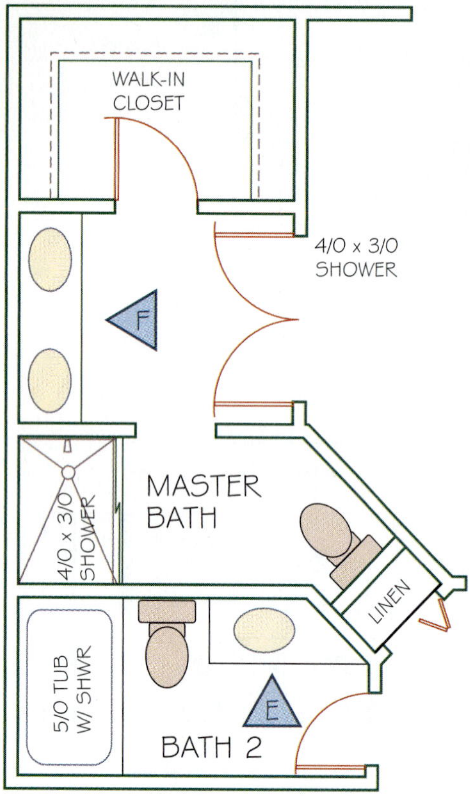

FIGURE 24.22 ■ Using the floor plan and elevation cabinet location symbol. *Courtesy Piercy & Barclay Designers, Inc. CPBD.*

with the same symbols labeled below the related elevations. A similar method may be used to relate cabinet construction details to the elevations. The detail identification symbol shown in Figure 24.25 is used to correlate a detail with the location from which it originated on the elevation. In some plans, small areas such as women's and men's rest rooms have interior elevations without specific orientation to the floor plan. The reason is that the relationship of floor plan to elevations is relatively obvious.

Various degrees of detailed representation may be needed for millwork, depending on its complexity and the design specifications. The sheet from a set of architectural working drawings shown in Figure 24.26 is an example of how millwork drawings can provide very clear and precise construction techniques.

Cabinet Layout

Cabinet layout is possible after the floor-plan drawings are complete. The floor-plan cabinet drawings are used to establish the cabinet elevation lengths. Cabinet elevations may be projected directly from the floor plan, or dimensions may be transferred from the floor plan to the elevations with a scale or dividers.

Kitchen Cabinet Layout

The cabinet elevations are drawn as though you were standing in the room looking directly at the cabinets. Cabinet elevations are two dimensional. Height and length are shown in an external view, and depth is shown where cabinets are cut with a cutting plane that is the line of sight. See Figure 24.27. For the kitchen shown in Figure 24.27 there are two elevations: one looking at the sink and dishwasher cabinet group, the other viewing the range, refrigerator, and adjacent cabinets.

Given the floor-plan drawing shown in Figure 24.27, draw the cabinet elevations as follows:

STEP 1 Make a copy of the cabinet floor-plan area, and tape the copy to the drafting table above the area where the elevations are to be drawn. Transfer length dimensions from the floor plan onto the blank drafting surface in the location where the desired elevation is to be drawn using construction lines, as shown in Figure 24.28. Some drafters prefer to transfer dimensions from the floor-plan drawings with a scale or dividers. As you gain experience, you will probably prefer to transfer dimensions. Notice that when the elevations are transferred from the floor plan, the scale is the same as the floor plan, 1/4″ = 1′–0″ (1:50 metric). Other scales are also used, such as 3/8″ = 1′–0″ or 1/2″ = 1′–0″, when additional clarity is needed to show small detail. When the cabinet elevation scale is different from the floor-plan scale, the dimensions may not be projected. In this case the dimensions are established from the floor plan and a scale change is implemented before transfer to the elevation drawing. In using CADD, the cabinets may be projected directly from the floor plan on their own layer. After drawing, the floor-plan layer may be turned off, leaving only the cabinet drawings on the screen. The cabinets may then be moved or scaled as needed.

STEP 2 Establish the room height from floor to ceiling. If the floor-to-ceiling height is 8′, then draw two lines 8′ apart with construction lines using the scale 1/4″ = 1′–0″. See Figure 24.29.

STEP 3 Establish the portion of the cabinetry where the line of sight cuts through base and upper cabinets and soffits. Use construction lines to draw lightly in these areas, or project on a construction layer in CADD, as shown in Figure 24.30.

STEP 4 Use construction lines to draw cabinet elevation features such as doors, drawers, and appliances. See Figure 24.31. Also draw background features, if any, such as backsplash, windows, and doors. Notice the rectangular shape that is drawn to the left of the dishwasher and below the countertop where the sink would be. This is a *false front,* designed to look like a drawer. An actual drawer cannot be placed here, as it would run into the sink. A false drawer front is used to establish

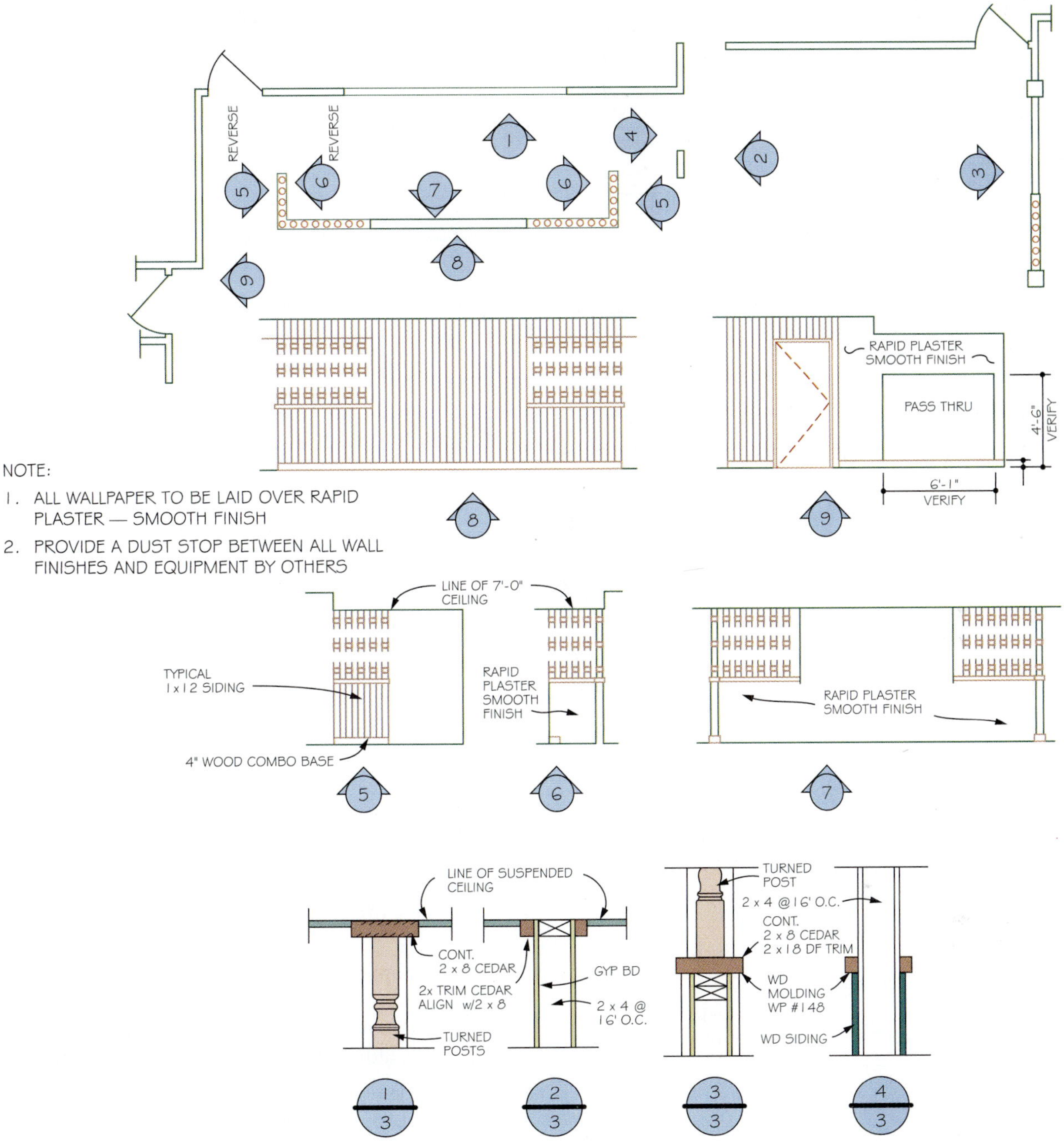

FIGURE 24.23 ■ Commercial floor plan with cabinet identification and location symbols. *Courtesy Ken Smith Structureform Masters, Inc.*

FIGURE 24.24 ■ Floor plan and elevation cabinet identification and location symbol.

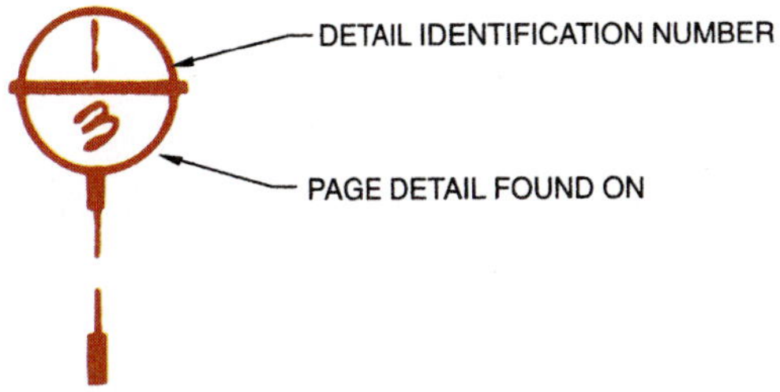

FIGURE 24.25 ■ Detail identification symbol.

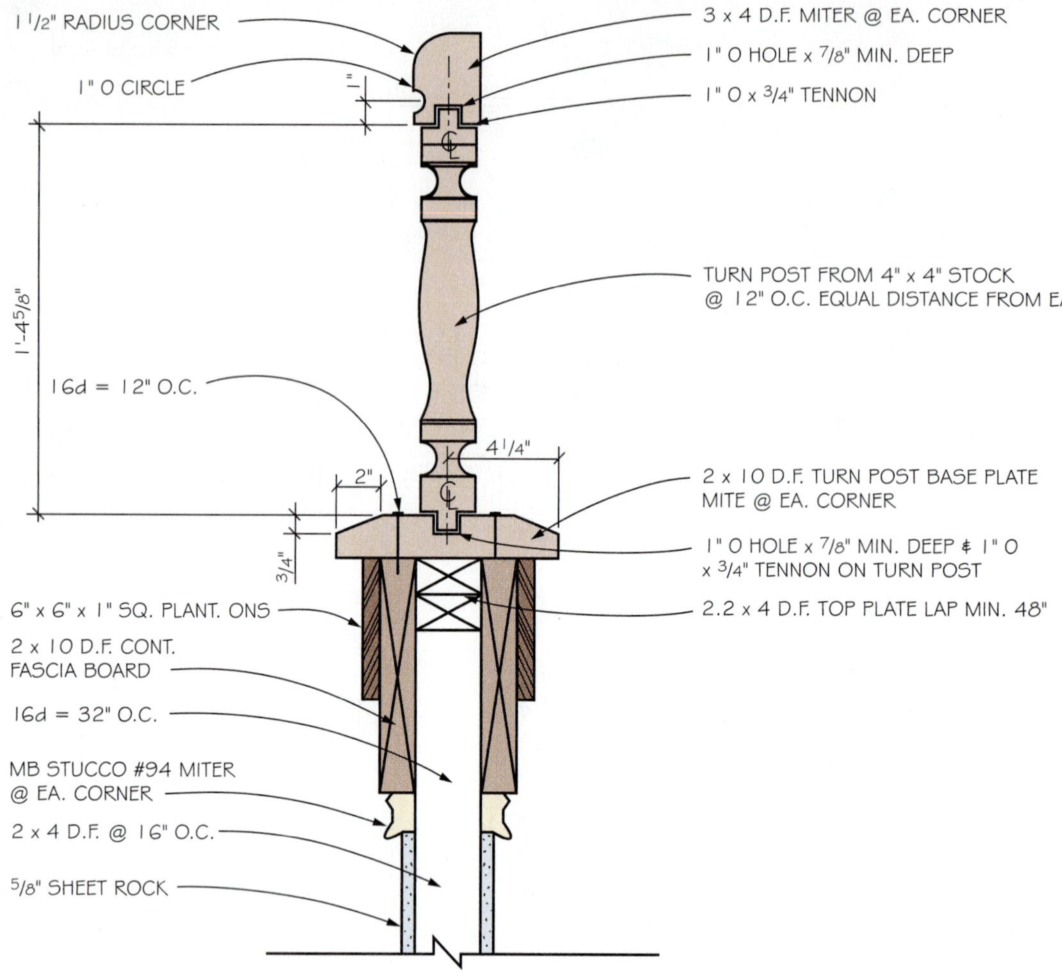

FIGURE 24.26 ■ Millwork and cabinet construction details. *Courtesy Structureform Masters, Inc.*

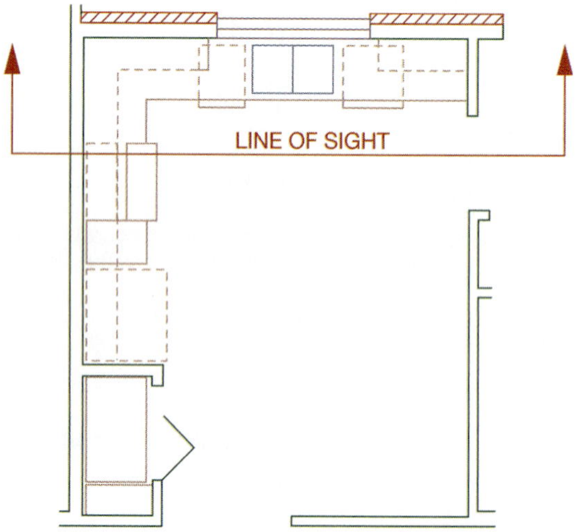

FIGURE 24.27 ■ Establishing the line of sight for cabinet elevations.

the appearance of a line of drawers just below the base cabinet top and before doors are used. Some designs use cabinet doors from the toe kick to the base cabinet top, although this method is not as common.

STEP 5 Darken all cabinet lines. Some drafters draw the outline of the ends of the cabinets and surrounding areas with a thick line. Draw dashed lines that represent hidden features, such as sinks and shelves, as seen in Figure 24.32.

STEP 6 Add dimensions and notes. See Figure 24.32. The dimensions shown in Figure 24.32 are given in feet and inches; it is standard practice in many offices to dimension cabinetry in inches or millimeters only.

Follow the same process for the remaining cabinet elevations. See Figure 24.33.

Bathroom Cabinet Layout

The same procedure may be used to draw bath vanity elevations, as shown in Figure 24.34, or other special cabinet or millwork elevations and details.

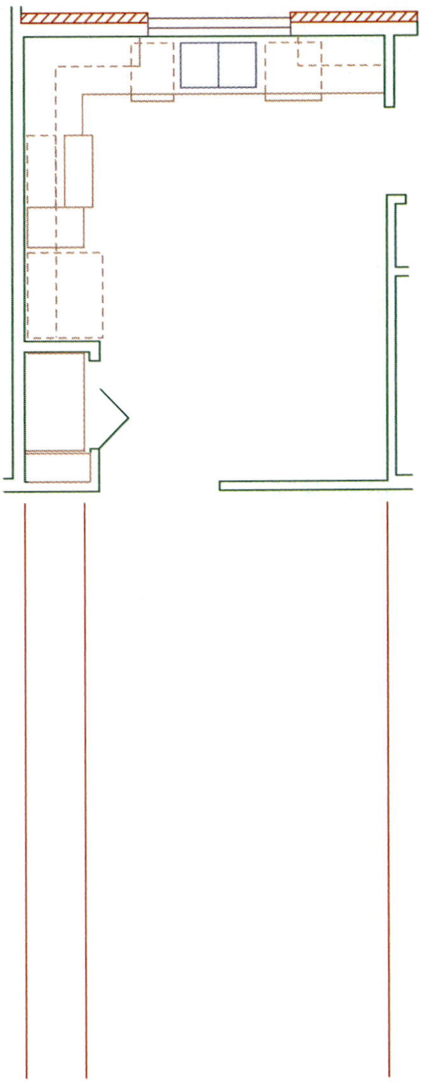

FIGURE 24.28 ■ Step 1: Transfer length dimensions from the floor plan to the cabinet elevation using construction lines.

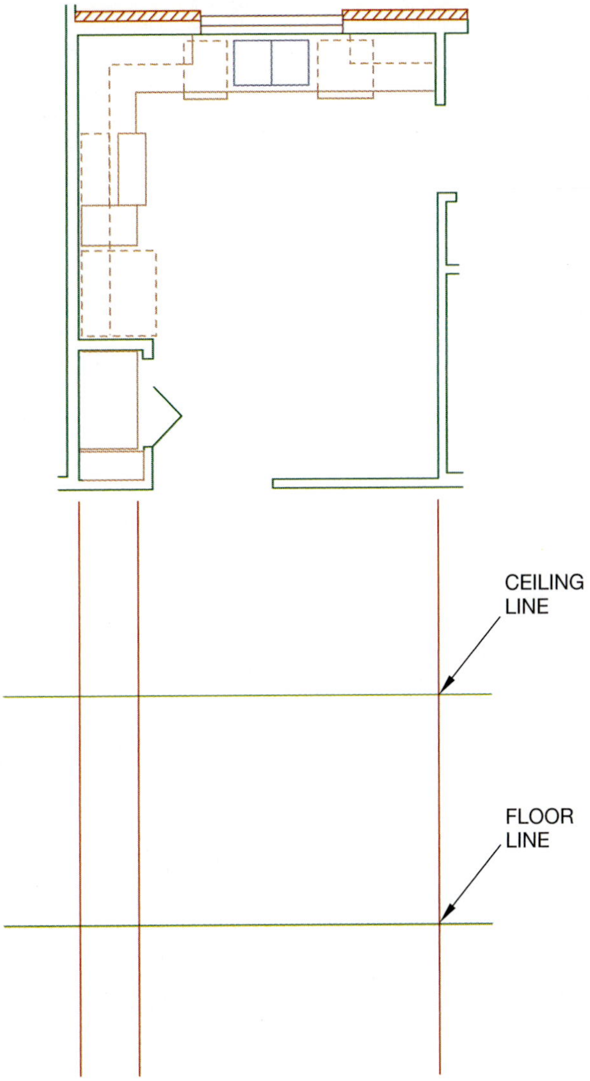

FIGURE 24.29 ■ Step 2: Establish the floor and ceiling lines using construction lines.

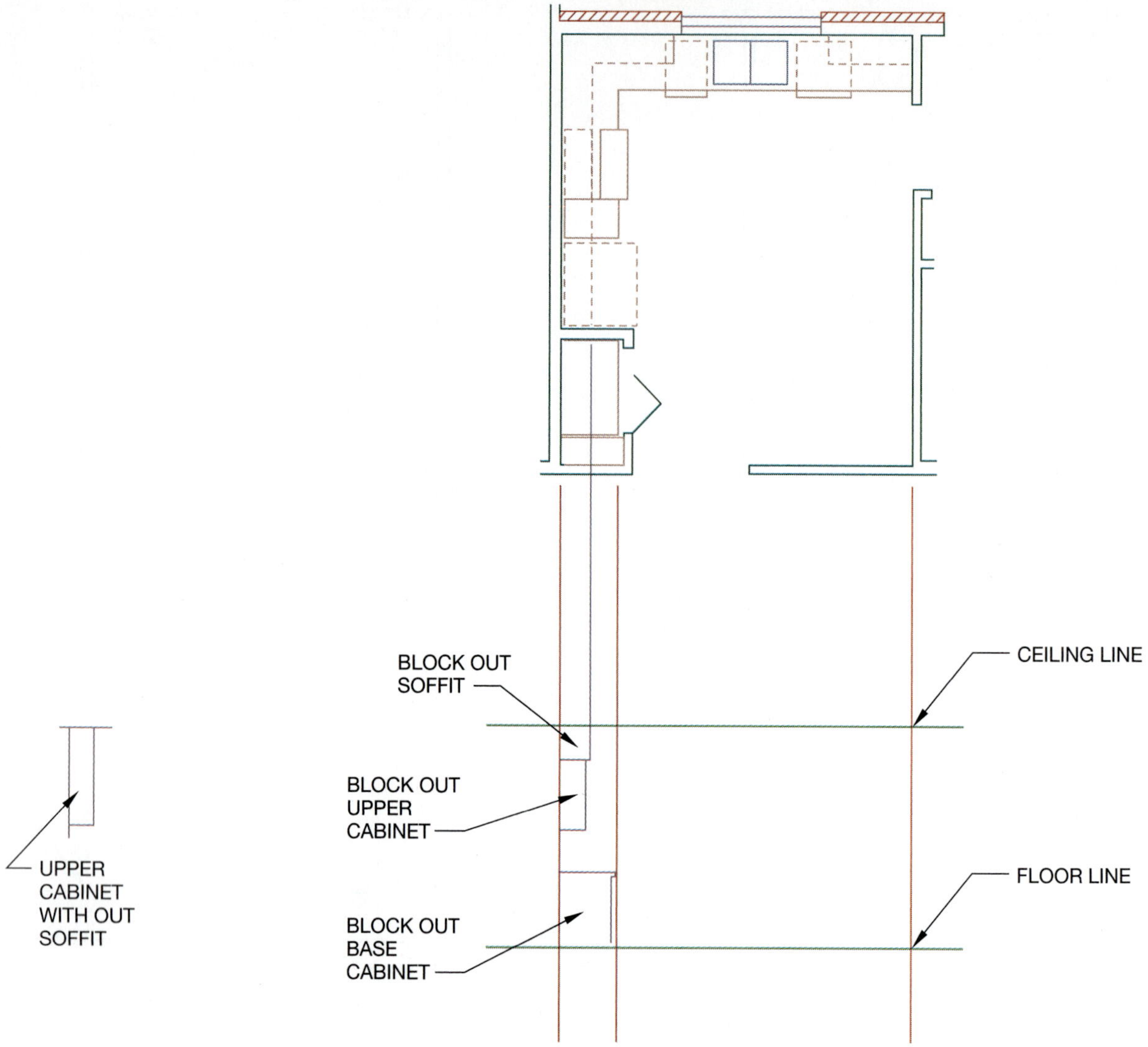

FIGURE 24.30 ■ Step 3: Establish base cabinets, upper cabinets, and soffits using construction lines.

(a)

(b)

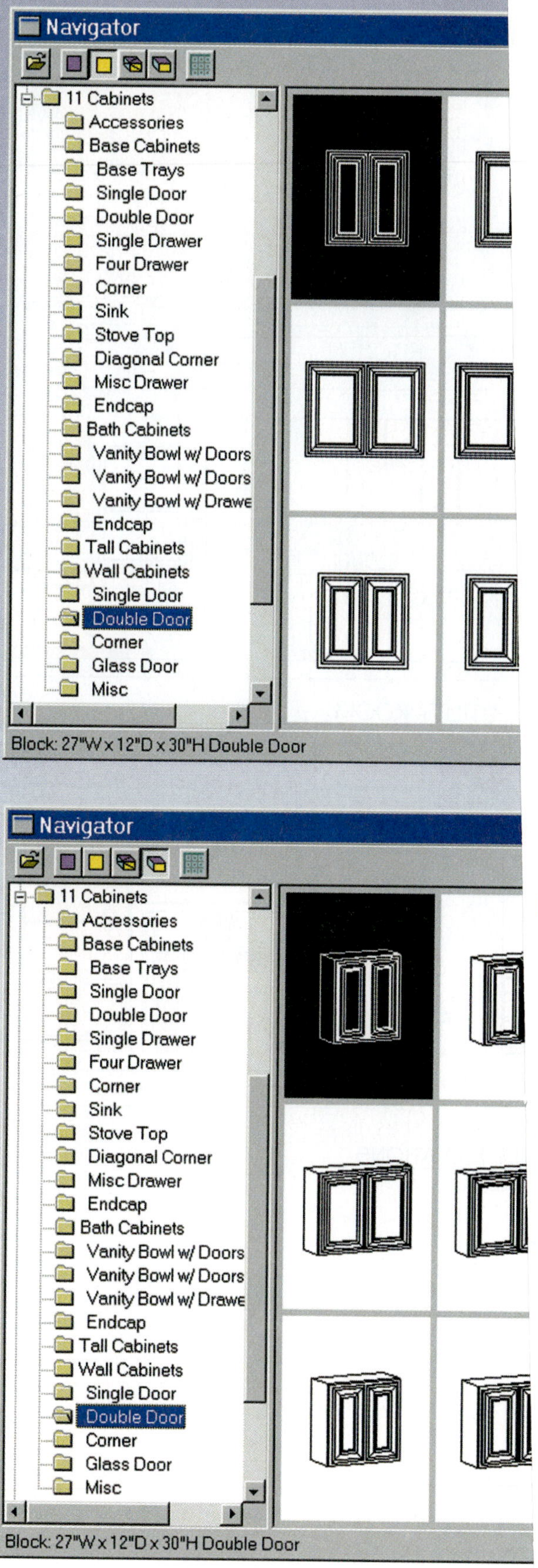

FIGURE 24.36 ■ (a) CADD menu options are available fo… 2D or 3D, as shown in (b). *Courtesy Visi…*

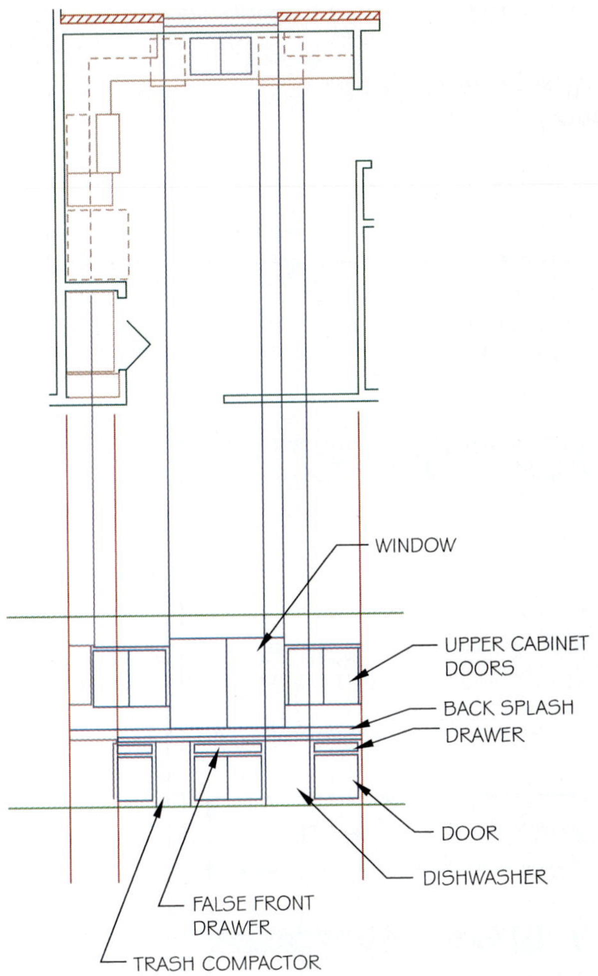

FIGURE 24.31 ■ Step 4: Use construction lines to establish cabinet elevation features.

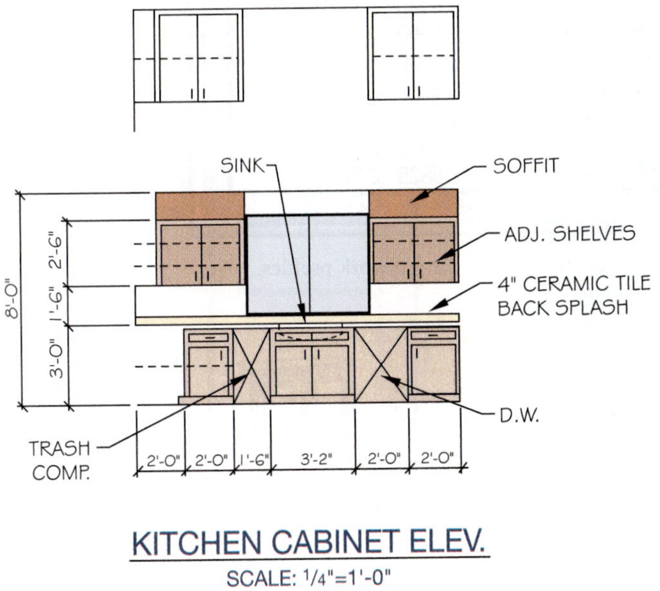

FIGURE 24.32 ■ Steps 5 and 6: Darken all cabinet lines, and add dimensions.

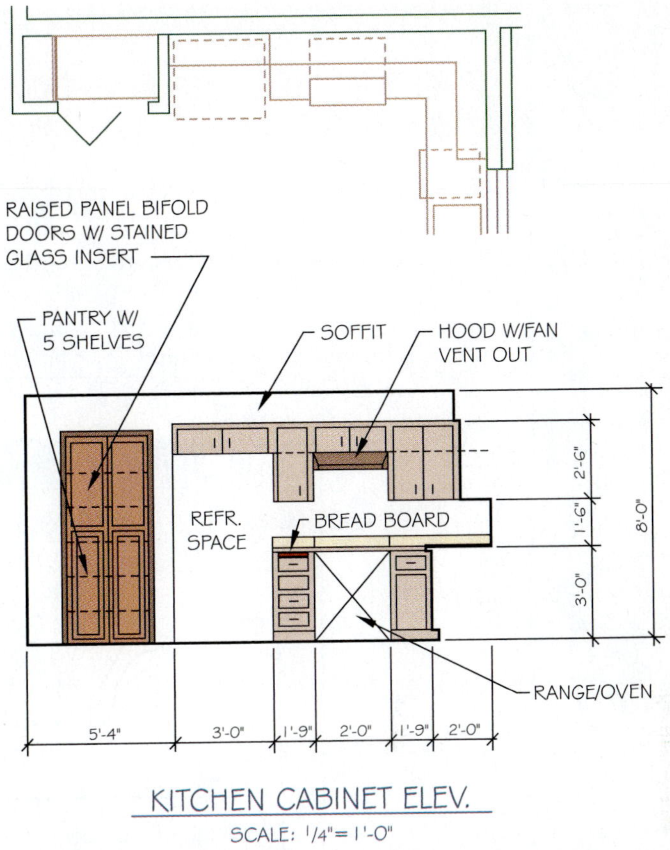

FIGURE 24.33 ■ Follow steps 1 through 6 to draw the other cabinet elevations.

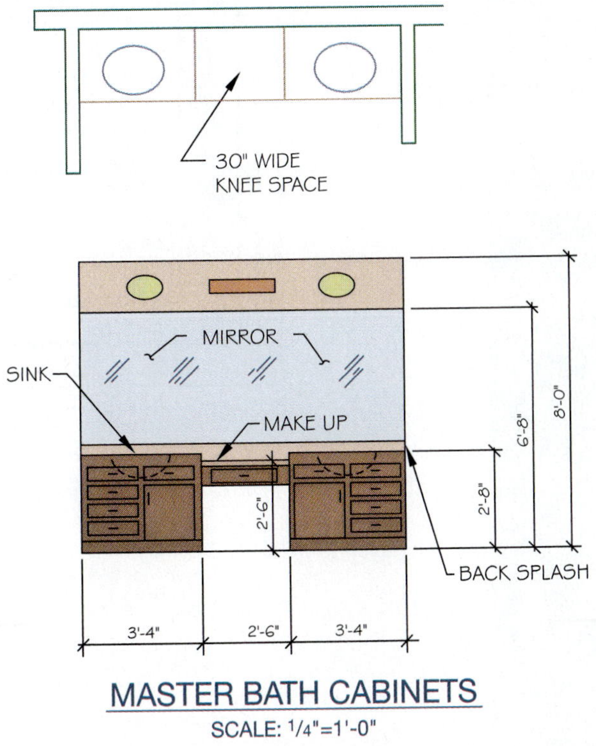

FIGURE 24.34 ■ Use cabinet layout steps 1 through 6 to draw all bath cabinet elevations.

CADD APPLICATIONS

DRAWING MILLW
ELEVATIONS WIT

CADD makes it easy to draw millwork d
tions with the use of custom menus libr
tain millwork profiles, as shown in Figu
CADD menus for cabinet floor plan
base and upper cabinets on the drawi
symbols for appliances and kitchen or l

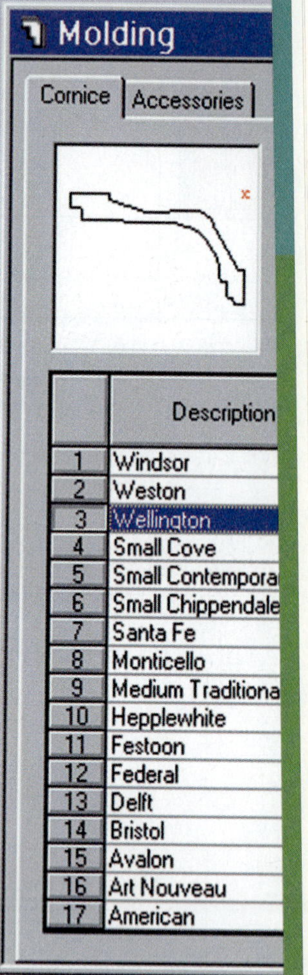

FIGURE 24.35 ■ CADD d
Courtesy

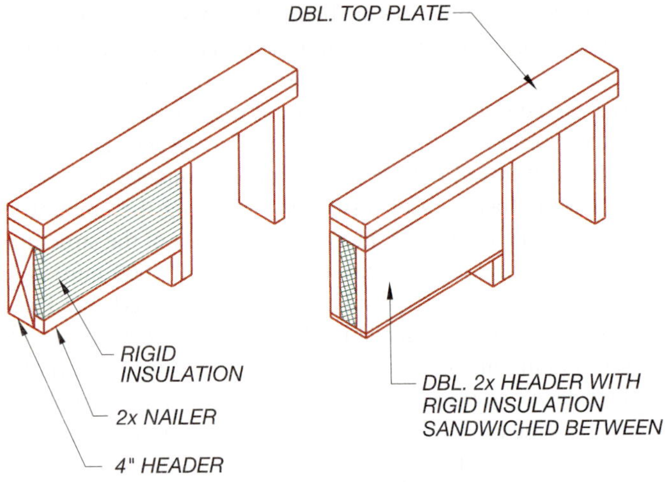

FIGURE 25.10 ■ Rigid insulation placed behind or between headers increases the insulation value of a header from R-4 to R-11.

sections using advanced framing techniques. Figure 25.10 shows examples of how insulated headers can be framed. Advanced framing methods also affect how the roof will intersect the walls. Figure 25.11 compares standard roof and wall intersection with those of advanced methods.

Structural insulated panels (SIPs) have been used for decades in the construction of refrigerated buildings, where thermal efficiency is important. SIPs have also been used in residential construction since the 1930s as part of research projects. Their popularity increased as energy costs rose through the 1980s, and they have become an economical method of construction in many parts of the country.

SIPs are composed of a continuous core of rigid foam insulation, which is laminated between two layers of structural board with an adhesive to form a single, solid panel; see Figure 25.12. Typically, expanded polystyrene (EPS) foam is used as the insulation material, and oriented strand board (OSB) is used for the outer shell of the panel. Panels with an EPS core and drywall skin are available for timber-framed structures. EPS retains its shape throughout the life of the structure, to offer uniform resistance to air infiltration and increased R values over a similar-size wood wall with batt insulation. Although values differ somewhat for each manufacturer, an R value of 4.35 per inch of EPS is common.

In addition to their energy efficiency, SIPs offer increased construction savings by combining framing, sheathing, and insulation procedures into one step during construction. Panels can be made in sizes ranging from 4′ through 24′ (1200-7300 mm) long and 8′ (2400 mm) high, depending on the manufacturer and design requirements. SIPs can be precut and custom-fabricated for use on the most elaborately shaped structures. Panels are quickly assembled on-site by fitting splines into routed grooves in the panel edges, as seen in Figure 25.13. Door and window openings can be precut during assembly or cut at the job site. A chase is typically installed in the panels to allow for electrical wiring.

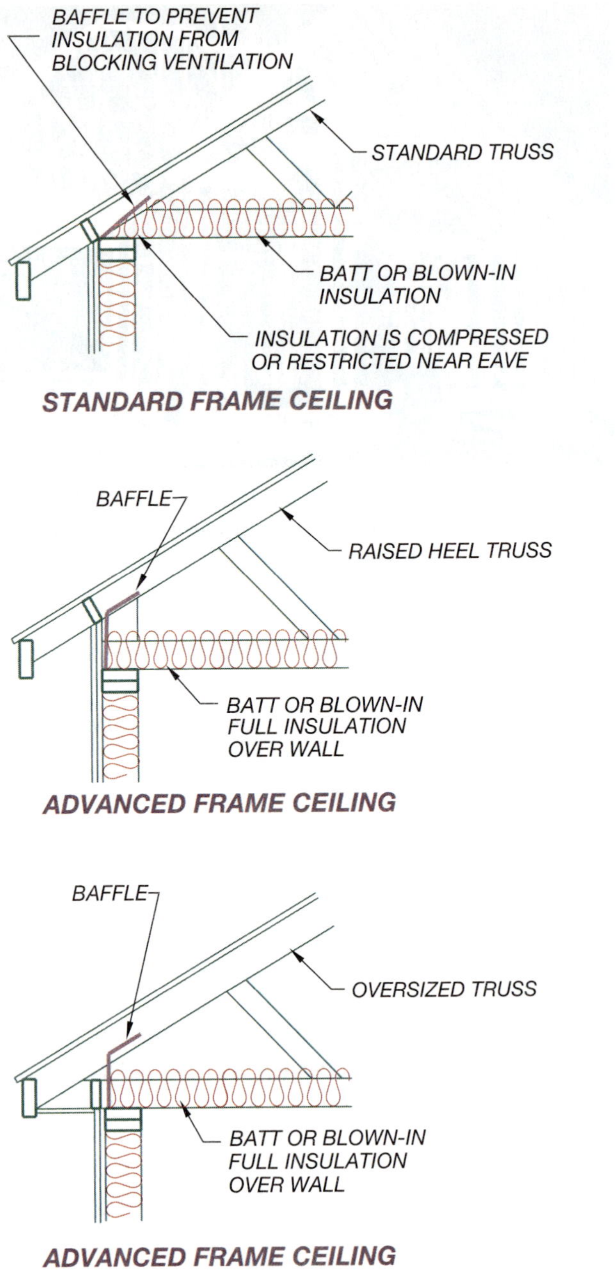

FIGURE 25.11 ■ By altering the bearing point of a truss or rafter, the ceiling insulation can be carried to the outer edge of the exterior wall and reduce heat loss.

POST-AND-BEAM FRAMING

Post-and-beam or *timber framing* construction places framing members at greater distances apart than platform methods do. In residential construction, posts and beams are usually spaced at 48″ (1200 mm) on center. Although this system uses less lumber than other methods, it requires larger members. Sizes vary, depending on the span, but beams of 4″ or 6″ (100-150 mm) wide are typically used. The subflooring and roofing over the beams are commonly 2 × 6, 2 × 8, or 1 1/8 (50 × 150, 50 × 200 or 28 mm) T&G plywood. Figure 25.14 shows typical construction members of

FIGURE 25.12 ■ Structural insulated panels are constructed of a continuous core of rigid foam insulation that is laminated between two layers of OSB. *Courtesy of Insulspan.*

FIGURE 25.13 ■ Structural insulated panels are joined to each other with tongue-and-groove seams. *Courtesy of Insulspan.*

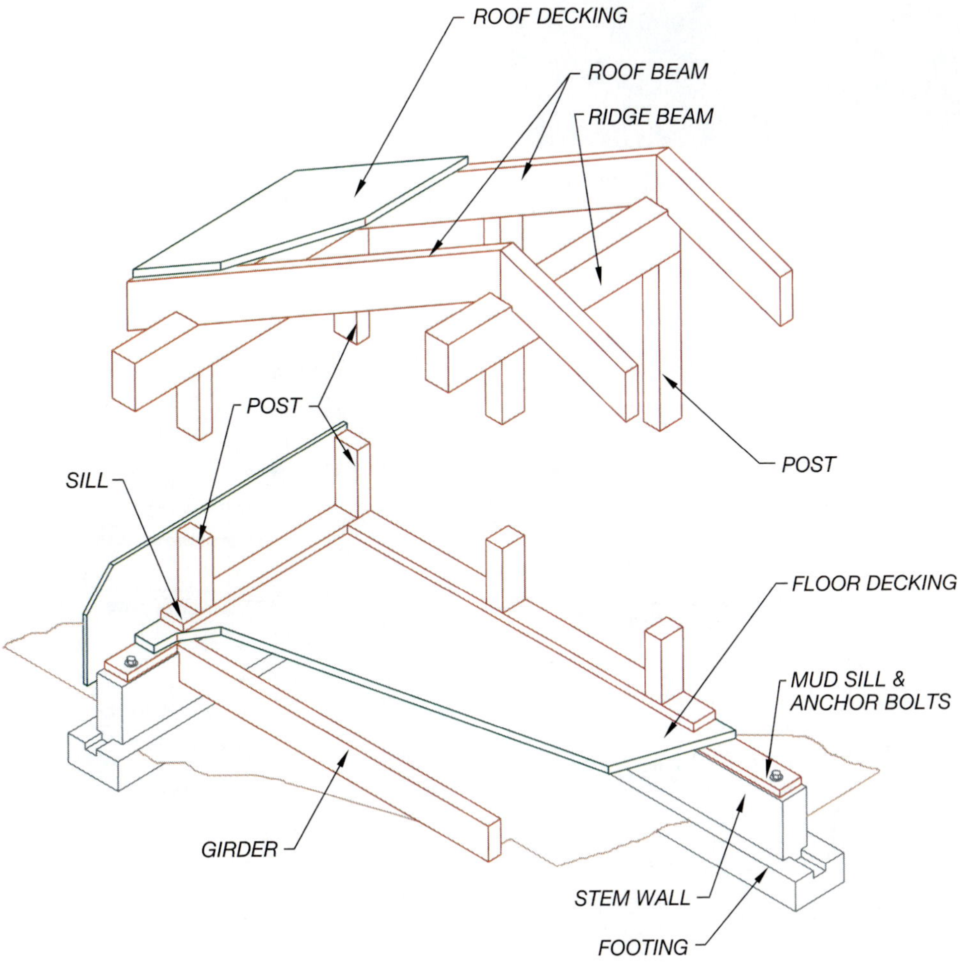

FIGURE 25.14 ■ Structural members of a post-and-beam framing system. In residential construction, supporting members are usually placed at 48″ o.c. In commercial uses, spacing may be 20′ or greater.

www.steelframingalliance.com	North American Steel Framing Alliance
www.polysteel.com	Polysteel Forms
www.concretehomes.com	Portland Cement Association (Concrete construction)
www.portcement.org	Portland Cement Association
www.timberpeg.com	Timberpeg Post and Beam Homes
www.southernpine.com	Southern Forest Products Association
www.spacejoist.com	SpaceJoist (Open web trusses)
www.ssina.com	Specialty Steel Industry of North America
www.steeljoist.org	Steel Joist Institute
www.masonrysociety.org	The Masonry Society
www.trimjoist.com	TrimJoist Engineered Wood Products
www.tjm.com	Trus-Joist MacMillan
www.wwpa.org	Western Wood Products Association
www.alhloghomes.com	Appalachian Log Homes
www.enercept.com	Enercept, Inc (SIPs)

CHAPTER 25

Framing Methods Test

DIRECTIONS

Answer the questions with short, complete statements or drawings as needed on an $8\frac{1}{2}'' \times 11''$ drawing sheet as follows:

1. Letter your name and the date at the top of the sheet.
2. Letter the question number and provide the answer. You do not need to write out the question.

QUESTIONS

Question 25–1 Explain two advantages of platform framing.

Question 25–2 Sketch a section view showing platform construction methods for a one-level house.

Question 25–3 Why is balloon framing not commonly used today?

Question 25–4 List five methods that can be used to frame a residence.

Question 25–5 Sketch and label a section showing a post-and-beam foundation system.

Question 25–6 How is brick typically used in residential construction?

Question 25–7 What factor dictates the reinforcing in masonry walls?

Question 25–8 Why would a post-and-beam construction roof be used with platform construction walls?

Question 25–9 What would be the R value for a wood wall 4″ wide?

Question 25–10 What is the typical spacing of posts and beams in residential construction?

Question 25–11 List three qualities that make steel framing a popular residential method.

Question 25–12 List four classifications of CMUs.

Question 25–13 Sketch two methods of framing a header with advance framing technology. Label each major component.

Question 25–14 What is a fire cut, and how is it used?

Question 25–15 How is moisture removed from a masonry cavity wall?

Question 25–16 What are two materials typically used to reinforce concrete block construction?

CHAPTER 26

Structural Components

INTRODUCTION

As with every other phase of drafting, construction drawings have their own terminology. The terms used in this chapter are basic for structural components of residential construction. These terms refer to floor, wall, and roof components.

FLOOR CONSTRUCTION

Two common methods of framing the floor system are conventional joist and post and beam. In some areas of the country it is common to use post-and-beam framing for the lower floor when a basement is not used and conventional framing for the upper floor. The project designer chooses the floor system or systems to be used. The size of the framing crew and the shape of the ground at the job site are the two main factors in the choice of framing methods.

Conventional Floor Framing

Conventional, or stick framing involves the use of members 2″ (50 mm) wide placed one at a time in a repetitive manner. Basic terms in this system are *mudsill, floor joist, girder,* and *rim joist.* Each can be seen in Figure 26.1 and throughout this chapter.

Floor joists can range in size from 2 × 6 (50 × 150) through 2 × 14 (50 × 350) and may be spaced at 12″, 16″, or 24″ (300, 400, or 600 mm) on center depending on the load, span, and size of joist to be used. A spacing of 16″ (400 mm) is most common. See Chapter 28 for a complete explanation of sizing joist. Because of the decreasing supply and escalating price of sawn lumber, several alternatives to floor joist have been developed. Four common substitutes are open web trusses, I joist, laminated veneer lumber, and steel joist.

The *mudsill,* or *base plate,* is the first of the structural lumber used in framing the home. The mudsill is the plate that rests on the masonry foundation and provides a base for all framing. Because of the moisture content of concrete and soil, the mudsill is required to be pressure-treated or made of foundation-grade redwood or cedar. Mudsills are set along the entire perimeter of the foundation and attached to the foundation with anchor bolts. The size of the mudsill is typically 2″ × 6″ (50 × 150 mm). Anchor bolts are required to be a minimum of 1/2 × 10″ (13 × 250 mm). Bolts 5/8 and 3/4″ (16 and 19 mm) in diameter are also common. Bolt spacing of 6′-0″ (1830 mm) is the maximum allowed by all codes and is common for most residential usage. Areas of a structure that are subject to lateral loads or uplifting will reflect bolts at a much closer spacing based on the engineer's load calculations. Section 9 will further discuss anchor bolts.

With the mudsills in place, the girders can be set to support the floor joists. *Girder* is another name for a beam. The girder is used to support the floor joists as they span across the foundation. Girders are usually sawn lumber, 4″ or 6″ (100 or 150 mm) wide, with the depth determined by the load to be supported and the span to be covered. Sawn members 2″ (50 mm) wide can

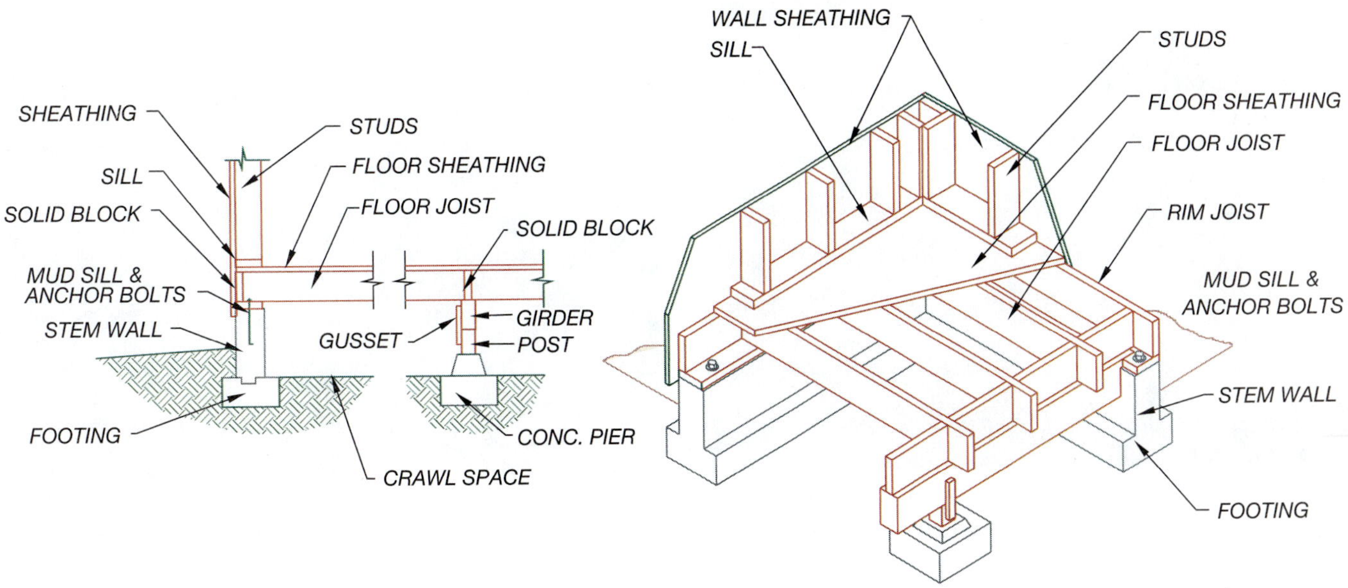

FIGURE 26.1 ■ Conventional, or stick floor framing at the concrete perimeter wall.

often be joined together to form a girder. If only two members are required to support the load, members can be nailed, according to the nailing schedule provided in the IRC. If more than three members must be joined to support a load, most building departments require them to be bolted together so the inspector can readily see that all members are attached. Another form of built-up beam is a *flitch beam,* which consists of steel plates bolted between wood members. Because they are labor-intensive and heavy, and because of the availability of engineered wood beams, flitch beams are becoming less popular. Chapter 29 introduces formulas that designers use to determine what beams size should be and whether built-up beams can be used.

In areas such as basements where a large open space is desirable, laminated wood beams called *glu-lam beams* are often used. They are made of sawn lumber that is glued together under pressure to form a beam stronger than its sawn lumber counterpart. Beam widths of 3 1/8″, 5 1/8″, and 6 3/4″ (80, 130, and 170 mm) are typical in residential construction. Although much larger sizes are available, depths typically range from 9″ through 16 1/2″ (229–419 mm) at 1 1/2″ (38 mm) intervals. See Figure 26.2. As the span of the beam increases, a *camber* or curve, is built in to help resist the tendency of the beam to sag when a load is applied. Glu-lams are often used where the beam will be left exposed because they do not drip sap and do not twist or crack, as a sawn beam does as it dries.

Engineered wood girders and beams are common throughout residential construction. Unlike glu-lam beams, which are laminated from sawn lumber such as a 2 × 4 or 2 × 6 (50 × 100 or 50 × 150 mm), parallel strand lumber (PSL) is laminated from veneer strips of fir and southern pine, which are coated with resin and then compressed and heated. Typical widths are 3 1/2″, 5 1/4,″ and 7″ (90, 130, and 180 mm). Depths range from 7″ through 18″ (180–460 mm). PSL beams have no crown or camber.

Laminated veneer lumber (LVL) is made from ultrasonically graded Douglas fir veneers, which are laminated with all grains parallel to each other with exterior grade adhesives under heat and pressure. LVL beams come in widths of 1–3/4″ (45 mm) and 3–1/2″ (90 mm) and depths ranging from 5 1/2″ through 18″ (140–460 mm). LVL beams offer performance and durability superior to that of other engineered products and often offer the smallest and lightest wood girder solution.

Steel girders such as those in Figure 26.3 are often used where foundation supports must be kept to a minimum and offer a tremendous advantage over wood for total load that can be supported on long spans. Steel beams are often the only type of beam that can support a specified load and still be hidden within the depth of the floor framing. They also offer the advantage of no expansion or shrinkage due to moisture content. Steel beams are named for their shape, depth, and weight. A steel beam with the designation W16 × 19 would represent a wide-flange steel beam with the shape of an I. The number 16 represents the approximate depth of the beam in inches, and the 19 represents the approximate weight of the beam in pounds per linear foot. Floor joists are usually set on top of the girder, as shown in Figure 26.1, but they also may be hung from the girder with joist hangers.

Posts are used to support the girders. As a general rule of thumb, a 4 × 4 (100 × 100 mm) post is typically used below a girder 4″ (100 mm) wide, and a post 6″ (150 mm) wide is typically used below a girder 6″ (150 mm) wide. Sizes can vary, depending on the load and the height of the post. LVL posts ranging in size from 3 1/2″ through 7″ (90–180 mm) are increasingly being used for their ability to support loads. A 1 1/2″ minimum bearing surface of (38 mm) is required to support a girder rest-

FIGURE 26.2 ■ Glu-lam beams span longer distances than sawn lumber and eliminate problems with twisting, shrinking, and splitting. *Courtesy Georgia Pacific Corporation.*

FIGURE 26.3 ■ Steel girders allow greater spans with fewer supporting columns for this residential foundation. *Courtesy Janice Jefferis.*

ing on a wood or steel support, and a 3″ (75 mm) bearing surface is required if the girder is resting on concrete or masonry.

Steel columns may be used in place of a wooden post, depending on the load to be transferred to the foundation. Because a wooden post will draw moisture out of the concrete foundation, it must rest on 55-lb felt. Sometimes an asphalt roofing shingle is used between the post and the girder. If the post is subject to uplift or lateral forces, a metal post base or strap may be specified by the engineer to attach the post firmly to the concrete. Figure 26.4 shows the steel connectors used to keep a post from lifting off the foundation. Forces causing uplift will be discussed further in Chapter 27.

Once the framing crew has set the support system, the floor joists can be set in place. *Floor joists* are the structural members used to support the subfloor, or rough floor. Floor joists usually span between the foundation and a girder, but, as shown in Figure 26.5, a joist may extend past its support. This extension is known as a *cantilever.*

Trusses that resemble an I are often referred to as I-joist are a high-strength, lightweight, cost-efficient alternative to sawn lumber. I-joists form a uniform size, have no crown, and do not shrink. They come in depths of 9 1/2″, 11 7/8″, 14″, and 18″ (240, 300, 360, and 460 mm). I-joists are able to span greater distances than comparable-sized sawn joists and are suitable for spans up to 40′ (12 200 mm) for residential uses. Figure 26.6 shows an example of a solid web truss with a plywood web.

Webs can be made from plywood or oriented strand board (OSB). OSB is increasingly being used to replace lumber products and is made from wood fibers arranged in a precisely controlled pattern; the fibers are coated with resin and then compressed and cured under heat. Holes can be placed in the web to allow for HVAC ducts and electrical requirements based on the manufacturer's specifications. Figure 26.6 shows a truss made with OSB.

FIGURE 26.4 ■ Steel straps are typically used to resist stress from lateral and uplift by providing a connection between the residential frame and the foundation. *Courtesy Simpson Strong-Tie Company, Inc.*

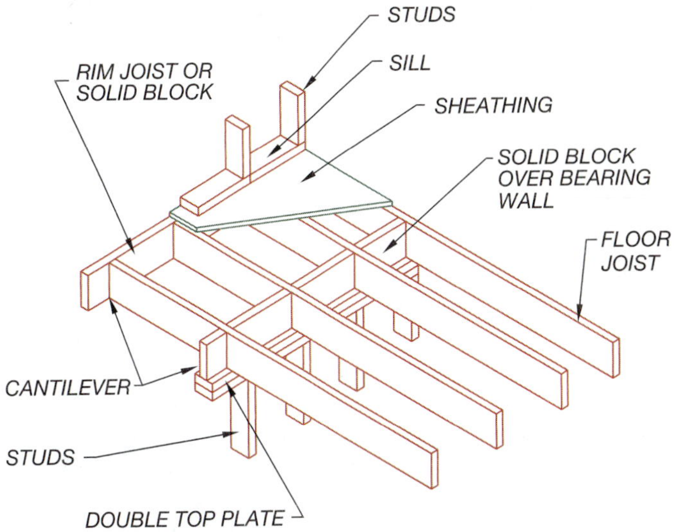

FIGURE 26.5 ■ A floor joist or beam that extends past its supporting member is cantilivered.

FIGURE 26.6 ■ Engineered lumber is a common material for residential construction. I-joist with a plywood web (center) are popular for their uniform size, light weight, and high strength. Oriented strand board (OSB) (top) is also a common web material for residential joist. Joists made of laminated veneer lumber (LVL) are used to support heavy loads. *Courtesy Georgia Pacific.*

Laminated veneer lumber (LVL) is made from ultrasonically graded Douglas fir veneers, which are laminated just as LVL beams are. LVL joists are 1 3/4″ (45 mm) wide and range in depths of 5 1/2″ through 18″ (140–460 mm). LVL joists offer superior performance and durability to other engineered products and are designed for single- or multi-span uses that must support heavy loads. Figure 26.6 shows an example of an LVL joist.

Figure 26.7 shows an example of a floor framed with engineered joist. Figure 26.8 shows typical floor construction using engineered floor joist. Depending on the loads to be supported, when the joist is supported by the sill, blocks may need to be placed beside each joist to provide additional support. Figure 26.9 shows another common method of supporting engineered floor joists. It is the responsibility of the drafter to verify specific construction requirements with the manufacturer. Most joist manufacturers supply disks of standard construction details that can be modified and inserted into the construction drawings. Figures 26.9*b* through 26.9*d* show examples of drawings supplied by an engineered joist manufacturer for use in designing engineered floor support.

FIGURE 26.7 ■ Engineered floor joist set in place to provide a residential floor system.

Open web floor trusses are a common alternative to using 2 × sawn lumber for floor joist. Open web trusses are typically spaced at 24″ (600 mm) o.c. for residential and 32″ (800 mm) for light commercial floors. Open web trusses are typically available for spans up to 38′ (11 590 mm) for residential floors. Figure 26.10*a* and 26.10*b* show examples of an open web floor truss. The horizontal members of the truss are called *top* and *bottom chords* respectively, and are typically made from 1.5″ × 3″ (40 × 75 mm) lumber laid flat. The diagonal members, called *webs,* are made of wood or tubular steel approximately 1″ (25 mm) in diameter. When drawn in section or details, the webs are typically represented with a bold center line drawn at a 45° angle. The exact angle is determined by the truss manufacturer and is unimportant to the detail. As the span or load on the joist increases, the size and type of material used are increased. Pairs of 1.5″ × 2.3″ (40 × 60 mm) laminated veneer lumber are used for the chords of many residential floor trusses. In drawing sections showing floor trusses, it is important to work with the manufacturer's details to find exact sizes and truss depth. Most

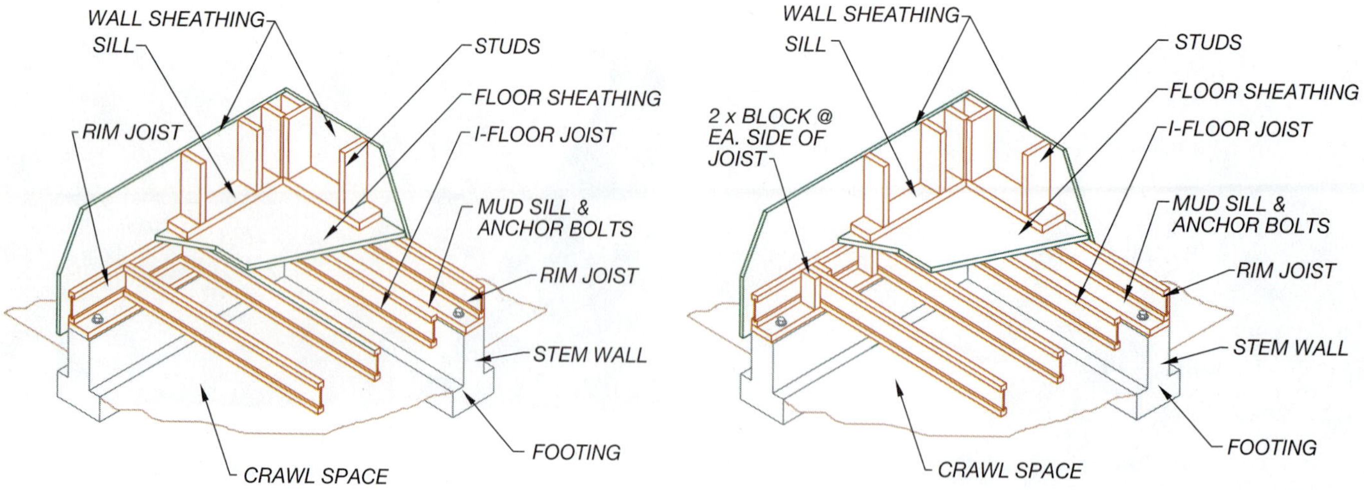

FIGURE 26.8 ■ Floor joist to sill connections using engineered lumber.

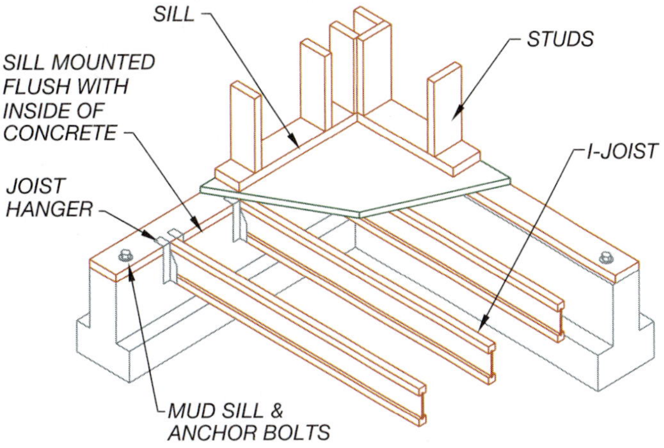

FIGURE 26.9(A) ■ Engineered floor joist can be supported using metal hangers.

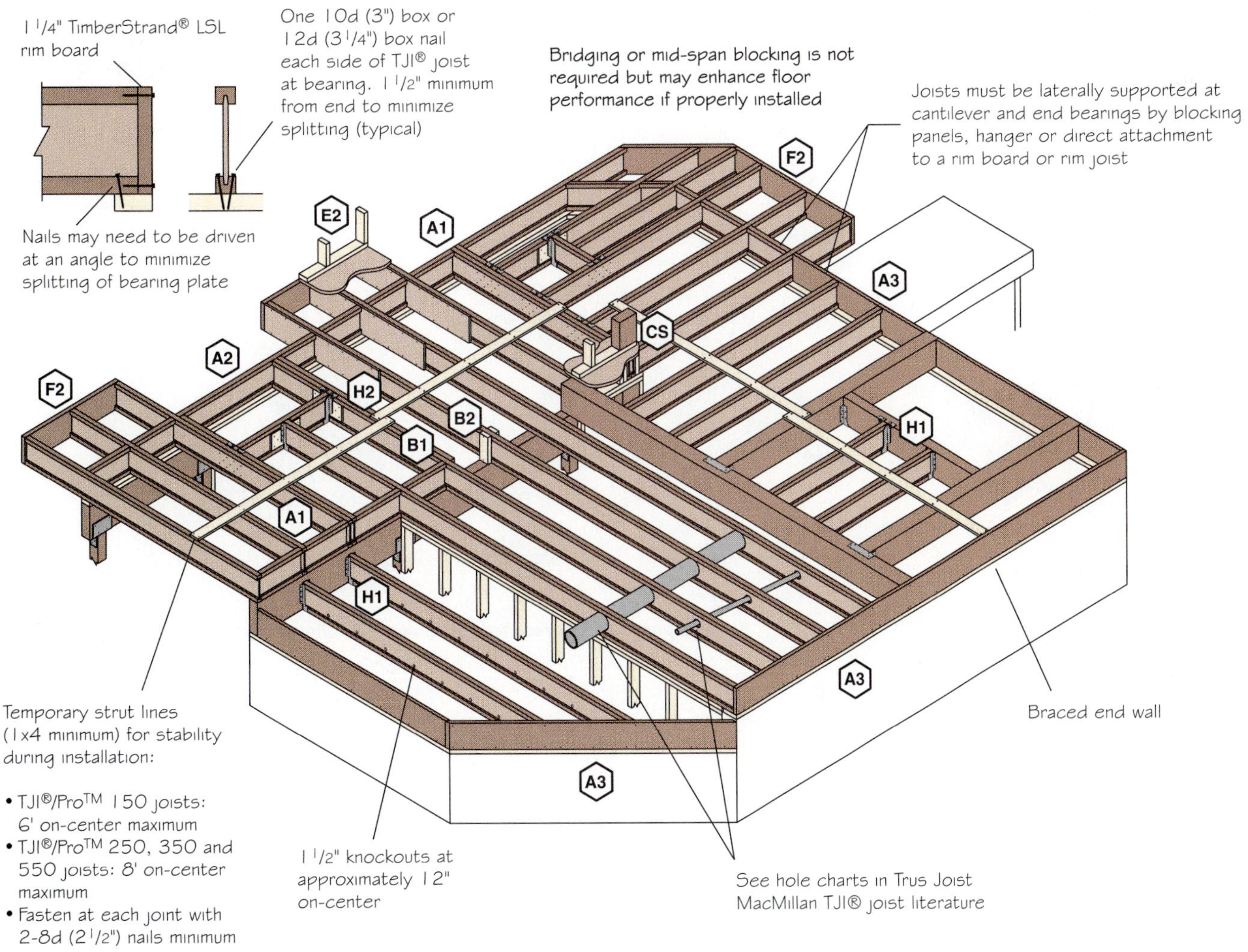

FIGURE 26.9(B) ■ Construction components of an engineered floor system. *Courtesy Trus Joist MacMillan.*

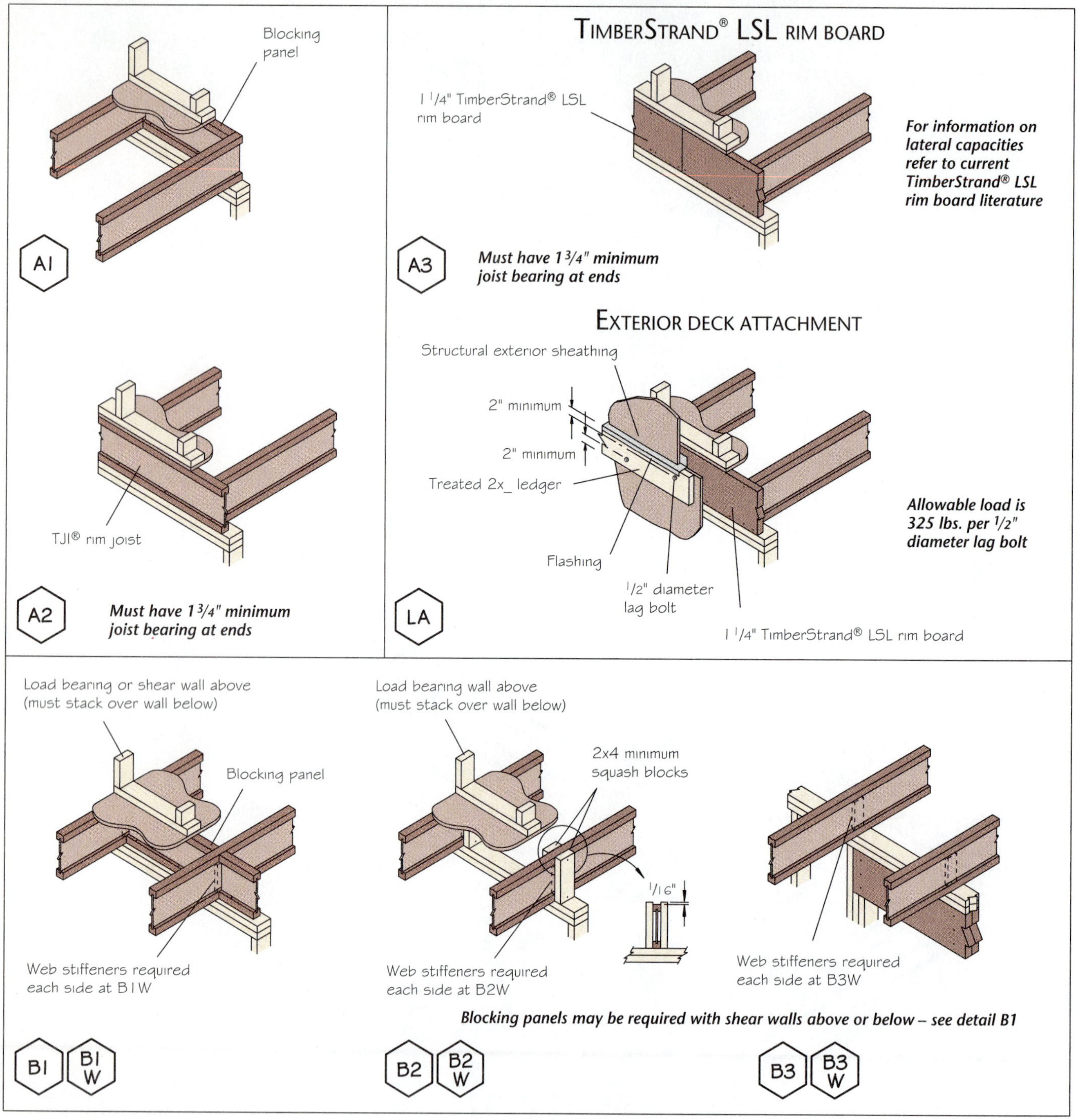

General Notes

MINIMUM BEARING LENGTH

- At joist ends: 1 3/4".
- At intermediate supports: 3 1/2".

BLOCKING PANELS, RIM BOARDS OR RIM JOIST

- Vertical load transfer at bearings must be checked for each application. Capacities of rim details shown are as follows:
  TJI® blocking............................2000 plf
  TJI® rim joist ....2000 plf
  TimberStrand® LSL–1 1/4"............4250 plf
- Bracing complying with the code shall be carried to the foundation.

FIGURE 26.9(C) ■ Common connection details of engineered floor joist. *Courtesy Trus Joist MacMillan.*

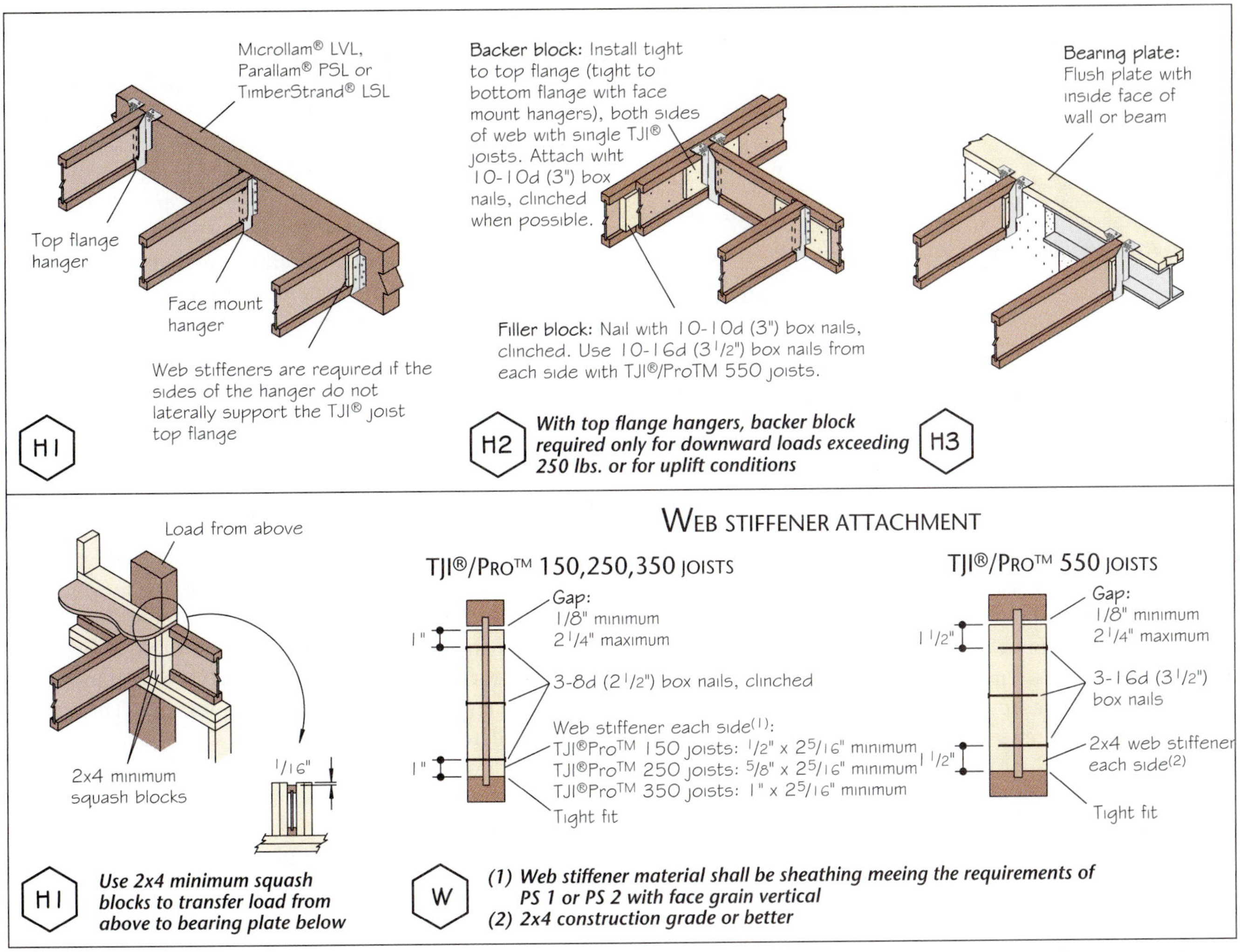

Microllam® LVL, Parallam® PSL or TimberStrand® LSL
Top flange hanger
Face mount hanger
Web stiffeners are required if the sides of the hanger do not laterally support the TJI® joist top flange
H1
Backer block: Install tight to top flange (tight to bottom flange with face mount hangers), both sides of web with single TJI® joists. Attach wiht 10-10d (3") box nails, clinched when possible.
Filler block: Nail with 10-10d (3") box nails, clinched. Use 10-16d (3 1/2") box nails from each side with TJI®/ProTM 550 joists.
H2
With top flange hangers, backer block required only for downward loads exceeding 250 lbs. or for uplift conditions
Bearing plate: Flush plate with inside face of wall or beam
H3
Load from above
2x4 minimum squash blocks
1/16"
H1
Use 2x4 minimum squash blocks to transfer load from above to bearing plate below
WEB STIFFENER ATTACHMENT
TJI®/PRO™ 150,250,350 JOISTS
Gap: 1/8" minimum 2 1/4" maximum
1"
3-8d (2 1/2") box nails, clinched
Web stiffener each side(1):
TJI®ProTM 150 joists: 1/2" x 2 5/16" minimum
TJI®ProTM 250 joists: 5/8" x 2 5/16" minimum
TJI®ProTM 350 joists: 1" x 2 5/16" minimum
Tight fit
TJI®/PRO™ 550 JOISTS
Gap: 1/8" minimum 2 1/4" maximum
1 1/2"
3-16d (3 1/2") box nails
2x4 web stiffener each side(2)
Tight fit
W
(1) Web stiffener material shall be sheathing meeing the requirements of PS 1 or PS 2 with face grain vertical
(2) 2x4 construction grade or better

FIGURE 26.9(D) ■ Common connection details of engineered floor joist. *Courtesy Trus Joist MacMillan.*

FIGURE 26.10(A) ■ Open web floor joist can be used to span large distances for residential designs. *Courtesy Leroy Cook.*

FIGURE 26.10(B) ■ Open web floor joist with metal webs.

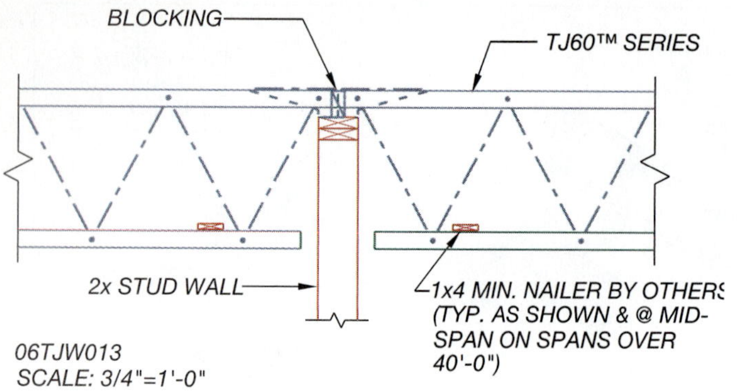

FIGURE 26.11 ■ Disk copies of truss details, supplied by most manufacturers, can be inserted into the working drawings. *Courtesy Trus Joist MacMillan.*

truss manufacturers supply disk copies of common truss connections. Figure 26.11 shows an example of a truss detail.

When steel joist are used to support the floor, a 6″ × 6″ × 54 mil L clip angle is bolted to the foundation to support the track that will support the floor joist as shown in Figure 26.12. Building codes require a minimum of 1/2″ bolts and clip angle at 6′–0″ o.c. to be used to attach the angle to the concrete. Eight #8 screws are required to attach the angle to the floor joist track. Floor joist are then bolted to the track using 2- #8 screws at each joist. Additional fasteners should be used if winds exceed 90 mph.

Floor Bracing

Because of the height-to-depth proportions of a joist, it will tend to roll over onto its side as loads are applied. To resist this tendency, a rim joist or blocking is used to support the joist. A *rim joist,* which is sometimes referred to as a band or header, is usually aligned with the outer face of the foundation and mudsill.

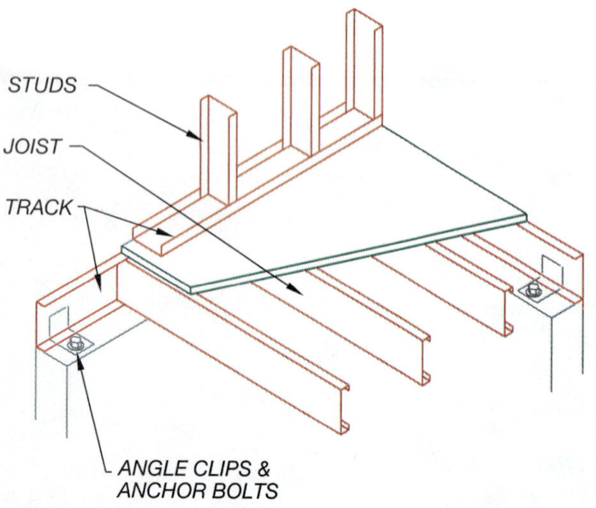

FIGURE 26.12 ■ Floor construction using steel floor joist.

Some framing crews set a rim joist around the entire perimeter and then end-nail the floor joists. An alternative to the rim joist is to use solid blocking at the sill between floor joists. Figure 26.13 shows the difference between a rim joist and solid blocking at an exterior wall.

Blocking in the floor system is typically used at the center of the joist span, as shown in Figure 26.14. Spans longer than 10′ (250 mm) must be blocked at center span to help transfer lateral forces from one joist to another and then to the foundation system. Another use of blocking is at the end of the floor joists at their bearing point. Figure 26.14 shows the use of solid blocking at an interior wall. Figure 26.15 shows an example of blocking for engineered floor joist. These blocks help keep the entire floor system rigid and are used in place of the rim joist. Blocking is also used to provide added support to the floor sheathing. Often walls are reinforced to resist lateral loads. These loads are, in turn, transferred to the foundation through the floor system and are resisted by a *diaphragm,* a rigid plate that acts similarly to a beam and can be found in the roof level, walls, or floor system. In a floor diaphragm, 2″ × 4″ (50 × 100 mm) blocking is typically laid flat between the floor joist to allow the edges of plywood panels to be supported. Nailing all edges of a plywood panel allows for approximately one-half to two times the design load to be resisted over unblocked diaphragms. The architect or engineer determines the size and spacing of nails and blocking required. Blocking is also used to reduce the spread of fire and smoke through the floor system.

Floor sheathing is installed over the floor joists to form the subfloor. The sheathing provides a surface for the base plate of the wall to rest on. Until the mid-1940s, 1 × 4s or 1 × 6s

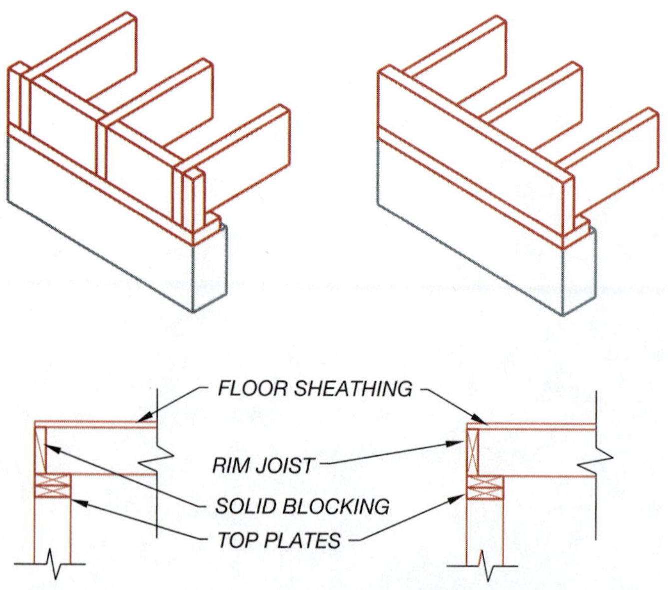

FIGURE 26.13 ■ Floor joist blocking at the center and end of the span. At midspan, solid or cross-blocking may be used. At the ends of the floor joists, solid blocking or a continuous rim joist may be used to provide stability.

FIGURE 26.14 ■ Solid blocking is placed between floor joist to help resist lateral loads.

FIGURE 26.15 ■ Blocking is used to provide stiffness to a floor system and to support plumbing and heating equipment. *Courtesy John Black.*

(25 × 100 or 25 × 150 mm) laid perpendicularly or diagonally to the floor joists were used for floor sheathing. Plywood and OSB are the most common materials used for floor sheathing. Depending on the spacing of the joist, plywood with a thickness of 1/2″, 5/8″, or 3/4″ (13, 16, or 19 mm) with an APA grade of EXP 1 or 2, EXT, STRUCT I-EXP-1, or STRUCT 1-EXT is used for sheathing. EXT represents exterior grade, STRUCT represents structural, and EXP represents exposure. Plywood also is printed with a number to represent the *span rating,* which represents the maximum spacing from center to center of supports. The span rating is listed as two numbers, such as 32/16, separated by a slash. The first number represents the maximum recommended spacing of supporters if the panel is used for roof sheathing and the long dimension of the sheathing is placed across three or more supports. The second number represents the maximum recommended spacing of supports if the panel is used for floor sheathing and the long dimension of the sheathing is placed across three or more supports. Plywood is usually laid so that the face grain on each surface is perpendicular to the floor joist. This provides a rigid floor system without having to block the edges of the plywood. Floor sheathing that will support ceramic tile is typically 32/16″ APA span rated 15/31″ (12 mm) or span rated 40/20″ and 19/32″ (15 mm) thick.

Waferboard or oriented strand board (OSB) is also a popular alternative to traditional plywood subfloors. OSB is made from three layers of small strips of wood that have been saturated with a binder and then compressed under heat and pressure. The exterior layers of strips are parallel to the length of the panel, and the core layer is laid in a random pattern.

Once the subfloor has been installed, an underlayment for the finish flooring is put down. The underlayment is not installed until the walls, windows, and roof are in place, making the house weathertight. The underlayment provides a smooth impact-resistant surface on which to install the finished flooring. Underlayment is usually 3/8″ or 1/2″ (9 or 13 mm) APA

underlayment GROUP 1, EXPOSURE 1 plywood, hardboard, or waferboard. Hardboard is typically referred to as medium- or high-density fiberboard (MDF or HDF) and is made from wood particles of various sizes that are bonded together with a synthetic resin under heat and pressure. The underlayment may be omitted if the holes in the plywood are filled. APA STURD-I-FLOOR rated plywood 19/32″ through 1 3/32″ (15–28 mm) thick can also be used to eliminate the underlayment.

Post-and-Beam Construction

Terms to be familiar with when working with post-and-beam floor systems include *girder, post, decking,* and *finished* floor. Each can be seen in Figure 26.16. Notice there are no floor joists with this system.

A mudsill is installed with post-and-beam construction just as with platform construction. Once set, the girders are also placed. With post-and-beam construction the girder is supporting the floor decking instead of floor joists. Girders are usually 4 × 6 (100 × 150 mm) beams spaced at 48″ (1200 mm) o.c., but the size and distance may vary depending on the loads to be supported. As with conventional methods, posts are used to support the girders. Typically a 4 × 4 (100 × 100 mm) is the minimum size used for a post, with a 4 × 6 (100 × 150 mm) post used to support joints in the girders. With the support system in place, the floor system can be installed. Figures 26.17*a* and 26.17*b* show the components of a post-and-beam floor system.

FIGURE 26.17(A) ■ Floors built with a post-and-beam system require no floor joist to support the floor sheathing. *Courtesy Saundra Powell.*

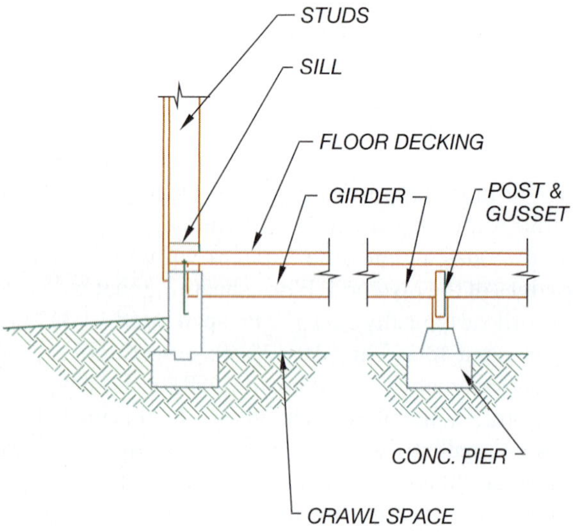

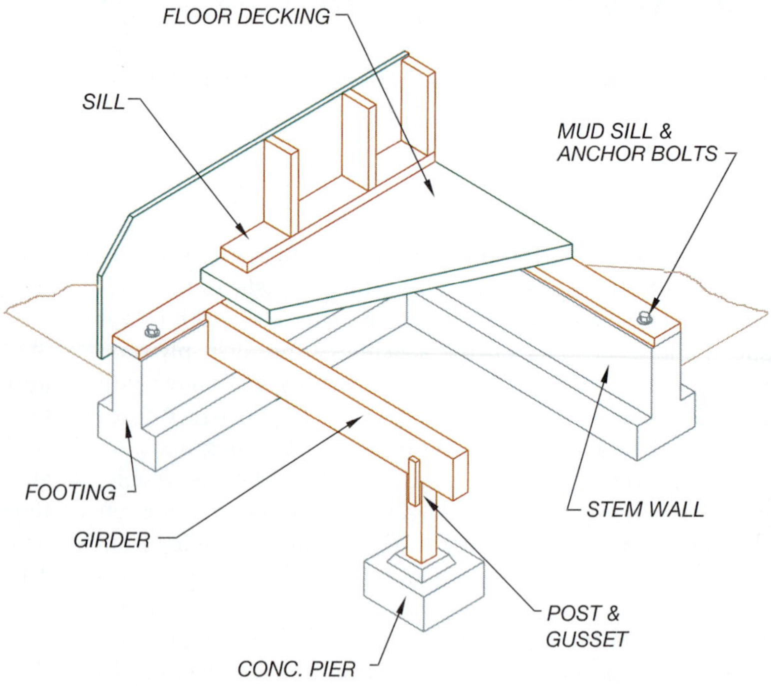

FIGURE 26.16 ■ Components of post-and-beam construction.

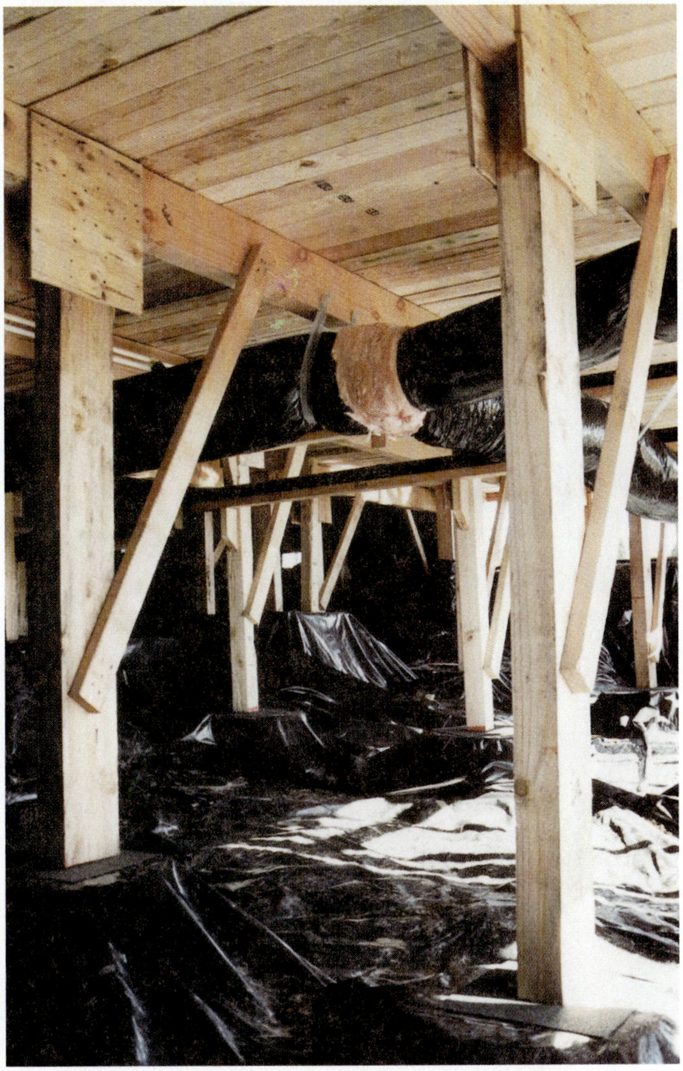

FIGURE 26.17(B) ■ The crawl space for a post and beam floor system. Plywood gusset plates are used to connect the post to the girder and a wood brace is used to resist lateral loads for tall post.

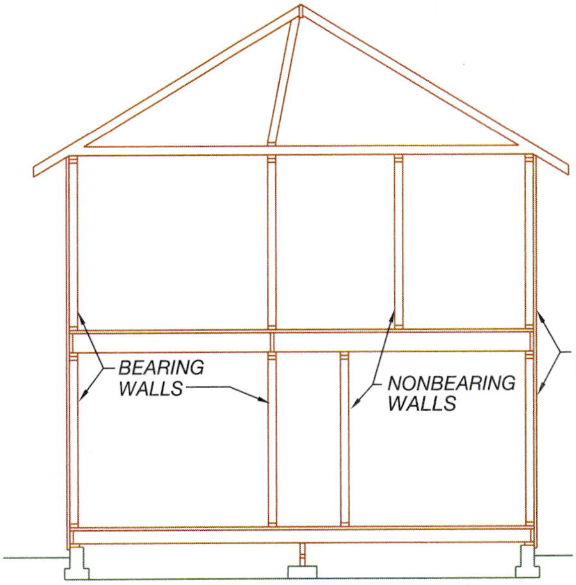

FIGURE 26.18 ■ Bearing and nonbearing walls. A bearing wall supports its own weight and the weight of floor and roof members. A nonbearing wall supports only its own weight. IRC allow ceiling weight to be supported on a wall and still be considered a nonbearing wall.

Decking is the material laid over the girders to form the subfloor. Typically decking is 2 × 6 or 2 × 8 (50 × 150 or 50 × 200 mm) tongue and groove (T&G) boards or 1 3/32″ (28 mm) APA rated 2-4-1 STURD-I-FLOOR EXP-1 plywood T&G floor sheathing. The decking is usually finished similarly to conventional decking with a hardboard overlay.

FRAMED WALL CONSTRUCTION

As a drafter you will be concerned with two types of walls, bearing and nonbearing. A bearing wall supports not only itself but also the weight of the roof or other floors constructed above it. A bearing wall requires some type of support under it at the foundation or lower floor level in the form of a girder or another bearing wall. Nonbearing walls are sometimes called partitions. A nonbearing wall serves no structural purpose. It is a partition used to divide rooms and could be removed without causing damage to the building. Bearing and nonbearing walls can be seen in Figure 26.18. In post-and-beam construction, any exterior walls placed between posts are nonbearing walls.

Bearing and nonbearing walls made of wood or engineered lumber are both constructed using a sole plate, studs, and a top plate. Each can be seen in Figure 26.19a, *b,* and *c.* The *sole,* or bottom, plate is used to help disperse the loads from the wall studs to the floor system. The sole plate also holds the studs in position as the wall is being tilted into place. *Studs* are the vertical framing members used to transfer loads from the top of the wall to the floor system. Typically 2 × 4 (50 × 100 mm) studs are spaced at 16″ (400 mm) o.c. and provide a nailing surface for the wall sheathing on the exterior side and the Sheetrock on the interior side. Studs 2 × 6 (50 × 150 mm) are often substituted for 2 × 4 studs (50 × 100 mm) to provide added resistance for lateral loads and to provide a wider area to install insulation. Engineered studs in 2 × 4 or 2 × 6 (50 × 100 or 50 × 150 mm) are also available in 8′, 9′, and 10′ (2400, 2700, and 3000 mm) lengths. Engineered studs are made from short sections of stud-grade lumber that have had the knots and splits removed. Sections of wood are joined together with 5/8″

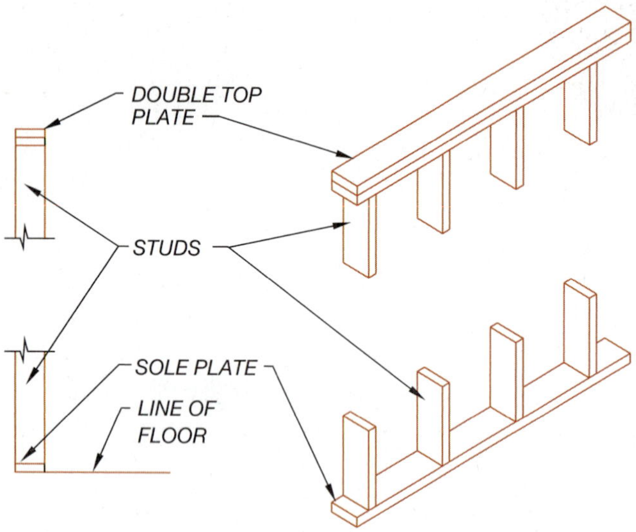

FIGURE 26.19(A) ■ Standard wall construction uses a double top plate on the top of the wall, a sole plate at the bottom of the wall, and studs.

FIGURE 26.19(C) ■ Common components for a steel framed wall system. The track is screwed to floor sheathing, and studs are screwed to the track.

FIGURE 26.19(B) ■ Common components for a wood framed wall system.

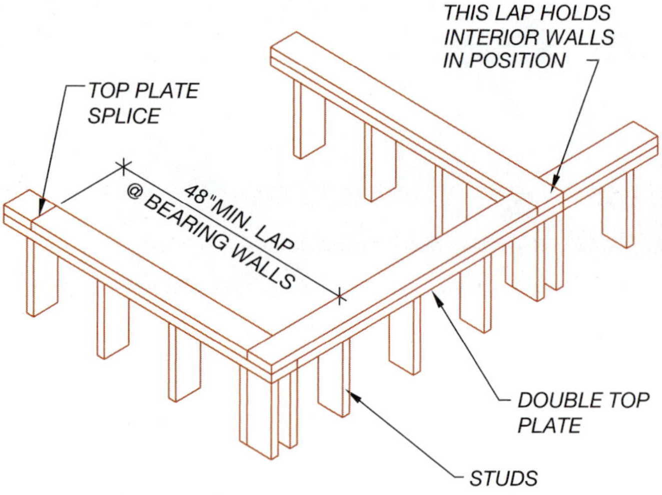

FIGURE 26.20 ■ Top plate laps are required to be 48″ minimum in bearing walls or held together with steel straps.

(16 mm) finger joints. The top plate is located on top of the studs and is used to hold the wall together. See Figure 26.20. Two top plates are required on bearing walls, and each must lap the other a minimum of 48″ (1200 mm). This lap distance provides a continuous member on top of the wall to keep the studs from separating. An alternative to the double top plate is to use one plate with a steel strap at each joint in the plate. The top plate also provides a bearing surface for the floor joists from an upper level or for the roof members. When steel studs are used, a track is placed above the studs and fastened with a minimum of 2-#8 screws to each stud. The size of the screws will vary depending on the gage of the track and the thickness of the steel studs. Figure 26.21 shows an example of a steel-framed wall.

OSB and plywood sheathing are primarily used as an insulator against the weather and also as a backing for the exterior siding. Sheathing may be considered optional, depending on your area of the country. When sheathing is used on exterior walls, it provides what is called *double-wall construction.* In *single-wall construction,* wall sheathing is not used, and the siding is attached over a vapor barrier such as Tyvek, Pinkwrap, Typar, or the traditional 15 # felt placed over the studs, as in Figure 26.22. The cost of the home and its location will have a great influence on whether wall sheathing is to be used. In areas where double-wall construction is required for weather or structural reasons, 3/8″ plywood is used on exterior walls as an underlayment. Plywood sheathing should be APA-rated EXP 1 or 2, EXT, STRUCT 1, EXT 1, or STRUCT 1 EXT. Many builders prefer to use 1/2″ (13 mm) OSB for sheathing in place of plywood.

FIGURE 26.21 ■ Steel studs and joist provide the support for the upper floor.

FIGURE 26.22 ■ The siding material can be placed directly over the vapor barrier and studs in what is referred to as single-wall construction. *Courtesy APA, The Engineered Wood.*

Figure 26.23*a* shows the use of plywood for wall sheathing. Figure 26.23*b* shows the use of OSB wall sheathing.

The design of the home may require the use of plywood sheathing for its ability to resist the tendency of a wall to twist or rack. Racking can be caused by wind or seismic forces. Plywood used to resist these forces is called a *shear panel*. See Figure 26.24 for an example of racking and how plywood can be used to resist this motion.

FIGURE 26.23(A) ■ Plywood is placed over the vapor barrier and studs to provide added protection from the elements and to provide resistance to lateral forces caused by winds and earthquakes.

FIGURE 26.23(B) ■ OSB is a common alternative for an underlayment for siding materials.

An alternative to plywood for shear panels is to use let-in braces. Figure 26.25 shows how a brace can be used to stiffen the studs. A notch is cut into the studs, and a 1 × 4 (25 × 100 mm) is laid flat in this notch at a 45° angle to the studs. If plywood siding rated APA Sturd-I-Wall is used, no underlayment or let-in braces are required.

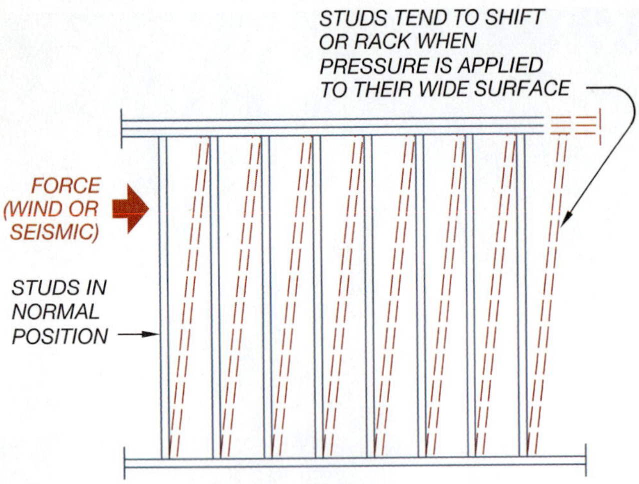

FIGURE 26.24 ■ Wall racking occurs when wind or seismic forces push the studs out of their normal position. Plywood panels can be used to resist these forces and keep the studs perpendicular to the floor. These plywood panels are called *shear* or braced wall panels.

Prior to the mid-1960s, blocking was common in walls to help provide stiffness. Blocking for structural or fire reasons is now no longer required unless a wall exceeds 10′ (3050 mm) in height. Blocking is often installed for a nailing surface for mounting cabinets and plumbing fixtures. Blocking is sometimes used to provide extra strength in some seismic zones.

In addition to the wall components mentioned, there are several other terms that the drafter needs to be aware of. These are terms that are used to describe the parts used to frame around an opening in a wall: *headers, subsill, trimmers, king studs,* and *jack studs*. Each can be seen in Figure 26.26. Figure 26.26*b* shows each component in a wall made of single-wall construction.

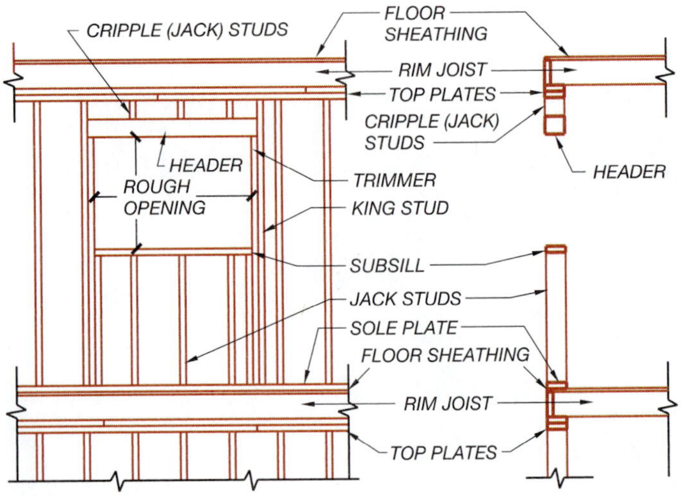

FIGURE 26.26(A) ■ Construction components at an opening.

FIGURE 26.25 ■ Let-in braces can sometimes be used instead of shear panels to resist wall racking. *Courtesy Gang-Nail Systems, Inc.*

FIGURE 26.26(B) ■ Framing for an opening in a wall includes the header, king stud, trimmer, sill, and jack studs.

A *header* in a wall is used over an opening such as a door or window. When an opening is made in a wall, one or more studs must be omitted. A header is used to support the weight that the missing studs would have carried. The header is supported on each side by a *trimmer.* Depending on the weight the header is supporting, double trimmers may be required. The trimmers also provide a nailing surface for the window and the interior and exterior finishing materials. A *king stud* is placed beside each trimmer and extends from the sill to the top plates. It provides support for the trimmers so that the weight imposed from the header can only go downward, not sideways.

Between the trimmers is a *subsill* located on the bottom side of a window opening. It provides a nailing surface for the interior and exterior finishing materials. *Jack studs,* or *cripples,* are studs that are not full height. They are placed between the sill and the sole plate and between a header and the top plates.

ROOF CONSTRUCTION

Roof framing includes both conventional and truss framing methods. Each has its own special terminology, but there are many terms that apply to both systems. These common terms will be described first, followed by the terms for conventional and truss framing methods.

Basic Roof Terms

Roof terms common to conventional and trussed roofs are *eave, cornice, eave blocking, fascia, ridge, sheathing, finishing roofing, flashing,* and *roof pitch dimensions.*

The *eave* is the portion of the roof that extends beyond the walls. The *cornice* is the covering that is applied to the eaves. Common methods for constructing the cornice can be seen in Figure 26.27. *Eave* or *bird blocking* is a spacer block placed between the rafters or truss tails at the eave. This block keeps the spacing of the rafters or trusses uniform and keeps small animals from entering the attic. It also provides a cap to the exterior siding, as seen in Figure 26.28.

A *fascia* is a trim board placed at the end of the rafter or truss tails and usually perpendicular to the building wall. It hides the truss or rafter tails and also provides a surface where the gutters

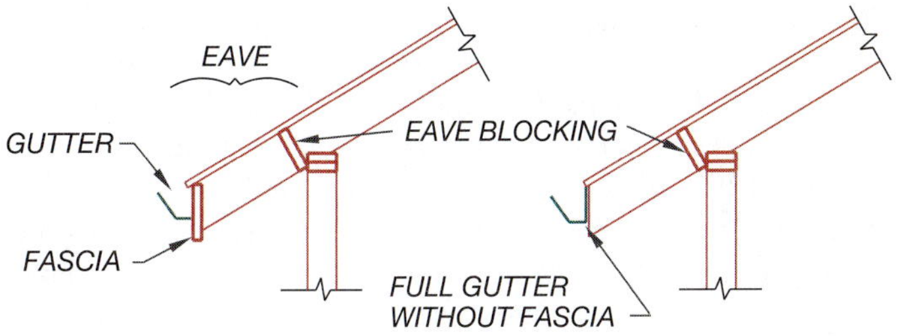

FIGURE 26.28 ■ Eave components.

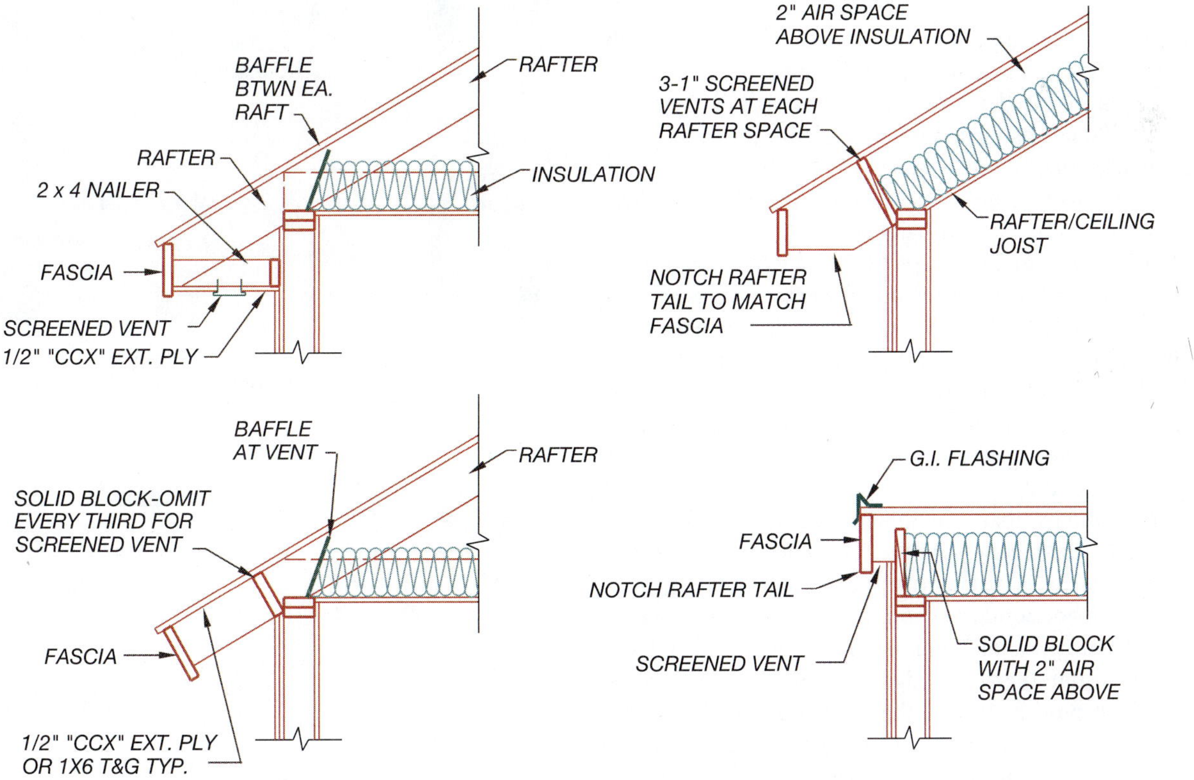

FIGURE 26.27 ■ Typical methods for constructing a cornice.

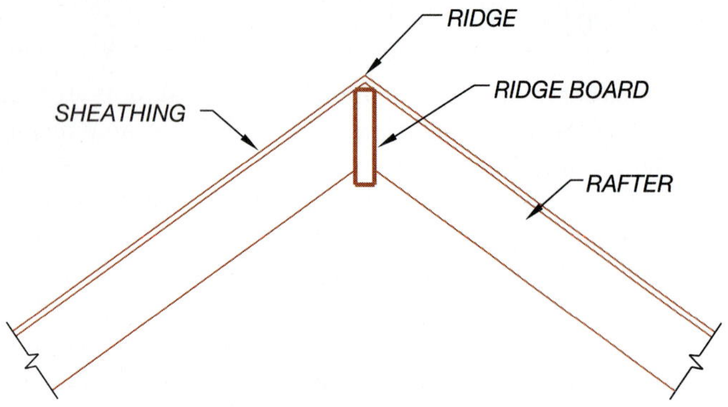

FIGURE 26.29 ■ Ridge construction.

may be mounted. The fascia can be made from either 1× or 2× material, depending on the need to resist warping. See Figure 26.28. The fascia is typically 2″ (50 mm) deeper than the rafter or truss tails. At the opposite end of the rafter from the fascia is the ridge. The *ridge* is the highest point of a roof and is formed by the intersection of the rafters or the top chords of a truss. See Figure 26.29.

Roof sheathing is similar to wall and floor sheathing in that it is used to cover the structural members. Roof sheathing may be either solid or skip. The area of the country and the finished roofing to be used will determine which type of sheathing is used. For solid sheathing, 1/2″ (13 mm) thick CDX plywood is generally used. CDX is the specification given by the American Plywood Association to designate standard-grade plywood. It provides an economical, strong covering for the framing, as well as an even base for installing the finished roofing. Common span ratings used for residential roofs are 24/16, 32/16, 40/20, and 48/24. Figure 26.30 shows plywood sheathing being applied to a conventionally framed roof. OSB is also used as a roof sheathing.

Skip sheathing is used with either tile or shakes. Typically 1 × 4s (25 × 100 mm) are laid perpendicular to the rafters with a 4″ (100 mm) space between pieces of sheathing. See Figure 26.31. A water-resistant sheathing must be used when the eaves are exposed to weather. This usually consists of plywood rated CCX or 1″ (25 mm) T&G decking. CCX is the specification for exterior use plywood. It is designated for exterior use because of the glue and the type of veneers used to make the panel.

The *finished roofing* is the weather protection system. Roofing might include built-up roofing, asphalt shingles, fiberglass shingles, cedar, tile, or metal panels. (See Chapter 20.) Flashing is generally 20 to 26 gage metal used at wall and roof intersections to keep water out.

Pitch, span, and overhang are dimensions that are needed to define the angle, or steepness, of the roof. Each can be seen in Figure 26.32. *Pitch,* used to describe the slope of the roof, is the ratio between the horizontal run and the vertical rise of the roof. The *run* is the horizontal measurement from the outside edge of the wall to the center line of the ridge. The *rise* is the vertical distance from the top of the wall to the highest point of the rafter being measured. Review Chapter 20 for a complete discussion of roof pitch. The *span* is the horizontal measurement between the inside edges of the supporting walls. (Chapter 27 will discuss span further.) The *overhang* is the horizontal measurement between the exterior face of the wall and the end of the rafter tail.

FIGURE 26.30 ■ Plywood with a span rating of 24/16 is typically placed over rafters to support the finish roofing material. *Courtesy Louisiana Pacific.*

FIGURE 26.31 ■ Skip or spaced sheathing is used under shakes and tile roofs. *Courtesy Saundra Powell.*

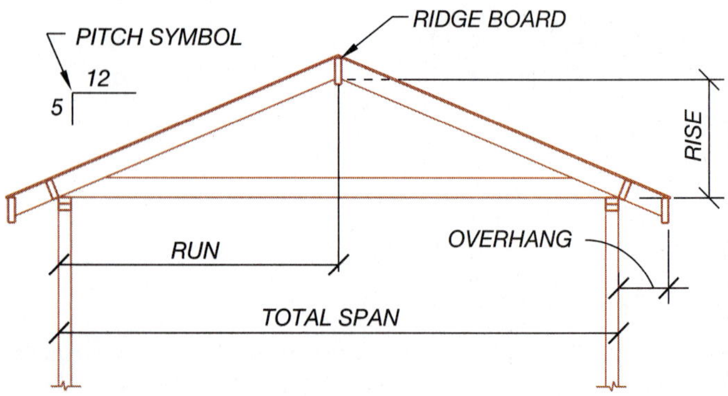

FIGURE 26.32 ■ Roof dimensions needed for construction.

FIGURE 26.33(A) ■ Engineered lumber used for rafters.

Conventionally Framed Roofs

Conventional, or stick, framing methods involve the use of wood members placed in repetitive fashion. Stick framing involves the use of members such as a ridge board, rafter, and ceiling joists. The *ridge board* is the horizontal member at the ridge, which runs perpendicular to the rafters. The ridge board is centered between the exterior walls when the pitch on each side is equal. The ridge board resists the downward thrust resulting from gravity trying to force the rafters into a V shape between the walls. The ridge board does not support the rafters but is used to align the rafters so that their forces are pushing against each other.

Rafters are the sloping members used to support the roof sheathing and finished roofing. Rafters are typically spaced at 24″ o.c. (600 mm), but 12″ and 16″ (300 and 400 mm) spacings are also used. Engineered rafters similar to Figure 26.33*a* are also used in residential construction. Figure 26.33*b* provides an example of the drawings supplied by an engineered joist manufacturer. Figures 26.33*c* and *d* show common construction connections of engineered rafters. There are various

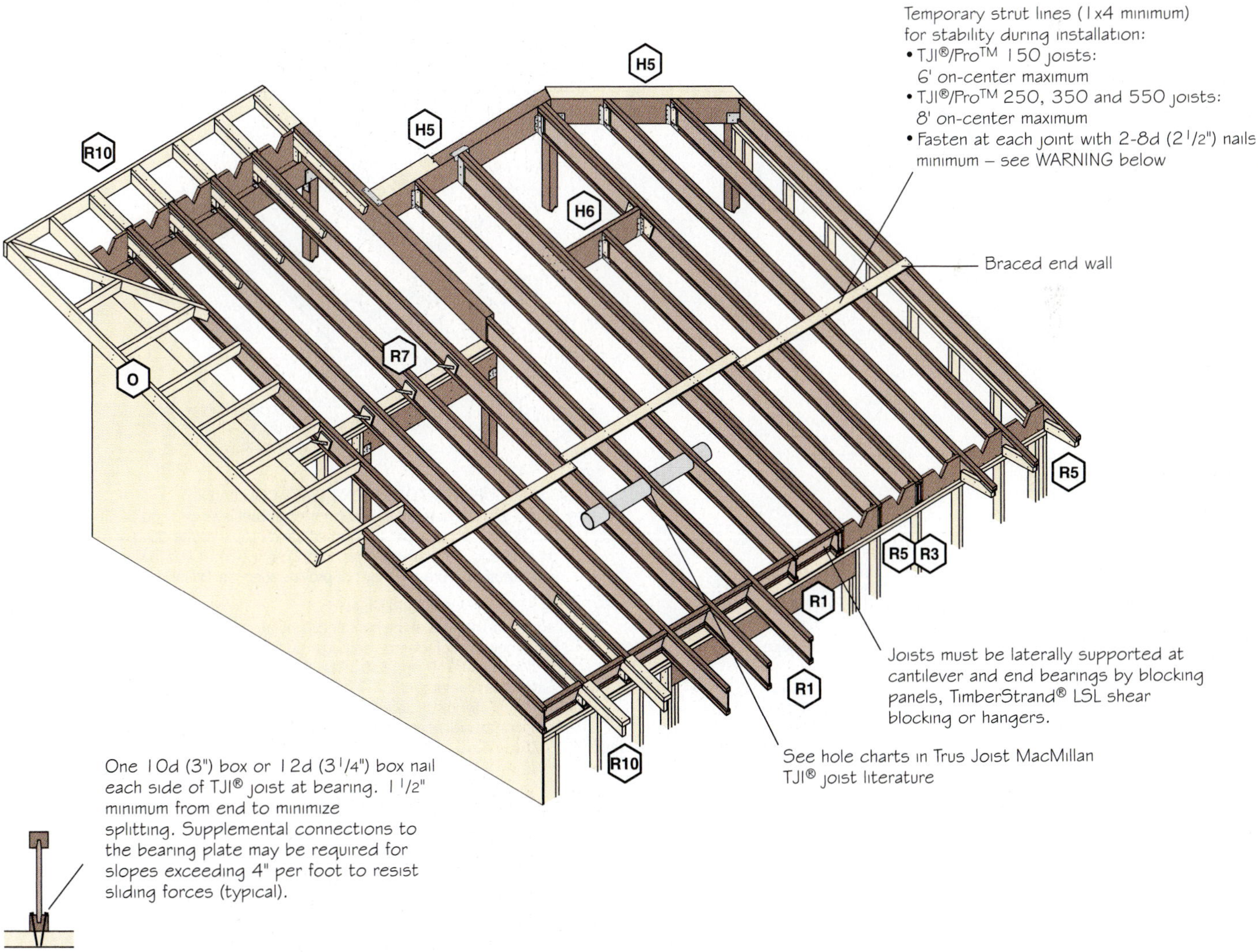

FIGURE 26.33(B) ■ Construction of a roof using engineered rafters. *Courtesy Trus Joist MacMillan.*

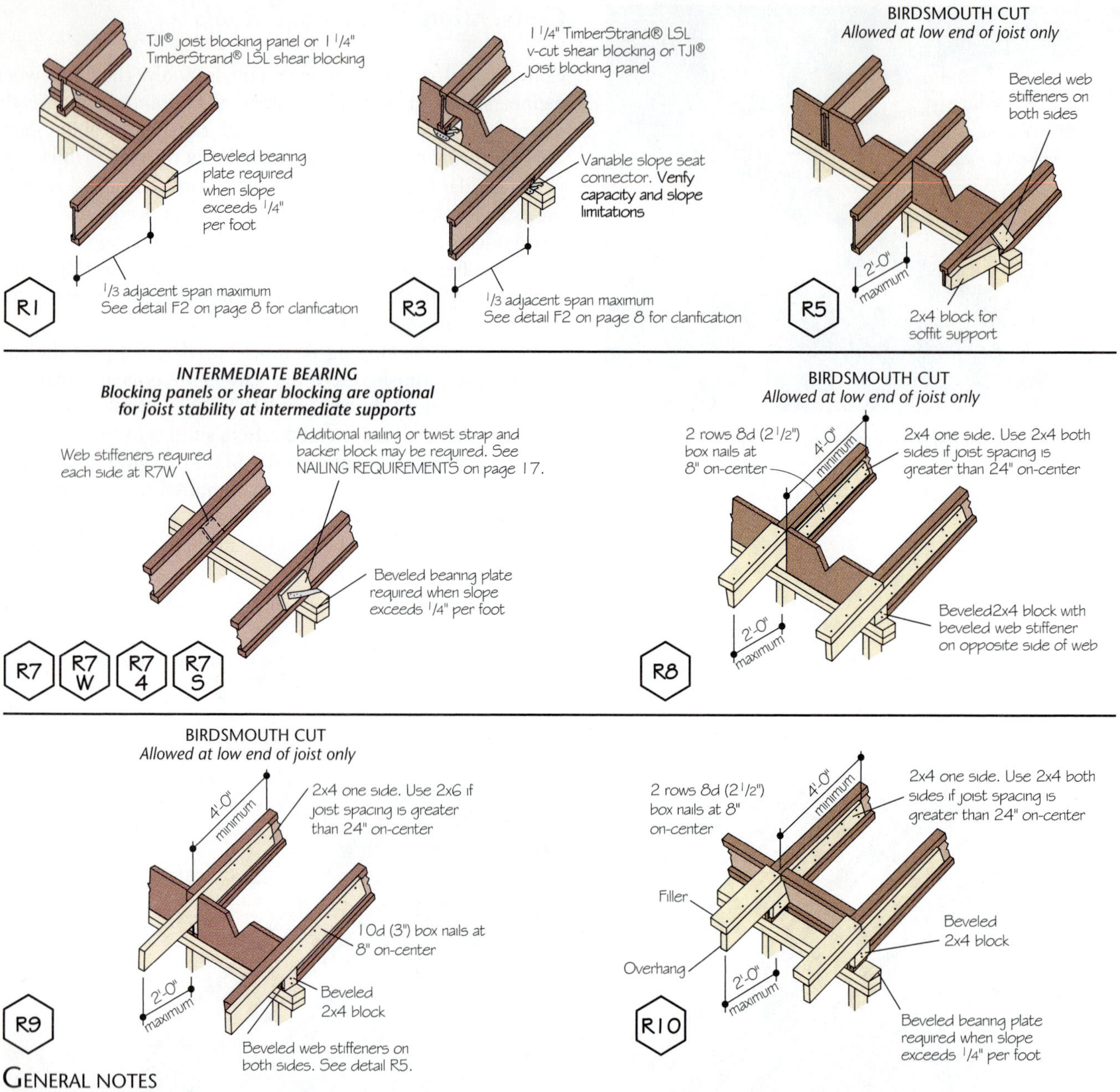

General Notes

MINIMUM BEARING LENGTH

- At joist ends: 1 3/4".
- At intermediate supports: 3 1/2".

SLOPE/BEVEL PLATE CRITERIA

- Unless otherwise noted, all details are valid to maximum 12" per foot slope.
- Supplemental connections to the bearing plate may be required for slopes exceeding 4" per foot to resist sliding forces.
- Wood bearing surfaces: Sloped bearing surface required when slope exceeds 1/4" per foot. This can be accomplished by using:
  - Beveled bearing plate.
  - Variable slope seat connector (verify connector capacity, see pages 20 and 21).
  - Birdsmouth cut (see detail BC). Allowed at low end of joist only.
- Hangers: Sloped seats and beveled web stiffeners required when slope exceeds 1/2" per foot.

LATERAL SUPPORT TO PREVENT JOIST ROLLOVER

- All roof joists must be laterally supported at cantilever and end bearings Use TJI®joist blocking panels, TimberStrand® LSL shear blocking or metal cross bracing. Attach metal cross bracing with 2-10d x 1 1/2" nails at each end. Metal cross bracing may not provide adequate lateral load transfer.

WEB STIFFENER REQUIREMENTS

- Required if the sides of the hanger do not laterally support the TJI® joist top flange or per footnotes on pages 20 and 21.
- TJI®/Pro™ 150, 250, and 350 joists: Required at all sloped hanger and birdsmouth cut locations.
- TJI®/Pro™ 550 joists: Required at all sloped hanger and birdsmouth cut locations, as well as any hanger locations where joist reaction exceeds 1475 lbs.

FIGURE 26.33(C) ■ Common construction details for engineered rafters. *Courtesy Trus Joist MacMillan.*

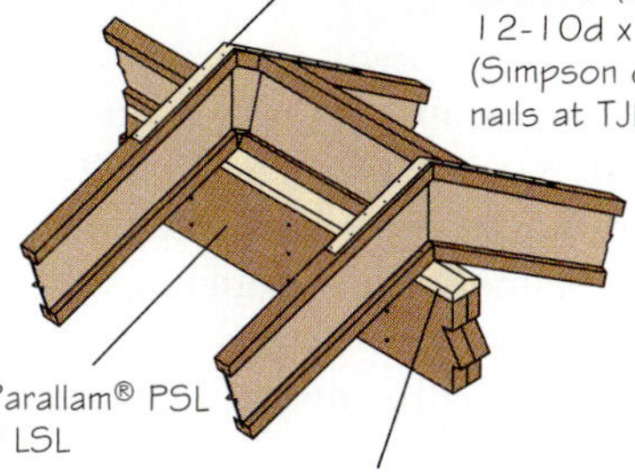
LSTA15 (Simpson or USP) strap with 12-10d x 1 1/2" nails, Use LSTA24 (Simpson or USP) strap with 18-10d x 1 1/2" nails at TJI®/Pro™ 550 joists.
Microllam® LVL, Parallam® PSL or TimberStrand® LSL
Double beveled bearing plate required when slope exceeds 1/4" per foot
R14

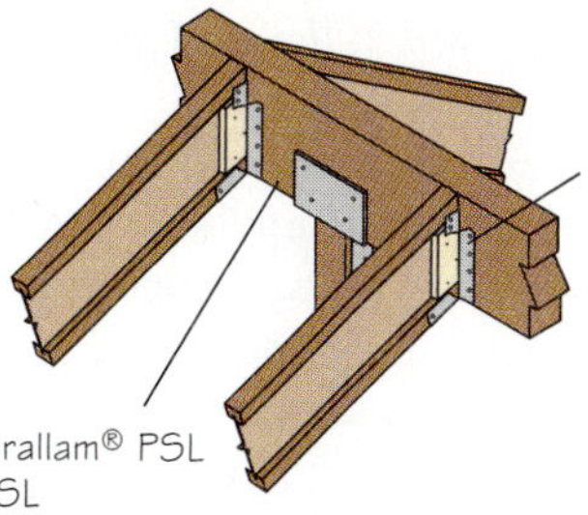
Simpson LSSU hanger with beveled web stiffeners. LSSU hangers allowed with 9 1/2", 11 7/8" and 14" TJI® joists only.
Microllam® LVL, Parallam® PSL or TimberStrand LSL
H5

Simpson LSTA15 strap with 12-10d x 1 1/2" nails may be required with hangers other than LSSU when slope exceeds 7"/12". Use LSTA24 strap with 18-10d x 1 1/2" nails at TJI® /Pro™ 550 joists.

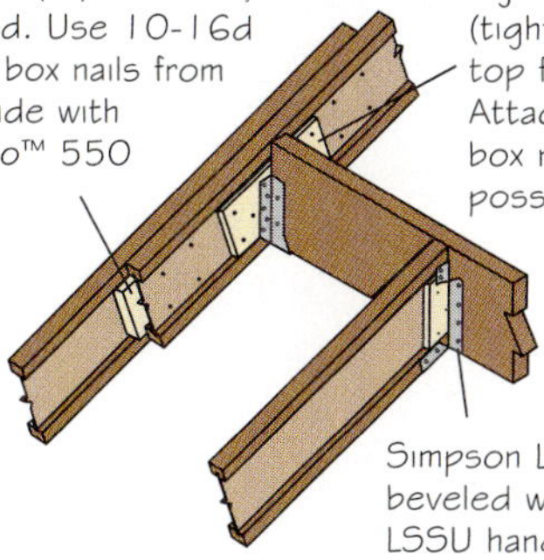
Filler block: Attach with 10-10d (3") box nails, clinched. Use 10-16d (3 1/2") box nails from each side with TJI®/Pro™ 550 joists.
Backer block: Install tight to bottom flange (tight to top flange with top flange hangers). Attach with 10-10d (3") box nails, clinched when possible.
Simpson LSSU hanger with beveled web stiffeners. LSSU hangers allowed with 9 1/2", 11 7/8" and 14" TJI® joists only.
H6

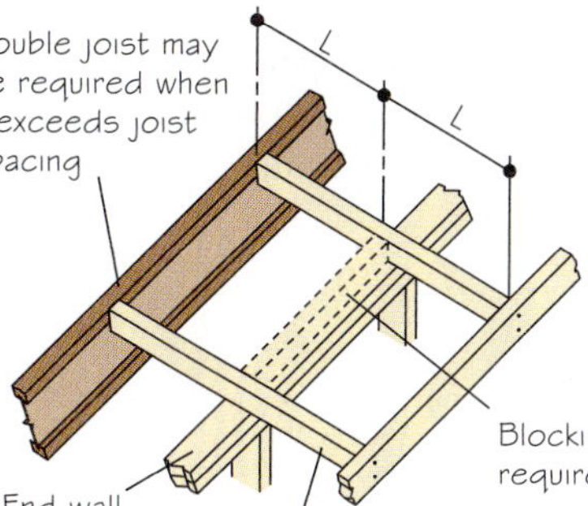
Double joist may be required when L exceeds joist spacing
L
L
Blocking as required
End wall
2x overhang. Notch around TJI® joist top flange.
O

BIRDSMOUTH CUT

Allowed at low end of joist only

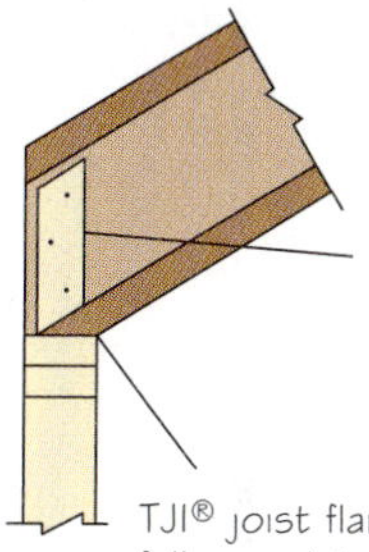
Beveled web stiffener each side of TJI® joist web
TJI® joist flange must bear fully on plate. Birdsmouth cut must not overhang inside face of plate.
BC

SHEAR BLOCKING AND VENTILATION HOLES ROOF ONLY

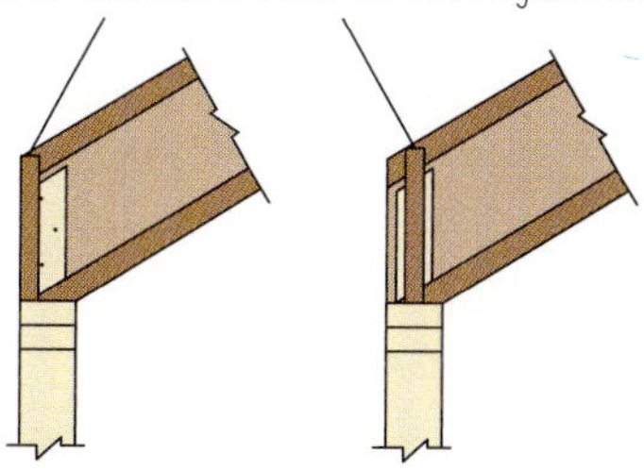
1 1/4" TimberStrand® LSL shear blocking (between joists) may be field trimmed to match joist depth at outer edge of wall or located on wall to match joist depth
SB

Vertical depth bearing of TJI® joists with high slopes (10"/12" to 12"/12") requires that TimberStrand® LSL shear blocking be one size deeper than the TJI® joist

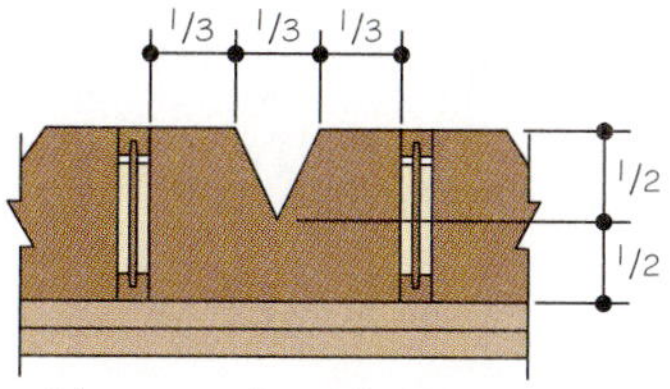
1/3
1/3
1/3
1/2
1/2

Maximum allowable V-cut

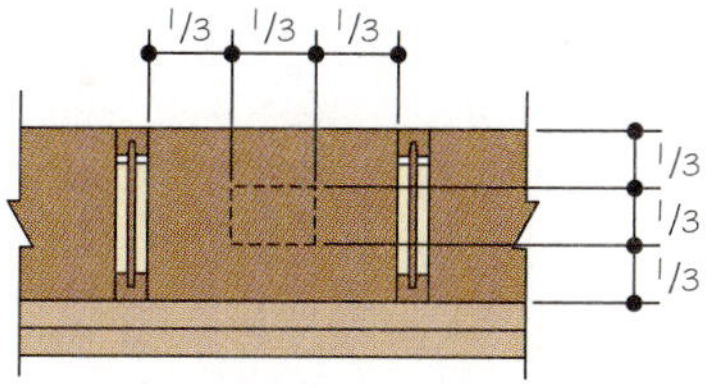
1/3
1/3
1/3
1/3
1/3
1/3

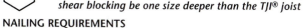

Allowed hole zone for round, square or rectangular holes

NAILING REQUIREMENTS

- **TJI® joists at end bearings:** 2-10d (3") box or 12d (3 1/4") box nails (1 each side), 1 1/2" minimum from end.
- **TJI® joists at intermediate bearings:**
  Roof slopes less than 4" per foot: 2-10d (3") box or 12d (3 1/4") box nails (1 each side). See detail R7.
  Roof slopes from 4" to 5" per foot: 4-10d (3") box or 12d (3 1/4") box nails (2 each side). See detail R74.
  Roof slopes greater than 5" per foot: 4-10d (3") box or 12d (3 1/4") box nails (2 each side) plus a twist strap and backer block. See detail R7S.
- **Blocking panels or shear blocking to bearing plate:**
  TJI® joist blocking panels: 10d (3") box nails at 6" on-center.
  TimberStrand® LSL shear blocking: Toenail with 10d (3") box nails at 6" on-center or 16d (3 1/2") box nails at 12" on-center.
  Shear transfer: Connections equivalent to decking nail schedule.

FILLER AND BACKER BLOCK SIZES

TJI®/Pro™	150	250		350		550	
DEPTH	9 1/2" or 11 7/8"	9 1/2" or 11 7/8"	14" or 16"	11 7/8"	14" or 16"	11 7/8"	14" or 16"
FILLER BLOCK (Detail H6)	1 1/8" net	2x6	2x8	2x6 +1/2" plywood	2x8 +1/2" plywood	2-2x6	2-2x8
BACKER BLOCK (Detail H6)	1/2" or 5/8"	5/8" or 3/4"	5/8" or 3/4"	1" net	1" net	2x6	2x8

If necessary, increase filler and backer block height for face mount hangers. Maintain 1/8" gap at top of joist; see detail W.
Filler and backer block dimensions should accommodate required nailing without splitting.

FIGURE 26.33(D) ■ Common construction details for engineered rafters. *Courtesy Trus Joist MacMillan.*

kinds of rafters: *common, hip, valley,* and *jack*. Each can be seen in Figure 26.34.

A *common rafter* is used to span and support the roof loads from the ridge to the top plate. Common rafters run perpendicular to both the ridge and the wall supporting them. The upper end rests squarely against the ridge board, and the lower end receives a bird's mouth notch and rests on the top plate of the wall. A *bird's mouth* is a notch cut in a rafter at the point where the rafter intersects a beam or bearing wall. This notch increases the contact area of the rafter by placing more rafter surface against the top of the wall, as shown in Figure 26.35.

Hip rafters are used when adjacent slopes of the roof meet to form an inclined ridge. The hip rafter extends diagonally across the common rafters and provides support to the upper end of the rafters. See Figure 26.36. The hip is inclined at the same pitch as the rafters. A *valley rafter* is similar to a hip rafter. It is inclined at the same pitch as the common rafters that it supports. Valley rafters get their name because they are located where adjacent roof slopes meet to form a valley. *Jack rafters* span from a wall to a hip or valley rafter. They are similar to a common rafter but span a shorter distance. Typically, a section will show only common rafters, with hip, valley, and jack rafters reserved for a very complex section.

Rafters tend to settle because of the weight of the roof and because of gravity. As the rafters settle, they push supporting walls outward. These two actions, downward and outward, require special members to resist them. These members are *ceiling joists, ridge bracing, collar ties, purlins,* and *purlin blocks and braces*. Each can be seen in Figure 26.37. When the ceiling joist are laid perpendicular to the rafters, a metal strap must be placed over the ridge to keep the rafters from separating, as shown in Figure 26.38.

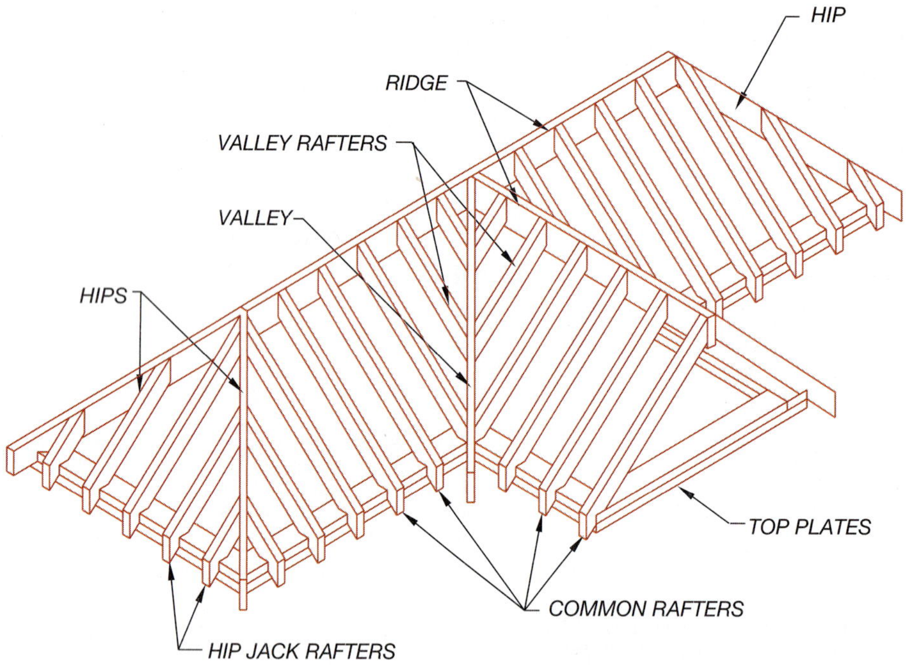

FIGURE 26.34 ■ Roof members in conventional construction.

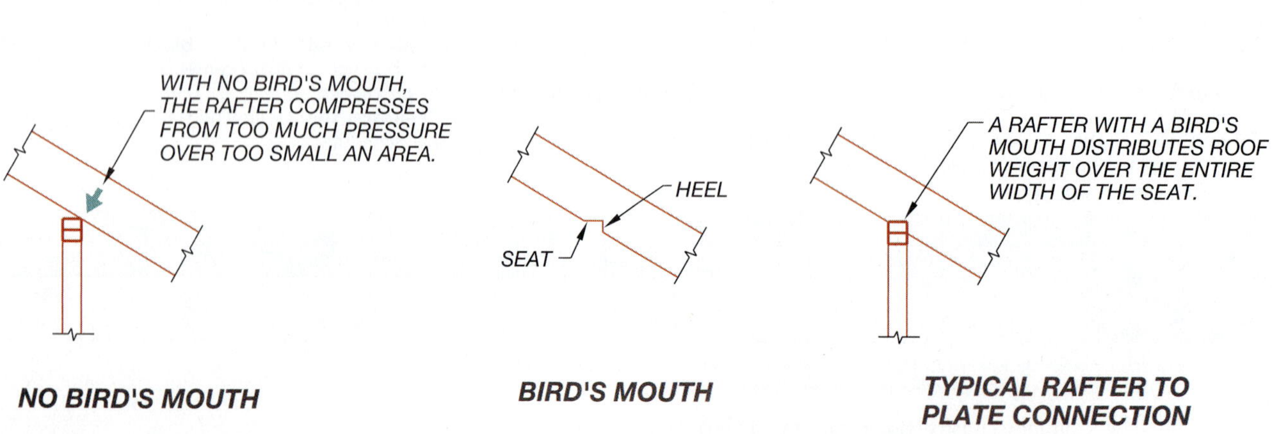

FIGURE 26.35 ■ A bird's mouth is a notch cut into the rafter to increase the bearing surface.

FIGURE 26.36 ■ A hip rafter is an inclined ridge board used to frame an exterior corner. *Courtesy Michelle Cartwright.*

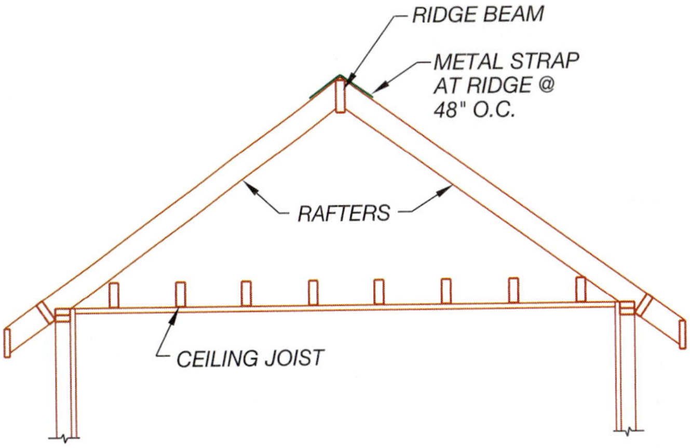

FIGURE 26.38 ■ When joist are perpendicular to the rafters, a metal strap can be used to connect the rafters to the ridge and resist the outward force of the rafters.

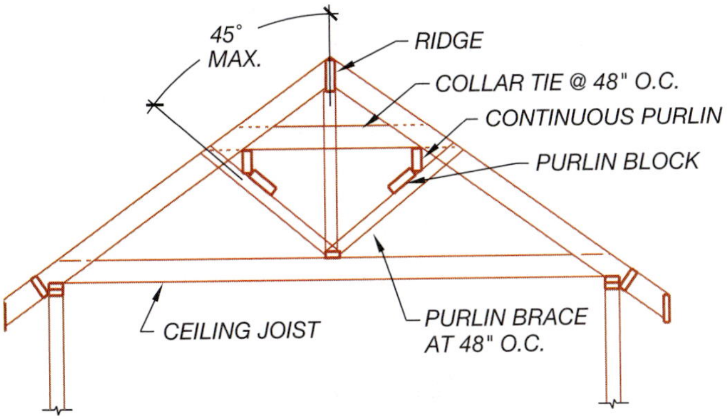

FIGURE 26.37 ■ Common roof supports.

Ceiling joists span between the top plates of bearing walls to resist the outward force placed on the walls from the rafters. The ceiling joists also support the finished ceiling. *Collar ties* are also used to help resist the outward thrust of the rafters. They are usually the same cross-section size as the rafter and placed in the upper third of the roof.

Ridge braces are used to support the downward action of the ridge board. The brace is usually a 2 × 4 (50 × 100) spaced at 48″ (1200 mm) o.c. maximum. The brace must be set at 45° maximum from vertical. A *purlin* is a brace used to provide support for the rafters as they span between the ridge and the wall. The purlin is usually the same cross-section size as the rafter and is placed below the rafter to reduce the span. As the rafter span is reduced, the size of the rafter can be reduced. See Chapter 28 for a further explanation of rafter sizes. *Purlin braces* are used to support the purlins. They are typically 2 × 4s (50 × 100s) spaced at 48″ (1220 mm) o.c. along the purlin and transfer weight from the purlin to a supporting wall. The brace is supported by an interior wall, or a 2 × 4 (50 × 100) lying across the ceiling joist. It can be installed at no more than 45° from vertical. A scrap block of wood is used to keep the purlin from sliding down the brace. When there is no wall to support the ridge brace, a strong back is added. A *strong back* is a beam placed over the ceiling joist to support the ceiling and roof loads. Figure 26.39 shows a strong back.

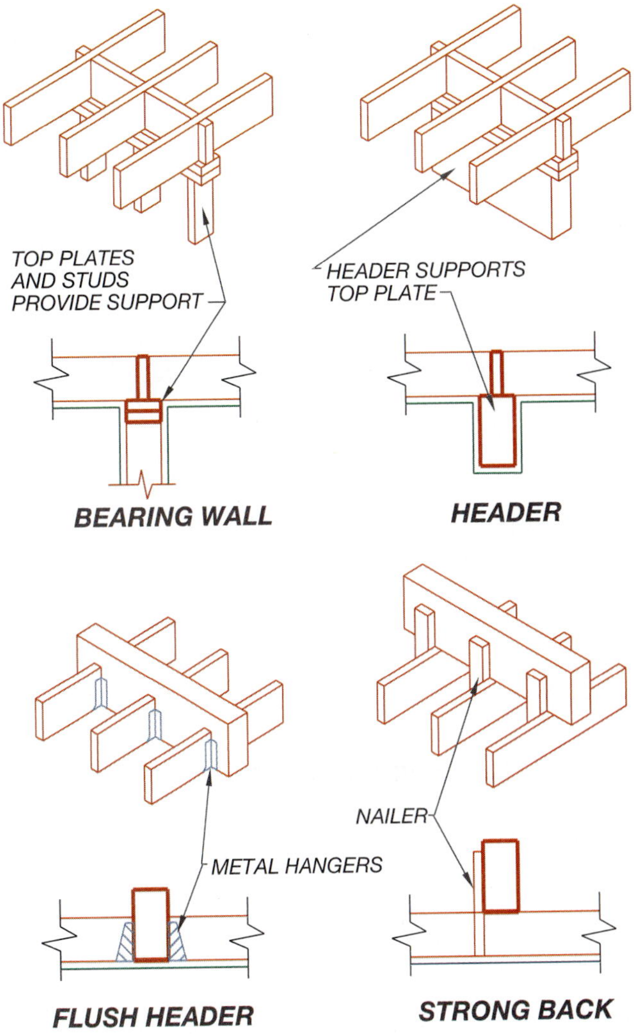

FIGURE 26.39 ■ Common methods of supporting loads include bearing walls, headers, flush headers, and the strongback. When a header is used, the loads rest on the header. With a flush header, the ceiling joist are supported by metal hangers. A piece of scrap wood is used to hang the ceiling joist to the strongback.

If a vaulted ceiling is to be represented, two additional terms will be used on the sections: *rafter/ceiling joist* and *ridge beam.* Both can be seen in Figure 26.40. A *rafter/ceiling joist,* or *rafter joist,* is a combination of rafter and ceiling joist. The rafter/ceiling joist is used to support both the roof loads and the finished ceiling. Typically a 2 × 12 (50 × 300 mm) rafter/ceiling joist is used to allow room for 10″ (250 mm) of insulation and 2″ (50 mm) of airspace above the insulation. The size of the rafter/ceiling joist must be determined by the load and span.

A *ridge beam* is used to support the upper end of the rafter/ceiling joist. Since there are no horizontal ceiling joists, metal joist hangers must be used to keep the rafters from separating from the ridge beam.

The final terms that you will need to be familiar with to draw a stick roof are *header* and *trimmer.* Both terms are used in wall construction, and they have a similar function when used as roof members. See Figure 26.41. A *header* at the roof level consists of two members nailed together to support rafters around an opening such as a skylight or chimney. *Trimmers* are two rafters nailed together to support the roofing on the inclined edge of an opening.

Truss Roof Construction

A *truss* is a component used to span large distances without intermediate supports. Residential trusses can be as short as 15′ (4570 mm) or as long as 50′ (15 200 mm). Trusses can be either prefabricated or job-built. Prefabricated trusses are commonly used in residential construction. Assembled at the truss company and shipped to the job site, the truss roof can quickly be set in place. A roof that might take two or three days to frame using conventional framing can be set in place in two or three hours using trusses, which are set in place by crane. The size of material used to frame trusses is smaller than with conventional framing. Typically, truss members need be only 2 × 4s (50 × 100 mm) set at 24″ (600 mm) o.c. The exact size of the truss members will be determined by an engineer working for the truss manufacturer. The drafter's responsibility when drawing a structure framed with trusses is to represent the span of the trusses on the framing plan and show the general truss shape and bearing points in the section drawings. Chapter 30 will introduce framing plans, and Chapter 35 will introduce sections.

A knowledge of truss terms is helpful in making these drawings. Terms that must be understood are *top chord, bottom chord, webs, ridge block,* and *truss clips.* Each is shown in Figure 26.42. The *top chord* serves a function similar to a

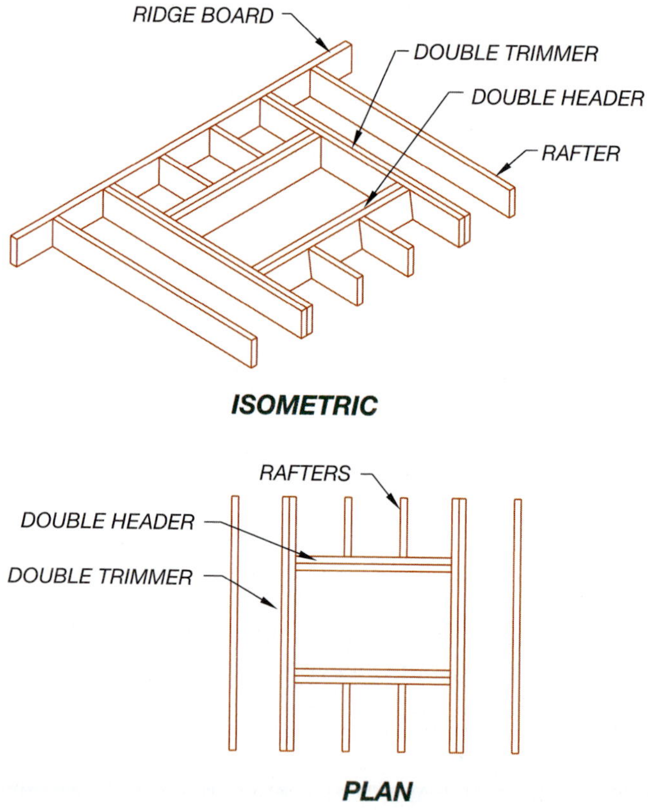

FIGURE 26.41 ■ Typical construction at roof opening.

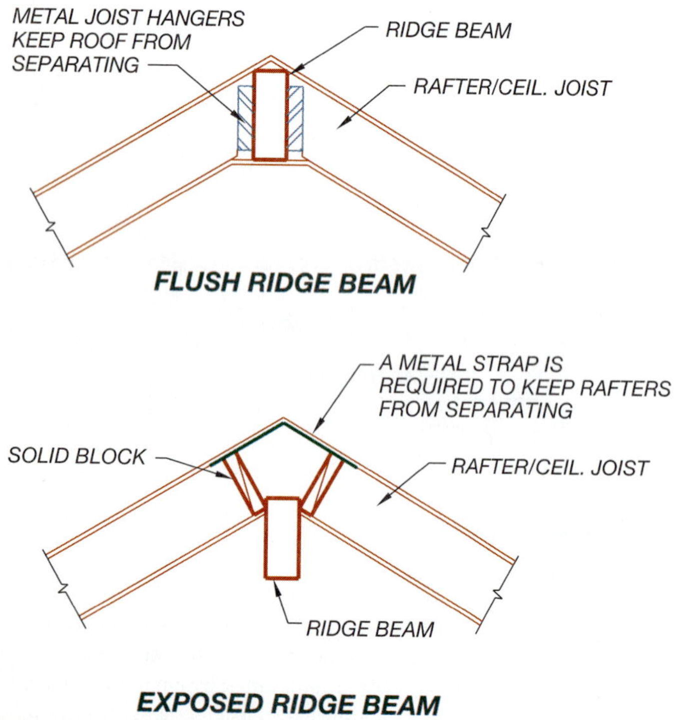

FIGURE 26.40 ■ Common connections between the ridge beam and rafters. The ridge may be exposed or hidden.

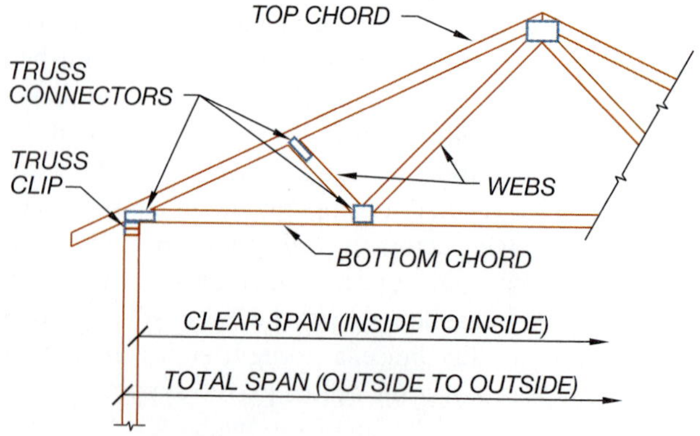

FIGURE 26.42 ■ Truss construction members.

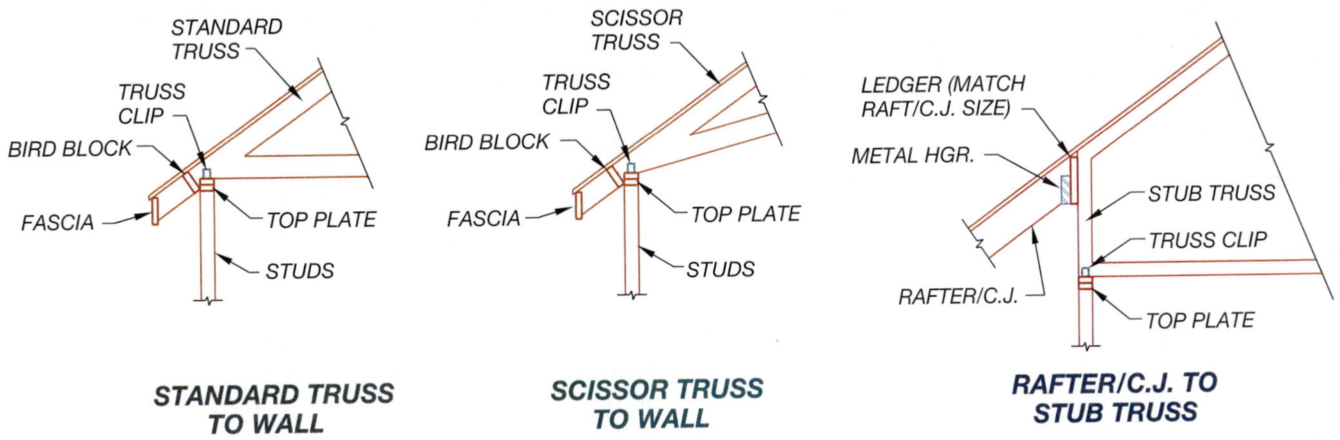

FIGURE 26.43 ■ Common truss connections normally shown in the sections. Notice that the top chord aligns with the outer face of the top plate. When detailing a scissor truss, the bottom chord is typically drawn a minimum of two pitches less than the top chord.

rafter. It is the upper member of the truss and supports the roof sheathing. The *bottom chord* serves a purpose similar to a ceiling joist. It resists the outward thrust of the top chord and supports the finished ceiling material. *Webs* are the interior members of the truss that span between the top and bottom chord. They are attached to the chords by the manufacturer using metal plate connectors. *Ridge blocks* are blocks of wood used at the peak of the roof to provide a nailing surface for the roof sheathing and as a spacer in setting the trusses into position. *Truss clips,* also known as *hurricane ties,* are used to strengthen the connection between the truss and top plate or header, which is used to support the truss. The truss clips transfer wind forces applied to the roof, which cause uplift down through the wall framing into the foundation. A block is also used where the trusses intersect a support. Common truss intersections with blocking and hurricane ties are shown in Figure 26.43.

The truss gains its strength from triangles formed throughout it. The shape of each triangle cannot be changed unless the length of one of the three sides is altered. The entire truss will tend to bend under the roof loads as it spans between its bearing points. Most residential trusses can be supported by a bearing point at or near each end of the truss. As spans exceed 40′ or as the shape of the bottom chord is altered, it may be more economical to provide a third bearing point at or near the center. For a two-point bearing truss, the top chords are in compression from the roof loads and tend to push out the heels and down at the center of the truss. The bottom chord is attached to the top chords and is in tension as it resists the outward thrust of the top chords. Webs closest to the center are usually stressed by tension, and the outer webs are usually stressed by compression. Figure 26.44 shows how the tendency of the truss to bend under the roof loads is resisted and the loads are transferred to the bearing points.

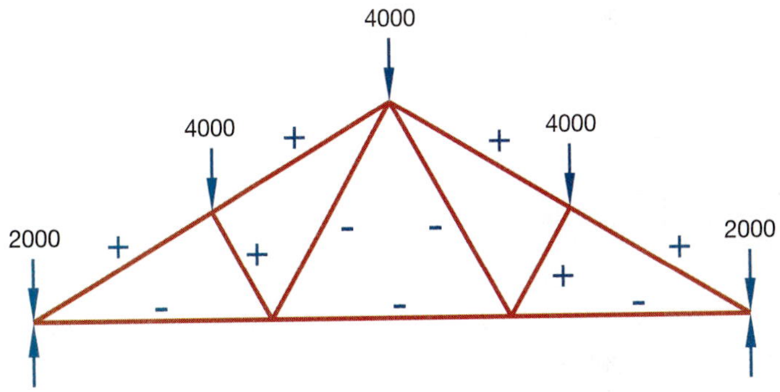

FIGURE 26.44 ■ Trusses are designed so that the weight to be supported is spread to the outer walls. This is done by placing some members in tension and some in compression. A member in compression is indicated by a plus sign (+), and one in tension is represented by a minus sign (−).

When they were introduced to the residential market, trusses were used primarily to frame gable roofs. Computers and sophisticated design software have enabled trusses to be manufactured easily in nearly any shape. Several of the common shapes and types of trusses available for residential roof construction can be seen in Figure 26.45. The truss most commonly used in residential construction is a *standard truss,* which is used to frame gable roofs. A *gable end wall truss* is used to form the exterior ends of the roof system and is aligned with the exterior side of the end walls of a structure. This truss is more like a wall than a truss, with the vertical supports typically spaced at 24″ (600 mm) o.c. to support exterior finishing material. A *girder truss* is used on houses with an L- or U-shaped roof where the roofs intersect. To form a girder truss, the manufacturer typically bolts two or three standard trusses together. The manufacturer determines the size and method of constructing the girder truss.

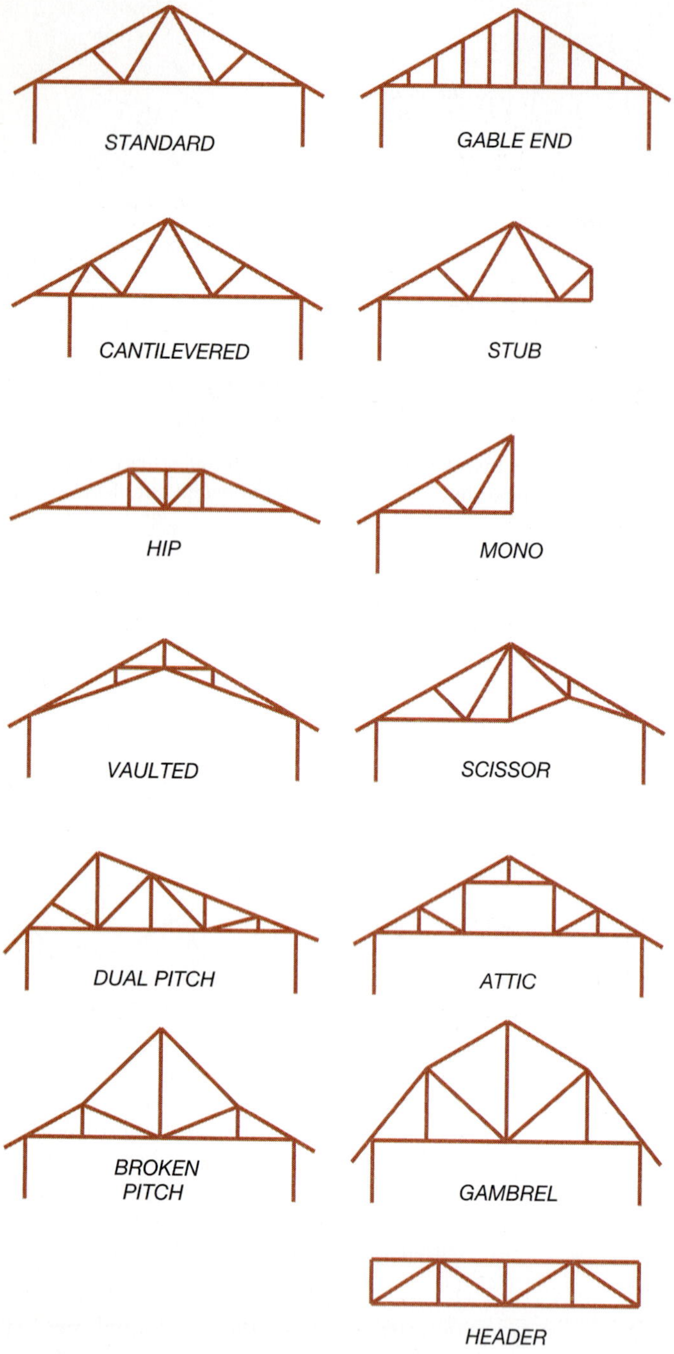

FIGURE 26.45 ■ Common types of roof trusses for residential construction.

The *cantilevered truss* is used where a truss must extend past its support to align with other roof members. Cantilevered trusses are typically used where walls jog to provide an interior courtyard or patio. A *stub truss* can be used where an opening will be provided in the roof or the roof must be interrupted. Rooms with glass skywalls, as in Figure 26.46, can often be framed using stub trusses. The shortened end of the truss can be supported by either a bearing wall, a beam, or by a header truss. A *header truss* has a flat top that is used to support stub trusses. The header truss has a depth to match that of the stub truss and is similar in function to a girder truss. The header truss spans between and is hung from two standard trusses. Stub trusses are hung from the header truss.

Hip trusses are used to form hip roofs. Each succeeding truss increases in height until the full height of the roof is achieved and standard trusses can be used. The height of each hip truss decreases as they get closer to the exterior wall, which is perpendicular to the ridge. Typically hip trusses must be 6′ (1500 mm) from the exterior wall to achieve enough height to support the roof loads. Figure 26.46 shows where hip trusses could be used to frame a roof. The exact distance will be determined by the truss manufacturer and will be further explained in Chapter 30. A *mono truss* is a single pitched truss, which can often be used in conjunction with hip trusses to form the external 6′ (1500 mm) of the hip. A mono truss is also useful in blending a one-level structure with a two-level structure, as seen in Figure 26.47.

If a vaulted roof is desired, vaulted or scissor trusses can be used. *Vaulted* and *scissor trusses* have inclined bottom chords. Typically there must be at least a two-pitch difference between the top and bottom chord. If the top chord is set at a 6/12 pitch, the bottom chord usually cannot exceed a 4/12 pitch. A section will need to be drawn to give the exact requirement for the vaulted portion of the truss, similar to Figure 26.48. If a portion of the truss needs to have a flat ceiling, it may be more economical to frame the lowered portion with conventional framing materials, a process often referred to as *scabbing on*. Although this process requires extra labor at the job site, it might eliminate a third bearing point near the center of the truss. Figure 26.49 shows an example of two areas with vaulted ceilings separated by a portion of a room with a flat ceiling.

METAL HANGERS

Metal hangers are used on floor, ceiling, and roof members, to keep structural members from separating. Figure 26.50 shows several common types of connectors that are used in light construction. These connectors keep beams from lifting off posts, keep posts from lifting off foundations, or hold one beam or joist to another.

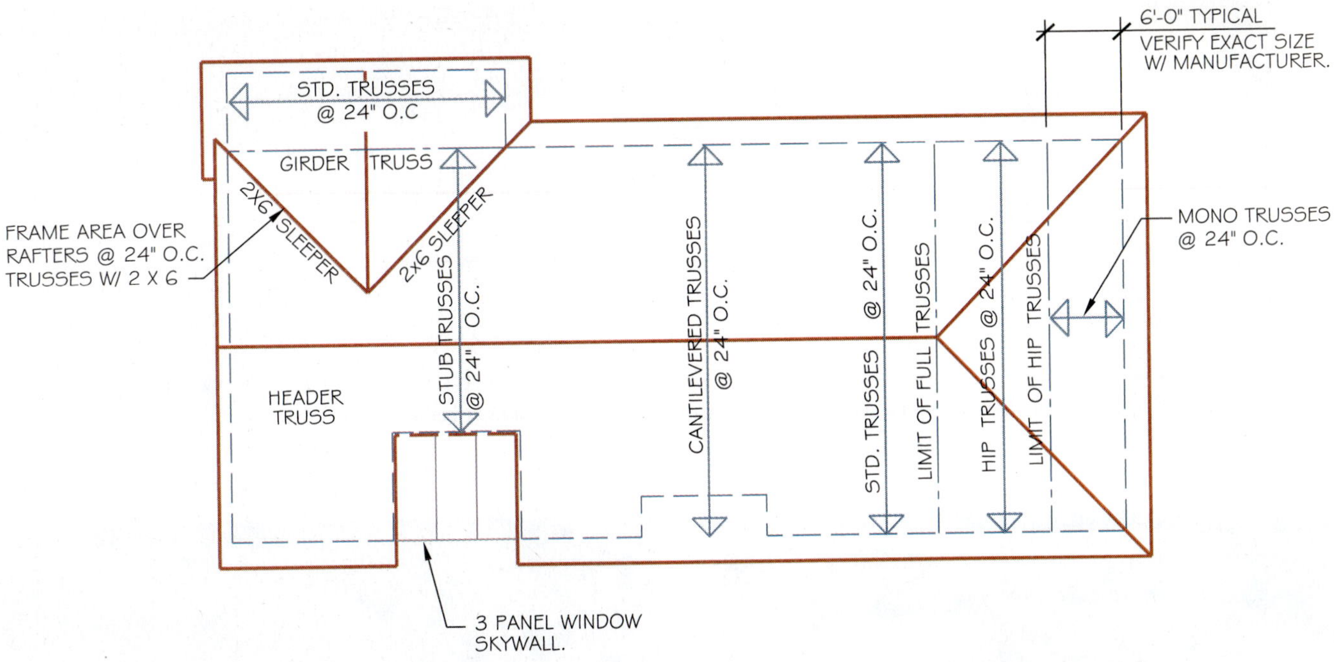

FIGURE 26.46 ■ Standard, girder, stub, cantilever, hip, and mono trusses can each be used to form different roof shapes.

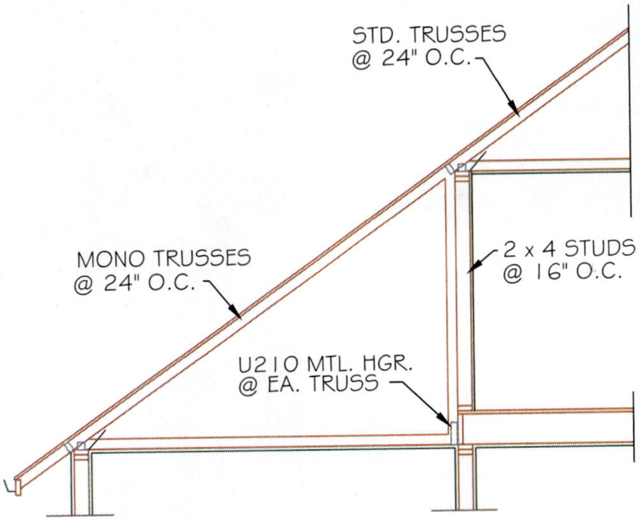

FIGURE 26.47 ■ A mono truss is often used to blend a one-story area into a two-story area.

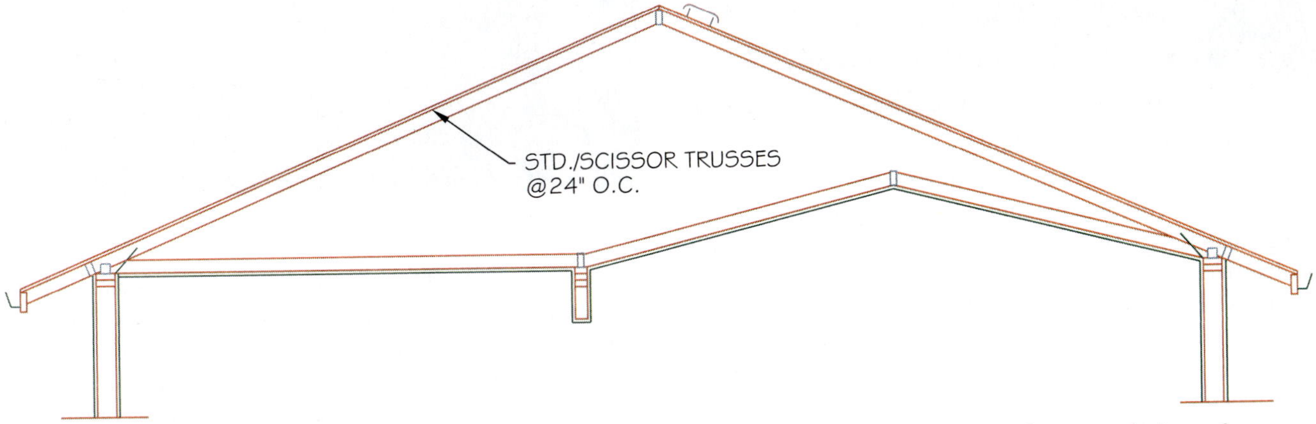

FIGURE 26.48 ■ A drawing of a standard/scissor truss is usually provided to show the manufacturer where to vault the roof.

Snow Loads

Snow loads may or may not be a problem in the area for which you are designing. You may be designing in an area where snow is something you dream about, not design for. If you are designing in an area where snow is something you shovel, it is also something you must allow for in your design. Because snow loads vary so greatly, the designer should consult the local building department to determine what amount of snow load should be considered. In addition to climatic variables, the elevation, wind frequency, duration of snowfall, and the exposure of the roof all influence the amount of live load design.

Dynamic Loads

Dynamic loads are those imposed on a structure from a sudden gust of wind or from an earthquake.

Wind Loads

Although wind design should be done only by a competent architect or engineer, a drafter must understand the areas of a structure that are subject to failure. Figure 27.1 shows a map of minimum basic wind speeds that should be designed for. Wind pressure creates wind loads on a structure. These loads vary greatly, depending on the location and the height of the structure above the ground. IRC defines four basic wind exposure categories based on ground surface irregularities of the job site. These include:

- Exposure A—A construction site in large city centers where 50 percent of the surrounding structures have a height of 70′ (21 300 mm) or greater for distance of 0.5 miles (0.8 km) or 10 times the building height.
- Exposure B—This wind exposure is the assumed design standard for the IRC. It includes urban, suburban,

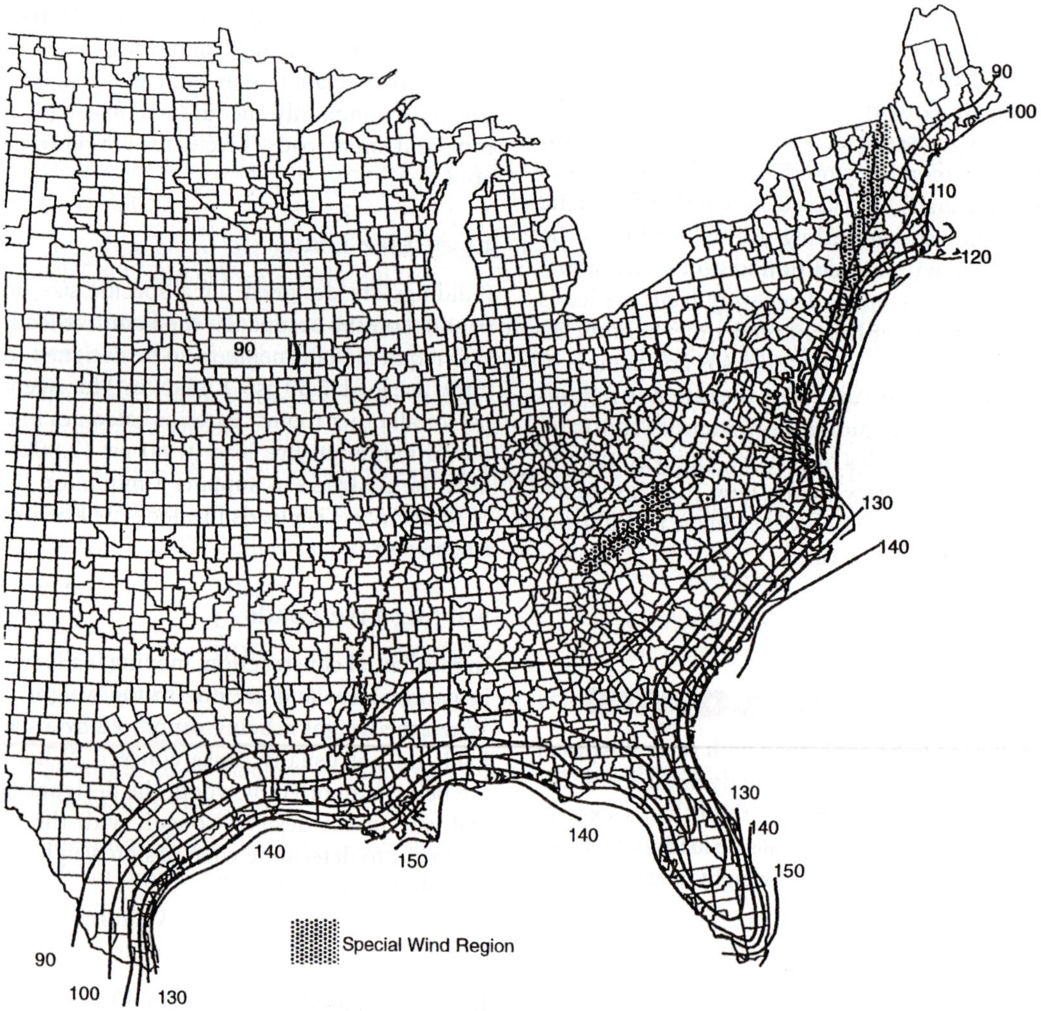

Location	V mph
Hawaii	105
Puerto Rico	145
Guam	170
Virgin islands	145
American Samoa	125

FIGURE 27.1 ■ Basic minimum wind speeds for the design of wind pressure on a residence. See the IRC for complete listings. *Reproduced from the* International Residential Code / 2000. *Copyright ©2000. Courtesy International Code Council, Inc.*

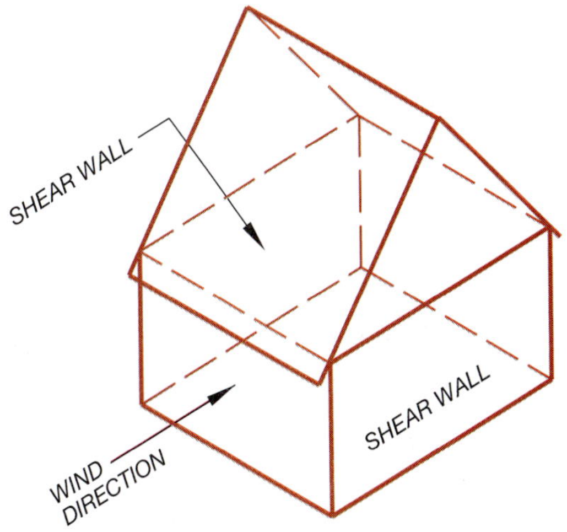

FIGURE 27.2 ■ Wind pressure on a wall will be resisted by the bolts connecting the wall sill to the foundation and by the roof structure. The wind is also resisted by shear walls, which are perpendicular to the wall under pressure.

wooded areas or other terrain with multiple closely spaced structures that are the size of single-family residence or larger.

- Exposure C—Open terrain with scattered obstructions with a height of less than 30′ (9100 mm) that extend more than 1500′ (457 200 mm) in any direction from the building site.
- Exposure D—Structures built on flat unobstructed sites within 1500′ (457 200 mm) of shorelines exposed to wind flowing over open bodies of water 1 mile (1.6 km) wide or larger. Sites built near shorelines in hurricane-prone regions are excluded from this category. Because such wide variations in wind speed can be encountered, the designer of the structure must rely on the local building department to provide information.

Figure 27.2 shows a simplified explanation of how winds can affect a house. Wind affects a wall just as it would the sail of a boat. With a boat, the desired effect is to move the boat. With a structure, this tendency to move must be resisted. The walls resisting the wind will tend to bow under the force of the wind pressure. The tendency can be resisted by roof and foundation members and perpendicular support walls. These supporting walls will tend to become parallelograms and collapse. The designer of the structure determines the anticipated wind speed and designs walls, typically referred to as *shear walls,* to resist this pressure.

In planning for loads from wind, prevailing wind direction cannot be assumed. Winds are assumed to act in any horizontal direction and will create a positive pressure on the windward side of a structure. A negative pressure on the leeward (downwind) side of the structure creates a partial vacuum. Design pressures for a structure are based on the total pressure a structure might be expected to encounter, which is equal to the sum of the positive and negative pressure. A design value of 30 psf (1.44kN/m^2) is common for residential projects, but this value should be verified with local building departments. Thirty psf (1.44kN/m^2) equals a wind speed of 108 miles per hour (mph). Other common values include:

PSF	N/M^2	MPH	KM/H	PSF	KN/M^2	MPH	KM/H
15	0.72	76	122	50	2.40	140	225
20	0.96	88	142	55	2.64	147	237
25	1.20	99	159	60	2.88	153	246
30	1.44	108	174	70	3.36	165	266
35	1.68	117	188	80	3.84	177	285
40	1.92	125	201	90	4.32	188	303
45	2.16	133	214	100	4.80	198	319

Areas subject to high winds from hurricanes or tornados should be able to withstand winds of 125 mph (201 km/h), with 3-second gusts as high as 160 mph (257 km/h). Using these values, the designer can determine existing wall areas and the resulting wind pressure that must be resisted. This information is then used to determine the size and spacing of anchor bolts and any necessary metal straps or ties needed to reinforce the structure.

Wind pressure is also critical to the design and placement of doors and windows. The design wind pressure will affect the size of the glazing area and the method of framing the rough openings for doors and windows. Openings in shear walls will reduce the effectiveness of the wall. Framing around openings must be connected to the frame and foundation to resist forces of uplift from wind pressure. *Uplift,* the tendency of members to move upward in response to wind or seismic pressure, is typically resisted by the use of steel straps or connectors that join the trimmer and king studs beside the window to the foundation or to framing members in the floor lever below. Wall areas with large areas of openings must also have design studies to determine the amount of wall area that will be required to resist the lateral pressure created by wind pressure. Figure 27.3 shows the relationship of openings to walls for shear resistance. As a general rule of thumb, a wall section 4′–0″ (1200 mm) is required for each 25 lineal feet (7600 mm) of wall. The amount of wall to be reinforced is based on the wall height, the seismic zone, the size of openings by the shear panel, and the method of construction used to reinforce the wall. Chapter 30 will explore methods of resisting the forces of shear using the prescriptive methods of the building codes.

Another problem created by wind pressure is the tendency of a structure to overturn. Structures built on pilings or other posted foundations allow wind pressure under the structure to exceed the pressure on the leeward side of the structure. Codes require that the resistance of the dead loads of a building to

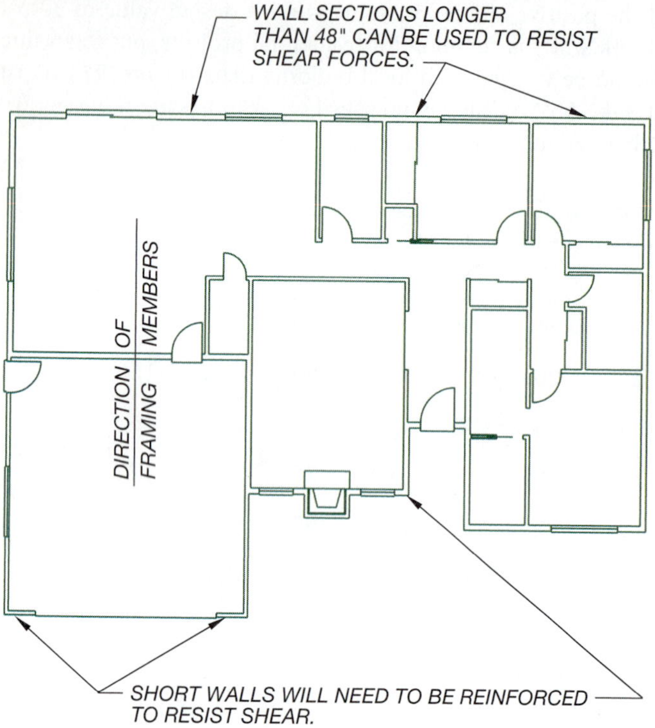

FIGURE 27.3 ■ Special provisions for shear walls are generally required where openings break up normal wall construction. For residential construction, 3/8–1/2″ (9.5–12.7 mm) plywood is typically used to reinforce the framing.

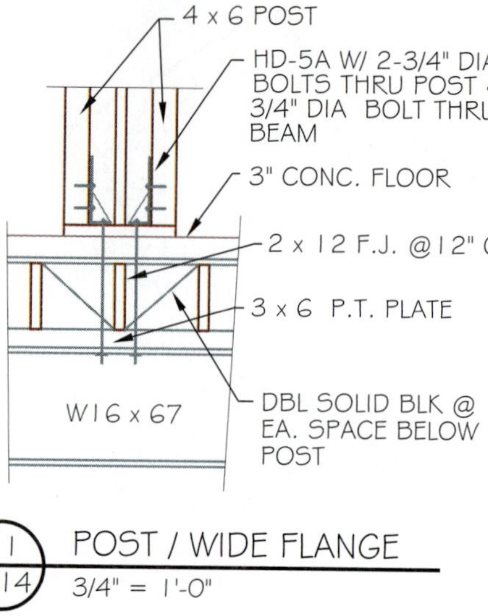

FIGURE 27.4 ■ Metal straps are used to transfer the loads from one floor through another and down to the foundation.

overturning be one and a half times the overturning effect of the wind. Metal ties are used to connect the walls to the floor, the floor joist to support beams, the support beams to supporting columns, and columns to the foundation. These ties are determined by an architect or engineer for each structure and are based on the area exposed to the wind and the wind pressure. Figure 27.4 shows an example of a detail designed by an engineer to resist the forces of uplift on a structure.

Seismic Loads

Seismic loads result from earthquakes. Figure 27.5 shows a seismic map of the United States and the risk in each area of damage from earthquakes. The IRC has specific requirements for seismic design based on the location and soil type of the building site and the building shape. Chapter 30 will explore building shape as it relates to construction materials. Chapter 31 will discuss soil types and their effect on structures. Six seismic zones are defined by IRC ranging from A through E. Zone A is the least prone to seismic damage and structures in zone E are most likely to be damaged by earthquakes. Zones A, B, and C do not require special construction methods. Special provisions are provided in the IRC for structures built in zones D1 and D2. Structures built in zone E are governed by the IBC.

Stress which results from an earthquake is termed a *seismic* load and is usually treated as a lateral load that involves the entire structure. Lateral loads created by wind affect only certain parts of a structure, while lateral forces created by seismic forces affect the entire structure. As the ground moves in varying directions at varying speeds, the entire structure is set in motion. Even after the ground comes to rest, the structure tends to wobble like Jell-O. Typically, structures fall to seismic forces in much the same way as they do to wind pressure. Intersections of roofs-to-wall, wall-to-floor and floor-to-foundation are critical to the ability of a structure to resist seismic forces. An architect or engineer should carefully design these connections. Typically, a structure must be fluid enough to move with the shock wave but so connected that individual components move as a unit and all units of a structure move as one. Figure 27.6 shows a detail of an engineer's design for a connection of a garage-door king stud designed to resist seismic forces. The straps that are attached to the wall cause the wall and foundation to move as one unit.

LOAD DESIGN

Once the floor plan and elevations have been designed, the designer can start the process of determining how the structure will resist the loads that will be imposed on it. The goal of load design is to achieve equilibrium between the structure and the forces that will act upon the structure. For the gravity loads pushing down on the structure, an opposite and equal force or *reaction* must resist the gravity loads. The structural members of the home form a load path to transfer gravity loads into the foun-

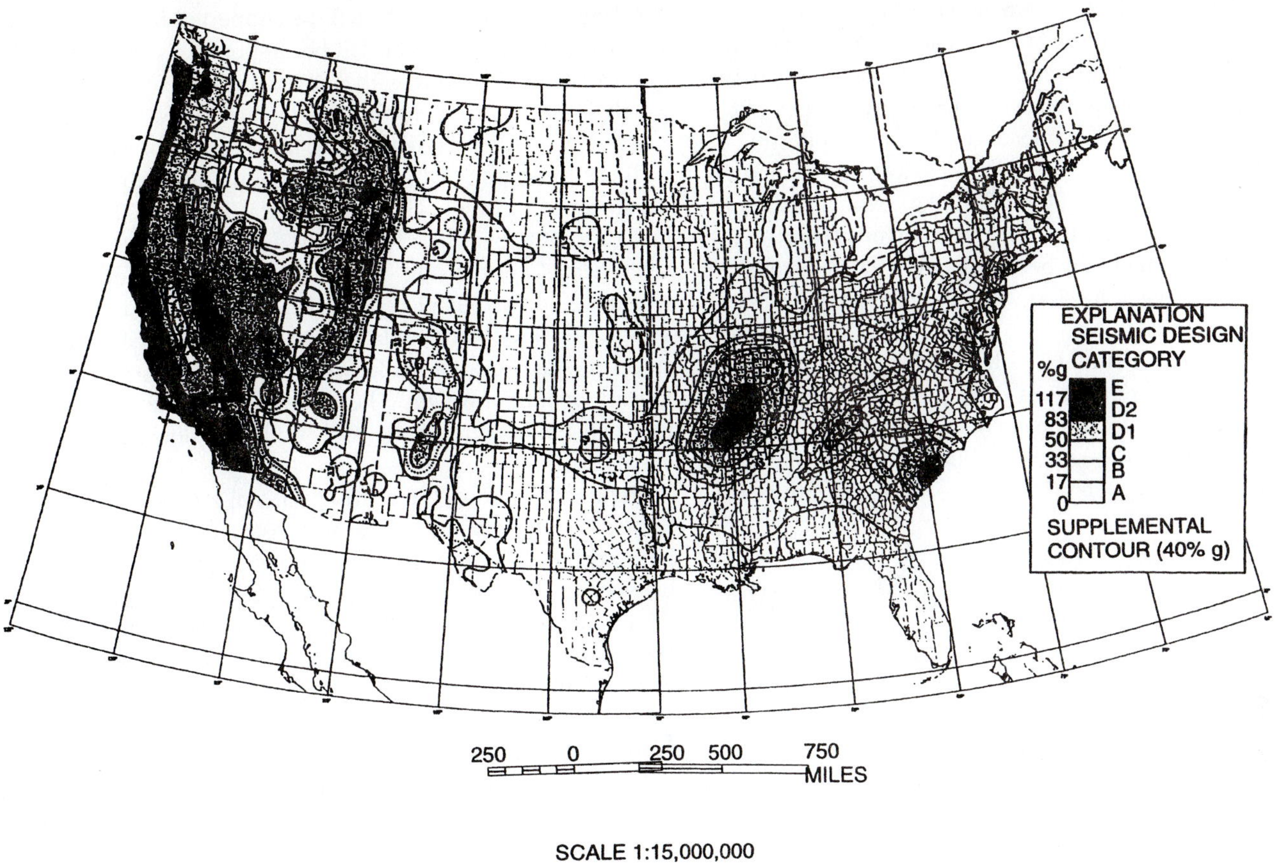

FIGURE 27.5 ■ Seismic zone map for site Class-D can be used to determine the potential risk of damage risk from earthquakes. *Reproduced from the* International Residential Code / 2000. *Copyright ©2000. Courtesy International Code Council, Inc.*

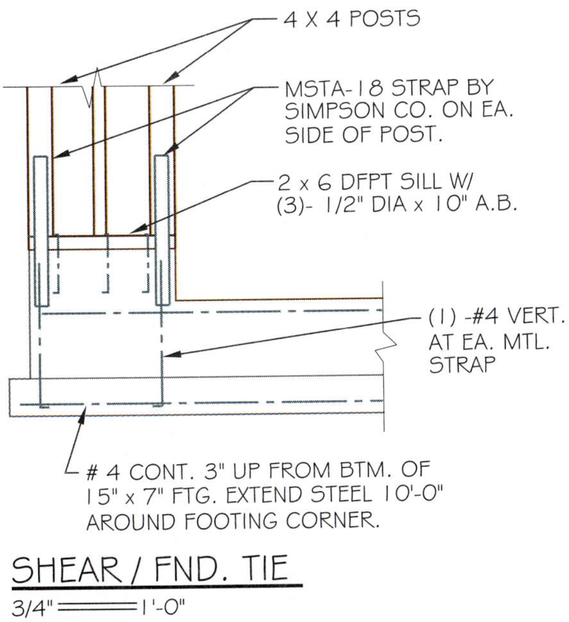

FIGURE 27.6 ■ Hold-down anchors are typically used to resist seismic forces and form a stable intersection between the floor and foundation.

dation and then into the soil. The *load path* is the route that is used to transfer the roof loads into the walls, then into the floor, and then to the foundation. The structural members must be of sufficient size to resist the loads that are above the member.

To determine the sizes of material required, it is always best to start at the roof and work down to the foundation. When you calculate from the top down, the loads will be accumulating; and when you work down to the foundation, you will have the total loads needed to size the footings.

As a beginning drafter you are not expected to design the structural components. In most offices, the structural design of even the simplest buildings is done by a designer or an engineer. The information in this chapter is a brief introduction into the size of structural members. The design of even simple beams can be very complex.

LOAD DISTRIBUTION

A beam may be simple or complex. A *simple beam* has a uniform load evenly distributed over the entire length of the beam and is supported at each end. With a simple beam, the load on it is

equally dispersed to each support. An individual floor joist, rafter, or truss or a wall may be thought of as a simple beam. For instance, the wall resisting the wind load in Figure 27.2 can be thought of as a simple beam because it spans between two supporting walls and has a uniform load. If a beam is supporting a uniformly distributed load of 10,000 lb (4536 kg), each supporting post would be resisting 5000 lb (2268 kg). A *complex beam* has a nonuniform load at any point of the beam, or supports that are not located at the end of the beam. Chapter 29 will introduce methods of determining beam sizes when the loads are not evenly distributed.

Figure 27.7 shows a summary of typical building weights based on minimum design values from the IRC. Figure 27.8 shows the bearing walls for a one-story structure framed with a truss roof system and a post-and-beam floor system. Remember that with a typical truss system, all interior walls are nonbearing. Half of the roof weight will be supported by the left wall, and half of the roof weight will be supported by the right wall. For a structure 32′ wide with 2′ overhangs, the entire roof will weigh 1440 lb (36′ × 40#). Each linear foot of wall will hold 720 lb (18 × 40# psf), which is half the total roof weight.

If the walls are 8′ tall, each wall will weigh 80 lb (8′ × 10#) per linear feet of wall. The foundation will hold 100 lb of floor load (2′ × 50#) per linear foot. Only 2′ of floor will be supported, since beams are typically placed at 48″ o.c. Half of the floor weight (2′ × 50#) is placed on the stem wall, and half of the weight (2′ × 50#) will be supported by the girder parallel to the stem wall. Each interior girder will support 200 lb (4′ × 50#) per linear foot of floor weight. The total weight on the stem wall per foot will be the sum of the roof, wall, and floor loads, which equals 900 lb.

Figure 27.9 shows the bearing walls of a two-level structure framed using western platform construction methods. This building has a bearing wall that is located approximately halfway between the exterior walls. For this type of building, one-half of the total building loads will be on the central bearing wall, and one-quarter of the total building loads will be on each exterior wall. Examine the building one floor at a time and see why.

MINIMUM DESIGN LOADS MEMBER	DL	LL	TOTAL
Floor (non-sleeping)	10	40	50
Floors (sleeping rooms)	10	30	40
Exterior balconies	10	60	70
Decks	10	40	50
Walls	10	—	—
Attics (without storage)	5	10	15
Attics (with storage)	10	10	20
Roofs (light coverings)	10	30	40
Roofs (tile)	25	20	45

FIGURE 27.7 ■ Typical loads for residential construction.

Upper Floor

At the upper floor level, the roof and ceiling loads are being supported. In a building 32′ × 15′, as shown in Figure 27.10, each rafter is spanning 16′. Of course the rafter is longer than 16′, but remember that the span is a horizontal measurement. If the loads are uniformly distributed throughout the roof, half of the weight of the roof will be supported at each end of the rafter. At the ridge, half of the total roof weight is being supported. At each exterior wall, one-quarter of the total roof load is being supported. At the ceiling and the other floor levels, the loading is the same. One-half of each joist is supported at the center wall and half at the outer wall.

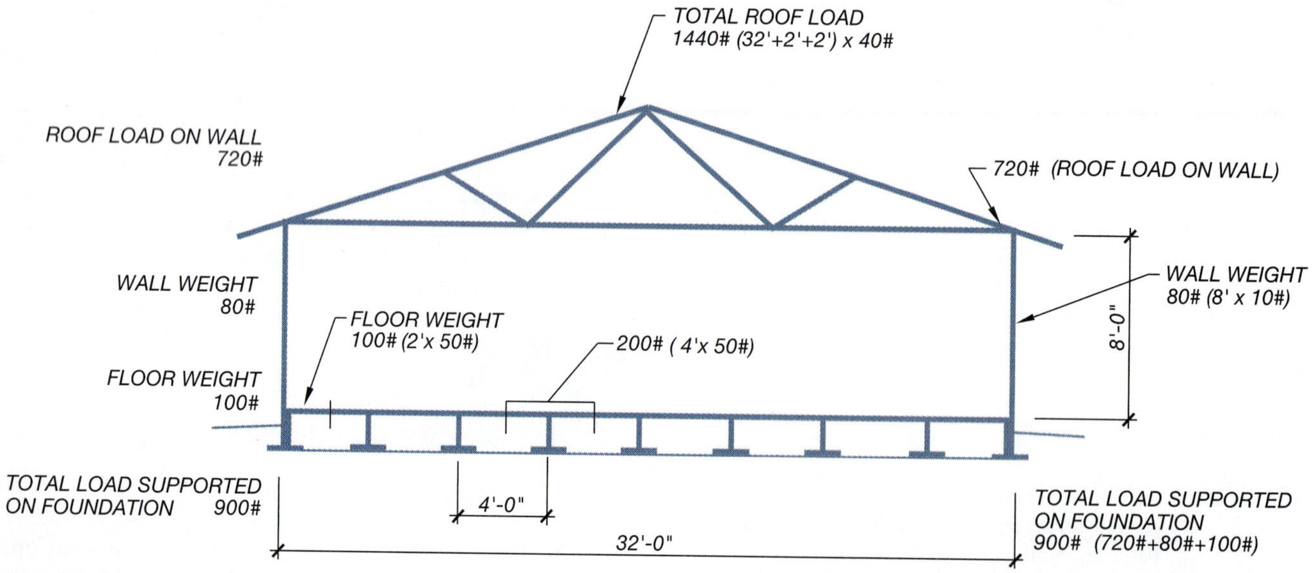

FIGURE 27.8 ■ Loads for a one-level structure framed with a truss roof with 2′ overhangs and a post-and-beam foundation system.

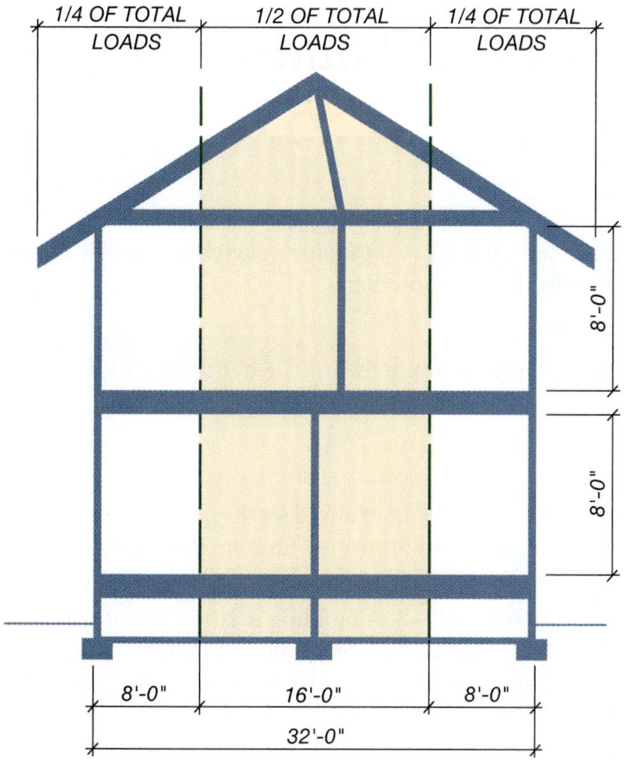

FIGURE 27.9 ■ Bearing walls and load distribution of a typical home.

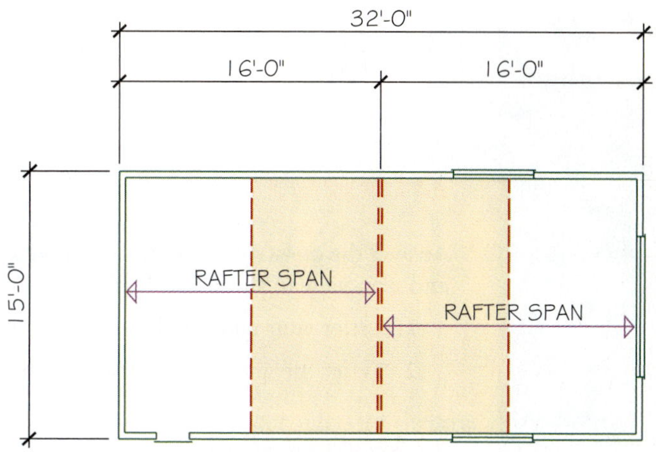

FIGURE 27.10 ■ Load distribution on a simple beam.

Using the loads from Figure 27.7, the weights being supported can be determined. To determine the total weight that a wall supports is a matter of determining the area being supported and multiplying by the weight.

Roof

The area being supported at an exterior wall is 15′ long by 10′ wide (8′ + 2′ overhang), which is 150 sq ft. The roof load equals the sum of the live and dead loads. Assume a live load of 30 psf. For the dead load assume the roof is built with asphalt shingles, 1/2″ ply, and 2 × 8 rafters for a dead load of approximately 8 psf. For simplicity 8 can be rounded up to 10 psf for the dead load. By adding the dead and live loads, you can determine that the total roof load is 40 psf. The total roof weight the wall is supporting is 6000 lb (150 × 40 lb = 6000 lb).

Because the center area in the example is twice as big, the weight will also be approximately twice as big. But just to be sure, check the total weight on the center wall. You should be using 15 × 16 × 40 lb for your calculations. The wall is 15′ long and holds 8′ of rafters on each side of the wall, for a total of 16′. Using a LL of 40 lb, the wall is supporting 9600 pounds.

Ceiling

The procedure for calculating a ceiling is the same as for a roof, but the loads are different. A typical loading pattern for a ceiling with storage is 30 lb. At the outer walls the formula would be 15 × 8 × 30 lb, or 3600 lb. The center wall would be holding 15 × 16 × 30 lb, or 7200 lb.

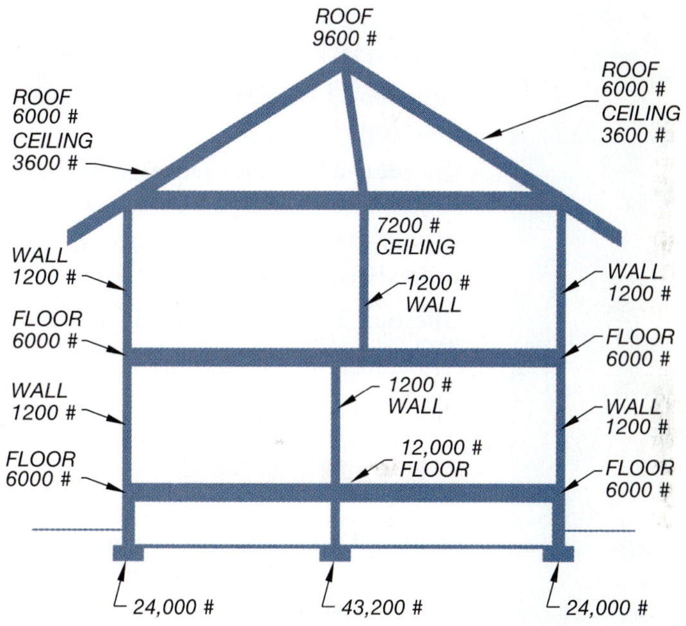

FIGURE 27.11 ■ Total loads supported by footings. Using the assumed loads multiplied by 15 (the length of the building) produces the total loads.

Lower Floor

Finding the weight of a floor is the same as finding the weight of a ceiling, but the loads are much greater. The LL for residential floors is 40 lb and the DL is 10 lb, for a total of 50 lb per sq ft.

Walls

The only other weight left to be determined is the weight of the walls. Generally, walls are 8′–0″ high and average about 10 lb psf. Determine the height of the wall and multiply by the length of the wall to find its area. Multiply the area by the weight per square foot, and you will have the total wall weight. Figure 27.11 shows the total weight that will be supported by the footings.

CEILING JOIST SPANS FOR COMMON LUMBER SPECIES
(Uninhabitable attics with limited storage, live load = 10 psf, L/Δ = 240)

CEILING JOIST SPACING (inches)	SPECIE AND GRADE		DEAD LOAD = 10 psf			
			2x4	2x6	2x8	2x10
			Maximum ceiling joist spans			
			(ft. - in.)	(ft. - in.)	(ft. - in.)	(ft. - in.)
12	Douglas fir-larch	SS	10-5	16-4	21-7	(a)
	Douglas fir-larch	#1	10-0	15-9	20-1	24-6
	Douglas fir-larch	#2	9-10	14-10	18-9	22-11
	Douglas fir-larch	#3	7-8	11-2	14-2	17-4
	Hem-fir	SS	9-10	15-6	20-5	(a)
	Hem-fir	#1	9-8	15-2	19-7	23-11
	Hem-fir	#2	9-2	14-5	18-6	22-7
	Hem-fir	#3	7-8	11-2	14-2	17-4
	Southern pine	SS	10-3	16-1	21-2	(a)
	Southern pine	#1	10-0	15-9	20-10	(a)
	Southern pine	#2	9-10	15-6	20-1	23-11
	Southern pine	#3	8-2	12-0	15-4	18-1
	Spruce-pine-fir	SS	9-8	15-2	19-11	25-5
	Spruce-pine-fir	#1	9-5	14-9	18-9	22-11
	Spruce-pine-fir	#2	9-5	14-9	18-9	22-11
	Spruce-pine-fir	#3	7-8	11-2	14-2	17-4
16	Douglas fir-larch	SS	9-6	14-11	19-7	25-0
	Douglas fir-larch	#1	9-1	13-9	17-5	21-3
	Douglas fir-larch	#2	8-9	12-10	16-3	19-10
	Douglas fir-larch	#3	6-8	9-8	12-4	15-0
	Hem-fir	SS	8-11	14-1	18-6	23-8
	Hem-fir	#1	8-9	13-5	16-10	20-8
	Hem-fir	#2	8-4	12-8	16-0	19-7
	Hem-fir	#3	6-8	9-8	12-4	15-0
	Southern pine	SS	9-4	14-7	19-3	24-7
	Southern pine	#1	9-1	14-4	18-11	23-1
	Southern pine	#2	8-11	13-6	17-5	20-9
	Southern pine	#3	7-1	10-5	13-3	15-8
	Spruce-pine-fir	SS	8-9	13-9	18-1	23-1
	Spruce-pine-fir	#1	8-7	12-10	16-3	19-10
	Spruce-pine-fir	#2	8-7	12-10	16-3	19-10
	Spruce-pine-fir	#3	6-8	9-8	12-4	15-0
19.2	Douglas fir-larch	SS	8-11	14-0	18-5	23-4
	Douglas fir-larch	#1	8-7	12-6	15-10	19-5
	Douglas fir-larch	#2	8-0	11-9	14-10	18-2
	Douglas fir-larch	#3	6-1	8-10	11-3	13-8
	Hem-fir	SS	8-5	13-3	17-5	22-3
	Hem-fir	#1	8-3	12-3	15-6	18-11
	Hem-fir	#2	7-10	11-7	14-8	17-10
	Hem-fir	#3	6-1	8-10	11-3	13-8
	Southern pine	SS	8-9	13-9	18-1	23-1
	Southern pine	#1	8-7	13-6	17-9	21-1
	Southern pine	#2	8-5	12-3	15-10	18-11
	Southern pine	#3	6-5	9-6	12-1	14-4
	Spruce-pine-fir	SS	8-3	12-11	17-1	21-8
	Spruce-pine-fir	#1	8-0	11-9	14-10	18-2
	Spruce-pine-fir	#2	8-0	11-9	14-10	18-2
	Spruce-pine-fir	#3	6-1	8-10	11-3	13-8
24	Douglas fir-larch	SS	8-3	13-0	17-1	20-11
	Douglas fir-larch	#1	7-8	11-2	14-2	17-4
	Douglas fir-larch	#2	7-2	10-6	13-3	16-3
	Douglas fir-larch	#3	5-5	7-11	10-0	12-3
	Hem-fir	SS	7-10	12-3	16-2	20-6
	Hem-fir	#1	7-6	10-11	13-10	16-11
	Hem-fir	#2	7-1	10-4	13-1	16-0
	Hem-fir	#3	5-5	7-11	10-0	12-3
	Southern pine	SS	8-1	12-9	16-10	21-6
	Southern pine	#1	8-0	12-6	15-10	18-10
	Southern pine	#2	7-8	11-0	14-2	16-11
	Southern pine	#3	5-9	8-6	10-10	12-10
	Spruce-pine-fir	SS	7-8	12-0	15-10	19-5
	Spruce-pine-fir	#1	7-2	10-6	13-3	16-3
	Spruce-pine-fir	#2	7-2	10-6	13-3	16-3
	Spruce-pine-fir	#3	5-5	7-11	10-0	12-3

For SI: 1 inch = 25.4 mm, 1 foot = 304.8 mm, 1 psf = 0.0479 kN/m^2.

a. Span exceeds 26 feet in length. Check sources for availability of lumber in lengths greater than 20 feet.

FIGURE 28.5 ■ This span table is suitable for determining ceiling joists with limited storage and a live load of 10 psf and a dead load of 10 psf. *Reproduced from the* International Residential Code / 2000. *Copyright © 2000. Courtesy International Code Council, Inc.*

TABLE R802.5.1(1)[a]
RAFTER SPANS FOR COMMON LUMBER SPECIES
(Roof live load=20 psf, ceiling not attached to rafters, L/Δ=180)

RAFTER SPACING (inches)	SPECIE AND GRADE	DEAD LOAD = 10 psf					DEAD LOAD = 20 psf				
		2x4	2x6	2x8	2x10	2x12	2x4	2x6	2x8	2x10	2x12
		Maximum rafter spans[b]									
		(ft. - in.)	(ft. - in.)	(ft. - in.)	(ft. - in.)	(ft. - in.)	(ft. - in.)	(ft. - in.)	(ft. - in.)	(ft. - in.)	(ft. - in.)
12	Douglas fir-larch SS	11-6	18-0	23-9	(b)	(b)	11-6	18-0	23-5	(b)	(b)
	Douglas fir-larch #1	11-1	17-4	22-5	(b)	(b)	10-6	15-4	19-5	23-9	(b)
	Douglas fir-larch #2	10-10	16-7	21-0	25-8	(b)	9-10	14-4	18-2	22-3	25-9
	Douglas fir-larch #3	8-7	12-6	15-10	19-5	22-6	7-5	10-10	13-9	16-9	19-6
	Hem-fir SS	10-10	17-0	22-5	(b)	(b)	10-10	17-0	22-5	(b)	(b)
	Hem-fir #1	10-7	16-8	21-10	(b)	(b)	10-3	14-11	18-11	23-2	(b)
	Hem-fir #2	10-1	15-11	20-8	25-3	(b)	9-8	14-2	17-11	21-11	25-5
	Hem-fir #3	8-7	12-6	15-10	19-5	22-6	7-5	10-10	13-9	16-9	19-6
	Southern pine SS	11-3	17-8	23-4	(b)	(b)	11-3	17-8	23-4	(b)	(b)
	Southern pine #1	11-1	17-4	22-11	(b)	(b)	11-1	17-3	21-9	25-10	(b)
	Southern pine #2	10-10	17-0	22-5	(b)	(b)	10-6	15-1	19-5	23-2	(b)
	Southern pine #3	9-1	13-6	17-2	20-3	24-1	7-11	11-8	14-10	17-6	20-11
	Spruce-pine-fir SS	10-7	16-8	21-11	(b)	(b)	10-7	16-8	21-9	(b)	(b)
	Spruce-pine-fir #1	10-4	16-3	21-0	25-8	(b)	9-10	14-4	18-2	22-3	25-9
	Spruce-pine-fir #2	10-4	16-3	21-0	25-8	(b)	9-10	14-4	18-2	22-3	25-9
	Spruce-pine-fir #3	8-7	12-6	15-10	19-5	22-6	7-5	10-10	13-9	16-9	19-6
16	Douglas fir-larch SS	10-5	16-4	21-7	(b)	(b)	10-5	16-0	20-3	24-9	(b)
	Douglas fir-larch #1	10-0	15-4	19-5	23-9	(b)	9-1	13-3	16-10	20-7	23-10
	Douglas fir-larch #2	9-10	14-4	18-2	22-3	25-9	8-6	12-5	15-9	19-3	22-4
	Douglas fir-larch #3	7-5	10-10	13-9	16-9	19-6	6-5	9-5	11-11	14-6	16-10
	Hem-fir SS	9-10	15-6	20-5	(b)	(b)	9-10	15-6	19-11	24-4	(b)
	Hem-fir #1	9-8	14-11	18-11	23-2	(b)	8-10	12-11	16-5	20-0	23-3
	Hem-fir #2	9-2	14-2	17-11	21-11	25-5	8-5	12-3	15-6	18-11	22-0
	Hem-fir #3	7-5	10-10	13-9	16-9	19-6	6-5	9-5	11-11	14-6	16-10
	Southern pine SS	10-3	16-1	21-2	(b)	(b)	10-3	16-1	21-2	(b)	(b)
	Southern pine #1	10-0	15-9	20-10	25-10	(b)	10-0	15-0	18-10	22-4	(b)
	Southern pine #2	9-10	15-1	19-5	23-2	(b)	9-1	13-0	16-10	20-1	23-7
	Southern pine #3	7-11	11-8	14-10	17-6	20-11	6-10	10-1	12-10	15-2	18-1
	Spruce-pine-fir SS	9-8	15-2	19-11	25-5	(b)	9-8	14-10	18-10	23-0	(b)
	Spruce-pine-fir #1	9-5	14-4	18-2	22-3	25-9	8-6	12-5	15-9	19-3	22-4
	Spruce-pine-fir #2	9-5	14-4	18-2	22-3	25-9	8-6	12-5	15-9	19-3	22-4
	Spruce-pine-fir #3	7-5	10-10	13-9	16-9	19-6	6-5	9-5	11-11	14-6	16-10
19.2	Douglas fir-larch SS	9-10	15-5	20-4	25-11	(b)	9-10	14-7	18-6	22-7	(b)
	Douglas fir-larch #1	9-5	14-0	17-9	21-8	25-2	8-4	12-2	15-4	18-9	21-9
	Douglas fir-larch #2	8-11	13-1	16-7	20-3	23-6	7-9	11-4	14-4	17-7	20-4
	Douglas fir-larch #3	6-9	9-11	12-7	15-4	17-9	5-10	8-7	10-10	13-3	15-5
	Hem-fir SS	9-3	14-7	19-2	24-6	(b)	9-3	14-4	18-2	22-3	25-9
	Hem-fir #1	9-1	13-8	17-4	21-1	24-6	8-1	11-10	15-0	18-4	21-3
	Hem-fir #2	8-8	12-11	16-4	20-0	23-2	7-8	11-2	14-2	17-4	20-1
	Hem-fir #3	6-9	9-11	12-7	15-4	17-9	5-10	8-7	10-10	13-3	15-5
	Southern pine SS	9-8	15-2	19-11	25-5	(b)	9-8	15-2	19-11	25-5	(b)
	Southern pine #1	9-5	14-10	19-7	23-7	(b)	9-3	13-8	17-2	20-5	24-4
	Southern pine #2	9-3	13-9	17-9	21-2	24-10	8-4	11-11	15-4	18-4	21-6
	Southern pine #3	7-3	10-8	13-7	16-0	19-1	6-3	9-3	11-9	13-10	16-6
	Spruce-pine-fir SS	9-1	14-3	18-9	23-11	(b)	9-1	13-7	17-2	21-0	24-4
	Spruce-pine-fir #1	8-10	13-1	16-7	20-3	23-6	7-9	11-4	14-4	17-7	20-4
	Spruce-pine-fir #2	8-10	13-1	16-7	20-3	23-6	7-9	11-4	14-4	17-7	20-4
	Spruce-pine-fir #3	6-9	9-11	12-7	15-4	17-9	5-10	8-7	10-10	13-3	15-5
24	Douglas fir-larch SS	9-1	14-4	18-10	23-4	23-4	8-11	13-1	16-7	20-3	23-5
	Douglas fir-larch #1	8-7	12-6	15-10	19-5	19-5	7-5	10-10	13-9	16-9	19-6
	Douglas fir-larch #2	8-0	11-9	14-10	18-2	18-2	6-11	10-2	12-10	15-8	18-3
	Douglas fir-larch #3	6-1	8-10	11-3	13-8	13-8	5-3	7-8	9-9	11-10	13-9
	Hem-fir SS	8-7	13-6	17-10	22-9	22-9	8-7	12-10	16-3	19-10	23-0
	Hem-fir #1	8-4	12-3	15-6	18-11	18-11	7-3	10-7	13-5	16-4	19-0
	Hem-fir #2	7-11	11-7	14-8	17-10	17-10	6-10	10-0	12-8	15-6	17-11
	Hem-fir #3	6-1	8-10	11-3	13-8	13-8	5-3	7-8	9-9	11-10	13-9
	Southern pine SS	8-11	14-1	18-6	23-8	23-8	8-11	14-1	18-6	22-11	(b)
	Southern pine #1	8-9	13-9	17-9	21-1	21-1	8-3	12-3	15-4	18-3	21-9
	Southern pine #2	8-7	12-3	15-10	18-11	18-11	7-5	10-8	13-9	16-5	19-3
	Southern pine #3	6-5	9-6	12-1	14-4	14-4	5-7	8-3	10-6	12-5	14-9
	Spruce-pine-fir SS	8-5	13-3	17-5	21-8	21-8	8-4	12-2	15-4	18-9	21-9
	Spruce-pine-fir #1	8-0	11-9	14-10	18-2	18-2	6-11	10-2	12-10	15-8	18-3
	Spruce-pine-fir #2	8-0	11-9	14-10	18-2	18-2	6-11	10-2	12-10	15-8	18-3
	Spruce-pine-fir #3	6-1	8-10	11-3	13-8	13-8	5-3	7-8	9-9	11-10	13-9

For SI: 1 inch = 25.4 mm, 1 foot = 304.8 mm, 1 psf = 0.0479 kN/m^2.

a. The tabulated rafter spans assume that ceiling joists are located at the bottom of the attic space or that some other method of resisting the outward push of the rafters on the bearing walls, such as rafter ties, is provided at that location. When ceiling joists or rafter ties are located higher in the attic space, the rafter spans shall be multiplied by the factors given below:

FIGURE 28.6 ■ This span table is suitable for determining rafters that do not support ceiling loads. Design loads include a live load of 20 psf and a dead load of either 10 psf or 20 psf. *Reproduced from the* International Residential Code / 2000. *Copyright © 2000. Courtesy International Code Council, Inc.*

H_C/H_R	Rafter Span Adjustment Factor
2/3 or greater	0.50
1/2	0.58
1/3	0.67
1/4	0.76
1/5	0.83
1/6	0.90
1/7.5 and less	1.00

where: H_C = Height of ceiling joists or rafter ties measured vertically above the top of the rafter support walls.

H_R = Height of roof ridge measured vertically above the top of the rafter support walls.

b. Span exceeds 26 feet in length. Check sources for availability of lumber in lengths greater than 20 feet.

FIGURE 28.6 ■ *(continued)*

interior finish. The column in Figure 28.6 with a load of 20 psf is suitable for supporting most tiles. Vendor catalogs should always be consulted to determine the required dead loads.

Once the live and dead loads have been determined, choose the appropriate table. Determine what size Hem-Fir rafter should be used to support a roof over a room 18′ wide framed with a gable roof. Assume the roof is supporting 235 lb. composition shingles with a live load of 20 psf, a dead load of 10 psf, and 24″ spacing. Since a gable roof is being framed, the actual horizontal span is only 9′–0″. Using the table in Figure 28.6, it can be seen that 2 × 6 HF #2 rafters spaced at 24″ o.c. are suitable for spans of up to 11′–7″. If the same room were to be framed with DFL #2 lumber, 2 × 6 members would be suitable for spans of up to 11′–9″. If the spacing is reduced to 16″ o.c., 2 × 4 DFL #2 rafters can be used for spans of up to 9′–10″. Even though they are legal, many framers will not use 2 × 4 members because they tend to split as they are attached to the wall top plates.

WORKING WITH ENGINEERED LUMBER

Determining the span of engineered lumber is similar to sizing sawn lumber, although span tables will vary slightly for each manufacturer. Most suppliers of engineered joists provide materials for determining floor joists and rafters. The balance of this chapter will explore material supplied by Trus Joist MacMillan.

Sizing Engineered Floor Joists

Figure 28.7 shows a floor span table for engineered floor joists by Trus Joist MacMillan. Notice that this actually consists of two tables. The left portion of the table describes spans based on code-allowed deflections. The right portion of the table describes spans based on the manufacturer's suggested deflection limits of 1/480. The increased deflection limits will produce an excellent floor system with little or no vibration or squeaking. To determine the joist size required to span 16′–0″ would require the following steps:

- Determine the deflection limit to be used. For this example L/360 will be used.
- Identify the loading condition to be used. For this example a 40 psf LL and a 10 psf DL will be used.
- Select the on-center spacing to be used. For this example 24″ o.c. will be used.
- Scan down the spacing column until a distance that exceeds the span is located. For this example, the first joist that meets the design criteria is the 11 7/8″ TJI/Pro250 joist with a span of 18′–8″.

Compare the same joist with the right half of the table. Using the stricter deflection limits, the TJI/Pro250 joist will span 16′–10″. If the spacing is changed to 16″ o.c., a 9 1/2″ TJI/Pro250 could be used with a maximum span of 16′–1″.

Supporting Cantilevered Floor Joists

Different blocking methods are required when engineered joists are used. The manufacturer's drawings should be consulted to reference required blocking on the framing plans. One common area where special support is required is for joists that cantilever. Figure 28.8 shows alternatives that can be used to resist joist deflection at cantilevers. Figure 28.9 shows the limits for the cantilever and the blocking options specified in Figure 28.8. Cantilever support can be determined using the tables in Figure 28.8; follow these steps:

- Use the first two columns to identify the TJI joist to be used. For this example, the 9 1/2″ TJI/Pro250 at 16″ o.c. from the previous example will be used.
- Locate the roof truss span that meets or exceeds the needed conditions. The 32′ span of the 250 portion of the 9 1/2″ depth will be used.
- Scan across the top of the table and find the roof total load column that meets or exceeds the required design condition. For this example, the 45 psf columns will be used.
- Select the spacing column that meets the design criteria. Find the required TJI span row (9 1/2″ = 250 – 32′ span) that intersects the total roof load/spacing column

MINIMUM CRITERIA PER CODE
L/360 LIVE LOAD DEFLECTION

Load	DEPTH	TJI®/Pro™	12" o.c.	16" o.c.	19.2" o.c.	24" o.c.
40 PSF LIVE LOAD / 10 PSF DEAD LOAD 12 PSF DEAD LOAD AT TJI®/Pro™ 550	9½"	150	18'-8"	17'-1"	16'-2"	14'-11"
		250	19'-6"	17'-10"	16'-10"	15'-8"
	11⅞"	150	22'-3"	20'-4"	18'-10"	15'-0"
		250	23'-3"	21'-3"	20'-0"	18'-8"(2)
		350	24'-10"	22'-8"	21'-4"	19'-11"(2)
		550	28'-2"	25'-8"	24'-2"	22'-6"
	14"	250	26'-5"	24'-1"	22'-9"(2)	18'-11"(2)
		350	28'-2"	25'-8"	24'-3"(2)	21'-4"(2)
		550	32'-0"	29'-1"	27'-5"	25'-6"(1)
	16"	250	29'-3"	26'-8"(2)	23'-8"(2)	18'-11"(2)
		350	31'-2"	28'-5"(2)	26'-8"(2)	21'-4"(2)
		550	35'-5"	32'-3"(1)	30'-4"(1)	26'-9"(1)(2)
40 PSF LIVE LOAD / 20 PSF DEAD LOAD 22 PSF DEAD LOAD AT TJI®/Pro™ 550	9½"	150	18'-8"	16'-8"	15'-3"	12'-6"
		250	19'-6"	17'-10"	16'-6"	13'-5"
	11⅞"	150	22'-3"	18'-10"	15'-8"	12'-6"
		250	23'-3"	20'-11"(2)	19'-1"(2)	15'-9"(2)
		350	24'-10"	22'-8"	20'-8"(2)	17'-9"(2)
		550	27'-10"	25'-4"	23'-11"	22'-3"(1)(2)
	14"	250	26'-5"	23'-2"(2)	19'-9"(2)	15'-9"(2)
		350	28'-2"	25'-1"(2)	22'-2"(2)	17'-9"(2)
		550	31'-7"	28'-9"	27'-1"(1)(2)	22'-5"(1)(2)
	16"	250	28'-11"(2)	23'-8"(2)	19'-9"(2)	15'-9"(2)
		350	31'-2"(2)	26'-8"(2)	22'-2"(2)	17'-9"(2)
		550	35'-0"	31'-10"(1)	28'-1"(1)(2)	22'-5"(1)(2)

IMPROVED PERFORMANCE SYSTEM
L/480 LIVE LOAD DEFLECTION

Load	DEPTH	TJI®/Pro™	12" o.c.	16" o.c.	19.2" o.c.	24" o.c.
40 PSF LIVE LOAD / 10 PSF DEAD LOAD 12 PSF DEAD LOAD AT TJI®/Pro™ 550	9½"	150	16'-11"	15'-5"	14'-7"	13'-7"
		250	17'-8"	16'-1"	15'-2"	14'-2"
	11⅞"	150	20'-1"	18'-4"	17'-4"	15'-0"
		250	21'-0"	19'-2"	18'-1"	16'-10"(2)
		350	22'-5"	20'-5"	19'-3"	17'-11"
		550	25'-6"	23'-2"	21'-10"	20'-3"
	14"	250	23'-10"	21'-9"	20'-6"(2)	18'-11"(2)
		350	25'-6"	23'-2"	21'-10"	20'-4"(2)
		550	28'-11"	26'-3"	24'-9"	23'-0"
	16"	250	26'-5"	24'-1"	22'-9"(2)	18'-11"(2)
		350	28'-2"	25'-8"	24'-2"(2)	21'-4"(2)
		550	32'-0"	29'-1"	27'-5"	25'-5"(1)
40 PSF LIVE LOAD / 20 PSF DEAD LOAD 22 PSF DEAD LOAD AT TJI®/Pro™ 550	9½"	150	16'-11"	15'-5"	14'-7"	12'-6"
		250	17'-8"	16'-1"	15'-2"	13'-5"
	11⅞"	150	20'-1"	18'-4"	15'-8"	12'-6"
		250	21'-0"	19'-2"	18'-1"(2)	15'-9"(2)
		350	22'-5"	20'-5"	19'-3"(2)	17'-9"(2)
		550	25'-6"	23'-2"	21'-10"	20'-3"(1)
	14"	250	23'-10"	21'-9"(2)	19'-9"(2)	15'-9"(2)
		350	25'-6"	23'-2"(2)	21'-10"(2)	17'-9"(2)
		550	28'-11"	26'-3"	24'-9"	22'-5"(1)(2)
	16"	250	26'-5"	23'-8"(2)	19'-9"(2)	15'-9"(2)
		350	28'-2"	25'-8"(2)	22'-2"(2)	17'-9"(2)
		550	32'-0"	29'-1"	27'-5"(1)(2)	22'-5"(1)(2)

- Long term deflection under dead load which includes the effect of creep, common to all wood members, has not been considered for any of the above applications. Shaded spans reflect initial dead load deflection exceeding 0.33", which may be unacceptable.

(1) Web stiffeners are required in hangers when the TJI®/Pro™ 550 joist span is greater than the spans shown in the following table:

TJI®/Pro™	40 PSF LIVE LOAD, 12 PSF DEAD LOAD				40 PSF LIVE LOAD, 22 PSF DEAD LOAD			
	12" o.c.	16" o.c.	19.2" o.c.	24" o.c.	12" o.c.	16" o.c.	19.2" o.c.	24" o.c.
550	Not Required	Not Required	28'-8"	22'-11"	Not Required	29'-10"	24'-10"	19'-10"

(2) Web stiffeners are required at intermediate supports of continuous span joists in conditions where the intermediate bearing width is less than 5 1/4" **and** the span on either side of the intermediate bearing is greater than the spans shown in the following table:

TJI®/Pro™	40 PSF LIVE LOAD, 10 PSF DEAD LOAD*				40 PSF LIVE LOAD, 20 PSF DEAD LOAD**			
	12" o.c.	16" o.c.	19.2" o.c.	24" o.c.	12" o.c.	16" o.c.	19.2" o.c.	24" o.c.
150	Web Stiffener Not Required				Web Stiffener Not Required			
250	Not Required	24'-3"	20'-2"	16'-1"	26'-11"	20'-2"	16'-9"	13'-5"
350	Not Required	27'-8"	23'-1"	18'-5"	30'-9"	23'-1"	19'-2"	15'-4"
550	Not Required			25'-8"	Not Required		26'-11"	21'-6"

*12 PSF Dead Load at TJI®/Pro™ 550 joists.

**22 PSF Dead Load at TJI®/Pro™ 550 joists.

GENERAL NOTES

Tables are based on:

- Assumed composite action with a single layer of appropriate span-rated glue-nailed wood sheathing for deflection only **(spans shall be reduced 5" when sheathing panels are nailed only)**.
- Uniformly loaded joists.
- Increase for repetitive member use has been included.
- Spans shown are clear disance between supports.
- Most restrictive of simple or multiple span.
- For loading conditions not shown, refer to PLF tables on page 11.

FIGURE 28.7 ■ Engineered floor joist span tables. *Courtesy Trus Joist MacMillan.*

DEPTH	TJI®/Pro™	ROOF TRUSS SPAN	ROOF TOTAL LOAD 35 PSF 16"	35 PSF 19.2"	35 PSF 24"	45 PSF 16"	45 PSF 19.2"	45 PSF 24"	55 PSF 16"	55 PSF 19.2"	55 PSF 24"
9½"	150	24'	0	0	1	0	1	1	1	1	X
		26'	0	0	1	0	1	X	1	1	X
		28'	0	0	1	1	1	X	1	X	X
		30'	0	1	1	1	1	X	1	X	X
		32'	0	1	1	1	1	X	1	X	X
		34'	0	1	X	1	X	X	X	X	X
		36'	0	1	X	1	X	X	X	X	X
	250	24'	0	0	1	0	0	1	0	1	1
		26'	0	0	1	0	1	1	1	1	X
		28'	0	0	1	0	1	1	1	1	X
		30'	0	0	1	0	1	1	1	1	X
		32'	0	0	1	1	1	X	1	1	X
		34'	0	1	1	1	1	X	1	X	X
		36'	0	1	1	1	1	X	1	X	X
11⅞"	150	26'	0	0	1	0	1	1	1	1	1
		28'	0	0	1	1	1	1	1	1	1
		30'	0	1	1	1	1	1	1	1	X
		32'	0	1	1	1	1	1	1	1	X
		34'	0	1	1	1	1	1	1	1	X
		36'	0	1	1	1	1	1	1	1	X
		38'	1	1	1	1	1	X	1	1	X
	250	26'	0	0	W	0	W	1	W	1	1
		28'	0	W	W	W	W	1	W	1	1
		30'	0	W	W	W	W	1	W	1	1
		32'	0	W	1	W	1	1	1	1	1
		34'	0	W	1	W	1	1	1	1	1
		36'	W	W	1	W	1	1	1	1	X
		38'	W	W	1	W	1	1	1	1	X
	350	26'	0	0	0	0	0	W	0	W	1
		28'	0	0	W	0	W	1	W	W	1
		30'	0	0	W	0	W	1	W	1	1
		32'	0	0	W	0	W	1	W	1	1
		34'	0	0	W	0	W	1	W	1	1
		36'	0	0	1	W	1	1	1	1	2
		38'	0	W	1	W	1	1	1	1	2
	550	26'	0	0	0	0	0	0	0	0	0
		28'	0	0	0	0	0	0	0	0	0
		30'	0	0	0	0	0	0	0	0	0
		32'	0	0	0	0	0	0	0	0	1
		34'	0	0	0	0	0	0	0	0	1
		36'	0	0	0	0	0	0	0	0	1
		38'	0	0	0	0	0	0	0	0	1

DEPTH	TJI®/Pro™	ROOF TRUSS SPAN	ROOF TOTAL LOAD 35 PSF 16"	35 PSF 19.2"	35 PSF 24"	45 PSF 16"	45 PSF 19.2"	45 PSF 24"	55 PSF 16"	55 PSF 19.2"	55 PSF 24"
14"	250	26'	0	0	W	0	W	1	W	1	1
		28'	0	W	W	W	W	1	W	1	1
		30'	0	W	1	W	W	1	W	1	1
		32'	0	W	1	W	1	1	1	1	1
		34'	0	W	1	W	1	1	1	1	1
		36'	W	W	1	W	1	1	1	1	2
		38'	W	W	1	W	1	1	1	1	2
	350	26'	0	0	W	0	W	W	W	W	1
		28'	0	0	W	0	W	1	W	W	1
		30'	0	0	W	0	W	1	W	1	1
		32'	0	0	W	0	W	1	W	1	1
		34'	0	W	W	W	W	1	W	1	1
		36'	0	W	1	W	1	1	W	1	2
		38'	0	W	1	W	1	1	1	1	2
	550	28'	0	0	0	0	0	0	0	0	0
		30'	0	0	0	0	0	0	0	0	0
		32'	0	0	0	0	0	0	0	0	W
		34'	0	0	0	0	0	0	0	0	W
		36'	0	0	0	0	0	0	0	0	1
		38'	0	0	0	0	0	W	0	0	1
		40'	0	0	0	0	0	W	0	0	1
16"	250	28'	0	W	W	W	W	1	W	1	1
		30'	0	W	1	W	W	1	W	1	1
		32'	0	W	1	W	1	1	1	1	1
		34'	0	W	1	W	1	1	1	1	1
		36'	W	W	1	W	1	1	1	1	2
		38'	W	W	1	W	1	1	1	1	2
		40'	W	W	1	1	1	1	1	1	2
	350	28'	0	0	W	0	W	1	W	W	1
		30'	0	0	W	0	W	1	W	1	1
		32'	0	0	W	0	W	1	W	1	1
		34'	0	W	W	W	W	1	W	1	1
		36'	0	W	1	W	1	1	W	1	2
		38'	0	W	1	W	1	1	1	1	2
		40'	0	W	1	W	1	1	1	1	2
	550	28'	0	0	0	0	0	0	0	0	0
		30'	0	0	0	0	0	0	0	0	W
		32'	0	0	0	0	0	0	0	0	W
		34'	0	0	0	0	0	0	0	0	W
		36'	0	0	0	0	0	0	0	0	W
		38'	0	0	0	0	0	W	0	0	W
		40'	0	0	0	0	0	W	0	W	1

GENERAL NOTES

Tables are based on:

- 15 psf roof dead load.
- 80 plf exterior wall load with 3'–0" maximum width window or door openings. For larger openings, or multiple 3'–0" width openings spaced less than 6'–0" on-center, additional joists beneath the opening's trimmers may be required.
- TimberStrand® LSL or spruce-pine-fir bearing plate or equivalent.
- Roof truss with 24" soffits.
- 3/4" reinforcement refers to 3/4" "Exposure 1" plywood or other 3/4" "Exposure 1" 48/24 rated sheathing that is cut to match the full depth of the TJI® joist. Install with face grain horizontal. Reinforcing member must bear fully on the wall plate. Minimum wall plate width is 3 1/2".
- For conditions beyond the scope of this table, use our TJ-Beam™ or TJ-Xpert™ software programs or contact your Trus Joist MacMillan representative for assistance.

LEGEND

0 *No reinforcement required.*

W *Web stiffener required each side of joist at bearing. See detail E1W.*

1 *3/4" x 48" reinforcement required on one side of joist (see detail E2) or double the joists (see detail E4). Do not use detail E4 with TJI®/Pro™ 550 joists.*

2 *3/4" x 48" reinforcement required on both sides of joint (see detail E3) or double the joists (see detail E4). Do not use detail E4 with TJI®/Pro™ 550 joists.*

X *Will not work. Reduce spacing of joists and recheck on table.*

FIGURE 28.8 ■ Engineered cantilevered floor joist span tables. *Courtesy Trus Joist MacMillan.*

NON-LOAD BEARING CANTILEVER DETAILS

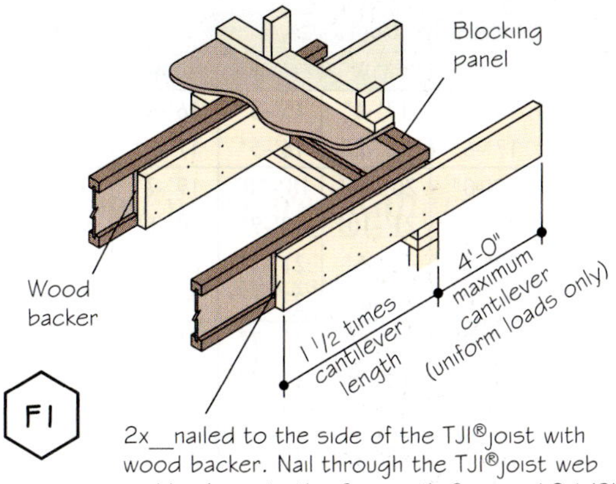

2x__nailed to the side of the TJI® joist with wood backer. Nail through the TJI® joist web and backer into the 2x__ with 2 rows 10d (3") common nails at 6" o.c. and clinch. Use 16d (3 1/2") nails with TJI®/Pro™ 550 joists.

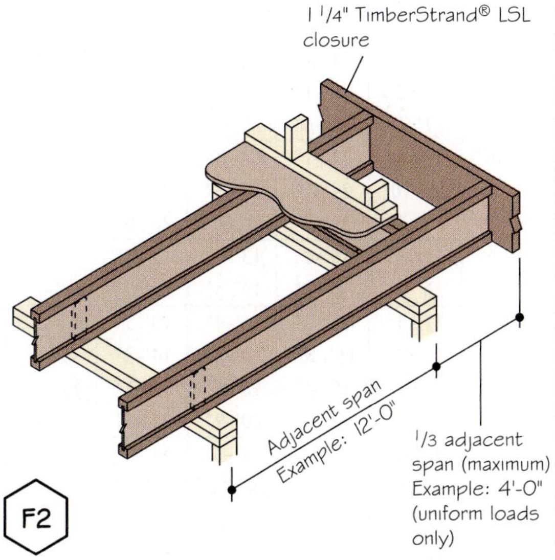

TJI® joists may be cantilevered up to 1/3 the adjacent span if not supporting concentrated loads on the cantilever. Cantilevers exceeding 4 feet may require special consideration. Refer to our TJ-Beam™ or TJ-Xpert™ software or contact your Trus Joist MacMillan representative for assistance.

LOAD BEARING CANTILEVER DETAILS

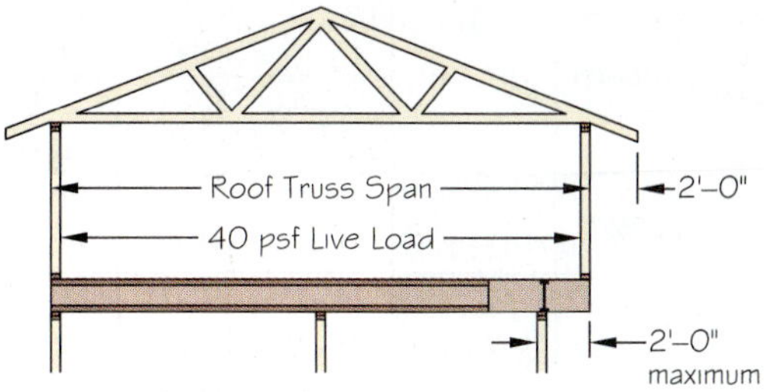

TJI® joists may be cantilevered up to a maximum of 2'-0" when supporting roof load, but may require reinforcement. **Consult tables on page 9 to determine required reinforcement.** See details below for methods of reinforcement.

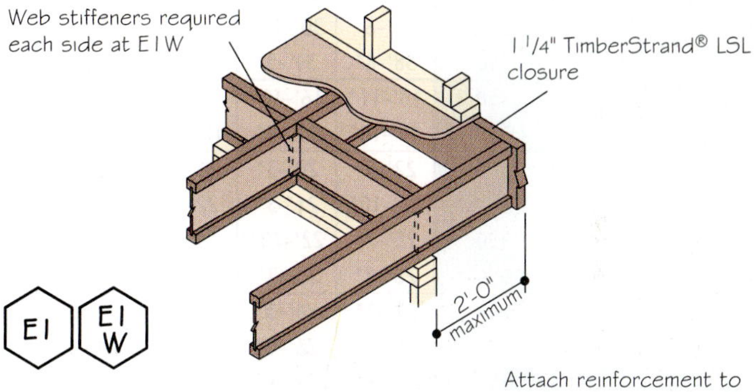

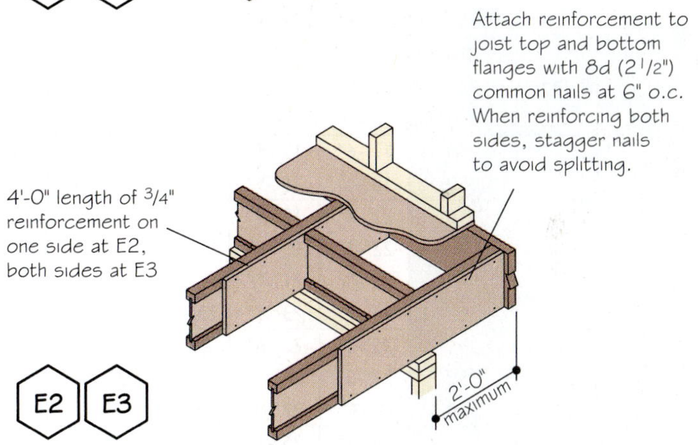

TJI®/Pro™ 150, 250 AND 350 JOISTS ONLY

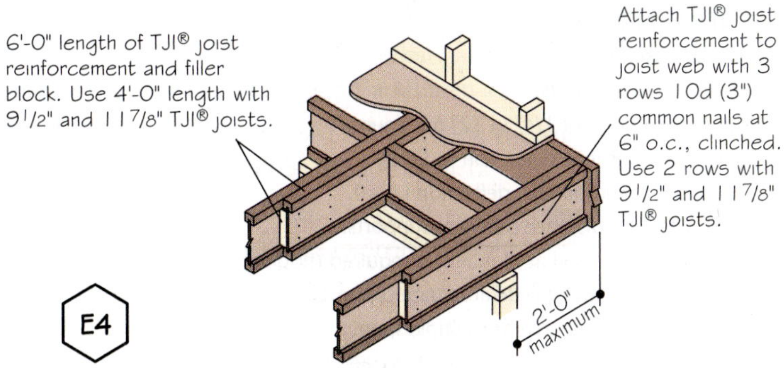

FIGURE 28.9 ■ Cantilevered floor joist details. *Courtesy Trus Joist MacMillan.*

Question 28–14 Use the tables presented in this chapter and determine the floor joist size required to span 17′–0″ using L/360 and L/480 using a spacing of 16″ o.c.

Question 28–15 What is the advantage of using a deflection value of L/480 if the building codes allow a greater deflection?

DIRECTIONS

Use the tables presented in this chapter to complete the following problems. Unless noted, use DFL #2 at standard spacing. Assume 30 lb. live load for all rafters.

PROBLEMS

Problem 28–1 A living room is 17′–9″ wide. What is the smallest size ceiling joist that could be used?

Problem 28–2 A contractor has bought a truckload of 2 × 6s 18′ long. Can they be used for ceiling joists with no attic storage spaced at 16″ o.c.?

Problem 28–3 What size rafter is needed for a home 28′ wide with a gable roof, 4/12 pitch with 235 lb. composition shingles?

Problem 28–4 What is the maximum span allowed using 2 × 10 rafters for a 3/12 pitch using built-up roofing?

Problem 28–5 Determine the rafter size to be used to span 14′ and support 20# dead load at a 6/12 pitch.

Problem 28–6 What size TJI should be used to span 17′–9″ if L/480 is to be used?

Problem 28–7 A 9 1/2″ TJI150 at 24″ o.c. will be used to cantilever 24″. The joist will support a floor load of 45 psf and a wall that supports a truss 28′ wide with 1′ overhangs. How should the cantilevered joist be reinforced?

Problem 28–8 If the joists in Problem 28–7 were placed at 16″ o.c., how should they be reinforced?

Problem 28–9 Engineered rafters are required to span 19′ and support 300# composition shingles on a 6/12 pitch in an area where no snow is expected. What size TJI should be used if 24″ spacing is desired?

Problem 28–10 Engineered rafters are required to span 17′–6″ and support a DL of 20 lbs. on a 5/12 pitch in an area where no snow is expected. What is the smallest size TJI that can be used? What is the smallest size rafter that can be used if 24″ spacing is desired?

CHAPTER 29

Determining Beam Sizes

INTRODUCTION

As you advance in your architectural skills, the need to determine the size of structural members will occur frequently. This chapter will introduce loads that must be considered, define structural lumber, and explore how the size of wood beam is determined. There are several skills that you will need to develop in order to determine the size of structural members easily. These skills include the ability to distinguish loading patterns on beams, recognize standard engineering symbols used in beam formulas, recognize common causes of beam failure and understand how to select beams to resist these tendencies.

LOADING AND SUPPORT PATTERNS OF BEAMS

There are two common ways to load and support a beam. Loads can be uniformly distributed over the entire span of a beam or can be concentrated in one small area of a beam. In Chapter 27, uniformly distributed loads were discussed. A concentrated load is one that is placed on only one area of a beam. A support post from an upper floor resting on a beam on a lower floor, or the weight of a car being transferred through a wheel into the floor system are examples of concentrated loads. Occasionally you will need to deal with *increasing loads* that result from triangular loading patterns. A beam used to frame a hip or valley supports a triangular load. The triangular load that results from a hip can be seen in Figure 29.1. The load starts at zero at the low end of the hip. The maximum load is at the high end of the hip, where it intersects the ridge.

Usually a beam is supported at each end. Common alternatives are to support a beam at the center of the span in addition to the ends or to provide a support at one end and near the other end. The type of beam that extends past the support is called a *cantilevered beam*. Examples of each type of load and the support system can be seen in Figure 29.2.

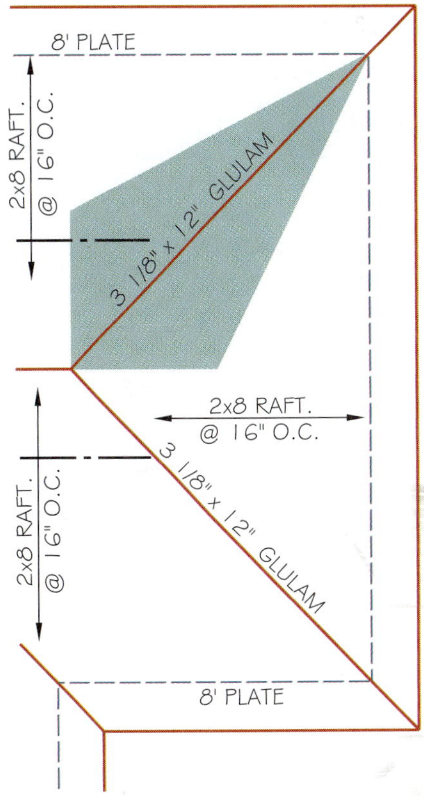

FIGURE 29.1 ■ The beam used to support a hip or valley supports a triangular load. The triangle is formed because half of the weight of each rafter is supported on the beam.

STRUCTURAL LUMBER

No matter how the loads are placed or what size the load is, all beams will bend to some extent. How and to what extent a beam will bend depends on its natural properties. These properties vary depending on the size of the member, the type of wood to be used, the moisture content of the wood, and defects in the wood.

Lumber Sizes

Lumber is described using a nominal size and a net size. *Nominal size* describes the width and depth of a piece of wood in whole inches such as a 6 × 10. The 6 represents the width and the 10 represents the depth of the beam. A 6 × 12 will shrink in size as it dries, and it will be reduced in size as it is smoothed with a planer. The final size of wood after planing is referred to as the *net* size. The net next size of a 6 × 10 is 5 1/2″ × 9 1/2″. The actual or net size is used in designing structural members. Common net sizes include widths of 1 1/2″, 3 1/2″, 5 1/2″ and 7 1/4″. The net depth will vary depending on the width. Common net sizes are:

2" MATERIAL		4" MATERIAL		6" MATERIAL	
1½" × 3½"	38 × 89	3½" × 3½"	89 × 89	5½" × 3½"	140 × 89
1½" × 5½"	38 × 140	3½" × 5½"	89 × 140	5½" × 5½"	140 × 140
1½" × 7¼"	38 × 184	3½" × 7¼"	89 × 184	5½" × 7½"	140 × 190
1½" × 9¼"	38 × 235	3½" × 9¼"	89 × 235	5½" × 9½"	140 × 241
1½" × 11¼"	38 × 286	3½" × 11¼"	89 × 286	5½" × 11½"	140 × 292
1½" × 13¼"	38 × 337	3½" × 13½"	89 × 343	5½" × 13½"	140 × 343

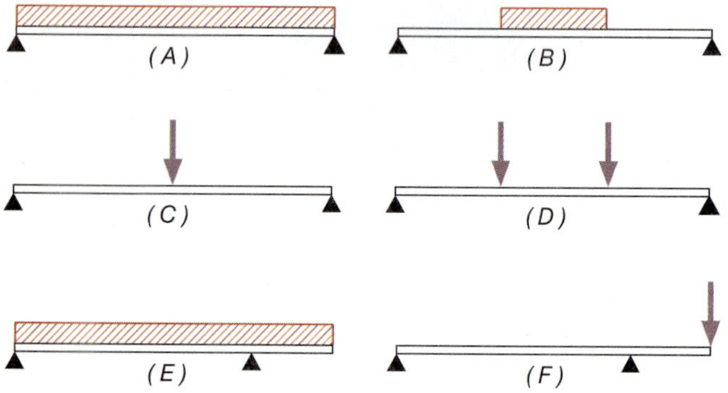

FIGURE 29.2 ■ Common loading patterns on a beam include the following: (*A*) simple beam with a uniformly distributed load; (*B*) beam with a partially distributed load at the center; (*C*) beam with a concentrated load at the center; (*D*) beam with two equal concentrated loads placed symmetrically; (*E*) cantilevered beam with a uniform load; (*F*) cantilevered beam with a concentrated load at the free end.

Lumber Grading

Most structural lumber is visually graded when it is sawn at a mill. With visual inspection methods, wood is evaluated by ASTM standards in cooperation with the U.S. Forest Products laboratory. Some structural lumber is tested nondestructively by machine and graded as *mechanically evaluated lumber* (MEL). The current standard for rating lumber is the NDS standard (National Design Specifications for Wood Construction), based on destructive testing of visually graded dimension lumber. The lumber guidelines can be found on the American Forest and Paper Association Web site. With NDS guidelines, wood is given a designation such as #1 to define its quality and a base value for the species. The base value is then adjusted, depending on the size of the lumber. The grading process takes into account what portion of the tree the member was sawn from and natural defects in the wood such as knots and checks. In addition to NDS lumber guidelines, LRFD (load reduction factor design standards) guidelines are being adopted by some municipalities. A wise designer should verify the current code and lumber standard that will govern the project. No matter how the wood is evaluated, common terms used to define it include dimension lumber, timbers, post and timbers and beams and stringers.

Dimension Lumber

Dimension lumber ranges in thickness from 2 to 4 inches (50–100 mm) and has moisture content of less than 19%. Common grades of dimension lumber include:

- Stud—2″–6″ (50–150 mm) width and a maximum length of 10′ (3000 mm)
- Select structural, No. 1 and better, No. 1, 2, & 3, Construction and Utility.

Timbers

Timbers is both a general classification for large sizes of structural lumber and the name of a specific grade and size: structural lumber that is at least 5″ (125 mm) thick. The two basic categories of timbers include post and timbers and beams and stringers.

- *Post and Timbers.* This classification refers to structural lumber that is 5 × 5 and larger with a thickness and width that are the same or within 2″ (50 mm) of each other such as a 6 × 6 or 6 × 8 beams.
- *Beams and Stringers.* This classification refers to structural lumber that is 5 × 5 and larger with a thickness 2″ (50 mm) or greater than the width such as a 6 × 10 or 6 × 12.

Properties of Lumber

In addition to the terms used to describe lumber, several key terms must be understood to determine the size of wood beams. These terms include the breadth, depth, area, extreme fiber stresses, neutral axis, moment of inertia, section modulus and grain. See Figure 29.3.

b = breadth of a rectangular beam in inches.
d = depth of a rectangular beam in inches.
A = area of the beam, determined using bd.

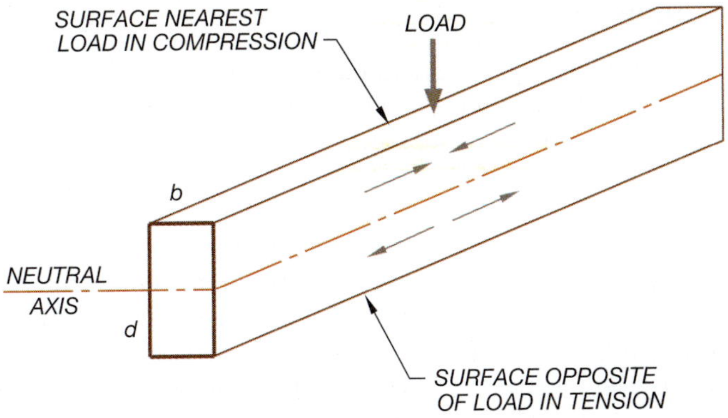

FIGURE 29.3 ■ Common properties of rectangular lumber. When a load is placed on the top of a beam, the surface nearest the load is placed in compression and the surface away from the beam is in tension.

Extreme fiber stresses = As a beam deflects from the load it supports, the surface of the beam supporting the load will be placed in compression. The bottom of the beam will be placed in tension.

Neutral axis = The axis formed where the forces of compression and tension in a beam reach equilibrium. It is formed at the midpoint between the forces of compression and tension. Since the upper and lower surfaces of the beam resist the most stress, holes can be drilled near the neutral axis without greatly affecting the strength of the beam. This chapter will explore the placement of loads on a beam and help determine the loads that must be resisted so that equilibrium can be achieved.

I = Moment of inertia is the sum of the products of each of the elementary areas of a beam multiplied by the square of their distance from the neutral axis of the cross section. This sounds technical, but a value for I is listed in tables that will be presented later in this chapter for use in deflection calculations. (See Figures 29.18 and 29.21.)

S = Section modulus is the moment of inertia divided by the distance from the neutral axis to the extreme fiber of the cross section. Once again, this sounds technical, but a value for S is listed in tables that will be presented later in this chapter for use in determining bending strength calculations. (See Figures 29.18 and 29.21.)

Grain = How a beam reacts will depend on the relationship of the load to the grain of the wood. Wood grain is composed of fibers that can be visualized as grains of rice that are aligned in the same direction. The wood fibers are strongest in their long direction. Common relationships of loads to grain include loads placed parallel and perpendicular to the grain.

- When a load is applied in the same direction as the fibers of a structural member, the force is said to be *parallel to the grain*. Loads that are parallel to the grain are represented by the symbol //.
- When a load is applied across the direction of the fibers of a structural member, the force is said to be *perpendicular to the grain* and is represented by the symbol ⊥. Figure 29.4 shows a force applied to a beam in each direction.

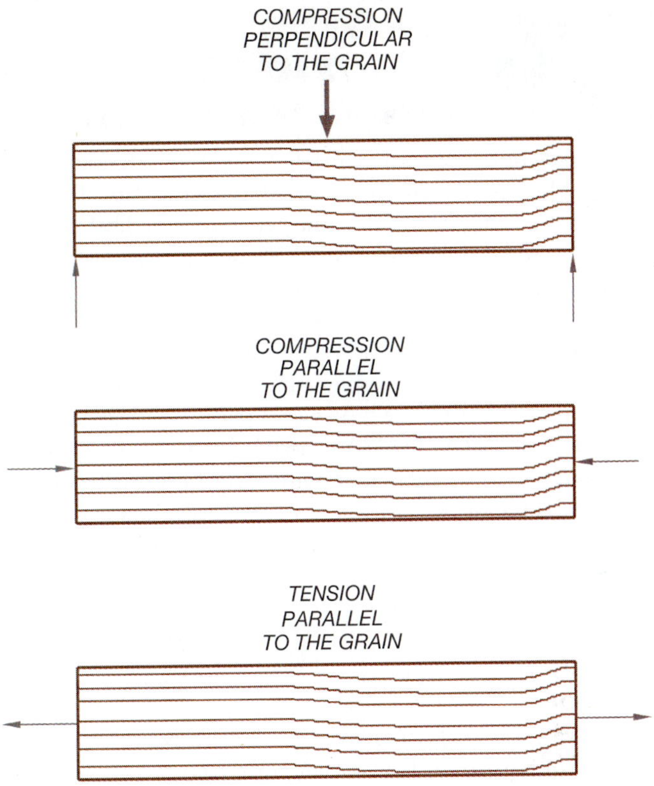

FIGURE 29.4 ■ Forces acting on horizontal members.

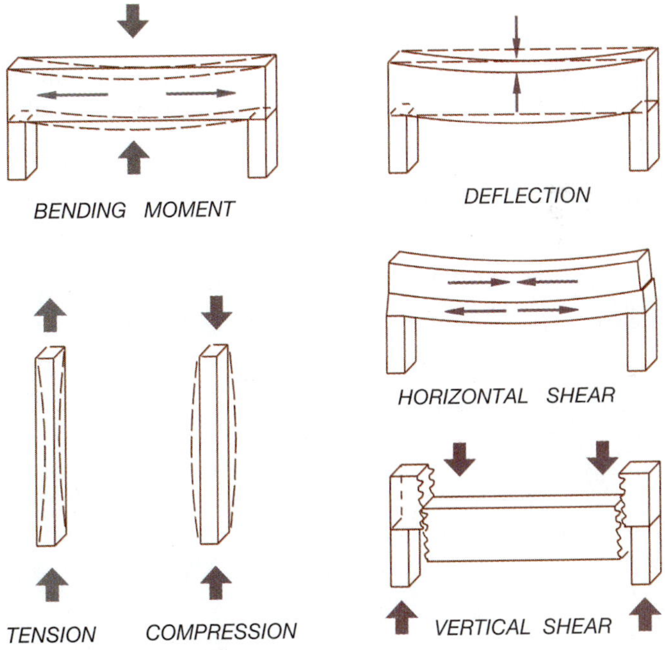

FIGURE 29.5 ■ Forces acting on structural members.

Loads typically affect structural members in five different ways: bending strength, deflection, horizontal shear, vertical shear, and bearing area. These forces will create stress that affects the fibers inside the beam. The forces of tension and compression from outside of a beam will also be considered. Each force can be seen in Figure 29.5.

LOADING REACTIONS OF WOOD MEMBERS

For every action there is an equal and opposite reaction. This is a law of physics that affects every structure ever built. There are several actions or stresses that must be understood before beam reactions can be considered. These stresses include fiber bending stress, deflection, horizontal shear, vertical shear, and compression. You do not need to know how these stresses are generated, only that they exist, to determine the size of a framing member using standard loading tables. Design loads have been discussed in earlier chapters. Review Figure 27.7 (page 548) for a partial list of design live (LL) and dead (DL) loads. These design loads will be useful in determining the load on a beam. Figure 29.6 shows a partial listing of base design values that are needed for the design of beams. This table includes a column of values for fiber bending stress (F_b), modulus of elasticity (E), and horizontal shear (F_v).

Fiber Bending Stress

The bending strength of a wood beam is measured in units of fiber bending stress. Earlier you read about extreme fiber stress (compression) that occurs on the beam surface supporting the load. Extreme fiber stress (tension) also occurs on the surface opposite the surface in compression. Engineers use a relationship of allowable fiber bending stress to the maximum bending stress to determine the required strength of a framing member. Fiber bending stresses are represented by the symbol F_b. The values listed in beam tables are the safe, allowable fiber bending.

Bending Strength

Bending strength is the determination of the beam strength required to resist the force applied to a beam measured in moments. The *bending moment* at any point of the beam is the measure of the tendency of the beam to bend due to the force acting on the beam. The magnitude of the bending moment varies throughout the length of the beam. The maximum bending stress occurs at the midpoint of a simple beam with a uniform load. The location for maximum bending moment in complex beams will be discussed later in this chapter. The letter M represents bending moment in engineering formulas. The relationship of the allowable extreme fiber bending stress (F_b) to the maximum bending moment (M) equals the required section modulus (S) of a beam. The design equation is $S = M/F_b$. The F_b base design value listed in psi for various lumber species can be determined using Figure 29.6. The base F_b value for 4× members varies depending on the depth of the beam. The base design value for 6× material remains fixed up to and including 6 × 12. Determining the required section modulus of a beam will provide one of the pieces of information needed to determine the size of beam needed to support a specific load over a given span.

Deflection

Deflection deals with the stiffness of a beam. It measures the tendency of a structural member to bend under a load and as a result of gravity. As a load is placed on a beam, the beam will sag between its supports. As the span increases, the tendency to deflect increases. Deflection rarely causes a beam to break but greatly affects the materials that the beam supports. When floor joists sag too much, Sheetrock will crack or doors and

DESIGN VALUES FOR BEAMS

SIZE, SPECIES, AND COMMERCIAL GRADE	EXTREME FIBER BENDING F_b			HORIZONTAL SHEAR F_v	MODULUS OF ELASTICITY (E)
DFL #2	Base Value	Modifier	Increased Value		
4 × 8	875	(1.3)	1138	85	1.6
4 × 10	875	(1.2)	1050	85	1.6
4 × 12	875	(1.1)	963	85	1.6
DFL #1					
6 × 8	1350			85	1.6
6 × 10	1350			85	1.6
6 × 12	1350			85	1.6
Reduce the base value of all members larger than 6 × 12 by $(12/d)1/9=C_f$.					
Hem-Fir					
6 × _	1050			70	1.3

FIGURE 29.6 ■ Partial listing of safe design values of common types of lumber used for beams. All values are based on National Design Standards for Wood Construction (NDS), published by the American Forest and Paper Association.

windows will stick. Two formulas will be used to determine deflection. The first is the legal limit of deflection that determines how much the building code will allow a specific beam to bend. Limits are set by code and are expressed as a ratio of the length of the beam in inches over a deflection value. Allowable deflection for a floor beam is represented by the ratio L/360. For a 10′ long floor beam, it is allowed to deflect 0.33″ (120″/360). The allowable deflection value defines the maximum amount of deflection (can sag). Other allowable deflections include L/240 (roofs supporting ceilings) and L/180 (roofs with no ceiling loads).

Modulus of Elasticity

Modulus of elasticity also deals with the stiffness of a structural member. It is a ratio of the amount a member will deflect in proportion to the applied load. The letter E is used in beam tables and design formulas to represent the value for the modulus of elasticity. The E value represents how much the member will deflect (the will sag value). For a beam to be safe, the E value (will sag) can not exceed the deflection value (can sag). Different formulas for varied loading conditions will be introduced throughout the balance of this chapter. These formulas consider the load and span of the beam, the modulus of elasticity, and the moment of inertia of the beam. Deflection is represented by the symbol Δ. The value for the modulus of elasticity is expressed as E for the species and grade of the beam to be used, and the moment of inertia is expressed as I.

Horizontal Shear

Horizontal shear is the tendency for the wood fibers to slide past each other at the neutral axis and fail along the length of the beam. Horizontal shear is a result of forces that affect the beam fibers parallel to the wood grain. Under severe bending pressure, adjacent wood fibers are pushed and pulled in opposite directions. The top portion of the beam nearest the load is in compression. The edge of the beam away from the load is under stress from tension. The line where the compression and tension forces meet is the point at which the beam will fail. Horizontal shearing stresses are greatest at the neutral surface. The maximum horizontal shear stress for a rectangular beam is one and one half times the average unit shear stress. The design formula is $v = (1.5)(V)/A$, where:

v = maximum unit horizontal shear stress in psi

V = total vertical shear in pounds

A = area of the structural member in square inches

Figure 29.6 lists the maximum allowable horizontal shear values in psi for major species of framing lumber. The value for v must be less than the maximum safe F_v value.

Horizontal shear has the greatest effect on beams with a relatively heavy load spread over a short span. Visualize a yardstick with supports at each end supporting a 2-lb. load at the center. The yardstick will bow but will probably not break. Move the supports toward the center of the beam so they are 1′ apart. The beam will not bend but will be more prone to failure from breaking from the stress of horizontal shear.

Vertical Shear

Vertical shear, represented by the letter V, is the tendency of a beam to fail perpendicular to the fibers of the beam from two opposing forces. Vertical shear will cause a beam to break and fall between its supporting posts. Vertical shear is rarely a design concern in residential construction, since a beam will first fail by horizontal shear.

Tension

In addition to the forces of tension that act within a beam, tension stresses attempt to lengthen a structural member. Tension is rarely a problem in residential and light commercial beam design. Tension is more likely to affect the intersection of beams than the beam itself. Beam details typically reflect a metal strap to join beams laid end to end, or a metal seat may be used to join two beams to a column.

Compression

Compression is the tendency to compress a structural member. Fibers of a beam tend to compress the portion of a beam resting on a column, but residential loads are usually not sufficient to cause a structural problem. Posts are another area where compression can be seen. Wood is strongest along the grain. The loads in residential design typically are not large enough to cause problems in the design of residential columns. However, the length of the post is important in relation to the load to be supported. As the length of the post increases, the post tends to bend rather than compress. Structures with posts longer than 10′ are often required by building departments to be approved by a structural engineer or architect. The forces of tension and compression will not be analyzed in this book.

METHODS OF BEAM DESIGN

Four methods of beam design are used by professionals: a computer program, a span computer, wood design books, and of course, the old-fashioned way of pencil, paper, and a few formulas.

Computer programs that will size beams are available for many personal and business computers. Each national wood distributor will supply programs that can determine spans, needed materials, and then print out all appropriate stresses. Software that will size beams made of wood, engineered lumber and steel is also available through private developers. Two popular programs include StruCalc 4 by Cascade Consulting Associates and Beam Chek by AC Software, Inc. Figure 29.7 shows an example of the prompts for solving a simple floor beam using StruCalc. Figure 29.8 shows the

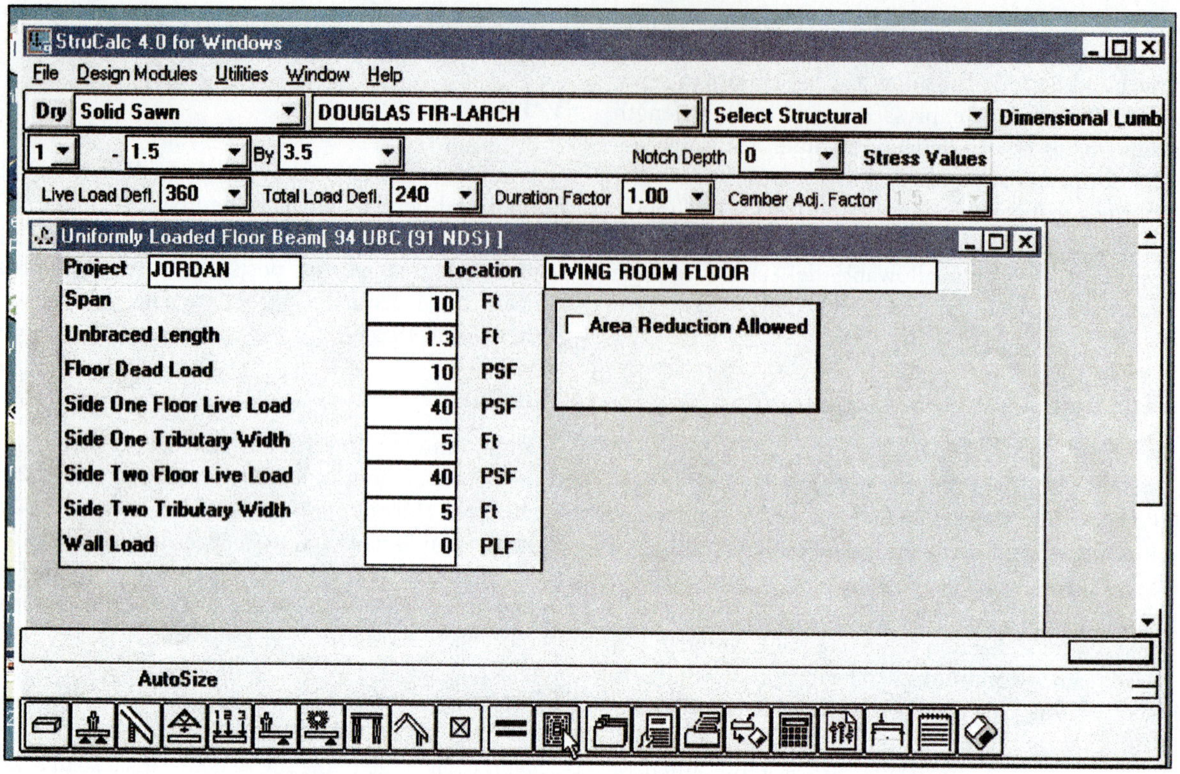

FIGURE 29.7 ■ Display for determining a beam that supports a floor load using the program StruCalc 4.0, developed by Cascade Consulting Engineers.

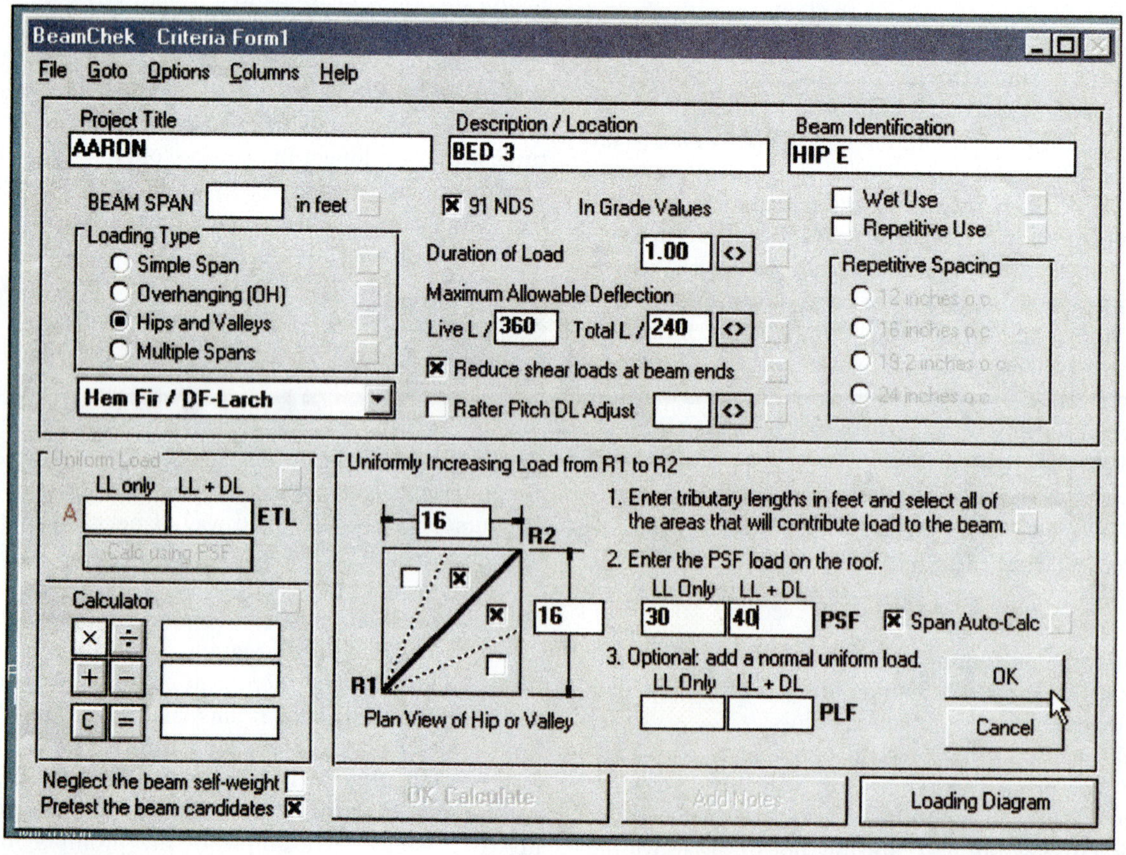

FIGURE 29.8 ■ Display for determining a hip beam using the program Beam Chek developed by AC Software, Inc.

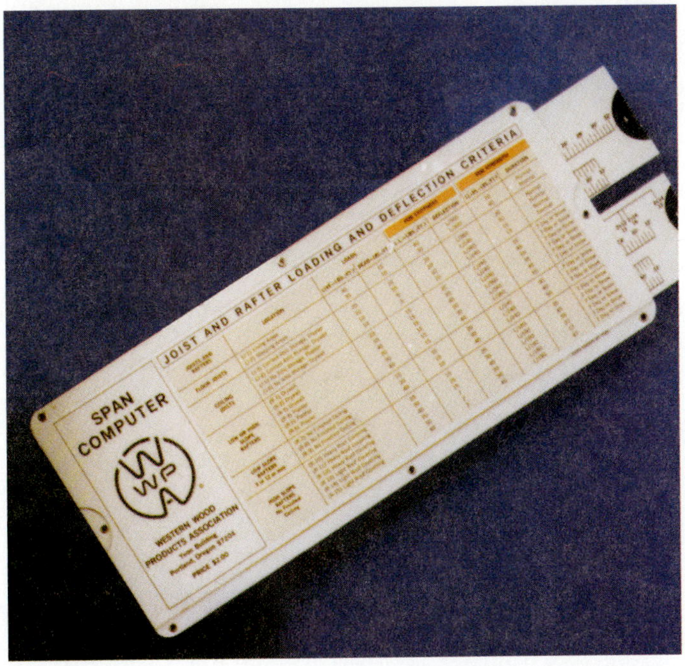

FIGURE 29.9 ■ Span computer used to determine joist and beam sizes. *Courtesy Western Wood Products Association.*

prompts for solving a hip using BeamChek. Web sites for each firm are listed at the end of the chapter. These programs typically ask for loading information and then proceed to determine the span size within seconds. Using a computer program to solve beam spans is extremely easy, but you do have to be able to answer questions that require an understanding of the basics of beam design and loading.

A span computer is very similar to a slide rule. Figure 29.9 shows a span computer made by the Western Wood Products Association. Beam sizes are determined by lining up values of the lumber to be used with a distance to be spanned. Western Wood Products also has a span calculator, named SpanMaster, available for sizing joists, rafters, and beams. Ordering information for each product can be found at the Western Wood Products Web site.

Another practical method of sizing beams is the use of books published by various wood associations. Two of the most common span books used in architectural offices are *Wood Structural Design Data* from the National Forest Products Association and *Western Woods Use Book* from the Western Wood Products Association. These books contain design information of wood members, standard formulas, and design tables that provide beam loads for a specific span. Figure 29.10 shows a partial listing for the 8′ span table. By following the instructions provided with the table, information on size, span, and loading patterns can be determined.

The final method of beam design is to use standard formulas to determine how the beam will be stressed. Figure 29.11 shows the formulas and the loading, shear, and moment diagrams for a simple beam. These are diagrams that can be drawn by the designer to determine where the maximum stress will occur. With a simple beam and a uniformly distributed load, the diagrams typically are not drawn, because the results remain constant.

NOTATIONS FOR FORMULAS

Standard symbols have been adopted by engineering and architectural communities to simplify the design of beams. Many symbols have been introduced throughout this chapter. Figure 29.12 gives a list of the notations that will be needed to design residential beams. Notice that some are written in uppercase letters and some in lowercase letters. Uppercase letters such as W and L represent measurements expressed in feet. W represents the total load in pounds per square feet; L represents the beam span in feet. Lowercase letters represent measurements expressed in inches. The letters b and d represent beam size in inches, and ℓ represents beam span in inches. As a beginning designer, you need not understand why the formulas work, but it is important that you know the notations to be used in the formulas.

SIZING WOOD BEAMS

Wood beams can be determined by following these steps:

1. Determine the area to be supported by the beam.
2. Determine the weight supported by 1 linear foot of beam.
3. Determine the reactions.
4. Determine the pier sizes.
5. Determine the bending moment.
6. Determine the horizontal shear.
7. Determine the deflection.

Determining the Area to Be Supported

To determine the size of a beam, the weight the beam is to support must be known. To find the weight, find the area the beam is to support and multiply by the total loads. Figure 29.13 shows a sample floor plan with a beam of undetermined size. The beam is supporting an area 10′–0″ long (the span). Floor joists are being supported on each side of the beam. Chapter 27 introduced methods of load dispersal. In this example, each joist on each side of the beam can be thought of as a simple beam. Half of the weight of each joist on side A will be supported by wall A, and half of the weight of joist A will be supported by the beam. Because the total length of the joist on side A is 10′, the beam will be supporting 5′ (half the length of each joist) of floor on side A. This 5′-wide area that the beam supports is referred to as a tributary area. In this case, it is tributary load A. On the right side of the beam, tributary load B would be 5′ wide (half of the

WOOD BEAMS—SAFE LOAD TABLES

Symbols used in the tables are as follows:

F_b = Allowable unit stress in extreme fiber in bending, psi.
W = Total uniformly distributed load, pounds
w = Load per linear foot of beam, pounds
F_v = Horizontal shear stress, psi, induced by load W
E = Modulus of elasticity, 100 psi, induced by load W for l/360 limit

Beam sizes are expressed as nominal sizes, inches, but calculations are based on net dimensions of S4S sizes.

SIZE OF BEAM		F_b									
		900	1000	1100	1200	1300	1400	1500	1600	1800	2000
8'–0" SPAN											
2 × 14	W	3291	3657	4023	4389	4754	5120	5486	5852	6583	7315
	w	411	457	502	548	594	640	685	731	822	914
	F_v	124	138	151	165	179	193	207	220	248	276
	E	489	543	597	652	706	760	815	869	978	1086
6 × 8	W	3867	4296	4726	5156	5585	6015	6445	6875	734	8593
	w	483	537	590	644	698	751	805	859	966	1074
	F_v	70	78	85	93	101	109	117	125	140	156
	E	864	960	1055	1152	1247	1343	1439	1535	1727	1919
4 × 10	W	3743	4159	4575	4991	5407	5823	6238	6654	7486	8318
	w	467	519	571	623	675	727	779	831	935	1039
	F_v	86	96	105	115	125	134	144	154	173	192
	E	700	778	856	934	1011	1089	1167	1245	1401	1556
3 × 12	W	3955	4394	4833	5273	5712	6152	6591	7031	7910	8789
	w	494	549	604	659	714	769	823	878	988	1098
	F_v	105	117	128	140	152	164	175	187	210	234
	E	576	640	704	768	832	896	959	1024	1151	1279
8 × 8	W	5273	5859	6445	7031	7617	8203	8789	9375	10546	11718
	w	659	732	805	878	952	1025	1098	1171	1318	1464
	F_v	70	78	85	93	101	109	117	125	140	156
	E	864	960	1055	1151	1247	1343	1439	1535	1727	1919
3 × 14	W	5486	6095	6705	7315	7924	8534	9143	9753	10972	12191
	w	685	761	838	914	990	1066	1142	1219	1371	1523
	F_v	124	138	151	165	179	193	207	220	248	276
	E	489	543	597	652	706	760	815	869	978	1086
4 × 12	W	5537	6152	6767	7382	7998	8613	9228	9843	11074	12304
	w	692	769	845	922	999	1076	1153	1230	1384	1538
	F_v	105	117	128	140	152	164	175	187	210	234
	E	576	640	704	768	832	896	960	1023	1151	1279
6 × 10	W	6204	6894	7583	8272	8962	9651	10341	11030	12409	13788
	w	775	861	947	1034	1120	1206	1292	1378	1551	1723
	F_v	89	98	108	118	128	138	148	158	178	197
	E	682	757	833	909	985	1061	1136	1212	1364	1515
3 × 16	W	7267	8075	8882	9690	10497	11305	12112	12920	14535	16150
	w	908	1009	1110	1211	1312	1413	1514	1615	1816	2018
	F_v	142	158	174	190	206	222	238	254	285	317
	E	424	472	519	566	613	660	708	755	849	944
4 × 14	W	7973	8859	9745	10631	11517	12403	13289	14175	15946	17718
	w	996	1107	1218	1328	1439	1550	1661	1771	1993	2214
	F_v	126	140	154	168	182	196	210	225	253	281
	E	480	533	586	640	693	746	800	853	960	1066

FIGURE 29.10 ■ Partial listing of a typical span table. Once W, w, F_b, E, and F_v are known, spans can be determined for a simple beam. See Appendix C for a complete listing. *From* Wood Structural Design Data. *Courtesy National Forest Products Association.*

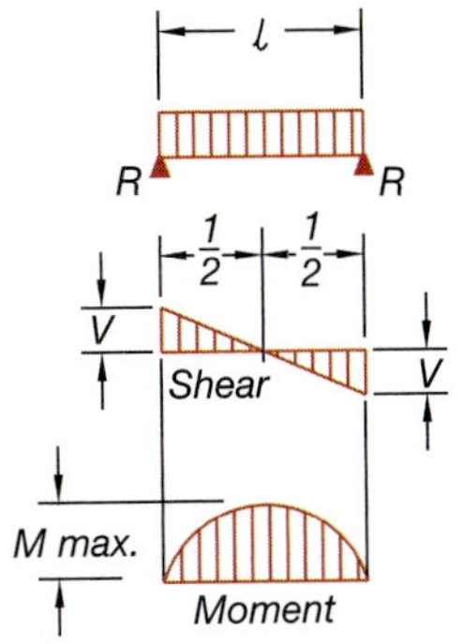

$$R = V = \frac{wl}{2}$$

$$M\ max.\ (at\ center) = \frac{wl^2}{8}$$

$$v = \frac{3V}{2bd}$$

$$\Delta\ max.\ (at\ center) = \frac{5wl^4}{384EI}$$

FIGURE 29.11 ■ Loading, shear, and moment diagrams for simple beam with uniform loads, showing where stress will affect a beam and the formulas for computing these stresses. *From* Wood Structural Design Data. *Courtesy National Forest Products Association.*

BEAM FORMULA NOTATIONS

b	=	breadth of beam in inches
d	=	depth of beam in inches
D	=	deflection due to load
E	=	modulus of elasticity
F_b	=	allowable unit stress in extreme fiber bending
F_v	=	unit stress in horizontal shear
I	=	moment of inertia of the section
l	=	span of beam in inches
L	=	span of beam in feet
M	=	bending or resisting moment
P	=	total concentrated load in pounds
S	=	section modulus
V=R	=	end reaction of beam
W	=	total uniformly distributed load in pounds
w	=	load per linear foot of beam in pounds

FIGURE 29.12 ■ Common notations used in beam formulas.

total span of 10′). A *tributary width* is defined as the accumulation of loads that are directed to a structural member. The tributary width will always be half the distance between the beam to be designed and the next bearing point. A second point to remember about tributary loads is that the total tributary width will always be half the total distance between the bearing points on each side of the beam, no matter where the beam is located. Figure 29.14 shows why this is true.

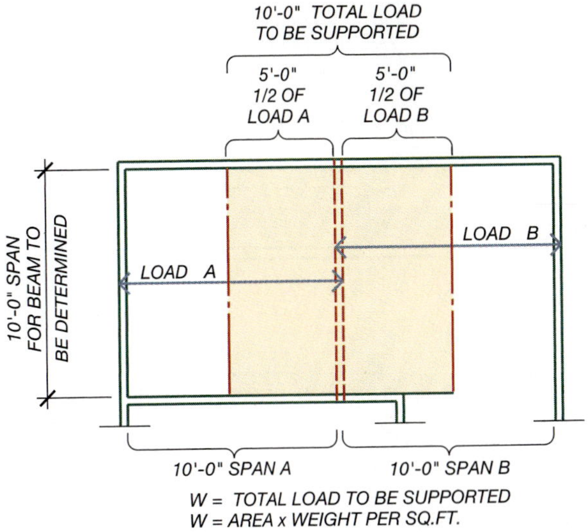

FIGURE 29.13 ■ Floor plan with a beam to be determined. This beam is supporting an area of 100 sq ft, with an assumed weight of 50 psf, W = 5000 lb.

Once the tributary width is known, the total area that the beam is supporting can be determined. Adding the width of tributary A + B and then multiplying this sum by the beam length measured in feet (represented by L) will determine the area that the beam is supporting. For the beam in Figure 29.13, the area would be (5′ + 5′)(10′) = 100 sq ft. Using the loads from Figure 27.7, the loads for floors can be determined. The live load for a floor is 40#, the dead load is 10#, and the total load is 50#. By multiplying the area by the load per square foot, the total load (W) can be determined. For this example, with an area of 100 sq ft and a total load of 50 psf, the total load is 5000#. The total load is represented by the letter W in beam formulas.

If the beam span (L) is to be determined using a computer program, the total live load and the total dead load that the beam is supporting may be required instead of the total load (W). For this example the total dead load is 10# × 100 sq ft, or 1000#. The total live load is 40# × 100 sq ft, or 4000#. Although this may seem obvious, the total live load and the total dead load should add up to the total load supported. Taking the time to check the obvious can help to eliminate human errors.

Determining Linear Weight

In some formulas only the weight per linear foot of beam is desired. This is represented by the letter w. In Figure 29.15, it can be seen that w is the product of the area to be supported (the total tributary width), multiplied by the weight per square foot. It can also be found by dividing the total weight (W) by the length of the beam. In our example, if the span (L) is 10′, w = 5000/10 or 500 lb.

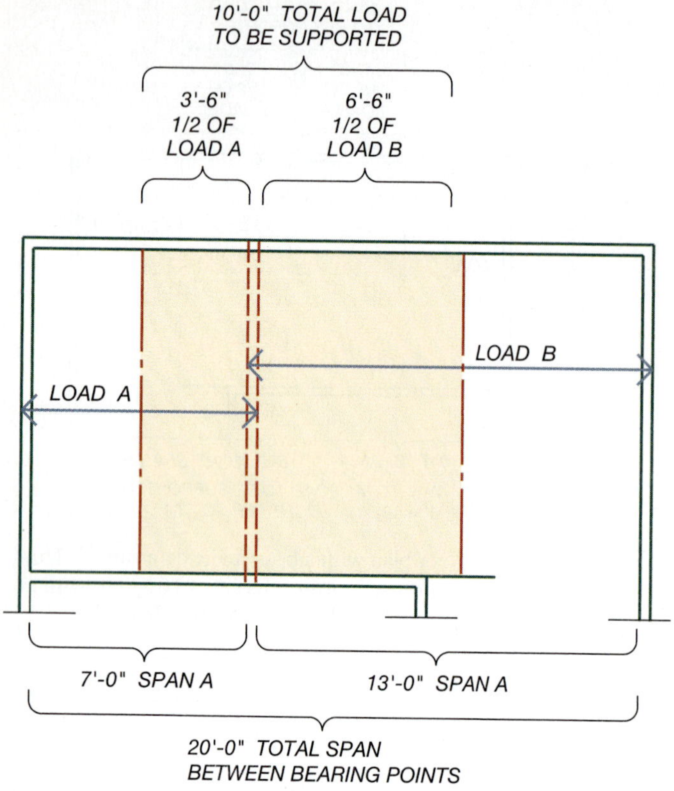

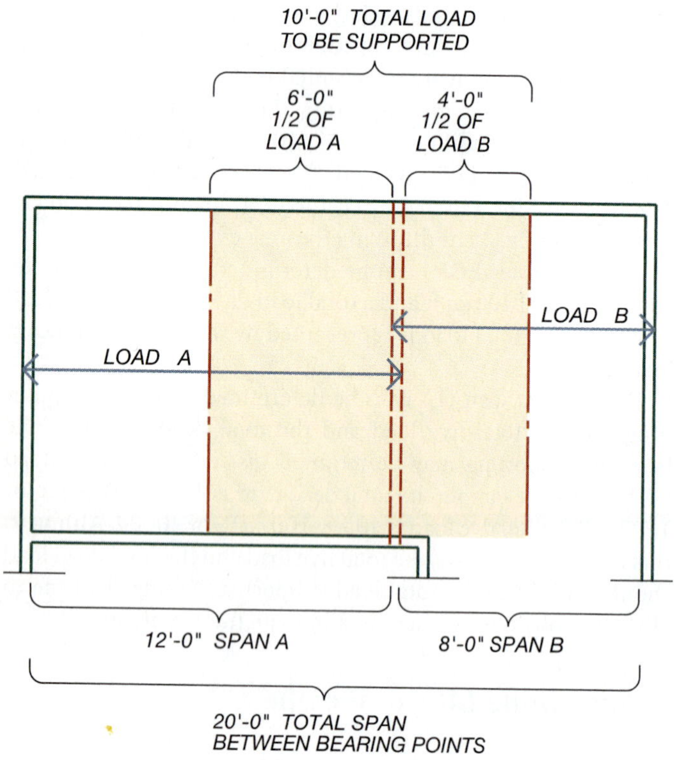

FIGURE 29.14 ■ The sum of the tributary widths is equal to half the total distance between the bearing point on each side of the beam being determined. With a total width of 20′, the tributary width supported by the beam will be 10′ no matter where the beam is placed.

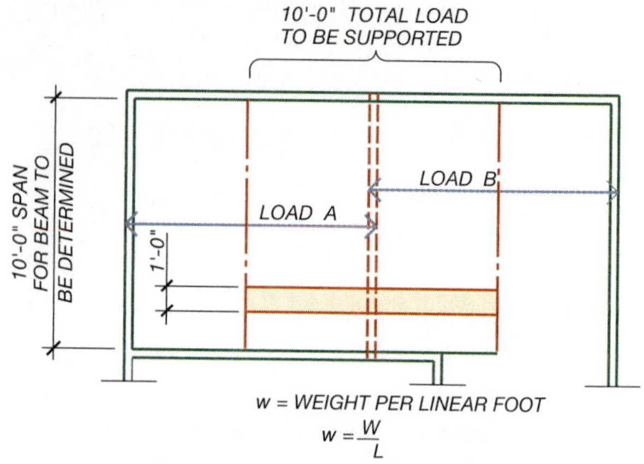

FIGURE 29.15 ■ Determining linear weights. For an area of 10 sq ft, with an assumed weight of 50 psf, w = 500 lb.

Determining Reactions

Even before the size of a beam is known, the supports for the beam can be determined. These supports for the beam are represented by the letter R, for reactions. The letter V is also sometimes used in place of the letter R. On a simple beam, half of the weight it is supporting (W) will be dispersed to each end. In the example, W = 5000 lb and R = 2500 lb. At this point W, w, and R have been determined (W = 5000 lb, w = 500 lb, and R = 2500 lb).

Determining Pier Sizes

The reaction is the load from the beam that must be transferred to the soil. The reaction from the beam will be transferred by a post to the floor and foundation system. Usually a concrete pier is used to support the loads at the foundation level. To determine the size of the pier, the working stress of the concrete and the bearing value of the soil must be taken into account.

Usually concrete with a working stress of 2000 to 3500 psi is used in residential construction. The working stress of concrete specifies how much weight in pounds can be supported by each square inch (psi) of concrete surface. If each square inch of concrete can support 2500 lb, only 1 sq in. of concrete would be required to support a load of 2500 lb. The bearing value of the soil must also be considered. The pier size is determined by dividing the load to be supported (R) by the soil pressure.

See Figure 29.16 for safe soil-loading values. Most building departments use 2000 psf for the assumed safe working value of soil.

Don't skim over these numbers. The concrete is listed in pounds per square inch. Soil is listed in pounds per square foot.

PRESUMPTIVE LOAD-BEARING VALUES OF FOUNDATION MATERIALS[a]

CLASS OF MATERIAL	LOAD-BEARING PRESSURE (pounds per square foot)
Crystalline bedrock	12,000
Sedimentary and foliated rock	4,000
Sandy gravel and/or gravel (GW and GP)	3,000
Sand, silty sand, clayey sand, silty gravel and clayey gravel (SW, SP, SM, SC, GM and GC)	2,000
Clay, sandy clay, silty clay, clayey silt, silt and sandy silt (CI, ML, MH and CH)	1,500[b]

For SI: 1 psf = 0.0479 kN/m^2.

a. When soil tests are required by Section R401.4, the allowable bearing capacities of the soil shall be part of the recommendations.

b. Where the building official determines that in-place soils with an allowable bearing capacity of less than 1,500 psf are likely to be present at the site, the allowable bearing capacity shall be determined by a soils investigation.

FIGURE 29.16 ■ Safe soil-bearing values. *Reproduced from the* International Residential Building Code /2000. *Copyright © 2000. Courtesy International Code Council, Inc.*

A pier supporting 2500 lb must be divided by 2000 lb (the soil-bearing value) to find the area of the pier needed. This will result in an area of 1.25 sq ft of concrete needed to support the load. See Figure 29.17 to determine the size of pier needed to obtain the proper soil support area.

Determining the Bending Moment

A *moment* is the tendency of a force to cause rotation about a certain point. In figuring a simple beam, W is the force and R is the point around which the force rotates. Determining the bending moment will calculate the size of the beam needed to resist the tendency for W to rotate around R.

To determine the size of the beam required to resist the force (W), use the following formula:

$$M = \frac{(w)(\ell^2)}{8}$$

PIER AREAS AND SIZES

ROUND PIERS	SQUARE PIERS
15″ DIA. = 1.23 SQ FT	15″ SQ = 1.56 SQ FT
18″ DIA. = 1.77 SQ FT	18″ SQ = 2.25 SQ FT
21″ DIA. = 2.40 SQ FT	21″ SQ = 3.06 SQ FT
24″ DIA. = 3.14 SQ FT	24″ SQ = 4.00 SQ FT
27″ DIA. = 3.97 SQ FT	27″ SQ = 5.06 SQ FT
30″ DIA. = 4.90 SQ FT	30″ SQ = 6.25 SQ FT
36″ DIA. = 7.07 SQ FT	36″ SQ = 9.00 SQ FT
42″ DIA. = 9.60 SQ FT	42″ SQ = 12.25 SQ FT

FIGURE 29.17 ■ Common pier areas and sizes. The area of the concrete pier can be determined by dividing the load to be supported by the soil-bearing pressure. Areas are shown for common pier sizes. Round piers are typically used for interiors, and square piers are generally used when piers are added to support the exterior foundation.

Once the moment is known, it can then be divided by the F_b of the wood to determine the size of beam required to resist the load ($S = M/F_b$). This whole process can be simplified by using the formula:

$$S = \frac{(3)(w)(L^2)}{(2)(F_b)}$$

Because F_b values for 4× members are size dependent the formula must be reworked with a new value if a 4× member is to be used. To determine if a 4 × 10 would be suitable, use the values for Douglas fir from Figure 29.6, and apply this formula to the beam span from Figure 29.13:

$$S = \frac{(3)(500)(100)}{(2)(1050)} = \frac{150,000}{2100} = 71.4 = S$$

This formula will determine the section modulus required to support a load of 5000 lb. Look at the table in Figure 29.18 to determine if a 4 × 10 is adequate. A beam must be selected that has an S value larger than the S value found. By examining the S column of the table, you will find that a 4 × 10 has an S value of only 49.9, so the beam is not adequate. To determine if a 4 × 12 is adequate, divide the top portion of the S formula by 2 × 963 (150,000/(2)963 = S = 78.4). A 4 × 12 is also inadequate. To determine what 6× is adequate, use an F_b value of 1350. The values of 963 and 1350 are determined using the table in Figure 29.6. Using the formula 150,000/2700 will require a minimum S value of 55.6. It can be seen in Figure 29.18 that a 6 × 8 has

PROPERTIES OF STRUCTURAL LUMBER						
NOMINAL SIZE	(b)(d)	(2)(b)(d)	S	A	I	(384)(E)(I) *
2 × 6	1.5 × 5.5	16.5	7.6	8.25	20.8	12,780
2 × 8	1.5 × 7.25	21.75	13.1	10.875	47.6	29,245
2 × 10	1.5 × 9.25	27.75	21.4	13.875	98.9	60,764
2 × 12	1.5 × 11.25	33.75	31.6	16.875	177.9	109,302
2 × 14	1.5 × 13.25	39.75	43.9	19.875	290.8	178,668
4 × 6	3.5 × 5.5	38.5	17.6	19.25	48.5	29,798
4 × 8	3.5 × 7.25	50.75	30.7	25.375	111.0	68,198
4 × 10	3.5 × 9.25	64.75	49.9	32.375	230.8	141,804
4 × 12	3.5 × 11.25	78.75	73.8	39.375	415.3	255,160
4 × 14	3.5 × 13.5	94.5	106.3	47.250	717.6	440,893
6 × 8	5.5 × 7.5	82.5	51.6	41.25	193.4	118,825
6 × 10	5.5 × 9.5	104.5	82.7	32.25	393.0	241,459
6 × 12	5.5 × 11.5	126.5	121.2	63.25	697.1	428,298
6 × 14	5.5 × 13.5	148.5	167.1	74.25	1127.7	692,859

*384 × E × I *values are listed in units per million, with* E *value assumed to be 1.6 for DFL.*

FIGURE 29.18 ■ Structural properties of wood beams. Columns (2)(b)(d) and (384)(E)(I) can be used in the horizontal shear and deflection formulas. These two columns are set up for Douglas fir no. 2. For different types of wood, use NDS values for these columns.

an F_b value of only 51.6, so it is not suitable. A 6 × 10 has an F_b value of 82.7, which is suitable to resist the forces of bending moment. The other formulas must now be used to verify that a 6 × 10 will be adequate for horizontal shear and deflection.

Determining Horizontal Shear Values

To compute the size of a beam required to resist the forces of horizontal shear, use the following formula:

$$F_v = \frac{3(V)}{(2)(b)(d)} = \# < 85 \text{ (safe design value for DFL)}$$

This will determine the minimal 2bd value required to resist the load. For a simple beam, V = R. To determine b and d, use the actual size of the beam determined for bending. For this example, use the 6 × 10 value. The actual size of a 6 × 10 is 5 1/2″ × 9 1/2″. Figure 29.18 shows the product (2bd) to be 104.5. These values can be inserted into the formula to find the shear on the beam.

$$F_v\ 5\ \frac{(3)(V)}{(2)(b)(d)}\ 5\ \frac{3\ 3\ 2500}{104.5}\ 5\ \frac{7500}{104.5}\ 5\ 71.7$$

This answer is meaningless unless you know the safe shear value for the type of wood in question. The safe limit for lumber in horizontal shear can be determined from the table in Figure 29.6. The value for DFL is 85. If the product of the formula is larger than the safe working stress, the beam fails. In the example above, 71.7 is less than the limit for Douglas fir. If the answer had been bigger than 85, the formula would have needed to be reworked by inserting a different beam 2bd value. The figures on the top of the formula will remain the same because these values are the result of the load, which has not changed.

Another way to determine F_v is to multiply V times 3 and then divide the F_v value for the lumber being used.

$$F_v = \frac{3V}{85} = \# < \text{(2bd) value for selected bm from Figure 29.18}$$

$$F_v = \frac{3\ 3\ 2500}{85} = \text{(minimum value needed. 2bd value must exceed the value.) } 88.23 = 6 \times 10$$

This will produce a minimum area required to resist the stress in horizontal shear. You have now determined W, w, R, V, S, and F_v. The last value to be determined is the stress for deflection.

Determining Deflection

Deflection is the amount of sag in a beam. Deflection limits are determined by building codes and are expressed as a fraction of an inch in relation to the span of the beam in inches. Limits set by the IRC are:

l/360 Floors and ceilings.

l/240 Roofs under 3/12 pitch, tile roofs, and vaulted ceilings

l/180 Roofs over 3/12 pitch

To determine deflection limits, the maximum allowable limit must first be known. Use the formula:

$$D_{max} = \frac{L \times 12}{360} = \text{maximum allowable deflection}$$

This formula requires the span in feet (L) to be multiplied by 12 (12 inches per foot) and then divided by 360 (the safe limit for floors and ceiling).

In the example used in this chapter, L equals 10′. The maximum safe limit would be:

$$D_{max} = \frac{L \times 12}{360} = \frac{10 \times 12}{360} = D = \frac{120}{360} = 0.33''$$

$$= \text{maximum allowable deflection}$$

The maximum amount the beam is allowed to sag is 0.33″. Now you need to determine how much the beam will actually sag under the load it is supporting.

The values for E and I need to be determined before the deflection can be determined. These values can be found in Figure 29.18. You will also need to know ℓ^3 and W, but these values are not available in tables. W has been determined. For our example:

$$\ell = 10 \times 12 = 120''$$

$$\ell^3 = (\ell)(\ell)(\ell) = (120)(120)(120) = 1{,}728{,}000 \text{ or } 1.728$$

Because the E value was reduced from 1,600,000 to 1.6, change the ℓ^3 value from 1,728,000 to 1.728. Each of these numbers is much easier to use for calculating.

To determine how much a beam will sag, use the following formula:

$$D = \frac{5(W)(\ell^3)}{(384)(E)(I)} \text{ or } D = 22.5\frac{(W)(\ell^3)}{(E)(I)}$$

The values determined thus far are W = 5000, w = 500, R = V = 2500, L = 10, ℓ = 120, ℓ^3 = 1.728 and I = 393 (see Figure 29.18, column I). Find also the value of (384)(E)(I) for a 6 × 10 beam from the table in Figure 29.18 (241,459) or use the (384)(E)(I) value from table 29.21. Now insert the values into the formula.

$$D = \frac{5(W)(\ell^3)}{384(E)(I)} = \frac{5 \times 5000 \times 1.728}{241{,}459} = \frac{43{,}200}{241{,}459} = 0.178''$$

Because 0.178″ is less than the maximum allowable deflection of 0.33″, a 6 × 10 beam can be used to support the loads.

REVIEW

In what may have seemed like an endless string of formulas, tables, and values, you have determined the loads and stresses on a 10′ long beam. Seven basic steps were required, as follows:

1. Determine the values for W, w, L, ℓ, ℓ^3 and R.
   W = area to be supported × weight (LL + DL)
   w = W/L
   L = span in feet
   ℓ = span in inches
   ℓ^3 = (l)(l)(l)
   R = V = W/2
2. Determine S:
   $$S = \frac{(3)(w)(L^2)}{(2)(F_b)}$$
3. Determine F_v:
   $$F_v = \frac{(3)(V)}{(2)(b)(d)} = \# < 85 \text{ or } \frac{(3)(V)}{85} = \# < (2)(b)(d)$$
4. Determine the value for D_{max}
   D = l/360 or l/240 or l/180
5. Determine D:
   $$D = \frac{(5)(W)(\ell^3)}{(384)(E)(I)}$$
6. Determine post supports: R = W/2
7. Determine piers size: R/soil bearing pressure

Figure 29.19 shows a floor plan with a ridge beam that needs to be determined. The beam will be SPF with no snow loads. Using the seven steps, determine the size of the beam.

STEP 1 Determine the values.
W = 10 × 12 × 40 = 4800 lb
w = W/L = 4800/12 = 400 lb
R = V = 2400
L = 12′
L^2 = 144
ℓ = 12′ × 12″ = 144
ℓ^3 = 2.986
F_b = 900 (See figure 29.20)
F_v = 65 max (see figure 29.20)
E = 1.2 max

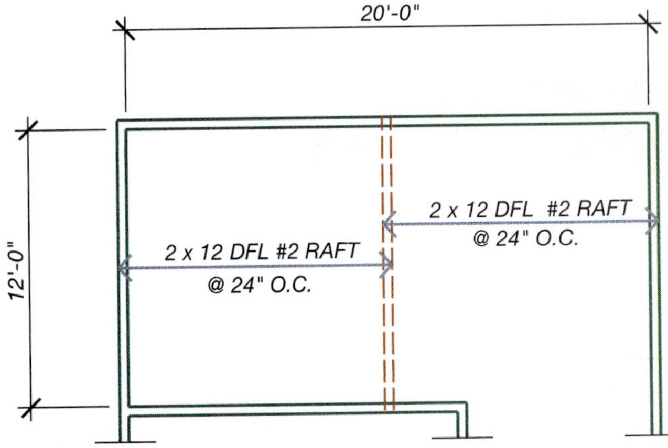

FIGURE 29.19 ■ Sample floor plan with a beam of undetermined size.

STEP 2 Determine the section modulus.

$$S = \frac{(3)(w)(L^2)}{(2)(F^b)} = \frac{3\ 3\ 400\ 3\ 144}{2\ 3\ 900} = \frac{172{,}800}{1800} = 96$$

Use a 4 × 14 or a 6 × 12 to determine F_v.

STEP 3 Determine horizontal shear. The value from Table 29.20 is 65.

$$4 \times 14\ F_v = \frac{(3)(V)}{(2)(b)(d)} = \frac{2\ 3\ 2400}{2\ 3\ 3.5\ 3\ 13.5} = \frac{7200}{94.5} = 76.19 > 65$$

Since the computed value is larger than the table value, the 4 × 14 beam is inadequate. Try a 6 × 12.

$$6 \times 12\ F_v = \frac{700}{126.5} = 56.92$$

Finally, this beam works. Now solve for deflection.

STEP 4 Determine D_{max}.

$$D_{max} = \frac{1}{180} = \frac{144}{180} = 0.8''$$

STEP 5 Determine D:

$$D = \frac{(5)(W)(l^3)}{(384)(E)(I)} = \frac{5\ 3\ 4800\ 3\ 2.986}{384\ 3\ 1.2\ 3\ 697}$$

$$= \frac{71{,}664}{321{,}178}$$

$$D = 0.223$$

STEP 6 Determine reactions:

$$R = W/2 = 2400$$

STEP 7 Determine piers. R/soil value (assume 2000 lb)

$$\frac{2400}{2000} = 1.2 \text{ sq ft}$$

According to the table (Figure 29.17) use either a 15″ diameter or 15″ square pier.

WOOD ADJUSTMENT FACTORS

To this point you've been introduced to the basics of solving simple beams using the NDF base design values for the particular species based on the size of the member. These values will provide an accurate but conservative description of how a specific member can be used. Building codes and the NDS guidelines allow the beam values to be adjusted based on conditions that will reflect the true use of the structural member. These reductions include repetitive use, load duration, moisture and temperature content, size, and shear stress.

Repetitive Use Factor

When a load is spread over several members, a reduction in the F_b value is allowed. The fiber bending value of lumber 2″ and 4″ thick placed repetitively can be applied if all of the following conditions are met:

- The structural members receiving the reductions must be placed at 24″ maximum spacing.
- A minimum of 3 members are used in the pattern.
- Each of the structural members are joined by sheathing or decking.

When all three conditions are met, an increase in the F_b value can be used. The increase is $F_b \times 1.15$.

Load Duration Factor

Tests have shown that a piece of lumber can carry greater maximum loads over a short period of time than over a long period of time. When a load is supported by a wood member for a short duration, the member will return to its original shape when the load is reduced. When the load is supported for a long period of time, the member will become permanently deformed. Design standards generally recognize several common duration periods that affect the maximum live load to be supported. These load factors include permanent, normal, two months, seven days, wind or earthquake, and impact. When two or more loads of different duration will be supported, the factors are not cumulative. Use the load duration factor that best represents the lifetime of the load. Because wind, earthquake, and impact loads can vary so widely, these should be based on local codes and considered with the supervision of a licensed professional.

LOAD	DESIGN FACTOR (APPLY TO SIZE ADJUSTED VALUE)
Permanent—A load supported for more than 10 years	0.90
Normal—Loads supported for up to 10 years	1.0
Two-month—Short-term loads such as snow on a roof	1.15
Seven-day–Short-term loads such as workers on a roof	1.25
One-day–Very short-term loads	1.33
Ten-minute—Wind and earthquake loads	1.6
Impact—Unexpected falling weight	2.0

Moisture and Temperature Content Factor

Wood that is exposed to extreme moisture and high temperatures will lose strength. Although this is generally not a problem with residential construction, building departments can specify that wood for decks and other uses exposed to long periods of moisture and high temperature be subject to wet use adjustments. Design adjustment factors for wet use include $F_b \times 0.85$, $F_c \times 0.67$, and $E \times 0.90$.

Size Factor

The depth of sawn lumber will affect how the member responds to a load. The size factor compensates for the natural properties of lumber. Values include those in the table below.

Shear Stress Factor

The values for $F_{v_{//}}$ that are listed in tables allow for structural members to have the maximum number of splits, checks, and shakes. When the actual size of the lumber to be used is known, adjustments can be made for the structural member. Designers typically use this adjustment when a desired beam is adequate for section bending and deflection but fails in horizontal shear. Horizontal shear values for 3″ and thicker lumber are established as if the lumber is split for the full length. When specific lengths of splits are known and no increase is expected, the following adjustments can be made to the base F_v value:

3″ AND THICKER LUMBER

WHEN LENGTH OF SPLIT ON WIDE FACE IS:	MULTIPLY TABULATED F_V VALUE BY:
No split	2.0
1/2 of narrow face	1.67
1 of narrow face	1.33
1 1/2 of narrow or more	1.00

SPAN TABLES

You have already been introduced to span tables. Now that you have sized a beam by solving for S, F_v, and E, you can better understand why span tables should be used whenever possible. The National Forest Products Association offers tables that are easy to use and accepted by most professionals as a design standard. Appendix D shows a partial listing of these tables for spans ranging from 6′–0″ through 16′–0″. Although shorter beams may be required in a residence, generally only the F_v value will need to be determined to find the beam size. Beams longer than 16′ are typically holding enough weight that a glu-lam beam should be used.

To use the information in Appendix A, W must be known. The type of wood must also be known so that the maximum

SIZE FACTORS C_F (APPLY TO DIMENSION LUMBER BASE VALUES)

GRADE	NOMINAL WIDTH (DEPTH)	Fb: 2 AND 3″ THICK NOMINAL	Fb: 4″ THICK NOMINAL	Ft	Fc//	OTHER PROPERTIES
	2″, 3″, & 4″	1.5	1.5	1.5	1.15	1.0
Select	5″	1.4	1.4	1.4	1.1	1.0
Structural	6″	1.3	1.3	1.3	1.1	1.0
No. 1 & Btr	8″	1.2	1.3	1.2	1.05	1.0
No. 1 & No. 2	10″	1.1	1.2	1.1	1.0	1.0
& No. 3	12″	1.0	1.1	1.0	1.0	1.0
	14″ & wider	0.9	1.0	0.9	0.9	1.0
Construction & Standard	2″, 3″, & 4″	1.0	1.0	1.0	1.0	1.0
Utility	2″ & 3″	0.4	—	0.4	0.6	1.0
	4″	1.0	1.0	1.0	1.0	1.0
Stud	2″, 3″, & 4″	1.1	1.1	1.1	1.05	1.0
	6″ & wider	1.0	1.0	1.0	1.0	1.0

values for F_b, F_v, and E can be used. In the beam solved in the last example, this value would be W = 4800#. If SPF is to be used, F_b = 900, F_v = 65, and E = 1.3. Because the beam is 12′ long, look in the 12′ span section of the table. Because F_b is 900, your answer will come from this column. Proceed down this column until you find a W value that is equal to or larger than the W to be supported (4800 lb). 4 × 14 (5315) and 6 × 12 (6061) will work. Check the F_v values for each. This number must be less than 65. In our example, both seem to fail, but *this is not the case.* Consider what the 6 × 12 column has told you. If you were to have a 12′-long beam, and it supported 6061 lb, the beam would be under 71 units of stress in F_v. The beam only needs to resist 4800 lb, not 6061. In this case, you still have to figure F_v by using the formula 3V/65 = 2bd. You determined this in the last example and know that a 6 × 12 will work. To determine if D is suitable, examine the D value for the 6 × 12. You will see the number 845. This represents 0.845 unit of stress. The maximum E value for SPF is 1.3. Since 0.845 is less than 1.3, the beam works.

Try solving the same beam for Douglas fir. The values to be used are W = 4800, F_b = 1350, F_v = 85, and E = 1.6. Enter the 1300 F_b column until a weight is found equal to or greater than 4800 lb. You could use a 4 × 12. Notice that F_v is 101. Because 101 is greater than 85, the beam appears to fail. Instead of working with the F_v formula, there is a shortcut. Because the beam only supports 4800 lb, go to the left of the W values until the W value is closer to 4800. As you move to the left in the W column, you'll find that at 4921 lb, the beam has F_v = 93. Since your beam holds only 4800 lb, you can continue to the left. At 4511 lb, the beam has an F_v of 85. 85 is the maximum F_v value, but at this point the 12′ beam cannot support the desired W of 4800.

Try a 4 × 14. Enter the F_b 1300 column. The maximum supported weight is 7678 lb. Using the W values for a 4 × 14, move to the left until you find a number close to 4800. The F_b 900 column is as close as you can get with a W value of 5315. A beam holding 5315 lb will have 84 units of F_v stress. A 4 × 14 beam is safe. Its E value is .720, less than 1.6, so E is safe also. Compare the values for a 6 × 10, and you will find that it, too, is safe. Enter the table in the 1300 F_b column, and you will find that the beam could support 5974 lb. Move to the left, and you will see that the weight to be supported falls between 4596 and 5055 lb. Each has an F_v value less than the maximum of 85. Staying in the 5055 column (which is more than our beam will have to support), the E value is 1.250, which is less than the maximum of 1.6 for Douglas fir.

Laminated Beams

Two beam problems have now been solved using the tables in this chapter. Both beams were chosen from standard dimensioned lumber. Often, glu-lams are used because of their superior strength. Using a glu-lam can greatly reduce the depth of a beam as compared with conventional lumber. Glu-lam beams are determined in the same manner as standard lumber, but different values are used based on values provided by the American Institute of Timber Construction. See Figure 29.20.

BASE DESIGN VALUES FOR BEAMS & STRINGERS

SPECIES AND COMMERCIAL GRADE	EXTREME FIBER BENDING F_b	HORIZONTAL SHEAR F_v	MODULUS OF ELASTICITY E
DFL #1	1350	85	1,600,000
24 F_b V4 DF/DF	2400	165	1,700,000
Hem Fir #1	1050	70	1,300,000
24 F_b E-2 HF/HF	2400	155	1,700,000
SPF #1	900	65	1,200,000
22F_b E-2 SP/SP	2200	200	1,700,000

BASE DESIGN VALUES FOR POSTS & TIMBERS

SPECIES AND COMMERCIAL GRADE	EXTREME FIBER BENDING F_b	HORIZONTAL SHEAR F_v	MODULUS OF ELASTICITY E
DLF #1	1200	85	1,600,000
Hem Fir #1	950	70	1,300,000
SPF #1	800	65	1,200,000

FIGURE 29.20 ■ Comparative values of common framing lumber with laminated beams of equal materials. *Values based on the Western Wood Products Association.*

PROPERTIES OF GLU-LAM BEAMS

SIZE (b)(d)	S	A	(2)(b)(d)	I	(384)(E)(I) *
3⅛ × 9.0	42.2	28.1	56.3	189.8	123,901
3⅛ × 10.5	57.4	32.8	65.6	301.5	196,819
3⅛ × 12.0	75.0	37.5	75.0	450.0	293,760
3⅛ × 13.5	94.9	42.2	84.4	640.7	418,249
3⅛ × 15.0	117.2	46.9	93.8	878.9	573,746
5⅛ × 9.0	69.2	46.1	92.0	311.3	203,217
5⅛ × 10.5	94.2	53.8	107.6	494.4	322,744
5⅛ × 12.0	123.0	61.5	123.0	738.0	481,766
5⅛ × 13.5	155.7	69.2	138.4	1,050.8	685,962
5⅛ × 15.0	192.2	76.9	153.8	1,441.4	940,946
5⅛ × 16.5	232.5	84.6	169.0	1,918.5	1,252,397
6¾ × 10.5	124.0	70.9	141.8	651.2	425,103
6¾ × 12.0	162.0	81.0	162.0	972.0	634.522
6¾ × 13.5	205.0	91.1	182.6	1,384.0	903,475
6¾ × 15.0	253.1	101.3	202.0	1,898.4	1,239,276
6¾ × 16.5	306.3	111.4	222.8	2,526.8	1,649,495

**All values for 384 × E × I are written in units per million. All E values are figured for Doug fir @ E = 1.7. Verify local conditions.*

FIGURE 29.21 ■ Structural properties of glu-lam beams. Columns (2)(b)(d) and (384)(E)(I) are set up for Douglas fir. If a different type of wood is to be used, you must use different values for these columns. *Values are based on the Western Wood Products Association.*

Figure 29.21 gives values for glu-lams made of Douglas fir. You will also need to consult the safe values for glu-lam beams. For beams constructed of Douglas fir, the values are $F_b = 2200$, $F_v = 165$, and $E = 1.7$.

Using Engineered Beams

Most manufacturers of engineered joists and rafters also supply beams made of LVL (laminated veneer lumber), PSL (parallel strand lumber) and LSL (laminated strand lumber). Figure 29.22 shows an example of a table for sizing Parallam beams made of PSL. To use this table, follow these steps:

- Determine the roof loading (snow duration and live and dead loads) and find the appropriate section of the table that represents the design loads.
- Find the house width from the appropriate loading section that meets or exceeds the span of the trusses.
- Locate the opening size in the "Rough Opening" column that meets or exceeds the required window or door rough opening.
- The beam listed at the intersection of the rough opening and the house width/roof load is the required header size.

Figure 29.23 shows an example of a table that can be used to determine the span of interior roof beams and Figure 29.24 shows an example of a table that can be used to determine the size of beams used to support floor loads. Because sizes vary with each manufacturer, specific standards and tables should be obtained from the beam supplier.

Design Properties

(1) For 12-inch depth. For others, multiply by $[\frac{12}{d}]^{0.111}$

(2) $F_{c\perp}$ shall not be increased for duration of load.

(3) 750 psi for all Eastern Species Parallam® PSL and 13/4" thick Western Species Parallam® PSL.

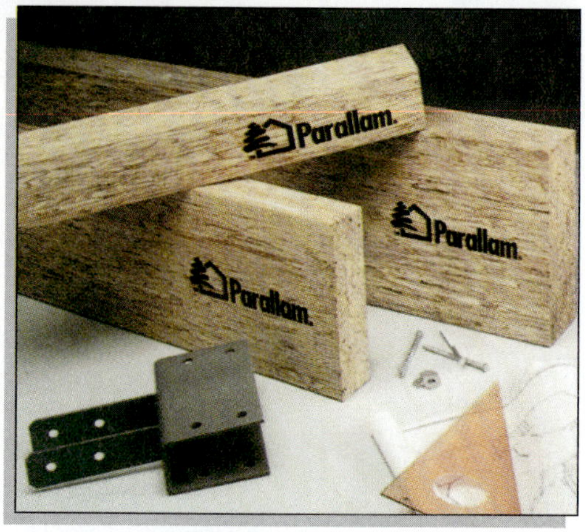

Allowable Design Properties (100% Load Duration)

1 3/4" 2.0E Parallam® PSL

DESIGN PROPERTY	DEPTH				
	9 1/4"	9 1/2"	11 1/4"	11 7/8"	14"
MOMENT (ft. lbs.)	6,210	6,530	8,985	9,950	13,58
SHEAR (lbs.)	3,130	3,215	3,805	4,020	4,735
MOMENT OF INERTIA (in^4)	115	125	210	245	400
WEIGHT (lbs./lin. ft.)	5.1	5.2	6.2	6.5	7.7

2 11/16" 2.0E Parallam® PSL

DESIGN PROPERTY	DEPTH						
	9 1/4"	9 1/2"	11 1/4"	11 7/8"	14"	16"	18"
MOMENT (ft. lbs.)	9,535	10,025	13,800	15,280	20,855	26,840	33,530
SHEAR (lbs.)	4,805	4,935	5,845	6,170	7,275	8,315	9,350
MOMENT OF INERTIA (in^4)	175	190	320	375	615	915	1,305
WEIGHT (lbs./lin. ft.)	7.8	8.0	9.5	10.0	11.8	13.4	15.1

5 1/4" 2.0E Parallam® PSL

DESIGN PROPERTY	DEPTH						
	9 1/4"	9 1/2"	11 1/4"	11 7/8"	14"	16"	18"
MOMENT (ft. lbs.)	18,625	19,585	26,955	29,855	40,740	52,430	65,495
SHEAR (lbs.)	9,390	9,645	11,420	12,055	14,210	16,240	18,270
MOMENT OF INERTIA (in^4)	345	375	625	735	1,200	1,790	2,550
WEIGHT (lbs./lin. ft.)	15.2	15.6	18.5	19.5	23.0	26.3	29.5

3 1/2" 2.0E Parallam® PSL

DESIGN PROPERTY	DEPTH						
	9 1/4"	9 1/2"	11 1/4"	11 7/8"	14"	16"	18"
MOMENT (ft. lbs.)	12,415	13,055	17,970	19,900	27,160	34,955	43,665
SHEAR (lbs.)	6,260	6,430	7,615	8,035	9,475	10,825	12,180
MOMENT OF INERTIA (in^4)	230	250	415	490	800	1,195	1,700
WEIGHT (lbs./lin. ft.)	10.1	10.4	12.3	13.0	15.3	17.5	19.7

7" 2.0E Parallam® PSL

DESIGN PROPERTY	DEPTH						
	9 1/4"	9 1/2"	11 1/4"	11 7/8"	14"	16"	18"
MOMENT (ft. lbs.)	24,830	26,115	35,940	39,805	54,325	69,905	87,325
SHEAR (lbs.)	12,520	12,855	15,225	16,070	18,945	21,655	24,360
MOMENT OF INERTIA (in^4)	460	500	830	975	1,600	2,390	3,400
WEIGHT (lbs./lin. ft.)	20.2	20.8	24.6	26.0	30.6	35.0	39.4

General Assumptions for Non-Treated Parallam® PSL

- Lateral support is required at all bearing points and along compression edge at intervals of 24" on-center or closer.
- Parallam® PSL beams are made without camber; therefore, in addition to complying with the deflection limits of the applicable building code, other considerations, such as long term deflection under sustained loads (including creep), ponding (positive drainage is essential) and aesthetics, must be evaluated.
- Roof members shall either be sloped for drainage or designed to account for load and deflection as specified in the applicable building code.
- Reductions applied in accordance with: 1994 UBC 1606, 1996 NBC 1606 and 1994 SBC 1604 for floor live load; 1994 UBC 1606, 1996 NBC 1607, and 1994 SBC 1604 for roof live load in non-snow (125%) conditions.
- 3 1/2" members may be two pieces of 1 3/4" or a single 3 1/2" width beam.
  5 1/4" members may be three pieces 1 3/4", one piece 1 3/4" with one piece 3 1/2", or a single 5 1/4" width beam.
  7" members may be two pieces 1 3/4" around one piece 3 1/2", two pieces 3 1/2", or a single 7" width beam.

FIGURE 29.22 ■ Design tables for solving exterior headers for a single-level residence using nontreated Parallam PSL beams. *Courtesy Trus Joist MacMillan.*

HEADERS SUPPORTING ROOF

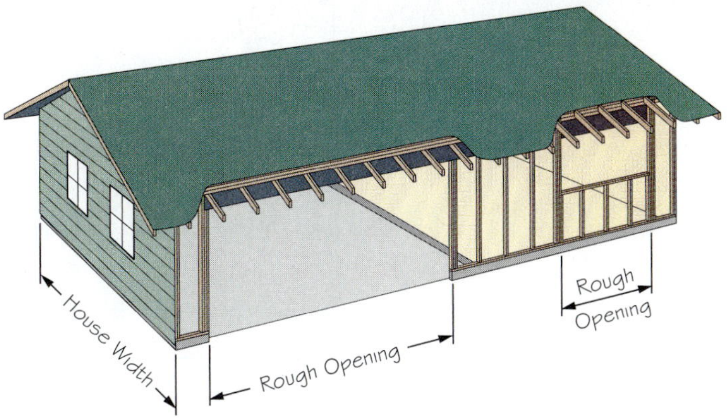

ROOF LOAD (psf)		HOUSE WIDTH	ROUGH OPENING 8'–0"	9'–3"	10'–0"	12'–0"	14'–0"	16'–3"	18'–3"
NON-SNOW AREA 125%	20LL + 15DL	24'–0"	3 1/2" x 9 1/4"	3 1/2" x 9 1/4"	3 1/2" x 9 1/4"	3 1/2" x 9 1/4"	3 1/2" x 11 1/4" 5 1/4" x 9 1/4"	3 1/2" x 11 1/4" 7" x 9 1/4"	3 1/2" x 14" 5 1/4" x 11 1/4"
		30'–0"	3 1/2" x 9 1/4"	3 1/2" x 9 1/4"	3 1/2" x 9 1/4"	3 1/2" x 9 1/4"	3 1/2" x 11 1/4" 5 1/4" x 9 1/4"	3 1/2" x 11 7/8" 5 1/4" x 11 1/4"	3 1/2" x 14" 5 1/4" x 11 7/8"
		36'–0"	3 1/2" x 9 1/4"	3 1/2" x 9 1/4"	3 1/2" x 9 1/4"	3 1/2" x 9 1/4"	3 1/2" x 11 1/4" 5 1/4" x 9 1/2"	3 1/2" x 14" 5 1/4" x 11 1/4"	3 1/2" x 14" 7" x 11 1/4"
	20LL + 20DL	24'–0"	3 1/2" x 9 1/4"	3 1/2" x 9 1/4"	3 1/2" x 9 1/4"	3 1/2" x 9 1/4"	3 1/2" x 11 1/4" 5 1/4" x 9 1/4"	3 1/2" x 11 7/8" 5 1/4" x 11 1/4"	3 1/2" x 14" 5 1/4" x 11 1/4"
		30'–0"	3 1/2" x 9 1/4"	3 1/2" x 9 1/4"	3 1/2" x 9 1/4"	3 1/2" x 9 1/4"	3 1/2" x 11 1/4" 5 1/4" x 9 1/4"	3 1/2" x 14" 5 1/4" x 11 1/4"	3 1/2" x 14" 7" x 11 1/4"
		36'–0"	3 1/2" x 9 1/4"	3 1/2" x 9 1/4"	3 1/2" x 9 1/4"	3 1/2" x 11 1/4" 5 1/4" x 9 1/4"	3 1/2" x 11 1/4" 7" x 9 1/4"	3 1/2" x 14" 5 1/4" x 11 1/4"	3 1/2" x 16" 5 1/4" x 14"
SNOW AREA 115%	25LL + 15DL	24'–0"	3 1/2" x 9 1/4"	3 1/2" x 9 1/4"	3 1/2" x 9 1/4"	3 1/2" x 9 1/4"	3 1/2" x 11 1/4" 5 1/4" x 9 1/4"	3 1/2" x 11 7/8" 5 1/4" x 11 1/4"	3 1/2" x 14" 5 1/4" x 11 7/8"
		30'–0"	3 1/2" x 9 1/4"	3 1/2" x 9 1/4"	3 1/2" x 9 1/4"	3 1/2" x 9 1/4"	3 1/2" x 11 1/4" 5 1/4" x 9 1/2"	3 1/2" x 14" 5 1/4" x 11 1/4"	3 1/2" x 14" 7" x 11 1/4"
		36'–0"	3 1/2" x 9 1/4"	3 1/2" x 9 1/4"	3 1/2" x 9 1/4"	3 1/2" x 11 1/4" 5 1/4" x 9 1/4"	3 1/2" x 11 7/8" 5 1/4" x 11 1/4"	3 1/2" x 14" 5 1/4" x 11 7/8"	3 1/2" x 16" 5 1/4" x 14"
	30LL + 15DL	24'–0"	3 1/2" x 9 1/4"	3 1/2" x 9 1/4"	3 1/2" x 9 1/4"	3 1/2" x 9 1/4"	3 1/2" x 11 1/4" 5 1/4" x 9 1/4"	3 1/2" x 14" 5 1/4" x 11 1/4"	3 1/2" x 14" 5 1/4" x 11 7/8"
		30'–0"	3 1/2" x 9 1/4"	3 1/2" x 9 1/4"	3 1/2" x 9 1/4"	3 1/2" x 11 1/4" 5 1/4" x 9 1/4"	3 1/2" x 11 1/4" 7" x 9 1/4"	3 1/2" x 14" 5 1/4" x 11 7/8"	3 1/2" x 16" 5 1/4" x 14"
		36'–0"	3 1/2" x 9 1/4"	3 1/2" x 9 1/4"	3 1/2" x 9 1/4"	3 1/2" x 11 1/4" 5 1/4" x 9 1/4"	3 1/2" x 14" 5 1/4" x 11 1/4"	3 1/2" x 16" 5 1/4" x 14"	3 1/2" x 16" 5 1/4" x 14"
	40LL + 15DL	24'–0"	3 1/2" x 9 1/4"	3 1/2" x 9 1/4"	3 1/2" x 9 1/4"	3 1/2" x 11 1/4" 5 1/2" x 9 1/4"	3 1/2" x 11 1/4" 7" x 9 1/4"	3 1/2" x 14" 5 1/4" x 11 7/8"	3 1/2" x 16" 5 1/4" x 14"
		30'–0"	3 1/2" x 9 1/4"	3 1/2" x 9 1/4"	3 1/2" x 9 1/4"	3 1/2" x 11 1/4" 5 1/4" x 9 1/4"	3 1/2" x 14" 5 1/4" x 11 1/4"	3 1/2" x 16" 5 1/4" x 14"	3 1/2" x 18" 5 1/4" x 14"
		36'–0"	3 1/2" x 9 1/4"	3 1/2" x 9 1/4"	3 1/2" x 9 1/2" 5 1/4" x 9 1/4"	3 1/2" x 11 7/8" 5 1/4" x 9 1/2"	3 1/2" x 14" 5 1/4" x 11 1/4"	3 1/2" x 16" 5 1/4" x 14"	5 1/4" x 16" 7" x 14"

GENERAL NOTES

Table is based on:

- Uniform loads
- Worst case of simple or continuous span. When sizing a continuous span application, use the longest span. Where ratio of short span to long span is less than 0.4, use the TJ-Beam™ software program or contact your Trus Joist MacMillan representative.
- Roof truss framing with 24" soffits
- Deflection criteria of L/240 live load and L/180 total load. All members 7 1/4" and less in depth are restricted to a maximum deflection of 5/16".

BEARING REQUIREMENTS

Minimum header support to be double trimmers (3" bearing).

In shaded areas, support headers with triple trimmers (4 1/2" bearing).

FIGURE 29.22 ■ (Continued)

Ridge Beams

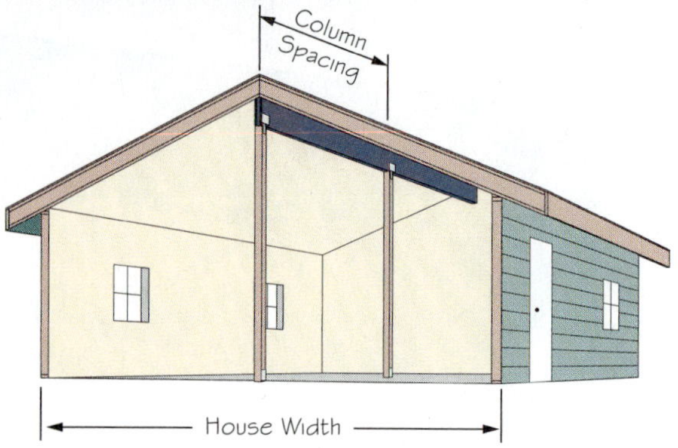

How to use this table

1. Determine the roof loading (live load, dead load and load duration factor) and find the appropriate section of the table.
2. Within that loading section, find the HOUSE WIDTH that meets or exceeds the sum of the spans for the supported roof framing on both sides of the beam.
3. Locate under COLUMN SPACING the span that meets or exceeds the required beam span.
4. Select Parallam® PSL beam size indicated in the appropriate cell of the table.

ROOF LOAD (psf)		HOUSE WIDTH	COLUMN SPACING						
			12'–0"	14'–0"	16'–0"	18'–0"	20'–0"	22'–0"	24'–0"
NON-SNOW AREA 125%	20LL + 15DL	24'–0"	3 1/2" x 9 1/4"	3 1/2" x 9 1/4"	3 1/2" x 11 1/4" 5 1/4" x 9 1/4"	3 1/2" x 11 7/8" 5 1/4" x 11 1/4"	3 1/2" x 14" 5 1/4" x 11 1/4"	3 1/2" x 14" 7" x 11 1/4"	3 1/2" x 16" 5 1/4" x 14"
		30'–0"	3 1/2" x 9 1/4"	3 1/2" x 11 1/4" 5 1/4" x 9 1/4"	3 1/2" x 11 1/4" 7" x 9 1/4"	3 1/2" x 14" 5 1/4" x 11 1/4"	3 1/2" x 14" 5 1/4" x 11 7/8"	3 1/2" x 16" 5 1/4" x 14"	3 1/2" x 18" 5 1/4" x 16"
		36'–0"	3 1/2" x 9 1/4"	3 1/2" x 11 1/4" 5 1/4" x 9 1/4"	3 1/2" x 11 7/8" 5 1/4" x 11 1/4"	3 1/2" x 14" 5 1/4" x 11 7/8"	3 1/2" x 16" 5 1/4" x 14"	3 1/2" x 16" 5 1/4" x 14"	3 1/2" x 18" 5 1/4" x 16"
	20LL + 20DL	24'–0"	3 1/2" x 9 1/4"	3 1/2" x 9 1/2" 5 1/4" x 9 1/4"	3 1/2" x 11 1/4" 5 1/4" x 9 1/2"	3 1/2" x 14" 5 1/4" x 11 1/4"	3 1/2" x 14" 5 1/4" x 11 7/8"	3 1/2" x 16" 5 1/4" x 14"	3 1/2" x 16" 5 1/4" x 14"
		30'–0"	3 1/2" x 9 1/4"	3 1/2" x 11 1/4" 5 1/4" x 9 1/4"	3 1/2" x 11 7/8" 5 1/4" x 11 1/4"	3 1/2" x 14" 5 1/4" x 11 1/4"	3 1/2" x 16" 5 1/4" x 14"	3 1/2" x 16" 5 1/4" x 14"	3 1/2" x 18" 5 1/4" x 16"
		36'–0"	3 1/2" x 9 1/4"	3 1/2" x 11 1/4" 5 1/4" x 9 1/4"	3 1/2" x 14" 5 1/4" x 11 1/4"	3 1/2" x 14" 5 1/4" x 11 7/8"	3 1/2" x 16" 5 1/4" x 14"	3 1/2" x 18" 5 1/4" x 16"	3 1/2" x 18" 5 1/4" x 16"
SNOW AREA 115%	25LL + 15DL	24'–0"	3 1/2" x 9 1/4"	3 1/2" x 9 1/2" 5 1/4" x 9 1/4"	3 1/2" x 11 1/4" 5 1/4" x 9 1/2"	3 1/2" x 14" 5 1/4" x 11 1/4"	3 1/2" x 14" 5 1/4" x 11 7/8"	3 1/2" x 16" 5 1/4" x 14"	3 1/2" x 18" 5 1/4" x 16"
		30'–0"	3 1/2" x 9 1/4"	3 1/2" x 11 1/4" 5 1/4" x 9 1/4"	3 1/2" x 11 7/8" 5 1/4" x 11 1/4"	3 1/2" x 14" 5 1/4" x 11 7/8"	3 1/2" x 16" 5 1/4" x 14"	3 1/2" x 16" 5 1/4" x 14"	3 1/2" x 18" 5 1/4" x 16"
		36'–0"	3 1/2" x 9 1/4"	3 1/2" x 11 1/4" 5 1/4" x 9 1/2"	3 1/2" x 14" 5 1/4" x 11 1/4"	3 1/2" x 14" 7" x 11 1/4"	3 1/2" x 16" 5 1/4" x 14"	3 1/2" x 18" 5 1/4" x 16"	5 1/4" x 18" 7" x 16"
	30LL + 15DL	24'–0"	3 1/2" x 9 1/4"	3 1/2" x 11 1/4" 5 1/4" x 9 1/4"	3 1/2" x 11 1/4" 7" x 9 1/4"	3 1/2" x 14" 5 1/4" x 11 1/4"	3 1/2" x 14" 7" x 11 1/4"	3 1/2" x 16" 5 1/4" x 14"	3 1/2" x 18" 5 1/4" x 16"
		30'–0"	3 1/2" x 9 1/4"	3 1/2" x 11 1/4" 5 1/4" x 9 1/4"	3 1/2" x 14" 5 1/4" x 11 1/4"	3 1/2" x 14" 5 1/4" x 11 7/8"	3 1/2" x 16" 5 1/4" x 14"	3 1/2" x 18" 5 1/4" x 16"	5 1/4" x 16"
		36'–0"	3 1/2" x 11 1/4" 5 1/4" x 9 1/4"	3 1/2" x 11 1/4" 7" x 9 1/4"	3 1/2" x 14" 5 1/4" x 11 1/4"	3 1/2" x 16" 5 1/4" x 14"	3 1/2" x 18" 5 1/4" x 14"	5 1/4" x 16" 7" x 14"	5 1/4" x 18" 7" x 16"
	40LL + 15DL	24'–0"	3 1/2" x 9 1/4"	3 1/2" x 11 1/4" 5 1/4" x 9 1/4"	3 1/2" x 14" 5 1/4" x 11 1/4"	3 1/2" x 14" 5 1/4" x 11 7/8"	3 1/2" x 16" 5 1/4" x 14"	3 1/2" x 18" 5 1/4" x 16"	3 1/2" x 18" 5 1/4" x 16"
		30'–0"	3 1/2" x 11 1/4" 5 1/4" x 9 1/4"	3 1/2" x 11 1/4" 7" x 9 1/4"	3 1/2" x 14" 5 1/4" x 11 1/4"	3 1/2" x 16" 5 1/4" x 14"	3 1/2" x 18" 5 1/4" x 16"	5 1/4" x 16"	5 1/4" x 18" 7" x 16"
		36'–0"	3 1/2" x 11 1/4" 5 1/4" x 9 1/4"	3 1/2" x 14" 5 1/4" x 11 1/4"	3 1/2" x 16" 5 1/4" x 14"	3 1/2" x 18" 5 1/4" x 14"	5 1/4" x 16" 7" x 14"	5 1/4" x 18" 7" x 16"	5 1/4" x 18"

General Notes

Table is based on:

- Uniform loads
- Worst case of simple or continuous span. When sizing a continuous span application, use the longest span. Where ratio of short span to long span is less than 0.4, use the TJ-Beam™ software program or contact your Trus Joist MacMillan representative.
- Deflection criteria of L/240 live load and L/180 total load.

BEARING REQUIREMENTS

Minimum beam support to be double trimmers (3" bearing) at ends and 7 1/2" bearing at intermediate supports of continuous spans.

In shaded areas, support beams with triple trimmers (4 1/2" bearing) at ends and 11 1/4" bearing at intermediate supports of continuous spans.

FIGURE 29.23 ■ Design tables for solving interior headers that support residential roof loads using nontreated Parallam PSL beams. *Courtesy Trus Joist MacMillan.*

FLOOR GIRDER BEAMS

HOW TO USE THIS TABLE

1. Determine the floor loading (live load and dead load) and find the appropriate section of the table.
2. Within that loading section, find the LENGTH OF SUPPORTED FLOOR FRAMING that meets or exceeds the sum of spans 1 and 2 for the supported floor joists. If floor joists are simple span (not continuous over the Parallam® PSL beam), then the LENGTH OF SUPPORTED FLOOR FRAMING may be taken as 80% of the sum of spans 1 and 2 of the floor joists.
3. Locate under COLUMN SPACING a span that meets or exceeds the required beam span.
4. Select Parallam® PSL beam size indicated in the appropriate cell of the table.

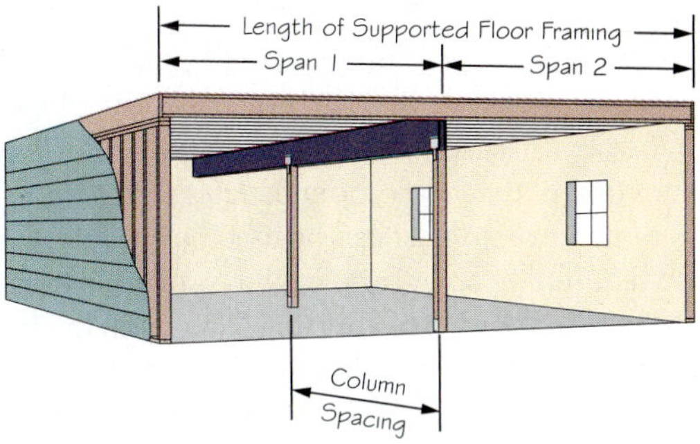

FLOOR LOAD (psf)	LENGTH OF SUPPORTED FLOOR FRAMING	COLUMN SPACING						
		12'-0"	14'-0"	16'-0"	18'-0"	20'-0"	22'-0"	24'-0"
40LL + 12DL	24'-0"	3 1/2" x 11 1/4" 5 1/4" x 9 1/2"	3 1/2" x 14" 5 1/4" x 11 1/4"	3 1/2" x 16" 5 1/4" x 14"	3 1/2" x 18" 5 1/4" x 14"	3 1/2" x 18" 5 1/4" x 16"	5 1/4" x 18" 7" x 16"	7" x 18"
	28'-0"	3 1/2" x 11 7/8" 5 1/4" x 11 1/4"	3 1/2" x 14" 5 1/4" x 11 7/8"	3 1/2" x 16" 5 1/4" x 14"	3 1/2" x 18" 5 1/4" x 16"	5 1/4" x 18" 7" x 16"	5 1/4" x 18" 7" x 16"	7" x 18"
	32'-0"	3 1/2" x 14" 5 1/4" x 11 1/4"	3 1/2" x 14" 7" x 11 1/4"	3 1/2" x 16" 5 1/4" x 14"	3 1/2" x 18" 5 1/4" x 16"	5 1/4" x 18" 7" x 16"	7" x 18"	7" x 18"
	36'-0"	3 1/2" x 14" 5 1/4" x 11 1/4"	3 1/2" x 16" 5 1/4" x 14"	3 1/2" x 18" 5 1/4" x 14"	3 1/2" x 18" 5 1/4" x 16"	5 1/4" x 18" 7" x 16"	7" x 18"	
	40'-0"	3 1/2" x 14" 5 1/4" x 11 1/4"	3 1/2" x 16" 5 1/4" x 14"	3 1/2" x 18" 5 1/4" x 16"	5 1/4" x 16"	5 1/4" x 18" 7" x 16"	7" x 18"	
	44'-0"	3 1/2" x 14" 5 1/4" x 11 7/8"	3 1/2" x 16" 5 1/4" x 14"	3 1/2" x 18" 5 1/4" x 16"	5 1/4" x 18" 7" x 16"	5 1/4" x 18"	7" x 18"	
40LL + 20DL	24'-0"	3 1/2" x 11 1/4" 7" x 9 1/4"	3 1/2" x 14" 5 1/4" x 11 1/4"	3 1/2" x 16" 5 1/4" x 14"	3 1/2" x 18" 5 1/4" x 16"	5 1/4" x 16"	5 1/4" x 18" 7" x 16"	7" x 18"
	28'-0"	3 1/2" x 11 7/8" 5 1/4" x 11 1/4"	3 1/2" x 14" 5 1/4" x 11 7/8"	3 1/2" x 16" 5 1/4" x 14"	3 1/2" x 18" 5 1/4" x 16"	5 1/4" x 18" 7" x 16"	5 1/4" x 18"	7" x 18"
	32'-0"	3 1/2" x 14" 5 1/4" x 11 1/4"	3 1/2" x 16" 5 1/4" x 14"	3 1/2" x 18" 5 1/4" x 14"	5 1/4" x 16" 7" x 14"	5 1/4" x 18" 7" x 16"	7" x 18"	
	36'-0"	3 1/2" x 14" 5 1/4" x 11 1/4"	3 1/2" x 16" 5 1/4" x 14"	3 1/2" x 18" 5 1/4" x 16"	5 1/4" x 16"	5 1/4" x 18" 7" x 16"	7" x 18"	
	40'-0"	3 1/2" x 14" 5 1/4" x 11 1/4"	3 1/2" x 18" 5 1/4" x 14"	5 1/4" x 16" 7" x 14"	5 1/4" x 18" 7" x 16"	7" x 18"	7" x 18"	
	44'-0"	3 1/2" x 16" 5 1/4" x 11 7/8"	5 1/4" x 14"	5 1/4" x 16" 7" x 14"	5 1/4" x 18" 7" x 16"	7" x 18"		

GENERAL NOTES

Table is based on:

- Uniform loads
- Worst case of simple or continuous span. When sizing a continuous span application, use the longest span. Where ratio of short span to long span is less than 0.4, use the TJ-Beam™ software program or contact your Trus Joist MacMillan representative.
- Deflection criteria of L/360 live load and L/240 total load

BEARING REQUIREMENTS

Minimum header support to be double trimmers (3" bearing) at ends and 7 1/2" bearing at intermediate supports of continuous spans.

In shaded areas, support beams with triple trimmers (4 1/2" bearing) at ends and 11 1/4" bearing at intermediate supports of continuous spans.

FIGURE 29.24 ■ Design tables for solving interior headers that support residential floor loads using nontreated Parallam PSL beams. *Courtesy Trus Joist MacMillan.*

SIMPLE BEAM WITH A LOAD CONCENTRATED AT THE CENTER

Often in light construction, load-bearing walls of an upper floor may not line up with the bearing walls of the floor below, as shown in Figure 29.25. This type of loading occurs when the function of the lower room dictates that no post be placed in the center. Since no post can be used on the lower floor, a beam will be required to span from wall to wall and be centered below the upper level post. This beam will have no loads to support other than the weight from the post.

In order to size the beam needed to support the upper area, use the formulas in Figure 29.26. Notice that a new symbol, P has been added to represent a point, or concentrated load in the formula used to determine the maximum bending moment (M). The point load is the sum of the reactions from each end of the beam.

Figure 29.27 shows a sample worksheet for this beam. Notice that the procedure used to solve for this beam is similar to the one used to determine a simple beam with uniform loads. With this procedure, you must determine the amount of point loads for the upper beams. The sketch shows the upper floor plan with each point load.

Once the point loads are known, determine M. Once the value for M has been determined, it can be used to find the required section modulus (S). Use the formula $M/F_b = S$. This will provide the S value for the beam. Try to use a beam that has a depth equal to the depth of the floor joist. Notice also in Figure 29.26 that there is no formula for determining the value for F_v in this or any of the following formulas. F_v is always determined using the formula $3V/(2)(b)(d)$ = required F_v value or $3V/\max F_v$ value = min 2bd value.

In Figure 29.27 three different beams have been selected that have the minimum required section modulus. Once the section modulus is known, the beam can be checked for horizontal shear and deflection. Because of the length of the beam, horizontal shear will not be a factor.

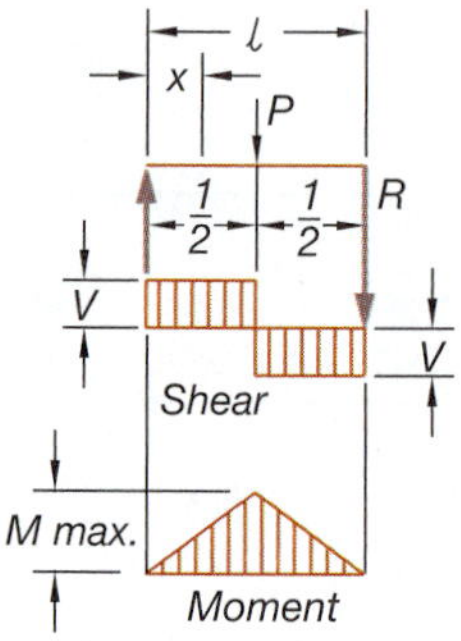

$R = V \quad = \frac{P}{2}$

$M \max. \text{ (at point of load)} = \frac{Pl}{4}$

$M_x \quad \left(\text{when } x < \frac{l}{2}\right) \quad = \frac{Px}{2}$

$\Delta \max. \text{ (at point of load)} = \frac{Pl^3}{48EI}$

$\Delta x \quad \left(\text{when } x < \frac{l}{2}\right) \quad = \frac{Px}{48EI}(3l^2 - 4x^2)$

FIGURE 29.26 ■ Shear and moment diagrams for a simple beam with a load concentrated at the center. *Data reproduced courtesy Western Wood Products Association.*

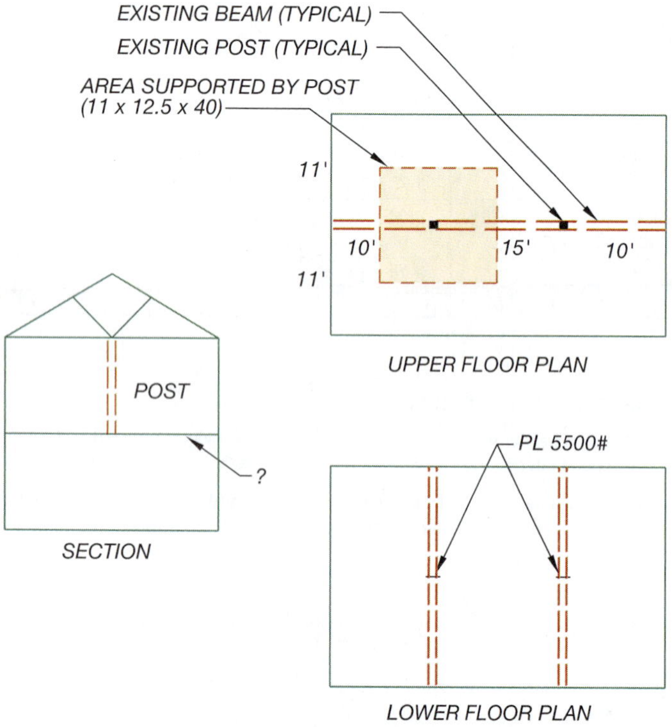

FIGURE 29.25 ■ An engineer's sketch to help determine loads on a beam, showing a load concentrated at the center.

22' BEAM @ LIVING RM.

P=W=5500# V=2750 L=22' l=264 l^3=18.399744

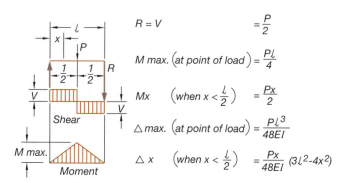

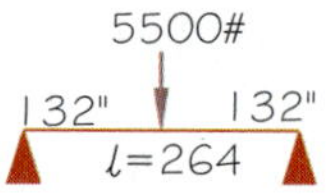

$M = \frac{(P)(l)}{4} = M = \frac{(5500)(264)}{4}$

$R_1=2750 \quad R_2=2750 \quad \frac{1,452,000}{4} = 363,000 = M$

$S = \frac{M}{fb} = \frac{363,000}{2200} = S = 165$ USE $5\frac{1}{8}$ x 15 (S = 192) OR $6\frac{3}{4}$ x 13.5 (S = 205)

$fv = \frac{(3)(V)}{165} = 2bd = \frac{(3)(2750)}{165} = \frac{8250}{165} = 50$ BOTH BM. OK

$D_{max.} = \frac{l}{360} = \frac{264}{360} = .73 \quad D = \frac{(P)(l^3)}{(48)(E)(I)} =$

$D = \frac{(5500)(18.4)}{(48 \times 1.7)(1384)} = \frac{101,200}{112,934} = .89 > .73$ = FAIL

? $5\frac{1}{8} \times 16\frac{1}{2}$ = I = 1441.4 = $\frac{101,200}{117,618}$ = .86 > .73 = FAIL

? $6\frac{3}{4} \times 15$ = I = 1898.4 = $\frac{101,200}{154,909}$ = .65 < .73 =

USE $6\frac{3}{4}$ x 15 fb 2200 GLU-LAM BEAM

FIGURE 29.27 ■ Worksheet for a beam with a load concentrated at the center. It is a good idea to include the location of the beam, a sketch of the loading, needed formulas, and the selected beam.

SIMPLE BEAM WITH A LOAD CONCENTRATED AT ANY POINT

Figure 29.28 shows a sample floor plan that will result in concentrated off-center loads at the lower floor level. Figure 29.29 shows the diagrams and formulas that would be used to determine the beams of the lower floor plan. Figure 29.30 shows the worksheet for this beam.

CANTILEVERED BEAM WITH A UNIFORM LOAD

This loading pattern typically occurs where floor joists of a deck cantilever past the supporting wall. One major difference of this type of loading is that the live loads of decks are almost double the normal floor live loads. Figure 29.31 shows the diagrams and formulas for a cantilevered beam or joist. Figure 29.32 shows the worksheet to determine the spacing of floor joist at a 3′ deck cantilever.

CANTILEVERED BEAM WITH A POINT LOAD AT THE FREE END

This type of beam results when a beam or joist is cantilevered and is supporting a point load. Floor joists that are cantilevered and support a wall and roof can be sized using the formulas in Figure 29.33. Figure 29.34 shows the worksheet for determining the size and spacing of floor joists to cantilever 2′ and support a bearing wall.

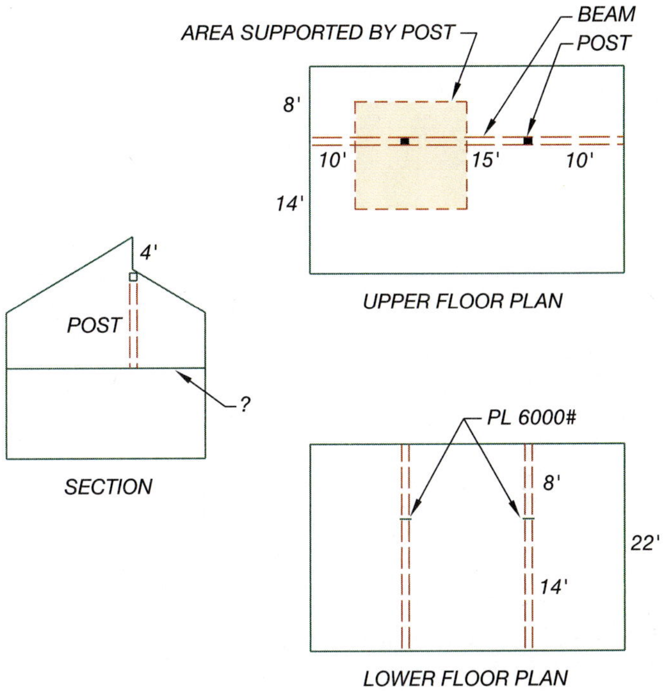

FIGURE 29.28 ■ An engineer's sketch for a beam with a load concentrated at any point on the beam.

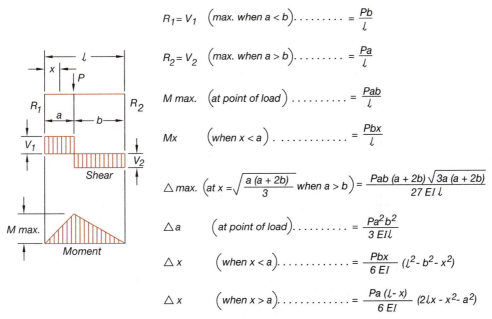

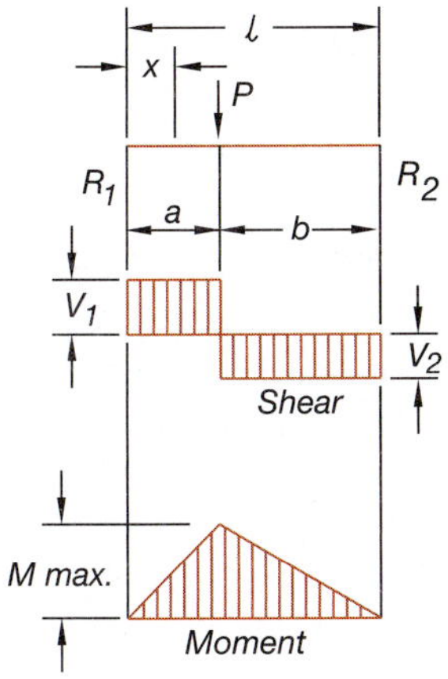

$$R_1 = V_1 \quad \left(\text{max. when } a < b\right) \ldots\ldots = \frac{Pb}{l}$$

$$R_2 = V_2 \quad \left(\text{max. when } a > b\right) \ldots\ldots = \frac{Pa}{l}$$

$$M\ \text{max.} \quad \left(\text{at point of load}\right) \ldots\ldots = \frac{Pab}{l}$$

$$M_x \quad \left(\text{when } x < a\right) \ldots\ldots = \frac{Pbx}{l}$$

$$\Delta\ \text{max.} \quad \left(\text{at } x = \sqrt{\frac{a(a+2b)}{3}} \text{ when } a > b\right) = \frac{Pab\,(a+2b)\sqrt{3a\,(a+2b)}}{27\,EI\,l}$$

$$\Delta a \quad \left(\text{at point of load}\right) \ldots\ldots = \frac{Pa^2b^2}{3\,EIl}$$

$$\Delta x \quad \left(\text{when } x < a\right) \ldots\ldots = \frac{Pbx}{6\,EI}\left(l^2 - b^2 - x^2\right)$$

$$\Delta x \quad \left(\text{when } x > a\right) \ldots\ldots = \frac{Pa\,(l - x)}{6\,EI}\left(2lx - x^2 - a^2\right)$$

FIGURE 29.29 ■ Shear and moment diagrams for a simple beam with a load concentrated at any point on the beam. *Courtesy Western Wood Products Association.*

22' SPAN @ LOWER FLOOR

5500#

a=96" | b=168" $W = (4 + 7)(5 + 7.5)(40\#) = W = 5500\#$

$l = 264$ $l = 264"$

R_1 R_2

$R_1 = \dfrac{(P)(b)}{l} = \dfrac{(5500)(168)}{264} = \dfrac{924,000}{264} = 3500\# = R_1$

$R_2 = P - R_1 = 5500 - 3500 = 2000 = R_2$

$M = \dfrac{(P)(a)(b)}{l} = \dfrac{(5500)(96)(168)}{264} = \dfrac{88,704,000}{264} = 336,000 = M$

$S = \dfrac{M}{fb} = \dfrac{336,000}{2200} = 152.7$ USE $5\frac{1}{8} \times 13\frac{1}{2}$
OR $6\frac{3}{4} \times 12$

$fv = \dfrac{(3)(V)}{165} = \dfrac{(3)(3500)}{165} = \dfrac{10,500}{165} = 63.6$ BOTH BM. OK

$D = \dfrac{l}{360} = \dfrac{264}{360} = .73$ max. $E = \dfrac{(P)(a)(b)(a + 2b)\sqrt{3a(a + 2b)}}{27(E)(l)(I)} =$

$D = \dfrac{(5500)(96)(168)(432)\sqrt{(288)(432)}}{(27)(1.7)(264)(I) = (12118)(I)} = \dfrac{13,516,525}{12,117.6\,(I)} =$

? $5\frac{1}{8} \times 13\frac{1}{2} = \dfrac{13,516,525}{12,733,174} = 106 > .73 =$ FAIL

? $5\frac{1}{8} \times 15 = \dfrac{13,516,525}{17,466,308} = .77 > .73 =$ FAIL

? $6\frac{3}{4} \times 15 = \dfrac{13,516,525}{23,004,051} = .58 < .73 =$ ok USE $6\frac{3}{4} \times 15$ fb 2200 GLU-LAM BEAM

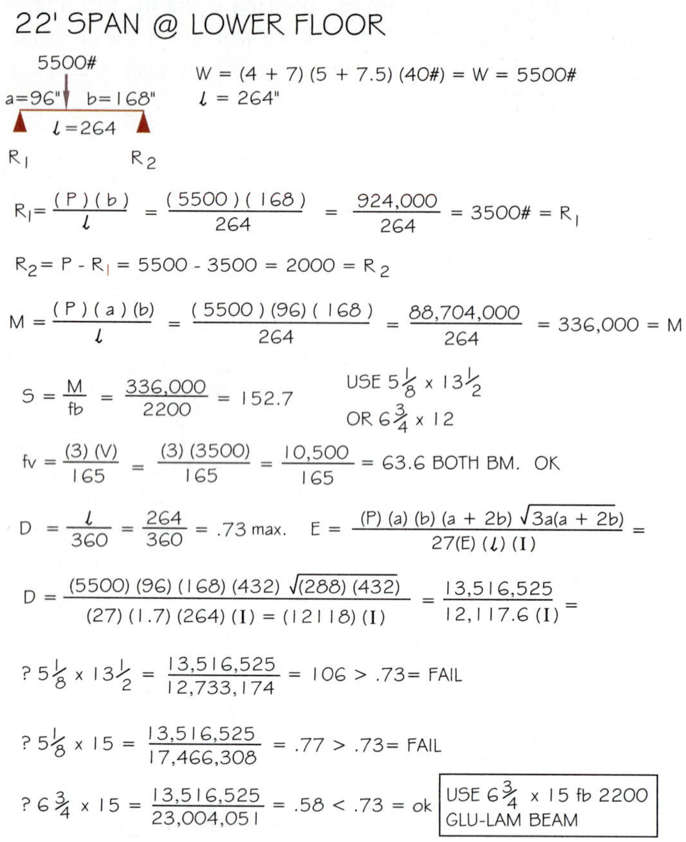

FIGURE 29.30 ■ Worksheet for a beam with a load concentrated off center.

FLOOR JOIST, 36" CANTILEVER

$W = 210\#$ $W = 210\#$ $w = 70\#$ $V = 210$

$l = 36"$ $L = 3'$ $l = 36"$

$l^2 = 1296$ $l^4 = 1.678,616\,(1.68)$

$M = \dfrac{(w)(l^2)}{2} = \dfrac{(70)(1296)}{2} = \dfrac{90,720}{2} = 45,360$

$S = \dfrac{M}{fb} = \dfrac{45,360}{1495} = 30.34$ USE 2 x 12 SS @ 12" O.C.

$2bd = \dfrac{(3)(V)}{fv} = \dfrac{(3)(210)}{85} = \dfrac{630}{85} = 7.4$ 2 x 12 OK

$D_{max.} = \dfrac{l}{360} = \dfrac{36}{360} = .1$ max.

$D = \dfrac{(w)(l^4)}{(8)(E)(I)} = \dfrac{(70)(1.68)}{(8)(1.6)(178)} = \dfrac{118}{2278} = .05 < .1$ OK

USE 2 x 12 F.J. DFL/SS @ 12" O.C.

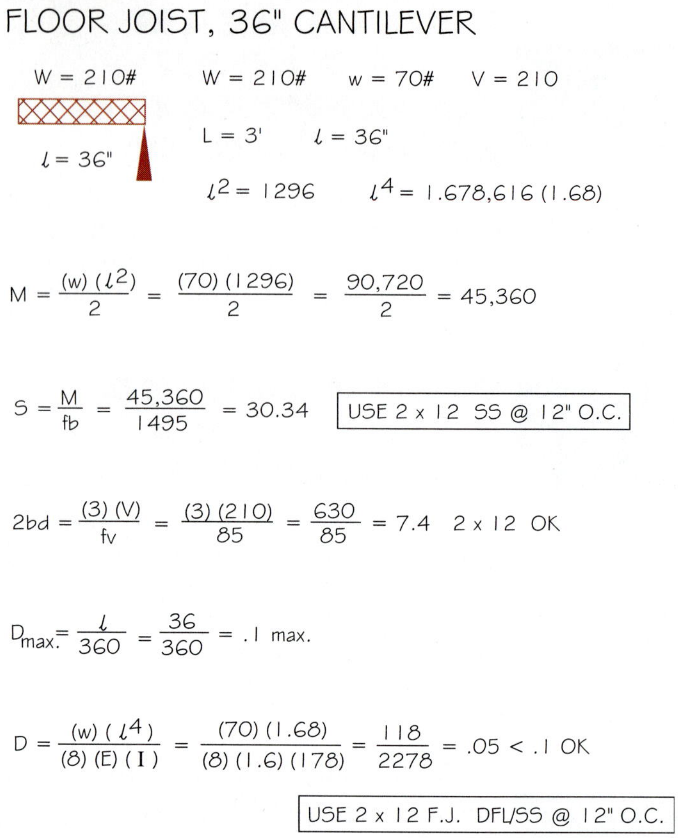

FIGURE 29.32 ■ Worksheet for a cantilevered joist with a uniform load.

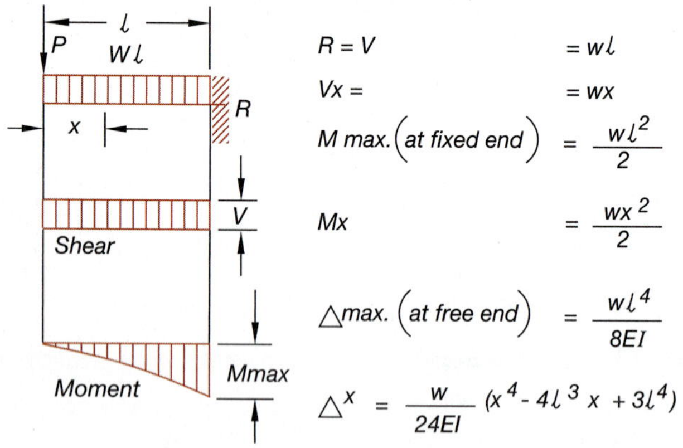

FIGURE 29.31 ■ Shear and moment diagrams for a cantilevered beam with a uniform load. *Data reproduced courtesy Western Wood Products Association.*

$R = V \qquad = wl$

$Vx = \qquad = wx$

$M\ max.\left(at\ fixed\ end\right) = \dfrac{wl^2}{2}$

$Mx \qquad = \dfrac{wx^2}{2}$

$\triangle max.\left(at\ free\ end\right) = \dfrac{wl^4}{8EI}$

$\triangle^x = \dfrac{w}{24EI}(x^4 - 4l^3 x + 3l^4)$

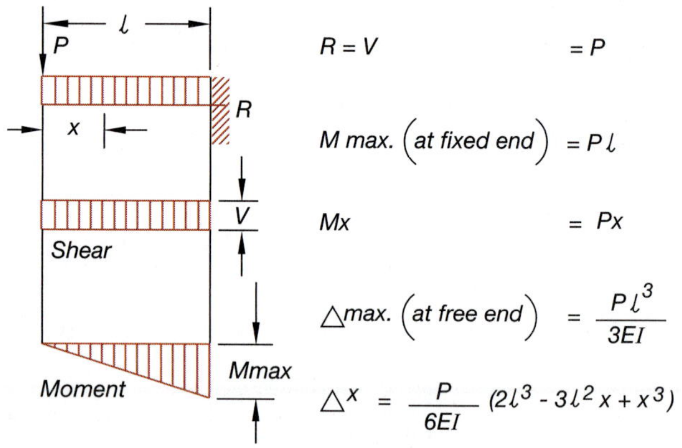

FIGURE 29.33 ■ Shear and moment diagrams for a cantilevered beam with a concentrated load at the free end. *Data reproduced courtesy Western Wood Products Association.*

$R = V \qquad = P$

$M\ max.\left(at\ fixed\ end\right) = Pl$

$Mx \qquad = Px$

$\triangle max.\left(at\ free\ end\right) = \dfrac{Pl^3}{3EI}$

$\triangle^x = \dfrac{P}{6EI}(2l^3 - 3l^2 x + x^3)$

24" CANTILEVER @ WALL

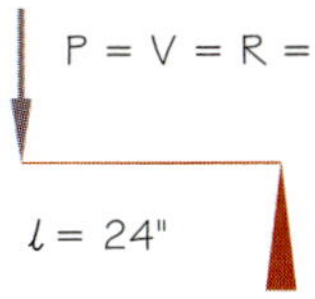

$$M = (P)(l) = (780)(24) = 18{,}720$$

$$S = \frac{M}{fb} = \frac{18{,}720}{1140} = 16.4$$

USE 2 x 10 F.J. @ 12" O.C.

$$fv = \frac{(3)(V)}{85} = \frac{(3)(780)}{85} = \frac{2380}{85} = 28 = 2bd$$

USE 2 x 10 FAIL
USE 2 x 12 F.J. OK

$$D_{max.} = \frac{l}{360} = \frac{24}{360} = .06 \text{ max.}$$

$$D = \frac{(P)(l^3)}{(3)(E)(I)} = \frac{(780)(.013824)}{(3)(1.6)(177.9)} = \frac{10.78}{853.92} = .013 < .06$$

USE 2 x 10 F.J. @ 12" O.C.

FIGURE 29.34 ■ Worksheet for a cantilevered beam with a concentrated load at the free end.

Chapter 29 Additional Reading

The following Web sites can be used as a resource to help you keep current on lumber manufacturers.

ADDRESS	COMPANY OR ORGANIZATION
www.afandpa.org	American Forest and Paper Association
www.aitc.glulam.org	American Institute of Timber Construction
www.awc.org	American Wood Council (NDF Standards)
www.beamchek.com	AC Software, Inc.
www.bc.com	Boise Cascade
www.bcewp.com	Boise Cascade engineered wood products
www.cwc.ca	Canadian Wood Council
www.gp.com	Georgia Pacific
www.inpa.org	International Wood Products Association
www.lp.com	Louisiana-Pacific
www.strucalc.com	StruCalc 4.0 developed by Cascade Consulting Engineers
www.sfpa.org	Southern Forest Products Association
www.tjm.com	Trus Joist MacMillan
www.wii.com	Willamette Industries
www.woodcom.com	Wood Industries Information Center
www.wwpa.org	Western Wood Products Association

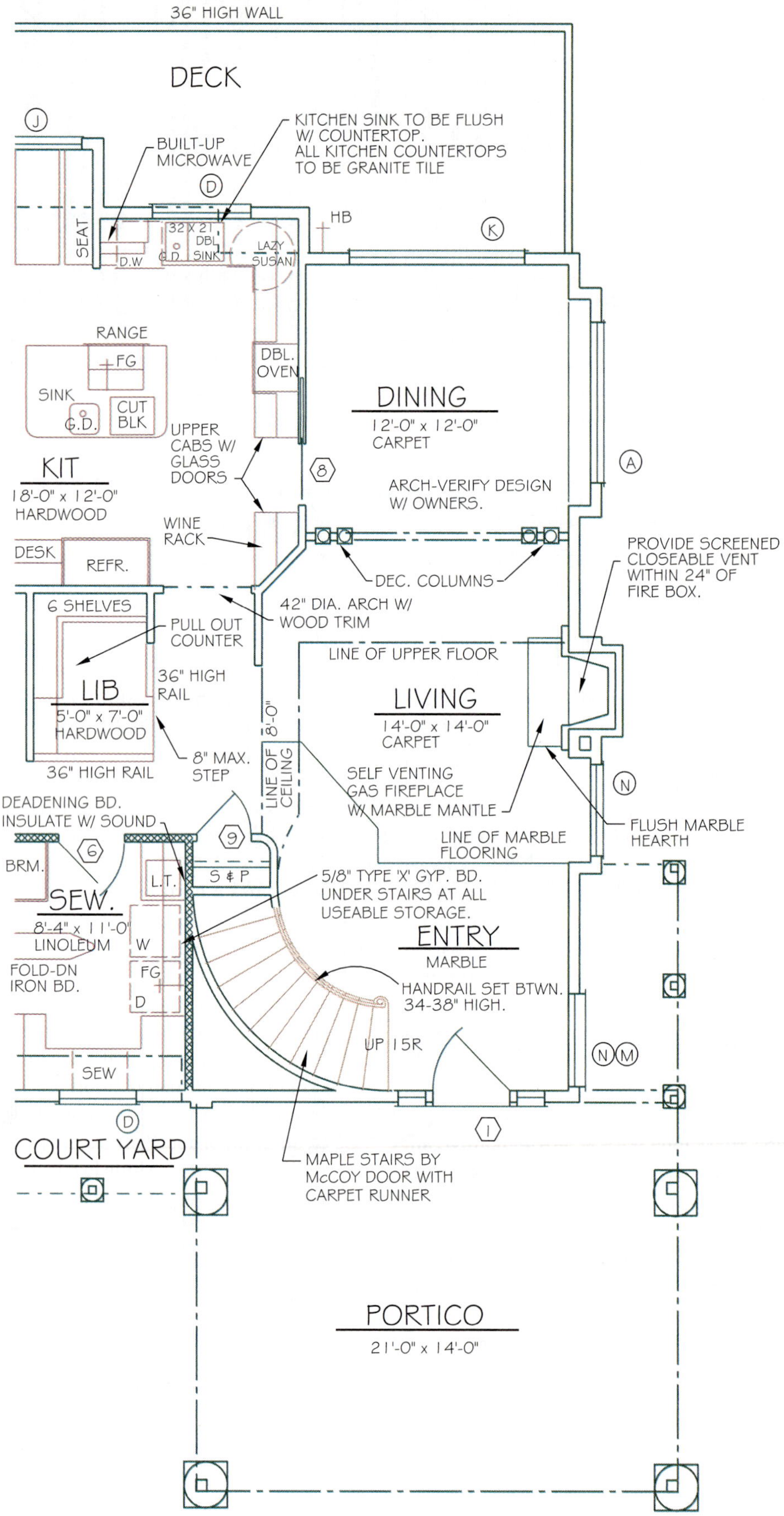

FIGURE 30.2 ■ For a detailed set of plans, typically only architectural information is placed on the floor plan. *Courtesy Residential Designs.*

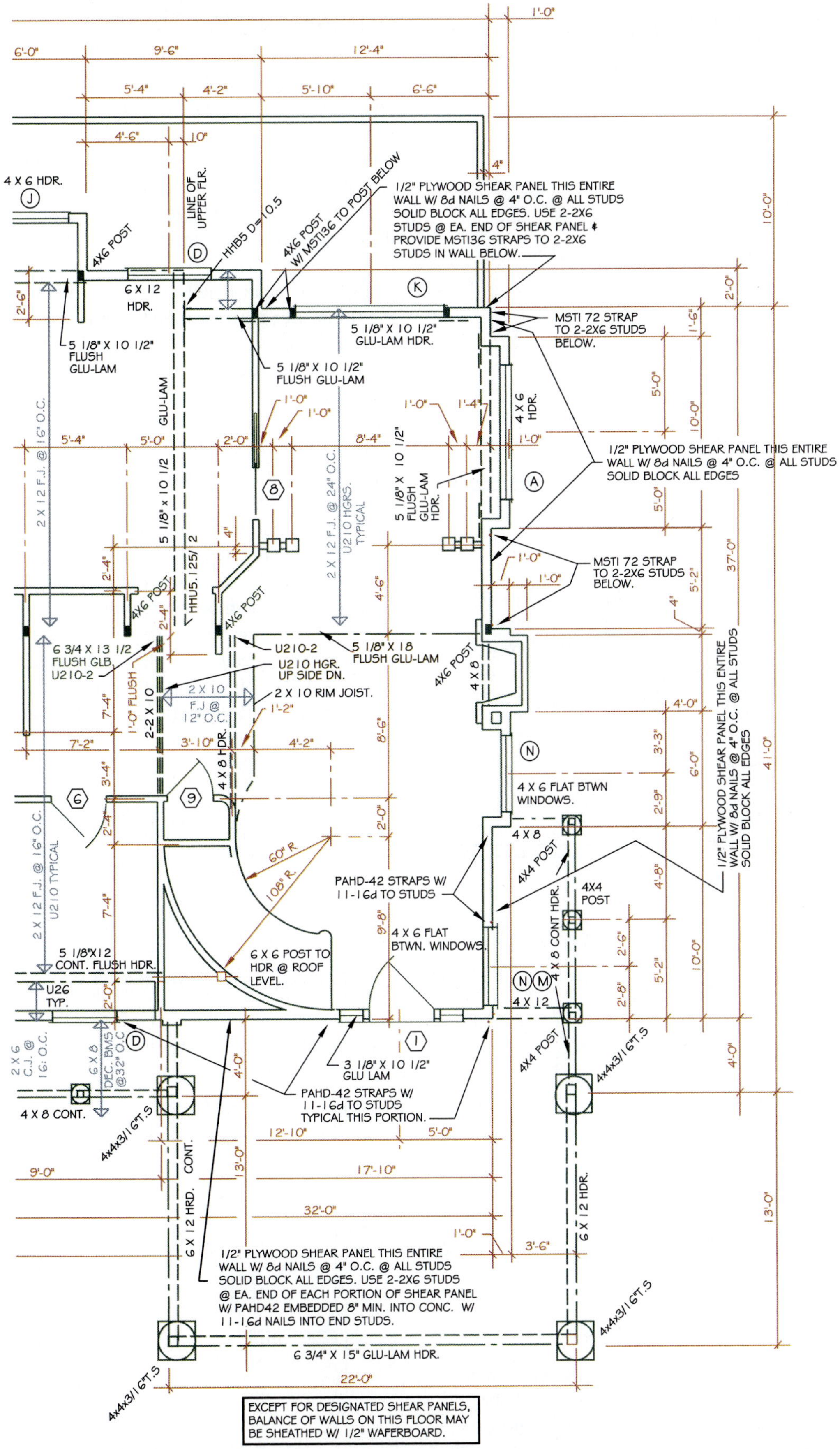

FIGURE 30.3 ■ The framing plan for the structure shown in Figure 30.2 contains only the structural information. *Courtesy Residential Designs.*

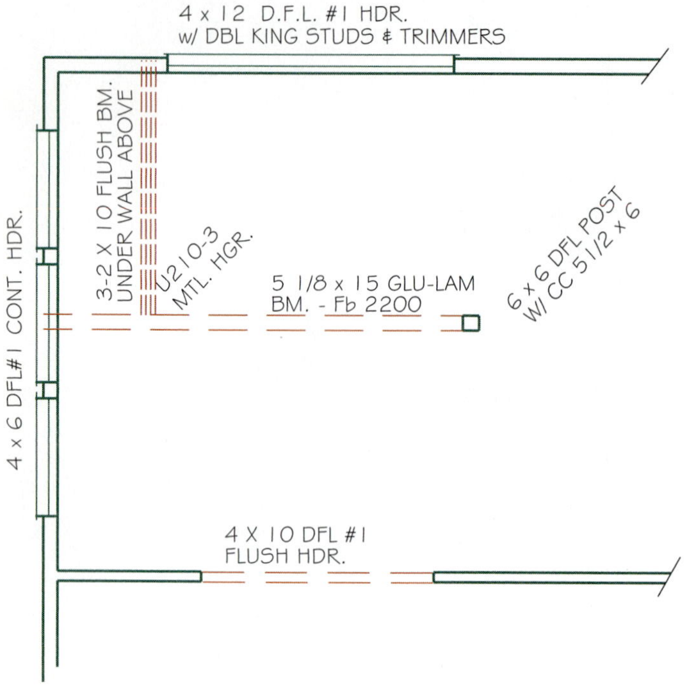

FIGURE 30.4 ■ Headers over a door or window are typically specified by note but not drawn. Framers realize that a door or window must have a header and will look for a note. Headers over wall opening or between walls must be drawn and specified.

Seismic and Wind Resistance

The severity of lateral loads will vary greatly, depending on the area of the country where you work. Typically plywood shear panels, extra blocking, metal angles, and metal connectors to tie posts of one level to members below them are common methods of resisting the forces caused by earthquakes or high winds. (See Chapter 25 for a review of these terms.) These members are determined by an engineer and placed on the framing plan by the drafter.

Plywood shear panels can be specified by a local note pointing to the area of the wall to be reinforced. If several walls are to receive shear panels, a note to explain the construction of the panel can be placed as a general note, with a shortened local note used merely to locate their occurrence. Horizontal metal straps may be required to strengthen connections of the top plate or to tie the header into wall framing. These straps can be represented by a line placed on the side of the wall with a note to specify the manufacturer, the size, and the required blocking. Where walls can be reinforced with a let-in brace, a bold diagonal line is usually placed on the wall.

Metal straps used to tie trimmers, king studs, or posts of one level to another can be represented on each layer of the framing plan by a bold line where they occur. This vertical strap should be shown and specified on the framing plan for each level being connected. The specification should include the manufacturer, size, and required nailing. Figure 30.6 shows how to represent shear panels, let-in braces, and metal straps. Straps are also

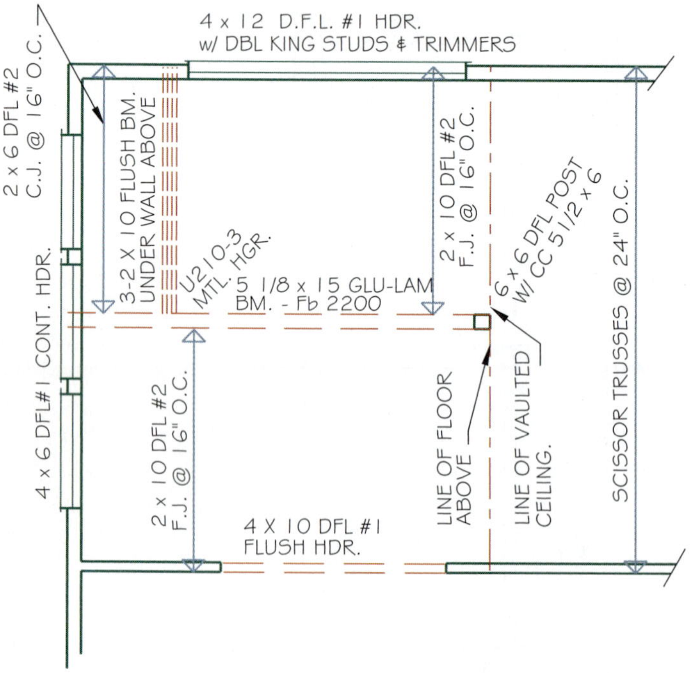

FIGURE 30.5 ■ Joist, rafters, and trusses are each represented by an arrow that shows the direction of the span, with a note to specify the type, size, and spacing of the framing member. Posts are often specified on an angle so that the notation is obvious.

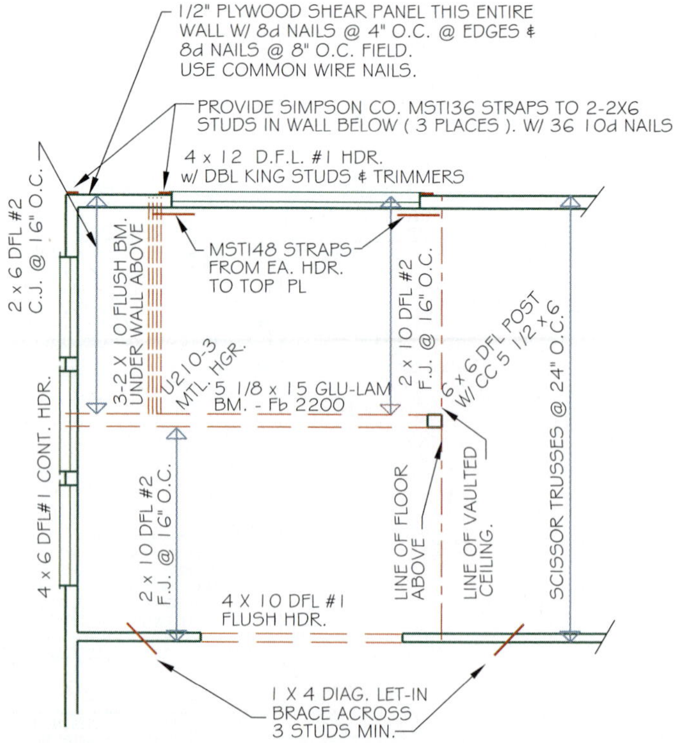

FIGURE 30.6 ■ Material for resisting lateral loads is specified on the framing plan by note. Bold lines can be used to represent metal straps.

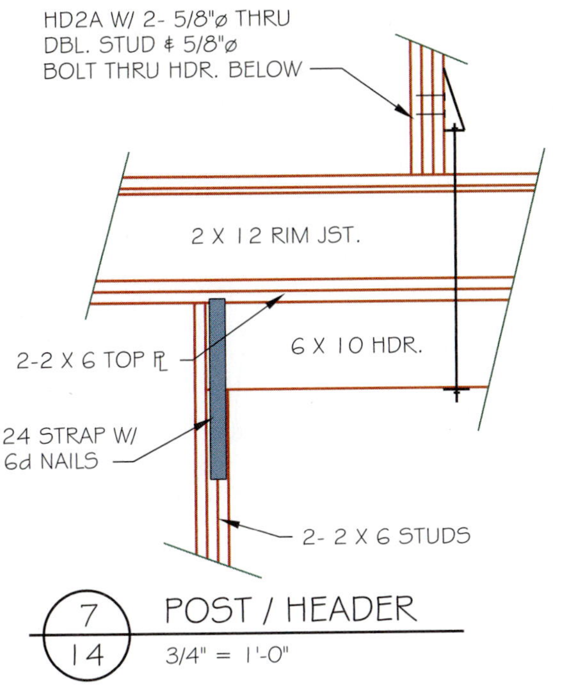

FIGURE 30.7 ■ Metal straps and hold-down anchors, which are specified on the framing, are typically detailed to show exact placement.

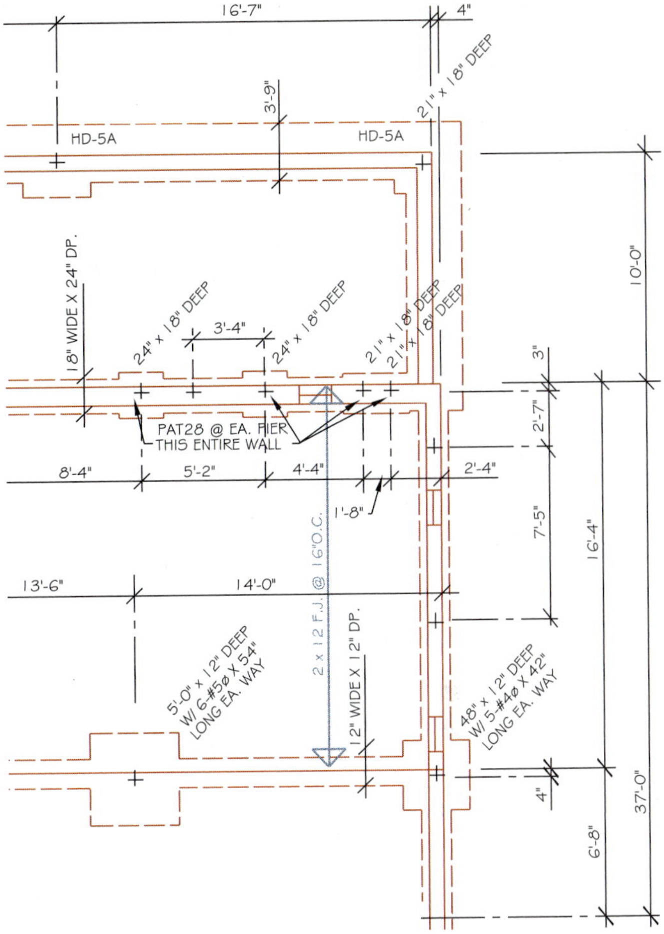

FIGURE 30.8 ■ Metal anchors must be located and specified on the foundation plan so that lateral and gravity loads are transferred into the concrete.

shown in sections and details, as in Figure 30.7, and they can be shown on the exterior elevations. When the lowest floor level is to be attached to the foundation level, the metal connector is often represented by a cross. Figure 30.8 shows common methods of representing these ties on the foundation plan.

Floors and roof members are often tied into the wall system to help resist lateral loads. In addition to the common nailing used to connect each member, metal angles are used to secure the truss or rafter to the top plate. Where lateral loads are severe, blocking is added between the trusses or rafters to keep these members from bending under lateral stress. These areas where joist or rafters are stiffened to resist bending from lateral loads are often referred to as a *diaphragm*. The size, location, and framing method are specified by an engineer. The angles and blocking for the diaphragm must be clearly specified on the framing plan. The angles are typically represented in a section or detail but are not drawn on the framing plan. They should be specified by a note that indicates manufacturer, model number, spacing, and which members are to be connected. Areas that are to receive special blocking should be indicated on the framing plan. Figure 30.9 shows how a diaphragm can be represented on a framing plan.

Resisting Lateral Loads Using Code Prescriptive Paths

Throughout this chapter it has been assumed that someone will be calculating the exact loads on each surface of a structure. An alternative to having an engineer design the method of resisting lateral loads is to use the prescriptive path provided by the building code. Prescriptive design methods provided by building codes are specific requirements for construction to ensure that a structure will be able to resist lateral loads. Prescriptive design methods are intended for fairly simple structures and tend to limit design alternatives. The prescriptive path limits ceiling heights, restricts where an opening can be placed in walls, and also limits wall placement. The prescriptive path of the IRC limits the design criteria for basic wind speeds below 130 mph. The IBC provides information for structures that must resist greater wind loads, but these codes are not the subject of this chapter. Homes that must resist greater winds should be designed under the supervision of an engineer. When prescriptive design methods are used, wall heights are limited to a maximum of 10 feet (3000 mm). Window and wall placement will be discussed throughout the balance of this chapter. Before applying the prescriptive path method, several key terms that are used by the codes must be understood.

Basic Terms of Prescriptive Design

The terms *braced wall lines* and *braced wall panels* are used to describe methods of resisting wall loads. Each exterior surface of a residence is considered a *braced wall line*. Figure 30.10 shows a simple structure with four exterior walls

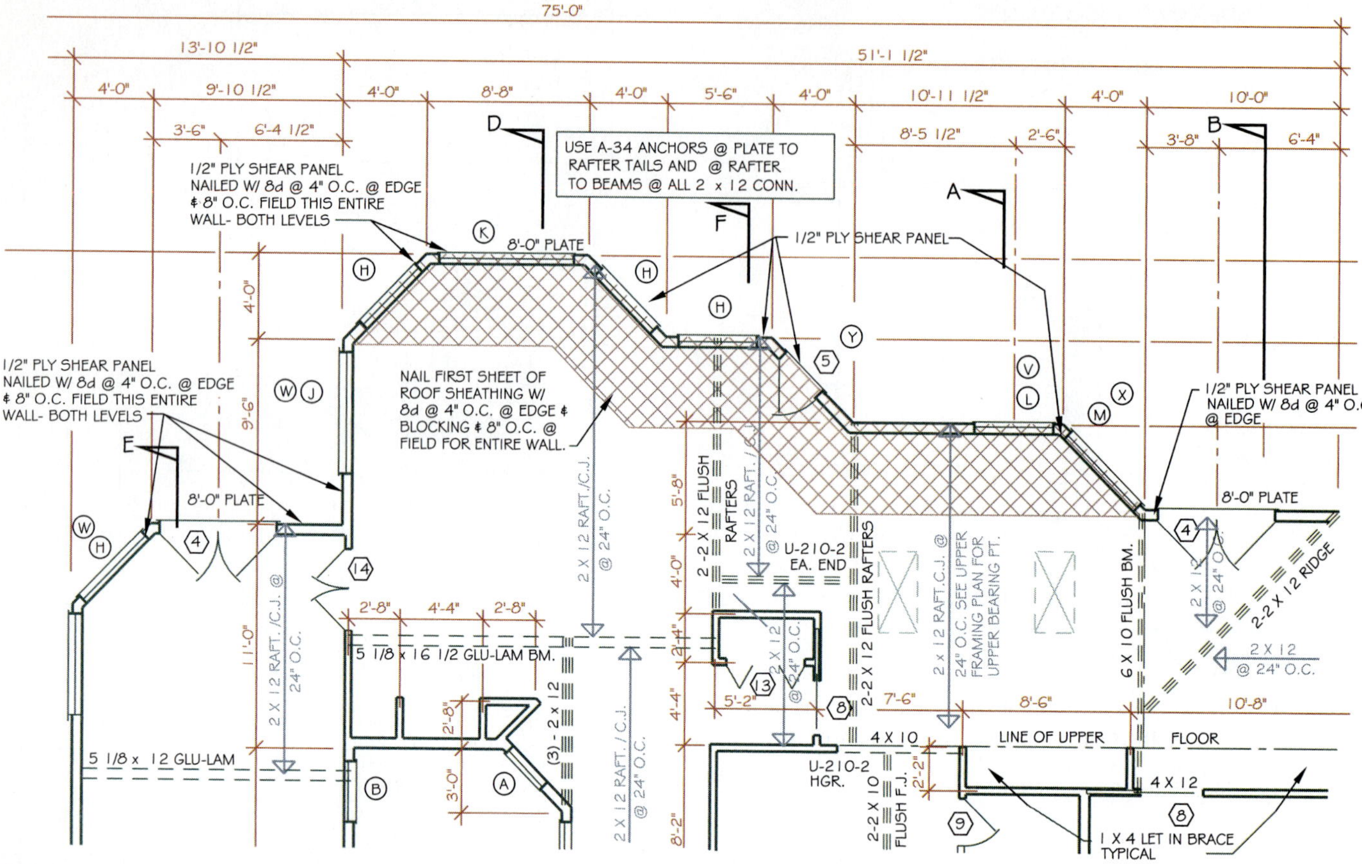

FIGURE 30.9 ■ Special nailing and reinforced connections between the roof and wall are used to resist lateral forces. This information must be shown on the framing plan. Because the rear face of the home has very few walls, the engineer is using the roof to keep the walls square. The walls support the weight of the roof, and the roof resists the lateral loads. *Courtesy David Jefferis.*

(braced wall lines). As wind is applied to the structure, the natural tendency is for the wall to change from a rectangle to a parallelogram. If wind force is applied to surface 1, wall lines 2 and 4 will be used to resist the force. Because the wind force can come from any direction, each surface must be reinforced to resist the force that may be applied to the adjoining surface.

Because very few homes are simple rectangles, the IRC allows some modifications to braced wall lines. These alterations include the following:

- A wall may be offset up to 48″ (1200 mm) and still be considered one plane provided that the total distance from each wall line is not greater than 8′ (2400 mm). See Figure 30.11. Offsets such as a recessed entry may not need reinforcing as long as the offset is less than 48″ (1200 mm).
- To use the prescriptive path design method, no more than 2 offsets in 25′ (7500 mm) are allowed. Offsets must be separated by a distance equal to or greater than the length of the wall offsets. See Figure 30.11.

Three methods are approved by most building departments for reinforcing a braced wall line. These methods include braced wall panels, alternative braced wall panels, and portal frames.

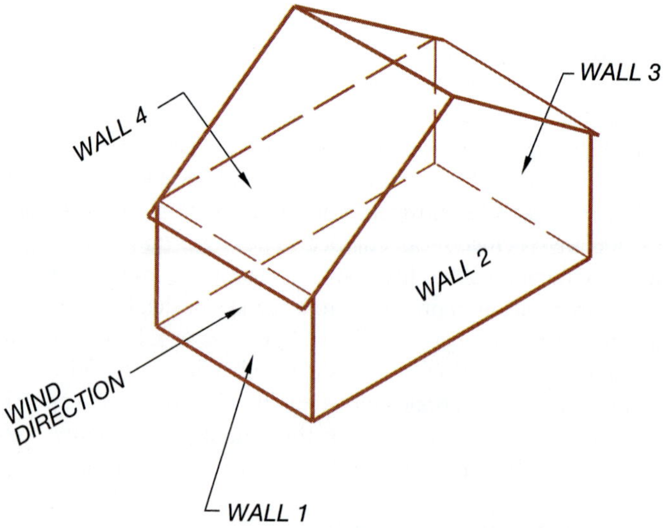

FIGURE 30.10 ■ A simple structure with four exterior walls (braced wall lines). As wind is applied to the structure, the natural tendency is for the wall to change from a rectangle into a parallelogram. If wind force is applied to surface 1, wall lines 2 and 4 will be used to resist the force.

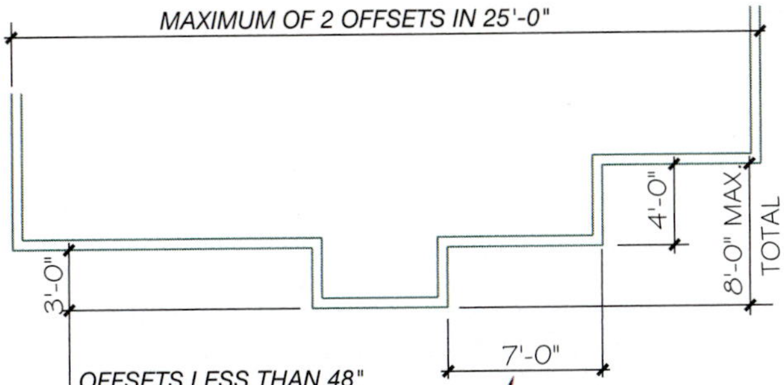

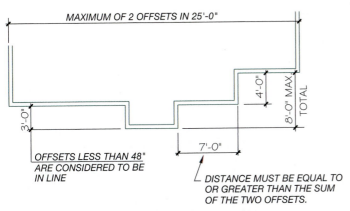

FIGURE 30.11A ■ The IRC allows braced wall lines with offsets to still be considered a single braced wall line, if three conditions are met: (1) The offset cannot be greater than 48″ (1200 mm). (2) No more than 2 offsets in 25′ (7500 mm) are allowed. (3) A distance equal or greater must separate each offset to the length of the wall offsets.

- A *braced wall panel* (BWP) is a method of braced wall line reinforcement that uses panels with a minimum length of 48″ (1200 mm) to resist lateral loads. The minimum length of braced panels can be reduced if the entire wall is sheathed. The size of the panel reduction is based on the wall height and the height of the wall opening. See Figure 30.11b.
- An *alternative braced wall panel* (ABWP) is a method of braced wall line reinforcement that uses panels with a minimum length of 2′–8″ (800 mm) to resist lateral loads.
- A *portal frame* (PF) is a method of braced wall line reinforcement that uses panels 22 1/2″ (570 mm) wide to resist lateral loads. Portal frames are not mentioned in the IRC, but are accepted by most building departments when based on APA specifications.

Components of Braced Wall Construction

Several methods are allowed by building codes for constructing the panel that will reinforce a braced wall line. IRC-approved construction methods for a braced wall panel include:

1. Nominal 1 × 4 (25 × 100) continuous diagonal let-in brace that is let into the top and bottom plates and a minimum of three studs. The brace cannot be placed at an angle greater than 60° or less than 45° from horizontal.
2. Diagonally placed wood boards with a net thickness of 5/8″ (16 mm) that are applied to studs spaced at a maximum of 24″ (600 mm) o.c.
3. Structural wood sheathing with a minimum thickness of 5/16″ (8 mm) over studs placed at 16″ (400 mm) o.c. or 3/8″ (9.5 mm) structural sheathing over studs spaced at 24″ (600 mm).
4. 1/2″ (12.7 mm) or 25/32″ (19.8 mm) structural fiberboard sheathing panels 4′ × 8′ (1200 × 2400 mm) applied vertically over studs placed at 16″ (406 mm) o.c.
5. Gypsum board 1/2″ (12.7 mm) thick × 4′ (1200 mm) wide over studs spaced at 24″ (600 mm).
6. Particleboard wall sheathing panels.
7. Portland cement plaster on studs placed at 16″ (400 mm) o.c.
8. Hardboard panel siding.

For these conditions to be allowable, each material must be applied using the connection methods listed in the fastener schedule shown in Figure 30.12. Other limitations include these:

- 1 × 4 (25 × 100) continuous diagonal let-in braces (method 1) are not allowed in seismic zones D1, D2, and E.
- Each panel must be a minimum of 48″ (1200 mm) long covering a minimum of three studs placed at 16″ (400 mm) o.c. or two studs placed at 24″ (600 mm) o.c. for methods 2, 3, 4, 6, 7, and 8.
- Each wall panel must be 96″ (2400 mm) long if gypsum board 1/2″ (12.7 mm) × 4′ (1200 mm) wide (method 5) is used on one wall face. The length can be reduced to 48″ (1200 mm) if the gypsum board is placed on each face of the stud wall.

If the walls are made rigid to keep their rectangular shape, a second tendency will occur. The edge of the wall panel nearest the load (the wind) will start to lift and rotate around the edge of the panel

LENGTH REQUIREMENTS FOR BRACED WALL PANELS IN A CONTINUOUSLY SHEATHED WALL[a,b]

LENGTH OF BRACED WALL PANEL (inches)			MAXIMUM OPENING HEIGHT NEXT TO THE BRACED WALL PANEL (% of wall height)
8-foot wall	9-foot wall	10-foot wall	
48	54	60	100%
32	36	40	85%
24	27	30	65%

For SI: 1 inch = 25.4 mm, 1 foot = 305 mm, 1 psf = 0.0479 kN/m².

a. Linear interpolation shall be permitted.

b. Full-height sheathed wall segments to either side of garage openings that support light frame roofs with roof covering dead loads of 3 psf or less shall be permitted to have a 4:1 aspect ratio.

FIGURE 30–11B ■ Length requirements for braced walls designed using prescriptive methods. *Reproduced from the* International Residential Code / 2000. *Copyright © 2000. Courtesy International Code Council, Inc.*

WALL BRACING

SEISMIC DESIGN CATEGORY OR WIND SPEED	CONDITION	TYPE OF BRACE[b,c]	AMOUNT OF BRACING[a,d,e]
Category A and B ($Ss \le 0.35g$ and $Sds \le 0.33g$) or 120 mph and less	One story Top of two or three story	Methods 1, 2, 3, 4, 5, 6, 7 or 8	Located at each end and at least every 25 feet on center but not less than 16% of braced wall line.
	First story of two story Second story of three story	Methods 1, 2, 3, 4, 5, 6, 7 or 8	Located at each end and at least every 25 feet on center but not less than 16% of braced wall line for Method 3 and 25% of braced wall line for Methods 2, 4, 5, 6, 7 or 8.
	First story of three story	Methods 2, 3, 4, 5, 6, 7 or 8	Minimum 48-inch-wide panels located at each end and at least every 25 feet on center but not less than 25% of braced wall line for method 3 and 35% of braced wall line for Methods 2, 4, 5, 6, 7 or 8.
Category C ($Ss \le 0.6g$ and $Sds \le 0.53g$) or less than 130 mph	One story Top of two or three story	Methods 1, 2, 3, 4, 5, 6, 7 or 8	Located at each end and at least every 25 feet on center but not less than 16% of braced wall line for Method 3 and 25% of braced wall line for Methods 2, 4, 5, 6, 7 or 8.
	First story of two story Second story of three story	Methods 2, 3, 4, 5, 6, 7 or 8	Located at each end and at least every 25 feet on center but not less than 30% of braced wall line for Method 3 and 45% of braced wall line for Methods 2, 4, 5, 6, 7 or 8.
	First story of three story	Methods 2, 3, 4, 5, 6, 7 or 8	Located at each end and at least every 25 feet on center but not less than 45% of braced wall line for Method 3 and 60% of braced wall line for Methods 2, 4, 5, 6, 7 or 8.
Category D_1 ($Ss \le 1.25g$ and $Sds \le 0.83g$) or less than 130 mph	One story Top of two or three story	Methods 2, 3, 4, 5, 6, 7 or 8	Located at each end and at least every 25 feet on center but not less than 20% of braced wall line for Method 3 and 30% of braced wall line for Methods 2, 4, 5, 6, 7 or 8.
	First story of two story Second story of three story	Methods 2, 3, 4, 5, 6, 7 or 8	Located at each end and not more than 25 feet on center but not less than 45% of braced wall line for Method 3 and 60% of braced wall line for Methods 2, 4, 5, 6, 7 or 8.
	First story of three story	Methods 2, 3, 4, 5, 6, 7 or 8	Located at each end and not more than 25 feet on center but not less than 60% of braced wall line for Method 3 and 85% of braced wall line for Method 2, 4, 5, 6, 7 or 8.
Category D_2 or less than 130 mph	One story Top of two story	Methods 2, 3, 4, 5, 6, 7 or 8	Located at each end and at least every 25 feet on center but not less than 25% of braced wall line for Method 3 and 40% of braced wall line for Methods 2, 4, 5, 6, 7 or 8.
	First story of two story	Methods 2, 3, 4, 5, 6, 7 or 8	Located at each end and not more than 25 feet on center but not less than 55% of braced wall line for Method 3 and 75% of braced wall line for Methods 2, 4, 5, 6, 7 or 8.
	Cripple walls	Method 3	Located at each end and not more than 25 feet on center but not less than 75% of braced wall line.

For SI: 1 inch = 25.4 mm, 1 foot = 304.8 mm, 1 psf = 0.0479 kN/m^2, 1 mile per hour = 1.609 km/hr.

a. Wall bracing amounts are based on a soil site class "D." Interpolation of bracing amounts between the S_{ds} values associated with the Seismic Design Categories shall be permitted when a site specific S_{ds} value is determined in accordance with Section 1613 of the *International Building Code*.

b. Foundation cripple wall panels shall be braced in accordance with Section R602.9.

c. Methods of bracing shall be as described in Section R602.10.2. The alternate braced wall panels described in Section R602.10.6 shall also be permitted.

d. The bracing amounts for Seismic Design Categories are based on a 15 psf wall dead load. For walls with a dead load of 8 psf or less, the bracing amounts shall be permitted to be multiplied by 0.85 provided that the adjusted bracing amount is not less than that required for the site's wind speed. The minimum length of braced panel shall not be less than required by Section R602.10.3.

e. When the dead load of the roof/ceiling exceeds 15 psf, the bracing amounts shall be increased in accordance with Section R301.2.2.4. Bracing required for a site's wind speed shall not be adjusted.

FIGURE 30.11C ■ Wall bracing methods for prescriptive design. *Reproduced from the* International Residential Code / 2000. *Copyright © 2000. Courtesy International Code Council, Inc.*

NAILING SCHEDULE

TAKEN FROM 1997 UBC TABLE 23-II-B-1 CONNECTION	NAILING (1)
1 JOIST TO SILL OR GIRDER, TOE NAIL	3- 8d
2 BRIDGING TO JOIST, TOENAIL EACH END	2- 8d
3 1 X 6 SUBFLOOR OR LESS TO EACH JOIST, FACE NAIL	2-8d
4 WIDER THAN 1 X 6 SUBFLOOR TO EA. JOIST, FACE NAIL	3-8d
5 2" SUBFLOOR TO JOIST OR GIRDER, BLIND AND FACE NAIL	2-16d
6 SOLE PLATE TO JOIST OR BLOCKING TYPICAL FACE NAIL SOLE PLATE TO JOIST OR BLOCKING, @ BRACED WALL PANELS	16d @ 16" O.C. 3- 16d PER 16"
7 TOP PLATE TO STUD, END NAIL	2-16d
8 STUD TO SOLE PLATE	4- 8d , TOENAIL OR 2- 16d , END NAIL
9 DOUBLE STUDS, FACE NAIL	16d @ 24" O.C.
10 DOUBLED TOP PLATES, TYPICAL FACE NAIL DOUBLE TOP PLATES, LAP SPLICE	16d @ 16" O.C. 8- 16d
11 BLOCKING BETWEEN JOIST OR RAFTERS TO TOP PLATE, TOENAIL	3- 8d
12 RIM JOIST TO TOP PLATE, TOENAIL	8d @ 6" O.C.
13 TOP PLATES, LAPS & INTERSECTIONS, FACE NAIL	2-16d
14 CONTINUOUS HEADER, TWO PIECES	16d @ 16" O.C. ALONG EACH EDGE
15 CEILING JOIST TO PLATE, TOE NAIL	3-8d
16 CONTINUOUS HEADER TO STUD, TOE NAIL	4-8d
17 CEILING JOIST, LAPS OVER PARTITIONS, FACE NAIL	3-16d
18 CEILING JOIST TO PARALLEL RAFTERS, FACE NAIL	3-16d
19 RAFTERS TO PLATE, TOE NAIL	3-8d
20 1" BRACE TO EA. STUD & PLATE FACE NAIL	2-8d
21 1" x 8" SHEATHING OR LESS TO EA. BEARING, FACE NAIL	2-8d
22 WIDER THAN 1" x 8" SHEATHING TO EACH BEARING, FACE NAIL	3-8d
23 BUILT-UP CORNER STUDS	16d @ 24" O.C.
24 BUILT-UP GIRDER AND BEAMS	20d @ 32" O.C. @ TOP & BTM. AND STAGGERED 2-20d @ ENDS & @ EACH SPLICE
25	2-16d 2" PLANKS @ EACH BEARING
26 WOOD STRUCTURAL PANELS AND PARTICLEBOARD(2) SUBFLOOR, ROOF AND WALL SHEATHING TO FRAMING: 19/32" - 3/4" 1/2" AND LESS 7/8 "- 1" 1 1/8"- 1 1/4" COMBINATION SUBFLOOR-UNDERLAYMENT TO FRAMING: 3/4" AND LESS 7/8" - 1" 1 1/8" - 1 1/4"	 8d COMMON OR 6d DEFORMED SHANK 6d COMMON OR DEFORMED SHANK 8d COMMON OR DEFORMED SHANK 10d COMMON OR 8d DEFORMED SHANK 6d DEFORMED SHANK 8d DEFORMED SHANK 10d COMMON OR 8d DEFORMED SHANK
27 PANEL SIDING TO FRAMING(2) 1/2" OR LESS CASING NAILS. 5/8"	 6d CORROSION-RESISTANT SIDING OR 8d CORROSION-RESISTANT SIDING OR
28 FIBERBOARD SHEATHING (3) 1/2" 25/32"	 No. 11 GA.(4) 6d COMMON NAILS No. 16 GA.(5) No. 11 GA.(4) 8d COMMON NAILS No. 16 GA.(5)
29. INTERIOR PANELING: 1/4" 25/32"	 4d (6) 6d (7)

1. COMMON OR BOX NAILS MAY BE USED WHERE OTHERWISE STATED.
2. NAILS SPACED @ 6" O.C. @ EDGES, 12" O.C. @ INTERMEDIATE SUPPORTS EXCEPT 6" @ ALL SUPPORTS WHERE SPANS ARE 48" OR MORE. FOR NAILING OF WOOD STRUCTURAL PANEL AND PARTICLEBOARD DIAPHRAGMS AND SHEAR WALLS SEE SPECIFIC NOTES ON DRAWINGS. NAILS FOR WALL SHEATHING MAY BE COMMON, BOX OR CASING.
3. FASTENERS SPACED 3" O.C. @ EXTERIOR EDGES & 6" O.C. @ INTERMEDIATE SUPPORTS.
4. CORROSION -RESISTANT ROOFING NAILS W/ 7/16" DIA. HEAD & 1 1/2" LENGTH FOR 1/2" SHEATHING & 1 3/4" LENGTH FOR 25/32" SHEATHING..
5. CORROSION-RESISTANT STAPLES W/ NOMINAL 7/16" CROWN & 1 1/8" LENGTH FOR 1/2" SHEATHING & 1 1/2" LENGTH FOR 25/32" SHEATHING..
6. PANEL SUPPORTS @ 16" (20" IF STRENGTH AXIS IN THE LONG DIRECTION OF THE PANEL, UNLESS OTHERWISE MARKED). CASING OR FINISH NAILS SPACED 6" ON PANEL EDGES, 12" @ INTERMEDIATE SUPPORTS.
7. PANEL SUPPORTS @ 24". CASING OR FINISH HAILS SPACED 6" ON PANEL EDGES, 12" AT INTERMEDIATE SUPPORTS.

WOOD STRUCTURAL PANEL ROOF SHEATHING NAILING SCHEDULE (1)

WIND REGION	NAILS	PANEL LOCATION	ROOF FASTENER ZONE (2) 1	2	3
			FASTENING SCHEDULE (INCHES ON CENTER) X 25.4 FOR MM		
GREATER THAN 90 MPH (145 km/h)	8d COMMON	PANEL EDGES (3)	6	6	4(4)
		PANEL FIELD	6	6	6(4)
GREATER THAN 80 MPH (129 km/h) TO 90 MPH (145 km/h)	8d COMMON	PANEL EDGES (3)	6	6	4
		PANEL FIELD	12	6	6
80 MPH (129 km/h) OR LESS	8d COMMON	PANEL EDGES (3)	6	6	6
		PANEL FIELD	12	12	12

1. APPLIES ONLY TO MEAN ROOF HEIGHTS UP TO 35 FT (10 700 MM). FOR MEAN ROOF HEIGHTS 35 FEET (10 700 MM), THE NAILING SHALL BE DESIGNED.
2. ROOF FASTENING ZONES ARE SHOWN BELOW.
3. EDGE SPACING ALSO APPLIES OVER ROOF FRAMING AT GABLE-END WALLS.
4. USE 8d RING-SHANK NAILS IN THIS ZONE IF MEAN ROOF HEIGHT IS GREATER THAN 25' (7600mm).

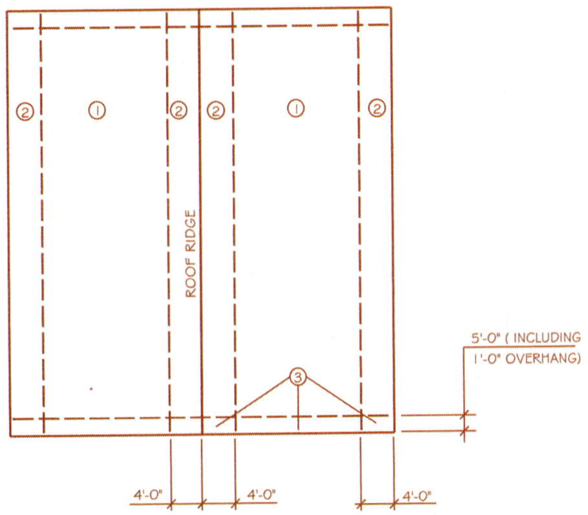

FIGURE 30.12 ■ Building codes provide specifications that regulate each connection specified on the framing plans. Providing a nailing schedule with the working drawings will eliminate having to specify nailing for each drawing. *Courtesy Residential Designs.*

that is farthest from the load. Figure 30.13 shows this tendency. In addition to specifying the material to cover the wall, building codes require several other types of information to be specified about the reinforced wall panel. Required information includes:

- Framing members to be used to reinforce the ends of the wall panel
- Methods of attaching the wall panel to the sill plate
- Methods of attaching the sill to the stem wall
- Methods of reinforcing the foundation that will support the wall panel

Because this is to be an introduction to reinforcing walls using prescriptive methods, each method will not be covered in depth. Remember that walls must be reinforced according to the governing building department and that research will be needed.

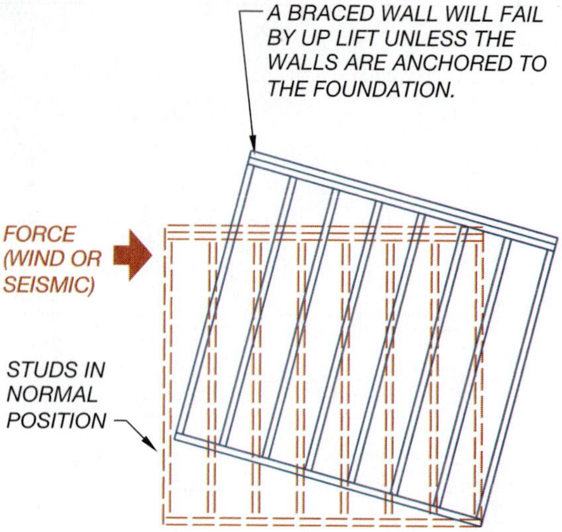
A BRACED WALL WILL FAIL BY UP LIFT UNLESS THE WALLS ARE ANCHORED TO THE FOUNDATION.
FORCE (WIND OR SEISMIC)
STUDS IN NORMAL POSITION

FIGURE 30.13 ■ If walls are made rigid to keep their rectangular shape, a second tendency will occur. The edge of the wall panel nearest the load (the wind) will start to lift and rotate around the edge of the panel that is farthest from the load.

Representing Wall Reinforcement on a Framing Plan

One of the requirements for designing using prescriptive methods is that the location of wall panels must be clearly marked on the framing drawings. When the code that governs your area is used, prescriptive construction methods can be summarized and given in a table to represent the materials that will be required to reinforce the walls and foundation. Figure 30.14

BRACED WALL SCHEDULE (FRAMING PLAN)

	MARK	WALL COVERING	EDGE NAILING	FIELD NAILING	PANEL SUPPORT	STRUCTURAL CONNECTORS
BRACED WALL PANEL	A	48" MIN. x 1/2" CD EXT. PLY 1 SIDE UNBLOCKED	8d @ 6" O.C	8d @ 12" O.C.	2 - 2 x 6 STUDS @ EA. END OF PANEL	PAHD42 TO CON
ALTERNATIVE BRACED WALL PANEL	B	2'-8" MIN. x 1/2" CD EXT. PLY 1 SIDE UNBLOCKED	8d @ 6" O.C	8d @ 12" O.C.	2 - 2 x 6 STUDS @ EA. END OF PANEL	HPAHD22 TO CO
PORTAL FRAME	C	CD EXT. PLY 1 SIDE -BLOCKED EDGE	2 ROWS OF 8d @ 3" O.C	8d @ 12" O.C.	4 x 4 POST @ EA. END OF PANEL	1-MSTA18 STRA EA. SIDE EA. END OF PANEL TO 4 x 12 MIN H (4 STRAPS MIN E END OF FRAME).
INTERIOR BWP	D	1/2" x 48" GYPSUM BOARD EACH SIDE OF WALL W/ 5d COOLER NAILS @ 7" O.C. MAXIMUM W/ BLOCKED EDG (1/2" x 96" LONG IF 1 SIDE ONLY).				

HOLD - DOWN SCHEDULE (FOUNDATION PLAN)

	MARK	HOLD - DOWN	CONNECTIONS	FOUNDATION REINFORCING
	0	NONE REQUIRED	———	DBL JOIST OR BEAM BELOW. SEE FOUNDATION PL/
BRACED WALL PANEL	1	PAHD42 STRAP	(16) 16d OR (3) 1/2"Ø M.B.	PROVIDE MIN. OF (2) 1/2"Ø A. B.
ALTERNATIVE BRACED WALL PANEL	2	HPAHD22 STRAP	(19) 16d OR (3) 1/2"Ø M.B.	PROVIDE (3) 1/2"Ø A.B. @ 1/5 POINTS OF SILL W/ 1 - #4 3" DN FROM TOP OF WALL & 1 - #4 3" UP FROM BOTTOM OF FOOTING. PROVIDE 1-#4 VERT @24" O.C. IN STEM WALL.
PORTAL FRAME	3	HTT22 STRAP & SSTB24 BOLT TO CONCRETE	(32) 16d SINKERS W/ 2" MAX OFFSET FROM POST FACE TO BOLT CENTERLINE	PROVIDE (3) 1/2"Ø A.B. @ MUDSILL W/ 2-2x6 SILL W/ 2 - #4 - 3" UP FROM BOTTOM OF 15" x 7" FTG. EXTEND STEEL 10'-0" AROUND FOOTING CORNER. PROVIDE 1 #4 VERT. @EA. SSTB24 CONNECTOR

FIGURE 30.14 ■ Tables can be established to specify the materials to be used to resist lateral shear. This table shows the material needed to resist lateral loads for a one-level residence. If a table is to be used, it should be near the framing plan. *Courtesy Michael Jefferis.*

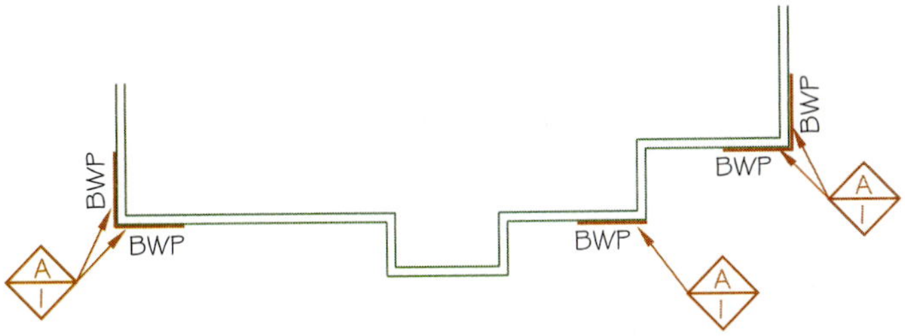

FIGURE 30.15 ■ Material for resisting lateral loads can be specified on the framing plan and referenced to a table similar to Figure 30.14.

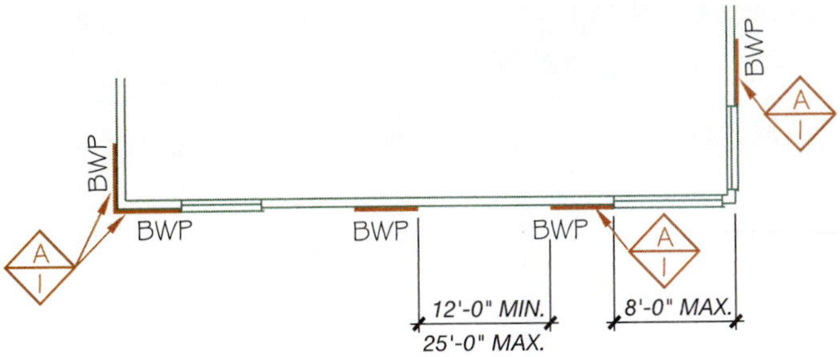

FIGURE 30.16 ■ Placement of braced wall and alternative braced wall panels to meet prescriptive code requirements.

shows an example of a table for specifying braced walls, alternative braced walls, and portal frames for a one-level home. The table will allow the required framing materials to be listed once and then specified in each required location by the use of a symbol. Figure 30.15 shows how lateral materials can be referenced on a framing plan.

Guidelines for Placement of Wall Reinforcement

The final stage of placing lateral bracing using prescriptive methods is to consider the actual placement of the reinforced wall panels. Although placement may vary depending on local conditions, the IRC requires the following standards for placement of reinforced wall panels:

Placement of BWP/ABWP Within a Braced Wall Line: See Figure 30.16.

- The reinforcement (the braced wall, alternative braced wall, or portal frame) must start within 8′ (2400 mm) of a braced wall line. The IRC allows the distance to be increased to 12.5′ (3810 mm) when additional bracing is provided.
- The maximum space between reinforcement panels is 25′ (7620 mm) o.c.

Placement of PF Within a Braced Wall Line: See Figure 30.17.

- Portal frames must be placed in pairs.
- A minimum distance of 8′ (2400 mm) between panels is required.
- The maximum distance between interior edges of portal frame panels is 25′ (7620 mm).

In addition to referencing portal frames on the framing plan, a detail should be added to the sections to describe construction methods. An example of a portal frame detail can be seen in Figure 8.46. Figures 30.18*a* and *b* show the steel that will be added to the foundation as a result of planning for lateral loads.

Placement of Braced Wall Lines: See Figure 30.19

A maximum space of 35′ (1036 cm) is allowed between braced wall lines to accommodate a single room. All other braced wall lines shall be a maximum of 25′ (7620 mm) o.c.

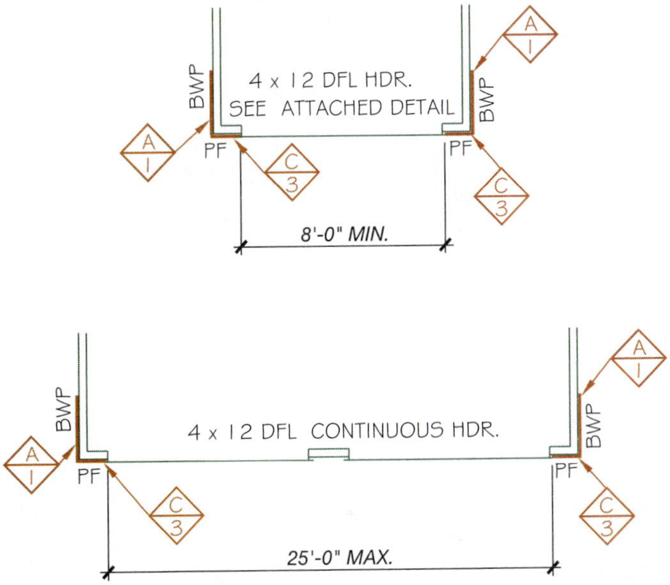

FIGURE 30–17 ■ Placement of portal frames to resist lateral loads using prescriptive design methods.

Support of Braced Wall Lines:

- An exterior braced wall line must be supported over a continuous footing.
- When floor joists are perpendicular to a braced wall line above, solid blocking or a continuous beam must be provided below the braced wall line.
- Floors with cantilevers or setbacks not exceeding four times the nominal depth of the floor joist may support braced wall panels if the following conditions are met:
  - Minimum joist size of 2 × 10 (50 × 250 mm) with a maximum spacing of 16″ (400 mm) o.c.
  - The ratio of back span to cantilever is a minimum of 2:1.
  - Floor joists at each end of a braced wall panel must be doubled.

FIGURE 30.18A ■ Metal straps are used to bind wood materials to concrete to resist uplift. These are the PAHD42 straps that are required by #1 of the table in Figure 30.14. *Courtesy Connie Willmon.*

FIGURE 30.18B ■ As the uplift stress increases, connections able to resist greater loads are required, such as these HD5A anchors specified by #4 of Figure 30.38*b*. *Courtesy Tim Taylor.*

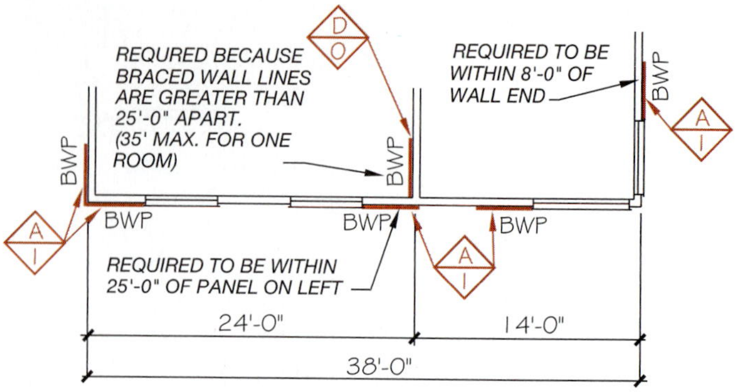

FIGURE 30.19 ■ Placement of braced wall lines using prescriptive design methods. When wall lines are more than 35′–0″ apart, an interior wall must be reinforced.

- A continuous rim joist is connected to ends of all cantilevered joists.
- Gravity loads carried at the end of the cantilevered joist are limited to uniform wall and roof load and the reactions from headers having a span of 8′ (2400 mm) or less.

- An interior braced wall line must be supported over a continuous footing at intervals of 50′ (1500 cm).
- In a building more than one story in height, all interior braced wall panels must be supported on a continuous footing unless:
  - Cripple wall height does not exceed 48″ (1200 mm).
  - First-floor braced wall panels are supported on double floor joists, continuous blocking, or floor beams.
  - The distance between braced lines does not exceed twice the building width parallel to the braced wall line.

Figure 30.20 shows a one-story residence with braced wall panels. Compare the drawing with the home in Figure 30.1. The residence in Figure 30.1 shows lateral support that was designed by calculating the actual wind loads that will affect the structure. Comparing that home with the results of using building codes prescriptive paths in Figure 30.20 will show that minor changes were required in the placement of some windows, and the addition of the storage area in the upper left corner of the home. These changes were required to provide adequate space for braced panels. As a general rule, using the prescriptive code may save the initial cost of engineering but require more material to resist the loads.

Dimensions

Chapter 15 introduced the process of placing dimensions on the floor plan. The process for placing dimensions on the framing plan is exactly the same. If a separate framing plan is to be drawn, the floor plan is usually not dimensioned; instead, all dimensions

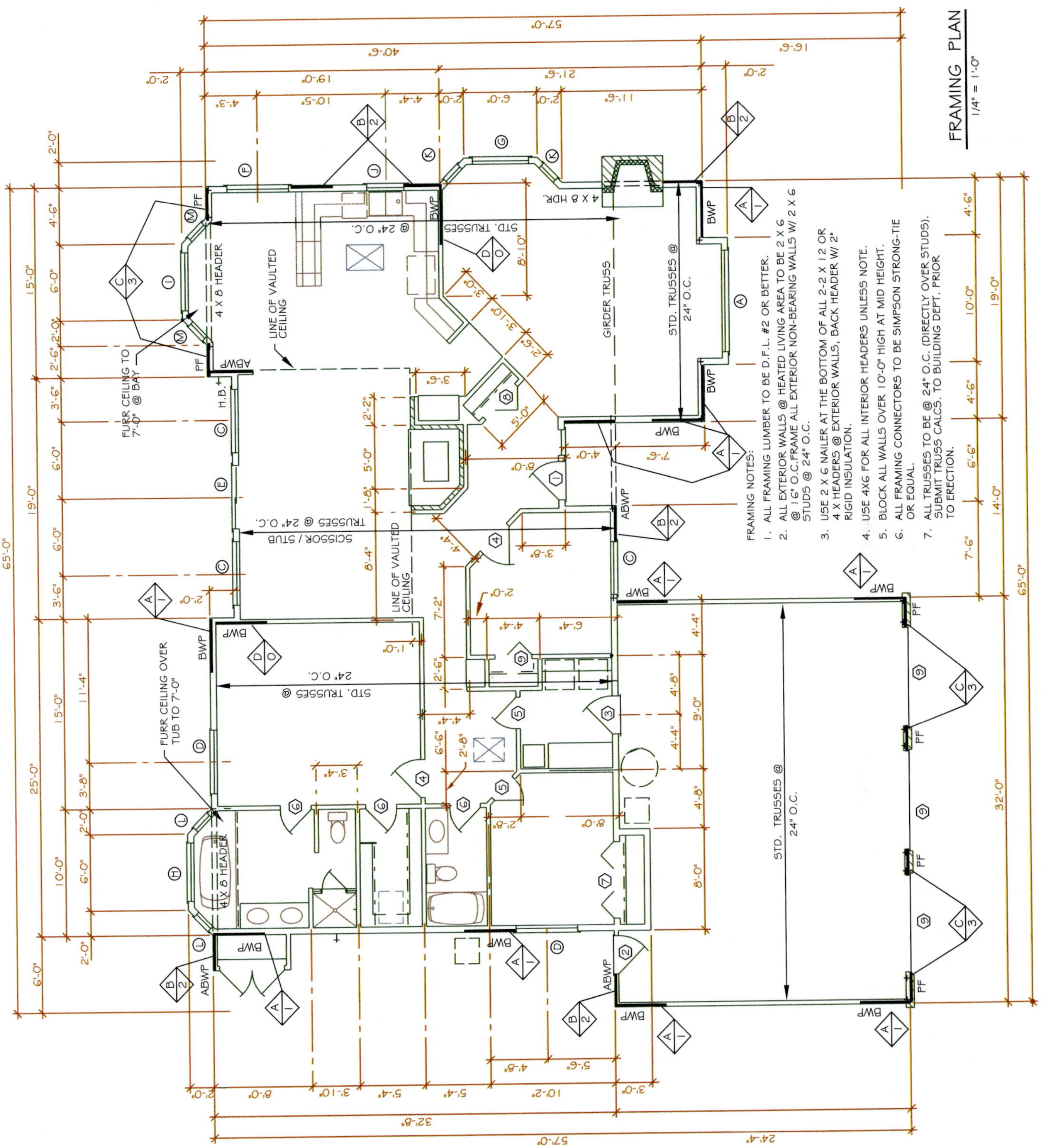
FRAMING PLAN
1/4" = 1'-0"
FRAMING NOTES:
1. ALL FRAMING LUMBER TO BE D.F.L. #2 OR BETTER.
2. ALL EXTERIOR WALLS @ HEATED LIVING AREA TO BE 2 X 6 @ 16" O.C.FRAME ALL EXTERIOR NON-BEARING WALLS W/ 2 X 6 STUDS @ 24" O.C.
3. USE 2 X 6 NAILER AT THE BOTTOM OF ALL 2-2 X 12 OR 4 X HEADERS @ EXTERIOR WALLS, BACK HEADER W/ 2" RIGID INSULATION.
4. USE 4X6 FOR ALL INTERIOR HEADERS UNLESS NOTE.
5. BLOCK ALL WALLS OVER 10'-0" HIGH AT MID HEIGHT.
6. ALL FRAMING CONNECTORS TO BE SIMPSON STRONG-TIE OR EQUAL.
7. ALL TRUSSES TO BE @ 24" O.C. (DIRECTLY OVER STUDS). SUBMIT TRUSS CALCS. TO BUILDING DEPT. PRIOR TO ERECTION.
STD. TRUSSES @ 24" O.C.
SCISSOR / STUB TRUSSES @ 24" O.C.
GIRDER TRUSS
LINE OF VAULTED CEILING
FURR CEILING TO 7'-0" @ BAY
FURR CEILING OVER TUB TO 7'-0"
4 X 8 HEADER
4 X 8 HDR.

FIGURE 30.20 ■ Representing the material to resist lateral loads using prescriptive design methods. The same home can be seen in Figure 30.1, where the loads were resisted by calculating the actual load on the structure.

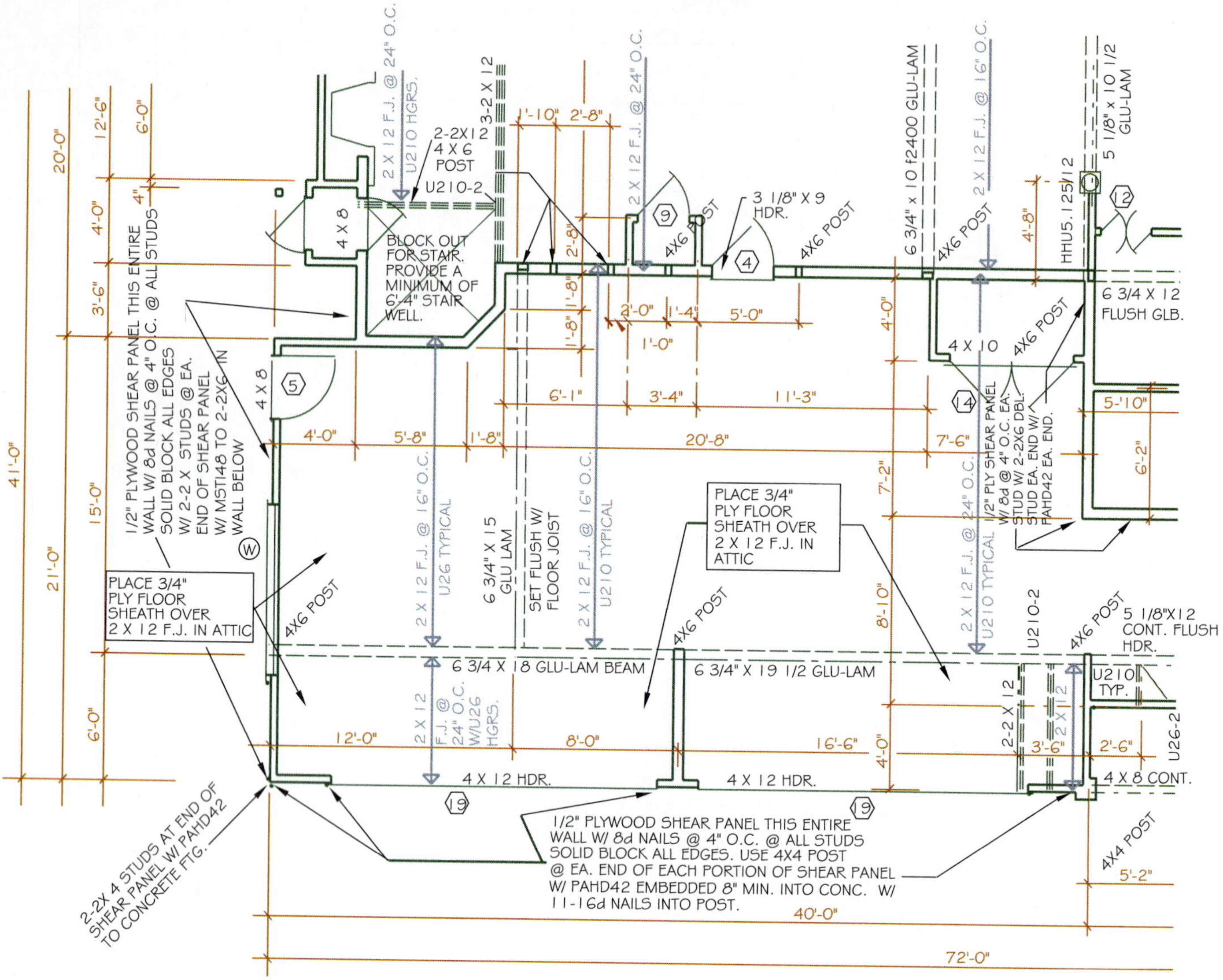

FIGURE 30.21 ■ The location of beams and post must be dimensioned on the framing plan. *Courtesy Aaron Jefferis.*

are placed on the framing plans. In addition to walls, doors, and windows, the location of framing members often needs to be dimensioned. Beams that span between an opening in a wall are located by the dimensions that locate the wall. Beams in the middle of a room must be located by dimensions.

Where joists extend past a wall, the length of the cantilever must be dimensioned. Joist size or type often varies when an upper level only partially covers another level. The limits of the placement of joists should also be dimensioned on a floor plan. Figure 30.21 shows how the location of structural members can be clarified for the framer.

Notes

The use of local notes to specify materials on the framing plan has been mentioned throughout this chapter. Many professionals also place general notes on the framing plan to ensure compliance with the code that governs construction. These framing notes can be placed on the framing plan, with the sections and details, or on a separate specifications page that is included with the working drawings. Figure 30.22 shows common notes that might be included with the framing plan.

Section References

The framing plans are used as reference maps to show where cross sections have been cut. Detail reference symbols are also placed on the framing plan to help the print reader understand material that is being displayed in the sections. (Chapter 34 will further explain section tags and their relationship to the sections and framing plans.) Figure 30.9 shows examples of section markers.

FRAMING NOTES:

NOTES SHALL APPLY TO ALL LEVELS. FRAMING STANDARDS ARE ACCORDING TO 2000 I.B.C.

1. ALL FRAMING LUMBER TO BE D.F.L. #2 MIN. ALL GLU-LAM BEAMS TO BE fb2400, V-4, DF/DF.
2. FRAME ALL EXTERIOR WALLS W/ 2 x 6 STUDS @ 16" O.C. FRAME ALL EXTERIOR NON-BEARING WALLS W/ 2 x 6 STUDS @ 24" O.C. USE 2 x 4 STUDS @ 16" O.C. FOR EXTERIOR UNHEATED WALLS AND FOR ALL INTERIOR WALLS UNLESS NOTED. ALL INT. HDRS TO BE 4 x 6 DFL#2 UNLESS NOTED.
3. ALL EXTERIOR HDRS. TO BE (2) 2x12 UNLESS NOTED. USE 2 x 6 NAILER AT THE BOTTOM OF ALL 2-2X12 OR 4 X HDR. @ EXTERIOR WALLS. BACK HDR .W/ 2" RIGID INSULATION.
4. PLYWOOD ROOF SHTG. TO BE 1/2" STD. GRADE 32/16 PLY. LAID PERP. TO RAFT. NAIL W/ 8 d'S @ 6" O.C. @ EDGES & 12" O.C. AT FIELD.
5. ALL SHEAR PANELS TO BE 1/2" PLY NAILED W/ 8 d'S @ 4" O.C. @ EDGE AND BLK. & 8 d'S @ 8" O.C. @ FIELD.
6. LET-IN BRACES TO BE 1 x 4 DIAG. BRACES @ 45 FOR ALL INTERIOR LOAD-BEARING WALLS. (CROSS 3 STUDS MIN).
7. BLOCK ALL WALLS OVER 10'-0" HIGH AT MID HEIGHT.
8. FLOOR SHTG TO BE 3/4" STD. GRADE T. & G. PLY. LAID PERP. TO FLR. JOIST. NAIL W/ 10 d'S @ 6" O.C. @ EDGES & BLK. & 12" O.C. @ FIELD. COVER W/ 3/8" HARDBOARD.
9. NOTCHES IN THE ENDS OF JOIST SHALL NOT EXCEED 1/4 OF THE JOIST DEPTH. HOLES DRILLED IN JOIST SHALL NOT BE IN THE UPPER OR LOWER 2" OF THE JOIST. THE DIA. OF HOLES DRILLED IN JOIST SHALL NOT EXCEED 1/3 THE DEPTH OF THE JOIST.
10. PROVIDE DOUBLE JOIST UNDER AND PARALLEL TO LOAD BEARING WALLS.
11. BLOCK ALL FLOOR JOIST AT SUPPORTED ENDS AND @ 10'-0" O.C. MAX. ACROSS SPAN.
12. ALL FRAMING CONNECTORS TO BE SIMPSON CO. OR EQUAL.

FIGURE 30.22 ■ General notes can be used to ensure that minimum standards established by building codes will be maintained.

COMPLETING A FRAMING PLAN

The order used to draw the framing plan depends on the method of construction and the level to be framed. A structure framed with trusses requires less detailing than the same plan framed with rafters and ceiling joists. Interior beams will be greatly reduced, if not entirely eliminated, when trusses are used. On multilevel structures, because a lower level is supporting more weight, beams tend to be shorter, requiring more and larger posts. Despite differences, framing plans also have similarities that can help in drafting the plan, no matter what level is to be drawn.

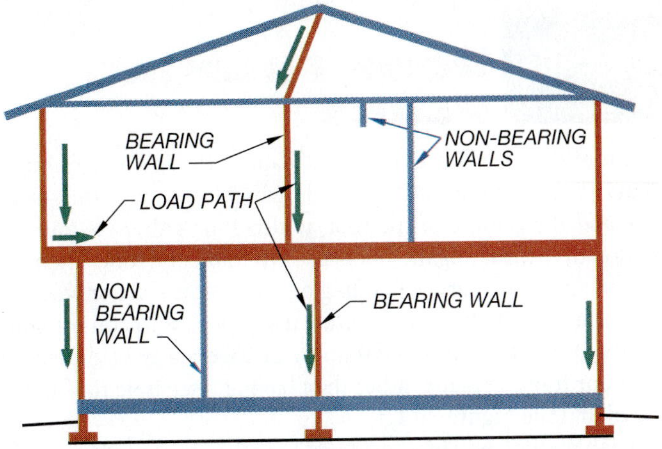

FIGURE 30.23 ■ Following the load path through a structure. Bearing walls are considered to be aligned as long as the distance of the offset does not exceed the thickness of the floor joist. If the offset of the bearing walls is greater than the thickness of the floor joist, the size of the joist must be determined by using the formulas in Chapter 28.

Once the walls have been located, dimensions can be placed to locate all walls and openings in them. Then individual framing members can be determined. Rafter direction is based on the shape of the roof, and its length is determined by the span for a specific size of the particular member. Rafter spans can be determined using the tables in Chapter 28. The process is similar if trusses are to be drawn, but the sizes of the trusses are determined by the manufacturer.

After the size and direction of the roof framing members are selected, loading bearing walls can be identified and headers selected for openings in the load-bearing walls. With the structural members specified, any dimensions required to locate structural members can be completed. Materials that can be specified with local notes can be added and the drawing completed by adding general notes.

Once the framing for the roof has been determined, framing for lower levels can be drawn. Bearing walls on the upper level need to have some method of supporting the loads as they are transferred downward. Support can be provided by stacking bearing walls or by transferring loads through floor joists to other walls. Figure 30.23 shows a simplified drawing of how loads can be transferred. Load bearing walls can be offset by the thickness of the floor joists and still be considered aligned. For example, if 2 × 10 (50 × 250 mm) floor joists are used, an upper wall could be offset 10″ (250 mm) from a lower wall and still be considered to be aligned. When floor joists are cantilevered or used to transfer vertical loads to lower walls that are not aligned, the tables in Chapter 28 cannot be used to determine joist sizes. These members need to be determined using the formulas in Chapter 29. Once the bearing walls for the lowest floor level have been determined, the foundation can be completed.

ROOF FRAMING PLANS

The roof plan was introduced in Chapter 21. The roof framing plan is similar, but in addition to showing the shape of the structure and the outline of the roof, it also shows the size and direction of the framing members used to frame the roof. The roof framing plan will vary depending on whether trusses or western platform construction methods are used. Depending on the simplicity of the structure, roof framing members may be drawn on the floor framing plans rather than having a separate roof framing plan. (See Figure 30.1.)

Each method will be considered for gable, hip, and Dutch hip roofs for the residence drawn in Chapter 16 using separate roof plans and framing plans. Each roof can be drawn by using the same steps to form the base drawing. If a roof plan has been drawn, it can be traced to form the base of the roof framing plan. To draw a roof framing plan, use construction lines for the following steps. Use the line quality described throughout this chapter and Chapter 21 to complete this drawing.

Roof Plan Base Drawing

STEP 1 Draw the outline of the exterior walls.

STEP 2 Draw any posts and headers required for covered porches.

STEP 3 Draw the limits of the overhangs. Unless your instructor provides different instructions, use 24″ (600 mm) overhangs at eaves and 12″ (300 mm) overhangs at gable end walls.

STEP 4 Locate all ridges.

STEP 5 Locate all hip or valleys formed by roof intersections.

STEP 6 Locate all roof openings, such as skylights and chimneys.

STEP 7 Draw all items from steps 1 through 6 using finished line quality. Your drawing should now resemble Figure 30.24.

Gable Roof Framing Plan with Truss Framing

STEP 8 Draw the boundaries of any areas to be vaulted. Assume the master bedroom will be vaulted. Provide dimensions to specify the limits.

STEP 9 Draw and label the arrows to indicate the framing members. Assume standard/scissor trusses over the master bedroom with standard roof trusses at 24″ o.c. for the balance of the upper level and over the garage. Specify 4 × 6 beams at 32″ o.c. at the entry porch. Support the entry beams on 4 × 8 beams on 4 × 4 post with PC44 post caps by Simpson Company or equal.

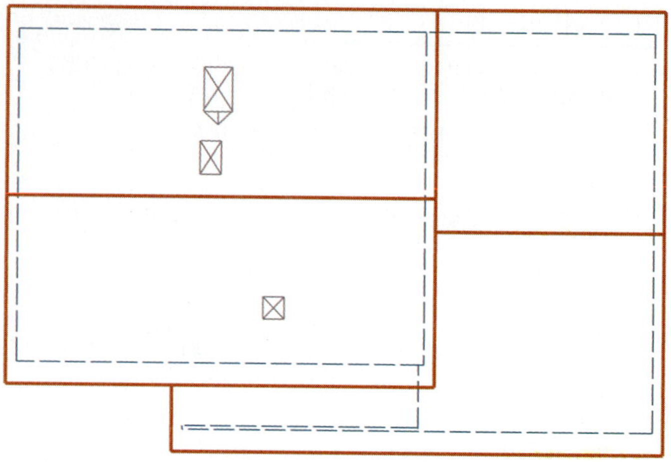

FIGURE 30.24 ■ The roof plan serves as a base for the framing plan.

STEP 10 Provide local notes to specify any necessary straps or metal connectors. For this residence, provide an ST6224 tie strap from the 4 × 8 header to the garage top plate.

STEP 11 Provide general notes to indicate any necessary framing material. Your drawing should now resemble Figure 30.25.

Gable Roof Framing Plan with Rafter Framing

Follow steps 1 through 7 to draw the base for this plan. The drawing should resemble Figure 30.24. Complete the plan for a roof framed using rafters, using the following steps:

STEP 1 Determine the location of interior walls, which are parallel to the ridge. Although the walls are not drawn on the framing plan, they can be used to support purlin braces.

STEP 2 Draw the boundaries of any areas to be vaulted. Assume the master bedroom will be vaulted. Provide dimensions to specify the limits.

STEP 3 Draw and label the arrows to indicate the framing members. Use the span tables from Chapter 28 to determine the required size. For this residence, DFL #2 will be used. Use 2 × 6 rafters at 24″ o.c. where possible. For this project the following framing will be used:

2 × 12 raft./c.j. over bedroom 1

4 × 14 DFL #1 exposed ridge beam

purlins at upper floor: 2 × 6

purlin over garage: 2 × 8

strongback over garage: 5 1/8 × 13 1/2 F_b 2400 glu-lam

beams at porch: 4 × 8

rafters at porch: 4 × 6 @ 32″ o.c.

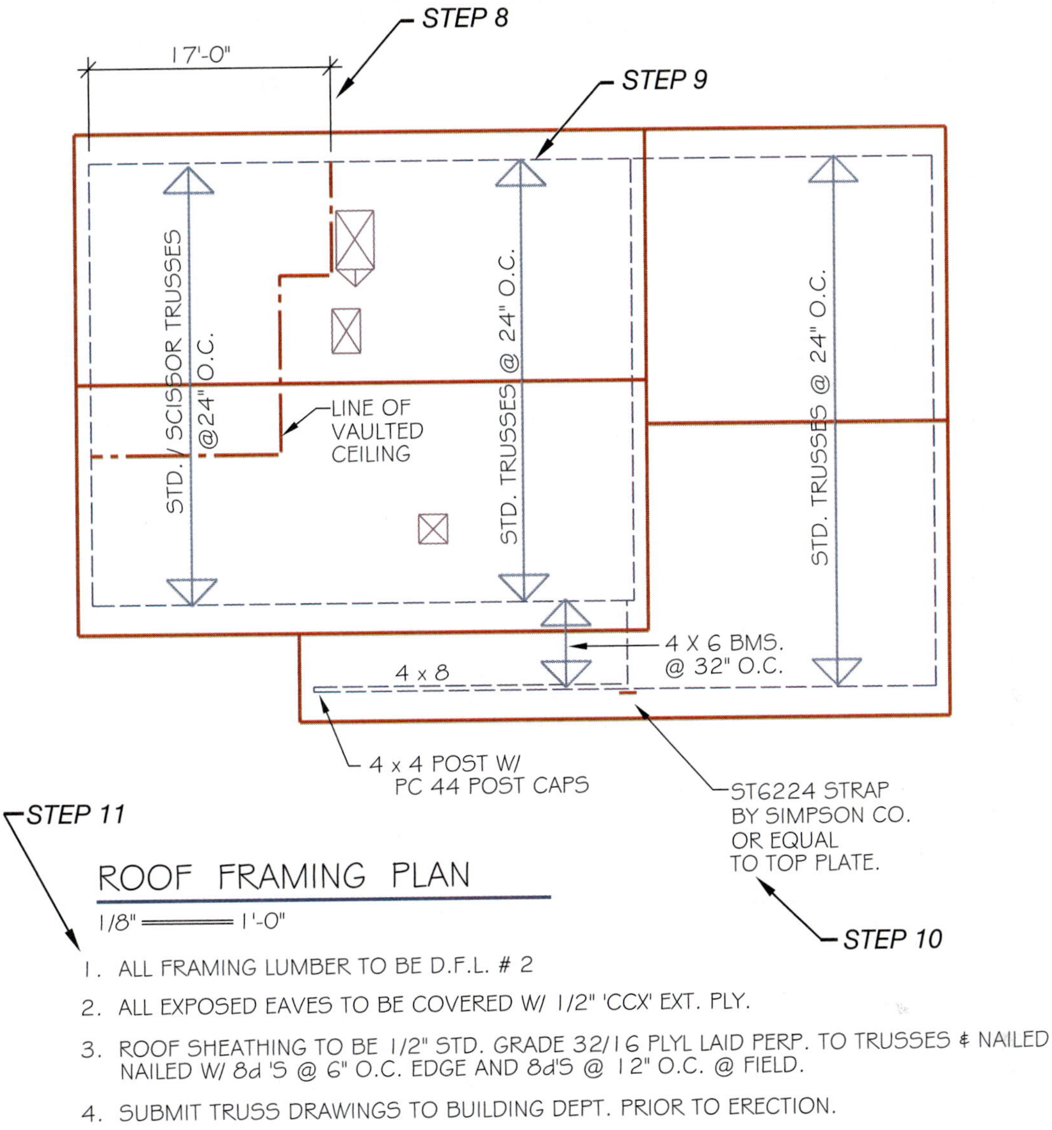

FIGURE 30.25 ■ Completed roof framing plan for a gable roof framed with trusses.

STEP 4 Provide purlins and strong backs at approximately the midpoint of rafter spans.

STEP 5 Provide local notes to specify any necessary straps or metal connectors. For this residence, provide an ST6224 tie strap from the 4 × 8 header to the garage top plate. Support the entry beams on 4 × 4 post with pcc44 post caps by Simpson Company or equal.

STEP 6 Provide general notes to indicate any necessary framing material. Your drawing should now resemble Figure 30.26.

Hip Roof Framing Plan with Truss Framing

Follow steps 1 through 7 to draw the base for this plan. The base drawing should resemble Figure 30.27.

STEP 1 Determine the limit of the standard trusses based on the intersection of the hips with the ridge.

STEP 2 Determine the limits of any required girder trusses. Because the roof of the structure has no perpendicular roofs, no girder trusses are required.

STEP 3 Draw the boundaries, and dimension the limits of any areas to be vaulted. Because the framing members in the master bedroom will be spanning in two different directions, the room will not be vaulted for this roof plan.

STEP 4 Determine the limits of the hip trusses. This distance is typically determined by the truss manufacturer and the pitch of the roof. For this plan assume 6′–0″ is required before hip trusses can be used.

STEP 5 Draw and label the arrows to indicate the framing members. Assume standard trusses at 24″ o.c. over the

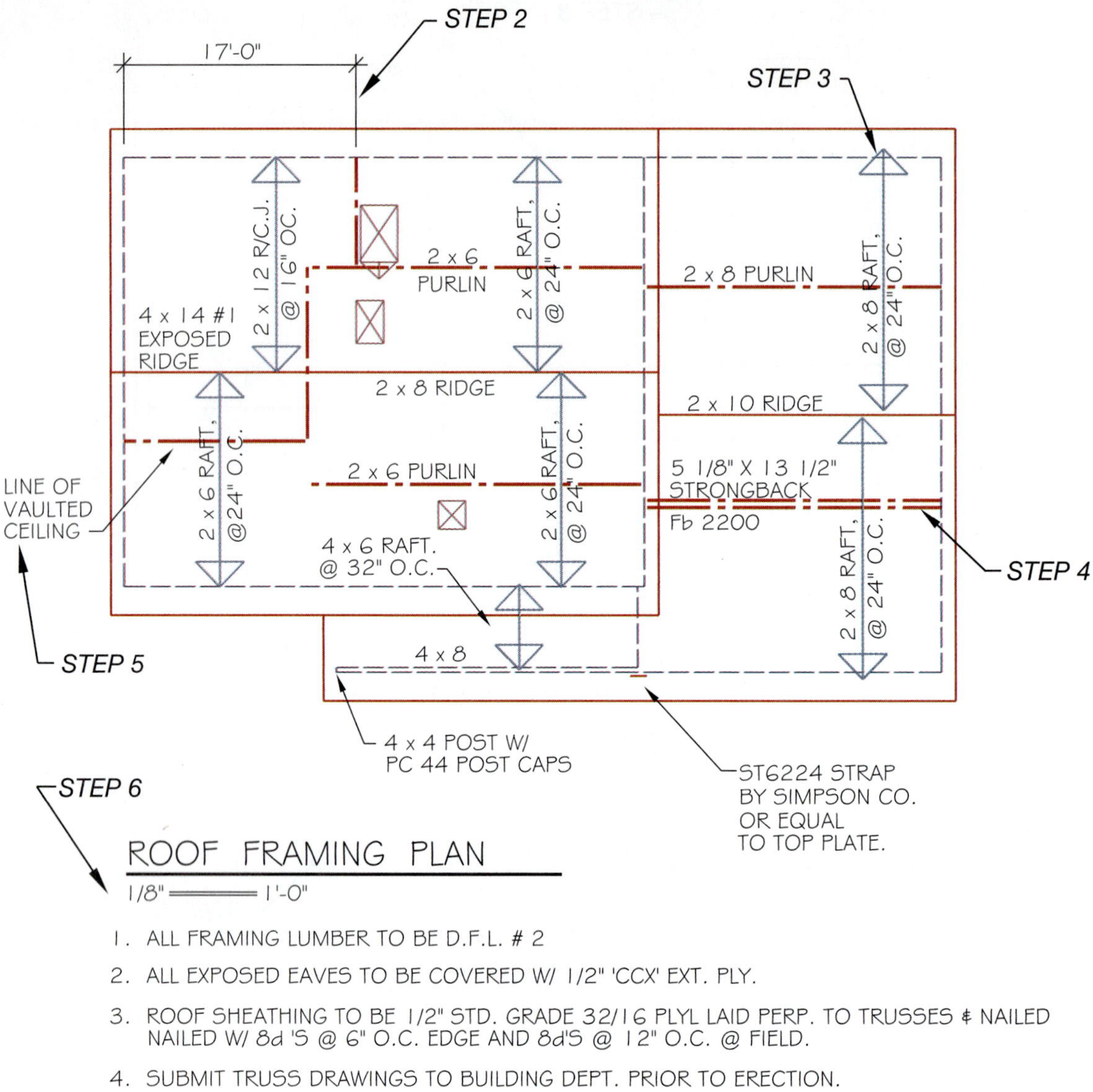

FIGURE 30.26 ■ Completed roof framing plan for a gable roof with rafters.

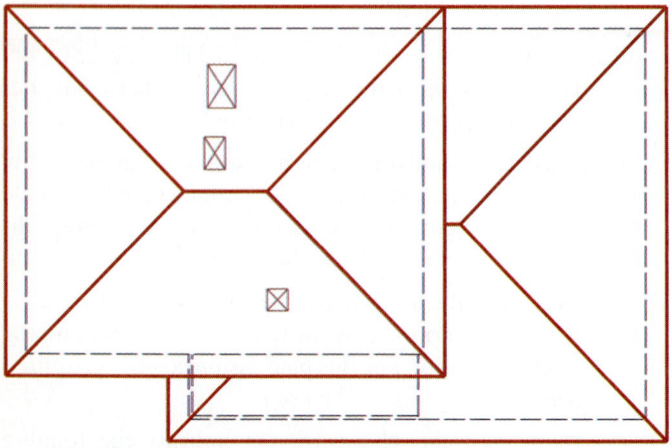

FIGURE 30.27 ■ The roof plan for a hip roof serves as a base for the roof framing plan.

entire upper floor and over the garage. Specify 4 × 6 beams at 32″ o.c. at the entry porch. Support the entry beams on 4 × 8 beams on 4 × 4 post with pc44 post caps by Simpson company or equal.

STEP 6 Provide local notes to specify any necessary straps or metal connectors. For this residence, provide an ST6224 tie strap from the 4 × 8 header to the garage top plate.

STEP 7 Provide general notes to indicate any necessary framing material. Your drawing should now resemble Figure 30.28.

Hip Roof Framing Plan with Rafter Framing

Follow steps 1 through 7 to draw the base for this plan. The drawing should resemble Figure 30.27. The layout of the roof will be similar to the layout of a stick framed gable roof.

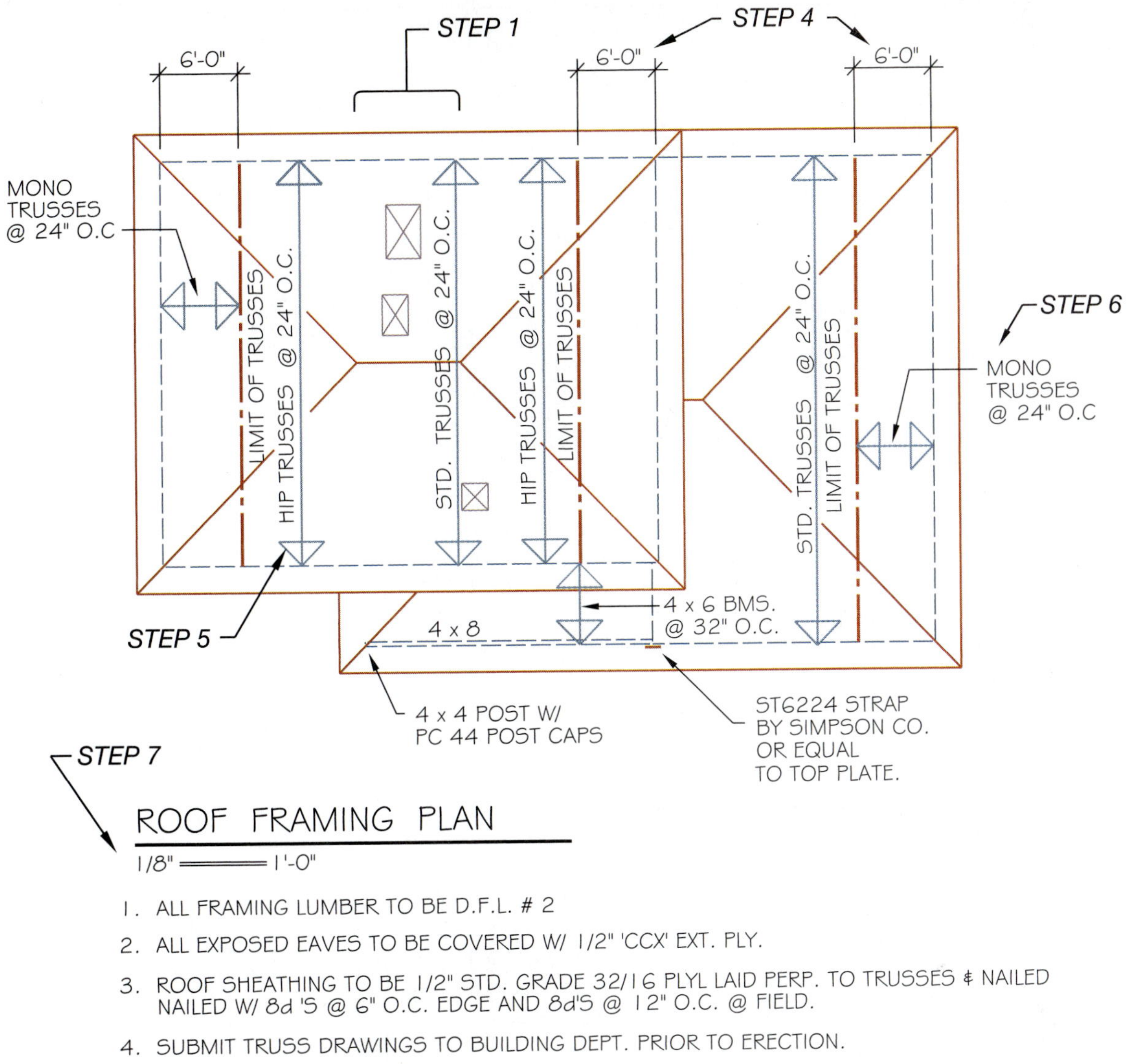

FIGURE 30.28 ■ Completed roof framing plan for a hip roof framed with trusses.

STEP 1 Determine the location of interior walls that are parallel to the ridge. These walls will support purlin braces.

STEP 2 Draw the boundaries of any vaulted areas.

STEP 3 Markers to show rafters can be drawn and labeled using the span tables in Chapter 28.

STEP 4 Locate purlins and strong backs where necessary based on rafter spans.

STEP 5 Provide local notes to specify any necessary straps or metal connectors.

STEP 6 Provide general notes to indicate any necessary framing material. Your drawing should now resemble Figure 30.29*a*. If a roof has a complicated shape, some companies will draw all framing members. Figure 30.29*b* shows a hip roof and all required framing members.

Dutch Hip Roof Framing Plan with Truss Framing

Follow steps 1 through 6 to draw the base for this plan. The base drawing should resemble Figure 30.30.

STEP 1 Determine the limit of the standard trusses based on the limits of the Dutch hip. For this plan, assume the gable end walls of the Dutch hip will be set 6′–0″ from the end walls.

STEP 2 Determine the location of any required girder trusses. Because the roof of the structure has no perpendicular roofs, no girder trusses will be required.

STEP 3 Draw the boundaries, and dimension the limits of any areas to be vaulted. Because the framing members in the master bedroom will be spanning in two different directions, the room will not be vaulted for this roof plan.

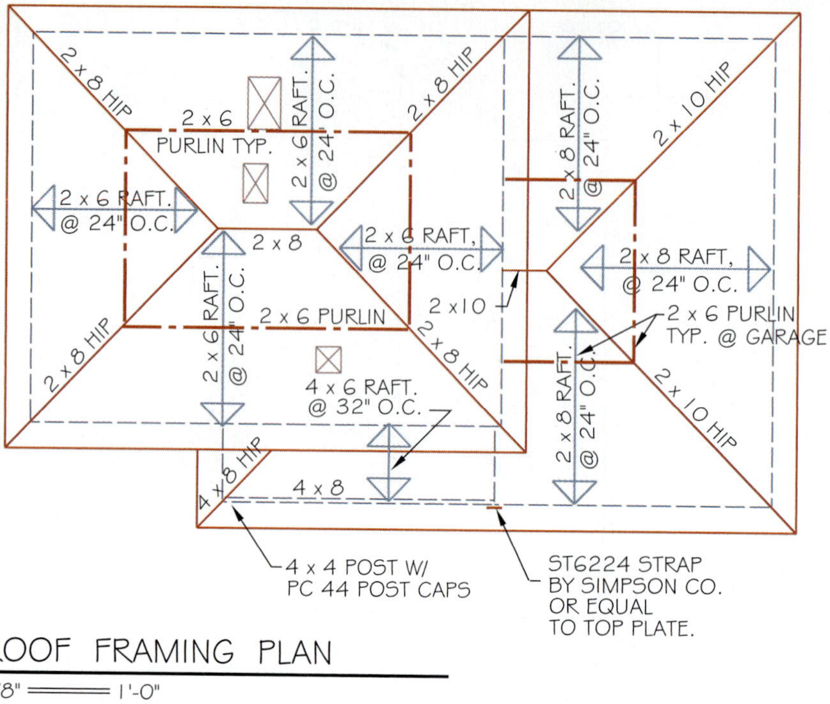

1. ALL FRAMING LUMBER TO BE D.F.L. #2
2. ALL EXPOSED EAVES TO BE COVERED W/ 1/2" 'CCX' EXT. PLY.
3. ROOF SHEATHING TO BE 1/2" STD. GRADE 32/16 PLY. LAID PERP. TO RAFT. NAILED W/ 8d'S @ 6" O.C. EDGE AND 8d'S @ 12" O.C. @ FIELD.
4. MAXIMUM SPANS:

2 x 8 RAFTERS	@ 24" O.C. =	11'-0"
" "	@ 16" O.C. =	13'-5"
" "	@ 12" O.C. =	15'-6"
2 x 6 RAFTERS	@ 24" O.C. =	11'-6"
" "	@ 16" O.C. =	14'-0"
" "	@ 12" O.C. =	16'-3"

A

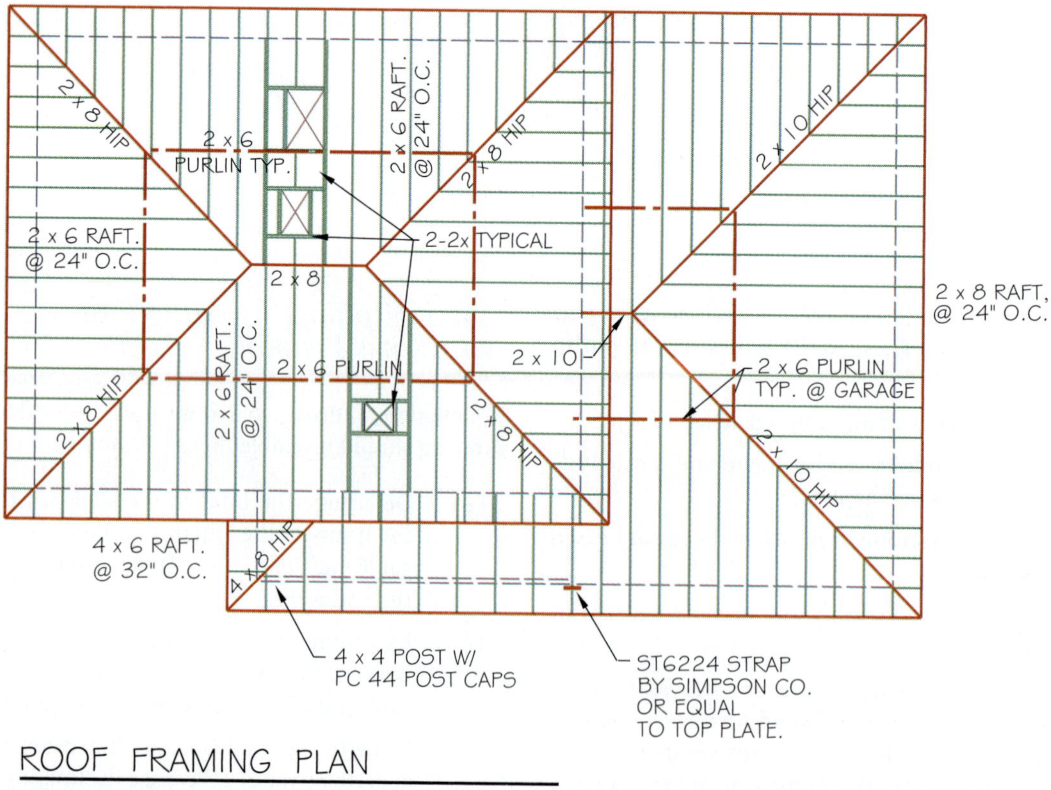

B

FIGURE 30.29 ■ (a) Completed roof framing plan for a hip roof framed with rafters. (b) Some offices show all framing members to eliminate any confusion. *Courtesy Jordan Jefferis.*

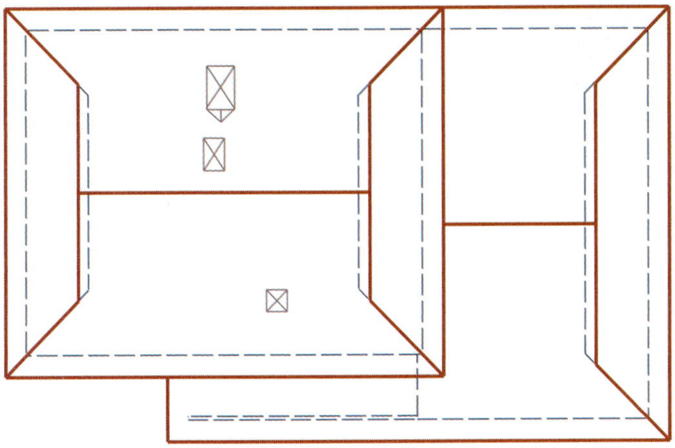

FIGURE 30.30 ■ The roof plan for a Dutch hip roof serves as a base for the roof framing plan.

STEP 4 Draw and label the arrows to indicate the framing members.

STEP 5 Provide local notes to specify any necessary straps or metal connectors.

STEP 6 Provide general notes to indicate any necessary framing material. Your drawing should now resemble Figure 30.31.

Dutch Hip Roof Framing Plan with Rafter Framing

Follow steps 1 through 7 for a truss roof framing plan to draw the base for this plan. The drawing should resemble Figure 30.30.

STEP 1 Lay out interior walls that are parallel to the ridge.

STEP 2 Draw the boundaries, and dimension the limits of any areas to be vaulted. Because the framing members in the master bedroom will be spanning in two different directions, the room will not be vaulted for this roof plan.

STEP 3 Draw and label the arrows to indicate the framing members. Use the span tables in Chapter 28 to determine the required size.

STEP 4 Provide purlins and strong backs where necessary based on rafter spans.

STEP 5 Add dimensions to locate all openings, overhangs, and purlins.

STEP 6 Provide local notes to specify any necessary straps or metal connectors.

STEP 7 Provide general notes to indicate any necessary framing material. Your drawing should now resemble Figure 30.32.

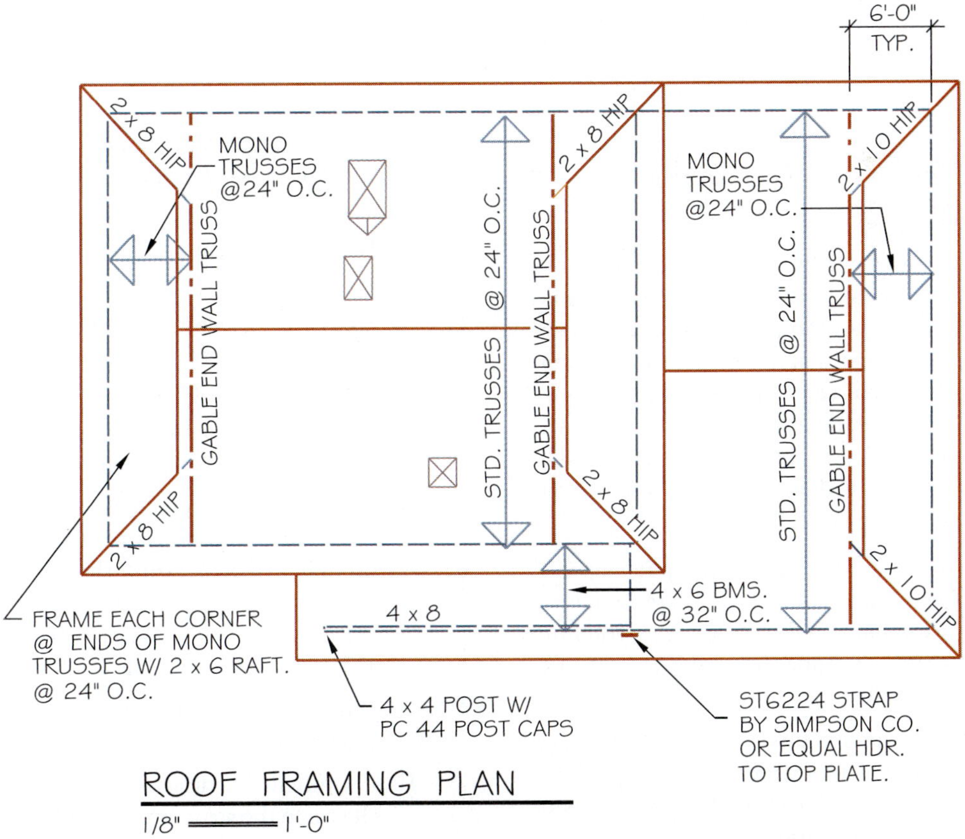

FIGURE 30.31 ■ Completed roof framing plan for a Dutch hip roof framed with trusses. Notes similar to those in Figure 30.28 should be included to complete the plan.

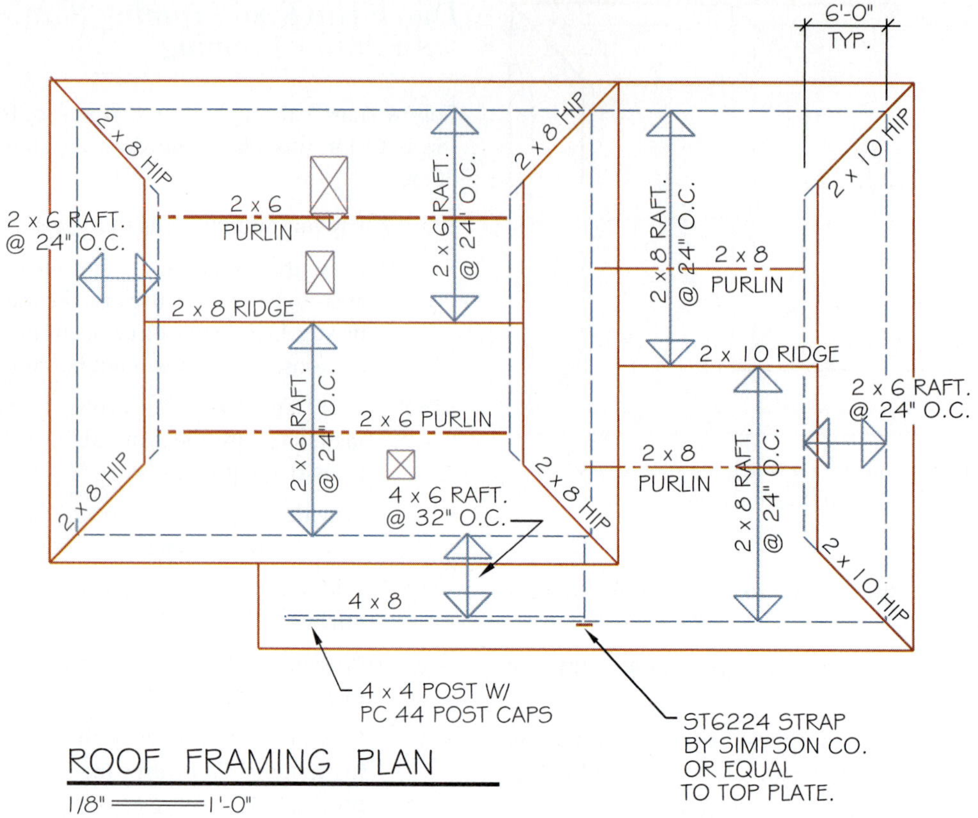

FIGURE 30.32 ■ Completed roof framing plan for a Dutch hip roof framed with rafters. Notes similar to those in Figure 30.28 should be included to complete the plan.

FLOOR FRAMING PLANS

The framing plan can be completed by using the base drawings for the floor plans. Ideally a Mylar copy of the floor plan should be made once the walls, doors, windows, and cabinets are drawn. If a copy of the floor plan was not made prior to completing the floor plan, the layout can be traced. If a computer program is being used, layers containing the walls, windows, doors, and cabinets can be displayed, with material unrelated to the framing plan frozen. The base drawing for the upper framing plan should resemble Figure 30.33.

STEP 1 If your floor plan was dimensioned, skip to step 9. If not, lay out all exterior dimension and extension lines using construction lines for each side of the structure. Establish lines for overall and major jogs from the exterior face of the wall.

STEP 2 Lay out dimensions and extension lines to locate each interior wall that intersects with an exterior wall. Interior walls will be located from center to center, with center lines for extension lines.

STEP 3 Lay out dimension and extension lines to locate exterior openings. Do not dimension doors in the garage on the framing plan.

STEP 4 Lay out dimension and extension lines to locate all interior walls that do not intersect with an exterior wall.

STEP 5 Place overall dimensions centered above each dimension line.

STEP 6 Specify dimensions that must be a specific size—for example, spaces for tubs, toilets, showers, and other manufactured items.

STEP 7 Specify dimensions for areas that are required to be a specific size—for example, hallways, closets, and stairwells.

STEP 8 Dimension all major jogs. Wherever possible, try to maintain modular sizes of the material being used. Although a distance from the edge of one wall to the center of the next wall may measure 12′–1″, 14′ joists would need to be purchased. Reducing the distance to 12′–0″ or less would allow 12′ joists to be used.

STEP 9 Dimension all remaining sizes, starting from the smallest rooms and going on to the largest.

STEP 10 Coordinate all dimensions. The lines of dimensions closest to the structure should add up to the total seen on the next line of dimensions. Dimensions on the interior of the structure should add up to match any exterior dimensions. Your framing plan should now resemble Figure 30.34.

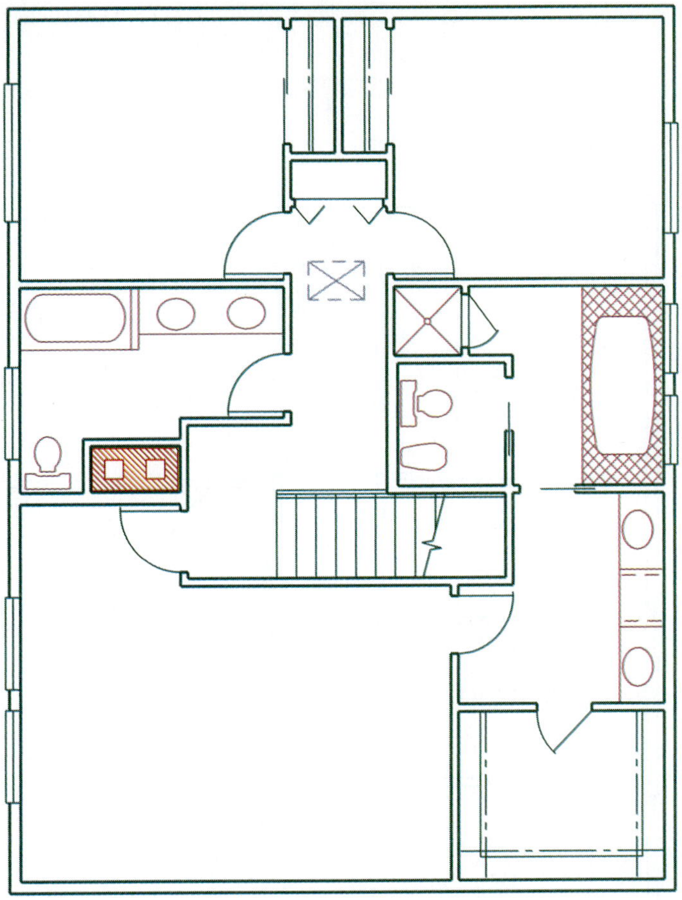

FIGURE 30.33 ■ The upper floor plan can be used as the base to draw the upper floor framing plan. On a trussed framed roof, this plan can be used in place of the roof framing plan.

STEP 11 Specify door and windows symbols, which were specified on the floor plan.

STEP 12 Draw the boundaries of any areas to be vaulted. Assume that the master bedroom will be vaulted. Provide dimensions to specify the limits.

STEP 13 Draw and label the arrows to indicate the framing members. Assume standard/scissor trusses over the master bedroom with standard roof trusses at 24″ o.c. for the balance of the upper level. If a roof other than a truss-framed gable is to be shown on this plan, review the steps used throughout the layout roof framing plan.

Note: If the roof is to be stick framed, only the members used to show the support for the ceiling are shown on the framing plan. If a roof framing plan was drawn, *do not repeat* the material shown in steps 12 through 13 on the upper floor framing plan.

STEP 14 Draw and specify all beams, headers, and posts.

STEP 15 Provide local notes to specify any materials needed for resisting lateral loads caused by wind or seismic forces.

Note: Because requirements vary widely for each area of the country, verify with your instructor materials that are typical for your area.

STEP 16 Provide local notes to specify any necessary straps or metal connectors.

STEP 17 Provide general notes to indicate any necessary framing material. Your drawing should now resemble Figure 30.35. If the upper level was framed using western platform construction techniques, the plan would resemble Figure 30.34.

The framing plan for the main floor can be completed following the same procedure that was used to draw the upper floor. Because the upper floor of the residence drawn in Chapter 16 does not completely cover the lower floor, part of the framing plan will show floor joists to support the upper floor, and part of the plan will show roof framing members. If a separate roof framing plan was drawn, all roof framing material will be omitted from this plan.

STEP 1 Dimension the framing plan following steps 1 through 9 of the upper framing plan.

STEP 2 Specify door and window symbols specified on the floor plan.

STEP 3 Draw the boundaries of any areas to be vaulted.

STEP 4 Draw and label the arrows to indicate the framing members. Use the tables in Chapter 28 to determine

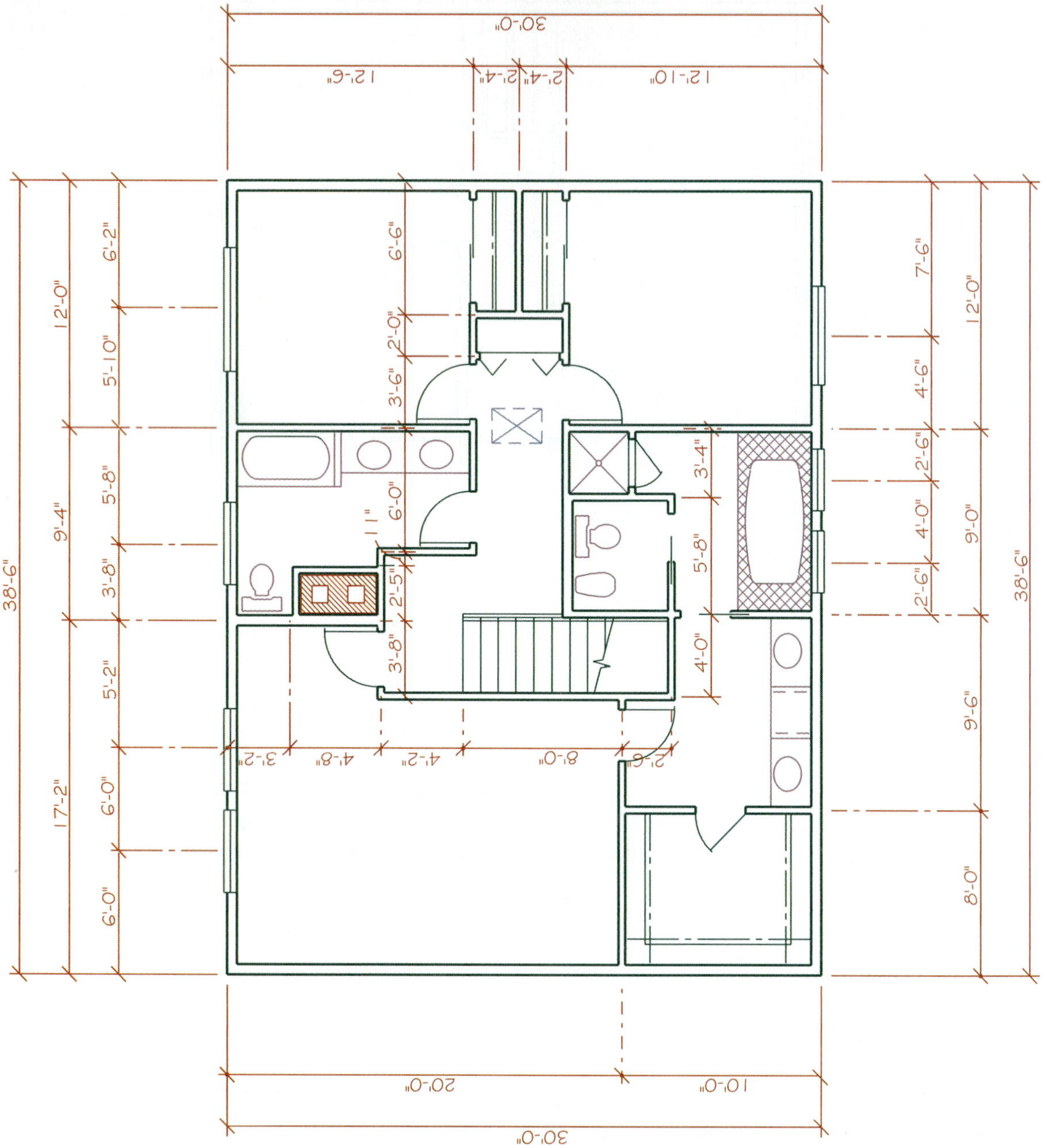

FIGURE 30.34 ■ Rather than placing dimensions on a floor plan, many offices use the framing plan to display all dimensions.

the sizes of members for species of framing lumber common to your area.

STEP 5 Using the tables and formulas in Chapter 29, draw and specify all beams, headers, and posts.

STEP 6 Provide local notes to specify any materials needed for resisting lateral loads caused by wind or seismic forces. Verify with your instructor materials that are common in your area.

STEP 7 Provide local notes to specify any necessary straps or metal connectors.

STEP 8 Provide general notes to indicate any necessary framing material. Your drawing should now resemble Figure 30.37.

If the upper level is framed with joists, the lower floor will be affected. Figure 30.38 shows the changes required to the lower-floor framing plan.

The information for framing the floor will be placed on the foundation plan. Procedures for showing the support for the lower plan will be introduced in Chapter 33. Because this plan has a partial basement, the drafter will have the choice of where the floor framing over the basement can be shown. One option is to show a separate framing plan for just the basement, with the floor for the basement and the balance of the structure shown on the foundation plan. This plan would be drawn using the steps for the main framing plan and would resemble Figure 30.39. Because the plan is relatively simple, many professional offices would show the material for framing the main floor over the basement on the foundation plan.

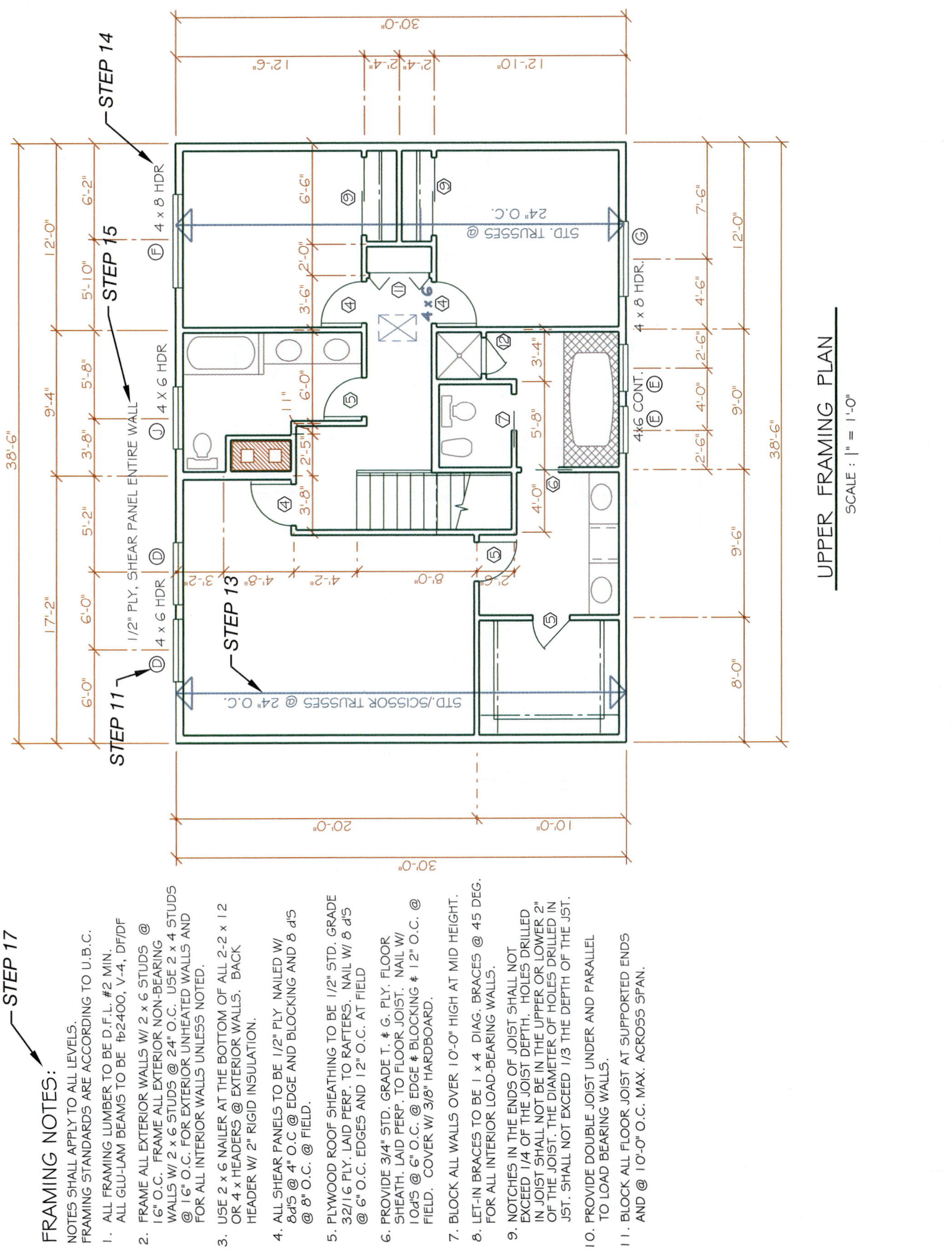

FRAMING NOTES:
NOTES SHALL APPLY TO ALL LEVELS.
FRAMING STANDARDS ARE ACCORDING TO U.B.C.
1. ALL FRAMING LUMBER TO BE D.F.L. #2 MIN. ALL GLU-LAM BEAMS TO BE fb2400, V-4, DF/DF
2. FRAME ALL EXTERIOR WALLS W/ 2 x 6 STUDS @ 16" O.C. FRAME ALL EXTERIOR NON-BEARING WALLS W/ 2 x 6 STUDS @ 24" O.C. USE 2 x 4 STUDS @ 16" O.C. FOR EXTERIOR UNHEATED WALLS AND FOR ALL INTERIOR WALLS UNLESS NOTED.
3. USE 2 x 6 NAILER AT THE BOTTOM OF ALL 2-2 x 12 OR 4 x HEADERS @ EXTERIOR WALLS. BACK HEADER W/ 2" RIGID INSULATION.
4. ALL SHEAR PANELS TO BE 1/2" PLY NAILED W/ 8d'S @ 4" O.C @ EDGE AND BLOCKING AND 8 d'S @ 8" O.C. @ FIELD.
5. PLYWOOD ROOF SHEATHING TO BE 1/2" STD. GRADE 32/16 PLY. LAID PERP. TO RAFTERS. NAIL W/ 8 d'S @ 6" O.C. EDGES AND 12" O.C. AT FIELD
6. PROVIDE 3/4" STD. GRADE T. & G. PLY. FLOOR SHEATH. LAID PERP. TO FLOOR JOIST. NAIL W/ 10d'S @ 6" O.C. @ EDGE & BLOCKING & 12" O.C. @ FIELD. COVER W/ 3/8" HARDBOARD.
7. BLOCK ALL WALLS OVER 10'-0" HIGH AT MID HEIGHT.
8. LET-IN BRACES TO BE 1 x 4 DIAG. BRACES @ 45 DEG. FOR ALL INTERIOR LOAD-BEARING WALLS.
9. NOTCHES IN THE ENDS OF JOIST SHALL NOT EXCEED 1/4 OF THE JOIST DEPTH. HOLES DRILLED IN JOIST SHALL NOT BE IN THE UPPER OR LOWER 2" OF THE JOIST. THE DIAMETER OF HOLES DRILLED IN JST. SHALL NOT EXCEED 1/3 THE DEPTH OF THE JST.
10. PROVIDE DOUBLE JOIST UNDER AND PARALLEL TO LOAD BEARING WALLS.
11. BLOCK ALL FLOOR JOIST AT SUPPORTED ENDS AND @ 10'-0" O.C. MAX. ACROSS SPAN.
STEP 17
STEP 11
STEP 13
STEP 14
STEP 15
STD. TRUSSES @ 24" O.C.
STD./SCISSOR TRUSSES @ 24" O.C.
4 x 8 HDR
4 X 6 HDR
1/2" PLY, SHEAR PANEL ENTIRE WALL
4x6 CONT.
4 x 8 HDR.
30'-0"
38'-6"
UPPER FRAMING PLAN
SCALE : 1" = 1'-0"

FIGURE 30.35 ■ Completed upper floor framing plan showing truss construction for the structure started in Chapter 16. Lateral bracing is based on the design specific method.

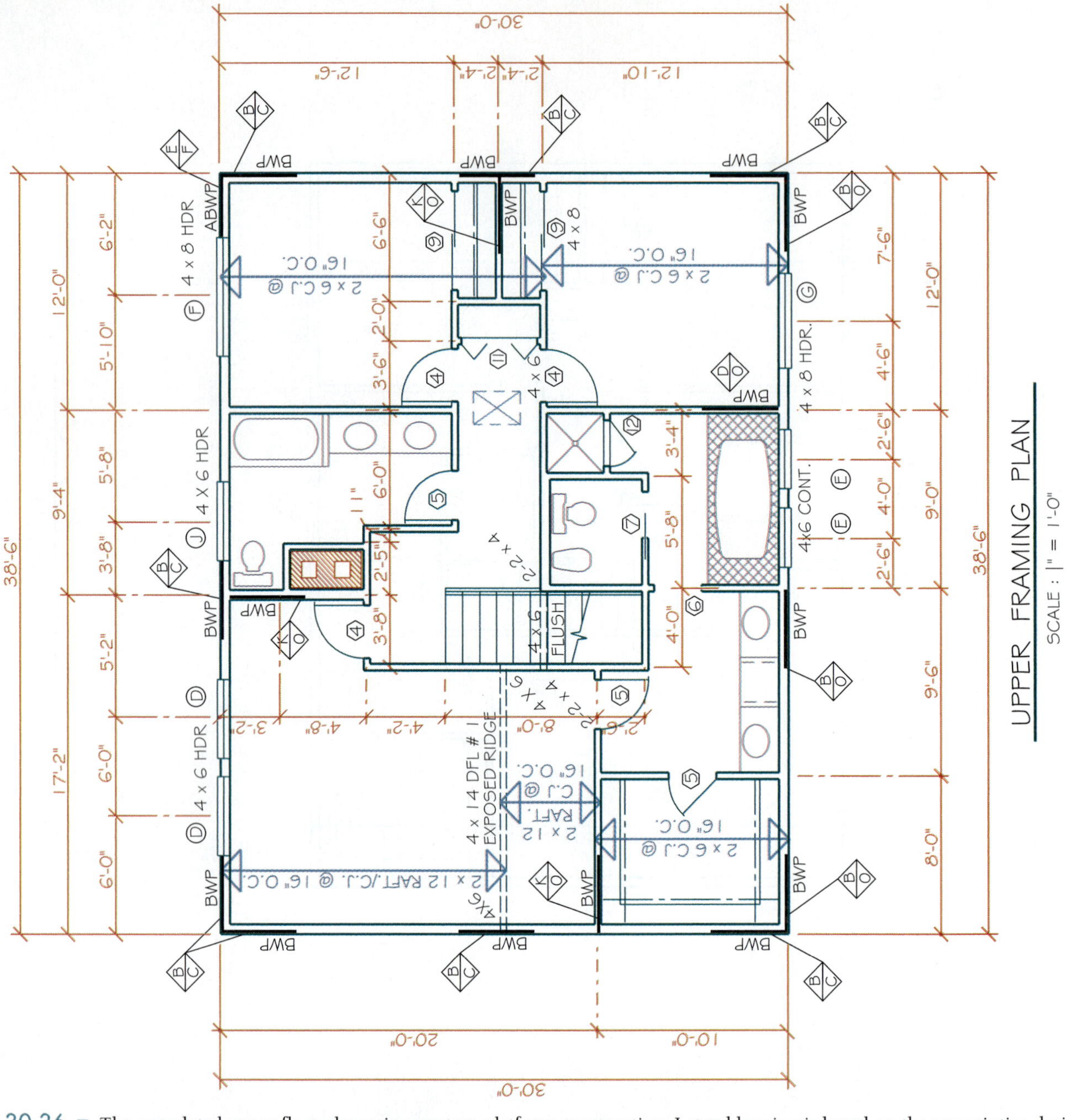

FIGURE 30.36 ■ The completed upper floor plan using western platform construction. Lateral bracing is based on the prescriptive design specific method.

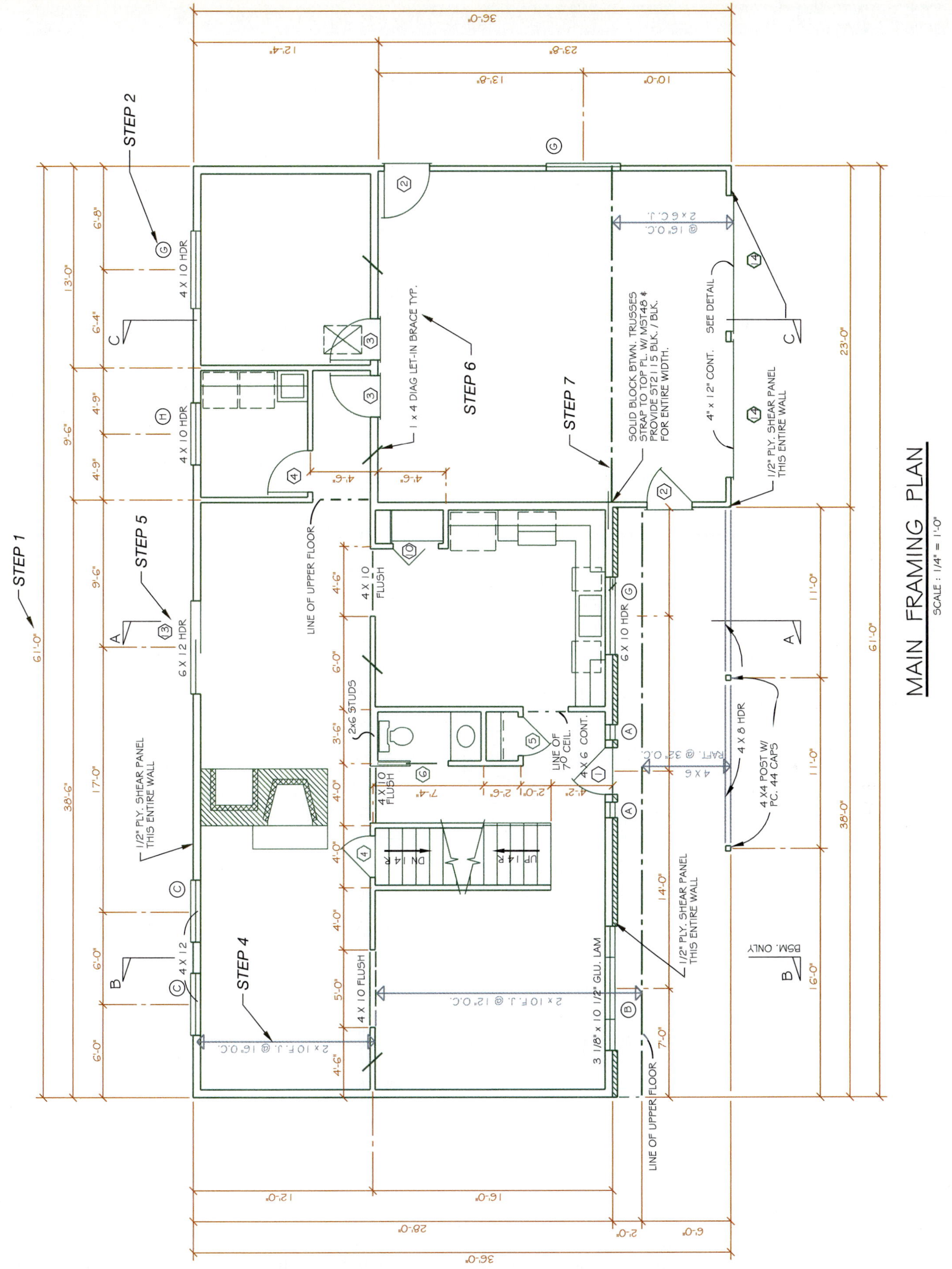

FIGURE 30.37 ■ Main floor framing plan for the residence started in Chapter 16. By separating the structural information from the architectural information, greater clarity is achieved. Lateral bracing is based on the design specific method.

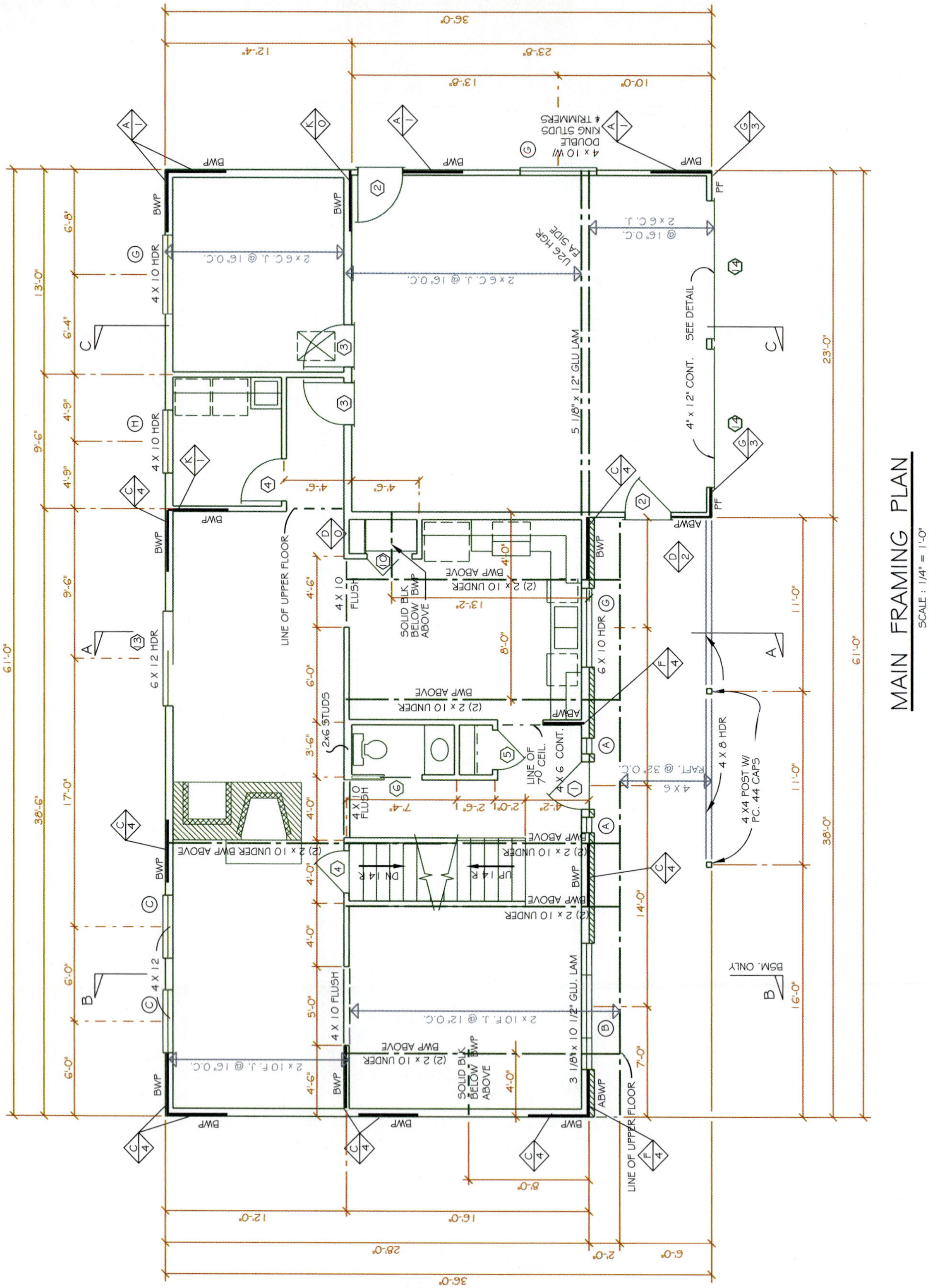

FIGURE 30.38(A) ■ Completed main floor plan using western platform construction for the entire level. Lateral bracing is based on the prescriptive design specific method.

BRACED WALL SCHEDULE

	MARK	WALL COVERING	EDGE NAILING	FIELD NAILING	PANEL SUPPORT	STRUCTURAL CONNECTORS
BWP 1 LEVEL	A	48" MIN. x 1/2" CD EXT. PLY 1 SIDE UNBLOCKED	8d @ 6" O.C	8d @ 12" O.C.	2 - 2 x 6 STUDS @ EA. END OF PANEL	PAHD42 TO CONC.
BWP UPPER LEVEL OF MULTI-LEVEL.	B	48" MIN. x 1/2" CD EXT. PLY 1 SIDE UNBLOCKED	8d @ 6" O.C	8d @ 12" O.C.	2 - 2 x 6 STUDS @ EA. END OF PANEL	MST37 UPPER TO LOWER POST
BWP LOWER LEVEL OF MULTI-LEVEL	C	48" MIN. x 1/2" CD EXT. PLY 2 SIDES UNBLOCKED	8d @ 3" O.C	8d @ 12" O.C.	4 x 6 POST @ EA. END OF PANEL	HD5A TO CONC. THRU DBL SILL
ABWP 1 LEVEL	D	2'-8" MIN. x 1/2" CD EXT. PLY 1 SIDE UNBLOCKED	8d @ 6" O.C	8d @ 12" O.C.	2 - 2 x 6 STUDS @ EA. END OF PANEL	HPAHD22 TO CONC.
ABWP UPPER OF MULTI-LEVEL.	E	2'-8" MIN. x 1/2" CD EXT. PLY 1 SIDE UNBLOCKED	8d @ 6" O.C	8d @ 12" O.C.	4 x 6 POST @ EA. END OF PANEL	MST37 UPPER TO LOWER POST
ABWP LOWER LEVEL OF MULTI-LEVEL	F	2'-8" MIN. x 1/2" CD EXT. PLY 2 SIDES UNBLOCKED	8d @ 3" O.C	8d @ 12" O.C.	2 - 2 x 6 STUDS @ EA. END OF PANEL	HD5A TO CONC. W/ (2) 3/4" M. B. TO POST & STTB20 TO CONCRETE
PORTAL FRAME ONE LEVEL	G	CD EXT. PLY 1 SIDE -BLOCKED EDGE	2 ROWS OF 8d @ 3" O.C	8d @ 12" O.C.	4 x 4 POST @ EA. END OF PANEL	1-MSTA18 STRAP EA. SIDE EA. END OF PANEL TO 4 x 12 MIN HDR. (4 STRAPS MIN EA. END OF FRAME).
PORTAL FRAME UPPER LEVEL OF MULTI-LEVEL	H	22 1/2" MIN. x 1/2" CD EXT. PLY 1 SIDE -BLOCKED EDGE	2 ROWS OF 8d @ 3" O.C	8d @ 12" O.C.	4 x 4 POST @ EA. END OF PANEL	HTT22 UPPER TO LOWER POST W/ (4) 1/2"M. B. EA POST @ 5/8" STUD BOLT THROUGH FLOOR.
PORTAL FRAME LOWER LEVEL OF MULTI-LEVEL.	J	22 1/2" MIN. x 1/2" CD EXT. PLY 2 SIDES -BLOCKED EDGE	2 ROWS OF 8d @ 3" O.C	8d @ 12" O.C.	4 x 4 POST @ EA. END OF PANEL	HD8A TO CONC. W/ (3) 7/8" M. B. TO POST & STTB28 TO CONCRETE
INTERIOR BRACED WALL	K	1/2" x 48" GYPSUM BOARD EACH SIDE OF WALL W/ 5d COOLER NAILS @ 7" O.C. MAXIMUM W/ BLOCKED EDGES PROVIDE 1/2" X 96" PANEL IF ONE SIDE ONLY.				

HOLD - DOWN SCHEDULE

MARK	HOLD - DOWN	CONNECTIONS	FOUNDATION REINFORCING
0	NONE REQUIRED	NONE REQUIRED	DBL. JOIST OR BEAM BELOW. SEE FRAMING/FND. PLAN
1	PAHD42 STRAP	(16) 16d OR (3) 1/2"Ø M.B.	PROVIDE MIN. OF (2) 1/2"Ø A. B.
2	HPAHD22 STRAP	(19) 16d OR (3) 1/2"Ø M.B.	PROVIDE (3) 1/2"Ø A.B. @ 1/5 POINTS OF SILL W/ 1 - #4 3" DN FROM TOP OF WALL & 1 - #4 3" UP FROM BOTTOM OF FOOTING. PROVIDE 1-#4 VERT @24" O.C. IN STEM WALL.
3	HTT22 STRAP & SSTB24 BOLT TO CONCRETE	(32) 16d SINKERS W/ 2" MAX OFFSET FROM POST FACE TO BOLT CENTERLINE	PROVIDE (3) 1/2"Ø A.B. @ MUDSILL W/ 2-2x6 SILL W/ 2 - #4 - 3" UP FROM BOTTOM OF 15" x 7" FTG. EXTEND STEEL 10'-0" AROUND FOOTING CORNER. PROVIDE 1 #4 VERT. @EA. SSTB24 CONNECTOR
4	HD5A	(2) 3/4" M. B. THRU POST	STTB20 W/ MIN. OF (3) A. B. THRU 2 - 2 x 6 SILL
5	HD8A	(4) 7/8" M. B. THRU POST	STTB28 W/ MIN. OF (3) A. B. THRU 2 - 2 x 6 SILL

FIGURE 30.38(B) ■ Braced wall schedule required for the framing plans in Figure 30.36 and 30.38*a*.

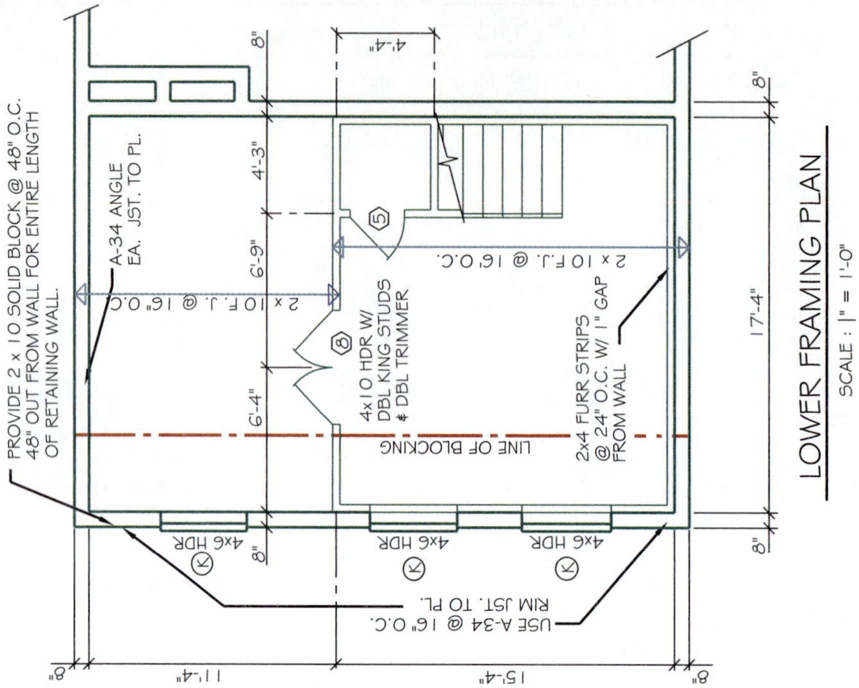

FIGURE 30.39 ■ The structural information for the basement can be shown on a separate framing plan or on the foundation plan, depending on the complexity of the structure. Because the load-bearing walls are concrete, no lateral load information is specified on this framing plan. Methods of attaching the connectors specified in Figure 30.38*a* will be specified on the foundation plan.

CHAPTER 30

Framing Plan Test

DIRECTIONS

Answer the following questions with short complete statements or drawings as needed. Answers should be based on the building code that governs your area. Use the tables in this chapter and Figure 30.36 to answer the following questions.

QUESTIONS

Question 30–1 What are the two major methods of determining lateral loads?

Question 30–2 Where should rafters be specified on the working drawing?

Question 30–3 How are the alternatives for locating trusses different from the alternatives for locating rafters?

Question 30–4 What are the options for locating ceiling joist on the construction drawings?

Question 30–5 You're looking at the main level framing plan of a two-story residence. What framing members would you expect to see specified?

Question 30–6 What is the maximum basic wind speed allowed by your building code for prescriptive wall design?

Question 30–7 An engineer has specified that a metal strap be used to tie a header to a top plate. How should the strap be represented, and what material should be specified on the framing plan?

Question 30–8 What is the maximum wall height allowed if prescriptive method of wall reinforcement is used?

Question 30–9 List the dimensional requirements for a portal frame.

Minimum width:

Minimum distance between panels:

Maximum distance between panels:

Question 30–10 How is a braced wall panel attached to the foundation?

Question 30–11 What foundation reinforcement is required for a portal frame?

Question 30–12 What is the most common method of constructing a braced wall panel in your area?

Question 30–13 What are three possible reactions walls might have as wind is applied perpendicular to the wall?

Question 30–14 What is the maximum distance between braced wall lines for a wind less than 85 mph?

Question 30–15 What nailing is required for interior wall braces using 1/2″ gypsum board?

Question 30–16 What size, method, and spacing are required if two studs will be nailed together to form a post?

Question 30–17 What size nails are required to attach the edge of 1/2″ plywood to trusses?

Question 30–18 What is the maximum distance two bearing walls can be offset and still be considered to be aligned?

Question 30–19 What is the maximum distance two walls can be offset and still considered to be the same braced wall line?

Question 30–20 Describe the requirements for offsets in a braced wall line.

FRAMING PLAN LAYOUT PROBLEM

Use information from the floor, roof plan, and elevations to draw the framing plan for the residence that was started in Chapter 16. Use the span tables in Chapter 28 to determine all rafter sizes and the formulas in Chapter 29 to determine all beam sizes. Verify with your instructor the risk of wind and seismic damage that should be considered in the design.

SECTION 9

Foundation Plans

31 CHAPTER

Foundation Systems

INTRODUCTION

All structures are required to have a foundation. The foundation provides a base to distribute the weight of the structure onto the soil. The weight, or load, must be evenly distributed over enough soil to prevent it from compressing the soil. In addition to resisting the load from gravity, the foundation must resist floods, winds, and earthquakes. Where flooding is a problem, the foundation system must be designed for the possibility that much of the supporting soil may be washed away. The foundation must also be designed to resist any debris that may be carried by floodwaters.

The forces of wind on a structure can cause severe problems for a foundation. The walls of a structure act as a large sail. If the structure is not properly anchored to the foundation, the walls can be ripped away by the wind. Wind tries to push a structure not only sideways but upward as well. Because the structure is securely fastened at each intersection, wind pressure is transferred into the foundation. Proper foundation design will resist this upward movement. Figure 31.1 shows an example of straps used to resist uplift.

Depending on the risk of seismic damage, special design may be required for a foundation. Although earthquakes cause both vertical and horizontal movement, it is the horizontal movement that causes the most damage to structures. The foundation system must be designed so that it can move with the ground yet keep its basic shape. Steel reinforcing and welded wire mesh are often required to help resist or minimize damage due to the movement of the earth.

FIGURE 31.1 ■ In addition to resisting the forces of gravity, a foundation must be able to resist the forces of uplift created by wind pressure acting on the structure. Anchor bolts and metal framing straps embedded in the concrete are two common methods of resisting the forces of nature. *Courtesy Tereasa Jefferis.*

SOIL CONSIDERATIONS

In addition to the forces of nature, the nature of the soil supporting the foundation must also be considered. The texture of the soil and the tendency of the soil to freeze will influence the design of the foundation system.

Soil Texture

The texture of the soil will affect its ability to resist the load of the foundation. Before a foundation can be designed for a structure, the bearing capacity of the soil must be known. This is a design value specifying the amount of weight a square foot of soil can support. The bearing capacity of soil depends on its composition and the moisture content. The International Residential Code provides five basic classifications for soil:

Crystalline bedrock	12,000 psf (574.8 kPa)
Sedimentary and foliated rock	4,000 psf (191.6 kPa)
Sandy gravel and/or gravel (SG and GP)	3,000 psf (143.7 kPa)
Sand, silty sand, clayey sand, silty gravel, and clayey gravel (SW, SP, SM, SC, GM and GC)	2,000 psf (95.8 kPa)
Clay, sandy clay, silty clay, clayey silt, silt, sandy silt (CI, ML, MH and CH)	1,500 psf (47.9 kPa)

The allowable bearing capacity will need to be determined by a soils engineer if the building department believes soil with an allowable bearing of less than 1,500 psf is likely to be present. A soil bearing pressure of 2,000 psf is the design value used for most stock home designs when the site conditions are not known.

Structures built on soils with low bearing capacity require footings that extend into stable soil or are spread over a wide area. Both options typically require the design to be approved by an engineer.

In residential construction, the type of soil can often be determined from the local building department. In commercial

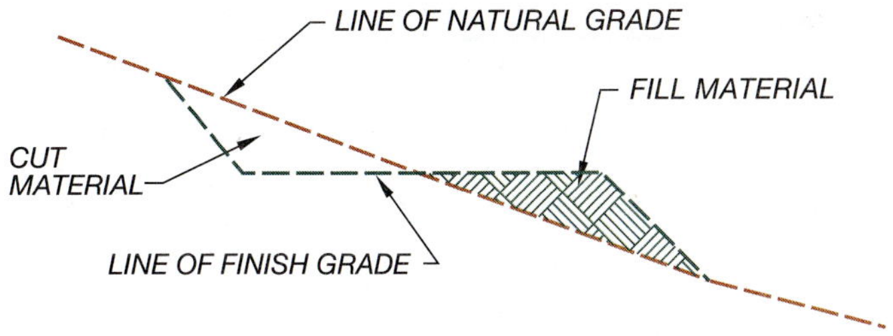

FIGURE 31.2 ■ As access roads are created, soil is often cut away and pushed to the side of the roadway, creating areas of fill. Unless the foundation extends into the natural grade, the structure will settle.

construction, a soils engineer is usually required to study the various types of soil at the job site and make recommendations for foundation design. The soil bearing values must be determined before a suitable material for the foundation can be selected. In addition to the texture, the tendency of freezing must also be considered.

The five common classifications are used to define natural, undisturbed soil; construction sites, however, often include soil that has been brought to the site or moved on the site. Soil that is placed over the natural grade is called *fill material*. Fill material is often moved to a lot when an access road is placed in a sloping site, as in Figure 31.2. After a few years, vegetation covers the soil and gives it the appearance of natural grade. Footings resting on fill material will eventually settle under the weight of a structure. All discussion of foundation depths in this text refers to the footing depth into the natural grade.

Compaction

Fill material can be compacted to increase its bearing capacity. Compaction is typically accomplished by vibrating, tamping, rolling, or adding a temporary weight. There are three major ways to compact soil.

1. Static force: A heavy roller presses soil particles together.
2. Impact forces: A ramming shoe strikes the ground repeatedly at high speed.
3. Vibration: High-frequency vibration is applied to soil through a steel plate.

Large job sites are typically compacted by mobile equipment. Small areas in the construction site are generally compacted by handheld mobile equipment. Granular soils are best compacted by vibration, and soils containing large amounts of clay are best compacted by force. Each of these methods will reduce air voids between grains of soil. Proper compaction lessens the effect of settling and increases the stability of the soil, which increases the load-bearing capacity. The effects of frost damage are minimized in compacted soil because penetration of water into voids in the soil is minimized.

Moisture content is the most important factor in efficient soil compaction because moisture acts as a lubricant to help soil particles move closer together. Compaction should be completed under the supervision of a geotechnical engineer or another qualified expert who understands the measurements of soil moisture. Compaction is typically accomplished in lifts of from 6″ through 12″ (150–300 mm). The soils or geotechnical engineer will specify the requirements for soils excavation and compaction. The drafter's job is to place the specifications of the engineer clearly on the plans.

Freezing

Don't confuse ground freezing with blizzards. Even in the warmer southern states, ground freezing can be a problem. Figure 31.3 shows the average frost penetration depths for the United States. A foundation should be built to a depth where the ground is not subject to freezing. Water in the soil expands as it freezes and then contracts as it thaws. Expansion and shrinking of the soil will cause heaving in the foundation. As the soil expands, the foundation can crack. As the soil thaws, water that cannot be absorbed by the soil can cause the soil to lose much of its bearing capacity, causing further cracking of the foundation. In addition to geographic location, the type of soil also affects freezing. Fine-grained soil is more susceptible to freezing because it tends to hold moisture.

A foundation must rest on stable soil so that the foundation does not crack. The designer will have to verify the required depth of foundations with the local building department. Once the soil bearing capacity and the depth of freezing are known, the type of foundation system to be used can be determined.

Water Content

The amount of water the foundation will be exposed to, as well as the permeability of the soil, must also be considered in the design of the foundation. As the soil absorbs water, it expands, causing the foundation to heave. In areas of the country with extended periods of rainfall, there is little variation in the soil moisture content; this minimizes the risk of heaving caused by soil expansion. Greater danger results from shrinkage during periods of decreased rainfall. In areas of the country that receive only minimal rainfall, followed by extended periods without moisture, soil shrinkage can cause severe foundation problems because of the greater moisture differential. To aid in the design of the foundation, each of the model code has included the Thornthwaite Moisture Index, which lists the amount of water that would be returned to the atmosphere by evaporation from the ground surface and transpiration if there was an unlimited supply of water to the plants and soil.

Concrete slab on grade is used primarily in dry areas from southern California to Texas, where the contrast between the dry soil under the slab and the damp soil beside the foundation creates a risk of heaving at the edge of the slab. If the soil be-

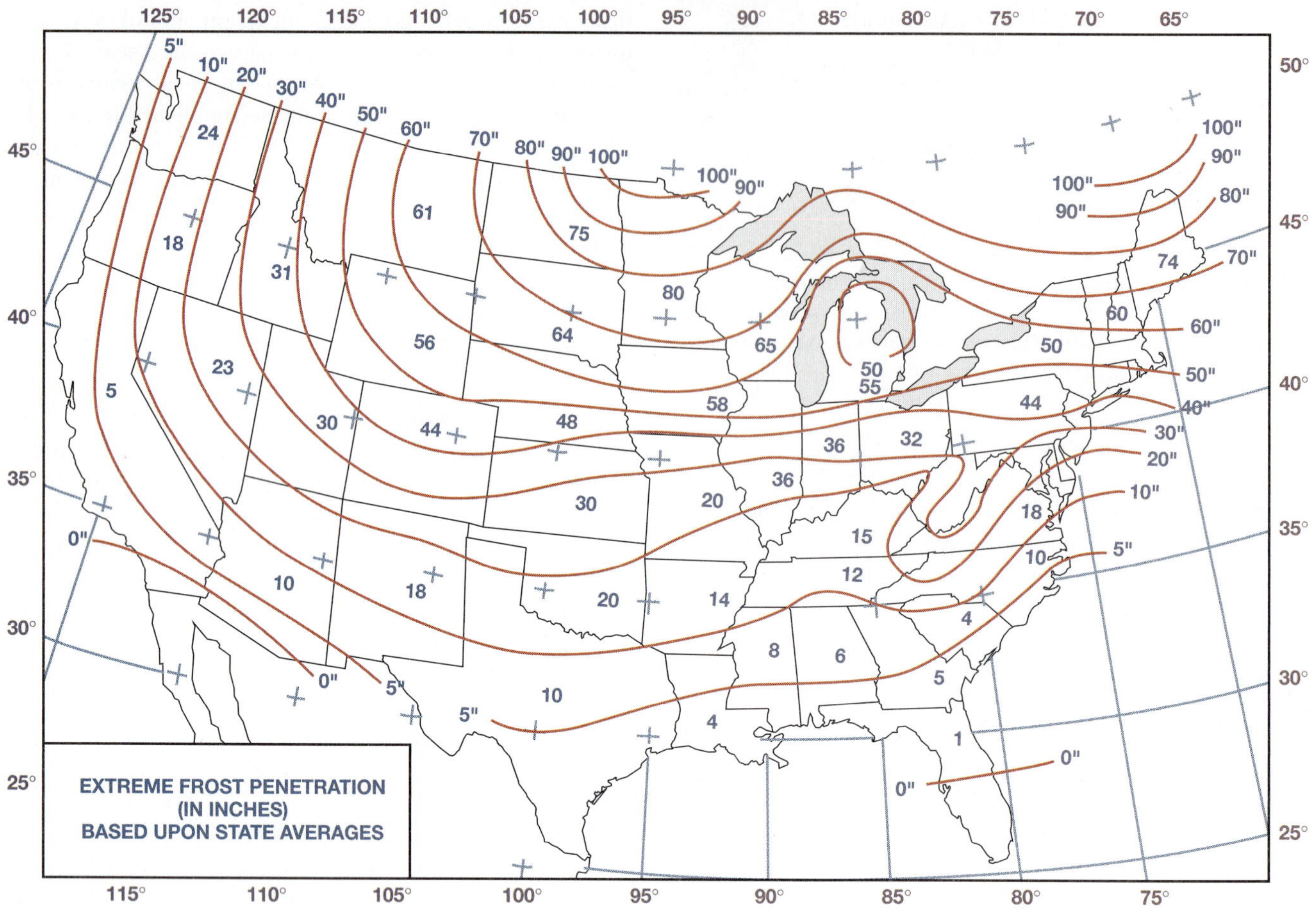

FIGURE 31.3 ■ Depth of freezing can greatly affect the type of foundation used for structure.

neath the slab experiences a change of moisture content after the slab is poured, the center of the slab can heave. Heaving can be resisted by proper drainage and reinforcing placed in the foundation and throughout the floor slab. The effects of soil moisture will be investigated in Chapter 33, which considers concrete slab construction.

Surface water and groundwater must be properly diverted from the foundation so that the soil can support the building load. Proper drainage also minimizes water leaks into the crawl space or basement, reducing mildew and rotting. Most codes require the finish grade to slope away from the foundation at a minimum slope of 6″ (152 mm) within the first 10′ (3048 mm). A 3 percent minimum slope is preferable for planted or grassy areas; a 1 percent slope is acceptable for paved areas.

Gravel or coarse-grained soils can be placed beside the foundation to increase percolation. In damp climates, a drain is often required beside the foundation at the base of a gravel bed to facilitate drainage. See Figure 31.4. As the amount of water in the soil surrounding the foundation is reduced, the lateral loads imposed on the foundation are also reduced. The pressure of the soil becomes increasingly important as the height of the foundation wall is increased. Foundation walls enclosing basements should be waterproofed, as in Figure 31.5. Asphaltic emulsion is often used to prevent water from penetration into the basement. Floor slabs below grade are required to be placed over a vapor barrier.

FIGURE 31.4 ■ A gravel bed and a drain divert water away from the foundation. *Courtesy Boccia Inc.*

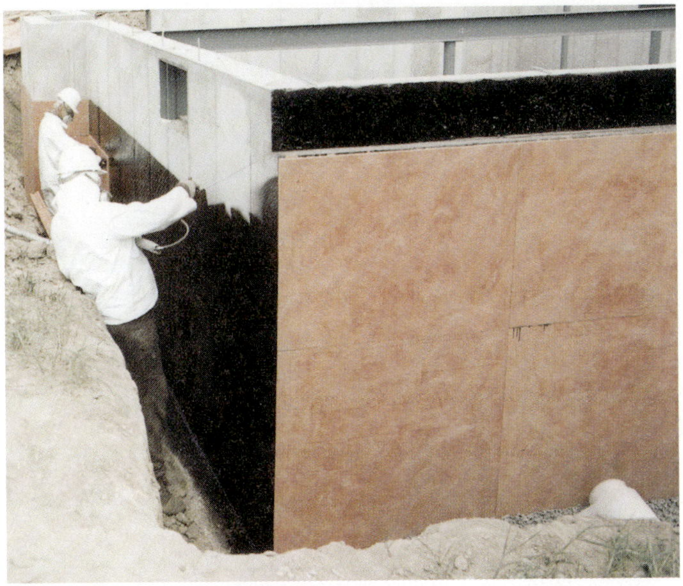

FIGURE 31.5 ■ A waterproof emulsion will reduce the risk of water penetrating the concrete wall. *Courtesy Boccia Inc.*

Radon

Structures built in areas of the country with high radon levels need to provide protection from this cancer-causing gas. The IRC and the EPA (Environmental Protection Agency) have mapped the continental United States by county and identified high-risk areas to radon. The buildup of radon can be reduced by making minor modifications to the gravel placed below basement slabs. A 4″ (100 mm) PVC vent can be placed in a minimum layer of 4″ (100 mm) gravel covered with 6 mil polyethylene. The plastic barrier should have a minimum lap of 12″ (300 mm) at intersections. Any joints, cracks, or penetrations in the floor slab must be caulked. The vent must run under the slab until it can be routed up through the framing system to an exhaust point, which is a minimum of 10′ (3000 mm) from other openings in the structure and 12″ (300 mm) above the roof. The system should include rough-in electrical wiring for future installation of a fan located in the vent stack and a system-failure warning device. This can usually be accomplished by placing an electrical junction box in an attic space for future installation of a fan.

TYPES OF FOUNDATIONS

The foundation is usually constructed of pilings, continuous footings, or grade beams.

Pilings

A *piling* foundation system uses beams placed between vertical supports, called a piling, to support structural loads. The columns may extend into the natural grade or may be supported on other material that extends into stable soil. Piling foundations are typically used:

- On steep hillside sites where it may not be feasible to use traditional excavating equipment.
- Where the load imposed by the structure exceeds the bearing capacity of the soil.
- On sites subject to flooding or other natural forces that might cause large amounts of soil to be removed.

Coastal property and sites near inland bodies of water subject to flooding often use a piling foundation to keep the habitable space of the structure above the floodplain level. Typically, a support beam is placed under or near each bearing wall. Beams are supported on a grid of vertical supports, which extend down to a more stable stratum of rock or dense soil. Beams can be steel, sawn or laminated wood, or prestressed concrete. Vertical supports may be concrete columns, steel tubes or beams, wood columns, or a combination of these materials.

On shallow pilings a hole can be bored, and poured concrete with steel reinforcing can be used. Figure 31.6 shows a detail for a poured-concrete footing. If the vertical support is required to extend deeper than 10′ (3000 mm), a pressure-treated wood timber or steel column can be driven into the soil. Figure 31.7 shows the layout of a residential floor

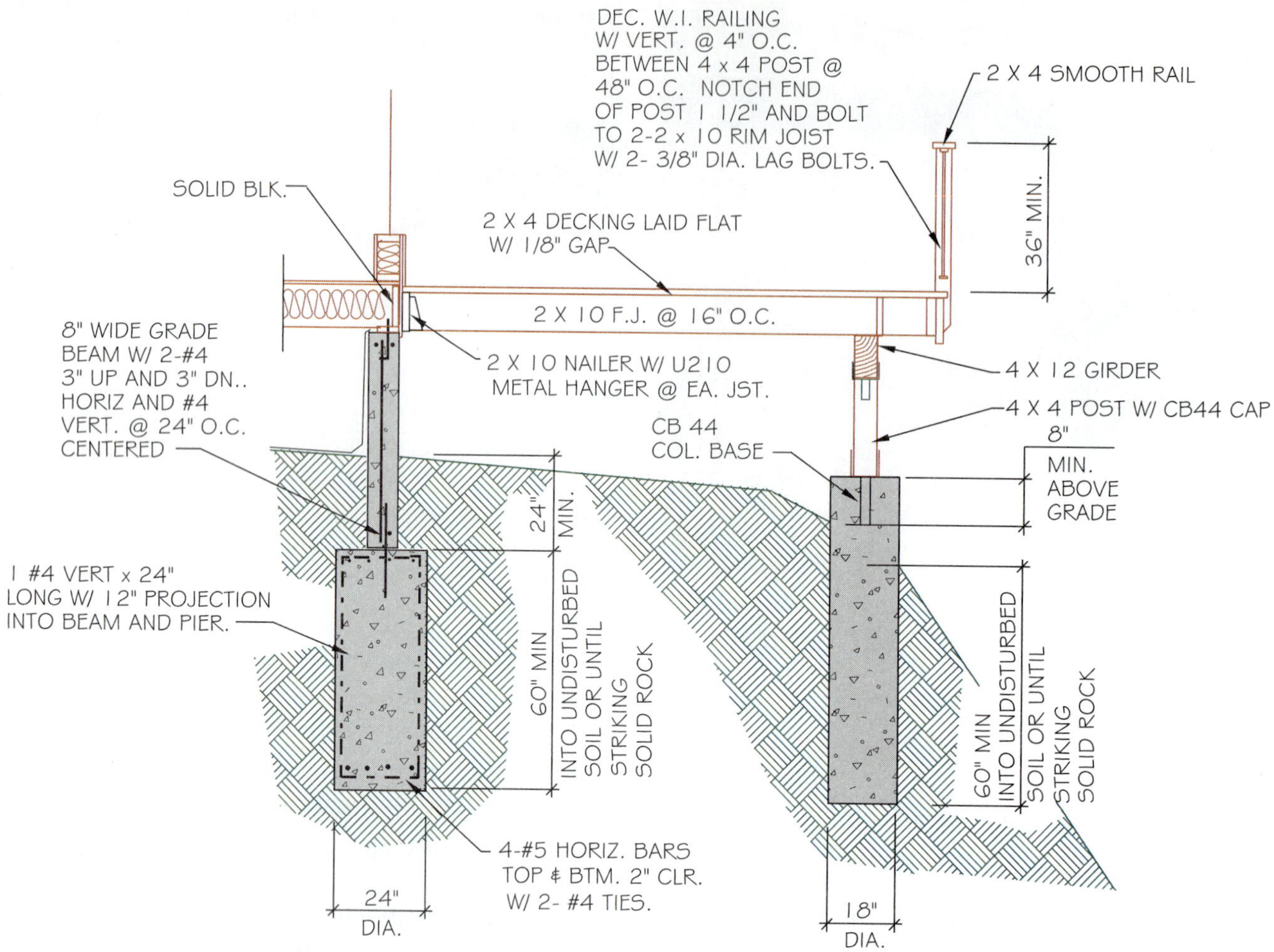

FIGURE 31.6 ■ Shallow pilings are often made of concrete with steel reinforcement. The size, type, and placement of the reinforcement will vary, depending on the load to be supported, the depth of the piling, and the engineer's design. The piling on the left is being used to support a grade beam, which is a concrete beam placed below the finished grade to span between pilings. The piling on the right is used to support wood post.

FIGURE 31.7 ■ A piling foundation is often used when a job site is too steep for traditional foundation methods. *Courtesy Sara Jefferis.*

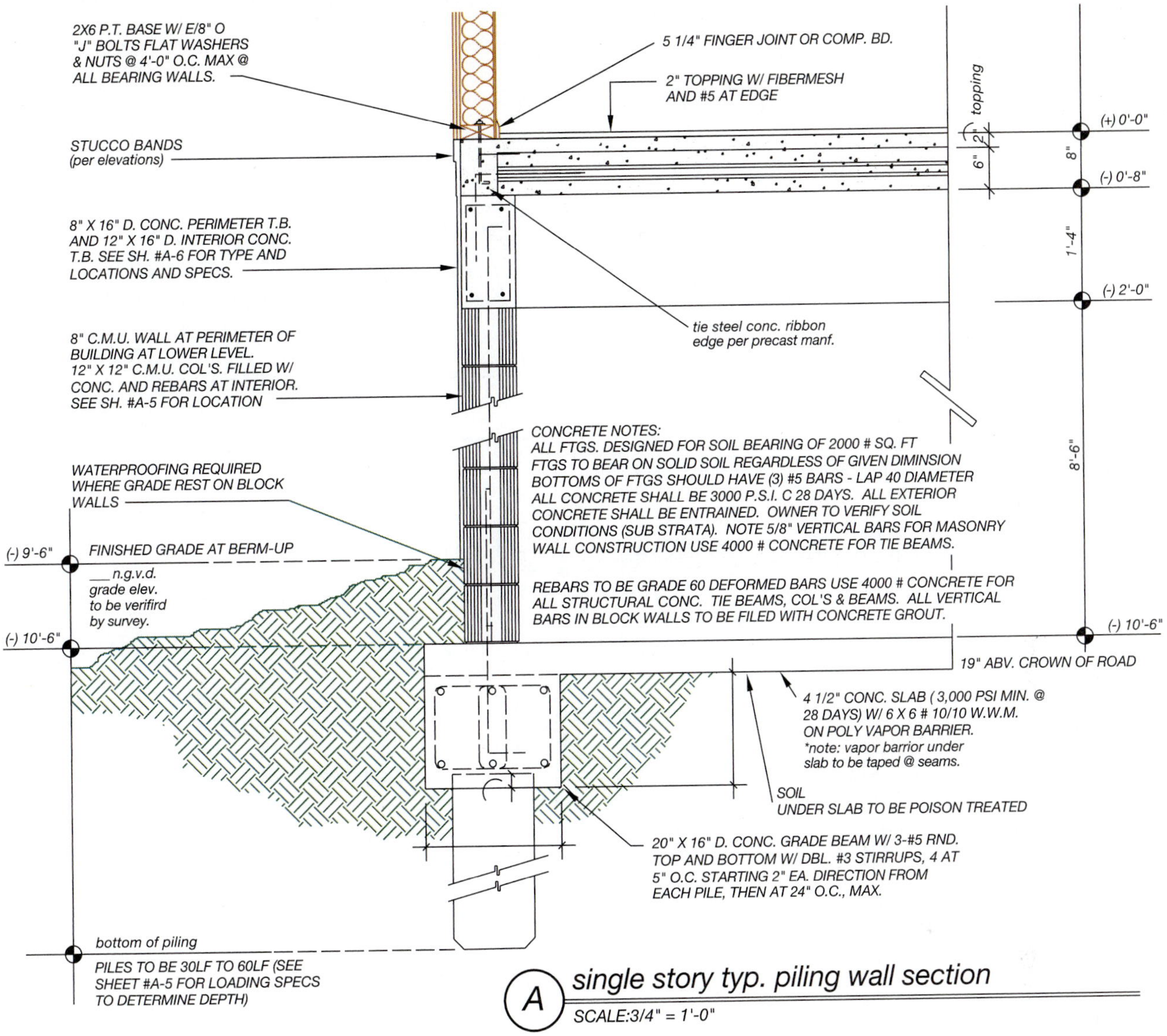

FIGURE 31.8 ■ Detail required to explain construction of a concrete piling and concrete floor intersection. *Courtesy Eric S. Brown Design Group.*

supported on a piling foundation. Figure 31.8 shows a concrete piling and vertical supports used to support a concrete floor system. Figure 31.9 shows the components of a piling plan for residence. In addition to the vertical columns and horizontal beams, diagonal steel cables or rods are placed between the columns to resist lateral and rotational forces. An engineer must design the piling and the connection of the pilings to the superstructure.

In addition to resisting gravity loads, a piling foundation must be able to resist forces from uplift, lateral force, and rotation. Figure 31.10 shows a detail of a concrete piling used to support steel columns, which in turn support the floor system of the residence; and a detail of a steel piling foundation, which is used to support steel columns above grade. Notice that the engineer has specified a system that uses braces approximately parallel to the ground to stabilize the top of the pilings against lateral loads. These braces would not be shown on the foundation plan but would be specified on the elevations, sections, and details.

Continuous or Spread Foundations

The most typical type of foundation used in residential and light commercial construction is a *continuous* or *spread foundation*. This consists of a footing and wall. The footing is the base of the foundation system and is used to displace the

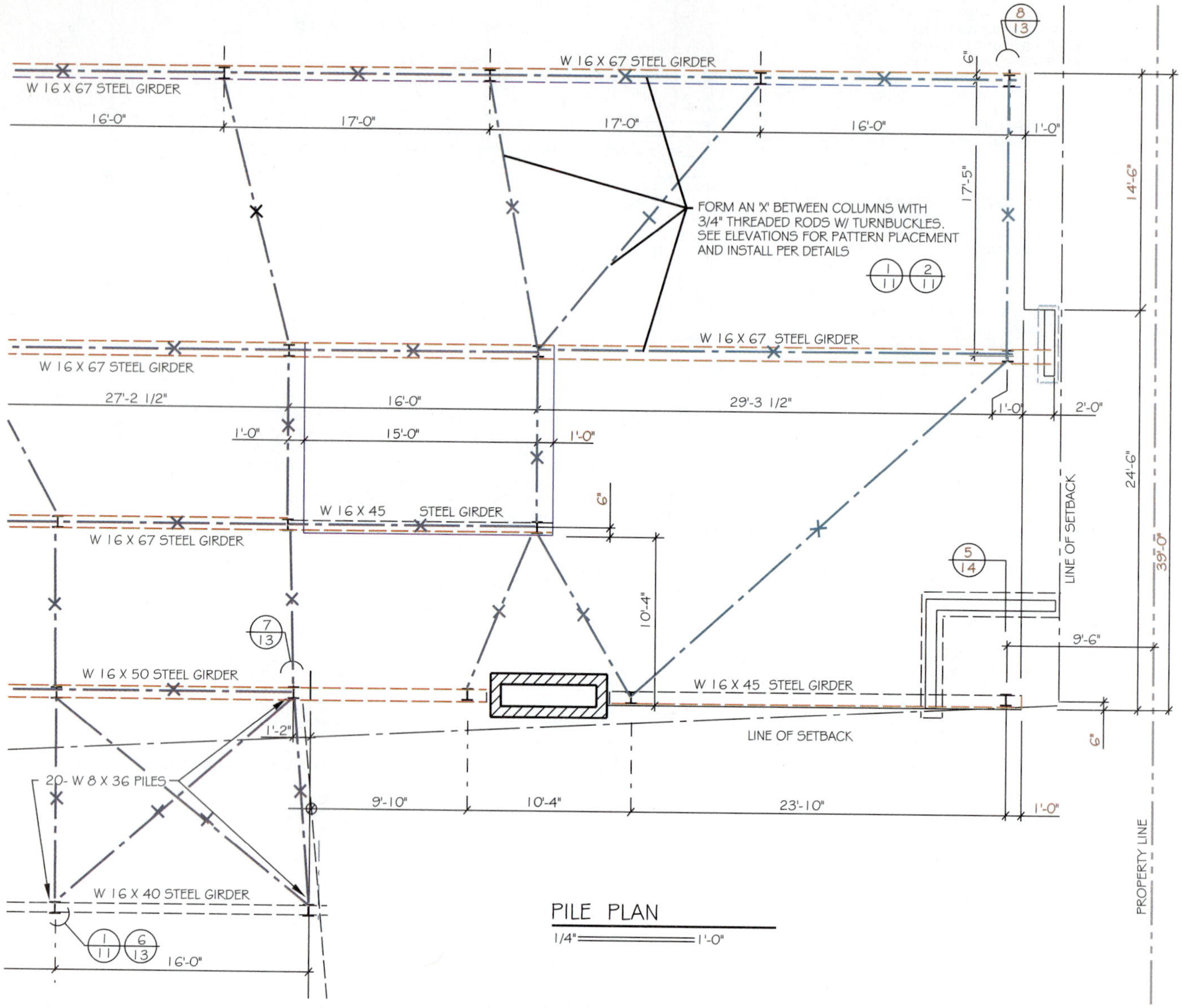

FIGURE 31.9 ■ The foundation plan for a structure supported on pilings shows the location of all pilings, the beams that span between the pilings, and any diagonal bracing required to resist lateral loads. *Courtesy Residential Designs.*

building loads over the soil. Figure 31.11 shows typical footings and how they are usually drawn on foundation plans. Footings are made of poured concrete and placed so that they extend below freezing level. Fully grouted masonry and wood foundations are also allowed by code. The size of the footing is based on the soil bearing value and the load to be supported. Figure 31.12 shows common footing sizes and depths required by the IRC. The strength of the concrete must also be specified; this is based on the location of the concrete in the structure and its chance of freezing. Figure 31.13 lists typical values.

For areas of soft soil or fill material, reinforcement steel is placed in the footing. Concrete is extremely durable when it supports a load and is compressed but very weak under tension.

If the footing is resting on fill material, the bottom of the concrete footing will bend. As the footing bends, the concrete will be under tension. Steel is placed near the bottom of the footing to resist the forces of tension in concrete. This reinforcing steel, or *rebar,* is not shown on the foundation plan but is specified in a note giving the size, quantity, and spacing of the steel. The size of rebar is represented by a number indicating

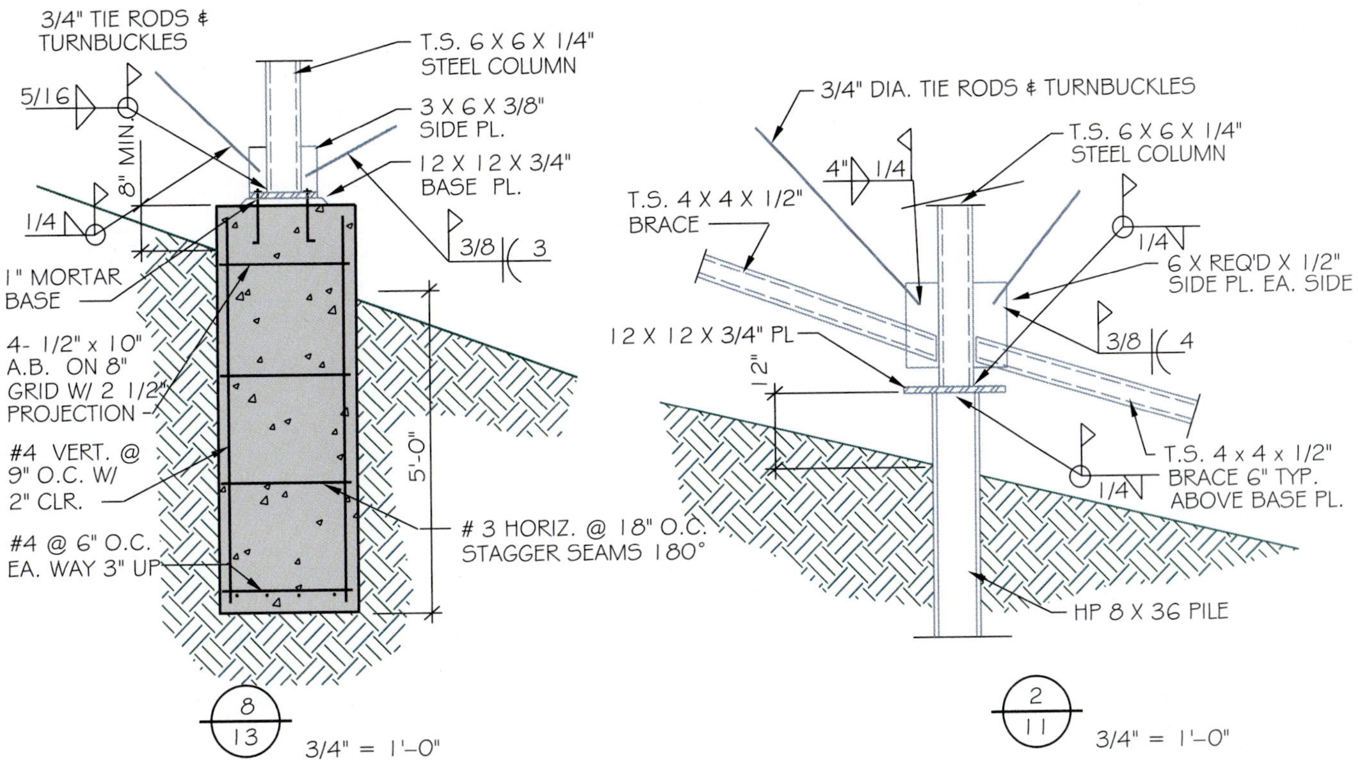

FIGURE 31.10 ■ The drafter draws details of the pilings to supplement the foundation plan to ensure that all of the engineer's specifications can be clearly understood. (a) The concrete piling on the left supports a steel column, (b) the piling on the right supports a steel column that extends up to the superstructure of the residence. Because of the loads to be supported, the depth of the pilings, and a severe risk of seismic damage, the engineer has specified horizontal steel tube bracing to the top of the pilings. *Courtesy Residential Designs.*

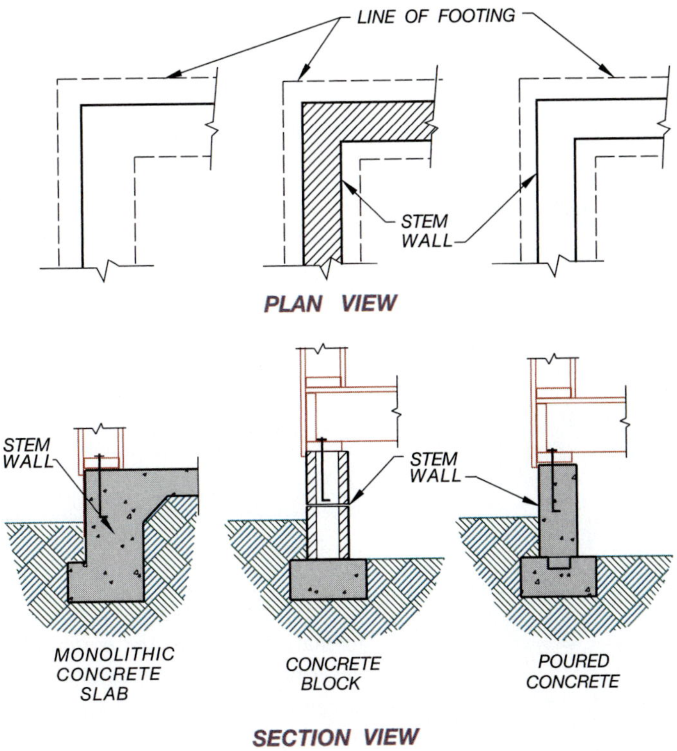

FIGURE 31.11 ■ A footing, used to spread building loads evenly into the soil, is represented on the foundation plan by hidden lines.

MINIMUM WIDTH OF CONCRETE OR MASONRY FOOTINGS (INCHES) (A)

LOAD-BEARING VALUE OF SOIL (PSF)	1,500	2,000	2,500	3,000	3,500	4,000
Conventional wood frame construction						
1-story	16	12	10	8	7	6
2-story	19	15	12	10	8	7
3-story	22	17	14	11	10	9
4-inch brick veneer over wood frame or 8-inch hollow concrete masonry						
1-story	19	15	12	10	8	7
2-story	25	19	15	13	11	10
3-story	31	23	19	16	13	12
8-inch solid or fully grouted masonry						
1-story	22	17	13	11	10	9
2-story	31	23	19	16	13	12
3-story	40	30	24	20	17	15

For SI: 1 inch = 25.4 mm, 1 pound per square foot = 47.88 Pa.

MINIMUM WIDTH OF STEM WALL (B)

Plain Concrete	Minimum Width
Walls less than 4′–6″ (1372 mm)	6″ (152 mm)
Walls greater than 4′–6″ (1372 mm)	7.5″ (191 mm)
Plain Masonry	
Solid grout or solid units	6″ (152 mm)
Non grouted units	8″ (203 mm)

MINIMUM FOOTING DEPTH (C)*

Minimum 1-story	6″ (152 mm)
Minimum 2-story	7″ (178 mm)
Minimum 3-story	8″ (203 mm)

*Values vary based on the type of soil. Verify local requirements with your building department.

MINIMUM FOOTING DEPTH INTO NATURAL GRADE (D)*

Minimum code value	12″ (305 mm)
Recommended 2-story	18″ (457 mm)
Recommended 3-story	24″ (610 mm)

*Values vary based on the frost depth and the type of soil. Verify local requirements with your building department.

(a)

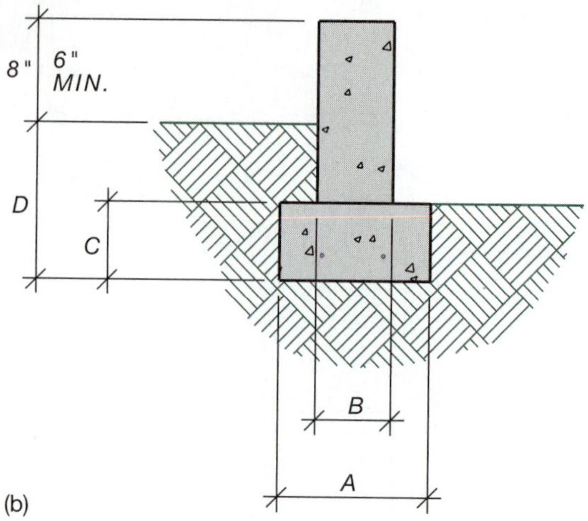

(b)

FIGURE 31.12 ■ Minimum footing requirements based on the International Residential Code / 2000. *Copyright @ 2000, Courtesy International Code Council, Inc.*

the approximate diameter of the bar in eighths of an inch. Sizes typically range from 1/4″ to 2 1/2″ diameter. A 1/4″ diameter rebar is 2/8″ or a #2 diameter; a 1/2″ diameter bar is a #4. IRC allows only rebar that is a #4 diameter or larger to be considered as reinforcing with the use of smaller steel considered as non-reinforced concrete. Steel sizes and numbers and their metric equivalent are:

REBAR SIZES

IMPERIAL SIZES		METRIC REBAR	
#3	3/8″	#10M	9.5 mm
#4	1/2″	#13M	12.7 mm
#5	5/8″	#16M	15.9 mm
#6	3/4″	#19M	19.1 mm
#7	7/8″	#22M	22.2 mm
#8	1″	#25M	25.4 mm
#9	1 1/8″	#29M	28.7 mm
#10	1 1/4″	#32M	32.3 mm
#11	1 7/16″	#36M	35.8 mm
#14	1 5/8″	#43M	43 mm
#18	2 1/2″	#57M	57.3 mm

Steel rebar may be smooth, but most uses of steel require deformations to help the concrete bind to the steel. Typical steel deformation patterns can be seen in Figure 31.14. Placed between the deformations are the identification marks of the steel manufacturer. The first letter identifies the mill that produced the rebar, followed by the bar size and the type of steel. The strength of the steel used for rebar can also vary. Common grades associated with residential construction are grades 40,

TYPE OR LOCATION OF CONCRETE	MINIMUM COMPREHENSIVE STRENGTH WEATHER POTENTIAL		
	NEGLIGIBLE	MODERATE	SEVERE
Basement walls and foundations not exposed to weather	2500	2500	2500
Basement slabs and interior slabs on grade, except garage floor slabs	2500	2500	2500
Basement walls, foundation walls, exterior walls, and other vertical concrete work exposed to weather	2500	3000	3000
Porches, concrete slabs and steps exposed to the weather, and garage floor slabs	2500	3000	3500

FIGURE 31.13 ■ Compressive strength of concrete, based on the International Residential Code /2000. *Copyright © 2000, Courtesy International Code Council, Inc.*

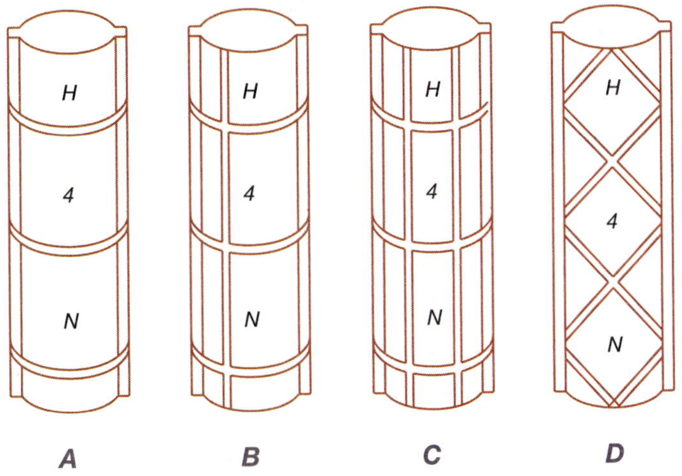

FIGURE 31.14 ■ Concrete construction is often strengthened by adding steel reinforcement to the face of the concrete member that is in tension. To increase the capacity of the concrete to bond to the steel, ribs or deformations are added to the bar. Between the deformations, identification marks are imprinted to designate the mill that produced the steel, the bar size, and the type of steel.

50, 60, and 75. Figure 31.15*a* shows the forms set for a reinforced footing, and Figure 31.15*b* shows the completed footing.

The material used to construct the foundation wall and the area in which the building is to be located will affect how the wall and footing are tied together. If the wall and footing are made at different times, a #4 bar must be placed at the top of the stem wall and a #4 bar must be placed near the bottom of the footing. A concrete slab resting on a footing must have a #4 bar at the top and bottom of the footing. The foundation wall and footing can

FIGURE 31.15(A) ■ Forms set in preparation for pouring the concrete for the footing. Horizontal and vertical steel has been placed to tie the stem wall the footing. *Courtesy of Tom Worcester.*

FIGURE 31.15(B) ■ Once the forms are removed from the footings, the forms can be set to place the stem wall. *Courtesy of Tom Worcester.*

also be strengthened by placing a *keyway* in the footing. The keyway is formed by placing a 2 × 4 (50 × 100 mm) in the top of the concrete footing while the concrete is still wet. Once the concrete has set, the 2 × 4 (50 × 100 mm) is removed, leaving a keyway in the concrete. When the concrete for the wall is poured, it will form a key by filling in the keyway. If a stronger bond is desired, steel is often used to tie the footing to the foundation wall. Both methods of attaching the foundation wall to the footing can be seen in Figure 31.16. Footing steel is not drawn on the foundation plan but is specified in a general note and shown in a footing detail similar to Figure 31.17.

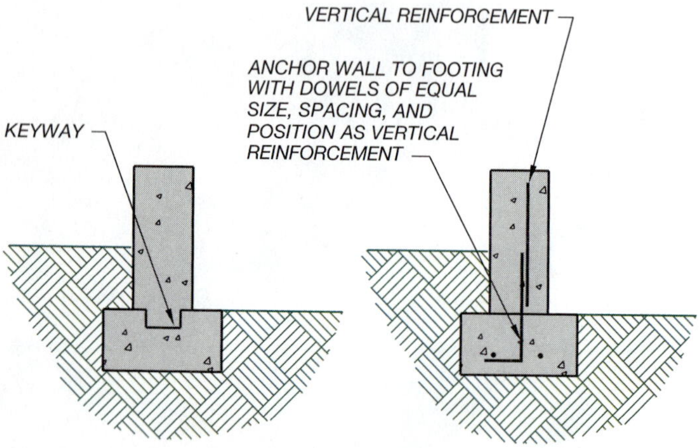

FIGURE 31.16 ■ The footing can be bonded to the foundation wall with a key or with steel.

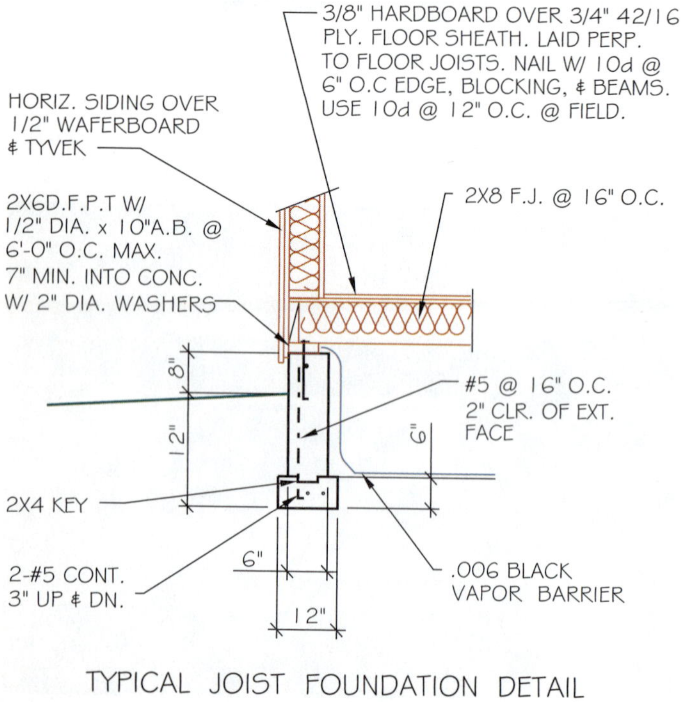

FIGURE 31.17 ■ When steel is required in the foundation, it is usually specified in a general note on the foundation plan and a detail.

Grade Beams

To provide added support for a foundation in unstable soil, a grade beam may be used in place of the foundation, as in Figure 31.6. The grade beam is similar to a wood beam that supports loads over a window. The grade beam is placed under the soil below the stem wall and spans between stable supports. The support may be stable soil or pilings. The depth and reinforcing required for a grade beam are determined by the load to be supported and are sized by an architect or engineer. A grade beam resembles a footing when drawn on a foundation plan. Steel reinforcing may be specified by notes and referenced to details rather than on the foundation plan.

Fireplace Footings

A masonry fireplace will need to be supported on a footing. IRC requires the footing to be a minimum of 12″ deep and extend 6″ past the face of the fireplace on each side. Figure 31.18 shows how fireplace footings are shown on the foundation plan.

Veneer Footings

If masonry veneer is used, the footing must be wide enough to provide adequate support for the veneer. Depending on the type

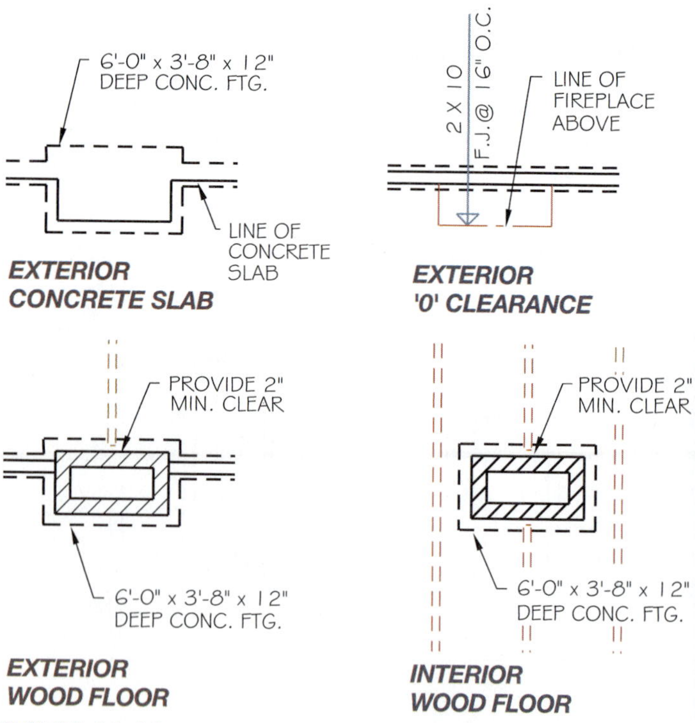

FIGURE 31.18 ■ A masonry fireplace is required to have a 12″ (300 mm) deep footing that extends 6″ (150 mm) past the face of the fireplace. A wood stove, a zero-clearance fireplace, or a gas fireplace is not required to have a footing, but the outline of the unit should be represented on the foundation plan if it extends beyond the foundation.

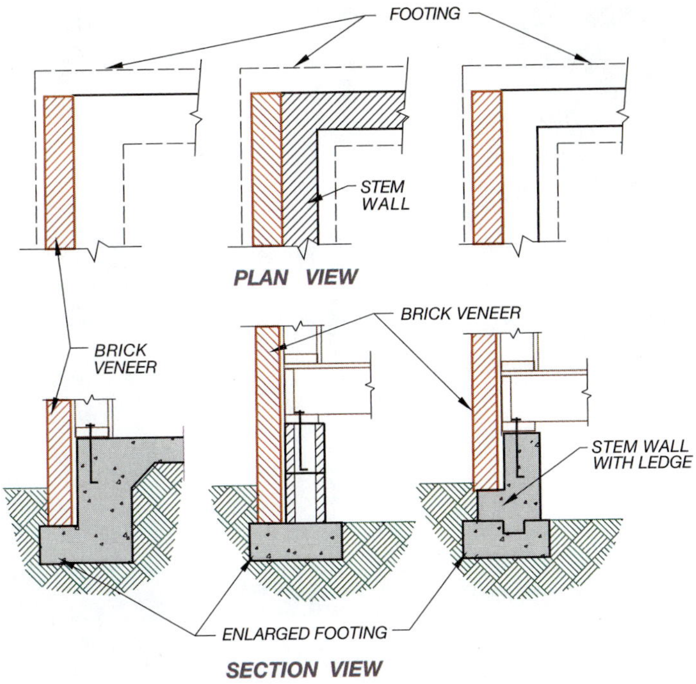

FIGURE 31.19 ■ When a masonry veneer is added to a wall, additional footing width is needed to provide support.

of veneer material, the footing is widened. Figure 31.19 shows common methods of providing footing support for veneer.

Foundation Wall

The foundation wall is the vertical wall that extends from the top of the footing up to the first-floor level of the structure, as shown in Figures 31.11, 31.16, and 31.19. The foundation wall is usually centered on the footing to help spread the loads being supported. The height of the wall must extend 6″ above the ground although 8″ (204 mm) is recommended. The height can be reduced to 4″ if masonry veneer is used. This height reflects the minimum distance that is required between wood-framing members and the grade. A width of 6″ (152 mm) is standard for plain concrete, and 8″ (203 mm) for plain masonry walls. The required width of the wall varies depending on the wall height and the type of soil. Figure 31.12 shows common wall dimensions. Common methods of forming the footing and stem wall can be seen in Figure 31.20. Figure 31.21 shows the forms for a poured concrete stem wall. Figure 31.22 shows the completed stem wall.

In addition to concrete block and poured concrete foundation walls, many companies are developing alternative methods of forming the stem wall. Blocks made of expanded polystyrene foam (EPS) or other lightweight materials can be stacked into the desired position and fit together with interlocking teeth. EPS block forms can be assembled in a much shorter time than traditional form work and remain in place to become part of the finished wall. Reinforcing steel can be set inside the block forms in patterns similar to traditional block walls. Once the forms are assembled, concrete can be pumped into the forms using any of the common methods of pouring. The finished wall has an R-value between R-22 and R-35, depending on the manufacturer. Figure 31.23 shows a foundation being built using EPS forms.

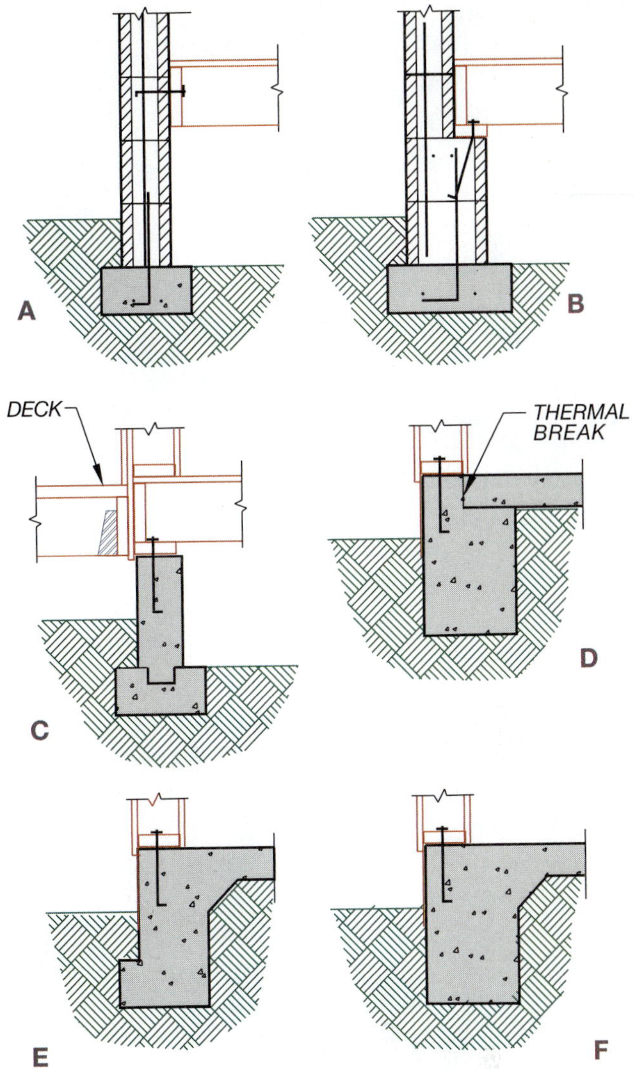

FIGURE 31.20 ■ Common methods of forming stem walls. (A) Concrete masonry units, with a pressure-treated ledger to support wood floor joist. (B) Floor joists supported on a pressure-treated sill with concrete masonry units. (C) Floor joists supported by pressure-treated sill with joist supporting a wood deck hung from a ledger. (D) Concrete floor slab supported on foundation with an isolation joint between the stem wall and slab to help the slab maintain heat. (E) Stem wall with projected footing. (F) Stem wall and foundation of equal width. Although more concrete is used, it can be formed quickly, saving time and material.

FIGURE 31.21 ■ Once the concrete is poured, anchor bolts are placed in preparation for placement of the wood sill. Once the floor has been placed, the soil can be placed against the stem wall to form the finish grade. *Courtesy Richard Schmitke.*

FIGURE 31.22 ■ With the wood sill held in place by the anchor bolts, framing of the floor can be started. *Courtesy Barbara Tooley.*

The top of the foundation wall must be level. When the building lot is not level, the foundation wall is often stepped. This helps reduce the material needed to build the foundation wall. As the ground slopes downward, the height of the wall is increased, as shown in Figure 31.24. Foundation walls may not step more than 24″ (610 mm) in one step. Wood framing between the wall and any floor being supported as Figure 31.25 may not be less than 14″ (356 mm) in height. The foundation wall will also change heights for a sunken floor. Figure 31.26 shows how a sunken floor can be represented on a foundation plan. Steel anchor bolts are placed in the top of the wall to secure the wood mudsill to the concrete. If concrete blocks are used, the cell in which the bolt is placed must be filled with

FIGURE 31.23 ■ Blocks made of expanded polystyrene foam provide both a permanent form for pouring the stem wall and insulation to prevent heat loss. *Courtesy American Polysteel Forms.*

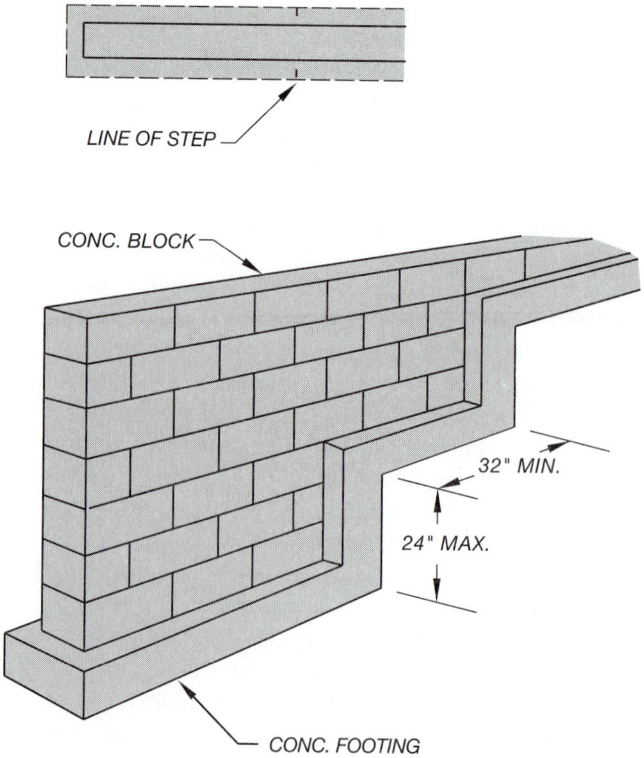

FIGURE 31.24 ■ The footing and foundation wall are often stepped on sloping lots to save material.

FIGURE 31.25 ■ Short studs called jack studs are often used to span between the stem wall and the floor, or between different floor levels. Jack studs must be a minimum of 14″ (350 mm) in length or the wall must be formed of solid wood. *Courtesy of Sam Griggs.*

grout. Figure 31.27 shows anchor bolts and lists minimum spacing requirements. The mudsill is required to be pressure-treated or to be made of some other water-resistant wood so that it will not absorb moisture from the concrete. Anchor bolts from the concrete extend through the mudsill. A 2″ (50 mm) round washer is placed over the bolt before the nut is installed to increase the holding power of the bolts. The mudsill and anchor bolt are not drawn on the foundation plan but are typically specified with a note, as in Figure 31.26.

The mudsill must also be protected from termites in many parts of the country. Among the most common methods of protection are metal caps between the mudsill and the wall, chemical treatment of the soil around the foundation, and chemically treated wood near the ground. The metal shield is not drawn on the foundation plan but is specified in a note near the foundation plan or in a foundation detail.

If the house is to have a wood flooring system, some method of securing support beams to the foundation must be provided. Typically a cavity or beam pocket is built into the foundation wall. The cavity provides a method of supporting the beam and helps tie the floor and foundation system together. A 3″ (75 mm) minimum bearing surface must be provided for the beam where it rests on concrete. A 1/2″ (13 mm) air space is allowed

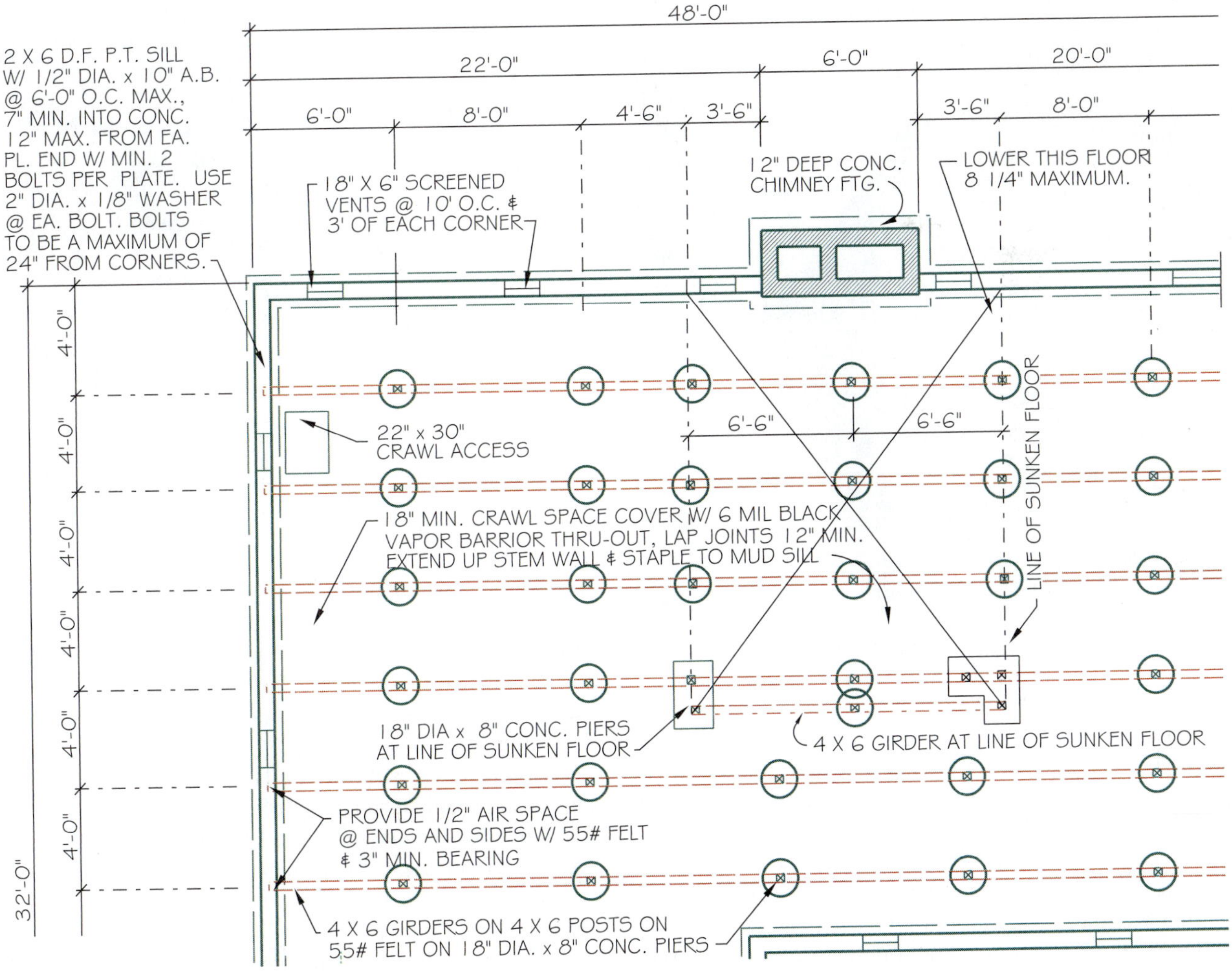

FIGURE 31.26 ■ Common components of a foundation plan are vents, crawl access, fireplace footing, sunken floor, beam pockets, stem wall, footings, and piers.

PLACEMENT OF ANCHOR BOLTS

Min. diameter	1/2″ ∅ (13 mm)
Min. depth into concrete or masonry	7″ (1478 mm)
Max. spacing (1-story)	6′–0″ (1829 mm)
Max. spacing (2-story) Zone D_1 and D_2	4′–0″ (1219 mm)

FIGURE 31.27A ■ Placement of anchor bolts.

FIGURE 31.27B ■ Anchor bolts are set in concrete to hold down the wood framing members. *Photo courtesy John Black.*

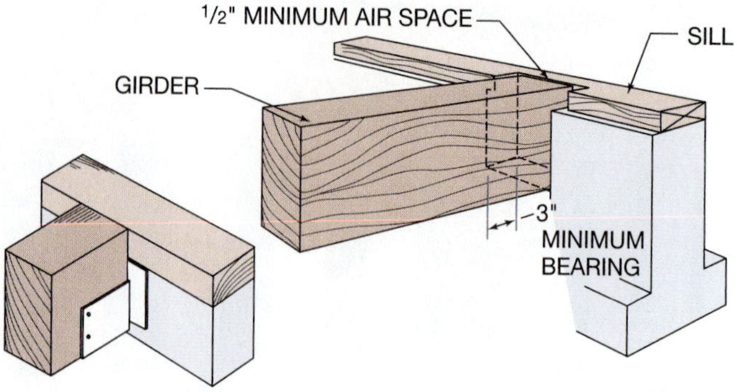

FIGURE 31.28 ■ A beam seat, or pocket, can be recessed into the foundation wall, or the beam may be supported by metal connectors. *Connectors courtesy Simpson Strong-Tie Company, Inc.*

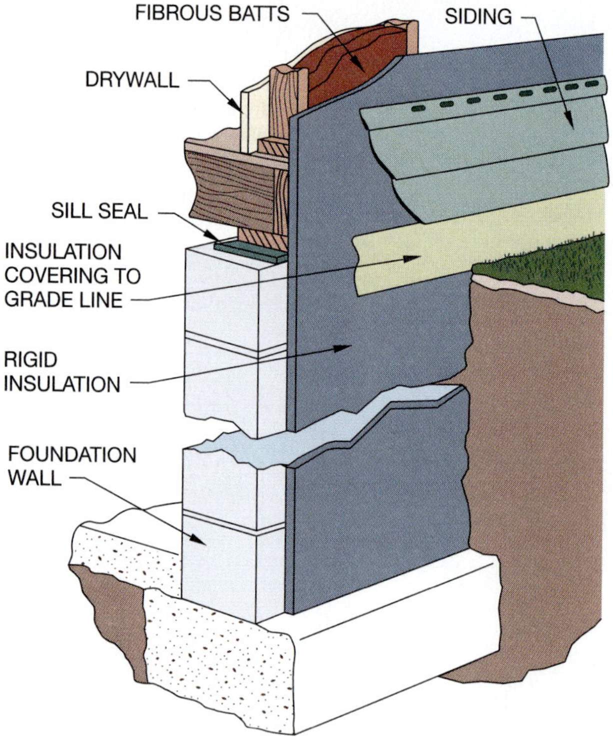

FIGURE 31.29 ■ Insulation is often placed on the foundation wall to help cut heat loss.

around the beam in the pocket for air circulation. Figure 31.28 shows common methods of beam support at the foundation wall. Air must also be allowed to circulate under the floor system. Vents must be provided that will provide 1 sq ft of ventilation for each 150 sq ft (0.67 m^2/100m^2) of crawl space. To supply ventilation under the floor, vents must be set into the foundation wall in addition to minimum size requirements, vents must be provided within 3′ 0″ (914 mm) of each corner to provide air current throughout the crawl space. If an access opening is not provided in the floor, an 18″ × 24″ (457 × 610 mm) opening will be needed in the foundation wall for access to the crawl space under the floor. Figure 31.26 shows how a crawl access, vents, and girder pockets are typically represented on a foundation plan.

When required by the IRC or if the foundation is for an energy-efficient structure, insulation is often added to the wall. Two-inch (50 mm) rigid insulation is used to protect the wall from cold weather, as with the wall shown in Figure 31.5. This can be placed on either side of the wall. If the insulation is placed on the exterior side of the wall, the wall will retain heat from the building. This placement does cause problems in protecting the insulation from damage. Figure 31.29 shows exterior insulation placement.

In addition to supporting the load of the structure, the foundation walls must also resist the lateral pressure of the soil against the wall. When the wall is over 24″ (600 mm) in height, vertical steel is added to the wall to help reduce tension. Horizontal steel is also added to foundation walls in some seismic zones to help strengthen the wall. If required, the steel is typically specified in a note.

Retaining Walls

Retaining or basement walls are primarily made of concrete blocks or poured concrete. Wood walls are used in some areas. The material used will depend on labor trends in your area. The material used will affect the height of the wall. If concrete blocks are used, the wall is approximately 12 blocks high from the top of the footing. If poured concrete is used, the wall will normally be 8′ (2438 mm) high from the top of the footing. This will allow 4 × 8 (1200 × 2400 mm) sheets of plywood to be used as forms for the sides of the wall.

Regardless of the material used, basement walls serve the same function as the shorter foundation walls. Because of the added height, the lateral forces acting on the side of these walls are magnified. As seen in Figure 31.30, lateral soil pressure bends the wall inward, thus placing the soil side of the wall in compression and the interior face of the wall in tension. To resist this tensile stress, steel reinforcing may be required by the building department. The seismic zone will affect the size and placement of the steel. Figure 31.31 shows common patterns of steel placement. Figure 31.32 shows a typical foundation detail used to represent the steel placement of a retaining wall. Figure 31.33 shows an example of steel placement for a poured concrete retaining wall.

The footing for a retaining wall is usually 16″ (400 mm) wide and either 8″ or 12″ (200 or 300 mm) deep. The depth depends on the weight to be supported. Steel is extended from the footing into the wall. At the top of the wall, anchor bolts are placed in the wall using the same method as for a foundation wall. Anchor bolts are placed much closer together for retaining walls. than for shorter walls. Bolt spacings of 24″ (610 mm) o.c. are common when floor joists are perpendicular to the wall, and at 32″ (813 mm) o.c. where joists are parallel to the retaining wall. It is very important that the wall and the floor system be securely tied together. The floor system is used to help strengthen the wall and resist soil pressure. Where seismic risk is great, metal angles are added to the anchor bolts to make the tie between the wall and the floor joist

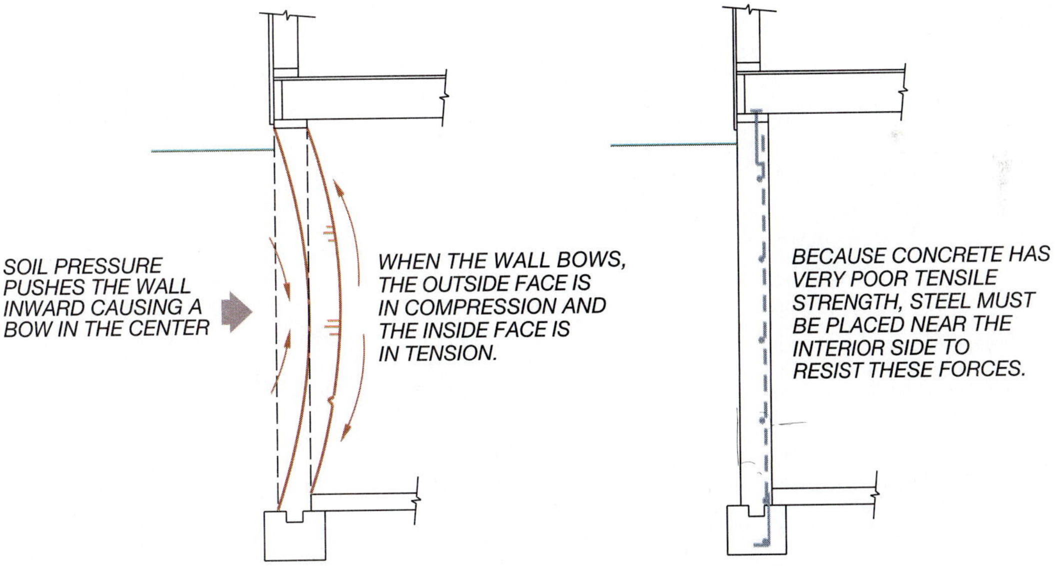

SOIL PRESSURE PUSHES THE WALL INWARD CAUSING A BOW IN THE CENTER
WHEN THE WALL BOWS, THE OUTSIDE FACE IS IN COMPRESSION AND THE INSIDE FACE IS IN TENSION.
BECAUSE CONCRETE HAS VERY POOR TENSILE STRENGTH, STEEL MUST BE PLACED NEAR THE INTERIOR SIDE TO RESIST THESE FORCES.

FIGURE 31.30 ■ Stresses acting on a retaining wall cause it to act as a simple beam spanning between each floor. The soil is the supported load causing the wall to bow.

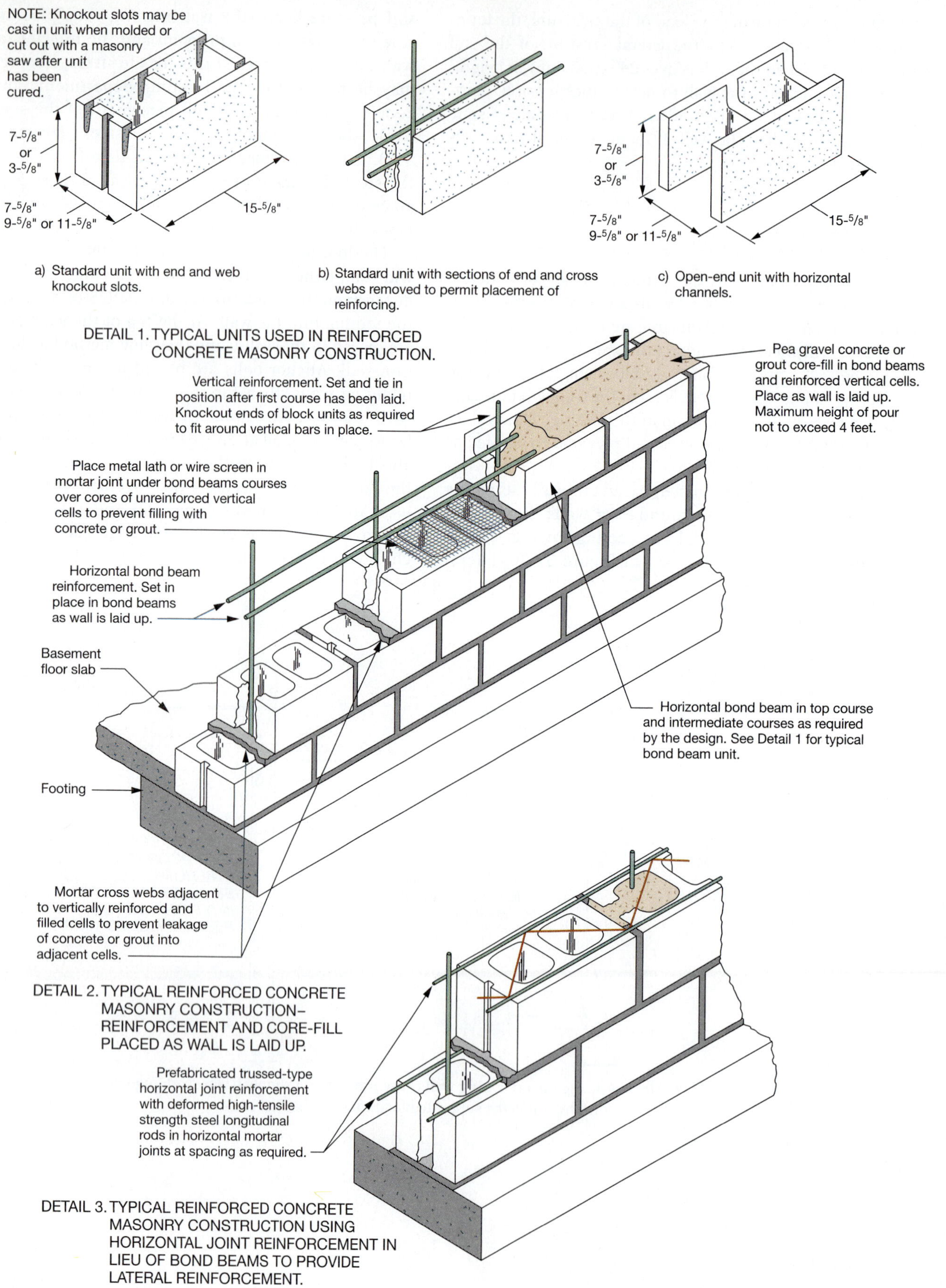
NOTE: Knockout slots may be cast in unit when molded or cut out with a masonry saw after unit has been cured.
7-5/8" or 3-5/8"
7-5/8" 9-5/8" or 11-5/8"
15-5/8"
a) Standard unit with end and web knockout slots.
b) Standard unit with sections of end and cross webs removed to permit placement of reinforcing.
7-5/8" or 3-5/8"
7-5/8" 9-5/8" or 11-5/8"
15-5/8"
c) Open-end unit with horizontal channels.
DETAIL 1. TYPICAL UNITS USED IN REINFORCED CONCRETE MASONRY CONSTRUCTION.
Vertical reinforcement. Set and tie in position after first course has been laid. Knockout ends of block units as required to fit around vertical bars in place.
Pea gravel concrete or grout core-fill in bond beams and reinforced vertical cells. Place as wall is laid up. Maximum height of pour not to exceed 4 feet.
Place metal lath or wire screen in mortar joint under bond beams courses over cores of unreinforced vertical cells to prevent filling with concrete or grout.
Horizontal bond beam reinforcement. Set in place in bond beams as wall is laid up.
Basement floor slab
Horizontal bond beam in top course and intermediate courses as required by the design. See Detail 1 for typical bond beam unit.
Footing
Mortar cross webs adjacent to vertically reinforced and filled cells to prevent leakage of concrete or grout into adjacent cells.
DETAIL 2. TYPICAL REINFORCED CONCRETE MASONRY CONSTRUCTION– REINFORCEMENT AND CORE-FILL PLACED AS WALL IS LAID UP.
Prefabricated trussed-type horizontal joint reinforcement with deformed high-tensile strength steel longitudinal rods in horizontal mortar joints at spacing as required.
DETAIL 3. TYPICAL REINFORCED CONCRETE MASONRY CONSTRUCTION USING HORIZONTAL JOINT REINFORCEMENT IN LIEU OF BOND BEAMS TO PROVIDE LATERAL REINFORCEMENT.

FIGURE 31.31A ■ Suggested construction details for reinforced concrete masonry foundation walls. *Courtesy National Concrete Masonry Association.*

REINFORCED CONCRETE AND MASONRY[a] FOUNDATION WALLS

MAXIMUM WALL HEIGHT (feet)	MAXIMUM UNBALANCED BACKFILL HEIGHT[c] (feet)	PLAIN CONCRETE MINIMUM NOMINAL WALL THICKNESS (inches)			PLAIN MASONRY[a] MINIMUM NOMINAL WALL THICKNESS (inches)		
		Soil classes[b]					
		GW, GP, SW and SP	GM, GC, SM, SM-SC and ML	SC, MH, ML-CL and inorganic CL	GW, GP, SW and SP	GM, GC, SM, SM-SC and ML	SC, MH, ML-CL and inorganic CL
5	4	6	6	6	6 solid[d] or 8	6 solid[d] or 8	6 solid[d] or 8
	5	6	6	6	6 solid[d] or 8	8	10
6	4	6	6	6	6 solid[d] or 8	6 solid[d] or 8	6 solid[d] or 8
	5	6	6	6	6 solid[d] or 8	8	10
	6	6	8[g]	8[g]	8	10	12
7	4	6	6	6	6 solid[d] or 8	8	8
	5	6	6	8[g]	6 solid[d] or 8	10	10
	6	6	8	8	10	12	10 solid[d]
	7	8	8	10	12	10 solid[d]	12 solid[d]
8	4	6	6	6	6 solid[d] or 8	6 solid[d] or 8	8
	5	6	6	8	6 solid[d] or 8	10	12
	6	8[h]	8	10	10	12	12 solid[d]
	7	8	10	10	12	12 solid[d]	Footnote e
	8	10	10	12	10 solid[d]	12 solid[d]	Footnote e
9	4	6	6	6	6 solid[d] or 8	6 solid[d] or 8	8
	5	6	8[g]	8	8	10	12
	6	8	8	10	10	12	12 solid[d]
	7	8	10	10	12	12 solid[d]	Footnote e
	8	10	10	12	12 solid[d]	Footnote e	Footnote e
	9	10	12	Footnote f	Footnote e	Footnote e	Footnote e

For SI: 1 inch = 25.4 mm, 1 foot = 304.8 mm, 1 psi = 6.895 Pa.

a. Mortar shall be Type M or S and masonry shall be laid in running bond. Ungrouted hollow masonry units are permitted except where otherwise indicated.

b. Soil classes are in accordance with the Unified Soil Classification System. Refer to Table R405.1.

c. Unbalanced backfill height is the difference in height of the exterior and interior finish ground levels. Where an interior concrete slab is provided, the unbalanced backfill height shall be measured from the exterior finish ground level to the top of the interior concrete slab.

d. Solid grouted hollow units or solid masonry units.

e. Wall construction shall be in accordance with Table R404.1.1(2) or a design shall be provided.

f. A design is required.

g. Thickness may be 6 inches, provided minimum specified compressive strength of concrete, f'_c, is 4,000 psi.

REINFORCED CONCRETE AND MASONRY[a] FOUNDATION WALLS

MAXIMUM WALL HEIGHT (feet)	MAXIMUM UNBALANCED BACKFILL HEIGHT[e] (feet)	MINIMUM VERTICAL REINFORCEMENT SIZE AND SPACING[b, c] FOR 8-INCH NOMINAL WALL THICKNESS		
		Soil classes[d]		
		GW, GP, SW and SP soils	GM, GC, SM, SM-SC and ML soils	SC, MH, ML-CL and inorganic CL soils
6	5	#4 @ 48" o.c.	#4 @ 48" o.c.	#4 @ 48" o.c.
	6	#4 @ 48" o.c.	#4 @ 40" o.c.	#5 @ 48" o.c.
7	4	#4 @ 48" o.c.	#4 @ 48" o.c.	#4 @ 48" o.c.
	5	#4 @ 48" o.c.	#4 @ 48" o.c.	#4 @ 40" o.c.
	6	#4 @ 48" o.c.	#5 @ 48" o.c.	#5 @ 40" o.c.
	7	#4 @ 40" o.c.	#5 @ 40" o.c.	#6 @ 48" o.c.
8	5	#4 @ 48" o.c.	#4 @ 48" o.c.	#4 @ 40" o.c.
	6	#4 @ 48" o.c.	#5 @ 48" o.c.	#5 @ 40" o.c.
	7	#5 @ 48" o.c.	#6 @ 48" o.c.	#6 @ 40" o.c.
	8	#5 @ 40" o.c.	#6 @ 40" o.c.	#6 @ 24" o.c.
9	5	#4 @ 48" o.c.	#4 @ 48" o.c.	#5 @ 48" o.c.
	6	#4 @ 48" o.c.	#5 @ 48" o.c.	#6 @ 48" o.c.
	7	#5 @ 48" o.c.	#6 @ 48" o.c.	#6 @ 32" o.c.
	8	#5 @ 40" o.c.	#6 @ 32" o.c.	#6 @ 24" o.c.
	9	#6 @ 40" o.c.	#6 @ 24" o.c.	#6 @ 16" o.c.

For SI: 1 inch = 25.4 mm, 1 foot = 304.8 mm.

a. Mortar shall be Type M or S and masonry shall be laid in running bond.

b. Alternative reinforcing bar sizes and spacings having an equivalent cross-sectional area of reinforcement per lineal foot of wall shall be permitted provided the spacing of the reinforcement does not exceed 72 inches.

c. Vertical reinforcement shall be Grade 60 minimum. The distance from the face of the soil side of the wall to the center of vertical reinforcement shall be at least 5 inches.

d. Soil classes are in accordance with the Unified Soil Classification System. Refer to Table R405.1.

e. Unbalanced backfill height is the difference in height of the exterior and interior finish ground levels. Where an interior concrete slab is provided, the unbalanced backfill height shall be measured from the exterior finish ground level to the top of the interior concrete slab.

FIGURE 31.31B ■ Required sizes of plain and reinforced concrete and masonry foundation walls. *Courtesy Reproduced from the International Residential Code / 2000. Copyright 2000, Courtesy International Code Council, Inc.*

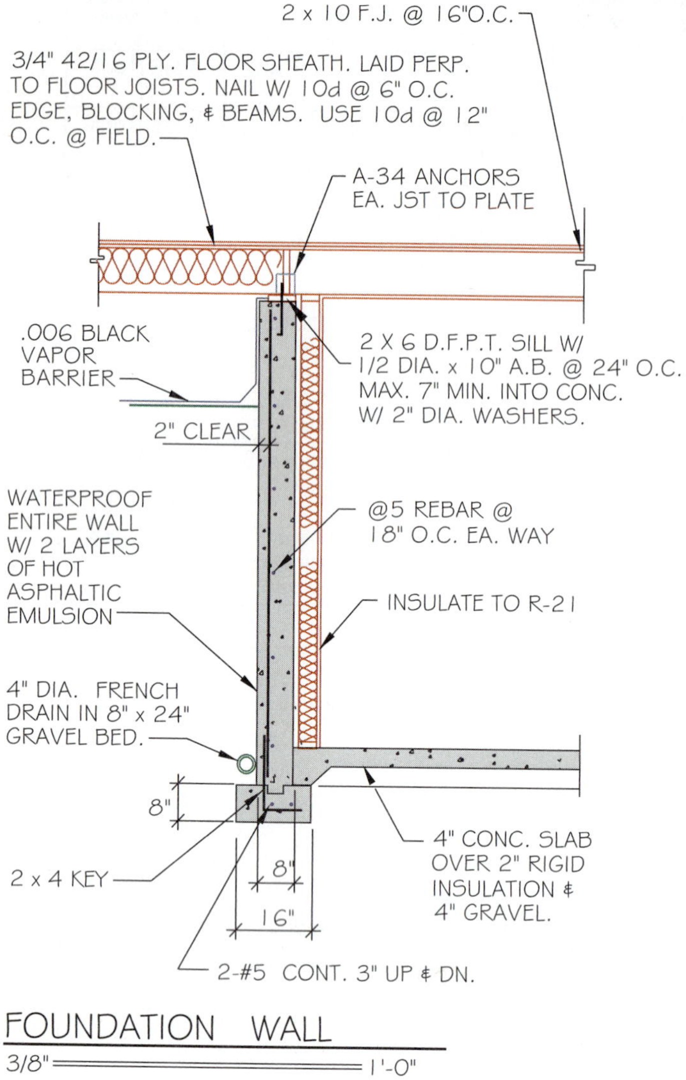

FIGURE 31.32 ■ The components of a concrete retaining wall are specified in a section or detail. Key components are the building material of the wall, reinforcing material, floor attachment, waterproofing, and drainage method. *Courtesy Residential Designs.*

FIGURE 31.33 ■ Reinforcing steel extends from the footing and ties to the wall steel to provide strength for the retaining wall. *Courtesy David Jefferis.*

extremely rigid. These connections are shown on the foundation plan, as seen in Figure 31.34.

To reduce soil pressure next to the footing, a drain is installed. The drain is set at the base of the footing to collect and divert water from the face of the wall. Figure 31.35 shows how the drain is placed. The area above the drain is filled with gravel so that subsurface water will percolate to the drain and away from the wall. Reducing the water content of the soil reduces the lateral pressure on the wall. The drain is not drawn on the foundation plan but must be specified in a note on the foundation plan, sections, and foundation details.

No matter what the soil condition, the basement wall must be protected to reduce moisture passing through the wall into the living area. The IRC specifies two levels of basement wall protection as damp proofing and water proofing. Each method must be applied to the exterior face of a foundation wall that surrounds habitable space. The protection must be installed from the top of the top of the footing to the grade. A masonry or concrete wall can be damp proofed by adding 3/8″ (9.5 mm) portland cement parging to the exterior side of the wall. The parging must then be protected by a bituminous coating, acrylic modified cement, or a coat of surface-bonding mortar. Materials used to waterproof a wall can also be applied. In areas with a high water table or other severe soil/water conditions, basement walls surrounding habitable space must be waterproofed. Concrete and mortar walls are waterproofed by adding a 2-ply hot-mopped felts, 6-mil polyethylene or 40-mil polymer-modified asphalt, 6-mil (0.15 mm) polyvinyl chloride, or 55# (25 kg) roll roofing. Additional provisions are provided by the IRC for ICF for wood foundation walls.

Adding windows to the basement can help cut down the moisture content of the basement. This will sometimes require adding a window well to prevent the ground from being pushed in front of the window. Figure 31.34 shows how a window well—or areaway, as it is sometimes called—can be drawn on the foundation plan.

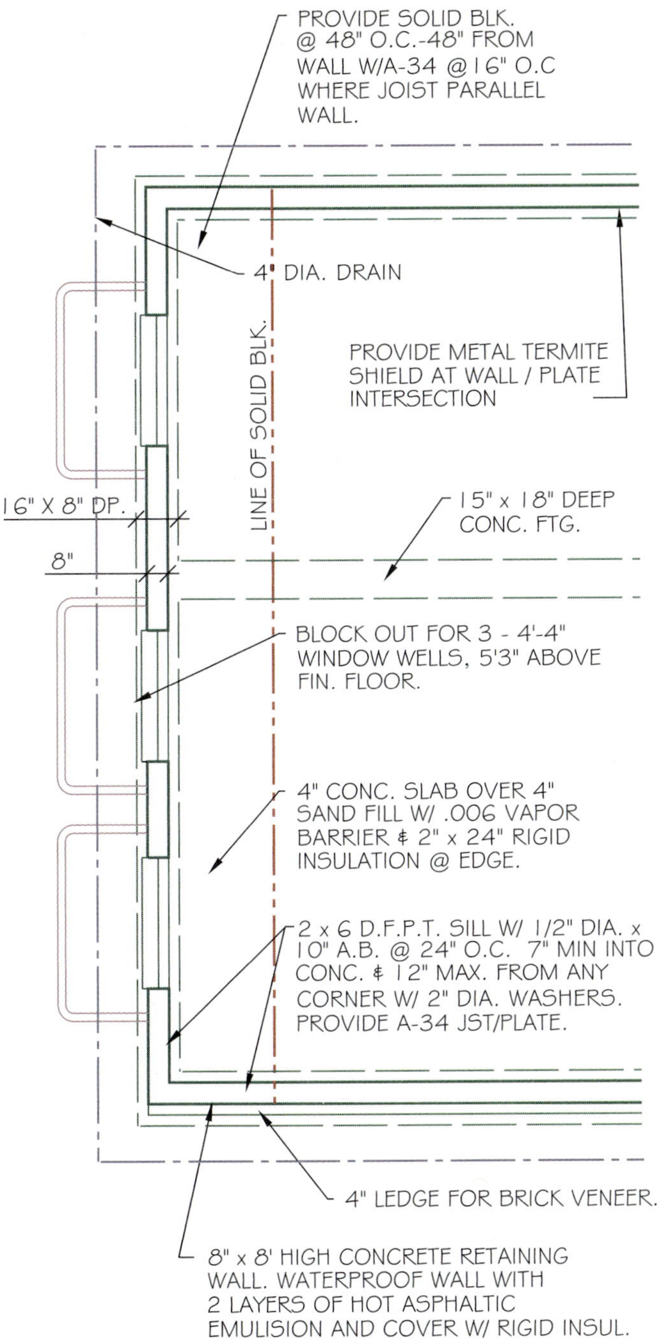

FIGURE 31.34 ■ Common elements that are shown on a foundation with a basement. If a window is to be placed in a full-height basement wall, a window well to restrain the soil around window is required. Blocking is placed between floor joists that are parallel to the retaining wall. The blocking provides rigidity to the floor system allowing lateral pressure from the wall to be transferred into the floor system.

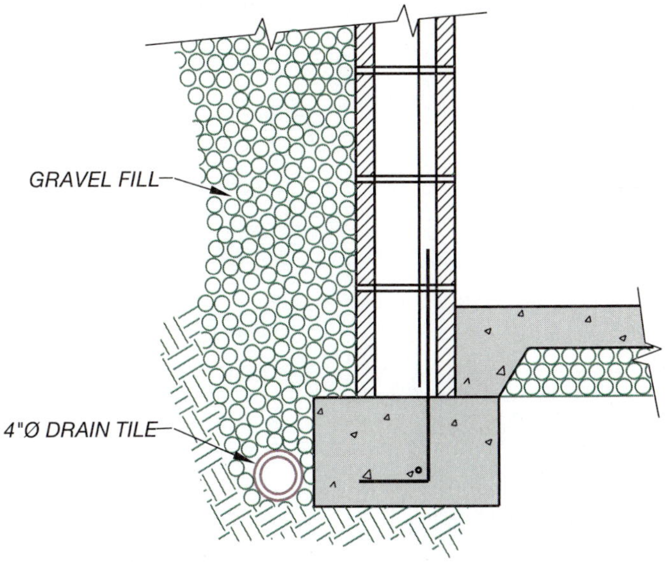

FIGURE 31.35 ■ Drain and gravel must be specified on foundation details.

Like a foundation wall, the basement wall needs to be protected from termites. The metal shield will not be drawn on the foundation plan but should be called out in a note.

Treated-Wood Basement Walls

Pressure-treated lumber can be used to frame both crawl space and basement walls. Treated-wood basement walls allow for easy installation of electrical wiring, insulation, and finishing materials.

Instead of a concrete foundation, a gravel bed is used to support the wall loads. Gravel is required to extend 4″ (100 mm) on each side of the wall and be approximately 8″ (200 mm) deep. A 2 × (50 mm) pressure-treated plate is laid on the gravel, and the wall is built of pressure-treated wood, as seen in Figure 31.36. Pressure-treated 1/2″ (13 mm) thick C-D grade plywood with exterior glue is laid perpendicular to the studs, covered with 6 mil (0.15 mm) polyethylene and sealed with an adhesive.

Restraining Walls

As seen in Figure 31.37, when a structure is built on a sloping site, the masonry wall may not need to be full height. Although less soil is retained than with a full-height wall, more problems are encountered. Figure 31.38 shows the tendencies for bending for this type of wall. Because the wall is not supported at the top by the floor, the soil pressure must be resisted through

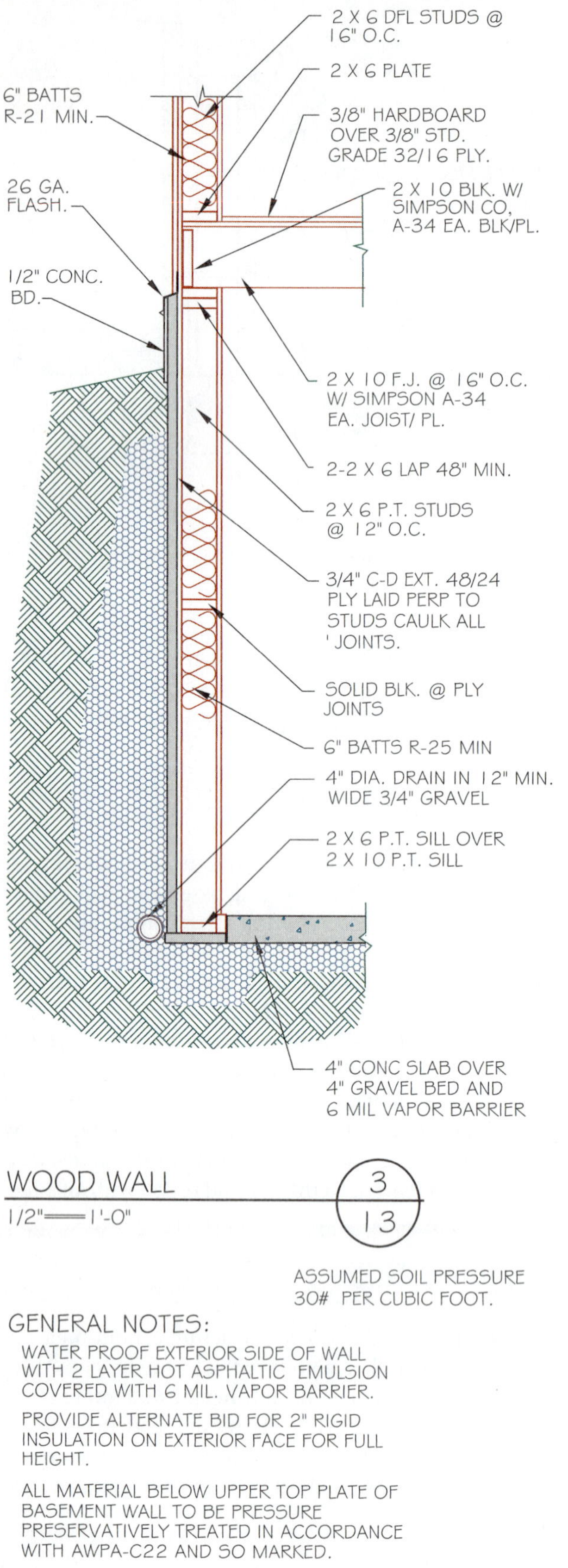

FIGURE 31.36 ■ Stem walls and basement retaining walls can be framed using pressure-treated wood with no foundation. The size and spacing of members are determined by the loads being supported and the height of the wall.

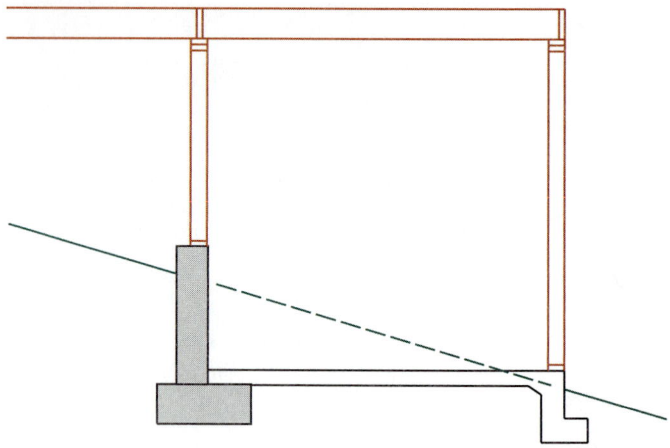

FIGURE 31.37 ■ An 8′ (2400 mm) high retaining wall is usually not required on sloping lots. A partial restraining wall with wood-framed walls above it is often used.

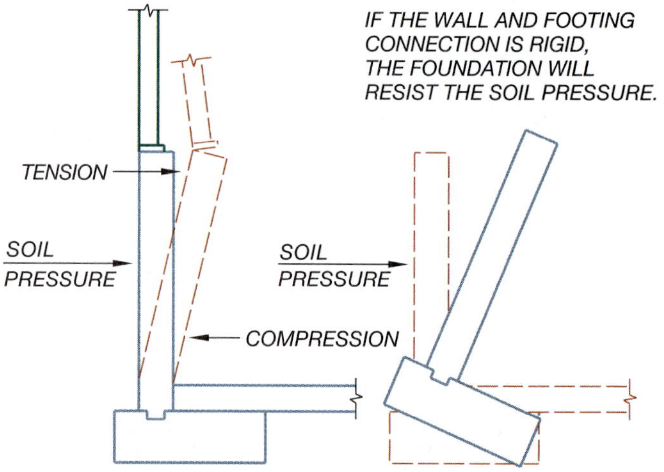

FIGURE 31.38 ■ When a wall is not held in place at the top, soil pressure will attempt to move the wall inward. The intersection of the wood and concrete walls is called a hinge point because of the tendency to move.

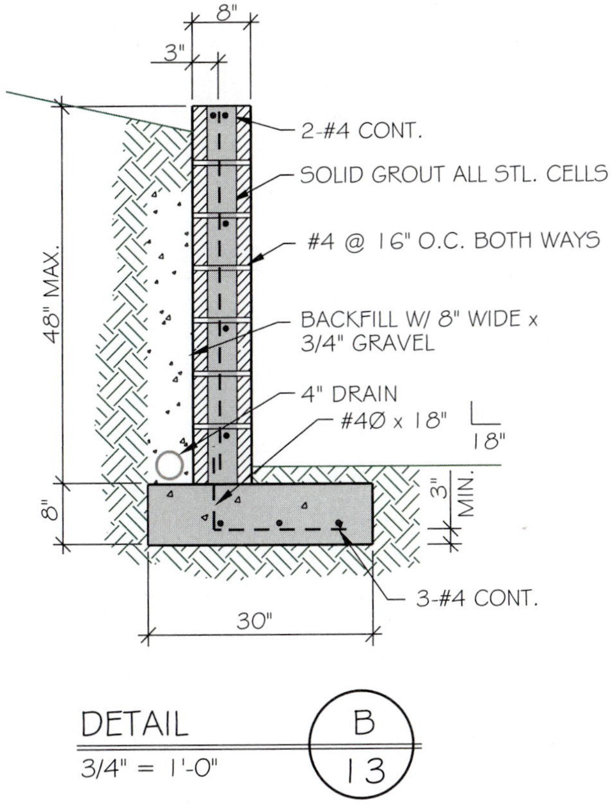

FIGURE 31.39 ■ Detail of a partial partial-height wall.

the footing. This requires a larger footing than for a full-height wall. If the wall and footing connection is rigid enough to keep the wall in position, the soil pressure will try to overturn the whole foundation. The extra footing width is required to resist the tendency to overturn. Figure 31.39 shows a detail that a drafter might be required to draw and how the wall would be represented on the foundation plan. Figure 31.40 shows how a restraining wall will be represented on a foundation plan. Depending on the slope of the ground being supported, a key may be required, as in Figure 31.41. A keyway on the top of the footing has been discussed. This key is added to the bottom of the footing to help keep it from sliding as a result of soil pressure against the wall. The key is not shown on the foundation plan but is shown in a detail of the wall.

Interior Supports

Foundation walls and footings are part of the foundation system supporting the exterior of the structure. Interior loads are supported on spot footings, or piers, as seen in Figure 31.42. Piers are formed by using either preformed or framed forms, or by excavation. Pier depth is generally required to match that of the footings. Although Figure 31.42 shows the piers above the finished grade, many states require the mass of the footing to be placed below grade to prevent problems with freezing and shifting from seismic activity. The placement of piers will be determined by the type of floor system to be used and will be discussed in Chapter 32. The size of the pier will depend on the load being supported and the soil bearing pressure. Piers are usually drawn on the foundation plan with dashed lines, as shown in Figure 31.43.

In addition to the exterior footings and interior piers, footings may also be required under braced walls. Braced walls and braced wall lines were introduced in Chapter 30. On the framing plan, when the distance between braced wall lines exceed 35′–0″ an interior braced wall line must be provided. A beam or double joist is required to support the interior braced wall line. At the foundation level, when the distance between the footings for braced wall lines exceeds 50′–0″, a continuous footing must be provided below the braced wall. A continuous footing must be provided below all multi-level braced walls in seismic zones D_1 and D_2.

A final consideration of foundation walls and footings is the placement of each by a garage door. For small openings such as a 36″ (900 mm) door, the footing is continued under the door opening. The stem wall is cut to allow access to the garage without having to step over the wall each time the garage is entered. For large opening such as the main garage door, in areas of the country with low seismic risk, the footing is not continuous across the door opening. In areas at risk of seismic activity, the footing must extend across the entire width of the door. The continuous footing helps the walls on each side of the door to act as one wall unit. The stem wall is cut to allow the concrete floor to cover the stem wall, providing a smooth entry into the garage.

Metal Connectors

Metal connectors are often used at the foundation level to resist stress from wind and seismic forces. Three of the most commonly used metal connectors are shown in Figure 31.44. The senior drafter or designer will determine the proper connector specify on the foundation plan. How the connector is used will determine how it is specified. Figure 31.45 shows how these connectors might be specified on the foundation plan.

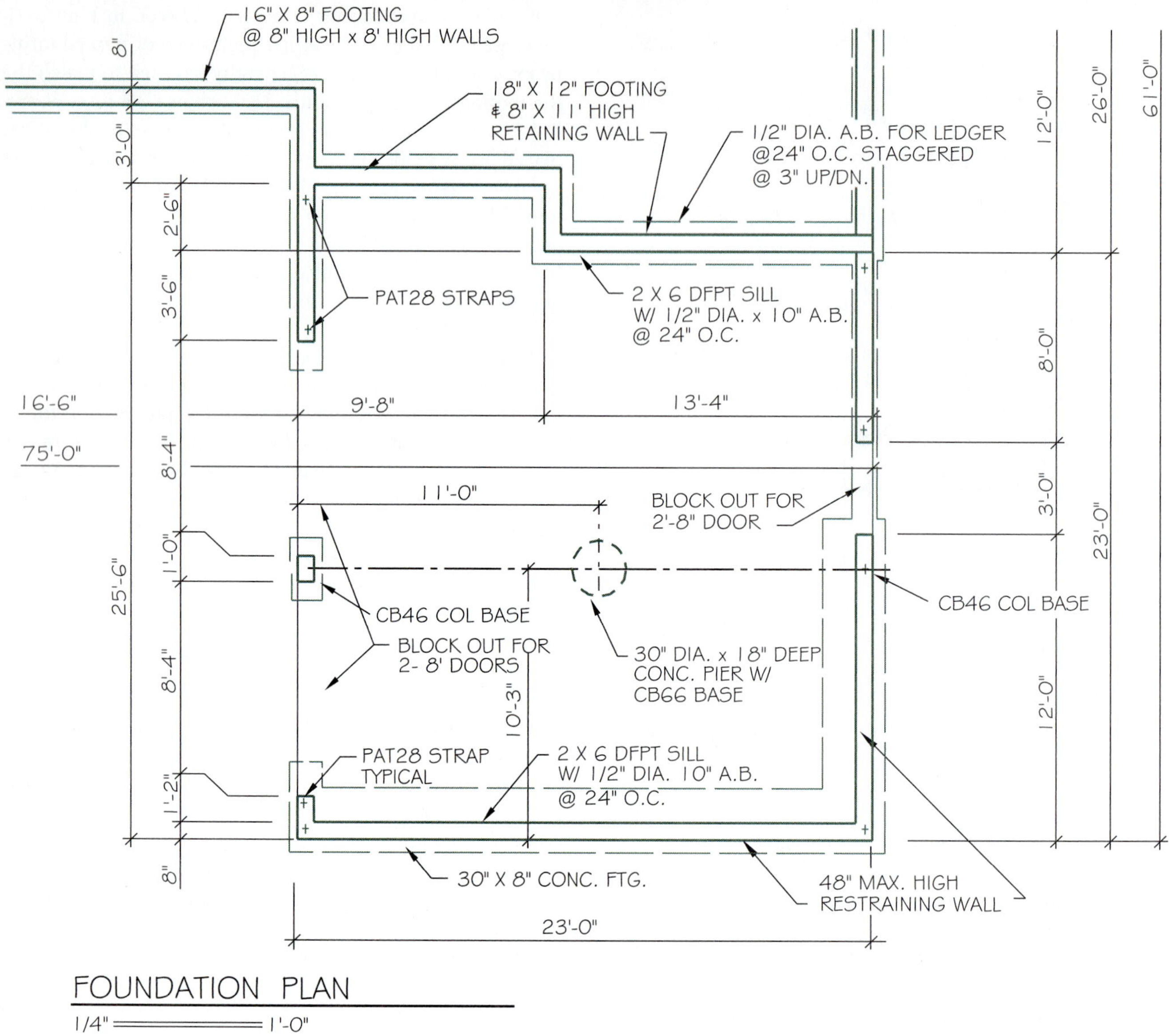

FIGURE 31.40 ■ Retaining and restraining walls are represented similarly to footings and stem walls. Because of the added height, the width is typically specified on the foundation plan, and the building components are specified in a detail or section.

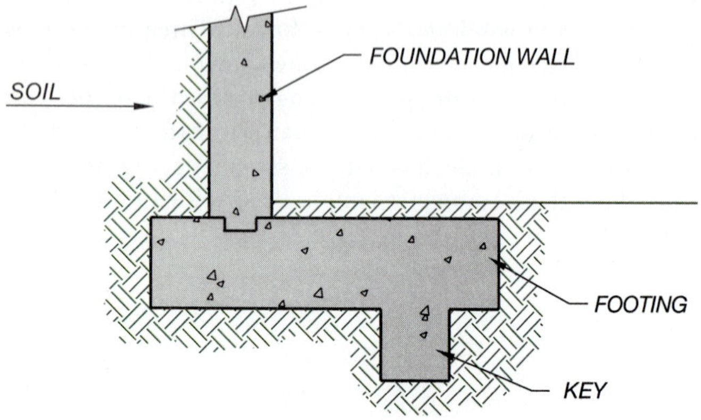

FIGURE 31.41 ■ Soil pressure will attempt to make the wall and footing slide across the soil. A key may be provided on the bottom of the footing to provide added surface area to resist sliding.

FIGURE 31.42 ■ Concrete piers can be formed by using preformed or framed forms or by excavating. Note the step in the footing to accommodate the slope of the site. *Courtesy Michelle Cartwright.*

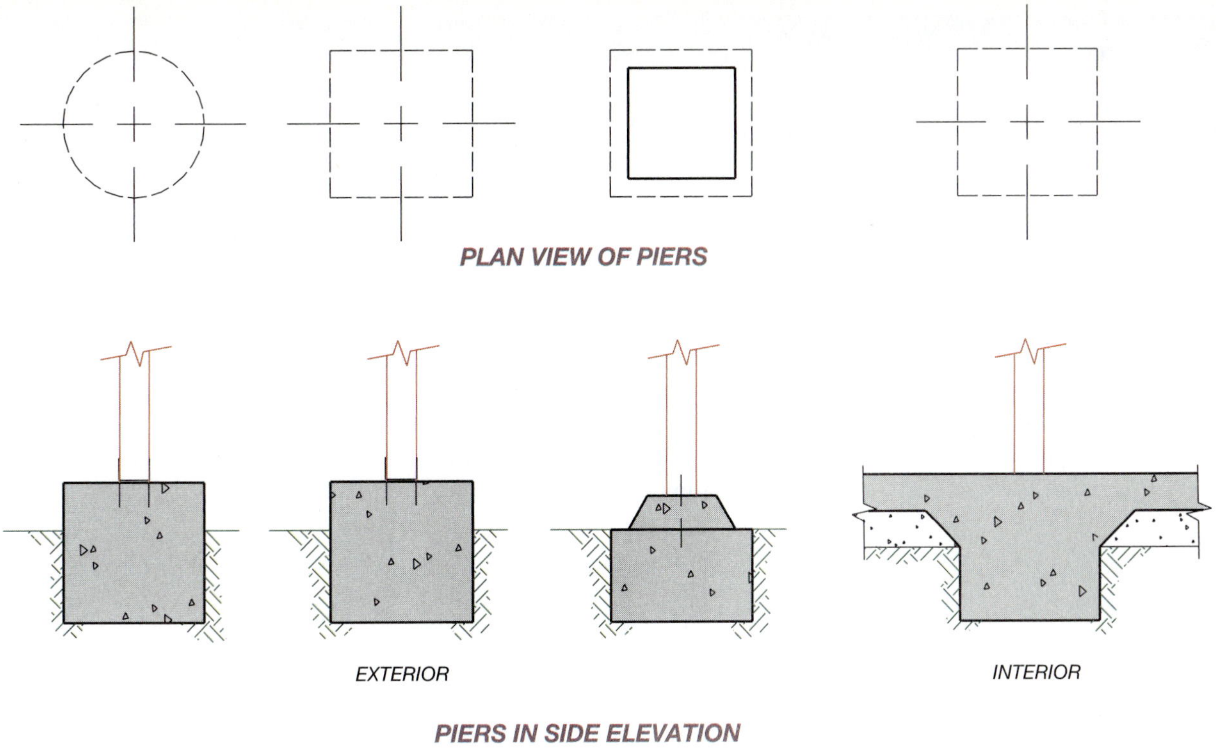

FIGURE 31.43 ■ Concrete piers are used to support interior loads. Codes require wood to be at least 6″ (152 mm) above grade, which means that piers must also extend 6″ (152 mm) above grade.

(a) (b) (c)

FIGURE 31.44 ■ Three common types of metal connectors used on the foundation system. Drafters are often required to use vendor information when drawing foundation plans. *Courtesy Simpson Strong-Tie Company, Inc.*

DIMENSIONING FOUNDATION COMPONENTS

This chapter has presented components of a foundation system and how they are drawn on the foundation plan. The manner in which they are dimensioned is equally important to the construction crew. The line quality for the dimension and leader lines is the same as was used on the floor plan. Jogs in the foundation wall are dimensioned using the same methods used on the floor or framing plan. Most of the dimensions for major shapes will be exactly the same as the corresponding dimensions on the floor plan. When the foundation plan is drawn using a computer, the foundation plan is typically placed in the same drawing file as the floor plan. This will allow the overall dimensions for the floor plan to be displayed on the foundation plan if different layers are used to place the dimensions. Layers such as BASEDIM, FLOORDIM, and FNDDIMEN will help differentiate between dimensions for the floor and foundation.

A different method is used to dimension the interior walls of a foundation than those used on a floor plan. Foundation walls are dimensioned from face to face rather than face to center, as on a floor plan. Footing widths are usually dimensioned from center to center. Each type of dimension can be seen in Figure 31.46.

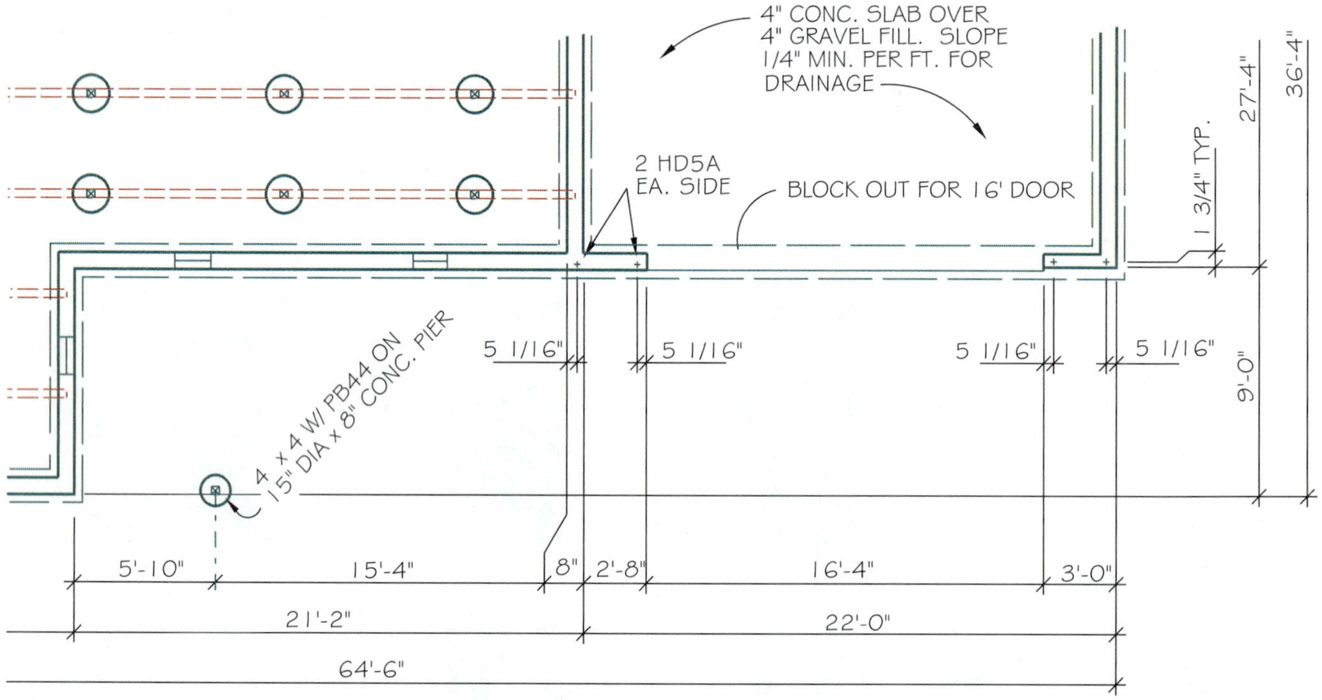

FIGURE 31.45 ■ Metal connectors are typically used to resist the forces of uplift, shear, flooding, and seismic stresses.

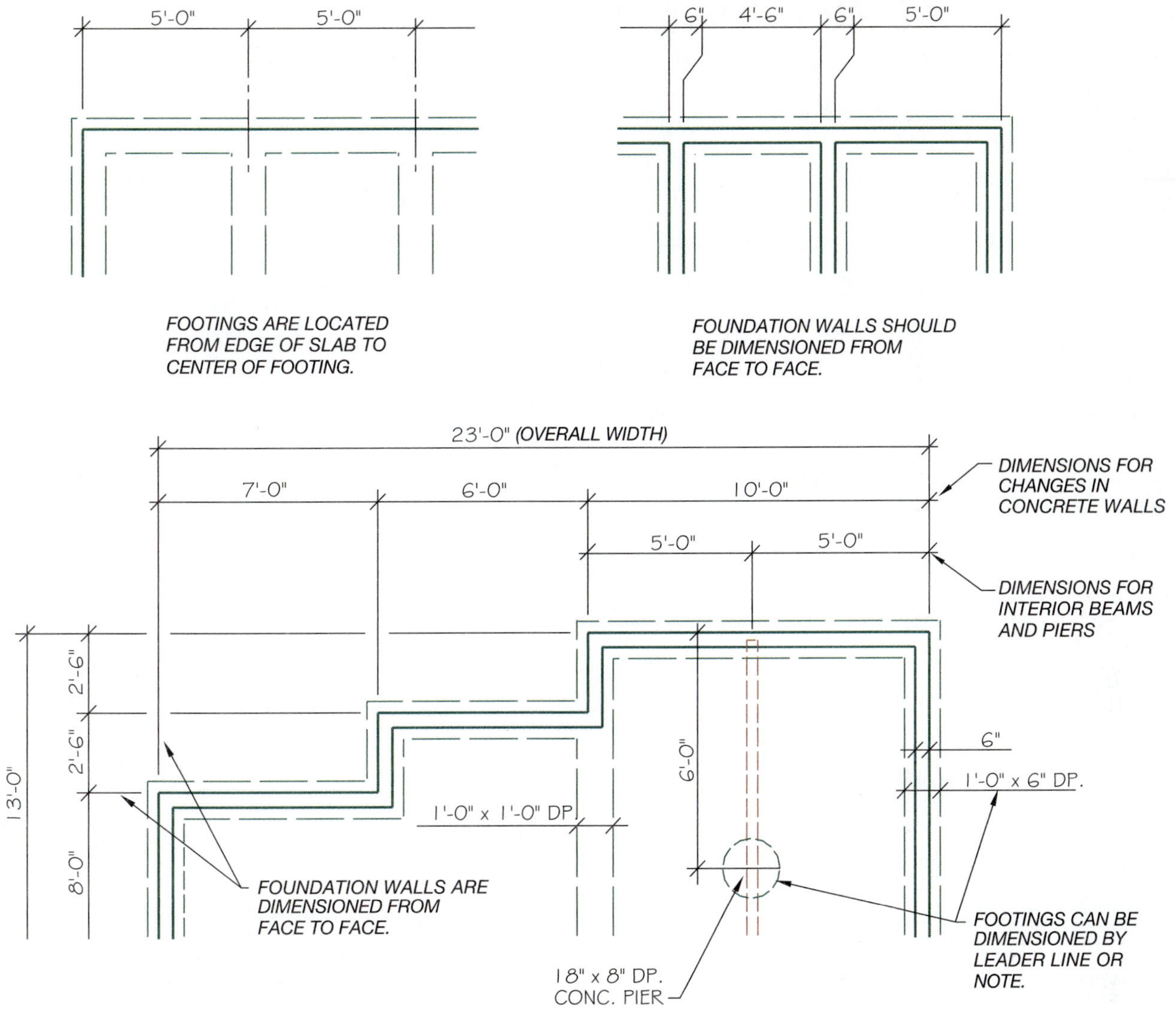

FIGURE 31.46 ■ Dimensioning techniques for foundation plans.

Chapter 31 Additional Reading

The following Websites can be used as a resource to help you keep current with changes in foundation materials.

ADDRESS	COMPANY OR ORGANIZATION
www.aci-int.org	American Concrete Institute International
www.bluemaxxaab.com	AAB Building Systems Inc. (ICFs)
www.afmcorp-epsfoam.com	AFM Corporation
www.boccia.org	Boccia, Inc.
www.eco-block.com	Eco-Block
www.portcement.org	Portland Cement Association
www.strongtie.com	Simpson Strong-Tie Connections
www.soils.org	Soil Science Society of America

CHAPTER

31 Foundation Systems Test

DIRECTIONS

Answer the questions with short, complete statements or drawings as needed.

1. Letter your name and the date at the top of the sheet.
2. Letter the question number and provide the answer. You do not need to write out the question.

QUESTIONS

Question 31–1 What are the major parts of the foundation system?

Question 31–2 What methods are used to determine the size of footings?

Question 31–3 List five forces that a foundation must withstand.

Question 31–4 How can the soil texture influence a foundation?

Question 31–5 List the major types of material used to build foundation walls.

Question 31–6 Why is steel placed in footings?

Question 31–7 Describe when a stepped footing might be used.

Question 31–8 What size footing should be used with a basement wall?

Question 31–9 What influences the size of piers?

Question 31–10 Describe the difference between a retaining and a restraining wall.

PROBLEMS

DIRECTIONS

Unless other instructions are given by your instructor, complete the following details for a one-level house. Use a scale of 3/4″ = 1′–0″ with proper line weight and quality and provide notes required to label typical materials. Provide dimensions based on common practice in your area. Omit insulation unless specified.

Problem 31–1 Show a typical foundation with poured concrete with a 2 × 4 key, 2 × 6 f.j. at 16″ o.c., and 2 × 4 studs at 24″ o.c.

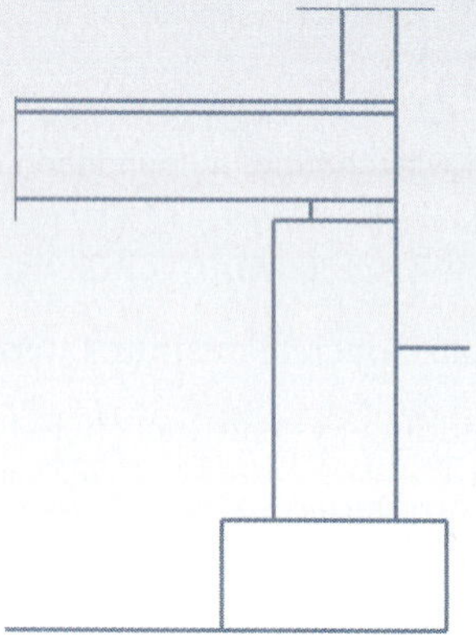

Problem 31–2 Show a typical post-and-beam foundation system. Show 4 × 6 girders @ 48″ o.c. parallel to the stem wall.

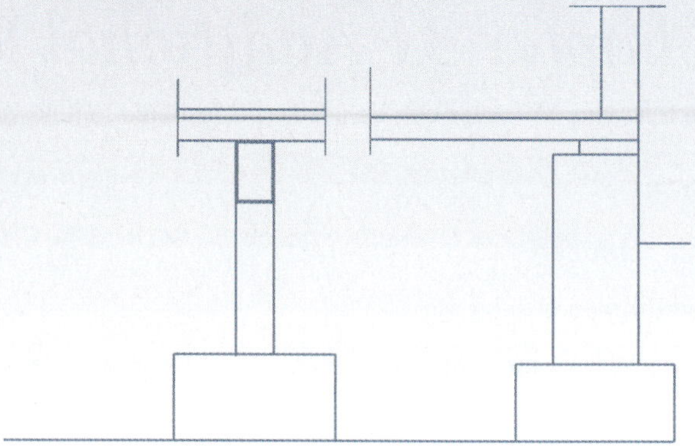

Problem 31–3 Draw a foundation showing a concrete floor system. Use #10 × 10-4″ × 4″ wwm 2″ dn. from top surface and 2 #4 bars centered in the footing 2″ up and down. Use 2 × 4 studs at 16″ o.c. for walls.

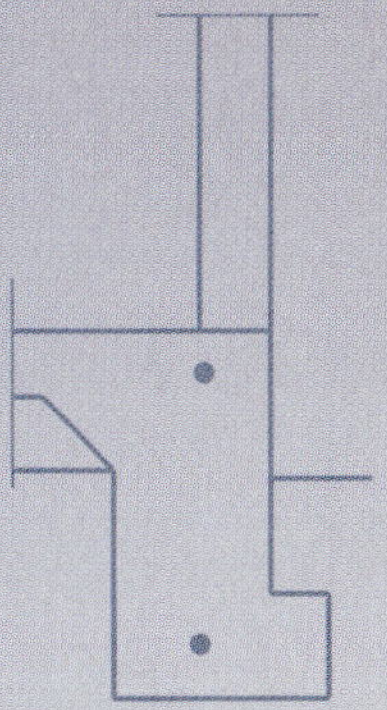

Problem 31–4 Draw a detail showing an interior footing for a concrete slab supporting a 2 × 4 stud-bearing wall. Anchor the wall with Ramset-type fasteners (or equal). Use the same slab reinforcing that was used in Problem 31–3.

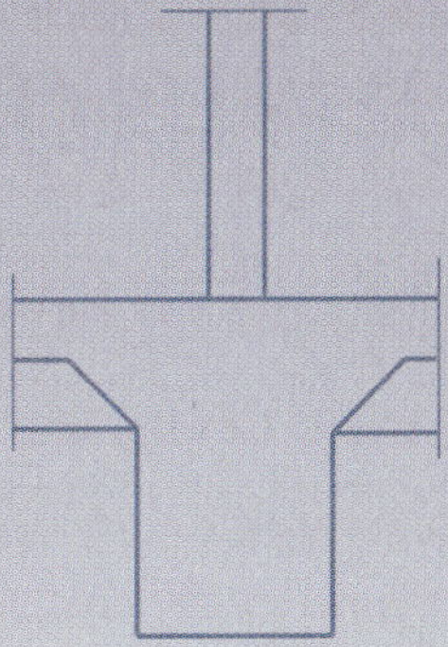

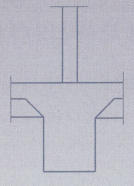

Problem 31–5 Design a retaining wall for an 8′ tall basement, using concrete blocks. Use 2 × 10 f.j. parallel to wall, w/wall, w/5/8″ dia. a.b. @ 32″ o.c. Show typical solid blocking and anchor rim joists with A35 anchors at 16″ o.c. Show #5 @ 18″ o.c. each way @ 2″ from tension side with 1-#5 in footing 2″ up. Use a scale of 1/2″ = 1′–0″.

Problem 31–6 Show a detail of the intersection of a 48″ high concrete restraining wall. Use a 30″ wide footing with 2-#5 cont. 3″ up @ 12″ o.c. and #5 @ 24″ o.c. in wall 2″ from tension side. Use 4″ drain in 8 × 30″ gravel bed. Use 2 × 6 sill with 5/8″ a.b. @ 24″ o.c. Use an 8′ ceiling with 2 × 10 f.j. @ 16″ o.c. Cantilever 15″ past wall w/1″ exterior stucco. Use a scale of 1/2″ = 1′–0″.

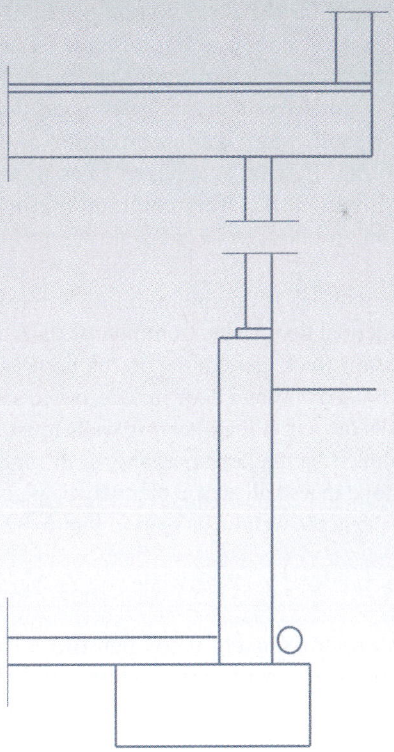

32 CHAPTER

Floor Systems and Foundation Support

INTRODUCTION

The foundation plan shows not only the concrete footings and walls but also the members that are used to form the floor. Two common types of floor systems are typically used in residential construction: floor systems with a crawl space or basement below the floor system, and floor systems built at grade level. Each has its own components and information that must be put on a foundation plan.

ON-GRADE FOUNDATIONS

A concrete slab is often used for the floor system of residential or commercial structures. A concrete slab provides a firm floor system with little or no maintenance and generally requires less material and labor than a conventional wood floor system. The floor slab is usually poured as an extension of the foundation wall and footing, in what is referred to as monolithic construction. See Figure 32.1. Other common methods of pouring the foundation and floor system are shown in Figure 32.2.

A 3 1/2″ (90 mm) concrete slab is the minimum thickness allowed by the IRC for residential floor slabs. Commercial slabs are often 5″ or 6″ (125–150 mm) thick, depending on the floor load to be supported. The slab is used only as a floor surface, not to support the weight of the walls or roof. If load-bearing walls must be supported, the floor slab must be thickened, as shown in Figure 32.3. If a load is concentrated in a small area, a pier may be placed under the slab to help disperse the weight, as seen in Figure 32.4.

Slab Joints

Concrete tends to shrink approximately 0.66″ per 100′ (16.7 mm/30480 mm) as the moisture in the mix hydrates and the concrete hardens. Concrete also continues to expand and shrink throughout its life, depending on the temperatures and the moisture in the supporting soil. This shrinkage can cause the floor slab to crack. To help control possible cracking, three types of joints can be placed in the slab: control, construction, and isolation joints. See Figure 32.5.

Control Joints

As the slab contracts during the initial drying process, the lower surface of the slab rubs against the soil, creating tensile stress. The friction between the slab and the soil causes cracking, which can be controlled by a *control* or *contraction joint.* Such a joint does not prevent cracking, but it does control where the cracks will develop in the slab. Control joints can be created by cutting the fresh concrete or sawing the concrete within 6 to 8 hours of placement. Joints are usually one-quarter of the slab depth. Because the

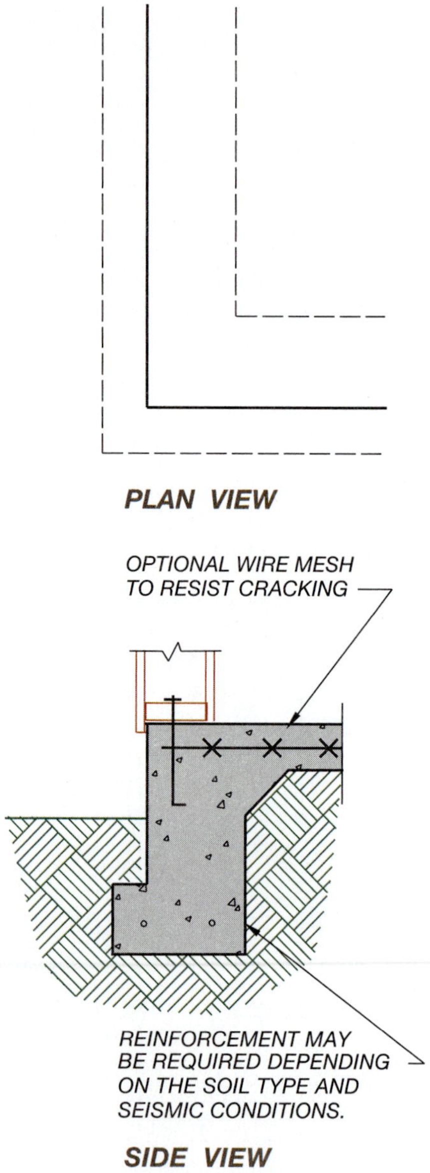
PLAN VIEW
OPTIONAL WIRE MESH
TO RESIST CRACKING
REINFORCEMENT MAY
BE REQUIRED DEPENDING
ON THE SOIL TYPE AND
SEISMIC CONDITIONS.
SIDE VIEW

FIGURE 32.1 ■ The foundation and floor system can often be constructed in one pour, saving time and money.

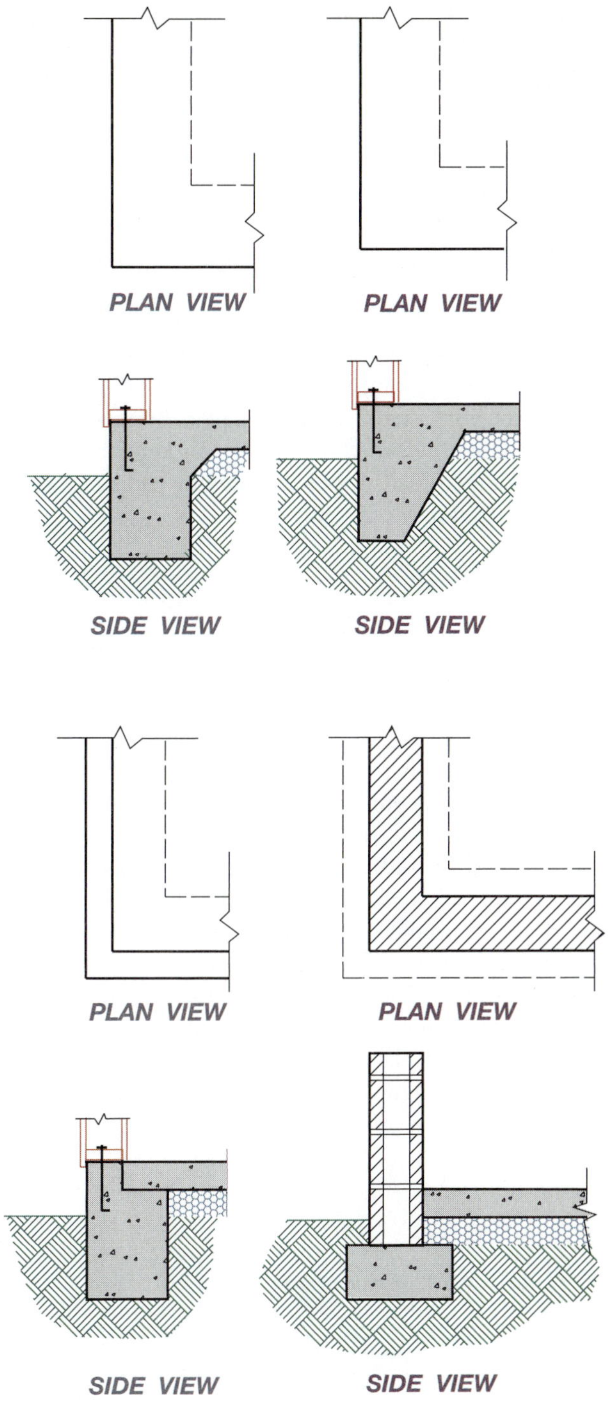
PLAN VIEW
PLAN VIEW
SIDE VIEW
SIDE VIEW
PLAN VIEW
PLAN VIEW
SIDE VIEW
SIDE VIEW

FIGURE 32.2 ■ Common foundation and slab intersections.

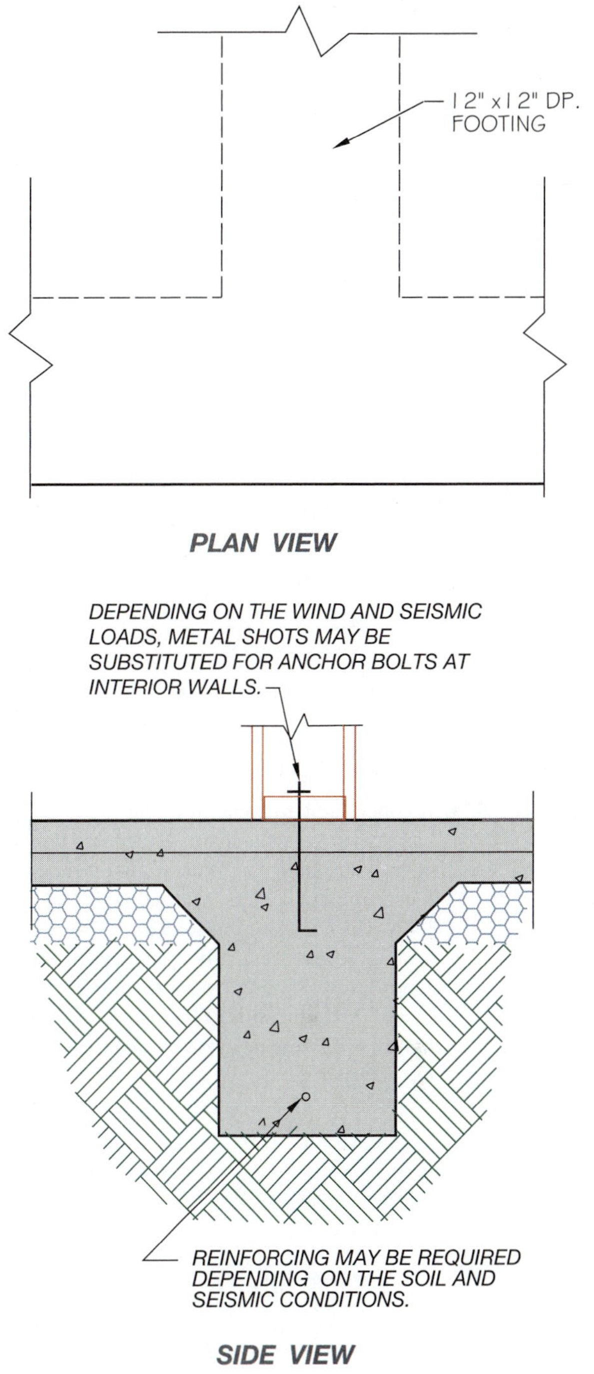
12" x 12" DP.
FOOTING
PLAN VIEW
DEPENDING ON THE WIND AND SEISMIC LOADS, METAL SHOTS MAY BE SUBSTITUTED FOR ANCHOR BOLTS AT INTERIOR WALLS.
REINFORCING MAY BE REQUIRED DEPENDING ON THE SOIL AND SEISMIC CONDITIONS.
SIDE VIEW

FIGURE 32.3 ■ Footing and floor intersections at an interior load-bearing wall.

slab has been weakened, any cracking due to stress will result along the joint. The American Concrete Institute (ACI) suggests that control joints be placed a distance in feet equal to about two and a half times the slab depth in inches. For a 4″ (100 mm) slab, joints would be placed at approximately 10′ (3000 mm) intervals. The locations of control joints are usually specified in note form for residential slabs. The spacing and method of placement can be specified in the general notes for the foundation plan.

Construction Joints

When concrete construction must be interrupted, a *construction joint* is used to provide a clean surface where work can be resumed. Because a vertical edge of one slab will have no bond to the next slab, a keyed joint is used to provide support between the two slabs. The key is typically formed by placing a beveled strip that is about one-fifth of the slab thickness and one-tenth of the slab thickness in width to the form used to mold the slab.

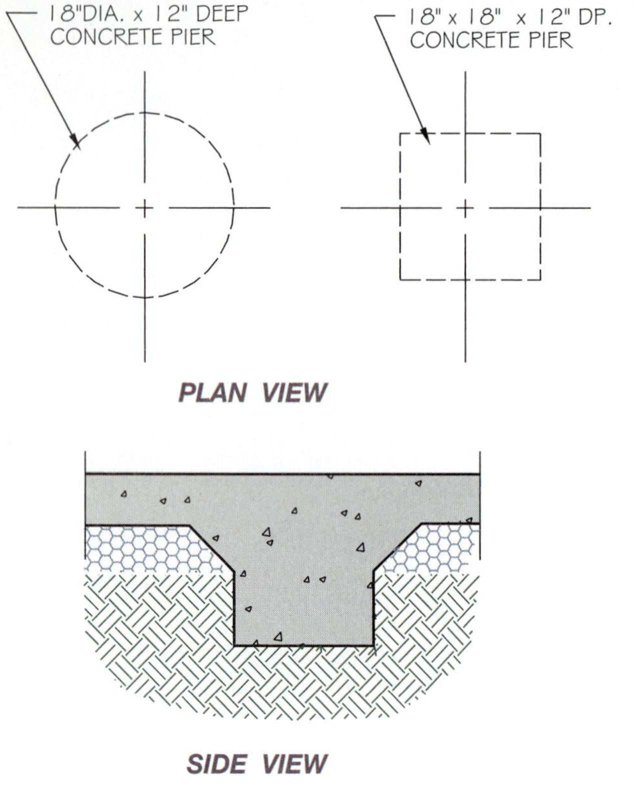

FIGURE 32.4 ■ A concrete pier is poured under loads concentrated in small areas. Piers are typically round or rectangular, depending on their location. Square piers are used when the exterior footing needs to be reinforced, or if the footing is so large that it must be framed with lumber. Round piers are used when the pier is placed inside the foundation. The load to be supported and the soil bearing capacity determine the size of each pier.

The method used to form the joint can be specified in note form on the foundation notes, but the crew placing the concrete will determine the location.

Isolation Joints

An *isolation* or *expansion joint* is used to separate a slab from an adjacent slab, wall, or column, or some other part of the structure. The joint prevents forces from an adjoining structural member from being transferred into the slab, causing cracking. Such a joint also allows for expansion of the slab caused by moisture or temperature. Isolation joints are typically between 1/4″ and 1/2″ (6–13 mm) wide. The location of isolation joints should be specified on the foundation plan. Because of the small size of residential foundations, isolation joints are not usually required. Chapter 46 discusses the drawing and specification of isolation joints in more detail.

Slab Placement

The slab may be placed above, below, or at grade level. Residential slabs are often placed above grade for hillside construction to provide a suitable floor for a garage. A platform made of wood or steel materials can be used to support a lightweight concrete floor slab. Concrete is considered lightweight depending on the amount of air that is pumped into the mixture during the manufacturing process. Above-grade residential slabs are typically supported by a wood-framed platform covered with plywood sheathing. Ribbed metal decks are typically used for heavier floors found in commercial construction. The components of an above-ground concrete floor are typically noted on a framing plan but not drawn. The foundation plan shows the columns and footings used to support the increased weight of the floor. An example of a framing plan to support an above-ground concrete slab can be seen in Figure 32.6. The foundation plan for the same

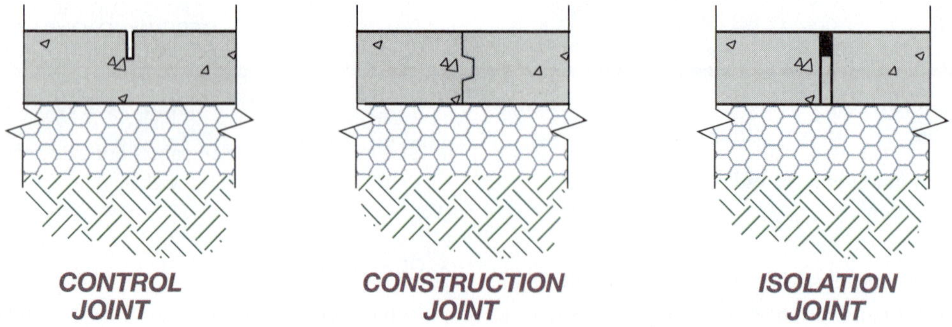

FIGURE 32.5 ■ Joints are placed in concrete to control cracking. A control joint is placed in the slab to weaken the slab and cause cracking to occur along the joint rather than throughout the slab. When construction must be interrupted, a construction joint is formed to increase bonding with the next day's pour. Isolation joints are provided to keep stress from one structural material from cracking another.

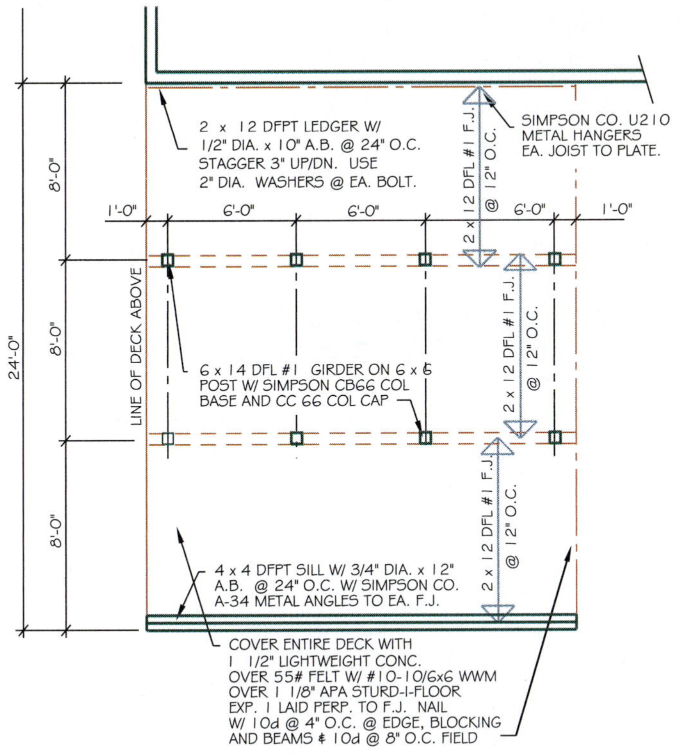

FIGURE 32.6 ■ An above-ground concrete floor is often used for a garage floor and driveway for hillside residential construction. The framing plan shows the framing of the platform used to support the concrete slab.

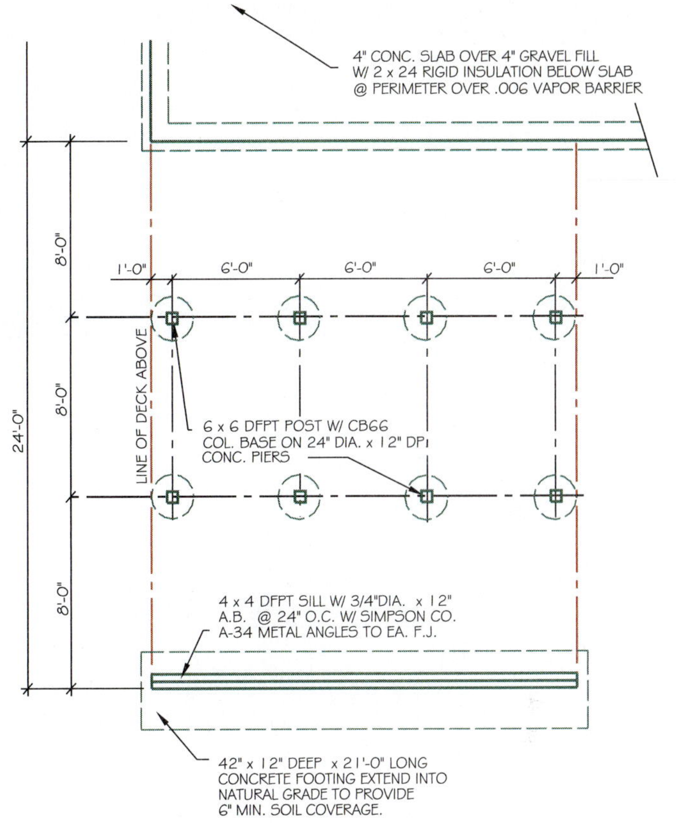

FIGURE 32.7 ■ The foundation plan for an above-ground concrete slab typically shows the support piers for the framing platform.

area can be seen in Figure 32.7. The construction process will also require details similar to those in Figure 32.8.

Slabs built below grade are most commonly used in basements. When used at grade, the slab is usually placed just above grade level. Most building codes require the top of the slab to be 8″ (203 mm) above the finish grade to keep structural wood away from the ground moisture.

Slab Preparation

When a slab is built at grade, approximately 8″ to 12″ inches (200–300 mm) of topsoil and vegetation is removed to provide a stable, level building site. Excavation usually extends about 5′ (1500 mm) beyond the building size to allow for the operation of excavating equipment needed to trench for the footings. Once forms for the footings have been set, fill material can be spread to support the slab. Most codes require the slab to be placed on a 4″ (102 mm)-minimum base of compacted sand or gravel fill. The area for which the building is designed will dictate the type of fill material that is typically used. The fill material provides a level base for the concrete slab and helps eliminate cracking in the slab caused by settling of the ground under the slab. The fill material is not shown on the foundation plan but is specified with a note. A typical note to specify the concrete and fill material might read:

> 4″ CONC. SLAB OVER .006 VISQUEEN OVER 1″ SAND FILL OVER 4″ COMPACTED GRAVEL FILL.

Slab Reinforcement

When the slab is placed on more than 4″ (102 mm) of uncompacted fill, welded wire fabric should be specified to help the slab resist cracking. Spacings and sizes of wires of welded wire fabric are identified by style and designations. A typical designation specified on a foundation plan might be: 6 × 12—W16 × W8, where:

- 6 = longitudinal wire spacing
- 12 = transverse wire spacing
- 16 = longitudinal wire size
- 8 = transverse wire size

The letter *W* indicates smooth wire. *D* can be used to represent deformed wire. Typically a steel mesh of number 10 wire in 6″ (150 mm) grids is used for residential floor slabs. A typical note to specify reinforced concrete and fill material might read:

> 4″ CONC. SLAB W/6 × 6 W12 × 12 WWM OVER .006 VISQUEEN OVER 1″ SAND FILL OVER 4″ COMPACTED GRAVEL FILL.

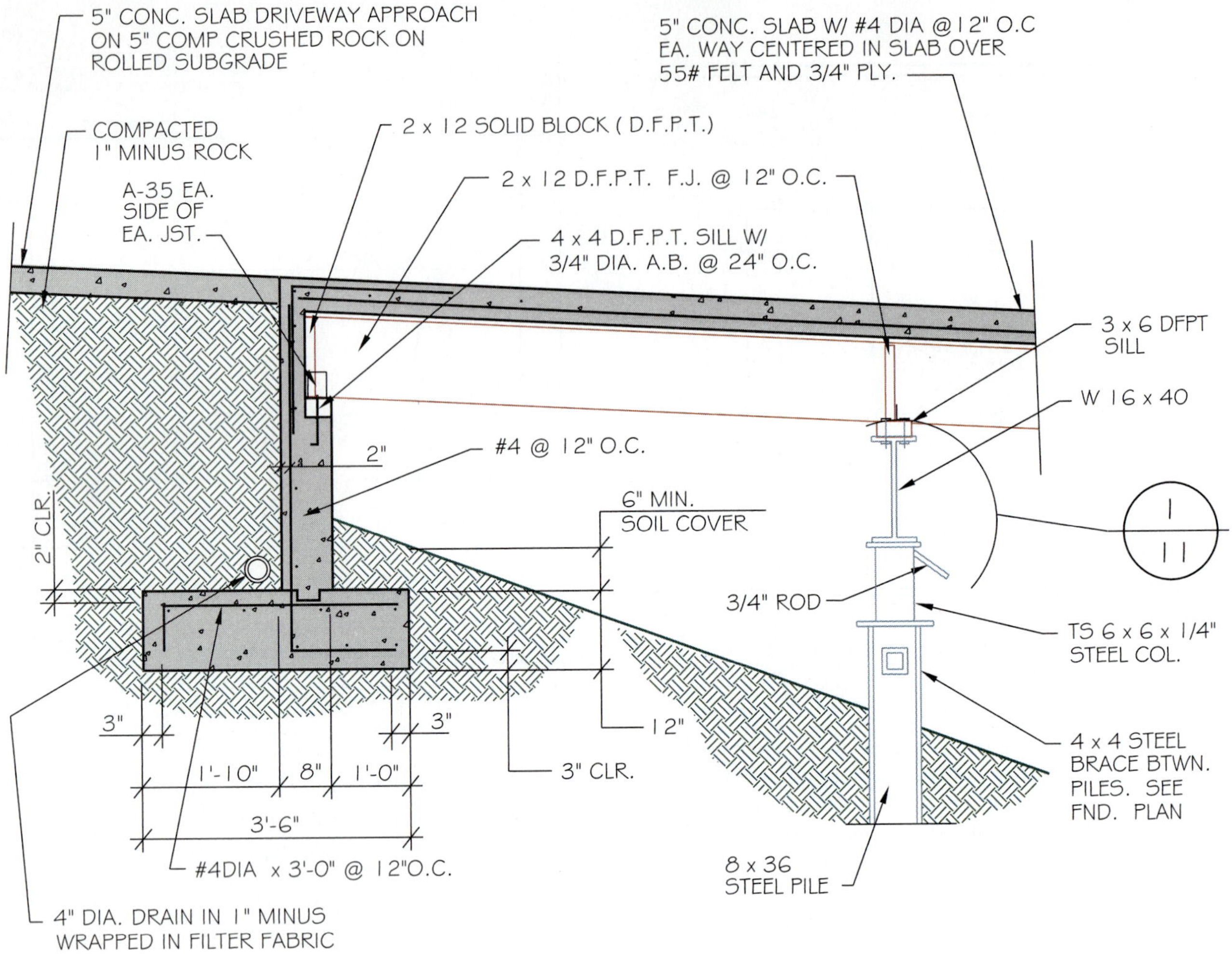

FIGURE 32.8 ■ In addition to the framing and foundation plans, details are used to clarify the concrete reinforcement.

FIGURE 32.9 ■ Welded wire mesh is often placed in concrete slabs to reduce cracking.

Figure 32.9 shows an example of the mesh used in concrete slabs.

Steel reinforcing bars similar to those shown in Figure 32.10 can be added to a floor slab to prevent bending of the slab due to expansive soil. While mesh is placed in a slab to limit cracking, steel reinforcement is placed in the concrete to prevent cracking due to bending. Steel reinforcing bars can be laid in a grid pattern near the surface of the concrete that will be in tension from bending. The placement of the reinforcement in the concrete is important to the effectiveness of the reinforcement. The amount of concrete placed around the steel is referred to as *coverage*. Proper coverage strengthens the bond between the steel and concrete and also protects the steel from corrosion if the concrete is exposed to chemicals, weather, or water. Proper coverage is also important to protect the steel from damage by fire. If steel is required to reinforce a residential concrete slab, an engineer will typically determine the size, spacing, coverage, and grade of bars to be used. As a general guideline to placing steel, the ACI recommends:

FIGURE 32.10 ■ Steel reinforcing is used in place of welded wire mesh when the slab is placed over fill material. *Courtesy Krista Herbel.*

- For concrete cast against and permanently exposed to earth (footings): 3″ (75 mm) minimum coverage.
- For concrete exposed to weather or earth such as basement walls: 2″ (50 mm) coverage for #6–18 bars and 1 1/2″ (38 mm) coverage for #5 and W31/D31 wire or smaller.
- For concrete not exposed to weather or in contact with ground: 1 1/2″ (38 mm) coverage for slabs, walls, and joists; 3/4″ (19 mm) coverage for #11 bars and smaller; and 1 1/2 ″ (38 mm) coverage for #14 and #18 bars.

Wire mesh and steel reinforcement is not shown on the foundation plan but is specified by a note similar to Figure 32.6.

Post-tensioned Concrete Reinforcement

The methods of reinforcement mentioned thus far assume that the slab will be poured over stable soil. Concrete slabs can often be poured over unstable soil by using a method of reinforcement known as *post-tensioning*. This method of construction was originally developed in the late 1950s for reinforcement of slabs that were to be poured at ground level and then lifted into place for multilevel structures. Adopted for the use of residential slabs, post-tensioning allows concrete slabs to be poured on grade over expansive soil. The technology has advanced sufficiently so that post-tensioning is now widely used even on stable soils.

For design purposes, a concrete slab can be considered as a wide, shallow beam that is supported by a concrete foundation at the edge. While a beam may sag at the center from loads and gravity, a concrete slab can either sag or bow, depending on the soil conditions. Soil at the edge of a slab will be exposed to more moisture than soil near the center of the slab. This differential in moisture content can cause the edges of the slab to heave, creating tension in the bottom portion of the slab and compression in the upper portion. Center lift conditions can result as the soil beneath the interior of the slab becomes wetter and expands, as the perimeter of the slab dries and shrinks, or as a combination of the two factors. As the center of the slab heaves, tension is created in the upper portion of the slab with compression in the lower portion. Because concrete is very poor at resisting stress from tension, the slab must be reinforced with a material such as steel that has high tensile strength. Steel tendons with anchors can be extended through the slab as it is poured. Usually between 3 and 10 days after the concrete has been poured, these tendons are stretched by hydraulic jacks, which place approximately 25,000 lb of force on each tendon. The tendon force is transferred to the concrete slab through anchorage devices at the ends of the tendons. This process creates an internal compressive force throughout the slab, increasing the ability of the slab to resist cracking and heaving. Post-tensioning usually allows for a thinner slab than normally would be required to span over expansive conditions, elimination of other slab reinforcing, and elimination of most slab joints.

Two methods of post-tensioning are typically used for residential slabs: flat slab and ribbed slab. The *flat slab* method uses steel tendons ranging in diameter from 3/8″ to 1/2″ (10–13 mm). Maximum spacings of tendons recommended by the Post Tensioning Institute (PTI) are:

3/8″ (9 mm) diameter: 5′–0″ (1524 mm) spacing.

7/16″ (11 mm) diameter: 6′–10″ (1829 mm) spacing.

1/2″ (13 mm) diameter: 9′–0″ (2743 mm) spacing.

The exact spacing and size of tendons must be determined by an engineer based on the loads to be supported and the strength and conditions of the soil. When required, the tendons can be represented on the foundation plan, as seen in Figure 32.11. Details will also need to be provided to indicate how the tendons will be anchored, as well as to show the exact locations of the tendons. Figure 32.12 is an example of a tendon detail. In addition to representing and specifying the steel throughout the floor system, the drafter will need to specify the engineer's requirements for the strength of the concrete at 28 days, the period when the concrete is to be stressed, as well as what strength the concrete should achieve before stressing.

A second method of post-tensioning an on-grade floor slab is with the use of concrete ribs or beams placed below the slab. These beams reduce the span of the slab over the soil and provide increased support. The width, depth, and spacing are determined by the engineer based on the strength and condition of the soil and the size of the slab. Figure 32.13 shows an example of a beam detail that a drafter would be required to draw

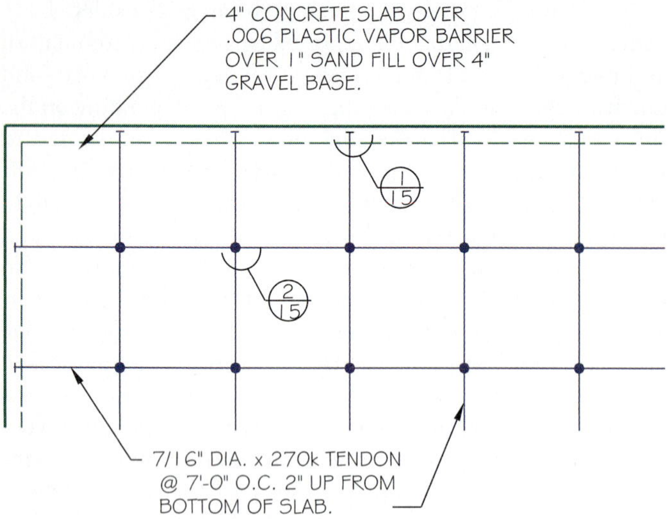

FIGURE 32.11 ■ When a concrete slab is post-tensioned, the tendons and anchors used to support the floor slab must be represented on the foundation plan.

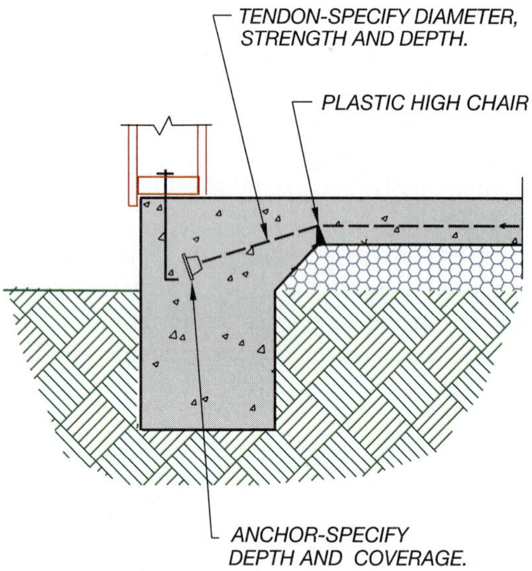

FIGURE 32.12 ■ Tendon details should be drawn by the drafter to reflect the design of the engineer.

to show the reinforcing specified by the engineer. Figure 32.14 is an example of how these beams could be shown on the foundation plan.

Moisture Protection

In most areas, the slab is required to be placed over 6 mil polyethylene sheet plastic to protect the floor from ground moisture. When a plastic vapor barrier is to be placed over gravel, a layer of sand should be specified to cover the gravel fill to avoid tearing the vapor barrier. An alternative is to use 55# rolled roofing in place of the plastic. The vapor barrier is not drawn on the foundation plan but is specified with a note on the foundation plan.

Slab Insulation

Depending on the risk of freezing, some municipalities require the concrete slab to be insulated to prevent heat loss. The insulation can be placed under the slab or on the outside of the stem wall. When it is placed under the slab, a 2 × 24″ (50 × 600 mm) minimum rigid insulation material should be used to insulate the slab. An isolation joint is typically provided between the stem wall and the slab to prevent heat loss through the stem wall. When placed on the exterior side of the stem wall, the insulation should extend past the bottom of the foundation. Care must be taken to protect exposed insulation on the exterior side of the wall. This can usually be done by placing a protective covering such as 1/2″ (13 mm) concrete board over the insulation. Figure 32.15 a and b show common methods of insulating concrete floor slabs. Insulation is not shown on the foundation plan but is represented by a note, as in Figure 32.16, and specified in sections and footing details.

Plumbing and Heating Requirements

Plumbing and heating ducts must be placed under the slab before the concrete is poured. On residential plans, plumbing is usually not shown on the foundation plan. Generally the skills of the plumbing contractor are relied on for the placement of re-

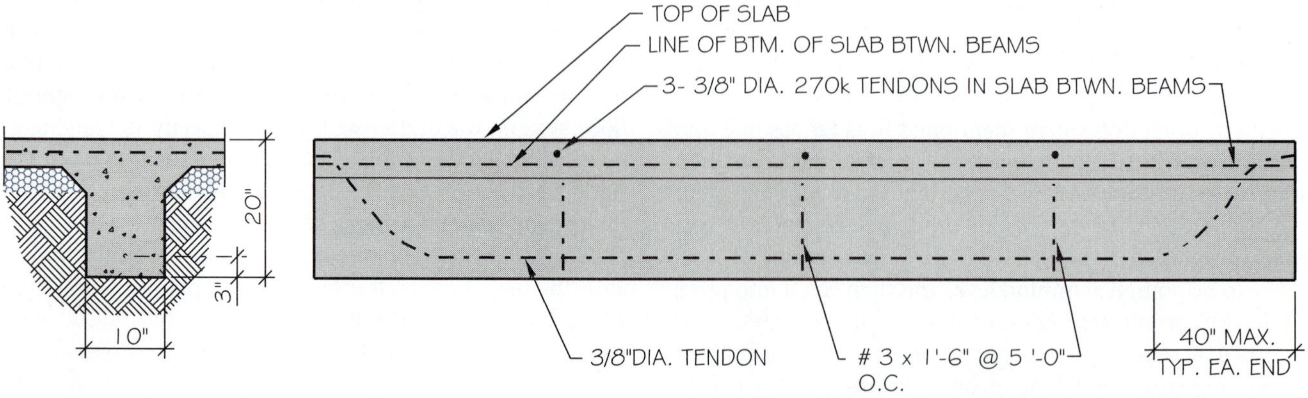

FIGURE 32.13 ■ Concrete beams can be placed below the floor slab to increase the effect of the slab tensioning. Details must be drawn to indicate how the steel will be placed in the beam, as well as how the beam steel will interact with the slab steel.

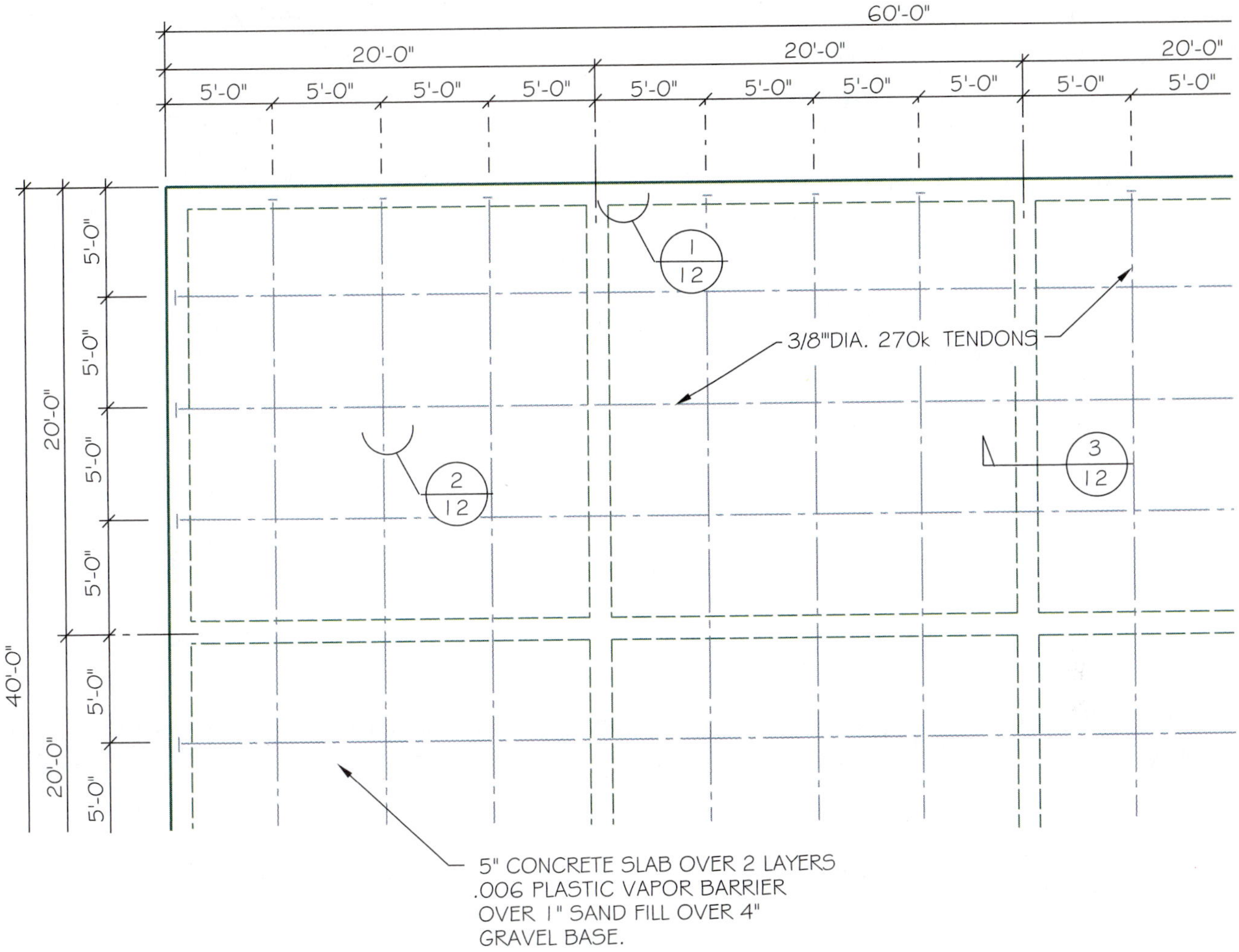

FIGURE 32.14 ■ Subslab beams must be located on the foundation plan using methods like those used to represent an interior footing.

quired utilities. Although piping runs are not shown, terminations such as floor drains are often shown and located on a concrete slab plan. Figure 32.16 shows how floor drains are typically represented. Figure 32.17 shows the placement of plumbing materials prior to pouring the slab. If heating ducts will be placed under the slab, they are usually drawn on the foundation plan as shown in Figure 32.18. Drawings for commercial construction usually include both a plumbing plan and a mechanical plan. Each will be presented in Chapter 44.

Changes in Floor Elevation

The floor level is often required to step down to meet the design needs of the client. A stem wall similar to Figure 32.19 is formed between the two floor levels and should match the required width for an exterior stem wall. The lower slab is typically thickened to a depth of 8″ (200 mm) to support the change in elevation. Steps between floors greater than 24″ (600 mm) should be designed as a retaining wall and detailed using the methods described in Chapters 31 and 36. Figure 32.16 shows how a lowered slab can be represented. The step often occurs at what will be the edge of a wall when the framing plan is complete. Great care must be taken to coordinate the dimensions of a floor plan with those of the foundation plan, so that walls match the foundation.

Common Components Shown on a Slab Foundation

Refer to Figure 32.16. Dimensions are needed for each of the following items to provide location information.

- Outline of slab
- Interior footing locations
- Changes in floor level
- Floor drains
- Exterior footing locations

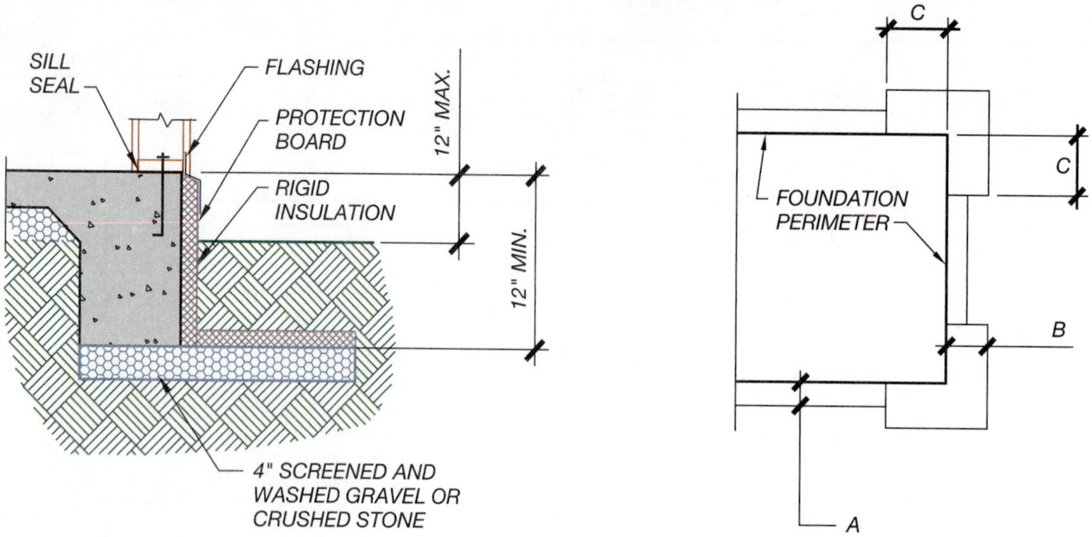

INSULATION DETAIL

HORIZONTAL INSULATION PLAN

FIGURE 32.15A ■ Insulation placement for frost protected footings in heated buildings. *Reproduced from the* International Residential Code / 2000. *Copyright © 2000. Courtesy International Code Council, Inc.*

MINIMUM INSULATION REQUIREMENTS FOR FROST-PROTECTED FOOTINGS IN HEATED BUILDINGS[A]

AIR FREEZING INDEX (°F-DAYS)[b]	VERTICAL INSULATION R-VALUE[c,d]	HORIZONTAL INSULATION R-VALUE[c,e]		HORIZONTAL INSULATION DIMENSIONS PER FIGURE R403.3(1) (INCHES)		
		ALONG WALLS	AT CORNERS	A	B	C
1,500 or less	4.5	NR	NR	NR	NR	NR
2,000	5.6	NR	NR	NR	NR	NR
2,500	6.7	1.7	4.9	12	24	40
3,000	7.8	6.5	8.6	12	24	40
3,500	9.0	8.0	11.2	24	30	60
4,000	10.1	10.5	13.1	24	36	60

For SI: 1 inch = 25.4 mm, °C = [(°F)-32]/1.8.

a. *Insulation requirements are for protection against frost damage in heated buildings. Greater values may be required to meet energy conservation standards. Interpolation between values is permissible.*

b. *See Figure R403.3(2) for Air Freezing Index values.*

c. *Insulation materials shall provide the stated minimum R-values under long-term exposure to moist, below-ground conditions in freezing climates. The following R-values shall be used to determine insulation thicknesses required for this application: Type II expanded polystyrene—2.4R per inch; Type IV extruded polystyrene—4.5R per inch; Type VI extruded polystyrene—4.5R per inch; Type IX expanded polystyrene—3.2R per inch; Type X extruded polystyrene—4.5R per inch. NR indicates that insulation is not required.*

d. *Vertical insulation shall be expanded polystyrene insulation or extruded polystyrene insulation.*

e. *Horizontal insulation shall be extruded polystyrene insulation.*

FIGURE 32.15B ■ Common alternatives of placing insulation for a concrete slab. *Reproduced from the* International Residential Code / 2000. *Copyright © 2000. Courtesy International Code Council, Inc.*

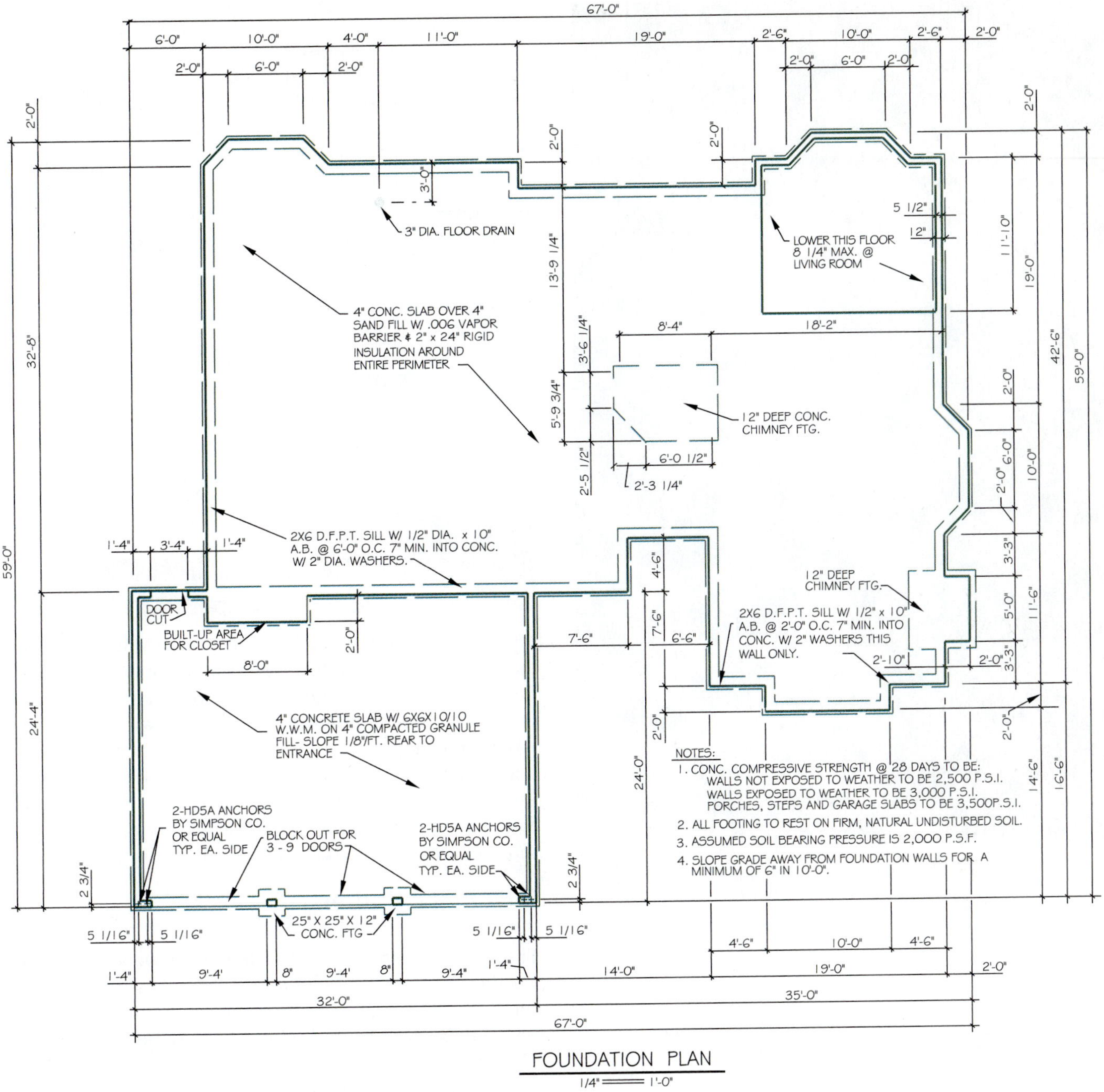

FIGURE 32.16 ■ A foundation plan for a home with a concrete slab floor system.

- Ducts for mechanical
- Metal anchors
- Patio slabs

Common Components Specified by Note Only

See Figure 32.16.

- Slab thickness and fill material
- Wire mesh
- Mudsill size
- Vapor barriers
- Pier sizes
- Assumed soil strength
- Reinforcing steel
- Anchor bolt size and spacing
- Insulation
- Slab slopes
- Concrete strength

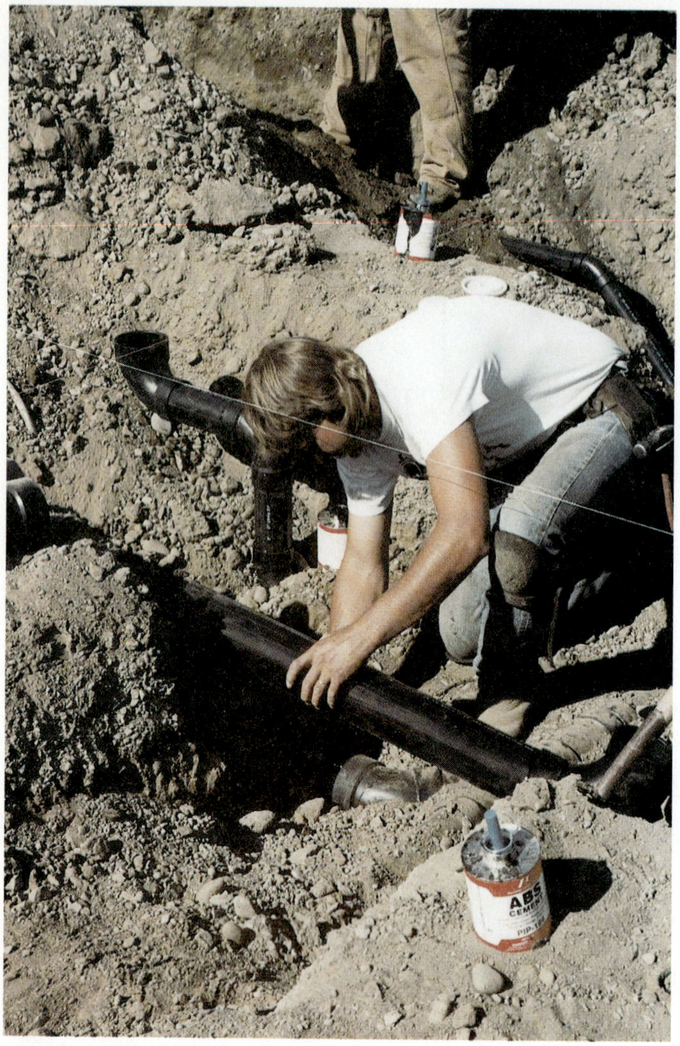

FIGURE 32.17 ■ Plumbing must be placed prior to preparing for the concrete slab. *Courtesy Judy Schmitke.*

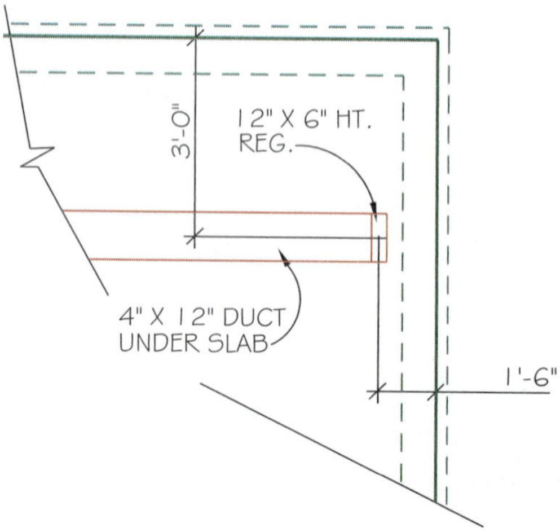

FIGURE 32.18 ■ HVAC ducts and registers that are to be located in the floor must be shown on the foundation plan.

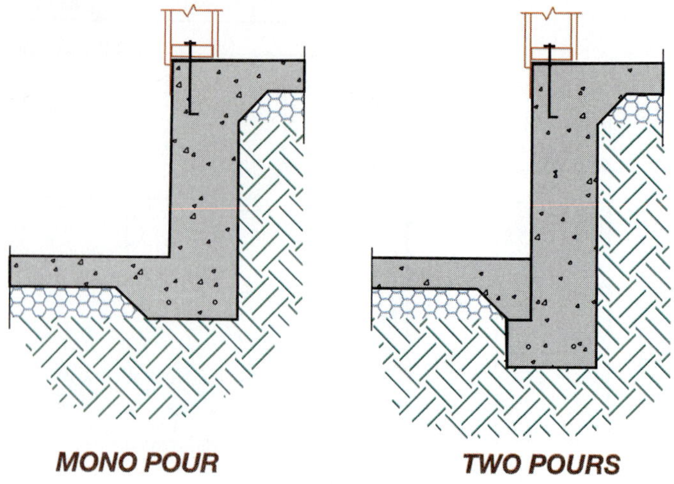

FIGURE 32.19 ■ A stem wall is created between floors when a change in elevation is required.

CRAWL SPACE FLOOR SYSTEMS

The crawl space is the area formed between the floor system and the ground. Building codes require a minimum of 18″ (457 mm) from the bottom of the floor to the ground and 12″ (305 mm) from the bottom of beams to the ground. Two common methods of providing a crawl space below the floor are the conventional method using floor joists and the post-and-beam system. An introduction to each system is needed to complete the foundation. Both floor systems were discussed in detail in relationship to the entire structure in Section 8.

Joist Floor Framing

The most common method of framing a wood floor is with wood members called floor joists. Floor joists are used to span between the foundation walls and are shown in Figure 32.20*a*. Sawn lumber ranging in size from 2 × 6 through 2 × 12 (50 × 150 to 50 × 300 mm) has traditionally been used to frame the floor platform. Contractors in many parts of the country are now framing with trusses or joists made from engineered lumber similar to Figure 32.20*b*. (These materials were discussed in Chapter 25.) The floor joists are usually placed at 16″ (400 mm) on center, but the spacing of joists may change depending on the span, the material used, and the load to be supported.

To construct a joist floor system, a pressure-treated sill is bolted to the top of the foundation wall with the anchor bolts that were placed when the foundation was formed. The floor joists can then be nailed to the sill. An alternative used to reduce the distance from the finished floor to the finish grade is to set the floor joists flush with the sill, as in Figure 32.21. This method requires the use of metal hangers to support the joists. With the floor joists in place, plywood floor sheathing is installed to provide a base for the finish floor. The size of the subfloor depends on the spacing of the floor joists and the floor

A

B

FIGURE 32.20 ■ (a) A floor framed with floor joists uses wood members ranging from 2 × 6 through 2 × 12 (50 × 150 to 50 × 300 mm) to span between supports. Solid blocking is placed between the joists at 10′–0″ (3000 mm) o.c. (b) Engineered lumber is a popular material for joist floor systems. Joists are typically placed at 16″ (400 mm) o.c. Depending on the load to be supported, the allowable deflection, and the span, a 12″ or 24″ (300 or 600 mm) spacing may be used. *Courtesy Leroy Cook.*

FIGURE 32.21 ■ Although floor joists are normally placed above the sill, they can be mounted flush to the sill to reduce the distance from the finish floor to the finish grade. The weight of the joist is transferred to the sill using joist hangers. *Courtesy Karen Griggs.*

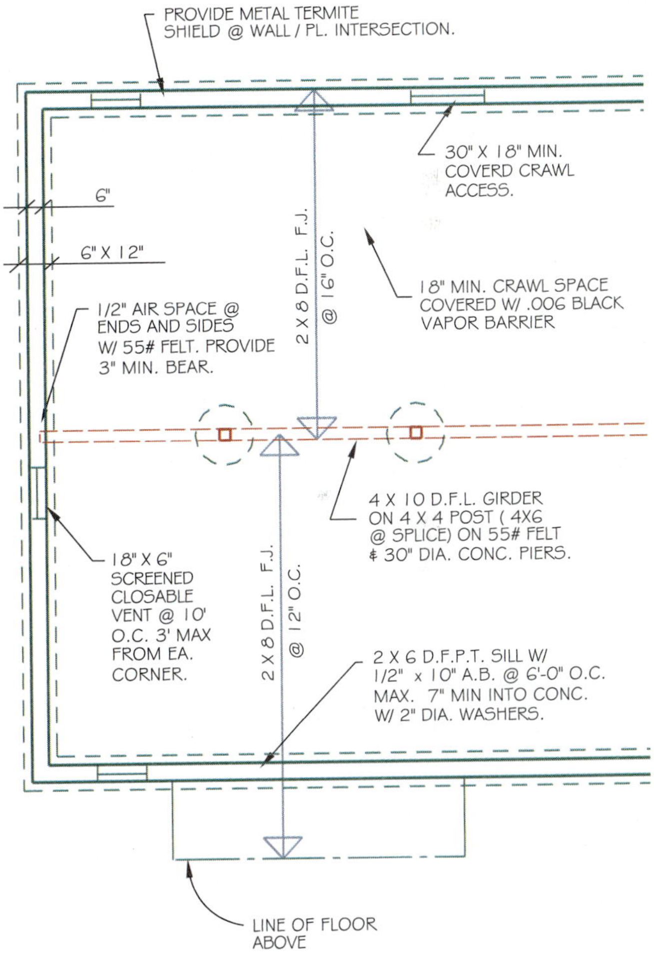

FIGURE 32.22 ■ Common methods of representing and specifying floor joist components in plan view.

loads that must be supported. The live load to be supported will also affect the thickness of the plywood to be used. The American Plywood Association (APA) has span tables for plywood from 7/16″ through 7/8″ (11.1–22.2 mm) to meet the various conditions that might be found in a residence. A subfloor can also be made from 1″ (25 mm) material such as 1 × 6 (25 × 150 mm) tongue-and-groove (T&G) lumber, although this method requires more labor. The floor sheathing is not represented on the foundation plan but is specified in a general note. Figure 32.22 shows common methods of drawing floor joists on the foundation plan.

When the distance between the foundation walls is too great for the floor joists to span, a girder is used to support the joists. A girder is a horizontal load-bearing member that spans between two or more supports at the foundation level. Depending on the load to be supported and the area you are in, either a wood, laminated wood, engineered wood product, or steel member may be

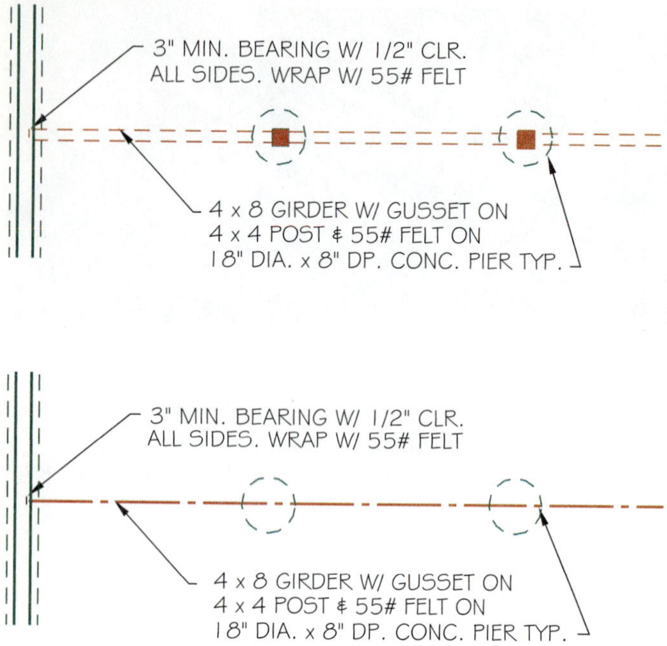

FIGURE 32.23 ■ Common methods of drawing girders in plan view.

FIGURE 32.24 ■ A gusset is used to attach the girder to the post so that each will move together during seismic activity. The gusset can be a scrap of wood, or a prefabricated metal connector. Post bracing may be required depending on the height of the post and the seismic zone.

used for the girder and for the support post. Girders may be drawn using a single bold line or pairs of thin lines similar to Figure 32.23. The girder is usually supported in a concrete beam pocket, where it intersects the foundation wall. Posts made of wood, laminated wood, engineered wood, or steel columns are used to support girders that span between the foundation walls. No matter what the material is, the girder must be attached to the post. When wood members are used, a gusset attaches the post and girder together, like the connection in Figure 32.24. A steel connector can also be used to make wood to wood connections or wood to steel connections. A steel girder may require no intermediate supports or can be welded to a steel column.

A concrete pier is placed under the intermediate girder supports to resist settling. The post must be attached to the pier in some seismic zones. A metal rod inserted into a pre-drilled hole in the post can be used in areas of low seismic risk. A metal strap or post base may be required where high wind or seismic risk is greater. Figure 32.25 shows common connections required for a joist floor system. Figure 32.26 shows methods of representing common floor system components on a foundation plan. Figure 32.27 shows a portion of the floor framing in place. Figure 32.22 shows methods of drawing the girders, posts, piers, and beam pocket on the foundation plan. Figure 32.26 shows a complete foundation plan using floor joists to support the floor.

Common Components Shown with a Joist Floor System

See Figure 32.26. Unless marked with an asterisk, all components on the following pages require dimensions to provide location information.

- Foundation walls
- Door openings in foundation wall
- Fireplace
- Floor joists
- Girders (bearing walls and floor support)
- Girder pockets
- Crawl access*
- Exterior footings
- Metal anchors
- Fireplace footings
- Outline of cantilevers
- Interior piers
- Changes in floor levels
- Vents for crawl space*

Common Components Specified by Note Only

See Figure 32.26.

- Floor joist size and spacing
- Anchor bolt size and spacing
- Insulation
- Crawl height

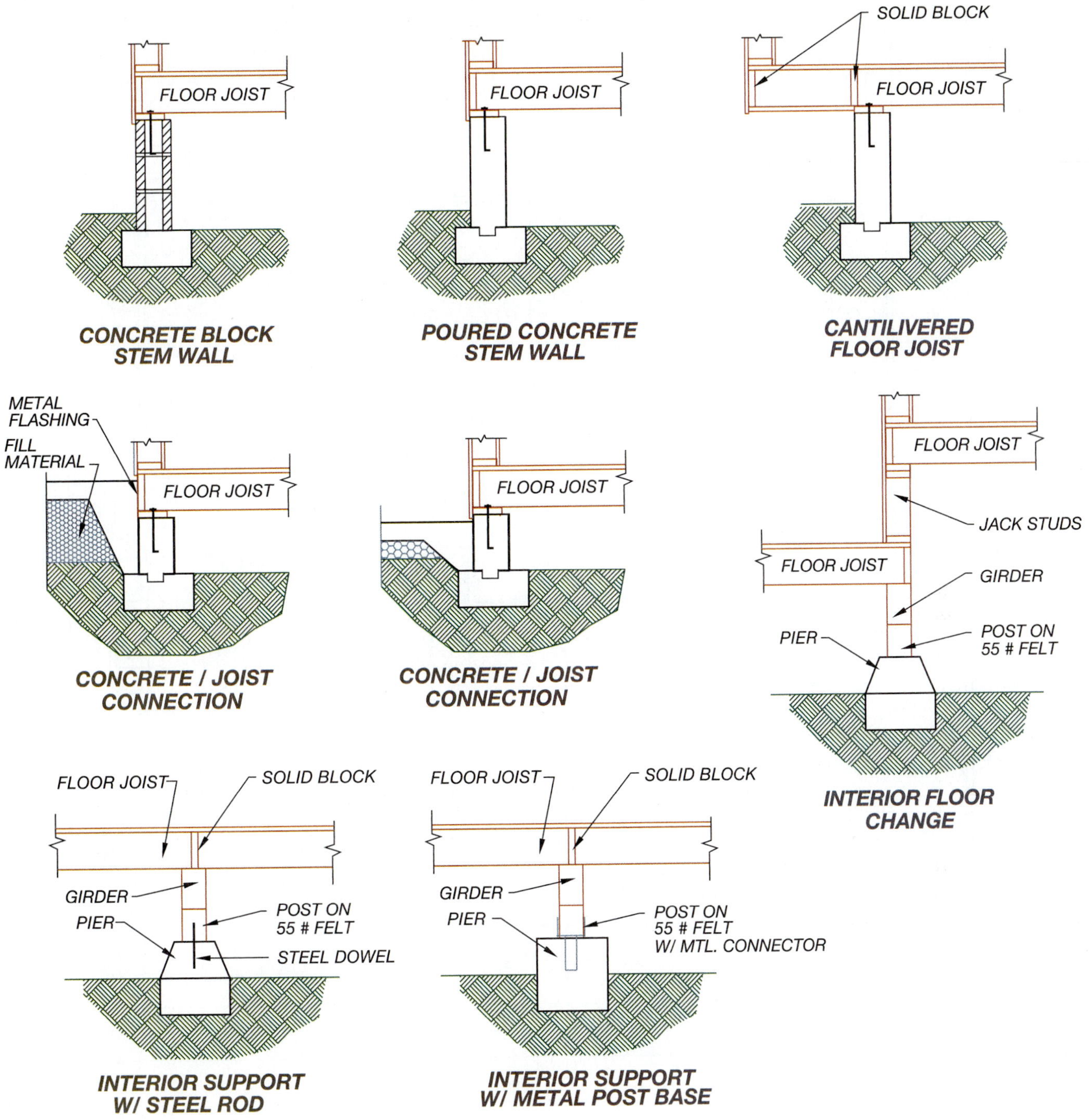

FIGURE 32.25 ■ Common connections for a joist floor system.

- Assumed soil strength
- Subfloor material
- Girder size
- Mudsill size
- Vapor barrier
- Concrete strength
- Wood type and grade

Post-and-Beam Floor Systems

A post-and-beam floor system is built using a standard foundation system. Rather than having floor joists span between the foundation walls, a series of beams are used to support the subfloor, as shown in Figure 32.28. Once the mudsill is bolted to the foundation wall, the beams are placed so that the top of each beam is flush with the top of the mudsill. The beams are usually placed at 48″ (1200 mm) on center, but the spacing can vary de-

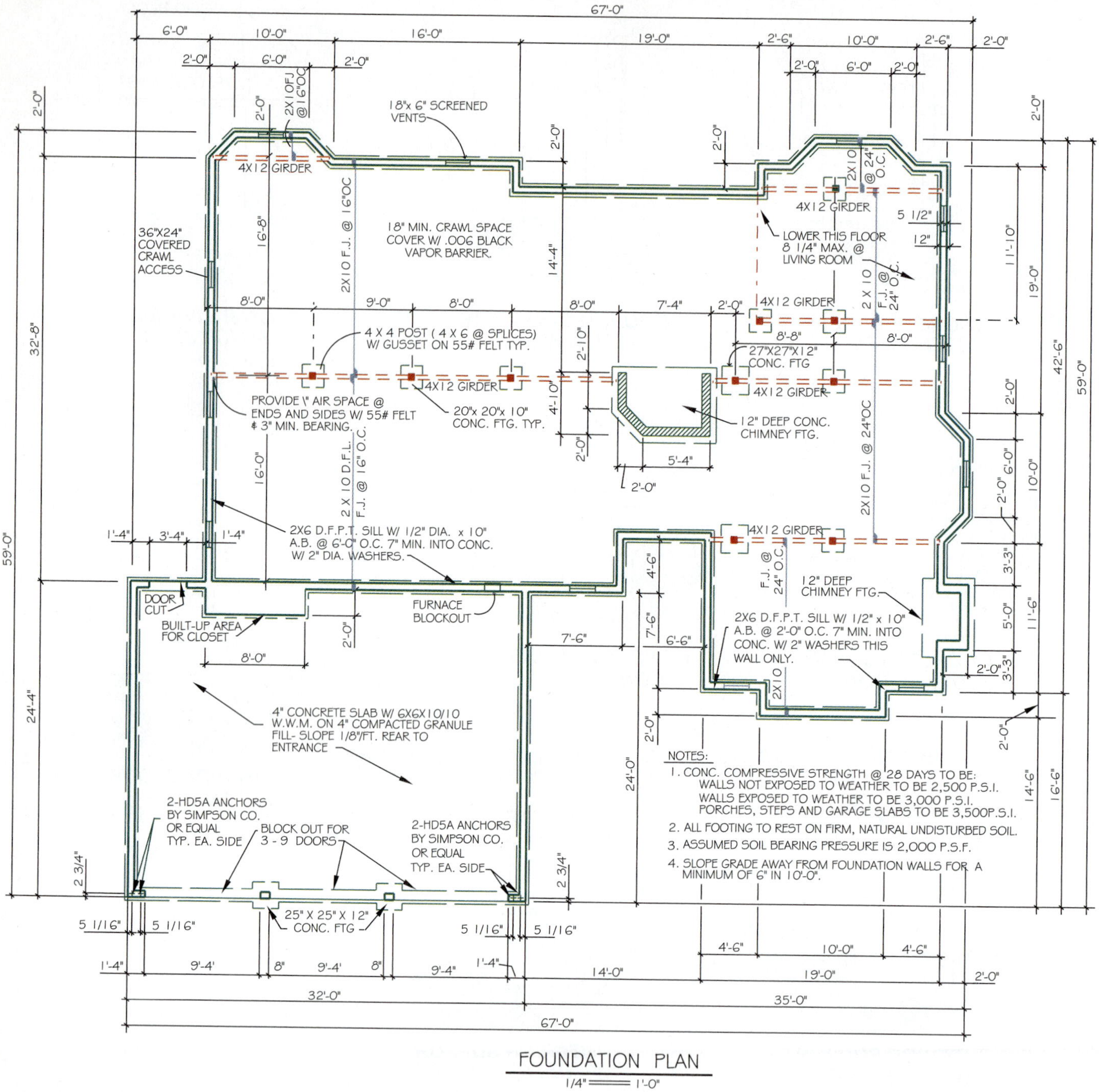

FIGURE 32.26 ■ Foundation plan for a home with a joist floor system.

pending on the size of the floor decking to be used. The area of the country where the structure is to be built affects how the subfloor is made. Generally material 2″ (50 mm) thick, such as 2 × 6 (50 × 150 mm) T & G boards laid perpendicular to the beams, are used for a subfloor. Plywood with a thickness of 1 1/8″ (28.5 mm) and an APA rating of STURD-I-FLOOR 2-4-1 with an exposure rating of EXP-1 can also be used to build a post-and-beam subfloor quickly and economically. When the subfloor is glued to the support beams, the strength and quality of the floor is greatly increased because squeaks, bounce, and nail popping are eliminated.

The beams are supported by wooden posts as they span between the foundation walls. Posts are usually placed at 8′ (2400 mm) on center but spacing can vary depending on the load to

FIGURE 32.27 ■ Placement of floor materials for a cantilevered bay.

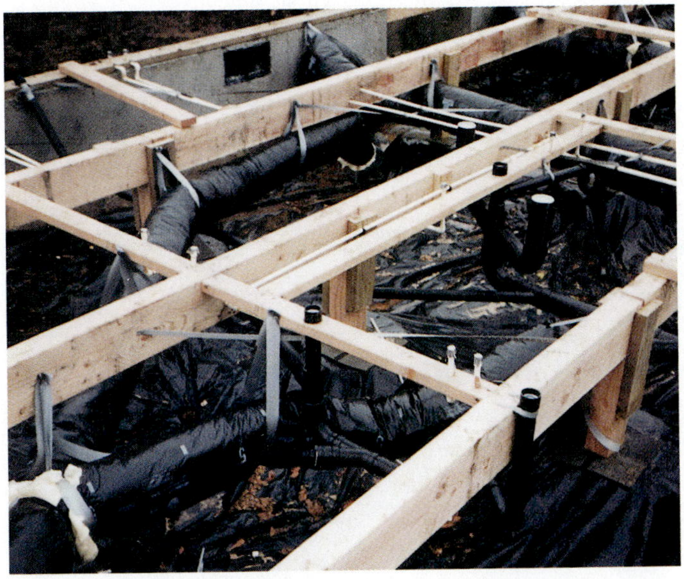

FIGURE 32.28 ■ Girders are usually placed at 48″ (1200 mm) o.c. with supports at 8′–0″ (2400 mm) o.c. Blocking is placed between the beams to support plumbing materials.

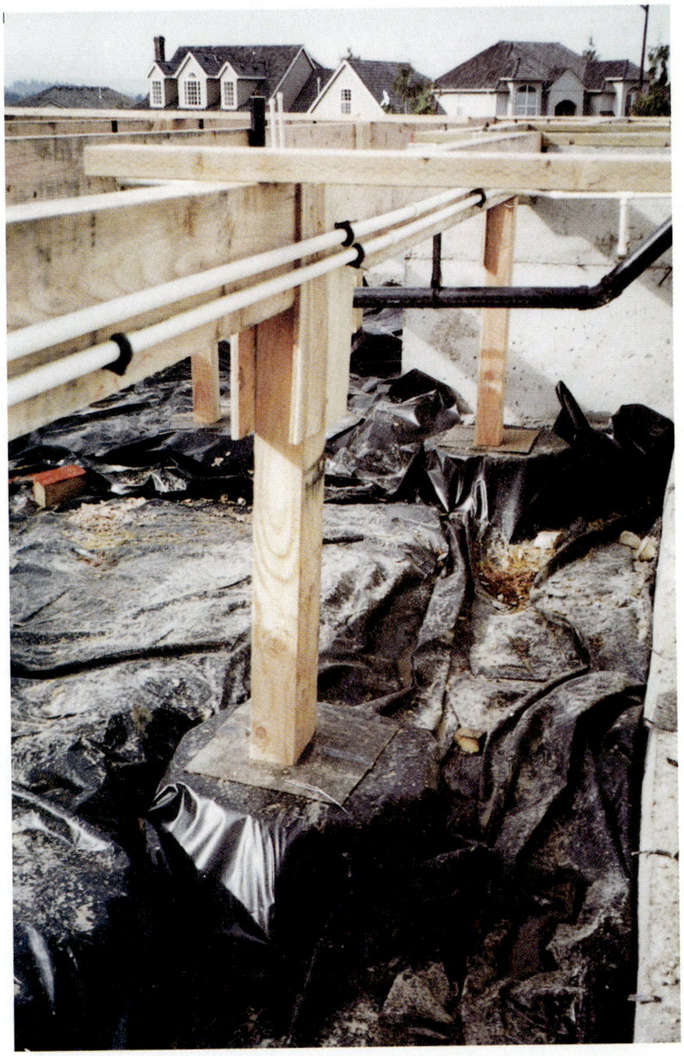

FIGURE 32.29 ■ Posts for a joist or post-and-beam system sit on a shingle or 55# felt placed above a concrete pier. The shingle is placed to stop moisture from being drawn into the post. A vapor barrier keeps moisture from passing from the soil into the crawl space.

be supported. Each post is supported by a concrete pier similar to Figure 32.29. Figure 32.30 shows common components of a post-and-beam floor system. Beams, posts, and piers are drawn on the foundation plan, as shown in Figure 32.23. Figure 32.31 shows a foundation plan with a post-and-beam floor system.

Common Components Shown for a Post-and-Beam System

See Figure 32.31. Unless marked with an asterisk, all components require dimensions to provide location information.

- Foundation walls
- Openings in walls for doors
- Fireplace
- Girders (beams)
- Girder pockets
- Crawl access*
- Exterior footings
- Metal anchors
- Fireplace footings
- Piers
- Changes in floor levels
- Vents for crawl space*

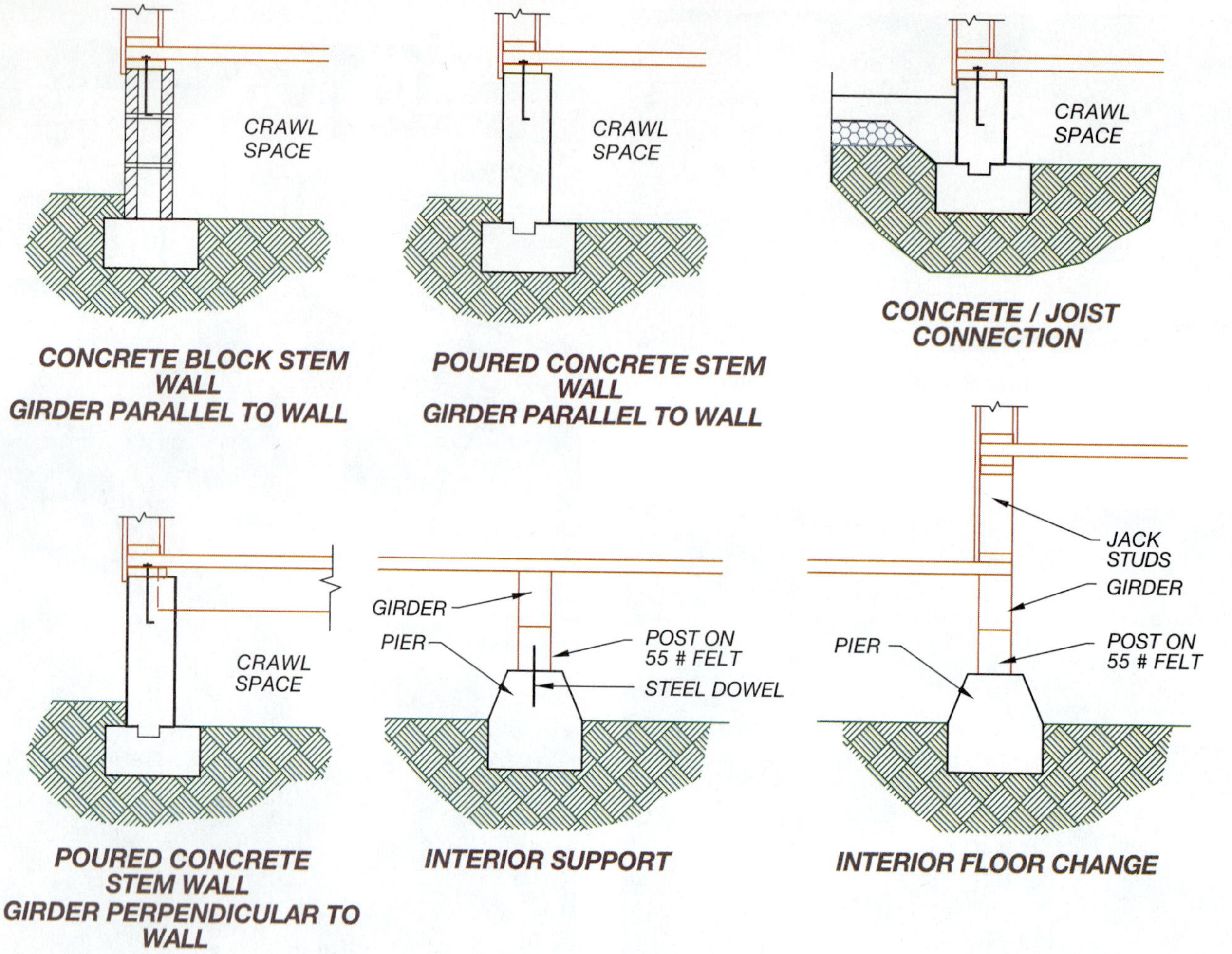

FIGURE 32.30 ■ Common connections for a post-and-beam floor system.

Common Components Specified by Note only

See Figure 32.31.

- Anchor bolt spacing
- Mudsill size
- Insulation
- Minimum crawl height
- Subfloor material
- Girder size
- Vapor barrier
- Concrete strength
- Wood type and grade
- Assumed soil strength

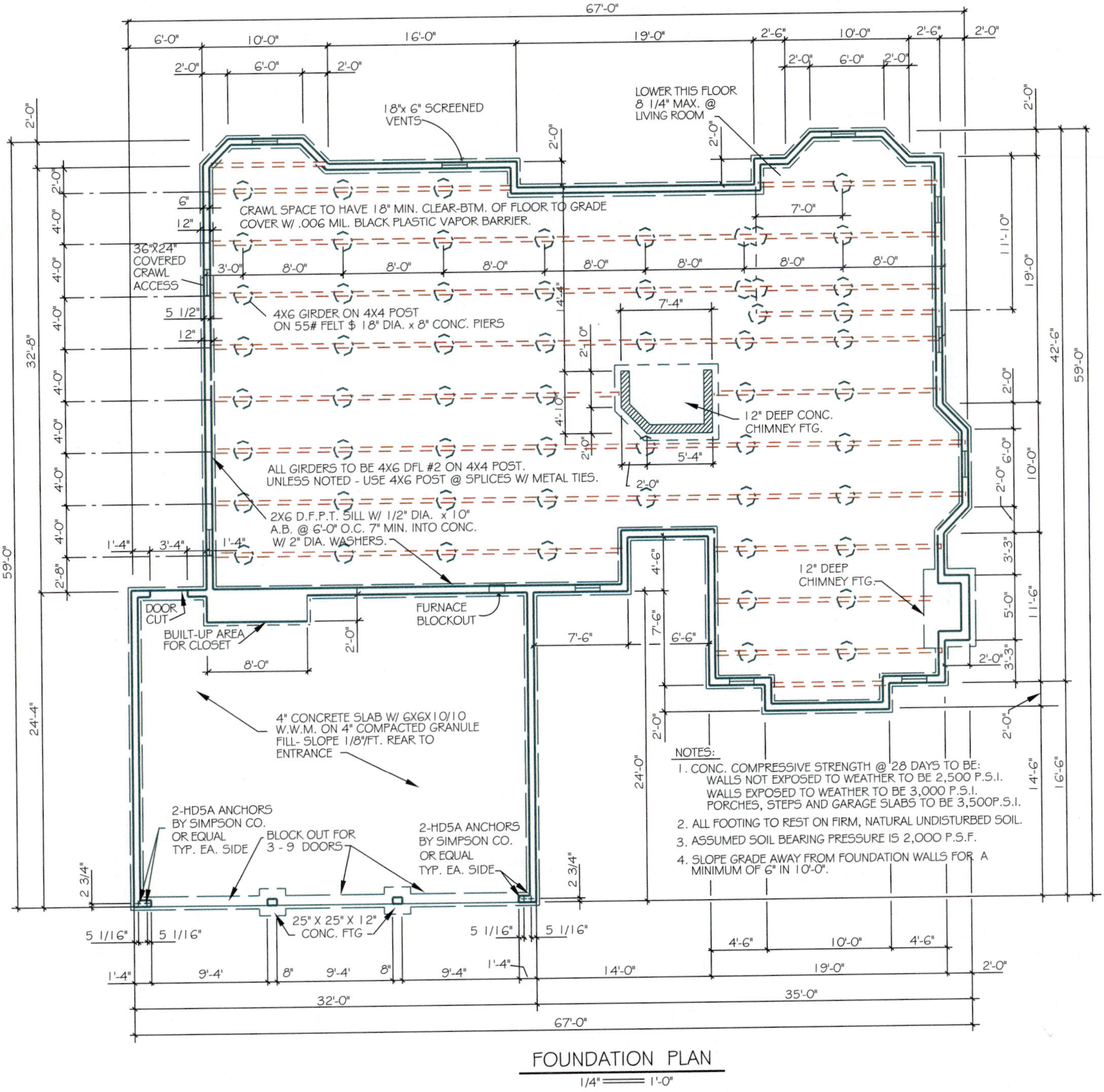

FIGURE 32.31 ■ Foundation plan for a home with a post-and-beam floor system.

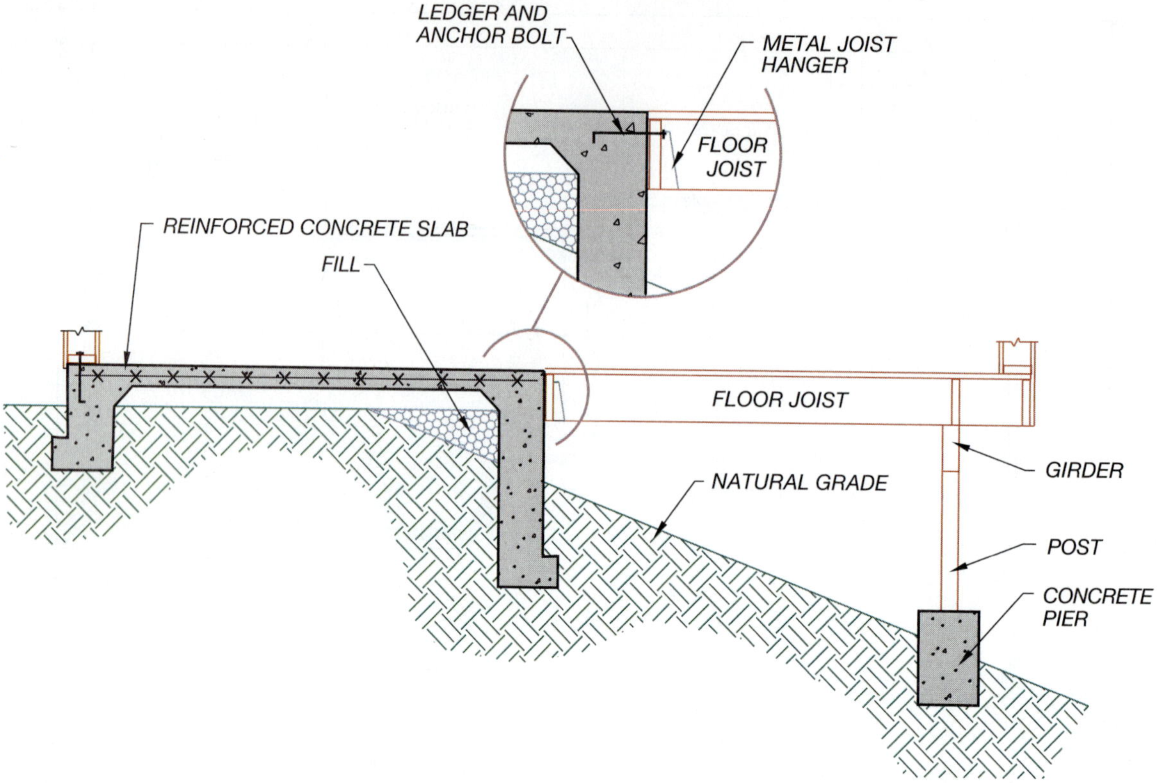

FIGURE 32.32 ■ Concrete slab and floor joist systems are often combined on sloping sites to help minimize fill material. A ledger is used to provide anchorage to floor joists where they intersect the concrete slab. Metal joist hangers are used to join the joists to the ledger.

Combined Floor Methods

Floor and foundation methods may be combined depending on the building site. This is typically done on partially sloping lots when part of a structure may be constructed with a slab and part of the structure with a joist floor system, as seen in Figure 32.32.

A residence with a basement is another example of construction that requires combined floor methods. A home with a partial basement will require the use of a concrete floor in the basement area, with either a joist or post-and-beam floor system over the crawl space. A joist floor for the crawl space is easier to match the floor over the basement area. Figure 32.33*a* shows a residence that has a partial basement. The right portion of the plan uses a joist floor system. The left side of the structure has a basement. Figure 32.33*b* shows a residence with a full basement. The entire basement can be constructed using the methods that were introduced earlier, when concrete floors were discussed. Figure 32.33*c* shows a residence with a full basement. The major difference between *b* and *c* is that the retaining wall does not totally enclose the basement. This type of basement is ideal for homes built on sloping sites. The basement is built on the low side of the site, allowing the lower walls to be constructed of wood. Notice on each side of the foundation that a retaining wall 48″ (1200 mm) high has been added. The wall allows the length of the wall 8′–0″ (2400 mm) high to be reduced, allowing the lower floor to be treated as an on-grade slab. Depending on your area, the drawings for the basement may require an engineer's stamp. Chapter 36 will explore the drawing needed to detail the basement walls.

One component typically used when floor systems are combined is a ledger. A ledger is used to provide support for floor joists and subfloor when they intersect the concrete. Unless felt is placed between the concrete and the ledger, the ledger must be pressure-treated lumber. The ledger can be shown on the foundation plan.

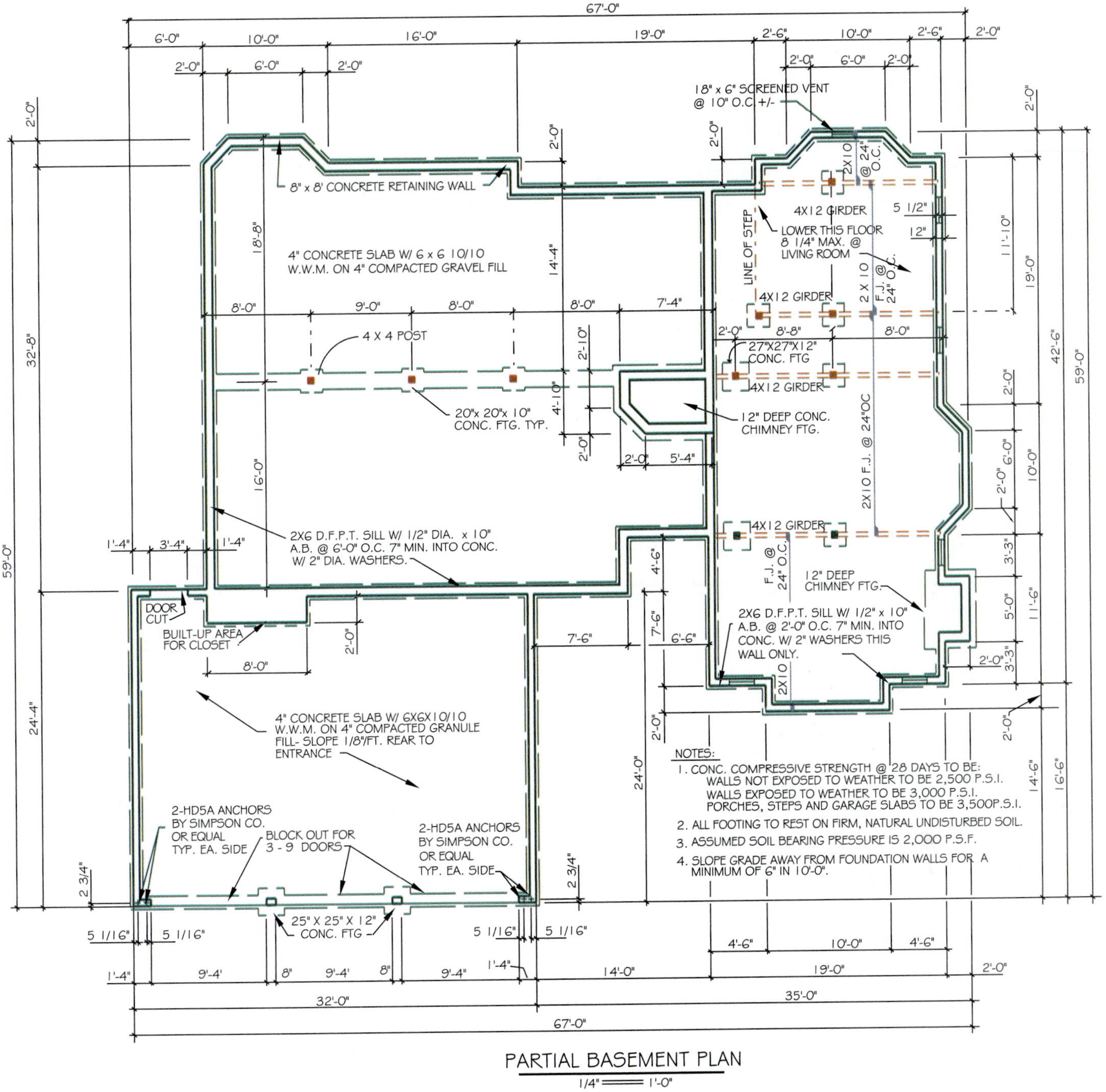

FIGURE 32.33(A) ■ Foundation plan for a home with a partial basement. The basement has a concrete floor, and the floor over the crawl space is supported by joists.

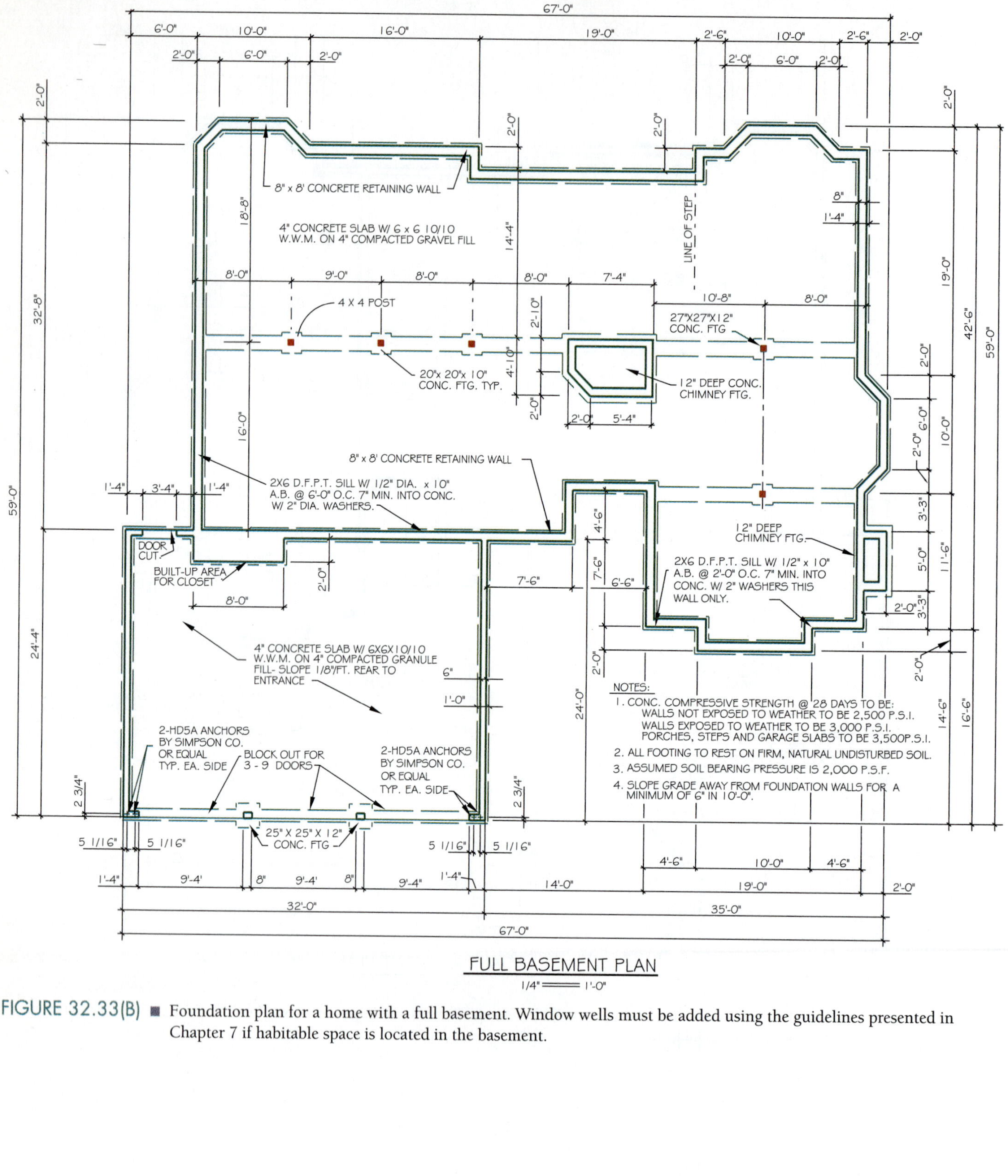

FIGURE 32.33(B) ■ Foundation plan for a home with a full basement. Window wells must be added using the guidelines presented in Chapter 7 if habitable space is located in the basement.

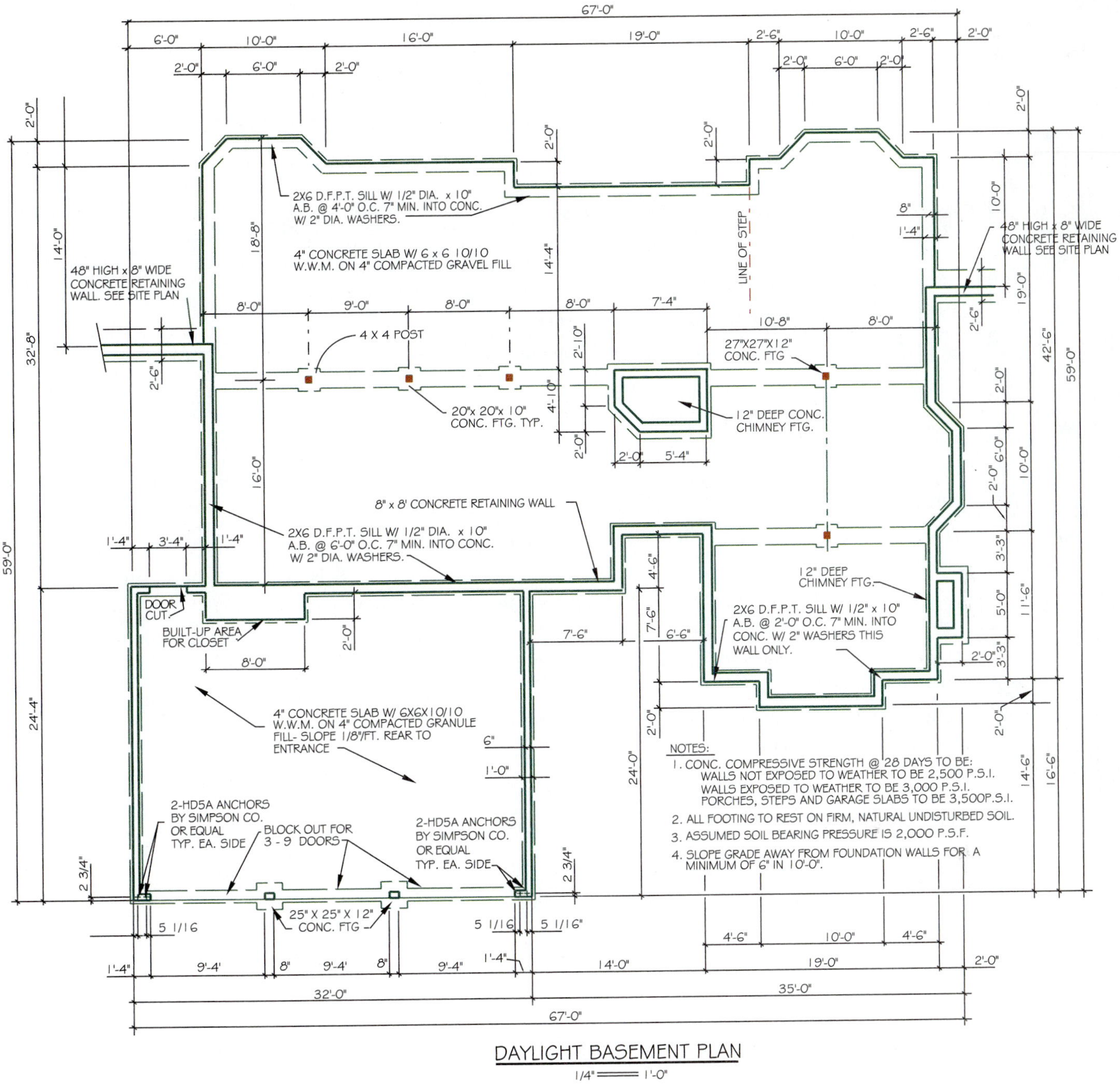

FIGURE 32.33(C) ■ A foundation plan for a home with a daylight basement. When a home is built on a sloping site, the low side of the basement can be constructed using slab-on-grade construction methods. The retaining walls 48″ (1200 mm) high on each side of the home allow standard wood construction to be used for lower walls.

CHAPTER 32

Floor Systems and Foundation Support Test

DIRECTIONS

Answer the questions with short, complete statements or drawings as needed.

1. Letter your name and the date at the top of the sheet.
2. Letter the question number and provide the answer. You do not need to write out the question.

QUESTIONS

Question 32–1 Why is a concrete slab called an on-grade floor system?

Question 32–2 What is the minimum thickness for a residential slab?

Question 32–3 Why are control joints placed in slabs?

Question 32–4 What is the minimum amount of fill placed under a slab?

Question 32–5 What thickness of vapor barrier is to be placed under a slab?

Question 32–6 What is the minimum height required in the crawl area?

Question 32–7 What is the purpose of a girder?

Question 32–8 How are floor joists attached to the foundation wall?

Question 32–9 How are girders supported at the foundation wall?

Question 32–10 What is a common spacing for beams in a post-and-beam floor?

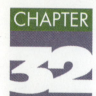

CHAPTER 33

Foundation Plan Layout

INTRODUCTION

The foundation plan is typically drawn at the same scale as the floor plan that it will support. Although the floor plan can be traced to obtain overall sizes, this practice can lead to major errors in the foundation plan. If you trace a floor plan that is slightly out of scale, you will reproduce the same errors in the foundation plan. A better method is to draw the foundation plan using the dimensions that are found on a print of the floor plan. If the foundation cannot be drawn using the dimensions on the floor plan, your floor plan may be missing dimensions or may contain errors. Great care needs to be taken with the foundation plan. If the foundation plan is not accurate, changes may be required that affect the entire structure.

This chapter includes guidelines for several types of foundation plans: concrete slab, joist construction, post-and-beam, and partial and full basements. Each is based on the floor plan that was used for examples in Section 4. Before attempting to draw a foundation plan, study the completed plan that follows each example so you will know what the finished drawing should look like.

The foundation plan can be drawn by dividing the work into the six stages: (1) foundation layout, (2) drawing foundation members, (3) drawing floor framing members, (4) dimensioning, (5) lettering, and (6) evaluation. As you progress through the drawing, you will use several types of line quality. For layout steps, construction lines in nonreproducible blue with a 6H lead will be best. When drawing finished-quality lines, use the following:

5 mm lead, #0 pen, or sharp 3H lead for thin lines

7 mm lead, #2 pen, or sharp H lead for bold lines

9 mm lead, #3 pen, or H lead and draw two parallel lines very close for very bold lines.

CONCRETE SLAB FOUNDATION LAYOUT

The following steps can be used to draw a foundation with a concrete slab floor system. Not all steps will be required for every house. When the concrete slab foundation is complete, it should resemble the plan shown in Figure 33.5. Use construction lines for steps 1 through 9. Each step can be seen in Figure 33.1.

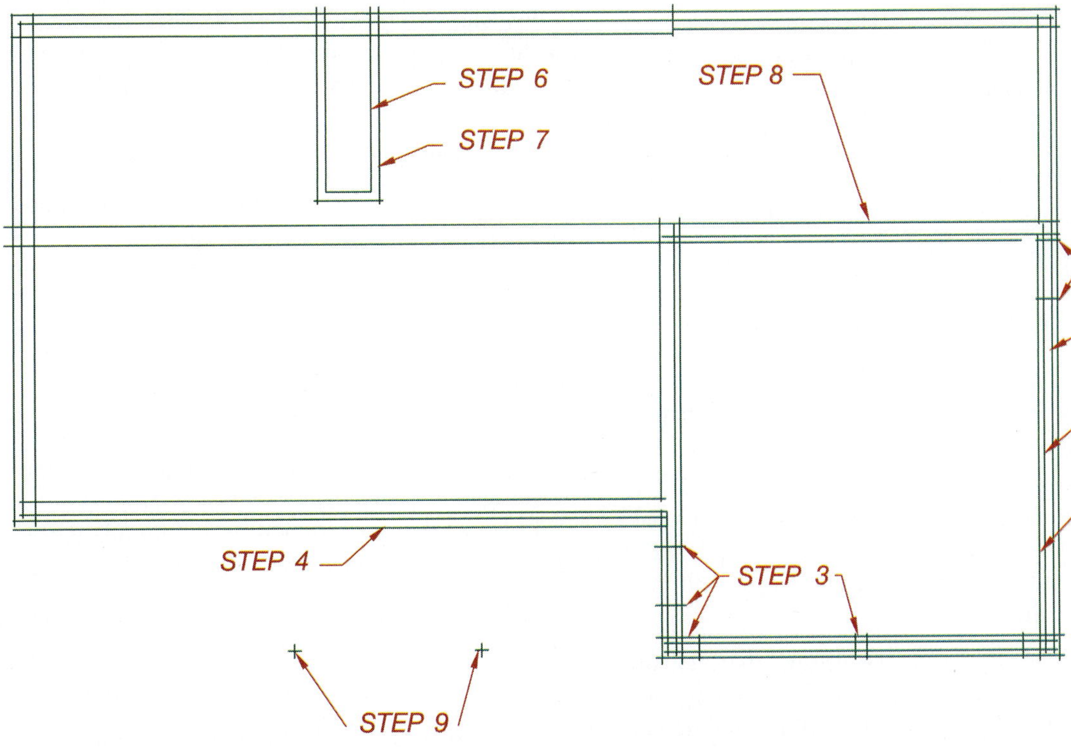

FIGURE 33.1 ■ Use construction lines to lay out the location of stem wall, footings, door locations, and the fireplace for a concrete slab foundation plan.

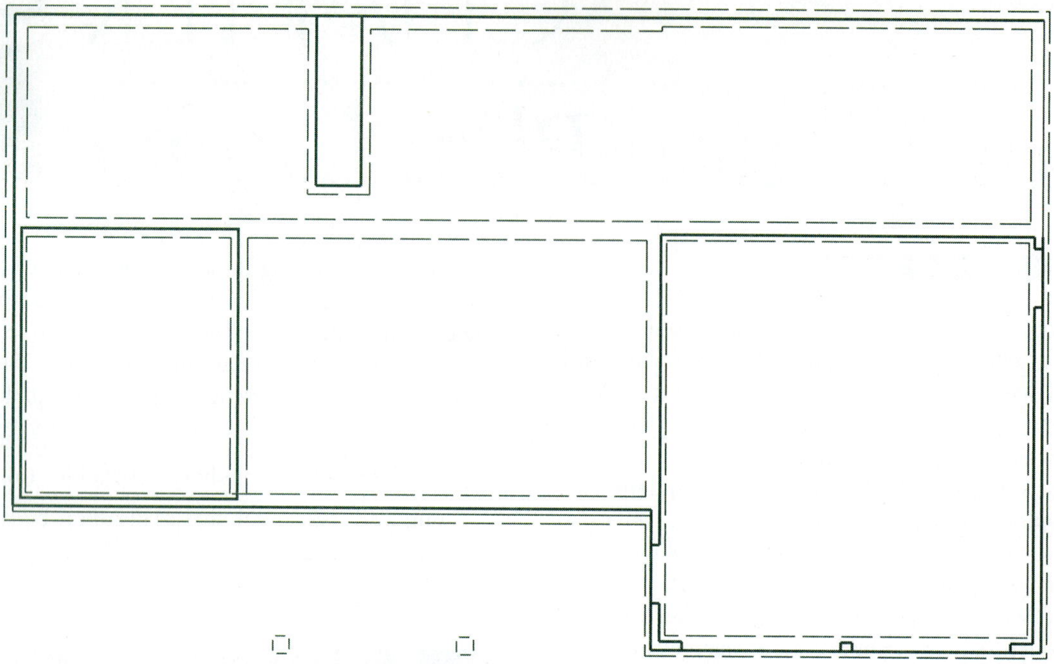

FIGURE 33.2 ■ Carefully darken all objects using the proper finished-line quality. Construction lines should be so light that they do not need to be erased.

STEP 1 Using the dimensions on your floor plan, lay out the exterior edge of the slab. The edge of the slab should match the exterior side of the exterior walls on the floor plan.

STEP 2 Draw the interior side of the stem wall around the slab at the garage area. See Figure 31.12a for a review of foundation dimensions.

STEP 3 Block out doors in the stem walls. Allow for the door size plus 4″ (100 mm).

STEP 4 Lay out a support ledge if brick veneer is to be used.

STEP 5 Lay out the exterior footing width.

STEP 6 Lay out the size of the fireplace based on measurements from the floor plan.

STEP 7 Lay out the fireplace footing so that it extends 6″ (150 mm) minimum beyond the face of the fireplace.

STEP 8 Lay out interior footings.

STEP 9 Lay out any exterior piers that might be required for decks or porches.

STEP 10 Darken all items that were drawn in steps 1–9. Use bold lines to draw steps 1–4 with finished-line quality. Use thin dashed lines to draw steps 5, 7, 8, and 9 with finished-line quality. Your drawing should now resemble the drawing in Figure 33.2.

See Figure 33.3 for steps 11 through 14. Because the following items are simple, they can be drawn without construction lines. These items may or may not be required, depending on your plan.

STEP 11 Draw changes in the floor levels.

STEP 12 Draw metal connectors and exterior piers.

STEP 13 Draw floor drains.

STEP 14 Draw heating registers if required.

You now have drawn all of the information that is required to represent the floor and foundation systems. Follow steps 15 through 20 to place the required dimensions on the drawing. Use thin lines for all extension and dimension lines. Your drawing should resemble Figure 33.4 when complete.

Draw extension and dimension lines to locate:

STEP 15 The overall size on each side of the foundation.

STEP 16 Jogs in the foundation walls.

STEP 17 Door openings in the stem wall.

STEP 18 Interior footings.

STEP 19 The fireplace.

STEP 20 Heating and plumbing materials.

The final drawing procedure is to place dimensions and specify the materials that are to be used. Figure 33.5 is an example of the notes that are required on the foundation. Use the following steps to complete the foundation plan.

STEP 21 Compute and neatly letter all dimensions in the appropriate location.

STEP 22 Neatly letter all required general notes.

STEP 23 Neatly letter all required local notes.

STEP 24 Place a title and scale under drawing.

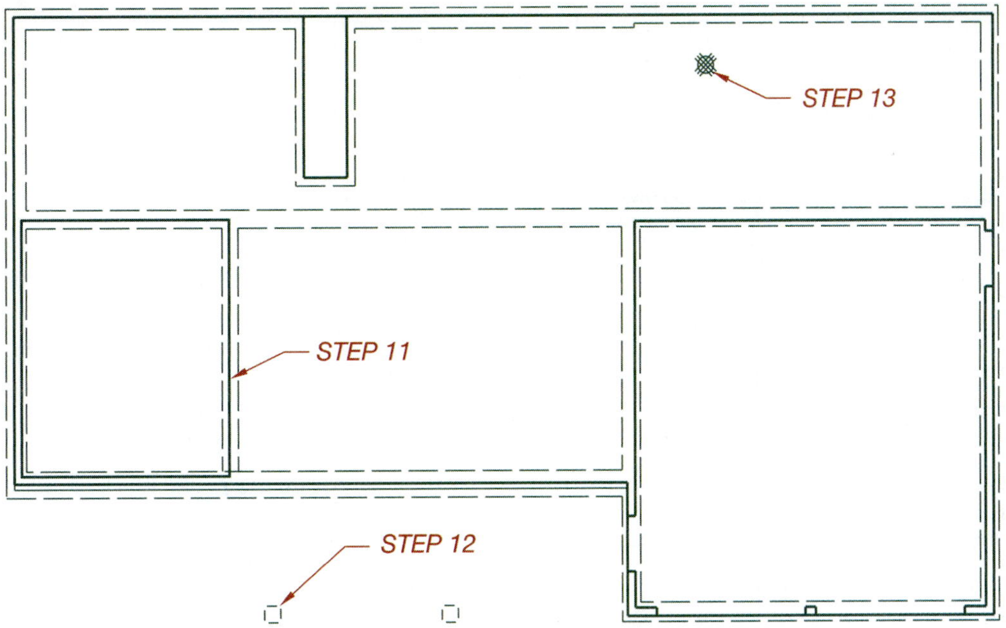

FIGURE 33.3 ■ Add any finishing materials such as lower slabs and plumbing, electrical, and HVAC material that will be below the slab.

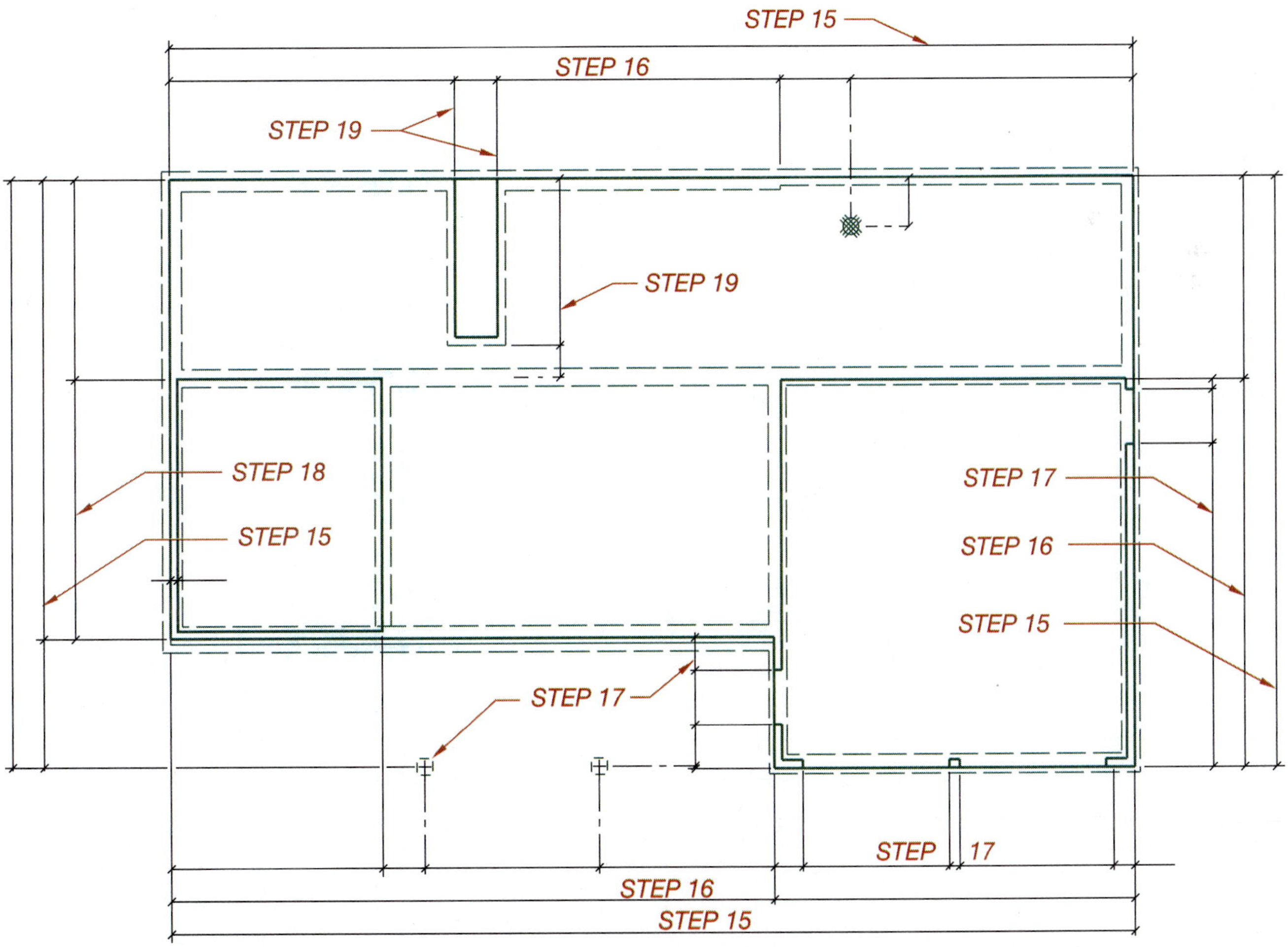

FIGURE 33.4 ■ Prepare the plan to be dimensioned by adding dimension and extension lines to describe the overall, major jogs, and each opening on each side of the foundation.

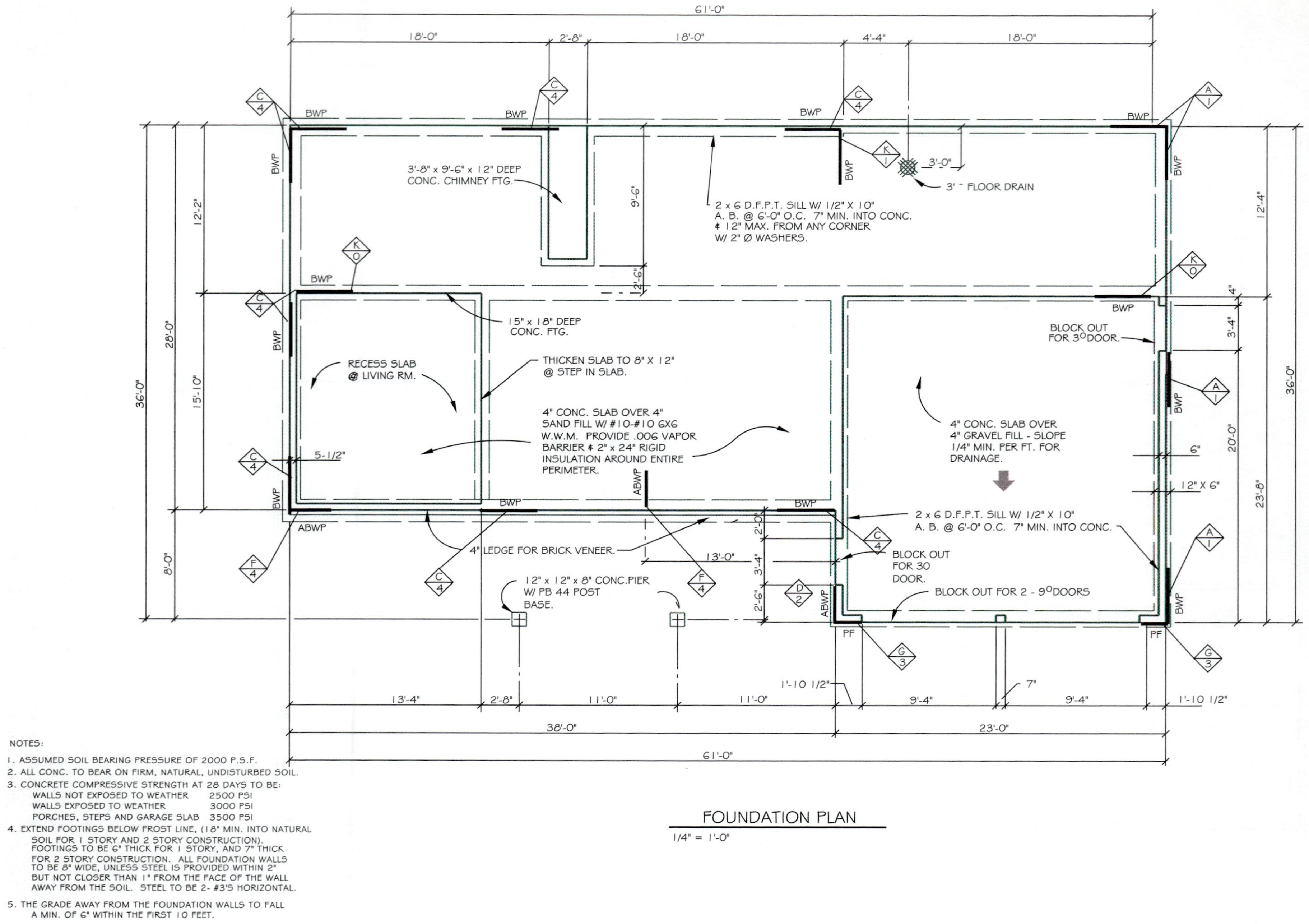
61'-0"
18'-0"
2'-8"
18'-0"
4'-4"
18'-0"
BWP
ABWP
12'-2"
28'-0"
36'-0"
15'-10"
8'-0"
12'-4"
20'-0"
23'-8"
3'-4"
4"
9'-6"
2'-6"
3'-0"
3'-8" x 9'-6" x 12" DEEP CONC. CHIMNEY FTG.
3' - FLOOR DRAIN
2 x 6 D.F.P.T. SILL W/ 1/2" X 10" A. B. @ 6'-0" O.C. 7" MIN. INTO CONC. & 12" MAX. FROM ANY CORNER W/ 2" Ø WASHERS.
15" x 18" DEEP CONC. FTG.
RECESS SLAB @ LIVING RM.
THICKEN SLAB TO 8" X 12" @ STEP IN SLAB.
4" CONC. SLAB OVER 4" SAND FILL W/ #10-#10 6X6 W.W.M. PROVIDE .006 VAPOR BARRIER & 2" x 24" RIGID INSULATION AROUND ENTIRE PERIMETER.
BLOCK OUT FOR 3⁰DOOR.
4" CONC. SLAB OVER 4" GRAVEL FILL - SLOPE 1/4" MIN. PER FT. FOR DRAINAGE.
6"
12" X 6"
2 x 6 D.F.P.T. SILL W/ 1/2" X 10" A. B. @ 6'-0" O.C. 7" MIN. INTO CONC.
5-1/2"
4" LEDGE FOR BRICK VENEER.
13'-0"
2'-0"
3'-4"
2'-6"
BLOCK OUT FOR 30 DOOR.
BLOCK OUT FOR 2 - 9⁰DOORS
12" x 12" x 8" CONC.PIER W/ PB 44 POST BASE.
PF
1'-10 1/2"
7"
13'-4"
2'-8"
11'-0"
11'-0"
9'-4"
9'-4"
1'-10 1/2"
38'-0"
23'-0"
61'-0"
FOUNDATION PLAN
1/4" = 1'-0"
NOTES:
1. ASSUMED SOIL BEARING PRESSURE OF 2000 P.S.F.
2. ALL CONC. TO BEAR ON FIRM, NATURAL, UNDISTURBED SOIL.
3. CONCRETE COMPRESSIVE STRENGTH AT 28 DAYS TO BE:
WALLS NOT EXPOSED TO WEATHER 2500 PSI
WALLS EXPOSED TO WEATHER 3000 PSI
PORCHES, STEPS AND GARAGE SLAB 3500 PSI
4. EXTEND FOOTINGS BELOW FROST LINE, (18" MIN. INTO NATURAL SOIL FOR 1 STORY AND 2 STORY CONSTRUCTION). FOOTINGS TO BE 6" THICK FOR 1 STORY, AND 7" THICK FOR 2 STORY CONSTRUCTION. ALL FOUNDATION WALLS TO BE 8" WIDE, UNLESS STEEL IS PROVIDED WITHIN 2" BUT NOT CLOSER THAN 1" FROM THE FACE OF THE WALL AWAY FROM THE SOIL. STEEL TO BE 2- #3'S HORIZONTAL.
5. THE GRADE AWAY FROM THE FOUNDATION WALLS TO FALL A MIN. OF 6" WITHIN THE FIRST 10 FEET.

FIGURE 33.5 ■ The foundation plan for a structure with a concrete slab is finished by adding dimensions and text.

STEP 25 Using a print of your plan, evaluate your drawing using the following checklist:

SLAB FOUNDATION PLAN CHECKLIST

CORRECT SYMBOL AND LOCATION	CORRECT STRUCTURAL MATERIALS
Outline of slabs	Proper footing size
Walls	Proper footing location
Footings and piers	Proper pier size
Doors	Proper pier location
Ductwork	
Plumbing	
Floor slopes	

REQUIRED LOCAL NOTES	REQUIRED GENERAL NOTES
Concrete slab thickness, fill and reinforcement	Soil bearing pressure Concrete strength
Veneer ledges	Anchor bolt size and spacing
Door blockouts	Vapor barriers
Fireplace footings	Slab insulation
Pier sizes	Reinforcement
Footing sizes	
Floor drains	
Heating registers	

FOUNDATION PLAN WITH JOIST CONSTRUCTION

The following steps (illustrated in Figure 33.6) can be used to draw a foundation plan showing continuous footings and floor joists. Use construction lines for steps 1 through 9.

STEP 1 Using the dimensions on your floor plan, lay out the exterior face of the foundation wall.

STEP 2 Determine the foundation wall thickness and lay out the interior face of the foundation wall. See Figure 31.12 for a review of foundation footing and wall dimensions.

STEP 3 Block out doors in the garage area. Typically the door size plus 4″ (100 mm) is provided as the foundation is formed.

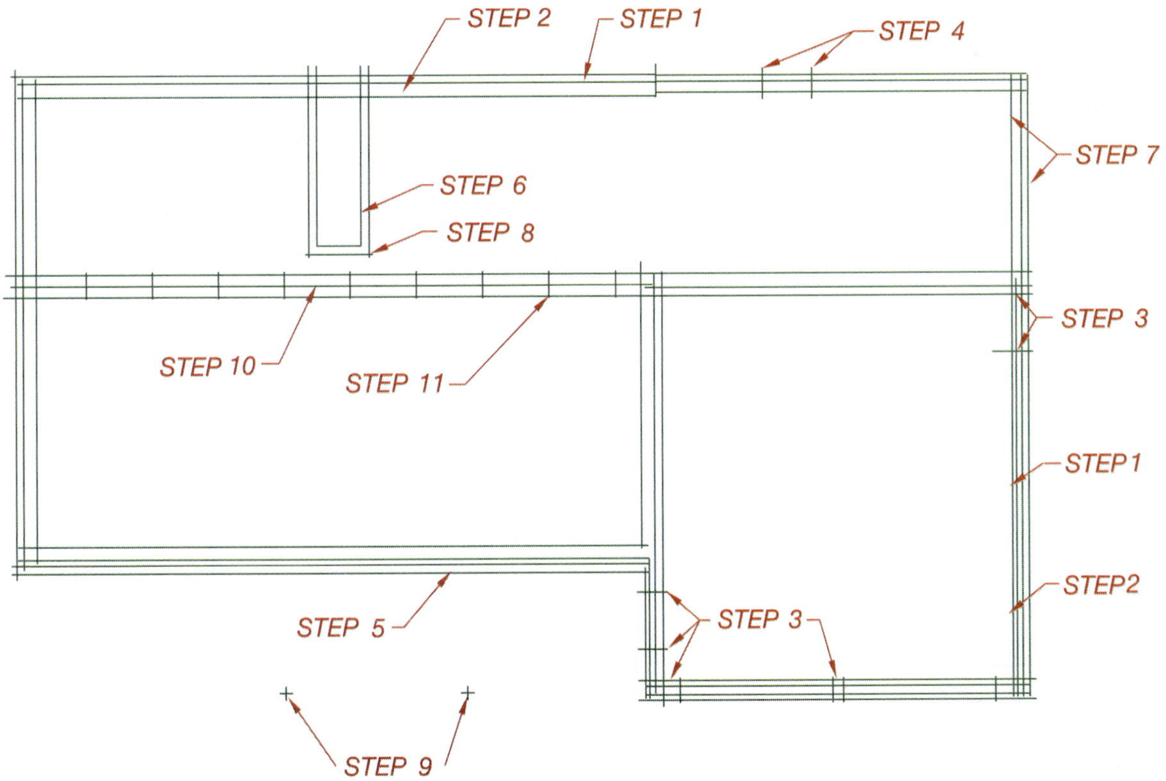

FIGURE 33.6 ■ Use construction lines to lay out the location of stem wall, footings, door locations, the fireplace, and girder locations.

STEP 4 Block out a crawl access.

STEP 5 Lay out a support ledge if you are using masonry veneer.

STEP 6 Lay out the size of the fireplace using the measurements on the floor plan.

STEP 7 Lay out the footing width around the perimeter walls.

STEP 8 Lay out the fireplace footing so it extends a minimum of 6″ (150 mm) past the face of the fireplace.

STEP 9 Lay out any exterior piers required for porches or deck support.

STEP 10 Lay out girder locations to support floor joists, load-bearing walls, and changes in floor elevation.

STEP 11 Locate the center of all support piers.

STEP 12 Darken all items drawn in steps 1–9. Use bold lines to represent the materials drawn in steps 1–6. Use thin dashed lines for steps 7–9. When you are finished with this step, your drawing should resemble Figure 33.7.

See Figure 33.8 for steps 13 through 18. The items in these steps are not drawn with construction lines. Because of their simplicity, these items can be drawn using finished lines and need not be traced.

STEP 13 Draw the girders, using thin dashed lines.

STEP 14 Draw arrows to represent the floor joist direction.

STEP 15 Draw the piers to support the girders.

STEP 16 Draw beam pockets.

STEP 17 Draw vents in the walls surrounding the crawl space. Use bold lines to represent the edges and thin lines to represent the vent.

STEP 18 Crosshatch the masonry chimney if concrete block or brick is used.

You now have drawn all of the information that is required to represent the floor and foundation systems. Follow steps 19 through 24 to dimension these items. Use thin lines for all extension and dimension lines. When complete, your drawing should resemble Figure 33.9.

Draw extension and dimension lines to locate:

STEP 19 The overall size of each side of the foundation.

STEP 20 The jogs in the foundation wall.

STEP 21 Door openings in the foundation wall.

STEP 22 The fireplace.

STEP 23 All girders and piers. Place extension lines on the outside of the foundation if possible.

STEP 24 Metal connectors if required.

All materials have now been drawn and located. The final drawing procedure is to place dimensions and specify the material to be used. This is done with general and local notes in a method similar to that used on the floor plan. Figure 33.10 shows

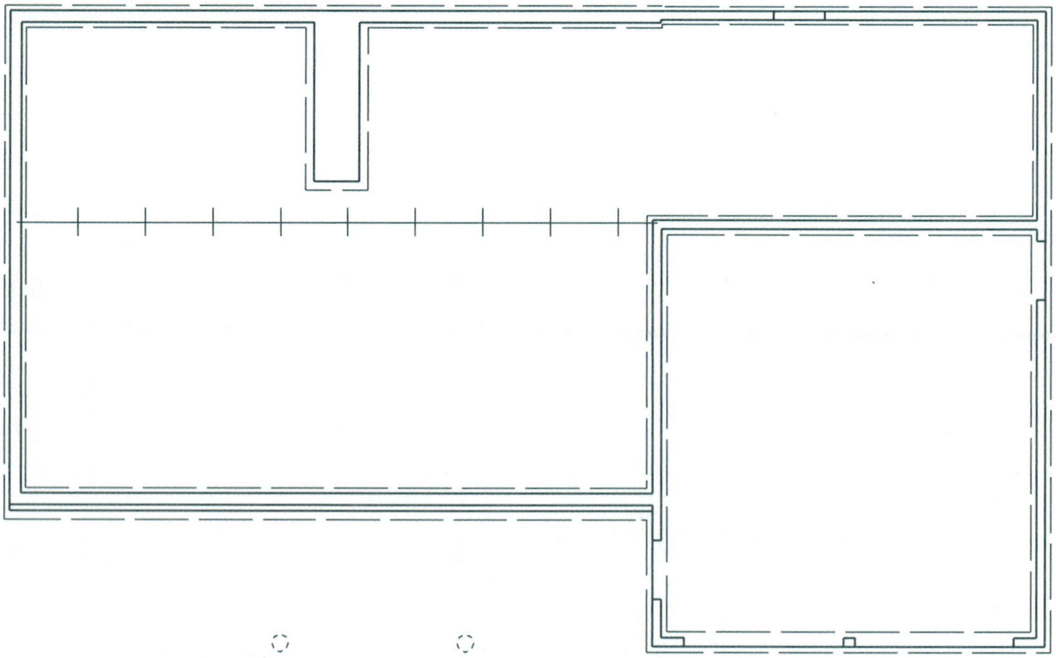

FIGURE 33.7 ■ Carefully darken all objects using the proper finished-line quality. Construction lines should be so light that they do not need to be erased.

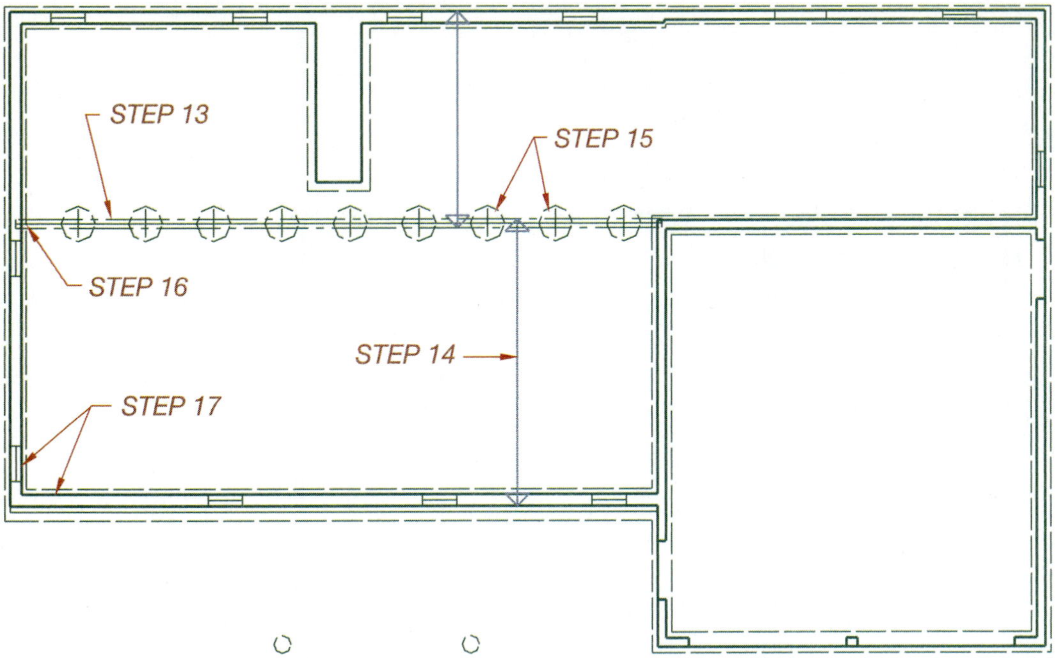

FIGURE 33.8 ■ Draw the girders, joists, and vents.

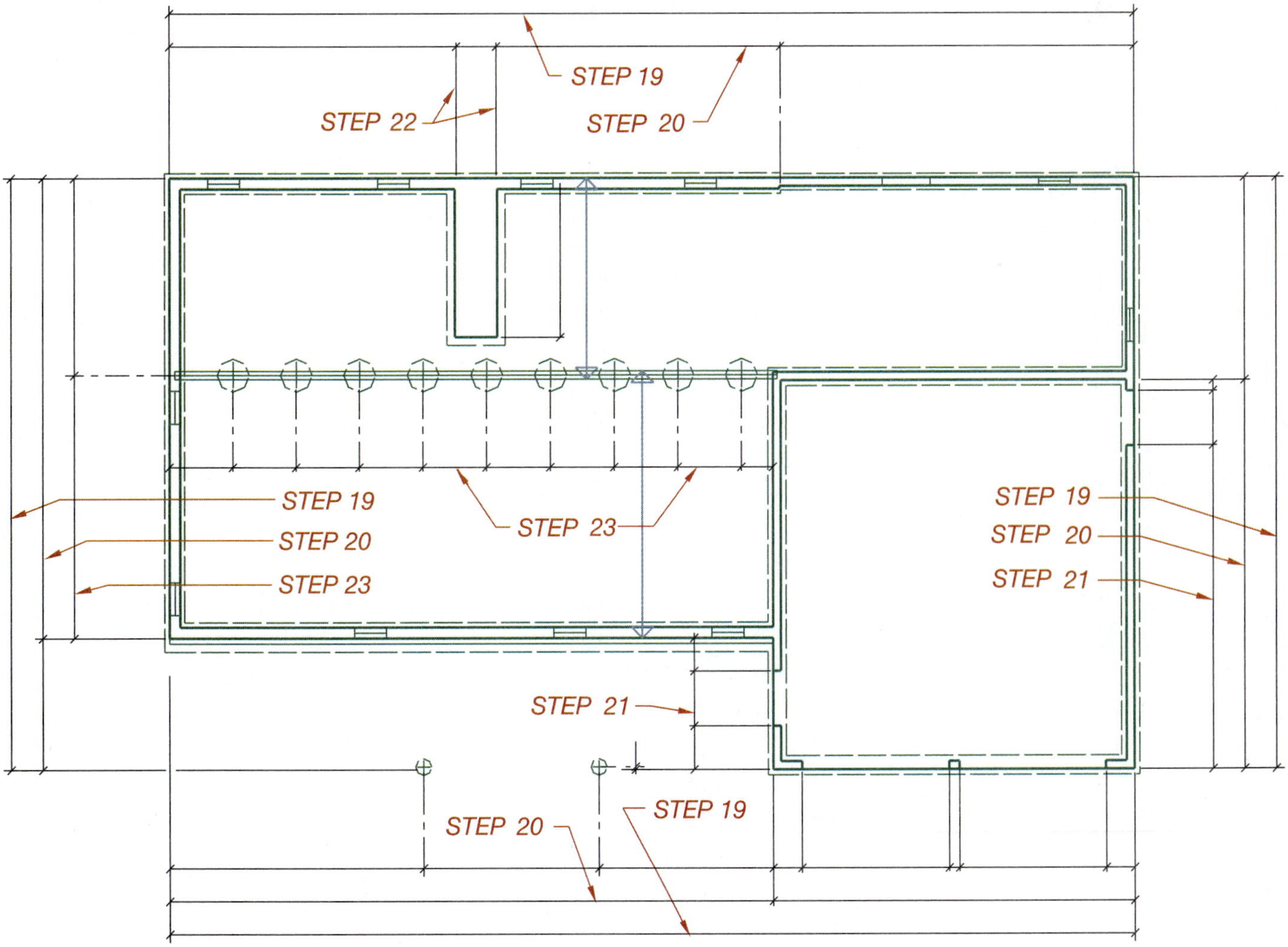

FIGURE 33.9 ■ Prepare the plan to be dimensioned by adding dimension and extension lines to describe the overall, major jogs, and each opening on each side of the foundation.

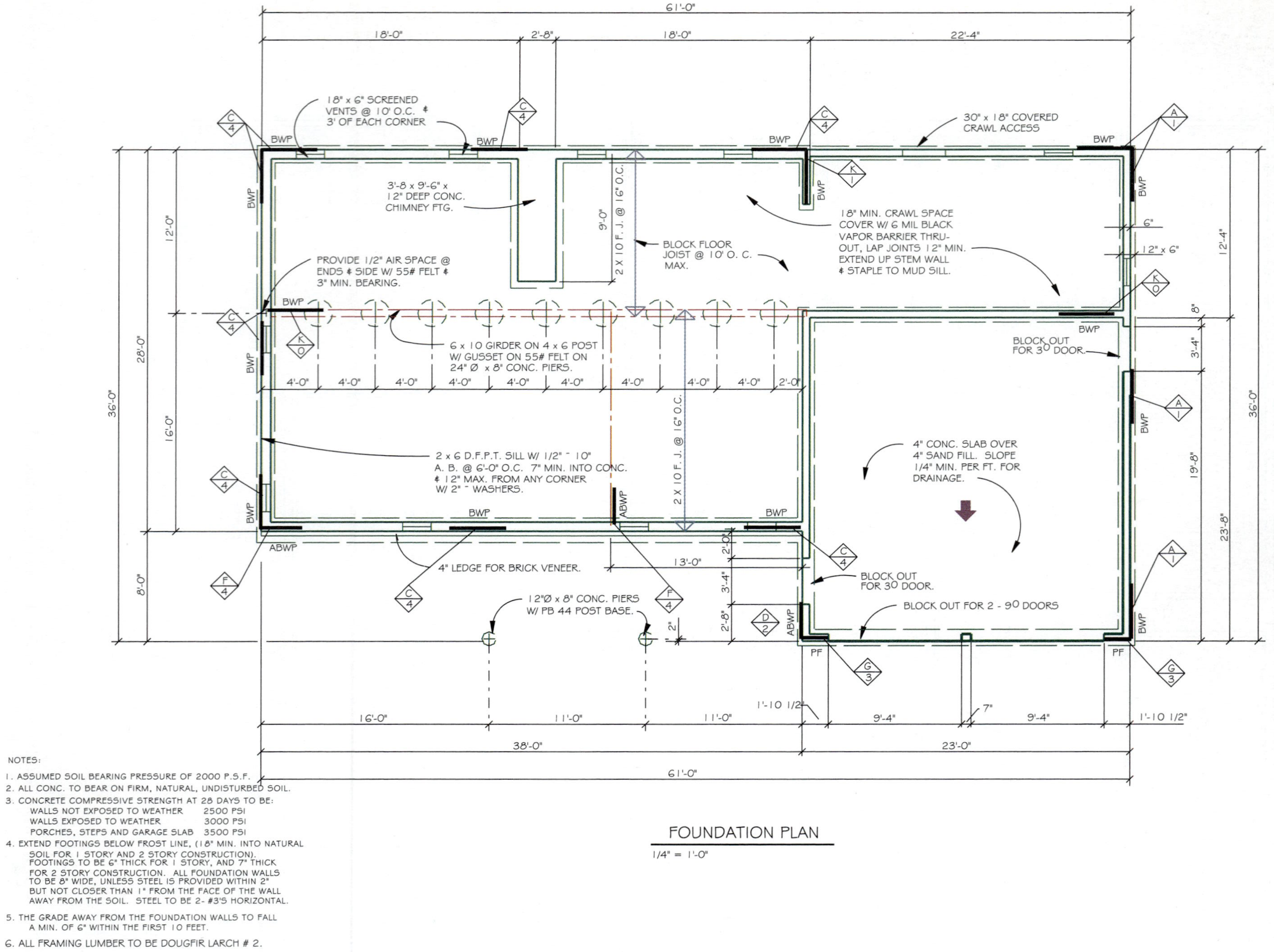
61'-0"
18'-0"
2'-8"
18'-0"
22'-4"
18" x 6" SCREENED VENTS @ 10' O.C. & 3' OF EACH CORNER
30" x 18" COVERED CRAWL ACCESS
BWP
ABWP
PF
3'-8 x 9'-6" x 12" DEEP CONC. CHIMNEY FTG.
9'-0"
2 X 10 F. J. @ 16" O.C.
BLOCK FLOOR JOIST @ 10' O. C. MAX.
18" MIN. CRAWL SPACE COVER W/ 6 MIL BLACK VAPOR BARRIER THRU-OUT, LAP JOINTS 12" MIN. EXTEND UP STEM WALL & STAPLE TO MUD SILL.
6"
12" x 6"
PROVIDE 1/2" AIR SPACE @ ENDS & SIDE W/ 55# FELT & 3" MIN. BEARING.
6 x 10 GIRDER ON 4 x 6 POST W/ GUSSET ON 55# FELT ON 24" Ø x 8" CONC. PIERS.
4'-0"
2'-0"
BLOCK OUT FOR 30 DOOR.
2 x 6 D.F.P.T. SILL W/ 1/2" - 10" A. B. @ 6'-0" O.C. 7" MIN. INTO CONC. & 12" MAX. FROM ANY CORNER W/ 2" - WASHERS.
4" CONC. SLAB OVER 4" SAND FILL. SLOPE 1/4" MIN. PER FT. FOR DRAINAGE.
4" LEDGE FOR BRICK VENEER.
13'-0"
12"Ø x 8" CONC. PIERS W/ PB 44 POST BASE.
BLOCK OUT FOR 2 - 90 DOORS
36'-0"
12'-0"
28'-0"
16'-0"
8'-0"
12'-4"
3'-4"
19'-8"
23'-8"
2'-8"
2"
16'-0"
11'-0"
1'-10 1/2"
9'-4"
7"
38'-0"
23'-0"
NOTES:
1. ASSUMED SOIL BEARING PRESSURE OF 2000 P.S.F.
2. ALL CONC. TO BEAR ON FIRM, NATURAL, UNDISTURBED SOIL.
3. CONCRETE COMPRESSIVE STRENGTH AT 28 DAYS TO BE:
WALLS NOT EXPOSED TO WEATHER 2500 PSI
WALLS EXPOSED TO WEATHER 3000 PSI
PORCHES, STEPS AND GARAGE SLAB 3500 PSI
4. EXTEND FOOTINGS BELOW FROST LINE, (18" MIN. INTO NATURAL SOIL FOR 1 STORY AND 2 STORY CONSTRUCTION). FOOTINGS TO BE 6" THICK FOR 1 STORY, AND 7" THICK FOR 2 STORY CONSTRUCTION. ALL FOUNDATION WALLS TO BE 8" WIDE, UNLESS STEEL IS PROVIDED WITHIN 2" BUT NOT CLOSER THAN 1" FROM THE FACE OF THE WALL AWAY FROM THE SOIL. STEEL TO BE 2- #3'S HORIZONTAL.
5. THE GRADE AWAY FROM THE FOUNDATION WALLS TO FALL A MIN. OF 6" WITHIN THE FIRST 10 FEET.
6. ALL FRAMING LUMBER TO BE DOUGFIR LARCH # 2.
FOUNDATION PLAN
1/4" = 1'-0"

FIGURE 33.10 ■ The foundation plan for a home with a joist floor system is finished by adding dimensions and text.

how the notes might appear in the foundation plan. Use the following steps as a guideline to completing the foundation plan.

STEP 25 Compute all dimensions and place them in their appropriate locations.

STEP 26 Neatly letter all required general notes.

STEP 27 Neatly letter all required local notes.

STEP 28 Place a title and scale under the drawing.

STEP 29 Use a print of your plan to evaluate your drawing using the following checklist.

FOUNDATION-PLAN CHECKLIST

CORRECT SYMBOL AND LOCATION	CORRECT STRUCTURAL MATERIALS
Walls	Proper footing sizes
Footings	Proper wall size
Crawl access	Proper beam placement
Vents	Proper beam size
Doors	Proper joist size and direction
Girders	Proper pier locations and sizes
Joist	
Piers	
REQUIRED LOCAL NOTES	
Joist size and spacing	Veneer ledges
Beam sizes	Door blockouts
Beam pockets	Fireplace footing size
Metal connectors	Vent size and spacing
Anchor bolts and mudsill	
Garage slab thickness	
Fill and slope direction	
REQUIRED GENERAL NOTES	REQUIRED DIMENSIONS
Soil bearing information	Overall
Concrete strength	Jogs
Crawl space covering	Openings
Framing lumber grade and species	Girder locations
	Pier locations
	Metal connectors

STANDARD FOUNDATION WITH POST-AND-BEAM FLOOR SYSTEM

When your drawing is complete, it should resemble the plan in Figure 33.14. Use construction lines for drawing steps 1 through 12. Each step can be seen in Figure 33.11.

STEP 1 Using the dimensions on your floor plan, lay out the exterior edge of the foundation wall.

STEP 2 Lay out the interior face of the foundation wall.

STEP 3 Block out doors in the garage area.

STEP 4 Block out for a crawl access.

STEP 5 Lay out the support ledge if masonry veneer is to be used.

STEP 6 Lay out the size of the fireplace using measurements from the floor plan.

STEP 7 Lay out the footing under the foundation walls.

STEP 8 Lay out the footing under the fireplace.

STEP 9 Lay out any exterior piers required to support porches and decks.

STEP 10 Lay out the center of each load-bearing wall.

STEP 11 Lay out all girders.

STEP 12 Lay out the center for the piers to support the girders.

See Figure 33.12 for step 13 through 15.

STEP 13 Darken all items in steps 1–12. Use bold lines to represent the material in steps 1–6. Use thin dashed lines to represent the material in steps 7–9 and 12.

STEP 14 Use pairs of thin dashed lines or a very bold center line to represent the girders.

The items to be drawn in these steps have not been drawn with construction lines. Because of their simplicity you can draw these items with finished lines.

STEP 15 Draw vents in the walls surrounding the crawl space.

STEP 16 Draw beam pockets. When you are finished with this step, your drawing should resemble the drawing in Figure 33.12.

STEP 17 Crosshatch materials formed using concrete blocks.

You now have drawn all of the information to represent the floor and foundation systems. Follow steps 18 through 23 to dimension these items. Use thin lines for extension and dimension lines. When complete, your foundation plan should resemble Figure 33.13.

Draw extension and dimension lines to locate:

STEP 18 The overall size of the foundation.

STEP 19 The jogs in the foundation walls.

STEP 20 All openings in the foundation walls except for vents.

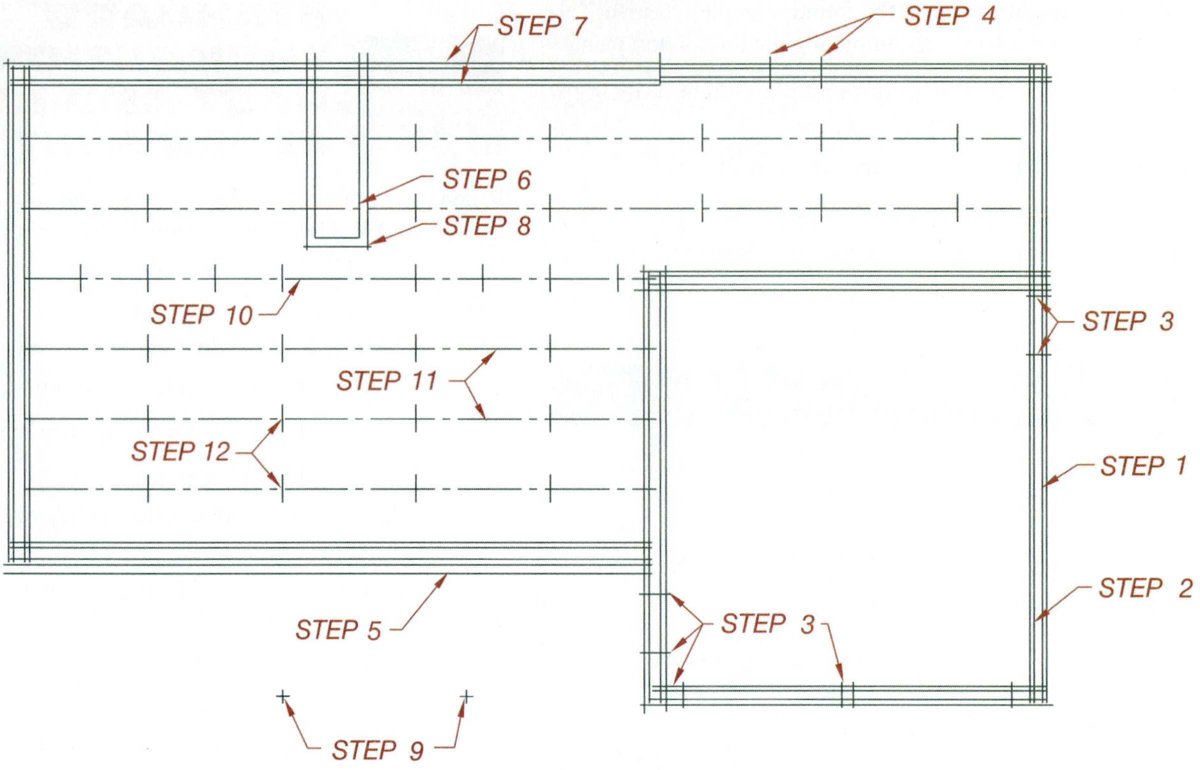

FIGURE 33.11 ■ Use construction lines to lay out the location of stem wall, footing, door locations, the fireplace, and girder and pier locations for a post-and-beam foundation plan.

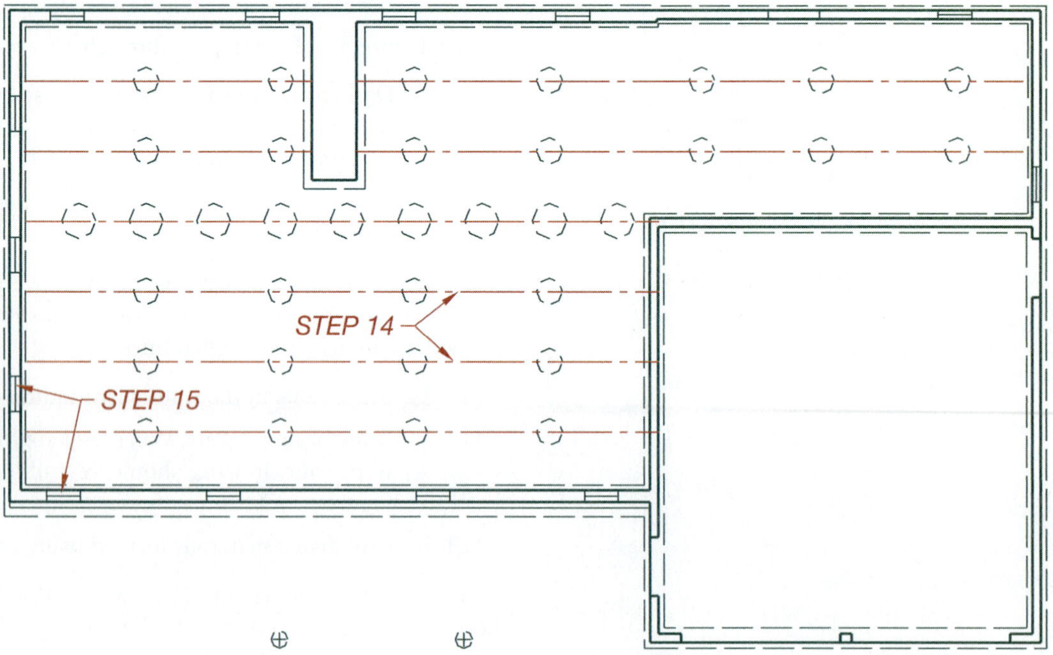

FIGURE 33.12 ■ Carefully darken all objects using the proper finished-line quality. Construction lines should be so light that they do not need to be erased.

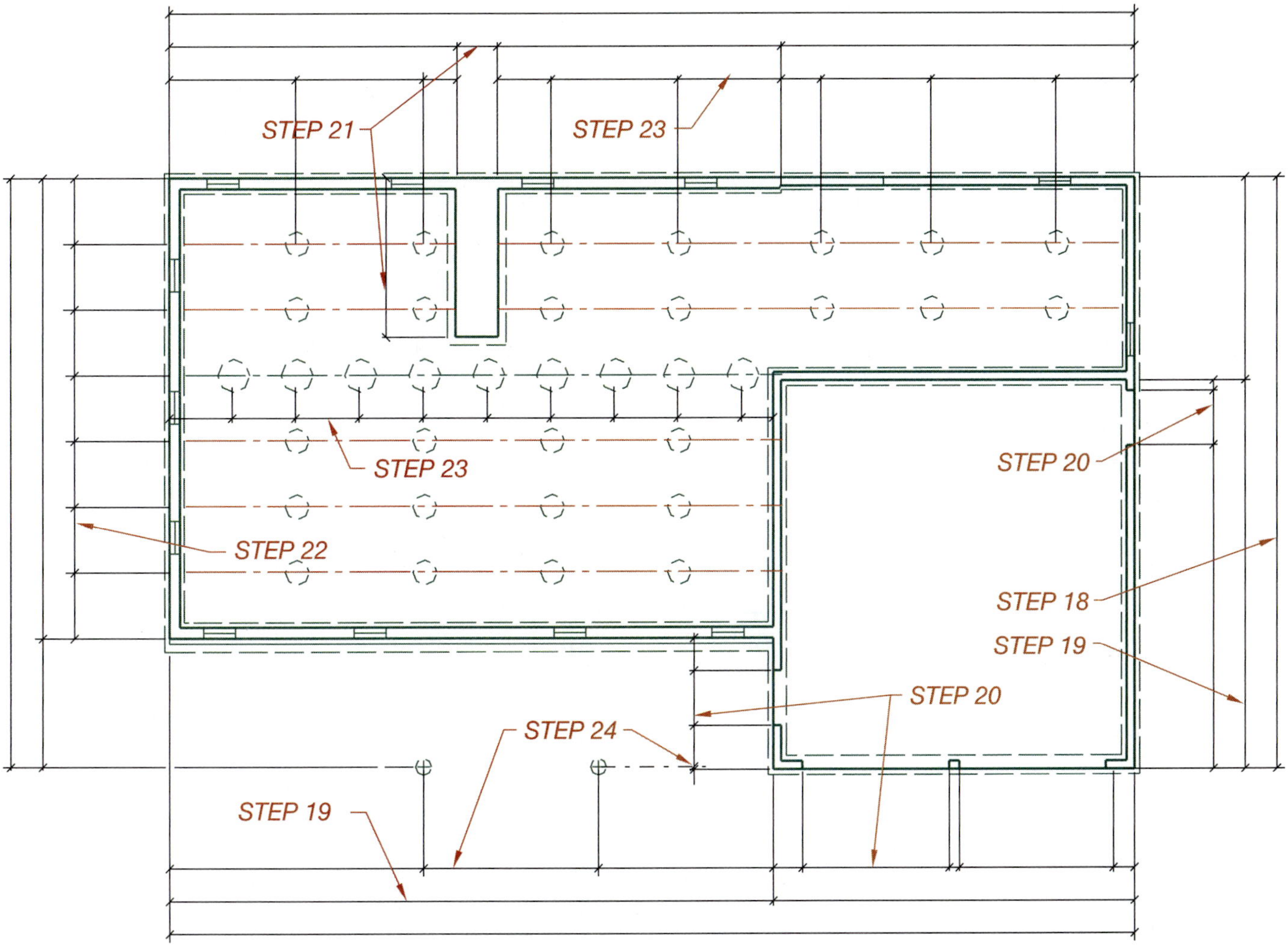

FIGURE 33.13 ■ Draw the girders and the piers with finished-line quality. Place dimension lines to describe the overall, major jogs, and each opening on each side of the foundation.

STEP 21 The fireplace.

STEP 22 All girders.

STEP 23 All piers.

STEP 24 All metal connectors.

All materials have now been drawn to represent the floor and foundation systems. The final drafting procedure is to place dimensions and specify the material to be used. This is done by the use of general and local notes in a method similar to what was used on the floor plan. Figure 33.14 shows how notes can be placed on the foundation plan. Complete the foundation plan using the following steps.

STEP 25 Compute and neatly letter dimensions in the appropriate place.

STEP 26 Neatly letter all required general notes.

STEP 27 Neatly letter all local notes.

STEP 28 Place a title and scale below the drawing.

STEP 29 Using a print of your plan, evaluate your drawing using the checklist on page 689.

COMBINATION SLAB AND CRAWL SPACE PLANS FOR A PARTIAL BASEMENT

A structure may require a combined slab and floor joist system. The foundation plan will have similarities to both a foundation with joist and a slab foundation system. Figure 33.19 is an

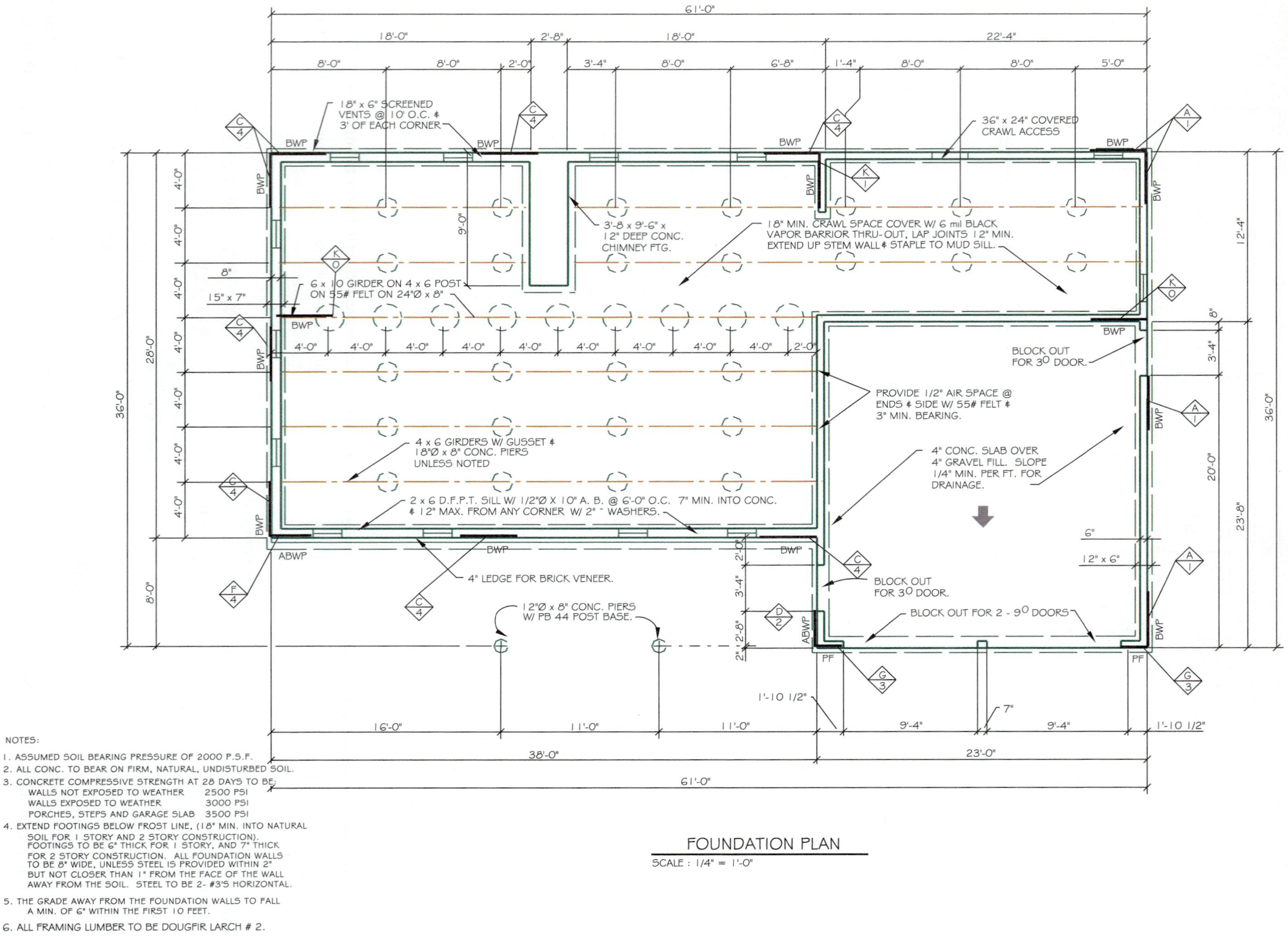

FIGURE 33.14 ■ The foundation plan for a home with a post-and-beam floor system is finished by adding dimensions and text.

POST-AND-BEAM CHECKLIST	
CORRECT SYMBOL AND LOCATION	CORRECT STRUCTURAL MATERIALS
Walls	Proper footing sizes
Footings	Proper wall size
Crawl access	Proper beam placement
Vents	Proper pier placement
Doors	Veneer ledges
Girders and piers Line of slabs	
Fireplace	
REQUIRED LOCAL NOTES	REQUIRED GENERAL NOTES
Beam size	Soil bearing values
Beam pockets	Concrete values
Metal connectors	Crawl space covering
Anchor bolts and mudsills	Crawl space insulation
Garage lab thickness, fill, slope direction	Framing lumber grade and species
Fireplace note	
Vent size and spacing	
Veneer ledges	
Door blockouts	
REQUIRED DIMENSIONS	
Overall	Girder locations
Jogs	Pier locations
Wall openings	Metal locations Fireplace location

example of a foundation plan with a basement slab and a crawl space with joist construction. The following steps can be used to draw the foundation plan. Use construction lines for steps 1 through 12. Each step can be seen in Figure 33.15.

STEP 1 Using the dimensions on your floor plan, lay out the exterior face of the foundation walls.

STEP 2 Determine the wall thickness and lay out the interior face of the foundation walls.

STEP 3 Block out door and window openings in the foundation walls.

STEP 4 Block out a crawl access.

STEP 5 Lay out a support ledge if masonry veneer is to be used.

STEP 6 Lay out the size of the fireplace based on the size drawn on the floor plan.

STEP 7 Lay out the footing under the fireplace.

STEP 8 Lay out the footing width under all foundation walls.

STEP 9 Lay out the footing width under all interior load-bearing walls.

STEP 10 Lay out any exterior piers required for porches and deck support.

STEP 11 Lay out girder locations to support floor and load-bearing walls.

STEP 12 Locate the center of all interior support piers.

STEP 13 Darken all items drawn in steps 1–10. Use bold lines for steps 1–6. Use thin dashed lines for items 7–10 and step 12. When complete, your drawing should resemble Figure 33.16.

See Figure 33.17 for steps 14 through 19. These items have not been drawn with construction lines. Each can be drawn using finished-quality lines and need not be traced.

STEP 14 Draw the girders using dashed lines.

STEP 15 Draw piers with thin dashed lines.

STEP 16 Draw vents in the crawl area only.

STEP 17 Draw windows with thin lines.

STEP 18 Draw window wells using thin lines.

STEP 19 Crosshatch materials if concrete blocks are used.

You now have drawn all of the information that is needed to represent the floor and foundation systems. Follow steps 20 through 23 to dimension these items. Use thin lines for all extension and dimension lines. When complete, your drawing should resemble Figure 33.18.

Draw extension and dimension lines to locate:

STEP 20 Overall size on each side of the foundation.

STEP 21 Jogs in the foundation walls.

STEP 22 All openings in the foundation walls except for vents.

STEP 23 All girders and piers.

All materials have now been drawn and located. The final drawing procedure is to place dimensions and specify the materials that were used. Materials may be specified with general and local notes. Figure 33.19 is an example of the notes that can be found on the foundation plan. Use the following steps as guidelines to complete the foundation plan.

STEP 24 Compute all dimensions and place them in the appropriate location.

STEP 25 Neatly letter all required general notes.

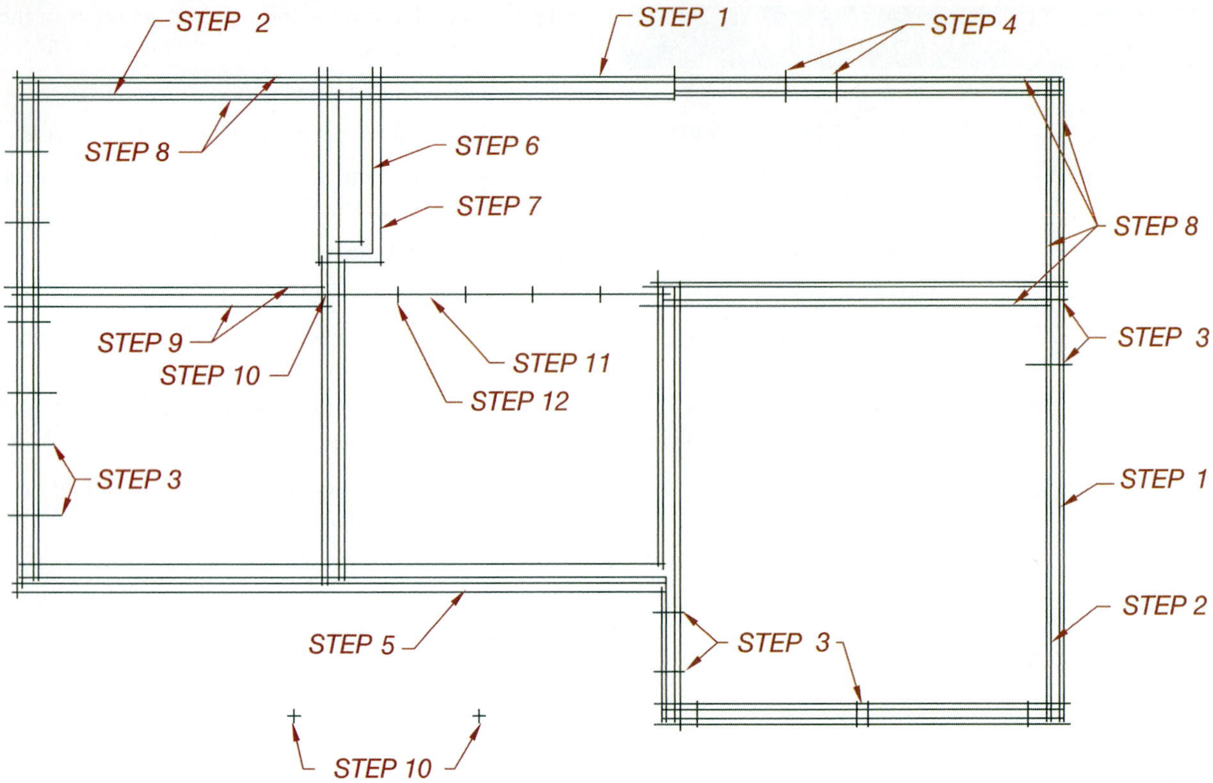

FIGURE 33.15 ■ Use construction lines to lay out the location of stem wall, retaining walls, footings, door locations, and the fireplace for a foundation plan with a partial basement.

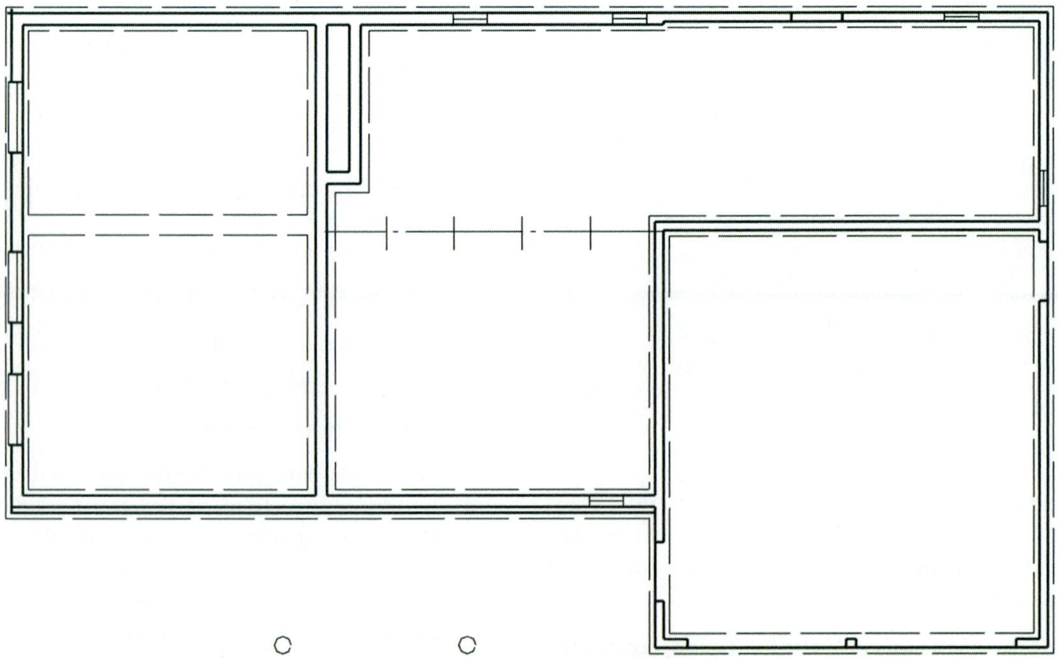

FIGURE 33.16 ■ Carefully darken all objects using the proper finished-line quality. Construction lines should be so light that they do not need to be erased.

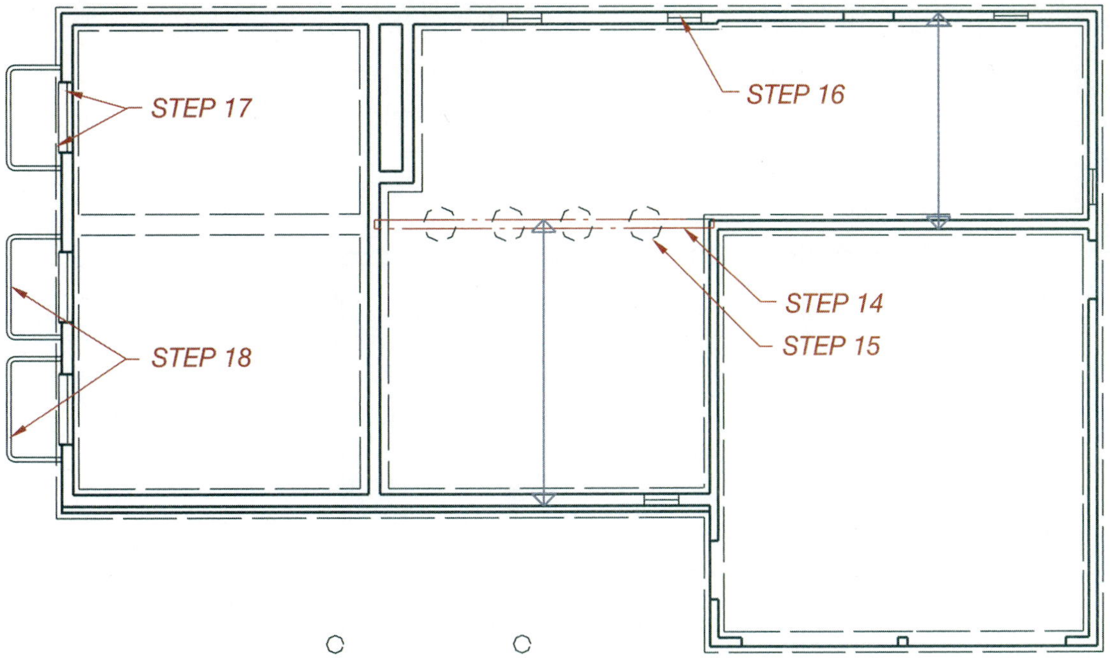

FIGURE 33.17 ■ Draw all girders, piers, windows, and window wells.

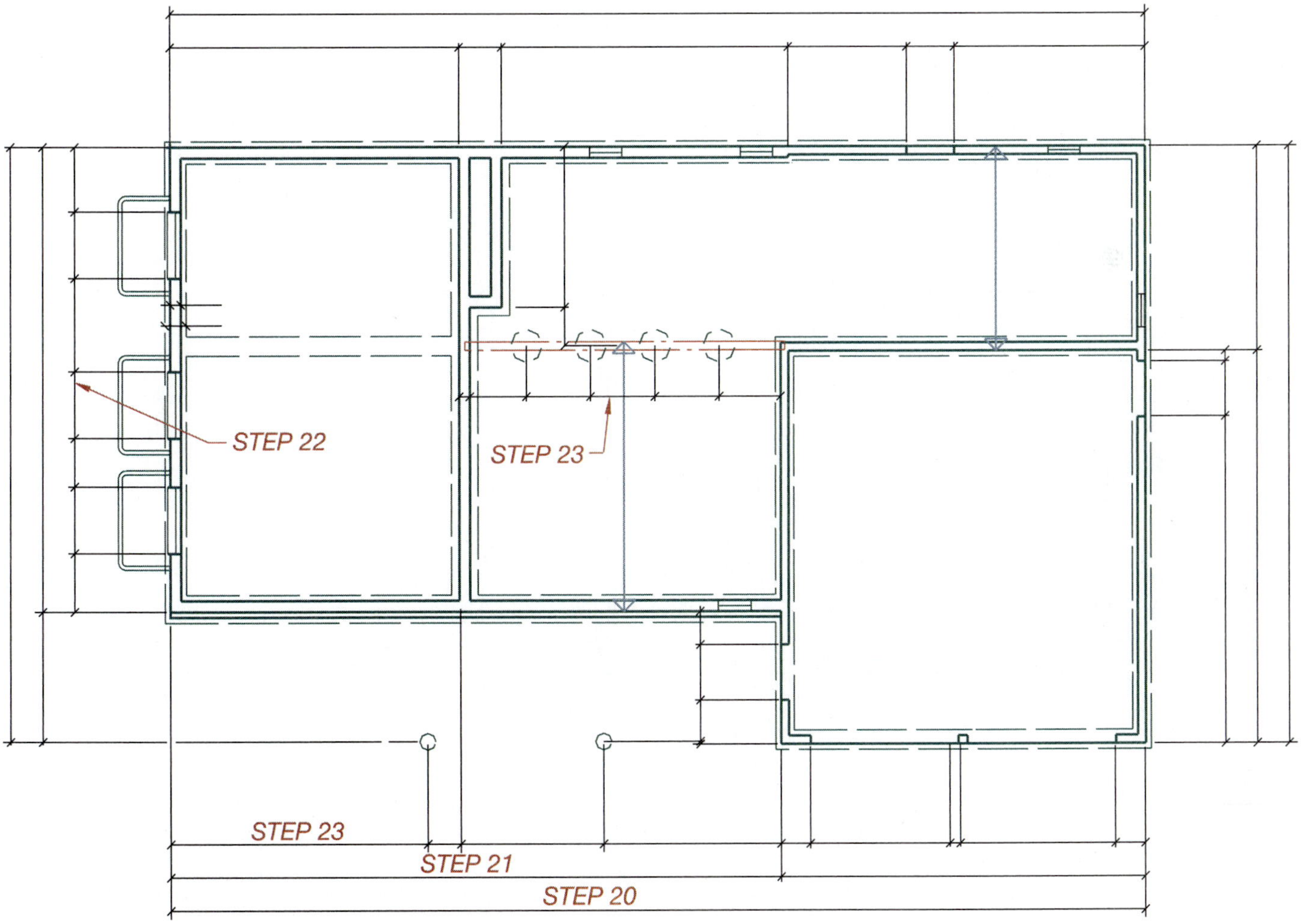

FIGURE 33.18 ■ Prepare the plan to be dimensioned by adding dimension and extension lines to describe the overall, major jogs and each opening on each side of the foundation.

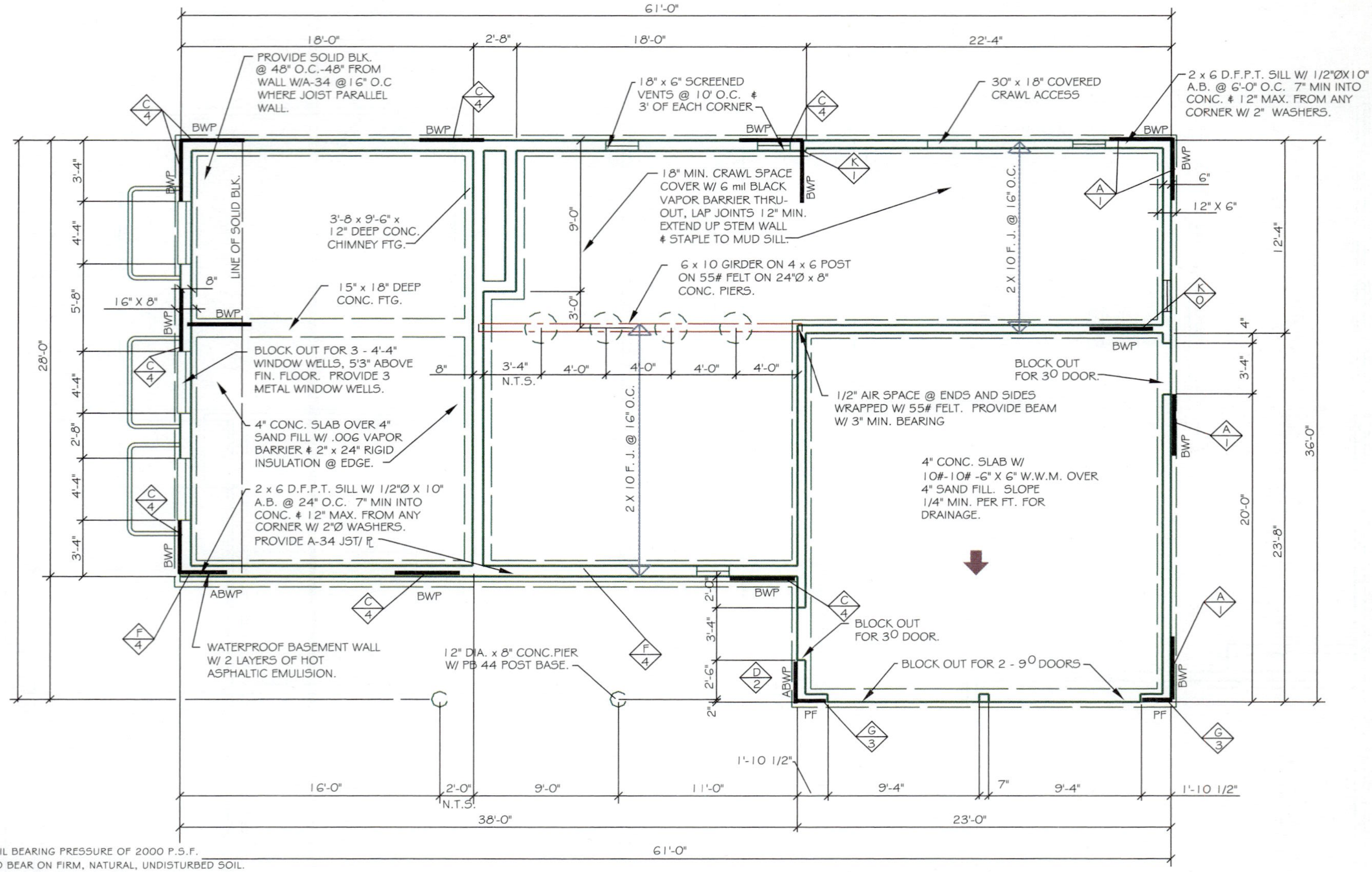

NOTES:

1. ASSUMED SOIL BEARING PRESSURE OF 2000 P.S.F.
2. ALL CONC. TO BEAR ON FIRM, NATURAL, UNDISTURBED SOIL.
3. CONCRETE COMPRESSIVE STRENGTH AT 28 DAYS TO BE:
   - WALLS NOT EXPOSED TO WEATHER 2500 PSI
   - WALLS EXPOSED TO WEATHER 3000 PSI
   - PORCHES, STEPS AND GARAGE SLAB 3500 PSI
4. EXTEND FOOTINGS BELOW FROST LINE, (18" MIN. INTO NATURAL SOIL FOR 1 STORY AND 2 STORY CONSTRUCTION). FOOTINGS TO BE 6" THICK FOR 1 STORY, AND 7" THICK FOR 2 STORY CONSTRUCTION. ALL FOUNDATION WALLS TO BE 8" WIDE, UNLESS STEEL IS PROVIDED WITHIN 2" BUT NOT CLOSER THAN 1" FROM THE FACE OF THE WALL AWAY FROM THE SOIL. STEEL TO BE 2- #3'S HORIZONTAL.
5. THE GRADE AWAY FROM THE FOUNDATION WALLS TO FALL A MIN. OF 6" WITHIN THE FIRST 10 FEET.

FOUNDATION PLAN

1/4" = 1'-0"

FIGURE 33.19 ■ The foundation plan with a crawl space and partial basement is finished by adding dimensions and text.

STEP 26 Neatly letter all required local notes.

STEP 27 Place a title and scale under the drawing.

STEP 28 Using a print of your plan, evaluate your drawing using the following checklist.

CHECKLIST FOR PARTIAL SLAB AND FLOOR JOIST FOUNDATION

CORRECT SYMBOL AND LOCATION	CORRECT STRUCTURAL MATERIALS
Walls	Footing size
Footings	Wall size
Crawl access	Beam placement
Vents	Beam size
Doors	Joist size
Girders and joist	Pier location
Piers	Beam location
Slabs	Pier sizes
Window wells	
REQUIRED LOCAL NOTES	**REQUIRED GENERAL NOTES**
Joist size and spacing	Soil bearing value
Beam size	Concrete strength
Beam pockets	Crawl space covering
Metal connectors	Framing lumber grade and species
Anchor bolts	Slab insulation
Mudsill	
Slab thickness	
Fill	
Slope direction of slab	
Window wells	

FULL BASEMENT

The foundation plan for a home with a full basement will have similarities to a slab foundation system. Figure 33.22 is an example of a foundation plan with a basement slab and a crawl space with joist construction. The following steps can be used to draw the foundation plan. Use construction lines for steps 1 through 12. Each step can be seen in Figure 33.20.

STEP 1 Using the dimensions on your floor plan, lay out the exterior face of the foundation walls.

STEP 2 Determine the wall thickness and lay out the interior face of the foundation walls.

STEP 3 Block out openings in the foundation walls.

STEP 4 Lay out a support ledge if masonry veneer is to be used.

STEP 5 Lay out the size of the fireplace based on the size drawn on the floor plan.

STEP 6 Lay out the footing under the fireplace.

STEP 7 Lay out the footing width under all foundation walls.

STEP 8 Lay out the footing width under all interior load-bearing walls.

STEP 9 Lay out any exterior piers required for porches and deck support.

STEP 10 Darken all items drawn in steps 1–9. Use bold lines for steps 1–6. Use thin dashed lines for items 7–9.

See Figure 33.21 for steps 11 through 13. These items have not been drawn with construction lines. Each can be drawn using finished-quality lines and need not be traced.

STEP 11 Draw window wells using thin lines.

STEP 12 Draw windows with thin lines.

STEP 13 Crosshatch any materials where concrete blocks are used. When complete, your drawing should resemble Figure 33.21.

You now have drawn all of the information that is needed to represent the foundation system. Follow steps 14 through 16 to dimension these items. Use thin lines for all extension and dimension lines. When complete, your drawing should resemble Figure 33.21.

Draw extension and dimension lines to locate:

STEP 14 Overall size on each side of the foundation.

STEP 15 Jogs in the foundation walls.

STEP 16 All openings in the foundation walls except for vents.

All materials have now been drawn and located. The final drawing procedure is to place dimensions and specify the materials that were used. Materials may be specified with general and local notes. Figure 33.22 is an example of the notes that can be found on the foundation plan. Use the following steps as guidelines to complete the foundation plan.

STEP 17 Compute all dimensions and place them in the appropriate location.

STEP 18 Neatly letter all required general notes.

STEP 19 Neatly letter all required local notes.

STEP 20 Place a title and scale under the drawing.

STEP 21 Using a print of your plan, evaluate your drawing using the checklists that were used to complete a partial slab/joist foundation.

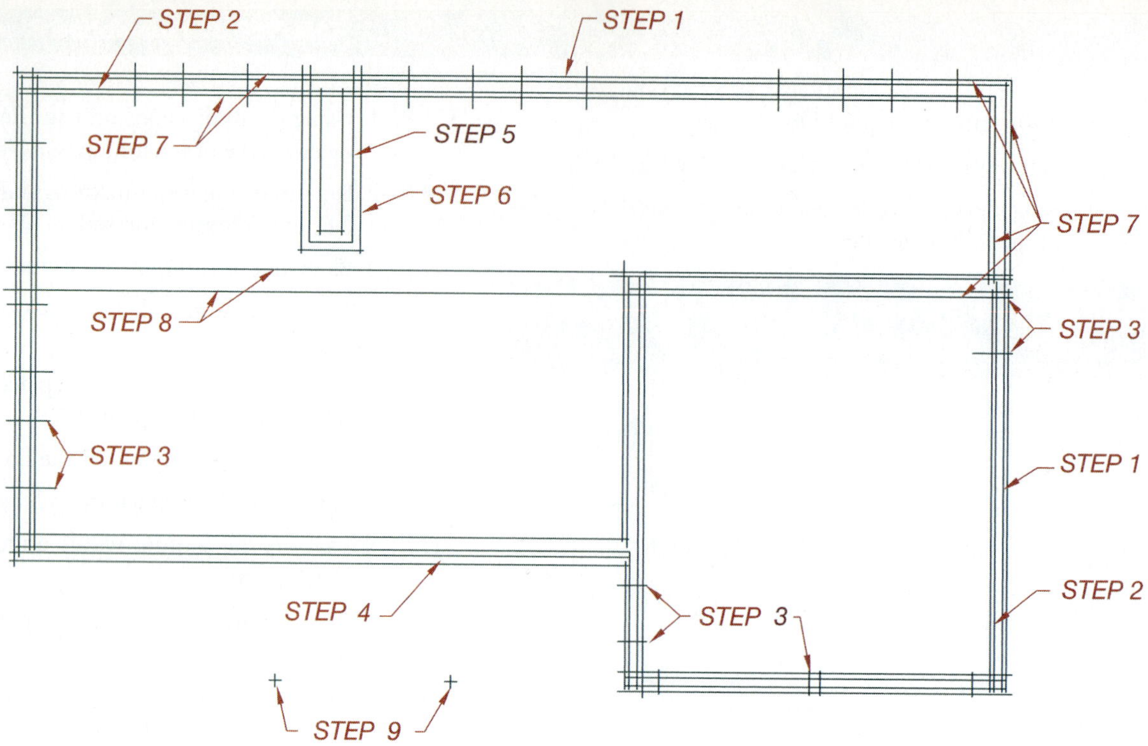

FIGURE 33.20 ■ Use construction lines to lay out the location of retaining walls, footings, door locations, and the fireplace for a full basement foundation plan.

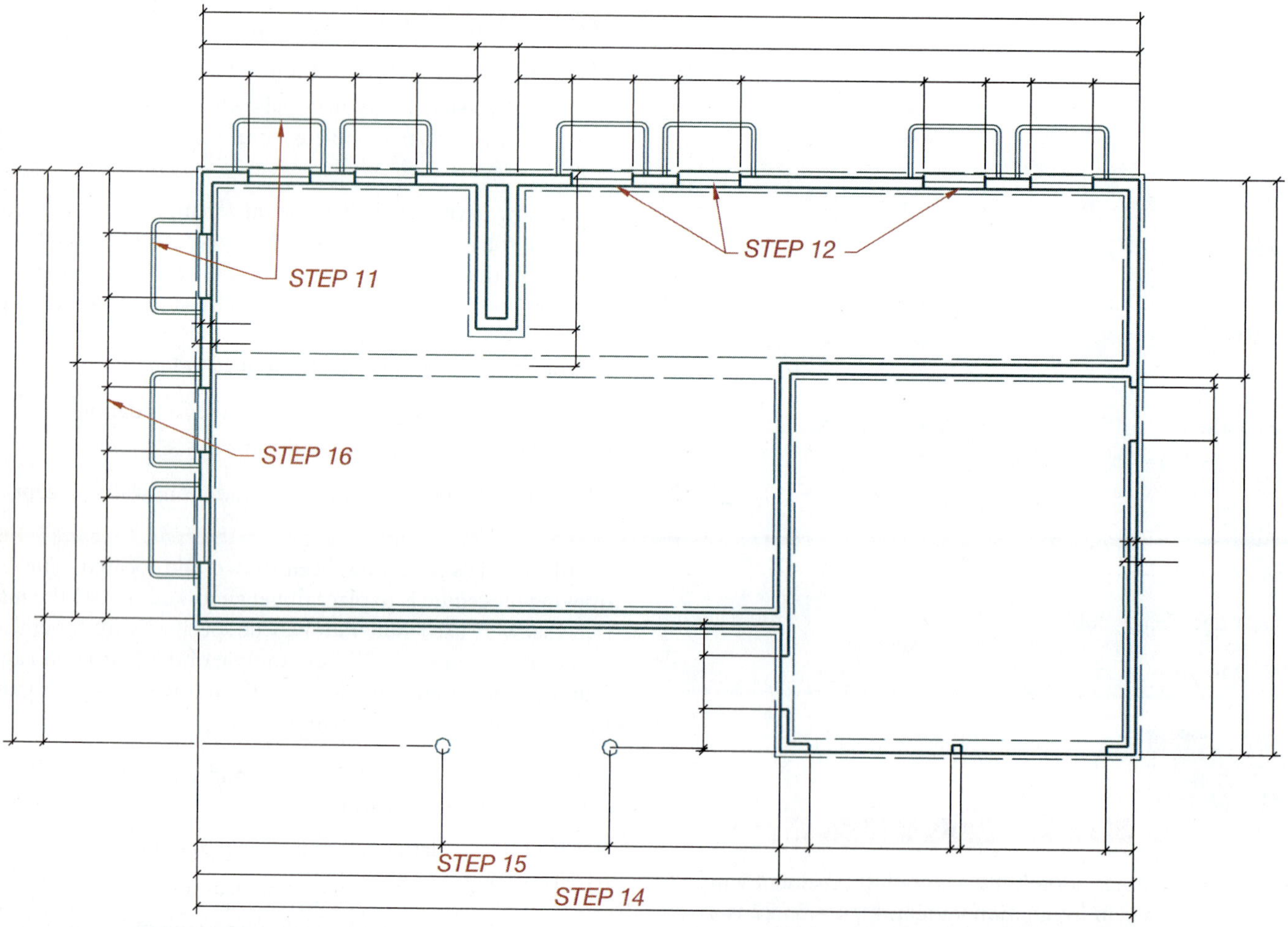

FIGURE 33.21 ■ Carefully darken all objects using the proper finished-line quality. Construction lines should be so light that they do not need to be erased. Add any finishing materials such as window wells. Add dimension and extension lines to describe the overall, major jogs, and opening on each side of the foundation.

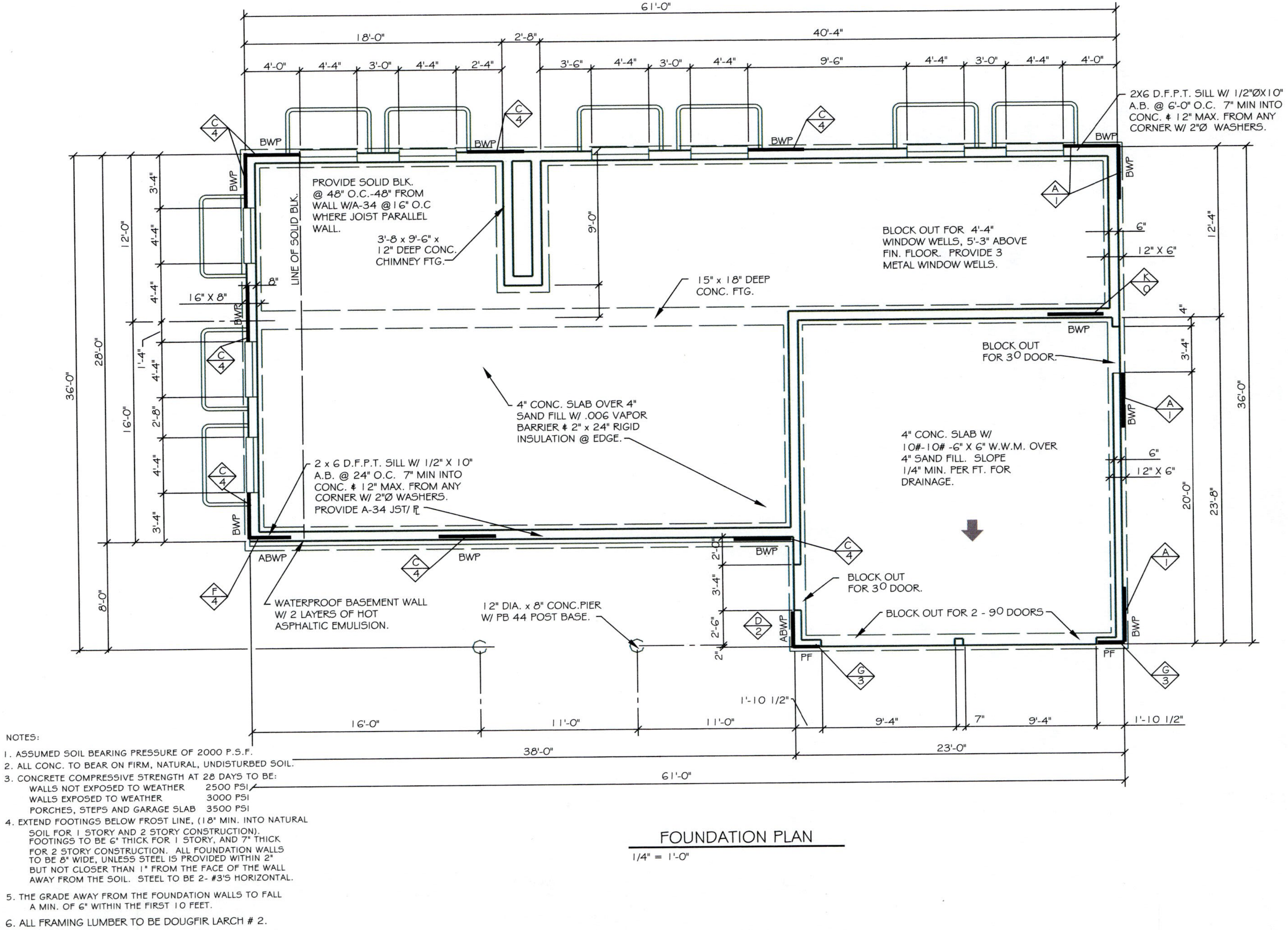
61'-0"
18'-0"
2'-8"
40'-4"
4'-0"
4'-4"
3'-0"
4'-4"
2'-4"
3'-6"
4'-4"
3'-0"
4'-4"
9'-6"
4'-4"
3'-0"
4'-4"
4'-0"
2X6 D.F.P.T. SILL W/ 1/2"ØX10" A.B. @ 6'-0" O.C. 7" MIN INTO CONC. & 12" MAX. FROM ANY CORNER W/ 2"Ø WASHERS.
BWP
PROVIDE SOLID BLK. @ 48" O.C.-48" FROM WALL W/A-34 @16" O.C WHERE JOIST PARALLEL WALL.
3'-8 x 9'-6" x 12" DEEP CONC. CHIMNEY FTG.
LINE OF SOLID BLK.
9'-0"
BLOCK OUT FOR 4'-4" WINDOW WELLS, 5'-3" ABOVE FIN. FLOOR. PROVIDE 3 METAL WINDOW WELLS.
6"
12" X 6"
15" x 18" DEEP CONC. FTG.
16" X 8"
8"
1'-4"
4'
3'-4"
BLOCK OUT FOR 3⁰ DOOR.
4" CONC. SLAB OVER 4" SAND FILL W/ .006 VAPOR BARRIER & 2" x 24" RIGID INSULATION @ EDGE.
4" CONC. SLAB W/ 10#-10# -6" X 6" W.W.M. OVER 4" SAND FILL. SLOPE 1/4" MIN. PER FT. FOR DRAINAGE.
2 x 6 D.F.P.T. SILL W/ 1/2" X 10" A.B. @ 24" O.C. 7" MIN INTO CONC. & 12" MAX. FROM ANY CORNER W/ 2"Ø WASHERS. PROVIDE A-34 JST/ ℄
36'-0"
28'-0"
12'-0"
16'-0"
8'-0"
12'-4"
20'-0"
23'-8"
ABWP
WATERPROOF BASEMENT WALL W/ 2 LAYERS OF HOT ASPHALTIC EMULISION.
12" DIA. x 8" CONC.PIER W/ PB 44 POST BASE.
BLOCK OUT FOR 3⁰ DOOR.
BLOCK OUT FOR 2 - 9⁰ DOORS
2'-0"
3'-4"
2'-6"
2"
PF
1'-10 1/2"
16'-0"
11'-0"
11'-0"
9'-4"
7"
9'-4"
1'-10 1/2"
38'-0"
23'-0"
61'-0"
NOTES:
1. ASSUMED SOIL BEARING PRESSURE OF 2000 P.S.F.
2. ALL CONC. TO BEAR ON FIRM, NATURAL, UNDISTURBED SOIL.
3. CONCRETE COMPRESSIVE STRENGTH AT 28 DAYS TO BE:
WALLS NOT EXPOSED TO WEATHER 2500 PSI
WALLS EXPOSED TO WEATHER 3000 PSI
PORCHES, STEPS AND GARAGE SLAB 3500 PSI
4. EXTEND FOOTINGS BELOW FROST LINE, (18" MIN. INTO NATURAL SOIL FOR 1 STORY AND 2 STORY CONSTRUCTION). FOOTINGS TO BE 6" THICK FOR 1 STORY, AND 7" THICK FOR 2 STORY CONSTRUCTION. ALL FOUNDATION WALLS TO BE 8" WIDE, UNLESS STEEL IS PROVIDED WITHIN 2" BUT NOT CLOSER THAN 1" FROM THE FACE OF THE WALL AWAY FROM THE SOIL. STEEL TO BE 2- #3'S HORIZONTAL.
5. THE GRADE AWAY FROM THE FOUNDATION WALLS TO FALL A MIN. OF 6" WITHIN THE FIRST 10 FEET.
6. ALL FRAMING LUMBER TO BE DOUGFIR LARCH # 2.
FOUNDATION PLAN
1/4" = 1'-0"

FIGURE 33.22 ■ The foundation plan for a full basement is finished by adding dimensions and text.

CADD APPLICATIONS

USING CADD TO DRAW FOUNDATION PLANS

Because you do not have the same kind of accuracy problems when working with CADD as you have with manual drafting, the CADD floor plan may be used as an accurate basis for drawing the foundation plan. Display the floor plan on a layer, and then begin the foundation drawing directly over the floor plan on another layer. Use layers with prefixes such as FNDWALLS, FNDFOOT, FNDJOIST, FNDTEXT, and FNDDIMEN to keep the foundation plan separate from the floor-plan files. By using the OSNAP command, the line representing the outer side of the stem walls can be drawn, using the walls of the floor plan as a guide. The OFFSET command can be used to lay out the thickness of the stem walls and footings. Corners can be adjusted by using the FILLET or TRIM command. The CHANGE command can be used to change the lines representing the footings from continuous to hidden. By following the step-by-step instructions for a particular foundation type, the plan can be completed.

Many of the dimensions used on the floor plan can also be used on the foundation plan. A layer such as BASEDIM can be used for placing dimensions required by the floor and foundation plans. General notes can be typed and stored as a WBLOCK and reused on future foundation plans. Many drafters also store lists of local notes required for a particular type of foundation as a WBLOCK and insert them into a drawing. Once inserted into the foundation plan, the notes can be moved to the desired position. When completed, the foundation can be stored separately from the floor plan to make plotting easier. Storing the foundation plan with the floor plan will save disk space, and proper use of layering can ease plotting. All foundation walls, bearing footings, and support beams will be in their correct locations. When you are finished drawing the foundation plan, turn OFF or FREEZE the floor plan layer to have a ready-to-plot foundation plan.

CHAPTER 33

Foundation-Plan Layout Test

DIRECTIONS

Answer the questions with short, complete statements.

1. Letter your name and the date at the top of the sheet.
2. Letter the question number and provide the answer. You do not need to write out the question.

QUESTIONS

Question 33–1 At what scale will the foundation be drawn?

Question 33–2 List five items that must be shown on a foundation plan for a concrete slab.

Question 33–3 What general categories of information must be dimensioned on a slab foundation?

Question 33–4 Show how floor joists are represented on a foundation plan.

Question 33–5 How large an opening should be provided in the stem wall for a garage door 8′–0″ wide?

Question 33–6 How much space should be provided for a 3′ entry door in a post-and-beam foundation? Explain your answer.

Question 33–7 How are the footings represented on a foundation plan?

Question 33–8 Show two methods of representing girders.

Question 33–9 What type of line quality is typically used to represent beam pockets?

Question 33–10 What are the disadvantages of tracing a print of the floor plan to lay out a foundation plan?

PROBLEMS

1. Draw a foundation plan that corresponds to the floor-plan problem from Chapter 16 that you have drawn. Draw the required foundation plan using the guidelines given in this chapter. Design a floor system that is suitable for the residence and your area of the country.
2. Draw the foundation plan using the same type and size of drawing material that you used for the floor plan.
3. Use the same scale that was used to draw the floor plan.
4. Refer to your floor plan to determine dimensions and position of load-bearing walls.
5. Refer to the text of this chapter and class lecture notes for complete information.
6. When your drawing is complete, turn in a print to your instructor for evaluation.

STEP 6 Lay out the bottom of the 4″ (100 mm) slab.

STEP 7 Lay out 4″ (100 mm) of fill material under the slab.

STEP 8 Lay out the width of the stem wall.

STEP 9 Lay out the width of the footings.

STEP 10 Lay out the 4″ (100 mm) wide ledge for brick veneer.

STEP 11 Block out the interior footings for load-bearing walls.

Wall Layout

With the foundation and floor system lightly drawn, proceed now to lay out the walls for the main level using construction lines. See Figure 35.2 for the layout of steps 12 through 18.

STEP 12 Locate the top of the top plates. Measure up 8′–0″ (2400 mm) above the floor. Draw a line to represent the top of the top plate. The walls can be drawn at 8′–0″ (2400 mm) high, but this is not their true height. The true height is the sum of the thickness of two top plates, the studs and the base plate. The plates are made from 2″ material, which is actually 1 1/2″ thick. The studs are milled in 88 5/8″ or 92 5/8″ lengths. Check with your instructor to see which size you should use. See Figure 35.3 to determine the true height of the walls. In a two-level house, this would also represent the bottom of the floor joist. In a one-level house this line will represent the bottom of the ceiling joist.

STEP 13 Locate the interior side of the exterior walls.

STEP 14 Locate the interior walls.

After locating the interior walls, make sure that any interior bearing walls line up over a footing. If they do not line up, there is a mistake in your measurements or math calculations or a conflict between the dimensions of the floor and foundation plans.

STEP 15 Lay out the tops of doors and windows. Headers for doors and windows are typically set at 6′–10″ (2080 mm) above the floor. Measure up from the top of the slab to locate the bottom of the header. The top will be drawn later.

LONG STUD (IN.)		SHORT STUD (IN.)
3	2 Top plates	3
92 5/8	Stud height	88 5/8
+ 1 1/2	Base plate	+ 1 1/2
97 1/8	Total height	93 1/8
8′–1 1/8″	Dimension to appear on section	7′–9 1/8″

FIGURE 35.3 ■ Determining floor to ceiling dimensions. Add the depth of the top plates, the base plate, and the height of the studs. Notice that the heights for the two common precut stud lengths are given.

STEP 16 Lay out the subsills. To establish the subsill location, you must know the window size. This can be found on either the floor plan or the window schedule. Measure down the required distance from the bottom of the header to establish the subsill location.

STEP 17 Lay out the 7′–0″ (2130 mm) ceilings for soffits in areas such as kitchens, baths, or hallways.

STEP 18 Lay out the patio post width and height.

Lay out the upper floor area following the same steps that were used for the lower level. See Figure 35.4 for steps 19 through 25.

STEP 19 Lay out exterior walls. Remember the 2′–0″ (600 mm) cantilever of the upper floor past the lower level walls.

STEP 20 Locate the depth of the floor joists.

STEP 21 Draw the plywood subfloor.

STEP 22 Lay out the ceiling 8′–0″ (2400 mm) above the floor.

STEP 23 Lay out the interior side of exterior walls.

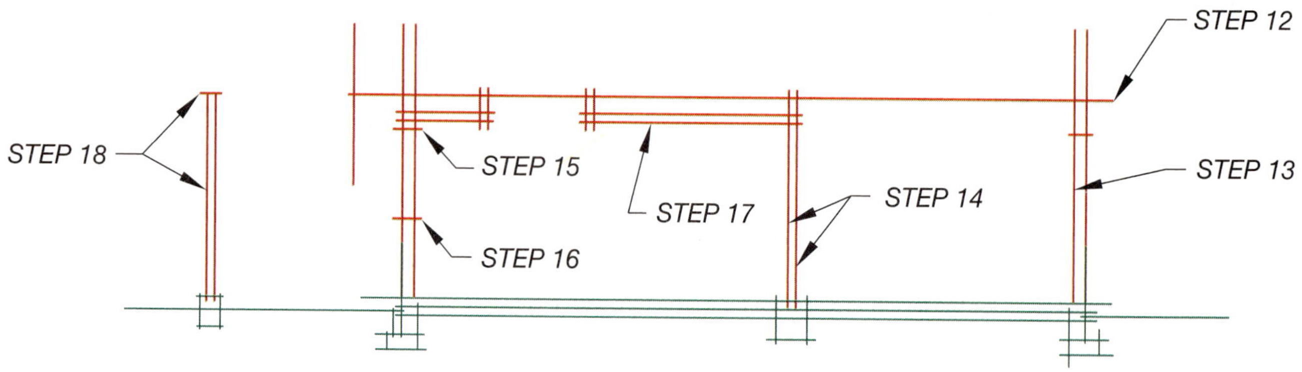

FIGURE 35.2 ■ Layout of the ceiling level.

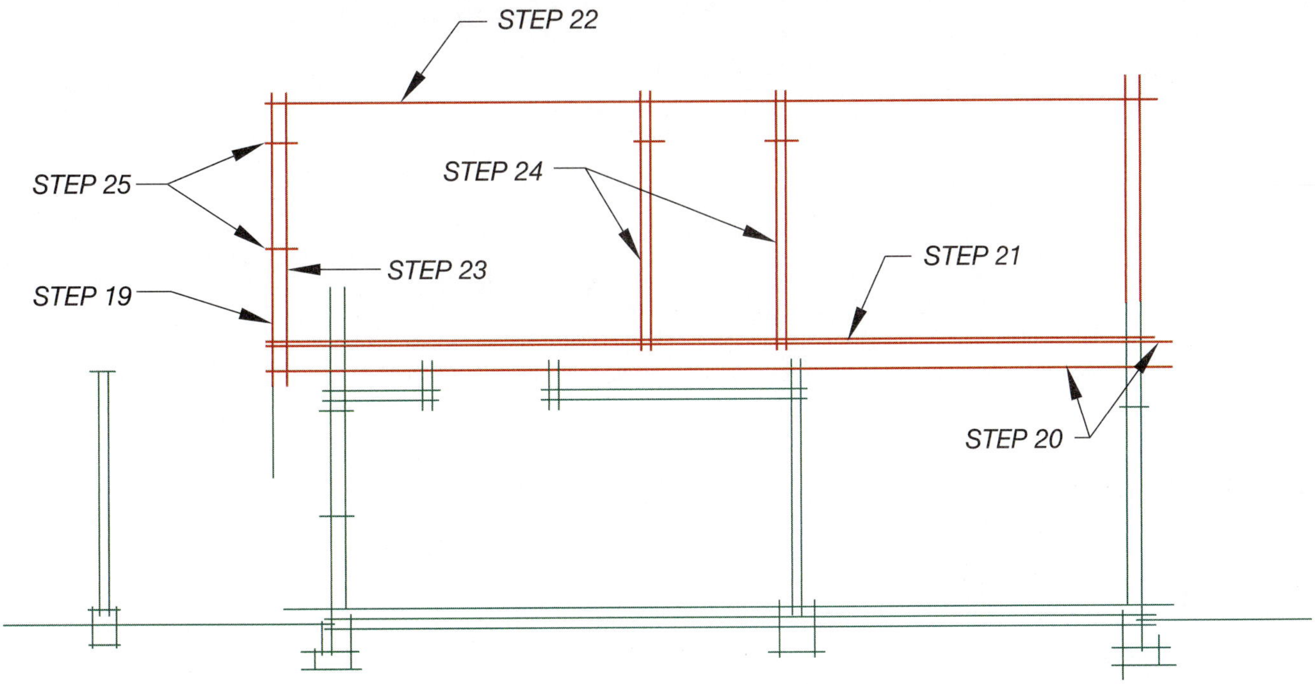

FIGURE 35.4 ■ Layout of the upper floor level.

STEP 24 Lay out the interior walls.

STEP 25 Lay out the windows.

Truss Roof Layout

Steps 26 through 33 can be seen in Figure 35.5. For the layout of other types of roofs, see Chapter 36.

STEP 26 Locate the ridge.

STEP 27 Locate the overhang on each side (see the roof plan).

STEP 28 Lay out the bottom side of the top chord. Starting at the intersection of the outer wall and the ceiling line, draw a line at a 5/12 pitch to match the pitch shown in the elevations. Do this for both sides. The lines should intersect at the ridge. A 5/12 pitch means that you measure up 5 units and over 12 units. A 5/12 pitch can be drawn at an angle of 22 1/2°.

STEP 29 Lay out the top side of the top chord. Assume chords to be 4″ (100 mm) deep.

STEP 30 Lay out the plywood roof sheathing.

STEP 31 Lay out the top side of the bottom chord.

STEP 32 Lay out the patio roof. Starting at the intersection of the inner wall and the ceiling, draw a line at a 5/12 pitch to represent the rafter.

STEP 33 Lay out the 1″ (25 mm) tongue-and-groove roof sheathing.

The entire section has now been outlined.

STAGE 3: FINISHED-QUALITY LINES—STRUCTURAL MEMBERS ONLY

When drawing with finished-quality lines, start at the roof and work down. This will help you keep the drawing clean. The steps followed will be divided into truss roof, walls, upper walls and floor, lower walls, and foundation.

Continuous members shown in section are traditionally drawn with a diagonal cross (X) placed in the member. Blocking is drawn with one diagonal line (/) through the member. This method of representing sectioned members is time-consuming to draw and may be difficult to read on small-scale sections. Rather than drawing symbols, the members can be drawn much more easily with different line qualities, as shown in Figure 35.6.

Three types of lines are used to draw structural material for the sections:

Thin. Use a 5-mm lead, a #0 pen, or a sharp 3H lead. If you are using AutoCAD, use a 0.010 line weight.

Bold. Use a 9-mm lead, a #2 pen, or an H lead with a wedge tip (0.024).

Very bold. Use a 9-mm lead, and draw two parallel lines with no white space between them, or use a #4 pen or an unpointed H lead (0.047).

As you read through these steps, you will be asked to draw items that have not been laid out. These are items that can be

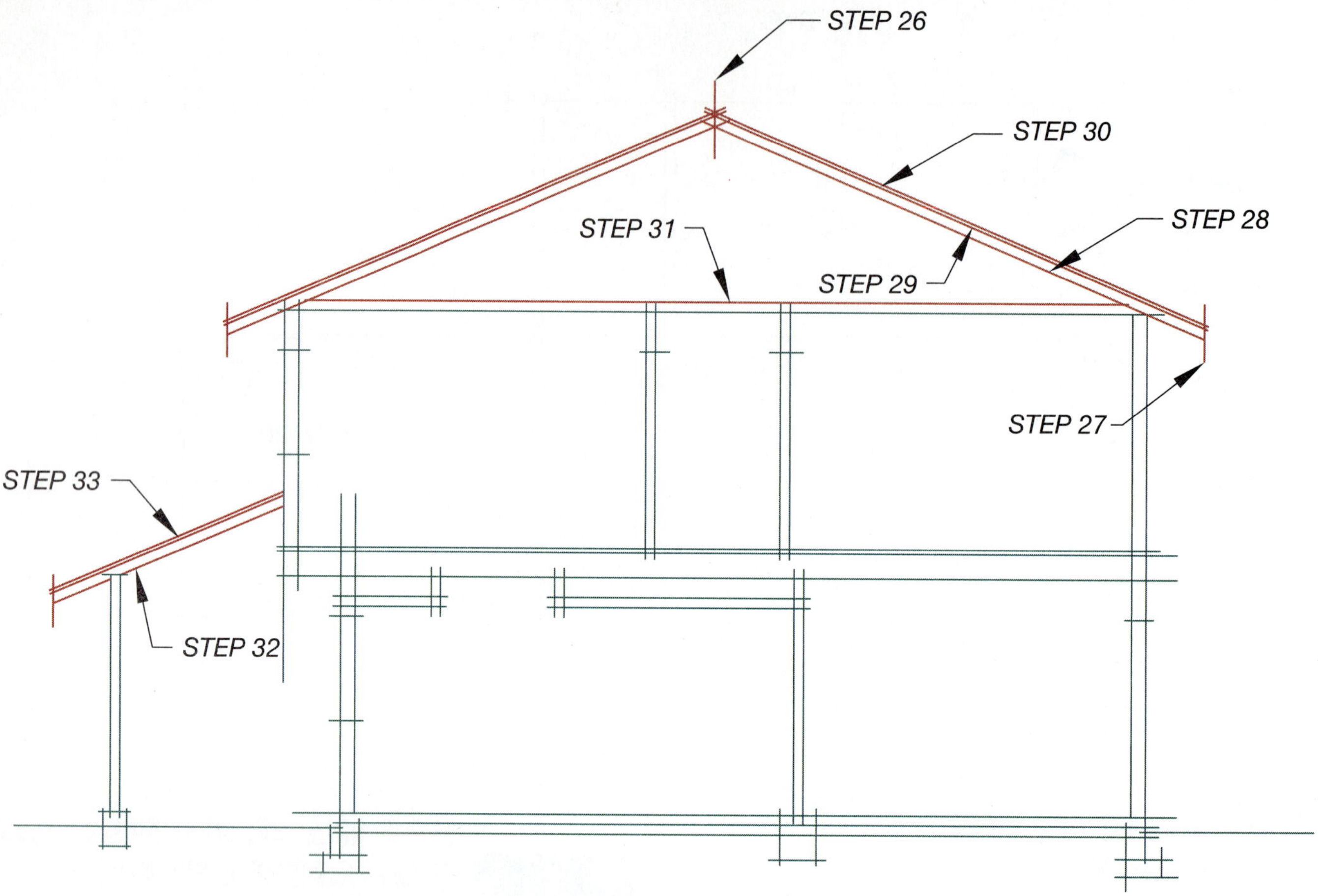

FIGURE 35.5 ■ Truss roof layout. The truss webs can be omitted, since the truss manufacturer will design and specify the required materials.

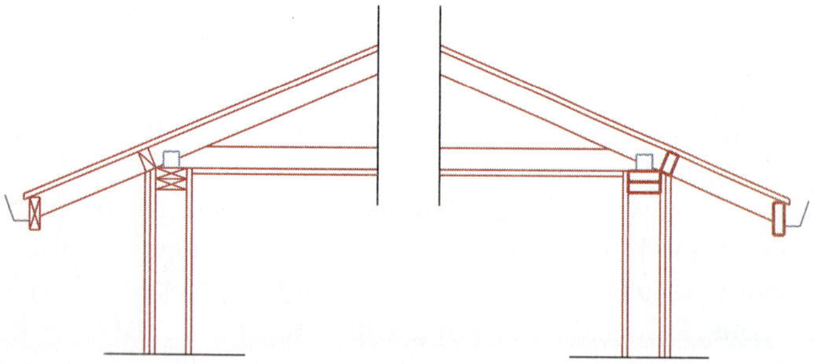

FIGURE 35.6 ■ Finished-quality lines for structural members. Lines to represent the blocking can be omitted for clarity when scales smaller than 3/4″ = 1′–0″ are used.

drawn by eye and need not be traced. Just be careful to keep 2″ (50 mm) members uniform in size so that they do not become 4″ (100 mm) members. When finished, your drawing should look like Figure 35.7.

Roof

To gain speed draw all parallel lines on one side of the roof before drawing the opposite side. Roof steps 1 through 7 are shown in Figure 35.8.

STEP 1 Use thin lines to draw the plywood sheathing.

STEP 2 Draw the top chord with thin lines.

STEP 3 Draw the eave blocking with bold lines.

STEP 4 Repeat steps 1 through 3 on the opposite side of the roof.

STEP 5 Draw the fascias with bold lines.

STEP 6 Draw the bottom chord with thin lines.

STEP 7 Draw the ridge blocking with a bold line.

Upper Walls and Floor

See Figure 35.9 for steps 8 through 18.

STEP 8 Draw all double top plates with bold lines.

STEP 9 Draw walls with thin lines.

STEP 10 Draw all headers with bold lines.

STEP 11 Draw all subsills with bold lines.

STEP 12 Draw all bottom plates with bold lines.

STEP 13 Draw exterior sheathing with thin lines.

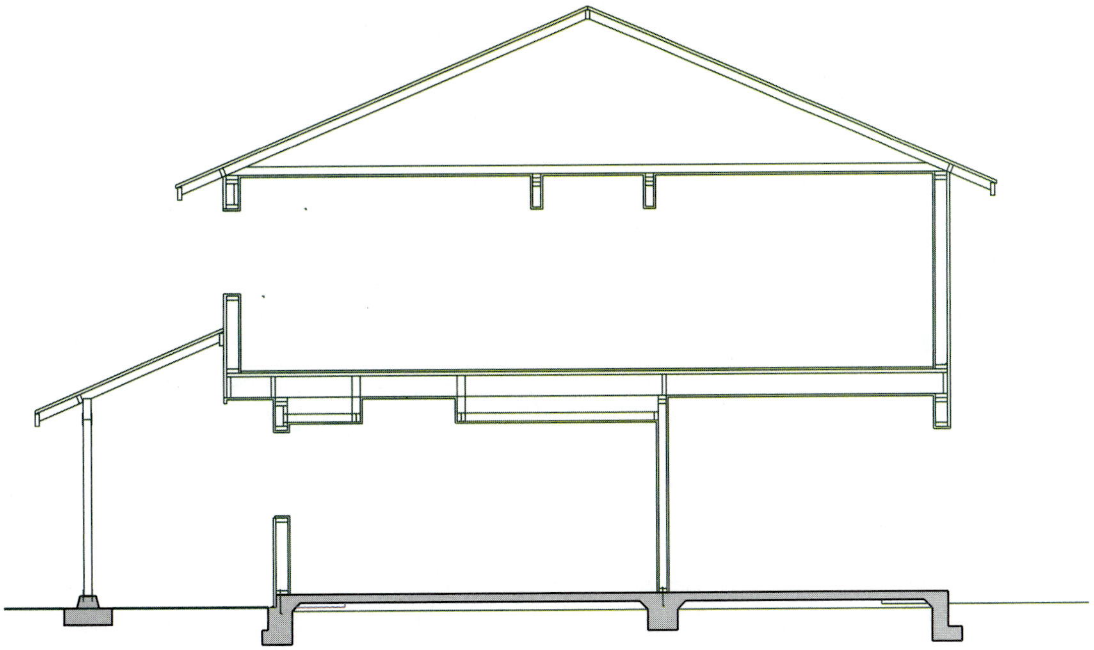

FIGURE 35.7 ■ All structural members drawn with finished-quality lines.

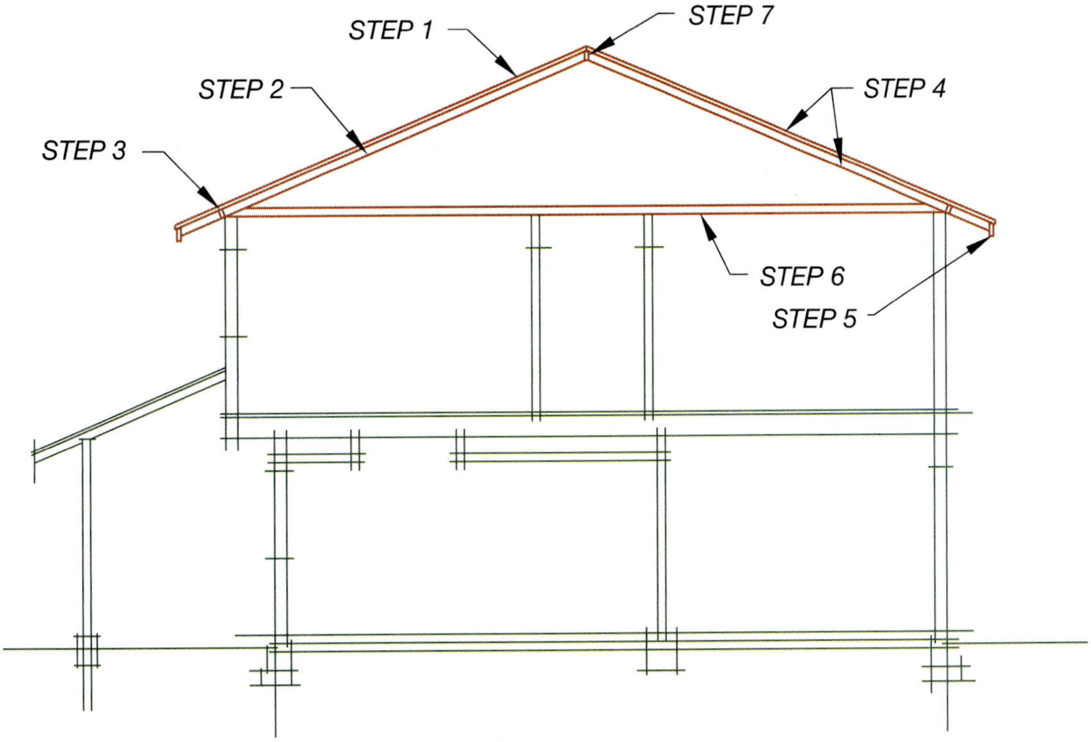

FIGURE 35.8 ■ Truss roof construction drawn with finished-quality lines.

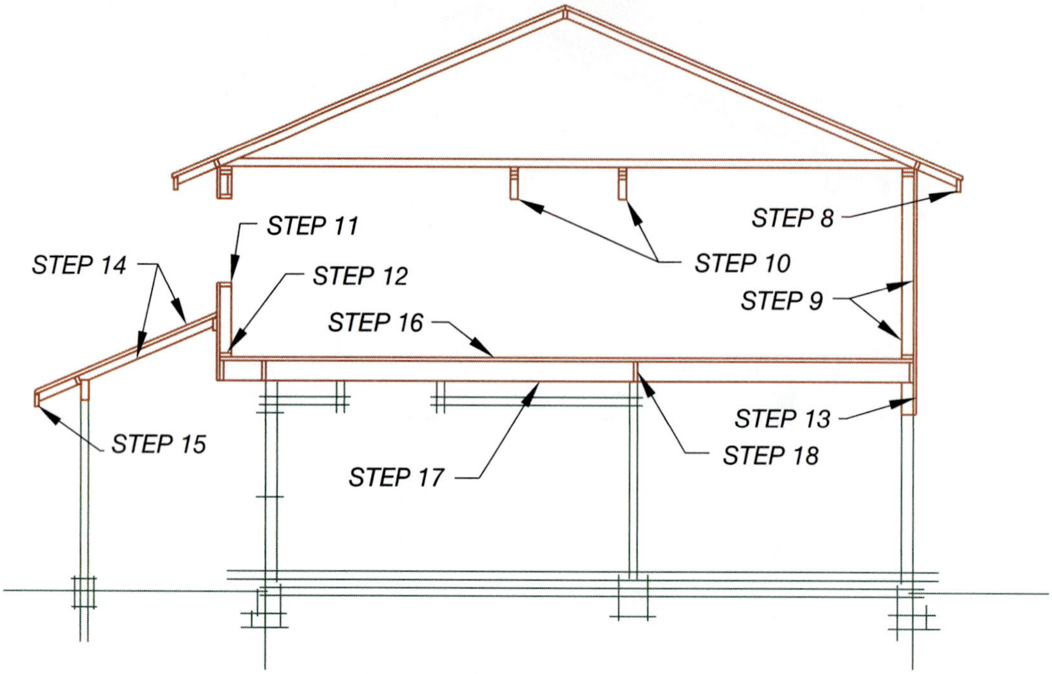

FIGURE 35.9 ■ Upper-floor structural members drawn with finished-quality lines.

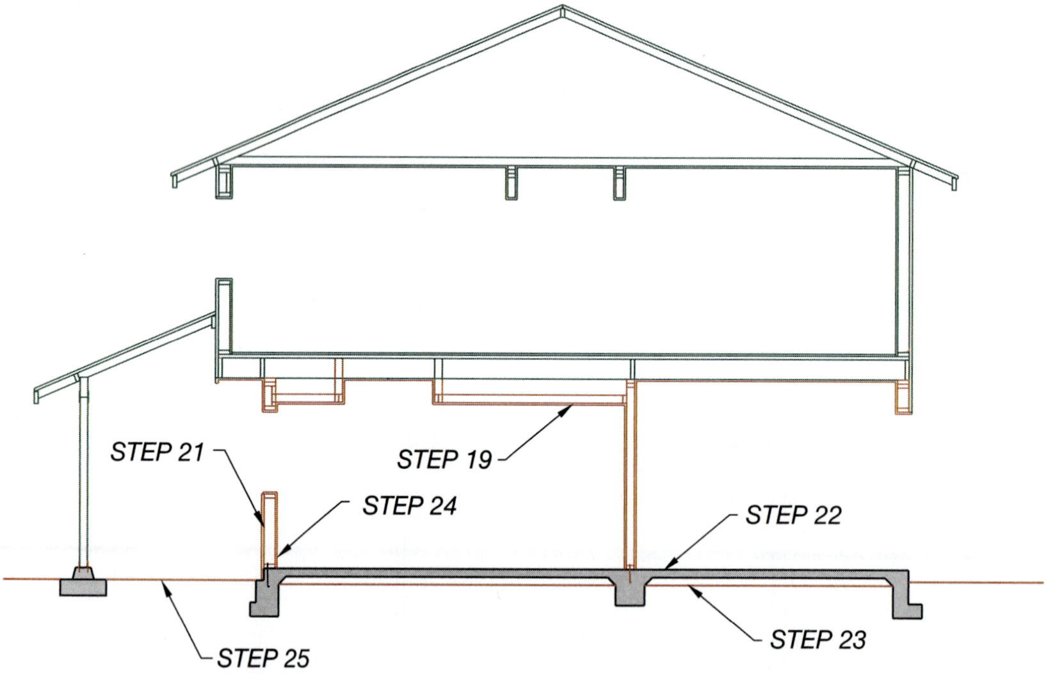

FIGURE 35.10 ■ Structural members of the main-floor level drawn with finished-quality lines.

STEP 14 Draw the porch rafters and decking with thin lines.

STEP 15 Draw fascia, block, header, and ledger for the porch roof with bold lines.

STEP 16 Draw plywood subfloor with thin lines.

STEP 17 Draw floor joists with thin lines.

STEP 18 Draw floor blocks with bold lines.

Lower Level Walls

See Figure 35.10 for steps 19 through 25.

STEP 19 Draw furred ceiling with thin lines.

STEP 20 Draw all blocks and ledgers for ceiling with bold lines.

STEP 21 Draw all exterior wall sheathing with thin lines.

Foundation

STEP 22 Draw the outline of the concrete using very bold lines.

STEP 23 Draw the 4″ (100 mm) fill material with a thin line.

STEP 24 Draw the mudsills and anchor bolts with bold lines.

STEP 25 Draw the finished grade with a very bold line.

All structural members are now drawn. Your drawing should resemble Figure 35.7.

STAGE 4: DRAWING FINISHING MATERIALS

The material drawn in this stage seals the exterior from the weather and finishes the interior. Start at the roof and work down to the foundation to keep the drawing clean. Use thin lines for this entire stage unless otherwise noted. See Figure 35.11 for steps 1 through 13.

Roof

STEP 1 Draw hurricane ties approximately 3″ (75 mm) square.

STEP 2 Draw baffles with a bold line.

STEP 3 Draw the finished ceiling.

STEP 4 Draw insulation in the ceiling approximately 10″ (250 mm) deep.

STEP 5 Draw ridge vents.

STEP 6 Draw gutters.

Walls and Floor

STEP 7 Draw the exterior siding.

STEP 8 Draw all windows.

STEP 9 Draw all interior finishes and insulation.

STEP 10 Draw and crosshatch the brick veneer.

STEP 11 Draw the hardboard underlayment. Note that the plywood subfloor extends under all walls, but the hardboard underlayment does not go under any walls.

STEP 12 Draw floor insulation.

STEP 13 Draw the metal connectors and flashing for the porch.

STAGE 5: DIMENSIONING

Now that the section is drawn, all structural members must be dimensioned so that the framing crew knows their vertical location. Figure 35.12 shows the needed dimensions. All leader lines should be thin, similar to those used on the floor and foundation plans. All lettering should be aligned. Place the required leader and dimension lines to locate the following dimensions:

STEP 1 Floor to ceiling. Place a dimension line from the top of the floor sheathing to the bottom of the floor or ceiling joist.

STEP 2 Specify the floor-to-floor dimensions.

STEP 3 Floor to bottom of bedroom windows, 44″ (1120 mm) maximum.

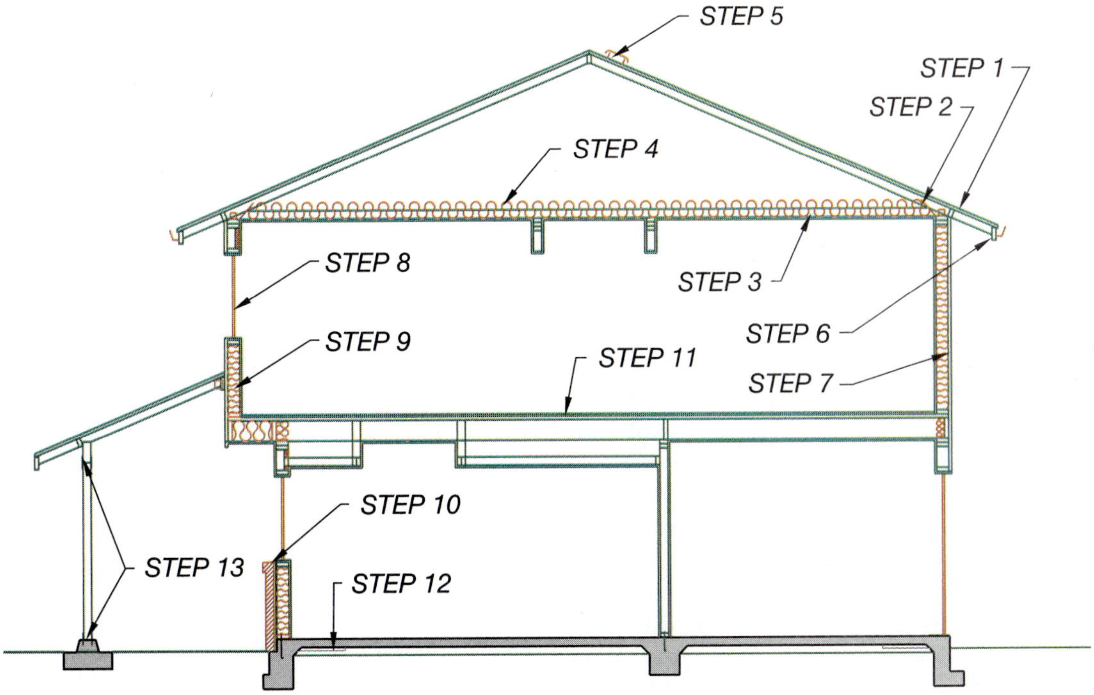

FIGURE 35.11 ■ Cross section with all the finishing materials for the roof, walls, and floors.

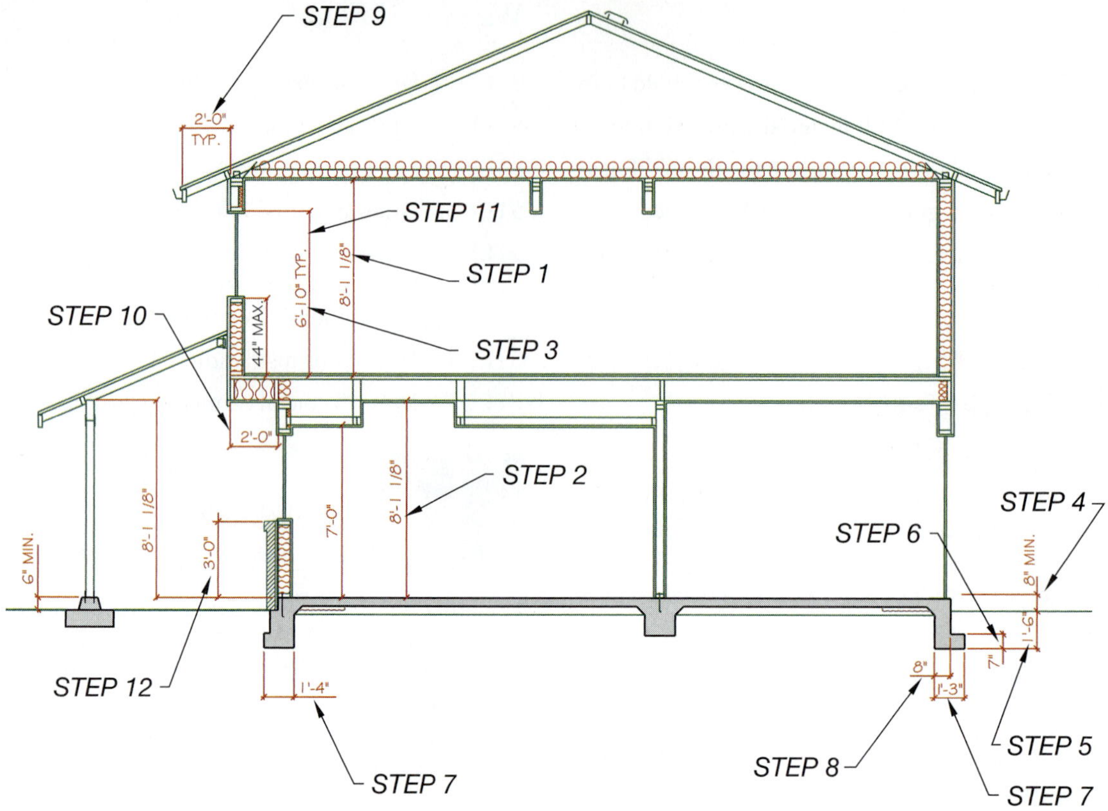

FIGURE 35.12 ■ Cross section with all of the required dimensions.

STEP 4 Finished grade to top of slab.

STEP 5 Depth of footing into grade.

STEP 6 Height of footings.

STEP 7 Width of footings.

STEP 8 Width of stem wall.

STEP 9 Eave overhangs.

STEP 10 Cantilevers.

STEP 11 Header heights from rough floor.

STEP 12 Height of brick veneer if necessary.

STEP 13 Lean back and relax! You're almost done.

STAGE 6: LETTERING NOTES

Everything that has been drawn and located must be explained. Place guidelines around the perimeter of the drawing and align the required notes on the guidelines, as seen in Figure 35.13. Doing so will help make your drawing neat and easy to read. Not all of the following notes will need to go on every section you draw. Typically, the primary section will be fully notated, and then other sections will have only construction notes. You will notice as you go through the list of general notes that some are marked (*). For the marked items, there are several options. Ask your instructor for help in determining which grade of material to use.

Remember that the following list is just a guideline. You, the drafter, will need to evaluate each section prior to placing the required notes. In an office setting your supervisor might give you a print with all of the notes that need to be placed on the original. As you gain confidence, you will probably be referred to a similar drawing for help in deciding which notes are needed. Eventually you will be able to draw and notate a section without any guidelines. The general notes that may appear on sections are as follows:

Roof

*1. RIDGE BLOCK, or 2 × ___ RIDGE BOARD.

2. SCREENED RIDGE VENTS @ 10′ O.C. ±. PROVIDE 1 SQ FT PER EACH 300 SQ FT OF ATTIC SPACE.

*3. 1/2″ STD GRADE 32/16 PLY ROOF SHEATH. LAID PERP. TO TRUSSES. NAIL W/ 8d @ 6″ O.C. @ EDGE AND 8d @ 12″ O.C. FIELD.

or

1 × 4 SKIP SHEATHING W/ 3 1/2″ SPACING.

*4. 235# (OR 300#) COMPO. ROOF SHINGLES OVER 15# FELT.

or

MED. CEDAR SHAKES OVER 15# FELT W/ 30# × 18″ WIDE FELT BETWEEN EACH COURSE W/ 10 1/2″ EXPOSURE.

or

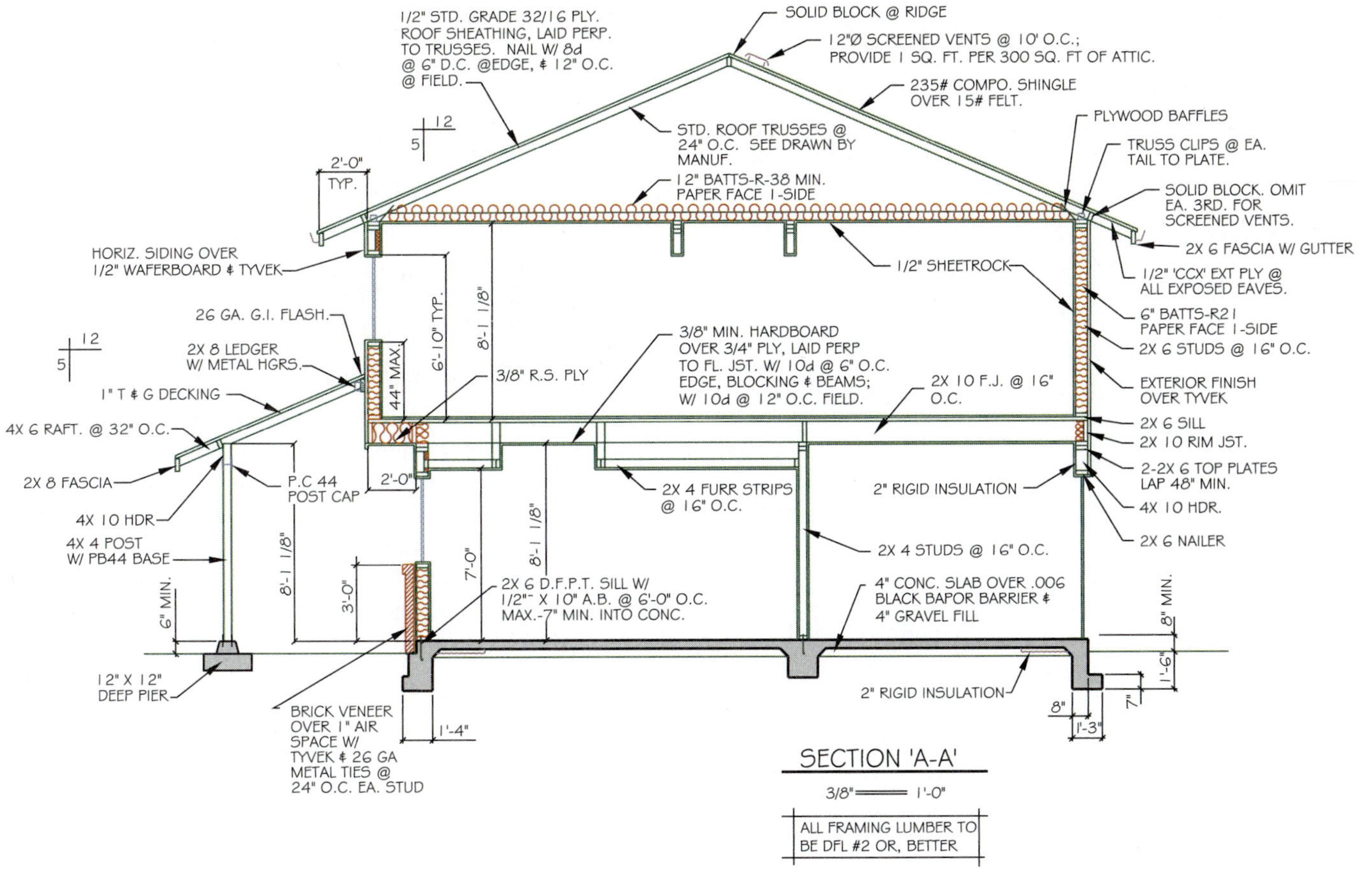

FIGURE 35.13 ■ Cross section completed with all structural and finishing material drawn, dimensioned, and lettered.

CONC. ROOF TILES BY (give manufacturer's name, and color weight of tiles and underlayment). INSTALL AS PER MANUF. SPECS.

5. METAL ROOFING BY (give the manufacturer's name, type of product, gage, and color). INSTALL AS PER MANUF. SPECS.
6. STD. ROOF TRUSSES @ 24″ O.C. SEE DRAW. BY MANUF.
7. __ BATTS, R-__ PAPER FACE @ HEATED SIDE.

or

__ BLOWN-IN INSULATION R-__ MIN.

8. BAFFLES AT EAVE VENTS.
9. SOLID BLOCK—OMIT EA. 3rd FOR SCREENED VENTS.
10. TRUSS CLIPS @ EA. TAIL TO PLATE.
11. __ × __ FASCIA W/ GUTTER.
12. 1/2″ 'CCX' EXT. PLY @ ALL EXPOSED EAVES.

or

1 × 4 T&G DECKING @ ALL EXPOSED EAVES.

Walls

*1. 2—2 × __ TOP PLATES, LAP 48″ MIN.

*2. 2 × __ STUDS @ ___″ O.C.

*3. 2 × __ SILL (Fill in the blank.)

*4. EXT. SIDING OVER 1/2″ WAFERBOARD & TYVEK.

or

EXT. SIDING OVER 3/8″ PLY & 15 FELT.

*5. 3 1/2 FIBERGLASS BATTS, R-11 MIN.—PAPER FACE ONE SIDE.

or

6″ FIBERGLASS BATTS, R-19, PAPER FACE ONE SIDE.

or

6″ FIBERGLASS BATTS R-21.

*6. __ × __ HEADER.

or

2—2 × 12 HDRS W/ 2 × 6 NAILER & 2″ RIGID INSULATION.

7. LINE OF INTERIOR FINISH.
8. SOLID BLOCK AT MID HEIGHT FOR ALL WALLS OVER 10′.
9. 5/8″ TYPE 'X' GYP. BD. FROM FLOOR TO BOTTOM OF ROOF SHEATH.

or

1/2″ GYP. BD. WALLS AND CEIL.

10. BRICK VENEER OVER 1″ AIR SPACE & TYVEK W/METAL TIES @ 24″ O.C. EA. STUD.

Upper Floor and Foundation

1. 3/8″ MIN. HARDBOARD UNDERLAYMENT.

*2. 3/4″ 42/16 PLY. FLOOR SHEATH. LAID PERP. TO FLOOR JOISTS. NAIL W/ 10d @ 6″ O.C. EDGE, BLOCKING, & BEAMS. USE 10d @ 12″ O.C. @ FIELD.

3. SOLID BLOCK @ 10′ O.C. MAX.

4. __ BATTS—R-__ MIN. or (R-21).

5. 2 × 6 PT. SILL W/1/2″ Ø × 10″ A.B. @ 6′–0″ O.C. MAX.—7″ MIN. INTO CONC. W/2″ DIA WASHERS.

6. 4″ CONC. SLAB OVER 4″ GRAVEL, 6 MIL VAPOR BARRIER, AND 2″ SAND FILL.

   or

   4″ CONC. SLAB OVER 4″ GRAVEL FILL OVER 55# FELT.

7. 2″ × 24″ RIGID INSULATION.

8. Give the section a title such as SECTION 'AA'.

9. Give the drawing a scale such as 3/8″ = 1′–0″.

10. List general notes near the title or near the title block.

    *ALL FRAMING LUMBER TO BE D.F.L. # 2 OR BETTER. (You may need to substitute a different type of wood.)

    SEE SECTIONS 'BB' & 'CC' FOR BALANCE OF NOTES.

Give your eyes and fingers a well-deserved rest. You now have 99 percent of the section complete.

STAGE 7: EVALUATE YOUR WORK

Don't assume that because you did the work and it took a long time, the drawing is complete. Run a print, and evaluate your work for accuracy and quality. Don't just compare it with someone else's drawing. Use the checklist, and make sure your section matches the material and location that you have specified on the floor and foundation plans. Your best chance of finding your own mistakes is to get away from your drawing for an hour or two before checking it.

CHAPTER 35

Section Layout Test

DIRECTIONS

Answer the questions with short, complete statements or drawings as needed.

1. Letter your name and the date at the top of the sheet.
2. Letter the question number and provide the answer. You do not need to write out the question.

QUESTIONS

Question 35–1 Define the following terms.

a. Rafter
b. Truss
c. Ceiling joist
d. Collar tie
e. Jack stud
f. Rim joist
g. Chord
h. Sheathing

Question 35–2 On a blank sheet of paper, sketch a section view of a conventionally framed roof showing all interior supports.

Question 35–3 Give the typical sizes of the following materials.

a. Mudsill
b. Stud height
c. Roof sheathing
d. Wall sheathing
e. Floor decking
f. Underlayment

Question 35–4 List three common scales for drawing sections, and tell when each is best used.

Question 35–5 List the seven stages of drawing a section.

Question 35–6 List two different types of drawings for which stock drawings are typically used.

CADD APPLICATIONS

DRAWING SECTIONS AND DETAILS WITH CADD

The steps for drawing sections presented in this chapter are used for both manual and computer-aided drafting. However, using the computer has some advantages: standard sections and details can be brought into the drawing from a menu library; placing notes on the section is easy with CADD. Also, some architectural CADD packages automatically draw a preliminary section from information you provide in relationship to the floor plan. This type of parametric design requires that an imaginary cutting plane line be placed through the floor plan in the desired section location. This is followed by computer prompts requesting information such as roof pitch and structural floor thicknesses. In programs of this type, the floor-plan walls are drawn with heights established, so these dimensions automatically convert to wall height information in the sectional view. All you do to complete the section is add material symbols, dimensions, and notes.

Some architectural CADD packages have template library menu overlays that contain typical section elements such as details, materials, and tags. After architects have used the CADD system awhile, they begin to save all typical or standard construction details as BLOCKS. These BLOCKS are commonly placed in a library manual for easy reference and can be called up and displayed at any time in any drawing. The standard details should be clearly labeled with an identification code for easy reference. Each CADD user should have a copy of the library reference manual, and every time a new detail is drawn, the reference manual should be updated with a drawing of the detail and the reference code. Figure 35.14 shows two standard details. Stock details can be merged to form a partial section. The partial section shown in Figure 35.15 was created by aligning the roof/wall detail directly above the post-and-beam detail. The line representing the top of the floor decking was offset to determine the location of the top of the wall. With the top of the wall located, the roof/wall detail was inserted into the required position. Many building departments will accept a partial section when an application is submitted for a building permit for a simple residence.

A partial section can be used to form a full section. The line that represents the outside edge of the stud can be offset a distance equal to the required width of the house. The partial section can then be mirrored to form the opposite side of the structure, and then lines from one side can be extended to the opposite side. The FILLET, EXTEND, and STRETCH commands of AutoCAD are excellent for connecting the two partial sections. Once the shell of the structure has been drawn, the interior wall and girders are added using the COPY and ARRAY commands. The location of the interior walls can be determined using the dimensions of the framing plan, and the girders are placed according to their location on the foundation plan. Figure 35.16 shows a full section created using the stock details from Figure 35.14.

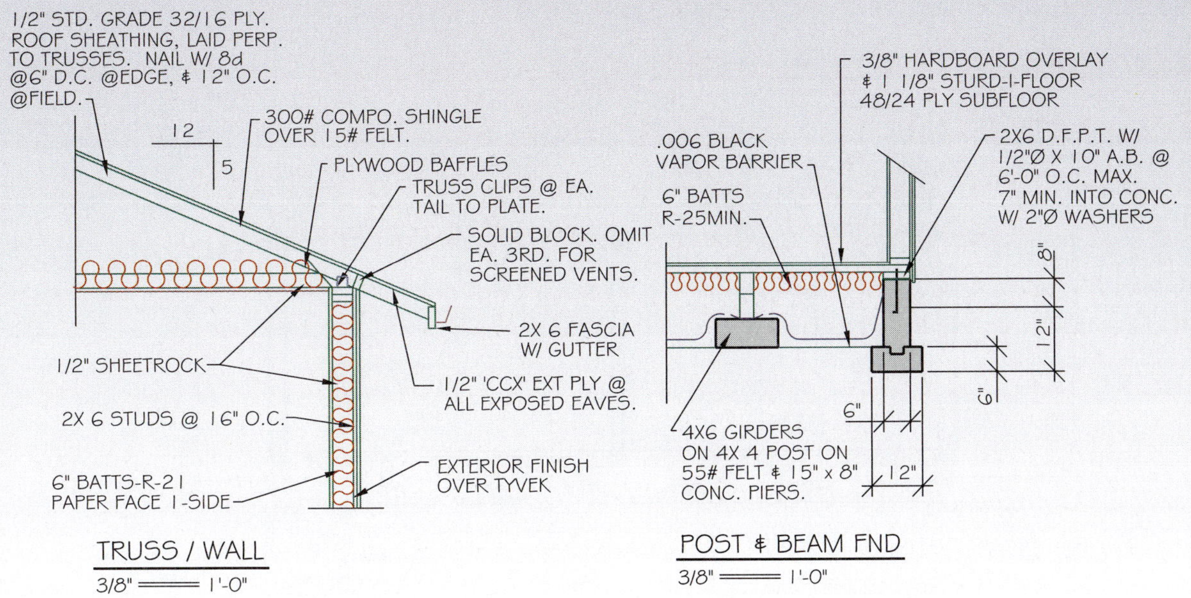

FIGURE 35.14 ■ CADD detail drawings can be saved as blocks and reused with other similar structures.

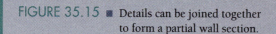

FIGURE 35.15 ■ Details can be joined together to form a partial wall section.

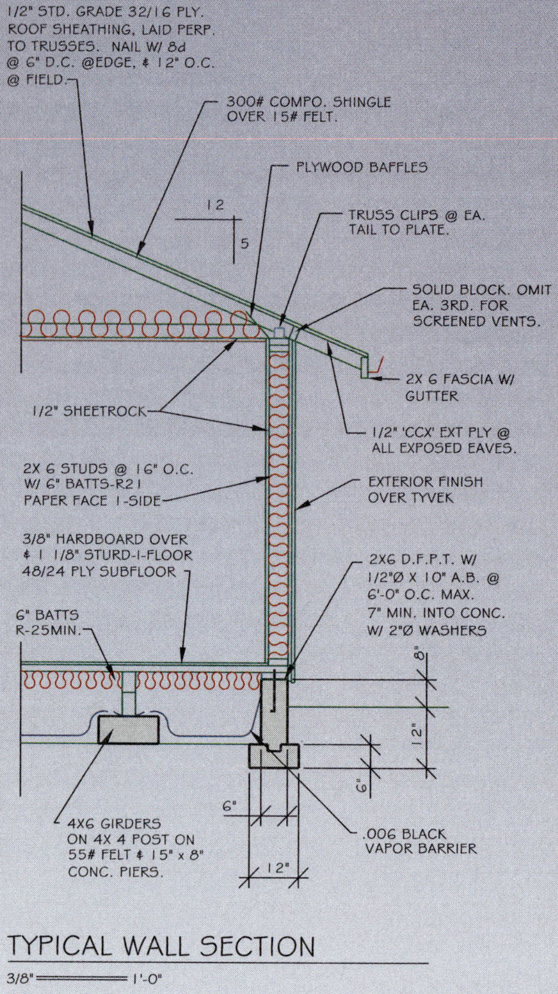

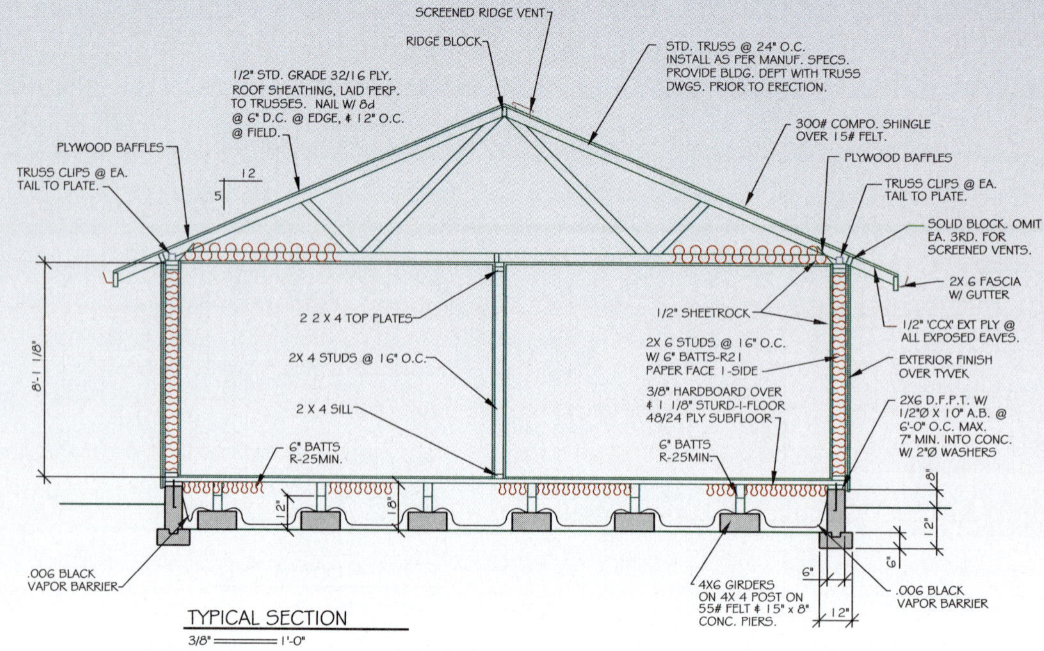

FIGURE 35.16 ■ A partial wall section can be used to form a full section.

CHAPTER 36

Alternative Layout Techniques

INTRODUCTION

Chapter 35 presented the basic methods for drawing a section with a concrete slab and truss roof. Those methods will not meet the needs of all drafters because of variations in the job site, the contractor's personal preference for framing, or the area of the country in which the house is to be built. This chapter provides you with other common construction methods to meet a variety of needs. Unless noted, all sizes given in the following steps are based on the minimum standards of the International Residential Code and should be compared with local standards.

FLOOR JOIST/FOUNDATION

Section 8 introduced methods of drawing a section with a concrete slab. This chapter will introduce methods of drawing a section with a joist floor system.

Joist Floor Layout

Use construction lines for steps 1 to 15. See Figure 36.1 for steps 1 to 5.

STEP 1 Lay out the width of the building.

STEP 2 Establish a baseline about 2″ (50 mm) above the border.

Before laying out steps 3 through 8, it may be necessary to review the basics of foundation design in Section 8. All sizes given in the following steps are based on the minimum standards of the International Residential Code (unless otherwise noted) and should be compared with local standards.

STEP 3 Measure up from the baseline the required thickness for a footing.

STEP 4 Locate the finished grade.

STEP 5 Locate the top of the stem wall. All wood should be 8″ minimum (200 mm) above finished grade.

See Figure 36.2 for steps 6 through 15.

STEP 6 Block out the width of the stem wall.

STEP 7 Block out ledge 4″ (100 mm) wide for brick veneer.

STEP 8 Block out the required width for footings. Center the footing below the stem walls.

STEP 9 Add 4″ (100 mm) in width to the footing to support the brick ledge.

STEP 10 Locate the depth for the 2×6 (50×150 mm) mudsill. The top of the mudsill also becomes the bottom of the floor joists.

STEP 11 Locate the depth of the floor joists.

STEP 12 Lay out the plywood subfloor. Draw a line as close to the floor joist as possible to represent the plywood while still leaving a gap between the plywood and the joist.

STEP 13 Lay out required girders beneath load-bearing walls or as needed on this basis of the joist spans.

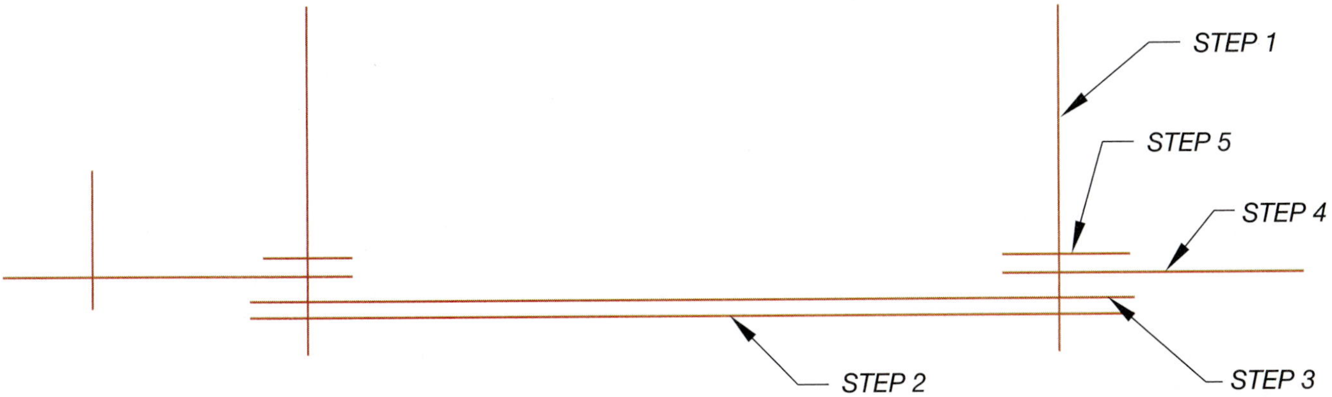

FIGURE 36.1 ■ Layout for the joist floor system using construction lines.

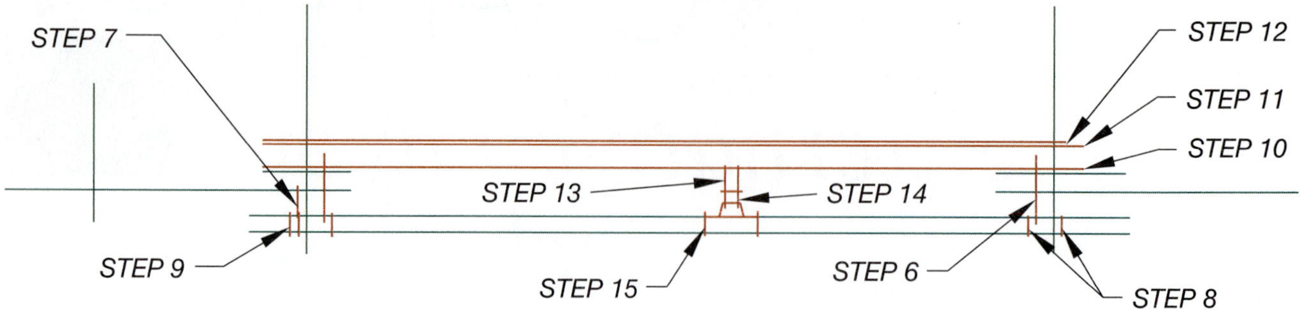

FIGURE 36.2 ■ Layout of the interior supports for a joist floor and foundation.

STEP 14 Lay out posts to support girders.

STEP 15 Lay out concrete piers centered under the post.

Finished Line Quality—Structural Materials

Unless otherwise noted, use the line weights and methods described in Chapter 35 to draw structural materials for a joist floor system and foundation with finished-line quality. See Figure 36.3 for steps 16 through 23.

STEP 16 Draw mudsills with bold lines.

STEP 17 Draw girders with bold lines.

STEP 18 Draw support post with thin lines.

STEP 19 Draw all exterior wall sheathing with thin lines.

STEP 20 Draw stem walls and footings with very bold lines.

STEP 21 Draw the piers with very bold lines.

STEP 22 Draw anchor bolts with bold lines.

STEP 23 Use a very bold line to draw the finished grade.

All structural material is now drawn.

Finished-Quality Lines—Finishing Materials

Use Figure 36.4 as a guide to add the finishing materials.

STEP 24 Draw the hardboard underlayment. Note that the plywood subfloor extends under all walls, but the hardboard underlayment does not go under any walls.

STEP 25 Draw floor insulation.

STEP 26 Draw the vapor barrier freehand, leaving a small space between the ground and the barrier.

STEP 27 Draw the metal connectors and flashing for the porch.

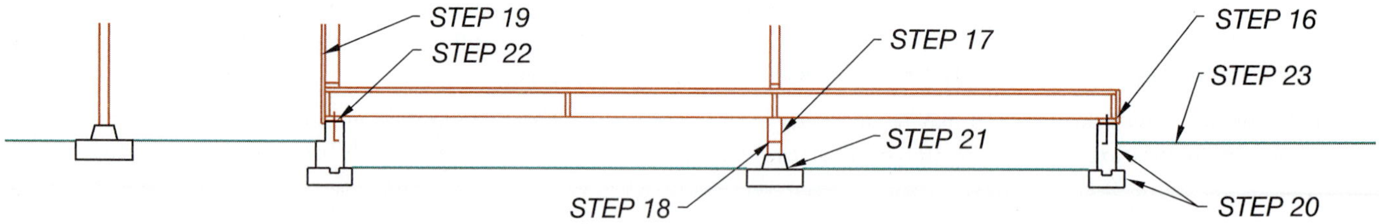

FIGURE 36.3 ■ Finished-quality lines on the structural members of a joist floor and foundation system. Varied line weights should be used to represent concrete, sectioned continuous materials, and the material that lies beyond the cutting plane.

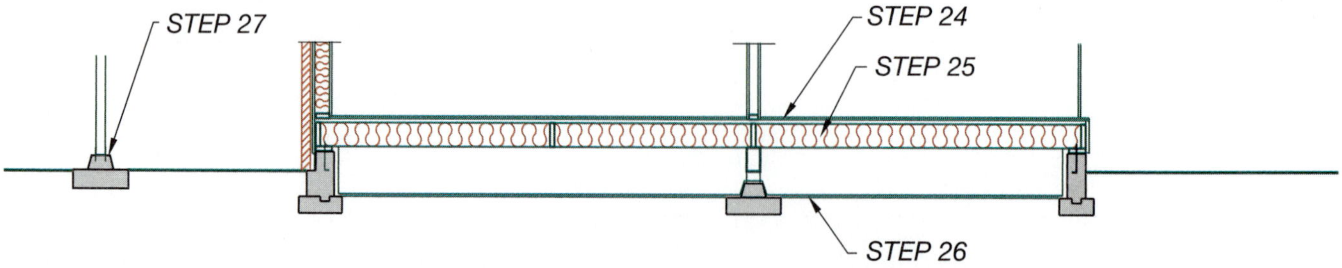

FIGURE 36.4 ■ Finish materials for a joist floor and foundation system include the floor overlay, insulation, metal connectors, and vapor barrier.

Dimensioning

All structural members must be dimensioned so that the framing crew knows their vertical location. Figure 36.5 shows the needed dimensions. Place the required leader and dimension lines to locate the following dimensions:

STEP 28 Finished grade to top of stem wall.

STEP 29 Depth of footing into grade.

STEP 30 Height of footing.

STEP 31 Width of footing.

STEP 32 Width of stem wall.

STEP 33 Depth of crawl space.

STEP 34 Bottom of girders to grade.

Floor and Foundation Notes

Letter the floor and foundation notes as shown in Figure 36.6. Typical notes should include the following:

1. 3/8″ MIN. HARDBOARD UNDERLAYMENT.
*2. 3/4″ 42/16 PLY. FLOOR SHEATH. LAID PERP. TO FLOOR JOISTS. NAIL W/ 10d @ 6″ O.C. EDGE, BLOCKING, & BEAMS. USE 10d @ 12″ O.C. @ FIELD.
3. SOLID BLOCK @ 10′ O.C. MAX.
4. __ BATTS—R-__ MIN. or (R-21).
*5. __ × __ GIRDERS (Fill in the blanks. See foundation plan to verify size.)
*6. 4 × 4 POST, 4 × 6 @ SPLICES, W/ GUSSET TO GIRDER ON 55# FELT ON __ × __ DEEP CONC. PIERS.
7. .006 BLACK VAPOR BARRIER or 55# ROLLED ROOFING.
*8. 2 × __ P.T. SILL W/ 1/2″ ∅ A.B. @ 6′–0″ O.C. MAX.—7″ MIN. INTO CONC. W-2″ ∅ WASHERS.
9. Give the section a title such as SECTION ‘AA’.
10. Give the drawing a scale such as 3/8″ = 1′–0″.
11. List general notes near the title or near the title block. *ALL FRAMING LUMBER TO BE D.F.L. #2 OR BETTER.

    (Instead of D.F.L. you may need to substitute a different type of wood.)

    SEE SECTIONS ‘BB’ & ‘CC’ FOR BALANCE OF NOTES.

Complete the section as in Chapter 35.

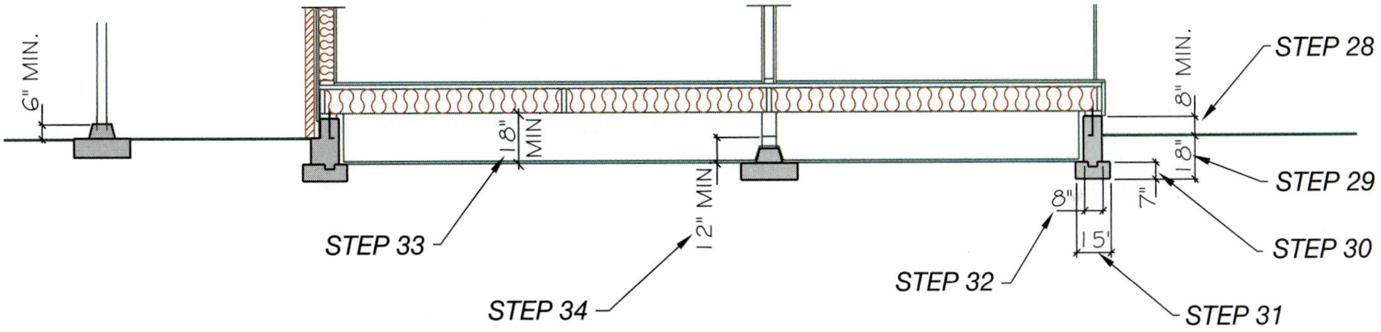

FIGURE 36.5 ■ Dimensions are required for a joist system to describe minimum clearances.

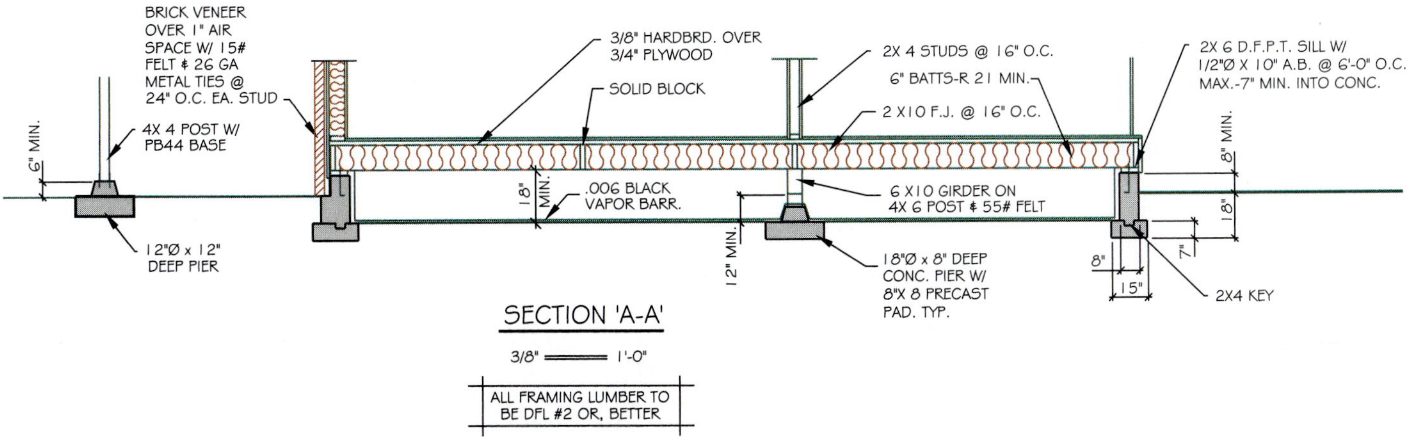

FIGURE 36.6 ■ Notes for a joist foundation system are required to describe all structural and finish materials. Some companies use generic notes for structural material such as joist and girders and refer the print reader to other plans for specific sizes. Instead of marking the girder as 6 × 10, a reference such a “Girder—see foundation plan” could be used.

POST-AND-BEAM FOUNDATIONS

Section 8 presented an introduction to post-and-beam construction. This chapter will describe post-and-beam construction as it applies to the foundation and lower floor.

Layout

Use construction lines for steps 1 through 14. See Figure 36.7 for steps 1 through 10.

STEP 1 Lay out the width of the building.

STEP 2 Lay out a baseline for the bottom of the footings.

STEP 3 Lay out the footing thickness.

STEP 4 Lay out the finished grade.

STEP 5 Locate the top of the stem wall.

STEP 6 Block out the width of the stem wall.

STEP 7 Block out the width of the footings, centered under the stem wall.

STEP 8 Block out a 4″ (100 mm) wide ledge for the brick veneer.

STEP 9 Add 4″ (100 mm) width to the footing to support the brick ledge.

STEP 10 Locate the 2 × 6″ (50 × 150 mm) mudsill.

See Figure 36.8 for steps 11 through 13.

STEP 11 Lay out the depth of the 2″ (50 mm) floor decking.

STEP 12 Lay out the girders and the support posts.

STEP 13 Lay out the concrete support piers centered under the posts. Lay out the width first for each pier, and then draw one continuous guideline to represent the top of all piers.

Finished-Quality Lines—Structural Material

For drawing structural members with finished-quality lines, use line quality as described in Chapter 35 unless otherwise noted. See Figure 36.9 for steps 14 through 22. You will notice that some items will be drawn that have not been laid out. These are items that are simple enough to draw by estimation.

STEP 14 With thin lines, draw the tongue-and-groove (T&G) decking. Do not show the actual T&G pattern. This is done only when details are shown at a larger scale such as 3/4″ = 1′–0″.

STEP 15 Draw the mudsills using bold lines.

STEP 16 Draw anchor bolts with bold lines.

STEP 17 Draw all posts with thin lines.

STEP 18 Draw all girders with bold lines.

STEP 19 Draw the stem walls and footings with very bold lines.

STEP 20 Draw piers with very bold lines.

STEP 21 Draw the post pads freehand with bold lines.

STEP 22 Draw the finished grade with very bold lines.

Finished-Quality Lines—Finishing Material

The materials in this section will be used to seal the exterior walls from the weather and to finish the interior. Use thin lines

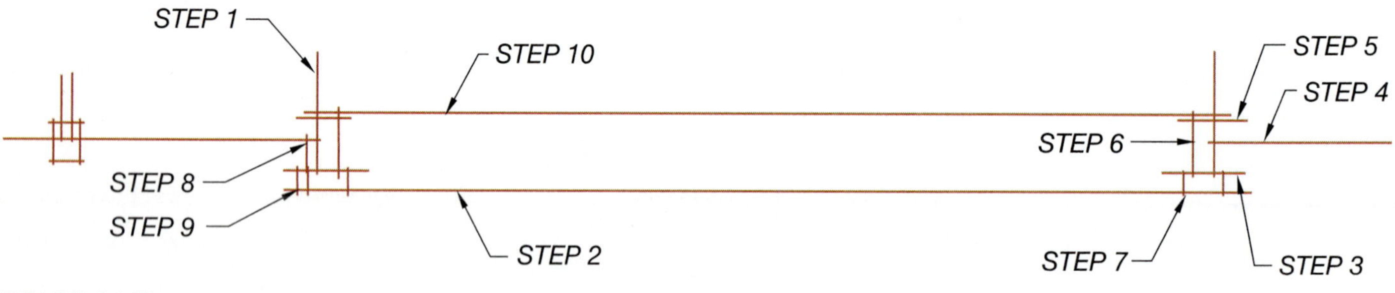

FIGURE 36.7 ■ Layout methods for a post-and-beam floor and foundation system.

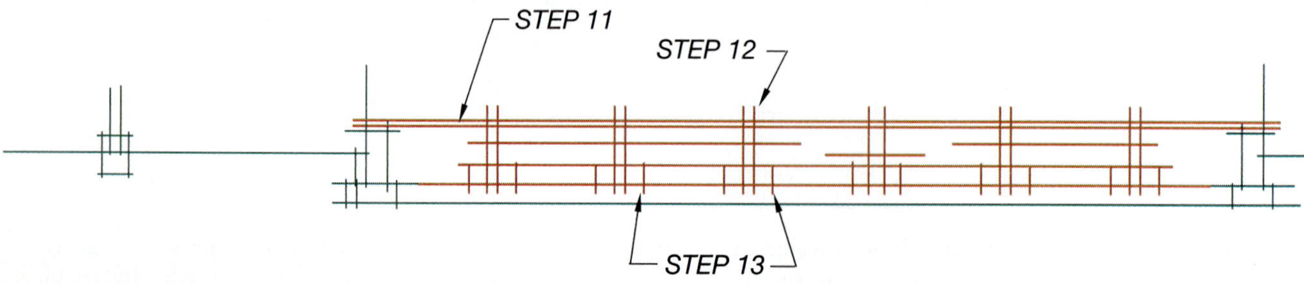

FIGURE 36.8 ■ Interior girders, post, and piers are located for a post-and-beam floor system.

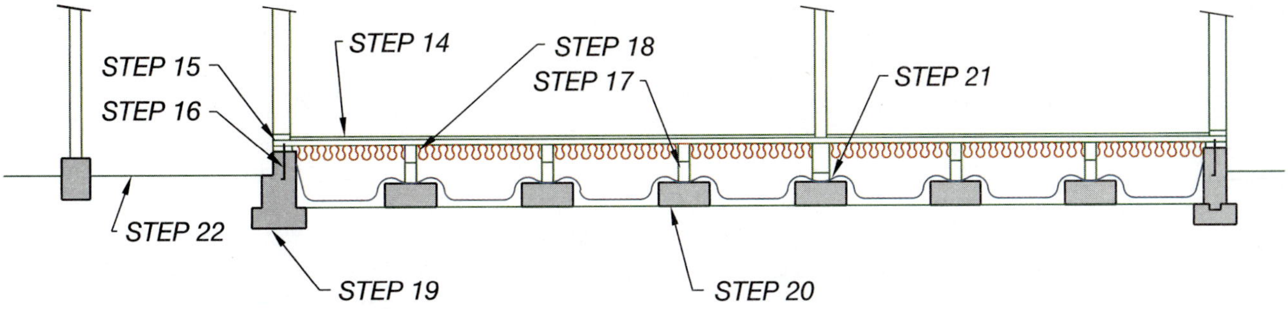

FIGURE 36.9 ■ Structural materials for post-and-beam floor system represented using finished-quality lines.

for each step, unless otherwise noted. See Figure 36.10 for steps 23 through 25.

STEP 23 Draw the hardboard subfloor from wall to wall.

STEP 24 Draw the insulation.

STEP 25 Draw the vapor barrier freehand.

Dimensioning

See Figure 36.11 for steps 26 through 32. Place the needed leader and dimension lines to locate the following dimensions:

STEP 26 Finished grade to the top of the stem wall.

STEP 27 Depth of the footing into grade.

STEP 28 Height of the footing.

STEP 29 Width of the footing.

STEP 30 Width of the stem wall.

STEP 31 Depth of the crawl space, 18″ (457 mm) minimum.

STEP 32 Bottom of girders to grade, 12″ (305 mm) minimum.

Notes

STEP 33 Letter the floor and foundation notes as shown in Figure 36.12. Typical notes should include the following.

a. 3/8″ HARDBOARD OVERLAY
b. 2 × 6 T&G DECKING or 1 1/8″ STURD·I· PLY
c. 6″ BATTS. R-19 MIN.
d. __ × __ GIRDERS
e. 4 × 4 POST (4 × 6 @ SPLICES) ON 55# FELT W/ GUSSET.

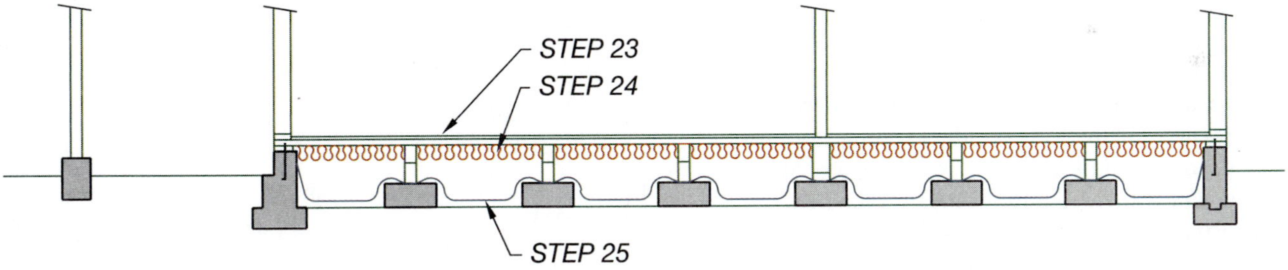

FIGURE 36.10 ■ Materials such as the floor overlay, insulation, metal connectors, and vapor barrier are added using finished-quality lines.

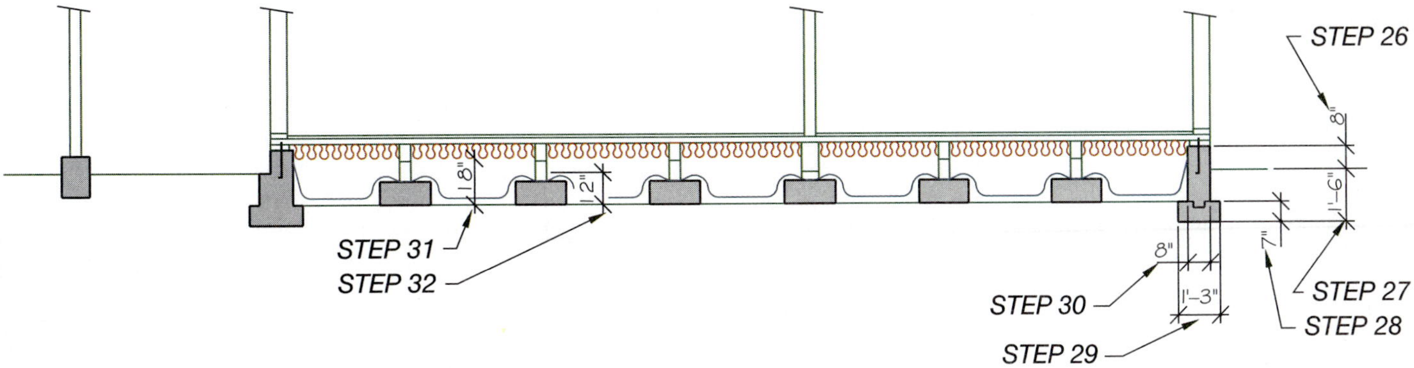

FIGURE 36.11 ■ Required dimensions for a post-and-beam floor and foundation system.

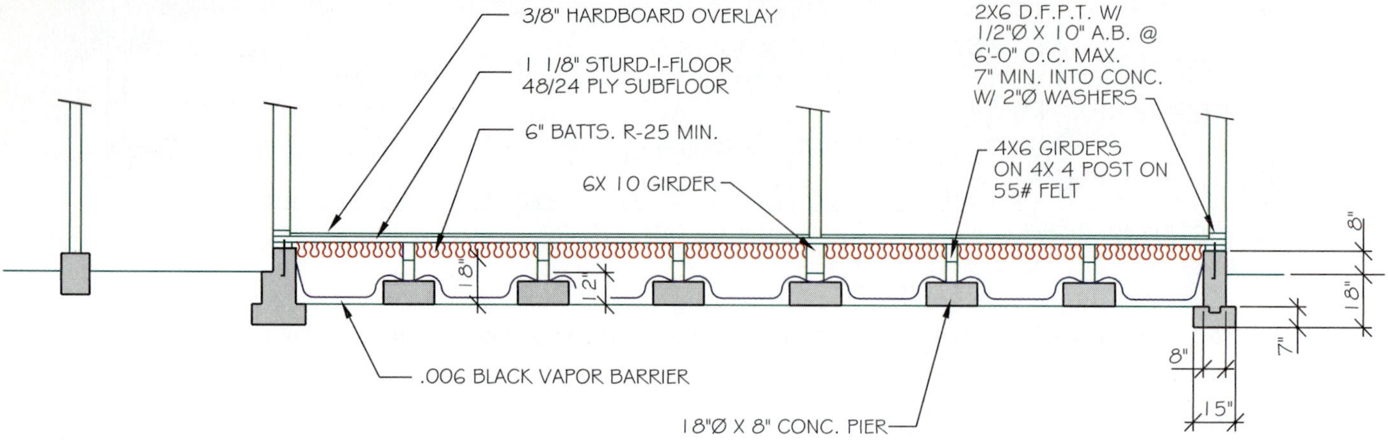

FIGURE 36.12 ■ Typical notes for a post-and-beam floor and foundation system.

f. __″ DIA × __″ DEEP CONC. PIERS

g. .006 BLACK VAPOR BARRIER OR 55# ROLLED ROOFING

h. 2 × 6 P.T. SILL W/ 1/2″ DIA × 10″ A.B. @ 6′–0″ O.C. MAX.—7″ MIN. INTO CONC. THRU 2″ DIA WASHERS.

STEP 34 Complete the section as in Chapter 35.

BASEMENT WITH CONCRETE SLAB

For the basement layout, use light construction lines. Depending on the seismic zone for which you are drawing, an engineer's drawing may be required for this kind of foundation. If so, you will need to follow engineers' design standards similar to those shown in Figure 36.13. Understanding engineers' design standards, or calculations (calcs), can be very frustrating. Calcs are generally divided into the areas of the item to be designed, the formulas to determine the size, and the solution.

The drafter does not need to understand the formulas that are used but must be able to convert the solution into a drawing. The solution in Figure 36.13 is the series of notes that are listed under the heading USE. For an entry-level drafter, the engineer will usually provide a sketch similar to the drawing in Figure 36.14 to explain the calculations. By using the written calculations and the sketch, a drafter can make a drawing similar to Figure 36.15.

The basement can be drawn by following steps 1 through 13. See Figure 36.16.

STEP 1 Lay out the width of the building.

STEP 2 Lay out a baseline for the bottom of the footings.

STEP 3 Measure up from the baseline for the footings. Footings are typically 8″ (200 mm) deep for a one-story building and 12″ (300 mm) deep for two or more stories.

STEP 4 Lay out the 4″ (100 mm) concrete slab.

STEP 5 Lay out the 2″ (50 mm) mudsill. The height of the basement will vary.

STEP 6 Lay out the wooden floor.

STEP 7 Lay out the finished grade 8″ (200 mm) below the mudsill.

STEP 8 Block out the width of the retaining wall. See Figure 31.31b for required wall thickness.

8' HIGH BEAM TYPE BLOCK RETAINING WALL

KENNETH D. SMITH - ARCHITECT
El Cajon, Ca. 92020

DESIGN TYPE- 8' BEAM TYPE BLOCK RETAINING WALL

M=(0.1283) (960) (8) = 985'#

TRY #5 - 16" O.C. VERT. PLACED 2" FROM INSIDE FACE OF 8" BLOCK

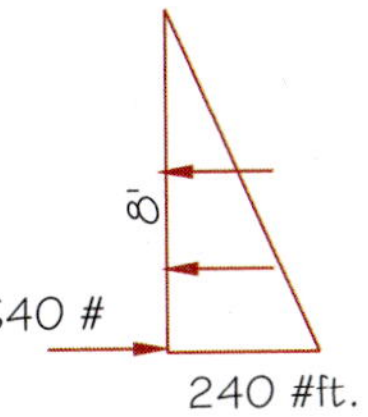

$$np = \frac{(43)(0.31)}{(16)(5.62)} = J=.861 \quad 2/KJ=.558$$

$$f_m = \frac{(1.33)(12)(985)(558)}{(16)(562)^2} = 174.5 \text{ p.s.i.}$$

$$f_s = \frac{(12)(1.33)(985)}{(0.31)(.861)(5.62)} = 10{,}500 \text{ p.s.i.} \qquad U = (11.0)\frac{16}{1.963} = 88.5 \text{ p.s.i.}$$

$$V = \frac{(1.33)(640)}{(16)(.861)(5.62)} = 11.0 \text{p.s.i.} \qquad ht = \frac{96}{8} = 12$$

$$f_a = (135)(.94) = 127 \text{ p.s.i.}$$

ALLOW FOR ROOF, 2nd FLOOR, & 1 st. FLOOR LOADING OF WALL FOR f_a

$$f_a = \frac{(40)(40)+(8)(32)+(2)(8)(16)+(4)(137)+(8)(63)+(8)(63)=(92)(5)}{(7.62)(12)}$$

f_a = 29.4 p.s.i.

MIN. HORIZ. STEEL= (0.007)(8)(96)=0.54in.^{2}

$$\frac{174.5}{225} + \frac{29.4}{127} = \boxed{1.00}$$

o.k.

USE:

- # 5 - VERT. @ 16" O.C. PLACED 2" FROM FACE OF BLOCK AWAY FROM SOIL.
- USE 8" GRADE 'A' CONC. BLOCK- SOLID GROUT ALL STEEL CELLS.
- INTERMEDIATE GRADE DEFORMED BARS LAP 40 DIA. @ SPLICES.
- DBL. ALL VERT. STEEL BESIDE OPENINGS & WALL ENDS & MATCH ALL DOWEL STEEL OUT OF FTGS. FOR SIZE & POSITION.
- USE 2-#4 - UNDER & OVER ALL OPENINGS (UNLESS NOTED OTHERWISE) & EXTEND 24" BEYOND OPENINGS. PLACE 1 1/2" - 2" UP FROM BTM. OF LINTEL IN INVERTED BOND BEAM BLOCK OR EQUAL & GROUT LINTEL CELLS SOLID.
- USE 4-#4 - CONT @ TOP OF WALL IN 8" X 16" BLOCK BOND BM. & 2-#4 - CONT. IN BOND BLOCK @ MIDHEIGHT (4'MAX.)
- GROUT AS PER SECTION 2103 & MORTAR AS PER TABLE 21-A TYPE 'S' 1997 U.B.C. W/ CENTERING BRACKET @ TOP & BTM.
- PROVIDE CLEANOUT HOLES @ BTM. OF STEEL CELLS WHERE WALLS ARE GROUTED IN MORE THAN 4' LIFTS.
- AT TOP OF WALLS, USE 4" X 3" X 1/4" X 3" LONG (OR EQUAL) ∠ W/ 3/4"Ø X 10" A.B. THRU 3" LEG & PLATE INTO WALL & 3/4" 3/4"Ø BOLT THRU 4" LEG @ JOIST (OR BLOCK WHERE ⊥ TO WALL W/ 2"Ø WASHERS AGAINST WOOD WHERE JOIST∥ WALL BLOCK OUT 4' @ 32" O.C. & USE 10d @ 4" O.C. FOR SUBFLOOR.
- WATERPROOF ENTIRE WALL & USE 4"Ø DRAIN TILE & GRAVEL @ BTM. OF WALL.
- DO NOT BACK FILL UNTIL FLOOR IS IN PLACE.

FIGURE 36.13 ■ A drafter is often required to work from engineers' calculations when drawing retaining walls. Calculations usually show the math formulas to solve a problem and the written solution to the problem. It is the written solution to the problem that the drafter must make into a drawing. Each of the materials in the "Use" portion of the calculations must be shown and specified in the section. *Courtesy Kenneth D. Smith and Associates, Inc.*

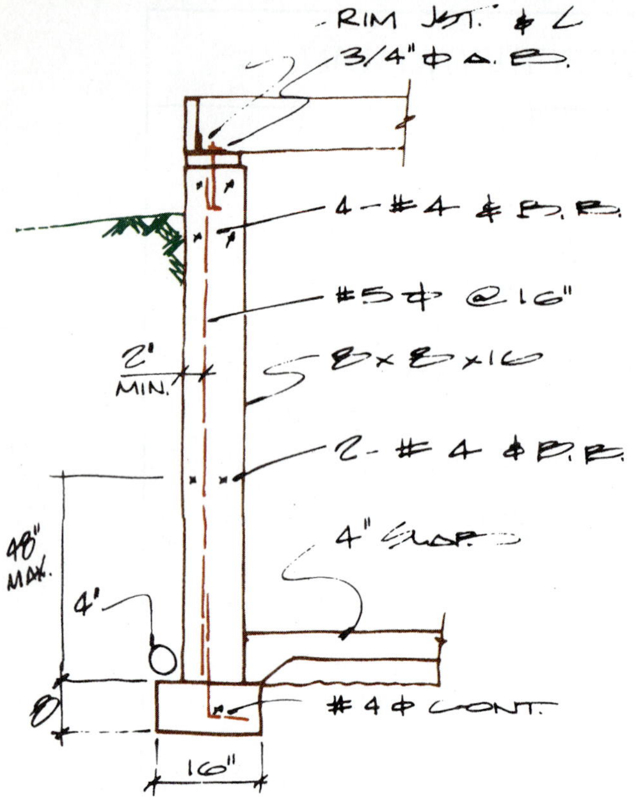

FIGURE 36.14 ■ A sketch is often given to the drafter to help explain the written calculations. Many of the written calculations appear on the sketch but are not written in proper form.

STEP 9 Block out the width of the footing. Footings for this type of wall are usually 16″ (400 mm) wide, centered under the retaining wall.

STEP 10 Lay out the 4″ (100 mm) ledge and add 4″ (100 mm) to the footing to support the brick veneer. An alternative to the ledge method is to form the wall 4″ (100 mm) out from the face of the exterior wall to support the brick. See Figure 36.17.

STEP 11 Draw a line to show the 4″ (100 mm) fill material under the slab.

STEP 12 Lay out the interior footings for bearing walls.

STEP 13 Lay out the steel reinforcing for the wall per Figure 36.13. This will vary greatly, depending on the seismic area you are in and how the wall is to be loaded. One typical placement pattern puts the vertical steel 2″ (51 mm) from the inside (tension) face, and the horizontal steel 18″ (457 mm) o.c. starting at the footing.

Finished-Quality Lines—Structural Material Only

See Figure 36.18 for steps 14 through 21.

STEP 14 Draw the plywood flooring with thin lines.

STEP 15 Draw the floor joists with thin lines. Draw blocking, and the floor joists that are perpendicular to the wall with bold lines.

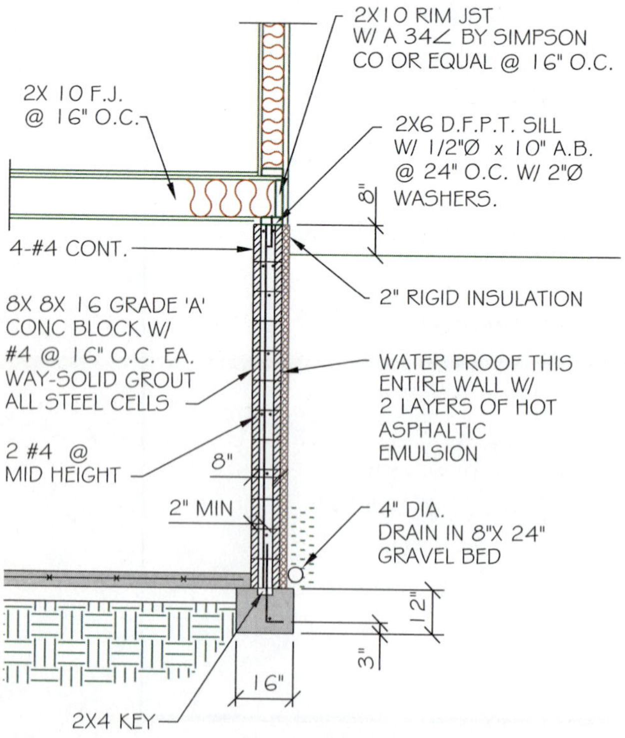

FIGURE 36.15(A) ■ A section of an 8′ (2400 mm) high concrete block retaining wall based on the specification shown in Figures 36.13 and 36.14. *Courtesy Dean Smith, Kenneth D. Smith Architects, Inc.*

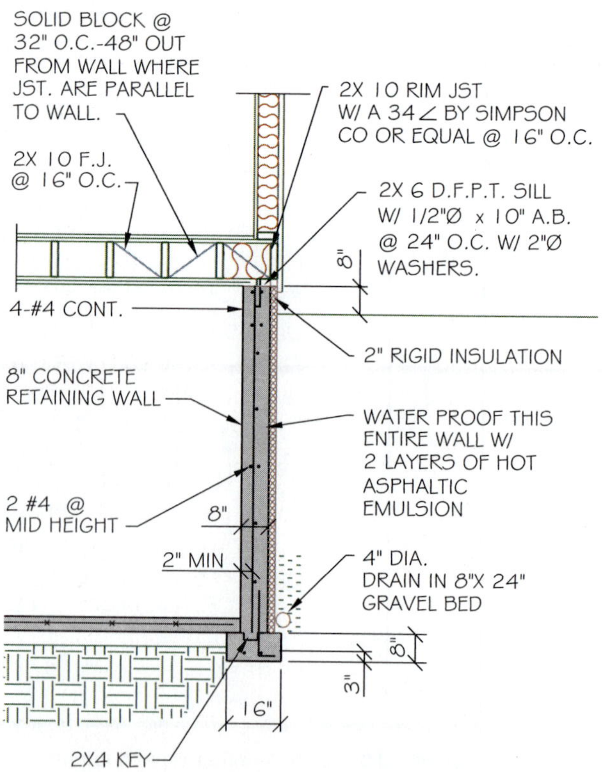

FIGURE 36.15(B) ■ A section of an 8′ (2400 mm) high poured concrete retaining wall with joist parallel to the retaining wall.

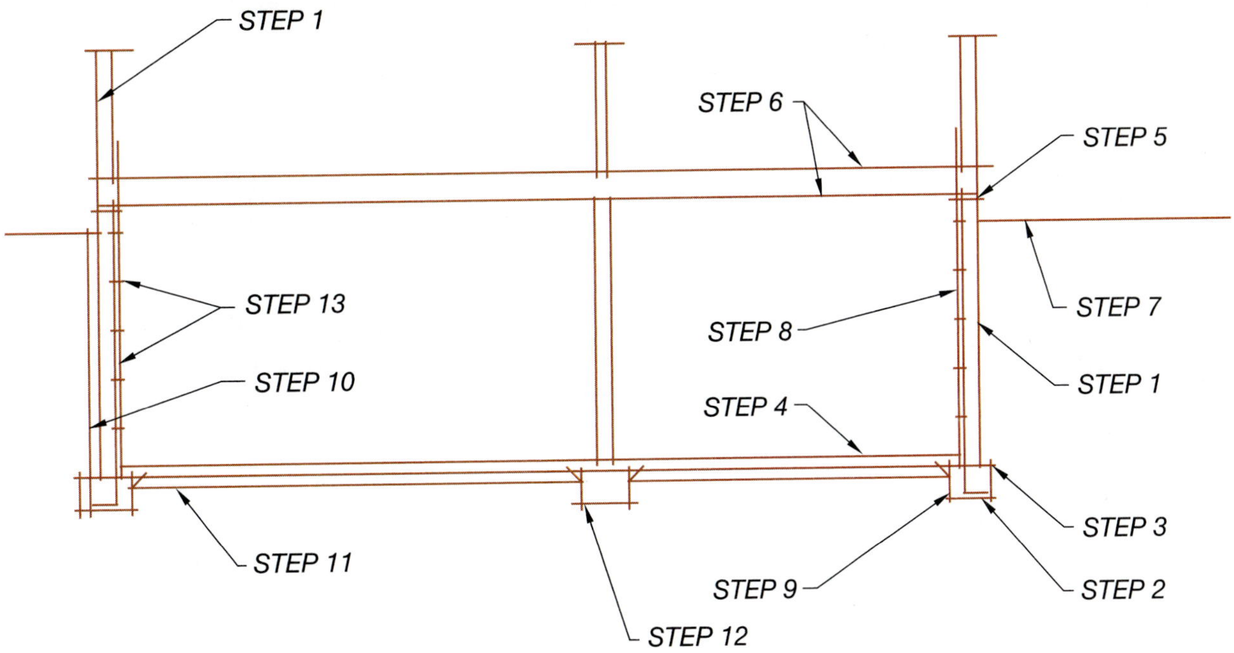

FIGURE 36.16 ■ The initial layout of a full basement with concrete block retaining walls.

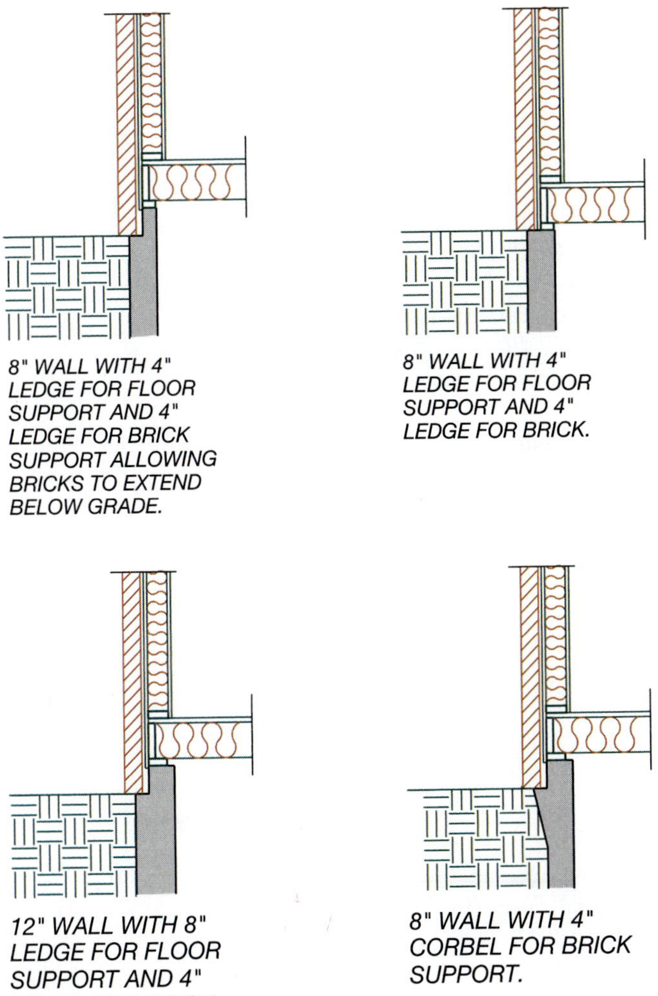

8" WALL WITH 4" LEDGE FOR FLOOR SUPPORT AND 4" LEDGE FOR BRICK SUPPORT ALLOWING BRICKS TO EXTEND BELOW GRADE.
8" WALL WITH 4" LEDGE FOR FLOOR SUPPORT AND 4" LEDGE FOR BRICK.
12" WALL WITH 8" LEDGE FOR FLOOR SUPPORT AND 4" LEDGE FOR BRICK.
8" WALL WITH 4" CORBEL FOR BRICK SUPPORT.

FIGURE 36.17 ■ Common methods used for supporting brick veneer.

STEP 16 Draw the mudsills and anchor bolts with bold lines.

STEP 17 Draw the walls with bold lines. If concrete blocks are used, draw the division lines of the blocks and crosshatch the blocks with thin lines.

STEP 18 Draw the footings with bold lines.

STEP 19 Draw the concrete slab with bold lines.

STEP 20 Draw the wall steel with bold dashed lines.

STEP 21 Complete the section as in Chapter 35.

Finishing Materials

See Figure 36.19 for steps 22 through 28.

STEP 22 Draw the floor insulation with thin lines.

STEP 23 Draw the finished grade with very bold lines.

STEP 24 Draw the 4″ (100 mm) diameter drain with thin lines.

STEP 25 Draw the 8″ × 24″ (200 × 600 mm) gravel bed at the base of the wall.

STEP 26 Draw the 4″ (100 mm) fill materials with thin lines.

STEP 27 Draw the brick veneer and the crosshatching with thin lines.

STEP 28 Draw the welded wire mesh with thin dashed lines.

Dimensions

See Figure 36.20 for steps 29 through 35. Place the needed leader and dimensions lines to locate the dimensions listed on page 729:

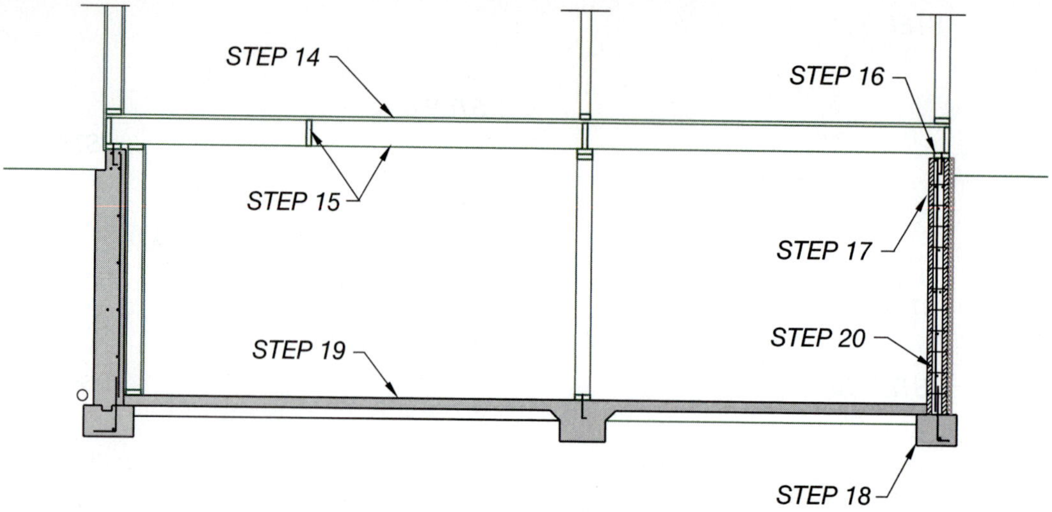

FIGURE 36.18 ■ Finished-quality lines of the structural material for a basement foundation system. The left wall is drawn showing a poured concrete wall. The right wall is drawn showing 8 × 8 × 16 (200 × 200 × 400) concrete blocks. The two wall systems are *not* used together but are shown here to illustrate construction methods.

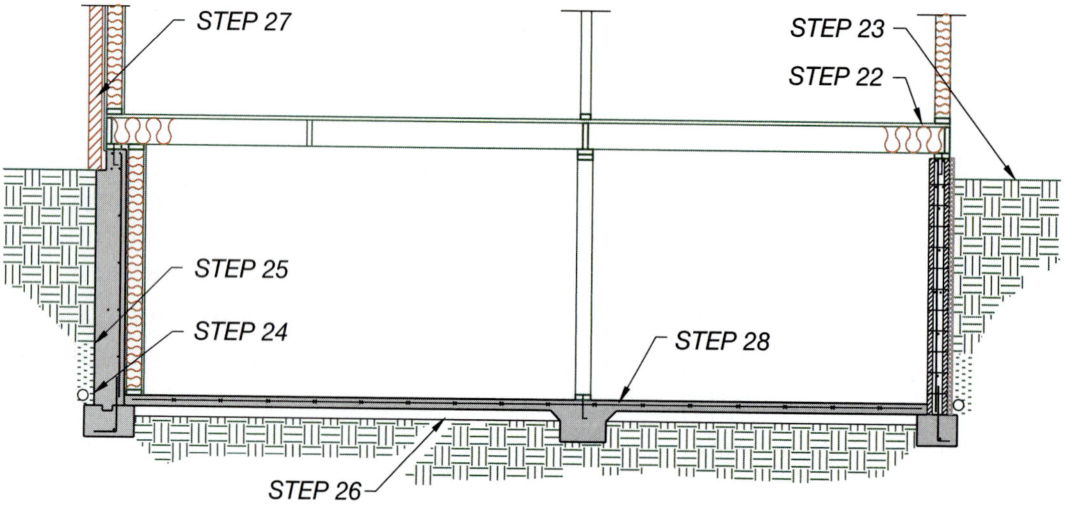

FIGURE 36.19 ■ Finishing materials for a basement foundation system.

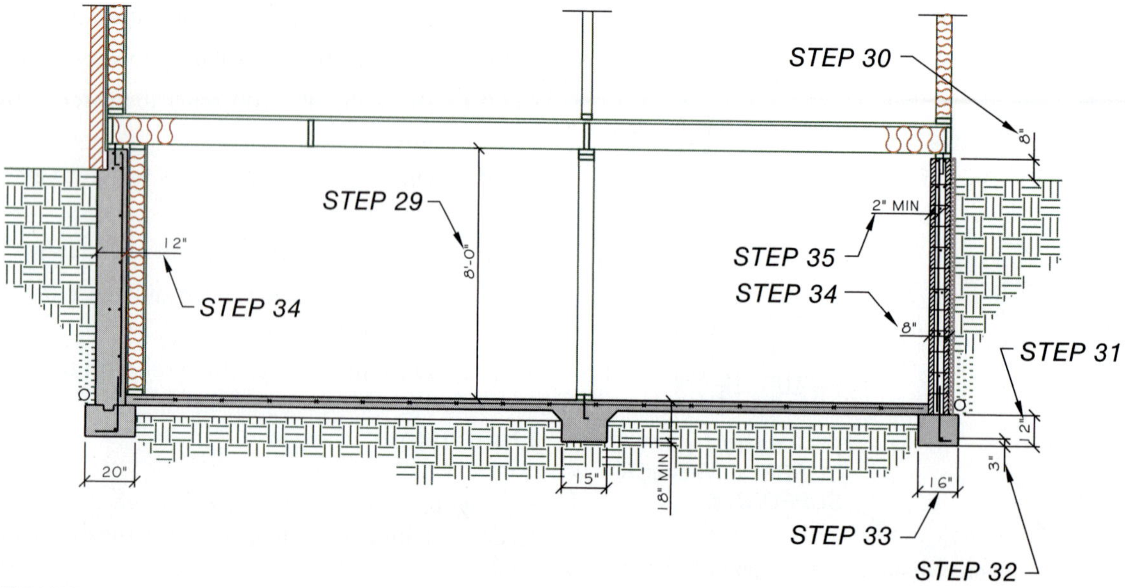

FIGURE 36.20 ■ Required dimensions for a basement foundation system.

STEP 29 Floor to ceiling.

STEP 30 Top of wall to finished grade.

STEP 31 Height of footing.

STEP 32 Footing steel to bottom of footing.

STEP 33 Width of footing.

STEP 34 Width of wall.

STEP 35 Edge of wall to vertical steel.

Notes

STEP 36 Letter the section notes as shown in Figure 36.21. Remember that these notes are only guidelines and may vary slightly because of local standards. Typical notes should include the following:

a. 2 × __ FLOOR JOIST @ __″ O.C.

b. SOLID BLOCK @ 48″ O.C.—48″ OUT FROM WALL WHERE JOISTS ARE PARALLEL TO WALL.

c. 2 × __ RIM JOIST W/A-34 ANCHORS BY SIMPSON CO. OR EQUAL.

d. 2 × 6 P.T. SILL W/ 1/2″ Ø × 10″ A.B. @ 24″ O.C. W/2″ Ø WASHERS.

e. 8″ CONC. WALL W/ #4 @ 18″ O.C. EA. WAY. 8 × 8 × 16 GRADE "A" CONC. BLOCKS W/ #4 @ 16″ O.C. EA. WAY—SOLID GROUT ALL STEEL CELLS.

f. WATERPROOF THIS WALL W/ HOT ASPHALTIC EMULSION.

g. 4″ Ø DRAIN IN 8″ × 24″ MIN. GRAVEL BED.

h. 2 × 4 KEY.

i. 4″ CONC. SLAB OVER 4″ GRAVEL FILL AND 55# ROLLED ROOFING.

or

4″ CONC. SLAB OVER 4″ SAND FILL AND 6 MIL VAPOR BARRIER, AND 2″ SAND FILL.

Note: Wire mesh is often placed in the concrete slab to help control cracking. This mesh is typically called out on the sections and foundation plan as 6″ × 6″—#10 × #10 W.W.M.

CONVENTIONAL (STICK) ROOF FRAMING

The two types of conventional roof framing presented are flat and vaulted ceilings. You will notice many similar features between the two styles.

Flat Ceilings

Use light construction lines to lay out the drawing. See Figure 36.22 for steps 1 through 7.

STEP 1 Place a vertical line to represent the ridge.

STEP 2 Locate each overhang.

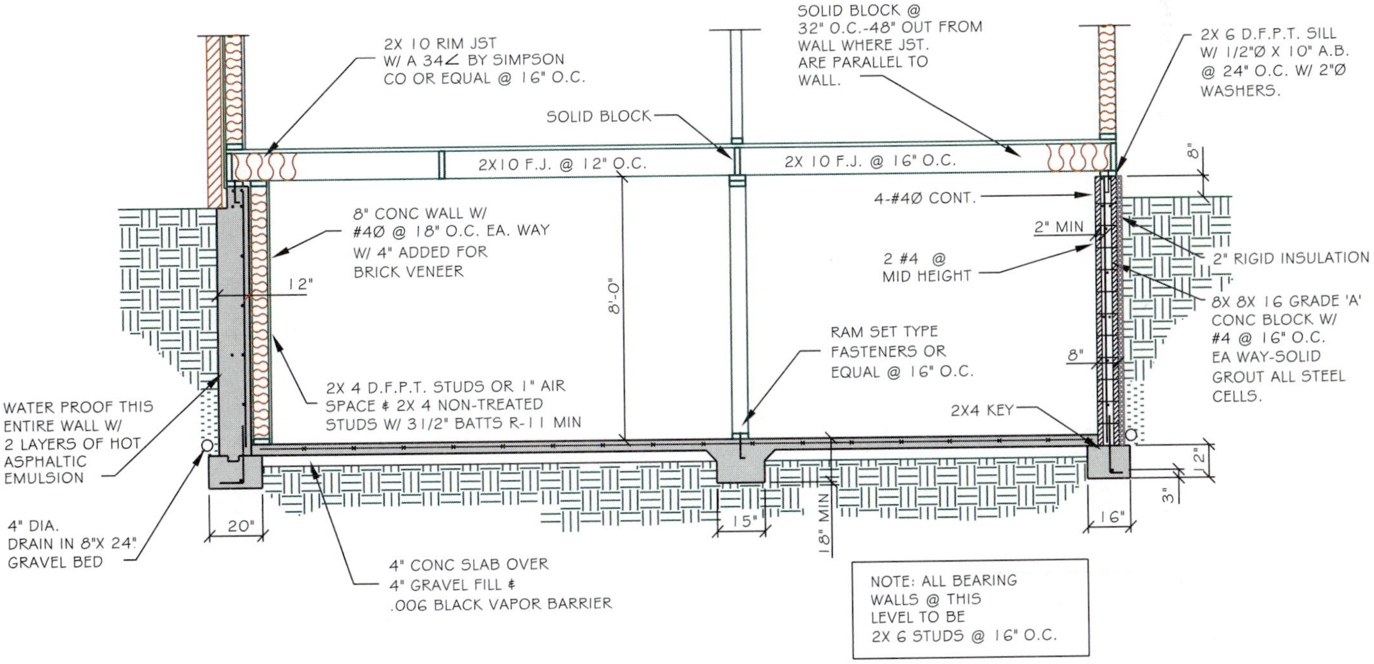

FIGURE 36.21 ■ Typical notes required to describe the materials for a full basement.

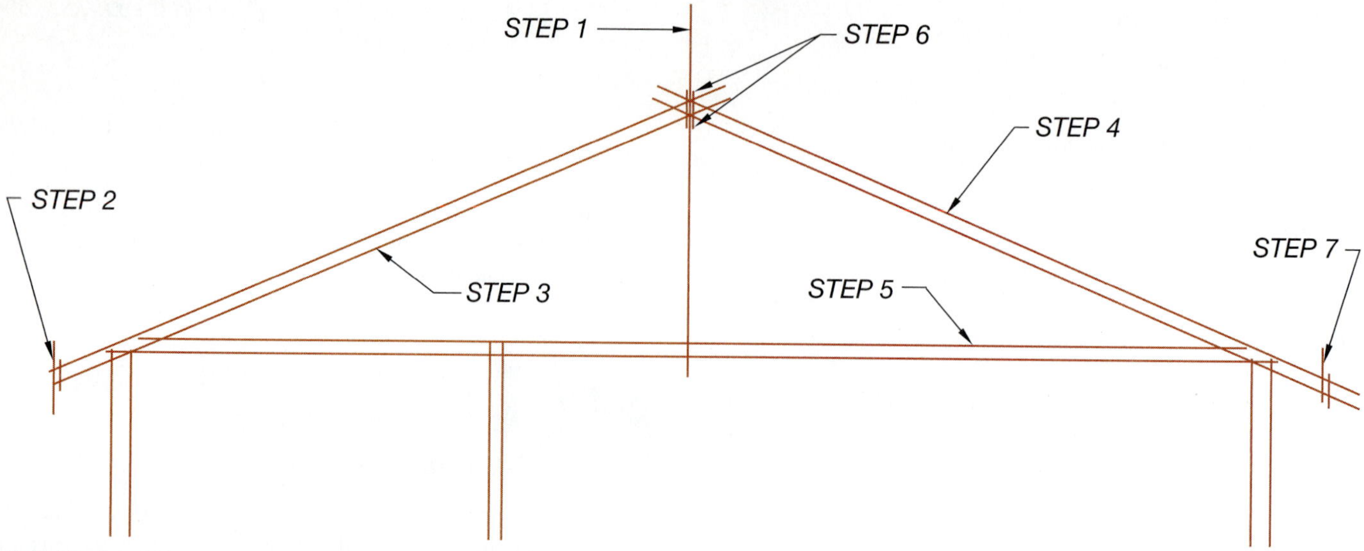

FIGURE 36.22 ■ Layout of a conventionally framed roof.

STEP 3 Lay out the bottom side of the rafter. Starting at the intersection of the ceiling and the interior side of the exterior wall, draw a line at a 5/12 pitch. Do this for both sides of the roof. Both lines should meet at the ridge line.

STEP 4 Lay out the top side of the rafters.

STEP 5 Lay out the top side of the ceiling joist.

STEP 6 Lay out the ridge board.

STEP 7 Lay out the fascias.

Lay Out the Interior Supports

See Figure 36.23 for steps 8 through 12.

STEP 8 Lay out the ridge brace and support.

STEP 9 Lay out the purlin braces and supports.

STEP 10 Lay out the strongbacks.

STEP 11 Lay out the purlins.

STEP 12 Lay out the collar ties.

Finished-Quality Lines—Structural Members Only

See Figure 36.24 for steps 13 through 20.

STEP 13 Draw the plywood sheathing with thin lines.

STEP 14 Draw the rafters with thin lines.

STEP 15 Draw the eave blocking with bold lines.

STEP 16 Repeat steps 13 through 15 for the opposite side of the roof.

STEP 17 Draw the fascias with bold lines.

STEP 18 Draw the strongbacks with bold lines.

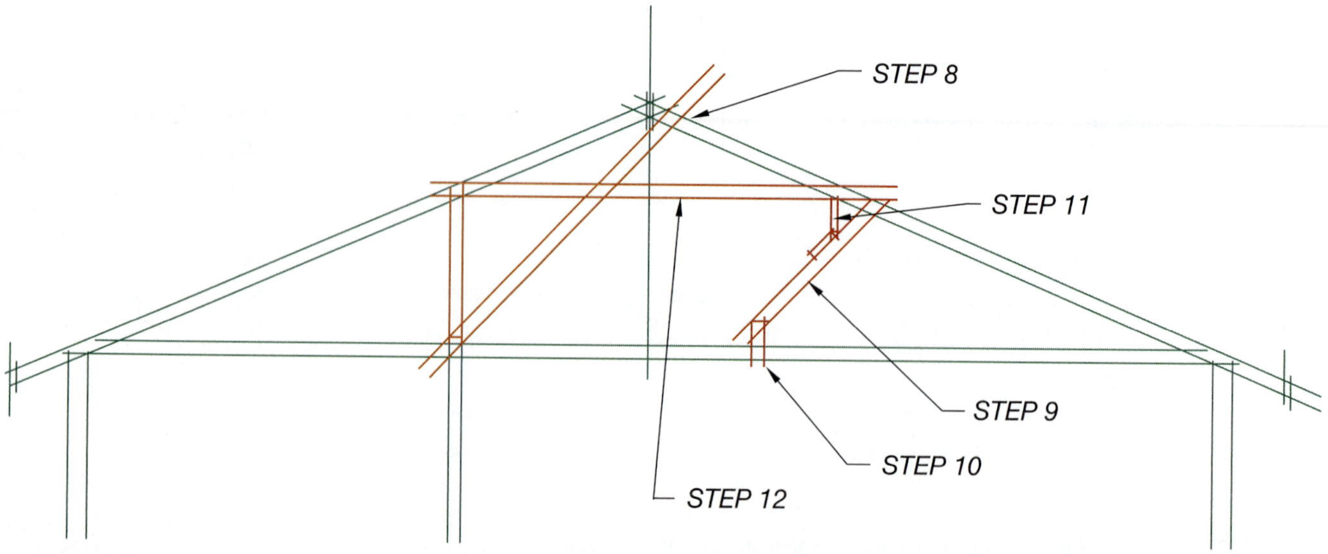

FIGURE 36.23 ■ Layout of interior supports for a conventionally framed roof. The strongback (step 10) can be used in place of a bearing wall or a flush header.

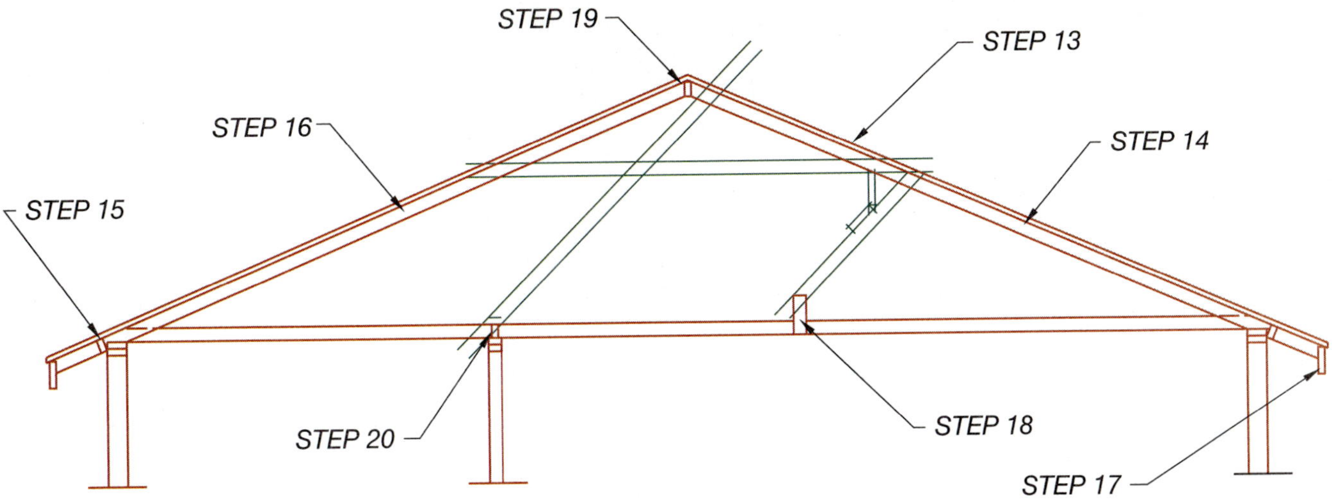

FIGURE 36.24 ■ Finished-quality lines of structural materials for a conventionally framed roof.

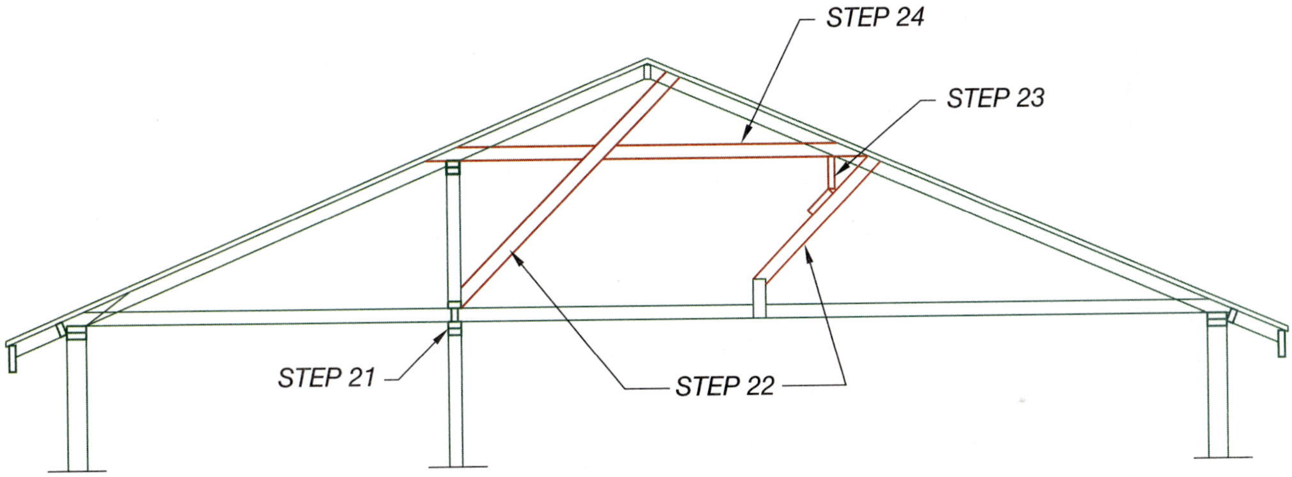

FIGURE 36.25 ■ Finished-quality lines represent the interior supports. The purlin, purlin brace, and collar ties may or may not be required, depending on the width of the structure and the rafter material and size.

STEP 19 Draw the ridge board with bold lines.

STEP 20 Draw blocking with bold lines.

Interior Support—Finished-Quality Lines

See Figure 36.25 for steps 21 through 24.

STEP 21 Draw all plates with bold lines.

STEP 22 Draw all braces with thin lines.

STEP 23 Draw all purlins with bold lines.

STEP 24 Draw the collar tie with thin lines.

Dimensions

Usually, no dimensions are required for the roof framing.

Finishing Materials

See Figure 36.26 for steps 25 through 29.

STEP 25 Draw baffles with bold lines.

STEP 26 Draw the finished ceiling with thin lines.

STEP 27 Draw joist hangers with thin lines.

STEP 28 Draw the 10″ thick insulation.

STEP 29 Draw the ridge vents.

Notes

See Chapter 35 for a complete listing of notes that apply to the roof. In addition to those notes, see Figure 36.27 for the notes needed to describe the interior supports. Roof notes should include the following:

a. 2 × __ RIDGE BOARD.
b. 2 × __ RAFTERS @ __″ O.C.
c. 2 × __ COLLAR TIE @ 48″ O.C. MAX.
d. 2 × __ PURLIN.
e. 2 × 4 BRACE @ 48″ O.C. MAX.

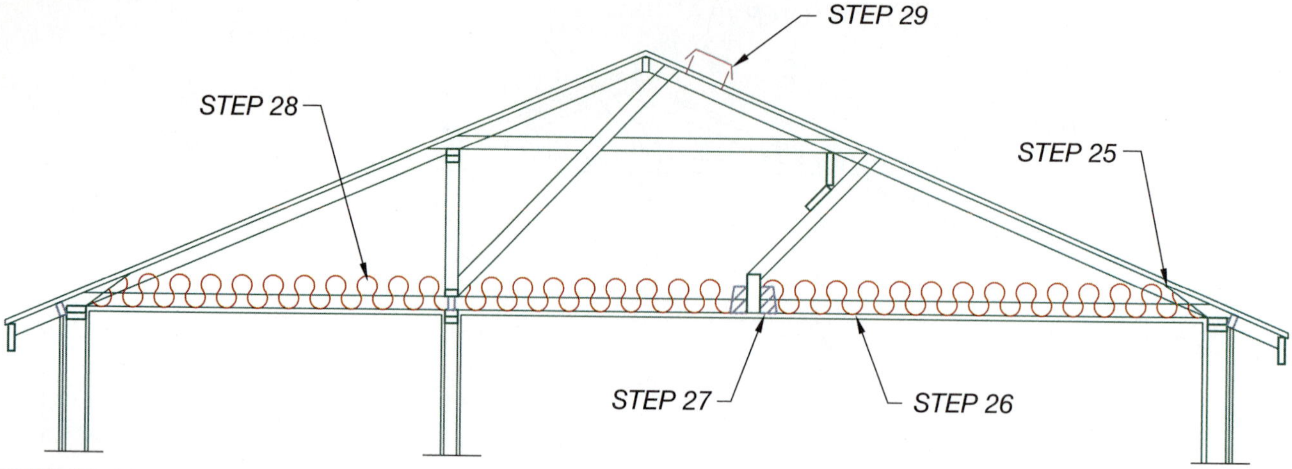

FIGURE 36.26 ■ Finishing materials for a conventionally framed roof include materials such as ridge vents, insulation, and metal hangers.

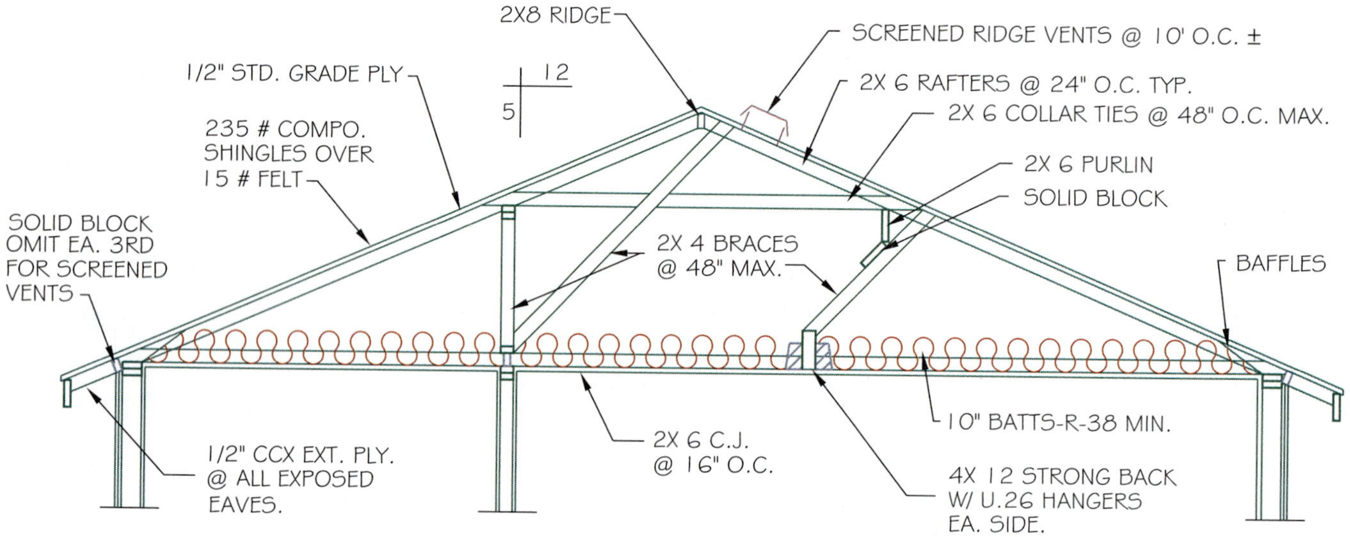

FIGURE 36.27 ■ Notes are required to describe all materials for a conventionally framed roof.

Vaulted Ceilings

If the entire roof level is to have vaulted ceilings, the process of laying it out is very similar to the procedure for laying out a stick roof with flat ceilings. Usually, though, part of the roof has 2 × 6 rafters with flat ceilings and a vaulted ceiling using 2 × 12 rafter/ceiling joists. In order to make the two roofs align on the outside, the plate must be lowered in the area having the vaulted ceilings. See Figure 36.28 for steps 1 through 7 of laying out the vaulted ceiling. Use construction lines for all steps.

STEP 1 Lay out the overhangs.

STEP 2 Lay out the bottom side of the 2 × 6 rafter.

STEP 3 Lay out the top side of the 2 × 6 rafter.

STEP 4 Measure down from the top side of the rafter drawn in step 3 to establish the bottom of the 2 × 12 rafter/ceiling joist. Notice that the 2 × 12 extends below the top plate of the wall.

STEP 5 Repeat steps 2 through 4 for the opposite side of the roof.

STEP 6 Lay out the ridge beam.

STEP 7 Lay out the fascias and the rafter tail cut for the 2 × 12 rafter/ceiling joists.

Finished-Quality Lines—Structural Members Only

See Figure 36.29 for steps 8 through 12.

STEP 8 Draw the plywood sheathing with thin lines.

STEP 9 Draw the rafter/ceiling joist with thin lines.

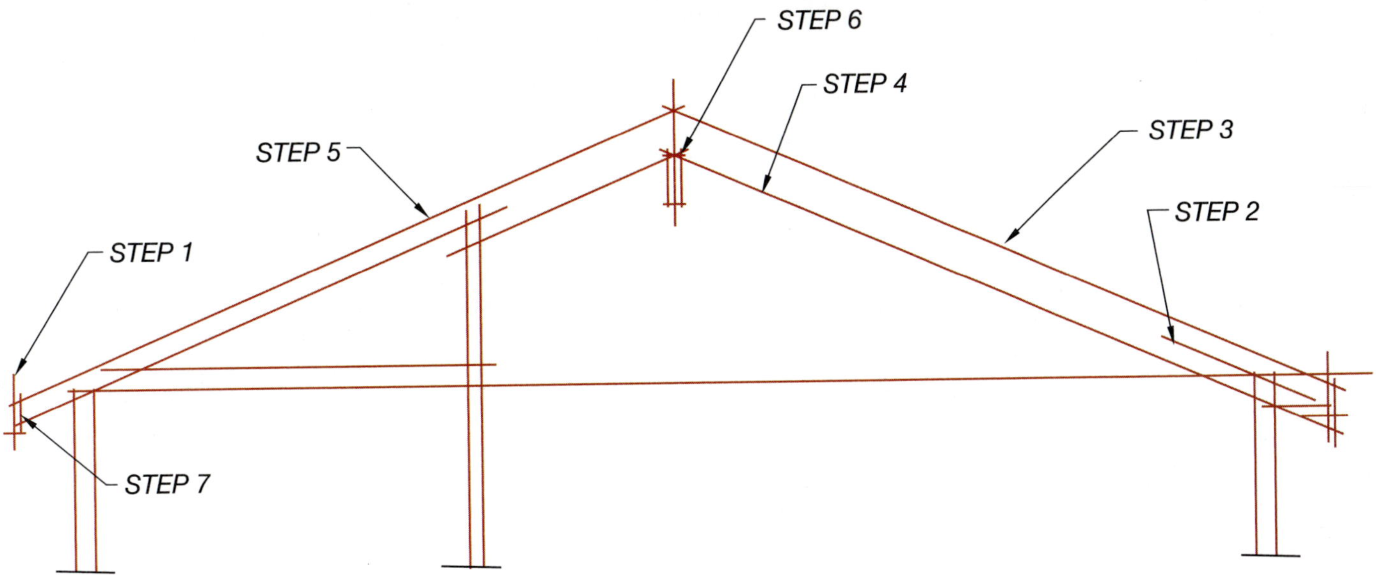

FIGURE 36.28 ■ Layout of a stick framed vaulted ceiling.

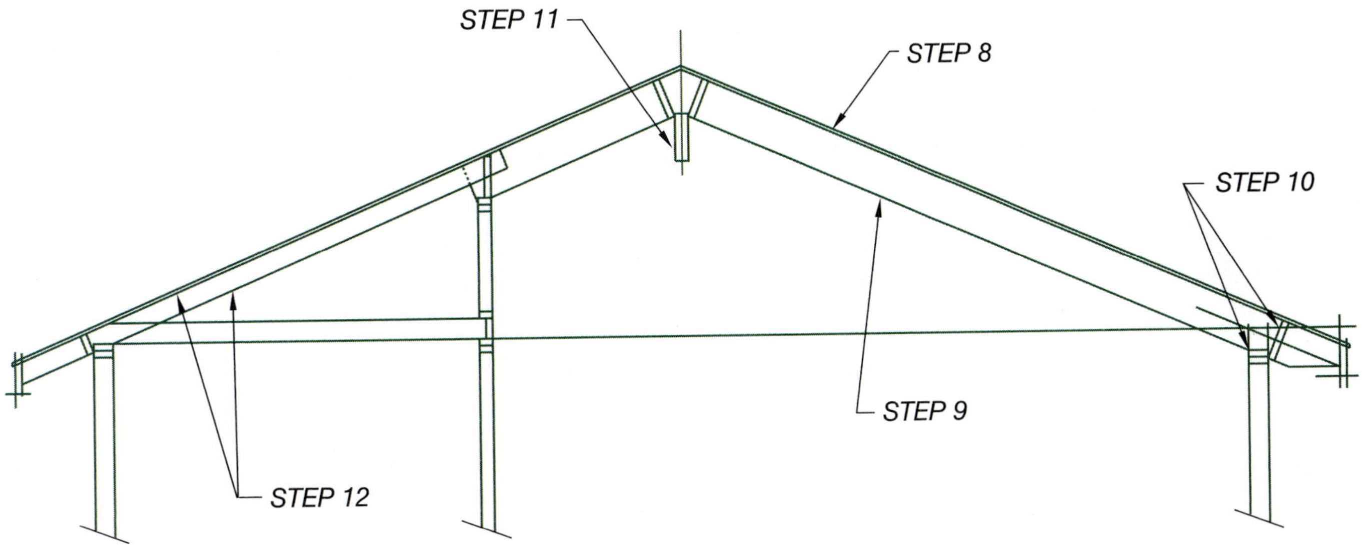

FIGURE 36.29 ■ Structural materials for a vaulted roof represented with finished-quality lines.

STEP 10 Draw all blocks, plates, and fascias with bold lines.

STEP 11 Draw the ridge beam with bold lines.

STEP 12 Repeat steps 8 through 10 for the opposite side of the roof.

Finishing Materials

See Figure 36.30 for steps 13 through 16.

STEP 13 Draw the baffles with bold lines.

STEP 14 Draw the insulation with thin lines.

STEP 15 Draw the ceiling interior finish with thin lines.

STEP 16 Draw the ridge vents and notches with thin lines.

STEP 17 Draw metal hangers and crosshatch with thin lines. Depending on the ridge beam option that you drew, joist hangers may not be required.

Dimensions

No dimensions are needed to describe the roof. You will need to give the floor-to-plate dimension. List this dimension as 7′–6″ ± VERIFY AT JOB SITE. By telling the framer to verify the height at the job site, you are alerting the framer to a problem. This is a problem that can be solved much more easily at the job site than at a CADD station.

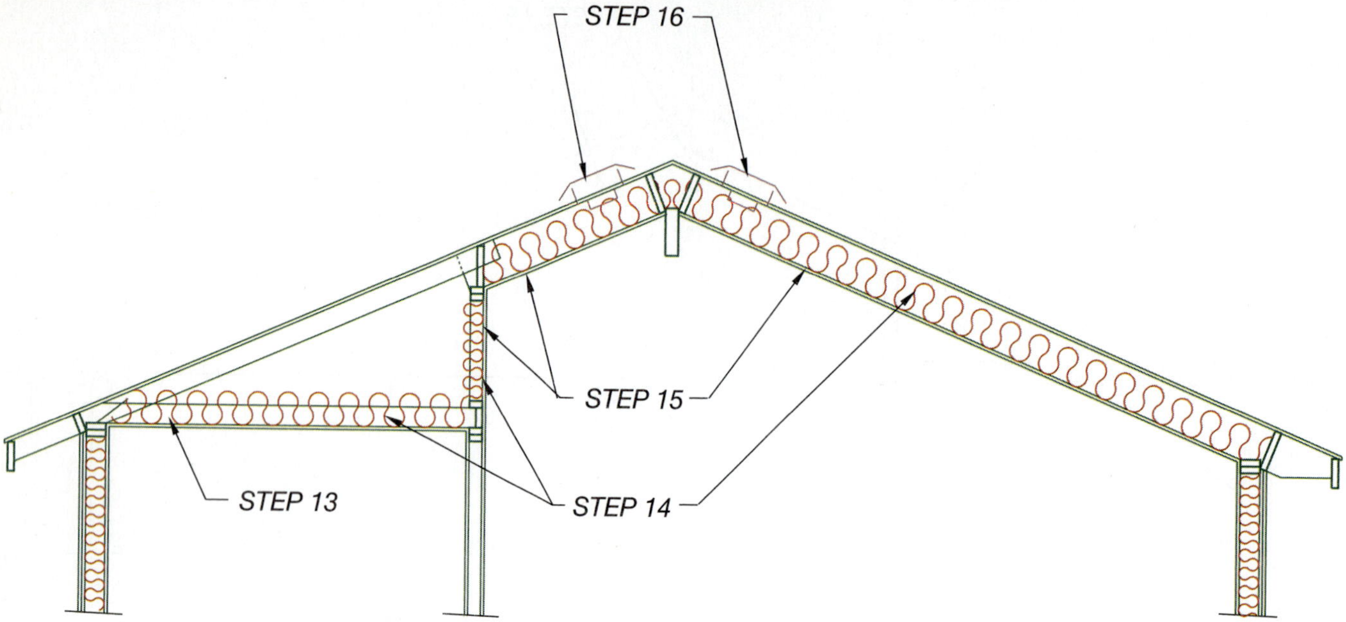

FIGURE 36.30 ■ Finishing materials for a vaulted roof include ridge vents, insulation and metal hangers.

Notes

See Chapter 35 for a complete listing of notes that apply to the roof. In addition to those notes, see Figure 36.31 for the notes needed to describe the roof completely. The notes should include the following:

a. __ × __ RIDGE BEAM.
b. 2 × __ RAFT./C.J. @ __″ O.C.
c. U-210 JST. HGR. BY SIMPSON CO. OR EQUAL.
d. SCREENED RIDGE VENTS @ EA. 3RD. SPACE. NOTCH RAFT. FOR AIRFLOW.
e. SOLID BLOCK W/ 3–1″ DIA SCREENED VENTS @ EA. SPACE.
f. NOTCH RAFTER TAILS AS REQD.
g. LINE OF INTERIOR FINISH.
h. __ BATTS—R-__MIN.

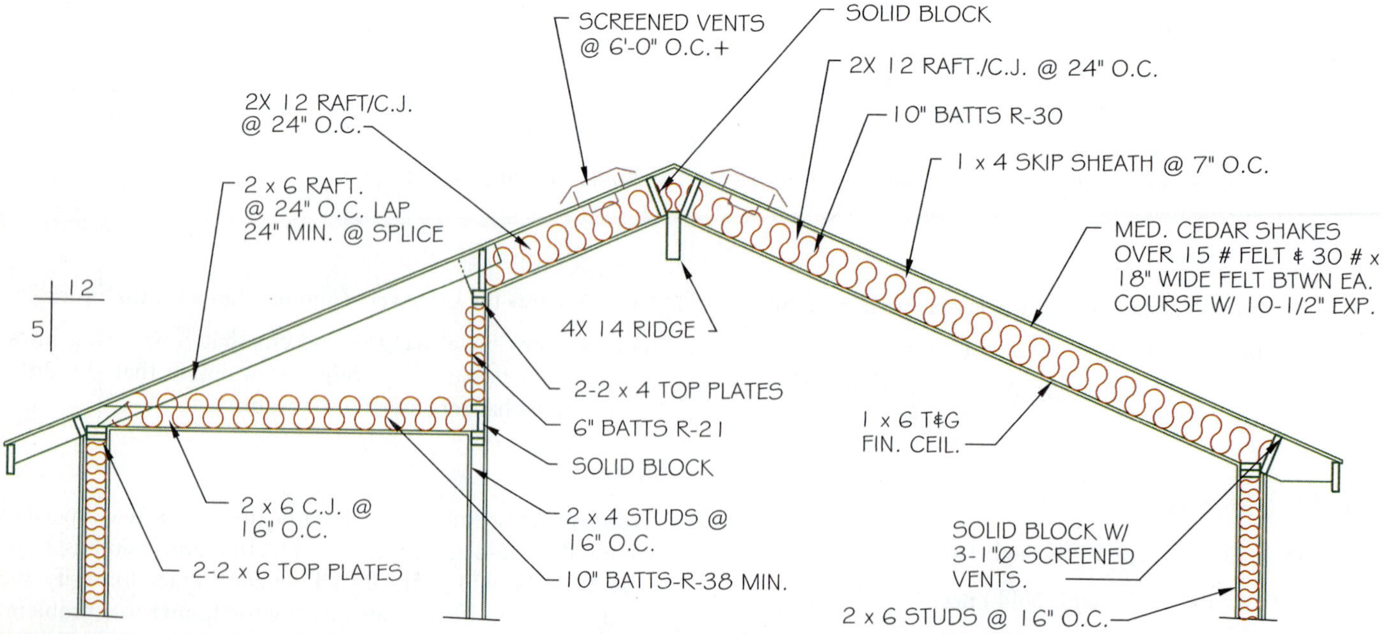

FIGURE 36.31 ■ Notes for a vaulted ceiling.

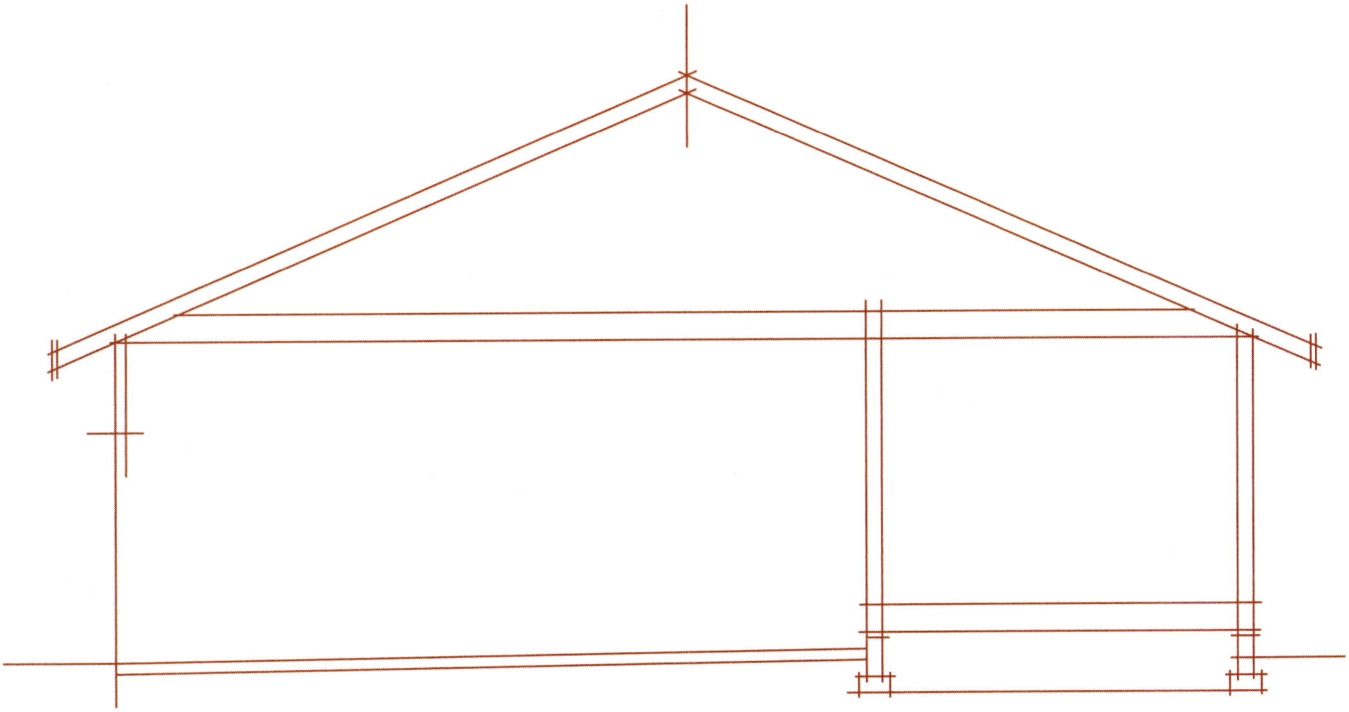

FIGURE 36.32 ■ Layout of a section drawn through the garage and the living area framed with a joist floor system.

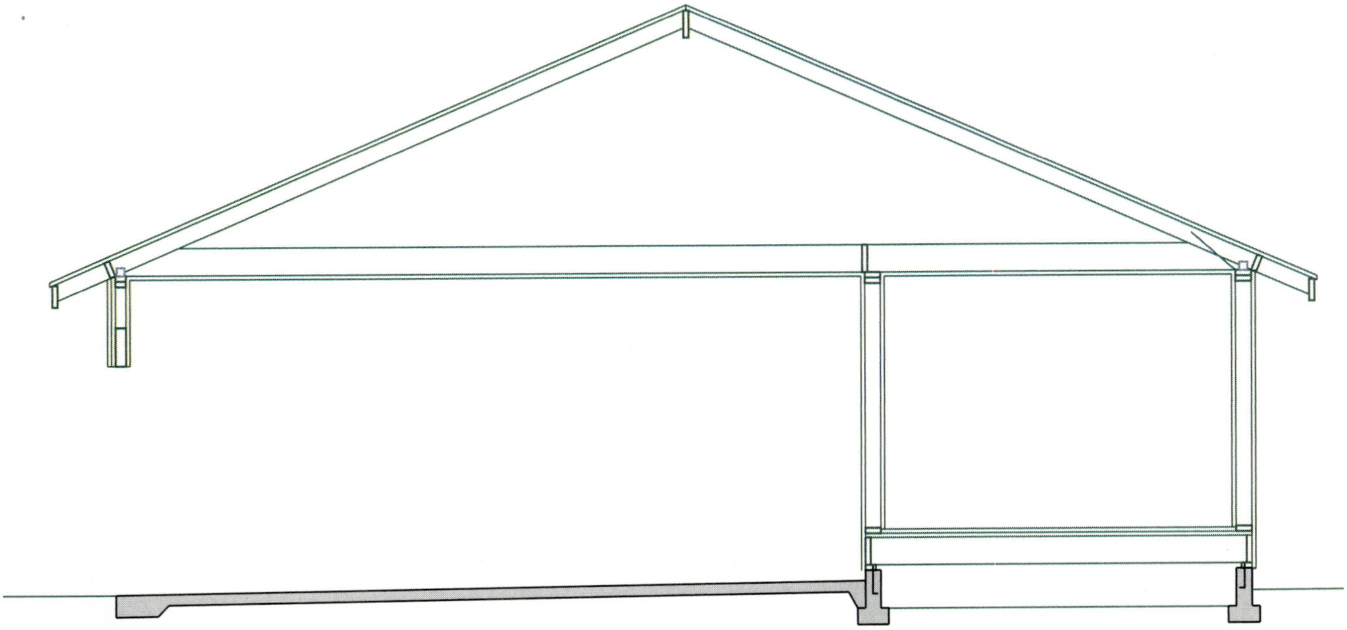

FIGURE 36.33 ■ Finished-quality lines for structural materials.

GARAGE/ RESIDENCE SECTION

Drawing this section is similar to drawing the sections just described, since it is a combination of several types of sections. The steps needed to draw this section can be seen in Figures 36.32 through 36.35.

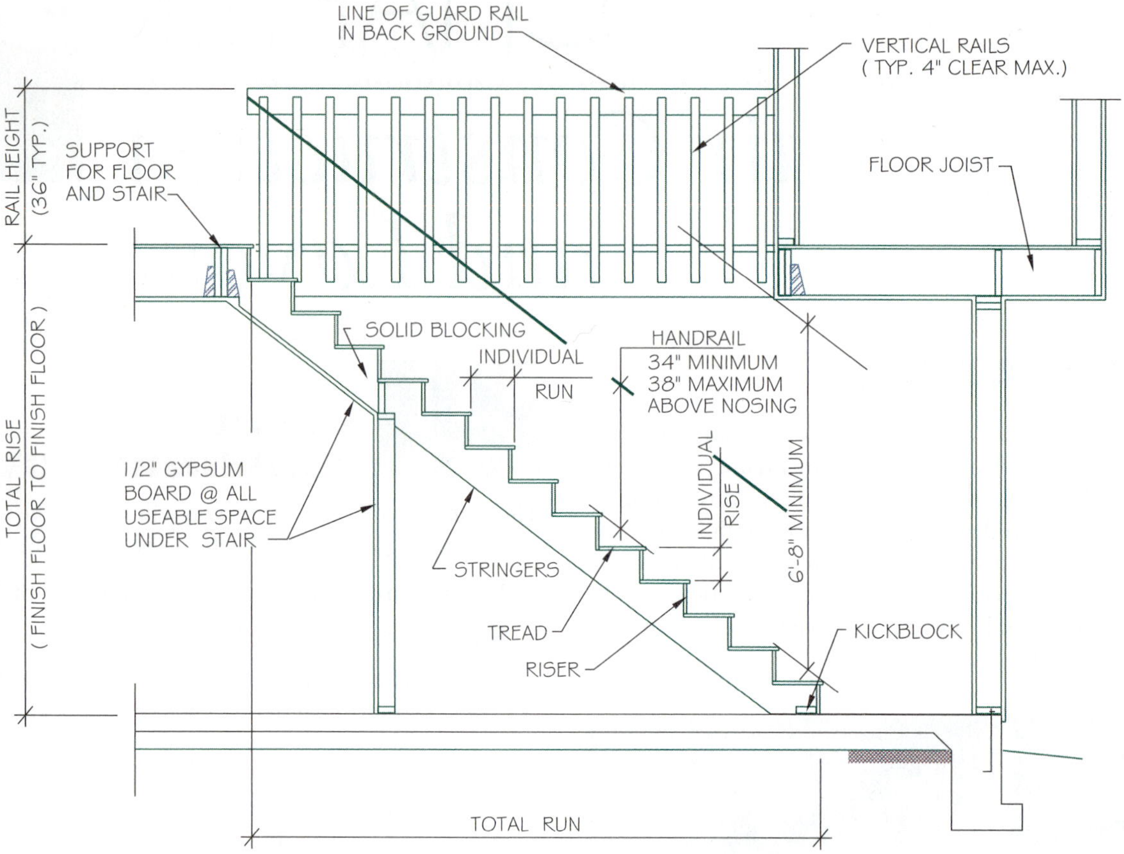

FIGURE 37.2 ■ Common stair terms.

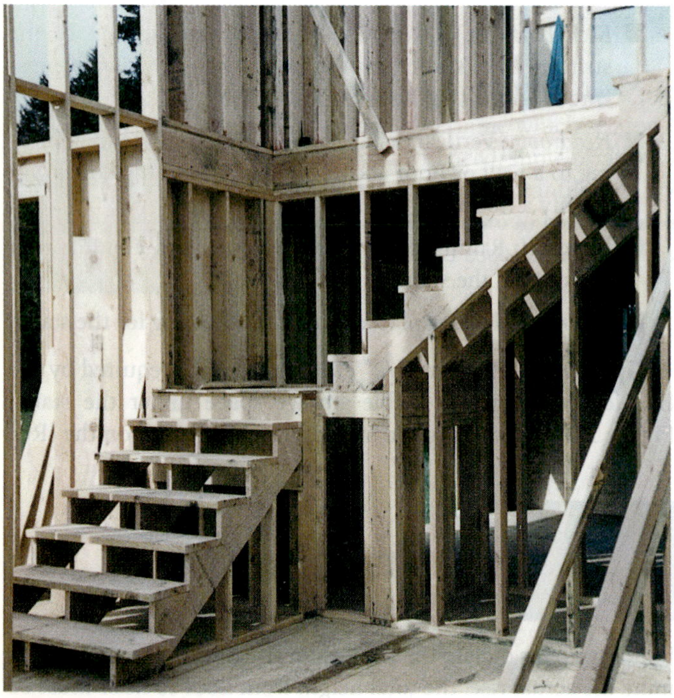

FIGURE 37.3 ■ Stairs with stringers in place, ready for the treads and risers. *Courtesy Angie Renner.*

STAIR TYPE	IRC
Straight stair	
Max. rise	7 3/4″ (197 mm)
Min. run	10″ (254 mm)
Min. headroom	6′–8″ (2032 mm)
Min. tread width	3′–0″ (914 mm)
Handrail height	34″ (864 mm) min./38″ (965 mm) max.
Guardrail height	36″ (914 mm) min.
Winders	
Min. tread depth	6″ (153 mm) min. @ edge/10″ (254 mm) @ 12″ (305 mm)
Spiral	
Min. width	26″ (660 mm)
Min. tread depth	7 1/2″ (190 mm) @ 12″ (305 mm)
Max. rise	9 1/2″ (241 mm)
Min. headroom	6′–6″ (1982 mm)

FIGURE 37.4 ■ Basic stair dimensions according to the IRC.

Step 1. Determine the total rise in inches.

19	3/4	plywood
235	9 1/4	floor joist
76	3	top plates
2353	92 5/8	studs
38	1 1/2	bottom plate
2721mm	107.125″	total rise

Step 2. From the number of risers required. Divide rise by 7 3/4″ [197 mm] is the maximum rise).

$$2721 \div 197 = 13.8 \qquad 107.125 \div 7\,3/4 = 13.8$$

Since you cannot have .8 risers, the number will be rounded up to 14 risers.

Step 3. Find the number of treads required.
Number of treads equal Rise − run.
14 − 1 = 13 treads required.

Step 4. Multiply the length of each tread by the number of treads to find the total run.

FIGURE 37.5 ■ Determining the rise and run needed for a flight of stairs.

the floor-to-ceiling height, the depth of the floor joist, and the depth of the floor covering. The total rise can then be divided by the maximum allowable rise to determine the number of steps required, as shown in Figure 37.5.

Once the required rise is determined, this information should be stored in your memory for future reference. Of the residential stairs you will lay out in your career as a drafter, probably 99 percent will have the same rise. So with a standard 8′–0″ (2430 mm) ceiling, you will always need 14 risers.

Once the rise is known, the required number of treads can be found easily, since there will always be one fewer tread than the number of risers. Thus, a typical stair for a house with 8′–0″ (2430 mm) ceilings will have 14 risers and 13 treads. If each tread is 10 1/2″ (267 mm) wide, the total run can be found by multiplying 10 1/2″ (267 mm) (the width) by 13 (the number of treads required). With this basic information, you are now ready to lay out the stairs. The layout for a straight stairway will be described first.

STRAIGHT STAIR LAYOUT

The straight-run stair is a common type of stair that will need to be drawn. It is a stair that goes from one floor to another in one straight run. An example of a straight-run stair for the residence shown in Chapter 16 can be seen in Figure 37.6. Figure 37.7 shows the stair section for the same house with the basement option. See Figure 37.8 for steps 1 through 4.

STEP 1 Lay out walls that may be near the stairs.

STEP 2 Lay out each floor level.

STEP 3 Lay out one end of the stairs. If no dimensions are available on the floor plans, scale your drawing.

STEP 4 Figure and lay out the total run of the stairs.

STEP 5 Lay out the required risers. See Figure 37.9.

STEP 6 Lay out the required treads. See Figure 37.9.

STEP 7 Lay out the stringer, and outline the treads and risers, as shown in Figure 37.10.

Drawing the Stairs with Finished-Quality Lines

Use a thin line to draw the stairs unless otherwise noted. See Figure 37.11 for steps 8 through 11.

STEP 8 Draw the treads and risers.

STEP 9 Draw the bottom side of the stringer.

STEP 10 Draw the upper stringer support or support wall where the stringer intersects the floor.

STEP 11 Draw metal hangers if there is no support wall.

See Figure 37.12 for steps 12 through 17.

STEP 12 Draw the kick block with bold lines.

STEP 13 Draw any intermediate support walls.

STEP 14 Draw solid blocking with bold lines.

STEP 15 Draw the 5/8″ (16 mm) type X gypsum board.

STEP 16 Draw any floors or walls that are over the stairs.

STEP 17 Draw the handrail.

Dimensions and Notes

See Figure 37.13 for steps 18 through 23. Place the required leader and dimension lines to locate the following needed dimensions:

STEP 18 Total rise

STEP 19 Total run

STEP 20 Rise

STEP 21 Run

STEP 22 Headroom

STEP 23 Handrail

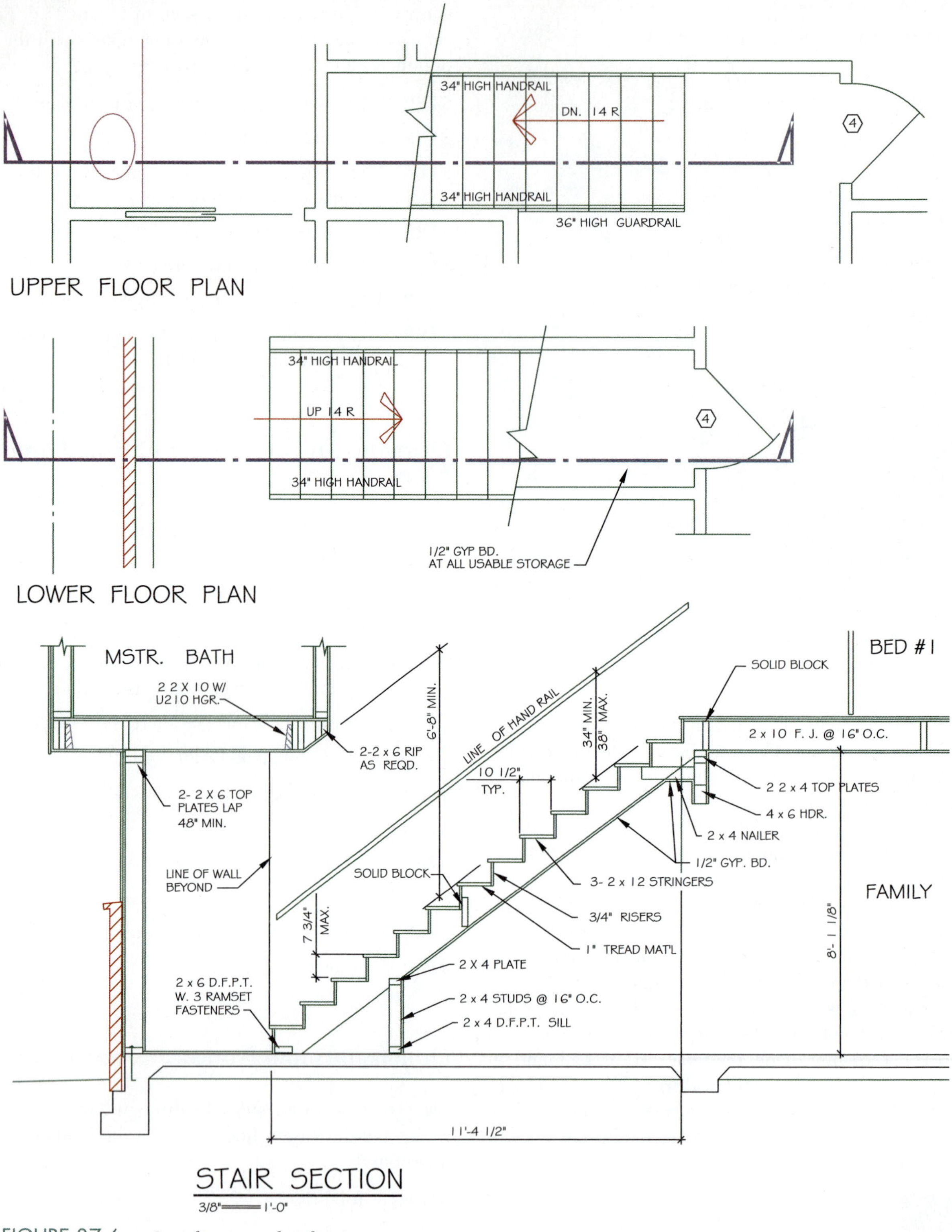

FIGURE 37.6 ■ Straight-run enclosed stairs.

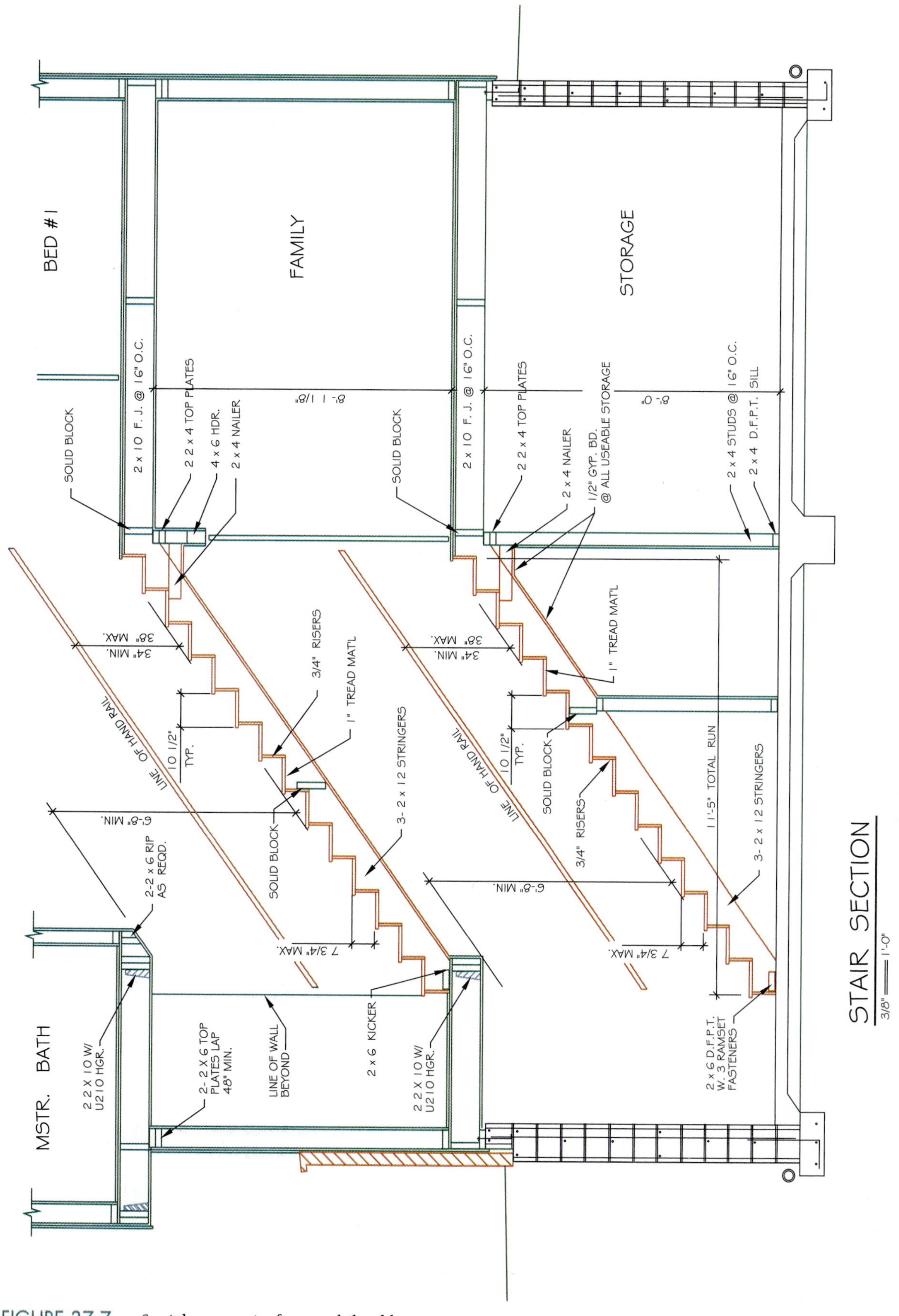

FIGURE 37.7 ■ Straight-run stairs for a multilevel home.

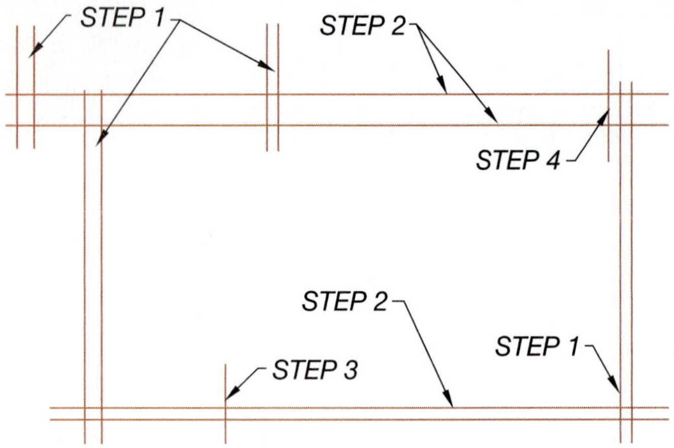

FIGURE 37.8 ■ Lay out walls, floor, and each end of the stairs.

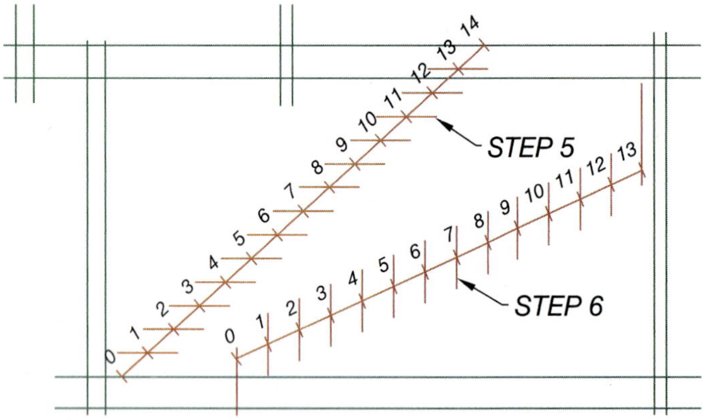

FIGURE 37.9 ■ Layout of the risers and treads.

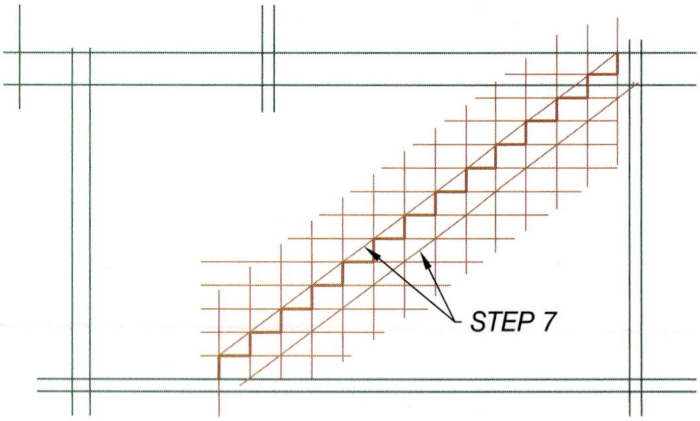

FIGURE 37.10 ■ Outline the treads, stringer, and risers.

See Figure 37.14 for typical notes that are placed on stair sections. Have your instructor specify local variations.

OPEN STAIRWAY LAYOUT

An open stairway is similar to a straight enclosed stairway. It goes from one level to the next in a straight run. The major difference is that with the open stair, there are no risers between

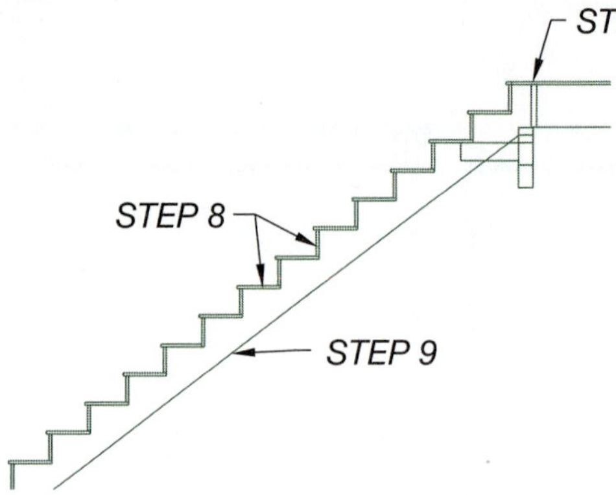

FIGURE 37.11 ■ Finished-quality lines for treads, risers, and the stringer.

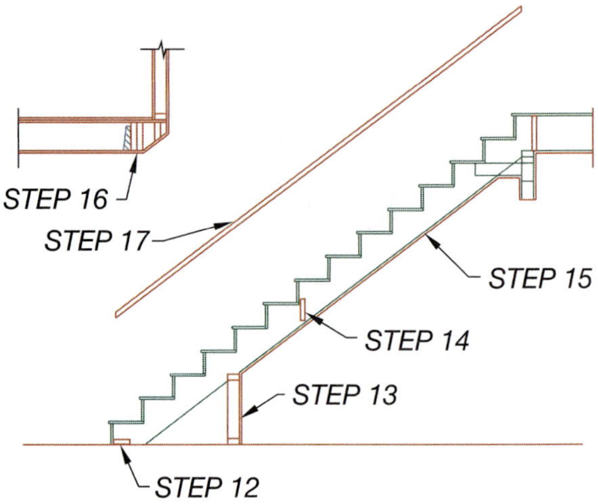

FIGURE 37.12 ■ Finishing materials include the kick block, support walls, blocking, fire-rated gypsum board, and handrails.

the treads. This allows for viewing from one floor to the next, creating an open feeling. See Figure 37.15 for an open stairway. See Figure 37.16 for steps 1, 2, and 3.

STEP 1 Lay out the stairs following steps 1 through 4 of the enclosed stair layout.

STEP 2 Lay out the 3 × 12 (75 × 300 mm) treads.

STEP 3 Lay out the 14″ (360 mm) deep stringer centered on the treads.

Finished-Quality Lines

Use thin lines for each step unless otherwise noted. See Figure 37.17 for steps 4 through 9.

STEP 4 Draw the treads with a bold line.

STEP 5 Draw the stringer.

STEP 6 Draw the upper stringer supports with bold lines.

STEP 7 Draw the metal hangers for the floor and stringer.

STEP 8 Draw any floors or walls that are near the stairs.

STEP 9 Draw the handrail.

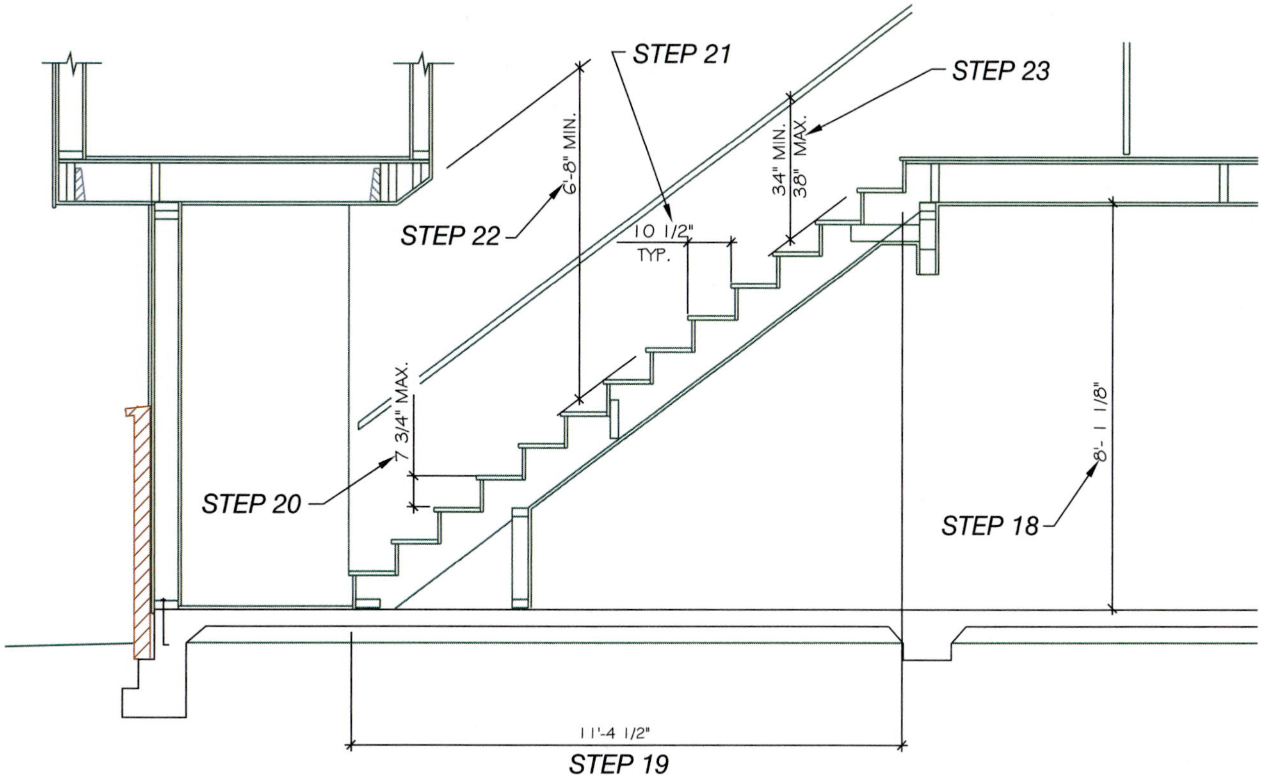

FIGURE 37.13 ■ Stair dimensions are required to show the individual and total rise and run, the location of the handrail, and the height between floors.

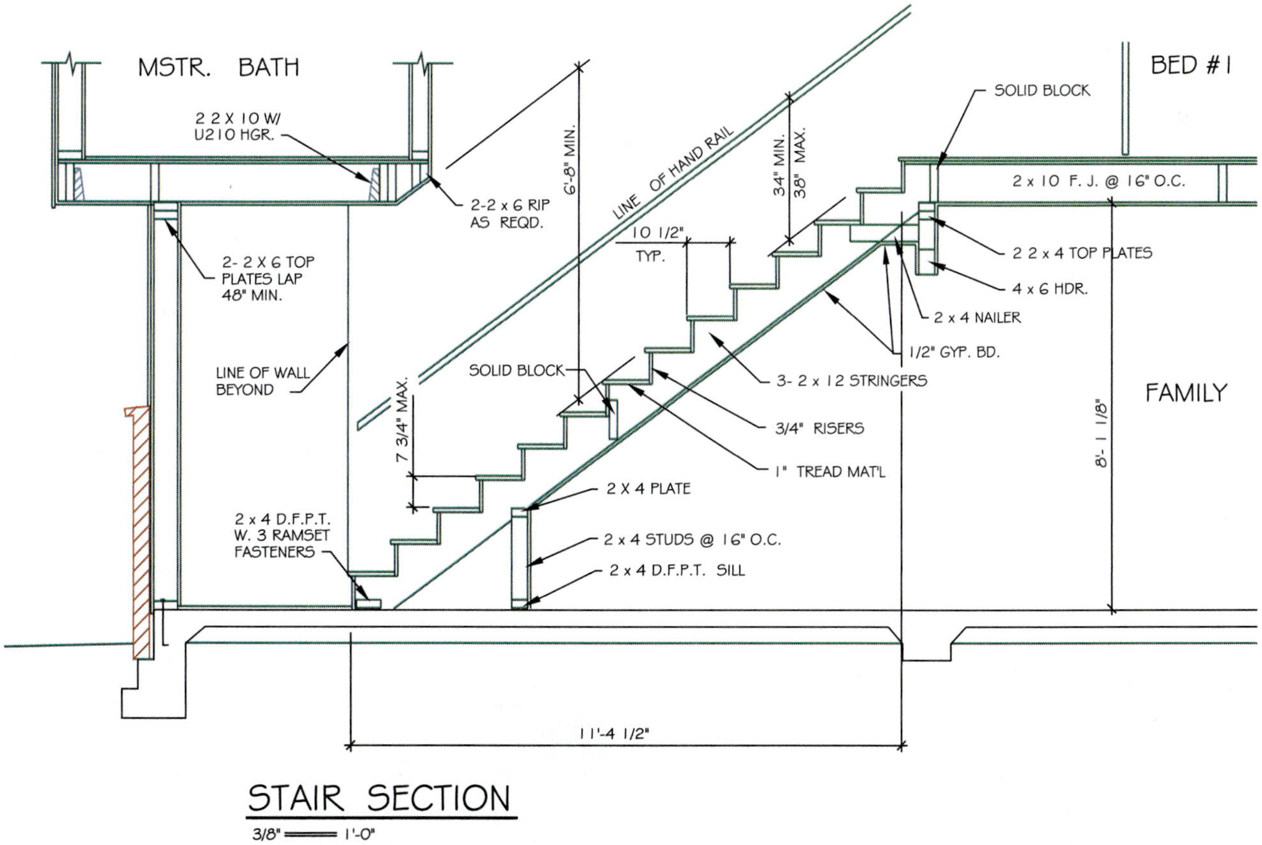

FIGURE 37.14 ■ The stair drawing is completed by adding notes to describe all materials.

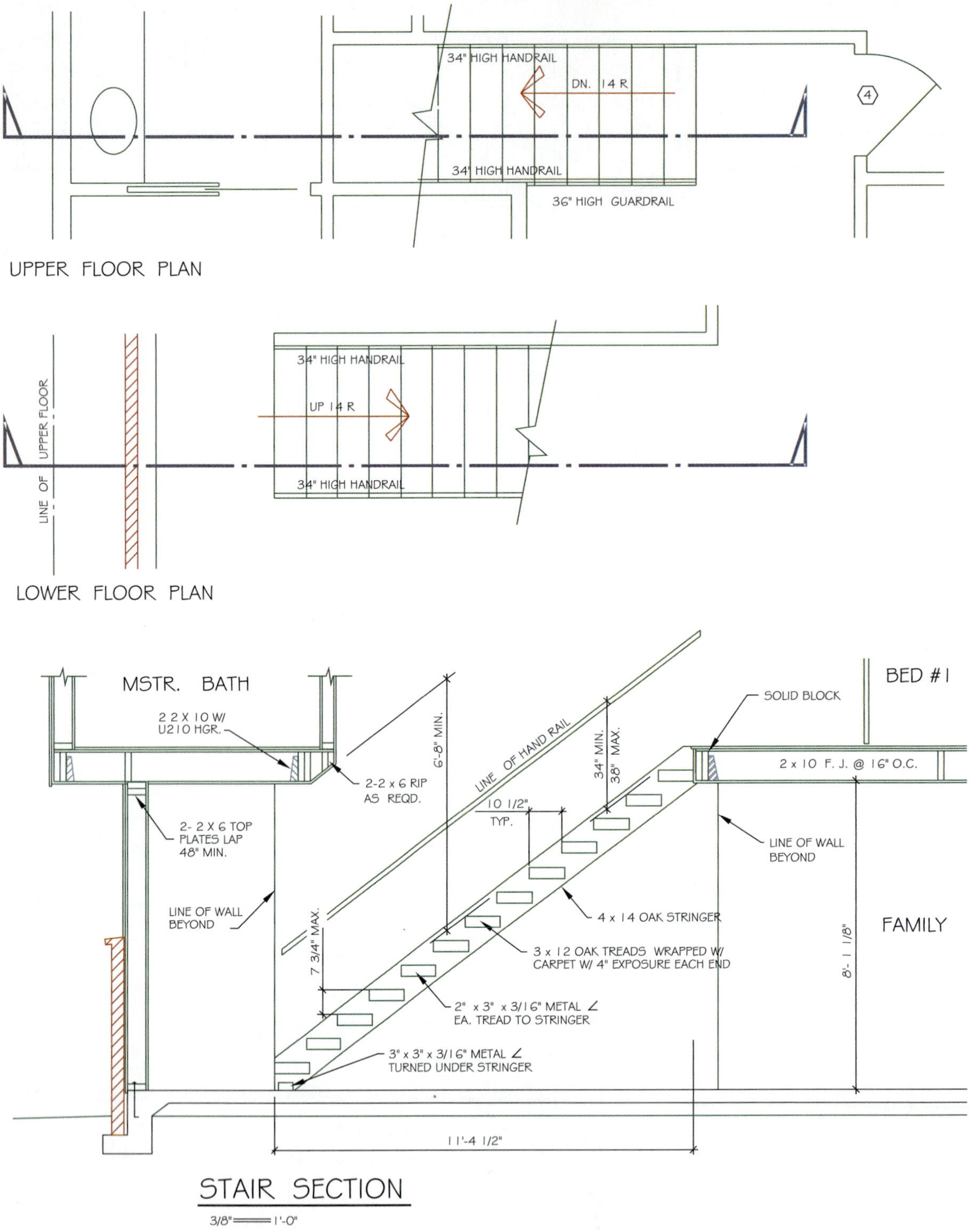

FIGURE 37.15 ■ Open stairway layout.

STEP 10 Place the required leader and dimension lines to provide the needed dimensions. See steps 18 through 24 of the enclosed stair layout for a guide to the needed dimensions. See Figure 37.18.

STEP 11 Place the required notes on the section. Use Figure 37.18 as a guide. Have your instructor specify local variations.

U-SHAPED STAIRS

The U-shaped stair is often used in residential design. Rather than going up a whole flight of steps in a straight run, this stair layout introduces a landing. The landing is usually located at

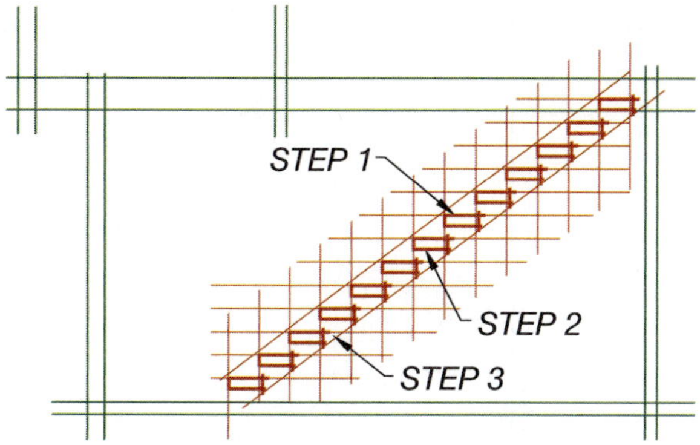

FIGURE 37.16 ■ Lay out the open tread stairs.

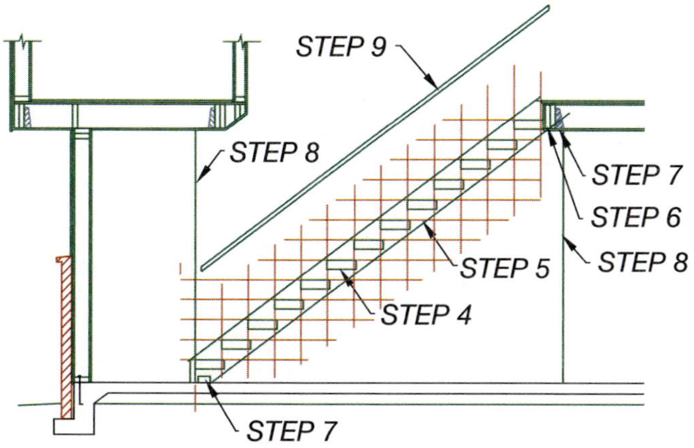

FIGURE 37.17 ■ Finished-quality lines for structural material on open stairs.

the midpoint of the run, but it can be offset, depending on the amount of room allowed for stairs on the floor plan. Figure 37.19 shows what a U-shaped stair looks like on the floor plan and in section.

The stairs may be either open or enclosed, depending on the location. The layout is similar to the layout of the straight-run stair but requires a little more planning in the layout stage because of the landing. Lay out the distance from the start of the stairs to the landing, based on the floor-plan measurements. Then proceed using a method similar to that used to draw the straight-run stair. See Figure 37.20 for help in laying out the section.

EXTERIOR STAIRS

It is quite common to need to draw sections of exterior stairs on multilevel homes. Figure 37.21 shows two different types of exterior stairs. Although there are many variations, these two options are common. Both can be laid out by following the procedure for straight run stairs.

There are some major differences in the finishing materials. Notice that there is no riser on the wood stairs, and the tread is thicker than the tread of an interior step. Usually the same material that is used on the deck is used for treads. In many parts of the country, a nonskid material should also be called for to cover the treads.

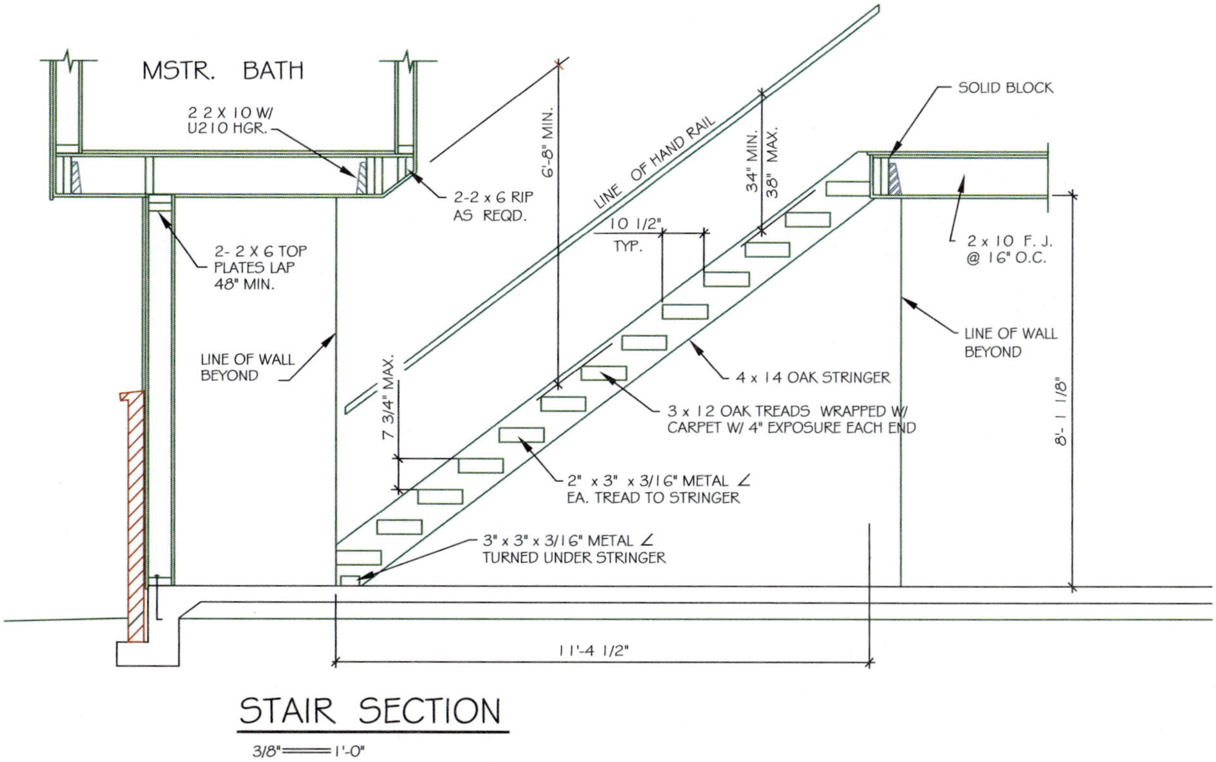

FIGURE 37.18 ■ Completed open stair with dimensions and notes added.

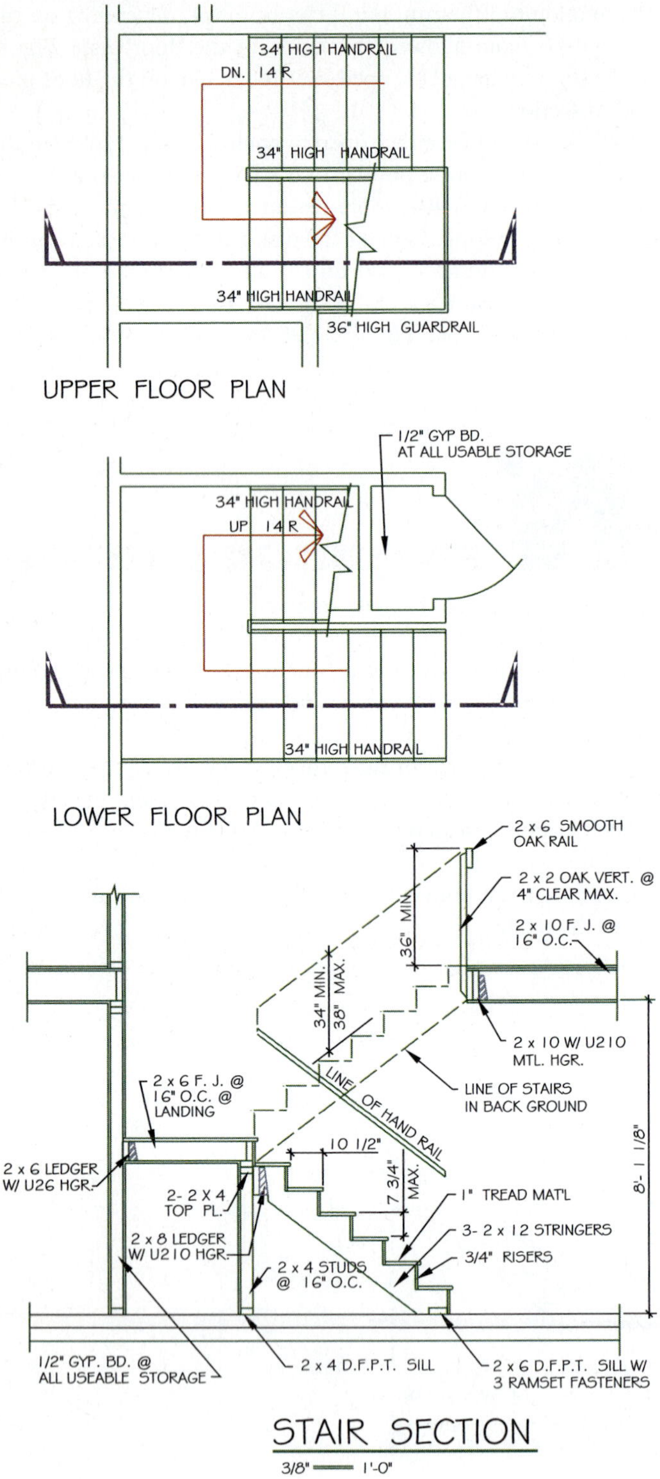

FIGURE 37.19 ■ U-shaped stair.

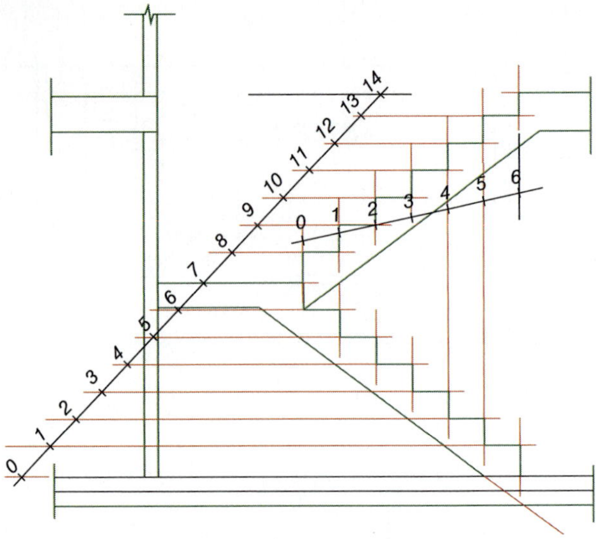

FIGURE 37.20 ■ Layout of U-shaped stair runs.

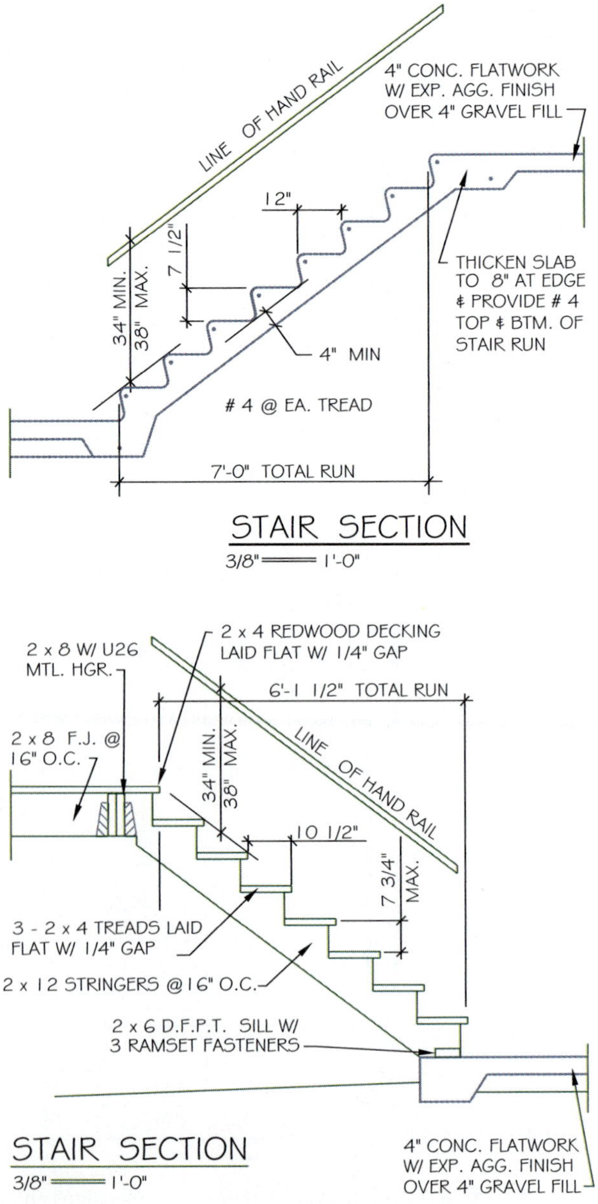

FIGURE 37.21 ■ Common types of exterior stairs.

The concrete stair can also be laid out by following the procedure for straight run stairs. Once the risers and run have been marked off, the riser can be drawn. Notice that the riser is drawn on a slight angle. It can be drawn at about 10° and not labeled. This is something the flatwork crew will determine at the job site, depending on their forms.

DRAWING STAIRS WITH CADD

A stair section can be drawn using a CAD system by following the manual layout procedure. Commands such as ARRAY, OFFSET, TRIM, and FILLET will quickly reproduce repetitive elements of the stair. The section is even easier to draw with a parametric CADD system. All you have to do is pick the starting point of the stairs, specify the number of steps required, give the stair width and direction, and give the total rise; the program then automatically calculates the rise of each step. The program asks you to provide handrails with or without bulusters and provides you with several options for handrail ends. After you have given the required information, the stair is drawn automatically.

There are also CADD stair detailing systems that reduce detailing from hours to minutes. The CADD program uses your specifications for type of stair construction, total rise, and total run. It then automatically calculates the individual rise and run. You specify the tread, riser type, thickness, stringer dimensions, and railing specifications. After all design variables are input, the computer automatically draws, completely dimensions, and labels the stair section.

Chapter 37 Additional Reading

The following Web sites can be used as a resource to help you keep current with changes in stair materials.

ADDRESS	COMPANY OR ORGANIZATION
www.arcways.com	Arcways, Inc.
www.stairwaysinc.com	Stairways, Inc.
www.southernstaircase.com	Southern Staircase
www.theironshop.com	The Iron Shop
www.yorkspiralstair.com	York Spiral Stairs

CHAPTER 37

Stair Construction and Layout Test

DIRECTIONS

Answer the questions with short, complete statements or drawings as needed.

1. Letter your name and the date at the top of the sheet.
2. Letter the question number and provide the answer. You do not need to write out the question.

QUESTIONS

Question 37–1 What is a tread?

Question 37–2 What is the minimum headroom required for a residential stair?

Question 37–3 What is the maximum individual rise of a step?

Question 37–4 What member is used to support the stairs?

Question 37–5 What spacing is required between the verticals of a railing?

Question 37–6 Describe the difference between a handrail and a guardrail.

Question 37–7 How many risers are required if the height between floors is 10′ (3048 mm)?

Question 37–8 Sketch three common stair types.

Question 37–9 If a run of 10″ (255 mm) is to be used, what will be the total run when the distance between floors is 9′ (2700 mm)?

Question 37–10 What is a common size for treads in an open-tread layout?

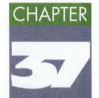

CHAPTER 38

Fireplace Construction and Layout

INTRODUCTION

Thought was given to the type and location of the fireplace when the floor plan was drawn. As the sections are being drawn, consideration must be given to the construction of the fireplace and chimney. The most common construction materials for a fireplace and chimney are masonry and metal. The metal, or zero-clearance, fireplace is manufactured and does not require a section to explain its construction. A metal fireplace is surrounded by wood walls that can be represented on the floor or framing plans. Figure 38.1 shows a metal fireplace installed in a wood-framed chase. The interior can be covered in masonry, stone, tile or some other noncombustible material, as in Figure 38.2. Notice in Figure 38.1 that a metal chimney has been provided. Metal fireplaces are available that do not require a chimney. Figure 38.3 shows a gas-burning fireplace that is vented through the wall.

FIGURE 38.1 ■ Metal fireplace insert encased with standard wood framing. *Courtesy Michael Jefferis.*

FIGURE 38.2 ■ Metal fireplace units can be covered in masonry, stone, tile, or some other noncombustible material. *Courtesy Leroy Cook.*

FIGURE 38.3 ■ Exterior view of a self-venting gas fireplace; it requires a vent similar to a dryer.

FIREPLACE TERMS

Fireplace construction has its own vocabulary that a drafter should be familiar with to draw a section. Figure 38.4 shows the major components of a fireplace and chimney.

The *fireplace opening* is the area between the side and top faces of the fireplace. The size of the fireplace opening should be given much consideration, since it is important for the appearance and operation of the fireplace. If the opening is too small, the fireplace will not produce a sufficient amount of heat. If the opening is too large, the fire could make a room too hot. Common design guidelines suggest that the opening be approximately 1/30 of the room area for small rooms, and 1/65 of the room area for large rooms. Figure 38.5 shows suggested fireplace opening sizes relative to room size. The ideal dimensions for a single-face fireplace have been determined to be 36″ (900 mm) wide and 26″ (660 mm) high. These dimensions may be varied slightly to meet the size of the brick or to fit other special dimensions of the room.

The *hearth* is the floor of the fireplace and consists of an inner and outer hearth. The inner hearth is the floor of the firebox. The hearth is made of fire-resistant brick and holds the burning fuel. An ash dump is usually located in the inner hearth of a masonry fireplace. The *ash dump* is an opening in the hearth that the ashes can be dumped into. The ash dump is covered with a small metal plate, which can be removed to provide access to the ash pit. The *ash pit* is the space below the fireplace where the ashes can be stored.

The outer hearth may be made of any noncombustible material. The material is usually selected to blend with other interior design features and may be brick, tile, marble, or stone. The outer hearth protects the combustible floor around the fireplace. Figure 38.6 shows the minimum sizes for the outer hearth.

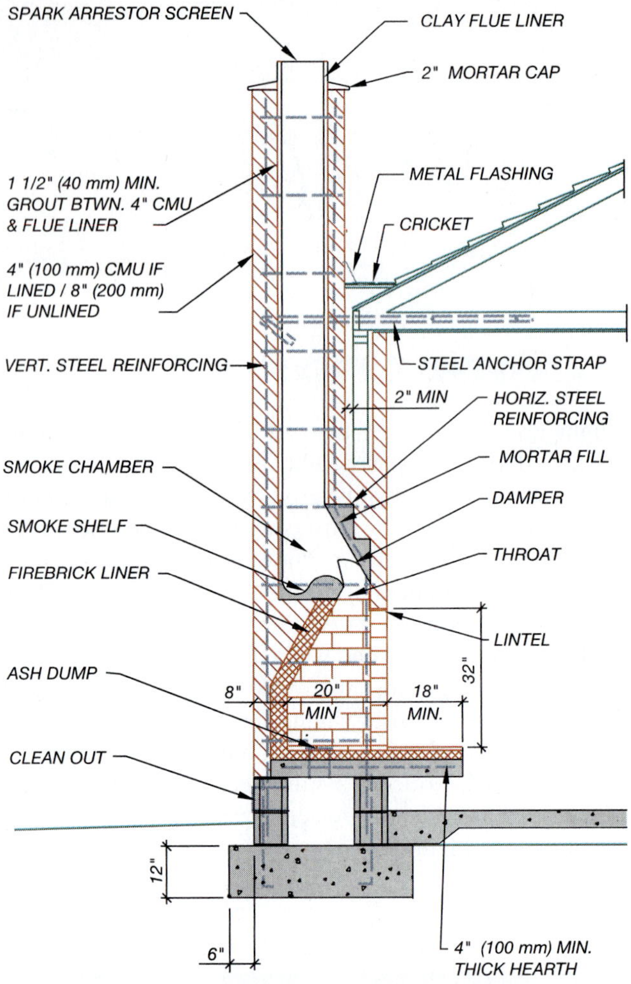

FIGURE 38.4 ■ Parts of a masonry fireplace and chimney.

SUGGESTED WIDTH OF FIREPLACE OPENINGS APPROPRIATE TO ROOM SIZE					
SIZE OF ROOM		IF IN SHORT WALL		IF IN LONG WALL	
(FT)	(MM)	(IN.)	(MM)	(IN.)	(MM)
10 × 14	(3000 × 4300)	24	(600)	24–32	(600–800)
12 × 16	(3600 × 4900)	28–36	(700–900)	32–36	(800–900)
12 × 20	(3600 × 6100)	32–36	(800–900)	36–40	(900–1000)
12 × 24	(3600 × 7300)	32–36	(800–900)	36–48	(900–1200)
14 × 28	(4300 × 8600)	32–40	(800–1000)	40–48	(1000–1200)
16 × 30	(4900 × 9100)	36–40	(900–1000)	48–60	(1200–1500)
20 × 36	(6100 × 11000)	40–48	(1000–1200)	48–72	(1200–1800)

FIGURE 38.5 ■ The size of the fireplace should be proportioned to the size of the room. This will give both a pleasing appearance and a fireplace that will not overheat the room.

GENERAL CODE REQUIREMENTS

ITEM	LETTER	UNIFORM BUILDING CODE* 1997
Hearth Slab Thickness	A	4″
Hearth Slab Width (Each side of opening)	B	8″ Fireplace opg. < 6 sq. ft. 12″ Fireplace opg. ≥ 6 sq. ft.
Hearth Slab Length (Front of opening)	C	16″ Firepl. opg. < 6 sq. ft. 20″ Firepl. opg. ≥ 6 sq. ft.
Hearth Slab Reinforcing	D	Reinforced to carry its own weight and all imposed loads.
Thickness of Wall or Firebox	E	10″ common brick or 8″ where a fireback lining is used. Jts. In fireback 1/4″ max.
Distance from Top of Opening to Throat	F	6″
Smoke Chamber Edge of Shelf Rear Wall—Thickness Front & Side Wall—Thickness	G	 6″ 8″
Chimney Vertical Reinforcing	**H	Four #4 full length bars for chimney up to 40″ wide. Add two #4 bars for each additional 40″ or fraction of width or each additional flue.
Horizontal Reinforcing	J	1/4″ ties at 18″ and two ties at each bend in vertical steel.
Bond Beams	K	No specified requirements. L.A. City requirements and good practice.
Fireplace Lintel	L	Incombustible material
Walls with Flue Lining	M	Brick with grout around lining 4″ min. from flue lining to outside face of chimney.
Walls with Unlined Flue	N	8″ solid masonry
Distance Between Adjacent Flues	O	4″ including flue liner
Effective Flue Area (Based on Area of Fireplace Opening)	P	Round lining—1/12 or 50 sq.in.min. Rectangular lining 1/10 or 64 sq. in. min. Unlined or lined with firebrick—1/8 or 100 sq. in. min.
Clearances Wood Frame Combustible Material Above Roof	R	1″ when outside of wall or 1/2″ gypsum board 1″ when entirely within structure 6″ min. to fireplace opening. 12″ from opening when material projecting more than 1/8 for ea. 1″ 2′ at 10′
Anchorage Strap Number Embedment into chimney Fasten to Bolts	S	 3/16″ × 1″ 2 12″ hooked around outer bar w/6″ ext. 4 joists Two 1/2″ dia.
Footing Thickness Width	T	 12″ min. 6″ each side of fireplace wall
Outside Air Intake	U	Optional
Glass Screen Door		Optional

Applies to Los Angeles County and Los Angeles City requirements.

***H EXCEPTION. Chimneys constructed of hollow masonry units may have vertical reinforcing bars spliced to footing dowels, provided that the splice is inspected prior to grouting of the wall.*

FIGURE 38.6(A) ■ General code requirements for fireplace and chimney construction. The letters in the second column will be helpful in locating specific items in Figure 38.6*b*.

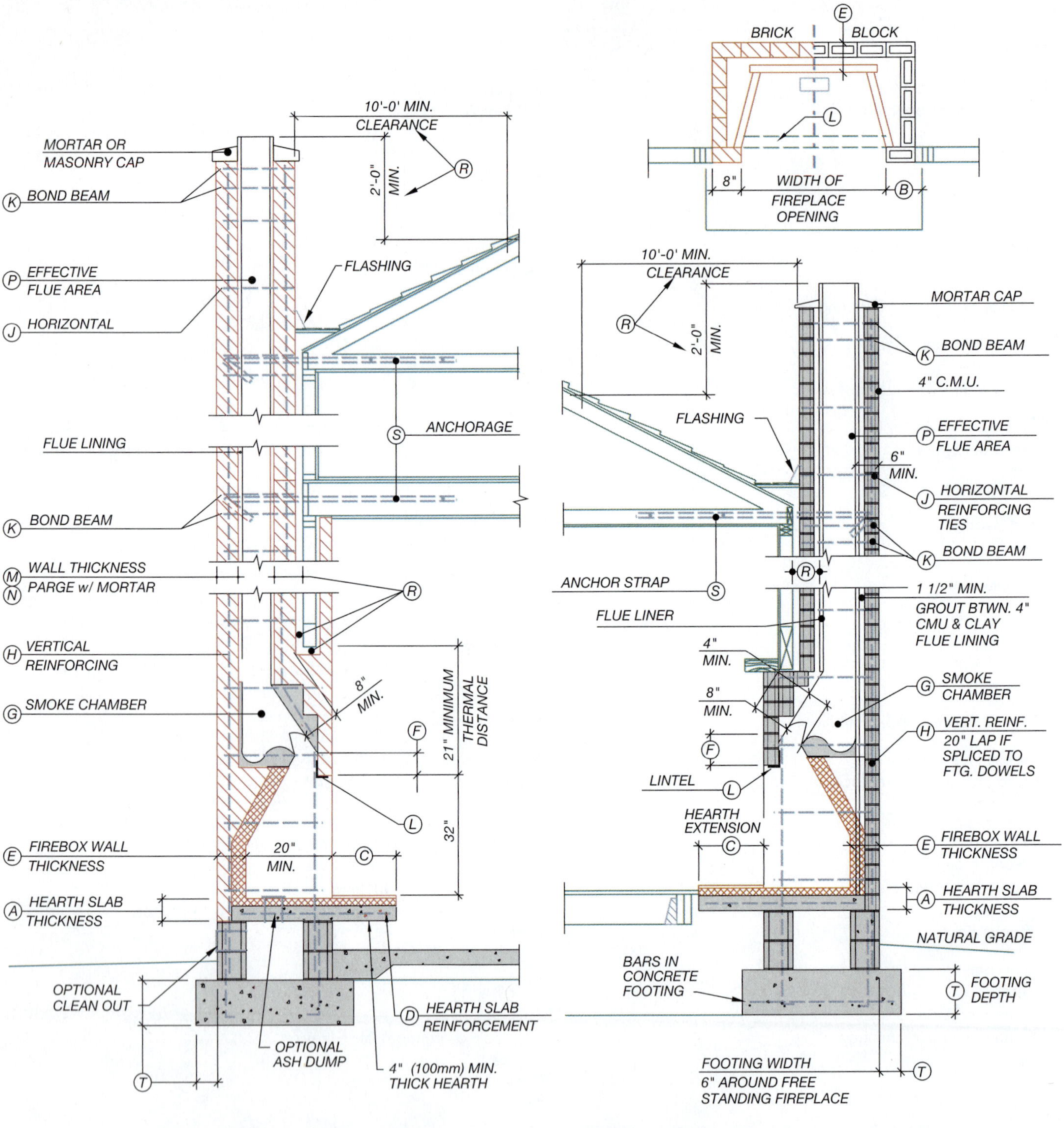

FIGURE 38.6(B) ■ Brick fireplace and chimney components. The circled letters refer to item references in Figure 38.6*a*.

The fireplace opening is the front of the firebox. The *firebox* is where the combustion of the fuel occurs. The side should be slanted slightly to radiate heat into the room. The rear wall should be inclined to provide an upward draft into the upper part of the fireplace and chimney. The firebox is usually constructed of fire-resistant brick set in fire-resistant mortar. Figure 38.6 shows minimum wall thickness for the firebox.

The firebox depth should be proportional to the size of the fireplace opening. Providing a proper depth ensures that smoke will not discolor the front face (breast) of the fireplace. With an opening of 36″ × 26″ (900 × 660 mm), a depth of 20″ (500 mm) should be provided for a single-face fireplace. Figure 38.7 lists recommended fireplace opening-to-depth proportions.

FIREPLACE TYPE	WIDTH OF OPENING (W) (IN.)	WIDTH OF OPENING (W) (MM)	HEIGHT OF OPENING (H) (IN.)	HEIGHT OF OPENING (H) (MM)	DEPTH OF OPENING (D) (IN.)	DEPTH OF OPENING (D) (MM)
Single Face	28	700	24	600	20	500
	30	760	24	600	20	500
	30	760	26	660	20	500
	36	900	26	660	20	500
	36	900	28	700	22	560
	40	1000	28	700	22	560
	48	1200	32	800	25	635
Two faces	34	865	27	685	23	585
adjacent "L"	39	990	27	685	23	585
or corner type.	46	1170	27	685	23	585
	52	1320	30	760	27	635
Two faces*	32	800	21	530	30	760
opposite	35	890	21	530	30	760
Look through	42	1070	21	530	30	760
	48	1200	21	530	34	865
Three face*	39	990	21	530	30	760
2 long, 1 short	46	1170	21	530	30	760
3-way opening	52	1320	21	530	34	865
Three face*	43	1090	27	685	23	585
1 long, 2 short	50	1270	27	685	23	585
3-way opening	56	1420	30	760	27	686

Fireplaces open on more than front and one end are* **not *recommended.*

FIGURE 38.7 ■ Guide for fireplace opening-to-depth proportions.

Above the fireplace opening is a lintel. The *lintel* is a reinforced masonry or steel angle that supports the fireplace face. The throat of a fireplace is the opening at the top of the firebox that opens into the chimney. The throat must be at least 8″ (200 mm) above the fireplace opening and must be a minimum of 4″ (100 mm) in depth. The throat should be closable when the fireplace is not in use. This is done by installing a damper. The *damper* extends the full width of the throat and is used to prevent heat from escaping up the chimney when the fireplace is not in use. When fuel is being burned in the firebox, the damper can be opened to allow smoke from the firebox into the smoke chamber of the chimney. The *smoke chamber* acts as a funnel between the firebox and the chimney. The shape of the smoke chamber should be symmetrical so that the chimney draft pulls evenly and creates an even fire in the firebox. The smoke chamber should be centered under the flue in the chimney and directly above the firebox. A *smoke shelf* is located at the bottom of the smoke chamber behind the damper. The smoke shelf prevents downdrafts from the chimney from entering the firebox.

The *chimney* is the upper extension of the fireplace and is built to carry off the smoke from the fire. The main components of the chimney are the flue, lining, anchors, cap, and spark arrester. The wall thickness of the chimney will be determined by the type of flue construction. Masonry chimneys are not allowed to change in size or shape within 6″ (150 mm) above or below any combustible floor, ceiling, or roof component penetrated by the chimney. Figure 38.6 shows the minimum wall thickness for chimneys.

The *flue* is the opening inside the chimney that allows smoke and combustion gases to pass from the firebox away from the structure. A flue may be constructed of normal masonry products or may be covered with a flue liner. The size of the flue must be proportional to the size of the firebox opening and the number of open faces of the fireplace. A flue that is too small will not allow the fire to burn well and will cause smoke to exit through the front of the firebox. A flue that is too large for the firebox will cause too great a draft through the house as the fire draws its combustion air. Flue sizes are generally required to equal either 1/8 or 1/10 of the fireplace opening. Figure 38.8 shows recommended areas for residential fireplaces.

A *chimney liner* is usually built of fire clay or terra-cotta. The liner is built into the chimney to provide a smooth surface to the flue wall and to reduce the width of the chimney wall. The smooth surface of the liner will help reduce the buildup of soot, which could cause a chimney fire. The *chimney cap* is the sloping surface on the top of the chimney. The slope prevents rain from collecting on the top of the chimney. The flue normally projects 2″ to 3″ (50–75 mm) above the cap so that water will

TYPE OF FIREPLACE	WIDTH OF OPENING W IN.	HEIGHT OF OPENING H IN.	DEPTH OF OPENING D IN.	AREA OF FIREPLACE OPENING FOR FLUE DETERMINATION SQ IN.	FLUE SIZE REQUIRED AT 1/10 AREA OF FIREPLACE OPENING	FLUE SIZE REQUIRED AT 1/8 AREA OF FIREPLACE OPENING
	28	24	20	672	8 1/2 × 13	8 1/2 × 17
	30	24	20	720	8 1/2 × 17	13″ round
	30	26	20	780	8 1/2 × 17	10 × 18
	36	26	20	936	10 × 18	13 × 17
	36	28	22	1008	10 × 18	10 × 21
	40	28	22	1120	10 × 18	10 × 21
	48	32	25	1536	13 × 21	10 × 21
	60	32	25	1920	17 × 21	21 × 21
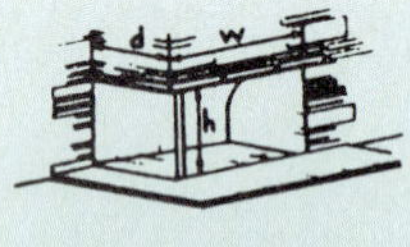	34	27	23	1107	10 × 18	10 × 21
	39	27	23	1223	10 × 21	13 × 21
	46	27	23	1388	10 × 21	13 × 21
	52	30	27	1884	13 × 21	17 × 21
	64	30	27	2085	17 × 21	21 × 21
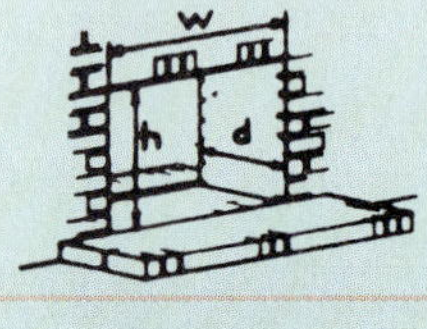	32	21	30	1344	13 × 17 or 10 × 21	17 × 17 or 13 × 21
	35	21	30	1470	17 × 17 or 13 × 21	17 × 21
	42	21	30	1764	17 × 21	17 × 21
	48	21	34	2016	17 × 21	21 × 21
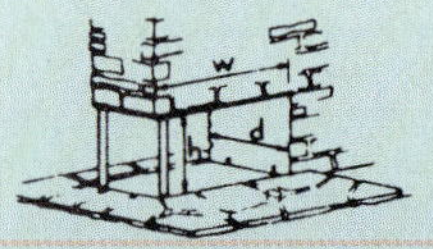	39	21	30	1638	13 × 21 or 17 × 17	17 × 21
	46	21	30	1932	17 × 21	21 × 21
	52	21	34	2184	17 × 21	21 × 21
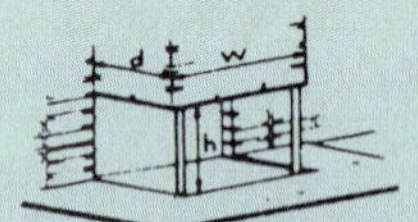	43	27	23	1782	13 × 21 or 17 × 17	17 × 21
	50	27	23	1971	17 × 21	17 × 21
	56	30	27	2490	21 × 21	21 × 21 or
	68	30	27	2850	13 × 21 & ▲ 10 × 21	2—10 × 21 ▲ 2—13 × 21 or 2—17 × 17 or ▲ 10 × 18 & 17 × 21

▲ *Rather than using 2 flue liners in chimney, the flue is often left unlined with 8″ masonry walls. Unlined flues must have a minimum area of 1/8 fireplace opening.*

FIGURE 38.8 ■ Recommended flue areas for residential fireplaces.

not run down the flue. The *chimney hood* is a covering that may be placed over the flue for protection from the elements. The hood can be made of either masonry or metal. The masonry cap is built so that openings allow for the prevailing wind to blow through the hood and create a draft in the flue. The metal hood can be rotated by wind pressure to keep the opening of the hood downwind and thus prevent rain or snow from entering the flue. A *spark arrester* is a screen placed at the top of the flue inside the hood to prevent combustibles from leaving the flue.

GOALS OF AN ENERGY-EFFICIENT FIREPLACE

In addition to adding charm and elegance to a room, a fireplace also needs to be functional. No matter what the material is, consideration must be given to the air that will be used by the fireplace. Three points to consider are how air will be supplied to the fire, how to use the heated air created by the fire, and how to minimize heated air escaping through the chimney.

Providing Air Supply

The construction methods introduced in Chapters 10 and 25 have greatly reduced the amount of air that enters the structure. When a fireplace is added to a relatively airtight residence, the fireplace can actually cool a room if proper precautions are not taken. Air must be provided to the fireplace for combustion, and air is expelled up the chimney after combustion. As the air is expelled after combustion, new air must be drawn into the fire. If the air for combustion is taken from the room being heated, a draft will be created. To reduce drafts and maximize the heat

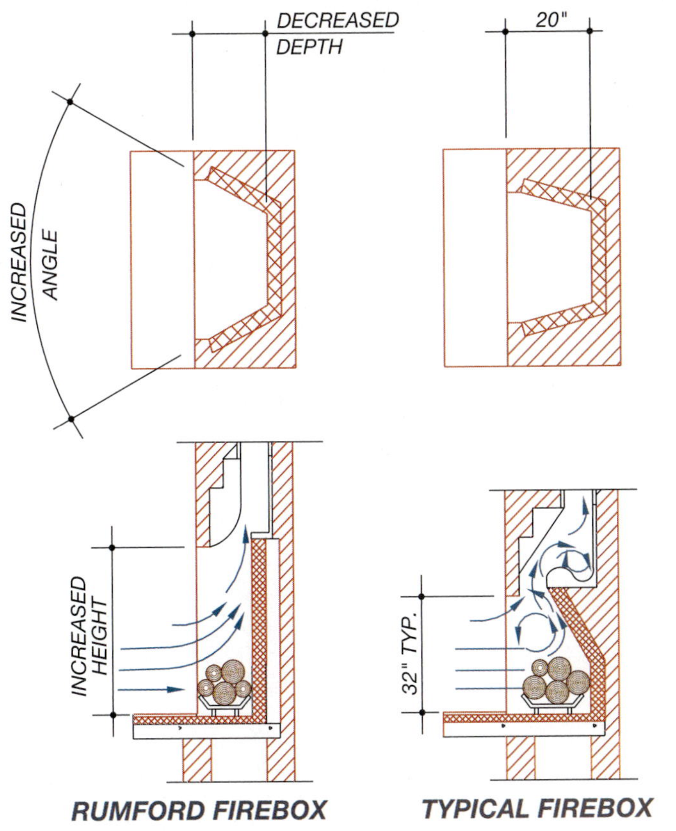

FIGURE 38.9 ■ Comparison of a Rumford-style firebox and traditional construction. The Rumford style, popular in the mid 1800s, has been reinvented to increase the amount of heated air that is returned to the room containing the fireplace.

produced by the fire, combustion air must be taken from an outside source. When the floor plan was completed, a note was placed on the plan reading:

PROVIDE A SCREENED, CLOSABLE VENT WITHIN 24″ OF THE FIREBOX.

Providing the note on the floor plan will help ensure that the fireplace meets the minimum building code requirements. It will also ensure that a draft is not created. As the fireplace section is drawn, the vent for air intake should be indicated. The vent should be placed on an exterior wall or in the back of the fireplace in a position that is high enough above the ground so that it will not be blocked by snow. Combustion kits are available that can be installed in the duct, to heat the outside air before it enters the firebox. Another alternative to drawing air from the outside is to draw combustion air from the cold air return ducts of the heating system. The HVAC contractor can calculate the needed size and provide a duct that takes the necessary cool air from inside the home to supply the fireplace.

Using the Heated Air Efficiently

Once the fire has heated the air in the firebox, that air must be used efficiently to heat the structure. Chapter 10 introduced passive solar heating and the use of a masonry mass to store heat. A masonry fireplace and chimney will provide mass to absorb, store, and radiate the heat back into the room long after the fire is out. When a fireplace is built on an outer wall, the mass of the fireplace is often outside the structure, to increase the usable floor space. With the chimney outside the structure, the interior face of the fireplace is the only surface that radiates heat back into the room. The amount of heat radiated into a room is increased as more of the fireplace and chimney mass is located inside the structure. Placing the fireplace and chimney totally within a structure will allow all surfaces of the masonry to radiate heat into the structure.

Altering the shape of the firebox will also increase the amount of air radiated back to the room. Figure 38.5 shows common sizes of fireplace openings. Increasing the width and height of the opening and decreasing the depth of the fireplace will increase the amount of heat in the room. This altered firebox shape is based on the Rumford design that was popular in the late 1800s. It has been reinvented and has become popular in many areas where it is permitted by building codes. Increasing the length of the side walls and decreasing the length of the rear walls will also increase the heat returned to the room. Figure 38.9 shows the difference between a Rumford-type firebox and traditional designs. A third method of increasing the heat returned to the room is a mechanical blower.

Decreasing Chimney Heat Loss

The traditional means of reducing heat loss up the chimney has been a damper. The damper is opened while the fire is burning and closed when the fire is out. Because the damper is not airtight, warm air from the room will rise and be drawn up the chimney. As the air in the chimney cools, it will start a reverse current, drawing cold air into the heated room. Placing a chimney on an outer wall will increase this tendency, since the chimney will be cooler. In very cold climates, the reverse convection current can be reduced by an additional damper at the top of the chimney.

CHIMNEY REINFORCEMENT

A minimum of four vertical reinforcing bars 1/2″ (13 mm) in diameter (#4) should be used in the chimney, extending from the foundation up to the top of the chimney. Typically these vertical bars are supported at 18″ (450 mm) intervals with 1/4″ (6 mm) horizontal rebar. Rebar is also installed when the vertical steel is bent for a change in chimney width. In addition to the reinforcement in the chimney, the chimney must be attached to the structure. This is typically done with *steel anchors* that connect

the fireplace to the framing members at each floor and ceiling level. Steel straps approximately 3/16″ × 1″ (5 × 25 mm) are embedded into the grout of the chimney or wrapped around the reinforcing steel and then bolted to the framing members.

DRAWING THE FIREPLACE SECTION

The fireplace section can be a valuable part of the working drawings. Although a fireplace drawing is required by most building departments when a home has a masonry fireplace and chimney, a drafter may not be required to draw a fireplace section. Because of the similarities of fireplace design, a fireplace drawing is often kept on file as a stock detail at many offices. When a house that has a fireplace is being drawn, the drafter needs only to get the stock detail from a file, make the needed copies, and attach them to the prints of the plans. A stock detail can be seen at the end of the chapter.

If you are required to draw a fireplace section, a print of the floor plans will be needed to help determine wall locations and the size of the fireplace. As the drawing for the fireplace is started, use construction lines. See Figure 38.10 for steps 1 through 4.

Layout of the Fireplace

STEP 1 Lay out the width of the fireplace: 20″ (500 mm) for the firebox and 8″ (200 mm) for the rear wall.

STEP 2 Lay out all walls, floors, and ceilings that are near the fireplace. Be sure to maintain the required minimum distance from the masonry to the wood as determined by the code in your area.

STEP 3 Lay out the foundation for the fireplace. The footing is typically 6″ (150 mm) wider on each side than the fireplace and 12″ (300 mm) deep.

STEP 4 Lay out the hearth in front of the fireplace. The hearth will vary in thickness, depending on the type of finishing material used. If the fireplace is being drawn for a house with a wood floor, the hearth will require a 4″ (100 mm) minimum concrete slab projecting from the fireplace base to support the finished hearth.

See Figure 38.11 for steps 5 through 8.

STEP 5 Lay out the firebox. Assume that a 36″ × 26″ × 20″ (900 × 660 × 500 mm) firebox will be used. See Figure 38.12 for guidelines for laying out the firebox.

STEP 6 Determine the size of flue required. A 36″ × 26″ opening has an area of 936 sq in. By examining Figure 38.13 you can see that a flue with an area of 91 sq in. is required. The flue walls can be constructed using either 8″ (200 mm) of masonry or 4″ (100 mm) of masonry and a clay flue liner. If a liner is used, a 13″ (330 mm) round liner is the minimum size required.

STEP 7 Lay out the flue with a liner. Draw the interior face of the liner. The thickness will be shown later.

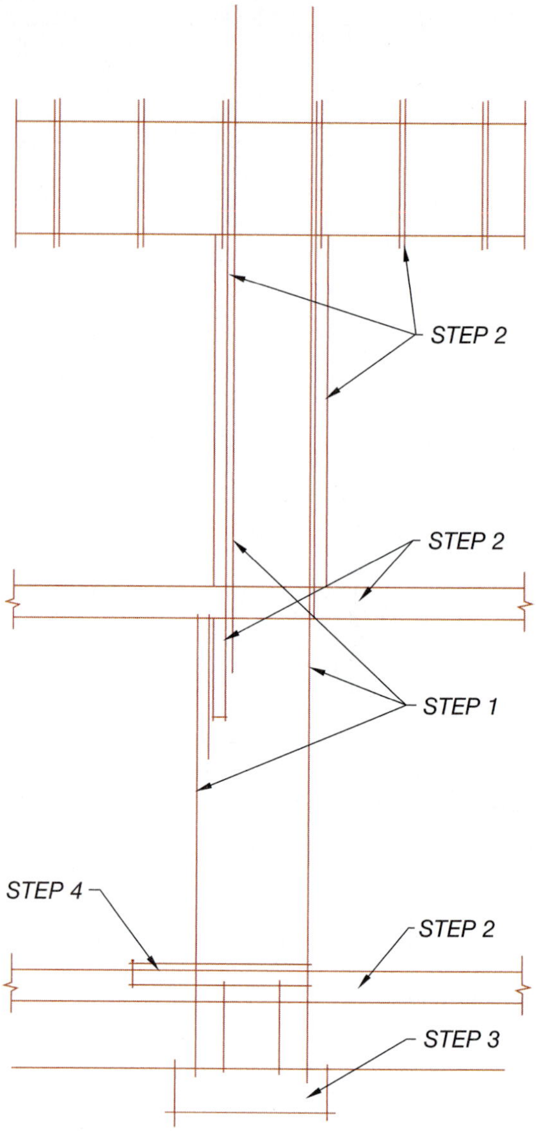

FIGURE 38.10 ■ Fireplace layout using construction lines.

STEP 8 Determine the height of the chimney. Most building codes require the chimney to project 24″ (600 mm) minimum above any construction within 10′ (3000 mm) of the chimney. See Figure 38.14.

Drawing with Finished-Quality Lines

STEP 9 Use bold lines to outline all framing materials, as shown in Figure 38.15.

See Figure 38.16 for steps 10 through 12.

STEP 10 Use bold lines to outline the masonry, flue liner, and hearth.

STEP 11 Use thin lines at a 45° angle to crosshatch the masonry.

STEP 12 Crosshatch the firebrick with lines at a 45° angle so that a grid is created.

See Figure 38.17 for steps 13 through 15.

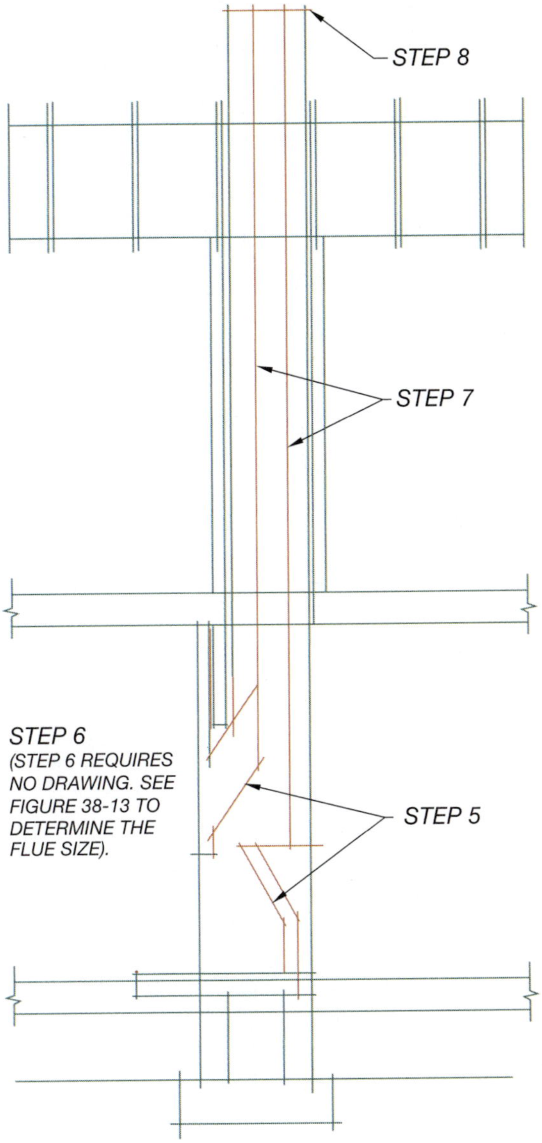

FIGURE 38.11 ■ Layout of the firebox and flue using construction lines.

STEP 13 Draw all reinforcing steel with bold lines. Check local building codes to determine what steel will be required.

STEP 14 Draw the lintel with a very bold line.

STEP 15 Draw the damper with a very bold line.

See Figure 38.18 for steps 16 and 17.

STEP 16 Dimension the drawing. Items that must be dimensioned include the following:

a. Height above roof
b. Width of hearth
c. Width and height of firebox
d. Width and depth of footings
e. Wood to masonry clearance

STEP 17 Place notes on the drawing using Figure 36.18 as a guide. Notes will vary, depending on the code that you follow and the structural material that has been specified on the floor, framing, and foundation plans.

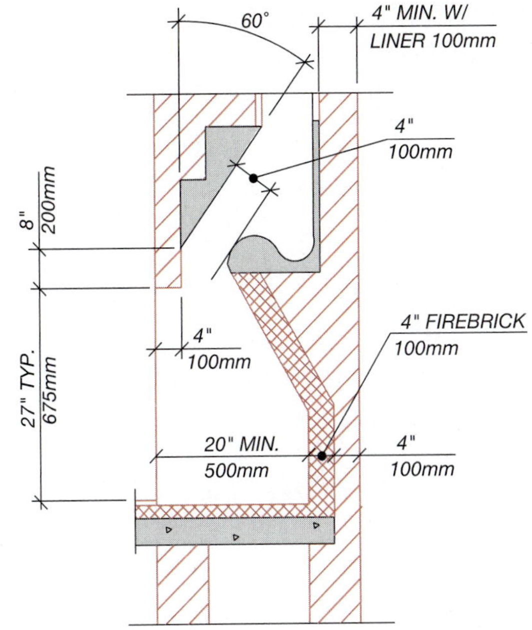

FIGURE 38.12 ■ Minimum sizes required for firebox layout.

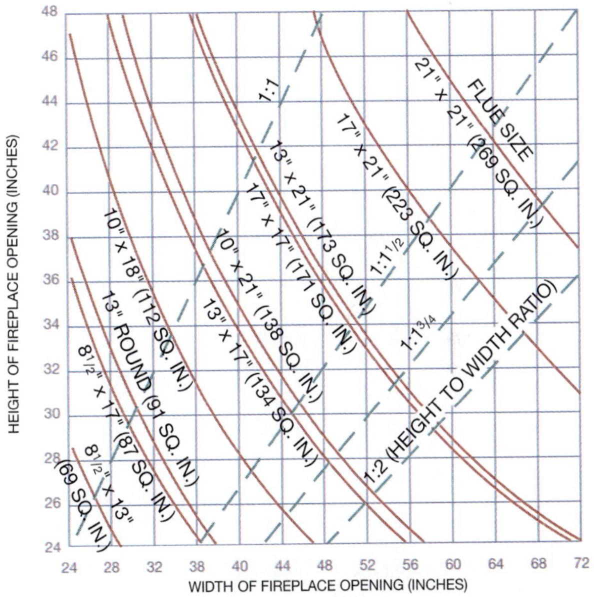

GRAPH TO DETERMINE THE PROPER FLUE SIZE FOR A SINGLE-FACE FIREPLACE, BY MATCHING THE WIDTH AND HEIGHT OF THE FIREPLACE OPENING, THE MINIMUM FLUE SIZE CAN BE DETERMINED. ONCE THE MINIMUM FLUE SIZE HAS BEEN DETERMINED, THE CHIMNEY SIZE CAN BE DETERMINED.

Nominal Dimension of Flue Lining	Actual Outside Dimensions of Flue Lining	Effective Flue Area	Max. Area of Fireplace Opening	Minimum Outside Dimension of Chimney
8 1/2" Round	8 1/2" Round	39 sq. in.	390 sq. in.	17" x 17"
8 1/2" x 13" oval	8 1/2" x 12 3/4"	69 sq. in.	690 sq. in.	17" x 21"
8 1/2" x 17" oval	8 1/2" x 16 3/4"	87 sq. in.	870 sq. in.	17" x 25"
13" Round	12 3/4" Round	91 sq. in.	1092 sq. in.	21" x 21"
10" x 18" oval	10" x 17 3/4"	112 sq. in.	1120 sq. in.	19" x 26"
10" x 21" oval	10" x 21"	138 sq. in.	1380 sq. in.	19" x 30"
13" x 17" oval	12 3/4" x 16 3/4"	134 sq. in.	1340 sq. in.	21" x 25"
13" x 21" oval	12 3/4" x 21"	173 sq. in.	1730 sq. in.	21" x 30"
17" x 17" oval	16 3/4" x 16 3/4"	171 sq. in.	1710 sq. in.	25" x 25"
17" x 21" oval	16 3/4" x 21"	223 sq. in.	2230 sq. in.	25" x 30"
21" x 21" oval	21" x 21"	269 sq. in.	2690 sq. in.	30" x 30"

FIGURE 38.13 ■ Common flue and chimney sizes based on the area of the fireplace opening.

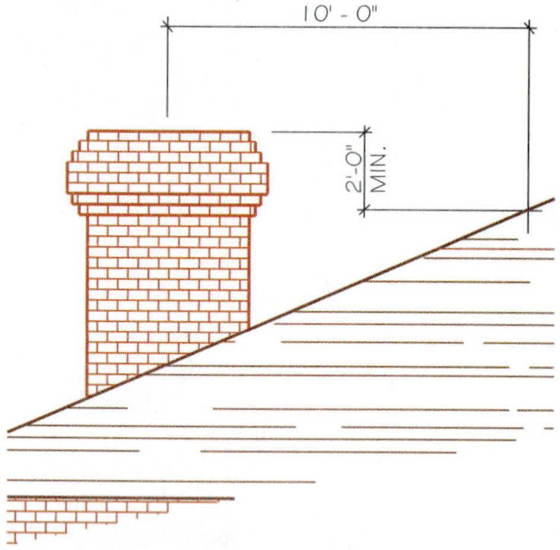

FIGURE 38.14 ■ The chimney is required to extend 2′ (600 mm) above any part of the structure that is within 10′.

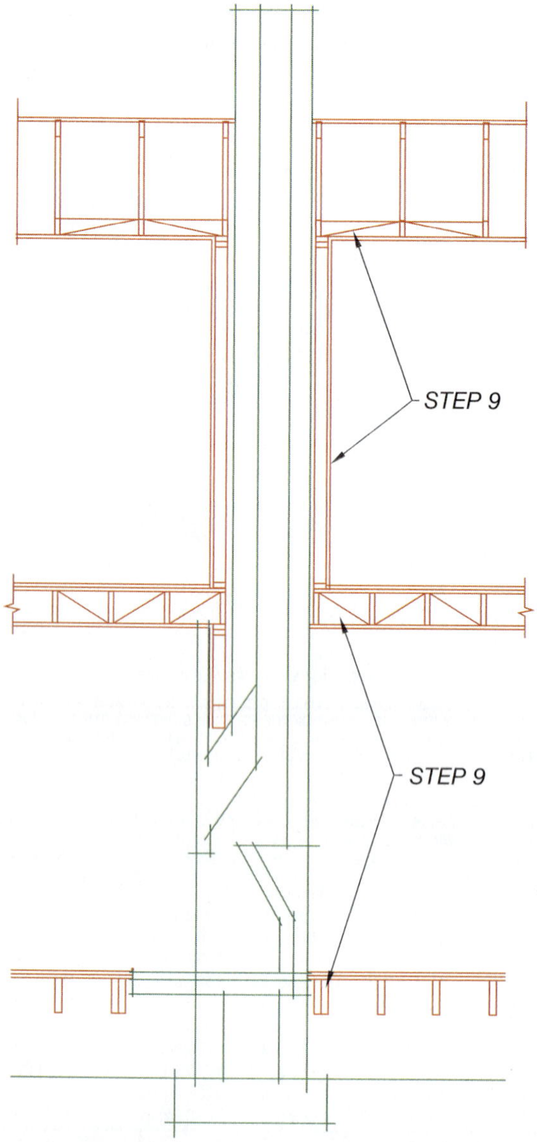

FIGURE 38.15 ■ Draw all framing members with finished-quality lines.

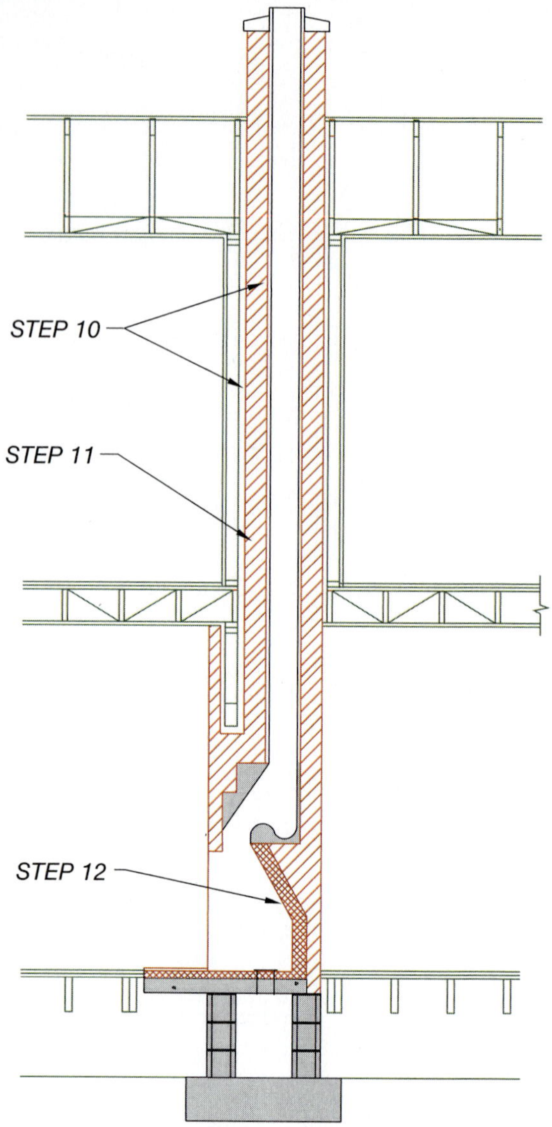

FIGURE 38.16 ■ Darken all masonry materials.

FIREPLACE ELEVATIONS

The fireplace elevation is the drawing of the fireplace as viewed from inside the home. Fireplace elevations typically show the size of the firebox and the material that will be used to decorate the face of the fireplace. Fireplace elevations can be drawn by using the following steps. See Figure 38.19 for steps 1 through 4.

Elevation Layout

STEP 1 Lay out the size of the wall that will contain the fireplace.

STEP 2 Lay out the width of the chimney.

STEP 3 Lay out the height and width of the hearth if a raised hearth is to be used.

STEP 4 Lay out the fireplace opening.

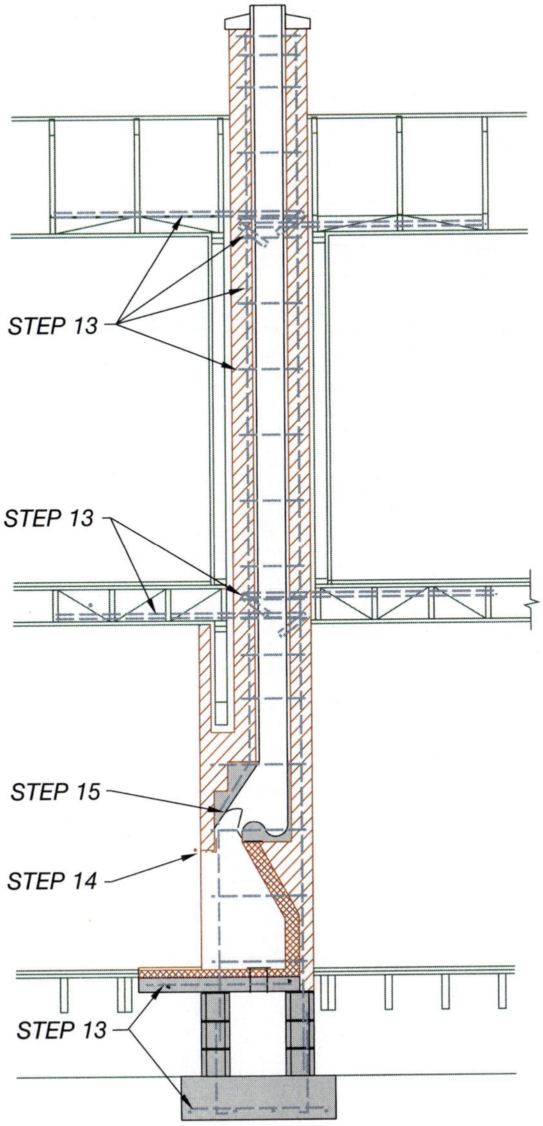

FIGURE 38.17 ■ Draw all reinforcing steel.

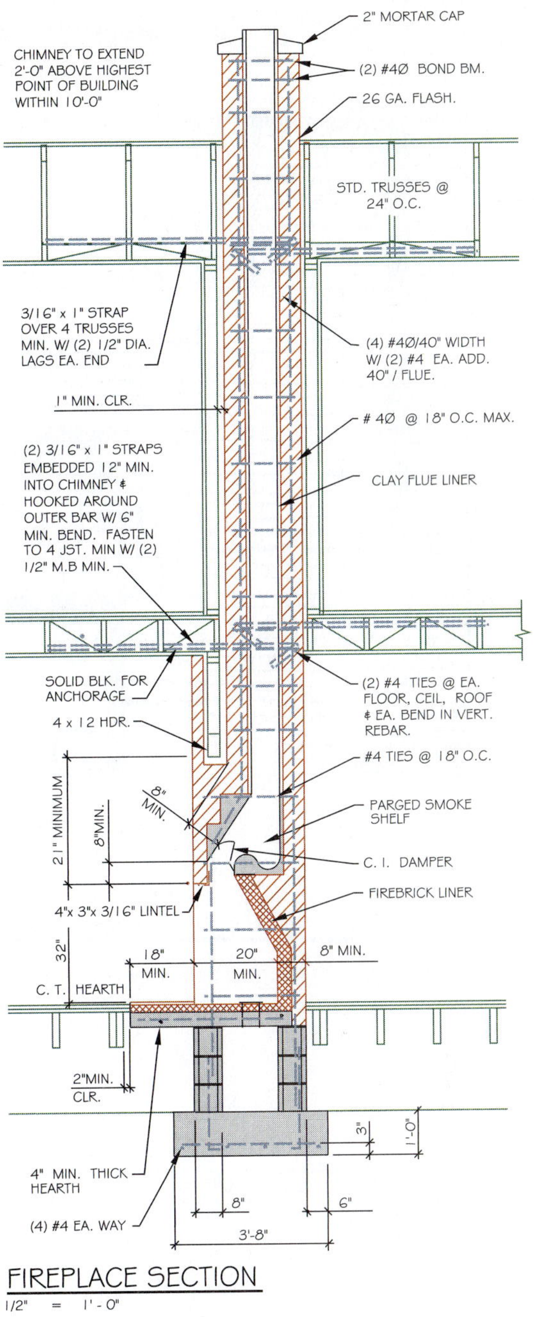

FIGURE 38.18 ■ Place the notes and dimensions.

Finished-Quality Lines

Usually the designer will provide the drafter with a rough sketch similar to Figure 38.20 to describe the finishing materials on the face of the fireplace. By using this sketch and the methods described in Chapter 22, the fireplace elevation can be drawn. See Figure 38.21. Once the finishing materials have been drawn, the elevation can be lettered and dimensioned as in Figure 38.22. Notes will need to be placed on the drawing to describe all materials. Dimensions will be required to describe:

1. Room height
2. Hearth height
3. Fireplace opening height and width
4. Mantel height

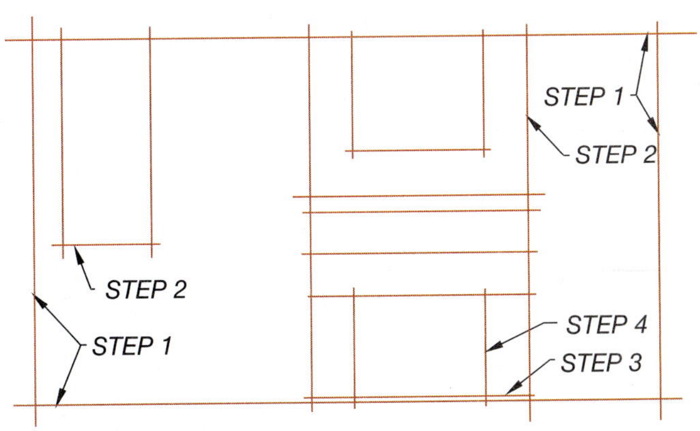

FIGURE 38.19 ■ Fireplace elevation layout.

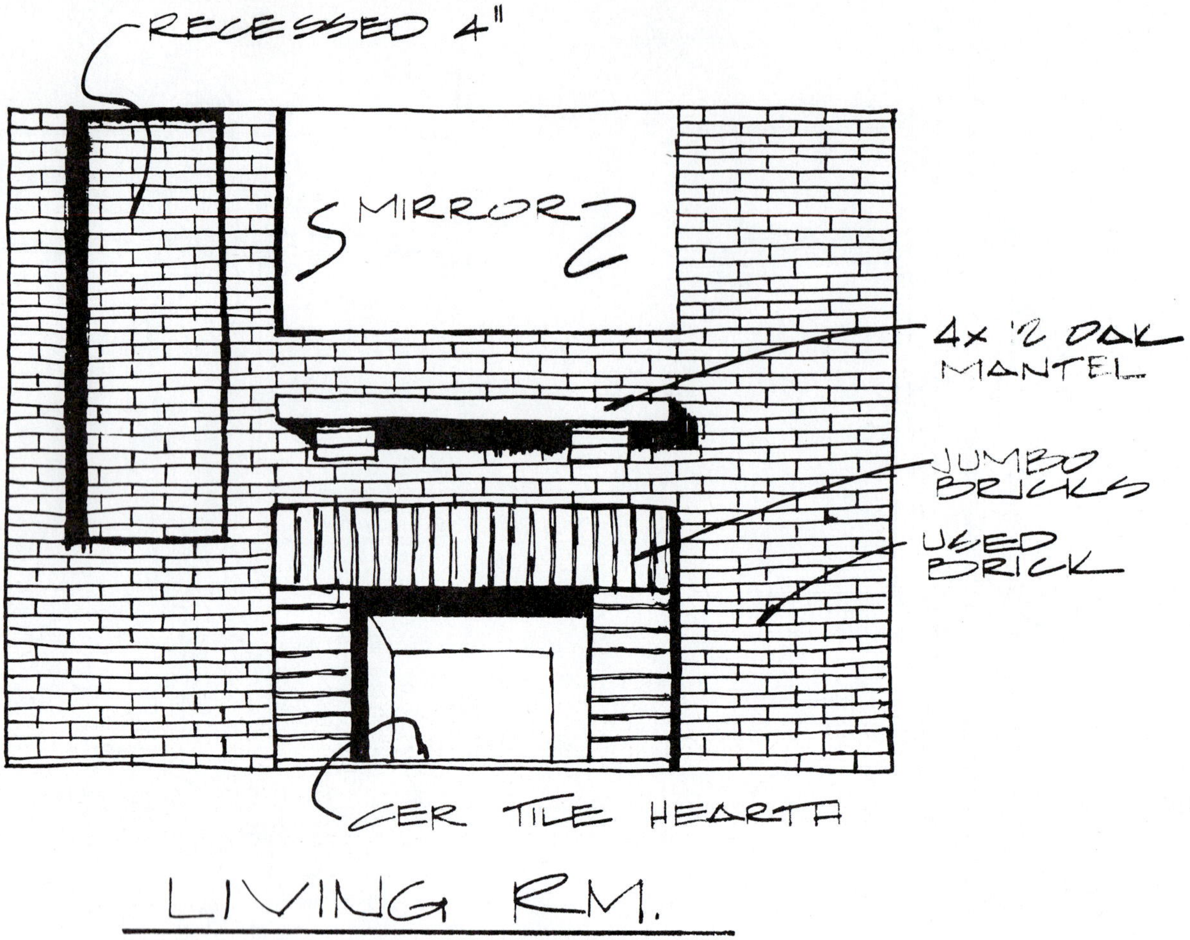

FIGURE 38.20 ■ The drafter can often use the designer's preliminary sketch as a guide when drawing the finishing materials for a fireplace elevation.

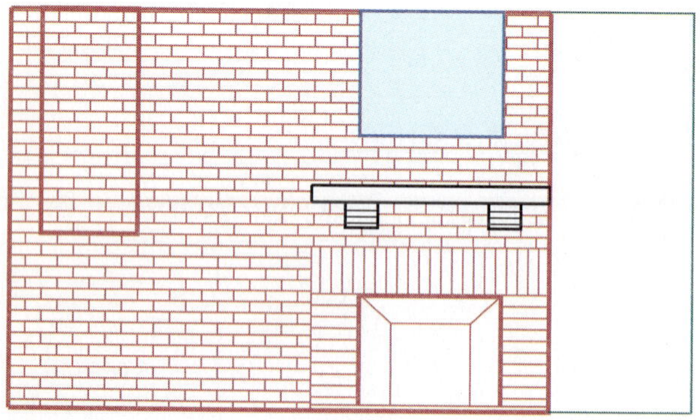

FIGURE 38.21 ■ Drawing the elevation with finished line quality.

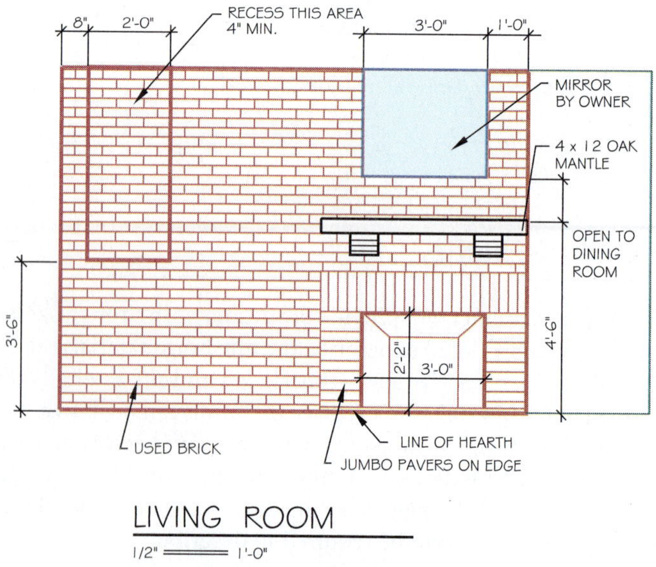

FIGURE 38.22 ■ Lettering and dimensions for a fireplace elevation.

CADD APPLICATIONS

DRAWING FIREPLACE SECTIONS AND ELEVATIONS WITH CADD

In many cases the architectural firm draws a number of standard fireplace sections and details similar to Figure 38.23. These typical sections are then saved in the CADD symbols library and called up when necessary to insert on a drawing. The standard sections may be completely dimensioned and noted, or the dimensions and specific notes may be added after the section is brought into the drawing.

Drawing fireplace elevations with CADD is easy because a fireplace elevation is normally either a fireplace box and a surrounding mantel, or a fireplace box and a masonry wall. In either case the job is simple. If you are drawing fireplace elevations with a surrounding mantel, draw a few standard applications and add one of them to the drawing at any time. Figure 38.24 shows a standard mantel in a fireplace elevation. With most CADD programs, these symbols are dragged onto the drawing, where the computer gives you a chance to scale the display up or down. This provides maximum flexibility for inserting the drawing. Once the standard mantel has been added to the drawing, it can be increased or decreased in size with commands such as SCALE, TRIM, or STRETCH. If the fireplace elevation is a masonry wall, all you do is define the area with lines and add elevation symbols such as stone or brick, as shown in Figure 38.25.

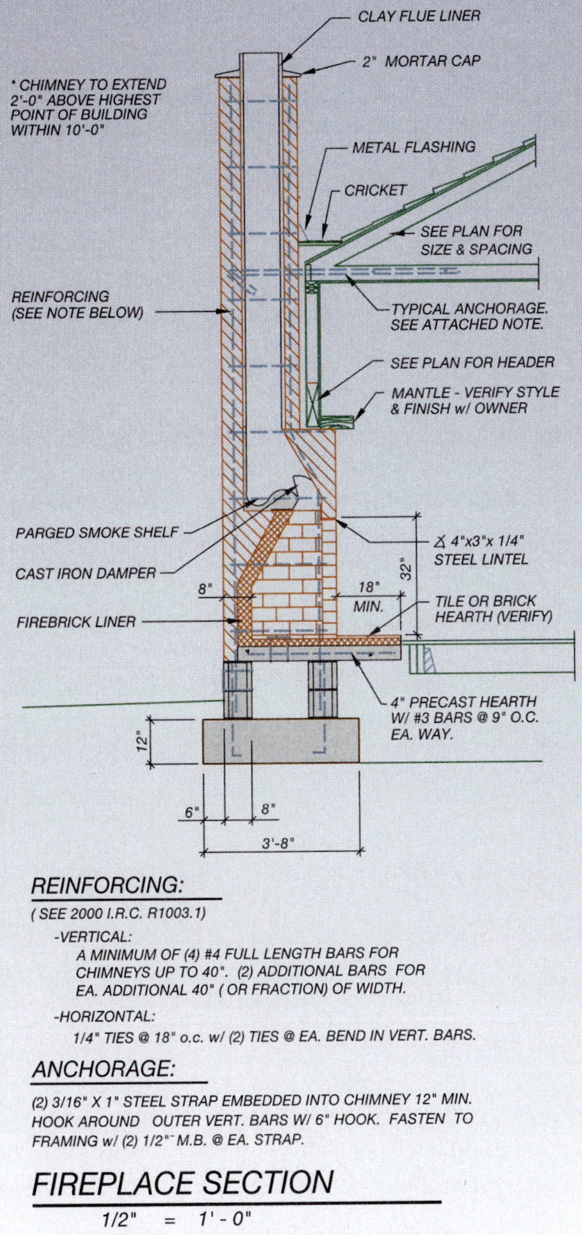

FIGURE 38.23 ■ A CADD fireplace section can be created as a stock detail.

FIGURE 38.24 ■ CADD mantel.

FIGURE 38.25 ■ CADD representation of a fireplace with a masonry front.

Chapter 38 Additional Reading

The following Web sites can be used as a resource to help you keep current with changes in fireplace materials.

ADDRESS	COMPANY OR ORGANIZATION
www.bia.org	Brick Industry Association
www.rumford.com	Buckley Rumford Fireplace
www.centralfireplace.com	Central Fireplace Gas Fireplace
www.fireplace-stove-insert.com	Directory of Fireplace Products
www.eleganceinstone.com	Elegance in Stone Fireplace Surrounds
www.fireplace-centre.com	The Fireplace Centre
www.arcat.com/arcatcos	Heat-N-Glow Fireplace Products, Inc.
www.maconline.org	Masonry Advisory Council
www.vermontcastings.com	Vermont Castings/Majestic Fireplaces

CHAPTER 38

Fireplace Construction and Layout Test

DIRECTIONS

Answer the questions with short, complete statements or drawings as needed.

1. Letter your name and the date at the top of the sheet.
2. Letter the question number and provide the answer. You do not need to write out the question.

QUESTIONS

Question 38–1 What purpose does a damper serve?

Question 38–2 What parts does the throat connect?

Question 38–3 What is the most common size for a fireplace opening?

Question 38–4 What is the required flue area for a fireplace opening of 44″ × 26″ (1118 × 660 mm)?

Question 38–5 What flues could be used for a fireplace opening of 1340 sq in. (864,514 mm^2)?

Question 38–6 How is masonry shown in cross section?

Question 38–7 Why is a fireplace elevation drawn?

Question 38–8 Why is fireplace information often placed in stock details?

Question 38–9 Where should chimney anchors be placed?

Question 38–10 How far should the hearth extend in front of the fireplace with an opening of 7 sq ft (0.65 m^2)?

SECTION 11

Architectural Rendering

39 CHAPTER

Presentation Drawings

INTRODUCTION

The term *presentation drawings* was introduced during the discussion of the design sequence in Chapter 1. As the residence was being designed, the floor plan and a rendering were drawn to help the owners understand the designer's ideas. Presentation drawings are the drawings that are used to convey basic design concepts from the design team to the owner or other interested persons. Presentation drawings are a very important part of public hearings and design reviews as a structure is studied by government and private agencies to determine its impact on the community. In residential architecture, presentation drawings are frequently used to show compliance with review board standards and to help advertise existing stock plans.

Your artistic ability, the type of drawing to be done, and the needs of the client will affect how the presentation drawings will be done and who will produce them. Some offices have an architectural illustrator do all presentation drawings. An illustrator combines the skills of an artist with the techniques of drafting. Other offices allow design drafters to make the presentation drawings. A drafter may be able to match the quality of an illustrator but not the speed. Many of the presentation drawings throughout this book would take an illustrator only a few hours to draw.

TYPES OF PRESENTATION DRAWINGS

The same types of drawings that are required for the working drawings can also be used to help present the design ideas. Because the owner, user, or general public may not be able to understand the working drawings fully, each can be drawn as a presentation drawing to present basic information. The most common types of presentation drawings are renderings, elevations, floor plans, plot plans, and sections.

Renderings

Renderings are the best type of presentation drawing for showing the shape or style of a structure. The term *rendering* can be used to describe an artistic process applied to a drawing. Each of the presentation drawings can be rendered using one of the artistic styles soon to be discussed. *Rendering* can also refer to a drawing created using the perspective layout method, which will be presented in Chapter 40. Although rendering is the artistic process used in a perspective drawing, the term is also often applied to the drawings themselves.

Figure 39.1 shows an example of a rendering. A rendering is used to present the structure as it will appear in its natural setting. Exterior renderings are typically drawn using two-point perspective. A rendering can also be very useful for showing the interior shape and layout of a room, as seen in Figure 39.2. Interior renderings are usually drawn using one-point perspective.

Elevations

A rendered elevation is often used as a presentation drawing to help show the shape of the structure. An example of a presentation elevation can be seen in Figure 39.3. This type of presentation drawing gives the viewer an accurate idea of the finished product, while the working drawing is aimed at giving the construction crew information about the materials that they will be installing. The rendered elevation shows the various changes in surface much better than the working elevation. Although the rendered elevation does not show the depth as well as a rendering does, it gives the viewer a clearer understanding of the project without requiring the time or money needed for a rendering. A rendered elevation usually includes all of the material shown

FIGURE 39.1 ■ The most common type of presentation drawing is a rendering, or perspective, drawing. The rendering presents an image of how the structure will look when complete. *Westbury, Plan 433 © 1994. Courtesy Stephen Fuller American Home Gallery.*

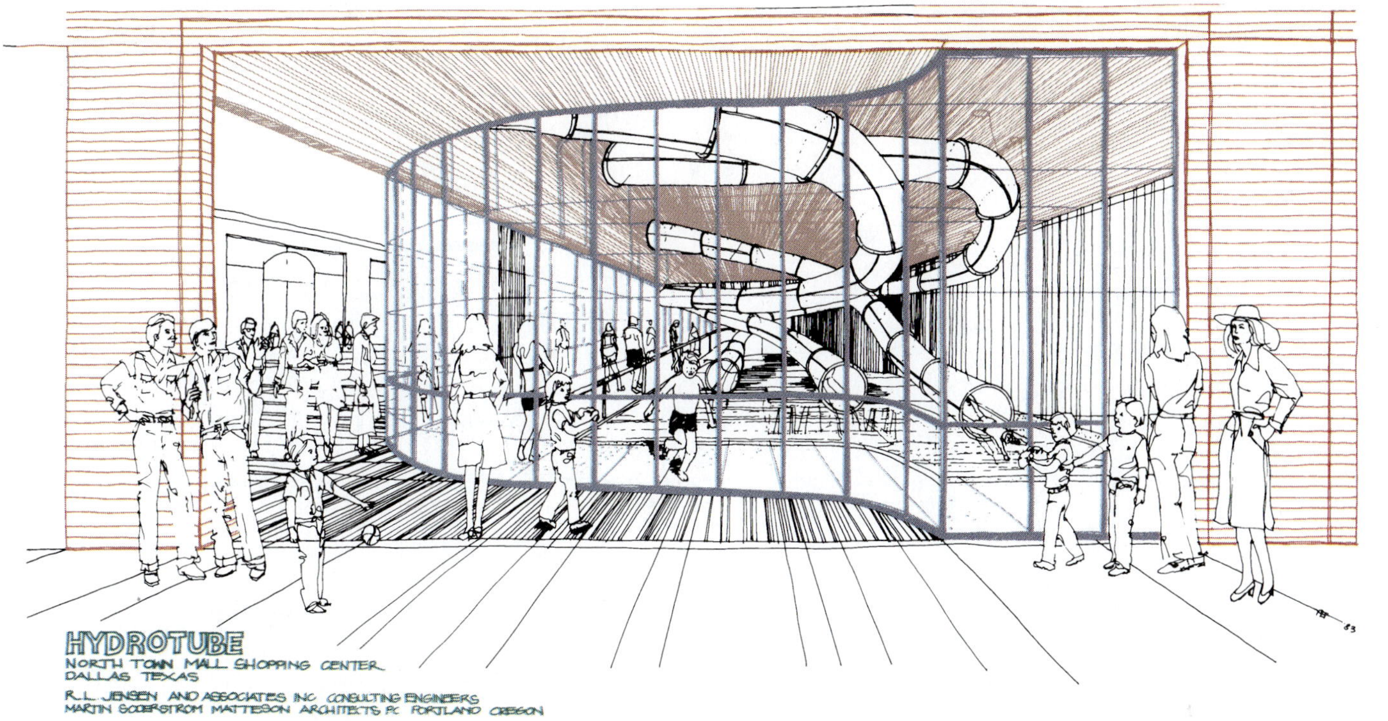

FIGURE 39.2 ■ Renderings are often helpful for showing the relationships of interior spaces. *Courtesy Paul Franks and Martin Soderstrom, Matteson Architects.*

FIGURE 39.3 ■ A presentation elevation helps to show the shape of the structure without taking as much time as a rendering. *Courtesy Design Basics, Inc.*

on a working elevation with the addition of shades and plants. Depending on the artistic level of the drafter, people and automobiles are often shown also.

Floor Plans

Floor plans are often used as presentation drawings to convey the layout of interior space. Similar to the preliminary floor plan in the design process, a presentation floor plan is used to show room relationships, openings such as windows and doors, and basic room sizes. Furniture and traffic patterns are also usually shown. Figure 39.4 provides an example of a presentation floor plan.

Site Plans

A rendered site plan is used to show how the structure will relate to the job site and to the surrounding area. The placement of the building on the site and the north arrow are the major items shown. As seen in Figure 39.5, streets, driveways, walkways, and plantings are usually shown. Although most of these items are shown on the working site plan, the presentation site plan shows this material more artistically than the site plan in the working drawings.

Sections

Sections are often part of the presentation drawings, to show vertical relationships within the structure. As seen in Figure 39.6, a section can be used to show the changes of floor or ceiling levels, vertical relationships, and sun angles. Working sections show these same items with emphasis on structural materials. The presentation sections may show some structural material, but the emphasis is on spatial relationships.

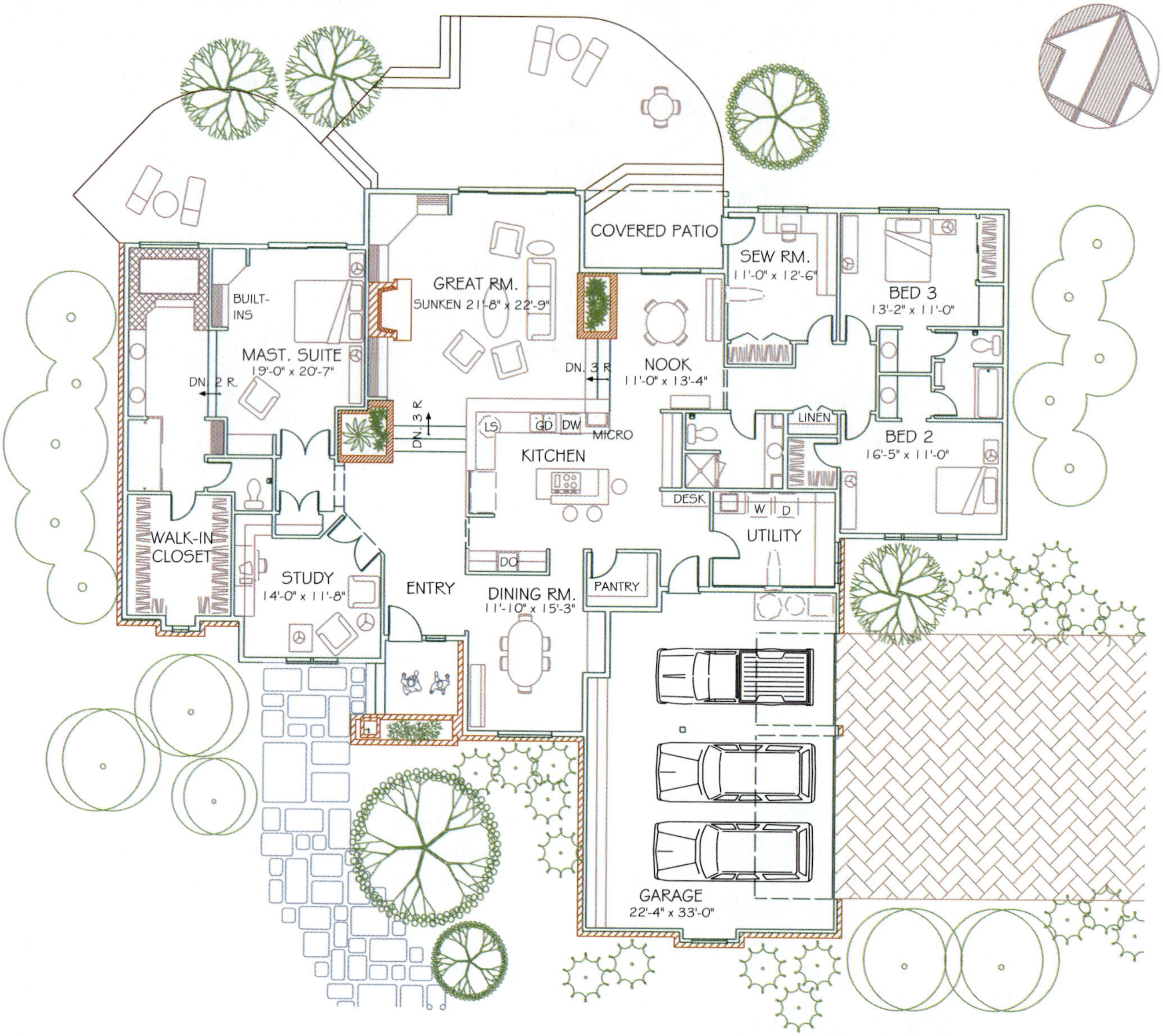

FIGURE 39.4 ■ A floor plan is often part of the presentation drawings and is used to show room relationships.

METHODS OF PRESENTATION

No matter what type of drawing is to be made, it can be drawn on any one of several different media. Drawings are made using different materials and line techniques.

Common Media

Common media for presentation drawings include sketch paper, vellum, Mylar, illustration board, and CADD. Sketch paper is often used for the initial layout stages of presentation drawings, and some drawings are even done in their finished state on sketch paper. Most drawing materials will adhere to sketch paper, but it is not very durable. Drafting vellum can be a durable drawing medium but will wrinkle when used with some materials. Vellum is best used with graphite or ink. Polyester film is often used in presentation drawings because of the ease of correcting errors. Polyester film provides both durability and a good-quality surface for making photographic reproductions. Ink is the most common element used on polyester film, but some types of watercolors and polyester leads can also be used.

Illustration board is extremely durable and is a suitable surface for all types of drawing materials. When the materials are to be photographed, a white board is usually used. Beige, gray, and light blue board also can give presentation drawings a pleasing appearance. Because of the difficulty of correcting mistakes, illustration board is not usually used by beginners. Until experience is gained, presentation drawings can be drawn on vellum and then mounted on illustration board.

When drawing on illustration board, the drafter must plan the arrangement of the entire layout carefully prior to placing

FIGURE 39.5 ■ A rendered plot plan is used to show how structures relate to the job site. *Courtesy Paul Franks.*

FIGURE 39.6 ■ Presentation sections are used to show vertical relationships. *Courtesy Home Planners, Inc.*

the first lines. Drawings are usually done on sketch paper first and then transferred to the illustration board. To transfer a drawing onto the board, lightly shade the back side of the sketch with a soft graphite lead, which will act as carbon paper when the original lines are traced to transfer them onto illustration board. To avoid smearing, graphite should be placed only over each line to be transferred, not over the entire back side of the drawing.

A pleasing effect can be achieved by combining media. A rendering can be drawn using overlay principles and combining vellum or polyester film and illustration board. For example, a structure may be drawn on illustration board and landscaping in color on polyester film. Because of the airspace between the polyester film and the illustration board, an illustration of depth is created.

Drawing Materials

Common materials used for presentation drawings are graphite, ink, colored pencil, felt-tip pens, watercolor, and CAD-generated drawings.

Because of the ease of correcting errors, graphite is a good material for the beginning drafter to experiment with. Common leads used for presentation work are 4H or 5H for layout work and 2H, H, F, HB, and 4B for drawing object lines. The choice of lead depends on the object to be drawn and the surface of the media being drawn upon. Figure 39.7 shows different qualities that can be achieved with graphite on vellum.

Because of its reproductive qualities, ink is commonly used for presentation drawings. Ink lines have a uniform density that can be easily reproduced photographically. Careful planning is required when working with ink to ensure that the drawing is properly laid out prior to inking and to determine the order in which the lines will be drawn. With proper planning, you can draw in one area while the ink dries in another area.

The complexity of a drawing affects the size of the pen points to be used. The technique to be used dictates which pen points will be used. Common points used for presentation work are numbers 0, 1, and 2. Many drafters and illustrators also add 000, 00, and number 4 points to get a greater variation in line contrast. Figure 39.8 shows examples of various line weights available for ink work. Figure 39.9 shows an elevation rendered in ink.

Color is often added to a presentation drawing by the use of colored pencils, markers, or pastels. The skill of the drafter and the use of the drawing will affect where the color is to be placed. Color is usually added to the original drawing when the illustration is to be reproduced. When the drawing is to be displayed, color is often placed on the print rather than on an original drawing. Best results are achieved when color is used to highlight a drawing rather than for every item in the drawing. The use of watercolors is common among professional illustrators because of the lifelike presentations that can be achieved. See Figure 39.10.

Line Techniques

Two techniques commonly used for drawing lines in presentation drawings are *mechanical* and *freehand.* They use the same methods for layout but vary in the method of achieving finished line quality. Figure 39.11 shows an example of a rendering drawn me-

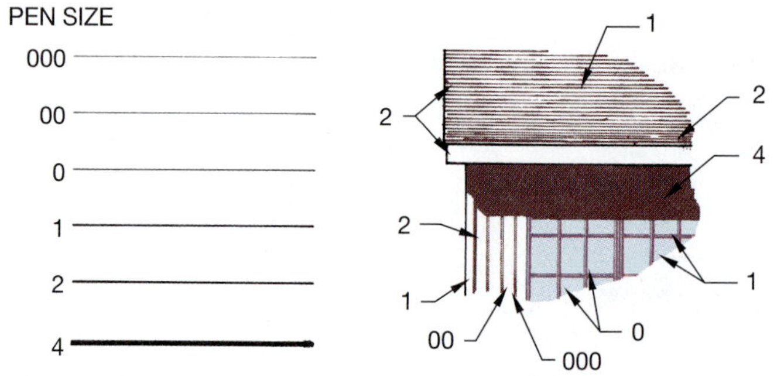

FIGURE 39.8 ■ Common pen point sizes used for presentation work. The size of the drawing and the amount of detail to be shown affect the size of point.

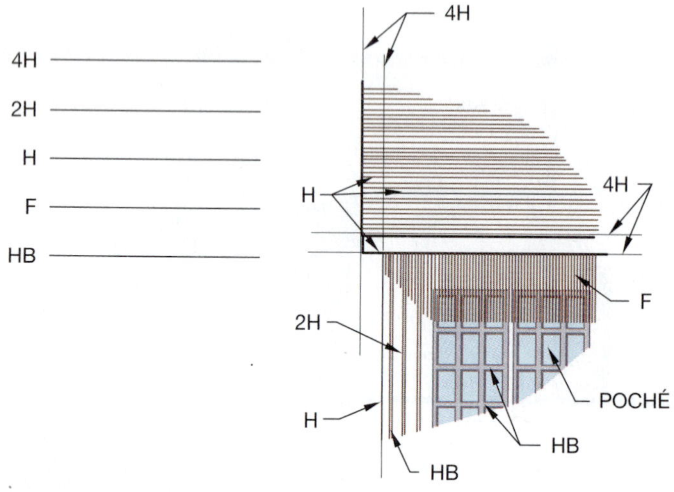

FIGURE 39.7 ■ Types of graphite lines that are typically used on renderings.

FIGURE 39.9 ■ Renderings drawn with ink and varied line widths can be used to represent varied materials. *Greenwood, Plan 542-b © 1996. Courtesy Stephen Fuller American Home Gallery.*

Banfield Hall, Plan No. 286
©1992 Stephen Fuller American Home Gallery, Ltd.

FIGURE 39.10 ■ Renderings drawn with opaque watercolor can be reproduced easily in color or black-and-white. *Banfield Hall, Plan 286 © 1992. Courtesy Stephen Fuller American Home Gallery.*

FIGURE 39.11 ■ Renderings drawn with tools to provide straight lines.

Lenox Retreat, Plan No. 146
©1988 Stephen Fuller American Home Gallery, Ltd.

FIGURE 39.12 ■ Presentation drawings are often drawn using freehand methods. *Lenox Retreat, Plan 146 © 1988. Courtesy Stephen Fuller American Home Gallery.*

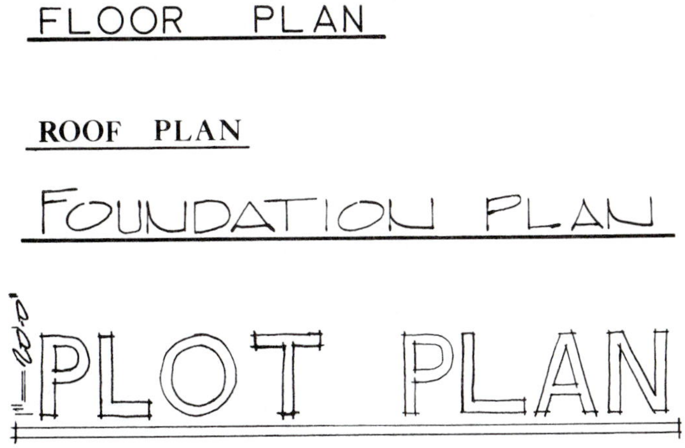

FIGURE 39.13 ■ Lettering can be drawn by using mechanical methods such as a LeRoy Lettering Guide or rub-ons, or with freehand techniques.

chanically using a straightedge to produce lines. If a more casual effect is desired, initial layout lines can be traced without using tools. Figure 39.12 shows an example of a freehand rendering.

In addition to line methods, many styles of lettering are also used on presentation drawings. Lettering may be placed with mechanical methods such as a lettering guide or rub-ons. Many illustrations are lettered with freehand lettering similar to Figure 39.13.

RENDERING PROCEDURE FOR DRAWINGS

Because of the number of steps required to make a perspective drawing, one- and two-point perspective will be covered in Chapters 40 and 41. Other drawings can be rendered as follows.

Elevations

Presentation elevations can be drawn by following the initial layout steps described in Chapter 22. The elevations are usually drawn at the same scale as the floor plan. Once the elevation is drawn with construction lines, the drafter should plan the elements to be included in the presentation. For our example, these will be limited to plants and shading. Care must be taken to ensure that the surroundings in the drawing do not overshadow the main structure.

Plants

Your artistic ability will determine how plants are to be drawn. Rub-on plants are a great time-saver, but they should not be your only means of representing plants because they are expensive and offer only a limited selection.

The method of drawing plantings should match the method used to draw the structure. No matter what method is used, plants are typically kept very simple. The area of the country where the structure will be built determines the types of plants

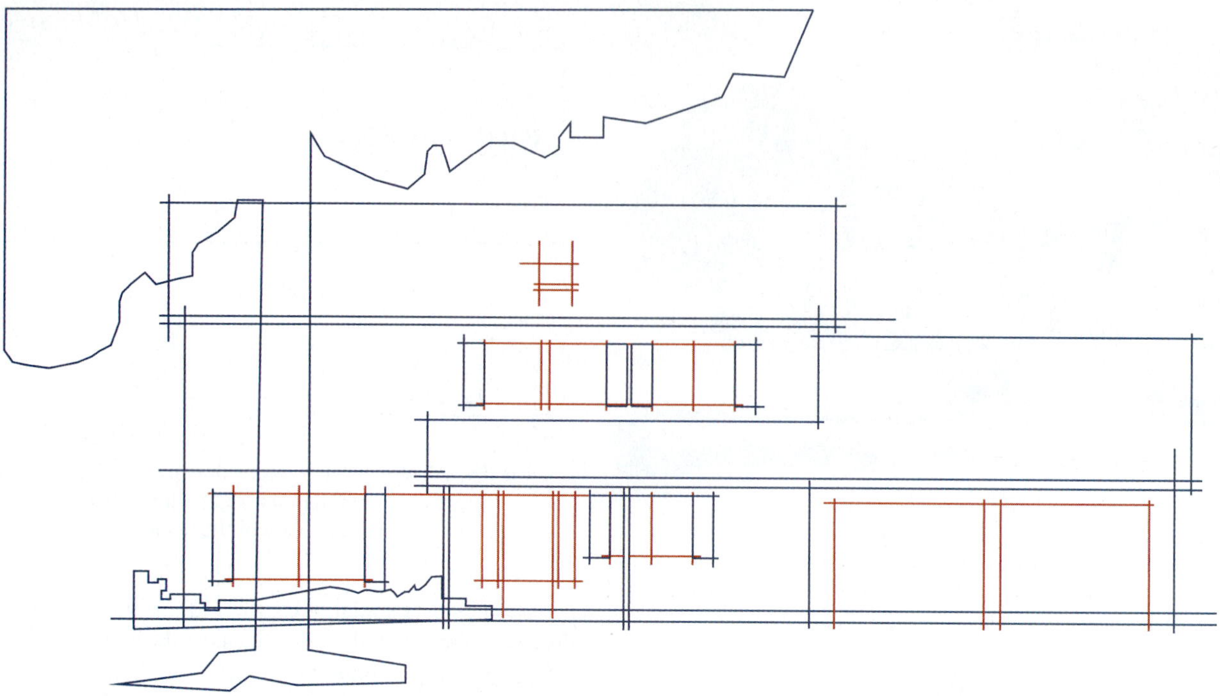

FIGURE 39.14 ■ With the elevation drawn with construction lines, trees in the foreground should be drawn. Be sure to draw trees that actually grow in the vicinity.

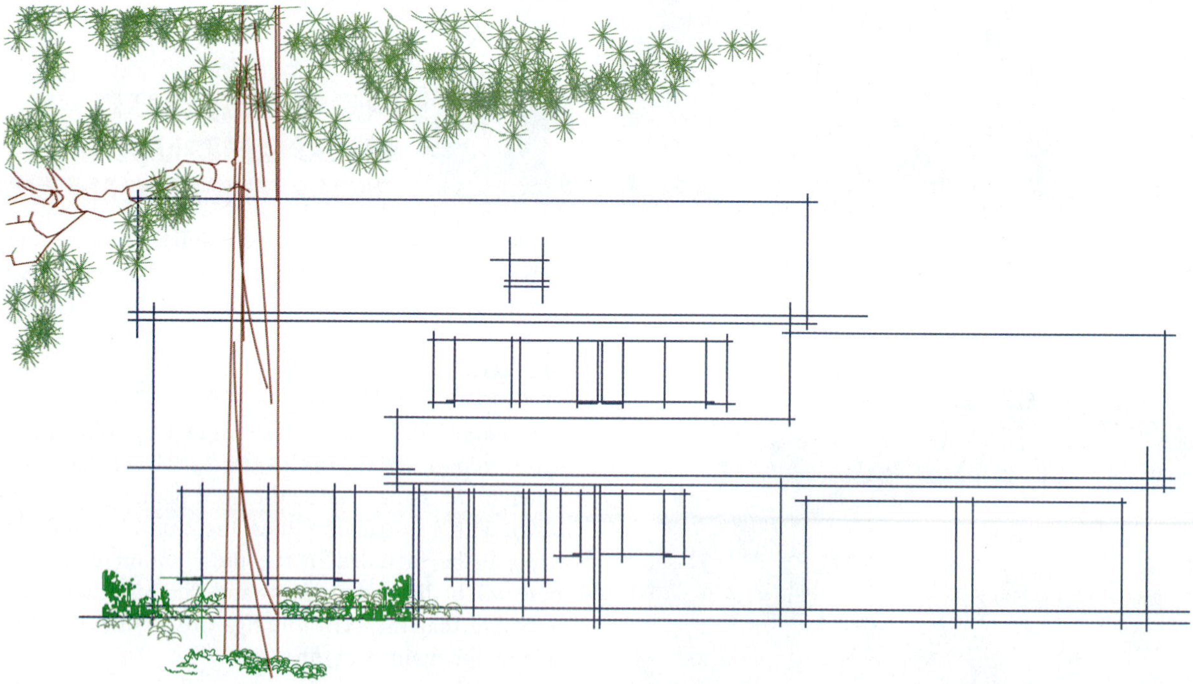

FIGURE 39.15 ■ Plantings that will be in the foreground should be drawn before drawing the elevation.

that are shown on the elevation. Common types of trees that can be placed on the elevations can be seen in the example in this chapter. Be sure to use plants that are typically seen in your area.

Start the presentation elevations using the same methods as on the working drawings. Sketch the area where the plants will be placed, as in Figure 39.14. Plants that will be in the foreground should be drawn prior to drawing the structure with finished-quality lines. Figure 39.15 shows the foreground planting in place. Once the foreground trees have been drawn, the material on the elevations can be drawn by following the same

FIGURE 39.16 ■ The presentation elevation can be completed using steps similar to those used to draw elevations in Chapter 23. For clarity, material in the background, such as a chimney, is often omitted.

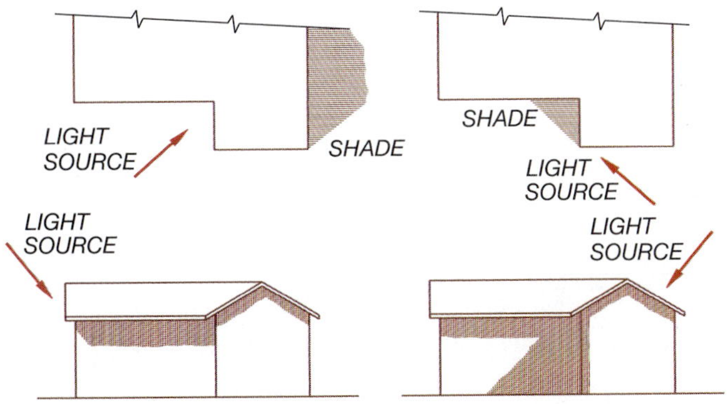

FIGURE 39.17 ■ The shape of the structure influences the selection of the light source. Select a source that will accent depth but will not dominate a surface.

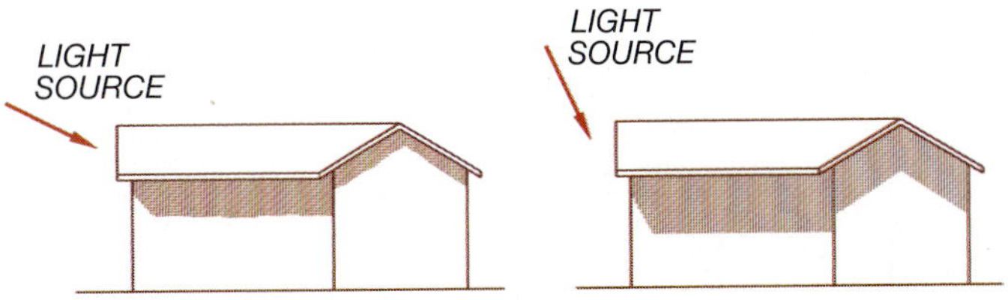

FIGURE 39.18 ■ As the light source is moved above the horizon line, the depth of the shadow will be increased.

steps as for the working drawings. The elevation should now resemble the drawing in Figure 39.16.

Shading

Shadows are often used on presentation drawings to show surface changes and help create a sense of realism. Before shadows can be drawn, a light source must be established. A light source should be selected that will give the best presentation. The source of light will be influenced by the shape of the building. This can be seen in Figure 39.17. Even though the sun may strike the structure as shown on the right, so little of the structure remains unshaded that it is hard to determine the appearance of the building. By moving the light source to the other side of the drawing, as seen at the left, more of the features of the building surface are identified.

Once the light source has been selected, the amount of shadow to be seen should be determined. Figure 39.18 shows the effect of moving the light source away from the horizon. Determine where shadows will be created and the amount of shade that will be created by projecting a line from the light source across the surface to be shaded, as in Figure 39.19. If you are not happy with the amount of shade, raise or lower the light source. Once the first shadow is projected, the proportions of the other shadows must be maintained. Be sure to use a very light line as you are lining out the location of shadows. If you are using CADD, freeze the layer containing the outline.

Shade can be drawn using several different methods. Figure 39.20 shows common methods of using graphite or ink. Shade

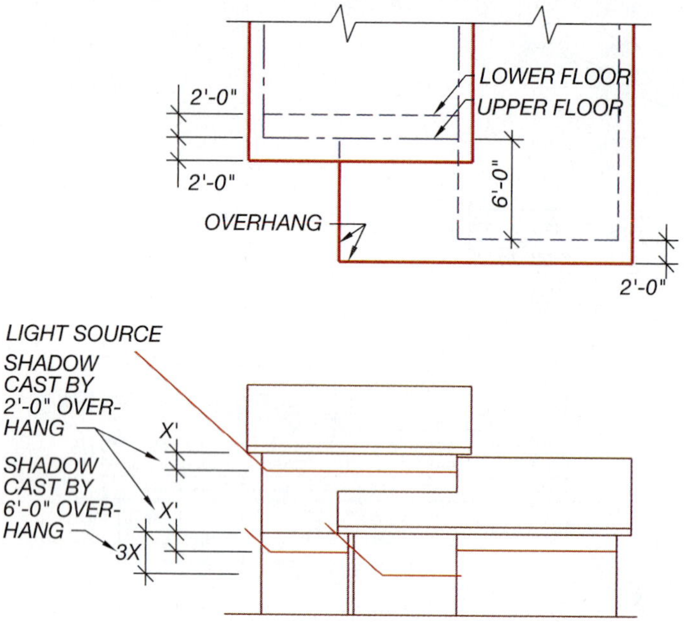

FIGURE 39.19 ■ Shadows are placed by determining a light source. The depth of the shadow can be determined by the illustrator. Select a depth that does not cover important information. Once the first shadow has been drawn, keep other shadows in proportion. Keep shadow outlines very light.

can be applied by CADD using a hatch pattern consisting of vertical parallel lines. Determine the method of shading you will be using, and then lay out the area where the shadows will be placed. The final procedure is to draw the small shadows created by changes in surface material. Figure 39.21 shows common areas that should receive shading.

With all material now drawn and rendered, basic materials should be specified. Complete specifications do not need to be given for the products to be used, but general specifications should be included. Figure 39.22 shows examples of notes that are often placed on the elevations.

Floor Plan

The use of the presentation floor plan will determine the scale at which it will be drawn. If the plan is to be mounted and displayed, the layout space will determine the scale to be used. When the plan will be reproduced, it is typically drawn at a scale of either 1/4″ = 1′–0″ or 1/8″ = 1′–0″ and then photographically reduced. The presentation floor plan can be drawn using many of the same steps that were described in Chapter 15.

Once the walls are drawn, the symbols for plumbing and electrical appliances can be drawn. Closets and storage areas are usually represented on the floor plan. These symbols can

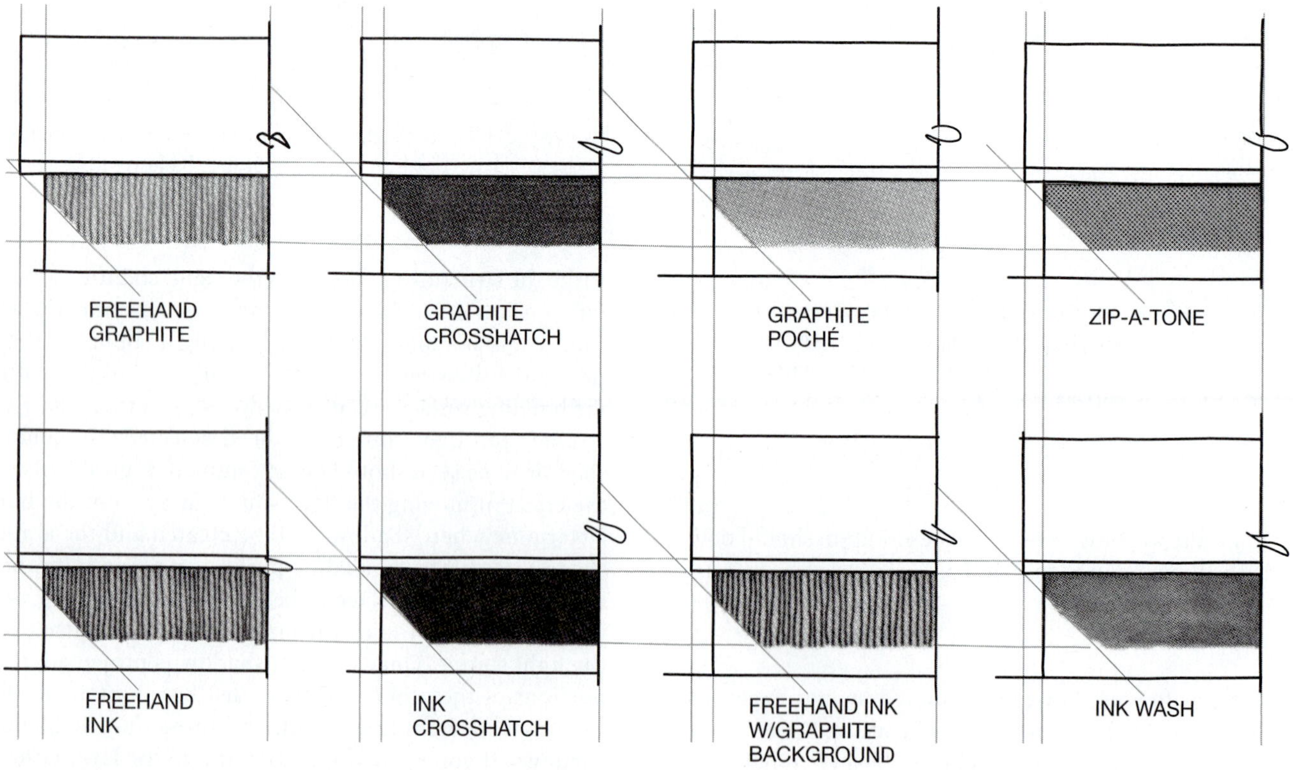

FIGURE 39.20 ■ Shading can be drawn using several methods.

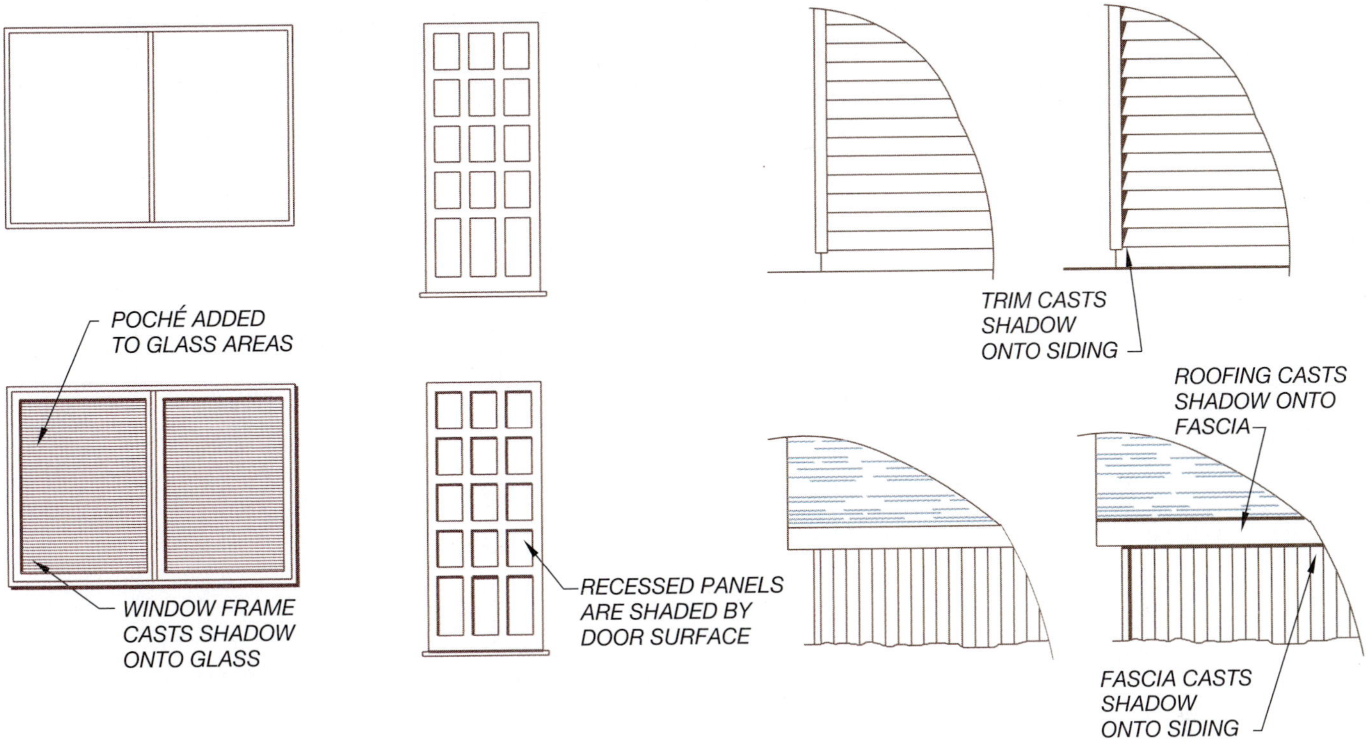

FIGURE 39.21 ■ Shading can affect the realism of a drawing. Shadows will be cast at each surface change.

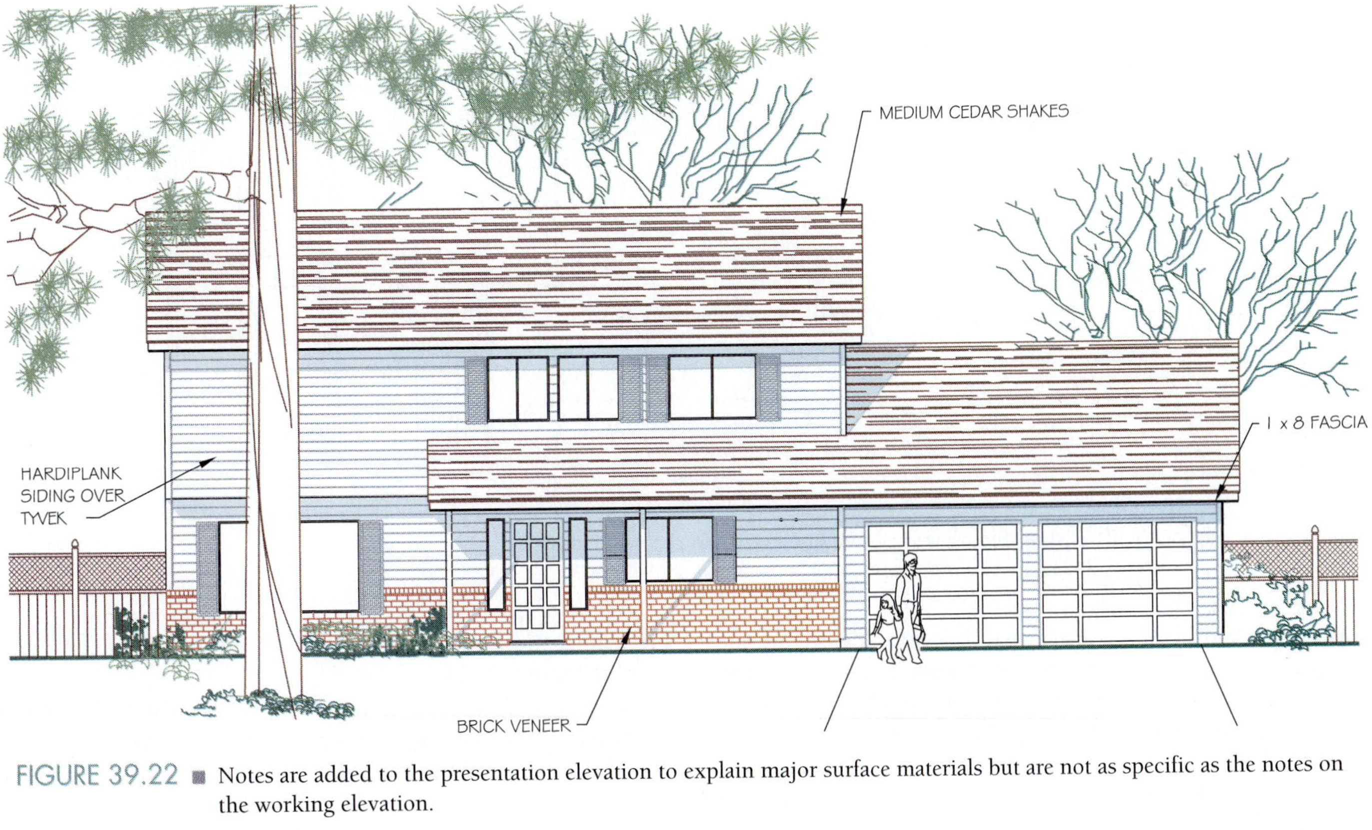

FIGURE 39.22 ■ Notes are added to the presentation elevation to explain major surface materials but are not as specific as the notes on the working elevation.

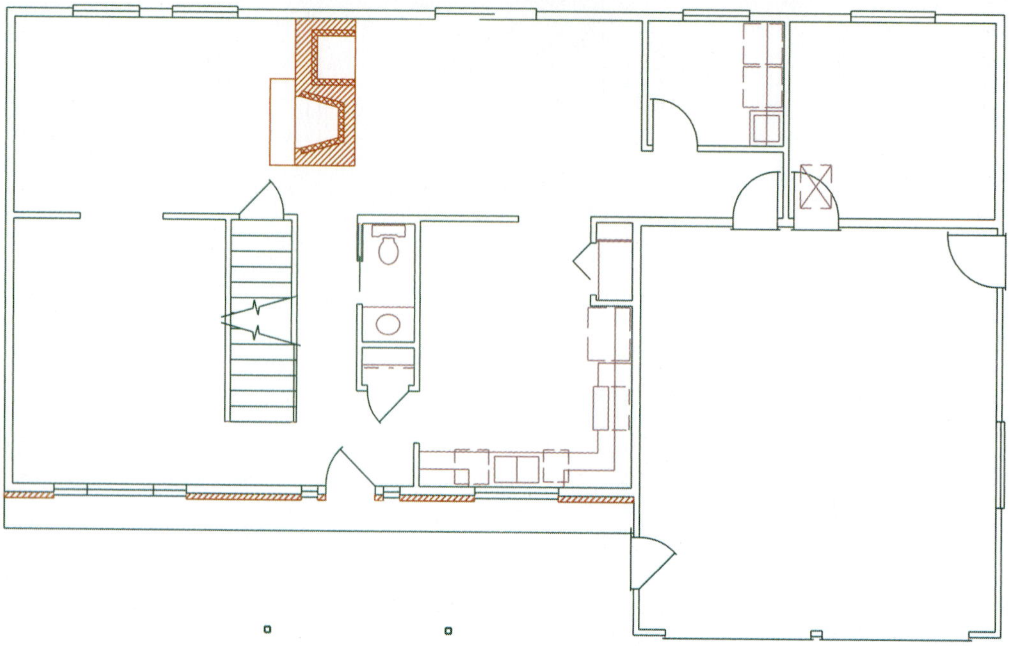

FIGURE 39.23 ■ With the walls, door, and windows drawn, the cabinets, stairs, and other interior features can be added to the presentation floor plan.

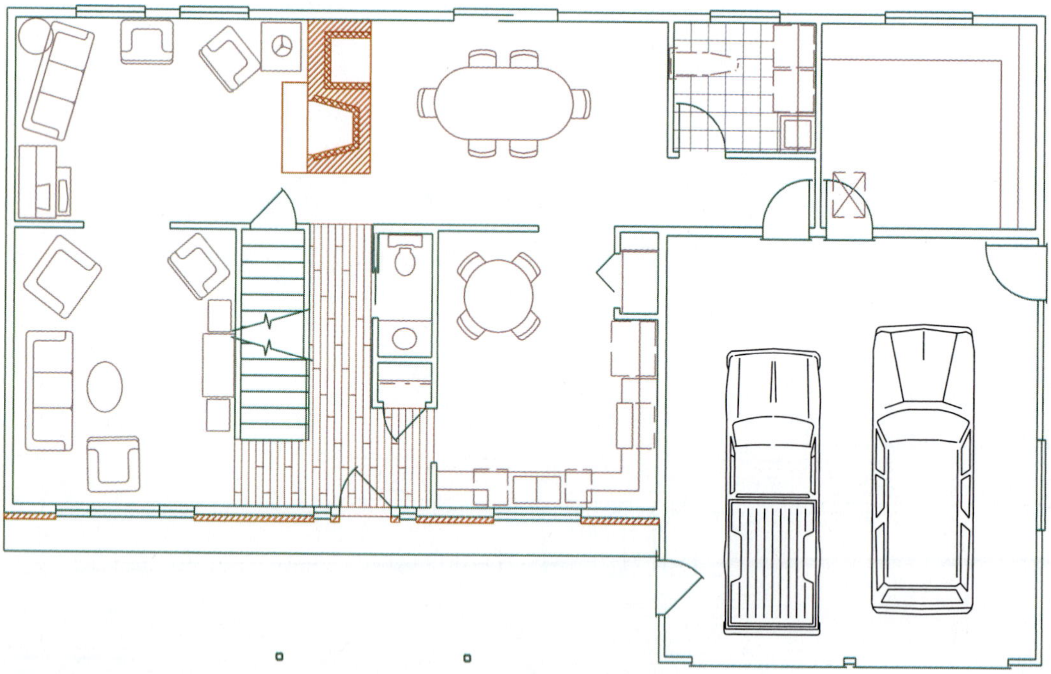

FIGURE 39.24 ■ Floor materials are drawn to help clarify the design.

be seen in Figure 39.23. Furniture is often shown on the floor plan to suggest possible living arrangements. Furniture can be sketched, drawn with templates, put in place with rub-ons, or inserted using pregenerated computer blocks. When brick, stone, or tile is used for a floor covering, it is often shown on the floor plan. A directional arrow may be shown on a floor plan to help orient the viewer to the position of the sun or to surrounding landmarks. The symbols for brick and tile floor covering have been added to the floor plan of Figure 39.24.

Written specifications on the floor plan are usually limited to room types and sizes, appliances, and general titles. Landscaping, decks, and walkways are usually represented on the floor plan to help show how the home will relate to the building site. Figure 39.25 shows the completed presentation floor plan for the plan that was drawn in Chapter 16.

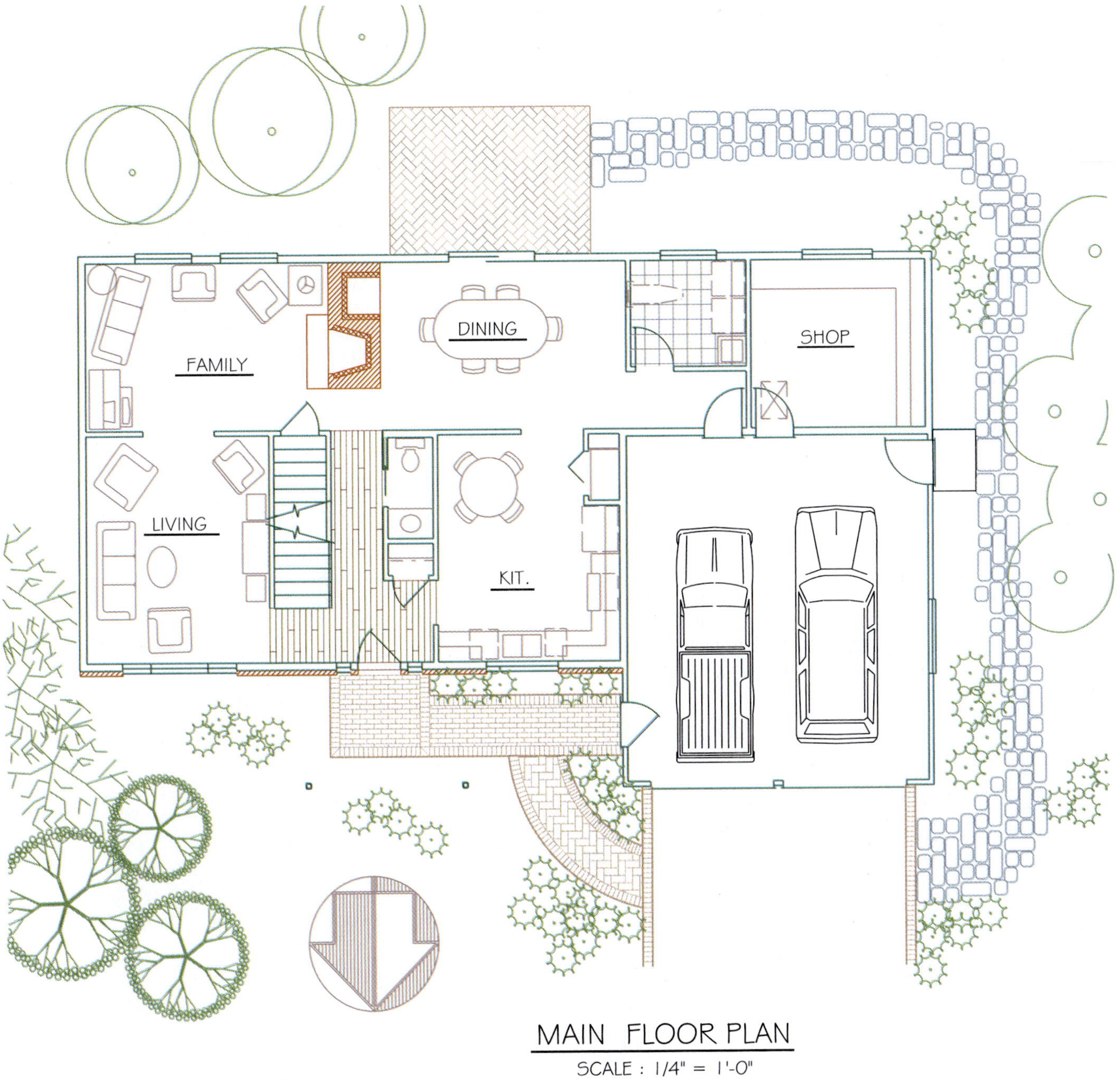

FIGURE 39.25 ■ The completed presentation floor plan is useful for showing the arrangement of interior space.

Site Plan

The site plan can be drawn by following many of the steps in Chapter 13. Once the lot and structure have been outlined, walks, driveways, decks, and pools should be drawn.

Once the structure is located, plantings and walkways are usually indicated on the plan, as shown in Figure 39.26. As with the other types of presentation drawings, basic sizes are specified on the site plan, as shown in Figure 39.27.

Sections

Section presentation drawings are made by using the methods described in Chapters 35 and 36. Sections are usually drawn at a scale similar to the presentation floor plan. The initial section layout for the house discussed in Chapter 16 can be seen in Figure 39.28. Lines of sun angles at various times of the year and lines of sight can be indicated on the section. Room names and ceiling heights are also specified on the section. Methods of presenting this type of information can be seen in Figure 39.29.

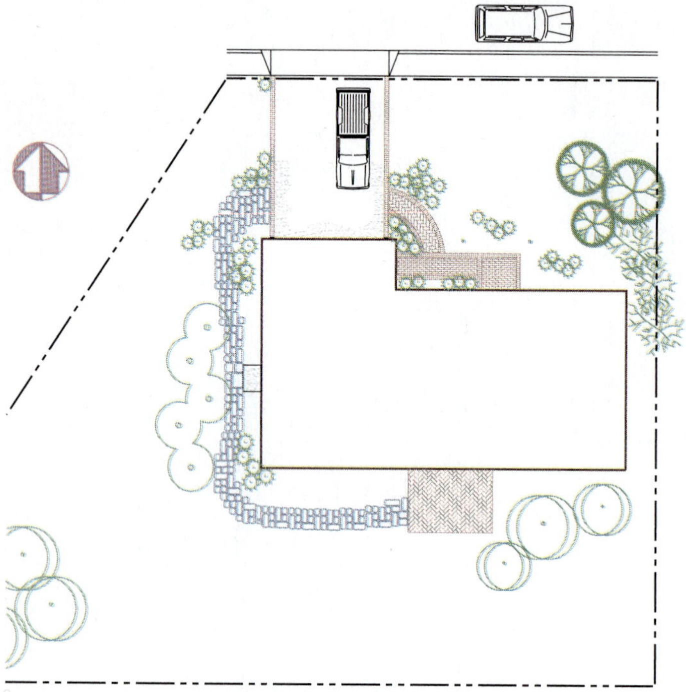

FIGURE 39.26 ■ Landscaping, decks, walks, and any other outdoor materials should be included on the presentation site plan to show how the structure will blend with its surroundings.

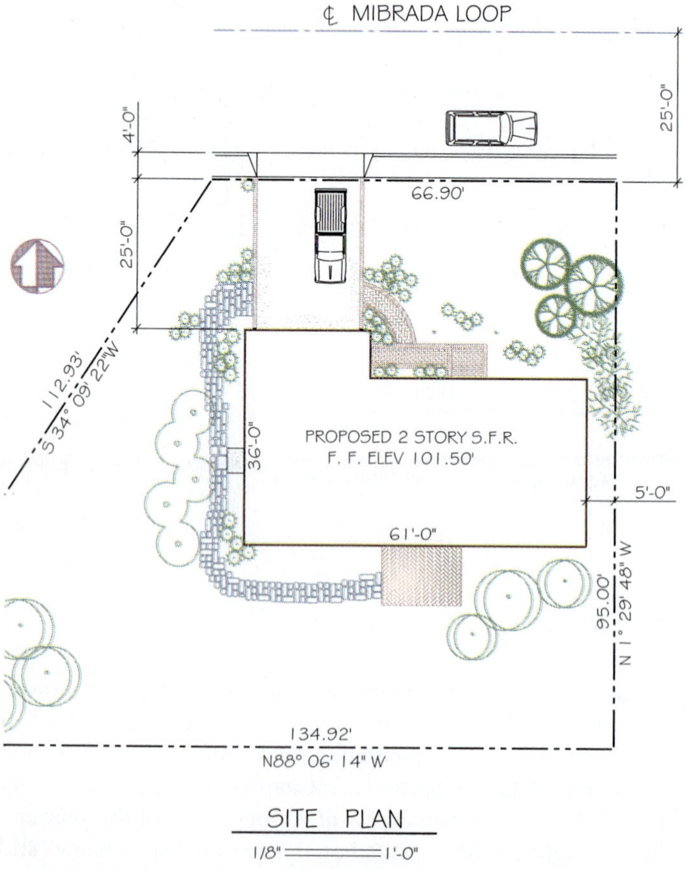

FIGURE 39.27 ■ Major items should be specified on the site plan to help the viewer better understand the project.

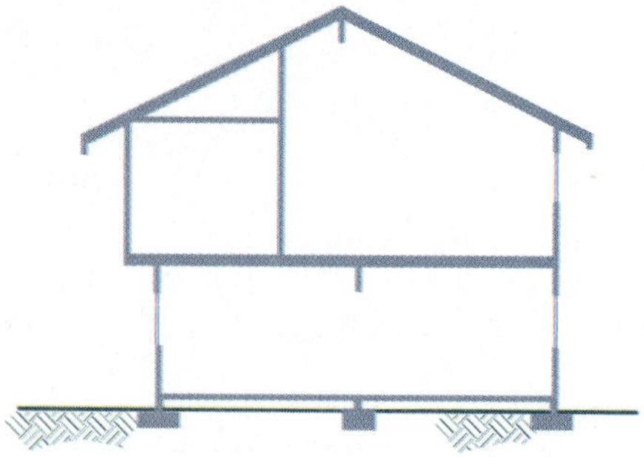

FIGURE 39.28 ■ The initial layout of the section can be drawn using procedures similar to those used for the layout of the working section in Chapters 35 and 36.

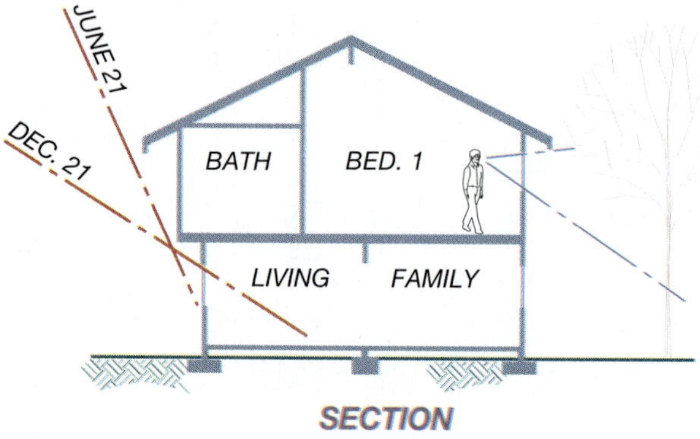

FIGURE 39.29 ■ The presentation section is often used to show sun angles and views. Label rooms so that the viewer will understand which areas are being shown.

USING CADD TO MAKE PRESENTATION DRAWINGS

When CADD is used to draw elevations (Chapters 22 and 23), the material representations look so realistic that they may also be used as presentation drawings. Most architectural drafters add materials such as trees and cars to the drawings, because with CADD these items can be placed on the drawing in seconds. You simply pick a symbol from a menu library and add it to the elevation in the desired location. This process is so easy that elevation drawings often contain such presentation features as landscaping and shading.

Preparing three-dimensional (3D) drawings is easy with CADD. Some of the powerful CADD packages allow you to generate a 3D drawing from the floor plan layout automatically. Wall and header heights are established as you draw the floor plan. When it is complete, you can view the drawing in 3D from any selected point in space. You can change the viewpoint until you find the view that displays the building best. Some architects have found that a combination of 3D CADD presentation of the building and the artistic addition of landscaping features creates beautiful presentation drawings in less time.

There are also CADD rendering programs available that let you turn 3D line drawings into realistically shaded pictures. You can create a drawing that looks like a photographer's image by adjusting camera and lighting locations. You can produce both color and black-and-white drawings with this process. Figure 39.30 shows a drawing rendered with computer software.

FIGURE 39.30 ■ Renderings using third party software create a realistic drawing using Auto CAD. *Courtesy Graphics Software, Inc.*

CADD APPLICATIONS

CHAPTER 39

Presentation Drawings Test

DIRECTIONS

Answer the questions with short complete statements or drawings as needed.

1. Letter your name and the date at the top of the sheet.
2. Letter the question number and provide the answer. You do not need to write out the question.

QUESTIONS

Question 39–1 What is the use of presentation drawings?

Question 39–2 What is the major difference between the information provided on a presentation drawing and that provided on a working drawing?

Question 39–3 What are some factors that might influence who will draw the presentation drawings?

Question 39–4 List and describe the major information found on each type of presentation plan.

Question 39–5 What type of drawing is best used to show the shape and style of a structure?

Question 39–6 List four different media that are often used for presentation drawings.

Question 39–7 Which drawing material is most often used for presentation drawings?

Question 39–8 What common methods are used to add color to a drawing?

Question 39–9 At what scale are sections usually drawn?

Question 39–10 What scale is typically used when a floor plan will be reproduced photographically?

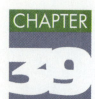

CHAPTER 40

Perspective Drawing Techniques

INTRODUCTION

All of the drawings that you have done so far have been orthographic projections, which have allowed the size and shape of a structure to be accurately presented. To develop a drawing similar to what is seen when looking at the project requires the use of perspective. Perspective drawing methods present structures very much as they appear in their natural setting. Figure 40.1 presents two elevations for a residence. This same residence can be seen in a perspective drawing in Figure 40.2, and in a photograph in Figure 40.3. As you can see, the perspective drawing resembles the photo much more than the working drawing resembles it.

TYPES OF PERSPECTIVE DRAWINGS

Three methods are used for perspective drawings: three-point, two-point, and one-point. The three-point method is primarily used to draw very tall multilevel structures, as in Figure 40.4, and will not be covered in this text.

Figure 40.2 is an example of two-point perspective. Notice that all horizontal lines merge into the horizon. With this method, at least two surfaces of the structure will be seen. Two-point perspective is generally used to present the exterior views of a structure, but the technique can also be used to present interior shapes. Figure 40.5 shows an example of two-point perspective.

One-point perspective is used to present interior space layouts or exterior views. The one-point method can also be used to present exterior views such as courtyards. Figure 40.6 shows an example of a home drawn as a one-point perspective. Notice that all lines merge into one central point.

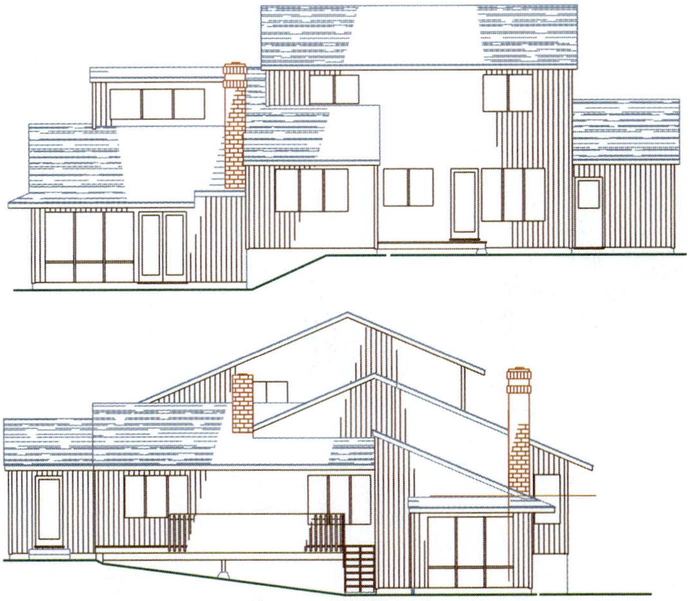

FIGURE 40.1 ■ Orthographic drawings are used to present information regarding the construction of a structure. Many people outside the construction field have a difficult time interpreting those drawings. *Courtesy Jeff's Residential Designs.*

FIGURE 40.2 ■ Two-point perspective is a drawing method that closely resembles a photograph. *Courtesy Jeff's Residential Designs.*

FIGURE 40.3 ■ Many people who are unable to interpret orthographic drawings are able to visualize a project when they see a photograph. *Courtesy Zachary Jefferis.*

FIGURE 40.4 ■ Three-point perspective is used to draw tall structures. *Courtesy Olympia & York Properties, Inc.*

PERSPECTIVE TERMS

In perspective drawing, six terms are used frequently: ground line, station point, horizon line, vanishing points, picture plane, and the true-height line. The relationship of these lines, points, and planes can be seen in Figure 40.7.

Ground Line

The *ground line* (G.L.) represents the horizontal surface at the base of the perspective drawing. It is this surface that gives the viewer a base from which to judge heights. The ground line is used as a base for making vertical measurements.

Station Point

The *station point* (S.P.) represents the position of the observer's eye. Figure 40.8 shows the theoretical positioning of the station point. The station point will be used in a manner similar to a vanishing point. Since the station point represents the location of the observer, all measurements of width will converge there.

Figure 40.9 shows the method used to find the horizontal location of a surface. A straight line is projected from each corner of the surface in the plan view to the station point. From the intersection of each line with the picture plane, another line is projected at 90° to the picture plane down to the area where the perspective will be drawn. When a surface extends past the picture plane (P.P.), a line is projected from the station point to the surface and then up to the picture plane, as shown in Figure 40.10.

FIGURE 40.5 ■ Two-point perspective is primarily used to present the exterior of a structure. *Courtesy Kenneth D. Smith & Associates, Inc.*

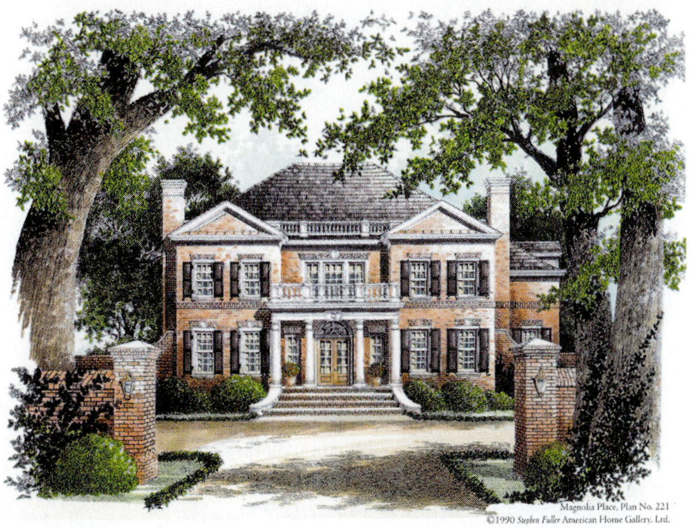

FIGURE 40.6 ■ One-point perspective is primarily used to present interior views of a structure, but it can also be used for exterior views. *Magnolia Place,* PLAN NO 221 © 1990. *COURTESY STEPHEN FULLER, INC.*

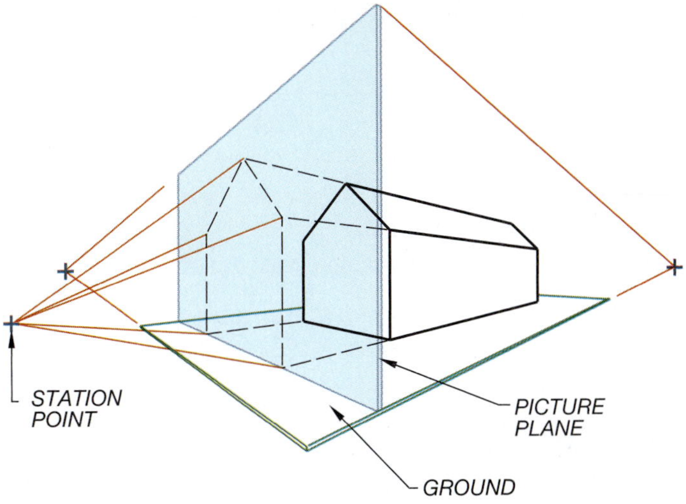

FIGURE 40.8 ■ The station point (S.P.) represents the location of the viewer. All points on the floor plan will converge at the station point.

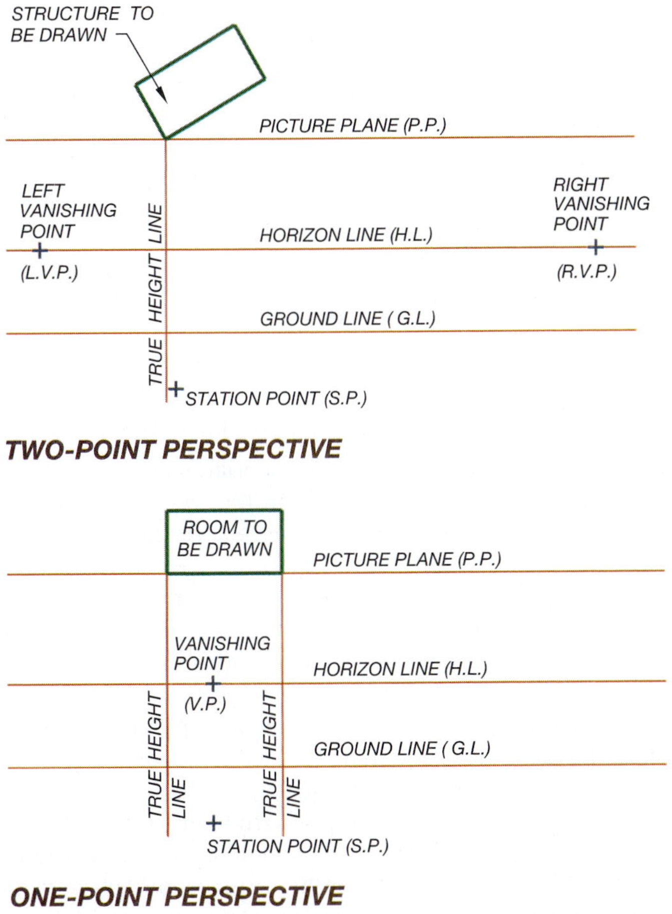

FIGURE 40.7 ■ In perspective drawing, six terms are used repeatedly: ground line, station point, horizon line, vanishing points, picture plane, and true-height line.

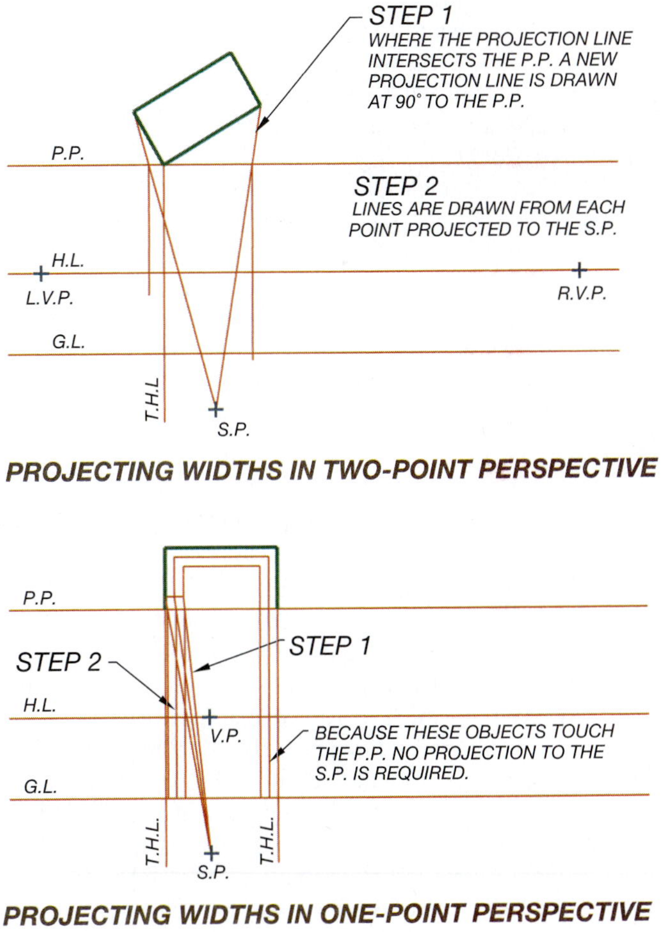

FIGURE 40.9 ■ The station point is used to project the width of the object to be drawn.

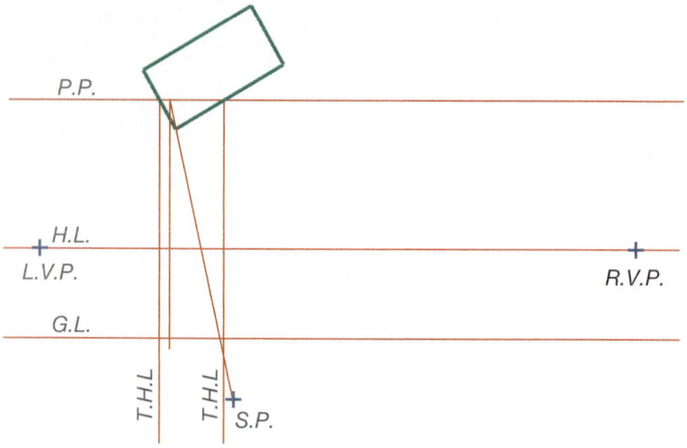

FIGURE 40.10 ■ When a structure extends past the picture plane, a line is projected from the station point through the point up to the picture plane. From this point on the picture plane, a line, perpendicular to the picture plane, is projected to the drawing area.

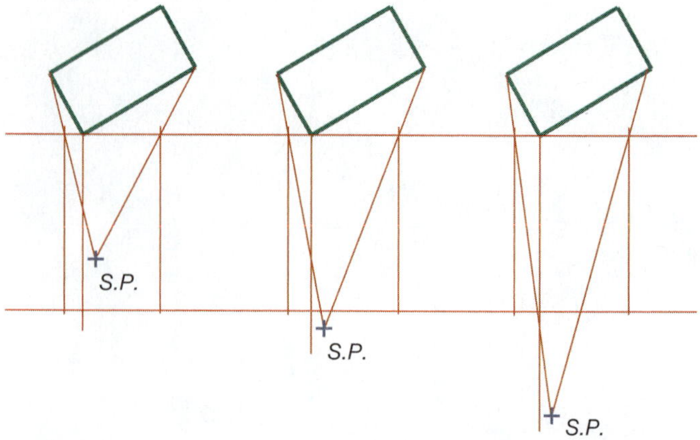

FIGURE 40.11 ■ As the distance between the picture plane and the station point is increased, the width of the drawing is also increased.

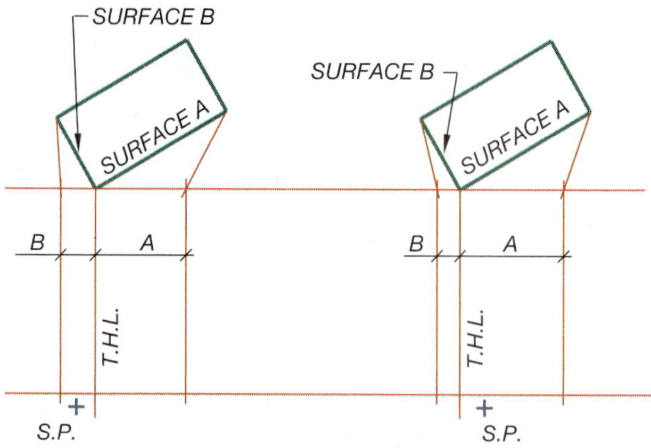

FIGURE 40.12 ■ As the station point is shifted to the left or right of the true-height line, the width of each surface is affected.

The location of the station point affects the total size of the finished drawing. Figure 40.11 shows the effect of moving the station point relative to the picture plane. As the distance between the station point and the picture plane is decreased, the width of the drawing will be decreased. Likewise, as the distance is increased between the station point and the picture plane, the perspective will become larger.

The station point can also affect the view of each surface to be presented in the perspective drawing. The station point can be placed anywhere on the drawing but is usually placed so that the structure will fit within a cone that is 30° wide. As the station point is moved in a horizontal direction, the width of each surface to be projected will be affected. Figure 40.12 shows the effect of moving the station point.

Horizon Line

The *horizon line* (H.L.) is drawn parallel to the ground line and represents the intersection of ground and sky. As the distance between the horizon line and ground line is varied, the view of the structure will be greatly affected. Figure 40.13 shows the differences that can be created as the horizon line is varied. The horizon line is usually placed at an eye level of between 5′ and 6′ (1500–1800 mm). It can be placed anywhere on the drawing, depending on the view desired. When it is placed above the highest point in the elevation, the viewer will be able to look down on the structure. Moving the horizon line above eye level can be very helpful if a roof with an intricate shape needs to be displayed. The horizon line can also be placed below the normal line of vision. This placement allows the viewer to see items that are normally hidden by eave overhangs. Avoid placing the horizon line at the top or bottom of the object to be drawn such as windows.

Vanishing Points

In one- or two-point perspective, the *vanishing point* (V.P.) or points will always be on the horizon line. As the structure is drawn, all horizontal lines will converge at the horizon line, as in Figure 40.14. In one-point perspective the vanishing point may be placed anywhere on the horizon line. Figure 40.15 shows the effect of changing the location of the vanishing point.

The vanishing points will be on the horizon line in two-point perspective, but their location will be determined by the location of the station point and the angle of the floor plan to the picture plane. Figure 40.16 shows how the vanishing points are established. With the floor plan and station point established, lines parallel to the floor plan can be projected from the station point. These lines are extended from the station point until they intersect the picture plane. A line at 90° to the picture plane is extended from this intersection down to the horizon line. The point where this line intersects the horizon line forms the vanishing point. Be sure that you project the line from the station point all the way to the picture plane. A common mistake in the layout is

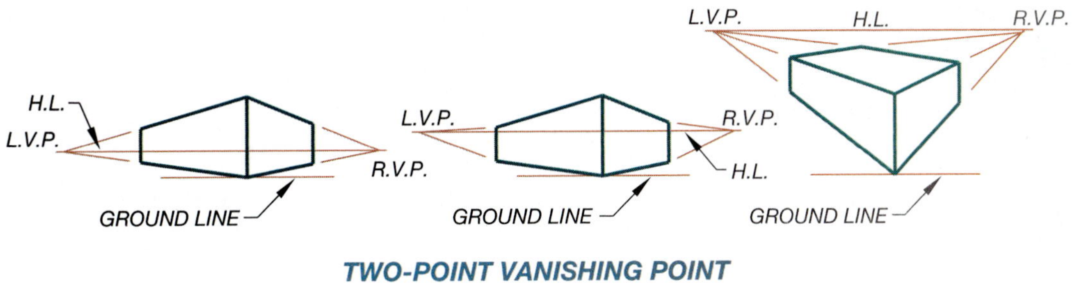

FIGURE 40.13 ■ Changing the distance from the horizon line to the ground line affects the view of the structure.

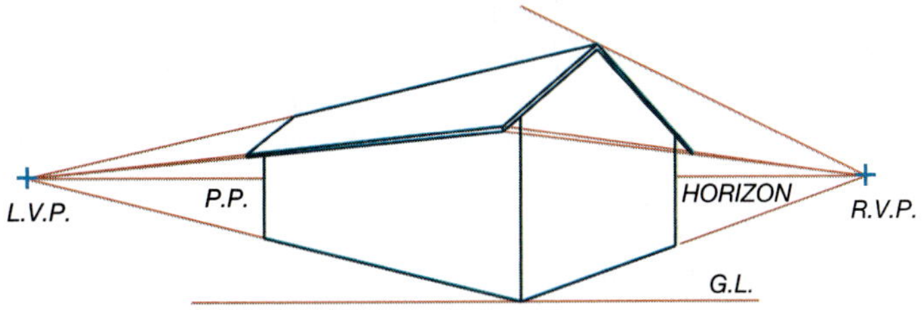

FIGURE 40.14 ■ In two-point perspective, all horizontal lines will converge at the vanishing points.

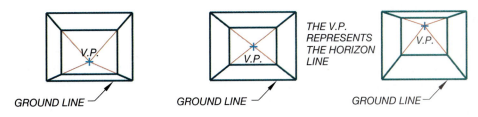

to project the angle directly onto the horizon line. This will result in a perspective that is greatly reduced in size (Figure 40.17).

Picture Plane

The *picture plane* (P.P.) represents the plane that the view of the object is projected onto. Figure 40.18 shows how the theory of the picture plane affects a perspective drawing. The picture plane is represented by a horizontal line in perspective drawing. It is a reference line on which the floor plan of the structure to be drawn is placed. Any point of the structure that is on the picture plane will be drawn in its true size in the perspective drawing. Parts of the structure that are above the picture plane will appear smaller in the perspective drawing, and parts of the structure that lie below the picture plane line will become larger. The relationship of the structure to the picture plane can be seen in Figure 40.17.

The relationship of the floor plan to the picture plane controls not only the height of the structure but also the amount of surface area that will be exposed. As a surface of the floor plan is rotated away from the picture plane, its length will be shortened in the perspective drawing. Figure 40.18 shows the effect of rotating the floor plan away from the picture plane.

The structure to be drawn can be placed at any angle to the picture plane. An angle of 45°, 30°, or 15° will be helpful in some of the projecting steps if you use a drafting machine. The

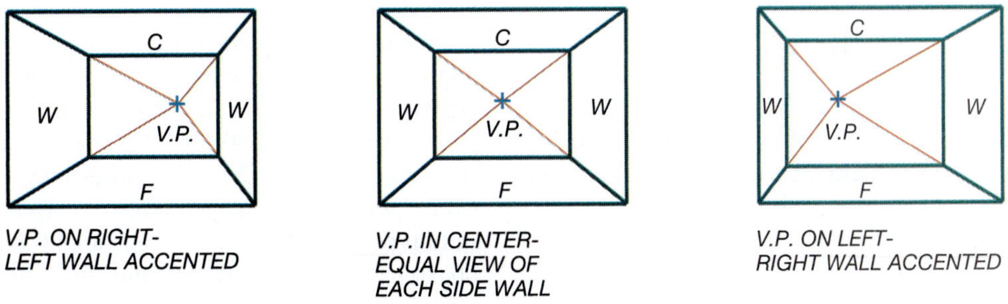

FIGURE 40.15 ■ In one-point perspective, the vanishing point may be placed anywhere. As the vanishing point is shifted from side to side, the shape of the walls is changed. The station point should always be kept below the vanishing point.

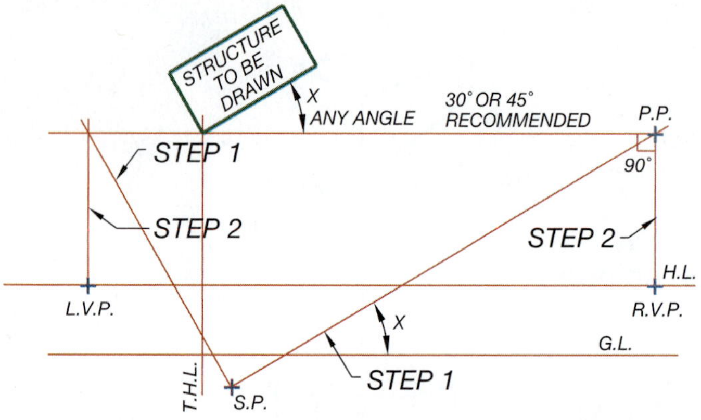

FIGURE 40.16 ■ The vanishing points in two-point perspective can be located with two easy steps. Start by projecting lines parallel to the structure from the station point to the picture plane. Where these lines intersect the picture plane, project a second line down to the horizon line at a 90° angle. Each vanishing point is located where the second projection line intersects the horizon line.

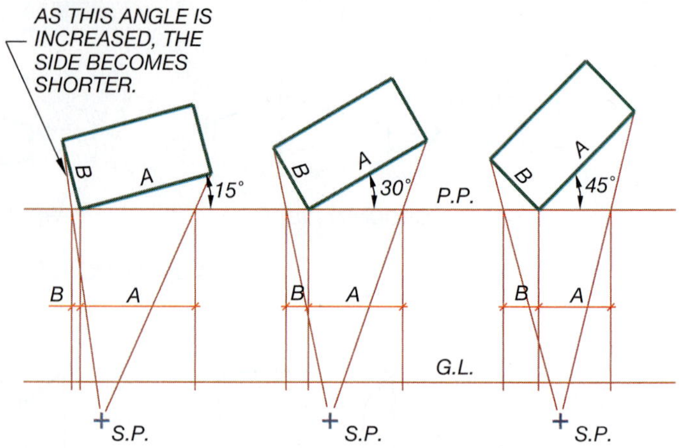

FIGURE 40.18 ■ As a surface of the plan view is rotated away from the picture plane, it will become foreshortened in the perspective view.

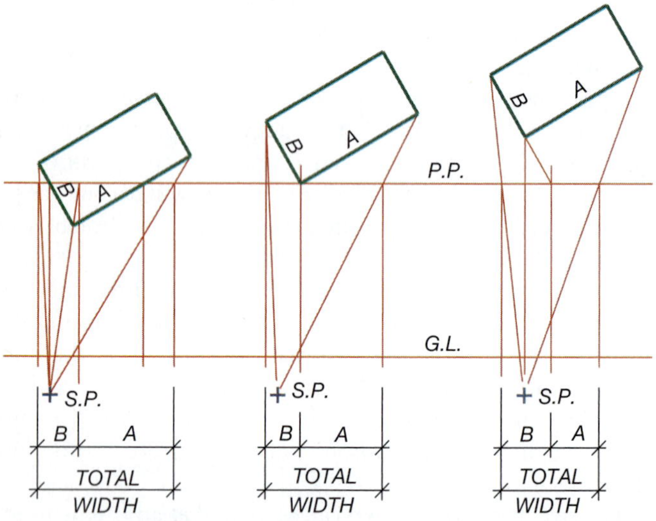

FIGURE 40.17 ■ Areas of the structure that extend past the picture plane will appear enlarged. When the structure is behind the picture plane, its size will be reduced.

key factor in choosing the angle should be the desired effect of the presentation drawing. If one side of the structure is very attractive and the other side is plain, choose an angle that will shorten the plain side.

True-Height Line

In the discussion of the picture plane, you learned that where the object to be drawn touches the picture plane is the only place where true height can be determined. Often in two-point perspective, only one corner of the structure touches the picture plane line. The line projected from that point is called the *true-height line* (T.H.L.). All other surfaces must have their height projected to this line and then projected to the vanishing point. To save time, an elevation is often used to project the height of the structure to the true-height line. This method can be seen in Figure 40.19.

Projecting true heights tends to be the most difficult aspect of the perspective drawing for the beginning illustrator. *All* heights can be projected from the elevation to the true-height line. Once projected to the true-height line, the height must be taken to the vanishing point, as shown in Figure 40.20. If the height is for a surface that is not on the picture plane, an additional step is required, as seen in Figure 40.21. This procedure requires the height to be projected to the true-height line, back toward the vanishing point, and then back to the true location.

Combining Effects

It takes plenty of experience to get over the initial confusion caused by the relationship of these points, lines, and planes. On your initial drawings, keep one corner of the floor plan on the picture plane. Experiment with changing the drawing size once you feel comfortable with the process of projecting heights. A typical layout for your first few drawings would be to keep the plan view at either a 30° or a 45° angle to the picture plane. Using an angle of less than 30° will usually place one of the vanishing points off your layout area. Keep the distance from the station point to the picture plane about twice the height of the object to be drawn and near the true-height line to avoid distortion of a surface. As you begin to feel confidence with the perspective method, start by changing one variable, then another, and then others as needed.

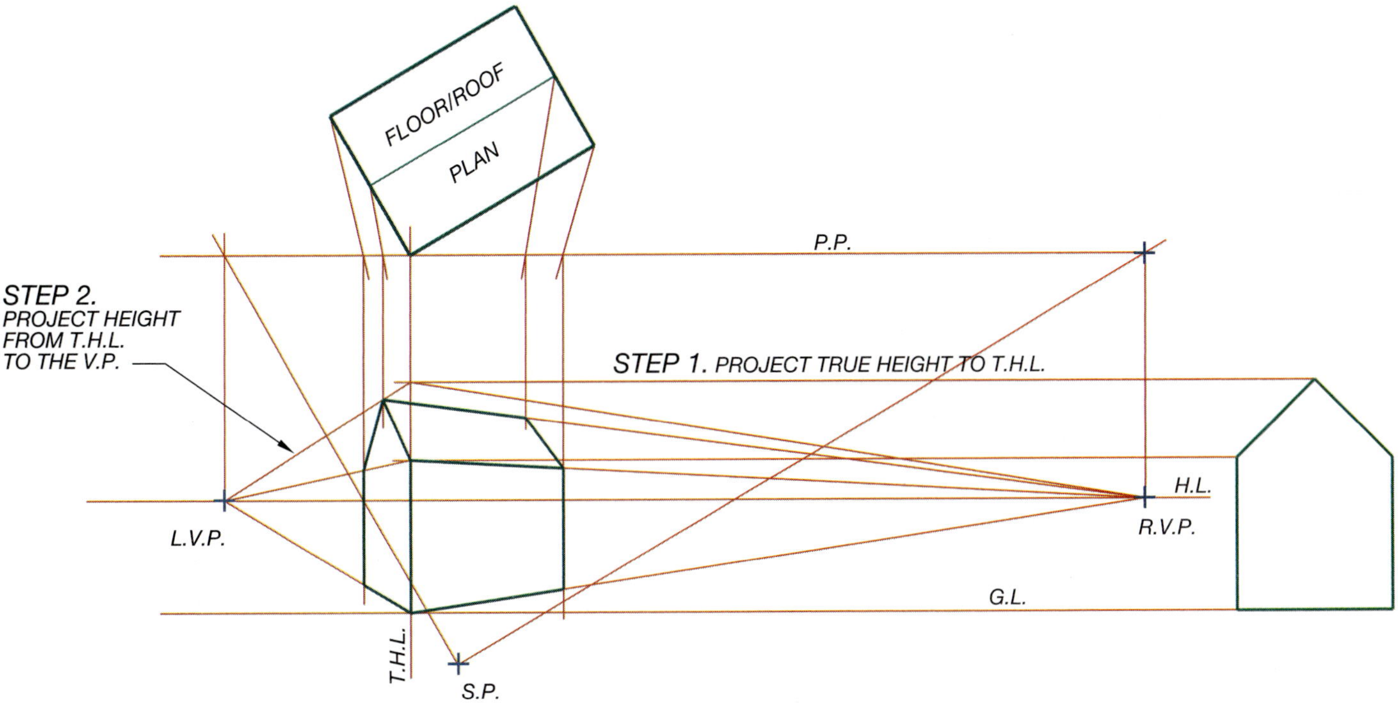

FIGURE 40.19 ■ Heights from an elevation are projected to the true-height line and then to the vanishing point.

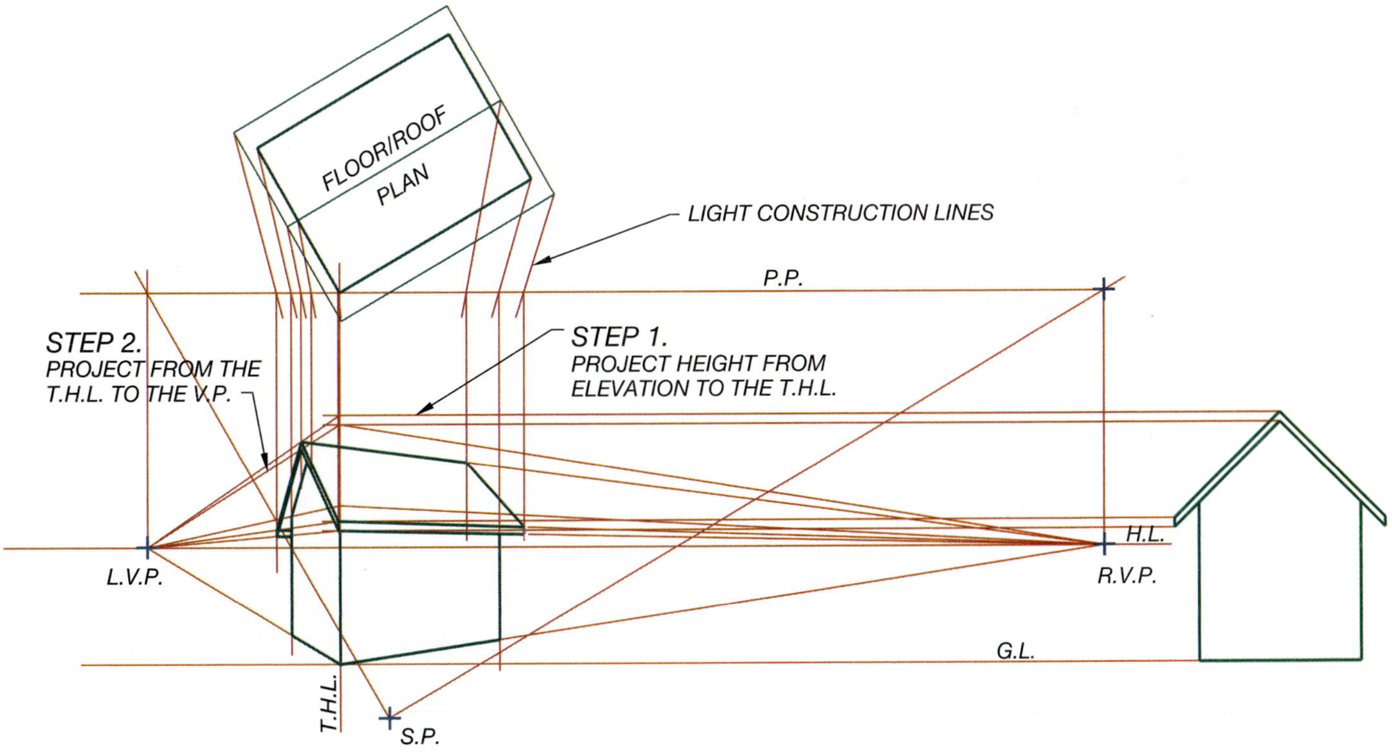

FIGURE 40.20 ■ All heights are projected initially to the true-height line and then to a vanishing point. The actual height is established where the height projection line intersects the location line from the plan view.

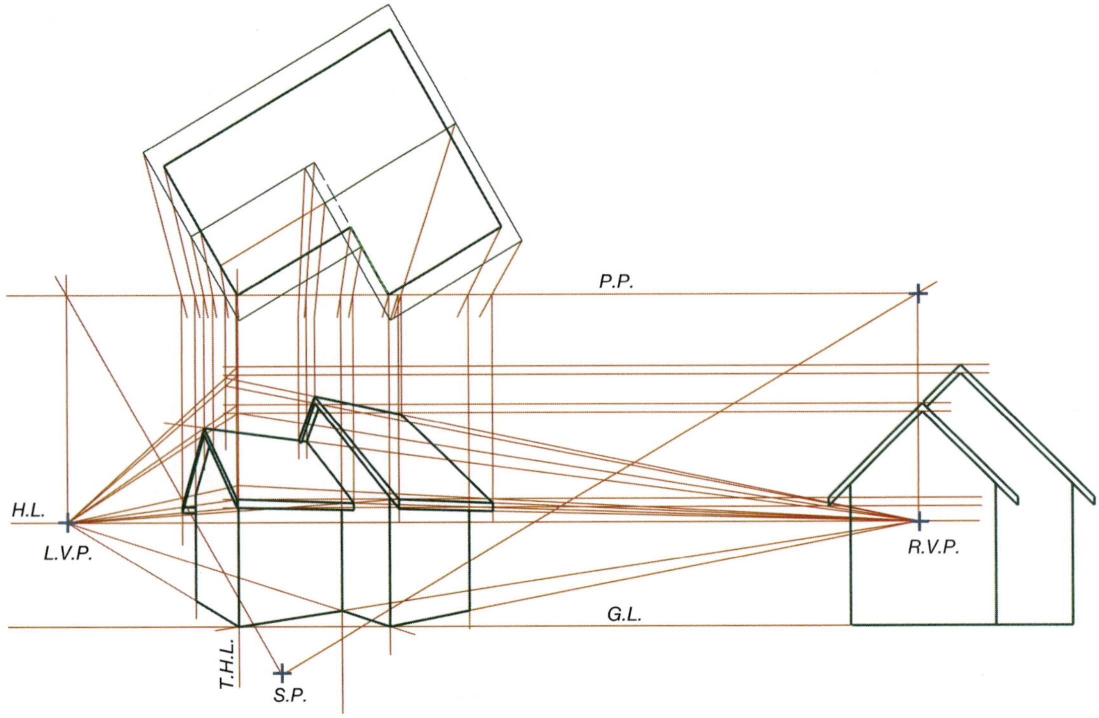

FIGURE 40.21 ■ When a surface does not touch the picture plane, it must be projected to a surface that does and then to the station point.

TWO-POINT PERSPECTIVE

There are numerous ways to draw two-point perspective, each with advantages and disadvantages. As you continue with your architectural training, you will be exposed to many other methods. Don't be afraid to experiment with a method. Keep in mind that this chapter is intended to be only an introduction to perspective methods.

Getting Started

To begin your drawing, you will need a print of your floor plan and an elevation. You will find it helpful to draw the outline of the roof on the print of the floor plan. You will also need a sheet of paper about 4′ (1200 mm) long and 2′ (600 mm) high on which to do the initial layout work. Butcher paper or the back of a sheet of blueprint paper provides a durable surface for drawing. Colored pencils can also help you keep track of various lines. Some students like to draw all roof lines in one color, all wall lines in a second color, and all window and door lines in a third color. The residence that was drawn in Chapter 16 will be used for our example here.

Setup of Basic Elements

See Figure 40.22 for steps 1 through 10.

STEP 1 Draw a line at the top of your paper to represent the picture plane.

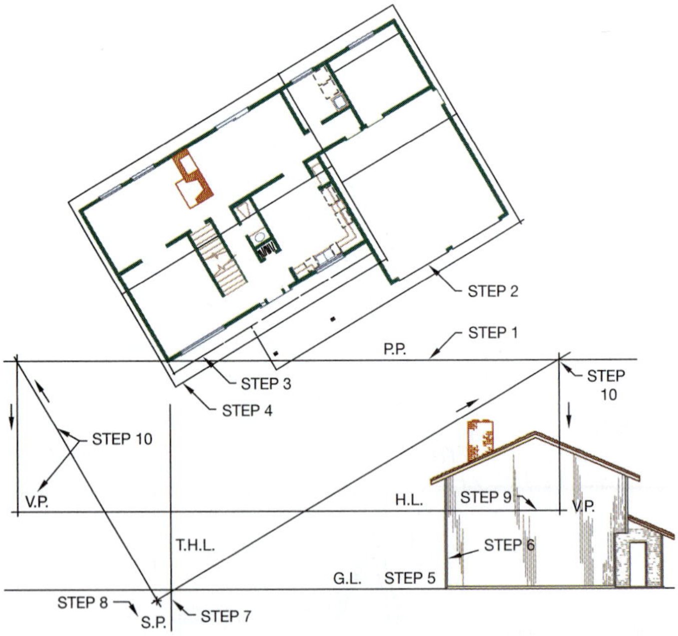

FIGURE 40.22 ■ Initial layout for two-point perspective. Establish the picture plane, horizon line, ground line, true-height line, station point, and vanishing points.

STEP 2 Tape a print of your floor plan on the paper so that a corner of the building touches the picture plane. For this example, a 30° angle between the long side of the floor plan and the picture plane was used.

STEP 3 Draw the line of any cantilevers for the upper floor.

STEP 4 Draw the outline of the roof on the floor plan.

STEP 5 Establish a ground line. The distance from the ground line to the picture plane should be greater than the height of the structure to be drawn, to avoid overlapping layout lines.

STEP 6 Tape a print of the elevation to the side of the drawing area with the ground of the elevation touching the ground line.

STEP 7 Establish the true-height line by projecting the intersection of the floor plan and the picture plane down at 90° from the picture plane.

STEP 8 Establish the station point. Keep the station point within an inch of the true height line, and a distance of about twice the height of the object to be drawn.

STEP 9 Establish the horizon line at about 6 scale ft above the ground.

STEP 10 Establish the vanishing points. Start at the station point and project lines parallel to the floor plan up to the picture plane. From the point where these lines intersect the picture plane, project a line down at 90° to the horizon line. The vanishing points are established where this line intersects the horizon line.

You are now ready to project individual points from the floor plan down to the area where the perspective drawing will be drawn. Start by determining the maximum size of the house.

STEP 11 Project lines from each corner of the roof down to the station point. Even though the lines converge on the station point, stop the lines just after they pass the picture plane. See Figure 40.23.

See Figure 40.24 for steps 12 through 14.

STEP 12 Project a line down from the picture plane to the drawing area to represent each corner of the roof.

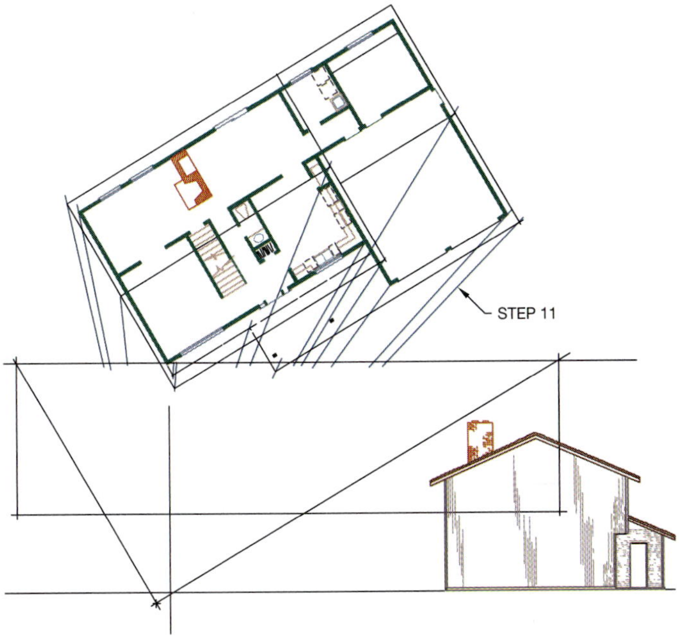

FIGURE 40.23 ■ Layout of the roof shape. Project corner locations from the plan to the station point.

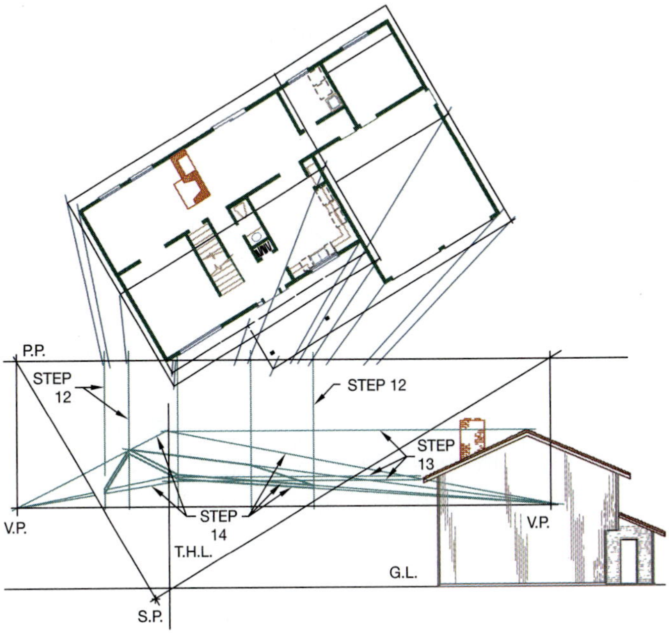

FIGURE 40.24 ■ The roof shape can be drawn by projecting the width and height lines. The roof plan must be drawn on the floor plan before it can be projected to the drawing area.

STEP 13 Project the height for the roof onto the true-height line.

STEP 14 Project the heights of the roof on the true-height line back to each vanishing point. As width and height lines intersect, you can begin to lay out the basic roof shape.

STEP 15 Continue to project height and width lines for each roof surface. Work from front to back and top to bottom. When all roof surfaces are projected, your drawing will resemble Figure 40.25.

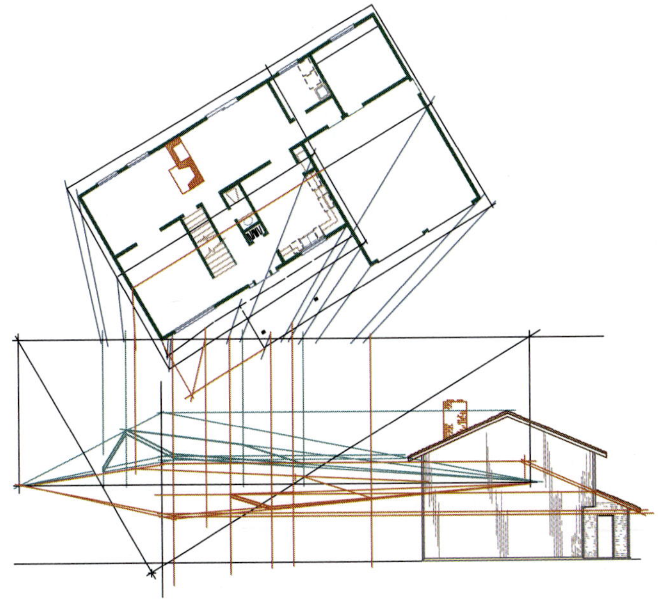

FIGURE 40.25 ■ Lay out the upper roof by projecting heights from the elevation to the true-height line and then back to the vanishing points. Corresponding points from the picture plane are projected down to the drawing area.

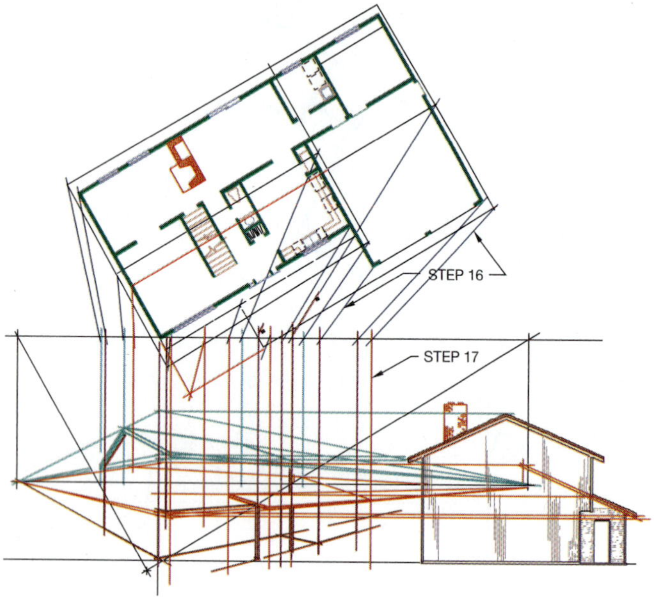

FIGURE 40.26 ■ The shape of the two-point perspective can be finished by connecting the height and width lines. Work from top to bottom and from front to back.

You now have the roof shape blocked out, plus a multitude of lines that may be making you tense. Don't panic. To keep track of the lines, use colored pencils for the walls' projection lines. See Figure 40.26 for steps 16 and 17.

STEP 16 Project a line from the corner of each wall to the station point, stopping at the picture plane.

STEP 17 Project a line from the picture plane to the drawing area to represent each wall.

STEP 18 Using a color different from that used for roof and wall projections, plot the outline for all windows and doors. Project the height from the elevations and the locations from the floor plan.

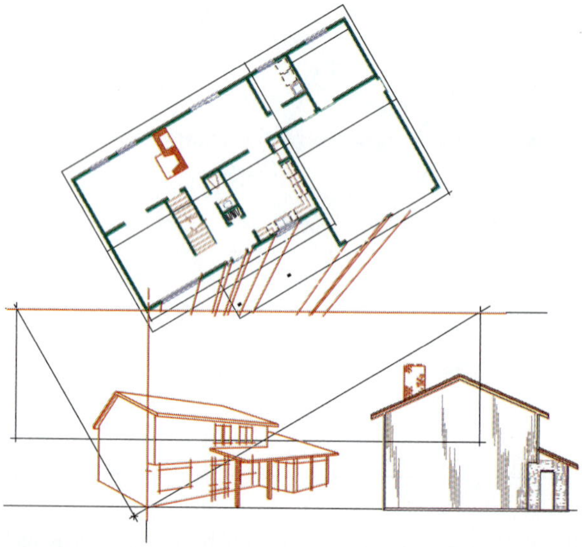

FIGURE 40.27 ■ The perspective drawing is now complete. Proceed to Chapter 41 for an explanation on how to render the drawing with finished-quality lines.

Your drawing should now resemble Figure 40.27, and it is now ready to be transformed into a rendered presentation drawing. The steps for completing this process will be discussed in the following chapter.

ONE-POINT PERSPECTIVE

Much of the process for creating a one-point perspective is similar to creating a two-point perspective. Differences in procedure will be explained step by step. To begin your drawing, you will need a sheet of butcher paper and a print of your floor plan. Depending on the complexity of the area to be drawn, you may need to draw an elevation. If you are working on a simple interior, you may be able to mark the heights on the true-height line rather than drawing an elevation. Depending on the size of the area you will be drawing, the floor plan may need to be enlarged. Often your drawing will seem very small if you start with a 1/4″ = 1′–0″ floor plan. Because kitchens are such a common item in presentation drawings, the kitchen from the residence in Chapter 16 will be drawn for our example. See Figure 40.28 for steps 1 through 7.

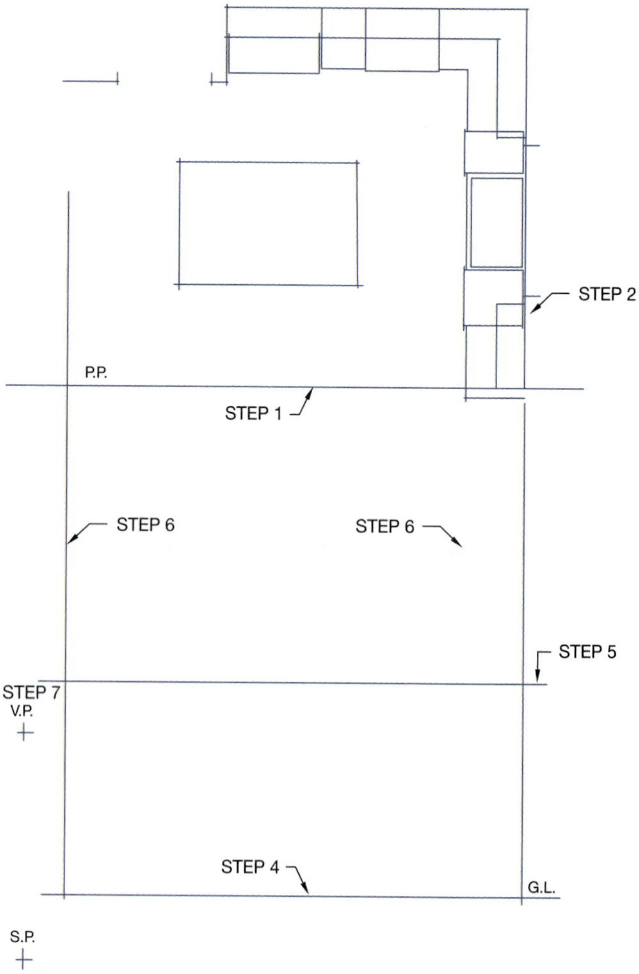

FIGURE 40.28 ■ Initial layout for one-point perspective. Start by taping a print of the floor plan to the picture plane.

STEP 1 Establish the picture plane.

STEP 2 Tape a print of the floor plan to the picture plane. Remember to redraw the floor plan at a larger scale.

STEP 3 Locate the station point using the same consideration as with the two-point method.

STEP 4 Establish the ground line. In one-point perspective, this line represents the floor.

STEP 5 Measure up from the ground line and establish the ceiling line. Notice that these lines will be parallel. All lines that are parallel to the picture plane will remain parallel in the drawing.

STEP 6 Establish the width of the drawing by projecting the points where the walls intersect the picture plane down to the drawing area.

STEP 7 Establish a vanishing point. The vanishing point can go at any height but should be kept above the station point.

See Figure 40.29 for steps 8 through 10.

STEP 8 Project lines from each of the room corners to the vanishing point to begin establishing the side walls, floor, and ceiling.

STEP 9 Project lines from each room corner of the plan to the station point. Extend the lines to just past the picture plane, to avoid construction lines in the area where the perspective will be drawn.

STEP 10 Lay out the floor, walls, and ceiling. These can be determined where the wall line intersects the lines that were projected from the floor and ceiling to the vanishing points.

See Figure 40.30 for steps 11 through 14.

STEP 11 Measure the height of the cabinets on the cabinet elevations in Chapter 24, and project these heights on either of the true-height lines. In one-point perspective, each side of the drawing is a true-height line.

STEP 12 Lay out the shape of the base cabinet. This can be done by projecting the width from the floor plan and measuring the height on the true-height line. Project the height back to the vanishing points.

STEP 13 Lay out the cabinet on the back wall by projecting the intersection point on the floor plan down onto the perspective.

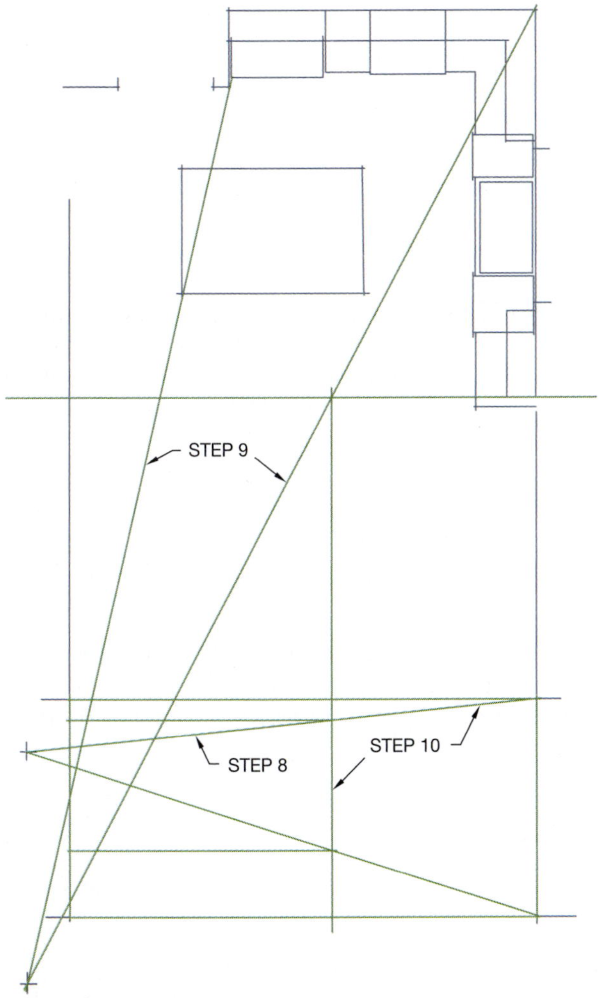

FIGURE 40.29 ■ Establishing the walls, floor, and ceiling.

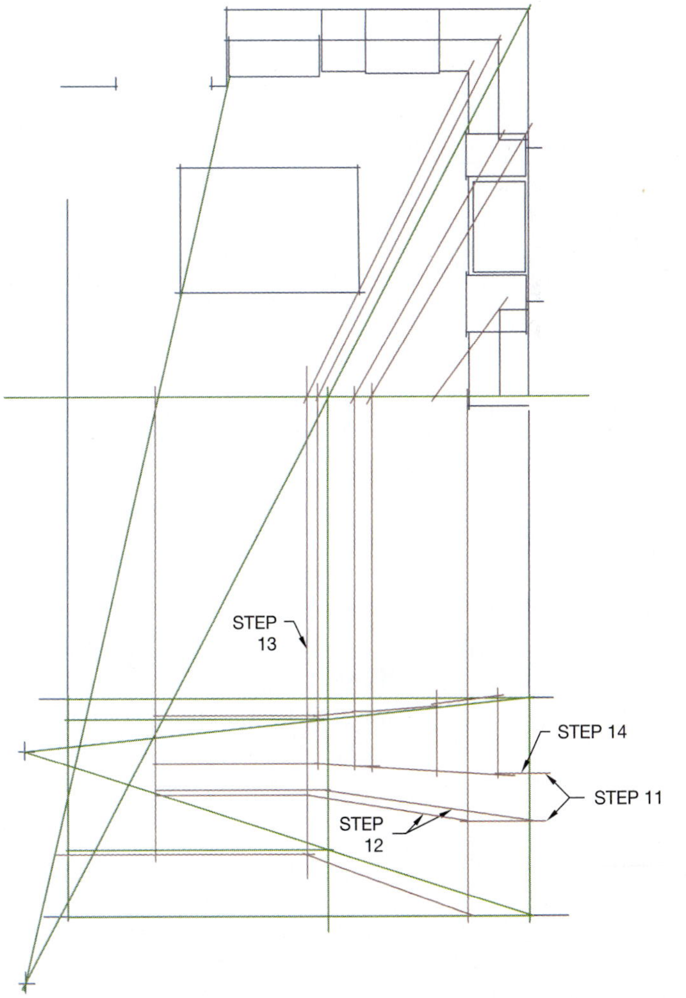

FIGURE 40.30 ■ Lay out the basic cabinet shapes by projecting each location from the floor plan.

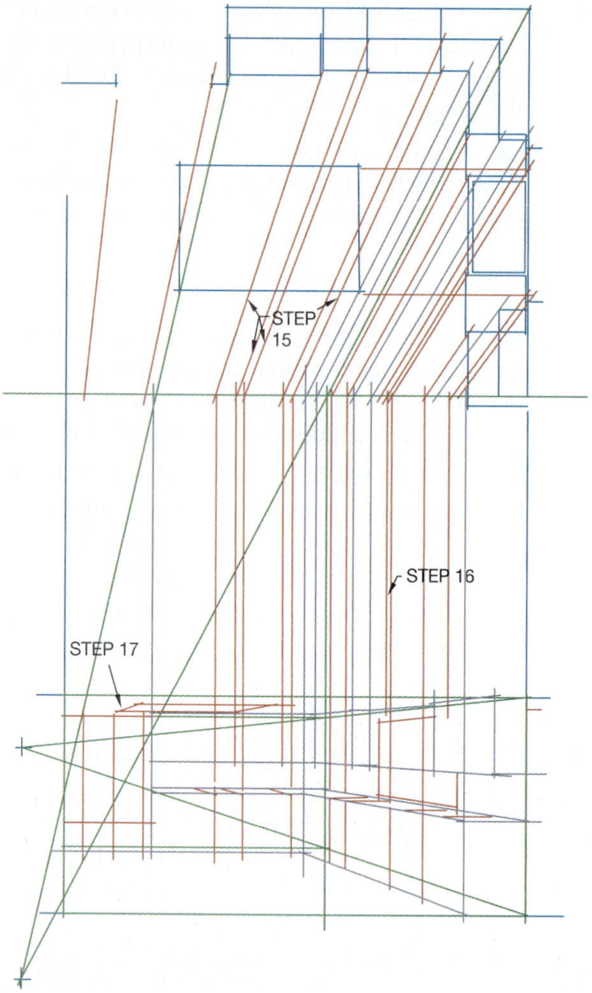

FIGURE 40.31 ■ Block out individual cabinets. Use drawings of the cabinet elevations to establish each location.

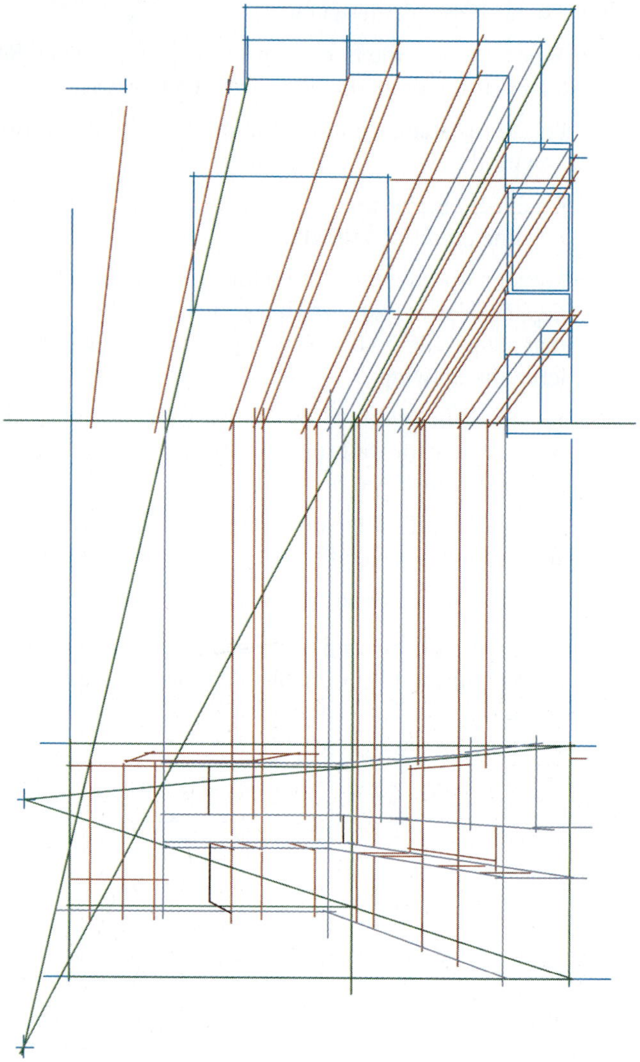

FIGURE 40.32 ■ Completed one-point-perspective drawing, ready to be rendered. See Chapter 41 for an explanation of how to render the drawing with finished-quality lines.

STEP 14 Lay out the upper cabinet following the same procedure used to lay out the base cabinet.

See Figure 40.31 for steps 15 through 17. Similar to the two-point perspective, your drawing is probably starting to get cluttered with projection lines. Don't hesitate to use multiple colors for projection lines.

STEP 15 Lay out the widths for all cabinets and appliances by projecting their widths to the station points and then down to the drawing. To determine heights, mark the true height on the true height line and then project the height back to the vanishing point.

STEP 16 Block out any windows.

STEP 17 Block out soffits, skylights, or other items in the ceiling.

You now have the basic materials drawn for the one-point perspective. Your drawing should resemble Figure 40.32. Chapter 41 will introduce you to the basics of rendering the perspective drawing.

USING CADD TO MAKE PERSPECTIVES

Preparing perspective drawings is easy with CADD. Most of the powerful CADD packages allow you to generate a perspective drawing automatically from the floor plan layout. Wall and header heights are established while you draw the floor plan. When the floor plan is complete, you can view the drawing in perspective from any selected point in space. You can change the viewpoint until you find the perspective view, lighting, colors, and materials that display the building best. Figure 40.33*a* and *b* show different views of the same structure. When you move the viewpoint, the CADD system automatically adjusts the picture plane, station point, and vanishing points, but the ground plane always remains the same. Interior perspectives may be drawn in the same manner, or you can create the drawing directly in 3-D. Some systems have a split screen that allows you to draw in plan view on one part of the screen while generating a 3-D view on the other.

FIGURE 40.33(A) ■ Creating two-point perspective with a computer program allows changes in the station and vanishing points to be made easily. *Courtesy Kenneth D. Smith & Associates, Inc.*

FIGURE 40.33(B) ■ By altering the horizon line, the view that is displayed is greatly altered. *Courtesy Kenneth D. Smith & Associates, Inc.*

Problem 40–5 Draw a one-point perspective of the following object at a scale of 1/2″ = 1′–0″. Draw a floor plan on the P.P. Set the G.L. 8″ below the P.P. Set the V.P. in the center of the drawing at a point above B. Set the S.P. 4″ directly below the V.P. Darken all object lines using graphite. Leave all construction lines.

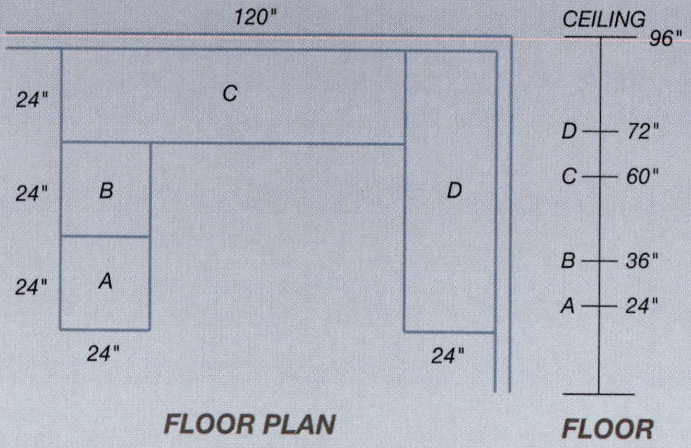

PROBLEM 40–5

Problem 40–6 Draw a one-point perspective of the following object at a scale of 1/2″ = 1′–0″. Draw the floor plan on the P.P. Establish your own G.L. and S.P. Set the V.P. in the left third of the drawing, at a point above D. Darken all object lines using graphite. Leave all construction lines. Redraw this object using steps 1 through 3. Set the V.P. at any point above the G.L. in the right third of the drawing. Darken all object lines using graphite. Leave all construction lines.

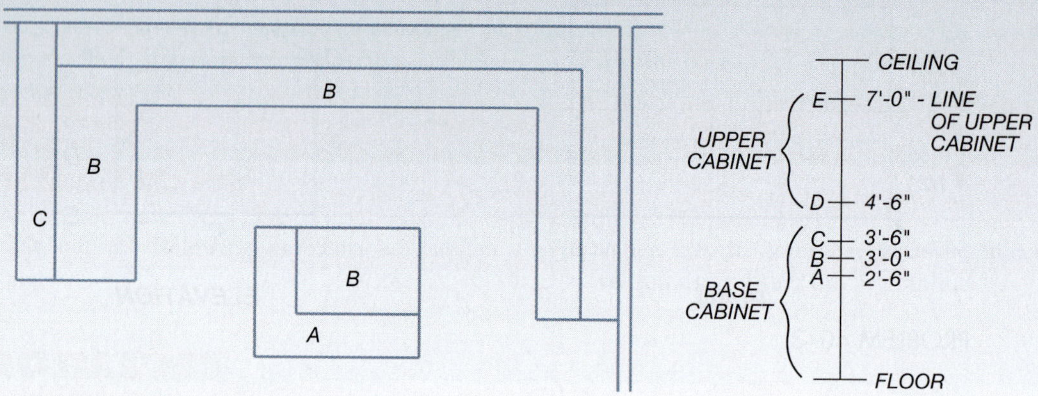

PROBLEM 40–6

Problem 40–7 Using a print of your floor plan and elevation, draw a two-point perspective of the residence that you have drawn for your project. Select a view that will not accent one surface greatly over the other. Leave all lines as construction lines. This drawing will be used in Chapter 41 for your rendering project.

Problem 40–8 Using a print of your floor plan and cabinet elevations, draw a one-point perspective of one of the rooms. Do all work on butcher paper. Leave all lines as construction lines. This drawing will be used in Chapter 41 for your rendering project.

CHAPTER 41

Rendering Methods for Perspective Drawings

INTRODUCTION

Figure 40.27 showed an example of a perspective drawing. It presented a better view of the structure than an orthographic drawing, but it still did not appear very realistic. In order to transform a perspective drawing into a realistic presentation of a structure, the drawing must be rendered. You have already been exposed to the process of rendering other types of presentation drawings. Similar materials and media are used to render perspective drawings. A rendered perspective drawing typically shows depth, shading, reflections, texture, and entourage, or surroundings.

More than any other type of presentation drawing, a rendered perspective can be an intimidating project for the beginning student. On other forms of presentation drawings, you will be able to use standard drafting tools to form the lines. When rendering perspective drawings, you will need to rely on your artistic ability. Don't panic if you feel you will never be able to draw a rendering. The techniques needed require lots of practice, but they can be mastered. Start by developing a scrapbook of renderings done by professional illustrators in various materials. Also collect photos of people, trees, and cars. Until you gain confidence in your artistic ability, you can use one of the photos or renderings from your scrapbook as a guide, or, if necessary, you could even trace it.

Your renderings will also be enhanced if you will take the time to notice and sketch the things around you. Pay special attention to trees, and to shadows that are cast by and onto buildings. You will notice that, depending on the location and intensity of sunlight, surfaces take on different appearances. Try to reproduce these effects in your sketches.

FIGURE 41.1 ■ A rendered perspective drawing can be used to show depth and texture. *Monte Vista Ranch,* PLAN NO. 363 © 1994. *Courtesy Stephen Fuller, American Home Gallery, Ltd.*

PRESENTING DEPTH

Figure 41.1 provides an example of how depth and surface texture can be shown. The perspective drawing itself gives an illusion of depth by presenting three surface areas in one view. In two-point perspective, a feeling of depth is often created at areas such as windows, doors, fascias, surface intersections, and trim boards. The feeling of depth can be created by using offsetting lines and by using different line weights. The same guidelines for showing depth should be used in one-point perspective. Cabinet doors, drawers, and appliances typically project past the face of a cabinet and should be shown on the perspective drawing.

SHADOWS

Shadows provide an excellent method of presenting depth. Shadows are cast anytime a surface blocks light from striking another surface. Figure 41.2 shows an example of the effect of shadows on a structure. The principles for projecting shadows in perspective are similar to the methods used in orthographic drawings. Before any shadows can be projected, a light source must be established, using the same guidelines that were used for presentation elevations. Keep the light source in a position such that major features of the structure will not be hidden.

Shadows are usually projected at an angle of 30°, 45°, or 50°. To save time in projecting, use a standard triangle or drafting machine setting. Figure 41.3 shows the effect of changing the angle of the light source.

Two principles affect the layout of shadows if the light source is drawn parallel to the picture plane.

1. A vertical wall or surface casts a shadow on the adjacent ground or horizontal surface in the *direction in which the light rays are traveling.*

FIGURE 41.2 ■ Shades and shadows are used to show depth. *Breckenridge Peak* , PLAN NO. 349 © 1995. *Courtesy Stephen Fuller, American Home Gallery, Ltd.*

2. The shadow on a plane, caused by a line parallel to the plane, forms a shadow line *parallel to the line.* These two lines converge at the same vanishing point.

Both principles can be seen in Figure 41.4. The vertical lines cast their shadows according to the first rule, and the horizontal lines cast a shadow according to the second rule. These same two principles can be applied as the shadows are cast onto other parts of the structure, as seen in Figure 41.5. When a structure has an overhang, the same principles can be applied to project the shaded areas.

You will often be required to project the height of a horizontal surface onto an inclined surface. This can be done by following the procedure shown in Figure 41.6. A similar problem is created when you are trying to project the shadow of an inclined surface onto a flat surface. Figure 41.7 shows how the shadow created by a roof can be projected onto the ground.

One of the problems of casting shadows parallel to the picture plane is that this method requires one of the two wall surfaces to be in the shade. Material can still be seen when it is in the shade, but it will not be as prominent as the same material in direct sunlight. Shading film, ink wash, and graphite shading are the most common methods of showing shaded areas. These same materials are used to show shadows, but they must be applied using a denser pattern. Methods similar to those for drawing shadows can be used on perspective drawings that were used on the rendered elevation.

If you would like to have both surfaces of a structure in direct sunlight, the light source will need to be moved so that it is no longer parallel to the picture plane. Figure 41.8 shows a rendering with the light source behind the viewer so that both sides are in sunlight.

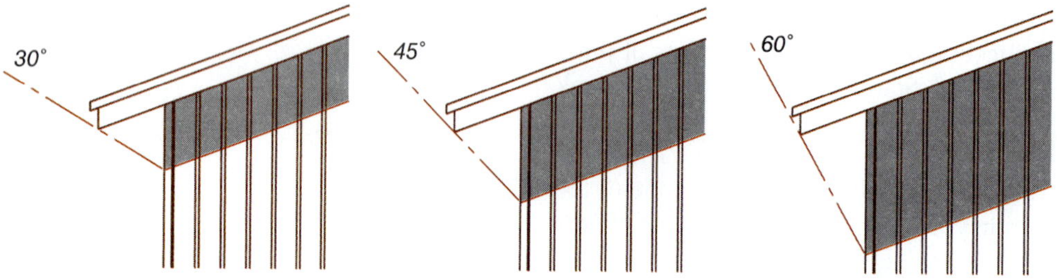

FIGURE 41.3 ■ As the angle of light becomes steeper, the length of the shadow becomes longer.

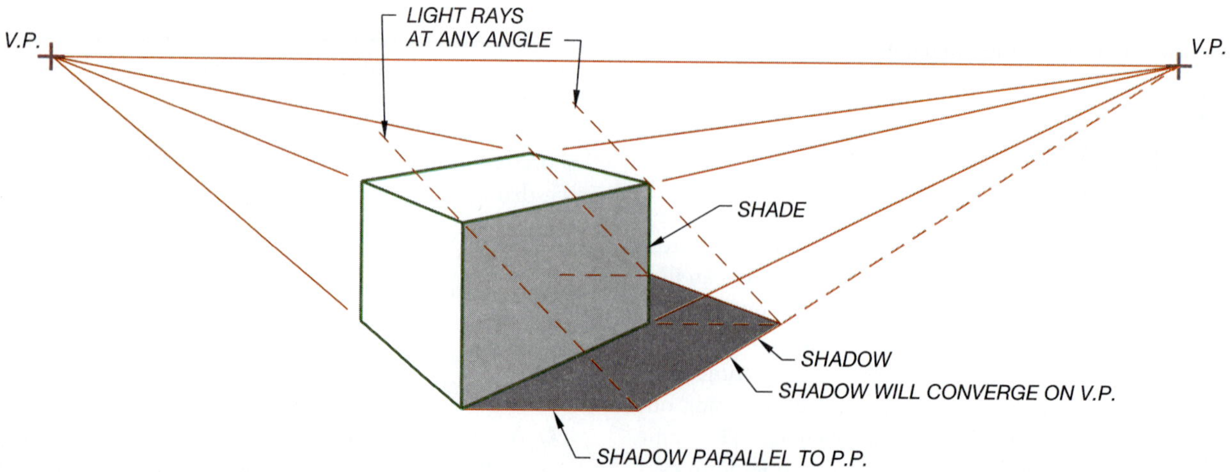

FIGURE 41.4 ■ A vertical surface casts a shadow on a horizontal surface in the direction that the light rays are traveling. The created shadow is parallel to the surface that created it and converges at the same vanishing point.

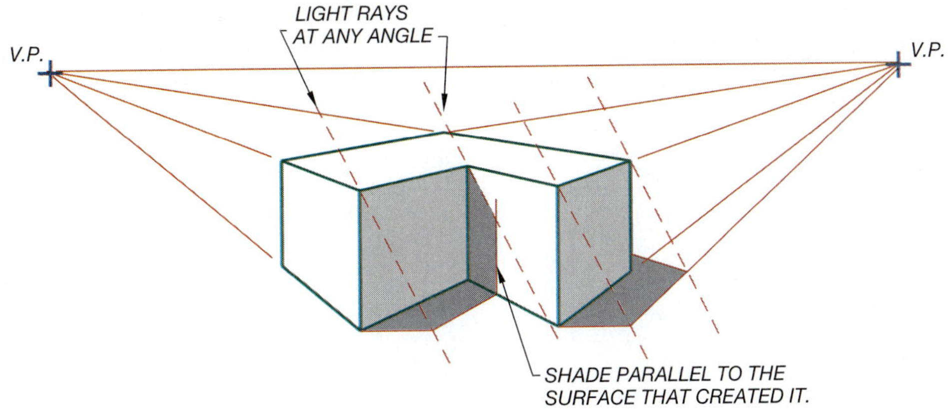

FIGURE 41.5 ■ Shadows cast onto a vertical wall are projected using the same methods used to project shadows onto the ground.

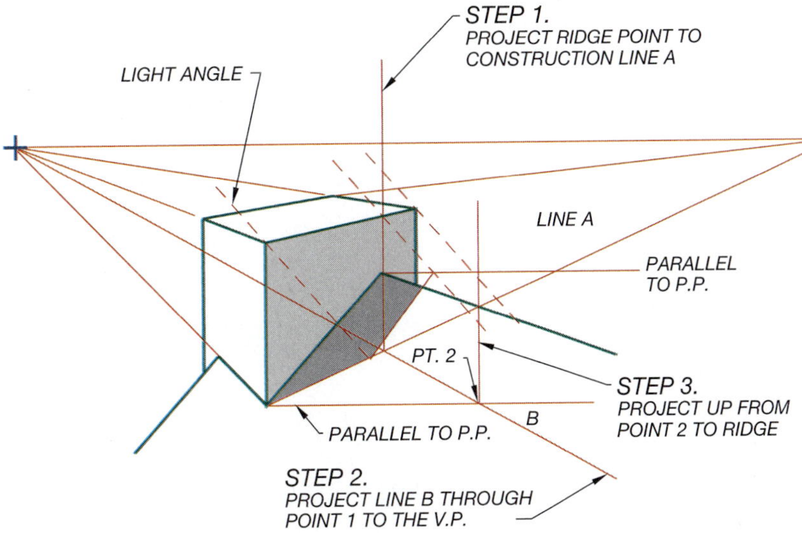

FIGURE 41.6 ■ Projecting shadows onto an angled surface.

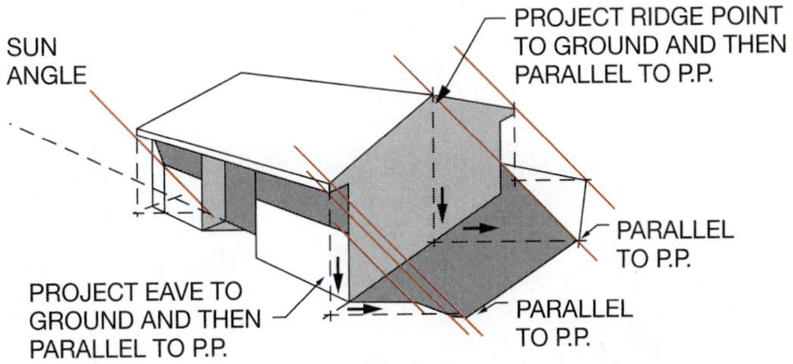

FIGURE 41.7 ■ Projecting shadows from an inclined roof onto a horizontal surface.

FIGURE 41.8 ■ The sun is often placed above and behind the viewer so that some shadows will be cast, but the majority of each surface will remain in sunlight. In this rendering, the garage is shaded to highlight the residence. *Santa Rosa Canyon*, PLAN NO. 358 © 1994. *Courtesy Stephen Fuller, American Home Gallery, Ltd.*

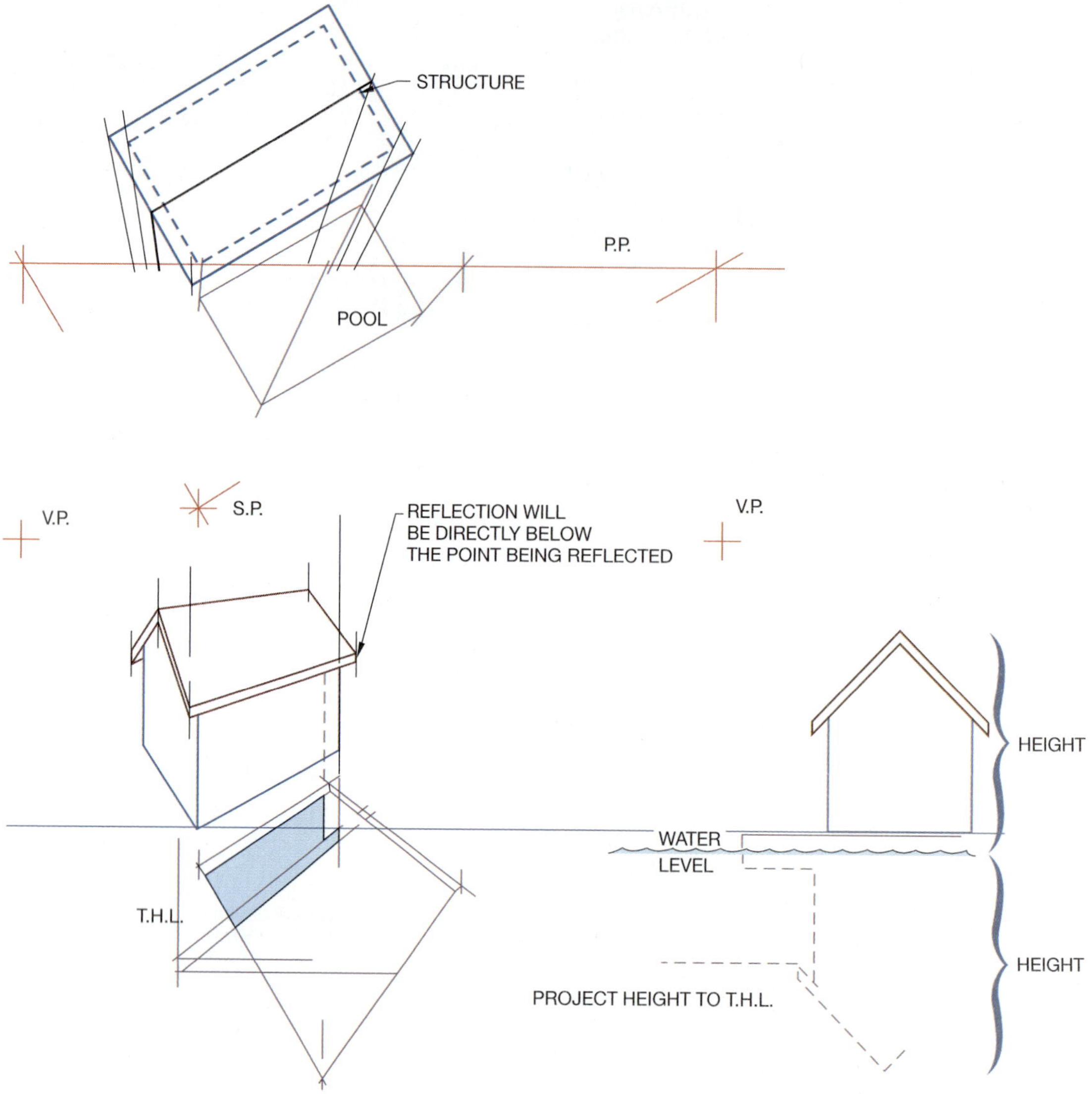

FIGURE 41.9 ■ Projecting reflections on water. The reflected image will fall directly below the point to be projected. The depth of the reflection will be equal to the distance of the point above the reflecting surface.

REFLECTIONS

Reflections in perspective drawings are typically caused by water or glass. In casting a reflection caused by water, the point to be projected will fall directly below that point and the depth of the reflection will be equal to the distance that the point is above the water's surface. Figure 41.9 shows how these two principles can be used to project a reflection. Figure 41.10 shows an example of rendering water.

Large areas of glass in a perspective drawing can often appear very dull unless rendered. Figure 41.11 shows common methods of rendering glass. Glass is often rendered solid black with small areas of white to indicate reflections. Glass can also be left unrendered, and all material in the background can be shown. A third alternative is to use an ink wash, a combination of water and ink applied with a small paintbrush that gives a gray

FIGURE 41.10 ■ Reflections on water.

FIGURE 41.11 ■ Glass can be rendered using several techniques. *Courtesy Kenneth D. Smith & Associates, Inc.*

tone rather than the solid black color of ink. Small areas of reflection are often left white or shaded with a soft graphite.

TEXTURE

An important part of the rendered perspective is to show the texture of the materials used to build the structure. Most materials can be drawn on the perspective drawing in styles similar to those used to draw an elevation. Although the material will appear similar, the actual lines used to represent the material should be different on a perspective drawing.

Roofing Materials

Shingles, tiles, and metal panels are the major types of materials that will be rendered on the roof. Horizontal construction lines that represent each course of shingles or tiles are usually drawn using tools and then traced using freehand techniques, as shown in Figure 41.12. Bar tiles are represented using methods similar to shingles. Curved tiles can be drawn using similar methods and represented as shown in Figure 41.13.

Siding or Paneling

Siding will be drawn using either horizontal or vertical lines. If horizontal siding is used, the lines representing the siding will merge at the vanishing points. When horizontal lap siding intersects another surface, a sawtooth shade pattern will result.

Vertical siding can be drawn using the methods used on the rendered elevation. For some types of siding, both thickness and texture must be drawn. Figure 41.14 shows common methods

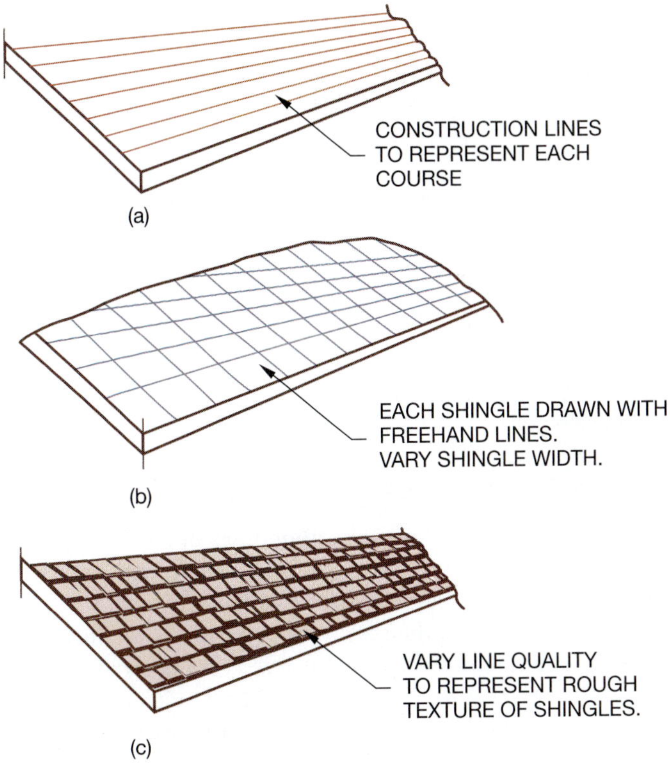
CONSTRUCTION LINES TO REPRESENT EACH COURSE
(a)
EACH SHINGLE DRAWN WITH FREEHAND LINES. VARY SHINGLE WIDTH.
(b)
VARY LINE QUALITY TO REPRESENT ROUGH TEXTURE OF SHINGLES.
(c)

FIGURE 41.12 ■ Drawing shingles. Shingles can be drawn by lightly drawing lines to the vanishing point to represent each course. Lines can then be drawn to represent each shingle. With each shingle drawn, the original horizontal lines should then be redrawn freehand to represent the rough texture of shingles.

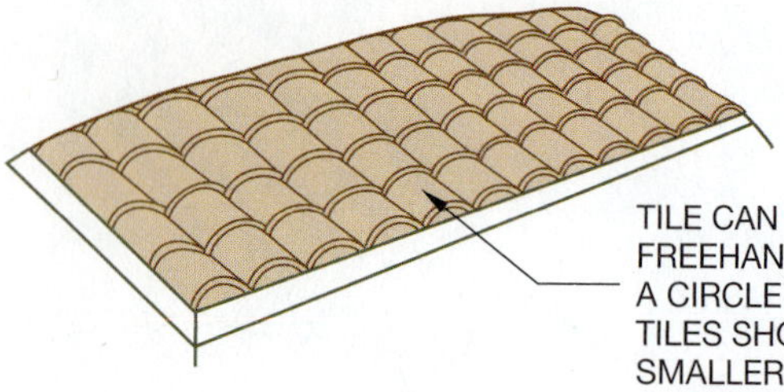
TILE CAN BE DRAWN
FREEHAND OR WITH
A CIRCLE TEMPLATE.
TILES SHOULD BE DRAWN
SMALLER AND FLATTER
AS THEY GET CLOSER
TO THE V.P.

FIGURE 41.13 ■ Drawing curved tiles is similar to the layout procedure for drawing shingles. Tiles should be drawn smaller as they get closer to the vanishing point.

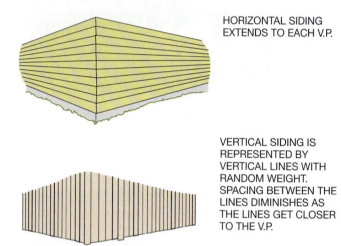

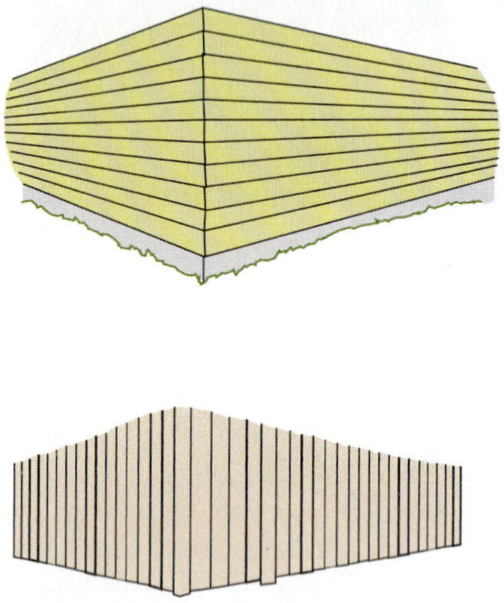
HORIZONTAL SIDING
EXTENDS TO EACH V.P.
VERTICAL SIDING IS
REPRESENTED BY
VERTICAL LINES WITH
RANDOM WEIGHT.
SPACING BETWEEN THE
LINES DIMINISHES AS
THE LINES GET CLOSER
TO THE V.P.

FIGURE 41.14 ■ Drawing horizontal or vertical siding.

of drawing siding. Paneling is usually drawn using a combination of vertical straightedge and freehand lines to represent the wood grain.

ENTOURAGE

Entourage is the term for the surroundings of a building. Surroundings consist of ground cover, trees, people, and cars. The entourage helps to create an attractive, realistic drawing. In addition, it helps define the scale of the drawing. The illustrator must be very careful that the entourage is not allowed to compete with the structure. The surroundings should accent rather than detract from the building. Figure 41.15 shows an example of entourage that blends well with the structure. Notice that rocks, plants, and trees help present the residence as it might appear but do not draw attention away from it.

Plants

Plants can be drawn by hand or applied as rub-ons. A wide variety of rub-on trees and plants are available for use on perspectives. If rub-ons are to be used, they should be placed on the drawing after the linework has been completed. This will help keep the

Sweetstream, Plan No. 256
©1992 Stephen Fuller American Home Gallery, Ltd.

FIGURE 41.15 ■ Entourage should be used to help present the structure in a lifelike setting. *Sweetstream,* PLAN NO. 256 © 1992. *Courtesy Stephen Fuller, American Home Gallery, Ltd.*

plants from being peeled off the illustration as you draw. Care must be taken to outline the shape of the plant so that the lines of the structure are not placed where the plants will go. If a diazo copy will be made, entourage can be placed on the back of the drawing. If the drawing will be reproduced photographically, the rub-ons must be placed on the front side of the drawing material.

Although rub-on plants may be convenient, every drafter should know how to draw entourage. Freehand entourage can best be drawn by spending time observing and sketching various kinds of trees and plants. The method used to draw the entourage should be consistent with the method used to draw the structure. A wide variety of plants can be seen by studying the rendered perspective drawings presented throughout this text. Practice drawing trees and plants that typically grow in your area until you gain confidence in your work. Keep trees in proportion to the structure. Heights for plants can be projected using methods similar to those used for the structure.

Another common element that should be included with plantings is a ground cover such as grass. Broad strokes of a soft lead are often used to represent grass and the shape of the ground, as shown in Figure 41.16. Ground cover provides a way to fill in the white space around a structure without distracting the viewer.

People

Many illustrators include people as part of their entourage. If properly drawn, people can help set the scale of a structure and give depth to a drawing. The heights of the people in your drawing should be projected from the vanishing point to the area where they will be placed, so that they will be consistent with the height of the structure. Figure 41.17 shows how people can be drawn in a rendering. As you study the people shown in the renderings throughout this text, you will notice that some of the figures have been outlined while others have been finely detailed.

The method used to draw people should be consistent with the scale of the drawing, and their location should not detract from the main focus of the drawing. Depending on your artistic ability, you may want to collect from magazines and clothes cat-

FIGURE 41.16 ■ Ground cover will often show individual plants. *West Paces Estate,* PLAN NO. 148 © 1988. *Courtesy Stephen Fuller, American Home Gallery, Ltd.*

alogs illustrations of people of various sizes in different poses. Such pictures could then be traced onto your rendering.

Furniture

For an interior rendering, furniture and people together help to provide a reference for scale. Furniture should be kept simple and consistent with the line quality of the drawing. Figure 41.18 shows how furniture can help draw the viewer into the drawing to create a sense of realism, rather than showing just a drawing of a room.

Cars

Cars can be a very attractive addition to a rendering but can easily overshadow the structure. Be careful not to detail a car so completely that the viewer is thinking about the car rather than the structure. Care must be taken in the placement of the car so that it blends with the surroundings. Drawing people near cars helps establish the scale of the entourage.

FIGURE 41.18 ■ Furniture helps show how a space can be used.

Content

You've been exposed to many of the elements that are typically placed in renderings, but you still must consider which elements should be used in your rendering and where these elements should be placed. The entourage should be consistent with the purpose of the drawing and should not obscure the structure. The following three guidelines are helpful in planning your drawing in order to maintain a proper relation between the structure and the entourage:

1. Use only entourage that is necessary to show the location, scale, and usage of the structure.
2. Draw the entourage with only enough detail to illustrate the desired material.
3. Never obscure structural elements with entourage.

FIGURE 41.17 ■ People and automobiles are often included in a drawing to give the viewer a sense of belonging. *Courtesy Kenneth D. Smith and Associates, Inc.*

PLANNING THE DRAWING

Before rendering a presentation drawing, do several sketches to determine the effectiveness of a layout. Consideration should be given to the placement of the building, people, landscaping, and contrasting values that will be created by the layout. Figure 41.19 is an example of a rendering that has made good use of contrast to help present a structure. Entourage has been used to create surfaces against which the structure can be contrasted.

The type of drawing that you are rendering should affect the type of balance that is used in your presentation. When you are making presentation drawings of residential projects, an informal method of balance will typically be most appropriate. Figure 41.20 shows an example of informal balance with the mass of the structure and the entourage offsetting each other.

FIGURE 41.19 ■ Entourage should be used to accent the structure. *Cantwell* PLAN NO. 432 © 1994. *Courtesy Stephen Fuller, American Home Gallery, Ltd.*

FIGURE 41.20 ■ Informal balance can be seen in this drawing: the trees are used to offset the mass of the structure. *Eastport Overlook* PLAN NO. 345, © 1994. *Courtesy Stephen Fuller, American Home Gallery, Ltd.*

TWO-POINT PERSPECTIVE RENDERING

See Figure 41.21 for steps 1 through 3.

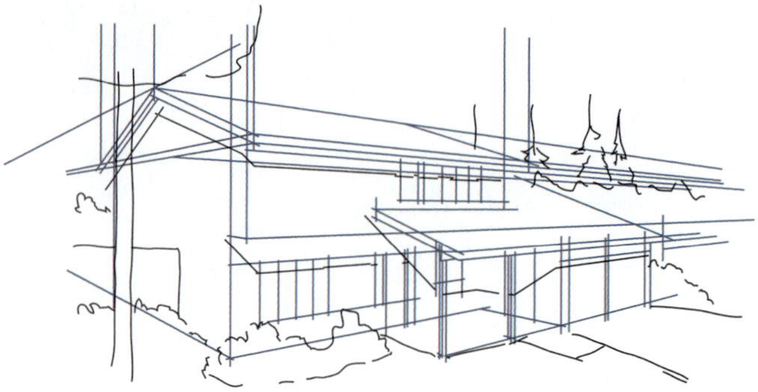

FIGURE 41.21 ■ With the perspective drawn with construction lines, entourage should be sketched in. Notice that features that are farthest from the viewer are often omitted to enhance the presentation.

STEP 1 Tape your perspective layout to your drawing surface. If you are using graphite and vellum, cover your layout with a thin sheet of sketch paper. This will keep the layout from being traced onto the back of your vellum as you trace the layout.

STEP 2 Lay out areas in the foreground where entourage will be placed.

STEP 3 Lay out areas that will be shaded.

See Figure 41.22 for steps 4 through 6.

FIGURE 41.22 ■ Draw the foreground entourage prior to drawing the structure. Background and ground cover can also be added.

STEP 4 Draw trees that will be placed in the foreground.

STEP 5 Draw people and cars that will be placed on the drawing. On residential renderings, people and cars are not usually drawn.

STEP 6 Draw entourage to help tie the structure to the ground.

FIGURE 41.23 ■ The structure can be rendered once the foreground entourage is drawn. Work from the top to bottom, and from front to back.

FIGURE 41.24 ■ Highlights are added to indicate surface changes.

STEP 7 Render the structure by first drawing material in the foreground; then work into the background. See Figure 41.23.

See Figure 41.24 for steps 8 through 10.

STEP 8 Draw major areas of shades and shadows.

STEP 9 Highlight areas in materials that will cast shadows.

STEP 10 Letter the job title or owner's name in an appropriate place.

ONE-POINT PERSPECTIVE RENDERING

One-point perspective can be drawn by following the same procedure as for two-point perspective. The entourage that will be shown for an interior view will be different from an exterior illustration; however, the interior rendering must be drawn in the same sequence as an exterior view. That is, draw the foreground entourage before drawing the cabinets or furniture in the background. See Figure 41.25.

FIGURE 41.25 ■ The perspective shown in Figure 40–32 is rendered to show depth and texture.

CADD APPLICATIONS

Rendering has been greatly altered by the use of computer software. Many companies provide software that will create a 3D image of a structure. Most programs will generate a wire frame drawing that presents the shape of the structure similarly to the drawings that were created in Chapter 40 using manual methods. The wire frame can be rotated, the vanishing points can be altered, or the station point can be relocated. Once a view is determined, surface materials can be added, shades and shadows can be cast, and entourage can be inserted. Many software programs that are used to create perspectives, can also be used to render drawings. Many programs are also available that can be used to render the wire frame. Figures 41.10, 41.11, 41.17, and 41.18 were rendered using CADD.

Chapter 41 Additional Reading

The following Web sites can be used as a resource to help you keep current with changes in rendering software.

ADDRESS	COMPANY OR ORGANIZATION
www.accurender.com	Accurender
www.chiefarch.com	Advanced Relational Technologies, Inc
www.ketiv.com	Ketiv Technologies
www.lightscape.com	Lightscape Technologies
www.peopleforpeople.com	People for People
www.solidbuilder.com	Eagle Point
www.softcad.com	SoftCAD International
www.softplan.com	SoftPlan Systems
www.visapp.com	Visual Applications

CHAPTER 41

Rendering Methods for Perspective Drawings Test

DIRECTIONS

Answer the questions with short, complete statements or drawings as needed.

1. Letter your name and the date at the top of the sheet.
2. Letter the question number and provide the answer. You do not need to write out the question.

QUESTIONS

Question 41–1 What two principles affect the layout of shade and shadows?

Question 41–2 Sketch an example of how shakes appear in a rendering.

Question 41–3 List three steps that could help improve a drafter's rendering technique.

Question 41–4 At what angles are shadows usually projected? Explain.

Question 41–5 If you would like to have both surfaces in two-point perspective in direct sunlight, where would you place the light source?

Question 41–6 What is entourage?

Question 41–7 List three guidelines for planning entourage.

Question 41–8 Sketch four examples of how glass can be rendered.

Question 41–9 Sketch examples of how depth can be shown at a window.

Question 41–10 Describe the elements to be considered in planning a rendering.

PROBLEMS

Problem 41–1 Using your two-point perspective drawing from Problem 40–7, render the residence using either graphite or ink. Draw entourage by hand; do not use rub-ons. Use the examples found in this text as guidelines for drawing your entourage.

Problem 41–2 Using your one-point perspective drawing from Problem 40–8, render the interior illustration using either graphite or ink. Draw entourage freehand; do not use rub-ons. Use the examples found in this text as guidelines for drawing your entourage.

SECTION 12

General Construction Specifications

42 CHAPTER

Construction Specifications

INTRODUCTION

Building plans, including all of the elements that make up a complete set of residential or commercial drawings, contain most general and some specific information about construction. However, it is very difficult to provide all of the required information on a set of plans. Schedules, as discussed in detail in Section 4, provide a certain amount of subordinate information. Information that cannot be provided clearly or completely on the drawing or in schedules is provided in construction specifications. Specifications are an integral part of any set of plans.

Most lenders have a format for giving residential construction specifications. The Federal Housing Administration (FHA) or the Federal Home Loan Mortgage Corporation (FHLMC) has a specification format entitled Description of Materials. This specifications form is used widely, as is or with revisions, by most residential construction lenders. The same form is used by the Farm Home Administration (FmHA) and by the Veterans Administration (VA). A completed FHA Description of Materials form for a typical structure is found on the Online Companion that accompanies this text. These plans, the construction specifications, and the building contract together become the legal documents for the construction project. These documents should be prepared very carefully in cooperation with the architect, client, and contractor. Any deviation from these documents should be approved by all three parties. When brand names are used, a clause specifying "or equivalent" may be added. This means that another brand equivalent in value to the one specified may be substituted with the construction supervisor's approval.

SPECIFICATIONS FOR COMMERCIAL CONSTRUCTION

Specifications are written documents that describe in detail the requirements for products, materials, and workmanship upon which the construction project is based. A *specification* is an exact statement describing the characteristics of a particular aspect of the project.

Specifications for commercial construction projects are often more complex and comprehensive than the documents for residential construction. Commercial project specifications may provide very detailed instructions for each phase of construction. Specifications may establish time schedules for the completion of the project. Also, in certain situations, the specifications include inspections in conjunction with or in addition to those required by a local jurisdiction.

Construction specifications often follow the guidelines of the individual architect or engineering firm although a common format has been established that is called *Master Format*™ and is published by the Construction Specifications Institute (CSI) and the Construction Specifications Canada (CSC). *Master Format* is a list of numbers and titles that are created for the organization of information into a standard sequence that relates to construction requirements, products, and activities. This format is called the "Master List of Numbers and Titles for the Construction Industry." The list is a standard filing system and format that ensures that construction specifications are prepared and filed in the same manner by every architect and contractor in the construction industry. *Master Format* is widely used in the United States and Canada and has been adopted for use by the U.S. Department of Defense. It includes the following general group of construction documents:

- Bidding Requirements
- Contract Requirements
- Facilities and Spaces
- Systems and Assemblies
- Construction Products and Activities—Specifications

"Bidding Requirements and Contract Requirements" are referred to as series zero because they begin with a 00 numbering system prefix. These documents are not specifications. They establish relationships, processes, and responsibilities for projects.

"Facilities and Spaces" and "Systems and Assemblies" are available for project-specific requirements. These areas do not have a *Master Format* numbering system or titles. However, they do have a *UniFormat*™ numbering system and title. *UniFormat* is an arrangement of construction information based on physical parts of a facility called systems and assemblies. It is intended to complement *Master Format*. *UniFormat* is set up in the following categories:

- Project Description
- A Substructure
- B Shell
- C Interiors

- D Services
- E Equipment and Furnishings
- F Special Construction and Demolition
- G Building Sitework
- Z General

"Project Description" provides an introduction to a project and does not have a letter designation. "General" (Z) is provided for expansion beyond the building construction. Categories A through Z are referred to as Level 1. The next subdivision of Level 1 is Level 2, which has a two-digit number. Further subdivisions are Level 3, with two more numbers followed by a period; and then Level 4, with numbers assigned by the user. For example, four levels in "Interiors" might read C1010.10 where C is Level 1, 10 is Level 2, 10 is Level 3, and .10 is Level 4. A Level 5 can also be created as a subset of Level 4 for presenting specialized design solutions. These numbers are displayed by placing another period after the Level 4 number. Level 1 and Level 2 of *UniFormat* are found on the Online Companion that accompanies this text.

"Construction Products and Activities—Specifications" are the core of *MasterFormat*. The specifications are established in 16 divisions. Subdivisions of each general *MasterFormat* division are identified with five-digit numbers. For example, one of the major categories is "Division 16 Electrical." One of Division 16 subsection is "16500 Lighting." Additional subsections are identified with dashed numbers, such as "–510 Interior Luminaries." The complete number down to this subtitle would be 16500–510. Level 1 and Level 2 of *MasterFormat* are found on the CD that accompanies this text.

CONSTRUCTION DOCUMENTS

Documents is a general term that refers to all drawings and written information related to a project. *Construction documents* are drawings and written documents prepared and assembled by architects and engineers for communicating the design of the project and administering the construction contract. The two major groups of construction documents are bidding requirements and contract documents. *Bidding requirements* are used to attract bidders and provide the procedures to be used for submitting bids. *Bidding documents* are the construction documents issued to bidders for the purpose of providing construction bids. *Contract documents* are the legal requirements that become part of the construction contract. Contract documents are where the construction drawings and specifications are found.

Construction Drawings

You have been studying and creating construction drawings while learning the concepts covered throughout this textbook. Drawings show lines and text for the purpose of providing information about the project. These drawings are a principal part of the set of construction documents. What drawings are needed will depend on the specific requirements of the construction project. The drawings for a small residential addition might fit on one or two pages; the drawings for a commercial building might be on a hundred or more pages. Drawings vary in how much information they show, depending on the use, the project phase, and the desired representation. In addition to plan views, elevations, sections and details, drawings can have schedules that have a detailed list of components, items, or parts to be furnished in the project.

Coordinating Drawings and Specifications

A complete set of construction documents contains drawings and specifications. One person or a unified team of people should coordinate drawings and specifications. The elements used in the drawings and specifications—such as symbols, abbreviations, and terminology—should be standardized to help avoid confusion. After the construction documents are prepared in a professional manner, the coordination and quality pledge must be conducted with effective communication.

The drawings should locate and identify materials and should include the assembly of components, dimensions, details, and diagrams. The drawings have local and specific notes, but notes should be used only to identify—not to describe—a material or part. Notes that are too detailed can obscure the drawing. Detailed written information should be placed in the specifications. Symbols that are used in the set of drawings should be represented as an approved standard and should be shown and labeled in a legend for reference purposes. Drawings do not need to be cross-referenced to the specifications. Drawings and specifications are combined to become the complete set of construction documents; this means that you do not have to provide notes that refer to the specifications when identifying an item on the drawings. The specifications are used to define the specific quality and type of material, equipment, and installation. The drawings provide quantity, capacity, location, and general written information in the form of notes, while the specifications clearly define items such as minimum requirements, physical properties, chemical composition, and installation procedures. Schedules are used on drawings to help simplify communication by providing certain items in a table format. The schedules can be placed on the drawings or in the specifications. The information found in schedules or on the drawings should not repeat the information found in the specifications. The drawings, schedules, and specifications must be carefully coordinated so the information is consistent. The *MasterFormat* divisions and sublevels can be used as a checklist to ensure that every required specification is included.

MasterFormat is an effective system for indexing large project specifications. Much of the *MasterFormat* cannot be effectively used in some light commercial or residential construction. When comprehensive specifications are required for residential construction, an abridged format of the CSI/CSC system may be used. This short form may include the major divisions and exclude the subdivisions.

CHAPTER 42

General Construction Specifications Test

DIRECTIONS

Answer the questions with short, complete statements or drawings as needed.

1. Letter your name and the date at the top of the sheet.
2. Letter the question number and provide the answer. You do not need to write out the question.

QUESTIONS

Question 42–1 Give a general definition of construction specifications.

Question 42–2 Identify the basic differences between residential and commercial specifications.

Question 42–3 List four factors that influence the specific requirements of minimum construction specifications established by local building officials.

Question 42–4 Using general terms, list the typical minimum construction requirements for the following categories:

a. Room dimensions
b. Light and ventilation
c. Foundation
d. Framing
e. Stairways
f. Roof
g. Chimney and fireplace
h. Thermal insulation and heating
i. Fire warning system

A partial example would be as follows:

4. d. Framing
    1. lumber grades
    2. beams bearing area

Question 42–5 Describe the format established by the Construction Specifications Institute (CSI) for construction specifications.

PROBLEM

Problem 42–1 Obtain a blank copy of the FHA Description of Materials form from your instructor or local FHA office. Using your set of architectural plans for the residence you have been drawing as a continuing problem, complete the FHA form. If you have more than one set of plans, select only one set or have your instructor select the set of plans to use.

CHAPTER 43

Construction Supervision Procedures

INTRODUCTION

The architect or designer may often be involved with several phases of preparation before construction of a project begins. The level of commitment may go beyond the preparation of plans to the complete supervision of the entire construction project. There are several items that any client may need before construction starts, depending on the complexity or unique requirements of the project. Architects are often involved with zone changes when necessary, specification preparation, building permit applications, bonding requirements, the client's financial statement, the lender's approval, and the building contractor's estimates and bid procurement.

LOAN APPLICATIONS

Loan applications vary depending on the requirements of the lender. Most applications for construction financing contain a variety of similar information. The construction appraisal requirements proposed by the FHA are, in part, as follows:

1. Plot plan (three copies per lot).
2. Prints (three copies per plan). Prints should include the following:

   Four elevations: front, rear, right side, and left side.

   Floor plan(s).

   Foundation plan.

   Wall section(s).

   Roof plan.

   Cross sections of exterior walls, stairs, etc.

   Cabinet detail including a cross section.

   Fireplace detail (manufacturer's detail if fireplace is prefabricated).

   Truss detail and engineering (include name and address of supplier).

   Heating plan. For forced-air, show size and location of ducts and registers plus cubic feet per minute (CFM) at each register, and furnace Btus.

   Location of wall units with watts and CFM at register.

3. Specifications (one copy per print). Areas commonly missed on specifications are the following:

   List all appliances with make and model number.

   List smoke detector with make and model number and "Direct Wire."

   If the project is landscaped, supply a typical detail for each plan type.

   List plumbing fixtures with make, model number, size, and color.

   List carpet by brand, style, and directory number.

   List carpet pad by brand, style, thickness, and directory number.

   Include name and address of manufacturer of trusses.

   Include name of manufacturer of fireplace and model number if applicable.

4. Public Utility District (PUD) or heating contractor's heat loss calculation (one per specification sheet). Room-by-room Btu heat loss count showing the total Btu heat loss for the dwelling as well as the total output of the heating system.
5. Proposed sale price.
6. Copy of earnest money agreement, blank or (for a presale) completed.
7. For a presale, the buyer as well as the builder must sign the specification sheet.

INDIVIDUAL APPRAISAL REQUIREMENTS

The subdivision should be FHA-approved, or a letter must accompany the submission with the following information:

1. Total number of lots in the subdivision.
2. Evidence of acceptance of streets and utilities by local authorities for continued maintenance.
3. One copy of the recorded plat and covenants. Covenants are legal agreements setting the conditions and restrictions for the property.

MASTER APPRAISAL REQUIREMENTS

A minimum of five lots for each plan type is required for a master. The subdivision must be FHA- or VA-approved and contain the following information:

1. Location map and a copy of the recorded plat and covenants.
2. A letter from the builder that includes the following information:
   a. The number of lots in the area owned by the builder and the number of lots proposed to be built upon.
   b. The number of homes presently under construction and the number completed and unsold.

If the builder has not worked with the FHA previously, an equal employment opportunity certificate and an affirmative fair housing marketing plan must be submitted for either a master or an individual appraisal request.

CHANGE ORDERS

Any physical change in the plans or specifications should be submitted to the FHA on FHA Form 2577. It should be noted whether the described change will be an increase or decrease in value and by what dollar amount. It must be signed by the lender, the builder, and, if the structure is sold, the purchaser, who must also sign prior to submission to the FHA.

Most lending institutions' applications are not as comprehensive as the FHA's; others may require additional information. The best practice is to research the lender to identify the needed information. A well-prepared set of documents and drawings usually stands a better chance of funding.

BUILDING PERMITS

The responsibility of completing the building permit application may fall to the architect, designer, or builder. The architect should contact the local building official to determine the process to be followed. Generally the building permit application is a basic form that identifies the major characteristics of the structure to be built, the legal description and location of the property, and information about the applicant. The application is usually accompanied by two sets of plans and up to five sets of plot plans. The fee for a building permit usually depends on the estimated cost of construction. The local building official determines the amount, which is based on a standard schedule at a given cost per square foot. The fees are often divided into two parts: a plan-check fee paid upon application and a building-permit fee paid when the permit is received. There are other permits and fees that may or may not be paid at this time. In some cases the mechanical, sewer, plumbing, electrical, and water permits are obtained by the general contractor or subcontractors. Water and sewer permits may be expensive, depending on the local assessments for these utilities.

CONTRACTS

Building contracts may be very complex documents for large commercial construction or short forms for residential projects. The main concern in the preparation of the contract is that all parties understand what will specifically be done, in what period of time, and for what reimbursement. The contract becomes an agreement between the client, general contractor, and architect. An example of a typical building contract is found on the Online Companion that accompanies this text.

It is customary to specify the date by which the project is to be completed. On some large projects, dates for completion of the various stages of construction are specified. Usually the contractor receives a percentage of the contract price for the completion of each stage of construction. Payments are typically made three or four times during construction. Another method is for the contractor to receive partial payment for the work done each month. Verify the method used with the lending agency.

The owner is usually responsible for having the property surveyed. The architect may be responsible for administering the contract. The contractor is responsible for construction and for the security of the site during the construction period.

Certain kinds of insurance are required during construction. The contractor is required to have liability insurance. This protects the contractor against being sued for accidents occurring on the site. The owner is required to have property insurance, including fire, theft, and vandalism. Workers' compensation is another form of insurance that provides income for contractors' employees if they are injured at work. The contractor must be licensed, bonded, and insured under the requirements of the state where the construction occurs.

The contract describes conditions under which it may be ended. A contract may be terminated if one party fails to comply with it, if one of the parties is disabled or dies, and for several other reasons.

There are two kinds of contracts in use for most construction—the fixed-sum and the cost-plus contract, each offering certain advantages and disadvantages.

Fixed-sum (sometimes called *lump-sum*) contracts are used most often. With a fixed-sum contract, the contractor agrees to complete the project for a certain amount of money. The greatest advantage of this kind of contract is that the owner knows in advance exactly what the cost will be. However, the contractor does not know what hidden problems may be encountered, and so the contractor's price must be high enough to cover unforseen circumstances, such as excessive rock in the excavation or sudden increases in the cost of materials.

A *cost-plus* contract is one in which the contractor agrees to complete the work for the actual cost, plus a percentage for

overhead and profit. The advantage of this type of contract is that the contractor does not have to allow for unforseen problems. A cost-plus contract is also useful when changes are to be made during the course of construction. The main disadvantage of this kind of contract is that the owner does not know exactly what the cost will be until the project is completed.

COMPLETION NOTICE

The *completion notice* is a document that should be posted in a conspicuous place on or adjacent to the structure. This legal document notifies all parties involved in the project that work has been substantially completed. There may be a very small part of work or cleanup to be done, but the project, for all practical purposes, is complete. The completion notice must be recorded in the local jurisdiction. Completion notices served several functions. Subcontractors and suppliers have a certain given period of time to file a claim or lien against the contractor or client to obtain reimbursement for labor or materials that have not been paid for. Lending institutions often hold a percentage of funds for a given period of time after the completion notice has been posted. It is important for the contractor to have this document posted so the balance of payment can be obtained. The completion notice is often posted in conjunction with a final inspection that includes the local building officials and the client. Some lenders require that all building inspection reports be submitted before payment is given, and some also require a private inspection by an agent of the lender. A sample completion notice is found on the Online Companion that accompanies this text.

BIDS

Construction bids are often obtained by the architect for the client. The purpose is to get the best price for the best work. Some projects require that work be given to the lowest bidder; other projects do not necessarily go to the lowest bid. In some situations, especially in the private sector, other factors are considered, such as an evaluation of the builder's history based on quality, ability to meet schedules, cooperation with all parties, financial stability, and license, bond, and insurance.

The bid becomes part of the legal documents for completion of the project. The legal documents may include plans, specifications, contracts, and bids. The following items are part of a total analysis of costs for residential construction. The architect and client should know clearly what the bid includes.

1. Plans
2. Permits, fees, specifications
3. Roads and road clearing
4. Excavation
5. Water connection (well and pump)
6. Sewer connection (septic)
7. Foundation, waterproofing
8. Framing, including materials, trusses, and labor
9. Fireplace, including masonry and labor
10. Plumbing, both rough and finished
11. Wiring, both rough and finished
12. Windows
13. Roofing, including sheet metal and vents
14. Insulation
15. Drywall or plaster
16. Siding
17. Gutters, downspouts, sheet-metal, and rain drains
18. Concrete flatwork and gravel
19. Heating
20. Garage and exterior doors
21. Painting and decorating
22. Trim and finish interior doors, including material and labor
23. Underlayment
24. Carpeting, including the amount of carpet and padding, and the cost of labor
25. Vinyl floor covering—amount and cost of labor
26. Formica
27. Fixtures and hardware
28. Cabinets
29. Appliances
30. Intercom or stereo system
31. Vacuum system
32. Burglar alarm
33. Weatherstripping and venting
34. Final grading, cleanup, and landscaping
35. Supervision, overhead, and profit
36. Subtotal of land costs
37. Financing costs

CONSTRUCTION INSPECTIONS

When the architect or designer is responsible for the supervision of the construction project, then it is necessary to work closely with the building contractor to get the proper inspections at the necessary times. Two types of inspections occur most frequently. The regularly scheduled code inspections that are required during specific phases of construction help ensure that the construction methods and materials meet local and national code requirements. The general intent of these inspections is to protect the safety of the occupants and the public. Another type of inspection is often conducted by the lender during

certain phases of construction. The purpose of these inspections is to ensure that the materials and methods described in the plans and specifications are being used. The lender has a valuable interest here. If the materials and methods are inferior or not of the standard expected, then the value of the structure may not be what the lender had considered when a preliminary appraisal was made. Another reason for these inspections, and probably the reason that the builder likes best, has to do with disbursement. Disbursement inspections may be requested at various times, such as monthly, or they may be related to a specific disbursement schedule: four times during construction, for example. The intended result of these inspections is the release of funds for payment of work completed.

When the architect or designer supervises the total construction, then he or she must work closely with the contractor to ensure that the project is completed in a timely manner. When a building project remains idle, the overhead costs, such as construction interest, begin to add up quickly. A contractor who bids a job high but builds quickly may be able to save money in the final analysis. Some overhead costs go on daily even when work has stopped or slowed. The supervisor should also have a good knowledge of scheduling so that inspections can be obtained at the proper time. If an inspection is requested when the project is not ready, the building official may charge a fee for excess time spent. Always try to develop a good rapport with building officials so each encounter goes as smoothly as possible.

CHAPTER 43 Construction Supervision Procedures Test

DIRECTIONS

Answer the questions with short, complete statements or drawings as needed on an 8 1/2″ × 11″ sheet of notebook paper, as follows:

1. Letter your name and the date at the top of the sheet.
2. Letter the question number and provide the answer. You do not need to write out the question.

QUESTIONS

Question 43–1 Define and list the elements of the following construction-related documents:

a. loan applications
b. contracts
c. building permits
d. completion notice
e. bids
f. change orders

Question 43–2 Outline and briefly discuss the required building construction inspections.

DIRECTIONS

1. All forms must be neatly hand-lettered or typed, unless otherwise specified by your instructor.

PROBLEMS

Problem 43–1 Copy and then complete the building permit application on page 817. Use this information.

- Project location address: 3456 Barrington Drive, your city and state
- Nearest cross street: Washington Street
- Subdivision name: Barrington Heights
- Township: 2S
- Range: 1E
- Section: 36
- Tax lot: 2400
- Lot size: 15000 sq ft

BUILDING PERMIT APPLICATION

Amount Due ________

TO BE FILLED IN BY APPLICANT

Project Location (Address) ________

Nearest Cross Street ________

Subdivision Name ________ Lot ________ Block ________

Township ________ Range ________ Section ________ Tax Lot ________

Lot Size ________ (Sq. Ft.) Building Area ________ (Sq. Ft.) Basement Area ________ (Sq. Ft.) Garage Area ________ (Sq. Ft.)

Stories ________ Bedrooms ________ Water Source ________ Sewage Disposal ________

Estimated Cost of Labor and Material ________

Plans and Specifications made by ________ accompany this application.

Owner's Name ________ Builder's Name ________

Address ________ Address ________

City ________ State ________ City ________ State ________

Phone ________ Zip ________ Phone ________ Zip ________

I certify that I am registered under the provisions of ORS Chapter 701 and my registration is in full force and effect. I also agree to build according to the above description, accompanying plans and specifications, the State of Oregon Building Code, and to the conditions set forth below.

________ APPLICANT ________ HOMEBUILDER'S REGISTRATION NO. ________ DATE

I agree to build according to the above description, accompanying plans and specifications, the State of Oregon Building Code, and to the conditions set foith below.

________ APPLICANT ________ DATE

- Building area: 2000 sq ft
- Basement area: None
- Garage area: 576 sq ft
- Stories: 1
- Bedrooms: 3
- Water source: Public
- Sewage disposal: Public
- Estimated cost of labor and materials: $88,500
- Plans and specifications made by: You
- Owner's name: Your teacher; address and phone may be fictitious
- Builder's name: You, your address, and your phone number
- You sign as applicant
- Homebuilder's registration number: Your social security number or another fictitious number
- Date: Today's date

Problem 43–2 Copy and then complete the building contract found on the Online Companion that accompanies this text. Use this information:

- Today's date.
- Name yourself as the contractor and your teacher as the owner. (Give complete first name, middle initial, and last name where names are required.)
- ARTICLE I:
  Construction of approximate 2000-square-foot house to be located at 3456 Barrington Drive, your city and state. Also known as Lot 8, Block 2, Barrington Heights, your county and state.
- Drawing and specifications prepared by you.
- All said work to be done under the direction of you.
- ARTICLE II:
  Commence work within 10 days and substantially complete on or before:
  Give a date four months from the date of the contract.
- ARTICLE III:
  $88,500
  Payable at the following times:
  1. One-third upon completion of the foundation.
  2. One-third upon completion of drywall.
  3. One-third 30 days after posting the completion notice.
- ARTICLE X:
  Insurance $250,000; $500,000; and $50,000.
- ARTICLE XVIII:
  Give the owner's name, address, and phone number. (A fictitious address and phone number may

CHANGE ORDER

Date: ____________________

Project Name: ____
____

Location: ____
____
____

Description of Change:

Additional Cost: ____________________

Reduction in cost: ____________________

Adjusted total project cost: ______________________________________

____
OWNER

____
BUILDER

____
LENDER

be used if preferred.) Give your name and address as the contractor.

- ARTICLE XX:
  This contract amount is valid for a period of 60 days. If, for reasons out of the contractor's control, construction has not begun by the end of this 60-day period, contractor has the right to rebid and revise the contract.

Problem 43–3 Copy and then complete the change order form above based on this information:

- Today's date
- Project Name: Your teacher's name
- Location: 3456 Barrington Drive, your city and state. Also known as Lot 8, Block 2, Barrington Heights, your county and state
- Description of change: Add a 24″ × 48″ insulated tempered flat glass skylight in the vaulted ceiling centered in the entry foyer.
- Additional cost: $850
- Adjusted total project cost: $89,350
- Signatures

Problem 43–4 Copy and then complete the completion notice (top half of form above double line only; the rest of the information is for official Notary Public and recording officer) found on the Online Companion that accompanies this text. Use this information:

- Today's date
- Owner or mortgagee: Your name
- P.O. address: Your address

SECTION 13

Commercial Drafting

44 CHAPTER

Building Codes and Commercial Design

INTRODUCTION

One of the biggest differences you will notice between residential and commercial drafting is that you will depend more on building codes. You were introduced to residential building codes in Chapter 7 and have seen their effect on construction throughout the book. In residential drafting you may have been able to use a shortened version of the code such as the 2000 International Residential Code for One- and Two-Family Dwellings. If so, commercial construction drafting will be your first introduction to the full building code. As you draw commercial projects such as the structure shown in Figure 44.1, the building codes will become much more influential. To be an effective drafter, you must be able to use properly the code that governs your area. Although there are many different codes in use throughout the country, most will be similar to the Uniform Building Code (UBC), the Basic National Building Code (BOCA), the Standard Building Code (SBC) or the International Building Code (IBC). These comprehensive codes have been developed over many years and are well researched. Each code is performance-based, describing the desired results with a wide latitude allowed to achieve those results. The 2000 IBC was still under development at the time of printing of this book.

If you haven't done so already, check with the building department in your area and determine the code that will cover your drafting. Typically you will spend approximately $90.00 to purchase a code book. This may seem like a large expenditure to you now, but it will be one of your most useful drafting tools. Your code book should be considered a drafting tool just as pencils, vellum, or your workstation are. If possible, purchase the loose-leaf version of the code rather than the hardcover edition. The hardcover edition may look nicer sitting on your shelf, but this will be one book that will not just sit. The loose-leaf edition will allow you to update your codes with yearly amendments, as well as state and local additions.

FIGURE 44.1 ■ Working on drawings for commercial structures such as the Honda Building in Quebec, Canada, will require extensive knowledge of the building code that governs the project. *Courtesy Trendstone.*

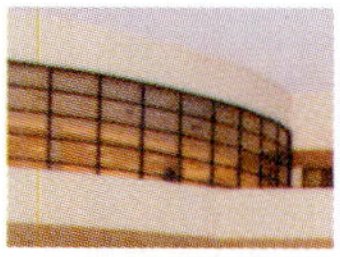

EXPLORING BUILDING CODES

Building codes influence every aspect of a construction project. Each national code consists of several related books that regulate design, electrical, energy, fire, mechanical, plumbing, structural, and zoning. This chapter will examine only portions of the Uniform Building Code that regulate the design of a project. Building codes do not dictate the style to be used for the structure, but they do regulate the location, size, and type of materials that can be used; these are based on the size of the building and on how it will be used. The project architect will initially make each of the code decisions that affect the project. Understanding the structure of the codes and how each aspect of the code defines the structure will make you a better drafter and will increase your ability to advance. To develop your skills, purchase a code book as a student, and don't hesitate to mark or place tabs on certain pages to identify key passages. You won't need to memorize the book, but you will need to refer to certain areas again and again. Page tabs will help you find needed tables or formulas quickly.

DETERMINING DESIGN CATEGORIES

To use a building code effectively during the initial design stage, you must determine classifications used to define a structure: the occupancy group, type of construction, floor area, height, building location and size, and the occupant load.

Occupancy Groups

The *occupancy group* specifies by whom or how the structure will be used. To protect the public adequately, buildings used for different purposes are designed to meet the hazards of each usage. Think of the occupancy group with which you are most familiar, the R occupancy. The R grouping not only covers the single-family residence but also includes duplexes, apartments, lodging areas, and hotels. This one group covers single-family residences and multistory apartments with hundreds of occupants. Obviously, the safety requirements for a multilevel apart-

ment with hundreds of occupants should be different from a single-family residence.

The code that you are using will affect how the occupancy is listed. The letter of the occupancy listing generally is the first letter of the word that it represents. Basic occupancy classifications for the model codes lump various uses into categories (classifications) such as:

CODE	USE
A	for Assembly-type uses such as theaters and churches
B	for nonhazardous, Business, such office buildings and buildings for service-type uses
E	for Educational buildings
F	for Factories—industrial
H	for Hazardous uses such as cabinet shops or areas where large volumes of flammable liquids are stored
I	for Institutional uses such as hospitals and jails
M	for Mercantile uses such as department stores and markets
R	for single-family or multifamily Residential
S	for Storage of moderate- and low-hazard materials
U	for Utility and miscellaneous (this does not appear in the SBC)

These uses may be further broken down into subcategories for factors such as the number of occupants or the degree of hazardous material used. Figure 44.2 shows a detailed listing of occupancy categories from the UBC.

Table 3-A—DESCRIPTION OF OCCUPANCIES BY GROUP AND DIVISION

GROUP AND DIVISION	SECTION	DESCRIPTION OF OCCUPANCY
A–1		A building or portion of a building having an assembly room with an occupant load of 1,000 or more and a legitimate stage.
A–2		A building or portion of a building having an assembly room with an occupant load of less than 1,000 and a legitimate stage.
A–2.1	303.1.1	A building or portion of a building having an assembly with an occupant load of 300 or more without a legitimate stage, including such buildings used for educational purposes and not classed as a Group E or Group B Occupancy.
A–3		Any building or portion of a building having an assembly room with an occupant load of less than 300 without a legitimate stage, including such buildings used for educational purposes and not classes as a Group E or Group B Occupancy.
A–4		Stadiums, reviewing stands and amusement park structures not included within other Group A Occupancies.
B	304.1	A building or structure, or a portion thereof, for office, professional or service-type transactions, including storage of records and accounts, and eating and drinking establishments with an occupant load of less than 50.
E–1		Any building used for educational purposes through the 12th grade by 50 or more persons for more than 12 hours per week or four hours in any one day.
E–2	305.1	Any building used for educational purposes through the 12th grade by less than 50 persons for more than 12 hours per week or four hours in any one day.
E–3		Any building or portion thereof used for day-care purposes for more than six persons.
F–1		Moderate-hazard factory and industrial occupancies include factory and industrial uses not classified as Group F, Division 2 Occupancies.
F–2	306.1	Low-hazard factory and industrial occupancies include facilities producing noncombustible or nonexplosive materials which during finishing, packing or processing do not involve a significant fire hazard.
H–1		Occupancies with a quantity of material in the building in excess of those listed in Table 3–D which present a high explosion hazard as listed in Section 307.1.1.
H–2	307.1	Occupancies with a quantity of material in the building in excess of those listed in Table 3–D which present a moderate explosion hazard or a hazard from accelerated burning as listed in Section 307.1.1.
H–3		Occupancies with a quantity of material in the building in excess of those listed in Table 3–D which present a high fire or physical hazard as listed in Section 307.1.1.
H–4		Repair garages not classified as Group S, Division 3 Occupancies.
H–5		Aircraft repair hangars not classified as Group S, Division 5 Occupancies and heliports.
H–6	307.1 and 307.11	Semiconductor fabrication facilities and comparable research and development areas when the facilities in which hazardous production materials are used, and the aggregate quantity of material is in excess of those listed in Table 3–D or 3–E.
H–7	307.1	Occupancies having quantities of materials in excess of those listed in Table 3–E that are health hazards as listed in Section 307.1.1.
I–1.1		Nurseries for the full-time care of children under the age of six (each accommodating more than five children), hospitals, sanitariums, nursing homes with nonambulatory patients and similar buildings (each accommodating more than five patients).
I–1.2	308.1	Health-care centers for ambulatory patients receiving outpatient medical care which may render the patient incapable of unassisted self-preservation (each tenant space accommodating more than five such patients).
I–2		Nursing homes for ambulatory patients, homes for children six years of age or over (each accommodating more than five persons).
I–3		Mental hospitals, mental sanitariums, jails, prisons, reformatories and buildings where personal liberties of inmates are similarly restrained.
M	309.1	A building or structure, or a portion thereof, for the display and sale of merchandise, and involving stocks of goods, wares or merchandise, incidental to such purposes and accessible to the public.
R–1	310.1	Hotels and apartment houses, congregate residences (each accommodating more than 10 persons).
R–3		Dwellings, lodging houses, congregate residences (each accommodating 10 or fewer persons).
S-1		Moderate hazard storage occupancies include buildings or portions of buildings used for storage of combustible maerials not classified as Group S, Division 2 or Group H Occupancies.
S-2	311.1	Low-hazard storage occupancies include buildings or portions of buildings used for storage of noncombustible materials.
S-3		Repair garages and parking garages not classified as Group S, Division 4 Occupancies.
S-4		Open parking garages.
S–5		Aircraft hangars and helistops.
U–1	312.1	Private garages, carports, sheds and agricultural buildings.
U–2		Fences over 6 feet (182.9 mm) high, tanks and towers.

FIGURE 44.2 ■ Common occupancy listings. *Reproduced from the 1997 edition of* The Uniform Building Code, *copyright © 1997, with permission of the publisher,* the *International Conference of Building Officials.*

Table 3–B Required Separation in Buildings of Mixed Occupancy[1] (Hours)

	A-1	A-2	A-2.1	A-3	A-4	B	E	F-1	F-2	H-2	H-3	H-4,5	H-6,7[2]	I	M	R-1	R-3	S-1	S-2	S-3	S-5	U-1[3]
A-1		N	N	N	N	3	N	3	3	4	4	4	4	3	3	1	1	3	3	4[5]	3	1
A-2			N	N	N	1	N	1	1	4	4	4	4	3	1	1	1	1	1	3[5]	1	1
A-2.1				N	N	1	N	1	1	4	4	4	4	3	1	1	1	1	1	3[5]	1	1
A-3					N	N	N	N	N	4	4	4	3	2	N	1	1	N	1	3[5]	1	1
A-4						1	N	1	1	4	4	4	4	3	1	1	1	1	1	3[5]	1	1
B							1	N[6]	N	2	1	1	1	2	N	1	1	N	1	1	1	1
E								1	1	4	4	4	3	1	1	1	1	1	1	3[5]	1	1
F-1									1	2	1	1	1	3	N[6]	1	1	N	1	1	1	1
F-2										2	1	2	1	2	1	1	1	N	1	1	1	1
H-1	NOT PERMITTED IN MIXED OCCUPANCIES. SEE CHAPTER 10.																					
H-2											1	1	2	4	2	4	4	2	2	2	2	1
H-3												1	1	4	1	3	3	1	1	1	1	1
H-4.5													1	4	1	3	3	1	1	1	1	1
H-6,7[2]														4	1	4	4	1	1	1	1	3
I															2	1	1	2	4	4[5]	3	1
M																1	1	1[4]	1[4]	1	1	1
R-1																	N	3	1	3[5]	1	1
R-3																		1	1	1	1	1
S-1																			1	1	1	1
S-2																				1	1	N
S-3																					1	1
S-4	OPEN PARKING GARAGES ARE EXCLUDED EXCEPT AS PROVIDED IN SECTION 311.2.																					
S-5																						N

N—No requirements for fire resistance.

[1]For detailed requirements and exceptions, see Section 302.4.

[2]For special provisions on highly toxic materials, see the Fire Code.

[3]For agricultural buildings, see also Appendix Chapter 3.

[4]See Section 310.2.1 for exception.

[5]Group S, Division 3 Occupancies used exclusively for parking or storage of pleasure-type motor vehicles and provided no repair or refueling is done may have the occupancy separation reduced one hour.

[6]For Group F, Division 1 woodworking establishments with more than 2,500 square feet, the occupancy separation shall be one hour.

FIGURE 44.3 ■ If a structure contains more than one occupancy grouping, the areas will need to be separated by protective construction methods to reduce the spread of fire. The fire rating of the separating construction can be determined by comparing the occupancy of each type of construction. *Reproduced from the 1997 edition of* The Uniform Building Code, *copyright © 1997, with permission of the publisher, the International Conference of Building Officials.*

A structure often has more than one occupancy group. Separation must be provided at the wall or floor dividing these different occupancies. Figure 44.3 shows a listing of required separations based on UBC requirements. If you are working on a structure that has a 10,000 sq ft office area over an enclosed parking garage, the two areas must be separated. By using the table in Figure 44.2, you can see that the office area is a B occupancy and the garage is an S4 occupancy. By using the table in Figure 44.3, you can determine that a floor with a one-hour rating must be provided between these two areas. The hourly rating is given to construction materials specifying the length of time that the item must resist structural damage that would be caused by a fire. Various ratings are assigned by the codes for materials ranging from nonrated to a four-hour rating. Whole chapters are assigned by each code to cover specific construction requirements for various materials. Chapter 7 of the UBC serves as an introduction to the fire resistance of materials.

Once the occupancy grouping is determined, turn to the section of the building code that introduces specifics of that occupancy grouping. Students often become frustrated with the amount of page turning that is required to find particular information in the codes. Often you will look in one section of the code only to be referred to another section of the code and then to another. It may be frustrating, but the design of a structure is often complicated and requires patience.

Table 6-A—TYPES OF CONSTRUCTION—FIRE-RESISTIVE REQUIREMENTS (In Hours)
For details, see occupancy section in Chapter 3, type of construction sections in this chapter and sections referenced in this table.

	TYPE I	TYPE II			TYPE III		TYPE IV	TYPE V	
	Noncombustible				Combustible				
BUILDING ELEMENT	Fire-resistive	Fire-resistive	1-Hr.	N	1-Hr.	N	H.T.	1-Hr.	N
1. Bearing walls—exterior	4 Sec. 602.3.1	4 Sec. 603.3.1	1	N	4 Sec. 604.3.1	4 Sec. 604.3.1	4 Sec. 605.3.1	1	N
2. Bearing walls—interior	3	2	1	N	1	N	1	1	N
3. Nonbearing walls—exterior	4 Sec. 602.3.1	4 Sec. 603.3.1	1 Sec. 603.3.1	N	4 Sec. 604.3.1	4 Sec. 604.3.1	4 Sec. 605.3.1	1	N
4. Structural frame[1]	3	2	1	N	1	N	1 or H.T.	1	N
5. Partitions—permanent	1[2]	1[2]	1[2]	N	1	N	1 or H.T.	1	N
6. Shaft enclosures[3]	2	2	1	1	1	1	1	1	1
7. Floors and floor-ceilings	2	2	1	N	1	N	H.T.	1	N
8. Roofs and roof-ceilings	2 Sec. 602.5	1 Sec. 603.5	1 Sec. 603.5	N	1	N	H.T.	1	N
9. Exterior doors and windows	Sec. 602.3.2	Sec. 603.3.2	Sec. 603.3.2	Sec. 603.3.2	Sec. 604.3.2	Sec. 604.3.2	Sec. 605.3.2	Sec. 606.3	Sec. 606.3
10. Stairway construction	Sec. 602.4	Sec. 603.4	Sec. 603.4	Sec. 603.4	Sec. 604.4	Sec. 604.4	Sec. 605.4	Sec. 606.4	Sec. 606.4

N—No general requirements for fire resistance. H.T.—Heavy timber.

[1]Structural frame elements in an exterior wall that is located where openings are not permitted or where protection of openings is required, shall be protected against external fire exposure as required for exterior bearing walls or the structural frame, whichever is greater.

[2]Fire-retardant-treated wood (see Section 207) may be used in the assembly, provided fire-resistance requirements are maintained. See Sections 602 and 603.

[3]For special provisions, see Sections 304.6, 306.6 and 711.

FIGURE 44.4 ■ Construction requirements for various parts of a structure, based on the UBC. *Reproduced from the 1997 edition of* The Uniform Building Code, *copyright © 1997, with permission of the publisher, the International Conference of Building Officials.*

Type of Construction

Once the occupancy of the structure has been determined, the type of construction used to protect the occupants can be determined. The type of construction will determine the kind of building materials that can or cannot be used. The kind of material used in construction will determine the ability of the structure to resist fire. Five general types of construction are typically specified by building codes and are represented by the numbers 1 through 5.

Construction of building types 1 and 2 requires the structural elements of the walls, floors, roofs, and exits to be constructed or protected by approved noncombustible materials such as steel, iron, concrete, and masonry. Construction of types 3, 4, and 5 can be steel, iron, concrete, masonry, or wood. In addition to specifying the structural framework, the type of construction will also dictate the material used for interior partitions, exit specifications, and a wide variety of other requirements for the building. Figure 44.4 shows the construction requirements for various areas of a structure, based on the type of construction. Specific areas of the code must now be researched to determine how these requirements can be met.

Building Area

Once the type of construction has been determined, the size of the structure can be determined. Figure 44.5 shows a listing of allowable floor area based on the type of construction to be used. These are basic square-foot sizes, which may be altered depending on the different construction techniques that can be used. If you are working on the plans for a multilevel office building with 20,000 sq ft per floor, this table can be used to determine the type of construction required. First determine the occupancy. By using Figure 44.2 you can find that the structure is a B occupancy. Enter Figure 44.5 on the right side using type V-N. For a Group B occupancy using type V-N construction, the structure would be limited to 8,000 sq ft. Type II F. R. or Type I F. R. construction methods would need to be used unless the structure is divided into smaller segments by fire-rated walls. You will now need to turn to the areas of the code that cover these types of construction to determine types of materials that could be used.

Determine the Height

The occupancy and type of construction will determine the maximum height of the structure. Figure 44.5 can also be used to determine the allowable height of a structure. The theoretical 20,000 sq ft office building could be 12 stories (or 160′) high if constructed of Type II F.R. construction materials, or of unlimited height if built with Type I materials. Remember that this height is based on building requirements governing fire and public safety. Zoning regulations for a specific area may further limit the height of the structure.

		TYPES OF CONSTRUCTION								
		I	II			III		IV	V	
		F.R.	F.R.	One-hour	N	One-hour	N	H.T.	One-hour	N
		Maximum Height (feet)								
TYPE OF CONSTRUCTION		UL	160 (48 768 mm)	65 (19 812 mm)	55 (16 764 mm)	65 (19 812 mm)	55 (16 764 mm)	65 (19 812 mm)	50 (15 240 mm)	40 (12 192 mm)
Use Group	Height/Area	Maximum Height (stories) and Maximum Area (sq. ft.) (× 0.0929 for m^2)								
A-1	H	UL	4	Not Permitted						
	A	UL	29,900							
A-2, 2.1[2]	H	UL	4	2	NP	2	NP	2	2	NP
	A	UL	29,900	13,500	NP	13,500	NP	13,500	10,500	NP
A-3, 4[2]	H	UL	12	2	1	2	1	2	2	1
	A	UL	29,900	13,500	9,100	13,500	9,100	13,500	10,500	6,000
B, F-1, M, S-1, S-3, S-5	H	UL	12	4	2	4	2	4	3	2
	A	UL	39,900	18,000	12,000	18,000	12,000	18,000	14,000	8,000
E-1, 2, 3[4]	H	UL	4	2	1	2	1	2	2	1
	A	UL	45,200	20,200	13,500	20,200	13,500	20,200	15,700	9,100
F-2, S-2	H	UL	12	4	2	4	2	4	3	2
	A	UL	59,900	27,000	18,000	27,000	18,000	27,000	21,000	12,000
H-1[5]	H	1	1	1	1	Not Permitted				
	A	15,000	12,400	5,600	3,700					
H-2[5]	H	UL	2	1	1	1	1	1	1	1
	A	15,000	12,400	5,600	3,700	5,600	3,700	5,600	4,400	2,500
H-3, 4, 5[5]	H	UL	5	2	1	2	1	2	2	1
	A	UL	24,800	11,200	7,500	11,200	7,500	11,200	8,800	5,100
H-6, 7	H	3	3	3	2	3	2	3	3	1
	A	UL	39,900	18,000	12,000	18,000	12,000	18,000	14,000	8,000
I-1.1, 1.2[6,10]	H	UL	3	1	NP	1	NP	1	1	NP
	A	UL	15,100	6,800	NP	6,800	NP	6,800	5,200	NP
I-2	H	UL	3	2	NP	2	NP	2	2	NP
	A	UL	15,100	6,800	NP	6,800	NP	6,800	5,200	NP
I-3	H	UL	2	Not Permitted[7]						
	A	UL	15,100							
R-1	H	UL	12	4	2[9]	4	2[9]	4	3	2[9]
	A	UL	29,900	13,500	9,100[9]	13,500	9,100[9]	13,500	10,500	6,000[9]
R-3	H	UL	3	3	3	3	3	3	3	3
	A	Unlimited								
S-4[3]	H	See Table 3-H								
	A									
U[8]	H	See Chapter 3								
	A									

A—Building area in square feet.
H—Building height in number of stories.
H.T.—Heavy timber.
NP—Not permitted.
N—No requirements for fire resistance.
F.R.—Fire resistive.
UL—Unlimited.

[1]For multistory buildings, see Section 504.2.
[2]For limitations and exceptions, see Section 303.2.
[3]For open parking garages, see Section 311.9.
[4]See Section 305.2.3.
[5]See Section 307.
[6]See Section 308.2.1 for exception to the allowable area and number of stories in hospitals, nursing homes and health-care centers.
[7]See Section 308.2.2.2.
[8]For agricultural buildings, see also Appendix Chapter 3.
[9]For limitations and exceptions, see Section 310.2.
[10]For Type II F.R., the maximum height of Group I, Division 1.1 Occupancies is limited to 75 feet (22 860 mm). For Type II, One-hour construction, the maximum height of Group I, Division 1.1 Occupancies is limited to 45 feet (13 716 mm).

FIGURE 44.5 ■ Basic allowable height and floor area of a structure, based on the UBC. Once the occupancy group is determined, the basic allowable floor area can be determined; this is based on the fire resistance of the construction materials to be used. Type I material has the highest resistance to fire. Type V-N construction is most susceptible to damage from fire. The allowable height in floors is above each listing of floor area. *Reproduced from the 1997 edition of* the Uniform Building Code, *copyright © 1997, with permission of the publisher, the International Conference of Building Officials.*

Building Location and Size

Zoning ordinances and how the structure will be used have a significant impact on where the building is located on the property. Where the building is located on the lot will, in turn, have an impact on the construction of the building. The location of the building in relationship to other buildings or the property lines is very important in regard to fire safety and will affect the requirements for the fire resistance of exterior walls and the openings in those walls. As the distance between two buildings is decreased, the probability that a fire in one will damage the other increases. Increasing the space between buildings also provides greater access for firefighting equipment. Each model code has provisions for increasing the allowable size of a building when large yards are provided. In determining floor area increases, a yard includes adjacent public streets, alleys, or courtyards, but yards on adjacent lots do not count. Figure 44.6 shows increases that can be applied to the base floor size, based on the number and size of the yards around the building. Increasing the size of a yard improves fire safety, and the code allows larger buildings to be built.

It may be beneficial to increase the size of a yard to avoid required fire-resistant construction. Figure 44.7 lists the required fire resistance for walls based on the size of the yard. The

SECTION 505—ALLOWABLE AREA INCREASES

505.1 General. The floor areas specified in Section 504 may be increased by employing one of the provisions in this section.

505.1.1 Separation on two sides. Where public ways or yards more than 20 feet (6096 mm) in width extend along and adjoin two sides of the building, floor areas may be increased at a rate of 1¼ percent for each foot (305 mm) by which the minimum width exceeds 20 feet (6096 mm), but the increase shall not exceed 50 percent.

515.1.2 Separation on three sides. Where public ways or yards more than 20 feet (6096 mm) in width extend along and adjoin three sides of the building, floor areas may be increased at a rate of 2½ percent for each foot (305 mm) by which the minimum width exceeds 20 feet (6090 mm), but the increase shall not exceed 100 percent.

515.1.3 Separation on all sides. Where public ways or yards more than 20 feet (6096) mm) in width extend on all sides of a building and adjoin the entire perimeter, floor areas may be increased at a rate of 5 percent for each foot (305 mm) by which the minimum exceeds 20 feet (6096 mm). Such increases shall not exceed 100 percent, except that greater increases shall be permitted for the following occupancies:

1. Group S, Division 1 aircraft storage hangars not exceeding one story in height.
2. Group S, Division 2 or Group F, Division 2 Occupancies not exceeding two stories in height.
3. Group H, Division 5 aircraft repair hangars not exceeding one story in height. Area increases shall not exceed 500 percent for aircraft repair hangars except as provided in Section 505.2.

505.2 Unlimited Area. The area of any one- or two-story building of Groups B; F, Division 1 or 2; M; S, Division 1, 2, 3, 4, or 5; and H, Division 5 Occupancies shall not be limited if the building is provided with an approved automatic sprinkler system throughout as specified in Chapter 9, and entirely surrounded and adjoined by public ways or yards not less than 60 feet (18 288 mm) in width.

The area of a Group S, Division 2 or Group F, Division 2 Occupancy in a one-story Type II, Type III One-hour or Type IV building shall not be limited if the building is entirely surrounded and adjoined by public ways or yards not less than 60 feet (18 288 mm) in width.

505.3 Automatic Sprinkler Systems. The areas specified in Table 5–13 and Section 504.2 may be tripled in one-story buildings and doubled in buildings of more than one story if the building is provided with an approved automatic sprinkler system throughout. The area increases permitted in this subsection may be compounded with that specified in Section 505.1.1., 505.1.2., or 505.1.3. The increases permitted in this subsection shall not apply when automatic sprinkler systems are installed under the following provisions:

1. Section 506 for an increase in allowable number of stories.
2. Section 904.2.6 for Group H, Divisions 1 and 2 Occupancies.
3. Substitution for one-hour fire-resistive construction pursuant to Section 508.
4. Section 402, Atria,

FIGURE 44.6 ■ The basic floor size of a structure can often be increased; this increase is based on the number and size of the yards surrounding the structure or by a sprinkler system. *Reproduced from the 1997 edition of* The Uniform Building Code, *copyright © 1997, with permission of the publisher, the International Conference of Building Officials.*

TABLE 5-A—EXTERIOR WALL AND OPENING PROTECTION BASED ON LOCATION ON PROPRIETY FOR ALL CONSTRUCTION TYPES[1,2,3]

Occupancy Group[4]	Construction Type	Exterior Walls: Bearing	Exterior Walls: Nonbearing	Openings[5]
		Distances are measured to property lines (see Section 503). × 304.8 for mm		
A-1	I-F.R. II-F.R.	Four-hour N/C	Four-hour N/C less than 5 feet Two-hour N/C less than 20 feet One-hour N/C less than 40 feet NR, N/C elsewhere	Not permitted less than 5 feet Protected less than 20 feet
A-1	II One-hour II-N III One-hour III-N IV-H.T. V One-hour V-N	Group A, Division 1 Occupancies are not allowed in these construction types.		
A-2 A-2.1 A-3 A-4	I-F.R. II-F.R. III One-hour IV-H.T.	Four-hour N/C	Four-hour N/C less than 5 feet Two-hour N/C less than 20 feet One-hour N/C less than 40 feet NR, N/C elsewhere	Not permitted less than 5 feet Protected less than 20 feet
A-2 A-2.1[2]	II One-hour	Two-hour N/C less than 10 feet One-hour N/C elsewhere	Same as bearing except NR, N/C 40 feet or greater	Not permitted less than 5 feet Protected less than 10 feet
	II-N III-N V-N	Group A, Divisions 2 and 2.1 Occupancies are not allowed in these construction types.		
	V One-hour	Two-hour less than 10 feet One-hour elsewhere	Same as bearing	Not permitted less than 5 feet Protected less than 10 feet
A-3	II One-hour	Two-hour N/C less than 5 feet One-hour N/C elsewhere	Same as bearing except NR, N/C 40 feet or greater	Not permitted less than 5 feet Protected less than 10 feet
	II-N	Two-hour N/C less than 5 feet One-hour N/C less than 20 feet NR, N/C elsewhere	Same as bearing	Not permitted less than 5 feet Protected less than 10 feet
	III-N	Four-hour N/C	Four-hour N/C less than 5 feet Two-hour N/C less than 20 feet One-hour N/C less than 40 feet NR, N/C elsewhere	Not permitted less than 5 feet Protected less than 20 feet
	V One-hour	Two-hour less than 5 feet One-hour elsewhere	Same as bearing	Not permitted less than 5 feet Protected less than 10 feet
	V-N	Two-hour less than 5 feet One-hour less than 20 feet NR elsewhere	Same as bearing	Not permitted less than 5 feet Protected less than 10 feet
A-4	II One-hour	One-hour N/C	Same as bearing except NR, NC 40 feet or greater	Protected less than 10 feet
	II-N	One-hour N/C less than 10 feet NR, N/C elsewhere	Same as bearing	Protected less than 10 feet
	III-N	Four-hour N/C	Four-hour N/C less than 5 feet Two-hour N/C less than 20 feet One-hour N/C less than 40 feet NR, N/C elsewhere	Not permitted less than 5 feet Protected less than 10 feet
	V One-hour	One-hour	Same as bearing	Protected less than 10 feet
	V-N	One-hour less than 10 feet NR elsewhere	Same as bearing	Protected less than 10 feet
B, F-1, M, S-1, S-3	I-F.R. II-F.R. III One-hour III-N IV-H.T.	Four-hour N/C less than 5 feet Two-hour N/C elsewhere	Four-hour N/C. less than 5 feet Two-hour N/C less than 20 feet One-hour N/C less than 40 feet NR, N/C elsewhere	Not permitted less than 5 feet Protected less than 20 feet

FIGURE 44.7 ■ The fire resistance of a wall or opening is regulated by its location and distance to the property line. *Reproduced from the 1997 edition of the* Uniform Building Code, *copyright © 1997 with permission of the publisher, the International Conference of Building Officials.*

TABLE 5-A—EXTERIOR WALL AND OPENING PROTECTION BASED ON LOCATION ON PROPRIETY FOR ALL CONSTRUCTION TYPES[1,2,3]

Occupancy Group[4]	Construction Type	Exterior Walls: Bearing	Exterior Walls: Nonbearing	Openings[5]
		Distances are measured to property lines (see Section 503). × 304.8 for mm		
B F-1 M S-1, S--3	II One-hour	One-hour N/C	Same as bearing except NR, N/C 40 feet or greater	Not permitted less than 5 feet Protected less than 10 feet
	II-N[3]	One-hour N/C less than 20 feet NR, N/C elsewhere	Same as bearing except NR, N/C 40 feet or greater	Not permitted less than 5 feet Protected less than 10 feet
	V One-hour	One-hour	Same as bearing	Not permitted less than 5 feet Protected less than 10 feet
	V-N	One-hour N/C less than 20 feet NR elsewhere	Same as bearing	Not permitted less than 5 feet Protected less than 10 feet
E-1 E-2 E-3	I-F.R. II-F.R. III One-hour III-N IV-H.T.	Four-hour N/C	Four-hour N/C less than 5 feet Two-hour N/C less than 20 feet One-hour N/C less than 40 feet NR, N/C elsewhere	Not permitted less than 5 feet Protected less than 20 feet
	II One-hour	Two-hour N/C less than 5 feet Two-hour N/C elsewhere	Same as bearing except NR, N/C 40 feet or greater	Not permitted less than 5 feet Protected less than 10 feet
E-1 E-2 E-3 cont.	II-N	Two-hour N/C less than 5 feet One-hour N/C less than 10 feet NR/NC elsewhere	Same as bearing except NR, N/C 40 feet or greater	Not permitted less than 5 feet Protected less than 10 feet
	V One-hour	Two-hour NC less than 5 feet One-hour elsewhere	Same as bearing	Not permitted less than 5 feet Protected less than 10 feet
	V-N	Two-hour less than 5 feet One-hour less than 10 feet NR elsewhere	Same as bearing	Not permitted less than 5 feet Protected less than 10 feet
F-2 S-2	I-F.R. II-F.R. III One-hour III-N IV-H.T.	Four-hour N/C less than 5 feet Two-hour N/C elsewhere	Four-hour N/C less than 5 feet Two-hour N/C less than 20 feet One-hour N/C less than 40 feet NR, N/C elsewhere	Not permitted less than 3 feet Protected less than 20 feet
	II One-hour	One-hour N/C	Same as bearing NR, N/C 40 feet or greater	Not permitted less than 5 feet Protected less than 10 feet
	II-N[3]	One-hour N/C less than 5 feet NR, N/C elsewhere	Sane as bearing	Not permitted less than 5 feet Protected less than 10 feet
	V One-hour	One-hour	Same as bearing	Not permitted less than 5 feet Protected less than 10 feet
	V-N	One-hour N/C less than 5 feet NR elsewhere	Same as bearing	Not permitted less than 5 feet Protected less than 10 feet
H-1[2,3]	I-F.R. II-F.R.	Four-hour N/C	NR N/C	Not restricted[3]
	II One-hour	One-hour N/C	NR N/C	Not restricted[3]
	II-N	NR N/C	Same as bearing	Not restricted[3]
	III One-hour III-N IV=H.T. V One-hour V-N	Group H, Division 1 Occupancies are not allowed in buildings of these construction types.		

FIGURE 44.7 ■ Continued

TABLE 5-A—EXTERIOR WALL AND OPENING PROTECTION BASED ON LOCATION ON PROPRIETY FOR ALL CONSTRUCTION TYPES[1,2,3]

Occupancy Group[4]	Construction Type	Exterior Walls: Bearing	Exterior Walls: Nonbearing	Openings[5]
		Distances are measured to property lines (see Section 503). × 304.8 for mm		
H-2[2,3] H-3[2,3] H-4[3] H-6 H-7	I-F.R. II-F.R. III One-hour III-N IV-H.T.	Four-hour N/C	Four-hour N/C less than 5 feet Two-hour N/C less than 10 feet One-hour N/C less than 40 feet NR, N/C elsewhere	Not permitted less than 5 feet Protected less than 20 feet
	II One-hour	Four-hour N/C less than 5 feet Two-hour N/C less than 10 feet	Four-hour N/C less than 5 feet Two-hour N/C less than 10 feet One-hour N/C elsewhere	Not permitted less than 5 feet Protected less than 20 feet One-hour N/C less than 20 feet NR, N/C elsewhere
	II-N	Four-hour N/C less than 5 feet Two-hour N/C less than 10 feet One-hour N/C less than 20 feet NR, N/C elsewhere	Same as bearing	Not permitted less than 5 feet Protected less than 20 feet
	V One-hour	Four-hour less than 5 feet Two-hour less than 10 feet One-hour elsewhere	Same as bearing	Not permitted less than 5 feet Protected less than 20 feet
	V-N	Four-hour less than 5 feet Two-hour less than 10 feet One-hour less than 20 feet NR elsewhere	Same as bearing	Not permitted less than 5 feet Protected less than 20 feet
H-5[2]	I-F.R. II-F.R. III One-hour III-N IV-H.T.	Four-hour N/C	Four-hour N/C less than 40 feet One-hour N/C less than 60 feet NR, N/C elsewhere	Protected less than 60 feet
	II One-hour	One-hour N/C	Same as bearing, except NR, N/C 60 feet or greater	Protected less than 60 feet
	II-N	One-hour N/C less than 60 feet NR elsewhere	Same as bearing	Protected less than 60 feet
	V One-hour	One-hour	Same as bearing	Protected less than 60 feet
	V-N	One-hour less than 60 feet NR elsewhere	Same as bearing	Protected less than 60 feet
I-1.1 I-1.2 I-2 I-3	I-F.R. II-F.R.	Four-hour N/C	Four-hour N/C less than 5 feet Two-hour N/C less than 20 feet One-hour N/C less than 40 feet NR, N/C elsewhere	Not permitted less than 5 feet Protected less than 20 feet
I-1.1 I-1.2 I-3[2]	II One-hour	Two-hour N/C less than 5 feet One-hour N/C elsewhere	Same as bearing, except NR, N/C 40 feet or greater	Not permitted less than 5 feet Protected less than 10 feet
	V One-hour	Two-hour less than 5 feet One-hour elsewhere	Same as bearing	Not permitted less than 5 feet Protected less than 10 feet
I-1.1 I-1.2 I-2 I-3	II-N III-N V-N	These occupancies are not allowed in buildings of these construction types.[6]		
I-3	IV-H.T.	Group I, Division 3 Occupancies are not allowed in buildings of this construction type.		
I-1.1 I-1.2 I-2 I-3	III One-hour	Four-hour N/C	Same as bearing except NR, N/C 40 feet or greater	Not permitted less than 5 feet Protected less than 20 feet
I-1.1 I-1.2 I-2	IV-H.T.	Four-hour N/C	Same as bearing, except NR, N/C 40 feet or greater	Not permitted less than 5 feet Protected less than 20 feet

FIGURE 44.7 ■ Continued

TABLE 5-A—EXTERIOR WALL AND OPENING PROTECTION BASED ON LOCATION ON PROPRIETY FOR ALL CONSTRUCTION TYPES[1,2,3]

Occupancy Group[4]	Construction Type	Exterior Walls: Bearing	Exterior Walls: Nonbearing	Openings[5]
		Distances are measured to property lines (see Section 503). × 304.8 for mm		
I-2	II One-hour	One-hour N/C	Same as bearing, except NR, N/C 40 feet or greater	Not permitted less than 5 feet Protected less than 10 feet
	V One-hour	One-hour	Same as bearing except	Not permitted less than 5 feet Protected less than 10 feet
R-1	I-F.R. II-F.R. III One-hour III-N IV-H.T.	Four-hour less than 3 feet Two-hour N/C elsewhere	Four-hour N/C less than 3 feet Two-hour N/C less than 20 feet One-hour N/C less than 40 feet NR, N/C elsewhere	Not permitted less than 3 feet Protected less than 20 feet
	II One-hour	One-hour N/C	Same as bearing except NR, N/C 40 feet or greater	Not permitted less than 5 feet
	II-N	One-hour N/C less than 5 feet NR, N/C elsewhere	Same as bearing except NR, N/C 40 feet or greater	Not permitted less than 5 feet
	V One-hour	One-hour	Same as bearing	Not permitted less than 5 feet
	V-N	One-hour less than 5 feet NR elsewhere	Same as bearing	Not permitted less than 5 feet
R-3	I-F.R. II-F.R. III One-hour III-N IV-H.T.	Four-hour N/C	Four-hour N/C less than 3 feet Two-hour N/C less than 20 feet One-hour N/C less than 40 feet NR, N/C elsewhere	Not permitted less than 3 feet Protected less than 20 feet
	II One-hour	One-hour N/C	Same as bearing except NR, N/C 40 feet or greater	Not permitted less than 3 feet
	II-N	One-hour N/C less than 3 feet NR, N/C elsewhere	Same as bearing	Not permitted less than 3 feet
	V One-hour	One-hour	Same as bearing	Not permitted less than 3 feet
	V-N	One-hour less than 3 feet NR elsewhere	Same as bearing	Not permitted less than 3 feet
S-4	I-F.R. II-F.R. III One-hour II-N[3]	One-hour N/C less than 10 feet NR, N/C elsewhere	Same as bearing	Not permitted less than 5 feet Protected less than 10 feet
	III One-hour III-N IV-H.T. V One-hour V-N	Group S, Division 4 open parking garages are not permitted in these types of construction.		
S-5	II One-hour	One-hour N/C	Same as bearing, except NR, N/C 40 feet or greater	Not permitted less than 5 feet Protected less than 20 feet
	III-N[3]	One-hour N/C less than 20 feet NR, N/C elsewhere	Same as bearing, except NR, N/C 40 feet or greater	Not permitted less than 5 feet Protected less than 20 feet
	V One-hour	One-hour	Same as bearing	Not permitted less than 5 feet Protected less than 20 feet
	V-N[5]	One-hour N/C less than 20 feet NR, elsewhere	Same as bearing	Not permitted less than 5 feet Protected less than 20 feet

FIGURE 44.7 ■ Continued

TABLE 5-A—EXTERIOR WALL AND OPENING PROTECTION BASED ON LOCATION ON PROPRIETY FOR ALL CONSTRUCTION TYPES[1,2,3]

Occupancy Group[4]	Construction Type	Exterior Walls: Bearing	Exterior Walls: Nonbearing	Openings[5]
		Distances are measured to property lines (see Section 503). × 304.8 for mm		
U-1[3]	I-F.R. II-F.R. III One-hour III-N IV-H.T.	Four-hour N/C	Four-hour N/C less than 3 feet Two-hour N/C less than 20 feet One-hour N/C less than 40 feet NR, N/C elsewhere	Not permitted less than 3 feet Protected less than 20 feet
	II One-hour	One-hour N/C	Same as bearing except NR, N/C 40 feet or greater	Not permitted less than 3 feet
	V One-hour	One-hour	Same as bearing	Not permitted less than 3 feet
	II-n[2]	One-hour N/C less than 3 feet[3] NR N/C elsewhere	Same as bearing	Not permitted less than 3 feet
	V-N	One-hour N/C less than 3 feet[3] NR elsewhere	Same as bearing	Not permitted less than 3 feet
U-2	All	Not regulated		

N/C—Noncombustible.
NR—Nonrated.
H.T.—Heavy timber.
F.R.—Fire resistive.
[1]See Section 503 for types of walls affected and requirements covering percentage of openings permitted in exterior walls. For walls facing streets, yards and public ways, see also Section 601.5.
[2]For additional restrictions see Chapter 3 and 6.
[3]For special provisions and exceptions, see also Section 503.4.
[4]See Table 3-A for a description of each occupancy type.
[5]Openings in exterior walls shall be protected by a fire assembly having at least a three-fourths-hour fire-protection rating.
[6]See Section 312.2.1, Exception 3.

FIGURE 44.7 ■ Continued

column on the far right side lists the fire resistance of openings (windows and doors) or completely prohibits them when yards become too small.

Increasing the Allowable Floor Area

The allowable increases of Figure 44.6 can be applied to the site plan in Figure 44.8. The proposed structure is to be a professional office building, two stories high, with 20,000 sq ft on each floor. Using the base square footage, it was found that only Type II-FR and Type I-FR construction have enough basic allowable area. These types of construction are relatively expensive; therefore, less costly options should be explored. According to Figure 44.6, the basic allowable area can be increased by installing fire sprinklers and by having large yards around the building. You will see from Figure 44.6 that the building size can be increased if it has at least 2 yards larger than 20′ (6100 mm). The building in Figure 44.8 has three yards larger than 20′, so a 3-yard increase can be applied. Subtract 20′ from 36′ (the smallest yard) and multiply the results by 2.5% to determine the basic allowable increase in building area. For this site, the basic allowable area can be increased by 40%. If a 2-yard increase is used, using the two street yards, a 50% increase can be achieved. Multiplying 1.25% by the yard plus the street width [(10′ + 60′) × 1.25] results in an 87.5% increase. The UBC restricts the increase to a maximum of 50% for 2 yards. Obviously, in this case the 2-yard increase should be used to achieve the maximum allowable building size.

Use Figure 44.6 to determine the available options. Multiplying the allowable area of a type V-N building (8,000 sq ft) by 1.5 gives an allowable area of 12,000 sq ft. With type V-1H construction, 21,000 sq ft are allowed (14,000 × 1.5) per floor. Another option for increasing the allowable area of the building is to install a fire sprinkler system. According to Figure 44.6, the size of a two-story building can be doubled if sprinklers are added to the structure. Each model code allows for doubling the allowable area of a two-story building or tripling the allowable area of a single-story building if it is fully protected by sprinklers. If the building is fully protected by a sprinkler system, even Type V-N construction can be used.

For a Type B occupancy it can be seen in Figure 44.6 that walls facing a yard smaller than 20′ (6100 mm) will be required fire-resist construction no matter what construction type is used. This means that the north wall of the building

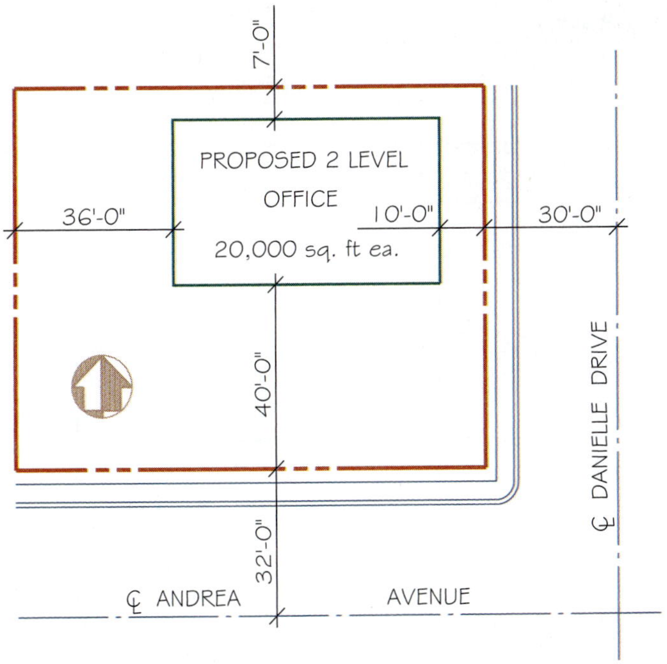

FIGURE 44.8 ■ The site plan for a proposed structure can be used to visualize the side yard setback and to determine the maximum building area. *Courtesy Dean Smith.*

will have to be fire-resistant. Openings in the north wall are allowed, but they must be fire-resistant (protected). Chapter 45 will introduce methods to achieve specific fire ratings. Because the street can be counted as part of the yard, the east wall is not required to be fire-resistant even though it is only 10′ (3050 mm) from the property line. Unprotected openings are allowed on the east wall because the street can be considered part of the required yard.

Determine the Occupant Load

You have determined how the structure will be used and established the allowable size. Now you must determine how many people may use the structure. The number of occupants in the building is determined by a ratio of the room area to the code specified area per person, as shown in Figure 44.9, or in some cases by the number of fixed seats in the room. Theaters and churches are examples of spaces in which the number of occupants is determined by the number of seats. In most other spaces, the area of the room is divided by the factor shown under "Occupant Load Factor" in Figure 44.9 to determine the number of occupants. In each occupancy, the intended size of the structure is divided by the occupant load factor to determine the occupant load. In a 20,000 sq ft office structure, the occupant load is 200. This is found by dividing the size (20,000 sq ft) by the occupant factor (100). (See listing No. 23.) The occupant load will be needed to determine the number of exits, the size and locations of doors, and many other construction requirements.

Exits

Exit Paths

After the basic exterior shape of the building has been determined, the interior layout can be designed. One of the most important steps in this process is to determine the exit path. The number and size of the exists will be determined in most cases by the number of occupants in the building. The center column of Figure 44.9 lists the UBC requirements for two exits. The other model codes require a minimum of two exits from a space, and then list exceptions based on each occupancy classification. All of the model codes require 3 exits if the building has 501 to 1000 occupants and 4 exits for 1001 and more.

The location of the exits is also an important factor to be considered. If more than one exit is required, each model code requires at least two of the exits to be separated by one-half of the diagonal of the area served. This separation ensures that a fire will not block two exits. If the exits are not reasonably separated, there is a chance that a fire might block multiple exits, denying people an escape route. Figure 44.10 shows how these requirements are applied. A second consideration of exit design is dead-end corridors. If the occupants of a building run down a long dead-end corridor, they will lose valuable time retracing their steps to get out of the building. Dead-end corridors over 20′ (6100 mm) in length are not allowed.

Determining Exits

A single-story building with 40,000 sq ft of office space and a 1000 sq ft conference room will be considered. Figure 44.9 requires that for the purpose of calculating exits you are to use 1 occupant for every 100 sq ft for office use, and 1 occupant for every 15 sq ft of conference room. This occupancy load is the same for each model code. Dividing 40,000 by 100 results in 400 occupants for the office area. Dividing 1000 by 15 results in 67 occupants, for a total of 467 occupants. The model codes require 0.2″ of exit width for every occupant (BOCA requires more in an unsprinkled building). Multiply 0.2″ by 467 requires 94″ (7′–10″) of exit width. This building requires only two exits, but two 3′–0″ doors will not provide the proper exit width. The code also states that the required width shall be divided approximately equally among the separate exits. This means that if you provide only 2 exits, a 3′–6″ door and a 4′–6″ door or a pair of 3′–0″ doors at the other may not be acceptable. Since the word *approximately* has been used in the code, it is advisable to check with the local building department for its interpretation on this matter.

There are minimum requirements in the codes for hall and door widths. Exit doors a minimum of 3′–0″ (900 mm) wide are required for most conditions. Most uses required a minimum hall width of 44″ (1100 mm), but a width of 36″ (900 mm) is acceptable for some limited uses. Stairways have required minimum widths and a required minimum width per occupant.

10-A—MINIMUM EGRESS REQUIREMENTS[1]

USE[2]	MINIMUM OF TWO EXITS OTHER THAN ELEVATORS ARE REQUIRED WHERE NUMBER OF OCCUPANTS IS AT LEAST	OCCUPANT LOAD FACTOR[3] (square feet) × 0.0929 for m^2
1. Aircraft hangars (no repair)	10	500
2. Auction rooms	30	7
3. Assembly areas, concentrated use (without fixed seats) Auditoriums Churches and chapels Dance floors Lobby accessory to assembly occupancy Lodge rooms Reviewing stands Stadiums	50	7
Waiting area	50	3
4. Assembly areas, less-concentrated use Conference rooms Dining rooms Drinking establishments Exhibit rooms Gymnasiums Lounges Stages	50	15
5. Bowling alley (assume no occupant load for bowling lanes)	50	4
6. Children's homes and homes for the aged	6	80
7. Classrooms	50	20
8. Congregate residences	10	200
9. Courtrooms	50	40
10. Dormitories	10	50
11. Dwellings	10	300
12. Exercising rooms	50	50
13. Garage, parking	30	200
14. Hospitals and sanitariums—		
Health-care center	10	80
Nursing homes		
Sleeping rooms	6	80
Treatment rooms	10	80
15. Hotels and apartments	10	200
16. Kitchen—commercial	30	200
17. Library reading room	50	50
18. Locker rooms	30	50
19. Malls (see Chapter 4)	—	—
20. Manufacturing areas	30	200
21. Mechanical equipment room	30	300
22. Nurseries for children (day care)	7	35
23. Offices	30	100
24. School shops and vocational rooms	50	50
25. Skating rinks	50	50 on the skating area; 15 on the deck
26. Storage and stock rooms	30	300
27. Stores—retail sales rooms		
Basements and ground floor	50	30
Upper floors	50	60
28. Swimming pools	50	50 for the pool area; 15 on the deck
29. Warehouses	30	500
30. All others	50	100

[1]Access to, and egress from, buildings for persons with disabilities shall be provided as specified in Chapter 11.

[2]For additional provisions on number of exits from Groups H and I Occupancies and from rooms containing fuel-fired equipment or cellulose nitrate, see Sections 1018, 1019 and 1020, respectively.

[3]This table shall not be used to determine working space requirements per person.

[4]Occupant load based on five persons for each alley, including 15 feet (4572 mm) of runway.

FIGURE 44.9 ■ The occupant load of a structure is determined by dividing the intended size of the structure by the occupant load factor. *Reproduced from the 1997 edition of* the Uniform Building Code, *copyright © 1997, with permission of the publisher, the International Conference of Building Officials.*

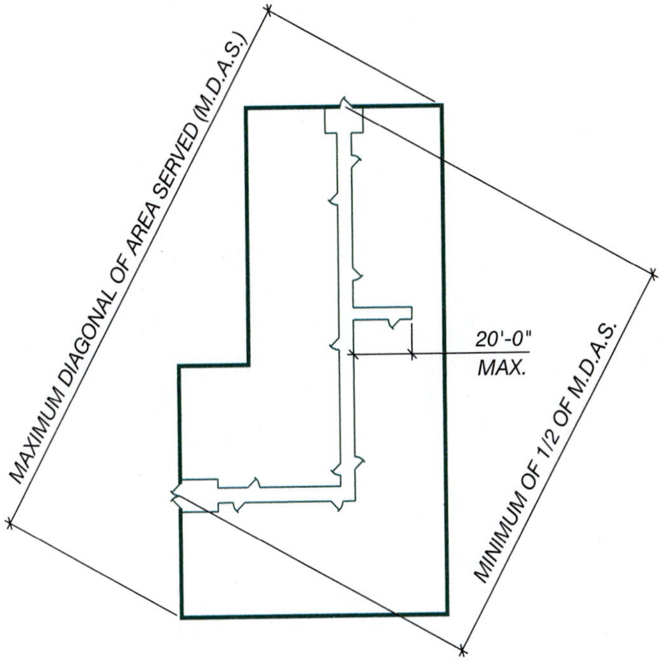

FIGURE 44.10 ■ If more than one exit from a structure is required, the distance between the exits must be equal to or greater than one-half of the area served (M.D.A.S.). *Courtesy Ginger Smith.*

USING THE CODES

Determining the five classifications of a building may seem senseless to you as a beginning drafter. Although these are procedures that the architect or engineer will perform in the initial design stage, the drafter must be aware of these classifications as basic drafting functions are performed on the project. Many of the problems that the drafter will need the code to solve will require knowledge of basic code limitations. Practice using the tables presented in this text as well as similar tables of the code that govern your area.

Once you feel comfortable determining the basic categories of a structure, study the chapter of your building code that presents basic building requirements for a structure. Chapters 7–9 of the UBC introduce fire protection, and Chapters 10–12 introduce occupant needs.

Once you feel at ease reading through the chapters dealing with general construction, work on the chapters of the code that deal with uses of different types of material. Start with the chapter that deals with the most common building material in your area. For most of you, this will be Chapter 23, which governs the use of wood. Although you will be familiar with the use of wood in residential structures, the building codes will open many new areas that were not required in residential construction.

Chapter 44 Additional Reading

The following Web sites can be used as a resource to help you keep current with changes in each of the major national building codes.

ADDRESS	COMPANY OR ORGANIZATION
■ www.bocai.org	1999 BOCA (Building Officials and Code Administrators International Inc).
■ www.icbo.org	1997 UBC (Uniform Building Code), written by the ICBO (International Conference of Building Officials).
■ www.sbcci.org	1999 SBC (Standard Building Code), written by the SBCCI (Southern Building Code Congress International, Inc.)
■ www.intlcode.org	The 2000 IBC (International Building Code), written by the International Code Council.

CHAPTER 44

Building Codes and Commercial Design Test

DIRECTIONS

Answer the questions with short, complete statements.

1. Letter your name and the date at the top of the sheet.
2. Letter the question number and provide the answer. You do not need to write out the question.

QUESTIONS

Question 44–1 What is the occupancy rating of a room used to train 50 or more junior college drafters?

Question 44–2 What fire rating would be required for a wall separating a single-family residence from a garage?

Question 44–3 What is the height limitation on an office building of type 2-N construction?

Question 44–4 What is the maximum allowable floor area of a public parking facility using type 4 construction methods?

Question 44–5 What fire resistance is required for the exterior walls of a structure used for full-time care of children that is within 5′ of a property line?

Question 44–6 What fire rating is required between a wall separating an A-1 occupancy from a B occupancy?

Question 44–7 What is the maximum height of a residence of type 2 through 5 construction?

Question 44–8 You are drafting a 12,000 sq ft retail sales store. Determine the occupancy, the least restrictive type of construction to be used, maximum height and square footage, and the occupant load for each possible level.

CHAPTER 45

Common Commercial Construction Materials

INTRODUCTION

As you work on commercial projects, you will be exposed to new types of structures and codes different from the ones you were using with residential drawing. You will also be using different types of materials from those used in residential construction. Most of the materials used in commercial construction can be used in residential construction, but they are not because of their cost and the cost of labor associated with each material. Wood is the exception. Common materials used in commercial construction are wood, concrete block, poured concrete, and steel.

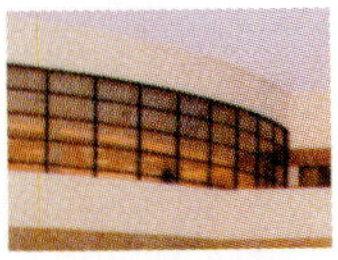

WOOD

Wood is used in many types of commercial buildings in a manner similar to its use in residential construction. The western platform system, which was covered in Chapter 25, is a common framing method for multifamily and office buildings as well as others. Heavy wood timbers are also used for some commercial construction. As a drafter, you will need to be familiar with both types of construction.

Platform Construction

Platform construction methods in commercial projects are similar to residential methods. Wood is not used at the foundation level but is a common material for walls and intermediate and upper-level floor systems. Trusses or truss joists are also common floor joist materials for commercial projects. Although each is used in residential construction, their primary usage is in commercial construction.

Walls

Wood is used to frame walls in many projects. The biggest difference in wood wall construction is not in the framing method but in the covering materials. Depending on the type of occupancy and the type of construction required, wood framed walls may require a special finish to achieve a required fire protection. Figure 45.1 shows several different methods of finishing a wood wall to achieve various fire ratings.

Roofs

Wood is also used to frame the roof system of many commercial projects. Joists, trusses, and panelized systems are the framing systems most typically used. Both joist and truss systems have been discussed in Sections 7 and 8. These systems usually allow

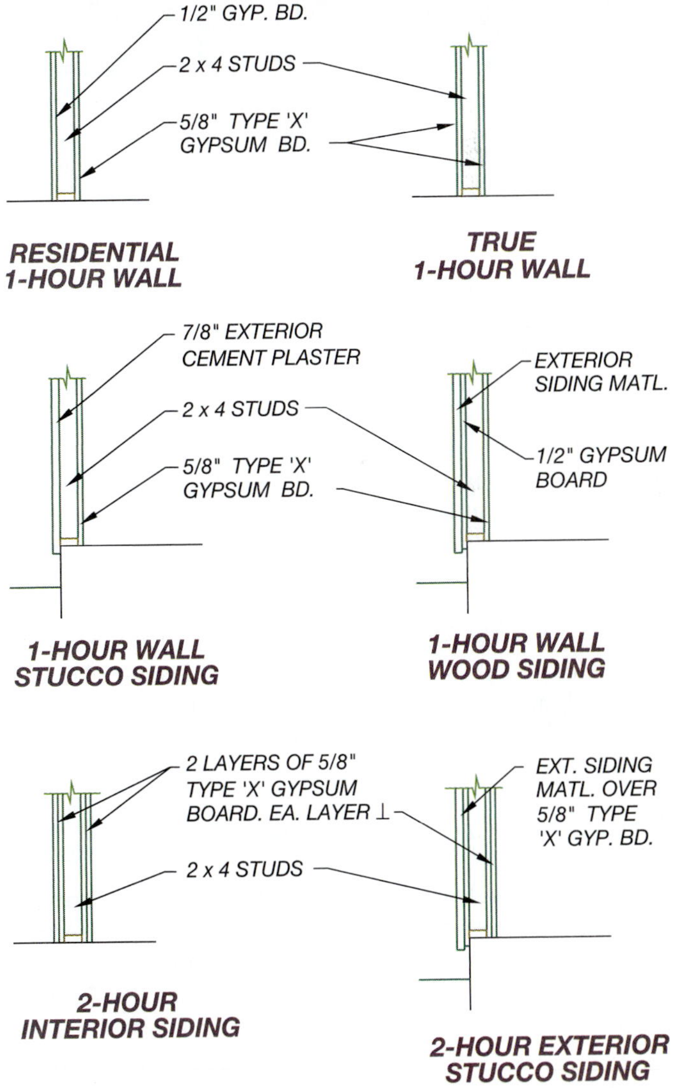

FIGURE 45.1 ■ Walls often require special treatment to achieve the needed fire rating for certain types of construction. (See Figure 44.3.) Consult your local building code for complete descriptions of coverings.

FIGURE 45.2 ■ Trusses are used throughout commercial construction to span large distances. Common spacings include 24″ (600 mm), 32″ (800 mm), and 48″ (1200 mm) o.c. *Courtesy Southern Forest Products Association.*

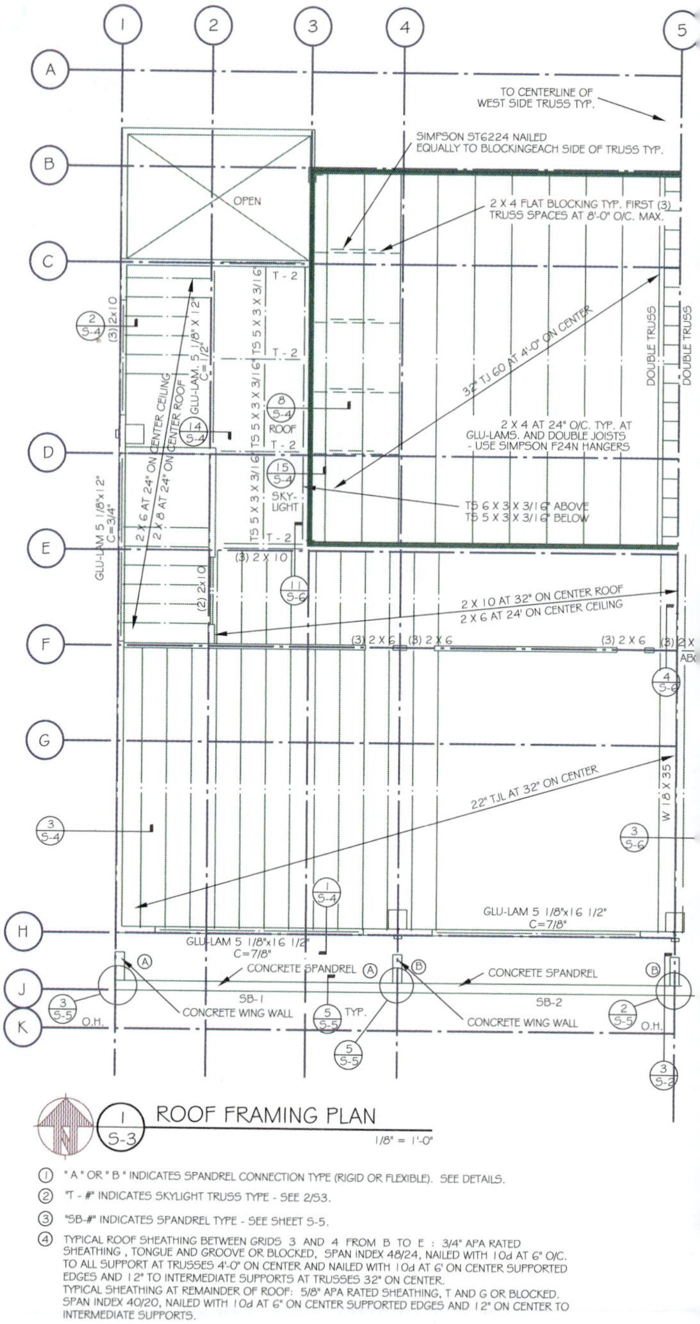

FIGURE 45.3 ■ Roof plan using trusses to span between supports. Lines are drawn to represent the span of each truss. A diagonal line is placed across the trusses to represent the extent of their placement. *Courtesy Van Domelen/Looijenga/McGarrigle/Knauf Consulting Engineers.*

the joists or trusses to be placed at 24″ or 32″ (600 or 800 mm) o.c. for commercial uses, as Figure 45.2. An example of representing a truss system can be seen in Figure 45.3. The panelized roof system typically uses beams placed approximately 20′ to 30′ (6100 to 9100 mm) apart. Smaller beams called purlins are then placed between the main beams, typically using an 8′ (2400 mm) spacing. Figure 45.4 shows a panelized roof framed with laminated beams and trusses. Joists that are 2″ or 3″ (50 or 75 mm) wide are then placed between these purlins at 24″ (600 mm) o.c. The roof is then covered with plywood sheathing. Figure 45.5 shows an example of a roof framing plan for a panelized roof system. The availability of materials, labor practices, and the use of the building determine which type of roof will be used.

Heavy Timber Construction

In addition to standard uses of wood in the western platform system, large wood members are sometimes used for the structural framework of a building. This method of construction consists of wood members that range from 5″ × 5″ through 12″ × 12″ (130 × 130 through 300 × 300 mm). Structural members larger than 12″ × 12″ (300 × 300 mm) are available but are not often used, for economic reasons. Heavy timber is used for both appearance and structural reasons. Heavy timbers have excellent structural and fire retardant qualities. In a fire, heavy wood members will char on their exterior surfaces but will maintain their structural integrity long after an equal-sized steel beam will have failed.

Representing Timber Members

The size and location of beams and posts are specified on the floor or framing plans. Posts are represented using lines similar to those used to represent walls. Post sizes are often specified using diagonal text, so that the specification will be clear. Beams are typically represented using dashed lines. Text

FIGURE 45.4 ■ This panelized roof is framed with laminated beams at 32′ o.c. and trusses placed at 8′–0″ o.c. and 2 × 6 members placed at 24″ o.c. *Courtesy Trus Joist MacMillan.*

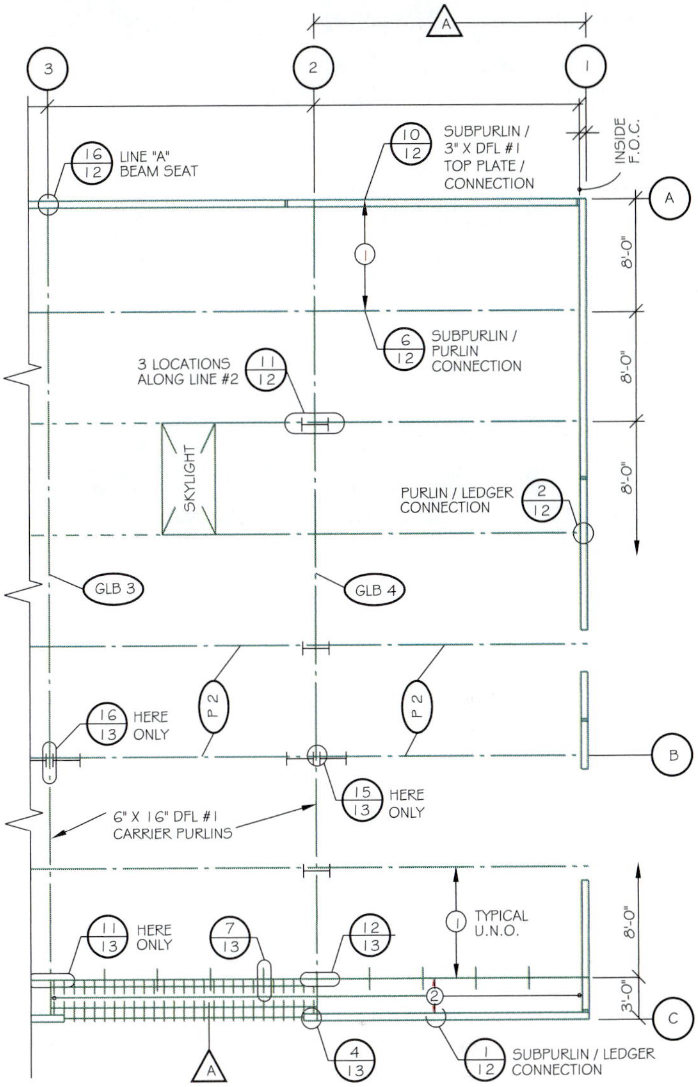

FIGURE 45.5 ■ Panelized roof framing plan. *Courtesy Kenneth D. Smith and Associates, Inc.*

to describe the beam is normally placed parallel to the beam. Figure 45.6 shows an example of specifying beams and post on a framing plan. Decking is typically specified in details and sections. Details showing the connection of post to foundation, post to beams at floor and roof levels, beam to beam connections, and decking to beams are typically required.

Laminated Beams

Because of the difficulty of producing large beams in long lengths from solid wood, large beams are often constructed from smaller members laminated together to form the larger beam. Laminated beams are a common material for buildings such as gymnasiums and churches that require large amounts of open space. Laminated beams are used for their appearance, their fire resistance, and the wide variety of shapes that can be created. Figure 45.7*a* shows examples of laminated arches. Three of the most common types of laminated beams are the single span, Tudor arch, and three-hinged arch. Examples of each can be seen in Figure 45.7*b*.

Single-span beams are often used in standard platform framing methods. These beams are typically referred to as glu-lams. Because of their increased structural qualities, a laminated beam can be used to replace a much larger sawed beam. A typical sawn beam has a fiber bending strength of 1350, compared with

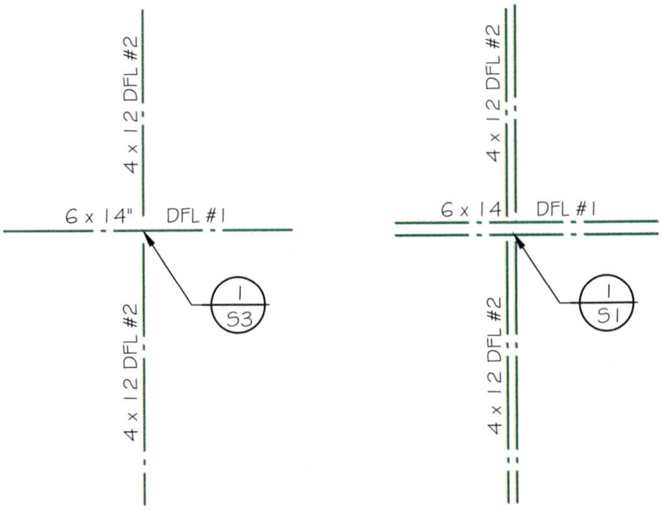

FIGURE 45.6 ■ Representing wood beams in plan view.

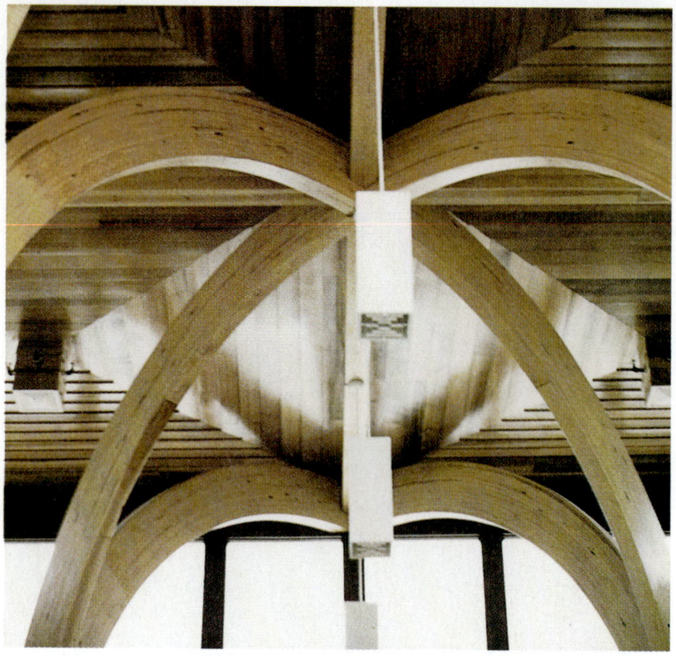

FIGURE 45.7(A) ■ Laminated beams come in a variety of sizes and shapes. *Courtesy American Institute of Timber Construction.*

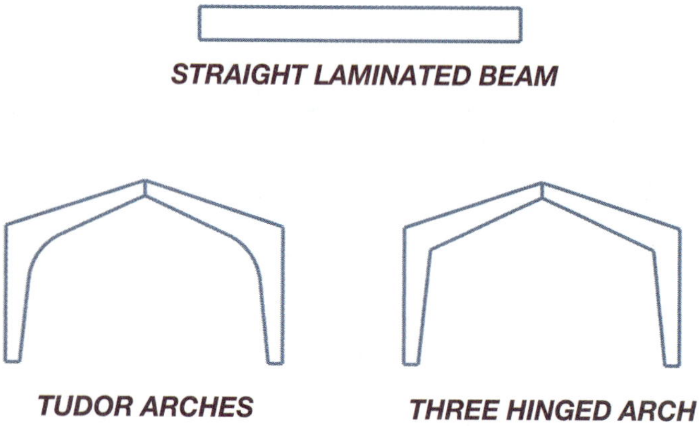

FIGURE 45.7(B) ■ Common shapes of laminated beams.

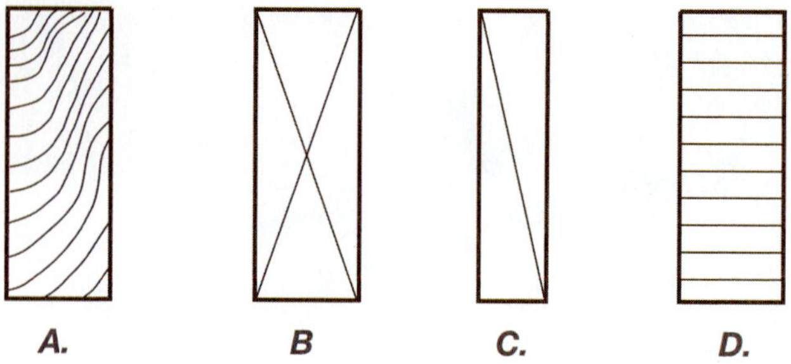

FIGURE 45.8 ■ Representing various types of beams in end view. (A) End grain; (B) continuous sawn beam; (C) noncontinuous beam or blocking; (D) laminated beam.

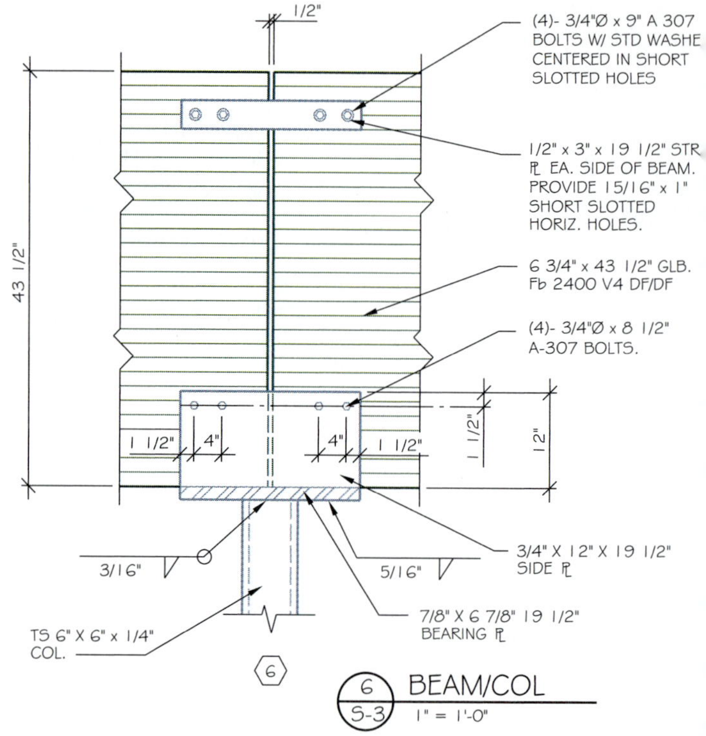

FIGURE 45.9 ■ When a prefabricated connector is unavailable in the proper size, a drafter will need to work from an engineer's sketches to draw the required connectors.

laminated beams with a bending strength of between 2000 and 2400. Laminated beams often have a built-in curve, or camber. The camber is designed into the beam to help resist the loads to be carried. The Tudor and three-hinged arch members are a post-and-beam system combined into one member.

Representing Beams

Beams drawn in end view can be represented using different methods, as shown in Figure 45.8. At a scale of 3/4″ = 1′–0″ or larger, the X is usually placed to help define the beam. The X is also used to represent the end view of a continuous member such as a beam. One diagonal line is used to represent the end view of noncontinuous members such as blocking placed between beams. Many offices use a thick outline around the perimeter to help distinguish beams. Thin lines are placed at 1 1/2″ o.c. in laminated beams to represent each lamination. When a side view of a laminated beam is shown, thin, light lines should be used to represent the lamination lines. Varied line weights will ensure that laminations are not confused with other materials. Figure 45.9 shows a detail of a laminated beam connection.

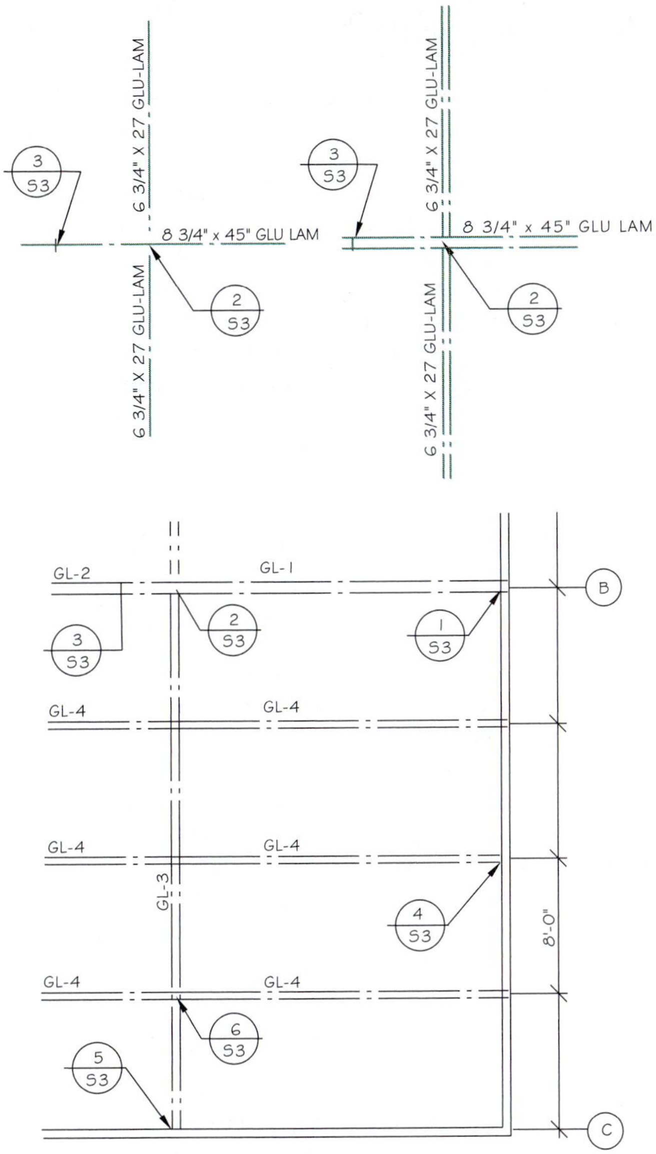

GLU-LAM BEAM SCHEDULE		
GL-1	8 3/4" x 45" GLU LAM	DF/HF 24fb-V5
GL-2	8 3/4" x 37" GLU LAM	DF/HF 24fb-V5
GL-3	6 3/4" X 27 GLU-LAM	DF/DF 20fb-V7
GL-4	5 1/8 x 15 GLU-LAM	DF/DF 20fb-V7

MANUFACTURER TO CERTIFY IN WRITING PRIOR TO INSTALLATION OF GLU-LAMINATED BEAMS THAT MINIMUM DESIGN VALUES HAVE BEEN PROVIDED. ALL BEAMS UNLESS NOTED SHALL BE Fv=165, Fc=450, fb AS NOTED, INDUSTRIAL APPEARANCE WITH EXTERIOR GLUE. PROVIDE "A.I.T.C. CERTIFICATE OF CONFORMANCE TO BUILDING DEPT. PRIOR TO ERECTION.

FIGURE 45.10 ■ Representing laminated beams in plan view. For a simple structure, the size of the beam can be specified beside the beam. A beam schedule can be used to simplify a drawing if several different sizes are required.

Laminated beams are represented in plan view just as sawn beams are, with the required text parallel to the beam. Depending on the complexity of the plan, a schedule may be used to list the specifications for laminated timbers. If a schedule is used, a beam symbol is placed near the beam so that appropriate material can be easily referenced. Notice in Figure 45.10 that many of the specifications are given in a general note. Sections and details may refer only to the beam size, grade, and material. Like sawn lumber, laminated beams will require many details to explain connection. Typically, details showing beam to beam, beam to post, truss or rafter to beam connections, and column to support connections are required.

CONCRETE BLOCK

In commercial construction, concrete blocks are used to form the wall system for many types of buildings (see Figure 45.11). Concrete blocks are durable and relatively inexpensive to install and maintain. Blocks are typically manufactured in 8 × 8 × 16 (200 × 200 × 400) modules.

FIGURE 45.11 ■ Concrete blocks are used to form the wall systems of many commercial structures. Placement of CMUs is aided by placement of scaffolding. *Courtesy Janice Jefferis.*

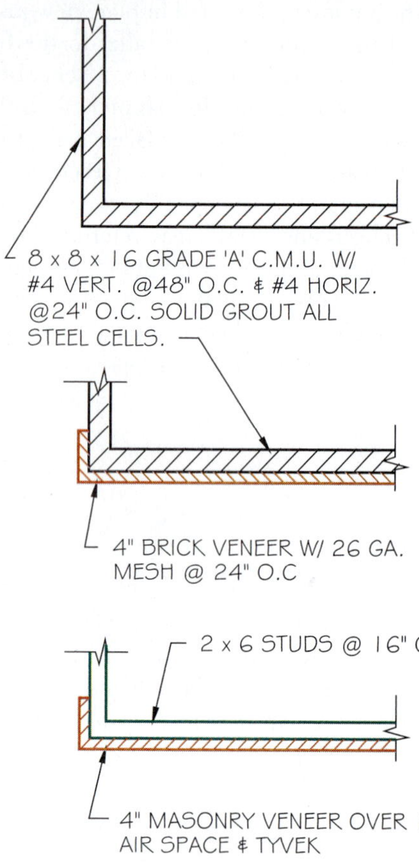

FIGURE 45.12 ■ Representing masonry products in plan view.

Sizes

The sizes listed are width, height, then length. Other common sizes are 4 × 8 × 16, 6 × 8 × 16, and 12 × 8 × 16 (100 × 200 × 400, 150 × 200 × 400, and 300 × 200 × 400). The actual size of the block is smaller than the nominal size so that mortar joints can be included. Although the person responsible for the design determines the size of the structure, it is important that the drafter be aware of the modular principles of concrete block construction. Lengths of walls, locations of openings in a wall, and heights of walls and openings must be based on the modular size of the block being used. Failure to maintain the modular layout can result in a tremendous increase in the cost of labor to cut and lay the blocks. To minimize cutting and labor cost, concrete block structures should be kept to its modular size. When the standard 8 × 8 × 16 (200 × 200 × 400 mm) CMU is used, structures that are an even number of feet long should have dimensions that end in 0″ or 8″. For instance structures that are 24′–0″ or 32′–8″ are modular. Dimensions for walls that are an odd number of feet should end with 4″. Walls that are 3′–4″, 9′–4″, and 15′–4″ are modular, but a wall that is 15′–8″ will require blocks to be cut.

Masonry Representation

Concrete blocks are represented in plan view as shown in Figure 45.12. Type, size, and reinforcement are specified on the plan views, with bold lines to represent the edges of the masonry. Thin lines for hatching are placed at a 45° angle to the edge of the wall. The *size* and location of concrete block walls must be dimensioned on the floor plan. Concrete block is dimensioned from edge to edge of walls, as seen in Figure 45.13. Openings in the wall are also located by dimensioning to the edge. Masonry is also shown in sections and details, but the scale of the drawing will affect the drawing method. At scales under 1/2″ = 1′–0″, the wall is typically drawn just as in plan view. At larger scales, details typically reflect the cells of the block. Figure 45.14 shows methods of representing concrete blocks in cross section.

When CMUs are required to support loads from a beam, a pilaster is placed in the wall to carry the loads. A pilaster is a thickened area of wall. Pilasters are also used to provide lateral support to the wall when the wall is required to span long distances. Figure 45.15 is an example of a detail that would be required to show the size and thickness of a pilaster.

When a wood floor or roof is to be supported by the block, a ledger similar to Figure 45.16 is bolted to the block. Holes are punched in the block to allow an anchor bolt to penetrate into the cell, where it is wired to the wall reinforcing and held firmly by filling the cell solid with grout. Figure 45.17 shows examples of details for connecting a mezzanine floor to a block wall. Care must be taken to tie the ledger bolts into the wall reinforcement, and to tie the floor joist to the ledger.

Steel Reinforcement

Masonry units are excellent at resisting forces from compression but are very weak at resisting forces in tension. Steel is excellent at resisting forces of tension but tends to buckle under forces of compression. The combination of these two materials forms an excellent unit for resisting great loads from lateral and vertical forces. Reinforced masonry structures are stable because the masonry, steel, grout, and mortar bond together effectively. Loads that create compression on the masonry are effectively transferred to the steel. If the steel is carefully placed, the forces will result in tension on the steel and will be safely resisted.

Reinforcing Bars

Steel reinforcing bars, called *rebar,* are used to reinforce masonry walls. Each code sets specific guidelines for the size and spacing of rebar based on the seismic zone of the construction site. Reinforcement will be specified by the engineer throughout the calculations and sketches and must be represented accurately by the drafter.

Reinforcing bars can be either smooth or deformed. Smooth bars are used primarily for joints in floor slabs. Because the bars are smooth, the concrete of the slab does not easily bond to the

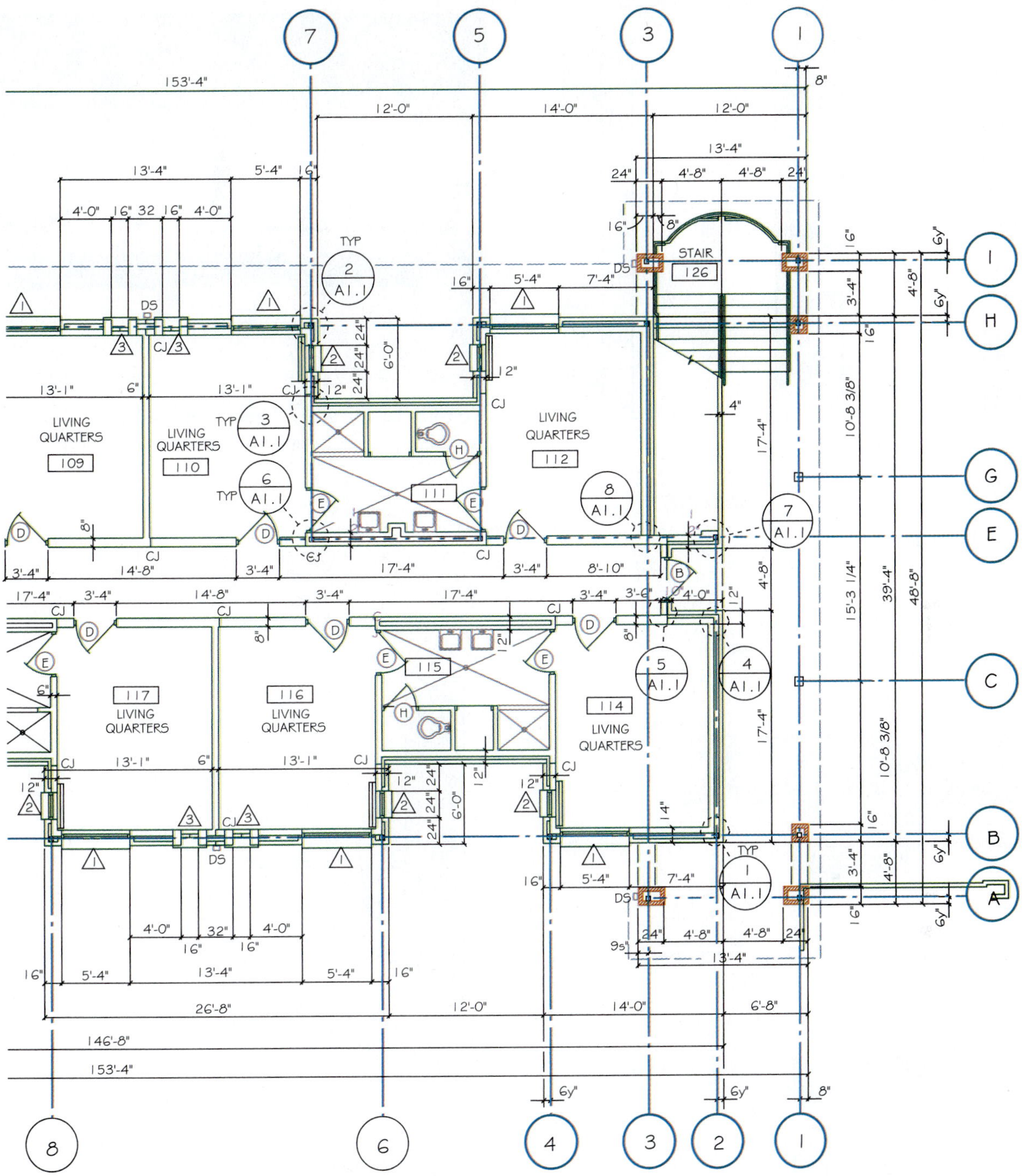

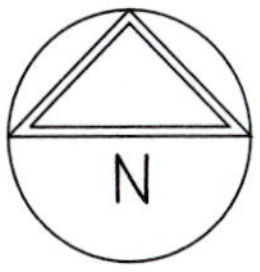

16
A1.1

FIRST FLOOR PLAN

SCALE: 1/8" = 1'-0"

FIRST FLOOR AREA = 4986 S.F 44 STUDENTS

TOTAL BUILDING AREA = 10,231 S.F. 1 FACULTY (PROCTOR)

45 TOTAL

GENERAL NOTES:

1. ALL WALLS ARE CONCRETE MASONRY UNITS UNLESS DESIGNATED AS BRICK:
2. SEE STRUCTURAL SHEETS FOR STRUCTURAL GRID DIMENSIONS.
3. SEE SHEET A1.3 KEYPLANS FOR MORE DETAIL, ENLARGEMENT, AND SECTION KEYS.
4. ALL EXPOSED MECHAINCAL & ELECTRICAL PENETRATIONS OF FLOORS, CEILINGS, AND/OR WALLS ARE TO RECEIVE "PENSIL 100 FIRESTOP SEALANT" BY GENERAL ELECTRIC.
5. CJ = CONTROL JOINT, SEE SPECIFIC DETAILS.

FIGURE 45.13 ■ Structural brick and concrete masonry units are dimensioned from exterior face to exterior face. The drafter must take great care to maintain modular sizes as walls and openings are dimensioned. *Courtesy G. Williamson Archer, A.I.A., Archer & Archer P.A. Architects.*

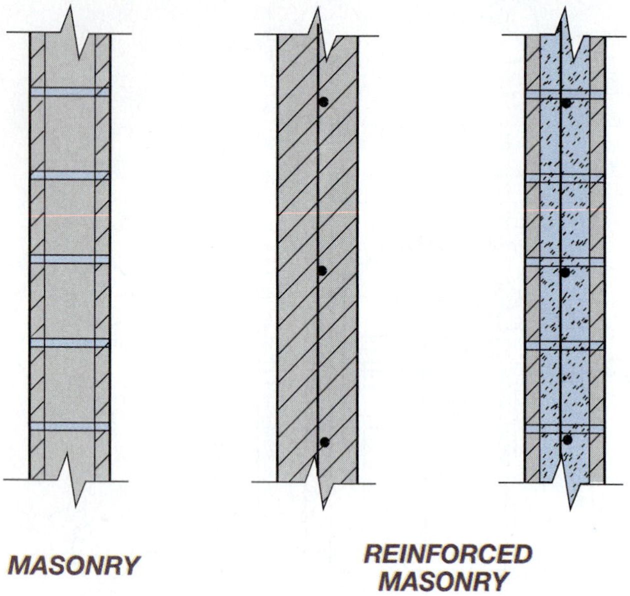

FIGURE 45.14 ■ Representing concrete blocks in section view.

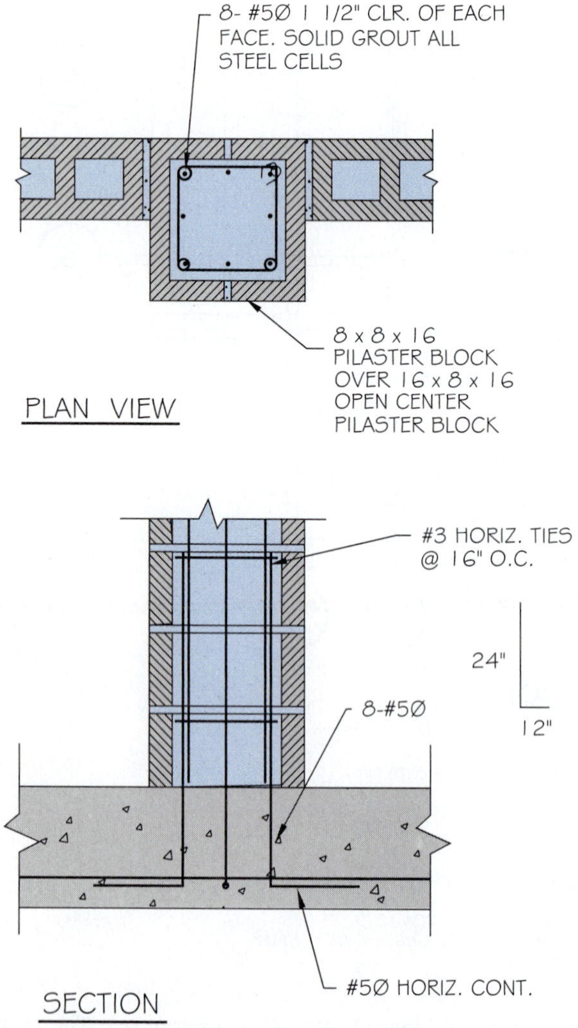

FIGURE 45.15 ■ A detail must be drawn to represent concrete block pilaster construction.

FIGURE 45.16 ■ A pressure-treated ledger is bolted to the wall to support the wood roof members. *Courtesy David Jefferis.*

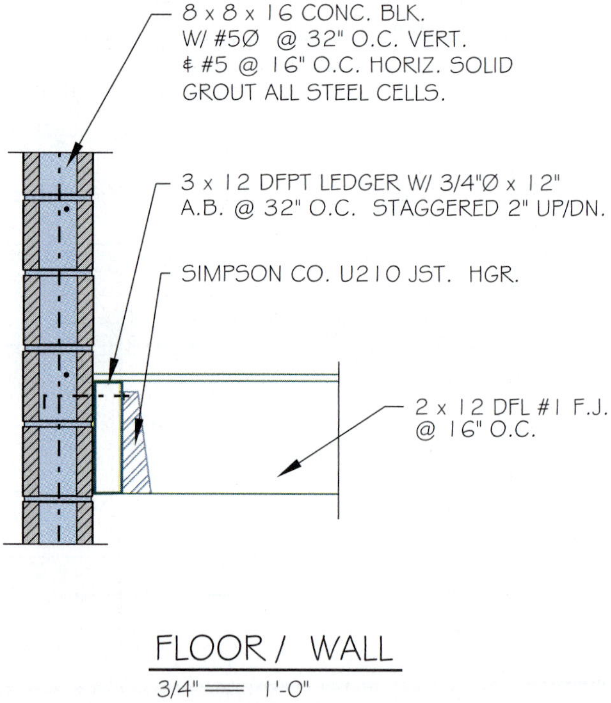

FIGURE 45.17 ■ A detail must be drawn for the intersection of each material. The drafter must represent each material clearly.

bar, allowing the concrete to expand and contract. Bars used for reinforcing masonry walls are usually deformed so that the concrete will bond more effectively with the bar and not allow slippage as the wall flexes under pressure. Deformations consist of small ribs that are placed around the surface of the bar. Deformed bars range in size from 3/8″ through 1 3/8″ diameter. Inch sizes do not readily convert to metric sizes. Bars are referenced on plans by a number rather than a size. A number represents the size of the steel in approximately 1/8″ increments. Common sizes of steel include:

BAR SIZE	NOMINAL DIAMETER IN INCHES
# 3	0.375 - 3/8
# 4	0.500 - 1/2
# 5	0.625 - 5/8
# 6	0.750 - 3/4
# 7	0.875 - 7/8
# 8	1.000 - 1
# 9	1.128 - 1 1/8
# 10	1.270 - 1 1/4
# 11	1.410 - 1 3/8
# 14	1.693 - 1 3/4
# 18	2.257 - 2 1/4

Steel Placement

Wall reinforcing is held in place by filling each masonry cell containing steel with grout. When exact locations within the cell are required by design, the steel is wired into position, so that it can't move as grout is added to the cell. The placement of steel varies with each application, but steel reinforcement is typically placed on the side of the wall that is in tension. Because a wall will receive pressure from each side, masonry wall construction usually has the steel centered in the wall cavity. For retaining walls, steel is normally placed near the side that is opposite the soil. The location of steel relative to the edge of the wall must be specified in the wall details.

Rebar Representation

The drafter will need to specify the quantity of bars, the bar size by number, the direction that the steel runs, and the grade on the drawings where steel is referenced. On drawings such as the framing plan, walls will be drawn, but the steel is not drawn. Steel specifications are generally included in the wall reference, as in Figure 45.13. When shown in section or detail, steel is represented by a bold line which can be solid or dashed, depending on office practice. Steel is represented by a solid circle when shown in end view. In detail, steel is often drawn at an exaggerated size so that it can be clearly seen. It should not be so small that it blends with the hatch pattern used for the grout, or so large that it is the first thing seen in the detail. Figure 45.17 shows how steel can be represented in detail.

Locating Steel

Dimensions will be required to show the location of the steel from the edge of the concrete. The engineer may provide a note in the calculations such as:

7- #5 horiz. @ 3″ up/dn.

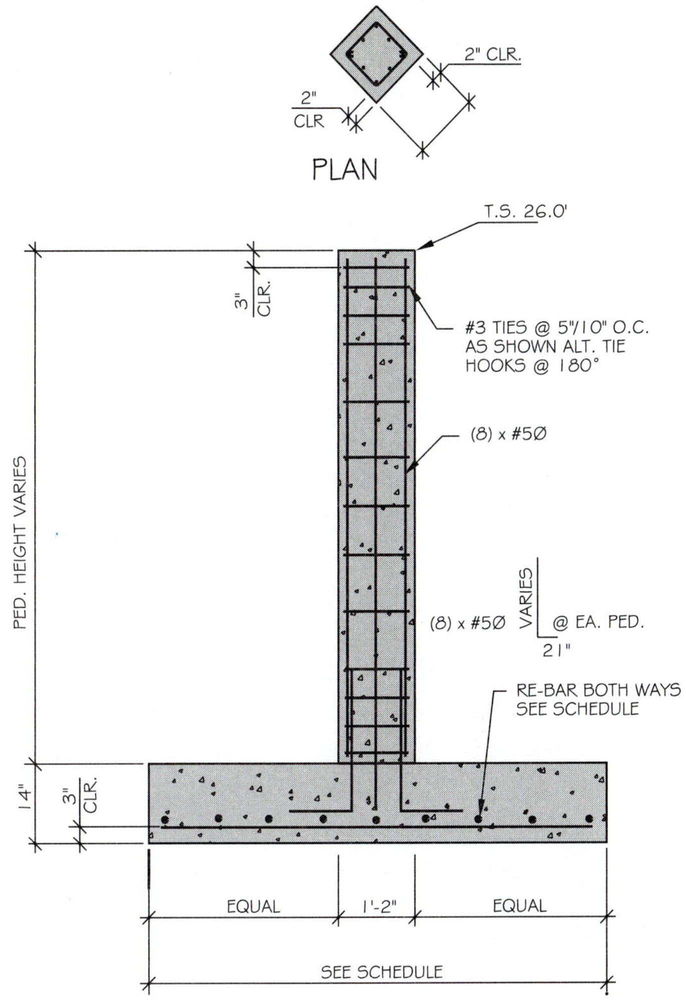

	PED HEIGHT	SIZE	REINFORCING STEEL
B-2	7.50	7'-0" ◻	#6 @ 10" O.C.
B-3	6.75	6'-6" ◻	#6 @ 12" O.C.
B-4	5.91	6'-6" ◻	#6 @ 12" O.C.
B-5	5.00	7'-0" ◻	#6 @ 10" O.C.

FIGURE 45.18 ■ Representing steel placement in detail.

Although this note could be placed on the drawings exactly as is, a better method can be seen in Figure 45.18. The quantity and size of the steel are specified in the detail, but the locations are placed using separate dimensions. Although this requires slightly more work on the part of the drafter, the more visual specifications will be less likely to be overlooked or misunderstood. Depending on the engineer and the type of stress to be placed on the masonry, the location may be given from edge of concrete to edge of steel or from edge of concrete to center of steel. To distinguish the edge of steel location, the term *clear* or CLR is added to the dimension. Figure 45.18 shows examples of each method of locating steel. In addition to the information placed in the details, steel will also be referenced in the written specifications. The written specifications will detail the grade and strength requirements for general areas of the structure such as walls, floors, columns, and retaining walls. Figure 45.19

CONCRETE BLOCK

1. DESIGN F'M = 1500 PSI FOR SOLID GROUTED WALLS, 1350 PSI FOR WALLS WITH REINFORCED CELLS ONLY GROUTED.
2. ALL CMU UNITS TO BE GRADE N TYPE 1 LIGHTWEIGHT UNITS PER ASTM C90 DRY TO INTERMEDIATE HUMIDITY CONDITION PER TABLE NO. 1.
3. MORTAR TO BE PER UBC TABLE 24A, TYPE S OR PER MANUFACTURER'S RECOMMENDATION
   TO REACH CORRECT STRENGTH. GROUT TO HAVE MINIMUM STRENGTH TO MEET DESIGN STRENGTH AND MADE WITH 3/8" MINUS AGGREGATE.
4. CONTRACTOR TO PREPARE AND TEST 3 GROUTED PRISMS AND 3 UNGROUTED PRISMS PER EVERY 5000 SQUARE FEET OF WALL, DURING CONSTRUCTION.
5. ALL WORK SHALL CONFORM TO SECTION 2402 THROUGH SECTION 2412 OF THE 1997 UNIFORM BUILDING CODE.
6. REINFORCING FOR MASONRY TO BE ASTM A615 GRADE 60 PLACED IN CENTERS OF CELLS AS FOLLOWS (UNLESS OTHERWISE INDICATED):
   A. VERTICAL: 1- #5 @ 48" O.C. PLUS 2- #5'S FULL HEIGHT EACH SIDE OF OPENINGS, UNLESS SHOWN OTHERWISE ON DRAWINGS.
   B. HORIZONTAL: 2- #4'S @ 48" O.C. (FIRST BOND BEAM 48" FROM GROUND FLOOR) PLUS 2- #4'S AT TOP OF WALL AND AT EACH INTERMEDIATE FLOOR LEVEL.
   C. LINTELS: LESS THAN 4'-0" WIDE---2- #4'S IN BOTTOM OF 8"DEEP LINTEL.
      4'-0" TO 6'-0" WIDE----2- #4'S IN BOTTOM OF 16" DEEP LINTEL.
   D. CORNERS AND INTERSECTIONS: 1- 24" X 24" CORNER BAR AT EACH BOND BEAM SAME SIZE AS HORIZONTAL REINFORCING.
7. GROUT ALL CELLS FULL AT EXTERIOR OR LOAD-BEARING WALLS. GROUT ALL CELLS THAT CONTAIN REINFORCING OR EMBEDDED ITEMS.
8. ELECTRICAL BOXES, CONDUIT AND PLUMBING SHALL NOT BE PLACED IN ANY CELL THAT CONTAINS REINFORCING.
9. SEE DRAWINGS FOR ADDITIONAL REINFORCING.

REINFORCING

1. ALL REINFORCING STEEL TO BE ASTM A615 GRADE 60, EXCEPT TIES, STIRRUPS, AND DOWELS TO MASONRY TO BE GRADE 40. WELDED WIRE FABRIC SHALL CONFORM TO ASTM A185 AND SHALL BE 6X6-W1.4XW1.4 WWF MATS.
2. FABRICATE AND INSTALL REINFORCING STEEL ACCORDING TO THE "MANUAL OF STANDARD PRACTICE FOR DETAILING REINFORCED CONCRETE STRUCTURES" ACI STANDARD 315.
3. PROVIDE 2'-0" X 2'-0" CORNER BARS TO MATCH HORIZONTAL REINFORCING IN POURED-IN-PLACE WALLS AND FOOTINGS AT ALL CORNERS AND INTERSECTIONS.
4. SPLICES IN WALL REINFORCING SHALL BE LAPPED 30 DIAMETERS (2'-0" MINIMUM) AND SHALL BE STAGGERED AT LEAST 4' AT ALTERNATE BARS.
5. ALL OPENINGS SMALLER THAN 30" X 30" THAT DISRUPT REINFORCING SHALL HAVE AN AMOUNT OF REINFORCING EQUAL TO THE AMOUNT DISRUPTED PLACED BOTH SIDES OF OPENING AND EXTENDING 2'-0" EACH SIDE OF OPENING.
6. PROVIDE THE FOLLOWING REINFORCING AROUND WALL OPENINGS LARGER THAN 30" X 30":
   A. (2) #5 OVER OPENING X OPENING WIDTH PLUS 2'-0" EACH SIDE.
   B. (2) #5 UNDER OPENING X OPENING WIDTH PLUS 2'-0" EACH SIDE.
   C. (2) #5 EACH SIDE OF OPENING X FULL STORY HEIGHT.
   D. PROVIDE 90 DEGREE HOOK FOR BARS AT OPENINGS IF REQUIRED EXTENSION PAST OPENING CANNOT BE OBTAINED.
7. PROVIDE (2) #4 CONTINUOUS BARS AT TOP AND BOTTOM AND AT DISCONTINUOUS ENDS OF ALL WALLS.
8. PROVIDE (2) #5 X OPENING DIMENSION PLUS 2'-0" EACH SIDE AROUND ALL EDGES OF OPENINGS LARGER THAN 15" X 15" IN STRUCTURAL SLABS AND PLACE (1) #4 X 4'-0" AT 45 DEGREES TO EACH CORNER.
9. PROVIDE DOWELS FROM FOOTINGS TO MATCH ALL VERTICAL WALL, PILASTER, AND COLUMN REINFORCING (POURED-IN-PLACE COLUMNS AND WALLS). LAP 30 DIAMETERS OR 2'-0" MINIMUM.
10 ALTERNATE ENDS OF BARS 12" IN STRUCTURAL SLABS WHENEVER POSSIBLE.
11 ALL WALL REINFORCING TO BE PLACED IN CENTER OF WALL UNLESS SHOWN OTHERWISE ON THE DRAWINGS.

FIGURE 45.19 ■ Written specifications for concrete block and steel reinforcing are often placed with the structural drawings. Standards notes are often stored in a library and inserted and modified as needed for each job. *Courtesy Van Domelen/ Looijenga/ McGarrigle/ Knauf Consulting Engineers.*

shows an example of steel specifications that accompany a small office structure.

POURED CONCRETE

Concrete is a common building material composed of sand and gravel bonded together with cement and water. One of the most common types of cement is portland cement. Portland cement contains pulverized particles of limestone, cement rock, oyster shells, silica sand, shale, iron ore, and gypsum. The gypsum determines the time required for the cement to set.

Concrete can be poured in place at the job site, formed off site and delivered ready to be erected into place, or formed at the job site and lifted into place. Your area of the country, the office that you work in, and the type of structure to be built dictate which of these concrete construction methods you will be using.

FIGURE 45.20 ■ Reinforcing steel is placed between wooden forms when concrete is cast in place. *Courtesy Michael Jefferis.*

Concrete Cast in Place

You have been exposed to the methods of cast-in-place concrete for residential foundations and retaining walls. Commercial applications are similar, but the size of the casting and the amount of reinforcing vary greatly. In addition to foundation and floor systems, concrete is often used for walls, columns, and floors above ground. Walls and columns are usually constructed by setting steel reinforcing in place and then surrounding it by wooden forms to contain the concrete, as in Figure 45.20. Once

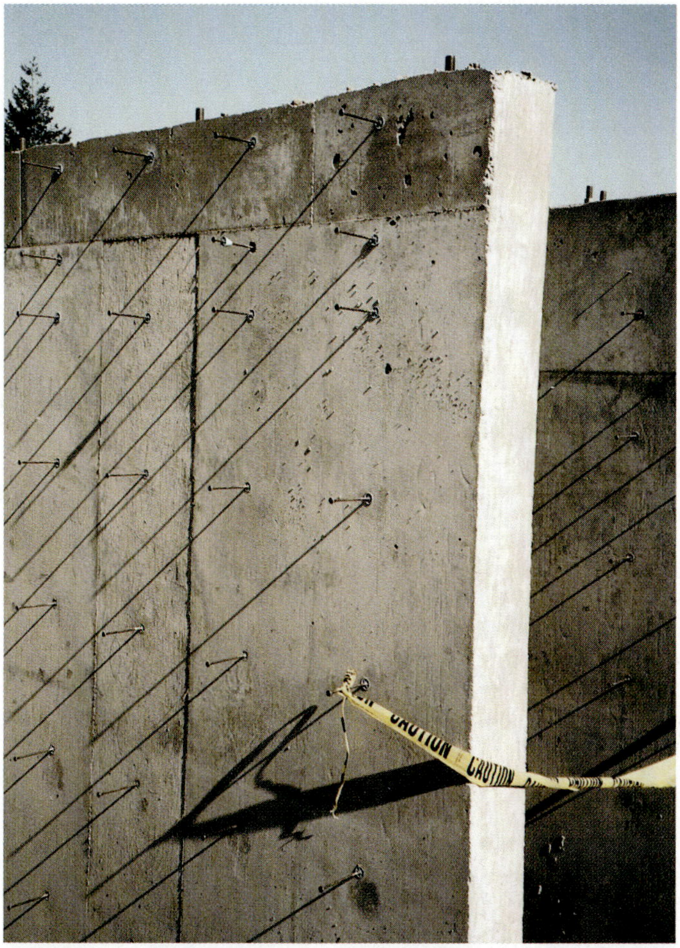

FIGURE 45.21 ■ Wood forms can be removed once the concrete has set. Metal form ties are removed as the concrete hardens. *Courtesy Tom Worcester.*

FIGURE 45.22 ■ Steel reinforcement for a concrete column. Notice that horizontal ties are placed around the vertical bars to keep the vertical bars from separating. *Courtesy Tereasa Jefferis.*

the concrete has been poured and allowed to set, the forms can be removed (see Figure 45.21). As a drafter, you will be required to draw details, showing not only sizes of the part to be constructed but also steel placement within the wall or column. This will typically consist of drawing the vertical steel and the horizontal ties. Ties are wrapped around the vertical steel to keep the column from separating when placed under a load, as in Figure 45.22. Figure 45.23 shows two examples of column reinforcing methods. Figure 45.24 shows the drawings required to detail the construction of a rectangular concrete column. Depending on the complexity of the object to be formed, the drafter may be required to draw the detail for the column and for the forming system.

Concrete is also used on commercial projects to form an above-grade floor. The floor slab can be supported by a steel deck or be entirely self-supporting. The steel deck system is typically used on structures constructed with a steel frame. Figure 44.25 shows an example of ribbed decking being placed. Two of the most common poured-in-place concrete floor systems are the ribbed and waffle floor methods.

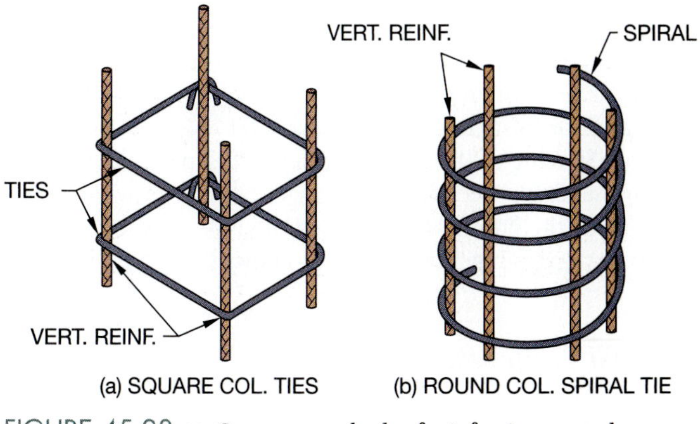

FIGURE 45.23 ■ Common methods of reinforcing poured concrete columns.

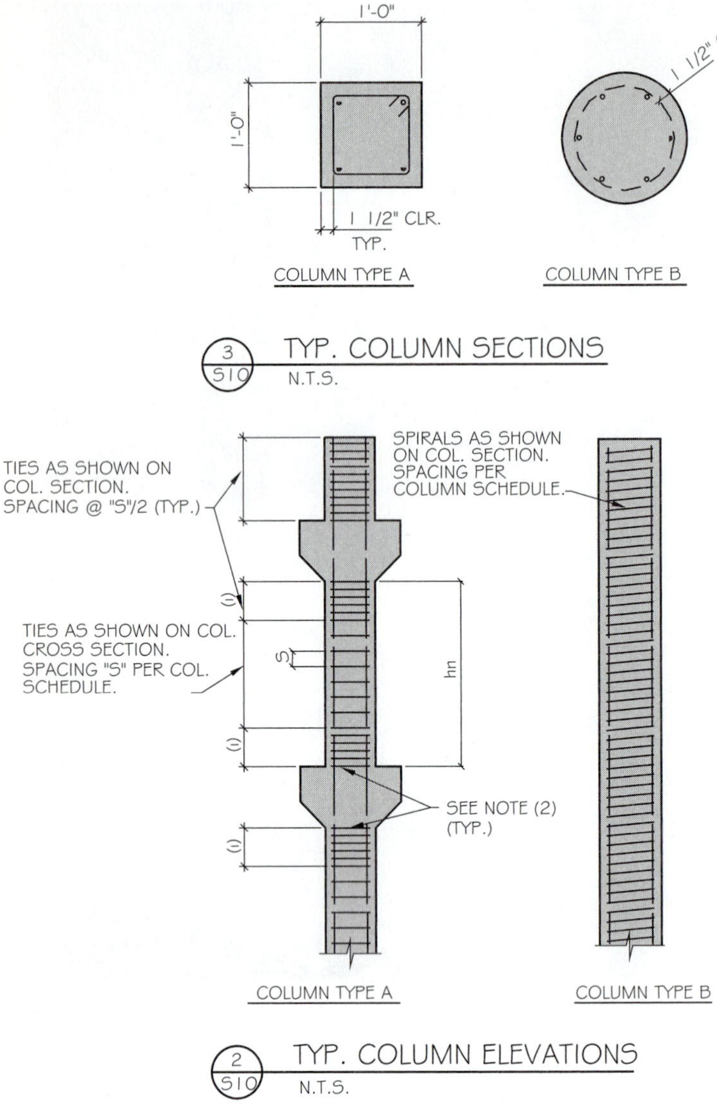

FIGURE 45.24 ■ Common methods of reinforcing poured concrete columns. The horizontal or spiral ties are attached to the vertical reinforcement with metal ties to resist movement as the concrete is placed. *Courtesy Van Domelen/Looijenga/McGarrigle/Knauf Consulting Engineers.*

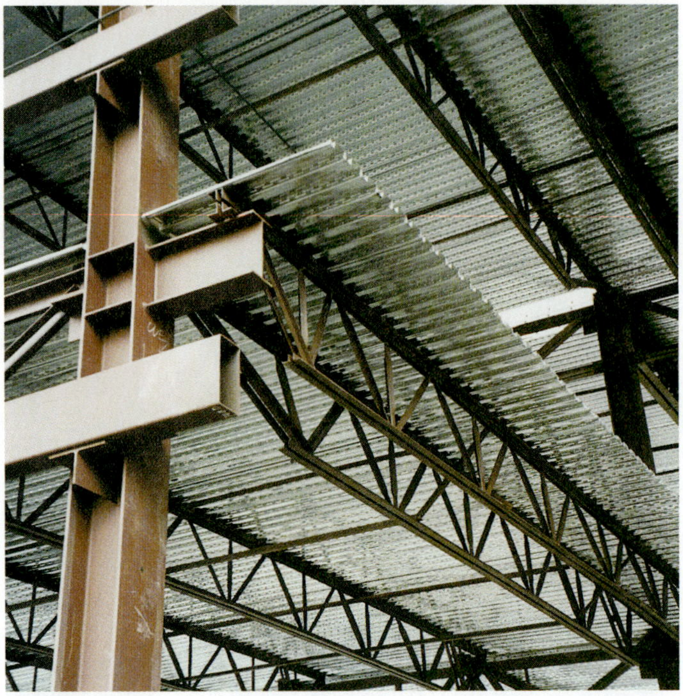

FIGURE 45.25 ■ Ribbed metal decking is used to support aboveground poured concrete slabs. *Courtesy Sara Jefferis.*

FIGURE 45.26 ■ Forms for a waffle floor system.

The ribbed system is used in many office buildings. The ribs serve as floor joists to support the slab but are actually part of the slab. Spacing of the ribs depends on the span and the reinforcing material. The waffle system is used to provide added support for the floor slab and is typically used in the floor system of parking garages. Figure 45.26 shows the forms for a waffle floor system.

Precast Concrete

Precast concrete construction consists of forming walls or other components off site and transporting the part to the job site. Figure 45.27 shows an example of a framing plan for a precast floor system. Figure 45.28*a* and Figure 45.28*b* show examples of precast floor panels. In addition to detailing how precast members will be constructed, drawings must also include methods of transporting and lifting the part into place. Precast parts (Figure 45.28) typically have an exposed metal flange so that the part can be connected to other parts. Figure 45.29 shows common details used for wall connections.

In addition to being precast, many concrete products are also prestressed. Concrete is prestressed by placing steel cables held in tension between the concrete forms while the concrete is poured around them. See Figure 45.30. Once the concrete has

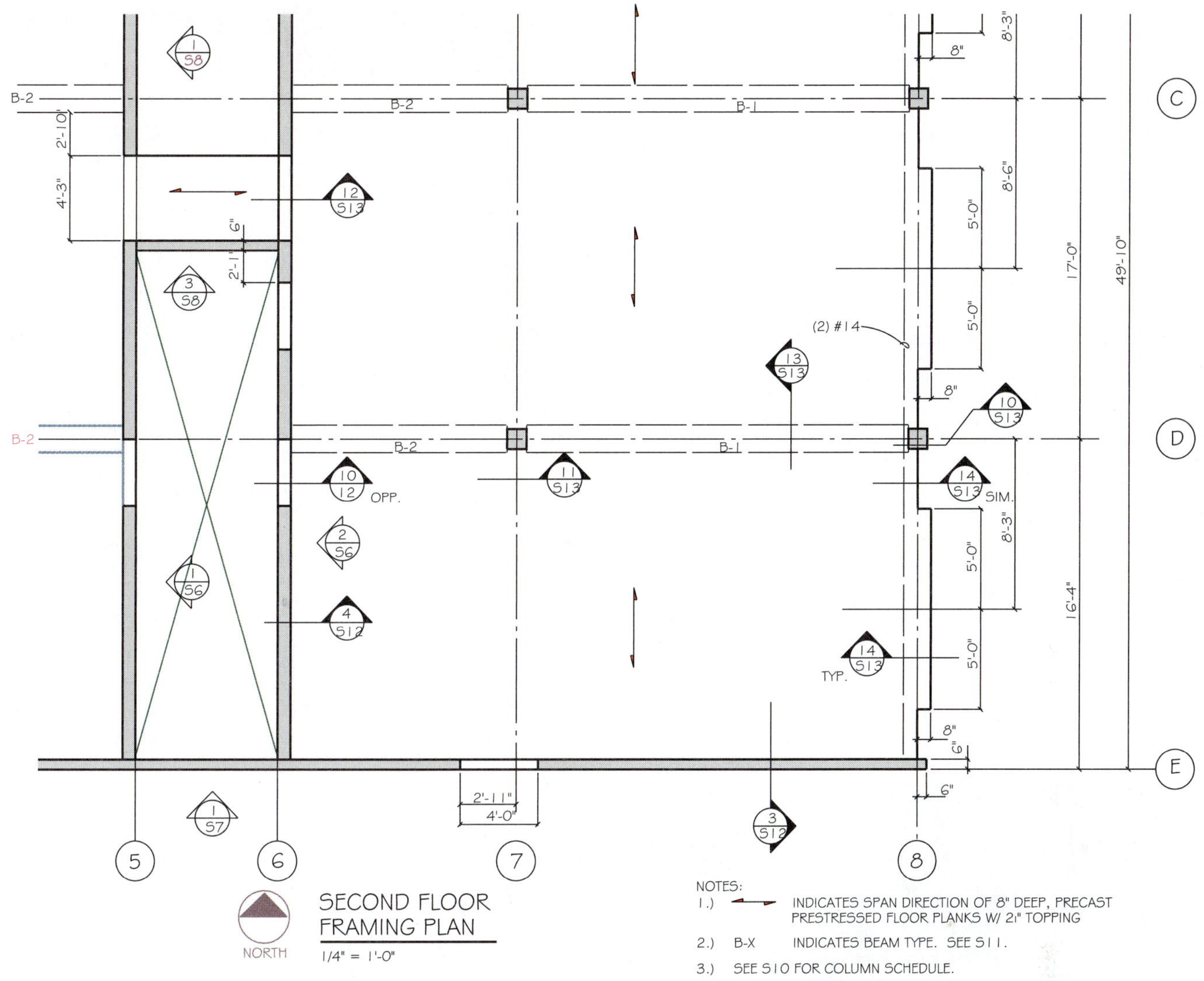

FIGURE 45.27 ■ A framing plan for a precast concrete structure shows the location of all concrete beams and panels. The plan serves as a reference map for detail markers. *Courtesy KPFF Consulting Engineers.*

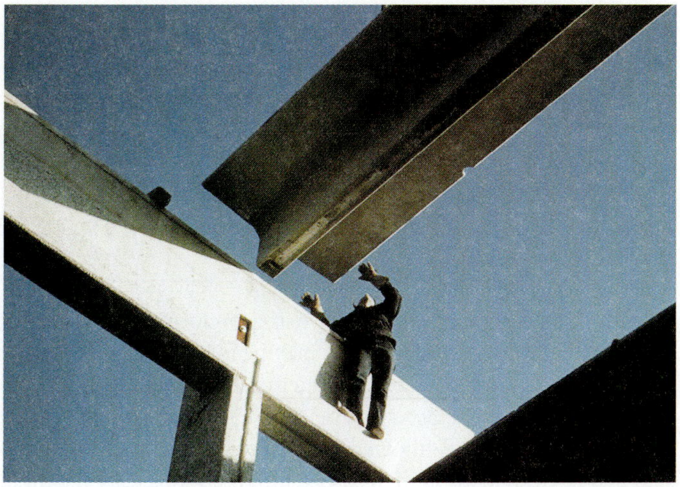

FIGURE 45.28(A) ■ Precast concrete members are cast off site and delivered to the site as needed. *Courtesy Portland Cement Association.*

FIGURE 45.28(B) ■ Precast hollow core slab units. *Courtesy Portland Cement Association.*

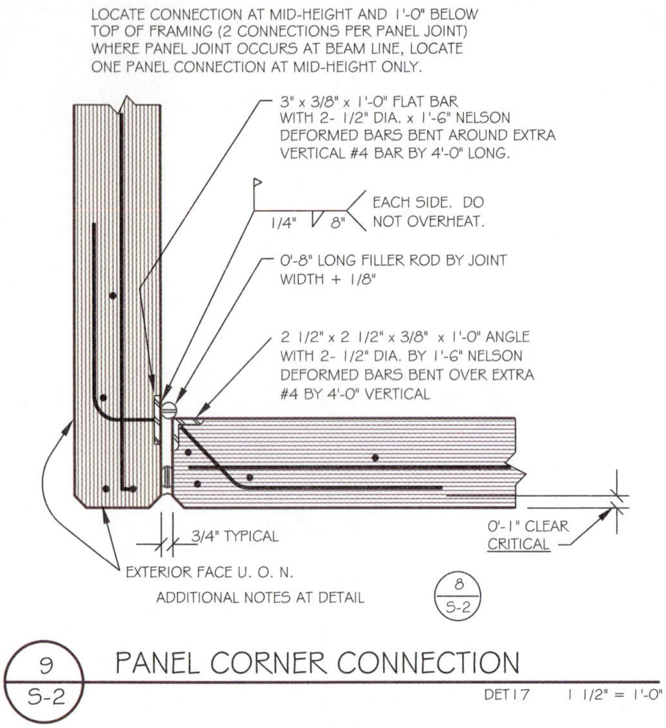

FIGURE 45.29 ■ Detail showing the connection of two precast wall panels. *Courtesy Van Domelen/Looijenga/McGarrigle/Knauf Consulting Engineers.*

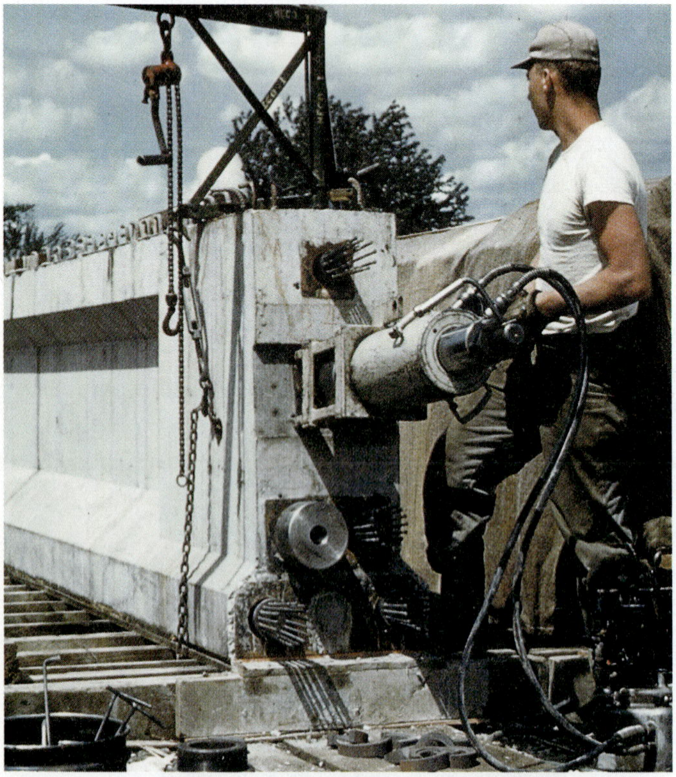

FIGURE 45.30 ■ Concrete for commercial projects is often reinforced by putting tension on steel cables placed in the concrete. The process is known as posttensioning when the cables are tensioned after the concrete has cured. *Courtesy Portland Cement Association.*

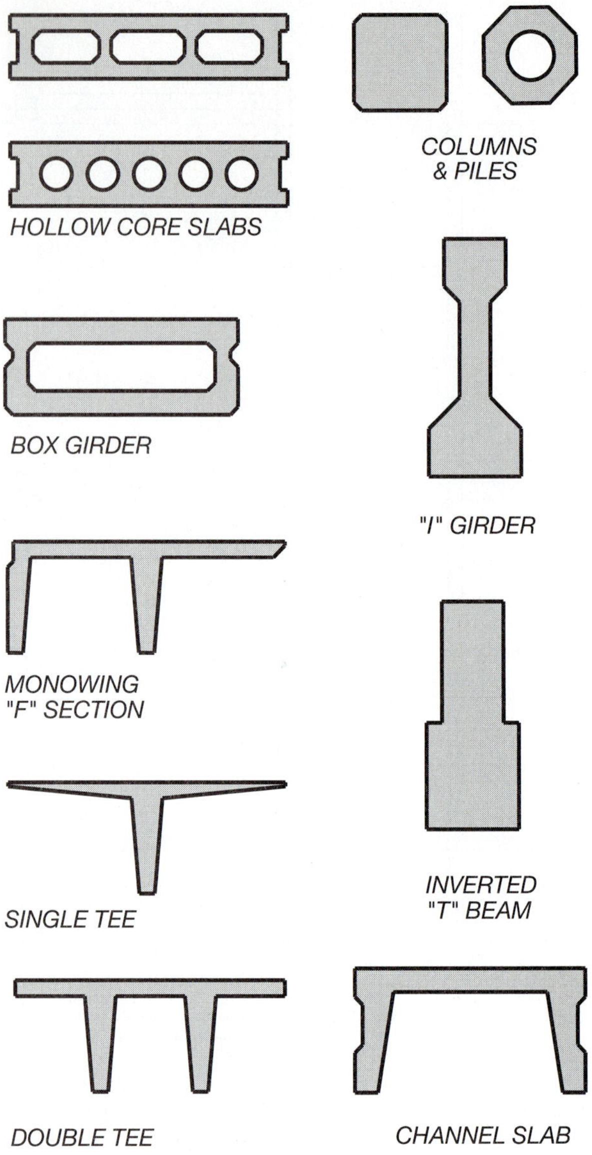

FIGURE 45.31 ■ Common precast concrete shapes.

hardened, the tension on the cables is released. As the cables attempt to regain their original shape, compression pressure is created within the concrete. The compression in the concrete helps prevent cracking and deflection and often allows the size of the member to be reduced. Figure 45.31 shows common shapes that are typically used in prestressed construction.

Tilt-Up

Tilt-up construction is a method using preformed wall panels, which are lifted into place. Panels may be formed at the job or off site. Forms for a wall are constructed in a horizontal position, and the required steel is placed in the form. Concrete is then poured around the steel and allowed to harden. Once the panel has reached its design strength, it can be lifted into place, as shown in Figure 45.32. In this type of construction, the drafter will usually draw a plan view to specify the panel locations, as in Figure 45.33. Figure 45.34 shows a typical steel placement drawing.

FIGURE 45.32 ■ Tilt-up concrete panels are used as an alternative to concrete block construction. Panels are poured at the job site and lifted into position once they have cured. *Courtesy Lynne Worcester.*

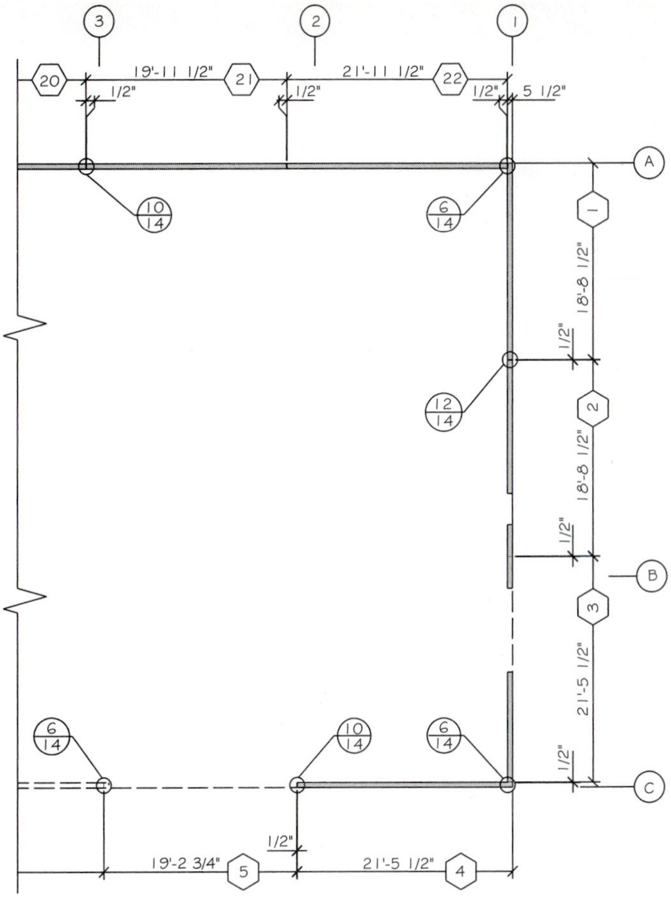

FIGURE 45.33 ■ A plan view is used to show the location of precast panels. Panels are also referenced to the exterior elevations. *Courtesy Kenneth D. Smith and Associates, Inc.*

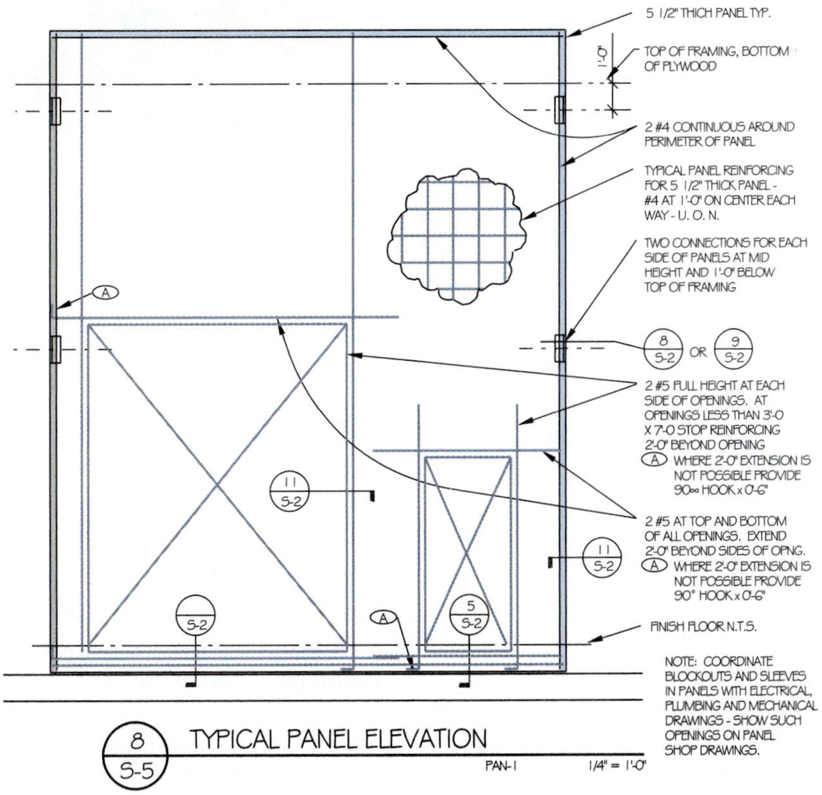

FIGURE 45.34 ■ Steel reinforcing in panels is shown in a panel elevation and referenced to a plan view. *Courtesy Van Domelen/Looijenga/McGarrigle/Knauf Consulting Engineers.*

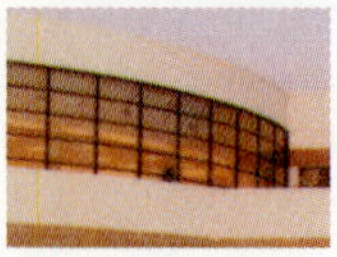

STEEL CONSTRUCTION

Steel construction can be divided into the three categories: lightweight structural members, prefabricated steel structures, and steel-framed structures. Lightweight structural members include studs, joists, open-web trusses, space frames, and decking.

Steel Studs

Prefabricated steel studs are used in many types of commercial structures to help meet the requirements of Types 1, 2, and 3 construction methods. Steel studs offer lightweight, noncombustible, corrosion-resistant framing for interior walls, and for load-bearing exterior walls up to four stories high. Steel members are available for use as studs or joists. Members are designed for rapid assembly and are predrilled for electrical and plumbing conduits. The standard 24″ (610 mm) spacing reduces the number of studs required by about one-third, compared with common studs spacing of 16″ (406 mm) o.c. Widths of studs range from 3 5/8″ to 10″ (93 to 254 mm) but can be manufactured in any width. The material used to produce studs ranges from 12 to 20 gage steel, depending on the loads to be supported. Figure 45.35 shows steel studs as they are typically used. Steel studs are mounted in a channel track at the top and bottom of the wall. This channel is similar to the top and bottom plates of a standard stud wall. Horizontal bridging is placed through the predrilled holes in the studs and then welded to the stud, serving a function similar to solid blocking in a stud wall. Figure 45.36 shows components of steel stud framing.

Stud Specifications

Section properties and steel specifications vary among manufactures, so the material to be specified on the plans will vary. Catalogs of specific vendors must be consulted to determine the structural properties of the desired stud. The architect or engineer will determine the size of stud to be used, but the drafter will often need to consult vendors' catalogs to completely specify the studs. As a minimum, the gage and usage are specified on the framing plan, details, and sections. If a manufacturer is to be specified, the callout will typically include the size, style, gage, and manufacturer of the stud. An example of a steel stud specification would read:

362SJ20 STEEL STUDS BY UNIMAST

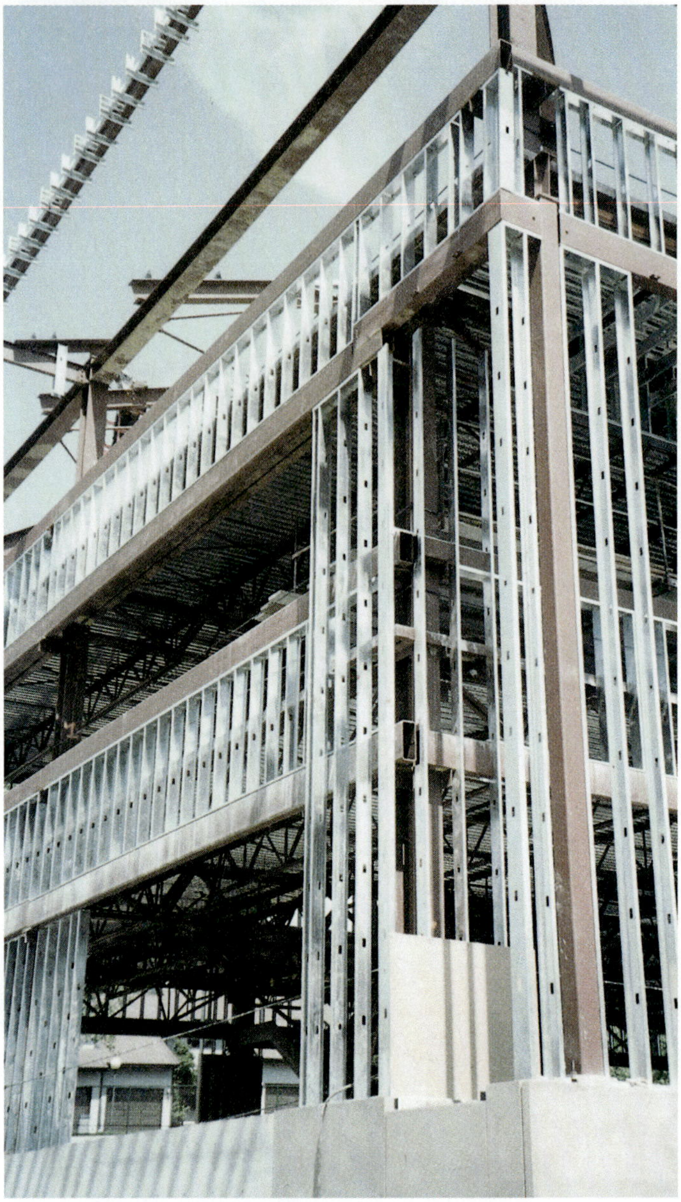

FIGURE 45.35 ■ Steel studs are often used where noncombustible construction is required. *Courtesy Sam Griggs.*

Steel Joists

Steel joists provide the same advantages over wood as steel studs. The gage and yield point for joists is similar to stud values. Many joists are manufactured so that a joist may be placed around another joist, or nested. *Nested joists* allow the strength of a joist to be greatly increased without increasing the size of the framing members. A nested joist is often used to support a bearing wall placed above the floor. Joists are available in lengths up to approximately 40′ (12,000 mm), depending on the manufacturer.

Open Web Steel Joists

Open web steel joists are manufactured in the same configurations as open web wood trusses. Open web steel joists offer the advantage of being able to support greater loads over longer spans than their wood counterparts with less dead load than solid members. Figure 45.25 shows an example of steel trusses used to support floor and roof loads. Open web joists are referred to by series number, which defines their use. The Steel Joist Institute (SJI) designations include K, LH, and DLH. All

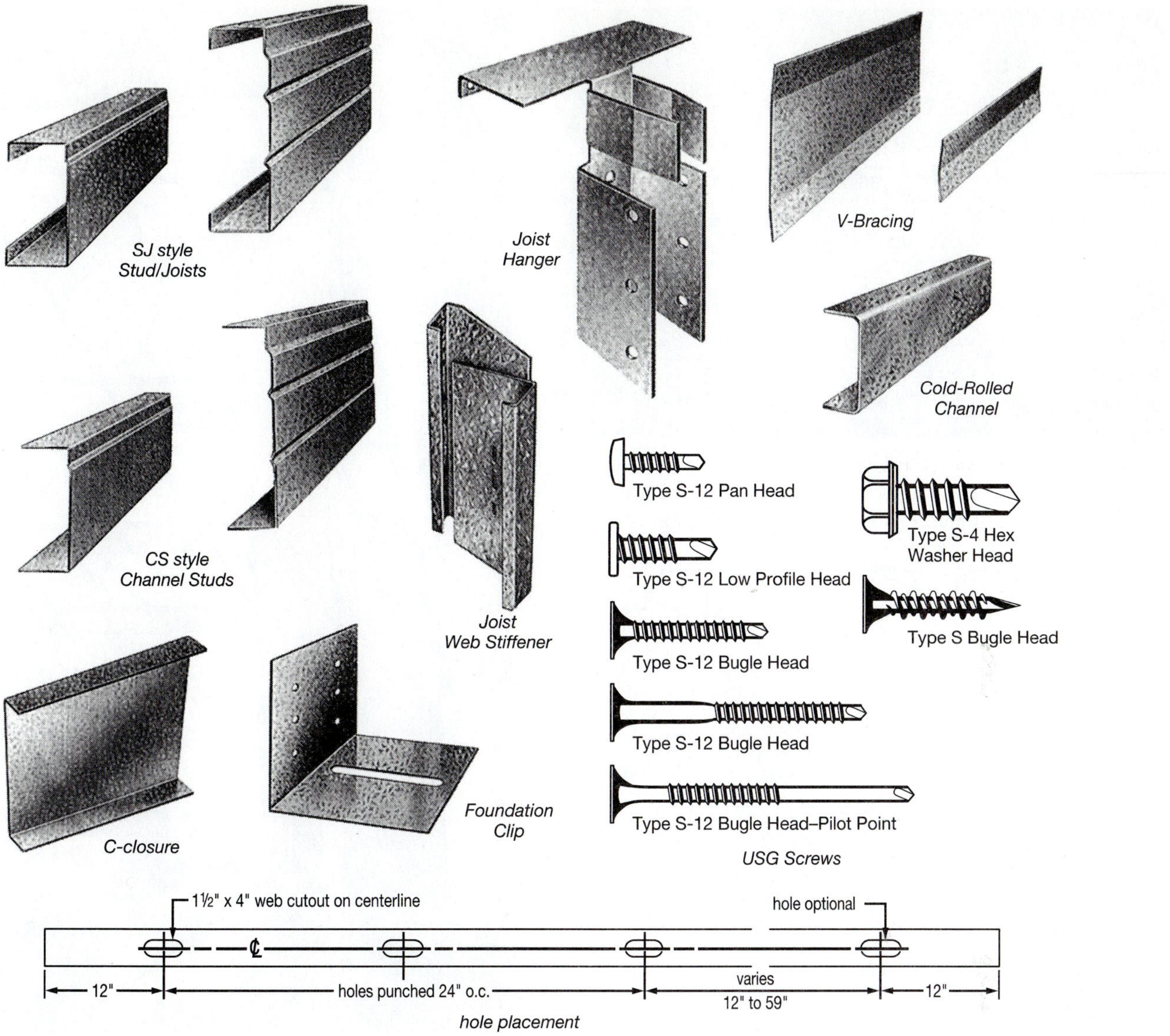

FIGURE 45.36 ■ Common components of steel stud construction. *Courtesy United States Gypsum Company.*

sizes listed below are based on SJI specifications for steel joists. The K-series joists are parallel web members which range in depth from 8″ through 30″ (200–760 mm) in 2″ (50 mm) increments for spans up to approximately 60′ (18,300 mm).

Steel joists are specified on the framing plans, structural details, and sections by providing the approximate depth, the series designation, the chord size, and the spacing. An open web steel joist specification would resemble:

14K4 OPEN WEB STEEL JOIST @ 32″ O.C.

Just as with a wood truss, span tables are available to determine safe working loads of steel joists. The engineer will determine the truss to be used, but the drafter will often be required to find information to complete the truss specification.

Space Frames

Figure 45.37 shows an example of the use of a space frame for supporting the exterior shell of a structure. Drafters working for the architect, the structural engineer, and the fabricator will be involved in the detailing of space frames.

Steel Decking

Corrugated steel decking is used for most lightweight floor systems supported by steel channels or open web joists. Decking is produced in a variety of shapes, depths, and gages. The three most common shapes are shown in Figure 45.38. Common depths include 1.5″, 3″, and 4.5″ (38, 75, and 115 mm), with lengths available up to 30′ (9140 mm). The engineer will determine the required rib depth, based on the load to be supported and the lateral load to be resisted. Steel decking is attached to the floor or roof framing system with screws or welds. Decking is usually covered with a minimum of 1 1/2″ (38 mm) of concrete above the ribs to provide the finished floor. Decking is specified in the project manual and in sections and details by giving the manufacturer, type of decking, depth, and gage. A typical specification would resemble:

Metal decking to be Verco, Type 'N',3″-20ga. galv. decking.

FIGURE 45.37 ■ Three-dimensional trusses form a lightweight metal support system referred to as a space frame. *Courtesy MERO Structures, Inc.*

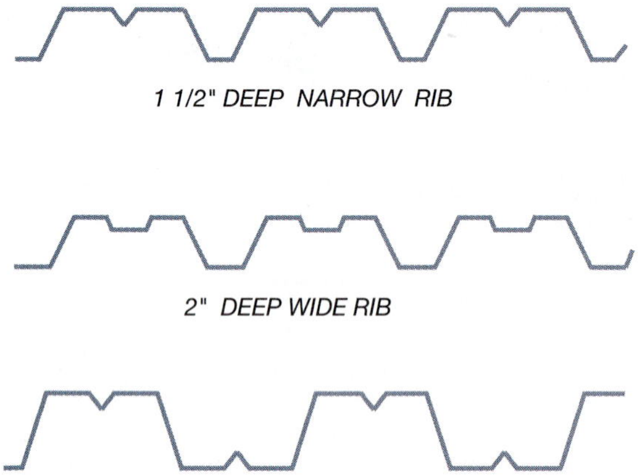

FIGURE 45.38 ■ Common corrugated decking shapes.

FIGURE 45.39 ■ A rigid frame structure consists of the wall and roof frame, girts, and metal decking siding. *Courtesy Aaron Jefferis.*

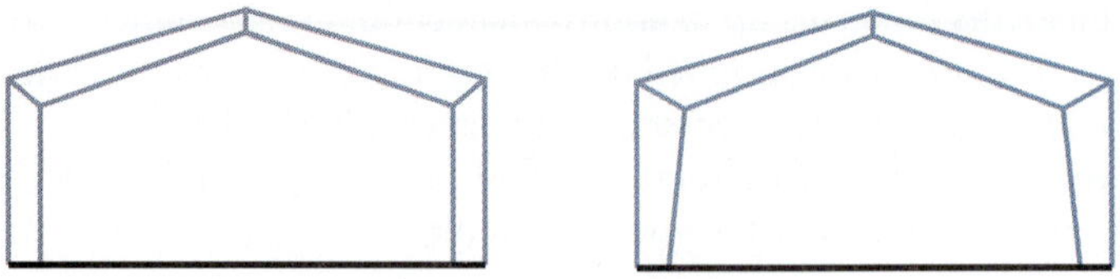

FIGURE 45.40 ■ Common shapes of rigid frame structures.

Prefabricated Steel Structures

Prefabricated or metal buildings have become common for commercial uses in many parts of the country. The frame, roof, and wall components are standardized and sold as modular units with given spans, wall heights, and lengths to reduce cost and eliminate wasted materials. Figure 45.39 shows a steel-framed structure being erected.

Typical Components

The structural system is made up of a frame that supports the walls, the roof, and all externally applied loads. Figure 45.40 shows the two most common frame shapes. The size of the members to be used is determined by engineers working for the fabricator. Tapering of members allows the minimum amount of material to be used, while still maintaining the required area to resist the loads to be supported. End plates are welded or bolted to each end of the vertical members. The plate at the base of the wall member is predrilled to allow for bolting the steel to the concrete support. The plate at the top of the frame is used to bolt the vertical to the inclined member. The vertical wall member is bolted to what will become the inclined roof member to form one rigid member. This frame member is similar to the three-hinged laminated arch discussed earlier in the chapter.

The wall system is made of horizontal girts attached to the vertical frame. Usually the girts are a channel or a Z which is

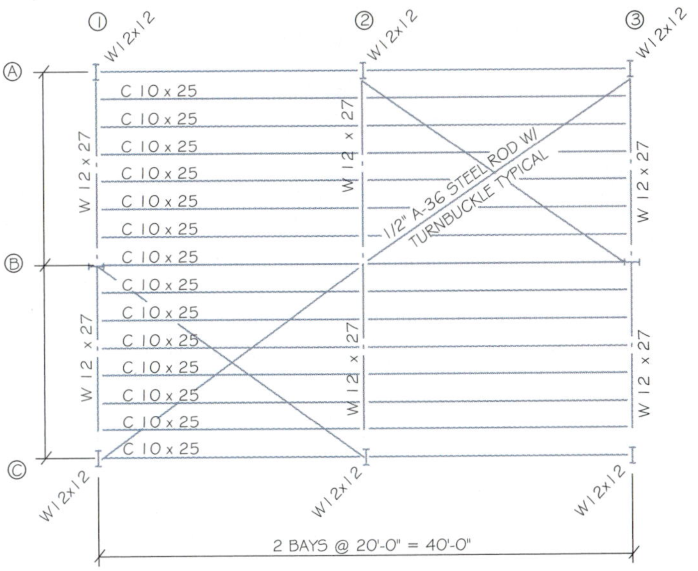

FIGURE 45.41 ■ Framing plan for a prefabricated metal structure.

welded to the frame. Girts can be seen in Figure 45.39. Metal siding is screwed to the girts to complete the wall. The roof is made with steel purlins that provide support for the roofing material. The spacing of the purlins will depend on the sheet metal material used to complete the roof. The roof and wall frames are both reinforced with steel cable X bracing to resist lateral loads.

Representation on Drawings

The engineering team working for the fabricator will complete the structural drawing required to build the structure. This would include drawings showing exact locations of all framing members and connections, as in Figure 45.41.

Steel-Framed Buildings

Steel-framed buildings require engineering and shop drawings similar to those used for concrete structures. As a drafter in an engineering or architectural firm, you will most likely be making engineering drawings similar to the one in Figure 45.42. An

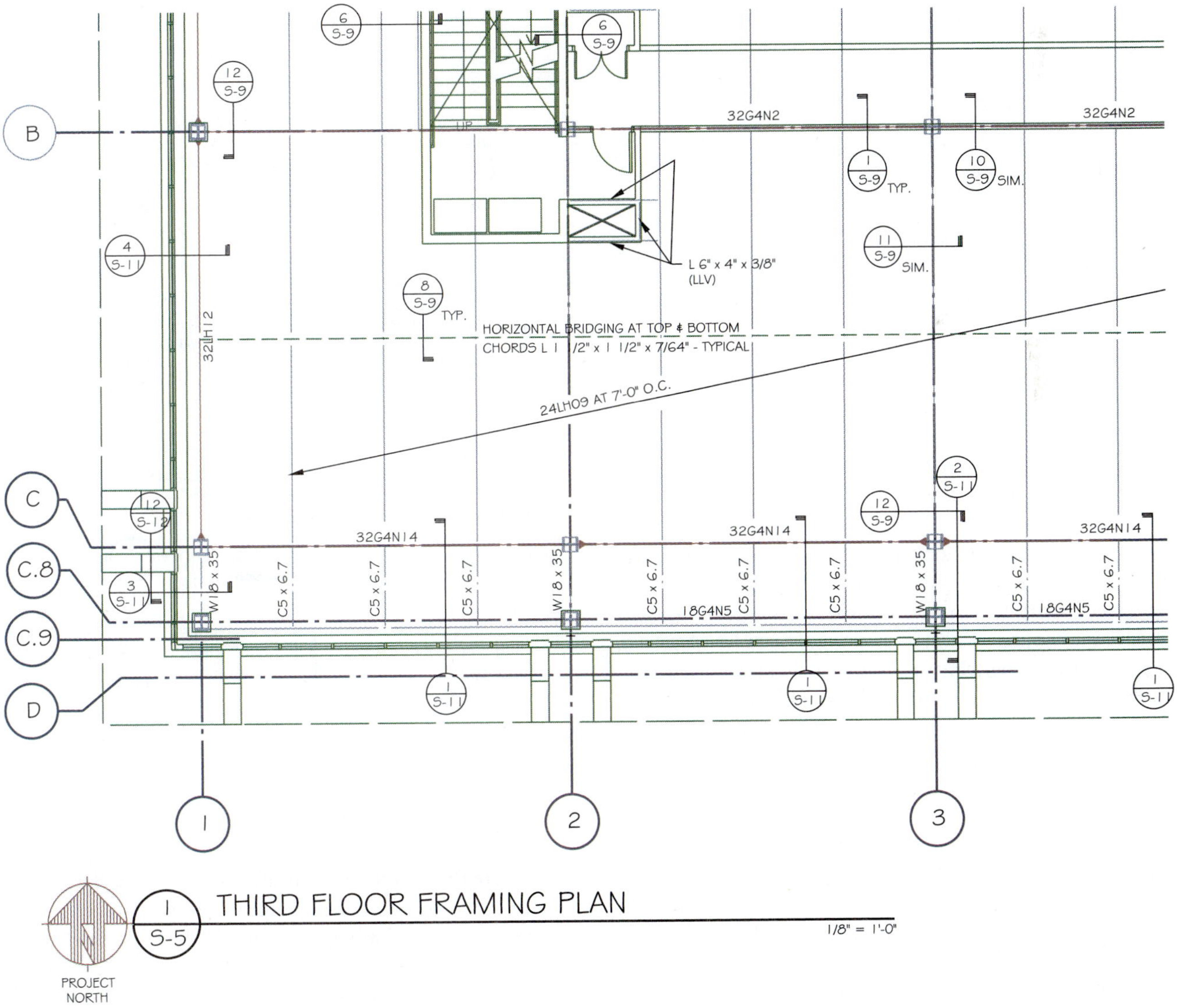

FIGURE 45.42 ■ Structural steel engineering drawings. *Courtesy Van Domelen/Looijenga/McGarrigle/Knauf Consulting Engineers.*

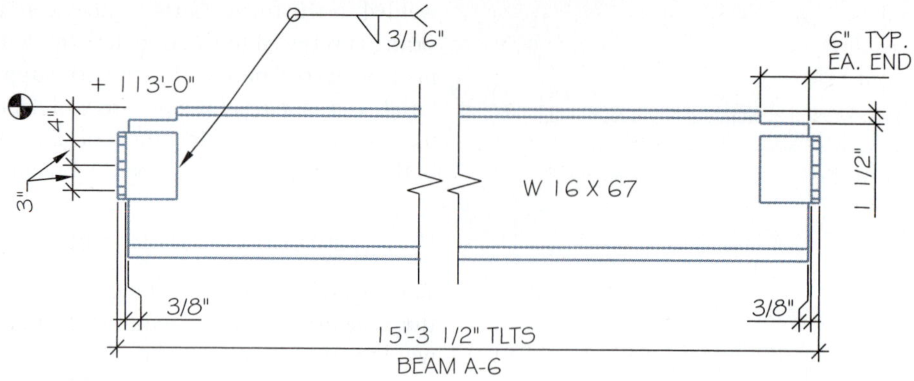

FIGURE 45.43 ■ Structural steel fabrication drawings are made for each steel component to specify exact size, drilling, and cutting locations.

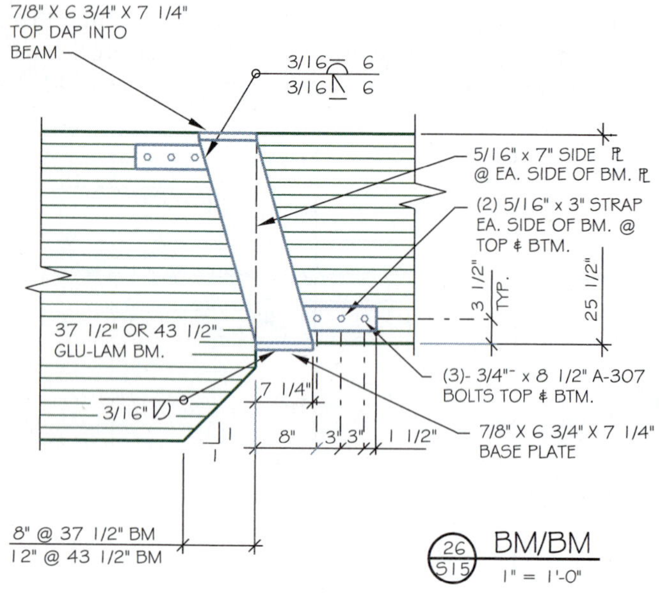

FIGURE 45.44 ■ Steel plates are often used to fabricate beam connectors. *Courtesy Jordan Jefferis.*

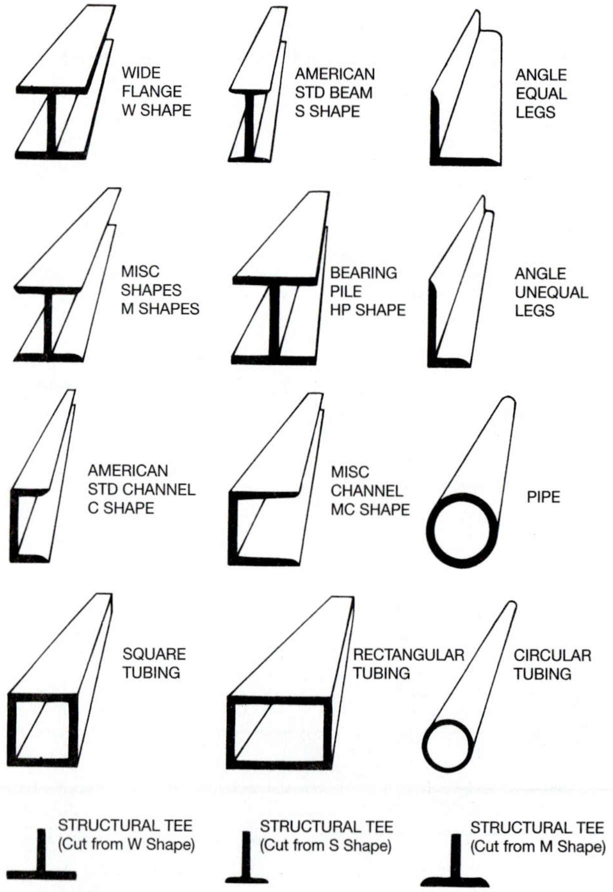

FIGURE 45.45 ■ Standard shapes of structural steel.

engineer or drafter working for the steel fabricator will typically develop the shop drawings, such as the one in Figure 45.43.

As a drafter working on steel-framed structures, you will need to become familiar with the *Manual of Steel Construction*, published by the American Institute of Steel Construction (AISC). In addition to your building code, this manual will be one of your prime references, since it will be helpful in determining dimensions and properties of common steel shapes.

Common Steel Products

Structural steel is typically identified as a plate, as a bar, or by its shape. Plates are flat pieces of steel of various thickness used at the intersection of different members. Figure 45.44 shows a steel connector that uses top, side, and bottom plates. Plates are typically specified on a drawing by giving the thickness, width, and length, in that order. The ℄ symbol is often used to specify plate material.

Bars are the smallest structural steel products. Bars are round, square, rectangular, or hexagonal when seen in cross section. They are often used as supports or braces for other steel parts.

Structural steel is produced in the shapes shown in Figure 45.45. M, S, and W are the names that are given to steel shapes

that have a cross-sectional area in the shape of the letter I. The three differ in the width of their flanges. The flange is the horizontal leg of the I shape, and the vertical leg is the web. In addition to varied flange widths, S-shape flanges also vary in depth.

Angles are structural steel components that have an L shape. The legs of the angle may be either equal or unequal in length but are usually equal in thickness. Channels have a squared C cross-sectional area and are represented by the letter C when they are specified in note form. Structural tees are cut from W, S, and M steel shapes by cutting the webs. Common designations are WT, ST, and MT.

Structural tubing is manufactured in square, rectangular, and round cross-sectional configurations. These members are used as columns to support loads from other members. Tubes are specified by the size of the outer wall followed by the thickness of the wall.

Representing Steel on Framing Plans

As with other methods of construction, plans are required to show the location and size of each steel column, girder, beam, and joist. An example of a steel framing plan can be seen in Figure 45.42. Columns are usually W-shape, but M and S are also used. Steel columns are shown on the framing plan by representing the proper shape, providing the location from center to center, and a specification for the size and strength to be used. TS tubes can be used for columns for heights below three stories depending on the loads to be supported. Steel studs are often used to frame between vertical supports but are not load-bearing when used in conjunction with steel columns.

Girders are normally formed from W shapes because of the ease of making connections. Intermediate beams may be either W, S, M, or C. Depending on the loads to be supported, girders can be built-up to increase the bearing capacity. A built-up girder can be composed of standard shapes or fabricated from steel plates. On the framing and floor plans, girders are usually represented by a single line placed between, but not touching, the columns that will be used for support. The size and spacing of the girders are typically specified by note. Steel joist and decking covered with lightweight concrete can be used to span between intermediate beams to form floor and roof levels. Reinforced concrete slabs are also used to span between intermediate support beams.

Representing Steel in Details

Details are required to show each intersection of structural members, and the method of connecting various materials to the steel frame. A common connection detail can be seen in Figure 45.46. Details showing the connections of structural materials will be referenced on the framing plan and sections. Joints for steel construction are either bolted or welded. Bolting and welding methods and symbols will be discussed later in this chapter.

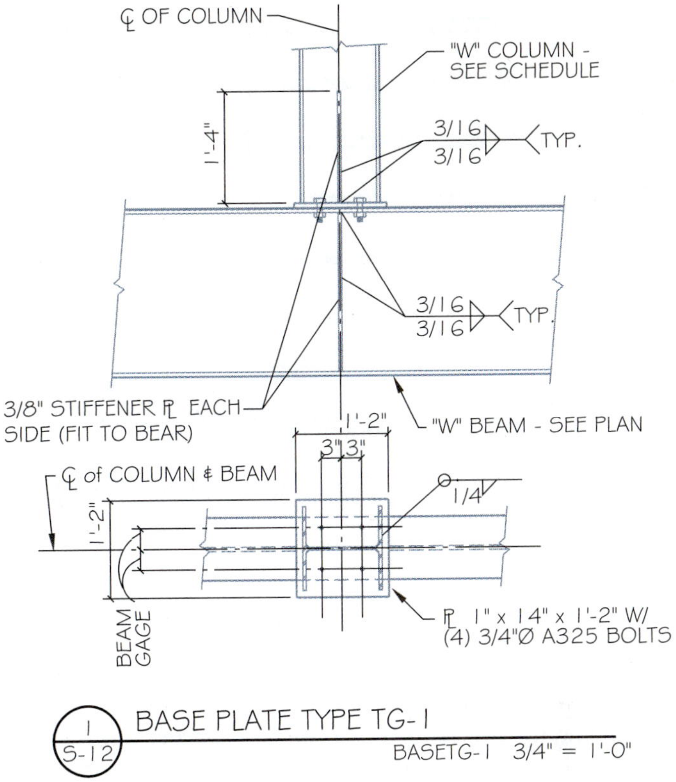

FIGURE 45.46 ■ Detail of a steel column and beam intersection, showing bolted and welded connections. *Courtesy Van Domelen/Looijenga/McGarrigle/Knauf Consulting Engineers.*

COMMON CONNECTION METHODS

The stress to be resisted and the materials to be connected will determine which connection method will be used. Common connection methods include nails, staples, power-driven studs, screws, metal connectors, bolts, and welds.

Nails

Nailing as it applies to residential construction was introduced in Chapter 30. Nails are the common connector for wood to wood members with a thickness of less than 1 1/2″ (40 mm). A nail is required by code to penetrate into the supporting member by half the depth of the supporting member. If two 2× (50×) members are being attached, a nail would be required to penetrate 3/4″ (20 mm) into the lower member. Thicker wood assemblies are normally bolted. The type and size of nails to be used, as well as the placement and nailing pattern, will all affect the joint. Each building code provides a schedule for specifying the method, quantity, and size of nailing to be used. A nailing schedule was introduced in Figure 30.12.

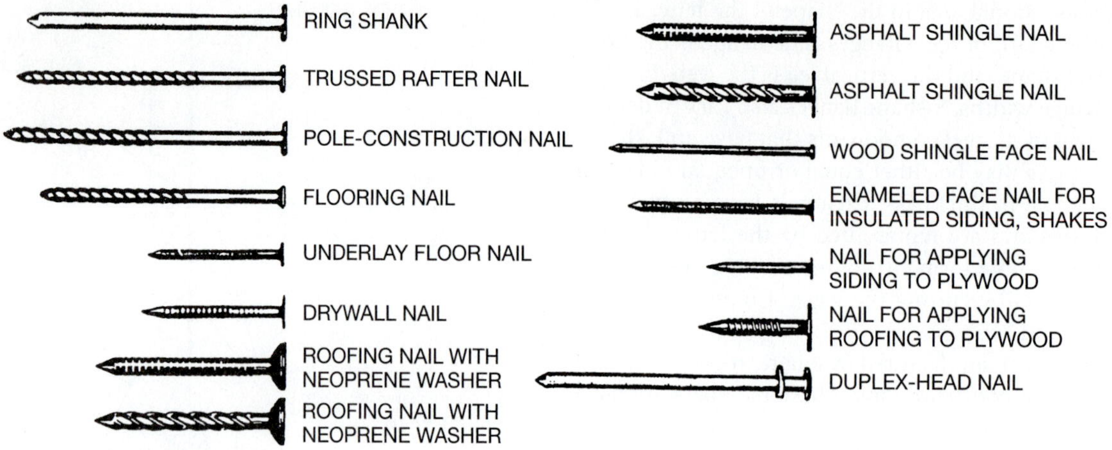

FIGURE 45.47 ■ Common types of nails used in construction.

Types of Nails

The most common nails specified by engineers in the calculations for framing include common, deformed, box, and spikes. Most nails are made from stainless steel, but copper and aluminum are also used. Figure 45.47 shows common types of nails used for construction. A *common* nail is typically used for most rough framing applications. *Box* nails are slightly thinner and have less holding power than common nails. Box nails are used because they generate less resistance in penetration and are less likely to split the lumber. Box nails come in sizes up to 16d. *Spikes* are nails larger than 20d. Nailing specifications are found throughout the written specifications, on framing plan and details.

Nail Sizes

Nail sizes are described by the term *penny,* which is represented by the symbol d. Standard nails range from 2d through 60d. *Penny* is a weight classification; it compares pounds per 1,000 nails. For instance, one thousand 8d nails weigh 8 pounds. Figure 45.48 shows common sizes of nails. The size, spacing, and quantity of nails to be used are typically specified in structural details.

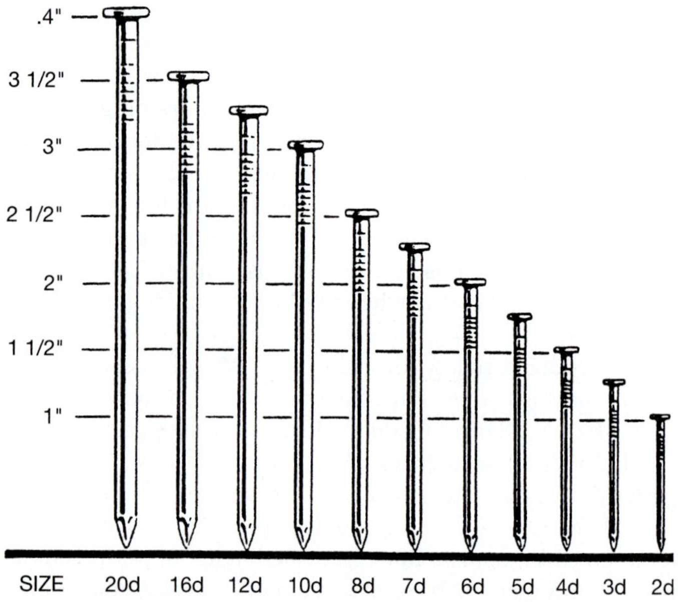

FIGURE 45.48 ■ Standard sizes of common and box nails.

Nailing Specifications

Nails smaller than 20d are typically specified by penny size and by spacing for continuous joints, such as attaching a plate to a floor system. An example of a nailing note found on a framing plan or detail is:

2 × 6 DFL SILL W/ 20d's @ 4″ O.C.

Nails are specified by a quantity and penny size for repetitive joints such as a joist to a sill. The specification on a section or detail would simply read 3-8d's and would point to the area in the detail where the nails will be placed. Depending on the scale of the drawing, the nails may or may not be shown. Specifications for spikes are given in a manner similar to those for nails, but the penny size is replaced by spike diameter.

Nailing Placement

Nailing placements are described by the manner in which they are driven into the members being connected. Common methods of driving nails, shown in Figure 45.49, include face, end, toe and blind nailing. The type of nailing to be used is determined by how accessible the head of the nail is during construction and by the type of stress to be resisted. The engineer will specify if nailing other than that recommended by the nailing schedule of the prevailing code is to be used. Nails that are required to resist shear are strongest when perpendicular to the grain. Nails that are placed parallel to the end grain, such as end nailing, are weakest and tend to pull out from as stress is applied. Common nailing placement includes:

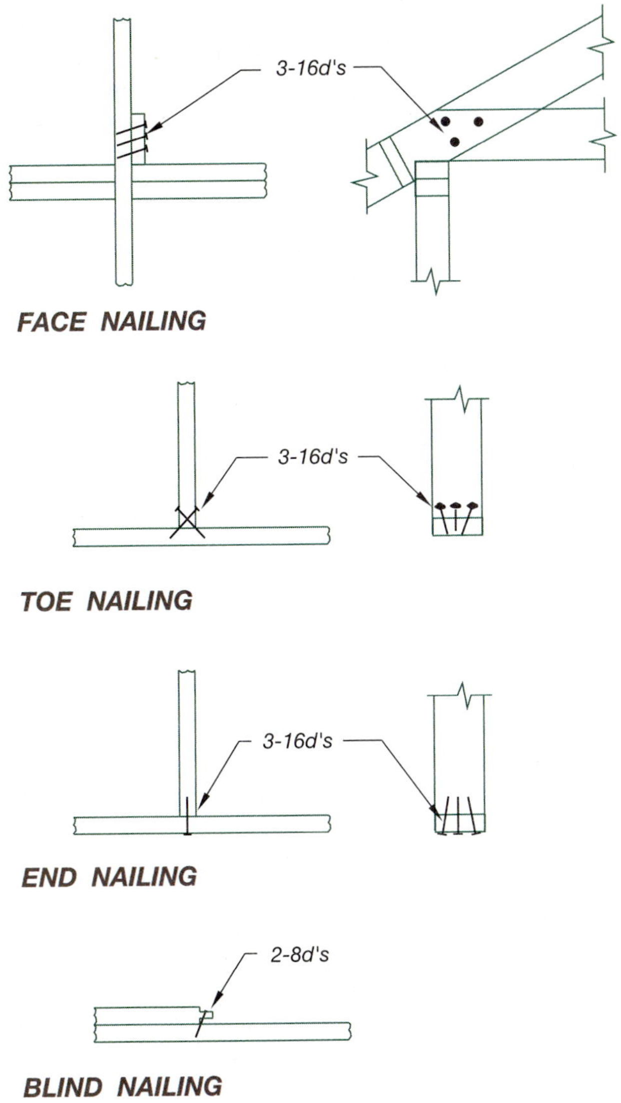

FIGURE 45.49 ■ Methods of placing nails.

- Face nailing—driving a nail through the face or surface of one board into the face of another. Face nailing is used to connect sheathing to rafters or studs, nail a plate to the floor sheathing, or to nail a let-in brace to a stud.
- End nailing—driving a nail through the face of one member into the end of another member. A plate is end-nailed into the studs as a wall is assembled.
- Toe nailing—driving a nail through the face of one board into the face of another. With face and end nailing, nails are driven in approximately 90° to the face. With toe nailing, the nail is driven in at approximately a 30° angle. Connections of rafters to top plates or a header to a trimmer are toe-nailed joints.
- Blind nailing—used where it is not desirable to see the nail head. Attaching wood flooring to the subfloor is done with blind nailing. Nails are driven at approximately a 45° angle through the tongue of tongue-and-groove flooring and hidden by the next piece of flooring.

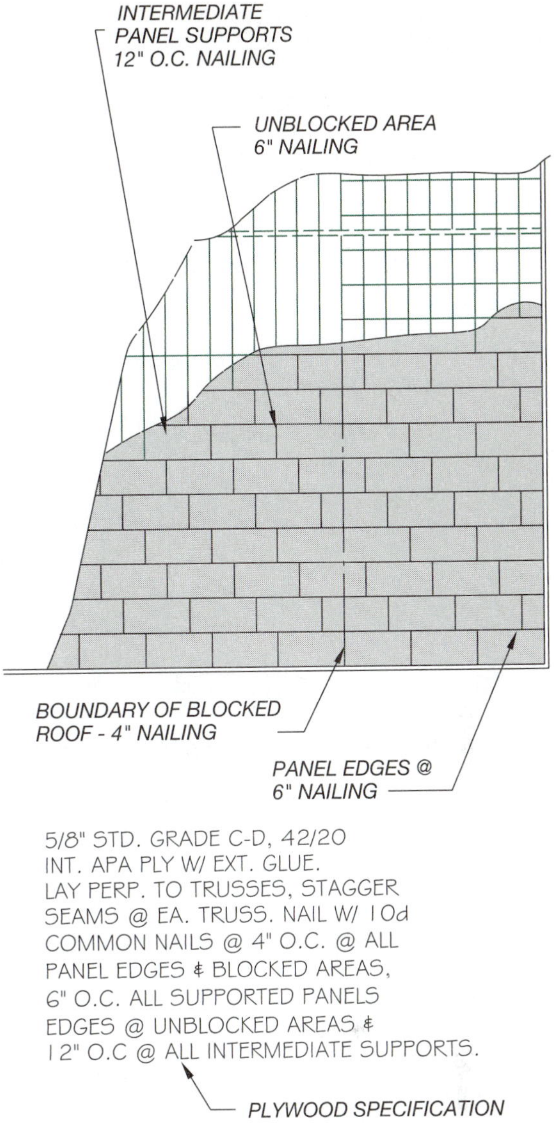

FIGURE 45.50 ■ Nail specification often refer to nail placement by edge, boundary, and field locations.

Nailing Patterns

Nail specifications for sheathing and other large areas of nailing often refer to nail placement along an edge, boundary, or field. Figure 45.50 shows an example of each location. Common placements include:

- Edge nailing—nails placed at the edge of a sheet of plywood.
- Field nailing—nails placed in the supports for a sheet of plywood, excluding the edges.
- Boundary nailing—nailing at the edge of a specific area of plywood.

Staples

Hammer-driven nails have been used for centuries. Power-driven nails have greatly increased the speed and ease in which nails can be inserted. As technology advanced, staples started to

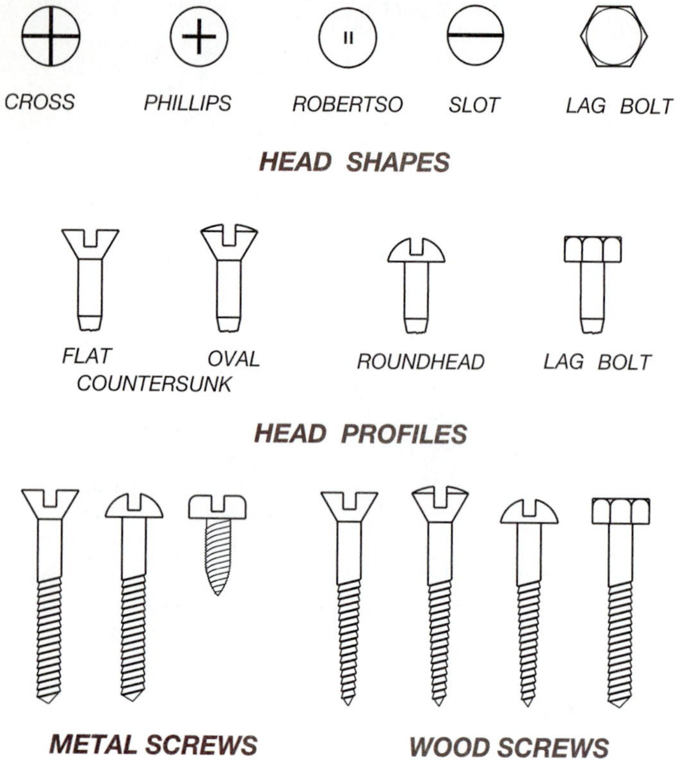

FIGURE 45.51 ■ Common types of screws used in construction.

replace nailing for some applications. Staples are most often used for connecting asphalt roofing and for attaching sheathing to roof, wall, and floor supports.

Power-Driven Studs

Metal studs can be used to anchor wood or metal to masonry. These studs range in diameter from 1/4″ through 1/2″ (6–13 mm) and in length from 3/4″ to 6″ (20 to 150 mm). *Power-driven studs* are made from heat-treated steel and inserted by a powder charge from a gunlike device. They are typically used where it would be difficult to insert the anchor bolts at the time the concrete is poured. Studs can also be used to join wood to steel construction.

Screws

Screws are often specified for use in wood connections that must be resistant to withdrawal. Three common screws are used throughout the construction industry. Each is identified by its head shape; see Figure 45.51. Common screws include:

- Flathead (countersunk) screw—specified in the architectural drawings for finish work where a nail head is not desirable.
- Roundhead screws—used at lumber connections where a head is tolerable. Roundhead screws are also used to connect lightweight metal to wood.
- Lag screws—have a hexagonal or square head that is designed to be tightened by a wrench rather than a screwdriver. Lag screws are used for lumber connections 1 1/2″ (40 mm) and thicker. Lag bolts are available with diameters ranging from 1/4″ to 1 1/4″ (6–30 mm). A washer is typically used with a lag bolt to guard against crushing wood fibers near the bolt. A pilot hole that is approximately 3/4 of the shaft diameter is often specified to reduce wood damage and increase the resistance to withdrawal.

Flathead and roundhead screws are designated by the gage, which specifies a diameter, by length in inches, and by head shape. A typical specification might be: #10 × 3″ F.H.W.S. (flathead wood screw). Lag screws are designated by their length and diameter. A typical specification might be:

5/8″ ∅ × 6″ LAG SCREW THROUGH 1 1/2″ ∅ WASHER.

Metal Framing Connectors

Premanufactured metal connectors by companies such as Simpson and Tyco are used at many wood connections to strengthen nailed connections. Joist hangers, post caps, post bases, and straps are some of the most common lightweight metal hangers used with wood construction. Each connector comes in a variety of gages of metal with sizes to fit a wide variety of lumber. Metal connectors are typically specified on the framing plans, sections, and details by listing the model number and type of connector. A metal connector specification for connecting 2-2 × 12 joists to a beam would resemble:

SIMPSON CO. HHUS212-2TF JST.HGR.

The supplier is typically specified in general notes and within the written specifications.

Depending on the connection, the nails or bolts used with the metal fastener may or may not be specified. If no specification is given, it is assumed that all nail holes in the connector will be filled. If bolts are to be used, they will normally be specified with the connector, based on the manufacturer's recommendations. Nails associated with metal connectors are typically labeled with the letter n instead of d. These nails are equal to their d counterpart, but the length has been modified by the manufacturer to fit the metal hanger.

If premanufactured connectors that can support the required loads are not available, the engineer must design a steel connector to be fabricated for a specific joint. These connectors are assembled in a shop and shipped to the job site. The drafter will need to specify the material, welds, and exact bolt locations. Figure 45.52 shows a fabricated metal connector.

Bolts and Washers

Bolts used in the construction industry include anchor bolts, carriage bolts, and machine bolts. Each can be seen in Figure 45.53. Washers are used under the head and nut for most bolting applications. *Washers* keep the bolt head and nut from pulling through the lumber and reduce damage to the lumber by spreading the stress from the bolt across more wood fibers. Typically a circular washer, specified by its diameter, is used. Common bolts used with commercial construction include anchor, carriage, machine and miscellaneous bolts.

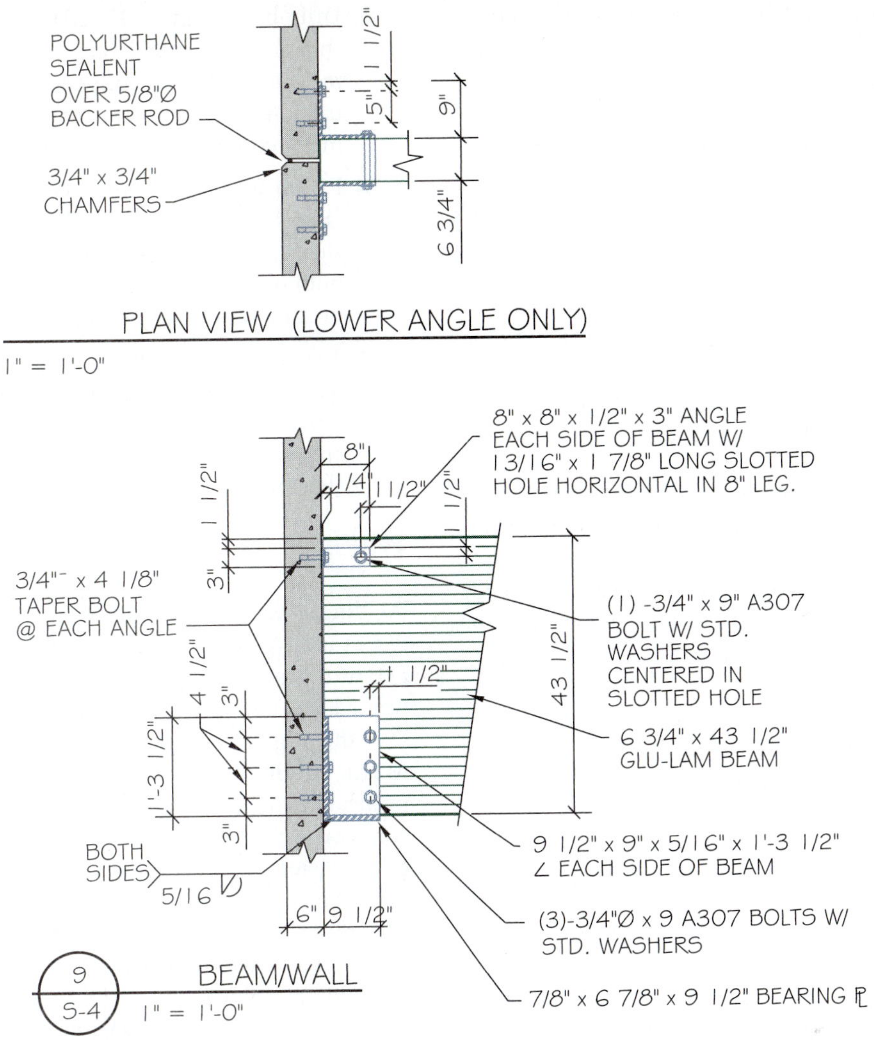

FIGURE 45.52 ■ Metal connectors can be fabricated from steel plates that are welded together. Beams are bolted to the connector.

Anchor Bolts

An anchor bolt is an L-shaped bolt, inserted into concrete, which is used to bolt lumber to the concrete (Figure 45.53A). The short leg of the L is inserted into the concrete to resist withdrawal. The upper end of the long leg is threaded to receive a nut. A 2″ (50 mm) washer is typically used with anchor bolts. Anchor bolts are represented in structural drawings with the letters **A.B.** along with a specification including the size of the member to be connected, the bolt diameter, length, embedment into concrete, spacing and washer size. A typical note might resemble:

2 × 6 DFPT SILL W/ 5/8″ Ø × 12″ A.B. @ 48″ O.C. W/ 2″ Ø WASHERS. PROVIDE 9″ MIN. EMBEDMENT.

When anchor bolts are used to attach lumber to the side of a concrete wall, a specification is typically given to stagger the bolts in relation to the edge of the lumber to be attached as well as the bolt spacing. The specification might resemble:

3 × 12 DFPT LEDGER W/ 3/4″ Ø × 12″ A.B. @ 32″ O.C. STAGGERED 3″ UP/DN WITH 2″ Ø WASHERS.

Carriage Bolts

A *carriage bolt* is used for connecting steel and other metal members as well as timber connections (Figure 45.53B). Carriage bolts have a rounded head with the lower portion of the shaft threaded. Directly below the head, at the upper end of the shaft, is a square shank. As the shank is pulled into the lumber, it will keep the bolt from spinning as the nut is tightened. Diameters range from 1/4″ to 1″ (6–25 mm), with lengths typically available to 12″ (300 mm). The specification for a carriage bolt will be similar to that for an anchor bolt, except for the designation of the bolt type.

Machine Bolts

A *machine bolt* is a bolt with a hexagonal head and a threaded shaft (Figure 45.53C). Machine bolts are divided into two classifications: unfinished or high-strength. Common bolts are used for attaching steel to steel, steel to wood, or wood to wood. Common bolts are often used in steel joints to provide a temporary connection while field welds are completed. Bolts are assumed to

be common unless high-strength bolts are specified. They are referred to with a note such as:

USE 3-3/4″ Ø × 8″ M.B. @ 3″ O.C. W/ 1 1/2″ Ø WASHERS.

The strength of bolts should be specified in a general note on the framing plans or on pages of details containing bolted connections. The engineer will determine the bolt locations based on the stress to be resisted. Common spacings are:

- 1 1/2″ (40 mm) in from an edge parallel to the grain.
- 3″ (75 mm) minimum from the edge when perpendicular to the grain.
- 1 1/2″ (40 mm) from the edge of steel members.
- 2″ (50 mm) from the edge of concrete.

Bolts used for connecting structural steel are referred to as *high-strength* bolts and come in three common compositions. These bolts are manufactured with an ASTM (American Society for Testing Materials) or Institute of Steel Construction Specifications grading number on the head. Common specifications used in commercial construction include:

A-325: High-strength bolts

A-490: High-strength bolts, medium carbon

A-441: High-strength, low-alloy steel bolts

A-242: Corrosion-resistant high-strength low-alloy steel bolts.

High-strength bolts are normally specified to be tightened with a pneumatic impact wrench.

Miscellaneous Bolts

Several other types of bolts are used in special circumstances. These include studs, drift bolts, expansion bolts, and toggle bolts.

- Stud—a bolt that has no head. A stud is welded to a steel beam so that a wood plate can be bolted to the beam.
- Drift bolt (Figure 45.53D)—a steel rod that has been threaded. Threaded rods can be driven into one wood member with another member bolted to the threaded protrusion. Threaded rods may also be used to span between metal connectors on two separate beams.
- Expansion bolts—bolts with a special expanding sleeve (Figure 45.53E). These bolts are designed so that the sleeve, once inserted into a hole, will expand to increase holding power. Expansion bolts are typically used for connecting lumber to masonry.
- Toggle bolt—a bolt that has a nut designed to expand when it is inserted through a hole, so that it cannot be removed. Toggle bolts (Figure 45.53F) are used where one end of the bolt may not be accessible because of construction parameters.

Welds

Welding is the method of providing a rigid connection between two or more pieces of steel. In welding, metal is heated to a temperature high enough to cause melting. The parts that are welded become one, with the welded joint actually stronger than the original material. Welding offers better strength, better weight distribution of supported loads, and a greater resistance to shear or rotational forces than a bolted connection. The most common welds in construction are shielded metal arc welding, gas tungsten arc welding, and gas metal arc welding. In each case, the components to be welded are placed in contact with each other, and the edges are melted to form a bond. Additional metal is also used to firm a sufficient bond.

Welds are typically specified in details as shown in Figure 45.54. A horizontal reference line is connected to the parts to be

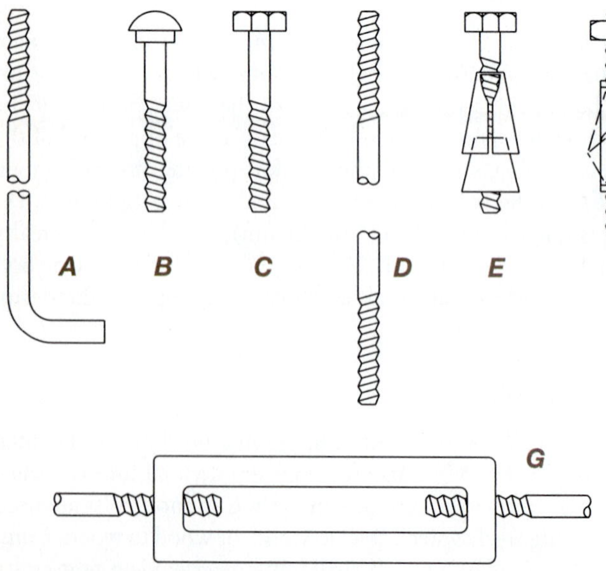

FIGURE 45.53 ■ Common bolts used in construction.

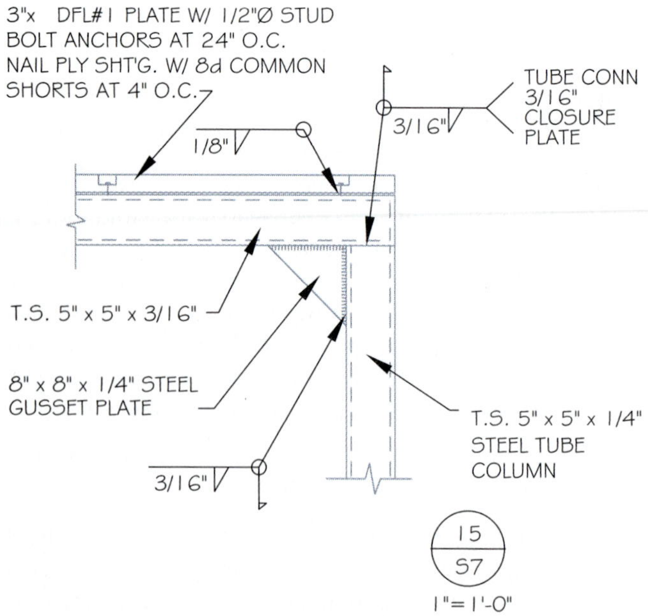

FIGURE 45.54 ■ Welds are specified in details by the use of specialized symbols that are referenced to the area to receive the weld.

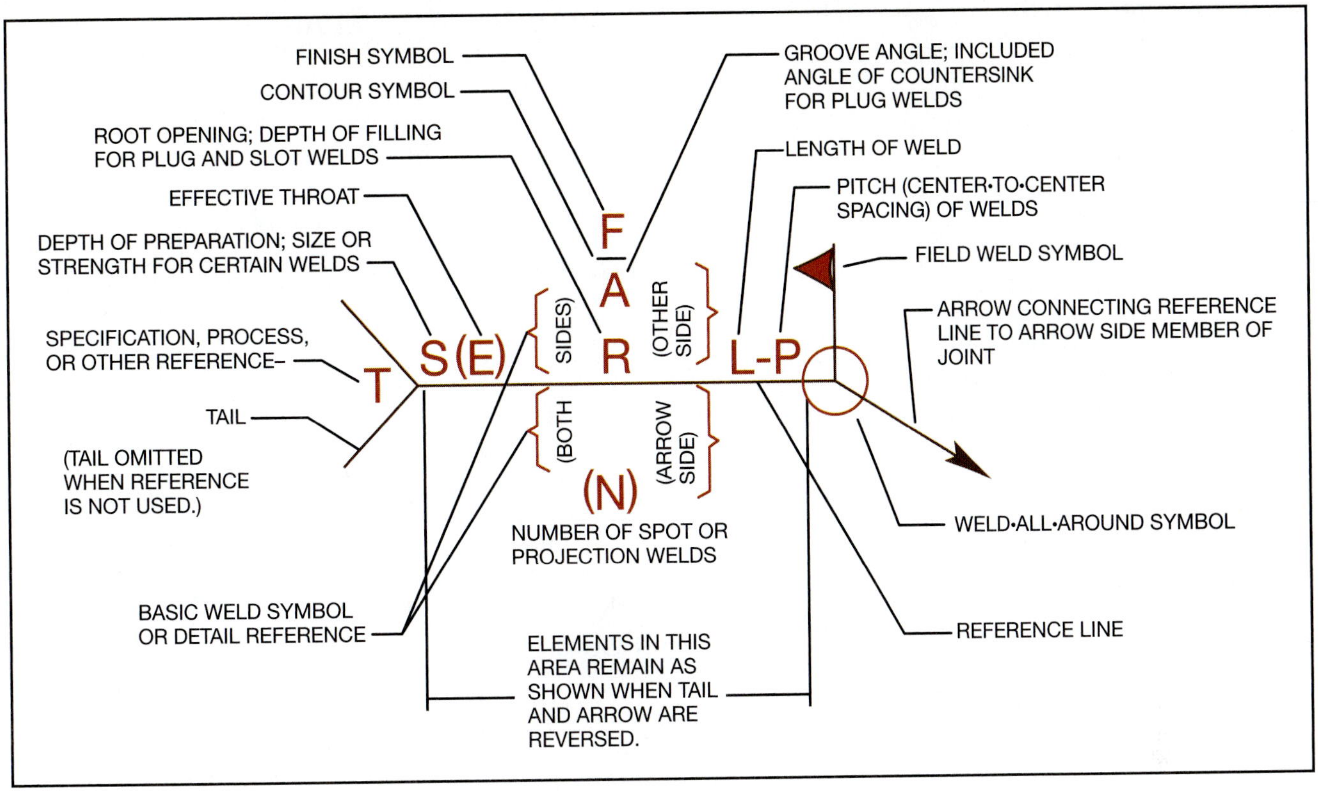

FIGURE 45.55 ■ Common locations of each element of a welding symbol. *Courtesy American Welding Society.*

welded by an inclined line with an arrow. The arrow touches the area to be welded. It is not uncommon to see a welding line bend to point into difficult-to-reach places, or to see more than one leader line extending from the reference line. Information about the type of weld, the location of the weld, the welding process, and the size and length of the weld is all specified on or near the reference line. Figure 45.55 shows a welding symbol and the proper location of information.

Welded Joints

The way that steel components intersect greatly influences the method used to weld the materials together and is often included in the specification. Common methods of arranging components to be welded are butt, lap, tee, outside corner, and edge joint. See Figure 45.56.

Types of Welds

The type of weld is associated with the weld shape or the type of groove in the metal components that will receive the weld, or both. Welds typically specified on structural and architectural drawings include fillet, groove, and plug welds. The method of joining steel and the symbol for each method can be seen in Figure 45.57.

- Fillet weld—The most common weld used in construction, a fillet weld (Figure 45.57A) is formed at the internal corner of two intersecting pieces of steel. The fillet can

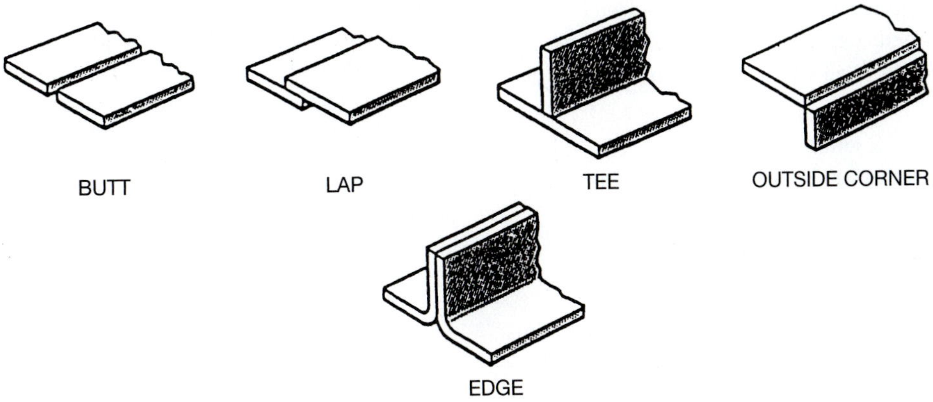

FIGURE 45.56 ■ Types of welded joints.

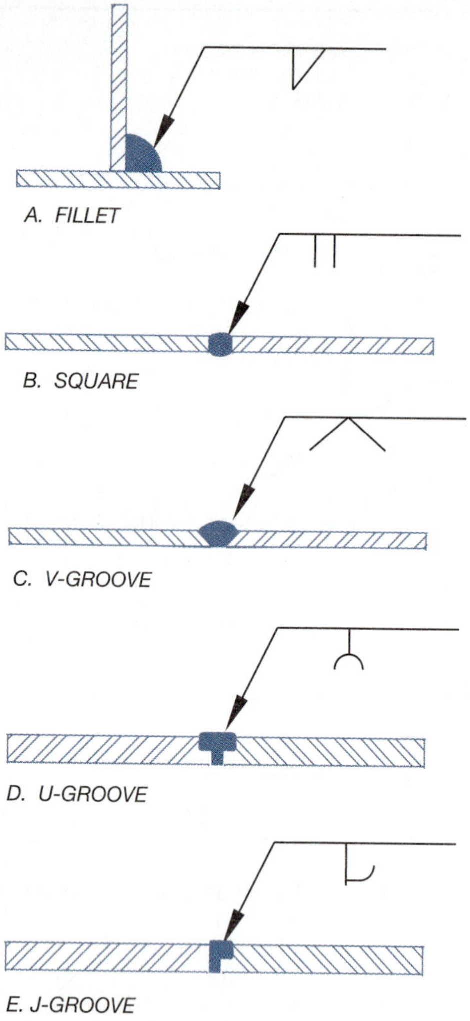

FIGURE 45.57 ■ The type of weld required by the engineer is represented by the shape of the material to be welded.

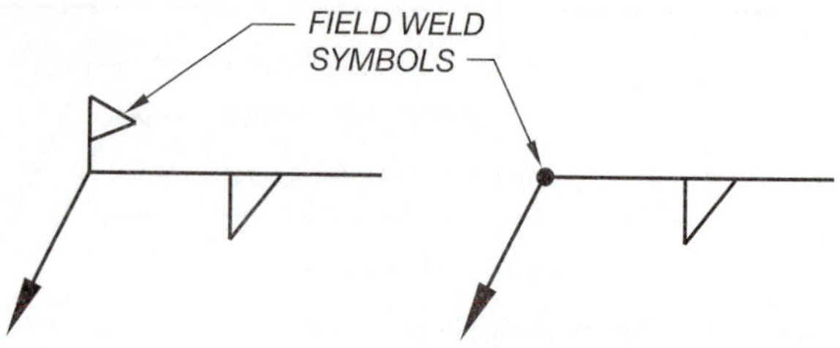

FIGURE 45.58 ■ The welding symbol can be used to represent on-site (field) or off-site welding.

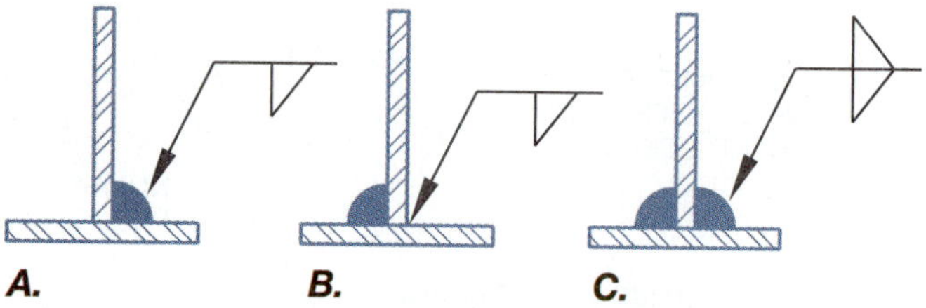

FIGURE 45.59 ■ The welding symbol can be used to describe the placement of the weld. (A) Symbol below the line places the weld on *this side.* (B) Symbol above the line places the weld on the *other side.* (C) Symbol on each side of the line places the weld on each side of the material.

be applied to one or both sides and can be continuous or a specified length and spacing.

- Square groove weld—applied when two pieces with perpendicular edges are joined end to end. The spacing between the two pieces of metal is called the *root opening.* The root opening is shown to the left of the symbol in Figure 45.57B.
- V-groove weld—applied when each piece of steel to be joined has an inclined edge which forms a V. The included angle is often specified, as well as the root opening. See Figure 45.57C.
- Beveled weld—created when only one piece of steel has a beveled edge. An angle for the bevel and the root opening is typically given.
- U-groove weld—created when the groove between the two mating parts forms a U. See Figure 45.57D.
- J-groove weld—results when one piece has a perpendicular edge and the other has a curved grooved edge. See Figure 45.57E. The included angle, the root opening, and the weld size are typically given for U and J-groove welds.

Welding Locations and Placements

Welds specified on structural drawings may be done away from the job site and then shipped ready to be installed. Large components that must be assembled at the job site are called *field-welded.* Two symbols used to refer to a field weld can be seen in Figure 45.58.

The placement of the welding symbol in relationship to the reference line is critical in explaining where the actual weld will take place. Figure 45.59 shows three examples and the effects of placing a fillet weld symbol on the reference. The distinction of symbol placement can be quite helpful to the drafter if adequate space for a symbol is not available on the proper side of the detail. The welding symbol can be placed on either side of the drawing, and the relationship to the reference line can be used to clarify the exact location. Options include:

- All-around—used for circular and rectangular parts to indicate that a weld is to be placed around the entire intersection. A circle placed at the intersection of the leader and reference line indicates that a feature is to be welded *all-around.* Figure 45.54 shows a column-beam intersection joined with welds requiring an all-around weld.
- Weld length and increment—If a weld does not surround the entire part, the length and spacing of the weld should be placed beside the weld symbol. The number preceding the weld symbols indicates the size of the weld. The number following the weld symbol indicates the length each weld is to be. The final size indicates the spacing of the weld along a continuous intersection of two mating parts.

USING CADD TO DRAW STRUCTURAL DETAILS

CADD has increased productivity in drawing structural details. One advantage is the use of standard details. Once a detail has been drawn, it may be automatically inserted into any drawing at any time. The detail may be inserted and used as is, or it may be inserted and modified to provide specific information that changes from one drawing to the next. Every time a CADD drafter draws a detail, it is saved in a drawing file for later use. With this process it is not necessary to render the same drawing more than once. CADD structural packages are available that provide a variety of standard structural details. Most of these packages are based on parametric design, which means that the standard structural detail may be altered by supplying the computer with data. For example, a set of structural stairs may be drawn automatically by providing the computer with total rise, number of risers, tread dimension, reinforcing size and spacing, and railing information.

Even for creating nonstandard original drawings, CADD is easier than manual drafting. Many CADD software packages have custom structural tablet menu overlays that provide standard structural shapes that may be inserted on the drawing at any time. This saves time, because the drafter need not draw these symbols individually with a pencil and manual template.

Chapter 45 Additional Reading

The following Web sites can be used as a resource to help you keep current with changes in commercial building materials.

ADDRESS	COMPANY OR ORGANIZATION
www.aisc.org	American Institute of Steel Construction
www.aitc-glulam.org	American Institute of Timber Construction
www.aws.org	American Welding Society
www.crsi.org	Concrete Reinforcing Steel Institute
www.dfi.org	Deep Foundation Institute
www.canammanac.com	Hambro (steel trusses)
www.industrial-fasteners.org	Industrial Fasteners Institute
www.mbma.com	Metal Building Manufacturers Association
www.ncma.com	National Concrete Masonry Association
www.precast.org	National Precast Concrete Associations
www.post-tenioning.org	Post Tensioning Institute
www.portcement.org	Portland Cement Association
www.steeljoist.org	Steel Joist Institute
www.spfa.org	Steel Plate Fabricators Association

CHAPTER 45 Common Commercial Construction Materials Test

DIRECTIONS

Answer the questions with short, complete statements.

1. Letter your name and the date at the top of the sheet.
2. Letter the question number and provide the answer. You do not need to write out the question.

QUESTIONS

Question 45–1 Under what conditions can wood be used to frame a wall that requires a two-hour fire rating?

Question 45–2 Why are the materials used to frame commercial buildings not typically used on residential projects?

Question 45–3 Describe how a panelized roof is framed.

Question 45–4 How does the F_b value for a DF/DF glu-lam beam compare with a 6X DFL #1 beam?

Question 45–5 Why is heavy timber a better construction material than steel in areas of high fire risk?

Question 45–6 Describe two types of drawings typically required for steel and concrete projects.

Question 45–7 How is concrete prestressed?

Question 45–8 How are precast concrete panels usually connected to each other?

Question 45–9 What is the recommended height limit for load-bearing steel studs?

Question 45–10 Sketch the proper symbol for a 3/16″ fillet weld, opposite side, field weld.

Question 45–11 List the nominal diameter of #4, #8, and #14 steel bars.

Question 45–12 What is a nested joist?

Question 45–13 How are members of a rigid frame structure connected?

Question 45–14 List five pieces of information that might be included in a bolt specification.

Question 45–15 What type of nail is most typically used for wood framing?

Question 45–16 Describe how rebar is represented in end-view in a section. How is it represented on the floor plan?

Question 45–17 List the names of the following bolts: A.B. C.B. H.S. M.B.

Question 45–18 How long is an 8-penny nail?

Question 45–19 What is the common location of steel reinforcement in a concrete block wall?

Question 45–20 What information is required to describe an open web steel joist?

Question 45–21 Use the nailing schedule in Chapter 30 to determine how a rafter will be connected to a top plate; to secure 3/4″ plywood to the floor joist.

Question 45–22 List the three possible locations for placing a weld and give the symbol that represents each location.

Question 45–23 Describe the difference between the terms *field, edge,* and *boundary.*

Question 45–24 What shapes are typically used to form a steel frame?

Question 45–25 What information is required to describe a steel stud?

Question 45–26 Why is a washer used in placing a bolt?

Question 45–27 List two types of reinforcing bars and explain the difference between them.

Question 45–28 What information is required, and where is the information placed to describe a steel decking?

Question 45–29 Sketch the symbol for a 1/8″ fillet weld applied all around a column which is applied at the job site.

Question 45–30 Explain methods for representing a sawn continuous beam, a noncontinuous beam, and a laminated beam.

Question 45–31 List five types of joints where welds can be applied.

Question 45–32 Sketch the symbol used for a 3/16″ × 3″ long V groove weld applied to this side.

Question 45–33 What are the most common methods of connecting large steel shapes to each other?

Question 45–34 Why are laminated beams used instead of steel or sawn beams?

Question 45–35 Explain the difference between representing poured concrete and representing concrete blocks in plan view.

Question 45–36 The following note is located on a sketch of a pier detail:

6 Ø @ 8″ O.C.E.W /T & B.

Sketch or describe what the steel would look like.

Question 45–37 Sketch the symbol for a 1/8″ field, fillet weld on both sides.

Question 45–38 Use the research material to determine the size of common glu-lam beam widths.

Question 45–39 Use the Internet to obtain material related to steel framing.

Question 45–40 Determine what code regulates commercial construction in your area.

46 CHAPTER

Commercial Construction Projects

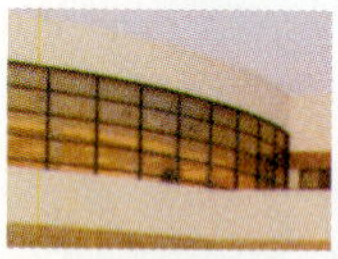

OFFICE PRACTICE

Although the design process is similar for residential and commercial projects, actual office practice is different. As a new employee, you will be given jobs that are typically not the office favorites. These might include making corrections or lettering drawings that were drawn by others on the staff. As your linework and lettering speed and quality improve, so will the variety of drafting projects that you work on. No matter what your drawing level, you will be required to research your drawing project. The two major tools that you will be using will be Sweets catalogs and the building code covering your area. Chapter 44 introduced you to building codes as they apply to commercial projects.

Sweets catalogs are a collection of vendors' brochures that are used in nearly every architectural and engineering office. In residential construction, a drafter may be required to conduct a limited amount of research. With commercial projects, because of the various materials that are available, the drafter usually spends several hours a week researching product information.

Figure 46.1 shows an example from the Simpson Strong-Tie catalog. This column cap (CC) is a common connector used to fasten beams to wood post. A common method of specifying this connection in the calculations might include only the use of a CC cap. The drafter would be required to determine the size of the post and the size of the beam to be supported, and then select the proper cap.

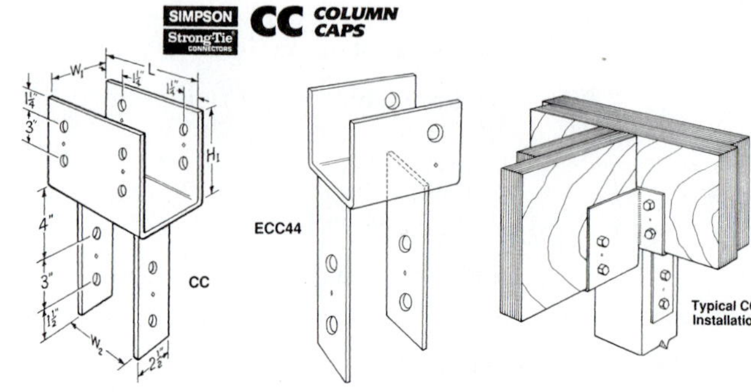

MODEL NO.	MATL	DIMENSIONS				FASTENERS		ALLOWABLE LOADS		
		W_1	W_2	L	H	BEAM	POST	UPLIFT (133)	DOWN (100) Fc⊥ @ 750 psi	DOWN (100) Fc⊥ @ 625 psi
CC44	7 ga	3 5/8	3 5/8	7	4	2-5/8 MB	2-5/8 MB	1220	18375	15310
CC46	7 ga	3 5/8	5 1/2	11	6 1/2	4-5/8 MB	2-5/8 MB	2330	28875	24060
CC64	7 ga	5 1/2	3 5/8	11	6 1/2	4-5/8 MB	2-5/8 MB	3660	45375	37810
CC66	7 ga	5 1/2	5 1/2	11	6 1/2	4-5/8 MB	2-5/8 MB	3660	45375	37810
CC6-7 1/8	7 ga	5 1/2	7 1/8	11	6 1/2	4-5/8 MB	2-5/8 MB	3660	45375	37810
CC7 1/8-4	3 ga	7 1/8	3 5/8	13	8	4-3/4 MB	2-3/4 MB	6260	68250	—
CC7 1/8-6	3 ga	7 1/8	5 1/2	13	8	4-3/4 MB	2-3/4 MB	6320	68250	—
CC7 1/8-7 1/8	3 ga	7 1/8	7 1/8	13	8	4-3/4 MB	2-3/4 MB	6320	68250	—
ECC44	7 ga	3 5/8	3 5/8	5 1/2	4	1-5/8 MB	2-5/8 MB	—	9185	7655
ECC46	7 ga	3 5/8	5 1/2	8 1/2	6 1/2	2-5/8 MB	2-5/8 MB	—	14435	12030
ECC64	7 ga	5 1/2	3 5/8	7 1/2	6 1/2	2-5/8 MB	2-5/8 MB	—	14435	12030
ECC66	7 ga	5 1/2	5 1/2	7 1/2	6 1/2	2-5/8 MB	2-5/8 MB	—	22685	18905

FIGURE 46.1 ■ Vendors' catalogs are needed to determine many components shown on details. This table from the Simpson Strong-Tie Company, can be used to determine the cap dimensions at post-to-beam details.

TYPES OF DRAWINGS

Most of the information to which you have been introduced while learning about residential architecture can be applied to commercial projects as well. There will be differences, of course, but the basic principles and procedures of drafting will be the same. Many of the differences will be discussed in this chapter.

Calculations

One of the prime differences between commercial and residential drafting is your need to work with engineers' calculations. No matter if you work for a designer, architect, or engineer, you will need to use a set of calculations as you draw the plans for a commercial project. In residential drafting, engineer's calcs are used to determine sizes of retaining walls and seismic and wind loads. In commercial drafting, a set of calculations is provided for the entire structure, detailing everything from beam sizes to the size and number of nails to use in a wall.

Although called "engineer's calcs," these calculations may be prepared by an engineer or an architect. By state law, commercial calculations are required to be signed by a licensed architect or engineer. Because of the complexity of many concrete and steel structures, architects typically design and coordinate the drafting of the project but hire a consulting engineer to design the structural members. The architect typically designs the structure and decides where structural columns and beams will be located, and an engineer determines the stress and the member required to resist this stress.

Figure 46.2 shows a page from an engineer's calcs. Most calcs are divided into three areas. The calc for a specific area usually begins with a statement of the problem. In Figure 46.2 the engineer is determining the size of the roof sheathing to be used

ENGINEER'S GOAL

ROOF SHEATHING - WAREHOUSE

$W_V = 25$ PSF

$$\frac{L}{W} = \frac{88658}{(2)(122)} = 363 \ \#1 \quad OR \quad \frac{59124}{(2)(391)} = 92 \ \#2$$

ASSUME TRUSSES @ 32" O.C.

CHECK 5/8" C-D 42/20 PLYWOOD

$w = 35$ PSF
$V = 425 > 363$ @ BLOCKED AREAS
$= 285 \ \#1$ UNBLOCKED
$= 215 \ \#2$ UNBLOCKED > 92 @#1

DESIGN METHOD

DETERMINE BLOCKING TERMINATION

$$x = \frac{(285)(160.5)}{363} = 126.1'$$

$d = 160.5 - 126 = 35.50 = > 37.33'$ FROM EA. END WALL

USE: 5/8" C-D, 42/20, INTERIOR APA PLYWOOD W/ EXTERIOR GLUE. PROVIDE BLOCKING @ 37.33' ADJACENT TO END WALLS. NAIL W/ 10 d COMMON NAILS @ 4" O.C. @ ALL PANEL EDGES @ BLOCKED AREAS, 6" O.C. @ ALL SUPPORTED PANEL EDGES @ UNBLOCKED AREAS ∉ 12"O.C. @ ALL INTERMEDIATE SUPPORTS.

THIS IS WHAT THE DRAFTER IS TO PUT ON THE WORKING DRAWINGS

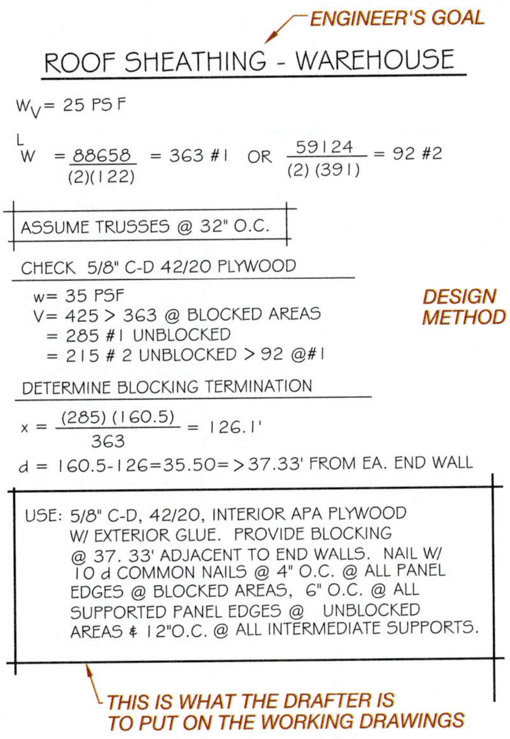

FIGURE 46.2 ■ Calculations for a structure typically include a problem to be solved, the mathematical solution, and the specification to be placed on the drawing by a drafter. *Courtesy Kenneth D. Smith, and Associates, Inc.*

CARRIER PURLINS:

LINE (3):
TA= 8 X 10 X 3 + 5X10=290#

$P_1 = (10+16) \ 8X20 = 1600 + 3200 = 4800$

$P_2 = (10+16) \ 8X \ 10 + (10+16) \ 5 \ X \ 10 = 2080 + 1300 = 3380$

$$R_B = \frac{(1600 + 3200) \ 11 + (1300+2080)3}{18.5} = 1162 + 2240 = 3402$$

$$R_C = \frac{(1600 + 3200) \ 7.5 + (1300+2080)15.5}{18.5} = 1738 + 3040 = 4778$$

$M = 3402 \ X \ 7.5 = 25515$ 6 X 16 S= 220

$$S = \frac{25515 \ X \ 12}{1.25 \ X \ 1300} = 188.4$$ 6 X 16 DFL #1

LINE (3):
TA= 8 X 10 X 2 + 5 X 10=260#

$P_1 = 4800$ (SEE ABOVE)

$P_2 = (10+16) \ 5X \ 20 = (1000 + 1600 = 2600$

$$R_B = \frac{(1600 + 3200) \ 11 + (1000+1600)3}{18.5} = 1114 + 2162 = 3276$$

$$R_C = \frac{(1600 + 3200) \ 7.5 + (1000+1600)15.5}{18.5} = 1486 + 2638 = 4124$$

$M = 3275 \ X \ 7.5 = 24562$ 6 X 16 DFL #1

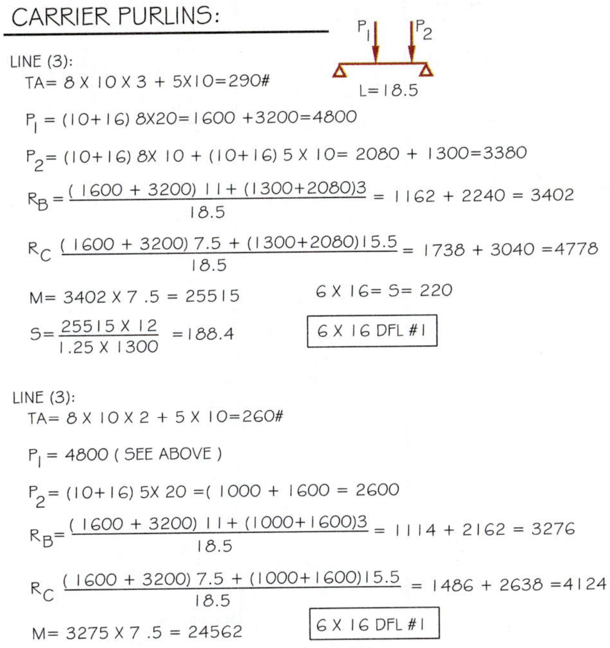

FIGURE 46.3 ■ An example of calculations to determine loading criteria at a beam connection over a column. The drafter draws the detail, using proper line and lettering quality. *Courtesy Kenneth D. Smith, and Associates, Inc.*

for a warehouse. The second area of the calcs is the mathematical formula used to determine the stress and the needed reaction to that stress. This area of the calculations is of little meaning to the inexperienced drafter. You will note that since the part of the calcs you will need is the solution area, the engineer has placed the solution to the math problem in a box for easy reference. This is the information that the drafter will place on the plans.

Most calculations start at the highest level of the structure and work down to the foundation level. This allows loads to be accumulated as formulas are worked out so that loads from a past solution can be used on a lower level. Figure 46.3 shows the calculations that were used to determine the loads on a column.

Figure 46.4 shows the drawing that the drafter produced from the calcs. Notice that some items on the drawing were not in the specifications. This is where experience of the construction process is used. On the basis of past drawing experience and knowledge of the construction process, the drafter is able to draw such items as the shim without having it specified. Depending on the engineer and the experience of the drafting team, the calcs may or may not contain sketches. If the engineer is specifying a common construction technique for a skilled drafting team, usually a sketch would not be required. The information needed to draw this detail was compiled from the foundation, grading, roof, and floor plans, as well as the sections and other areas of the calcs.

Floor Plans

Commercial projects are similar to residential projects in their components and use similar types of drawings to display basic

SIMPSON HW412 W/14 N2ON NAILS

3/16"PLY SHIM EA. SIDE OF GLU-LAM ∉ HAND NAILED W/ 4 ROWS OF (8) 10d COMMON

SIMPSON MST27 TIE STRAP @ EA. SIDE W/ (30)-16d

6 x 16 DFL#1

4 x 12 DFL #2

SIMPSON A35

SIMPSON CC5 1/4-8 COL. CAP

GLU-LAM POST

5 1/8 x 31 1/2 GLU-LAM 8"

16 / S-3 BEAM / BEAM / POST
1" = 1'-0"

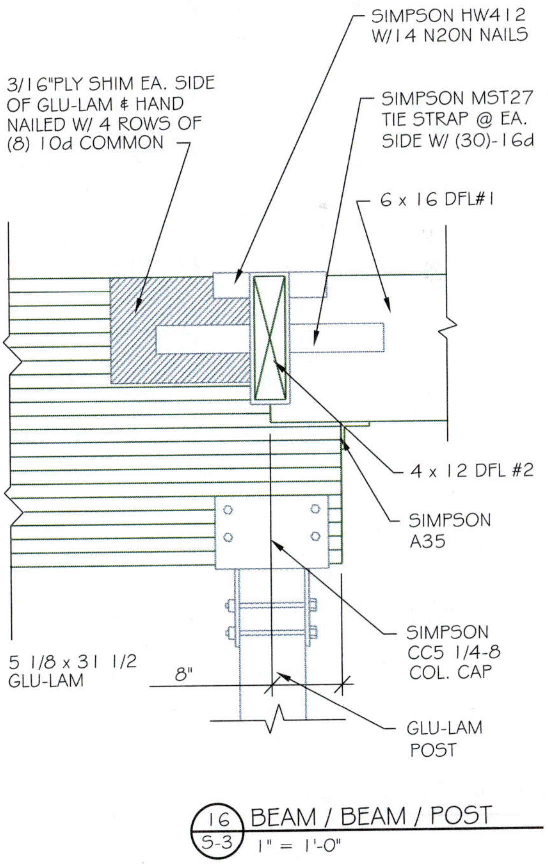

FIGURE 46.4 ■ Detail drawn to convey the information in the engineer's calculations shown in Figure 46.3. *Courtesy Dean Smith, Kenneth D. Smith, and Associates, Inc.*

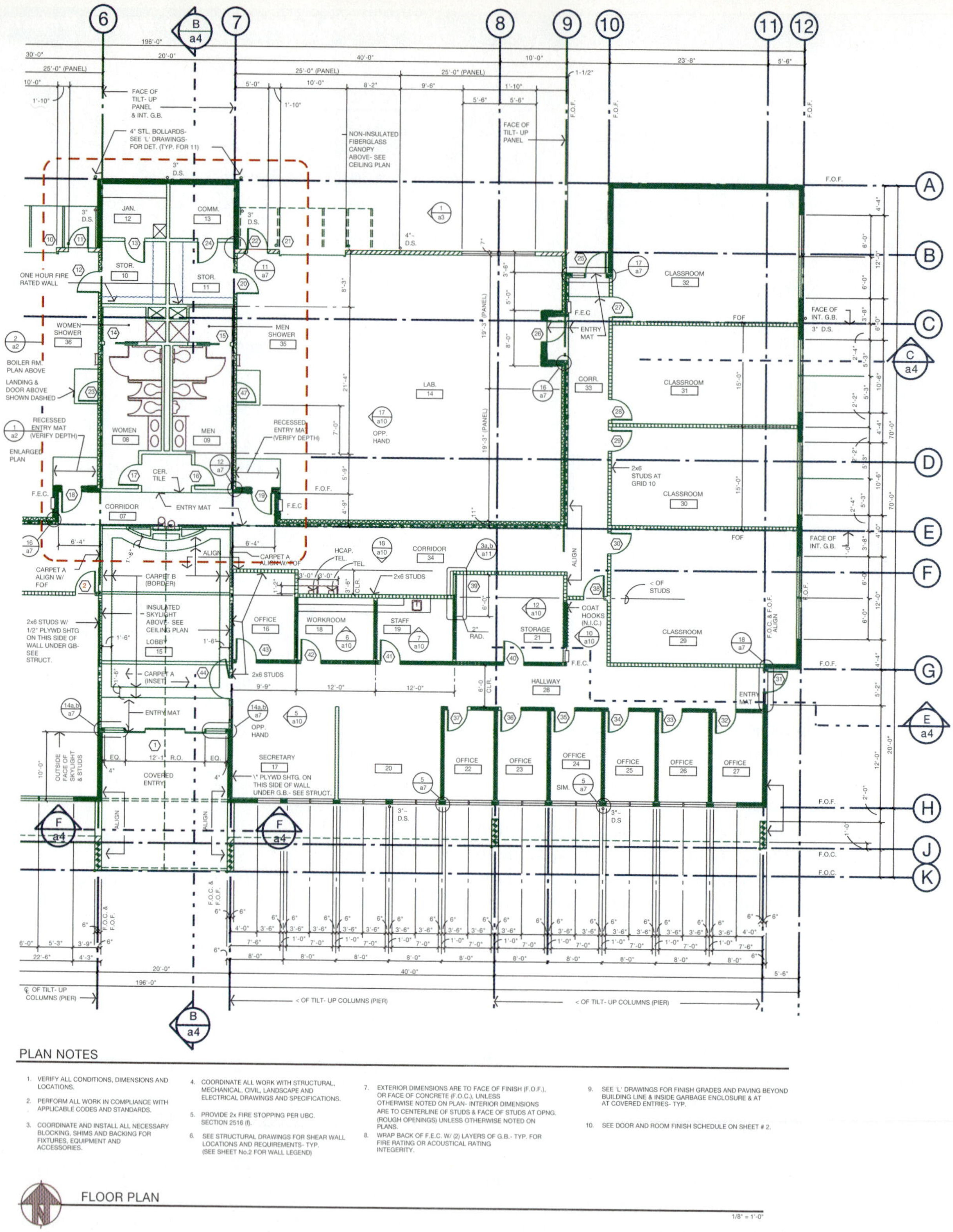

FIGURE 46.5 ■ The floor plan for a structure shows all wall and opening locations, as well as a reference map to help the print reader understand the relationship of other drawings to the floor plan. *Courtesy Architects Barrentine, Bates, & Lee, AIA.*

information. The floor plan of a commercial project is used to show the locations of materials in the same way that a residential floor plan does. The difference in the floor plan comes in the type of material being specified. Figure 46.5 shows a floor plan for college classrooms.

Because most commercial projects are so complex, specific areas of the structure are often clarified by grid reference markers. Figure 46.5 shows grid references 6 through 12 and A through K. The print reader can use these references to relate each drawing and detail back to the floor plan. Commercial floor plans are usually larger than residential plans to clarify complicated areas. Figure 46.6 shows an enlarged view of the bath areas shown in Figure 46.5. To keep the floor plans uncluttered and clear, schedules, notes, and symbols are used

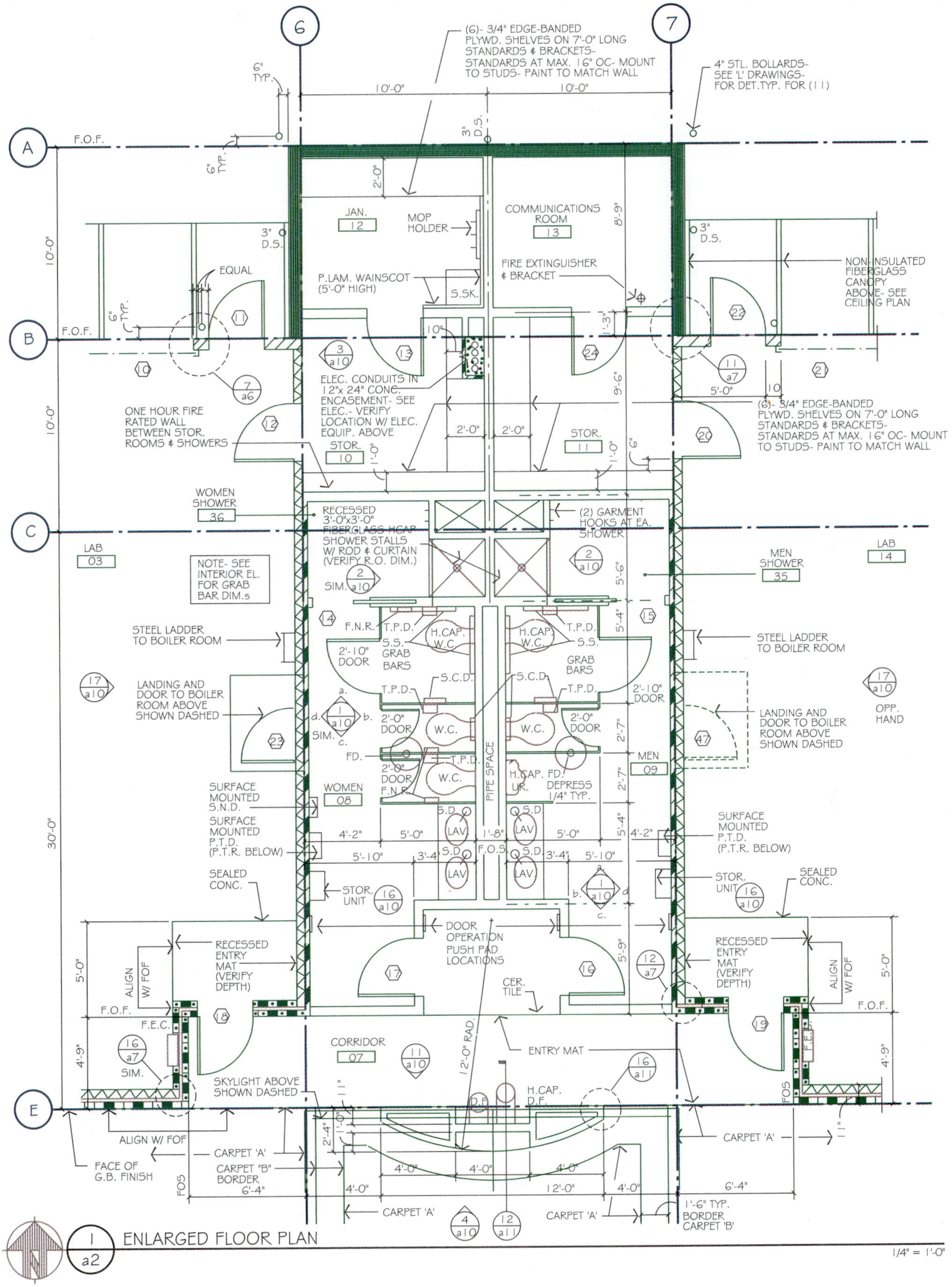

FIGURE 46.6 ■ Portions of a floor plan are often enlarged for clarity. This drawing was provided to clarify the floor plan shown in Figure 46.5. The grids can be used to correlate the two drawings. *Courtesy Architects Barrentine, Bates & Lee, AIA.*

ROOM FINISH SCHEDULE

NO.	ROOM NAME	FLOOR	BASE	WALLS N	WALLS E	WALLS S	WALLS W	CEILING	HEIGHT	REMARKS
01	LOUNGE	4	1	1	1	1	1	1/6	VARIES	
02	CORRIDOR	1/2	1	1	1	1	1	3	7'-6" / 9'-0"	
03	LAB	1/6	1	3	1/3	1/3	3	4	VARIES	BASE AT G.B. WALLS ONLY
04	VOCAT. CLASSROOM	2	1	1	1	1	1	2	9'-0"	
05	CLASSROOM	2	1	1	1	1	1	2	9'-0"	
06	CLASSROOM	2	1	1	1	1	1	2	9'-0"	
07	CORRIDOR	1/5	1		1	*	1	3	7'-6"	* ENTRY KIOSK
08	WOMEN	5	2	2	2	2	2	1	7'-6" / 8'-8"	
09	MEN	5	2	2	2	2	2	1	7'-6" / 8'-8"	
10	STORAGE	6	1	1	1	1	1/4	1	8'-8"	
11	STORAGE	6	1	1	1/4	1	1	1	8'-8"	
12	JANITOR	6	1	2	2/	2/5	2	1	VARIES	
13	COMMUN. ROOM	6	1	1	1	1	1	1	VARIES	
14	LAB	1/6	1	1/3	1/3	1/3	1/3	4	VARIES	BASE AT G.B. WALLS ONLY
15	LOBBY	2/3/1	1	* 1	1	1	1	6	VARIES	* ENTRY KIOSK
16	OFFICE	3	1	1	1	1	1	2	9'-0"	
17	SECRETARY	2	1	1	1	1	1	1	9'-0"	
18	WORKROOM	4	1	1	1	1	1	2	9'-0"	
19	STAFF	4	1	1	1	1	1	2	9'-0"	
20	LIBRARY	2	1	1	1	1	1	1	9'-0"	
21	STORAGE	4	1	1	1	1	1	2	9'-0"	
22	OFFICE	3	1	1	1	1	1	2	9'-0"	
23	OFFICE	3	1	1	1	1	1	2	9'-0"	
24	OFFICE	3	1	1	1	1	1	2	9'-0"	
25	OFFICE	3	1	1	1	1	1	2	9'-0"	
26	OFFICE	3	1	1	1	1	1	2	9'-0"	
27	OFFICE	3	1	1	1	1	1	2	9'-0"	
28	HALLWAY	1/2	1	1	1	1	1	1	9'-0"	
29	CLASSROOM	2	1	1	1	1	1	2	9'-0"	
30	CLASSROOM	2	1	1	1	1	1	2	9'-0"	
31	CLASSROOM	2	1	1	1	1	1	2	9'-0"	
32	CLASSROOM	2	1	1	1	1	1	2	9'-0"	
33	CORRIDOR	1/2	1	1	1	1	1	3	9'-0" / 7'-6"	
34	CORRIDOR	2	1	1	1	1	1	3	9'-0" / 7'-6"	
35	MEN SHOWER	5	2	2	2	2	2	1	8'-0"	
36	WOMEN SHOWER	5	2	2	2	2	2	1	8'-0"	
37	BOILER ROOM	6	1	4	4	2	2	5	VARIES	
38	ELECTRICAL ROOM	6	1	4	2	2	4	5	VARIES	

ROOM FINISH KEY

FLOORS

1. ENTRY MAT
2. CARPET 'A'
3. CARPET 'B'
4. SHEET VINYL
5. CERAMIC TILE (COLOR No.1)
6. SEALED CONC.

BASE

1. 4" COVED RUBBER BASE
2. 5" COVED CER. TILE (SEE INTERIOR ELEVATIONS) THINSET OVER WR. G.B. (COLOR No.2)

NOTES

* SEE INT. ELEVATIONS

— NO WORK REQUIRED

WALLS

1. GYPSUM BOARD PAINT
2. WATER RESISTANT GYPSUM BOARD PAINT
3. EXPOSED CONCRETE- ELASTOMERIC PAINT
4. EXPOSED CONCRETE W/ SEALER
5. 5'-0" HIGH P.LAM. WAINSCOT OVER W.R. G.B. W/ METAL EDGES

CEILINGS

1. GYPSUM BOARD ON 2x FRAMING AT 16" OC.(1 HOUR FIRE RATED & WATER RESISTANT G.B. AT ROOMS 08, 09, 35 & 36)
2. 2'x4' SUSP. ACOUSTIC TILE (NON-RATED)
3. GYPSUM BOARD ON EACH SIDE OF 2x CEILING JOISTS AT 24" OC. (ONE HOUR FIRE RATED AT CORRIDORS)
4. EXPOSED ROOF STRUCTURE- PAINT
5. (2) LAYERS OF 5/8" GYPSUM BOARD OVER MET. HAT CHANNELS AT 16" OC.- (ONE HOUR FIRE RATED)
6. INSULATED FIBERGLASS SKYLIGHT

7/a11

WALL LEGEND

EXTERIOR WALLS

1. — EXTERIOR INSULATION & FINISH SYSTEM (EIFS.) OVER 1/2" PLYWD. WALL SHTG., OVER 2x6 STUDS AT 16" OC. W/ R-19 FOILFACE FIBERGLASS INSUL. & 5/8" GYP. BOARD INTERIOR FINISH (CONTINUE DOUBLE LAYER OF FINISH FROM INTERIOR WALLS AT OFFSETS IN EXTERIOR WALLS) — 14/a5
2. — EXTERIOR INSULATION & FINISH SYSTEM (EIFS.) OVER 5-1/2" TILT-UP CONC. WALLS W/ PAINTED EXPOSED CONC. ON GARBAGE ENCLOSURE SIDE ONLY- (8'-10" HIGH WALL) — 19/a6
3. — EXTERIOR INSULATION & FINISH SYSTEM (EIFS.) OVER 5-1/2" TILT- UP CONC. WALLS W/ SEALED OR PAINTED INTERIOR FINISH. — 8/a5
4. — 12" TILT- UP CONC. COLUMNS- PAINT ALL EXPOSED SURFACES — 1/a5, 2/a5, 3/a5

INTERIOR WALLS

5. — 2x4 STUDS AT 16" OC. W/ 3" ACOUSTIC INSUL. IN STUD CAVITY - G.B. ON ONE SIDE (LEAVE AIR SPACE BETWEEN STUDS AND/OR TILT- UP CONC. WALL) EXTEND FROM FLOOR LINE UP TO ROOF FRAMING (ONE HOUR FIRE RATED CONSTRUCTION) — 7/a5
6. — 5-1/2" TILT- UP CONC. WALL W/ EXTERIOR INSUL. & FINISH SYSTEM (AT EXTERIOR) ABOVE ALL LOW ROOF AREAS-SEAL OR PAINT ALL EXPOSED CONC. SURFACES — 6/a5
7. — 2x4 STUDS AT 16" OC. W/ 3" ACOUSTIC INSUL.- (2) LAYERS OF G.B. ON BOTH SIDES EXTEND FROM FLOOR LINE UP TO ROOF FRAMING (END SECOND LAYER OF G.B. 6" ABOVE CEILING LINE)- 2x6 STUDS AS NOTED ON FLOOR PLAN (ONE HOUR FIRE RATED CONSTRUCTION AT LOBBY (15) & CORRIDORS) — 19/a11
8. — 2x4 STUDS AT 16" OC. W/ 3" ACOUSTIC INSUL.- G.B. ON BOTH SIDES OF WALL EXTEND FROM FLOOR LINE UP TO FLOOR FRAMING, UNLESS OTHERWISE NOTED (ONE HOUR FIRE RATED AS NOTED ON PLANS) — 12/a7
9. — 2x4 STUDS AT 16" OC. W/ 3" ACOUSTIC INSUL.- (1/2" RESELEINT CHANNELS AT 16" OC. W/ G.B. ON ONE SIDE & (2) LAYERS OF G.B. ON OPPOSITE SIDE- EXTEND FROM FLOOR LINE UP TO 6" ABOVE CEILING LINE — 16a,b/a11
10. — METAL HAT CHANNELS AT 16" OC. FURRING OVER CONC. WALL OR CONC. CONDUIT ENCASEMENT W/ G.B. FINISH- EXTEND FROM FLOOR LINE UP TO FLOOR FRAMING — 12/a7
11. — 2x4 STUDS AT 16" OC. W/ 3" ACOUSTIC INSUL. & (2) LAYERS OF G.B. ON ONE SIDE (LEAVE AIR SPACE BETWEEN STUDS) EXTEND FROM FLOOR LINE UP TO ROOF FRAMING (ONE HOUR FIRE RATED CONSTRUCTION) — 16/a7

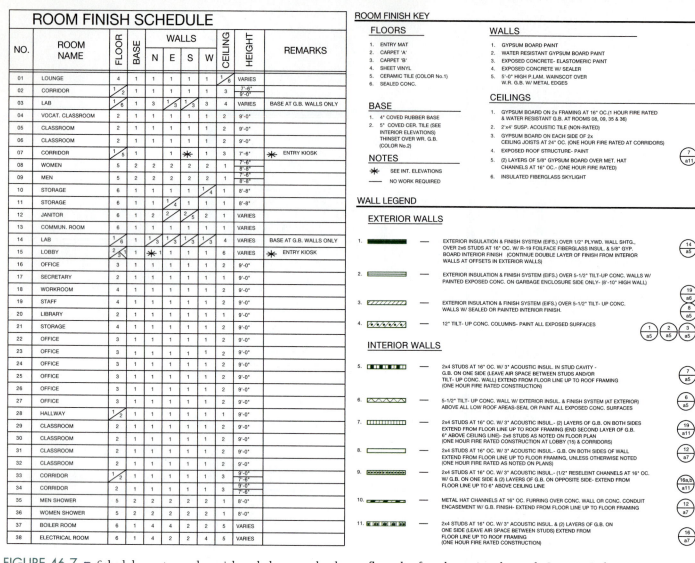

FIGURE 46.7 ■ Schedules, notes, and special symbols are used to keep a floor plan from becoming cluttered. *Courtesy Architects Barrentine, Bates & Lee, AIA.*

extensively throughout commercial drawings. Figure 46.7 shows the finish schedule for the structure in Figure 46.5.

Electrical Plans

On a set of residential plans, an electrical plan is drawn to provide information about the locations of outlets, switches, and light fixtures, and other related information, for the electrician. A similar electrical plan is provided for a set of commercial drawings. Information for the electrical plans is usually specified by the owner or by major tenants. This information is given to an electrical engineer, who will specify how the circuitry is to be installed. The engineer typically marks the needed information on a print of the floor plan. This print is then given to a drafter who transfers the information to an electrical plan. Because of the complexity of the electrical needs of commercial projects, a separate plan is drawn to show power outlets, as in Figure 46.8 along with a plan for lighting, as in Figure 46.9. A plan is typically provided for roof-mounted or other exterior equipment. Schematic drawings may also be provided by the electrical engineer and must be incorporated into the working drawings. See Figure 46.10.

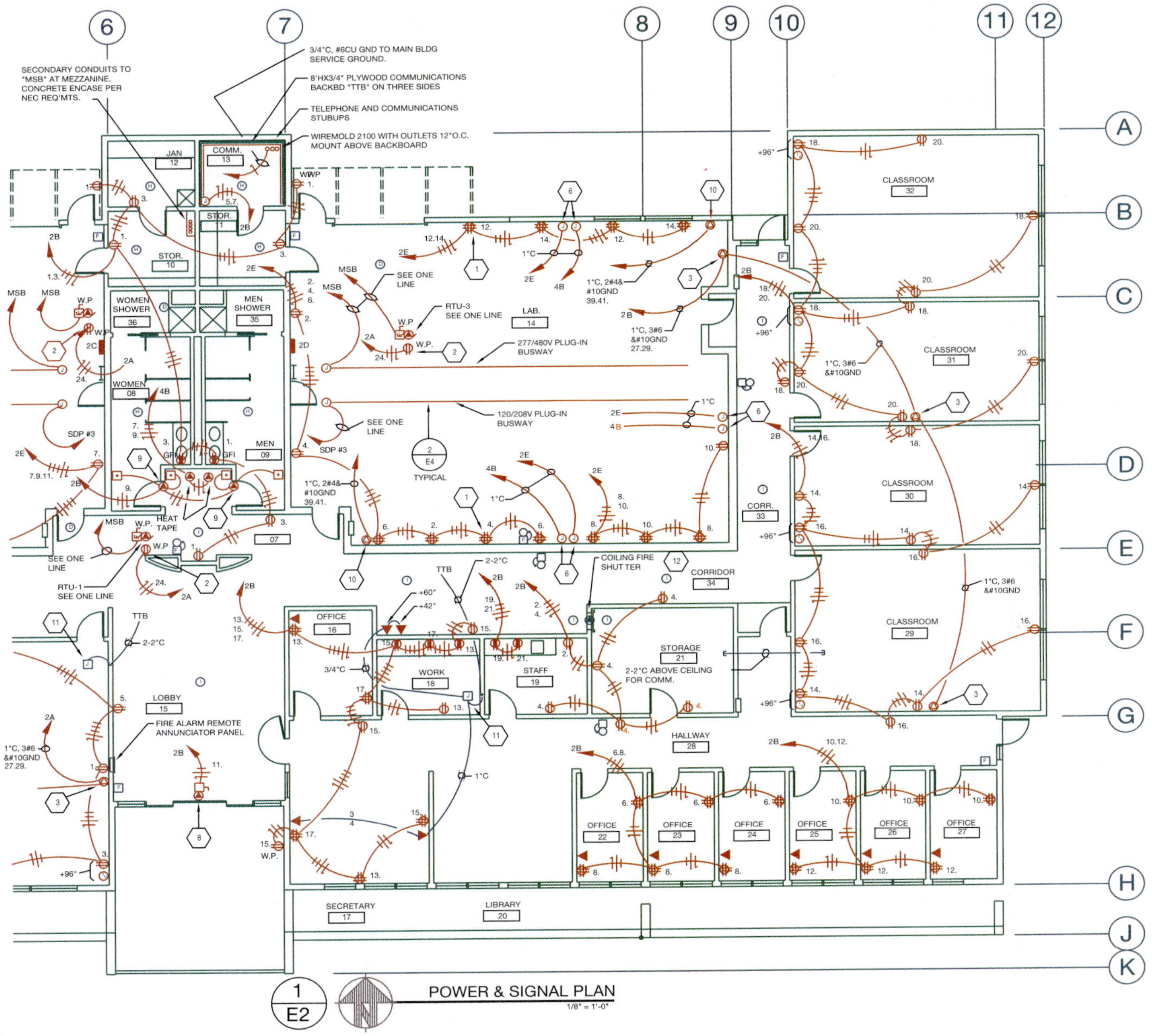

PLAN NOTES

1. ALL OUTLETS AND JUNCTION BOXES ON THIS WALL TO BE MOUNTED AT +48" AFF.
2. RECEPTACLE LOCATED ON ROOF. MOUNT ON EQUIPMENT ENCLOSURE.
3. PROVIDE RECEPTACLE FOR "SPIDER BOX", HUBBELL #SR50.
4. GENERAL NOTE: PROVIDE TWO HUBBELL #SB101 "SPIDER BOXES" AECH WITH #SCB50 CABLE AND PLUG TO MATCH #SR50 RECEPTACLE.
5. PROVIDE AITKEN #QHL326 INFRARED HTR. W/ (2)1600WATT QUARTZ LAMPS, 277V. STEM SUSPEND AT MINIMUM 6" FROM CEILING. INSTALL PER MANUFACTURERS RECOMMENDATIONS. SEE DETAIL #3/E4 FOR CONTROL SCHEMATIC.
6. PROVIDE FLUSH J-BOX WITH CONDUIT AND PULL STRING TO INDICATED PANEL FOR FUTURE USE. PROVIDE LABEL ON J-BOX COVER WITH PANEL DESIGNATION.
7. PROVIDE TWO-PIECE METAL RACEWAY, WIREMOLD #6000 SERIES WITH OUTLETS AS REQUIRED BY OWNER. PROVIDE TWO STACKED LENGTHS AT LOCATIONS AS SHOWN ON PLAN (1)POWER & (1)COMMUNICATIONS. MOUNT AT HEIGHT SPECIFIED BY OWNER.
8. CONNECT AUTO DOOR OPERATOR, 120V. COORDINATE REQUIREMENTS WITH EQUIP. SUPPLIER.
9. CONNECT HANDICAP ACCESS DOOR OPERATOR AND ASSOCIATED PUSH PADS. COORDINATE REQUIREMENTS WITH EQUIPMENT SUPPLIER.
10. PROVIDE 60A/2P RECEPTACLE WITH GROUND FOR WELDER, MATCH EQUIPMENT CORD CAP.
11. PROVIDE 12" X 12" X 4"D J-BOX WITH COVER FOR COMMUNICATIONS CABLING. MOUNT ABOVE ACCESSIBLE CEILING.
12. CONNECT TIME DELAY RELEASE DEVICE ASSOCIATED WITH COILING SHUTTER LOCATED ABOVE CEILING OF STORAGE 21. COORDINATE REQUIREMENTS WITH EQUIPMENT SUPPLIER.

FIGURE 46.8 ■ A plan is drawn to show power supplies. *Courtesy Architects Barrentine, Bates & Lee, AIA, and Rob Connell of MFIA, Inc., Consulting Engineers.*

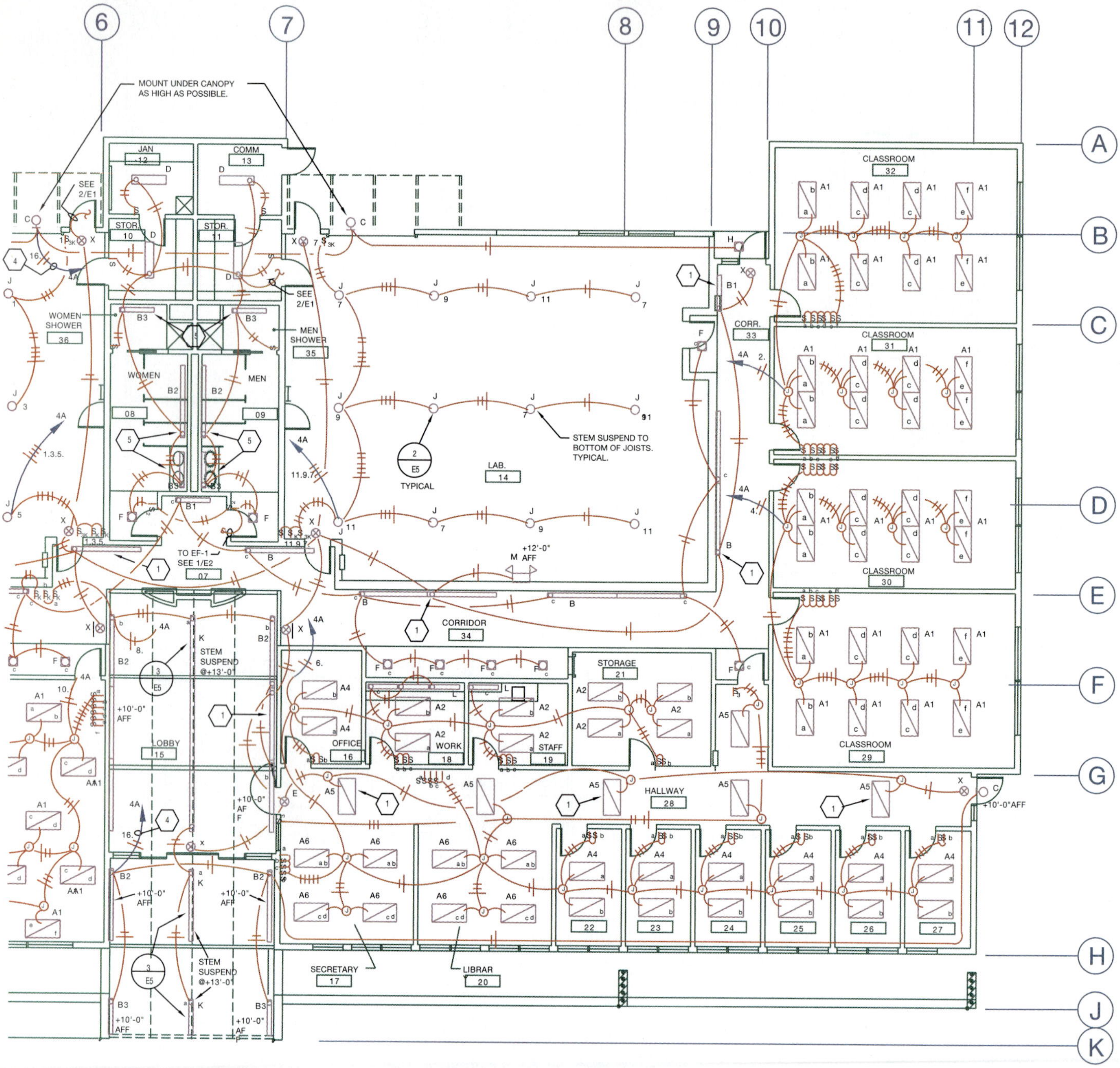

1
E1

LIGHTING PLAN
1/8" = 1'-0"

PLAN NOTES

1. PROVIDE BODINE B-50 EMERGENCY BATTERY BALLAST IN THIS FIXTURE FOR ALL LAMPS. CONCEAL TEST SWITCH AND CHARGE LIGHT BEHIND LENSE.
2. GENERAL NOTE: CORNER MOUNT TYPE "B" & "B1" FIXTURES AT CEILING UNLESS OTHERWISE NOTED.
3. LIGHT FIXTURES AT COVERED ENTRY ARE TO BE PROVIDED WITH "0" DEGREE COLD WEATHER BALLAST.
4. CONNECT TO DESIGNATED PANEL VIA EXTERIOR LIGHTING CONTROL "ELC".
5. MOUNT FIXTURE ON WALL AT +7'-0" ABOVE FINISHED FLOOR.
6. PROVIDE BODINE B-50 EMERGENCY BATTERY BALLAST IN THIS FIXTURE. CONNECT TO UNSWITCHED LEG OF CIRCUIT.

FIGURE 46.9 ■ Lighting equipment and fixtures are shown using an overlay of the floor plan. *Courtesy Architects Barrentine, Bates & Lee, AIA; and Rob Connell of MFIA, Inc., Consulting Engineers.*

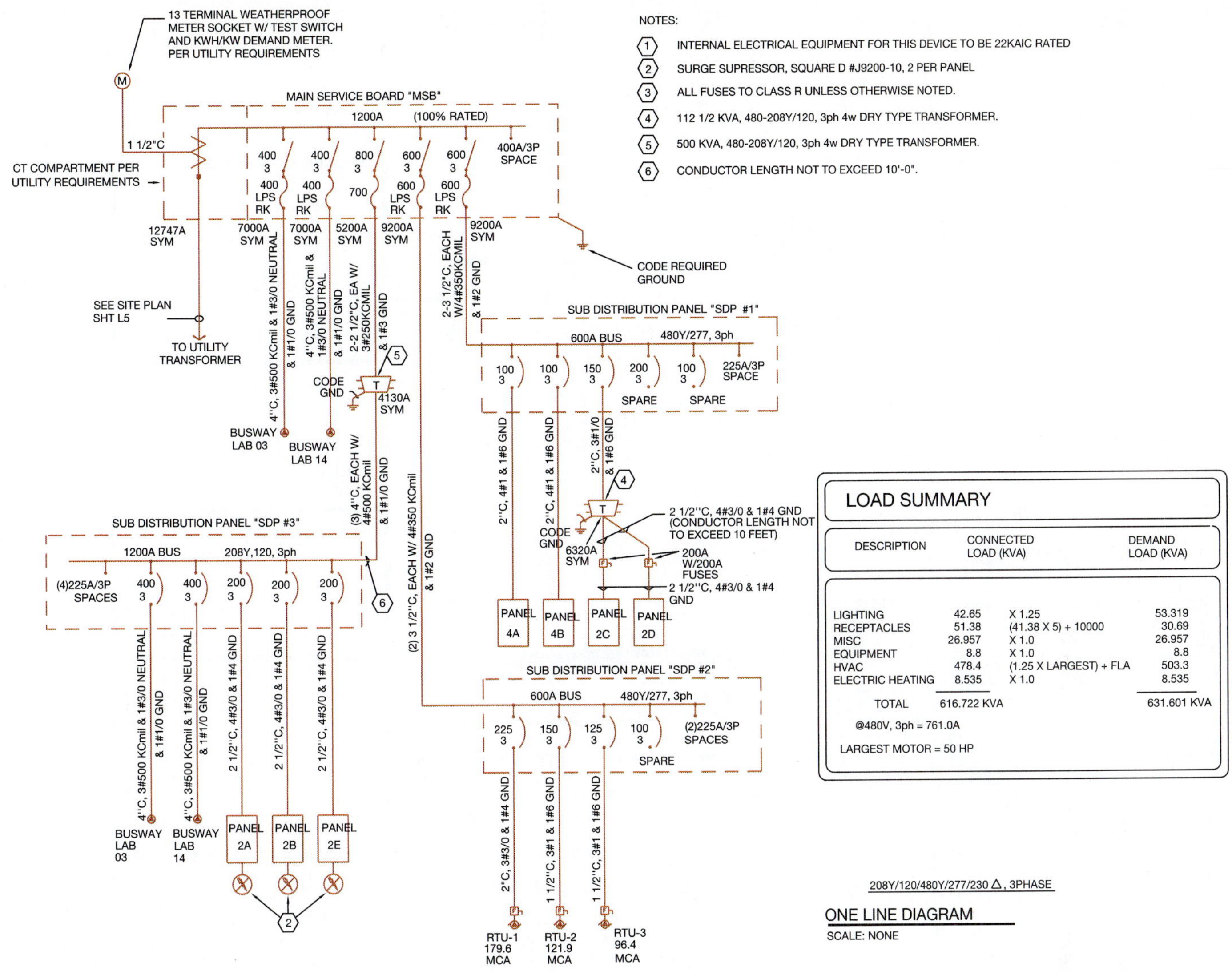

FIGURE 46.10 ■ A drafter is often required to draw schematic diagrams to help explain electrical needs. *Courtesy Architects Barrentine, Bates & Lee, AIA; and Rob Connell of MFIA, Inc., Consulting Engineers.*

Reflected Ceiling Plan

Acoustical ceilings, typically used in commercial projects, are supported by wires in what is called a *t-bar system*. A reflected ceiling plan shows how the ceiling tiles are to be placed. The plan also shows the location of light panels or ceiling-mounted heat registers. Figure 46.11 shows a reflected ceiling plan. The drafter must work with information provided by both the electrical and the mechanical engineers to draw this plan properly.

Mechanical Plans

Mechanical plans are used in the same manner for commercial projects as for residential projects. These plans show the location of heating and cooling equipment as well as the duct runs. As with other plans, the mechanical contractor marks all equipment sizes and duct runs on a print, and a drafter redraws the information on the mechanical plans. Figure 46.12 shows a mechanical plan. Figure 46.13 and 46.14 show details that supplement the plan view.

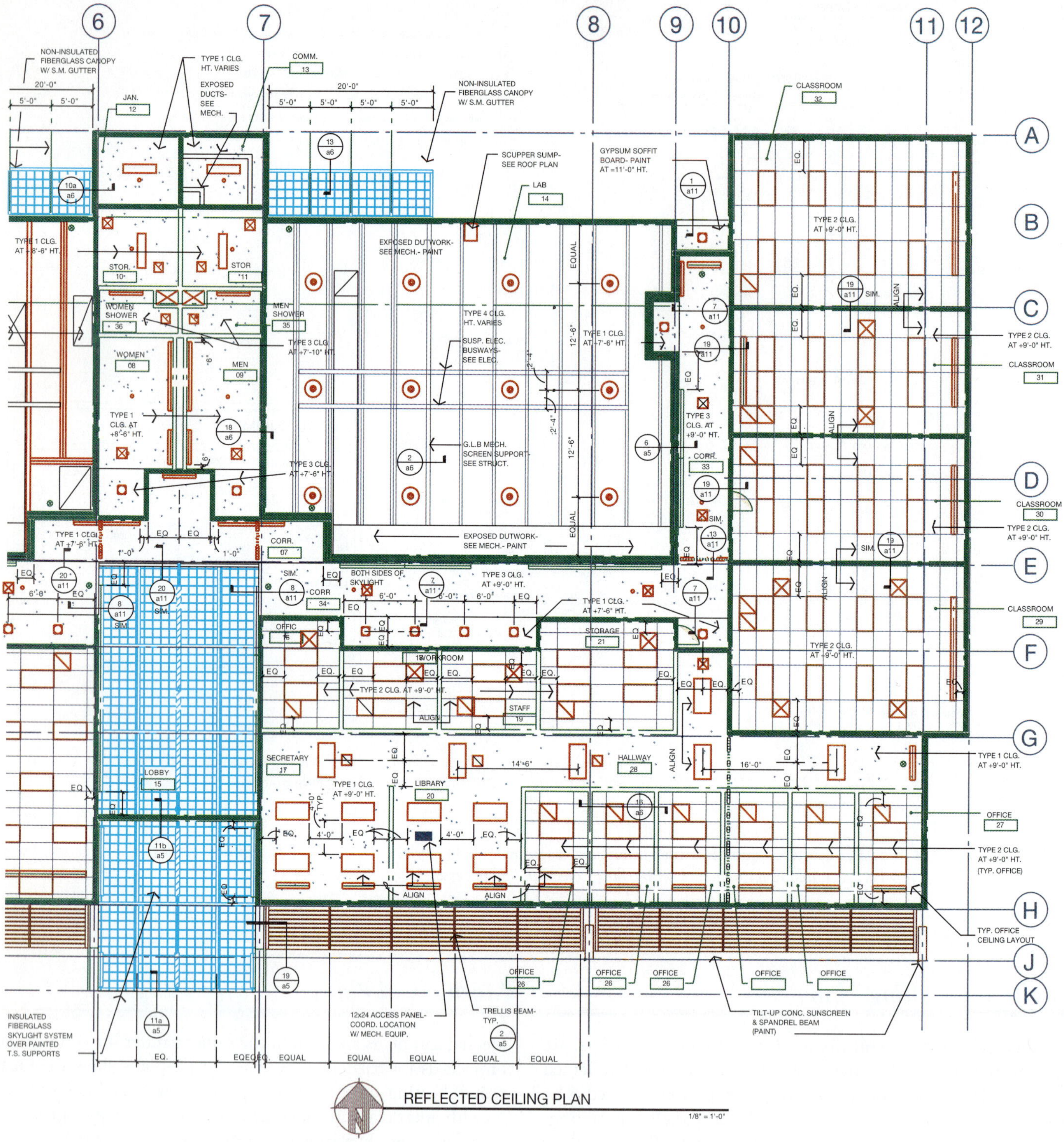

FIGURE 46.11 ■ The reflected ceiling plan for the structure shown in Figure 46.5. The ceiling plan is used to show the layout for the suspended ceiling system. *Courtesy Architects Barrentine, Bates & Lee, AIA.*

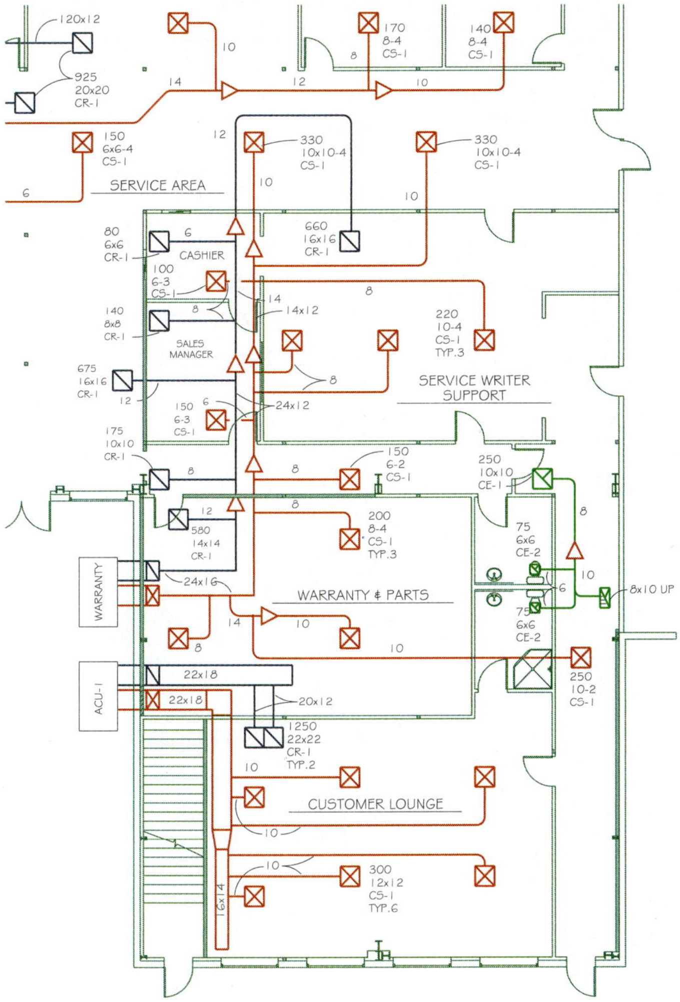

FIGURE 46.12 ■ A mechanical plan is drawn to show heating and air-conditioning requirements.

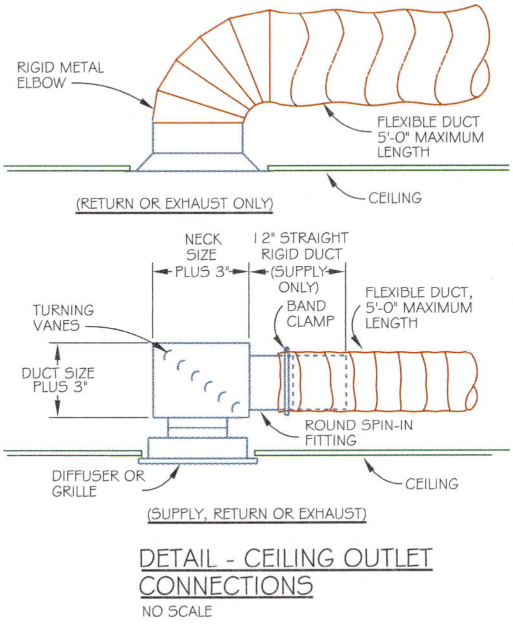

FIGURE 46.13 ■ Details are required to show how HVAC equipment is installed. *Courtesy Manfull Curtis Consulting Engineering.*

GENERAL NOTES

1 INFORMATION PERTAINING TO EXISTING HVAC EQUIPMENT, MEDICAL GAS AND VACUUM, AND PLUMBING PIPING, FIXTURES, ITEMS, ETC., SHOWN ON THESE DRAWINGS HAS BEEN TAKEN FROM VARIOUS RECORD DRAWINGS AND SOME INVESTIGATION BUT ALL INFORMATION HAS NOT BEEN VERIFIED AT THE SITE. VERIFY ALL EXISTING CONDITIONS WHERE POSSIBLE, RELATIVE TO SCOPE OF WORK, PRIOR TO SUBMITTING BID.

2 SOME CONCEALED DUCTS, PIPING, AND OTHER ITEMS HAVE BEEN SHOWN IN AN ASSUMED LOCATION BUT NOT VERIFIED. CONTRACTOR SHALL VERIFY.

3 PRIOR TO INSTALLATION OF ANY NEW PIPING, EXPOSE ALL POINTS OF CONNECTION AND VERIFY EXACT SIZE, LOCATION, ELEVATION, AND TYPE OF MATERIAL. DETERMINE THAT EXISTING PIPE IS CORRECT SIZE AND MATERIAL AND THAT THE ELEVATION IS LOW ENOUGH TO ACHIEVE REQUIRED SLOPE BEFORE NEW PIPING IS CONNECTED. IF NOT, NOTIFY THE ARCHITECT AND DO NOT PROCEED WITH INSTALLATION UNTIL INSTRUCTED TO DO SO.

4 EXISTING DUCTS AND PIPING ENCOUNTERED DURING CONSTRUCTION WHICH ARE NO LONGER REQUIRED SHALL BE REMOVED AND BE DISPOSED OF AS SPECIFIED. CAP PIPING IN A CONCEALED LOCATION AND AS CLOSE TO SERVING MAIN AS POSSIBLE TO LIMIT CAPPED DEAD END RUNS TO 2'-0" MAXIMUM FOR ALL SERVICES, EXCEPT AS NOTED.

5 PRIOR TO DISCONNECTING, REMOVING, OR CAPPING EXISTING PIPING, VERIFY THAT IT DOES NOT SERVE EXISTING FIXTURES OR EQUIPMENT TO REMAIN. IF IT DOES, LEAVE SYSTEM INTACT OR CONNECT TO CLOSEST NEW SERVICES. RECORD ON AS-BUILT DRAWINGS.

6 REMOVE EXISTING FIXTURE, EQUIPMENT, AND RELATED PIPING, FITTINGS, AND APPURTENANCES WHERE INDICTATED AND AS REQUIRED. CAP SERVICE PIPING IN A CONCEALED LOCATION. EXISTING MECHANICAL EQUIPMENT, PLUMBING FIXTURES, AND APPURTENANCES WHICH ARE REMOVED AND NOT RELOCATED OR REINSTALLED SHALL BE DISPOSED OF AS SPECIFIED.

7 ALL DEVIATIONS FROM CONTRACT DRAWINGS, INCLUDING EXISTING, SHALL BE RECORDED ON AS-BUILT DRAWINGS.

8 REFER TO ARCHITECTURAL DOCUMENTS FOR EXACT LOCATION AND HEIGHT OF ALL PLUMBING FIXTURES AND MEDICAL GAS OUTLETS. COORDINATE WITH OTHER TRADES.

9 BECAUSE OF THE SMALL SCALE OF THE DRAWING, IT IS NOT POSSIBLE TO INDICATE ALL PIPING RUNS, OFFSETS, FITTINGS, VALVES, AND ACCESSORIES WHICH MAY BE REQUIRED. CAREFULLY INVESTIGATE CONDITIONS SURROUNDING INSTALLATION OF THE WORK AND FURNISH NECESSARY FITTINGS, VALVES, TRAPS, ETC. , WHICH MAY BE REQUIRED TO COMPLETE THE INSTALLATION IN A SATISFACTORY AND CODE APPROVED MANNER.

10 RELOCATE TO NEAREST ACCEPTABLE LOCATION AND RECONNECT EXISTING PIPING REQUIRED TO REMAIN WHICH INTERFERES WITH NEW CONSTRUCTION.

11 ALL SERVICE DISRUPTIONS SHALL BE SCHEDULED AT LEAST 24 HOURS IN ADVANCE WITH USER STAFF AND ENGINEERING SERVICES.

12 LOCATION AND ROUTING OF SERVICES FOR SOME EXISTING FIXTURES IS UNKNOWN. VERIFY IN FIELD AND REMOVE AND CAP AS REQUIRED.

13 EXISTING MECHANICAL EQUIPMENT, PLUMBING FIXTURES, AND APPURTENANCES WHICH ARE TO BE FURNISHED BY OWNER, RELOCATED AND/OR REINSTALLED UNDER THIS CONTRACT SHALL BE COMPLETELY CLEANED AND INSPECTED. IF REPAIR IS REQUIRED TO PUT THEM IN SATISFACTORY WORKING CONDITION, NOTIFY OWNER BEFORE REINSTALLING.

FIGURE 46.14 ■ A drafter will need to update standard notes for each HVAC project. *Courtesy Manfull Curtis Consulting Engineering.*

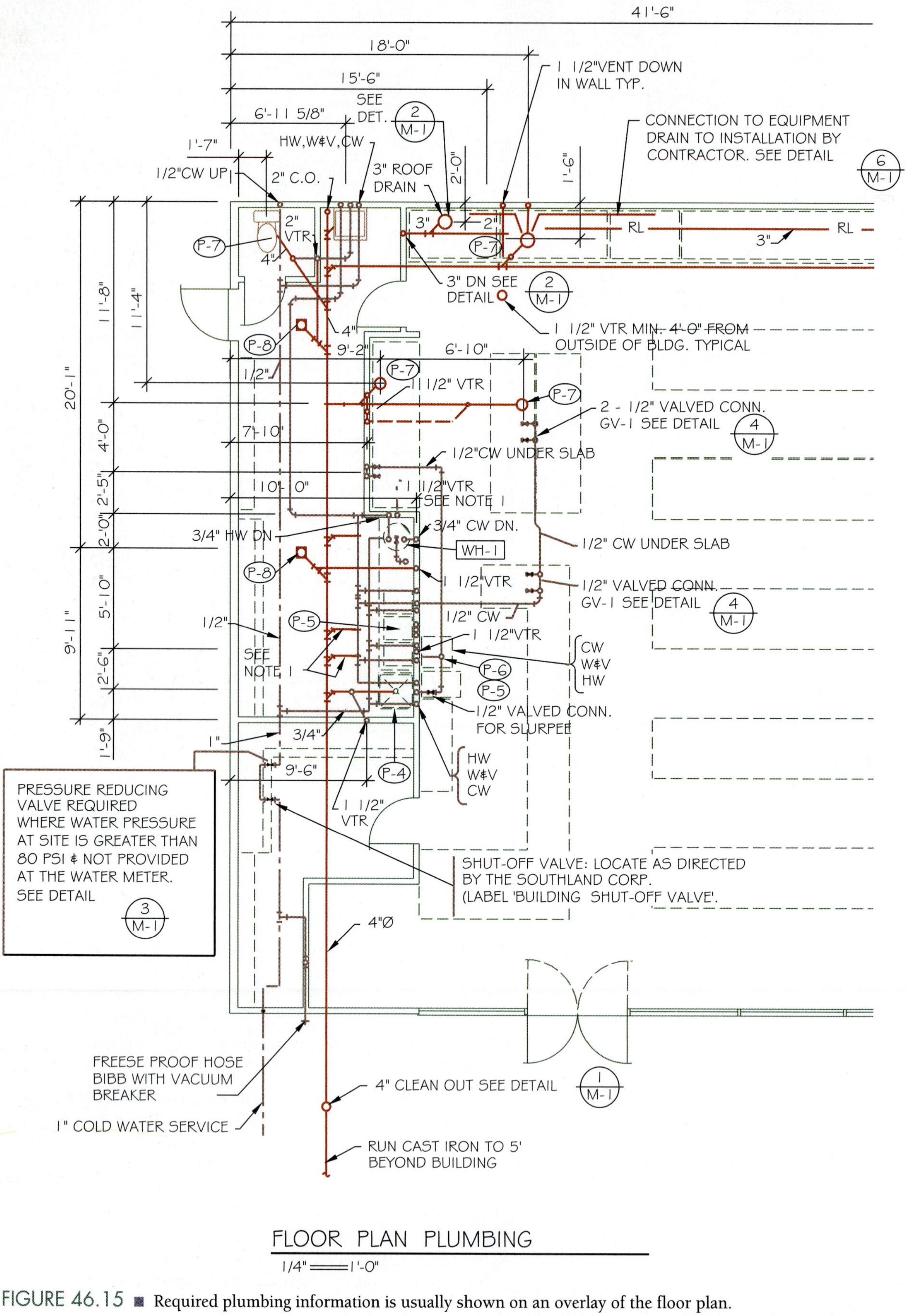

FIGURE 46.15 ■ Required plumbing information is usually shown on an overlay of the floor plan.

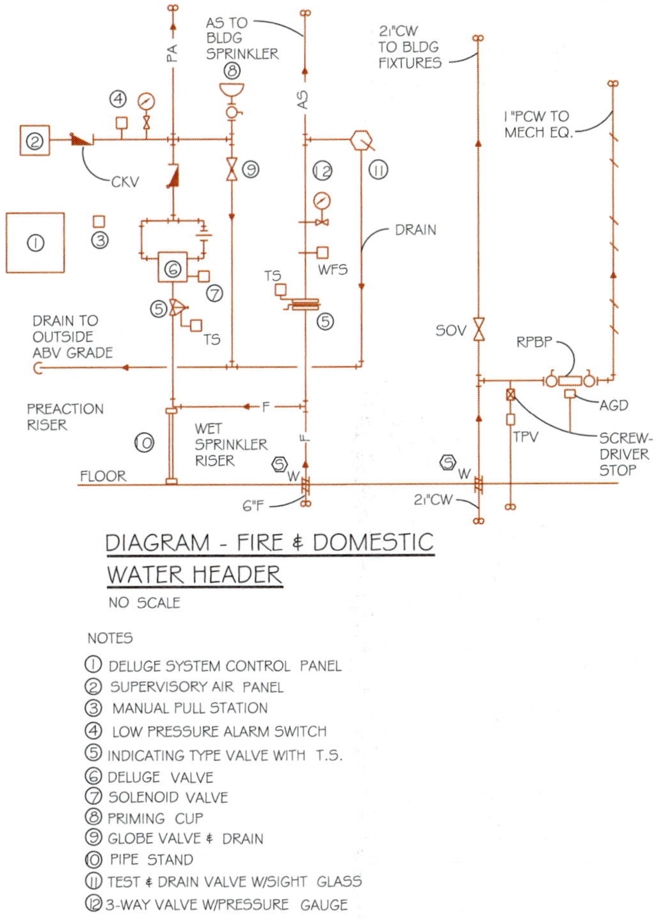

FIGURE 46.16(A) ■ A schematic plumbing diagram will be designed by an engineer and completed by a drafter. *Courtesy Manfull Curtis Consulting Engineering.*

Plumbing Plan

As with other plans that are generated from the floor plan, the information for the plumbing plan is supplied by the engineer and redrawn by a drafter. Depending on the complexity of the project, the plumbing plan may be divided into sewer and fresh water plans. Figure 46.15 shows a single plumbing plan that combines both systems. Notice that this plan shows far more detail and dimensions than the plans used for residential projects. Because of the complexity of commercial projects, more details are usually placed on the plumbing plan. Figure 46.16 shows line drawings and details that the drafter was required to place on the plan view.

Foundation Plans

Foundations in commercial work are typically concrete slabs because of their durability and low labor cost. As seen in Figure 46.17, the foundation for a commercial project is similar to that of a residential project. The size of the footings will vary, but the same type of stresses that affect a residence affect a commercial project. They differ only in their magnitude. In commercial projects it is common to have both a slab and a foundation plan. The slab plan, or slab-on-grade plan, as it is sometimes called, is a plan view of the construction of the floor system and specifies the size and location of concrete pours. A slab-on-grade plan can be seen in Figure 46.18. The foundation plan is used to show below-grade concrete work. A foundation plan can be seen in Figure 46.19.

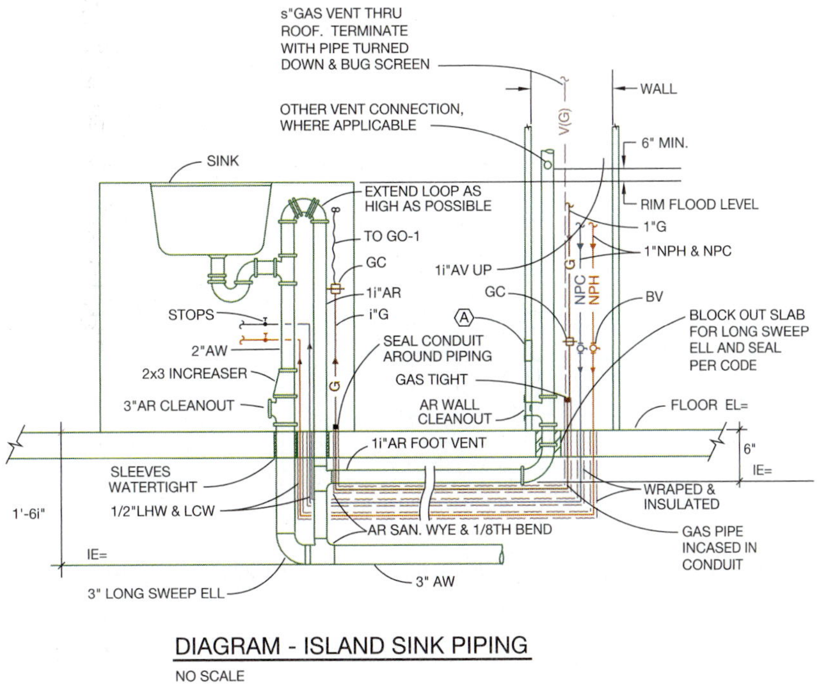

FIGURE 46.16(B) ■ Details are often required for commercial plumbing drawings. *Courtesy Manfull Curtis Consulting Engineering.*

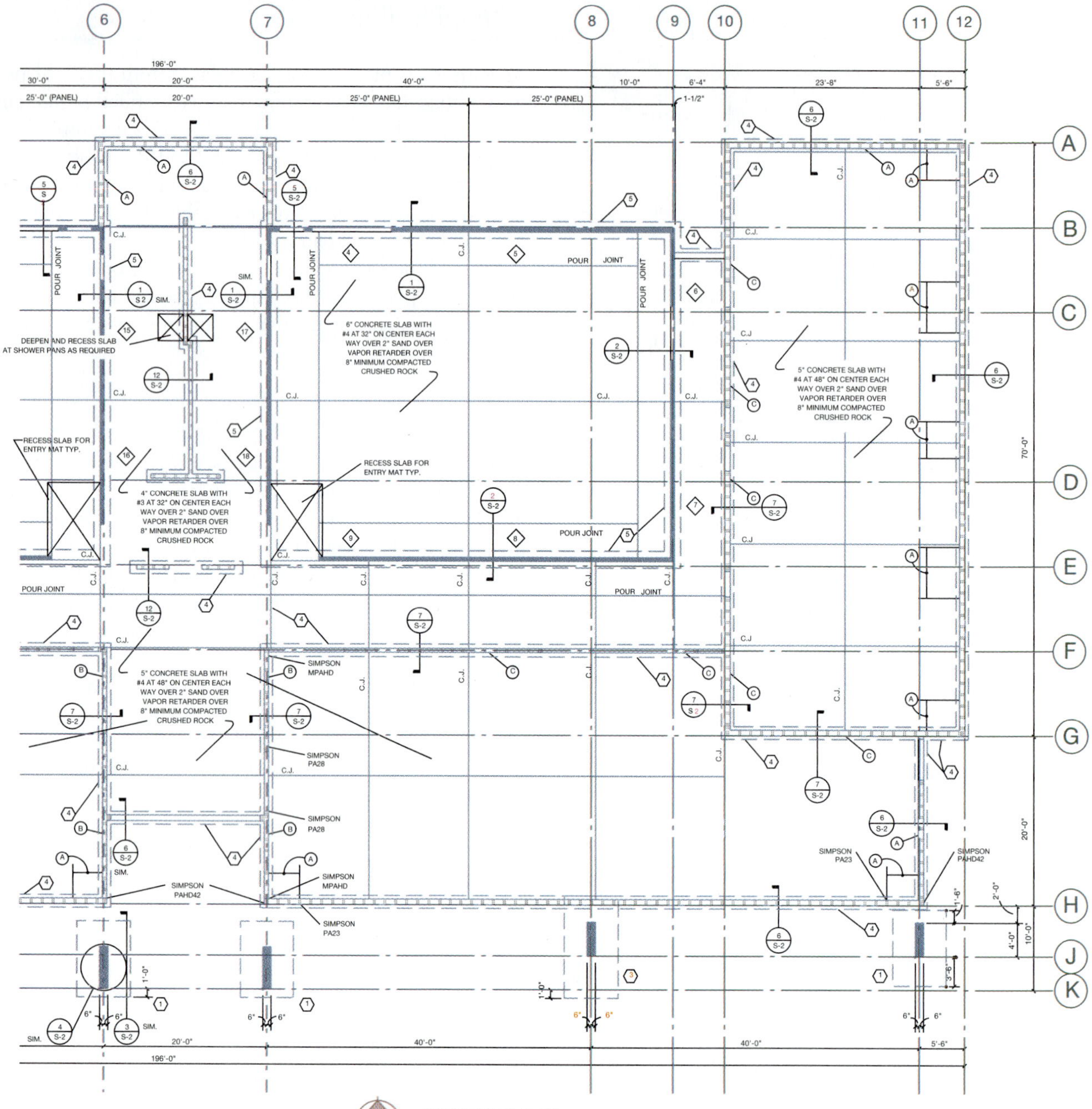

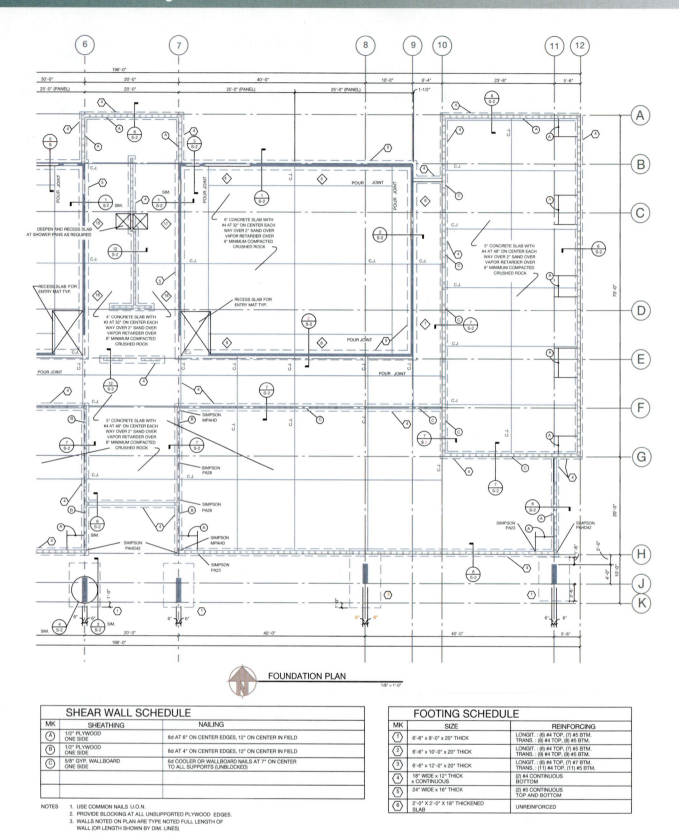

FOUNDATION PLAN
1/8" = 1'-0"

SHEAR WALL SCHEDULE		
MK	SHEATHING	NAILING
(A)	1/2" PLYWOOD ONE SIDE	8d AT 6" ON CENTER EDGES, 12" ON CENTER IN FIELD
(B)	1/2" PLYWOOD ONE SIDE	8d AT 4" ON CENTER EDGES, 12" ON CENTER IN FIELD
(C)	5/8" GYP. WALLBOARD ONE SIDE	6d COOLER OR WALLBOARD NAILS AT 7" ON CENTER TO ALL SUPPORTS (UNBLOCKED)

NOTES
1. USE COMMON NAILS U.O.N.
2. PROVIDE BLOCKING AT ALL UNSUPPORTED PLYWOOD EDGES.
3. WALLS NOTED ON PLAN ARE TYPE NOTED FULL LENGTH OF WALL (OR LENGTH SHOWN BY DIM. LINES)

FOOTING SCHEDULE		
MK	SIZE	REINFORCING
1	6'-6" x 9'-0" x 20" THICK	LONGIT. : (6) #4 TOP, (7) #5 BTM. TRANS. : (8) #4 TOP, (8) #5 BTM.
2	6'-6" x 10'-0" x 20" THICK	LONGIT. : (6) #4 TOP, (7) #5 BTM. TRANS. : (9) #4 TOP, (9) #5 BTM.
3	6'-6" x 12'-0" x 20" THICK	LONGIT. : (6) #4 TOP, (7) #7 BTM. TRANS. : (11) #4 TOP, (11) #5 BTM.
4	18" WIDE x 12" THICK x CONTINUOUS	(2) #4 CONTINUOUS BOTTOM
5	24" WIDE x 16" THICK	(2) #5 CONTINUOUS TOP AND BOTTOM
6	2'-0" X 2'-0" X 18" THICKENED SLAB	UNREINFORCED

FIGURE 46.17 ■ A concrete slab foundation is typically used for commercial structures. *Courtesy Architects Barrentine, Bates & Lee, AIA.*

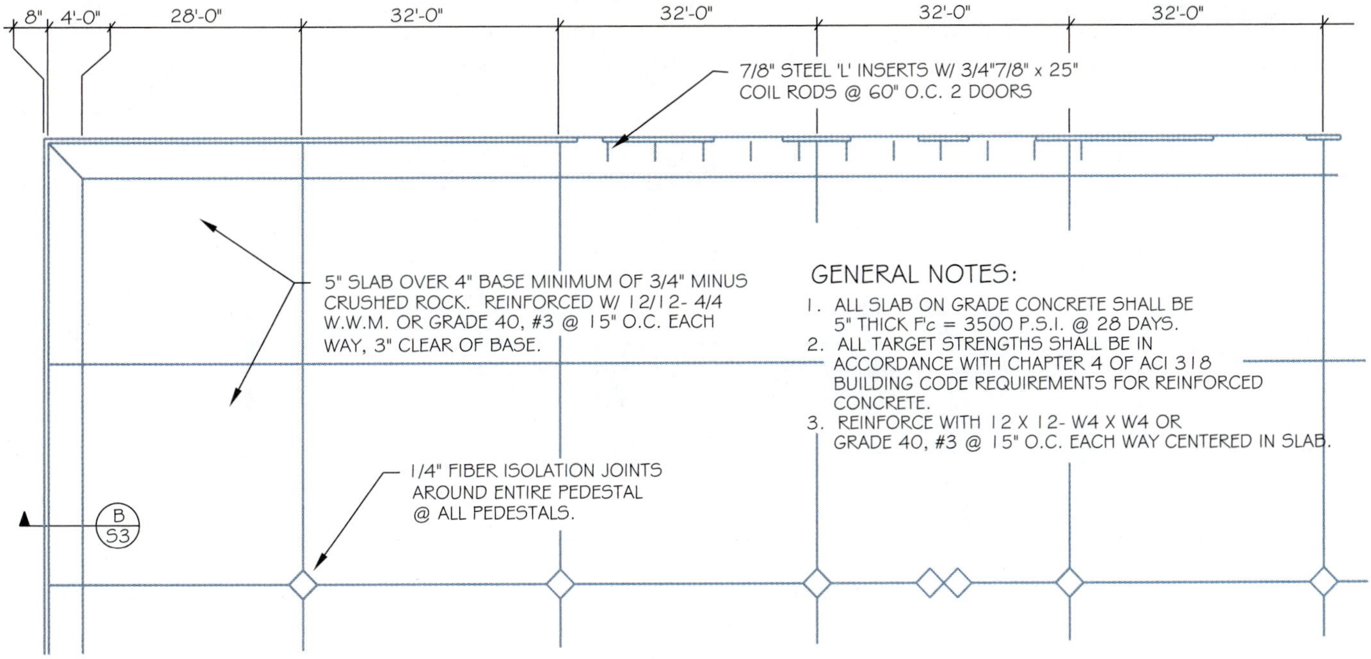

FIGURE 46.18 ■ A slab-on-grade plan shows the size and location of all concrete pours. *Courtesy Gisela Smith, Kenneth D. Smith and Associates, Inc.*

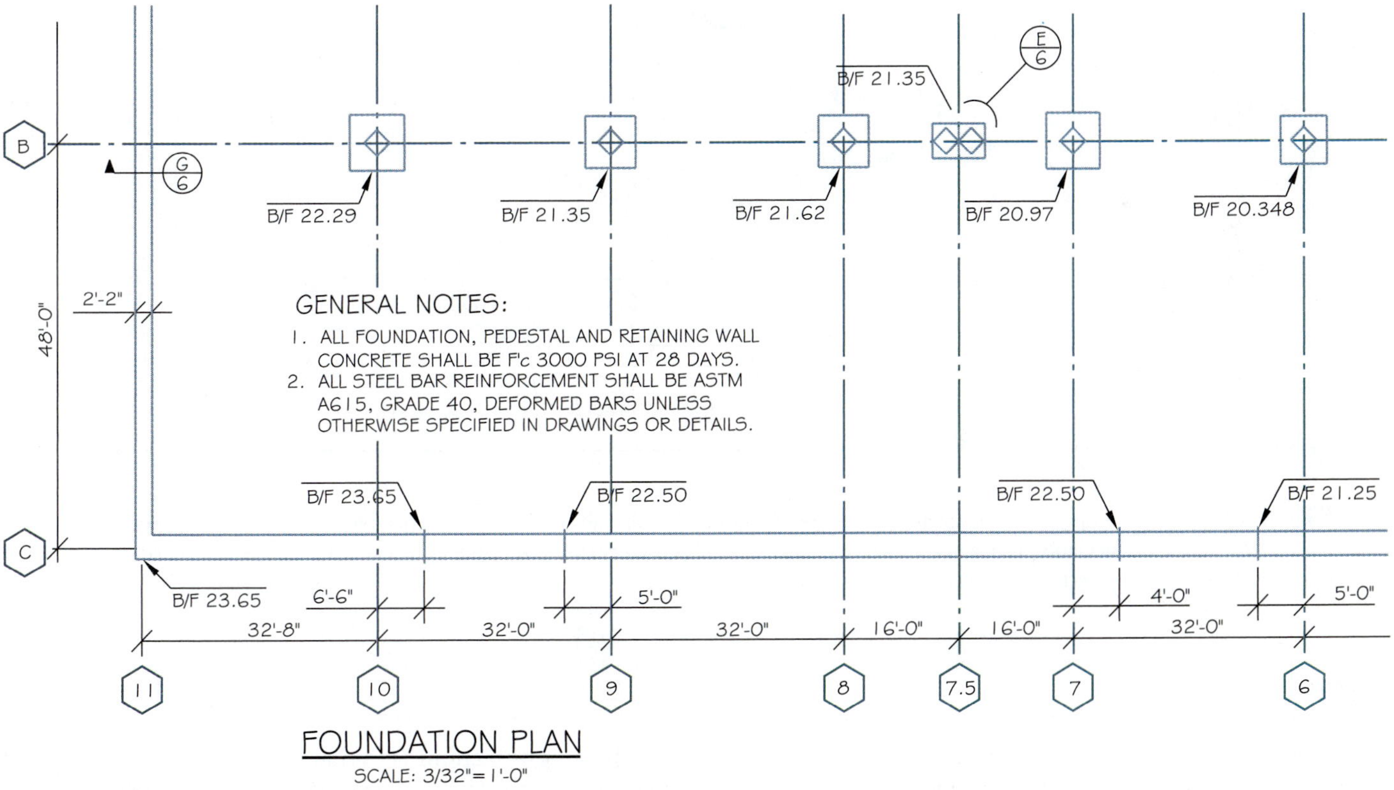

FIGURE 46.19 ■ A foundation plan is used to show below-grade concrete work. *Courtesy Ginger Smith, Kenneth D. Smith and Associates, Inc.*

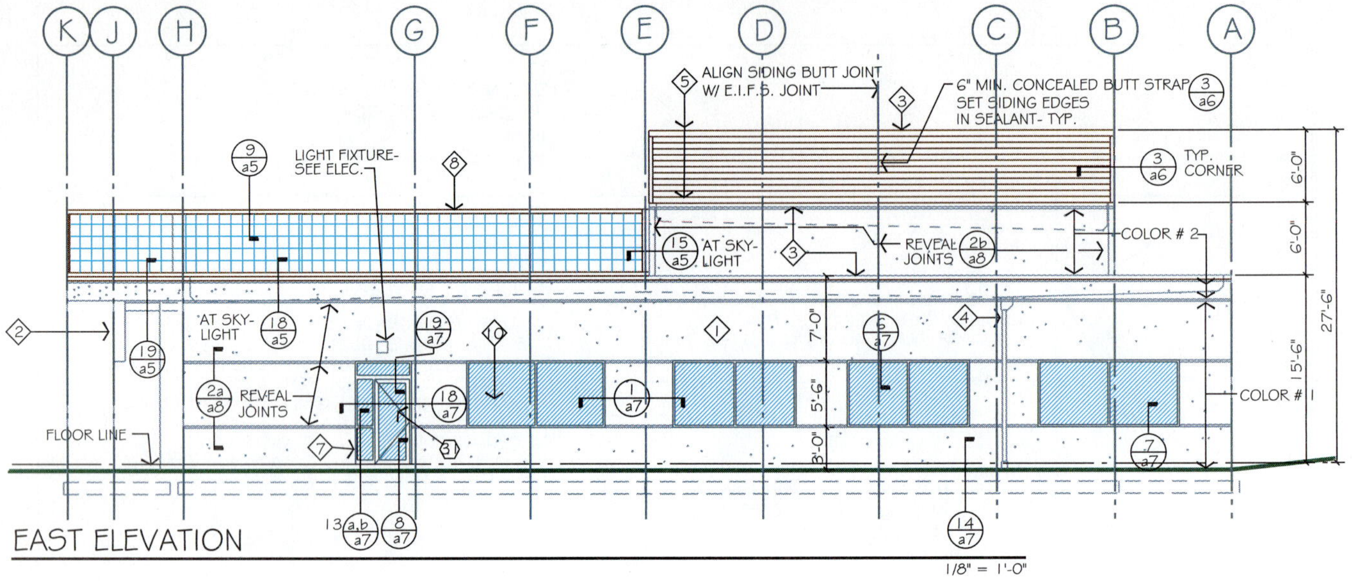

FIGURE 46.20 ■ Elevations for commercial projects are similar to the elevations for residential projects. This elevation is the east side of the floor plan shown in Figure 46.5. *Courtesy Architects Barrentine, Bates & Lee, AIA.*

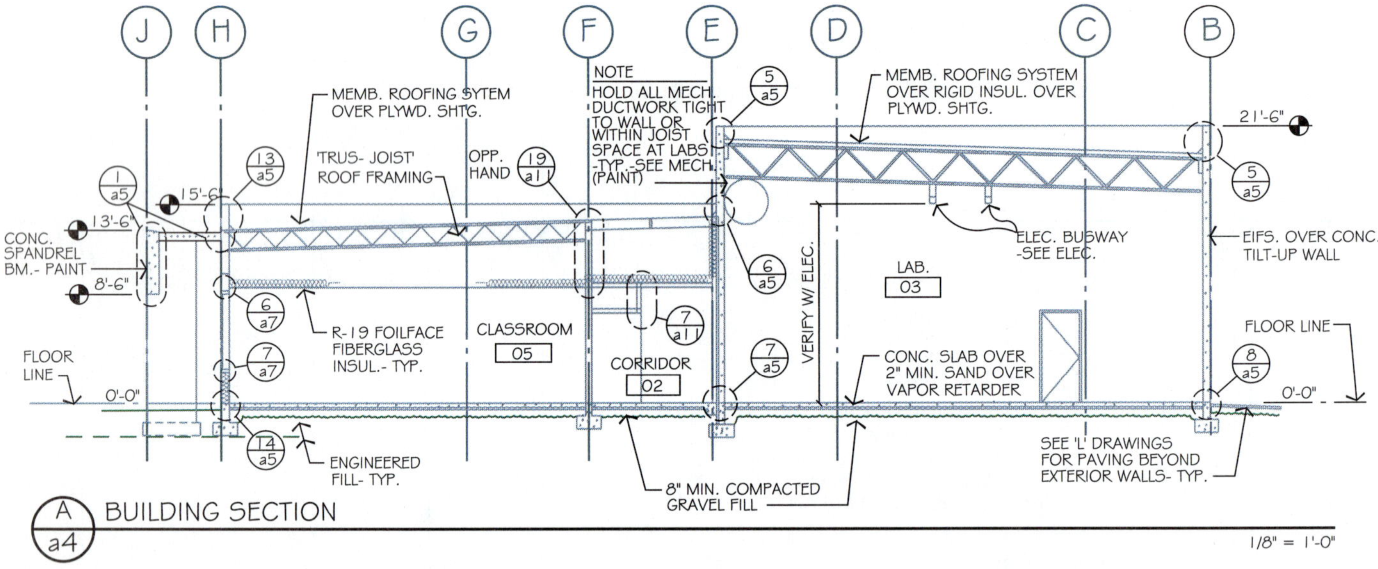

FIGURE 46.21 ■ Sections for commercial projects are often drawn at a small scale to show major types of construction. Specific information is usually shown in details. This section is referenced to the floor plan in Figure 46.5. *Courtesy Architects Barrentine, Bates & Lee, AIA.*

Elevations

Elevations for commercial projects are very similar to the elevations required for a residential project. Figure 46.20 shows a commercial elevation. Notice that the elevation shows the same types of information as a residential elevation but uses much more detail. Because many commercial structures are so large, commercial projects are sometimes drawn at a scale of 1/8″ = 1′–0″ or smaller and have very little detail added. Material is clarified through the use of separate details.

Sections

The drawing of sections is similar for commercial projects and residential projects, but the sections often have a different use in commercial projects. In residential projects, sections are used to show major types of construction. This use is also made of sections on commercial plans, but the sections are used primarily as a reference map of the structure. Sections are usually drawn at a scale of 1/4″ = 1′–0″ or smaller, and details of specific intersections are referenced to the sections. Figure 46.21

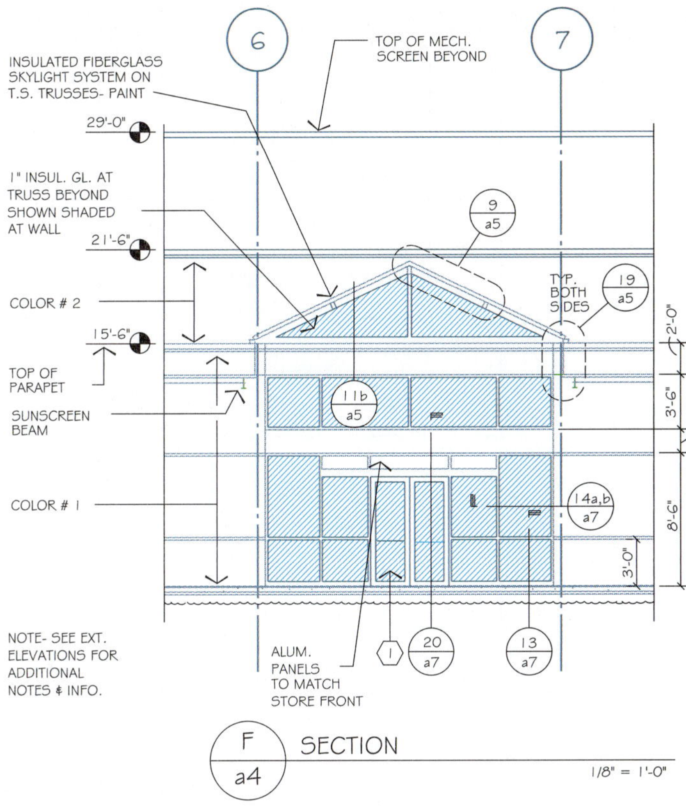

FIGURE 46.22 ■ Partial sections are often used to show construction information. *Courtesy Architects Barrentine, Bates & Lee, AIA.*

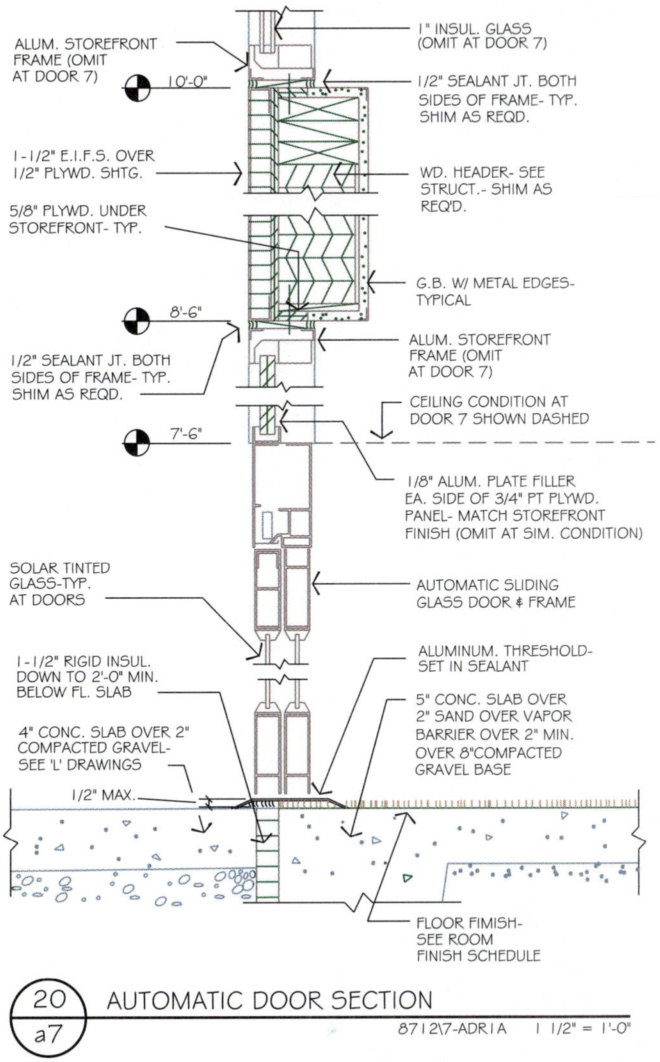

FIGURE 46.23 ■ Drawing details is one of the most common jobs performed by drafters. This detail is referenced to the section in Figure 46.22. *Courtesy Architects Barrentine, Bates & Lee, AIA.*

shows an example of this type of section. Another common method is to draw several partial sections, as in Figure 46.22.

Details

On a residential plan, you may draw details of a few special connections or of stock items such as fireplaces or footings. In commercial projects, drawing and editing details is one of the primary jobs of the drafter. Because there are so many variables in construction techniques and materials, details are used to explain a specific area of construction. Figure 46.23 shows one of the details that was required to supplement the section.

Roof Plans

Roof plans and roof drainage plans are typically used more in commercial projects than in residential work. Because the roof system is usually flat and contains concentrated loads from mechanical equipment, a detailed placement of beams and interior supports is usually required. Two of the most common types of roof systems used in commercial projects are truss and panelized systems. The use of trusses was discussed in Chapter 25. A panelized roof consists of large beams supporting smaller beams, which in turn support 4 × 8 roofing panels. The system was discussed in Chapter 45. The roof plan can be seen in Figure 46.24, and the roof framing plan can be seen in Figure 46.25.

Interior Elevations

The kind of interior elevations drawn for a commercial project will depend on the type of project to be drawn. In an office or warehouse building, cabinets are minimal and usually are not drawn. For an office setting, an interior designer may be used to coordinate and design interior spaces. On projects such as a restaurant, the drafter probably will work closely with the supplier of the kitchen equipment to design an area that will suit both the users and the equipment. This requires coordinating the floor plan with the interior elevations. Figure 46.26 shows an elevation for the bathroom in Figure 46.6.

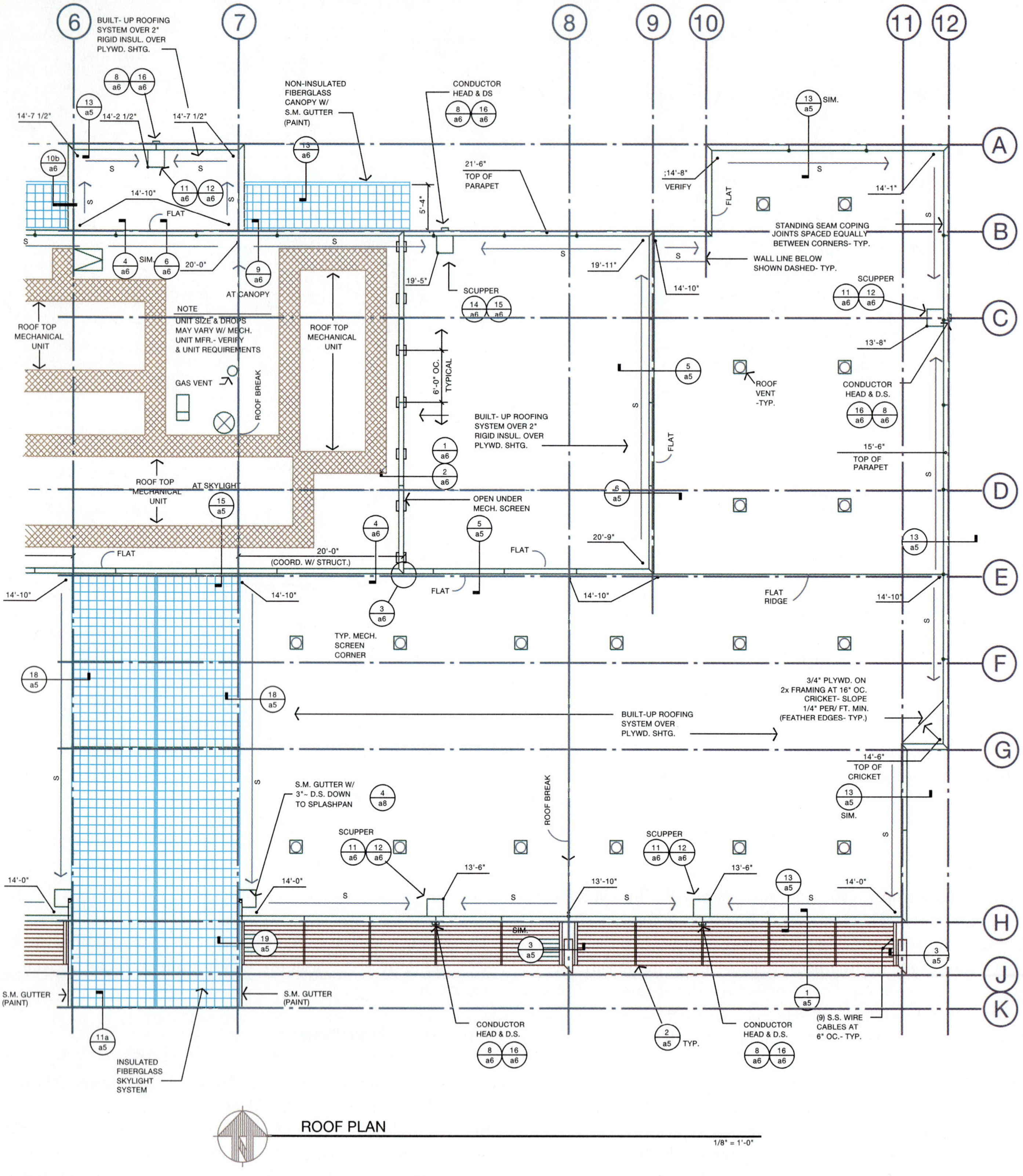

FIGURE 46.24 ■ A roof plan is used to show construction components at the roof level. *Courtesy Architects Barrentine, Bates & Lee, AIA.*

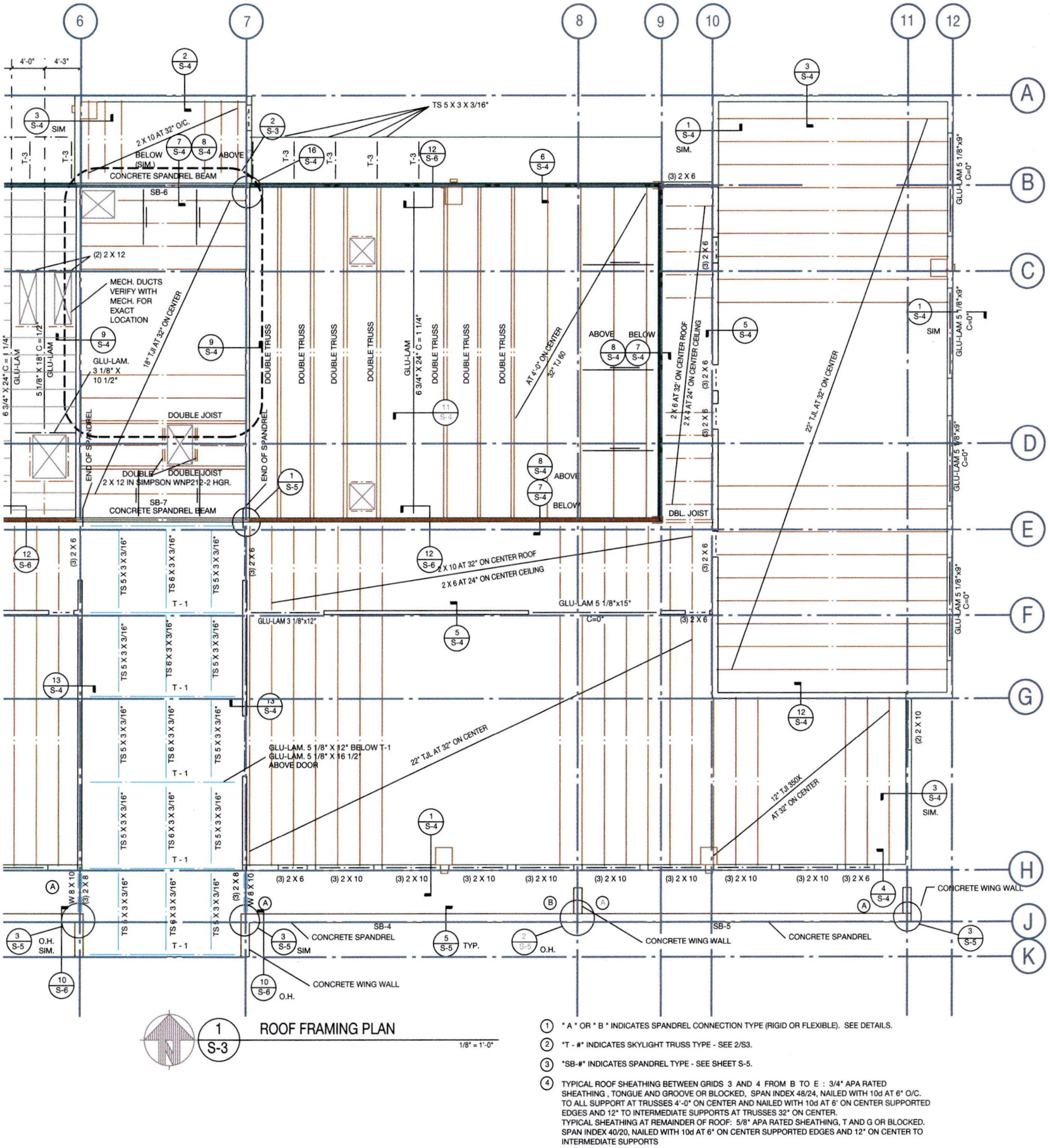

FIGURE 46.25 ■ A roof framing plan is used to show structural components at the roof level. *Courtesy Architects Barrentine, Bates & Lee, AIA.*

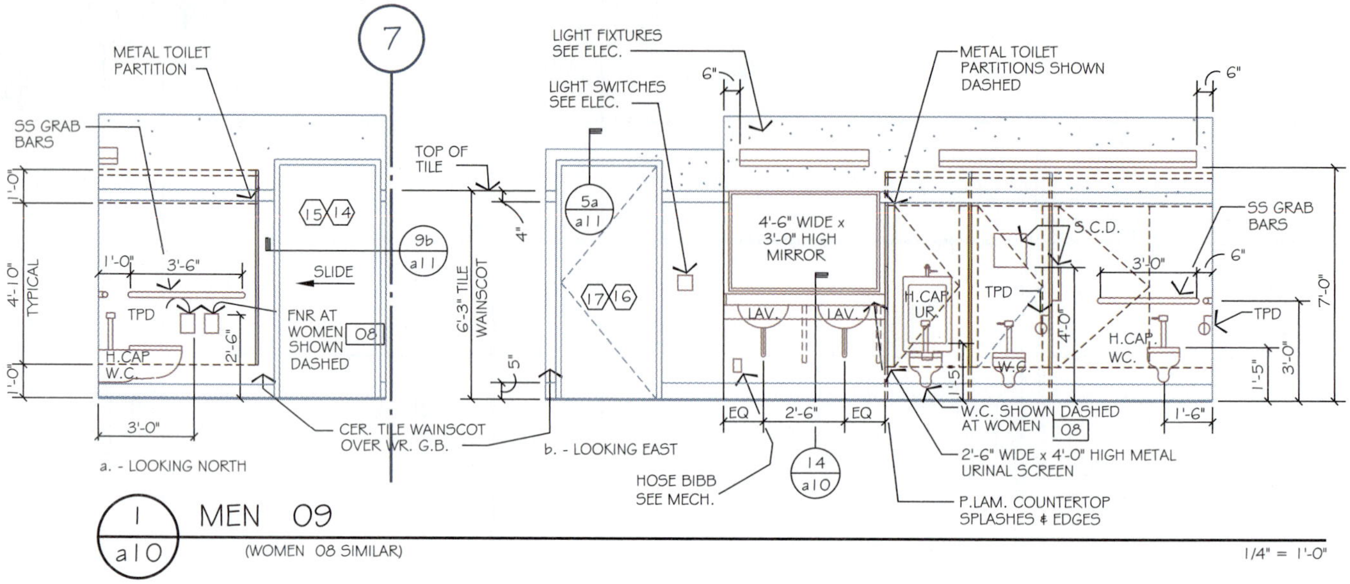

FIGURE 46.26 ■ In addition to structural details, the drafter may be required to draw interior elevations and details. These elevations are referenced to the enlarged floor plan in Figure 46.6. *Courtesy Architects Barrentine, Bates & Lee, AIA.*

Site Plans

The *site plan* for a commercial project resembles the site plan for a residential project. One major difference is the need to draw parking spaces, driveways, curbs, and walkways. In commercial projects, the site plan is often drawn as part of the preliminary design study. Parking spaces are determined by the square footage of the building and how the building will be used. Another major consideration is locating all existing utilities. Water lines and meters, sewer laterals and manholes, and underground electric and communication services adjacent to the property must be shown. Figure 46.27 shows a site plan for an office building.

Grading Plan

A *grading plan* shows the existing soil contours and any changes that must be made to the site to accommodate the structure and related improvements. Finish-grade contours representing 1′ and 5′ (300 and 1525 mm) intervals are typically shown, along with retaining walls and underground drainage facilities. Contours also reflect surface depressions used to divert water away from the structure to drainage grates. A drafter working for a civil engineer will then translate field notes into a grading plan, as in Figure 46.28.

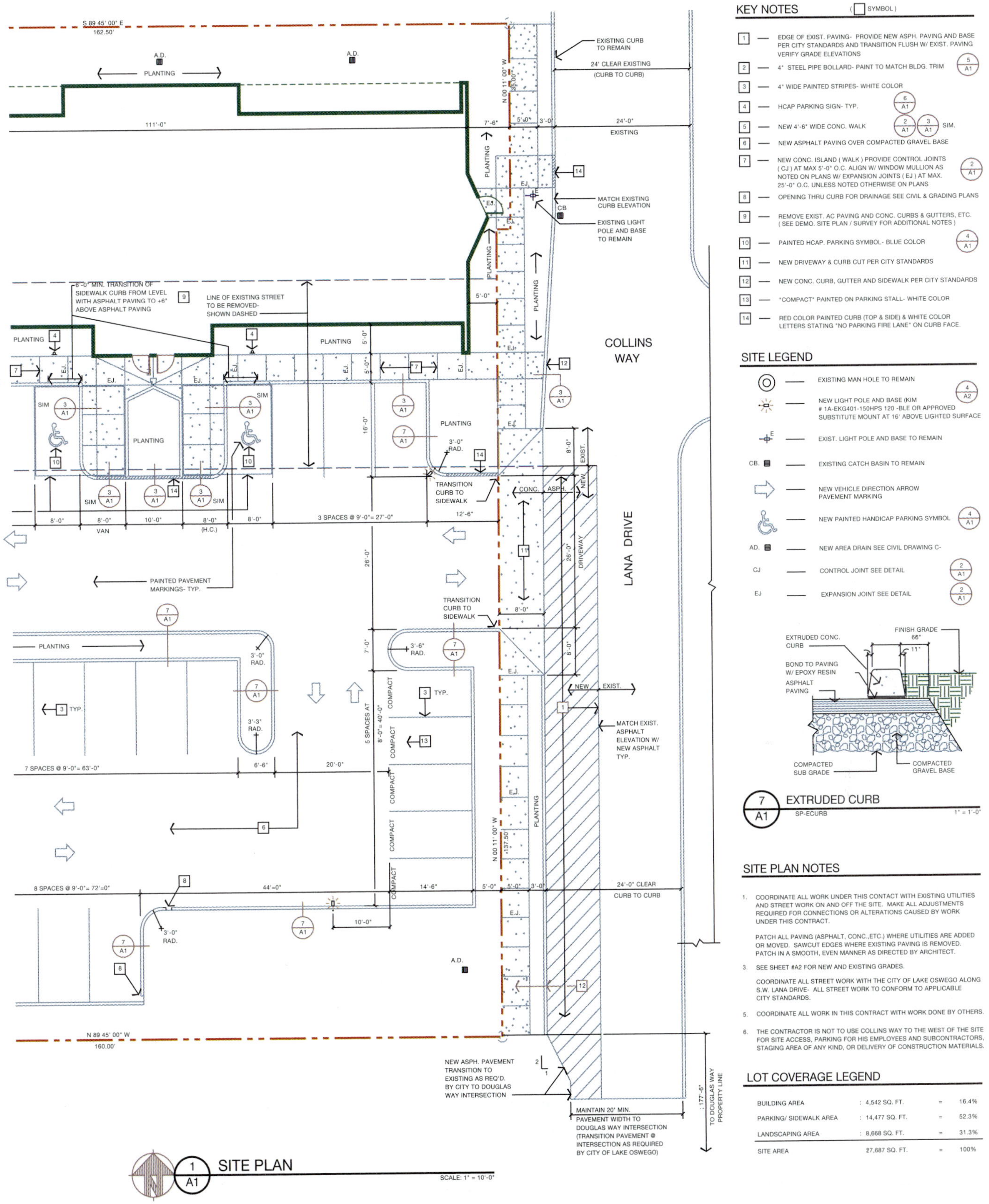

FIGURE 46.27 ■ A site plan for a commercial project is similar to a residential plan. Information about parking is a major addition to a commercial plan. *Courtesy Architects Barrentine, Bates & Lee, AIA.*

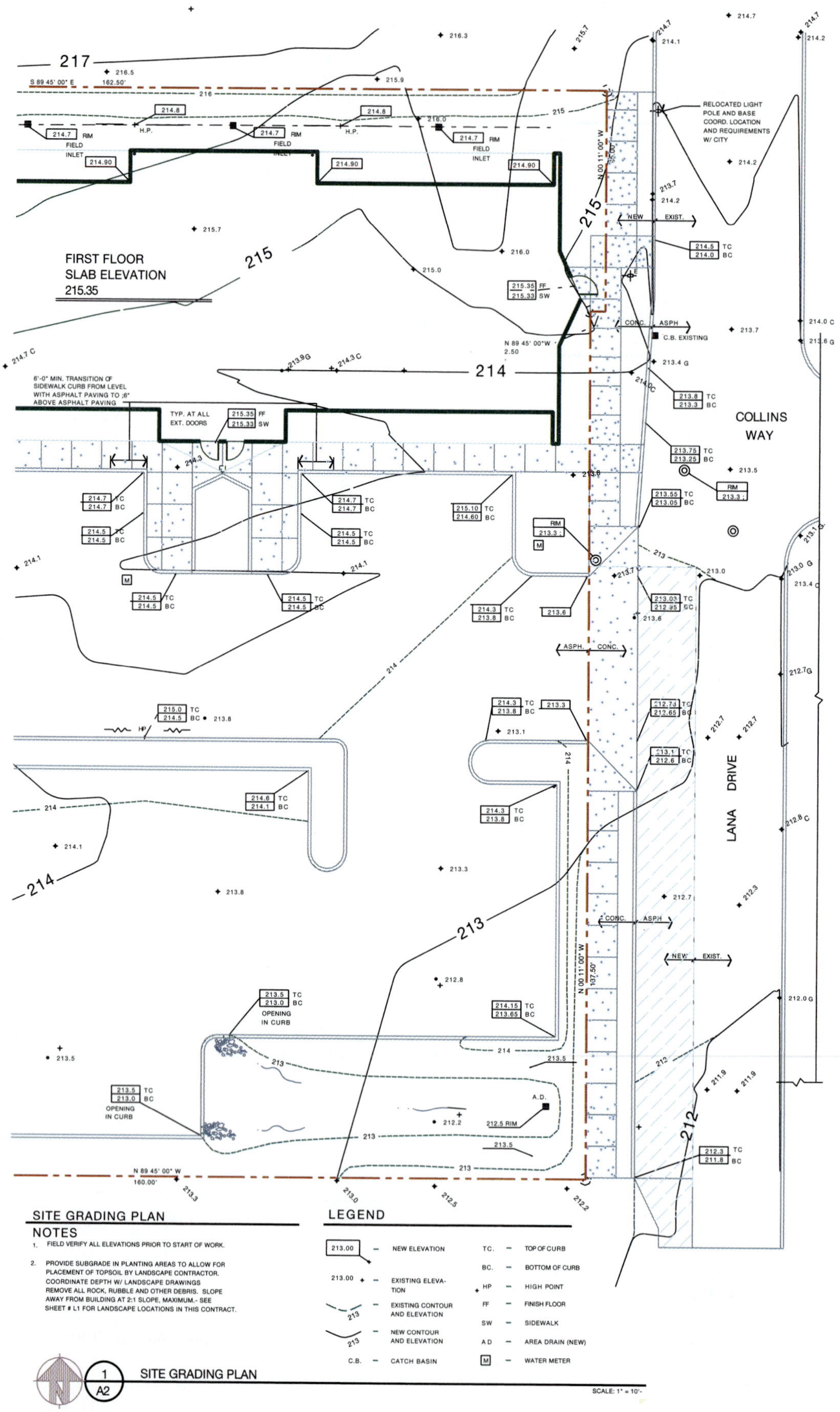

FIGURE 46.28 ■ Grading information is often placed on an overlay of the site plan to show cut and fill requirements of a job site. *Courtesy Architects Barrentine, Bates & Lee, AIA.*

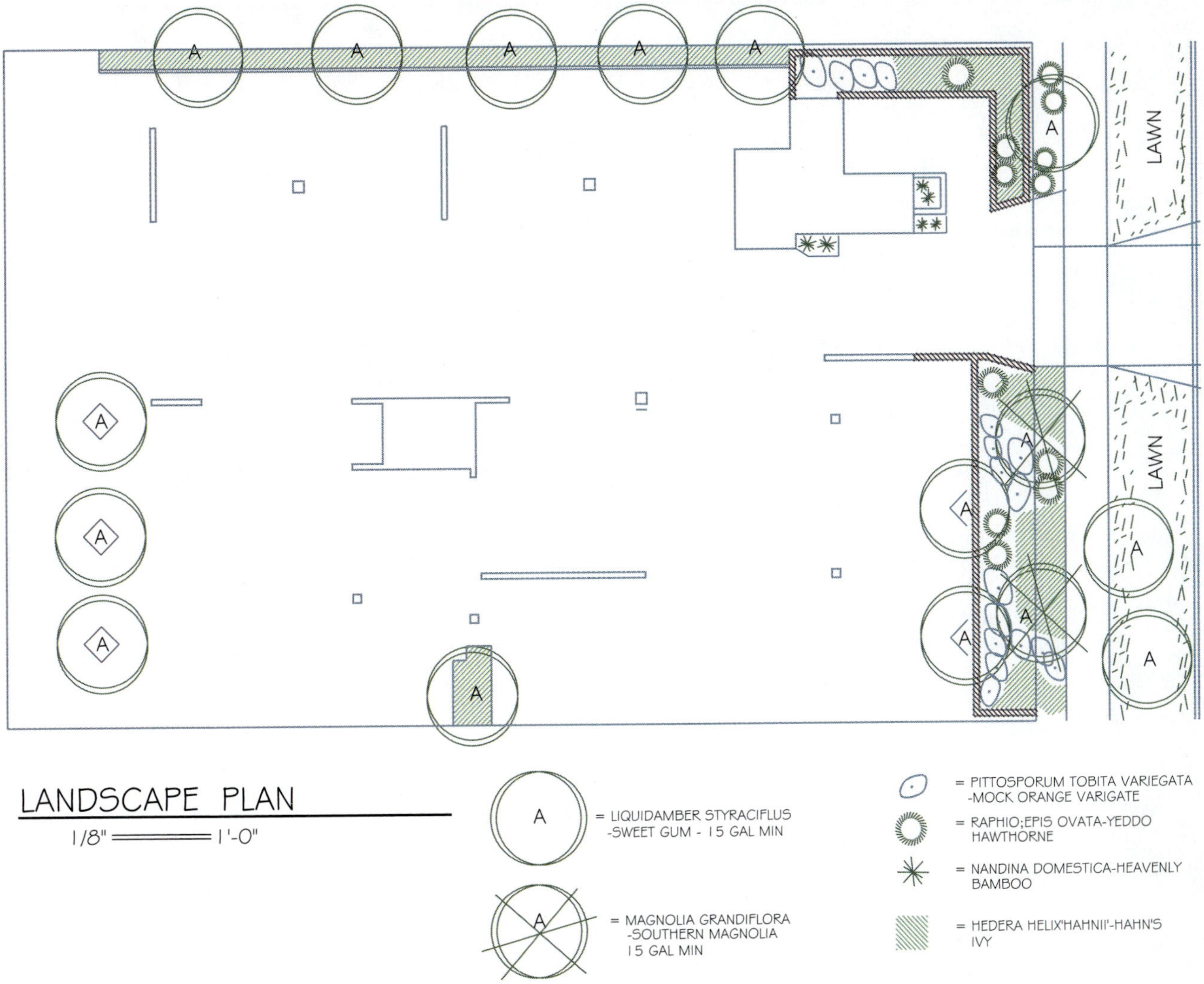

FIGURE 46.29 ■ A planting or landscape plan is drawn as an overlay to the site plan to show how the site will be planted and maintained. *Courtesy Janice Jefferis.*

Landscape Plan

The size of a project determines how the landscape plan is drawn. On a small project, a drafter may be given the job of drawing the landscape plan. When this is the case, the drafter must determine suitable plants for the area and the job site and then specify the plants by their proper Latin name on the plan. On typical commercial projects, a copy of the site plan is given to a landscape architect, who will specify the size, type, and location of plantings. A method of maintaining the plantings is also typically shown on the plan, as in Figure 46.29.

CADD APPLICATIONS

USING CADD FOR COMMERCIAL PROJECTS

In most commercial architectural projects, the architect works with consultants in the fields of mechanical, electrical, and structural engineering; plumbing systems; and the planning of interior design and space. These professionals use background drawings supplied by the architect to create their individual designs and plans. Background drawings are a subset of the information contained in the architect's floor plans.

Background drawings sent to individual consultants may contain different information. For example, the mechanical, electrical, and plumbing engineers need background drawings with sinks, toilets, showers, and utilities shown; these items may be omitted, to avoid confusion, on the structural background drawing. In multistory architectural projects, the background drawings may be used to coordinate floor-to-floor stacking relationships.

Multifamily housing projects provide a perfect application for CADD. Most of these projects have three or more unit plans, which may be copied, mirrored, and rotated to create a complete multifamily plan. After the unit plan is drawn, it can be edited using mirror or copy, command, or rotated to create the building floor plan.

In using CADD for architectural projects, it is important to relate one drawing to the next for accurate overlay purposes. The datum point method relates all plan drawings in the set through a common point. The datum point may be any convenient point on the drawing. Many drafters select the lower left corner of the drawing, represented by X and Y coordinates 0,0, as the datum point. Then the lower right corner of each additional drawing is coordinated the same way. The relationship of the floor plans is established relative to the datum point, so each plan is aligned, one over the other.

CHAPTER 46

Commercial Construction Projects Test

DIRECTIONS

Answer the questions with short, complete statements or drawings.

1. Letter your name and the date at the top of the sheet.
2. Letter the question number and provide the answer. You do not need to write out the question.

QUESTIONS

Question 46–1 List three types of offices in which you might be able to work as a drafter.

Question 46–2 List two major research tools that you will be using as a commercial drafter.

Question 46–3 What is a reflected ceiling plan?

Question 46–4 What are engineer's calculations?

Question 46–5 What two types of plumbing plans may be drawn for a commercial project?

Question 46–6 Describe two methods of drawing sections in commercial drawings.

Question 46–7 List and describe two types of plans that are usually based on the site plan.

Question 46–8 What are the three parts of a set of calcs? How do they affect a drafter?

Question 46–9 Describe three plans that can be used to describe site-related material.

Question 46–10 Explain the difference between two types of plans that are used to describe work done at or below the finished grade.

47 CHAPTER

Structural Drafting

INTRODUCTION

This chapter introduces basic concepts and drawing methods expected of an entry-level drafter in an architect's or engineer's office. The project example at the end of this chapter will be working drawings for a concrete tilt-up warehouse.

The project has two main objectives: to coordinate various plans and to draw coordinated details that correspond to different areas of the building.

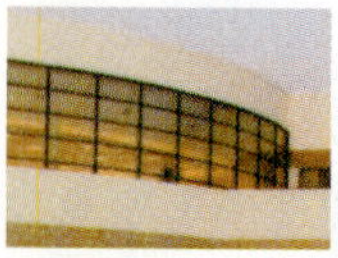

PLAN THE DRAWING

Before you begin drawing, search all plan views for similarities. Start at the upper level and work to the lowest. This should show you the loads that must be supported as your work progresses. As you work from top to bottom, you should see beams in the roof supported by columns on the floor plan, pedestals on the slab plan, and footings at the foundation. Had you begun at the foundation level, it might not have been clear why a given pier was in a particular location. As you work on the various plans, you should also be studying details that relate to the plan. Details will give you a better understanding, which should result in faster drawing production.

As you draw, you may not be able to get your questions answered as quickly as they arise. This is typical of work situations; it is often difficult to find an engineer to answer your questions. Using a nonreproducible blue pencil, keep a list of your questions right on your drawing.

PROCEDURE—FROM CALCULATIONS TO WORKING DRAWING

As you work on projects for an architect or engineer, you will be given a set of calculations, sketches, and similar drawings that have been done by the office. The calculations are the mathematical solutions to particular problems. Calculations are printed following the related problem. As a drafter, you will be given a set of specifications and will be expected to determine the goal of the engineer. You will then translate the results into working drawings. Typically, this will mean skipping over the math work. For the project you will be working on, the math has been omitted and only the material to be placed on the drawings has been shown.

Sketches an engineer provides are tools to help solve calculations. A drafter is expected to use the sketch as a guide to help determine size and material to be used.

> **Note:** *The sketches in this chapter are to be used as a guide only.* As you progress through the project, you will find that some parts of a drawing do not match things that have already been drawn. It is your responsibility to coordinate the material that you draw with other drawings for this project. If you are unsure of what to draw, consult your instructor.

In addition to the calculations and sketches for a specific job, a drafter is expected to study similar jobs and details for common elements that can be used in the new drawing. This includes many common notes as well as common connection methods. A drafter would also be expected to consult vendors' catalogs for specific details of prefabricated material.

Drawing Layout

There is no one totally correct layout for this project. Usually a project will be drawn so details and plans that relate to a given construction crew will be grouped together. Therefore, the roofer will never see the foundation plan. Often this technique of grouping details and plans requires material to be specified on plans more than once. As you lay out this project, two common methods might be used.

1. Group all plan views together, then all roof details, all concrete details, etc., in the same order that the plan views are presented.
2. Group a plan view with all related details; for instance, show the roof plan and then the roof details; the slab plan and then the slab details.

The layout format will not really be important until you have the plan views complete and you start to layout details. By that time, you should have a much better understanding of the project.

One frustration associated with this project and this type of drawing is that it cannot be completed without interruptions. You will be required to draw the basic drawing, stop, then solve a related problem at another level. You will be working on several drawings at the same time. The good news is that all drawings belong to the same project. It is common in an office to switch from one job to another several times a day, depending on an engineer's or a client's appointments. On this project, you will be expected to shift

from one drawing to another, keep them all coordinated, and still keep your sanity.

Material

Drafters in structural firms often work with several different drawing materials: ink on vellum or Mylar, graphite on vellum, polyester on vellum, any combination of these, or CADD. The plan views, elevations, and panel elevations are well suited to CADD drafting because there are so many layering possibilities. Verify the drawing medium with your instructor whenever you start a drawing.

The material is not as important as the end result. Excellent line quality reflecting consistent width and density is a must for a structural detailer. Many of the details in this text are similar to what you will be drawing. The details feature several different line weights, as should your drawings.

DETAIL COORDINATION

As you draw each detail, provide room for a title, scale, and detail marker under the detail. The title will typically tell the function of the drawing, such as BEAM/WALL CONNECTION, and should be neatly printed using lettering 1/4″ (6 mm) high. The scale should be placed under the title in lettering 1/8″ (3 mm) high. A line should separate the two, and there should be a detail reference circle on the right end of the line. The reference circle should be about 3/4″ (19 mm) in diameter with the line passing through the center of the circle. There will eventually be a detail number on the top of the line. On the bottom of the line, place the page number where the detail was drawn. Do not fill in the circles until all of the drawing is complete.

Throughout the plans, you will be referring to a specific detail location for more information. Typically, circles 3/8″–1/2″ (9–13 mm) in diameter are used so the reference circle does not totally dominate the plan.

LETTERING

Lettering is an important skill to be mastered, even in this age of computers. You will probably get your initial job interview solely on the basis of your lettering. Aside from this minor detail, excellent lettering is essential for easy-to-use drawings. The best structural lettering has simple shapes, is easy to read, and is quick to produce. Lettering must be uniform in size, angle, and spacing if you are to advance in the field. Practice lettering that has uniform angles of vertical strokes, uniform spacing, and proper placement of notes. Place notes so they require the shortest leader line without getting in the way of the drawing. Notes are often placed in the drawing or detail. Be careful to keep the vertical strokes of letters away from lines in the drawing.

ORDER OF PRECEDENCE

This project may be unlike others you have done. Not only does it have sketches and engineer's calculations (calcs); **it also has some very large errors.** Most are so obvious (at the beginning) that you will have no trouble finding them. They get tougher as you gain more knowledge. The errors are here to help you learn to think as a drafter. If you think you have found an error, do not make changes in the drawings until you have discussed them with the engineer. (In this case, discuss them with your instructor.

> Because engineers are not perfect, you sometimes need to sort through conflicting information to solve a problem. Usually though, the engineer is correct, and you will find that you have misinterpreted the information.

To sort through conflicting information, use the following order of precedence.

1. Written changes by the engineer (your instructor) as change orders.
2. Verbal changes given by the engineer (your instructor).
3. Engineer's calcs.
4. Sketches by engineer.
5. Lecture notes and sketches.
6. Your own decision (should it come to this, please check with your instructor).

OCCUPANCY

The space between grid 1 and grid 7.5 will be used for manufacturing wood cabinets including storage of stains, varnish, and glue. The space between grid 7.5 and grid 11 will be used for sales and office space. Before you start drawing, review Chapter 44 and determine the occupancy of each area, the least restrictive types of construction to be used, maximum height, square footage, and occupant load of each portion.

CHAPTER 47

Structural Drafting Problems

DIRECTIONS

Students do not usually have adequate time to draw the entire set of plans required to construct this building. Your instructor may wish to add or delete drawings or have you work individually or in teams. Verify with your instructor what problems will be drawn.

Problem 47–1 Roof Drainage Plan.

Goal The purpose of this drawing is to show the elevations of the roof supports. By changing the elevation of supports, you can control the flow of water and direct it to downspouts.

Problem The building on which you will work will be nearly as large as a football field. Therefore, you have a large area that collects water. If the water is not quickly drained from the roof, it must be treated as a live load, and the beam sizes must be increased. This plan will show the slope of structural members and the roof, and it will show overflow drain locations. Use the sketch as a guide.

Method At a scale of 1/16″=1′0″ draw:

1. The outline of the entire structure with walls 6″ wide. Determine size of building from other sketches.
2. The center of the glu-lams at grid B.
3. Four roof drains, overflow drains, and scuppers equally spaced on grids A and C; assume drains to be 8″ in diameter.
4. Dimension as per sketch.
5. Letter all elevations at drain.
6. Specify drains, overflows, and scuppers.
7. General notes: (1) Roof and overflow drains to be general-purpose type with nonferrous domes and 4″ diameter outlets. (2) Overflow drains to be set with inlet 2″ above drain inlet and shall be independently connected to drain lines. (3) Scuppers to be 4″ high × 7″ wide with 4″ rectangular corrugated downspouts. Provide a 6 × 9 conductor head at the top of down spouts.

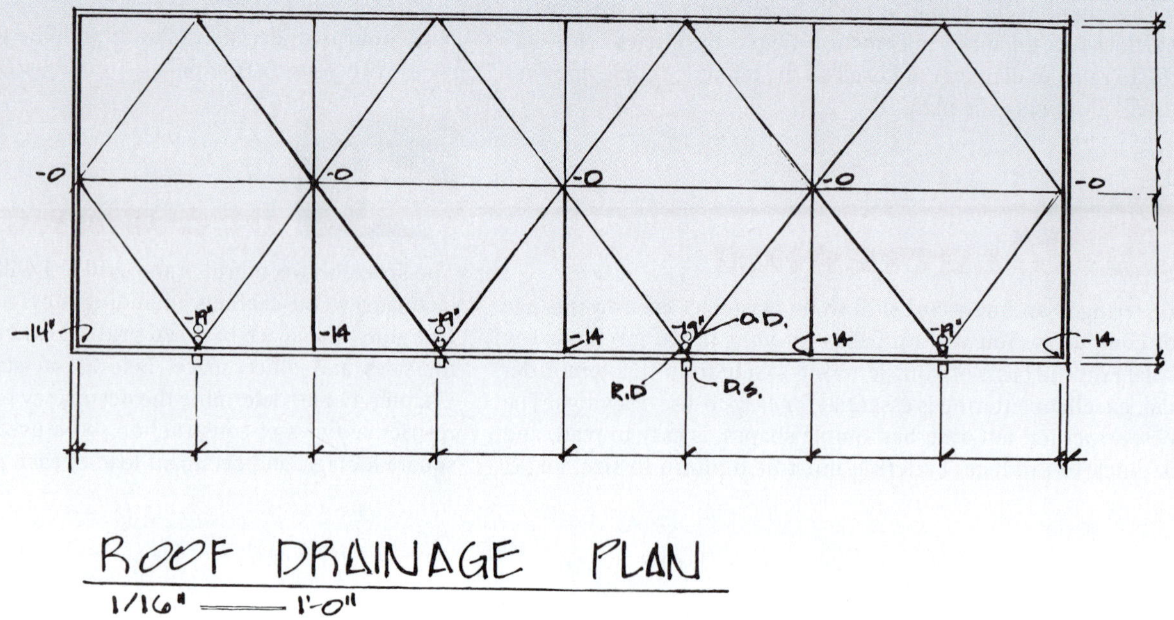

PROBLEM 47–1

Problem 47–2 Roof Framing Plan.

Goal The purpose of this drawing is to show roof framing used to support the finished roofing. This structure will have a series of laminated beams, which extend through the center of the structure at grid B from grid 1 through 11. It will also have a wood ledger bolted to each wall at grid A and C. Trusses will span between the ledger and the glu-lams. Blocking will be shown at each end of the roof between the trusses. Above the trusses will be sheets of plywood. Four skylights will be equally placed in each bay. This plan will also show how the separation wall intersects the beams.

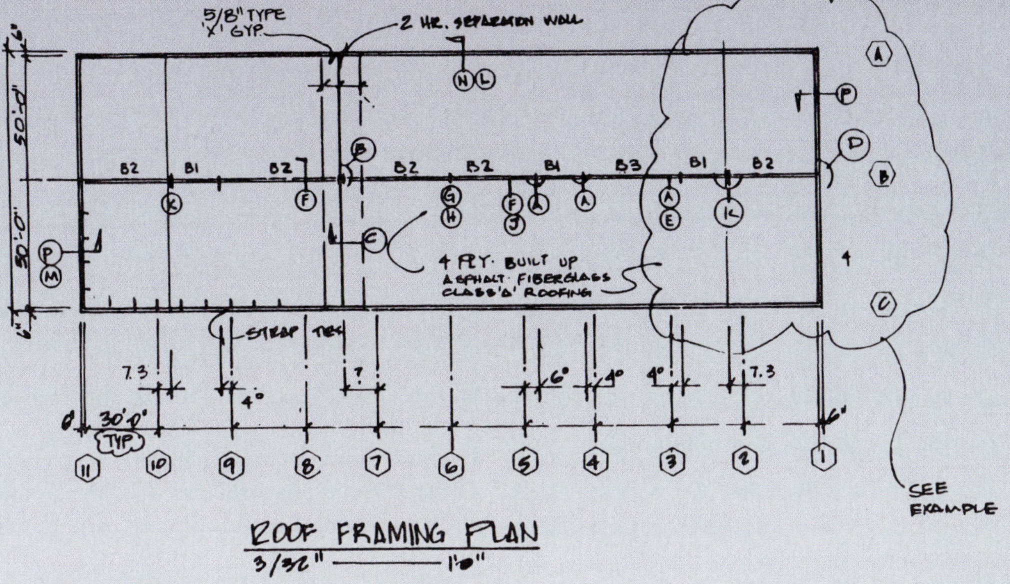

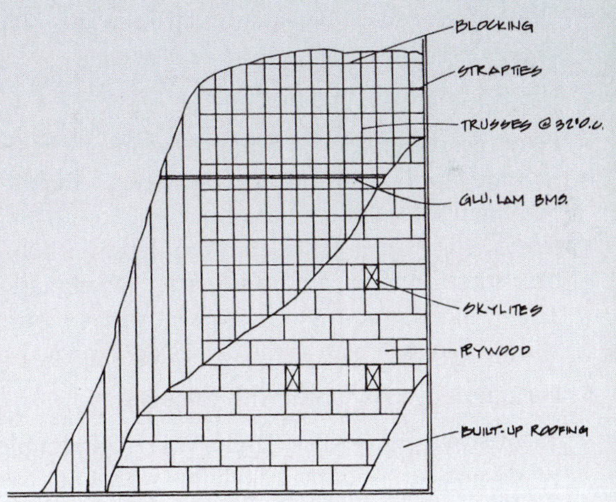

ROOF SHEATHING

w_V = 25 PSI

$w_L = \frac{88658}{(2)(122)} = 363$ #/l

$= \frac{59124}{(2)(321)} = 92$ #/l

ASSUME TRUSSES @ 32" O.C.

CHECK }", C-D 42/20 PLYWOOD

w= 35 PSF 25 PSF

V= 425 #/l BLOCKED > 363 #/l

= 285 #/l UNBLOCKED

= 215 #/l UNBLOCKED > 92 #/l

DETERMINE BLOCKING TERMINATION

$x = \frac{(285)(160.5)}{363} = 126.01'$ d= 160.5-126= 35.50= 37.33 from end of wall

USE 5/8" C-D, 42/20, INTERIOR APA PLY W/ EXT. GLUE. LAY PERP. TO TRUSSES, STAGGER SEAMS AT EACH TRUSS. NAIL W/ 10d COMMON NAILS @ 4" O.C. @ ALL PANEL EDGES AND BLOCKED AREAS, @ 6" O.C. @ ALL SUPPORTED PANEL EDGES @ UNBLOCKED AREAS & @ 12" O.C. @ ALL INTERMEDIATE SUPPORTS.

ROOF BEAMS

BEAM 1 = DF/HF 24f-V5 6 3/4 x25 1/2 GLU- LAM

BEAM 2= DF/DF 20f- V7 6 3/4 x43 1/2 GLU-LAM

BEAM 3= DF/DF 20f- V7 6 3/4 x 37 1/2 GLU-LAM

ROOF TRUSSES

W_L =25 PSF W_Δ= 15 PSF ASSUME TRUSSES @ 32"O.C.

$\ell = \frac{122-1}{2} = 60.50'$ w= (2.67) (25 + 15) =107 #/l

$R = \frac{(107)(60.5)}{2} = 3237$ $\Delta = \frac{(60.50)(12)}{240}$ 3.03"

CHECK 45" TJ/60 TRUSS JOIST

w=109 #l > 107 #l R= 4800# > 3237

$\Delta \frac{(107)(60.50)^4}{(308000)(45-2.3)}$ 2 =2.55" < 3.03"

USE 45" TJ/60 TRUSS JOISTS AT 32" O.C.

PROBLEM 47–2

Problem The building on which you are working is approximately 300′ long × 100′ wide × 27′ high. The walls will act like a sail. To keep the structure rigid, the roof framing will be connected to the walls to form a rigid connection. In addition, a diaphragm will be built into each end of the structure. As the walls bend from wind or seismic pressure, the roof will shift in the direction of the pressure. As the pressure is decreased, the walls will return to their original shape. The rigid connections ensure that when the walls spring back to their original position, the roof will also return to its original position. If the connections at this level fail, the roof will fall from its supports. The roof framing plan is used to show how and where these items are located. The two ends of this structure are to be symmetrical. Use the sketches of the roof as a guide.

Method Use a scale of 3/32″ = 1′–0″. If you're unsure of what you are drawing, look through the roof framing and truss details.

1. Lay out and draw the walls.
2. Lay out and draw the locations of beams.
3. Dimension and label all grids.
4. Lay out truss, blocking, skylights, strap ties, and 4 × 8 plywood as per sketch and roof details.
5. Draw plywood over blocking and trusses and strap ties and skylights.
6. Draw 2-hour wall and roof area.
7. Letter general notes: (1) Beam sizes. Specify beam size and type by beam, or provide a beam schedule. See engineer's calcs for size of beams 1–3. (2) Bay referencing. Use 1/4″ hex and print letter or number centered in hex. (3) Do not place detail markers on your plan at this time. These markers should be added to each plan as the details are drawn.
8. Local notes (verify notes with calculations).
    1. Bristolite 3069-A.S.-DD-CC/WTH-C.P. double domed curb-mounted skylights with 4 per bay (80 total).
    2. 45″ TJ/60 Trus joist at 32″ o.c.
    3. 3 × 4 dfl std. and better solid blocking at 48″ o.c. approximately 36′ out from each end wall.
    4. Simpson MST 27 strap ties at 8′–0″ o.c. for entire perimeter. See Detail ???. (Place detail marker in note for future reference).
    5. 5/8″ C-D 42/20 . . . . (See engineer's calcs for complete specifications and provide required information on drawing.)

Problem 47–3 Slab on Grade Plan.

Goal This drawing will provide directions for pouring the concrete floor. Major items shown on this plan include the floor slab and control joints, pedestal footings to support steel columns, loading docks, and doors.

Problem The concrete slab is so large that the concrete crew will pour it in stages. This will account for the 30 × 25 grids. These control joints will also serve to minimize cracking. The 4′ wide strip at the perimeter is to allow for movement of the concrete walls. The small squares at grid B are to allow for movement of the slab from the steel columns, which support the roof.

Method Your goals on this drawing are to determine door locations in the walls, slab control joints, and L connectors which will support the slab where it is over fill. Use a scale of 3/32″ = 1′–0″. Use the sketch as a guide for your drawing.

1. Lay out grids A and C and 1–11.
2. Lay out the exterior walls. Keep the drawing in the upper right corner of the page to allow for possible detail placement around the plan.
3. Lay out door openings using sketches of the panel elevations for locations.
4. Locate the loading docks. See slab details for the wall thickness. Locate the dock so that there is an equal amount of space at each side of the door

and the dock walls. This is not the same as centering the dock on the center of the doors.

5. Lay out control joints at grid B, grids 2–10, 4′ perimeter, and pedestals. See wall details for the pedestal size. Draw at an appropriate scale.
6. Draw walls and control joints. Use varied line weight to distinguish between the two.
7. Draw the structural connectors using bold lines.
8. Dimension as per sketch.
9. Label the grids, elevations, and notes.
10. General notes: (1) All slab on grade concrete shall be 5″ thick F′c = 3500 p.s.i. @ 28 days. (2) All target strengths shall be in accordance with Chapter 4 of ACI 318 Building Code Requirements for reinforced concrete. (3) Reinforce with 12 × 12- w4 × w4 or grd 40, #3 @ 15″ o.c. each way centered in slab.
11. Local notes:

ON SKETCH	NOTE SHOULD READ
Doweled joints . . .	Doweled joints w/#5 × 15″ smooth dowels @ 12″ o.c.
1/4″ joint . . .	1/4″ fiber isolation joints around entire pedestal.
5″ slab. . .	5″ slab over 4″ base minimum of 3/4″ minus crushed rock. Reinforce w/ 12/ 12-4/4 w.w.m. or grade 40, # 3 @ 15″ o.c. each way, 3″ clear of base.
L connectors . . .	3/4″ dia. ′L′ structural connection inserts w/ 3/4″ diameter × 25″ coil rods @ 5′–7″ o.c. Provide 2″ min. rod penetration into inserts 2 3/4″ from top of slab.

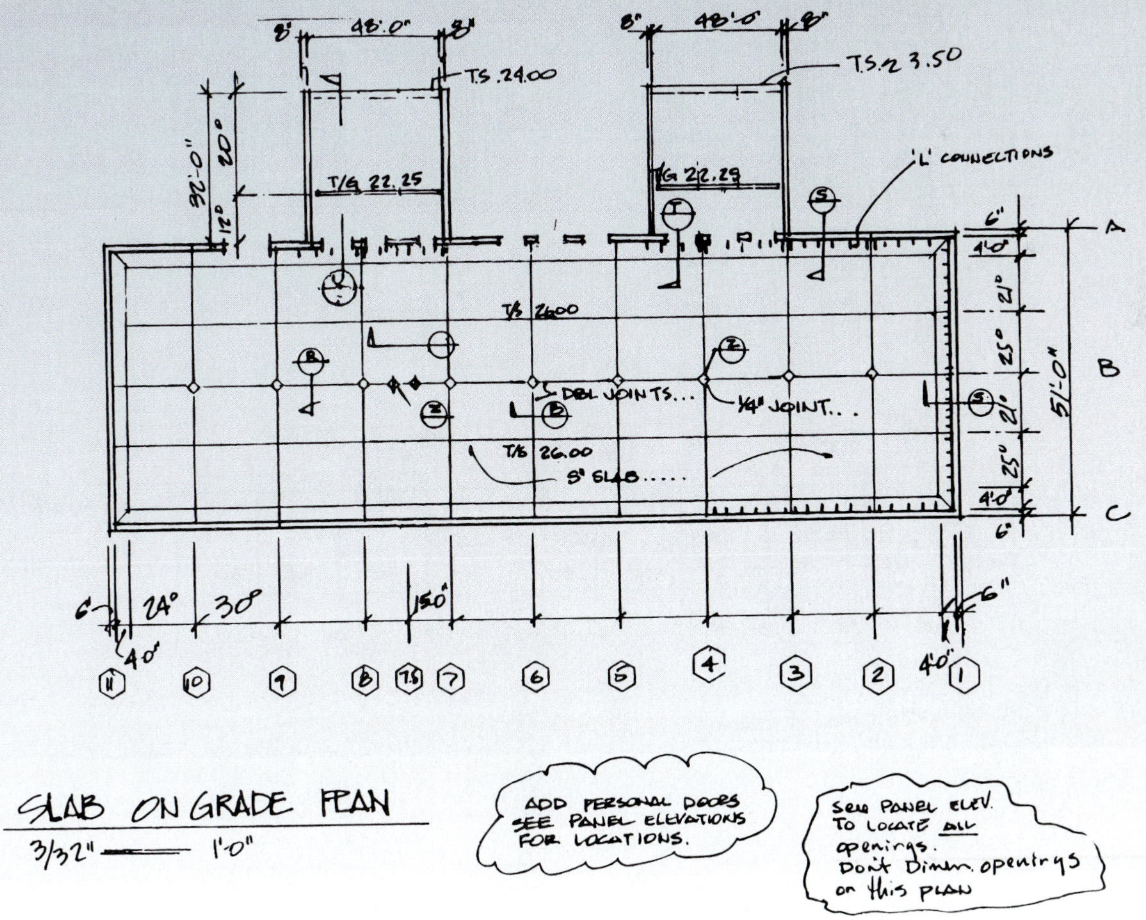

PROBLEM 47–3

Problem 47–4 Foundation Plan.

Goal The purpose of this drawing is to show concrete supports for the walls and columns. Supports will consist of a continuous footing at the perimeter of the structure and individual piers placed under the pedestals that support the columns.

Problem This drawing will show the size and location of the footings, as well as any change in their elevation. All elevations should be given using the proper symbol.

Method Use a scale of 3/32″ = 1′–0″ for this drawing. Place the drawing in the upper right-hand corner of the page. Use the sketch of the foundation as a guide.

1. Lay out grids A–C, and 1–11.
2. Lightly lay out the exterior face of the concrete walls and loading docks. This is for your reference only, since only the retaining walls will be darkened on this plan.
3. Determine the size of the exterior footing, and lay it out in the proper location. Refer to the foundation details for size information.
4. Draw the location of all elevation changes in the footing.
5. Determine the size of the pedestal footings. Since all footings are close in size, given the scale of the drawing, lay out all footings at the *average* size.
6. Draw all footings.
7. Dimension drawing as per sketch.
8. Label all grids, elevations, and notes.
9. General notes. (1) All foundation, pedestal, and retaining wall concrete shall be F′c 3000 PSI at 28 days. (2) All steel reinforcement shall be ASTM A615, Grade 40, deformed bars unless otherwise specified in drawings or details.
10. Local notes. None required.

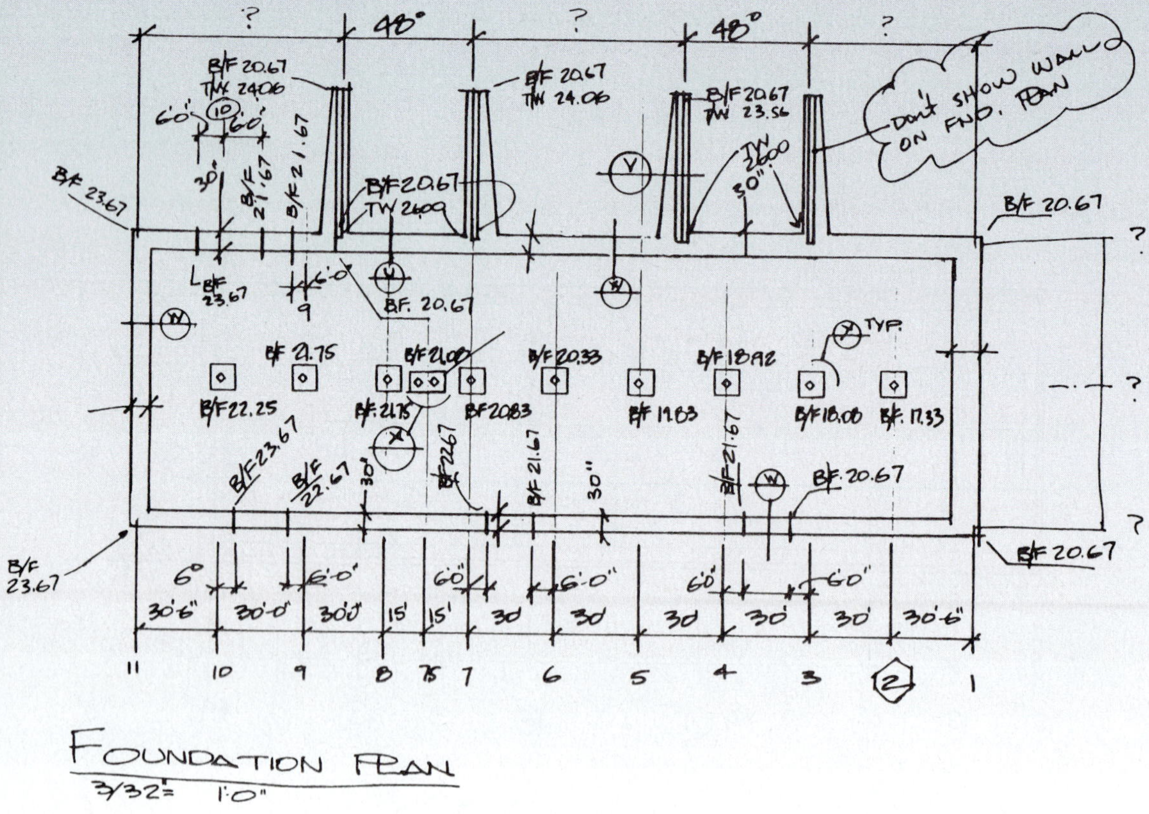

PROBLEM 47–4

Problem 47–5 Roof Beam Details.

Goal The purpose of these details is to show how the glu-lam beams are attached in end-to-end connections, to columns, and to the wall at grids 1 and 11. Use the sketches and calcs as a guide for your drawings.

- DETAIL A: BEAM TO BEAM. Use of a saddle to hang 25.5″ beam from the larger beams.
- DETAIL B: BEAM TO WALL. Beams supported on columns with 2-hour wall between. Gyp. bd. to remain unbroken.
- DETAIL C: WALL TO ROOF. Similar to "B" but shows roof between beam and concrete walls.
- DETAIL D: BEAM TO WALL. Top and side views to show plates for beam support.

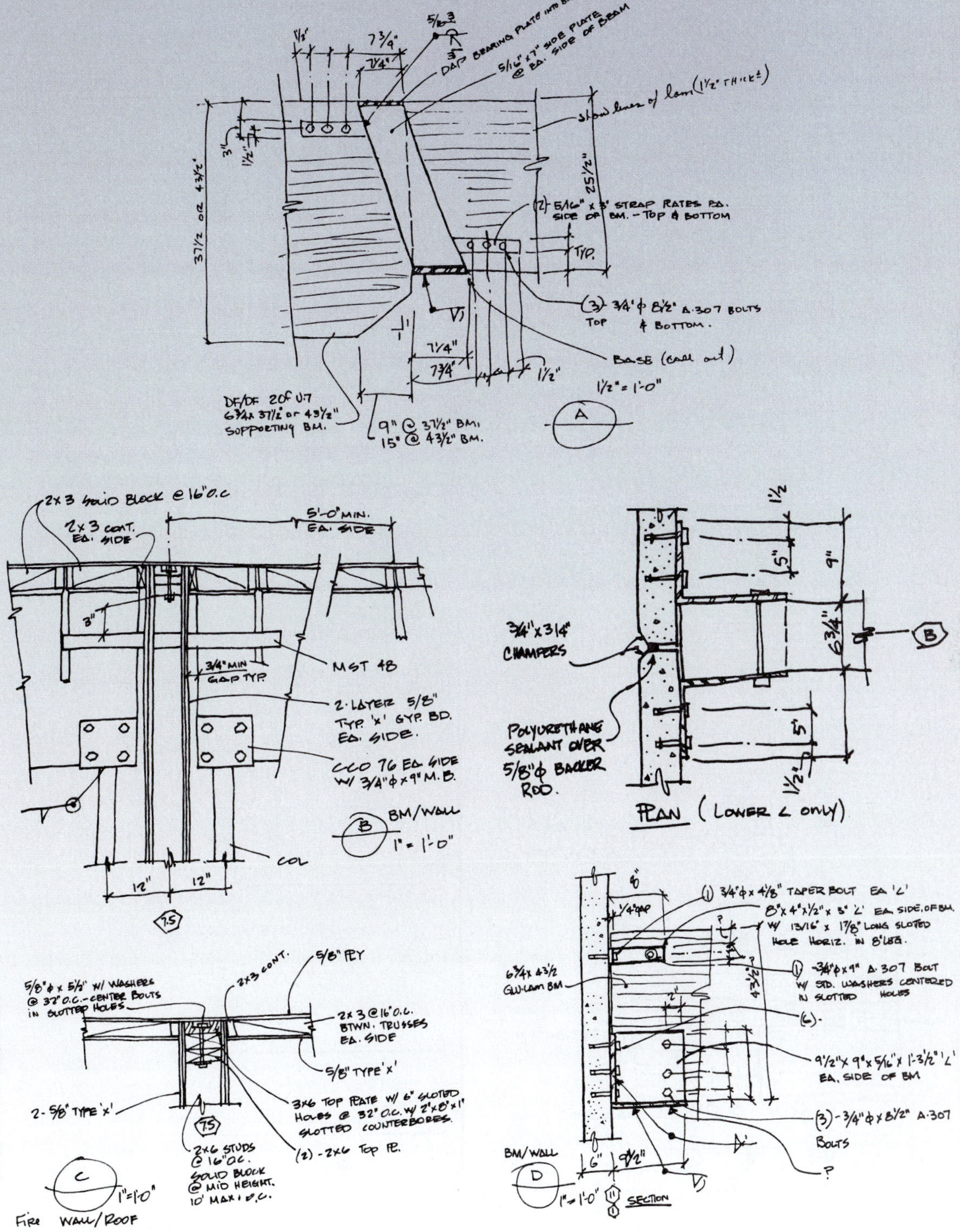

PROBLEM 47–5A

- DETAILS E AND F: BEAM TO COLUMN. Both show bearing-plate connections to steel column.
- DETAIL G: BEAM TO COLUMN. Intersection of two beams over a steel column. All other beam intersections are in midspan and are not over a column.

Method These drawings should be done in two stages. Layout and final drawing with lettering and dimensions. Use the calcs, the sketches, and material from vendor's brochures or Sweets catalogs.

1. Determine the location of details. Details should be drawn either with the roof drainage and framing plans or on a separate sheet. In addition, determine in what order the details will be presented. Place the details in an order that reflects their relationship to the building, not the order in which the engineer thought of them.
2. Lightly block out each detail, allowing approximately 15 minutes per detail. Given this brief amount of time, draw only major features, such as beam sizes, metal brackets, and bolt centerlines.
3. Draw beams with finish-line quality. Use several line types or colors. Draw steel plates and brackets as boldest lines and beam outlines slightly thinner, followed by bolts or other connectors, followed by glu-lam lines.
4. Dimension all information from the sketch or calcs. Interrupt lam lines as required for easy reading.
5. Place weld information on details as required. Be sure you understand what you are connecting before you order a weld. Also, be mindful of symbols for "this side" or "opposite side."
6. Notes. Place notes in or near the drawing as required. Similar information in two details can be covered in one note carefully placed between the two details.
7. Place detail markers, title, and scale below each detail. Do not place any information in the circle yet.

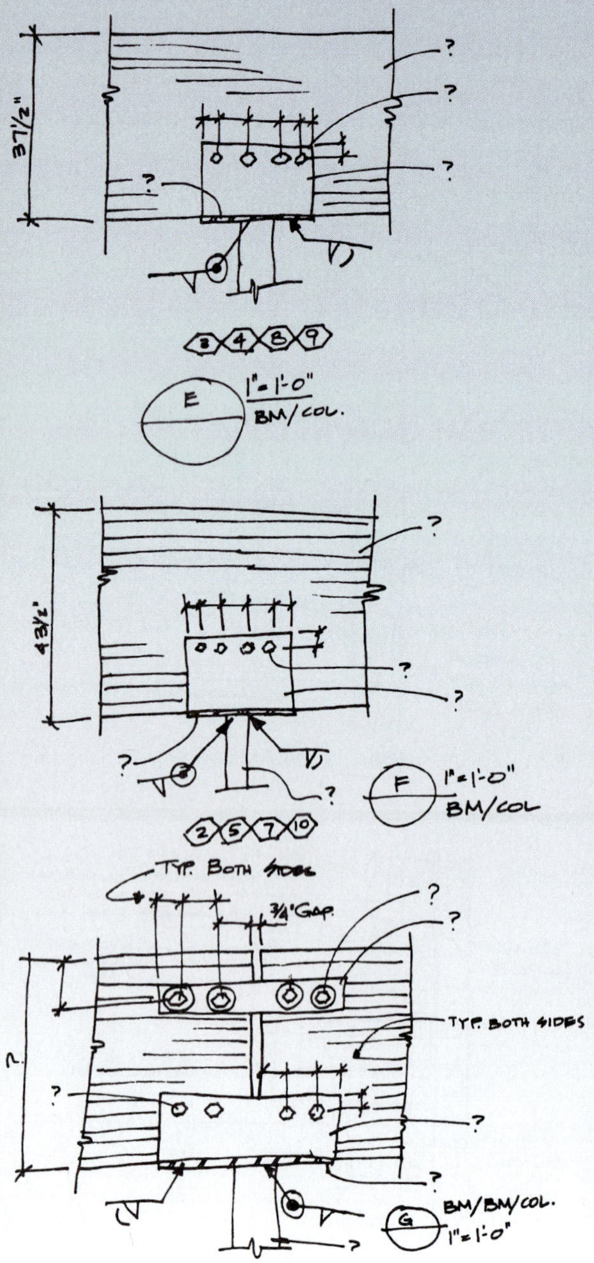

PROBLEM 47–5B

ROOF / BEAM-BEAM

ASSUMPTIONS:
ALL BOLTS TO BE 3/4" DIA. @ 3" O.C. UNLESS NOTED
PROVIDE 1 1/2" MIN. PLATE EDGE TO BOLT CENTER TYP. UNLESS NOTED
ALL BOLTS TO BE 3 1/2" FROM TOP AND BOTTOM OF BEAMS.
ALL WELDS AT BM CONNECTIONS TO BE 5/16" UNLESS NOTED.
PROVIDE 3/8" FILLET WELD @ COL / PL. UNLESS NOTED

HINGE CONNECTORS

USE: 5/16" SIDE PLs w/ 7/8" x 6 7/8" x 7 1/4" BEARING PL's @ TOP AND BOTTOM. USE 5/16" SIDE STRAPS W/ (3) 3/4" DIA. BOLTS.

BEAM / TILT UP WALL

USE 9 1/2 x 1'-3 1/2" x 5/16 FABRICATED SIDE ANGLES W/ 7/8" x 6 7/8" x 9 1/2" BEARING PL. USE (6) 3/4" DIA. x 4 1/8" TAPERBOLTS @ 5" O.C. EA. SIDE OF ANGLE & (4) 3/4" DIA. BOLTS @ 5" O.C. THRU

BEAM / FIREWALL

USE: 5/8" TYPE 'X' GYP. BD. MIN 5' EACH SIDE OF WALL. VERIFY TRUSS LOCATION AND DIMEN. DETAIL ACCORDINGLY.
PLACE MST 48 @ 3" MAX. FROM BOTTOM OF PL.

BEAM / COLUMN (3,4,8,9)

USE: 5/16 x 10 x 22 SIDE PL. W/ 7/8" x 6 7/8" x 22 BEARING PL
(4) 3/4" DIA. BOLTS @ 6 1/4" O.C.
USE A TS 6 x 6 x 1/4" COLUMN W/ 1/4" FILLET TO PL.

BEAM / COLUMN (2,5,7,10)

5/16" x 10 x 24 1/2" SIDE PL. W/ 7/8" x 6 7/8" x 24 1/2" BEARING PL. W/
(4) 3/4" DIA. BOLTS @ 7" O.C.
USE A TS 6 x 6 x 5/16 COLUMN W/ 1/4" FILLET TO PL.

BEAM / COLUMN (6)

5/16 x 10 x 19 SIDE PL. W/ 7/8" x 6 7/8" 19 1/2" BEARING PL.
(4) 3/4" DIA. BOLTS
TOP > TO BE 1/2" x 19 1/2" x 3 STRAP PL. W/ 13/16 x 1" SHORT SLOTTED HORIZ. HOLES (EA. SIDE OF BEAM) W/ (4) 3/4" DIA. W/ STD WASHERS CENTERED EA. SIDE. USE A TS 6 x 6 x 5/16 COLUMN W/ 1/4" FILLET TO PL.

Problem 47–6 Truss Details.

Goal The purpose of these details is to show how the trusses are connected in end-to-end connections, to beams, to ledgers, and to the wall at grids 1 and 11 at A and C.

- DETAIL H: Basic truss-to-beam connection. Information in this detail should be reflected in all roof details.
- DETAIL I: Remember that I, O, and Q are never used in the final detail callouts.
- DETAIL J: Truss-to-beam at a column. Braces are added from the truss to the glu-lam so that the beam will not roll off the column as a result of wind pressure against the walls at grid A and C.
- DETAIL K: Truss-to-truss connection at the inner edge of the roof diaphragm. This bolting will, in effect, make two trusses function as one. Make use of vendor specs for sizes.
- DETAIL L: Ledger splice at grid A and C.
- DETAIL M: Ledger splice at grid 1 and 11.
- DETAIL N: Truss-to-ledger connection at A and C, perpendicular.
- DETAIL P: Truss-to-ledger connection at 1 and 11, parallel.

Method These drawings should be done in two stages. Layout and final drawing with lettering and dimensions. Use the calcs and the vendor's material to supply needed information.

1. Determine the location of details. Details should be drawn with the beam details or on a separate sheet. Place the details in an order that reflects their relationship to the building.
2. Lay out each detail, allowing approximately 15 minutes per detail. Assume the web is 1″ diameter aluminum placed at a 45° angle starting 6″ from the end of the truss. The bottom chord of the truss *never* touches the beam. Also, carefully study the truss diagram provided by the vendor. The top chord, which supports the entire load on the truss, never touches the beam. The truss is supported by a metal connector.
3. Draw beams with finish-line quality. Use several line types or colors. Draw steel plates and brackets

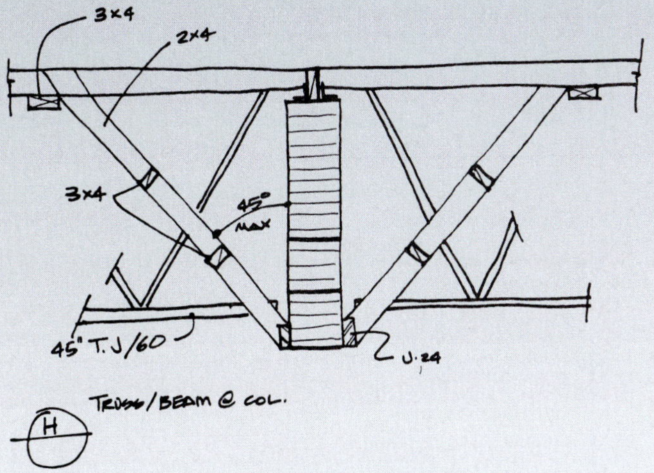

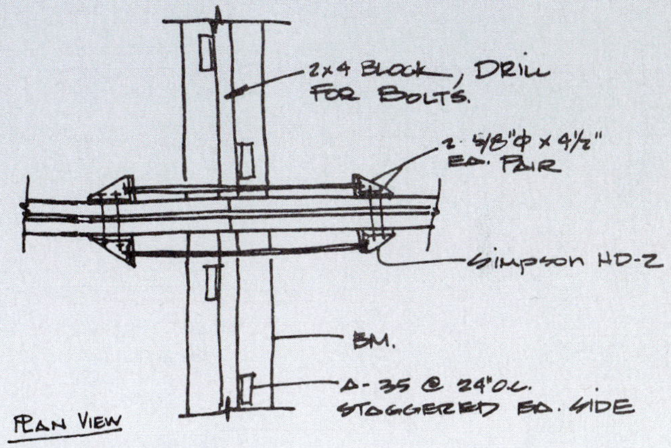

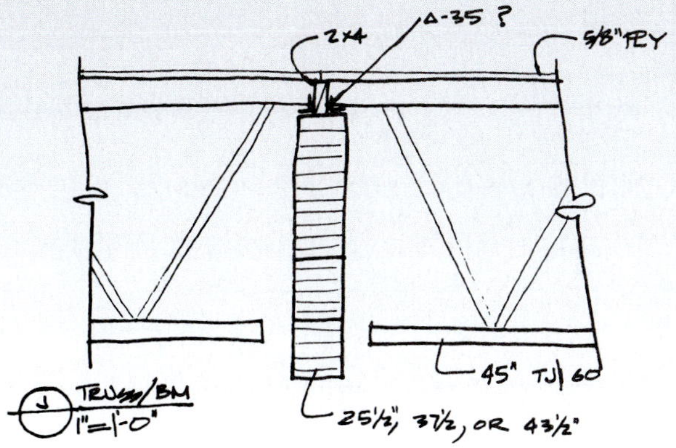

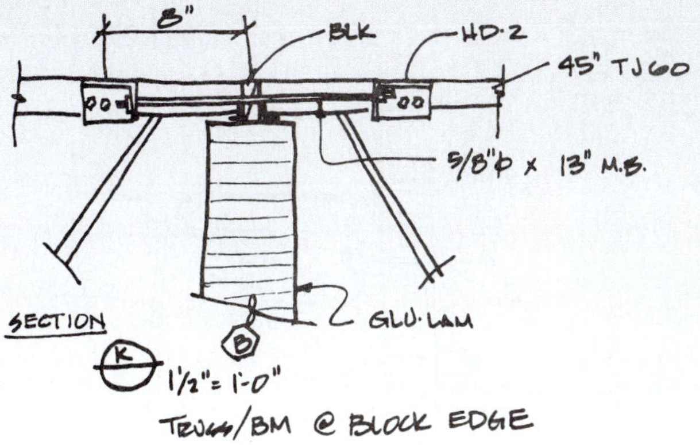

PROBLEM 47–6A

as boldest lines and beam outlines slightly thinner, followed by bolts or other connectors, followed by glu-lam lines.

4. Dimension all information from the sketch or calcs.
5. Place weld information on details as required.
6. Notes. Place notes in or near the drawing as required.
7. Place detail markers and title and scale below each detail. Do not yet place any information in the circle, however.

TRUSS / BEAM DETAILS

ROOF LEDGERS: USE 4x12 D.F.L. # 2

LEDGER SPLICES: USE (2) 1/2" x 3 STRAP PL. EXTENDING 4 BOLTS ON EACH SIDE OF SPLICE. USE 3/4" DIA. x 8" BOLTS @ 20" O.C. W/ 2 ROWS @ 3 1/2" O.C.

LEDGER SPLICE/ NON BEARING WALLS (1 & 11):
USE (3) SIMPSON MST48 STRAP TIES CENTERED OVER SPLICE @ 3 1/2" O.C.- 1 3/4" DOWN FROM LEDGER TOP. PROVIDE (4) 3/4" x 8" BOLTS THRU 1/4" x 3" WASHERS CENTERED BETWEEN PL's.

TRUSSES: USE 45" DEEP TJ/60 TRUSSJOIST @ 32" O.C.
ASSUME: 2.3" DEEP TOP AND BOTTOM CHORD.
1" DIA. ALUM. WEBS @ 45 DEGREES; 6" IN FROM END OF TOP CHORD. BTM. CHORD TO BE 2" MIN. CLEAR OF BM.

TRUSS/BM CONNECTION:
PROVIDE 2x4 SOLID BLOCKING BTWN. TRUSSES OVER BEAM. ATTACH TO GLU-LAM W/ SIMPSON CO. A-35 @ 32" O.C. @ EA. BLOCK ALTERNATE SIDES & PROVIDE (2) EA. BLK. NAIL W/ n8 ALL HOLES.

TRUSS-BM/COLUMN: PROVIDE 2x4 BRACE ALONG BEAM FOR (2) TRUSS SPACES EA. SIDE OF COLUMN. USE 3x4 SOLID BLOCK BTWN. BRACES. NAIL BLOCK TO BRACE W/ (5)-16d NAILS. PROVIDE 3x4X10' NAILER @ TRUSSES. NAIL W/ (2)-16d'S EA. TRUSS. USE SIMPSON U-24 HGR. 2" UP FOM BTM. OF BEAM TO BRACE. SET BRACE AT 45 DEGREES MAX. FROM BEAM.

TRUSS-BM @ BLOCKING EDGE: USE (2) SIMPSON HD-2 HOLD DOWNS EITHER SIDE OF SPLICE W/ (2)-5/8" x 4 1/2"Ø BOLTS THRU TRUSS, AND 5/8" x 15"Ø BOLTS ACROSS SPLICE.

TRUSS/WALL (A & C): USE 4x12 DFL #2 TREATED LEDGER W/ 3/4" x 8" BOLTS @ 48" O.C. THRU 1/4"x 3" WASHERS. SOLID BLOCK BTWN. TRUSSES @ LEDGER W/ 2x4 BLOCKS W/ (2) SIMPSON A-35 EA. BLK. PROVIDE 3/4" x 3" x 4" PL. ABOVE PLY. SHEATH. BOLT TO WALL W/ (1) 3/4" x 4 1/8"Ø BOLTS AT EA. 3rd TRUSS. WELD MST 27 STRAP TO PL. W/ 1/8" FILLET WELD AND NAIL TO TRUSSES W/ n8 EA. HOLE.

TRUSS/WALL (1, 11):
USE 4x12 DFL TREATED LEDGER W/ 3/4"x 8" BOLTS @ 6'-0" O.C. THRU 1/8" x 3" WASHERS. USE 3 x 4 PL. (SEE ABOVE) @ EA. 3rd TRUSS W/ MST STRAP. PROVIDE 3x4 SOLID BLOCK @ 48" O.C. FOR 37.3' OUT FROM WALL (SEE PLYWOOD DIAPHRAM SPECS.) CONNECT BLOCK TO TRUSS W/ SIMPSON Z-38 HGR.

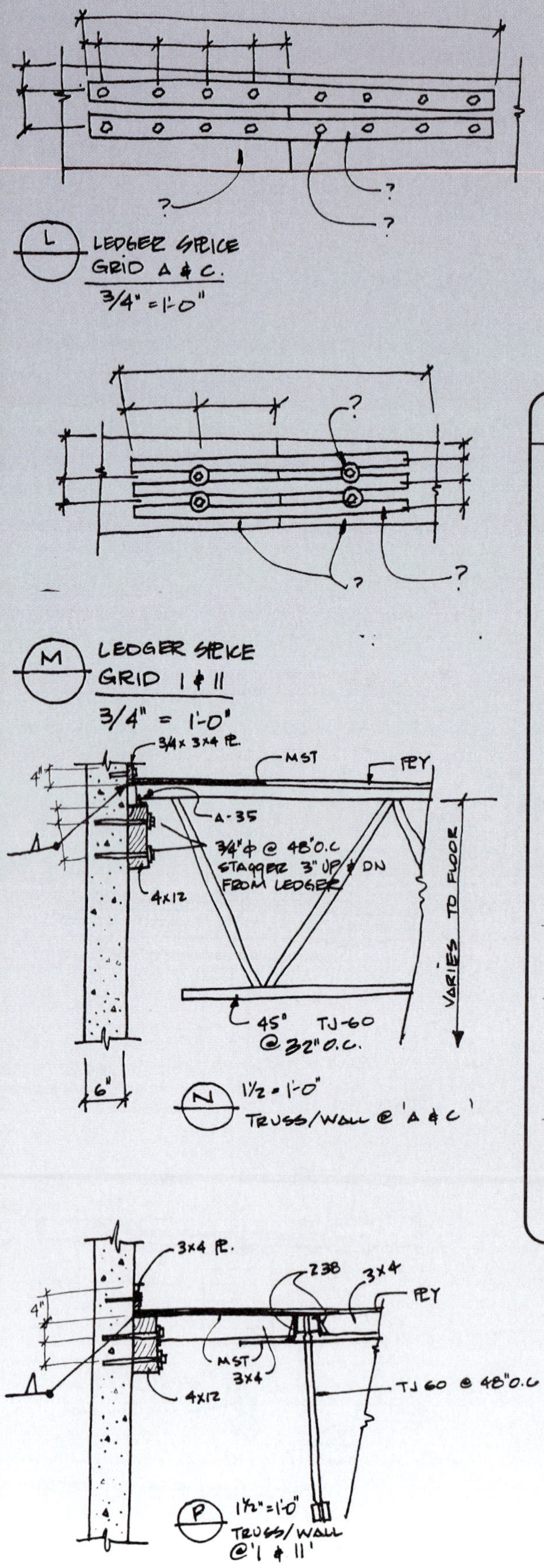

PROBLEM 47–6B

Problem 47–7 Slab Details.

Goal These drawings will provide information required to complete the floor slab, showing intersections in the concrete slab and door joints, and elevations.

Method Use a scale as indicated on each detail. Place details on the same sheet as the slab, on grade plan, or on a separate sheet with all other concrete details.

1. Determine in what order the details will be presented.
2. Lay out each detail, allowing approximately 15 minutes per detail.
3. Draw all lines. Be careful that all rebar is drawn with the same type of line quality. Draw mesh as a thin line with the "X" spaced at 4″ o.c.
4. Dimension as per sketches.
5. Place detail markers on the slab plan and section.
6. Label as required.

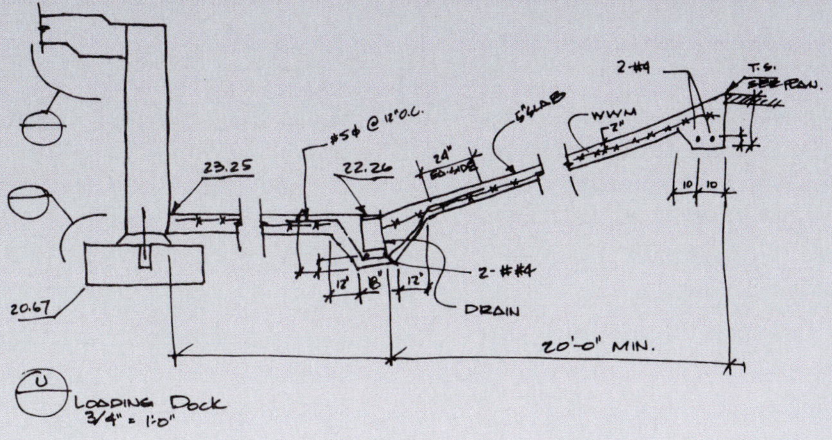

PROBLEM 47–7A

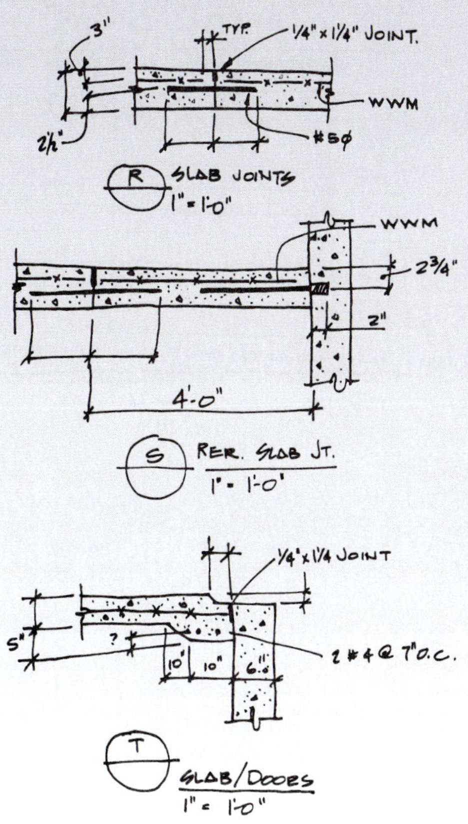

SLAB/ DOCK

SLAB:

ALL SLAB ON GRADE CONC. TO BE F'c = 3500 PSI @ 28 DAYS. ALL TARGET STRENGTHS SHALL BE IN ACCORDANCE W/ CHAPTER 4 OF ACI 318 BUILDING CODE REQUREMENTS FOR REINFORCED CONCRETE. SLABS TO BE 5" THICK UNLESS NOTED W/ WWF 12x12 -W4xW4 OR GRD 40 # 3 @ 15" O.C. EA. WAY CENTERED IN SLAB.

JOINTS:

PROVIDE 1/4" FIBER INSOLATION JOINTS W/ NEOPRENE JOINT SEALANT OVER 3/8" BACKER BEAD AT ALL CONTROL JOINTS AND AROUND ALL FOUNDATION PEDESTALS.
PROVIDE # 5 x 15" LONG SMOOTH DOWELLS @ 12"O.C. @ ALL CONTROL JOINTS UNLESS NOTED (SEE PER. JTS.) COVER W/ PAPER SLEEVES ON EA SIDE OF JOINT.
KEEP WWM 2" MIN. CLEAR OF JOINT. KEEP # 5's 2 1/2" CLR. OF BOTTOM OF SLAB.

PERIMETER JOINTS:

SAME AS ABOVE EXCEPT: ALL SMOOTH DOWELLS TO BE # 5x 30" LONG.

PROVIDE 3/4" DIA. 'L' STRUCTURAL CONNECTION INSERTS W/ 3/4" DIA. X 25" LONG COIL RODS @ 5'-7" O.C. PROVIDE 2" MIN. ROD PENETRATION INTO INSERTS.

DOORS:

PROVIDE 3" WIDE x 3/4" x 3/4" CHAMFER IN SLAB AT ALL DOORS. THICKEN SLAB AT ALL DOORS TO 10" AND PROVIDE (2) #4 BARS 2" CLR OF BTM. OF SLAB.

LOADING DOCK:

PROVIDE 12' WIDE BASE PLATFORM W/ 1/4" x 12" SLOPE TO DRAIN. DRAIN TO BE 8" x 12" DEEP POLYESTER CONC DRAIN W/ C.I. GRATE. THICKEN SLAB @ GRATE TO 18" DP. W/ 2 # 4 @ 15" O.C. 3" CLR OF BOT. OF FTG.

PROVIDE 20" MIN SLOPE IN DOCK. THICKEN SLAB TO 10" @ ENTRY W/ (2) #4 DIA. @ 6" O.C. 3" CLR. OF BTM OF FTG.

PROBLEM 47–7B

Problem 47–8 Foundation Details.

Goal These details will provide information about the concrete at the foundation level, showing either the exterior foundation or the pedestals.

Method Use a scale as indicated on each detail. Place details on the same sheet as the foundation plan or on a separate sheet with all other concrete details. Use the sketches and the calcs as a guide for your drawing.

1. Determine in what order the details will be presented.
2. Determine the missing dimensions from the calculations prior to layout work.
3. Lay out each detail. Be sure to allow room for schedules by each detail where they are required. Be careful that all rebar is drawn with the same type of line quality as used in the slab details.
4. Dimension as per sketches.
5. Place detail markers on the foundation plan and section.
6. Label as required.

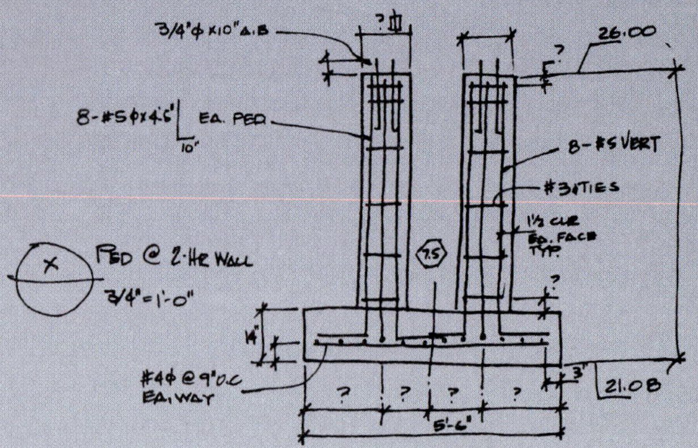

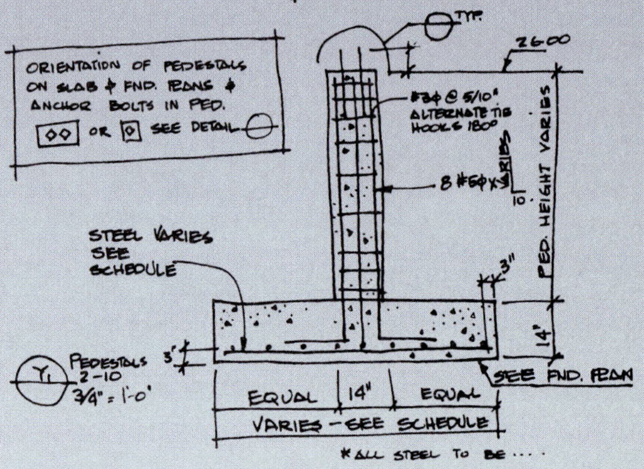

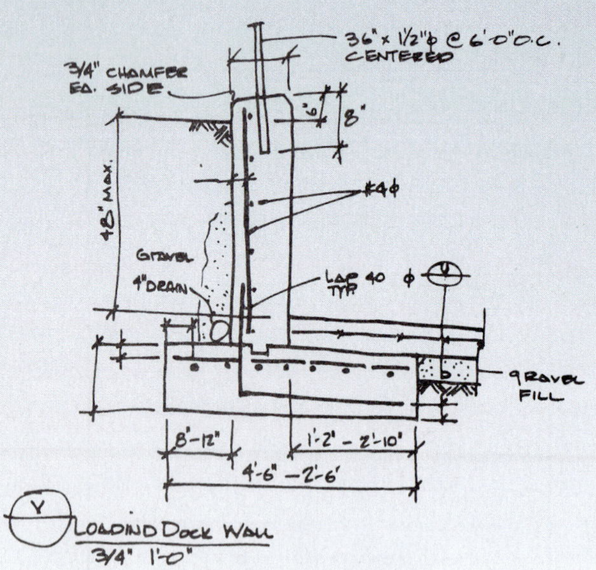

FOOTING	PEDESTAL HEIGHT	SIZE	REINFORCING STEEL	BTM. OF FTG.
B-2	7.50	7'-0" ϕ	# ϕ @ o.c.	17.33
B-3				
B-4				
B-5				
B-6				
B-7				
B-7.5				
B-8				
B-9				
B-10				
ALL STEEL TO BE GRADE ..... UNLESS NOTED				

Y2

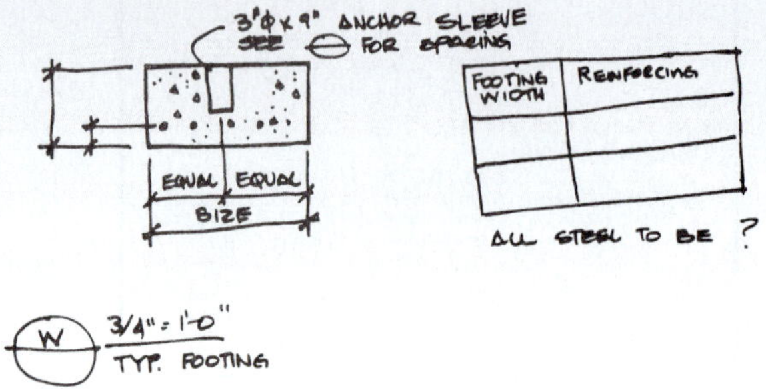

PROBLEM 47–8A

FOUNDATION

ASSUME:
ALL FOUNDATION, PEDESTAL, AND RETAINING WALL CONC. TO BE F'c =3000 PSI @ 28 DAYS.

ALL STEEL BAR REINFORCEMENT SHALL BE ASTM A615, GRADE 40, DEFORMED BARS UNLESS OTHERWISE SPECIFIED IN DRAWINGS OR DETAILS.

PEDESTALS: USE 14" W/ (8)- GRADE 40 #5 VERT. RE-BAR 1 1/2" CLR. OF PEDESTAL FACE (HEIGHT VARIES, SEE FOUNDATION PLAN AND COMPLETE SCHEDULE). USE # 3 HORIZ. TIES W/ TOP AND BTM. (3) TIES @ 5" O.C. & INTERMEDIATES @ 10" O.C. HORIZ. TIES TO BE WITHIN 1 1/2" OF TOP AND BOTTOM. PROVODE (4) 3/4" x 10" A.B. IN 6 1/2" GRID W/ 2 1/2" BOLT PROJECTION.

PEDESTALS: (@ 7.5) SIMILAR TO TYPICAL PEDESTALS (ABOVE) UNLESS NOTED. PROVIDE # 3 HORIZ TIES (2) @ 5"O.C. WITHIN TOP OF 1 1/2" OF PEDESTAL, AND BALANCE @ 10"O.C. W/ LAST TIE WITHIN 1 1/2" OF BTM OF PED. USE 5'-6" x 3'-6" x 18" DP. FTG.

PEDESTAL FOUNDATIONS: (@ B-6) USE A 6" x 14" DP. W/GRD. 40, # 5 @10"O.C. EA.WAY 3" CLEAR OF BASE

PEDESTAL FOUND. (@ 2,5,7,10) USE 7' x 14" DP. W/ GRD. 40, # 6 @ 10"O.C. EA.WAY 3" CLR. OF BASE.

PEDESTAL FOUND. (@ 3,4,8,9,) USE 6.5' x 14" DP. W/ GRADE 40, # 6 @ 12"O.C. EACH WAY 3" CLEAR OF BASE.

FOUND.- LOAD BEAR:
36" x 14" DP. W/ (3) GRD. 40,#5@ 9"O.C. 3" CLR. OF FTG. BTM.

FOUND.- NON-BEAR: (1& 11) USE 26" x 14" DP. FOOTING W/ (2) GRD. 40, #5 @ 12"O.C. 3" CLR. OF FTG. BTM.

RETAINING WALL: USE 8" THICK WALL W/GRD. 40, #4 @ 10" O.C. EA. WAY, 2" CLR. OF SOIL SIDE OF WALL. USE 14" THICK FTG x 4.5' WIDE W/ GRADE 40, #4 @ 10" O.C. EA. WAY, 2" CLR OF TOP OF FTG. PROVIDE #5 @ 10" O.C. EA. WAY 3" CLR. OF BTM OF FTG. W/ #5 @ @10" O.C. 'L' SHAPED BAR W/ 24" MIN. PROJECTION INTO FOOTING. USE #4 CONT. @ "L". FOOTING TO TAPER FROM 8"-12" PROJECTION FROM WALL ON SOIL SIDE AND BTWN. 1'-2"- 2'-10" ON DOCK SIDE. PROVIDE 4" DIA. DRAIN IN 2" x 24" MIN. 3/4" MINUS GRAVEL BED.

PROBLEM 47–8B

Problem 47–9 Typical cross section.

Goal The purpose of this drawing is to show the vertical relationship of structural members. It functions as a reference map and is not intended to provide a detailed explanation of the connections. Most of the details that will be drawn will be referenced on this drawing.

Method Using a scale of 3/16″ = 1′–0″, draw the section. Do not draw the entire section.

1. Place a section marker in the same location on the roof drainage, roof framing, slab, and foundation plans before starting the drawing.
2. Establish baselines to represent the top of the slab floor and the wall at C and the beams at B.
3. Determine the elevations of the footings at the wall and pedestal from the foundation plan. Your detail must match your reference marker, not your neighbor's or the engineer's sketch. The location of the reference marker will affect your wall size.
4. Establish vertical heights at B and C, based on panel elevation.
5. Lay out the ledger, trusses, and beams.
6. Draw all items. Be careful to distinguish between concrete and wood members.
7. Dimension as per the sketch.
8. Label grids and local notes.

ON PROBLEM	NOTE SHOULD READ
5/8″ ply . . .	See note on roof calcs.
T.J.I. . . .	45″ TJ/60 trus joists at 24″ o.c.
6″ walls . . .	6″ tilt up conc. walls/
Roofing . . .	4 ply built-up asphaltic fiberglass class 'A' roofing.
Beam . . .	6 3/4″ × 25.5, 37.5 or 43.5 glu-lam beam

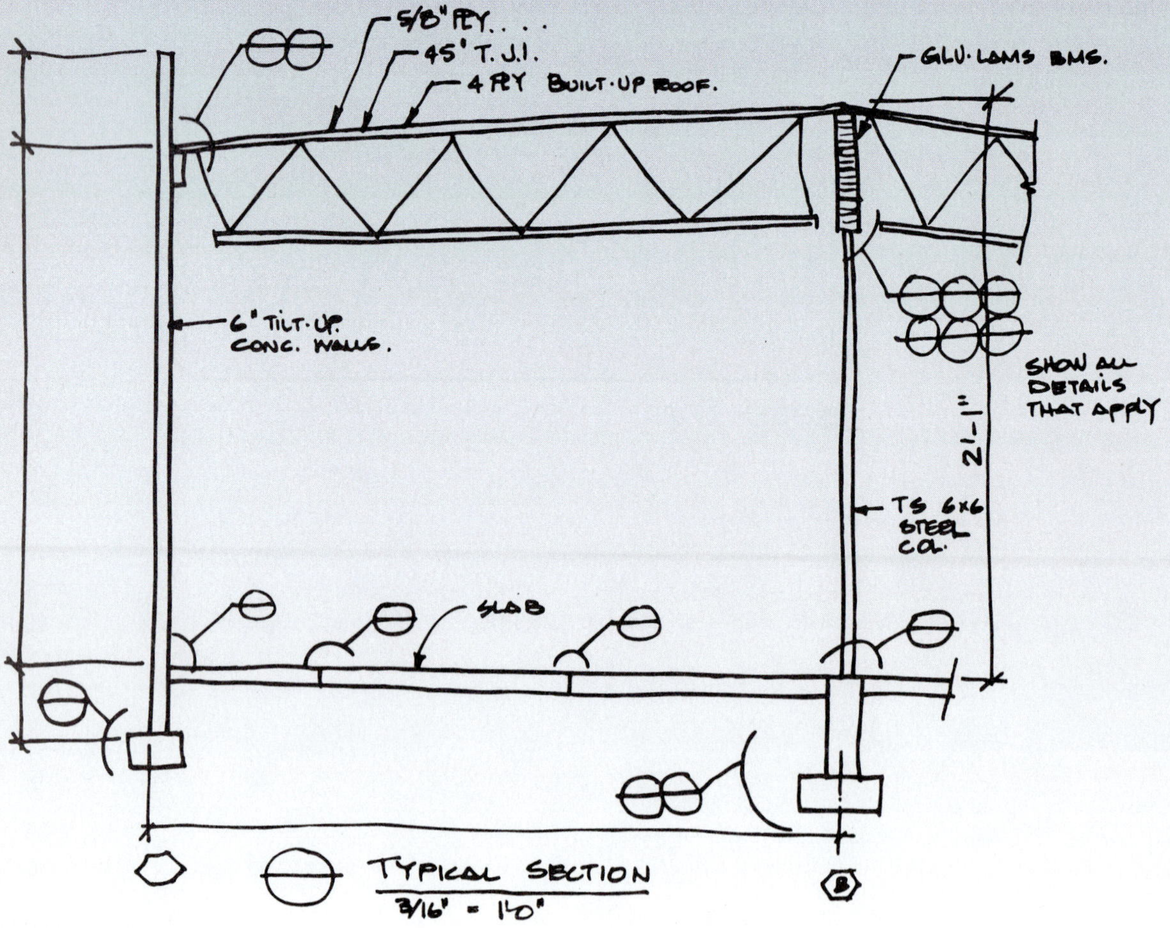

PROBLEM 47–9

Problem 47–10 Exterior Elevations.

Goal These drawings will provide an external view of the building to be constructed.

Method Typically, an elevation of each side of a structure would be drawn. Verify with your instructor which elevations will be drawn for your project. Use a scale of 3/32″ = 1′–0″ and follow the procedures outlined in Chapters 22 and 23. For best use of a page, place the elevations in the upper portion of a sheet to allow room for other drawings. If you are using CADD, compare the elevations and the panel elevations so both drawings can be done in layers.

Layout

1. Draw a reference line to represent the slab. This line will never be on the finished drawing but will help locate the bottom of all doors.
2. Establish each end of the structure and each grid.
3. Establish the top of the walls using information on the typical cross sections and details.
4. Establish finish grade elevations.
5. Lay out all door locations using the locations shown on the panel elevation sketches.
6. Locate loading docks according to the foundation plan.
7. Determine the locations of three strips, and sign as per sketch.
8. Draw the grade and loading dock using bold lines.
9. Draw the outline of the structure.
10. Draw grid lines as two thin parallel lines.
11. Draw doors.
12. Draw stripes and sign.
13. Draw concrete symbols on some of the walls.
14. Label as per sketch.

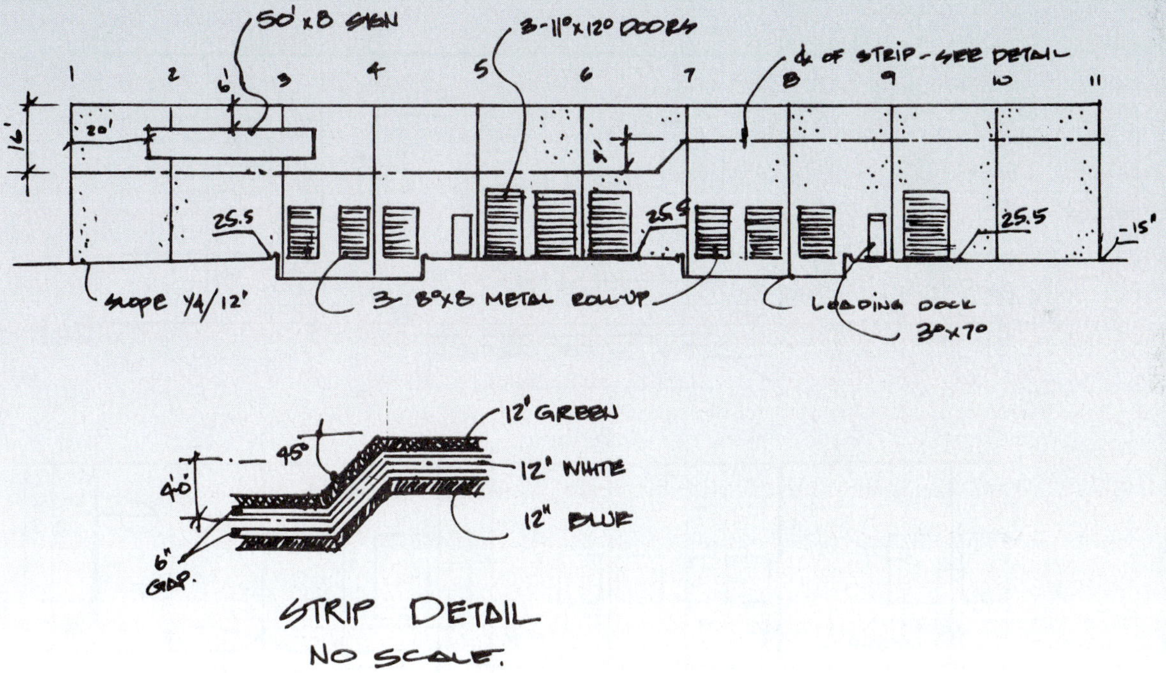

PROBLEM 47–10

Problem 47–11 Panel Elevations.

Goal These drawings will provide an internal view of the walls to be constructed, showing wall shape and openings.

Method Typically, an elevation of each panel of a structure would be drawn. Verify with your instructor if more than the north face will be drawn for your project. Use a scale of 3/32″ = 1′–0″. Because of the length of the structure, the sketch of the north panel elevations is shown on two separate drawings. Your drawings should reflect the true layout of the structure and include all information specified in the calcs. Place this drawing on the same sheet as the elevations. Use a print of the elevations to speed the layout, or set this up as a layer to be used on the exterior elevations if using CADD.

Layout

1. Draw a reference line to represent the slab. This line will never be on the finished drawing but will help locate the bottom of each panel.
2. Establish each end of the structure and each grid.
3. Establish the top of the walls using information on the sections and elevations.
4. Determine the foundation elevations and lay out the top of the footing.
5. Lay out all door locations.
6. Draw the outline of the panels.
7. Draw the grid lines.
8. Draw door openings.
9. Provide dimensions to describe all wall shapes and all door openings. Keep size location separate from location dimension.
10. Print all notes.

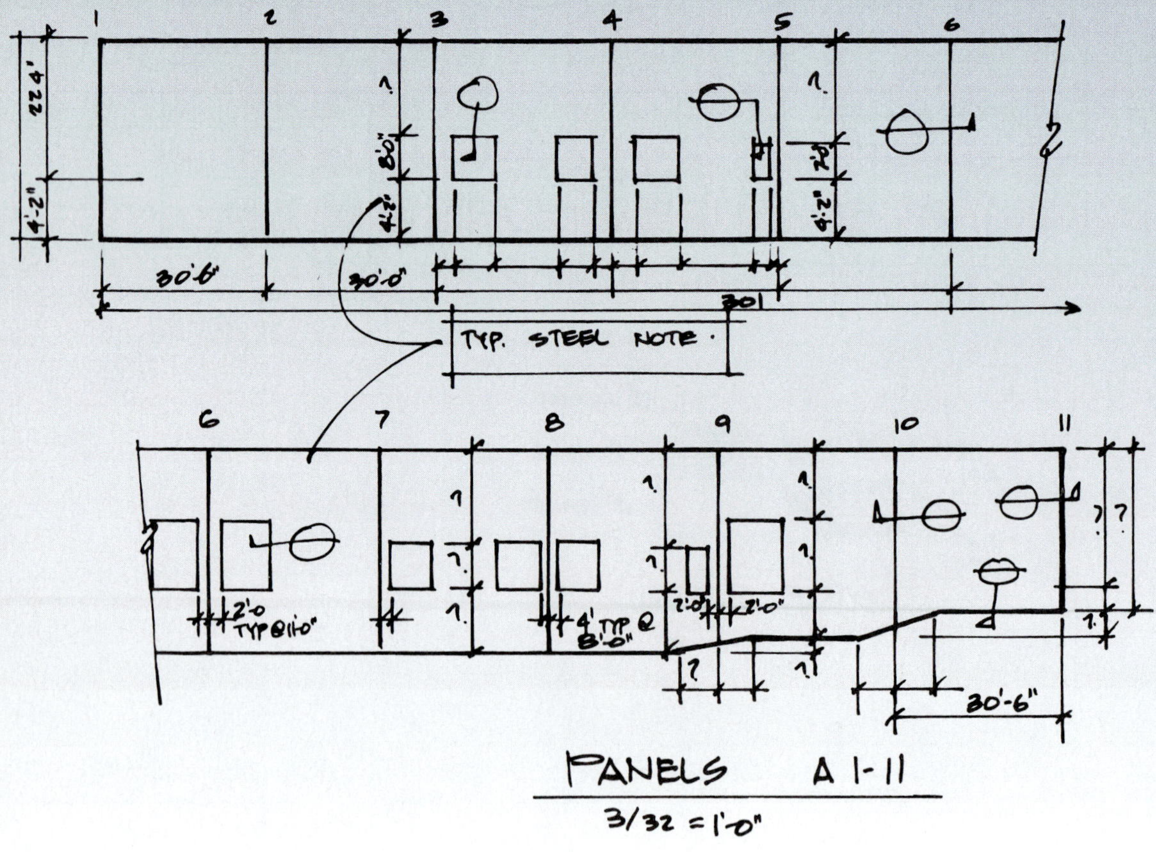

PROBLEM 47–11

Problem 47–12 Panel Details.

Goal The purpose of these drawings is to show steel placement around any openings in the tilt-up walls. Steel could be shown on the panel elevations, but more clarity is gained by providing details.

Method A detail will be provided for each size opening, to show typical steel placement. Be aware that all steel specified by the doors is in addition to any steel specified in the general note for the walls. Typical wall steel is usually only specified in a note and not shown. This allows special steel patterns to be shown and easily specified. Use a scale of 1/4″ = 1′–0″ minimum. Place these details on the same sheet as the panel elevations. Reference these details to the section, and slab on grade plan.

1. Draw the door opening allowing approximately 7′–0″ on all four sides.
2. Lay out steel placements based on dimensions specified on the panel elevation sketches and calcs.
3. Draw the steel much bolder than the opening outline.
4. Dimension each opening as per the sketch.
5. Label all steel.

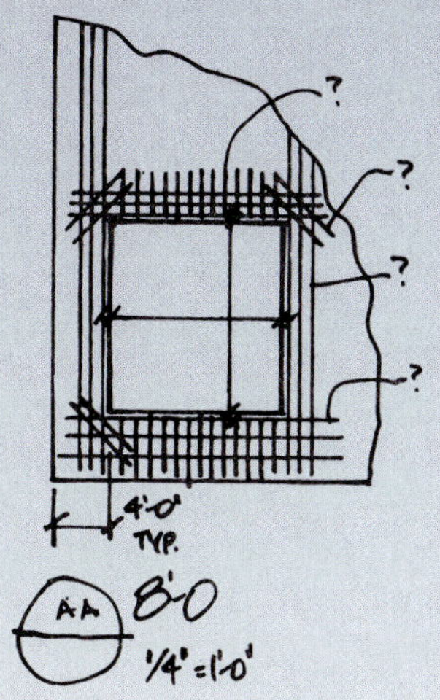

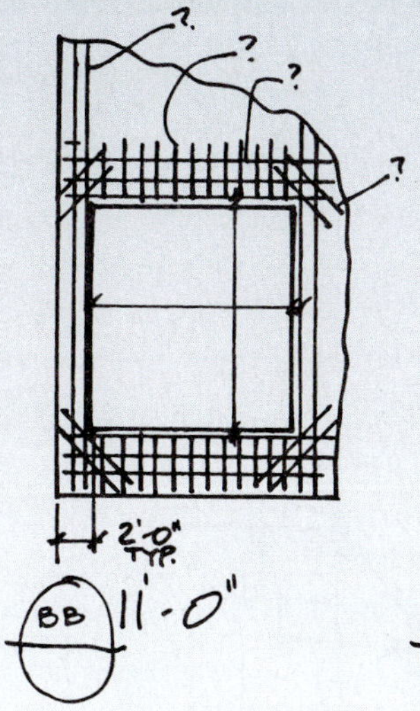

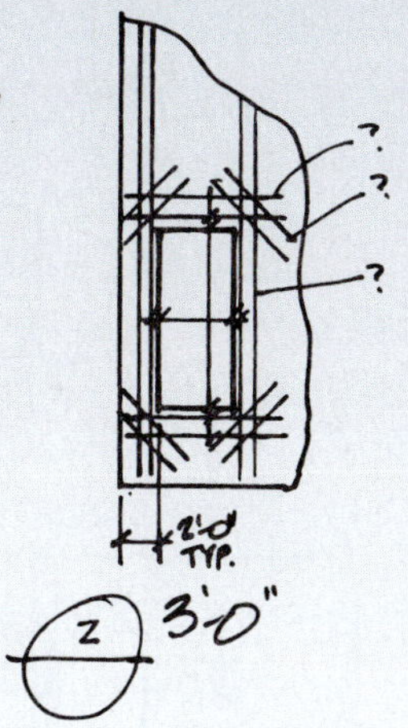

WALL PANELS

PANELS: ALL WALLS TO BE 6" THICK, 4000 PSI CONCRETE W/ GRADE 60-# 5 @ 10"O.C. VERT. & GRADE 40-# 4 @ 12" O.C. HORIZONTALLY CENTERED IN WALL. EXTEND ALL WALL STEEL TO WITHIN 2" OF TOP AND BOTTOM.

ALL PANELS TO HAVE EXPOSED AGGREGATE FINISHED UNLESS NOTED

PROVIDE (2) GRADE 40, # 5 x 4'-0" 45 DEGREE DIAG. STEEL TYP. TOP AND BOTTOM OF ALL DOORS. TIE FIRST DIAG. TO INTERSECTION OF HORIZ. AND VERT. STEEL. W/ SECOND BAR AT 8" O.C.

PANEL/PANEL: USE (2) 3/4" x 2 1/2" COIL INSERTS THRU 1" HOLES IN 11"h x 10"w x 3/8" INSERTS TO BE 1 1/2" MIN. FROM PL. EDGES, 8"O.C. AND 3" MIN. FROM EDGE OF CONC. PANEL.

PROVIDE 3 x 13 x 3/8" PL. INSET FLUSH INTO CONC. 3" MIN. FROM PANEL EDGE W/ (2) 3/4" x " HEADED CONC. ANCHORS @ 10" O.C. 1 1/2" MIN. FROM PL. EDGES. WELD TO PL. W/ 1/8" FILLET ALL AROUND @ FIELD. PROVIDE 1 1/2" MIN. LAP OF 11 x 10 PL. OVER 3x13 PL. & WELD W/ 1/4" FILLET WELD.

ALL CONNECTOR PL. TO BE @ 5'-0" O.C. MAX. 12" MIN. FROM TOP AND BTM OF PANELS.

PANEL CORNERS: SIMILAR TO PANEL/PANEL JOINT UNLESS NOTED. USE 3/8" x 5" x 3" x 11" PL.

ALL CORNER PL"s TO BE @ 6'-0" O.C. MAX. 12" MIN. FROM TOP AND BTM. OF PANEL EDGES.

PANEL/ FND: 7/8" DIA. x 14" STRUCTURAL CONNECTOR BY RICHMOND OR EQUAL, 8" MIN. INTO FND. WITH 5" MIN WALL PENETRATION. SET IN NON-SHRINK GROUT IN 3 x 9 ANCHOR SLEEVE IN FOUNDATION. USE:

6	PER	PANEL	@	GRID	11
9	"	"	@	"	1
8	"	"	@	"	C
8	"	"	@	"	1-4
7	"	"	@	"	4-7
6	"	"	@	"	7-11

11'-0" DOOR: IN ADDITION TO WALL STEEL PROVIDE:
(2) GRADE 60, # 6 @ 8"O.C. 1" CLEAR OF EACH FACE VERT. FOR FULL HEIGHT OF WALL.
(3) GRADE 60, # 4 @ 12" O.C. HORIZ. ABOVE AND BELOW DOORS, EXTEND 24" MIN EA. SIDE OF DOOR.

ABOVE AND BELOW DOOR PROVIDE GRD. 40, # 4 @ 12"O.C VERT. EXTEND VERT. 36" ABOVE AND BELOW DOOR..

**** ALL STEEL PATTERNS TO START WITHIN 1 1/2" OF WALL OPENINGS.

8'-0" DOOR: IN ADDITION TO WALL STEEL PROVIDE:
(3) GRADE 40, # 5 @ 8" O.C. @ 1" CLR. OF EACH FACE VERT. FOR FULL HEIGHT OF WALL.

(3) GRADE 60, # 4 @ 12" O.C. HORIZ. ABOVE AND BELOW DOORS, EXTEND 24" MIN EA. SIDE OF DOOR.
ABOVE AND BELOW DOOR PROVIDE GRD. 40, # 4 @ 12" O.C. VERT. EXTEND VERT. 36" ABOVE AND BELOW DOOR..

**** ALL STEEL PATTERNS TO START WITHIN 1" OF WALL OPENINGS.

3'-0" DOOR: IN ADDITION TO WALL STEEL PROVIDE:

(2) GRADE 60, # 6 @ 6" O.C. 1" CLR. OF EA. FACE VERT. FOR FULL HEIGHT OF WALL.

(2) GRD 40, # 5 x 7'-0" TOP AND BTM. @ 8" O.C.

PROBLEM 47–12

Problem 47–13 Wall Details.

Goal The purpose of the wall details is to aid in the erection and connection of the concrete wall panels and to show connections between the steel columns and pedestals.

Method Use the scale as indicated on each sketch. Place the details with the panel elevations, with a section, or with other concrete details.

1. Determine in what order the details will be presented. Place the details in an order that reflects their relationship to the building.
2. Lightly block out each detail. Draw only major items such as concrete, steel connector, and column outlines.
3. Draw details with finish-line quality. Use several line types or colors to distinguish clearly concrete, steel plates, and bolts.
4. Dimension all information based on sketches and calcs. Do not share dimension locations of bolts.
5. Place all weld symbols as required.
6. Place notes as required. Notes can be shared between details.
7. Place detail markers, title, and scale below each detail. Do not place information in the circle.

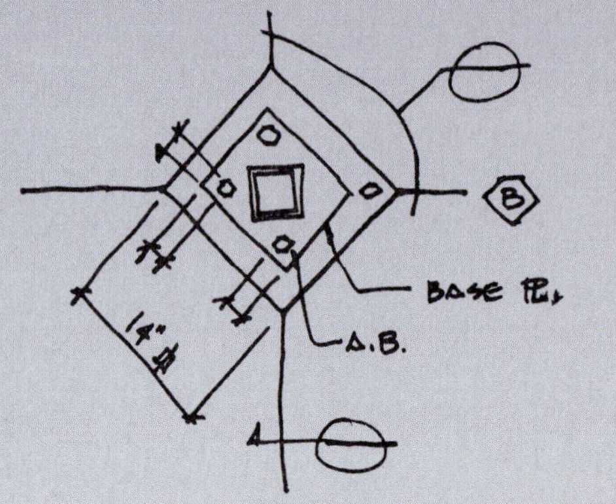

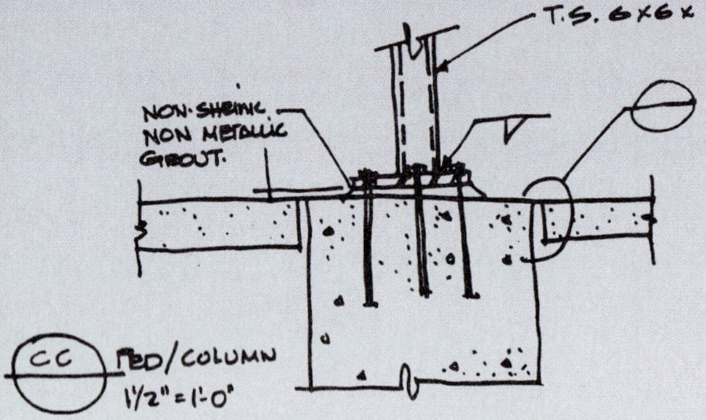

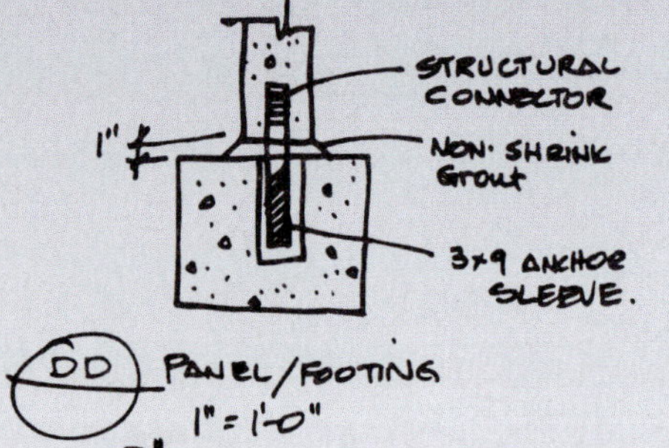

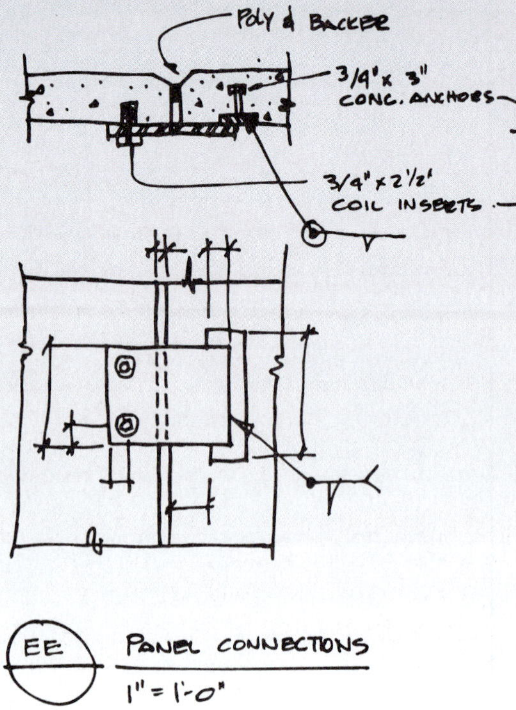

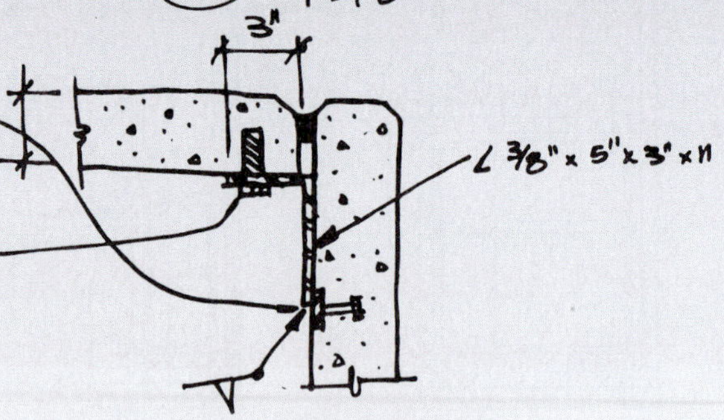

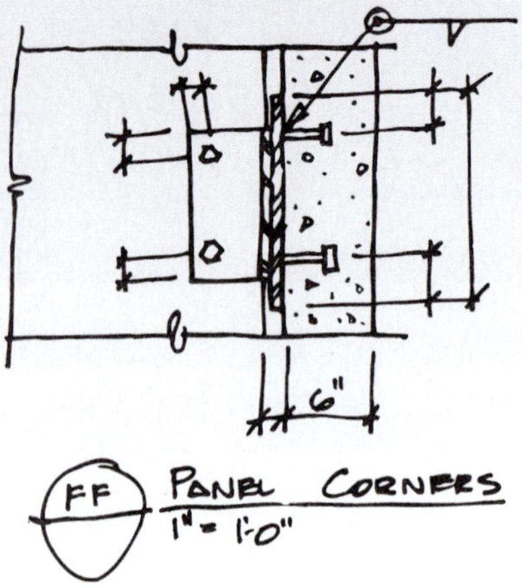

PROBLEM 47–13

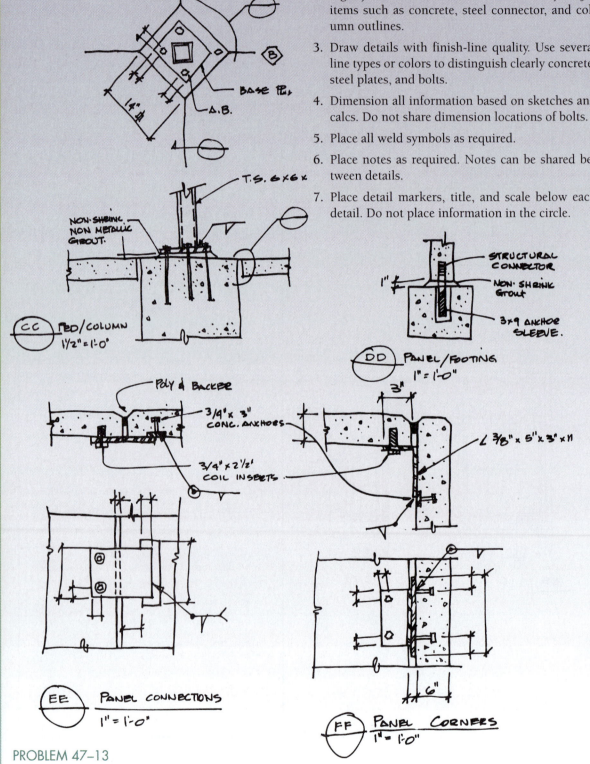

Problem 47–14 Project Coordination.

Goal Up to this point, you have drawn plan views, elevations, sections, and details. These drawings must now be placed in a logical presentation order and be referenced to each other.

Method

1. Place all drawings in order. Suggested order: roof plan, roof details, slab plan, slab details, etc., through to the foundation level, sections, and elevations. OR: place all plan views from drainage to foundation in descending order, sections, elevations, panels, and details.
2. Once the page order has been determined, place a page number in the title block. Structural drawings will always reflect the total number of pages in the title block as well as the actual page being viewed. In the title block, the numbers 2 of 7 would be placed on the second of seven pages.
3. Assign a page number to all details. All details drawn on page 1 would have a 1 placed in the lower half of the detail reference circle. All details drawn on page 2 would have a 2 placed in the lower half of the reference circle. This pattern should be repeated until all details have a page number.
4. Assign a reference letter to all details. Two methods may be used. Step A is the same for each method.
   A. Starting on page 1, label each detail A, B, C, etc., until all details on page 1 have been assigned a letter. Start in one corner and work across or up and down the page. Do not skip all over the page as you assign reference letters.
   B. On page 2, assign each detail a letter starting with the letter A, and work through the alphabet. Use the same pattern that was used on page 1. Repeat this process for all pages with details. OR
   B. On page 2, instead of starting over at A, use the letter that would follow the last letter used on the preceding detail. Never use the letters I, O, or Q. After detail Z, use AA, BB, etc.
   C. After all details have been referenced, make sure that each detail referenced on the plan views and sections, etc., has the correct reference symbol. Some details will be referenced on more than one plan. Most will be on the section.

A APPENDIX

Metric in Construction

The following tables are courtesy *Metric in Construction Newsletter,* published by the Construction Metrication Council of National Institute of Building Sciences, Washington, D.C., fax: 202-289-1092. E-mail: bbrenner@nibs.org Internet: www.nibs.org

▼ UNIT CONVERSION TABLES

SI SYMBOLS AND PREFIXES

Quantity	Unit	Symbol
BASE UNITS		
Length	Meter	m
Mass	Kilogram	kg
Time	Second	s
Electric current	Ampere	A
Thermodynamic temperature	Kelvin	K
Amount of substance	Mole	mol
Luminous intensity	Candela	cd
SI SUPPLEMENTARY UNITS		
Plane angle	Radian	rad
Solid angle	Steradian	sr

SI PREFIXES

Multiplication Factor	Prefix	Symbol
1 000 000 000 000 000 000 = 10^{18}	exa	E
1 000 000 000 000 000 = 10^{15}	peta	P
1 000 000 000 000 = 10^{12}	tera	T
1 000 000 000 = 10^{9}	giga	G
1 000 000 = 10^{6}	mega	M
1 000 = 10^{3}	kilo	k
100 = 10^{2}	hecto	h
10 = 10^{1}	deka	da
0.1 = 10^{-1}	deci	d
0.01 = 10^{-2}	centi	c
0.001 = 10^{-3}	milli	m
0.000 001 = 10^{-6}	micro	μ
0.000 000 001 = 10^{-9}	nano	n
0.000 000 000 000 = 10^{-12}	pico	p
0.000 000 000 000 000 = 10^{-15}	femto	f
0.000 000 000 000 000 000 = 10^{-18}	atto	a

CONVERSION FACTORS

To convert	to	multiply by
	LENGTH	
1 mile (U.S. statute)	km	1.609 344
1 yd	m	0.9144
1 ft	m	0.3048
	mm	304.8
1 in	mm	25.4
	AREA	
1 mile^2 (U.S. statute)	km^2	2.589 998
1 acre (U.S. survey)	ha	0.404 6873
	m^2	4,046,873
1 yd^2	m^2	0.836 1274
1 ft^2	m^2	0.092 903 04
1 in^2	mm^2	645.16
	VOLUME, MODULUS OF SECTION	
1 acre ft	m^3	1,233.489
1 yd^3	m^3	0.764 5549
100 board ft	m^3	0.235 9737
1 ft^3	m^3	0.028 316 85
	L (dm^3)	28.3168
1 in^3	mm^3	16 387.06
	mL (cm^3)	16.3871
1 barrel (42 U.S. gallons)	m^3	0.158 9873
	(FLUID) CAPACITY	
1 gal (U.S. liquid)*	L**	3.785 412
1 qt (U.S. liquid)	mL	946.3529
1 pt (U.S. liquid)	mL	473.1765
1 fl oz (U.S.)	mL	29.5735
1 gal (U.S. liquid)	m^3	0.003 785 412
*1 gallon (UK) approx. 1.2 gal (U.S.)		
**1 liter approx. 0.001 cubic meters		
	SECOND MOMENT OF AREA	
1 in^4	mm^4	4,162,314
	m^4	4,162,314 × 10^{-7}
	PLANE ANGLE	
1° (degree)	rad	0.017 453 29
	mrad	17.453 29
1′ (minute)	urad	290.8882
1″ (second)	urad	4.848 137

CONVERSION FACTORS—(Continued)

To convert	to	multiply by
	VOLUME RATE OF FLOW	
1 ft^3/s	m^3/s	0.028 316 85
1 ft^3/min	L/s	0.471 9474
1 gal/min	L/s	0.063 0902
1 gal/min	m^3/min	0.0038
1 gal/h	mL/s	1.051 50
1 million gal/d	L/s	43.8126
1 acre ft/s	m^3/s	1233.49
	TEMPERATURE INTERVAL	
1° F	°C or K	0.555 556 5⁄9°C = 5⁄9K
	MASS	
1 ton (short)	metric ton	0.907 185
	kg	907.1847
1 lb	kg	0.453 5924
1 oz	g	28.349 52
1 long ton (2,240 lb)	kg	1,016.047
	MASS PER UNIT AREA	
1 lb/ft^2	kg/m^2	4.882 428
1 oz/yd^2	g/m^2	33.905 75
1 oz/ft^2	g/m^2	305.1517
	DENSITY (MASS PER UNIT VOLUME)	
1 lb/ft^3	kg/m^3	16.01846
1 lb/yd^3	kg/m^3	0.593 2764
1 ton/yd^3	t/m^3	1.186 553
	FORCE	
1 tonf (ton-force)	kN	8.896 44
1 kip (1,000 lbf)	kN	4.448 22
1 lbf (pound-force)	N	4.448 22
	MOMENT OF FORCE, TORQUE	
1 lbf · ft	N · m	1.355 818
1 lbf · in	N · m	0.112 9848
1 tonf · ft	kN · m	2.711 64
1 kip · ft	kN · m	1.355 82
	FORCE PER UNIT LENGTH	
1 lbf/ft	N/m	14.5939
1 lbf/in	N/m	175.1268
1 tonf/ft	kN/m	29.1878
	PRESSURE, STRESS, MODULUS OF ELASTICITY (FORCE PER UNIT AREA) (1 Pa = 1 N/m^2)	
1 $tonf/in^2$	MPa	13.7895
1 $tonf/ft^2$	kPa	95.7605
1 kip/in^2	MPa	6.894 757
1 lbf/in^2	kPa	6.894 757
1 lbf/ft^2	Pa	47.8803
Atmosphere	kPa	101.3250
1 inch mercury	kPa	3.376 85
1 foot (water column at 32°F.)	kPa	2.988 98

CONVERSION FACTORS—(Continued)

To convert	to	multiply by
	WORK, ENERGY, HEAT (1J = 1N · m = 1W · s)	
1 kWh (550ft · lbf/s)	MJ	3.6
1 Btu (Int. Table)	kJ	1.055 056
	J	1,055.056
1 ft · lbf	J	1.355 818
	COEFFICIENT OF HEAT TRANSFER	
1 Btu/(ft^2 · h · °F)	W/(m^2 · K)	5.678 263
	THERMAL CONDUCTIVITY	
1 Btu/(ft · h · °F)	W/(m · K)	1.730 735
	ILLUMINANCE	
1 lm/ft^2 (footcandle)	lx (lux)	10.763 91
	LUMINANCE	
1 cd/ft^2	cd/m^2	10.7639
1 foot lambert	cd/m^2	3.426 259
1 lambert	kcd/m^2	3.183 099

CONVERSION AND ROUNDING

The conversion of inch-pound units to metric is an important part of the metrication process. But conversion can be deceptively simple because most measurements have implied, not expressed, tolerances and many products (like 2-by-4s) are designated in rounded, easy-to-remember "nominal" sizes, not actual ones. For instance, if anchor bolts are to be embedded in masonry to a depth of 8 inches, what should this depth be in millimeters? A strict conversion (using 1 inch = 25.4 mm) results in an exact dimension of 203.2 mm. But this implies an accuracy of 0.1 mm (1/254 inch) and a tolerance of ± 0.05 mm (1/508 inch), far beyond any reasonable measure for field use. Similarly, 203 mm is overly precise, implying an accuracy of 1 mm (about 1/25 inch) and a tolerance of ± 0.5 mm (about 1/50 inch). As a practical matter, ± 3 mm (1/8 inch) is well within the tolerance for setting anchor bolts. Applying ± 3 mm to 203.2 mm, the converted dimension should be in the range of 200 mm to 206 mm. Metric measuring devices emphasize 10 mm increments and masons work on a 200 mm module, so the selection of 200 mm would be a convenient dimension for masons to use in the field. Thus, a reasonable metric conversion for 8 inches, *in this case*, is 200 mm.

B APPENDIX

Layer Titles

Layer Names	Comments	Civil Site Plan	Site Details	Foundation Plan	Architectural Floor Plan	Reflected Ceiling Plan	Finish Plan	Roof/Framing Plan	Elevations	Sections	Schedules and Details	HVAC Plan	Power Plan	Lighting Plan	Plumbing Plan
Partial CAD File List		C8713-1.DWG	C8713-2.DWG	S8713-1.DWG	A8713-2.DWG			A8713-3.DWG	A8713-4.DWG		A8713-9.DWG	M8713-1.DWG			
A-WALL	Walls				●	○	○					○	○	○	○
A-WALL-EXST	Existing walls				●										
A-WALL-DEMO	Demolished walls				●										
A-WALL-HEAD	Door headers					●						○		○	
A-WALL-PATT	Wall poche				●	○	○					○	○	○	○
A-DOOR	Doors				●	○	○					○	○	○	○
A-DOOR-EXST	Existing doors				●										
A-DOOR-DEMO	Demolished doors				●										
A-DOOR-IDEN	Door numbers, hardware group				●										
A-GLAZ	Windows				●	○	○					○	○	○	○
A-GLAZ-EXST	Existing windows				●										
A-GLAZ-DEMO	Demolished windows				●										
A-FLOR-CASE	Casework				●		○						○		○
A-FLOR-IDEN	Room numbers, names, tags				●	○	○					○	○	○	○
A-FLOR-OVHD	Overhead platform				●										
A-FLOR-PATT	Floor pavers						●								
A-EQPM	General equipment				●								○		○
A-CLNG	General ceiling linework					●						○		○	
A-CLNG-GRID	Ceiling 2'x2' tiles					●								○	
A-CLNG-OPEN	Ceiling /roof penetrations					●						○		○	
A-CLNG-PATT	Drywall ceiling poche					●						○		○	
A-CLNG-TEES	Ceiling main tees					●								○	
A-ROOF	Roof plan line work							●							
A-ROOF-GUTT	Gutters							●							
A-ROOF-PATT	Shingles hatch and stipple poche							●							
A-ELEV	Building elevations line work								●						
A-ELEV-OTLN	Building outlines								●	○					
A-ELEV-NOTE	Building elevations reference notes								●						
A-ELEV-SYMB	Building elevations drafting symbols								●						
A-ELEV-DIMS	Building elevations dimensions								●						
A-ELEV-PATT	Building elevations texture								●						
A-ELEV-TTLB	Building elevations title block info								●						
A-SECT	Wall sections line work									●					
A-SECT-NOTE	Wall sections reference notes									●					
A-SECT-SYMB	Wall sections drafting symbols									●					
A-SECT-DIMS	Wall sections dimensions									●					
A-SECT-PATT	Wall sections hatching									●					
A-SECT-TTLB	Wall sections title block info									●					
A-DETL	Construction details line work										●				
A-DETL-NOTE	Construction details reference notes										●				
A-DETL-SYMB	Construction details drafting symbols										●				
A-DETL-DIMS	Construction details dimensions										●				
A-DETL-PATT	Insulation										●				
A-DETL-TTLB	Construction details title block info										●				
A-SHBD	Architectural border and title block	●	○		○	○	○	○	○	○	○				
A-PLCL-NOTE	Reflected ceiling plan reference notes					●									
A-PLCL-SYMB	Reflected ceiling plan drafting symbols					●									
A-PLCL-DIMS	Reflected ceiling plan dimensions					●									
A-PLCL-TTLB	Reflected ceiling plan title block info					●									
A-MFN-NOTE	Floor plan reference notes				●										
A-MFN-SYMB	Floor plan drafting symbols				●										
A-MFN-DIMS	Floor plan dimensions				●										
A-MFN-TTLB	Floor plan title block info				●										
A-PLFN-NOTE	Finish plan reference notes						●								
A-PLFN-SYMB	Finish plan drafting symbols						●								
A-PLFN-DIMS	Finish plan dimensions						●								
A-PLFN-TTLB	Finish plan title block info						●								
A-PROF-NOTE	Roof plan reference notes							●							
A-PROF-SYMB	Roof plan drafting symbols							●							
A-PROF-DIMS	Roof plan dimensions							●							
A-PROF-TTLB	Roof plan title block info							●							
A-SCHD-TEXT	Schedules text and line work										●				
S-FNDN	General foundation linework			●											
S-SLAB	General slab linework			●											
S-FRAM	Framing plan beams, joists							●							
S-SHBD	Structural sheet border and title block			●											
S-PFND-NOTE	Foundation plan reference notes			●											
S-PFND-SYMB	Foundation plan drafting symbols			●											
S-PFND-DIMS	Foundation plan dimensions			●											
S-PFND-TTLB	Foundation plan title block info			●											

Layer Names	Comments	Civil Site Plan	Site Details	Foundation Plan	Architectural Floor Plan	Reflected Ceiling Plan	Finish Plan	Roof/Framing Plan	Elevations	Sections	Schedules and Details	HVAC Plan	Power Plan	Lighting Plan	Plumbing Plan
Partial CAD File List		C8713-1.DWG	C8713-2.DWG	S8713-1.DWG	A8713-2.DWG			A8713-3.DWG	A8713-4.DWG		A8713-9.DWG	M8713-1.DWG			
M-EXHS-DUCT	Exhaust system ductwork											●			
M-HVAC-CDFF	HVAC ceiling diffusers											●			
M-HVAC-DUCT	HVAC ductwork											●			
M-HVAC-EQPM	HVAC equipment											●			
M-HOTW-PIPE	Hot water piping											●			
M-CWAT-PIPE	Chilled water piping											●			
M-SHBD	Mechanical sheet border and title block											●	○	○	○
M-PHVA-NOTE	HVAC plan reference notes											●			
M-PHVA-SYMB	HVAC plan drafting symbols											●			
M-PHVA-DIMS	HVAC plan dimensions											●			
M-PHVA-TTLB	HVAC plan title block info											●			
P-DOMW-EQPM	Domestic hot & cold water equipment														●
P-DOMW-PIPE	Domestic hot & cold water piping														●
P-SANR-PIPE	Sanitary piping														●
P-SANR-FIXT	Plumbing fixtures				●		○						○		○
P-SANR-FLDR	Floor drains														●
P-PPLM-NOTE	Plumbing plan reference notes														●
P-PPLM-SYMB	Plumbing plan drafting symbols														●
P-PPLM-DIMS	Plumbing plan dimensions														●
P-PPLM-TTLB	Plumbing plan title block info														●
E-LITE	Light fixtures					●								○	
E-LITE-EXST	Existing light fixtures					●								○	
E-LITE-SWCH	Lighting switches													●	
E-LITE-CIRC	Lighting circuits													●	
E-POWR-WALL	Power wall outlets and receptacles												●		
E-POWR-PANL	Power panels												●		
E-POWR-CIRC	Power circuits												●		
E-POWR-NUMB	Power circuit numbers												●		
E-PPOW-NOTE	Power plan reference notes												●		
E-PPOW-SYMB	Power plan drafting notes												●		
E-PPOW-DIMS	Power plan dimensions												●		
E-PPOW-TTLB	Power plan title block info												●		
E-PLIT-NOTE	Lighting plan reference notes													●	
E-PLIT-SYMB	Lighting plan drafting symbols													●	
E-PLIT-DIMS	Lighting plan dimensions													●	
E-PLIT-TTLB	Lighting plan title block info													●	
C-BLDG	New building footprint and poche	●													
C-BLDG-DEMO	Existing building to be removed	●													
C-BLDG-EXST	Existing building footprints	●													
C-BLDG-PATT	New building area poche	●													
C-DETL	Site details line work		●												
C-DETL-NOTE	Site details reference notes		●												
C-DETL-SYMB	Site details drafting symbols		●												
C-DETL-DIMS	Site details dimensions		●												
C-DETL-TTLB	Site details title block info		●												
C-PKNG	New parking lot stripes and precast curbs	●													
C-PKNG-DEMO	Existing parking lots to be removed	●													
C-PKNG-PATT	New parking lot poche	●													
C-PROP	Property and setback lines	●		○	○			○					○		○
C-ROAD	New roads	●		○	○			○					○		○
C-ROAD-EXST	Existing roads	●													
C-ROAD-PATT	New roads poche	●													
C-PSIT-NOTE	Site plan reference notes	●													
C-PSIT-SYMB	Site plan drafting symbols	●													
C-PSIT-DIMS	Site plan dimensions	●													
C-PSIT-TTLB	Site plan title block info	●													
L-SITE-DEMO	Existing fencing to be removed	●													
L-WALK	New walks and steps	●			○		○	○					○		○
L-WALK-EXST	Existing walks and steps	●			○		○	○					○		○
L-WALK-PATT	New walks and steps poche	●			○										

Courtesy American Institute of Architects

APPENDIX C

Wood Beam Safe Load Table

This appendix shows a partial listing of span tables from *Wood Structural Design Data,* 1986 edition, courtesy of National Forest Products Association. Once W, F_b, E, and F_v are known, spans can be determined for a simple beam of spans that are normally found in a residence. Refer to *Wood Structural Design Data* for a complete listing of all lumber sizes.

WOOD BEAM SAFE LOAD TABLE

SIZE OF BEAM		F_b 900	1,000	1,100	1,200	1,300
6′–0″						
4 x 4	W	714	793	873	952	1,032
	w	119	132	145	158	172
	F_v	43	48	53	58	63
	E	1,388	1,542	1,697	1,851	2,005
4 x 6	W	1,764	1,960	2,156	2,352	2,548
	w	294	326	359	392	424
	F_v	68	76	84	91	99
	E	883	981	1,079	1,175	1,276
4 x 8	W	3,066	3,406	3,747	4,088	4,428
	w	511	567	624	616	738
	F_v	90	100	110	120	130
	E	670	744	819	893	968
6 x 6	W	2,772	3,081	3,389	3,697	4,005
	w	462	513	564	616	667
	F_v	68	76	84	91	99
	E	883	981	1,079	1,178	1,276
6 x 8	W	5,156	5,729	6,302	6,875	7,447
	w	859	954	1,050	1,145	1,241
	F_v	93	104	114	125	135
	E	648	719	792	864	935

SIZE OF BEAM		F_b 900	1,000	1,100	1,200	1,300
7′–0″						
4 x 6	W	1,512	1,680	1,848	2,016	2,184
	w	216	240	264	288	312
	F_v	58	65	72	78	85
	E	1,030	1,145	1,259	1,374	1,489
4 x 8	W	2,628	2,920	3,212	3,504	3,796
	w	375	417	458	500	542
	F_v	77	86	94	103	112
	E	782	868	955	1,042	1,129
4 x 10	W	4,278	4,753	5,228	5,704	6,179
	w	611	679	746	814	882
	F_v	99	110	121	132	143
	E	612	681	749	817	885
6 x 6	W	2,376	2,640	2,904	3,169	3,433
	w	339	377	414	452	490
	F_v	58	65	72	78	85
	E	1,030	1,145	1,259	1,374	1,489
6 x 8	W	4,419	4,910	5,401	5,892	6,383
	w	631	701	771	841	911
	F_v	80	89	98	107	116
	E	755	839	924	1,007	1,091

WOOD BEAM SAFE LOAD TABLE

SIZE OF BEAM		F_b 900	1,000	1,100	1,200	1,300
8′–0″						
4 x 6	W	1,323	1,470	1,617	1,764	1,911
	w	165	183	202	220	238
	F_v	51	57	63	68	74
	E	1,178	1,309	1,439	1,570	1,701
4 x 8	W	2,299	2,555	2,810	3,066	3,321
	w	287	319	351	383	415
	F_v	67	75	83	90	98
	E	893	993	1,092	1,191	1,291
4 x 10	W	3,743	4,159	4,575	4,991	5,407
	w	467	519	571	623	675
	F_v	86	96	105	115	125
	E	700	778	856	934	1,011
4 x 12	W	5,537	6,152	6,767	7,382	7,998
	w	692	769	845	922	999
	F_v	105	117	128	140	152
	E	576	640	703	768	832
4 x 14	W	7,973	8,859	9,745	10,361	11,517
	w	996	1,107	1,218	1,328	1,439
	F_v	126	140	154	168	182
	E	480	533	586	640	693
6 x 6	W	2,079	2,310	2,541	2,772	3,003
	w	259	288	317	346	375
	F_v	51	57	63	68	74
	E	1,178	1,309	1,439	1,570	1,701
6 x 8	W	3,867	4,296	4,726	5,156	5,585
	w	483	537	590	644	698
	F_v	70	78	86	93	101
	E	864	960	1,055	1,152	1,247
6 x 10	W	6,204	6,894	7,583	8,272	8,962
	w	775	861	947	1,034	1,120
	F_v	89	98	108	118	128
	E	682	757	833	909	985
6 x 12	W	9,092	10,102	11,112	12,122	13,133
	w	1,136	1,262	1,389	1,515	1,641
	F_v	107	119	131	143	155
	E	563	626	688	751	813
6 x 14	W	12,529	13,921	15,314	16,706	18,098
	w	1,566	1,740	1,914	2,088	2,262
	F_v	126	140	154	168	182
	E	480	533	586	640	693

SIZE OF BEAM		F_b 900	1,000	1,100	1,200	1,300
9′–0″						
4 x 8	W	2,044	2,271	2,498	2,725	2,952
	w	227	252	277	302	328
	F_v	60	67	73	80	87
	E	1,005	1,117	1,228	1,340	1,452
4 x 10	W	3,327	3,697	4,066	4,436	4,806
	w	369	410	451	492	534
	F_v	77	85	94	102	111
	E	788	875	963	1,050	1,138
4 x 12	W	3,921	5,468	6,015	6,562	7,109
	w	546	607	668	729	789
	F_v	93	104	114	125	135
	E	648	720	792	863	936
4 x 14	W	7,087	7,875	8,662	9,450	10,237
	w	787	875	962	1,050	1,137
	F_v	112	125	137	150	162
	E	540	600	660	720	780
6 x 8	W	3,437	3,819	4,201	4,583	4,965
	w	381	424	466	509	551
	F_v	62	69	76	83	90
	E	972	1,079	1,187	1,295	1,403
6 x 10	W	5,515	6,128	6,740	7,353	7,966
	w	612	680	748	817	885
	F_v	79	87	96	105	114
	E	767	852	937	1,023	1,108
6 x 12	W	8,081	8,979	9,877	10,775	11,673
	w	897	997	1,097	1,197	1,297
	F_v	95	106	117	127	138
	E	633	704	774	845	915
6 x 14	W	11,137	12,375	13,612	14,850	16,087
	w	1,237	1,375	1,512	1,650	1,787
	F_v	112	125	137	150	162
	E	540	600	659	720	779

WOOD BEAM SAFE LOAD TABLE

SIZE OF BEAM		F_b 900	1,000	1,100	1,200	1,300
10′-0″						
4 x 10	W	2,994	3,327	3,660	3,992	4,325
	w	299	332	366	399	432
	F_v	69	77	84	92	100
	E	875	972	1,070	1,167	1,264
4 x 12	W	4,429	4,921	5,414	5,906	6,398
	w	442	492	541	590	639
	F_v	84	93	103	112	121
	E	720	800	880	960	1,040
4 x 14	W	6,378	7,087	7,796	8,505	9,213
	w	637	708	779	850	921
	F_v	101	112	123	135	146
	E	600	666	733	800	866
6 x 8	W	3,093	3,437	3,781	4,125	4,468
	w	309	343	378	412	446
	F_v	56	62	68	75	81
	E	1,079	1,199	1,319	1,439	1,559
6 x 10	W	4,963	5,515	6,066	6,618	7,169
	w	496	551	606	661	716
	F_v	71	79	87	95	102
	E	852	947	1,042	1,136	1,231
6 x 12	W	7,273	8,081	8,890	9,698	10,506
	w	727	808	889	969	1,050
	W	86	95	105	115	124
	E	704	782	860	939	1,017
6 x 14	W	10,023	11,137	12,251	13,365	14,478
	w	1,002	1,113	1,225	1,336	1,447
	F_v	101	112	123	135	146
	E	600	666	733	800	866

SIZE OF BEAM		F_b 900	1,000	1,100	1,200	1,300
11′-0″						
4 x 8	W	1,672	1,858	2,044	2,229	2,415
	w	152	168	185	202	219
	F_v	49	54	60	65	71
	E	1,228	1,365	1,502	1,638	1,775
4 x 10	W	2,722	3,024	3,327	3,629	3,932
	w	247	274	302	329	357
	F_v	63	70	77	84	91
	E	963	1,070	1,177	1,284	1,391
4 x 12	W	4,026	4,474	4,921	5,369	5,816
	w	366	406	447	488	528
	F_v	76	85	93	102	110
	E	792	879	968	1,055	1,143
4 x 14	W	5,798	6,443	7,087	7,731	8,376
	w	527	585	644	702	761
	F_v	92	102	112	122	132
	E	660	733	806	880	953
6 x 8	W	2,812	3,125	3,437	3,750	4,062
	w	255	284	312	340	369
	F_v	51	56	62	68	73
	E	1,187	1,319	1,451	1,585	1,715
6 x 10	W	4,512	5,013	5,515	6,016	6,518
	w	410	455	501	546	592
	F_v	64	71	79	86	93
	E	937	1,042	1,146	1,250	1,350
6 x 12	W	6,612	7,347	8,081	8,816	9,551
	w	601	667	734	801	868
	F_v	78	87	95	104	113
	E	774	860	946	1,033	1,119
6 x 14	W	9,112	10,125	11,137	12,150	13,162
	w	828	920	1,012	1,104	1,196
	F_v	92	102	112	122	132
	E	659	733	806	880	953

WOOD BEAM SAFE LOAD TABLE

SIZE OF BEAM		F_b 900	1,000	1,100	1,200	1,300
		12'–0"				
4 x 8	W	1,533	1,703	1,873	2,044	2,214
	w	127	141	156	170	184
	F_v	45	50	55	60	65
	E	1,340	1,489	1,638	1,787	1,936
4 x 10	W	2,495	3,772	3,050	3,327	3,604
	w	207	231	254	277	300
	F_v	57	64	70	77	83
	E	1,050	1,167	1,284	1,401	1,517
4 x 12	W	3,691	4,101	4,511	4,921	5,332
	w	307	341	375	410	444
	F_v	70	78	85	93	101
	E	863	960	1,055	1,151	1,247
4 x 14	W	5,315	5,906	6,496	7,087	7,678
	w	442	492	541	590	639
	F_v	84	93	103	112	121
	E	720	800	880	960	1,040
6 x 8	W	2,578	2,864	3,151	3,437	3,723
	w	214	238	262	286	310
	F_v	46	52	57	62	67
	E	1,295	1,439	1,583	1,727	1,871
6 x 10	W	4,136	4,596	5,055	5,515	5,974
	w	344	383	421	459	497
	F_v	59	65	72	79	85
	E	1,023	1,136	1,250	1,364	1,477
6 x 12	W	6,061	6,734	7,408	8,081	8,755
	w	505	561	617	673	729
	F_v	71	79	87	95	103
	E	845	939	1,033	1,126	1,220
6 x 14	W	8,353	9,281	10,209	11,137	12,065
	w	696	773	850	928	1,005
	F_v	84	93	103	112	121
	E	720	800	880	960	1,040

SIZE OF BEAM		F_b 900	1,000	1,100	1,200	1,300
		13'–0"				
4 x 8	W	1,415	1,572	1,729	1,886	2,044
	w	108	120	133	145	157
	F_v	41	46	51	55	60
	E	1,452	1,613	1,775	1,936	2,097
4 x 10	W	2,303	2,559	2,815	3,071	3,327
	w	177	196	216	236	255
	F_v	53	59	65	71	77
	E	1,138	1,264	1,391	1,517	1,644
4 x 12	W	3,407	3,786	4,164	4,543	4,921
	w	262	291	320	349	378
	F_v	64	72	79	86	93
	E	936	1,039	1,143	1,247	1,351
4 x 14	W	4,906	5,451	5,997	6,542	7,087
	w	377	419	461	503	545
	F_v	77	86	95	103	112
	E	780	866	953	1,039	1,126
6 x 8	W	2,379	2,644	2,908	3,173	3,437
	w	183	203	223	244	264
	F_v	43	48	52	57	62
	E	1,403	1,559	1,715	1,871	2,027
6 x 10	W	3,818	4,242	4,666	5,091	5,515
	w	293	326	358	391	424
	F_v	54	60	66	73	79
	E	1,108	1,231	1,354	1,477	1,601
6 x 12	W	5,595	6,216	6,838	7,460	8,081
	w	430	478	526	573	621
	F_v	66	73	81	88	95
	E	915	1,017	1,119	1,220	1,322
6 x 14	W	7,710	8,567	9,424	10,280	11,137
	w	593	659	724	790	856
	F_v	77	86	95	103	112
	E	779	866	953	1,039	1,126

WOOD BEAM SAFE LOAD TABLE

14'-0"

SIZE OF BEAM		F_b 900	1,000	1,100	1,200	1,300
4 x 8	W	1,314	1,460	1,606	1,752	1,898
	w	93	104	114	125	135
	F_v	38	43	47	51	56
	E	1,564	1,737	1,911	2,085	2,259
4 x 10	W	2,139	2,376	2,614	2,852	3,089
	w	152	169	186	203	220
	F_v	49	55	60	66	71
	E	1,225	1,362	1,498	1,634	1,770
4 x 12	W	3,164	3,515	3,867	4,218	4,570
	w	226	251	276	301	326
	F_v	60	66	73	80	87
	E	1,008	1,119	1,231	1,343	1,455
4 x 14	W	4,556	5,062	5,568	6,075	6,581
	w	325	361	397	433	470
	F_v	72	80	88	96	104
	E	840	933	1,026	1,119	1,213
6 x 8	W	2,209	2,455	2,700	2,946	3,191
	w	157	175	192	210	227
	F_v	40	44	49	53	58
	E	1,511	1,679	1,847	2,015	2,183
6 x 10	W	3,545	3,939	4,333	4,727	5,121
	w	253	281	309	337	365
	F_v	50	56	62	67	73
	E	1,193	1,326	1,458	1,591	1,724
6 x 12	W	5,195	5,772	6,350	6,927	7,504
	w	371	412	453	494	536
	F_v	61	68	75	82	88
	E	986	1,095	1,205	1,314	1,424
6 x 14	W	7,159	7,955	8,750	9,546	10,341
	w	511	568	625	681	738
	F_v	72	80	88	96	104
	E	839	933	1,026	1,119	1,213

15'-0"

SIZE OF BEAM		F_b 900	1,000	1,100	1,200	1,300
4 x 8	W	1,226	1,362	1,499	1,635	1,771
	w	81	90	99	109	118
	F_v	36	40	44	48	52
	E	1,675	1,862	2,048	2,234	2,420
4 x 10	W	1,996	2,218	2,440	2,661	2,883
	w	133	147	162	177	192
	F_v	46	51	56	61	66
	E	1,313	1,459	1,605	1,751	1,897
4 x 12	W	2,953	3,281	3,609	3,937	4,265
	w	196	218	240	262	284
	F_v	56	62	68	75	81
	E	1,079	1,199	1,319	1,439	1,559
4 x 14	W	4,252	4,725	5,197	5,670	6,142
	w	283	315	346	378	409
	F_v	67	75	82	90	97
	E	900	1,000	1,099	1,199	1,299
6 x 8	W	2,062	2,291	2,520	2,750	2,979
	w	137	152	168	183	198
	F_v	37	41	45	50	54
	E	1,619	1,799	1,979	2,159	2,339
6 x 10	W	3,309	3,676	4,044	4,412	4,779
	w	220	245	269	294	318
	F_v	47	52	58	63	68
	E	1,278	1,421	1,563	1,705	1,847
6 x 12	W	4,849	5,387	5,926	6,465	7,004
	w	323	359	395	431	466
	F_v	57	63	70	76	83
	E	1,056	1,173	1,291	1,408	1,526
6 x 14	W	6,682	7,425	8,167	8,910	9,652
	w	445	495	544	594	643
	F_v	67	75	82	90	97
	E	900	1,000	1,099	1,199	1,299

WOOD BEAM SAFE LOAD TABLE

SIZE OF BEAM		F_b 900	1,000	1,100	1,200	1,300
			16′–0″			
4 x 8	W	1,149	1,277	1,405	1,533	1,660
	w	71	79	87	95	103
	F_v	33	37	41	45	49
	E	1,787	1,986	2,184	2,383	2,582
4 x 10	W	1,871	2,079	2,287	2,495	2,703
	w	116	129	142	155	168
	F_v	43	48	52	57	62
	E	1,401	1,556	1,712	1,868	2,023
4 x 12	W	2,768	3,076	3,383	3,691	3,999
	w	173	192	211	230	249
	F_v	52	58	64	70	76
	E	1,151	1,279	1,407	1,535	1,663
4 x 14	W	3,986	4,429	4,872	5,315	5,758
	w	249	276	304	332	359
	F_v	63	70	77	84	91
	E	960	1,066	1,173	1,279	1,386
6 x 8	W	1,933	2,148	2,363	2,578	2,792
	w	120	134	147	161	174
	F_v	35	39	42	46	50
	E	1,727	1,919	2,111	2,303	2,495
6 x 10	W	3,102	3,447	3,791	4,196	4,481
	w	193	215	236	258	280
	F_v	44	49	54	59	64
	E	1,364	1,515	1,667	1,818	1,970
6 x 12	W	4,546	5,051	5,556	6,061	6,566
	w	284	315	347	378	410
	F_v	53	59	65	71	77
	E	1,126	1,252	1,377	1,502	1,627
6 x 14	W	6,264	6,960	7,657	8,353	9,049
	w	391	435	478	522	565
	F_v	63	70	77	84	91
	E	960	1,066	1,173	1,279	1,386

SIZE OF BEAM	F_b 900	1,000	1,100	1,200	1,300

APPENDIX D

COUNTING THE COST

In Chapter 1 you were introduced to the cost of a set of plans and the cost of construction. Once the plans are completed, the drawings are submitted to several bidders for bids. Contractors complete the estimate for construction cost based on their knowledge of current material, labor, and economic factors. Companies such as R.S. Means and the Dodge Report published by F.W. Dodge provide contractors with up to date costs on products, manufacturers, materials, and supplies for the construction industry. An architect will generally oversee the bidding process, but the client must make the final decision on which contractor to hire. Construction cost should not be the only factor in determining the builder, since bids can vary based on:

- Differences in design details and materials unless plans and specifications are very specific;
- Builders may have varied experience with certain home styles or features, site conditions, or specific types of materials and equipment;
- Cost of construction loan cost, insurance, permits and inspections, and utility hook-up fees may or may not be included in the bid.
- Quality and the speed of the project based on the number of other on-going projects.
- Varied administrative charges, overhead and profit margin.

Although drafters are not involved in cost estimating unless they work directly for an estimator, an understanding of the role of the contractor and the subcontractors in determining the cost of the project will help you advance within architectural and engineering offices. Keep in mind that the following costs are averages, and that actual cost will vary for each project based on the area of the country, the labor supply, the time of year, and the economy of the area. The following costs and fees are based on a home to be constructed with a value of $321,550.

Construction Costs:	Approx. % of const. cost	Bid
Permits	5.00	14,600
Temp. power/excavation/backfill	4.00	11,700
Footings & foundations (matl. & labor)	2.40	7,000
Framing material	9.30	27,125
Framing labor	9.30	27,125
Roofing	1.50	4,370
Windows & sliders	3.90	11,370
Rough plumbing	3.40	9,915
Rough electrical	3.70	10,800
Fireplace	2.50	7,290
Masonry veneer/chimney	2.47	7,200

Construction Costs:	Approx. % of const. cost	Bid
Basement/garage floor/flatwork	1.50	4,370
Siding (matl. & labor)	4.30	12,540
Heating & air conditioning (matl. & labor)	3.75	10,930
Insulation	1.85	5,395
Drywall/tape/texture	4.70	13,705
Septic/sewer	0.75	2,190
Water hook-up	0.30	875
Downspouts & gutter	0.75	2,190
Exterior painting	1.30	3,790
Interior painting	1.60	4,660
Paneling & wall paper	0.62	1,810
Cabinets/hardware	9.50	27,700
Linoleum/tile/vinyl	2.90	8,450
Finish plumbing	2.75	8,020
Garage door/openers	0.98	2,860
Finish electrical	1.54	4,490
Carpets	2.80	8,150
Hardwood flooring	1.25	3,640
Interior/exterior doors & trims	5.00	14,580
Appliances	1.60	4,660
Decks/rails/misc. extras	2.15	6,270
Clean-up	0.61	1,780
Total construction amount	100.00	291,550
Operating expenses/profit		30,000

General Fees:	
Construction cost	321,550
Architect's design fees	29,000
Landscaping/sprinklers	25,000
Total construction cost	375,550
Misc. closing fees (loan fees, title charges, recording fees, reserves)	9,825
Land purchase price	72,900
Total cost	458,275
20% down + fees	101,480
80% mortgage	358,760
Interest on mortgage	500,502*
Average monthly interest	1,390
Monthly payment	2,387
Total cost to owner over 30 years	958,777**

**Interest assumed to be 7% on $458,760 mortgage*

**Not including property taxes and assumes no prepayments.

Construction rates supplied by Walt Townsend of Townsend Construction

Financial information supplied by Michael Jefferis

ABBREVIATIONS

Drafters and designers use many abbreviations to conserve space. Using standard abbreviations ensures that drawings are interpreted accurately.

Here are three guidelines for proper use:

1. Abbreviations are in capital letters.
2. A period is used only when the abbreviation may be confused with a word.
3. Several words use the same abbreviation; use is defined by the location.

access panel	**ap**
acoustic	**ac**
acoustic plaster	**ac pl**
actual	**act**
addition	**add**
adhesive	**adh**
adjustable	**adj**
aggregate	**aggr**
air conditioning	**a.c.**
alternate	**alt**
alternating current	**ac**
aluminum	**alum**
American Institute of Architects	A.I.A.
American Institute of Building Designers	A.I.B.D.
American Institute of Steel Construction	A.I.S.C.
American Institute of Timber Construction	A.I.T.C.
American National Standards Institute	A.N.S.I.
American Plywood Association	A.P.A.
American Society of Civil Engineers	A.S.C.E.
American Society of Heating, Refrigerating, and Air Conditioning Engineers	A.S.H.R.A.E.
American Society of Landscape Architects	A.S.L.A.
American Society for Testing and Materials	A.S.T.M.
amount	**amt**
ampere	**amp**
anchor bolt	**a.b.**
angle	$\angle$
approximate	**approx**
approved	**appd**
architectural	**arch**
area	**a**
asbestos	**asb.**
asphalt	**asph**
asphaltic concrete	**asph conc**
at	@
automatic	**auto**
avenue	**ave**
average	**avg**
balcony	**balc**
base	**b**
basement	**basm**
Basic National Building Code	BOCA
batten	**batt.**
bathroom	**b**
bathtub	**bt.**
beam	**bm**
bearing	**br**
bedroom	**br.**
benchmark	**B.M.**
bending moment	**M**
better	**btr**
between	**btwn**
beveled	**bev**
bidet	**bdt**
block	**blk**
blocking	**blkg**
blower	**blo**
board	**bd.**
board feet	**bd ft**
both sides	**b.s.**

both ways	b.w.
bottom	btm
bottom of footing	b.f.
boulevard	blvd
brass	br
brick	brk
British thermal unit	Btu
bronze	brz
broom closet	bc
building	bldg
building line	bL
built-in	blt-in
buzzer	buz
by	×
cabinet	cab.
cast concrete	c conc
cast iron	c.i.
catalog	cat
catch basin	C.B.
caulking	calk
ceiling	clg
ceiling diffuser	c.d.
ceiling joist	c.j. or ceil. jst
cement	cem
center	ctr
center to center	c-c
center line	℄
centimeter	cm
ceramic	cer
chamfer	cham
channel	c
check	chk.
cinder block	cin blk
circle	cir
circuit	cir
circuit breaker	cir bkr
class	cl
cleanout	c.o.
clear	clr
coated	ctd
cold water	c.w.
column	col
combination	comb
common	com
composition	comp
computer-aided drafting	CAD
concrete	conc
concrete masonry unit	c.m.u.
conduit	cnd
construction	const
Construction Standards Institute	CSI
continuous	cont
contractor	contr
control joint	c.j.
copper	cop
corridor	corr
corrugate	corr.
countersink	csk.
courses	c.
cubic	cu.
cubic feet	cu ft
cubic feet per minute	cfm
cubic inch	cu in
cubic yard	cu yd
damper	dpr.
dampproofing	dp
dead load	dl
decibel	db
decking	dk
deflection	d.
degree	° or deg
design	dsgn
detail	det.
diagonal	diag
diameter	ϕ or dia
diffuser	dif.
dimension	dimen
dining room	dr
dishwasher	d/w
disposal	disp
ditto	″ or do
division	div
door	dr.
double	dbl
double hung	d.h.
Douglas fir	df
down	dn
downspout	d.s.
drain	d
drawing	dwg
drinking fountain	d.f.
dryer	d.
drywall	D.W.
each	ea
each face	E.F.
each way	E.W.
east	e
elbow	el
electrical	elect
elevation	elev
enamel	enam
engineer	engr
entrance	ent.
Environmental Protection Agency	EPA
Equal	eq
Equipment	equip
estimate	est
excavate	exc
exhaust	exh
existing	exist
expansion joint	exp jt

exposed	expo
extension	extn
exterior	ext.
fabricate	fab
face brick	f.b.
face of studs	f.o.s.
Fahrenheit	F
Federal Housing Administration	F.H.A.
feet / foot	′ or ft
feet per minute	fpm
finished	fin
finished floor	fin fl
finished grade	fin gr
finished opening	f.o.
firebrick	fbrk.
fire hydrant	F.H.
fireproof	f.p.
fixture	fix.
flammable	flam
flange	flg.
flashing	fl.
flexible	flex
floor	flr.
floor drain	f.d.
floor joist	fl jst
floor sink	f.s.
fluorescent	fluor
folding	fldg
foot	(′) ft
footcandle	fc
footing	ftg
foot pounds	ft lb
forced air unit	f.a.u.
foundation	fnd
front	fnt
full size	fs
furnace	furn
furred ceiling	fc
future	fut
gallon	gal
galvanized	galv
galvanized iron	g.i.
gage	ga.
garage	gar
gas	g.
girder	gird
glass	gl.
glue laminated	glu-lam
grade	gr
grade beam	grbm
grating	grtg
gravel	gvl
grille	gr
ground	gnd
ground fault circuit interrupter	gfci
grout	gt
gypsum	gyp
gypsum board	gyp bd
hardboard	hdb
hardware	hdw
hardwood	hdwd
head	hd.
header	hdr.
heater	htr.
heating	htg.
heating/ventilating/air conditioning	hvac
height	ht.
hemlock	hem
hemlock-fir	hem-fir
hollow core	h.c.
horizontal	horiz.
horsepower	h.p.
hose bibb	h.b.
hot water	h.w.
hot water heater	h.w.h.
hundred	c
illuminate	illum.
incandescent	incan.
inch	″ or in.
inch pounds	in. lb.
incinerator	incin.
inflammable	infl.
inside diameter	i.d.
inside face	i.f.
inspection	insp.
install	inst.
insulate	ins.
insulation	insul.
interior	int.
International Code Council	ICC
International Conference of Building Officials	ICBO
International Residential Code	IRC
iron	i
jamb	jmb.
joint	jt.
joist	jst.
junction	jct.
junction box	J-box
kiln dried	k.d.
kilowatt	kW
kilowatt hour	kWh
Kip (1,000 lb)	K
kitchen	kit.
knockout	k.o.
laboratory	lab.
laminated	lam.

landing	**ldg.**
laundry	**lau**
lavatory	**lav**
length	**lgth**
level	**lev**
light	**lt**
linear feet	**lin ft**
linen closet	**l cl**
linoleum	**lino.**
live load	**LL**
living room	**liv**
long	**lg**
louver	**lv**
lumber	**lum.**
machine bolt	**m.b.**
manhole	**m.h.**
manufacturer	**manuf**
marble	**mrb.**
masonry	**mas.**
material	**matl.**
maximum	**max.**
mechanical	**mech.**
medicine cabinet	**m.c.**
medium	**med**
membrane	**memb**
metal	**mtl**
meter	**m**
mile	**mi**
minimum	**min**
minute	**(′) min.**
mirror	**mirr.**
miscellaneous	**misc**
mixture	**mix**
model	**mod**
modular	**mod.**
molding	**mldg**
motor	**mot**
mullion	**mull**
National Association of Home Builders	**N.A.H.B.**
National Bureau of Standards	**N.B.S.**
natural	**nat.**
natural grade	**nat. gr.**
noise reduction coefficient	**n.r.c.**
nominal	**nom**
not applicable	**n.a.**
not in contract	**N.I.C.**
not to scale	**N.T.S.**
number	**# or no**
obscure	**obs**
on center	**O.C.**
opening	**opg**
opposite	**opp**
ounce	**oz**

outside diameter	**O.D.**
outside face	**O.F.**
overhead	**ovhd.**
painted	**ptd.**
pair	**pr.**
panel	**pnl.**
parallel	**// or par.**
part	**pt.**
partition	**part.**
pavement	**pvmt.**
penny	**d**
per	**/**
perforate	**perf.**
perimeter	**per.**
permanent	**perm.**
perpendicular	**⊥ or perp.**
pi (3.1416 . . .)	**π**
plaster	**pls.**
plasterboard	**pls. bd.**
plastic	**plas.**
plate	**℄ or pl.**
platform	**plat.**
plumbing	**plmb.**
plywood	**ply.**
polished	**pol.**
polyethelyne	**poly.**
polyvinyl chloride	**pvc**
position	**pos.**
pound	**# or lb**
pounds per square foot	**psf**
pounds per square inch	**psi**
precast	**prcst.**
prefabricated	**prefab.**
preferred	**pfd.**
preliminary	**prelim.**
pressure treated	**p.t.**
property	**prop.**
pull chain	**p.c.**
pushbutton	**p.b.**
quality	**qty.**
quantity	**qty.**
radiator	**rad.**
radius	**r or rad.**
random length and width	**r l & w**
range	**r.**
receptacle	**recp.**
recessed	**rec.**
redwood	**rdwd.**
reference	**ref.**
refrigerator	**refr.**
register	**reg.**
reinforcing	**reinf.**
reinforcing bar	**rebar**
reproduce	**repro.**

required	reqd.
resistance moment	rm
return	ret.
revision	rev.
ridge	rdg.
riser	ris.
roof	rf.
roof drain	R.D.
roofing	rfg.
room	rm.
rough	rgh.
rough opening	r.o.
round	ϕ or rd
safety	saf.
sanitary	san.
scale	sc.
schedule	sch.
screen	scrn.
screw	scr.
second	sec.
section	sect.
select	sel.
select structural	sel. st.
self-closing	s.c.
service	serv.
sewer	sew.
sheathing	shtg.
sheet	sht.
shower	sh.
side	s.
siding	sdg.
sill cock	s.c.
similar	sim.
single hung	s.h.
socket	soc.
soil pipe	s.p.
solid block	sol. blk.
solid core	S.C.
Southern Building Code	S.B.C.
Southern pine	SP
Southern Pine Inspection Bureau	S.P.I.B.
specifications	specs.
spruce-pine-fir	SPF
square	□ or sq
square feet	♯ or sq ft
square inch	♯ or sq in.
stainless steel	sst.
standpipe	st. p.
standard	std.
steel	stl.
stirrup	stir.
stock	stk.
storage	sto.
storm drain	S.D.
street	st.
structural	str.
structural clay tile	S.C.T.
substitute	sub.
supply	sup.
surface	sur.
surface four sides	S4S
surface two sides	S2S
suspended ceiling	susp. clg.
switch	sw.
symbol	sym.
symmetrical	sym.
synthetic	syn.
system	sys.
tangent	tan.
tar and gravel	t & g
tee	t
telephone	tel.
television	tv.
temperature	temp.
terra-cotta	t.c.
terrazzo	tz.
thermostat	thrm.
thickness	thk.
thousand	m
thousand board feet	MBF
threshold	thr.
through	thru
toilet	tol.
tongue and groove	t & g
top of wall	t.w.
total	tot.
tread	tr.
tubing	tub.
typical	typ.
Underwriters' Laboratories, Inc.	U.L.
unfinished	unfin.
Uniform Building Code	UBC
United States Department of Housing and Urban Development	H.U.D.
urinal	ur.
utility	util.
V-joint	v-jt.
valve	v
vanity	van.
vapor barrier	v.b.
vapor proof	vap. prf.
ventilation	vent.
vent pipe	vp
vent stack	v.s.
vent through roof	v.t.r.
vertical	vert.
vertical grain	vert. gr.
vinyl	vin.
vinyl asbestos tile	v.a.t.
vinyl base	v.b.

vinyl tile . . . . **v.t.**
vitreous . . . . **vit.**
vitreous clay tile . . . . **v.c.t.**
volt . . . . **v**
volume . . . . **vol.**

wainscot . . . . **wsct.**
wall vent . . . . **w.v.**
washing machine . . . . **wm**
waste stack . . . . **w.s.**
water closet . . . . **w.c.**
water heater . . . . **w.h.**
waterproof . . . . **w.p.**
watt . . . . **W**
weather stripping . . . . **ws**
weatherproof . . . . **wp.**
weep hole . . . . **wh**
weight . . . . **wt.**
welded wire fabric . . . . **w.w.f.**
west . . . . **w**
white pine . . . . **wp**
wide flange . . . . **W or** WF
width . . . . **w**
window . . . . **wdw.**
with . . . . **w**
without . . . . **w/o**
wood . . . . **wd**
wrought iron . . . . **w.i.**

yard . . . . **yd**
yellow pine . . . . **yp**

zinc . . . . **zn.**

GLOSSARY

Accessibility Ability to go in, out, and through a building and its rooms with ease regardless of disability.

Acoustics The science of sound and sound control.

Active solar energy System that uses mechanical devices to absorb, store, and use solar heat.

Adobe Heavy clay used in many southwestern states to make sun-dried bricks.

Aggregate Stone, gravel, cinder, or slag used as one of the components of concrete.

Air duct A pipe, typically made of sheet metal, that carries air from a source such as a furnace or air conditioner to a room within a structure.

Air trap A "U"-shaped pipe placed in wastewater lines to prevent backflow of sewer gas.

Air-dried lumber Lumber that has been stored in yards or sheds for a period of time after cutting. Building codes typically assume a 19 percent moisture content when determining joist and beams of air-dried lumber.

Alcove A small room adjoining a larger room often separated by an archway.

Ampere (amps) A measure of electrical current.

Anchor A metal tie or strap used to tie building member to each other.

Anchor bolt A threaded bolt used to fasten wooden structural members to masonry.

Angle iron A structural piece of steel shaped to form a 90-degree angle.

Appraisal The estimated value of a piece of property.

Apron The inside trim board placed below a windowsill. The term is also used to apply to a curb around a driveway or parking area.

Architect A licensed professional who designs commercial and residential structures.

Areaway A subsurface enclosure to admit light and air to a basement. Sometimes called a window well.

Asbestos A mineral that does not burn or conduct heat; it is usually used for roofing material.

Ash dump An opening in the hearth into which ashes can be dumped.

Ash pit An area in the bottom of the firebox of a fireplace to collect ash.

Ashlar masonry Squared masonry units laid with a horizontal bed joint.

Asphalt An insoluble material used for making floor tile and for waterproofing walls and roofs.

Asphalt shingle Roof shingles made of asphalt-saturated felt and covered with mineral granules.

Asphaltic concrete A mixture of asphalt and aggregate which is used for driveways.

Assessed value The value assigned by governmental agencies to determine the taxes to be assessed on structures and land.

Atrium An inside courtyard of a structure which may be either open at the top or covered with a roof.

Attic The area formed between the ceiling joists and rafters.

Awning window A window that is hinged along the top edge.

Backfill Earth, gravel, or sand placed in the trench around the footing and stem wall after the foundation has cured.

Baffle A shield, usually made of scrap material, to keep insulation from plugging eave vents. Also used to describe wind- or sound-deadening devices.

Balance A principle of design dealing with the relationship between the various areas of a structure as they relate to an imaginary centerline.

Balcony A deck or patio that is above ground level.

Balloon framing A building construction method that has vertical wall members that extend uninterrupted from the foundation to the roof.

Band joist A joist set at the edge of the structure that runs parallel to the other joist (also called a rim joist).

Banister A hand rail beside a stairway.

Base cabinets Cabinets in a kitchen or bathroom which sit on the floor.

Base course The lowest course in brick or concrete masonry unit construction.

Baseline A reference line.

Baseboard The finish trim where the wall and floor intersect, or an electric heater that extends along the floor.

Basement A level of a structure that is built either entirely below grade level (full basement) or partially below grade level (daylight basement).

Batt Blanket insulation usually made of fiberglass to be used between framing members.

Batten A board used to hide the seams when other boards are joined together.

Batter board A horizontal board used to form footings.

Bay A division of space within a building usually divided by beams or columns.

Beam A horizontal structure member that is used to support roofs or wall loads (often called a header).

Beamed ceiling A ceiling that has support beams that are exposed to view.

Bearing plate A support member, often a steel plate, used to spread weight over a larger area.

Bearing wall A wall that supports vertical loads in addition to its own weight.

Benchmark A reference point used by surveyors to establish grades and construction heights.

Bending One of three major forces acting on a beam. It is the tendency of a beam to bend or sag between its supports.

Bending moment A measure of the forces that cause a beam to break by bending. Represented by (M).

Beveled siding Siding that has a tapered thickness.

Bibb An outdoor faucet which is threaded so that a hose may be attached. (Represented by H.B. on floor plans).

Bill of material A part of a set of plans that lists all of the material needed.

Bird block A block placed between rafters to maintain a uniform spacing and to keep animals out of the attic.

Bird's mouth A notch cut into a rafter to provide a bearing surface where the rafter intersects the top plate.

Blind nailing Driving nails in such a way that the heads are concealed from view.

Blocking Framing members, typically wood, placed between joist, rafters, or studs to provide rigidity (also called bridging).

Board and batten A type of siding using vertical boards with small wood strips (battens) used to cover the joints of the boards.

Board foot The amount of wood contained in a piece of lumber 1-in. thick by 12 in. wide by 12 in. long.

Bond The mortar joint between two masonry units, or a pattern in which masonry units are arranged.

Bond beam A reinforced concrete beam used to strengthen masonry walls.

Bottom chord The lower, usually horizontal, member of a truss.

Box beam A hollow built-up structural unit.

Braced wall line Each exterior surface of a residence.

Braced wall panel A method of reinforcing a braced wall line using 48″ wide panels to resist lateral loads.

Breaker An electrical safety switch that automatically opens the circuit when excessive amperage occurs in the circuit.

Breezeway A covered walkway with open sides between two different parts of a structure.

Bridging Cross blocking between horizontal members used to add stiffness. Also called blocking.

Broker A representative of the seller in property sales.

Btu British thermal unit. A unit used to measure heat.

Building code Legal requirements designed to protect the public by providing guidelines for structural, electrical, plumbing, and mechanical areas of a structure.

Building line An imaginary line determined by zoning departments to specify on which area of a lot a structure may be built (also known as a setback).

Building paper A waterproofed paper used to prevent the passage of air and water into a structure.

Building permit A permit to build a structure issued by a governmental agency after the plans for the structure have been examined and the structure is found to comply with all building code requirements.

Built-up beam A beam built to smaller members that are bolted or nailed together.

Built-up roof A roof composed of three or more layers of felt, asphalt, pitch, or coal tar.

Bullnose Rounded edges of cabinet trim.

Butt joint The junction where two members meet in a square-cut joint; end to end, or edge to edge.

Buttress A projection from a wall often located below roof beams to provide support to the roof loads and to keep long walls in the vertical position.

Cabinet work The interior finish woodwork of a structure, especially cabinetry.

CADD Computer-aided design and drafting.

Cant strip A small built-up area between two intersecting roof shapes to divert water.

Cantilever Projected construction that is fastened at only one end.

Carport A covered automobile parking structure that is typically not fully enclosed.

Carriage The horizontal part of a stair stringer that supports the tread.

Casement window A hinged window that swings outward.

Casing The metal, plastic, or wood trim around a door or a window.

Catch basin An underground reservoir for water drained from a roof before it flows to a storm drain.

Cathedral window A window with an upper edge that is parallel to the roof pitch.

Caulking A soft, waterproof material used to seal seams and cracks in construction.

Cavity wall A masonry wall formed with two wythes with an air space between each face.

Ceiling joist The horizontal member of the roof which is used to resist the outward spread of the rafters and to provide a surface on which to mount the finished ceiling.

Cement A powder of alumina, silica, lime, iron oxide, and magnesia pulverized and used as an ingredient in mortar and concrete.

Central heating A heating system in which heat is distributed throughout a structure from a single source.

Cesspool An underground catch basin for the collection and dispersal of sewage.

Chair rail Molding placed horizontally on the wall at the height where chair backs would otherwise damage the wall.

Chamfer A beveled edge formed by removing the sharp corner of a piece of material.

Channel A standard form of structural steel with three sides at right angles to each other forming the letter "C".

Chase A recessed area of column formed between structural members for electrical, mechanical, or plumbing materials.

Check Lengthwise cracks in a board caused by natural drying.

Check valve A valve in a pipe that permits flow in only one direction.

Chimney An upright structure connected to a fireplace or furnace that passes smoke and gases to outside air.

Chimney cap The sloping surface on the top of the chimney.

Chord The upper and lower members of a truss which are supported by the web.

Cinder block A block made of cinder and cement used in construction.

Circuit The various conductors, connections, and devices found in the path of electrical flow from the source through the components and back to the source.

Circuit breaker A safety device which opens and closes an electrical circuit.

Clapboard A tapered board used for siding that overlaps the board below it.

Cleanout A fitting with a removable plug that is placed in plumbing drainage pipelines to allow access for cleaning out the pipe.

Clearance A clear space between building materials to allow for airflow or access.

Clerestory A window or group of windows that are placed above the normal window height, often between two roof levels.

Code A performance-based description of the desired results with wide latitude allowed to achieve the results.

Collar ties Horizontal ties between rafters near the ridge to help resist the tendency of the rafters to separate.

Colonial A style of architecture and furniture adapted from the American colonial period.

Column A vertical structural support, usually round and made of steel.

Common wall The partition that divides two different dwelling units.

Complex beam Beam with a non-uniform load at any point on it and has supports that are not located at its end.

Compression A force that crushes or compacts.

Computer aided drafting Using a computer as a drafting aid.

Computer input Information placed into a computer.

Computer mainframe The main computer that controls many smaller desk terminals.

Computer output Information which is displayed or printed by a computer.

Computer software Programs used to control a computer.

Concentrated load A load centralized in a small area. The weight supported by a post results in a concentrated load.

Concrete A building material made from cement, sand, gravel, and water.

Concrete blocks Blocks of concrete that are precast. The standard size is 8 × 8 × 16.

Condensation The formation of water on a surface when warm air comes in contact with a cold surface.

Conductor Any material that permits the flow of electricity. A drainpipe that diverts water from the roof (a downspout).

Conduit A metal, fiber pipe, or tube used to enclose one or more electrical conductors.

Conduit A bendable pipe or tubing used to encase electrical wiring.

Construction joint A joint used when concrete construction must be interrupted and is used to provide a clean surface when work can be resumed.

Construction loan A mortgage loan to provide cash to the builder for material and labor that is typically repaid when the structure is completed.

Continuous beam A single beam that is supported by more than two supports.

Contours A line that represents land formations.

Contractor The manager of a construction project, or one specific phase of it.

Control joint An expansion joint in a masonry wall formed by raking mortar from the vertical joint.

Control point survey A survey method that establishes elevations that is recorded on a map.

Convenience outlet An electrical receptacle though which current is drawn for the electrical system of an appliance.

Coping A masonry cap placed on top of a block wall to protect it from water penetration.

Corbel A ledge formed in a wall by building out successive courses of masonry.

Cornice The part of the roof that extends out from the wall. Sometimes referred to as the eave.

Counterflash A metal flashing used under normal flashing to provide a waterproof seam.

Course A continuous row of building material such as shingles, stone, or brick.

Court An unroofed space surrounded by walls.

Crawl space The area between the floor joists and the ground.

Cricket A diverter built to direct water away from an area of a roof where it would otherwise collect, such as behind a chimney.

Cripple A wall stud that is cut at less than full length.

Cross bracing Boards fastened diagonally between structural members, such as floor joists, to provide rigidity.

Crosshatch Thin lines drawn about 1/16″ apart, typically at 45 degrees, to show masonry in plain view or wood that has been sectioned.

Cul-de-sac A dead-end street with no outlet which provides a circular turn-around.

Culvert Underground passageways for water, usually part of a drainage system.

Cupola A small structure built above the main roof level to provide light or ventilation.

Cure The process of concrete drying to its maximum design strength, usually taking 28 days.

Curtain wall An exterior wall which provides no structural support.

Damper A moveable plate that controls the amount of draft for a woodstove, fireplace, or furnace.

Datum A reference point for starting a survey.

Dead load The weight of building materials or other immovable objects in a structure.

Deadening A material used to control the transmission of sound.

Deck Exterior floor that is supported on at least two opposing sides by adjoining structures, piers, or posts.

Decking A wood material used to form the floor or roof, typically used in 1 inch and 2 inch thicknesses.

Deflection The tendency of a structural member to bend under a load and as a result of gravity.

Density The number of people allowed to live in a specific area of land or to work in a specific area of a structure.

Depreciation Loss of monetary value.

Designer A person who designs buildings but is not licensed as an architect.

Details Enlargements of specific areas of a structure that are drawn where several components intersect or where small members are required.

Diaphragm A rigid plate that acts similar to a beam and can be found in the roof level, walls, or floor system.

Digitizer An input tool for computer aided drafting used to draw on a flat bed plotter. The device translates images to numbers for transmission to the computer.

Dimension line Line showing the length of the dimension and terminates at the related extension line with slashes, arrowheads, or dots.

Dimension lumber Lumber ranging in thickness from 2 to 4 inches and having moisture content of less than 19 percent.

Disk Computer storage units that store information or programs used to run the computer.

Distribution panel Panel where the conductor from the meter base is connected to individual circuit breakers, which are connected to separate circuits for distribution to various locations throughout the structure.

Diverter A metal strip used to divert water.

Documents A general term that refers to all drawings and written information related to a project.

Dormer A structure which projects from a sloping roof to form another roofed area. This new area is typically used to provide a surface to install a window.

Double hung A type of window in which the upper and lower halves slide past each other to provide an opening at the top and bottom of the window.

Downspout A pipe which carries rainwater from the gutters of the roof to the ground.

Drafter A person who makes drawings and details from another person's creations.

Drain A collector for a pipe that carries water.

Dressed lumber Lumber that has been surfaced by a planing machine to give the wood a smooth finish.

Dry rot A type of wood decay caused by fungi that leaves the wood a soft powder.

Dry well A shallow well used to disperse water from the gutter system.

Drywall An interior wall covering installed in large sheets made from gypsum board.

Ducts Pipes, typically made of sheet metal, used to conduct hot or cold air of the HVAC system.

Duplex outlet A standard electrical convenience outlet with two receptacles.

Dutch door A type of door that is divided horizontally in the center so that each half of the door may be opened separately.

Dutch hip A type of roof shape that combines features of a gable and a hip roof.

Dynamic loads The loads imposed on a structure from a sudden gust of wind or from an earthquake.

Easement An area of land that cannot be built upon because it provides access to a structure or to utilities, such as power or sewer lines.

Eave The lower part of the roof that projects from the wall (also see cornice).

Egress A term used in building codes to describe access.

Elastic limit The extent to which a material can be bent and still return to its original shape.

Elbow An "L"-shaped plumbing pipe.

Elevation The height of a specific point in relation to another point. The exterior views of a structure.

Ell An extension of the structure at a right angle to the main structure.

Eminent domain The right of a government to condemn private property so that it may be obtained for public use.

Enamel A paint that produces a hard, glossy, smooth finish.

Engineer A licensed professional who applies mathematical and scientific principles to the design and construction of structures.

Entourage The surroundings of a rendered building consisting of groundcover, trees, people, and automobiles.

Equity The value of real estate in excess of the balance owed on the mortgage.

Ergonomics The study of human space and movement needs as they relate to a given work area, such as a kitchen.

Excavation The removal of soil for construction purposes.

Expansion joint A joint installed in concrete construction to reduce cracking and to provide workable areas.

Extension line Line showing the extent of a dimension that are generally thin, dark, crisp lines.

Fabrication Work done on a structure away from the job site.

Facade The exterior covering of a structure.

Face brick Brick that is used on the visible surface to cover other masonry products.

Face grain The pattern in the visible veneer of plywood.

Fascia A horizontal board nailed to the end of rafters or trusses to conceal their ends.

Federal Housing Administration (FHA) A governmental agency that insures home loans made by private lending institutions.

Felt A tar-impregnated paper used for water protection under roofing and siding materials. Sometimes used under concrete slabs for moisture resistance.

Fiber bending stress The measurement of structural members used to determine their stiffness. (Fb)

Fiberboard Fibrous wood products that have been pressed into a sheet. Typically used for the interior construction of cabinets and for a covering for the subfloor.

Fill Material used to raise an area for construction. Typically gravel or sand is used to provide a raised, level building area.

Filled insulation Insulation material that is blown or poured into place in attics and walls.

Fillet weld A weld between two surfaces that butt at 90 degrees to each other with the weld filling the inside corner.

Finished lumber Wood that has been milled with a smooth finish suitable for use as trim and other finish work.

Finished size Sometimes called the dressed size, the finished size represents the actual size of lumber after all milling operations and is typically about 1/2 inch smaller than the nominal size, which is the size of lumber before planing.

Nominal Size (inches)	Finished Size (inches)
1	3/4
2	1 1/2
4	3 1/2
6	5 1/2
8	7 1/2
10	9 1/2
12	11 1/2
14	13 1/2

Fire cut An angular cut on the end of a joist or rafter that is supported by masonry. The cut allows the wood member to fall away from the wall without damaging a masonry wall when wood is damaged by fire.

Fire door A door used between different types of construction which has been rated as being able to withstand fire for a certain amount of time.

Fire rated A rating given to building materials to specify the amount of time the material can resist damage caused by fire.

Fire wall A wall constructed of materials resulting in a specified time that the wall can resist fire before structural damage will occur.

Firebox The combustion chamber of the fireplace where the fire occurs.

Firebrick A refractory brick capable of withstanding high temperatures and used for lining fireplaces and furnaces.

Fireproofing Any material that is used to cover structural materials to increase their fire rating.

Fire-stop Blocking placed between studs or other structural members to resist the spread of fire.

Fitting A standard pipe used for joining two or more sections of pipe.

Fixed window Window that is designed without hinges so it cannot be opened.

Flagstone Flat stones used typically for floor and wall coverings.

Flashing Metal used to prevent water leaking through surface intersections.

Flat roof A roof with a minimal roof pitch, usually about 1/4″ per 12″.

Floor plan Architectural drawing of a room or building as if seen from above.

Floor plug A 110-convenience outlet located in the floor.

Flue A passage inside of the chimney to conduct smoke and gases away from a firebox to outside air.

Flue liner A terra-cotta pipe used to provide a smooth flue surface so that unburned materials will not cling to the flue.

Footing The lowest member of a foundation system used to spread the loads of a structure across supporting soil.

Footing form The wooden mold used to give concrete its shape as it cures.

Foundation The system used to support a building's loads and made up to stem walls, footings, and piers. The term is used in many areas to refer to the footing.

Frame The structural skeleton of a building.

French door Exterior or interior doors that have glass panels and swings into a room.

Frost line The depth to which soil will freeze.

Furring Wood strips attached to structural members that are used to provide a level surface for finishing materials when different sized structural members are used.

Gable A type of roof with two sloping surfaces that intersect at the ridge of the structure.

Gable end wall The triangular wall that is formed at each end of a gable roof between the top plate of the wall and the rafters.

Galvanized Steel products that have had zinc applied to the exterior surface to provide protection from rusting.

Gambrel A type of roof formed with two planes on each side of the ridge. The lower pitch is steeper than the upper portion of the roof.

Geothermal system Heating and cooling system which uses the constant, moderate temperature of the ground to provide space heating and cooling, or domestic hot water, by placing a heat exchanger in the ground, or in wells, lakes, rivers, or streams.

Girder A horizontal support member at the foundation level.

Glued-laminated beam (glu-lam) A structural member made up of layers of lumber that are glued together.

Grade The designation of the quality of a manufactured piece of wood.

Grading The moving of soil to effect the elevation of land at a construction site.

Gravel stop A metal strip used to retain gravel at the edge of built-up roofs.

Gravity Uniform force that affects all structures due to the gravitational force from the earth.

Green lumber Lumber that has not been kiln-dried and still contains moisture.

Ground An electrical connection to the earth by means of a rod.

Ground fault circuit interrupter (gfci or gfi) A 110-convenience outlet with a built-in circuit breaker. GFCI outlets are to be used within 5′0″ of any water source.

Ground line A line which represents the horizontal surface at the base of a perspective drawing.

Grout A mixture of cement, sand, and water used to fill joints in masonry and tile construction.

Guardrail A horizontal protective railing used around stairwells, balconies, and changes of floor elevation greater than 30 inches.

Gusset A metal or wood plate used to strengthen the intersection of structural members.

Gypsum board An interior finishing material made of gypsum and fiberglass and covered with paper which is installed in large shapes.

Half-timber A frame construction method where spaces between wood members are filled with masonry.

Hanger A metal support bracket used to attach two structural members.

Hardboard Sheet material formed of compressed wood fibers.

Head The upper portion of a door or window frame.

Header A horizontal structural member used to support other structural members over openings, such as doors and windows.

Header course A horizontal masonry course with the end of each masonry unit exposed.

Headroom The vertical clearance over a stairway.

Hearth The fire resistant floor within and extending a minimum of 18 inches in front of the firebox.

Heartwood The inner core of a tree trunk.

Heat pump A unit designed to produce forced air for heating and cooling.

Hip The exterior edge formed by two sloping roof surfaces.

Hip roof A roof shape with four sloping sides.

Hopper window Window that is hinged at the bottom and swings inward.

Horizon line A line drawn parallel to the ground line which represents the intersection of ground and sky.

Horizontal shear One of three major forces acting on a beam, it is the tendency of the fibers of a beam to slide past each other in horizontal direction. (Fv)

Hose bibb A water outlet that is threaded to receive a hose.

Humidifier A mechanical device that controls the amount of moisture inside of a structure.

Hydroelectric power Electricity generated by the conversion of the energy created by falling water.

I Beam The generic term for a wide flange or American standard steel beam with a cross section in the shape of the letter "I".

Indirect lighting Mechanical lighting that is reflected off a surface.

Insertion point Location where text is started when using a computer program.

Insulated concrete form (ICF) An energy efficient wall framing system in which poured concrete is placed in polystyrene forms that are left in place to create a super-insulated wall.

Insulation Material used to restrict the flow of heat, cold, or sound from one surface to another.

International Residential Code (IRC) A national building code for one- and two-family dwellings.

Isometric A drawing method which enables three surfaces of an object to be seen in one view, with the base of each surface drawn at 30 degrees to the horizontal plane.

Jack rafter A rafter which is cut shorter than the other rafters to allow for an opening in the roof.

Jack stud A wall member which is cut shorter than other studs to allow for an opening such as a window (also called a cripple stud).

Jalousie A type of window made of thin horizontal panels that can be rotated between the open and closed position.

Jamb The vertical members of a door or window frame.

Joist A horizontal structural member used in repetitive patterns to support floor and ceiling loads.

Junction box A box that protects electrical wiring splices in conductors or joints in runs.

Kick block A block used to keep the bottom of the stringer from sliding on the floor when downward pressure is applied to the stringer.

Kiln A heating unit for the removal of moisture from wood.

Kiln dried A method of drying lumber in a kiln or oven. Kiln dried lumber has a reduced moisture content when compared to lumber that has been air-dried.

King stud A full-length stud placed at the end of a header.

Kip Used in some engineering formulas to represent 1,000 pounds.

Knee wall A wall of less than full height.

Knot A branch or limb of a tree that is cut through in the process of manufacturing lumber.

Lally column A vertical steel column that is used to support floor or foundation loads.

Laminated Several layers of material that have been glued together under pressure.

Landing A platform between two flights of stairs.

Lateral Sideways motion in a structure caused by wind or seismic forces.

Lath Wood or sheet metal strips which are attached to the structural frame to support plaster.

Lattice A grille made by criss-crossing strips of material.

Lavatory A bathroom sink, or a room which is equipped with a washbasin.

Layers In CADD, details of a design or different drafting information that is separated so that one type of information is over the other.

Leach lines Soil absorbent field used to disperse liquid material from a septic system.

Ledger A horizontal member which is attached to the side of wall members to provide support for rafters or joists.

Lien A monetary claim on property.

Lintel A horizontal steel member used to provide support for masonry over an opening.

Lisp A programming language used to customize CADD software.

Live load The load from all movable objects within a structure including loads from furniture and people. External loads from snow and wind are also considered live loaded.

Load bearing wall A support wall that holds floor or roof loads in addition to its own weight.

Load path The route that is used to transfer the roof loads into the walls, then into the floor, and then to the foundation.

Lookout A beam used to support eave loads.

Louver An opening with horizontal slats to allow for ventilation.

Low-e glass Low-emissivity glass that has a transparent coating on its surface that acts as a thermal mirror.

Mansard A four-sided, steep-sloped roof.

Mantel A decorative shelf above the opening of a fireplace.

Market value The amount that property can be sold for.

Master format A list of numbers and titles that are created to organize information into a standard sequence that relates to construction requirements, products, and activities.

Mesh A metal reinforcing material placed in concrete slabs and masonry walls to help resist cracking.

Metal ties A manufactured piece of metal for joining two structural members together.

Metal wall ties Corrugated metal strips used to bond brick veneer to its support wall.

Millwork Finished woodwork that has been manufactured in a milling plant. Examples are window and door frames, mantels, moldings and stairway components.

Mineral wool An insulating material made of fibrous foam.

Modular cabinet Prefabricated cabinets that are constructed in specific sizes called modules.

Modulas of elasticity (e) The degree of stiffness of a beam.

Module A standardized unit of measurement.

Moisture barrier Typically a plastic material used to restrict moisture vapor from penetrating into a structure.

Molding Decorative strips, usually made of wood, used to conceal the seam in other finishing materials.

Moment The tendency of a force to rotate around a certain point.

Monolithic Concrete construction created in one pour.

Monument A boundary marker set to mark property corners.

Mortar A combination of cement, sand, and water used to bond masonry units together.

Mortgage A contract pledging the sale of property upon full payment of purchase price.

Mortgagee The lender of the money to the mortgagor for the purchase of property.

Mortgagor The buyer of property who is paying off the mortgage to the mortgagee.

Moving loads Loads that are not stationary such as those produced by automobiles and construction equipment.

Mud room A room or utility entrance where soiled clothing can be removed before entering the main portion of the residence.

Mudsill The horizontal wood member that rests on concrete to support other wood members.

Mullion A horizontal or vertical divider between sections of a window.

Multiview A drawing that shows the top, fronts, and right side views of an object.

Muntin A horizontal or vertical divider within a section of a window.

Nailer A wood member bolted to concrete or steel members to provide a nailing surface for attaching other wood members.

Net size Final size of wood after planing.

Neutral axis The axis formed where the forces of compression and tension in a beam reach equilibrium.

Newel The end post of a stair railing.

Nominal size An approximate size achieved by rounding the actual material size to the nearest larger whole number.

Nonferrous metal Metal, such as copper or brass, that contains no iron.

Nosing The rounded front edge of a tread which extends past the riser.

Obscure glass Glass that is not transparent.

Occupancy group A specification that indicates by whom or how the structure will be used.

On center A measurement taken from the center of one member to the center of another member.

Orientation The locating of a structure on property based on the location of the sun, prevailing winds, view, and noise.

Orthographic projection Any projection of an object's features onto an imaginary plane called a plane of projection.

Outlet An electrical receptacle which allows for current to be drawn from the system.

Outrigger A support for roof sheathing and the fascia which extends past the wall line perpendicular to the rafters.

Outsourcing The practice of sending parts of the project out to subcontractors for completion.

Overhang The horizontal measurement of the distance the roof projects from a wall.

Overlay drafting The practice of drawing a structure in several layers which must be combined to produce the finished drawing.

Pad An isolated concrete pier.

Parapet A portion of wall that extends above the edge of the roof.

Parging A thin coat of plaster used to smooth a masonry surface.

Parquet flooring Wood flooring laid to form patterns.

Partition An interior wall.

Party wall A wall dividing two adjoining spaces such as apartments or offices.

Passive solar system System that uses natural, architectural means to radiate, store, and radiate solar heat.

Patio Ground-level exterior entertaining area that is made of concrete, stone, brick, or treated wood.

Penny The length of a nail represented by the letter "d"; "16d" is read as "sixteen penny."

Percolation test Test used to determine if the soil could accommodate a septic system.

Perspective A drawing method that provides the illusion of depth by the use of vanishing points.

Photodrafting The use of photography to produce a base drawing on which additional drawings can be made.

Pictorial drawing A drawing that shows an object in a three-dimensional format.

Picture plane The plane onto which the view of the object is projected.

Pier A concrete or masonry foundation support.

Pilaster A reinforcing column built into or against a masonry wall.

Piling A vertical foundation support driven into the ground to provide support on stable soil or rock.

Pitch A description of roof angle comparing the vertical rise to the horizontal run.

Plank Lumber which is 1 1/2 to 2 1/2 inches in thickness.

Plaster A mix of sand, cement, and water used to cover walls and ceilings.

Plat A map of an area of land which shows the boundaries of individual lots.

Platform framing A building construction method where each floor acts as a platform in the framing.

Plenum An air space for transporting air from the HVAC system.

Plot A parcel of land.

Plotter An output device used in computer aided drafting to draw lines and symbols.

Plumb True vertical.

Plywood Wood composed of three or more layers with the grain of each layer placed at 90 degrees to each other and bonded with glue.

Poche A shading method using graphite applied with a soft tissue in a rubbing motion.

Pocket door A door that slides into a pocket that has been built into the wall.

Point-of-beginning Fixed location on a plot of land where the survey begins.

Porch A covered entrance to a structure.

Portico A roof supported by columns instead of walls.

Portland cement A hydraulic cement made of silica, lime and aluminum that has become the most common cement used in the construction industry because of its strength.

Post A vertical wood structural member usually 4 × 4 or larger.

Post tensioning Concrete slabs that are poured over unstable soil by using a method of reinforcement.

Precast A concrete component that has been cast in a location other than the one in which it will be used.

Prefabricated Buildings or components that are built away from the job site and transported ready to be used.

Presentation drawing Drawings used to convey basic design concepts.

Prestressed A concrete component that is placed in compression as it is cast to help resist deflection.

Prevailing winds Direction from which the wind most frequently blows in a given area of the country.

Principal The original amount of money loaned before interest is applied.

Profile Vertical section of the surface of the ground, and/or of underlying earth that is taken along any desired fixed line.

Program A set of instructions which controls the functions of a computer.

Proportion A principle of design that deals with size and shape of areas and their relationship to one another.

Puck A pointing device with multiple buttons used to issue commands or pick items from a menu.

Purlin A horizontal roof member which is laid perpendicular to rafters to help limit deflection.

Purlin brace A support member which extends from the purlin down to a load-bearing wall or header.

Quad A courtyard surrounded by the walls of buildings.

Quarry tile An unglazed, machine-made tile.

Quarter round Wood molding that has the profile of one-quarter of a circle.

Rabbet A rectangular groove cut on the edge of a board.

Radial survey A survey method used to locate property corners, structures, natural features, and elevation points.

Radiant barrier Material made of aluminum foil with backing and is used to stop heat from radiating through the attic.

Radiant heat Heat emitted from a particular material such as brick, electric coils, or hot water pipe without use of air movement.

Radon A naturally occurring radioactive gas, usually found in a basement, that breaks down into compounds that are carcinogenic when inhaled over long periods of times.

Rafter The inclined structural member of a roof system designed to support roof loads.

Rafter/ceiling joist An inclined structural member which supports both the ceiling and the roof materials.

Rake joint A recessed mortar joint.

Reaction The upward forces acting at the supports of a beam.

Rebar Reinforcing steel used to strengthen concrete.

Reference bubble A symbol used to designate the origin of details and sections.

Register An opening in a duct for the supply of heated or cooled air.

Reinforced concrete Concrete that has steel rebar placed in it to resist tension.

Relative humidity The amount of water vapor in the atmosphere compared to the maximum possible amount at the same temperature.

Rendering An artistic process applied to drawings to add realism.

Restraining wall A masonry wall supported only at the bottom by a footing that is designed to resist soil loads.

Retaining wall A masonry wall supported at the top and bottom, designed to resist soil loads.

R-factor A unit of thermal resistance applied to the insulating value of a specific building material.

Rheostat An electrical control device used to regulate the current reaching a light fixture. A dimmer switch.

Rhythm A principle of design related to the repetitive elements of a design that brings order and uniformity to it.

Ribbon A structural wood member framed into studs to support joists or rafters.

Ridge The uppermost area of two intersecting roof planes.

Ridge board A horizontal member that rafters are aligned against to resist their downward force.

Ridge brace A support member used to transfer the weight from the ridge board to a bearing wall or beam. The brace is typically spaced at 48 inches O.C. and may not exceed a 45-degree angle from vertical.

Rim joist A joist at the perimeter of a structure that runs parallel to the other floor joist.

Rise The amount of vertical distance between one tread and another.

Riser The vertical member of stairs between the treads.

Roll roofing Roofing material of fiber or asphalt that is shipped in rolls.

Roof drain A receptacle for removal of roof water.

Rough floor The subfloor, usually plywood, which serves as a base for the finished floor.

Rough hardware Hardware used in construction, such as nails, bolts, and metal connectors, which will not be seen when the project is complete.

Rough in To prepare a room for plumbing or electrical additions by running wires or piping for a future fixture.

Rough lumber Lumber that has not been surfaced but has been trimmed on all four sides.

Rough opening The unfinished opening between framing members allowed for doors, windows, or other assemblies.

Rowlock A pattern for laying masonry units so that the end of the unit is exposed.

Run The horizontal distance of a set of steps or the measurement describing the depth of one step.

R-value Measurement of thermal resistance used to indicate the effectiveness of insulation.

Saddle A small gable-shaped roof used to divert water from behind a chimney.

Sash An individual frame around a window.

Scab A short member that overlaps the butt joint of two other members used to fasten those members.

Scale A measuring instrument used to draw materials at reduced size.

Schedule A written list of similar components such as windows and doors.

Scratch coat The first coat of stucco which is scratched to provide a good bonding surface for the second coat.

Seasoning The process of removing moisture from green lumber by either air (natural) or kiln drying.

Section A type of drawing showing an object as if it had been cut through to show interior construction.

Seismic Earthquake related forces.

Septic tank A tank in which sewage is decomposed by bacteria and dispersed by drain tiles.

Service connection The wires that run to a structure from a power pole or transformer.

Setback The minimum distance required between the structure and the property line.

Shake A hand-split wooden roof shingle.

Shear The stress that occurs when two forces from opposite directions are acting on the same member. Shearing stress tends to cut a member just as scissors cut paper.

Shear panel A wall panel designed by an engineer to resist wind or seismic forces.

Sheathing A covering material placed over walls, floors and roofs which serves as a backing for finishing materials.

Shim A piece of material used to fill a space between two surfaces.

Shiplap A siding pattern of overlapping rabbeted edges.

Sill A horizontal wood member placed at the bottom of walls and openings in walls.

Simple beam Beam with a uniform load evenly distributed over its entire length and supported at each end.

Site orientation Placement of a structure on a property with certain environmental and physical factors taken into consideration.

Skylight An opening in the roof to allow light and ventilation that is usually covered with glass or plastic.

Slab A concrete floor system typically poured at ground level.

Sleepers Strips of wood placed over a concrete slab in order to attach other wood members.

Smoke chamber The portion of the chimney located directly over the firebox that acts as a funnel between the firebox and the chimney.

Smoke shelf A shelf located at the bottom of the smoke chamber to prevent down-drafts from the chimney from entering the firebox.

Soffit A lowered ceiling, typically found in kitchens, halls, and bathrooms to allow for recessed lighting or HVAC ducts.

Soil stack The main vertical waste-water pipe.

Soil vent Vent which runs up the wall and vents out the roof, allowing vapor to escape and ventilating the system.

Solar heat Heat that comes from energy generated from sunlight.

Solar panel Solar cells that convert sunlight into electricity.

Solarium Glassed-in porch designed on the south side of a house that is in direct exposure to the sun's rays.

Soldier A masonry unit laid on end with its narrow surface exposed.

Sole plate The plate placed at the bottom of a wall.

Spackle The covering of sheetrock joints with joint compound.

Span The horizontal distance between two supporting members.

Spark arrester A screen placed at the top of the flue to prevent combustibles from leaving the flue.

Specifications An exact statement describing the characteristics of a particular aspect of the project.

Splice Two similar members that are joined together in a straight line usually by nailing or bolting.

Split-level A house that has two levels, one about half a level above or below the other.

Square An area of roofing covering 100 square feet.

Stack A vertical plumbing pipe.

Stair well The opening in the floor where a stair will be framed.

Station point The position of the observer's eye in a perspective drawing.

Stations Numbers given to the horizontal lines of a grid survey.

Stile A vertical member of a cabinet, door, or decorative panel.

Stirrup A "U"-shaped metal bracket used to support wood beams.

Stock Common sizes of building materials.

Stop A wooden strip used to hold windows in place.

Stress A live or dead load acting on a structural member. Stress results as the fibers of a beam resist an external force.

Stressed-skin panel A hollow, built-up member typically used as a beam.

Stretcher A course of masonry laid horizontally with the end of the unit exposed.

Stringer The inclined support member of a stair that supports the risers and treads.

Stucco A type of plaster made from Portland cement, sand, water, and a coloring agent that is applied to exterior walls.

Stud The vertical framing member of a wall which is usually 2 × 4 or 2 × 6 in size.

Stylus A pen-like device used for transferring information to a computer by writing, moving, or picking on an electronic tablet.

Subfloor The flooring surface that is laid on the floor joint and serves as a base layer for the finished floor.

Subsill A sill located between the trimmers and bottom side of a window opening. It provides a nailing surface for interior and exterior materials.

Sump A recessed area in a basement floor to collect water so that it can be removed by a pump.

Surfaced lumber Lumber that has been smoothed on at least one side.

Survey map Map of a property showing its size, boundaries, and topography.

Swale A recessed area formed in the ground to help divert ground water away from a structure.

Tamp To compact soil or concrete.

Temporary loads The loads that must be supported by a structure for a limited time.

Tensile strength The resistance of a material or beam to the tendency to stretch.

Tension Forces that cause a material to stretch or pull apart.

Termite shield A strip of sheet metal used at the intersection of concrete and wood surfaces near ground level to prevent termites from entering the wood.

Terra-cotta Hard-baked clay typically used as a liner for chimneys.

Terrain Characteristic of the land on which a proposed structure is to be built.

Thermal conductor A material suitable for transmitting heat.

Thermal resistance Represented by the letter "R", resistance measures the ability of a material to resist the flow of heat.

Thermostat A mechanical device for controlling the output of HVAC units.

Threshold The beveled member directly under a door.

Throat The narrow opening to the chimney that is just above the firebox. The throat of a chimney is where the damper is placed.

Tilt-up A method of construction in which concrete walls are cast in a horizontal position and then lifted into place.

Timber Lumber with a cross-sectional size of 4 × 6 inches or larger.

Toenail Nails driven into a member at an angle.

Tongue and groove A joint where the edge of one member fits into a groove in the next member.

Topography Physical description of land surface showing its variation in elevation and location of features such as rivers, lakes, or towns.

Traffic flow Route that people follow as they move from one area to another.

Transom A window located over a door.

Trap A "U"-shaped pipe below plumbing fixtures which holds water to prevent odor and sewer gas from entering the fixture.

Tread The horizontal member of a stair on which the foot is placed.

Tributary width The accumulation of loads that are directed to a structural member. It is always half the distance between the beam to be designed and the next bearing point.

Trimmer Joist or rafters that are used to frame an opening in a floor, ceiling, or roof.

Truss A prefabricated or job-built construction member formed of triangular shapes used to support roof or floor loads over long spans.

Ultimate strength The unit stress within a member just before it breaks.

Uniformat An arrangement of construction information based on physical parts of a facility called systems and assemblies.

Unit stress The maximum permissible stress a structural member can resist without failing.

Unity A principle of design related to the common design or decorating pattern that ties a structure together.

Uplift The tendency of structural members to move upward due to wind or seismic pressure.

Valley The internal corner formed between two intersecting roof structures.

Vanishing point The point at which an extended line intersects the horizon line.

Vapor barrier Material that is used to block the flow of water vapor into a structure. Typically 6 mil (.0006 inch) black plastic.

Vault An inclined ceiling area.

Vellum Drafting paper that is specially designed to accept pencil or ink.

Veneer A thin outer covering or nonload bearing masonry face material.

Vent pipe Pipes that provide air into the waste lines to allow drainage by connecting each plumbing fixture to the vent stack.

Vent stack A vertical pipe of a plumbing system used to equalize pressure within the system and to vent sewer gases.

Ventilation The process of supplying and removing air from a structure.

Vertical shear A stress acting on a beam, that causes a beam to drop between its supports.

Vestibule A small entrance or lobby.

Virtual reality A computer simulated world that appears to be a real world.

Wainscot Paneling applied to the lower portion of a wall.

Wallboard Large flat sheets of gypsum, typically 1/2 or 5/8 inch thick, used to finish interior walls.

Warp Variation from true shape.

Waterproof Material or a type of construction that prevents the absorption of water.

Weather strip A fabric or plastic material placed along the edges of doors, windows, and skylights to reduce air infiltration.

Webs Interior members of the truss that span between the top and bottom chord.

Weep hole An opening in the bottom course of a masonry to allow for drainage.

Weld A method of providing a rigid connection between two or more pieces of steel.

Work triangle The triangular area created in the kitchen by drawing a line from the sink, to the refrigerator, and to the cooking area.

Wythe A single unit thickness of a masonry wall.

Zoning An ordinance that regulates the location, size, and type of a structure in a building zone.

INDEX